THE UNIVERSITY OF ARIZONA SPACE SCIENCE SERIES

RICHARD P. BINZEL, GENERAL EDITOR

Protostars and Planets VI
Henrik Beuther, Ralf S. Klessen, Cornelis P. Dullemond, and
Thomas Henning, editors, 2014, 914 pages

Comparative Climatology of Terrestrial Planets
Stephen J. Mackwell, Amy A. Simon-Miller, Jerald W. Harder,
and Mark A. Bullock, editors, 2013, 610 pages

Exoplanets
S. Seager, editor, 2010, 526 pages

Europa
Robert T. Pappalardo, William B. McKinnon,
and Krishan K. Khurana, editors, 2009, 727 pages

The Solar System Beyond Neptune
M. Antonietta Barucci, Hermann Boehnhardt, Dale P. Cruikshank,
and Alessandro Morbidelli, editors, 2008, 592 pages

Protostars and Planets V
Bo Reipurth, David Jewitt, and Klaus Keil, editors, 2007, 951 pages

Meteorites and the Early Solar System II
D. S. Lauretta and H. Y. McSween, editors, 2006, 943 pages

Comets II
M. C. Festou, H. U. Keller,
and H. A. Weaver, editors, 2004, 745 pages

Asteroids III
William F. Bottke Jr., Alberto Cellino, Paolo Paolicchi,
and Richard P. Binzel, editors, 2002, 785 pages

TOM GEHRELS, GENERAL EDITOR

Origin of the Earth and Moon
R. M. Canup and K. Righter, editors, 2000, 555 pages

Protostars and Planets IV
Vincent Mannings, Alan P. Boss,
and Sara S. Russell, editors, 2000, 1422 pages

Pluto and Charon
S. Alan Stern and David J. Tholen, editors, 1997, 728 pages

**Venus II—Geology, Geophysics, Atmosphere,
and Solar Wind Environment**
S. W. Bougher, D. M. Hunten,
and R. J. Phillips, editors, 1997, 1376 pages

Protostars and Planets VI

Protostars and Planets VI

Edited by

Henrik Beuther
Ralf S. Klessen
Cornelis P. Dullemond
Thomas Henning

With the assistance of

Renée Dotson

With 250 collaborating authors

THE UNIVERSITY OF ARIZONA PRESS
Tucson

in collaboration with

LUNAR AND PLANETARY INSTITUTE
Houston

About the front cover:

Herschel far-infrared image of the Carina Nebula (blue = 70 µm, green = 160 µm, red = 250 µm). This complex network of gas and dust filaments forms a cosmic nursery where thousands of stars are born. Courtesy of ESA/PACS/SPIRE/Thomas Preibisch, Universitäts-Sternwarte München, Ludwig-Maximilians-Universität München, Germany.

About the back cover:

Three-dimensional magneto-hydrodynamical simulation of a protoplanetary disk. The forming planet has opened a gap in the disk and still continues accreting material from the protoplanetary disk. Based on simulations by Ana Uribe, Hubert Klahr, and Thomas Henning (2013, *Astrophysical Journal, 769,* 97).

The University of Arizona Press
in collaboration with the Lunar and Planetary Institute
© 2014 The Arizona Board of Regents
All rights reserved
∞ This book is printed on acid-free, archival-quality paper.
Manufactured in the United States of America

21 20 19 18 17 16   7 6 5 4 3 2

Library of Congress Cataloging-in-Publication Data

Protostars and planets VI / edited by Henrik Beuther, Ralf S. Klessen, Cornelis P. Dullemond, and Thomas Henning ; with the assistance of Renee Dotson ; with 250 collaborating authors.
     pages cm — (The University of Arizona space science series)
  Proceedings of a conference held in Heidelberg, Germany, July 15–20, 2013.
  Includes bibliographical references and index.
  ISBN 978-0-8165-3124-0 (cloth : alk. paper)
  1. Protostars—Congresses. 2. Planets—Origin—Congresses. 3. Molecular clouds—Congresses. 4.
Stars—Formation—Congresses. I. Beuther, Henrik. II. Series: University of Arizona space science series.
  QB806.P78 2014
  523.2'4—dc23
                            2014025163

Contents

PART II: DISK FORMATION AND EVOLUTION

PART III: PLANETARY SYSTEMS — SEARCH, FORMATION, AND EVOLUTION

PART IV: ASTROPHYSICAL CONDITIONS FOR LIFE

Scientific Advisory Committee

Philippe André
Javier Ballesteros-Paredes
Isabelle Baraffe
Alan Boss
John Bradley
Nuria Calvet
Gael Chauvin
Therese Encrenaz
Guido Garay

Tristan Guillot
Nader Haghighipour
Shigeru Ida
Ray Jayawardhana
Willy Kley
Alexander Krot
Katharina Lodders
Karl Menten
Michael Meyer

Alessandro Morbidelli
Ralph Pudritz
Bo Reipurth
Dimitar Sasselov
Motohide Tamura
Ewine van Dishoeck
Stephane Udry
Alycia Weinberger

List of Contributing Authors

List of Contributing Authors
(continued)

Kley W.   667
Knutson H.   739
Kokubo E.   595
Kóspál A.   387
Krasnopolsky R.   173
Kraus A.   497
Kraus S.   497
Krivov A. V.   521
Krot A. N.   809
Kruijssen J. M. D.   291
Krumholz M. R.   3, 149, 243
Lammer H.   883
Laughlin G. P.   715
Lebedev S.   451
Lesur G.   411
Li H.-B.   101
Li Z.-Y.   101, 173, 243
Lin D.   691
Lis D. C.   835
Lodato G.   387
Longmore S. N.   291
Lunine J. I.   835
Macintosh B.   715
Mac Low M.-M.   3
Madhusudhan N.   739
Mahadevan S.   715
Mamajek E. E.   219
Martin P.   125
Marty B.   363
Masset F.   667
Matt S. P.   433
Matthews B. C.   521
Maury A.   173
Mayer L.   643
McKee C. F.   77, 149
Meeus G.   317
Megeath S. T.   195
Meru F.   643
Mohanty S.   433
Molinari S.   125
Moore T.   125
Morbidelli A.   595, 787
Mordasini C.   691
Moreaux E.   53
Morishima R.   595

Mousis O.   363, 859
Muzerolle J.   497
Myers P. C.   195
Najita J.   497
Nakamura F.   243
Natta A.   339
Nayakshin S.   643
Naylor T.   219
Nelson R.   667
Nisini B.   451
Nomura H.   317
Nordlund Å.   77
Noriega-Crespo A.   125
Novak G.   101
Öberg K. I.   363
O'D. Alexander C. M.   809
Offner S. S. R.   53, 195
Ormel C.   547
Ostriker E. C.   3
Paardekooper S.-J.   667
Padoan P.   77
Papaloizou J.   667
Pascucci I.   475
Petaev M. I.   809
Piétu V.   317
Pilat-Lohinger E.   883
Pineda J.   27
Pizzarello S.   859
Plume R.   125
Podolak M.   643
Pontoppidan K. M.   363
Poteet C. A.   195
Pudritz R. E.   27, 173
Qi C.   317
Rafikov R.   619
Raga A.   451
Rathborne J.   291
Rauer H.   883
Ray T. P.   451
Raymond S. N.   595, 787
Reipurth B.   267
Ribas I.   883
Ricci L.   339
Rickman H.   547
Robert F.   859

Rodríguez L. F.   267
Romanova M. M.   387
Sahlmann J.   715
Salyk C.   363
Scholz A.   433
Semenov D.   317, 859
Shang H.   173
Skinner S. L.   387
Soderblom D. R.   219
Sotin C.   763
Spohn T.   571
Sridharan T. K.   101
Stassun K. G.   267, 433
Stephan T.   809
Stolte A.   149, 291
Stutz A. M.   195
Tan J. C.   149
Tanaka H.   547
Tang K. S.   101
Testi L.   125, 291, 339
Tobin J. J.   195
Tokovinin A.   267
Trieloff M.   571
Turner N. J.   411
van Dishoeck E. F.   835
Vázquez-Semadeni E.   3, 125
Veras D.   787
Vorobyov E. I.   195, 387
Wakelam V.   317
Walsh K. J.   595
Wardle M.   411
Ward-Thompson D.   27
Whitworth A. P.   53
Williams J. P.   339
Wilner D. J.   339
Wood B. E.   883
Wyatt M. C.   521
Yee J. C.   715
Zanni C.   433
Zavagno A.   125
Zhang Q.   243
Zhu Z.   387, 497
Zinnecker H.   267

Preface

The Protostars and Planets book and conference series has been a long-standing tradition that commenced with the first meeting led by Tom Gehrels and held in Tucson, Arizona, in 1978. The goal then, as it still is today, was to bridge the gap between the fields of star and planet formation as well as the investigation of planetary systems and planets. As Tom Gehrels stated in the preface to the first Protostars and Planets book, "Cross-fertilization of information and understanding is bound to occur when investigators who are familiar with the stellar and interstellar phases meet with those who study the early phases of solar system formation." The central goal remained the same for the subsequent editions of the books and conferences Protostars and Planets II in 1984, Protostars and Planets III in 1990, Protostars and Planets IV in 1998, and Protostars and Planets V in 2005, but has now been greatly expanded by the flood of new discoveries in the field of exoplanet science.

The original concept of the Protostars and Planets series also formed the basis for the sixth conference in the series, which took place on July 15–20, 2013. It was held for the first time outside of the United States in the bustling university town of Heidelberg, Germany. The meeting attracted 852 participants from 32 countries, and was centered around 38 review talks and more than 600 posters. The review talks were expanded to form the 38 chapters of this book, written by a total of 250 contributing authors. This Protostars and Planets volume reflects the current state-of-the-art in star and planet formation, and tightly connects the fields with each other. It is structured into four sections covering key aspects of molecular cloud and star formation, disk formation and evolution, planetary systems, and astrophysical conditions for life. All poster presentations from the conference can be found at *www.ppvi.org*.

In the eight years that have passed since the fifth conference and book in the Protostars and Planets series, the field of star and planet formation has progressed enormously. The advent of new space observatories like Spitzer and more recently Herschel have opened entirely new windows to study the interstellar medium, the birthplaces of new stars, and the properties of protoplanetary disks. Millimeter and radio observatories, in particular interferometers, allow us to investigate even the most deeply embedded and youngest protostars. Complementary to these observational achievements, novel multi-scale and multi-physics theoretical and numerical models have provided new insights into the physical and chemical processes that govern the birth of stars and their planetary systems. Sophisticated radiative transfer modeling is critical in order to better connect theories with observations.

Since the last Protostars and Planets volume, more than 1000 new extrasolar planets have been identified and there are thousands more waiting to be verified. Such a large database allows for the first time a statistical assessment of the planetary properties as well as their evolution pathways. These investigations show the enormous diversity of the architecture of planetary systems and the properties of planets. High-contrast imaging at short and long wavelengths has resolved protoplanetary disks and associated planets, and transit spectroscopy is a new tool that allows us to study even the physical properties of extrasolar planetary atmospheres.

The understanding of our own solar system has also progressed enormously since 2005. For instance, the sample-return Stardust mission has provided direct insight into the composition of comets and asteroids, and has demonstrated the importance of mixing processes in the early solar system. And much more is now known about the origin and role of short-lived nuclides at these stages of the solar system.

For generations of astronomers, the Protostars and Planets volumes have served as an essential resource for our understanding of star and planet formation. They are used by students to dive into new topics, and they are much valued by experienced researchers as a comprehensive overview of the field with all its interactions. We hope that you will enjoy reading (and learning from) this book as much as we do.

The organization of the Protostars and Planets conference was carried out in close collaboration between the Max Planck Institute for Astronomy and the Center for Astronomy of the University Heidelberg, with generous support from the German Science Foundation. This volume is a product of effort and care by many people. First and foremost, we want to acknowledge the 250 contributing authors, as it is only due to their expertise and knowledge that such a comprehensive review compendium in all its depth and breadth is possible. The Protostars and Planets VI conference and this volume was a major undertaking, with support and contributions by many people and institutions. We like to thank the members of the Scientific Advisory Committee who selected the 38 teams and chapters out of more than 120 submitted proposals. Similarly, we are grateful to the reviewers, who provided valuable input and help to the chapter authors. The book would also not have been possible without the great support of Renée Dotson and other staff from USRA's Lunar and Planetary Institute, who handled the detailed processing of all manuscripts and the production of the book, and of Allyson Carter and other staff from the University of Arizona Press. We are also grateful to Richard Binzel, the General Editor of the Space Science Series, for his constant support during the long process, from the original concept to this final product. Finally, we would like to express a very special thank you to the entire conference local organizing committee, and in particular, Carmen Cuevas and Natali Jurina, for their great commitment to the project and for a very fruitful and enjoyable collaboration.

Henrik Beuther
Ralf Klessen
Cornelis Dullemond
Thomas Henning
Heidelberg, June 2014

Part I:
Molecular Clouds and Star Formation

Dobbs C. L., Krumholz M. R., Ballesteros-Paredes J., Bolatto A. D., Fukui Y., Heyer M., Low M-M. M., Ostriker E. C., and Vázquez-Semadeni E. (2014) Formation of molecular clouds and global conditions for star formation. In *Protostars and Planets VI* (H. Beuther et al., eds.), pp. 3–26. Univ. of Arizona, Tucson, DOI: 10.2458/azu_uapress_9780816531240-ch001.

Formation of Molecular Clouds and Global Conditions for Star Formation

Clare L. Dobbs
University of Exeter

Mark R. Krumholz
University of California, Santa Cruz

Javier Ballesteros-Paredes
Universidad Nacional Autónoma de México

Alberto D. Bolatto
University of Maryland, College Park

Yasuo Fukui
Nagoya University

Mark Heyer
University of Massachusetts

Mordecai-Mark Mac Low
American Museum of Natural History

Eve C. Ostriker
Princeton University

Enrique Vázquez-Semadeni
Universidad Nacional Autónoma de México

Giant molecular clouds (GMCs) are the primary reservoirs of cold, star-forming molecular gas in the Milky Way and similar galaxies, and thus any understanding of star formation must encompass a model for GMC formation, evolution, and destruction. These models are necessarily constrained by measurements of interstellar molecular and atomic gas and the emergent, newborn stars. Both observations and theory have undergone great advances in recent years, the latter driven largely by improved numerical simulations, and the former by the advent of large-scale surveys with new telescopes and instruments. This chapter offers a thorough review of the current state of the field.

1. INTRODUCTION

Stars form in a cold, dense, molecular phase of the interstellar medium (ISM) that appears to be organized into coherent, localized volumes or clouds. The star formation history of the universe, the evolution of galaxies, and the formation of planets in stellar environments are all coupled to the formation of these clouds, the collapse of unstable regions within them to stars, and the clouds' final dissipation. The physics of these regions is complex, and descriptions of cloud structure and evolution remain incomplete and require continued exploration. Here we review the current status of observations and theory of molecular clouds, focusing on key advances in the field since *Protostars and Planets V* (*Reipurth et al.,* 2007).

The first detections of molecules in the ISM date from the 1930s, with the discovery of CH and CN within the diffuse interstellar bands (*Swings and Rosenfeld,* 1937; *McKellar,* 1940) and later the microwave lines of OH (*Weinreb et al.,* 1963), NH_3 (*Cheung et al.,* 1968), water vapor (*Cheung et al.,* 1969), and H_2CO (*Snyder et al.,* 1969). Progress accelerated in the 1970s with the first measurements of molecular hydrogen (*Carruthers*, 1970) and the ^{12}CO J=1–0 line at 2.6 mm (*Wilson et al.*, 1970) and the continued development of millimeter-wave instrumentation and facilities.

The first maps of CO emission in nearby star-forming regions and along the Galactic plane revealed the unexpectedly large spatial extent of giant molecular clouds (GMCs) (*Kutner et al.*, 1977; *Lada*, 1976; *Blair et al.*, 1978; *Blitz and Thaddeus*, 1980) and their substantial contribution to the mass budget of the ISM (*Scoville*, 1975; *Gordon and Burton*, 1976; *Burton and Gordon*, 1978; *Sanders et al.*, 1984). Panoramic imaging of ^{12}CO emission in the Milky Way from both the northern and southern hemispheres enabled the first complete view of the molecular gas distribution in the Galaxy (*Dame et al.*, 1987, 2001) and the compilation of GMC properties (*Solomon et al.*, 1987; *Scoville et al.*, 1987). Higher-angular-resolution observations of optically thin tracers of molecular gas in nearby clouds revealed a complex network of filaments (*Bally et al.*, 1987; *Heyer et al.*, 1987), and high-density tracers such as NH_3, CS, and HCN revealed the dense regions of active star formation (*Myers*, 1983; *Snell et al.*, 1984). Since this early work, large, millimeter-filled aperture [Institut de Radioastronomie Millimétrique (IRAM) 30 m, Nobeyama Radio Observatory (NRO) 45 m], interferometric [Berkeley-Illinois-Maryland Association (BIMA), Owens Valley Radio Observatory (OVRO), Plateau de Bure (PdBI)], and submillimeter [Caltech Submillimeter Observatory (CSO), James Clark Maxwell Telescope (JCMT)] facilities have provided improved sensitivity and the ability to measure higher excitation conditions. Observations to date have identified ~200 distinct interstellar molecules (*van Dishoeck and Blake*, 1998; *Müller et al.*, 2005), and the last 40 years of observations using these molecules have determined a set of cloud properties on which our limited understanding of cloud physics is based.

Theoretically, the presence of molecular hydrogen in the ISM was predicted long before the development of large-scale CO surveys (e.g., *Spitzer*, 1949). In the absence of metals, formation of H_2 by gas phase reactions catalyzed by electrons and protons is extremely slow, but dust grains catalyze the reaction and speed it up by orders of magnitude. As a result, H_2 formation is governed by the density of dust grains, gas density, and the ability of hydrogen atoms to stick to dust grains and recombine (*van de Hulst*, 1948; *McCrea and McNally*, 1960; *Gould and Salpeter*, 1963; *Hollenbach and Salpeter*, 1971). The ISM exhibits a sharp transition in molecular fraction from low to high densities, typically at 1–100 cm^{-3} (or $\Sigma \sim$ 1–100 M$_\odot$ pc^{-2}), dependent mostly on the ultraviolet (UV) radiation field and metallicity (*van Dishoeck and Black*, 1986; *Pelupessy et al.*, 2006; *Glover and Mac Low*, 2007a; *Dobbs et al.*, 2008; *Krumholz et al.*, 2008, 2009a; *Gnedin et al.*, 2009). This dramatic increase in H_2 fraction represents a change to the regime where H_2 becomes self-shielding. Many processes have been invoked to explain how atomic gas reaches the densities ($\geq$100 cm^{-3}) required to become predominantly

molecular (see section 3). Several mechanisms likely to govern ISM structure became apparent in the 1960s: cloud-cloud collisions (*Oort*, 1954; *Field and Saslaw*, 1965), gravitational instabilities (e.g., *Goldreich and Lynden-Bell*, 1965a), thermal instabilities (*Field*, 1965), and magnetic instabilities (*Parker*, 1966; *Mouschovias*, 1974). At about the same time, *Roberts* (1969) showed that the gas response to a stellar spiral arm produces a strong spiral shock, likely observed as dust lanes and associated with molecular gas. Somewhat more recently, the idea of cloud formation from turbulent flows in the ISM has emerged (*Ballesteros-Paredes et al.*, 1999), as well as colliding flows from stellar feedback processes (*Koyama and Inutsuka*, 2000).

The nature of GMCs, their lifetime, and whether they are virialized remains unclear. Early models of cloud-cloud collisions required very long-lasting clouds [100 million years (m.y.)] in order to build up more massive GMCs (*Kwan*, 1979). Since then, lifetimes have generally been revised downward. Several observationally derived estimates, including up to the present date, have placed cloud lifetimes at around 20–30 m.y. (*Bash et al.*, 1977; *Blitz and Shu*, 1980; *Fukui et al.*, 1999; *Kawamura et al.*, 2009; *Miura et al.*, 2012), although there have been longer estimates for molecule-rich galaxies (*Tomisaka*, 1986; *Koda et al.*, 2009) and shorter estimates for smaller, nearby clouds (*Elmegreen*, 2000; *Hartmann et al.*, 2001).

In the 1980s and 1990s, GMCs were generally thought to be supported against gravitational collapse, and in virial equilibrium. Magnetic fields were generally favored as a means of support (*Shu et al.*, 1987). Turbulence would dissipate unless replenished, while rotational support was found to be insufficient (e.g., *Silk*, 1980). More recently these conclusions have been challenged by new observations, which have revised estimates of magnetic field strengths downward, and new simulations and theoretical models that suggest that clouds may in fact be turbulence-supported, or that they may be entirely transient objects that are not supported against collapse at all. These questions are all under active discussion, as we review below.

In section 2, we describe the main new observational results and corresponding theoretical interpretations. These include the extension of the Schmidt-Kennicutt relation to other tracers, notably H_2, as well as to much smaller scales, e.g., those of individual clouds. Section 2 also examines the latest results on GMC properties, both within the Milky Way and in external galaxies. Compared to the data that were available at the time of *Protostars and Planets V* (*Reipurth et al.*, 2007), CO surveys offer much higher resolution and sensitivity within the Milky Way, and are able to better cover a wider range of environments beyond it. In section 3, we discuss GMC formation, providing a summary of the main background and theory, while reporting the main advances in numerical simulations since *Protostars and Planets V*. We also discuss progress on calculating the conversion of atomic to molecular gas, as well as CO chemistry. Section 4 describes the various scenarios for the evolution of GMCs, including the revival of globally collapsing clouds as a viable theoretical model, and examines the role of different forms of stellar feedback as internal and external sources of cloud motions. Then in section 5 we relate the star-forming properties in GMCs to these different scenarios. Finally, in section 6 we look forward to what we can expect between now and Protostars and Planets VII.

2. OBSERVED PROPERTIES OF GIANT MOLECULAR CLOUDS

2.1. Giant Molecular Clouds and Star Formation

Molecular gas is strongly correlated with star formation on scales from entire galaxies (*Kennicutt*, 1989, 1998; *Gao and Solomon*, 2004; *Saintonge et al.*, 2011b), to kiloparsec and subkiloparsec regions (*Wong and Blitz*, 2002; *Bigiel et al.*, 2008; *Rahman et al.*, 2012; *Leroy et al.*, 2013b), to individual GMCs (*Evans et al.*, 2009; *Heiderman et al.*, 2010; *Lada et al.*, 2010, 2012). These relations take on different shapes at different scales. Early studies of whole galaxies found a power-law correlation between total gas content (Hɪ plus H_2) and star formation rate (SFR) with an index $N \sim 1.5$ (*Kennicutt*, 1989, 1998). These studies include galaxies that span a very large range of properties, from dwarfs to ultraluminous infrared (IR) galaxies, so it is possible that the physical underpinnings of this relation are different in different regimes. Transitions with higher critical densities such as HCN (1–0) and higher-J CO lines (*Gao and Solomon*, 2004; *Bayet et al.*, 2009; *Juneau et al.*, 2009; *García-Burillo et al.*, 2012) also show power-law correlations but with smaller indices; the index appears to depend mostly on the line critical density, a result that can be explained through models (*Krumholz and Thompson*, 2007; *Narayanan et al.*, 2008a,b).

Within galaxies the SFR surface density, Σ_{SFR}, is strongly correlated with the surface density of molecular gas as traced by CO emission and only very weakly, if at all, related to atomic gas. The strong correlation with H_2 persists even in regions where atomic gas dominates the mass budget (*Schruba et al.*, 2011; *Bolatto et al.*, 2011). The precise form of the SFR-H_2 correlation is a subject of study, with results spanning the range from superlinear (*Kennicutt et al.*, 2007; *Liu et al.*, 2011; *Calzetti et al.*, 2012) to approximately linear (*Bigiel et al.*, 2008; *Blanc et al.*, 2009; *Rahman et al.*, 2012; *Leroy et al.*, 2013b), to sublinear (*Shetty et al.*, 2013). Because CO is used to trace H_2, the correlation can be altered by systematic variations in the CO to H_2 conversion factor, an effect that may flatten the observed relation compared to the true one (*Shetty et al.*, 2011; *Narayanan et al.*, 2011, 2012).

The SFR-H_2 correlation defines a molecular depletion time, $\tau_{dep}(H_2) = M(H_2)/SFR$, which is the time required to consume all the H_2 at the current SFR. A linear SFR-H_2 correlation implies a constant $\tau_{dep}(H_2)$, while superlinear (sublinear) relations yield a timescale $\tau_{dep}(H_2)$ that decreases (increases) with surface density. In regions where CO emission is present, the mean depletion time over kiloparsec scales is $\tau_{dep}(H_2) = 2.2$ gigayears (G.y.) with ±0.3 dex scatter, with some dependence on the local conditions (*Leroy et al.*, 2013b). *Saintonge et al.* (2011a) find that, for entire galaxies, $\tau_{dep}(H_2)$ decreases by a factor of ~3 over 2 orders of magnitude increase in the SFR surface. *Leroy et al.* (2013b) show that the kiloparsec-scale measurements within galaxies are consistent with this trend, but that $\tau_{dep}(H_2)$ also correlates with the dust-to-gas ratio. For normal galaxies, using a CO-to-H_2 conversion factor that depends on the local dust-to-gas ratio removes most of the variation in $\tau_{dep}(H_2)$.

On scales of a few hundred parsecs, the scatter in $\tau_{dep}(H_2)$ rises significantly (e.g., *Schruba et al.*, 2010; *Onodera et al.*, 2010) and the SFR-H_2 correlation breaks down. This is partially a manifestation of the large dispersion in SFR per unit mass in individual GMCs (*Lada et al.*, 2010), but it is also a consequence of the timescales involved (*Kawamura et al.*, 2009; *Kim et al.*, 2013). Technical issues concerning the interpretation of the tracers also become important on the small scales (*Calzetti et al.*, 2012).

On sub-GMC scales there are strong correlations between star formation and extinction, column density, and volume density. The correlation with volume density is very close to that observed in ultraluminous IR galaxies (*Wu et al.*, 2005). Some authors have interpreted these data as implying that star formation only begins above a threshold column density of $\Sigma_{H_2} \sim 110$–130 $M_\odot$ yr^{-1} or volume density $n \sim 10^{4-5}$ cm^{-3} (*Evans et al.*, 2009; *Heiderman et al.*, 2010; *Lada et al.*, 2010, 2012). However, others argue that the data are equally consistent with a smooth rise in SFR with volume or surface density, without any particular threshold value (*Krumholz and Tan*, 2007; *Narayanan et al.*, 2008a,b; *Gutermuth et al.*, 2011; *Krumholz et al.*, 2012; *Burkert and Hartmann*, 2013).

2.2. Giant Molecular Clouds as a Component of the Interstellar Medium

Molecular clouds are the densest, coldest, highest column density, highest extinction component of the interstellar medium. Their masses are dominated by molecular gas (H_2), with a secondary contribution from He (~26%), and a varying contribution from Hɪ in a cold envelope (e.g., *Fukui and Kawamura*, 2010) and interclump gas detectable by Hɪ self-absorption (*Goldsmith and Li*, 2005). Most of the molecular mass in galaxies is in the form of molecular clouds, with the possible exception of galaxies with gas surface densities substantially higher than that of the Milky Way, where a substantial diffuse H_2 component exists (*Papadopoulos et al.*, 2012a,b; *Pety et al.*, 2013; *Colombo et al.*, 2014).

Molecular cloud masses range from ~10^2 $M_\odot$ for small clouds at high Galactic latitudes (e.g., *Magnani et al.*, 1985) and in the outer disk of the Milky Way (e.g., *Brand and Wouterloot*, 1995; *Heyer et al.*, 2001) up to giant ~10^7 $M_\odot$ clouds in the central molecular zone of the Galaxy (*Oka et al.*, 2001). The measured mass spectrum of GMCs (see section 2.3) implies that most of the molecular mass resides in the largest GMCs. Bulk densities of clouds are $\log[n_{H_2}/cm^{-3}] = 2.6 \pm 0.3$ (*Solomon et al.*, 1987; *Roman Duval et al.*, 2010), but clouds have inhomogenous density distributions with large contrasts (*Stutzki et al.*, 1988). The ratio of molecular to stellar mass in galaxies shows a strong trend with galaxy color from high in blue galaxies [10% for near-ultraviolet (NUV), r ~ 2) to low in red galaxies ($\leq$0.16% for NUV, r $\geq$ 5) (*Saintonge et al.*, 2011b). The typical molecular to atomic ratio in galaxies where both HI and H_2 are detected is $R_{mol} \equiv M_{H_2}/M_{HI} \approx 0.3$ with scatter of $\pm$0.4 dex. The large scatter reflects the fact that the atomic and molecular masses are only weakly correlated, and in contrast with the molecular gas to stellar mass fraction, the ratio R_{mol} shows only weak correlations with galaxy properties such as color (*Leroy et al.*, 2005; *Saintonge et al.*, 2011b).

In terms of their respective spatial distributions, in spiral galaxies H_2 is reasonably well described by an exponential profile with a scale length $\ell_{CO} \approx 0.2\ R_{25}$, rather smaller than the optical emission (*Young et al.*, 1995; *Regan et al.*, 2001; *Leroy et al.*, 2009; *Schruba et al.*, 2011), where R_{25} is the 25th magnitude isophotal radius for the stellar light distribution. In contrast, HI shows a nearly flat distribution with typical maximum surface density $\Sigma_{HI,max} \sim 12\ M_\odot\ pc^{-2}$ [similar to the HI column seen toward solar neighborhood clouds (*Lee et al.*, 2012)]. Galaxy centers are the regions that show the most variability and the largest departures from these trends (*Regan et al.*, 2001; *Bigiel and Blitz*, 2012). At low metallicities the HI surface density can be much larger, possibly scaling as $\Sigma_{HI,max} \sim Z^{-1}$ (*Fumagalli et al.*, 2010; *Bolatto et al.*, 2011; *Wong et al.*, 2013). In spiral galaxies the transition between the atomic and molecular-dominated regions occurs at R ~ 0.4 R_{25} (e.g., *Leroy et al.*, 2008). The CO emission also shows much more structure than the HI on the small scales (*Leroy et al.*, 2013a). In spirals with well-defined arms (NGC 6946, M 51, NGC 628) the interarm regions contain at least 30% of the measured CO luminosity (*Foyle et al.*, 2010), but at fixed total gas surface density R_{mol} is very similar for arm and interarm regions, suggesting that arms act mostly to collect gas rather to directly trigger H_2 formation (*Foyle et al.*, 2010) (see Fig. 1). We discuss the relationship between HI and H_2 in more detail in section 3.3.

2.3. Statistical Properties of Giant Molecular Clouds

Statistical descriptions of GMC properties have provided insight into the processes that govern their formation and evolution since large surveys first became possible in the 1980s (see section 1). While contemporary observations are more sensitive and feature better angular resolution and sampling than earlier surveys, identification of clouds within position-position-velocity (PPV) data cubes remains a significant problem. In practice, one defines a cloud as a set of contiguous voxels in a PPV data cube of CO emission above a surface brightness threshold. Once a cloud is defined, one can compute global properties such as size,

velocity dispersion, and luminosity (*Williams et al.*, 1994; *Rosolowsky and Leroy*, 2006). While these algorithms have been widely applied, their reliability and completeness are difficult to evaluate (*Ballesteros-Paredes and Mac Low*, 2002; *Pineda et al.*, 2009; *Kainulainen et al.*, 2009b), particularly for surveys of ^{12}CO and ^{13}CO in the Galactic plane that are subject to blending of emission from unrelated clouds. The improved resolution of modern surveys helps reduce these problems, but higher surface brightness thresholds are required to separate a feature in velocity-crowded regions. High resolution can also complicate the accounting, as the algorithms may identify cloud substructure as distinct clouds. Moreover, even once a cloud is identified, deriving masses and mass-related quantities from observed CO emission generally requires application of the CO-to-H_2 conversion factor or the H_2 to ^{13}CO abundance ratio, both of which can vary within and between clouds in response to local conditions of UV irradiance, density, temperature, and metallicity (*Bolatto et al.*, 2013; *Ripple et al.*, 2013). Millimeter-wave interferometers can resolve large GMC complexes in nearby galaxies but must also account for missing flux from an extended component of emission.

Despite these observational difficulties, there are some robust results. Over the mass range M > $10^4\ M_\odot$ where it can be measured reliably, the cloud mass spectrum is well-fit by a power-law $dN/dM \sim M^{-\gamma}$ [cumulative distribution function $N(> M) \sim M^{-\gamma+1}$], with values $\gamma < 2$ indicating that most of the mass is in large clouds. For GMCs in the Milky Way, γ is consistently found to be in the range 1.5 to 1.8 (*Solomon et al.*, 1987; *Kramer et al.*, 1998; *Heyer et al.*, 2001; *Roman Duval et al.*, 2010) with the higher value likely biased by the inclusion of cloud fragments identified as distinct clouds. Giant molecular clouds in the Magellanic clouds exhibit a steeper mass function overall, and specifically for massive clouds (*Fukui et al.*, 2008; *Wong et al.*, 2011). In M33, γ ranges from 1.6 in the inner regions to 2.3 at larger radii (*Rosolowsky and Blitz*, 2005; *Gratier et al.*, 2012).

In addition to clouds' masses, we can measure their sizes and thus their surface densities. The *Solomon et al.* (1987) catalog of inner Milky Way GMCs, updated to the current Galactic distance scale, shows a distribution of GMCs surface densities $\Sigma_{GMC} \approx 150^{+95}_{-70}\ M_\odot\ pc^{-2}$ ($\pm 1\sigma$ interval), assuming a fixed CO-to-H_2 conversion factor $X_{CO} = 2 \times 10^{20}\ cm^{-2}\ (K\ km\ s^{-1})^{-1}$, and including the He mass (*Bolatto et al.*, 2013). *Heyer et al.* (2009) reobserved these clouds in ^{13}CO and found $\Sigma_{GMC} \sim 40\ M_\odot\ pc^{-2}$ over the same cloud areas, but concluded that this is likely at least a factor of 2 too low due to non-long-term evolution (LTE) and optical depth effects. *Heiderman et al.* (2010) find that ^{13}CO can lead to a factor of 5 underestimate. A reanalysis by *Roman Duval et al.* (2010) shows $\Sigma_{GMC} \sim 144\ M_\odot\ pc^{-2}$ using the ^{13}CO rather than the ^{12}CO contour to define the area. Measurements of surface densities in extragalactic GMCs remain challenging, but with the advent of the Atacama Large Millimeter Array (ALMA) the field is likely to evolve quickly. For a sample of nearby galaxies,

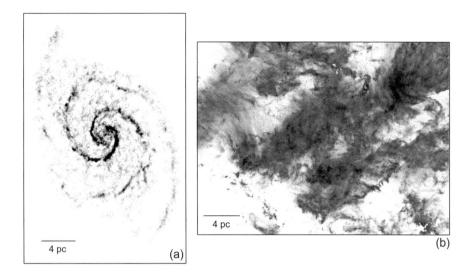

Fig. 1. (a) CO J=1–0 image of M51 from *Koda et al.* (2009) showing that the largest cloud complexes are distributed in spiral arms, while smaller GMCs lie both in and between spiral features. **(b)** An image of velocity-integrated CO J=1–0 emission from the Taurus molecular cloud from *Narayanan et al.* (2008c) illustrating complex gas motions within clouds.

many of them dwarfs, *Bolatto et al.* (2008) find $\Sigma_{GMC} \approx$ 85 $M_\odot$ pc^{-2}. Other recent extragalactic surveys find roughly comparable results, $\Sigma_{GMC} \sim$ 40–170 $M_\odot$ pc^{-2} (*Rebolledo et al.,* 2012; *Donovan Meyer et al.,* 2013).

Giant molecular cloud surface densities may prove to be a function of environment. The PdBI Arcsecond Whirlpool Survey (PAWS) survey of M 51 finds a progression in surface density (*Colombo et al.,* 2014), from clouds in the center ($\Sigma_{GMC} \sim$210 $M_\odot$ pc^{-2}), to clouds in arms ($\Sigma_{GMC} \sim$ 185 $M_\odot$ pc^{-2}), to those in interarm regions ($\Sigma_{GMC} \sim$ 140 $M_\odot$ pc^{-2}). *Fukui et al.* (2008), *Bolatto et al.* (2008), and *Hughes et al.* (2010) find that GMCs in the Magellanic clouds have lower surface densities than those in the inner Milky Way ($\Sigma_{GMC} \sim$ 50 $M_\odot$ pc^{-2}). Because of the presence of extended H_2 envelopes at low metallicities (section 2.6), however, this may underestimate their true molecular surface density (e.g., *Leroy et al.,* 2009). Even more extreme variations in Σ_{GMC} are observed near the Galactic center and in more extreme starburst environments (see section 2.7).

In addition to studying the mean surface density of GMCs, observations within the Galaxy can also probe the distribution of surface densities within GMCs. For a sample of solar neighborhood clouds, *Kainulainen et al.* (2009a) use IR extinction measurements to determine that probability density functions (PDFs) of column densities are lognormal from $0.5 < A_V < 5$ (roughly 10–100 $M_\odot$ pc^{-2}), with a power-law tail at high column densities in actively star-forming clouds. Column density images derived from dust emission also find such excursions (*Schneider et al.,* 2012, 2013). *Lombardi et al.* (2010), also using IR extinction techniques, find that, although GMCs contain a wide range of column densities, the mass M and area A contained within a specified extinction threshold nevertheless obey the *Larson* (1981) M $\propto$ A relation, which implies constant column density.

Finally, we warn that *all* column density measurements are subject to a potential bias. Giant molecular clouds are identified as contiguous areas with surface brightness values or extinctions above a threshold typically set by the sensitivity of the data. Therefore, pixels at or just above this threshold comprise most of the area of the defined cloud, and the measured cloud surface density is likely biased toward the column density associated with this threshold limit. Note that there is also a statistical difference between "mass-weighed" and "area-weighed" Σ_{GMC}. The former is the average surface density that contributes most of the mass, while the latter represents a typical surface density over most of the cloud extent. Area-weighed Σ_{GMC} tend to be lower, and although perhaps less interesting from the viewpoint of star formation, they are also easier to obtain from observations.

In addition to mass and area, velocity dispersion is the third quantity that we can measure for a large sample of clouds. It provides a coarse assessment of the complex motions in GMCs that may be responsible for the observed structure in GMCs shown in Fig. 1. *Larson* (1981) identified scaling relationships between velocity dispersion and cloud size suggestive of a turbulent velocity spectrum, and a constant surface density for clouds. Using more sensitive surveys of GMCs, *Heyer et al.* (2009) found a scaling relation that extends the Larson relationships such that the one-dimensional velocity dispersion σ_v depends on the physical radius, R, and the column density Σ_{GMC}, as shown in Fig. 2. The points follow the expression

$$\sigma_v = 0.7 \left(\Sigma_{GMC} / 100 \ M_\odot \ pc^{-2} \right)^{1/2} \left(R/1 \ pc \right)^{1/2} \ km \ s^{-1} \quad (1)$$

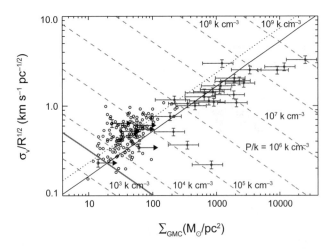

Fig. 2. The variation of $\sigma_v/R^{1/2}$ with surface density, Σ_{GMC}, for Milky Way GMCs from *Heyer et al.* (2009) (open circles) and massive cores from *Gibson et al.* (2009) (filled circles on cross-hatch lines). For clarity, a limited number of error bars are displayed for the GMCs. The horizontal error bars for the GMCs convey lower limits to the mass surface density derived from ^{13}CO. The vertical error bars for both datasets reflect a 20% uncertainty in the kinematic distances. The horizontal error bars for the massive cores assume a 50% error in the $C^{18}O$ and N_2H^+ abundances used to derive mass. The solid and dotted black lines that cut across the figure from the lower left to the upper right show loci corresponding to gravitationally bound and marginally bound clouds respectively. Lines of constant turbulent pressure are illustrated by the dashed lines from the upper left to the lower right, and the mean thermal pressure of the local ISM is shown as the heavy solid line in the lower left corner of the figure.

More recent compilations of GMCs in the Milky Way (*Roman Duval et al.*, 2010) have confirmed this result, and studies of local group galaxies (*Bolatto et al.*, 2008; *Wong et al.*, 2011) have shown that it applies to GMCs outside the Milky Way as well. Equation (1) is a natural consequence of gravity playing an important role in setting the characteristic velocity scale in clouds, either through collapse (*Ballesteros-Paredes et al.*, 2011b) or virial equilibrium (*Heyer et al.*, 2009). Unfortunately, one expects only factor of $\sqrt{2}$ differences in velocity dispersion between clouds that are in free-fall collapse or in virial equilibrium (*Ballesteros-Paredes et al.*, 2011b), making it extremely difficult to distinguish between these possibilities using observed scaling relations. Concerning the possibility of pressure-confined but mildly self-gravitating clouds (*Field et al.*, 2011), Fig. 2 shows that the turbulent pressures, $P = \rho\sigma_v^2$, in observed GMCs are generally larger than the mean thermal pressure of the diffuse ISM (*Jenkins and Tripp*, 2011), so these structures must be confined by self-gravity.

As with column density, observations within the Galaxy can also probe internal velocity structure. *Brunt* (2003), *Heyer and Brunt* (2004), and *Brunt et al.* (2009) used principal component analysis of GMC velocity fields to investigate the scales on which turbulence in molecular clouds could be driven. They found no break in the velocity dispersion-size relation, and reported that the second principle component has a "dipole-like" structure. Both features suggest that the dominant processes driving GMC velocity structure must operate on scales comparable to or larger than single clouds.

2.4. Dimensionless Numbers: Virial Parameter and Mass-to-Flux Ratio

The virial theorem describes the large-scale dynamics of gas in GMCs, so ratios of the various terms that appear in it are a useful guide to what forces are important in GMC evolution. Two of these ratios are the virial parameter, which evaluates the importance of internal pressure and bulk motion relative to gravity, and the dimensionless mass-to-flux ratio, which describes the importance of magnetic fields compared to gravity. Note, however, that neither of these ratios accounts for potentially important surface terms (e.g., *Ballesteros-Paredes et al.*, 1999).

The virial parameter is defined as $\alpha_G = M_{virial}/M_{GMC}$, where $M_{virial} = 5\sigma_v^2 R/G$ and M_{GMC} is the luminous mass of the cloud. For a cloud of uniform density with negligible surface pressure and magnetic support, $\alpha_G = 1$ corresponds to virial equilibrium and $\alpha_G = 2$ to being marginally gravitationally bound, although in reality $\alpha_G > 1$ does not strictly imply expansion, nor does $\alpha_G < 1$ strictly imply contraction (*Ballesteros-Paredes*, 2006). Surveys of the Galactic plane and nearby galaxies using ^{12}CO emission to identify clouds find an excellent, near-linear correlation between M_{virial} and the CO luminosity, L_{CO}, with a coefficient implying that (for reasonable CO-to-H_2 conversion factors) the typical cloud virial parameter is unity (*Solomon et al.*, 1987; *Fukui et al.*, 2008; *Bolatto et al.*, 2008; *Wong et al.*, 2011). Virial parameters for clouds exhibit a range of values from $\alpha_G \sim 0.1$ to $\alpha_G \sim 10$, but typically α_G is indeed ~1. *Heyer et al.* (2009) reanalyzed the *Solomon et al.* (1987) GMC sample using ^{13}CO J=1–0 emission to derive cloud mass and found a median $\alpha_G = 1.9$. This value is still consistent with a median $\alpha_G = 1$, since excitation and abundance variations in the survey lead to systematic underestimates of M_{GMC}. A cloud catalog generated directly from the ^{13}CO emission of the BU-FCRAO Galactic Ring Survey resulted in a median $\alpha_G = 0.5$ (*Roman Duval et al.*, 2010). Previous surveys (*Dobashi et al.*, 1996; *Yonekura et al.*, 1997; *Heyer et al.*, 2001) tended to find higher α_G for low-mass clouds, possibly a consequence of earlier cloud-finding algorithms preferentially decomposing single GMCs into smaller fragments (*Bertoldi and McKee*, 1992).

The importance of magnetic forces is characterized by the ratio M_{GMC}/M_{cr}, where $M_{cr} = \Phi/(4\pi^2 G)^{1/2}$ and Φ is the magnetic flux threading the cloud (*Mouschovias and Spitzer*, 1976; *Nakano*, 1978). If $M_{GMC}/M_{cr} > 1$ (the supercritical case), then the magnetic field is incapable of providing the requisite force to balance self-gravity, while if $M_{GMC}/M_{cr} < 1$ (the subcritical case), the cloud can be supported against self-gravity by the magnetic field. Initially subcritical volumes can become supercritical through ambipolar diffusion (*Mouschovias, 1987; Lizano and Shu, 1989*). Evaluating whether a cloud is sub- or supercritical is challenging. Zeeman measurements of the OH and CN lines offer a direct measurement of the line of sight component of the magnetic field at densities $\sim 10^3$ and $\sim 10^5$ cm^{-3}, respectively, but statistical corrections are required to account for projection effects for both the field and the column density distribution. *Crutcher (2012)* provides a review of techniques and observational results, and report a mean value $M_{GMC}/M_{cr} \approx 2$–3, implying that clouds are generally supercritical, although not by a large margin.

2.5. Giant Molecular Cloud Lifetimes

The natural time unit for GMCs is the free-fall time, which for a medium of density ρ is given by $\tau_{ff} = [3\pi/(32G\rho)]^{1/2} = 3.4(100/n_{H_2})^{1/2}$ m.y., where n_{H_2} is the number density of H_2 molecules, and the mass per H_2 molecule is 3.9×10^{-24} g for a fully molecular gas of cosmological composition. This is the timescale on which an object that experiences no significant forces other than its own gravity will collapse to a singularity. For an object with $\alpha_G \approx 1$, the crossing timescale is $\tau_{cr} = R/\sigma \approx 2\tau_{ff}$. It is of great interest how these natural timescales compare to cloud lifetimes and depletion times.

Scoville et al. (1979) argue that GMCs in the Milky Way are very long-lived ($>10^8$ yr) based on the detection of molecular clouds in interarm regions, and *Koda et al.* (2009) apply similar arguments to the H_2-rich galaxy M 51. They find that, while the largest GMC complexes reside within the arms, smaller ($<10^4$ M$_\odot$) clouds are found in the interarm regions, and the molecular fraction is large ($>75\%$) throughout the central 8 kpc (see also *Foyle et al., 2010*). This suggests that massive GMCs are rapidly built up in the arms from smaller, preexisting clouds that survive the transit between spiral arms. The massive GMCs fragment into the smaller clouds upon exiting the arms, but have column densities high enough to remain molecular (see section 3.4). Since the time between spiral arm passages is ~ 100 m.y., this implies a similar cloud lifetime $\tau_{life} \gtrsim 100$ m.y. $\gg \tau_{ff}$. Note, however, this is an argument for the mean lifetime of a H_2 molecule, not necessarily for a single cloud. Furthermore, these arguments do not apply to H_2-poor galaxies like the Large Magellanic Cloud (LMC) and M 33.

Kawamura et al. (2009) (see also *Fukui et al., 1999; Gratier et al., 2012*) use the NANTEN survey of ^{12}CO J=1–0 emission from the LMC, which is complete for clouds with mass $>5 \times 10^4$ M$_\odot$, to identify three distinct cloud types that are linked to specific phases of cloud evolution. Type I clouds are devoid of massive star formation and represent the earliest phase. Type II clouds contain compact H II regions, signaling the onset of massive star formation. Type III clouds, the final stage, harbor developed stellar clusters and H II regions. The number counts of cloud types

indicate the relative lifetimes of each stage, and age-dating the star clusters found in type III clouds then makes it possible to assign absolute durations of 6, 13, and 7 m.y. for types I, II, and III respectively. Thus the cumulative GMC lifetime is $\tau_{life} \sim 25$ m.y. This is still substantially greater than τ_{ff}, but by less so than in M 51.

While lifetime estimates in external galaxies are possible only for large clouds, in the solar neighborhood it is possible to study much smaller clouds, and to do so using timescales derived from the positions of individual stars on the HR diagram. *Elmegreen* (2000), *Hartmann et al.* (2001), and *Ballesteros-Paredes and Hartmann* (2007), examining a sample of solar neighborhood GMCs, note that their HR diagrams are generally devoid of post T Tauri stars with ages of ~ 10 m.y. or more, suggesting this as an upper limit on τ_{life}. More detailed analysis of HR diagrams, or other techniques for age-dating stars, generally points to age spreads of at most ~ 3 m.y. (*Reggiani et al., 2011; Jeffries et al., 2011*).

While the short lifetimes inferred for Galactic clouds might at first seem inconsistent with the extragalactic data, it is important to remember that the two datasets are probing essentially nonoverlapping ranges of cloud mass and length scale. The largest solar neighborhood clouds that have been age-dated via HR diagrams have masses $<10^4$ M$_\odot$ (the entire Orion cloud is more massive than this, but the age spreads reported in the literature are only for the few thousand M$_\odot$ central cluster), below the detection threshold of most extragalactic surveys. Since larger clouds have, on average, lower densities and longer free-fall timescales, the difference in τ_{life} is much larger than the difference in τ_{life}/τ_{ff}. Indeed, some authors argue that τ_{life}/τ_{ff} may be ~ 10 for Galactic clouds as well as extragalactic ones (*Tan et al., 2006*).

2.6. Star Formation Rates and Efficiencies

We can also measure star formation activity within clouds. We define the star formation efficiency or yield, ε_*, as the instantaneous fraction of a cloud's mass that has been transformed into stars, $\varepsilon_* = M_*/(M_* + M_{gas})$, where M_* is the mass of newborn stars. In an isolated, non-accreting cloud, ε_* increases monotonically, but in an accreting cloud it can decrease as well. *Krumholz and McKee* (2005), building on classical work by *Zuckerman and Evans* (1974), argue that a more useful quantity than ε_* is the star formation efficiency per free-fall time, defined as $\varepsilon_{ff} = \dot{M}_*/(M_{gas}/\tau_{ff})$, where $\dot{M}_*$ is the instantaneous SFR. This definition can also be phrased in terms of the depletion timescale introduced above: $\varepsilon_{ff} = \tau_{ff}/\tau_{dep}$. One virtue of this definition is that it can be applied at a range of densities ρ, by computing $\tau_{ff}(\rho)$ then taking M_{gas} to be the mass at a density $\geq \rho$ (*Krumholz and Tan, 2007*). As newborn stars form in the densest regions of clouds, ε_* can only increase as one increases the density threshold used to define M_{gas}. It is in principle possible for ε_{ff} to both increase and decrease, and its behavior as a function of density encodes important information about how star formation behaves.

Within individual clouds, the best available data on ε_* and ε_{ff} come from campaigns that use the Spitzer Space Telescope to obtain a census of young stellar objects with excess IR emission, a feature that persists for 2–3 m.y. of pre-main-sequence evolution. These are combined with cloud masses and surface densities measured by millimeter dust emission or IR extinction of background stars. For a set of five star-forming regions investigated in the Cores to Disks Spitzer Legacy program, *Evans et al.* (2009) found $\varepsilon_* = 0.03–0.06$ over entire GMCs, and $\varepsilon_* \sim 0.5$ considering only dense gas with n $\sim 10^5$ cm^{-3}. On the other hand, $\varepsilon_{ff} \approx 0.03–0.06$ regardless of whether one considers the dense gas or the diffuse gas, due to a rough cancellation between the density dependence of M_{gas} and τ_{ff}. *Heiderman et al.* (2010) obtain comparable values in 15 additional clouds from the Gould's Belt Survey. *Murray* (2011) find significantly higher values of $\varepsilon_{ff} = 0.14–0.24$ for the star clusters in the Galaxy that are brightest in WMAP free emission, but this value may be biased high because it is based on the assumption that the molecular clouds from which those clusters formed have undergone negligible mass loss despite the clusters' extreme luminosities (*Feldmann and Gnedin,* 2011).

At the scale of the Milky Way as a whole, recent estimates based on a variety of indicators put the Galactic SFR at ≈ 2 M$_\odot$ yr^{-1} (*Robitaille and Whitney,* 2010; *Murray and Rahman,* 2010; *Chomiuk and Povich,* 2011), within a factor of ~ 2 of earlier estimates based on groundbased radio catalogs (e.g., *McKee and Williams,* 1997). In comparison, the total molecular mass of the Milky Way is roughly 10^9 M$_\odot$ (*Solomon et al.,* 1987), and this, combined with the typical free-fall time estimated in the previous section, gives a galaxy-average $\varepsilon_{ff} \sim 0.01$ (see also *Krumholz and Tan,* 2007; *Murray and Rahman,* 2010).

For extragalactic sources one can measure ε_{ff} by combining SFR indicators such as Hα, UV, and IR emission with tracers of gas at a variety of densities. As discussed above, observed H$_2$ depletion times are τ_{dep}(H$_2$) ≈ 2 G.y., whereas GMC densities of n$_H \sim 30–1000$ cm^{-3} correspond to free-fall times of $\sim 1–8$ m.y., with most of the mass probably closer to the smaller value, since the mass spectrum of GMCs ensures that most of the mass is in the large clouds, which tend to have lower densities. Thus $\varepsilon_{ff} \sim 0.001–0.003$. Observations using tracers of dense gas (n $\sim 10^5$ cm^{-3}) such as HCN yield $\varepsilon_{ff} \sim 0.01$ (*Krumholz and Tan,* 2007; *García-Burillo et al.,* 2012); given the errors, the difference between the HCN and CO values is not significant. As with the *Evans et al.* (2009) clouds, higher-density regions subtend smaller volumes and comprise smaller masses. ε_{ff} is nearly constant because M_{gas} and $1/\tau_{ff}$ both fall with density at about the same rate.

Figure 3 shows a large sample of observations compiled by *Krumholz et al.* (2012), which includes individual Galactic clouds, nearby galaxies, and high-redshift galaxies, covering a very large range of mean densities. They find that all the data are consistent with $\varepsilon_{ff} \sim 0.01$, albeit with considerable scatter and systematic uncertainty. Even with the uncertainties, however, it is clear that $\varepsilon_{ff} \sim 1$ is strongly ruled out.

2.7. Giant Molecular Clouds in Varying Galactic Environments

One gains useful insight into GMC physics by studying their properties as a function of environment. Some of the most extreme environments, such as those in starbursts or metal-poor galaxies, also offer unique insights into astrophysics in the primitive universe, and aid in the interpretation of observations of distant sources.

Galactic centers, which feature high metallicity and stellar density, and often high surface densities of gas and star formation, are one unusual environment to which we have observational access. The properties of the bulge, and presence of 1 bar, appear to influence the amount of H$_2$ in the center (*Fisher et al.,* 2013). Central regions with high Σ_{H_2} preferentially show reduced τ_{dep}(H$_2$) compared to galaxy averages (*Leroy et al.,* 2013b), suggesting that central GMCs convert their gas into stars more rapidly. Reduced τ_{dep}(H$_2$) is correlated with an increase in CO (2–1)/(1–0) ratios, indicating enhanced excitation (or lower optical depth). Many galaxy centers also exhibit a superexponential increase in CO brightness, and a drop in CO-to-H$_2$ conversion factor [*Sandstrom et al.* (2013), which reinforces the short τ_{dep}(H$_2$) conclusion]. On the other hand, in our own Galactic center, *Longmore et al.* (2013) show that there are

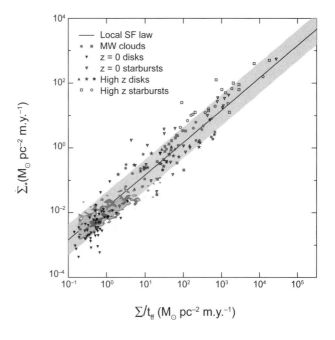

Fig. 3. SFR per unit area vs. gas column density over freefall time (*Krumholz et al.,* 2012). Different shapes indicate different data sources and different types of objects: circles and squares are Milky Way clouds, filled triangles are unresolved z = 0 galaxies, open triangles are unresolved z = 0 starbursts, filled symbols are unresolved z > 1 disk galaxies, and open symbols are unresolved z > 1 starburst galaxies. Contours show the distribution of kpc-sized regions within nearby galaxies. The black line is $\varepsilon_{ff} = 0.01$, and the gray band is a factor of 3 range around it.

massive molecular clouds that have surprisingly little star formation, and depletion times $\tau_{dep}(H_2)$ ~1 G.y. comparable to disk GMCs (*Kruijssen et al., 2013*), despite volume and column densities orders of magnitude higher (see the chapter by Longmore et al. in this volume).

Obtaining similar spatially resolved data on external galaxies is challenging. *Rosolowsky and Blitz* (2005) examined several very large GMCs (M ~ 10^7 M$_\odot$, R ~ 40–180 pc) in M 64. They also find a size-linewidth coefficient somewhat larger than in the Milky Way disk, and, in ^{13}CO, high surface densities. Recent multi-wavelength, high-resolution ALMA observations of the center of the nearby starburst NGC 253 find cloud masses M ~ 10^7 M$_\odot$ and sizes R ~ 30 pc, implying $\Sigma_{GMC} \geq 10^3$ M$_\odot$ pc^{-2} (Leroy et al., 2014, in preparation). The cloud linewidths imply that they are self-gravitating.

The low-metallicity environments of dwarf galaxies and outer galaxy disks supply another fruitful laboratory for study of the influence of environmental conditions. Because of their proximity, the Magellanic Clouds provide the best locations to study metal-poor GMCs. Owing to the scarcity of dust at low metallicity (e.g., *Draine et al., 2007*), the abundances of H_2 and CO in the ISM are greatly reduced compared to what would be found under comparable conditions in a higher metallicity galaxy (see the discussion in section 3.3). As a result, CO emission is faint, only being present in regions of very high column density (e.g., *Israel et al., 1993; Bolatto et al., 2013*, and references therein). Despite these difficulties, there are a number of studies of low-metallicity GMCs. *Rubio et al.* (1993) reported GMCs in the Small Magellanic Clouds (SMC) exhibit sizes, masses, and a size-linewidth relation similar to that in the disk of the Milky Way. However, more recent work suggests that GMCs in the Magellanic Clouds are smaller and have lower masses, brightness temperatures, and surface densities than typical inner Milky Way GMCs, although they are otherwise similar to Milky Way clouds (*Fukui et al., 2008; Bolatto et al., 2008; Hughes et al., 2010; Muller et al., 2010; Herrera et al., 2013*). Magellanic Cloud GMCs also appear to be surrounded by extended envelopes of CO-faint H_2 that are ~30% larger than the CO-emitting region (*Leroy et al., 2007, 2009*). Despite their CO faintness, though, the SFR-H_2 relation appears to be independent of metallicity once the change in the CO-to-H_2 conversion factor is removed (*Bolatto et al., 2011*).

3. GIANT MOLECULAR CLOUD FORMATION

We now turn to the question of how GMCs form. The main mechanisms that have been proposed, which we discuss in detail below, are converging flows driven by stellar feedback or turbulence (section 3.1), agglomeration of smaller clouds (section 3.2), gravitational instability (section 3.3) and magnetogravitational instability (section 3.4), and instability involving differential buoyancy (section 3.5). All these mechanisms involve converging flows, i.e., $\nabla \times$ **u** < 0, where **u** is the velocity, which may be continuous or

intermittent. Each mechanism, however, acts over different sizes and timescales, producing density enhancements of different magnitudes. Consequently, different mechanisms may dominate in different environments, and may lead to different cloud properties. In addition to these physical mechanisms for gathering mass, forming a GMC involves a phase change in the ISM, and this too may happen in a way that depends on the large-scale environment (section 3.6).

Two processes that we will not consider as cloud formation mechanisms are thermal instabilities (TI) (*Field, 1965*) and magneto-rotational instabilities (MRI) (*Balbus and Hawley, 1991*). Neither of these by themselves are likely to form molecular clouds. Thermal instabilities produce the cold (100 K) atomic component of the ISM, but the cloudlets formed are ~parsecs in scale, and without the shielding provided by a large gas column this does not lead to 10–20 K molecular gas. Furthermore, TI does not act in isolation, but interacts with all other processes taking place in the atomic ISM (*Vázquez-Semadeni et al., 2000; Sánchez-Salcedo et al., 2002; Piontek and Ostriker, 2004; Kim et al., 2008*). Nevertheless, the formation of the cold atomic phase aids GMC formation by enhancing both shock compression and vertical settling, as discussed further in section 3.2. The MRI is not fundamentally compressive, so again it will not by itself lead to GMC formation, although it can significantly affect the development of large-scale gravitational instabilities that are limited by galactic angular momentum (*Kim et al., 2003*). Magneto-rotational instabilities or TI may also drive turbulence (*Koyama and Inutsuka, 2002; Kritsuk and Norman, 2002; Kim et al., 2003; Piontek and Ostriker, 2005, 2007; Inoue and Inutsuka, 2012*), and thereby aid cloud agglomeration and contribute to converging flows. However, except in regions with very low SFR, the amplitudes of turbulence driven by TI and MRI are lower than those driven by star formation feedback.

3.1. Localized Converging Flows

Stellar feedback processes such as the expansion of H$_{II}$ regions (*Bania and Lyon, 1980; Vázquez-Semadeni et al., 1995; Passot et al., 1995*) and supernova blast waves (*Mc-Cray and Kafatos, 1987; Gazol-Patiño and Passot, 1999; de Avillez, 2000; de Avillez and Mac Low, 2001; de Avillez and Breitschwerdt, 2005; Kim et al., 2011; Ntormousi et al., 2011*) can drive converging streams of gas that accumulate to become molecular clouds, either in the Galactic plane or above it, or even after material ejected vertically by the local excess of pressure due to the stars/supernovae falls back into the plane of the disk. Morphological evidence for this process can be found in large-scale extinction maps of the Galaxy (see Fig. 4), and recent observations in both the Milky Way and the LMC confirm that MCs can be found at the edges of supershells (*Dawson et al., 2011, 2013*).

Locally — on scales up to ~100 pc — it is likely that these processes play a dominant role in MC formation, since on these scales the pressure due to local energy sources is typically P/k_B ~ 10^4 K cm^{-3}, which exceeds the mean pressure of the ISM in the solar neighborhood (*Draine*, 2011). The mass of MCs created by this process will be defined by the mean density ρ_0 and the velocity correlation length, L, of the converging streams; L is less than the disk scale height H for local turbulence. For solar neighborhood conditions, where there are low densities and relatively short timescales for coherent flows; this implies a maximum MC mass of a few times 10^4 $M_\odot$. Converging flows driven by large-scale instabilities can produce higher-mass clouds (see below).

A converging flow is not by itself sufficient to form a MC; the detailed initial velocity, density, and magnetic field structure must combine with TI to produce fast cooling (e.g., *McCray et al.*, 1975; *Bania and Lyon*, 1980; *Vázquez-Semadeni et al.*, 1995; *Hennebelle and Pérault*, 1999; *Koyama and Inutsuka*, 2000; *Audit and Hennebelle*, 2005; *Heitsch et al.*, 2006; *Vázquez-Semadeni et al.*, 2007). This allows rapid accumulation of cold, dense atomic gas, and thus promotes molecule formation. The accumulation of gas preferentially along field lines [perhaps due to magneto-Jeans instability (MJI) — see section 3.4] also increases the mass to flux ratio, causing a transition from subcritical gas to supercritical (*Hartmann et al.*, 2001; *Vázquez-Semadeni et al.*, 2011). Thus the accumulation of mass from large-scale streams, the development of a molecular phase with negligible thermal support, and the transition from magnetically subcritical to supercritical all happen essentially simultaneously, at a column density of ~10^{21} cm^{-2} (*Hartmann et al.*, 2001) (see also section 3.6), allowing simultaneous molecular cloud and star formation.

This mechanism for GMC formation naturally explains the small age spread observed in MCs near the Sun (see section 2.5), since the expected star formation timescale in

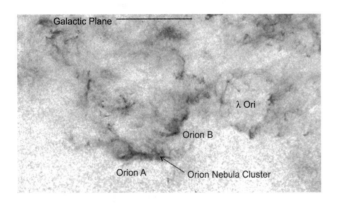

Fig. 4. Extinction map toward the Orion-Monoceros molecular complex. Different features are located at different distances, but at the mean distance of Orion, the approximate size of the A and B clouds, as well as the λ Ori ring, is on the order of 50 pc. Note the presence of shells at a variety of scales, presumably due to OB stars and/or supernovae. The λ Ori ring surrounds a 5-m.y.-old stellar cluster, and it is thought to have been produced by a supernova [data from *Rowles and Froebrich* (2009), as adapted by L. Hartmann (2013, in preparation)].

these models is the thickness of the compressed gas (~few parsecs) divided by the inflow velocity (~few kilometers squared). In addition to the overall age spread, the shape of the stellar age distribution produced by this mechanism is consistent with those observed in nearby MCs: Most of the stars have ages of 1–3 m.y., and only few are older. While some of these regions may be the result of contamination by nonmembers (*Hartmann*, 2003), the presence of some older stars might be the result of a few stars forming prior to the global, but hierarchical and chaotic contraction concentrates most of the gas and forms most of the stars (*Hartmann et al.*, 2012). This model is also consistent with the observed morphology of solar neighborhood clouds, since their elongated structures would naturally result from inflow compression.

3.2. Spiral-Arm-Induced Collisions

While localized converging flows can create clouds with masses up to ~10^4 $M_\odot$, most of the molecular gas in galaxies is found in much larger clouds (section 2.3). Some mechanism is required to either form these large clouds directly, or to induce smaller clouds to agglomerate into larger ones. Although not yet conclusive, there is some evidence of different cloud properties in arms compared to interarm regions, which could suggest different mechanisms operating (*Colombo et al.*, 2014). It was long thought that the agglomeration of smaller clouds could not work because, for clouds moving with observed velocity dispersions, the timescale required to build a 10^5–10^6-$M_\odot$ cloud would be >100 m.y. (*Blitz and Shu*, 1980). However, in the presence of spiral arms, collisions between clouds become much more frequent, greatly reducing the timescale (*Casoli and Combes*, 1982; *Kwan and Valdes*, 1983; *Dobbs*, 2008). Figure 5 shows a result from a simulation where GMCs predominantly form from spiral-arm-induced collisions. Even in the absence of spiral arms, in high-surface-density galaxies, *Tasker and Tan* (2009) find that collisions are frequent enough (every ~0.2 orbits) to influence GMC properties, and possibly star formation (*Tan*, 2000). The frequency and success of collisions is enhanced by clouds' mutual gravity (*Kwan and Valdes*, 1987; *Dobbs*, 2008), but is suppressed by magnetic fields (*Dobbs and Price*, 2008).

This mechanism can explain a number of observed features of GMCs. Giant molecular associations in spiral arms display a quasiperiodic spacing along the arms (*Elmegreen and Elmegreen*, 1983; *Efremov*, 1995), and *Dobbs* (2008) show that this spacing can be set by the epicyclic frequency imposed by the spiral perturbation, which governs the amount of material that can be collected during a spiral arm passage. A stronger spiral potential produces more massive and widely spaced clouds. The stochasticity of cloud-cloud collisions naturally produces a power-law GMC mass function (*Field and Saslaw*, 1965; *Penston et al.*, 1969; *Taff and Savedoff*, 1973; *Handbury et al.*, 1977; *Kwan*, 1979; *Hausman*, 1982; *Tomisaka*, 1984), and the power-law indices produced in modern hydrodynamic simulations agree well with observations (*Dobbs, 2008; Dobbs et al., 2011a; Tasker and Tan, 2009*), provided the simulations also include a subgrid feedback recipe strong enough to prevent runaway cloud growth (*Dobbs et al., 2011a; Hopkins et al., 2011*). A third signature of either local converging flows or agglomeration is that they can produce clouds that are counterrotating compared to the galactic disk (*Dobbs, 2008; Dobbs et al., 2011a; Tasker and Tan, 2009*), consistent with observations showing that retrograde clouds are as common as prograde ones (*Blitz, 1993; Phillips, 1999; Rosolowsky et al., 2003; Imara and Blitz, 2011; Imara et al., 2011*). Clouds formed by agglomeration also need not be gravitationally bound, although, again, stellar feedback is necessary to maintain an unbound population (*Dobbs et al., 2011b*).

3.3. Gravitational Instability

An alternative explanation for massive clouds is that they form in a direct, top-down manner, and one possible mechanism for this to happen is gravitational instability. Axisymmetric perturbations in single-phase infinitesimally thin gas disks with effective sound speed c_{eff}, surface density Σ, and epicyclic frequency κ can occur whenever the Toomre parameter $Q \equiv \kappa c_{eff}/(\pi G \Sigma) < 1$. For the nonaxisymmetric case, however, there are no true linear (local) instabilities because differential rotation ultimately shears any wavelet into a tightly wrapped trailing spiral with wavenumber $k \propto t$ in which pressure stabilization (contributing to the dispersion relation as $k^2 c_{eff}^2$) is stronger than self-gravity (contributing as $-2\pi G \Sigma_{gas}|k|$). Using linear perturbation

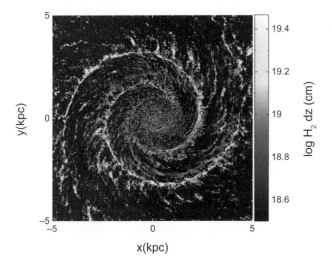

Fig. 5. A snapshot is shown from a simulation of the gas in a spiral galaxy, including a fixed spiral potential, gas heating and cooling, self-gravity, and stellar feedback (from *Dobbs and Pringle, 2013*). The GMCs in this simulation predominantly form by cloud-cloud collisions in the spiral arms. Darker shades of gray show the total gas column density, while the lighter shades show the H_2 fraction integrated through the disk.

theory, *Goldreich and Lynden-Bell* (1965b) were the first to analyze the growth of low-amplitude shearing wavelets in gaseous disks; the analogous "swing amplification" process for stellar disks was studied by *Julian and Toomre* (1966). Spiral-arm regions, because they have higher gaseous surface densities, are more susceptible than interarm regions to self-gravitating fragmentation, and this process has been analyzed using linear theory (e.g., *Elmegreen, 1979; Balbus and Cowie, 1985; Balbus, 1988*). Magnetic effects are particularly important in spiral-arm regions because of the reduced shear compared to interarm regions; this combination leads to a distinct process, the MJI (see section 3.4).

Nonaxisymmetric disturbances have higher Q thresholds for growth (i.e., amplification is possible at lower Σ) than axisymmetric disturbances. Because the instabilities are nonlinear, numerical simulations are required to evaluate these thresholds. *Kim and Ostriker* (2001) found, for local thin-disk simulations and a range of magnetization, that the threshold is at Q = 1.2–1.4. Vertical thickness of the disk dilutes self-gravity, which tends to reduce the critical Q below unity, but magnetic fields and the contribution of stellar gravity provide compensating effects, yielding $Q_{crit} \sim 1.5$ (*Kim et al., 2002, 2003; Kim and Ostriker, 2007; Li et al., 2005*).

In the absence of a preexisting "grand design" stellar spiral pattern, gravitational instability in the combined gas-star disk can lead to flocculent or multi-armed spirals, as modeled numerically by, e.g., *Li et al.* (2005), *Kim and Ostriker* (2007), *Robertson and Kravtsov* (2008), *Tasker and Tan* (2009), *Dobbs et al.* (2011a), *Wada et al.* (2011), and *Hopkins* (2012). In cases where a grand design is present (e.g., when driven by a tidal encounter), gas flowing through the spiral pattern supersonically experiences a shock, which raises the density and can trigger gravitational collapse (*Roberts, 1969*); numerical investigations of this include, e.g., *Kim and Ostriker* (2002, 2006), *Shetty and Ostriker* (2006), and *Dobbs et al.* (2012) (see also section 3.4 for a discussion of magnetic effects).

Clouds formed by gravitational instabilities in single-phase gas disks typically have masses $\sim 10\times$ the two-dimensional Jeans mass $M_{J,2D} = c_{eff}^4/G^2\Sigma \sim 10^8\ M_\odot(c_{eff}/7\ \mathrm{km\ s^{-1}})^4(\Sigma/M_\odot\ \mathrm{pc^{-2}})^{-1}$. The gathering scale for mass is larger than the two-dimensional Jeans length $L_{J,2D} = c_{eff}^2/(G\Sigma)$ in part because the fastest-growing scale exceeds $L_{J,2D}$ even for infinitesimally thin disks, and this increases for thick disks; also, the cloud formation process is highly anisotropic because of shear. At moderate gas surface densities, $\Sigma < 100\ M_\odot\ \mathrm{pc^{-2}}$, as found away from spiral-arm regions, the corresponding masses are larger than those of observed GMCs when $c_{eff} \sim 7\ \mathrm{km\ s^{-1}}$, comparable to large-scale mean velocity dispersions for the atomic medium. The absence of such massive clouds is consistent with observations, indicating that Q values are generally above the critical threshold (except possibly in high redshift systems), such that spiral-arm regions at high Σ and/or processes that reduce c_{eff} locally are required to form observed GMCs via self-gravitating instability. While many analyses and simulations assume a single-phase medium, the diffuse ISM from which GMCs form is in fact a multi-phase medium, with cold clouds surrounded by a warmer intercloud medium. The primary contribution to the effective velocity dispersion of the cold medium is turbulence. This turbulence can dissipate due to cloud-cloud collisions, as well as large-scale flows in the horizontal direction, and flows toward the mid-plane from high latitude. Turbulent dissipation reduces the effective pressure support, allowing instability at lower Σ (*Elmegreen*, 2011). In addition, the local reduction in c_{eff} enables gravitational instability to form lower-mass clouds. Simulations that include a multi-phase medium and/or feedback from star formation (which drives turbulence and also breaks up massive GMCs) find a broad spectrum of cloud masses, extending up to several $\times 10^6$ $M_\odot$ (e.g., *Wada et al.*, 2000; *Shetty and Ostriker*, 2008; *Tasker and Tan*, 2009; *Dobbs et al.*, 2011a; *Hopkins*, 2012).

3.4. Magneto-Jeans Instability

In magnetized gas disks, another process, now known as MJI, can occur. This was first investigated using linear perturbation theory for disks with solid-body rotation by *Lynden-Bell* (1966), and subsequently by *Elmegreen* (1987), *Gammie* (1996), and *Kim and Ostriker* (2001) for general rotation curves. Magneto-Jeans instability occurs in low-shear disks at any nonzero magnetization, whereas swing amplification (and its magnetized variants) described in section 3.3 requires large shear. In MJI, magnetic tension counteracts the Coriolis forces that can otherwise suppress gravitational instability in rotating systems, and low shear is required so that self-gravity takes over before shear increases k to the point of stabilization. *Kim and Ostriker* (2001) and *Kim et al.* (2002) have simulated this process in uniform, low-shear regions as might be found in inner galaxies.

Because low shear is needed for MJI, it is likely to be most important either in central regions of galaxies or in spiral arms (*Elmegreen*, 1994), where compression by a factor of Σ/Σ_0 reduces the local shear as d ln Ω/d ln R ~ Σ/Σ_0–2. Given the complex dynamics of spiral-arm regions, this is best studied with MHD simulations (*Kim and Ostriker*, 2002, 2006; *Shetty and Ostriker*, 2006), which show that MJI can produce massive, self-gravitating clouds either within spiral arms or downstream, depending on the strength of the spiral shock. Clouds that collapse downstream are found within overdense spurs. The spiral arms maintain their integrity better in magnetized models that in comparison unmagnetized simulations. Spacings of spiral-arm spurs formed by MJI are several times the Jeans length $c_{eff}^2/(G\Sigma_{arm})$ at the mean arm gas surface density Σ_{arm}, consistent with observations of spurs and giant HII regions that form "beads on a string" in grand design spirals (*Elmegreen and Elmegreen*, 1983; *La Vigne et al.*, 2006). The masses of gas clumps formed by MJI are typically ~10^7 $M_\odot$ in simulations with c_{eff} = 7 km s^{-1}, comparable to the most massive observed giant molecular cloud complexes and consistent with expectations from linear theory. Simulations of MJI in multi-phase disks with feedback-driven turbulence have not yet been conducted, so there are no predictions for the full cloud spectrum.

3.5. Parker Instability

A final possible top-down formation mechanism is Parker instability, in which horizontal magnetic fields threading the ISM buckle due to differential buoyancy in the gravitational field, producing dense gas accumulations at the mid-plane. Because the most unstable modes have wavelengths ~$(1–2)2\pi H$, where H is the disk thickness, gas could in principle be gathered from scales of several hundred parsecs to produce quite massive clouds (*Mouschovias*, 1974; *Mouschovias et al.*, 1974; *Blitz and Shu*, 1980). However, Parker instabilities are self-limiting (e.g., *Matsumoto et al.*, 1988), and in single-phase media saturate at a factor of a few density enhancements at the mid-plane (e.g., *Basu et al.*, 1997; *Santillán et al.*, 2000; *Machida et al.*, 2009). Simulations have also demonstrated that three-dimensional dynamics, which enhance reconnection and vertical magnetic flux redistribution, result in end states with relatively uniform horizontal density distributions (e.g., *Kim et al.*, 1998, 2003). Some models suggest that spiral-arm regions may be favorable for undular modes to dominate over interchange ones (*Franco et al.*, 2002), but simulations including self-consistent flow through the spiral arm indicate that vertical gradients in horizontal velocity (a consequence of vertically curved spiral shocks) limit the development of the undular Parker mode (*Kim and Ostriker*, 2006). While Parker instability has important consequences for vertical redistribution of magnetic flux, and there is strong evidence for the formation of magnetic loops anchored by GMCs in regions of high magnetic field strength such as the Galactic center (*Fukui et al.*, 2006; *Torii et al.*, 2010), it is less clear if Parker instability can create massive, highly overdense clouds.

For a medium subject to TI, cooling of overdense gas in magnetic valleys can strongly enhance the density contrast of structures that grow by Parker instability (*Kosiński and Hanasz*, 2007; *Mouschovias et al.*, 2009). However, simulations to date have not considered the more realistic case of a preexisting cloud/intercloud medium in which turbulence is also present. For a nonturbulent medium, cold clouds could easily slide into magnetic valleys, but large turbulent velocities may prevent this. In addition, whatever mechanisms drive turbulence may disrupt the coherent development of large-scale Parker modes. An important task for future modeling of the multi-phase, turbulent ISM is to determine whether Parker instability primarily redistributes the magnetic field and alters the distribution of low-density coronal gas, or if it also plays a major role in creating massive, bound GMCs near the mid-plane.

3.6. Conversion of Atomic Hydrogen to Molecular Hydrogen, and Carbon$^+$ to Carbon Monoxide

Thus far our discussion of GMC formation has focused on the mechanisms for accumulating high-density gas. However, the actual observable that defines a GMC is usually CO emission, or in some limited cases other tracers of

H_2 (e.g., *Bolatto et al.,* 2011). Thus we must also consider the chemical transition from atomic gas, where the hydrogen is mostly HI and carbon is mostly C^+, to molecular gas characterized by H_2 and CO. The region over which this transition occurs is called a photodissociation or photon-dominated region (PDR). Molecular hydrogen molecules form (primarily) on the surfaces of dust grains, and are destroyed by resonant absorption of Lyman- and Werner-band photons. The equilibrium H_2 abundance is controlled by the ratio of the far ultraviolet (FUV) radiation field to the gas density, so that H_2 becomes dominant in regions where the gas is dense and the FUV is attenuated (*van Dishoeck and Black,* 1986; *Black and van Dishoeck,* 1987; *Sternberg,* 1988; *Krumholz et al.,* 2008, 2009a; *Wolfire et al.,* 2010). Once H_2 is abundant, it can serve as the seed for fast ion-neutral reactions that ultimately culminate in the formation of CO. Nonetheless, CO is also subject to photodissociation, and it requires even more shielding than H_2 before it becomes the dominant C repository (*van Dishoeck and Black,* 1988).

The need for high extinction to form H_2 and CO has two important consequences. One, already alluded to in section 2, is that the transition between HI and H_2 is shifted to higher surface densities in galaxies with low metallicity and dust abundance and thus low extinction per unit mass (*Fumagalli et al.,* 2010; *Bolatto et al.,* 2011; *Wong et al.,* 2013). A second is that the conversion factor X_{CO} between CO emission and H_2 mass increases significantly in low-metallicity galaxies (*Bolatto et al.,* 2013, and references therein), because the shift in column density is much greater for CO than for H_2. Carbon monoxide and H_2 behave differently because self-shielding against dissociation is much stronger for H_2 than for CO, which must rely on dust shielding (*Wolfire et al.,* 2010). As a result, low-metallicity galaxies show significant CO emission only from high-extinction peaks of the H_2 distribution, rather than from the bulk of the molecular material.

While the chemistry of GMC formation is unquestionably important for understanding and interpreting observations, the connection between chemistry and dynamics is considerably less clear. In part, this is due to numerical limitations. At the resolutions achievable in galactic-scale simulations, one can model H_2 formation only via subgrid models that are either purely analytic (e.g., *Krumholz et al.,* 2008, 2009a,b; *McKee and Krumholz,* 2010; *Narayanan et al.,* 2011, 2012; *Kuhlen et al.,* 2012; *Jaacks et al.,* 2013), or that solve the chemical rate equations with added "clumping factors" that are tuned to reproduce observations (e.g., *Robertson and Kravtsov,* 2008; *Gnedin et al.,* 2009; *Pelupessy and Papadopoulos,* 2009; *Christensen et al.,* 2012). Fully-time-dependent chemodynamical simulations that do not rely on such methods are restricted to either simple low-dimensional geometries (*Koyama and Inutsuka,* 2000; *Bergin et al.,* 2004) or to regions smaller than typical GMCs (*Glover and Mac Low,* 2007a,b, 2011; *Glover et al.,* 2010; *Clark et al.,* 2012; *Inoue and Inutsuka,* 2012).

One active area of research is whether the HI to H_2 or

C^+ to CO transitions are either necessary or sufficient for star formation. For CO the answer appears to be "neither." The nearly fixed ratio of CO surface brightness to SFR we measure in solar metallicity regions (see section 2.1) drops dramatically in low-metallicity ones (*Gardan et al.,* 2007; *Wyder et al.,* 2009; *Bolatto et al.,* 2011, 2013), strongly suggesting that the loss of metals is changing the carbon chemistry but not the way stars form. Numerical simulations suggest that, at solar metallicity, CO formation is so rapid that even a cloud undergoing free-fall collapse will be CO-emitting by the time there is substantial star formation (*Hartmann et al.,* 2001; *Heitsch and Hartmann,* 2008). Conversely, CO can form in shocks even if the gas is not self-gravitating (*Dobbs et al.,* 2008; *Inoue and Inutsuka,* 2012). Moreover, formation of CO does not strongly affect the temperature of molecular clouds, so it does not contribute to the loss of thermal support and onset of collapse (*Krumholz et al.,* 2011; *Glover and Clark,* 2012a). Thus it appears that CO accompanies star formation, but is not causally related to it.

The situation for H_2 is much less clear. Unlike CO, the correlation between star formation-H_2 correlation appears to be metallicity independent, and is always stronger than the star formation-total gas correlation. Reducing the metallicity of a galaxy at fixed gas surface density lower both the H_2 abundance and SFR, and by nearly the same factor (*Wolfe and Chen,* 2006; *Rafelski et al.,* 2011; *Bolatto et al.,* 2011). This still does not prove causality, however. *Krumholz et al.* (2011) and *Glover and Clark* (2012a) suggest that the explanation is that both H_2 formation and star formation are triggered by shielding effects. Only at large extinction is the photodissociation rate suppressed enough to allow H_2 to become dominant, but the same photons that dissociate H_2 also heat the gas via the grain photoelectric effect, and gas only gets cold enough to form stars where the photoelectric effect is suppressed. This is, however, only one possible explanation of the data.

A final question is whether the chemical conditions in GMCs are in equilibrium or non-equilibrium. The gas-phase ion-neutral reactions that lead to CO formation have high rates even at low temperatures, so carbon chemistry is likely to be in equilibrium (*Glover et al.,* 2010; *Glover and Mac Low,* 2011; *Glover and Clark,* 2012b). The situation for H_2 is less clear. The rate coefficient for H_2 formation on dust grain surfaces is quite low, so whether gas can reach equilibrium depends on the density structure and the time available. Averaged over ~100-pc size scales and at metallicities above ~1% of solar, *Krumholz and Gnedin* (2011) find that equilibrium models of *Krumholz et al.* (2009a) agree very well with time-dependent ones by *Gnedin et al.* (2009), and the observed metallicity-dependence of the HI-H_2 transition in external galaxies is also consistent with the predictions of the equilibrium models (*Fumagalli et al.,* 2010; *Bolatto et al.,* 2011; *Wong et al.,* 2013). However, *Mac Low and Glover* (2012) find that equilibrium models do not reproduce their simulations on ~1–10-pc scales. Nonetheless, observations of solar neighborhood clouds on such scales appear to be consistent with equlibrium (*Lee et al.,* 2012). All models agree, however, that non-equilibrium effects must become dominant at metallicities below ~1–10% of solar, due to the reduction in the rate coefficient for H_2 formation that accompanies the loss of dust (*Krumholz,* 2012; *Glover and Clark,* 2012b).

4. STRUCTURE, EVOLUTION, AND DESTRUCTION

Now that we have sketched out how GMCs come into existence, we consider the processes that drive their internal structure, evolution, and eventual dispersal.

4.1. Giant Molecular Cloud Internal Structure

Giant molecular clouds are characterized by a very clumpy, filamentary structure (see the chapters by André et al. and Molinari et al. in this volume), which can be produced by a wide range of processes, including gravitational collapse (e.g., *Larson,* 1985; *Nagai et al.,* 1998; *Curry,* 2000; *Burkert and Hartmann,* 2004), non-self-gravitating supersonic turbulence (*Padoan et al.,* 2001), and colliding flows plus thermal instability (*Audit and Hennebelle,* 2005, *Vázquez-Semadeni et al.,* 2006); all these processes can be magnetized or unmagnetized. Some recent observational studies argue that the morphology is consistent with multi-scale infall (e.g., *Galván-Madrid et al.,* 2009; *Schneider et al.,* 2010; *Kirk et al.,* 2013), but strong conclusions will require quantitative comparison with a wide range of simulations.

One promising approach to such comparisons is to develop statistical measures that can be applied to both simulations and observations, either two-dimensional column density maps or three-dimensional position-position-velocity cubes. *Padoan et al.* (2004a,b) provide one example. They compare column density PDFs produced in simulations of sub-Alfvénic and super-Alfvénic turbulence, and argue that observed PDFs are better fit by the super-Alfvénic model. This is in some disagreement with observations showing that magnetic fields remain well-ordered over a wide range of length scales [see the recent review by *Crutcher* (2012) as well as the chapter by H.-B. Li et al. in this volume]. The need for super-Alfvénic turbulence in the simulations may arise from the fact that they did not include self-gravity, thus requiring stronger turbulence to match the observed level of structure (*Vázquez-Semadeni et al.,* 2008). Nevertheless, the *Padoan et al.* (2004a,b) results probably do show that magnetic fields cannot be strong enough to render GMCs sub-Alfvénic.

A second example comes from *Brunt* (2010), *Brunt et al.* (2010 a,b), and *Price et al.* (2011), who use the statistics of the observed two-dimensional column-density PDF to infer the underlying three-dimensional volume-density PDF, and in turn use this to constrain the relationship between density variance and Mach number in nearby molecular clouds. They conclude from this analysis that a significant fraction of the energy injection that produces turbulence must be in compressive rather than solenoidal modes. Various authors (*Kainulainen et al.,* 2009a; *Kritsuk et al.,* 2011; *Ballesteros-Paredes et al.,* 2011a; *Federrath and Klessen,* 2013) also argue that the statistics of the density field are also highly sensitive to the amount of star formation that has taken place in a cloud, and can therefore be used as a measure of evolutionary state.

4.2. Origin of Nonthermal Motions

As discussed in section 2.3, GMCs contain strong nonthermal motions, with the bulk of the energy in modes with size scales comparable to the cloud as a whole. For a typical GMC density of ~100 cm^{-3}, temperature of ~10 K, and bulk velocity of ~1 km s^{-1}, the viscous dissipation scale is ~10^{12} cm (~0.1 AU), implying that the Reynolds number of these motions is ~10^9. Such a high value of the Reynolds number essentially guarantees that the flow will be turbulent. Moreover, since the bulk velocity greatly exceeds the sound speed, the turbulence must be supersonic, although not necessarily super-Alfvénic. *Zuckerman and Evans* (1974) proposed that this turbulence would be confined to small scales, but modern simulations of supersonic turbulence indicate that the power is mostly on large scales. It is also possible that the linewidths contain a significant contribution from coherent infall, as we discuss below.

While turbulence is expected, the deeper question is why the linewidths are so large in the first place. Simulations conducted over the last ~15 years have generally demonstrated that, in the absence of external energy input, turbulence decays in ~1 crossing time of the outer scale of the turbulent flow (*Mac Low et al.,* 1998; *Mac Low,* 1999; *Stone et al.,* 1998; *Padoan and Nordlund,* 1999), except in the case of imbalanced MHD turbulence (*Cho et al.,* 2002). Thus the large linewidths observed in GMCs would not in general be maintained for long periods in the absence of some external input. This problem has given rise to a number of proposed solutions, which can be broadly divided into three categories: global collapse, externally driven turbulence, and internally driven turbulence.

4.2.1. Global collapse scenario. The global collapse scenario, first proposed by *Goldreich and Kwan* (1974) and *Liszt et al.* (1974), and more recently revived by *Vázquez-Semadeni et al.* (2007, 2009), *Heitsch and Hartmann* (2008), *Heitsch et al.* (2008, 2009), *Ballesteros-Paredes et al.* (2011a,b), and *Hartmann et al.* (2012) as a nonlinear version of the hierarchical fragmentation scenario proposed by *Hoyle* (1953), offers perhaps the simplest solution: The linewidths are dominated by global gravitational collapse rather than random turbulence. This both provides a natural energy source (gravity) and removes the need to explain why the linewidths do not decay, because in this scenario GMCs, filaments, and clumps are not objects that need to be supported, but rather constitute a hierarchy of stages in a global, highly inhomogeneous collapse flow, with each stage accreting from its parent (*Vázquez-Semadeni et al.,* 2009).

Investigations of this scenario generally begin by considering an idealized head-on collision between two single-phase, warm, diffuse gas streams, which might be caused by either local feedback or large-scale gravitational instability (cf. section 3). (Simulations of more realistic glancing collisions between streams already containing dense clumps have yet to be performed.) The large-scale compression triggers the formation of a cold cloud, which quickly acquires many Jeans masses because the warm-cold phase transition causes a factor of ~100 increase in density and a decrease by the same factor in temperature. Thus the Jeans mass, $M_J \propto \rho^{-1/2}T^{3/2}$, decreases by a factor ~10^4 (e.g., *Vázquez-Semadeni,* 2012). The cloud therefore readily fragments into clumps (*Heitsch et al.,* 2005; *Vázquez-Semadeni et al.,* 2006), and the ensemble of clumps becomes gravitationally unstable and begins an essentially pressure-free collapse. It contracts first along its shortest dimension (*Lin et al.,* 1965), producing sheets and then filaments (*Burkert and Hartmann,* 2004; *Hartmann and Burkert,* 2007; *Vázquez-Semadeni et al.,* 2007, 2010, 2011; *Heitsch and Hartmann,* 2008; *Heitsch et al.,* 2009). Although initially the motions in the clouds are random and turbulent, they become ever-more infall-dominated as the collapse proceeds. However, because these motions have a gravitational origin, they naturally appear virialized (*Vázquez-Semadeni et al.,* 2007; *Ballesteros-Paredes et al.,* 2011b). Accretion flows consistent with the scenario have been reported in several observational studies of dense molecular gas (e.g., *Galván-Madrid et al.,* 2009; *Schneider et al.,* 2010; *Kirk et al.,* 2013), but observations have yet to detect the predicted inflows at the early, large-scale stages of the hierarchy. These are difficult to detect because it is not easy to separate the atomic medium directly connected to molecular clouds from the general HI in the galaxy, and because the GMCs are highly fragmented, blurring the inverse p-Cygni profiles expected for infall. In fact, *Heitsch et al.* (2009) show that the CO line profiles of chaotically collapsing clouds match observations of GMCs.

While this idea elegantly resolves the problem of the large linewidths, it faces challenges with respect to the constraints imposed by the observed 20–30-m.y. lifetimes of GMCs (section 2.5) and the low rates and efficiencies of star formation (section 2.6). We defer the question of SFRs and efficiencies to section 5. Concerning GMC lifetimes, a semi-analytic model by *Zamora-Avilés et al.* (2012) for the evolution of the cloud mass and SFR in this scenario shows agreement with the observations within factors of a few. Slightly smaller timescales (~10 m.y.) are observed in numerical simulations of cloud buildup that consider the evolution of the molecular content of the cloud (*Heitsch and Hartmann,* 2008), although these authors considered substantially smaller cloud masses and more dense flows. Simulations considering larger cloud masses (several 10^4 M$_\odot$) exhibit evolutionary timescales ~20 m.y. (*Colín et al.,* 2013), albeit no tracking of the molecular fraction was performed there. Simulations based on the global collapse scenario have not yet examined the formation and

evolution of clouds of masses above 10^5 M$_\odot$, comparable to those studied in extragalactic observations. Moreover, in all these models the lifetime depends critically on the duration and properties of the gas inflow that forms the clouds, which is imposed as a boundary condition. Self-consistent simulations in which the required inflows are generated by galactic-scale flows also remain to be performed.

4.2.2. External driving scenario. The alternative possibility is that the large linewidths of GMCs are dominated by random motions rather than global collapse. This would naturally explain relatively long GMC lifetimes and (as we discuss in section 5) low SFRs, but in turn raises the problem of why these motions do not decay, giving rise to a global collapse. The external driving scenario proposes that this decay is offset by the injection of energy by flows external to the GMC. One obvious source of such external energy is the accretion flow from which the cloud itself forms, which can be subject to nonlinear thin-shell instability (*Vishniac,* 1994) or oscillatory overstability (*Chevalier and Imamura,* 1982) that will drive turbulence. *Klessen and Hennebelle* (2010) point out that only a small fraction of the gravitational potential energy of material accreting onto a GMC would need to be converted to bulk motion before it is dissipated in shocks in order to explain the observed linewidths of GMCs, and semi-analytic models by *Goldbaum et al.* (2011) confirm this conclusion. Numerical simulations confirm that cold dense layers confined by the ram pressure of accretion flows indeed are often turbulent (*Hunter et al.,* 1986; *Stevens et al.,* 1992; *Walder and Folini,* 2000; *Koyama and Inutsuka,* 2002; *Audit and Hennebelle,* 2005; *Heitsch et al.,* 2005; *Vázquez-Semadeni et al.,* 2006), although numerical simulations consistently show that the velocity dispersions of these flows are significantly smaller than those observed in GMCs unless the flows are self-gravitating (*Koyama and Inutsuka,* 2002; *Heitsch et al.,* 2005; *Vázquez-Semadeni et al.,* 2007, 2010). This can be understood because the condition of simultaneous thermal and ram pressure balance implies that the Mach numbers in both the warm and cold phases are comparable (*Banerjee et al.,* 2009).

While accretion flows are one possible source of energy, there are also others. Galactic-scale and kiloparsec-scale simulations by *Tasker and Tan* (2009), *Tasker* (2011), *Dobbs et al.* (2011a,b, 2012), *Dobbs and Pringle* (2013), *Hopkins et al.* (2012), and *Van Loo et al.* (2013) all show that GMCs are embedded in large-scale galactic flows that subject them to continuous external buffeting — cloud-cloud collisions, encounters with shear near spiral arms, etc. — even when the cloud's mass is not necessarily growing. These external motions are particularly important for the most massive clouds, which preferentialy form via large-scale galactic flows, and can drive turbulence in them over a time significantly longer than τ_{ff}. This mechanism seems particularly likely to operate in high-surface density galaxies where the entire ISM is molecular and thus there is no real distinction between GMCs and other gas, and in fact seems to be required to explain the large velocity dispersions observed in high-redshift galaxies (*Krumholz and Burkert,* 2010).

4.2.3. Internal driving scenario. The internally driven scenario proposes that stellar feedback internal to a molecular cloud is responsible for driving turbulence, explaining the large linewidths seen, in conjunction with externally driven turbulence in the very rare clouds without significant star formation [e.g., the so-called Maddalena's cloud (*Williams et al.,* 1994)]. There are a number of possible sources of turbulent driving, including H II regions, radiation pressure, protostellar outflows, and the winds of main-sequence stars. (Supernovae are unlikely to be important as an internal driver of turbulence in GMCs in most galaxies because the stellar evolution timescale is comparable to the crossing timescale, although they could be important as an external driver, for the dispersal of GMCs, and in starburst galaxies; see below.) *Matzner* (2002) and *Dekel and Krumholz* (2013) provide useful summaries of the momentum budgets associated with each of these mechanisms, and they are discussed in much more detail in the chapter by Krumholz et al. in this volume.

H II regions are one possible turbulent driver. *Krumholz et al.* (2006) and *Goldbaum et al.* (2011) conclude from their semi-analytic models that H II regions provide a power source sufficient to offset the decay of turbulence, that most of this power is distributed into size scales comparable to the size of the cloud, and that feedback power is comparable to accretion power. The results from simulations are less clear. *Gritschneder et al.* (2009) and *Walch et al.* (2012) find that H II regions in their simulations drive turbulence at velocity dispersions comparable to observed values, while *Dale et al.* (2005, 2012, 2013) and *Colín et al.* (2013) find that H II regions rapidly disrupt GMCs with masses up to $\sim 10^5 \, M_\odot$ within less than 10 m.y. [consistent with observations showing that >10-m.y.-old star clusters are usually gas-free (*Leisawitz et al.,* 1989; *Mayya et al.,* 2012)], but do not drive turbulence. The origin of the difference is not clear, as the simulations differ in several ways, including the geometry they assume, the size scales they consider, and the way that they set up the initial conditions.

Nevertheless, in GMCs where the escape speed approaches the 10 km s^{-1} sound speed in photoionized gas, H II regions can no longer drive turbulence nor disrupt the clouds, and some authors have proposed that radiation pressure might take over (*Thompson et al.,* 2005; *Krumholz and Matzner,* 2009; *Fall et al.,* 2010; *Murray et al.,* 2010; *Hopkins et al.,* 2011, 2012). Simulations on this point are far more limited, and the only ones published so far that actually include radiative transfer (as opposed to a subgrid model for radiation pressure feedback) are those of *Krumholz and Thompson* (2012, 2013), who conclude that radiation pressure is unlikely to be important on the scales of GMCs. Figure 6 shows a result from one of these simulations.

Stellar winds and outflows can also drive turbulence. Winds from hot main-sequence stars have been studied by *Dale and Bonnell* (2008) and *Rogers and Pittard* (2013), who find that they tend to ablate the clouds but not drive

significant turbulence. On the other hand, studies of protostellar outflows find that on small scales they can maintain a constant velocity dispersion while simultaneously keeping the SFR low (*Cunningham et al.,* 2006; *Li and Nakamura,* 2006; *Matzner,* 2007; *Nakamura and Li,* 2007; *Wang et al.,* 2010). On the other hand, these studies also indicate that outflows cannot be the dominant driver of turbulence on ~ 10–100-pc scales in GMCs, both because they lack sufficient power, and because they tend to produce a turbulent power spectrum with a distinct bump at ~ 1-pc scales, in contrast to the pure power law usually observed.

Whether any of these mechanisms can be the dominant source of the large linewidths seen in GMCs remains unsettled. One important caveat is that only a few of the simulations with feedback have included magnetic fields, and *Wang et al.* (2010) and *Gendelev and Krumholz* (2012) show (for protostellar outflows and H II regions, respectively) that magnetic fields can dramatically increase the ability of internal mechanisms to drive turbulence, because they provide an effective means of transmitting momentum into otherwise difficult-to-reach portions of clouds.

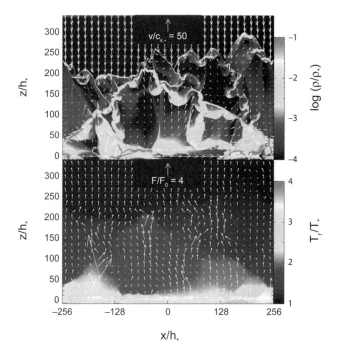

Fig. 6. Time slice from a radiation-hydrodyanmic simulation of a molecular cloud with a strong radiation flux passing through it. In the top panel, different shades show density and arrows show velocity; in the bottom panel, different shades show temperature and arrows show radiation flux. Density and temperature are both normalized to characteristic values at the cloud edge; velocity is normalized to the sound speed at the cloud edge, and flux to the injected radiation flux. The simulation demonstrates that radiation pressure-driven turbulence is possible, but also that the required radiation flux and matter column density are vastly in excess of the values found in real GMCs. Figure taken from *Krumholz and Thompson* (2012).

Magneto-hydrodynamic simulations of feedback in GMCs are clearly needed.

4.3. Mass Loss and Disruption

As discussed in section 2.5, GMCs are disrupted long before they can turn a significant fraction of their mass into stars. The question of what mechanism or mechanisms are responsible for this is closely tied to the question of the origin of GMC turbulence discussed in the previous section, as each proposed answer to that question imposes certain requirements on how GMCs must disrupt. In the global collapse scenario, disruption must occur in less than the mean-density free-fall time to avoid excessive star formation. Recent results suggest a somewhat slower collapse in flattened or filamentary objects (*Toalá et al.,* 2012; *Pon et al.,* 2012), but disruption must still be fast. In the externally driven or internally driven turbulence scenarios disruption can be slower, but must still occur before the bulk of the material can be converted to stars. Radiation pressure and protostellar outflows appear unlikely to be responsible for the same reasons (discussed in the previous section) that they cannot drive GMC-scale turbulence. For main-sequence winds, *Dale and Bonnell* (2008) and *Rogers and Pittard* (2013) find that they can expel mass from small, dense molecular clumps, but it is unclear if the same is true of the much larger masses and lower density scales that characterize GMCs.

The remaining stellar feedback mechanisms, photo-ionization and supernovae, are more promising. Analytic models have long predicted that photoionization should be the primary mechanism for ablating mass from GMCs (e.g., *Field,* 1970; *Whitworth,* 1979; *Cox,* 1983; *Williams and McKee,* 1997; *Matzner,* 2002; *Krumholz et al.,* 2006; *Goldbaum et al.,* 2011), and numerical simulations by *Dale et al.* (2012, 2013) and *Colín et al.* (2013) confirm that photoionization is able to disrupt GMCs with masses of $\sim 10^5$ $M_\odot$ or less. More massive clouds, however, may have escape speeds too high for photoionization to disrupt them unless they suffer significant ablation first.

Supernovae are potentially effective in clouds of all masses, but are in need of further study. Of cataloged Galactic supernova remnants, 8% (and 25% of X-ray emitting remnants) are classified as "mixed morphology," believed to indicate interaction between the remnant and dense molecular gas (*Rho and Petre,* 1998). This suggests that a nonnegligible fraction of GMCs may interact with supernovae. Because GMCs are clumpy, this interaction will differ from the standard solutions in a uniform medium (e.g., *Cioffi et al.,* 1988; *Blondin et al.,* 1998), but theoretical studies of supernova remnants in molecular gas have thus far focused mainly on emission properties (e.g., *Chevalier,* 1999; *Chevalier and Fransson,* 2001; *Tilley et al.,* 2006) rather than the dynamical consequences for GMC evolution. Although some preliminary work (*Kovalenko and Korolev,* 2012) suggests that an outer shell will still form, with internal clumps accelerated outward when they are overrun by the expanding shock front (cf. *Klein et al.,* 1994; *Mac Low et al.,* 1994), complete simulations of supernova remnant expansion within realistic GMC environments are lacking. Obtaining a quantitative assessment of the kinetic energy and momentum imparted to the dense gas will be crucial for understanding GMC destruction.

5. REGULATION OF STAR FORMATION IN GIANT MOLECULAR CLOUDS

Our discussion thus far provides the framework to address the final topic of this review: What are the dominant interstellar processes that regulate the rate of star formation at GMC and galactic scales? The accumulation of GMCs is the first step in star formation, and large-scale, top-down processes appear to determine a cloud's starting mean density, mass-to-magnetic-flux ratio, Mach number, and boundedness. But are these initial conditions retained for times longer than a cloud dynamical time, and do they affect the formation of stars within the cloud? If so, how stars form is ultimately determined by the large-scale dynamics of the host galaxy. Alternatively, if the initial state of GMCs is quickly erased by internally driven turbulence or external perturbations, then the regulatory agent of star formation lies instead on small scales within individual GMCs. In this section, we review the proposed schemes and key GMC properties that regulate the production of stars.

5.1. Regulation Mechanisms

Star formation occurs at a much lower pace than its theoretical possible free-fall maximum (see section 2.6). Explaining why this is so is a key goal of star formation theories. These theories are intimately related to the assumptions made about the evolutionary path of GMCs. Two theoretical limits for cloud evolution are a state of global collapse with a duration $\sim \tau_{\rm ff}$ and a quasisteady state in which clouds are supported for times $\gg \tau_{\rm ff}$.

In the global collapse limit, one achieves low SFRs by having a low net star formation efficiency ε_* over the lifetime $\sim \tau_{\rm ff}$ of any given GMC, and then disrupting the GMC via feedback. The mechanisms invoked to accomplish this are the same as those invoked in section 4.2.3 to drive internal turbulence: photoionization and supernovae. Some simulations suggest this these mechanisms can indeed enforce low ε_*: *Vázquez-Semadeni et al.* (2010) and *Colín et al.* (2013), using a subgrid model for ionizing feedback, find that $\varepsilon_* \lesssim 10\%$ for clouds up to $\sim 10^5$ $M_\odot$, and *Zamora-Avilés et al.* (2012) find that the evolutionary timescales produced by this mechanism of cloud disruption are consistent with those inferred in the LMC (section 2.5). On the other hand, it remains unclear what mechanisms might be able to disrupt $\sim 10^6$ $M_\odot$ clouds.

If clouds are supported against large-scale collapse, then star formation consists of a small fraction of the mass "percolating" through this support to collapse and form stars. Two major forms of support have been considered:

magnetic (e.g., *Shu et al.,* 1987; *Mouschovias,* 1991a,b) and turbulent (e.g., *Mac Low and Klessen,* 2004; *Ballesteros-Paredes et al.,* 2007). While dominant for over two decades, the magnetic support theories, in which the percolation was allowed by ambipolar diffusion, are now less favored (although see *Mouschovias et al.,* 2009; *Mouschovias and Tassis,* 2010) due to growing observational evidence that molecular clouds are magnetically supercritical (section 2.4). We do not discuss these models further.

In the turbulent support scenario, supersonic turbulent motions behave as a source of pressure with respect to structures whose size scales are larger than the largest scales of the turbulent motions (the "energy-containing scale" of the turbulence), while inducing local compressions at scales much smaller than that. A simple analytic argument suggests that, regardless of whether turbulence is internally or externally driven, its net effect is to increase the effective Jeans mass as $M_{J,turb} \propto v_{rms}^2$, where v_{rms} is the rms turbulent velocity (*Mac Low and Klessen,* 2004). Early numerical simulations of driven turbulence in isothermal clouds (*Klessen et al.,* 2000; *Vázquez-Semadeni et al.,* 2003) indeed show that, holding all other quantities fixed, raising the Mach number of the flow decreases the dimensionless SFR ε_{ff}. However, this is true only as long as the turbulence is maintained; if it is allowed to decay, then raising the Mach number actually raises ε_{ff}, because in this case the turbulence simply accelerates the formation of dense regions and then dissipates (*Nakamura and Li,* 2005). Magnetic fields, even those not strong enough to render the gas subcritical, also decrease ε_{ff} (*Heitsch et al.,* 2001; *Vázquez-Semadeni et al.,* 2005; *Padoan and Nordlund,* 2011; *Federrath and Klessen,* 2012).

To calculate the SFR in this scenario, one can idealize the turbulence level, mean cloud density, and SFR as quasistationary, and then attempt to compute ε_{ff}. In recent years, a number of analytic models have been developed to do so (*Krumholz and McKee,* 2005; *Padoan and Nordlund,* 2011; *Hennebelle and Chabrier,* 2011) [see *Federrath and Klessen* (2012) for a useful compilation, and for generalizations of several of the models]. These models generally exploit the fact that supersonic isothermal turbulence produces a PDF with a lognormal form (*Vazquez-Semadeni,* 1994), so that there is always a fraction of the mass at high densities. The models then assume that the mass at high densities (above some threshold), M_{hd}, is responsible for the instantaneous SFR, which is given as SFR = M_{hd}/τ, where τ is some characteristic timescale of the collapse at those high densities.

In all these models ε_{ff} is determined by other dimensionless numbers: the rms turbulent Mach number $\mathcal{M}$, the virial ratio α_G, and (when magnetic fields are considered) the magnetic β parameter; the ratio of compressive to solenoidal modes in the turbulence is a fourth possible parameter (*Federrath et al.,* 2008; *Federrath and Klessen,* 2012). The models differ in their choices of density threshold and timescale (see the chapter by Padoan et al. in this volume), leading to variations in the predicted dependence of ε_{ff} on $\mathcal{M}$, α_G, and β. However, all the models produce

$\varepsilon_{ff} \sim 0.01$–$0.1$ for dimensionless values comparable to those observed. *Federrath and Klessen* (2012) and *Padoan et al.* (2012) have conducted large campaigns of numerical simulations where they have systematically varied $\mathcal{M}$, α_G, and β, measured ε_{ff}, and compared to the analytic models. *Padoan et al.* (2012) give their results in terms of the ratio t_{ff}/t_{dyn} rather than α_G, but the two are identical up to a constant factor (*Tan et al.,* 2006). In general they find that ε_{ff} decreases strongly with α_G and increases weakly with $\mathcal{M}$, and that a dynamically significant magnetic field (but not one so strong as to render the gas subcritical) reduces ε_{ff} by a factor of ~3. Simulations produce $\varepsilon_{ff} \sim 0.01$–$0.1$, in general agreement with the range of analytic predictions.

One can also generalize the quasistationary turbulent support models by embedding them in time-dependent models for the evolution of a cloud as a whole. In this approach one computes the instantaneous SFR from a cloud's current state (using one of the turbulent support models or based on some other calibration from simulations), but the total mass, mean density, and other quantities evolve in time, so that the instantaneous SFR does too. In this type of model, a variety of assumptions are necessarily made about the cloud's geometry and about the effect of the stellar feedback. *Krumholz et al.* (2006) and *Goldbaum et al.* (2011) adopt a spherical geometry and compute the evolution from the virial theorem, assuming that feedback can drive turbulence that inhibits collapse. As illustrated in Fig. 7, they find that most clouds undergo oscillations around equilibrium before being destroyed at final SFEs ~ 5–10%. The models match a wide range of observations, including the distributions of column density, linewidth-size relation, and cloud lifetime. In constrast, *Zamora-Avilés et al.* (2012) (see also Fig. 7) adopt a planar geometry [which implies longer free-fall times than in the spherical case (*Toalá et al.,* 2012)] and assume that feedback does not drive turbulence or inhibit contraction. With these models they reproduce the SFRs seen in low- and high-mass clouds and clumps, and the stellar age distributions in nearby clusters. As shown in Fig. 7, the overall evolution is quite different in the two models, with the *Goldbaum et al.* (2011) clouds undergoing multiple oscillations at roughly fixed Σ and M_*/M_{gas}, while the *Zamora-Avilés et al.* (2012) model predicts a much more monotonic evolution. Differentiating between these two pictures will require a better understanding of the extent to which feedback is able to inhibit collapse.

5.2. Connection Between Local and Global Scales

Extragalactic star formation observations at large scales average over regions several times the disk scale height in width, and over many GMCs. As discussed in section 2.1, there is an approximately linear correlation between the surface densities of SFR and molecular gas in regions where $\Sigma_{gas} \lesssim 100$ $M_\odot$ pc^{-2}, likely because observations are simply counting the number of GMCs in a beam. At higher Σ_{gas}, the volume-filling factor of molecular material approaches unity, and the index N of the correlation

$\Sigma_{SFR} \propto \Sigma_{gas}^N$ increases. This can be due to increasing density of molecular gas leading to shorter gravitational collapse and star-formation timescales, or because higher total gas surface density leads to stronger gravitational instability and thus faster star formation. At the low values of Σ_{gas} found in the outer disks of spirals (and in dwarfs), the index N is also greater than unity. This does not necessarily imply that there is a cut-off of Σ_{SFR} at low gas surface densities, although simple models of gravitational instability in isothermal disks can indeed reproduce this result (*Li et al.*, 2005), but instead may indicate that additional parameters beyond just Σ_{gas} control Σ_{SFR}. In outer disks, the ISM is mostly diffuse atomic gas and the radial scale length of Σ_{gas} is quite large [comparable to the size of the optical disk (*Bigiel and Blitz,* 2012)]. The slow fall-off of Σ_{gas} with radial distance implies that the sensitivity of Σ_{SFR} to other parameters will become more evident in these regions. For example, a higher surface density in the old stellar disk appears to raise Σ_{SFR} (*Blitz and Rosolowsky,* 2004, 2006; *Leroy et al.,* 2008), likely because stellar gravity confines

the gas disk, raising the density and lowering the dynamical time. Conversely, Σ_{SFR} is lower in lower-metallicity galaxies (*Bolatto et al.,* 2011), likely because lower dust shielding against UV radiation inhibits the formation of a cold, star-forming phase (*Krumholz et al.,* 2009b).

Feedback must certainly be part of this story. Recent large-scale simulations of disk galaxies have consistently pointed to the need for feedback to prevent runaway collapse and limit SFRs to observed levels (e.g., *Kim et al.,* 2011; *Tasker,* 2011; *Hopkins et al.,* 2011, 2012; *Dobbs et al.,* 2011a; *Shetty and Ostriker,* 2012; *Agertz et al.,* 2013). With feedback parameterizations that yield realistic SFRs, other ISM properties (including turbulence levels and gas fractions in different HI phases) are also realistic [see above and also *Joung et al.* (2009) and *Hill et al.* (2012)]. However, it is still also an open question as to whether feedback is the entire story for the large-scale SFR. In some simulations (e.g., *Ostriker and Shetty,* 2011; *Dobbs et al.,* 2011a; *Hopkins et al.,* 2011, 2012; *Shetty and Ostriker,* 2012; *Agertz et al.,* 2013), the SFR on $\geq$100-pc scales is mainly set by the time required for gas to become gravitationally unstable on large scales and by the parameters that control stellar feedback, and is insensitive to the parameterization of star formation on $\lesssim$parsec scales. In other models the SFR is sensitive to the parameters describing both feedback and small-scale star formation [e.g., ε_{ff} and H_2 chemistry (*Gnedin and Kravtsov,* 2010, 2011; *Kuhlen et al.,* 2012, 2013)].

Part of this disagreement is doubtless due to the fact that current simulations do not have sufficient resolution to include the details of feedback, and in many cases they do not even include the required physical mechanisms (e.g., radiative transfer and ionization chemistry). Instead, they rely on subgrid models for momentum and energy injection by supernovae, radiation, and winds, and the results depend on the details of how these mechanisms are implemented. Resolving the question of whether feedback alone is sufficient to explain the large-scale SFRs of galaxies will require both refinement of the subgrid feedback models using high-resolution simulations, and comparison to observations in a range of environments. In at least some cases, the small-scale simulations have raised significant doubts about popular subgrid models (e.g., *Krumholz and Thompson,* 2012, 2013).

A number of authors have also developed analytic models for large-scale SFRs in galactic disks. *Krumholz et al.* (2009b) propose a model in which the fraction of the ISM in a star-forming molecular phase is determined by the balance between photodissociation and H_2 formation, and the SFR within GMCs is determined by the turbulence-regulated star formation model of *Krumholz and McKee* (2005). This model depends on assumed relations between cloud complexes and the properties of the interstellar medium on large scales, including the assumption that the surface density of cloud complexes is proportional to that of the ISM on kiloparsec scales, and that the mass fraction in the warm atomic ISM is negligible compared to the mass in cold atomic and molecular phases. *Ostriker et al.* (2010)

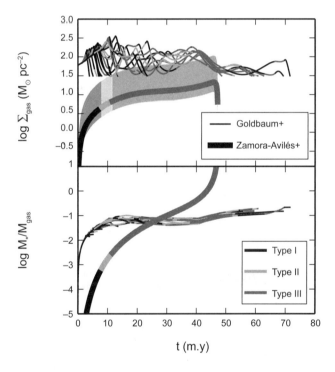

Fig. 7. Predictions for the large-scale evolution of GMCs using the models of *Goldbaum et al.* (2011) (thin lines, each line corresponding to a different realization of a stochastic model) and *Zamora-Avilés et al.* (2012) (thick line). The top panel shows the gas surface density. The minimum in the *Goldbaum et al.* (2011) models is the threshold at which CO dissociates. For the planar *Zamora-Avilés et al.* (2012) model, the thick line is the median and the shaded region is the 10–90% range for random orientation. The bottom panel shows the ratio of instantaneous stellar to gas mass. Type classifications shown are based on *Kawamura et al.* (2009) (see also section 2.5), computed based on the Hα and V-band luminosities of the stellar populations.

and *Ostriker and Shetty* (2011) have developed models in which star formation is self-regulated by feedback. In these models, the equilibrium state is found by simultaneously balancing ISM heating and cooling, turbulent driving and dissipation, and gravitational confinement with pressure support in the diffuse ISM. The SFR adjusts to a value required to maintain this equilibrium state. Numerical simulations by *Kim et al.* (2011) and *Shetty and Ostriker* (2012) show that ISM models including turbulent and radiative heating feedback from star formation indeed reach the expected self-regulated equilibrium states. However, as with other large-scale models, these simulations rely on subgrid feedback recipes whose accuracies have yet to be determined. In all these models, in regions where most of the neutral ISM is in gravitationally bound GMCs, Σ_{SFR} depends on the internal state of the clouds through the ensemble average of $\varepsilon_{ff}/\tau_{ff}$. If GMC internal states are relatively independent of their environments, this would yield values of $\langle\varepsilon_{ff}/\tau_{ff}\rangle$ that do not strongly vary within a galaxy or from one galaxy to another, naturally explaining why $\tau_{dep}(H_2)$ appears to be relatively uniform, ~2 G.y. wherever $\Sigma_{gas} \le 100$ $M_\odot$ pc^{-2}.

Many of the recent advances in understanding large-scale star formation have been based on disk galaxy systems similar to our own Milky Way. Looking to the future, we can hope that the methods being developed to connect individual star-forming GMCs with the larger-scale ISM in local "laboratories" will inform and enable efforts in high-redshift systems, where conditions are more extreme and observational constraints are more challenging.

6. LOOKING FORWARD

6.1. Observations

Systematic surveys and detailed case studies will be enriched by the expansion in millimeter-submillimeter capabilities over the next decade. These will contribute in two main modes: through cloud-scale observations in other galaxies, and in expanding the study of clouds in our own Milky Way. The increased sensitivities of ALMA and the NOrthern Extended Millimeter Array (NOEMA) will sample smaller scales at larger distances, resolving GMC complexes and investigating the physical state and formation mechanisms of GMCs in a variety of extragalactic environments. The smaller interferometers like the Combined Array for Research in Millimeter-wave Astronomy (CARMA) and the Submillimeter Array (SMA) will likely focus on systematic mapping of large areas in the Milky Way or even external galaxies. At centimeter wavelengths, the recently upgraded Jansky Very Large Array (JVLA) brings new powerful capabilities in continuum detection at 7 mm to study cool dust in disks, as well as the study of free-free continuum, molecular emission, and radio recombination lines in our own galaxy and other galaxies. Single-dish millimeter-wave facilities equipped with array receivers and continuum cameras such as the IRAM

30-m, Nobeyama Radio Observatory (NRO) 45-m, Large Millimeter Telescope (LMT) 50-m, and future Cerro Chajnator Atacama Telescope (CCAT) facility will enable fast mapping of large areas in the Milky Way, providing the much-needed large-scale context to high-resolution studies. These facilities, along with the Atacama Pathfinder Experiment (APEX) and NANTEN2 telescopes, will pursue multi-transition/multi-scale surveys sampling star-forming cores and the parent cloud material simultaneously and providing valuable diagnostics of gas kinematics and physical conditions.

There is hope that some of the new observational capabilities will break the theoretical logjam described in this review. At present, it is not possible for observations to distinguish between the very different cloud formation mechanisms proposed in section 3, nor between the mechanisms that might be responsible for controlling cloud density and velocity structure, and cloud disruption (section 4), nor between various models for how star formation is regulated (section 5). There is reason to hope that the new data that will become available in the next few years will start to rule out some of these models.

6.2. Simulations and Theory

The developments in numerical simulations of GMCs since *Protostars and Planets V* (*Reipurth et al.,* 2007) will continue over the next few years to Protostars and Planets VII. There is a current convergence of simulations toward the scales of GMCs, as well as GMC-scale physics. Galaxy simulations are moving toward ever smaller scales, while individual cloud simulations seek to include more realistic initial conditions. At present, galactic models are limited by the resolution required to adequately capture stellar feedback, internal cloud structure, cloud motions, and shocks. They also do not currently realize the temperatures or densities of GMCs observed, nor typically include magnetic fields. Smaller-scale simulations, on the other hand, miss larger-scale dynamics such as spiral shocks, shear, and cloud-cloud collisions, which will be present either during the formation or throughout the evolution of a molecular cloud. Future simulations, which capture the main physics on GMC scales, will provide a clearer picture of the evolution and lifetimes of GMCs. Furthermore, they will be more suitable for determining the role of different processes in driving turbulence and regulating star formation in galaxies, in conjunction with analytic models and observations.

Continuing improvements in codes, including the use of moving mesh codes in addition to adaptive mesh refinement (AMR) and smoothed-particle hydrodynamics (SPH) methods, will also promote progress on the description of GMCs, and allow more consistency checks between different numerical methods. More widespread and further development of chemodynamical modeling will enable the study of different tracers in cloud and galaxy simulations. In conjunction with these techniques, synthetic observations, such as HI and CO maps, will become increasingly important for comparing the results of numerical models and simulations, and testing whether the simulations are indeed viable representations of galaxies and clouds.

Acknowledgments. Support for this work was provided by the European Research Council through the FP7 ERC starting grant project LOCALSTAR (C.L.D.); the NSF through grants AST09-55300 (M.R.K.), AST11-09395 (M.-M.M.L.), AST09-08185 (E.C.O.), AST09-55836 (A.D.B.), and AST10-09049 (M.H.); NASA through ATP grant NNX13AB84G (M.R.K.), Chandra award number GO2-13162A (M.R.K.) issued by the Chandra X-ray Observatory Center, which is operated by the Smithsonian Astrophysical Observatory for and on behalf of the National Aeronautics Space Administration under contract NAS8-03060, and Hubble Award Number 13256 (M.R.K.) issued by the Space Telescope Science Institute, which is operated by the Association of Universities for Research in Astronomy, Inc., under NASA contract NAS 5-26555; a Research Corporation for Science Advancement Cottrell Scholar Award (A.D.B.); an Alfred P. Sloan Fellowship (M.R.K.); the hospitality of the Aspen Center for Physics, which is supported by the National Science Foundation Grant PHY-1066293 (M.R.K.); and CONACYT grant 102488 (E.V.S.).

REFERENCES

Agertz O. et al. (2013) *Astrophys. J., 770,* 25.
Audit E. and Hennebelle P. (2005) *Astron. Astrophys., 433,* 1.
Balbus S. A. (1988) *Astrophys. J., 324,* 60.
Balbus S. A. and Cowie L. L. (1985) *Astrophys. J., 297,* 61.
Balbus S. A. and Hawley J. F. (1991) *Astrophys. J., 376,* 214.
Ballesteros-Paredes J. (2006) *Mon. Not. R. Astron. Soc., 372,* 443.
Ballesteros-Paredes J. and Hartmann L. (2007) *Rev. Mexicana Astron. Astrofis., 43,* 123.
Ballesteros-Paredes J. and Mac Low M.-M. (2002) *Astrophys. J., 570,* 734.
Ballesteros-Paredes J. et al. (1999) *Astrophys. J., 515,* 286.
Ballesteros-Paredes J. et al. (2007) In *Protostars and Planets V* (B. Reipurth et al., eds.), pp. 63–80. Univ. of Arizona, Tucson.
Ballesteros-Paredes J. et al. (2011a) *Mon. Not. R. Astron. Soc., 416,* 1436.
Ballesteros-Paredes J. et al. (2011b) *Mon. Not. R. Astron. Soc., 411,* 65.
Bally J. et al. (1987) *Astrophys. J. Lett., 312,* L45.
Banerjee R. et al. (2009) *Mon. Not. R. Astron. Soc., 398,* 1082.
Bania T. M. and Lyon J. G. (1980) *Astrophys. J., 239,* 173.
Bash F. N. et al. (1977) *Astrophys. J., 217,* 464.
Basu S. et al. (1997) *Astrophys. J. Lett., 480,* L55.
Bayet E. et al. (2009) *Mon. Not. R. Astron. Soc., 399,* 264.
Bergin E. A. et al. (2004) *Astrophys. J., 612,* 921.
Bertoldi F. and McKee C. F. (1992) *Astrophys. J., 395,* 140.
Bigiel F. and Blitz L. (2012) *Astrophys. J., 756,* 183.
Bigiel F. et al. (2008) *Astron. J., 136,* 2846.
Black J. H. and van Dishoeck E. F. (1987) *Astrophys. J., 322,* 412.
Blair G. N. et al. (1978) *Astrophys. J., 219,* 896.
Blanc G. A. et al. (2009) *Astrophys. J., 704,* 842.
Blitz L. (1993) In *Protostars and Planets III* (E. H. Levy and J. I. Lunine, eds.), pp. 125–161. Univ. of Arizona, Tucson.
Blitz L. and Rosolowsky E. (2004) *Astrophys. J. Lett., 612,* L29.
Blitz L. and Rosolowsky E. (2006) *Astrophys. J., 650,* 933.
Blitz L. and Shu F. H. (1980) *Astrophys. J., 238,* 148.
Blitz L. and Thaddeus P. (1980) *Astrophys. J., 241,* 676.
Blondin J. M. et al. (1998) *Astrophys. J., 500,* 342.
Bolatto A. D. et al. (2008) *Astrophys. J., 686,* 948.
Bolatto A. D. et al. (2011) *Astrophys. J., 741,* 12.
Bolatto A. D. et al. (2013) *Annu. Rev. Astron. Astrophys., 51,* 207.
Brand J. and Wouterloot J. G. A. (1995) *Astron. Astrophys., 303,* 851.
Brunt C. M. (2003) *Astrophys. J., 584,* 293.
Brunt C. M. (2010) *Astron. Astrophys., 513,* A67.
Brunt C. M. et al. (2009) *Astron. Astrophys., 504,* 883.

Brunt C. M. et al. (2010a) *Mon. Not. R. Astron. Soc., 405,* L56.
Brunt C. M. et al. (2010b) *Mon. Not. R. Astron. Soc., 403,* 1507.
Burkert A. and Hartmann L. (2004) *Astrophys. J., 616,* 288.
Burkert A. and Hartmann L. (2013) *Astrophys. J., 773,* 48.
Burton W. B. and Gordon M. A. (1978) *Astron. Astrophys., 63,* 7.
Calzetti D. et al. (2012) *Astrophys. J., 752,* 98.
Carruthers G. (1970) *Astrophys. J. Lett., 161,* L81.
Casoli F. and Combes F. (1982) *Astron. Astrophys., 110,* 287.
Cheung A. C. et al. (1968) *Phys. Rev. Lett., 25,* 1701.
Cheung A. C. et al. (1969) *Nature, 221,* 626.
Chevalier R. A. (1999) *Astrophys. J., 511,* 798.
Chevalier R. A. and Fransson C. (2001) *Astrophys. J. Lett., 558,* L27.
Chevalier R. A. and Imamura J. N. (1982) *Astrophys. J., 261,* 543.
Cho J. et al. (2002) *Astrophys. J., 564,* 291.
Chomiuk L. and Povich M. S. (2011) *Astron. J., 142,* 197.
Christensen C. et al. (2012) *Mon. Not. R. Astron. Soc., 425,* 3058.
Cioffi D. F. et al. (1988) *Astrophys. J., 334,* 252.
Clark P. C. et al. (2012) *Mon. Not. R. Astron. Soc., 424,* 2599.
Colín P. et al. (2013) *Mon. Not. R. Astron. Soc., 435,* 1701.
Colombo D. et al. (2014) *Astrophys. J., 784,* 3.
Cox D. P. (1983) *Astrophys. J. Lett., 265,* L61.
Crutcher R. M. (2012) *Annu. Rev. Astron. Astrophys., 50,* 29.
Cunningham A. J. et al. (2006) *Astrophys. J., 653,* 416.
Curry C. L. (2000) *Astrophys. J., 541,* 831.
Dale J. E. and Bonnell I. A. (2008) *Mon. Not. R. Astron. Soc., 391,* 2.
Dale J. E. et al. (2005) *Mon. Not. R. Astron. Soc., 358,* 291.
Dale J. E. et al. (2012) *Mon. Not. R. Astron. Soc., 424,* 377.
Dale J. E. et al. (2013) *Mon. Not. R. Astron. Soc., 430,* 234.
Dame T. M. et al. (1987) *Astrophys. J., 322(1),* 706.
Dame T. M. et al. (2001) *Astrophys. J., 547(1),* 792.
Dawson J. R. et al. (2011) *Astrophys. J., 728,* 127.
Dawson J. R. et al. (2013) *Astrophys. J., 763,* 56.
de Avillez M. A. (2000) *Mon. Not. R. Astron. Soc., 315,* 479.
de Avillez M. A. and Breitschwerdt D. (2005) *Astron. Astrophys., 436,* 585.
de Avillez M. A. and Mac Low M.-M. (2001) *Astrophys. J. Lett., 551,* L57.
Dekel A. and Krumholz M. R. (2013) *Mon. Not. R. Astron. Soc., 432,* 455.
Dobashi K. et al. (1996) *Astrophys. J., 466(1),* 282.
Dobbs C. L. (2008) *Mon. Not. R. Astron. Soc., 391,* 844.
Dobbs C. L. and Price D. J. (2008) *Mon. Not. R. Astron. Soc., 383,* 497.
Dobbs C. L. and Pringle J. E. (2013) *Mon. Not. R. Astron. Soc., 432,* 653.
Dobbs C. L. et al. (2008) *Mon. Not. R. Astron. Soc., 389,* 1097.
Dobbs C. L. et al. (2011a) *Mon. Not. R. Astron. Soc., 417,* 1318.
Dobbs C. L. et al. (2011b) *Mon. Not. R. Astron. Soc., 413,* 2935.
Dobbs C. L. et al. (2012) *Mon. Not. R. Astron. Soc., 425,* 2157.
Donovan Meyer J. et al. (2013) *Astrophys. J., 772,* 107.
Draine B. T. (2011) *Physics of the Interstellar and Intergalactic Medium.* Princeton Univ., Princeton.
Draine B. T. et al. (2007) *Astrophys. J., 663(2),* 866.
Efremov Y. N. (1995) *Astron. J., 110,* 2757.
Elmegreen B. G. (1979) *Astrophys. J., 231,* 372.
Elmegreen B. G. (1987) *Astrophys. J., 312,* 626.
Elmegreen B. G. (1994) *Astrophys. J., 433,* 39.
Elmegreen B. G. (2000) *Astrophys. J., 530,* 277.
Elmegreen B. G. (2011) *Astrophys. J., 737,* 10.
Elmegreen B. G. and Elmegreen D. M. (1983) *Mon. Not. R. Astron. Soc., 203,* 31.
Evans N. J. I. et al. (2009) *Astrophys. J. Suppl., 181(2),* 321.
Fall S. M. et al. (2010) *Astrophys. J. Lett., 710,* L142.
Federrath C. and Klessen R. S. (2012) *Astrophys. J., 761,* 156.
Federrath C. and Klessen R. S. (2013) *Astrophys. J., 763,* 51.
Federrath C. et al. (2008) *Astrophys. J. Lett., 688,* L79.
Feldmann R. and Gnedin N. Y. (2011) *Astrophys. J. Lett., 727,* L12.
Field G. et al. (2011) *Mon. Not. R. Astron. Soc., 416(1),* 710.
Field G. B. (1965) *Astrophys. J., 142,* 531.
Field G. B. (1970) *Mem. Soc. Roy. Sci. Liège, 19,* 29.
Field G. B. and Saslaw W. C. (1965) *Astrophys. J., 142,* 568.
Fisher D. B. et al. (2013) *Astrophys. J., 764(2),* 174.
Foyle K. et al. (2010) *Astrophys. J., 725(1),* 534.
Franco J. et al. (2002) *Astrophys. J., 570,* 647.

Fukui Y. and Kawamura A. (2010) *Annu. Rev. Astron. Astrophys.,* 48, 547.
Fukui Y. et al. (1999) *Publ. Astron. Soc. Japan, 51,* 745.
Fukui Y. et al. (2006) *Science, 314,* 106.
Fukui Y. et al. (2008) *Astrophys. J.Suppl., 178(1),* 56.
Fumagalli M. et al. (2010) *Astrophys. J., 722,* 919.
Galván-Madrid R. et al. (2009) *Astrophys. J., 706,* 1036.
Gammie C. F. (1996) *Astrophys. J., 462,* 725.
Gao Y. and Solomon P. M. (2004) *Astrophys. J., 606(1),* 271.
García-Burillo S. et al. (2012) *Astron. Astrophys., 539,* A8.
Gardan E. et al. (2007) *Astron. Astrophys., 473,* 91.
Gazol-Patiño A. and Passot T. (1999) *Astrophys. J., 518,* 748.
Gendelev L. and Krumholz M. R. (2012) *Astrophys. J., 745,* 158.
Gibson D. et al. (2009) *Astrophys. J., 705,* 123.
Glover S. C. O. and Clark P. C. (2012a) *Mon. Not. R. Astron. Soc.,* 421, 9.
Glover S. C. O. and Clark P. C. (2012b) *Mon. Not. R. Astron. Soc.,* 426, 377.
Glover S. C. O. and Mac Low M.-M. (2007a) *Astrophys. J. Suppl.,* 169, 239.
Glover S. C. O. and Mac Low M.-M. (2007b) *Astrophys. J., 659,* 1317.
Glover S. C. O. and Mac Low M.-M. (2011) *Mon. Not. R. Astron. Soc.,* 412, 337.
Glover S. C. O. et al. (2010) *Mon. Not. R. Astron. Soc., 404,* 2.
Gnedin N. Y. and Kravtsov A. V. (2010) *Astrophys. J., 714,* 287.
Gnedin N. Y. and Kravtsov A. V. (2011) *Astrophys. J., 728,* 88.
Gnedin N. Y. et al. (2009) *Astrophys. J., 697,* 55.
Goldbaum N. J. et al. (2011) *Astrophys. J., 738,* 101.
Goldreich P. and Kwan J. (1974) *Astrophys. J., 189,* 441.
Goldreich P. and Lynden-Bell D. (1965a) *Mon. Not. R. Astron. Soc.,* 130, 125.
Goldreich P. and Lynden-Bell D. (1965b) *Mon. Not. R. Astron. Soc.,* 130, 125.
Goldsmith P. F. and Li D. (2005) *Astrophys. J., 622,* 938.
Gordon M. A. and Burton W. B. (1976) *Astrophys. J., 208(1),* 346.
Gould R. J. and Salpeter E. E. (1963) *Astrophys. J., 138,* 393.
Gratier P. et al. (2012) *Astron. Astrophys., 542,* A108.
Gritschneder M. et al. (2009) *Astrophys. J. Lett., 694,* L26.
Gutermuth R. A. et al. (2011) *Astrophys. J., 739(2),* 84.
Handbury M. J. et al. (1977) *Astron. Astrophys., 61,* 443.
Hartmann L. (2003) *Astrophys. J., 585,* 398.
Hartmann L. and Burkert A. (2007) *Astrophys. J., 654,* 988.
Hartmann L. et al. (2001) *Astrophys. J., 562,* 852.
Hartmann L. et al. (2012) *Mon. Not. R. Astron. Soc., 420,* 1457.
Hausman M. A. (1982) *Astrophys. J., 261,* 532.
Heiderman A. et al. (2010) *Astrophys. J., 723(2),* 1019.
Heitsch F. and Hartmann L. (2008) *Astrophys. J., 689,* 290.
Heitsch F. et al. (2001) *Astrophys. J., 547,* 280.
Heitsch F. et al. (2005) *Astrophys. J. Lett., 633,* L113.
Heitsch F. et al. (2006) *Astrophys. J., 648,* 1052.
Heitsch F. et al. (2008) *Astrophys. J., 683,* 786.
Heitsch F. et al. (2009) *Astrophys. J., 704,* 1735.
Hennebelle P. and Chabrier G. (2011) *Astrophys. J. Lett., 743,* L29.
Hennebelle P. and Pérault M. (1999) *Astron. Astrophys., 351,* 309.
Herrera C. N. et al. (2013) *Astron. Astrophys., 554,* 91.
Heyer M. H. and Brunt C. M. (2004) *Astrophys. J. Lett., 615,* L45.
Heyer M. H. et al. (1987) *Astrophys. J., 321,* 855.
Heyer M. H. et al. (2001) *Astrophys. J., 551(2),* 852.
Heyer M. et al. (2009) *Astrophys. J., 699,* 1092.
Hill A. S. et al. (2012) *Astrophys. J., 750,* 104.
Hollenbach D. and Salpeter E. E. (1971) *Astrophys. J., 163,* 155.
Hopkins P. F. (2012) *Mon. Not. R. Astron. Soc., 423,* 2016.
Hopkins P. F. et al. (2011) *Mon. Not. R. Astron. Soc., 417,* 950.
Hopkins P. F. et al. (2012) *Mon. Not. R. Astron. Soc., 421,* 3488.
Hoyle F. (1953) *Astrophys. J., 118,* 513.
Hughes A. et al. (2010) *Mon. Not. R. Astron. Soc., 406(3),* 2065.
Hunter J. H. Jr. et al. (1986) *Astrophys. J., 305,* 309.
Imara N. and Blitz L. (2011) *Astrophys. J., 732,* 78.
Imara N. et al. (2011) *Astrophys. J., 732,* 79.
Inoue T. and Inutsuka S.-i. (2012) *Astrophys. J., 759,* 35.
Israel F. P. et al. (1993) *Astron. Astrophys., 276,* 25.
Jaacks J. et al. (2013) *Astrophys. J., 766,* 94.

Jeffries R. D. et al. (2011) *Mon. Not. R. Astron. Soc., 418,* 1948.
Jenkins E. B. and Tripp T. M. (2011) *Astrophys. J., 734,* 65.
Joung M. R. et al. (2009) *Astrophys. J., 704,* 137.
Julian W. H. and Toomre A. (1966) *Astrophys. J., 146,* 810.
Juneau S. et al. (2009) *Astrophys. J., 707,* 1217.
Kainulainen J. et al. (2009a) *Astron. Astrophys., 508,* L35.
Kainulainen J. et al. (2009b) *Astron. Astrophys., 497,* 399.
Kawamura A. et al. (2009) *Astrophys. J. Suppl., 184(1),* 1.
Kennicutt R. C. J. (1989) *Astrophys. J., 344,* 685.
Kennicutt R. C. J. (1998) *Astrophys. J., 498,* 541.
Kennicutt R. C. J. et al. (2007) *Astrophys. J., 671(1),* 333.
Kim C.-G. et al. (2008) *Astrophys. J., 681,* 1148.
Kim C.-G. et al. (2011) *Astrophys. J., 743,* 25.
Kim J. et al. (1998) *Astrophys. J. Lett., 506,* L139.
Kim J.-H. et al. (2013) *Astrophys. J., 779,* 8.
Kim W.-T. and Ostriker E. C. (2001) *Astrophys. J., 559,* 70.
Kim W.-T. and Ostriker E. C. (2002) *Astrophys. J., 570,* 132.
Kim W.-T. and Ostriker E. C. (2006) *Astrophys. J., 646,* 213.
Kim W.-T. and Ostriker E. C. (2007) *Astrophys. J., 660,* 1232.
Kim W.-T. et al. (2002) *Astrophys. J., 581,* 1080.
Kim W.-T. et al. (2003) *Astrophys. J., 599,* 1157.
Kirk H. et al. (2013) *Astrophys. J., 766,* 115.
Klein R. I. et al. (1994) *Astrophys. J., 420,* 213.
Klessen R. S. and Hennebelle P. (2010) *Astron. Astrophys., 520,* A17.
Klessen R. S. et al. (2000) *Astrophys. J., 535,* 887.
Koda J. et al. (2009) *Astrophys. J. Lett., 700,* L132.
Kosiński R. and Hanasz M. (2007) *Mon. Not. R. Astron. Soc., 376,* 861.
Kovalenko I. and Korolev V. (2012) Conference poster, available online at *http://zofia.sese.asu.edu/_evan/cosmicturbulence posters/ Kovalenko poster.pdf*.
Koyama H. and Inutsuka S.-I. (2000) *Astrophys. J., 532,* 980.
Koyama H. and Inutsuka S.-I. (2002) *Astrophys. J. Lett., 564,* L97.
Kramer C. et al. (1998) *Astron. Astrophys., 329,* 249.
Kritsuk A. G. and Norman M. L. (2002) *Astrophys. J. Lett., 569,* L127.
Kritsuk A. G. et al. (2011) *Astrophys. J. Lett., 727,* L20.
Kruijssen J. M. D. et al. (2013) *ArXiv e-prints,* arXiv:1306.5285.
Krumholz M. R. (2012) *Astrophys. J., 759,* 9.
Krumholz M. R. and Burkert A. (2010) *Astrophys. J., 724,* 895.
Krumholz M. R. and Gnedin N. Y. (2011) *Astrophys. J., 729,* 36.
Krumholz M. R. and Matzner C. D. (2009) *Astrophys. J., 703,* 1352.
Krumholz M. R. and McKee C. F. (2005) *Astrophys. J., 630,* 250.
Krumholz M. R. and Tan J. C. (2007) *Astrophys. J., 654,* 304.
Krumholz M. R. and Thompson T. A. (2007) *Astrophys. J., 669,* 289.
Krumholz M. R. and Thompson T. A. (2012) *Astrophys. J., 760,* 155.
Krumholz M. R. and Thompson T. A. (2013) *Mon. Not. R. Astron. Soc.,* 434, 2329.
Krumholz M. R. et al. (2006) *Astrophys. J., 653,* 361.
Krumholz M. R. et al. (2008) *Astrophys. J., 689,* 865.
Krumholz M. R. et al. (2009a) *Astrophys. J., 693,* 216.
Krumholz M. R. et al. (2009b) *Astrophys. J., 699,* 850.
Krumholz M. R. et al. (2011) *Astrophys. J., 731,* 25.
Krumholz M. R. et al. (2012) *Astrophys. J., 745,* 69.
Kuhlen M. et al. (2012) *Astrophys. J., 749,* 36.
Kuhlen M. et al. (2013) *Astrophys. J., 776,* 34.
Kutner M. L. et al. (1977) *Astrophys. J., 215(1),* 521.
Kwan J. (1979) *Astrophys. J., 229,* 567.
Kwan J. and Valdes F. (1983) *Astrophys. J., 271,* 604.
Kwan J. and Valdes F. (1987) *Astrophys. J., 315,* 92.
La Vigne M. A. et al. (2006) *Astrophys. J., 650,* 818.
Lada C. (1976) *Astrophys. J. Suppl., 32(1),* 603.
Lada C. J. et al. (2010) *Astrophys. J., 724(1),* 687.
Lada C. J. et al. (2012) *Astrophys. J., 745(2),* 190.
Larson R. B. (1981) *Mon. Not. R. Astron. Soc., 194,* 809.
Larson R. B. (1985) *Mon. Not. R. Astron. Soc., 214,* 379.
Lee M.-Y. et al. (2012) *Astrophys. J., 748,* 75.
Leisawitz D. et al. (1989) *Astrophys. J. Suppl., 70,* 731.
Leroy A. et al. (2005) *Astrophys. J., 625(2),* 763.
Leroy A. et al. (2007) *Astrophys. J., 658(2),* 1027.
Leroy A. et al. (2008) *Astron. J., 136,* 2782.
Leroy A. K. et al. (2009) *Astron. J., 137(6),* 4670.
Leroy A. K. et al. (2013a) *Astrophys. J. Lett., 769(1),* L12.
Leroy A. K. et al. (2013b) *Astron. J., 146,* 19.

Li Y. et al. (2005) *Astrophys. J., 626*, 823.
Li Z.-Y. and Nakamura F. (2006) *Astrophys. J. Lett., 640*, L187.
Lin C. C. et al. (1965) *Astrophys. J., 142*, 1431.
Liszt H. S. et al. (1974) *Astrophys. J., 190*, 557.
Liu G. et al. (2011) *Astrophys. J., 735(1)*, 63.
Lizano S. and Shu F. (1989) *Astrophys. J., 342(1)*, 834.
Lombardi M. et al. (2010) *Astron. Astrophys., 519*, L7.
Longmore S. N. et al. (2013) *Mon. Not. R. Astron. Soc., 429(2)*, 987.
Lynden-Bell D. (1966) *The Observatory, 86*, 57.
Mac Low M. (1999) *Astrophys. J., 524*, 169.
Mac Low M. et al. (1998) *Phys. Rev. Lett., 80*, 2754.
Mac Low M.-M. and Glover S. C. O. (2012) *Astrophys. J., 746*, 135.
Mac Low M.-M. and Klessen R. S. (2004) *Rev. Mod. Phys., 76*, 125.
Mac Low M.-M. et al. (1994) *Astrophys. J., 433*, 757.
Machida M. et al. (2009) *Publ. Astron. Soc. Japan, 61*, 411.
Magnani L. et al. (1985) *Astrophys. J., 295*, 402.
Matsumoto R. et al. (1988) *Publ. Astron. Soc. Japan, 40*, 171.
Matzner C. D. (2002) *Astrophys. J., 566*, 302.
Matzner C. D. (2007) *Astrophys. J., 659*, 1394.
Mayya Y. D. et al. (2012) *EPJ Web Conf., 19*, 08006.
McCray R. and Kafatos M. (1987) *Astrophys. J., 317*, 190.
McCray R. et al. (1975) *Astrophys. J., 196*, 565.
McCrea W. H. and McNally D. (1960) *Mon. Not. R. Astron. Soc., 121*, 238.
McKee C. F. and Krumholz M. R. (2010) *Astrophys. J., 709*, 308.
McKee C. F. and Williams J. P. (1997) *Astrophys. J., 476*, 144.
McKellar A. (1940) *Publ. Astron. Soc. Pac., 52(1)*, 187.
Miura R. E. et al. (2012) *Astrophys. J., 761*, 37.
Mouschovias T. (1987) In *Physical Processes in Interstellar Clouds* (G. E. Morfill and M. Scholer, eds.), pp. 453–489. Proceedings of the NATO Advanced Study Institute, NATO Science Series C, Vol. 210, Reidel, Dordrecht.
Mouschovias T. C. (1974) *Astrophys. J., 192*, 37.
Mouschovias T. C. (1991a) In *The Physics of Star Formation and Early Stellar Evolution* (C. J. Lada and N. D. Kylafis, eds.), p. 61. NATO ASIC Proc. 342, Kluwer, Dordrecht.
Mouschovias T. C. (1991b) In *The Physics of Star Formation and Early Stellar Evolution* (C. J. Lada and N. D. Kylafis, eds.), p. 449. NATO ASIC Proc. 342, Kluwer, Dordrecht.
Mouschovias T. and Spitzer L. (1976) *Astrophys. J., 210(1)*, 326.
Mouschovias T. C. and Tassis K. (2010) *Mon. Not. R. Astron. Soc., 409*, 801.
Mouschovias T. C. et al. (1974) *Astron. Astrophys., 33*, 73.
Mouschovias T. C. et al. (2009) *Mon. Not. R. Astron. Soc., 397*, 14.
Muller E. et al. (2010) *Astrophys. J., 712(2)*, 1248.
Müller H. S. P. et al. (2005) *J. Molec. Structure, 742*, 215.
Murray N. (2011) *Astrophys. J., 729*, 133.
Murray N. and Rahman M. (2010) *Astrophys. J., 709*, 424.
Murray N. et al. (2010) *Astrophys. J., 709*, 191.
Myers P. C. B. P. J. (1983) *Astrophys. J., 266(1)*, 309.
Nagai T. et al. (1998) *Astrophys. J., 506*, 306.
Nakamura F. and Li Z.-Y. (2005) *Astrophys. J., 631*, 411.
Nakamura F. and Li Z.-Y. (2007) *Astrophys. J., 662*, 395.
Nakano T. (1978) *Publ. Astron. Soc. Japan, 30(1)*, 681.
Narayanan D. et al. (2008a) *Astrophys. J., 684*, 996.
Narayanan D. et al. (2008b) *Astrophys. J. Lett., 681*, L77.
Narayanan G. et al. (2008c) *Astrophys. J. Suppl., 177*, 341.
Narayanan D. et al. (2011) *Mon. Not. R. Astron. Soc., 418*, 664.
Narayanan D. et al. (2012) *Mon. Not. R. Astron. Soc., 421*, 3127.
Ntormousi E. et al. (2011) *Astrophys. J., 731*, 13.
Oka T. et al. (2001) *Astrophys. J., 562(1)*, 348.
Onodera S. et al. (2010) *Astrophys. J. Lett., 722(2)*, L127.
Oort J. H. (1954) *Bull. Astron. Inst. Netherlands, 12*, 177.
Ostriker E. C. and Shetty R. (2011) *Astrophys. J., 731*, 41.
Ostriker E. C. et al. (2010) *Astrophys. J., 721*, 975.
Padoan P. and Nordlund A. (1999) *Astrophys. J., 526*, 279.
Padoan P. and Nordlund A. (2011) *Astrophys. J., 730*, 40.
Padoan P. et al. (2001) *Astrophys. J., 553*, 227.
Padoan P. et al. (2004a) *Phys. Rev. Lett., 92(19)*, 191102.
Padoan P. et al. (2004b) *Astrophys. J. Lett., 604*, L49.
Padoan P. et al. (2012) *Astrophys. J. Lett., 759*, L27.
Papadopoulos P. P. et al. (2012a) *Mon. Not. R. Astron. Soc., 426*, 2601.

Papadopoulos P. P. et al. (2012b) *Astrophys. J., 751*, 10.
Parker E. N. (1966) *Astrophys. J., 145*, 811.
Passot T. et al. (1995) *Astrophys. J., 455*, 536.
Pelupessy F. I. and Papadopoulos P. P. (2009) *Astrophys. J., 707*, 954.
Pelupessy F. I. et al. (2006) *Astrophys. J., 645*, 1024.
Penston M. V. et al. (1969) *Mon. Not. R. Astron. Soc., 142*, 355.
Pety J. et al. (2013) *Astrophys. J., 779*, 43.
Phillips J. P. (1999) *Astron. Astrophys. Suppl., 134*, 241.
Pineda J. E. et al. (2009) *Astrophys. J. Lett., 699*, L134.
Piontek R. A. and Ostriker E. C. (2004) *Astrophys. J., 601*, 905.
Piontek R. A. and Ostriker E. C. (2005) *Astrophys. J., 629*, 849.
Piontek R. A. and Ostriker E. C. (2007) *Astrophys. J., 663*, 183.
Pon A. et al. (2012) *Astrophys. J., 756*, 145.
Price D. J. et al. (2011) *Astrophys. J. Lett., 727*, L21.
Rafelski M. et al. (2011) *Astrophys. J., 736*, 48.
Rahman N. et al. (2012) *Astrophys. J., 745(2)*, 183.
Rebolledo D. et al. (2012) *Astrophys. J., 757*, 155.
Regan M. W. et al. (2001) *Astrophys. J., 561(1)*, 218.
Reggiani M. et al. (2011) *Astron. Astrophys., 534*, A83.
Reipurth B. et al., eds. (2007) *Protostars and Planets V*, Univ. of Arizona, Tucson.
Rho J. and Petre R. (1998) *Astrophys. J. Lett., 503*, L167.
Ripple F. et al. (2013) *Mon. Not. R. Astron. Soc., 431(2)*, 1296.
Roberts W. W. (1969) *Astrophys. J., 158*, 123.
Robertson B. E. and Kravtsov A. V. (2008) *Astrophys. J., 680*, 1083.
Robitaille T. P. and Whitney B. A. (2010) *Astrophys. J. Lett., 710*, L11.
Rogers H. and Pittard J. M. (2013) *Mon. Not. R. Astron. Soc., 431*, 1337.
Roman Duval J. et al. (2010) *Astrophys. J., 723*, 492.
Rosolowsky E. and Blitz L. (2005) *Astrophys. J., 623(2)*, 826.
Rosolowsky E. and Leroy A. (2006) *Publ. Astron. Soc. Pac., 118(1)*, 590.
Rosolowsky E. et al. (2003) *Astrophys. J., 599*, 258.
Rowles J. and Froebrich D. (2009) *Mon. Not. R. Astron. Soc., 395*, 1640.
Rubio M. et al. (1993) *Astron. Astrophys., 271*, 9.
Saintonge A. et al. (2011a) *Mon. Not. R. Astron. Soc., 415(1)*, 61.
Saintonge A. et al. (2011b) *Mon. Not. R. Astron. Soc., 415(1)*, 32.
Sánchez-Salcedo F. J. et al. (2002) *Astrophys. J., 577*, 768.
Sanders D. et al. (1984) *Astrophys. J., 276(1)*, 182.
Sandstrom K. M. et al. (2013) *Astrophys. J., 777*, 5.
Santillán A. et al. (2000) *Astrophys. J., 545*, 353.
Schneider N. et al. (2010) *Astron. Astrophys., 520*, A49.
Schneider N. et al. (2012) *Astron. Astrophys., 540(1)*, L11.
Schneider N. et al. (2013) *Astrophys. J., 766(2)*, L17.
Schruba A. et al. (2010) *Astrophys. J., 722(2)*, 1699.
Schruba A. et al. (2011) *Astron. J., 142(2)*, 37.
Scoville N. Z. S. P. M. (1975) *Astrophys. J. Lett., 199(2)*, L105.
Scoville N. Z. et al. (1979) In *The Large-Scale Characteristics of the Galaxy* (W. B. Burton, ed.), pp. 277–282. IAU Symp. 84, Reidel, Dordrecht.
Scoville N. Z. et al. (1987) *Astrophys. J. Suppl., 63*, 821.
Shetty R. and Ostriker E. C. (2006) *Astrophys. J., 647*, 997.
Shetty R. and Ostriker E. C. (2008) *Astrophys. J., 684*, 978.
Shetty R. and Ostriker E. C. (2012) *Astrophys. J., 754*, 2.
Shetty R. et al. (2011) *Mon. Not. R. Astron. Soc., 412(3)*, 1686.
Shetty R. et al. (2013) *Mon. Not. R. Astron. Soc., 430(1)*, 288.
Shu F. H. et al. (1987) *Annu. Rev. Astron. Astrophys., 25*, 23.
Silk J. (1980) In *Star Formation* (A. Maeder and L. Martinet, eds.), p. 133. Saas-Fee Advanced Course 10, Geneva Observatory.
Snell R. L. et al. (1984) *Astrophys. J., 276*, 625.
Snyder L. E. et al. (1969) *Phys. Rev. Lett., 22*, 679.
Solomon P. M. et al. (1987) *Astrophys. J., 319*, 730.
Spitzer L. Jr. (1949) *Astrophys. J., 109*, 337.
Sternberg A. (1988) *Astrophys. J., 332*, 400.
Stevens I. R. et al. (1992) *Astrophys. J., 386*, 265.
Stone J. M. et al. (1998) *Astrophys. J. Lett., 508*, L99.
Stutzki J. et al. (1988) *Astrophys. J., 332*, 379.
Swings P. and Rosenfeld L. (1937) *Astrophys. J., 86*, 483.
Taff L. G. and Savedoff M. P. (1973) *Mon. Not. R. Astron. Soc., 164*, 357.
Tan J. C. (2000) *Astrophys. J., 536*, 173.
Tan J. C. et al. (2006) *Astrophys. J. Lett., 641*, L121.
Tasker E. J. (2011) *Astrophys. J., 730*, 11.
Tasker E. J. and Tan J. C. (2009) *Astrophys. J., 700*, 358.

Thompson T. A. et al. (2005) *Astrophys. J., 630,* 167.
Tilley D. A. et al. (2006) *Mon. Not. R. Astron. Soc., 371,* 1106.
Toalá J. A. et al. (2012) *Astrophys. J., 744,* 190.
Tomisaka K. (1984) *Publ. Astron. Soc. Japan, 36,* 457.
Tomisaka K. (1986) *Publ. Astron. Soc. Japan, 38,* 95.
Torii K. et al. (2010) *Publ. Astron. Soc. Japan, 62,* 1307.
van de Hulst H. C. (1948) *Harvard Obs. Monogr., 7,* 73.
van Dishoeck E. F. and Black J. H. (1986) *Astrophys. J. Suppl., 62,* 109.
van Dishoeck E. F. and Black J. H. (1988) *Astrophys. J., 334,* 771.
van Dishoeck E. F. and Blake G. A. (1998) *Annu. Rev. Astron. Astrophys., 36,* 317.
Van Loo S. et al. (2013) *Astrophys. J., 764,* 36.
Vázquez-Semadeni E. (1994) *Astrophys. J., 423,* 681.
Vázquez-Semadeni E. (2012) *ArXiv e-prints,* arXiv:1208.4132.
Vázquez-Semadeni E. et al. (1995) *Astrophys. J., 441,* 702.
Vázquez-Semadeni E. et al. (2000) *Astrophys. J., 540,* 271.
Vázquez-Semadeni E. et al. (2003) *Astrophys. J. Lett., 585,* L131.
Vázquez-Semadeni E. et al. (2005) *Astrophys. J. Lett.,* 630, L49.
Vázquez-Semadeni E. et al. (2006) *Astrophys. J., 643,* 245.
Vázquez-Semadeni E. et al. (2007) *Astrophys. J., 657,* 870.
Vázquez-Semadeni E. et al. (2008) *Mon. Not. R. Astron. Soc., 390,* 769.
Vázquez-Semadeni E. et al. (2009) *Astrophys. J., 707,* 1023.
Vázquez-Semadeni E. et al. (2010) *Astrophys. J., 715,* 1302.
Vázquez-Semadeni E. et al. (2011) *Mon. Not. R. Astron. Soc., 414,* 2511.
Vishniac E. T. (1994) *Astrophys. J., 428,* 186.
Wada K. et al. (2000) *Astrophys. J., 540,* 797.
Wada K. et al. (2011) *Astrophys. J., 735,* 1.
Walch S. K. et al. (2012) *Mon. Not. R. Astron. Soc., 427,* 625.
Walder R. and Folini D. (2000) *Astrophys. Space Sci., 274,* 343.
Wang P. et al. (2010) *Astrophys. J., 709,* 27.
Weinreb S. et al. (1963) *Nature, 200,* 829.
Whitworth A. (1979) *Mon. Not. R. Astron. Soc., 186,* 59.
Williams J. P. and McKee C. F. (1997) *Astrophys. J., 476,* 166.
Williams J. P. et al. (1994) *Astrophys. J., 428,* 693.
Wilson R. W. et al. (1970) *Astrophys. J. Lett., 161,* L43.
Wolfe A. M. and Chen H.-W. (2006) *Astrophys. J., 652,* 981.
Wolfire M. G. et al. (2010) *Astrophys. J., 716,* 1191.
Wong T. and Blitz L. (2002) *Astrophys. J., 569,* 157.
Wong T. et al. (2011) *Astrophys. J. Suppl., 197,* 16.
Wong T. et al. (2013) *Astrophys. J. Lett., 777,* L4.
Wu J. et al. (2005) *Astrophys. J. Lett., 635,* L173.
Wyder T. K. et al. (2009) *Astrophys. J., 696,* 1834.
Yonekura Y. et al. (1997) *Astrophys. J. Suppl., 110,* 21.
Young J. S. et al. (1995) *Astrophys. J. Suppl., 98,* 219.
Zamora-Avilés M. et al. (2012) *Astrophys. J., 751,* 77.
Zuckerman B. and Evans N. J. (1974) *Astrophys. J. Lett., 192,* L149.

André P., Di Francesco J., Ward-Thompson D., Inutsuka S-I., Pudritz R. E., and Pineda J. (2014) From filamentary networks to dense cores in molecular clouds: Toward a new paradigm for star formation. In *Protostars and Planets VI* (H. Beuther et al., eds.), pp. 27–51. Univ. of Arizona, Tucson, DOI: 10.2458/azu_uapress_9780816531240-ch002.

From Filamentary Networks to Dense Cores in Molecular Clouds: Toward a New Paradigm for Star Formation

Philippe André
Laboratoire d'Astrophysique de Paris-Saclay

James Di Francesco
National Research Council of Canada

Derek Ward-Thompson
University of Central Lancashire

Shu-ichiro Inutsuka
Nagoya University

Ralph E. Pudritz
McMaster University

Jaime Pineda
University of Manchester and European Southern Observatory

Recent studies of the nearest star-forming clouds of the Galaxy at submillimeter wavelengths with the Herschel Space Observatory have provided us with unprecedented images of the initial and boundary conditions of the star-formation process. The Herschel results emphasize the role of interstellar filaments in the star-formation process and connect remarkably well with nearly a decade's worth of numerical simulations and theory that have consistently shown that the interstellar medium (ISM) should be highly filamentary on all scales, and star formation is intimately related to self-gravitating filaments. In this review, we trace how the apparent complexity of cloud structure and star formation is governed by relatively simple universal processes — from filamentary clumps to galactic scales. We emphasize two crucial and complementary aspects: (1) the key observational results obtained with Herschel over the past three years, along with relevant new results obtained from the ground on the kinematics of interstellar structures; and (2) the key existing theoretical models and the many numerical simulations of interstellar cloud structure and star formation. We then synthesize a comprehensive physical picture that arises from the confrontation of these observations and simulations.

1. INTRODUCTION

The physics controlling the earliest phases of star formation is not yet well understood. Improving our global understanding of these phases is crucial for gaining insight into the general inefficiency of the star-formation process, the global rate of star formation on galactic scales, the origin of stellar masses, and the birth of planetary systems.

Since *Protostars and Planets V* (*Reipurth et al.,* 2007) seven years ago, one area that has seen the most dramatic advances has been the characterization of the link between star formation and the structure of the cold interstellar medium (ISM). In particular, extensive studies of the nearest star-forming clouds of our Galaxy with the Herschel Space Observatory have provided us with unprecedented images of the initial and boundary conditions of the star-formation process (e.g., Fig. 1). The Herschel images reveal an intricate network of filamentary structures in every interstellar cloud. The observed filaments share common properties, such as their central widths, but only the densest filaments contain prestellar cores, the seeds of future stars. Overall, the Herschel data, as well as other observations from, e.g., near-infrared (IR) extinction studies, favor a scenario in which interstellar filaments and prestellar cores represent two key steps in the star-formation process. First, large-scale supersonic flows compress the gas, giving rise to a universal web-like filamentary structure in the ISM. Next, gravity takes over and controls the further fragmentation

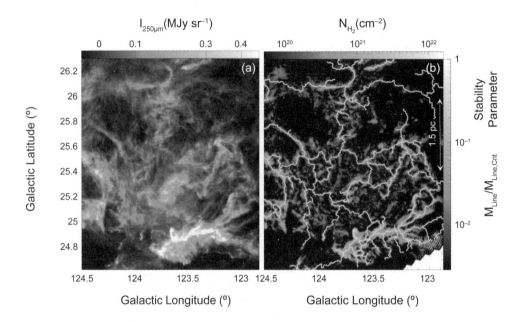

Fig. 1. (a) Herschel /SPIRE 250-μm dust continuum map of a portion of the Polaris flare translucent cloud (e.g., *Miville-Deschênes et al.,* 2010; *Ward-Thompson et al.,* 2010). **(b)** Corresponding column density map derived from Herschel data (e.g., *André et al.,* 2010). The contrast of the filaments has been enhanced using a curvelet transform (cf. *Starck et al.,* 2003). The skeleton of the filament network identified with the DisPerSE algorithm (*Sousbie,* 2011) is shown by the gray lines. A similar pattern is found with other algorithms such as *getfilaments* (*Men'shchikov,* 2013). Given the typical width ~0.1 pc of the filaments (*Arzoumanian et al.,* 2011) (see Fig. 5 below), this column density map is equivalent to a map of the mass per unit length along the filaments (see gray scale on right).

of filaments into prestellar cores and ultimately protostars.

The new observational results connect remarkably well with nearly a decade's worth of numerical simulations and theory that have consistently shown the ISM should be highly filamentary on all scales, and star formation is intimately connected with self-gravitating filaments (e.g., Fig. 10). The observations set strong constraints on models for the growth of structure in the ISM, leading to the formation of young stellar populations and star clusters. Numerical simulations now successfully include turbulence, gravity, a variety of cooling processes, magneto-hydrodynamics (MHD), and most recently, radiation and radiative feedback from massive stars. These numerical advances have been essential in testing and developing new insights into the physics of filaments and star formation, including the formation, fragmentation, and further evolution of filaments through accretion, and the central role of filaments in the rapid gathering of gas into cluster-forming, dense regions.

2. UNIVERSALITY OF THE FILAMENTARY STRUCTURE OF THE COLD INTERSTELLAR MEDIUM

2.1. Evidence of Interstellar Filaments Prior to Herschel

The presence of parsec-scale filamentary structures in nearby interstellar clouds and their potential importance for star formation have been pointed out by many authors for more than three decades. For example, *Schneider and Elmegreen* (1979) discussed the properties of 23 elongated dark nebulae, visible on optical plates and showing evidence of marked condensations or internal globules, which they named "globular filaments." High-angular resolution observations of Hɪ toward the Riegel-Crutcher cloud by *McClure-Griffiths et al.* (2006) also revealed an impressive network of aligned Hɪ filaments. In this case, the filaments were observed in Hɪ self-absorption (HISA) but with a column density ($A_V < 0.5$ mag) that is not enough to shield CO from photodissociation. These Hɪ filaments appear aligned with the ambient magnetic field, suggesting they are magnetically dominated. Tenuous CO filaments were also observed in diffuse molecular gas by *Falgarone et al.* (2001) and *Hily-Blant and Falgarone* (2007).

Within star-forming molecular gas, two nearby complexes were noted to have prominent filamentary structure in both CO and dust maps: the Orion A cloud (e.g., *Bally et al.,* 1987; *Chini et al.,* 1997; *Johnstone and Bally,* 1999) and the Taurus cloud (e.g., *Abergel et al.,* 1994; *Mizuno et al.,* 1995; *Hartmann,* 2002; *Nutter et al.,* 2008; *Goldsmith et al.,* 2008). Other well-known examples include the molecular clouds in Musca-Chamaeleon (e.g., *Cambrésy,* 1999), Perseus (e.g., *Hatchell et al.,* 2005), and S106 (e.g., *Balsara et al.,* 2001). More distant "infrared dark clouds" (IRDCs) identified at mid-IR wavelengths with the Infrared Space Observatory (ISO), Midcourse Space Experiment (MSX), and Spitzer (e.g., *Pérault et al.,* 1996; *Egan et al.,* 1998; *Peretto and Fuller,* 2009; *Miettinen and Harju,*

2010), some of which are believed to be the birthplaces of massive stars (see the chapter by Tan et al. in this volume), also have clear filamentary morphologies. Collecting and comparing observations available prior to Herschel, *Myers* (2009) noticed that young stellar groups and clusters are frequently associated with dense "hubs" radiating multiple filaments made of lower column density material.

For the purpose of this review, we will define a "filament" as any elongated ISM structure with an aspect ratio larger than ~5–10 that is significantly overdense with respect to its surroundings. The results of Galactic imaging surveys carried out with the Herschel Space Observatory (*Pilbratt et al.,* 2010) between late 2009 and early 2013 now demonstrate that such filaments are truly ubiquitous in the cold ISM (e.g,. section 2.2), present a high degree of universality in their properties (e.g., section 2.5), and likely play a key role in the star-formation process (see section 6).

2.2. Ubiquity of Filaments in Herschel Imaging Surveys

Herschel images provide key information on the structure of molecular clouds over spatial scales ranging from the sizes of entire cloud complexes (≥10 pc) down to the sizes of individual dense cores (<0.1 pc). While many interstellar clouds were already known to exhibit large-scale filamentary structures long before (cf. section 2.1), one of the most spectacular early findings from Herschel continuum observations was that the cold ISM [e.g., individual molecular clouds, giant molecular clouds (GMCs), and the Galactic plane] is highly structured with filaments pervading clouds (e.g., see *André et al.,* 2010; *Men'shchikov et al.,* 2010; *Molinari et al.,* 2010; *Henning et al.,* 2010; *Motte et al.,* 2010). These ubiquitous structures were seen by Herschel for the first time due to its extraordinary sensitivity to thermal dust emission both at high resolution and over (larger) scales previously inaccessible from the ground.

Filamentary structure is omnipresent in every cloud observed with Herschel, irrespective of its star-forming content (see also the chapter by Molinari et al. in this volume). For example, Fig. 1a shows the 250-μm continuum emission map of the Polaris Flare, a translucent, non-star-forming cloud (*Ward-Thompson et al.,* 2010; *Miville-Deschênes et al.,* 2010). Figure 1b shows a Herschel-derived column density map of the same cloud, appropriately filtered to emphasize the filamentary structure in the data. In both figures, filaments are clearly seen across the entire cloud, although no star formation has occurred. This omnipresent structure suggests the formation of filaments precedes star formation in the cold ISM and is tied to processes acting within the clouds themselves.

2.3. Common Patterns in the Organization of Filaments

The filaments now seen in the cold ISM offer clues about the nature of the processes in play within molecular clouds.

First, we note in the clouds shown in Fig. 1 and Fig. 2 (i.e., Polaris and Taurus), as well as other clouds, that filaments are typically very long, with lengths of ~1 pc or more, and up to several tens of parsecs in the case of some IRDCs (e.g., *Jackson et al.,* 2010; *Beuther et al.,* 2011). Despite small-scale deviations, filaments are in general quite linear (unidirectional) over their lengths, with typically minimal overall curvature and no sharp changes in overall direction. Moreover, many (although not all) filaments appear co-linear in direction to the longer extents of their host clouds. Indeed, some clouds appear globally to be filamentary but also contain within themselves distinct populations of (sub-)filaments [e.g., IC 5146 or NGC 6334; see *Arzoumanian et al.* (2011) and *Russeil et al.* (2013)]. What is striking about these characteristics is that they persist from cloud to cloud, although presumably filaments (and their host clouds) are three-dimensional objects with various orientations seen in projection on the sky. Nevertheless, these common traits suggest filaments originate from processes acting over the large scales of their host cloud (e.g., large turbulent modes).

What controls the organization of filaments? *Hill et al.* (2011) noted how filament networks varied within Vela C, itself quite a linear cloud. They found filaments were arranged in more varied directions (i.e., disorganized) within "nests" in outer, lower column density locations, but filaments appeared more unidirectional within "ridges" in inner, higher column density locations. Indeed, Hill et al. found that a greater concentration of mass within filaments is seen in ridges vs. nests. Hill et al. argued (from comparing column density probability functions) that these differences were due to the relative influences of turbulence and gravity in various locations within Vela C, with the former and latter being dominant in nests and ridges respectively. Similar trends can be seen in other clouds. For example, we note in Fig. 1 the relatively disorganized web-like network of filaments seen in the generally lower column density Polaris Flare, whose filaments are likely unbound and where turbulence likely dominates. On the other hand, we note in Fig. 2 the dominant and more unidirectional B211/B213 filament in Taurus, which, as judged from Herschel data, is dense enough to be self-gravitating (*Palmeirim et al.,* 2013).

Although direct observational constraints on the magnetic field inside molecular clouds remain scarce, an emerging trend is that dense (self-gravitating) filaments tend to be *perpendicular* to the direction of the local magnetic field, while low-density (unbound) filaments or striations tend to be *parallel* to the local magnetic field (cf. Fig. 3) (*Peretto et al.,* 2012; *Palmeirim et al.,* 2013; Cox et al., in preparation) [see also the related chapter by H.-B. Li et al. in this volume and caveats discussed by *Goodman et al.* (1990)].

2.4. Density and Temperature Profiles of Filaments

Detailed analysis of resolved filamentary column density profiles derived from Herschel data (e.g., *Arzoumanian et al.,* 2011; *Juvela et al.,* 2012; *Palmeirim et al.,* 2013) (Fig. 4) suggests that the shape of filament radial profiles

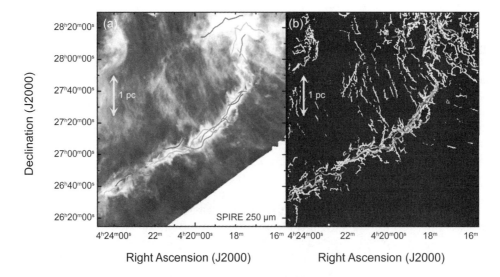

Fig. 2. See Plate 1 for color version. **(a)** Herschel/SPIRE 250-μm dust continuum image of the B211/B213/L1495 region in Taurus (*Palmeirim et al.,* 2013). The curves display the velocity-coherent "fibers" identified by *Hacar et al.* (2013) using $C^{18}O(1-0)$ observations. **(b)** Fine structure of the Herschel/SPIRE 250-μm dust continuum emission from the B211/B213 filament obtained by applying the multi-scale algorithm *getfilaments* (*Men'shchikov,* 2013) to the 250-μm image shown in **(a)**. Note the faint striations perpendicular to the main filament and the excellent correspondence between the small-scale structure of the dust continuum filament and the bundle of velocity-coherent fibers traced by *Hacar et al.* (2013) in $C^{18}O$ [same colored curves as in **(a)**].

is quasi-universal and well-described by a Plummer-like function of the form (cf. *Whitworth and Ward-Thompson,* 2001; *Nutter et al.,* 2008; *Arzoumanian et al.,* 2011)

$$\rho_p(r) = \frac{\rho_c}{\left[1 + (r/R_{flat})^2\right]^{p/2}} \qquad (1a)$$

for the density profile, equivalent to

$$\Sigma_p(r) = A_p \frac{\rho_c R_{flat}}{\left[1 + (r/R_{flat})^2\right]^{\frac{p-1}{2}}} \qquad (1b)$$

for the column density profile, where ρ_c is the central density of the filament, R_{flat} is the radius of the flat inner region, $p \approx 2$ is the power-law exponent at large radii ($r \gg R_{flat}$), and A_p is a finite constant factor that includes the effect of the filament's inclination angle to the plane of sky. Note that the density structure of an isothermal gas cylinder in hydrostatic equilibrium follows equation (1) with $p = 4$ (*Ostriker,* 1964). These recent results with Herschel on the radial density profiles of filaments, illustrated in Fig. 4, now confirm on a very strong statistical basis similar findings obtained on just a handful of filamentary structures from, e.g., near-IR extinction and groundbased submillimeter continuum data (e.g., *Lada et al.,* 1999; *Nutter et al.,* 2008; *Malinen et al.,* 2012).

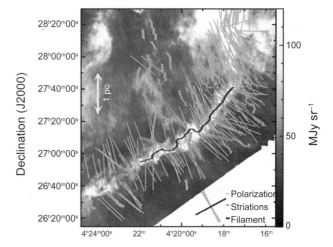

Fig. 3. See Plate 2 for color version. Display of optical and infrared polarization vectors from, e.g., *Heyer et al.* (2008) and *Chapman et al.* (2011) tracing the magnetic field orientation, overlaid on the Herschel/SPIRE 250-μm image of the B211/B213/L1495 region in Taurus (*Palmeirim et al.,* 2013) (see Fig. 2a). The plane-of-the-sky projection of the magnetic field appears to be oriented perpendicular to the B211/B213 filament and roughly aligned with the general direction of the striations overlaid as gray lines in the top left of the figure. A very similar pattern is observed in the Musca cloud (Cox et al., in preparation).

One possible interpretation for why the power-law exponent of the density profile is p ≈ 2 at large radii, and not p = 4 as in the isothermal equilibrium solution, is that dense filaments are not strictly isothermal but better described by a polytropic equation of state, $P \propto \rho^\gamma$ or $T \propto \rho^{\gamma-1}$ with $\gamma \leq 1$ (see *Palmeirim et al.*, 2013) (Fig. 4b). Models of polytropic cylindrical filaments undergoing gravitational contraction indeed have density profiles scaling as $\rho \propto r^{-\frac{2}{2-\gamma}}$ at large radii

(*Kawachi and Hanawa*, 1998; *Nakamura and Umemura*, 1999). For γ values close to unity, the model density profile thus approaches $\rho \propto r^{-2}$, in agreement with the Herschel observations. However, filaments may be more dynamic systems undergoing accretion and flows of various kinds (see section 4 and Fig. 9 below), which we address in section 5.

2.5. Quasi-Universal Inner Width of Interstellar Filaments

Remarkably, when averaged over the length of the filaments, the diameter $2 \times R_{flat}$ of the flat inner plateau in the radial profiles is a roughly constant ~0.1 pc for all filaments, at least in the nearby clouds of Gould's belt (cf. *Arzoumanian et al.*, 2011). For illustration, Fig. 5 shows that interstellar filaments from eight nearby clouds are characterized by a narrow distribution of inner full width at half maximum (FWHM) widths centered at about 0.1 pc. In contrast, the range of filament column densities probed with Herschel spans 2 to 3 orders of magnitude (see Fig. 3 of *Arzoumanian et al.*, 2011).

The origin of this quasi-universal inner width of interstellar filaments is not yet well understood (see discussion in section 6.5). A possible hint is that it roughly matches the scale below which the *Larson* (1981) power-law linewidth vs. size relation breaks down in molecular clouds (cf. *Falgarone et al.*, 2009), and a transition to "coherence" is observed between supersonic and subsonic turbulent gas motions [cf. *Goodman et al.* (1998) and section 4.2 below].

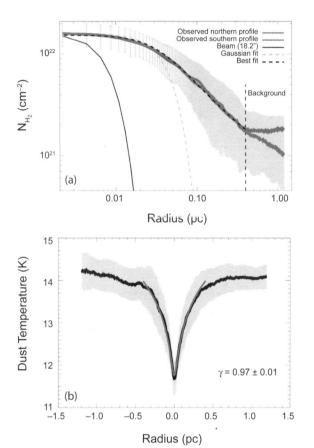

(a)

(b)

$\gamma = 0.97 \pm 0.01$

Fig. 4. See Plate 3 for color version. **(a)** Mean radial column density profile observed perpendicular to the B213/B211 filament in Taurus (*Palmeirim et al.*, 2013), for both the northern and southern part of the filament. The yellow area shows the (±1σ) dispersion of the distribution of radial profiles along the filament. The inner solid curve shows the effective 18" HPBW resolution (0.012 pc at 140 pc) of the Herschel column density map used to construct the profile. The dashed black curve shows the best-fit Plummer model (convolved with the 18" beam) described in equation (1) with p = 2.0 ± 0.4 and a diameter $2 \times R_{flat} = 0.07 \pm 0.01$ pc, which matches the data very well for r ≤ 0.4 pc. **(b)** Mean dust temperature profile measured perpendicular to the B213/B211 filament in Taurus. The solid gray curve shows the best polytropic model temperature profile obtained by assuming $T_{gas} = T_{dust}$ and that the filament has a density profile given by the Plummer model shown in **(a)** (with p = 2) and obeys a polytropic equation of state, $P \propto \rho^\gamma$ [and thus $T(r) \propto \rho(r)^{(\gamma-1)}$]. This best fit corresponds to a polytropic index $\gamma = 0.97 \pm 0.01$ (see *Palmeirim et al.*, 2013, for further details).

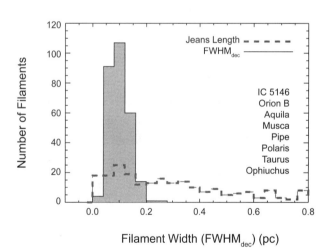

Fig. 5. Histogram of deconvolved FWHM widths for a sample of 278 filaments in eight nearby regions of the Gould belt, all observed with Herschel (at resolutions ranging from ~0.01 pc to ~0.04 pc) and analyzed in the same way (Arzoumanian et al., in preparation) [see *Arzoumanian et al.* (2011) for initial results on a subsample of 90 filaments in three clouds]. The distribution of filament widths is narrow with a median value of 0.09 pc and a standard deviation of 0.04 pc (equal to the bin size). In contrast, the distribution of Jeans lengths corresponding to the central column densities of the filaments (dashed histogram) is much broader.

This may suggest that the filament width corresponds to the sonic scale below which interstellar turbulence becomes subsonic in diffuse, non-star-forming molecular gas (cf. *Padoan et al.,* 2001). For the densest (self-gravitating) filaments, which are expected to radially contract with time (see section 5.1), this interpretation is unlikely to work, however; we speculate in section 6.5 that *accretion* of background cloud material (cf. section 4) plays a key role.

2.6. Converging Filaments and Cluster Formation

As illustrated in Fig. 6 and discussed in detail in section 3.2 and section 6.1, another key result from Herschel is the direct connection between filament structure and the formation of cold dense cores. Filament intersections can provide higher column densities at the location of intersection, enhancing star formation even further. Indeed, localized mass accumulation due to filament mergers may provide the conditions necessary for the onset of clustered star formation. Intersecting filaments as a preferred environment for massive star and cluster formation have been discussed by *Myers* (2009, 2011). For example, *Peretto et al.* (2012) revealed the "hub-filament" structure (see *Myers,* 2009) of merged filaments within the B59 core of the Pipe Nebula cloud. The Pipe Nebula is generally not star-forming, with the sole exception of that occurring in B59 itself. Indeed, star formation may have only been possible in the otherwise dormant Pipe Nebula due to the increased local density (dominance of gravity) in the B59 hub resulting from filament intersection. Merging filaments may also influence the formation of higher-mass stars. For example, merging structures may be dominated by a single massive filament that is being fed by smaller adjacent subfilaments (cf. Figs. 2 and 3). *Hill et al.* (2011) and *Hennemann et al.* (2012) found examples of such merged filaments in the RCW36 and DR21 ridges of Vela C and Cygnus X, respectively [see also *Henshaw et al.* (2013) for an IRDC example]. At those locations, more massive protostellar candidates or high-mass stars are found, possibly since merged filaments have again yielded locally higher densities. Intersecting filaments may provide the very high densities needed locally for cluster formation. In the Rosette molecular cloud, *Schneider et al.* (2012) found a high degree of coincidence between high column densities, filament intersections, and the locations of embedded IR clusters throughout that cloud.

3. DENSE CORE PROPERTIES

3.1. Core Definition and Core-Finding Algorithms

Conceptually, a dense core is an individual fragment or local overdensity that corresponds to a local minimum in the gravitational potential of a molecular cloud. A starless core is a dense core with no associated protostellar object. A prestellar core may be defined as a dense core that is both starless and gravitationally bound. In other words, a prestellar core is a self-gravitating condensation of gas and

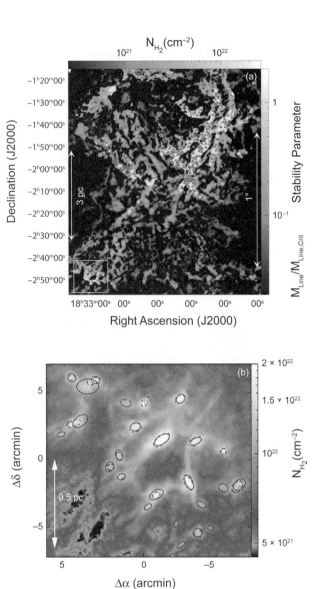

Fig. 6. See Plate 4 for color version. **(a)** Column density map of a subfield of the Aquila star-forming region derived from Herschel data (*André et al.,* 2010). The contrast of the filaments has been enhanced using a curvelet transform (cf. *Starck et al.,* 2003). Given the typical width ~0.1 pc of the filaments (*Arzoumanian et al.,* 2011) (see Fig. 5), this map is equivalent to a map of the mass per unit length along the filaments. The areas where the filaments have a mass per unit length larger than half the critical value $2c_s^2/G$ (cf. *Inutsuka and Miyama,* 1997) (section 5.1) and are thus likely gravitationally unstable are highlighted in white. The bound prestellar cores identified by *Könyves et al.* (2010) are shown as small triangles. **(b)** Close-up column density map of the area shown as a square box on the left. The black ellipses mark the major and minor FWHM sizes of the prestellar cores found with the source extraction algorithm *getsources* (*Men'shchikov et al.,* 2012); four protostellar cores are also shown by star symbols. The dashed ellipses mark the FHWM sizes of the sources independently identified with the *csar* algorithm (*J. Kirk et al.,* 2013). The effective resolution of the image is ~18″ or ~0.02 pc at d ~ 260 pc.

dust within a molecular cloud, which may potentially form an individual star (or system) by gravitational collapse (e.g., *Ward-Thompson et al.*, 1994, 2007; *André et al.*, 2000; *Di Francesco et al.*, 2007). A protostellar core is a dense core within which a protostar has formed. While in general the gravitational potential cannot be inferred from observations, it turns out to be directly related to the observable column density distribution for the postshock, filamentary cloud layers produced by supersonic turbulence in numerical simulations of cloud evolution (*Gong and Ostriker*, 2011). In practical terms, this means that one can define a dense core as the immediate vicinity of a local maximum in observed column density maps such as those derived from Herschel imaging (see Fig. 6 for examples). [In more mathematical terms, the projection of a dense core onto the plane of sky corresponds to a "descending two-manifold" (cf. *Sousbie*, 2011) associated to a local peak P in column density, i.e., the set of points connected to P by integral lines following the gradient of the column density distribution.] One may use significant breaks in the gradient of the column density distribution around the core peak to define the core boundaries. This is clearly a difficult task, however, unless the core is relatively isolated and the instrumental noise in the data is negligible (as is often the case with Herschel observations of nearby clouds).

Prestellar cores are observed at the bottom of the hierarchy of interstellar cloud structures and depart from the *Larson* (1981) self-similar scaling relations [see *Heyer et al.* (2009) for a recent reevaluation of these relations]. They are the smallest units of star formation (e.g., *Bergin and Tafalla*, 2007). To first order, known prestellar cores have simple, convex (and not very elongated) shapes, and their density structures approach that of Bonnor-Ebert (BE) isothermal spheroids bounded by the external pressure exerted by the parent cloud (e.g., *Johnstone et al.*, 2000; *Alves et al.*, 2001; *Tafalla et al.*, 2004). These BE-like density profiles do not imply that prestellar cores are necessarily in hydrostatic equilibrium, however; they are also consistent with dynamic models (*Ballesteros-Paredes et al.*, 2003).

Since cores are intrinsically dense, they can be identified observationally within molecular clouds as compact objects of submillimeter/millimeter continuum emission (or optical/IR absorption). In these cases, the higher densities of cores translate into higher column densities than their surroundings, aiding their detection. Cores can also be identified in line emission, using transitions excited by cold, dense conditions from molecules that do not suffer from freeze-out in those conditions, e.g., low-level transitions of NH_3 and N_2H^+. (Line identification of cores can be problematic if the lines used have high-optical depths and are self-absorbed, like CO. Hyperfine lines of NH_3 and N_2H^+, however, tend to have low optical depths.) In general, the high-mass sensitivities of continuum observations suggest such data are the superior means of detecting cores in molecular clouds. Of course, such emission is influenced by both the column density *and* the temperature of the emitting dust. Moreover, it can be sometimes difficult to disentangle objects along the

line-of-sight (los) without kinematic information, possibly leading to false detections. Even with kinematic information, los superpositions of multiple objects can complicate the derivation of reliable core properties (cf. *Gammie et al.*, 2003). Cores can also be difficult to disentangle if the resolution of the observations is too low and cores are blended. Most of these problems can be largely mitigated with sufficiently high-resolution data; in general, being able to resolve down to a fraction of 0.1 pc (e.g., ~0.03 pc or less) appears to be adequate.

Apart from mapping speed, two key advantages of Herschel's broadband imaging for core surveys are that (1) dust continuum emission is largely bright and optically thin at far-IR/submillimeter wavelengths toward cores and thus directly traces well their column densities (and temperatures), and (2) the ~18″ half-power beamwidth (HPBW) angular resolution of Herschel at $\lambda = 250$ μm, corresponding to ~0.03 pc at a distance d = 350 pc, is sufficient to resolve cores.

In practice, systematic core extraction in wide-field far-IR or submillimeter dust continuum images of highly structured molecular clouds is a complex problem. There has been much debate over which method is the best for identifying extended sources such as dense cores in submillimeter surveys (see, e.g., *Pineda et al.*, 2009; *Reid et al.*, 2010). The problem can be conveniently decomposed into two subtasks: (1) source/core detection, and (2) source/core measurement. In the presence of noise and background cloud fluctuations, the detection of sources/cores reduces to identifying statistically significant intensity/column density peaks based on the information provided by finite-resolution continuum images. The main problem in the measurement of detected sources/cores is finding the spatial extent or "footprint" of each source/core. Previously, submillimeter continuum maps obtained from the ground were spatially filtered (due to atmosphere) and largely monochromatic, so simpler methods [e.g., the eye; *Sandell and Knee* (2001), *clumpfind* (*Williams et al.*, 1994; *Johnstone et al.*, 2000), and *gaussclumps* (*Stutzki and Güsten*, 1990; *Motte et al.*, 1998)] were utilized to identify cores within those data.

Herschel continuum data require more complex approaches, as they are more sensitive than previous maps, retaining information on a wider range of scales. In addition, Herschel continuum data can include up to six bands and have a resolution depending linearly on wavelength. To meet this challenge, the new *getsources* method was devised by *Men'shchikov et al.* (2010, 2012) and used to extract cores in the Herschel Gould Belt Survey (HGBS) data. Two alternative methods that have also been used on Herschel data include *csar* (*J. Kirk et al.*, 2013), a conservative variant of the well-known segmentation routine *clumpfind*, and *cutex* (*Molinari et al.*, 2011), an algorithm that identifies compact sources by analyzing multi-directional second derivatives and performing "curvature" thresholding in a monochromatic image. Comparison of the three methods on the same data shows broad agreement (see, e.g., Fig. 6b), with some differences seen in the case of closely packed groups of cores embedded in a strong background. Generally speaking, a core can be considered a robust detection if it is independently found by more than one algorithm. One merit of the *csar* method is that it can also retain information pertaining to the tree of hierarchical structure within the field, making it similar to other recent dendrogram (*Rosolowsky et al.,* 2008) or substructure codes (*Peretto and Fuller,* 2009).

Once cores have been extracted from the maps, the Herschel observations provide a very sensitive way of distinguishing between protostellar cores and starless cores based on the respective presence or absence of point-like 70-μm emission. Flux at 70 μm can trace very well the internal luminosity of a protostar (e.g., *Dunham et al.,* 2008), and Herschel observations of nearby (d < 500 pc) clouds have the sensitivity to detect even candidate "first hydrostatic cores" (cf. *Pezzuto et al.,* 2012), the very first and lowest-luminosity (~0.01–0.1 $L_\odot$) stage of protostars (e.g., *Larson,* 1969; *Saigo and Tomisaka,* 2011; *Commerçon et al.,* 2012) (see also the chapter by Dunham et al. in this volume).

The Herschel continuum data can also be used to divide the sample of starless cores into gravitationally bound and unbound objects based on the locations of the cores in a mass vs. size diagram (such as the diagram in Fig. 8a) and comparison of the derived core masses with local values of the Jeans or BE mass (see *Könyves et al.,* 2010).

3.2. Spatial Distribution of Dense Cores

As dense cores emit the bulk of their luminosity at far-IR and submillimeter wavelengths, Herschel mapping observations are ideally suited for taking a deep census of such cold objects in nearby molecular cloud complexes. Furthermore, since the maps of the HGBS essentially cover the entire spatial extent of nearby clouds, they provide an unbiased and unprecedented view of the spatial distribution of dense cores within the clouds. The HGBS results show that more than 70% of the prestellar cores identified with Herschel are located within filaments. More precisely, bound prestellar cores and deeply embedded protostars (e.g., class 0 objects) are primarily found in filaments with column densities $N_{H_2} \gtrsim 7 \times 10^{21}$ cm^{-2}. To illustrate, Fig. 6 shows the locations of bound cores identified by *Könyves et al.* (2010) on a column density map of the Aquila Rift cloud derived from the Herschel data (see also *André et al.,* 2010). As can be plainly seen, bound cores are located predominantly in dense filaments (shown in white in Fig. 6a) with supercritical masses per unit length $M_{line} > M_{line,crit}$, where $M_{line,crit} = \frac{2c_s^2}{G}$ is the critical line mass of a nearly isothermal cylindrical filament [see *Inutsuka and Miyama* (1997) and section 5.1] and c_s is the isothermal sound speed. [Throughout this chapter, by supercritical or subcritical filament, we will mean a filament with $M_{line} > M_{line,crit}$ or $M_{line} < M_{line,crit}$, respectively. This notion should not be confused with the concept of a magnetically supercritical or subcritical cloud/core (e.g., *Mouschovias,* 1991).] Interestingly, the median projected spacing ~0.09 pc observed between the prestellar cores of Aquila roughly matches

the characteristic ~0.1 pc inner width of the filaments (see section 2.5). Overall the spacing between Herschel cores is not periodic, although hints of periodicity are observed in a few specific cases (see, e.g., Fig. 8b). Only a small (<30%) fraction of bound cores are found unassociated with any filament or only associated with subcritical filaments. In the L1641 molecular cloud mapped by the HGBS in Orion A, *Polychroni et al.* (2013) report that only 29% of the prestellar cores lie off the main filaments and that these cores tend to be less massive than those found on the main filaments of Orion A. The remarkable correspondence between the spatial distribution of compact cores and the most prominent filaments [Fig. 6a and *Men'shchikov et al.* (2010)] suggests that *prestellar dense cores form primarily by cloud fragmentation along filaments.* Filaments of significant column density (or mass per unit length) are more likely to fragment into self-gravitating cores that form stars. We will return to this important point in more detail in section 6.

3.3. Lifetimes of Cores and Filaments

Observationally, a rough estimate of the lifetime of starless cores can be obtained from the number ratio of cores with and without embedded young stellar objects (YSOs) in a given population (cf. *Beichman et al.,* 1986), assuming a median lifetime of ~2 × 10^6 yr for class II YSOs (cf. *Evans et al.,* 2009). Using this technique, *Lee and Myers* (1999) found that the typical lifetime of starless cores with average volume density ~10^4 cm^{-3} is ~10^6 yr. By considering several samples of isolated cores spanning a range of core densities, *Jessop and Ward-Thompson* (2000) subsequently established that the typical core lifetime decreases as the mean volume density in the core sample increases (see also *Kirk et al.,* 2005).

Core statistics derived from recent HGBS studies of nearby star-forming clouds such as the Aquila complex are entirely consistent with the pre-Herschel constraints on core lifetimes (see *Ward-Thompson et al.,* 2007, and references therein). Figure 7a shows a plot of estimated lifetime vs. average volume density, similar to that introduced by *Jessop and Ward-Thompson* (2000), but for the sample of starless cores identified with Herschel in the Aquila complex (*Könyves et al.,* 2010). As can be seen, the typical lifetime of cores denser than ~10^4 cm^{-3} on average remains ~10^6 yr. Indeed, all estimated core lifetimes lie between one free-fall time (t_{ff}), the timescale expected in free-fall collapse, and 10 × t_{ff}, roughly the timescale expected for highly subcritical cores undergoing ambipolar diffusion (e.g., *Mouschovias,* 1991). Interestingly, Fig. 7a suggests that prestellar cores denser than ~10^6 cm^{-3} on average evolve essentially on a free-fall timescale. Note, however, that statistical estimates of core lifetimes in any given region are quite uncertain since it is assumed all observed cores follow the same evolutionary path and that the core/star formation rate is constant. Nevertheless, the fact that similar results are obtained in different regions suggests that the timescales given by a

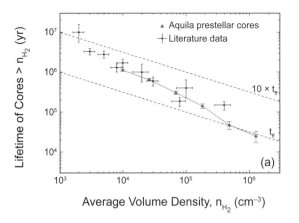
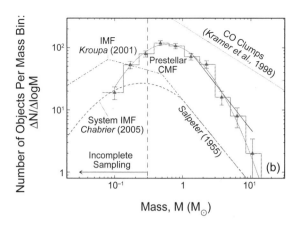

Fig. 7. See Plate 5 for color version. **(a)** Plot of estimated lifetime against minimum average core density (after *Jessop and Ward-Thompson,* 2000) for the population of prestellar and starless cores identified with Herschel in the Aquila cloud complex (blue triangles; Könyves et al., in preparation) and literature data [black crosses; see *Ward-Thompson et al.* (2007) and references therein]. Note the trend of decreasing core lifetime with increasing average core density. For comparison, the dashed lines correspond to the free-fall timescale (t_{ff}) and a rough approximation of the ambipolar diffusion timescale ($10 \times t_{ff}$). **(b)** Core mass function (histogram with error bars) of the ~500 candidate prestellar cores identified with Herschel in Aquila (*Könyves et al.,* 2010; *André et al.,* 2010). The IMF of single stars [corrected for binaries; e.g., *Kroupa* (2001)], the IMF of multiple systems (e.g., *Chabrier,* 2005), and the typical mass spectrum of CO clumps (e.g., *Kramer et al.,*1998) are shown for comparison. A lognormal fit to the observed CMF is superimposed; it peaks at ~0.6 $M_\odot$, close to the Jeans mass within marginally critical filaments at T ⋅ 10 K (see section 6.3).

plot such as Fig. 7a are at least approximately correct (i.e., within factors of ~2–3).

It is not clear yet whether density is the only parameter controlling a core's lifetime or whether mass also plays an additional role. Several studies suggest that massive prestellar cores (i.e., precursors to stars >8 $M_\odot$), if they exist at all, are extremely rare with lifetimes shorter than the free-fall timescale (e.g., *Motte et al.,* 2007; *Tackenberg et al.,* 2012). These findings may be consistent with the trend seen in Fig. 7a and the view that massive stars can only form from very dense structures. Alternatively, they may indicate that core evolution is mass dependent (see *Hatchell and Fuller,* 2008) or that massive protostars acquire the bulk of their mass from much larger scales than a single prestellar core (see *Peretto et al.,* 2013) (section 4.3).

The above lifetime estimates for low-mass prestellar cores, coupled with the result that a majority of the cores lie within filaments (section 3.2), suggest that dense, star-forming filaments such as the B211/B213 filament in Taurus (cf. Fig. 2) survive for at least ~10^6 yr. A similar lifetime can be inferred for lower-density, non-star-forming filaments (cf. Fig. 1) from a typical sound-crossing time ≥5 × 10^5 yr, given their characteristic width ~0.1 pc (section 2.5) and nearly sonic internal velocity dispersion (section 4.4 and Fig. 9b).

3.4. Core Mass Functions

Using *getsources*, about 200 (class 0 and class I) protostars were identified in the Herschel images of the whole (~11 deg^2) Aquila cloud complex (*Bontemps et al.,* 2010; *Maury et al.,* 2011), along with more than 500 starless

cores ~0.01–0.1 pc in size (see Fig. 6 for some examples). On a mass vs. size diagram, most (>60%) Aquila starless cores lie close to the loci of critical Bonnor-Ebert spheres, suggesting that the cores are self-gravitating and prestellar in nature (*Könyves et al.,* 2010). The core mass function (CMF) derived for the entire sample of >500 starless cores in Aquila is well fit by a lognormal distribution and closely resembles the initial mass function (IMF) [Fig. 7b (*Könyves et al.,* 2010; *André et al.,* 2010)]. The similarity between the Aquila CMF and the *Chabrier* (2005) system IMF is consistent with an essentially one-to-one correspondence between core mass and stellar system mass ($M_{\star sys} = \varepsilon_{core}M_{core}$). Comparing the peak of the CMF to the peak of the system IMF suggests that the efficiency ε_{core} of the conversion from core mass to stellar system mass is between ~0.2 and ~0.4 in Aquila.

The first HGBS results on core mass functions therefore confirm the existence of a close similarity between the prestellar CMF and the stellar IMF, using a sample ~2–9× larger than those of earlier groundbased studies (cf. *Motte et al.,* 1998; *Johnstone et al.,* 2000; *Stanke et al.,* 2006; *Alves et al.,* 2007; *Enoch et al.,* 2008). The efficiency factor ε_{core} ~ 30% may be attributable to mass loss due to the effect of outflows during the protostellar phase (*Matzner and McKee,* 2000). More work is needed, however, to derive a reliable prestellar CMF at the low-mass end and fully assess the potential importance of subtle observational biases (e.g., background-dependent incompleteness and blending of unresolved groups of cores). The results from the full HGBS will also be necessary to characterize fully the nature of the CMF-IMF relationship as a function of environment.

Furthermore, as pointed out by, e.g., *Ballesteros-Paredes et al.* (2006), establishing a *physical* link between the CMF and the IMF is not a straightforward task.

The findings based on early analysis of the Herschel data nevertheless tend to support models of the IMF based on precollapse cloud fragmentation such as the gravoturbulent fragmentation picture (e.g., *Larson,* 1985; *Klessen and Burkert,* 2000; *Padoan and Nordlund,* 2002; *Hennebelle and Chabrier,* 2008). Independently of any model, the Herschel observations suggest that one of the keys to the problem of the origin of the IMF lies in a good understanding of the formation mechanism of prestellar cores. In addition, further processes, such as rotational subfragmentation of prestellar cores into binary/multiple systems during collapse (e.g., *Bate et al.,* 2003; *Goodwin et al.,* 2008) and competitive accretion at the protostellar stage (e.g., *Bate and Bonnell,* 2005), may also play a role and, e.g., help populate the low- and high-mass ends of the IMF, respectively (see the chapter by Offner et al. in this volume).

3.5. Observational Evidence of Core Growth

Herschel observations have uncovered a large population of faint, unbound starless cores (*Ward-Thompson et al.,* 2010; *André et al.,* 2010), not seen in earlier submillimeter continuum data, that may be the precursors to the better-characterized population of prestellar cores.

As an example, Fig. 8a (Plate 6a) shows the mass vs. size relation for sources found in the Herschel data of the S1 and S2 regions of Taurus. The blue triangles represent the starless and prestellar cores, red stars are cores with embedded protostars and black stars are T Tauri stars. Prior to Herschel, cores seen in the submillimeter continuum and CO cores were found to lie in very different areas of the mass-size plane (*Motte et al.,* 2001). It was not known if this difference was due to selection effects, or whether there were two distinct (but possibly related) populations of objects, i.e., low-density non-star-forming starless cores (cf. *Ward-Thompson et al.,* 2010) and higher-density prestellar cores (*Ward-Thompson et al.,* 1994, 2007).

Figure 8 shows that Herschel cores in the S1/S2 region of Taurus lie in the region of mass-size diagram partly overlapping the CO cores, partly in the region of the groundbased continuum cores and partly in between (see also *Kirk et al.,* 2013). This rules out the possibility that there are two unrelated populations of objects. Instead, at least a fraction of CO clumps and lower density cores may evolve into prestellar cores by accreting additional cloud mass and migrating across the diagram of Fig. 8a, before collapsing to form protostars. Likewise, *Belloche et al.* (2011) argued that a fraction (~20–50%) of the unbound starless cores found in their Large APEX Bolometer Camera (LABOCA) observations of Chamaeleon may be still growing in mass and turn prestellar in the future.

Given that a majority of cores lie within filaments, we speculate that this process of core growth occurs primarily through filamentary accretion (cf. Fig. 8b). Interestingly,

core growth through filamentary accretion may have been detected toward the starless core L1512 in Taurus (*Falgarone et al.,* 2001) [see also *Schnee et al.* (2007) for TMC-1C], and is seen in a number of numerical models. *Balsara et al.* (2001) were among the first to discuss accretion via filaments and showed that their model was consistent with observations of S106. *Smith et al.* (2011, 2012) described accretion flows from filaments onto cores, and *Gómez and Vázquez-Semadeni* (2013) discussed the velocity field of the filaments and their environment, and how the filaments can refocus the accretion of gas toward embedded cores.

Figure 8a also shows tracks of core evolution in the mass vs. size diagram proposed by *Simpson et al.* (2011). In this picture, cores evolve by accreting mass quasistatically and by maintaining Bonnor-Ebert equilibrium. In this phase, they move diagonally to the upper right in the mass vs. size diagram. Upon reaching Jeans instability, a core collapses to form a protostar, moving leftward in the diagram. However, protostellar cores lie beyond the Jeans limit, in the upper left part of the diagram, near the resolution limit of the observations (they have been given an additional 0.3 $M_\odot$ to account for mass already accreted onto the protostar).

To substantiate this evolutionary picture, *Simpson et al.* (2011) showed that cores in L1688 (Ophiuchus) showing weak signs of infall [i.e., through blue-asymmetric double-peaked line profiles (see *Evans,* 1999)] appeared close to, or beyond, the Jeans mass line while cores exhibiting characteristic infall signatures all appeared beyond it. In L1688, 40% of the cores beyond the Jeans mass line showed signs of infall; 60% if those exhibiting possible blue-asymmetric line profiles were included.

Using this diagram, an approximate final stellar mass could be derived for any given core by following a Bonnor-Ebert track diagonally upward to the Jeans mass line. From that point on in the evolution, an efficiency for the collapse of the core must be assumed to reach an estimate of the final stellar mass. If a core continues to accrete while collapsing, its evolutionary track will also move slightly upward after crossing the Jeans mass and turning left. Timescale arguments suggest, however, that accretion of background material cannot play a major role after the onset of protostellar collapse, at least for low-mass protostars (cf. *André et al.,* 2007). High-mass protostars may differ significantly from their low-mass counterparts since there is mounting evidence of parsec-scale filamentary accretion onto massive protostellar cores (e.g., *Peretto et al.,* 2013).

4. KINEMATICS OF FILAMENTS AND CORES

4.1. Evidence of Velocity-Coherent Structures

Molecular line observations of several regions have also shown the presence of filamentary structures. Single-dish observations of a few regions in the nearby Taurus cloud using molecular lines revealed that several filaments are identified in velocity, although in most cases they overlap on the plane of sky (*Hacar and Tafalla,* 2011; *Hacar et*

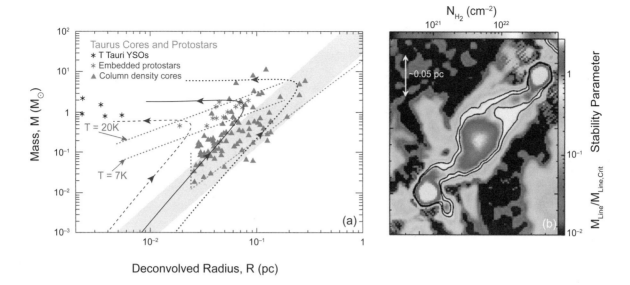

Fig. 8. See Plate 6 for color version. **(a)** Mass vs. size diagram for the cores identified with Herschel in the Taurus S1/S2 region (*J. Kirk et al.,* 2013). The two solid straight lines mark the loci of critically self-gravitating isothermal Bonnor-Ebert spheres at T = 7 K and T = 20 K, respectively, while the gray band shows the region of the mass-size plane where low-density CO clumps are observed to lie (cf. *Elmegreen and Falgarone,* 1996). The dotted curve indicates both the mass sensitivity and the resolution limit of the *J. Kirk et al.* (2013) study. Model evolutionary tracks from *Simpson et al.* (2011) are superposed as black curves (dashed, solid, and dotted for ambient pressures $P_{ext}/k_B = 10^6$ K cm^{-3}, 10^5 K cm^{-3}, and 10^4 K cm^{-3}, respectively), in which starless cores (filled triangles) first grow in mass by accretion of background material before becoming gravitationally unstable and collapsing to form protostars (star symbols) and T Tauri stars (darker star symbols). **(b)** Blow-up Herschel column density map of a small portion of the Taurus B211/B213 filament (cf. Fig. 2a) showing two protostellar cores on either side of a starless core (Palmeirim et al., in preparation). The contrast of filamentary structures has been enhanced using a curvelet transform (see *Starck et al.,* 2003). Note the bridges of material connecting the cores along the B211/B213 filament, suggesting that the central starless core may still be growing in mass through filamentary accretion, a mechanism seen in some numerical simulations (cf. *Balsara et al.,* 2001) (section 5).

al., 2013). In these cases, the centroid velocity along the filaments does not vary much, suggesting that the filaments are coherent (i.e., "real") structures. For L1495/B211/B213 in Taurus, the average density was derived from line data, and used to estimate that the thickness of the structure along the line of sight is consistent with a cylinder-like structure instead of a sheet seen edge-on (*Li and Goldsmith,* 2012). Also, some of these filaments contain dense cores that share the kinematic properties of the associated filaments, suggesting that these cores form through filament fragmentation.

An excellent example is the L1495/B211 filament, observed both in dust continuum by the HGBS (*Palmeirim et al.,* 2013) (see Fig. 2a) and in various lines by *Hacar et al.* (2013). These data allow for a direct comparison between properties derived from each tracer. In the Herschel data, *Palmeirim et al.* (2013) identified this filament as a single, thermally supercritical structure >5 pc in length. In the molecular line data, on the other hand, *Hacar et al.* (2013) identified 35 different filamentary components with typical lengths of 0.5 pc, with usually a couple of components overlapping on the plane of sky but having distinct velocities. The masses per unit length (line masses) of these components or "fibers" are close to the stability value (see section 5.1). Interestingly, most of the line-identified fibers can also be

seen in the Herschel dust continuum observations when large-scale emission is filtered out, enhancing the contrast of small-scale structures in the data (cf. Fig. 2b). The *Hacar et al.* (2013) results suggest that some supercritical Herschel-identified filaments consist of bundles of smaller and more stable fibers. Other examples of thermal fibers include the narrow ones revealed by high-angular resolution (6″) NH$_3$ observations of the B5 region in Perseus (*Pineda et al.,* 2011).

4.2. Transition to Coherence

Dense cores in general exhibit velocity dispersions where nonthermal motions (e.g., turbulence) are smaller than thermal motions (*Myers,* 1983). *Goodman et al.* (1998) and *Caselli et al.* (2002) coined the term "coherent core" to describe the location where nonthermal motions are subsonic (or at most transonic) and roughly constant. *Goodman et al.* (1998) further showed that the lower-density gas around cores, as traced by OH and C^{18}O, have supersonic velocity dispersions that decrease with size, as expected in a turbulent flow. Meanwhile, NH$_3$ line emission, tracing dense core gas, shows a nearly constant, nearly thermal width. Therefore, a transition must occur at some point between turbulent gas and more quiescent gas.

More than a decade later, *Pineda et al.* (2010) obtained a large, sensitive NH_3 map toward B5 in Perseus. This map showed that the sonic region within B5 is surrounded by NH_3 emission displaying supersonic turbulence. The transition between subsonic and supersonic turbulence (e.g., an increase in the gas velocity dispersion by a factor of 2) occurs in less than a beam width (<0.04 pc). The typical size of the sonic region is ~0.1 pc, similar to the width of filaments seen by Herschel (see section 2.5).

Transitions in turbulence levels have also been observed in other clouds [e.g., L1506 in Taurus (*Pagani et al.*, 2010)] and other parts of the Perseus molecular cloud (Pineda et al., in preparation). Both fields with lower levels of star-formation activity (e.g., L1451 and B5) and more active fields (e.g., L1448 and IC348-SW) show well-defined sonic regions completely surrounded by more turbulent gas when mapped in NH_3 at high resolution. Moreover, the transition between sonic and more turbulent gas motions is sharp (less than a beam width) in all fields, regardless of their environment or level of star-formation activity.

From these results, it is clear that regions of subsonic turbulence mark the locations where material is readily available for star-formation. It is very likely that these sonic regions are formed in the central parts of the filaments identified with Herschel. If the sonic regions are massive enough, further substructures, like the isothermal fibers observed in B5, could develop. It should also be stressed that these ~0.1-pc sonic regions are much larger than the milliparsec-scale structures in which subsonic interstellar turbulence is believed to dissipate (*Falgarone et al.*, 2009).

4.3. Subfilaments and Low-Density Striations

The material within filaments is not static. In fact, several lines of evidence have shown that (1) material is being added to filaments from striations, and (2) filaments serve as highly efficient routes for feeding material into hubs or ridges where clustered star formation is ongoing.

In the low-mass case, the Taurus cloud presents the most striking evidence for striations along filaments. CO maps tracing low-density gas show the clear presence of striations in B211/B213 as low-level emission located away from the denser filament (*Goldsmith et al.*, 2008). These striations are also prominent in Herschel dust continuum maps of the region from the HGBS (*Palmeirim et al.*, 2013) (see Figs. 2 and 3). These results suggest that material may be funneled through the striations onto the main filament. The typical velocities expected for the infalling material in this scenario are ~0.5–1 km s⁻¹, which are consistent with the kinematical constraints from the CO observations.

In denser cluster-forming clumps like the Serpens South protocluster in Aquila or the DR21 ridge in Cygnus X, filaments are observed joining into a central hub where star formation is more active (*H. Kirk et al.*, 2013, see Fig. 9a; see also *Schneider et al.*, 2010). Single-dish line observations show classical infall profiles along the filaments, and infall rates of 10^{-4}–10^{-3} $M_\odot$ yr⁻¹ are derived. Moreover,

velocity gradients toward the central hub suggest that material is flowing along the filaments, at rates similar to the current local star-formation rate (SFR).

More recently, Atacama Large Millimeter Array (ALMA) interferometric observations of N_2H^+ emission from a more distant IRDC, SDC335, hosting high-mass star formation, show the kinematic properties of six filaments converging on a central hub (*Peretto et al.*, 2013). Velocity gradients along these filaments are seen, suggesting again that material is being gathered in the central region through the filaments at an estimated rate of 2.5×10^{-3} $M_\odot$ yr⁻¹. This accretion rate is enough to double the mass of the central region in a free-fall time.

4.4. Internal Velocity Dispersions

Given the wide range of column densities observed toward Herschel-identified filaments (see section 2.5), can turbulent motions provide effective stability to filaments across the entire range? Naively, Larson's relation ($\sigma_v \propto R^{0.5}$) can give a rough estimate of the level of turbulent motions within filaments.

Heyer et al. (2009) used the Five College Radio Astronomy Observatory (FCRAO) Galactic Ring Survey ^{13}CO (1–0) data to study line widths across a wide range of environments. They focused on the kinematical properties of whole molecular clouds, which likely include filaments like those seen with Herschel. They found that the velocity dispersion is dependent on the column density, a deviation from Larson's relation. For example, filaments with higher column density have a higher level of turbulent motions than is expected from Larson's relation. These linewidth estimates, however, were not carried out toward specific filaments, and a more focused study on filaments was needed to properly assess the role of nonthermal motions in the dynamical stability of filaments.

Recently, *Arzoumanian et al.* (2013) presented molecular line measurements of the internal velocity dispersions in 46 Herschel-identified filaments. Figure 9b shows that the level of turbulent motions is almost constant and does not dominate over thermal support for thermally subcritical and nearly critical filaments. On the other hand, a positive correlation is found between the level of turbulent motions and the filament column density for thermally supercritical filaments. This behavior suggests that thermally subcritical filaments are gravitationally unbound with transonic line-of-sight velocity dispersions, but thermally supercritical filaments are in approximate virial equipartition (but not necessarily virial equilibrium).

Moreover, it may be possible to glean aspects of filament evolution from plots like Fig. 9b. For example, once a filament becomes thermally supercritical (i.e., self-gravitating) it will contract and increase its central column density. Meanwhile, the accretion of material onto the filament (e.g., through striations or subfilaments) will increase its internal velocity dispersion and virial mass per unit length. Following this line of reasoning, *Arzoumanian et*

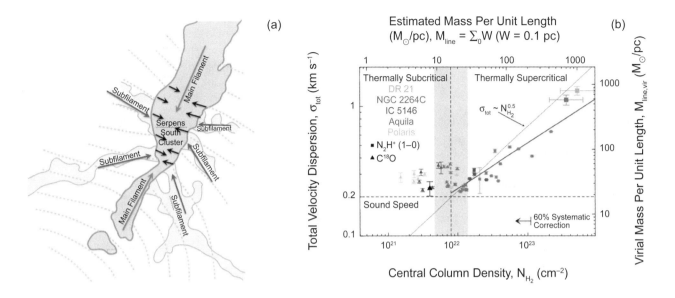

Fig. 9. See Plate 7 for color version. **(a)** Sketch of the typical velocity field inferred from line observations (e.g., *H. Kirk et al., 2013*) within and around a supercritical filament [here Serpens-South in the Aquila complex; adapted from *Sugitani et al.* (2011)]: Two long arrows mark longitudinal infall along the main filament; short black arrows indicate radial contraction motions; and other arrows mark accretion of background cloud material through striations or subfilaments, along magnetic field lines (dotted curves from Sugitani et al.). **(b)** Total (thermal + nonthermal) velocity dispersion vs. central column density for a sample of 46 filaments in nearby interstellar clouds (*Arzoumanian et al., 2013*). The horizontal dashed line shows the value of the thermal sound speed ~0.2 km s^{-1} for T = 10 K. The vertical gray band marks the border zone between thermally subcritical and thermally supercritical filaments where the observed mass per unit length M_{line} is close to the critical value $M_{line,crit} \sim 16$ $M_\odot$/pc for T = 10 K. The black straight line shows the best power-law fit $\sigma_{tot} \propto N_{H_2}^{0.36 \pm 0.14}$ to the data points corresponding to supercritical filaments.

al. (2013) speculated that this accretion process may allow supercritical filaments to maintain approximately constant inner widths while evolving. We will return to this point in section 6.5.

5. THEORETICAL MODELS

5.1. Basic Physics of Self-Gravitating Filaments in the Interstellar Medium

Theoretical treatments of filaments in molecular clouds have focused on several different kinds of possible filament states: (1) equilibria, (2) collapsing and fragmenting systems that follow from unstable equilibria, (3) equilibria undergoing considerable radial accretion, and finally (4) highly dynamical systems for which equilibrium models are not good descriptions. Indeed, all these aspects of filamentary cloud states may be at play in different clouds or at different times and regions in the same cloud. The fact that star-forming cores are observed preferentially in filaments that are predicted to be gravitationally unstable on the basis of equilibrium models (cf. Fig. 6a) implies that some aspects of basic equilibrium theory are relevant in real clouds. We turn first to these results.

Early papers on filament structure, such as the classic solution for a self-gravitating isothermal filament (*Stodolk-*

iewicz, 1963; *Ostriker,* 1964), assumed that filaments are in cylindrical hydrostatic equilibrium.

If we multiply Poisson's equation for self-gravity in a cylinder of infinite length by r on both sides and integrate from the center to the outermost radius r = R, we obtain

$$R \left. \frac{d\Phi}{dr} \right|_{r=R} = 2G \int_0^R 2\pi r \rho dr \equiv 2GM_{line} \qquad (2)$$

where Φ denotes gravitational potential and M_{line} is the mass per unit length (or line mass) defined as $M_{line} = \int_0^R 2\pi r \rho dr$. This quantity remains constant in a change of the cylinder radius where $\rho \propto R^{-2}$. Thus, the self-gravitational force per unit mass $F_{g,cyl}$ is

$$F_{g,cyl} = \left. \frac{d\Phi}{dr} \right|_{r=R} = 2 \frac{GM_{line}}{R} \propto \frac{1}{R} \qquad (3)$$

On the other hand, if we denote the relation between pressure P and density ρ as $P = K\rho^{\gamma eff}$, the pressure gradient force per unit mass F_p scales as

$$F_p = \frac{1}{\rho} \frac{\partial P}{\partial r} \propto R^{1-2\lambda_{eff}} \qquad (4)$$

Therefore, we have

$$\frac{F_p}{F_{g,cyl}} \propto R^{2-2\lambda_{eff}} \tag{5}$$

If $\gamma_{eff} > 1$, the pressure gradient force will dominate self-gravity for a sufficiently small outer radius R. In contrast, if $\gamma_{eff} \leq 1$, radial collapse will continue indefinitely once self-gravity starts to dominate over the pressure force. Therefore, we can define the critical ratio of specific heats for the radial stability of a self-gravitating cylinder as $\gamma_{crit,cyl} = 1$. In the case of $\gamma_{eff} = 1$, and under a sufficiently small external pressure, a cylinder will be in hydrostatic equilibrium only when its mass per unit length has the special value for which $F_p = F_{g,cylinder}$. Therefore, we can define a critical mass per unit length for an isothermal cylinder

$$M_{line,crit} \equiv \int_0^\infty 2\pi\rho_4(r)rdr = \frac{2c_s^2}{G} \tag{6}$$

where $\rho_4(r)$ is given by equation (1) in section 2.4 with p = 4. Note that, when γ_{eff} is slightly less than 1 (as observed; cf. Fig. 4b), the maximum line mass of *stable* cylinders is close to the critical value $M_{line,crit}$ for an isothermal cylinder.

Using similar arguments, the critical γ_{eff} is found to be $\gamma_{crit,sphere} = 4/3$ for a sphere and $\gamma_{crit,sheet} = 0$ for a sheet. The thermodynamical behavior of molecular clouds corresponds to $\gamma_{eff} \leq 1$ (e.g., *Koyama and Inutsuka*, 2000) (see also Fig. 4b for the related anticorrelation between dust temperature and column density). Therefore, the significance of filamentary geometry can be understood in terms of ISM thermodynamics. For a sheet-like cloud, there is always an equilibrium configuration since the internal pressure gradient can always become strong enough to halt the gravitational collapse independently of its initial state (e.g., *Miyama et al.*, 1987; *Inutsuka and Miyama*, 1997). In contrast, the radial collapse of an isothermal cylindrical cloud cannot be halted and no equilibrium is possible when the mass per unit length exceeds the critical value $M_{line,crit}$. Conversely, if the mass per unit length of the filamentary cloud is less than $M_{line,crit}$, gravity can never be made to dominate by increasing the external pressure, so that the collapse is always halted at some finite cylindrical radius.

Another way of describing this behavior is in terms of the effective gravitational energy per unit length of a filament, denoted $\mathcal{W}$, which takes a simple form (*Fiege and Pudritz*, 2000)

$$\mathcal{W} = -\int \rho r \frac{d\Phi}{dr} d\mathcal{V} = -M_{line}^2 G \tag{7}$$

where $\mathcal{V}$ is the volume per unit length (i.e., the cross-sectional area) of the filament. $\mathcal{W}$ is the gravitational term in the virial equation. It is remarkable in that it remains constant during any radial contraction caused by an increased external pressure. The scaling was first derived by *McCrea*

(1957). If a filament is initially in equilibrium, the effective gravitational energy can never be made to dominate over the pressure support by squeezing it (*McCrea*, 1957; *Fiege and Pudritz*, 2000). Thus, filaments differ markedly from isothermal spherical clouds, which can always be induced to collapse by a sufficient increase in external pressure (e.g., *Bonnor*, 1956; *Shu*, 1977).

The critical mass per unit length $M_{line,crit} \approx 16 \, M_\odot \, pc^{-1} \times (T_{gas}/10 \, K)$, as originally derived, depends only on gas temperature T_{gas}. This expression can be readily generalized to include the presence of nonthermal gas motions. In this case, the critical mass per unit length becomes $M_{line,vir} = 2\sigma_{tot}^2/G$, also called the virial mass per unit length, where $\sigma_{tot} = \sqrt{c_s^2 + \sigma_{NT}^2}$ is the total one-dimensional velocity dispersion including both thermal and nonthermal components (*Fiege and Pudritz*, 2000). Clearly, both the equation of state of the gas and filament turbulence will play a role in deciding this critical line mass.

Magnetic fields also play an important role in filamentary equilibria. The most general magnetic geometry consists of both poloidal as well as toroidal components. The latter component arises quite generally in oblique shocks as well as in shear flows, both of which impart a sense of local spin to the gas and twisting of the poloidal field lines (see the chapter by H.-B. Li et al. in this volume). Toroidal fields also are generated during the collapse of magnetized cores (e.g., *Mouschovias and Paleologou*, 1979) and carry off substantial fluxes of angular momentum (see the chapter by Z.-Y. Li et al. in this volume). The general treatment of the virial theorem for magnetized gas (*Fiege and Pudritz*, 2000) shows that the poloidal field that threads a filament provides a pressure that helps support a filament against gravity. On the other hand, the toroidal field contribution to the total magnetic energy is destabilizing; it squeezes down on the surface of the filament and assists gravity. Thus the total magnetic energy can be negative in regions where the toroidal field dominates, and positive where the poloidal field dominates. *Fiege and Pudritz* (2000) showed that $M_{line,vir}$ is modified in the case of magnetized filaments, i.e., $M_{line,vir}^{mag} = M_{line,vir}^{unmag} \times (1 - \mathcal{M}/|\mathcal{W}|)^{-1}$, where $\mathcal{M}$ is the magnetic energy per unit length (positive for poloidal fields and negative for toroidal fields). While molecular gas in dense regions typically has $|\mathcal{M}|/|\mathcal{W}| \leq 1/2$ (*Crutcher*, 1999, 2012), there are large variations over these measurements. For the typical case, $M_{line,vir}^{mag}$ differs from $M_{line,vir}^{unmag} \equiv 2\sigma_{tot}^2/G$ by less than a factor of 2. Radial profiles for toroidal isothermal models follow an $\rho \propto r^{-2}$ form, in agreement with observations (cf. section 2.4).

The essence of a virial theorem analysis of filaments can be demonstrated in a diagram where the ratio of observed surface pressure to total turbulent pressure $(P_S/\langle P \rangle)$ of a filament is plotted against the ratio of its observed mass per unit length to the critical value $(M_{line}/M_{line,vir})$ (see *Fiege and Pudritz*, 2000). The virial theorem for filaments gives $P_S/\langle P \rangle = 1 - \{(M_{line}/M_{line,vir}) \times [1 - (\mathcal{M}/|\mathcal{W}|)]\}$. In this diagram, purely hydrodynamic filaments can be easily discriminated from filaments with net positive or negative magnetic energy. Analysis by *Contreras et al.* (2013) of a collection of filaments in the ATLASGAL catalog of 870-μm images as well as ^{13}CO data revealed that most of the filaments fell in the part of this virial diagram where toroidal fields dominate (see Fig. 3 of *Contreras et al.,* 2013). In addition, work by *Hernandez and Tan* (2011) on several IRDC filaments indicated the presence of large surface pressures $(P_S/\langle P\rangle) > 1$, suggesting that these systems may be quite out of equilibrium.

Even moderate fields can have a very strong effect on the formation of unstable regions in filaments (*Tilley and Pudritz,* 2007; *Li et al.,* 2010). Figure 10b shows that rather flattened filaments form in three-dimensional MHD simulations because the magnetic field strongly affects the turbulent flow (*Li et al.,* 2010), forcing material to flow and accumulate along field lines. The magnetic Jeans mass, $M_J = (B_0/2\pi G^{1/2}\rho_0^{2/3})^3$, exceeds the thermal Jeans mass in this case, suppressing the formation of intermediate-mass stars but leaving the more massive stars relatively unaffected.

The fragmentation properties of filaments and sheets differ from those of spheroidal clouds in that there is a preferred scale for gravitational fragmentation that directly scales with the scale height of the filamentary or sheet-like medium (e.g., *Larson,* 1985) (see also section 5.3). In the spherical case, the largest possible mode (i.e., overall collapse of the medium) has the fastest growth rate so that global collapse tends to overwhelm the local collapse of finite-sized density perturbations, and fragmentation is generally suppressed in the absence of sufficiently large initial density enhancements (e.g., *Tohline,* 1982). It is also well known that spherical collapse quickly becomes strongly centrally concentrated (*Larson,* 1969; *Shu,* 1977), which tends to produce a single central density peak rather than several condensations (e.g., *Whitworth et al.,* 1996). In contrast, sheets have a natural tendency to fragment into filaments (e.g., *Miyama et al.,* 1987) and filaments with masses per unit length close to $M_{line,crit}$ a natural tendency to fragment into spheroidal cores (e.g., *Inutsuka and Miyama,* 1997). The filamentary geometry is thus the most favorable configuration for small-scale perturbations to collapse locally and grow significantly before global collapse overwhelms them (*Pon et al.,* 2011, 2012; *Toalá et al.,* 2012). To summarize, theoretical considerations alone emphasize the potential importance of filaments for core and star formation.

5.2. Origin of Filaments: Various Formation Models

Since the mid 1990s, simulations of supersonic turbulence have consistently shown that gas is rapidly compressed into a hierarchy of sheets and filaments (e.g., *Porter et al.,* 1994; *Vázquez-Semadeni,* 1994; *Padoan et al.,* 2001; see Fig. 10). Furthermore, when gravity is added into turbulence simulations, the denser gas undergoes gravitational collapse to form stars (e.g., *Ostriker et al.,* 1999; *Ballesteros-Paredes et al.,* 1999; *Klessen and Burkert,* 2000; *Bonnell et al.,* 2003; *Mac Low and Klessen,* 2004; *Tilley and Pudritz,* 2004; *Krumholz et al.,* 2007). There are many

sources of supersonic turbulent motions in the ISM out of which molecular clouds can arise, i.e., galactic spiral shocks in which most GMCs form, supernovae, stellar winds from massive stars, expanding Hɪɪ regions, radiation pressure, cosmic-ray streaming, Kelvin-Helmholtz and Rayleigh-Taylor instabilities, gravitational instabilities, and bipolar outflows from regions of star formation (*Elmegreen and Scalo,* 2004). An important aspect of supersonic turbulent shocks is that they are highly dissipative. Without constant replenishment, they damp within a cloud crossing time.

Filaments are readily produced in the complete absence of gravity. Simulations of turbulence often employ a spectrum of plane waves that are random in direction and phase. As is well known, the crossing of two planar shock wave fronts is a line: the filament (e.g., *Pudritz and Kevlahan,* 2013). The velocity field in the vicinity of such a filament shows a converging flow perpendicular to the filament axis as well as turbulent vortices in the filament wake. *Hennebelle* (2013) reported that filaments can be formed by the velocity shear that is ubiquitous in magnetized turbulent media (see also *Passot et al.,* 1995). Others have attributed filaments to being stagnation regions in turbulent media, where filaments are considered to be rather transient structures (e.g., *Padoan et al.,* 2001). *Li et al.* (2010) have shown that filaments are formed preferentially perpendicular to the magnetic field lines in strongly magnetized turbulent clouds (see also *Basu et al.,* 2009). An important and observationally desirable aspect of supersonic turbulence is that it produces a hierarchical structure that can be described by a lognormal density distribution (*Vázquez-Semadeni,* 1994). A lognormal arises whenever the probability of each new density increment in the turbulence is independent of the previous one. In a shocked medium, the density at any point is the product of the shock-induced, density jumps and multiplicative processes of this kind rapidly converge to produce lognormals as predicted by the central limit theorem (*Kevlahan and Pudritz,* 2009).

As the mass accumulates in a filament, it becomes gravitationally unstable and filamentary flows parallel to the filament axis take place as material flows toward a local density enhancement (e.g., *Banerjee et al.,* 2006). Since the advent of sink particles to trace subregions that collapse to form stars, simulations have consistently shown that there is a close association between sink particles and filaments (e.g., *Bate et al.,* 2003). The mechanism of the formation of such fragments in terms of gravitational instabilities above a critical line mass does not seem to have been strongly emphasized in the simulation literature until recently, however.

Simulations show that as gravity starts to become more significant, it could be as important as turbulence in creating filamentary web structures and their associated "turbulent" velocity fields in denser collapsing regions of clouds. Simulations of colliding gas streams show that the resulting dense clouds are always far from equilibrium even though a simplistic interpretation of the virial theorem might indicate otherwise. The point is that as clouds build

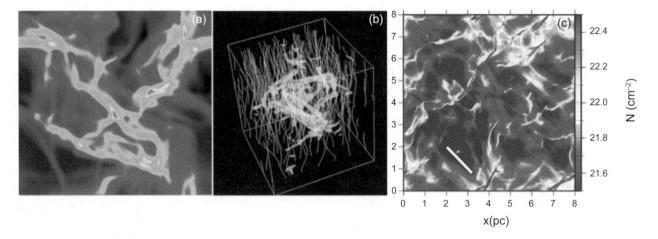

Fig. 10. See Plate 8 for color version. **(a)** Formation of filaments and cores from shock compression in numerical simulations of isothermal supersonic turbulence with rms Mach number ~10 (*Padoan et al.*, 2001). **(b)** Three-dimensional view of flattened filaments resulting from mass accumulation along magnetic field lines in the MHD model of *Li et al.* (2010). The magnetic field favors the formation of both low- and high-mass stars compared to intermediate-mass stars, due to fragmentation of high-density filaments and global clump collapse, respectively. **(c)** Face-on column density view of a shock-compressed dense layer in numerical MHD simulations by Inutsuka et al. (in preparation). The gray scale (in cm^{-2}) is shown on the right. The mean magnetic field is in the plane of the layer and its direction is shown by a white bar. Note the formation of dense, magnetized filaments whose axes are almost perpendicular to the mean magnetic field. Fainter "striation"-like filaments can also be seen that are almost perpendicular to the dense filaments.

up, the global gravitational self-energy of the system $|E_g|$ grows to the point where gravitational collapse begins. The kinetic energy resulting from the collapse motions E_k tracks this gravitational term in such a way that $|E_g| \simeq 2E_k$ (e.g., *Vázquez-Semadeni et al.*, 2007; *Ballesteros-Paredes et al.*, 2011), mimicking virial equilibrium. Thus, gravitational collapse in media that have a number of Jeans masses will have multiple centers of collapse, and the velocity fields that are set up could be interpreted as turbulence [see *Tilley and Pudritz* (2007) for magnetic analog]. Such systems are never near equilibrium, however.

As noted, simulations of colliding streams produce clouds that are confined to a flattened layer. We therefore consider the fragmentation of a dense shell created by one-dimensional compression (e.g., by an expanding HII region, an old supernova remnant, or collision of two clouds) in more detail — specifically the formation of filaments by self-gravitational fragmentation of sheet-like clouds. As discussed in section 5.1, since the critical ratio of specific heats $\gamma_{crit,sheet} = 0$, sheet-like configurations are stable against compression. Here, the compressing motions are in the direction of the thickness and these can always be halted by the resulting increase in (central) pressure of the sheet, justifying the analysis of quasi-equilibrium sheet-like clouds.

Self-gravitational instability of a sheet-like equilibrium is well known. The critical wavelength for linear perturbations on the sheet is a few times its thickness. Sheet-like clouds are unstable to perturbations whose wavelengths are larger than the critical wavelength, and the most unstable wavelength is about twice the critical wavelength. The growth timescale is on the order of the free-fall timescale. The

detailed analysis by *Miyama et al.* (1987) on the nonlinear growth of unstable perturbations shows that the aspect ratio of dense regions increases with time, resulting in the formation of filaments. Thus, sheet-like clouds are expected to break up into filaments whose separations are about twice the critical wavelength (i.e., several times the thickness of the sheet). If the external pressure is much smaller than the central pressure, the thickness of the sheet is on the order of the Jeans length ($\lambda_J \sim c_s t_{ff}$). In contrast, if the external pressure is comparable to the central pressure, the thickness of the sheet can be much smaller than the Jeans length. As pointed out by *Myers* (2009), the fragmentation properties expected for compressed sheet-like clouds resemble the frequently observed "hub-filament systems" (see also *Burkert and Hartmann*, 2004).

Magnetic fields that are perpendicular to the sheet tend to stabilize the sheet. Indeed, if the field strength is larger than the critical value ($B_{crit} = 4\pi^2 G\Sigma^2$, where Σ is the surface density of the sheet), the sheet is stable against gravitational fragmentation (*Nakano and Nakamura*, 1978). In contrast, if the magnetic field is in the plane of the sheet, the stabilizing effect is limited, but the direction of the field determines the directions of fragmentation (*Nagai et al.*, 1998). If the external pressure is much smaller than the central pressure of the sheet, the sheet will fragment into filaments whose axis directions are perpendicular to the mean direction of magnetic field lines. In this case, the masses per unit length of the resulting filaments are expected to be about twice the critical value ($M_{line} \sim 2M_{line,crit}$). Such filaments may be the dense filaments observed with Herschel. On the other hand, if the external pressure is comparable to the central pressure in the sheet, the resulting filaments will be parallel to the mean magnetic field lines. In this case, the masses per unit length of the resulting filaments are smaller than the critical value ($M_{line} < M_{line,crit}$). Such filaments may be the "striations" seen with Herschel.

Numerical simulations of the formation of molecular clouds through one-dimensional compression are now providing interesting features of more realistic evolution (*Inoue and Inutsuka,* 2008, 2009, 2012). Figure 10c shows a snapshot of the face-on view of a non-uniform molecular cloud compressed by a shock wave travelling at 10 km s^{-1} (Inutsuka et al., in preparation). The magnetic field lines are mainly in the dense sheet of compressed gas. Many dense filaments are created, with axes perpendicular to the mean magnetic field lines. We can also see many faint filamentary structures that mimic observed "striations" and are almost parallel to the mean magnetic field lines. Those faint filaments are feeding gas onto the dense filaments, just as envisioned in section 4.4.

Although further analysis is required to give a conclusive interpretation of the observations, a simple picture based on one-dimensional compression of a molecular cloud may be a promising direction for understanding sheet and filament formation.

5.3. Fate of Dense Filaments

As shown earlier in section 3.2 and Fig. 6a, dense filaments with supercritical masses per unit length harbor starforming cores. Here, we theoretically outline what happens in such dense filaments.

Self-gravitating cylindrical structures are unstable to perturbations whose wavelengths are sufficiently large. For simplicity, we first focus on unmagnetized equilibrium filaments with isothermal or polytropic equation of state (*Inutsuka and Miyama,* 1992). The general behavior of the growth rate does not depend much on the polytropic index. A cylinder is unstable for wavelengths larger than about twice its diameter, and the most unstable wavelength is about 4× its diameter. The growth timescale is somewhat longer than the free-fall timescale. *Nagasawa* (1987) and *Fiege and Pudritz* (2000) investigated the effect of an axial magnetic field on the stability of the isothermal equilibrium filament. The stabilizing effect of an axial field is strong in the case where the line mass of the filament is much less than the critical value, while the effect saturates and does not qualitatively change the character of the instability for a filament of nearly critical line mass. An axial field slightly increases the critical line mass and the most unstable wavelength. In contrast, the inclusion of a helical magnetic field complicates the character of the instability (*Nakamura et al.,* 1993; *Fiege and Pudritz,* 2000). In the presence of a very strong toroidal field, for instance, the nature of the instability changes from gravitational to magnetic (*Fiege and Pudritz,* 2000).

The evolution of massive, self-gravitating isothermal filaments was studied in detail by *Inutsuka and Miyama* (1992, 1997). They found that unless a large nonlinear

perturbation exists initially, a supercritical filament mainly collapses radially without fragmentation. Thus, the characteristic length scale, and hence the mass scale, is determined by the fragmentation of the filament at the bouncing epoch of the radial collapse. The bouncing of the radial collapse is expected when compressional heating starts to overwhelm radiative cooling, resulting in a significant increase in the gas temperature ($\gamma_{eff} > 1$). The characteristic density, ρ_{crit}, of this breakdown of isothermality was studied with radiation hydrodynamics simulations by *Masunaga et al.* (1998) and its convenient expression for filamentary clouds was derived analytically by *Masunaga and Inutsuka* (1999).

Once the radial collapse of a massive filament has been halted by an increase in temperature at higher density or some other mechanism, longitudinal motions of fragmentation modes are expected to become significant. These motions may create a number of dense cores located along the main axis of the filament, just as observed (e.g., Figs. 6a and 8b). Note that the characteristic core mass resulting from fragmentation in this case is much smaller than the mass corresponding to the most unstable mode of the cylinder before the radial collapse. The resulting core masses are expected to depend on the initial amplitude distribution of density perturbations along the filament axis. Indeed, *Inutsuka* (2001) employed the so-called Press-Schechter formalism, a well-known method in cosmology (*Press and Schechter,* 1974), to predict the mass function of dense cores as a function of the initial power spectrum of fluctuations in mass per unit length. Figure 11 shows the typical time evolution of the core mass function. The evolution starts from initial perturbations of mass per unit length whose spectrum has a power-law index of −1.5, i.e., close to the

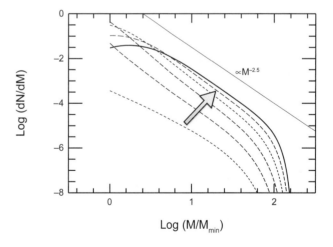

Fig. 11. Evolution of the mass function of dense cores produced by filament fragmentation in the case of $\delta(k) \propto k^{-1.5}$ (*Inutsuka,* 2001). The curves correspond to the snapshots at $t/t_g = 0$ (dashed), 2, 4, 6, 8, and 10 (solid), where t_g is the free-fall timescale, and the arrow indicates the time sequence. M_{min} corresponds to the most unstable wavelength of the quasistatic filament. The thin straight line corresponds to dN/dM $\propto M^{-2.5}$.

Kolmogorov index of –1.67. This theoretical work suggests that if we know the power spectrum of fluctuations in line mass along filaments before gravitational fragmentation, we can predict the mass function of the resulting dense cores. Thus, it is of considerable interest to measure such power spectra for filaments that have not yet produced many cores (see section 6.3).

5.4. Roles of Feedback Effects

Filamentary organization of the cloud plays a significant role in how important feedback effects can be.

To understand the effect of radiative feedback from massive stars, consider the condition for having insufficient ionizing radiative flux from some mass M undergoing an accretion flow to prevent the stalling or collapse of an HII region. For spherical symmetry, *Walmsley* (1995) derived a simple formula for the critical accretion rate $\dot{M}_{crit}$ onto a source of mass M and luminosity L above which the ionizing photons from the source would be absorbed by the accreting gas and cause the HII region to contract: $\dot{M}_{crit} > (4\pi LGMm_H^2/\alpha_B)^{1/2}$ where α_B is the hydrogenic recombination (n > 1) coefficient. Filamentary accretion offers a way out of this dilemma since it is intrinsically asymmetric. Ionizing flux can still escape the forming cluster in directions not dominated by the filamentary inflow (*Dale and Bonnell*, 2011). Simulations show that the ionizing flux from forming massive clusters escape into the low-density cavities created by infall onto the filaments (*Dale and Bonnell*, 2011). Such simulations have the character of a network of filaments and a "froth" of voids. The filamentary flows themselves are far too strong to be disrupted by the radiation field. *Dale and Bonnell* (2011) (see their Fig. 12) find that the filamentary distribution of the dense star-forming gas (about 3% of the mass) is not disrupted in their simulation of a 10^6-$M_\odot$ clump seeded with supersonic Kolmogorov turbulence. Figure 12 shows the distribution of stars, filaments, voids, and low-density ionized gas that fills the cavities in these simulations. The authors conclude that photoionization has little effect in disrupting the strong filamentary accretion flows that are building the clusters, although this seems to contradict the results of other work (e.g., *Vázquez-Semadeni et al.*, 2010; *Colín et al.*, 2013). More work is needed to clarify the situation.

The feedback from outflows in a filament-dominated cloud can also have significant effects on stellar masses. Simulations by *Li et al.* (2010) show that outflows typically reduce the mass of stars by significant factors (e.g., from 1.3 to 0.5 $M_\odot$). Going further, *Wang et al.* (2010) assumed a simplified model for outflows in which the momentum injected from the sink particle into the surrounding gas $\Delta P = P_*\Delta M$ every time the sink particle has gained a mass of ΔM. The energy and momentum are injected in two highly collimated bipolar jets. The outflows from the cluster region are collectively powerful enough in these simulations to break up the very filaments that are strongly feeding material into the cluster. This severing of the filamentary accretion flow slows up the rate for cluster formation. In regions that are

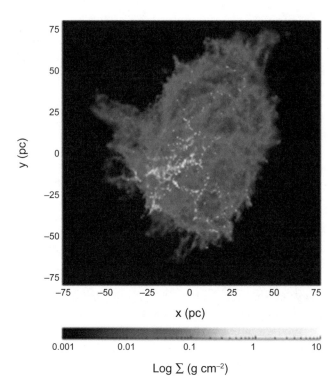

Fig. 12. See Plate 9 for color version. Distribution of stars, cold gas filaments, voids, and the low-density ionized gas that fills the cavities in the SPH numerical simulations of *Dale and Bonnell* (2011), in which a massive stellar cluster forms at the junction of a converging network of filaments, from the gravitational collapse of a turbulent molecular cloud.

sufficiently massive, however, the trickle of material that manages to avoid being impacted by these outflows can still feed the formation of massive stars.

Fall et al. (2010) (see their Fig. 2) show that outflows play an important role for lower-mass clouds and clusters (M < 10^4 $M_\odot$). For higher masses and column densities $\Sigma > 0.1$ g cm^{-2}, however, radiation pressure seems to dominate over the feedback effects. It will be interesting to see if this scaling still holds for filament-dominated clouds.

6. CONFRONTING OBSERVATIONS AND THEORY

6.1. Column-Density Threshold for Prestellar Cores

As already mentioned (e.g., section 3.2), prestellar cores identified with Herschel are preferentially found within the *densest filaments*, i.e., those with column densities exceeding $\sim 7 \times 10^{21}$ cm^{-2} (*André et al.*, 2010) (Fig. 6a). In the Aquila region, for instance, the distribution of background cloud column densities for the population of prestellar cores shows a steep rise above $N_{H_2}^{back} \sim 5 \times 10^{21}$ cm^{-2} (cf. Fig. 13a) and is such that $\sim$90% of the candidate bound cores are found above a background column density $N_{H_2}^{back} \sim 7 \times 10^{21}$ cm^{-2}, corresponding to a background visual extinction $A_V^{back} \sim 8$.

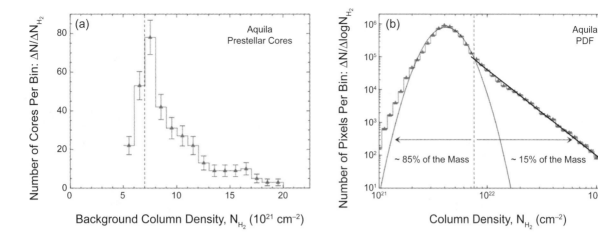

Fig. 13. (a) Distribution of background column densities for the candidate prestellar cores identified with Herschel in the Aquila Rift complex (cf. *Könyves et al., 2010*). The vertical dashed line marks the column density or extinction threshold at $N_{H_2}^{back} \sim 7 \times 10^{21}$ cm^{-2} or $A_V^{back} \sim 8$ (also corresponding to $\Sigma_{gas}^{back} \sim 130\ M_\odot$ pc^{-2}). **(b)** Probability density function of column density in the Aquila cloud complex, based on the column density image derived from Herschel data (*André et al., 2011; Schneider et al., 2013*). A lognormal fit at low-column densities and a power-law fit at high column densities are superimposed. The vertical dashed line marks the same column density threshold as in **(a)**.

The Herschel observations of Aquila therefore strongly support the existence of a column density or visual extinction threshold for the formation of prestellar cores at $A_V^{back} \sim$ 5–10, which had been suggested earlier based on ground-based studies of, e.g., the Taurus and Ophiuchus clouds (cf. *Onishi et al., 1998; Johnstone et al., 2004*). Interestingly, a very similar extinction threshold at $A_V^{back} \sim 8$ is independently observed in the spatial distribution of YSOs with Spitzer (*Heiderman et al., 2010; Lada et al., 2010*). In the Polaris flare cirrus, the Herschel observations are also consistent with such an extinction threshold since the observed background column densities are all below $A_V^{back} \sim 8$ and there are no or at most a handful of examples of bound prestellar cores in this cloud (e.g., *Ward-Thompson et al., 2010*). More generally, the results obtained with Herschel in nearby clouds suggest that a fraction $f_{pre} \sim 15\%$ of the total gas mass above the column density threshold is in the form of prestellar cores.

6.2. Theoretical Interpretation of the Threshold

These Herschel findings connect very well with theoretical expectations for the gravitational instability of filaments (cf. section 5.1) and point to an *explanation* of the star-formation threshold in terms of the filamentary structure of molecular clouds. Given the typical width $W_{fil} \sim$ 0.1 pc measured for filaments (*Arzoumanian et al., 2011*) (see Fig. 5) and the relation $M_{line} \approx \Sigma_0 \times W_{fil}$ between the central gas surface density Σ_0 and the mass per unit length M_{line} of a filament (cf. Appendix A of *André et al., 2010*), the threshold at $A_V^{back} \sim 8$ or $\Sigma_{gas}^{back} \sim 150\ M_\odot$ pc^{-2} corresponds to within a factor of $\ll 2$ to the critical mass per unit length $M_{line,crit} = 2c_s^2/G \sim 16\ M_\odot$ pc^{-1} of nearly isothermal, long cylinders [see section 5.1 and *Inutsuka and Miyama*

(1997)] for a typical gas temperature T $\sim$ 10 K. Thus, the core-formation threshold approximately corresponds to the *threshold above which interstellar filaments are gravitationally unstable* (*André et al., 2010*). Prestellar cores tend to be observed only above this threshold (cf. Fig. 6a and Fig. 13a) because they form out of a filamentary background and only filaments with $M_{line} > M_{line,crit}$ are able to fragment into self-gravitating cores.

The column-density threshold for core and star formation within filaments is not a sharp boundary but a smooth transition for several reasons. First, observations only provide information on the *projected* column density $\Sigma_{obs} = \frac{1}{\cos i} \Sigma_{int}$ of any given filament, where i is the inclination angle to the plane of sky and Σ_{int} is the intrinsic column density of the filament (measured perpendicular to the long axis). For a population of randomly oriented filaments with respect to the plane of sky, the net effect is that Σ_{obs} *overestimates* Σ_{int} by a factor $\langle \frac{1}{\cos i} \rangle = \frac{\pi}{2} \sim 1.57$ on average. Although systematic, this projection effect remains small and has little impact on the global classification of observed filamentary structures as supercritical or subcritical.

Second, there is a (small) spread in the distribution of filament inner widths of about a factor of 2 on either side of 0.1 pc (*Arzoumanian et al., 2011*) (cf. Fig. 5), implying a similar spread in the intrinsic column densities corresponding to the critical filament mass per unit length $M_{line,crit}$.

Third, interstellar filaments are not all exactly at T = 10 K and their internal velocity dispersion sometimes includes a small nonthermal component, σ_{NT}, which must be accounted for in the evaluation of the critical or virial mass per unit length $M_{line,vir} = 2\sigma_{tot}^2/G$ (see section 5.1) (*Fiege and Pudritz, 2000*). The velocity dispersion measurements of *Arzoumanian et al.* (2013) (Fig. 9b) confirm that there is a critical threshold in mass per unit length above which interstellar filaments are self-gravitating and below which they are unbound, and that the position of this threshold lies around ~16–32 $M_\odot$ pc^{-1}, i.e., within a factor of 2 of the thermal value of the critical mass per unit length $M_{line,crit}$ for T = 10 K. The results shown in Fig. 9b emphasize the role played by the thermal critical mass per unit length $M_{line,crit}$ in the evolution of filaments. Combined with the Herschel findings summarized above and illustrated in Figs. 6a and 13a, they support the view that the gravitational fragmentation of filaments may control the bulk of core formation, at least in nearby Galactic clouds.

6.3. Filament Fragmentation and the Core Mass Function/Initial Mass Function Peak

Since most stars appear to form in filaments, the fragmentation of filaments at the threshold of gravitational instability is a plausible mechanism for the origin of (part of) the stellar IMF. We may expect local collapse into spheroidal protostellar cores to be controlled by the classical Jeans criterion M ≥ M_{BE} where the Jeans or critical Bonnor-Ebert mass M_{BE} (e.g., *Bonnor,* 1956) is given by

$$M_{BE} \sim 1.3 c_s^4 / G^2 \Sigma_{cl} \qquad (8)$$

or

$$M_{BE} \sim 0.5 M_\odot \times \left(T_{eff} / 10K \right)^2 \times \left(\Sigma_{cl} / 160 M_\odot pc^{-2} \right)^{-1}$$

If we consider a quasi-equilibrium isothermal cylindrical filament on the verge of *global* radial collapse, it has a mass per unit length equal to the critical value $M_{line,crit}$ = $2c_s^2/G$ (~16 $M_\odot$ pc^{-1} for T_{eff} ~ 10 K) and an effective diameter $D_{flat,crit}$ = $2c_s^2/G\Sigma_0$ (~0.1 pc for T_{eff} ~ 10 K and Σ_0 ~ 160 $M_\odot$ pc^{-2}). A segment of such a cylinder of length equal to $D_{flat,crit}$ contains a mass $M_{line,crit} \times D_{flat,crit}$ = $4c_s^4/G^2\Sigma_0$ ~ 3 ×M_{BE} (~1.6 $M_\odot$ for T_{eff} ~ 10 K and Σ_0 ~ 160 $M_\odot$ pc^{-2}) and is thus locally Jeans unstable. Since local collapse tends to be favored over global collapse in the case of filaments (e.g., *Pon et al.,* 2011) (see section 5.1), gravitational fragmentation into spheroidal cores is expected to occur along supercritical filaments, as indeed found in both numerical simulations (e.g., *Bastien et al.,* 1991; *Inutsuka and Miyama,* 1997) and Herschel observations (see Fig. 6a and section 6.1). Remarkably, the peak of the prestellar CMF at ~0.6 $M_\odot$ as observed in the Aquila cloud complex (cf. Fig. 7b) corresponds very well to the Bonnor-Ebert mass M_{BE} ~ 0.5 $M_\odot$ within marginally critical filaments with $M_{line} \approx M_{line,crit}$ ~ 16 $M_\odot$ pc^{-1} and surface densities $\Sigma \approx \Sigma_{gas}^{crit}$ ~ 160 $M_\odot$ pc^{-2}. Likewise, the median projected spacing ~0.09 pc observed between the prestellar cores of Aquila (cf. section 3.2) roughly matches the thermal Jeans length within marginally critical filaments. All of this is consistent with the idea that gravitational fragmentation is the dominant physical mechanism generating prestellar cores within interstellar filaments. Furthermore, a typical prestellar core mass of ~0.6 $M_\odot$ translates into a charac-

teristic star or stellar system mass of ~0.2 $M_\odot$, assuming a typical efficiency ε_{core} ~ 30% (cf. section 3.4).

A subtle point needs to be made about the median spacing and mass of prestellar cores, however. Namely, the values observed with Herschel are a factor of ≥4 smaller than the characteristic core spacing and core mass predicted by linear stability analysis for infinitely long, ~0.1-pc-wide self-gravitating isothermal cylinders (see section 5.3). In their high-resolution IR extinction study of the filamentary IRDC G11, *Kainulainen et al.* (2013) recently noted a similar mismatch between the typical spacing of small-scale (≤0.1 pc) cores and the predictions of cylinder fragmentation theory. They were able to show, however, that the typical spacing of the larger-scale (≥0.5 pc), lower-density substructures identified along the G11 filament was in reasonably good agreement with the global fragmentation properties expected for a self-gravitating cylinder (see Fig. 14). This work suggests that on large scales the fragmentation properties of a self-gravitating filament are dominated by the most unstable mode of the global structure, while on small scales the fragmentation properties depend primarily on local conditions and the local Jeans/Bonnor-Ebert criterion applies.

In any event, the Herschel results tend to support the *Larson* (1985) interpretation of the peak of the IMF in terms of the typical Jeans mass in star-forming clouds. Overall, the Herschel findings suggest that the gravitational fragmentation of supercritical filaments produces the peak of the prestellar CMF, which, in turn, may account for the lognormal "base" (cf. *Bastian et al.,* 2010) of the IMF.

It remains to be seen whether the bottom end of the IMF and the Salpeter power-law slope at the high-mass end can be also explained by filament fragmentation. Naively, gravitational fragmentation should produce a narrow prestellar CMF, sharply peaked at the median thermal Jeans mass. Note, however, that a small (~25%) fraction of prestellar cores do not appear to form along filaments (section 3.2). Furthermore, a Salpeter power-law tail at high masses may result from filament fragmentation if turbulence has generated an appropriate field of initial density fluctuations within the filaments in the first place (cf. *Inutsuka,* 2001).

Addressing the high-mass power-law tail of the IMF, *Inutsuka* (2001) has shown that if the power spectrum of initial density fluctuations along the filaments approaches $P(k) \equiv |\delta_k|^2 \propto k^{-1.5}$ then the CMF produced by gravitational fragmentation evolves toward $dN/dM \propto M^{-2.5}$ (see Fig. 11), similar to the Salpeter IMF ($dN/dM_\star \propto M_\star^{-2.35}$). Interestingly, the power spectrum of column density fluctuations along the filaments observed with Herschel in nearby clouds is typically $P(k) \propto k^{-1.6}$, which is close to the required spectrum (Roy et al., in preparation).

Alternatively, a CMF with a Salpeter power-law tail may result from the gravitational fragmentation of a population of filaments with a distribution of supercritical masses per unit length. Observationally, the supercritical filaments observed as part of the HGBS do seem to have a power-law distribution of masses per unit length $dN/dM_{line} \propto M_{line}^{-2.2}$ above ~20 $M_\odot$ pc^{-1} (Arzoumanian et al., in preparation).

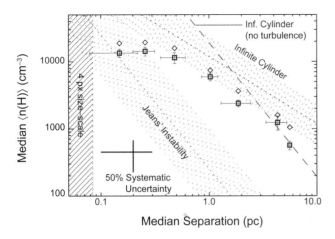

Fig. 14. See Plate 10 for color version. Median density of significant substructures as a function of separation in the high-(2″ or 0.035 pc)-resolution column density map of the massive, filamentary IRDC G11.11-0.12 by *Kainulainen et al.* (2013). The left dotted line shows the relation expected from standard Jeans' instability arguments. The right dotted line shows the relation predicted by the linear stability analysis of self-gravitating isothermal cylinders. Note how the structure of G11 is consistent with global cylinder fragmentation on scales ≥0.5 pc and approaches local Jeans' fragmentation on scales ≥0.2 pc. From *Kainulainen et al.* (2013.)

Since the width of the filaments is roughly constant ($W_{fil} \sim$ 0.1 pc), the mass per unit length is directly proportional to the central surface density, $M_{line} \sim \Sigma \times W_{fil}$. Furthermore, the total velocity dispersion of these filaments increases roughly as $\sigma_{tot} \propto \Sigma^{0.5}$ (*Arzoumanian et al.,* 2013) (see Fig. 9b), which means that their effective temperature (including thermal and nonthermal motions) scales roughly as $T_{eff} \propto \Sigma$. Hence $M_{BE} \propto \Sigma \propto M_{line}$, and the observed distribution of masses per unit length directly translates into a power-law distribution of Bonnor-Ebert masses $dN/dM_{BE} \propto M_{BE}^{-2.2}$ along supercritical filaments, which is also reminiscent of the Salpeter IMF.

6.4. A Universal Star Formation Law Above the Threshold?

The realization that, at least in nearby clouds, prestellar core formation occurs primarily along gravitationally unstable filaments of roughly constant width $W_{fil} \sim 0.1$ pc may also have implications for our understanding of star formation on global Galactic and extragalactic scales. Remarkably, the critical mass per unit length of a filament, $M_{line,crit} = 2c_s^2/G$, depends only on gas temperature (i.e., $T \sim 10$ K for the bulk of molecular clouds, away from the immediate vicinity of massive stars) and is modified by only a factor on the order of unity for filaments with realistic levels of magnetization (*Fiege and Pudritz,* 2000) (see section 5.1). These simple conditions may set a quasi-

universal threshold for star formation in the cold ISM of galaxies at $M_{line,crit} \sim 16$ $M_\odot$ pc^{-1} in terms of filament mass per unit length, or $M_{line,crit}/W_{fil} \sim 160$ $M_\odot$ pc^{-2} in terms of gas surface density, or $M_{line,crit}/W_{fil}^2 \sim 1600$ $M_\odot$ pc^{-3} in terms of gas density (i.e., a number density $n_{H_2} \sim 2 \times 10^4$ cm^{-3}).

While further work is needed to confirm that the width of interstellar filaments remains close to $W_{fil} \sim 0.1$ pc in massive star-forming clouds beyond the Gould belt, we note here that recent detailed studies of the RCW 36 and DR 21 ridges in Vela-C (d ~ 0.7 kpc) and Cygnus X (d ~ 1.4 kpc) are consistent with this hypothesis (*Hill et al.,* 2012; *Hennemann et al.,* 2012). As already pointed out in section 6.2, the threshold should be viewed as a smooth transition from non-star-forming to star-forming gas rather than as a sharp boundary. Furthermore, the above threshold corresponds to a *necessary* but not automatically *sufficient* condition for widespread star formation within filaments. In the extreme environmental conditions of the central molecular zone near the Galactic center, for instance, star formation appears to be largely suppressed above the threshold (*Longmore et al.,* 2013). In the bulk of the Galactic disk where more typical environmental conditions prevail, however, we may expect the above threshold in filament mass per unit length or (column) density to provide a fairly good selection of the gas directly participating in star formation.

Recent near- and mid-IR studies of the star formation rate (SFR) as a function of gas surface density in both Galactic and extragalactic cloud complexes (e.g., *Heiderman et al.,* 2010; *Lada et al.,* 2010) show that the SFR tends to be linearly proportional to the mass of dense gas above a surface density threshold $\Sigma_{gas}^{th} \sim 120$–130 $M_\odot$ pc^{-2} and drops to negligible values below Σ_{gas}^{th} [see *Gao and Solomon* (2004) for external galaxies]. Note that this is essentially the *same* threshold as found with Herschel for the formation of prestellar cores in nearby clouds (cf. section 6.1 and Figs. 13a and 9b). Moreover, the relation between the SFR and the mass of dense gas (M_{dense}) above the threshold is estimated to be SFR = 4.6×10^{-8} M$_\odot$ yr^{-1} × $(M_{dense}/M_\odot)$ in nearby clouds (*Lada et al.,* 2010), which is close to the relation SFR = 2×10^{-8} $M_\odot$ yr^{-1} × $(M_{dense}/M_\odot)$ found by *Gao and Solomon* (2004) for galaxies (see Fig. 15).

Both of these values are very similar to the SFR per unit solar mass of dense gas SFR/$M_{dense} = f_{pre} \times \varepsilon_{core}/t_{pre} \sim 0.15 \times 0.3/10^6 \sim 4.5 \times 10^{-8}$ yr^{-1} that we may derive based on Herschel results in the Aquila cloud complex. For this estimate, we consider that only a fraction $\times f_{pre} \sim 15\%$ of the gas mass above the column density threshold is in the form of prestellar cores (cf. section 6.1), that the local star-formation efficiency at the level of an individual core is $\varepsilon_{core} \sim 30\%$ (cf. section 3.4), and that the typical lifetime of the Aquila cores is $t_{pre} \sim 10^6$ yr (cf. section 3.3) (Könyves et al., in preparation). Despite relatively large uncertainties, the agreement with the extragalactic value of *Gao and Solomon* (2004) is surprisingly good, implying that there may well be a quasi-universal "star-formation law" converting the gas of dense filaments into stars above the threshold (Fig. 15) (see also *Lada et al.,* 2012).

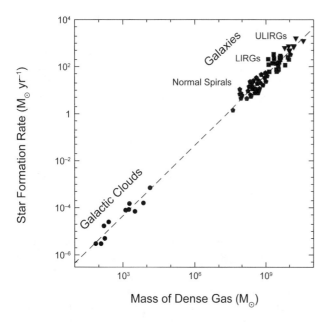

Fig. 15. Relation between star-formation rate (SFR) and mass of dense gas (M_{dense}) for local molecular clouds (*Lada et al., 2010*) and external galaxies (*Gao and Solomon, 2004*). M_{dense} is the mass of dense gas above the star-formation threshold ($n_{H_2} \sim 2 \times 10^4$ cm^{-3}; see text). The dashed line going through the data points represents the linear relation SFR = 4.5×10^{-8} M$_\odot$ yr^{-1} × (M_{dense}/M$_\odot$) inferred from the Herschel results on prestellar cores in the Aquila cloud complex (see text and Könyves et al., in preparation). Adapted from *Lada et al.* (2012).

6.5. Origin of the Characteristic Inner Width of Filaments

The fact that the same ~0.1-pc width is measured for low-density, subcritical filaments suggests that this characteristic scale is set by the physical process(es) producing the filamentary structure. Furthermore, at least in the case of diffuse, gravitationally unbound clouds such as Polaris (Fig. 1), gravity is unlikely to be involved. As mentioned in section 5.2, large-scale compression flows (turbulent or not) and the dissipation of the corresponding energy provide a potential mechanism. In the picture proposed by *Padoan et al.* (2001), for instance, the dissipation of turbulence occurs in shocks, and interstellar filaments correspond to dense, postshock stagnation gas associated with compressed regions between interacting supersonic flows. One merit of this picture is that it can account qualitatively for the ~0.1-pc width. The typical thickness of shock-compressed structures resulting from supersonic turbulence in the ISM is indeed expected to be roughly the sonic scale of the turbulence, i.e., ~0.1 pc in diffuse interstellar gas (cf. *Larson, 1981; Federrath et al., 2010*; discussion in *Arzoumanian et al., 2011*). Direct evidence of the role of large-scale compressive flows has been found with Herschel in the Pipe Nebula in the form of filaments with asymmetric column density profiles that most likely result from compression

by the winds of the nearby Sco OB2 association (*Peretto et al., 2012*).

A more complete picture, proposed by *Hennebelle* (2013), is that filaments result from a combination of turbulent compression and shear [see *Hily-Blant and Falgarone* (2009) for an observed example of a diffuse CO filament in Polaris corresponding to a region of intense velocity shear]. Interestingly, the filament width is comparable to the cutoff wavelength $\lambda_A \sim 0.1$ pc × $\left(\frac{B}{10\ \mu G}\right) \times \left(\frac{n_{H_2}}{10^3\ cm^{-3}}\right)^{-1}$ below which MHD waves cannot propagate in the primarily neutral gas of molecular clouds (cf. *Mouschovias, 1991*), if the typical magnetic field strength is B ~ 10 μG (*Crutcher, 2012*). Hence the tentative suggestion that the filament width may be set by the dissipation mechanism of MHD waves due to ion-neutral friction (*Hennebelle, 2013*). Alternatively, the characteristic width may also be understood if interstellar filaments are formed as quasi-equilibrium structures in pressure balance with a typical ambient ISM pressure $P_{ext}/k_B \sim$ 2–5 × 10^4 K cm^{-3} (*Fischera and Martin, 2012*; Inutsuka et al., in preparation). Clearly, more work is needed to clarify the origin of the width of subcritical filaments.

That star-forming, supercritical filaments also maintain roughly constant inner widths ~0.1 pc while evolving (*Arzoumanian et al., 2011*) (Figs. 4 and 5) is even more surprising at first sight. Indeed, supercritical filaments are unstable to radial collapse and are thus expected to undergo rapid radial contraction with time (see section 5.1). The most likely solution to this paradox is that supercritical filaments are *accreting* additional background material while contracting. The increase in velocity dispersion with central column density observed for supercritical filaments (*Arzoumanian et al., 2013*) (see Fig. 9b) is indeed suggestive of an increase in (virial) mass per unit length with time.

As mentioned in section 4.3, more direct observational evidence of this accretion process for supercritical filaments exists in several cases in the form of low-density striations or subfilaments seen perpendicular to the main filaments and apparently feeding them from the side (see Figs. 2, 3, and 9a). Accretion onto dense filaments is also seen in numerical simulations (*Gómez and Vázquez-Semadeni, 2013*). This process supplies gravitational energy to supercritical filaments that is then converted into turbulent kinetic energy (cf. *Heitsch et al., 2009; Klessen and Hennebelle, 2010*) and may explain the observed increase in velocity dispersion with column density ($\sigma_{tot} \propto \Sigma_0^{0.5}$; cf. Fig. 9b). Indeed, the fine substructure and velocity-coherent "fibers" observed within Herschel supercritical filaments [cf. Fig. 2 and *Hacar et al.* (2013)] may possibly be the manifestation of accretion-driven "turbulence." The central diameter of such accreting filaments is expected to be on the order of the effective Jeans length $D_{J,eff} \sim 2\sigma_{tot}^2/G\Sigma_0$, which *Arzoumanian et al.* (2013) have shown to remain close to ~0.1 pc. Hence, through accretion of parent cloud material, supercritical filaments may keep roughly constant inner widths and remain in rough virial balance while contracting (see *Heitsch, 2013a,b; Hennebelle and André, 2013*). This process may effectively prevent the global (radial) collapse of supercritical filaments and thus favor their fragmentation into cores (e.g., *Larson, 2005*), in agreement with the Herschel results (see Figs. 6a and 8b).

7. CONCLUSIONS: TOWARD A NEW PARADIGM FOR STAR FORMATION?

The observational results summarized in sections 2–4 provide key insight into the first phases of the star-formation process. They emphasize the role of filaments and support a scenario in which the formation of prestellar cores occurs in two main steps. First, the dissipation of kinetic energy in large-scale MHD flows (turbulent or not) appears to generate ~0.1-pc-wide filaments in the ISM. Second, the densest filaments fragment into prestellar cores by gravitational instability above the critical mass per unit length $M_{line,crit} \approx 16\ M_\odot\ pc^{-1}$, equivalent to a critical (column) density threshold $\Sigma_{gas}^{crit} \sim 160\ M_\odot\ pc^{-2}$ ($A_V^{crit} \sim 8$) or $n_{H_2}^{crit} \sim 2 \times 10^4\ cm^{-3}$.

In contrast to the standard gravoturbulent fragmentation picture (e.g., *Mac Low and Klessen, 2004*), in which filaments are present but play no fundamental role, our proposed paradigm relies heavily on the unique properties of filamentary geometry, such as the existence of a critical mass per unit length for nearly isothermal filaments. That the formation of filaments in the diffuse ISM represents the first step toward core and star formation is suggested by the filaments *already* being pervasive in a gravitationally unbound, non-star-forming cloud such as Polaris (cf. Fig. 1) (*Hily-Blant and Falgarone, 2007; Men'shchikov et al., 2010; Miville-Deschênes et al., 2010*). Hence, many interstellar filaments are not produced by large-scale gravity and their formation must precede star formation.

The second step appears to be the gravitational fragmentation of the densest filaments with supercritical masses per unit length ($M_{line} \geq M_{line,crit}$) into prestellar cores (cf. section 6.2). In active star-forming regions such as the Aquila complex, most of the prestellar cores identified with Herschel are indeed concentrated within supercritical filaments (cf. Fig. 6a). In contrast, in non-star-forming clouds such as Polaris, all the filaments have subcritical masses per unit length and very few (if any) prestellar cores and no protostars are observed (cf. Fig. 1).

The present scenario may explain the peak for the prestellar CMF and the base of the stellar IMF (see section 6.3 and Fig. 7b). It partly accounts for the general inefficiency of the star-formation process since, even in active star-forming complexes such as Aquila (Fig. 6), only a small fraction of the total gas mass (~15% in the case of Aquila; see Fig. 13b) is above of the column density threshold, and only a small fraction $f_{pre} \sim 15\%$ of the dense gas above the threshold is in the form of prestellar cores (see section 6.1). Therefore, the vast majority of the gas in a GMC (~98% in the case of Aquila) does not participate in star formation at any given time (see also *Heiderman et al., 2010; Evans, 2011*). Furthermore, the fact that essentially the same "star-formation law" is observed above the column density

threshold in both Galactic clouds and external galaxies (see section 6.4; Fig. 15) (*Lada et al., 2012*) suggests the star-formation scenario sketched above may well apply to the ISM of other galaxies.

The results reviewed in this chapter are very encouraging as they tentatively point to a unified picture of star formation on GMC scales in both Galactic clouds and external galaxies. Much more work would be needed, however, to fully understand the origin of the characteristic width of interstellar filaments and to determine whether the same ~0.1-pc width also holds beyond the clouds of the Gould belt. In particular, confirming and refining the scenario proposed here will require follow-up observations to constrain the dynamics of the filaments imaged with Herschel as well as detailed comparisons with numerical simulations of molecular cloud formation and evolution. Both ALMA and the NOrthern Extended Millimeter Array (NOEMA) will be instrumental in testing whether this scenario based on Herschel results in nearby Galactic clouds is truly universal and applies to the ISM of all galaxies.

Acknowledgments. We thank D. Arzoumanian, J. Kirk, V. Könyves, and P. Palmeirim for useful discussions and help with several figures. We also thank H. Beuther, E. Falgarone, and the referee for constructive comments. Ph.A. is partially supported by the European Research Council under the European Union's Seventh Framework Programme (Grant Agreement No. 291294) and by the French National Research Agency (Grant No. ANR-11-BS56-0010). R.E.P. is supported by a Discovery grant from the National Science and Engineering Research Council of Canada.

REFERENCES

Abergel A. et al. (1994) *Astrophys. J. Lett., 423,* L59.
Alves J. F. et al. (2001) *Nature, 409,* 159.
Alves J. F. et al. (2007) *Astron. Astrophys., 462,* L17.
André P. et al. (2000) In *Protostars and Planets IV* (V. Mannings et al., eds.), p. 59. Univ. of Arizona, Tucson.
André P. et al. (2007) *Astron. Astrophys., 472,* 519.
André Ph. et al. (2010) *Astron. Astrophys., 518,* L102.
André Ph. et al. (2011) In *Computational Star Formation* (J. Alves et al., eds.), p. 255. IAU Symp. 270, Cambridge Univ., Cambridge.
Arzoumanian D. et al. (2011) *Astron. Astrophys., 529,* L6.
Arzoumanian D. et al. (2013) *Astron. Astrophys., 553,* A119.
Ballesteros-Paredes J. et al. (1999) *Astrophys. J., 527,* 285.
Ballesteros-Paredes J. et al. (2003) *Astrophys. J., 592,* 188.
Ballesteros-Paredes J. et al. (2006) *Astrophys. J., 637,* 384.
Ballesteros-Paredes J. et al. (2011) *Mon. Not. R. Astron. Soc., 411,* 65.
Bally J. et al. (1987) *Astrophys. J. Lett., 312,* L45.
Balsara D. et al. (2001) *Mon. Not. R. Astron. Soc., 327,* 715.
Banerjee R. et al. (2006) *Mon. Not. R. Astron. Soc., 373,* 1091.
Bastian N. et al. (2010) *Annu. Rev. Astron. Astrophys., 48,* 339.
Bastien P. et al. (1991) *Astrophys. J., 378,* 255.
Basu S. et al. (2009) *New Astron., 14,* 483.
Bate M. R. and Bonnell I. A. (2005) *Mon. Not. R. Astron. Soc., 356,* 1201.
Bate M. R. et al. (2003) *Mon. Not. R. Astron. Soc., 339,* 577.
Beichman C. A. et al. (1986) *Astrophys. J., 307,* 337.
Belloche A. et al. (2011) *Astron. Astrophys., 535,* A2.
Bergin E. A. and Tafalla M. (2007) *Annu. Rev. Astron. Astrophys., 45,* 339.
Beuther H. et al. (2011) *Astron. Astrophys., 533,* A17.
Bonnell I. A. et al. (2003) *Mon. Not. R. Astron. Soc., 343,* 413.
Bonnor W. B. (1956) *Mon. Not. R. Astron. Soc., 116,* 351.
Bontemps S. et al. (2010) *Astron. Astrophys., 518,* L85.
Burkert A. and Hartmann L. (2004) *Astrophys. J., 616,* 288.

Cambrésy L. (1999) *Astron. Astrophys., 345,* 965.
Caselli P. et al. (2002) *Astrophys. J., 572,* 238.
Chabrier G. (2005) In *The Initial Mass Function 50 Years Later* (E. Corbelli et al., eds.) p. 41. Astrophys. Space Sci. Library, Vol. 327, Springer-Verlag, Berlin.
Chapman N. L. et al. (2011) *Astrophys. J., 741,* 21.
Chini R. et al. 1997) *Astrophys. J. Lett., 474,* L135.
Colín P. et al. (2013) *Mon. Not. R. Astron. Soc., 435,* 1701.
Commerçon B. et al. (2012) *Astron. Astrophys., 545,* A98.
Contreras et al. (2013) *Mon. Not. R. Astron. Soc., 433,* 251.
Crutcher R. M. (1999) *Astrophys. J., 520,* 706.
Crutcher R. M. (2012) *Annu. Rev. Astron. Astrophys., 50,* 29.
Dale J. E. and Bonnell I. A. (2011) *Mon. Not. R. Astron. Soc., 414,* 321.
Di Francesco J. et al. (2007) In *Protostars and Planets V* (B. Reipurth et al., eds.), pp. 669–684. Univ. of Arizona, Tucson.
Dunham M. M. et al. (2008) *Astrophys. J. Suppl., 179,* 249.
Egan M. P. et al. (1998) *Astrophys. J. Lett., 494,* L199.
Elmegreen B. G. and Falgarone E. (1996) *Astrophys. J., 471,* 816.
Elmegreen B. G. and Scalo J. (2004) *Annu. Rev. Astron. Astrophys., 42,* 211.
Enoch M. L. et al. (2008) *Astrophys. J., 684,* 1240.
Evans N. J. (1999) *Annu. Rev. Astron. Astrophys., 37,* 311.
Evans N. J. (2011) In *Computational Star Formation* (J. Alves et al., eds.), p. 25. IAU Symp. 270, Cambridge Univ., Cambridge.
Evans N. J. et al. (2009) *Astrophys. J. Suppl., 181,* 321.
Fall S. M. et al. (2010) *Astrophys. J. Lett., 710,* L142.
Falgarone E. et al. (2009) *Astron. Astrophys., 507,* 355.
Falgarone E. et al. (2001) *Astrophys. J., 555,* 178.
Federrath C. et al. (2010) *Astron. Astrophys., 512,* A81.
Fiege J. D. and Pudritz R. E. (2000) *Mon. Not. R. Astron. Soc., 311,* 85.
Fischera J. and Martin P. G. (2012) *Astron. Astrophys., 542,* A77.
Gammie C. F. et al. (2003) *Astrophys. J., 592,* 203.
Gao Y. and Solomon P. (2004) *Astrophys. J., 606,* 271.
Goldsmith P. F. et al. (2008) *Astrophys. J., 680,* 428.
Gómez G. C. and Vázquez-Semadeni E. (2013) *ArXiv e-prints,* arXiv:astro-ph/1308.6298.
Gong H. and Ostriker E. C. (2011) *Astrophys. J., 729,* 120.
Goodman A. A. et al. (1990) *Astrophys. J., 359,* 363.
Goodman A. A. et al. (1998) *Astrophys. J., 504,* 223.
Goodwin S. P. et al. (2008) *Astron. Astrophys., 477,* 823.
Hacar A. and Tafalla M. (2011) *Astron. Astrophys., 533,* A34.
Hacar A. et al. (2013) *Astron. Astrophys., 554,* A55.
Hartmann L. (2002) *Astrophys. J., 578,* 914.
Hatchell J. and Fuller G. A. (2008) *Astron. Astrophys., 482,* 855.
Hatchell J. et al. (2005) *Astron. Astrophys., 440,* 151.
Heiderman A. et al. (2010) *Astrophys. J., 723,* 1019.
Heitsch F. (2013a) *Astrophys. J., 769,* 115.
Heitsch F. (2013b) *Astrophys. J., 776,* 62.
Heitsch F. et al. (2009) *Astrophys. J., 704,* 1735.
Hennebelle P. (2013) *Astron. Astrophys., 556,* A153.
Hennebelle P. and André Ph. (2013) *Astron. Astrophys., 560,* A68.
Hennebelle P. and Chabrier G. (2008) *Astrophys. J., 684,* 395.
Hennemann M. et al. (2012) *Astron. Astrophys., 543,* L3.
Henning Th. et al. (2010) *Astron. Astrophys., 518,* L95.
Henshaw J. D. et al. (2013) *Mon. Not. R. Astron. Soc., 428,* 3425.
Hernandez A. K. and Tan J. C. (2011) *Astrophys. J., 730,* 44.
Heyer M. et al. (2008) *Astrophys. J., 680,* 420.
Heyer M. et al. (2009) *Astrophys. J., 699,* 1092.
Hill T. et al. (2011) *Astron. Astrophys., 533,* A94.
Hill T. et al. (2012) *Astron. Astrophys., 548,* L6.
Hily-Blant P. and Falgarone E. (2007) *Astron. Astrophys., 469,* 173.
Hily-Blant P. and Falgarone E. (2009) *Astron. Astrophys., 500,* L29.
Inoue T. and Inutsuka S. (2008) *Astrophys. J., 687,* 303.
Inoue T. and Inutsuka S. (2009) *Astrophys. J., 704,* 161.
Inoue T. and Inutsuka S. (2012) *Astrophys. J., 759,* 35.
Inutsuka S. (2001) *Astrophys. J. Lett., 559,* L149.
Inutsuka S. and Miyama S. M. (1992) *Astrophys. J., 388,* 392.
Inutsuka S. and Miyama S. M. (1997) *Astrophys. J., 480,* 681.
Jackson J. M. et al. (2010) *Astrophys. J. Lett., 719,* L185.
Jessop N. E. and Ward-Thompson D. (2000) *Mon. Not. R. Astron. Soc., 311,* 63.
Johnstone D. and Bally J. (1999) *Astrophys. J. Lett., 510,* L49.
Johnstone D. et al. (2000) *Astrophys. J., 545,* 327.
Johnstone D. et al. (2004) *Astrophys. J. Lett., 611,* L45.
Juvela M. et al. (2012) *Astron. Astrophys., 541,* A12.
Kainulainen J. et al. (2013) *Astron. Astrophys., 557,* A120.

Kawachi T. and Hanawa T. (1998) *Publ. Astron. Soc. Japan, 50,* 577.
Kevlahan N. and Pudritz R. E. (2009) *Astrophys. J., 702,* 39.
Kirk H. et al. (2013) *Astrophys. J., 766,* 115.
Kirk J. M. et al. (2005) *Mon. Not. R. Astron. Soc., 360,* 1506.
Kirk J. M. et al. (2013) *Mon. Not. R. Astron. Soc., 432,* 1424.
Klessen R. S. and Burkert A. (2000) *Astrophys. J. Suppl., 128,* 287.
Klessen R. S. and Hennebelle P. (2010) *Astron. Astrophys., 520,* A17.
Könyves V. et al. (2010) *Astron. Astrophys., 518,* L106.
Koyama H. and Inutsuka S. (2000) *Astrophys. J., 532,* 980.
Kramer C. et al. (1998) *Astron. Astrophys., 329,* 249.
Kroupa P. (2001) *Mon. Not. R. Astron. Soc., 322,* 231.
Krumholz M. R. et al. (2007) *Astrophys. J., 656,* 959.
Lada C. J. et al. (1999) *Astrophys. J., 512,* 250.
Lada C. J. et al. (2010) *Astrophys. J., 724,* 687.
Lada C. J. et al. (2012) *Astrophys. J., 745,* 190.
Larson R. B. (1969) *Mon. Not. R. Astron. Soc., 145,* 271.
Larson R. B. (1981) *Mon. Not. R. Astron. Soc., 194,* 809.
Larson R. B. (1985) *Mon. Not. R. Astron. Soc., 214,* 379.
Larson R. B. (2005) *Mon. Not. R. Astron. Soc., 359,* 211.
Lee C. W. and Myers P. C. (1999) *Astrophys. J. Suppl., 123,* 233.
Li D. and Goldsmith P. F. (2012) *Astrophys. J., 756,* 12.
Li Z.-Y. et al. (2010) *Astrophys. J. Lett., 720,* L26.
Longmore S. N. et al. (2013) *Mon. Not. R. Astron. Soc., 429,* 987.
Mac Low M.-M. and Klessen R. S. (2004) *Rev. Mod. Phys., 76,* 125.
Malinen J. et al. (2012) *Astron. Astrophys., 544,* A50.
Masunaga H. et al. (1998) *Astrophys. J., 495,* 346.
Masunaga H. and Inutsuka S. (1999) *Astrophys. J., 510,* 822.
Matzner C. D. and McKee C. F. (2000) *Astrophys. J., 545,* 364.
Maury A. et al. (2011) *Astron. Astrophys., 535,* A77.
McClure-Griffiths N. M. et al. (2006) *Astrophys. J., 652,* 1339.
McCrea W. H. (1957) *Mon. Not. R. Astron. Soc., 117,* 562.
Men'shchikov A. (2013) *Astron. Astrophys., 560,* A63.
Men'shchikov A. et al. (2010) *Astron. Astrophys., 518,* L103.
Men'shchikov A. et al. (2012) *Astron. Astrophys., 542,* A81.
Miettinen O. and Harju J. (2010) *Astron. Astrophys., 520,* A102.
Miville-Deschênes M.-A. et al. (2010) *Astron. Astrophys., 518,* L104.
Miyama S. M. et al. (1987) *Prog. Theor. Phys., 78,* 1273.
Mizuno A. et al. (1995) *Astrophys. J. Lett., 445,* L161.
Molinari S. et al. (2010) *Astron. Astrophys., 518,* L100.
Molinari S. et al. (2011) *Astron. Astrophys., 530,* A133.
Motte F. et al. (1998) *Astron. Astrophys., 336,* 150.
Motte F. et al. (2001) *Astron. Astrophys., 372,* L41.
Motte F. et al. (2007) *Astron. Astrophys., 476,* 1243.
Motte F. et al. (2010) *Astron. Astrophys., 518,* L77.
Mouschovias T. Ch. (1991) *Astrophys. J., 373,* 169.
Mouschovias T. Ch. and Paleologou E. V. (1979) *Astrophys. J., 230,* 204.
Myers P. C. (1983) *Astrophys. J., 270,* 105.
Myers P. C. (2009) *Astrophys. J., 700,* 1609.
Myers P. C. (2011) *Astrophys. J., 735,* 82.
Nagai T. et al. (1998) *Astrophys. J., 506,* 306.
Nagasawa M. (1987) *Prog. Theor. Phys., 77,* 635.
Nakano T. and Nakamura T. (1978) *Publ. Astron. Soc. Japan, 30,* 671.
Nakamura F. and Umemura M. (1999) *Astrophys. J., 515,* 239.
Nakamura F. et al. (1993) *Publ. Astron. Soc. Japan, 45,* 551.
Nutter D. et al. (2008) *Mon. Not. R. Astron. Soc., 384,* 755.
Onishi T. et al. (1998) *Astrophys. J., 502,* 296.
Ostriker E. C. et al. (1999) *Astrophys. J., 513,* 259.
Ostriker J. (1964) *Astrophys. J., 140,* 1056.
Padoan P. and Nordlund A. (2002) *Astrophys. J., 576,* 870.
Padoan P. et al. (2001) *Astrophys. J., 553,* 227.
Pagani L. et al. (2010) *Astron. Astrophys., 512,* A3.
Palmeirim P. et al. (2013) *Astron. Astrophys., 550,* A38.
Passot T. et al. (1995) *Astrophys. J., 455,* 536.
Pérault M. et al. (1996) *Astron. Astrophys., 315,* L165.
Peretto N. and Fuller G. A. (2009) *Astron. Astrophys., 505,* 405.
Peretto N. et al. (2012) *Astron. Astrophys., 541,* A63.
Peretto N. et al. (2013) *Astron. Astrophys., 555,* A112.
Pezzuto S. et al. (2012) *Astron. Astrophys., 547,* A54.
Pilbratt G. L. et al. (2010) *Astron. Astrophys., 518,* L1.
Pineda J. et al. (2009) *Astrophys. J. Lett., 699,* L134.
Pineda J. et al. (2010) *Astrophys. J. Lett., 712,* L116.
Pineda J. et al. (2011) *Astrophys. J. Lett., 739,* L2.
Polychroni D. et al. (2013) *Astrophys. J. Lett., 777,* L33.
Pon A. et al. (2011) *Astrophys. J., 740,* 88.
Pon A. et al. (2012) *Astrophys. J., 756,* 145.
Porter D. et al. (1994) *Phys. Fluids, 6,* 2133.

Press W. and Schechter P. (1974) *Astrophys. J., 187,* 425.
Pudritz R. E. and Kevlahan N. K.-R. (2013) *Philos. Trans. R. Soc. Ser. A, 371,* 20120248.
Reid M. A. et al. (2010) *Astrophys. J., 719,* 561.
Reipurth B. et al., eds. (2007) *Protostars and Planets V,* Univ. of Arizona, Tucson.
Rosolowsky E. W. et al. (2008) *Astrophys. J., 679,* 1338.
Russeil D. et al. (2013) *Astron. Astrophys., 554,* A42.
Saigo K. and Tomisaka K. (2011) *Astrophys. J., 728,* 78.
Sandell G. and Knee L. B. G. (2001) *Astrophys. J. Lett., 546,* L49.
Schneer S. et al. (2007) *Astrophys. J., 671,* 1839.
Schneider N. et al. (2013) *Astrophys. J. Lett., 766,* L17.
Schneider N. et al. (2010) *Astron. Astrophys., 520,* A49.
Schneider N. et al. (2012) *Astron. Astrophys., 540,* L11.
Schneider S. and Elmegreen B. G. (1979) *Astrophys. J. Suppl., 41,* 87.
Shu F. (1977) *Astrophys. J., 214,* 488.
Simpson R. J. et al. (2011) *Mon. Not. R. Astron. Soc., 417,* 216.
Smith R. J. et al. (2011) *Mon. Not. R. Astron. Soc., 411,* 1354.
Smith R. J. et al. (2012) *Astrophys. J., 750,* 64.
Sousbie T. (2011) *Mon. Not. R. Astron. Soc., 414,* 350.
Stanke T. et al. (2006) *Astron. Astrophys., 447,* 609.
Starck J. L. et al. (2003) *Astron. Astrophys., 398,* 785.
Stodolkiewicz J. S. (1963) *Acta Astron., 13,* 30.

Stutzki J. and Güsten R. (1990) *Astrophys. J., 356,* 513.
Sugitani K. et al. (2011) *Astrophys. J., 734,* 63.
Tackenberg J. et al. (2012) *Astron. Astrophys., 540,* A113.
Tafalla M. et al. (2004) *Astron. Astrophys., 416,* 191.
Tilley D. A. and Pudritz R. E. (2004) *Mon. Not. R. Astron. Soc., 353,* 769.
Tilley D. A. and Pudritz R. E. (2007) *Mon. Not. R. Astron. Soc., 382,* 73.
Toalá J. A. et al. (2012) *Astrophys. J., 744,* 190.
Tohline J. E. (1982) *Fund. Cosmic Phys., 8,* 1.
Vázquez-Semadeni E. (1994) *Astrophys. J., 423,* 681.
Vásquez-Semadeni E. et al. (2007) *Astrophys. J., 657,* 870.
Vázquez-Semadeni E. et al. (2010) *Astrophys. J., 715,* 1302.
Walmsley M. (1995) *Rev. Mex. Astron. Astrofis. Ser. Conf., 1,* 137.
Wang P. et al. (2010) *Astrophys. J., 709,* 27.
Ward-Thompson D. et al. (1994) *Mon. Not. R. Astron. Soc., 268,* 276.
Ward-Thompson D. et al. (2007) In *Protostars and Planets V* (B. Reipurth et al., eds.), pp. 33–46. Univ. of Arizona, Tucson.
Ward-Thompson D. et al. (2010) *Astron. Astrophys., 518,* L92.
Whitworth A. P. and Ward-Thompson D. (2001) *Astrophys. J., 547,* 317.
Whitworth A. P. et al. (1996) *Mon. Not. R. Astron. Soc., 283,* 1061.
Williams J. P. et al. (1994) *Astrophys. J., 428,* 693.

Offner S. S. R., Clark P. C., Hennebelle P., Bastian N., Bate M. R., Hopkins P. F., Moraux E., and Whitworth A. P. (2014) The origin and universality of the stellar initial mass function. In *Protostars and Planets VI* (H. Beuther et al., eds.), pp. 53–75. Univ. of Arizona, Tucson, DOI: 10.2458/azu_uapress_9780816531240-ch003.

The Origin and Universality of the Stellar Initial Mass Function

Stella S. R. Offner
Yale University
(now at University of Massachusetts, Amherst)

Paul C. Clark
Universität Heidelberg

Patrick Hennebelle
Le Service d'Astrophysique et le Laboratoire AIM-Centre d'Etude de Saclay

Nathan Bastian
Liverpool John Moores University

Matthew R. Bate
University of Exeter

Philip F. Hopkins
California Institute of Technology

Estelle Moraux
Institut de Planétologie et d'Astrophysique de Grenoble

Anthony P. Whitworth
Cardiff University

We review current theories for the origin of the stellar initial mass function (IMF) with particular focus on the extent to which the IMF can be considered universal across various environments. To place the issue in an observational context, we summarize the techniques used to determine the IMF for different stellar populations, the uncertainties affecting the results, and the evidence for systematic departures from universality under extreme circumstances. We next consider theories for the formation of prestellar cores by turbulent fragmentation and the possible impact of various thermal, hydrodynamic, and magneto-hydrodynamic (MHD) instabilities. We address the conversion of prestellar cores into stars and evaluate the roles played by different processes: competitive accretion, dynamical fragmentation, ejection and starvation, filament fragmentation and filamentary accretion flows, disk formation and fragmentation, critical scales imposed by thermodynamics, and magnetic braking. We present explanations for the characteristic shapes of the present-day prestellar core mass function (CMF) and the IMF and consider what significance can be attached to their apparent similarity. Substantial computational advances have occurred in recent years, and we review the numerical simulations that have been performed to predict the IMF directly and discuss the influence of dynamics, time-dependent phenomena, and initial conditions.

1. INTRODUCTION

Measuring the stellar initial mass function (IMF), and understanding its genesis, is a central issue in the study of star formation. It also has a fundamental bearing on many other areas of astronomy, e.g., modeling the microphysics of galactic structure and evolution, and tracking the buildup of the heavy elements that are essential for planet formation and for life. Since it appears that the formation and architecture of a planetary system strongly depends on the mass of the host star, and — for stars in clusters — on feedback effects from nearby massive stars, an understanding of the IMF is critical on this count as well.

Many theoretical ideas have been proposed to explain the origin of the IMF. Current popular ideas can be roughly divided into two opposing categories: models that are deterministic, and models that are stochastic. Simply put, the debate is one of nature (cores) vs. nurture (dynamics).

Theories postulating a direct mapping between the core mass function (CMF) and the IMF are predicated on the idea that stellar masses are inherited from the distribution of natal core masses (*Padoan and Nordlund*, 2002; *Hennebelle and Chabrier*, 2008, 2009; *Oey*, 2011; *Hopkins*, 2012b). In such models the origin of the IMF is essentially deterministic since they assume that stellar mass is accreted from a local reservoir of gas. Models for the CMF are based on the ubiquitous presence of energetic, turbulent motions within molecular clouds, which naturally create a distribution of density and velocity fluctuations. Proponents of such models appeal to the striking similarity between the CMF and IMF shapes. If the mapping between the CMF and IMF is sufficiently simple and cores evolve in relative isolation from one another, then the problem of predicting the IMF reduces to understanding the CMF.

At the opposite extreme, some models propose that final stellar masses are completely independent of initial core masses, and thus any similarity between the IMF and CMF is coincidental (*Bonnell et al.*, 2001b; *Bate et al.*, 2003; *Clark et al.*, 2007). Models that emphasize the importance of dynamical interactions and stochastic accretion, e.g., those based on the idea of stars competing for gas, fall in this category. The dynamical view of star formation has its roots in early, simple numerical studies. *Larson* (1978) was the first to perform numerical simulations of the collapse of a cloud to form a group of protostars and emphasized that a mass spectrum of objects could be obtained "at least in part by accretion processes and the competition between different accreting objects." Both *Larson* (1978) and *Zinnecker* (1982) developed simple analytic models, finding that mass spectra of the form $dN/d \log M \approx M^{-1}$ (resembling the high-mass end of the IMF) could be produced by protostars competing for the accretion of gas from their parent cloud. This pioneering work was neglected for close to two decades until more advanced numerical simulations of the formation of groups of protostars could be performed.

Advances in computational power have subsequently increased the scope and utility of numerical simulations of star formation. Current state-of-the-art calculations of cluster formation may include a variety of physical effects as well as resolution down to astronomical unit scales. A variety of parameter studies have enabled the exploration of the IMF as a function of initial conditions and environment. Although still physically incomplete, these simulations provide tantalizing clues on the origin of the IMF in addition to a wealth of secondary metrics such as gas kinematics, accretion disk properties, stellar velocities, and stellar multiplicities, which provide further constraints on IMF theories.

Meanwhile, challenges persist for the observational determination of the IMF. While observations of resolved stellar populations support a universal stellar IMF, statistics are sparse for both brown dwarfs and very high-mass stars, stellar multiplicity is often unresolved, and masses depend significantly on uncertain stellar evolutionary models. Recent observations are extending beyond local environments and the Milky Way by employing a variety of techniques. Persuasive evidence for IMF variations has been presented for extreme regions such as the galactic center, giant ellipticals, and dwarf galaxies (*van Dokkum and Conroy*, 2010; *Cappellari et al.*, 2012; *Lu et al.*, 2013; *Geha et al.*, 2013). Although statistical and observational uncertainties remain formidable, these observations provide fresh impetus for general and predictive theoretical models.

The apparent Galactic IMF invariance and the categorical differences between current theories motivate a number of important, open questions that we will discuss in this chapter. What sets the characteristic stellar mass and why is it apparently invariant in the local universe? Do high-mass stars form differently than low-mass stars and, if so, what implications does this have for their relative numbers? How sensitive is a star's mass to its initial core mass? To what extent do dynamical interactions between forming stars influence the IMF? Finally, what physical effects impact stellar multiplicity and how does this impinge upon star formation efficiency? We will address these questions in the context of current theories and discuss the prognosis for verification and synthesis of the theoretical ideas given observational constraints.

We begin by reviewing the observational determination of the IMF in section 2. In section 3 we discuss the form of the CMF and theories for its origin. Next, section 4 considers the possible link between the CMF and IMF. We discuss simulations of forming clusters in section 5 and evaluate the role of dynamics, time-dependent phenomena, and initial conditions in producing the IMF. We summarize the status of open questions and present suggestions for future work in section 6.

2. OBSERVATIONAL DETERMINATION OF THE INITIAL MASS FUNCTION

2.1. Forms of the Initial Mass Function

When interpreting observations and models of the IMF, it is common to compare them with standard analytical forms that have been put forward in the literature. For example, studies focusing on stars more massive than a few solar masses commonly adopt a power-law distribution of the form $dN \propto M^{-\alpha} dM$. This was the form originally adopted by *Salpeter* (1955), who determined $\alpha = 2.35$ by fitting observational data. More than 55 years later, this value is still considered the standard for stars above 1 $M_\odot$. However, this function diverges as it approaches zero, so it is clear that there must be a break or turnover in the IMF at low masses. This was noted by *Miller and Scalo* (1979), who found that the IMF was well approximated by a lognormal distribution between 0.1 $M_\odot$ and ~30 $M_\odot$, where there is a clear flattening below 1 $M_\odot$. However, current data suggest that a lognormal underpredicts the number of massive stars (>20 $M_\odot$) due to the turndown above this mass, relative to a continuous power law (see, e.g., *Bastian et al.*, 2010).

More modern forms often adopt a lognormal distribution at low masses and a power-law tail above a solar mass (*Chabrier,* 2003, 2005). A very similar approach, although analytically different, is to represent the full mass range of the IMF as a series of power laws, such as in *Kroupa* (2001, 2002). Above ~0.2 $M_\odot$, the multiple-power-law segments and lognormal with a power-law tail at high masses agree very well, as shown in Fig. 1. However, the form of the IMF at the low-mass end is still relatively uncertain and subject to ongoing debate (e.g., *Thies and Kroupa,* 2007, 2008; *Kroupa et al.,* 2013). The philosophical difference between these two forms reduces to whether one believes that star formation is a continuous process or whether there are distinct physical processes that dominate in certain regimes, such as at the stellar/substellar boundary.

Other proposed functional forms include a truncated exponential where $dN \propto M^{-\alpha}\{1-\exp[(-M/M_p)^{-\beta}]\}dM$ (*de Marchi and Paresce,* 2001; *Parravano et al.,* 2011); the Pareto-Levy family of stable distributions, which includes the Gaussian (*Cartwright and Whitworth,* 2012); and a lognormal joined with power laws at both low and high masses (*Maschberger,* 2013). It is usually not possible to discriminate between these forms on the basis of observations due to significant uncertainties. We note that Fig. 1 shows only the proposed IMF functions and not the error associated with the derivations of these fits from the data. In all cases shown, the authors fit the galactic field IMF over a restricted mass range and then extrapolated their functional fit to the whole mass range. Observers and theoreticians now use these functional forms without any error bars to evaluate how well their data (or models) are consistent within their uncertainties.

2.2. Initial Mass Function Universality in the Milky Way

2.2.1. Uncertainties in the initial mass function determination from resolved populations. The basic observational method to derive the IMF consists of three steps. First, observers measure the luminosity function (LF) of a complete sample of stars that lie within some defined volume. Next, the LF is converted into a present-day mass function (PDMF) using a mass-magnitude relationship. Finally, the PDMF is corrected for the star-formation history, stellar evolution, galactic structure, cluster dynamical evolution, and binarity to obtain the individual star IMF. None of these steps is straightforward, and many biases and systematic uncertainties may be introduced during the process. A detailed discussion of these limitations at low masses and in the substellar regime is given by *Jeffries* (2012) and *Luhman* (2012).

When obtaining the LF, defining a complete sample of stars for a given volume can be challenging. Studies of field stars from photometric surveys, either wide and relatively shallow (e.g., *Bochanski et al.,* 2010) or narrow but deep (e.g., *Schultheis et al.,* 2006), are magnitude limited and need to be corrected for the Malmquist bias. Indeed, when

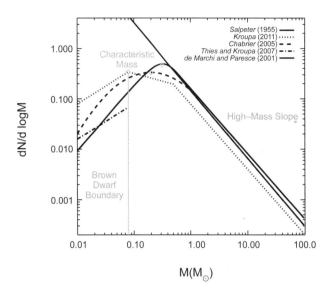

Fig. 1. Initial mass function (IMF) functional forms proposed by various authors from fits to Galactic stellar data. With the exception of the Salpeter slope, the curves are normalized such that the integral over mass is unity. When comparing with observational data, the normalization is set by the total number of objects as shown in Fig. 2.

distances are estimated from photometric or spectroscopic parallax, the volume limited sample inferred will be polluted by more distant bright stars due to observational uncertainties and intrinsic dispersion in the color-magnitude relation. This effect can be fairly significant and change the slope of the luminosity function by more than 10%. Studies of nearby stars, for which the distance is determined from trigonometric parallax, are not affected by this bias, but they are affected by the *Lutz and Kelker* (1973) bias, in which the averaged measured parallax is larger than the true parallax, if the parallax uncertainties exceed 10%. These issues reduce the studies within 20 pc to a completeness level of ~80% (*Reid et al.,* 2007), which results in small number statistics, especially at low masses. Cluster studies do not have these problems since a complete sample can in principle be obtained from the complete cluster census. However, secure membership cannot be assessed from photometry alone and proper motion measurements as well as spectroscopic follow-up are often necessary. In addition, the remaining contamination by field stars may be large if cluster properties such as age and proper motion are similar to the field.

For young star-forming regions, differential extinction may be a significant problem. This is usually taken into account by limiting the sample up to a given A_V. However, imposing a limit may exclude the more central regions, which are often more extincted, from the determination of the LF. If the stellar spatial distribution is not the same at all masses and, in particular, if mass segregation is present, excluding high-A_V sources may introduce a strong bias. Such an effect can be tested by simulating a fake cluster as in *Parker et al.* (2012).

Even with no extinction, mass segregation must be taken into account because the cluster LF may be different in the center and in the outskirts. The best way to account for this issue is to cover an area larger than the cluster extent. However, in very rich and distant clusters, this may not be sufficient. Photometric surveys are blind in the vicinity of very bright stars since they are not able to detect nearby faint objects, and crowding can result in blended stars. Both effects may produce a strong bias in the LF. All the above considerations clearly show that incompleteness is often a major problem, one that is more severe at low masses, and should be treated carefully when determining the LF.

Once the LF has been obtained, the luminosities must be converted into masses. Different methods may be used, which convert either absolute magnitude, bolometric luminosities, or effective temperature, and which use either empirical or theoretical relationships. For main-sequence stars with accurate parallax and photometry, dynamical masses can be measured in binary systems to give an empirical mass-magnitude relationship (e.g., *Delfosse et al.,* 2000) that can then be used directly. While there is almost no scatter and there is good agreement with evolutionary models when using near-infrared magnitude, this is clearly not the case in the V-band due to opacity uncertainties and metallicity dependence, especially at lower masses.

Another issue to consider is the age dependence of the mass-luminosity relation and the difficulty of estimating stellar ages unless the stars are coeval. The population of field stars has a large age spread, and many of the oldest stars have already left the main sequence. In this blended population, the simplest assumption is that all field stars are approximately the age of the universe. In this case, stars more massive than 0.8 $M_\odot$ have already left the main sequence while younger low-mass stars ($M < 0.1\ M_\odot$), which take a considerable amount of time to contract, are still on the pre-main sequence (e.g., *Baraffe et al.,* 1998). In the substellar domain, the situation is more complicated because brown dwarfs never reach the main sequence. They never initiate hydrogen fusion but simply cool with time, which leads to a degeneracy between mass, effective temperature, and age. In principle, position in the Hertzsprung-Russell (HR) diagram could determine both mass and age, but different evolutionary models lead to large differences in the deduced masses. The discrepancy between the *Burrows et al.* (1997) and *Chabrier et al.* (2000) models is mainly around 10% but can be as large as 100% (*Konopacky et al.,* 2010). Uncertainties in the models may be due to the formation/disappearance of atmospheric clouds, equation of state, assumptions for the initial conditions and accretion (*Hosokawa et al.,* 2011; *Baraffe et al.,* 2012), treatment of convection, and effect of magnetic fields (*Chabrier and Küker,* 2006). We also note that even if a model is very accurate, a typical measurement uncertainty of 100 K in the effective temperature may correspond to a factor of 2 difference in mass (see, e.g., *Burrows et al.,* 1997).

The last step in the process, which consists of deriving the IMF from the PDMF, is also subject to many biases as it often relies on various models and assumptions for the star-formation history, Galactic structure, and stellar evolution. For example, short-lived massive stars are underrepresented in the field population since they evolve to become white dwarfs or neutron stars. Consequently, an estimate of the star-formation history is required to transform the field PDMF to the IMF above 0.8 $M_\odot$. Moreover, the low-mass stars are older on average and are more widely distributed above and below the Galactic plane. This must be corrected for, especially for nearby surveys that do not explore the full Galactic disk scale height. In contrast to the field population, clusters contain stars of similar age, distance, and initial composition. This is advantageous since we expect the PDMF of young clusters to be the IMF. However, below 10 million years (m.y.), reliable masses are difficult to obtain due to high and variable extinction, reddening, uncertain theoretical models (e.g., *Baraffe et al.,* 2009; *Hosokawa et al.,* 2011), and possible age spread. At older ages the PDMF is more robust since it is less affected by these uncertainties; however, it may no longer resemble the IMF.

As discussed above, mass segregation introduces a radial dependence to the mass function (MF) that must be taken into account if the survey does not include the entire cluster. This is true for all clusters of any age. Indeed, many young clusters appear mass segregated, whereby the most massive stars are more concentrated toward the center (e.g., *Hillenbrand and Hartmann,* 1998), indicating that mass segregation may be primordial or can occur dynamically on a very short timescale (e.g., *Allison et al.,* 2010). More importantly, secular evolution due to two-body encounters leads to a preferential evaporation of lower-mass members, which shifts the peak of the MF toward higher masses (*de Marchi et al.,* 2010). However, N-body numerical simulations suggest this effect is not important when the cluster is younger than its relaxation time: $t_{relax} = NR_h/(8\sigma_V \ln N)$ (e.g., *Binney and Tremaine,* 1987), where N is the number of star systems in the cluster, R_h is the half mass radius, and σ_V is the velocity dispersion.

Finally, another important limitation is that binaries are not resolved in most of these studies. Consequently, the cluster MFs as well as the field MF obtained from deep surveys correspond to the system MF, not the individual star MF. Correction for binarity can be done for the field stellar MF but not for the substellar domain, since the binary properties of field brown dwarfs are fairly uncertain. There is usually no attempt to correct for binarity in clusters, mainly to avoid introducing additional bias since binary properties may depend on age and environment.

2.2.2. The field initial mass function. Since the seminal study of *Salpeter* (1955), the Galactic field stellar IMF is usually described as a power law with $\alpha = 2.35$ above 1 $M_\odot$, although it is occasionally claimed to be steeper at higher masses (see, e.g., *Scalo,* 1986). Between 0.1 $M_\odot$ and 0.8 $M_\odot$, the PDMF corresponds to the IMF and is now relatively well constrained thanks to recent studies based on local stars with Hipparcos parallax (*Reid et al.,* 2002) and the much larger sample of field stars with less accurate distances (*Bochanski et al.,* 2010; *Covey et al.,* 2008; *Deacon et al.,* 2008). The results of these studies are in reasonable agreement for M < 0.6 M$_\odot$ and suggest the single-star IMF at low masses is well described by a power law with $\alpha \sim 1.1$. The IMF obtained by *Bochanski et al.* (2010) suggests a deficit of stars with M > 0.6 M$_\odot$ and is better fit by either a lognormal with a peak mass around 0.18 M$_\odot$ and $\sigma = 0.34$ or a broken power law. However, the evidence for a peak is weak on the basis of this mass range alone, and the perceived deficit may be due to neglecting binary companions in systems with total mass above 0.8 M$_\odot$ (*Parravano et al.,* 2011). Although the statistical uncertainties are much smaller for extended samples than for small local samples of field stars, the systematic errors are larger, yielding similar final uncertainties in both cases. Overall, most results are converging on a field IMF in the stellar domain that has a Salpeter-like slope above ~1 M$_\odot$ and flattens significantly below. This distribution can be equally well represented by segmented power laws (*Kroupa,* 2001) or by a lognormal function with peak mass of 0.15–0.25 M$_\odot$ (*Chabrier,* 2005).

In the substellar domain the shape of the Galactic disk IMF is more uncertain. Instead of converting a LF directly to a MF, which is hampered by the brown dwarf age dependence in the mass-luminosity relationship, the IMF is usually estimated by predicting the LF via Monte-Carlo simulations and assuming an IMF functional form and a star-formation history (e.g., *Burgasser,* 2004). Recent results from the Wide-field Infrared Survey Explorer (WISE) (*Kirkpatrick et al.,* 2012) and the United Kingdom Infra-Red Telescope (UKIRT) Infrared Deep Sky Survey (UKIDSS) (*Burningham et al.,* 2013; *Day-Jones et al.,* 2013) using this approach indicate that the substellar IMF extends well below 10 M$_{Jup}$ and may be represented by a power law with $-1 < \alpha < 0$, although it is still consistent with a lognormal shape due to the large uncertainties.

2.2.3. The initial mass function in young clusters. Several studies have been carried out in young open clusters with ages of 30–150 m.y. to determine the IMF across the substellar boundary (0.03 M$_\odot \lesssim$ M $\lesssim$ 0.3 M$_\odot$). Since the comprehensive review by *Jeffries* (2012), IMF determinations have been obtained using the UKIDSS Galactic cluster survey, which includes IC 4665 (*Lodieu et al.,* 2011), the Pleiades (*Lodieu et al.,* 2012a), and α Per (*Lodieu et al.,* 2012b). The results agree well with previous studies, indicating that the system IMF is well represented by a power law with $\alpha \simeq 0.6$ in the mass range 0.03–0.5 M$_\odot$ or a lognormal with a peak mass around 0.2–0.3 M$_\odot$ and $\sigma \simeq$ 0.5, consistent with *Chabrier* (2005). Results obtained in star-forming regions [ONC, ρ-Oph, NGC 2024, NGC 1333, IC 348, σ Ori, Cha I, NGC 6611, λ Ori, Upper Sco; see references in *Bastian et al.* (2010) and *Jeffries* (2012); for more recent studies, see *Lodieu* (2013), *Alves de Oliveira et al.* (2012, 2013), *Peña Ramírez et al.* (2012), *Scholz et al.* (2012), *Muzic et al.* (2011, 2012), *Geers et al.* (2011), and *Bayo et al.* (2011)] are also in remarkable agreement in the same mass range (see Fig. 2). The only clear exception

is Taurus, which shows an excess of 0.6–0.8 M$_\odot$ stars and peaks at higher masses (*Luhman et al.,* 2009). However, note that an agreement between the system IMFs does not necessarily imply agreement between the individual star IMFs, since binary properties may be different.

At lower masses, the lognormal form may remain a good parameterization, although this is currently debated, since an excess of low-mass brown dwarfs has been reported in Upper Sco (*Lodieu,* 2013) and in σ Ori (*Peña Ramírez et al.,* 2012) (see Fig. 2). These excesses may result from uncertainties in the mass-luminosity relation at very low masses and young ages.

Obtaining measurements of the IMF at high stellar masses (M $\simeq$ 3–100 M$_\odot$) is also challenging. Recent studies since the review by *Bastian et al.* (2010) include the young, massive clusters NGC 2264 (*Sung and Bessell,* 2010), Tr 14 and Tr 16 (*Hur et al.,* 2012), and NGC 6231 (*Sung et al.,* 2013). These regions exhibit a high-mass slope that is consistent with Salpeter, although the IMF is found to be slightly steeper for NGC 2264 where $\alpha = 2.7 \pm 0.1$. In contrast, the MF appears to vary spatially in the much

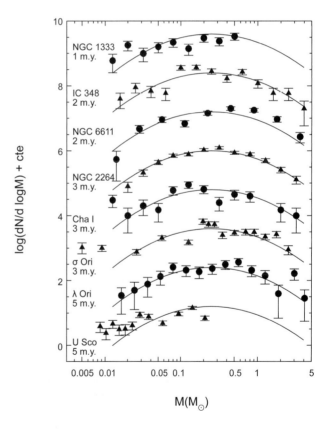

Fig. 2. Recent IMF estimates for eight star-forming regions: NGC 1333 (*Scholz et al.,* 2012), IC 348 (*Alves de Oliveira et al.,* 2013), NGC 6611 (*Oliveira et al.,* 2009), NGC 2264 (*Sung and Bessell,* 2010), Cha I (*Luhman,* 2007), σ Ori (*Peña Ramírez et al.,* 2012), λ Ori (*Bayo et al.,* 2011), and Upper Sco (*Lodieu,* 2013). The error bars represent the Poisson error for each data point. The solid lines are not a fit to the data but the lognormal form proposed by *Chabrier* (2005) for the IMF, normalized to best follow the data.

denser and more massive NGC 3603 and Westerlund 1 clusters, as well as in clusters within the Galactic center such as the Arches, Quintuplet, and young nuclear star clusters. The high-mass slopes of these extreme regions appear to be consistently flatter than the standard IMF in the inner part of the cluster and steepen toward the outskirts. This is a strong indication that these clusters are mass segregated despite their youth. While the global IMF of the Arches is found to be consistent with a Salpeter IMF (*Habibi et al.*, 2013), the IMF of Westerlund 1 is controversial since *Lim et al.* (2013) report a flatter IMF while *Gennaro et al.* (2011) report a normal IMF. In the Quintuplet cluster, *Hussmann et al.* (2012) derived an IMF that is flatter than Salpeter, but they only investigated the central part of the cluster. In the young nuclear cluster in the Galactic center, *Lu et al.* (2013) found a MF slope of $\alpha = 1.7 \pm 0.2$, which is again flatter than Salpeter. They claim that this region is one of the few environments where significant deviations have been found from an otherwise near-universal IMF.

2.3. Initial Mass Function Universality in the Local Universe

With high-resolution imaging provided by the Hubble Space Telescope (HST), we can resolve individual stars out to a few milliparsec. This technique has proven to be extremely powerful for determining the star-formation histories of galaxies, and in some cases, the stellar IMF can be directly constrained. In general, these studies are limited to the high-mass portion of the IMF ($M > 1 \ M_\odot$) simply due to detection limits; however, a few recent studies have attempted to push significantly past this limit. It is beyond the scope of this review to discuss the plethora of studies that have measured the PDMF/IMF through resolved stellar populations. Instead, we highlight just a few recent studies and refer the reader to *Bastian et al.* (2010) for a more thorough compilation. [For an alternative discussion of observational variations, see the IMF review by *Kroupa et al.* (2013).]

Increasing distance amplifies the potential caveats and biases present in studies of local regions, e.g., crowding, photometric errors, unresolved binaries, and mass segregation. *Weisz et al.* (2013) have supplied an overview of these plus additional biases and potential pitfalls and have provided the statistical tools to account for them. These authors have collated observations of 89 young clusters and star-forming regions from the literature and have recalculated the index of the MF (assuming that it is a power law) given the statistical and observational uncertainties. They find the best-fit α of 2.46 with a 1σ dispersion of 0.34, which is similar to Salpeter. Additionally, the authors warn that deriving a physical upper stellar limit to the IMF is difficult and fraught with biases and, consequently, could not be determined from their collected sample.

Hubble Space Telescope Wide Field Camera 3 (WFC3) observations of the young, $\sim 10^5$-$M_\odot$ cluster Westerlund 1 have measured its MF down to $\sim 0.1 \ M_\odot$. While mass

segregation is certainly present and the inner regions are somewhat incomplete, the overall MF is consistent with the expectation of *Chabrier* (2005), i.e., a peak around 0.2–0.3 $M_\odot$ and a power law at higher masses with a nearly Salpeter slope (Andersen et al., in preparation).

Using extremely deep HST images, a few recent studies have attempted to exceed the 1-$M_\odot$ boundary in extragalactic studies. For example, *Kalirai et al.* (2013) have studied a single HST field in the Small Magellanic Clouds (SMC) and have been able to constrain the MF from 0.37 $M_\odot$ to 0.93 $M_\odot$. They find that the data are well represented by a power law with index of 1.9. This is near the index implied by the IMF determined by *Chabrier* (2005) for an upper mass of 0.8–0.9 $M_\odot$. Surprisingly, a single index for the IMF was able to reproduce their observations over this mass range, i.e., they did not find evidence for a flattening at lower masses as expected for the Chabrier IMF. For example, at 0.5 $M_\odot$, the expected slope is 1.5, which is within the observational uncertainties. Hence, the results presented in *Kalirai et al.* (2013) only differ from the *Chabrier* (2005) expectation below $\sim 0.45 \ M_\odot$ and then by only 1–2σ. This highlights the difficulty of fitting a single power-law index to the data when a smooth distribution with a continuously changing slope is expected.

We conclude that currently there is no strong evidence (more than 2σ) for IMF variations within local galaxies as determined from studies of resolved stellar populations.

2.4. Extragalactic Determinations of the Initial Mass Function

While techniques to infer the IMF in extragalactic environments are necessarily more indirect than those used locally, the ability to sample more extreme environments provides a promising avenue to discover systematic variations. We refer the reader to *Bastian et al.* (2010) for a summary of extragalactic IMF studies that were published before 2010. These authors concluded that the majority of observations were consistent with a normal IMF and that many of the claims of systematic IMF variations were potentially due to the assumptions and/or complications, such as extinction corrections and systematic offsets in star-formation rate indicators, in the methods used.

Since 2010, there has been a flurry of IMF studies focusing on ancient early type (elliptical) galaxies. These studies have employed two main techniques: Either they dynamically determine the mass of galaxies in order to compare their mass-to-light ratios to the expectations of stellar population synthesis (SPS) models (e.g., *Cappellari et al.*, 2012), or they use gravity-sensitive integrated spectral features to determine the relative numbers of low-mass ($M < 0.5 \ M_\odot$) stars (e.g., *van Dokkum and Conroy*, 2010). Contrary to previous studies that found an overabundance of high-mass stars, i.e., a "top-heavy" IMF, in the high-redshift progenitors of these systems (e.g., *Davé*, 2008), the above studies have suggested an overabundance of low-mass stars, i.e., a "bottom-heavy" IMF, which is consistent with a Salpeter slope (or steeper) down to the hydrogen-burning limit. However, it is important to note that while both types of studies find systematic variations, namely that galaxies with higher-velocity dispersion have more bottom-heavy IMFs, the reported variations are much smaller than had been previously suggested. The most recent results are within a factor of ~2–3 in total mass for a given amount of light.

One potential caveat is that the above analyses depend on comparing observed properties with expectations of SPS models. While a large amount of work has gone into developing such models, i.e., by investigating stellar evolution and atmospheres, it is necessary to calibrate the models with known stellar populations. For old populations, this is generally done with globular clusters (e.g., *Maraston et al.,* 2003; *Conroy and Gunn,* 2010). However, there is a deficit of resolved globular clusters in the Galaxy and Magellanic Clouds that reach the relatively high metallicity ($Z \geq 1/2\ Z_{\odot}$) of elliptical galaxies. Hence, it is necessary to extrapolate the SPS models outside the calibrated regime. Using the higher-metallicity globular cluster population of M31, *Strader et al.* (2011) tested the validity of SPS extrapolations and found systematic offsets between the observations and SPS model predictions. Interestingly, the offset goes in the opposite direction to that inferred from the dynamical studies of elliptical galaxies, which may imply larger IMF variations.

We conclude that while elliptical galaxies show intriguing evidence for systematic IMF variations, which are potentially due to their unique formation environment, uncertainties underlying the current techniques require further study to confirm these potential systematic variations.

3. FORM AND ORIGIN OF THE CORE MASS FUNCTION

Observations of the IMF necessarily occur after the natal gas has formed stars or been dispersed. However, examining the properties of molecular clouds and the star formation within provides clues for the initial conditions of star formation and how gas collapses and produces the IMF.

Within molecular clouds, observers find dense condensations, often containing embedded protostars, which appear to have masses distributed like the IMF. A natural possibility is that these "cores" serve as gas reservoirs from which stars accrete most of their mass. If stellar masses are indeed largely determined by the amount of gas in their parent cores, then understanding the origin and distribution of core masses becomes a fundamental part of understanding the IMF.

Before continuing we define several important terms that we use throughout the chapter. Henceforth, a *core* refers to a blob of gas that will form one to a few stars, i.e., a single star system. A *prestellar core* is one that is expected to undergo collapse but is not yet observed to contain a protostar. In contrast, a *clump* refers to a larger collection of gas, which will likely form a cluster of stars. A *star-forming clump* specifically indicates a clump that is observed to contain several forming stars, while *CO clumps* are simply large interstellar clouds that are observed in CO. The exact meaning of these terms varies widely in the literature, and we discuss the observational challenges of core and clump definition further in section 4.

The possible importance of cores in the star-formation process leads to a central question: What determines the mass reservoir that is eventually accreted by a star and when is it determined (before the star forms or concomitant with it)? A natural possibility is that this reservoir is broadly traced by the population of observed, dense prestellar cores. In this scenario, observationally identified cores constitute a large part of the final stellar mass. This does not preclude gas expulsion by winds, division of mass into binary and multiple systems, and further core accretion from the surrounding medium during collapse. These processes are all expected to have some influence but are assumed to be secondary in importance to the core masses. We dedicate this section to discussing observations and theories for the CMF.

3.1. Observational Determination of the Core Mass Function

The first determination of the CMF was achieved 15 years ago using millimeter continuum dust emission in nearby molecular clouds, namely ρ Oph (*Motte et al.,* 1998; *Johnstone et al.,* 2000) and Serpens (*Testi and Sargent,* 1998). In these studies, cores were identified from the emission data using either a wavelet analysis (*Langer et al.,* 1993; *Starck et al.,* 1995) or "CLUMPFIND" (*Williams et al.,* 1994), a procedure that separates objects along minimum flux boundaries. About 30–50 objects were identified in each case with masses spanning a wide range (~0.1–10 $M_{\odot}$). These condensations appear sufficiently dense and compact to be gravitationally bound. The inferred core radial densities suggest that they are similar to Bonnor-Ebert spheres with a constant density in the central region. All three studies found a similar core mass spectrum. Below 0.5 $M_{\odot}$, the mass distribution scales as $dN/dM \propto M^{-1.5}$, while above ~0.5 $M_{\odot}$, the distribution is significantly steeper and consistent with a power law $dN/dM \propto M^{-\alpha}$ where $\alpha \simeq 2$–2.5. The authors cautioned that the low-mass part of the distribution was likely affected by insufficient sensitivity and resolution. The inferred high-mass distribution is significantly steeper than the mass spectrum inferred for CO clumps, which have $\alpha \simeq 1.7$ (*Heithausen et al.,* 1998; *Kramer et al.,* 1998). The authors noted that the steep mass spectrum was reminiscent of the Salpeter IMF slope and proposed that the observed cores were indeed the reservoirs that would contribute most of the eventual stellar masses. More recent observations by *Enoch et al.* (2007) found similar results for the CMFs of the nearby molecular clouds Perseus and Ophiuchus but a shallower mass distribution for Serpens ($\alpha \simeq 1.6$).

The CMF has also been inferred in the Pipe Nebula (*Alves et al.,* 2007; *Román-Zúñiga et al.,* 2010) using near-infrared extinction maps produced from the Two Micron All-Sky Survey (2MASS) point source catalogs (*Lombardi et al.*, 2006). This technique nicely complements the results based on dust emission because it is independent of dust temperature. The cores identified in the Pipe Nebula are all starless and much less dense than the cores previously observed in emission, suggesting that they may be at a very early evolutionary stage. For masses larger than about 3 $M_\odot$, the mass spectrum is again very similar to the Salpeter IMF, but at lower masses it displays a flat profile with a possible peak around 1 $M_\odot$. Altogether the mass distribution has a shape similar to the Chabrier and Kroupa IMFs but is shifted toward larger masses by a factor of ~3. *Nutter and Ward-Thompson* (2007) obtained similar results in Orion. The CMF offset from the IMF has often been interpreted as a core-to-star conversion efficiency on the order of 1/3.

The most recent determination of the CMF has been obtained for the Aquila rift cloud complex using Herschel data (*André et al.*, 2010; *Könyves et al.*, 2010). Since more than 500 cores have been detected, the statistics are about 1 order of magnitude better than in the previous studies. Most of the cores appear to be gravitationally bound, meaning that the estimated core masses are comparable to or larger than the local Jeans mass, $M_J \simeq c_s^3/(\rho^{1/2}G^{3/2})$, where c_s is the sound speed and ρ is the gas density. The Herschel data confirm the conclusions of the previous studies: For high masses the CMF slope is comparable to Salpeter, while at low masses the CMF turns over with a peak located at ~1 $M_\odot$. The CMF has also been inferred in the Polaris cloud, which is much less dense than Aquila and is non-star-forming. Interestingly, the identified cores do not appear to be self-gravitating, but the core mass spectrum still resembles a lognormal, yet peaks at a much smaller value [$\simeq 2 \times 10^{-2}$ $M_\odot$ (*André et al.*, 2010)]. In summary, the observed CMF in star-forming regions can generally be fitted with a lognormal and resembles the IMF but shifted to higher masses by a factor of 3.

Finally, we note that cores within infrared dark clouds (IRDCs) also have a mass distribution that is similar to the IMF shape. *Peretto and Fuller* (2010) derived a clump mass spectrum from the Spitzer Galactic Legacy Infrared Mid-Plane Survey Extraordinaire (GLIMPSE) data catalog of IRDCs (*Peretto and Fuller,* 2009). They found that while the mass spectrum of the IRDCs themselves was similar to the mass distribution inferred for CO clumps ($\alpha \simeq 1.7$), the mass spectrum of the embedded structures was steeper and comparable to the Salpeter IMF (although see also *Ragan et al.*, 2009; *Wilcock et al.*, 2012). Unfortunately, their observational completeness limit did not allow a determination of the distribution peak. The discrepant slopes of core and clump distributions may indicate a transition scale, which we discuss below in the context of theoretical models.

Fifteen years of investigation and increasingly high-quality data have reinforced the similarity between the shapes of the CMF and the IMF at large masses. At lower masses, a change in the slope of the mass spectrum around a few solar masses appears to be a common feature of many studies. However, a definitive demonstration of such a peak will require better observational completeness limits.

3.2. Analytical Theories for the Core Mass Function

If the distribution of stellar masses is in some way inherited from the distribution of core masses, then theories predicting the CMF are an important first step in understanding the IMF. At this point, we stress that the theories presented below attempt to infer the mass of the gas reservoir that eventually ends up in the stars rather than the masses of cores as defined observationally. Indeed, these latter are transient objects, whose masses are time-dependent, which are difficult to properly model using a statistical approach. In some circumstances, such as in especially compact and clustered regions, discrete cores are difficult to identify observationally. However, individual "reservoirs" (collapsing regions of gas from which an embedded protostar will accrete most of its mass) can still be well-defined theoretically. In the following section, for convenience we use "CMF" to refer to the mass distribution of the stellar mass reservoirs even though these objects are likely somewhat different than observed cores.

Turbulence is ubiquitous in the interstellar medium and appears to be intimately bound to the process of star formation (*Mac Low and Klessen*, 2004; *Elmegreen and Scalo*, 2004; *Scalo and Elmegreen*, 2004). Most theories of the CMF are predicated on the dual effects of turbulence. Supersonic turbulent motions can both disperse gas and provide support against gravity (*Bonazzola et al.*, 1987). In the interstellar medium, turbulent support appears to be extremely important, since turbulent motions approximately balance gravity within giant molecular clouds (GMCs) and other cold, dense regions where the thermal pressure is negligible (*Evans*, 1999). However, this support cannot simply be treated as hydrostatic "pressure." The same turbulent motions necessarily generate compressions, shocks, and rarefactions; since these effects are multiplicative, the central limit theorem naturally leads to a lognormal density distribution whose width increases with the Mach number, $\mathcal{M}$ [see *Vazquez-Semadeni* (1994) and *Padoan et al.* (1997), although some non-lognormal corrections are discussed in *Hopkins* (2013b)]. Because density fluctuations grow exponentially while velocities grow linearly, the presence of supersonic turbulence will always produce dense, local regions that are less stable against self-gravity, even if turbulent motions globally balance gravity.

Padoan et al. (1997) presented one of the first models to combine the effects of density fluctuations and gravity to predict a CMF. Assuming a lognormal density probability density function (PDF) and isothermal gas, they estimated the number of regions in which the density would be sufficiently high so that gravity overwhelmed thermal support. They further assumed these regions should be analogous to observed cores. Their method naturally predicts a lognormal-like CMF with a peak and a power-law extension at high masses. In this calculation, the peak of the IMF comes from the lognormal density PDF, which stiffly drops at high density. The slope of the distribution at high masses reflects the counting of thermal Jeans masses in a medium that follows a lognormal density distribution. The result is similar to observed CMFs, although both the high- and low-mass slopes predicted appear to be shallower than those observed. *Padoan et al.* (1997) noted two main limitations to their derivation. First, they neglected local turbulent support. Second, gravitational instability, as well as the actual mass of a region, depends not just on the density and temperature but also on the size scale. *Padoan and Nordlund* (2002) extended this method by adopting the simple heuristic assumption that the size of each region followed the characteristic size scale of postshock gas, which could be computed given some $\mathcal{M}$. They argued that this more naturally led to a power-law high-mass slope.

Inutsuka (2001) was the first to approach the problem of the CMF with the formalism from *Press and Schechter* (1974), which is used within cosmology to calculate the MF of dark matter halos. *Hennebelle and Chabrier* (2008, 2009) then extended this previous work in more detail. As in *Press and Schechter* (1974), *Hennebelle and Chabrier* (2008) assumed the gas density is a Gaussian random field, although in the context of turbulent clouds the field becomes log-density rather than linear-density. If the density power spectrum is known, the density can be evaluated for all scales, resolving the second limitation above. The fluctuations on all scales can be compared to some threshold to assess gravitational boundedness and thus calculate the mass spectrum of self-gravitating turbulent fluctuations above some simple overdensity. By considering the importance of turbulent as well as thermal support in defining self-gravitating regions, *Hennebelle and Chabrier* (2008) predicted a CMF with a peak, a Salpeter-like power law at high masses, and a rapidly declining ("stiff") distribution at low masses. This multi-scale approach provides one solution for the "cloud-in-cloud" problem identified by *Bond et al.* (1991). This problem arises because small-scale structures are generally embedded within larger ones, which complicates structure counting. The original formulation by *Press and Schechter* (1974) suffered from this problem and led them to overestimate the number of structures by a factor of 2 (*Bond et al.*, 1991). The *Hennebelle and Chabrier* (2008) approach also accounts for turbulent fragmentation but not for gravitationally induced fragmentation, which is much more difficult to include. *Hennebelle and Chabrier* (2008) found that their prediction for non-self-gravitating masses agreed well with the observed mass spectrum of CO clumps but was less consistent with the IMF-like observed CMF.

Hopkins (2012a,b) extended this approach even further to encompass the entire galactic disk and include rotational support. This revised approach invokes "excursion set theory," which is essentially the theory of the statistics of random walks. Excursion set theory was first applied to cosmology by *Bond et al.* (1991) to make the *Press and Schechter* (1974) approach mathematically rigorous and solve the cloud-in-cloud problem [for a review of its current applications in cosmology, see *Zentner* (2007)]. In short, imagine constructing a statistical realization of the density field. In the context of star formation, one can use the density power spectrum

of supersonic turbulent gas to correctly include contributions from flows on different scales. *Hopkins* (2012a,b) showed that the density variance vs. scale, which is the most important quantity needed to construct the density field, naturally follows from the largest Mach numbers in the galactic disk. Next, one picks a random point in the field and smooths the gas density around that point in a "window" of some arbitrary radius. As this radius increases from zero to some arbitrarily large value, the mean density in the window can be compared to some threshold governing the self-gravitating collapse of the gas. The threshold may be computed by including thermal, turbulent, magnetic, and rotational support and varies as a function of the size scale. If the field crosses this threshold at any point, then there is both a minimum and a maximum "window radius" in which the smoothed field is above the threshold. Within this window, the field may cross back and forth multiple times. Excursion set theory naturally resolves the cloud-in-cloud problem by explicitly predicting substructures inside larger structures. And indeed, because of the very large dynamic range of turbulence in galaxies, self-gravitating objects inherently include self-gravitating substructures on a wide range of scales.

Hopkins defined the "first crossing" as the largest "smoothing scale" for which a structure is self-gravitating and demonstrated that the predicted mass spectrum and properties such as the size-mass and linewidth-size relations of the objects agreed very well with those of molecular clouds. At the other extreme, on small scales where thermal support becomes important, there is always a minimum self-gravitating scale, the "last-crossing," within a structure. *Hopkins* (2012a) showed that the distribution of structures defined by the last crossing agreed well with the CMF. In fact, under the right circumstances, the predicted last-crossing distribution reduces to the *Hennebelle and Chabrier* (2008) formulation. Figure 3 illustrates the predictions of the various theories and shows the corresponding mass functions. Because the turbulent cascade is set by the global galactic disk properties, this approach naturally predicts an invariant CMF/IMF at high masses within the Milky Way (*Hopkins*, 2013c). *Hopkins* (2013d) noted that this formalism also predicts the clustering of cores, which is analogous to galaxy clustering. Consequently, these statistics provide a quantitative explanation for the large-scale clustering in star formation.

Statistical tools, such as excursion set theory, are the theoretical analog of new observational metrics such as "dendrograms"; both seek to describe the statistical hierarchy of self-gravitating structures as a function of scale (*Rosolowsky et al.*, 2008b; *Goodman et al.*, 2009). Comparing theory with observations requires the additional step of smoothing the observed gas on various scales and accounting for projection effects, which may be challenging (e.g., *Beaumont et al.*, 2013). However, by constructing the statistics of self-gravitating objects defined on their smallest/largest self-gravitating scales, it is possible to rigorously define the observational analog of last/first crossings. One can imagine that even more powerful constraints can be obtained by using these methods to directly compare the

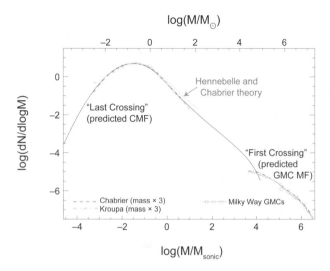

Fig. 3. Core mass spectrum and GMC mass spectrum calculated by *Hopkins* (2012b) as a function of stellar mass and sonic mass. The gray dotted curve shows the CMF inferred by *Hennebelle and Chabrier* (2008), while the lines indicate structures defined by the last (gray dashed) and first crossing (dot-dashed) of the self-gravitating barrier. The *Chabrier* (2005) and *Kroupa* (2001) fits to the observed IMF are shown by the gray dashed and light-gray dot-dashed lines, respectively. Observed Milky Way GMCs are indicated by the solid gray line with circles.

behavior of the density and velocity fields as a function of smoothing scale (e.g., by examining power spectra, covariance, the spatial correlation function of self-gravitating regions on different scales, or the rate of "back-and-forth" crossings as a function of scale). This would allow observations and theory to compare to the entire dynamic range of turbulent fragmentation using all the information in the field rather than focusing on some, ultimately arbitrary, definition of a "core" and a "cloud."

In summary, these theories provide a natural link between the CMF, GMC MF, and star cluster MF for a wide range of systems. This link derives from the inherently scale-free physics of turbulent fluctuations plus gravity. The most robust features of these models are the Salpeter-like slope ($\alpha \sim 2.2$–2.4), which emerges generically from scale-free processes, and the lognormal-like turnover, which emerges from the central limit theorem. The location of the turnover, in any model where turbulent density fluctuations are important, is closely tied to the sonic length, which is the scale, R_s, below which thermal or magnetic support dominates over turbulence. Below this scale density fluctuations become rapidly damped and cannot produce new cores. As a result, the location of the peak, and the rapidity of the CMF turnover below it, depend dramatically on the gas thermodynamics (*Larson*, 2005; *Jappsen et al.*, 2005). Unlike what is often assumed, however, the dependence on thermodynamics displayed by turbulent models and simulations with driven turbulence is not the same as a Jeans-mass dependence (see *Hennebelle and Chabrier*,

2009). The difference between the scaling of the Jeans mass, $M_J \simeq c_s^3/(\rho^{1/2} G^{3/2})$, and the sonic mass, $M_s \simeq v_{turb}^2 R_s/G$, can be dramatic when extrapolating to extreme regions such as starbursts where the turbulent velocity, v_{turb}, is large (*Hopkins,* 2013c). Consequently, a better understanding of the role of thermodynamics in extreme environments is needed.

Other processes such as magnetic fields, feedback from protostars, and intermittency of turbulence may also be important and are not always considered in these models (see section 5 and the chapter by Krumholz et al. in this volume). For example, supersonic turbulence is intermittent and spatially correlated, so both velocity and density structures can exhibit coherency and non-Gaussian features, including significant deviations from lognormal density distributions (*Federrath et al.,* 2010; *Hopkins,* 2013b). In addition, the models above generally assume a steady-state system, where in fact time-dependent effects may be critical (see section 5). Certainly these effects are necessary to consider when translating the CMF models into predictions of star-formation rates (*Clark et al.,* 2007; *Federrath and Klessen,* 2012; and the chapter by Padoan et al. in this volume). Preliminary attempts to incorporate time-dependence into the models (*Hennebelle and Chabrier,* 2011, 2013; *Hopkins,* 2013a) are worth exploring in more detail in future work.

3.3. The Core Mass Function Inferred from Numerical Simulations

A variety of numerical simulations have been performed to study the origin of the CMF and the IMF (see section 5 for a discussion of cluster simulations). Here, we consider a subset of numerical simulations that focus on modeling turbulence and the formation of cores. The most important assumptions in these simulations are the explicit treatment of self-gravity and turbulent driving, since these two processes lie at the very heart of the analytic models described above. Other physical processes such as magnetic fields and the equation of state are also very important, but we temporarily ignore these for the purpose of comparing the analytic models to simulations. A more technical but severe issue is the definition of the object mass used to compute the mass spectrum. First, the exact definition of a core can be important, and second, "sink" particles are often used to mimic protostars or cores that accrete from the surrounding gas. Given that these sinks cannot self-consistently model all the relevant physical processes, they are usually intermediate between a core and a star, although when feedback is properly accounted for, they may more closely resemble a star (see section 5.2). Another crucial issue is numerical resolution. Broadly speaking, the IMF covers about 4 orders of magnitude in mass, which corresponds to 5–7 orders of magnitude in spatial scales assuming a typical mass-size relation $M_{CORE} \propto L^{1-2}$ (e.g., *Larson,* 1981). Since most numerical simulations span less than 5 orders of magnitude, the mass spectrum they compute is necessarily limited.

The CMF has been determined in driven turbulence simulations without self-gravity, where prestellar cores are identified as regions that would have been gravitationally bound if gravity had been included. The exact cores identified depend on the types of support considered. *Padoan et al.* (2007) performed both hydrodynamical and magneto-hydrodynamical (MHD) simulations without gravity. They found that when only thermal support was considered in the core determination, the mass spectrum was too steep with $\alpha = 3$. When magnetic pressure support was included in the core definition, the mass spectrum became shallower and $\alpha \simeq 2$–2.5, which is similar to the Salpeter exponent. *Schmidt et al.* (2010) performed similar simulations using both solenoidal and compressive random forcing and including thermal and turbulent support. They also found that when only thermal support was used to assess core boundedness, the mass spectrum was too steep with $\alpha \simeq 3$. When turbulent support was included, the mass spectrum again became shallower and the slope approached the Salpeter slope. *Schmidt et al.* (2010) also demonstrated that the position of the distribution peak noticeably depends on the turbulent forcing. The peak shifts toward smaller masses for compressive forcing. This occurs because as the gas experiences more compression, smaller collapsing regions are generated; the sonic mass (and Jeans mass in those dense regions) became smaller because the fraction of compressive motions increased. This effect is also predicted by analytic models. Schmidt et al. performed quantitative comparisons with both the predictions by *Hennebelle and Chabrier* (2008) and the hydrodynamical predictions by *Padoan et al.* (1997) and found good agreement.

Simulations that include both turbulence and self-gravity have been performed by a variety of groups (see section 5). Of particular interest here, *Smith et al.* (2009) used simulations to investigate the CMF, the IMF, and the correlation between them. They obtained the CMF by defining prestellar cores on the basis of either the gravitational potential or thermal and turbulent support and defined the IMF as the distribution of sink masses. Smith et al. found that both the simulated CMF and the particle MF resembled the observed IMF. To investigate the correlation between the two, they identified the core in which a sink particle formed (the parent core) and tracked the relationship between the sink masses and parent cores in time. They found that the correlation remained very good for about three free-fall times, indicating that sink masses, at least for early times, are determined by their parent core. We discuss this work further and its relevance for understanding the link between the CMF and IMF in section 4.

4. THE LINK BETWEEN THE CORE MASS FUNCTION AND THE INITIAL MASS FUNCTION

Star-forming molecular clouds are observed to contain ensembles of cores whose mass distribution closely resembles the stellar IMF. As discussed in section 3, theoretical models of turbulent fragmentation predict distributions of core masses that also resemble the IMF. At first glance, we would appear to be done: The theoretical core mass function resembles the observed CMF, which in turn resembles the stellar IMF.

However, the picture is not so simple. If the shape of the IMF derives directly from the shape of the CMF, several conditions must hold. First, all observed cores must be truly "prestellar," i.e., bound structures destined to condense into stars. Second, cores must not alter their mass by either accretion or mergers, or, if they do, they must do so in a self-similar fashion such that the shape of the CMF is preserved. Third, all cores must have the same star-formation efficiency. Fourth, if cores undergo internal fragmentation, they must do so in a self-similar fashion. Fifth, all cores must condense into stars at the same rate; otherwise, cores that condense more slowly will be overrepresented in the CMF. Figure 4 illustrates what happens when each of these five conditions is violated. Finally, it remains to be shown whether the cores produced by turbulent fragmentation theory can sensibly be identified with observed cores.

4.1. Inferring the Intrinsic Properties of Prestellar Cores

4.1.1. Masses and thermal support. While observational studies of the CMF have tried to ensure that the cores identified are truly prestellar, there are many uncertainties. Since a prestellar core is by definition destined to form a single star or multiple system, an important quantity is the estimated ratio of the core mass to Jeans mass, which is commonly used by observers to determine whether a core is gravitationally bound. Unfortunately, this fundamental quantity is extremely sensitive to the derived temperature of the core, going as

$$\frac{M_{CORE}}{M_{JEANS}} \sim 0.05 \left(\frac{\pi}{\Omega^3} \right)^{\frac{1}{4}} \left(\frac{G\bar{m}F_\lambda D}{c\kappa_\lambda} \right)^{\frac{3}{2}} \frac{\lambda^6}{k_B^3 T_{DUST}^3} \qquad (1)$$

where F_λ is the monochromatic flux from the core, D its distance, Ω the solid-angle enclosed by the core's boundary, T_{DUST} the dust temperature, κ_λ the monochromatic mass-opacity (due to dust, but normalized per unit mass of dust and gas), k_B the Boltzmann constant, G the universal gravitational constant, $\bar{m}$ the mean gas-particle mass, and c the speed of light. Observational estimates of this ratio are very uncertain, primarily because κ_λ and T_{DUST} are heavily degenerate (*Shetty et al.,* 2009a,b). The situation has been improved by Herschel (*André et al.,* 2010). By combining Herschel multi-wavelength data with a careful statistical analysis, it is possible to break these degeneracies (*Kelly et al.,* 2012; *Launhardt et al.,* 2013). However, estimates still typically assume that the dust and gas have the same temperature, which is only likely above a number density of ~10^5 cm^{-3} when the dust and gas thermally couple (e.g., *Glover and Clark,* 2012). Consequently, the question of whether an observed core truly belongs in the prestellar CMF (i.e., $M_{CORE}/M_{JEANS} > 1$) can be fraught.

(a) Not all cores are "prestellar." Here we show the emerging IMF that could arise if the low-mass cores in the CMF are transient "fluff."

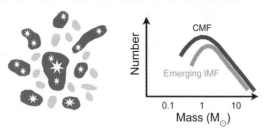

(b) Core growth is not self-similar. Here we show the emerging IMF that could arise if, say, only the low-mass cores in the CMF are still accreting.

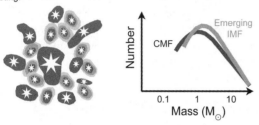

(c) Varying star formation efficiency (SFE). Here we show the emerging IMF that could arise if the high-mass cores in the CMF have a lower SFE than their low-mass siblings.

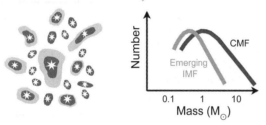

(d) Fragmentation is not self-similar. Here we show the emerging IMF that could arise if the cores in the CMF fragment are based on the number of initial Jeans masses they contain.

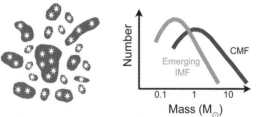

(e) Varying embedded phase timescale. Here we show the emerging IMF that could arise if the low-mass cores in the CMF finish before the high-mass cores.

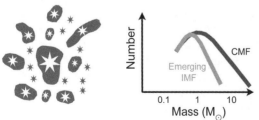

Fig. 4. Schematics illustrating what would happen if the conditions necessary for the CMF to map to the IMF were to be violated (see section 4).

4.1.2. Non-thermal support. Constraints on the kinematics of cores can be obtained from observations of molecular line profiles (e.g., *André et al.,* 2007), but the kinematic information is normally convolved with the effects of in-homogeneous abundances and excitation conditions, finite resolution, superposition of emission from disparate regions along the line of sight, or optical thickness (concealment of regions along the line of sight). In addition, spectra only probe the radial component of the velocity. These uncertainties aside, the linewidths measured from observations of N_2H^+, NH_3, and HCO^+ toward dense cores in Perseus and Ophiuchus suggest that the internal velocities, while generally subsonic, can contribute a significant amount of energy (*André et al.,* 2007; *Johnstone et al.,* 2010; *Schnee et al.,* 2012). Consequently, some of the cores identified purely on the basis of dust emission or extinction may actually be unbound, transient objects. *Enoch et al.* (2008) demonstrated that this is likely to affect only the lowest mass-bins of their CMF, and then only by a small amount, but they note that the statistics are still small.

Similar constraints pertain to estimates of the magnetic field in cores. Field estimates are obtained using either the Zeeman effect, which only probes the line-of-sight component, or the Chandrasekhar Fermi conjecture, which can be used to estimate the transverse component. *Crutcher et al.* (2009) presented statistical arguments to suggest that magnetic fields are not able to support prestellar cores against gravity. However, observing magnetic field strengths in cores is challenging (see the chapter by Li et al. in this volume for more discussion).

4.1.3. Extracting cores from column density maps. Any evaluation of the CMF inherently depends on the procedure used to identify cores and map their boundaries. Different algorithms[e.g., GaussClumps (*Stutzki and Guesten,* 1990); ClumpFind (*Williams et al.,* 1994); dendrograms (*Rosolowsky et al.,* 2008b)], even when applied to the same observations, do not always identify the same cores, and when they do, they sometimes assign widely different masses. Interestingly, different methods for extracting cores usually find similar CMFs even though there may be a poor correspondence between individual cores [for example, the CMFs derived for Ophiuchus by *Motte et al.* (1998) and by *Johnstone et al.* (2000)]. A similar problem arises in the analysis of simulations (*Smith et al.,* 2008). *Pineda et al.* (2009) have shown that the number and properties of cores extracted often depend critically on the values of algorithmic parameters. Therefore, the very existence of the cores that contribute to an observed CMF should be viewed with caution, particularly at the low-mass end where the sample may be incomplete (*André et al.,* 2010).

4.2. Phenomenology of Core Growth, Collapse, and Fragmentation

In the theory of turbulent fragmentation, cores form in layers assembled at the convergence of large-scale flows or in shells swept up by expanding nebulae such as HII regions, stellar wind bubbles, and supernova remnants. Numerical simulations indicate that these cores are delivered by a complex interplay between shocks, thermal instability, self-gravity, and magnetic fields, moderated by thermal and chemical microphysics and radiation transfer (*Bhattal et al.,* 1998; *Klessen,* 2001; *Clark and Bonnell,* 2005, 2006; *Vázquez-Semadeni et al.,* 2007; *Padoan et al.,* 2007; *Hennebelle et al.,* 2008; *Banerjee et al.,* 2009; *Heitsch et al.,* 2011; *Walch et al.,* 2012). Due to the anisotropies introduced by shocks, magnetic fields, and self-gravity, these layers and shells tend to break up first into filaments from which cores then condense, in particular at the places where filaments intersect. However, all cores do not form stars; some disperse without forming stars and others may be assimilated by larger cores.

The main focus of star formation is those cores that become gravitationally unstable. A number of authors have studied collapsing cores in numerical simulations (*Klessen et al.,* 2000; *Klessen,* 2001; *Tilley and Pudritz,* 2004; *Clark and Bonnell,* 2005, 2006; *Vázquez-Semadeni et al.,* 2007; *Smith et al.,* 2009). These studies show that very-low-mass objects ($M_{CORE} < 0.5$ $M_{\odot}$) tend not to be self-gravitating but are instead transient structures with excess turbulent and thermal energy. In contrast, the first gravitationally unstable cores to form in these simulations are always around the mean thermal Jeans mass, $M_{CORE} \sim 1$ $M_{\odot}$. Structure decompositions of these clouds yield MFs that resembles the CMF even though most of the identified "cores" are unbound and transient.

Studies that include sink particles and employ Lagrangian fluid methods have investigated the mapping between the masses of prestellar cores at the point they become bound and the final masses of the stars that form within them (*Clark and Bonnell,* 2005; *Smith et al.,* 2009). To date, this has been done with smoothed particle hydrodynamics (SPH), but a similar analysis could be performed using tracer particles in Eulerian fluid codes. As previously mentioned in section 3 there appears to be a good correlation between core and star masses at early times, but the correlation erodes after several core dynamical times. *Chabrier and Hennebelle* (2010) reanalyzed the *Smith et al.* (2009) data to quantify the CMF/IMF correlation by comparing the distributions with a simple Gaussian distribution. They found that the level of mismatch between stellar masses and the parent cores can be reproduced with a modest statistical dispersion on the order of one-third of the core mass. However, the final mismatch between stellar masses and their natal core masses in these studies could in part be due to the dense initial conditions employed, the lack of feedback to halt accretion, and limited dynamic range. More recent simulations follow the assembly of cores from typical GMC densities (e.g., *Glover and Clark,* 2012) or from even lower densities typical of the interstellar medium (e.g., *Hennebelle et al.,* 2008; *Clark et al.,* 2012), and these may provide a more useful framework within which to evaluate the core-formation process. Such studies have the advantage of capturing the transition between the line-

cooling and dust-cooling regimes, and the characteristic stellar mass may depend upon the conditions of the gas at this transition (*Larson,* 2005; *Jappsen et al.,* 2005).

Numerical calculations suggest that the flow of material onto cores is highly anisotropic, where the largest and most fertile cores acquire much of their mass from the filaments in which they are embedded (e.g., *Whitworth et al.,* 1995; *Smith et al.,* 2011). Moreover, some of the material flowing into a core may already have formed stars before it arrives at the center (*Girichidis et al.,* 2012). These stars can then continue to grow by competitive accretion while cold, dense gas remains (*Bonnell et al.,* 1997; *Delgado-Donate et al.,* 2003). Stars may become detached from the gas reservoir if they are ejected through dynamical interactions with other stars (*Reipurth and Clarke,* 2001). Even if the material flowing into a core has not yet condensed into a star when it arrives, the anisotropic inflow manifests itself as turbulence in the core, and therefore creates a substructure that may help the core to fragment (*Goodwin et al.,* 2004; *Walch et al.,* 2012; *Offner et al.,* 2010).

However, early simulations and most of those above do not include MHD effects or radiative feedback (see also section 5). When MHD (*Tilley and Pudritz,* 2007; *Hennebelle and Teyssier,* 2008), radiative feedback (*Krumholz et al.,* 2007, 2009; *Cunningham et al.,* 2011), or both (*Price and Bate,* 2007; *Commerçon et al.,* 2011; *Peters et al.,* 2011; *Myers et al.,* 2013) are included, the efficiency of fragmentation is reduced, and the number of protostars formed by a core may be reduced. Magnetic torques may also act to redistribute angular momentum, which affects fragmentation (*Basu and Mouschovias,* 1995).

When a protostar forms in a core it is normally attended by an accretion disk, which may fragment in some circumstances to form (usually low-mass) companions (e.g., *Bonnell and Bate,* 1994). Early simulations indicated that this could provide an important channel for forming brown dwarfs and very-low-mass H-burning stars (*Bate et al.,* 2002; *Stamatellos and Whitworth,* 2009a,b). Magneto-hydrodynamic simulations have cast some doubt on whether sufficiently massive disks can form given observed mass-to-flux ratios, because magnetic torques efficiently remove angular momentum. However, misalignment of the field relative to the angular momentum vector may provide an avenue for massive disks to form (e.g., *Joos et al.,* 2012; *Seifried et al.,* 2012; *Krumholz et al.,* 2013). In addition, turbulence, Parker-like instabilities, and non-ideal effects may be important and must be included in simulations to resolve issue of massive disk formation.

Even if massive disks do form, it is critical to consider radiative feedback from the central star, and any other stars that form. If a star accretes continuously from its disk, it has a high-accretion luminosity, and the consequent radiative heating stabilizes its disk against fragmentation (see section 5 and the chapter by Krumholz et al. in this volume). On the other hand, if the accretion is episodic, with most of the accretion energy being released in short bursts, and if the time between bursts is sufficiently long, then disk fragmentation can still occur and appears to produce low-mass stars in the numbers and with the statistical properties observed (*Stamatellos et al., 2012*).

4.3. High-Mass Cores and the High-Mass Initial Mass Function

McKee and Tan (2002) proposed that high-mass star formation occurs via the roughly monolithic collapse of high-mass cores in a scaled-up version of low-mass star formation. This implies that any mapping between the CMF and IMF should extend to higher masses, as suggested by *Oey* (2011). A few $\simeq$10-$M_\odot$ prestellar cores have been reported in the literature and included in CMF determinations (*Nutter and Ward-Thompson, 2007; Alves et al., 2007*); however, it is usually unclear whether these more massive objects will go on to form a single star system or are clumps, which will form a cluster of stars. While several class 0 candidates (i.e., cores with a detected embedded source) with masses greater than 50 $M_\odot$ have been found (*Bontemps et al., 2010; Palau et al., 2013; Zhang et al., 2013a*), there is no unequivocal high-mass prestellar core. It is unclear whether this implies these objects never exist, and thus the CMF/IMF mapping does not extend to such high masses, or whether such cores are extremely short lived, as proposed by *Hatchell and Fuller* (2008).

Infrared dark clouds have often been suggested as the best candidates for high-mass prestellar cores given their high masses and column densities. However, their clumpy interiors (*Ragan et al., 2009*) and subvirial velocity dispersions (*Ragan et al., 2012*) suggest that these objects may be similar to local and lower-mass regions of star formation such as Ophiuchus (*André et al., 2007*). If so, these objects are more likely the progenitors of entire clusters (*Rathborne et al., 2006*). More recent studies have shown that some parts of IRDC G035.39-00.33 may be in virial equilibrium (*Hernandez et al., 2012*) and thus consistent with the scenario proposed by *McKee and Tan* (2002). Further interferometric studies of the interiors of these objects are needed and could be performed with the Atacama Large Millimeter Array (ALMA).

It remains unclear whether the observed high-mass class 0 candidates contain a single massive protostellar system or a deeply embedded protocluster, such as those discussed in section 5. Extensive radiative transfer modeling [along the lines of *Zhang et al.* (2013b) and *Offner et al.* (2012)] may be able to help distinguish between these scenarios. Furthermore, class 0 cores are not necessarily isolated objects; they are embedded within larger star-forming regions. Observations of local regions suggest that the relative motion of low-mass cores is small (e.g., *André et al., 2007*); however, in more clustered regions cores could be undergoing accretion and merging as described above.

4.4. Statistical Constraints

As mentioned above, the statistical mapping of the CMF to the IMF requires that on average cores fragment self similarly and have a roughly constant efficiency that is independent of mass. A simple self-similar mapping appears to be possible because a very simple model can reproduce binary statistics, such as binary frequencies and mass ratios, provided that the typical core forms four or five stars and has a high-star-formation efficiency (*Holman et al., 2013*). A less-restrictive mapping between the CMF and IMF would entail that the most abundant stars, $\bar{M} \sim 0.2\ M_\odot$, mainly form from the most abundant prestellar cores, $\bar{M}_{CORE} \sim 1\ M_\odot$, while the most massive stars form only from the most massive cores (*Goodwin et al., 2008; Oey, 2011*).

Initial estimates of the efficiency with which core gas forms stars are based on the premise that most cores produce a single star. Observationally, the CMF appears to have a peak at around 1 $M_\odot$ (e.g., *Alves et al., 2007; Nutter and Ward-Thompson, 2007; Enoch et al., 2008; André et al., 2010*), which suggests an efficiency of ~30%. These early estimates have subsequently been increased to ~50% to allow for the formation of binary systems. This estimate is consistent with the theoretical predictions of *Matzner and McKee* (2000), who look at the impact of the generalized X-wind model (*Shu et al., 1995*) on a protostellar core. They predict efficiencies in the range 20–70%, depending on the geometry of the core, and stress that this effect should be insensitive to the mass of both the protostar and its parent core. In addition, efficiencies of around 50% have been claimed in numerical studies of the collapse of magnetized, rotating prestellar cores (*Machida et al., 2009; Machida and Matsumoto, 2012; Price et al., 2012*). These studies capture the protostellar outflows that arise both from the magnetic-tower structure, which dominates the polar regions of the core (*Lynden-Bell, 1996*), and from the centrifugal acceleration that is powered in the protostellar disk (*Blandford and Payne, 1982; Pudritz and Norman, 1983*). While these results are encouraging for the CMF/IMF mapping, it should be stressed that both the theoretical and numerical predictions quoted here rely on idealized geometries; the final efficiencies in more realistic, turbulent cores, such as those studied by *Seifried et al.* (2012), are still unknown. Finally, *Holman et al.* (2013) argued that the efficiency might be much higher and perhaps even larger than 100% if cores continue accreting while forming stars.

Although efficiencies must be taken into account when mapping the CMF to the IMF, the observed CMF peak is also quite uncertain. CMF peaks have been reported to be ~0.02 $M_\odot$ in Polaris (*André et al., 2010*); ~0.1 $M_\odot$ in ρ Oph (*Motte et al., 1998*); $\simeq$2 $M_\odot$ in the Pipe Nebula (*Alves et al., 2007*); and ~1 $M_\odot$ in Orion, Aquila, and the combined populations of ρ Oph, Serpens, and Perseus (*Nutter and Ward-Thompson, 2007; Enoch et al., 2008; André et al., 2010*). *Goodwin et al.* (2008) noted that the peak of the CMF appears to depend on the source distance, which could occur if blending creates bias in more distant regions. As discussed in section 3.1, the different peak values may be due to different core definitions (e.g., the inclusion of low-density unbound objects), but they may also due to observational incompleteness, which is often near the peak position.

Additional constraints on the CMF-IMF mapping are imposed by considering the time evolution of the CMF. For example, *Clark et al.* (2007) stressed that cores must have at least a Jeans mass to form stars. Assuming a roughly constant temperature, this presents two options: Either the cores all have the same Jeans mass (~0.1 $M_\odot$), and so cores contain different numbers of Jeans masses, or the cores each have a few Jeans masses such that the low-mass cores are denser than high-mass cores. The latter scenario is more likely to result in the self-similar mapping between the CMF and IMF discussed above, since it avoids the problem of an extremely Jeans-unstable core fragmenting to form a large cluster. However, this scenario also implies that cores likely evolve on different timescales, and low-mass cores form protostars more quickly than their high-mass counterparts. The emerging stellar MF should therefore be biased toward lower masses than the full IMF, which is not evident in nearby regions. If different-mass stars form on different timescales, *McKee and Offner* (2010) showed that the protostellar PDMF will be distinct from one in which stars form on the same timescale. Since the masses of embedded protostars are difficult if not impossible to measure, it may be possible to constrain the IMF through the distribution of protostellar luminosities, which provides a promising indirect means to probe the underlying mode of star-formation times and accretion rates.

The observational evidence for differences between prestellar and protostellar core mass distributions is ambiguous. *Hatchell and Fuller* (2008) showed that the mass distribution of prestellar and protostellar cores in Perseus were markedly different: The former resembles the IMF, while the latter is strongly peaked at around 8 $M_\odot$. One interpretation of these observations is that high-mass cores evolve faster than low-mass cores. Given the criterion for Jeans instability, it is not obvious how this could be the case. Alternatively, *Hatchell and Fuller* (2008) noted that these results could imply that most low-mass cores are transient. As discussed above, such a picture is broadly consistent with what is found in simulations and would imply that the observed CMF is seriously contaminated by unbound clumps. Such a scenario has also been used to explain the chemistry of low-mass cores in L673 (*Morata et al.,* 2005). In contrast, *Enoch et al.* (2008) measured the combined CMF of ρ Oph, Serpens, and Perseus and demonstrated that the prestellar and protostellar CMFs have statistically similar distributions. Since the *Hatchell and Fuller* (2008) sample contains less than half the number of cores of the *Enoch et al.* (2008) sample, the apparent differences may be purely statistical (see also the chapter by Dunham et al. in this volume for discussion of the challenges of observing and classifying protostars).

The Press-Schechter and excursion-set IMF models discussed in section 3 potentially provide a solution to the timescale issue. The models envisage that low-mass cores are more likely to be disrupted by turbulence than high-mass cores — an effect that *Hennebelle and Chabrier* (2008) refer to as "turbulent dispersion." This process acts to lower the overall star-formation rate in the low-mass cores, thus helping to counteract the fact that low-mass cores may, individually, collapse faster than their higher-mass counterparts. Again, this would imply that not all the observed cores are truly "prestellar." They also find that the cores that make up the bound CMF have similar densities yet still follow an IMF-like distribution. While this potentially provides a resolution to the apparent timescale problem, it does not address the question of why a 10-$M_\odot$ core containing many Jeans masses would only form a small number of objects. One solution to this puzzle may be supplied by the addition of more complex physics, which is discussed further in the following section.

5. FORMATION OF CLUSTERS

In our Galaxy, many stars are observed to form in groups, associations, or clusters (*Lada and Lada,* 2003; *Bressert et al.,* 2010), and the IMF is usually measured by observing these systems because they provide ensembles of stars with similar ages. Thus far, our theoretical discussion has focused on the dense cores that are observed to be the sites of star formation, tackling their formation, properties, and whether and how they produce stars. We have largely considered them to evolve in isolation from each other with each producing one or perhaps a few stars. However, particularly in dense star-forming regions, the formation of a cluster and the formation of stars occur simultaneously. Thus, the formation of stellar clusters is a dynamical and time-dependent process, where protostars may interact with each other and the evolving cloud via many different physical mechanisms. Due to this time dependence and inherent complexity, numerical simulations are essential to explore the role of each mechanism in producing the IMF.

In the mid-1990s, interest in "competitive accretion" as a mechanism for producing a mass spectrum was revived by studies using SPH with sink particles representing protostars (*Bonnell et al.,* 1997). Simulations of protostellar "seeds" embedded in and accreting from a gas reservoir demonstrated that the initial mass spectrum was relatively unimportant (see also *Zinnecker,* 1982), but that differential accretion naturally produced both a mass spectrum and mass segregation, with the more massive stars located near the cluster center as is often observed in young clusters. These early studies identified the importance of the stellar spatial distribution and dynamical interactions between protostars. In some cases, dynamical interactions were able to eject stars from the cloud and thus halt accretion. Subsequent papers found that the high-mass end of the mass spectrum was steeper than $dN/d \log M \approx M^{-1}$ due to changes in the dynamics of accretion and mass segregation (*Bonnell et al.,* 2001b,a). Rather than begin with protostellar "seeds," *Klessen et al.* (1998) studied the fragmentation of structured gas clouds to form protostars and followed their subsequent competitive accretion. They found a lognormal protostellar mass spectrum resembling the observed IMF and again noted the role of dynamical interactions in terminating accretion.

These early studies laid the foundation for the research performed over the last decade, which has included increasingly more physical processes. In this section we review numerical work investigating the roles of initial conditions, protostellar interactions, and feedback on the IMF.

5.1. Role of Initial Conditions in Shaping the Initial Mass Function

A variety of different initial conditions and physical effects have been explored in simulations of star cluster formation. A number of these find good agreement with the observed IMF, while others highlight interesting variations. However, we stress that currently no simulations of cluster formation include all the important physics, produce sufficient stellar statistics, and reach sufficiently high resolution to resolve close binaries. Moreover, simulations that follow the accretion and dynamical evolution of stars employ sink particles as a numerical convenience in order to limit the resolution of gravitational collapse and reduce computational expense. Consequently, the results summarized here are best considered suggestive "numerical experiments."

5.1.1. Cloud initialization. Simulations typically adopt one of three main approaches to the problem of cloud initialization. The first approach is to consider an isolated cloud, typically with an initially analytic density distribution (e.g., *Girichidis et al.,* 2011). Cloud structure may be seeded by applying a random velocity field at the initial time. A second approach is to consider a piece of a molecular cloud by adopting periodic boundary conditions (e.g., *Klessen and Burkert,* 2000; *Offner et al.,* 2008a). In the latter case, the large-scale turbulent energy cascade can be modeled by continuous turbulent driving (*Mac Low,* 1999). Both of these approaches dispense with the issue of molecular cloud formation and instead initialize with mean observed molecular cloud properties. Both of these approaches produce a CMF and/or IMF similar to observations (e.g., *Bate,* 2012; *Krumholz et al.,* 2012) (see Fig. 5), which suggests the IMF may be fairly insensitive to the details of cloud formation in numerical simulations. A third approach directly models the cloud formation. For example, some simulations begin with the collision of two opposing atomic streams of gas (*Koyama and Inutsuka,* 2002; *Vázquez-Semadeni et al.,* 2007; *Heitsch et al.,* 2008b; *Hennebelle et al.,* 2008; *Banerjee et al.,* 2009). This setup is predicated on the idea that some molecular clouds form from coherent gas streams (see the chapter by Dobbs et al. in this volume for further discussion of cloud formation mechanisms). Alternatively, simulations may begin with a full or partial galactic disk that forms clouds self-consistently by including interstellar medium physics. While this achieves more realistic and self-consistent initial cloud conditions, current resolution is not sufficient to follow the formation of individual stars.

5.1.2. Turbulent properties. Most cloud simulations include some degree of supersonic turbulence. The details of this turbulence have been widely varied, reflecting both current observational uncertainties and underlying theoretical

disagreement. Over the last decade simulations have typically adopted one of two treatments: Either turbulent motions are continually driven to maintain a fixed turbulent energy or the turbulence is allowed to freely decay after initialization. Arguments in favor of the former appeal to the very low star-formation rate in clouds (*Krumholz and Tan,* 2007), the continuing cascade of energy from larger scales, and the approximate gravitational and kinetic energy equipartition in clouds (*Krumholz et al.,* 2006). However, approaches applying random velocity perturbations in lieu of kinematic stellar feedback are not physical (see section 5.3). Alternatively, some turbulence may be regenerated through gravitational collapse (*Huff and Stahler,* 2007; *Field et al.,* 2008; *Klessen and Hennebelle,* 2010; *Robertson and Goldreich,* 2012). Both approaches have been successful in reproducing IMF characteristics (e.g., *Bate,* 2012; *Krumholz et al.,* 2012). Clouds with excess turbulent motions, i.e., gravitationally unbound clouds, have lower star-formation efficiencies and produce IMFs that are flatter than observed (*Clark et al.,* 2008).

The turbulent outer driving scale is generally assumed to be comparable to the simulated cloud size on the basis that turbulence is generated during the cloud-formation process and cascades smoothly to small scales (but see also *Swift and Welch,* 2008). *Klessen* (2001) demonstrated that driven turbulence injected on small scales produces a CMF/IMF that is too flat compared to observations. In the case of decaying turbulence, variations in the initial turbulent power spectrum slope apparently have little effect on the IMF (*Bate,* 2009b). *Girichidis et al.* (2011) found that the resultant IMFs were also relatively insensitive to whether the initial turbulence was solenoidal (divergence-free), compressive, or some combination of the two (see Fig. 5).

One universal characteristic of turbulent simulations, independent of turbulence details, is the production of filamentary structure (e.g., *Heitsch et al.,* 2008a; *Collins et al.,* 2012). Star-forming cores often appear at regular intervals along filaments (*Men'shchikov et al.,* 2010; *Offner et al.,* 2008b; *Smith et al.,* 2012), and gas may flow freely along them, supplying additional material to forming stars (see the chapter by André et al. in this volume for additional discussion of filaments). However, it remains to be seen whether the observed filamentary structure is simply a consequence of turbulence or serves some more fundamental role in the determination of the IMF.

5.1.3. Initial density profiles. Simulations have explored a range of underlying analytic density profiles, e.g., $\rho(r) \propto r^{-\beta}$ with β ranging from 0 to 2. Most recently, *Girichidis et al.* (2011) and *Girichidis et al.* (2012) found that turbulent, centrally condensed density profiles form a more massive star and fewer lower-mass stars, whereas turbulent, flat density profiles produce more distributed clusters. The different density profiles also yielded demonstratively different IMFs, where the IMFs produced by uniform and Bonnor-Ebert density distributions more closely resembled the observed IMF (see Fig. 5).

5.1.4. Magnetic field strength. The degree of magnetization and the relative importance of turbulent and magnetic

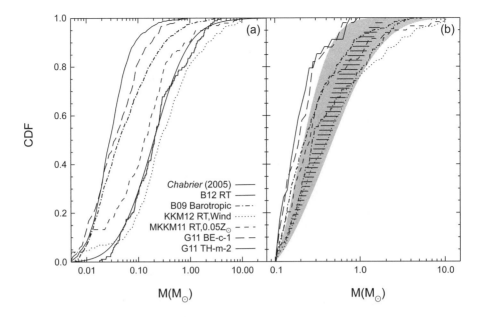

Fig. 5. Cumulative distribution functions (CDFs) for stellar masses from the simulations BE-c-1 (Bonnor-Ebert sphere with decaying compressive turbulence), TH-m-2 (uniform density with decaying mixed turbulence) [both from *Girichidis et al.* (2011)], simulation with 0.05 $Z_\odot$ and radiative feedback (*Myers et al.,* 2011), simulation with radiative feedback and protostellar outflows (*Krumholz et al.,* 2012), simulation with decaying turbulence and a barotropic EOS (*Bate,* 2009a), and simulation similar to B09 but including radiative transfer (*Bate,* 2012). **(a)** The CDFs are shown with the *Chabrier* (2005) single-star IMF for a mass range of 0.005–100 $M_\odot$. **(b)** The same CDFs but computed only for the mass range 0.1–15 $M_\odot$. The shaded regions show the fits and associated error for the observed IMF at the half-mass cluster radius for Westerlund 1 [solid gray (Andersen et al., in preparation)] and for the combined data of the Pleiades (*Moraux et al.,* 2003), IC 4665 (*de Wit et al.,* 2006), and Blanco 1 [horizontal lines (*Moraux et al.,* 2007)]. The fits of the observational data are performed for the ranges 0.15–15 $M_\odot$ and 0.045–2 $M_\odot$, respectively.

energy on different scales remain controversial, but there is consensus that magnetic support can have a dramatic impact on molecular cloud evolution. Strong magnetization provides pressure support that reduces the core and star-formation rate (*Heitsch et al.,* 2001; *Price and Bate,* 2009; *Dib et al.,* 2010; *Padoan and Nordlund,* 2011; *Padoan et al.,* 2012). *Li et al.* (2010) showed that magnetic fields can also reduce the characteristic stellar mass by increasing the number of low-mass stars formed in filaments and reducing the number of intermediate-mass stars formed from turbulent compressions. The authors attributed this difference to the suppression of intermediate-mass stars from converging flows; magnetic support inhibited isolated collapse and instead stars formed predominantly within magnetized filaments, which tended to produce smaller-mass stars. An anticorrelation between characteristic stellar mass and magnetic field strength was also demonstrated in simulations of star formation in the galactic center (*Hocuk et al.,* 2012). In principle, a non-ideal MHD treatment would allow the field to diffuse from magnetically subcritical regions; however, few simulations have included such effects and the impact on the shape of the IMF is unclear. *Vázquez-Semadeni et al.* (2011) found that including ambipolar diffusion had little effect on the star-formation rate in their simulations. Larger cloud simulations that produce more stars and include non-ideal effects are necessary to fully understand the effect of magnetic fields on the IMF.

5.1.5. Thermal assumptions. The role of thermal physics in shaping the IMF remains an open question. If the characteristic stellar mass is due to turbulent fragmentation, then thermal physics may play only a minor role (see section 3). Alternatively, thermal physics, by setting the characteristic Jeans mass, may set the peak of the IMF (*Jappsen et al.,* 2005; *Larson,* 2005; *Bate and Bonnell,* 2005; *Bonnell et al.,* 2006; *Klessen et al.,* 2007). Simulations have shown that the peak of the IMF is sensitive to the adiabatic index, γ, of the gas and the critical density at which the gas transitions from being isothermal (efficient cooling) to adiabatic (inefficient cooling). For a polytropic equation of state, $P = K\rho^\gamma$, the characteristic Jeans mass can be expressed as $M_J = \pi^{5/2}/6(K/G)^{3/2}\gamma^{3/2}\rho^{(3/2)\gamma-2}$, where ρ is the gas density, K is a constant, and G is the gravitational constant (*Jappsen et al.,* 2005). Proponents of this model argue that while turbulent and magnetic support may delay gravitational collapse, such support will eventually dissipate or diffuse, leaving thermal pressure alone to oppose gravity. In either case, the peak cannot be too sensitive to the cloud environment without producing large variations, which is hard to reconcile with observations of a generally invariant IMF. However, if the gas can be described by a barotropic equation of state with some fixed critical density, as suggested by *Larson* (2005), the IMF peak mass may be set independently of the cloud Jeans mass and thus partially decouple the IMF from the cloud environment (*Bonnell et al.*, 2006). If only thermal physics is important and if the effective equation of state depends only on fundamental atomic parameters and does not also depend strongly on metallicity, then this could explain an invariant IMF.

More detailed thermal physics can be considered by including heating and cooling by dust, molecules, and atomic lines in place of a fixed equation of state. *Myers et al.* (2011) found that simulations including radiative transfer produced only weak variations in the IMF even as the metallicity varied by a factor of 100. Using a different approach, *Glover and Clark* (2012) performed simulations including heating and cooling computed via a reduced chemical network. These showed that the star-formation rate is also relatively insensitive to the gas metallicity for variations over a factor of 100. Similarly, *Dopcke et al.* (2011) found that dust cooling enabled small-scale fragmentation and hence the formation of low-mass stars for metallicities $Z \geq 10^{-5} Z_\odot$. Altogether, these studies help explain the observed IMF invariance over environments with a range of metallicities.

5.2. Role of Protostellar Feedback and Interactions

It is possible that the protostars themselves play a part in setting the IMF since they may interact with each other and the molecular cloud as the cluster forms. If they play the dominant role, this may explain why the IMF is not observed to vary greatly with environment (see section 2); the generation of the IMF may be a self-regulating process rather than depending sensitively on initial conditions.

5.2.1. Dynamical interactions. There are many ways by which protostars may affect the cloud in which they form and each other. As mentioned previously, dynamical interactions between accreting protostars can produce a spectrum of masses because the accretion rate of any particular protostar depends on its location, the surrounding gas density, and its speed. Typically, slow-moving protostars located near the cloud center, where the overall gravitational potential of the cloud tends to funnel the gas, accrete at higher rates than those in the outskirts of the cluster or those that have been involved in dynamical interactions that have given them higher speeds. However, the accretion rate of any particular protostar also tends to be highly variable (*Bonnell et al.*, 1997; *Klessen*, 2001). Dynamical interactions can be an effective mechanism for stopping accretion (by ejecting protostars from the cloud or increasing their speeds so that their accretion rate becomes negligible), and theories of the IMF can be constructed based on accreting protostars with probabilistic stopping times (*Basu and Jones*, 2004; *Bate and Bonnell*, 2005; *Myers*, 2009) in which accretion may be terminated due to dynamical interactions, outflow or radiative feedback, and other mechanisms. As mentioned in section 5.1, however, hydrodynamical simulations of competitive accretion alone result in IMFs whose characteristic stellar mass depends on the initial Jeans mass in the cloud (*Bate and Bonnell*, 2005; *Jappsen et al.*, 2005; *Bonnell et al.*, 2006) and typi-

cally produce too many brown dwarfs (*Bate*, 2009a). *Bate and Bonnell* (2005) argue that the dependence on the initial Jeans mass comes about within competitive accretion models because the characteristic stellar mass is given by the product of the typical accretion rate, ($\dot{M} \propto c_s^3/G$, where c_s is the sound speed) and the typical dynamical interaction timescale ($t \propto \rho^{-1/2}$), which terminates accretion. This is problematic because the IMF is not observed to vary greatly with initial conditions.

5.2.2. Radiative heating. The effects of radiation transfer and radiative feedback from protostars has been investigated in star-cluster calculations following three different approaches (*Bate*, 2009c; *Offner et al.*, 2009b; *Urban et al.*, 2010). These studies found that the inclusion of radiative heating in the vicinity of protostars dramatically reduced the amount of fragmentation, particularly of massive circumstellar disks. In turn this reduced the rate of dynamical interactions between protostars. Since in the competitive accretion process, brown dwarfs and low-mass stars are usually formed as protostars that have their accretion terminated quickly after they have formed by dynamical interactions, radiative feedback also reduced the proportion of brown dwarfs to stars. Figure 5 illustrates that an overabundance of low-mass objects is one of the main areas of disagreement between some simulations and observations. Potentially of even more importance, however, is the discovery that radiative feedback also removes the dependence of the mass spectrum on the initial Jeans mass of the clouds (*Bate*, 2009c). *Bate* (2009c) presented a qualitative analytical argument for how radiative feedback weakens the dependence of the IMF on the initial conditions. For example, consider two clouds with different initial densities. In the absence of protostellar heating, the Jeans mass and length are both smaller in the high-density cloud, so fragmentation will tend to produce more stars, which have smaller masses and are closer together, than in the low-density cloud. However, protostellar heating is most effective on small scales, so when protostellar heating is included in the higher-density cloud, it raises the effective Jeans mass and length by a larger factor (resulting in fewer stars with greater masses) than when it is included in the lower-density cloud (where the stars would already form further apart). Thus, the dependence on the initial density is weakened. More recently, *Krumholz* (2011) took this argument further to link the characteristic mass of the IMF with fundamental constants. Regardless of whether these analytic arguments portray the full picture or not, subsequent radiation hydrodynamics simulations have shown that excellent agreement with the observed IMF can be obtained, along with many other statistical properties of stellar systems (*Bate*, 2012, *Krumholz et al.*, 2012).

On the other hand, in very dense molecular clouds that produce massive stars, radiative feedback can be too effective at inhibiting fragmentation. *Krumholz et al.* (2011) showed that radiative feedback could lead to an overheating problem in which, after an initial cluster of protostars is formed, further fragmentation is prevented while the protostars continue to accrete. In this case, the characteristic stellar mass continually increases in time, leading to a top-heavy IMF. When combined with a strong magnetic field (mass-to-flux ratio ~2), *Commerçon et al.* (2011) and *Myers et al.* (2013) both demonstrate that radiative feedback and magnetic support could almost completely prevent fragmentation of such initial conditions, leading to a single massive star or binary rather than a cluster. However, the details of fragmentation appear to be sensitive to a combination of the initial cloud conditions and the code applied. For example, when modeling the collapse of high-mass cores, *Hennebelle et al.* (2011) found that the effects of radiation in reducing fragmentation are modest in comparison to magnetic fields. Likewise, work investigating radiative heating, including ionization from high-mass sources and using a long-characteristics radiative transfer approach, found a more modest impact by radiation on fragmentation (*Peters et al., 2010a,b, 2011*). *Peters et al.* (2010b) modeled a 1000-$M_\odot$ clump, finding that fragmentation was mainly suppressed in the inner ~1000 AU and otherwise continued unabated within the dense filaments feeding the central region. In any case, current two- and three-dimensional radiation-hydrodynamic methods adopt approximations of the radiative transfer equations. To better understand the effects of radiative feedback on the IMF in clusters, more work is required to implement methods that accurately follow the propagation of multi-frequency, angle-dependent radiation and achieve accuracy in both the optically thick and thin limits.

5.2.3. Kinematic feedback: Outflows and ionization. Another form of feedback is kinetic feedback via protostellar jets and outflows, stellar winds, and supernovae. All of these likely contribute to the turbulent motions within clouds and affect the efficiency of star formation (e.g., *Li and Nakamura*, 2006), but the level of the different contributions is unclear (see the chapter by Krumholz et al. in this volume). Calculations to determine how kinetic feedback may contribute to the IMF have only recently started to be performed, and in all cases to date the outflows have been added in a prescribed manner rather than generated self-consistently. *Dale and Bonnell* (2008) investigated the effects of stellar winds in hydrodynamical simulations of cluster formation. They found that outflows slowed the star-formation process, but had little effect on the IMF except at the high-mass end where they disrupted the accretion of high-mass protostars. *Wang et al.* (2010) performed MHD simulations with and without outflows and came to the same conclusion — outflows slowed the star-formation rate (by a factor of 2) and reduced the accretion rates of the most massive stars by disrupting the filaments that feed them and slowing the global collapse of the cloud. *Li et al.* (2010) investigated the effects of magnetic fields and outflows on the IMF, finding that although the general forms of the mass spectrums were the same in all calculations, strong magnetic fields reduced the characteristic stellar mass by a factor of 4 over hydrodynamical simulations. Adding outflows to the MHD simulation decreased the characteristic mass by

another factor of 2. *Hansen et al.* (2012) performed the first cluster formation calculations to include both radiative feedback and outflows. Although only one-third of the material accreted by a protostar was ejected in each outflow, the characteristic mass of the stars was reduced by two-thirds relative to calculations without outflows. They also pointed out that because the protostellar accretion rates were lower, radiative feedback was less important. *Krumholz et al.* (2012) expanded on this, showing that the combination of outflows and turbulence reduced the overheating problem. The protostellar population produced by their calculations with turbulence and outflows was in statistical agreement with the observed IMF.

Finally, ionizing feedback by massive stars can both trigger fragmentation by compressing dense cores and suppress star formation by destroying dense molecular gas and disrupting accretion flows (*Dale et al.*, 2005, 2007; *Walch et al.*, 2013). Thus far, however, no discernible effect on the IMF has been found (*Dale and Bonnell*, 2012). However, it is important to note that the ionizing radiation included in these calculations only determines whether the gas is ionized and hot ($T \approx 10^4$ K) or molecular and cold ($T \approx 10$ K). Heating of the cold molecular gas by nonionizing photons is not included. Its inclusion may lead to a change in the IMF similar to that discussed in the previous section. *Peters et al.* (2010b) do include ionization and heating by nonionizing photons but do not make a prediction for the IMF.

5.3. Other Predictions and Observables

5.3.1. Gas kinematics. At early times, protostars are often too dim and too enshrouded by dust and gas to make observational determinations about their spatial distribution, velocities, and masses. However, observations of dense gas kinematics provide spatial and dynamical clues about the initial conditions and properties of protostars. For example, the turbulent linewidths of low-mass cores are approximately sonic, where protostellar cores do not possess significantly broader widths than starless cores (*Kirk et al.*, 2007; *André et al.*, 2007; *Rosolowsky et al.*, 2008a). In addition, the relative velocity offsets between the dense gas and core envelopes are small, suggesting that protostars form from gas that does not move ballistically with respect to its surroundings (*Kirk et al.*, 2007, 2010; *André et al.*, 2007), which disfavors a dynamical scenario at very early times.

Simulations can be compared directly to these metrics through synthetic observations using radiative transfer. Both *Ayliffe et al.* (2007) and *Offner et al.* (2008b) found reasonable agreement between simulated and observed linewidths for different numerical parameters. *Offner et al.* (2009a) showed that protostellar dispersions in simulations of turbulent fragmentation are initially small and motions are well-correlated with the surrounding gas. *Rundle et al.* (2010) verified this for star-forming cores in simulations where protostars later underwent competitive accretion. Molecular line modeling by *Smith et al.* (2012, 2013) explored the evolution of line profiles for higher mass cores and filaments, where observations are often messier and more difficult to interpret than for local low-mass star-forming cores. Additional work compares simulated dust continuum observations of forming protostars with observations (*Kurosawa et al.*, 2004; *Offner et al.*, 2012). Such detailed synthetic observations are needed to constrain protostellar masses, gas morphologies, and outflow characteristics, which are essential metrics for distinguishing between IMF theories. Future work requires additional quantitative synthetic observations in order to provide more discriminating observational comparisons.

5.3.2. Cluster evolution and mass segregation. Stars may significantly dynamically interact during and after formation. Understanding their initial spatial distribution and subsequent evolution is necessary to accurately account for dynamical effects when measuring the IMF in clusters. The stellar spatial distribution may also present some clues for distinguishing between different theoretical models. For example, in the competitive accretion picture, the most massive stars form in a clustered environment and are naturally mass segregated at birth (*Bonnell et al.*, 2001b). In this picture instances of isolated massive stars are generally stars ejected from a cluster via dynamical interactions. In the turbulent core model, the massive cores that are predicted to form massive star systems are more likely to be found in the cloud center, where the highest column density gas is located (*Krumholz and McKee*, 2008). However, massive stars are not physically precluded from forming in isolation (*Tan et al.*, 2013). The observational evidence for isolated massive star formation is obfuscated by small number statistics and the large distances to massive targets. However, young, isolated massive stars appear to exist. *Bressert et al.* (2012) and *Oey et al.* (2013) located a number of isolated massive stars that do not appear to have been dynamically ejected from a cluster. *Lamb et al.* (2010) found that a couple of apparently isolated massive stars are in fact runaways, while several more appear to be in very sparse clusters. Using a sample of 22 candidate runaway O-type stars, *de Wit et al.* (2005) concluded that only ~4% of Galactic O-type stars form outside clusters. Explaining the frequency of isolated massive stars and the mass distributions of small clusters is an ongoing challenge for theoretical models.

5.3.3. The multiplicity of protostellar and stellar systems. One important, intermediate step between the CMF and single-star IMF is the frequency and mass dependence of multiple star systems. Resolving close binary systems and accurately counting wide binary systems remains an observational challenge. However, there is convincing evidence that younger systems have higher multiplicity than field stars (*Duchêne and Kraus*, 2013). This is not surprising since higher-order multiple star systems naturally dynamically decay with time. As shown in Fig. 6, simulations with sufficient resolution have successfully reproduced the multiplicity fraction of star systems (*Bate*, 2009a, 2012; *Krumholz et al.*, 2012). Note that Fig. 6 compares with the field multiplicity fraction, and it is not obvious why simulations should agree well with the field population while the cluster is gas-dominated and not dynamically relaxed. However, if the cluster environment is initially dense, *Moeckel and Bate* (2010) used N-body simulations to demonstrate that the multiplicity is nearly constant while the remaining gas (modeled by a background potential) is dispersed. Likewise, *Parker and Reggiani* (2013) found that the shape of the mass ratio distribution is preserved during cluster relaxation because binary disruption is mass-independent. The distribution of multiples must also be reconciled with the stabilizing effect of magnetic fields (*Commerçon et al.*, 2011; *Myers et al.*, 2013), which appears to have a deleterious impact on small-scale fragmentation and early multiplicity. In principle, the distribution of separations and mass ratios should provide strong constraints on simulations (*Bate*, 2012) but are usually neglected. Future studies are required to determine how these properties depend on initial conditions and additional physics. More pointedly, future work must determine what such multiplicity statistics imply about the efficiency of star formation in dense gas and the origin of the IMF (*Holman et al.*, 2013).

6. SUMMARY

Despite many lingering areas of debate surrounding the origin of the IMF, a few key areas enjoy consensus. The presence of supersonic turbulence plays an essential role in seeding structures that go on to spawn the stellar mass distribution. Another fundamental area of physics is radiative transfer. A proper thermal treatment and heating from protostars tends to self-regulate the occurrence of additional fragmentation within cores and, when included in cluster simulations, reduces the numbers of substellar objects and the number of stars per core, resulting in IMFs that agree well with the observed Galactic IMF. While the exact importance of magnetic effects remains poorly constrained, magnetization appears to have similar role in reducing fragmentation, while also creating outflows that entrain and eject core material, thus influencing the characteristic stellar mass.

In contrast, debate continues surrounding several other fundamental questions related to the IMF origin. For example, do the details and properties of the initial self-gravitating structures, "cores," have a unique and essential role in producing the IMF? The essence of the debate is one of nature vs. nurture: Do the initial dense cores play a greater role in determining stellar mass, or does the environment and dynamical interactions dominate? This issue is obfuscated by the subtleties of defining cores observationally and connecting them to theoretical models. In recent years, simulations have shown that the combined influence of radiative and magnetic effects reduce fragmentation and hence dynamical interactions in the early stages of star formation, which makes stellar masses more similar to the masses of their parent cores; however, simulations have also shown that a core should not be viewed as a static object

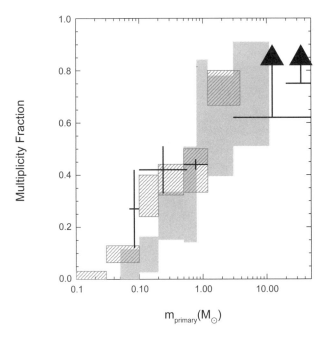

Fig. 6. Multiplicity fraction as a function of primary stellar mass. The solid shaded blocks indicate the multiplicity and associated statistical uncertainty for a simulation with radiative and kinematic feedback from forming protostars (*Krumholz et al., 2012*). The striped blocks show the multiplicity and statistical uncertainty for a simulation with radiative transfer and decaying turbulence (*Bate, 2012*). The black crosses represent observational results, with the horizontal width indicating the mass range for the observations and the vertical range showing the stated uncertainty. The two highest mass observational data points are lower limits. From left to right, the data shown are taken from *Basri and Reiners* (2006) and *Allen* (2007) (shown as a single combined point), *Fischer and Marcy* (1992), *Raghavan et al.* (2010), *Preibisch et al.* (1999), and *Mason et al.* (2009).

but as one that can gain mass on a dynamical timescale. Consequently, we conclude that both nature (cores) and nurture (dynamics) are involved to some extent in the origin of the IMF. Future studies are needed to quantify this issue, especially with respect to connecting simulations more directly with observations.

An additional debate revolves around the meaningfulness of the similarity between the CMF and IMF, which seems to suggest a gas-to-star efficiency factor ≥ 0.3. This efficiency likely results from some combination of multiplicity, mass loss from outflows, and dynamics; however, the interplay between these processes is highly nonlinear and exactly how each depends on core and star masses remains poorly understood. Future full-physics, parsec-sized simulations of star formation with subastronomical unit resolution are necessary to disentangle these effects.

The fidelity of observations in both local and extreme regions has increased during the last decade, underscoring the apparent universality of the IMF across a wide range of environments to within observational uncertainty. This invariance suggests the IMF has minimal dependence on cluster size, cluster boundedness, and metallicity. Instances of proposed variation such as within the galactic center, giant ellipticals, and dwarf galaxies are tantalizing but tenuous. If the formation of stars is relatively insensitive to initial conditions and depends strongly on self-regulating fundamental physics as simulations and CMF models suggest, then this would account for IMF invariance across a range of environments. Proposed theoretical solutions for the origin of IMF variance in extreme environments are still in their infancy. Future work is needed to provide more robust constraints on observed low-mass stellar and core MFs, natal conditions, and secondary characteristics such as multiplicity, which are necessary pieces of the theoretical puzzle of the IMF origin.

Acknowledgments. The authors thank the referees, R. Pudritz and R. Klessen, for thoughtful comments that improved the review, along with M. Krumholz, A. Myers, P. Girichidis, and M. Andersen for sharing their data. Support for S.S.R.O. was provided by NASA through Hubble Fellowship grant HF-51311.01 awarded by the Space Telescope Science Institute, which is operated by the Association of Universities for Research in Astronomy, Inc., for NASA, under contract NAS 5-26555. P.C.C. is supported by grant CL 463/2-1, part of the DFG priority program 1573 "Physics of the Interstellar Medium," and acknowledges financial support from the Deutsche Forschungsgemeinschaft (DFG) via SFB 881"The Milky Way System" (subprojects B1 and B2). P.H. has received funding from the European Research Council under the European Community's Seventh Framework Programme (FP7/2007-2013 Grant Agreement no. 306483). N.B. acknowledges partial support from a Royal Society University Research Fellowship. E.M. acknowledges funding from the Agence Nationale de la Recherche contract ANR-2010- JCJC-0501-1 (DESC). A.P.W. thanks the UK STFC for support through grant PP/E000967/1.

REFERENCES

Adams T. et al. (2002) *Mon. Not. R. Astron. Soc., 333(3)*, 547.
Allen P. R. (2007) *Astrophys. J., 668*, 492.
Allison R. J. et al. (2010) *Mon. Not. R. Astron. Soc., 407*, 1098.
Alves J. et al. (2007) *Astron. Astrophys., 462*, L17.
Alves de Oliveira C. et al. (2012) *Astron. Astrophys., 539*, 151.
Alves de Oliveira C. et al. (2013) *Astron. Astrophys., 549*, A123.
André P. et al. (2007) *Astron. Astrophys., 472*, 519.
André P. et al. (2010) *Astron. Astrophys., 518*, L102.
Ayliffe B. A. et al. (2007) *Mon. Not. R. Astron. Soc., 374*, 1198.
Banerjee R. et al. (2009) *Mon. Not. R. Astron. Soc., 398*, 1082.
Baraffe I. et al. (1998) *Astron. Astrophys., 337*, 403.
Baraffe I. et al. (2009) *Astrophys. J. Lett., 702(1)*, L27.
Baraffe I. et al. (2012) *Astrophys. J., 756(2)*, 118.
Basri G. and Reiners A. (2006) *Astron. J., 132*, 663.
Bastian N. et al. (2010) *Annu. Rev. Astron. Astrophys., 48*, 339.
Basu S. and Jones C. E. (2004) *Mon. Not. R. Astron. Soc., 347*, L47.
Basu S. and Mouschovias T. C. (1995) *Astrophys. J., 452*, 386.
Bate M. R. (2009a) *Mon. Not. R. Astron. Soc., 392*, 590.
Bate M. R. (2009b) *Mon. Not. R. Astron. Soc., 397*, 232.
Bate M. R. (2009c) *Mon. Not. R. Astron. Soc., 392*, 1363.
Bate M. R. (2012) *Mon. Not. R. Astron. Soc., 419*, 3115.
Bate M. R. and Bonnell I. A. (2005) *Mon. Not. R. Astron. Soc., 356*, 1201.
Bate M. R. et al. (2002) *Mon. Not. R. Astron. Soc., 332*, L65.
Bate M. R. et al. (2003) *Mon. Not. R. Astron. Soc., 339*, 577.
Bayo A. et al. (2011) *Astron. Astrophys., 536*, A63.

Beaumont C. N. et al. (2013) *Astrophys. J., 777,* 173.
Bhattal A. S. et al. (1998) *Mon. Not. R. Astron. Soc., 297,* 435.
Binney J. and Tremaine S. (1987) *Galactic Dynamics.* Princeton Univ., Princeton. 747 pp.
Blandford R. D. and Payne D. G. (1982) *Mon. Not. R. Astron. Soc., 199,* 883.
Bochanski J. J. et al. (2010) *Astron. J., 139(6),* 2679.
Bonazzola S. et al. (1987) *Astron. Astrophys., 172,* 293.
Bond J. R. et al. (1991) *Astrophys. J., 379,* 440.
Bonnell I. A. and Bate M. R. (1994) *Mon. Not. R. Astron. Soc., 269,* L45.
Bonnell I. A. et al. (1997) *Mon. Not. R. Astron. Soc., 285,* 201.
Bonnell I. A. et al. (2001a) *Mon. Not. R. Astron. Soc., 324,* 573.
Bonnell I. A. et al. (2001b) *Mon. Not. R. Astron. Soc., 323,* 785.
Bonnell I. A. et al. (2006) *Mon. Not. R. Astron. Soc., 368,* 1296.
Bontemps S. et al. (2010) *Astron. Astrophys., 524,* A18.
Bressert E. et al. (2010) *Mon. Not. R. Astron. Soc., 409,* L54.
Bressert E. et al. (2012) *Astron. Astrophys., 542,* A49.
Burgasser A. J. (2004) *Astrophys. J. Suppl., 155(1),* 191.
Burningham B. et al. (2013) *Mon. Not. R. Astron. Soc., 433,* 457–497.
Burrows A. et al. (1997) *Astrophys. J., 491(2),* 856.
Cappellari M. et al. (2012) *Nature, 484,* 485.
Cartwright A. and Whitworth A. P. (2012) *Mon. Not. R. Astron. Soc., 423,* 1018.
Chabrier G. (2003) *Publ. Astron. Soc. Pac., 115,* 763.
Chabrier G. (2005) In *The Initial Mass Function 50 Years Later* (E. Corbelli et al., eds.), pp. 41–50. Astrophys. Space Sci. Library, Vol. 327, Kluwer, Dordrecht.
Chabrier G. and Hennebelle P. (2010) *Astrophys. J. Lett., 725,* L79.
Chabrier G. and Küker M. (2006) *Astron. Astrophys., 446, 3,* 1027.
Chabrier G. et al. (2000) *Astrophys. J., 542(1),* 464.
Clark P. C. and Bonnell I. A. (2005) *Mon. Not. R. Astron. Soc., 361, 2.*
Clark P. C. and Bonnell I. A. (2006) *Mon. Not. R. Astron. Soc., 368,* 1787.
Clark P. C. et al. (2007) *Mon. Not. R. Astron. Soc., 379,* 57.
Clark P. C. et al. (2008) *Mon. Not. R. Astron. Soc., 386,* 3.
Clark P. C. et al. (2012) *Mon. Not. R. Astron. Soc., 424,* 2599.
Collins D. C. et al. (2012) *Astrophys. J., 750,* 13.
Commerçon B. et al. (2011) *Astrophys. J. Lett., 742,* L9.
Conroy C. and Gunn J. E. (2010) *Astrophys. J., 712,* 833.
Covey K. R. et al. (2008) *Astron. J., 136(5),* 1778.
Crutcher R. M. et al. (2009) *Astrophys. J., 692,* 844.
Cunningham A. J. et al. (2011) *Astrophys. J., 740,* 107.
Dale J. E. and Bonnell I. A. (2008) *Mon. Not. R. Astron. Soc., 391,* 2.
Dale J. E. and Bonnell I. A. (2012) *Mon. Not. R. Astron. Soc., 422,* 1352.
Dale J. E. et al. (2005) *Mon. Not. R. Astron. Soc., 358,* 291.
Dale J. E. et al. (2007) *Mon. Not. R. Astron. Soc., 377,* 535.
Davé R. (2008) *Mon. Not. R. Astron. Soc., 385,* 147.
Day-Jones A. C. et al. (2013) *Mon. Not. R. Astron. Soc., 430(2),* 1171.
Deacon N. R. et al. (2008) *Astron. Astrophys., 486(1),* 283.
Delfosse X. et al. (2000) *Astron. Astrophys., 364,* 217.
Delgado-Donate E. J. et al. (2003) *Mon. Not. R. Astron. Soc., 342,* 926.
de Marchi G. and Paresce F. (2001) In *Astronomische Gesellschaft Meeting Abstracts* (E. R. Schielicke, ed.), p. 551. Astronomische Gesellschaft, Heidelberg.
de Marchi G. et al. (2010) *Astrophys. J., 718(1),* 105.
de Wit W. J. et al. (2005) *Astron. Astrophys., 437,* 247.
de Wit W. J. et al. (2006) *Astron. Astrophys., 448,* 189.
Dib S. et al. (2010) *Astrophys. J., 723,* 425.
Dopcke G. et al. (2011) *Astrophys. J. Lett., 729,* L3.
Duchêne G. and Kraus A. (2013) *Annu. Rev. Astron. Astrophys., 51,* 269.
Elmegreen B. G. and Scalo J. (2004) *Annu. Rev. Astron. Astrophys., 42,* 211.
Enoch M. L. et al. (2007) *Astrophys. J., 666,* 982.
Enoch M. L. et al. (2008) *Astrophys. J., 684,* 1240.
Evans N. J. II (1999) *Annu. Rev. Astron. Astrophys., 37,* 311.
Federrath C. and Klessen R. S. (2012) *Astrophys. J., 761,* 156.
Federrath C. et al. (2010) *Astron. Astrophys., 512,* A81+.
Field G. B. et al. (2008) *Mon. Not. R. Astron. Soc., 385,* 181.
Fischer D. A. and Marcy G. W. (1992) *Astrophys. J., 396,* 178.
Geers V. et al. (2011) *Astrophys. J., 726(1),* 23.
Geha M. et al. (2013) *Astrophys. J., 771,* 29.
Gennaro M. et al. (2011) *Mon. Not. R. Astron. Soc., 412(4),* 2469.
Girichidis P. et al. (2011) *Mon. Not. R. Astron. Soc., 413,* 2741.
Girichidis P. et al. (2012) *Mon. Not. R. Astron. Soc., 420,* 613.
Glover S. C. O. and Clark P. C. (2012) *Mon. Not. R. Astron. Soc., 426,* 377.

Goodman A. A. et al. (2009) *Nature, 457,* 63.
Goodwin S. P. et al. (2004) *Astron. Astrophys., 414,* 633.
Goodwin S. P. et al. (2008) *Astron. Astrophys., 477,* 823.
Habibi M. et al. (2013) *Astron. Astrophys., 556,* A26.
Hansen C. E. et al. (2012) *Astrophys. J., 747,* 22.
Hatchell J. and Fuller G. A. (2008) *Astron. Astrophys., 482,* 855.
Heithausen A. et al. (1998) *Astron. Astrophys., 331,* L65.
Heitsch F. et al. (2001) *Astrophys. J., 547,* 280.
Heitsch F. et al. (2008a) *Astrophys. J., 674,* 316.
Heitsch F. et al. (2008b) *Astrophys. J., 683,* 786.
Heitsch F. et al. (2011) *Mon. Not. R. Astron. Soc., 415,* 271.
Hennebelle P. and Chabrier G. (2008) *Astrophys. J., 684,* 395.
Hennebelle P. and Chabrier G. (2009) *Astrophys. J., 702,* 1428.
Hennebelle P. and Chabrier G. (2011) *Astrophys. J. Lett., 743,* L29.
Hennebelle P. and Chabrier G. (2013) *Astrophys. J., 770,* 150.
Hennebelle P. and Teyssier R. (2008) *Astron. Astrophys., 477,* 25.
Hennebelle P. et al. (2008) *Astron. Astrophys., 486,* L43.
Hennebelle P. et al. (2011) *Astron. Astrophys., 528,* A72.
Hernandez A. K. et al. (2012) *Astrophys. J. Lett., 756,* L13.
Hillenbrand L. A. and Hartmann L. W. (1998) *Astrophys. J., 492,* 540.
Hocuk S. et al. (2012) *Astron. Astrophys., 545,* A46.
Holman K. et al. (2013) *Mon. Not. R. Astron. Soc., 432,* 3534.
Hopkins P. F. (2012a) *Mon. Not. R. Astron. Soc., 423,* 2016.
Hopkins P. F. (2012b) *Mon. Not. R. Astron. Soc., 423,* 2037.
Hopkins P. F. (2013a) *Mon. Not. R. Astron. Soc., 430,* 1653.
Hopkins P. F. (2013b) *Mon. Not. R. Astron. Soc., 430,* 1880.
Hopkins P. F. (2013c) *Mon. Not. R. Astron. Soc., 433,* 170.
Hopkins P. F. (2013d) *Mon. Not. R. Astron. Soc., 428,* 1950.
Hosokawa T. et al. (2011) *Astrophys. J., 738,* 140.
Huff E. M. and Stahler S. W. (2007) *Astrophys. J., 666,* 281.
Hur H. et al. (2012) *Astron. J., 143(2),* 41.
Hussmann B. et al. (2012) *Astron. Astrophys., 540,* A57.
Inutsuka S.-I. (2001) *Astrophys. J. Lett., 559,* L149.
Jappsen A.-K. et al. (2005) *Astron. Astrophys., 435,* 611.
Jeffries R. D. (2012) In *Low Mass Stars and the Transition Stars/Brown Dwarfs — Evry Schatzman School on Stellar Physics XXIII* (C. Reylé et al., eds.), pp. 45–89. EAS Publ. Series, Vol. 57, Cambridge Univ., Cambridge.
Johnstone D. et al. (2000) *Astrophys. J., 545,* 327.
Johnstone D. et al. (2010) *Astrophys. J., 711,* 655.
Joos M. et al. (2012) *Astron. Astrophys., 543,* A128.
Kalirai J. S. et al. (2013) *Astrophys. J., 763,* 110.
Kelly B. C. et al. (2012) *Astrophys. J., 752,* 55.
Kirk H. et al. (2007) *Astrophys. J., 668,* 1042.
Kirk H. et al. (2010) *Astrophys. J., 723,* 457.
Kirkpatrick D. J. et al. (2012) *Astrophys. J., 753(2),* 156.
Klessen R. S. (2001) *Astrophys. J. Lett., 550,* L77.
Klessen R. S. and Burkert A. (2000) *Astrophys. J. Suppl., 128,* 287.
Klessen R. S. and Hennebelle P. (2010) *Astron. Astrophys., 520,* A17.
Klessen R. S. et al. (1998) *Astrophys. J. Lett., 501,* L205.
Klessen R. S. et al. (2000) *Astrophys. J., 535,* 887.
Klessen R. S. et al. (2007) *Mon. Not. R. Astron. Soc., 374,* L29.
Konopacky Q. M. et al. (2010) *Astrophys. J., 711(2),* 1087.
Könyves V. et al. (2010) *Astron. Astrophys., 518,* L106.
Koyama H. and Inutsuka S.-I. (2002) *Astrophys. J. Lett., 564,* L97.
Kramer C. et al. (1998) *Astron. Astrophys., 329,* 249.
Kroupa P. (2001) *Mon. Not. R. Astron. Soc., 322,* 231.
Kroupa P. (2002) *Science, 295,* 82.
Krumholz M. R. (2011) *Astrophys. J., 743,* 110.
Krumholz M. R. and McKee C. F. (2008) *Nature, 451,* 1082.
Krumholz M. R. and Tan J. C. (2007) *Astrophys. J., 654,* 304.
Krumholz M. R. et al. (2006) *Astrophys. J., 653,* 361.
Krumholz M. R. et al. (2007) *Astrophys. J., 656,* 959.
Krumholz M. R. et al. (2009) *Science, 323,* 754.
Krumholz M. R. et al. (2011) *Astrophys. J., 740,* 74.
Krumholz M. R. et al. (2012) *Astrophys. J., 754,* 71.
Krumholz M. R. et al. (2013) *Astrophys. J. Lett., 767,* L11.
Kurosawa R. et al. (2004) *Mon. Not. R. Astron. Soc., 351,* 1134.
Lada C. J. and Lada E. A. (2003) *Annu. Rev. Astron. Astrophys., 41,* 57.
Lamb J. B. et al. (2010) *Astrophys. J., 725,* 1886.
Langer W. D. et al. (1993) *Astrophys. J. Lett., 408,* L45.
Larson R. B. (1978) *Mon. Not. R. Astron. Soc., 184,* 69.
Larson R. B. (1981) *Mon. Not. R. Astron. Soc., 194,* 809.
Larson R. B. (2005) *Mon. Not. R. Astron. Soc., 359,* 211.
Launhardt R. et al. (2013) *Astron. Astrophys., 551,* A98.
Li Z.-Y. and Nakamura F. (2006) *Astrophys. J. Lett., 640,* L187.

Li Z.-Y. et al. (2010) *Astrophys. J. Lett., 720,* L26.
Lim B. et al. (2013) *Astron. J., 145(2),* 46.
Lodieu N. (2013) *Mon. Not. R. Astron. Soc., 431(4),* 3222.
Lodieu N. et al. (2011) *Astron. Astrophys., 532,* A103.
Lodieu N. et al. (2012a) *Mon. Not. R. Astron. Soc., 422(2),* 1495.
Lodieu N. et al. (2012b) *Mon. Not. R. Astron. Soc., 426(4),* 3403.
Lombardi M. et al. (2006) *Astron. Astrophys., 454,* 781.
Lu J. R. et al. (2013) *Astrophys. J., 764,* 155.
Luhman K. L. (2007) *Astrophys. J. Suppl., 173,* 104.
Luhman K. L. (2012) *Annu. Rev. Astron. Astrophys., 50,* 65.
Luhman K. L. et al. (2009) *Astrophys. J., 703(1),* 399.
Lutz T. E. and Kelker D. H. (1973) *Publ. Astron. Soc. Pac., 85,* 573.
Lynden-Bell D. (1996) *Mon. Not. R. Astron. Soc., 279,* 389.
Mac Low M.-M. (1999) *Astrophys. J., 524,* 169.
Mac Low M.-M. and Klessen R. S. (2004) *Rev. Mod. Phys., 76,* 125.
Machida M. N. and Matsumoto T. (2012) *Mon. Not. R. Astron. Soc., 421,* 588.
Machida M. N. et al. (2009) *Astrophys. J. Lett., 699,* L157.
Maraston C. et al. (2003) *Astron. Astrophys., 400,* 823.
Maschberger T. (2013) *Mon. Not. R. Astron. Soc., 429,* 1725.
Mason B. D. et al. (2009) *Astron. J., 137,* 3358.
Matzner C. D. and McKee C. F. (2000) *Astrophys. J., 545,* 364.
McKee C. F. and Offner S. S. R. (2010) *Astrophys. J., 716,* 167.
McKee C. F. and Tan J. C. (2002) *Nature, 416,* 59.
Men'shchikov A. et al. (2010) *Astron. Astrophys., 518,* L103.
Miller G. E. and Scalo J. M. (1979) *Astrophys. J. Suppl., 41,* 513.
Moeckel N. and Bate M. R. (2010) *Mon. Not. R. Astron. Soc., 404,* 721.
Morata O. et al. (2005) *Astron. Astrophys., 435,* 113.
Moraux E. et al. (2003) *Astron. Astrophys., 400,* 891.
Moraux E. et al. (2007) *Astron. Astrophys., 471,* 499.
Motte F. et al. (1998) *Astron. Astrophys., 336,* 150.
Muzic K. et al. (2011) *Astrophys. J., 732(2),* 86.
Muzic K. et al. (2012) *Astrophys. J., 744(2),* 134.
Myers A. T. et al. (2011) *Astrophys. J., 735,* 49.
Myers A. T. et al. (2013) *Astrophys. J., 766,* 97.
Myers P. C. (2009) *Astrophys. J., 706,* 1341.
Nutter D. and Ward-Thompson D. (2007) *Mon. Not. R. Astron. Soc., 374,* 1413.
Oey M. S. (2011) *Astrophys. J. Lett., 739,* L46.
Oey M. S. et al. (2013) *Astrophys. J., 768,* 66.
Offner S. S. R. and McKee C. F. (2011) *Astrophys. J., 736,* 53.
Offner S. S. R. et al. (2008a) *Astrophys. J., 686,* 1174.
Offner S. S. R. et al. (2008b) *Astron. J., 136,* 404.
Offner S. S. R. et al. (2009a) *Astrophys. J. Lett., 704,* L124.
Offner S. S. R. et al. (2009b) *Astrophys. J., 703,* 131.
Offner S. S. R. et al. (2010) *Astrophys. J., 725,* 1485.
Offner S. S. R. et al. (2012) *Astrophys. J., 753,* 98.
Oliveira J. M. et al. (2009) *Mon. Not. R. Astron. Soc., 392,* 1034.
Padoan P. and Nordlund Å. (2002) *Astrophys. J., 576,* 870.
Padoan P. and Nordlund Å. (2011) *Astrophys. J., 730,* 40.
Padoan P. et al. (1997) *Mon. Not. R. Astron. Soc., 288,* 145.
Padoan P. et al. (2007) *Astrophys. J., 661,* 972.
Padoan P. et al. (2012) *Astrophys. J. Lett., 759,* L27.
Palau A. et al. (2013) *Astrophys. J., 762,* 120.
Parker R. J. and Reggiani M. M. (2013) *Mon. Not. R. Astron. Soc., 432,* 2378.
Parker R. J. et al. (2012) *Mon. Not. R. Astron. Soc., 426(4),* 3079.
Parravano A. et al. (2011) *Astrophys. J., 726(1),* 27.
Peña Ramírez K. et al. (2012) *Astrophys. J., 754(1),* 30.
Peretto N. and Fuller G. A. (2009) *Astron. Astrophys., 505,* 405.
Peretto N. and Fuller G. A. (2010) *Astrophys. J., 723,* 555.
Peters T. et al. (2010a) *Astrophys. J., 711,* 1017.
Peters T. et al. (2010b) *Astrophys. J., 725,* 134.
Peters T. et al. (2011) *Astrophys. J., 729,* 72.
Pineda J. E. et al. (2009) *Astrophys. J. Lett., 699,* L134.
Preibisch T. et al. (1999) *New Astron., 4,* 531.
Press W. H. and Schechter P. (1974) *Astrophys. J., 187,* 425.
Price D. J. and Bate M. R. (2007) *Mon. Not. R. Astron. Soc., 377,* 77.

Price D. J. and Bate M. R. (2009) *Mon. Not. R. Astron. Soc., 398,* 33.
Price D. J. et al. (2012) *Mon. Not. R. Astron. Soc., 423,* L45.
Pudritz R. E. and Norman C. A. (1983) *Astrophys. J., 274,* 677.
Ragan S. E. et al. (2009) *Astrophys. J., 698,* 324.
Ragan S. E. et al. (2012) *Astrophys. J., 746,* 174.
Raghavan D. et al. (2010) *Astrophys. J. Suppl., 190,* 1.
Rathborne J. M. et al. (2006) *Astrophys. J., 641,* 389.
Reid I. N. et al. (2002) *Astron. J., 124(5),* 2721.
Reid I. N. et al. (2007) *Astrophys. J. Suppl., 133(6),* 2825.
Reipurth B. and Clarke C. (2001) *Astrophys. J., 122,* 432.
Robertson B. and Goldreich P. (2012) *Astrophys. J. Lett., 750,* L31.
Román-Zúñiga C. G. et al. (2010) *Astrophys. J., 725,* 2232.
Rosolowsky E. W. et al. (2008a) *Astrophys. J. Suppl., 175,* 509.
Rosolowsky E. W. et al. (2008b) *Astrophys. J., 679,* 1338.
Rundle D. et al. (2010) *Mon. Not. R. Astron. Soc., 407,* 986.
Salpeter E. E. (1955) *Astrophys. J., 121,* 161.
Scalo J. M. (1986) *Fund. Cosmic Phys., 11,* 1.
Scalo J. and Elmegreen B. G. (2004) *Annu. Rev. Astron. Astrophys., 42,* 275.
Schmidt W. et al. (2010) *Astron. Astrophys., 516,* A25.
Schnee S. et al. (2012) *Astrophys. J., 755,* 178.
Scholz A. et al. (2012) *Astrophys. J., 744(1),* 6.
Schultheis M. et al. (2006) *Astron. Astrophys., 447(1),* 185.
Seifried D. et al. (2012) *Mon. Not. R. Astron. Soc., 423,* L40.
Shetty R. et al. (2009a) *Astrophys. J., 696,* 2234.
Shetty R. et al. (2009b) *Astrophys. J., 696,* 676.
Shu F. H. et al. (1995) *Astrophys. J. Lett., 455,* L155.
Smith R. J. et al. (2008) *Mon. Not. R. Astron. Soc., 391,* 1091.
Smith R. J. et al. (2009) *Mon. Not. R. Astron. Soc., 396,* 830.
Smith R. J. et al. (2011) *Mon. Not. R. Astron. Soc., 411,* 1354.
Smith R. J. et al. (2012) *Astrophys. J., 750,* 64.
Smith R. J. et al. (2013) *Astrophys. J., 771,* 24.
Stamatellos D. and Whitworth A. P. (2009a) *Mon. Not. R. Astron. Soc., 392,* 413.
Stamatellos D. and Whitworth A. P. (2009b) *Mon. Not. R. Astron. Soc., 400,* 1563.
Stamatellos D. et al. (2012) *Mon. Not. R. Astron. Soc., 427,* 1182.
Starck J.-L. et al. (1995) In *Astronomical Data Analysis Software and Systems IV* (R. A. Shaw et al., eds.), p. 279. ASP Conf. Series 77, Astronomical Society of the Pacific, San Francisco.
Strader J. et al. (2011) *Astron. J., 142,* 8.
Stutzki J. and Guesten R. (1990) *Astrophys. J., 356,* 513.
Sung H. and Bessell M. S. (2010) *Astron. J., 140(6),* 2070.
Sung H. et al. (2013) *Astron. J., 145(2),* 37.
Swift J. J. and Welch W. J. (2008) *Astrophys. J. Suppl., 174,* 202.
Tan J. C. et al. (2013) *ArXiv e-prints,* arXiv:1310.2710.
Testi L. and Sargent A. I. (1998) *Astrophys. J. Lett., 508,* L91.
Thies I. and Kroupa P. (2007) *Astrophys. J., 671,* 767.
Thies I. and Kroupa P. (2008) *Mon. Not. R. Astron. Soc., 390,* 1200.
Tilley D. A. and Pudritz R. E. (2004) *Mon. Not. R. Astron. Soc., 353,* 769.
Tilley D. A. and Pudritz R. E. (2007) *Mon. Not. R. Astron. Soc., 382,* 73.
Urban A. et al. (2010) *Astrophys. J., 710,* 1343.
van Dokkum P. G. and Conroy C. (2010) *Nature, 468,* 940.
Vazquez-Semadeni E. (1994) *Astrophys. J., 423,* 681.
Vázquez-Semadeni E. et al. (2007) *Astrophys. J., 657,* 870.
Vázquez-Semadeni E. et al. (2011) *Mon. Not. R. Astron. Soc., 414,* 2511.
Walch S. et al. (2012) *Mon. Not. R. Astron. Soc., 419,* 760.
Walch S. et al. (2013) *Mon. Not. R. Astron. Soc., 435,* 917.
Wang P. et al. (2010) *Astrophys. J., 709,* 27.
Weisz D. R. et al. (2013) *Astrophys. J., 762,* 123.
Whitworth A. P. et al. (1995) *Mon. Not. R. Astron. Soc., 277,* 727.
Wilcock L. A. et al. (2012) *Mon. Not. R. Astron. Soc., 422,* 1071.
Williams J. P. et al. (1994) *Astrophys. J., 428,* 693.
Zentner A. R. (2007) *Intl. J. Mod. Phys. D, 16,* 763.
Zhang Y. et al. (2013a) *Astrophys. J., 767,* 58.
Zhang Y. et al. (2013b) *Astrophys. J., 766,* 86.
Zinnecker H. (1982) *New York Acad. Sci. Ann., 395,* 226.

Padoan P., Federrath C., Chabrier G., Evans N. J. II, Johnstone D, Jørgensen J. K., McKee C. F., and Nordlund Å. (2014) The star formation rate of molecular clouds. In *Protostars and Planets VI* (H. Beuther et al., eds.), pp. 77–100. Univ. of Arizona, Tucson, DOI: 10.2458/azu_uapress_9780816531240-ch004.

The Star Formation Rate of Molecular Clouds

Paolo Padoan
University of Barcelona

Christoph Federrath
Monash University

Gilles Chabrier
Ecole Normale Supérieure de Lyon

Neal J. Evans II
The University of Texas at Austin

Doug Johnstone
University of Victoria

Jes K. Jørgensen
University of Copenhagen

Christopher F. McKee
University of California, Berkeley

Åke Nordlund
University of Copenhagen

We review recent advances in the analytical and numerical modeling of the star formation rate in molecular clouds and discuss the available observational constraints. We focus on molecular clouds as the fundamental star formation sites, rather than on the larger-scale processes that form the clouds and set their properties. Molecular clouds are shaped into a complex filamentary structure by supersonic turbulence, with only a small fraction of the cloud mass channeled into collapsing protostars over a free-fall time of the system. In recent years, the physics of supersonic turbulence has been widely explored with computer simulations, leading to statistical models of this fragmentation process, and to the prediction of the star formation rate as a function of fundamental physical parameters of molecular clouds, such as the virial parameter, the root mean square (rms) Mach number, the compressive fraction of the turbulence driver, and the ratio of gas to magnetic pressure. Infrared space telescopes, as well as groundbased observatories, have provided unprecedented probes of the filamentary structure of molecular clouds and the location of forming stars within them.

1. INTRODUCTION

Understanding and modeling the star formation rate (SFR) is a central goal of a theory of star formation. Cosmological simulations of galaxy formation demonstrate the impact of SFR models on galaxy evolution (*Agertz et al.,* 2013), but they neither include star formation self-consistently, nor its feedback mechanisms. They must rely instead on suitable subgrid-scale models to include the effects of star formation and feedback.

The SFR has been a fundamental problem in astrophysics since the 1970s, when it was shown that the gas depletion time in our Galaxy is much longer than any characteristic free-fall time of star-forming gas (*Zuckerman and Palmer,* 1974; *Williams and McKee,* 1997). The same problem applies to individual molecular clouds (MCs), where the gas depletion time is much longer than the free-fall time as well (e.g., *Evans et al.,* 2009). It has been argued that for local clouds (*Krumholz and Tan,* 2007), as well as for disk and starburst galaxies at low and high redshift (*Krumholz et al.,* 2012a; *Federrath,* 2013b), the gas depletion time is always on the order of 100 free-fall times.

The first solution to the SFR problem was proposed under the assumption that MCs are supported against gravity by relatively strong magnetic fields, with star formation resulting from the contraction of otherwise subcritical cores by ambipolar drift (*Shu et al.,* 1987). The relevant timescale for star formation would then be the ambipolar drift time, much longer than the free-fall time in MCs. By accounting for the effect of photoionization on the coupling of gas and magnetic field, *McKee* (1989) derived a model for the SFR that predicted gas-depletion times consistent with the observations.

When it became possible to carry out relatively large three-dimensional simulations of magneto-hydrodynamic (MHD) turbulence in the 1990s, the focus of most SFR studies gradually moved from the role of magnetic fields to the role of supersonic turbulence, as described in a number of reviews (*Scalo and Elmegreen,* 2004; *Elmegreen and Scalo,* 2004; *Mac Low and Klessen,* 2004; *Ballesteros-Paredes et al.,* 2007; *McKee and Ostriker,* 2007). *Padoan and Nordlund* (1999) showed that the Zeeman-splitting measurements of the magnetic field in MCs were likely to indicate a mean magnetic field strength much weaker than previously assumed (see also *Lunttila et al.,* 2008, 2009; *Bertram et al.,* 2012). Based on a body of recent observational results, the review by *McKee and Ostriker* (2007) concludes that MCs are mostly supercritical instead of subcritical, in which case the ambipolar-drift time cannot be the solution to the SFR problem.

Current simulations of star formation by supersonic turbulence show a high sensitivity of the SFR to the virial parameter, defined as twice the ratio of turbulent kinetic energy to gravitational energy (*Padoan and Nordlund,* 2011; *Federrath and Klessen,* 2012; *Padoan et al.,* 2012), and thus to the ratio between the free-fall time and the turbulence crossing time, as suggested by analytical models based on the turbulent fragmentation paradigm (*Krumholz and Mc-Kee,* 2005; *Padoan and Nordlund,* 2011; *Hennebelle and Chabrier,* 2011; *Federrath and Klessen,* 2012; *Hennebelle and Chabrier,* 2013). Interestingly, the models and simulations also show that the magnetic field is still needed to make the SFR as low as observed in MCs, even if the SFR is not controlled by ambipolar drift.

Most studies of the SFR published after *Protostars and Planets V* (*Reipurth et al.,* 2007) have focused on the role of turbulence, so this review revolves around supersonic turbulence as well. Besides the theoretical ideas and the numerical simulations, we discuss the observations that are used to constrain the SFR in MCs and have a potential to test the theory. Extragalactic observations can be used to constrain SFR models as well (*Krumholz et al.,* 2012a; *Federrath,* 2013b). They present the advantage of a greater

variety of star formation environments relative to local MCs, but they are not as detailed as studies of nearby clouds. Although the extragalactic field is rapidly evolving and very promising, the observational side of this review is limited to studies of local clouds, and we only briefly discuss the issue of relating the local studies to the extragalactic literature. For more information on this topic, we refer the reader to the recent review by *Kennicutt and Evans* (2012), and to the chapter by Dobbs et al. in this volume.

We start by reviewing observational estimates of the SFR in MCs in section 2, with a critical discussion of the methods and their uncertainties. We also contrast the methods based on the direct census of stellar populations in MCs with indirect ones used to determine the SFR on larger scales in extragalactic studies.

We then review the theoretical models in section 3, focusing on those that account for the turbulent nature of the interstellar medium, and discuss their differences, limitations, and potential for further development. The theory relies on statistical results from numerical studies of supersonic MHD turbulence, particularly on the probability distribution of gas density, whose dependence on fundamental physical parameters has been recently quantified in several studies. Besides the distribution of gas density, the key ingredient in the theory is the concept of a critical density for star formation. However, MCs are characterized by highly nonlinear turbulent motions, producing shocks and filaments that are often described as fractal structures, casting doubt on such a threshold density concept. We discuss how the various models justify this approximation, and how the observations support the idea of a critical density.

Turbulence simulations have been used to directly derive the SFR, by introducing self-gravity and sink particles to trace gravitationally unstable and star-forming gas (e.g., *Federrath et al.,* 2010b). Over the last two years, these simulations have been employed in vast parameter studies that can be used to test and constrain theoretical models. In section 4, we offer a critical view of numerical methods and experiments and we summarize the most important findings from the comparison between simulations and theories.

In section 5 we compare both theory and simulations to the observational estimates of the SFR. We draw conclusions and outline future directions for both models and observations in section 6.

2. OBSERVATIONS

Over the last decade, infrared studies from groundbased telescopes, such as 2MASS (*Kleinmann,* 1992), and from space with the Spitzer Space Telescope (*Werner et al.,* 2004) and Herschel Space Observatory (*Pilbratt et al.,* 2010) have provided extended surveys of the populations of young stellar objects (YSOs) and their evolutionary stages, as well as their distributions within their parental clouds. Together with large-scale extinction and submillimeter continuum maps, which provide measurements of the cloud mass and column density, these observations allow direct estimates of the star formation rates and efficiencies in different cloud regions.

In writing this chapter, the authors realized that many misunderstandings were generated because of different conceptions of what a MC is. The obvious case is the "cloud in the computer" vs. the "cloud imaged by observers," but even the latter is subject to definitional issues because of evolving sensitivity and observing techniques. We thus begin the next subsection with a discussion of techniques for measuring clouds and conclude with the evolving observational definition of a MC.

2.1. Measuring Cloud Mass and Surface Density: Methods and Uncertainties

The distribution of mass in nearby MCs has been determined primarily from extinction mapping in the visible and infrared, using Spitzer and 2MASS data, and from groundbased submillimeter surveys of dust continuum emission. These observations allow the surface densities and masses to be derived without assumptions about abundance and excitation of gas tracers, such as CO and its rarer isotopologs. The estimation of the total cloud mass does, however, rely on a consistency in the dust to gas ratio within and across clouds. These techniques have thus largely replaced maps of CO for nearby clouds, although CO is still required to uncover cloud kinematics. For more distant clouds, maps of CO and other species are still used, and we discuss the issues arising for cloud definition and properties in section 2.3.

The advantages of the extinction-mapping technique are twofold. First, the availability and sensitivity of large-area optical and infrared detectors allow large regions to be observed efficiently. Also, only the extinction properties of the intervening dust are required to convert from extinction to column density. The major disadvantages are the lack of resolution available, unless the infrared observations are extremely deep, and the inability to measure very high column densities where the optical depth in the infrared becomes too large to see background stars.

Extinction maps from optical data have provided valuable measures of cloud extents, masses, and surface densities for modest extinctions (e.g., *Cambrésy*, 1999; *Dobashi et al.*, 2005), and the application to the near-infrared has extended these techniques to larger extinctions (e.g., *Lombardi and Alves*, 2001). The extinction of individual background stars are measured and converted to a column density of intervening gas and dust assuming appropriate properties for dust in MCs. Studies of the extinction law in nearby clouds show that the dust is best represented by models with ratios of total to selective extinction, $R_V = 5.5$ (*Chapman et al.*, 2009; *Ascenso et al.*, 2012, 2013). For the nearby clouds imaged with Spitzer, 2×10^4 to 1×10^5 background stars were identified (*Evans et al.*, 2009); when added to the 2MASS data base, extinctions up to $A_V = 40$ mag can be measured with spatial resolution of about 270″ (*Heiderman et al.*, 2010). The conversion from A_V to mass surface density is

$$\Sigma_{gas}\left(g\ cm^{-2}\right) = \mu m_H \left[1.086 C_{ext}\left(V\right)\right]^{-1} A_V \qquad (1)$$

where $\mu = 1.37$. $C_{ext}(V)$ is the extinction per column density of H nucleons, $N(H) = N(HI) + 2N(H_2)$ (*Draine*, 2003) (see online tables at *http://www.astro.princeton.edu/~draine/dust/dustmix.html*). Two different grain models are available for $R_V = 5.5$. For normalized Case A grains, the newer models, $C_{ext}(V) = 6.715 \times 10^{-22}$ cm^2, and $\Sigma_{gas} = 15 A_V\ M_\odot$ pc^{-2}. For Case B grains, $C_{ext}(V) = 4.88 \times 10^{-22}$ cm^2, and $\Sigma_{gas} = 21 A_V\ M_\odot$ pc^{-2}. The Case B grains match observations well (*Ascenso et al.*, 2013), but have some theoretical issues (B. T. Draine, personal communication). Following *Heiderman et al.* (2010) we adopt the normalized Case A grain model, noting the possibility that all masses and surface densities are about 40% higher. Note that $C_{ext} = 4.896 \times 10^{-22}$ cm^2 (normalized Case A) or 5.129×10^{-22} cm^2 (Case B) for $R_V = 3.1$, so use of the diffuse interstellar medium (ISM) conversions will result in higher estimates of Σ_{gas}. These values may apply to regions with $A_V < 2$ mag, but *Ascenso et al.* (2013) finds no clear change from $R_V = 5.5$.

Groundbased submillimeter mapping of dust continuum emission provides an alternative, but it requires assumptions about both the dust emissivity properties at long wavelengths and the dust temperature distribution along the line of sight, usually taken to be constant. In addition, the groundbased submillimeter maps lose sensitivity to large-scale structure. The resolution, however, can be significantly higher than for extinction mapping, approximately tens of arcseconds rather than hundreds, and the submillimeter emission remains optically thin even at extreme column densities. For a description of planned maps with SCUBA-2, see *Johnstone et al.* (2005) and *Ward-Thompson et al.* (2007b).

More recently, Herschel has mapped a large fraction of the nearby MCs from the far-infrared through the submillimeter. The resulting spectral energy distributions add information about the mean dust temperature along the line of sight and thus yield more accurate column density maps of MCs (e.g., *Sadavoy et al.*, 2012, 2013; *Kirk et al.*, 2013; *Schneider et al.*, 2013). The spatial resolution of Herschel lies between the standard extinction map scale and the groundbased submillimeter maps. We can expect a rich harvest of results once these surveys are fully integrated.

2.2. Mass and Surface Density of Clouds

The three largest clouds in the c2d project were completely mapped at 1.1 mm using Bolocam on the Caltech Submillimeter Observatory (CSO) (*Enoch et al.*, 2006, 2007; *Young et al.*, 2006). Additional maps were made at 850 μm with SCUBA for Ophiuchus (*Johnstone et al.*, 2004), Perseus (*Hatchell and Fuller*, 2008; *Kirk et al.*, 2006), and Orion (*Johnstone and Bally*, 1999, 2006). Smaller clouds and regions were mapped at 350 μm by *Wu et al.* (2007), providing higher-resolution data at those wavelengths than was possible with Herschel. A uniform reprocessing of all SCUBA data provides a database for many individual regions (*di Francesco et al.*, 2006).

On larger scales, we now have surveys of the Galactic Plane at millimeter/submillimeter wavelengths, notably the Bolocam Galactic Plane Survey (BGPS) (*Aguirre et al.*, 2011; *Rosolowsky et al.*, 2010; *Ginsburg et al.*, 2013) and the APEX Telescope Large Area Survey of the Galaxy (ATLASGAL) projects (*Schuller et al.*, 2009; *Contreras et al.*, 2013).

The primary product of these blind submillimeter surveys was a catalog of relatively dense structures. Although dust continuum emission is related to dust temperature and column density throughout the cloud, the loss of sensitivity to large-scale emission for groundbased instruments results in an effective spatial filtering. This yields a particular sensitivity to small-scale column density enhancements and thus picks out structures with high volume density. For nearby clouds, millimeter sources correspond to cores, the sites of individual star formation (*Enoch et al.*, 2006), whereas Galactic Plane surveys mostly pick up clumps, the sites of cluster formation (*Dunham et al.*, 2011). This can also be seen in the fraction of cloud mass identified through the submillimeter mapping. In nearby clouds the submillimeter sources account for a few percent of the cloud mass (*Johnstone et al.*, 2004; *Kirk et al.*, 2006), while at larger distances the fraction of cloud mass observed appears to be much higher, reaching tens of percentages, and the slope of the clump mass function flattens (*Kerton et al.*, 2001; *Muñoz et al.*, 2007).

For the nearby clouds, Spitzer data were used to distinguish protostellar cores from starless cores, almost all of which are probably prestellar (i.e., gravitationally bound), using the definitions of *di Francesco et al.* (2007) and *Ward-Thompson et al.* (2007a). This distinction allowed a clarification of the properties of prestellar cores. Three results are particularly salient for this review.

First, the core mass function was consistent with a picture in which core masses map into stellar masses with a core-to-star efficiency of $\varepsilon = 0.25$ to 0.4 (*Enoch et al.*, 2008; *Sadavoy et al.*, 2010; *Alves et al.*, 2007; *André et al.*, 2010). Second, the timescale for prestellar cores with mean densities above about 10^4 cm^{-3} to evolve into protostellar cores is about 0.5 million years (m.y.) (*Enoch et al.*, 2008). Third, prestellar cores are strongly concentrated to regions of high extinction. For example, 75% of the cores lie above $A_V = 8$, 15, and 23 mag in Perseus, Serpens, and Ophiuchus, respectively (*Johnstone et al.*, 2004; *Kirk et al.*, 2006; *Enoch et al.*, 2007). These were some of the first quantitative estimates of the degree to which star formation is concentrated to a small area of the clouds characterized by high surface densities. In contrast, less than 20% of the area and 38% of the cloud mass lies above $A_V = 8$ mag (*Evans et al.*, 2014).

For 29 nearby clouds mapped by Spitzer, Evans et al. have measured mean surface densities above the $A_V = 2$ contour. These have a mean value of $\Sigma_{gas} = 79 \pm 22$ M$_\odot$ pc^{-2}. For maps of 12 nearby clouds that go down to $A_V = 0.5$ mag

with 0.1-pc resolution, mean $A_V = 1.6 \pm 0.7$ mag for starforming clouds and $A_V = 1.5 \pm 0.7$ mag if five non-starforming clouds are included (*Kainulainen et al.*, 2009). Clearly, the mean surface densities are heavily influenced by how low an extinction contour is included. These mean extinctions would correspond to 23 to 26 M$_\odot$ pc^{-2} with the adopted extinction law, but other extinction laws could raise these values by about 40%.

2.3. Dust Versus CO Observations

The images of clouds from extinction or Herschel images of dust emission are qualitatively different from earlier ones. At low extinction levels, clouds appear wispy, windswept, and diffuse, more like cirrus clouds in our atmosphere than like cumulus clouds. The Herschel images tracing higher extinction regions reveal a strong theme of filaments and strands with high contrast against the more diffuse cloud (see chapter by André et al. in this volume). These modern images of cloud properties must be contrasted with older images, which still dominate the mental images of MCs for many astronomers. The very definition of MC depends on the technique used to measure it. For the nearby clouds, optical and near-infrared imaging define clouds to well below $A_V = 1$, and it is only at $A_V < 1$ that lognormal column density distributions become apparent, as discussed in later sections, while their SFR properties were generally measured down to only $A_V = 2$, as discussed in the next section. In contrast, the maps of MCs based on maps of CO probe down to a particular value of antenna temperature or integrated intensity, and the conversion to gas surface density depends on that limiting value.

The most detailed study comparing extinction and CO at low levels is that by *Pineda et al.* (2010), which combined deep extinction mapping with 0.14-pc resolution with extensive CO maps (*Goldsmith et al.*, 2008). CO emission can be seen down to $A_V \sim 0.1$ mag, but the conversion from CO emission to mass surface density rises rapidly as A_V decreases below 1 mag. The mass in areas where ^{13}CO is not detected is roughly equal to the mass where it was detected. The net result is a mean surface density of 39 M$_\odot$ pc^{-2}, based on the total cloud mass and area (*Pineda et al.*, 2010).

For more distant clouds, or for clouds at low Galactic longitude and latitude, extinction and submillimeter maps cannot easily separate the cloud from the general Galactic field without additional kinematic data. The main tracer of clouds across the Galaxy has been maps of CO and isotopologs. One of the earliest and most influential studies of cloud properties was by *Solomon et al.* (1987), and this study remains the enduring image in the mental toolkit of many theorists and nearly all extragalactic astronomers. By defining clouds at a threshold of 3 K or higher in T_R^* and assuming that the virial theorem applied, *Solomon et al.* (1987) derived a mean surface density for the inner Galaxy clouds of $\Sigma_{gas} = 170$ M$_\odot$ pc^{-2}, a value still quoted by many astronomers. When scaled to the newer distance to the

Galactic Center, this value becomes 206 $M_\odot$ pc^{-2} (*Heyer et al.*, 2009). The *Solomon et al.* (1987) work was, however, based on severely undersampled maps, and the selection criteria favored "warm" clouds with active star formation. It has been superseded by better sampled and deeper maps of ^{13}CO by *Heyer et al.* (2009) and *Roman-Duval et al.* (2010). *Heyer et al.* (2009) found that the median surface density of inner Galaxy MCs is 42 $M_\odot$ pc^{-2} within the extrapolated 1-K contour of CO, but noted that abundance variations could raise the value to 80–120 $M_\odot$ pc^{-2}. *Roman-Duval et al.* (2010) found a mean Σ_{gas} = 144 $M_\odot$ pc^{-2} within the 4σ contour of ^{13}CO, which traced gas with a median $A_V >$ ~7 mag. These values are thus measuring a fraction of the cloud with elevated surface densities, if those clouds are like the local clouds. *Roman-Duval et al.* (2010) find that Σ_{gas} declines for Galactocentric radii beyond 6.6 kpc to the lower values seen in nearby clouds. These values depend on assumptions about CO abundance and excitation. The definition and properties of clouds appear to depend on their location in the Galaxy as well as on technique and sensitivity.

Extensive maps in molecular lines of the nearby clouds have also allowed studies of the dynamical properties of the clouds and locations of outflows (e.g., *Ridge et al.* 2006). These may also be used to compare to detailed simulations of line profiles from turbulent clouds.

2.4. Measuring the Star Formation Rate in Molecular Clouds: Methods and Uncertainties

Methods for measuring the SFRs in general have been reviewed and tabulated by *Kennicutt and Evans* (2012). The most direct measures of the SFRs in nearby MCs are based on counting of YSOs and assigning a timescale to the observed objects. Recent surveys of MCs with Spitzer provide a quite complete census of YSOs with infrared excesses (*Evans et al.*, 2009; *Dunham et al.*, 2013). There is, however, considerable uncertainty in the low luminosity range, where contamination by background objects becomes severe. Tradeoffs between completeness and reliability are inevitable. The YSO identifications in *Evans et al.* (2009) emphasized reliability; other methods have suggested 30–40% more YSOs (e.g., *Kryukova et al.* 2012; *Hsieh and Lai*, 2013). Optical photometry and spectroscopy also provided valuable follow-up data for the later stages (e.g., *Oliveira et al.*, 2009; *Spezzi et al.*, 2010, 2008). Although we may expect updates as the Herschel surveys of the Gould belt clouds are fully analyzed, initial results from the Orion clouds suggest that the percentage of sources identifiable only with Herschel data is about 5% (*Stutz et al.*, 2013).

The selection of YSOs by infrared excess means that older objects are not counted, but it has the advantage that the lifetime of infrared excess has been studied extensively by counting the fraction of stars with infrared excess in clusters of different ages. The half-life of infrared excess is determined to be t_{excess} = 2 ± 1 m.y. (e.g., *Mamajek*, 2009), with the main source of uncertainty being the choice of pre-

main-sequence (PMS) evolutionary tracks. All the measures of SFR ($\dot{M}_*$) discussed later depend on the assumed half-life of infrared excesses. Increases in the ages of young clusters (see chapter by Soderblom et al. in this volume) will decrease the estimates proportionally. The third element needed is the mean mass of a YSO, for which we assume a fully sampled system initial mass function (IMF), with $\langle M_* \rangle$ = 0.5 $M_\odot$ (*Chabrier*, 2003). The nearby clouds are not forming very massive stars, so they are not sampling the full IMF; this effect would cause an overestimate of $\dot{M}_*$.

Putting these together

$$\langle \dot{M}_* \rangle = N(YSOs)\langle M_* \rangle / t_{excess} \qquad (2)$$

This equation has been used to compute $\dot{M}_*$ for 20 nearby clouds (*Evans et al.*, 2009; *Heiderman et al.*, 2010), with the assumed t_{excess} = 2 m.y. being the major source of uncertainty [for a numerical study investigating the effect of varying t_{excess}, see *Federrath and Klessen* (2012)]. From these measures of $\dot{M}_*$, one can compute surface densities of star formation rate, Σ_{SFR}, efficiencies, defined by SFE = M(YSO)/[M(YSO) + M(cloud)] [see, e.g., the definition in *Federrath and Klessen* (2013)], rates per mass (often referred to as efficiencies in extragalactic work), defined by $\dot{M}_*/M_{gas}$, and their reciprocals, the depletion times, t_{dep}. Star formation rates determined from star counting are the most reliable, but are available only for nearby clouds (e.g., *Heiderman et al.*, 2010; *Lada et al.*, 2010) reaching out to Orion (*Megeath et al.*, 2012). However, even the Orion clouds are not fully representative of the regions of massive star formation in the inner galaxy or in other galaxies.

Other methods of measuring $\dot{M}_*$ are mostly taken from extragalactic studies, where star counting is impractical, and their application to the Galaxy is tricky. Among the methods listed in *Kennicutt and Evans* (2012), those that are not too sensitive to extinction, such as 24-μm emission, total far-infrared emission, and thermal radio continuum, may be useful also in localized regions of the Galactic Plane. *Vutisalchavakul and Evans* (2013) tested a number of these and found that none worked well for nearby regions where star counts are available. They noted that all the extragalactic methods assume a well-sampled IMF and well-evolved cluster models, so their failure in nearby regions is not surprising. They did find consistency in estimated $\dot{M}_*$ between thermal radio and total far-infrared methods for regions with total far-infrared luminosity greater than $10^{4.5}$ $L_\odot$. As long as the overall SFR in a region is dominated by massive, young clusters, the measures that are sensitive to only the massive stars work to within a factor of 2 (e.g., *Chomiuk and Povich*, 2011).

In summary, methods using YSO counting are reasonably complete in nearby (d < 500 pc) clouds, and SFRs, averaged over 2 m.y., are known to within a factor of about 2. For more distant clouds, where YSO counting is not practical, indirect measures can fail by orders of magnitude unless the region in question has a sufficiently well-sampled IMF to satisfy the assumptions in models used to convert tracers into masses of stars.

2.5. Mass in Young Stars and the Star Formation Rate in Molecular Clouds

The Spitzer/c2d survey summarized by *Evans et al.* (2009) mapped seven nearby star-forming regions: Perseus; Ophiuchus; Lupus I, III, and IV; Chamaeleon II, and Serpens. Additional data for JHK bands from the 2MASS project provided critical information for sorting YSOs from background stars and galaxies in Spitzer surveys [*Harvey et al.* (2007) for c2d and Gould belt; *Rebull et al.* (2010) for Taurus; and *Gutermuth et al.* (2009) for Orion and nearby clusters].

Using the near-infrared extinction maps, these studies reveal a relatively small scatter in the star formation rate per unit area (Σ_{SFR}) of the c2d clouds. The values of Σ_{SFR} are in the range 0.65–3.2 $M_\odot$ m.y.$^{-1}$ pc^{-2}, with an average of 1.6 $M_\odot$ m.y.$^{-1}$ pc^{-2}. Efficiencies per 2 m.y., the characteristic timescale for YSOs that can be detected and characterized at these wavelengths, are generally small, 3–6%. This is similar to the star formation efficiency (SFE) inferred from 2MASS studies of the Perseus, Orion A and B, and MonR2 MCs (*Carpenter,* 2000). In the smaller group L673, *Tsitali et al.* (2010) likewise find SFE = 4.6%. In contrast, *Lada and Lada* (2003) concluded that the SFE is larger for embedded clusters than for MCs as a whole. Typical values for embedded clusters are in the range 10–30%, with lower values usually found for less-evolved clusters.

These estimates largely differ because of the scales probed. For example, considering the larger-scale Ophiuchus and Perseus clouds, *Jørgensen et al.* (2008) found that the SFE was strongly dependent on the column density and regions considered: The younger clusters such as L1688, NGC 1333, and IC 348 showed values of 10–15%, contrasting with the SFE of a few percent found on larger scales (*Jørgensen et al.,* 2008; *Evans et al.,* 2009; *Federrath and Klessen,* 2013). Likewise, *Maury et al.* (2011) utilized the Herschel Space Observatory and IRAM 30-m observations to infer Σ_{SFR} = 23 $M_\odot$ m.y.$^{-1}$ pc^{-2} for the Serpens South cluster (see also *Gutermuth et al.,* 2008). Gutermuth et al. noted that the YSO surface density of the Serpens South cluster was >430 pc^{-2} within a circular region with a 0.2-pc radius, corresponding to a higher $\Sigma_{SFR} \approx$ 700 $M_\odot$ m.y.$^{-1}$ pc^{-2} — again likely reflecting the density of material there. Zooming in on the material just associated with dense cores, the gas + dust mass becomes comparable to the mass of the YSOs — an indication that the efficiency of forming stars is high once material is in sufficiently dense cores (*Enoch et al.,* 2008; *Jørgensen et al.,* 2008).

In particular, *Federrath and Klessen* (2013) measured the SFE in 29 clouds and cloud regions with a new technique based on the column-density power spectrum. They find that the SFE increases from effectively zero in large-scale Hı clouds and non-star-forming clouds to typical star-forming MCs (SFE = 1–10%) to dense cores (SFE > 10%), where the SFE must eventually approach the core-to-star efficiency, $\varepsilon \approx$ 0.3–0.7. Undoubtedly, the SFE is not constant over time within a cloud. The observed SFE must depend

both on the properties of a cloud and on its the evolutionary state. The above qualitative SFE sequence encapsulates both of these effects.

In systematic comparisons between clouds or substructures of similar scales, differences are also seen: *Lada et al.* (2010) combined the data for a sample of MCs and found a strong (linear) correlation between the SFR and the mass of the cloud at extinctions above $A_K \approx$ 0.8 mag ($A_V \approx$ 7.5 mag). *Heiderman et al.* (2010) and *Gutermuth et al.* (2011) examined the relation between the SFR and local gas surface density in subregions of clouds and likewise found relations between the two: *Gutermuth et al.* (2011) found that $\Sigma_{SFR} \propto \Sigma_{gas}^2$, while *Heiderman et al.* (2010) found a relation with a steeper increase (faster than squared) at low surface densities and a less steep increase (linear) at higher surface densities. One difference between these studies was the inclusion of regions forming more massive stars: While the study by *Gutermuth et al.* (2011) includes SFRs based on direct counts for a number of clusters forming massive stars, the *Heiderman et al.* (2010) study is based on the far-infrared luminosities and HCN column densities for a sample of distant, more active, star-forming regions, which may underestimate the true SFRs (*Heiderman et al.,* 2010; *Vutisalchavakul and Evans,* 2013).

Another result of these studies is the presence of a star formation threshold in surface density: *Heiderman et al.* (2010) and *Lada et al.* (2010) find evidence for a characteristic extinction threshold of $A_V \approx$ 7–8 mag separating regions forming the great majority of stars from those where star formation is rare. Similar thresholds have previously been suggested, e.g., on the basis of mapping of $C^{18}O$ emission (*Onishi et al.,* 1998) and (sub)millimeter continuum maps (e.g., *Johnstone et al.,* 2004; *Kirk et al.,* 2006; *Enoch et al.,* 2007) of a number of the nearby star-forming clouds. With our adopted conversion (see section 2.1), the extinction threshold translates to gas surface density of about 120 ± 20 $M_\odot$ pc^{-2}, but other conversions would translate to about 160 $M_\odot$ pc^{-2} (see, e.g., *Lada et al.,* 2013).

All these methods measure the SFR for relatively nearby star-forming regions where individual YSOs can be observed and characterized. An alternative on larger scales is to measure the SFR using the free-free emission, e.g., from the Wilkinson Microwave Anisotropy Probe (WMAP). The free-free emission is expected to be powered by the presence of massive stars through the creation of HII regions. Using this method and by comparing to the giant molecular cloud (GMC) masses from CO line maps, *Murray* (2011) find varying star formation efficiencies ranging from 0.2% to 20% with an average of 8%. However, as the author notes, the sample is biased toward the most luminous sources of free-free emission, potentially selecting the most actively star-forming regions in our Galaxy.

Observationally, the SFR scales with the cloud mass and, with a much reduced scatter, on the mass of "dense" gas (*Lada et al.,* 2010, 2012). Figure 1 shows the log of the SFR/mass vs. the mass, using either the mass of dense gas (A_V > 8 mag), or the full cloud mass, down to $A_V \sim$

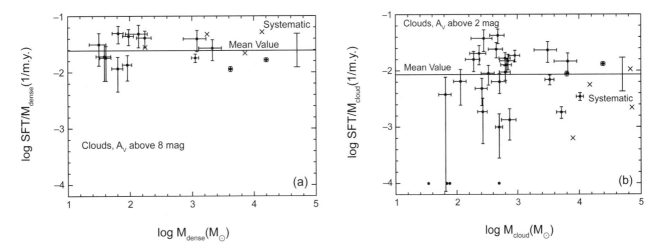

Fig. 1. **(a)** The log of the SFR per mass of dense gas vs. the log of the mass of dense gas and **(b)** the log of the SFR per total cloud mass vs. the log of the total cloud mass. The filled circles are based on the c2d and Gould belt clouds (*Evans et al.,* 2014), while the crosses represent Orion A, Orion B, Taurus, and the Pipe, taken from *Lada et al.* (2010). While there are some differences in identification and selection of YSOs, they are small. In **(a)**, the extinction contour defining the dense gas is A_V = 8 mag for clouds taken from *Evans et al.* (2014) and A_K = 0.8 mag for those taken from *Lada et al.* (2010). In **(b)**, the extinction contour defining the cloud is usually A_V = 2 mag for clouds taken from *Evans et al.* (2014) and A_K = 0.1 mag for those taken from *Lada et al.* (2010). Uncertainties about observables, including cloud distance, have been propagated for the c2d and Gould belt clouds; the requisite information is not available for the "Lada" clouds. The four points plotted at –4 on the y axis are clouds with no observed star formation. The horizontal lines show the mean values for the c2d and Gould belt clouds and the error bars represent the likely systematic uncertainties, dominated by those in the SFR.

2 mag. The large scatter in the star formation rate per unit total gas mass and the much smaller scatter per unit mass of dense gas are clearly illustrated. The average values for the c2d plus Gould belt clouds are (*Evans et al.,* 2014) t_{dep} = 201 ± 240 m.y. for clouds, 47 ± 24 m.y. for dense gas; t_{ff} = 1.47 ± 0.58 m.y. for clouds, 0.71 ± 0.38 m.y. for dense gas; $SFR_{ff} \equiv t_{ff}/t_{dep}$ = 0.018 ± 0.014 for clouds, 0.018 ± 0.008 for dense gas.

Removing this first-order dependence by looking at variations in SFR per mass of either cloud or dense gas vs. other, nondimensional parameters is probably the best way forward. In doing so, we need to be clear about the entity in which we are comparing the SFRs. It could be for the whole cloud, defined by either extinction or CO emission, most of which lies well below the dense gas threshold, or it could refer only to the denser gas, for which the dispersion is much smaller. The latter might best be thought of as the star-forming portion of the cloud, which is often called a clump (*Williams et al.,* 2000).

2.6. Comparison with Extragalactic Results

As many surveys of the Milky Way are becoming available, it is increasingly interesting to put our Galaxy in the context of galactic star formation studies. For other galaxies, the standard way to look at star formation is via the Kennicutt-Schmidt (KS) relation (*Kennicutt,* 1998a,b). Generalizing earlier suggestions (*Schmidt,* 1959, 1963), involving a relation

between star formation and gas volume densities, *Kennicutt* (1998b) found a relation between more readily measured surface densities of the form

$$\Sigma_{SFR} \propto \Sigma_{gas}^N \qquad (3)$$

with N ≈ 1.4 providing a good fit over many orders of magnitude for averages over whole galaxies.

Studies of resolved star formation over the face of nearby galaxies reveal a more complex picture with a threshold around Σ_{gas} = 10 $M_\odot$ pc^{-2}, roughly coincident with the transition from atomic to molecular-gas-dominated interstellar media, and a linear dependence on the surface density of molecular gas, Σ_{mol}, above that threshold (*Bigiel et al.,* 2010). Another transition around Σ_{mol} = 100 $M_\odot$ pc^{-2}, where the transition from normal to starburst galaxies is found, has been suggested (*Genzel et al.,* 2010; *Daddi et al.,* 2010), but this transition depends on interpretation of the CO observations, particularly the uncertainty in the extrapolation of the total column density. The existence of this second transition is consistent with relations between star formation and dense gas measured by HCN emission found earlier by *Gao and Solomon* (2004) and extended to clumps forming massive stars in the Milky Way by *Wu et al.* (2005). Together with the evidence for a sharp increase in SFR above about Σ_{gas} = 120 $M_\odot$ pc^{-2} in local clouds, these results led *Kennicutt and Evans* (2012) to suggest a second transition to a more rapid form of star formation around $\Sigma_{mol} \approx 100–300$ M$_\odot$ pc^{-2}, depending on the physical conditions.

When the nearby clouds are plotted on the KS relation, all but the least active lie well above the relation for whole galaxies or the linear relation found in nearby galaxies (*Evans et al.*, 2009). They are more consistent with the relation for starburst galaxies and the dense gas relations noted above (*Heiderman et al.*, 2010). *Federrath and Klessen* (2012), *Krumholz et al.* (2012a), and *Federrath* (2013b) provide suggestions to resolve this discrepancy between extragalactic and Milky Way SFRs.

3. THEORY

The complexity of turbulence, especially in the supersonic regime and in the case of magnetized and self-gravitating gas, precludes the development of fully analytical theories of the dynamics of MCs. However, the nonlinear chaotic behavior of turbulent flows results in a macroscopic order that can be described with universal statistics. These statistics are flow properties that only depend on fundamental nondimensional parameters expressing the relative importance of various terms in the underlying equations, such as the Reynolds and Mach numbers. Once established, scaling laws and probability density functions (PDFs) of turbulent flow variables can be applied to any problem involving turbulent flows with known parameters. Furthermore, because of the chaotic nature of turbulent flows, their steady-state statistics cannot even depend on the initial conditions. Because of this universality of turbulence, the statistical approach to the dynamics of MCs is a powerful tool to model the MC fragmentation induced by supersonic turbulence (with the obvious caveat that each MC is an individual realization of a turbulent flow, with specific local forces and possibly significant deviations from isotropy and steady state).

For many astrophysical problems involving supersonic turbulence, the turbulent fragmentation, i.e., the creation of a highly nonlinear density field by shocks, is the most important effect of the turbulence. One can then tackle those problems by focusing on the end result, the statistics of density fluctuations, largely ignoring the underlying dynamics. The PDF of gas density in supersonic turbulence is then the statistics of choice. This is especially true in modeling star formation, because the density enhancements, under characteristic MC conditions, are so large that the turbulence alone can seed the gravitational collapse of compressed regions directly into the very nonlinear regime. In other words, much of the fragmentation process is directly controlled by the turbulence (the nonlinear coupling of the velocity field on different scales).

These considerations explain the choice of some members of the numerical community, starting in the 1990s, to focus the research on idealized numerical experiments of supersonic turbulence, using periodic boundary conditions, isothermal equation of state, random driving forces, and, notably, no self-gravity (e.g., *Mac Low et al.*, 1998; *Padoan and Nordlund*, 1999; *Padoan et al.*, 2004; *Kritsuk et al.*,

2007; *Federrath et al.*, 2010a). These idealized simulations have stimulated important theoretical advances and better understanding of supersonic turbulence (e.g., *Falkovich et al.*, 2010; *Aluie*, 2011, 2013; *Galtier and Banerjee*, 2011; *Kritsuk et al.*, 2013). The setup of these experiments (although with the inclusion of self-gravity and sink particles) are described in section 4. Here we briefly summarize the most important numerical results regarding the gas density PDF (section 3.1). We then introduce the concept of a critical density for star formation (section 3.2), and discuss its application to the prediction of the SFR (section 3.3).

3.1. Probability Density Function of Gas Density in Supersonic Turbulence

The gas density PDF was first found to be consistent with a lognormal function in three-dimensional simulations of compressible homogeneous shear flows by *Blaisdell et al.* (1993), in two-dimensional simulations of supersonic turbulence by *Vazquez-Semadeni* (1994), and in three-dimensional simulations of supersonic turbulence by *Padoan et al.* (1997b). The lognormal function is fully determined by its first- and second-order moments, the mean and the standard deviation, as it is defined by a trivial transformation of the Gaussian distribution: If the PDF of $\tilde{\rho} = \rho/\rho_0$ is lognormal (ρ_0 is the mean density), then the PDF of $s = \ln(\tilde{\rho})$ is Gaussian

$$p(s)\,ds = \frac{1}{\left(2\pi\sigma_s^2\right)^{1/2}}\exp\left[-\frac{\left(s-s_0\right)^2}{2\sigma_s^2}\right]ds \qquad (4)$$

where $s_0 = -\sigma_s^2/2$. *Padoan et al.* (1997a) and *Nordlund and Padoan* (1999) found that the standard deviation of the density scales linearly with the Mach number, $\sigma_{\tilde{\rho}} = b\mathcal{M}_s$, where $\mathcal{M}_s \equiv \sigma_v/c_s$, and $b \approx 1/2$. This relation and the value of b can be derived from a simple model based on a single shock (*Padoan and Nordlund*, 2011). In terms of the logarithmic density, s, this is equivalent to

$$\sigma_s^2 = \ln\left[1 + b^2\mathcal{M}_s^2\right] \qquad (5)$$

The lognormal nature of the PDF is valid only in the case of an isothermal equation of state (at low densities, time-averaging may be necessary to cancel out deviations due to large-scale expansions and retrieve the full lognormal shape). Strong deviations from the lognormal PDF can be found with a (non-isothermal) polytropic equation of state (*Passot and Vázquez-Semadeni*, 1998; *Scalo et al.*, 1998; *Nordlund and Padoan*, 1999). However, *Glover et al.* (2010) and *Micic et al.* (2012) find nearly lognormal PDFs when a detailed chemical network including all relevant heating and cooling processes is taken into account. Further simulations tested the validity of equation (5) and derived more precise values of b (e.g., *Ostriker et al.*, 1999; *Klessen*, 2000; *Glover and Mac Low*, 2007; *Kritsuk et al.*, 2007; *Federrath et al.*, 2008a, 2010a; *Schmidt et al.*, 2009; *Price et al., 2011; Konstandin et al., 2012; Federrath, 2013a).* Other numerical studies considered the effect of magnetic fields (e.g., *Ostriker et al., 2001; Lemaster and Stone, 2008; Padoan and Nordlund, 2011; Molina et al., 2012),* and the effect of gravity (e.g., *Klessen, 2000; Federrath et al., 2008b; Collins et al., 2011; Kritsuk et al., 2011b; Cho and Kim, 2011; Collins et al., 2012; Federrath and Klessen, 2013; Kainulainen et al., 2013).*

The value of b depends on the ratio of compressional to total power in the driving force, with b ≈ 1 for purely compressive driving, and b ≈ 1/3 for purely solenoidal driving (*Federrath et al., 2008a, 2010a*). It has also been found that the PDF can be very roughly described by a lognormal also in the case of magnetized turbulence, with a simple modification of equation (5) derived by *Padoan and Nordlund* (2011) and *Molina et al.* (2012)

$$\sigma_s^2 = \ln\left[1 + b^2 \mathcal{M}_s^2 \beta/(\beta+1)\right] \qquad (6)$$

where β is the ratio of gas to magnetic pressures, $\beta = 8\pi\rho c_s^2/B^2 = 2c_s^2/v_A^2 = 2\mathcal{M}_A^2/\mathcal{M}_s^2$ for isothermal gas, $v_A \equiv B/\sqrt{4\pi\rho}$ is the Alfvén velocity, and $\mathcal{M}_A \equiv \sigma_v/v_A$ is the Alfvénic Mach number. This relation simplifies to the nonmagnetized case (equation (5)) as β → ∞. The derivation accounts for magnetic pressure, but not magnetic tension. In other words, it neglects the anisotropy of MHD turbulence that tends to align velocity and magnetic field. The anisotropy becomes important for a large magnetic field strength, and the relation in equation (6) is known to break down for trans- or sub-Alfvénic turbulence, requiring a more sophisticated model.

It has been found that self-gravity does not affect the density PDF at low and intermediate densities, and only causes the appearance of a power-law tail at large densities, reflecting the density profile of collapsing regions (*Klessen, 2000; Dib and Burkert, 2005*). *Kritsuk et al.* (2011b) have used a very deep adaptive mesh refinement (AMR) simulation covering a range of scales from 5 pc to 2 AU (*Padoan et al., 2005*), where individual collapsing cores are well resolved, and the density PDF develops a power law covering approximately 10 orders of magnitude in probability. By comparing the density profiles of the collapsing cores with the power-law high-density tail of the density PDF, they demonstrated that the PDF tail is consistent with being the result of an ensemble of collapsing regions. They point out that a spherically symmetric configuration with a power-law density profile, $\rho = \rho_0(r/r_0)^{-n}$ has a power-law density PDF with slope m given by m = −3/n, and a projected density PDF with slope p given by p = −2/(n−1) (see also *Tassis et al., 2010; Girichidis et al., 2014*). They find that in their simulation the power-law exponent decreases over time, and may be nearly stationary at the end of the simulation, when it reaches a value m = −1.67 for the density PDF, and p = −2.5 for the projected density PDF. Both values give n = −1.8, consistent with similarity solutions of the collapse of isothermal spheres (*Whitworth and Summers, 1985*). *Collins et al.* (2011) find a very similar slope, m = −1.64, also

in the case of magnetized turbulence, and *Federrath and Klessen* (2013) show that this result does not depend on the Mach number or on the ratio of compressive to total power, consistent with being primarily the effect of self-gravity in collapsing regions. The slope has a slight dependence on the turbulence Alfvénic Mach number, as the absolute value of m increases with increasing magnetic field strength (*Collins et al., 2012*).

3.2. Critical Density for Star Formation

Current theories assume that MCs are supersonically turbulent and that prestellar cores arise as gravitationally unstable density fluctuations in the turbulent flow. As explained in section 3.3, they derive the SFR as the mass fraction above an effective critical density for star formation, ρ_{crit}, which can be computed by assuming that the density PDF is lognormal, as discussed in section 3.1. The derivation of such a critical density is therefore a crucial step in current SFR models. *Federrath and Klessen* (2012) have summarized the various treatments of the critical density in recent theories of the SFR by *Krumholz and McKee* (2005) (KM), *Padoan and Nordlund* (2011) (PN), and *Hennebelle and Chabrier* (2011, 2013) (HC). Table 1 summarizes the main differences in those theories, in particular the different choice of density threshold.

One of the fundamental dimensionless parameters in the theories is the virial parameter, given by the ratio of turbulent and gravitational energies. Because of the difficulty of estimating the virial parameter in non-idealized systems, we distinguish between the theoretical virial parameter, $\alpha_{vir,T} = 2E_{kin}/E_{grav}$, and the observational one

$$\alpha_{vir,O} = 5\sigma_{v,1D}^2 R/GM \qquad (7)$$

where E_{kin} and $\sigma_{v,1D}$ include thermal effects. The two virial parameters are equal for an isothermal, spherical cloud of constant density (*Bertoldi and McKee, 1992; Federrath and Klessen, 2012*); for centrally concentrated clouds, $\alpha_{vir,O} > \alpha_{vir,T}$. *Federrath and Klessen* (2012) find that the fractal structure (e.g., *Scalo, 1990; Federrath et al., 2009*) of the clouds and the assumed boundary conditions (particularly periodic vs. nonperiodic) can lead to order-of-magnitude differences between $\alpha_{vir,O}$ and $\alpha_{vir,T}$. In the absence of a magnetic field, isolated clouds with $\alpha_{vir,T} < 1$ cannot be in equilibrium and will collapse. Including the effects of surface pressure, the critical value of $\alpha_{vir,T}$ for collapse is somewhat greater than 1 (*Bertoldi and McKee, 1992*). The corresponding critical value for $\alpha_{vir,O}$ depends on the density distribution; e.g., for a thermally supported Bonnor-Ebert sphere, it is 2.054. In the following, we refer to the virial parameter on the cloud diameter scale L = 2R as α_{cl}.

KM assumed that low-mass prestellar cores are thermally supported, so that in a turbulent medium the Jeans length of a critical core is proportional to the one-dimensional sonic length, $\lambda_{J,crit} = \ell_{s,1D}/\phi_x$, where ϕ_x is a numerical factor that was assumed to be on the order of unity. They derived the

TABLE 1. Six analytical models for the star formation rate per free-fall time.

Analytic Model	Free-Fall Time Factor	Critical Density $\rho_{crit}/\rho_0 = \exp(s_{crit})$	SFR$_{ff}$
KM	1	$(\pi^2/45)\,\phi_x^2 \times \alpha_{cl}\mathcal{M}_s^2\,(1+\beta^{-1})^{-1}$	$\varepsilon/(2\phi_t)\,\{1 + \mathrm{erf}\,[(\sigma_s^2-2s_{crit})/(8\sigma_s^2)^{1/2}]\}$
PN	$t_{ff}\,(\rho_0)/t_{ff}\,(\rho_{crit})$	$(0.067)\,\theta^{-2} \times \alpha_{cl}\mathcal{M}_s^2\,f(\beta)$	$\varepsilon/(2\phi_t)\,\{1 + \mathrm{erf}\,[(\sigma_s^2-2s_{crit})/(8\sigma_s^2)^{1/2}]\}\,\exp\,[(1/2)s_{crit}]$
HC	$t_{ff}\,(\rho_0)/t_{ff}\,(\rho)$	$(\pi^2/5)\,y_{cut}^{-2} \times \alpha_{cl}\mathcal{M}_s^2\,(1+\beta^{-1}) + \tilde{\rho}_{crit,turb}$	$\varepsilon/(2\phi_t)\,\{1 + \mathrm{erf}\,[(\sigma_s^2-2s_{crit})/(2\sigma_s^2)^{1/2}]\}\,\exp\,[(3/8)\sigma_s^2]$
Multi-ff KM	$t_{ff}\,(\rho_0)/t_{ff}\,(\rho)$	$(\pi^2/5)\,\phi_x^2 \times \alpha_{cl}\mathcal{M}_s^2\,(1+\beta^{-1})^{-1}$	$\varepsilon/(2\phi_t)\,\{1 + \mathrm{erf}\,[(\sigma_s^2-2s_{crit})/(2\sigma_s^2)^{1/2}]\}\,\exp\,[(3/8)\sigma_s^2]$
Multi-ff PN	$t_{ff}\,(\rho_0)/t_{ff}\,(\rho)$	$(0.067)\,\theta^{-2} \times \alpha_{cl}\mathcal{M}_s^2\,f(\beta)$	$\varepsilon/(2\phi_t)\,\{1 + \mathrm{erf}\,[(\sigma_s^2-2s_{crit})/(2\sigma_s^2)^{1/2}]\}\,\exp\,[(3/8)\sigma_s^2]$
Multi-ff HC	$t_{ff}\,(\rho_0)/t_{ff}\,(\rho)$	$(\pi^2/5)\,y_{cut}^{-2} \times \alpha_{cl}\mathcal{M}_s^2\,(1+\beta^{-1})$	$\varepsilon/(2\phi_t)\,\{1 + \mathrm{erf}\,[(\sigma_s^2-2s_{crit})/(2\sigma_s^2)^{1/2}]\}\,\exp\,[(3/8)\sigma_s^2]$

Note that the critical density for the KM and multi-ff KM models were derived in *Krumholz and McKee* (2005) based on the one-dimensional velocity dispersion, while *Federrath and Klessen* (2012) used the three-dimensional velocity dispersion, which changes the coefficient from $(\pi^2/45)\phi_x^2$ to $(\pi^2/5)\phi_x^2$. The function f(β), entering the critical density in the PN and multi-ff PN models, is given in *Padoan and Nordlund* (2011). The added turbulent contribution $\tilde{\rho}_{crit,turb}$ in the critical density of the HC model is given in *Hennebelle and Chabrier* (2011), *Federrath and Klessen* (2012), and equation (12). For further details, see *Federrath and Klessen* (2012), from which this table was taken.

sonic length using a linewidth-size relation of the form $\sigma_{v,1D} \propto r^q$. In their numerical evaluation, they adopted $q \simeq 1/2$, consistent with both simulations (e.g., *Padoan*, 1995; *Kritsuk et al.*, 2007; *Federrath et al.*, 2010a; *Federrath*, 2013a) and observations (e.g., *Ossenkopf and Mac Low*, 2002; *Heyer and Brunt*, 2004; *Padoan et al.*, 2006, 2009; *Roman-Duval et al.*, 2011). As a result, they obtained

$$\rho_{crit,KM} = \left(\pi^2\phi_x^2/45\right)\alpha_{cl}\mathcal{M}_{s,cl}^2\rho_{cl} \qquad (8)$$

where α_{cl} is the observational virial parameter for the cloud, ρ_{cl} is the mean density of the cloud, and $\mathcal{M}_{s,cl}$, like all Mach numbers in this paper, is the three-dimensional Mach number. They estimated the factor ϕ_x to be 1.12 from comparison with the simulations of *Vázquez-Semadeni et al.* (2003). If the sonic length is defined based on the three-dimensional velocity dispersion, then the coefficient in equation (8) changes from $(\pi^2\phi_x^2/45)$ to $(\pi^2\phi_x^2/5)$ (*Federrath and Klessen,* 2012). The KM result for ρ_{crit} is a generalization of the result of *Padoan* (1995) that $\rho_{core,crit} \simeq \mathcal{M}_{s,cl}^2\rho_{cl}$.

In PN, the critical density was obtained by requiring that the diameter of a critical Bonnor-Ebert sphere be equal to the characteristic thickness of shocked layers (identified as the size of prestellar cores), inferred from isothermal shock conditions (MHD conditions when the effect of the magnetic field is included)

$$\rho_{crit,PN} \simeq 0.067\,\theta_{int}^{-2}\alpha_{cl}\mathcal{M}_{s,cl}^2\rho_{cl} \qquad (9)$$

where θ_{int} = integral-scale radius/cloud diameter ≈ 0.35. PN did not assume a velocity-size relation, so their result does not depend on a specific value of q. However, for a reasonable value of q = 1/2, the critical density in KM agrees with that in PN to within a factor of 2.

The virial parameter can also be expressed as $\alpha_{cl} \propto \sigma_{v,cl}^2/\rho_{cl}R_{cl}^2$. Thus, if the exponent of the line width-size relation is q = 1/2, the critical density from equations (8) and (9) becomes independent of cloud density, virial parameter, and Mach number

$$n_{H,crit} \approx (5-10)\times10^4\,\mathrm{cm}^{-3}\left(T_{cl}/10\,\mathrm{K}\right)^{-1} \qquad (10)$$

assuming $\sigma_{v,1D} = 0.72\,\mathrm{km\,s^{-1}\,pc^{-1/2}}(R/1\,\mathrm{pc})^{1/2}$ (*McKee and Ostriker,* 2007). This previously overlooked result suggests a possible theoretical explanation for the nearly constant threshold density for star formation suggested by observations (see sections 2 and 5.3). However, it must be kept in mind that $n_{H,crit}$ varies as the square of the coefficient in the linewidth-size relation, so regions with different linewidth-size relations should have different critical densities. It should also be noted that magnetic fields tend to reduce the critical density by an uncertain amount (see below), and that the HC formulation of the critical density does not yield a constant value.

If the core is turbulent, McKee et al. (in preparation) show that the critical density can be expressed as

$$\rho_{core,crit} \simeq \left(\frac{\sigma_{v,cl}}{\sigma_{v,core}}\right)^2\frac{\alpha_{cl}}{\alpha_{crit}}\rho_{cl} \qquad (11)$$

The numerical factor $\alpha_{crit} = O(1)$ must be determined from theory or simulation, but the virial parameter of the cloud, α_{cl}, can be determined from observations. Cores that are turbulent can be subject to further fragmentation, but this is suppressed by protostellar feedback (e.g., *Myers et al.,* 2013). For cores that are primarily thermally supported $(\sigma_{v,1D,core} \simeq c_s)$, this expression has the same form as the KM and PN results above; if the core is turbulent, the critical density is reduced.

HC include the effects of turbulence in prestellar cores, but they argue that their model does not have a critical density for star formation. That is true of their derivation of the stellar IMF, where cores of different densities are allowed to collapse. For example, large turbulent cores with density lower than that of thermally supported cores are at the origin of massive stars, as in *McKee and Tan* (2003). However, when integrating the mass spectrum of gravitationally unstable density fluctuations to derive the SFR,

HC face the problem of choosing the largest mass cutoff in the integral, corresponding to the most massive star, which introduces an effective critical density. They chose to extend the integral up to a fraction, y_{cut}, of the cloud size, L_{cl}. Their critical density is thus the density such that the Jeans length (including turbulent support) is equal to $y_{cut}L_{cl}$

$$\rho_{crit,HC} = \left[\left(\frac{\pi^2}{5} \right) y_{cut}^{-2} \mathcal{M}_{s,cl}^{-2} + \left(\frac{\pi^2}{15} y_{cut}^{-1} \right) \right] \alpha_{cl} \rho_{cl} \quad (12)$$

where the first term is the contribution from thermal support, and the second term the contribution from turbulent support. The second term is larger than the first one if $\mathcal{M}_{s,cl} > (3/y_{cut})^{1/2}$, and can be obtained from equation (11) by noting that, for q = 1/2, one has $\mathcal{M}_{s,core,cut}^2/\mathcal{M}_{s,cl}^2 = R_{core,cut}/R_{cl} = y_{cut}$ for the maximum core mass; the second term of equation (12) then follows by taking $\alpha_{core} = 15/\pi^2$, similar to KM. HC assume that y_{cut} is a constant, independent of the properties of the cloud; the physical justification for this has yet to be provided. In summary, the critical density for KM and PN is defined for thermally supported cores, whereas that for HC is for the turbulent core corresponding to the most massive star.

Besides affecting the density PDF in a complex way (see section 3.1), magnetic fields can alter the critical density. Equation (11) remains valid, but the values of the virial parameters change. *McKee and Tan* (2003) estimated that $\alpha_{vir,0}$ is reduced by a factor of $(1.3 + 1.5/\mathcal{M}_A^2)$ in the presence of a magnetic field, provided that the gas is highly supersonic. If the core is embedded in a cloud with a similar value of $\mathcal{M}_A$ and the cloud is bound, then the magnetic field reduces $\alpha_{vir,0}$ for both the core and the cloud, and the net effect of the field is small. *Hopkins* (2012) finds that the critical density depends on the ratio of $(c_s^2 + \sigma^2 + v_A^2)$ on the cloud and core scales, so the field does not have a significant effect if $\mathcal{M}_A$ is about the same on the two scales. By contrast, *Federrath and Klessen* (2012) include magnetic fields in the KM derivation of ρ_{crit} by generalizing the comparison of Jeans length and sonic length to the comparison of the magnetic Jeans length with the magneto-sonic length. This reduces the KM value by a factor of $(1 + \beta^{-1})$, where β is the ratio of thermal to magnetic pressure, in agreement with the PN model (see Table 1). For the HC model, instead, *Federrath and Klessen* (2012) find that the critical density increases with increasing magnetic field strength, because it depends inversely on $\mathcal{M}_{s,cl}$ (see equation (12)) rather than directly, as in the KM and PN models. These contradictory results on the effect of the magnetic field are discussed in more detail in McKee et al. (in preparation).

3.3. Star Formation Rate

The dimensionless star formation rate per free-fall time, SFR_{ff}, is defined as the fraction of cloud (or clump) mass converted into stars per cloud mean free-fall time, $t_{ff,0} = \sqrt{3\pi/32G\rho_0}$, where ρ_0 is the mean density of the cloud

$$SFR_{ff}(t) = \frac{\dot{M}_*(t)}{M_{cl}(t)} t_{ff,0} \quad (13)$$

where all quantities are in principle time-dependent.

Detailed comparisons between the KM, PN, and HC models have been presented in *Hennebelle and Chabrier* (2011) and in *Federrath and Klessen* (2012) and will only be summarized here. A more general formalism to derive the SFR has been developed by *Hopkins* (2013), but as this author does not provide explicit formulae for the SFR, it is difficult to include here the results of his model.

The common point between these models is that they all rely on the so-called turbulent fragmentation scenario for star formation, with star formation resulting from a field of initial density fluctuations imprinted by supersonic turbulence. The models assume that gravitationally unstable density fluctuations are created by the supersonic turbulence and thus follow the turbulence density PDF (equation (4)). The SFR is then derived as the integral of the density PDF above a critical density (the HC model is actually derived from the integral of the mass spectrum of gravitationally unstable density fluctuations, but, after some algebra and simplifying assumptions, it is reduced to an integral of the density PDF above a critical density). The differences between the SFR models result from the choice of the critical density (see section 3.2 and Table 1), at least when the same general formulation for the SFR is used for all models, as first proposed by *Hennebelle and Chabrier* (2011) and further explored by *Federrath and Klessen* (2012)

$$SFR_{ff} = \frac{\varepsilon}{\phi t} \int_{\tilde{\rho}_{crit}}^{\infty} \frac{t_{ff,0}}{t_{ff}(\tilde{\rho})} p(\tilde{\rho}) d\tilde{\rho} \quad (14)$$

where $p(\tilde{\rho})d\tilde{\rho} = \tilde{\rho}p(s)ds$, and $p(s)ds$ is given by equation (4), with the variance following the relations in equations (5) or (6). Notice that, in the HC model, the relations in equations (5) or (6) include a scale dependence reflecting the property of the turbulence velocity power spectrum, $\sigma_s(R) = \sigma_{s,0}f(R)$ [e.g., equation (5) in *Hennebelle and Chabrier* (2008)]. As mentioned below, this point is of importance when calculating the SFR.

In equation (14), the PDF is divided by the free-fall time of each density, because unstable fluctuations are assumed to turn into stars in a free-fall time. In steady state, the shape of the density PDF should be constant with time and thus the turbulent flow must replenish density fluctuations of any amplitude. This may require a time longer that the free-fall time at a given density, hence the replenishment factor, ϕ_t, is introduced. It may in general be a function of $\tilde{\rho}$, but given our poor understanding of the replenishment process, it is simply assumed to be constant. *Hennebelle and Chabrier* (2011) suggested using the turbulence turnover time as an estimate of the replenishment time. They showed (in their appendix) that this choice yields $\phi_t \approx 3$. KM assumed that the replenishment time was on the order of the global free-fall time, not the local one, and their factor ϕ_t was defined accordingly. PN assumed that the replenishment time was on the order of the free-fall time of the critical density, and did not introduce a parameter ϕ_t to parameterize the uncertainty of this choice. Fitting each theory to numerical simulations, *Federrath and Klessen* (2012) find that $\phi_t \approx$ 2–5 in the best-fit models (see Fig. 3).

Note that the factor $t_{ff,0}/t_{ff}(\rho)$ appears inside the integral in equation (14) because gas with different densities has different free-fall times, which must be taken into account in the most general case (see *Hennebelle and Chabrier, 2011*). Previous estimates for SFR_{ff} either used a factor $t_{ff,0}/t_{ff,0} = 1$ (*Krumholz and McKee*, 2005), or a factor $t_{ff,0}/t_{ff}(\rho_{crit})$ (*Padoan and Nordlund*, 2011), both of which are independent of density and were thus pulled out of the general integral, equation (14). We refer to the latter models as "single free-fall" models, while equation (14) is a "multi-free-fall" model, but we can still distinguish three different density thresholds $s_{crit} = \ln(\tilde{\rho}_{crit})$ based on the different assumptions in the KM, PN, and HC theories. The general solution of equation (14) is

$$SFR_{ff} = \frac{\varepsilon}{2\phi_t} \exp\left(\frac{3}{8}\sigma_s^2\right) \left[1 + \mathrm{erf}\left(\frac{\sigma_s^2 - s_{crit}}{\sqrt{2\sigma_s^2}}\right)\right] \quad (15)$$

and depends only on s_{crit} and σ_s, which in turn depend on the four basic, dimensionless parameters α_{vir}, $\mathcal{M}_s$, b, and β, as shown and derived for all models in *Federrath and Klessen* (2012) (see Table 1 for a summary of the differences between each theoretical model for the SFR). It must be stressed that analytically integrating equation (14) to yield equation (15) is possible only if the variance of the PDF does not entail a scale dependence. In the HC formalism, this corresponds to what these authors refer to as their "simplified" multi-free-fall theory [section 2.4 of *Hennebelle and Chabrier* (2011)], while their complete theory properly accounts for the scale dependence.

The coefficient ε accounts for two efficiency factors: the fraction of the mass with density larger than ρ_{crit} that can actually collapse (a piece of very dense gas may be too small or too turbulent to collapse), and the fraction that ends up as actual stars (the so called core-to-star efficiency). The first factor should already be accounted for in the HC derivation of equation (14), and therefore should not be incorporated in ε. However, given the simplifications in the derivation, and the uncertainty in choosing the extreme of integration of the mass spectrum mentioned above (section 3.2), it may be necessary to allow for a fraction of this efficiency factor to be included in ε. The second factor (the core-to-star efficiency) is partly constrained by observations as well as analytical calculations and simulations, suggesting a value $\varepsilon \simeq 0.3$–0.7 (e.g., *Matzner and McKee*, 2000; *Federrath and Klessen*, 2012, 2013). As in the case of ϕ_t, ε is assumed to be a constant, even if, in principle, it may have a density dependence. To be consistent with

their physical meaning, the values of these two parameters should satisfy the conditions $\phi_t \geq 1$ and $\varepsilon \leq 1$.

As first proposed by *Hennebelle and Chabrier* (2011), the KM and PN models should be extended based on equation (14), which is easily done by inserting the KM and PN expressions for ρ_{crit} in that equation. The original KM and PN models were equivalent to a simplified version of equation (14), where one substitutes $t_{ff}(\tilde{\rho})$ with $t_{ff,0}$ (KM) or $t_{ff}(\tilde{\rho}_{crit})$ (PN). The effect of this extension is significant in the case of the KM model, because it introduces a dependence of SFR_{ff} on $\mathcal{M}_s$ that is otherwise missing, but rather small in the case of the PN model, where that dependence was already present. *Federrath and Klessen* (2012) showed that this extension from a single-free-fall to a multi-free-fall theory of the SFR does improve the match of the KM and PN models with their simulations (see section 4.4).

4. SIMULATIONS

Numerical simulations where physical conditions may be easily varied and controlled can greatly contribute, both qualitatively and quantitatively, to our understanding of the most important factors influencing the star formation rate. In this section, we first discuss numerical methods, then present results from large parameter studies.

4.1. Numerical Methods

Star formation rates measured in numerical experiments should be primarily influenced by the fundamental numerical parameters, such as Mach numbers, virial numbers, types of external driving, etc. However, shortcomings in the numerical representation of turbulence, on the one hand, and specific details (or lack thereof) in the recipes used to implement accretion onto "sink particles" representing the real stars, on the other, may also influence the results significantly.

Some of the general shortcomings, particularly in relation to the need to reproduce a turbulent cascade covering a wide range of scales, were reviewed in *Klein et al.* (2007). Since then, comparisons of numerical codes have been presented by *Kitsionas et al.* (2009) for the case of decaying supersonic HD turbulence, and by *Kritsuk et al.* (2011a) for the case of decaying supersonic MHD turbulence. Price and *Federrath* (2010) compared grid and particle methods for the case of driven, supersonic HD turbulence.

From the point of view of the current discussion, the bottom line of these comparisons is that the effective resolution of the various grid-based codes used in the most recent SFR measurements are quite similar, and that a much larger range of resolution results from the use (or not) of AMR than from the use of different codes. For problems where direct comparisons can be made (mainly HD problems), grid and particle-based codes also show similar results. Even though these problems do not have analytical solutions that may be used for verification, the general behavior of both HD and MHD turbulence is well represented by the codes that have been used to empirically measure the SFR.

An important point to keep in mind, however, is that both the *Kitsionas et al.* (2009) and *Kritsuk et al.* (2011a) comparisons, as well as *Federrath et al.* (2010a) and *Federrath et al.* (2011a), illustrate that turbulent structures are resolved well only down to a scale of 10–30 grid cells; smaller scales are, to quote *Kitsionas et al.* (2009), "significantly affected by the specific implementation of the algorithm for solving the equations of hydrodynamics." These effects, as well as the relative similarity with which different codes are affected, are well illustrated by Fig. 3 of *Kritsuk et al.* (2011a), which shows a drop in compensated velocity power spectra sets about 1 order of magnitude below the Nyquist wave number, i.e., at about 20 cells per wavelength, in essentially all codes. Even the world's highest-resolution simulation of supersonic turbulence to date, with 40,963 grid cells, only yields a rather limited inertial range (*Federrath*, 2013a). The possible influence of the magnetic Prandtl number, effectively close to unity in numerical experiments, while typically much larger in the ISM, should also be kept in mind (*Federrath et al.*, 2011b; *Schober et al.*, 2012; *Bovino et al.*, 2013; *Schleicher et al.*, 2013).

In numerical experiments, the SFR is measured using "sink particles" that can accumulate gas from their surroundings. The use of sink particles is convenient, because one can directly measure the conversion of gas mass into stellar mass. It is also inevitable, since it would otherwise be very difficult to handle the mass that undergoes gravitational collapse. The sink-particle technique goes back to *Boss and Black* (1982), who used a single, central "sink cell" in a grid code. Since then, sink-particle methods have been described and discussed by a number of authors (e.g., *Bate et al.*, 1995; *Krumholz et al.*, 2004; *Jappsen et al.*, 2005; *Federrath et al.*, 2010b).

Bate et al. (1995) applied a sink-particle technique to the modeling of accretion in protobinary systems using an SPH method, with a series of criteria for deciding when to dynamically form new sink particles. The most important criteria are that the density exceeds a certain, fixed threshold, and that the total energy is negative; i.e., the gas is bound. To decide whether to accrete onto existing sink particles they used a fixed accretion radius, within which an SPH particle would be accreted if it was gravitationally bound, with an angular momentum implying a circular orbit less than the accretion radius, and with auxiliary conditions imposed to resolve ambiguous cases.

Krumholz et al. (2004) applied a similar method in the context of an AMR grid code (Orion). To decide when to create new sink particles they used primarily the local Jeans length, which is equivalent to a density threshold criterion. Sink particles were created when a Jeans length was resolved with less than four grid cells, thus combining the sink-particle creation criterion with the Truelove criterion (*Truelove et al.*, 1997), which prevents artificial gravitational fragmentation. To handle the creation of very close sink particles they merged all sink particles with distances below a given accretion radius. To decide how rapidly to accrete onto existing sink particles they used a method based primarily on the Bondi-Hoyle accretion formula, complemented by taking the effects of residual angular momentum into account.

Federrath et al. (2010b) implemented sink particles in a version of the Flash code (*Fryxell et al.*, 2000; *Dubey et al.*, 2008), using a similar series of criteria as in *Bate et al.* (1995) to guard against spurious creation in shocks (for example). They checked the resulting behavior in the AMR case against the behavior in the original SPH case. They demonstrated that omitting some of the sink-creation criteria may result in significantly overestimating the number of stars created, as well as overestimating the SFR.

Krumholz et al. (2004), *Federrath et al.* (2010b), and Haugbølle et al. (in preparation) seem to agree on the fact that the global SFR in simulations is not as sensitive to details of the sink-particle recipes as the number and mass distribution of sink particles. It appears that the rate of accretion is determined at some relatively moderate density level, well below typical values adopted as thresholds for sink-particle creation, and that various auxiliary criteria can result in the formation of a larger or smaller number of sinks, without affecting the global SFR very much.

An important class of effects that are largely missing from current simulations aimed at measuring the SFR consists of (direct and indirect) effects of stellar feedback. Bipolar outflows, for example, both reduce the final mass of the stars (a "direct" effect), and feed kinetic energy back into the ISM, thus increasing the virial number and reducing the SFR (an "indirect" effect).

As discussed elsewhere in this chapter, both comparisons of the core mass function with the stellar IMF as well as comparisons of numerical simulations with observations indicate that on the order of half of the mass is lost in bipolar outflows, i.e., leading to a core-to-star efficiency $\varepsilon \approx 0.3$–0.7 (*Matzner and McKee*, 2000; *Federrath and Klessen*, 2012). In the near future, it may become possible to model the outflow in sufficient detail to measure the extent to which accretion is diverted into outflows, using local zoom-in with very deep AMR simulations, which are able to resolve the launching of outflows in the immediate neighborhood of sink particles (*Seifried et al.*, 2012; *Price et al.*, 2012; *Nordlund et al.*, 2013).

The indirect effect of outflow feedback has been modeled by adopting assumptions about the mass and momentum loading of bipolar outflows (e.g., *Nakamura and Li*, 2008; *Wang et al.*, 2010; *Cunningham et al.*, 2011). When attempting to quantify the importance of the feedback, it is important to adopt realistic values of the virial number. If the virial number of the initial setup is small, the relative importance of the feedback from outflows is increased. A number of studies of MCs have found very extended inertial ranges (e.g., *Ossenkopf and Mac Low*, 2002; *Heyer and Brunt*, 2004; *Padoan et al.*, 2006; *Roman-Duval et al.*, 2011), and relatively smooth velocity power spectra without bumps or features at high wavenumbers (e.g., *Padoan et al.*, 2009; *Brunt et al.*, 2009), suggesting that local feedbacks are not a dominant energy source, compared to the inertial energy cascade from very large scales.

Finally, a few comments on the importance of feedback from stellar radiation (see the chapters by Krumholz et al. and Offner et al. in this volume). As is obvious from a number of well-known Hubble images, and also from simple back-of-the-envelope estimates, the UV radiation from newborn massive stars can have dramatic effects on the surrounding ISM, ionizing large ISM bubbles and cooking away the outer layers of cold, dusty MCs. Such effects have recently started to become incorporated in models of star formation (e.g., *Peters et al.*, 2010; *Commerçon et al.*, 2011; *Kuiper et al.*, 2011; *Krumholz et al.*, 2012b; *Bate*, 2012). In a series of papers, *Dale et al.* (2012a,b, 2013a,b) have explored the effect of H$_{II}$ regions on the SFR. They find that the ionization feedback reduces the SFR by an amount that depends on the virial parameter and total mass of the cloud. To derive a realistic measure of the effect of feedbacks, one needs to make sure that other factors that affect the SFR have realistic levels as well. For example, most numerical studies of the effect of massive star feedback have neglected magnetic fields, which are known to reduce the SFR by a factor of 2 or 3 (*Padoan and Nordlund*, 2011; *Federrath and Klessen*, 2012) (see also section 4.3) even without any stellar feedback. Furthermore, as real clouds are generally not isolated, inertial driving from larger scales (e.g., gas accretion onto a cloud) (see, e.g., *Klessen and Hennebelle*, 2010; *Schmidt et al.*, 2013) may also reduce the SFR.

4.2. Setups of Numerical Experiments

Besides the differences in numerical methods discussed in the previous section, numerical studies of star formation may also differ significantly with respect to their boundary and initial conditions. Ideally, one would like to model the whole process starting from cloud formation in galaxy simulations (e.g., *Tasker and Bryan*, 2006; *Dobbs et al.*, 2006, see also the chapter by Dobbs et al. in this volume) and large-scale colliding flows (*Heitsch et al.*, 2005; *Vázquez-Semadeni et al.*, 2006; *Hennebelle et al.*, 2008; *Banerjee et al.*, 2009) down to the formation of individual stars. However, this is beyond the current capabilities of supercomputers. Due to their local nature (typically ~1–10 pc), star formation simulations do not couple the formation of prestellar cores and their collapse with the large-scale processes driving the ISM turbulence, such as the evolution of supernova (SN) remnants, and the formation and evolution of GMCs. In setting up such simulations, one can choose either to completely ignore the effect of the larger scales (e.g., *Clark et al.*, 2005; *Bate*, 2009; *Bonnell et al.*, 2011; *Girichidis et al.*, 2011), or to model it with external forces (e.g., *Klessen et al.*, 2000; *Heitsch et al.*, 2001; *Padoan and Nordlund*, 2011; *Padoan et al.*, 2012; *Federrath and Klessen*, 2012).

When driving forces are included, it is usually done with periodic boundary conditions and with a random, large-scale force generated in Fourier space. This artificial random force is supposed to mimic a turbulent cascade of energy starting at a much larger scale, but its effect may be different from that of the true driving forces of the ISM, such as SNs, spiral arm compression, expanding radiation fronts, winds, jets, outflows, and other mechanisms. The SFR in the simulations depends on whether a driving force is included or not (e.g., *Klessen et al.*, 2000; *Vázquez-Semadeni et al.*, 2003). Without an external force, the turbulence decays rapidly (*Padoan and Nordlund*, 1997; *Stone et al.*, 1998; *Mac Low et al.*, 1998; *Padoan and Nordlund*, 1999; *Mac Low*, 1999) and star formation proceeds at a very high rate, while if the turbulence is continuously driven, a lower SFR can be maintained for an extended time. Both types of numerical setups have drawbacks. If external forces are neglected and star formation proceeds very rapidly, unphysical initial conditions may strongly affect the results; if the turbulence is driven, initial conditions are forgotten over time, but the results may depend on the artificial force.

Large differences between simulations are also found in their initial conditions. These can be taken to be rather artificial, e.g., with stochastic velocity fields inconsistent with supersonic turbulence apart from their power spectrum (e.g., *Clark et al.*, 2008; *Bate*, 2009), or self-consistently developed by driving the turbulence over many dynamical times prior to the inclusion of self-gravity (e.g., *Offner et al.*, 2009; *Padoan and Nordlund*, 2011; *Padoan et al.*, 2012; *Federrath and Klessen*, 2012). In some studies, initial conditions are taken to follow ad hoc density profiles or velocity power spectra with the purpose of studying the effect of those choices (e.g., *Bate and Bonnell*, 2005; *Girichidis et al.*, 2011).

4.3. Results from Large Parameter Studies

Some of the first numerical studies testing the effects of the turbulence, its driving scale, and the magnetic field strength on the SFR were carried out by *Klessen et al.* (2000), *Heitsch et al.* (2001), and *Vázquez-Semadeni et al.* (2003). Recent high-resolution simulations by *Padoan and Nordlund* (2011), *Padoan et al.* (2012), and *Federrath and Klessen* (2012), covering the largest range of MC parameters available to date, demonstrate that the SFR is primarily determined by four dimensionless parameters: (1) the virial parameter, $\alpha_{vir} = 2E_{kin}/E_{grav}$; (2) the sonic Mach number, $\mathcal{M}_s = \sigma_v/c_s$; (3) the turbulence driving parameter, b, in equations (5) and (6); and (4) the plasma $\beta = \langle P_{gas} \rangle / \langle P_{mag} \rangle$, or the Alfvénic Mach number $\mathcal{M}_A$, defined as $\mathcal{M}_A = \sigma_v/v_A = \sigma_v/(\langle B^2 \rangle/\langle 4\pi\rho \rangle)^{1/2}$.

These parameter studies adopted periodic boundary conditions, an isothermal equation of state, and large-scale random driving. The turbulence was driven without self-gravity for several dynamical times, to reach a statistical steady state. Self-gravity was then included, and collapsing regions above a density threshold, ρ_{max}, and satisfying other conditions (see section 4.1) were captured by sink particles.

Padoan and Nordlund (2011) carried out the first large parameter study of the SFR using the Stagger-code to run simulations on uniform numerical grids of 500^3 and 1000^3 computational points. They created the sink particles only based on a density threshold, $\rho_{max} = 8000\langle\rho\rangle$, large enough to avoid noncollapsing density fluctuations created by shocks in the turbulent flow. They explored two different values of $\mathcal{M}_s$, two different values of β, and seven different values of α_{vir}. Their simulations showed that SFR$_{ff}$ can be reduced by the magnetic field by approximately a factor of 3, that in the nonmagnetized case SFR$_{ff}$ increases with increasing values of $\mathcal{M}_s$, and that, both with and without magnetic fields, SFR$_{ff}$ decreases with increasing values of α_{vir}, confirming the predictions of their model.

In their following work, *Padoan et al.* (2012) analyzed an even larger parameter study, based on 45 AMR simulations with the Ramses code, with a maximum resolution equivalent to $32,768^3$ computational points. Thanks to the very large dynamic range, they could adopt a threshold density for the creation of sink particles of $\rho_{max} = 10^5\langle\rho\rangle$, much larger than in the uniform-grid simulations. The creation of a sink particle also required that the cell was at a minimum of the gravitational potential, and that the velocity divergence was negative. They explored two values of the sonic Mach number, $\mathcal{M}_s = 10$ and 20; four values of the initial Alfvénic Mach number, $\mathcal{M}_A = 1.25, 5, 20$, and 33; and seven values of the virial parameter, in the approximate range $0.2 < \alpha_{vir} < 20$. Their results are presented in Fig. 2, and can be summarized in three points: (1) SFR$_{ff}$ decreases exponentially with increasing t_{ff}/t_{dyn} ($\propto \alpha_{vir}^{1/2}$), but is insensitive to changes in $\mathcal{M}_s$ (in the range $10 \leq \mathcal{M}_s \leq 20$), for constant values of t_{ff}/t_{dyn} and $\mathcal{M}_A$. (2) Decreasing values of $\mathcal{M}_A$ (increasing magnetic field strength) reduce SFR$_{ff}$, but only to a point, beyond which SFR$_{ff}$ increases with a further decrease of $\mathcal{M}_A$. (3) For values of $\mathcal{M}_A$ characteristic of star-forming regions, SFR$_{ff}$ varies with $\mathcal{M}_A$ by less than a factor of 2. Therefore, *Padoan et al.* (2012) proposed a simple law for the SFR depending only on t_{ff}/t_{dyn}, based on the empirical fit to the minimum SFR$_{ff}$: SFR$_{ff} \approx \varepsilon \exp(-1.6\, t_{ff}/t_{dyn})$ (dashed line in Fig. 2), where ε is the core-to-star formation efficiency.

These results were confirmed and extended in a third parameter study by *Federrath and Klessen* (2012), based on 34 uniform grid simulations with the Flash code, using a resolution of up to 512^3 computational points (plus one AMR run with maximum resolution equivalent to 1024^3 computational points). Six of their runs include a magnetic field, covering the range $1.3 \leq \mathcal{M}_A \leq 13$, but all with approximately the same sonic Mach number, $\mathcal{M}_s \approx 10$, so the lack of dependence of SFR$_{ff}$ on $\mathcal{M}_s$ found by *Padoan et al.* (2012) could not be verified. On the other hand, the 28 runs without magnetic fields span a wide range of values of sonic Mach number, $2.9 \leq \mathcal{M}_s \leq 52$, which allowed *Federrath and Klessen* (2012) to confirm the analytical and numerical result of *Padoan and Nordlund* (2011), that in the nonmagnetized case SFR$_{ff}$ increases with increasing $\mathcal{M}_s$. *Federrath and Klessen* (2012) also studied the effect of varying b (the ratio of compressible to total energy in the turbulence driving), and carried out a systematic comparison of their simulations with the predictions of SFR models and observations (see sections 4.4 and 5.1, respectively).

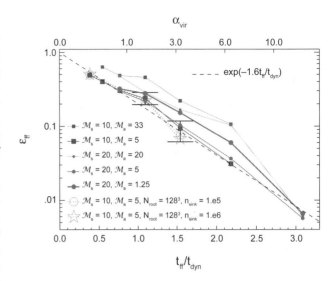

Fig. 2. SFR per free-fall time, SFR$_{ff}$, vs. t_{ff}/t_{dyn} (bottom abscissa) and α_{vir} (top abscissa). The symbols for each series of runs, where only the strength of gravity is changed and $\mathcal{M}_s$ and $\mathcal{M}_A$ are kept constant, are connected by a line to better distinguish each series. The two error bars give the mean of SFR$_{ff}$, plus and minus the rms values, for each group of five 32^3-root-grid runs with identical parameters ($\mathcal{M}_s \approx 10$ and $\mathcal{M}_A \approx 5$), but different initial conditions. The dashed line is an approximate exponential fit to the minimum value of SFR$_{ff}$ vs. t_{ff}/t_{dyn}. From *Padoan et al.* (2012), reproduced with permission of AAS.

Federrath and Klessen (2012) found that both $\mathcal{M}_s$ and b can introduce order-of-magnitude variations in SFR$_{ff}$, in the absence of magnetic fields. Increasing b and $\mathcal{M}_s$ produces a wider density PDF (see equations (5) and (6)) and thus pushes a larger fraction of gas above the critical density for star formation, increasing SFR$_{ff}$ (a larger $\mathcal{M}_s$ results in a larger critical density, but also in a shorter free-fall time at such density). For purely compressive driving, *Federrath and Klessen* (2012) found a 4× higher SFR$_{ff}$, when increasing $\mathcal{M}_s$ from 5 to 50. For fixed $\mathcal{M}_s = 10$, which is a reasonable average Mach number for Milky Way clouds, they found that purely compressive (curl-free, b = 1) driving yields about 10× higher SFR$_{ff}$ compared to purely solenoidal (divergence-free, b = 1/3) driving. The increase of SFR$_{ff}$ for compressive driving is caused by the denser structures (filaments and cores) produced with such driving, which are more gravitationally bound than the structures produced with purely solenoidal driving.

Increasing the magnetic field strength and thus decreasing β reduces SFR$_{ff}$. Numerical simulations by *Padoan and Nordlund* (2011), *Padoan et al.* (2012), and *Federrath and Klessen* (2012) quantify the effect of the magnetic field and find a maximum reduction of the SFR by a factor of 2–3 in strongly magnetized, trans- to sub-Alfvénic turbulence compared to purely hydrodynamic turbulence. This is a relatively small change in SFR$_{ff}$ compared to changes induced by α_{vir}, $\mathcal{M}_s$, and b, in the absence of magnetic fields.

4.4. Comparison with Theoretical Models

To compare them with numerical results, we separate the theories of the SFR into six cases (see section 3.3), named "KM," "PN," "HC," and "multi-ff KM," "multi-ff PN," "multi-ff HC," following the notations in *Hennebelle and Chabrier* (2011) and *Federrath and Klessen* (2012). The first three represent the original analytical derivations by *Krumholz and McKee* (2005), *Padoan and Nordlund* (2011), and *Hennebelle and Chabrier* (2011), while the last three are all based on the multi-free-fall expression of equation (14). Notice that the HC models, here and in *Federrath and Klessen* (2012), refer to the aforementioned simplified version of the HC model without a scale dependence in the density variance, which yields equation (15). In case of the complete theory, the integral of equation (14) can no longer be performed analytically. Besides this important point, the difference among the multi-free-fall models is the expression for the critical density (the lower limit of the integral in equation (14)). Table 1 summarizes the differences among all six theories following *Federrath and Klessen* (2012), who also extended all theoretical models to include the effects of magnetic pressure (except PN, where this dependence was already included).

The comparison between the theoretical and the numerical SFR$_{ff}$ is shown in Figs. 3a (KM, PN, HC) and 3b (multi-free-fall KM, PN, HC). Any of the multi-free-fall theories provides significantly better fits to the numerical simulations than the single-free-fall models ($\chi^2_{red} = 1.3$ and 1.2 for multi-ff KM and PN; $\chi^2_{red} = 5.7$ and 1.8 for KM and PN). Note that HC and multi-ff HC are both multi-ff models, but *Hennebelle and Chabrier* (2011) defined them this way to distinguish their model where the critical density includes both thermal and turbulent support (called HC) from the one where only thermal support is included (called multi-ff HC) (see Table 1 and section 3.2).

The simplified HC model, although originally designed as a multi-free-fall model, does not fit the simulations as well as the multi-ff KM and multi-ff PN models. This is likely related to the definition of the critical density in HC, where it scales with $\mathcal{M}_s^{-2}$, while in KM and PN it scales with $\mathcal{M}_s^2$ [see section 3.2, Table 1, and the derivation in *Federrath and Klessen* (2012)]. A caveat of the present comparison is that the density PDF in the models is assumed to be perfectly lognormal. Accounting for deviations of the PDF in future semianalytical models may further improve the agreement with the simulations.

Figure 3 shows that the multi-free-fall KM and PN theories provide good fits to the numerical simulations. However, some data points in Fig. 3 deviate from the theoretical predictions by a factor of 2–3. The simulations that deviate the most have relatively low numerical resolution. A careful resolution study of these models shows that they converge to the diagonal line in Fig. 3. Notably, simulation models 32, 33, and 34 with numerical resolutions of 256^3, 512^3, and 1024^3 grid cells (in the upper right corner of Fig. 3) clearly converge to the diagonal line with increas-

ing resolution. The same is true for the MHD models 19 and 20 with 256^3 and 512^3 resolution in the inset of Fig. 3. This convergence with increasing resolution supports the conclusion that the multi-free-fall theories for the SFR provide a good match to numerical simulations where the interplay of self-gravity and supersonic turbulence is the primary process controlling the SFR.

5. THEORY AND SIMULATIONS VERSUS OBSERVATIONS

5.1. Model Predictions Versus Observed Star Formation Rate

A comparison of simulations from *Federrath and Klessen* (2012) with observations of Σ_{SFR} in Milky Way clouds by *Heiderman et al.* (2010) is shown in Fig. 4. The observational data are from Galactic observations of clouds and YSOs identified in the Spitzer cores-to-disks (c2d) and Gould's belt (GB) surveys (*Evans et al.*, 2009) of massive dense clumps (*Gao and Solomon*, 2004; *Wu et al.*, 2005, 2010) and of the Taurus MC (*Goldsmith et al.*, 2008; *Pineda et al.*, 2010; *Rebull et al.*, 2010). Simulation data are shown as contours for SFE = 1% (inner contour line) and SFE = 10% (outer contour line).

The simulation data from *Federrath and Klessen* (2012) shown as contours in Fig. 4 were transformed into the observational space by measuring Σ_{gas} and Σ_{SFR} with a method as close as possible to what observers do to infer Σ_{SFR}-to-Σ_{gas} relations (see, e.g., *Bigiel et al.*, 2008; *Heiderman et al.*, 2010), including the effects of telescope beam smoothing. This was done by computing two-dimensional projections of the gas column density, Σ_{gas}, and the sink particle column density, Σ_{SF}, along each coordinate axis: x, y, z. Each of these two-dimensional maps was smoothed to a resolution $(N_{res}/8)^2$ for a given three-dimensional numerical resolution $N_{res}^3 = 128^3$–1024^3 grid cells [for a complete list of simulations and their parameters, see Table 2 in *Federrath and Klessen* (2012)], such that the size of each pixel in the smoothed maps slightly exceeds the sink particle diameter (which is 5 grid cells). We then search for pixels with a sink particle column density greater than zero, $\Sigma_{SF} > 0$, and extract the corresponding pixel in the gas column density map, which gives Σ_{gas} in units of $M_\odot$ pc^{-2} for that pixel. The SFR column density is computed by taking the sink particle column density Σ_{SF} of the same pixel and dividing it by a characteristic timescale for star formation, t_{SF}, which yields $\Sigma_{SFR} = \Sigma_{SF}/t_{SF}$ in units of $M_\odot$ yr^{-1} kpc^{-2}. The simplest choice for t_{SF} is a fixed star formation time, $t_{SF} = 2$ m.y., based on an estimate of the elapsed time between star formation and the end of the Class II phase (e.g., *Evans et al.*, 2009; *Covey et al.*, 2010). This is also the t_{SF} adopted by *Lada et al.* (2010) and *Heiderman et al.* (2010) to convert YSO counts into an SFR column density (see section 2.4), so we used it here as the standard approach. However, *Federrath and Klessen* (2012) also experimented with two other choices for t_{SF}, which gave similar results,

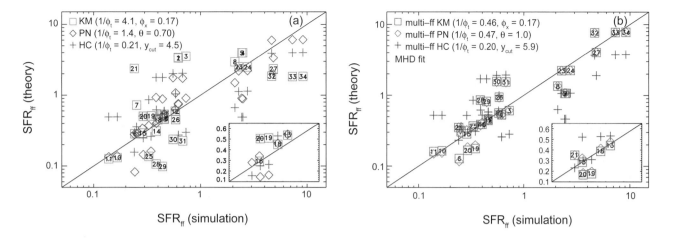

Fig. 3. Comparison of SFR$_{ff}$ (theory) with SFR$_{ff}$ (simulation). **(a)** Original KM (boxes), PN (diamonds), and HC (crosses) theories; **(b)** multi-free-fall version of each theory defined in *Hennebelle and Chabrier* (2011). The multi-free-fall prescription is superior to all single-free-fall models and provides good fits to the numerical simulations (insets show blowups of the MHD simulations; the x-range of the insets is identical to the y-range). The multi-ff KM and multi-ff PN models agree to within a factor of 3 with any of the numerical simulations over the 2 orders of magnitude in SFRs tested. The simulation number in Table 2 in *Federrath and Klessen* (2012) is given in the boxes for each SFR$_{ff}$ (simulation). From *Federrath and Klessen* (2012), reproduced with permission of AAS.

considering the large dispersion of the observational data.

The *Heiderman et al.* (2010) sample of SFR column densities for Galactic clouds in Fig. 4 covers different evolutionary stages of the clouds, such that a single SFE for the whole sample is unlikely. However, we can reasonably assume SFEs in the range 1% to 10% (*Evans et al.,* 2009; *Federrath and Klessen,* 2013). For a typical SFE = 3%, *Federrath and Klessen* (2012) find a best-fit match of their simulations to the *Heiderman et al.* (2010) sample by scaling the simulations to a core-to-star efficiency ε = 0.5, in agreement with theoretical results (*Matzner and McKee,* 2000), with observations of jets and outflows (*Beuther et al.,* 2002), and with independent numerical simulations that concentrate on the collapse of individual cloud cores (e.g., *Wang et al.,* 2010; *Seifried et al.,* 2012).

The simulation data in Fig. 4 are also consistent with the Milky Way cloud samples in *Lada et al.* (2010) and *Gutermuth et al.* (2011), showing that Σ_{SFR} can vary by more than 1 order of magnitude for any given Σ_{gas}. These variations can be explained by variations in α_{vir} (*Hennebelle and Chabrier,* 2011; *Padoan et al.,* 2012), $\mathcal{M}_s$ (*Renaud et al.,* 2012; *Federrath,* 2013b), b (*Federrath and Klessen,* 2012), and β (*Padoan and Nordlund,* 2011; *Federrath and Klessen,* 2012) from cloud to cloud. Besides these physical effects primarily related to the statistics of self-gravitating turbulence, the Σ_{SFR}-Σ_{gas} relation is also somewhat affected by observational issues [e.g., telescope resolution (see *Heiderman et al.,* 2010; *Federrath and Klessen,* 2012)] and by variations in the metallicity (*Krumholz et al.,* 2012a). The latter effect introduces uncertainties by a factor of about 2 (*Glover and Clark,* 2012).

Figure 5 compares the SFR obtained with the complete HC formalism for four different cloud sizes, R$_c$ = 0.5, 2, 5,

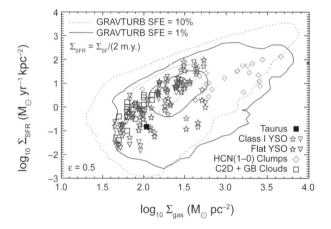

Fig. 4. Star formation rate column density Σ_{SFR} vs. gas column density Σ_{gas} for Milky Way clouds compiled in *Heiderman et al.* (2010) (symbols) and in the GRAVTURB simulations by *Federrath and Klessen* (2012) (contours). Individual data points: Taurus — filled black box (data from *Goldsmith et al.,* 2008; *Pineda et al.,* 2010; *Rebull et al.,* 2010); Class I YSOs and flat YSOs — stars, with upper limits shown as downward-pointing triangles; HCN (1–0) clumps — golden diamonds (data from *Gao and Solomon,* 2004; *Wu et al.,* 2005, 2010); and c2d + GB clouds —boxes (data from *Evans et al.,* 2009). Contours show data from numerical simulations by *Federrath and Klessen* (2012) for a star formation efficiency SFE = 1% (inner contour line) and SFE = 10% (outer contour line). The thick contours enclose 50% of all (Σ_{gas}, Σ_{SFR}) simulation pairs, while the thin contours enclose 99%. All simulation data were scaled to a local core-to-star efficiency of ε = 0.5 (*Matzner and McKee,* 2000), providing the best fit to the observational data. From *Federrath and Klessen* (2012), reproduced with permission of AAS.

and 20 pc, with the data from *Heiderman et al.* (2010) and *Gutermuth et al.* (2011), for the case of both isothermal and non-isothermal gas, for $y_{cut} = 0.25$ and $b = 0.5$ (approximate equipartition between solenoidal and compressive modes of turbulence). For such typical Milky Way MC conditions, the agreement is fairly reasonable. The observed scatter, with no one-to-one correspondence between the SFR and the surface density, is well explained by the strong dependence of the SFR upon the cloud's size/mass, besides the effect of the turbulence parameters. The theory also adequately reproduces the observational SFR values of *Gutermuth et al.* (2011) (bracketed areas) for the adequate (large) cloud sizes, down to the very low-density regime, $\Sigma \sim 30$ $M_{\odot}$ pc^{-2}. This supports the suggestion that there is no real step-function threshold condition for star formation and that the latter can occur even in rather low-density regions, although at a much lower rate.

In Fig. 5, the HC model is applied assuming a density-size relation, $\rho \propto R_c^{-0.7}$, which determines the dependence on Σ_{gas} seen in the theoretical curves. This density-size relation is confirmed by observations where clouds are defined by constant ^{13}CO brightness contours (*Roman-Duval et al.*, 2010). However, when clouds are defined by constant surface density contours, one gets a different density-size relation, $\rho \propto R_c^{-1.0}$ (*Lombardi et al.*, 2010), which changes the position of the theoretical curves of Fig. 5 slightly (the curves would be a bit more vertical and closer to one another). This highlights the importance of adopting unambiguous definitions of MCs in both theoretical models and observational works, or rather of deriving theoretical models that do not depend on MC definitions.

5.2. Density Probability Density Function in Molecular Clouds

Several studies have tried to compare the PDF predicted by supersonic turbulence with measurements in MCs. The PDF of a MC was first shown to be consistent with a lognormal function by *Padoan et al.* (1997a). They used the extinction measurements of IC 5146 by *Lada et al.* (1994), specifically their σ_{A_V}–AV relation, to constrain the shape and standard deviation of the three-dimensional density PDF, and the density power spectrum. They concluded that the density PDF was consistent with a lognormal, that the standard deviation was $\sigma_{\bar{\rho}} = 5.0 \pm 0.5$, that the slope of the density power spectrum was -2.6 ± 0.5, and that density fluctuations had to reach at least a scale of 0.03 pc. Given the Mach number derived for IC 5146, $\mathcal{M}_s \approx 10$, these results were all in good agreement with their numerical simulations of super-Alfvénic turbulence, showing that $\sigma_{\bar{\rho}} = b\mathcal{M}_s$, with $b \approx 0.5$.

Brunt et al. (2010b) analytically derived the relation between the variance of the three-dimensional density field and the observed variance and power spectrum of the projected density. Applied to the Taurus MC, this method gave a density variance yielding $b = 0.48^{+0.15}_{-0.11}$, similar to the case of IC 5146, and consistent with that of turbulence simulations driven with a mixture of solenoidal and compressive modes (*Brunt,* 2010; *Price et al.,* 2011).

The relation between the Mach numbers and the variance of the projected density can be derived directly from simulations. *Burkhart and Lazarian* (2012) found that the variance, σ_S, of the logarithm of the column density, S =

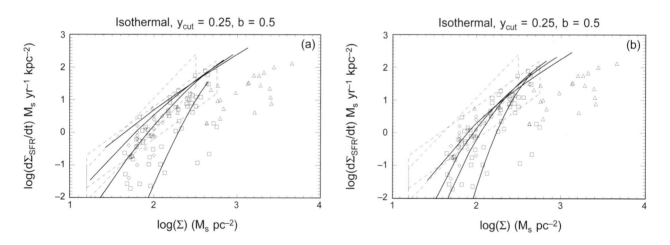

Fig. 5. Star formation rate per unit area, Σ_{SFR}, as a function of gas surface density, Σ_{gas}. Solid lines are results obtained with the complete HC SFR theory (*Hennebelle and Chabrier,* 2013) for cloud sizes $R_c = 20$, 5, 2, and 0.5 pc, from left to right, in the case of **(a)** isothermal and **(b)** nonisothermal gas, for $y_{cut} = 0.25$ and $b = 0.5$, and assuming a density-size relation, $\rho \propto R_c^{-0.7}$. Each curve is obtained by varying the normalization of the density-size relation (for details, see *Hennebelle and Chabrier,* 2013). The data correspond to observational determinations by *Heiderman et al.* (2010) for massive clumps (triangles) and molecular cloud YSOs (diamonds + squares), and by *Gutermuth et al.* (2011) (bracketed areas). From *Hennebelle and Chabrier* (2013), reproduced with permission of AAS.

$\ln(\Sigma/\Sigma_0)$ (where Σ_0 is the mean column density), depends on the sonic Mach number, and follows the relation $\sigma_S^2 = A \ln [1 + b^2 \mathcal{M}_s^2]$, where $A \approx 0.11$, for a range of values of Mach numbers, $0.4 \leq \mathcal{M}_s \leq 8.8$ and $0.3 \leq \mathcal{M}_A \leq 3.2$. They also used that relation to map the Mach number in the Small Magellanic Cloud (*Burkhart et al.,* 2010). By comparing with equation (5), one can see that the ratio of density and column density variances is independent of the Mach number, $\sigma_S^2/\sigma_{\mathcal{S}}^2 = A$. In a subsequent paper, *Burkhart et al.* (2013) studied the role of radiative transfer effects on the derivation of the column density variance using molecular emission lines. They showed that the parameter A is quite sensitive to the optical depth of the observed line, so extinction maps, rather than molecular emission lines, are the method of choice to directly constrain the density PDF, unless opacities and molecular abundances are known. Similar conclusions about the uncertainties in deriving column density PDFs in MCs with CO emission lines were also reached by *Goodman et al.* (2009), through a comparison of column density measured with dust emission and extinction. They also found that the relation between the mean and the variance of the projected density were consistent with a lognormal distribution.

Besides deriving the density variance, a more ambitious goal is to derive the whole PDF using MC observations. This was first done by *Padoan et al.* (1999), who compared a simulation of supersonic and super-Alfvénic turbulence with a ^{13}CO survey of the Perseus region. Radiative transfer effects were properly accounted for, because the comparison was done using synthetic observations of the simulation, generated with a three-dimensional non-local thermodynamic equilibrium (LTE) radiative transfer code (*Padoan et al.,* 1998). However, besides radiative transfer, the interpretation of molecular emission lines would also require a careful study of molecular abundances, in order to probe a more extended density range. The interpretation of extinction maps is more straightforward, at least for MCs that are far enough from the Galactic Plane to avoid confusion, and close enough to guarantee a large number of background stars. Moreover, *Brunt et al.* (2010a) developed a statistical method to reconstruct the volumetric (three-dimensional) density PDF from observations of the column (two-dimensional) density PDF and power spectrum.

Extinction maps were used by *Kainulainen et al.* (2009) to derive column density PDFs of many nearby clouds. They found that the PDFs are consistent with a lognormal function at all densities in non-star-forming clouds. In regions of active star formation, the PDF is lognormal at intermediate and low densities, and a power law at high densities, consistent with the results of supersonic turbulence with self-gravity. Other studies based on extinction maps used the cumulative PDF of column density, and found again deviations from the lognormal at high extinctions in regions of active star formation (*Froebrich and Rowles,* 2010; *Kainulainen et al.,* 2013). The relation between extinction-based cumulative PDF of column density and the SFR was derived even more explicitly by *Lada et al.* (2010), who found a linear correlation between the SFR and the mass of gas with extinction $A_K > 0.8$ mag.

Recently, far-infrared dust emission maps from Herschel have also been used to derive the PDF of column density in MCs. For example, *Schneider et al.* (2012, 2013) have shown that PDFs from different regions of the Rosette MC deviate from the lognormal, although the PDF tails are not well described by single power laws. Cumulative column density PDFs from Herschel maps are found to agree with cumulative PDFs from near- and mid-infrared extinction maps, and are therefore a powerful probe of the structure of infrared dark clouds (*Kainulainen and Tan,* 2013).

5.3. Star Formation Thresholds in Molecular Clouds

As already mentioned in section 2.5, studies of nearby clouds, based on a variety of methods (near-infrared dust extinction, submillimeter dust emission, or molecular emission lines) have demonstrated the existence of a column density threshold for star formation of approximately 120 M$_\odot$ pc^{-2}, below which star formation appears to be rare. This threshold should not be interpreted as a strict step-function condition for star formation (presumably, the SFE continues to increase at larger Σ_{gas}, up to a maximum equal to ε, the core-to-star efficiency, when the characteristic Σ_{gas} of cores is reached; see section 2.5). The idea of the threshold is to exclude the large mass of the cloud that is currently not forming stars. Understanding how to stop this mass from eventually forming stars lies at the heart of the SFE problem for clouds and galaxies. The correlation between the mass fraction of gas with surface density above the threshold and the SFR/area found by *Heiderman et al.* (2010) and *Lada et al.* (2010) may become even stronger at even larger column densities (*Gutermuth et al.,* 2011). This is suggested, e.g., by the increase in the SFE toward smaller (denser) scales (*Federrath and Klessen,* 2013), with the mass in dense cores being comparable to the mass in YSOs (*Enoch et al.,* 2008; *Jørgensen et al.,* 2008).

It is nevertheless interesting to contrast the observed threshold with the critical volume density of the SFR models, ρ_{crit}. Indeed, *Lada et al.* (2010) have argued that the column density threshold they found could also be interpreted as a volume density threshold of approximately 10^4 cm^{-3}, and the equivalent mean volume density within the column density threshold for the c2d + GB clouds is about 6000 cm^{-3} (*Evans et al.,* 2014). The mass per unit length threshold for star formation found by *André et al.* (2010) in MC filaments probed by Herschel can also be viewed both as a column or volume density threshold, because the filaments they select have a well-defined characteristic thickness, independent of their surface density (*Arzoumanian et al.,* 2011, 2013).

We have demonstrated in section 3.2 that the general expression for the theoretical critical density, equation (11), as well as the equivalent expressions from the KM model, equation (8) and the PN model, equation (9), yield a constant value, $n_{H,crit} = 3.0 \times 10^4$ cm^{-3} (see equation (10)). This is not true for the HC model, where the critical density, given by equation (12), depends on the cloud size (the larger the cloud, the smaller the critical density), even after assuming a standard linewidth-size relation. We conclude that, in the case of the KM and PN models, one would expect observational signatures of a star formation threshold approximately independent of cloud properties, at least within a MC sample following approximately the same velocity-size relation. Based on the HC model, instead, one would expect the threshold to be approximately the same only for MCs of the same size, and to increase toward smaller cloud sizes.

The observed threshold of *Lada et al.* (2010), 10^4 cm^{-3}, is a few times smaller than the above value of $n_{H,crit}$. However, as mentioned above, it is not a strict step-function condition for star formation, and it is to be expected that the correlation between the SFR and the gas mass fraction above the threshold would become increasingly stronger as the observed threshold is increased toward the theoretical value. If the theoretical value were reached, then the SFR per unit mass should be approximately the mass fraction above $n_{H,crit}$ divided by the free-fall time of $n_{H,crit}$, multiplied by the local core-to-star efficiency. In other words, most of the gas denser than $n_{H,crit}$ should be collapsing or about to start to do so.

The prediction for high-mass star-forming cores would probably be different than for local MCs. The cores in the samples of *Plume et al.* (1997) and *Shirley et al.* (2003), for example, have rms velocities well in excess of the Larson relations for nearby clouds. Assuming that their virial parameter is not very different from that of nearby clouds, their larger rms Mach number at a given size (their warmer temperature probably does not compensate the increased linewidth) would imply a larger value of $n_{H,crit}$ than in nearby clouds. On the other hand, if the HC model were correct, the threshold density in these regions should not increase with the Mach number.

6. CONCLUSIONS AND FUTURE DIRECTIONS

6.1. Summary and Conclusions

Progress on theory, simulations, and observations of star formation has been substantial since *Protostars and Planets V* (*Reipurth et al.*, 2007). We have summarized this progress, focusing on the observational and theoretical estimates of the SFR in individual clouds, with some reference to larger-scale star formation on galaxy scales.

Observationally (section 2) we now have extinction maps of nearby clouds, defining cloud structures, surface densities, and masses, with far less uncertainty than was inherent in molecular line maps. Most of the cloud mass lies at low surface densities, well below 100 M$_\odot$ pc^{-2}. The surface densities follow a lognormal PDF at low extinctions (A$_V$ < few mag), but clouds with active star formation show power-law tails. Maps of dust continuum emission show that the denser parts of clouds are highly filamentary,

with dense cores lying along filaments containing only a very small fraction of the cloud mass. We also have quite complete censuses of YSOs in all the larger nearby (d < 500 pc) MCs, allowing robust estimates of SFRs averaged over the half-life of infrared excesses (~2 m.y.). These are quite low compared to what would be expected if clouds were collapsing at the free-fall rate, calculated from the mean cloud density, and making stars with unity efficiency; SFR$_{ff}$ ~ 0.01 for the clouds as a whole, but the scatter is large. If attention is focused on regions above a surface density "threshold" of about 120 M$_\odot$ pc^{-2}, the dispersion in SFR per mass of this "dense" gas is much smaller, essentially consistent with observational uncertainties, and the mean SFR$_{ff}$ doubles to ~0.02. Thus, most of the mass of nearby clouds is quite inactive in star formation. The mean depletion time of the nearby clouds is about 140 m.y.; if only 10% of the mass forms clouds that form stars before dissipating, the galaxy-scale depletion time of about 1–2 G.y. can result from these characteristic values of local efficiency: Star formation may thus be seen as the ultimate outcome of an "inefficient hierarchy."

The fact that star formation is slow compared to free-fall times, and ultimately inefficient (mass depletion times are much greater than structure lifetimes), may be understood from current theories (section 3) as essentially a consequence of an excess of the turbulent kinetic plus magnetic energies over the gravitational energy of a cloud. This understanding has come about as a consequence of numerical efforts that resulted in large parameter studies based on many simulations (section 4). A number of formalisms addressing the problem of estimating the SFR in clouds with magnetized turbulence have been developed. Their predictions have been checked against both idealized simulations (section 4) of such clouds, and against observations (section 5). While formulations differ in detail, the differences have been narrowed down and clarified in recent papers, as discussed in this review.

As indicated by the theoretical models (section 3), and illustrated in practice by numerical simulations (section 4), in addition to the virial ratio there are a number of other factors that also influence the SFR, among them the sonic and Alfvénic Mach numbers, and the relative importance of compressive and incompressive components of the random forces driving the turbulence. These parameters are all included in the current theoretical models for the SFR, and the best theories can successfully predict the SFR within a factor of 2–3 (Fig. 3), over at least 2 orders of magnitude in SFR and for a wide range of parameter space, which is very encouraging given the simplified assumptions in the analytical theories (most importantly, the current assumption of a purely lognormal PDF together with a volume density threshold for star formation). The importance of radiative feedback from early type stars is also broadly appreciated and accepted, as is the possible importance of kinetic feedback from low-mass stars (see the chapter by Krumholz et al. in this volume). These latter aspects are, however, very difficult to incorporate into the analytical theories, and are also nontrivial and computationally costly to include in numerical simulations. We thus expect further developments of both analytical and numerical models addressing these issues in the near future.

6.2. Future Directions

An important future step will be to apply what we have learned here about the SFR in the Milky Way to distant galaxies at low and high redshift, including starburst galaxies. Extragalactic studies focus primarily on Kennicutt-Schmidt-type relations, i.e., measuring the SFR column density, Σ_{SFR}, as a function of gas column density, Σ_{gas} (e.g., *Kennicutt,* 1998b; *Bigiel et al.,* 2008; *Kennicutt and Evans,* 2012), as in Fig. 4.

In a recent theoretical effort with comparison to observations, *Krumholz et al.* (2012a) suggested a unification of star formation in the Milky Way and in distant galaxies at low and high redshift into a universal law where Σ_{SFR} is about 1% of the gas collapse rate, Σ_{gas}/t_{ff}. This model is purely empirical, but fits observations with Σ_{SFR} varying over 5 orders of magnitude quite well. Yet a physical foundation of this empirical model is still missing; it must of course ultimately be a result of the physical laws governing star formation that we have discussed in this review, with the apparent uniformity a result of a corresponding uniformity of the particular combination of physical properties that determine the SFR. The remaining scatter in Σ_{SFR} around the proposed universal law is likely to reflect, in addition to an unavoidable observational scatter, local statistical fluctuations in physical properties.

Adding more observational data, including all local cloud and YSO measurements from *Heiderman et al.* (2010), *Gutermuth et al.* (2011), the central molecular zone (*Yusef-Zadeh et al.,* 2009), and the Small Magellanic Cloud (*Bolatto et al.,* 2011), in addition to the low- and high-redshift disk and starburst galaxies, *Federrath* (2013b) confirms the relatively tight correlation of Σ_{SFR} with Σ_{gas}/t_{ff} (compared to Σ_{gas} only) and also suggests a likely contribution to the scatter in this correlation. *Federrath* (2013b) included simulation data with different forcing of the turbulence, sonic Mach numbers ranging from 5–50, and varying magnetic field strength in the *Krumholz et al.* (2012a) framework, showing that a significant fraction of the scatter might be explained by variations in the properties of the turbulence, in particular by variations in the sonic Mach number of the star-forming clouds. Applying the theoretical models for the local SFR derived and discussed in this chapter, *Federrath* (2013b) shows that they are consistent with the simulation data and with the notion that the scatter seen in the observations can be explained by variations in the parameters of the turbulence. To test this suggestion, it will be necessary to measure the parameters of the turbulence, in particular the virial parameter, the sonic and Alfvén Mach numbers, the driving parameter b, and the star formation efficiency of the star-forming portions of all the objects contributing to the Σ_{SFR}-Σ_{gas}/t_{ff} relation.

Specific tasks for future studies include:

1. Clarify definitions of MCs, both from the point of view of observations alone, and, more importantly, for establishing a better basis for comparing simulations with observations. This will become easier when using larger-scale simulations, where clouds can be selected using essentially the same methods as in observations.

2. Investigate the dense-gas mass fraction of star-forming clouds as a direct test of theoretical predictions (irrespective of a determination of the SFR). This could include a detailed numerical and observational study of the relation between the critical density of SFR theories, and the density where the PDF departs from the lognormal.

3. Compare numerical simulations with observations avoiding the (necessarily highly uncertain) "inversion" of observations into fundamental parameters, and focusing instead on the use of statistical "fingerprints" that can be readily constructed from observations and simulations alike. These may include spatial correlation functions of cores or YSOs, as well as column-density PDFs.

4. Determine how to relate the SFR from counting YSOs in local clouds to indirect methods for more distant clouds and to global averages, using extragalactic methods. In such an effort, which aims at "calibrating" the extragalactic methods, the use of "forward analysis" based on numerical simulations (constructing synthetic observations from simulations) may constitute a useful tool.

5. Extend studies of the SFR vs. gas properties to regions forming massive stars and clusters. These typically are denser and more turbulent and may reveal the importance of parameters like Mach number and turbulent forcing. Because they are more distant, different techniques are needed to study both the cloud structure and the star formation properties. From the modeling perspective they also pose challenges, because chemistry and radiative energy transfer is needed for a quantitative comparison with observations.

6. Explore even larger scales, for a greater variety of SF conditions, to avoid modeling star-forming clouds as isolated systems with artificial boundary and initial conditions, and to study the role of large-scale feedbacks and driving. This will be achieved with higher-resolution extragalactic observations adequately interpreted (*Kruijssen and Longmore,* 2014), and with multiscale star-formation simulations (whole galaxies or galactic fountains) that account for realistic driving forces such as galactic dynamics, SN explosions, stellar radiation, and stellar outflows (*Mac Low and Klessen,* 2004).

Ultimately, the point of view taken in most of this review (as well as in the research papers reviewed), namely to explore and understand how the SFR depends on fundamental physical properties of the ISM, needs to be combined with the complementary view: asking how these fundamental properties of the ISM are, in turn, affected by star formation, and to what extent they are also modulated and affected by other external factors, such as galactic dynamics and density waves.

Given that we already know, from numerical simulations and theoretical modeling consistent with the simulations, that the SFR is a very steep function of the virial parameter, in particular at SFR values as low as those that are observed, one conclusion is practically unavoidable: There must be a strong feedback from the average SFR level back to the physical properties that, in turn, determine the SFR. If one assumed, hypothetically, that the SFR was entirely controlled by external factors, with, e.g., the normalization of the turbulent cascade (the constant factor in Larson's velocity relation) being entirely determined by galactic dynamics, with no influence of the feedback from star formation, that would be tantamount to claiming that the rather universal observed SFR_{ff} of ~1% was a coincidental outcome of these external agents, maintaining a sufficiently uniform virial parameter for the SFR to come out right (or else manipulating other variables as well to serendipitously compensate for virial parameter variations). This appears much less likely than assuming that the feedback from stars closes a "servo loop," which keeps the SFR at a value corresponding to consistency between the integrated feedback (SN explosions, outflows, and radiation) generated by star formation, and the distribution of physical parameters (particularly the virial parameter) that these feedbacks can sustain.

External factors such as galactic dynamics are surely present, but star formation must play an important role for this servo loop to be effective. A possible problem with this idea is that the feedback is much stronger and different in regions forming massive stars, yet the star formation rate per free-fall time is similar. However, the existence of a pervasive inertial range in velocity dispersion, covering a large range of scales, shows that, whatever the ultimate source of the driving, the mechanical energy is distributed rather efficiently in space, thus being able to also power low-mass star formation. The relative uniformity of the SFR may be just another consequence of this. Determining the ratio of contributions from stellar feedback relative to the total driving, and the resulting quasistable state of the feedback loop, via a combination of numerical and theoretical modeling, with the eventual success gauged by the consistency with observations, must indeed be considered as the ultimate goal of research into star formation and the SFR. When the fundamental properties of the ISM, such as the (distribution of) virial parameters, Mach numbers, and driving modes, come out self-consistently from such modeling with essentially no free parameters, then that goal has been reached.

Acknowledgments. We thank R. Klessen, an anonymous referee, P. Hennebelle, A. Kritsuk, and M. Krumholz for reading the manuscript and providing useful comments. P.P. is supported by the FP7-PEOPLE-2010-RG grant PIRG07-GA-2010-261359. Simulations by P.P. were carried out on the NASA/Ames Pleiades supercomputer, and under the PRACE project pra50751 running on SuperMUC at the LRZ (project ID pr86li). C.F. acknowledges support from the Australian Research Council for a Discovery Projects Fellowship (Grant DP110102191). N.J.E. was supported by NSF Grant AST-1109116 to the University of Texas at Austin. The research of C.F.M. is supported in part by NSF grant AST-1211729 and NASA grant NNX13AB84G. D.J. is supported by the National Research Council of Canada and by a Natural Sciences and Engineering Research Council of Canada (NSERC) Discovery Grant. J.K.J. is supported by a Lundbeck Foundation Junior Group Leader Fellowship. Research at Centre for Star and Planet Formation was funded by the Danish National Research Foundation and the University of Copenhagens Programme of Excellence. Supercomputing time at Leibniz Rechenzentrum (PRACE projects pr86li, pr89mu, and project pr32lo) at Forschungszentrum Jülich (project hhd20) and at DeIC/KU in Copenhagen are gratefully acknowledged.

REFERENCES

Agertz O. et al. (2013) *Astrophys. J., 770,* 25.
Aguirre J. E. et al. (2011) *Astrophys. J. Suppl., 192,* 4.
Aluie H. (2011) *Phys. Rev. Lett., 106(17),* 174502.
Aluie H. (2013) *Phys. D Nonlinear Phenom., 247,* 54.
Alves J. et al. (2007) *Astron. Astrophys., 462,* L17.
André P. et al. (2010) *Astron. Astrophys., 518,* L102.
Arzoumanian D. et al. (2011) *Astron. Astrophys., 529,* L6.
Arzoumanian D. et al. (2013) *Astron. Astrophys., 553,* A119.
Ascenso J. et al. (2012) *Astron. Astrophys., 540,* A139.
Ascenso J. et al. (2013) *Astron. Astrophys., 549,* A135.
Ballesteros-Paredes J. et al. (2007) In *Protostars and Planets V* (B. Reipurth et al., eds.), pp. 63–80. Univ. of Arizona, Tucson.
Banerjee R. et al. (2009) *Mon. Not. R. Astron. Soc., 398,* 1082.
Bate M. R. (2009) *Mon. Not. R. Astron. Soc., 392,* 590.
Bate M. R. (2012) *Mon. Not. R. Astron. Soc., 419,* 3115.
Bate M. R. and Bonnell I. A. (2005) *Mon. Not. R. Astron. Soc., 356,* 1201.
Bate M. R. et al. (1995) *Mon. Not. R. Astron. Soc., 277,* 362.
Bertoldi F. and McKee C. F. (1992) *Astrophys. J., 395,* 140.
Bertram E. et al. (2012) *Mon. Not. R. Astron. Soc., 420,* 3163.
Beuther H. et al. (2002) *Astron. Astrophys., 383,* 892.
Bigiel F. et al. (2008) *Astron. J., 136,* 2846.
Bigiel F. et al. (2010) *Astron. J., 140,* 1194.
Blaisdell G. A. et al. (1993) *J. Fluid Mech., 256,* 443.
Bolatto A. D. et al. (2011) *Astrophys. J., 741,* 12.
Bonnell I. A. et al. (2011) *Mon. Not. R. Astron. Soc., 410,* 2339.
Boss A. P. and Black D. C. (1982) *Astrophys. J., 258,* 270.
Bovino S. et al. (2013) *New J. Phys., 15(1),* 013055.
Brunt C. M. (2010) *Astron. Astrophys., 513,* A67.
Brunt C. M. et al. (2009) *Astron. Astrophys., 504,* 883.
Brunt C. M. et al. (2010a) *Mon. Not. R. Astron. Soc., 405,* L56.
Brunt C. M. et al. (2010b) *Mon. Not. R. Astron. Soc., 403,* 1507.
Burkhart B. and Lazarian A. (2012) *Astrophys. J. Lett., 755,* L19.
Burkhart B. et al. (2010) *Astrophys. J., 708,* 1204.
Burkhart B. et al. (2013) *Astrophys. J., 771,* 122.
Cambrésy L. (1999) *Astron. Astrophys., 345,* 965.
Carpenter J. M. (2000) *Astron. J., 120,* 3139.
Chabrier G. (2003) *Publ. Astron. Soc. Pac., 115,* 763.
Chapman N. L. et al. (2009) *Astrophys. J., 690,* 496.
Cho W. and Kim J. (2011) *Mon. Not. R. Astron. Soc., 410,* L8.
Chomiuk L. and Povich M. S. (2011) *Astron. J., 142,* 197.
Clark P. C. et al. (2005) *Mon. Not. R. Astron. Soc., 359,* 809.
Clark P. C. et al. (2008) *Mon. Not. R. Astron. Soc., 386,* 3.
Collins D. C. et al. (2011) *Astrophys. J., 731,* 59.
Collins D. C. et al. (2012) *Astrophys. J., 750,* 13.
Commerçon B. et al. (2011) *Astrophys. J. Lett., 742,* L9.
Contreras Y. et al. (2013) *Astron. Astrophys., 549,* A45.
Covey K. R. et al. (2010) *Astrophys. J., 722,* 971.
Cunningham A. J. et al. (2011) *Astrophys. J., 740,* 107.
Daddi E. et al. (2010) *Astrophys. J. Lett., 714,* L118.
Dale J. E. et al. (2012a) *Mon. Not. R. Astron. Soc., 427,* 2852.
Dale J. E. et al. (2012b) *Mon. Not. R. Astron. Soc., 424,* 377.
Dale J. E. et al. (2013a) *Mon. Not. R. Astron. Soc., 431,* 1062.
Dale J. E. et al. (2013b) *Mon. Not. R. Astron. Soc., 430,* 234.
Dib S. and Burkert A. (2005) *Astrophys. J., 630,* 238.
di Francesco J. et al. (2006) In *Revealing the Molecular Universe: One Antenna is Never Enough* (D. C. Backer et al., eds.), p. 275. ASP Conf. Series 356, Astronomical Society of the Pacific, San Francisco.
di Francesco J. et al. (2007) In *Protostars and Planets V* (B. Reipurth et al., eds.), pp. 17–32. Univ. of Arizona, Tucson.

Dobashi K. et al. (2005) *Publ. Astron. Soc. Japan, 57,* 1.
Dobbs C. L. et al. (2006) *Mon. Not. R. Astron. Soc., 371,* 1663.
Draine B. T. (2003) *Annu. Rev. Astron. Astrophys., 41,* 241.
Dubey A. et al. (2008) In *Numerical Modeling of Space Plasma Flows* (N. V. Pogorelov et al., eds.), p. 145. ASP Conf. Series 385, Astronomical Society of the Pacific, San Francisco.
Dunham M. K. et al. (2011) *Astrophys. J., 741,* 110.
Dunham M. M. et al. (2013) *Astron. J., 145,* 94.
Elmegreen B. G. and Scalo J. (2004) *Annu. Rev. Astron. Astrophys., 42,* 211.
Enoch M. L. et al. (2006) *Astrophys. J., 638,* 293.
Enoch M. L. et al. (2007) *Astrophys. J., 666,* 982.
Enoch M. L. et al. (2008) *Astrophys. J., 684,* 1240.
Evans N. J. II et al. (2009) *Astrophys. J. Suppl., 181,* 321.
Evans N. J. II et al. (2014) *Astrophys. J., 782,* 114.
Falkovich G. et al. (2010) *J. Fluid Mech., 644,* 465.
Federrath C. (2013a) *Mon. Not. R. Astron. Soc., 436,* 1245.
Federrath C. (2013b) *Mon. Not. R. Astron. Soc., 436,* 3167.
Federrath C. and Klessen R. S. (2012) *Astrophys. J., 761,* 156.
Federrath C. and Klessen R. S. (2013) *Astrophys. J., 763,* 51.
Federrath C. et al. (2008a) *Astrophys. J. Lett., 688,* L79.
Federrath C. et al. (2008b) *Phys. Scrip. T, 132(1),* 014025.
Federrath C. et al. (2009) *Astrophys. J., 692,* 364.
Federrath C. et al. (2010a) *Astron. Astrophys., 512,* A81.
Federrath C. et al. (2010b) *Astrophys. J., 713,* 269.
Federrath C. et al. (2011a) *Astrophys. J., 731,* 62.
Federrath C. et al. (2011b) *Phys. Rev. Lett., 107(11),* 114504.
Froebrich D. and Rowles J. (2010) *Mon. Not. R. Astron. Soc., 406,* 1350.
Fryxell B. et al. (2000) *Astrophys. J. Suppl., 131,* 273.
Galtier S. and Banerjee S. (2011) *Phys. Rev. Lett., 107(13),* 134501.
Gao Y. and Solomon P. M. (2004) *Astrophys. J., 606,* 271.
Genzel R. et al. (2010) *Mon. Not. R. Astron. Soc., 407,* 2091.
Ginsburg A. et al. (2013) *Astrophys. J. Suppl., 208,* 14.
Girichidis P. et al. (2011) *Mon. Not. R. Astron. Soc., 413,* 2741.
Girichidis P. et al. (2014) *Astrophys. J., 781,* 91.
Glover S. C. O. and Clark P. C. (2012) *Mon. Not. R. Astron. Soc., 426,* 377.
Glover S. C. O. and Mac Low M.-M. (2007) *Astrophys. J., 659,* 1317.
Glover S. C. O. et al. (2010) *Mon. Not. R. Astron. Soc., 404,* 2.
Goldsmith P. F. et al. (2008) *Astrophys. J., 680,* 428.
Goodman A. A. et al. (2009) *Astrophys. J., 692,* 91.
Gutermuth R. A. et al. (2008) *Astrophys. J. Lett., 673,* L151.
Gutermuth R. A. et al. (2009) *Astrophys. J. Suppl., 184,* 18.
Gutermuth R. A. et al. (2011) *Astrophys. J., 739,* 84.
Harvey P. et al. (2007) *Astrophys. J., 663,* 1149.
Hatchell J. and Fuller G. A. (2008) *Astron. Astrophys., 482,* 855.
Heiderman A. et al. (2010) *Astrophys. J., 723,* 1019.
Heitsch F. et al. (2001) *Astrophys. J., 547,* 280.
Heitsch F. et al. (2005) *Astrophys. J. Lett., 633,* L113.
Hennebelle P. and Chabrier G. (2008) *Astrophys. J., 684,* 395.
Hennebelle P. and Chabrier G. (2011) *Astrophys. J. Lett., 743,* L29.
Hennebelle P. and Chabrier G. (2013) *Astrophys. J., 770,* 150.
Hennebelle P. et al. (2008) *Astron. Astrophys., 486,* L43.
Heyer M. H. and Brunt C. M. (2004) *Astrophys. J. Lett., 615,* L45.
Heyer M. et al. (2009) *Astrophys. J., 699,* 1092.
Hopkins P. F. (2012) *Mon. Not. R. Astron. Soc., 423,* 2016.
Hopkins P. F. (2013) *Mon. Not. R. Astron. Soc., 430,* 1653.
Hsieh T.-H. and Lai S.-P. (2013) *Astrophys. J. Suppl., 205,* 5.
Jappsen A.-K. et al. (2005) *Astron. Astrophys., 435,* 611.
Johnstone D. and Bally J. (1999) *Astrophys. J. Lett., 510,* L49.
Johnstone D. and Bally J. (2006) *Astrophys. J., 653,* 383.
Johnstone D. et al. (2004) *Astrophys. J. Lett., 611,* L45.
Johnstone D. et al. (2005) In *Protostars and Planets V,* Abstract #8485, Lunar and Planetary Institute, Houston.
Jørgensen J. K. et al. (2008) *Astrophys. J., 683,* 822.
Kainulainen J. and Tan J. C. (2013) *Astron. Astrophys., 549,* A53.
Kainulainen J. et al. (2009) *Astron. Astrophys., 508,* L35.
Kainulainen J. et al. (2013) *Astron. Astrophys., 553,* L8.
Kennicutt R. C. and Evans N. J. (2012) *Annu. Rev. Astron. Astrophys., 50,* 531.
Kennicutt R. C. Jr. (1998a) *Annu. Rev. Astron. Astrophys., 36,* 189.
Kennicutt R. C. Jr. (1998b) *Astrophys. J., 498,* 541.
Kerton C. R. et al. (2001) *Astrophys. J., 552,* 601.
Kirk H. et al. (2006) *Astrophys. J., 646,* 1009.
Kirk J. M. et al. (2013) *Mon. Not. R. Astron. Soc., 432,* 1424.

Kitsionas S. et al. (2009) *Astron. Astrophys., 508,* 541.
Klein R. I. et al. (2007) In *Protostars and Planets V* (B. Reipurth et al., eds.), pp. 99–116. Univ. of Arizona, Tucson.
Kleinmann S. G. (1992) In *Robotic Telescopes in the 1990s* (A. V. Filippenko, ed.), pp. 203–212. ASP Conf. Series 34, Astronomical Society of the Pacific, San Francisco.
Klessen R. S. (2000) *Astrophys. J., 535,* 869.
Klessen R. S. and Hennebelle P. (2010) *Astron. Astrophys., 520,* A17.
Klessen R. S. et al. (2000) *Astrophys. J., 535,* 887.
Konstandin L. et al. (2012) *Astrophys. J., 761,* 149.
Kritsuk A. G. et al. (2007) *Astrophys. J., 665,* 416.
Kritsuk A. G. et al. (2011a) *Astrophys. J., 737,* 13.
Kritsuk A. G. et al. (2011b) *Astrophys. J. Lett., 727,* L20.
Kritsuk A. G. et al. (2013) *Mon. Not. R. Astron. Soc., 436,* 3247–3261.
Kruijssen J. M. D. and Longmore S. N. (2014) *Mon. Not. R. Astron. Soc., 439,* 3239.
Krumholz M. R. and McKee C. F. (2005) *Astrophys. J., 630,* 250.
Krumholz M. R. and Tan J. C. (2007) *Astrophys. J., 654,* 304.
Krumholz M. R. et al. (2004) *Astrophys. J., 611,* 399.
Krumholz M. R. et al. (2012a) *Astrophys. J., 745,* 69.
Krumholz M. R. et al. (2012b) *Astrophys. J., 754,* 71.
Kryukova E. et al. (2012) *Astron. J., 144,* 31.
Kuiper R. et al. (2011) *Astrophys. J., 732,* 20.
Lada C. J. and Lada E. A. (2003) *Annu. Rev. Astron. Astrophys., 41,* 57.
Lada C. J. et al. (1994) *Astrophys. J., 429,* 694.
Lada C. J. et al. (2010) *Astrophys. J., 724,* 687.
Lada C. J. et al. (2012) *Astrophys. J., 745,* 190.
Lada C. et al. (2013) *Astrophys. J., 778,* 133.
Lemaster M. N. and Stone J. M. (2008) *Astrophys. J. Lett., 682,* L97.
Lombardi M. and Alves J. (2001) *Astron. Astrophys., 377,* 1023.
Lombardi M. et al. (2010) *Astron. Astrophys., 519,* L7.
Lunttila T. et al. (2008) *Astrophys. J., 686,* L91.
Lunttila T. et al. (2009) *Astrophys. J. Lett., 702,* L37.
Mac Low M.-M. (1999) *Astrophys. J., 524,* 169.
Mac Low M.-M. and Klessen R. S. (2004) *Rev. Mod. Phys., 76,* 125.
Mac Low M.-M. et al. (1998) *Phys. Rev. Lett., 80,* 2754.
Mamajek E. E. (2009) In *Exoplanets and Disks: Their Formation and Diversity* (T. Usuda et al., eds.), pp. 3–10. AIP Conf. Proc. 1158, American Institute of Physics, College Park, Maryland. Matzner C. D. and McKee C. F. (2000) *Astrophys. J., 545,* 364.
Maury A. J. et al. (2011) *Astron. Astrophys., 535,* A77.
McKee C. F. (1989) *Astrophys. J., 345,* 782.
McKee C. F. and Ostriker E. C. (2007) *Annu. Rev. Astron. Astrophys., 45,* 565.
McKee C. F. and Tan J. C. (2003) *Astrophys. J., 585,* 850.
Megeath S. T. et al. (2012) *Astron. J., 144,* 192.
Micic M. et al. (2012) *Mon. Not. R. Astron. Soc., 421,* 2531.
Molina F. Z. et al. (2012) *Mon. Not. R. Astron. Soc., 423,* 2680.
Muñoz D. J. et al. (2007) *Astrophys. J., 668,* 906.
Murray N. (2011) *Astrophys. J., 729,* 133.
Myers A. T. et al. (2013) *Astrophys. J., 766,* 97.
Nakamura F. and Li Z.-Y. (2008) *Astrophys. J., 687,* 354.
Nordlund Å. K. and Padoan P. (1999) In *Interstellar Turbulence* (J. Franco and A. Carraminana, ed.), p. 218. Cambridge Univ., Cambridge.
Nordlund Å. et al. (2013) In *Exploring the Formation and Evolution of Planetary Systems* (B. Matthews and J. Graham, eds.), IAU Symposium 299, Cambridge Univ., Cambridge.
Offner S. S. R. et al. (2009) *Astrophys. J., 703,* 131.
Oliveira I. et al. (2009) *Astrophys. J., 691,* 672.
Onishi T. et al. (1998) *Astrophys. J., 502,* 296.
Ossenkopf V. and Mac Low M.-M. (2002) *Astron. Astrophys., 390,* 307.
Ostriker E. C. et al. (1999) *Astrophys. J., 513,* 259.
Ostriker E. C. et al. (2001) *Astrophys. J., 546,* 980.
Padoan P. (1995) *Mon. Not. R. Astron. Soc., 277,* 377.
Padoan P. and Nordlund A. (1997) *ArXiv e-prints,* arXiv:astro-ph/9706176.
Padoan P. and Nordlund Å. (1999) *Astrophys. J., 526,* 279.
Padoan P. and Nordlund Å. (2011) *Astrophys. J., 730,* 40.
Padoan P. et al. (1997a) *Astrophys. J., 474,* 730.
Padoan P. et al. (1997b) *Mon. Not. R. Astron. Soc., 288,* 145.
Padoan P. et al. (1998) *Astrophys. J., 504,* 300.
Padoan P. et al. (1999) *Astrophys. J., 525,* 318.
Padoan P. et al. (2004) *Phys. Rev. Lett., 92(19),* 191102.
Padoan P. et al. (2005) *Astrophys. J. Lett., 622,* L61.
Padoan P. et al. (2006) *Astrophys. J. Lett., 653,* L125.

Padoan P. et al. (2009) *Astrophys. J. Lett., 707,* L153.
Padoan P. et al. (2012) *Astrophys. J. Lett., 759,* L27.
Passot T. and Vázquez-Semadeni E. (1998) *Phys. Rev. E, 58,* 4501.
Peters T. et al. (2010) *Astrophys. J., 711,* 1017.
Pilbratt G. L. et al. (2010) *Astron. Astrophys., 518,* L1.
Pineda J. L. et al. (2010) *Astrophys. J., 721,* 686.
Plume R. et al. (1997) *Astrophys. J., 476,* 730.
Price D. J. and Federrath C. (2010) *Mon. Not. R. Astron. Soc., 406,* 1659.
Price D. J. et al. (2011) *Astrophys. J. Lett., 727,* L21.
Price D. J. et al. (2012) *Mon. Not. R. Astron. Soc., 423,* L45.
Rebull L. M. et al. (2010) *Astrophys. J. Suppl., 186,* 259.
Reipurth B. et al., eds. (2007) *Protostars and Planets V.* Univ. of Arizona, Tucson.
Renaud F. et al. (2012) *Astrophys. J. Lett., 760,* L16.
Ridge N. A. et al. (2006) *Astron. J., 131,* 2921.
Roman-Duval J. et al. (2010) *Astrophys. J., 723,* 492.
Roman-Duval J. et al. (2011) *Astrophys. J., 740,* 120.
Rosolowsky E. et al. (2010) *Astrophys. J. Suppl., 188,* 123.
Sadavoy S. I. et al. (2010) *Astrophys. J., 710,* 1247.
Sadavoy S. I. et al. (2012) *Astron. Astrophys., 540,* A10.
Sadavoy S. I. et al. (2013) *Astrophys. J., 767,* 126.
Scalo J. (1990) In *Physical Processes in Fragmentation and Star Formation* (R. Capuzzo-Dolcetta et al., eds.), pp. 151–176. Astrophys. Space Sci. Library, Vol. 162, Kluwer, Dordrecht.
Scalo J. and Elmegreen B. G. (2004) *Annu. Rev. Astron. Astrophys., 42,* 275.
Scalo J. et al. (1998) *Astrophys. J., 504,* 835.
Schleicher D. R. G. et al. (2013) *New J. Phys., 15(2),* 023017.
Schmidt M. (1959) *Astrophys. J., 129,* 243.
Schmidt M. (1963) *Astrophys. J., 137,* 758.
Schmidt W. et al. (2009) *Astron. Astrophys., 494,* 127.
Schmidt W. et al. (2013) *Mon. Not. R. Astron. Soc., 431,* 3196.
Schneider N. et al. (2012) *Astron. Astrophys., 540,* L11.
Schneider N. et al. (2013) *Astrophys. J. Lett., 766,* L17.
Schober J. et al. (2012) *Phys. Rev. E, 85(2),* 026303.
Schuller F. et al. (2009) *Astron. Astrophys., 504,* 415.
Seifried D. et al. (2012) *Mon. Not. R. Astron. Soc., 422,* 347.
Shirley Y. L. et al. (2003) *Astrophys. J. Suppl., 149,* 375.
Shu F. H. et al. (1987) *Annu. Rev. Astron. Astrophys., 25,* 23.
Solomon P. M. et al. (1987) *Astrophys. J., 319,* 730.
Spezzi L. et al. (2008) *Astrophys. J., 680,* 1295.
Spezzi L. et al. (2010) *Astron. Astrophys., 513,* A38.
Stone J. M. et al. (1998) *Astrophys. J. Lett., 508,* L99.
Stutz A. M. et al. (2013) *Astrophys. J., 767,* 36.
Tasker E. J. and Bryan G. L. (2006) *Astrophys. J., 641,* 878.
Tassis K. et al. (2010) *Mon. Not. R. Astron. Soc., 408,* 1089.
Truelove J. K. et al. (1997) *Astrophys. J. Lett., 489,* L179.
Tsitali A. E. et al. (2010) *Astrophys. J., 725,* 2461.
Vázquez-Semadeni E. (1994) *Astrophys. J., 423,* 681.
Vázquez-Semadeni E. et al. (2003) *Astrophys. J. Lett., 585,* L131.
Vázquez-Semadeni E. et al. (2006) *Astrophys. J., 643,* 245.
Vutisalchavakul N. and Evans II N. J. (2013) *Astrophys. J., 765,* 129.
Wang P. et al. (2010) *Astrophys. J., 709,* 27.
Ward-Thompson D. et al. (2007a) In *Protostars and Planets V* (B. Reipurth et al., eds.), pp. 33–46. Univ. of Arizona, Tucson.
Ward-Thompson D. et al. (2007b) *Publ. Astron. Soc. Pac., 119,* 855.
Werner M. W. et al. (2004) *Astrophys. J. Suppl., 154,* 1.
Whitworth A. and Summers D. (1985) *Mon. Not. R. Astron. Soc., 214,* 1.
Williams J. P. and McKee C. F. (1997) *Astrophys. J., 476,* 166.
Williams J. P. et al. (2000) In *Protostars and Planets IV* (V. Mannings et al., eds.), p. 97. Univ. of Arizona, Tucson.
Wu J. et al. (2005) *Astrophys. J. Lett., 635,* L173.
Wu J. et al. (2007) *Astron. J., 133,* 1560.
Wu J. et al. (2010) *Astrophys. J. Suppl., 188,* 313.
Young K. E. et al. (2006) *Astrophys. J., 644,* 326.
Yusef-Zadeh F. et al. (2009) *Astrophys. J., 702,* 178.
Zuckerman B. and Palmer P. (1974) *Annu. Rev. Astron. Astrophys., 12,* 279.

Li H.-B., Goodman A., Sridharan T. K., Houde M., Li Z.-Y., Novak G., and Tang K. S. (2014) The link between magnetic fields and cloud/star formation. In *Protostars and Planets VI* (H. Beuther et al., eds.), pp. 101–123. Univ. of Arizona, Tucson, DOI: 10.2458/azu_uapress_9780816531240-ch005.

The Link Between Magnetic Fields and Cloud/Star Formation

Hua-bai Li
Max Planck Institute for Astronomy and The Chinese University of Hong Kong

Alyssa Goodman and T. K. Sridharan
Harvard-Smithsonian Center for Astrophysics

Martin Houde
University of Western Ontario and California Institute of Technology

Zhi-Yun Li
University of Virginia

Giles Novak
Northwestern University

Kwok Sun Tang
The Chinese University of Hong Kong

The question of whether magnetic fields play an important role in the processes of molecular cloud and star formation has been debated for decades. Recent observations have revealed a simple picture that may help illuminate these questions: Magnetic fields have a tendency to preserve their orientation at all scales that have been probed — from 100-pc-scale intercloud media down to subparsec-scale cloud cores. This ordered morphology has implications for the way in which self-gravity and turbulence interact with magnetic fields: Both gravitational contraction and turbulent velocities should be anisotropic due to the influence of dynamically important magnetic fields. Such anisotropy is now observed. Here we review these recent observations and discuss how they can improve our understanding of cloud/star formation.

1. INTRODUCTION

How molecular clouds form and then fragment into filaments, cores, protostellar disks, and finally stars is still mysterious. The existence of large-scale magnetic fields (B fields) in the interstellar medium (ISM) was first shown by *Hiltner* (1951) and *Hall* (1951) nearly 65 years ago. On the cloud-formation scale (10^3–10^2 pc), B-field strength is expected to determine whether a cloud can rotate with the angular momentum inherited from galactic shear and/or turbulence. The fragmentation of a cloud (e.g., the formation of filamentary structures; see the chapter by André et al. in this volume) also strongly depends on the cloud B-field strength. For protostellar disk formation, the debate is centered on whether magnetic braking can prevent the formation of large rotationally supported disks at early times (see the chapter by Z.-Y. Li et al. in this volume). At almost all stages of cloud/star formation, the roles played by B fields remain highly controversial, in large part as a result of the lack of observational constraints.

This situation is gradually improving, owing to data obtained in some decade-long surveys of B-field strength [Zeeman measurements (e.g., *Troland and Crutcher*, 2008)] or morphologies [submillimeter polarimetry from, e.g., *Dotson et al.* (2010) and *Matthews et al.* (2009); infrared (IR) polarimetry from, e.g., *Clemens et al.* (2012)] of molecular clouds having been released recently. In this paper, we review a series of relatively recent surveys (23 surveys are selected and divided into 14 categories) (Fig. 1) that lead us to a picture in which magnetic fields play a variety of roles at different points during the star-formation process. A significant portion of this chapter is devoted to the magnetic topology problem raised in *Protostars and Planets III* (*McKee et al.,* 1993): How does the B-field topology evolve as molecular clouds form out of the ISM and as cores within the cloud contract to form stars? After 20 years, recent surveys have finally shed some light on this problem.

We have organized the following discussion by the scales involved in each process: cloud formation, filament

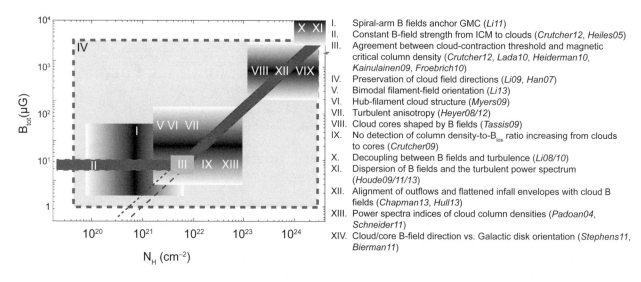

I.   Spiral-arm B fields anchor GMC (*Li11*)
II.   Constant B-field strength from ICM to clouds (*Crutcher12, Heiles05*)
III. Agreement between cloud-contraction threshold and magnetic critical column density (*Crutcher12, Lada10, Heiderman10, Kainulainen09, Froebrich10*)
IV.   Preservation of cloud field directions (*Li09, Han07*)
V.   Bimodal filament-field orientation (*Li13*)
VI.   Hub-filament cloud structure (*Myers09*)
VII. Turbulent anisotropy (*Heyer08/12*)
VIII. Cloud cores shaped by B fields (*Tassis09*)
IX.   No detection of column density-to-B_{los} ratio increasing from clouds to cores (*Crutcher09*)
X.   Decoupling between B fields and turbulence (*Li08/10*)
XI.   Dispersion of B fields and the turbulent power spectrum (*Houde09/11/13*)
XII. Alignment of outflows and flattened infall envelopes with cloud B fields (*Chapman13, Hull13*)
XIII. Power spectra indices of cloud column densities (*Padoan04, Schneider11*)
XIV. Cloud/core B-field direction vs. Galactic disk orientation (*Stephens11, Bierman11*)

Fig. 1. Twenty-three recent surveys (noted by the leading authors and years) are discussed here in 14 categories. The rectangles in the plot show the limits of the typical B-field strengths and column densities the observations probed.

formation, core formation, and protostellar disk formation are discussed in sections 2–5 respectively. In section 6, we synthesize the results from various observations and try to explain their discrepancy.

2. MOLECULAR CLOUD FORMATION

The formation of molecular clouds is poorly understood. While some cloud formation models suggest that a large-scale galactic magnetic field is irrelevant at the scale of molecular clouds (e.g., *Dobbs,* 2008), because the turbulence and rotation of a cloud may randomize the orientation of its B field (Fig. 2b), other models (e.g., *Shetty and Ostriker,* 2006) have envisioned a galactic B field strong enough to channel cloud accumulation and fragmentation (Fig. 2a). Recent observations have shed light on this debate.

2.1. Observation I: Spiral-Arm B Fields Anchor Giant Molecular Clouds

A measurement of the field direction in individual clouds and comparison to the spiral arms should determine which model is correct, but this is difficult to perform in the Milky Way because the arms cannot be observed due to the edge-on view of the Galactic disk. Furthermore, state-of-the-art instruments are not sufficiently efficient to probe the cloud B fields from a face-on galaxy with the conventional cloud B-field tracer: polarization of dust thermal emission. Thus *Li and Henning* (2011) tried to probe cloud fields in M33 with the polarization of CO-line emission (due to the Goldreich-Kylafis effect), which is much stronger than thermal dust emission. One problem with the Goldreich-Kylafis effect (*Goldreich and Kylafis,* 1981; *Cortes et al.,* 2005) is that a line polarization can be either perpendicular or parallel to the local B-field direction projected on the sky. Their argument that the 90° ambiguity can still be statistically

useful is as follows: An intrinsically flat field distribution, as happens when the turbulence is super-Alfvénic or the cloud is rotating, will still be flat with this ambiguity. On the other hand, an intrinsically single-peaked Gaussian-like field distribution, as happens when the turbulence is sub-Alfvénic, will either stay single-peaked or split into two peaks approximately 90° apart.

M33 is the nearest face-on galaxy with pronounced optical spiral arms. The subcompact configuration of the Submillimeter Array (SMA) offers a linear spatial resolution of ~15 parsec at 230 GHz (the frequency of the CO J=2–1 transition) at the distance of M33 (900 kpc). *Li and Henning* (2011) picked the six most massive clouds from M33 for their strong CO-line emission. The distribution of the offsets between the CO polarization and the local arm directions clearly shows double peaks (Fig. 3). The distribution can be fitted by a double-Gaussian function with the two peaks lying at −1.9° ± 4.7° and 91.1° ± 3.7° and a standard deviation of 20.7° ± 2.6°. This indicates that the mean field directions are well defined and highly correlated with the spiral arms, which is consistent with the scenario that galactic B fields can exert tension forces strong enough to resist cloud rotation (Fig. 2a).

2.2. Observation II: Constant B-Field Strength from Intercloud Media to Clouds

In a recent review of molecular cloud magnetic field measurements, *Crutcher* (2012) summarized the Zeeman measurements from the past decade in a B_{los} (line-of-sight field strength)-vs.-N_H (H column density) plot. We show this plot in Fig. 4 and highlight the column density ranges of intercloud media (ICM) traced by HI data. Based on a Bayesian analysis, *Crutcher et al.* (2010b) concluded that the two most probable scenarios for ICM B-field strength are (1) constant strength around 10 µG and (2) any strength

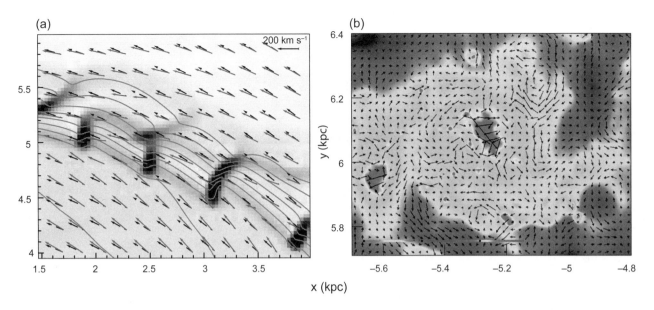

Fig. 2. Two competing scenarios of cloud formation. **(a)** A patch from a global galaxy simulation (*Shetty and Ostriker,* 2006). The solid vectors show the instantaneous gas velocity in the frame rotating with the spiral potential. The dotted vectors show the initial velocities (pure circular motion). The solid blue lines show B-field orientations. The grayscale stands for the relative surface density. The B fields of the spiral arm are only slightly twisted in the molecular cloud complexes (dark elongated regions), and in turn the field tension is strong enough to hinder the cloud rotation. **(b)** A similar simulation (*Dobbs et al.,* 2008) but the well-developed cloud rotation has produced tidal tails extending from the GMC, and the B fields (vectors) follow the rotation and have lost the "memory" of the galactic field direction. A color version of this figure is available at *http://arxiv.org/pdf/1404.2024v1.pdf*.

between 0 and 10 μG with a median of 6 μG. Most importantly, the B-field strength remains relatively constant in both cases over column densities from the ICM to the lower-density regions in clouds. Independent analysis from *Heiles and Troland* (2005) on the same data shown in the ICM zone of Fig. 4 also concluded that there was a B-field strength median of ~6 μG, independent of column density. This is interpreted as evidence that gas can only accumulate along the B fields during cloud formation (*Crutcher et al.,* 2010b; *Crutcher,* 2012), again consistent with Fig. 2a.

Lazarian et al. (2012) suggested an alternative scenario: The B fields were compressed during the cloud-formation process, but the turbulence-enhanced "reconnection diffusion" of B fields is efficient enough to keep the field strength independent of density (however, see the discussion of Observation III in section 2.3).

2.3. Observation III: Agreement Between Cloud-Contraction Threshold and Magnetic Critical Column Density

Both Observations I and II support the strong-B-field scenario (Fig. 2a) that cloud masses are first accumulated along B fields. Only after a cloud has accumulated enough gas to reach the magnetic critical density can gravitational contraction happen in all directions. In Fig. 4, for roughly $N_H > 10^{21}$ cm^{-2}, the B-field strength increases with N_H, which implies that self-gravity is able to compress the field lines after accumulating adequate mass along the fields. The

slanted solid line in Fig. 4, B (μG) = 3.8 × 10^{-21} N_H (cm^{-2}) represents the so-called magnetic critical condition (*Crutcher,* 2012; *Nakano and Nakamura,* 1978): Regions above the line can, in principle, be supported against their self-gravity by the magnetic forces alone. Since the cloud mass is accumulated along the B fields, the cloud shape should be more sheet-like (e.g., *Shetty and Ostriker,* 2006) instead of spherical, so statistically the observed column density should be twice the value to calculate the criticality (the column density observed with a sight line aligned with the B field) due to projection effects (*Shu et al.,* 1999). Taking the projection effects into account, we add to Fig. 4 the corrected magnetic critical condition

$$B(\mu G) = 1.9 \times 10^{-21} N_{H,crit} (cm^{-2}) \qquad (1)$$

Assuming an equipartition between turbulent and magnetic energies (upper limit of turbulent energy of sub-Alfvénic clouds), and magnetic virial equilibrium, 2T + M + U = 0 [where T, M, and U are, respectively, kinetic, magnetic and gravitational potential energies (*McKee et al.,* 1993)], the critical condition becomes

$$B(\mu G) = 1.1 \times 10^{-21} N_{H,crit} (cm^{-2}) \qquad (2)$$

Equations (1) and (2) give the lower and upper limits of the critical column density. For B = 6–10 μG, the critical column density range is $N_H = [3.2, 9.1] \times 10^{21}$ cm^{-2} (Fig. 4).

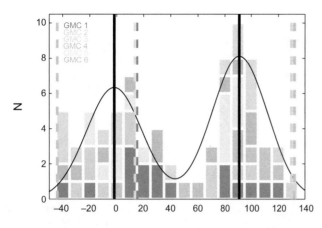

Offset of CO Polarization from Local Arm Direction (°)

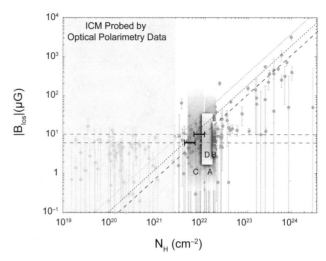

N_H (cm^{-2})

Fig. 3. Distribution of the CO polarization-spiral arm offsets (*Li and Henning*, 2011). The offsets are from the difference between the orientations of CO polarization and local arms in M33. Contributions from different GMCs are distinguished by the colors. The distribution can be fitted by a double-Gaussian function with a separation of ~90°, suggesting that the cloud B-field directions are correlated with the arm directions. The dashed lines show directions of synchrotron polarization, which are not relevant to the discussion here. A color version of this figure is available at *http://arxiv.org/pdf/1404.2024v1.pdf.*

Fig. 4. The agreement between the magnetic critical density and gravitational contraction threshold. On top of the plot of line-of-sight field strength (B_{los}) vs. column density (N_H) (*Crutcher*, 2012), we indicate the intercloud media (ICM) by the light shaded zone, which is also the typical column density traced by optical polarimetry data (Observations IV and V). The field strength is quite constant from the ICM to lower-density regions of the clouds (*Crutcher et al.*, 2010b). The two horizontal lines mark 10 µG and 6 µG, respectively (*Crutcher et al.*, 2010b), and the slanted solid line is the theoretical magnetic critical condition from *Crutcher* (2012). Applying a projection-effect correction (*Shu et al.*, 1999) to it, we obtain the short-dashed line. Assuming an equipartition condition between magnetic and turbulent energies, we obtain the upper limit of the critical condition (slanted long-dashed line). The two "H"-shaped symbols mark the range of possible critical column densities for B = 10 and 6 µG respectively. The emprical star formation thresholds, A_V = 7.3 ± 1.8 mag from *Lada et al.* (2010) and A_V = 8.1 ± 0.9 mag from *Heiderman et al.* (2010), are shown by the zones labled A and B. Zones C (*Kainulainen et al.*, 2009) and D (*Froebrich and Rowles*, 2010) show where the observed column density PDF turns from log-normal to power-law-like. A color version of this figure is available at *http://arxiv.org/pdf/1404.2024v1.pdf.*

An indicator of gravitational contraction is the shape of a probability density function (PDF) of cloud column densities. Numerical simulations show that this PDF of nongravitating clouds is log-normal for both sub- and super-Alfvénic clouds (e.g., *Li P.-S. et al.*, 2008; *Collins et al.*, 2012). When self-gravity is also accounted for, the PDF of high-density regions, where gravitational energy dominates other forms of energy, deviates from the log-normal function that describes the low-density PDF (e.g., *Nordlund and Padoan*, 1999). This log-normal type PDF and the deviation are indeed observed (*Kainulainen et al.*, 2009; *Froebrich and Rowles*, 2010) and the transition point, cloud-contraction threshold (N_H = [3.7, 9.4] × 10^{21} cm^{-2}), is comparable to the critical column density (N_H = [3.2, 9.1] × 10^{21} cm^{-2}) (*Li et al.*, 2013). We note that the empirical star-formation threshold, N_H = [1.0, 1.7] × 10^{22} cm^{-2} (*Heiderman et al.*, 2010; *Lada et al.*, 2010), is a bit larger than the cloud-contraction threshold (Fig. 4).

3. FILAMENT FORMATION

Recent observational studies, especially those exploiting the Herschel data (*André et al.*, 2010; *Molinari et al.*, 2010; *Henning et al.*, 2010; *Hill et al.*, 2011; *Ragan et al.*, 2012), have supported the decades-old "bead string" scenario of star formation (*Schneider and Elmegreen*, 1979; *Mizuno et al.*, 1995; *Nagahama et al.*, 1998). In this picture, molecular clouds first form filaments (parsec- to tens-of-parsecs-long

"strings"), which then further fragment into dense cores ("beads"). This scenario emphasizes the significance of filamentary structures as a critical step in star formation.

3.1. Models

However, the formation mechanism behind filamentary clouds is still not understood. One popular model for filament formation is shock compression due to stellar feedback, supernovae, or turbulence (e.g., *Padoan et al.*, 2001; *Hartmann et al.*, 2001; *Arzoumanian et al.*, 2011). However, this model is in contradiction with the fact that molecular clouds commonly show long filaments parallel with each other (*Myers*, 2009). Stellar feedback (see, e.g., Fig. 3 of *Hartmann et al.*, 2001) and isotropic super-Alfvénic turbulence (see, e.g., Fig. 2 of *Padoan et al.,* 2001) cannot explain these parallel cloud filaments naturally. There are two other possible mechanisms to form filamentary clouds that both require dynamically dominant B fields. These are B-field channeled gravitational contraction (e.g., *Nakamura and Li,* 2008) and anisotropic sub-Alfvénic turbulence (*Stone et al.,* 1998). In the former mechanism, the Lorentz force causes gas to contract significantly more in the direction along the field lines than perpendicular to the field lines, if the gas pressure is not strong enough to support the cloud against self-gravity along the B field (*Mouschovias,* 1976). This contraction will result in flattened structures, which look elongated in the sky (e.g., *Nakamura and Li,* 2008). When there are multiple contraction centers, the gas will end up in parallel filaments. Sub-Alfvénic anisotropic turbulence has the opposite effect: Turbulent pressure extends the gas distribution more in the direction along the field lines and leads to filaments aligned with the B field (see Fig. 2 of *Stone et al.,* 1998; *Li P.-S. et al.,* 2008). This means that the competition between gravitational and turbulent pressures in a medium dominated by B fields will shape the cloud to be elongated either parallel or perpendicular to the B fields.

The presence of ordered dynamically important B fields required in the scenarios discussed above has recently received significant observational support.

3.2. Observation IV: Preservation of Cloud Field Directions

It is hard to tell whether a cloud is globally super- or sub-Alfvénic by directly measuring the field strength. This is because both the conventional methods, i.e., Zeeman measurements (e.g., *Troland and Crutcher,* 2008) and the Chandrasekhar-Fermi method (e.g., *Crutcher et al.,* 2004), measure only the strength of some field components and have significant uncertainty in their estimates (e.g., *Hildebrand et al.,* 2009; *Mouschovias and Tassis,* 2009, 2010; *Crutcher et al.,* 2010a; *Houde et al.,* 2009).

A strategy to tackle this problem is to see whether cloud B fields are ordered (with well-defined mean direction) or are random. It takes only a slightly super-Alfvénic turbulence to make a B-field morphology random in numerical simulations. The transition from ordered to random field morphologies is quite sensitive to the Alfvénic Mach number (M_A, the ratio of turbulent to Alfvénic velocity). For example, *Falceta-Gonçalves et al.* (2008) showed that B-field morphologies are ordered for $M_A = 0:7$ but random for $M_A = 2$ (no value in between was shown; see discussion in section 6.2). However, mapping the B-field morphology of a whole cloud using the polarization of thermal dust emission was extremely time consuming. The Submillimeter Polarimeter for Antarctic Remote Observations (SPARO) team (*Novak,* 2006) published the first four molecular cloud B-field maps with ~5-arcmin resolution (*Li et al.,* 2006). Each cloud took about a month of telescope time from the Antarctic station. The field morphologies they observed are

either shaped by the shells of HII bubbles or quite ordered.

Other more efficient techniques had been developed to survey cloud fields. The plane-of-the-sky B-field orientations in ICM and in cloud cores can be measured using optical and submillimeter polarimetry respectively within a reasonable observation time (Fig. 5), and polarization archives have been built for both cases. *Li et al.* (2009) compared the fields from 25 cloud cores (subparsec-scale), from the Hertz (*Dotson et al.,* 2010) and SCUpol (*Curran and Chrysostomou,* 2007) archives, to their surrounding ICM (hundreds-of-parsec-scale) fields direction inferred by the *Heiles* (2000) optical polarization archive and found a significant correlation (Fig. 6): 90% of the offsets between core and ICM B fields are less than 45°. The probability for obtaining this correlation from two random distributions is less than 0.01%. This result agrees with *Li et al.* (2006). Compared to the cloud simulations in the literature, only sub-Alfvénic ones present a similar picture.

A similar idea is to compare B_{los} directions from cloud cores and from the ICM. Zeeman measurements of OH maser lines can probe core B_{los}, and Faraday rotation measurements of pulsars probe Galactic B_{los}. Comparing the two datasets, *Han and Zhang* (2007) concluded that B fields in the clouds still "remember" the directions of Galactic ICM B-field directions. In principle, Zeeman measurements of emission probe three-dimensional field morphology for regions with $n(H_2) > 10^5$ cm^{-3}, but the interpretation of emission polarization is not straightforward (*Crutcher,* 2012; *Vlemmings et al.,* 2011).

Nowadays, a survey of cloud B fields has been made possible by state-of-the-art instruments, and their early results look consistent with SPARO and Fig. 6. These surveys use either more efficient submillimeter polarimeters, e.g., the Balloon-borne Large Aperture Submillimeter Telescope (BLAST)-Pol (*Fissel et al.,* 2010) and Planck (*Fauvet et al.,* 2012), or use infrared (IR) polarimetry to fill the gap between cloud cores and ICM, e.g., Mimir (*Clemens et al.,* 2007) and SIRPOL (*Kandori et al.,* 2006). Some examples are shown in Fig. 7.

3.3. Observation V: Bimodal Filament-Field Orientation

Observations I through IV tell us that molecular clouds are threaded by B fields in well-defined directions, which means that cloud B fields are not randomized by self-gravity or turbulence during the cloud and core-formation processes. They are also not compressed in regions with $N_H < 5 \times 10^{21}$ cm^{-2} that comprise most of the volume of a molecular cloud (*Kainulainen et al.,* 2009; *Lada et al.,* 2010). This ordered B field should, in turn, channel turbulent and gravitational gas motion, such that the resulting cloud shapes are elongated in directions either parallel or perpendicular to the local ICM B fields. This theoretical model prediction was discussed in *Protostars and Planets III* (*Heiles et al.,* 1993), where, based on the study of Taurus, Ophiuchus, and Perseus (*Goodman et al.,* 1990), no significant alignment was found

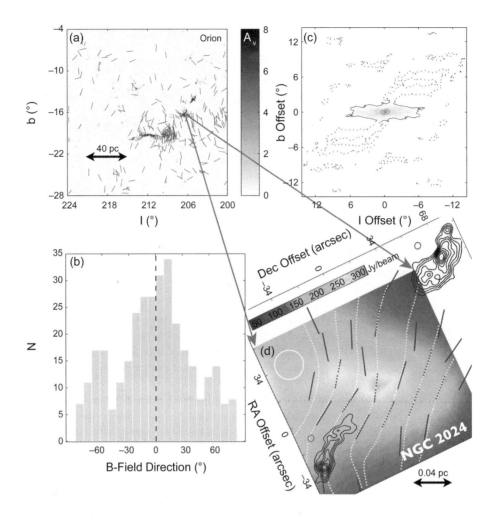

Fig. 5. (a) A_V maps from *Dobashi* (2011) in Galactic coordinates (degrees) overlapped with optical polarization from *Heiles* (2000). **(b)** Distributions of the ICM B-field orientations inferred from the optical polarimetry data shown in **(a)**, measured counterclockwise from Galactic north in degrees. The dashed line shows the Stokes mean of all the field detections in the map. **(c)** The autocorrelation of the A_V map shown in **(a)**. The coordinates are the offsets in degrees. The contour direction gives the cloud elongation. There are clearly several parallel contours, and they are almost perpendicular to the mean field direction from **(b)**. A survey of the correlation between cloud elongation (filaments) and mean field direction is given in Fig. 8 (Observation V). **(d)** Closeup of the NGC 2024 core in (a). False-color map: dust thermal emission at 350 μm (*Dotson et al.,* 2010) with a 20-arcsec resolution (white circle), centered on the FIR-3 core. Vectors are the magnetic field directions inferred from the polarimetry data, and dotted lines are magnetic field lines based on the polarimetry data (*Li et al.,* 2010). Note that the field direction in **(a)** is very close to the field direction here, regardless of the very different scales and densities. A survey of the correlation between ICM and core B fields is given in Fig. 6 (Observation IV). Also note that the B field is almost perpendicular to the core elongation; more discussion is given in Observation VIII. Contours are HCO+(3–2) (*Li et al.,* 2010). Note that the dense dust ridge is perpendicular to the mean field direction and that the HCO+ streaks are parallel with the fields. Similar structures are seen at cloud (parsec) scale and are called hub-filament structure (Observation VI) (*Myers,* 2009). A color version of this figure is available at *http://arxiv.org/pdf/1404.2024v1.pdf*.

between the field orientations and cloud elongation. *Li et al.* (2013) revisited this issue using a larger sample and came to a different conclusion. Their results are summarized below.

Cloud elongation and their local ICM B fields can be probed respectively by extinction measurement [A_V (*Dobashi,* 2011)] and optical polarization data (*Heiles,* 2000), as illustrated in Fig. 5. *Li et al.* (2013) looked into these two archival datasets and compared cloud elongation and ICM B fields from 13 regions along the Gould belt. Their result is summa-

rized in Fig. 8: All pairs of mean fields and cloud directions fall within 30° from being either parallel or perpendicular to each other. The probability for obtaining this correlation from two random distributions is less than 0.6%. Monte Carlo simulations and Bayesian analyses are performed to study the typical range of the three-dimensional offset from parallel and perpendicular alignments: The 95% confidential range is 0° to 20°. This indicates that ICM B fields are strong enough to guide gravitational contraction to form flat

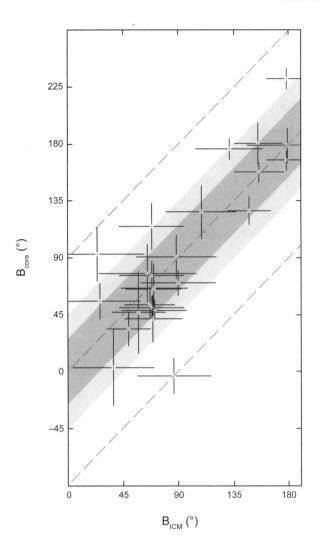

Fig. 6. Correlation between B-field directions of cloud cores (B$_{core}$) and surrounding intercloud media (B$_{ICM}$) (*Li et al.,* 2009). The directions are measured from north to south in J2000 coordinates, increasing counterclockwise. The bars indicate the interquartile ranges (IQRs) of the B-field directions. The mean of the IQRs from all the ICM regions is approximately 52° (dark-shaded area). About 70% of the core/ICM pairs deviate from perfect parallelism by less than 26°. Nearly 90% of the B$_{core}$ values are more nearly parallel than perpendicular to B$_{ICM}$ (dark- and light-shaded together).

condensations perpendicular to them (e.g., *Nakamura and Li,* 2008) and strong enough to channel turbulent flows to result in filaments aligned with them (e.g., *Stone et al.,* 1998; *Li P.-S. et al.,* 2008).

For cloud cores, recent SMA polarimetry survey (Zhang et al., in preparation) also observed a trend of bimodal alignment between core elongation and core B fields. A different trend for cloud cores is also observed; see the discussion of Observation VIII in section 4.1.

3.4. Observation VI: Hub-Filament Cloud Structure

Figure 8 covers all the nearby clouds examined by *My-*

ers (2009), who concluded that the clouds can be often described by a "hub-filament" morphology: The clouds have high-density elongated "hubs," which host most of the star formation in the clouds and lower-density "parallel filaments" directed mostly along the short axes of the hubs. *Myers* (2009) suggested that the parallel filaments are due to layer fragmentation. The main result of Observation V, i.e., that elongated/filamentary structures tend to be aligned either parallel or perpendicular to the ambient ICM B fields, suggests that the gas layers in this scenario must host ordered and dynamically dominant B fields, as shown by *Nagai et al.* (1998). Note that the two types of B-field-regulated filaments discussed in Observation V can also explain the hub-filament structures. The two types of filaments may form at the same time, with the denser and more massive filaments (hubs) perpendicular to the B field and finer filaments in the vicinity aligned with the field [see observations from, e.g., *Palmeirim et al.* (2013) and *Goldsmith et al.* (2008) and simulations from *Nakamura and Li* (2008) and *Price and Bate* (2008)], which is identical to the hub-filament structure described by *Myers* (2009). In this scenario, hubs and filaments should always be perpendicular to each other, but their sky projections are not necessarily so. This means that at least some of the exceptions, i.e., nonperpendicularity (e.g., Fig. 9 of *Myers,* 2009), can be explained by projection effects. The hub-filament system can be an explanation (besides rotation during contraction) of the larger differences between cloud directions defined by different column densities (e.g., Aquila in Fig. 8).

The hub-filament system could be a self-similar structure. For example, the Herschel Space Observatory resolved part of the Pipe nebula with 0.5-arcmin resolution and showed that a hub from *Myers* (2009) can fragment into a network of perpendicular filaments (*Peretto et al.,* 2012). The network is aligned with the mean B-fields direction. An Observation V-type analysis of the Herschel Gould belt data will be of interest. *Li et al.* (2011) observed one of these core regions (NGC 2024; Fig. 5) with 3-arcsec resolution and found filaments perpendicular to the core (aligned with the B field), i.e., a hub-filament structure. A survey of core vicinities with high angular resolution and sensitivity as performed by *Li et al.* (2011) is necessary to tell whether NGC 2024 is a special case or not.

3.5. Observation VII: Turbulent Anisotropy

Given that cloud fields are ordered, turbulent velocities should be anisotropic. Turbulent energy cascades more easily in the direction perpendicular to the mean field than in the direction aligned with the field (i.e., along B fields the velocity should be more coherent and less turbulent), if the field is dynamically important compared to the turbulence. This is because the development of turbulent eddies will be suppressed in the direction parallel to the field. This phenomenon should be more prominent at smaller scales, where turbulent energy is lower and field curvature (and thus tension) will be larger if the field is bent. An analytic model of anisotropic,

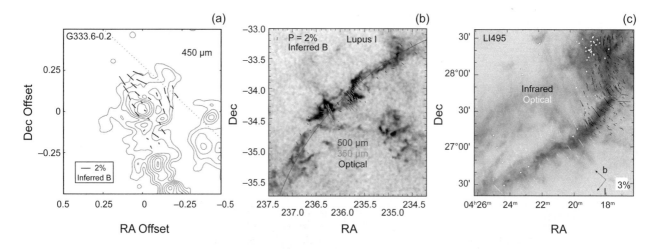

Fig. 7. Examples of cloud B-field morphologies. Vectors show B-field directions probed by different wavelengths. **(a)** Sub-millimeter intensity contours overlapped by submillimeter polarimetry inferred B fields (*Li et al.,* 2006). **(b)** Inferred B fields on top of the A_V map. Optical and submillimeter data are aligned (*Matthews et al,* 2013). **(c)** Grayscale shows relative A_V. Optical and IR data are aligned (*Chapman et al.,* 2011). A color version of this figure is available at *http://arxiv.org/pdf/1404.2024v1.pdf.*

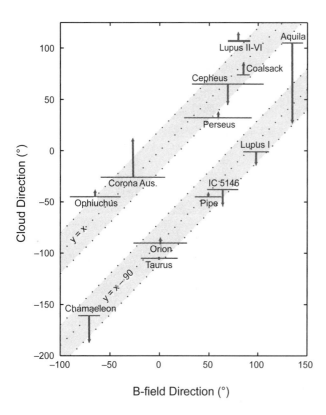

Fig. 8. Cloud elongations vs. B-field directions (*Li et al.,* 2013). The range of directions of a cloud traced by the different column densities is shown by the vertical arrow, where the tail and head represent the directions of the whole cloud and higher densities (1/5 peak value) respectively. The width (42°) of the shaded regions in the X and Y directions is equal to the averaged value of the 13 IQRs (horizontal error bars) of the B-field directions. The filaments and B fields tend to be either aligned or perpendicular to each other.

incompressible magneto-hydrodynamic (MHD) turbulence was first proposed by *Goldreich and Sridhar* (1995, 1997). *Cho and Vishniac* (2000), *Maron and Goldreich* (2001), and *Cho et al.* (2002) numerically verified the anisotropy predicted by the Goldreich-Sridhar model. Similar anisotropy was also observed in compressible MHD simulations from *Cho and Lazarian* (2002) and *Vestuto et al.* (2003).

Heyer et al. (2008) first observed turbulence anisotropy in a molecular cloud by showing that the ^{12}CO (1–0) turbulence velocity is more coherent in the direction along the B field (Fig. 9). The region they studied is roughly ($2° \times 2°$) centered at 4h50m0s, 27°0'0" (J2000) in the Taurus Molecular Cloud. A_V is less than 2 mag in this region (*Schlegel et al.,* 1998), and thus the conclusion from this work is restricted to the cloud envelope. In the same region, they also observed that the ^{12}CO emission exhibits "filaments" (see Observation V) that are aligned along the magnetic field direction. B-field channeled turbulent anisotropy is also seen with ^{12}CO (1–0) spectra from Ophiuchus (Ji and Li, in preparation), where the direction with the least velocity dispersion tends to align with the B field. *Heyer and Brunt* (2012) extended the study of *Heyer et al.* (2008) to higher density using ^{13}CO (1–0) lines, which traces A_V = 4–10 mag and exhibits little evidence for anisotropy.

Heyer and Brunt (2012) interpret the observation as the turbulence transiting from sub- to super-Alfvénic in regions with higher column density. However, we note that A_V = 4 mag is measured from the "hub" (see Observation VI) and happens to be the cloud-contraction threshold of TMC (*Kainulainen et al.,* 2009) (Observation III). It means that, over this density limit, contraction velocities in all directions dominate the turbulent velocity so turbulent anisotropy is less clearly observable. Moreover, super-Alfvénic turbulence in regions with A_V = 4–10 mag also contradicts

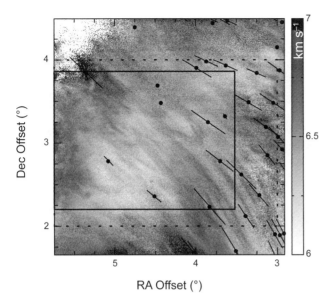

Fig. 9. Image from *Heyer et al.* (2008) showing ^{12}CO velocity centroid overlapped with B-field directions inferred from the optical polarimetry data from *Heiles* (2000). The velocity streaks are aligned with the B-field direction. The solid line box and dotted line box show the areas within which the velocity anisotropy and B-field direction are estimated in *Heyer et al.* (2008).

Observation IV: B fields in densities below and above this density range are aligned. *Li et al.* (2011) observed the B-field-aligned velocity anisotropy with ^{12}CO (7–6) emission from a region of NGC 2024 with the mean $A_V = 7$ mag, after carefully examining the line profiles away from regions with contraction/expansion signatures. If the angle dispersion of a polarimetry map stems from turbulence, the angle dispersion should also be anisotropic if the turbulence is. This is exactly what *Chitsazzadeh et al.* (2012) found for OMC-1, which has A_V above 30 mag (*Scandariato et al.*, 2011). Given the available data, it is hard to tell whether there is a density limit for turbulent anisotropy.

4. CLOUD CORE FORMATION

By cloud cores, we mean structures like in Fig. 5d, typically with linear scales of subparsec and mean density $n_H = 10^{5-6}$ cm^{-3}. They are nurseries of star formation.

4.1. Observation VIII: Cloud Cores Shaped by B Fields

Given that cloud fields and ICM fields are aligned, flux freezing should make a gravitational contraction anisotropic, and (somewhat) flattened high-density regions perpendicular to the mean field direction should be expected, as observed in the NGC 2024 core region (Fig. 5). Figure 5d covers the same region as *Crutcher* (1999) analyzed for NGC 2024. He estimated that the mass-to-flux ratio here is highly su-

percritical (4.6× of the critical value), the highest among the 15 core regions ($N_H > 10^{23}$) with reliable detections in Fig. 4. Even so, the magnetic field still clearly channels the contraction to form an elongated core with the long axis perpendicular to the field (Fig. 5).

Observation of this kind of correlation, however, is not always straightforward, because the real core flatness is not always observable due to the projection effect. Observations suggest that core projections are not generally spherical (e.g., *Benson and Myers,* 1989) and that the most probable core shapes are nearly oblate (e.g., *Jones et al.,* 2001). If the shortest axis of an oblate core indeed orients close to the mean field direction, then the closer the line of sight is to perpendicular to the shortest axis, the better the alignment that should be observed between the field projection and the short axis of the core projection. In other words, the more elongated the core projection, the better the alignment should be. This is exactly the trend *Tassis et al.* (2009) observed from 32 cloud cores (Fig. 10) with the Caltech Submillimeter Observatory (CSO).

4.2. Observation IX: No Detection of Column Density-to-B_{los} Ratio Increasing from Cloud to Cores

The proposal that B fields can regulate star formation through ambipolar diffusion (AD) has been put forth for more than five decades (*Mestel and Spitzer,* 1956; *Shu et al.,* 1999; *Ciolek and Mouschovias,* 1993), but observers had been unable to even try to test it until the pioneer work of *Crutcher et al.* (2009). If AD plays a major role in core formation, an increase in mass-to-flux ratio from clouds to cores should be expected. With OH Zeeman measurements, *Crutcher et al.* (2009) used column density-to-B_{los} ratios to estimate mass-to-flux ratios. Comparing the mass-to-flux ratios between cores and surrounding envelopes from four regions (B217-2, L1544, B1, and L1448), they reported that the ratios of cores are smaller than the envelopes for all cases. They thus concluded that the result is more in agreement with the super-Alfvénic cloud simulation from *Lunttila et al.* (2008), which is without AD and can produce cores with smaller column density-to-B_{los} ratios compared to their envelopes.

However, other than AD, there are uncontrolled factors in this experiment that also affect column density-to-B_{los} ratios. These factors make it difficult for this experiment to discriminate or support any core formation mechanism. In the following paragraphs we discuss these factors.

4.2.1. Factor 1: B_{los} reversal. Starless cores are usually more quiescent than their envelopes (*Benson and Myers,* 1989; *Goodman et al.,* 1998; *Barranco and Goodman,* 1998). The stronger the turbulence, the larger the B-field dispersion (*Chandrasekhar and Fermi,* 1953), and the more field reversal might happen within a telescope beam to reduce the mean B_{los} measured by the Zeeman effect. The stronger turbulence in envelopes can bias the column density-to-B_{los} ratios toward the higher end. For sub-Alfvénic turbulence, B-field direction has a Gaussian-like dispersion

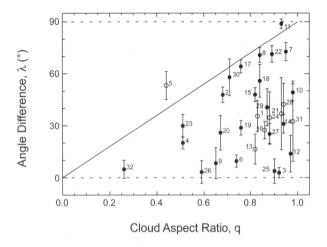

Fig. 10. The alignment between the B fields and the short axes of cores (*Tassis et al., 2009*). λ is the angle difference between the sky projection of the core B field and the short axis of the core projection. q is defined as the short-to-long axis ratio of a core projection. The model suggested by *Tassis et al.* (2009) is that cores forming under dynamically important B fields will be flattened such that the core short axes will align with the B fields in three dimensions. This geometry is most easily observed (small λ) when the line of sight is close to perpendicular to the B fields or, in other words, when the flattened core is viewed edge-on (when q is small). The upper limit of λ indeed increases with q, i.e., most data points are below the solid line (λ/90° = q). The probability of getting only 2 out of 32 data points above the solid line by chance is only ∼10^{-7}.

and the field reversal within a beam depends on the angle between the line of sight and the mean field direction. For example, a Gaussian B-field distribution with a STD = 30° and a mean direction 60° from the line of sight will have ∼17% of the B_{los} being flipped. This observational bias is illustrated by the sub-Alfvénic simulation shown in Fig. 11 (*Bertram et al., 2012*), where the column density-to-B_{los} ratios from the cores and from the envelopes are compared along various lines of sight.

By definition, mass-to-flux ratios should be estimated by column density-to-B_{los} ratios measured along the B-field direction (Z axis in Fig. 11). But *Crutcher et al.* (2009) proposed that the ratio (R) between the column density-to-B_{los} ratios from a core and from its envelope should be independent of lines of sight, assuming that B fields from the core and from the envelope are aligned. Based on Observation IV, this seems to be a fair assumption. However, it ignores the fact that larger misalignment between B fields and the line of sight produces more B_{los} flips to disperse the measurements of B_{los}. With a line of sight aligned with the mean field (no B_{los} reversal), Fig. 11 shows that all cores have mass-to-flux ratios *larger* than their envelopes. Along other lines of sight, field reversal can significantly lower the B_{los} of many envelopes and give R < 1. Figure 11 shows that sub-Alfvénic turbulence is enough to flip B_{los};

super-Alfvénic turbulence (*Lunttila et al.,* 2008) is not a necessary condition for observing R < 1. *Bertram et al.* (2012) also performed super-Alfvénic simulations and obtained results similar to Fig. 11, only with more R < 1. But in super-Alfvénic cases the distribution of green data is not too different from red and blue data (*Bertram et al., 2012*) due to tangled B-field lines, which disagrees with Observation IV. Sub-Alfvénic turbulence introduces structures to B fields without changing the mean direction, which can produce the R scattering in Fig. 11 as well as the multi-scale field alignment shown in Fig. 6.

4.2.2. Factor 2: Anisotropic gravitational contraction. Figure 11 also shows that R > 1 is possible even without AD. While mass is accumulated more along the B-field lines due to anisotropic contraction (Observation VIII) or anisotropic turbulent shocks (Observation VII) channeled by dynamically important B fields, one should observe larger column density-to-B_{los} ratios of cores (R > 1) even without AD. This is because the column density increases faster than field strength does in cores due to the anisotropic gas contraction/compression mainly along the field lines (*Li et al., 2011*). Only a compression perpendicular to the field lines can increase the B-field strength. Like factor 1, the larger the offset between the line of sight and the B field, the stronger the factor 2.

4.2.3. Factor 3: Reconnection diffusion. Lazarian et al. (2012) proposed that turbulent eddies can bring B-field lines to cross each other and get reconnected, and this reconnection can cause field diffusion with an efficiency increasing as turbulent energy increases. If cores are more quiescent than their envelopes, the latter will have a higher reconnection diffusion efficiency and thus higher mass-to-flux ratio (R < 1).

While AD and factor 2 tend to make R > 1, factors 1 and 3 tend to make R < 1. The competition between these factors can produce a spectrum of R. This is already seen in Fig. 11 (blue and red data points) even without AD. Note that only the green data points in Fig. 11 are the ratios between the mass-to-flux ratios from cores and from envelopes, the quantities *Crutcher et al.* (2009) planned to measure. In reality, however, the column density-to-B_{los} ratios, which are the values *Crutcher et al.* (2009) used to estimate mass-to-flux ratios, will give biased Rs, like the blue or red data points in Fig. 11. Bearing in mind that B_{los} may reverse from position to position to cause the scattering in the Zeeman measurements, *Mouschovias and Tassis* (2009) reanalyzed the data from *Crutcher et al.* (2009) (although their reason for reversal is not turbulence). They obtained a much higher possibility for R > 1 compared to the analysis from *Crutcher et al.* (2009), who assumed that the variation in their Zeeman measurements are mainly due to observation errors (*Crutcher et al., 2010a*).

An interesting fact about the simulations from *Bertram et al.* (2012) is that the mean of all Rs is always larger than 1, regardless of the magnetic Mach number. This is different from the results of the four observations from *Crutcher et al.* (2009). Whether this discrepancy is physical or due to

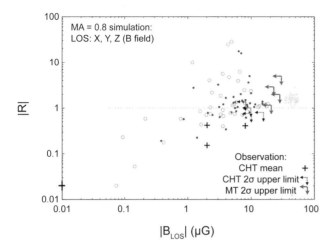

Fig. 11. Distribution of clumps from the sub-Alfvénic simulation (*Bertram et al.*, 2012) and from observations (*Crutcher et al.*, 2009) (CHT). The simulated data is projected along different line-of-sight directions (X, Y, and Z axes) to compare with observations. Plotted is the absolute value of R, the ratio between the column density-to-B_{los} ratios from the cores and those from the envelopes, against the absolute value of the average of B_{los}. Only the column density-to-B_{los} ratios measured along the mean B field (Z axis; light "+" symbols) are equivalent to the mass-to-flux ratios by definition. The initial magnetic field strength is almost unchanged when observed along the B field, meaning that the B field is not tangled and turbulence only introduces field dispersion. This field dispersion can be observed as B_{los} reversal along other lines of sight and produces a large dispersion of R compared to those observed along the B field. Observations from *Crutcher et al.* (2009) are also plotted, with the 2σ upper limits from *Crutcher et al.* (2009) and from *Mouschovias and Tassis* (2009) (MT). The latter authors analyzed the same set of observational data with different assumptions (see text). A color version of this figure is available at *http://arxiv.org/pdf/1404.2024v1.pdf*.

the small observation sample size remains a question. On the other hand, super-Alfvénic simulations from *Lunttila et al.* (2008) indeed produce ~80% of the clumps with R < 1. However, their |R| tends to be inversely proportional to B_{los}, which is again different from the trend of the four observations (Fig. 11) and different from the super-/sub-Alfvénic simulations from *Bertram et al.* (2012). While more work is needed to better understand simulations and observations, one thing is obvious: With or without AD, |R| can be either larger or less than unity.

5. PROTOSTELLAR DISK FORMATION

The competition between angular momentum and magnetic braking, as shown in Fig. 2, should also take place at the scale of protostellar disk formation. Compared to the scale of cloud formation (Fig. 2), here the angular momentum is weaker and magnetic braking is stronger.

Galactic shear can be ignored at this scale and turbulence is also much weaker based on Larson's Law, and B fields are stronger based on Fig. 4. Yet, unlike Observation I, here we know that angular momentum wins the competition in many cases where disks are observed (e.g., *Murillo and Lai*, 2013; *Tobin et al.*, 2012), so how it can win becomes a mystery.

Due to the lack of instrumental sensitivity, observations of B fields on scales of starless cores to disks is far behind the development of related theories. The Atacama Large Millimeter/submillimeter Array (ALMA) should offer adequate resolution and sensitivity to probe the B fields of protostellar disks and the dense cores in which they are embedded. Observational constraints on competing numerical models are extremely important, because the models of disk formation are not converging. In some models, efficient B-field diffusion (e.g., *Li Z.-Y. et al.*, 2011) is indispensable for solving the so-called magnetic braking catastrophe, i.e., the B-field tension force will prevent disk formation if field lines are well coupled to the dense cores. Other groups, for example, claim that turbulence (e.g., *Seifried et al.*, 2012) or mass rearrangement from the core to the pseudo-disk (*Machida et al.*, 2011) can mitigate the effect of magnetic braking and thus enable disk formation. Numerical (artificial) effects have not been ruled out as a cause for the differences between these various models. The chapter by Z.-Y. Li et al. in this volume has a detailed review of the disk formation theories. Here, we review recent observations, focusing on the role played by B fields in disk formation.

5.1. Observation X: Decoupling Between B Fields and Turbulence, a New Model of Ion-Neutral Linewidth Difference

For the scales of galactic disks and molecular clouds, flux freezing (the perfect coupling between gas and magnetic field lines) is a good approximation, based on the MHD induction equation. Indeed, this is why weak fields are expected to follow cloud rotation and turbulence, while strong fields will channel turbulence and suppress cloud rotation (Fig. 2). Moving toward small scales, however, the MHD induction equation also suggests decoupling between turbulent eddies and B fields [when the magnetic Reynolds number, R_M, is approximately below unity (*Li and Houde*, 2008)]. This may help to solve the magnetic braking catastrophe, if this decoupling scale, L′, is comparable to the scales of protostellar disks (a few hundred astronomical units) and turbulence at this scale is the main energy source of disk angular momentum. Below the decoupling scale, ions and neutrals should have different turbulent velocity (energy) spectra (Fig. 12), because ions are still coupled with field lines due to the Lorentz force. The identification of L′ is challenging because it could be on the order of milliparsecs (*Li and Houde*, 2008), which is near the resolution limit of current radio and submillimeter telescopes. However, since all the turbulent eddies within a telescope beam contribute to the observed linewidth (velocity dispersion), the decoupling should still affect the linewidth even if their beam size is larger than L′. In other words, the linewidths of coexistent ions and neutrals should be different due to neutral gas-field decoupling. Indeed, many other factors — e.g., opacities, spatial distributions, hyperfine structures and outflows — can affect linewidths, but these nonmagnetic factors can be ruled out by carefully choosing the regions of observation and the ion/neutral pairs. HCO⁺/HCN and H¹³CO⁺/H¹³CN can be used for this purpose (*Li and Houde, 2008; Hezareh et al.,* 2010), because the ion species generally have larger opacity, slightly more extended distribution, and more unresolved hyperfine structures (*Li et al.,* 2010); all these factors suggest a wider linewidth, but ion spectra are systematically narrower (*Houde et al.,* 2000a,b; 2002). Most importantly, *Li et al.* (2010) have shown that the linewidth difference between HCO⁺ and HCN is proportional to the B-field strength (Fig. 12). This definitely cannot be explained by any reason other than decoupling between B fields and small turbulent eddies. The decoupling is called turbulent ambipolar diffusion (TAD) in *Li and Houde* (2008) to distinguish it from AD, the decoupling between B fields and gravitationally contracting gas (Observation IX).

By setting RM (the ratio between the advection term and diffusion terms of the induction equation) equal to unity, it can be shown that (*Li and Houde,* 2008)

$$B_{pos}^2 \approx 4\pi n_i \mu v_i V_n' L' \qquad (3)$$

where B_{pos}, n_i, v_i, and μ are, respectively, the plane-of-sky B-field strength, the ion density, the collision rate between an ion and the neutrals, and the reduced mass for such collisions. V_n' is the neutral turbulent velocity at L'. *Li and Houde* (2008) showed how V_n' and L' can be derived from the position-position-velocity data cube with, most importantly, the linewidth difference between coexistent ions and neutrals. In this way, L′ is estimated to be on the order of a few hundred astronomical units (about 1 mpc), the size of a typical protostellar disk. So TAD must, to some degree, play a role in mitigating the magnetic braking catastrophe if disk angular momentum originates from turbulence. The most direct way to test the above TAD model (Fig. 12) is to probe turbulent velocity spectra of HCO⁺ and HCN below the TAD scale, where the model predicts that the ions should have a steeper slope. With the SMA, we have observed turbulent velocity spectra down to the scale of 0.01 pc, where ions and neutrals still share the same slope. With ALMA, it will be possible to probe turbulent velocity spectra well below 1 mpc. This also allows direct measurements of V_n' and L', which can be used to estimate B-field strength with equation (3) above.

While both models try to explain the ion/neutral linewidth difference, the model discussed here (*Li and Houde,* 2008) is independent from the model introduced earlier by *Houde et al.* (2000). *Li and Houde* (2008) relate linewidth difference to B_{pos} through turbulence spectra and the decoupling scale (L′). These turbulence properties are not treated in the earlier model, which relates the linewidth difference to the B-field orientation. The two models are sometimes confused in literature (e.g., *Crutcher,* 2012). The observation that the coexistent ions and neutrals should have different turbulent velocities (linewidths) is also supported by simulations from various groups (e.g., *Tilley and Balsara,* 2010, 2011; *Li et al.,* 2012). Although these simulations can reproduce the ion/neutral linewidth difference, the detailed results, e.g., decoupling scale and slopes of ion/neutral turbulent velocity spectra, vary from code to code. Turbulent ambipolar diffusion simulations are challenging; all these codes must either adopt some kind of approximation or sacrifice resolution in order to realize a simulation within a reasonable CPU time.

5.2. Observation XI: Dispersion of Magnetic Fields and the Turbulent Power Spectrum

The aforementioned TAD decoupling scale between the magnetic field (ions) and the neutral component of the gas can also be inferred through polarization maps. This is accomplished by the application and generalization of techniques of analysis well known in turbulence studies of polarization data. More precisely, something akin to a structure function of the second order often applied to turbulent velocity fields (*Frisch,* 1995) can be transposed to polarization angles (*Dotson et al.,* 1996; *Falceta-Gonçalves et al.,* 2008; *Hildebrand et al.,* 2009; *Houde et al.,* 2009, 2011, 2013). In other words, if we define by $\Delta\Phi(\ell)$ the difference between two polarization angles separated by a distance ℓ on a polarization map, then we can introduce the angular dispersion function $1-\langle\cos[\Delta\Phi(\ell)]\rangle$, where the average $\langle\cdots\rangle$ is performed on all pairs of polarization angles for a given distance ℓ. It is straightforward to verify that the dispersion function is closely related to the structure function of the polarization angles $\langle\Delta\Phi^2(\ell)\rangle$ through

$$1-\langle\cos[\Delta\Phi(\ell)]\rangle \simeq \frac{1}{2}\langle\Delta\Phi^2(\ell)\rangle \qquad (4)$$

when $\Delta\Phi(\ell) \ll 1$ (*Houde et al.,* 2013). Although for the purpose of the following discussion the dispersion or structure functions could be used interchangeably, the advantage in using the dispersion function is its close connection to the power spectrum (see below).

With the assumption that the magnetic field can be modeled as the sum of turbulent ($\mathbf{B}_t$; i.e., zero-mean and random) and ordered ($\mathbf{B}_0$) components, it is easy to show that the structure (or the dispersion) function is also the sum of corresponding turbulent and ordered functions (*Houde et al.,* 2013). In other words

$$\langle\Delta\Phi^2(\ell)\rangle = \langle\Delta\Phi_t^2(\ell)\rangle + \langle\Delta\Phi_0^2(\ell)\rangle \qquad (5)$$

Given that the turbulent and ordered components of the field are usually characterized by different length scales, with the ordered component being of a larger scale, they can

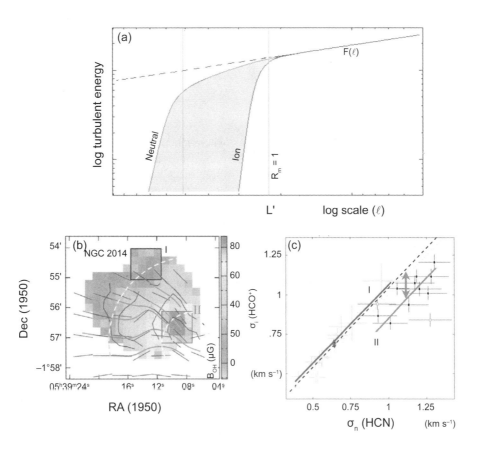

Fig. 12. **(a)** Illustration of the difference between the ion and neutral velocity spectra, from *Li and Houdo* (2008). This is a log-log plot of the turbulent energy vs. scale size in arbitrary units. The decoupling between neutrals and B fields happens at scales smaller than L′, where $R_M < 1$. The energy spectrum in the inertial range is common for both species, while ions and neutrals have different spectra at scales smaller than L′. The square of the velocity dispersion (σ) measured at a particular scale (beam size) is proportional to the integral of the energy spectrum over all scales smaller than the beam size. At scales larger than L′, the observed difference between the two velocity dispersion spectra is proportional to the shaded area. **(b)** Example of the correlation between ion/neutral velocity dispersion difference and B-field strength. The different shades show B_{los} in NGC 2024, estimated by OH Zeeman measurements (*Crutcher et al.,* 1999). The vectors indicate the magnetic field directions inferred from 100-µm polarimetry data (*Crutcher et al.,* 1999). The solid curved lines are the magnetic field lines (*Li et al.,* 2010). Velocity dispersions from the boxed regions (I and II) are shown in **(c)**. **(c)** HCO⁺ (3–2) vs. HCN (3–2) velocity dispersions (σ_i and σ_n). The maps of HCO⁺ (3–2) from the boxed regions (zones I and II) are shown in Fig. 5d. The light and dark shades are for the weak field (zone I) and strong field (zone II) in **(b)** respectively. The dashed line is X = Y. Note that the region with the stronger field has the larger difference, on average (shown by the double-headed arrow), between σ_i and σ_n (*Li et al.,* 2010). A color version of this figure is available at *http://arxiv.org/pdf/1404.2024v1.pdf.*

be cleanly separated without any assumption on the exact morphology of the ordered (large-scale) magnetic field. This is a significant improvement on previous analyses of polarization maps, where the determination of the magnetic field dispersion [to be used, for example, in the so-called Chandrasekhar-Fermi equation (*Chandrasekhar and Fermi,* 1953)] necessitated modeling the large-scale magnetic field with a particular shape (e.g., an hourglass). Since it is highly likely that this description of the large-scale magnetic field with a predefined function will not, in general, perfectly fit the true nature of the field, any error thus introduced in the analysis will be propagated in subsequent calculations.

An example of a dispersion function analysis applied to low-mass star formation simulations taken from *Hen-*

nebelle et al. (2011) is shown in Fig. 13. It is shown how the turbulent component of the dispersion function (i.e., the turbulent autocorrelation function; symbols, Fig. 13b) can be extracted from the data (symbols, Fig. 13a) by simply fitting the component due to the large-scale magnetic field by a Taylor series in a range where we expect contributions from the turbulent magnetic field to vanish.

It is important to note that the overall width of the autocorrelation function results not only from the intrinsic correlation length of the magnetized turbulence but also from the size of the beam defining the resolution of the data. Still, the excess of the autocorrelation width beyond the beam's contribution is a measure of the turbulence correlation length. Although the problem is further com-

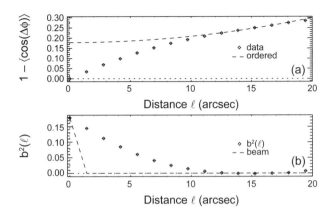

Fig. 13. Angular dispersion function of a low-mass star formation simulation taken from *Hennebelle et al.* (2011). **(a)** The total dispersion function (i.e., the sum of the turbulent and ordered components) is represented by the symbols, while a Taylor series fit to data located at $\ell > 13''$ (broken curve) accounts for the contribution of the ordered component to the total dispersion function (shifted up by the constant level of the turbulent component in the fitting range). **(b)** Subtraction of the data from the Taylor series fit [from **(a)**] yields the normalized magnetic field autocorrelation function (symbols), which is broadened by the finite resolution of the data (i.e., the beam size; broken curve).

plicated by the fact that the signal is integrated through the line of sight and across the area subtended by the telescope beam, it was shown by *Houde et al.* (2009) that the turbulent correlation length and the relative level of magnetic turbulent energy can be readily determined when the underlying observations are realized at sufficiently high spatial resolution (i.e., when the beam width is narrower than the width of the turbulent autocorrelation function, as is the case in Fig. 13b). The knowledge of these parameters is especially important to the application of the Chandrasekhar-Fermi equation for estimating the magnetic field strength B_{pos}. This is because it then becomes possible to account for the number of turbulent cells contained within the column of gas probed by the observations and correct for the aforementioned signal integration that averages down the turbulent dispersion function. For example, *Houde et al.* (2009) performed a dispersion analysis on a 350-μm Submillimeter High Angular Resolution Camera II Polarimeter (SHARP) map of the OMC-1 molecular cloud (*Vaillancourt et al.*, 2008). They determined the correlation length of the magnetized turbulence to be 16 mpc and the number of turbulent cells probed by the telescope beam to be approximately 21, yielding a plausible value of $B_{pos} = 760$ μG. CN($N = 1 \rightarrow 0$) Zeeman measurements yielded a line-of-sight field strength of approximately 360 μG in this source; see *Crutcher et al.* (1999).

Because it is the inverse of the spectral width of the turbulent power spectrum, the turbulent correlation length is to some extent linked to the TAD decoupling scale. As was mentioned in the previous section, this is the scale of

turbulent energy dissipation that characterizes the upper end (in k-space) of the power spectrum. This spectrum is connected to the dispersion function through a simple Fourier transform

$$\langle \mathbf{\bar{B}}_t \cdot \mathbf{\bar{B}}_t (\ell) \rangle \Leftrightarrow R_t (k) \| H(k) \|^2 \qquad (6)$$

where $\langle \mathbf{\bar{B}}_t \cdot \mathbf{\bar{B}}_t (\ell) \rangle$ is the measured autocorrelation function of the turbulent magnetic field, $R_t (k)$ is the intrinsic turbulent power spectrum, and $H(k)$ is the spectral profile of the telescope beam. The turbulent power spectrum associated with the autocorrelation of Fig. 13 is shown in Fig. 14. Multiplying the resulting power spectrum (symbols, Fig. 14a) by $2\pi k$ yields a one-dimensional spectrum (solid curve) that, for example, can be fitted to a power law in its inertial range (broken curve). In this example, the log-log plot of the one-dimensional spectrum (symbols, Fig. 14b) reveals a (numerical) dissipation scale at $\log (k/2\pi) \simeq -0.7$ arcsec^{-1}. *Houde et al.* (2011) have applied this spectral analysis to three high-resolution polarization maps obtained with the SMA (for Orion KL, IRAS 16293, and NGC 1333 IRAS 4A), and although the spectra obtained are not as cleanly resolved as the one shown in Fig. 13, a dissipation scale was clearly detected at approximately 10 mpc for Orion KL. This scale is most likely associated with TAD, as discussed in the previous section and first detected by *Li and Houde* (2008) in M17 using comparisons of HCO$^+$ and HCN linewidths.

5.3. Observation XII: Alignment of Outflows and Flattened Infall Envelopes with Cloud B Fields?

Observationally, we have not yet fully characterized the relationship between the symmetry axes of structures in the protostellar environment, such as bipolar outflows and flattened infall envelopes, and the orientation of the local cloud B field. However, new instrumentation is improving the situation, leading to new tests for disk-formation models such as those reviewed at the beginning of section 5. If B fields in cloud cores are sufficiently strong, one expects collapse to proceed primarily along field lines, leading to infall envelopes that are flattened along the direction of the B field (Observation VIII). Such magnetically shaped flattened infalling gas structures [a.k.a. "pseudodisks" (*Galli and Shu*, 1993; *Allen et al.*, 2003)] have characteristic scales of several thousands of astronomical units for low-mass protostars. If cloud cores have their rotation axes aligned with the B field due to magnetic braking (*Mouschovias and Paleologou*, 1979) then the smaller-scale Keplerian disks should also have their symmetry axes parallel to the B field. We can test for such an alignment by comparing B-field orientation with outflow axis. Not all simulations of magnetized core collapse predict good alignment between core B field and Keplerian disk axis (e.g., see *Joos et al.*, 2012).

Tests for alignment between protostellar structures and B fields have employed various techniques; here we restrict the discussion to submillimeter- or millimeter-wave

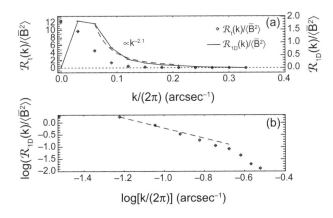

Fig. 14. **(a)** The turbulent power spectrum (symbols) calculated from the Fourier transform of the autocorrelation function shown in Fig. 13b; this spectrum is multiplied by $2\pi k$ to obtain the one-dimensional (Kolmogorov-type) power spectrum (solid curve). **(b)** Log-log plot of the one-dimensional turbulent power spectrum (symbols). Both panels show a power-law fit to the inertial range of the one-dimensional turbulent power spectrum (broken curve); the (numerical) dissipation scale is seen in **(b)** at log $(k/2\pi) \simeq -0.7$.

polarimetry. *Curran and Chrysostomou* (2007) carried out submillimeter polarimetry using the James Clerk Maxwell Telescope (JCMT) Submillimetre Common-User Bolometer Array (SCUBA) for a sample of 16 high-mass star-forming regions and compared mean B-field directions with outflow axes. The projected misalignment angles were nearly uniformly distributed between 0° and 90°, suggesting the absence of any correlation, as confirmed with a Kolgomorov-Smirnov test yielding p = 0.849. The authors pointed out that for outflows lying nearly parallel to the line of sight it would be difficult to observe any intrinsic alignment that might exist between outflow and field. However, when they omitted from their analysis all outflows having inclination angles below 45° (i.e., those oriented relatively close to the line of sight), the distribution was still consistent with a random one (p = 0.572).

Wolf et al. (2003) and *Davidson et al.* (2011) carried out similar types of observational studies for the case of nearby, low-mass cores, pointing out that such targets have generally simpler geometries, are often more isolated, and most importantly are the types of objects most often simulated by theorists. *Wolf et al.* (2003) used SCUBA/JCMT polarimetry to map four targets and *Davidson et al.* (2011) used the SHARP submillimeter polarimeter at CSO (*Li H. et al.*, 2008) to study three sources. Both results are suggestive of an alignment between field and outflow, at least for the younger sources, but neither study reported statistical tests for an overall alignment. Two very recent studies of nearby low-mass cores that did carry out such tests are SHARP/CSO polarimetry by *Chapman et al.* (2013) on 10,000-AU scales, and Combined Array for Research in Millimeter-wave Astronomy (CARMA) interferometric 1.3-mm polarimetry by *Hull et al.* (2013) on 1000-AU

scales. *Chapman et al.* (2013) restricted themselves to single class 0 sources while *Hull et al.* (2013) included multiples as well as class I targets.

The survey of *Chapman et al.* (2013) included seven targets, and they compared position angles of mean B field, flattened infall envelope ("pseudodisk"), and outflow. Following the example of *Curran and Chrysostomou* (2007), they made use of the estimated outflow inclination angles. They found a tight alignment between the position angles of pseudodisk minor axes and outflow axes, and they exploited this by using the outflow inclination angles as proxies for the unknown pseudodisk symmetry axis inclination angles. Using all these angles, they tested for correlations, in three-dimensional space, between B field and pseudodisk symmetry axis and between B-field and outflow axis, finding evidence for positive correlations in both cases, with 95% and 96% confidence, respectively. As a further test, they combined polarimetry data from the subset of their sources having inclination angle below 45°, first rotating each map according to its apparent pseudodisk axis, to form a source-averaged magnetic field map having improved signal-to-noise ratio. This map, shown in Fig. 15, shows that the overall direction of the source-averaged B field is roughly perpendicular to the pseudodisk plane, as expected from their reported correlation between pseudodisk symmetry axis and B field. Similar results were obtained when the individual maps were rotated according to their outflow position angles rather than referencing to the pseudodisks.

The interferometric polarimetry survey of *Hull et al.* (2013) included 16 sources, and they compared the position angles of mean B field and outflow. The projected misalignment angles were found to range from 0° to 90° with a nearly uniform distribution, consistent with a random alignment. The Kolgomorov-Smirnov test yielded p = 0.64. Inclination angles were not considered.

How can we reconcile the *Chapman et al.* (2013) and *Hull et al.* (2013) surveys? One possibility is that the inclusion in the CARMA survey of the (more complex) multiples and (more evolved) class I sources obscures the correlation. This is suggested by the fact that the *Hull et al.* (2013) results do not seem random if we consider only the four relatively simple and young sources that are included in both surveys. L1157 and L1448-IRS2 show very good alignment between field and outflow/pseudodisk in both surveys, and Serp-FIR1 can be discounted as its outflow lies nearly parallel to the line of sight. This leaves L1527, where on the scales of the SHARP/CSO maps the B field is consistent with a pinched poloidal field aligned with the outflow axis, while on the smaller scales probed by CARMA the field is perpendicular to the outflow, suggesting a transition of field directions from large to small scales. This transition is demonstrated in numerical simulations of disks and outflows (e.g., Fig. 15). The simulation in Fig. 15 (*Tomisaka*, 2011) has the core B field, disk minor axis, and outflow well aligned, and the B field is perpendicular to the disk on 10,000-AU scales (Figs. 15a,c) but can be at any direction on 1000-AU scales (Figs. 15b,d). This is consistent

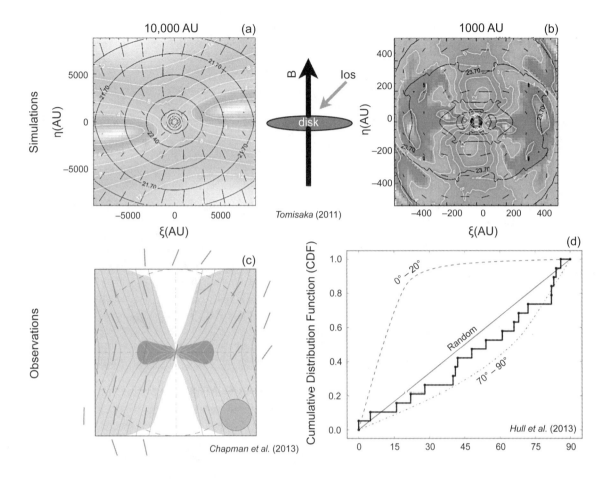

Fig. 15. (a),(b) Simulations: B-field directions inferred from the polarization of thermal dust emission (*Tomisaka*, 2011) based on a MHD and radiation-transfer simulation. The simulation results in a disk perpendicular to the initial B field, and bipolar outflows driven by magnetic pressure. The figure shows the projections along a line of sight 45° from the initial B-field direction. The relative column density is shown by the white contours, and the dark vectors and contours show the field directions and polarization degrees. **(a)** Shown here, on a scale comparable to *Chapman et al.* (2013), are field directions roughly perpendicular to the pseudodisk. **(b)** Closeup of **(a)** showing the central 1000-AU region, which is comparable to *Hull et al.* (2013). The field morphology on the 1000-AU scale starts to be affected by the disk rotation and can be oriented at any direction; i.e., the mean field direction on this scale has nothing to do with the mean field direction of the cloud core. **(c)** Observations at 10,000 AU: Source-average observed B-field directions (bars) superposed on model magnetic field lines (solid lines) and gas density (light and dark shading), adapted from *Chapman et al.* (2013). The model is from *Allen et al.* (2003). The observed source-average B-field map has a mean direction that is nearly vertical and shows hints of a pinch, in accordance with the model. **(d)** Observations at 1000 AU: A B-field survey from *Hull et al.* (2013) shows no correlation between the mean field and the inferred disk orientation. A color version of this figure is available at *http://arxiv.org/pdf/1404.2024v1.pdf.*

with both surveys. Zooming in to the central 300-AU scale, *Tomisaka* (2011) sees toroidal fields. Because of this field complexity at small scales due to outflows and disk rotation, B-field directions on 1000-AU scales are not representative of core fields and thus cannot put constraints on whether cloud fields and disks should be aligned (e.g., *Tomisaka*, 2011) or unaligned (e.g., *Joos et al.*, 2012).

The morphology of gas density and B field evident in Fig. 15 is suggestive of B fields guiding protostellar collapse, just as was the hourglass B field discovered earlier in the class 0 protobinary NGC 1333 IRAS 4A by *Girart et al.* (2006). However, we should not discount the possibility

that feedback may influence the appearance of such maps. Is the field controlling the collapse or vice versa? In the case of NGC 1333 IRAS 4A, the SHARP/CSO map of *Attard et al.* (2009) shows that B fields continue smoothly to larger scales (see the polarization vectors in Fig. 17), indicating that indeed they guide the collapse in this source. But for the *Chapman et al.* (2013) result of Fig. 15, data on larger scales have yet to be compiled. In this case, near-IR polarimetry (*Clemens et al.*, 2007) may provide a way forward. In the longer run, large area submillimeter polarimetry from the stratosphere (*Dowell et al.*, 2010; *Pascale et al.*, 2012) will be combined with ALMA polarimetry to yield multi-scale maps for scores of targets.

Improvements will also come from the development of new analysis techniques. One troublesome effect is the "bias toward 90°." To understand this, consider taking the average of a set of measured B-field position angles lying in the interval 0° to 180°. If the distribution is broad, there will be a tendency for this average to lie closer to 90° rather than to 0° or 180°. For example, note that all 16 of the well-separated independent clouds studied by *Curran and Chrysostomou* (2007) have mean B-field angles lying between 45° and 135°. The strength of this bias varies from study to study, but it is always present unless removed. A simple method for removing it is to use the Equal Weight Stokes Mean statistic (*Li et al., 2006*) in place of a straight mean.

6. DISCUSSION

B-field observations are difficult and undoubtedly involve many uncertainties. However, one should not give up hope, because many useful arguments can still be made despite the uncertainties. For example, Observation I properly treats the ambiguity of the CO-line polarization direction. Also, although there is no way to precisely control the region probed by optical polarization data, contamination from the lines of sight should not "create" the correlations seen in Figs. 6 and 8; if anything, the contamination does the opposite. Many factors indeed contribute to emission linewidths, but how many are even arguably the possible cause of the ion/neutral linewidth difference that is correlated with B-field strength (Observation X)?

In the following, we discuss critically some of the "evidence" that has been used to argue against the strong-field scenario: (1) Does data scattering under the magnetic critical line in the B_{los} vs. N_H plot (Fig. 4) imply super-Alfvénic turbulence? (2) Does the ~2/3 slope in the B_{los} vs. n_H plot imply isotropic core contraction? (3) Are the optical polarization angle dispersions (Figs. 6 and 8) too large for dynamically dominant B fields? (4) Do submillimeter "polarization holes" imply that high-density cores are not probed by submillimeter polarimetry? (5) Is there observational evidence for super-Alfvénic turbulence?

6.1. Zeeman Measurement Scattering and Mass-to-Flux Ratio

The four cores and envelopes in Observation IX (from Perseus and Taurus molecular clouds) are mostly within regions with $A_V > 4$ mag, which is very likely to be higher than the critical column density (see Fig. 4). The cloud contraction thresholds defined by the observed column density PDFs of these two clouds (*Kainulainen et al., 2009*) are indeed lower than $A_V = 4$ mag. Thus, it is possible that self-gravity forms the four cores without help from either AD or super-Alfvénic turbulence. This agrees with the suggestion from *Elmegreen* (2007) that core evolution starts close and stays close to the magnetically critical state. Whether these

cores can further fragment with or without AD is another story, because subregions have smaller mass-to-flux ratios (mass $\propto r^3$ and flux $\propto r^2$, where r is the scale).

A criticism of this magnetic-criticality-hypothesis is that data points in Fig. 4 are consistent with B_{total} widely scattered under the critical line, not just slightly below it (*Crutcher et al., 2010b*), and this is seen in simulations assuming weak B fields. However, with the lessons learned from many of the observations discussed here, we should review the way the data scattering in Fig. 4 is interpreted.

First, the efforts to explain Observation IX improved our understanding of Zeeman measurements. While the interpretation of Observation IX had been very controversial (e.g., *Mouschovias and Tassis, 2009; Crutcher et al., 2010a*), various groups (e.g., *Lunttila et al., 2008, Bertram et al., 2012, Mouschovias and Tassis, 2009, Bourke et al., 2001*) seem to reach one agreement: B_{los} reversals will cause Zeeman measurements to underestimate B_{los}, which can explain Observation IX.

Second, mass contraction along B fields (Observations II and VIII) will not increase the total mass in the system, but will increase the column density observed, as long as the line of sight is not aligned with the B fields (*Li et al., 2011*)

Both of these effects will increase column density-to-B_{los} ratios and should be considered when interpreting data in Fig. 4. Given the same intrinsic B-field strength, the same total mass, and even the same non-zero offset between the line of sight and the mean B field, the appearance of Fig. 4 can still vary. Turbulence brings a data point *downward* by dispersing (not necessarily tangling; see Observation IX) the B-field direction, and gravitational contraction/turbulence compression channeled by B fields moves a data point toward the *right* in Fig. 4. Observing an oblate core with B-field direction close to the symmetry axis (as suggested by Observation VIII) with different lines of sight also situates data points at different positions in Fig. 4: While B_{los} certainly moves down as the line of sight moves away from the core axis, also note that the column density will increase as our view becomes more edge-on to the oblate core. All these effects will cause column density-to-B_{los} ratios to *overestimate* mass-to-flux ratios and cause the data points in Fig. 4 to be scattered *under* the magnetically critical line. So even if B fields are dynamically dominant, data points should still be scattered under the critical line. For the same reason, "R" is scattered in Fig. 11.

6.2. Observation VIII Versus the 2/3 Slope of the B_{los} Versus n_H Plot

Observation VIII also supports the idea that core evolution should stay close to the magnetically critical state (*Elmegreen, 2007*). If the cores are highly supercritical, gravitational contraction should be isotropic and their shape should not correlate with B-field directions, as seen in Observation VIII. However, *Crutcher et al.* (2010b) concluded that a dynamically important B field during core formation is inconsistent with the Zeeman measurements from clouds and cloud cores (based on OH and CN data respectively), because the upper limit of the B_{los} vs. n_H plot has a slope ~2/3. This slope implies that gravitational contraction is isotropic. One possible explanation for this discrepancy is that the OH measurements are from a dark cloud survey, while the CN data is mostly from massive cluster-forming clumps in giant molecular clouds (GMCs). Since it is unlikely for dense cores of nearby dark clouds to evolve into massive cluster-forming clumps, using the slope to infer an isotropic collapse is questionable.

Also worth noticing is that isotropic contraction should result in radial-like B-field morphologies. However, this is rarely observed (e.g., see Figs. 5d, 6, 10, and 17). Even the "hour-glass"-shaped field morphologies (e.g., Fig. 17) are not often observed.

6.3. Optical Polarization Direction Angle Dispersion

The orderliness of B fields can be used as an indication of the relative strength between B fields and turbulence. Based on numerical MHD simulations, field dispersion is sensitive to the Alfvénic Mach number (M_A; Fig. 16); even slightly super-Alfvénic turbulence is enough to significantly distort or even randomize (Fig. 16) field directions. The telescope beam average may lower the dispersion (Obser-

vation XI), but cannot artificially introduce a well-defined mean field direction to a random field. So the fields shown in Figs. 6, 7, and 8 cannot be super-Alfvénic (random). Moreover, the projected two-dimensional B-field dispersion is usually larger than the intrinsic three-dimensional field dispersion (Fig. 16).

Following *Chandrasekhar and Fermi* (1953), we can estimate the lower limit of B-field direction dispersions for super-Alfvénic turbulence and compare this with the standard devision dispersion (STD) shown in Figs. 5 and 7 (on average 31°). Assuming that the STD of B-field direction (σ, observed with a line of sight perpendicular to the mean field) is completely due to gas turbulence, Chandrasekhar and Fermi (CF) derived the relation (CF relation): σ (radians) = $[4\pi\rho]^{1/2}v/B$, where B, ρ, and v are, respectively, the B-field strength (Gauss), density (g cm^{-3}) and line-of-sight turbulent velocity (centimeters per second). They used the small angle approximation, and numerical simulations (e.g., *Ostriker et al.*, 2001) indicate that the CF relation is a good approximation only when $\sigma < 25°$. The average optical polarization STD is 31°, which is too large to use the CF relation. *Falceta-Gonçalves et al.* (2008) improved the relation by replacing σ with tan (σ) and numerically showed that this new relation is applicable for larger σ. Setting $M_A = 1$, i.e., v = $B/(12\pi\rho)^{1/2}$, the improved CF relation gives $\sigma = 30°$, which is almost identical to the observed value.

However, it is not only M_A that affects the observed angle dispersion. Projection effects (Fig. 16) and nonturbulent structures of the B field also contribute to the angle dispersion. Nonturbulent B-field structures include, e.g., those caused by stellar feedback and Galactic B-field structures (intercloud B fields in Figs. 6 and 8 spread out over hundreds of parsecs along a line of sight). Observation XI introduced new analytical tools for removing nonturbulent structures in the polarimetry data, but these have not been applied to the optical polarization data. With all these nonturbulent factors that can significantly increase the B-field dispersion, the B-field dispersion (31°) is only nearly equivalent to the trans-Alfvénic condition assuming turbulence is the only force that can deviate field directions. This suggests that the turbulence is sub-Alfvénic.

6.4. Submillimeter "Polarization Holes"

It is generally observed that the degree of polarization from thermal dust emission decreases with increasing column density. An example is shown in Fig. 17. This phenomenon is called "polarization holes." A straightforward and popular interpretation is that increasing density will reduce the dust grain alignment efficiency; *Padoan et al.* (2001), for example, suggested that grains are not aligned for $A_v > 3$ mag. However, this simple scenario fails to explain the other general observation that was found later with interferometers: Zooming in on polarization holes with interferometers reveals polarization in a similar direction but with *much* larger polarization fractions, some

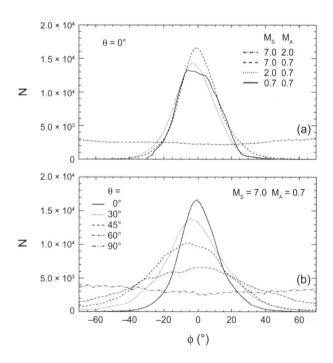

Fig. 16. Simulated submillimeter polarization direction angle (ϕ) dispersion (*Falceta-Gonçalves et al., 2008*). θ is how far the line of sight is from being perpendicular to the mean field direction; M_S and M_A are sonic Mach number and Alfvénic Mach number. **(a)** Super-Alfvénic ($M_A > 1$) turbulence results in random polarization direction. **(b)** Illustration of the projection effect: ϕ dispersion increases with θ.

even larger than those from the low-density sight lines in the single-dish map (Fig. 17). Since interferometers filter out regions that are less dense and more extended, lower polarization should have been observed by interferometers if grain alignment is "turned off" in high-density regions.

Assuming constant grain alignment efficiency, increasing dispersion of B-field direction angles can also lower the polarization fraction. Whether this idea can explain polarization holes depends on whether B-field direction dispersion increases with column density. This seems plausible, because, for a typical clump as shown in Fig. 17, larger column density usually implies larger line-of-sight

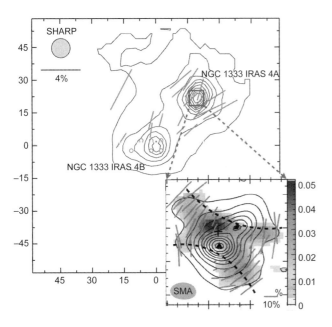

Fig. 17. Low-grain alignment efficiency cannot explain polarization holes. The main figure illustrates the polarization holes in NGC 1333 observed by SHARP/CSO (*Attard et al.,* 2009). Dust emission flux is shown as contours. The vector lengths are proportional to the polarization fractions and the circles are where the polarization is too low to be detected by SHARP. In the upper-left corner are the single-dish beam size and a sample vector with the length of 4% polarization. The coordinates are offsets in arcseconds. Inset: *Girart et al.* (2006) used the SMA to focus on one of the polarization holes and revealed much larger degrees of polarization compared to the single-dish observation; note the sample vector of 10% polarization in the lower-right corner. This certainly goes against the idea that higher density will lower the grain alignment efficiency. The synthesized SMA beam is shown in the lower-left corner. Dust emission flux is shown as contours and the color indicates polarized flux. Note that the vectors in both maps show the submillimeter polarization directions. The inferred B-field directions are orthogonal to the vectors. The dashed lines in the SMA map trace two B-field lines, which appear as an hourglass shape. The B-field direction from the SHARP map is roughly aligned with the short axis of the core (see Observation VIII) and is also aligned with the axis of the hourglass.

dimensions, which should host a larger velocity dispersion based on Larson's law. Then the CF relation implies that B-field direction dispersion is proportional to the velocity dispersion and therefore column density (assuming field strength roughly proportional to the square root of the density). *Li et al.* (2009) indeed observed that polarization fractions decrease with increasing line-of-sight velocity dispersions in DR21; the correlation is even higher than between polarization fraction/column density.

Here we test this idea using the simulations from *Ostriker et al.* (2001). First we select a subregion (Fig. 18) of the beta = 0.01 case, which contains the densest filaments in the simulation. In the X-Y plane, for example, we study the correlation between column densities and B-field direction angle dispersions after averaging along Z. The result is shown in Fig. 18; a clear positive correlation is seen. Note that here no changes in grain alignment efficiency are involved. Simply the B-field dispersion along the line of sight is enough to cause polarization holes.

6.5. Observational Support of Super-Alfvénic Turbulence?

While the observations we reviewed point toward a picture of sub-Alfvénic turbulence, there are other observations that are used to support the scenario of super-Alfvénic clouds. Here we briefly discuss these observations and explain why these claims are based on assumptions that may not be fulfilled.

6.6. Observation XIII: Power Spectra Indices of Cloud Column Densities

One analysis commonly used in support for super-Alfvénic clouds is the column density power spectra suggested by *Padoan et al.* (2004). Their simulations show that Alfvénic flows provide a power law index for column density of 2.25, while highly super-Alfvénic clouds have an index around 2.7. Comparing with these simulations, they concluded that Perseus, Taurus, and Rosetta are all super-Alfvénic, because the power-law indices of their ^{13}CO maps are around 2.75.

However, there are several caveats to the above conclusion. First, the simulated power-law index depends on the exact set-up of the simulations. For example, cloud simulations of *Collins et al.* (2012) show much shallower spectra of column densities, and the power-law indices are not correlated with magnetic Mach numbers. Second, the range of the observed power law indices is between 2 and 3; see Table I of *Schneider et al.* (2011), for example. Third, different tracers can result in very different power-law indices. For example, the indices of Perseus, Taurus, and Rosetta are 2.16, 2.20, and 2.55, respectively, when dust extinction is used to trace the column densities (*Schneider et al.,* 2011). *Schneider et al.* (2011) concluded that the indices probed by dust extinction are usually significantly lower than those probed by CO. With careful comparison of extinction, thermal emission, and CO maps of Perseus,

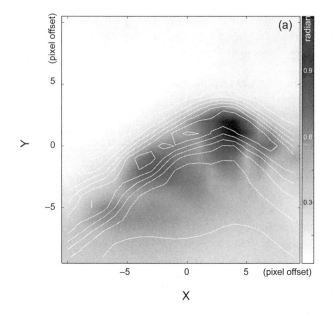

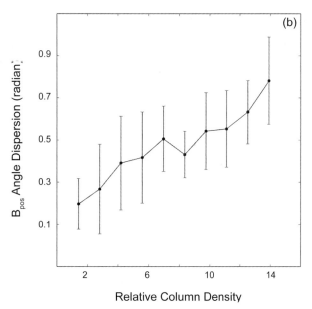

Fig. 18. An illustration of how polarization holes may be caused by B-field structures in a sub-Alfvénic cloud. **(a)** *Ostriker et al.* (2001) simulated a sub-Alfvénic ($\beta = 0.01$) cloud and here we study the densest filament (pixel coordinates: x = 227–247; y = 50–70; z = 237–250) in this simulation. The contours show the mean density along each line of sight; the values of the highest contour and the spacing between adjacent contours are respectively 140× and 20× the uniform density in the initial condition. The grayscale shows the B_{pos} angle dispersion along each line of sight. B_{pos} refers to the plane-of-sky (X-Y plane) component, which will affect the polarization. The higher the angle dispersion, the lower the polarization fraction, assuming constant grain alignment efficiency. **(b)** B_{pos} dispersion vs. column density based on data in **(a)**. The B_{pos} dispersion indeed increases with column density, and thus polarization holes at high column density are expected even with a constant grain alignment efficiency.

Goodman et al. (2009) concluded that dust is a better tracer of column density than CO, because it has no problems of threshold density, opacity, and chemical depletion.

Furthermore, as we have seen in Observation V, filamentary structures are equivalent to anisotropic autocorrelation functions (which is how the filament directions are defined) and thus anisotropic power spectra. The averaged index in this case depends on how the filamentary structure is projected on the sky. These considerations put in doubt the conclusion that empirical column density power laws support super-Alfvénic states of molecular clouds.

6.7. Observation XIV: Cloud/Core B-Field Direction Versus Galactic Disk Orientation

As another argument against the strong-field scenario, *Stephens et al.* (2011) and *Bierman et al.* (2011) used the fact that cloud or core fields are not aligned with the Galactic disk to conclude that cloud B fields must have decoupled from the Galactic B fields. Their argument relies on the assumption that Galactic fields are largely aligned with the disk plane; however, this is not the case at the scale of cloud accumulation length, as we will show in the following.

Figure 6 in *Stephens et al.* (2011) shows an angle distribution of almost all the polarimetry detections from the *Heiles* (2000) catalog, and the distribution clearly peaks in the direction of the Galactic disk plane. Note that this plot contains stars from distances of 140 pc to several kiloparsecs, and thus shows the (Stokes) mean B fields from various scales because the polarization of a star samples the entire sight line (Fig. 19). As a result, one cannot establish from their plot whether the B-field coherence happens at every scale or only at certain scale ranges.

To distinguish between the two possibilities, in Fig. 19, we plot similar polarization distributions but only for stars with distances within 100-pc bins centered at 100, 300, 700, 1500, and 2500 pc in distance respectively. We also use the optical data archive of *Heiles* (2000). We exclude data for which the ratio of the polarization level to its uncertainty is less than 2. The numbers of stars in each distance range are, from nearest to farthest, 1072, 339, 116, 82, and 51. At 100-pc scale the distribution is very flat, i.e., Galactic B fields can have any direction. As shown in Fig. 19, the so-called coherent Galactic B field only appears at scales above 700 pc, where structures at smaller scales are averaged out.

Also shown in Fig. 19 (with a dashed line) is the distribution of the B-field directions from 52 cloud cores at parsec to subparsec scales from *Stephens et al.* (2011). They concluded that the core B fields must have decoupled from the Galactic B fields, because the direction distribution of the core fields is not as peaked as their Fig. 6. However, as the accumulation length of even a GMC is only ~400 pc (*Williams et al.*, 2000), the core B fields are not expected to be related to alactic B-field structures larger than 400 pc. In fact, the distributions of the core fields and the Galactic fields at 100–300-pc scales are very similar in Fig. 19. With the same archives, Observation IV studied the core

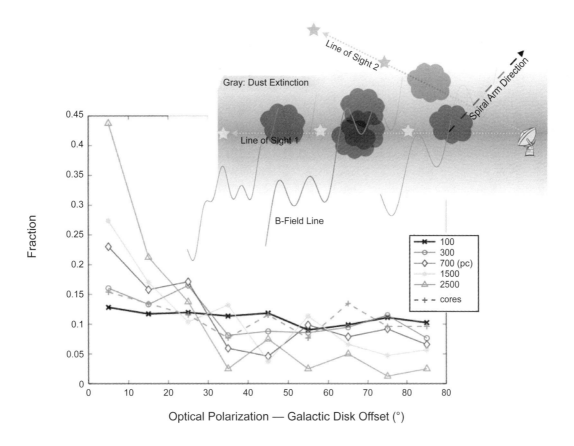

Fig. 19. We sample the stars with reliable polarization detections from 100-pc bins centered at a distance of 100, 300, 700, 1500, and 2500 pc. The plots show that the galactic B field is more coherent at larger scales (above 700 pc), but is almost random at scales near 100 pc. Also plotted is the distribution of the B fields from cloud cores (dashed line) at parsec to subparsec scales, probed with thermal dust emission (*Dotson et al., 2010*). The core field distribution is very similar to that of galactic fields at 100–300-pc scales, the size of the accumulation length of a typical molecular cloud. Inset, top: An illustration of the fact that galactic B fields (smooth lines) follow spiral arms (dark dashed arrow) and anchor clouds (Observations I and IV), but have rich structures perpendicular to the galactic disk. Compared to line-of-sight 1, line-of-sight 2 passes through less galactic mass other than one particular cloud. Stars with larger distance provide averaged field directions corresponding to larger scales. We sample the stars with reliable polarization detections from 100-pc bins centered at distance 100, 300, 700, 1500, and 2500 pc. The plots show that the Galactic B-field is more coherent at larger scales (above 700 pc), but is almost random at scales near 100 pc. Also plotted is the distribution of the B-fields from cloud cores (dashed line) at parsec to subparsec scales, probed with thermal dust emission (*Dotson et al., 2010*). The core field distribution is very similar to that of Galactic fields at 100–300-pc scales, the size of the accumulation length of a typical molecular cloud. A color version of this figure is available at *http://arxiv.org/pdf/1404.2024v1.pdf*.

fields and the polarization within 100–200 pc (accumulation length) of each core, and showed a significant correlation (Fig. 6). This means that the structures of Galactic B fields at the scale of cloud formation are preserved in the cores.

7. SUMMARY

Recent B-field surveys and other related observations are discussed (Fig. 1). Observations I (the correlation between cloud fields and spiral arms) and II (the constant ICM field strength) show that galactic B fields are strong enough to hinder cloud rotation and channel gravitational contraction and imprint their directions onto molecular clouds. Only after the accumulated mass reaches a critical

value defined by the galactic B-field strength (~10 μG) can the cloud also contract in a direction perpendicular to the field lines and increase the field strength (Observation III). The cloud contraction density threshold can be reached by accumulation along B fields; AD is not necessary for cloud contraction. However, observational proof (or lack thereof) of the existence of AD is very challenging (Observation IX). Observation IV shows that the galactic field directions anchor into clouds all the way down to cloud cores. This B-field direction will make the ISM anisotropic in density profiles (Observations V, VI, and VIII) and in velocity profiles of turbulence (Observation VII). The fact that cloud cores are flattened with the short axes oriented close to the B fields (Observation VIII) implies that, although cores might form under the supercritical state (Observation III), they cannot be so supercritical that the contraction becomes isotropic.

While a number of mechanisms have been proposed to explain how protostellar disks can overcome magnetic braking (see the chapter by Z.-Y. Li et al. in this volume), observations of the B-field effects on disk formation are still at a very early stage. To our knowledge, two types of observations have been tried: (1) Probing the B-field-turbulence decoupling scale: Two methods (Observations X and XI) agree on a value of several milliparsecs, which is comparable to the scale of disk formation. (2) Studying core field-outflow alignments: Assuming outflow directions as an indication of disk orientations (because outflow directions are easier to identify), Observation XII is aimed at studying whether core fields have any effect on disk orientations. However, we note that, like turbulent flows and gravitational contraction, outflows and disk rotation also interact with core B fields. One cannot know for sure whether the outflows and/or disk rotation have destroyed the alignment or created fake alignments, especially on relatively small (100–1000 AU) scales. Multiple-scale observations, similar to Observation IV, to determine whether B-field directions are changed in the outflows, disks, and vicinities are critical for these kinds of experiments.

At the end (section 6), we show that the observations used to criticize the strong-field scenario are not without their own difficulties upon close examination. ALMA should be able to further test the strong-field scenario (Observations IV–VIII) in regions with even higher densities.

Acknowledgments. H.L. appreciates the support from the Deutsche Forschungs-gemeinschaft priority program 1573 ("Physics of the Interstellar Medium"). Z.Y.L. is supported in part by NASA NNX10AH30G, NNX14AB38G, and NSF AST1313083.

REFERENCES

Allen A. et al. (2003) *Astrophys. J., 599,* 363.
Andrè Ph. et al (2010) *Astron. Astrophys., 518,* 102.
Arzoumanian D. et al. (2011) *Astron. Astrophys., 529,* 6.
Attard M. et al. (2009) *Astrophys. J., 702,* 1584.
Barranco J. and Goodman A. (1998) *Astrophys. J., 504,* 207.
Benson P. and Myers P. (1989) *Astrophys. J. Suppl., 71,* 89.
Bertram E. et al. (2012) *Mon. Not. Roy. Astron. Soc., 420,* 3163.
Bierman E. et al. (2011) *Astrophys. J., 741,* 81.
Bourke T. et al. (2001) *Astrophys. J., 554,* 916.
Chandrasekhar S. and Fermi E (1953) *Astrophys. J., 118,* 113.
Chapman N. et al. (2011) *Astrophys. J., 741,* 21.
Chapman N. et al. (2013) *Astrophys. J., 770,* 151.
Chitsazzadeh S. et al. (2012) *Astrophys. J., 749,* 45.
Cho J. and Lazarian A. (2002) *Phys. Rev. Lett., 88,* 245001.
Cho J. and Vishniac E. T. (2000) *Astrophys. J., 539,* 273.
Cho J. et al. (2002) *Astrophys. J., 564,* 291.
Ciolek G. and Mouschovias T. (1993) *Astrophys. J., 418,* 774.
Clemens D. et al. (2007) *Publ. Astron. Soc. Pac., 119,* 1385.
Clemens D. et al. (2012) *Astrophys. J. Suppl., 200,* 21.
Collins D. et al. (2012) *Astrophys. J., 750,* 13–18.
Cortes P. et al. (2005) *Astrophys. J., 628,* 780.
Crutcher R. (2012) *Annu. Rev. Astron. Astrophys., 50,* 29.
Crutcher R. et al. (1999) *Astrophys. J. Lett., 514,* L121.
Crutcher R. et al. (2004) *Astrophys. J., 600,* 279.
Crutcher R. et al. (2009) *Astrophys. J., 692,* 844–855.
Crutcher R. et al. (2010a) *Mon. Not. R. Astron. Soc., 402,* 64.
Crutcher R. M. et al. (2010b) *Astrophys. J., 725,* 466-479.
Curran R. and Chrysostomou A. (2007) *Mon. Not. R. Astron. Soc., 382,* 699.
Davidson J. et al. (2011) *Astrophys. J., 732,* 97.
Dobashi K. (2011) *Publ. Astron. Soc. Japan, 63,* 1.
Dobbs C. G (2008) *Mon. Not. R. Astron. Soc., 391,* 844.
Dotson J. (1996) *Astrophys. J., 470,* 566.
Dotson J. L. et al. (2010) *Astrophys. J. Suppl., 186,* 406.
Dowell D. et al. (2010) In *Ground-Based and Airborne Instrumentation for Astronomy III* (I. S. McLean et al., eds.), 77356H. SPIE Conf. Series 7735, Bellingham, Washington, DOI: 10.1117/12.857842.
Elmegreen B. (2007) *Astrophys. J., 668,* 1064.
Falceta-Gonçalves D. et al. (2008) *Astrophys. J., 679,* 537.
Fauvet L. et al. (2012) *Astron. Astrophys., 540,* 122.
Fissel L. et al. (2010) In *The Balloon-Borne Large-Aperture Submillimeter Telescope for Polarimetry: BLAST-Pol* (W. S. Holland and J. Zmuidzinas, eds.), p. 9. SPIE Conf. Series 7741, Bellingham, Washington.
Frisch U. (1995) *Turbulence: The Legacy of A. N. Kolmogorov.* Cambridge Univ., Cambridge. 296 pp.
Froebrich D. and Rowles J. (2010) *Mon. Not. R. Astron. Soc., 406,* 1350–1357.
Galli D. and Shu F. (1993) *Astrophys. J., 417,* 220.
Girart J. et al. (2006) *Science, 313,* 812.
Goldreich P. and Kylafis N. (1981) *Astrophys. J., 243,* 75.
Goldreich P. and Sridhar H. (1995) *Astrophys. J., 438,* 763.
Goldreich P. and Sridhar H. (1997) *Astrophys. J., 485,* 680.
Goldsmith P. et al. (2008) *Astrophys. J., 680,* 428.
Goodman A. et al. (1990) *Astrophys. J., 359,* 363.
Goodman A. et al. (1998) *Astrophys. J., 504,* 223.
Goodman A. et al. (2009) *Astrophys. J., 692,* 91.
Hall J. (1951) *Astrophys. J., 56,* 40.
Han J. L. and Zhang J. S. (2007) *Astron. Astrophys., 464,* 609.
Hartmann L. et al. (2001) *Astrophys. J., 562,* 852.
Heiderman A. et al. (2010) *Astrophys. J., 723,* 1019–1037.
Heiles C. (2000) *Astron. J., 119,* 923.
Heiles C and Troland T. (2005) *Astrophys. J., 624,* 773.
Heiles C. et al. (1993) In *Protostars and Planets III* (E. H. Levy and J. I. Lunine, eds.), p. 279. Univ. of Arizona, Tucson.
Hennebelle P. et al. (2011) *Astron. Astrophys., 528,* 72.
Henning Th. et al. (2010) *Astron. Astrophys., 518,* 95.
Heyer M. H. and Brunt C. M. (2012) *Mon. Not. R. Astron. Soc., 420,* 1562–1569.
Heyer M. et al. (2008) *Astrophys. J., 680,* 420.
Hezareh T. et al. (2010) *Astrophys. J., 720,* 603.
Hildebrand R. et al. (2009) *Astrophys. J., 696,* 567.
Hill T. et al. (2011) *Astron. Astrophys., 533,* A94.
Hiltner W. (1951) *Astrophys. J., 114,* 241.
Houde M. et al. (2000a) *Astrophys. J., 537,* 245.
Houde M. et al. (2000b) *Astrophys. J., 536,* 857.
Houde M. et al. (2002) *Astrophys. J., 569,* 803.
Houde M. et al. (2009) *Astrophys. J., 706,* 1504.
Houde M. et al. (2011) *Astrophys. J., 733,* 109.
Houde M. et al. (2013) *Astrophys. J., 766,* 49.
Hull C. et al. (2013) *Astrophys. J., 768,* 159.
Jones C. et al. (2001) *Astrophys. J., 551,* 387.
Joos M. et al. (2012) *Astron. Astrophys., 543,* 128.
Kainulainen J. et al. (2009) *Astron. Astrophys., 508,* 35–39.
Kandori R. et al. (2006) In *Ground-Based and Airborne Instrumentation for Astronomy* (I. S. McLean and M. Iye, eds.), p. 159. SPIE Conf. Series 6269, Bellingham, Washington.
Lada C. et al. (2010) *Astrophys. J., 724,* 687.
Lazarian A. et al. (2012) *Astrophys. J., 757,* 154.
Li H.-b. and Henning T. (2011) *Nature, 479,* 499–501.
Li H.-b. and Houde M. (2008) *Astrophys. J., 677,* 1151.
Li H.-b. et al. (2006) *Astrophys. J., 648,* 340.
Li H. et al. (2008) *Applied Optics, 47,* 422.
Li H.-b. et al. (2009) *Astrophys. J., 704,* 891.
Li H.-b. et al. (2010) *Astrophys. J., 718,* 905.
Li H.-b. et al. (2011) *Mon. Not. R. Astron. Soc., 411,* 2067–2075.
Li H.-b. et al. (2013) *Mon. Not. R. Astron. Soc., 436,* 3707.
Li P.-S. et al. (2008) *Astrophys. J., 684,* 380–394.
Li P.- S. et al. (2012) *Astrophys. J., 744,* 73.
Li Z.-Y. et al. (2011) *Astrophys. J., 738,* 180.

Lunttila T. et al. (2008) *Astrophys. J., 686,* 91–94.

Machida M. et al. (2011) *Publ. Astron. Soc. Japan, 63,* 555.

Maron J. and Goldreich P. (2001) *Astrophys. J., 554,* 1175.

Matthews B. et al. (2009) *Astrophys. J. Suppl., 182,* 143–204.

Matthews T. G. et al. (2013) *ArXiv e-prints,* arXiv:1307.5853.

McKee C. et al. (1993) In *Protostars and Planets III* (E. H. Levy and J. I. Lunine, eds.) p. 327. Univ. of Arizona, Tucson.

Mestel L. and Spitzer L. Jr. (1956) *Mon. Not. R. Astron. Soc., 116,* 503.

Mizuno A. et al. (1995) *Astron. Astrophys., 445,* 161.

Molinari S. et al. (2010) *Astron. Astrophys., 518,* 100.

Mouschovias T. (1976) *Astrophys. J., 207,* 141.

Mouschovias T. and Paleologou E. (1979) *Astrophys. J., 230,* 204.

Mouschovias T. and Tassis K. (2009) *Mon. Not. R. Astron. Soc., 400,* 15.

Mouschovias T. and Tassis K. (2010) *Mon. Not. R. Astron. Soc., 409,* 801.

Murillo N. and Lai S.-P. (2013) *Astrophys. J., 764,* 15.

Myers P. (2009) *Astrophys. J., 700,* 1609.

Nagahama T. et al. (1998) *Astron. J., 116,* 336.

Nagai T. et al. (1998) *Astrophys. J., 506,* 306.

Nakamura F. and Li Z. (2008) *Astron. J., 687,* 354.

Nakano T. and Nakamura T. (1978) *Publ. Astron. Soc. Japan, 30,* 681.

Nordlund A. K. and Padoan P. (1999) In *Interstellar Turbulence, Proceedings of the 2nd Guillermo Haro Conference* (J. Franco and A. Carraminana, eds.), pp. 218–223. Cambridge Univ., Cambridge.

Novak G. (2006) *SPIE Newsroom,* 27 September 2006, DOI: 10.1117/2.1200609.0373.

Ostriker E. C. et al. (2001) *Astrophys. J., 546,* 980.

Padoan P. et al. (2001) *Astrophys. J., 553,* 227.

Padoan P. et al. (2004) *Astrophys. J., 604,* 49.

Palmeirim P. et al. (2013) *Astron. Astrophys., 550,* 38.

Pascale E. et al. (2012) In *Ground-Based and Airborne Telescopes IV* (L. M. Stepp et al., eds.), p. 15. SPIE Conf. Series 8444, Bellingham, Washington.

Peretto N. et al. (2012) *Astron. Astrophys., 541,* 63.

Price D. and Bate M. (2008) *Mon. Not. R. Astron. Soc., 385,* 1820.

Ragan S. et al. (2012) *ArXiv e-prints,* arXiv:1207.6518R.

Scandariato G. et al. (2011) *Astron. Astrophys., 533,* 38.

Schlegel D. et al. (1998) *Astrophys. J., 50,* 525.

Schneider N. et al. (2011) *Astron. Astrophys., 529,* 1.

Schneider S. and Elmegreen B. (1979) *Astrophys. J. Suppl., 41,* 87.

Seifried D. et al. (2012) *Mon. Not. R. Astron. Soc., 423,* 40.

Shetty R. and Ostriker E. (2006) *Astrophys. J., 647,* 997.

Shu F. et al. (1999) In *The Origin of Stars and Planetary Systems* (C. J. Lada and N. D. Kylafis, eds.), p. 193. Kluwer, Dordrecht.

Stephens I. et al. (2011) *Astrophys. J., 728,* 99.

Stone J. et al. (1998) *Astrophys. J., 508,* 99.

Tassis K. et al. (2009) *Mon. Not. R. Astron. Soc., 399,* 1681.

Tilley D. and Balsara D. (2010) *Astrophys. J., 406,* 1201.

Tilley D. and Balsara D. (2011) *Mon. Not. R. Astron. Soc., 415,* 3681.

Tobin J. et al. (2012) *Nature, 942,* 83.

Tomisaka K. (2011) *Publ. Astron. Soc. Japan, 63,* 147.

Troland T. and Crutcher R. (2008) *Astrophys. J., 680,* 457–465.

Vaillancourt J. et al. (2008) *Astrophys. J., 679,* 25.

Vestuto J. G. et al. (2003) *Astrophys. J., 590,* 858.

Vlemmings W. et al. (2011) *Astron. Astrophys., 529,* 95.

Williams J. et al. (2000) In *Protostars and Planets IV* (V. Mannings et al., eds.), p. 97. Univ. of Arizona, Tucson.

Wolf S. et al. (2003) *Astrophys. J., 592,* 233.

Molinari S., Bally J., Glover S., Moore T., Noreiga-Crespo A., Plume R., Testi L., Vázquez-Semadeni E., Zavagno A., Bernard J.-P., and Martin P. (2014) The Milky Way as a star formation engine. In *Protostars and Planets VI* (H. Beuther et al., eds.), pp. 125–148. Univ. of Arizona, Tucson, DOI: 10.2458/azu_uapress_9780816531240-ch006.

The Milky Way as a Star Formation Engine

Sergio Molinari
Istituto Nazionale di Astrofisica

John Bally
University of Colorado at Boulder

Simon Glover
Universität Heidelberg

Toby Moore
Liverpool John Moores University

Alberto Noriega-Crespo
California Institute of Technology and Space Telescope Science Institute

René Plume
University of Calgary

Leonardo Testi
European Southern Observatory, Istituto Nazionale di Astrofisica, and Excellence Cluster Universe

Enrique Vázquez-Semadeni
Universidad Nacional Autonoma de Mexico

Annie Zavagno
Aix Marseille Université

Jean-Philippe Bernard
Institut de Recherche en Astrophysique et Planétologie

Peter Martin
University of Toronto

The cycling of material from the interstellar medium (ISM) into stars and the return of stellar ejecta into the ISM is the engine that drives the galactic ecology in normal spirals. This ecology is a cornerstone in the formation and evolution of galaxies through cosmic time. There remain major observational and theoretical challenges in determining the processes responsible for converting the low-density, diffuse components of the ISM into dense molecular clouds, forming dense filaments and clumps, fragmenting them into stars, expanding OB associations and bound clusters, and characterizing the feedback that limits the rate and efficiency of star formation. This formidable task can be attacked effectively for the first time thanks to the synergistic combination of new global-scale surveys of the Milky Way from infrared (IR) to radio wavelengths, offering the possibility of bridging the gap between local and extragalactic star-formation studies. The Herschel Space Observatory Galactic Plane Survey (Hi-GAL) survey, with its five-band 70–500-μm full Galactic Plane mapping at 6″–36″ resolution, is the keystone of a set of continuum surveys that include the Galactic Legacy Infrared Mid-Plane Survey Extraordinaire (GLIMPSE)(360)+MIPSGAL@Spitzer, Wide-field Infrared Survey Explorer (WISE), Midcourse Space Experiment (MSX), APEX Telescope Large Area Survey of the Galaxy (ATLASGAL)@Atacama Pathfinder EXperiment (APEX), Bolocam Galactic Plane Survey (BGPS)@Caltech Submillimeter Observatory (CSO), and CORNISH@Very Large Array (VLA). This suite enables us to measure the Galactic distribution and physical properties of dust on all scales and in all components of the ISM from diffuse clouds to filamentary complexes and hundreds of thousands of dense clumps. A complementary suite of spectroscopic surveys in various atomic and molecular tracers is providing the chemical fingerprinting of dense clumps and filaments, as well as essential kinematic information to derive distances and thus transform panoramic data into a three-dimensional representation. The latest results emerging from these Galaxy-scale surveys are reviewed. New insights into cloud formation and evolution, filaments and their relationship to channeling gas onto gravitationally-bound clumps, the properties of these clumps, density thresholds for gravitational collapse, and star and cluster formation rates are discussed.

1. INTRODUCTION

Phase changes in the Galactic interstellar medium (ISM) are to a large extent controlled by the formation of massive stars. The cycling of the ISM from mostly neutral atomic (HI) clouds into molecular (H_2) clouds, traced by low-excitation species such as OH and CO, leads to the formation of dense, self-gravitating clumps and cores traced by high-density species such as NH_3, CS, HCN, HCO^+, and N_2H^+, and by other even higher dipole-moment molecules in regions where stars form.

Dust continuum emission in the mid-infrared (IR), far-IR, submillimeter, and millimeter ranges of the spectrum reveal progressively cooler and higher column density dust associated with all phases of the ISM and with the high-density material in protostellar envelopes and disks. Dust extinction measurements in the ultraviolet (UV), visual, near-IR, and mid-IR enable detection of progressively higher column densities of dust located in front of background stellar and diffuse emission sources.

Feedback from young stars limits the rate and efficiency of star formation by generating turbulence and disrupting the parent clouds. In giant molecular clouds (GMCs), feedback, usually dominated by the most massive young stellar objects (YSOs), tends to disrupt the parent cloud by the time 2–20% of its mass has formed stars. Ionization and shocks produced by massive stars, associations, and clusters convert the remaining gas into the 10^4–10^8 K photoionized and shock-heated phases of the ISM-HII regions traced by H and He recombination lines and free-free emission, hot superbubbles traced by X-ray emission, and 0.1- to 1-kpc-scale supershells that eventually cool, condense, and reform the cool ~20 to 10^4 K HI phase. Compression by shocks and gravity leads to the formation of new GMCs.

In the solar vicinity, atoms cycle through this loop on a timescale of 50–100 million years (m.y.). A mean star-formation efficiency (SFE) of 5% per GMC implies that atoms on average pass through this loop about 20× before being incorporated into a star. This galactic ecology is modulated by large-scale processes such as spiral arms and the central bar of the Milky Way. While star formation depletes the ISM at a rate of ~2 $M_\odot$ yr^{-1}, infall of gas from the Local Group supplements it at a highly uncertain rate of 0.1 to 1 $M_\odot$ yr^{-1}. The balance between star formation, which sequesters matter for the main-sequence lifetime of stars, and recycling from stellar winds and dying stars, supplemented by infall from the Local Group, determines the timescale on which the Galactic ISM is depleted. Large-scale Galactic Plane surveys of the last decade provide the data required to flesh out the details of this galactic ecology.

The Milky Way, a mildly barred, gas-rich spiral galaxy, supports the most active star formation in the Local Group. The Galactic disk contains about 1–3 × 10^9 $M_\odot$ of H_2 and about 2–6 × 10^9 $M_\odot$ of HI (*Combes*, 1991; *Kalberla and Kerp*, 2009). The molecular disk becomes prominent from Galactocentric radius R_{gal} ~ 3 kpc, beyond the central bar, with surface density peaking at about R_{gal} ~4–6 kpc (often

referred to as the molecular ring). It declines exponentially toward larger radii but can be traced well beyond R_{gal} ~10 kpc. The HI disk extends to beyond R_{gal} ~20 kpc. The disk ISM is dominated by a four-arm spiral pattern consisting of two major and two minor spiral arms and a variety of interarm spurs. The Sun is currently located in an interarm region between the Sagittarius and Perseus spiral arms, near a spur that extends from a distance of several kiloparsecs in the direction of Cygnus to at least 1 kpc beyond Orion in the opposite direction (*Xu et al.*, 2013). In molecular tracers such as CO, the arm-interarm contrast in the molecular ring is around 3:1, but in the outer Galaxy beyond the solar circle at $R_{gal} \approx 8.5$ kpc the contrast is much higher, approaching a value of 40:1 toward the Perseus arm.

Although Galactic rotation is well described by circular motions with orbit speeds ranging from 200 to 250 km s^{-1} from R_{gal} ~1 kpc to beyond 20 kpc, substantial radial motions are seen in gas tracers toward the Galactic Center and anticenter. The line of sight toward the Galactic Center shows three clearly defined spiral arms, distinguished by their negative radial velocities with respect to the local standard of rest (LSR, defined by the mean motion of stars in the solar vicinity). The innermost is the so-called 3-kpc arm in the molecular ring ($V_{LSR} \approx$ –60 km s^{-1}), next is the Scutum arm ($V_{LSR} \approx$ –35 km s^{-1}), and then closest (~2 kpc) is the Sagittarius arm ($V_{LSR} \approx$ –15 km s^{-1}). Radial streaming motions on the order of 10 to 30 km s^{-1} are also seen toward the Perseus arm and the far outer arm beyond that. These motions might reflect the gravitational potential well depth of the spiral arms.

The inner edge of the molecular ring at R_{gal} ~ 3 kpc lies near the outer radius of the Galactic bulge and bar. The central 3-kpc region of the Galaxy is dominated by the stellar bar whose major axis is inclined by 20° to 40° with respect to our line of sight (*Binney et al.*, 1991; *Morris and Serabyn*, 1996). Within 0.5 kpc of the nucleus lies the Central Molecular Zone (CMZ), containing ~10% of the molecular gas in the Galaxy. The CMZ clouds are 1 to 2 orders of magnitude denser than GMCs in the molecular ring and an order of magnitude more turbulent.

2. PHYSICAL PROPERTIES OF THE MILKY WAY INTERSTELLAR MEDIUM FROM DUST AND GAS TRACERS

With the advent of photography and multi-color photometry in the late 1800s and early 1900s, the presence of interstellar dust was recognized by its extinction and reddening effect on the light from distant stars. Evidence for interstellar gas followed in the 1930s with the detection of narrow absorption lines of sodium and potassium and a few simple molecules such as CH and CH^+. The discovery of the 21-cm line of HI in the 1940s led to the first all-sky surveys of the ISM and the recognition that the Galaxy contains more than 10^9 $M_\odot$ of gas. The detection of absorption and thermal emission from OH, NH_3, CO, and many other molecules in the 1960s and 1970s led to the discovery of molecular clouds and the recognition that GMCs were the birth sites of stars.

The last 30 years have seen an impressive set of IR, millimeter, and radio surveys both in spectroscopy and in continuum bands that have covered the Galactic Plane. Table 1 gives a representative list. These have increased exponentially the amount of information available but at the same time have exposed the complexity of the puzzle of the galactic ecology.

2.1. Spectral Line Surveys

Detector technology has to a large extent dictated the development of Galactic Plane surveys. From the late 1970s until the late 1990s, heterodyne receivers dominated, with a number of ever more detailed spectral line surveys of

the entire sky in HI, portions of the Galactic Plane in CO and selected regions such as Orion, Sgr B2, and other starforming complexes, as well as a few post-main-sequence stars in a range of other molecular species.

Since the early 1970s, CO emission has been the most commonly used tracer of molecular gas in the Galaxy. The CO molecule is the most abundant species in molecular gas next to H_2 and He. Its small (~ 0.1 Debye) dipole-moment results in bright CO emission widespread along the Galactic Plane. Its large abundance ($\sim 10^{-4} \times$ that of H_2), high opacity, and consequent radiative trapping enables CO to trace molecular gas at H_2 densities above $\sim 10^2$ cm^{-3}. The first comprehensive survey in the CO J=1–0 line (*Dame et al.,* 1987) covered the Galactic Plane over a latitude range of 10°–20° with a grid resolution of 0.5°, and parts of the Galactic Plane

TABLE 1. List of most representative surveys covering the Galactic Plane.

Survey Facilities	λ or lines	Survey Notes
Groundbased		
Columbia/CfA	CO, ^{13}CO	9–25′ resolution (*Dame et al.,* 2001)
DRAO/ATCA/VLA	HI-21 cm OH/Hα-RRL/1–2-GHz continuum 5-GHz continuum	IGPS: unbiased HI-21 cm 255° ≤ l ≤ 357° and 18° ≤ l ≤ 147° (*McClure-Griffiths et al.,* 2001; *Gibson et al.,* 2000; *Stil et al.,* 2006) + THOR: unbiased HI-21 cm/OH/Hα-RRLs/1–2 GHz continuum 15° ≤ l ≤ 67° (Beuther et al., in preparation) + CORNISH: 5-GHz continuum 10° ≤ l ≤ 65° (*Hoare et al.,* 2012)
FCRAO 14 m	CO, ^{13}CO	55″ resolution. Galactic Ring Survey (*Jackson et al.,* 2006) + Outer Galaxy Survey (*Heyer et al.,* 1998)
Mopra 22 m	CO, ^{13}CO, N_2H+, ($NH_3 + H_2O$) maser, HCO+/H^{13}CO+ + others	HOPS: (*Walsh et al.,* 2011; *Purcell et al.,* 2012), MALT90: ~2000 clumps 20° ≥ l ≥ –60° (*Foster et al.,* 2013), Southern GPS CO: unbiased 305° ≤ l ≤ 345° (*Burton et al.,* 2013), ThrUMMS: unbiased 300° ≤ l ≤ 358° (*Barnes et al.,* 2013), CMZ: (*Jones et al.,* 2012, 2013)
Parkes	CH$_3$OH maser	Methanol Multi-Beam Survey (*Green et al.,* 2009)
NANTEN/ NANTEN2	CO, ^{13}CO, C^{18}O	NGPS: unbiased, 200° ≤ l ≤ 60° (*Mizuno and Fukui,* 2004) + NASCO: unbiased in progress, 160° ≤ l ≤ 80°
CSO 10 m	1.3 mm continuum	Bolocam Galactic Plane Survey (BGPS), 33″ (*Aguirre et al.,* 2011)
APEX 12 m	870 μm continuum	ATLASGAL, 60° ≥ l ≥ –80° (*Schuller et al.,* 2009)
Spaceborne		
IRAS	2, 25, 60 and 100 μm continuum	3–5′, 96% of the sky
MSX	8.3, 12.1, 14.7, 21.3 μm continuum	Full Galactic Plane (*Price et al.,* 2001)
WISE	3.4, 4.6, 11, 22 μm continuum	All-sky (*Wright et al.,* 2010)
Akari	65, 90, 140, 160 μm continuum	All-sky (*Ishihara et al.,* 2010)
Spitzer	3.6, 4.5, 6, 8, 24 μm continuum	GLIMPSE+GLIMPSE360: Full Galactic Plane (*Benjamin et al.,* 2003; *Benjamin and GLIMPSE360 Team,* 2013) + MIPSGAL, 63° ≥ l ≥ –62° (*Carey et al.,* 2009)
Planck	350, 550, 850, 1382, 2098, 3000, 4285, 6820, 10^4 μm continuum	All-sky, resolution ≥5′ (*Planck Collaboration et al.,* 2013a)
Herschel	70, 160, 250, 350, 500 μm continuum	Hi-GAL: Full Galactic Plane (*Molinari et al.,* 2010a)

with a resolution of 9′, revealing the large-scale spatial and velocity distribution of CO-emitting molecular gas. *Dame et al.* (2001) combined spectra from 31 subsurveys to cover 4°–10° in latitude with 9′–15′ angular resolution. The first all-sky image of CO emission was produced from Planck data through careful analysis, detector by detector, of Planck High Frequency Instrument (HFI) continuum images (*Planck Collaboration et al.*, 2013b). Velocity-integrated maps of the three lowest rotational transitions of CO were produced at ~5′ resolution. In the regions of overlap, the agreement between the *Dame et al.* (2001) and the Planck products is excellent. The Planck CO images reveal many new high-latitude CO-emitting clouds.

Higher-angular-resolution (<few arcminutes) CO surveys cover selected portions of the Galactic Plane. The Five College Radio Astronomy Observatory (FCRAO) 14-m telescope surveyed the Northern Galactic Plane with a resolution of ~45″ (*Heyer et al.*, 1998; *Roman-Duval et al.*, 2010). Data covering the remaining portions of the Northern Galactic Plane were obtained before FCRAO was closed, but have not yet been published. Selected regions of the southern sky have been obtained with the NANTEN2, Atacama Submillimeter Telescope Experiment (ASTE), Atacama Pathfinder EXperiment (APEX), and Mopra telescopes. Although CO is an excellent tracer of H_2, most of the gas traced by this low-excitation species is not directly associated with star formation.

Detailed studies of selected nearby molecular clouds such as those in Taurus, Ophiuchus, Perseus, and Orion, as well as of clouds in more distant portions of the Galactic Plane, demonstrated that high-dipole moment (~2 to 4 Debye) species such as NH_3, CS, HCN, HCO^+, and N_2H^+ are more closely associated with dense, self-gravitating clumps and cores associated with the formation of clusters and individual stars. Recently, the Mopra telescope surveyed the southern Milky Way in the 1.3-cm NH_3 and H_2O maser lines [H_2O southern galactic Plane Survey (HOPS)] (*Walsh et al.*, 2011). Mopra is also conducting surveys of dense gas tracers around 3-mm [Millimetre Astronomy Legacy Team 90GHz (MALT90)] (*Foster et al.*, 2011, 2013), CH, and CO isotopologs [Galactic Census of High and Medium Mass Protostars (CHaMP)] (*Barnes et al.*, 2011).

2.2. Continuum Surveys

The Infrared Astronomical Satellite (IRAS) survey in the mid-1990s provided the first far-IR all-sky survey in the dust continuum. The development of focal plane arrays in the IR to far-IR wavelength region and submillimeter bolometer arrays in the late 1990s led to a series of surveys from spaceborne facilities that include the Infrared Space Observatory (ISO), Midcourse Space Experiment (MSX), and Cosmic Background Explorer (COBE). The last decade has seen a number of large continuum surveys with Spitzer, Akari, the Wide-field Infrared Survey Explorer (WISE), and, most recently, Herschel and Planck. The 12-m APEX telescope was used for the APEX Telescope Large Area

Survey of the Galaxy (ATLASGAL) survey of the Southern Galactic Plane at 870 μm (*Schuller et al.*, 2009). The 10-m Caltech Submillimeter Observatory (CSO) was used to obtain the 1.1-mm BGPS of the Northern Galactic Plane at 1300 μm with a 33″ beam (*Aguirre et al.*, 2011). The 15-m James Clark Maxwell Telescope (JCMT) is being used for 850-μm and 450-μm SCUBA2 surveys of the Galactic Plane and several nearby Gould belt clouds.

A major breakthrough in the study of massive star and cluster formation was the discovery of IR dark clouds (IRDCs), which trace the highest column density clumps and cores in GMCs. Unbiased extractions of IRDCs from the MSX maps produced catalogs of several thousand clouds (*Simon et al.*, 2006). Variations in the mid-IR background, however, can mimic an IRDC without dense and cold intervening material. When a more refined approach was used to extract IRDCs from the higher-quality Spitzer/Infrared Array Camera (IRAC) images of the Galactic Legacy Infrared Mid-Plane Survey Extraordinaire (GLIMPSE) survey (*Benjamin et al.*, 2003), *Peretto and Fuller* (2009) found twice as many clouds as *Simon et al.* (2006), yet did not confirm many of Simon et al.'s IRDCs. It has been suggested that IRDCs are the high-mass analogs of low-mass prestellar cores. Early radio spectroscopic follow-up (e.g., *Pillai et al.*, 2006) suggested that star formation in IRDCs is ongoing. *Peretto and Fuller* (2009) showed that as many as 70% of their IRDCs were indeed associated with a 24-μm compact source from the Spitzer/MIPSGAL Survey (*Carey et al.*, 2009), again indicating ongoing star formation.

Dust continuum emission at far-IR, submillimeter, and millimeter wavelengths provides a robust tracer of the dust column density and mean grain temperature along the line of sight, independent of the phase of the associated ISM. Surveys with Spitzer and WISE have mapped most of the Galactic Plane at 3.6, 4.5, 6, 8, and 24 μm with resolutions of a few to 10″. The Herschel Space Observatory Galactic Plane Survey (Hi-GAL) and its successor extensions (*Molinari et al.*, 2010b) have mapped the entire Galactic Plane in five bands at 70, 160, 250, 350, and 500 μm with angular resolutions of about 6″ to 36″.

Distances must be determined to extract physical quantities from continuum data. Some dust continuum morphological features are associated with objects for which distance estimates exist. However, for the majority of continuum sources there are no such associations. Continuum surveys of the Galactic Plane must be combined with complementary heterodyne surveys that provide measurements of the radial velocity of gas associated with dust continuum emission or absorption features. Most current distance determination methods rely on kinematic distances deduced from an axisymmetric Galactic rotation model combined with a variety of methods to resolve the kinematic distance ambiguity (KDA) for objects located within the solar circle. However, systematic streaming motions and local peculiar velocities limit the precision of such distance estimates (*Xu et al.*, 2013).

Radial velocity measurements can be used to associate clumps and features for which there are no direct distance measurements with sources for which distance estimates exist. The Hi-GAL group working on distances (*Russeil et al., 2011*) used the radial velocities of dust clumps and linked them to nearby sources (HII regions, stellar groups, or star-forming regions) for which reliable distance estimates exist. This "bootstrapping" approach, together with the use of near-IR extinction mapping information to help resolve the KDA, has provided approximate distance estimates for tens of thousands of Hi-GAL sources (Elia et al., in preparation).

The key challenge for kinematic distance estimates is resolution of the KDA. Distance from the Galactic midplane, the presence of an associated IRDC, a 21-cm "HI narrow self-absorption" (HINSA) feature at the LSR velocity of the dust clump (produced by cold atomic hydrogen in and around the molecular cloud), and the density of foreground stars can be used to resolve the KDA. Association with a high-contrast IRDC, the presence of HINSA, significant displacement from the Galactic mid-plane, and a low density of foreground stars in the near-IR favor placement at the near-kinematic distance. Absence of an associated IRDC and HINSA (because the 21-cm absorption is filled in by intervening warm HI emission), along with a high density of foreground stars in the Two Micron All-Sky Survey (2MASS), Spitzer, or WISE images, and location close to the mid-plane favor the far-kinematic distance.

Ellsworth-Bowers et al. (2013) have developed an automated statistical distance determination tool that ingests the Galactic latitude, radial velocity, and Spitzer 8-μm images of the region, and uses an *a priori* model of Galactic diffuse 8-μm emission to resolve the KDA using "distance-probability distribution functions." This method can be generalized to incorporate star counts and reddening toward dust emission features and the presence or absence of HINSA to provide a product of probabilities for resolving the KDA.

Higher-precision distance assignments will require precision parallax measurements toward associated sources, determination of radial streaming motions, and a full three-dimensional vector model of Galactic rotation. In a few years, visual wavelength parallax determinations with the Gaia mission will provide distances to sources and so might help with distance determination in relatively nearby regions. The Bar and Spiral Structure Legacy (BeSSeL) survey (*Brunthaler et al., 2011*) will provide direct parallax determinations toward about 400 Galactic sources exhibiting centimeter-wavelength maser emission, with a precision of ~3% to 10%. BeSSeL will provide direct trigonometric distances throughout the Galaxy because, unlike the Gaia measurements of stars, radio masers are not subject to extinction.

3. OVERVIEW OF GIANT MOLECULAR CLOUDS ON THE LARGE SCALE

The mass distribution of GMCs can be described by a power-law $dN/dM \propto M^{\alpha}$ with a cutoff at about $10^{6.5}$ $M_{\odot}$ for the Galactic disk (*Rosolowsky,* 2005). Because $\alpha = -1.5 \pm 0.1$, massive clouds dominate the total mass of H_2.

Giant molecular clouds exhibit supersonic internal motions on scales larger than individual star-forming cores (~0.01 to 0.05 pc). At $R_{gal} > 1$ kpc, tracers such as CO reveal linewidths on the order of 1 to 10 km $^{-1}$ in gas of mean density 10^2–10^3 cm^{-3} (*Roman-Duval et al.,* 2010). Dense clumps and cores with $n(H_2) > 10^4$ cm^{-3}, traced by high-dipole-moment dense gas tracers, typically cover only a few percent of the projected surface area of GMCs traced by CO emission. However, in the Central Molecular Zone (CMZ) such dense gas tracers have linewidths of 10 to over 30 km s^{-1} and cover most of the projected surface area traced by CO, indicating that CMZ clouds have 1 to 2 orders of magnitude larger mean densities (e.g., *Morris and Serabyn,* 1996).

Heyer et al. (2009), using FCRAO ^{13}CO data, found a dynamic range of nearly 2 orders of magnitude in the surface density of the GMCs (a few 10 $M_{\odot}$ pc^{-2} to a few 1000 $M_{\odot}$ pc^{-2}). Their generalization of the classic (*Larson,* 1981) linewidth-size scaling relation is $\sigma(\ell) \propto \Sigma^{1/2} \ell^{1/2}$. Massive star-forming clumps also appear to follow this relation (*Ballesteros-Paredes et al.,* 2011).

Molecular clouds have internal filamentary structure, first seen in deep photographs of nearby dark clouds (*Barnard,* 1907) and extensively explored in CO maps, e.g., in Orion and Taurus (*Bally et al.,* 1987a; *Narayanan et al.,* 2008; *Kirk J. et al.,* 2013). Widespread filamentary IRDCs, seen in silhouette against the bright mid-IR background of the Galactic Plane (*Egan et al.,* 1998), are of heightened interest because IRDCs are the densest, highest-column-density parts of GMCs most closely associated with star formation (*Carey et al.,* 2000). Now, thanks to large-scale dust emission surveys with Herschel, it has become apparent that filaments are a ubiquitous feature of the ISM at all size scales. For example, Figs. 1 and 2 show Herschel Hi-GAL images of filamentary structure in the W3–5 and $\ell = 111°$ complexes, respectively.

3.1. Formation

Theories of GMC formation have evolved enormously since the discovery of CO. Many of the key considerations are discussed in the review by *McKee and Ostriker* (2007). The chapter by Dobbs et al. in this volume also covers the formation of molecular clouds and so here we will simply highlight a few recent developments.

There is growing evidence that GMCs, and the star-forming clumps within them, form in a multi-step process. The first step is the concentration of the diffuse nonmolecular ISM, with a mean density of 1 cm^{-3}, into HI superclouds and related giant molecular associations (GMAs) with mean densities of 10 cm^{-3} (*Elmegreen and Elmegreen,* 1987). Giant molecular associations are characterized by 10^6–4×10^7 $M_{\odot}$ HI halos surrounding GMC complexes at 1- to 2-kpc intervals along the major spiral arms of the Galaxy (*Dame et al.,* 1986). For example, in the Perseus arm of the outer Milky Way, GMAs are found at approximately 2-kpc intervals at longitudes l ~ 110° to 112° associated

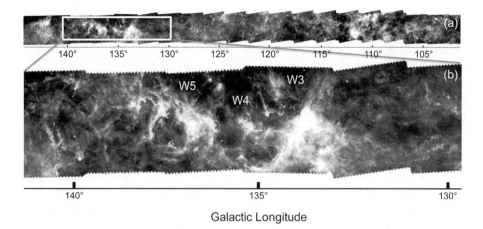

Fig. 1. **(a)** Hi-GAL composite of a portion of the Perseus Arm from data at 70 μm, 160 μm, and 350 μm. **(b)** The W5, W4, and W3 GMA and giant Hɪɪ region complex near l = 134°. The box outlines this region in **(a)**.

with NGC 7538 (*Fallscheer et al.,* 2013); l ~ 132° to 136° associated with the W3 (*Rivera-Ingraham et al.,* 2013), W4, and W5 giant Hɪɪ regions; and l ~ 172° to 180° associated with the Sh2-235 Hɪɪ region. Multi-wavelength Herschel imaging brings out the spectacular structures in such complexes, as illustrated in Figs. 1 (W3–W5) and 2 (NGC 7538). The GMAs form by the combined effects of self-gravity, the gravitational potential of the Galactic plane, and compression by spiral arms (*Elmegreen,* 1994); they are gravitationally bound with internal motions dominated by turbulence (*Dame et al.,* 1986).

The subsequent steps concern the development of structures of higher density. The second is the formation of dense sheets in post-shock layers where supersonic turbulence or the shells and bubbles produced by massive-star feedback produce converging and colliding flows (e.g., *Audit and Hennebelle,* 2005). The third step is the cooling of the post-shock layers and conversion of predominantly atomic gas into molecular gas. Fourth is the formation of dense filaments from sheets. Fifth is the fragmentation of filaments and sheets into star-forming clumps. Recent theoretical developments and observational evidence concerning filaments and dense clumps formation are discussed in sections 4.1 and 4.2. At each step, the density is increased by about 1 order of magnitude, thus reaching $n(H) \sim 10^6$ cm^{-3} in clumps and cores.

It is not yet clear at which step the resulting substructure becomes recognizable as CO-bright molecular clouds. As discussed below, gravity and supersonic convergence can compress the CO-dark H_2 to sufficient density to produce and excite CO, producing CO-bright structures only shortly before the onset of star formation.

3.1.1. Carbon-monoxide-free hydrogen and dark gas as fuel for giant molecular cloud formation. Aperture synthesis observations of nearby spiral galaxies (*Allen et al.,* 2004; *Heiner et al.,* 2009) comparing their CO and Hɪɪ region distributions and radial velocities show that most of the gas

traced by the 21-cm emission line is located downstream from spiral arms, active sites of star formation, and Hɪɪ regions. These results provide evidence that this Hɪ might be primarily a photodissociation product. In this picture, GMCs would primarily form from Hɪ clouds through CO-dark molecular gas rather than directly from Hɪ clouds.

Pringle et al. (2001) hypothesized a component of the ISM in which hydrogen is mostly molecular but that is CO-free and not traced by bright CO emission. There is growing observational evidence for such "dark molecular gas" (*Wolfire et al.,* 2010). Comparison of the total hydrogen column density using gamma-ray (*Grenier et al.,* 2005; *Abdo et al.,* 2010) or dust emission (e.g., *Planck Collaboration et al.,* 2011) with maps of 21-cm Hɪ and CO-bright H_2 show that there is significant dark H_2 surrounding most nearby molecular clouds. The mass of such CO-free dark gas might be comparable to the mass of H_2 traced by CO emission. The GOT C$^+$ survey of the Milky Way disk in the Cɪɪ 158-μm line also provides evidence for a substantial amount of CO-dark H_2 (*Pineda et al.,* 2013). Additionally, *Allen et al.* (2012) obtained deep 18-cm observations of OH emission in a portion of the Perseus arm, finding abundant OH in regions that do not show detectable CO emission.

Hydrogen can reach a high-equilibrium abundance in the ISM in gas flows that have lower volume and surface densities than those typical of GMCs (e.g., *Dobbs et al.,* 2008). For instance, *Clark et al.* (2012) demonstrate that the formation of CO-bright regions in the gas occurs only around 2 m.y. before the onset of star formation. In contrast, H_2-dominated but CO-free regions form much earlier. The timescales might be long enough that Hɪ and CO-dark gas can co-exist, as evidenced by the massive Hɪ envelopes surrounding the Perseus (*Lee et al.,* 2012) and Taurus molecular clouds (*Heiner and Vázquez-Semadeni,* 2013). The level of dissociating UV flux is not sufficient to explain all this Hɪ as a photodissociation product, suggesting that these Hɪ envelopes might be converting into CO-dark gas.

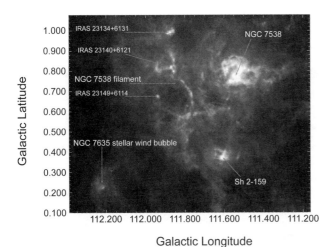

Fig. 2. Hi-GAL composite of the GMC complex associated with the NGC 7538 HII region at l = 111 in the Perseus Arm, from data at 70 μm, 160 μm, and 350 μm, in Galactic coordinates.

3.1.2. Conditioning the interstellar medium reservoir. Large-scale gravitational instabilities in the Galactic disk can drive GMC formation. Stability of a disk can be quantified in terms of the Toomre parameter (*Toomre*, 1964) $Q \equiv c_s \kappa / \pi G \Sigma$, where c_s is the sound-speed of the gas; κ is the epicyclic frequency, which is typically of the same order of magnitude as Ω, the rotational frequency of the disk; and Σ is the surface density of the disk. A pure gas disk is gravitationally unstable wherever $Q < 1$. The analysis for a disk that contains both gas and stars is more complex (e.g., *Elmegreen,* 2011a), but instability still requires a Toomre parameter on the order of unity.

The question of whether the Milky Way is Toomre stable has been examined recently by *Kruijssen et al.* (2013). They conclude that in the CMZ the gas is marginally unstable, with $Q \sim 1$, only within the central so-called 100-pc ring (see section 3.3) revealed by Herschel (*Molinari et al.,* 2011). On larger scales in the disk, they find stability for $R_{gal} < 4$ kpc. *Kruijssen et al.* (2013) also find that the disk is highly stable at $R_{gal} > 20$ kpc, due to its low surface density, and although metallicity-dependent variations in the CO-to-H_2 conversion factor could lead to instability, it is unlikely to make enough of a difference to render the disk unstable.

It is therefore plausible that on the largest scales, gravitational instability helps to gather together the gas required for GMC formation. However, GMCs probably do not form directly from this instability: The growth timescale is long (*Elmegreen,* 2011a), and the applicability of the Toomre analysis (which assumes an infinitely thin disk) to scales smaller than the disk scale height is also questionable. It is far more plausible that gravitational instability promotes GMC formation indirectly, by enhancing non-axisymmetric features such as spiral arms that help to gather the diffuse gas together (e.g., *Dobbs et al.,* 2012) and by directly driving turbulence in this gas (e.g., *Wada et al.,* 2002). Both

mechanisms can induce the converging flows that constitute the last stage of compression necessary to form GMCs.

Another form of mass assembly is bubbles created by massive star feedback. Giant HII regions and the collective effects of multiple supernovae in OB associations can sweep up kiloparsec-scale shells that blow out of the disk and massive, slowly expanding rings of gas in the Galactic Plane. The 21-cm surveys of *Hartmann and Burton* (1997) reveal hundreds of HI supershells in the Galaxy surrounding OB associations (*Koo et al.,* 1992; *Ehlerová and Palouš,* 2013). These 0.1- to 1-kpc-scale features surround bubbles of hot plasma traced by extremely-low-density HII regions and X-ray emission. They appear to be powered by the combined impact of photoionization, stellar winds, and supernova explosions. *McCray and Kafatos* (1987) proposed that gravitational instabilities in the supershells swept up by expanding superbubbles would first occur on the mass scale of GMCs in the solar vicinity. As superbubbles sweep up the surrounding ISM, the resulting shell expansion velocities decrease. Once the local velocity dispersion of a given patch of the shell drops below the gravitational escape velocity, that region becomes subject to gravitational instability. Both small and large length and mass scales are stable, giving rise to a critical intermediate scale, which in the solar vicinity is similar to the observed scales of GMCs.

Regardless of whether it is ultimately gravity or stellar feedback that is more important for assembling diffuse gas into dense clouds, the behavior of the gas on small scales is likely to be rather similar. Because GMC formation requires large amounts of gas to be gathered into one place, GMCs will typically form at the stagnation points of converging flows of gas, and the details of their formation will have little dependence on how these flows are driven.

3.1.3. Large- and small-scale converging flows. The potential importance of converging or "colliding" flows as a GMC formation mechanism was first stressed by *Ballesteros-Paredes et al.* (1999) and *Hartmann et al.* (2001). The earliest one-dimensional models (e.g., *Hennebelle and Pérault,* 1999; *Koyama and Inutsuka,* 2000) showed that the collision of streams of thermally stable warm gas could increase their density sufficiently to render the gas thermally unstable (*Field,* 1965). The resulting thermal instability leads to a large increase in the gas density from a value ~ 1 cm^{-3}, characteristic of the warm neutral medium (WNM), to ~ 100 cm^{-3}, characteristic of the cold neutral medium (CNM). At the same time, the temperature of the gas falls by 2 orders of magnitude, from $T \sim 6000$ K in the WNM to $T \sim 60$ K in the CNM.

Subsequent two-dimensional and three-dimensional modeling demonstrated that the thermal instability occurring in the post-shock gas layer naturally produces turbulence with a velocity dispersion similar to the sound speed in the unshocked gas (e.g., *Koyama and Inutsuka,* 2002; *Heitsch et al.,* 2005; *Vázquez-Semadeni et al.,* 2006; *Banerjee et al.,* 2009). Thermal instability can create dense structures directly via the isobaric form of the instability. However, most of the resulting small clumps are not self-gravitating

(*Koyama and Inutsuka,* 2002; *Heitsch et al.,* 2005; *Glover and Mac Low,* 2007), and instead it is the ensemble of clumps produced by a coherent converging flow that might become unstable (*Vázquez-Semadeni et al.,* 2007).

An important feature of the clouds formed in colliding flows is that they typically undergo global gravitational collapse primarily in the directions perpendicular to the flow (see, e.g., *Vázquez-Semadeni et al.,* 2007). Such collapse naturally leads to the formation of filaments. The nonlinear development of the thermal instability also proceeds by forming a network of filaments (see section 4.1) at whose intersections clumps (see section 4.2) are subsequently formed (*Audit and Hennebelle,* 2005; *Vázquez-Semadeni et al.,* 2006; *Heitsch et al.,* 2008).

3.2. Lessons from the Solar Vicinity

Currently, the Sun appears to be located inside a superbubble in an interarm spur of gas and dust between two major spiral arms of the Galaxy (*Xu et al.,* 2013). Most of the ISM in the solar vicinity is expanding with a mean velocity of 2–5 km s^{-1}, discovered in HI and more apparent in molecular gas (e.g., *Lindblad et al.,* 1973; *Dame et al.,* 2001). At the center of "Lindblad's ring" is the 50-m.y.-old Cas-Tau group, a "fossil" OB association (*Blaauw,* 1991), whose most massive members have evolved and become supernovae. The kiloparsec-scale superbubble produced by the Cas-Tau group blew out of the ISM both above and below the Galactic Plane. Consistent with ballistic trajectories of dense clouds entrained in or formed from supershells, HI surveys reveal complexes of infalling gas (*Stark et al.,* 1992; *Kuntz and Danly,* 1996).

The nearest OB associations, such as Scorpius-Centaurus, Per OB2, Orion OB1, Lac OB1b, and the B and A stars that trace the so-called "Gould belt" of nearby young and inter-mediate-aged stars, are all associated with Lindblad's ring (*Lesh,* 1968; *de Zeeuw et al.,* 1999), evidence for secondary star formation in clouds that condensed from the ancient, expanding, and tidally sheared supershell (see Figs. 5 and 6 in *Perrot and Grenier,* 2003). These nearby OB associations have spawned their own superbubbles with sizes ranging from 100 to 300 pc. The remaining GMCs in each of these regions are embedded within their respective superbubbles. Thus most of the star-forming molecular clouds in the solar vicinity have been extensively impacted and processed by massive star feedback from nearby OB associations.

Figure 3, after *Bally* (2010) and updated using Planck data (*Planck Collaboration et al.,* 2013a), shows the eastern part of the Orion-Eridanus superbubble in detail. The northern portion of the Orion A cloud hosts the most active site of ongoing star formation within 500 pc of the Sun. The Orion A cloud is cometary with its northern end located toward the projected interior of the superbubble. The Orion Nebula and the OMC1 core, the closest site of active high-mass star formation, is in the center of the degree-long, high-density integral shaped filament (ISF) (see section 4.1.2) (*Johnstone and Bally,* 1999, 2006).

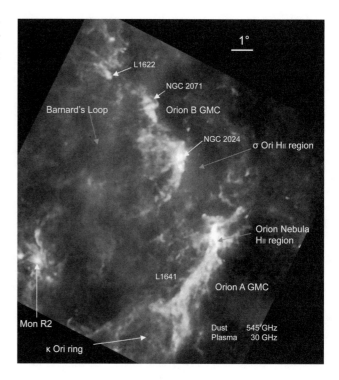

Fig. 3. Planck view of the 10^5 M$_\odot$ Orion A and B GMCs, traced by dust continuum emission at 545 GHz, and the eastern portion of the Orion-Eridanus superbubble, in free-free emission at 30 GHz. The cometary Orion A cloud faces the center of the Orion-Eridanus bubble. The Orion Nebula and the OMC1/2/3 clumps are the most active sites of star formation in the solar vicinity.

3.3. Lessons from the Central Molecular Zone

Extending from longitudes $\sim$–5° to +5° (Fig. 4), the CMZ provides a unique opportunity to investigate the properties of dense gas in a circumnuclear environment containing a supermassive black hole (SMBH). Approximately 10% of the Milky Way's ISM lies within 0.5 kpc of the Galactic Center's SMBH. The CMZ contains the most extreme GMCs and star-forming regions in the Galaxy.

The CMZ contains 2–6 × 10^7 M$_\odot$ of molecular gas with average densities greater than 10^4 cm^{-3}, sufficient to excite most high-dipole-moment molecules such as NH$_3$, CS, HCN, HCO$^+$, and HNCO everywhere in the CMZ (*Dahmen et al.,* 1998; *Ferrière et al.,* 2007; *Jones et al.,* 2012). The CMZ GMCs differ from GMCs in the Galactic disk in several ways: (1) They have 1 to 2 orders of magnitude higher mean density. (2) High-dipole-moment molecules are excited throughout the volume of CMZ GMCs, not just in isolated clumps and cores as in Galactic disk GMCs. (3) Central Molecular Zone GMCs have larger linewidths (typically 15 to more than 50 km s^{-1}) than disk GMCs (typically 2 to 5 km s^{-1}). Large linewidths are an indication of large-amplitude internal motions. Indeed, *Huettemeister et al.* (1998) found extensive SiO emission from the CMZ clouds, indicating the presence of shocks sufficiently strong

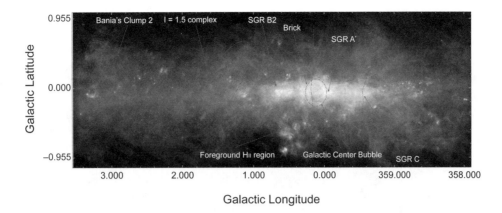

Fig. 4. The CMZ in a composite image combining Spitzer 24 μm with Herschel 70 μm and 500 μm. Noticeable features are identified in the figure; the Sofue-Handa lobe structure (see text) is delimited by the two nearly vertical dashed lines at l ~ 0.2°, and l ~ –0.5°.

to sputter dust grains to produce a significant gas-phase abundance of SiO. Molecular gas in the CMZ is not only dense, but also warm, and able to excite high-J transitions of CO, as well as a variety of high-dipole-moment molecules (*Martin et al.,* 2004).

The orbital dynamics of the gas is critical to the evolution. Non-self-intersecting orbits form two families: the x1, orbits whose major axes are aligned with the bar, and the x2 orbits, which tend to be elongated orthogonal to the bar. The x1 orbits become more "cuspy" as their semimajor axes shrink, eventually becoming self-intersecting. Clouds on such orbits will collide, resulting in loss of orbital angular momentum and settling onto the more compact x2 family of orbits. While the x1 orbits are populated by HI and some molecular clouds, the x2 orbits contain the densest and most turbulent molecular clouds in the Galaxy. Almost all star formation in this gas, heating the warm dust as seen in Fig. 4, occurs within about 100 pc of the nucleus. The CMZ gas is asymmetrically located with respect to the dynamic center of the Galaxy. Two-thirds of this emission is at positive longitudes and has positive radial velocities; only one-third is at negative longitudes and has negative radial velocities (*Bally et al.,* 1987b, 2010; *Jones et al.,* 2012). Remarkably, the majority of 24-μm sources in Fig. 4 are located at negative Galactic longitudes, not on the side of the nucleus containing the bulk of the dense CMZ gas and associated cold dust. *Molinari et al.* (2011) found that the coolest dust and the highest column density gas are located in an *infinity-sign-shaped* ring with a radius of about 100 pc (Fig. 5).

Other notable features are marked in these figures. The Sgr B2 molecular cloud at l = 0.8° contains over 10^6 $M_\odot$ of molecular gas and hosts what might be the most luminous site of massive star formation in the Galaxy. It is located adjacent to the older Sgr B1 region at the high-longitude end of the 100-pc ring. Sgr C, located at the negative longitude end of the ring, hosts the second most massive site of star formation in the CMZ.

The prominent oval of warm dust (left of center Fig. 5b) is a very young 30-pc-diameter superbubble [the "Galactic Center Bubble" (GCB)] whose interior contains the massive, 3- to 5-m.y.-old Arches and Quintuplet clusters. The Sgr A region and the SMBH are located outside the GCB, and thus the GCB might be powered mostly by massive stars. The GCB is the smallest and brightest member of a set of nested bubbles emerging from the central 100-pc region of our Galaxy. The Sofue-Handa lobe (*Sofue and Handa,* 1984) is a degree-scale bubble traced by free-free emission blowing out of this region. The relationship between these features and the recently recognized kiloparsec-scale Fermi-LAT-Planck bubble (*Su et al.,* 2010; *Planck Collaboration et al.,* 2013c) remains unclear. Is this bubble powered by the merging of superbubbles fed by dying massive stars in the CMZ, fed by occasional active galactic nucleus (AGN) activity of the SMBH, or a combination of these mechanisms?

The compact, high-column-density clump of cold dust located to the left of the GCB ("Brick" in Figs. 4 and 5) is the most extreme IRDC in the Galaxy (see the chapter by Longmore et al. in this volume). This clump is located in the 100-pc ring, has a mass greater than 10^5 $M_\odot$ within a radius of 3 pc, yet shows no evidence of any star formation (*Longmore et al.,* 2012). It is part of a string of massive clouds stretching from a location near Sgr A* to Sgr B2, including several with various degrees of ongoing star formation, possibly triggered by a close passage near the SMBH (*Longmore et al.,* 2013a).

Wide-field mapping of the CMZ in the dust continuum, high-J CO lines, the CI line, and a variety of dense gas tracers will shed light on a number of critical questions about star formation, the ISM, and its interactions with the SMBH. Does nuclear star formation shield the central black hole from growth? Or does occasional flaring of the SMBH regulate the state of the ISM and star formation in the CMZ? What is the star formation rate (SFR) in the CMZ? Does it follow the Schmidt-Kennicutt relation, or is the SFR per unit mass of dense gas lower in the CMZ than in galactic disks (*Longmore et al.,* 2013b; *Kruijssen et al.,* 2013)?

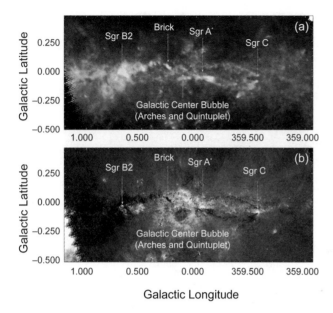

Fig. 5. Maps of background-subtracted **(a)** dust column density and **(b)** dust temperature from Hi-GAL images (*Molinari et al.,* 2011) of the inner 2°.2 of the CMZ. Large regions of cold dust are associated with Sgr B2 (left), Sgr C (right), and the 20 km s^{-1} cloud (right of center). The warm GCB is to the left of Sgr A.

4. CONVERTING GIANT MOLECULAR CLOUDS TO DENSE STRUCTURES

It is well established that only a small fraction of the gas in GMCs is dense enough to be involved in star formation. This is apparent from CS surveys in high-mass star-forming regions (e.g., *Plume et al., 1992, 1997*), as well as from large-scale gas and dust mapping of nearby low-mass star-forming regions (e.g., *Goldsmith et al., 2008*).

Herschel surveys allow investigation of the properties of IRDCs because the cold and dense dust shines brightly in the submillimeter continuum. Visual inspection by *Wilcock et al.* (2012) of the Hi-GAL survey in the 300° ≤ l ≤ 330° region of the Galactic Plane suggests that only 50% of the *Peretto and Fuller* (2009) IRDCs are bona fide cold and dense structures, while the rest are regions devoid of background mid-IR emission. Nevertheless, the Peretto and Fuller column densities derived from the mid-IR opacity provide independent measurements that can be compared with estimates obtained from Herschel's complete far-IR photometric coverage (*Elia et al.,* 2013; Elia et al., in preparation; *Schisano et al.,* 2014).

Unbiased surveys allow statistically significant estimates of the lifetimes and relative durations of the various stages of evolution of clouds, clumps, and cores. For example, *Tackenberg et al.* (2012) use a portion of the ATLASGAL survey to estimate the timescale for massive starless clumps using association to GLIMPSE sources, obtaining quite short (6–8 × 10^4 y) timescales. *Wilcock et al.* (2012) analyze a portion of the Hi-GAL survey and find that only 18% of

the Peretto and Fuller bona fide IRDCs in the 300° ≤ l ≤ 330° range appear not to be associated with either 8- or 24-μm counterparts.

An interesting approach has been adopted by *Battersby et al.* (2011), in which Hi-GAL column density and temperature maps in the l = 30° and l = 59° fields were compared to companion maps in which each pixel is populated with a variety of star-formation indicators, including 8- and 24-μm emission, association with methanol masers, and extended green objects suggestive of the presence of outflows. A pixel-to-pixel comparison showed that pixels with more star-formation indicators have higher column densities and higher temperatures with $T_{dust} > 20$ K.

4.1. Filaments

Surveys with Herschel reveal that the densest regions in molecular clouds are organized in complex filamentary structures (*Molinari et al.,* 2010a). As discussed above and by André et al. (see the chapter in this volume), nearby clouds such as Taurus and Ophiuchus are filamentary and Hi-GAL reveals similar structures on much larger scales (several to tens of parsecs) along the Galactic Plane. The ISM is permeated by a web of filamentary structures from very low column density "cirrus" emission to gravitationally "supercritical" clouds, and from the small scales of star-forming cores to scales of tens to hundreds of parsecs in the Galactic Plane.

Photodetector Array Camera and Spectrometer (PACS) and Spectral and Photometric Imaging Receiver (SPIRE) maps at 70–500 μm showed that the Aquila Rift and the Polaris Flare contain an extensive network of filaments, many of which are several parsecs in length (*André et al.,* 2010). In IC 5146, *Arzoumanian et al.* (2011) show that the filaments have radial density profiles that fall off as r$^{-1.5}$ to r$^{-2.5}$ and mean full width at half maximum (FWHM) widths of 0.10 ± 0.03 pc. This is much narrower than the distribution of Jeans lengths, which range from ~0.02 to 0.3 pc. Because the filament widths are within a factor of 2 of the sonic scale at which the turbulent velocity dispersion equals the sound speed, they suggest that large-scale turbulence could be responsible for forming the filaments, whereas gravity then drives their evolution by accreting material from the surroundings. This idea seems to be at least circumstantially supported by *Chapman et al.* (2011), whose polarization map of the L1495/B213 filament shows a number of subfilaments oriented perpendicular to the main filament, but aligned with the magnetic field direction as measured in the surrounding material. These subfilaments might be related to mass accretion onto the main filament (*Palmeirim et al.,* 2013), which in turn might help to form the dense cores within (*Hacar and Tafalla,* 2011).

Fischera and Martin (2012a,b) discuss equilibrium models of filaments as isothermal self-gravitating infinite cylinders in pressure equilibrium with the ambient medium. These offer an analytical and quantitative explanation of the radial profiles and the width-column-density relation.

However, there is a maximum mass line density $M_{lin,max}$ beyond which there is no equilibrium solution (often referred to as "critical"). If the filaments have only thermal support, then at 10 K $M_{lin,max} \sim 16\ M_\odot$ pc^{-1}, corresponding to a maximum column density of $N_{H_2,max} \sim 10^{22}$ cm^{-2} or $A_{V,max} \sim 10$. Note that if there were in addition *turbulent* support, as seems relevant to accreting filaments, then a quasistatic state might exist with higher M_{lin}. Filaments close to or exceeding $M_{lin,max}$ are subject to fragmentation into self-gravitating prestellar cores (*Fischera and Martin*, 2012b). Given the derived dust temperature in these filaments, and assuming that the gas temperature is similar, *André et al.* (2010) find that 60% of the bound (self-gravitating) prestellar cores in these regions are located in filaments with $M_{lin} > M_{lin,max}$, indicating a strong association of supercritical filaments and star formation. On the other hand, while filaments with $M_{lin} < M_{lin,max}$ interestingly contain cores, few of these cores have strong self-gravity.

Herschel has shown that filamentary structures are organized into complex interconnected branches and also loops. Examples include the Vela region (*Hill et al.*, 2011) and DR21 (*Schneider et al.*, 2010a). The first results from the Hi-GAL survey showed widespread filamentary structures on larger scales in a $2° \times 2°$ region toward $l = 59°$ (*Molinari et al.*, 2010a). Like *André et al.* (2010), they find that dense cores/clumps tend to be associated with the filaments. They find that the column densities of the filaments lie in the range 10^{21} cm$^{-2} \leq N(H_2) \leq 10^{22}$ cm^{-2}, corresponding to $1 \leq A_V \leq 10$. This range is lower than found for local clouds like the Aquila Rift (distance 260 pc) and the Polaris Flare (150 pc) and could be due in part to the greater distances of the filaments in the $l = 59°$ region ($\sim$2–8 kpc) and the resulting beam dilution (at these distances; Hi-GAL detects mostly clumps rather than cores).

More rigorous detection and quantitative assessment of filaments is now underway using a variety of tools based on a variety of morphological analyses of multi-scale diffuse emission. One of these approaches is based on the analysis of the Hessian matrix computed over large-scale column-density maps (but see also *Arzoumanian et al.*, 2011), and it has been applied by *Schisano et al.* (2014) over an $8° \times 2°$ Hi-GAL column-density map of the Galactic Plane over the longitude range $l = 217°$–$224°$. They identified over 500 filaments (with $\sim$2000 branches) with lengths from 1.5 to 9 pc and column densities from 10^{19} to 10^{22} cm^{-2} (see Fig. 6). Again, these column densities are lower than those cited by *André et al.* (2010) for two possible reasons: beam dilution caused by the greater distances to these filaments (500 pc–3.5 kpc) and the fact that *Schisano et al.* (2014) use the average column density rather than the column density at the spine of the filaments. This latter effect tends to decrease the quoted column density by a factor of 4. Focusing on only the nearest filaments, they find average lengths, widths, and column densities to be 2.6 pc, 0.26 pc, and 9.5×10^{20} cm^{-2}, respectively. Nevertheless, these filaments are still wider and less dense than those in *Arzoumanian et al.* (2011) by a factor of 2–3, and so not all are demonstrably supercritical.

The analysis of *Schisano et al.* (2014) also deals with the physical properties of what they call the individual "branches" of the filamentary networks, i.e., the subportions of filamentary structures limited by the nodal points created by the intersections. They found that the branches containing protostellar clumps have $M_{lin} \sim 20$–30 $M_\odot$ pc^{-1} (arguably supercritical), whereas branches with only prestellar clumps have $M_{lin} \sim 4\ M_\odot$ pc^{-1}. Although "prestellar" clumps are simply those with no counterpart in the Hi-GAL 70-μm band, the absence of associated 70-μm emission is an indication of considerably lower rates of star formation than found in protostellar clumps containing detectable 70-μm sources.

4.1.1. Filament formation mechanisms. There are a number of physical mechanisms that can form filamentary structure. These include gravity, compressive flows in supersonic turbulence and colliding sheets, production of tails by shadowing by dense clumps (the pillars of M16 are a well-known example), fragmentation of expanding shells and supershells, generation of fingers through Rayleigh-Taylor (RT) instability, and magnetic fields.

4.1.1.1. Gravity: *Lin et al.* (1965) demonstrated that spheroidal gas clouds tend to flatten as they collapse, yielding sheet-like structures, and sheets collapse into filaments (see also *Larson*, 1985). To produce sheets and filaments, the mass of the collapsing cloud must be much larger than the Jeans mass, because in clouds with masses close to M_J, isotropic pressure forces tend to keep the clouds quasispherical.

4.1.1.2. Supersonic turbulence: Both purely hydrodynamic (HD) and magneto-hydrodynamic (MHD) supersonic turbulence can lead to filament formation (e.g., *Padoan et al.*, 2001). Supersonic turbulence creates a complex network of shocked-layers confined by the ram pressure of the convergent parts of the flow. The supersonic collision between shock-formed sheets produces denser filaments. If these shock-compressed layers become sufficiently massive, their self-gravity can lead them to collapse. Otherwise, they will tend to dissolve on a turbulent crossing time. The Mach number of the turbulence determines the density contrast ρ_2/ρ_1 between pre-and post-shocked gas: For HD shocks, this scales as $\mathcal{M}^2$, where the Mach number $\mathcal{M} = V/c_s$. For MHD shocks, $\rho_2/\rho_1 \approx \mathcal{M}$ (*Padoan and Nordlund*, 2002), with $\mathcal{M}$ being given by V/c_A in this case. Here, c_s is the sound speed and $c_A = B/(4\pi\rho)^{1/2}$ is the Alfvén speed. Highly supersonic turbulence therefore yields high-density contrasts, while transonic or subsonic turbulence creates only small perturbations that then must be amplified by some other mechanism, such as gravity. The mass accumulated in post-shock layers depends on the correlation length of the turbulent velocity field; larger correlation lengths imply higher mass accumulation layers.

Both gravity and supersonic turbulence formation mechanisms work more readily in cold and dense gas with a low sound (or Alfvén) speed and small Jeans mass. Therefore, the thermal evolution of the gas plays an important role. Rapid and efficient cooling can lead to nonlinear development of thermal instability that can enhance filament formation (e.g.,

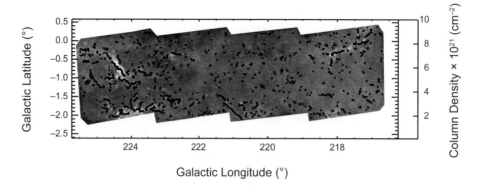

Fig. 6. Column density map from Hi-GAL images in the Galactic longitude range $217° \leq l \leq 224°$ (*Elia et al.,* 2013). The thick black lines delineate the main spine of each detected filament (branches are neglected for clarity). Detected filaments are those that show a contrast greater than 3× the standard deviation of the column density computed locally.

Vázquez-Semadeni et al., 2000, 2006; *Heitsch et al.,* 2005). Which of these processes (gravity, turbulence, or thermal instability) is primarily responsible for the formation of the filaments observed in the ISM remains undetermined; plausibly, all three processes could play important roles. Nevertheless, recent simulations (e.g., *Gómez and Vázquez-Semadeni,* 2013) suggest that large-scale gravitational collapse plays the dominant role here.

4.1.1.3. *Shadowing and cometary clouds:* Many H$_{II}$ regions exhibit cometary clouds with long filamentary tails pointing away from the direction of ionization. For example, Fig. 1 shows several cometary clouds and filamentary tails in the W4 and W5 giant H$_{II}$ regions. Figure 3 shows that the entire Orion A cloud is cometary with streamers trailing away from the dense ISF containing the Orion Nebula Cluster (ONC) and the OMC1, 2, and 3 cloud cores, which together have spawned over 2000 YSOs in the last few million years.

4.1.1.4. *Fragmentation of expanding shells and supershells:* Bubbles created by the non-ionizing radiation field of late B and A stars and H$_{II}$ regions as well as superbubbles and supernova remnants can sweep up ambient gas into expanding shells (see section 3.1.2). Limb brightening of the shells at the edges can make them look filamentary. Hi-GAL data reveals many approximately circular or semicircular filaments such as shown in Fig. 2. Additionally, various hydrodynamic instabilities, such as the Vishniac or RT instability or the overrunning of dense clumps, can lead to filamentary structure. The presence of a regular entrained magnetic field can also confine the motion of the charged component, thereby breaking the symmetry to produce filamentary structure.

4.1.1.5. *Rayleigh-Taylor instability:* Rayleigh-Taylor instabilities occur when a dense fluid overlies a light fluid, or when there is an acceleration of a light fluid though a dense layer. When the shock fronts associated with expanding giant H$_{II}$ regions or superbubbles reach radii comparable to or larger than the scale height of the Galactic gas layer, they can accelerate as they encounter an exponentially decreasing density profile. Additionally, ionization fronts propagating in a density profile that decreases faster than r^{-2} tend to become RT unstable. The RT instability creates long fingers of dense material oriented parallel to the pressure or density gradient, with the appearance of elongated, filamentary, cometary clouds. Dense clouds located more than a few scale heights above the Galactic mid-plane tend to become elongated orthogonal to the plane due to the effect of superbubbles breaking out of the Galactic gas layer.

4.1.1.6. *Magnetic fields:* In the Taurus clouds, interstellar polarization of background stars reveals that the minor filaments that run orthogonal to the dense Taurus filaments (*Goldsmith et al.,* 2008) are parallel to the direction of the field over most of the Taurus constellation. This morphology suggests that the charged component of the gas associated with the minor filaments is magnetically confined. The charged component is free to stream along the magnetic field lines and prevented from crossing field lines, instead gyrating about the field. Ion-neutral collisions effectively couple the neutral atoms and molecules to the field.

4.1.2. *Filament substructure.* Filaments are observed to have compact substructures consisting of clumps and cores (*Schneider and Elmegreen,* 1979; *Myers,* 2009; *André et al.,* 2010; *Molinari et al.,* 2010b), as expected at least for those that are strongly self-gravitating (with M_{line} near or exceeding $M_{line,max}$), but are actually present in most.

Figure 7 shows a clear relationship between the masses of the clumps and M_{line} of the filaments hosting them. M_{line} ranges over more than 2 orders of magnitude and the relationship extends well above $M_{line,max}$ (*Schisano et al.,* 2014). This correlation might indicate that a spectrum of filament mass line densities is produced by large-scale turbulence or gravitational collapse, with clumps almost immediately fragmenting so that their mass distribution would tend to carry memory of the gravitational state of the parent filament.

Alternatively, the correlation might result from evolution of filaments and clumps toward larger masses as accretion continues from the surrounding environment. Observations suggest continuous flow onto filaments as well as from filaments onto cores (*Schneider et al.,* 2010a; *Kirk H. et al.,* 2013). Numerical simulations of nonequilibrium collapsing clouds show similar behavior (*Gómez and Vázquez-Semadeni,* 2013). In such dynamical evolution, filaments fragment into denser clumps that collapse on shorter timescales than the filaments because of different dimensionality: Collapse timescales for finite filaments are longer than the clump free-fall time by a factor depending on their aspect ratio (*Toalá et al.,* 2012; *Pon et al.,* 2012). This implies that the clumps begin forming stars before the filaments have finished funneling the entire reservoir of gas onto the cores.

The kinematics of the ISF in the Orion A molecular cloud that contains the ONC (Fig. 3) exhibits this behavior. The ~7-pc (600)-long and 0.2-pc-wide ISF exhibits a velocity jump of 2–3 km s^{-1} at the position of the OMC1 cloud core located immediately behind the Orion Nebula. The southern half of the ISF has a radial velocity of $V_{LSR} \approx 8$ km s^{-1}, while just north of OMC1 V_{LSR} jumps to about 10 km s^{-1}; V_{LSR} increases to over 12 km s^{-1} near the northern end of the filament adjacent to the NGC 1977 H II region (*Ikeda et al.,* 2007). Despite the unknown projection effects, this velocity jump near the location of the Orion Nebula appears to be similar to the escape velocity with respect to the mass of the OMC1 cloud core and the ONC at a radius of 1 pc from the cluster core, about 3 km s^{-1}. Thus, it seems likely that material in the ISF is being drawn along its length by the gravity of the ONC and the adjacent OMC1 core. If this picture is correct, then the smaller OMC2 and OMC3 clusters might be dragged into the ONC, helping to fuel continued growth of the stellar population there. The timescale for such a merger

is roughly 1 m.y., longer than the crossing time of the ONC (<0.3 m.y.) or the OMC1 core (<0.05 m.y.).

The histograms of M_{line} in Fig. 8 show systematic differences depending on the presence of prestellar to protostellar clumps. If the filament branches with prestellar clumps (dashed line) are to evolve into the branches with protostellar clumps (solid line), then they would have to accumulate enough mass for an order of magnitude increase in M_{line} within only the prestellar core lifetime of a few × 10^5 yr, channeling a fraction of this material onto the clumps, thus increasing their mass as well. This would imply extremely high-accretion rates on the order of ~10^{-3}/10^{-4} M$_\odot$ pc^{-1} yr^{-1} from the surrounding environment and along the filament branches. Orion's ISF has a mass on the order of 3–5 × 10^3 M$_\odot$. An accretion timescale of 1 m.y. implies an accretion rate into the ONC of about 3–5 × 10^{-3} M$_\odot$ yr^{-1}. Such rates, although high, seem consistent with recent observations of the SDC335.579-0.272 cloud by *Peretto et al.* (2013). In this scenario, filaments such as the ISF are long-lived entities that become denser and more massive over time.

Schisano et al. (2014) find no correlation between M_{line} of a branch and the clump formation efficiency within that branch. However, the L/M ratio of protostellar clumps increases with M_{line}, suggesting that the increased M_{line} in branches containing protostellar clumps results in an increased SFR. Contrary to previous proposals, M_{line} might not always trace the evolution of a filament. When filaments form from large-scale turbulence, many might be transient objects, with only those that are strongly self-gravitating forming stars.

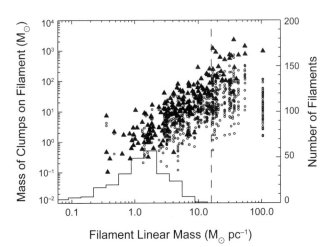

Fig. 7. Mass of the clumps located in detected filaments in the l = 217°–224° Galactic Plane region (*Schisano et al.,* 2014) (Fig. 6) vs. M_{line} of the hosting filament. Small circles are individual clump masses, while triangles represent the total mass in clumps on that filament. The distribution of M_{line} for filaments without clumps is also reported with the solid line histogram (refer to right axis for unit scale). The vertical dashed line marks $M_{line,max} \sim 16$ M$_\odot$ pc^{-1} for T ~ 10 K.

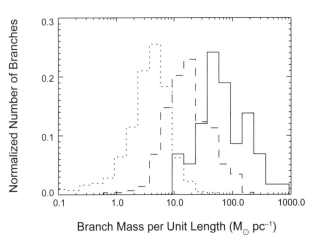

Fig. 8. Normalized histograms of M_{line} of filament branches detected by *Schisano et al.* (2014): with protostellar clumps (solid line), with prestellar clumps (dashed line), and without clumps (dotted line). Each histogram is normalized to its integral to emphasize intrinsic shape differences. Among the latter, those with higher M_{line} are part of more complex filaments that host both prestellar and protostellar clumps. The histograms have been normalized to the total number of filaments to emphasize intrinsic differences between the distributions.

4.2. Clumps

4.2.1. Observed properties. Clumps and cores have been the subject of star-formation studies for decades (see, e.g., *Lada et al.,* 2007, or the chapters by Andre et al. and Offner et al. in this volume). In nearby clouds, the core mass spectrum closely resembles the stellar initial mass function (IMF) (*Motte et al.,* 1998; *Alves et al.,* 2007), leading to arguments that it determines the IMF. Such detailed studies are limited to low-mass objects, less than about 10 $M_\odot$. Estimates toward more distant massive clumps are less definitive due to the relatively limited spatial resolution available until now for large samples (*Shirley et al.,* 2003; *Reid and Wilson,* 2005; *Beltrán et al.,* 2006).

Herschel-based studies by *Schisano et al.* (2014) show that denser clumps are preferentially found on filamentary structures along the Galactic Plane (see Fig. 9), similar to what is found at smaller scales in Aquila (*André et al.,* 2010) or toward Orion A (*Polychroni et al.,* 2013). Furthermore, the denser and more massive clumps are preferentially located at the intersection of filaments (*Myers,* 2009; *Schneider et al.,* 2012), indicating that such extreme clumps form from the funneling of gas along filaments.

Unbiased surveys of the Galactic Plane such as BGPS, ATLASGAL, and Hi-GAL are revolutionizing our understanding of the formation and star-forming evolution of dense protocluster-forming clumps. The ATLASGAL survey (*Schuller et al.,* 2009; *Contreras et al.,* 2013) has been matched to the methanol multi-beam (MMB) survey (*Urquhart et al.,* 2013a), identifying 577 submillimeter continuum clumps with maser emission in the longitude range $280° \leq l \leq 20°$ (within 1.5° of the Galactic mid-plane). Ninety-four percent of methanol masers are associated with submillimeter dust emission and are preferentially associated with the most massive clumps. These clumps are centrally condensed, with envelope structures that appear to be scale-free, and the mean offset of the maser from the peak submillimeter column density is $0'' \pm 4''$. Assuming a Kroupa IMF and a SFE of ~30%, they found that over two-thirds of the maser-associated clumps are likely to form stellar clusters with masses ≥ 20 $M_\odot$ and that almost all clumps satisfy the empirical mass-size criterion for massive star formation (*Kauffmann and Pillai,* 2010). *Urquhart et al.* (2013a) find that the SFE is significantly reduced in the Galactic Center region compared to the rest of the survey area where it is broadly constant and that there is a significant drop in the massive SFR density in the outer Galaxy. *Longmore et al.* (2013b) compared the absolute SFR derived from Wilkinson Microwave Anisotropy Probe (WMAP) free-free emission with the surface density of dense gas and dust from HOPS and Hi-GAL, also deducing a much lower star formation efficiency toward the CMZ.

A further association of ATLASGAL and UC H II regions detected by the CORNISH survey of the inner Galactic Plane (*Purcell et al.,* 2012; *Hoare et al.,* 2012; *Urquhart et al.,* 2013b) identified a lower envelope of 0.05 g cm^{-2} for the surface density of molecular clumps hosting mas-

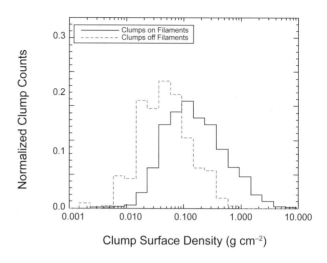

Fig. 9. Clump surface density distribution for clumps located within filaments and or not. From *Schisano et al.* (2014).

sive star formation and found that the mass of the most massive embedded star is closely correlated with the mass of the associated clump. They find little evolution in the structure of the molecular clumps between the methanol-maser and UC H II region phases. The value above is significantly lower than the theoretical prediction of 1 g cm^{-2} as a surface density threshold for massive star formation originally proposed by *Krumholz and McKee* (2008) and below the value of 0.2 g cm^{-2} recently proposed by *Butler and Tan* (2012) analyzing GLIMPSE counterparts toward a sample of IRDCs.

The 1.1-mm BGPS (*Aguirre et al.,* 2011; *Ginsburg et al.,* 2013) identified more than 8500 clumps in the Northern Galactic Plane (*Rosolowsky et al.,* 2010). Bolocam Galactic Plane Survey observations of the CMZ can be used to derive a clump mass spectrum for the central 500-pc region (*Bally et al.,* 2010). Using various assumptions about the dust temperature to derive masses, *Bally et al.* (2010) deduce clump mass spectra with a slope $\alpha = 2.4$ to 2.7 (dN/dM $\propto$ $M^{-\alpha}$) in the mass range 10^2 to 10^4 $M_\odot$, somewhat steeper than found for nearby low-mass cores or the stellar IMF.

Motivated by the detection of the 10^5 $M_\odot$ CMZ cloud dubbed the "Brick," which appears to be nearly devoid of star formation despite its high mass and density, *Ginsburg et al.* (2012) searched in a region excluding the CMZ for massive clumps with sufficient mass to form a young massive cluster (YMC) with a stellar mass of about 10^4 $M_\odot$, similar to the masses of the most massive Galactic clusters such as NGC 3603 or Westerlund 1. Under the assumption that the SFE is about 30%, *Ginsburg et al.* (2012) mined the BGPS clump catalog for clumps with masses larger than 5000 $M_\odot$, also requiring the radius to be smaller than 2.5 pc. The latter criterion was based on the arguments presented by *Bressert et al.* (2012) that to form a massive, gravitationally bound cluster, the gravitational escape speed from the cloud had to exceed the sound speed in photoionized plasma, about 10 km s^{-1}. All of the 18 massive candidate protocluster clumps were found to be actively forming massive stars or clusters based on their IR properties. Most of these massive clumps are associated with well-known giant H$_{II}$ region/star-forming complexes such as W43, W49, and W51. *Ginsburg et al.* (2012) concluded that the duration of any quiescent "prestellar" or "precluster" phase of massive protocluster clumps (MPCCs) must have a duration less than about 0.5 m.y. Thus, massive clusters in the Galactic disk might be assembled from smaller already actively star-forming units, rather than arising from an isolated massive MPCC (see the chapter by Longmore et al. in this volume).

Hi-GAL will greatly increase the number of dense clumps detected. From the first release of Hi-GAL photometric compact sources catalogs (Molinari et al., in preparation), more than 5×10^5 individual band-merged entries can be combined for the inner Galaxy only. Downselection to sources with at least three detections in adjacent bands delivers an initial catalog of nearly 100,000 compact sources in the range $+67° \geq 1 \geq -70°$; for nearly 60,000 of them a distance could be tentatively assigned, allowing the estimation of sizes ranging from 0.1 to above 1 pc and masses up to and in excess of 10^5 M$_\odot$ (Elia et al., in preparation). Dust temperatures range from 7 to 40 K with a median ~12 K, below the value of 20 K that is often adopted to derive masses from submillimeter data (*Urquhart et al.*, 2013a) and in better agreement with the ammonia temperature (*Wienen et al.*, 2012) measured toward 862 dense ATLASGAL 870-μm clumps. Preliminary studies of the clump mass functions in two very different Hi-GAL fields, toward the near tip of the Galactic bar and toward the Vul OB1 region (l = 59°), show very similar slopes (above the completeness limits) between 1.8 and 1.9 but very different mass scales (*Olmi et al.*, 2013). The much higher sensitivity of Hi-GAL with respect to BGPS or ATLASGAL, and its higher spatial resolution for λ ≤ 250 μm, enable larger statistical samples for clumps that may be hosting massive star formation. Figure 10 reports the mass-radius relationship for Hi-GAL clumps limited to the lines of sight toward the tips of the Galactic bar (*Veneziani et al.*, 2014). Clumps extend over areas of the plot where conditions are suited for massive star formation according to a variety of prescriptions. The gray solid and horizontal dashed areas are the loci with surface densities higher than 1 g cm^{-2} (*Krumholz and McKee*, 2008) and 0.2 g cm^{-2} (*Butler and Tan*, 2012), respectively.

Larger statistical samples also allow improved searches for potential precursors of YMCs, which, according to *Bressert et al.* (2012), should populate the vertical dashed area in Fig. 10. While none of the clumps revealed at the location of the tips of the bar seem to fulfill this criterion, other regions of the plane seem to contain such candidates. Figure 11 shows surface density vs. mass for the nearly 60,000 Hi-GAL clumps in the inner Galaxy for which distance estimates exist (Elia et al., in preparation), in the context of other structures as reported by *Tan* (2005). The clumps lie in the area familiar for "Galactic clumps," and extend beyond it toward surface density above 5 g cm^{-2} and

mass well above the few 10^4 M$_\odot$, to the right of the dashed line marking bounded ionized gas (*Bressert et al.*, 2012), typical of progenitors of the most massive Galactic clusters. For example, using the same mass criteria as *Ginsburg et al.* (2012) but with a more stringent threshold of 5 g cm^{-2} in mean clump surface density, more than 100 objects are selected in Hi-GAL (Ginsburg et al., in preparation).

4.2.2. Theory. Until recently, most researchers considered roundish clumps and cores in isolation as the typical sites of star formation (see, e.g., the reviews by *Beuther et al.*, 2007; *di Francesco et al.*, 2007; *Lada et al.*, 2007; *Ward-Thompson et al.*, 2007) and modeled them as thermally supported, equilibrium Bonnor-Ebert spheres, perhaps undergoing oscillations. However, since *Protostars and Planets V* (*Reipurth et al.*, 2007), it has been realized that cores are highly clustered, and most of them are embedded in the filaments from which they accrete (*Myers*, 2009; *André et al.*, 2010; *Molinari et al.*, 2010b; *Arzoumanian et al.*, 2011; *Kirk H. et al.*, 2013).

In turbulent models of GMCs, transient density fluctuations produced by shocks with sufficient compression and mass to exceed the Jeans mass result in gravitationally bound clumps and cores, while simultaneously providing support for the clouds as a whole (see, e.g., the review by *Ballesteros-Paredes et al.*, 2007, and references therein). However, recent simulations of molecular cloud formation and evolution including self-gravity (*Vázquez-Semadeni et al.*, 2007, 2009, 2011; *Heitsch and Hartmann*, 2008; *Banerjee et al.*, 2009) suggest that clouds experience global, hierarchical, and chaotic gravitational contraction, so that no support is really necessary.

The transformation of the diffuse ISM involves multiple stages of collapse and fragmentation. The cores in this scenario constitute the last stage of a mass cascade (*Field et al.*, 2008) from the large, H$_I$ supercloud scale, through GMCs, down to clumps and individual protostellar cores (*Clark and Bonnell*, 2005; *Vázquez-Semadeni et al.*, 2007, 2009). Collapse occurs along the smallest dimension first, forming sheets, then filaments, and finally clumps, as is expected for pressure-free structures (*Lin et al.*, 1965). This condition applies to GMCs, because their masses are much larger than their Jeans mass. Clumps and cores in this scenario are parts of dynamic rather than hydrostatic structures.

4.3. Triggered Star Formation

Star formation can be triggered by dynamical processes such as protostellar outflows, far-ultraviolet (FUV)-driven bubbles, expansion of H$_{II}$ regions, supernova remnants, superbubbles, etc. (*Elmegreen*, 1998, 2011b; *Deharveng et al.*, 2010). Triggering seems to be associated with a local increase of the SFR and SFE. Estimates of the percentage of YSOs formed by triggering in the Milky Way lie between 25% and 50% (*Snider et al.*, 2009; *Deharveng et al.*, 2010; *Thompson et al.*, 2012).

In general, any agent that results in converging flows or compression on sufficiently large scales to overcome

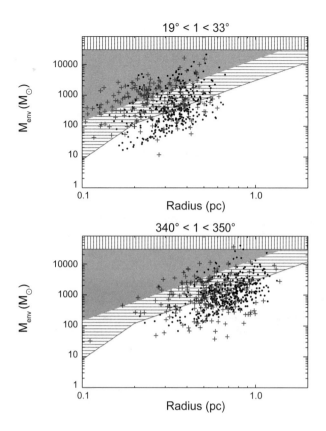

Fig. 10. Mass vs. radius for the Hi-GAL dense clumps in two ranges of Galactic longitude containing the tips of the Galactic bar from *Veneziani et al.* (2014). Protostellar (i.e., with 70-μm counterpart, crosses) and prestellar (i.e., without 70-μm counterpart, small circles) clumps at the location of the tips of the bar are reported. See text for a description of the colored and shaded areas.

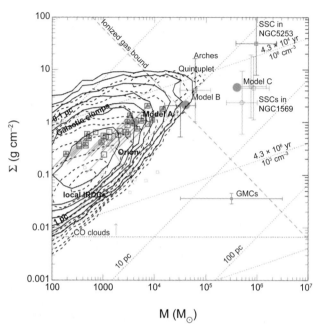

Fig. 11. Surface density vs. mass diagram from *Tan* (2005), overplotting the source density contours for the nearly 60,000 Hi-GAL detected clumps in the $67° \geq 1 \geq 19°$, and $-10° \leq 1 \leq -70°$, for which a distance estimate could be made (Elia et al., in preparation). Thick contours are for protostellar clumps, while dashed contours indicate prestellar clumps (i.e., undetected at 70 μm); source number density is computed in bins of 0.1 in decimal logarithm, and contour levels are 5, 10, 20, 40, 70, 100, 200, 400, 700 and 1000 sources per bin. Typical locations of a variety of structures are also indicated. Diagonal contours of constant radius and hydrogen number density or free-fall timescale are shown with dotted lines. The minimum surface density for CO clouds in the local Galactic UV radiation field is shown, as are typical parameters of GMCs. The condition for ionized gas to remain bound is indicated by the dashed line. Locations for a selection of IRDCs and dense star-forming clumps (squares) are reported [see *Tan* (2005) for a detailed description]. Several massive clusters are also indicated.

internal pressure or turbulent support can create conditions for the onset of gravitational instability. Observations show that the highest SFRs occur in regions of extreme pressure and density encountered in galaxy collisions and mergers. The processes act on different scales ranging from small-scale (~parsec) compression of globules, intermediate-scale (~1 to 10 pc) triggering by expanding FUV-driven bubbles such as NGC 2023 in Orion and extreme ultraviolet (EUV)-powered HII regions, to larger-scale triggering by expanding superbubbles.

Models of triggered star formation were motivated more than a half-century ago by observations of age sequences in nearby OB associations (*Blaauw*, 1991). For example, the 10–15-m.y.-old subgroup Ori OB1 is located in the northwest portion of Orion. A sequence of ever younger groups [the 1c, 1b, and 1d subgroups (*Brown et al.*, 1994)] extends toward the southeast part of Orion toward the Orion Nebula, which marks the location of the youngest subgroup.

Spatial segregation of age sequences prompted models of triggered, self-propagating star formation in which young second-generation stars form in the shock-compressed layers produced by D-type ionization fronts propagating into

clouds with speeds of a few kilometers per second (*Elmegreen and Lada*, 1977). If the second-generation stars are massive, their own HII regions continue the process, forming third-generation stars. However, if the clouds have large internal density variations, the D-type shock could lack the coherence required to collect sufficient mass to allow formation of massive stars.

Triggering can occur through compression of preexisting clumps and cores by application of an external pressure. For example, because the pressure exerted by ionization tends to be orthogonal to the ionization front, clouds and globules overrun by HII regions can be compressed. In principle, application of any external pressure can lead to collapse. However, if the pressure jump is too large, the momentum transfer to the denser medium being overrun causes it to be disrupted rather than compressed. Triggering is most likely when the shocks associated with the external pressure have speeds below the escape speed of the cloud. In this scenario, second-generation stars are expected to be younger than the first generation and to have random velocities with respect to them, because they retain the velocity dispersion of the preexisting cores.

In the collect and collapse mode (*Elmegreen and Lada,* 1977), expanding bubbles sweep up dense shells of gas from the surrounding medium. As such shells accumulate and decelerate, they can become subject to local gravitational instability and collapse, resulting in star formation. In this scenario, the younger second-generation stars are expected to be moving away from the older generation with a speed of 1–10 km s^{-1}. If the medium is relatively uniform, second-generation stars formed earlier from the decelerating shell should have larger velocities and so be located farther from the original stars than those formed later, resulting in an inverted age sequence.

Triggering is an attractive theoretical idea, but the association of YSOs with bubble rims does not necessarily imply that their formation was triggered. The observed morphologies and age sequences could have arisen from spontaneous star formation without the action of the adjacent bubble. The propagation velocities of ionization fronts and shocks decrease and stall when they encounter dense gas. Thus, even for spontaneous, untriggered star formation, bubbles produced by adjacent star formation will run up against other cores and clumps, giving the impression of triggering.

Proof of triggering in the Galactic context requires precise measurements of stellar age and proper motion. Chronometry needs to constrain stellar ages with a time resolution better than the age separation; evidence for age differences using near-IR high-resolution spectroscopy (*Martins et al.,* 2010) are still not conclusive. Proper motions with a precision of a few kilometers per second are needed to differentiate preexisting star formation from triggered star formation. For the nearby star-forming complexes, the Gaia mission should soon produce the required proper motion data.

4.3.1. Recent observational results. Figure 1 shows young stars and star clusters at the periphery of the giant HII regions in the W3, W4, and W5 complex. The rims of these giant HII regions are surrounded by dense cores, pillars, and hundreds of YSOs identified by Spitzer. In the largest and oldest portion of this complex, it has been shown that the formation of these objects was triggered by the expansion of the 3- to 5-m.y.-old W5 (*Koenig et al.,* 2008a,b; *Koenig and Allen,* 2011; *Deharveng et al.,* 2012). For the highest-column-density, most massive and luminous young clumps found by Herschel in W3, *Rivera-Ingraham et al.* (2013) suggest an active/dynamic "convergent constructive feedback" scenario to account for the formation of a cluster with decreasing age and increasing system/source mass toward the innermost regions and creation of an environment suitable for the formation of a Trapezium-like system.

The Spitzer satellite has revealed many thousands of bubble-shaped ionized regions in the Galaxy. The Milky Way Project (*Simpson et al.,* 2012), combined with other surveys such as the Red MSX Source (RMS) (*Urquhart et al.,* 2008), led to a statistical study of massive star formation associated with IR bubbles (*Kendrew et al.,* 2012). About 67 ± 3% of the massive young stellar objects (MYSOs) and (ultra-)compact HII regions in the RMS survey are associated with bubbles and 22 ± 2% of massive young stars might have formed as a result of feedback from expanding HII regions (see also *Thompson et al.,* 2012). A similar result was obtained previously by *Deharveng et al.* (2010) using Spitzer-GLIMPSE, Spitzer-MIPSGAL, and ATLASGAL surveys, indicating that about 25% of the bubbles might have triggered the formation of massive objects.

Alexander et al. (2013) and *Kerton et al.* (2013) combined Herschel images with HI and radio continuum data to probe bubbles and MYSOs at their periphery. Spectral energy distributions of YSOs derived from Herschel combined with the spatial distribution of sources shows age gradients away from the bubble center in regions such as the Rosette (*Schneider et al.,* 2010b) and N49 (*Zavagno et al.,* 2010).

Hi-GAL has also been used to assess the role of triggering. Hi-GAL sources observed at the edges of HII regions were examined. To determine if these sources and the HII regions are physically associated, we used velocity information from different kinematical surveys [including the MALT90 survey (*Foster et al.,* 2011)], as well as pointed observations. In particular, we have produced velocity maps toward 300 HII regions (mosaicing the individual MALT90 fields for large regions). Preliminary results (Zavagno et al., in preparation) show that the velocity fields observed around HII regions using molecular line data agree well with the velocities obtained for the ionized gas from optical and radio recombination line surveys, implying that much of the molecular gas with embedded sources is associated physically with the ionized regions. However, especially in the inner Galaxy, other molecular components are observed with different velocities along the line of sight, and our images show that young sources are also observed toward these components. These sources seen in projection near ionized regions might not be physically associated with them and so their formation would have nothing to do with the ionized region and its evolution. Cross-correlation with Hi-GAL distance information will help in sorting out these critical cases (Zavagno et al., in preparation).

4.3.2. Recent theoretical developments. Numerical models of the ionization of molecular clouds (*Mellema et al.,* 2006; *Arthur et al.,* 2011; *Tremblin et al.,* 2012a,b) show that compression and photo-erosion produce pillars, elephant trunks, globules, and shells of swept-up, shock-compressed dense gas that can form stars. Radiation hydrodynamical simulations show the formation of pillars from the curvature fluctuations resulting from turbulence in the dense shell. Ionization of the lower density gas behind the dense pillar heads produces cometary globules (*Haworth and Harries,* 2012; *Haworth et al.,* 2013). These models predict distinctive velocity fields that can be compared to observations. *Haworth et al.* (2013) developed molecular line diagnostics of triggered star formation using synthetic observations. *Walch et al.* (2013) use smoothed-particle hydrodynamics (SPH) simulations to explore the effects of O stars on a molecular cloud, finding that the structure of swept-up molecular gas can be either shell-like or dominated by pillars and globules, depending on the fractal dimension of the cloud.

5. STAR-FORMATION THRESHOLDS AND RATES

5.1. Star-Formation Thresholds

5.1.1. Observational studies of local clouds. Is there a column density threshold for star formation in local molecular clouds? Does the SFR increase with density or column density? *Onishi et al.* (1998) mapped the Taurus molecular cloud in $C^{18}O$ and established that prestellar cores (or class 0 protostars) were found only in regions with H_2 column densities greater than $N(H_2) \sim 8 \times 10^{21}$ cm^{-2}, corresponding to a visual extinction $A_V \sim 8.5$. On the other hand, *Hatchell et al.* (2005) found that in the Perseus cloud the number of cores is a steeply increasing function of the $C^{18}O$ integrated intensity (and therefore column density). However, $C^{18}O$ is photodissociated at low column densities, becomes optically thick at high column densities, and is depleted in the coldest, densest regions. Thus, dust is a better tracer of the column density. *Johnstone et al.* (2004) mapped the Ophiuchus cloud at 850 µm and report that dense cores and class-0 protostars were found only in regions with $A_V > 7$, indicating an extinction threshold for the formation of dense cores. *Heiderman et al.* (2010) reexamined this issue using a sample of YSOs drawn from the Spitzer c2d survey (*Evans et al., 2009*) and found a steep dependence of the SFR on the surface density, finding a column density threshold ~130 M$_\odot$ pc^{-2} (cf. Fig. 12), corresponding to $A_V \sim 8.5$. *Lada et al.* (2010) found that the number of YSOs is uncorrelated with the total mass of a cloud, but correlates with the mass above a K-band extinction of 0.8 mag (corresponding to $A_V \sim 7.3$), again consistent with a column density threshold.

Gutermuth et al. (2011) found a steep dependence of the SFR on the column density, $\Sigma_{SFR} \propto \Sigma_{gas}^2$, rather than a threshold and *Burkert and Hartmann* (2013) also argued that the data are more consistent with a steep, continuous increase of the SFR surface density with gas column density. From a first analysis of the initial Hi-GAL clump sample, Molinari et al. (in preparation) find SFR values for individual clumps that are in fairly good agreement with YSOs, clouds, and HCN clumps as reported by *Heiderman et al.* (2010) (see contours in Fig. 12); no clear evidence for a sharp column density threshold is found. Clumps that are considered "prestellar" based on the absence of a 70-µm counterpart, and which are not used to estimate the SFR, have surface densities that are similar to the "protostellar" clumps (the histogram in Fig. 12). They likely represent clumps that are on the verge of forming stars.

These studies demonstrate that there is a strong dependence of the SFR on column density, but whether/how the SFR continues to increase above some threshold remains uncertain.

Is there a threshold column density for the appearance of clumps and cores on filaments? A threshold of $A_V \sim 8$ is reported for the appearance of bound cores on filaments in the Aquila region by *André et al.* (2010) (see also the chapter by André et al. in this volume). In the more distant large filamentary complex in the outer Galaxy studied by *Schisano et al.* (2014), no threshold is apparent. In particular, Fig. 7 shows a wide range of mass line densities M_{line} for filaments with bound clumps [according to Larson's criteria as applied by *Elia et al.* (2013)]. However, the M_{line} distribution for filaments without clumps (histogram) is offset to lower values.

5.1.2. Theoretical models. Two classes of model can explain column density thresholds. In the first, the column density directly affects the likelihood of star formation:

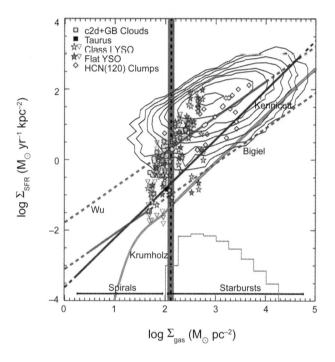

Fig. 12. Surface density of the SFR vs. mass surface density after *Heiderman et al.* (2010), including data from many studies thereby referenced. For comparison, the contours represent the distribution of the values for the Hi-GAL protostellar clumps (Elia et al., in preparation) classified as protostellar based on the presence of 70-µm emission. Because no SFR estimate is available for Hi-GAL prestellar clumps, their distribution in clump surface density is represented by the outlined histogram. The ranges of gas surface density for spiral and circumnuclear starburst galaxies in the *Kennicutt* (1998) sample are denoted by the gray horizontal lines. The gray shaded vertical band denotes the threshold surface density Σ_{th} of 129 ± 14 M$_\odot$ pc^{-2} (see text). Several Σ_{SFR} vs. Σ_{gas} prescriptions are shown; see *Heiderman et al.* (2010) for more details.

Star formation is strongly suppressed in gas with a column density below the threshold value. One plausible mechanism is photoelectric heating of the low extinction gas by the interstellar radiation field (ISRF). In turbulent clouds, the complex cloud structure enables radiation to penetrate deeply into the clouds, and the transition from warm, unshielded gas to cooler, shielded gas occurs only when the line-of-sight extinction exceeds $A_V \sim 8$. *Clark and Glover* (2013) show that such a model naturally explains the correlation between the mass of gas with $A_K > 0.8$ and the SFR found by *Lada et al.* (2010). In their simulations, gas with a line-of-sight extinction $A_K > 0.8$ is heated only by cosmic rays, resulting in temperatures lower than in the low-extinction regions, which are heated and partially ionized by the ISRF. Therefore, the Jeans mass in the high-extinction regions is considerably lower, making gravitational collapse and star formation more likely.

The second class of models is based on a dependence on the *volume* density, n_{th}, and minimum size, L_{min}. Only regions with column densities exceeding $n_{th}L_{min}$ can form stars. The existence of a volume density threshold is motivated by observations that show that a strong correlation exists between the amount of dense gas and the SFR (see, e.g., *Kennicutt and Evans,* 2012, and references therein). However, simple analytical models based on this correlation tend to predict a continuous dependence of the SFR on the gas density, albeit perhaps steep, rather than the existence of a threshold density (*Krumholz et al.,* 2012; *Burkert and Hartmann,* 2013).

5.2. Star-Formation Rates

5.2.1. Observations. The RMS survey has provided statistically significant determinations of the timescales and luminosity functions of massive YSOs and UC HII regions (*Mottram et al.,* 2011; *Davies et al.,* 2011). Both the radio-quiet MYSO phase, which could correspond to the swollen, cool, rapidly accreting phase predicted by *Hosokawa et al.* (2010), and the UC HII-region phase have timescales of one to a few $\times$ 10^5 yr, depending on the luminosity. The luminosity functions of the two phases are different, with UC HII regions detected up to luminosities of $\sim 10^6$ $L_\odot$ but no MYSOs above 10^5 $L_\odot$, which could be due to rapid evolution of the latter because the MYSO lifetimes become comparable to the Kelvin-Helmholz timescale, as predicted in the evolutionary models of *Molinari et al.* (2008). *Davies et al.* (2011) found that luminosity functions and spatial distributions are consistent with accelerating accretion rates as the MYSO grows in mass (predicted by turbulent core and competitive accretion models). Their results rule out models with constant or decreasing accretion rates, implying a global average Galactic SFR of 1.5–2 $M_\odot$ yr^{-1}. For comparison, the free-fall estimate of the SFR obtained by dividing the total molecular mass in the Galaxy by the free-fall time at the mean density of the molecular gas is a few $\times$ 10^2 $M_\odot$ yr^{-1}. Thus, star formation on average in the Galaxy is highly inefficient, with only $\sim 1\%$ of the molecular gas mass being converted to stars in a free-fall time. Equivalently, the depletion timescale of the molecular gas is ~ 1 gigayears (G.y.) in the Milky Way, similar to values inferred for nearby galaxies (see, e.g., *Leroy et al.,* 2013).

5.2.2. Theory. Models produced in recent years to explain the observed inefficiency of star formation can be grouped into two broad classes (see also the chapters by Dobbs et al. and Padoan et al. in this volume): (1) "cloud-support" models, in which the SFR is regulated by turbulence that globally maintains the cloud in quasi-equilibrium conditions while inducing local compressions that cause a small fraction of cloud's mass to collapse per global free-fall time; and (2) "cloud-collapse" models, in which the clouds are assumed to be undergoing collapse on all scales, with small-scale collapses occurring earlier than larger-scale ones, and with an increasing SFR, until the clouds are eventually destroyed by stellar feedback before much of their mass is turned into stars.

A common assumption in all models is that the clouds are internally turbulent and isothermal and therefore characterized by a log-normal density probability density function (PDF) (*Vazquez-Semadeni,* 1994; *Padoan et al.,* 1997; *Passot and Vázquez-Semadeni,* 1998). The densest regions in the cloud have shorter free-fall times and therefore collapse earlier if they occur in connected regions with more than one Jeans mass. Fluctuations with sufficiently high density will collapse on much shorter timescales than the whole cloud, producing the "instantaneous" SFR within the cloud. The models differ in the way they choose the threshold densities defining the instantaneously collapsing material, in the timescale associated with this instantaneous SF, and in the global physical conditions in the cloud.

In the cloud-support models (*Krumholz and McKee,* 2005; *Hennebelle and Chabrier,* 2011; *Padoan and Nordlund,* 2011), the clouds are in a stationary state and no time dependence is considered. It is interesting to calculate the fraction of the total mass that is converted into stars over a cloud's free-fall time, ε_{ff} [this has been called the "star formation rate per free-fall time" by *Krumholz and McKee* (2005), although by definition it is an efficiency achieved on the free-fall timescale]. To determine ε_{ff}, the cloud is assumed to be in virial equilibrium and to satisfy the Larson linewidth-size relationship. The various models then differ in their choice of the characteristic densities and associated timescales.

Krumholz and McKee (2005) assume that the timescale is close to the global cloud free-fall time, and that the density is that whose corresponding thermal Jeans length equals the sonic scale of the turbulence. The resulting Jeans masses are ~ 0.1 to 2 $M_\odot$ (for T = 10–40 K), comparable to the mean masses of stars. Note, however, that numerical simulations of isothermal, driven turbulence do not confirm the existence of the required hypothetical clumps that are simultaneously subsonic and super-Jeans (*Vázquez-Semadeni et al.,* 2008). *Padoan and Nordlund* (2011) instead assume that the relevant timescale is the free-fall time at the threshold density, and that this density is determined by the magnetic shock jump conditions and the magnetic critical mass for collapse. Finally, *Hennebelle and Chabrier* (2011) assume that the density threshold is such that its turbulent (i.e., including the turbulent velocity dispersion as a source of pressure) Jeans length is a specified fraction of the system scale, implying that this threshold is scale-dependent because the turbulent motions in general exhibit a scale dependence described by the turbulent energy spectrum. For the timescale, these authors assume that each density above the threshold collapses on its own free-fall time, a feature they term "multifree-fall." They also propose multi-free-fall versions of the *Krumholz and McKee* (2005) and *Padoan and Nordlund* (2011) theories.

In general, the turbulent-support models predict the dependence of ε_{ff} on various parameters of the clouds, such as the rms turbulent Mach number $\mathcal{M}$, the virial parameter α (the ratio of kinetic to gravitational energy in the cloud), and, when magnetic fields are considered, the magnetic β parameter (the ratio of thermal to magnetic pressure). *Federrath and Klessen* (2012) show that the multi-free-fall versions of these models all predict $\varepsilon_{ff} \sim 1$–10%, in qualitative agreement with observations.

On the other hand, cloud-collapse models (*Zamora-Avilés et al.*, 2012) assume that the clouds are in free-fall and therefore that both their density and instantaneous star-forming mass fraction are time dependent. These authors calibrate the threshold density for instantaneous SF by comparing to the simulations of *Vázquez-Semadeni et al.* (2010), take the associated timescale as the free-fall time at the threshold density, and assume that the evolution ends when the cloud is finally destroyed. They find that the SFR is generally increasing and compares favorably with the evolutionary timescales for GMCs in the Large Magellanic Cloud (LMC) (*Kawamura et al.*, 2009) and the stellar-age histograms of *Palla and Stahler* (2000, 2002). Moreover, *Zamora-Avilés and Vázquez-Semadeni* (2013) also find that the time-averaged SFR as a function of cloud mass compares favorably with the relation found by *Gao and Solomon* (2004).

An intermediate class of models has been presented by *Krumholz et al.* (2006) and *Goldbaum et al.* (2011). They compute the time-dependent virial balance of the clouds assuming spherical symmetry and that the expansion of HII regions drives supporting turbulence in the clouds, whereas winds produced by massive stars escape the clouds. Their models predict that the clouds undergo an oscillatory stage that is short for low-mass clouds and longer (several freefall times) for more massive clouds.

A crucial difference between the various models is the assumed effect of stellar feedback on the clouds, which ranges from simply driving turbulence to entirely evaporating them. Recent numerical simulations have addressed this issue, finding that while on short timescales (~1 m.y.) stellar ionizing feedback indeed drives turbulence into the clouds (*Gritschneder et al.*, 2009; *Walch et al.*, 2012), the final effect on timescales ~10 m.y. is the destruction of the star-forming site for clouds up to 10^5 M$_\odot$ (*Dale et al.*, 2012, 2013; *Colín et al.*, 2013).

5.3. Cloud Disruption

The low SFE in gravitationally unbound molecular clouds is a consequence of their dissolution in a crossing time, $L/\sigma \sim 3$–10 m.y. for 10–30-pc clouds obeying Larson's relationships. Bound clouds require energy inputs to be disrupted and dissociated. Feedback from forming stars is the most likely agent. For instance, feedback from low-mass (M < few M$_\odot$) stars is dominated by their outflows (*Li and Nakamura*, 2006; *Wang et al.*, 2010; *Nakamura and Li*, 2011). Although stars of all masses generate such flows during formation, in more massive stars other mechanisms inject greater amounts of energy after accretion stops. Intermediate-mass YSOs (2 < M < 8 M$_\odot$) produce copious nonionizing radiation that heats adjacent cloud surfaces to temperatures of 10^2 to over 6×10^3 K, raising the sound speed to 1–6 km s^{-1} and dramatically increasing pressure in the heated layer. The resulting gas expansion can drive shocks into surrounding gas, accelerating it. Chaotic cloud structure results in chaotic motions. Late B and A stars are thus an important source of heating, motion, and turbulence generation through the action of their photon-dominated regions. High-mass (M > 8 M$_\odot$) stars ionize their environments, creating 10^4 K HII regions whose pressure accelerates surrounding gas.

The growth of HII regions and their feedback on the structure of molecular clouds has been studied by many different groups (see, e.g., *Franco et al.*, 1994; *Matzner*, 2002; *Walch et al.*, 2012). Small molecular clouds are efficiently disrupted by ionizing radiation on short timescales, but the largest Galactic GMCs have escape velocities that are significantly higher than the sound speed of photoionized gas, and are therefore not disrupted (see, e.g., *Krumholz et al.*, 2006; *Dale et al.*, 2012)

In massive GMCs radiation pressure from absorbed or scattered photons could be important (*Murray et al.*, 2010). For single scattering or absorption, the momentum transferred to surrounding gas is $\dot{P} \sim L\tau/c \sim 1.3 \times 10^{29} \, \tau \, L_6$ g cm s^{-2} where τ is the optical depth and L_6 is the total embedded bolometric luminosity in units of 10^6 L$_\odot$. Because stars generally form in the densest parts of clouds, this momentum transfer contributes to gas dispersal. The effectiveness of this form of feedback depends on both how much radiation is produced by the stars and what fraction of this radiation is trapped by the cloud (*Krumholz and Thompson*, 2013). Radiation can leak out through cavities in highly structured clouds, lowering its effectiveness as a feedback mechanism; this is the case for most Galactic GMCs.

During their accretion phase, all stars produce powerful MHD-powered bipolar outflows whose momentum injection rate is 10× to 100× L/c. When massive stars form, their outflows are likely to clear channels through which radiation can leak out of the parent clump, decreasing the feedback from radiation pressure. Models of the effects of radiation pressure on star-forming clouds show that in low-mass clouds radiative trapping is small and that photoionization is more important than radiation pressure for destroying clouds (*Agertz et al.,* 2013). Radiative trapping could be larger in more massive clouds, although *Krumholz and Thompson* (2013) argue that RT instability of pressure-driven gas dramatically reduces its effective optical depth (but see *Kuiper et al.,* 2013).

Feedback from outflows, FUV, ionizing radiation, and radiation pressure produce comparable momentum injection rates. In turn, these are similar to momentum dissipation rates for typical parameters. This points to a dynamic balance between dissipation and injection. Furthermore, the low SFE in typical GMCs suggests that once the SFE reaches 2–20%, feedback wins and disrupts the cloud. Failure of feedback could lead to high SFE and formation of bound clusters (*Bressert et al.,* 2012).

OB stars that eventually explode as type II supernovae have lifetimes of 3–30 m.y., longer than GMC lifetimes. Stellar winds, radiation pressure, and the supernovae generated have powerful feedback on clouds, even those at dozens or hundreds of parsecs from the explosion site as observations of the local superbubbles show. Supernovae probably play the dominant role in driving turbulence on large scales in the ISM (*Mac Low and Klessen,* 2004).

5.4. Global Mapping of Thresholds and Star-Formation Rates

An assessment of the available estimates of the present-day SFR in the Milky Way was carried out by *Chomiuk and Povich* (2011) using a common IMF and stellar models applied to measured free-free emission rates and mid-IR YSO counts. Their median result is 1.9 ± 0.4 $M_\odot$ yr^{-1}. The GLIMPSE and MIPSGAL surveys (*Robitaille et al.,* 2008) result in a modeled SFR ~ 0.68–1.45 $M_\odot$ yr^{-1} (*Robitaille and Whitney,* 2010); the main uncertainty arises from contamination by asymptotic giant branch (AGB) stars. Using the classic method of measuring the Lyman continuum luminosity from the free-free radio continuum of HII regions, *Murray and Rahman* (2010) use the WMAP data to find SFR = 1.3 ± 0.2 $M_\odot$ yr^{-1} for a Salpeter IMF, and values between 0.9 and 2.2 $M_\odot$ yr^{-1} for a range of IMF slopes. This may be a lower bound since it not clear that the analysis of WMAP free-free emission properly accounts for Lyman continuum lost from the Galaxy that leaks out of chimneys to illuminate the Lockman layer.

Kennicutt and Evans (2012) present a review connecting the Galactic star formation with extragalactic systems. *Evans et al.* (2009) and *Lada et al.* (2010) found that for a given surface density in nearby molecular clouds the extragalactic Schmidt-Kennicutt relationship underestimates the SFR, because studies of the SFR in Galactic GMCs tend to ignore the surface density from extensive, non-star-forming gas; when this is included, the Galactic SFRs fit the extragalactic relationship. An example of this is shown in Fig. 12, where *Heiderman et al.* (2010) plot local clouds and Galactic clumps onto the extragalactic star formation relationship. The figure shows an order of magnitude range in SFR at any given gas surface density.

For individual protostellar clumps cataloged by Elia et al. (in preparation) using Hi-GAL data, we can derive the SFR and mass per unit area. The distribution of these values is overlaid as contours in Fig. 12.

Herschel provides a wealth of data on the formation rate of massive stars in different Galactic environments. For example, in four square degrees at longitudes l = 30°(near the Scutum arm tangent) and at l = 59° (*Russeil et al.,* 2011; *Veneziani et al.,* 2013), Hi-GAL found average rates of $1.6 \pm 0.7 \times 10^{-2}$ and $7.9 \pm 3.6 \times 10^{-4}$ $M_\odot$ yr^{-1} respectively. In the rich massive star-forming region G305, *Faimali et al.* (2012) found rates of 0.01–0.02 $M_\odot$ yr^{-1}. In the l = 217°–224° region situated in the outer Galaxy at Galactocentric distances greater than 10 kpc, *Elia et al.* (2013) found a value of 2.6×10^{-3} $M_\odot$ yr^{-1}. *Veneziani et al.* (2014) analyze the star-formation content at the tips of the Galactic bar using the same method as in *Veneziani et al.* (2013), but also incorporating a statistical correction to predicted luminosities to account for the fact that these clumps host protoclusters rather than single massive forming stars; they find values $\sim 1.5 \times 10^{-2}$ $M_\odot$ yr^{-1} kpc^{-2}, at both sides of the bar. Globally combining all the massive objects in the preliminary Hi-GAL catalog, limited to the $+67° \geq l \geq -70°$ range and excluding the inner ~25° of the CMZ, Molinari et al. (in preparation) find a total SFR ~1.6 $M_\odot$ yr^{-1}, which can be considered a lower limit especially because only a fraction (although significant) of the Galaxy is considered.

The conversion efficiency of molecular gas into dense potentially star-forming clumps and into stellar luminosity has been examined by *Eden et al.* (2012) and *Moore et al.* (2012) as a function of position, both in and out of spiral-arms. There is little or no increase in SFE associated with the Scutum-tangent region and the near end of the Galactic bar, which includes the W43 complex. Investigation of the Galactic distribution of the ~1650 embedded, young massive objects in the RMS survey, one-third of which are above the completeness limit of $\sim 2 \times 10^4$ $L_\odot$ (*Urquhart et al.,* 2013a), shows that the distribution of massive star formation in the Milky Way is correlated with a model of four spiral arms. The increase in SFR density in spiral arms is due to the pile-up of clumps, rather than to a higher specific rate of conversion of clumps into stars (a higher SFE). That the source and luminosity surface densities of RMS sources are correlated with the surface density of the molecular gas suggests that the massive star-formation rate per unit molecular mass is approximately constant across the Galaxy. The total luminosity of the embedded massive-star population is found to be $\sim 0.8 \times 10^8$ $L_\odot$, 30% of which is associated with the 10 most active star-forming complexes.

Given the enormous reservoir of dense H_2 in the CMZ, a high SFR might be expected there. Local relationships (*Lada et al.,* 2012; *Krumholz,* 2012) extrapolated to Galactic Center conditions predict 0.18–0.74 $M_\odot$ yr^{-1} (based on extinction) or 0.14–0.4 $M_\odot$ yr^{-1} (based on volume density). Surprisingly, however, *Yusef-Zadeh et al.* (2009) and *Immer et al.* (2012) found 0.08–0.14 $M_\odot$ yr^{-1}, also consistent with the Schmidt-Kennicutt relationship. Using WMAP free-free emission as a tracer of star formation within 500 pc of the Galactic Center (about 1 kpc^{-2}), *Longmore et al.* (2013b) found even lower values, 3.6×10^{-3} to 1.6×10^{-2} M$_\odot$ yr^{-1}, again in disagreement with predictions.

6. CONCLUSIONS

The avalanche of new IR to radio continuum and spectroscopic imaging of the Galactic Plane is a goldmine that we are just starting to exploit, and still more are desirable. Current quadrant-scale CO spectroscopic surveys of the third and fourth quadrants only deliver superarcminute resolution. Subarcminute CO surveys are required; this effort is indeed starting to be assembled (see Table 1) but more is needed. Additional high-spatial-resolution surveys in molecular tracers like NH$_3$ and other high-density tracers are also needed. Considerable potential also resides in post-Herschel unbiased arcsecond-resolution continuum surveys in the far-IR and submillimeter, as well as in the radio where sensitivities of current unbiased surveys are still not adequate for detailed Galactic studies. These will be within reach of future facilities such as the Space Infrared Telescope for Cosmology and Astrophysics (SPICA), the Cornell Caltech Atacama Telescope (CCAT), the Millimetron project, and the Square Kilometre Array (SKA) pathfinders, while ALMA and JWST will be able to concentrate on selected samples of star-forming complexes. Large-scale subarcminute-resolution polarization surveys in the far-IR and submillimeter are badly needed to ascertain the role of magnetic fields throughout the cloud-filament-clump-core evolutionary sequence.

In our opinion, however, the real challenge and opportunities lie in an "industrial revolution" of the science analysis methodologies in Galactic astronomy. Current approaches are largely inadequate to execute a timely and effective scientific exploitation of complete Galactic Plane surveys covering several hundreds of square degrees. Progress is needed in tools for pattern recognition and extraction of clumps, filaments, bubbles, and sheets in highly variable background conditions, for robust construction of spectral energy distributions (SEDs), for evolutionary classification of compact clumps and cores, and for three-dimensional vector modeling of Galactic rotation incorporating streaming motions, to name a few. Toward building a Galactic Plane knowledge base, we envision these technical developments integrated within a framework of data mining, machine learning, and interactive three-dimensional visualization, in order to transfuse the astronomer's know-how into a set of automated and supervised tools with decision-making capabilities.

REFERENCES

Abdo A. A. et al. (2010) *Astrophys. J., 710,* 133.
Agertz O. et al. (2013) *Astrophys. J., 770,* 25.
Aguirre J. E. et al. (2011) *Astrophys. J. Suppl., 192,* 4.
Alexander M. et al. (2013) *Astrophys. J., 770,* 1.
Allen R. J. et al. (2004) *Astrophys. J., 608,* 314.
Allen R. J. et al. (2012) *Astron. J., 143,* 97.
Alves J. et al. (2007) *Astron. Astrophys., 462,* L17.
André P. et al. (2010) *Astron. Astrophys., 518,* L102.
Arthur S. J. et al. (2011) *Mon. Not. R. Astron. Soc., 414,* 1747.
Arzoumanian D. et al. (2011) *Astron. Astrophys., 529,* L6.
Audit E. and Hennebelle P. (2005) *Astron. Astrophys., 433,* 1.
Bally J. et al. (1987a) *Astrophys. J. Lett., 312,* L45.
Bally J. et al. (1987b) *Astrophys. J. Suppl., 65,* 13.
Bally J. et al. (2010) *Astrophys. J., 721,* 137.
Banerjee R. et al. (2009) *Mon. Not. R. Astron. Soc., 398,* 1082.
Barnard E. E. (1907) *Astrophys. J., 25,* 218.
Barnes P. J. et al. (2011) *Astrophys. J. Suppl., 196,* 12.
Battersby C. et al. (2011) *Astron. Astrophys., 535,* A128.
Beltrán M. T. et al. (2006) *Astron. Astrophys., 447,* 221.
Benjamin R. A. and GLIMPSE360 Team (2013) *AAS Meeting #222,* Abstract #303.03.
Benjamin R. A. et al. (2003) *Publ. Astron. Soc. Pac., 115,* 953.
Beuther H. et al. (2007) In *Protostars and Planets V* (B. Reipurth et al., eds.), pp. 165–180. Univ. of Arizona, Tucson.
Binney J. et al. (1991) *Mon. Not. R. Astron. Soc., 252,* 210.
Blaauw A. (1991) In *The Physics of Star Formation and Early Stellar Evolution* (C. J. Lada and N. D. Kylafis, eds.), p. 125, NATO ASIC Proc. 342.
Bressert E. et al. (2012) *Astrophys. J. Lett., 758,* L28.
Brown A. G. A. et al. (1994) *Astron. Astrophys., 289,* 101.
Brunthaler A. et al. (2011) *Astron. Nachr., 332,* 461.
Burkert A. and Hartmann L. (2013) *Astrophys. J., 773,* 48.
Burton M. G. et al. (2013) *Publ. Astron. Soc. Australia, 30,* 44.
Butler M. J. and Tan J. C. (2012) *Astrophys. J., 754,* 5.
Carey S. J. et al. (2000) *Astrophys. J. Lett., 543,* L157.
Carey S. J. et al. (2009) *Publ. Astron. Soc. Pac., 121,* 76.
Chapman N. L. et al. (2011) *Astrophys. J., 741,* 21.
Chomiuk L. and Povich M. S. (2011) *Astron. J., 142,* 197.
Clark P. and Bonnell I. (2005) *Mon. Not. R. Astron. Soc., 361,* 2.
Clark P. C. and Glover S. C. O. (2013) *ArXiv e-prints,* arXiv:1306.5714.
Clark P. C. et al. (2012) *Mon. Not. R. Astron. Soc., 424,* 2599.
Colín P. et al. (2013) *Mon. Not. R. Astron. Soc., 435,* 1701.
Combes F. (1991) *Annu. Rev. Astron. Astrophys., 29,* 195.
Contreras Y. et al. (2013) *Astron. Astrophys., 549,* A45.
Dahmen G. et al. (1998) *Astron. Astrophys., 331,* 959.
Dale J. E. et al. (2012) *Mon. Not. R. Astron. Soc., 424,* 377.
Dale J. E. et al. (2013) *Mon. Not. R. Astron. Soc., 430,* 234.
Dame T. M. et al. (1986) *Astrophys. J., 305,* 892.
Dame T. M. et al. (1987) *Astrophys. J., 322,* 706.
Dame T. M. et al. (2001) *Astrophys. J., 547,* 792.
Davies B. et al. (2011) *Mon. Not. R. Astron. Soc., 416,* 972.
de Zeeuw P. T. et al. (1999) *Astron. J., 117,* 354.
Deharveng L. et al. (2010) *Astron. Astrophys., 523,* 6.
Deharveng L. et al. (2012) *Astron. Astrophys., 546,* 74.
di Francesco J. et al. (2007) In *Protostars and Planets V* (B. Reipurth et al., eds.), pp. 17–32. Univ. of Arizona, Tucson.
Dobbs C. L. et al. (2008) *Mon. Not. R. Astron. Soc., 389,* 1097.
Dobbs C. L. et al. (2012) *Mon. Not. R. Astron. Soc., 425,* 2157.
Eden D. J. et al. (2012) *Mon. Not. R. Astron. Soc., 422,* 3178.
Egan M. P. et al. (1998) *Astrophys. J. Lett., 494,* L199.
Ehlerová S. and Palouš J. (2013) *Astron. Astrophys., 550,* A23.
Elia D. et al. (2013) *Astrophys. J., 772,* 45.
Ellsworth-Bowers T. P. et al. (2013) *Astrophys. J., 770,* 39.
Elmegreen B. G. (1994) *Astrophys. J., 433,* 39.
Elmegreen B. G. (1998) In *Origins* (C. E. Woodward et al., eds.), p. 150. ASP Conf. Ser. 148, Astronomical Society of the Pacific, San Francisco.
Elmegreen B. G. (2011a) *Astrophys. J., 737,* 10.
Elmegreen B. G. (2011b) *Triggered Star Formation,* EAS Publ. Series, Vol. 51, Cambridge Univ., Cambridge.
Elmegreen B. G. and Elmegreen D. M. (1987) *Astrophys. J., 320,* 182.
Elmegreen B. G. and Lada C. J. (1977) *Astrophys. J., 214,* 725.
Evans N. J. II et al. (2009) *Astrophys. J. Suppl., 181,* 321.
Faimali A. et al. (2012) *Mon. Not. R. Astron. Soc., 426,* 402.
Fallscheer C. et al. (2013) *Astrophys. J., 773,* 102.
Federrath C. and Klessen R. S. (2012) *Astrophys. J., 761,* 156.
Ferrière K. et al. (2007) *Astron. Astrophys., 467,* 611.
Field G. B. (1965) *Astrophys. J., 142,* 531.
Field G. B. et al. (2008) *Mon. Not. R. Astron. Soc., 385,* 181.
Fischera J. and Martin P. G. (2012a) *Astron. Astrophys., 547,* A86.

Fischera J. and Martin P. G. (2012b) *Astron. Astrophys., 542,* A77.
Foster J. B. et al. (2011) *Astrophys. J. Suppl., 197,* 25.
Foster J. B. et al. (2013) *Publ. Astron. Soc. Australia, 30,* 38.
Franco J. et al. (1994) *Astrophys. J., 436,* 795.
Gao Y. and Solomon P. M. (2004) *Astrophys. J., 606,* 271.
Gibson S. J. et al. (2000) *Astrophys. J., 540,* 851.
Ginsburg A. et al. (2012) *Astrophys. J. Lett., 758,* L29.
Ginsburg A. et al. (2013) *Astrophys. J. Suppl., 208,* 14.
Glover S. C. O. and Mac Low M.-M. (2007) *Astrophys. J. Suppl., 169,* 239.
Glover S. C. O. and Mac Low M.-M. (2011) *Mon. Not. R. Astron. Soc., 412,* 337.
Goldbaum N. J. et al. (2011) *Astrophys. J., 738,* 101.
Goldsmith P. F. et al. (2008) *Astrophys. J., 680,* 428.
Gómez G. C. and Vázquez-Semadeni E. (2013) *Astrophys. J., 428,* 186.
Green J. A. et al. (2009) *Mon. Not. R. Astron. Soc., 392,* 783.
Grenier I. A. et al. (2005) *Science, 307,* 1292.
Gritschneder M. et al. (2009) *Astrophys. J. Lett., 694,* L26.
Gutermuth R. A. et al. (2011) *Astrophys. J., 739,* 84.
Hacar A. and Tafalla M. (2011) *Astron. Astrophys., 533,* A34.
Hartmann D. and Burton W. B. (1997) *Atlas of Galactic Neutral Hydrogen,* Cambridge Univ., Cambridge. 248 pp.
Hartmann L. et al. (2001) *Astrophys. J., 562,* 852.
Hatchell J. et al. (2005) *Astron. Astrophys., 440,* 151.
Haworth T. J. and Harries T. J. (2012) *Mon. Not. R. Astron. Soc., 420,* 562.
Haworth T. J. et al. (2013) *Mon. Not. R. Astron. Soc., 431,* 3470.
Heiderman A. et al. (2010) *Astrophys. J., 723,* 1019.
Heiner J. S. and Vázquez-Semadeni E. (2013) *Mon. Not. R. Astron. Soc., 429,* 3584.
Heiner J. S. et al. (2009) *Astrophys. J., 700,* 545.
Heitsch F. and Hartmann L. (2008) *Astrophys. J., 689,* 290.
Heitsch F. et al. (2005) *Astrophys. J. Lett., 633,* L113.
Heitsch F. et al. (2008) *Astrophys. J., 674,* 316.
Hennebelle P. and Chabrier G. (2011) *Astrophys. J. Lett., 743,* L29.
Hennebelle P. and Pérault M. (1999) *Astron. Astrophys., 351,* 309.
Heyer M. et al. (2009) *Astrophys. J., 699,* 1092.
Heyer M. H. et al. (1998) *Astrophys. J. Suppl., 115,* 241.
Hill T. et al. (2011) *Astron. Astrophys., 533,* 94.
Hoare M. G. et al. (2012) *Publ. Astron. Soc. Pac., 124,* 939.
Hosokawa T. et al. (2010) *Astrophys. J., 721,* 478.
Huettemeister S. et al. (1998) *Astron. Astrophys., 334,* 646.
Ikeda N. et al. (2007) *Astrophys. J., 665,* 1194.
Immer K. et al. (2012) *Astron. Astrophys., 548,* A120.
Ishihara D. et al. (2010) *Astron. Astrophys., 514,* A1.
Jackson J. M. et al. (2006) *Astrophys. J. Suppl., 163,* 145.
Johnstone D. and Bally J. (1999) *Astrophys. J. Lett., 510,* L49.
Johnstone D. and Bally J. (2006) *Astrophys. J., 653,* 383.
Johnstone D. et al. (2004) *Astrophys. J. Lett., 611,* L45.
Jones P. A. et al. (2012) *Mon. Not. R. Astron. Soc., 419,* 2961.
Jones P. A. et al. (2013) *Mon. Not. R. Astron. Soc., 433,* 221.
Kalberla P. M. W. and Kerp J. (2009) *Annu. Rev. Astron. Astrophys., 47,* 27.
Kauffmann J. and Pillai T. (2010) *Astrophys. J. Lett., 723,* L7.
Kawamura A. et al. (2009) *Astrophys. J. Suppl., 184(1),* 1.
Kendrew S. et al. (2012) *Astrophys. J., 755,* 71.
Kennicutt R. C. Jr. (1998) *Astrophys. J., 498,* 541.
Kennicutt R. C. Jr. and Evans N. J. II (2012) *Annu. Rev. Astron. Astrophys., 50,* 531.
Kerton C. R. et al. (2013) *Astron. J., 145,* 78.
Kirk H. et al. (2013a) *Astrophys. J., 766,* 115.
Kirk J. M. et al. (2013b) *Mon. Not. R. Astron. Soc., 432,* 1424.
Koenig X. P. and Allen L. E. (2011) *Astrophys. J., 726,* 18.
Koenig X. P. et al. (2008a) *Astrophys. J., 688,* 1142.
Koenig X. P. et al. (2008b) *Astrophys. J. Lett., 687,* L37.
Koo B. et al. (1992) *Astrophys. J., 390,* 108.
Koyama H. and Inutsuka S.-I. (2000) *Astrophys. J., 532,* 980.
Koyama H. and Inutsuka S.-I. (2002) *Astrophys. J. Lett., 564,* L97.
Kruijssen J. M. D. et al. (2013) *ArXiv e-prints,* arXiv:1303.6286.
Krumholz M. R. (2012) *Astrophys. J., 759,* 9.
Krumholz M. R. and McKee C. F. (2005) *Astrophys. J., 630,* 250.
Krumholz M. R. and McKee C. F. (2008) *Nature, 451,* 1082.
Krumholz M. R. and Thompson T. A. (2013) *Mon. Not. R. Astron. Soc., 434,* 2329.
Krumholz M. R. et al. (2006) *Astrophys. J., 653,* 361.
Krumholz M. R. et al. (2012) *Astrophys. J., 745,* 69.

Kuiper R. et al. (2013) In *The Labyrinth of Star Formation* (D. Stamatellos et al., eds.), Astrophys. Space Sci. Library, Vol. 36, Kluwer, Dordrecht.
Kuntz K. D. and Danly L. (1996) *Astrophys. J., 457,* 703.
Lada C. J. et al. (2007) In *Protostars and Planets V* (B. Reipurth et al., eds.), pp. 3–15. Univ. of Arizona, Tucson.
Lada C. J. et al. (2010) *Astrophys. J., 724,* 687.
Lada C. J. et al. (2012) *Astrophys. J., 745,* 190.
Larson R. B. (1981) *Mon. Not. R. Astron. Soc., 194,* 809.
Larson R. B. (1985) *Mon. Not. R. Astron. Soc., 214,* 379.
Lee M.-Y. et al. (2012) *Astrophys. J., 748,* 75.
Leroy A. et al. (2013) *Astron. J., 146,* 19.
Lesh J. R. (1968) *Astrophys. J. Suppl., 17,* 371.
Li Z.-Y. and Nakamura F. (2006) *Astrophys. J. Lett., 640,* L187.
Lin C. C. et al. (1965) *Astrophys. J., 142,* 1431.
Lindblad P. O. et al. (1973) *Astron. Astrophys., 24,* 309.
Longmore S. N. et al. (2012) *Astrophys. J., 746,* 117.
Longmore S. N. et al. (2013a) *Mon. Not. R. Astron. Soc., 433,* L15.
Longmore S. N. et al. (2013b) *Mon. Not. R. Astron. Soc., 429,* 987.
Mac Low M.-M. and Klessen R. S. (2004) *Rev. Mod. Phys., 76,* 125.
Martin C. L. et al. (2004) *Astrophys. J. Suppl., 150,* 239.
Martins F. et al. (2010) *Astron. Astrophys., 510,* 32.
Matzner C. D. (2002) *Astrophys. J., 566,* 302.
McClure-Griffiths N. M. et al. (2001) *Astrophys. J., 551,* 394.
McCray R. and Kafatos M. (1987) *Astrophys. J., 317,* 190.
McKee C. F. and Ostriker E. C. (2007) *Annu. Rev. Astron. Astrophys., 45,* 565.
Mellema G. et al. (2006) *Astrophys. J., 647,* 397.
Mizuno A. and Fukui Y. (2004) In *Milky Way Surveys: The Structure and Evolution of Our Galaxy* (D. Clemens et al., eds.), p. 59. ASP Conf. Ser. 317, Astronomical Society of the Pacific, San Francisco.
Molinari S. et al. (2008) *Astron. Astrophys., 481,* 345.
Molinari S. et al. (2010a) *Astron. Astrophys., 518,* L100.
Molinari S. et al. (2010b) *Publ. Astron. Soc. Pac., 122,* 314.
Molinari S. et al. (2011) *Astrophys. J. Lett., 735,* L33.
Moore T. J. T. et al. (2012) *Mon. Not. R. Astron. Soc., 426,* 701.
Morris M. and Serabyn E. (1996) *Annu. Rev. Astron. Astrophys., 34,* 645.
Motte F. et al. (1998) *Astron. Astrophys., 336,* 150.
Mottram J. C. et al. (2011) *Astrophys. J. Lett., 730,* L33.
Murray N. and Rahman M. (2010) *Astrophys. J., 709,* 424.
Murray N. et al. (2010) *Astrophys. J., 709,* 191.
Myers P. C. (2009) *Astrophys. J., 700,* 1609.
Nakamura F. and Li Z.-Y. (2011) *Astrophys. J., 740,* 36.
Narayanan G. et al. (2008) *Astrophys. J. Suppl., 177,* 341.
Olmi L. et al. (2013) *Astron. Astrophys., 551,* 111.
Onishi T. et al. (1998) *Astrophys. J., 502,* 296.
Padoan P. and Nordlund Å. (2002) *Astrophys. J., 576,* 870.
Padoan P. and Nordlund Å. (2011) *Astrophys. J., 730,* 40.
Padoan P. et al. (1997) *Mon. Not. R. Astron. Soc., 288,* 145.
Padoan P. et al. (2001) In *From Darkness to Light: Origin and Evolution of Young Stellar Clusters* (T. Montmerle and P. André, eds.), p. 279. ASP Conf. Ser. 243, Astronomical Society of the Pacific, San Francisco.
Palla F. and Stahler S. W. (2000) *Astrophys. J., 540,* 255.
Palla F. and Stahler S. W. (2002) *Astrophys. J., 581,* 1194.
Palmeirim P. et al. (2013) *Astron. Astrophys., 550,* A38.
Passot T. and Vázquez-Semadeni E. (1998) *Phys. Rev. E, 58,* 4501.
Peretto N. and Fuller G. A. (2009) *Astron. Astrophys., 505,* 405.
Peretto N. et al. (2013) *Astron. Astrophys., 555,* 112.
Perrot C. A. and Grenier I. A. (2003) *Astron. Astrophys., 404,* 519.
Pillai T. et al. (2006) *Astron. Astrophys., 450,* 569.
Pineda J. L. et al. (2013) *Astron. Astrophys., 554,* A103.
Planck Collaboration et al. (2011) *Astron. Astrophys., 536,* A19.
Planck Collaboration et al. (2013a) *ArXiv e-prints,* arXiv:1303.5062.
Planck Collaboration et al. (2013b) *ArXiv e-prints,* arXiv:1303.5073.
Planck Collaboration et al. (2013c) *Astron. Astrophys., 554,* A139.
Plume R. et al. (1992) *Astrophys. J. Suppl., 78,* 505.
Plume R. et al. (1997) *Astrophys. J., 476,* 730.
Polychroni D. et al. (2013) *Astrophys. J. Lett., 777,* L33.
Pon A. et al. (2012) *Astrophys. J., 756,* 145.
Price S. D. et al. (2001) *Astron. J., 121,* 2819.
Pringle J. E. et al. (2001) *Mon. Not. R. Astron. Soc., 327,* 663.
Purcell C. R. et al. (2012) *Mon. Not. R. Astron. Soc., 426,* 1972.
Reid M. A. and Wilson C. D. (2005) *Astrophys. J., 625,* 891.
Reipurth B. et al., eds. (2007) *Protostars and Planets V,* Univ. of Arizona, Tucson.

Rivera-Ingraham A. et al. (2013) *Astrophys. J., 766,* 85.
Robitaille T. P. and Whitney B. A. (2010) *Astrophys. J. Lett., 710,* L11.
Robitaille T. P. et al. (2008) *Astron. J., 136,* 2413.
Roman-Duval J. et al. (2010) *Astrophys. J., 723,* 492.
Rosolowsky E. (2005) *Publ. Astron. Soc. Pac., 117,* 1403.
Rosolowsky E. et al. (2010) *Astrophys. J. Suppl., 188,* 123.
Russeil D. et al. (2011) *Astron. Astrophys., 526,* 151.
Schisano E. et al. (2014) *Astrophys. J.,* in press.
Schneider N. et al. (2010a) *Astron. Astrophys., 520,* 49.
Schneider N. et al. (2010b) *Astron. Astrophys., 518,* L83.
Schneider N. et al. (2012) *Astron. Astrophys., 540,* L11.
Schneider S. and Elmegreen B. G. (1979) *Astrophys. J. Suppl., 41,* 87.
Schuller F. et al. (2009) *Astron. Astrophys., 504,* 415.
Shirley Y. et al. (2003) *Astrophys. J. Suppl., 149,* 375.
Simon R. et al. (2006) *Astrophys. J., 653,* 1325.
Simpson R. J. et al. (2012) *Mon. Not. R. Astron. Soc., 424,* 2442.
Snider K. D. et al. (2009) *Astrophys. J., 700,* 506.
Sofue Y. and Handa T. (1984) *Nature, 310,* 568.
Stark A. A. et al. (1992) *Astrophys. J. Suppl., 79,* 77.
Stil J. et al. (2006) *Astron. J., 132,* 1158.
Su M. et al. (2010) *Astrophys. J., 724,* 1044.
Tackenberg J. et al. (2012) *Astron. Astrophys., 540,* 113.
Tan J. C. (2005) In *Cores to Clusters: Star Formation with Next Generation Telescopes* (M. S. Kumar et al., eds.), p. 87. Astrophys. Space Sci. Library, Vol. 324, Kluwer, Dordrecht.
Thompson M. A. et al. (2012) *Mon. Not. R. Astron. Soc., 421,* 408.
Toalá J. A. et al. (2012) *Astrophys. J., 744,* 190.
Toomre A. (1964) *Astrophys. J., 139,* 1217.
Tremblin P. et al. (2012a) *Astron. Astrophys., 538,* A31.
Tremblin P. et al. (2012b) *Astron. Astrophys., 546,* 33.
Urquhart J. S. et al. (2008) In *Massive Star Formation: Observations Confront Theory* (H. Beuther et al., eds.), p. 381. ASP Conf. Ser. 387, Astronomical Society of the Pacific, San Francisco.

Urquhart J. S. et al. (2013a) *Mon. Not. R. Astron. Soc., 431,* 1752.
Urquhart J. S. et al. (2013b) *Mon. Not. R. Astron. Soc., 435,* 400.
Vázquez-Semadeni E. (1994) *Astrophys. J., 423,* 681.
Vázquez-Semadeni E. et al. (2000) *Astrophys. J., 540,* 271.
Vázquez-Semadeni E. et al. (2006) *Astrophys. J., 643,* 245.
Vázquez-Semadeni E. et al. (2007) *Astrophys. J., 657,* 870.
Vázquez-Semadeni E. et al. (2008) *Mon. Not. R. Astron. Soc., 390,* 769.
Vázquez-Semadeni E. et al. (2009) *Astrophys. J., 707,* 1023.
Vázquez-Semadeni E. et al. (2010) *Astrophys. J., 715,* 1302.
Vázquez-Semadeni E. et al. (2011) *Mon. Not. R. Astron. Soc., 414,* 2511.
Veneziani M. et al. (2013) *Astron. Astrophys., 549,* 130.
Veneziani M. et al. (2014) *Astrophys. J.,* in press.
Wada K. et al. (2002) *Astrophys. J., 577,* 197.
Walch S. K. et al. (2012) *Mon. Not. R. Astron. Soc., 427,* 625.
Walch S. et al. (2013) *Mon. Not. R. Astron. Soc., 435,* 917.
Walsh A. J. et al. (2011) *Mon. Not. R. Astron. Soc., 416,* 1764.
Wang P. et al. (2010) *Astrophys. J., 709,* 27.
Ward-Thompson D. et al. (2007) In *Protostars and Planets V* (B. Reipurth et al., eds.), pp. 33–46. Univ. of Arizona, Tucson.
Wienen M. et al. (2012) *Astron. Astrophys., 544,* 146.
Wilcock L. A. et al. (2012) *Mon. Not. R. Astron. Soc., 424,* 716.
Wolfire M. G. et al. (2010) *Astrophys. J., 716,* 1191.
Wright E. L. et al. (2010) *Astron. J., 140,* 1868.
Xu Y. et al. (2013) *Astrophys. J., 769,* 15.
Yusef-Zadeh F. et al. (2009) *Astrophys. J., 702,* 178.
Zamora-Avilés M. and Vázquez-Semadeni E. (2013) *ArXiv e-prints,* arXiv:1308.4918.
Zamora-Avilés M. et al. (2012) *Astrophys. J., 751,* 77.
Zavagno A. et al. (2010) *Astron. Astrophys., 518,* L101.

Tan J. C., Beltrán M. T., Caselli P., Fontani F., Fuente A., Krumholz M. R., McKee C. F., and Stolte A. (2014) Massive star formation. In *Protostars and Planets VI* (H. Beuther et al., eds.), pp. 149–172. Univ. of Arizona, Tucson, DOI: 10.2458/azu_uapress_9780816531240-ch007.

Massive Star Formation

Jonathan C. Tan
University of Florida

Maria T. Beltrán
Istituto Nazionale di Astrofisica–Osservatorio Astrofisico di Arcetri

Paola Caselli
Max-Planck-Institute for Extraterrestrial Physics

Francesco Fontani
Istituto Nazionale di Astrofisica–Osservatorio Astrofisico di Arcetri

Asunción Fuente
Observatorio Astronómico Nacional

Mark R. Krumholz
University of California, Santa Cruz

Christopher F. McKee
University of California, Berkeley

Andrea Stolte
Universität Bonn

The enormous radiative and mechanical luminosities of massive stars impact a vast range of scales and processes, from the reionization of the universe, to the evolution of galaxies, to the regulation of the interstellar medium, to the formation of star clusters, and even to the formation of planets around stars in such clusters. Two main classes of massive star formation theory are under active study: core accretion and competitive accretion. In core accretion, the initial conditions are self-gravitating, centrally concentrated cores that condense with a range of masses from the surrounding, fragmenting clump environment. They then undergo relatively ordered collapse via a central disk to form a single star or a small-N multiple. In this case, the prestellar core mass function has a similar form to the stellar initial mass function. In competitive accretion, the material that forms a massive star is drawn more chaotically from a wider region of the clump without passing through a phase of being in a massive, coherent core. In this case, massive star formation must proceed hand in hand with star cluster formation. If stellar densities become very high near the cluster center, then collisions between stars may also help to form the most massive stars. We review recent theoretical and observational progress toward understanding massive star formation, considering physical and chemical processes, comparisons with low and intermediate-mass stars, and connections to star cluster formation.

1. INTRODUCTION

Across the universe, massive stars play dominant roles in terms of their feedback and their synthesis and dispersal of heavy elements. Achieving a full theoretical understanding of massive star formation is thus an important goal of contemporary astrophysics. This effort can also be viewed as a major component of the development of a general theory of star formation that seeks to explain the birth of stars of all masses and from all varieties of star-forming environments.

Two main classes of theory are under active study: core accretion and competitive accretion. In core accretion, extending "standard" low-mass star formation theory (*Shu et al.,* 1987), the initial conditions are self-gravitating, centrally concentrated cores of gas that condense with a range of masses from a fragmenting *clump* (i.e., protocluster) environment. These cores then undergo gravitational collapse via a central disk, to form a single star or small-N multiple. The prestellar core (PSC) mass function (CMF) has a shape similar to the stellar initial mass function (IMF). In competitive accretion, gas that forms a massive star is drawn chaotically from a wider region of the clump, without ever being in a massive, coherent, gravitationally bound, starless core. Also, a forming massive star is always surrounded by a swarm of low-mass protostars. Competitive accretion is sometimes said to lead naturally to the IMF (*Bonnell et al.,* 2001, 2007): The total mass of massive stars must then be a small fraction of the total stellar mass formed from the clump. If the density of protostars congregating near the cluster center becomes sufficiently high, then stellar collisions may also assist in forming the most massive stars.

Recent advances in theoretical/numerical modeling of massive star formation involve inclusion of more physical processes, like radiation pressure, magnetic fields and protostellar outflows. Observationally, progress has resulted from telescopes such as Spitzer, Herschel, the Stratospheric Observatory for Infrared Astronomy (SOFIA), the Atacama Large Millimeter/submillimeter Array (ALMA), and the Very Large Array (VLA). Galactic plane surveys have yielded large samples of candidate massive protostars and their birth clouds.

This review aims to summarize massive star formation research, focusing on developments since the reviews of *Beuther et al.* (2007), *Zinnecker and Yorke* (2007), and *McKee and Ostriker* (2007). We do not discuss formation of the first stars, which are thought to have been massive (e.g., *Bromm,* 2013). Given the complexity of massive star formation, detailed comparison of theoretical predictions with observational results is needed for progress in understanding which accretion mechanism(s) is relevant and which physical and chemical processes are important. We thus first overview basic observed properties of massive star-forming regions (section 1.1), which set boundary conditions on theoretical models. Next we present a theoretical overview of physical processes likely involved in forming massive stars (section 2), including the different accretion models, protostellar evolution and feedback, and results from numerical simulations. We then focus on observational results on the earlier, i.e., initial condition (section 3), and later, i.e., accretion (section 4), stages of massive star formation. Here we discuss astrochemical modeling, as well as general comparisons of massive star formation with intermediate-/low-mass star formation. The relation of massive star formation to star cluster formation is examined in section 5. We conclude in section 6.

1.1. The Birth Environments of Massive Stars

The basic physical properties of regions observed to be forming or have formed massive stars, i.e., gas clumps and young star clusters, are shown in Fig. 1, plotting mass surface density, $\Sigma = M/(\pi R^2)$, of the structure vs. its mass, M. Stars, including massive stars, form in molecular gas, which is mostly found in giant molecular clouds (GMCs) with $\Sigma \sim$ 0.02 g cm^{-2}. Note 1.0 g cm$^{-2} \equiv 4790$ $M_\odot$ pc^{-2}, for which $N_H = 4.27 \times 10^{23}$ cm^{-2} (assuming $n_{He} = 0.1$ n_H so mass per H is $\mu_H = 2.34 \times 10^{-24}$ g) and visual extinction is $A_V = (N_H/2.0 \times 10^{21}$ cm$^{-2})$ mag $= 214$ mag. However, star formation is seen to be localized within star-forming clumps within GMCs, which typically have $\Sigma_{cl} \sim 0.1$–1 g cm^{-2}. Some massive systems, usually already formed star clusters, have Σ up to $\sim$30 g cm^{-2}.

In terms of Σ_{cl} (in g cm^{-2}) and $M_{cl,3} = M_{cl}/(1000$ $M_\odot)$, the radius and (H number) density of a spherical clump are

$$R_{cl} = 0.258 M_{cl,3}^{1/2} \Sigma_{cl}^{-1/2} \text{ pc} \tag{1}$$

$$\bar{n}_{H,cl} = 4.03 \times 10^5 \Sigma_{cl}^{3/2} M_{cl,3}^{-1/2} \text{ cm}^{-3} \tag{2}$$

Gas clumps massive enough to form a cluster of mass $M_{*cl} \sim 500$ $M_\odot$, i.e., with median expected maximum stellar mass $\sim$30 $M_\odot$ (for Salpeter IMF from 0.1 to 120 $M_\odot$), are thus $\sim$0.3 pc in size (if $\Sigma_{cl} \sim 1$ g cm^{-2} and efficiency $\varepsilon_{*cl} \equiv M_{*cl}/M_{cl} \sim 0.5$), only moderately larger than the $\cdot$ 0.1 pc sizes of well studied low-mass starless cores in regions such as Taurus (*Bergin and Tafalla,* 2007). However, mean densities in such clumps are at least 10 times larger.

2. THEORETICAL OVERVIEW

2.1. Physical Processes in Self-Gravitating Gas

The importance of self-gravity in a cloud of mass M and radius R can be gauged by the virial parameter

$$\alpha_{vir} \equiv 5\sigma^2 R/(GM) = 2aE_K/E_G \tag{3}$$

where σ is one-dimensional mass-averaged velocity dispersion, $a \equiv E_G/(3$ $GM^2/[5$ $R])$ is the ratio of gravitational energy, E_G (assuming negligible external tides), to that of a uniform sphere, and E_K is the kinetic energy (*Bertoldi and McKee,* 1992). Often α_{vir} is set as 2 E_K/E_G, with the advantage of clearly denoting bound ($E_K < E_G$) and virialized ($E_K = 0.5$ E_G) clouds, but the disadvantage that a is difficult to observe. For spherical clouds with a power-law density distribution, $\rho \propto r^{-k_\rho}$, a rises from 1 to 5/3 as k_ρ goes from 0 (uniform density) to 2 (singular isothermal sphere). A cloud in free-fall has $\alpha_{vir} \to 2a$ from below as time progresses. The cloud's escape velocity is $v_{esc} = (10/\alpha_{vir})^{1/2}\sigma$.

The velocity dispersion in a cloud is thus given by

$$\sigma \equiv (\pi \alpha_{vir} G \Sigma R/5)^{1/2} \tag{4}$$

where we have used the identity symbol to emphasize that this follows from the definition of α_{vir}. Clouds that are gravitationally bound with $\alpha_{vir} \sim 1$ and that have similar

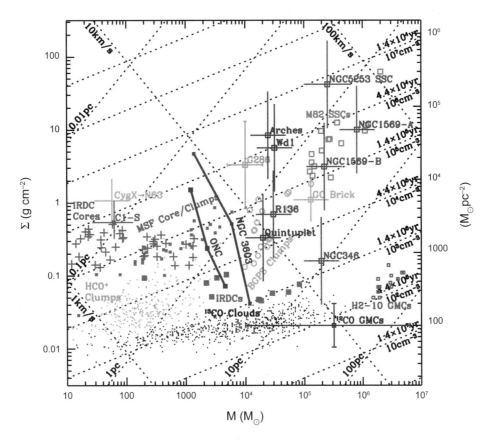

Fig. 1. The environments of massive star formation. Mass surface density, $\Sigma \equiv M/(\pi R^2)$, is plotted vs. mass, M. Dotted lines of constant radius, R, H number density, n_H (or free-fall time, $t_{ff} = [3\pi/(32G\rho)]^{1/2}$), and escape speed, $v_{esc} = (10/\alpha_{vir})^{1/2}\sigma$, are shown. Stars form from molecular gas, which in the Galaxy is mostly organized into GMCs. Typical ^{12}CO-defined GMCs have $\Sigma \sim 100$ $M_\odot$ pc^{-2} (*Solomon et al.,* 1987) [see *Tan et al.* (2013a) for detailed discussion of the methods for estimating Σ for the objects plotted here], although denser examples have been found in Henize 2-10 (*Santangelo et al.,* 2009). The ^{13}CO-defined clouds of *Roman-Duval et al.* (2010) are indicated, along with HCO$^+$ clumps of *Barnes et al.* (2011), including G286.21+0.17 (*Barnes et al.,* 2010). Along with G286, the BGPS clumps (*Ginsburg et al.,* 2012) and the Galactic center "Brick" (*Longmore et al.,* 2012) are some of the most massive, high-Σ gas clumps known in the Milky Way. Ten example infrared dark clouds (IRDCs) (*Kainulainen and Tan,* 2013) and their internal core/clumps (*Butler and Tan,* 2012) are shown, including the massive, monolithic, highly deuterated core C1-S (*Tan et al.,* 2013b). CygX-N63, a core with similar mass and size as C1-S, appears to be forming a single massive protostar (*Bontemps et al.,* 2010; *Duarte-Cabral et al.,* 2013). The IRDC core/clumps overlap with massive star-forming (MSF) core/clumps (*Mueller et al.,* 2002). Clumps may give rise to young star clusters, like the ONC (e.g., *Da Rio et al.,* 2012) and NGC 3603 (*Pang et al.,* 2013) (radial structure is shown from core to half-mass, $R_{1/2}$, to outer radius), or even more massive examples, e.g., Westerlund 1 (*Lim et al.,* 2013), Arches (*Habibi et al.,* 2013), and Quintuplet (*Hussmann et al.,* 2012) (shown at $R_{1/2}$), that are in the regime of "super star clusters" (SSCs), i.e., with $M_* \gtrsim 10^4$ $M_\odot$. Example SSCs in the Large Magellanic Cloud (LMC) [R136 (*Andersen et al.,* 2009)] and Small Magellanic Cloud (SMC) [NGC 346 (*Sabbi et al.,* 2008)] display a wide range of Σ, but no evidence of IMF variation (section 5.2). Even more massive clusters can be found in some dwarf irregular galaxies, such as NGC 1569 (*Larsen et al.,* 2008) and NGC 5253 (*Turner and Beck,* 2004), and starburst galaxy M82 (*McCrady and Graham,* 2007).

surface densities then naturally satisfy a line-width size (LWS) relation $\sigma \propto R^{1/2}$ (*Larson,* 1985), consistent with observations (*McKee and Ostriker,* 2007; *Heyer et al.,* 2009). *McKee et al.* (2010) termed this the virialized LWS relation for $\alpha_{vir} = 1$, as suggested by *Heyer et al.* (2009). By contrast, the standard turbulence-dominated LWS relation, $\sigma = \sigma_{pc}R_{pc}^{1/2}$, where $\sigma_{pc} \simeq 0.72$ km s^{-1} in the Galaxy (*McKee and Ostriker,* 2007), is independent of Σ. The virialized LWS relation applies for mass surface densities $\Sigma > \Sigma_{LWS} = (5/[\pi G])(\sigma_{pc}^2/1 \text{ pc}) \simeq 0.040$ g cm^{-2}. Since regions of

massive star formation have column densities substantially greater than this, they lie above the turbulence dominated LWS relation, as observed (*Plume et al.,* 1997).

Evaluating equation (4) for massive star-forming regions

$$\sigma = 1.826\alpha_{vir}^{1/2}\left(M_3 \Sigma\right)^{1/4} \text{ km s}^{-1} \qquad (5)$$

yields supersonic motions, as the isothermal sound speed $c_{th} = 0.188(T/10 \text{ K})^{1/2}$ km s^{-1} and $T \lesssim 30$ K for gas not too close to a massive star. These motions cannot be primarily infall, since clump infall (section 4.1) and star-formation rates per free-fall time are quite small (*Krumholz and Tan,* 2007). It is therefore likely that regions of massive star formation are dominated by supersonic turbulence. If so, then the gas inside a clump of radius R_{cl} will obey a LWS relation $\sigma = (R/R_{cl})^q \sigma_{cl}$ with $q \leq 1/2$ (*Matzner,* 2007), so

$$\alpha_{vir} = (5/\pi)\left(\sigma_{cl}^2/\left[R_{cl}G\Sigma\right]\right) = (\Sigma_{cl}/\Sigma)\alpha_{vir,cl} \qquad (6)$$

At a typical point in the clump, the density is less than average, so α_{vir} of a subregion of size R is $> (R_{cl}/R)\,\alpha_{vir,cl}$. Even if the clump is bound, the subregion is not. However, for $q = 1/2$, a subregion compressed so that $\Sigma \geq \Sigma_{cl}$ has $\alpha_{vir} \lesssim \alpha_{vir,cl}$ and is bound if the clump is; for $q < 1/2$, extra compression is needed to make a bound subregion.

Isothermal clouds more massive than the critical mass, M_{cr}, cannot be in hydrostatic equilibrium and will collapse. In this case, M_{cr} is termed the Bonnor-Ebert mass, given by

$$M_{cr} = M_{BE} = \bar{\mu}_{cr} c_{th}^3 \left(G^3\bar{\rho}\right)^{-1/2} = \bar{\mu}_{cr} c_{th}^4 \left(G^3\bar{P}\right)^{-1/2} \qquad (7)$$

where $\bar{P}$ is mean pressure in the cloud, and $\bar{\mu}_{cr} = 1.856$ (*McKee and Holliman,* 1999) (M_{cr} can also be expressed in terms of density at the cloud's surface; then $\bar{\mu}_{cr} = 1.182$). For nonmagnetic clouds this relation can be generalized to an arbitrary equation of state by replacing c_{th}^2 with σ^2. One can show that $\bar{\mu}_{cr}$ corresponds to a critical value of the virial parameter, $\alpha_{vir,cr} = 5(3/[4\pi])^{1/3}\bar{\mu}_{cr}^{-2/3}$; clouds with $\alpha_{vir} < \alpha_{vir,cr}$ will collapse. For example, equilibrium isothermal clouds have $\alpha > \alpha_{vir,cr} = 2.054$.

The critical mass associated with magnetic fields can be expressed in two ways (e.g., *McKee and Ostriker,* 2007): as $M_\Phi = \Phi/(2\pi G^{1/2})$, where Φ is the magnetic flux, or as $M_B = M_\Phi^3/M^2$, which can be rewritten as

$$M_B = \left(9/\left[16\sqrt{\pi}\right]\right)v_A^3 \left(G^3\bar{\rho}\right)^{-1/2} \qquad (8)$$

where v_A is the Alfvén velocity. Magnetically critical clouds have $M = M_\Phi = M_B$. Most regions with Zeeman observations are magnetically supercritical (*Crutcher,* 2012), i.e., $M > M_\Phi$, so that the field, on its own, cannot prevent collapse.

Magnetized isothermal clouds have $M_{cr} \simeq M_{BE} + M_\Phi$, and since most GMCs are gravitationally bound (e.g., *Roman-Duval et al.,* 2010; *Tan et al.,* 2013a), magnetized, and turbulent, they are expected to have $M \simeq M_{cr} \simeq 2\,M_\Phi$ (*McKee,* 1989). Magnetically subcritical clouds can evolve to being supercritical by flows along field lines and/or by ambipolar diffusion. In a quiescent medium, the ambipolar diffusion time is about 10 free-fall times (*Mouschovias,* 1987); this timescale is reduced in the presence of turbulence (e.g., *Fatuzzo and Adams,* 2002; *Li et al.,* 2012b). Lazarian and collaborators (e.g., *de Gouveia Dal Pino et al.,* 2012, and references therein) have suggested super-Alfvénic turbulence drives rapid reconnection that can efficiently remove magnetic flux from a cloud.

Self-gravitating clouds in virial equilibrium have a mean total pressure (thermal, turbulent, and magnetic) that is related to the total Σ via [*McKee and Tan,* 2003 (*MT03*)]

$$\bar{P} \equiv \left(3\pi/20\right)f_g\phi_{goem}\phi_B\alpha_{vir}G\Sigma^2 \qquad (9)$$

where f_g is the fraction of total mass surface density in gas (as opposed to stars), ϕ_{geom} is an order of unity numerical factor that accounts for the effect of nonspherical geometry, $\phi_B \simeq 1.3 + 3/(2\,\mathcal{M}_A^2)$ accounts for the effect of magnetic fields, and $\mathcal{M}_A = \sqrt{3}(\sigma/v_A)$ is the Alfvén Mach number. A magnetized cloud with the same total pressure and surface density as a nonmagnetic cloud will therefore have a virial parameter that is smaller by a factor of ϕ_B. Clouds that are observed to have small α_{vir} (e.g., *Pillai et al.,* 2011) (see also section 3.1) are therefore either in the very early stages of gravitational collapse or are strongly magnetized.

The characteristic time for gravitational collapse is the free-fall time. For a spherical cloud, this is

$$\bar{t}_{ff} = \left(3\pi/[32G\bar{\rho}]\right)^{1/2} = 6.85\times10^4 M_3^{1/4}\Sigma^{-3/4}\ \text{yr} \qquad (10)$$

where the bar on t_{ff} indicates that it is given in terms of the mean density of the mass M. The free-fall velocity is

$$v_{ff} = \left(2GM/r\right)^{1/2} = 5.77\left(M_3\Sigma\right)^{1/4}\ \text{km s}^{-1} \qquad (11)$$

An isothermal filament with mass/length $m_\ell > 2\,c_{th}^2/G = 16.4\,(T/10\ \text{K})\ M_\odot\ \text{pc}^{-1}$ cannot be in equilibrium and will collapse. Its free-fall time and velocity are $(1/2)\,(G\bar{\rho})^{-1/2}$ and $2\,[Gm_\ell\,\ln(r_0/r)]^{1/2} = 1.3[m_{\ell,100}\,\ln(r_0/r)]^{1/2}\ \text{km s}^{-1}$, with r_0 the initial radius of collapsing gas and $m_{\ell,100} = m_\ell/100\ M_\odot\ \text{pc}^{-1}$. Infall velocities much less than this indicate either collapse has just begun or that it is quasistatic.

2.2. Formation Mechanisms

A key parameter in both core and competitive accretion is the characteristic accretion rate in a cloud with $M \geq M_{cr}$

$$\dot{m}_{ff} = M/\bar{t}_{ff} = \left(8G/\sqrt{\pi}\right)^{1/2}\left(M\Sigma\right)^{3/4} \qquad (12)$$

$$= 1.46\times10^{-2}\left(M_3\Sigma\right)^{3/4}\ M_\odot\ \text{yr}^{-1} \qquad (13)$$

In competitive accretion models, the star-forming clump undergoes global, typically free-fall, collapse, so this is the characteristic accretion rate in the entire forming cluster. In core accretion models, this is the characteristic accretion rate to the central star and disk in the core, with the accreted gas then supplied to just one or a few protostars. The properties of the surrounding clump are assumed to be approximately constant during the formation of the star.

The corresponding accretion time, $t_{acc} \propto M/\dot{m}_{ff} \propto \bar{t}_{ff} \propto M^{1/4}$, is a weak function of mass for clouds of a given Σ. Note that a singular isothermal sphere has $\rho \propto r^{-2}$ so its collapse leads to $\dot{m}_{ff} \propto (M\Sigma)^{3/4} = \text{const}$ (*Shu,* 1977).

2.2.1. Core accretion. The principal assumption of core accretion models is that the initial conditions for intermediate and massive star formation are gravitationally bound cores, scaled up in mass from the low-mass examples known to form low-mass stars. Different versions of these models invoke varying properties of the cores, including their expected densities, density profiles, sources of internal pressure, and dynamical states. A distinguishing feature of these models is that the prestellar CMF is hypothesized to be similar in shape to the stellar IMF, with stellar masses being $m_* = \varepsilon_c M_c$, where $\varepsilon_c \sim 0.5$, perhaps set by protostellar outflow feedback (*Matzner and McKee, 2000*) (see section 2.4). This feature of some kind of one-to-one correspondence between the CMF and IMF is an underlying assumption of recent theories of the IMF, which predict the CMF based on the conditions needed to form bound cores in a turbulent medium (e.g., *Padoan and Nordlund, 2007*; cf. *Clark et al., 2007*).

There are at least two main differences between low- and high-mass star formation: First, for sufficiently massive stars, the Kelvin-Helmholtz time can be less than the accretion time, so the star accretes while on the main sequence (*Kahn, 1974*). Second, cores forming massive stars are large enough that internal turbulence can dominate thermal motions (*Myers and Fuller, 1992*; *Caselli and Myers, 1995*). Extending the work of these authors, *McKee and Tan (2002; MT03)* developed the turbulent core model, based on the assumptions that the internal pressure is mostly nonthermal, in the form of turbulence and/or magnetic fields, and that the initial core is reasonably close to internal virial equilibrium, so that its structure can be approximated as a singular polytropic sphere. Also, approximate pressure equilibrium with the surrounding clump is assumed, which thus normalizes the size, density, and velocity dispersion of a core of given mass to Σ_{cl}. *MT03* focused on the case in which $\rho \propto r^{-k_\rho}$ with $k_\rho = 1.5$, similar to observed values (section 3.1); for this case, $\Sigma_c = 1.22 \Sigma_{cl}$. For example, the core radius, given by equation (1) with core properties in place of those of the clump, can be expressed in terms of core mass and clump surface density: $R_c = 0.074 (M_{c,2}/\Sigma_{cl})^{1/2}$ pc.

The characteristic accretion rate in core accretion models is given by equation (12). In the *Shu (1977)* model, based on collapse of a singular isothermal sphere, the actual accretion rate is $0.38 \, \dot{m}_{ff}$. This result ignores the contraction needed to create the sphere. *Tan and McKee (2004)* argued (in the context of primordial star formation, but similar reasoning may apply locally) that it was more reasonable to include the formation phase of the collapsing cloud using one of *Hunter's (1977)* subsonic collapse solutions, which has an accretion rate 2.6 times larger and gives an accretion rate onto the star + disk system of $\dot{m}_{*d} \simeq \dot{m}_{ff}$. For collapse that begins from a marginally stable Bonnor-Ebert sphere, $\dot{m}_{*d}$ is initially $\gg \dot{m}_{ff}$, but then falls to about the *Shu (1977)* rate. For the turbulent core model, the dependence of the accretion rate on $M\Sigma$ can be reexpressed in terms of the current value of the idealized collapsed mass that has been supplied to the central disk in the zero-feedback limit, M_{*d} (note, the actual protostar plus disk mass accretion rate is

$\dot{m}_{*d} = \varepsilon_{*d}\dot{M}_{*d}$ and the integrated protostar plus disk mass is $m_{*d} = \bar{\varepsilon}_{*d}M_{*d}$), and Σ_{cl}. For $k_\rho = 1.5$ and allowing for the effects of magnetic fields (*MT03*), this gives

$$\dot{m}_{*d} = 1.37 \times 10^{-3} \varepsilon_{*d} \left(M_{c,2} \Sigma_{cl}\right)^{3/4} \left(M_{*d}/M_c\right)^{1/2} \, M_\odot \, yr^{-1} \quad (14)$$

for $\varepsilon_{*d} = 1$, this corresponds to $\dot{m}_{*d} = 0.64 \, \dot{m}_{ff}$. If the disk mass is assumed to be a constant fraction, f_d, of the stellar mass, then the actual accretion rate to the protostar is $\dot{m}_* = (1/[1 + f_d])\dot{m}_{*d}$. A value of $f_d \simeq 1/3$, i.e., a relatively massive disk, is expected in models where angular momentum transport is due to moderately self-gravitating disk turbulence and larger-scale spiral density waves.

Two challenges faced by core accretion are: (1) What prevents a massive core, perhaps containing $\sim 10^2$ Jeans masses, from fragmenting into a cluster of smaller stars? This will be addressed in section 2.4. (2) Where are the accretion disks expected around forming single and binary massive stars? Disks have been discovered around some massive stars, but it has not been shown that they are ubiquitous (section 4.2).

2.2.2. Competitive accretion. Competitive accretion (*Bonnell et al., 2001*) involves protostars accreting ambient clump gas at a rate

$$\dot{m}_{*d} = \pi \rho_{cl} v_{rel} r_{acc}^2 \quad (15)$$

where v_{rel} is the relative velocity of stars with respect to clump gas, ρ_{cl} is the local density, and r_{acc} is the accretion radius. Two limits for r_{acc} were proposed: (1) Gas-dominated regime (set by tidal radius): $r_{acc} \simeq r_{tidal} = 0.5[m_*/M_{cl}(R)]^{1/3}R$, where R is the distance of the star from the clump center; and (2) star-dominated regime (set by Bondi-Hoyle accretion radius): $r_{acc} \simeq r_{BH} = 2Gm_*/(v_{rel}^2 + c_s^2)$. The star-dominated regime was suggested to be relevant for massive star formation — the stars destined to become massive being those that tend to settle to protocluster centers, where high ambient gas densities are maintained by global clump infall. The accretion is assumed to be terminated by stellar feedback or by fragmentation-induced starvation (*Peters et al., 2010b*).

In addition to forming massive stars, *Bonnell et al. (2001, 2007)* proposed that competitive accretion is also responsible for building up the IMF for $m_* \gtrsim M_{BE}$. These studies have since been developed to incorporate additional physics (see section 2.5) and include comparisons to both the IMF and binary properties of the stellar systems (*Bate, 2012*).

Bonnell et al. (2004) tracked the gas that joined the massive stars in their simulation, showing it was initially widely distributed throughout the clump, so the final mass of the star did not depend on the initial core mass present when it first started forming. Studies of the gas cores seen in simulations exhibiting competitive accretion have been carried out by *Smith et al. (2011, 2013)*, with nonspherical, filamentary morphologies being prevalent, along with total accretion being dominated by that accreted later from beyond the original core volume. Other predictions of the competitive accretion scenario are relatively small accretion disks, with chaotically varying orientations, which would also be reflected in protostellar outflow directions. Massive stars would always be observed to form at the center of a cluster in which the stellar mass was dominated by low-mass stars.

As competitive accretion is "clump-fed," we express the average accretion rate of a star of final mass m_{*f} via

$$\langle \dot{m}_{*d} \rangle = \varepsilon_{ff} \dot{m}_{ff} m_{*f} / (\varepsilon_{cl} M_{cl})$$
$$\rightarrow 1.46 \times 10^{-4} \varepsilon_{ff,0.1} \frac{m_{*f,50}}{\varepsilon_{cl,0.5}} \frac{\Sigma_{cl}^{3/4}}{M_{cl,3}^{1/4}} M_\odot \, yr^{-1} \quad (16)$$

(see also *Wang et al.*, 2010), where ε_{ff} is the star formation efficiency per free-fall time and ε_{cl} is the final star formation efficiency from the clump. *Krumholz and Tan* (2007) estimated $\varepsilon_{ff} \simeq 0.04$ in the ONC. The average accretion rate (4.6×10^{-5} $M_\odot$ yr^{-1}) of the most massive star (46.4 $M_\odot$) in the *Wang et al.* (2010) simulation with outflow feedback ($\Sigma_{cl} = 0.08$ g cm^{-2}, $M_{cl} = 1220$ $M_\odot$, $\varepsilon_{cl} = 0.18$, $\varepsilon_{ff} = 0.08$) agrees with equation (16) to within 10%. This shows that a major difference between the competitive accretion model of massive star formation and the core accretion model is that its average accretion rate to the star is much smaller (cf. equation 14).

This low rate of competitive accretion was noted before in the context of accretion from a turbulent medium with $\alpha_{vir} \sim 1$, as is observed in most star-forming regions (*Krumholz et al.*, 2005a). *Bonnell et al.* (2001) came to essentially the same conclusion by noting competitive accretion would not be fast enough to form massive stars, if stars were virialized in the cluster potential (i.e., high v_{rel}). They suggested efficient star formation ($\varepsilon_{ff} \sim 1$) occurs in regions of global gravitational collapse with negligible random motions. *Wang et al.* (2010), by including the effects of protostellar outflows and moderately strong magnetic fields that slowed down star cluster formation, found massive star formation via competitive accretion occurred relatively slowly over about 1 million years (m.y.) (equation (16)). Accretion to the clump center was fed by a network of dense filaments, even while the overall clump structure remained in quasivirial equilibrium. As discussed further in section 2.5, these results may depend on the choice of initial conditions, such as the degree of magnetization and/or use of an initially smooth density field, which minimizes the role of turbulence.

Another challenge for competitive accretion is the effect of feedback. *Edgar and Clarke* (2004) noted that radiation pressure disrupts dusty Bondi-Hoyle accretion for protostellar masses and ≥ 10 $M_\odot$. Protostellar outflows, such as those included by *Wang et al.* (2010), also impede local accretion to a star from some directions around the accretion radius. This issue is examined further in section 2.4.

In summary, the key distinction between competitive and core accretion is whether competitive, "clump-fed" accre-

tion of gas onto stars, especially intermediate and massive stars, dominates over that present in the initial PSC. In core accretion, the PSC will likely gain some mass via accretion from the clump, but it will also lose mass due to feedback; the net result is that the mass of the PSC will be $\gtrsim m_{*f}$. In competitive accretion, the PSC mass is $\ll m_{*f}$. Of course, reality may be somewhere between these extremes, or might involve different aspects. We note that an observational test of this theoretical distinction requires that it be possible to identify PSCs that may themselves be turbulent. As discussed in section 2.5, to date no simulations have been performed with self-consistent initial conditions and with the full range of feedback. Such simulations will be possible in the near future and should determine whether massive PSCs can form in such an environment, as required for core accretion models, or whether low-/intermediate-mass stars can accrete enough mass by tidally truncated Bondi accretion to grow into massive stars.

2.2.3. Protostellar collisions. *Bonnell et al.* (1998) proposed massive stars may form (i.e., gain significant mass) via protostellar collisions, including those resulting from the hardening of binaries (*Bonnell and Bate*, 2005). This model was motivated by the perceived difficulty of accreting dusty gas onto massive protostars — merging protostars are optically thick and therefore unaffected by radiation pressure feedback. Note that such protostellar collisions are distinct from those inferred to be driven by binary stellar evolution (*Sana et al.*, 2012). Universal formation of massive stars via collisions would imply massive stars always form in clusters. Indeed, for collisional growth to be rapid compared to stellar evolution timescales requires cluster environments of extreme stellar densities, $\gtrsim 10^8$ pc^{-3} (i.e., $n_H \geq 3 \times 10^9$ cm^{-3}) (e.g., *Moeckel and Clarke*, 2011; *Baumgardt and Klessen*, 2011), never yet observed (Fig. 1). *Moeckel and Clarke* (2011) find that when collisions are efficient, they lead to runaway growth of one or two extreme objects, rather than smoothly filling the upper IMF. Thus collisional growth appears to be unimportant in typical massive star-forming environments.

2.3. Accretion Disks and Protostellar Evolution

In both core and competitive accretion, the angular momentum of the gas is expected to be large enough that most accretion to the protostar proceeds via a disk. Here angular momentum is transferred outward via viscous torques resulting from the magneto-rotational instability (MRI) or gravitational instability, which produces spiral arms and, if strong enough, fragmentation to form a binary or higher-order multiple (*Kratter et al.*, 2010). For the turbulent core model, an upper limit for the size of the disk forming in a core of rotational energy $\beta_{rot} = E_{rot}/|E_{grav}|$ is evaluated by assuming conservation of angular momentum of gas streamlines inside the sonic point of the flow. Then the disk radius, r_d, is a fraction of the initial core size: For a 60 $M_\odot$, $\beta_{rot} = 0.02$ core forming in a clump with $\Sigma_{cl} = 1$ g cm^{-2}, we have $r_d = 57.4$, 102 AU when $m_* = 8$, 16 $M_\odot$

(*Zhang et al.,* 2014) (see Figs. 2 and 3). However, magnetic braking of the accretion flow may make the disk much smaller (see the chapter by Z.-Y. Li et al. in this volume). Disks around massive protostars also arise in competitive accretion models (e.g., *Bate,* 2012), but are likely to be smaller than in core accretion models.

Angular momentum may also be transferred via torques associated with a large-scale magnetic field threading the disk that couples to a disk wind (*Blandford and Payne,* 1982; *Königl and Pudritz,* 2000). The final accretion to the protostar may be mediated by a strong stellar B-field, as proposed for X-wind models around low-mass protostars (*Shu et al.,* 2000). For massive stars, the required field strengths would need to be ≥kG (*Rosen et al.,* 2012). Or, the disk may continue all the way in to the protostellar surface, in which case one expects high (near breakup) initial rotation rates of massive stars. However, such high rates are typically not observed and the necessary spin down would require either stronger B-fields or longer disk lifetimes than those inferred from observations (*Rosen et al.,* 2012).

The evolution of the protostar depends on its rate of accretion of mass, energy, and angular momentum from the disk. Since the dynamical time of the star is short compared to the growth time, this process is typically modeled as a sequence of equilibrium stellar structure calculations (e.g., *Palla and Stahler,* 1991; *Hosokawa et al.,* 2010; *Kuiper and Yorke,* 2013; *Zhang et al.,* 2014). Most models developed so far have been for nonrotating protostars (see *Haemmerlé et al.,* 2013, for an exception). A choice must also be made for the protostellar surface boundary condition: photospheric or nonphotospheric. In the former, accreting material is able to radiate away its high internal energy that has just been produced in the accretion shock, while in the latter the gas is optically thick (i.e., the photosphere is at a larger radius than the protostellar surface). At a given mass, the protostar will respond to advecting more energy by having a larger size. If accretion proceeds through a disk, this is usually taken to imply photospheric boundary conditions (cf. *Tan and McKee,* 2004). In the calculation shown in Fig. 2, transition from quasispherical, nonphotospheric accretion is made to photospheric at $m_* \lesssim 0.1\ M_\odot$ based on an estimate of when outflows first affect the local environment. Subsequent evolution is influenced by D-burning, "luminosity-wave" swelling, and contraction to the zero age main sequence (ZAMS) once the protostar is older than its current Kelvin-Helmholtz time. This explains why protostars with higher accretion rates, i.e., in higher Σ_{cl} clumps, reach the ZAMS at higher masses. Protostars may still accrete along the ZAMS. The high temperatures and ionizing luminosities of this phase are a qualitative difference from lower-mass protostars, especially leading to ionization of the outflow cavity and eventually the core envelope (section 2.4).

Radiative transfer (RT) calculations are needed to predict the multiwavelength appearance and total spectral energy distribution (SED) of the protostar (e.g., *Robitaille et al.,* 2006; *Molinari et al.,* 2008; *Johnston et al.,* 2011; *Zhang*

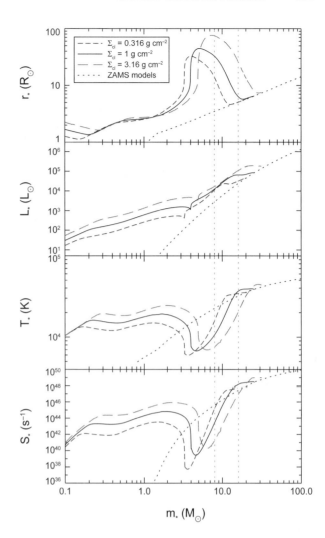

Fig. 2. Evolution of a massive protostar forming from a 60-$M_\odot$ core in $\Sigma_{cl} \simeq 0.3$, 1, 3 g cm^{-2} clumps. Top to bottom: radius, luminosity (including accretion), surface temperature, and Hionizing luminosity (*Zhang et al.,* 2014; see also *Hosokawa et al.,* 2010). Dotted lines show the ZAMS.

and Tan, 2011; *Zhang et al.,* 2013a, 2014). The luminosity of the protostar and disk are reprocessed by the surrounding gas and dust in the disk, envelope, and outflow cavity. Figure 3 shows example models of the density, velocity, and temperature structure of a massive protostar forming inside a 60 $M_\odot$ core embedded in a $\Sigma_{cl} = 1$ g cm^{-2} environment at two stages, when $m_* = 8$ and 16 $M_\odot$, zooming from the inner 100 to 10^3 to 10^4 AU (*Zhang et al.,* 2014). One feature of these models is that they self-consistently include the evolution of the protostar from the initial starless core in a given clump environment, including rotating infall envelope, accretion disk, and disk-wind-driven outflow cavities, that gradually open up as the wind momentum flux becomes more powerful (section 2.4). These models are still highly idealized, being axisymmetric about a single protostar. A real source would most likely be embedded in a forming cluster that includes other protostars in the vicinity.

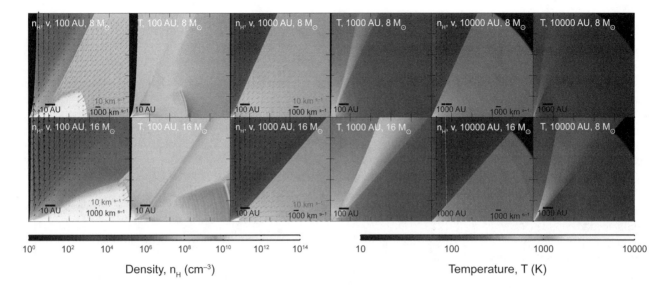

Fig. 3. Density and temperature profiles for a massive protostellar core (*Zhang et al.,* 2014), when the central protostar (at bottom left of each panel) reaches $m_* = 8\ M_\odot$ (top row) and $16\ M_\odot$ (bottom row). The disk mid-plane coincides with the x-axis, the outflow/rotation axis with the y-axis. The core has initial mass $M_c = 60\ M_\odot$ and rotational to gravitational energy ratio of $\beta_{rot} = 0.02$ and is embedded in a clump with mean surface density $\Sigma_{cl} = 1$ g cm^{-2}. At each stage, three pairs of box sizes are shown (left to right, 100, 10^3, 10^4 AU). Overlaid on density plots are arrows showing infall/outflow velocities (arrow length scale is 10/1000 km s^{-1}, respectively).

Similar continuum RT calculations have yet to be made for competitive accretion. We anticipate they would show more disordered morphologies and have smaller masses and Σs of gas in the close vicinity of the protostar than core accretion models, which may affect their SEDs at a given evolutionary stage, i.e., value of m_*. These potential SED differences are worth exploring, especially at the stages when ionization becomes important for creating hypercompact (HC) and ultracompact (UC) HII regions (section 2.4).

2.4. Feedback Processes During Accretion

Massive protostars are much more luminous and hotter than low-mass protostars so, all else being equal, one expects radiative feedback (i.e., thermal heating, dissociation/ ionization of hydrogen, radiation pressure on dust) to be more important. The same is true for mechanical feedback from stellar winds (i.e., those from the stellar surface) and protostellar outflows (magneto-centrifugally driven flows powered by accretion). Alternatively, if massive stars tend to form in denser, more highly pressurized environments, then feedback will have a harder time disrupting accretion. For core accretion models, a major goal of feedback studies is to estimate the star-formation efficiency from a core of a given initial mass and density (or surrounding clump pressure). For both core and competitive accretion, an important goal is to determine whether there are processes that lead to IMF truncation at some maximum mass. Feedback may also affect the ability of a core to fragment to form a binary and the efficiency of a clump to fragment into a cluster. Feedback also produces observational signatures, such as outflow cavities, HII regions, and excitation of masers,

which all serve as diagnostics of massive star formation. A general review of feedback is given in the chapter by Krumholz et al. in this volume. Here we discuss processes directly relevant to massive star formation.

For massive stars to form from massive cores, a mechanism is needed to prevent the core from fragmenting. *Krumholz and Mckee* (2008) suggested this may be due to radiative feedback from surrounding lower-mass protostars that have high accretion luminosities if they are forming in a high-pressure clump. This model predicts a minimum Σ for clumps to form massive stars, $\Sigma_{cl} \gtrsim 1$ g cm^{-2}. On the other hand, *Kunz and Mouschovias* (2009) and *Tan et al.* (2013b) invoked a non-feedback-related mechanism of magnetic field support to allow massive cores to resist fragmentation. This does not require a minimum Σ_{cl} threshold, but does require that there be relatively strong, ~mG, B-fields in at least some parts of the clump, so that the core mass is set by the magnetic critical mass. Simulations confirm that magnetic fields can suppress fragmentation (section 2.5). The observational evidence for a whether there is a Σ_{cl} threshold for massive star formation is discussed in section 4.7.

Once a massive protostar starts forming, but before contraction to the ZAMS, the dominant feedback is expected to be due to protostellar outflows (see also section 4.3). As a consequence of their extraction of angular momentum, these magneto-centrifugally-launched disk- and/or X-winds tend to have mass flow rates $\dot{m}_w = f_w \dot{m}_*$ with $f_w \sim 0.1-0.3$ and terminal velocities $v_w \sim v_K(r_0)$, where v_K is the Keplerian speed in the disk at the radius, r_0, of the launching region. The total outflow momentum flux can be expressed as $\dot{p}_w = \dot{m}_w v_w = f_w \dot{m}_* v_w \equiv \phi_w\ \dot{m}_* v_K(r_*)$: *Najita and Shu* (1994) X-wind models have a dimensionless parameter

$\phi_w \simeq 0.6$. An implementation of the *Blandford and Payne* (1982) disk-wind model has $\phi_w \simeq 0.2$, relatively independent of m_* (*Zhang et al.*, 2013a). Outflows are predicted to be collimated with $dp_w/d\Omega = (p_w/4\pi)[\ln(2/\theta_0)(1 + \theta_0^2 - \mu^2)]^{-1}$ (*Matzner and McKee*, 1999), where $\mu = \cos\theta$, θ is the angle from outflow axis, and $\theta_0 \sim 10^{-2}$ is a small angle of the core of the outflow jet. *Matzner and McKee* (2000) found star formation efficiency from a core due to such outflow momentum feedback of $\bar{\varepsilon}_{*d} \sim 0.5$. For the protostars in Fig. 2, $\bar{\varepsilon}_{*d} \simeq 0.45$, 0.57, and 0.69 for $\Sigma_{cl} = 0.3$, 1, 3 g cm^{-2} respectively, indicating protostellar outflow feedback may set a relatively constant formation efficiency from low- to high-mass cores.

The protostar's luminosity heats its surroundings, mostly via absorption by dust, which at high densities ($n_H \gtrsim 10^5$ cm^{-3}) is well-coupled thermally to the gas (*Urban et al.*, 2010). Dust is destroyed at $T \gtrsim 1500$ K, i.e., at $\lesssim 10$ AU for models in Fig. 3. Hot core chemistry (section 4.5) is initiated for temperatures $\gtrsim 100$ K. Thermal heating reduces subsequent fragmentation in the disk (see section 2.5).

As the protostar grows in mass and settles toward the main sequence, the temperature and H-ionizing luminosity begin to increase. The models in Fig. 3 have H-ionizing photon luminosities of 2.9×10^{43} s^{-1} and 1.6×10^{48} s^{-1} when $m_* = 8$ M$_\odot$ and 16 M$_\odot$, respectively. A portion of the inner outflow cavity will begin to be ionized — an "outflow-confined" H II region (*Tan and McKee*, 2003). Inner, strongly bound parts of the infall envelope that are unaffected by outflows could also confine the H II region (*Keto*, 2007). The H II region structure is detectable via radio continuum observations of thermal bremsstrahlung emission (section 4.4). Its extent depends sensitively on the density and dust content of the gas. Feedback from the H II region is driven by its high temperature, $\sim 10^4$ K, which sets up a pressure imbalance at the ionization front boundary with neutral gas. Since the MHD-outflow momentum flux is likely to dominate over the H II region thermal pressure, ionization feedback will only begin to be effective once the entire outflow cavity is ionized and ionization fronts start to erode the core infall envelope (cf. *Peters et al.*, 2011). Once the core envelope is mostly cleared, leaving equatorial accretion and a remnant accretion disk, the diffuse ionizing radiation field that is processed by the disk atmosphere can photoevaporate the disk (*Hollenbach et al.*, 1994). This process has been invoked by *McKee and Tan* (2008) to shut off accretion of the first stars around ~ 100–200 M$_\odot$ (see also *Hosokawa et al.*, 2011; *Tanaka et al.*, 2013), but its role in present-day massive star formation is unclear, especially given the presence of dust that can absorb ionizing photons.

Radiation pressure acting on dust has long been regarded as a potential impediment to massive star formation (*Larson and Starrfield*, 1971; *Kahn*, 1974; *Wolfire and Cassinelli*, 1987). However, as long as the accretion flow remains optically thick, e.g., in a disk, then there does not seem to be any barrier to forming massive stars (*Nakano*, 1989; *Jijina and Adams*, 1996; *Yorke and Sonnhalter*, 2002; *Krumholz et al.*, 2009; *Kuiper et al.*, 2010a, 2011; *Tanaka*

and Nakamoto, 2011). Outflows also reduce the ability of radiation pressure to terminate accretion, since they provide optically thin channels through which the radiation can escape (*Krumholz et al.*, 2005b). This contributes to the "flashlight effect" (*Yorke and Sonnhalter*, 2002; *Zhang et al.*, 2013a), leading to factors of several variation in the bolometric flux of a protostar depending on viewing angle. Numerical simulations of radiation pressure feedback are summarized in section 2.5.

A potential major difference between core and competitive accretion is their ability to operate in the presence of feedback. As discussed above, core accretion to a disk is quite effective at resisting feedback: Gas comes together into a self-gravitating object before the onset of star formation. Competitive accretion of ambient gas from the clump may be more likely to be disrupted by feedback. Considering the main feedback mechanism for low-mass stars, we estimate the ram pressure associated with a MHD (X- or disk-)wind of mass-loss rate $\dot{m}_w$ and velocity v_w as $P_w = \rho_w v_w^2 = f_\theta \dot{m}_w v_w/(4\pi r^2) = f_\theta \phi_w \dot{m}_* v_K(r_*)/(4\pi r^2)$, where $f_\theta \equiv 0.1 f_{\theta,0.1}$ is the factor by which the momentum flux of the wide-angle component of the wind is reduced from the isotropic average and where we have normalized, conservatively, to parameter values implied by disk-wind or X-wind models [e.g., the fiducial *Matzner and McKee* (1999) distribution has a minimum $f_\theta \simeq 0.2$ at $\theta = 90°$]. Evaluating P_w at $r_{BH} = 2$ Gm_*/σ^2 (appropriate for competitive accretion from a turbulent clump) around a protostar of current mass $m_* \equiv m_{*,1}$ M$_\odot$, adopting accretion rates from equation (16) and setting $r_* \equiv r_{*,3} 3$ R$_\odot$ (Fig. 2), we find the condition for the clump mean pressure to overcome the ram pressure of the wind, $\bar{P}_{cl} > P_w(r_{BH})$

$$\Sigma_{cl} > 11.7 \left(\frac{f_{\theta,0.1} \phi_{w,0.1} \alpha_{vir} \varepsilon_{ff,0.1}}{\phi_B \phi_{geom} \varepsilon_{cl,0.5}} \right)^4 \frac{M_{cl,3}^3 m_{*f,1}^4}{r_{*,3}^2 m_{*,1}^6} \text{ g cm}^{-2} \quad (17)$$

Thus in most clumps shown in Fig. 1, $\bar{P}_{cl}$ is too weak to confine gas inside the Bondi radius in the presence of such outflows. Note that here m_* is the mass scale at which feedback is being considered, while m_{*f} parameterizes the accretion rate needed to form a star of final mass m_{*f}. Equation (17) shows that MHD-wind feedback generated by the accretion rates expected in competitive accretion severely impacts accretion over most of the Bondi sphere, especially if the mass scale at which competitive accretion starts, following initial core accretion, is small ($m_* \sim 1$ M$_\odot$). We suggest simulations have so far not fully resolved the effects of MHD-wind feedback on competitive accretion and that this feedback may lead to a major reduction in its efficiency.

2.5. Results from Numerical Simulations

Numerical simulations have long been a major tool for investigating massive star formation. Today, the majority use either the Lagrangian technique smoothed particle hydrodynamics (SPH) (e.g., *Lucy*, 1977) or the Eulerian technique adaptive mesh refinement (AMR) (*Berger and Oliger,*

1984), both of which provide high dynamic range, allowing collapse to be followed over orders of magnitude in length scale in general geometries. Both code types include self-gravity, hydrodynamics, and sink particles to represent stars (e.g., *Bate et al.,* 1995; *Krumholz et al.,* 2004).

Probably the most significant advance in simulations since Protostars and Planets V has been the addition of extra physical processes. For SPH, there are implementations of magnetohydrodynamics (*Price and Monaghan,* 2004), flux-limited diffusion (FLD) for RT of dust-reprocessed non-ionizing radiation (*Whitehouse and Bate,* 2004), and ray-tracing for ionizing RT (*Dale et al.,* 2005; *Bisbas et al.,* 2009), the latter specifically used to study massive star formation. AMR codes include an even broader range of physics, all of which have been brought to bear on massive star formation: MHD (e.g., *Fryxell et al.,* 2000; *Li et al.,* 2012a), FLD for non-ionizing radiation (*Krumholz et al.,* 2007b; *Commerçon et al.,* 2011b), ray-tracing for ionizing and (in restricted circumstances) non-ionizing radiation (*Peters et al.,* 2010a), and protostellar outflows (*Wang et al.,* 2010; *Cunningham et al.,* 2011). More sophisticated RT schemes than pure FLD or ray-tracing are also available in some non-adaptive grid codes (e.g., *Kuiper et al.,* 2010b).

While this is progress, a few caveats are in order. First, no code yet includes all these physical processes; e.g., ORION includes MHD, dust-reprocessed radiation and outflows, but not ionizing radiation, while FLASH has MHD and ionizing radiation, but not outflows or dust-reprocessed radiation. Second, some physical processes have only been studied in isolation by a single code and their relative importance is unclear. Examples include imperfect thermal coupling between gas and dust (*Urban et al.,* 2010), dust coagulation and drift relative to gas (*Suttner et al.,* 1999), and ambipolar drift (*Duffin and Pudritz,* 2008).

Still, the advances in simulation technique have yielded some important general conclusions. First, concerning fragmentation, hydrodynamics-only simulations found that collapsing gas clouds invariably fragmented into stars with initial masses of ~0.1 $M_\odot$ (*Bonnell et al.,* 2004; *Dobbs et al.,* 2005), implying formation of massive stars would have to arise via subsequent accretion onto these fragments. However, *Krumholz et al.* (2007a, 2012) showed that adding non-ionizing, dust-reprocessed radiative feedback suppresses this behavior, as the first few stars to form heat the gas around them via their accretion luminosities, raising the Jeans mass and preventing much fragmentation. Similarly, *Hennebelle et al.* (2011) showed that magnetic fields also inhibit fragmentation, and *Commerçon et al.* (2011a) and *Myers et al.* (2013) (Fig. 4) have combined magnetic fields and radiation to show that the two together suppress fragmentation much more effectively than either one alone.

Second, massive star feedback is not very effective at halting accretion. Photoionization can remove lower-density gas, but dense gas that is already collapsing onto a massive protostar is largely self-shielding and is not expelled by ionizing radiation (*Dale et al.,* 2005; *Peters et al.,* 2010a,b, 2011). As for radiation pressure, two-dimensional simula-

tions with limited resolution (~100 AU) generally found that it could reverse accretion, thus limiting final stellar masses (*Yorke and Sonnhalter,* 2002). However, higher-resolution two-dimensional and three-dimensional simulations find that radiation pressure does not halt accretion since, in optically thick flows, the gas is capable of reshaping the radiation field and beaming it away from the bulk of the incoming matter. This beaming can be provided by radiation Rayleigh-Taylor fingers (e.g., *Krumholz et al.,* 2009; *Jiang et al.,* 2013; cf. *Kuiper et al.,* 2012), by an optically thick disk (*Kuiper et al.,* 2010a, 2011; *Kuiper and Yorke,* 2013), or by an outflow cavity (*Krumholz et al.,* 2005b; *Cunningham et al.,* 2011).

While there is general agreement on the two points above, simulations have yet to settle the question of whether stars form primarily via competitive or core accretion, or some hybrid of the two. Resolving this requires simulations large enough to form an entire star cluster, rather than focusing on a single massive star. *Bonnell et al.* (2004) and *Smith et al.* (2009) tracked the mass that eventually ends up in massive stars in their simulations of cluster formation, concluding that it is drawn from a ~1-pc cluster-sized region rather than a single well-defined ~0.1-pc core, and that there are no massive bound structures present. However, these simulations lacked radiative feedback or B-fields, and thus likely suffer from overfragmentation. *Wang et al.* (2010) performed simulations including outflows and magnetic fields. They found the most massive star in their simulations ultimately draws its mass from a ~1-pc-sized reservoir comparable to the size of its parent cluster, consistent with competitive accretion, but that the flow on large scales is mediated by outflows, preventing onset of rapid global collapse. As described in section 2.2.2, the average accretion rate to the massive star is relatively low in this simulation compared to the expectations of core accretion, which may be possible to test observationally (section 4). As discussed in section 2.4, outflow feedback on the scale of the Bondi accretion radius may be important in further limiting the rate of competitive accretion and so far has not been well resolved.

Peters et al. (2010a,b) simulated massive cluster formation with direct (i.e., not dust-reprocessed) radiation and B-fields, starting from smooth, spherical, slowly rotating initial conditions. They found massive stars draw mass from large but gravitationally bound regions, but that the mass flow onto these stars is ultimately limited by fragmentation of the accreting gas into smaller stars. *Girichidis et al.* (2012) extended this result to more general geometries. *Krumholz et al.* (2012) conducted simulations of cluster formation including radiation and starting from an initial condition of fully developed turbulence. They found massive stars do form in identifiable massive cores, with several tens of solar masses within ~0.01 pc. Core mass is not conserved in a Lagrangian sense, as gas flows in or out, but they are nonetheless definable objects in an Eulerian sense.

These contradictory results likely have several origins. One is initial conditions (e.g., *Girichidis et al.,* 2012). Those lacking any density structure and promptly undergoing

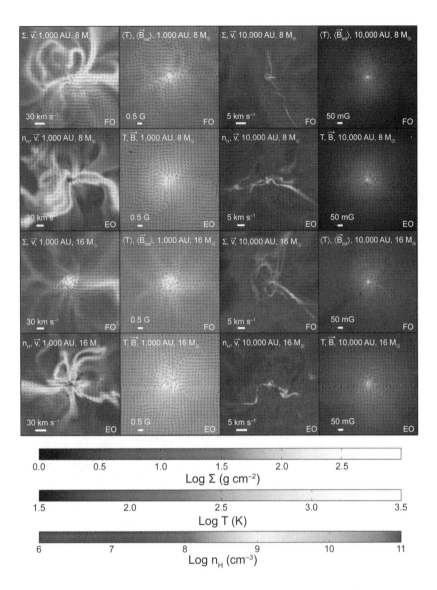

Fig. 4. Simulation of massive star formation including MHD and radiation pressure feedback (*Myers et al.,* 2013) from an initial core with M_c = 300 $M_\odot$ ≃ 2 M_Φ (i.e., twice the magnetic critical mass), Σ_c = 2 g cm^{-2}, and σ = 2.3 km s^{-1} (so that α_{vir} = 2.1; turbulence decays, leading to global collapse). Protostellar outflow feedback (e.g., *Wang et al.,* 2010; *Cunningham et al.,* 2011) is not included. The top row shows a face-on (FO) view of the accretion disk centered on the protostar when it has a mass of 8 $M_\odot$: From left to right are mass surface density, Σ, and mean velocity, $\langle\vec{v}\rangle$, in a 10^3-AU box; mass-weighted temperature, $\langle T\rangle$, and total magnetic field strength, $\langle\vec{B}_{tot}\rangle$, in a 10^3-AU box; then the same two figures but for a 10^4-AU box. The second row shows edge-on (EO) views of this structure, with slices in a plane containing the protostar of, from left to right, H number density, n_H, and velocity, $\vec{v}$, of a 10^3-AU square; temperature, T, and in-plane component of magnetic field, $\vec{B}$, of a 10^3-AU square; then the same two figures but for a 10^4-AU square. The third row repeats the first row, but now for a 16-$M_\odot$ protostar, and the fourth row repeats the second row for this protostar. With an initially turbulent core, the accretion flow is relatively disordered (cf. Fig. 3).

global collapse (e.g., *Bonnell et al.,* 2004) tend to find there are no bound, massive structures that can be identified as the progenitors of massive stars, while those either beginning from saturated turbulence (e.g., *Krumholz et al.,* 2012) or self-consistently producing it via feedback (e.g., *Wang et al.,* 2010) do contain structures identifiable as massive cores. Another issue is the different range of included physical mechanisms, with none of the published cluster simula-

tions combining dust-reprocessed radiation and magnetic fields — shown to be so effective at suppressing fragmentation on smaller scales. A final issue may simply be one of interpretation, with SPH codes tending to focus on the Lagrangian question of where the individual mass elements that make up a massive star originate, while Eulerian codes focus on the presence of structures at a particular point in space regardless of the paths of individual fluid elements.

3. OBSERVATIONS OF INITITAL CONDITIONS

3.1. Physical Properties of Starless Cores and Clumps

Initial conditions in core accretion models are massive starless cores, with $\Sigma \sim 1$ g cm^{-2}, similar to the Σs of their natal clump. For competitive accretion, massive stars are expected to form near the centers of the densest clumps. Thus to test these scenarios, methods are needed to study high Σ and volume density ($n_H \sim 10^6$ cm^{-3}), compact ($r_c \sim 0.1$ pc, i.e., 7″ at 3 kpc), and potentially very cold (T $\sim$ 10 K) structures. Recently, many studies of initial conditions have focused on infrared dark clouds (IRDCs): regions with such high Σ that they appear dark at mid-infrared (MIR) (~10 μm) and even up to far-infrared (FIR) (~100 μm) wavelengths. Indeed, selection of cores that may be starless often involves checking for the absence of a source at 24 or 70 μm.

3.1.1. Mass surface densities, masses, and temperatures. One can probe Σ structures via MIR extinction mapping of IRDCs, using diffuse Galactic background emission from warm dust. Spitzer Infrared Array Camera (IRAC) [e.g., Galactic Legacy Infrared Mid-Plane Survey Extraordinaire (GLIMPSE) (*Churchwell et al.,* 2009)] 8-μm images resolve down to 2″ and can probe to $\Sigma \sim 0.5$ g cm^{-2} [e.g., *Butler and Tan,* 2009, 2012 (*BT12*); *Peretto and Fuller,* 2009; *Ragan et al.,* 2009]. The method depends on the 8-μm opacity per unit total mass, $\kappa_{8\mu m}$ [*BT12* use 7.5 cm^2 g^{-1} based on the moderately coagulated thin ice mantle dust model of *Ossenkopf and Henning* (1994)], but is independent of dust temperature, T_d. Allowance is needed for foreground emission, best measured by finding "saturated" intensities toward independent, optically thick cores (*BT12*). Only differences in Σ relative to local surroundings are probed, so the method is insensitive to low-Σ IRDC environs. This limitation is addressed by combining near-infrared (NIR) and MIR extinction maps (*Kainulainen and Tan,* 2013). Even with careful foreground treatment, there are ~30% uncertainties in κ and thus Σ and, adopting 20% kinematic distance uncertainties, a 50% uncertainty in mass. Ten IRDCs studied in this way are shown in Fig. 1.

The high-resolution Σ maps derived from Spitzer images allow measurement of core and clump structure. Parameterizing density structure as $\rho \propto r^{-k_\rho}$ and looking at 42 peaks in their Σ maps, *BT12* found $k_\rho \simeq 1.1$ for "clumps" (based on total Σ profile) and $k_\rho \simeq 1.6$ for "cores" (based on Σ profile after clump envelope subtraction). These objects, showing total Σ, are also plotted in Fig. 1. A Σ map of one of these core/clumps is shown in Fig. 5. *Tan et al.* (2013b) used the fact that some of these cores are opaque at 70 μm to constrain $T_d \lesssim 13$ K. *Ragan et al.* (2009) measured an IRDC core/clump mass function, $dN/dM \propto M^{-\alpha_{cl}}$ with $\alpha_{cl} \simeq 1.76 \pm 0.05$ from 30 to 3000 M$_\odot$, somewhat shallower than that of the Salpeter stellar IMF ($\alpha_* \simeq 2.35$).

The Σ of these clouds can also be probed by the intensity, S_ν/Ω of FIR/millimeter dust emission, requiring T_d and κ_ν. For optically thin RT and black-body emission, $\Sigma = 4.35 \times 10^{-3}([S_\nu/\Omega]/[MJy/sr])\kappa_{\nu,0.01}^{-1}\lambda_{1.2}^3[\exp(0.799T_{d,15}^{-1}\lambda_{1.2}^{-1})-1]$

g cm^{-2}, where $\kappa_{\nu,0.01} \equiv \kappa_\nu/[0.01$ cm^2/g$)]^{-1}$, $\lambda_{1.2} \equiv \lambda/1.2$ mm, and $T_{d,15} \equiv T_d/15$ K. A common choice of κ_ν is again that predicted by the moderately coagulated thin ice mantle dust model of *Ossenkopf and Henning* (1994), with opacity per unit dust mass of $\kappa_{1.2mm,d} = 1.056$ cm^2 g^{-1}. A gas-to-refractory-component dust-mass ratio of 141 is estimated by *Draine* (2011), so $\kappa_{1.2mm} = 7.44 \times 10^{-3}$ cm^2 g^{-1}. Uncertainties in κ_ν and T_d now contribute to Σ; e.g., *Tan et al.* (2013b) adopt κ uncertainties of 30% and $T_d = 10 \pm 3$ K, leading to a factor of ~2 uncertainties in 1.3-mm-derived Σs. *Rathborne et al.* (2006) studied 1.2-mm emission at 11″ resolution in 38 IRDCs, finding core/clumps with ~10 to 10^4 M$_\odot$ (for $T_d = 15$ K). In their sample of 140 sources they found $dN/dM \propto M^{-\alpha_{cl}}$, with $\alpha_{cl} \simeq 2.1 \pm 0.4$.

Herschel observations of dust emission at 70 to 500 μm allow simultaneous measurement of T_d and Σ at ~20−30″ resolution, and numerous studies have been made of IRDCs (e.g., *Peretto et al.,* 2010; *Henning et al.,* 2010; *Beuther et al.,* 2010a; *Battersby et al.,* 2011; *Ragan et al.,* 2012). For MIR-dark regions, *Battersby et al.* (2011) derived a median $\Sigma \simeq 0.2$ g cm^{-2}, but with some values extending to ~5 g cm^{-2}. The median T_d of regions with $\Sigma \geq 0.4$ g cm^{-2} was 19 K, but the high-Σ tail had $T_d \sim$10 K.

Interferometric studies have probed millimeter dust emission at higher resolution. "Clumps" are often seen to contain substructure, i.e., a population of "cores." Core mass function measurements have been attempted; e.g., *Beuther and Schilke* (2004; see also *Rodón et al.,* 2012) observed IRAS 19410+2336, finding $\alpha_c = 2.5$ from M$_c \sim$ 1.7 to 25 M$_\odot$ (but with few massive cores). While the similarity of CMF and IMF shapes is intriguing, there are

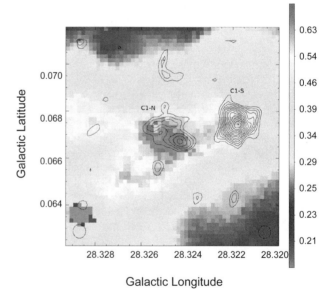

Fig. 5. Candidate massive starless cores, C1-S and C1-N, traced by N$_2$D$^+$(3-2) (contours), observed by ALMA (*Tan et al.,* 2013b). Background shows MIR Σ map (g cm^{-2}). C1-S has ~60 M$_\odot$. The high value of [N$_2$D$^+$]/[N$_2$H$^+$] ~ 0.1 (Kong et al., in preparation) is a chemical indicator that C1-S is starless.

caveats, e.g., whether cores are resolved, whether they are PSCs rather than non-star-forming over densities or already star-forming cores, the possibility of mass-dependent lifetimes of PSCs (*Clark et al.,* 2007), and binary/multiple star formation from PSCs.

Some massive (~60 $M_\odot$) cores, e.g., IRDC G28.34+0.06 P1 (*Zhang et al.,* 2009), Cygnus X N63 (*Bontemps et al.,* 2010) [note that recent detection of a bipolar outflow indicates a protostar is forming in this source (*Duarte-Cabral et al.,* 2013)], and IRDC C1-S (*Tan et al.,* 2013b) (Figs. 1 and 5), have apparently monolithic, centrally concentrated structures with little substructure, even though containing many (~100) Jeans masses. This suggests fragmentation is being inhibited by a nonthermal mechanism, i.e., magnetic fields. *Tan et al.* (2013b) estimate ~1-mG field strengths are needed for the mass of C1-S to be set by its magnetic critical mass, given its density of $\bar{n}_H \simeq 6 \times 10^5$ cm^{-3}.

Many molecular lines have been observed from IRDCs. Using integrated molecular line intensities to derive Σ is possible in theory, but common species like CO are frozen out from the gas phase (see below), and other species still present have uncertain and likely spatially varying abundances. Nevertheless, if the astrochemistry is understood, then species that are expected to become relatively abundant in the cold, dense conditions of starless cores, such as deuterated N-bearing molecules (section 3.2), can be used to identify PSCs, distinguishing them from the surrounding clump.

Infrared dark cloud gas temperatures of 10–20 K have been derived from NH$_3$ inversion transitions (e.g., *Pillai et al.,* 2006; *Wang et al.,* 2008; *Sakai et al.,* 2008; *Chira et al.,* 2013).

3.1.2. Magnetic fields. Polarization of dust continuum emission is thought to arise from alignment of nonspherical grains with B-fields and is thus a potential probe of plane-of-sky projected field morphology and, with greater uncertainty, field strength. The correlated orientation of polarization vectors with the orientations of filaments, together with the correlated orientations of polarization vectors of dense cores with their lower density surroundings (see chapter by H. Li et al. in this volume), suggests B-fields play some role in the formation of dense cores. However, these polarization results are typically for relatively nearby molecular clouds, such as Taurus, Pipe Nebula, and Orion, and only a few, lower-resolution studies have been reported for IRDCs (*Matthews et al.,* 2009).

Line-of-sight B-field strengths can be derived from Zeeman splitting of lines from molecules with an unpaired electron, such as OH, which probes lower-density envelopes, and CN, which traces denser gas. Unfortunately, measurement of Zeeman splitting in CN is very challenging observationally, requiring bright lines, and the reported measurements in massive star-forming regions are all toward already star-forming cores (section 4). From the results of *Falgarone et al.* (2008) as summarized by *Crutcher* (2012), at densities $n_H \gtrsim 300$ cm^{-3}, $B_{max} \simeq 0.44(n_H/10^5$ cm$^{-3})^{0.65}$ mG, with a distribution of B flat between 0 and B_{max}. Such field strengths, if present in massive starless cores like C1-S (Fig. 5), could support them against rapid free-fall collapse

and fragmentation.

3.1.3. Kinematics and dynamics. Measurement of cloud kinematics requires molecular line tracers, but again one faces the problem of being unsure about which parts of the cloud along the line of sight are being probed by a given tracer. The kinematics of ionized and neutral species can differ due to magnetic fields (*Houde et al.,* 2009). The spectra of molecular tracers of IRDCs, such as ^{13}CO, C^{18}O, N$_2$H$^+$, NH$_3$, HCN, HCO$^+$, and CCS, show linewidths ~0.5–2 km s^{-1}, i.e., consistent with varying degrees of supersonic turbulence (e.g., *Wang et al.,* 2008; *Sakai et al.,* 2008; *Fontani et al.,* 2011). In studying the kinematics of IRDC G035.3900.33, *Henshaw et al.* (2013) have shown it breaks up into a few distinct filamentary components separated by up to a few kilometers per second, and it is speculated these may be in the process of merging. Such a scenario may be consistent with the detection of widespread (greater than parsec-scale) SiO emission, a shock tracer, by *Jiménez-Serra et al.* (2010) along this IRDC. However, in general it is difficult to be certain about flow geometries from only line-of-sight velocity information. While infall/converging flow signatures have been claimed via inverse P-Cygni profiles in star-forming cores and clumps (section 4), there are few such claims in starless objects (*Beuther et al.,* 2013a). The L1544 PSC has ~8 $M_\odot$ and an infall speed of $\simeq 0.1$ km s^{-1} on scales of 10^3 AU — subsonic and $\ll v_{ff}$ (*Keto and Caselli,* 2010).

Given a measurement of cloud velocity dispersion, σ, the extent to which it is virialized can be assessed, but with the caveat that the amount of B-field support is typically unknown. Comparing ^{13}CO-derived σs with MIR + NIR extinction masses, *Kainulainen and Tan* (2013) found $\bar{\alpha}_{vir} =$ 1.9. *Hernandez et al.* (2012) compared MIR + NIR extinction masses with C^{18}O-derived σs and surface pressures in strips across IRDC G035.3900.33, finding results consistent with virial equilibrium (*Fiege and Pudritz,* 2000).

For starless cores, *Pillai et al.* (2011) studied the dynamics of cold cores near UC HII regions using NH$_2$D-derived σ and 3.5-mm emission to measure mass, finding $\bar{\alpha}_{vir} \sim$ 0.3. *Tan et al.* (2013b) measured mass and Σ from both MIR + NIR extinction and millimeter dust emission to compare predictions of the turbulent core accretion model (including surface pressure confinement and Alfvén Mach number $\mathcal{M}_A = 1$ magnetic support) with observed σ, derived from N$_2$D$^+$. In six cores they found a mean ratio of observed to predicted velocity dispersions of 0.81 ± 0.13. However, for the massive monolithic core C1-S, they found a ratio of 0.48 ± 0.17, which at face value implies subvirial conditions. However, virial equilibrium could apply if the magnetic fields were stronger so that $\mathcal{M}_A \simeq 0.3$ rather than 1, requiring $B \simeq 1.0$ mG. *Sánchez-Monge et al.* (2013c) used NH$_3$-derived mass and σ to find $\alpha_{vir} \sim 10$ for several tens of mostly low-mass starless cores, which would suggest they are unbound. However, they also found a linear correlation of M with virial mass $M_{vir} \equiv \alpha_{vir}M$, only expected if cores are self-gravitating, so further investigation of the accuracy of the absolute values of α_{vir} is needed.

3.2. Chemical Properties of Starless Cores and Clumps

Infrared dark cloud chemical properties resemble those of low-mass dense cores (e.g., *Vasyunina et al.*, 2012; *Miettinen et al.*, 2011; *Sanhueza et al.*, 2013), with widespread emission of NH_3 and N_2H^+ (e.g., *Zhang et al.*, 2009; *Henshaw et al.*, 2013). In the Nobeyama survey of *Sakai et al.* (2008), no CCS was detected, suggesting the gas is chemically evolved, i.e., atomic carbon is mostly locked into CO.

3.2.1. Carbon monoxide freeze-out. Carbon monoxide is expected to freeze-out from the gas phase onto dust grains when $T_d \le 20$ K (e.g., *Caselli et al.*, 1999). The CO depletion factor, $f_D(CO)$, is defined as the ratio of the expected CO column density given a measured Σ [assuming standard gas phase abundances, e.g., $n_{CO}/n_{H_2} = 2 \times 10^{-4}$ (*Lacy et al.*, 1994)] to the observed CO column density. *Miettinen et al.* (2011) compared CO(1-0) and (2-1) observations with Σ derived from FIR/millimeter emission, finding no evidence for depletion. *Hernandez et al.* (2012) compared NIR- and MIR-extinction-derived Σ with $C^{18}O(2-1)$ and (1-0) to map f_D in IRDC G035.39-00.33, finding widespread depletion with $f_D \sim 3$. *Fontani et al.* (2012) compared $C^{18}O(3-2)$ with FIR/millimeter-derived Σ in 21 IRDCs and found $\overline{f}_D \sim 30$, perhaps due to CO(3-2) tracing higher-density (shorter depletion timescale) regions. On the other hand, *Zhang et al.* (2009) found $f_D \sim 10^2-10^3$ in IRDC G28.34+0.06 P1 and P2 by comparing $C^{18}O(2-1)$ to Σ from FIR/millimeter emission.

3.2.2. Deuteration. Freeze-out of CO and other neutrals boosts the abundance of (ortho-)H_2D^+ and thus the deuterium fractionation of other species left in the gas phase (*Dalgarno and Lepp*, 1984). Low-mass starless cores on the verge of star formation, i.e., PSCs, show an increase in $D_{frac}(N_2H^+) \equiv N(N_2D^+)/N(N_2H^+)$ to ≥ 0.1 (*Crapsi et al.*, 2005). High (ortho-)H_2D^+ abundances are also seen (*Caselli et al.*, 2008). In the protostellar phase, $D_{frac}(N_2H^+)$ and $N(H_2D^+)$ decrease as the core envelope is heated (*Emprechtinger et al.*, 2009; see chapter by Ceccarelli et al. in this volume).

To see if these results apply to the high-mass regime, *Fontani et al.* (2011) selected core/clumps, both starless and those associated with later stages of massive star formation, finding (1) the average $D_{frac}(N_2H^+)$ in massive starless core/clumps located in quiescent environments tends to be as large as in low-mass PSCs (~0.2), and (2) the abundance of N_2D^+ decreases in core/clumps that either harbor protostars or are starless but externally heated and/or shocked (see also *Chen et al.*, 2011; *Miettinen et al.*, 2011).

HCO^+ also becomes highly deuterated, but as CO freezes out, formation rates of both HCO^+ and DCO^+ drop, so DCO^+ is not as good a tracer of PSCs as N_2D^+. Deuteration of NH_3 is also high (≥ 0.1) in starless regions of IRDCs, but, in contrast to N_2D^+, can remain high in the protostellar phase (e.g., *Pillai et al.*, 2011), perhaps since NH_2D and NH_3 also form in grain mantles, unlike N_2H^+ and N_2D^+, so abundant NH_2D can result from mantle evaporation. DNC/HNC are different, with smaller $D_{frac}(HNC)$ in colder, earlier-stage cores (*Sakai et al.*, 2012).

3.2.3. Ionization fraction. The ionization fraction, x_e, helps set the ambipolar diffusion timescale, t_{ad}, and thus perhaps the rate of PSC contraction. Observations of the abundance of molecular ions like $H^{13}CO^+$, DCO^+, N_2H^+, and N_2D^+ can be used to estimate x_e (*Dalgarno*, 2006). *Caselli et al.* (2002) measured $x_e \sim 10^{-9}$ in the central regions of PSC L1544, implying $t_{ad} \simeq t_{ff}$. In massive starless cores, *Chen et al.* (2011) and *Miettinen et al.* (2011) found $x_e \sim 10^{-8}-10^{-7}$, implying either larger cosmic-ray ionization rates or lower densities than in L1544. However, accurate estimates of x_e require detailed chemical modeling, currently lacking in the above studies, as well as knowledge of core structure — typically not well constrained. Core B-fields can also affect low-energy cosmic-ray penetration, potentially causing variation in cosmic-ray ionization rate (*Padovani and Galli*, 2011).

3.3. Effect of Cluster Environment

The cluster environment may influence the physical and chemical properties of PSCs due to, e.g., warmer temperatures, enhanced turbulence, and (proto-)stellar interactions. Surveys of cores in cluster regions have started to investigate this issue, but have so far mostly targeted low-mass star-forming regions, like Perseus (e.g., *Foster et al.*, 2009). These studies find cores have higher kinetic temperatures (~15 K) than isolated low-mass cores (~10 K). In spite of turbulent environments, cores have mostly thermal line widths.

Studies of protoclusters containing an intermediate- or high-mass forming star [e.g., IRAS 05345+3157 (*Fontani et al.*, 2008), G28.34+0.06 (*Wang et al.*, 2008), and W43 (*Beuther et al.*, 2012)] have shown starless cores can have supersonic internal motions and $D_{frac}(NH_3, N_2H^+)$ values similar to low-mass star-forming regions. *Sánchez-Monge et al.* (2013c) analyzed VLA NH_3 data of 15 intermediate-/high-mass star-forming regions, finding 73 cores, classified as quiescent starless, perturbed starless, or protostellar. The quiescent starless cores have smaller line widths and gas temperatures (1.0 km s^{-1}; 16 K) than perturbed starless (1.4 km s^{-1}; 19 K) and protostellar (1.8 km s^{-1}; 21 K) cores. Still, even the most quiescent starless cores possess significant nonthermal components, contrary to the cores in low-mass star-forming regions. A correlation between core temperature and incident flux from the most massive star in the cluster was seen. These findings suggest the initial conditions of star formation vary depending on the cluster environment and/or proximity of massive stars.

3.4. Implications for Theoretical Models

The observed properties of PSCs, including their dependence on environment, constrain theoretical models of (massive) star formation. For example, the massive (~60 $M_\odot$), cold ($T_d \sim 10$ K), highly deuterated, monolithic starless core shown in Fig. 5 contains many Jeans masses, has modestly supersonic line widths, and requires relatively strong, ~mG magnetic fields if it is in virial equilibrium. More generally, the apparently continuous, power-law distribution of the shape of the low- to high-mass starless CMF implies fragmentation of dense molecular gas helps to shape the eventual stellar IMF. Improved observations of the PSC mass function (e.g., as traced by cores showing high deuteration of N_2H^+) are needed to help clarify this connection.

4. OBSERVATIONS OF THE ACCRETION PHASE

4.1. Clump and Core Infall Envelopes

Infall motions can be inferred from spectral lines showing an inverse P-Cygni profile. This results from optically thick line emission from a collapsing cloud with a relatively smooth density distribution and centrally peaked excitation temperature. The profile shows emission on the blue-shifted side of line center (from gas approaching us on the cloud's farside) and self-absorption at line center and on the red-shifted side. Detection of a symmetric optically thin line profile from a rarer isotopolog helps confirm infall is being seen, rather than just independent velocity components. Infall to low-mass protostars is seen via spectral lines tracing densities above $\sim 10^4$ cm^{-3} showing such inverse P-Cygni profiles (e.g., *Mardones et al., 1997*).

Infall in high-mass protostellar cores is more difficult to find, given their typically larger distances and more crowded environments. It can also be difficult to distinguish core from clump infall. Single-dish observations of HCN, CS, HCO$^+$, CO, and isotopologs (e.g., *Wu and Evans, 2003; Wu et al., 2005b; Fuller et al., 2005; Barnes et al., 2010; Chen et al., 2010; López-Sepulcre et al., 2010; Schneider et al., 2010; Klaassen et al., 2012; Peretto et al., 2013*) reveal evidence of infall on scales ~ 1 pc, likely relevant to the clump/protocluster. Derived infall velocities and rates range from ~ 0.2 to 1 km s^{-1} and $\sim 10^{-4}$–10^{-1} M$_\odot$ yr^{-1}. However, these rates are very uncertain. *López-Sepulcre et al. (2010)* suggest they may be upper limits as the method assumes most clump mass is infalling, whereas the self-absorbing region may be only a lower-density outer layer.

Clump infall times, $t_{infall} \equiv M_{cl}/\dot{M}_{infall}$ can be compared to t_{ff}. For example, *Barnes et al. (2010)* measured $\dot{M}_{infall} \sim 3 \times 10^{-2}$ M$_\odot$ yr^{-1} in G286.21+0.17, with $M_{cl} \sim 10^4$ M$_\odot$ and $R_{cl} \simeq 0.45$ pc (Fig. 1). Thus $t_{infall}/t_{ff} \simeq 3.3 \times 10^5$ yr/5.0 $\times 10^4$ yr = 6.7. Note that this clump has the largest infall rate out of ~ 300 surveyed by *Barnes et al. (2011)*. Similar results hold for the $\sim 10^3$ M$_\odot$ clumps NGC 2264 IRS 1 and 2 (*Williams and Garland, 2002*), with t_{infall}/t_{ff} = 14 and 8.8, respectively. For the central region of SDC335 studied by *Peretto et al. (2013)*, with M_{cl} = 2600 M$_\odot$, R_{cl} = 0.6 pc and $\dot{M}_{infall}$ = 2.5 $\times 10^{-3}$ M$_\odot$ yr^{-1} (including a boosting factor of 3.6 to account for accretion outside observed filaments), then t_{infall}/t_{ff} = 7.0. This suggests clump/cluster assembly is gradual, allowing establishment of approximate pressure equilibrium (*Tan et al., 2006*).

On the smaller scales of protostellar cores, for bright embedded continuum sources, infall is inferred from red-shifted line profiles seen in absorption against the continuum (the blue-shifted inverse P-Cygni emission profile is difficult to distinguish from the continuum). In a few cases, this red-shifted absorption is observed in NH$_3$ at centimeter wavelengths against free-free emission of an embedded HC H II region [G10.62–0.38 (*Sollins et al., 2005*); note *Keto (2002)* has also reported ionized gas infall in this source (section 4.4); G24.78+0.08 A1 (*Beltrán et al., 2006*)]. In other cases, it is observed with millimeter interferometers in CN, C^{34}S, and ^{13}CO against core dust continuum emission [W51 N (*Zapata et al., 2008*), G19.61–0.23 (*Wu et al., 2009*), G31.41+0.31 (*Girart et al., 2009*), and NGC 7538 IRS1 (*Beuther et al., 2013b*)]. *Wyrowski et al. (2012)* saw absorption of rotational NH$_3$ transitions against FIR dust emission with SOFIA. For the interferometric observations, infall on scales of $\sim 10^3$ AU is traced. Infall speeds are a few kilometers per second; $\dot{M}_{infall} \sim 10^{-3}$–$10^{-2}$ M$_\odot$ yr^{-1}. *Goddi et al. (2011a)* used CH$_3$OH masers in AFGL 5142 to infer $\dot{M}_{infall} \sim 10^{-3}$ M$_\odot$ yr^{-1} on 300-AU scales. The above results indicate collapse of cores, in contrast to clumps, occurs rapidly, i.e., close to free-fall rates.

Dust continuum polarization is observed toward some massive protostars to infer B-field orientations (e.g., *Tang et al., 2009; Beuther et al., 2010b; Sridharan et al., 2014*). *Girart et al. (2009)* observed a relatively ordered "hourglass" morphology in G31.41+0.31, suggesting contraction has pinched the B-field. However, since the region studied is only moderately supercritical [$\Sigma \sim 5$ g cm^{-2} and plane of sky B ~ 2.5 mG (Frau et al., personal communication)], the field may still be dynamically important, e.g., in transferring angular momentum and suppressing fragmentation.

4.2. Accretion Disks

In core accretion models, the infall envelope is expected to transition from near-radial infall to gradually greater degrees of rotational support, until near-circular orbits are achieved in a disk. If the disk is massive, then one does not expect a Keplerian velocity field. Also, massive moderately gravitationally unstable disks may form large-scale, perhaps lopsided, spiral arms that may give the disk an asymmetric appearance (*Krumholz et al., 2007c*). Disk gravitational instability is a likely mechanism to form binaries or small-N multiples. Once the infall envelope has dispersed, either by feedback or exhaustion via accretion, then a remnant, lower-mass, near-Keplerian disk may persist for a time, until it also dissipates via feedback or accretion.

One of the simplest methods by which detection of accretion disks has been claimed is via imaging of a flattened NIR-extinction structure surrounding a NIR source (e.g., *Chini et al., 2004; Preibisch et al., 2011*). The latter authors report a 5500-AU-diameter disk of 2 M$_\odot$ around a 10–15 M$_\odot$ star. However, in the absence of kinematic confirmation from molecular line observations, one must also consider the possibility of chance alignment of the source with a non-rotationally-supported filamentary dust lane.

Hot and warm dust in close (≤ 100 AU) proximity to the protostar, likely in a disk or outflow cavity wall, can sometimes be inferred from NIR or MIR interferometry (e.g., *Kraus et al.,* 2010; *de Wit et al.,* 2011; *Boley et al.,* 2013). Most of the 24 MIR sources studied by Boley et al. show deviations from spherical symmetry, but it is difficult to tell if these are due primarily to disks or outflows.

For methods tracing kinematics, there are ~10 examples of "rotating toroids" in which velocity gradients traced by, e.g., $C^{34}S$, HDO, $H^{18}O$, or CH_3CN have been seen on $\geq$few × 1000 AU scales that are perpendicular to protostellar outflows (section 4.3) emerging from "hot cores" (section 4.5) (*Cesaroni et al.,* 2007; *Beltrán et al.,* 2011). Most are probably in the process of forming B stars, such as AFGL 2591–VLA3 (*Wang et al.,* 2012), but the sample also includes the UC HII regions G10.62–0.38 and G29.96–0.02 (*Beltrán et al.,* 2011), with $m_* \geq 15\ M_\odot$. The disk reported by *Wang et al.* (2012) appears to have sub-Keplerian kinematics, together with an expanding component perhaps driven by the outflow. Keplerian rotation has been claimed for IRAS 20126+4104 (*Cesaroni et al.,* 2005) and G35.20–0.74N (*Sánchez-Monge et al.,* 2013a) (Fig. 6). However, in the latter, where a $r_d \geq 2500$ AU disk is inferred from an arc of centroid positions of sequential velocity channels of CH_3CN observed with ~1000 AU resolution, there is misalignment of the projected rotation axis with the N-S MIR and ionized jet, thought to define the outflow axis (*Zhang et al.,* 2013b) (see section 4.3). On smaller scales, usually in nearby, lower-mass and -luminosity (~$10^4\ L_\odot$) systems, there is also evidence of flattened structures with kinematic gradients perpendicular to outflows [e.g., Cep A HW2 (*Patel et al.,* 2005), IRAS 16547-4247 (*Franco-Hernández et al.,* 2009), and IRAS 18162-2048 (*Fernández-López et al.,* 2011, *Carrasco-González et al.,* 2012)].

Near-infrared spectroscopic observations of CO(2–0) bandhead emission, sometimes emerging via scattered light through outflow cavities, can provide information about protostellar and disk photospheres, where temperatures are ~2000–5000 K, i.e., scales $\lesssim$ a few astronomical units (e.g., *Bik and Thi,* 2004; *Davies et al.,* 2010; *Ilee et al.,* 2013). With spectral resolutions $\geq 10^4$, disk kinematics can begin to be resolved. In Ilee et al.'s study, all 20 sources can be fit with a Keplerian model. For radio source I in the Orion Kleinmann-Low (KL) region (e.g., *Menten and Reid,* 1995; *Plambeck et al.,* 2013), photospheric temperatures ~4500 K are inferred (*Morino et al.,* 1998; *Testi et al.,* 2010). *Hosokawa and Omukai* (2009) modeled this as emission from a very large ~100-$R_\odot$ protostar (swollen by high accretion rates), while *Testi et al.* (2010) preferred accretion disk models.

There are claims of massive protostellar accretion disks based on methanol masers (e.g., *Pestalozzi et al.,* 2009). However, characterization of disks by this method is hampered by the uncertain excitation conditions and nonlinear nature of the maser emission, together with possible confusion with outflow motions. In Cep A HW2 methanol masers appear to trace the outflow (*Torstensson et al.,* 2011). Note that Zeeman splitting of these maser lines allow B-field strengths (~20 mG) and morphologies (perpendicular to the disk) to be measured (*Vlemmings et al.,* 2010).

Spatially and kinematically well-resolved observations via thermal line emission from massive protostar disks remain lacking. This is not surprising if disk diameters are typically $\leq$1000 AU (section 2.3), i.e., $\leq$0.5″ at 2 kpc. The high angular resolution to be achieved by ALMA in the coming years should provide breakthrough capabilities in this area.

4.3. Protostellar Outflows

Collimated, bipolar protostellar outflows (see also the chapter by Frank et al. in this volume) have been observed from massive protostars, mostly via CO and HCO^+, and their isotopologs (e.g., *Beuther et al.,* 2002; *Wu et al.,* 2005a; *Garay et al.,* 2007; *López-Sepulcre et al.,* 2009). Correlations are seen between outflow power, force, and mass loss rate, with bolometric luminosity over a range L ~ 0.1–$10^6\ L_\odot$. This suggests outflows from massive protostars are driven in the same way as those from low-mass protostars, namely via magneto-centrifugal X- or disk-winds [momentum from radiation pressure, ~L/c, is far too small (*Wu et al.,* 2005a)].

Based on a tentative trend inferred from several sources, *Beuther and Shepherd* (2005; see also *Vaidya et al.,* 2011) proposed a scenario in which outflow collimation decreases with increasing protostellar mass, perhaps due to the increasing influence of quasispherical feedback (winds, ionization, or radiation pressure). However, such evolution is also expected in models of disk-wind breakout from a self-gravitating core (*Zhang et al.,* 2014) (see Fig. 3).

Study of SiO may help disentangle "primary" outflow (i.e., material launched from the disk) from "secondary"

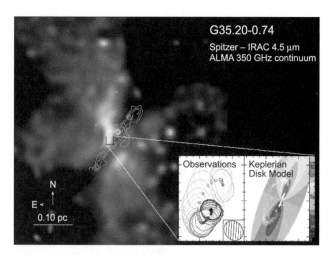

Fig. 6. G35.20–0.74N massive protostar (*Sánchez-Monge et al.,* 2013a). Large-scale image of 4.5 μm emission, expected to trace outflow cavities, with contours showing 850 μm continuum observed with ALMA. Left inset shows CH_3CN(19-18) K = 2 emission peaks (solid circles; outer circle is 50% contour) from a two-dimensional Gaussian fit channel by channel (velocity scale on right). Right inset compares these emission peaks with a Keplerian model.

outflow (i.e., swept-up core/clump material). Silicon monoxide may trace the primary outflow more directly, since its abundance is likely enhanced for the part of the disk-wind (and all the X-wind) launched from inside the dust destruction radius. However, SiO may also be produced in internal shocks in the outflow or at external shocks at the cavity walls. The single-dish survey of *López-Sepulcre et al.* (2011) found a decrease of the SiO luminosity with increasing luminosity-to-mass ratio in massive protostars (however, see *Sánchez-Monge et al.,* 2013b). Interferometric SiO observations, necessary to resolve the structure of massive protostellar outflows, are relatively few and mostly focused on sources with $L < 10^5$ $L_\odot$ [e.g., AFGL 5142 (*Hunter et al.,* 1999), IRAS 18264–1152 (*Qiu et al.,* 2007), IRAS 18566+0408 (*Zhang et al.,* 2007), G24.78+0.08 (*Codella et al.,* 2013), and IRAS 17233–3606 (*Leurini et al.,* 2013)]. These have traced well-collimated jets, with collimation factors similar to those from low-mass protostars. For higher-mass protostars with $L > 10^5$ $L_\odot$, interferometric SiO observations are scarcer and collimation results inconclusive. *Sollins et al.* (2004) mapped the shell-like UC HII region G5.89−0.39 with the SMA, finding a collimated SiO outflow, but the resolution was insufficient to distinguish the multiple outflows later detected in CO by *Su et al.* (2012). On the other hand, for IRAS 23151+5912 (*Qiu et al.,* 2007) and IRAS 18360–0537 (*Qiu et al.,* 2012), the molecular outflows traced by SiO are not well collimated and are consistent with ambient gas being entrained by an underlying wide-angle wind. Vibrationally excited SiO maser emission is thought to trace a wide-angle bipolar disk wind on scales of 10–100 AU around the massive protostar source I in Orion KL (e.g., *Greenhill et al.,* 2013).

Thermal (bremsstrahlung) radio jets should become prominent when the protostar contracts toward the main sequence (i.e., for $m_* \gtrsim 15$ $M_\odot$), causing its H-ionizing luminosity to increase dramatically. Shock ionization, including at earlier stages, is also possible (*Hofner et al.,* 2007). The primary outflow will be the first gas ionized, so centimeter continuum and radio recombination lines (RRLs) can trace massive protostar outflows. Elongated, sometimes clumpy, thermal radio continuum sources are observed around massive protostars, e.g., G35.20–0.74N (*Gibb et al.,* 2003) and IRAS 16562-3959 (*Guzmán et al.,* 2010). Many unresolved HC HII regions (section 4.4) may be the central parts of ionized outflows, since the emission measure is predicted to decline rapidly with projected radius (*Tan and McKee,* 2003). Synchrotron radio jets are seen from some massive protostars, e.g., W3(H_2O) (*Reid et al.,* 1995) and HH 80–81 (*Carrasco-González et al.,* 2010), but why some have synchrotron emission while most others are thermal is unclear.

Outflows also manifest themselves via cavities cleared in the core. Cavity walls, as well as exposed disk surface layers, experience strong radiative, and possibly shock, heating, which drive astrochemical processes that liberate and create particular molecular species that can serve as further diagnostics of outflows, such as water and light hydrides (see section 4.5) and maser activity (e.g., H_2O, CH_3OH) (e.g.,

Ellingsen et al., 2007; *Moscadelli et al.,* 2013). High-J CO lines are another important tracer of this warm, dense gas (*Fuente et al.,* 2012; *Yıldız et al.,* 2013; *San José-García et al.,* 2013). Densities and temperatures of outflowing gas in AFGL 2591 have been measured by CO and other highly excited linear rotors (*van der Wiel et al.,* 2013).

Given the high extinctions, outflow cavities can be the main escape channel for MIR (and even some FIR) radiation, thus affecting source morphologies. *De Buizer* (2006) proposed that this explains the 10 and 20 μm appearance of G35.20–0.74N, where only the northern outflow cavity that is inclined toward us (and is aligned with the northern radio jet) is prominent in groundbased imaging. *Zhang et al.* (2013b) observed this source with SOFIA at wavelengths up to ~40 μm and detected the fainter counter jet. Comparing with core accretion RT models, they estimated the protostar has $m_* \sim 20$–34 $M_\odot$, embedded in a core with $M_c = 240$ $M_\odot$, in a clump with $\Sigma_{cl} \simeq 0.4$–1 g cm^{-2}.

While much MIR emission from outflow cavities is due to thermal heating of cavity walls, in the NIR a larger fraction is emitted from the protostar and inner disk/outflow, reaching us directly or via scattering. The Brγ line and rovibrational H_2 lines in the NIR can reveal information about the inner outflow (e.g., *Cesaroni et al.,* 2013). Polarization vectors from scattered light can help localize the protostar: For example, in Orion KL at 4 μm (*Werner et al.,* 1983), these vectors point to a location consistent with radio source I.

The Orion KL region also serves as an example of the rare class of "explosive" outflows. Forming the inner part of the outflow from KL is a wide-angle flow containing "bullets" of NIR H_2 and Fe line emission (*Allen and Burton,* 1993; *Bally et al.,* 2011). Their spectra and kinematics yield a common age of ~10^3 yr. A 10^4-yr-old example of such a flow has been claimed by *Zapata et al.* (2013) in DR21. The KL outflow has been interpreted as being due to tidally enhanced accretion and thus outflow activity from the close (~few × 10^2 AU) passage near source I of the Becklin-Neugebauer (BN) runaway star, itself ejected from interaction with the θ^{1}C binary in the Trapezium (*Tan,* 2004). Becklin-Neugebauer's ejection from θ^{1}C has left a distinctive dynamical fingerprint on θ^{1}C, including recoil motion, orbital binding energy, and eccentricity — properties unlikely (probability ≤10^{-5}) to arise by chance (*Chatterjee and Tan,* 2012). This scenario attributes the "explosive" outflow as being the perturbed high-activity state of a previously normal massive protostellar outflow (akin to an FU Orionis outburst, but triggered by an external encounter rather than an internal disk instability). Alternatively, *Bally and Zinnecker* (2005) and *Goddi et al.* (2011b) proposed BN was ejected by source I, which must then be a hard binary or have suffered a merger. This would imply much closer, disruptive dynamical interactions involving the massive protostar(s) at source I (*Bally et al.,* 2011). In either scenario, recent perturbation of gas on ~10^2–10^3 AU scales around source I has occurred, likely affecting observed hot core complexity (*Beuther et al.,* 2006) and interpretation of maser disk and outflow kinematics (*Greenhill et al.,* 2013).

4.4. Ionized Gas

Observationally, HC and UC H$_{II}$ regions are defined to have sizes <0.01 pc and <0.1 pc, respectively (*Beuther et al.*, 2007; *Hoare et al.*, 2007). They have rising radio spectral indices, due to thermal bremsstrahlung emission from ~10^4 K plasma. A large fraction of HC H$_{II}$ regions show broad (FWHM $\geq$ 40 km s^{-1}) RRLs (*Sewilo et al.*, 2011). Studies of Brackett series lines in massive protostars also show broad lines, perhaps consistent with disk or wind kinematics (*Lumsden et al.*, 2012). Demographics of the UC H$_{II}$ region population imply a lifetime of ~10^5 yr (*Wood and Churchwell*, 1989; *Mottram et al.*, 2011), much longer than the expansion time at the ionized gas sound speed, so a confinement or replenishment mechanism is needed.

The above observational classification may mix different physical states that are expected theoretically during massive star formation (section 2.3 and section 2.4). An outflow-confined H$_{II}$ region (*Tan and McKee*, 2003) is expected first, appearing as a radio jet that gradually opens up as the entire primary-outflow-filled cavity is ionized. Together with outflow feedback, ionization should then start to erode the core infall envelope, driving a photoevaporative flow. Strongly bound parts of the core may become ionized yet continue to accrete (*Keto*, 2007), as inferred in G10.62–0.38 by *Keto* (2002) and W51e2 by *Keto and Klaassen* (2008). Eventually, remnant equatorial accretion may continue to feed a disk that is subject to photoevaporation (section 2.4).

Since massive star ionizing luminosities vary by factors of ~100 from B to O stars (Fig. 2), H$_{II}$ region sizes will also vary. So while, in general, one expects earlier phases of accretion to be associated with smaller H$_{II}$ regions, it is possible that some UC H$_{II}$ regions still harbor accreting massive protostars, while some HC H$_{II}$ regions host non-accreting, already-formed B stars in dense clump environments.

Using the Red MSX Source (RMS) survey (*Lumsden et al.*, 2013) and comparing with main-sequence lifetimes, *Mottram et al.* (2011) derived lifetimes of radio-quiet (RQ) massive protostars (likely the accretion phase before contraction to the ZAMS; Fig. 2) and "compact" (including UC) H$_{II}$ regions as a function of source luminosity. Radio-quiet massive protostars have lifetimes $\simeq$5 × 10^5 yr for L $\simeq$ 10^4 L$_\odot$, declining to $\simeq$1 × 10^5 yr for L $\simeq$ 10^5 L$_\odot$. No RQ massive protostars were seen with L $\gg$ 10^5 L$_\odot$, consistent with Fig. 2: By this luminosity most protostars should have contracted to the ZAMS and thus become "radio-loud" (for $\Sigma_{cl} \lesssim 3$ g cm^{-2}). The "compact" H$_{II}$ regions have lifetimes $\simeq$3 × 10^5 yr (independent of L). *Davies et al.* (2011) extended this work to show that the data favor models in which the accretion rate to massive protostars increases with time, as expected in the fiducial turbulent core accretion model (*MT03*) with k$_\rho$ = 1.5 and with accretion rates appropriate for Σ_{cl} ~ 1. However, their derived trend of increasing accretion rates is also compatible with the competitive accretion model of *Bonnell et al.* (2001).

4.5. Astrochemistry of Massive Protostars

Massive protostars significantly affect the chemical composition of their surroundings. First, they heat gas and dust, leading to sublimation of icy mantles formed during the cold PSC phase (e.g., *Charnley et al.*, 1992; *Caselli et al.*, 1993) — the hot core phase. Second, they drive outflows that shock the gas, enabling some reactions with activation energies and endothermicities to proceed (e.g., *Neufeld and Dalgarno*, 1989). Knowledge of chemical processes is vital to understand the regions traced by the various molecular lines and thus to study the structure and dynamics of the gas surrounding the protostar, i.e., its infall envelope, disk, and outflow (e.g., *Favre et al.*, 2011; *Bisschop et al.*, 2013), fundamental to test formation theories. Unfortunately, the chemistry in these regions is based heavily on poorly known surface processes, so it is important to keep gathering high-sensitivity and spectral/angular resolution data to constrain astrochemical theory, as well as laboratory data to lessen the uncertainties in the rate and collisional coefficients required in astrochemical and RT codes.

The majority of chemical models of these early stages of massive protostellar evolution do not take into account shocks and focus on three main temporal phases (see review by *Herbst and van Dishoeck*, 2009): (1) Cold phase (T ~ 10 K), before protostar formation, where freeze-out, surface hydrogenation/deuteration, and gas-phase ion-neutral reactions are key processes. The main constituents of icy mantles are H$_2$O, followed by CO and CO$_2$, and then by CH$_3$OH, NH$_3$, and CH$_4$ (*Öberg et al.*, 2011), as O, C, N, and CO are mainly hydrogenated, since H is by far the fastest element on the surface at such low temperatures. (2) Warm-up phase, when the protostar starts to heat the surroundings and temperatures gradually increase from ~10 to $\geq$100 K (e.g., *Viti et al.*, 2004). (3) Hot core phase, when all mantles are sublimated and only gas phase chemistry proceeds (e.g., *Brown et al.*, 1988). The warm-up phase is thought to be critical for surface formation of complex molecules (*Garrod and Herbst*, 2006). *Öberg et al.* (2013) claim their observations of N- and O-bearing organics toward high-mass star-forming region NGC 7538 IRS9 are consistent with the onset of complex chemistry at 25–30 K. At these dust temperatures, hydrogen atoms evaporate, while heavier species and molecules can diffuse more quickly within the mantles and form species more complex than in the cold phase. *Garrod* (2013) showed glycine can form in ice mantles at temperatures between 40 and 120 K and be detected in hot cores with ALMA. *Aikawa et al.* (2012) coupled a comprehensive gas-grain chemical network with one-dimensional hydrodynamics, also showing that complex molecules are efficiently formed in the warm-up phase. Early complex molecule formation in the cold phase, perhaps driven by cosmic-ray-induced UV photons or dust heating (*Bacmann et al.*, 2012), may need to be included in the above models.

Herschel has discovered unexpected chemistry in massive star-forming regions. Light hydrides such as OH$^+$ and H$_2$O$^+$, never observed before, have been detected in absorption and weak emission toward W3 IRS5, tracing outflow cavity walls heated and irradiated by protostellar UV radiation (e.g., *Benz et al., 2010; Bruderer et al., 2010*).

Water abundance and kinematics have been measured toward several massive star-forming regions, with different hot cores showing a variety of abundance levels. *Neill et al.* (2013) found abundances of ~6 × 10^{-4} toward Orion KL (making H$_2$O the predominant repository of O) and a relatively large HDO/H$_2$O ratio (~0.003), while *Emprechtinger et al.* (2013) measured 10^{-6} H$_2$O abundance and HDO/H$_2$O ~ 2 × 10^{-4} in NGC 6334 I. Thus, evaporation of icy mantles may not always be complete in hot cores, unlike the assumption made in astrochemical models. It also suggests that the level of deuteration is different in the bulk of the mantle compared to the upper layers that are first returned to the gas phase (*Kalvāns and Shmed,* 2013; *Taquet et al.,* 2013), or that these two hot cores started from slightly different initial dust temperatures, which may highly affect water deuteration (*Cazaux et al.,* 2011). Water vapor abundance has also been found to be low in the direction of outflows [7 × 10^{-7} (*van der Tak et al.,* 2010)], again suggesting ice mantles are more resistant to destruction in shocks than previously thought. The low water abundances in the outer envelopes of massive protostars [e.g., 2 × 10^{-10} in DR21 (*van der Tak et al.,* 2010), 8 × 10^{-8} in W43-MM1 (*Herpin et al.,* 2012)] are likely due to water being mostly in solid form.

4.6. Comparison to Lower-Mass Protostars

Some continuities and similarities were already noted between high- and low-mass starless cores, including the continuous, power-law form of the CMF (although work remains to measure the prestellar CMF) and the chemical evolution of CO freeze-out and high deuteration of certain species. Differences include massive starless cores having larger nonthermal internal motions, although these are also being found in lower-mass cores in high-mass star-forming regions (*Sánchez-Monge et al.,* 2013c). Massive cores may also tend to have higher Σs (section 4.7), whereas low-mass cores can be found with a wider range down to lower values.

For low-mass protostars, an evolutionary sequence from PSCs to pre-main-sequence stars was defined by *Lada* (1991) and *André* (1995): class 0, I, II, and III objects based on SEDs. As described above, an equivalent sequence for massive protostars is not well established. Core and competitive accretion models predict different amounts and geometries of dense gas and dust in the vicinity of the protostar. Even for core accretion models (and also for low-mass protostars), the SED will vary with viewing angle for the same evolutionary stage. For a given mass accretion rate and model for the evolution of the protostar and its surrounding disk and envelope, the observed properties of the system can be calculated. Examples of such models include those of *Robitaille et al.* (2006), *Molinari et al.* (2008), and *Zhang et al.* (2014) (see Figs. 2 and 3).

For massive protostars one expects a "radio-quiet" phase before contraction to the ZAMS. A growing region, at first confined to the disk and outflow, is heated to ≥100 K, exhibiting hot core chemistry. Protostellar outflows are likely to have broken out of the core, and be gradually widening the outflow cavities. Up to this point, the evolution of lower-mass protostars is expected to be qualitatively similar. Contraction to the ZAMS leads to greatly increased H-ionizing luminosities and thus a "radio-loud" phase, corresponding to HC or UC HII regions. Hot core chemistry will be more widespread, but there will also now be regions (perhaps confined to the outflow and disk surfaces) that are exposed to high FUV radiation fields. Stellar winds from the ZAMS protostar should become much stronger than those from low-mass protostars, especially on crossing the "bistability jump" at T$_*$ ≃ 21,000 K (*Vink et al.,* 2001).

How do observed properties of lower-mass protostars compare with massive ones? Helping address this are studies of "intermediate-mass" protostars with L ~ 100–10^4 L$_\odot$ and sharing some characteristics of their more massive cousins (e.g., clustering, creation of photodissociation regions). Many are closer than 1 kpc, allowing determination of the physical and chemical structures at similar spatial scales as in well-studied low-mass protostars.

For disks, while there are examples inferred to be present around massive protostars (section 4.2), information is lacking about their resolved structure or even total extent, making comparison with lower-mass examples difficult. Most massive protostellar disks appear to have sizes ≤10^3 AU. If disk size is related to initial core size, then one expects r$_d$ ∝ R$_c$ ∝ (M$_c$/Σ$_{cl}$)$^{1/2}$, so larger disks resulting from more massive cores may be partly counteracted if massive cores tend to be in higher Σ$_{cl}$ clumps. If stronger magnetization is needed to support more massive cores, then this may lead to more efficient magnetic braking during disk formation. Survival of remnant disks around massive stars may be inhibited by more efficient feedback (section 2.4).

Comparison of outflow properties was discussed in section 4.3, noting the continuity of outflow force and mass loss rates with L. Similar collimation factors are also seen, at least up to L ~ 10^5 L$_\odot$. The rare "explosive" outflows may affect massive protostars more than low-mass ones, but too few examples are known to draw definitive conclusions.

Equivalent regions exhibiting aspects of hot core chemistry, e.g., formation of complex organic molecules, have been found around low-mass [e.g., IRAS 16293–2422 (*Cazaux et al.,* 2003)] and intermediate-mass [e.g., IC 1396N (*Fuente et al.,* 2009), NGC 7129 (*Fuente et al.,* 2012), IRAS 22198+6336, and AFGL 5142 (also discussed in section 4.1 and section 4.3) (*Palau et al.,* 2011)] protostars. Around low-mass protostars these regions appear richer in O-bearing molecules like methyl-formate (CH$_3$OOCH) and poorer in the N-bearing compounds. They have high deuteration fractions (e.g., *Demyk et al.,* 2010), unlike hot cores around massive protostars. The situation in intermediate-mass protostars is mixed: IRAS 22198+6336 and AFGL 5142 are richer in oxygenated molecules, while NGC 7129 is richer in N-species. The recent detection of the vibrationally excited lines of CH$_3$CN and HC$_3$N in this hot core also points to a higher gas temperature (*Fuente et al.*, in preparation). *Palau et al.* (2011) discussed that these hot cores can encompass two different types of regions, inner accretion disks and outflow shocks, helping to explain the observed diversity.

Finally, the stellar IMF for $m_* \geq 1$ $M_\odot$ appears well-described by a continuous and universal power law (see section 5), with no evidence of a break that might evidence a change in the physical processes involved in star formation.

In summary, many aspects of the star formation process appear to either be very similar or vary only gradually as a function of protostellar mass. While some of these properties remain to be explored at the highest masses, we conclude that the bulk of existing observations support a common star formation mechanism from low to high masses.

4.7. Conditions for Massive Star Formation

Do clumps that form massive stars require a threshold Σ or other special properties? *López-Sepulcre et al.* (2010) found an increase in outflow detection rate from 56% to 100% when bisecting their clump sample by a threshold of $\Sigma_{cl} = 0.3$ g cm^{-2}. This is a factor of a few smaller than the threshold predicted by *Krumholz and McKee* (2008) from protostellar heating suppression of fragmentation of massive cores (section 2.4), and thus consistent within the uncertainties in deriving Σ. However, this clump sample contains a mixture of IR-dark and bright objects spanning this threshold, whereas one expects protostellar heating to be associated with IR-bright objects. *Longmore et al.* (2011) estimated the low-mass stellar population needed to be responsible for the observed temperature structure in the fragmenting clump G8.68–0.37, concluding it is too large compared to that allowed by the clump's bolometric luminosity. *BT12* found relatively low values of $\Sigma_{cl} \sim 0.3$ g cm^{-2} in their IRDC sample and advocated magnetic suppression of fragmentation. *Kauffmann et al.* (2010) found that three clouds that contain massive star formation (Orion A, G10.15–0.34, and G11.11–0.12) satisfy $M(r) \geq 870\ r_{pc}^{4/3}\ M_\odot$, while several clouds not forming massive stars do not. This empirical condition is equivalent to $\Sigma \geq 0.054\ M_3^{-1/2}$ g cm^{-2}, i.e., a relatively low threshold that may apply to the global clump even though massive stars form in higher Σ peaks. In summary, more work is needed to better establish if there are minimum threshold conditions for massive star formation. This is difficult since once one is sure a massive star is forming, it will have altered its environment. Thus it may be more fruitful to study the formation requirements of massive PSCs, although there are currently very few examples (section 3).

5. RELATION TO STAR CLUSTER FORMATION

5.1. Clustering of Massive Star Formation

de Wit et al. (2005) studied Galactic field O stars, concluding that the fraction born in isolation was low (4 ± 2%). *Bressert et al.* (2012) have found a small number of O stars that appear to have formed in isolation in the 30 Dor region of the LMC, while *Selier et al.* (2011) and *Oey et al.* (2013) have presented examples in the SMC. For the Galactic sample, the low fraction of "isolated-formation" O stars could be modeled by extrapolating a stochastically sampled power-law initial cluster mass function (ICMF) down to very low masses, including "clusters" of single stars. Such a model suggests that massive star formation is not more clustered than lower-mass star formation and that the "clustering" of star formation does not involve a minimum threshold of cluster mass or density (see also *Bressert et al.*, 2010).

The question of whether massive stars tend to form in the central regions of clumps/clusters is difficult to answer. Observationally, there is much evidence for the central concentration of massive stars within clusters (e.g., *Hillenbrand*, 1995; *Qiu et al.*, 2008; *Kirk and Myers*, 2012; *Pang et al.*, 2013; *Lim et al.*, 2013). For clusters, like the ONC, where a substantial fraction of the initial gas clump mass has formed stars, dynamical evolution leading to mass segregation during star cluster formation may overwhelm any signature of primordial mass segregation (e.g., *Bonnell and Davies*, 1998; *Allison and Goodwin*, 2011; *Maschberger and Clarke*, 2011), especially if cluster formation extends over many local dynamical times (*Tan et al.*, 2006). This problem is even more severe in gas-free, dynamically older systems like NGC 3603, Westerlund 1, and the Arches.

Earlier phase studies are needed. *Kumar et al.* (2006) searched 2MASS images for clusters around 217 massive protostar candidates, finding 54. Excluding targets most affected by Galactic plane confusion, the detection rate was ~60%. *Palau et al.* (2013) studied 57 (millimeter-detected) cores in 18 "protoclusters," finding quite low levels of fragmentation and relatively few associated NIR/MIR sources.

Do massive stars tend to form earlier, later, or contemporaneously with lower-mass stars? In turbulent core accretion (*MT03*), the formation times of stars from their cores show a weak dependence with mass, $t_{*f} \propto m_{*f}^{1/4}$, and the overall normalization is short compared to the global cluster formation time, if that is spread out over at least a few free-fall times. Competitive accretion models (e.g., *Wang et al.*, 2010) involve massive stars gaining their mass gradually over the same timescales controlling global clump evolution, suggesting that massive stars would form later than typical low-mass stars. Systematic uncertainties in young stellar age estimates (see chapter by Soderblom et al. in this volume) make this a challenging question to answer. In the ONC, *Da Rio et al.* (2012) have shown there is a real age spread of a few million years, i.e., at least several mean free-fall times, but no evidence for a mass-age correlation. Massive stars are forming today in the ONC, i.e., source I. If the runaway stars μ Col and AE Aur, together with the resulting binary, ι Ori, were originally in the ONC about 2.5 m.y. ago (*Hoogerwerf et al.*, 2001), then massive stars appear to have formed contemporaneously with the bulk of the cluster population.

5.2. Initial Mass Function and Binarity of Massive Stars

The massive star IMF and its possible variation with environment are potential tests of formation mechanisms. The IMF is constrained by observations of massive stars in young clusters, especially "super star clusters" (SSCs) with $M_* \gtrsim 10^4 M_\odot$, where effects of incomplete statistical sampling are reduced. For example, for IMF $dN/dm_* \propto m_*^{-\alpha_*}$ and $\alpha_* = 2.35$ (Salpeter) from $m_{*l} = 0.1 M_\odot$ to upper truncation mass of $m_{*u} = 100$ or $1000 M_\odot$, the median expected maximum stellar mass in a cluster with $M_* = 10^4 M_\odot$ is 83.6, 226 $M_\odot$, respectively (e.g., *McKee and Williams,* 1997). For $M_* = 10^5 M_\odot$ it is 98.0, 692 $M_\odot$, respectively.

Estimates of m_{*u} range from $\simeq 150 M_\odot$ (e.g., *Figer,* 2005) to $\simeq 300 M_\odot$ (*Crowther et al.,* 2010), with uncertainties due to crowding, unresolved binarity, extinction corrections, and the NIR magnitude-mass relation. Also, a limiting m_{*u} arising from star formation may occasionally appear to be breached by mergers or mass transfer in binary systems (e.g., *Banerjee et al.,* 2012; *Schneider et al.,* 2014). It is not yet clear if m_{*u} is set by local processes [e.g., ionization or radiation pressure feedback (section 2.4, section 2.5), rapid mass loss due to stellar instability] or by the cluster environment (e.g., *Weidner et al.,* 2013; however, see *Krumholz,* 2014).

Deriving initial stellar masses can thus require modeling stellar evolution, including the effects of rotation, mass loss, and binary mass transfer (e.g., *Sana et al.,* 2012; *de Mink et al.,* 2013; *Schneider et al.,* 2014). Dynamical evolution in clusters leads to mass segregation and ejection of stars, further complicating IMF estimation from observed mass functions (MFs) of either current or initial stellar masses.

Many attempts have been made to derive MFs in SSCs. For Westerlund 1, the most massive young cluster in the Galaxy, *Lim et al.* (2013) find $\alpha_* = 1.8 \pm 0.1$ within $r < 2.8$ pc over mass range $5 < m_*/M_\odot < 85$, and an even shallower slope of $\alpha_* = 1.5$ if the statistically incomplete highest-mass bins are excluded. A similar slope of $\alpha_* = 1.9 \pm 0.15$ for $1 < m_*/M_\odot < 100$ is measured for proper motion members in the central young cluster of NGC 3603, with an even shallower slope of $\alpha_* = 1.3 \pm 0.3$ found in the cluster core for the intermediate- to high-mass stars [$4 < m_*/M_\odot < 100$ (*Pang et al.,* 2013)]. For R136 in 30 Doradus in the LMC, *Andersen et al.* (2009) find $\alpha_* = 2.2 \pm 0.2$ for $1 < m_{*f}/M_\odot < 20$ and a radial coverage of 3 to 7 pc. However, the cluster core remains poorly resolved and its MF uncertain. In NGC 346 in the SMC, *Sabbi et al.* (2008) find $\alpha_* = 2.43 \pm 0.18$ for $0.8 < m_{*f}/M_\odot < 60$.

Environmental conditions of temperature, cosmic-ray flux, magnetization, and orbital shear are all higher in the Galactic Center, so one may expect IMF variations (e.g., *Morris and Serabyn,* 1996). In the Quintuplet cluster core, $r < 0.5$ pc, the present-day MF is found to be $\alpha_* = 1.7 \pm 0.2$ for $4 < m_{*f}/M_\odot < 48$ (*Hussmann et al.,* 2012). As in Westerlund 1 and NGC 3603, a steepening of the IMF slope with distance from cluster center is observed with $\alpha_* = 2.1 \pm 0.2$ for radii 1.2 to 1.8 pc, close to the expected tidal radius (B. Hussmann, personal communication). The core of the Arches cluster also exhibits a relatively shallow MF, but the combined MF slope out to the tidal radius is $\alpha_* = 2.5 \pm 0.2$ for $15 < m_{*f}/M_\odot < 80$ (*Habibi et al.,* 2013).

Detailed N-body simulations have been carried out to model the Arches cluster (e.g., *Harfst et al.,* 2010). The excellent match between the radial variation of the MF in these simulations and the observed increase in the MF slope with radius provide strong evidence that the steepening is caused by dynamical mass segregation alone. By analogy, the relatively shallow slopes observed in the central regions of all the above young, massive clusters are likely influenced, and possibly completely caused, by internal dynamical evolution of these clusters on timescales as short as 1–3 m.y., within the current ages of these clusters.

The most extreme star-forming environment resolved to date is the young nuclear cluster surrounding the supermassive black hole SgrA* in the center of the Milky Way. If the effects of increased tidal shear and temperatures cause an increase in the Jeans mass, it should most likely be observed in this environment. Previous studies suggested a slope as shallow as $\alpha_* = 0.45 + 0.3$ for $m_* > 10 M_\odot$ and with a truncation of $m_{*u} \simeq 30 M_\odot$, and hence proposed the most extreme stellar MF observed in a resolved population to date (*Bartko et al.,* 2010). Many of the young stars in the nuclear cluster are in an elongated disk-like structure (e.g., *Paumard et al.,* 2006), and optimizing for the inclusion of young disk candidates as members of the cluster revises this picture. Employing Keck spectroscopy along the known disk of young stars, *Lu et al.* (2013) found a slope of $\alpha_* = 1.7 \pm 0.2$ from detailed Bayesian modeling to derive individual stellar masses. While still flatter than the Salpeter slope, this result is now in agreement with the shape of the MFs found in the central regions of all other young, massive clusters outside this very extreme environment. The effects of mass segregation and ejection for altering the observed MF are not very well known. Modulo these uncertainties, there is no evidence for IMF variation in the Galactic Center compared to other massive clusters.

In summary, the massive young clusters resolved to date exhibit somewhat shallow present-day MFs in their cluster cores, with a steepening of the MF observed toward larger radii. Numerical simulations suggest that the central top-heavy mass distribution can be explained by mass segregation, and is not evidence for a deviating IMF in the high-mass regime. The fact that there is little or no variation of the shape of the high-mass IMF from NGC 346 to the Arches or Westerlund 1 suggests that the process of massive star formation has a very weak dependence on density, which varies by two to three orders of magnitude between these clusters (Fig. 1). This implies stellar collisions are not important for forming massive stars in these environments, in agreement with theoretical estimates of collision rates by, e.g., *Moeckel and Clarke* (2011). Predictions of the dependence of the IMF with density are needed from simulations and models of core and competitive accretion.

The binary properties of massive stars have been discussed extensively by *Zinnecker and Yorke* (2007) (see also *Sana et al.,* 2012; *de Mink et al.,* 2013). They are more likely to be in binary or multiples than lower-mass stars. For stars in cluster centers, these properties may also have been affected by dynamical evolution via interactions (e.g., *Parker et al.,* 2011; *Allison and Goodwin,* 2011), which tend to harden and increase the eccentricity of binary orbits and can also lead to ejection of runaway stars. For example, such an interaction has been proposed to explain the properties of the θ^1 C binary near the center of the ONC (*Chatterjee and Tan,* 2012). Thus, one should be cautious using the observed binary properties of massive stars to constrain massive star formation theories, with attention ideally focused on objects that are either very young (i.e., still forming) or relatively isolated in lower-density environments.

6. CONCLUSIONS AND FUTURE OUTLOOK

It is a challenge to understand the wide variety of interlocking physical and chemical processes involved in massive star formation. Still, significant theoretical progress is being made in modeling these processes, both individually and in combination in numerical simulations. However, such simulations still face great challenges in being able to adequately resolve the scales and processes that may be important, including MHD-driven outflows, radiative feedback, and astrochemistry. There are also uncertainties in how to initialize these simulations. Accurate prediction of the IMF, including massive stars, of a cluster forming under given environmental conditions remains a distant goal.

Close interaction with observational constraints is essential. Here rapid progress is also being made and, with the advent of ALMA, this should only accelerate. One challenge is development of the astrochemical sophistication needed to decipher the rich variety of diagnostic tracers becoming available for both pre- and protostellar phases. Determination of prestellar core mass functions and resolution of massive protostellar accretion disks, including possible binary formation, and outflows are important goals.

Core and competitive accretion theories are being tested by both simulation and observation. Core accretion faces the challenges of understanding fragmentation properties of magnetized, turbulent gas, following development of accretion disks and outflows from collapsing cores, and assessing the importance of external interactions in crowded cluster environments. Competitive accretion is also challenged by theoretical implementation of realistic feedback from MHD outflows and by observations of massive starless cores, together with apparent continuities and similarities of the star-formation process across protostellar mass and luminosity distributions. There is much work to be done!

Acknowledgments. We thank S. Kong, A. Myers, Á. Sánchez-Monge, and Y. Zhang for figures and discussions, and H. Beuther, M. Butler, P. Clark, N. Da Rio, M. Hoare, P. Klaassen, P. Kroupa, W. Lim, and an anonymous referee for discussions.

REFERENCES

Aikawa Y. et al. (2012) *Astrophys. J., 760,* 40.
Allen D. A. and Burton M. G. (1993) *Nature, 363,* 54.
Allison R. J. and Goodwin S. P. (2011) *Mon. Not. R. Astron. Soc., 415,* 1967.
Andersen M. et al. (2009) *Astrophys. J., 707,* 1347.
André P. (1995) *Astrophys. Space Sci., 224,* 29.
Bacmann A. et al. (2012) *Astron. Astrophys., 541,* L12.
Bally J. and Zinnecker H. (2005) *Astron. J., 129,* 2281.
Bally J. et al. (2011) *Astrophys. J., 727,* 113.
Banerjee S. et al. (2012) *Mon. Not. R. Astron. Soc., 426,* 1416.
Barnes P. J. et al. (2010) *Mon. Not. R. Astron. Soc., 402,* 73.
Barnes P. J. et al. (2011) *Astrophys. J. Suppl., 196,* 12.
Bartko H. et al. (2010) *Astrophys. J., 708,* 834.
Battersby C. et al. (2011) *Astron. Astrophys., 535,* 128.
Bate M. R. (2012) *Mon. Not. R. Astron. Soc., 419,* 3115.
Bate M. R. et al. (1995) *Mon. Not. R. Astron. Soc., 277,* 362.
Baumgardt H. and Klessen R. S. (2011) *Mon. Not. R. Astron. Soc., 413,* 1810.
Beltrán M. T. et al. (2006) *Nature, 443,* 427.
Beltrán M. T. et al. (2011) *Astron. Astrophys., 525,* 151.
Benz A. O. et al. (2010) *Astron. Astrophys., 521,* L35.
Berger M. J. and Oliger J. (1984) *J. Comp. Phys., 53,* 484.
Bergin E. A. and Tafalla M. (2007) *Annu. Rev. Astron. Astrophys., 45,* 339.
Bertoldi F. and McKee C. F. (1992) *Astrophys. J., 395,* 140.
Beuther H. and Schilke P. (2004) *Science, 303,* 1167.
Beuther H. and Shepherd D. S. (2005) In *Cores to Clusters* (M. Kumar et al.), pp. 105–119. Springer, New York.
Beuther H. et al. (2002) *Astron. Astrophys., 383,* 892.
Beuther H. et al. (2006) *Astrophys. J., 636,* 323.
Beuther H. et al. (2007) In *Protostars and Planets V* (B. Reipurth et al.), pp. 165–180. Univ. of Arizona, Tucson.
Beuther H. et al. (2010a) *Astron. Astrophys., 518,* L78.
Beuther H. et al. (2010b) *Astrophys. J. Lett., 724,* L113.
Beuther H. et al. (2012) *Astron. Astrophys., 538,* 11.
Beuther H. et al. (2013a) *Astron. Astrophys., 553,* 115.
Beuther H. et al. (2013b) *Astron. Astrophys., 558,* 81.
Bik A. and Thi W. F. (2004) *Astron. Astrophys., 427,* L13.
Bisbas T. G. et al. (2009) *Astron. Astrophys., 497,* 649.
Bisschop S. E. et al. (2013) *Astron. Astrophys., 552,* 122.
Blandford R. D. and Payne D. G. (1982) *Mon. Not. R. Astron. Soc., 199,* 883.
Boley P. A. et al. (2013) *Astron. Astrophys., 558,* 24.
Bonnell I. A. and Bate M. R. (2005) *Mon. Not. R. Astron. Soc., 362,* 915.
Bonnell I. A. and Davies M. B. (1998) *Mon. Not. R. Astron. Soc., 295,* 691.
Bonnell I. A. et al. (1998) *Mon. Not. R. Astron. Soc., 298,* 93.
Bonnell I. A. et al. (2001) *Mon. Not. R. Astron. Soc., 323,* 785.
Bonnell I. A. et al. (2004) *Mon. Not. R. Astron. Soc., 349,* 735.
Bonnell I. A. et al. (2007) In *Protostars and Planets V* (B. Reipurth et al.), pp. 149–164. Univ. of Arizona, Tucson.
Bontemps S. et al. (2010) *Astron. Astrophys., 524,* 18.
Bressert E. et al. (2010) *Mon. Not. R. Astron. Soc., 409,* L54.
Bressert E. et al. (2012) *Astron. Astrophys., 542,* 49.
Bromm V. (2013) *Rept. Prog. Phys., 76,* 11.
Brown P. D. et al. (1988) *Mon. Not. R. Astron. Soc., 231,* 409.
Bruderer S. et al. (2010) *Astron. Astrophys., 521,* L44.
Butler M. J. and Tan J. C. (2009) *Astrophys. J., 696,* 484.
Butler M. J. and Tan J. C. (2012) *Astrophys. J., 754,* 5.
Carrasco-González C. et al. (2010) *Science, 330,* 1209.
Carrasco-González C. et al. (2012) *Astrophys. J. Lett., 752,* L29.
Caselli P. and Myers P. C. (1995) *Astrophys. J., 446,* 665.
Caselli P. et al. (1993) *Astrophys. J., 408,* 548.
Caselli P. et al. (1999) *Astrophys. J. Lett., 523,* L165.
Caselli P. et al. (2002) *Astrophys. J., 565,* 344.
Caselli P. et al. (2008) *Astron. Astrophys., 492,* 703.
Cazaux S. et al. (2003) *Astrophys. J. Lett., 593,* L51.
Cazaux S. et al. (2011) *Astrophys. J. Lett., 741,* L34.
Cesaroni R. et al. (2005) *Astron. Astrophys., 434,* 1039.
Cesaroni R. et al. (2007) In *Protostars and Planets V* (B. Reipurth et al., eds.), p. 197–212. Univ. of Arizona, Tucson.
Cesaroni R. et al. (2013) *Astron. Astrophys., 549,* 146.
Charnley S. B. et al. (1992) *Astrophys. J. Lett., 399,* L71.
Chatterjee S. and Tan J. C. (2012) *Astrophys. J., 754,* 152.

Chen H.-R. et al. (2011) *Astrophys. J., 743,* 196.
Chen X. et al. (2010) *Astrophys. J., 710,* 150.
Chini R. et al. (2004) *Nature, 429,* 155.
Chira R.-A. et al. (2013) *Astron. Astrophys., 552,* 40.
Churchwell E. et al. (2009) *Publ. Astron. Soc. Pacific, 121,* 213.
Clark P. C. et al. (2007) *Mon. Not. R. Astron. Soc., 379,* 57.
Codella C. et al. (2013) *Astron. Astrophys., 550,* 81.
Commerçon B. et al. (2011a) *Astrophys. J. Lett., 742,* L9.
Commerçon B. et al. (2011b) *Astron. Astrophys., 529,* 35.
Crapsi A. et al. (2005) *Astrophys. J., 619,* 379.
Crowther P. A. et al. (2010) *Mon. Not. R. Astron. Soc., 408,* 731.
Crutcher R. M. (2012) *Annu. Rev. Astron. Astrophys., 50,* 29.
Cunningham A. J. et al. (2011) *Astrophys. J., 740,* 107.
Da Rio N. M. et al. (2012) *Astrophys. J., 748,* 14.
Dale J. E. et al. (2005) *Mon. Not. R. Astron. Soc., 358,* 291.
Dalgarno A. (2006) *Proc. Natl. Acad. Sci., 103,* 12269.
Dalgarno A. and Lepp S. (1984) *Astrophys. J. Lett., 287,* L47.
Davies B. et al. (2010) *Mon. Not. R. Astron. Soc., 402,* 1504.
Davies B. et al. (2011) *Mon. Not. R. Astron. Soc., 416,* 972.
De Buizer J. M. (2006) *Astrophys. J. Lett., 642,* L57.
de Gouveia Dal Pino E. et al. (2012) *Phys. Script., 86,* 018401.
de Mink S. E. et al. (2013) *Astrophys. J., 764,* 166.
Demyk K. et al. (2010) *Astron. Astrophys., 517,* 17.
de Wit W. J. et al. (2005) *Astron. Astrophys., 437,* 247.
de Wit W. J. et al. (2011) *Astron. Astrophys., 526,* L5.
Dobbs C. L. et al. (2005) *Mon. Not. R. Astron. Soc., 360,* 2.
Draine B. T. (2011) *Physics of the Interstellar and Intergalactic Medium.* Princeton Univ., Princeton.
Duarte-Cabral A. et al. (2013) *Astron. Astrophys., 558,* 125.
Duffin D. F. and Pudritz R. E. (2008) *Mon. Not. R. Astron. Soc., 391,* 1659.
Edgar R. and Clarke C. (2004) *Mon. Not. R. Astron. Soc., 349,* 678.
Ellingsen S. P. et al. (2007) In *Astrophysical Masers and Their Environments* (J. M. Chapman and W. A. Baan, eds.), p. 213. IAU Symp. 242, Cambridge Univ., Cambridge.
Emprechtinger M. et al. (2009) *Astron. Astrophys., 493,* 89.
Emprechtinger M. et al. (2013) *Astrophys. J., 765,* 61.
Falgarone E. et al. (2008) *Astron. Astrophys., 487,* 247.
Fatuzzo M. and Adams F. C. (2002) *Astrophys. J., 570,* 210.
Favre C. et al. (2011) *Astron. Astrophys., 532,* 32.
Fernández-López M. et al. (2011) *Astron. J., 142,* 97.
Fiege J. D. and Pudritz R. E. (2000) *Mon. Not. R. Astron. Soc., 311,* 85.
Figer D. F. (2005) *Nature, 434,* 192.
Fontani F. et al. (2008) *Astron. Astrophys., 477,* L45.
Fontani F. et al. (2011) *Astron. Astrophys., 529,* L7.
Fontani F. et al. (2012) *Mon. Not. R. Astron. Soc., 423,* 2342.
Foster J. B. et al. (2009) *Astrophys. J., 696,* 298.
Franco-Hernández R. et al. (2009) *Astrophys. J., 701,* 974.
Fryxell B. et al. (2000) *Astrophys. J. Suppl., 131,* 273.
Fuente A. et al. (2009) *Astron. Astrophys., 507,* 1475.
Fuente A. et al. (2012) *Astron. Astrophys., 540,* 75.
Fuller G. A. et al. (2005) *Astron. Astrophys., 442,* 949.
Garay G., Mardones D. et al. (2007) *Astron. Astrophys., 463,* 217.
Garrod R. T. (2013) *Astrophys. J., 765,* 60.
Garrod R. T. and Herbst E. (2006) *Astron. Astrophys., 457,* 927.
Gibb A. G. et al. (2003) *Mon. Not. R. Astron. Soc., 339,* 198.
Ginsburg A. et al. (2012) *Astrophys. J. Lett., 758,* L29.
Girart J. et al. (2009) *Science, 324,* 1408.
Girichidis P. et al. (2012) *Mon. Not. R. Astron. Soc., 420,* 613.
Goddi C. et al. (2011a) *Astron. Astrophys., 535,* L8.
Goddi C. et al. (2011b) *Astrophys. J., 728,* 15.
Greenhill L. J. et al. (2013) *Astrophys. J. Lett., 770,* L32.
Guzmán A. E. et al. (2010) *Astrophys. J., 725,* 734.
Habibi M. et al. (2013) *Astron. Astrophys., 556,* 26.
Haemmerlé L. et al. (2013) *Astron. Astrophys., 557,* 112.
Harfst S. et al. (2010) *Mon. Not. R. Astron. Soc., 409,* 628.
Hennebelle P. et al. (2011) *Astron. Astrophys., 528,* 72.
Henning Th. et al. (2010) *Astron. Astrophys., 518,* L95.
Henshaw J. et al. (2013) *Mon. Not. R. Astron. Soc., 428,* 3425.
Herbst E. and van Dishoeck E. F. (2009) *Annu. Rev. Astron. Astrophys., 47,* 427.
Hernandez A. K. et al. (2012) *Astrophys. J. Lett., 756,* L13.
Herpin F. et al. (2012) *Astron. Astrophys., 542,* 76.
Heyer M. H. et al. (2009) *Astrophys. J., 699,* 1092.
Hillenbrand L. A. et al. (1995) *Astron. J., 109,* 280.
Hoare M. G. et al. (2007) In *Protostars and Planets V* (B. Reipurth et al.), pp. 181–196. Univ. of Arizona, Tucson.

Hofner P. et al. (2007) *Astron. Astrophys., 465,* 197.
Hollenbach D. et al. (1994) *Astrophys. J., 428,* 654.
Hoogerwerf R. et al. (2001) *Astron. Astrophys., 365,* 49.
Hosokawa T. and Omukai K. (2009) *Astrophys. J., 691,* 823.
Hosokawa T. et al. (2010) *Astrophys. J., 721,* 478.
Hosokawa T. et al. (2011) *Science, 334,* 1250.
Houde M. et al. (2009) In *Submillimeter Astrophysics and Technology: A Symposium Honoring Thomas G. Phillips* (D. Lis et al., eds.), p. 265. ASP Conf. Series 417, Astronomical Society of the Pacific, San Francisco.
Hunter C. (1977) *Astrophys. J., 218,* 834.
Hunter T. et al. (1999) *Astrophys. J., 118,* 477.
Hussmann B. et al. (2012) *Astron. Astrophys., 540,* 57.
Ilee J. D. et al. (2013) *Mon. Not. R. Astron. Soc., 429,* 2960.
Jiang Y.-F. et al. (2013) *Astrophys. J., 763,* 102.
Jijina J. and Adams F. C. (1996) *Astrophys. J., 462,* 874.
Jiménez-Serra I. et al. (2010) *Mon. Not. R. Astron. Soc., 406,* 187.
Johnston K. G. et al. (2011) *Mon. Not. R. Astron. Soc., 415,* 2952.
Kahn F. D. (1974) *Astron. Astrophys., 37,* 149.
Kainulainen J. and Tan J. C. (2013) *Astron. Astrophys., 549,* 53.
Kalvāns J. and Shmeld I. (2013) *Astron. Astrophys., 554,* 111.
Kauffmann J. et al. (2010) *Astrophys. J., 716,* 433.
Keto E. (2002) *Astrophys. J., 568,* 754.
Keto E. (2007) *Astrophys. J., 666,* 976.
Keto E. and Caselli P. (2010) *Mon. Not. R. Astron. Soc., 402,* 1625.
Keto E. and Klaassen P. D. (2008) *Astrophys. J. Lett., 678,* L109.
Kirk H. and Myers P. C. (2012) *Astrophys. J., 745,* 131.
Klaassen P. D. et al. (2012) *Astron. Astrophys., 538,* 140.
Königl A. and Pudritz R. E. (2000) In *Protostars and Planets IV* (V. Mannings et al., eds.), p. 759. Univ. of Arizona, Tucson.
Kratter K. M. et al. (2010) *Astrophys. J., 708,* 1585.
Kraus S. et al. (2010) *Nature, 466,* 339.
Krumholz M. R. (2014) *Phys. Rept.,* in press, arXiv:1402.0867.
Krumholz M. R. and McKee C. F. (2008) *Nature, 451,* 1082.
Krumholz M. R. and Tan J. C. (2007) *Astrophys. J., 654,* 304.
Krumholz M. R. et al. (2005a) *Nature, 438,* 332.
Krumholz M. R. et al. (2005b) *Astrophys. J. Lett., 618,* L33.
Krumholz M. R. et al. (2007a) *Astrophys. J., 656,* 959.
Krumholz M. R. et al. (2007b) *Astrophys. J., 667,* 626.
Krumholz M. R. et al. (2007c) *Astrophys. J., 665,* 478.
Krumholz M. R. et al. (2009) *Science, 323,* 754.
Krumholz M. R. et al. (2012) *Astrophys. J., 754,* 71.
Kuiper R. and Yorke H. W. (2013) *Astrophys. J., 763,* 104.
Kuiper R. et al. (2010a) *Astrophys. J., 722,* 1556.
Kuiper R. et al. (2010b) *Astron. Astrophys., 511,* 81.
Kuiper R. et al. (2011) *Astrophys. J., 732,* 20.
Kuiper R. et al. (2012) *Astron. Astrophys., 537,* 122.
Kumar M. S. N. et al. (2006) *Astron. Astrophys., 449,* 1033.
Kunz M. W. and Mouschovias T. Ch. (2009) *Mon. Not. R. Astron. Soc., 399,* L94.
Lacy J. H. et al. (1994) *Astrophys. J. Lett., 428,* L69.
Lada C. J. (1991) In *The Physics of Star Formation and Early Stellar Evolution* (C. Lada and N. Kylafis, eds.), p. 329. Kluwer, Dordrecht.
Larsen S. S. et al. (2008) *Mon. Not. R. Astron. Soc., 383,* 263.
Larson R. B. (1985) *Mon. Not. R. Astron. Soc., 214,* 379.
Larson R. B. and Starrfield S. (1971) *Astron. Astrophys., 13,* 190.
Leurini S. et al. (2013) *Astron. Astrophys., 554,* 35.
Li P. S. et al. (2012a) *Astrophys. J., 745,* 139.
Li P. S. et al. (2012b) *Astrophys. J., 744,* 73.
Lim B. et al. (2013) *Astron. J., 145,* 46.
Longmore S. N. et al. (2011) *Astrophys. J., 726,* 97.
Longmore S. N. et al. (2012) *Astrophys. J., 746,* 117.
López-Sepulcre A. et al. (2009) *Astron. Astrophys., 499,* 811.
López-Sepulcre A. et al. (2010) *Astron. Astrophys., 517,* 66.
López-Sepulcre A. et al. (2011) *Astron. Astrophys., 526,* L2.
Lu J. R. et al. (2013) *Astrophys. J., 764,* 155.
Lucy L. B. (1977) *Astron. J., 82,* 1013.
Lumsden S. L. et al. (2012) *Mon. Not. R. Astron. Soc., 424,* 1088.
Lumsden S. L. et al. (2013) *Astrophys. J. Suppl., 208,* 11.
Mardones D. et al. (1997) *Astrophys. J., 489,* 719.
Maschberger Th. and Clarke C. J. (2011) *Mon. Not. R. Astron. Soc., 416,* 541.
Matthews B. C. et al. (2009) *Astrophys. J. Suppl., 182,* 143.
Matzner C. D. (2007) *Astrophys. J., 659,* 1394.
Matzner C. D. and McKee C. F. (1999) *Astrophys. J. Lett., 526,* L109.
Matzner C. D. and McKee C. F. (2000) *Astrophys. J., 545,* 364.

McCrady N. and Graham J. R. (2007) *Astrophys. J., 663*, 844.
McKee C. F. (1989) *Astrophys. J., 345*, 782.
McKee C. F. and Hollimann J. H. (1999) *Astrophys. J., 522*, 313.
McKee C. F. and Ostriker E. C. (2007) *Annu. Rev. Astron. Astrophys., 45*, 565.
McKee C. F. and Tan J. C. (2002) *Nature, 416*, 59.
McKee C. F. and Tan J. C. (2003) *Astrophys. J., 585*, 850.
McKee C. F. and Tan J. C. (2008) *Astrophys. J., 681*, 771.
McKee C. F. and Williams J. P. (1997) *Astrophys. J., 476*, 144.
McKee C. F. et al. (2010) *Astrophys. J., 720*, 1612.
Menten K. M. and Reid M. J. (1995) *Astrophys. J. Lett., 445*, L157.
Miettinen O. et al. (2011) *Astron. Astrophys., 534*, 134.
Moeckel N. and Clarke C. J. (2011) *Mon. Not. R. Astron. Soc., 410*, 2799.
Molinari S., Pezzuto S. et al. (2008) *Astron. Astrophys., 481*, 345.
Morino J.-I. et al. (1998) *Nature, 393*, 340.
Morris M. and Serabyn E. (1996) *Annu. Rev. Astron. Astrophys., 34*, 645.
Moscadelli L. et al. (2013) *Astron. Astrophys., 549*, 122.
Mottram J. C. et al. (2011) *Astrophys. J. Lett., 730*, L33.
Mouschovias T. Ch. (1987) In *Physical Processes in Interstellar Clouds* (G. Morfill and M. Scholer, eds.), p. 453. Reidel, Dordrecht.
Mueller K. E. et al. (2002) *Astrophys. J. Suppl., 143*, 469.
Myers A. T. et al. (2013) *Astrophys. J., 766*, 97.
Myers P. C. and Fuller G. A. (1992) *Astrophys. J., 396*, 631.
Najita J. R. and Shu F. H. (1994) *Astrophys. J., 429*, 808.
Nakano T. (1989) *Astrophys. J., 345*, 464.
Neill J. et al. (2013) *Astrophys. J., 770*, 142.
Neufeld D. A. and Dalgarno A. (1989) *Astrophys. J., 340*, 869.
Öberg K. I. et al. (2011) *Astrophys. J., 740*, 109.
Öberg K. I. et al. (2013) *Astrophys. J., 771*, 95.
Oey M. S. et al. (2013) *Astrophys. J., 768*, 66.
Ossenkopf V. and Henning T. (1994) *Astron. Astrophys., 291*, 943.
Padoan P. and Nordlund A. (2007) *Astrophys. J., 661*, 972.
Padovani M. and Galli D. (2011) *Astron. Astrophys., 530*, 109.
Palau A. et al. (2011) *Astrophys. J. Lett., 743*, L32.
Palau A. et al. (2013) *Astrophys. J., 762*, 120.
Palla F. and Stahler S. W. (1991) *Astrophys. J., 375*, 288.
Pang X. et al. (2013) *Astrophys. J., 764*, 73.
Parker R. J. et al. (2011) *Mon. Not. R. Astron. Soc., 418*, 2565.
Patel N. A. et al. (2005) *Nature, 437*, 109.
Paumard T. et al. (2006) *Astrophys. J., 643*, 1011.
Peretto N. and Fuller G. A. (2009) *Astron. Astrophys., 505*, 405.
Peretto N. et al. (2010) *Astron. Astrophys., 518*, L98.
Peretto N. et al. (2013) *Astron. Astrophys., 555*, 112.
Pestalozzi M. R. et al. (2009) *Astron. Astrophys., 501*, 999.
Peters T. et al. (2010a) *Astrophys. J., 711*, 1017.
Peters T. et al. (2010b) *Astrophys. J., 725*, 134.
Peters T. et al. (2011) *Astrophys. J., 729*, 72.
Pillai T. et al. (2006) *Astron. Astrophys., 450*, 569.
Pillai T. et al. (2011) *Astron. Astrophys., 530*, 118.
Plambeck R. L. et al. (2013) *Astrophys. J., 765*, 40.
Plume R. et al. (1997) *Astrophys. J., 476*, 730.
Preibisch T. et al. (2011) *Astron. Astrophys., 530*, 40.
Price D. J. and Monaghan J. J. (2004) *Mon. Not. R. Astron. Soc., 348*, 123.
Qiu K. et al. (2007) *Astrophys. J., 654*, 361.
Qiu K. et al. (2008) *Astrophys. J., 685*, 1005.
Qiu K. et al. (2012) *Astrophys. J., 756*, 170.
Ragan S. E. et al. (2009) *Astrophys. J., 698*, 324.
Ragan S. E. et al. (2012) *Astron. Astrophys., 547*, 49.
Rathborne J. M. et al. (2006) *Astrophys. J., 641*, 389.
Reid M. J. et al. (1995) *Astrophys. J., 443*, 238.
Robitaille T. P. et al. (2006) *Astrophys. J. Suppl., 167*, 256.
Rodón J. A. et al. (2012) *Astron. Astrophys., 545*, 51.
Roman-Duval J. et al. (2010) *Astrophys. J., 723*, 492.
Rosen A. L. et al. (2012) *Astrophys. J., 748*, 97.
Sabbi E. et al. (2008) *Astron. J., 135*, 173.
Sakai T. et al. (2008) *Astrophys. J., 678*, 1049.
Sakai T. et al. (2012) *Astrophys. J., 742*, 140.
San José-García I. et al. (2013) *Astron. Astrophys., 553*, 125.
Sana H. et al. (2012) *Science, 337*, 444.
Sánchez-Monge Á. et al. (2013a) *Astron. Astrophys., 552*, L10.

Sánchez-Monge Á. et al. (2013b) *Astron. Astrophys., 552*, L10.
Sánchez-Monge Á. et al. (2013c) *Mon. Not. R. Astron. Soc., 432*, 3288.
Sanhueza P. et al. (2013) *Astrophys. J., 773*, 123.
Santangelo G. et al. (2009) *Astron. Astrophys., 501*, 495.
Schneider F. R. N. et al. (2014) *Astrophys. J., 780*, 117.
Schneider N. et al. (2010) *Astron. Astrophys., 520*, 49.
Selier R. et al. (2011) *Astron. Astrophys., 529*, 40.
Sewilo M. et al. (2011) *Astrophys. J. Suppl., 194*, 44.
Shu F. H. (1977) *Astrophys. J., 214*, 488.
Shu F. H. et al. (1987) *Annu. Rev. Astron. Astrophys., 25*, 23.
Shu F. H. et al. (2000) In *Protostars and Planets IV* (V. Mannings et al., eds.), p. 789. Univ. of Arizona, Tucson.
Smith R. J. et al. (2009) *Mon. Not. R. Astron. Soc., 400*, 1775.
Smith R. J. et al. (2011) *Mon. Not. R. Astron. Soc., 411*, 1354.
Smith R. J. et al. (2013) *Astrophys. J., 771*, 24.
Sollins P. et al. (2004) *Astrophys. J. Lett., 616*, L35.
Sollins P. et al. (2005) *Astrophys. J. Lett., 624*, L49.
Solomon P. M. et al. (1987) *Astrophys. J., 319*, 730.
Sridharan T. K. et al. (2014) *Astrophys. J. Lett., 783*, L31.
Su Y.-N. et al. (2012) *Astrophys. J. Lett., 744*, L26.
Suttner G. et al. (1999) *Astrophys. J., 524*, 857.
Tan J. C. (2004) *Astrophys. J. Lett., 607*, L47.
Tan J. C. and McKee C. F. (2003) *ArXiv preprints,* arXiv:astro-ph/0309139.
Tan J. C. and McKee C. F. (2004) *Astrophys. J., 603*, 383.
Tan J. C. et al. (2006) *Astrophys. J. Lett., 641*, L121.
Tan J. C. et al. (2013a) In *Molecular Gas, Dust, and Star Formation* (T. Wong and J. Ott, eds.), p. 19. IAU Symp. 292, Cambridge Univ., Cambridge.
Tan J. C. et al. (2013b) *Astrophys. J., 779*, 76.
Tanaka K. E. I. and Nakamoto T. (2011) *Astrophys. J. Lett., 739*, L50.
Tanaka K. E. I. et al. (2013) *Astrophys. J., 773*, 155.
Tang Y-W. et al. (2009) *Astrophys. J., 700*, 251.
Taquet V. et al. (2013) *Astron. Astrophys., 550*, 127.
Testi L. et al. (2010) *Astron. Astrophys., 522*, 44.
Torstensson K. J. E. et al. (2011) *Astron. Astrophys., 529*, 32.
Turner J. L. and Beck S. C. (2004) *Astrophys. J. Lett., 602*, L85.
Urban A. et al. (2010) *Astrophys. J., 710*, 1343.
Vaidya B. et al. (2011) *Astrophys. J., 742*, 56.
van der Tak F. F. S. et al. (2010) *Astron. Astrophys., 518*, L107.
van der Wiel M. H. D. et al. (2013) *Astron. Astrophys., 553*, 11.
Vasyunina T. et al. (2012) *Astrophys. J., 751*, 105.
Vink J. S. et al. (2001) *Astron. Astrophys., 369*, 574.
Viti S. et al. (2004) *Mon. Not. R. Astron. Soc., 354*, 1141.
Vlemmings W. et al. (2010) *Mon. Not. R. Astron. Soc., 404*, 134.
Wang K.-S. et al. (2012) *Astron. Astrophys., 543*, 22.
Wang P. et al. (2010) *Astrophys. J., 709*, 27.
Wang Y. et al. (2008) *Astrophys. J. Lett., 672*, L33.
Weidner C. et al. (2013) *Mon. Not. R. Astron. Soc., 434*, 84.
Werner M. W. et al. (1983) *Astrophys. J. Lett., 265*, L13.
Whitehouse S. C. and Bate M. R. (2004) *Mon. Not. R. Astron. Soc., 353*, 1078.
Williams J. P. and Garland C. A. (2002) *Astrophys. J., 568*, 259.
Wolfire M. G. and Cassinelli J. (1987) *Astrophys. J., 319*, 850.
Wood D. O. S. and Churchwell E. (1989) *Astrophys. J., 340*, 265.
Wu J. and Evans N. J. (2003) *Astrophys. J. Lett., 592*, L79.
Wu Y. et al. (2005a) *Astron. J., 129*, 330.
Wu Y. et al. (2005b) *Astrophys. J. Lett., 628*, L57.
Wu Y. et al. (2009) *Astrophys. J. Lett., 697*, L116.
Wyrowski F. et al. (2012) *Astron. Astrophys., 542*, L15.
Yıldız U. A. et al. (2013) *Astron. Astrophys., 556*, 89.
Yorke H. and Sonnhalter C. (2002) *Astrophys. J., 569*, 846.
Zapata L. A. (2008) *Astrophys. J. Lett., 479*, L25.
Zapata L. A. et al. (2013) *Astrophys. J. Lett., 765*, L29.
Zhang Q. et al. (2007) *Astrophys. J., 470*, 269.
Zhang Q. et al. (2009) *Astrophys. J., 696*, 268.
Zhang Y. and Tan J. C. (2011) *Astrophys. J., 733*, 55.
Zhang Y. et al. (2013a) *Astrophys. J., 766*, 86.
Zhang Y. et al. (2013b) *Astrophys. J., 767*, 58.
Zhang Y. et al. (2014) *Astrophys. J., 788*, 166.
Zinnecker H. and Yorke H. (2007) *Annu. Rev. Astron. Astrophys., 45*, 481.

Li Z.-Y., Banerjee R., Pudritz R. E., Jørgensen J. K., Shang H., Krasnopolsky R., and Maury A. (2014) The earliest stages of star and planet formation: Core collapse, and the formation of disks and outflows. In *Protostars and Planets VI* (H. Beuther et al., eds.), pp. 173–194. Univ. of Arizona, Tucson, DOI: 10.2458/azu_uapress_9780816531240-ch008.

The Earliest Stages of Star and Planet Formation: Core Collapse, and the Formation of Disks and Outflows

Zhi-Yun Li
University of Virginia

Robi Banerjee
Universität Hamburg

Ralph E. Pudritz
McMaster University

Jes K. Jørgensen
Copenhagen University

Hsien Shang and Ruben Krasnopolsky
Academia Sinica

Anaëlle Maury
Harvard-Smithsonian Center for Astrophysics

The formation of stars and planets are connected through disks. Our theoretical understanding of disk formation has undergone drastic changes in recent years, and we are on the brink of a revolution in disk observation enabled by the Atacama Large Millimeter Array (ALMA). Large rotationally supported circumstellar disks, although common around more evolved young stellar objects (YSOs), are rarely detected during the earliest, "class 0" phase; however, a few excellent candidates have been discovered recently around both low- and high-mass protostars. In this early phase, prominent outflows are ubiquitously observed; they are expected to be associated with at least small magnetized disks. Whether the paucity of large Keplerian disks is due to observational challenges or intrinsically different properties of the youngest disks is unclear. In this review, we focus on the observations and theory of the formation of early disks and outflows and their connections with the first phases of planet formation. Disk formation — once thought to be a simple consequence of the conservation of angular momentum during hydrodynamic core collapse — is far more subtle in magnetized gas. In this case, the rotation can be strongly magnetically braked. Indeed, both analytic arguments and numerical simulations have shown that disk formation is suppressed in the strict ideal magnetohydrodynamic (MHD) limit for the observed level of core magnetization. We review what is known about this "magnetic braking catastrophe," possible ways to resolve it, and the current status of early disk observations. Possible resolutions include non-ideal MHD effects (ambipolar diffusion, Ohmic dissipation, and the Hall effect), magnetic interchange instability in the inner part of protostellar accretion flow, turbulence, misalignment between the magnetic field and rotation axis, and depletion of the slowly rotating envelope by outflow stripping or accretion. Outflows are also intimately linked to disk formation; they are a natural product of magnetic fields and rotation and are important signposts of star formation. We review new developments on early outflow generation since *Protostars and Planets V* (*Reipurth et al.,* 2007). The properties of early disks and outflows are a key component of planet formation in its early stages and we review these major connections.

1. OVERVIEW

This review focuses on the earliest stages of star and planet formation, with an emphasis on the origins of early disks and outflows, and conditions that characterize the earliest stages of planet formation.

The importance of disks is obvious. They are the birthplace of planets, including those in our solar system. Nearly 1000 exoplanets have been discovered to date (*http://exoplanet.eu*) (see the chapters by Chabrier et al. and Helled et al. in this volume). The prevalence of planets indicates that disks must be common at least at some point in time around Sun-like stars. Observations show that this is indeed the case.

Direct evidence for circumstellar disks around young stellar objects (YSOs) first came from the Hubble Space Telescope (HST) observations of the so-called Orion "proplyds" (*O'Dell and Wen*, 1992; *McCaughrean and O'Dell*, 1996), where the disks are seen in silhouette against the bright background. More recently, with the advent of millimeter and submillimeter arrays, there is now clear evidence from molecular line observations that some protoplanetary disks have Keplerian velocity fields, indicating rotational support [e.g., Atacama Large Millimeter Array (ALMA) observations of TW Hydra; see section 2 on observations of early disks]. Indirect evidence, such as protostellar outflows and infrared excess in spectral energy distribution, indicates that the majority of, if not all, low-mass, Sun-like stars pass through a stage with disks, in agreement with the common occurrence of exoplanets.

Theoretically, disk formation — once thought to be a trivial consequence of the conservation of angular momentum during hydrodynamic core collapse — is far more subtle in magnetized gas. In the latter case, the rotation can be strongly magnetically braked. Indeed, disk formation is suppressed in the strict ideal magnetohydrodynamic (MHD) limit for the observed level of core magnetization; the angular momentum of the idealized collapsing core is nearly completely removed by magnetic braking close to the central object. How is this resolved?

We review what is known about this so-called "magnetic braking catastrophe" and its possible resolutions (section 3). Important processes to be discussed include non-ideal MHD effects (ambipolar diffusion, Ohmic dissipation, and the Hall effect), magnetic interchange instability in the inner part of protostellar accretion flow, turbulence, misalignment between the magnetic field and rotation axis, and depletion of the slowly rotating envelope by outflow stripping or accretion. We then turn to a discussion of the launch of the earliest outflows, and show that two aspects of such outflows — the outer magnetic "tower" and the inner centrifugally driven disk wind that have dominated much of the discussion and theory of early outflows — are actually two regimes of the same unified MHD theory (section 4). In section 5, we discuss the earliest phases of planet formation in such disks, which are likely quite massive and affected by the angular momentum transport via both strong spiral waves and powerful outflows. We synthesize the results in section 6.

2. OBSERVATIONS

2.1. Dense Cores

Dense cores are the basic units for the formation of at least low-mass stars. Their properties determine the characteristics of the disk, outflow, and planets — the byproducts of the star-formation process. Particularly important for the formation of the outflow and disk (and its embedded planets) are the core rotation rate and magnetic field strength. We begin our discussion of observations with these two key quantities, before moving on to early outflows and disks.

Rotation rate is typically inferred from the velocity gradient measured across a dense core (*Goodman et al.*, 1993; *Caselli et al.*, 2002). Whether the gradient traces true rotation or not is still under debate (*Bergin and Tafalla*, 2007; *Dib et al.*, 2010). For example, infall along a filament can mimic rotation signature (*Tobin et al.*, 2012a) [for synthetic line emission maps from filament accretion and their interpretation, see also *Smith et al.* (2012, 2013)]. If the gradient does trace rotation, then the rotational energy of the core would typically be a few percent of the gravitational energy. Such a rotation would not be fast enough to prevent the dense core from gravitational collapse. It is, however, more than enough to form a large, 10^2-AU-sized circumstellar disk, if angular momentum is conserved during the core collapse.

Magnetic fields are observed in the interstellar medium on a wide range of scales (see chapter by H.-B. Li et al. in this volume). Their dynamical importance relative to gravity is usually measured by the ratio of the mass of a region to the magnetic flux threading the region. On the core scale, the field strength was characterized by *Troland and Crutcher* (2008), who carried out an OH Zeeman survey of a sample of dense cores in nearby dark clouds. They inferred a median value $\lambda_{los} \approx 4.8$ for the dimensionless mass-to-flux ratio [in units of the critical value $(2\pi G^{1/2})^{-1}$ (*Nakano and Nakamura*, 1978)]. Geometric corrections should reduce the ratio to a typical value of $\lambda \approx 2$ (*Crutcher*, 2012). It corresponds to a ratio of magnetic to gravitational energy of tens of percent, since the ratio is given roughly by λ^{-2}. Such a field is not strong enough to prevent the core from collapsing into stars. It is, however, strong enough to dominate the rotation in terms of energy, and therefore is expected to strongly affect disk formation (section 3).

2.2. Early Outflows

Jets and outflows are observed during the formation of stars over the whole stellar spectrum, from brown dwarfs (e.g., *Whelan et al.*, 2005, 2012) to high-mass stars (e.g., *Motogi et al.*, 2013; *Carrasco-González et al.*, 2010; *Qiu et al.*, 2011, 2008, 2007; *Zhang et al.*, 2007), strongly indicating a universal launching mechanism at work (see section 4 and also the chapter by Frank et al. in this volume).

Young brown dwarfs have optical forbidden line spectra similar to those of low-mass young stars that are indicative of outflows (*Whelan et al.,* 2005, 2006). The inferred outflow speeds on the order of 40–80 km s^{-1} are somewhat lower than those of young stars (*Whelan et al.,* 2007; *Joergens et al.,* 2013). *Whelan et al.* (2012) suggest that brown dwarf outflows can be collimated and episodic, just as their low-mass star counterparts. There is some indication that the ratio of outflow and accretion rates, $\dot{M}_{out}/\dot{M}_{accr}$, is higher for young brown dwarf and very low-mass stars (*Comerón et al.,* 2003; *Whelan et al.,* 2009; *Bacciotti et al.,* 2011) than for the classical T Tauri stars (e.g., *Hartigan et al.,* 1995; *Sicilia-Aguilar et al.,* 2010; *Fang et al.,* 2009; *Ray et al.,* 2007). Whether this is generally true remains to be established.

Since planetary systems like the jovian system with its Galilean moons are thought to be built up from planetary subdisks (*Mohanty et al.,* 2007), one would naturally expect them to launch outflows as well (see, e.g., *Machida et al.,* 2006; *Liu and Schneider,* 2009). There is, however, no direct observational evidence yet for such circumplanetary disk-driven outflows.

At the other end of the mass spectrum, there is now evidence that outflows around young massive stars can be highly collimated, even at relative late evolutionary stages (e.g., *Carrasco-González et al.,* 2010; *Rodríguez et al.,* 2012; *Chibueze et al.,* 2012; *Palau et al.* 2013). For example, interferometric observations reveal that the young, luminous (~10^5 L$_\odot$) object IRAS 19520+2759 drives a well-collimated CO outflow, with a collimation factor of 5.7 (*Palau et al.,* 2013). It appears to have evolved beyond a central B-type object, but still drives a collimated outflow, in contrast with the expectation that massive YSO outflows decollimate as they evolve in time (*Beuther and Shepherd,* 2005). Interestingly, an H II region has yet to develop in this source. It may be quenched by protostellar accretion flow (*Keto,* 2002, 2003; *Peters et al.,* 2010, 2011) or absent because the central stellar object is puffed up by rapid accretion [and thus not hot enough at the surface to produce ionizing radiation (*Hosokawa et al.,* 2010)].

2.3. Early Disks

From an observational point of view, the key question to address is when rotationally supported circumstellar disks are first established and become observable. It is clear that, after approximately 0.5 million years (m.y.) (*Evans et al.,* 2009), gaseous Keplerian (protoplanetary) disks are present on the scales of ~100–500 AU around both low- and intermediate-mass stars ["T Tauri" and "Herbig Ae" stars, respectively — or class II YSOs (see *Dutrey et al.,* 2007)]. Whether they are present at earlier times requires high-resolution studies of the youngest protostars, e.g., class 0 and class I objects.

One way to constrain the process of disk formation is to study the rotation rates on difference scales (see the caveats in inferring rotation rate in section 2.1). From larger to smaller scales, a clear progression in kinematics is evident [Fig. 1, reproduced from *Belloche* (2013)]: At large distances from the central protostar and in prestellar cores, the specific angular momentum decreases rapidly toward smaller radii, implying that the angular velocity is roughly constant (e.g., *Goodman et al.,* 1993; *Belloche et al.,* 2002). Observations of objects in relatively late stages of evolution suggest that the specific angular momentum tends to a constant value ($v_{rot} \propto r^{-1}$) between ~10^2 and 10^4 AU, as expected from conservation of angular momentum under infall. A rotationally supported disk is expected to show increasing specific angular momentum as a function of radius (Keplerian rotation, $v_{rot} \propto r^{-0.5}$). To characterize the properties of disks being formed, the task at hand is to search for the location where the latter two regimes, Keplerian disk with $v_{rot} \propto r^{-0.5}$ and the infalling envelope with $v_{rot} \propto r^{-1}$, separate.

2.3.1. Techniques. Observationally the main challenge in revealing the earliest stages of the circumstellar disks is the presence of the larger-scale protostellar envelopes during the class 0 and I stages, which reprocess most of the emission from the central protostellar object itself at shorter wavelengths and easily dominate the total flux at longer wavelengths. The key observational tools at different wavelength regimes are described below (for convenience we define the near-infrared as $\lambda < 3$ μm, mid-infrared as 3 μm $< \lambda < 50$ μm, far-infrared as 50 μm $< \lambda < 250$ μm, and (sub)millimeter as 250 μm $< \lambda < 4$ mm):

1. Mid-infrared: At mid-infrared wavelengths the observational signatures of young stars are dominated by the balance between the presence of warm dust and degree of extinction on small scales — and in particular, highly sensitive observations with the Spitzer Space Telescope in 2003–2009 have been instrumental in this field. One of the key results relevant to disk formation during the embedded stages is that simple infalling envelope profiles (e.g., $\rho \propto r^{-2}$ or $r^{-1.5}$) cannot extend unmodified to less than approximately a few hundred astronomical unit scales (*Jørgensen et al.,* 2005b; *Enoch et al.,* 2009): The embedded protostars are brighter in the mid-infrared than expected from such profiles, which can, for example, be explained if the envelope is flattened or has a cavity on small scales. Indeed, extinction maps reveal that protostellar environments are complex on 10^3-AU scales (*Tobin et al.,* 2010) and in some cases show asymmetric and filamentary structures that complicate the canonical picture of formation of stars from the collapse of relatively spherical dense cores.

2. Far-infrared: At far-infrared wavelengths the continuum emission comes mainly from thermal dust grains with temperatures of a few tens of Kelvins. This wavelength range is accessible almost exclusively from space only. Consequently, observatories such as the Infrared Space Observatory (ISO) and Herschel Space Observatory (*Pilbratt et al.,* 2010) have been the main tools for characterizing protostars there. At the time of writing, surveys from the Herschel Space Observatory are starting to produce large samples of deeply embedded protostellar cores that can be followed up by other instruments. The observations at far-infrared wavelengths provide important information about the peak of the luminosity of the embedded protostars and

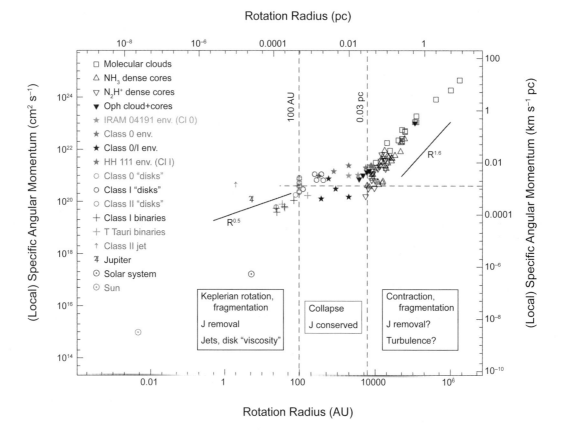

Fig. 1. See Plate 11 for color version. Progression of specific angular momentum as function of scale and/or evolutionary stage of YSOs. From *Belloche* (2013).

the distribution of low surface brightness dust — but due to the limitation in angular resolution, provide less information on the few hundred astronomical unit scales of disks around more evolved YSOs.

3. (Sub)millimeter: The (sub)millimeter wavelengths provide a unique window on the thermal radiation from the cooler dust grains on small scales (*André et al.,* 1993; *Chandler and Richer,* 2000). Aperture synthesis observations at these wavelengths resolve scales down to ~100 AU or better in nearby star-forming regions. The flux from the thermal dust continuum emission is strongly increasing with frequency ν as $F_\nu \propto \nu^2$ or steeper, making these wavelengths ideally suited for detecting dense structures while discriminating from possible free-free emission. Likewise the high-spectral resolution for a wide range of molecular rotational transitions can be tailored to study the structure (e.g., temperature and kinematics) of the different components in the protostellar systems (*Jørgensen et al.,* 2005a).

2.3.2. Millimeter continuum surveys. The main observational tool for understanding disk formation is millimeter surveys using aperture synthesis techniques that probe how the matter is distributed on the few hundred astronomical unit scales. The use of such a technique to address this question goes back to *Keene and Masson* (1990) and *Terebey et al.* (1993), with the first larger arcsecond-scale surveys appearing in the early 2000s (e.g., *Looney et al.,*

2000) and detailed radiative transfer modeling appearing about the same time (e.g., *Hogerheijde and Sandell,* 2000; *Harvey et al.,* 2003; *Looney et al.,* 2003; *Jørgensen et al.,* 2004, 2005a).

The general conclusion from these studies is that, in most cases, both the large- and small-scale continuum emission cannot be reproduced by a single analytical model of a simple, axisymmetric envelope. In some cases (e.g., *Brown et al.,* 2000; *Jørgensen et al.,* 2005a; *Enoch et al.,* 2009), these structures are well resolved on a few hundred astronomical unit scales. Some noteworthy counterexamples where no additional dust continuum components are required include L483 (*Jørgensen,* 2004), L723 (*Girart et al.,* 2009), and three of the nine class 0 protostars in Serpens surveyed by *Enoch et al.* (2011). Typically the masses for individual objects derived in these studies agree well with each other once similar dust opacities and temperatures are adopted. *Jørgensen et al.* (2009) compared the dust components for a sample of 18 embedded class 0 and I protostars and did not find an increase in mass of the modeled compact component with bolometric temperature as one might expect from the growth of Keplerian disks. *Jørgensen et al.* (2009) suggested that this could reflect the presence of the rapid formation of disk-like structure around the most deeply embedded protostars, although the exact kinematics of those around the most deeply embedded (class 0) sources were unclear. An unbiased survey of embedded protostars in Serpens with envelope masses larger than 0.25 $M_\odot$ and luminosities larger than 0.05 $L_\odot$ by *Enoch et al.* (2011) finds similar masses for the compact structures around the sources in that sample. Generally these masses are found to be small relative to the larger-scale envelopes in the class 0 stage on 10,000-AU scales — but still typically 1–2 orders of magnitude larger than the mass on similar, few hundred astronomical unit scales extrapolated from the envelope.

Still, the continuum observations do not provide an unambiguous answer to what these compact components represent. By the class I stage some become the Keplerian disks surrounding the protostars. The compact components around the class 0 protostars could be the precursors to these Keplerian disks. However, it is unlikely that such massive rotationally supported disks could be stably supported given the expected low stellar mass for the class 0 protostars: They should be prone to fragmentation (see section 5).

An alternative explanation for the compact dust emission detected in interferometric continuum observations may be the presence of "pseudo-disks." In the presence of magnetic fields, torsional Alfvén waves in twisted field lines carry away angular momentum, preventing the otherwise natural formation of large rotationally supported disks. However, strong magnetic pinching forces deflect infalling gas toward the equatorial plane to form a flattened structure — the "pseudo-disk" (*Galli and Shu*, 1993; *Allen et al.*, 2003; *Fromang et al.*, 2006). Unlike Keplerian disks observed at later stages, this flattened inner envelope is not supported by rotational motions, but can be partially supported by magnetic fields. Observationally disentangling disks and pseudodisks is of paramount importance since accretion onto the protostar proceeds very differently through a rotationally supported disk or a magnetically induced pseudo-disk.

Maury et al. (2010) compared the results from an Institut de Radioastronomie Millimétrique (IRAM) Plateau de Bure study of five class 0 to synthetic model images from three numerical simulations — in particular focusing on binarity and structure down to scales of ~50 AU. The comparison shows that magnetized models of protostar formation including pseudo-disks agree better with the observations than, e.g., pure hydrodynamical simulation in the case of no initial perturbation or turbulence. With turbulence, magnetized models can produce small disks (~100 AU) (*Seifried et al.*, 2013), which might still be compatible with the observations by *Maury et al.* (2010). The compact continuum components could also represent the "magnetic walls" modeled by *Tassis and Mouschovias* (2005), although in some sources excess unresolved emission remains unaccounted for in this model (*Chiang et al.*, 2008).

2.3.3. Kinematics. A number of more evolved class I YSOs show velocity gradients that are well fit by Keplerian profile (e.g., *Brinch et al.*, 2007; *Lommen et al.*, 2008; *Jørgensen et al.*, 2009) [see also *Harsono et al.* (2014) and *Lee* (2011) for evidence of Keplerian disks around class I object TMC1A and HH 111, respectively]. Generally the problem in studying the kinematics on disk scales in embed-

ded protostars is that many of the traditional line tracers are optically thick in the larger-scale envelope. *Jørgensen et al.* (2009), for example, showed that the emission from the (sub) millimeter transitions of HCO^+ would be optically thick on scales of ~100 AU for envelopes with masses larger than 0.1 $M_\odot$. An alternative is to trace less-abundant isotopologs. Recently one embedded protostar, L1527, was found to show Keplerian rotation in $^{13}CO/C^{18}O$ 2–1 (*Tobin et al.*, 2012b) [see also *Murillo et al.* (2013) for VLA1623A]; this result was subsequently strengthened by ALMA observations (N. Ohashi, personal communication; see also *http://www.almasc.org/upload/presentations/DC-05.pdf*). Another example of Keplerian motions is found in the protostellar binary L1551-NE (a borderline class 0/I source), in $^{13}CO/C^{18}O$ 2–1 (*Takakuwa et al.*, 2012). Large circumstellar disks are inferred around a number of high-mass YSOs, including G31.41+0.31 (*Cesaroni et al.*, 1994), IRAS 20126+4104 (*Cesaroni et al.*, 1997), IRAS 18089-1732 (*Beuther et al.*, 2004), IRAS 16547-4247 (*Franco-Hernández et al.*, 2009) and IRAS 18162-2048 (*Fernández-López et al.*, 2011). Whether they are rotationally supported remains uncertain.

Brinch et al. (2009) investigated the dynamics of the deeply embedded protostar NGC 1333-IRAS2A in subarcsecond-resolution observations of HCN (4–3) and the same line of its isotopolog $H^{13}CN$. Through detailed line radiative transfer modeling they showed that the ~300-AU compact structure seen in dust continuum was in fact dominated by infall rather than rotation.

Pineda et al. (2012) presented some of the first ALMA observations of the deeply embedded protostellar binary IRAS 16293-2422. These sensitive observations revealed a velocity gradient across one component in the binary, IRAS 16293A, in lines of the complex organic molecule methyl formate. However, this velocity gradient does not reflect Keplerian rotation and does not require a central mass beyond the envelope mass enclosed on the same scale. This is also true for the less-dense gas traced by lines of the rare $C^{17}O$ and $C^{34}S$ isotopologs in extended Submillimeter Array (eSMA) observations (0.5″ resolution) (*Favre et al.*, 2014).

The above observations paint a picture of complex structure of the material on small scales around low-mass protostellar systems. They raise a number of important potential implications, which we discuss next.

2.4. Implications and Outlook

2.4.1. Protostellar mass evolution. An important constraint on the evolution of protostars is how the bulk mass is transported and accreted from the larger scales through the circumstellar disks onto the central stars (section 5). An important diagnostic from the above observations is to compare the disk masses — either from dust continuum or line observations (taking into account the caveats about the dust properties and/or chemistry) to stellar masses inferred, e.g., from the disk dynamical profiles. *Jørgensen et al.* (2009) compared these quantities for a sample of predominantly class I YSOs with well-established disks; an updated version of this figure including recent measurements of dynamical masses for additional sources (*Tobin et al.,* 2012b; *Takakuwa et al.,* 2012) is shown in Fig. 2. These measurements are compared to standard semi-analytic models for collapsing rotating protostars (*Visser et al.,* 2009); these models typically underestimate the stellar masses relative to the disk masses. These simple models of collapse of largely spherical cores are likely inapplicable on larger scales where filamentary structures are sometimes observed (*Tobin et al.,* 2010; *Lee et al.,* 2012). Still, this comparison illustrates a potential avenue to explore when observations of a large sample of embedded protostars become available in the ALMA era; it provides direct measures of the accretion rates that more sophisticated numerical simulations need to reproduce.

2.4.2. Grain growth. Multi-wavelength continuum observations in the millimeter and submillimeter are also interesting for studying the grain properties close to the newly formed protostars. In more evolved circumstellar disks a flattened slope of the spectral energy distribution at millimeter or longer wavelengths is taken as evidence that significant grain growth has taken place (for a review, see, e.g., *Natta et al.,* 2007). On larger scales of protostellar envelope where the continuum emission is optically thin, interstellar medium (ISM)-like dust would result in submillimeter spectral slopes of 3.5–4. The more compact dust components observed at a few hundred astronomical unit scales have lower spectral indices of 2.5–3.0 (*Jørgensen et al.,* 2007; *Kwon et al.,* 2009; *Chiang et al.,* 2012), either indicating that some growth of dust grains to millimeter sizes has occurred or that the compact components are optically thick. At least in a few cases where the dust emission is clearly resolved, the inferred spectral slopes are in agreement with those observed for more-evolved T Tauri stars, indicating that dust rapidly grows to millimeter sizes (*Ricci et al.,* 2010). Assessing the occurrence of grain growth during the embedded stages is important for not only understanding the formation of the seeds for planetesimals (section 5), but also evaluating the non-ideal MHD effects in magnetized core collapse and disk formation, since they depend on the ionization level, which in turn is strongly affected by the grain size distribution (section 3).

2.4.3. Chemistry. The presence of disk-like structures on 100-AU scales may also have important implications for the chemistry in those regions. The presence of a disk may change the temperature, allowing molecules to freeze out again that would otherwise stay in the gas phase. Water and its isotopologs are a particularly clear example of these effects: For example, in observations of the $H_2^{18}O$ isotopolog toward the centers of a small sample of class 0 protostars, *Jørgensen and van Dishoeck* (2010) and *Persson et al.* (2012) found lower H_2O abundances than expected in the typical gas phase, consistent with a picture in which a significant fraction of the material at scales ≤100 AU has temperatures lower than ~100 K. This complicates the interpretation of the chemistry throughout the envelope. For example, it makes extrapolations of envelope physical and abundance structures from larger scales invalid (*Visser et al.,* 2013). An ongoing challenge is to better constrain the physical structure of the protostellar envelopes and disks on these scales. This must be done before significant progress can be made on understanding the initial conditions for chemical evolution in protoplanetary disks. The challenge is even more formidable for massive stars, although *Isokoski et al.* (2013) did not find any chemical differentiation between massive stars with and without disk-like structures.

2.4.4. Linking observations and theory. Observations suggest that, in principle, there is typically more than enough angular momentum on the core scale to form large, 10^2-AU-scale rotationally supported disks. The common presence of fast jets around deeply embedded protostars implies that the formation of such disks has begun early in the process of star formation. There is, however, currently little direct evidence that large, well-developed, Keplerian disks are prevalent around class 0 protostars, as one may naively expect based on angular momentum conservation during hydrodynamic core collapse. The paucity of large, early disks indicates that disk formation is not as straightforward as generally expected. The most likely reason is that star-forming cores are observed to be significantly magnetized, and magnetic fields are known to interact strongly with rotation. They greatly affect, perhaps even control, disk formation, as we show next.

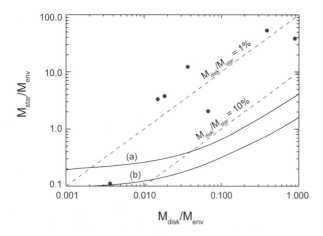

Fig. 2. Updated version of Fig. 18 from *Jørgensen et al.* (2009). Predicted stellar mass, M_{star}, vs. disk mass, M_{disk}, both measured relative to the envelope mass M_{env} in the models of *Visser et al.* (2009) with $\Omega_0 = 10^{-14}$ s^{-1} and c_s of **(a)** 0.19 km s^{-1} and **(b)** 0.26 km s^{-1} (solid lines). The sources for which stellar, disk, and envelope masses are measured are shown with dots (*Jørgensen et al.,* 2009; *Takakuwa et al.,* 2012; *Tobin et al.,* 2012b). The dashed lines indicate M_{disk}/M_{star} ratios of 1% (upper) and 10% (lower).

3. THEORY OF MAGNETIZED DISK FORMATION

How disks form has been a long-standing theoretical problem in star formation. Early work on this topic was reviewed by *Bodenheimer* (1995) and *Boss* (1998). Both reviews listed a number of unsolved problems. Topping both lists was the effect of the magnetic field, which turns out to present a formidable obstacle to disk formation. Substantial progress has been made in recent years in overcoming this obstacle, especially through Ohmic dissipation, field rotation misalignment, and turbulence. This progress will be summarized below.

3.1. Magnetic Braking Catastrophe in Ideal Magnetohydrodynamic Limit

The basic difficulty with disk formation in magnetized dense cores can be illustrated using analytic arguments in the strict ideal MHD limit where the magnetic field lines are perfectly frozen into the core material. In this limit, as a finite amount of mass is accreted onto the central object (the protostar), a finite amount of magnetic flux will be dragged into the object as well. The magnetic flux accumulated at the center forms a magnetic split monopole, with the field lines fanning out radially (*Galli et al.*, 2006; see sketch in their Fig. 1). As a result, the magnetic field strength increases rapidly with decreasing distance to the center, as $B \propto r^{-2}$. The magnetic energy density, which is proportional to the field strength squared, increases with decreasing radius even more rapidly, as $E_B \propto r^{-4}$. This increase is faster than, for example, the energy density of the accretion flow, which can be estimated approximately from spherical free-fall collapse as $E_{ff} \propto r^{-5/2}$. As the infalling material approaches the central object, it will become completely dominated by the magnetic field sooner or later. The strong magnetic field at small radii is able to remove all the angular momentum in the collapsing flow, leading to the so-called "magnetic braking catastrophe" for disk formation (*Galli et al.*, 2006). The braking occurs naturally in a magnetized collapsing core, because the faster-rotating matter that has already collapsed closer to the rotation axis remains connected to the more slowly rotating material at larger (cylindrical) distances through field lines. The differential rotation generates a field-line twist that brakes the rotation of the inner, faster-rotating part and transports its angular momentum outward [see *Mouschovias and Paleologou* (1979, 1980) for analytic illustrations of magnetic braking].

The catastrophic braking of disks in magnetized dense cores in the ideal MHD limit was also found in many numerical as well as semi-analytic calculations. *Krasnopolsky and Königl* (2002) were the first to show semi-analytically, using the so-called "thin-disk" approximation, that the formation of rotationally supported disks (RSDs) can be suppressed if the efficiency of magnetic braking is large enough. However, the braking efficiency was parameterized rather than computed self-consistently. Similarly, *Dapp and Basu* (2010) and *Dapp et al.* (2012) demonstrated that, in the absence of any magnetic diffusivity, a magnetic split monopole is produced at the center and RSD formation is suppressed, again under the thin-disk approximation.

Indirect evidence for potential difficulty with disk formation in ideal MHD simulations came from *Tomisaka*

(2000), who studied the (two-dimensional) collapse of a rotating, magnetized dense core using a nested grid under the assumption of axisymmetry. He found that, while there is little magnetic braking during the phase of runaway core collapse leading up to the formation of a central object, once an outflow is launched, the specific angular momentum of the material at the highest densities is reduced by a large factor (up to ~10^4) from the initial value. The severity of the magnetic braking and its deleterious effect on disk formation were not fully appreciated until *Allen et al.* (2003) explicitly demonstrated that the formation of a large, numerically resolvable, rotationally supported disk was completely suppressed in two dimensions by a moderately strong magnetic field (corresponding to a dimensionless mass-to-flux ratio λ of several) in an initially self-similar, rotating, magnetized, singular isothermal toroid (*Li and Shu*, 1996). They identified two key ingredients behind the efficient braking during the accretion phase: (1) concentration of the field lines at small radii by the collapsing flow, which increases the field strength, and (2) the fanning out of field lines due to equatorial pinching, which increases the lever arm for magnetic braking.

The catastrophic braking that prevents the formation of RSDs during the protostellar accretion phase has been confirmed in several subsequent two-dimensional and three-dimensional ideal MHD simulations (*Mellon and Li*, 2008; *Hennebelle and Fromang*, 2008; *Duffin and Pudritz*, 2009; *Seifried et al.*, 2012a; *Santos-Lima et al.*, 2012). *Mellon and Li* (2008), in particular, formulated the disk formation problem in the same way as *Allen et al.* (2003), by adopting a self-similar rotating, magnetized, singular isothermal toroid as the initial configuration. Although idealized, the adopted initial configuration has the advantage that the subsequent core collapse should remain self-similar. The self-similarity provides a useful check on the correctness of the numerically obtained solution. They found that the disk formation was suppressed by a field as weak as $\lambda = 13.3$.

Hennebelle and Fromang (2008) carried out three-dimensional simulations of the collapse of a rotating dense core of uniform density and magnetic field into the protostellar accretion phase using an ideal MHD adaptive mesh refinement (AMR) code (RAMSES). They found that the formation of a RSD is suppressed as long as λ is on the order of 5 or less. However, the $\lambda = 5$ case in *Price and Bate*'s (2007) smoothed particle magnetohydrodynamics (SPMHD) simulations appears to have formed a small disk (judging from the column density distribution). It is unclear whether the disk is rotationally supported or not, since the disk rotation rate was not given in the paper. Furthermore, contrary to the grid-based simulations, there appears little, if any, outflow driven by twisted field lines in the smoothed particle hydrodynamics (SPH) simulations, indicating that the efficiency of magnetic braking is underestimated [prominent outflows are produced, however, in other SPMHD simulations (e.g., *Bürzle et al.*, 2011; *Price et al.*, 2012)]. Another apparently discrepant result is that of *Machida et al.* (2011). They managed to form a large, 10^2-AU scale, rotationally supported disk for a very strongly magnetized core of λ = 1 even in the ideal MHD limit (their model 4) using a nested grid and sink particle. This contradicts the results from other simulations and semi-analytic calculations.

To summarize, both numerical simulations and analytic arguments support the notion that, in the ideal MHD limit, catastrophic braking makes it difficult to form rotationally supported disks in (laminar) dense cores magnetized to a realistic level (with a typical λ of a few). In what follows, we will explore the potential resolutions that have been proposed in the literature to date.

3.2. Non-Ideal Magnetohydrodynamic Effects

Dense cores of molecular clouds are lightly ionized [with a typical electron fractional abundance on the order of 10^{-7} (*Bergin and Tafalla*, 2007)]. As such, the magnetic field is not expected to be perfectly frozen into the bulk neutral material. There are three well-known non-ideal MHD effects that can in principle break the flux-freezing condition that lies at the heart of the magnetic braking catastrophe in the strict ideal MHD limit. They are ambipolar diffusion, the Hall effect, and Ohmic dissipation (for a review, see *Armitage,* 2011). Roughly speaking, in the simplest case of an electron-ion-neutral medium, both ions and electrons are well tied to the magnetic field in the ambipolar diffusion regime. In the Hall regime, electrons remain well tied to the field, but not ions. At the highest densities, both electrons and ions are knocked off the field lines by collisions before they finish a complete gyration; in such a case, Ohmic dissipation dominates. This simple picture is complicated by dust grains, whose size distribution in dense cores is relatively uncertain (see section 2.4), but that can become the dominant charge carriers. Under typical cloud conditions, ambipolar diffusion dominates over the other two effects at densities typical of cores (e.g., *Nakano et al.*, 2002; *Kunz and Mouschovias,* 2010). It is the most widely studied non-ideal MHD effect in the context of core formation and evolution in the so-called "standard" picture of low-mass star formation out of magnetically supported clouds (*Nakano,* 1984; *Shu et al.,* 1987; *Mouschovias and Ciolek,* 1999). We will first concentrate on this effect.

Ambipolar diffusion enables the magnetic field lines that are tied to the ions to drift relative to the bulk neutral material. In the context of disk formation, its most important effect is to redistribute the magnetic flux that would have been dragged into the central object in the ideal MHD limit to a circumstellar region where the magnetic field strength is greatly enhanced (*Li and McKee,* 1996). Indeed, the enhanced circumstellar magnetic field is strong enough to drive a hydromagnetic shock into the protostellar accretion flow (*Li and McKee,* 1996; *Ciolek and Königl,* 1998; *Contopoulos et al.* 1998; *Krasnopolsky and Königl,* 2002; *Tassis and Mouschovias,* 2007; *Dapp et al.,* 2012). *Krasnopolsky and Königl* (2002) showed semi-analytically, using the one-dimensional thin-disk approximation, that disk formation may be suppressed in the strongly magnetized post-shock

region if the magnetic braking is efficient enough. The braking efficiency, parameterized in *Krasnopolsky and Königl* (2002), was computed self-consistently in the two-dimensional (axisymmetric) simulations of *Mellon and Li* (2009), which were performed under the usual assumption of ion density proportional to the square root of neutral density. Three-dimensional simulations of ambipolar diffusion were performed by *Duffin and Pudritz* (2009) using a specially developed, single-fluid AMR code (*Duffin and Pudritz,* 2008), as well as by a two-fluid SPH code (*Hosking and Whitworth,* 2004). *Mellon and Li* (2009) found that ambipolar diffusion does not weaken the magnetic braking enough to allow rotationally supported disks to form for realistic levels of cloud core magnetization and cosmic-ray ionization rate. In many cases, the magnetic braking is even enhanced. These findings were strengthened by *Li et al.* (2011), who computed the ion density self-consistently using the simplified chemical network of Nakano and collaborators (*Nakano et al.* 2002; *Nishi et al.* 1991), which includes dust grains. An example of their simulations is shown in Fig. 3. It shows clearly the rapid slowdown of the infalling material near the ambipolar diffusion-induced shock (located at a radius of $\sim 10^{15}$ cm in this particular example) and the nearly complete braking of the rotation in the post-shock region, which prevents the formation of a RSD. The suppression of RSD is also evident from

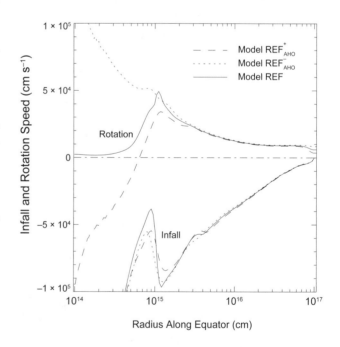

Fig. 3. Infall and rotation speeds along the equatorial plane of a collapsing rotating, magnetized dense core during the protostellar mass accretion phase for three representative models of *Li et al.* (2011). Model REF (solid lines) includes only ambipolar diffusion, whereas the other two include all three non-ideal MHD effects, especially the Hall effect. The initial magnetic field and rotation axis are in the same direction in one model (REF+, dashed lines) and in opposite directions in the other (REF−, dotted).

the fast, supersonic infall close to the central object. We should note that RSD formation may still be possible if the cosmic-ray ionization rate can be reduced well below the canonical value of 10^{-17} s^{-1} (*Mellon and Li,* 2009), through, for example, the magnetic mirroring effect, which may turn a large fraction of the incoming cosmic rays back before they reach the disk-forming region (*Padovani et al.,* 2013).

As the density increases, the Hall effect tends to become more important (the exact density for this to happen depends on the grain size distribution). It is less explored than ambipolar diffusion in the star-formation literature. A unique feature of this effect is that it can actively increase the angular momentum of a collapsing, magnetized flow through the so-called "Hall spin-up" (*Wardle and Ng,* 1999). In the simplest case of electron-ion-neutral fluid, the spin-up is caused by the current carriers (the electrons) moving in the azimuthal direction, generating a magnetic torque through field twisting; the toroidal current is produced by gravitational collapse, which drags the poloidal field into a pinched, hourglass-like configuration. The Hall spin-up was studied numerically by *Krasnopolsky et al.* (2011) and semi-analytically by *Braiding and Wardle* (2012a,b). *Krasnopolsky et al.* (2011) showed that a rotationally supported disk can form even in an initially *nonrotating* core, provided that the Hall coefficient is large enough. Interestingly, when the direction of the initial magnetic field in the core is flipped, the disk rotation is reversed. This reversal of rotation is also evident in Fig. 3, where the Hall effect spins up the nearly nonrotating material in the post-ambipolar diffusion shock region to highly supersonic speeds, but in different directions depending on the field orientation. The Hall effect, although dynamically significant, does not appear capable of forming a rotationally supported disk under typical dense core conditions according to *Li et al.* (2011). This inability is illustrated in Fig. 3, where the equatorial material collapses supersonically on the 10^2-AU scale even when the Hall effect is present.

Ohmic dissipation becomes the dominant non-ideal MHD effect at high densities (e.g., *Nakano et al.,* 2002). It has been investigated by different groups in connection with disk formation. *Shu et al.* (2006) studied semi-analytically the effects of a spatially uniform resistivity on the magnetic field structure during the protostellar mass accretion phase. They found that, close to the central object, the magnetic field decouples from the collapsing material and becomes more or less uniform. They suggested that a rotationally supported disk may form in the decoupled region, especially if the resistivity is higher than the classic (microscopic) value. This suggestion was confirmed by *Krasnopolsky et al.* (2010) (see also *Santos-Lima et al.,* 2012), who found numerically that a large, 10^2-AU-scale, Keplerian disk can form around a 0.5-$M_\odot$ star, provided that the resistivity is on the order of 10^{19} cm^2 s^{-1} or more; such a resistivity is significantly higher than the classic (microscopic) value over most of the density range relevant to disk formation.

Machida and Matsumoto (2011) and *Machida et al.* (2011) studied disk formation in magnetized cores including

only the classic value of resistivity estimated from *Nakano et al.*'s (2002) numerical results. The former study found that a relatively small, 10-AU-scale, rotationally supported disk formed within a few years after the formation of the stellar core. Inside the disk, the density is high enough for magnetic decoupling to occur due to Ohmic dissipation. This work was extended to much later times by *Machida et al.* (2011), who included a central sink region in the simulations. They concluded that the small RSD can grow to large, 10^2-AU size at later times, especially after the most of the envelope material has fallen onto the disk and the central object. A caveat, pointed out by *Tomida et al.* (2013) (see also *Dapp and Basu,* 2010), is that they used a form of induction equation that is, strictly speaking, inappropriate for the nonconstant resistivity adopted in their models; it may generate magnetic monopoles that are subsequently cleaned away using the Dedner's method (*Dedner et al.,* 2002). This deficiency was corrected in *Tomida et al.* (2013), who carried out radiative MHD simulations of magnetized core collapse to a time shortly (~1 yr) after the formation of the second (protostellar) core. They found that the formation of a (small, astronomical-unit-scale) rotationally supported disk was suppressed by magnetic braking in the ideal MHD limit but was enabled by Ohmic dissipation at this early time; the latter result is in qualitative agreement with *Machida and Matsumoto* (2011) and *Machida et al.* (2011), although it remains to be seen how the small disks in *Tomida et al.*'s (2013) simulations evolve further in time.

Dapp and Basu (2010) studied the effects of Ohmic dissipation on disk formation semi-analytically, using the "thin-disk" approximation for the mass distribution and an approximate treatment of magnetic braking. The approximations enabled them to follow the formation of both the first and second core. They found that a small, subastronomical-unit, rotationally supported disk was able to form soon after the formation of the second core in the presence of Ohmic dissipation; it was suppressed in the ideal MHD limit, in agreement with the later three-dimensional simulations of *Tomida et al.* (2013). This work was extended by *Dapp et al.* (2012) to include a set of self-consistently computed charge densities from a simplified chemical network and ambipolar diffusion. They showed that their earlier conclusion that a small, subastronomical-unit-scale, RSD is formed through Ohmic dissipation holds even in the presence of a realistic level of ambipolar diffusion. This conclusion appears reasonably secure in view of the broad agreement between the semi-analytic work and numerical simulations. When and how such disks grow to the much larger, 10-AU-scale, size deserve to be explored more fully.

3.3. Magnetic Interchange Instabilities

The formation of a large-scale RSD in a magnetized core is made difficult by the accumulation of magnetic flux near the accreting protostar. As discussed earlier, this is especially true in the presence of a realistic level of ambipolar diffusion, which redistributes the magnetic flux that would have been dragged into the central object to the circumstellar region [Ohmic dissipation has a similar effect; see *Li et al.* (2011) and *Dapp et al.* (2012)]. The result of the flux redistribution is the creation of a strongly magnetized region close to the protostar where the infall speed of the accreting flow is slowed down to well below the free-fall value (i.e., it is effectively held up by magnetic tension against the gravity of the central object), at least in two dimensions (assuming axisymmetry). It has long been suspected that such a magnetically supported structure would become unstable to interchange instabilities in three dimensions (*Li and McKee*, 1996; *Krasnopolsky and König*, 2002). Recent three-dimensional simulations have shown that this is indeed the case.

Magnetic interchange instability in a protostellar accretion flow driven by flux redistribution was first studied in detail by *Zhao et al.* (2011). They treated the flux redistribution through a sink particle treatment: When the mass in a cell is accreted onto a sink particle, the magnetic field is left behind in the cell (see also *Seifried et al.*, 2011; *Cunningham et al.*, 2012); it is a crude representation of the matter-field decoupling expected at high densities [on the order of 10^{12} cm or higher (*Nakano et al.*, 2002; *Kunz and Mouschovias*, 2010)]. The decoupled flux piles up near the sink particle, leading to a high magnetic pressure that is released through the escape of field lines along the directions of least resistance. As a result, the magnetic flux dragged into the decoupling region near the protostar along some azimuthal directions is advected back out along other directions in highly magnetized, low-density, expanding regions. Such regions are termed decoupling-enabled magnetic structure (DEMS) by *Zhao et al.* (2011); they appear to be present in the formally ideal MHD simulations of *Seifried et al.* (2011), *Cunningham et al.* (2012), and *Joos et al.* (2012) as well.

Krasnopolsky et al. (2012) improved upon the work of *Zhao et al.* (2011) by including two of the physical processes that can lead to magnetic decoupling: ambipolar diffusion and Ohmic dissipation. They found that the basic conclusion of *Zhao et al.* (2011), that the inner part of the protostellar accretion flow is driven unstable by magnetic flux redistribution, continues to hold in the presence of realistic levels of non-ideal MHD effects (see Fig. 4 for an illustrative example). The magnetic flux accumulated near the center is transported outward not only diffusively by the microscopic non-ideal effects, but also advectively through the bulk motions of the strongly magnetized expanding regions (the DEMS) generated by the instability. The advective flux redistribution in three dimensions lowers the field strength at small radii compared to the two-dimensional (axisymmetric) case where the instability is suppressed. It makes the magnetic braking less efficient and the formation of a RSD easier in principle. In practice, the magnetic interchange instability does not appear to enable the formation of rotationally supported disks by itself, because the highly magnetized DEMS that it creates remain trapped at relatively small distances from the protostar by the proto-

stellar accretion flow (see Fig. 4); the strong magnetic field inside the DEMS blocks the accretion flow from rotating freely around the center object to form a complete disk.

3.4. Magnetic Field-Rotation Misalignment

Misalignment between the magnetic field and rotation axis as a way to form large RSDs has been explored extensively by Hennebelle and collaborators (*Hennebelle and Ciardi*, 2009; *Ciardi and Hennebelle*, 2010; *Joos et al.* 2012; see also *Machida et al.*, 2006; *Price and Bate*, 2007; *Boss and Keiser*, 2013). The misalignment is expected if the angular momenta of dense cores are generated through turbulent motions (e.g., *Burkert and Bodenheimer*, 2000; *Seifried et al.*, 2012b; *Myers et al.*, 2013; *Joos et al.*, 2013). Plausible observational evidence for it was recently uncovered by *Hull et al.* (2013) using the Combined Array for Research in Millimeter-wave Astronomy (CARMA), who found that the distribution of the angle between the magnetic field on the 10^3-AU scale and the bipolar outflow axis (taken as a proxy for the rotation axis) is consistent with being random. If true, it would imply that in half the sources the two axes are misaligned by an angle greater than $60°$. *Joos et al.* (2012) found that such a large misalignment enables the formation of RSDs in moderately magnetized dense cores with a dimensionless mass-to-flux ratio λ of ~3–5; RSD formation is suppressed in such cores if the magnetic field and rotation axis are less misaligned (see

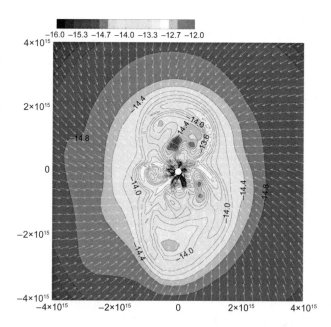

Fig. 4. An example of the inner protostellar accretion flow driven unstable by magnetic flux redistribution (from *Krasnopolsky et al.*, 2012). Plotted are the distribution of logarithm of density (in units of g cm⁻³) and velocity field on the equatorial plane (length in centimeters). The expanding, low-density regions near the center are the so-called DEMS that are strongly magnetized. They present a formidable barrier to disk formation.

Fig. 5). They attributed the disk formation to a reduction in the magnetic braking efficiency induced by large misalignment. In more strongly magnetized cores with $\lambda \leq 2$, RSD formation is suppressed independent of the misalignment angle, whereas in very weakly magnetized cores RSDs are formed for all misalignment angles.

Based on the work of *Hull et al.* (2013) and *Joos et al.* (2012), *Krumholz et al.* (2013) estimated that the field rotation misalignment may enable the formation of large RSDs in ~10–50% of dense cores. If the upper range is correct, the misalignment would go a long way toward solving the problem of excessive magnetic braking in protostellar disk formation.

Li et al. (2013) carried out simulations similar to those of *Joos et al.* (2012), except for the initial conditions. They confirmed the qualitative result of *Joos et al.* (2012) that the field-rotation misalignment is conducive to disk formation. In particular, large misalignment weakens the strong outflow in the aligned case and is a key reason behind the formation of RSDs in relatively weakly magnetized cores. For more strongly magnetized cores with $\lambda \leq 4$, RSD formation is suppressed independent of the degree of misalignment. This threshold value for the mass-to-flux ratio is about a factor of 2 higher than that obtained by *Joos et al.* (2012). The difference may come, at least in part, from the different initial conditions adopted: uniform density with a uniform magnetic field for *Li et al.* (2013) and a centrally condensed density profile with a non-uniform but unidirectional field for *Joos et al.* (2012); the magnetic braking is expected to

be more efficient at a given (high) central density for the former initial configuration, because its field lines would become more pinched, with a longer lever arm for braking. Whether there are other factors that contribute significantly to the above discrepancy remains to be determined.

If the result of *Li et al.* (2013) is correct, then a dense core must have both a large field-rotation misalignment *and* a rather weak magnetic field in order to form a RSD. This dual requirement would make it difficult for the misalignment alone to enable disk formation in the majority of dense cores, which are typically rather strongly magnetized according to *Troland and Crutcher* (2008) (with a median mass-to-flux ratio of $\lambda \sim 2$). In a more recent study, *Crutcher et al.* (2010) argued, based on Bayesian analysis, that a fraction of dense cores could be very weakly magnetized, with a dimensionless mass-to-flux ratio λ well above unity [see *Bertram et al.* (2012) for additional arguments for weak field, including field reversal]. However, since the median mass-to-flux ratio remains unchanged for the different distributions of the total field strength assumed in *Crutcher*'s (2012) Bayesian analysis, it is unlikely for the majority of dense cores to have λ much greater than the median value of 2. For example, *Li et al.* (2013) estimated the fraction of dense cores with $\lambda > 4$ at ~25%. There is also concern that the random distribution of the field-rotation misalignment angle found by *Hull et al.* (2013) on the 10^3-AU scale may not be representative of the distribution on the larger core scale. Indeed, *Chapman et al.* (2013) found that the field orientation on the core scale (measured using a single dish telescope) is within ~30° of the outflow axis for three of the four sources in their sample (see also *Davidson et al.,* 2011); the larger angle measured in the remaining source may be due to projection effects because its outflow axis lies close to the line of sight. If the result of *Chapman et al.* (2013) is robust and if the outflow axis reflects the rotation axis, dense cores with large misalignment between the magnetic and rotation axes would be rare. In such a case, it would be even less likely for the misalignment to be the dominant mechanism for disk formation.

3.5. Turbulence

Turbulence is a major ingredient for star formation (see reviews by, e.g., *Mac Low and Klessen,* 2004; *McKee and Ostriker,* 2007). It can generate local angular momentum by shear flows and form highly asymmetric dense cores [see results from Herschel observations (e.g., *Menshchikov et al.,* 2010; *Molinari et al.,* 2010)]. There is increasing evidence that it also promotes RSD formation. *Santos-Lima et al.* (2012) contrasted the accretion of turbulent and laminar magnetized gas onto a preexisting central star, and found that a nearly Keplerian disk was formed in the turbulent but not laminar case (see Fig. 6). The simulations were carried out at a relatively low resolution [with a rather large cell size of 15.6 AU; this was halved, however, in *Santos-Lima et al.* (2013), who found similar results], and turbulence was driven to a root mean square (rms) Mach

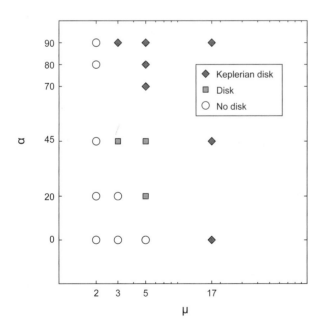

Fig. 5. Parameter space for disk formation according to *Joos et al.* (2012). The parameter α is the angle between the magnetic field and rotation axis, and μ is the dimensionless mass-to-flux ratio of the dense core (denoted by λ in the rest of the article). Diamonds denote disks with Keplerian rotation profile, squares those with flat rotation curves, and circles the cases with no significant disk.

number of ~4, which may be too large for low-mass cores. Nevertheless, the beneficial effect of turbulence on disk formation is clearly demonstrated. They attributed the disk formation to the turbulence-induced outward diffusion of magnetic flux, which reduces the strength of the magnetic field in the inner, disk-forming, part of the accretion flow. Similar results of disk formation in turbulent cloud cores are presented by *Seifried et al.* (2012b) and *Seifried et al.* (2013), although these authors attribute their findings to different mechanisms. They argued that the turbulence-induced magnetic flux loss is limited well outside their disks, based on the near constancy of an approximate mass-to-flux ratio computed on a sphere several times the disk size (with a 500-AU radius). They proposed instead that the turbulence-induced tangling of field lines and strong local shear are mainly responsible for the disk formation: The disordered magnetic field weakens the braking and the shear enhances rotation. Similarly, *Myers et al.* (2013) also observed formation of a nearly Keplerian disk in their radiative MHD simulation of a turbulent massive (300 $M_{\odot}$) core, although they refrained from discussing the origin of the disk in detail since it was not the focus of their investigation.

Seifried et al. (2013) extended their previous work to include both low-mass and high-mass cores and both subsonic and supersonic turbulence. They found disk formation in all cases. Particularly intriguing is the formation of rotationally dominated disks in the low-mass, subsonically turbulent cores. They argued that, as in the case of massive core with supersonic turbulence of *Seifried et al.* (2012b), such disks are not the consequence of turbulence-induced magnetic flux loss, although such loss appears quite severe on the disk scale, which may have contributed to the long-term survival of the formed disk. While disks appear to form only at sufficiently high mass-to-flux ratios $\lambda \geq 10$ in ordered magnetic fields, disks in the turbulent MHD simulations form at much lower and more realistic values of λ. *Seifried et al.* (2012b) found that λ increases gradually in the vicinity of the forming disk, which may have more to do with the growing accreting mass relative to the magnetic flux than with dissipative effects of the magnetic field by turbulent diffusion or reconnection.

Joos et al. (2013) investigated the effects of turbulence of various strengths on disk formation in a core of intermediate mass (5 $M_{\odot}$). They found that an initially imposed turbulence has two major effects. It produces an effective diffusivity that enables magnetic flux to diffuse outward, broadly consistent with the picture envisioned in *Santos-Lima et al.* (2012, 2013). It also generates a substantial misalignment between the rotation axis and magnetic field direction [an effect also seen in *Seifried et al.* (2012b) and *Myers et al.* (2013)]. Both of these effects tend to weaken magnetic braking and make disk formation easier. If the turbulence-induced magnetic diffusion is responsible, at least in part, for the disk formation, then numerical effects would be a concern. In the ideal MHD limit, the diffusion presumably comes from turbulence-enhanced reconnections due to finite grid resolution. Indeed, *Joos et al.* (2013) reported that their simulations did not appear to be fully converged, with disk masses differing by a factor up to ~2 in higher-resolution simulations. The situation is further complicated by numerical algorithms for treating magnetic field evolution, especially those relying on divergence cleaning, which could introduce additional artificial magnetic diffusion. To make further progress, it would be useful to determine when and how the reconnections occur and exactly how they lead to the magnetic diffusion that are apparent in the simulations of *Joos et al.* (2013), *Santos-Lima et al.* (2012), Li et al. (in preparation), and perhaps *Seifried et al.* (2012b, 2013).

3.6. Other Mechanisms

The magnetic braking catastrophe in disk formation would disappear if the majority of dense cores are nonmagnetic or only weakly magnetized [$\lambda \geq 5$ or even $\lambda \geq 10$ (see, e.g., *Hennebelle and Ciardi*, 2009; *Ciardi and Hennebelle*, 2010; *Seifried et al.*, 2011)]. However, such weakly magnetized cloud cores are rather unlikely. Although the recent study by *Crutcher et al.* (2010) indicates that some cloud cores might be highly supercritical, they are certainly not the majority. Furthermore, consider, for example, a typical core of 1 $M_{\odot}$ in mass and 10^4 cm^{-3} in H_2 number density. To have a dimensionless mass-to-flux ratio $\lambda \geq 5$, its field strength must be B ≤ 4.4 μG, less than the median field strength inferred for the atomic cold neutral medium (CNM) [~6 μG (*Heiles and Troland*, 2005)], which is unlikely. We therefore expect the majority of dense cores to have magnetic fields corresponding to $\lambda \leq 5$ [in agreement with *Troland and Crutcher* (2008)], which are strong enough to make RSD formation difficult.

Another proposed solution is the depletion of the protostellar envelope. The slowly rotating envelope acts as a brake on the more rapidly rotating material closer to the central ob-

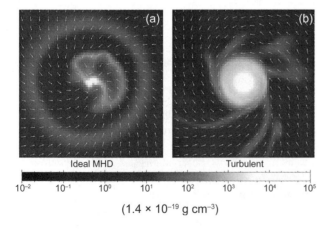

Fig. 6. Accretion of rotating, magnetized material onto a preexisting central object **(a)** with and **(b)** without turbulence (adapted from *Santos-Lima et al.,* 2012). The formation of a nearly Keplerian disk is clearly suppressed in the laminar case [**(a)**], by excessive magnetic braking, but is enabled by turbulence [**(b)**].

ject that is magnetically connected to it. Its depletion should promote RSD formation. Indeed, *Machida et al.* (2010) found that envelope depletion is conducive to the formation of large RSDs toward the end of the main accretion phase. This is in line with the expectation of *Mellon and Li* (2008), who envisioned that most of the envelope depletion is achieved through wind stripping rather than accretion, as would be the case if the star-formation efficiency of individual cores is relatively low [say, ~1/3 (*Alves et al.*, 2007)]. Given the ubiquity of fast outflows, their effects on envelope depletion and disk formation should be investigated in more detail.

3.7. Summary and Outlook

The formation of rotationally supported disks turns out to be much more complicated than envisioned just a decade ago. This is because star-forming dense cores are observed to be rather strongly magnetized in general [although the magnetization in a fraction of them can be rather weak; see *Crutcher et al.* (2010) and discussion above], with a magnetic energy typically much higher than the rotational energy. The field strength is further amplified by core collapse, which tends to concentrate the field lines in the region of disk formation close to the protostar. Both analytic calculations and numerical simulations have shown that the collapse-enhanced (ordered) magnetic field can prevent the RSD formation through catastrophic magnetic braking in the simplest case of ideal MHD limit, aligned magnetic field and rotation axis, and no turbulence. Ambipolar diffusion, the Hall effect, and magnetic interchange instabilities have profound effects on the dynamics of the inner protostellar accretion flow, but they do not appear capable of forming RSDs by themselves under typical conditions. Ohmic dissipation, on the other hand, can enable the formation of at least small, astronomical-unit-scale disks at early times. How such disks evolve in the presence of the instabilities and other non-ideal MHD effects remains to be quantified. Magnetic field rotation misalignment is conducive to disk formation, but it is unlikely to enable the formation of RSDs in the majority of dense cores, because of the dual requirement of both a relatively weak magnetic field and a relatively large tilt angle, which may be uncommon on the core scale. Turbulence appears to facilitate RSD formation in a number of numerical simulations. It is possible that the turbulence-enhanced numerical reconnection plays a role in the appearance of RSDs in these formally ideal MHD simulations, although turbulence by itself could reduce braking efficiency. The possible role of reconnection needs to be better understood and quantified.

4. EARLY OUTFLOWS

4.1. Introduction

Generally, low-mass YSOs are accompanied by highly collimated optical jets (see, e.g., *Cabrit et al.*, 1997; *Reipurth and Bally*, 2001), whereas high-mass stars are often obscured, hard to observe, and until recently, thought to drive much less collimated outflows (e.g., *Shepherd and Churchwell*, 1996a,b; *Beuther and Shepherd*, 2005) (see also discussion below). Nevertheless, there is strong evidence that the underlying launching process is based on the same physical mechanism, namely the magneto-rotational coupling: Magnetic fields anchored to an underlying rotor (e.g., an accretion disk) will carry along gas that will be flung outward [the same mechanism could also apply to galactic jets (e.g., *Pudritz et al.*, 2007, 2009)]. For example, *Guzmán et al.* (2012) concluded that collimated thermal radio jets are associated with high-mass YSOs, although for a relatively short time (~4×10^4 yr).

With the seminal theoretical work by *Blandford and Payne* (1982) and *Pudritz and Norman* (1983) the idea of magneto-centrifugally driven jets was first established. It was shown that magnetic fields anchored to the disk around a central object can lift off gas from the disk surface. A magneto-centrifugally driven jet will be launched if the poloidal component of magnetic field is inclined with respect to the rotation axis by more than 30°. Numerical simulations of Keplerian accretion disks threaded by such a magnetic field have shown that these jets are self-collimated and accelerated to high velocities (e.g., *Fendt and Camenzind*, 1996; *Ouyed and Pudritz*, 1997). This driving mechanism, where the launching of the jet is connected to the underlying accretion disk, predicts that jets rotate and carry angular momentum off the disk. The first plausible observational confirmation came from HST detection of rotational signatures in the optical jet of DG Tauri (*Bacciotti et al.*, 2002) [see, however, *Soker* (2005), *Cerqueira et al.* (2006), and *Fendt* (2011) for different interpretations of the observation]. Further evidence is provided by ultraviolet (UV) (*Coffey et al.*, 2007) and infared observations (*Chrysostomou et al.*, 2008).

The most viable mechanism to launch jets and wider angle outflows from accretion disks around YSOs is the coupling through magnetic fields where the gas from the disk surface is accelerated by the Lorentz force. Generally, this force can be divided into a magnetic tension and a magnetic pressure term. In an axisymmetric setup, the magnetic tension term is responsible for the magneto-centrifugal acceleration and jet collimation via hoop stress. The magnetic pressure can also accelerate gas off the underlying disk. These magnetic pressure-driven winds are sometimes known as *magnetic twist* (*Shibata and Uchida*, 1985), *plasma gun* (*Contopoulos*, 1995), or *magnetic tower* (*Lynden-Bell*, 2003).

Protostellar disks around YSOs themselves are the result of gravitational collapse of molecular cloud cores (see section 3). Since the molecular clouds are permeated by magnetic fields of varying strength and morphology (see, e.g., *Crutcher et al.*, 1999; *Beck*, 2001; *Alves et al.*, 2008), there should be a profound link between the collapse and magneto-rotationally driven outflows.

There are still many unresolved problems concerning jets and outflows. These include the details of the jet launching, the driving of molecular outflows, the efficiency of outflows around massive protostars, the influence of outflow feedback on star formation, and how efficient they are in clearing off the envelope material around the YSO. Shedding light on the last problem will also help determine whether there is a clear physical link between the core mass function (CMF) and initial mass function (IMF) (see the chapter by Offner et al. in this volume).

In this section we summarize our knowledge of early jets and outflows and discuss open questions.

4.2. Jet Launching and Theoretical Modeling

Self-consistent modeling of jet launching is a challenging task, especially during the earliest phases of star formation, when the core collapse has to be modeled at the same time. The most practical approach to study the self-consistent jet launching during the collapse of self- gravitating gas is through direct numerical simulation. Even then, the large dynamical range of length scales (from the 10-AU molecular core to the subastronomical-unit protostellar disk) and timescales (from the initial free-fall time of 10^5 yr to the orbital time of one year or less) require expensive AMR or SPH simulations that include magnetic fields. Furthermore, non-ideal MHD effects such as Ohmic dissipation and ambipolar diffusion complicate the calculations (see section 3).

One of the first collapse simulations in which outflows are observed was done more than a decade ago by *Tomisaka* (1998) with an axisymmetric nested grid technique. These simulations of magnetized, rotating, cylindrical cloud cores showed that a strong toroidal field component builds up, which eventually drives a bipolar outflow. Subsequently, a number of collapse simulations from different groups and different levels of sophistication were performed (e.g., *Tomisaka,* 2002; *Boss,* 2002; *Allen et al.,* 2003; *Matsumoto and Tomisaka,* 2004; *Hosking and Whitworth,* 2004; *Machida et al.,* 2004, 2005a,b; *Ziegler,* 2005; *Machida et al.,* 2006; *Banerjee and Pudritz,* 2006, 2007; *Machida et al.,* 2007, 2008; *Price and Bate,* 2007; *Hennebelle and Fromang,* 2008; *Duffin and Pudritz,* 2009; *Mellon and Li,* 2009; *Commerçon et al.,* 2010; *Bürzle et al.,* 2011; *Seifried et al.,* 2012a). Despite the diversity in numerical approach (e.g., AMR vs. SPH simulations) and initial problem setup, all simulations enforce the same general picture, that magnetically launched outflows are a natural outcome of magnetized core collapse. The details of the outflows generated depend, of course, on the initial parameters such as the degree of core magnetization and the core rotation rate.

4.2.1. Outflow driving. Traditionally, there is a clear distinction between outflows driven by centrifugal acceleration (*Blandford and Payne,* 1982; *Pudritz and Norman,* 1986; *Pelletier and Pudritz,* 1992) or the magnetic pressure gradient (*Lynden-Bell,* 1996, 2003). A frequently used quantity to make the distinction is the ratio of the toroidal to poloidal magnetic field, B_ϕ/B_{pol} (e.g., *Hennebelle and Fromang,* 2008). If this ratio is significantly above 1, the outflow is often believed to be driven by the magnetic pressure. However, the consideration of B_ϕ/B_{pol} alone can be misleading, as in centrifugally driven flows this value can be as high as

10 (*Blandford and Payne,* 1982). Close to the disk surface one can check the inclination of the magnetic field lines with respect to the vertical axis. The field lines have to be inclined by more than 30° for centrifugal acceleration to work (*Blandford and Payne,* 1982). Although this criterion is an exact solution of the ideal, stationary, and axisymmetric MHD equations for an outflow from a Keplerian disk, its applicability is limited to the surface of the disk. A criterion to determine the driving mechanism above the disk was used by *Tomisaka* (2002) comparing the centrifugal force F_c and the magnetic force F_{mag}. By projecting both forces on the poloidal magnetic field lines, it can be determined which force dominates the acceleration. For the outflow to be driven centrifugally, F_c has to be larger than F_{mag}. However, for this criterion to be self-consistent the gravitational force and the fact that any toroidal magnetic field would reduce the effect of F_c have to be taken into account.

In *Seifried et al.* (2012a), a general criterion was derived to identify centrifugally driven regions of the outflows and to differentiate those from magnetic pressure-driven outflows. The derivation assumed a stationary axisymmetric flow, which leads to a set of constraint equations based on conservation laws along magnetic field lines (see also *Blandford and Payne,* 1982; *Pudritz and Norman,* 1986; *Pelletier and Pudritz,* 1992). This criterion is applicable throughout the entire outflow.

The general condition for outflow acceleration is (*Seifried et al.,* 2012a)

$$\partial_{pol}\left(\frac{1}{2}v_\phi^2 + \Phi - \frac{v_\phi}{v_{pol}}\frac{1}{4\pi}\frac{B_\phi B_{pol}}{\rho} + \frac{1}{4\pi}\frac{B_\phi^2}{\rho}\right) < 0 \qquad (1)$$

where ∂_{pol} denotes the derivative along the poloidal magnetic field. It describes all regions of gas acceleration including those dominated by the effect of B_ϕ. This general outflow criterion should be compared to the case of centrifugal acceleration where $B_\phi = 0$, i.e. in the case with no resulting Lorentz force along the poloidal field line

$$\frac{r}{z}\frac{1}{GM}\left(\frac{v_\phi^2}{r^2}(r^2+z^2)^{3/2} - GM\right)\bigg/\left(\frac{B_z}{B_r}\right) > 1 \qquad (2)$$

where M, r, and z are the mass of the central object, the cylindrical radius, and distance along the z axis, respectively. Using both equations, one can distinguish between regions dominated by centrifugal acceleration and those by the toroidal magnetic pressure. Note that equation (2) does not assume a Keplerian disk, hence it is also applicable to early-type sub-Keplerian configurations.

An example of those outflow criteria is shown in Fig. 7, where one can see that the centrifugally launched region is narrower and closer to the rotation axis but faster than the outer part of the outflow. Generally, such early type outflows are driven by both mechanisms, i.e., by magnetic pressure and magneto-centrifugal forces, but the centrifugal launching should become more dominant while the underlying disk evolves toward a more stable Keplerian configuration.

Another, indirect, support for the outflow generation mechanism involving magnetic driving comes from a recent numerical study by *Peters et al.* (2012). Their simulations, which include feedback from ionizing radiation from massive protostars, show pressure-driven bipolar outflows reminiscent of those observed around massive stars. But detailed analysis through synthetic CO maps show that the pressure-driven outflows are typically too weak to explain the observed ones [e.g., outflows in G5.89 (*Puga et al.,* 2006; *Su et al.,* 2012)]. The failure suggests that a mechanism other than ionizing radiation must be found to drive the massive outflows. Since massive star-forming regions are observed to be significantly magnetized (e.g., *Girart et al.,* 2009; *Tang et al.,* 2009), the magnetic field is a natural candidate for outflow driving: The observed massive outflows might be driven magnetically by the massive stars themselves, or by the collection of lower- and intermediate-mass stars in the young massive cluster.

4.2.2. Outflow collimation. A general finding of self-consistent numerical simulations is that the degree of outflow collimation is time dependent and depends on the initial field strength. At very early stages (10^3–10^4 yr) outflows in typically magnetized, massive cores (with mass-to-flux ratios of $\lambda \leq 5$) are found to be poorly collimated with collimation factors of 1–2 instead of 5–10, still in agreement with observations of outflows around most young massive protostellar objects (e.g., *Ridge and Moore,* 2001; *Torrelles et al.,* 2003; *Wu et al.,* 2004; *Sollins et al.,* 2004; *Surcis et al.,* 2011). It is suggestive that during the earliest stage, i.e., before the B1–B2-type phase of the scenario described by *Beuther and Shepherd* (2005), the outflows are rather poorly collimated except in case of an unusually weak magnetic field. In their further evolution, however, the collimation will increase quickly due to the development of a fast, central jet coupled to the buildup of a Keplerian disk. Therefore it might be problematic to directly link the evolutionary stage of the massive YSO to the collimation of the observed outflow, as suggested by *Beuther and Shepherd* (2005). Additional

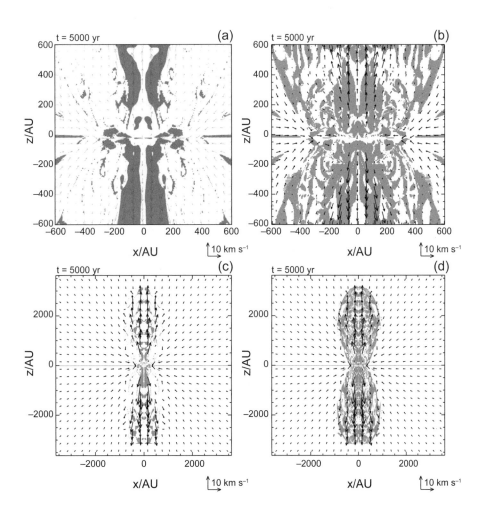

Fig. 7. Application of the outflow criteria derived in *Seifried et al.* (2012a). The left panels show that centrifugal acceleration works mainly close to the z-axis up to a height of about 800 AU, which agrees very well with the region where the highest velocities are found (equation (2)). The general criterion (equation (1)) is more volume filling and also traces regions in the outer parts. From *Seifried et al.* (2012a).

difficulties for correlating ages of YSOs and the collimation of outflows arise from the fragmentation of massive disks. For instance, circumsystem outflows (from around binaries or higher multiples) from large sub-Keplerian disks at early stages are possible. But those outflows are often uncollimated and might even show spherical morphologies due to fragmentation of the highly unstable accretion disk (e.g., *Peters et al.*, 2011; *Seifried et al.*, 2012a). Although it seems likely that the subsequent outflows from around single massive protostars should be collimated, the evidence from numerical simulations of clustered star formation showing the self-consistent launching of such outflows is still missing due to the lack of resolution.

Direct confirmation of those evolutionary scenarios is difficult, as one would need independent information of the age of the YSOs and details of the magnetization of the environment (see also discussion in *Ray*, 2009). Such observations are hard to obtain and therefore rather rare. However, there is an interesting observation that supports the picture of very early stage, poorly collimated outflows successively collimating over time. Observing two spatially adjacent, massive protostars in the star-forming region W75N, *Torrelles et al.* (2003) and *Surcis et al.* (2011) find the younger of the two has a spherical outflow, whereas the more-evolved protostar has a well-collimated outflow. Due to their close proximity to each other, they should have similar environmental conditions. Therefore the difference should rather be a consequence of different evolutionary stages, where the younger, poorly collimated outflow is possibly only a transient.

4.3. Feedback by Jets and Outflows

As mentioned earlier, jets and outflows are already present at very early stages of star formation. Hence, their influence on the subsequent evolution within star-forming regions may not be neglected. In particular, in cluster-forming regions, outflows are believed to influence or even regulate star formation (see also chapters by Krumholz et al. and Frank et al. in this volume), as originally proposed by *Norman and Silk* (1980). Since then a number of numerical simulations tried to address this issue (e.g., *Li and Nakamura*, 2006; *Banerjee et al.*, 2007, 2009; *Nakamura and Li*, 2007; *Wang et al.*, 2010; *Li et al.*, 2010; *Hansen et al.*, 2012), but with different outcomes. Detailed single jet simulations demonstrated that the jet power does not couple efficiently to the ambient medium and is not able to drive volume-filling supersonic turbulence (*Banerjee et al.*, 2007). This is because the bow shock of a highly collimated jet and developed jet instabilities mainly excite subsonic velocity fluctuations. Similarly, the simulations by *Hansen et al.* (2012) showed that protostellar outflows do not significantly affect the overall cloud dynamics, at least in the absence of magnetic fields. Otherwise, the results from simulations of star-cluster-forming regions (e.g., *Li and Nakamura*, 2006; *Nakamura and Li*, 2007; *Wang et al.*, 2010; *Li et al.*, 2010) clearly show an impact of outflows on

the cloud dynamics, accretion rates, and star-formation efficiency. But this seems to be only effective if rather strong magnetic fields and a high amount of initial turbulence are present in those cloud cores. The jet energy and momentum are better coupled to a turbulent ambient medium than a laminar one (*Cunningham et al.*, 2009).

4.4. Future Research on Outflows Around Protostars

Undoubtedly, jets and outflows from YSOs are strongly linked through the magnetic field to both the disk and surrounding envelope. Deciphering the strength and morphological structure of magnetic fields of jets and outflows will be a key to gain a better understanding of these exciting phenomena. Unfortunately, there are very few direct measurements of magnetic field strengths in YSO jets to date. The situation should improve with the advent of new radio instruments (see also *Ray*, 2009). For example, the Magnetism Key Science Project of the Low-Frequency Array (LOFAR) (*http://www.lofar.org*) plans to spatially resolve the polarized structure of protostellar jets to examine their magnetic field structure and investigate the impact of the field on the launching and evolution of protostellar jets. Prime targets would be the star-forming regions of Taurus, Perseus, and Cepheus Flare molecular clouds with subarcsecond resolution. The forthcoming Square Kilometer Array (SKA) (*http://www.skatelescope.org*) will also offer unprecedented sensitivity to probe the small-scale structure of outflows around protostars (*Aharonian et al.*, 2013).

5. CONNECTING EARLY DISKS TO PLANET FORMATION

As we have seen, the first 10^5 yr in the life of a protostellar disk are witness to the accretion of the bulk of the disk mass and the rapid evolution of its basic dynamics, as well as the most vigorous phase of its outflow activity. This is also the period when the basic foundations of the star's planetary system are laid down. Giant planet formation starts either by rapid gravitationally driven fragmentation in the more distant regions of massive disks (e.g., *Mayer et al.*, 2002; *Rafikov*, 2009), or by the formation of rocky planetary cores that over longer (million years) timescales will accrete massive gaseous envelopes (e.g., *Pollack et al.*, 1996) (see also the chapter by Helled et al. in this volume). Terrestrial planet formation is believed to occur as a consequence of the oligarchic collisional phase that is excited by perturbations caused by the appearance of the giant planets (see the chapter by Raymond et al. in this volume). There are several important connections between the first phases of planet formation and the properties of early protostellar disks. We first give an overview of some of the essential points, before focusing on two key issues.

At the most fundamental level, the disk mass is central to the character of both star and planet formation. Most of a star's mass is accreted through its disk, while at the same time, giant planets must compete for gas from the same gas reservoir. Sufficiently massive disks, roughly a tenth of the stellar mass, can in these early stages generate strong spiral waves that drive rapid accretion onto the central star. Such disks are also prone to fragmentation. Early disk masses exceeding 0.01 $M_\odot$ provide a sufficient gas supply to quickly form jovian planets (*Weidenschilling*, 1977). The lifetime of protostellar disks, known to be in the range 3–10 m.y., provides another of the most demanding constraints on massive planet formation. Detailed studies using the Spitzer Space Telescope indicate that 80% of gas disks around stars of less than 2 $M_\odot$ have dissipated by 5 m.y. after their formation (*Carpenter et al.*, 2006; *Hernández et al.*, 2008, 2010).

As already discussed, protostellar outflows are one of the earliest manifestations of star formation. Class 0 sources are defined as having vigorous outflows and this implies that magnetized disks are present at the earliest times. Magnetic fields that thread such disks are required for the outflow launching. These fields also have a strong quenching effect on the fragmentation of disks, which has consequences for the gravitational instability picture of planet formation. How early does Keplerian behavior set in? Some simulations (e.g., *Seifried et al.*, 2012b) suggest that, even within the first few 10 yr, the disk has already become Keplerian (section 3), making the launch of centrifugally driven jets all the more efficient. Before this, it is possible that angular momentum transport by spiral waves is significant.

Stars form as members of star clusters and this may have an effect on disk properties and therefore upon aspects of planet formation. Observations show that as much as 90% of the stars in the galactic disk originated in embedded young clusters (*Gutermuth et al.*, 2009). The disk fraction of young stars in clusters such as λ Orionis (*Hernández et al.*, 2010) and other clusters such as Upper Scorpius (*Carpenter et al.*, 2006) is similar to that of more dispersed groups, indicating that the dissipative timescale for disks is not strongly affected by how clustered the star-formation process is. We note in passing that the late time dissipation of disks is controlled both by photoevaporation, which is dominated by the far-ultraviolet (FUV) and X-ray radiation fields of the whole cluster and not of the host star, as well as by the clearing of holes in disks by multiple giant planets (see the chapter by Espaillat et al. in this volume). Calculations show that FUV radiation fields, produced mainly by massive stars, would inhibit giant planet formation in one-third to two-thirds of planetary systems, depending on the dust attenuation. However, this photoevaporation affects mainly the outer regions of disks, leaving radii out to 35 AU relatively unscathed (*Holden et al.*, 2011).

Rocky planetary cores on the order of 10 $M_\oplus$ are essential for the core accretion picture of giant planets. This process must take place during the early disk phase — the first 10^5 yr or so (see the chapter by Johansen et al. in this volume) in order to allow enough time for the accretion of a gas envelope. Therefore the appearance of planetesimals out of which such giant cores are built must also be quite rapid and is another important part of the first phases of giant planet formation in early protostellar disks. It is important to realize therefore that the various aspects of non-ideal MHD discussed in the context of disk formation in section 3, being dependent on grain properties, take place in a rapidly evolving situation wherein larger grains settle, agglomerate, and go on to pebble formation.

Finally, a major factor in the development of planetary systems in early disks arises from the rapid migration of the forming planetary cores. As is well known, the efficient exchange of planetary orbital angular momentum with a gaseous disk by means of Lindblad and co-rotation resonances leads to very rapid inward migration of small cores on 10^5-yr timescales (*Ida and Lin*, 2008) (see also the chapter by Benz et al. in this volume). One way of drastically slowing such migration is by means of planet traps — regions of zero net torque on the planet that occur at disk (*Masset et al.*, 2006; *Matsumura et al.*, 2009; *Hasegawa and Pudritz*, 2011, 2012). Inhomogeneities in disks, such as dead zones, ice lines, and heat transitions regions (from viscous disk heating to stellar irradiation domination), form special narrow zones where growing planets can be trapped. The early appearance of disk inhomogeneities and such planet traps encodes the basic initial architecture of forming planetary systems. We now turn to a couple of these major issues in more detail.

5.1. Fragmentation in Early Massive Disks

Early disks are highly time dependent, with infall of the core continuously delivering mass and angular momentum to the forming disk. Moreover, the central star is still forming by rapid accretion of material through the disk. Depending on the infall rate, the disk may or may not be self-gravitating (see below). Disk properties and masses can be measured toward the end of this accretion phase in the class I sources. A CARMA survey of 10 class I disks carried out by *Eisner* (2012) as an example showed that only a few class I disks exceed 0.1 $M_\odot$, and the range of masses from <0.01 to >0.1 $M_\odot$ exceeds that of disks in the class II phase. This is a tight constraint on the formation of massive planets and already suggests that the process must have been well on its way before even the class I state has been reached.

Two quantities that control the gravitational stability of a hydrodynamic disk are the ratio of the disk mass to the total mass of the system $\mu = M_d/(M_d + M_*)$ and the Toomre instability parameter $Q = c_s\kappa/\pi G\Sigma_d$, where $\Sigma_d = M_d/(2\pi R_d^2)$ is the surface mass density of the disk. Heating and cooling of the disk as well as its general evolution alter Q and infall onto the disk changes μ (*Kratter et al.*, 2008). The mass balance in an early disk will depend upon the efficiency of angular momentum transport, which is widely believed to be governed by two mechanisms: gravitational torque as well as the magneto-rotational instability (MRI). Angular momentum transport through the disk by these agents can be treated as "effective" viscosities: α_{GI} and $\alpha_{MRI} \simeq 10^{-2}$. These drive the total disk viscosity $\alpha = \alpha_{GI}(Q, \mu) + \alpha_{MRI}$.

The data produced by numerical simulations can be used to estimate α_{GI} in terms of Q and μ (e.g., simulations of *Vorobyov and Basu,* 2005, 2006). The resulting accretion rate onto the central star can be written in dimensionless form as $\dot{M}_*/M_d\Omega$, where the epicyclic frequency has been replaced by the angular frequency Ω of the disk. The accompanying Fig. 8 shows the accretion rate through the disks in this phase space. The figure shows that the greatest part of this disk Q-μ phase space is dominated by MRI transport rather than by gravitational torques from spiral waves. Evolutionary tracks for accreting stars can be computed in this diagram, which can be used to trace the evolution of the disks. Low-mass stars will have lower values of μ, MRI-dominated evolution, masses on the order of 30% of the system mass, and typical outer radii on the order of 50 AU. High-mass stars by comparison are predicted to have high values of $\mu \simeq 0.35$ and an extended period of local fragmentation as the accretion rates peak, as well as a disk outer edge at 200 AU.

The fragmentation of disks is markedly affected by the presence of significant magnetic fields. One of the main effects of a field is to modify the Toomre criterion. Because part of the action of the threading field in a disk is to contribute a supportive magnetic pressure, the Toomre Q parameter is modified with the Alfvén velocity v_A (the typical propagation speed of a transverse wave in a magnetic field); $Q_M = (c_s^2 + v_A^2)^{1/2}\kappa/\pi G\Sigma_d$. For typical values of the mass to flux ratio, magnetic energy densities in disk are comparable to thermal or turbulent energy densities.

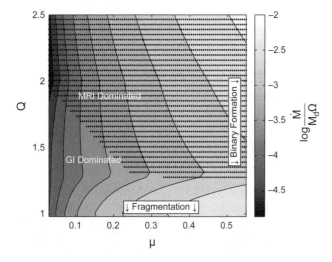

Fig. 8. Contours of the dimensionless accretion rate $\dot{M}_*/(M_d\Omega)$ from the disk onto the star from both transport components of the model. The lowest contour level is $10^{-4.8}$, and subsequent contours increase by 0.3 dex. The effect of each transport mechanism is apparent in the curvature of the contours. The MRI causes a mild kink in the contours across the Q = 2 boundary and is more dominant at higher disk masses due to the assumption of a constant disk turbulence parameter. Adapted from *Kratter et al.* (2008).

Seifried et al. (2011) find that the magnetic suppression of disk fragmentation occurs in most of their models. Even the presence of ambipolar diffusion of the disk field does not significantly enhance the prospects for fragmentation (*Duffin and Pudritz,* 2009).

Gravitational fragmentation into planets or low-mass companions requires that Q ~ 1; however, this is not sufficient. Fragments must also cool sufficiently rapidly, as was first derived by *Gammie* (2001) and generalized by *Kratter and Murray-Clay* (2011). The condition for sufficiently rapid cooling depends, in turn, upon how the disk is heated. The inner regions of disks are dominated by viscous heating, which changes into dominant radiative heating by irradiation of the disk by the central star, in the outer regions (*Chiang and Goldreich,* 1997). The transition zone between these two regions, which has been called a "heat transition radius" (*Hasegawa and Pudritz,* 2011) is of importance, both in regard to gravitation fragmentation (*Kratter and Murray-Clay,* 2011) as well as the theory of planet traps. This radius occurs where the heating of the surface of the disk by irradiation by the central star balances the heating of the disk at the midplane by viscous heating. Stellar irradiation dominates viscous heating if the temperature T exceeds a critical value: $T > [(9/8)(\alpha\Sigma/\sigma)(k/\mu)\tau_R\Omega]^{1/3}$, where τ_R is the Rosseland mean opacity, α is the viscosity parameter, σ is the Stefan-Boltzmann constant, k is the Boltzmann's constant, and Ω is the orbital angular frequency.

The long-term survival of fragments depends upon three different forces: gas pressure, shearing in the disk, and mutual interaction and collisions (*Kratter and Murray-Clay,* 2011). The role of pressure in turn depends upon how quickly the gas can cool, as well as its primary source of energy. Fragmentation in the viscously heated regime — which is where giant planets may typically form — can occur if the ratio of the cooling time to the dynamical time β is sufficiently small for a gas with adiabatic index γ: $\beta < ([4/9\gamma(\gamma-1)]\alpha_{sat}^{-1}$, where the saturated value of the viscous α parameter refers to the turbulent amplitude that can be driven by gravitational instability. Infall plays an important role in controlling this fragmentation. Generally, a higher infall rate $\dot{M}$ drives the value of Q downward, as seen in Fig. 8. Rapid infall will therefore tend to drive disks closer to instability, and perhaps even on to fragmentation. Irradiated disks, by contrast, have a harder time fragmenting. The basic point here is that while gravitationally driven turbulence can be dissipated to maintain Q = 1, irradiated disks do not have this property. Once an irradiated disk moves into a critical Q regime, they are more liable to fragment since there is no intrinsic self-regulatory mechanism for maintaining disk temperature near a critical value. The results indicate that irradiated disks can be driven by infall to fragment at lower accretion rates onto the disk.

Spiral arms compress the disk gas and are the most likely sites for fragmentation. A key question is what are the typical masses of surviving fragments. Recent numerical simulations of self-gravitating disks, without infall, have gone much farther into the nonlinear regimes to follow the fragmentation into planet-scale objects (e.g., *Boley et al.,* 2010; *Rogers and Wadsley,* 2012). Using realistic cooling functions for disks, *Rogers and Wadsley* (2012) have simulated self-gravitating disks and have found a new criterion for the formation of bound fragments. Consider a patch of a disk that has been compressed into a spiral arm of thickness l_1 (see Fig. 9). This arm will form gravitationally bound fragments if l_1 lies within the Hill radius $H_{Hill} = [G\Sigma_1 l_1^2 / 3\Omega^2]^{1/3}$ of the arm, or $(l_1/2H_{Hill}) < 1$.

This criterion also addresses the ability of shear to prevent the fragmentation of the arm. Numerical simulations verify that this criterion describes the survival of gravitationally induced fragments in the spiral arms. The results have been applied to disks around A stars and show that fragments of masses 15 M_{Jup} can form and survive at large distances on the order of 95 AU from the central star, in this radiation-heating-dominated regime. Brown dwarf scale masses seem to be preferred.

Finally, we note that clumps formed in the outer parts of disks may collapse very efficiently, perhaps on as little as thousand-year timescales. This suggests that clumps leading to the formation of giant planets could collapse quickly and survive transport to the interior regions of the disk (*Galvagni et al.,* 2012). We turn now to planetary transport through disks.

5.2. Planet Traps and the Growth and Radial Migration of Planets

The survival of planetary cores in early disks faces another classic problem arising from the exchange of orbital angular momentum between the low-mass plan-

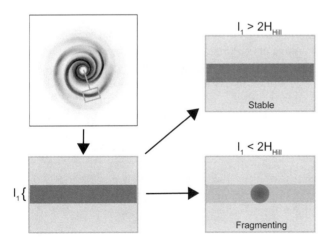

Fig. 9. The Hill criterion for spiral arm fragmentation: If a piece of the spiral arm of width l_1 lies within its own Hill thickness, then that section of the arm is free to collapse and fragmentation takes place. If such a section of the spiral arm lies outside its own Hill thickness, then shear stabilizes the arm and fragmentation does not take place. Adapted from *Rogers and Wadsley* (2012).

etary core and its surrounding gaseous disk. In these early phases, protoplanetary cores can raise significant wakes or spiral waves to the interior and exterior radial regions of the disk. These waves in turn exert torques back on the planet, resulting in its migration. For homogeneous disks with smooth, decreasing density and temperature profiles, the inner wake (which transfers angular momentum to the protoplanet, driving it outward) is slightly overcome by the outer wake (which extracts orbital angular momentum from the protoplanet, driving it inward). This results in a net torque that results in the rapid inward (type I) migration of the planet. As is well known, Monte Carlo population synthesis calculations on protoplanets in evolving accreting disks find that protoplanetary cores migrate into the center of the disk within 10^5 yr (*Ida and Lin,* 2005, 2008) — the timescale characterizing early disks.

Rapid type I migration must be slowed down by at least a factor of 10 to make it more compatible with the lifetime of the disk (see the chapter by Benz et al. in this volume). As noted, this can be achieved in disks at density and temperature inhomogeneities at whose boundaries migration can be rapidly slowed, or even stopped. We have already encountered two of these in other contexts, namely dead zones and heat transitions, wherein there is a change in density and in disk temperature gradient respectively. Heat transitions were shown to be planet traps by *Hasegawa and Pudritz* (2011). A third type of trap is the traditional ice line, wherein the temperature transition gives rise to a change in disk opacity with a concomitant fluctuation in the density (*Ida and Lin,* 2008). *Hasegawa and Pudritz* (2011, 2012) showed that protoplanetary cores can become trapped at these radii.

Dead zone edges act to stop planetary migration, as has been shown in theory (*Masset et al.,* 2006) and numerical simulations (*Matsumura et al.,* 2009). A region that is starved of ionization, which occurs where disk column densities are high enough to screen out ionizing cosmic rays, has large Ohmic dissipation, which prevents the operation of the MRI instability. This region is known as a dead zone because it is unable to generate the MHD turbulence necessary to sustain a reasonable viscosity α_{MRI} (*Gammie,* 1996). A dead zone is most likely to be present in the inner regions of disks in their later stages. During the earliest stages of disk evolution, when the disks are the most massive, these would extend out to 10 AU or so (*Matsumura and Pudritz,* 2006). It is necessary to transport away disk angular momentum to allow an ongoing large accretion flow that is measured for young stars. Older models supposed that a thin surface layer is sufficiently well ionized to support MRI turbulence. Recent simulations show, however, that such MRI active surface layers may not occur (*Bai and Stone,* 2013). The inclusion of all three of the non-ideal effects discussed in section 3 (Ohmic resistivity, Hall effect, and ambipolar diffusion) strongly changes the nature of MHD instability in vertically stratified disks. Going from the dense mid-plane to the surface of the disk, the dominant dissipation mechanism changes: from Ohmic resistivity at the mid-plane, to the Hall effect at mid-scale heights, to ambipolar diffusion in the surface layers. The results show that ambipolar diffusion shuts down the MRI even in the surface layers in the presence of a net (nonzero) magnetic flux through the disk. Instead of the operation of an MRI in the dead zone, a strong MHD disk wind is launched, and this carries off the requisite disk angular momentum very efficiently (for a review, see *Pudritz et al.*, 2007) (section 4 of this paper).

Planetary cores accrete most of their mass while moving along with such traps. As the disk accretion rate falls in an evolving disk, the traps move inward at different rates. At early times, with high accretion rates, the traps are widely separated. As the disk accretion rate falls from a high of 10^{-6} $M_\odot$ yr^{-1} to lower values at later times, the traps slowly converge at small disk radii, which likely initiates planet-planet interactions.

In summary, this section has shown that the properties of early magnetized disks as characterized by high infall rates and disk masses, as well as powerful outflows, can strongly influence the early phases of planet formation and migration.

6. SYNTHESIS

Both observational and theoretical studies of disk formation are poised for rapid development. Observationally, existing dust continuum surveys of deeply embedded "class 0" objects indicate that a compact emission component apparently distinct from the protostellar envelope is often present. Whether this component is a rotationally supported disk (RSD) or not is currently unclear in general. With unprecedented sensitivity and resolution, ALMA should settle this question in the near future.

On the theoretical side, recent development has been spurred largely by the finding that magnetic braking is so efficient as to prevent the formation of a RSD in laminar dense cores magnetized to realistic levels in the ideal MHD limit — the so-called "magnetic braking catastrophe." Although how exactly this catastrophe can be avoided remains unclear, two ingredients emerge as the leading candidates for circumventing the excessive braking: Ohmic dissipation and complex flow pattern (including turbulence, misalignment of magnetic field and rotation axis, and possibly irregular core shapes), the former through the decoupling of magnetic field lines from the bulk neutral matter at high densities, and the latter through, at least in part, turbulence/misalignment-induced magnetic reconnection. It is likely that both ingredients play a role, with Ohmic dissipation enabling a dense, small (perhaps astronomical-unit-scale) RSD to form early in the protostellar accretion phase, and flow complexity facilitating the growth of the disk at later times by weakening the magnetic braking of the lower-density protostellar accretion flow.

The above hybrid picture, although probably not unique, has the virtue of being at least qualitatively consistent with the available observational and theoretical results. The small Ohmic-dissipation-enabled disk can in principle drive powerful outflows that are a defining characteristic of class 0 sources from close to the central object where the magnetic field and matter are well coupled due to thermal ionization of alkali metals. During the class 0 phase, it may have grown sufficiently in mass to account for the compact component often detected in interferometric continuum observations, but not so much in size as to violate the constraint that the majority of the compact components remain unresolved to date. The relatively small size of early disks could result from magnetic braking.

A relatively small early disk could also result from a small specific angular momentum of the core material to begin with. There is a strong need to determine more systematically the magnitude and distribution of angular momentum in prestellar cores through detailed observations. Another need is to determine the structure of the magnetic fields on the 10^2–10^3-AU scale that is crucial to disk formation. For example, detection of magnetic field twisting would be direct evidence for magnetic braking. With the polarization capability coming online soon, ALMA is expected to make progress on this observational front.

On the theory front, there is a strong need to carry out simulations that combine non-ideal MHD effects with turbulence and complex initial conditions on the core scale, including magnetic field-rotation misalignment. This will be technically challenging to do, but is required to firm up disk-formation scenarios such as the hybrid one outlined above.

All the evidence and theory shows that the formation of outflows is deeply connected with the birth of magnetized disks during gravitational collapse. Early outflows and later higher-speed jets may be two aspects of a common underlying physical picture in which acceleration is promoted both by toroidal magnetic field pressure on larger scales as well as centrifugal "fling" from smaller scales. Simulations, theory, and observations are converging on the idea that the collapse and outflow phenomenon is universal, covering the full range of stellar mass scales from brown dwarfs to massive stars. Finally, the earliest stages of planet formation take place on the very same timescales as disks are formed and outflows are first launched. While little is yet known about this connection, it is evident that this must be important. Many aspects of planet formation are tied to the properties of early disks.

In summary, firm knowledge of disk formation will provide a solid foundation for understanding the links between early disks, outflows, and planets, opening the way to discovering the deep connections between star and planet formation.

Acknowledgments. Z.-Y.L. is supported in part by NASA NNX10AH30G, NNX14AB38G, and NSF AST1313083, R.B. by the Deutsche Forschungsgemeinschaft (DFG) via the grant BA 3706/1-1, R.E.P. by a Discovery Grant from the National Science and Engineering Research Council (NSERC) of Canada, and J.K.J. by a Lundbeck Foundation Junior Group Leader Fellowship as well as Centre for Star and Planet Formation, funded by the Danish National Research Foundation and the University of Copenhagen's Programme of Excellence.

REFERENCES

Aharonian F. et al. (2013) *ArXiv e-prints*, arXiv:1301.4124.
Allen A. et al. (2003) *Astrophys. J., 599*, 363.
Alves F. O. et al. (2008) *Astron. Astrophys., 486*, L13.
Alves J. et al. (2007) *Astron. Astrophys., 462*, L17.
André P. et al. (1993) *Astrophys. J., 406*, 122.
Armitage P. J. (2011) *Annu. Rev. Astron. Astrophys., 49*, 195.
Bacciotti F. et al. (2002) *Astrophys. J., 576*, 222.
Bacciotti F. et al. (2011) *Astrophys. J. Lett., 737*, L26.
Bai X.-N. and Stone J. M. (2013) *Astrophys. J., 769*, 76.
Banerjee R. and Pudritz R. E. (2006) *Astrophys. J., 641*, 949.
Banerjee R. and Pudritz R. E. (2007) *Astrophys. J., 660*, 479
Banerjee R. et al. (2007) *Astrophys. J., 668*, 1028.
Banerjee R. et al. (2009) In *Protostellar Jets in Context* (K. Tsinganos et al., eds.), p. 421. Springer-Verlag, Berlin.
Beck R. (2001) *Space Sci. Rev., 99*, 243.
Belloche A. (2013) In *Role and Mechanisms of Angular Momentum Transport During the Formation and Early Evolution of Stars* (P. Hennebelle and C. Charbonnel, eds.), p. 25. Cambridge Univ., Cambridge.
Belloche A. et al. (2002) *Astron. Astrophys., 393*, 927.
Bergin E. A. and Tafalla M. (2007) *Annu. Rev. Astron. Astrophys., 45*, 339.
Bertram E. et al. (2012) *Mon. Not. R. Astron. Soc., 420*, 3163.
Beuther H. and Shepherd D. (2005) In *Cores to Clusters: Star Formation with Next Generation Telescopes* (M. S. N. Kumar et al., eds.), p. 105. Springer, New York.
Beuther H. et al. (2004) *Astrophys. J. Lett., 616*, L23.
Blandford R. D. and Payne D. G. (1982) *Mon. Not. R. Astron. Soc., 199*, 883.
Bodenheimer P. (1995) *Annu. Rev. Astron. Astrophys., 33*, 199.
Boley A. C. et al. (2010) *Icarus, 207*, 509.
Boss A. P. (1998) In *Origins* (C. E. Woodward et al., eds.), pp. 314–326. ASP Conf. Ser. 148, Astronomical Society of the Pacific, San Francisco.
Boss A. P. (2002) *Astrophys. J., 568*, 743.
Boss A. P. and Keiser S. A. (2013) *Astrophys. J., 764*, 136.
Braiding C. R. and Wardle M. (2012a) *Mon. Not. R. Astron. Soc., 422*, 261.
Braiding C. R. and Wardle M. (2012b) *Mon. Not. R. Astron. Soc., 427*, 3188.
Brinch C. et al. (2007) *Astron. Astrophys., 475*, 915.
Brinch C. et al. (2009) *Astron. Astrophys., 502*, 199.
Brown D. W. et al. (2000) *Mon. Not. R. Astron. Soc., 319*, 154.
Burkert A. and Bodenheimer P. (2000) *Astrophys. J., 543*, 822.
Bürzle F. et al. (2011) *Mon. Not. R. Astron. Soc., 417*, L61.
Cabrit S. et al. (1997) In *Herbig-Haro Flows and the Birth of Stars* (B. Reipurth and C. Bertout, eds.), p. 163. Springer-Verlag, Berlin.
Carpenter J. M. et al. (2006) *Astrophys. J. Lett., 651*, L49.
Carrasco-González C.et al. (2010) *Science, 330*, 1209.
Caselli P. et al. (2002) *Astrophys. J., 572*, 238.
Cerqueira A. H. et al. (2006) *Astron. Astrophys., 448*, 231.
Cesaroni R. et al. (1997) *Astron. Astrophys., 325*, 725.
Cesaroni R. et al. (1994) *Astrophys. J. Lett., 435*, L137.
Chandler C. J. and Richer J. S. (2000) *Astrophys. J., 530*, 851
Chapman N. L. et al. (2013) *Astrophys. J., 770*, 151.
Chiang E. I. and Goldreich P. (1997) *Astrophys. J., 490*, 368.
Chiang H.-F. et al. (2008) *Astrophys. J., 680*, 474.
Chiang H.-F. et al. (2012) *Astrophys. J., 756*, 168.
Chibueze J. O. et al. (2012) *Astrophys. J., 748*, 146.
Chrysostomou A. et al. (2008) *Astron. Astrophys., 482*, 575.
Ciardi A. and Hennebelle P. (2010) *Mon. Not. R. Astron. Soc., 409*, L39.
Ciolek G. E. and Königl A. (1998) *Astrophys. J., 504*, 257.
Coffey D. et al. (2007) *Astrophys. J., 663*, 350.
Comerón F. et al. (2003) *Astron. Astrophys., 406*, 1001.
Commerçon B. et al. (2010) *Astron. Astrophys., 510*, L3.
Contopoulos I. et al. (1998) *Astrophys. J., 504*, 247.
Contopoulos J. (1995) *Astrophys. J., 450*, 616.
Crutcher R. M. (2012) *Annu. Rev. Astron. Astrophys., 50*, 29.
Crutcher R. M. et al. (1999) *Astrophys. J. Lett., 514*, L121.
Crutcher R. M. et al. (2010) *Astrophys. J., 725*, 466.
Cunningham A. J.et al. (2009) *Astrophys. J., 692*, 816.
Cunningham A. J. et al. (2012) *Astrophys. J., 744*, 185.
Dapp W. B. and Basu S. (2010) *Astron. Astrophys., 521*, L56.
Dapp W. B. et al. (2012) *Astron. Astrophys., 541*, A35.

Davidson J. A. et al. (2011) *Astrophys. J., 732*, 97.
Dedner A. et al. (2002) *J. Comp. Phys., 175*, 645.
Dib S. et al. (2010) *Astrophys. J., 723*, 425.
Duffin D. F. and Pudritz R. E. (2008) *Mon. Not. R. Astron. Soc., 391*, 1659.
Duffin D. F. and Pudritz R. E. (2009) *Astrophys. J. Lett., 706*, L46.
Dutrey A. et al. (2007) In *Protostars and Planets V* (B. Reipurth et al., eds.), pp. 495–506. Univ. of Arizona, Tucson.
Eisner J. A. (2012) *Astrophys. J., 755*, 23.
Enoch M. L. et al. (2011) *Astrophys. J. Suppl., 195*, 21.
Enoch M. L. et al. (2009) *Astrophys. J., 707*, 103.
Evans N. J. II et al. (2009) *Astrophys. J. Suppl., 181*, 321.
Fang M. et al. (2009) *Astron. Astrophys., 504*, 461.
Favre C. et al. (2014) *Astrophys. J., 790*, 55.
Fendt C. (2011) *Astrophys. J., 737*, 43.
Fendt C. and Camenzind M. (1996) *Astrophys. Lett. Comm., 34*, 289.
Fernández-López M. et al. (2011) *Astron. J., 142*, 97.
Franco-Hernández R. et al. (2009) *Astrophys. J., 701*, 974.
Fromang S. et al. (2006) *Astron. Astrophys., 457*, 371.
Galli D. and Shu F. H. (1993) *Astrophys. J., 417*, 220.
Galli D. et al. (2006) *Astrophys. J., 647*, 374.
Galvagni M. et al. (2012) *Mon. Not. R. Astron. Soc., 427*, 1725.
Gammie C. F. (1996) *Astrophys. J., 457*, 355.
Gammie C. F. (2001) *Astrophys. J., 553*, 174.
Girart J. M. et al. (2009) *Astrophys. J., 694*, 56.
Goodman A. A. et al. (1993) *Astrophys. J., 406*, 528.
Gutermuth R. A. et al. (2009) *Astrophys. J. Suppl., 184*, 18.
Guzmán A. E. et al. (2012) *Astrophys. J., 753*, 51.
Hansen C. E. et al. (2012) *Astrophys. J., 747*, 22.
Harsono D. et al. (2014) *Astron. Astrophys., 562*, 77.
Hartigan P. et al. (1995) *Astrophys. J., 452*, 736.
Harvey D. W. A. et al. (2003) *Astrophys. J., 583*, 809.
Hasegawa Y. and Pudritz R. E. (2011) *Mon. Not. R. Astron. Soc., 417*, 1236.
Hasegawa Y. and Pudritz R. E. (2012) *Astrophys. J., 760*, 117.
Heiles C. and Troland T. H. (2005) *Astrophys. J., 624*, 773.
Hennebelle P. and Ciardi A. (2009) *Astron. Astrophys., 506*, L29.
Hennebelle P. and Fromang S. (2008) *Astron. Astrophys., 477*, 9.
Hernández J. et al. (2008) *Astrophys. J., 686*, 1195.
Hernández J. et al. (2010) *Astrophys. J., 722*, 1226.
Hogerheijde M. R. and Sandell G. (2000) *Astrophys. J., 534*, 880.
Holden L. et al. (2011) *Publ. Astron. Soc. Pac., 123*, 14.
Hosking J. G. and Whitworth A. P. (2004) *Mon. Not. R. Astron. Soc., 347*, 1001.
Hosokawa T. et al. (2010) *Astrophys. J., 721*, 478.
Hull C. L. H. et al. (2013) *Astrophys. J., 768*, 159.
Ida S. and Lin D. N. C. (2005) *Astrophys. J., 626*, 1045.
Ida S. and Lin D. N. C. (2008) *Astrophys. J., 685*, 584.
Isokoski K. et al. (2013) *Astron. Astrophys., 554*, 100.
Joergens V. et al. (2013) *Astron. Nachr., 334*, 159.
Joos M. et al. (2012) *Astron. Astrophys., 543*, A128.
Joos M. et al. (2013) *Astron. Astrophys., 554*, A17.
Jørgensen J. K. (2004) *Astron. Astrophys., 424*, 589.
Jørgensen J. K. and van Dishoeck E. F. (2010) *Astrophys. J. Lett., 710*, L72.
Jørgensen J. K. et al. (2004) *Astron. Astrophys., 413*, 993.
Jørgensen J. K. et al. (2005a) *Astrophys. J., 632*, 973.
Jørgensen J. K. et al. (2005b) *Astrophys. J. Lett., 631*, L77.
Jørgensen J. K. et al. (2007) *Astrophys. J., 659*, 479.
Jørgensen J. K. et al. (2009) *Astron. Astrophys., 507*, 861.
Keene J. and Masson C. R. (1990) *Astrophys. J., 355*, 635.
Keto E. (2002) *Astrophys. J., 580*, 980.
Keto E. (2003) *Astrophys. J., 599*, 1196.
Krasnopolsky R. and Königl A. (2002) *Astrophys. J., 580*, 987.
Krasnopolsky R. et al. (2010) *Astrophys. J., 716*, 1541.
Krasnopolsky R. et al. (2011) *Astrophys. J., 733*, 54.
Krasnopolsky R. et al. (2012) *Astrophys. J., 757*, 77.
Kratter K. M. et al. (2008) *Astrophys. J., 681*, 375.
Kratter K. M. and Murray-Clay R. A. (2011) *Astrophys. J., 740*, 1.
Krumholz M. R. et al. (2013) *Astrophys. J. Lett., 767*, L11.
Kunz M. W. and Mouschovias T. C. (2010) *Mon. Not. R. Astron. Soc., 408*, 322.
Kwon W. et al. (2009) *Astrophys. J., 696*, 841.
Lee C.-F. (2011) *Astrophys. J., 741*, 62.
Lee K. et al. (2012) *Astrophys. J., 761*, 171.
Li Z.-Y. and McKee C. F. (1996) *Astrophys. J., 464*, 373.

Li Z.-Y. and Nakamura F. (2006) *Astrophys. J. Lett., 640,* L187.
Li Z.-Y. and Shu F. H. (1996) *Astrophys. J., 472,* 211.
Li Z.-Y. et al. (2010) *Astrophys. J. Lett., 720,* L26.
Li Z.-Y. et al. (2011) *Astrophys. J., 738,* 180.
Li Z.-Y. et al. (2013) *Astrophys. J., 774,* 82.
Liu J. and Schneider T. (2009) *Eos Trans. AGU, 90(52),* Fall Meet. Suppl., Abstract P32C-04.
Lommen D. et al. (2008) *Astron. Astrophys., 481,* 141.
Looney L. W. et al. (2000) *Astrophys. J., 529,* 477.
Looney L. W. et al. (2003) *Astrophys. J., 592,* 255.
Lynden-Bell D. (1996) *Mon. Not. R. Astron. Soc., 279,* 389.
Lynden-Bell D. (2003) *Mon. Not. R. Astron. Soc., 341,* 1360.
Machida M. N. and Matsumoto T. (2011) *Mon. Not. R. Astron. Soc., 413,* 2767.
Machida M. N. et al. (2004) *Mon. Not. R. Astron. Soc., 348,* L1.
Machida M. N. et al. (2005a) *Mon. Not. R. Astron. Soc., 362,* 369.
Machida M. N. et al. (2005b) *Mon. Not. R. Astron. Soc., 362,* 382.
Machida M. N. et al. (2006) *Astrophys. J. Lett., 649,* L129.
Machida M. N. et al. (2007) *Astrophys. J., 670,* 1198.
Machida M. N. et al. (2008) *Astrophys. J., 676,* 1088.
Machida M. N. et al. (2010) *Astrophys. J., 724,* 1006.
Machida M. N. et al. (2011) *Publ. Astron. Soc. Japan, 63,* 555.
Mac Low M.-M. and Klessen R. S. (2004) *Rev. Mod. Phys., 76,* 125.
Masset F. S. et al. (2006) *Astrophys. J., 642,* 478.
Matsumoto T. and Tomisaka K. (2004) *Astrophys. J., 616,* 266.
Matsumura S. and Pudritz R. E. (2006) *Mon. Not. R. Astron. Soc., 365,* 572.
Matsumura S. et al. (2009) *Astrophys. J., 691,* 1764.
Maury A. J. et al. (2010) *Astron. Astrophys., 512,* A40.
Mayer L. et al. (2002) *Science, 298,* 1756.
McCaughrean M. J. and O'Dell C. R. (1996) *Astron. J., 111,* 1977.
McKee C. F. and Ostriker E. C. (2007) *Annu. Rev. Astron. Astrophys., 45,* 565.
Mellon R. R. and Li Z.-Y. (2008) *Astrophys. J., 681,* 1356.
Mellon R. R. and Li Z.-Y. (2009) *Astrophys. J., 698,* 922.
Menshchikov A. et al. (2010) *Astron. Astrophys., 518,* L103.
Mohanty S. et al. (2007) *Astrophys. J., 657,* 1064.
Molinari S. et al. (2010) *Astron. Astrophys., 518,* L100.
Motogi K. et al. (2013) *Mon. Not. R. Astron. Soc., 428,* 349.
Mouschovias T. C. and Ciolek G. E. (1999) In *The Origin of Stars and Planetary Systems* (C. J. Lada and N. D. Kylafis, eds.), p. 305. Springer-Verlag, Berlin.
Mouschovias T. C. and Paleologou E. V. (1979) *Astrophys. J., 230,* 204.
Mouschovias T. C. and Paleologou E. V. (1980) *Astrophys. J., 237,* 877.
Murillo N. M. et al. (2013) *Astron. Astrophys., 560,* A103.
Myers A. T. (2013) *Astrophys. J., 766,* 97.
Nakamura F. and Li Z.-Y. (2007) *Astrophys. J., 662,* 395.
Nakano T. (1984) *Fund. Cosmic Phys., 9,* 139.
Nakano T. and Nakamura T. (1978) *Publ. Astron. Soc. Japan, 30,* 671.
Nakano T. et al. (2002) *Astrophys. J., 573,* 199.
Natta A. et al. (2007) In *Protostars and Planets V* (B. Reipurth et al., eds.), pp. 767–781. Univ. of Arizona, Tucson.
Nishi R. et al. (1991) *Astrophys. J., 368,* 181.
Norman C. and Silk J. (1980) *Astrophys. J., 238,* 158.
O'Dell C. R. and Wen Z. (1992) *Astrophys. J., 387,* 229.
Ouyed R. and Pudritz R. E. (1997) *Astrophys. J., 482,* 712.
Padovani M. et al. (2013) *Astron. Astrophys.,* in press, arXiv:1310.2158.
Palau A. et al. (2013) *Mon. Not. R. Astron. Soc., 428,* 1537.
Pelletier G. and Pudritz R. E. (1992) *Astrophys. J., 394,* 117.
Persson M. V. et al. (2012) *Astron. Astrophys., 541,* A39.
Peters T. et al. (2010) *Astrophys. J., 711,* 1017.
Peters T. et al. (2011) *Astrophys. J., 729,* 72.
Peters T. et al. (2012) *Astrophys. J., 760,* 91.
Pilbratt G. L. et al. (2010) *Astron. Astrophys., 518,* L1.
Pineda J. E. et al. (2012) *Astron. Astrophys., 544,* L7.
Pollack J. B. et al. (1996) *Icarus, 124,* 62.
Price D. J. and Bate M. R. (2007) *Astrophys. Space Sci., 311,* 75.
Price D. J. et al. (2012) *Mon. Not. R. Astron. Soc., 423,* 45.
Pudritz R. E. and Norman C. A. (1983) *Astrophys. J., 274,* 677.

Pudritz R. E. and Norman C. A. (1986) *Astrophys. J., 301,* 571.
Pudritz R. E. et al. (2007) In *Protostars and Planets V* (B. Reipurth et al., eds.), pp. 277–294. Univ. of Arizona, Tucson.
Pudritz R. E. et al. (2009) In *Structure Formation in Astrophysics* (G. Chabrier, ed.), p. 84. Cambridge Univ., Cambridge.
Puga E. et al. (2006) *Astrophys. J., 641,* 373.
Qiu K. et al. (2007) *Astrophys. J., 654,* 361.
Qiu K. et al. (2008) *Astrophys. J., 685,* 1005.
Qiu K. et al. (2011) *Astrophys. J. Lett., 743,* L25.
Rafikov R. R. (2009) *Astrophys. J., 704,* 281.
Ray T. P. (2009) *Rev. Mex. Astron. Astrofis. Ser. Conf., 36,* 179.
Ray T. et al. (2007) In *Protostars and Planets V* (B. Reipurth et al., eds.), pp. 231–244. Univ. of Arizona, Tucson.
Reipurth B. and Bally J. (2001) *Annu. Rev. Astron. Astrophys., 39,* 403.
Reipurth B. et al., eds. (2007) *Protostars and Planets V.* Univ. of Arizona, Tucson.
Ricci L. et al. (2010) *Astron. Astrophys., 521,* 66.
Ridge N. A. and Moore T. J. T. (2001) *Astron. Astrophys., 378,* 495.
Rodríguez T. et al. (2012) *Astrophys. J., 755,* 100.
Rogers P. D. and Wadsley J. (2012) *Mon. Not. R. Astron. Soc., 423,* 1896.
Santos-Lima R. et al. (2012) *Astrophys. J., 747,* 21.
Santos-Lima R. et al. (2013) *Mon. Not. R. Astron. Soc., 429,* 3371.
Seifried D. et al. (2011) *Mon. Not. R. Astron. Soc., 417,* 1054.
Seifried D. et al. (2012a) *Mon. Not. R. Astron. Soc., 422,* 347.
Seifried D. et al. (2012b) *Mon. Not. R. Astron. Soc., 423,* L40.
Seifried D. et al. (2013) *Mon. Not. R. Astron. Soc., 432,* 3320.
Shepherd D. S. and Churchwell E. (1996a) *Astrophys. J., 457,* 267.
Shepherd D. S. and Churchwell E. (1996b) *Astrophys. J., 472,* 225.
Shibata K. and Uchida Y. (1985) *Publ. Astron. Soc. Japan, 37,* 31.
Shu F. H. et al. (1987) *Annu. Rev. Astron. Astrophys., 25,* 23.
Shu F. H. et al. (2006) *Astrophys. J., 647,* 382.
Sicilia-Aguilar A. et al. (2010) *Astrophys. J., 710,* 597.
Smith R. J. et al. (2013) *Astrophys. J., 771,* 24.
Smith R. J. et al. (2012) *Astrophys. J., 750,* 64.
Soker N. (2005) *Astron. Astrophys., 435,* 125.
Sollins P. K. et al. (2004) *Astrophys. J. Lett., 616,* L35.
Su Y.-N. et al. (2012) *Astrophys. J. Lett., 744,* L26.
Surcis G. et al. (2011) *Astron. Astrophys., 527,* A48.
Takakuwa S. et al. (2012) *Astrophys. J., 754,* 52.
Tang Y.-W. et al. (2009) *Astrophys. J., 700,* 251.
Tassis K. and Mouschovias T. C. (2005) *Astrophys. J., 618,* 783.
Tassis K. and Mouschovias T. C. (2007) *Astrophys. J., 660,* 388.
Terebey S. et al. (1993) *Astrophys. J., 414,* 759.
Tobin J. J. et al. (2010) *Astrophys. J., 712,* 1010.
Tobin J. J. et al. (2012a) *Astrophys. J., 748,* 16.
Tobin J. J. et al. (2012b) *Nature, 492,* 83.
Tomida K. et al. (2013) *Astrophys. J., 763,* 6.
Tomisaka K. (1998) *Astrophys. J. Lett., 502,* L163.
Tomisaka K. (2000) *Astrophys. J. Lett., 528,* L41.
Tomisaka K. (2002) *Astrophys. J., 575,* 306.
Torrelles J. M. et al. (2003) *Astrophys. J. Lett., 598,* L115.
Troland T. H. and Crutcher R. M. (2008) *Astrophys. J., 680,* 457.
Visser R. et al. (2009) *Astron. Astrophys., 495,* 881.
Visser R. et al. (2013) *Astrophys. J., 769,* 19.
Vorobyov E. I. and Basu S. (2005) *Astrophys. J. Lett., 633,* L137.
Vorobyov E. I. and Basu S. (2006) *Astrophys. J., 650,* 956.
Wang P. et al. (2010) *Astrophys. J., 709,* 27.
Wardle M. and Ng C. (1999) *Mon. Not. R. Astron. Soc., 303,* 239.
Weidenschilling S. J. (1977) *Astrophys. Space Sci., 51,* 153.
Whelan E. T. et al. (2005) *Nature, 435,* 652.
Whelan E. T. et al. (2006) *New Astron. Rev., 49,* 582.
Whelan E. T. et al. (2007) *Astrophys. J. Lett., 659,* L45.
Whelan E. T. et al. (2009) *Astrophys. J. Lett., 691,* L106.
Whelan E. T. et al. (2012) *Astrophys. J., 761,* 120.
Wu Y. et al. (2004) *Astron. Astrophys., 426,* 503.
Zhang Q. et al. (2007) *Astron. Astrophys., 470,* 269.
Zhao B. et al. (2011) *Astrophys. J., 742,* 10.
Ziegler U. (2005) *Astron. Astrophys., 435,* 385.

Dunham M. M., Stutz A. M., Allen L. E., Evans N. J. II, Fischer W. J., Megeath S. T., Myers P. C., Offner S. S. R., Poteet C. A., Tobin J. J., and Vorobyov E. I. (2014) The evolution of protostars: Insights from ten years of infrared surveys with Spitzer and Herschel. In *Protostars and Planets VI* (H. Beuther et al., eds.), pp. 195–218. Univ. of Arizona, Tucson, DOI: 10.2458/azu_uapress_9780816531240-ch009.

The Evolution of Protostars: Insights from Ten Years of Infrared Surveys with Spitzer and Herschel

Michael M. Dunham
Yale University and Harvard-Smithsonian Center for Astrophysics

Amelia M. Stutz
Max Planck Institute for Astronomy

Lori E. Allen
National Optical Astronomy Observatory

Neal J. Evans II
The University of Texas at Austin

William J. Fischer and S. Thomas Megeath
The University of Toledo

Philip C. Myers
Harvard-Smithsonian Center for Astrophysics

Stella S. R. Offner
Yale University

Charles A. Poteet
Rensselaer Polytechnic Institute

John J. Tobin
National Radio Astronomy Observatory

Eduard I. Vorobyov
University of Vienna and Southern Federal University

Stars form from the gravitational collapse of dense molecular cloud cores. In the protostellar phase, mass accretes from the core onto a protostar, likely through an accretion disk; and it is during this phase that the initial masses of stars and the initial conditions for planet formation are set. Over the past decade, new observational capabilities provided by the Spitzer Space Telescope and Herschel Space Observatory have enabled wide-field surveys of entire star-forming clouds with unprecedented sensitivity, resolution, and infrared wavelength coverage. We review resulting advances in the field, focusing both on the observations themselves and the constraints they place on theoretical models of star formation and protostellar evolution. We also emphasize open questions and outline new directions needed to further advance the field.

1. INTRODUCTION

The formation of stars occurs in dense cores of molecular clouds, where gravity finally overwhelms turbulence, magnetic fields, and thermal pressure. This review assesses the current understanding of the early stages of this process, focusing on the evolutionary progression from dense cores up to the stage of a star and disk without a surrounding envelope: the protostellar phase of evolution.

Because of the obscuration at short wavelengths by dust and the consequent reemission at longer wavelengths, protostars are best studied at infrared and radio wavelengths, using both continuum emission from dust and spectral lines. At the time of the last Protostars and Planets conference, only initial Spitzer results were reported and Herschel had not yet launched. For this review, we can incorporate much more complete Spitzer results and report on some initial Herschel results. We focus this review on the progress made with these facilities toward answering several key questions about protostellar evolution, as outlined below.

How do surveys find protostars and distinguish them from background galaxies, asymptotic giant branch (AGB) stars, and more evolved star plus disk systems? What is the resulting census of protostars and young stars in nearby molecular clouds? Various identification methods have been employed and are compared in section 2, leading to a much firmer census from relatively unbiased surveys of clouds in the Gould belt and Orion.

How do we distinguish various stages in protostellar evolution? Theoretically, a collapsing core initially forms a first hydrostatic core, which quickly collapses again to form a true protostar, referred to as stage 0. As material moves from the surrounding core to the disk and onto the star, the stellar mass eventually exceeds the core mass, and the system becomes a stage I object. We discuss in section 2 the observational correlatives to these stages (called classes) and the reliability of using observational signatures to distinguish between the stages, while various candidates for the elusive first hydrostatic core are considered in section 4.

How long do each of the observational classes last? Ideally, we would ask this question about the stages, but the issues addressed in section 2 restrict current information about timescales to the classes. With the large surveys now available and with careful removal of contaminants, we can now estimate durations of the observational classes (sections 2 and 4).

What is the luminosity distribution of protostars (section 2)? What histories of accretion are predicted by various theoretical and semi-empirical models, and how do these compare to the observations (section 3)? Does accretion onto the star vary smoothly or is it more episodic? This question is addressed in significant detail in the chapter by Audard et al. in this volume. Here we address it with respect to the luminosity distribution in section 3, and other observational constraints are discussed in section 5. While some results of monitoring are becoming available, most information on the protostellar accretion or luminosity variations comes from indirect proxies such as outflow patterns and chemical effects with timescales appropriate to reflect the luminosity evolution.

How does the disk form and evolve during the protostellar phase, and how is this affected by the accretion history? This question is also addressed from a theoretical perspective in the chapter by Z.-Y. Li et al. in this volume, so we focus in section 6 on the observational evidence for and properties of disks in the protostellar phases.

What is the microscopic (chemical and mineralogical) evolution of the infalling material? Spectroscopy has provided information on the nature of the building blocks of comets and planetesimals as they fall toward the disk (section 7).

What is the effect of the star-forming environment? Protostars are found in a wide range of environments, ranging from isolated dense cores to low-density clusters of cores in a clump, to densely clustered regions like Orion. These issues are addressed in section 8.

The organization of this review chapter is motivated by first providing a very broad overview of the protostars that have been revealed by Spitzer and Herschel surveys (section 2), followed by a general theoretical overview of the protostellar mass accretion process (section 3). The rest of the review then focuses on specialized topics related to protostellar evolution, including the earliest stages of evolution (section 4), the evidence for episodic mass accretion in the protostellar stage (section 5), the formation and early evolution of protostellar disks (section 6), the evolution of infalling material (section 7), and the role of environment (section 8).

2. PROTOSTARS REVEALED BY INFRARED SURVEYS

Within the nearest 500 pc, low-mass young stellar objects (YSOs) can be detected by modern instruments across most of the initial mass function (IMF) over a wide range of wavelengths. Due to the emission from surrounding dust in envelopes and/or disks, YSOs are best identified at infrared wavelengths where all but the most embedded protostars are detectable (see section 4). Older pre-main-sequence stars that have lost their circumstellar dust are not revealed in the infrared but can be identified in X-ray surveys (e.g., *Audard et al.*, 2007; *Winston et al.*, 2010; *Pillitteri et al.*, 2013).

Several large programs to detect and characterize YSOs have been carried out with the Spitzer Space Telescope (*Werner et al.*, 2004), which operated at 3–160 μm during its cryogenic mission. These include "From Molecular Cores to Planet-Forming Disks" (c2d) (*Evans et al.*, 2003, 2009), which covered seven large, nearby molecular clouds and ~100 isolated dense cores; the Spitzer Gould Belt Survey (*Dunham et al.*, 2013), which covered 11 additional nearby clouds; and Spitzer surveys of the Orion (*Megeath et al.*, 2012) and Taurus (*Rebull et al.*, 2010) molecular clouds. These same regions have recently been surveyed by a number of key programs with the far-infrared Herschel Space Observatory (*Pilbratt et al.*, 2010), which observed at 55–670 μm, including the Herschel Gould's Belt (GB) Survey (*André et al.*, 2010) and Herschel Orion Protostar Survey (HOPS) (*Fischer et al.*, 2013; *Manoj et al.*, 2013; *Stutz et al.*, 2013). The wavelength coverage and resolution offered by Herschel allow a more precise determination of protostellar properties than ever before. Many smaller Spitzer and Herschel surveys have also been performed and their results will be referred to when relevant.

The results from these large surveys differ due to a combination of sample selection, methodology, sensitivity, resolution, and genuine differences in the star-formation environments. While the last of these is of major scientific interest and is touched upon in section 8, definitive evidence for differences in star formation caused by environmental factors awaits a rigorous combining of the samples and analysis with uniform techniques, which is beyond the scope of this review. Here, we highlight the broad similarities and differences in the surveys.

2.1. Identification

Young stellar objects are generally identified in the infrared by their red colors relative to foreground and background stars. Reliable identification is difficult since many other objects have similar colors, including star-forming galaxies, active galactic nuclei (AGN), background AGB stars, and knots of shock-heated emission. Selection criteria based on position in various color-color and color-magnitude diagrams have been developed to separate YSOs from these contaminants and are described in detail by *Harvey et al.* (2007) for the c2d and GB surveys and by *Gutermuth et al.* (2009), *Megeath et al.* (2009, 2012), and *Kryukova et al.* (2012) for the Orion survey.

Once YSOs are identified, multiple methods are employed to pick out the subset that are protostars still embedded in and presumed to be accreting from surrounding envelopes. For the c2d and GB clouds, *Evans et al.* (2009) and *Dunham et al.* (2013) defined as protostars those objects with at least one detection at $\lambda \geq 350$ μm, with the rationale being that (sub)millimeter detections in typical surveys of star-forming regions trace surrounding dust envelopes but not disks due to the relatively low mass sensitivities of these surveys (see *Dunham et al.*, 2013). On the other hand, *Kryukova et al.* (2012) identified protostars in the c2d, Taurus, and Orion clouds using a set of color and magnitude criteria designed to pick out objects with the expected red colors of protostars, but they did not require (sub)millimeter detections. A comparison of the two samples shows general agreement except for a large excess of faint protostars in the *Kryukova et al.* (2012) sample.

This discrepancy in the number of faint protostars may be due to incompleteness in the *Dunham et al.* (2013) sample, whose requirement of a (sub)millimeter detection may have introduced a bias against the lowest-mass cores and thus the lowest-mass (and luminosity) protostars. It may also be due to unreliability of some detections by *Kryukova et al.* (2012), whose corrections for contamination from galaxies, edge-on disks, and highly extinguished class II YSOs is somewhat uncertain, especially at the lowest luminosities where contamination steeply rises. Both effects likely contribute. Resolution of this discrepancy should be possible in the near future once complete results from the Herschel and James Clerk Maxwell Telescope Submillimetre Common-User Bolometer Array (SCUBA)-2 Legacy GB surveys are available, since together they will fully characterize the population of dense cores in all the GB clouds with sensitivities below 0.1 M$_\odot$. This general trade-off of completeness vs. reliability remains the subject of ongoing study, with *Hsieh and Lai* (2013) recently presenting a new method of identifying YSOs in Spitzer data

that may increase the number of YSOs by ~30% without sacrificing reliability. We anticipate continued developments in the coming years.

2.2. Classification

The modern picture of low-mass star formation resulted from the merger of an empirical classification scheme with theoretical work on the collapse of dense, rotating cores (e.g., *Shu et al.*, 1987; *Adams et al.*, 1987; *Wilking et al.*, 1987; *Kenyon et al.*, 1993), and was reviewed by *White et al.* (2007) and *Allen et al.* (2007) at the last Protostars and Planets conference. In this picture there are three stages of evolution after the initial starless core. In the first, the protostellar stage, an infalling envelope feeds an accreting circumstellar disk. In the second, the envelope has dissipated, leaving a protoplanetary disk that continues to accrete onto the star. In the third, the star has little or no circumstellar material, but it has not yet reached the main sequence. Defining S_λ as the flux density at wavelength λ, *Lada* (1987) used the infrared spectral index

$$\alpha = \frac{d \log(\lambda S_\lambda)}{d \log \lambda} \qquad (1)$$

to divide objects into three groups — class I, II, and III — meant to correspond to these three stages. *Greene et al.* (1994) added a fourth class, the "flat-SED" sources (see section 2.4), leading to the following classification system:

Class I: $\alpha \geq 0.3$
Flat-SED: $-0.3 \leq \alpha < 0.3$
Class II: $-1.6 \leq \alpha < -0.3$
Class III: $\alpha < -1.6$

Class 0 objects were later added as a fifth class for protostars too deeply embedded to detect in the near-infrared but inferred through other means [i.e., outflow presence (*Andre et al.*, 1993)]. They were defined observationally as sources with a ratio of submillimeter to bolometric luminosity (L_{smm}/L_{bol}) greater than 0.5%, where L_{smm} is calculated for $\lambda \geq 350$ μm. Figure 1 shows example spectral energy distributions (SEDs) for a starless core and for class 0, class I, and flat-SED sources, as well as several other types of objects that will be discussed in subsequent sections. For the first time, Spitzer and Herschel surveys of star-forming regions have provided truly complete spectral coverage for most protostars.

Due to the effects of inclination, aspherical geometry, and foreground reddening, there is not a one-to-one correspondence between observational class and physical stage (e.g., *Whitney et al.*, 2003a,b; *Robitaille et al.*, 2006; *Crapsi et al.*, 2008). To avoid confusion, *Robitaille et al.* (2006) proposed the use of stages 0, I, II, and III when referring to the physical or evolutionary status of an object and the use of "class" only when referring to observations. We follow this recommendation throughout this review.

While the original discriminant between class 0 and I protostars is L_{smm}/L_{bol} this quantity has historically been difficult to calculate because it requires accurate submillimeter photometry. Another quantity often used in its place is the bolometric temperature (T_{bol}): the effective temperature of a blackbody with the same flux-weighted mean frequency as the observed SED (*Myers and Ladd,* 1993). T_{bol} begins near 20 K for deeply embedded protostars (*Launhardt et al.,* 2013) and eventually increases to the effective temperature of a low-mass star once all the surrounding core and disk material has dissipated. *Chen et al.* (1995) proposed the following class boundaries in T_{bol}: 70 K (class 0/I), 650 K (class I/II), and 2800 K (class II/III).

With the sensitivity of Spitzer, class 0 protostars are routinely detected in the infrared, and class I sources by α are both class 0 and I sources by T_{bol} (*Enoch et al.,* 2009). Additionally, sources with flat α have T_{bol} consistent with class I or class II, extending roughly from 350 K to 950 K, and sources with class II and III α have T_{bol} consistent with class II, implying that T_{bol} is a poor discriminator between α-based classes II and III (*Evans et al.,* 2009).

T_{bol} may increase by hundreds of Kelvins, crossing at least one class boundary, as the inclination ranges from edge-on to pole-on (*Jorgensen et al.,* 2009; *Launhardt et al.,* 2013; *Fischer et al.,* 2013). Thus, many class 0 sources by T_{bol} may in fact be stage I sources, and vice versa. Far-infrared and submillimeter diagnostics have a superior ability to reduce the influence of foreground reddening and inclination on the inferred protostellar properties. At such wavelengths, foreground extinction is sharply reduced and observations probe the colder, outer parts of the envelope that are less optically thick and thus where geometry is less important. Flux ratios at $\lambda \geq 70$ µm respond primarily to envelope density, pointing to a means of disentangling these effects and developing more robust estimates of evolutionary stage (*Ali et al.,* 2010; *Stutz et al.,* 2013). Along these lines, several authors have recently argued that L_{smm}/L_{bol} is a better tracer of underlying physical stage than T_{bol} (*Young and Evans,* 2005; *Dunham et al.,* 2010a; *Launhardt et al.,* 2013).

Recent efforts have vastly expanded the available 350-µm data for protostars via, e.g., the Herschel GB survey (see the chapter by André et al. in this volume), several Herschel key programs (e.g., *Launhardt et al.,* 2013; *Green et al.,* 2013b), and groundbased observations (e.g., *Wu et al.,* 2007). These efforts have enabled accurate calculation of L_{smm}/L_{bol} for large samples. Figure 2 compares classification via L_{smm}/L_{bol} and T_{bol} for the c2d, GB, and HOPS protostars. While there are methodology differences in the details of how L_{smm} is calculated for the c2d+GB (*Dunham et al.,* 2013) and HOPS (*Stutz et al.,* 2013) protostars that must be resolved in future studies, the two classification methods agree 81% of the time (counting T_{bol} class II and L_{smm}/L_{bol} class I as agreement). In a similar analysis of nine isolated globules, *Launhardt et al.* (2013) did not find such a clear agreement between T_{bol} and L_{smm}/L_{bol} classification, although methodology and possibly environmental differences are substantial (*Launhardt et al.,* 2013). Between the increased availability of submillimeter data and the evidence that L_{smm}/L_{bol} is a better tracer of underlying physical stage, we suggest using L_{smm}/L_{bol} rather than T_{bol}

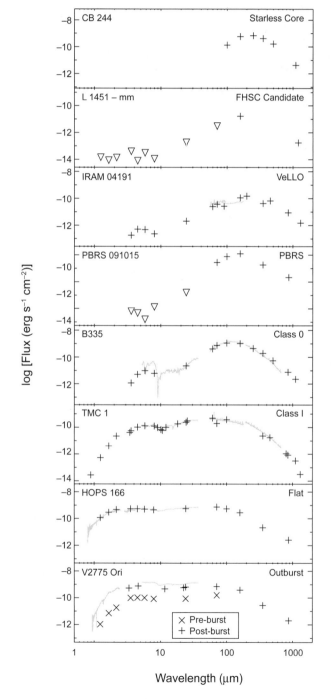

Fig. 1. SEDs for a starless core (*Stutz et al.,* 2010; *Launhardt et al.,* 2013), a candidate first hydrostatic core (*Pineda et al.,* 2011), a very low-luminosity object (*Dunham et al.,* 2008; *Green et al.,* 2013b), a PACS bright red source (*Stutz et al.,* 2013), a class 0 protostar (*Stutz et al.,* 2008; *Launhardt et al.,* 2013; *Green et al.,* 2013b), a class I protostar (*Green et al.,* 2013b), a flat-SED source (*Fischer et al.,* 2010), and an outbursting class I protostar (*Fischer et al.,* 2012). The + and × symbols indicate photometry, triangles denote upper limits, and gray lines show spectra.

as the primary tracer of the evolutionary status of protostars. However, the concept of monotonic evolution through the observational classes breaks down if accretion is episodic (see sections 3 and 5). Instead, protostars will move back and forth across class boundaries as their accretion rates and luminosities change (*Dunham et al.,* 2010a).

Table 1 lists the numbers of YSOs classified via α for the c2d+GB, Orion, and Taurus surveys. Since portions of Orion suffer from incompleteness, we present data for subregions where the counts are most complete (L1630 and L1641). Additionally, due to the high extinction toward Orion, we present for this region both the total number of flat-SED sources as well as the number that are likely reddened stage II objects based on analysis of longer-wavelength data. Not included are new protostars in Orion discovered by Herschel (see section 4.2.3); including these increases the total number by only 5–8%, emphasizing that Spitzer surveys missed relatively few protostars.

Table 1 also gives lifetimes for the protostellar (class 0 + I) and flat-SED objects. These are calculated under the following set of assumptions: (1) Time is the only variable, (2) star formation is continuous over at least the assumed class II lifetime, and (3) the class II lifetime is 2 million years (m.y.) (see the chapter by Soderblom et al. in this volume). This lifetime is best thought of as a "half-life"

rather than an absolute lifetime. Averaged over all surveys, we derive a protostellar lifetime of ~0.5 m.y. While the flat-SED lifetime by this analysis is ~0.4 m.y., this class may be an inhomogeneous collection of objects that are not all YSOs at the end of envelope infall (section 2.4). In all cases these are lifetimes of *observed classes* rather than *physical stages*, since the latter are not easily observable quantities. Indeed, some recent theoretical studies suggest the true lifetime of the protostellar stage may be shorter than the lifetime for the class 0 + I protostars derived here (*Offner and McKee,* 2011; *Dunham and Vorobyov,* 2012).

2.3. Protostellar Luminosities

Figure 3 plots L_{bol} vs. T_{bol} [a "BLT" diagram (*Myers and Ladd,* 1993)] for the protostars in the c2d+GB and Orion surveys, along with the separate L_{bol} and T_{bol} distributions. The 220 c2d+GB protostars are the YSOs in those surveys with submillimeter detections. The 332 Orion protostars are the 317 identified by *Megeath et al.* (2012) that were detected in HOPS 70-μm observations, plus an additional 15 that were newly discovered by HOPS [section 4.2.3 (*Stutz et al.,* 2013)]. Bolometric properties were calculated by trapezoidal integration under the available SEDs with no extrapolation to shorter or longer wavelengths and no corrections for foreground extinction. As found in previous work (*Kenyon et al.,* 1990), the luminosity distribution extends over several orders of magnitude, but this is now the case for hundreds of sources, and the distributions extend to even lower luminosities. Explanations for these broad distributions will be evaluated in section 3.

Statistics for the L_{bol} distributions appear in Table 2. Since the flux completeness limits of the c2d, GB, and HOPS surveys can be a function of position in regions with diffuse emission, and the relationship between source fluxes and L_{bol} depends on distance, source evolutionary status,

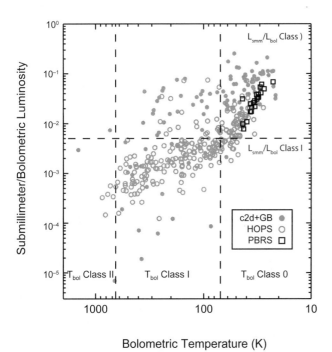

Fig. 2. Comparison of L_{smm}/L_{bol} and T_{bol} for the protostars in the c2d, GB, and HOPS surveys. The PBRS (section 4.2.3) are the 18 Orion protostars that have the reddest 70–24-μm colors, 11 of which were discovered with Herschel. The dashed lines show the class boundaries in T_{bol} from *Chen et al.* (1995) and in L_{smm}/L_{bol} from *Andre et al.* (1993). Protostars generally evolve from the upper right to the lower left, although the evolution may not be monotonic if accretion is episodic.

TABLE 1. YSO numbers and lifetimes.

	c2d+GB	L1630*	L1641†	Taurus
Numbers				
Class 0 + I‡	384	51	125	26
Flat	259	48	131	22
Flat§	—	30	74	—
Class II	1413	243	559	125
Lifetimes¶				
Class 0 + I (m.y.)	0.54	0.42	0.45	0.42
Flat (m.y.)	0.37	0.40	0.47	0.35
Class 0 + I (m.y.)**	—	0.37	0.39	—
Flat (m.y.)**	—	0.13	0.18	—

*Omitting regions of high nebulosity.
†Omitting the Orion Nebula region.
‡Not including new Herschel sources.
§Number of previous row that are likely reddened class II.
¶Assuming a class II lifetime of 2 m.y. (see text).
**Counting sources in fourth row (§) as class II.

local strength of the external radiation field, and total core mass available to be heated externally, it is difficult to derive an exact completeness limit in L_{bol}. *Dunham et al.* (2013) derive an approximate completeness limit of 0.05 $L_{\odot}$ for all but the most distant cloud in the c2d and GB surveys (IC 5146), which increases to 0.2 $L_{\odot}$ when IC 5146 is included. A full analysis of the HOPS completeness limit is still under investigation (*Stutz et al.*, in preparation).

The c2d+GB distribution extends over 3 orders of magnitude. Except for a statistically significant excess of low-luminosity class I sources, the distributions are generally similar for class 0 and I. Looking at the same clouds, *Kryukova et al.* (2012) find a distribution shifted to lower luminosities, with a mean (median) of 0.66 (0.63) $L_{\odot}$. The discrepancy between these two studies is partially resolved by updating the empirical relationship adopted by *Kryukova et al.* (2012) to calculate L_{bol}, but some discrepancy remains (*Dunham et al.*, 2013). As discussed above, there are concerns with both the completeness of *Dunham et al.* (2013) and the reliability of *Kryukova et al.* (2012).

Including new Herschel sources, the HOPS distribution extends over 4 orders of magnitude, with 13 protostars that are more luminous than the most luminous c2d+GB source. These sources increase the mean values of the HOPS luminosity distribution for both the class 0 and I subsamples by factors of 6 and 2.4, respectively, and also increase the median class 0 luminosity by a factor of 2.5 compared to the c2d+GB value. The degree to which these changes can be attributed to environmental differences as opposed to

incompleteness effects in the various samples is currently under investigation (*Stutz et al.*, in preparation).

2.4. Refining the Timeline of Protostellar Evolution

Determining when (and how quickly) envelopes dissipate is a major constraint for theories of how mass accretes onto protostars (see section 3). One key metric is the relative numbers (and thus implied durations) of class 0 and I protostars, which should be sensitive to the accretion history. We list the numbers of each in Table 3, using T_{bol} for classification since not all of the c2d and GB protostars have sufficient data for reliable L_{smm}/L_{bol} calculations. We find that 30% of protostars are in class 0 based on T_{bol} (*Enoch et al.*, 2009; *Dunham et al.*, 2013; *Fischer et al.*, 2013), implying a class 0 lifetime of 0.15 m.y., but interpreting these results is complicated by the lack of a one-to-one correspondence between class and stage. The combination of ubiquitous high-sensitivity far-infrared and submillimeter observations and improved modeling possible in the Herschel and Atacama Large Millimeter Array (ALMA) eras will lead to a better understanding of the relative durations of these stages.

In this review we have quoted a derived protostellar duration of ~0.5 m.y. However, the true duration remains uncertain. On one hand, some recent observations suggest that up to 50% of class I YSOs may be highly reddened stage II objects (*van Kempen et al.*, 2009; *Heiderman et al.*, 2010), a possibility also discussed by *White et al.* (2007). However, the exact fraction is uncertain and may have been overestimated by these studies. On the other hand, the nature of the flat-SED sources is unclear and has not yet been well determined in the literature. Some of them may represent objects in transition between stage I and II (e.g., *Calvet et al.*, 1994), but determining their exact nature is critical for determining the duration of the protostellar stage.

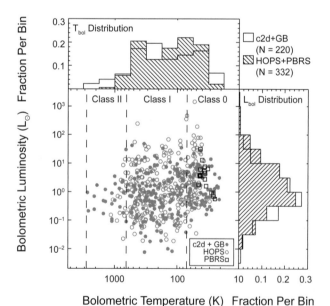

Fig. 3. The combined bolometric luminosity vs. temperature (BLT) diagram for the 552 protostars identified in the c2d, GB, and Spitzer Orion/HOPS surveys. BLT properties are measured by integrating under the observed SEDs. Marginal distributions of T_{bol} and L_{bol} are shown as histograms. The full population of protostars spans more than four orders of magnitude in luminosity.

TABLE 2. Protostellar L_{bol} statistics ($L_{\odot}$).

	c2d+GB	HOPS
Range	0.01–69	0.02–1440
Mean (median)	4.3 (1.3)	14 (1.2)
Class 0 mean (median)	4.5 (1.4)	27 (3.5)
Class I mean (median)	3.8 (1.0)	9.3 (1.0)

Note: Classes are by T_{bol}. See text for details on which sources are counted as protostars and included here.

TABLE 3. Protostellar numbers by T_{bol}.

	c2d+GB	HOPS
Class 0	63	93
Class I	132	222
Fraction of class 0*	0.32	0.30
Lifetime of class 0 (m.y.)†	0.16	0.15

*Ratio of class 0 to class 0 + I.

†Ratio multiplied by the 0.5-m.y. class 0 + I lifetime.

Future work on this front is clearly needed. Finally, all the timescales discussed in this chapter scale directly with the timescale for the class II phase, which we have assumed to be 2 m.y. (see the chapters in this volume on the ages of young stars by Soderblom et al. and on transitional disks by Espaillat et al. for further discussion of this timescale).

3. PROTOSTELLAR ACCRETION

The fundamental problem of star formation is how stars accrete their mass. In this section we give a general theoretical overview of how stars gain their masses and discuss the challenges of modeling protostellar luminosities. Understanding protostellar properties and evolution depends upon both the macrophysics of the core environment and the microphysics that drives gas behavior close to the protostar. After discussing protostellar luminosities (section 3.1), we describe three categories of models: those that depend on core properties (section 3.2), those that are based on the core environment and feedback (section 3.3), and those focused on accretion disk evolution (section 3.4). This separation is mainly for clarity since actual protostellar accretion is determined by a variety of nonlinear and interconnected physical processes that span many orders of magnitude in density and scale.

3.1. The Protostellar Luminosity Problem

The luminosity of a protostar provides an indirect measure of two quantities: the instantaneous accretion rate and the protostellar structure. Unfortunately, the luminosity contributions of each are difficult to disentangle. Over the past two decades various theories have attempted and failed to explain observed protostellar luminosites, which are found to be generally ~10× less luminous than expected (*Kenyon et al.*, 1990; *Kenyon and Hartmann*, 1995; *Young and Evans*, 2005; *Evans et al.*, 2009). This discrepancy became known as the "protostellar luminosity problem" and can be stated as follows. The total protostellar luminosity is given by

$$L_p = L_{phot} + f_{acc}\frac{Gm\dot{m}}{r} \qquad (2)$$

where L_{phot} is the photospheric luminosity generated by deuterium burning and Kelvin-Helmholtz contraction, m is the protostellar mass, $\dot{m}$ is the instantaneous accretion rate, r is the protostellar radius, and f_{acc} is the fraction of energy radiated away in the accretion shock. For accretion due to gravitational collapse (e.g., *Stahler et al.*, 1980), $\dot{m} \sim (c_s^2 + v_A^2 + v_t^2)^{3/2}/G \simeq 10^{-5}\ M_\odot$ yr^{-1}, where c_s, v_A, and v_t are the sound, Alfvén, and turbulent speeds, respectively. Given a typical protostellar radius r = 3.0 $R_\odot$ (*Stahler*, 1988; *Palla and Stahler*, 1992), the accretion luminosity of a 0.25-$M_\odot$ protostar is ~25 $L_\odot$. This is many times the observed median value and is only a lower limit to the true problem since it neglects the contributions from L_{phot} and from external heating by the interstellar radiation field.

However, equation (2) contains several poorly constrained quantities. The accretion rate onto a protostar is due to a combination of infalling material driven by gravitational collapse on large scales and the transport of material through an accretion disk on small scales. The disk properties are set by the core properties and the rate of infall (e.g., *Vorobyov*, 2009a; *Kratter et al.*, 2010a, see also the chapter by Z.-Y. Li et al. in this volume), and significant theoretical debate continues on the relationship between infall through the core and accretion onto the protostar, both instantaneously and averaged over time. Although it is difficult to directly measure $\dot{m}$, estimates for T Tauri stars based on infrared emission lines suggest $\dot{m} \lesssim 10^{-7}\ M_\odot$ yr^{-1} (e.g., *Muzerolle et al.*, 1998). Based on observational estimates for the embedded lifetime (see section 2), it is not possible for protostars to reach typical stellar masses without significantly higher average accretion during the embedded phase. While similar measurements of infrared emission lines in the protostellar phase are especially difficult, some observations suggest similarly low rates for class I objects (e.g., *Muzerolle et al.*, 1998; *White et al.*, 2007). Presently it is unclear if such observations are representative of all protostars or instead biased toward those objects already near the end of the protostellar stage and thus detectable in the near-infrared. A higher accretion rate is inferred in at least one protostar from its current accretion luminosity and mass derived from the Keplerian velocity profile of its disk (*Tobin et al.*, 2012). Mass infall rates derived from millimeter molecular line observations are also typically higher (e.g., *Brinch et al.*, 2009; *Mottram et al.*, 2013), although such rates are only available for some sources and are highly model dependent.

The intrinsic stellar parameters are also poorly constrained. While main-sequence stellar evolution is relatively well understood, pre-main-sequence evolution, especially during the first few million years, is much less well constrained. Consequently, both L_{phot} and r are uncertain and depend on the properties of the first core, past accretion history, accretion shock physics, and the physics of stellar interiors (*Baraffe et al.*, 2009; *Hosokawa et al.*, 2011; *Baraffe et al.*, 2012).

The luminosity problem was first identified by *Kenyon et al.* (1990), who also proposed several possible solutions, including slow and episodic accretion. In the slow accretion scenario, the main protostellar accretion phase lasts longer than a freefall time, which was then typically assumed to be ~0.1 m.y. Current observations suggest a protostellar duration of ~0.5 m.y. (section 2), which helps to alleviate the problem. In the episodic accretion scenario, accretion is highly variable and much of the mass may be accreted in statistically rare bursts of high accretion, observational evidence for which is discussed in section 5 and in the chapter by Audard et al. in this volume. One additional solution concerns the radiative efficiency of the accretion shock. If some of the shock kinetic energy is absorbed by the star or harnessed to drive outflows such that $f_{acc} < 1$ (e.g., *Ostriker and Shu*, 1995), the radiated energy would be reduced. Recent models applying combinations of these solutions have made significant progress toward reconciling theory and observation, as discussed in the remainder of this section.

3.2. Core-Regulated Accretion

A variety of theoretical models have been proposed for the gravitational collapse of dense cores. These models predict quantities such as the core density profile, gas velocities, and, most crucially, the rate of mass infall to the core center. If the accretion rate of a forming protostar is identical to the gas infall rate, as expected in models where disks either efficiently transfer mass onto the protostar or do not form at all, then these models also predict protostellar accretion rates. Numerical simulations of forming clusters suggest that accretion rates predicted by theoretical models are on average comparable to the stellar accretion rate (e.g., *Krumholz et al.*, 2012). In this section and in section 3.3, we discuss models that assume that the accreting gas is efficiently channeled from 0.1-pc to 0.1-AU scales, equivalent to focusing on the time-averaged accretion rate. Models where the instantaneous and time-averaged accretion rates may diverge significantly are discussed in section 3.4.

Figure 4 illustrates the accretion rate vs. time for a 1-$M_\odot$ star for a variety of theoretical models. Most core-regulated accretion models are based on the assumption that gas collapses from a local dense reservoir on the order of ~0.1 pc, i.e., a "core." The collapse of an isothermal, constant density sphere including thermal pressure is the simplest model with a self-similar solution. *Larson* (1969) and *Penston* (1969)

separately calculated the resulting accretion rate to be $\dot{m}$ = $46.9 c_s^3/G$ = 7.4×10^{-5} $(T/10K)^{3/2} M_\odot$ yr^{-1}, where c_s and T are the thermal sound speed and temperature, respectively. The self-similar solution for the collapse of a centrally condensed, isothermal sphere was computed by *Shu* (1977), who found $\dot{m}$ = 0.975 c_s^3/G, a factor of ~50 less than the Larson-Penston solution. In both cases, accretion does not depend on the initial mass, which leads to the prediction that accretion rate is independent of the instantaneous protostellar and final stellar mass and implies that the only environmental variable that affects accretion is the local gas temperature.

The above models consider only thermal pressure and gravity. However, cores are observed to be both magnetized and somewhat turbulent. The most massive clumps (e.g., *Barnes et al.*, 2011), which are possible progenitors of massive protostars, have turbulent linewidths several times the thermal linewidth. *McKee and Tan* (2003) proposed a "turbulent core" model to account for the higher column densities and turbulent linewidths of high-mass cores. In this model, $\dot{m} \propto m^{1/2} m_f^{3/4}$, where m and m_f are the instantaneous and final protostellar masses, respectively.

Significant numerical work has been devoted to modeling the formation of star clusters (see the chapter by Offner et al. in this volume). Many of these simulations present a very dynamical picture in which protostellar accretion rates vary as a function of location within the global gravitational potential and protostars "compete" with one another for gas (*Bonnell et al.*, 2001). The most massive stars form in the center of the gravitational potential where they can accrete at high rates. Protostars accrete until the gas is either completely accreted or dispersed, leading to a constant accretion time for all stars that is proportional to the global free-fall time. Analytically, this suggests $\dot{m} \propto m^{2/3} m_f$ (*Bonnell et al.*, 2001; *McKee and Offner*, 2010).

All core-regulated accretion models then fall somewhere between the limits of constant accretion rate and constant star-formation time. *McKee and Offner* (2010) propose hybrid models that include both a turbulent and a thermal component ("two-component turbulent core") or a competitive and a thermal component ("two-component competitive accretion"). Additional models have sought to analytically include rotation (*Terebey et al.*, 1984), nonzero initial velocities (*Fatuzzo et al.*, 2004), and magnetic fields (*Adams and Shu*, 2007). For example, *Adams and Shu* (2007) modify the isothermal sphere collapse problem to consider ambipolar diffusion. They derive accretion rates that are enhanced by a factor of 2–3 relative to the nonmagnetized case.

Declining accretion rates that fall by an order of magnitude or more from their peak are produced in some numerical simulations (*Vorobyov and Basu*, 2008; *Offner et al.*, 2009). *McKee and Offner* (2010) model this decline by imposing a "tapering" factor, $(1-t/t_f)$, such that accretion declines and terminates at some specified formation time, t_f. Infall could also be variable due to turbulence or magnetic effects (e.g., *Tassis and Mouschovias*, 2005), but such scenarios have not been well studied theoretically and are observationally difficult to constrain.

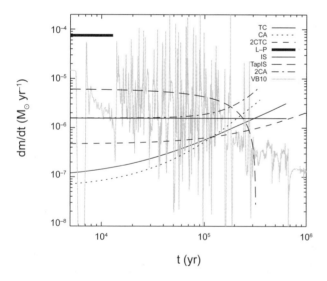

Fig. 4. Various model accretion rates as a function of time for a star of final mass 1 $M_\odot$. The models are the turbulent core (TC), competitive accretion (CA), two-component turbulent core (2CTC), Larson-Penston (L-P), isothermal sphere (IS), tapered isothermal sphere (TapIS), two-component accretion (2CA), and a simulated accretion history from *Vorobyov and Basu* (2010a). The TC, CA, and 2CTC are computed assuming a mean formation time of $\langle t_f \rangle$ = 0.44 m.y.

3.3. Feedback-Regulated Accretion

Most stars are born in cores in extended molecular clouds that are surrounded by lower-density filamentary gas and other recently formed protostars (e.g., *Bergin and Tafalla,* 2007; *Evans et al.,* 2009). These core environments provide mass for accretion from beyond the core, and "stellar feedback" in the form of ionizing radiation, winds, and outflows may also disperse star-forming gas. In this case the core environment is important, and the initial core mass is not sufficient to predict the final protostar mass. Instead the protostar mass depends on its accretion history and the competition between infall and dispersal.

Early models of such feedback-regulated accretion concentrated on the competition between accretion and outflows (*Norman and Silk,* 1980). *Basu and Jones* (2004) proposed a lognormal distribution of initial core masses, whose accretion rates are proportional to their mass, and accretion durations follow a waiting-time distribution. For this distribution, the probability density that accretion endures for t and then stops between t and t + dt is $(1/\tau)e^{-t/\tau}$, where τ is the mean accretion duration. A similar distribution of accretion durations was proposed to describe ejections by small multiple systems (*Bate and Bonnell,* 2005).

Recently, *Myers* (2010) proposed a feedback-regulated model that accounts for protostellar masses that follow the IMF. The basic ideas of the model are (1) protostars accrete from core-clump condensations, (2) the duration of accretion is the most important factor in setting the final mass of a protostar, and (3) accretion durations vary due to a combination of ejections, dispersal by stellar feedback, accretion competition, and exhaustion of initial gas. The waiting-time distribution describes this combination.

In dense clusters, a promising direction for protostar mass models is to treat the mass accretion rate as a function of mass or time, rather than as due to the collapse of a particular initial configuration. This approach allows discrimination between various IMF-forming accretion models (see also section 3.2). *Myers* (2009b, 2010) formulate a two-component accretion (2CA) rate having a constant, thermal component and a mass-dependent component (see Fig. 4). The 2CA accretion rate is similar to that of the two-component turbulent core model. *Myers* (2010) combined this 2CA model with an explicit distribution of accretion durations to derive the protostellar mass distribution and showed that it closely resembles the stellar IMF.

3.4. Disk-Regulated Accretion

Since most core mass likely accretes through a disk (section 6), it is important to determine how this mass is redistributed within the disk and transported onto the star. Two main processes of mass and angular momentum transport in disks have been proposed: viscous torques due to turbulence triggered by the magneto-rotational instability (MRI) (*Balbus and Hawley,* 1991) and gravitational torques induced by gravitational instability (GI) (*Lin and*

Pringle, 1987; *Laughlin and Bodenheimer,* 1994). Which of these mechanisms dominates and the efficiency with which the disk transports mass onto the protostar (and thus the degree to which the instantaneous infall and accretion rates are coupled) are currently debated and may depend on the details of local physical conditions, including the core properties and the infall rate. Numerical models that circumvent the complicated physics of the MRI and GI and treat both as a local viscous transport mechanism have been developed based on the *Shakura and Sunyaev* (1973) α-parameterization. These α-disk models have been successful in describing many aspects of disk physics and accretion (e.g., *Kratter et al.,* 2008; *Zhu et al.,* 2009), although their applicability may be limited for fairly massive disks (*Vorobyov,* 2010).

The MRI is known to operate only if the ionization fraction is sufficiently high to couple the magnetic field with the gas (*Blaes and Balbus,* 1994). Protostellar disks are sufficiently cold and dense that known ionization sources (thermal, X-rays, cosmic rays) may fail to provide the ionization needed to sustain the MRI, particularly in the disk mid-plane. If this is the case, *Gammie* (1996) proposed that the MRI may be active only in a layer near the surface, an idea known as "layered accretion." Various disk models suggest that the so-called dead zone, wherein the MRI is largely suppressed, may occupy a significant fraction of the disk volume at astronomical unit scales (see review by *Armitage,* 2011).

On the other hand, analytic models and numerical hydrodynamics simulations indicate that the physical conditions in protostellar disks may be favorable for the development of GI (*Toomre,* 1964; *Lin and Pringle,* 1987; *Laughlin and Bodenheimer,* 1994). Whenever the destabilizing effect of self-gravity becomes comparable to the stabilizing effects of pressure and shear, spiral density waves develop. If heating due to GI is balanced by disk cooling, the disk settles into a quasisteady state in which GI transports angular momentum outward, allowing mass to accrete onto the central star (*Lodato and Rice,* 2004). In this picture, gravitational torques alone are sufficient to drive accretion rates consistent with observations of intermediate- and upper-mass T Tauri stars (*Vorobyov and Basu,* 2007, 2008). However, for the very-low-mass disks around low-mass stars and brown dwarfs, GI is likely to be suppressed and, consequently, viscous mass transport due to MRI alone may explain the accretion rates of these objects (*Vorobyov and Basu,* 2009).

The degree of GI depends on a number of factors including the angular momentum of the parent core, infall rate, disk mass, and amount of radiative heating from the central protostar (*Kratter et al.,* 2010a; *Offner et al.,* 2010; *Vorobyov and Basu,* 2010a; *Stamatellos et al.,* 2011). If the disk is relatively massive and the local disk cooling time is faster than the dynamical time, sections of spiral arms can collapse into bound fragments and lead to a qualitatively different mode of disk evolution. In this mode, disk accretion is an intrinsically variable process due to disk fragmentation, non-axisymmetric structure, and gravitational torques

(*Vorobyov,* 2009b). Fragments that form in the disk outer regions are quickly driven inward due to the loss of angular momentum via gravitational interaction with the spiral arms (*Vorobyov and Basu,* 2005, 2006; *Baruteau et al.,* 2011; *Cha and Nayakshin,* 2011). As they accrete onto the protostar, clumps trigger luminosity bursts similar in magnitude to FU-Orionis-type or EX-Lupi-type events (*Vorobyov and Basu,* 2005, 2006, 2010a; *Machida et al.,* 2011a).

This burst mode of accretion mostly operates in the embedded phase of protostellar evolution when the continuing infall of gas from the parent core triggers repetitive episodes of disk fragmentation. The mass accretion onto the burgeoning protostar is characterized by short ($\lesssim$100–200 yr) bursts with accretion rate $\dot{m} \gtrsim$ a few $\times$ 10^{-5} $M_\odot$ yr^{-1} alternated with longer (10^3–10^4 yr) quiescent periods with $\dot{m} \lesssim 10^{-6}$ $M_\odot$ yr^{-1} (*Vorobyov and Basu,* 2010a). After the parent core is accreted or dispersed, any remaining fragments may trigger final accretion bursts or survive to form wide-separation planets or brown dwarfs (*Vorobyov and Basu,* 2010b; *Kratter et al.,* 2010b; *Vorobyov,* 2013).

Gravitational fragmentation is one of many possible mechanisms that can generate accretion and luminosity bursts (see the chapter by Audard et al. in this volume). Among the other mechanisms, a combination of MRI and GI has been the most well studied (*Zhu et al.,* 2009, 2010; *Martin et al.,* 2012). In this scenario, GI in the outer disk transfers gas to the inner subastronomical unit region where it accumulates until the gas density and temperature reach values sufficient for thermal ionization to activate the MRI. The subsequent enhanced angular momentum transport triggers a burst of accretion. The GI + MRI burst mechanism may act in disks that are not sufficiently massive to trigger disk fragmentation, but the details of the MRI are still poorly understood. Independent of their physical origin, luminosity bursts have important implications for disk fragmentation, accretion, and theoretical models of star formation.

3.5. Comparison Between Models and Observations

Here, we consider direct comparisons between observations and several theoretical models.

3.5.1. Protostellar luminosities. In order to consider the luminosity problem, *McKee and Offner* (2010) developed an analytic formalism for the present-day protostellar mass function (PMF). The PMF depends on the instantaneous protostellar mass, final mass, accretion rate, and average protostellar lifetime. Given some model for protostellar luminosity as a function of protostellar properties, *Offner and McKee* (2011) then derived the present-day protostellar luminosity function (PLF). Since the PLF depends only on observable quantities such as the protostellar lifetime and on a given theoretical model for accretion, the PLF can be used to directly compare star-formation theories with observations.

Offner and McKee (2011) computed the predicted PLFs for a variety of models and parameters, including the iso-

thermal sphere, turbulent core, competitive accretion, and two-component turbulent core models. Figure 5 compares the observed protostellar luminosity distribution with some of these predicted PLFs where the accretion rate is allowed to taper off as the protostar approaches its final mass. Models in which the accretion rate depends on the final mass (such as the turbulent core or competitive accretion models) naturally produce a broad distribution of luminosities. Offner and McKee also found that the theoretical models actually produce luminosities that are *too dim* compared to observations, given an average formation time of $\langle t_f \rangle \sim 0.5$ m.y. and allowing for episodic accretion. They concluded that a star-formation time of $\langle t_f \rangle \simeq 0.3$ m.y. provides a better match to the mean and median observed luminosities.

On the other hand, given that numerical simulations of disk evolution indicate that protostellar accretion may be an intrinsically variable process, the wide spread in the observed protostellar luminosity distribution may result from large-scale variations in the protostellar accretion rate (*Dunham et al.,* 2010a). This idea was further developed by *Dunham and Vorobyov* (2012), who used numerical hydrodynamic simulations of collapsing cores coupled with radiative transfer calculations to compare the model and observed properties of young embedded sources in the c2d clouds. They showed that gravitationally unstable disks with accretion rates that both decline with time and feature short-term variability and episodic bursts can reproduce the

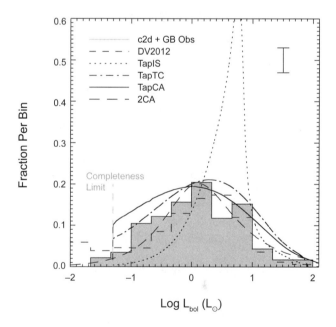

Fig. 5. Distribution of extinction-corrected protostellar luminosities from the c2d+GB surveys (shaded), predicted from disk simulations [dash (*Dunham and Vorobyov,* 2012)], TapIS (dots), tapered TC (dot-dash), tapered CA [dot-dot-dash (*Offner and McKee,* 2011)], and 2CA [long dash (*Myers,* 2012)]. The tapered models adopt a completeness limit of 0.05 $L_\odot$. Typical observational uncertainties are shown in the upper right.

full spread of observations, including very-low-luminosity objects. As shown in Fig. 5, accretion variability induced by GI and disk fragmentation can thus provide a reasonable match to the observed protostellar luminosity distribution and resolve the long-standing luminosity problem.

Finally, the distribution of protostellar masses can be obtained for the case of feedback-regulated accretion by combining the 2CA model with an explicit distribution of accretion times. *Myers* (2011, 2012) used the corresponding distribution of masses and accretion rates to compute the predicted PLF and found that it is in reasonable agreement with the observed protostellar luminosity distribution in nearby clouds.

3.5.2. Ages of young clusters. Models that specify accretion durations can also be tested against age estimates of young clusters. Such models can predict as a function of time the number of cluster members that are protostars, since they are still accreting, and the number that are pre-main-sequence stars (PMS), since they have stopped accreting. Application of the 2CA model to embedded cluster members identified as protostars or PMS indicate typical cluster ages of 1–3 m.y., in good agreement with estimates from optical and infrared spectroscopy and pre-main-sequence evolutionary tracks. This method can be used to date obscured young subclusters that are inaccessible to optical spectroscopy (*Myers*, 2012).

3.5.3. Future work. The models described above provide tangible, diverse resolutions to the luminosity problem, including mass-dependent and highly variable accretion histories. Future observational work should concentrate on constraining the magnitude and timescales of protostellar variability to assess its effects on protostellar luminosities and, ultimately, its importance in the mass-accretion process. On the theoretical front, high-resolution global-disk simulations, which include protostellar heating, magnetization, and ionization, are needed to improve our understanding of disk-regulated accretion variability. Improved theoretical understanding of outflow launching and evolution is needed to interpret observed outflow variability and how it correlates with accretion. Additional synthetic observations of models that take into account the complex, asymmetric morphologies of accreting protostars and the effects of external heating are required to properly compare to observed luminosities. Finally, dust evolution and chemical reaction networks determine the distribution of species, which is the lens through which we perceive all observational results. Many theoretical studies gloss over or adopt simplistic chemical assumptions, which undermines direct comparison between theory and observation.

4. THE EARLIEST OBSERVABLE STAGES OF PROTOSTELLAR EVOLUTION

The identification of young stars in the very earliest stages of formation is motivated by the goals of studying the initial conditions of the dense gas associated with star formation before modification by feedback and of deter-

mining the properties of cores when collapse begins. The earliest theoretically predicted phases are the first hydrostatic core and stage 0 protostars. Observational classes of young objects include class 0 sources, very-low-luminosity objects (VeLLOs) (section 4.2.1), candidate first hydrostatic cores (FHSC) (section 4.2.2), and PACS bright red sources (PBRS) (section 4.2.3). The challenge is to unambiguously tie observations to theory, a process complicated by optical depth effects, geometric (inclination) degeneracies, and intrinsically faint and/or very red observed SEDs. We begin with a theoretical overview of the earliest stages of evolution and then follow with observational anchors provided by Spitzer, Herschel, and other facilities.

4.1. Theoretical Framework

Once a dense molecular cloud core begins to collapse, the earliest object that forms is the FHSC. First predicted by *Larson* (1969), the FHSC exists between the starless and protostellar stages of star formation and has not yet been unambiguously identified by observations. The FHSC forms once the central density increases to the point where the inner region becomes opaque to radiation [$\rho_o \gtrsim 10^{-13}$ g cm^{-3} (*Larson*, 1969)], rendering the collapse adiabatic rather than isothermal. This object continues to accrete from the surrounding core and both its mass and central temperature increase with time. Once the temperature reaches ~2000 K, the gravitational energy liberated by accreting material dissociates H_2, preventing the temperature from continuing to rise to sufficiently balance gravity. At this point the second collapse is initiated, leading to the formation of the second hydrostatic core, more commonly referred to as the protostar.

The current range in predicted FHSC properties are summarized in Table 4. The significant variation in predicted properties is largely driven by different prescriptions and assumptions about magnetic fields (e.g., *Commerçon et al.*, 2012a), rotation (e.g., *Saigo and Tomisaka*, 2006; *Saigo et al.*, 2008), and accretion rates. Rotation may produce a flattened, disk-like morphology for the entire FHSC (*Saigo*

TABLE 4. Predicted properties of FHSCs.

Property	Range	References
Maximum mass ($M_\odot$)	0.04–0.05	[1,2,3,4]
	0.01–0.1*	
Lifetime (k.y.)	0.5–50	[1,3,4,6,7]
Internal luminosity ($L_\odot$)	10^{-4}–10^{-1}	[2,3]
Radius (AU)	~5	[2]
	$\geq$10–20*	[5,6]

*With the effects of rotation included.

References: [1] *Boss and Yorke* (1995); [2] *Masunaga et al.* (1998); [3] *Omukai* (2007); [4] *Tomida et al.* (2010); [5] *Saigo and Tomisaka* (2006); [6] *Saigo et al.* (2008); (7) *Commerçon et al.* (2012a).

and Tomisaka, 2006; Saigo et al., 2008), implying that disks may actually form before protostars with masses larger than the protostars themselves (*Bate, 2011; Machida and Matsumoto, 2011*).

Studies of FHSC SEDs have shown that the emitted radiation is completely reprocessed by the surrounding envelope, with SEDs characterized by emission from 10 to 30 K dust and no observable emission below ~20–50 μm (*Boss and Yorke, 1995; Masunaga et al., 1998; Omukai, 2007; Saigo and Tomisaka, 2011*). Emission profiles in various molecular species and simulated ALMA continuum images (*Saigo and Tomisaka, 2011; Tomisaka and Tomida, 2011; Commerçon et al., 2012b; Aikawa et al., 2012*) demonstrate the critical value of additional observational constraints beyond continuum SEDs. *Machida et al.* (2008) showed that first cores drive slow (~5 km s^{-1}) outflows with wide opening angles while *Price et al.* (2012) showed that, under certain assumptions, first core outflows may show strong collimation. Future theoretical attention is needed. Indeed, despite the dire observational need, no clear and unambiguous predictions for how a first core can be differentiated from a very young protostar have yet been presented.

4.2. Observations

Here we outline recent observational developments in the identification of the youngest sources, focusing on discoveries of three classes of observationally defined objects mentioned above: very-low-luminosity objects, candidate FHSCs, and PACS bright red sources (PBRS).

4.2.1. Very-low-luminosity objects. Prior to the launch of Spitzer, dense cores were identified as protostellar or starless based on the presence or absence of an associated Infrared Astronomical Satellite (IRAS) detection. Beginning with *Young et al.* (2004), Spitzer c2d observations revealed a number of faint protostars in cores originally classified as starless. This led to the definition of a new class of objects called VeLLOs: protostars embedded in dense cores with internal luminosities $L_{int} \leq 0.1\ L_{\odot}$ (*di Francesco et al., 2007*), where L_{int} excludes the luminosity arising from external heating by the interstellar radiation field. A total of 15 VeLLOs have been identified in the c2d regions (*Dunham et al., 2008*), with 6 the subject of detailed observational and modeling studies (*Young et al., 2004; Dunham et al., 2006; Bourke et al., 2006; Lee et al., 2009b; Dunham et al., 2010b; Kauffmann et al., 2011*).

The above studies have postulated three explanations for VeLLOs, whose very low luminosities require very low protostellar masses and/or accretion rates: (1) extremely young protostars with very little mass yet accreted, (2) older protostars observed in quiescent periods of a cycle of episodic accretion, and (3) protobrown dwarfs. The properties of the host cores and outflows driven by VeLLOs vary greatly from source to source, suggesting that VeLLOs (which are defined observationally) do not correspond to a single evolutionary stage and are instead a heterogeneous mixture of all three possibilities listed above. While the

relatively strong mid-infrared detections guarantee that none are FHSCs, at least some VeLLOs are consistent with being extremely young class 0 protostars just beyond the end of the first core stage (see section 4.2.2).

4.2.2. Candidate first hydrostatic cores. Nine other objects embedded within cores originally classified as starless have been detected and identified as candidate FHSCs. These objects have been revealed through faint mid-infrared detections of compact sources below the sensitivities of the large Spitzer surveys (*Belloche et al., 2006; Enoch et al., 2010*), (sub)millimeter detections of molecular outflows driven by "starless" cores (*Chen et al., 2010; Pineda et al., 2011; Schnee et al., 2012; Chen et al., 2012; Murillo and Lai, 2013*), and far-infrared detections indicating the presence of warm dust heated by an internal source (*Pezzuto et al., 2012*). However, we caution that even the very existence of some of these objects remains under debate (e.g., Schmalzl et al., 2013, in preparation).

Two significant questions have emerged: (1) Are any candidates true first cores? (2) How many "starless" cores are truly starless? The answer to the first question is not yet known; none clearly stand out as the best candidate(s) for a bona fide FHSC, with arguments for and against each. Nonetheless, it is extremely unlikely that all are FHSCs. Six of the nine candidate FHSCs are located in the Perseus molecular cloud. *Enoch et al.* (2009) identified 66 protostars in Perseus. Assuming that the duration of the protostellar stage is ~0.5 m.y. (section 2) and a first core lifetime of 0.5–50 k.y., as quoted above, we expect that there should be between 0.07 and 7 FHSCs in Perseus. Thus, unless the very longest lifetime estimates are correct, at least some (and possibly all) are very young second cores (protostars). Such objects would also be consistent with the definition of a VeLLO, emphasizing that these classes of objects are defined observationally and are not necessarily mutually exclusive in terms of evolutionary stage.

The number of "starless" cores that are truly starless is found by combining results for VeLLOs and candidate FHSCs. Combined, *Dunham et al.* (2008) and *Schnee et al.* (2012) find that approximately 18–38% of cores classified as starless prior to the launch of Spitzer in 2003 in fact harbor low-luminosity sources, although the exact statistics are still quite uncertain.

4.2.3. PACS bright red sources (PBRS). Using HOPS Herschel 70-μm imaging, *Stutz et al.* (2013) searched for new protostars in Orion that were too faint at λ ≤ 24 μm to be identified by Spitzer. They found 11 new objects with 70-μm and 160-μm emission that were either faint [m(24) > 7 mag] or undetected at λ ≤ 24 μm. In addition, they found seven previously Spitzer identified protostars with equally red colors (log $\lambda F_\lambda(70)/\lambda F_\lambda(24) > 1.65$). These 18 PBRS are the reddest known protostars in Orion A and Orion B. Figure 6 shows five representative PBRS.

Although the emission at λ ≤ 24 μm is faint for all PBRS, some are detected at 3.6–8.0 μm with Spitzer. These detections may be due to scattered light and shocked gas in outflow cavities, indicating the presence of outflows. A

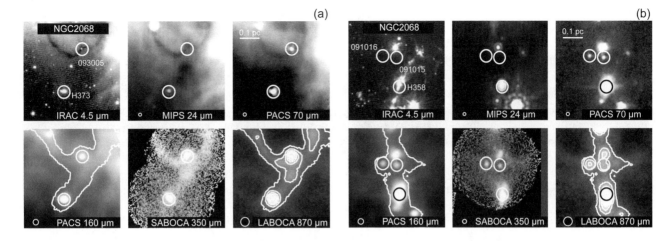

Fig. 6. 4' × 4' images of five PBRS at the indicated wavelengths. **(a)** PBRS 093005 and H373. Source 093005 is the reddest PBRS and therefore the reddest embedded object known in Orion, and lies at the intersection of three filaments seen in both absorption (8 μm, 24 μm, and 70 μm) and in emission (350 μm and 870 μm). **(b)** PBRS 091015, 019016, and H358. Figure adapted from *Stutz et al.* (2013).

total of eight PBRS are undetected by Spitzer at 24 μm, leading to a lower limit for their $\lambda F_\lambda(70)/\lambda F_\lambda(24)$ color. At $\lambda \geq 70$ μm all PBRS have SEDs that are well characterized by modified blackbodies (see Fig. 1 for a representative PBRS SED). The mean 70-μm flux of the PBRS is similar to that of the rest of the HOPS sample of protostars, with a comparable but somewhat smaller spread in values.

The overall fraction of PBRS to protostars in Orion is ~5%. Assuming PBRS represent a distinct phase in protostellar evolution (see further discussion of this assumption in section 4.3), and further assuming a constant star-formation rate and a protostellar duration of ~0.5 m.y. (section 2), the implied duration of the PBRS phase is 25 k.y., averaging over all Orion regions. However, the spatial distribution of the PBRS displays a striking non-uniformity compared to the distribution of normal HOPS protostars: Only 1% of the protostars in Orion A are PBRS, compared to ~17% in Orion B. Whether this large variation in the spatial distribution indicates a recent burst of star formation or environmental differences remains to be determined.

Basic parameters of interest include L_{bol}, T_{bol}, and L_{smm}/L_{bol}, as well as modified blackbody fits to the long-wavelength SEDs. The PBRS L_{bol} and T_{bol} distributions are shown in Fig. 3; L_{bol} spans a typical range compared to other protostars, but T_{bol} is restricted to very low values ($T_{bol} \leq 44$ K). Similarly, L_{smm}/L_{bol} values for PBRS occupy the extreme high end of the distribution for all protostars in Orion (*Stutz et al.,* 2013) (see Fig. 2). Modified blackbody fits to the thermal portions of their SEDs yield PBRS envelope mass estimates in the range of ~0.2–1.0 $M_\odot$. Radiative transfer models confirm that the 70-μm detections are inconsistent with externally heated starless cores and instead require the presence of compact internal objects. While the overall fraction of PBRS relative to protostars in Orion is small (~5%, see above), they are significant for their high envelope densities. Indeed, the PBRS have

70/24 colors consistent with very high envelope densities, near the expected class 0/I division or higher. The above evidence points toward extreme youth, making PBRS one of the few observational constraints we have on the earliest phases of protostellar evolution, all at a common distance with a striking spatial distribution within Orion.

4.3. Synthesis of Observations

With the discoveries of VeLLOs, candidate FHSCs, and PBRS, these large Spitzer and Herschel surveys of star-forming regions have expanded protostellar populations to objects that are both less luminous and more deeply embedded than previously known. These three object types are defined observationally, often based on serendipitous discoveries, and are not necessarily mutually exclusive in terms of the physical stages of their constituents. All are likely heterogeneous samples encompassing true FHSCs and protostars of varying degrees of youth. Indeed, some of the lowest-luminosity PBRS may in fact be FHSCs, and others with low luminosities are consistent with the definition of VeLLOs. Furthermore, at least some, and possibly all, of the objects identified as candidate FHSCs may not be true FHSCs but instead young protostars that have already evolved beyond the end of the FHSC stage. Such objects would be consistent with the definition of VeLLOs since they all have low luminosities, and some may be detected as PBRS once Herschel observations of the GB clouds are published. Integrating these objects into the broader picture of protostellar evolution is of great importance and remains a subject of ongoing study. Further progress in characterizing the evolutionary status of these objects and using them as probes of the earliest stages of star formation depend on specific theoretical predictions for distinguishing between FHSCs and very young protostars followed by observations that test such predictions.

5. PROTOSTELLAR ACCRETION BURSTS AND VARIABILITY

There is a growing body of evidence that the protostellar accretion process is variable and punctuated by short bursts of very rapid accretion (the "episodic accretion" paradigm). Indeed, optical variability was one of the original, defining characteristics of a young stellar object (*Joy*, 1945; *Herbig*, 1952). We provide here a summary of the observational evidence for episodic mass accretion in the protostellar stage.

5.1. Variability, Bursts, and Flares in Protostars

There are two general classes of outbursting young stars known, FU Orionis-type objects (FUors) and EX Lupi-type objects (EXors), which differ in their outburst amplitudes and timescales (see the chapter by Audard et al. in this volume). As an example of an outbursting source, Fig. 1 shows pre- and post-outburst SEDs of V2775 Ori, an outbursting protostar in the HOPS survey area (*Fischer et al.*, 2012). Based on the current number of known outbursting young stars, and assuming that all protostars undergo repeated bursts, *Offner and McKee* (2011) estimate that approximately 25% of the total mass accreted during the protostellar stage does so during such bursts. Some of the outbursting young stars detected to date are clearly still in the protostellar stage of evolution (e.g., *Fischer et al.*, 2012; *Green et al.*, 2013a; see also the chapter by Audard et al. in this volume), but whether or not all protostars undergo large-amplitude accretion bursts remains an open question.

Infrared monitoring campaigns offer the best hope for answering this question (e.g., *Johnstone et al.*, 2013), although we emphasize that not all detected variability is due to accretion changes; detailed modeling is necessary to fully constrain the variability mechanisms (e.g., *Flaherty et al.*, 2012, 2013). *Carpenter et al.* (2001, 2002) studied the near-infrared variability of objects in the Orion A and Chamaeleon I molecular clouds and identified numerous variable stars in each. *Morales-Calderón et al.* (2011) presented preliminary results from the Young Stellar Object Variability project (YSOVAR), a Spitzer warm mission program to monitor young clusters in the mid-infrared, in which they identified over 100 variable protostars. Once published, the final results of the program should provide robust statistics on protostellar variability (*Morales-Calderón et al.*, 2011; Rebull et al., 2013, in preparation). *Wolk et al.* (2013) found a variability fraction of 84% among YSOs in Cygnus OB7 with two-year near-infrared monitoring. *Megeath et al.* (2012) found that 50% of the YSOs in L1641 in Orion exhibited mid-infrared variability, with a higher fraction among the protostars alone compared to all YSOs, in multi-epoch Spitzer data spanning ~6 months. *Billot et al.* (2012) presented Herschel far-infrared monitoring of 17 protostars in Orion and found that 8 (40%) show >10% variability on timescales from 10 to 50 days, which they attributed to accretion variability. Finally, *Findeisen et al.* (2013) report results from the Palomar Transient Fac-

tory optical monitoring of the North American and Pelican Nebulae. By monitoring at optical wavelengths they are very incomplete to the protostars in these regions, but all three of the protostars they detect show variability.

Another approach is to compare observations from different telescopes, which provides longer time baselines but introduces uncertainties from differing band passes and instrument resolutions. Three such studies have recently been published, although none had sufficient statistics to restrict their analysis to only protostars. *Kóspál et al.* (2012) compared Infrared Space Observatory (ISO) and Spitzer spectra for 51 young stars and found that 79% show variability of at least 0.1 mag, and 43% show variability of at least 0.3 mag, with all the variability existing on timescales of about a year or shorter. *Scholz* (2012) compared 2Micron All Sky Survey (2MASS) and United Kingdom Infrared Telescope Archive (UKIRT) Infrared Deep Sky Survey (UKIDDS) near-infrared photometry with a time baseline of 8 yr for 600 young stars and found that 50% show >2σ variability and 3% show >0.5-mag variability, with the largest amplitudes seen in the youngest star-forming regions. Based on their statistics they derived an interval of at least 2000–2500 yr between successive bursts. *Scholz et al.* (2013) compared Spitzer and Wide-field Infrared Survey Explorer (WISE) photometry with 5-yr time baselines for 4000 young stars and identified 1–4 strong burst candidates with >1-mag increases between the two epochs. Based on these statistics they calculated a typical interval of 10,000 yr between bursts.

At present, neither direct monitoring campaigns nor comparison of data from different telescopes at different epochs offer definitive statistics on protostellar variability or the role variability plays in shaping the protostellar luminosity distribution. However, they do demonstrate that variability is common among all YSOs, and protostars in particular, and they do offer some of the first statistical constraints on variability and burst statistics. We anticipate continued progress in this field in the coming years.

5.2. Accretion Variability Traced by Outflows

Molecular outflows are driven by accretion onto protostars and are ubiquitous in the star-formation process (see the chapter by Frank et al. in this volume). Any variability in the underlying accretion process should directly correlate with variability in the ejection process. Indeed, many outflows show clumpy structure that can be interpreted as arising from separate ejection events. Although such clumpy structures can also arise from the interaction between outflowing gas and a turbulent medium even in cases where the underlying ejection process is smooth (*Offner et al.*, 2011), many outflows with such structure also display kinematic evidence of variability. In particular, outflow clumps in molecular CO gas, which are often spatially coincident with near-infrared H_2 emission knots, are found at higher velocities than the rest of the outflowing gas, and within these clumps the velocities follow Hubble laws (increasing velocity with increasing distance from the protostar). This creates distinct structures in position velocity diagrams called "Hubble wedges" by *Arce and Goodman* (2001), which are consistent with theoretical expectations for prompt entrainment of molecular gas by an episodic jet (e.g., *Arce and Goodman,* 2001, and references therein). One of the most recent examples of this phenomenon is presented by *Arce et al.* (2013) for the HH 46/47 outflow using ALMA data and is shown in Fig. 7. The timescales of the episodicity inferred from the clump spacings and velocities range from less than 100 yr to greater than 1000 yr (e.g., *Bachiller et al.,* 1991; *Lee et al.,* 2009a; *Arce et al.,* 2013).

Additional evidence for accretion variability comes from the integrated properties of molecular outflows. In a few cases, low-luminosity protostars drive strong outflows, implying higher *time-averaged* accretion rates traced by the integrated outflows than the *current* rates traced by the protostellar luminosities (*Dunham et al.,* 2006, 2010b; *Lee et al.,* 2010; *Schwarz et al.,* 2012). As discussed by *Dunham et al.* (2010b), the amount by which the accretion rates must have decreased over the lifetimes of the outflows are too large to be explained solely by the slowly declining accretion rates predicted by theories lacking short-timescale variability and bursts (see section 3).

5.3. Chemical Signatures of Variable Accretion

If the accretion onto the central star is episodic, driving substantial changes in luminosity, there can be observable effects on the chemistry of the infalling envelope and these can provide clues to the luminosity history. Several authors have recently explored these effects using chemical evolution models, as discussed in more detail in the chapter on episodic accretion by Audard et al. in this volume. A particular opportunity to trace the luminosity history exists in the absorption spectrum of CO_2 ice; observations of the 15.2-μm CO_2 ice absorption feature toward low-luminosity protostars with Spitzer spectroscopy have provided strong evidence for past accretion and luminosity bursts in these objects (*Kim et al.,* 2012) (see section 7).

6. THE FORMATION AND EVOLUTION OF PROTOSTELLAR DISKS

6.1. Theoretical Overview

The formation of a circumstellar disk is a key step in the formation of planets and/or binary star systems, where in this chapter "disk" refers to a rotationally supported, Keplerian structure. Disks must form readily during the star-formation process as evidenced by their ubiquity during the T Tauri phase (see Fig. 14 of *Hernández et al.,* 2007). In order to form a disk during the collapse phase, the infalling material must have some specific angular momentum (*Cassen and Moosman,* 1981); disks may initially start small and grow with time. The formation of the disk is not dependent on whether the angular momentum derives from initial cloud rotation (*Cassen and Moosman,* 1981;

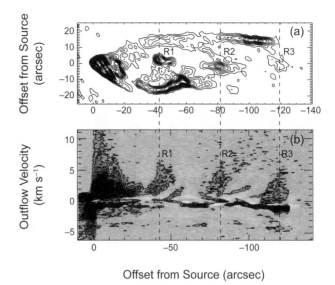

Fig. 7. **(a)** Integrated ^{12}CO (1–0) emission of the red-shifted lobe of the outflow driven by the HH 46/47 protostellar system based on the data presented by *Arce et al.* (2013), with the positions of three clump-like structures labeled. **(b)** Position-velocity diagram along the axis of the red-shifted lobe of the same outflow. The three clumps clearly exhibit higher velocities than the rest of the outflowing gas, with each clump exhibiting its own Hubble law.

Terebey et al., 1984) or the turbulent medium in which the core formed (e.g., *Offner et al.,* 2010), but the subsequent growth of the disk will depend on the distribution of angular momentum. After disk formation, most accreting material must be processed through the disk. Moreover, models for accretion bursts and outflows depend on the disk playing a dominant role (see section 3 and the chapters by Z.-Y. Li et al., Audard et al., and Frank et al. in this volume). Finally, disk rotation, when detectable, allows the only means of direct determination of protostellar masses (see the chapter by Dutrey et al. in this volume), since spectral types are unavailable for most protostars that are too deeply embedded to detect with optical and/or near-infrared spectroscopy, and mass is only one of several parameters that determines the luminosity of a protostar (see section 3.1).

Numerical hydrodynamic models of collapsing, rotating clouds predict the formation of massive and extended disks in the early embedded stages of protostellar evolution (*Vorobyov,* 2009a), with the disk mass increasing with the protostellar mass (*Vorobyov,* 2011b). However, these and earlier studies neglected the role of magnetic fields. *Allen et al.* (2003) presented ideal magneto-hydrodynamic (MHD) simulations examining the effects of magnetic braking and showed that collapsing material drags the magnetic field inward and increases the field strengths toward smaller radii. The magnetic field is anchored to the larger-scale envelope and molecular cloud, enabling angular momentum to be removed from the collapsing inner envelope, preventing the formation of a rotationally supported disk.

This finding is referred to as the "magnetic braking catastrophe" and had been verified in the ideal MHD limit analytically by *Galli et al.* (2006) and in further numerical simulations by *Mellon and Li* (2008). These studies concluded that the magnetic braking efficiency must be reduced in order to enable the formation of rotationally supported disks. Non-ideal MHD simulations have shown that Ohmic dissipation can enable the formation of only very small disks (R ~ 10 $R_\odot$) that are not expected to grow to hundred-astronomical-unit scales until after most of the envelope has dissipated in the class II phase (*Dapp and Basu,* 2010; *Machida et al.,* 2011b). Disks with greater radial extent and larger masses can form if magnetic fields are both relatively weak and misaligned relative to the rotation axis (*Joos et al.,* 2012), and indeed observational evidence for such misalignments has been claimed by *Hull et al.* (2013). *Krumholz et al.* (2013) estimated that 10–50% of protostars might have large (R > 100 AU), rotationally supported disks by applying current observational constraints on the strength and alignment of magnetic fields to the simulations of *Joos et al.* (2012).

6.2. Observations of Protostellar Disks

The earliest evidence for the existence of protostellar disks comes from models fit to the unresolved infrared and (sub)millimeter continuum SEDs of protostars (e.g., *Adams and Shu,* 1986; *Kenyon et al.,* 1993; *Butner et al.,* 1994; *Calvet et al.,* 1997; *Whitney et al.,* 2003b; *Robitaille et al.,* 2006; *Young et al.,* 2004). Models including circumstellar disks provided good fits to the observed SEDs, and in many cases, disks were required in order to obtain satisfactory fits. However, SED fitting is often highly degenerate between disk and inner core structure, and even in the best cases does not provide strong constraints on disk sizes and masses, the two most relevant quantities for evaluating the significance of magnetic braking. Thus in the following sections we concentrate on observations that do provide constraints on these quantities, with a particular focus on (sub)millimeter interferometric observations.

6.2.1. Class I disks. Due to complications with separating disk and envelope emission, a characterization of class I disks on par with class II disks (e.g., *Andrews et al.,* 2010) has not yet been possible. Several approaches have been taken toward characterizing class I disks, including focusing on SEDs, scattered light imaging, and millimeter emission. Analysis of millimeter emission requires invoking models to disentangle disk and envelope emission.

Analysis of millimeter emission requires invoking models to disentangle disk and envelope emission. This approach was carried out by *Jorgensen et al.* (2009), who found 10 class I disks with M ~ 0.01 $M_\odot$, although the disks were not resolved. *Eisner* (2012) modeled the disk and envelope properties of eight class I protostars in Taurus combining the SEDs, millimeter imaging, and near-infrared scattered light imaging (see also *Eisner et al.,* 2005). The median radius of the sample was found to be 250 AU and

the median disk mass was 0.01 $M_\odot$, but the derived parameters for each source depended strongly on the weighting of the model fits. They also did not resolve the disks, with only ~1″ resolution.

The most definitive constraints on class I disks have come from the detailed modeling of edge-on disks resolved in both near-infrared scattered light and millimeter dust emission. Two recent examples of such work include studies of the edge-on class I protostars IRAS 04302+2247 and CB 26. IRAS 04302+2247 was found to have a disk with R ~ 300 AU and M ~ 0.07 $M_\odot$ (*Wolf et al.,* 2008; *Gräfe et al.,* 2013). CB 26 has a disk with R ~ 200 AU and a possible inner hole of 45 AU cleared by an undetected binary companion (*Sauter et al.,* 2009). These representative cases illustrate that relatively large, massive disks may be common in the class I stage.

Sensitive molecular line observations are also enabling the kinematic structure of class I disks to be examined. A number of class I disks have been found to be rotationally supported (e.g., *Jorgensen et al.,* 2009; *Brinch et al.,* 2007; *Lommen et al.,* 2008; *Takakuwa et al.,* 2012; *Hara et al.,* 2013). Thus far, the number of class I protostellar mass measurements is only 6, and their masses range between 0.37 $M_\odot$ and 2.5 $M_\odot$. The disk masses are typically less than 10% of the stellar masses, modulo uncertainties in calculating mass from dust emission and modeling systematics. *Jorgensen et al.* (2009) pointed out possible discrepancies between models of disk formation from the simple picture of an infalling, rotating envelope (*Terebey et al.,* 1984). Semianalytic models of protostellar collapse from *Visser et al.* (2009) and hydrodynamic models of *Vorobyov* (2009a, 2011b) seem to suggest much more massive disks (up to 40–60% that of the star) than revealed by observations.

6.2.2. Class 0 disks. Observations of class 0 disks face challenges similar to the class I disks, but with the added complication of more massive envelopes responsible for ~90% of the emission at millimeter wavelengths (*Looney et al.,* 2000). The SEDs of class 0 protostars are dominated by the far-infrared component and the near-infrared is generally dominated by scattered light from the outflow cavity. Much like the class I systems, it is impossible to garner any constraints of the disk properties in class 0 systems from the SEDs alone.

B335 was one of the first class 0 systems to be examined in great detail and found to have a disk with R < 60 AU and M ~ 0.0014 $M_\odot$ (*Harvey et al.,* 2003); while that study assumed a distance of 250 pc, we have scaled these numbers to the more recent estimates that place B335 at ~150 pc (*Stutz et al.,* 2008). Among other class 0 protostars, HH 211 shows extended structure perpendicular to the outflow that may indicate a circumstellar disk (*Lee et al.,* 2009a). On the other hand, the class 0 protostar L1157-mm shows no evidence of resolved disk structure on any scale with resolutions as fine as 0.3″, implying any disk present must have R < 40 AU and M < 0.004–0.025 $M_\odot$ (*Chiang et al.,* 2012), and ALMA observations of IRAS 16293 2422 revealed only a very small disk with R ~ 20 AU (*Chen et al.,* 2013; *Zapata et al.,* 2013).

Jørgensen et al. (2007, 2009) coupled a submillimeter survey of 10 class 0 protostars with models for envelope emission and found class 0 disks with M ~ 0.01 M$_\odot$ (with significant scatter), and no evidence of disk mass growth between class 0 and I sources. On the other hand, *Maury et al.* (2010) did not find any evidence for disk structures >100 AU in a millimeter survey of five class 0 protostars at subarcsecond resolution (linear resolution <100 AU).

The clearest evidence of class 0 disks thus far is found toward the protostars L1527 IRS and VLA 1623A. For L1527, high-resolution scattered light and (sub)millimeter imaging observations found strong evidence of an edge-on disk (*Tobin et al.*, 2012) (see Fig. 8). Furthermore, molecular line observations were found to trace Keplerian rotation, implying a protostellar mass of 0.19 ± 0.04 M$_\odot$. The mass of the surrounding envelope is ~5× larger than the protostar, and the disk has R = 70–125 AU and M ~ 0.007 M$_\odot$ (*Tobin et al.*, 2012, 2013). L1527 IRS is classified as class 0 based on its SED. While its edge-on nature can bias this classification (section 2), it does have a more massive and extended envelope than a typical class I protostar in Taurus and is consistent with the stage 0 definition of M$_*$ < M$_{env}$. For VLA 1623A, *Murillo and Lai* (2013) and *Murillo et al.* (2013) identified a confirmed disk via detection of Keplerian rotation, with R ~ 150 and M ~ 0.02 M$_\odot$.

6.3. Summary of Protostellar Disks

Current observations clearly indicate that there are large (R > 100 AU) disks present in the class I phase but their properties have been difficult to detail en masse thus far. There is evidence for large disks in *some* class 0 systems (L1527 IRS, VLA 1623, and HH 211), but more observations with sufficient resolution to probe scales where disks dominate the emission are needed for a broader characterization.

Efforts to detect more disk structures in the nearby Perseus molecular cloud are underway and have found two strong candidates for class 0 disks (Tobin et al., in preparation).

Despite remarkable progress in characterizing disks around protostars in the last decade, several puzzles remain. First, if disks form via a process similar to the rotating collapse model, the observed disk masses are about an order of magnitude lower than theoretical estimates for a given protostellar mass (*Vorobyov*, 2009a; *Visser et al.*, 2009; *Yorke and Bodenheimer*, 1999). Second, models considering magnetic braking seem to suggest small disks until the class II phase, unless there are misaligned fields; however, observations seem to indicate at least some large disks in the class 0 and I phases. Finally, it is not clear if there is any observational evidence for magnetic braking during collapse, as *Yen et al.* (2013) recently showed that angular momentum is conserved in two class 0 systems.

Moving forward, ALMA will provide key observations for determining the presence and properties of class 0 disks, offering significant improvements over the resolution and sensitivity limits of current interferometers. By detecting Keplerian rotation, ALMA holds the promise to enable mass measurements for large numbers of protostars, possibly even directly observing for the first time the protostellar mass function (*McKee and Offner*, 2010). At the same time, the upgraded Very Large Array (VLA) has very high sensitivity to dust continuum emission at 7 mm and may also contribute to the characterization of protostellar disks.

7. THE EVOLUTION OF INFALLING MATERIAL

Silicate dust and ices are the building blocks of comets and, ultimately, Earth-like planets, and both undergo significant chemical and structural (crystallinity) changes during

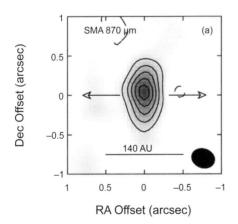 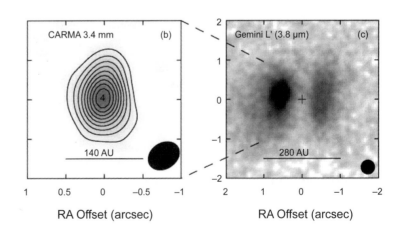

Fig. 8. Images of the edge-on disk around the protostar L1527 from *Tobin et al.* (2012). High-resolution images of L1527 showing the disk in dust continuum emission and scattered light are shown at wavelengths of **(a)** 870 µm from the Submillimeter Array (SMA), **(b)** 3.4 mm from Combined Array for Research in Millimeter-wave Astronomy (CARMA), and **(c)** 3.8 µm from Gemini. The Gemini image is shown on a larger scale. The sub/millimeter images are elongated in the direction of the dark lane shown in **(c)**, consistent with an edge-on disk in this class 0 protostellar system. The outflow direction is indicated by arrows in **(a)**. The cross in **(c)** marks the central position of the disk from the SMA images.

the protostellar stage. Understanding how these materials evolve as they flow from molecular cores to protostellar envelopes and disks is an essential component in determining how matter is modified in planet-forming disks. While most of this chapter focuses on macroscopic protostellar evolution, we now examine the microscopic evolution of solid-state matter surrounding protostars.

Mid-infrared spectra toward low-mass protostars contain a wealth of solid-state absorption features due to the vibrational modes of silicate dust and many molecular ice species (e.g., *Boogert et al., 2008; Furlan et al., 2008*). These materials are inherited from the dense interstellar medium, and are present in the infalling envelopes and accretion disks surrounding protostars. Within such environments, radiative and mechanical processing by ultraviolet (UV) photons and accretion shocks may sublimate or modify the properties of dust and ices. Indeed, theoretical disk evolutionary models of low-mass protostars predict that infalling ices, with sublimation temperatures greater than T ≈ 50 K, remain on grain surfaces after reaching the accretion disk (*Visser et al., 2009*), suggesting that at least a fraction of cometary ices are of protostellar origin. Furthermore, the formation of crystalline silicates are predicted to have occurred during the earliest phases of disk formation, when a fraction of the amorphous silicate dust falls close to the central protostar and is subsequently heated to temperatures of T = 800–1200 K (*Dullemond et al., 2006; Visser and Dullemond, 2010*).

While it is generally assumed that solids present in the infalling envelopes of low-mass protostars are mostly pristine without significant processing, the Spitzer Space Telescope has now revealed that such matter shows clear evidence for thermal processing. In this section, we review the current observational evidence for high- to low-temperature processing of silicates and CO_2 ice, and their implications for both protostellar evolution and the inventory of matter delivered to planet-forming disks.

7.1. Thermal Processing of Amorphous Silicates

Prior to the launch of the Spitzer Space Telescope, little knowledge of the silicate dust surrounding low-mass protostars existed. Because observations with ISO were sensitivity-limited to the most luminous, massive protostars, detailed spectral modeling was possible for only a few high-mass protostars (*Demyk et al., 1999, 2000*). Similar to the silicate composition of the diffuse interstellar medium, which is thought to be almost entirely amorphous in structure (*Kemper et al., 2005*), only a small fraction (1–2%) of the silicate dust mass present in protostellar envelopes was predicted to be in crystalline form during the pre-Spitzer era.

The high sensitivity of the Infrared Spectrograph (IRS) onboard Spitzer permitted the routine detection of crystalline silicates emission features toward T Tauri stars (e.g., *Sargent et al., 2009; Olofsson et al., 2010*) and comets (e.g., *Lisse et al., 2007; Kelley and Wooden, 2009*). Crystalline silicate emission features are also detected toward the

Serpens protostellar binary SVS 20 (*Ciardi et al., 2005*); however, a recent evaluation of its Spitzer-IRS spectrum suggests that SVS 20 may be a T Tauri star possessing a flattened, settled disk (*Oliveira et al., 2010*).

In contrast, more than 150 low-mass protostars have been studied in the literature, and their Spitzer-IRS spectra exhibit 10- and 20-μm amorphous silicate absorption features that are almost always characterized by broad, smooth profiles, lacking any superimposed narrow structure. However, the one exception is HOPS 68, a low-mass embedded protostar situated in the Orion Molecular Cloud complex (*Poteet et al., 2011*). The Spitzer-IRS spectrum toward HOPS 68 exhibits narrow *absorption* features of forsterite (Fig. 9), indicating that a significant fraction (≤17%) of amorphous silicates within its infalling envelope have experienced strong thermal processing (T ≥ 1000 K). Although the mechanisms responsible for such processing are still not fully understood, it is proposed that amorphous silicates were annealed or vaporized within the warm inner region of the disk and/or envelope and subsequently transported outward by entrainment in protostellar outflows. Alternatively, an *in situ* formation by shock processing in the outflow working surface may be responsible for the production of crystalline silicates within the envelope of HOPS 68.

The detection of crystalline silicates in cometary material necessitates a process for transporting thermally processed silicates to the cold outer regions of planet-forming disks (e.g., *Ciesla, 2011*) or requires an *in situ* formation route to be present at large radial distances (e.g., *Vorobyov, 2011a; Nayakshin et al., 2011*). The detection of crystalline silicates toward HOPS 68 demonstrates that at least in one case, thermally processed silicates may be delivered to the outer accretion disk by infall from the protostellar envelope. Of course, HOPS 68 may be a unique case. However, on the basis of its peculiar SED, it is argued that HOPS 68 has a highly flattened envelope structure that provides a line of sight to the inner region without passing through the intervening cold, outer envelope (*Poteet et al., 2011, 2013*). Thus, HOPS 68 may provide a rare glimpse into the thermally processed inner regions of protostellar envelopes.

7.2. Thermal Processing of Interstellar Carbon Dioxide Ice

Interstellar ices serve as excellent tracers of the thermal history of protostellar environments (*Boogert and Ehrenfreund, 2004*). As material falls from the outer envelope to the accretion disk, grains are subject to elevated temperatures while traversing the inner envelope region, resulting in the crystallization of hydrogen-rich ices or the sublimation of CO-rich ices (e.g., *Visser et al., 2009*). Moreover, protostellar feedback mechanisms, such as outflow-induced shocks and episodic accretion events, may also be responsible for the crystallization and sublimation of interstellar ices (*Bergin et al., 1998; Lee, 2007*).

Carbon dioxide ice has proven to be a sensitive diagnostic of the thermal history toward protostars. Pure CO_2

ice may be produced by (1) segregation of CO_2 from hydrogen-rich ice mixtures (e.g., $H_2O:CH_3OH:CO_2$ and $H_2O:CO_2$) or (2) thermal desorption of CO from a $CO:CO_2$ ice mixture at temperatures of T = 30–60 K and T = 20–30 K, respectively (*Ehrenfreund et al.*, 1998; *Öberg et al.*, 2009; *van Broekhuizen et al.*, 2006). Spectroscopically, thermally processed mixtures of hydrogen- and CO-rich CO_2 ices produce a double-peaked substructure in the 15.2-μm CO_2 ice bending mode profile, characteristic of pure, crystalline CO_2 ice (Fig. 9).

The presence of pure CO_2 ice is generally interpreted as a result of thermal processing of pristine icy grains in cold, quiescent regions of dense molecular clouds. To date, ~100 high-resolution Spitzer-IRS spectra toward low-mass protostars have been described in the literature, and the majority of the CO_2 is found in hydrogen- and CO-rich ices typical of molecular cloud cores (*Pontoppidan et al.*, 2008; *Zasowski et al.*, 2009; *Cook et al.*, 2011; *Kim et al.*, 2012). However, nearly 40% of the spectra show some evidence for the presence of pure CO_2 ice, suggesting that low-mass protostars have thermally processed inner envelopes. Among these, relatively large abundances (~15% of the total CO_2 ice column density) of pure CO_2 ice have been detected toward a sample of low-luminosity (0.08 $L_\odot \leq L \leq$ 0.69 $L_\odot$) embedded protostars (*Kim et al.*, 2012). Because the present environments of low-luminosity protostars do not provide the thermal conditions needed to produce pure CO_2 in their inner envelopes, a transient phase of higher luminosity must have existed sometime in the past (*Kim*

et al., 2012). The presence and abundance of pure CO_2 ice toward these low-luminosity protostars may be explained by episodic accretion events, in which pure CO_2 is produced by distillation of a $CO:CO_2$ mixture during each high-luminosity transient phase. Other accretion models have not yet been tested and may also explain the observed presence of pure CO_2 ice toward low-luminosity protostars; however, continuous accretion models with monotonically increasing luminosity cannot reproduce the observed abundances of CO_2 ice and $C^{18}O$ gas (*Kim et al.*, 2012).

The highest level of thermally processed CO_2 ice is found toward the moderately luminous (1.3 $L_\odot$) protostar HOPS 68 (*Poteet et al.*, 2013). Its CO_2 ice spectrum reveals an anomalous 15.2-μm bending mode profile that indicates little evidence for the presence of unprocessed ice. Detailed profile analysis suggests that 87–92% of its CO_2 ice is sequestered as spherical, CO_2-rich mantles, while typical interstellar ices are dominated by irregularly shaped, hydrogen-rich CO_2 mantles (e.g., *Pontoppidan et al.*, 2008). The nearly complete absence of unprocessed ices along the line of sight to HOPS 68 is best explained by a highly flattened envelope structure, which lacks cold absorbing material in its outer envelope, and possesses a large fraction of material within its inner (10 AU) envelope region. Moreover, it is proposed that the spherical, CO_2-rich ice mantles formed as volatiles rapidly froze out in dense gas, following an energetic but transient event that sublimated primordial ices within the inner envelope region of HOPS 68. The mechanism responsible for the sublimation is proposed to be either an episodic accretion event or shocks in the interaction region between the protostellar outflow and inner envelope. Presently, it is unknown if such feedback mechanisms are also responsible for the presence of crystalline silicates toward HOPS 68.

8. DOES ENVIRONMENT MATTER?

Low-mass stars form in a wide range of environments: isolated globules, small groups, and rich clusters of low- and high-mass stars (e.g., *Gutermuth et al.*, 2009; *Launhardt et al.*, 2013), with no apparent dichotomy between clustered and isolated star formation (*Bressert et al.*, 2010). This range of environments provides a natural laboratory for studying low-mass star formation in different physical conditions, with the goal of elucidating how those physical conditions may guide the formation of the stars. In this section, we explore how the environment affects the incidence of protostars and their properties. We define local environment as the region outside the immediate core/envelope, i.e., beyond 10,000–20,000 AU (*Enoch et al.*, 2008). Thus, the environmental conditions would include the properties of the gas in the surrounding molecular cloud or the density of young stars in the vicinity of the protostar, but would exclude the local molecular core/envelope as well as companions in a multiple star system.

Many of the environmental conditions have yet to be measured over the spatial extent of the Spitzer and Herschel

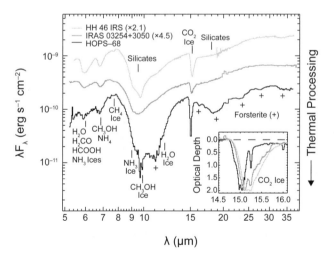

Fig. 9. Spitzer-IRS spectra of low-mass protostars, ordered by increasing evidence for thermal processing, with solid-state features indicated. Unlike HOPS 68 (black), the spectra of HH 46 IRS (gray) and IRAS 03254+3050 (dark gray) show no signs for the presence of crystalline silicates (forsterite). *Insert:* Arbitrarily scaled CO_2 ice optical depth spectra of protostars, exhibiting various degrees of ice processing that is evident by their double-peaked substructure. Data are presented in *Boogert et al.* (2008), *Pontoppidan et al.* (2008), and *Poteet et al.* (2011, 2013).

surveys. Thus we focus on three measures of the environment: (1) the gas column density, (2) cloud geometry, and (3) the surface density of YSOs, to address four specific questions.

8.1. Does the Incidence and Density of Protostars Depend on the Local Gas Column Density?

A combination of (sub)millimeter surveys for cores and extinction maps constructed from 2MASS and Spitzer photometry of background stars have enabled an examination of the incidence of molecular cores as a function of the column density of gas smoothed over spatial scales of 0.2 pc to 1 pc (see also the chapter by Padoan et al. in this volume for a similar discussion in the context of star-formation laws). Millimeter continuum surveys of the Ophiuchus and Serpens clouds show that the cores are found in environments where the column densities exceed $A_V = 7$ mag (*Johnstone et al.*, 2004; *Enoch et al.*, 2008), consistent with the model of photoionization-regulated star formation in magnetized clouds (*McKee*, 1989). However, in the Perseus cloud, *Enoch et al.* (2008) found that 25% of cores are located at $A_V = 7$ mag (however, see *Kirk et al.*, 2006), and *Hatchell et al.* (2005) showed that the probability of finding a submillimeter core in the Perseus cloud increased continuously with the integrated intensity of the $C^{18}O$ (1 → 0) line to the third power. In total, these results indicate a rapid rise in the incidence of cores with column density, but with a small number of cores detected below $A_V = 7$ mag.

A similar result comes from the examination of Spitzer-identified infrared protostars. In a combined sample of seven molecular clouds, *Gutermuth et al.* (2011) and *Masiunas et al.* (2012) showed that the surface density of dusty YSOs (i.e., class 0/I *and* class II objects) increases with the second to third power of the gas column density. *Heiderman et al.* (2010) also found a steep rise in the surface density of c2d and GB protostars with the column density of gas. Although Spitzer identified candidate protostars at column densities as low as 60 $M_\odot$ pc^{-2} [$A_V = 3$ mag (*Gutermuth et al.*, 2011)], the protostars are preferentially found in much denser environments, with a rapid, power-law-like rise in the density of protostars with increasing gas column density. Currently, it is not clear whether the rapid rise in the incidence of cores and density of protostars with column density is due to the inhibition of star formation in regions of low column density, as argued by the photoionization-regulated star-formation model, or a nonlinear dependence of the density of protostars on the gas column density, such as that predicted by Jeans fragmentation (*Larson*, 1985).

8.2. What is the Connection Between Cloud Geometry and the Spatial Distribution of Protostars?

The elongated and filamentary nature of interstellar clouds has been evident since the earliest optical observations of dark clouds (*Barnard*, 1907). With improvements

in sensitivity and resolution, it has become clear that star-forming molecular clouds are complex filamentary networks, with filamentary structure on scales from hundreds of parsecs to hundreds of astronomical units (*Lynds*, 1962; *Schneider and Elmegreen*, 1979; *Myers*, 2009a; *Tobin et al.*, 2010; *André et al.*, 2010; *Molinari et al.*, 2010). Large-scale filaments harbor star-forming cores and protostars, especially at bends, branch points, or hubs that host groups or clusters of cores and protostars (*Myers*, 2009a; *Schneider et al.*, 2012). Filaments may both accrete gas and channel its flow along their length (*Csengeri et al.*, 2011); recent observations of one young cluster found that flow onto and along filaments may feed gas to the central, cluster-forming hub (*Kirk et al.*, 2013). The filamentary structure of star-forming gas in Orion and the corresponding distribution of protostars is apparent in Fig. 10.

Self-gravitating filaments may fragment with a characteristic spacing (*Larson*, 1985; *Inutsuka and Miyama*, 1992, 1997). Indeed, the spacings determined from the initial temperature, surface density, and length of observed filaments predict the approximate number of low-mass stars in the Taurus complex and other complexes that lack rich clusters (*Hartmann*, 2002; *Myers*, 2011). A relationship between the spacing of protostars with gas column density is clearly apparent in Fig. 10, which shows recent HOPS observations of two filamentary regions within the Orion cloud. The OMC2/3 region is directly north of the Orion Nebula and is considered a northern extension of the Orion Nebula Cluster (ONC) (*Peterson and Megeath*, 2008; *Megeath et al.*, 2012). In contrast, L1641S is a more quiescent region in the southern part of the L1641 cloud (*Carpenter*, 2000; *Allen and Davis*, 2008). Although both regions host multi-parsec filaments with mass to length ratios exceeding the limit for a stable, thermally supported filament (*André et al.*, 2010; *Fischera and Martin*, 2012), the average column density calculated above a cutoff of $N(H_2) = 3 \times 10^{21}$ cm^{-2} is twice as high in OMC2/3: in $N(H_2) = 1.4 \times 10^{22}$ cm^{-2} OMC2/3 as compared to 7.6×10^{21} cm^{-2} in L1641S. The density of protostars is correspondingly higher in the OMC2/3 region, with the typical protostellar spacing being 6500 AU and 14,000 AU in OMC2/3 and L1641, respectively.

8.3. What Fraction of Protostars May Be Interacting?

With most protostars concentrated in regions of high gas column density and high stellar surface density, interactions between protostars may be common. *Winston et al.* (2007) found that the protostars in the Serpens main cluster have a median projected separation of 8000 AU [corrected to the revised distance for Serpens of 429 pc (*Dzib et al.*, 2011)]; smaller than the typical diameter of cores of 10,000–20,000 AU (*Enoch et al.*, 2008). They also found that the low relative velocities of the protostars and high gas density of the surrounding cloud core are consistent with the competitive accretion models of *Bonnell and Bate* (2006), where the protostars compete for gas from a common reservoir. The projected spacing of the protostars in OMC2/3 is also smaller than the typical diameters of cores (Fig. 10) (also see *Takahashi et al.*, 2013; *Li et al.*, 2013). Megeath et al. (in preparation) examined the spacing between the Spitzer-identified protostars in the Orion molecular clouds. They found that 11% and 23% of the protostars are within a projected separation of 5000 and 10,000 AU from another protostar, respectively. Thus, sources separated by ≤10,000 AU would potentially be able to interact through the collisions of their envelopes and tidal forces. However, since the actual separations are larger than the measured projected separations, these percentages should be considered upper limits. On the other hand, dynamical motions in such highly clustered regions (e.g., *Parker and Meyer*, 2012) may increase the percentage of sources that interact.

8.4. Do the Properties of Protostars Depend on Their Environment?

In their study of the protostellar luminosity distributions (PLDs) in nine nearby molecular clouds, *Kryukova et al.* (2012) searched for variations between the high- and low-density environments within a single cloud. By dividing protostars into two samples based on the local surface density of YSOs, they compared PLDs for protostars located in crowded, high-density regions to those in regions of low densities. In the Orion molecular clouds, they found the PLDs of the high- and low-stellar-density regions had a very low probability of being drawn from the same parent distribution, with the PLD for the protostars in denser environments biased to higher luminosities. Thus, within the Orion molecular clouds, the luminosities of the protostars appear to depend on the surface density of the YSOs in the surrounding environment. Unfortunately, the smaller numbers of protostars found in the other eight molecular clouds precluded a definitive comparison of the high- and low-density regions in those clouds.

One weakness of the *Kryukova et al.* (2012) study is the use of an empirical relationship to extrapolate from the infrared to bolometric luminosity. However, using the full SEDs from HOPS (Fig. 1) to directly determine the bolometric luminosity of the protostars in Fig. 10, the median L_{bol} is 6.2 $L_{\odot}$ in the densely clustered OMC2/3 region and 0.7 $L_{\odot}$ in the lower-density L1641S region. This corroborates the result of *Kryukova et al.* (2012) by showing the more densely clustered OMC2/3 region has systematically higher-luminosity protostars. Furthermore, it supports an interconnected relationship between the column density of the gas, the luminosities of the protostars, and the surface density of the protostars: Filaments with higher gas column density will have both a higher density of protostars and more luminous protostars. In the near future, Herschel and groundbased (sub)millimeter imaging of protostellar environments should revolutionize our understanding of how the properties of protostars depend on their environment, and thereby place strong constraints on models of protostellar evolution.

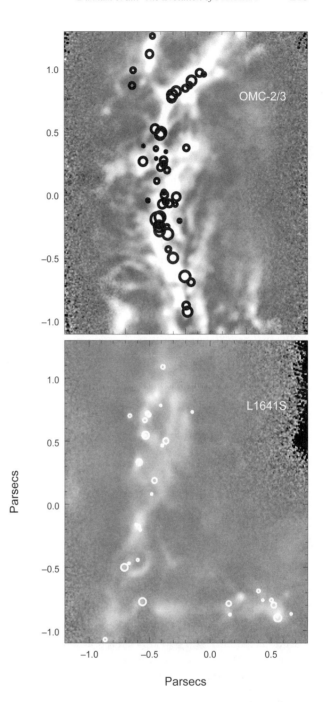

Fig. 10. Atacama Pathfinder Experiment (APEX)/Large APEX BOlometer CAmera (LABOCA) 870-μm maps of two regions in the Orion molecular clouds: OMC2/3 and L1641S (Stanke et al., in preparation). They are plotted on the same spatial scale with the same intensity scale for the 870-μm intensity. Overlaid are the position of the protostars; the size of the symbol is proportional to the log of the luminosity determined from the HOPS data.

9. SUMMARY

We have reported on recent progress in finding, identifying, and characterizing protostars in nearby clouds using infrared surveys and follow-up studies. The coverage, sensitivity, and resolution of Spitzer and Herschel have revealed objects of very low ($L_{int} < 0.1\ L_{\odot}$) luminosity (VeLLOs), objects with very red SEDs (PBRS), and candidates for first hydrostatic cores (FHSCs). Based on the full census of YSOs found in the c2d, GB, Orion, and Taurus surveys, classified as in *Greene et al.* (1994), and an assumed lifetime for class II objects of 2 m.y., we infer class 0 + I lifetimes of 0.42–0.54 m.y., with the longer estimate applying to the GB clouds. While some differences in identification and classification remain, we have attempted to minimize them.

The luminosity distribution is very broad. After reviewing models of evolution, focusing on the accretion histories, we compared the luminosity distributions from the models to the observations; either mass-dependent or episodic accretion models can match the data, but the isothermal sphere, constant accretion rate model cannot. Evidence of protostellar luminosity variability, outflow episodicity, and ice evolution all support some form of episodic accretion. The remaining questions are how extreme the variations can be, what the effect of variations are, and how much of the mass of the protostar is accreted during episodes of high accretion. Challenges for the future are to constrain short-term (decade-scale) variability as a possible clue to longer-term (e.g., 10^3 yr) variability, and to ultimately determine the relative importance of stochastic (episodic) and secular processes in building the final masses of stars.

The first stage of star formation, the FHSC, remains elusive. There are a number of candidates, but more theoretical work is needed to pin down the expected characteristics and to understand the evolutionary status of the other novel objects, such as PBRS and VeLLOs.

The existence and properties of disks in the protostellar phase are beginning to yield to observational scrutiny, and ALMA will provide a major advance. Current evidence suggests early formation of relatively massive disks, although only a small number of detections have been made to date. Theories without magnetic fields predict more massive disks than have been seen, while theories with rotation axes and magnetic fields aligned struggle to create disks at all. Further theoretical exploration is needed to understand the relationship of mass infall rate from the core, its transition to the disk, and the processes that allow it to accrete onto the growing star. Typical stage II disks process matter about 100× more slowly than matter falls in during stage I, so faster processes are needed in stage I to avoid mass buildup and instabilities.

Material falling onto the disk undergoes substantial chemical and, in some cases, mineralogical evolution. Dust grains grow and may become crystallized, while ice in mantles evolves chemically. Distillation of pure CO_2 ice provides clues to the luminosity evolution. The state of ice and gas arriving at the disk set the stage for later chemical evolution during stage II.

The complete surveys reported here reveal the location of star formation within molecular clouds. Prestellar cores and protostars are highly concentrated in regions of high surface, and presumably volume, density (see the chapter by Padoan et al. in this volume). The protostars in the very crowded environments of Orion appear to be systematically more luminous, and may be interacting in the regions of highest protostellar density. The "clumps" hosting cluster formation are generally very filamentary, and rich clusters prefer the nexus of filaments for their maternity wards.

The future is bright. The Atacama Large Millimeter Array will allow studies of unprecedented detail of the density, temperature, and velocity fields in infalling envelopes (e.g., *Pineda et al.,* 2012). Studies of embedded disks will clarify their masses and sizes, and detection of Keplerian motions will finally constrain the masses of the growing stars. The James Webb Space Telescope will allow deeper studies of the shorter (infrared) wavelengths of deeply embedded objects, and the Stratospheric Observatory for Infrared Astronomy (SOFIA) will provide spectroscopic data on the brighter sources. On the theoretical side, modelers should use physically realistic calculations to predict self-consistently all the observations: the IMF and its realization in different regions; the structure, velocity field, and chemistry of infalling envelopes; the luminosity distribution; protostellar disk sizes and masses; the diversity of disks emerging in the stage II phase; variability measurements; protostellar masses [measured with Keplerian disks with ALMA and the NOrthern Extended Millimeter Array (NOEMA), the successor to the Plateau de Bure Interferometer (PdBI)]; and environmental dependencies.

Acknowledgments. The authors thank R. Launhardt, J. Green, and H. G. Arce for providing data for Fig. 1 and Fig. 7, and the anonymous referee for comments that have improved this review. This review is based primarily on observations made with the Spitzer Space Telescope and Herschel Space Observatory. Spitzer is operated by the Jet Propulsion Laboratory, California Institute of Technology under a contract with NASA, and Herschel is a European Space Agency (ESA) space observatory with science instruments provided by European-led Principal Investigator consortia and with important participation from NASA. This review has made use of NASA's Astrophysics Data System Bibliographic Services. M.M.D., S.T.M., and W.J.F. acknowledge support from NASA through awards issued by JPL/Caltech, and M.M.D. acknowledges NSF support through grant AST-0845619 to Yale University. The work of A.M.S. was supported by the Deutsche Forschungsgemeinschaft priority program 1573 ("Physics of the Interstellar Medium"). N.J.E. acknowledges support from the NSF through grant AST-1109116 to the University of Texas at Austin. C.A.P. acknowledges support provided by the NASA Astrobiology Institute through contract NNA09DA80A. J.T. acknowledges support provided by NASA through Hubble Fellowship grant #HST-HF-51300.01-A awarded by the Space Telescope Science Institute, which is operated by the Association of Universities for Research in Astronomy, Inc., for NASA, under contract NAS 5-26555. E.I.V. performed numerical simulations on the SHARCNET, ACEnet, and VSC-2 scientific computer clusters, and acknowledges support from RFBR grant 130200939.

REFERENCES

Adams F. C. and Shu F. H. (1986) *Astrophys. J., 308,* 836.
Adams F. C. and Shu F. H. (2007) *Astrophys. J., 671,* 497.
Adams F. C. et al. (1987) *Astrophys. J., 312,* 788.

Aikawa Y. et al. (2012) *Astrophys. J., 760*, 40.

Ali B. et al. (2010) *Astron. Astrophys., 518*, L119.

Allen A. et al. (2003) *Astrophys. J., 599*, 363.

Allen L. E. and Davis C. J. (2008) In *Handbook of Star Forming Regions, Volume I: The Northern Sky ASP Monograph Publications* (B. Reipurth, ed.), p. 621. ASP Conf. Series 4, Astronomical Society of the Pacific, San Francisco.

Allen L. et al. (2007) In *Protostars and Planets V* (B. Reipurth et al., eds.), pp. 361–376. Univ. of Arizona, Tucson.

Andre P. et al. (1993) *Astrophys. J., 406*, 122.

André P. et al. (2010) *Astron. Astrophys., 518*, L102.

Andrews S. M. et al. (2010) *Astrophys. J., 723*, 1241.

Arce H. G. and Goodman A. A. (2001) *Astrophys. J., 554*, 132.

Arce H. G. et al. (2013) *Astrophys. J., 774*, 39.

Armitage P. J. (2011) *Annu. Rev. Astron. Astrophys., 49*, 195.

Audard M. et al. (2007) *Astron. Astrophys., 468*, 379.

Bachiller R. et al. (1991) *Astron. Astrophys., 251*, 639.

Balbus S. A. and Hawley J. F. (1991) *Astrophys. J., 376*, 214.

Baraffe I. et al. (2009) *Astrophys. J. Lett., 702*, L27.

Baraffe I. et al. (2012) *Astrophys. J., 756*, 118.

Barnard E. E. (1907) *Astrophys. J., 25*, 218.

Barnes P. J. et al. (2011) *Astrophys. J. Suppl., 196*, 12.

Baruteau C. et al. (2011) *Mon. Not. R. Astron. Soc., 416*, 1971.

Basu S. and Jones C. E. (2004) *Mon. Not. R. Astron. Soc., 347*, L47.

Bate M. R. (2011) *Mon. Not. R. Astron. Soc., 417*, 2036.

Bate M. R. and Bonnell I. A. (2005) *Mon. Not. R. Astron. Soc., 356*, 1201.

Belloche A. et al. (2006) *Astron. Astrophys., 454*, L51.

Bergin E. A. and Tafalla M. (2007) *Annu. Rev. Astron. Astrophys., 45*, 339.

Bergin E. A. et al. (1998) *Astrophys. J., 499*, 777.

Billot N. et al. (2012) *Astrophys. J. Lett., 753*, L35.

Blaes O. M. and Balbus S. A. (1994) *Astrophys. J., 421*, 163.

Bonnell I. A. and Bate M. R. (2006) *Mon. Not. R. Astron. Soc., 370*, 488.

Bonnell I. A. et al. (2001) *Mon. Not. R. Astron. Soc., 323*, 785.

Boogert A. C. A. and Ehrenfreund P. (2004) In *Astrophysics of Dust* (A. N. Witt et al., eds.), p. 547. ASP Conf. Series 309, Astronomical Society of the Pacific, San Francisco.

Boogert A. C. A. et al. (2008) *Astrophys. J., 678*, 985.

Boss A. P. and Yorke H. W. (1995) *Astrophys. J. Lett., 439*, L55.

Bourke T. L. et al. (2006) *Astrophys. J. Lett., 649*, L37.

Bressert E. et al. (2010) *Mon. Not. R. Astron. Soc., 409*, L54.

Brinch C. et al. (2007) *Astron. Astrophys., 461*, 1037.

Brinch C. et al. (2009) *Astron. Astrophys., 502*, 199.

Butner H. M. et al. (1994) *Astrophys. J., 420*, 326.

Calvet N. et al. (1994) *Astrophys. J., 434*, 330.

Calvet N. et al. (1997) *Astrophys. J., 481*, 912.

Carpenter J. M. (2000) *Astron. J., 120*, 3139.

Carpenter J. M. et al. (2001) *Astron. J., 121*, 3160.

Carpenter J. M. et al. (2002) *Astron. J., 124*, 1001.

Cassen P. and Moosman A. (1981) *Icarus, 48*, 353.

Cha S.H. and Nayakshin S. (2011) *Mon. Not. R. Astron. Soc., 415*, 3319.

Chen H. et al. (1995) *Astrophys. J., 445*, 377.

Chen X. et al. (2010) *Astrophys. J., 715*, 1344.

Chen X. et al. (2012) *Astrophys. J., 751*, 89.

Chen X. et al. (2013) *Astrophys. J., 768*, 110.

Chiang H.F. et al. (2012) *Astrophys. J., 756*, 168.

Ciardi D. R. et al. (2005) *Astrophys. J., 629*, 897.

Ciesla F. J. (2011) *Astrophys. J., 740*, 9.

Commerçon B. et al. (2012a) *Astron. Astrophys., 545*, A98.

Commerçon B. et al. (2012b) *Astron. Astrophys., 548*, A39.

Cook A. M. et al. (2011) *Astrophys. J., 730*, 124.

Crapsi A. et al. (2008) *Astron. Astrophys., 486*, 245.

Csengeri T. et al. (2011) *Astrophys. J. Lett., 740*, L5.

Dapp W. B. and Basu S. (2010) *Astron. Astrophys., 521*, L56.

Demyk K. et al. (1999) *Astron. Astrophys., 349*, 267.

Demyk K. et al. (2000) In *ISO Beyond the Peaks: The 2nd ISO Workshop on Analytical Spectroscopy* (A. Salama et al., eds.), p. 183. ESA SP-456, Noordwijk, The Netherlands.

di Francesco J. et al. (2007) In *Protostars and Planets V* (B. Reipurth et al., eds.), pp. 17–32. Univ. of Arizona, Tucson.

Dullemond C. P. et al. (2006) *Astrophys. J. Lett., 640*, L67.

Dunham M. M. and Vorobyov E. I. (2012) *Astrophys. J., 747*, 52.

Dunham M. M. et al. (2006) *Astrophys. J., 651*, 945.

Dunham M. M. et al. (2008) *Astrophys. J. Suppl., 179*, 249.

Dunham M. M. et al. (2010a) *Astrophys. J., 710*, 470.

Dunham M. M. et al. (2010b) *Astrophys. J., 721*, 995.

Dunham M. M. et al. (2013) *Astron. J., 145*, 94.

Dzib S. et al. (2011) *Rev. Mex. Astron. Astrofis. Ser. Conf., 40*, 231–232.

Ehrenfreund P. et al. (1998) *Astron. Astrophys., 339*, L17.

Eisner J. A. (2012) *Astrophys. J., 755*, 23.

Eisner J. A. et al. (2005) *Astrophys. J., 635*, 396.

Enoch M. L. et al. (2008) *Astrophys. J., 684*, 1240.

Enoch M. L. et al. (2009) *Astrophys. J., 692*, 973.

Enoch M. L. et al. (2010) *Astrophys. J. Lett., 722*, L33.

Evans N. J. II et al. (2003) *Publ. Astron. Soc. Pac., 115*, 965.

Evans N. J. II et al. (2009) *Astrophys. J. Suppl., 181*, 321.

Fatuzzo M. et al. (2004) *Astrophys. J., 615*, 813.

Findeisen K. et al. (2013) *Astrophys. J., 768*, 93.

Fischer W. J. et al. (2010) *Astron. Astrophys., 518*, L122.

Fischer W. J. et al. (2012) *Astrophys. J., 756*, 99.

Fischer W. J. et al. (2013) *Astron. Nachr., 334*, 53.

Fischera J. and Martin P. G. (2012) *Astron. Astrophys., 542*, A77.

Flaherty K. M. et al. (2012) *Astrophys. J., 748*, 71.

Flaherty K. M. et al. (2013) *Astron. J., 145*, 66.

Furlan E. et al. (2008) *Astrophys. J. Suppl., 176*, 184.

Galli D. et al. (2006) *Astrophys. J., 647*, 374.

Gammie C. F. (1996) *Astrophys. J., 457*, 355.

Gräfe C. et al. (2013) *Astron. Astrophys., 553*, A69.

Green J. D. et al. (2013a) *Astrophys. J., 772*, 117.

Green J. D. et al. (2013b) *Astrophys. J., 770*, 123.

Greene T. P. et al. (1994) *Astrophys. J., 434*, 614.

Gutermuth R. A. et al. (2009) *Astrophys. J. Suppl., 184*, 18.

Gutermuth R. A. et al. (2011) *Astrophys. J., 739*, 84.

Hara C. et al. (2013) *Astrophys. J., 771*, 128.

Hartmann L. (2002) *Astrophys. J., 578*, 914.

Harvey D. W. A. et al. (2003) *Astrophys. J., 596*, 383.

Harvey P. et al. (2007) *Astrophys. J., 663*, 1149.

Hatchell J. et al. (2005) *Astron. Astrophys., 440*, 151.

Heiderman A. et al. (2010) *Astrophys. J., 723*, 1019.

Herbig G. H. (1952) *J. R. Astron. Soc. Can., 46*, 222.

Hernández J. et al. (2007) *Astrophys. J., 662*, 1067.

Hosokawa T. et al. (2011) *Astrophys. J., 738*, 140.

Hsieh T.H. and Lai S.P. (2013) *Astrophys. J. Suppl., 205*, 5.

Hull C. L. H. et al. (2013) *Astrophys. J., 768*, 159.

Inutsuka S.I. and Miyama S. M. (1992) *Astrophys. J., 388*, 392.

Inutsuka S.I. and Miyama S. M. (1997) *Astrophys. J., 480*, 681.

Johnstone D. et al. (2004) *Astrophys. J. Lett., 611*, L45.

Johnstone D. et al. (2013) *Astrophys. J., 765*, 133.

Joos M. et al. (2012) *Astron. Astrophys., 543*, A128.

Jørgensen J. K. et al. (2007) *Astrophys. J., 659*, 479.

Jorgensen J. K. et al. (2009) *Astron. Astrophys., 507*, 861.

Joy A. H. (1945) *Astrophys. J., 102*, 168.

Kauffmann J. et al. (2011) *Mon. Not. R. Astron. Soc., 416*, 2341.

Kelley M. S. and Wooden D. H. (2009) *Planet. Space Sci., 57*, 1133.

Kemper F. et al. (2005) *Astrophys. J., 633*, 534.

Kenyon S. J. and Hartmann L. (1995) *Astrophys. J. Suppl., 101*, 117.

Kenyon S. J. et al. (1990) *Astron. J., 99*, 869.

Kenyon S. J. et al. (1993) *Astrophys. J., 414*, 676.

Kim H. J. et al. (2012) *Astrophys. J., 758*, 38.

Kirk H. et al. (2006) *Astrophys. J., 646*, 1009.

Kirk H. et al. (2013) *Astrophys. J., 766*, 115.

Kóspál Á. et al. (2012) *Astrophys. J. Suppl., 201*, 11.

Kratter K. M. et al. (2008) *Astrophys. J., 681*, 375.

Kratter K. M. et al. (2010a) *Astrophys. J., 708*, 1585.

Kratter K. M. et al. (2010b) *Astrophys. J., 710*, 1375.

Krumholz M. R. et al. (2012) *Astrophys. J., 754*, 71.

Krumholz M. R. et al. (2013) *Astrophys. J. Lett., 767*, L11.

Kryukova E. et al. (2012) *Astron. J., 144*, 31.

Lada C. J. (1987) In *Star Forming Regions* (M. Peimbert and J. Jugaku, eds.), pp. 1–17. IAU Symp. 115, Cambridge Univ., Cambridge.

Larson R. B. (1969) *Mon. Not. R. Astron. Soc., 145*, 271.

Larson R. B. (1985) *Mon. Not. R. Astron. Soc., 214*, 379.

Laughlin G. and Bodenheimer P. (1994) *Astrophys. J., 436*, 335.

Launhardt R. et al. (2013) *Astron. Astrophys., 551*, A98.

Lee C. F. et al. (2009a) *Astrophys. J., 699*, 1584.

Lee C. W. et al. (2009b) *Astrophys. J., 693*, 1290.

Lee J. E. (2007) *J. Korean Astron. Soc., 40*, 83.

Lee J. E. et al. (2010) *Astrophys. J. Lett., 709*, L74.

Li D. et al. (2013) *Astrophys. J. Lett., 768*, L5.

Lin D. N. C. and Pringle J. E. (1987) *Mon. Not. R. Astron. Soc.,* *225,* 607.

Lisse C. M. et al. (2007) *Icarus, 191,* 223.

Lodato G. and Rice W. K. M. (2004) *Mon. Not. R. Astron. Soc.,* *351,* 630.

Lommen D. et al. (2008) *Astron. Astrophys., 481,* 141.

Looney L. W. et al. (2000) *Astrophys. J., 529,* 477.

Lynds B. T. (1962) *Astrophys. J. Suppl., 7,* 1.

Machida M. N. and Matsumoto T. (2011) *Mon. Not. R. Astron. Soc.,* *413,* 2767.

Machida M. N. et al. (2008) *Astrophys. J., 676,* 1088.

Machida M. et al. (2011a) *Astrophys. J., 729,* 42.

Machida M. N. et al. (2011b) *Publ. Astron. Soc. Japan, 63,* 555.

Manoj P. et al. (2013) *Astrophys. J., 763,* 83.

Martin R. G. et al. (2012) *Mon. Not. R. Astron. Soc., 423,* 2718.

Masiunas L. C. et al. (2012) *Astrophys. J., 752,* 127.

Masunaga H. et al. (1998) *Astrophys. J., 495,* 346.

Maury A. J. et al. (2010) *Astron. Astrophys., 512,* A40.

McKee C. F. (1989) *Astrophys. J., 345,* 782.

McKee C. F. and Offner S. S. R. (2010) *Astrophys. J., 716,* 167.

McKee C. F. and Tan J. C. (2003) *Astrophys. J., 585,* 850.

Megeath S. T. et al. (2009) *Astron. J., 137,* 4072.

Megeath S. T. et al. (2012) *Astron. J., 144,* 192.

Mellon R. R. and Li Z. Y. (2008) *Astrophys. J., 681,* 1356.

Molinari S. et al. (2010) *Astron. Astrophys., 518,* L100.

Morales-Calderón M. et al. (2011) *Astrophys. J., 733,* 50.

Mottram J. C. et al. (2013) *Astron. Astrophys., 558,* A126.

Murillo N. M. and Lai S.P. (2013) *Astrophys. J. Lett., 764,* L15.

Murillo N. M. et al. (2013) *Astron. Astrophys., 560,* A103.

Muzerolle J. et al. (1998) *Astron. J., 116,* 2965.

Myers P. C. (2009a) *Astrophys. J., 700,* 1609.

Myers P. C. (2009b) *Astrophys. J., 706,* 1341.

Myers P. C. (2010) *Astrophys. J., 714,* 1280.

Myers P. C. (2011) *Astrophys. J., 735,* 82.

Myers P. C. (2012) *Astrophys. J., 752,* 9.

Myers P. C. and Ladd E. F. (1993) *Astrophys. J. Lett., 413,* L47.

Nayakshin S. et al. (2011) *Mon. Not. R. Astron. Soc., 416,* L50.

Norman C. and Silk J. (1980) *Astrophys. J., 238,* 158.

Öberg K. I. et al. (2009) *Astron. Astrophys., 505,* 183.

Offner S. S. R. and McKee C. F. (2011) *Astrophys. J., 736,* 53.

Offner S. S. R. et al. (2009) *Astrophys. J., 703,* 131.

Offner S. S. R. et al. (2010) *Astrophys. J., 725,* 1485.

Offner S. S. R. et al. (2011) *Astrophys. J., 743,* 91.

Oliveira I. et al. (2010) *Astrophys. J., 714,* 778.

Olofsson J. et al. (2010) *Astron. Astrophys., 520,* A39.

Omukai K. (2007) *Publ. Astron. Soc. Japan, 59,* 589.

Ostriker E. C. and Shu F. H. (1995) *Astrophys. J., 447,* 813.

Palla F. and Stahler S. W. (1992) *Astrophys. J., 392,* 667.

Parker R. J. and Meyer M. R. (2012) *Mon. Not. R. Astron. Soc.,* *427,* 637.

Penston M. V. (1969) *Mon. Not. R. Astron. Soc., 144,* 425.

Peterson D. E. and Megeath S. T. (2008) In *Handbook of Star Forming Regions, Volume I: The Northern Sky ASP Monograph Publications* (B. Reipurth, ed.), p. 590. ASP Conf. Series 4, Astronomical Society of the Pacific, San Francisco.

Pezzuto S. et al. (2012) *Astron. Astrophys., 547,* A54.

Pilbratt G. L. et al. (2010) *Astron. Astrophys., 518,* L1.

Pillitteri I. et al. (2013) *Astrophys. J., 768,* 99.

Pineda J. E. et al. (2011) *Astrophys. J., 743,* 201.

Pineda J. E. et al. (2012) *Astron. Astrophys., 544,* L7.

Pontoppidan K. M. et al. (2008) *Astrophys. J., 678,* 1005.

Poteet C. A. et al. (2011) *Astrophys. J. Lett., 733,* L32.

Poteet C. A. et al. (2013) *Astrophys. J., 766,* 117.

Price D. J. et al. (2012) *Mon. Not. R. Astron. Soc., 423,* L45.

Rebull L. M. et al. (2010) *Astrophys. J. Suppl., 186,* 259.

Robitaille T. P. et al. (2006) *Astrophys. J. Suppl., 167,* 256.

Saigo K. and Tomisaka K. (2006) *Astrophys. J., 645,* 381.

Saigo K. and Tomisaka K. (2011) *Astrophys. J., 728,* 78.

Saigo K. et al. (2008) *Astrophys. J., 674,* 997.

Sargent B. A. et al. (2009) *Astrophys. J. Suppl., 182,* 477.

Sauter J. et al. (2009) *Astron. Astrophys., 505,* 1167.

Schnee S. et al. (2012) *Astrophys. J., 745,* 18.

Schneider N. et al. (2012) *Astron. Astrophys., 540,* L11.

Schneider S. and Elmegreen B. G. (1979) *Astrophys. J. Suppl., 41,* 87.

Scholz A. (2012) *Mon. Not. R. Astron. Soc., 420,* 1495.

Scholz A. et al. (2013) *Mon. Not. R. Astron. Soc., 430,* 2910.

Schwarz K. R. et al. (2012) *Astron. J., 144,* 115.

Shakura N. I. and Sunyaev R. A. (1973) *Astron. Astrophys., 24,* 337.

Shu F. H. (1977) *Astrophys. J., 214,* 488.

Shu F. H. et al. (1987) *Annu. Rev. Astron. Astrophys., 25,* 23.

Stahler S. W. (1988) *Astrophys. J., 332,* 804.

Stahler S. W. et al. (1980) *Astrophys. J., 241,* 637.

Stamatellos D. et al. (2011) *Astrophys. J., 730,* 32.

Stutz A. M. et al. (2008) *Astrophys. J., 687,* 389.

Stutz A. M. et al. (2010) *Astron. Astrophys., 518,* L87.

Stutz A. M. et al. (2013) *Astrophys. J., 767,* 36.

Takahashi S. et al. (2013) *Astrophys. J., 763,* 57.

Takakuwa S. et al. (2012) *Astrophys. J., 754,* 52.

Tassis K. and Mouschovias T. C. (2005) *Astrophys. J., 618,* 783.

Terebey S. et al. (1984) *Astrophys. J., 286,* 529.

Tobin J. J. et al. (2010) *Astrophys. J., 712,* 1010.

Tobin J. J. et al. (2012) *Nature, 492,* 83.

Tobin J. J. et al. (2013) *Astrophys. J., 771,* 48.

Tomida K. et al. (2010) *Astrophys. J. Lett., 725,* L239.

Tomisaka K. and Tomida K. (2011) *Publ. Astron. Soc. Japan, 63,* 1151.

Toomre A. (1964) *Astrophys. J., 139,* 1217.

van Broekhuizen F. A. et al. (2006) *Astron. Astrophys., 451,* 723.

van Kempen T. A. et al. (2009) *Astron. Astrophys., 498,* 167.

Visser R. and Dullemond C. P. (2010) *Astron. Astrophys., 519,* A28.

Visser R. et al. (2009) *Astron. Astrophys., 495,* 881.

Vorobyov E. I. (2009a) *Astrophys. J., 692,* 1609.

Vorobyov E. I. (2009b) *Astrophys. J., 704,* 715.

Vorobyov E. I. (2010) *New Astron., 15,* 24.

Vorobyov E. I. (2011a) *Astrophys. J. Lett., 728,* L45.

Vorobyov E. I. (2011b) *Astrophys. J., 729,* 146.

Vorobyov E. I. (2013) *Astron. Astrophys., 552,* 129.

Vorobyov E. I. and Basu S. (2005) *Astrophys. J. Lett., 633,* 137.

Vorobyov E. I. and Basu S. (2006) *Astrophys. J., 650,* 956.

Vorobyov E. I. and Basu S. (2007) *Mon. Not. R. Astron. Soc., 381,* 1009.

Vorobyov E. I. and Basu S. (2008) *Astrophys. J. Lett., 676,* 139.

Vorobyov E. I. and Basu S. (2009) *Astrophys. J., 703,* 922.

Vorobyov E. I. and Basu S. (2010a) *Astrophys. J., 719,* 1896.

Vorobyov E. I. and Basu S. (2010b) *Astrophys. J. Lett., 714,* 133.

Werner M. W. et al. (2004) *Astrophys. J. Suppl., 154,* 1.

White R. J. et al. (2007) In *Protostars and Planets V* (B. Reipurth et al., eds.), pp. 117–132. Univ. of Arizona, Tucson.

Whitney B. A. et al. (2003a) *Astrophys. J., 591,* 1049.

Whitney B. A. et al. (2003b) *Astrophys. J., 598,* 1079.

Wilking B. A. et al. (1987) *Bull. Am. Astron. Soc., 19,* 1092.

Winston E. et al. (2007) *Astrophys. J., 669,* 493.

Winston E. et al. (2010) *Astron. J., 140,* 266.

Wolf S. et al. (2008) *Astrophys. J. Lett., 674,* L101.

Wolk S. J. et al. (2013) *Astrophys. J., 773,* 145.

Wu J. et al. (2007) *Astron. J., 133,* 1560.

Yen H. W. et al. (2013) *Astrophys. J., 772,* 22.

Yorke H. W. and Bodenheimer P. (1999) *Astrophys. J., 525,* 330.

Young C. H. and Evans N. J. II (2005) *Astrophys. J., 627,* 293.

Young C. H. et al. (2004) *Astrophys. J. Suppl., 154,* 396.

Zapata L. A. et al. (2013) *Astrophys. J. Lett., 764,* L14.

Zasowski G. et al. (2009) *Astrophys. J., 694,* 459.

Zhu Z. et al. (2009) *Astrophys. J., 694,* 1045.

Zhu Z. et al. (2010) *Astrophys. J., 713,* 1134.

Soderblom D. R., Hillenbrand L. A., Jeffries R. D., Mamajek E. E., and Naylor T. (2014) Ages of young stars. In *Protostars and Planets VI*
(H. Beuther et al., eds.), pp. 219–241. Univ. of Arizona, Tucson, DOI: 10.2458/azu_uapress_9780816531240-ch010.

Ages of Young Stars

David R. Soderblom
Space Telescope Science Institute

Lynne A. Hillenbrand
California Institute of Technology

Rob D. Jeffries
Keele University

Eric E. Mamajek
University of Rochester

Tim Naylor
University of Exeter

Determining the sequence of events in the formation of stars and planetary systems and their
timescales is essential for understanding those processes, yet establishing ages is fundamentally
difficult because we lack direct indicators. In this review we discuss the age challenge for
young stars, specifically those less than ~100 m.y. old. Most age determination methods that
we discuss are primarily applicable to groups of stars but can be used to estimate the age of
individual objects. A reliable age scale is established above 20 m.y. from measurement of the
lithium depletion boundary (LDB) in young clusters, and consistency is shown between these
ages and those from the upper-main-sequence (UMS) and main-sequence (MS) turn-off — if
modest core convection and rotation is included in the models of higher-mass stars. Other
available methods for age estimation include the kinematics of young groups, placing stars in
Hertzsprung-Russell diagrams (HRDs), pulsations and seismology, surface gravity measurement, rotation and activity, and Li abundance. We review each of these methods and present
known strengths and weaknesses. Below ~20 m.y., both model-dependent and observational
uncertainties grow, the situation is confused by the possibility of age spreads, and no reliable
absolute ages yet exist. The lack of absolute age calibration below 20 m.y. should be born in
mind when considering the lifetimes of protostellar phases and circumstellar material.

1. MOTIVATION AND SCOPE

We know the timing of key events in the early history
of our solar system from meteorites because we can apply
radiometric dating techniques to samples in a laboratory.
From this we infer the Sun's age, a very precise one, but
even the Sun itself does not directly reveal its age in its
visible properties. The techniques for age-dating stars, and
thereby other planetary systems, are of much lower precision. Establishing accurate or even precise ages for stars
remains difficult, especially for young stars at the very
stages where many interesting things, including planet formation and subsequent dynamical evolution, are happening.

1.1. What Does "Young" Mean?

For the purposes of this review we consider "young"
to be anything with an age (τ) below about 100 million
years (m.y.), i.e., the age of the Pleiades and younger. For
a coeval sample of stars, the higher stellar masses at this
age will have begun core hydrogen burning and the highest
masses will even have exhausted theirs. The lower stellar
masses will be still in the pre-main-sequence (PMS) phase
of radial contraction, which can take ~500 m.y. for objects
just above the brown dwarf limit.

In the context of understanding protostars and planets,
we attempt here to establish timescales and ordering of

events independently of the phenomena being studied. So, for instance, we note the presence of circumstellar material as suggestive of youth, but what we really want to do is to study the time evolution of circumstellar material and to bring to bear all applicable independent information. We cannot yet fully succeed in this goal, but progress is being made.

1.2. Goals in Considering Age Estimation Methods

We wish, in order of increasing difficulty, to accomplish the following: (1) Broadly categorize objects into wide age bins such as "very young" or less than 5-m.y.-old stars generally associated with star-forming regions; "young PMS" stars, which for solar mass are generally less than 30 m.y. old; and "young field" stars, generally less than a few hundred million years old. (2) Ascertain the correct ordering of phenomena, events, and ages. (3) Reliably detect differences in age, either for individual stars or for groups. (4) Derive absolute ages, which requires not only sound methodology but also a definition of τ_0, the evolutionary point at which age starts.

1.3. Recent Reviews

Ages and timescales have always been implicit in the study of stars, and in recent years there have been several discussions of this subject: (1) *Soderblom* (2010) wrote a general review of stellar ages and methods that emphasizes MS stars but includes a section on PMS objects. Therein, age-dating methodologies are characterized as we do here into fundamental, semifundamental, model-dependent, and empirical categories. (2) The Soderblom review grew out of IAU Symposium 258 ("The Ages of Stars"), which was held in Baltimore in 2008. The proceedings (*Mamajek et al.,* 2009) include reviews and contributions on PMS and young stars; in particular, see *Jeffries* (2009), *Hillenbrand* (2009), and *Naylor et al.* (2009). (3) More recently, *Jeffries* (2012) has discussed age spreads in star-forming regions. (4) *Preibisch* (2012) has presented a thorough discussion of color magnitude diagrams (CMDs) and Hertzsprung-Russell diagrams (HRDs) for PMS stars and how ages are estimated from those diagrams.

There are also several chapters in the present volume that relate closely to our topic, particularly the chapters by Bouvier et al., Dunham et al., and Alexander et al.

2. CRITICAL QUESTIONS ABOUT AGES

1. Can we establish reliable and consistent ages for young stars that are independent of the phenomena we wish to study (section 8)?

2. What is the age scale for PMS and young stars and what inherent uncertainties affect it? Can we construct a recommended age scale as a reference (section 3.1)?

3. We see scatter and dispersion in the HRDs and CMDs of PMS groups. Is that evidence for an age spread? Or can

that plausibly be accounted for by unappreciated physics such as variable and differing accretion rates (section 6)?

4. Are age classifiers such as abundant Li infallible, or is it possible for a PMS star to have little or no Li (section 5.3)?

5. Can the velocities of stars in groups reliably establish kinematic ages (section 3.2)?

6. What is the chemical composition of young stellar populations, and how do uncertainties in composition affect stellar age estimation (section 7)?

7. What can be done to improve PMS ages over the next decade (section 8)?

We will discuss individual methods and their precision and accuracy, as well as areas of applicability, weaknesses, and strengths. We note that it is not strictly necessary for us to decide upon an absolute zero point to the age scale. This is a matter of both physical and philosophical debate (section 6.2), but all estimates of stellar age presently have uncertainties that exceed any uncertainty in the definition of this zero point.

3. SEMIFUNDAMENTAL AGE TECHNIQUES

The only fundamental age in astrophysics is that of the Sun because the physics involved (decay of radioactive isotopes in meteorites) is completely understood and all necessary quantities can be measured. Semifundamental methods require assumptions, but ones that appear to be well founded and that are straightforward. Because the models that predict the lithium depletion boundary (LDB) at the very-low-mass end of cluster CMDs are physically simple, more so than for, say, cluster turn-offs, we recommend first establishing an age scale by considering clusters that have ages from this technique. Kinematic dating methods are also semifundamental in that they use a simple method with few assumptions, but, as we discuss, those assumptions may be invalid and the ages are problematic.

3.1. The Lithium Depletion Boundary

As PMS stars age and contract toward the zero age main sequence (ZAMS), their core temperatures rise. If the star is more massive than $\simeq 0.06$ $M_\odot$, the core temperature will eventually become high enough (~3 MK) to burn Li (^{7}Li) in proton capture reactions (*Basri et al.,* 1996; *Chabrier et al.,* 1996; *Bildsten et al.,* 1997; *Ushomirsky et al.,* 1998). Pre-MS stars reach this Li-burning temperature on a mass-dependent timescale and, since the temperature dependence of the nuclear reactions is steep, and the mixing timescale in fully convective PMS stars is short, total Li depletion throughout the star occurs rapidly. This creates a sharp, age-dependent, LDB between stars at low luminosities retaining all their initial Li and those at only slightly higher luminosities with total Li depletion.

The LDB technique is "semifundamental" because it relies on well-understood physics and is quite insensitive to variations in assumed opacities, metallicity, convective efficiency, equation of state, and stellar rotation. *Burke et al.* (2004) conducted experiments with theoretical models, varying the input physics within plausible limits, and found that absolute LDB ages have "1σ" theoretical uncertainties ranging from 3% at 200 m.y. to 8% at 20 m.y. Comparisons of LDB ages predicted by a variety of published evolutionary models show agreement at the level of ≤10% across a similar age range. This model insensitivity applies only to ages determined from the LDB luminosity; the predicted T_{eff} at the LDB is much more sensitive to the treatment of convection and atmospheric physics. L_{bol} is also much easier to measure than T_{eff}. Uncertainties in empirical bolometric corrections have a much smaller effect on derived ages than typical T_{eff} uncertainties from spectral types. In summary, the luminosity of the LDB is a good absolute age indicator, but the temperature of the LDB is not (*Jeffries*, 2006).

Lithium depletion boundary ages can be determined for coeval groups by determining the luminosity at which the transition from depleted to undepleted Li occurs; an example is shown in Fig. 1. This transition is expected to take <5% of the stellar age, so the exact definition of "Li-depleted" is not crucial. This is useful since determining accurate Li abundances for cool stars (T_{eff} < 4000 K) is difficult. Instead, one can rely on measuring the pseudo-equivalent width (EW) of the strong Li I 6708 Å resonance feature. In an undepleted cool star EW(Li) ≃ 0.6 Å, falling to <0.25 Å in a similar star that has burned >99% of its Li (*Palla et al.*, 2007).

Lithium depletion boundary ages have been measured for nine clusters between 21 m.y. and 132 m.y. (see Table 1); the published uncertainties are combinations of theoretical uncertainties and observational uncertainties associated with estimating the luminosity of the LDB. Contributing to the latter are distance estimates, reddening, and bolometric corrections (for stellar luminosity not in the band observed), but these are usually small compared with the difficulty in locating the LDB in sparse datasets and where the low-mass stars may be variable and may be in unresolved binary systems. In younger clusters, any age spread of more than a few million years might also blur the otherwise sharp LDB (see section 6).

In Table 1, we homogeneously reevaluate these LDB ages using the LDB locations (and uncertainty), cluster distances, and reddening given in the original LDB papers, but combined uniformly with the bolometric corrections used in *Jeffries and Oliveira* (2005) and the evolutionary models of *Chabrier and Baraffe* (1997). Age uncertainties include those due to LDB location, an assumed 0.1-mag systematic uncertainty in both color and magnitude calibrations (usually negligible), and, presented separately, the systematic absolute age uncertainty estimated by *Burke et al.* (2004). These results show that LDB age uncertainties have yet to attain the floor set by levels of theoretical understanding. There is scope for improvement, particularly in defining the position of the LDB, photometric calibration of cool stars, and better estimates of cluster distances.

The LDB technique is limited by both physical and practical constraints to clusters of age 20–200 m.y. Measuring the strength of the Li I 6708 Å feature at sufficient spectral resolution (R ≥ 3000) and signal-to-noise ratio to distinguish Li depletion in very faint, cool stars is challenging. Table 1 lists the apparent Cousins I magnitude of the LDB in the clusters where this has been possible. A photometric survey for candidates is required, and subsequent spectroscopy must be capable of distinguishing cluster members from nonmembers or risk confusing Li-depleted members with older, unrelated field stars. Although the relationship between LDB luminosity and age does become shallower at older ages, leading to an increase in (fractional) age uncertainty, it is the faintness of the LDB that leads to the upper age limit on its applicability. At ≃200 m.y., the LDB is at $L_{bol}/L_{\odot} \simeq 5 \times 10^{-4}$. The nearest clusters with ages ≥200 m.y. are at distances of ~300 pc or more, so finding the LDB would entail good intermediate-resolution spectroscopy of groups of stars with I > 20, beyond the realistic grasp of current 8–10-m telescopes. A lower limit to the validity of the technique is ≃20 m.y. and arises because at 10 m.y. some models do not predict significant Li destruction, and at ages of 10–20 m.y., the difference in LDB ages derived from different models is 20–30%.

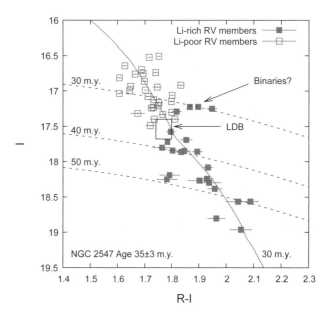

Fig. 1. Locating the LDB in the color-magnitude diagram of NGC 2547 (adapted from *Jeffries and Oliveira*, 2005). The plot shows three dashed loci of constant luminosity, corresponding to LDB ages of 30, 40, and 50 m.y. The solid locus is a low-mass isochrone at 30 m.y. The points represent cluster members that are found to possess a strong (undepleted) Li I 6708 Å line or not, and there is a reasonably sharp transition between these categories. A box marks the likely location of the LDB in this cluster and yields an age of 35.4 ± 3.3 m.y. (see Table 1), but its precise position is made harder to judge by the presence of probable unresolved binary systems that appear overluminous for their color.

TABLE 1. LDB ages compared with ages determined from upper-main-sequence fitting
using models both with and without convective overshoot.

Cluster	I_{LDB} (mag)	LDB Age (m.y.)	Ref.	M_{bol} (mag)	Homogeneous LDB Age (m.y.)	Mermilliod MS Age (m.y.)	Overshoot MS Age (m.y.)	Ref.
β Pic MG		21 ± 4	[1]	8.28 ± 0.54	20.3 ± 3.4 ± 1.7			[2]
NGC 1960	18.95 ± 0.30	22 ± 4	[3]	8.57 ± 0.33	23.2 ± 3.3 ± 1.9	<20	$26.3^{+3.2}_{-5.2}$	[11]
IC 4665	16.64 ± 0.10	28 ± 5	[4]	8.78 ± 0.34	25.4 ± 3.8 ± 1.9	36 ± 5	41 ± 12	[12]
NGC 2547	17.54 ± 0.14	35 ± 3	[5]	9.58 ± 0.20	35.4 ± 3.3 ± 2.2		48^{+14}_{-21}	[13]
IC 2602	15.64 ± 0.08	46^{+6}_{-5}	[6]	9.88 ± 0.17	40.0 ± 3.7 ± 2.5	36 ± 5	44^{+18}_{-16}	[13]
IC 2391	16.21 ± 0.07	50 ± 5	[7]	10.31 ± 0.16	48.6 ± 4.3 ± 3.0	36 ± 5	45 ± 5	[14]
α Per	17.70 ± 0.15	90 ± 10	[8]	11.27 ± 0.21	80 ± 11 ± 4	51 ± 7	80	[15]
Pleiades	17.86 ± 0.10	125 ± 8	[9]	12.01 ± 0.16	126 ± 16 ± 4	78 ± 9	120	[15]
Blanco 1	18.78 ± 0.24	132 ± 24	[10]	12.01 ± 0.29	126 ± 23 ± 4		115 ± 16	[10]

Columns 2–4 list the apparent I magnitude of the LDB, the published LDB, age and the source paper. Columns 5 and 6 give a bolometric magnitude and LDB age that have been homogeneously reevaluated using the locations of the LDB from the original papers, the evolutionary models of *Chabrier and Baraffe* (1997), and bolometric corrections used in *Jeffries and Oliveira* (2005). The error estimates include uncertainty in the LDB location, distance modulus, a calibration error of 0.1 mag, and then separately, a physical absolute uncertainty estimated from *Burke et al.* (2004). The last three columns give an upper main sequence age from *Mermilliod* (1981) using models with no convective overshoot, followed by literature age estimate using models with moderate convective overshoot.

References: [1] *Binks and Jeffries* (2013), I mag not available, M_{bol} calculated from K_{LDB}; [2] most massive member is A6V, hence no UMS age; [3] *Jeffries et al.* (2013); [4] *Manzi et al.* (2008); [5] *Jeffries and Oliveira* (2005); [6] *Dobbie et al.* (2010); [7] *Barrado y Navascués et al.* (2004); [8] *Stauffer et al.* (1999); [9] *Stauffer et al.* (1998); [10] *Cargile et al.* (2010); [11] *Bell et al.* (2013); [12] *Cargile and James* (2010); [13] *Naylor et al.* (2009); [14] derived by E. Mamajek using data from *Hauck and Mermilliod* (1998) and isochrones from *Bertelli et al.* (2009); [15] *Ventura et al.* (1998).

Although possible in only a few clusters, the LDB method is well placed to calibrate other age-estimation techniques used in those same clusters. Its usefulness in estimating ages for individual stars is limited; the detection of (undepleted) Li in a low-mass star at a known distance and luminosity, or less accurately with an estimated T_{eff}, will give an upper limit to its age. Conversely, the lack of Li in a similar star places a lower limit to its age. This can be useful for finding low-mass members of nearby, young moving groups or estimating the ages of field L dwarfs (*Martín et al.*, 1999), although might be confused by a weakening of the 6708-Å line in low-gravity, very-low-mass objects (*Kirkpatrick et al.*, 2008).

An example of how LDB ages can calibrate other techniques is provided by a comparison with ages determined for the same clusters from the main-sequence turn-off (MSTO) and UMS. These are derived by fitting photometric data with stellar evolutionary models, but depend on a number of uncertain physical ingredients including the amount of convective core overshooting and the stellar rotation rate (e.g., *Maeder and Meynet, 1989; Meynet and Maeder, 2000*) (see section 4.1). Upper MS/MSTO ages are listed, where available, in Table 1 using models with no convective core overshoot (*Mermilliod, 1981*) and using models with moderate core overshoot [primarily the models of *Schaller et al.* (1992) with 0.2 pressure scale heights of overshoot]. The comparison is shown in Fig. 2. As pointed out by *Stauffer et al.* (1998), cluster ages determined from models with no core overshoot are about 30% younger than LDB ages, but there is better agreement if moderate over-shooting is included. Recent models incorporating rotation for high-mass stars show that this can mimic the effects of overshooting in extending MS lifetimes (*Ekström et al., 2012*). Hence LDB ages are consistent with evolutionary models that incorporate overshooting, rotation, or a more moderate quantity of both, and comparison with LDB ages alone is unlikely to disentangle these. A homogeneous reanalysis of the UMS and MSTO ages for clusters with LDB ages, using uniform models and fitting techniques, would be valuable.

Below is a summary of the LDB method:

Pro: The LDB method involves few assumptions, and these appear to be on solid physical ground, making the method more reliable than others for clusters in the age range 20–200 m.y.

Pro: LBD ages depend only weakly on stellar composition and are insensitive to observational uncertainties.

Pro: LDB observations require minimal analysis or interpretation: The Li feature is either clearly present or it is not.

Pro: Age errors for this method appear to be ~10–20%, but could be lowered to ~5% with better observations.

Con: Detecting the presence or absence of Li at the LDB means acquiring spectra of moderate resolution for extremely faint objects, limiting its use to a few clusters. It is unlikely that many more LDB ages will become available without larger telescopes.

Con: The age range for which the LDB method is applicable (~20–200 m.y.) does not extend far past the ZAMS and does not cover star-forming regions.

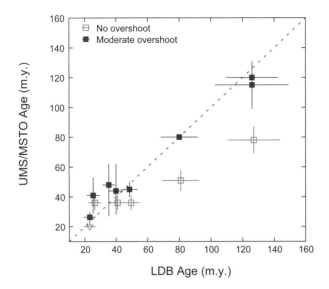

Fig. 2. A comparison of cluster LDB ages with ages determined from the upper-main-sequence (UMS) and main-sequence turn-off (MSTO) using models without core convective overshoot and with a moderate amount of core overshoot (about 0.2 pressure scale heights). Data and sources are from Table 1.

Con: The LDB method can only be applied to groups of stars of similar age, and provides only limits on the ages of individual low-mass stars.

3.2. Kinematic Ages

Kinematic ages theoretically can be derived from either proper motion or three-dimensional velocity data for members of unbound, expanding associations by measuring either the group's expansion rate ("expansion ages") or estimating the epoch of smallest volume ("traceback ages"). Additionally, some authors have calculated "flyby ages" by calculating the time of minimum separation between stellar groups, or individual stars and stellar groups, in the past. The twentieth-century work on kinematic ages is summarized in *Brown et al.* (1997), while *Soderblom* (2010) provides a review on more recent efforts.

Kinematic age techniques provide the hope of stellar age estimation independent of the many issues related to stellar evolution models; in particular, they promise to yield absolute ages at <20 m.y. where the LDB technique is unavailable. However, there are practical difficulties that have made the three common classifications of expansion ages unreliable. Historically, kinematic ages were calculated for OB subgroups, entire OB associations, and the entire Gould's belt complex. More recently, kinematic ages have been estimated for nearby young associations like the TW Hya and β Pic groups (e.g., *de la Reza et al.,* 2006; *Ortega et al.,* 2002a). Kinematic techniques do not provide useful information on age spreads within star-formation episodes, although runaways' ages have the potential to

place lower limits on ages if ejected stars and birth sites can be unambiguously matched.

3.2.1. OB associations. Brown et al. (1997) discussed kinematic age estimates for OB subgroups based on proper motions alone. Table 1 of *Brown et al.* (1997) shows that rarely is there good agreement between either expansion ages or traceback ages vs. ages derived via evolutionary tracks ("nuclear ages"). In most instances, kinematic ages were significantly shorter than nuclear ages. Through simple N-body simulations of unbound associations using realistic input parameters, *Brown et al.* (1997) demonstrated that calculating reliable kinematic ages of expanding associations, via either expansion rates or kinematic traceback, could not be estimated using proper motions alone because of their uncertainties; good radial velocities and parallaxes are needed.

3.2.2. Nearby associations. The availability of the Hipparcos astrometric dataset provided the opportunity to attempt the estimation of kinematic ages using three-dimensional velocities. The primary focus of these studies has been on newly discovered young associations within 100 pc (e.g., *Zuckerman and Song,* 2004), with the primary targets being the TW Hya Association (TWA) and β Pic Moving Group (BPMG). Some authors have elected to consider the kinematic ages of TWA and BPMG to be so well determined and model independent that they have been used to judge the veracity of other age-dating techniques (e.g., *Song et al.,* 2012); however, the veracity of the kinematic ages of TWA and BPMG are worth further scrutiny. We spend some time discussing both groups given the potential of their kinematic ages to act as benchmarks for other age-dating techniques.

3.2.3. TW Hya association. The most widely cited isochronal age for the TWA is that of *Webb et al.* (1999): 10 m.y. *Makarov and Fabricius* (2001) estimated an expansion age of 8.3 m.y.; however, the majority of stars used in the analysis were later shown spectroscopically to not be bona fide TWA members (*Song et al.,* 2003). Later, a traceback age of 8.3 ± 0.8 m.y. for TWA was calculated by *de la Reza et al.* (2006), who simulated the past orbits of four TWA members using Hipparcos astrometry. However, as de la Reza et al. admit, the traceback age for TWA hinged critically upon the membership of one system – TWA 19 – that turns out to be much more distant [d ≃ 92 ± 11 pc (*van Leeuwen,* 2007)] than other TWA members [d ≃ 56 pc (*Weinberger et al.,* 2013)], and its position, proper motion, distance, and age are much more commensurate with membership in the ~16-m.y.-old Lower Centaurus-Crux association (*de Zeeuw et al.,* 1999; *Mamajek et al.,* 2002; *Song et al.,* 2012).

A subsequent calculation by *Makarov et al.* (2005) calculated an expansion age for TWA of 4.7 ± 0.6 m.y. by including two additional young stars (HD 139084 and HD 220476) in the analysis with three well-established members (TWA 1, 4, 11). However, this age calculation appears to be unreliable, as HD 220476 is a mid-G ZAMS (not PMS) star in terms of absolute magnitude and spectroscopic surface gravity (e.g., *Gray et al.,* 2006), and HD 139084 appears to be a noncontroversial β Pic group member (*Zuckerman and Song,* 2004).

Other attempts at calculating a kinematic age for TWA have been unsuccessful. Examining trends of radial velocity vs. distance, *Mamajek* (2005) estimated a lower limit on the expansion age of TWA of 10 m.y. (95% confidence limit). The extensive trigonometric astrometric survey by *Weinberger et al.* (2012) was unable to discern any kinematic signature among the TWA members that might lead to a calculable kinematic age. It is odd that the first two studies that published well-defined kinematic ages both estimated the same age [in agreement with the isochronal age published by *Webb et al.* (1999)], despite serious issues with regard to inclusion of nonmembers in both analyses. We conclude that no reliable kinematic age has yet been determined for TWA.

3.2.4. β Pic moving group. Zuckerman et al. (2001) estimated an isochronal age of 12^{+8}_{-4} m.y. for BPMG. Traceback ages of 11–12 m.y. for BPMG have been calculated by *Ortega et al.* (2002a,b) and *Song et al.* (2003). *Ortega et al.* (2002a) simulated the orbits of the entire BPMG membership list from *Zuckerman et al.* (2001), and showed that their positions were most concentrated 11.5 m.y. ago (which they adopt as the age), and that at birth the BPMG members spanned a region 24 pc in size. *Ortega et al.* (2002b) later revised their traceback analysis, and derived an age of 10.8 ± 0.3 m.y. for BPMG. *Song et al.* (2003) added new Hipparcos stars to the membership of BPMG, and found that "excepting a few outliers . . . all [BPMG] members were confined in a smaller space about ~12 m.y. ago." *Song et al.* (2003) included ~20 BPMG stars that appeared to be close together ~12 m.y. ago, but rejected half a dozen other systems that showed deviant motion.

While these results seem self-consistent, they are inconsistent with the new LDB age of 21 ± 4 m.y. determined by *Binks and Jeffries* (2013), and the question should be raised as to whether the inclusion or exclusion of individual members impacted the age derived from the traceback analysis. *Makarov* (2007) found a wide distribution of flyby ages between individual β Pic members and the group centroid, consistent with ages of 22 ± 12 m.y. A new analysis of the kinematics of the BPMG by E. Mamajek (in preparation), using revised Hipparcos astrometry for the BPMG membership from the review by *Zuckerman and Song* (2004), was unable to replicate the *Ortega et al.* (2002a,b) and *Song et al.* (2003) kinematic age results. The expansion rate of the BPMG stars in U, V, and W velocity components vs. Galactic coordinates results in an ill-defined expansion age of 25^{+14}_{-7} m.y. More surprisingly, it was found that BPMG was not appreciably smaller ~12 m.y. ago, and fewer than 20% of BPMG members had their nearest "flyby" of the BPMG centroid during the interval 8–15 m.y. ago. Statements about the size of the group reflected dispersions measured using 68% intervals, hence they are reflecting the overall behavior of the stars, and therefore not subject to rejection of individual members because they did not produce a desired outcome (past convergence).

One is left with two interpretations: Either a small subset of BPMG stars are kinematically convergent in the past (and the current membership lists are heavily contaminated with interlopers), or the current BPMG membership lists contain mostly bona fide members, but that the group is not convergent in the past as previously thought. Either way, the kinematic ages for BPMG appear to be unreliable.

3.2.5. Flyby ages. Several studies have attempted to estimate "flyby ages" of stellar groups by calculating when in the past certain groups were most proximate, or when individual stars were closest to a group. The relation between the flyby times and group ages is ambiguous. The turbulent nature of molecular clouds may give way to random motions in star-forming complexes, and that means that tracing the bulk motions of various star-formation episodes back in time may lead to minimum separation times that are unrelated to the epoch of star formation. A few intriguing cases have appeared in the literature that deserve consideration. *Mamajek et al.* (2000) noted that the newly found η Cha and ε Cha groups were in close proximity in the past, and a Galactic orbit simulation by *Jilinski et al.* (2005) set their minimum separation at only a few parsecs 6.7 m.y. ago — similar to the isochronal ages of both groups. However, subsequent investigations have convincingly shown that the η Cha and ε Cha groups have significantly different ages based on considerations of color-magnitude positions and surface gravity indices (*Lawson et al.,* 2009), and a more recent kinematic analysis found that the groups were not much closer in the past than their current separation of ~30 pc (*Murphy et al.,* 2013). At present, there appears to be no *reliable* kinematic age for the η Cha and ε Cha groups.

Another interesting flyby age was calculated by *Ortega et al.* (2007), who demonstrated that the AB Doradus Moving Group (ABDMG) and the Pleiades were in close (~40 pc) proximity 119 ± 20 m.y. ago, commensurate with modern ages for both groups (*Stauffer et al.,* 1998; *Barrado y Navascués et al.,* 2004; *Luhman et al.,* 2005; *Barenfeld et al.,* 2013). Hence both could have formed in the same OB association (*Luhman et al.,* 2005). However, the uncertainties in the velocities of both ABDMG and the Pleiades are at the ~0.5 km s^{-1} level, so the minimum uncertainty in their three-dimensional positions ~120 m.y. ago is ~60 pc per coordinate, and hence the uncertainty in their mutual separation is ~100 pc.

From consideration of the properties of OB associations as a whole (see upcoming discussion on ages of stellar associations), it is unclear whether one could ever reliably determine a kinematic age to better than ~20 m.y. accuracy for two subgroups (in this case ABDMG and Pleiades) that are assumed to have formed in the same OB association over scales of ~10^2 pc. One would need to demonstrate that the groups formed in a very small volume, and one would need astrometry far superior than that delivered by Hipparcos to do so. Subsequent work by *Barenfeld et al.* (2013) has shown that a nonnegligible fraction of ABDMG "stream" members are chemically heterogenous, and hence could have formed from multiple birth sites unrelated to the AB Dor "nucleus." We conclude that the co-location of ABDMG and Pleiades ~120 m.y. ago is intriguing, but it is unclear whether the actual flyby age is sufficiently solid to reliably test other age techniques.

3.2.6. Runaway ages. Runaway stars are those ejected from binaries after supernovae explosions or from the decay of higher multiple systems. In some cases their origins may be traceable to a particular star-forming region or cluster, the traceback time giving their minimum age. The classic examples are AE Aur, μ Col, and ι Ori, two individual O/B stars and an O/B binary that *Hoogerwerf et al.* (2001) suggested were ejected from the ONC 2.5 m.y. ago, hence star formation was ongoing at least as long ago as this. However, ι Ori appears to be near the center of, and co-moving with, its own, dense, ~4–5-m.y.-old cluster NGC 1980 (*Alves and Bouy,* 2012; *Pillitteri et al.,* 2013), perhaps casting doubt on the original location of the runaway stars and calling into question any conclusions regarding the ONC. Without a population of runaway stars securely identified with a particular birth location it will be difficult to bring traceback ages to bear on the question of mean ages or age spreads. Lower-mass runaways may be more plentiful, but will be harder to find and generally have smaller peculiar velocities (*Poveda et al.,* 2005; *O'Dell et al.,* 2005). Precise distances and proper motions from Gaia may open up this avenue of research and will clearly assist in the vital task of locating the origin of runaway stars.

3.2.7. Inherent uncertainties in ages of association members. There are inherent astrophysical uncertainties in the ages of members of stellar associations that can arise, particularly in the context of kinematics. The assumption of single mean ages for members of large stellar associations is likely not a good one for groups larger in scale than typical embedded clusters and molecular cloud cores (*Evans et al.,* 2009). Significant velocity substructure is detected within giant molecular clouds (*Larson,* 1981), and significant age differences are seen among subgroups of OB associations (*Briceño et al.,* 2007). The combination of molecular cloud properties and observed properties of young stellar objects conspire to produce a characteristic timescale for the duration of star formation τ_{SF} over a region of length scale ℓ

$$\tau_{SF} \sim \ell_{pc}^{1/2} \quad \text{(m.y.)} \tag{1}$$

The relation comes from consideration of the observational data for star formation over scales of 0.1–10^3 pc (*Elmegreen and Efremov,* 1996) and the characteristic timescales for molecular clouds over similar scales (*Larson,* 1981). Hence the modeling of "bursts" of star formation, and the implicit or explicit assumption of coevality of a stellar group, should take into account empirical limits on the duration of star formation in a molecular cloud complex over a certain length scale. When adopting a *mean* age $\bar{\tau}$ for a member of an extended stellar association, one should naively predict a lower limit on the age precision for the star if the association's star-forming region was of size ℓ. The fractional age

precision for an individual group member when adopting a mean group age can be estimated as

$$\varepsilon = \frac{\delta\tau}{\bar{\tau}} \sim \frac{\tau_{SF}}{\tau} \sim \tau_{m.y.}^{-1} \ell_{pc}^{1/2} \tag{2}$$

For example, the nearest OB subgroup Lower Centaurus-Crux (LCC) has a mean age of ~17 m.y. and covers $\ell \sim 50$ pc in size (*Mamajek et al.,* 2002; *Pecaut et al.,* 2012), so we would naively predict a limit to the age precision of $\varepsilon \sim 50\%$ when adopting a mean group age for an individual member. Indeed, after taking into account the scatter in isochronal ages due to the effects of observational errors, the age spread in LCC has been inferred to be on the order of ~10 m.y. (*Mamajek et al.,* 2002; *Pecaut et al.,* 2012). It follows that adopting mean ages for members of entire associations that were larger than tens of parsec in size becomes untenable (age errors $\varepsilon \sim 100\%$) as one predicts $\delta\tau \sim \bar{\tau}$ — unless one can convincingly demonstrate that the group was kinematically confined to a small region in the past. This is difficult to do due to substantial uncertainties in present-day velocities, but may become possible by combining more precise Gaia astrometry with precise radial velocities.

Below is a summary of kinematic methods:

Pro: Methods are independent of stellar astrophysics.

Pro: Gaia should provide precise astrometry for many faint members of young groups, and that will enable the determination of more statistically sound kinematic ages, and for more distant groups.

Con: Kinematic ages derived using proper motions alone have been shown to be unreliable. Accurate radial velocities and parallaxes are required.

Con: Recent traceback analyses appear to suffer from some degree of subjectivity in the inclusion or exclusion of individual group members (especially for the TWA and BPMG groups).

Con: Some traceback ages have not held up when improved astrometric data comes available.

Con: It is unclear whether any reliable, repeatable, kinematic mean age for a *group* has ever been determined.

Con: Unless it can be shown that a kinematic group traces back to a very small volume, there is no reason to suppose that these stars are coeval.

4. MODEL-DEPENDENT METHODS

4.1. Placing Pre-Main-Sequence Stars in Hertzsprung-Russell Diagrams

This section discusses the methodology for comparing PMS evolutionary models with observations of young stars, including the effects of extinction/reddening, photometric variability, and ongoing accretion, which make the task more challenging than comparable exercises for open and globular clusters.

4.1.1. Models and classifications of models. The theory of PMS evolution requires an appreciation of the physics that governs the radial gradients of density, pressure, temperature, and mass within stellar interiors, as PMS stars globally contract over time. Our understanding of radiative transfer and consideration of the relevant energy sources (gravitational and light-element nuclear burning) and opacity sources (atomic and molecular gas and possibly dust as well for the coolest stars) further leads us to predictions of observables.

Models such as those by *D'Antona and Mazzitelli* (1997), *Baraffe et al.* (1998), *Siess et al.* (2000), *Yi et al.* (2003), *Demarque et al.* (2004), *Dotter et al.* (2008), and *Tognelli et al.* (2011) provide the radii, luminosities, and effective temperatures of stars of given mass at a given time. These models may differ in their inputs regarding the equation of state, opacities, convection physics, outer boundary condition of the stellar interior, and treatment of atmospheres. Additional physics such as fiducial initial conditions, accretion outbursts, ongoing accretion, rotation, and magnetic fields are also involved (e.g., *Palla and Stahler,* 1999; *Baraffe et al.,* 2002, 2009, 2012; *Tout et al.,* 1999; *Hartmann et al.,* 1998). From any of these evolutionary models, isochrones can be produced in either the natural plane of the theory ($L/L_\odot$ or g vs. T_{eff}) or in any color-magnitude or color-color diagram used by observers. As shown by, for example, *Hillenbrand et al.* (2008), the differences between ages predicted by the various theory groups increases toward younger ages and toward lower masses; there is <0.1–0.3 dex systematic variance in predicted ages at spectral type G2, but 0.25–0.6 dex at K6 among the models cited above. For an example of isochrone differences, see Fig. 3.

4.1.2. Transformations and empirical errors. If the theory is transformed into the observational plane, it must be done carefully and systematically to ensure that the theory and the data are in the same photometric system. Alternately, the observations can be transformed into the theoretical plane, requiring similarly detailed attention. Empirical requirements are a stellar spectral type or a direct estimate of the stellar temperature using model atmospheres, and at least two bands of photometry that enable a reddening estimate via comparison to the unreddened color expected for a star of the same spectral type or temperature. While traditional spectral typing was done in the "classical MK" region from 4000 to 5000 Å, the significant extinction toward star-forming regions led to the subsequent development of spectral sequences in the red optical and then the near-infrared (IR) (summarized in *Gray and Corbally,* 2009).

In practice, locating young PMS stars in the HRD involves many challenges. One is that the effects of changing surface gravity as stars contract to the MS often are not considered with the rigor they deserve. The mass-dependent surface gravity evolution over the tens to hundreds of million-year timescales that it takes high- to low-mass stars to reach the MS affects temperatures, colors, and bolometric corrections.

An intermediate temperature scale has been advocated by *Luhman* (1999) for stars a few million years old having

spectral types later than about M4, motivated in part by a desire to match the isochrones of *Baraffe et al.* (1998), which could be brought about by assuming the stars to be warmer. Intermediate color scales in various bands have been investigated by *Lawson et al.* (2009), *Da Rio et al.* (2010a), *Scandariato et al.* (2012), and *Pecaut and Mamajek* (2013). The differences from the MS are not necessarily systematic. For example, some colors are redder toward lower surface gravity down to some spectral type, such as M3 for (V–I), and then become bluer than MS values, continuing to later types. There is also a need for consideration of intermediate bolometric corrections rather than the broad application of MS relations, which are increasingly incorrect toward later-spectral-type young PMS stars. Figure 4 shows an example of intrinsic color and bolometric correction differences on transformation of the same evolutionary tracks.

Another challenge in placing young stars in the HRD is the extinction correction. Reddening toward young populations is often differential, i.e., spatially variable across a star-forming region with some or most of the reddening effect possibly arising in the local circumstellar environment itself. Thus the extinction corrections must be performed on a star-by-star basis. Furthermore, a wavelength range must be found for determination of the reddening correction and application of the bolometric correction, which is dominated by stellar (as opposed to strongly contaminated by circumstellar) emission.

Many young stars exhibit excess flux due to either (or both) accretion luminosity, which peaks in the ultraviolet (UV) but can extend through the entire optical wavelength

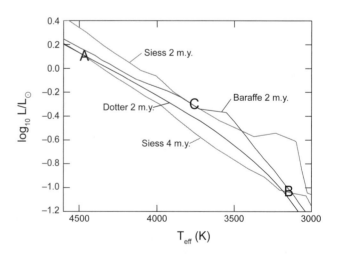

Fig. 3. An illustration of the difference in age obtained by using different interior models. Stars at points A or B would be given an age of 2 m.y. using the interior models of *Dotter et al.* (2008) and *Baraffe et al.* (1998) but 4 m.y. using the isochrones of *Siess et al.* (2000). A star at point C would be given an age of 2 m.y. using the isochrones of *Siess et al.* (2000) and *Baraffe et al.* (1998) but 1.3 m.y. on the isochrones of *Dotter et al.* (2008) (isochrone not shown).

range out to the Y band, and disk emission, which peaks in the mid- to far-IR but can extend as short as the I band. The appropriate wavelength range for de-reddening appears to be in the blue optical (B, V) for earlier-type stars, in the red optical (I, Y) for late-type stars, and possibly in the near-IR (J) for young brown dwarfs. An alternate approach is to use high-dispersion spectroscopy to determine a specific veiling value (the wavelength-dependent excess continuum flux due to accretion relative to the stellar photospheric flux) in a particular band, and use it to veiling-correct the broadband photometry before de-reddening or application of a bolometric correction.

The presence of disks and accretion adds a further complication in the form of photometric variability. This occurs at the level of a few percent for more active young spotted stars, up to the 5–30% levels typical of stars with disks that are accreting. Unless the true nature of the variability is understood, i.e., whether it is caused mostly by accretion effects or mostly by high-latitude dust obscuration events, it is not clear whether the right choice for placing stars in the HRD would utilize the bright-state vs. the faint-state magnitudes from a data stream. For disks with either large inclination or scale height, the observed photometry may be dominated by scattered light. This can render the stars fainter than they would be if seen more directly, and also leads to underestimates of the extinction.

The minimum statistical errors on just the basic parameters required from observations for HRD placement are ~50–200 K in T_{eff} and ~0.01–0.1 mag on measured photometry, along with ~10% uncertainty in mean cluster distance. An additional systematic error is the usually unknown multiplicity status of individual sources observed in seeing-limited conditions; this has an effect that depends on the binary mass ratio with maximum amplitude of 100% in luminosity overestimate for an equal mass/age system (see, e.g., the simulations at two ages in Fig. 5), modulo any error due to differential extinction.

Likely additional random errors for the youngest stars in star-forming regions where disk effects may complicate derivations include a typical ~0.05 mag in average photometric variability and ~0.3–1 mag in visual extinction determination if optical colors are used, or 2–5 mag if IR colors are used (not including a possible systematic effect due to the form of the reddening law in environments known to be exhibiting grain growth). In practice, the effective extinction errors are somewhat lower since HRDs are generally made by applying bolometric corrections to I-band or J-band photometry. For PMS stars there is also a 0.05–0.1 mag error from intrinsic color uncertainty, another 0.05–0.1 mag from the bolometric correction uncertainty, plus an additional <10% uncertainty on individual stellar distances due to cluster depth (at 150 pc, assuming the clusters are as deep as they are wide on the sky, although less for more distant star-forming regions).

The above numbers suggest ~0.2–0.3 dex in random error alone, driven by the error in the extinction correction, to which the systematic distance and binary errors would need to be added. If this is the case, summation of all error

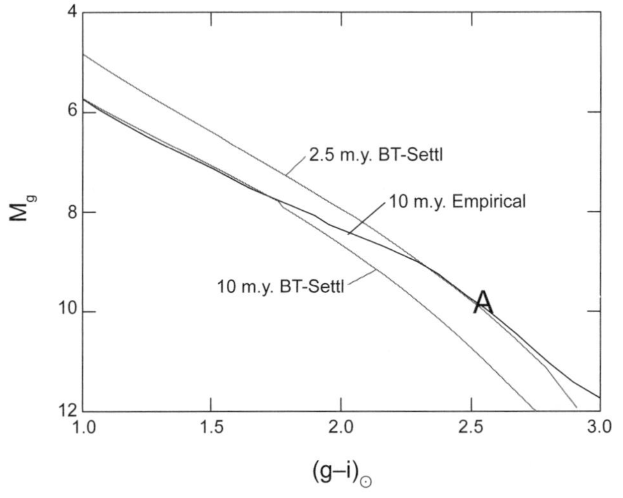

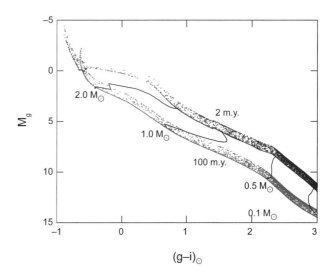

Fig. 4. An illustration of the difference in age obtained by using different color-effective temperature relationships. A star at point A would be assigned an age of 2.5 m.y. using the interior models of *Dotter et al.* (2008) and the BT-Settl model atmospheres of *Allard et al.* (2011), but 10 m.y. using the same *Dotter et al.* (2008) interiors with the semi-empirical color-effective temperature and bolometric corrections used in *Bell et al.* (2013).

Fig. 5. A simulated color-magnitude diagram for clusters that are 2 and 100 m.y. old. The masses are sampled from a Kroupa mass function (*Dabringhausen et al.,* 2008), with the prescription for binaries as in *Bell et al.* (2013). The interior models used are (from high to low mass) *Schaller et al.* (1992), *Dotter et al.* (2008), and *Baraffe et al.* (1998) and the model atmospheres from *Castelli and Kurucz* (2004) (hot stars) and *Allard et al.* (2011) (cool stars). We have chosen to join points between the models to minimize the dislocations, although they are still (just) visible.

sources is roughly consistent with, although arguably somewhat less than (*Burningham et al.,* 2005), the 0.2–0.6 dex root mean square (rms) spread in luminosity that can be measured for most young clusters, a topic we discuss in detail in section 6. From similar considerations, *Hartmann* (2001), *Reggiani et al.* (2011), and *Preibisch* (2012) each estimate <0.1–0.15 dex as the typical luminosity error, in which case the observed luminosity spreads are much greater than that attributed to random error. However, as illustrated recently by *Manara et al.* (2013), there are large trade-offs between accounting for accretion and extinction in analyzing observed spectra and colors. Thus the propagation of random errors may grossly underestimate the true luminosity uncertainties of heavily extincted accreting stars, which can be subject to large systematics. These issues possibly account for the factors of several variation among authors in the extinctions and luminosities reported for the same young stars. For populations that are more distant than the closest regions and/or for which differential extinction and the uncertainties induced by accretion/disk effects are smaller, the errors in HRD placement would be lower, <0.1–0.15 dex uncertainty in luminosity to accompany the 50–200 K uncertainty in temperature.

Simulations in, e.g., *Hillenbrand et al.* (2008), *Naylor et al.* (2009), *Preibisch* (2012), and *Bell et al.* (2013) quantify how, when the multiplicity effect dominates, stars appear systematically more luminous and hence younger in the low-mass PMS HRD. To quantify, considering most of the applicable error terms, *Preibisch* (2012) reports simulating that a true 2-m.y.-old star had inferred ages ranging over 1.2–2.2 m.y. (1σ) around a mean age of 1.7 m.y., while a true 5-m.y.-old star had inferred ages ranging over 2.7–5.5 m.y. (1σ) around a mean age of 4.1 m.y. Derived stellar ages in studies that do not account for multiplicity and known sources of error would appear young by 15–20%, but also would scatter by factors of nearly 50%. At older ages, the dominating systematic binary effect corresponds to a much larger age difference than at the younger ages, given the spacing of the isochrones. In the youngest regions where the extinction and disk effects discussed above dominate over the multiplicity effect, the luminosity errors may be more symmetric and hence the ages not necessarily biased low.

4.1.3. Star-forming region and cluster results. Figure 6 illustrates empirically how inferred stellar luminosities evolve from very young star-forming regions associated with dense gas, like the Orion Nebula Cluster, to older dispersing young associations lacking molecular gas, like Upper Sco, to open clusters like α Per and the Pleiades. As expected based on the error considerations discussed above, the luminosity spreads are larger and more symmetric at younger ages but settle down to the predicted ±0.1–0.2 dex of scatter at older ages, with the asymmetric signature of the binary effects apparent in most clusters older than ~15 m.y. Median luminosity plots like these can provide an age ordering to clusters and star-forming regions — as long as the distances are accurately known

and the samples unbiased. For example, it is possible to rank in increasing age (although not necessarily distinguish the ages of individual members, due to the apparent luminosity spreads), e.g., NGC 2024 and Mon R2, followed by the ONC, then Taurus and NGC 2264, IC 348, and σ Ori, Lupus, and Chamaeleon, TW Hydra and Upper Sco, UCL/LCC, IC 2602, and IC 2391, α Per, and finally the Pleiades cluster. Due to the general luminosity decline with age for low-mass PMS stars, this ranking seems meaningful even if the absolute ages remain highly debated, and even though the run of luminosity with effective temperature for any given cluster is not well matched by modern theoretical isochrones.

Performing the same exercise in color-magnitude diagrams, instead of the HRD, *Mayne et al.* (2007) and *Bell et al.* (2013) concluded that empirical tuning of bolometric corrections and temperature scales to particular models is necessary (see also *Stauffer et al.,* 2003) in order to derive absolute ages. Age ranking avoids this problem, and that of luminosity spreads due to multiplicity. Combining *Mayne et al.* (2007) and *Bell et al.* (2013) we find that the ONC, NGC 6611, NGC 6530, IC 5146, and NGC 2244 are indistinguishable in age, but are the youngest clusters. Similarly, there is a somewhat older group composed of σ Ori, NGC 2264, IC 348, and Cep OB3b. Beyond this age a ranking is possible for individual clusters, with λ Ori next, followed by NGC 2169, NGC 2362, NGC 7160, χ Per, and finally NGC 1960 at ~20 m.y.

In addition to testing whether empirical cluster sequences match theoretical models tied to stellar age, comparisons can

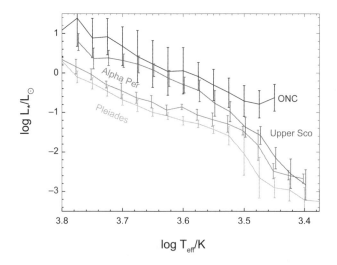

Fig. 6. Median luminosity and luminosity dispersion exhibited by four example benchmark clusters: the Orion Nebula cluster, Upper Sco, α Per, and Pleiades. Error bars indicate the empirical 1σ spread in luminosity with the red and magenta bars artificially offset along the abscissa for clarity. Stars were placed in the HRD using available spectral types and optical photometry from the literature, and the median luminosities calculated at each 0.025 dex bin in temperature.

be made of the coevality in HRDs of spatially resolved PMS binaries (e.g., *Hartigan and Kenyon,* 2003; *Kraus and Hillenbrand,* 2009) as well as of the radii inferred from double-line eclipsing PMS binaries. Such eclipsing systems along with astrometric systems also enable fundamental mass determinations for comparison with theory (e.g., *Gennaro et al.,* 2012). Given that there are still systematic discrepancies between models and fundamental masses in the PMS phase (e.g., *Hillenbrand and White,* 2004), the ages predicted by these same models likely also suffer systematic errors.

Below is a summary of isochrone placement for PMS stars:

Pro: Stellar luminosity changes rapidly and systematically during PMS contraction.

Con: Many factors influence the apparent luminosity of a PMS star, with not all of them known to us. In addition, the resulting change in luminosity with time may not be monotonic.

Con: The differences in published models are large.

Con: As a result of the above points, absolute ages from PMS isochrones are not yet credible.

Con: Placing individual stars in an HRD requires good estimates of T_{eff} and L_{bol}, both of which can be affected by extinction and non-uniform reddening, and also circumstellar material and accretion.

Pro: When the above effects are small, CMDs may be sufficient, and indeed more precise given the quantification of T_{eff} inherent in spectral typing.

Con: Unresolved binaries add uncertainty to luminosities.

Con: Estimating errors is difficult because of poorly understood factors that lead from the observations to modeled quantities.

Con: The effects of observational uncertainties increase the age errors toward the ZAMS as the evolution slows and the isochrone density increases.

4.2. Main-Sequence Evolution To and Beyond the Turnoff

Although the majority of stars in young clusters and star-forming regions are still in their PMS phase, a small number of the more massive stars will have reached the MS, and indeed the most massive will have evolved beyond it. The position of these stars in either the CMD or HRD gives us more age diagnostics that are based on very different physics from PMS evolution. Terminology is important here. While the star is still undergoing core hydrogen burning it is (by definition) evolving from the ZAMS to the terminal-age main sequence (TAMS), and an age derived from this gradual cooling and increase in luminosity is best described as a MS age, or sometimes a nuclear age. Once a star reaches the turn-off, the evolution is rapid, and so a further possible age diagnostic is the luminosity of the top of the MS — a turn-off age. Finally, the position of a star in its post-MS evolution can be yield useful ages, although the paucity of these high-mass stars can cause difficulties.

Overlaying main-sequence isochrones on the upper main sequence can yield useful constraints on the age of a cluster

(*Sung et al.,* 1999), allowing one to match both the position of the turn-off and the main-sequence evolution. However, more precise ages with statistically robust uncertainties can be obtained by using modern CMD fitting techniques (e.g., *Naylor et al.,* 2009).

A clear advantage of this part of the CMD/HRD is that the model atmospheres are well understood, lacking the complications of molecular opacities. The potential problems lie in the interior models, and involve enhanced convective overshoot (whether by rotation or otherwise) and mass loss. Stellar rotation will induce better mixing, enabling a star to burn hydrogen longer, but it also makes a star less luminous in its early MS evolution (*Ekström et al.,* 2012), and so the overall position of the isochrones changes little. As illustrated in *Pecaut et al.* (2012), the calculated age for Upper Sco changes by only 10% between rotating and nonrotating isochrones. Similar arguments apply to enhanced convective overshoot, with a similarly small effect (compare Figs. 4 and 5 of *Maeder and Mermilliod,* 1981). Mass loss also tends to move the masses along the isochrone, and again has only a small effect on the overall shape and position (see the discussion in *Naylor et al.,* 2009). However, the position of the turn-off does change significantly with rotation, a given luminosity corresponding to an age of ~30% younger at 10 m.y. for the rotating case.

In summary, main-sequence fitting seems to offer a method where the model dependence is small, but the model dependence for the turn-off is stronger. However, the movement from the ZAMS to the TAMS is subtle, and requires either well-calibrated photometry, or excellent conversion into the HRD. Furthermore, right at the top of the main-sequence, which is the most age-sensitive part, is exactly where the mass function can push us into small number statistics, which also makes the technique vulnerable to uncertain extinction and binarity.

Lyra et al. (2006) showed that for the range 10–150 m.y., MS ages and PMS ages agree. However, *Naylor et al.* (2009) showed that for clusters younger than 10 m.y. the MS ages were a factor of 2 larger than those derived from PMS isochrones. *Pecaut et al.* (2012) showed that there is excellent agreement between the isochronal ages (~11 m.y.) for Upper Sco members that were post-MS (Antares), MS B-type stars, and PMS AFG-type stars; lower-mass stars, however, trend toward younger ages. *Bell et al.* (2013) show reconciliation of the MS and PMS ages for several well-studied clusters, with a general movement toward higher ages than most previous studies. Reasonable concordance between MS, PMS, and LDB ages has been demonstrated for the oldest (22 ± 4 m.y.) cluster in the Bell et al. sample (*Jeffries et al.,* 2013).

Below is a summary of isochrone placement for MS stars:

Pro: Post-ZAMS models are in better agreement with one another and with the data, compared to PMS models.

Con: Main-sequence evolution is slow and higher-mass stars rare, leading to errors in inferred age that are driven mostly by the observational uncertainties and small number statistics.

4.3. Surface Gravity Diagnostics

The gravities of stars of the same temperature, but spanning ages from 1 to 100 m.y., can be different by 1 dex in surface gravity, depending on the temperature range considered (e.g., *Dotter et al.*, 2008). This will lead to differences in both the spectra and the colors that in principle could be used to determine the gravity and hence, in combination with isochrones, the age. However, the differences in color due to gravity for a given T_{eff} are modest for temperatures above 3000 K. Furthermore, surface gravity effects would have to be disentangled from reddening effects when interpreting observed colors.

Spectroscopic measures of surface gravity are more promising and have the key advantage of distance independence. Specifically, there is a "triangular" shape of the H-band spectra of low-gravity late-M-type stars (*Lucas et al.*, 2001; *Allers et al.*, 2007) that, although used as a method of selecting young stars, is not a viable route to precise ages. Other gravity-sensitive lines include the K-band NaI 2.206-μm line (*Takagi et al.*, 2011) and the J-band KI doublets at 1.17 and 1.25 μm (e.g., *Slesnick et al.*, 2004).

The optical Na 8190-Å doublet is now widely used for mid- and late-M-type stars. Line strength is driven by both gravity and temperature, so two-dimensional classification is required (e.g., *Slesnick et al.*, 2006; *Lawson et al.*, 2009). *Schlieder et al.* (2012) compare data with theoretical predictions and conclude that although this line, like the IR lines, clearly can be an effective diagnostic for identifying young stars, it may not be particularly useful for accurate age-dating of individual stars. However, it appears to do a good job of sorting cluster samples (*Lawson et al.*, 2009) with a resolution of a few million years, and fitting of model atmospheres by *Mentuch et al.* (2008) leads to an association age ranking, from youngest to oldest, as η Cha, TW Hya, β Pic, Tuc/Hor, and AB Dor. *Prisinzano et al.* (2012) have examined the log g sensitivity of the CaI lines around 6100 Å, which appear useful for late G-, K-, and M-type stars. Bluer surface gravity sensitive lines such as the MgI b triplet are generally not practical for faint, extinct young PMS stars.

4.4. Pulsations and Seismology

Solar-like p-mode oscillations have now been detected in hundreds of stars with the ultraprecise photometry of the Convection, Rotation and planetary Transits (CoRoT) and Kepler missions (e.g., *Chaplin and Miglio*, 2013). These include many evolved stars, for which it is now possible to distinguish stars ascending the red giant branch from those descending, even though they occupy identical regions in the HRD (*Bedding et al.*, 2011). Observations of stellar oscillations can reveal interior properties that constrain model fits much more precisely than is possible from only surface information.

The p-mode oscillations have also been detected in stars near 1 $M_\odot$, but the detections are mostly for older stars more massive than the Sun (*Chaplin and Miglio*, 2013). When detected, these oscillation frequencies provide significant physical constraints (e.g., density vs. radius) on stellar models, enabling ages to be calculated to as precise as 5–10% if spectra have been obtained to establish the star's composition.

Similar oscillations have not been detected in PMS stars, although there have been few attempts. The primary reason is the inherent high level of photometric variability in very young stars that is related to their high activity levels (*Chaplin et al.*, 2011). This leads to a noise level that inhibits detecting the oscillations even if they are present. However, some of the more massive PMS stars cross the instability strip in the HRD as they approach the MS. These intermediate-mass stars include δ Scutis and some B stars that oscillate via the κ mechanism (e.g., *Ruoppo et al.*, 2007; *Alecian et al.*, 2007). Pulsations of this kind have been seen in some stars of NGC 2264 (*Zwintz et al.*, 2013), and may be capable of providing model-dependent ages that are more precise than those from the HRD as they are distance independent and insensitive to extinction.

Pulsations are also predicted for low-mass stars and brown dwarfs via the ε mechanism (*Palla and Baraffe*, 2005), although thus far the amplitudes are not well constrained theoretically and only upper limits are available from observations (*Cody and Hillenbrand*, 2011).

Below is a summary of seismology for PMS stars:

Lower-mass stars, which form the vast majority of PMS stars, appear to be too noisy to exhibit detectable oscillations. Some higher-mass stars fall in the instability strip and so detailed observations of them can provide physical constraints that help lead to more precise ages.

4.5. Projected Stellar Radii

Young stars with convective envelopes are magnetically active and often rapidly rotating (section 5.1; see also the chapter by Bouvier et al. in this volume). These properties combine to produce light curves that are rotationally modulated by cool starspots, from which rotational periods, P_{rot}, can be estimated. The projected equatorial velocity of a star, v sin i, where i is the inclination of the spin axis to the line of sight, can be estimated using high-resolution spectroscopy to measure rotational broadening. If these quantities are multiplied together, the projected radius of a star is $(R/R_\odot)$ sin i = 0.02(P_{rot}/day)(v sin i/km s^{-1}). Although i is unknown, this formula gives the minimum radius of the star and because PMS stars become smaller as they move toward the ZAMS, this minimum radius can then be compared with isochrones of R vs. T_{eff} to give a model-dependent upper limit to the age.

This method is independent of distance and insensitive to extinction and binarity (as long as one star is much brighter), but requires time series photometry, a spectrum with sufficient resolution to resolve the rotational broadening and an estimate of T_{eff}. For field stars this can be a useful technique to confirm their youth, with a range of applicability equal to the time taken to reach the ZAMS at any particular spectral type (e.g., *Messina et al.,* 2010). In large groups of stars where spin axis orientation can be assumed random, the same technique offers a way of statistically determining the average radius or distribution of radii (see section 6). The main disadvantage of the method is that while it offers a geometric radius determination, independent of any luminosity estimate, interpreting this in terms of an age is still entirely model dependent. Furthermore, there are strong indications that the radii of young, magnetically active stars are significantly larger than predicted by current models, and this could lead to a systematic error (*López-Morales,* 2007; *Jackson et al.,* 2009).

4.6. Location and Size of the Radiative-Convective Gap

It has long been known that there is a discontinuity in the CMDs of young clusters at an age-dependent mass. This is illustrated in the 2-m.y. model of Fig. 5, where there is a gap at (g–i) = 0 with MS stars to the blue, and the PMS stars to the red (see *Walker,* 1956, for an observational version). The gap is created because there is a rapid evolution between the two sequences, as the stars move from the (largely) convective pre-MS, developing radiative cores (a progression along the Henyey tracks). Various authors have recognized this resulting gap, calling it the "H feature" (*Piskunov and Belikov,* 1996), the "pre-main-sequence transition" (*Stolte et al.,* 2004), and the "radiative-convective" (R-C) gap (*Mayne et al.,* 2007). Both *Gregory et al.* (2012) and *Mayne* (2010) discuss the gap in its wider context, but for age determination its importance stems from the fact that its position and size vary as a function of age. It becomes almost imperceptible after 13 m.y. (see the figures in *Mayne et al.,* 2007).

The position of the gap in absolute magnitude can be used as an age indicator since it produces a dip in the luminosity function. This is particularly useful in the extragalactic context where distances may be known from other information, where this method has been pioneered by *Cignoni et al.* (2010), but has also been applied to Galactic clusters (*Piskunov et al.,* 2004).

The real potential power of the R-C gap, however, is that at a given age its width measured in either color or magnitude space is independent of both the distance and the extinction. The practical problem is observationally defining the edges of the gap, as there are issues with both field star contamination and the fact that there clearly are a few stars in the gap itself (*Rochau et al.,* 2010). Given a clean sample, perhaps from Gaia, one may be able to solve this problem using two-dimensional fitting to model isochrones (see *Naylor and Jeffries,* 2006), but whether the models can correctly follow the rapid changes in stellar structure through this phase remains to be tested. An alternative may be to construct empirical isochrones.

5. EMPIRICAL METHODS

All the empirical methods for determining stellar ages are inherently circular in their reasoning, especially for PMS stars. Indeed, the general trends and variations in properties are often the very focus of investigations.

5.1. Rotation and Activity in Pre-Main-Sequence Stars

Using the observed rotation rates of PMS stars to estimate their ages conflicts with one of our primary goals, which is to establish ages independently of the phenomena being studied. The origin and evolution of angular momentum — and its surface observable (the rotation period or v sin i) — is one of the major areas of study in the field of star formation and early stellar evolution (see the chapter by Bouvier et al. in this volume).

Nevertheless, we ask here if our knowledge of rotation and how it changes is sufficient to invert the problem to estimate an age. And if the process will not work for a single star, is there some sample size for which a median or average rotation is a clear function of age?

The essence of the situation is illustrated in Fig. 7, which is taken from *Gallet and Bouvier* (2013). The different lines show various models that are not of interest here. Instead, note the range of observed rotation periods for solar-mass stars in a number of young clusters. First, in any one cluster the range is 1.5–2.5 dex. Second, this spread is not "noise" to which Gaussian statistics apply because the values are not concentrated toward an average value but instead are spread fairly uniformly in log P_{rot}. Third, the data for any one cluster encompasses a range of masses, but for these young stars P_{rot} varies little with mass (or color), so that does not matter. Fourth, and most important here, for the first ~100 m.y. or so there is no clear trend at all in mean or median rotation. Even taking 10 or 20 stars to average would not lead to a useful result. In terms of rotation, a group of stars at 10 m.y. looks very much like a similar group at 100 m.y.

This lack of distinct change in the distributions of P_{rot} values is in part due to the systematic decrease in stellar radii over the same age range as these stars approach the ZAMS: Angular momentum loss is balanced by the star's shrinkage to keep P_{rot} roughly constant. At lower masses it seems that angular momentum loss may not be so efficient. *Henderson and Stassun* (2012) point out that between 1 m.y. and 10 m.y. the fraction of slowly rotating M stars in young clusters rapidly diminishes. This empirical finding will not yield reliable ages for individual M dwarfs, because a slowly rotating object could be either very young or quite old. In groups of (presumed coeval) M dwarfs, unless distances are unavailable, PMS isochrones in the HRD/CMD are likely to lead to more precise ages.

"Activity" refers to various nonthermal emissions that arise in the upper atmospheres of stars. The easiest to

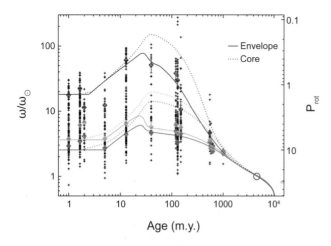

Fig. 7. Observed rotation rates of stars near 1 $M_\odot$ in young clusters (*Gallet and Bouvier,* 2013) (see also the chapter by Bouvier et al. in this volume). The different lines show several models that are not of relevance here. The plus symbols show observed rotation periods. The top, middle, and lower diamonds represent the 90th percentile, the 25th percentile, and the median for each cluster's distribution of P_{rot}. Note the very large spread (1.5 to 2.5 dex at any one age) and the lack of a clear trend for at least the first ~100 m.y.

measure are the emission reversals in the cores of the Ca II H and K lines, but in T Tauri stars in particular, Hα emission is prominent and is a defining characteristic of the class. Additional features are seen in UV wavelengths, and coronal emission in X-rays is often used as a means of identifying very young stars in a field because the X-rays are able to penetrate the surrounding material better than other wavelengths.

Activity in PMS stars is highly variable (as it is even for our Sun), at a level that masks any underlying age trend. In addition, the activity seen in young stars tends to saturate (e.g., *Berger et al.,* 2010), so there is little variation in activity with age up to ~200 m.y. or so, even in the mean for groups of stars. There is even some evidence for "supersaturation" in which extremely active stars exhibit less emission in activity signatures such as X-rays compared with somewhat less active stars (*Berger et al.,* 2010).

Using activity to estimate age even for MS stars can be problematic. There is a well-defined relation in the mean (e.g., *Soderblom et al.,* 1991) but substantial scatter within coeval groups and disagreement between members of binaries (*Mamajek and Hillenbrand,* 2008).

Below is a summary of rotation and activity for PMS stars:

The inherent dynamic range of rotation and activity at any one age for PMS stars equals or exceeds any broad trend in average rotation rates and activity levels with time. In addition, activity saturates at the high rotation rates seen in PMS stars and the early rotation history appears to be tied to the presence of disks. Both rotation and activity may be useful age indicators for older PMS and MS stars, depending on the mass [e.g., ~0.5 gigayears (G.y.)] and older for solar-mass stars). Ages from seismology provided by Kepler are helping to calibrate these relations.

5.2. Accretion and Circumstellar Disks

At the earliest visible ages, most stars have near- to mid-IR excesses diagnostic of warm dust at 0.1–10 AU, which is thought to arise from a primordial circumstellar disk, and show signs of gas accretion. Observations in young clusters and associations then show how spectral energy distributions (SEDs) evolve and the fraction of stars exhibiting these signatures declines. Various age-dependent relationships have been proposed. The fraction of stars with excesses at near-IR wavelengths halves in about 3 m.y. and becomes close to zero after 10 m.y. (*Haisch et al.,* 2001; *Hillenbrand,* 2005; *Dahm and Hillenbrand,* 2007; *Hernández et al.,* 2008; *Mamajek et al.,* 2009). The median SED of class II PMS stars appears to show a monotonic progression with age (e.g., *Alves and Bouy,* 2012). Likewise, the fraction of PMS stars showing accretion-related Hα emission declines from around 60% at the youngest ages, with an exponential timescale of ~2–3 m.y., and becomes very small at ≥10 m.y. (*Jayawardhana et al.,* 2006; *Jeffries et al.,* 2007; *Mamajek et al.,* 2009; *Fedele et al.,* 2010). In principle, these relationships, calibrated with fiducial clusters, could be inverted to obtain rough age estimates for stellar groups, or to rank the ages of populations with measured disk diagnostics.

In practice, there are complications that must be considered. First, the diagnostics used must be defined and measured in the same way in the calibrating clusters and using stars of similar mass. There is evidence that the fraction of stars with IR excess is both wavelength dependent and mass dependent (longer wavelengths probe dust at a larger radii and provide greater contrast with cool photospheres; lower-mass stars have lower mass disks) (*Carpenter et al.,* 2006; *Hernández et al.,* 2010). There are also differing methods of diagnosing active accretion that could alter the accretion disk frequency (*Barrado y Navascués and Martín,* 2003; *White and Basri,* 2003). It also appears that disk dispersal timescales increase with decreasing stellar mass (*Kennedy and Kenyon,* 2009; *Mamajek et al.,* 2009; *Luhman and Mamajek,* 2012). Second, an individual star may have a disk lifetime from <1 m.y. to ~15 m.y. It is also possible that different clusters at the same age have different disk frequencies (*Mamajek et al.,* 2009). The mechanisms by which disks disperse and accretion decays are poorly understood and must involve parameters other than age, such as initial disk mass, exposure to external UV radiation, planetesimal formation, etc. (*Adams et al.,* 2004; *Williams and Cieza,* 2011). Until their relative importance is established, age estimates for groups of stars, and especially age estimates for individual stars, based on disk properties are at risk of significant error. For example, a group of stars emerging from a molecular cloud that all possess near-IR excesses may be very young, or perhaps have been shielded from the disruptive influence of external UV radiation. Third, some diversity of disk/accretion diagnostics may be due to disk geometry or accretion variability. This should not be an issue for large statistical samples but is of serious concern for individual stars. Finally, disk frequency can only be considered a secondary age indicator at best. It requires calibration using clusters with ages determined by other methods and shares any inaccuracy with those methods. As we have seen in previous sections, no absolute age scale has been established on the timescales relevant for disk dispersal.

Mamajek et al. (2009) quantified the time evolution of the protoplanetary disk fraction in a group using the simple relation $f = \exp(-\tau/\tau_{disk})$, where $f = n/N$ = number of stars with accretion disks n divided by total number of stars N. With the data then available, τ_{disk} was estimated to be ~2.5 m.y., but this depends on the ages adopted for the calibrating clusters, and recent work suggests it could be a factor of 2 larger (e.g., *Pecaut et al.*, 2012; *Bell et al.*, 2013). There also appears to be real cosmic scatter in disk fractions at given age, and as a function of stellar mass and multiplicity. The approximate *disk fraction age* of a group is thus

$$\tau = -\tau_{disk} \ln f = \tau_{disk} (\ln N - \ln n) \qquad (3)$$

If one can quantify the intrinsic scatter σ_τ in the exponential decay constant τ_{disk} (i.e., "cosmic" scatter in the disk fraction due to the many factors that affect disk depletion), then the uncertainty in the age of a sample determined using disk fraction would be $\sigma \simeq \sqrt{(\ln f)^2 \, \sigma_\tau^2 + \tau^2/n}$ assuming Poisson noise from the counting of disked stars. As an example, consider the η Cha cluster, where *Sicilia-Aguilar et al.* (2009) find that $n = 4$ of $N = 18$ members are accretors ($f = 22\%$). Adopting a decay timescale $\tau = 2.5$ m.y., intrinsic scatter $\sigma_\tau = 1$ m.y. (to be measured), one would estimate an age for the η Cha cluster of 3.8 ± 2.0 m.y.; i.e., the disk fraction gives an age estimate with a precision of ~50%, but this precision becomes poorer when f or n are small. This age is consistent with published estimates (*Fang et al.*, 2013a).

Below is a summary of circumstellar disks as an age indicator:

There is a clear trend in protoplanetary disk fraction vs. time that could be used as a rough age indicator, but better precision is likely to be available from other methods (e.g., PMS isochrones). Its utility may be confined to instances where representative censuses of accretors and class III objects complete to some mass can be made (e.g., joint IR/X-ray surveys) and distances are uncertain. More work is needed to understand and quantify the factors responsible for the cosmic scatter in disk fraction as a function of age (and mass, multiplicity, environment, etc.).

5.3. Lithium Abundances

Lithium depletion in stars with $M > 0.4$ $M_\odot$ is more complex than in lower-mass stars (see section 3.1). As the core temperature rises during PMS contraction, the opacity falls and a radiative core develops, pushing outward to include an increasing mass fraction. During a short temporal window, stars ≤ 1 $M_\odot$ can deplete Li in their cores and convectively mix Li-depleted material to the surface before the convection zone base temperature falls below the Li-burning threshold and photospheric Li depletion halts. Lithium depletion commences after a few million years in solar-type stars and photospheric depletion should cease after ~15 m.y., before the ZAMS, with a large fraction of the initial Li remaining. In contrast, the convection zone base temperature does not decrease quickly enough to prevent total depletion for stars of $0.4 < M/M_\odot < 0.6$ within 30 m.y. (e.g., *Baraffe et al.*, 1998; *Siess et al.*, 2000). At masses >1 $M_\odot$, the radiative core develops before Li depletion starts and little PMS Li depletion can occur. These processes lead to predicted isochrones of photospheric Li depletion that are smoothly varying functions of mass or T_{eff} that could be used to estimate stellar ages.

Ages based on Li depletion among stars with $0.4 < M/M_\odot < 1.2$ (or equivalently late-F to early-M stars) are not "semifundamental" in the same way as the LDB technique. Different evolutionary models make vastly different predictions; Li depletion is exquisitely sensitive to uncertain conditions at the base of the shrinking convection zone and hence highly dependent on assumed interior and atmospheric opacities, the metallicity and He abundance, and most importantly, the convective efficiency (*Pinsonneault*, 1997; *Piau and Turck-Chièze*, 2002). It is also clear that "standard" convective mixing during the PMS is not the only process responsible for Li depletion; the decrease in Li abundances observed between ZAMS stars in the Pleiades and similar stars in older clusters and the Sun is probably due to rotationally induced mixing processes or gravity waves (*Chaboyer et al.*, 1995; *Charbonnel and Talon*, 2005).

Until stellar interior models are greatly improved, Li depletion in low-mass stars cannot provide accurate absolute ages. Indeed, Li depletion has chiefly been used to investigate the uncertain physics above using clusters of "known" age. However, imperfect evolution models do not preclude using comparative Li depletion as an empirical age estimator or as a means of ranking stars in age order. Clusters of stars with ages determined by other techniques (LDB, MSTO) can calibrate empirical Li isochrones. There is a clear progression of Li depletion with age between the youngest objects in star-forming regions, through young clusters like IC 2391/2602, α Per, and the Pleiades (Fig. 8) to older clusters like the Hyades and Praesepe [sources of such data can be found in *Sestito and Randich* (2005)]. There are (at least) three physical sources of uncertainty in using these isochrones: (1) The initial Li abundance of a star is not known, but assumed close to meteoritic. The possible scatter may be estimated from Galactic chemical evolution models or from early G stars in several young clusters at a similar age (e.g., *Bubar et al.*, 2011). It is unlikely to have a dispersion of more than around 0.1 dex among young population I stars in the solar neighborhood (*Lambert and Reddy*, 2004). (2) There is a scatter in Li abundance at a given T_{eff} in most young clusters. This is modest among G stars but grows to 1 dex or more in K stars (e.g., *Soderblom et al.*, 1993; *Randich et al.*, 2001). The scatter may be due to mass loss, rotation-induced mixing, inhibition of convection by magnetic fields, or some other phenomenon (e.g., *Ventura et al.*, 1998; *Sackmann and Boothroyd*, 2003), but for determining the age of a particular star it is a major nuisance. (3) Lithium depletion should be sensitive to metallicity and He abundance, particularly at lower masses. The range and availability of precise metallicities among available calibrating clusters is too limited to quantify this.

The first effect listed above is only relevant for G stars. The cosmic dispersion in initial Li may be significant compared to levels of Li depletion at ≤100 m.y. The second effect will be more important in K and M stars where Li is depleted by 1–2 dex in 100 m.y. An individual star can be assigned an age but with an associated uncertainty that depends on the dispersion of Li abundance [or EW(Li), see

below] in the coeval calibrating datasets. The third effect means that at present, ages can only be properly estimated for stars with close-to-solar metallicity.

Efforts to estimate stellar ages using Li depletion have focused on its use as a relatively crude, but distance-independent, indicator of youth or as a means of supporting age determinations for isolated field stars or kinematic "moving groups," where other age indicators are unavailable and HRDs cannot easily be constructed (*Jeffries*, 1995; *Barrado y Navascués*, 2006; *Mentuch et al.*, 2008; *Brandt et al.*, 2013). The measurement of Li abundance in FGK stars requires high-resolution (R ≥ 10,000) spectroscopy of the Li I 6708 Å feature and the ability to estimate T_{eff} and log g from spectra or photometry. The abundances, derived via model atmospheres, are sensitive to T_{eff} uncertainties and also to uncertain non-long-term-evolution (NLTE) effects (e.g., *Carlsson et al.*, 1994). If comparison with theoretical models is not required, then it makes little sense to compare data and empirical isochrones in the Li abundance, T_{eff} plane, where the coordinates have correlated uncertainties. A better comparison is made in the EW(Li) color plane (appropriately corrected for reddening), although there remain issues about the comparability of EW(Li) measured using different methods of integration, continuum estimation, and dealing with rotational broadening and blending when the Li feature is weak or there is accretion veiling in very young stars.

Figure 8 illustrates how the effectiveness of Li depletion as an age indicator is dependent on spectral type. At very young ages (<10 m.y.) little or no Li depletion is expected and the observations are consistent with that. Between 10 and 50 m.y., Li depletion is rapid in late K and M dwarfs but barely gets started in hotter stars. Age values (as opposed to limits) in this range can only be estimated from EW(Li) for stars with these spectral types or for groups including such stars. M dwarfs with Li (and below the LDB) must be younger than 50 m.y. Conversely, M dwarfs without discernible Li are older than 10 m.y. Individual M dwarfs cannot be age dated much more precisely than about ±20 m.y. due to the wide dispersion of EW(Li) among the M dwarfs of clusters at 40–50 m.y. (*Randich et al.*, 2001) and the lack of data for calibrating clusters at 10–40 m.y. At older ages the focus moves to K dwarfs with longer Li depletion timescales; these can be used to estimate ages in the range 50 m.y. (where some undepleted K dwarfs are still found) to ~500 m.y. (where all K dwarfs have completely depleted their Li). Within this range, the likely age uncertainties for one star are ~0.5 dex, due to the spread in Li depletion observed in calibrating clusters, but there may also be a tail of extreme outliers and close binarity can also confuse the issue. The uniform analysis of clusters in *Sestito and Randich* (2005) shows that the Li abundance drops by only ~0.2–0.3 dex over the first 200 m.y. of the life of an early G star. With an observed dispersion 0.1–0.2 dex, and similar uncertainties associated with metallicity and initial Li, Li depletion cannot be used

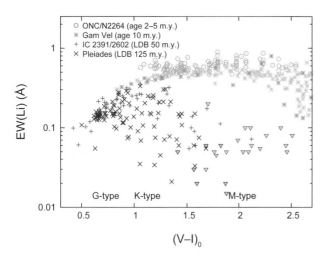

Fig. 8. The EW of the Li I 6708 Å feature vs. intrinsic color for groups of stars of fiducial ages. This demonstrates the empirical progression of Li depletion with age. There are several notable features: (1) The very young stars (2–10 m.y.) have EW(Li) consistent with zero depletion. (2) Rapid depletion commences in M dwarfs between 10 and 50 m.y. (3) Depletion occurs on longer timescales for K dwarfs, but a large scatter develops. (4) Little depletion occurs in the first 100 m.y. in G dwarfs. Data are taken from *Soderblom et al.* (1993), *Jones et al.* (1996), *Randich et al.* (1997, 2001), *Sergison et al.* (2013), and the Gaia-ESO survey (*Gilmore et al.*, 2012).

to estimate reliable ages for young G stars.

Below is a summary for Li depletion:

Pro: The primary Li feature at 6708 Å is easy to detect and measure in PMS stars, having an EW of 100 mÅ or more and lying near the peak of CCD sensitivity.

Pro: The broad trend of declining Li abundance with age is strong for PMS stars, particularly late K and M dwarfs.

Pro: PMS stars at <10 m.y. all appear to be Li-rich. There may be exceptions (e.g., *White and Hillenbrand*, 2005), but only in small numbers (see section 6).

Con: Detecting Li in a star requires spectra of good resolution and signal-to-noise. Only one feature at 6708 Å is generally visible.

Con: Lithium depletion is not understood physically; additional theoretical ingredients are likely to be required. Models can reproduce the solar abundance (e.g., *Charbonnel and Talon*, 2005) but there are no other old stars with well-determined Li abundances and fundamental parameters to constrain the theory.

Con: Converting an observed Li EW to an abundance is very temperature sensitive, can require a NLTE correction, and may be influenced by surface inhomogeneities such as starspots (*Soderblom et al.*, 1993). This and the item above prevent any absolute age determination from Li measurements.

Con: Substantial and poorly understood scatter exists in Li for stars of the same age, particularly below 1 $M_\odot$, which diminishes its effectiveness as an empirical age indicator.

Con: The rate of Li depletion is too low to be useful at ages ≤100 m.y. and M ≥ 1 $M_\odot$.

6. AGE SPREADS AMONG PRE-MAIN-SEQUENCE GROUPS

There is considerable debate about how long it takes to form a cluster of stars. Some argue that the dissipation of supersonic turbulence in molecular clouds leads to rapid collapse and star formation on a freefall time [<1 m.y. for a cluster like the ONC (*Ballesteros-Paredes and Hartmann*, 2007; *Elmegreen*, 2007)]. Others take the view that turbulence is regenerated or the collapse moderated by magnetic fields and that star formation occurs over 5–10 free-fall times (*Tan et al.*, 2006; *Mouschovias et al.*, 2006; *Krumholz and Tan*, 2007).

At some level and over some spatial scale, an age spread must exist. If we look at a region containing many star-forming "events" taking place over a large volume, then we expect some dispersion in their star-forming histories (see section 3.2). How can we measure the age spreads, and what are the appropriate sizes to consider?

The coexistence of class I, II, and III PMS stars (i.e., in progressively advanced evolutionary stages) in the same cluster/cloud/region is sometimes taken as evidence for an age spread, under the assumption that disk evolution is monotonic. Others attribute this to variation in disk dissipation timescales. An order-of-magnitude spread in luminosity

at a given temperature is almost ubiquitous in the HRDs of young clusters and star-forming regions with mean age <10 m.y. (an example for the Orion Nebula Cluster is shown in Fig. 9). The interpretation of this luminosity dispersion has important consequences for age estimation. If it is taken as evidence for an age spread, then that spread typically has FWHM ≃ 1 dex in log τ, would suggest a star-forming epoch lasting ≥10 m.y. (*Palla and Stahler*, 2000; *Huff and Stahler*, 2006), and would call into question the use of very young clusters as fiducial "points" to investigate early evolutionary processes. Others argue that luminosity estimates could be inaccurate due to observational uncertainties or

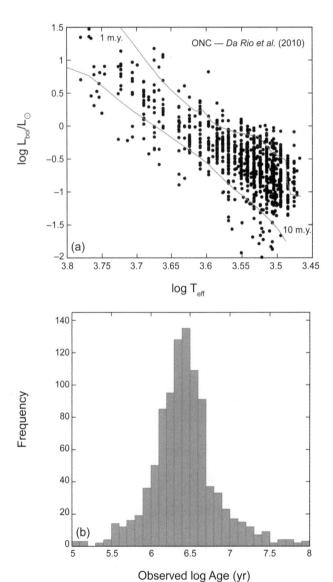

Fig. 9. **(a)** The HRD of the Orion Nebula cluster from the catalog compiled by *Da Rio et al.* (2010a). The loci are isochrones at 1 and 10 m.y. from the models of *Siess et al.* (2000). **(b)** The inferred distribution of log age (in years) for these stars.

astrophysical nuisance parameters such as binarity, accretion, and variability (see section 8.1) or even that the relationship between HRD position and age is scrambled at early ages by variable accretion history or magnetic activity. If so, then an age determined from the HRD for an individual star might be in error by factors of a few, and have an uncertain systematic bias.

6.1. Observed Age Spreads and Observational Uncertainties

The luminosity of a low-mass PMS star declines as $\tau^{-2/3}$ at almost constant T_{eff}. The typical Gaussian dispersion in log L (e.g., in Fig. 9) is $\simeq$0.25–0.3 dex, with a consequent pseudo-Gaussian dispersion of inferred log τ of $\simeq$0.4 dex. It has been realized for some time (see discussion in section 8.1) (e.g., *Hartmann*, 2001; *Hillenbrand*, 2009) that a mixture of uncertainties, both observational and astrophysical, might contribute significantly to this dispersion. Much recent work has focused on attempting to assess and simulate these uncertainties in an effort to establish what fraction of the luminosity dispersion is attributable to a genuine age spread, in some cases concluding that this fraction is large (*Hillenbrand et al.*, 2008; *Slesnick et al.*, 2008; *Preibisch*, 2012), in other cases quite small (*Da Rio et al.*, 2010b; *Reggiani et al.*, 2011).

The conclusions from the studies above are critically dependent on an accurate assessment of the observational uncertainties and their (probably non-Gaussian) distribution, and also on establishing the bona fide cluster membership of stars contributing to the age distribution. There is often a small (few percent) tail of very low luminosity objects (see Fig. 9) with implied ages a lot older than 10 m.y. It is possible that some of these objects are not genuine cluster members, but unassociated field stars (a classic, lingering case of this are the sources HBC 353–357 associated for many decades with Taurus, but more likely members of the background Perseus complex, raising their luminosity). Even if they share mean cluster kinematics, it is possible they were captured during cluster formation (*Pflamm-Altenburg and Kroupa*, 2007). Another possibility that seems borne out for some fraction of these "older" objects is that they are young objects with circumstellar disks that are viewed primarily through scattered light (*Slesnick et al.*, 2004; *Guarcello et al.*, 2010). Finally, it should be remembered that what is viewed on the sky could be a projection of two or more separate star-forming regions with different ages, and this would be misinterpreted as requiring an extended cluster formation timescale (e.g., *Alves and Bouy*, 2012). The young Galactic super star cluster NGC 3603 is an excellent example of the confusion that can arise. *Beccari et al.* (2010) show CMDs from the cluster and surrounding area, implying age spreads of more than 20 m.y. However, from a much smaller "core" region where extinction is uniform and disks may have photoevaporated, and with a proper-motion selected sample, *Kudryavtseva et al.* (2012) claim that star formation is "instantaneous." This cautions

that one must consider the size of the region being observed, ensure that membership of the cluster is secure, and carefully assess the observational uncertainties. There is some contrary evidence from multiwavelength studies that some star-forming regions show age spreads of ~2 m.y. (*Wang et al.*, 2011; *Bik et al.*, 2012).

6.2. Luminosity Spreads Without Age Spreads?

When does age start and what does $\tau = 0$ mean? The interpretation of a luminosity spread as due to an age spread implicitly relies on the veracity of PMS isochrones in the HRD or at least on the assumption that there is a one-to-one mapping between luminosity and age. In fact, the age of a PMS star deduced from the HRD will depend on the definition of $\tau = 0$ and the initial conditions will remain important to any age estimate until they are erased after a Kelvin-Helmholtz (K-H) timescale. *Stahler* (1988) has advocated the use of a "birthline," where PMS deuterium (D) burning defines a core mass-radius relationship. However, it has long been appreciated that early accretion can change this relationship, providing a spread in birthlines that is then propagated for the first few million years of the PMS lifetime (*Mercer-Smith et al.*, 1984; *Hartmann et al.*, 1997; *Tout et al.*, 1999; *Baraffe et al.*, 2002).

New observations and theoretical calculations have suggested that accretion onto embedded protostars can be at very high rates ($\dot{M} \geq 10^{-5}$ $M_\odot$ yr^{-1}, but interspersed with longer periods of much lower accretion rates (e.g., *Vorobyov and Basu*, 2006, 2010; *Enoch et al.*, 2009; *Evans et al.*, 2009). Models of early evolution incorporating high levels of accretion during the class I PMS phase have been presented by *Baraffe et al.* (2009), *Hosokawa et al.* (2011), and *Baraffe et al.* (2012). *Baraffe et al.* (2009) presented models of "cold accretion," where a negligible fraction, $\alpha \sim 0$, of the accretion kinetic energy is absorbed by the star. In such circumstances, and where the accretion timescale is much shorter than the K-H timescale, the response of the star is to contract and appear in the HRD, mimicking a much older (and smaller) star, and where it will remain for a K-H timescale of ~10 m.y.

On the other hand, if a significant fraction ($\alpha \geq 0.2$) of the accretion energy is absorbed, the star swells and becomes more luminous but, because its K-H timescale becomes much smaller, it quickly returns to follow the appropriate non-accreting isochrone. Variations in α and the accretion rate (or more precisely the amount of material accreted on a short timescale) would scatter a 1-m.y. coeval population in the HRD between 1 and 10 m.y. non-accreting isochrones. *Hosokawa et al.* (2011) present similar models, concluding that the effects discussed by *Baraffe et al.* (2009) apply only to PMS stars seen with $T_{eff} > 3500$ K once accretion has ceased; since luminosity spreads are observed at lower temperatures as well, then it is possible that inferred age spreads are genuine. *Baraffe et al.* (2012) reconciled these views to some extent, finding that for stars with a small final mass (and T_{eff}), initial seed protostellar mass was an important parameter, and at smaller values than assumed by *Hosokawa et al.* (2011), cold accretion did lead to significant post-accretion luminosity spreads in stars with differing accretion histories. Both ages and masses from the HRD would be overestimated using non-accreting isochrones. Theoretical arguments are hampered by the lack of very detailed three-dimensional numerical simulations of the formation of the protostellar core and the accretion process.

Another possible contributor to the dispersion in the HRD is that magnetic fields or significant spot coverage will inhibit the flow of energy out of the star and may change the luminosity and radius significantly and hence also T_{eff} (*Spruit and Weiss*, 1986; *Chabrier et al.*, 2007; *Feiden and Chaboyer*, 2012). Such anomalies have been measured in close, magnetically active, eclipsing binary systems, where T_{eff} may be decreased by 10% or more. It is possible that the effects may be more significant in fully convective PMS stars (*Jackson et al.*, 2009; *MacDonald and Mullan*, 2013), but no definitive investigation of magnetic activity or spottiness vs. HRD position has been performed.

6.3. Testing Age Spreads Independently of the Hertzsprung-Russell Diagrams

The reality of significant (~10 m.y.) age spreads can be tested independently of the HRD using other age indicators. Younger stars ought to have larger radii, lower gravities, lower core temperatures, and, for a given angular momentum, slower rotation rates than older stars of similar T_{eff} in the same cluster. *Jeffries* (2007) modeled the distribution of projected stellar radii (see section 4.5) in the ONC, demonstrating that a single age failed to match the broad spread of R sin i and that a more consistent match was found by assuming the radii were those given by HRD position. This implies a genuine factor of ~3 spread in radii at a given T_{eff} and that contributions to the luminosity dispersion from observational uncertainties and variability are small. While this does support the notion of a large luminosity dispersion, it does not necessarily indicate a wide age spread.

Evolutionary models featuring efficient convection or large mixing length begin to exhibit surface Li depletion at $\simeq$10 m.y. at M ~ 0.5 M$_\odot$ (see section 5.3). A number of authors find stars at these or even cooler spectral types that appear to have lost Li despite being members of young clusters/associations like the ONC and Taurus. Some of these stars also show evidence for accretion disks, requiring a careful veiling correction (*White and Hillenbrand*, 2005; *Palla et al.*, 2007), but in others the weak Li line is unambiguous (*Sacco et al.*, 2007). The ages estimated from the HRD are generally younger than implied by the level of Li depletion, and the masses lower — a feature that may persist to older ages (*Yee and Jensen*, 2010).

These Li-depleted stars support the notion of a (very) wide age spread, but represent at most a few percent of the populations of their respective clusters. Yet this Li age test may not be independent of the accretion history of the

star: *Baraffe and Chabrier* (2010) suggest that the same early accretion that would lead to smaller luminosities and radii, and apparently older ages compared to non-accreting models, will also lead to higher core temperatures and significant Li depletion. *Sergison et al.* (2013) found a modest correlation between the strength of Li absorption and isochronal age in the ONC and NGC 2264, but no examples of the severely Li-depleted objects that *Baraffe and Chabrier* (2010) suggest would be characteristic of the large accretion rates required to significantly alter the position of a PMS star in the HRD.

Littlefair et al. (2011) examined the relationship between rotation and age implied by CMD position within individual clusters. Pre-main-sequence contraction should lead to significant spin-up in older stars, or if the "disk locking" mechanism is effective [for stars that keep their inner disks for that long (*Koenigl*, 1991)], rotation periods might remain roughly constant. In all the examined clusters the opposite correlation was found — the apparently "young" objects were faster rotating than their "older" siblings — questioning whether position in the CMD truly reflects the stellar age and possibly finding an explanation in the connection between accretion history and present-day rotation.

The presence or lack of a disk or accretion acts as a crude age indicator (see section 5.2). In the scenario where most stars are born with circumstellar disks, and where disk signatures decay monotonically (on average) over timescales of only a few million years, any age spread greater than this should lead to clear differences in the age distributions of stars with and without disks. Some observations match this expectation (*Bertout et al.*, 2007; *Fang et al.*, 2013b), but many do not and the age distributions of stars with and without accretion signatures cannot be distinguished (*Dahm and Simon*, 2005; *Winston et al.*, 2009; *Rigliaco et al.*, 2011). *Jeffries et al.* (2011) performed a quantitative analysis of stars in the ONC with ages given by *Da Rio et al.* (2010a) and circumstellar material diagnosed with Spitzer IR results. The age distributions of stars with and without disks were indistinguishable, suggesting that any age spread must be smaller than the median disk lifetime.

Kinematic expansion is unlikely to constrain any age spread, but there are examples of runaway stars that apparently originated from star-forming regions. Even with excellent kinematic data, we would need several runaways from the same region in order to draw any conclusions about age spreads.

Pulsations among intermediate-mass stars in the instability strip (see section 4.4) have potential for diagnosing age spreads if good pulsation data for a number of stars in the same cluster can be obtained. While not independent of the evolutionary models, the ages determined in a seismology analysis are not affected by the same observational uncertainties as the HRD and can be considerably more precise (e.g., *Zwintz et al.*, 2013).

Below is a summary on age spreads:

The issue of whether significant age spreads exist is not settled either observationally or theoretically and the degree of coevality may vary in different star-forming environments. It is clear that the age implied by the HRD for an individual star in a young (≤10 m.y.) cluster should not be taken at face value, especially if that star has circumstellar material. At best there are large observational uncertainties and at worst the inferred ages are meaningless, almost entirely dependent on the accretion history of the star. It is possible that mean isochronal ages for a group of stars holds some validity but could be subject to an uncertain systematic bias.

7. THE EFFECTS OF COMPOSITION

There are several additional factors related to the composition of stars that influence age determinations because they affect both observations and models. In particular, what initial composition should be adopted for young (<100-m.y.-old) stars in the solar vicinity? Is the present-day solar photospheric or estimated protosolar abundance mix adequate?

Evolutionary tracks require mass fractions of H, He, and metals — and of course a choice of individual mass fractions for the "metal" elements. To first order, the young stars in the solar vicinity appear to be "solar" in composition. Unfortunately, it is still unclear exactly what the solar composition is, as there is disagreement between the solar metal fraction derived using comparing solar spectra to new three-dimensional stellar atmosphere models and that derived using helioseismological sound speed profiles (*Antia and Basu*, 2006; *Asplund et al.*, 2009; *Caffau et al.*, 2010). As we noted above, LDB ages are not sensitive to changes in the composition of models and thus those ages are robust.

7.1. Helium Abundances

Over the past decade it has become clear from the CMDs of some globulars that apparently similar stars can have a range of He abundances. Helium is important in determining the structure and hence evolution of a star, yet it remains stubbornly difficult to measure observationally. The general assumption is that other stars have the same He as the Sun, or that there is a general trend in Galactic evolution of He increasing with [Fe/H] over time. It needs to be recognized that uncertainty in He adds fundamental uncertainty to most aspects of PMS stars, including their ages.

Recently there has grown a general consensus that the protosolar He mass fraction was in the range $Y_\odot^{init} \simeq$ 0.27–0.28 (*Grevesse and Sauval*, 1998; *Asplund et al.*, 2009; *Lodders et al.*, 2009; *Serenelli and Basu*, 2010). Few He abundances measurements are available for nearby young <100-m.y.-old stars. Spectra of young B-type stars in nearby clusters shows that their He abundances are more or less similar to the modern photosphere, albeit with high dispersion, which appears to be due to rotationally induced mixing (*Huang and Gies*, 2006). Table 9 of *Nieva and Przybilla* (2012) shows a useful modern comparison between abundances of He, C, N, O, Ne, Mg, Si, and Fe among

nearby B stars, other young samples of stars, the interstellar medium, and the Sun. Save for Ne, there is fairly good agreement between the inferred abundances for B-type stars in the solar vicinity and the solar photospheric abundances from *Asplund et al.* (2005). Solar He abundances have also been inferred for members of the Pleiades (*Southworth et al.*, 2005; *An et al.*, 2007) and other nearby very young clusters (*Mathys et al.*, 2002; *Southworth et al.*, 2004; *Southworth and Clausen*, 2007; *Alecian et al.*, 2007).

7.2. Metallicity

Metallicities of PMS stars can be determined from observations, but with lower precision and accuracy than for MS stars. Complications may arise from the effects of starspots and plage regions and NLTE affects effective spectral line formation (*Stauffer et al.*, 2003; *Viana Almeida et al.*, 2009; *Bubar et al.*, 2011). The majority of young stellar groups investigated that are near the Sun appear to have solar metallicity. Metallicities consistent with [Fe/H] ≃ 0.0 have been measured among nearby young open clusters (*Randich et al.*, 2001; *Chen et al.*, 2003). Star-forming regions also appear to have metallicities consistent with solar [Fe/H] ≃ 0.00 (*Padgett*, 1996; *D'Orazi et al.*, 2009, 2011; *Viana Almeida et al.*, 2009), or perhaps slightly less at ≃–0.08 (*Santos et al.*, 2008). The cloud-to-cloud metallicity dispersion appears to be low, ~0.03 dex among 6 nearby star-forming regions (*Santos et al.*, 2008) and ~0.06 dex among 11 nearby young associations (*Viana Almeida et al.*, 2009). There are claims of some young nearby clusters having [Fe/H] upward of ~0.2 dex (*Monroe and Pilachowski*, 2010). For 15 open clusters within 1 kpc of the Sun from the compilation of *Chen et al.* (2003), the mean metallicity is [Fe/H] = –0.02 ± 0.04. The rms dispersion is ±0.16 dex; however, this is likely an upper limit to the real scatter as the metallicities come from many studies, and the size of the uncertainties is not listed. The results seem consistent with most nearby young stars having metallicity near solar, with 1σ dispersion <0.1 dex.

We conclude that the input protostellar composition for evolutionary models for <100-m.y.-old stellar populations in the solar vicinity should probably have Z/X similar to the modern-day solar photosphere (hence [Fe/H] ≃ 0.0), and He mass fraction similar to protosolar. A good starting point for models of nearby young stellar populations might be the recommended "present-day cosmic matter in the solar neighborhood" abundances from *Przybilla et al.* (2008), which are X = 0.715, Y = 0.271, Z = 0.014, Z/X = 0.020.

8. CONCLUSIONS AND RECOMMENDATIONS

8.1. Age Techniques Across the Hertzsprung-Russell Diagram

Table 2 categorizes the applicability of the different methods discussed above in various domains of stellar mass and age. The individual techniques within each cell are listed in order of reliability in the opinion of the present authors.

TABLE 2. Useful age-dating methods for various mass and age ranges in the H-R diagram.

	1–10 m.y.	~10–100 m.y.	>100 m.y.
<0.1 $M_\odot$	isochrones, gravity, R sin i, seismology?	LDB, isochrones, gravity	LDB, isochrones
0.1–0.5 $M_\odot$	isochrones, gravity, R sin i, disks	isochrones, Li, gravity	rotation/activity
0.5–2.0 $M_\odot$	isochrones, disks	Li	rotation/activity
>2.0 $M_\odot$	isochrones, seismology, R-C gap	isochrones, seismology	isochrones

8.2. General Comments

We have summarized the methods known to us by which one might estimate the age of a young star or group. Two conclusions are noteworthy.

First, we feel that the evidence supports using the age scale established by the LDB (section 3.1) for young clusters with age >20 m.y. The number of clusters with LDB ages is modest, but they span a broad range of ages and the LDB scale avoids the problems and uncertainties that have arisen in computing models for intermediate-mass stars at the MSTO points of young clusters. Adopting the LDB ages implies that a moderate amount of convective core overshoot or rotation needs to be included in models of intermediate-mass stars, a conclusion now supported by the analysis of oscillations, detected by Kepler, in stars slightly more massive than the Sun (*Silva Aguirre et al., 2013*).

Second, below 20 m.y. there is no well-defined absolute age scale. This should be considered carefully before using, for example, the median disk lifetime and ratios of different protostellar classes to estimate the lifetimes of protostellar phases. Kinematic ages ought to provide a timeline independent of any stellar physics uncertainties, but — frustratingly — attempts to estimate the ages of groups of very young stars through their kinematics appear to consistently fail (see section 3.2).

What does work? All the empirical methods (rotation, activity, Li, IR excesses) either have inherently large scatters at any one age, to a degree that exceeds any age trend, or they are the very properties we hope to study as essential aspects of PMS evolution. Placement of stars and associations in CMDs and HRDs remains fundamental in this field, despite the many and known difficulties. Pre-main-sequence isochronal ages can have good precision and can allow groups of stars to be ranked, but they have model-dependent systematic uncertainties of at least a factor of 2 below 10 m.y. Ages from the UMS or MSTO are probably more reliable, but are often less precise because there are fewer high-mass stars.

As we asked at the beginning, can we, in fact, establish reliable and consistent ages for young stars that are independent of the phenomena being studied? Our answer is probably "yes" for ages >20 m.y., but has to be "no" for younger objects, tempered by some progress, and with expectations of that progress continuing. The biggest improvement by far that we can anticipate is Gaia. All of stellar astrophysics has great expectations for the success of that mission, and it can seem as though all hope is vested in it. Nevertheless, the expectations are driven by realistic estimates of Gaia's performance. For PMS studies, Gaia will make it possible to measure directly the precise distances to individual stars in star-forming regions out to Orion and beyond. That obviates uncertainties due to extinction and reddening and allows the full three-dimensional structure of those regions to be seen. Gaia should detect more and lower-mass runaway stars, and its precise astrometry, when combined with good radial velocities, should reveal precise three-dimensional velocities that could lead to accurate kinematic ages.

Related to Gaia, the Gaia-European Southern Observatory (ESO) spectroscopic survey (*Gilmore et al., 2012*) will provide large, homogeneous datasets in ~30 young clusters, yielding uniformly determined abundances (Fe, Li, and others), along with RVs more precise than possible with Gaia (<0.5 km s^{-1}), with which to determine membership.

8.3. Recommendations and a Final Thought

To theorists and modelers: Please include the PMS in your published evolutionary tracks (most now do), please produce very dense grids of evolutionary tracks and isochrones by mass and age, and please carefully consider assumptions about composition.

To observers: When estimating ages of stars using evolutionary tracks, we recommend that you also try your technique on a few test stars to make sure that your results make sense. As a case in point, old field M dwarfs can have luminosities and effective temperatures that make them appear to be PMS and <100 m.y. This is because evolutionary tracks have difficulty predicting the MS among the low-mass stars. Age-dating groups of stars (clusters, associations, multiple stars) will most likely always be more accurate than age-dating individual stars (save perhaps special cases using asteroseismology).

An idea: There should be a D equivalent of the LDB, and because D burns at lower temperatures, it occurs earlier and in lower-mass objects. Stars of mass 0.1 $M_\odot$ deplete their D in just 2 m.y., but it takes 20 m.y. to reach D depletion in a 0.02-$M_\odot$ brown dwarf (*Chabrier et al., 2000*). Hence, for a coeval group of stars and brown dwarfs with an age in this interval, there should be a luminosity below which D is present, but above which it is absent. Detecting and measuring D in stars is challenging, but M-type spectra have molecular bands (perhaps HDO, CrD) that might be amenable to this, and the goal is not a precise measurement of D/H but instead just detection of the isotope. This could then potentially yield absolute ages below 20 m.y., although the definition of t = 0, the possibility of age spreads, the role of initial conditions, and the early or ongoing accretion (of D-rich material) would certainly make interpretation challenging.

Acknowledgments. E.E.M. acknowledges support from NSF grant AST-1008908. The careful reading by the referee was appreciated.

REFERENCES

Adams F. C. et al. (2004) *Astrophys. J., 611,* 360.
Alecian E. et al. (2007) *Astron. Astrophys., 473,* 181.
Allard F. et al. (2011) In *16th Cambridge Workshop on Cool Stars, Stellar Systems, and the Sun* (C. Johns-Krull et al., eds.), p. 91. ASP Conf. Series 448, Astronomical Society of the Pacific, San Francisco.
Allers K. N. et al. (2007) *Astrophys. J., 657,* 511.
Alves J. and Bouy H. (2012) *Astron. Astrophys., 547,* A97.
An D. et al. (2007) *Astrophys. J., 655,* 233.
Antia H. M. and Basu S. (2006) *Astrophys. J., 644,* 1292.
Asplund M. et al. (2005) In *Cosmic Abundances as Records of Stellar Evolution and Nucleosynthesis* (T. G. Barnes III and F. N. Bash, eds.), p. 25. ASP Conf. Series 336, Astronomical Society of the Pacific, San Francisco.
Asplund M. et al. (2009) *Annu. Rev. Astron. Astrophys., 47,* 481.
Ballesteros-Paredes J. and Hartmann L. (2007) *Rev. Mexicana Astron. Astrofis., 43,* 123.
Baraffe I. and Chabrier G. (2010) *Astron. Astrophys., 521,* A44.
Baraffe I. et al. (1998) *Astron. Astrophys., 337,* 403.
Baraffe I. et al. (2002) *Astron. Astrophys., 382,* 563.
Baraffe I. et al. (2009) *Astrophys. J. Lett., 702,* L27.
Baraffe I. et al. (2012) *Astrophys. J., 756,* 118.
Barenfeld S. A. et al. (2013) *Astrophys. J., 766,* 6.
Barrado y Navascués D. (2006) *Astron. Astrophys., 459,* 511.
Barrado y Navascués D. and Martín E. L. (2003) *Astron. J., 126,* 2997.
Barrado y Navascués D. et al. (2004) *Astrophys. J., 614,* 386.
Basri G. et al. (1996) *Astrophys. J., 458,* 600.
Beccari G. et al. (2010) *Astrophys. J., 720,* 1108.
Bedding T. R. et al. (2011) *Nature, 471,* 608.
Bell C. P. M. et al. (2013) *Mon. Not. R. Astron. Soc., 434,* 806.
Berger E. et al. (2010) *Astrophys. J., 709,* 332.
Bertelli G. et al. (2009) *Astron. Astrophys., 508,* 355.
Bertout C. et al. (2007) *Astron. Astrophys., 473,* L21.
Bik A. et al. (2012) *Astrophys. J., 744,* 87.
Bildsten L. et al. (1997) *Astrophys. J., 482,* 442.
Binks A. S. and Jeffries R. D. (2013) *ArXiv e-prints,* arXiv:1310.2613.
Brandt T. D. et al. (2013) *ArXiv e-prints,* arXiv:1305.7264.
Briceño C. et al. (2007) In *Protostars and Planets V* (B. Reipurth et al., eds.), pp. 345–360. Univ. of Arizona, Tucson.
Brown A. G. A. et al. (1997) *Mon. Not. R. Astron. Soc., 285,* 479.
Bubar E. J. et al. (2011) *Astron. J., 142,* 180.
Burke C. J. et al. (2004) *Astrophys. J., 604,* 272.
Burningham B. et al. (2005) *Mon. Not. R. Astron. Soc., 363,* 1389.
Caffau E. et al. (2010) *Astron. Astrophys., 514,* A92.
Cargile P. A. and James D. J. (2010) *Astron. J., 140,* 677.
Cargile P. A. et al. (2010) *Astrophys. J. Lett., 725,* L111.
Carlsson M. et al. (1994) *Astron. Astrophys., 288,* 860.
Carpenter J. M. et al. (2006) *Astrophys. J. Lett., 651,* L49.
Castelli F. and Kurucz R. L. (2004) *ArXiv e-prints,* arXiv:astro-ph/0405087.
Chaboyer B. et al. (1995) *Astrophys. J., 441,* 876.
Chabrier G. and Baraffe I. (1997) *Astron. Astrophys., 327,* 1039.
Chabrier G. et al. (1996) *Astrophys. J. Lett., 459,* L91.
Chabrier G. et al. (2000) *Astrophys. J. Lett., 542,* L119.
Chabrier G. et al. (2007) *Astron. Astrophys., 472,* L17.
Chaplin W. J. and Miglio A. (2013) *Annu. Rev. Astron. Astrophys., 51,* 353.
Chaplin W. J. et al. (2011) *Astrophys. J. Lett., 732,* L5.

Charbonnel C. and Talon S. (2005) *Science, 309,* 2189.
Chen Y. Q. et al. (2003) *Astrophys. J., 591,* 925.
Cignoni M. et al. (2010) *Astrophys. J. Lett., 712,* L63.
Cody A. M. and Hillenbrand L. A. (2011) *Astrophys. J., 741,* 9.
Dabringhausen J. et al. (2008) *Mon. Not. R. Astron. Soc., 386,* 864.
Dahm S. E. and Hillenbrand L. A. (2007) *Astron. J., 133,* 2072.
Dahm S. E. and Simon T. (2005) *Astron. J., 129,* 829.
D'Antona F. and Mazzitelli I. (1997) *Mem. Soc. Astron. Ital., 68,* 807.
Da Rio N. et al. (2010a) *Astrophys. J., 722,* 1092.
Da Rio N. et al. (2010b) *Astrophys. J., 723,* 166.
de la Reza R. et al. (2006) *Astron. J., 131,* 2609.
Demarque P. et al. (2004) *Astrophys. J. Suppl., 155,* 667.
de Zeeuw P. T. et al. (1999) *Astron. J., 117,* 354.
Dobbie P. D. et al. (2010) *Mon. Not. R. Astron. Soc., 409,* 1002.
D'Orazi V. et al. (2009) *Astron. Astrophys., 501,* 973.
D'Orazi V. et al. (2011) *Astron. Astrophys., 526,* A103.
Dotter A. et al. (2008) *Astrophys. J. Suppl., 178,* 89.
Ekström S. et al. (2012) *Astron. Astrophys., 537,* A146.
Elmegreen B. G. (2007) *Astrophys. J., 668,* 1064.
Elmegreen B. G. and Efremov Y. N. (1996) *Astrophys. J., 466,* 802.
Enoch M. L. et al. (2009) *Astrophys. J., 692,* 973.
Evans N. J. II et al. (2009) *Astrophys. J. Suppl., 181,* 321.
Fang M. et al. (2013a) *Astron. Astrophys., 549,* A15.
Fang M. et al. (2013b) *Astrophys. J. Suppl., 207,* 5.
Fedele D. et al. (2010) *Astron. Astrophys., 510,* A72.
Feiden G. A. and Chaboyer B. (2012) *Astrophys. J., 761,* 30.
Gallet F. and Bouvier J. (2013) *Astron. Astrophys., 556,* A36.
Gennaro M. et al. (2012) *Mon. Not. R. Astron. Soc., 420,* 986.
Gilmore G. et al. (2012) *The Messenger, 147,* 25.
Gray R. O. and Corbally J. C., eds. (2009) *Stellar Spectral Classification.* Princeton Univ., Princeton. 616 pp.
Gray R. O. et al. (2006) *Astron. J., 132,* 161.
Gregory S. G. et al. (2012) *Astrophys. J., 755,* 97.
Grevesse N. and Sauval A. J. (1998) *Space Sci. Rev., 85,* 161.
Guarcello M. G. et al. (2010) *Astron. Astrophys., 521,* A18.
Haisch K. E. Jr. et al. (2001) *Astrophys. J. Lett., 553,* L153.
Hartigan P. and Kenyon S. J. (2003) *Astrophys. J., 583,* 334.
Hartmann L. (2001) *Astron. J., 121,* 1030.
Hartmann L. et al. (1997) *Astrophys. J., 475,* 770.
Hartmann L. et al. (1998) *Astrophys. J., 495,* 385.
Hauck B. and Mermilliod M. (1998) *Astron. Astrophys. Suppl., 129,* 431.
Henderson C. B. and Stassun K. G. (2012) *Astrophys. J., 747,* 51.
Hernández J. et al. (2008) *Astrophys. J., 686,* 1195.
Hernández J. et al. (2010) *Astrophys. J., 722,* 1226.
Hillenbrand L. A. (2005) *ArXiv e-prints,* arXiv:0511.0083.
Hillenbrand L. A. (2009) In *The Ages of Stars* (E. E. Mamajek et al., eds.), pp. 81–94. IAU Symp. 258, Cambridge Univ., Cambridge.
Hillenbrand L. A. and White R. J. (2004) *Astrophys. J., 604,* 741.
Hillenbrand L. A. et al. (2008) In *14th Cambridge Workshop on Cool Stars, Stellar Systems, and the Sun* (G. van Belle, ed.), p. 200. ASP Conf. Series 384, Astronomical Society of the Pacific, San Francisco.
Hoogerwerf R. et al. (2001) *Astron. Astrophys., 365,* 49.
Hosokawa T. et al. (2011) *Astrophys. J., 738,* 140.
Huang W. and Gies D. R. (2006) *Astrophys. J., 648,* 591.
Huff E. M. and Stahler S. W. (2006) *Astrophys. J., 644,* 355.
Jackson R. J. et al. (2009) *Mon. Not. R. Astron. Soc., 399,* L89.
Jayawardhana R. et al. (2006) *Astrophys. J., 648,* 1206.
Jeffries R. D. (1995) *Mon. Not. R. Astron. Soc., 273,* 559.
Jeffries R. D. (2006) In *Chemical Abundances and Mixing in Stars in the Milky Way and Its Satellites* (S. Randich and L. Pasquini, eds,), p. 163. Springer-Verlag, Berlin.
Jeffries R. D. (2007) *Mon. Not. R. Astron. Soc., 381,* 1169.
Jeffries R. D. (2009) In *The Ages of Stars* (E. E. Mamajek et al., eds.), pp. 95–102. IAU Symp. 258, Cambridge Univ., Cambridge
Jeffries R. D. (2012) In *Star Clusters in the Era of Large Surveys* (A. Moitinho and J. Alves, eds.), p. 163. Springer-Verlag, Berlin.
Jeffries R. D. and Oliveira J. M. (2005) *Mon. Not. R. Astron. Soc., 358,* 13.
Jeffries R. D. et al. (2007) *Mon. Not. R. Astron. Soc., 376,* 580.
Jeffries R. D. et al. (2011) *Mon. Not. R. Astron. Soc., 418,* 1948.
Jeffries R. D. et al. (2013) *Mon. Not. R. Astron. Soc., 434,* 2438.
Jilinski E. et al. (2005) *Astrophys. J., 619,* 945.
Jones B. F. et al. (1996) *Astron. J., 112,* 186.
Kennedy G. M. and Kenyon S. J. (2009) *Astrophys. J., 695,* 1210.

Kirkpatrick J. D. et al. (2008) *Astrophys. J., 689*, 1295.
Koenigl A. (1991) *Astrophys. J. Lett., 370*, L39.
Kraus A. L. and Hillenbrand L. A. (2009) *Astrophys. J., 704*, 531.
Krumholz M. R. and Tan J. C. (2007) *Astrophys. J., 654*, 304.
Kudryavtseva N. et al. (2012) *Astrophys. J. Lett., 750*, L44.
Lambert D. L. and Reddy B. E. (2004) *Mon. Not. R. Astron. Soc., 349*, 757.
Larson R. B. (1981) *Mon. Not. R. Astron. Soc., 194*, 809.
Lawson W. A. et al. (2009) *Mon. Not. R. Astron. Soc., 400*, L29.
Littlefair S. P. et al. (2011) *Mon. Not. R. Astron. Soc., 413*, L56.
Lodders K. et al. (2009) *Landolt-Börnstein*, p. 44. Springer-Verlag, Berlin.
López-Morales M. (2007) *Astrophys. J., 660*, 732.
Lucas P. W. et al. (2001) *Mon. Not. R. Astron. Soc., 326*, 695.
Luhman K. L. (1999) *Astrophys. J., 525*, 466.
Luhman K. L. and Mamajek E. E. (2012) *Astrophys. J., 758*, 31.
Luhman K. L. et al. (2005) *Astrophys. J. Lett., 628*, L69.
Lyra W. et al. (2006) *Astron. Astrophys., 453*, 101.
MacDonald J. and Mullan D. J. (2013) *Astrophys. J., 765*, 126.
Maeder A. and Mermilliod J. C. (1981) *Astron. Astrophys., 93*, 136.
Maeder A. and Meynet G. (1989) *Astron. Astrophys., 210*, 155.
Makarov V. V. (2007) *Astrophys. J. Suppl., 169*, 105.
Makarov V. V. and Fabricius C. (2001) *Astron. Astrophys., 368*, 866.
Makarov V. V. et al. (2005) *Mon. Not. R. Astron. Soc., 362*, 1109.
Mamajek E. E. (2005) *Astrophys. J., 634*, 1385.
Mamajek E. E. and Hillenbrand L. A. (2008) *Astrophys. J., 687*, 1264.
Mamajek E. E. et al. (2000) *Astrophys. J., 544*, 356.
Mamajek E. E. et al. (2002) *Astron. J., 124*, 1670.
Mamajek E. E. et al., eds. (2009) *The Ages of Stars*. IAU Symp. 258, Cambridge Univ., Cambridge.
Manara C. F. et al. (2013) *Astron. Astrophys., 558*, A114.
Manzi S. et al. (2008) *Astron. Astrophys., 479*, 141.
Martín E. L. et al. (1999) *Astron. J., 118*, 2466.
Mathys G. et al. (2002) *Astron. Astrophys., 387*, 890.
Mayne N. J. (2010) *Mon. Not. R. Astron. Soc., 408*, 1409.
Mayne N. J. et al. (2007) *Mon. Not. R. Astron. Soc., 375*, 1220.
Mentuch E. et al. (2008) *Astrophys. J., 689*, 1127.
Mercer-Smith J. A. et al. (1984) *Astrophys. J., 279*, 363.
Mermilliod J. C. (1981) *Astron. Astrophys., 97*, 235.
Messina S. et al. (2010) *Astron. Astrophys., 520*, A15.
Meynet G. and Maeder A. (2000) *Astron. Astrophys., 361*, 101.
Monroe T. R. and Pilachowski C. A. (2010) *Astron. J., 140*, 2109.
Mouschovias T. C. et al. (2006) *Astrophys. J., 646*, 1043.
Murphy S. J. et al. (2013) *Mon. Not. R. Astron. Soc., 435*, 1325.
Naylor T. and Jeffries R. D. (2006) *Mon. Not. R. Astron. Soc., 373*, 1251.
Naylor T. et al. (2009) In *The Ages of Stars* (E. E. Mamajek et al., eds.), pp. 103–110. IAU Symp. 258, Cambridge Univ., Cambridge.
Nieva M.-F. and Przybilla N. (2012) *Astron. Astrophys., 539*, A143.
O'Dell C. R. et al. (2005) *Astrophys. J. Lett., 633*, L45.
Ortega V. G. et al. (2002a) *Astrophys. J. Lett., 575*, L75.
Ortega V. G. et al. (2002b) *Astrophys. J. Lett., 575*, L75.
Ortega V. G. et al. (2007) *Mon. Not. R. Astron. Soc., 377*, 441.
Padgett D. L. (1996) *Astrophys. J., 471*, 847.
Palla F. and Baraffe I. (2005) *Mem. Soc. Astron. Ital., 76*, 229.
Palla F. and Stahler S. W. (1999) *Astrophys. J., 525*, 772.
Palla F. and Stahler S. W. (2000) *Astrophys. J., 540*, 255.
Palla F. et al. (2007) *Astrophys. J. Lett., 659*, L41.
Pecaut M. J. and Mamajek E. E. (2013) *Astrophys. J., 208*, 9.
Pecaut M. J. et al. (2012) *Astrophys. J., 746*, 154.
Pflamm-Altenburg J. and Kroupa P. (2007) *Mon. Not. R. Astron. Soc., 375*, 855.
Piau L. and Turck-Chièze S. (2002) *Astrophys. J., 566*, 419.
Pillitteri I. et al. (2013) *Astrophys. J., 768*, 99.
Pinsonneault M. (1997) *Annu. Rev. Astron. Astrophys., 35*, 557.
Piskunov A. E. and Belikov A. N. (1996) *Astron. Lett., 22*, 466.
Piskunov A. E. et al. (2004) *Mon. Not. R. Astron. Soc., 349*, 1449.

Poveda A. et al. (2005) *Astrophys. J. Lett., 627*, L61.
Preibisch T. (2012) *Res. Astron. Astrophys., 12*, 1.
Prisinzano L. et al. (2012) *Astron. Astrophys., 546*, A9.
Przybilla N. et al. (2008) *Astrophys. J. Lett., 688*, L103.
Randich S. et al. (1997) *Astron. Astrophys., 323*, 86.
Randich S. et al. (2001) *Astron. Astrophys., 372*, 862.
Reggiani M. et al. (2011) *Astron. Astrophys., 534*, A83.
Rigliaco E. et al. (2011) *Astron. Astrophys., 525*, A47.
Rochau B. et al. (2010) *Astrophys. J. Lett., 716*, L90.
Ruoppo A. et al. (2007) *Astron. Astrophys., 466*, 261.
Sacco G. G. et al. (2007) *Astron. Astrophys., 462*, L23.
Sackmann I.-J. and Boothroyd A. I. (2003) *Astrophys. J., 583*, 1024.
Santos N. C. et al. (2008) *Astron. Astrophys., 480*, 889.
Scandariato G. et al. (2012) *Astron. Astrophys., 545*, A19.
Schaller G. et al. (1992) *Astron. Astrophys. Suppl., 96*, 269.
Schlieder J. E. et al. (2012) *Astron. J., 143*, 114.
Serenelli A. M. and Basu S. (2010) *Astrophys. J., 719*, 865.
Sergison D. J. et al. (2013) *Mon. Not. R. Astron. Soc., 434*, 966.
Sestito P. and Randich S. (2005) *Astron. Astrophys., 442*, 615.
Sicilia-Aguilar A. et al. (2009) *Astrophys. J., 701*, 1188.
Siess L. et al. (2000) *Astron. Astrophys., 358*, 593.
Silva Aguirre V. et al. (2013) *Astrophys. J., 769*, 141.
Slesnick C. L. et al. (2004) *Astrophys. J., 610*, 1045.
Slesnick C. L. et al. (2006) *Astron. J., 131*, 3016.
Slesnick C. L. et al. (2008) *Astrophys. J., 688*, 377.
Soderblom D. R. (2010) *Annu. Rev. Astron. Astrophys., 48*, 581.
Soderblom D. R. et al. (1991) *Astrophys. J., 375*, 722.
Soderblom D. R. et al. (1993) *Astron. J., 106*, 1059.
Song I. et al. (2003) *Astrophys. J., 599*, 342.
Song I. et al. (2012) *Astron. J., 144*, 8.
Southworth J. and Clausen J. V. (2007) *Astron. Astrophys., 461*, 1077.
Southworth J. et al. (2004) *Mon. Not. R. Astron. Soc., 349*, 547.
Southworth J. et al. (2005) *Astron. Astrophys., 429*, 645.
Spruit H. C. and Weiss A. (1986) *Astron. Astrophys., 166*, 167.
Stahler S. W. (1988) *Astrophys. J., 332*, 804.
Stauffer J. R. et al. (1998) *Astrophys. J. Lett., 499*, L199.
Stauffer J. R. et al. (1999) *Astrophys. J., 527*, 219.
Stauffer J. R. et al. (2003) *Astron. J., 126*, 833.
Stolte A. et al. (2004) *Astron. J., 128*, 765.
Sung H. et al. (1999) *Mon. Not. R. Astron. Soc., 310*, 982.
Takagi Y. et al. (2011) *Publ. Astron. Soc. Japan, 63*, 677.
Tan J. C. et al. (2006) *Astrophys. J. Lett., 641*, L121.
Tognelli E. et al. (2011) *Astron. Astrophys., 533*, A109.
Tout C. A. et al. (1999) *Mon. Not. R. Astron. Soc., 310*, 360.
Ushomirsky G. et al. (1998) *Astrophys. J., 497*, 253.
van Leeuwen F. (2007) *Astron. Astrophys., 474*, 653.
Ventura P. et al. (1998) *Astron. Astrophys., 331*, 1011.
Viana Almeida P. et al. (2009) *Astron. Astrophys., 501*, 965.
Vorobyov E. I. and Basu S. (2006) *Astrophys. J., 650*, 956.
Vorobyov E. I. and Basu S. (2010) *Astrophys. J., 719*, 1896.
Walker M. F. (1956) *Astrophys. J. Suppl., 2*, 365.
Wang Y. et al. (2011) *Astron. Astrophys., 527*, A32.
Webb R. A. et al. (1999) *Astrophys. J. Lett., 512*, L63.
Weinberger A. J. et al. (2012) *Am. Astron. Soc. Meeting Abstracts #219*, 337.10.
Weinberger A. J. et al. (2013) *Astrophys. J., 762*, 118.
White R. J. and Basri G. (2003) *Astrophys. J., 582*, 1109.
White R. J. and Hillenbrand L. A. (2005) *Astrophys. J. Lett., 621*, L65.
Williams J. P. and Cieza L. A. (2011) *Annu. Rev. Astron. Astrophys., 49*, 67.
Winston E. et al. (2009) *Astron. J., 137*, 4777.
Yee J. C. and Jensen E. L. N. (2010) *Astrophys. J., 711*, 303.
Yi S. K. et al. (2003) *Astrophys. J. Suppl., 144*, 259.
Zuckerman B. and Song I. (2004) *Annu. Rev. Astron. Astrophys., 23(42)*, 685.
Zuckerman B. et al. (2001) *Astrophys. J. Lett., 562*, L87.
Zwintz K. et al. (2013) *Astron. Astrophys., 552*, A68.

Krumholz M. R., Bate M. R., Arce H. G., Dale J. E., Gutermuth R., Klein R. I., Li Z.-Y., Nakamura F., and Zhang Q. (2014)
Star cluster formation and feedback. In *Protostars and Planets VI* (H. Beuther et al., eds.), pp. 243–266. Univ. of Arizona, Tucson,
DOI: 10.2458/azu_uapress_9780816531240-ch011.

Star Cluster Formation and Feedback

Mark R. Krumholz
University of California, Santa Cruz

Matthew R. Bate
University of Exeter

Héctor G. Arce
Yale University

James E. Dale
Excellence Cluster "Universe" and Ludwig-Maximillians-University

Robert Gutermuth
University of Massachusetts, Amherst

Richard I. Klein
University of California, Berkeley and Lawrence Livermore National Laboratory

Zhi-Yun Li
University of Virginia

Fumitaka Nakamura
National Astronomical Observatory of Japan

Qizhou Zhang
Harvard-Smithsonian Center for Astrophysics

Stars do not generally form in isolation. Instead, they form in clusters, and in these clustered environments, newborn stars can have profound effects on one another and on their parent gas clouds. Feedback from clustered stars is almost certainly responsible for a number of otherwise puzzling facts about star formation: (1) it is an inefficient process that proceeds slowly when averaged over galactic scales; (2) most stars disperse from their birth sites and dissolve into the galactic field over timescales ≪1 gigayears (G.y.); and (3) newborn stars follow an initial mass function (IMF) with a distinct peak in the range 0.1–1 $M_\odot$, rather than an IMF dominated by brown dwarfs. In this review, we summarize current observational constraints and theoretical models for the complex interplay between clustered star formation and feedback.

1. INTRODUCTION

1.1. Why is Feedback Essential?

Newborn stars have profound effects on their birth environments, and any complete theory for star formation must include them. Perhaps the best argument for this statement is an image such as Fig. 1, which shows 30 Doradus, the largest HII region in the Local Group, powered by the cluster NGC 2070 and its 2400 OB stars (*Parker,* 1993). The figure illustrates several of the routes by which young stars can influence their surroundings. Observations reveal extensive 8-µm emission around the outer rim of the structure shown, marking where gas has been warmed by far ultraviolet (UV)

radiation from young stars. On the inner rim of the cavity-like structures, Hα emission indicates that ionizing radiation has converted the ISM to a warm ionized phase. Finally, the seemingly-empty cavities are bright in X-ray emission produced by a ~10^7-K phase created by shocks in the fast winds launched by O stars. The entire region is expanding at ~25 km s^{-1} (*Chu and Kennicutt,* 1994). Any theory for how NGC 2070 arrived at its present state must address the role played by these processes, and several others such as protostellar outflow feedback and radiative heating by infrared (IR) light. These are not visually apparent in Fig. 1, but can be seen clearly in other regions, and are perhaps equally important.

In addition to the visual impression provided by Fig. 1 and similar observations, there are a number of more subtle

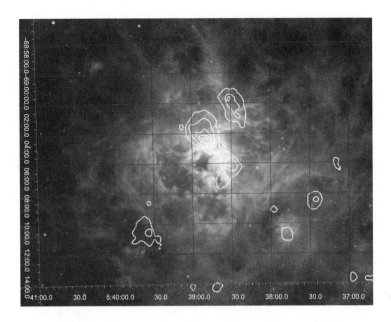

Fig. 1. Image of 30 Doradus: 8 μm, Hα, and 0.5–8 keV X-rays. White contours show $^{12}CO(1–0)$ emission. Figure taken from *Lopez et al.* (2011).

but equally compelling arguments for the importance of stellar feedback for cluster formation. The first, which dates back to the seminal work of *Zuckerman and Evans* (1974), is based on comparing the molecular mass M_{gas} to the star-formation rate $\dot{M}_*$ — either in an entire galaxy, or in a smaller region defined by a specified volume or column density threshold — to deduce a characteristic depletion time $t_{dep} = M_{gas}/\dot{M}_*$. Note that this should not be confused with the timescale over which star formation in a particular cloud takes place, as diagnosed, for example, by the stellar age spread (see the chapter by Soderblom et al. in this volume). The two are identical only if star formation proceeds until all the gas is converted to stars, and in general, the depletion time is 1–2 orders of magnitude larger.

While Zuckerman and Evans applied this technique to the low-density molecular gas traced by low-J CO emission, and numerous modern studies have done the same (see the chapters by Dobbs et al. and Padoan et al. in this volume), it has also become possible in the last 10 years to perform the same analysis for tracers of the denser regions from which clusters presumably form. Techniques for studying such regions include the low-J lines of heavy rotor molecules such as HCN, HCO^+, and CS (which have critical densities $\geq 10^4$ cm^{-3}), thermal emission from cold dust at submillimeter wavelengths, and dust extinction at near-IR wavelengths. The consensus from such studies is that the depletion time is always ~1–3 orders of magnitude longer than the free-fall time $t_{ff} \sim 1/\sqrt{G\rho}$, where ρ is the characteristic density selected by the tracer (*Krumholz and Tan,* 2007; *Evans et al.,* 2009; *Juneau et al.,* 2009; *Krumholz et al.,* 2012a; *Federrath,* 2013). In contrast, numerical simulations of star cluster formation that do not include any form of feedback generally produce $t_{dep} \sim t_{ff}$ (e.g., *Klessen and*

Burkert, 2000, 2001; *Bate et al.,* 2003; *Bonnell et al.,* 2003).

Magnetic fields alone are unlikely to prevent this outcome. Observations suggest that the median cloud is magnetically supercritical by a factor of 2 (*Crutcher,* 2012, and references therein), and simulations indicate that a magnetic field of this strength only decreases the star-formation rate by a factor of a few compared to the purely hydrodynamic case (*Price and Bate,* 2009; *Padoan and Nordlund,* 2011; *Federrath and Klessen,* 2012) (see also the chapter by Padoan et al. in this volume). Reduction of the star-formation rate by feedback, perhaps in conjunction with magnetic fields, is a prime candidate to resolve this problem.

A second, closely related argument has to do with the fraction of stars found in bound clusters. Most regions of active star formation are much denser than the field (e.g., *Gutermuth et al.,* 2009). We will refer to these regions as clusters, defined roughly as suggested by *Lada and Lada* (2003): a collection of physically-related stars within which the stellar mass density is ≫1 $M_\odot$ pc^{-3} [compared to ~0.01 in the field near the Sun (*Holmberg and Flynn,* 2000)], and where the total number of stars is greater than several tens. Such regions are typically ~1–10 pc in size, and, at least when they are young, also contain gas with a mass density greatly exceeding that of the stars. We do not require, as do some authors interested primarily in N-body dynamics (e.g., *Portegies Zwart et al.,* 2010), that the stars in question be gravitationally bound, or old enough to be dynamically relaxed; the former condition is often impossible to evaluate in clusters that are still embedded in their natal gas clouds, while the latter necessarily excludes the phase of formation in which we are most interested.

The argument for the importance of feedback can be made by observing that, while almost all regions of active star formation qualify as clusters by this definition, by an age of ~10–100 million years (m.y.), only a few percent of stars remain part of clusters with stellar densities noticeably above that of the field (e.g., *Silva-Villa and Larsen,* 2011; *Fall and Chandar,* 2012). In a cluster of N stars with a crossing time t_{cross} (typically ~0.1–1 m.y. in observed clusters), two-body evaporation does not become important until an age of $(10N/\ln N)t_{cross}$ (*Binney and Tremaine,* 1987). This is ~100–1000 m.y. even for a modest cluster of N = 1000. Thus a gravitationally bound cluster will not disperse on its own over the timescale demanded by observations. However, simulations of star cluster formation that do not include feedback have a great deal of difficulty reproducing this outcome (*Bate et al.,* 2003; *Bonnell et al.,* 2003). Instead, they invariably produce bound clusters.

One might think that this problem could be avoided by positing that most stars form in gravitationally unbound gas clouds. However, such a model suffers from two major problems. First, as discussed in the chapter by Dobbs et al. in this volume, recent surveys of molecular clouds find that their typical virial ratios are $\alpha_G \approx 1$, whereas $\alpha_G > 2$ is required to render a cloud unbound. Second, *Clark et al.* (2005) find that even unbound clouds (they consider one with $\alpha_G = 4$) leave most of their stars in bound clusters. A high virial ratio means that a cloud produces a number of smaller, mutually unbound clusters rather than a single large one, but each subcluster is still internally bound and has >1000 stars. These would survive too long to be consistent with observations. Stellar feedback represents the most likely way out of this problem, as the dispersal of gas by feedback can reduce the star-formation efficiency to the point where few stars remain members of bound clusters.

A final argument for the importance of feedback comes from the problem of explaining the origin of the stellar initial mass function (IMF). As we discuss below, simulations that do not include radiative feedback tend to have problems reproducing the observed IMF, while those including it do far better (also see the chapter by Offner et al. in this volume). In the following sections, we discuss the role of feedback in solving each of these problems, and highlight both the successes and failures of current models for its operation.

1.2. A Taxonomy of Feedback Mechanisms

Before discussing individual feedback mechanisms in detail, it is helpful to lay out some categories that can be used to understand them. Although many such taxonomies are possible, we choose to break feedback mechanisms down into three categories: (1) momentum feedback, (2) "explosive" feedback, (3) and thermal feedback.

Momentum feedback is, quite simply, the deposition of momentum into star-forming clouds so as to push on the gas, drive turbulent motions within it, and, if the feedback is strong enough, to unbind it entirely. The key feature of momentum feedback, which distinguishes it from explosive feedback, is the role of radiative energy loss. The dense, molecular material from which stars form, or even the

less-dense gas of the atomic interstellar medium (ISM), is extremely efficient at radiative cooling. As a result, when stars inject energy into the ISM, it is often the case that the energy is then radiated away on a timescale that is short compared to the dynamical time of the surrounding cloud. In this case the amount of energy delivered to the cloud matters little, and the effectiveness of the feedback is instead determined by the amount of momentum that is injected, since this cannot be radiated away. As we discuss below, protostellar outflows and (probably) radiation pressure are forms of momentum feedback.

In contrast, explosive feedback occurs when stars heat gas so rapidly, and to such a high temperature, that it is no longer able to cool on a cloud dynamical timescale. In this case at least some of the energy added to the cloud is not lost to radiation, and feedback is accomplished when the hot, overpressured gas expands explosively and does work on the surrounding cold molecular material. To understand the distinction between explosive and momentum-driven feedback, consider a point source injecting a wind of material into a uniform, cold medium, and sweeping up an expanding shell of material of mass M_{sh} and radius r_{sh}. If the wind is launched with mass flux $\dot{M}_w$ at velocity v_w, and in the process of sweeping up the shell there are no radiative losses (the extreme limit of the explosive case), then after a time t the kinetic energy of the shell is $M_{sh}\dot{r}_{sh}^2 \sim \dot{M}_w v_w^2 t$. On the other hand, in the case of momentum feedback where energy losses are maximal, the momentum of the shell is $M_{sh}\dot{r}_{sh} \sim \dot{M}_w v_w t$, and its kinetic energy is $M_{sh}\dot{r}_{sh}^2 \sim \dot{M}_w v_w \dot{r}_{sh}t$. Thus without radiative losses, the kinetic energy of the shell at equal times is larger by a factor of ~$v_w/\dot{r}_{sh}$. This is not a small number: In the example of 30 Doradus (Fig. 1), the measured velocity of the shell is ~25 km s^{-1} (*Chu and Kennicutt,* 1994), while typical launching velocities for O-star winds are >1000 km s^{-1}. Thus, when it operates, explosive feedback can be very effective. Winds from hot main-sequence stars, photoionizing radiation, and supernovae are all forms of explosive feedback.

Our final category is thermal feedback, which describes feedback mechanisms that do not necessarily cause the gas to undergo large-scale flows, but do alter its temperature. This is significant because the temperature structure of interstellar gas is strongly linked to how it fragments, and thus to the production of the IMF. Non-ionizing radiation is the main form of thermal feedback.

2. MOMENTUM FEEDBACK

2.1. Protostellar Outflows

2.1.1. Theory. Protostellar outflows are observed to be an integral part of star formation. Outflows eject a significant amount of mass from the regions around newborn stars, thereby helping to set the most important quantity for individual stars, their mass. Collectively, the outflows inject energy and momentum into their surroundings, modifying the environment in which the stars form (*Norman and Silk,*

1980; *McKee*, 1989; *Shu et al.*, 1999). Here we focus on this interaction, leaving the question of wind launching mechanisms to the chapter by Frank et al. in this volume. A key issue for outflows, as for all feedback mechanisms, is their momentum budget per unit mass of stars formed (a quantity with units of velocity), which we denote V_{out}. Note that V_{out} is not the velocity with which an individual outflow is launched; it is the total momentum carried by the outflows divided by the total mass of stars formed. It is therefore smaller than the velocity of an individual outflow by a factor equal to the ratio of the mass injected into the outflow divided by the stellar mass formed. A number of authors have attempted to estimate V_{out} from both theoretical models of outflow launching and from observed scaling relationships between outflow momenta and stellar properties. *Matzner and McKee* (2000) and *Matzner* (2007) estimate $V_{out} = 20$–40 km s^{-1}. The bulk of the momentum is produced by low-mass stars rather than massive ones, because outflow launch speeds scale roughly with the escape speeds from stellar surfaces, and the escape speeds from high-mass stars are not larger than those from low-mass stars by enough to compensate for the vastly greater mass contained in low-mass stars.

Protostellar outflow feedback is expected to be especially important wherever a large number of stars form close together in both space and time. The paradigmatic object for this type of feedback is the low-mass protocluster NGC 1333, where molecular line and infrared observations reveal numerous outflows packed closely together (*Knee and Sandell*, 2000; *Walawender et al.*, 2005, 2008; *Curtis et al.*, 2010; *Plunkett et al.*, 2013). The significance of outflow feedback in cluster formation can be illustrated using a simple estimate. Let the mass of the stars in a cluster be M_*, so the total momentum injected into the cluster-forming clump is $M_* V_{out}$. This momentum is in principle enough to move all the clump material (of total mass M_c) by a speed of

$$v \sim SFE \times V_{out} \sim 5 \text{ km s}^{-1} \left(\frac{SFE}{0.2} \right) \left(\frac{V_{out}}{25 \text{ km s}^{-1}} \right) \quad (1)$$

where $SFE = M_*/M_c$ is the star-formation efficiency of the clump. For typical parameters, this speed is significantly higher than the velocity dispersion of low-mass protoclusters such as NGC 1333. If all the momenta from the outflows were to be injected simultaneously, they would unbind the clump completely. If they are injected gradually, they may maintain the turbulence in the clump against dissipation and keep the stars forming at a relatively low rate over several free-fall times. This slow star formation over an extended period is consistent with the simultaneous presence of objects in all evolutionary stages, from prestellar cores to evolved class III objects that have lost most of their disks. The latter objects should be at least a few million years old, several times the typical free-fall time of the dense clumps that form NGC 1333-like clusters (*Evans et al.*, 2009).

Numerical simulations have demonstrated that outflows can indeed drive turbulence in a cluster-forming clump and maintain star formation well beyond one free-fall time. *Li and Nakamura* (2006) simulated magnetized cluster formation with outflow feedback assuming that stellar outflows are launched isotropically, and showed that the outflows drive turbulence that keeps the cluster-forming clump in quasi-equilibrium. The same conclusion was reached independently by *Matzner* (2007), who studied outflow-driven turbulence analytically. *Nakamura and Li* (2007) showed that collimated outflows are even more efficient in driving turbulence than spherical ones, because they reach large distances and larger-scale turbulence tends to decay more slowly. *Banerjee et al.* (2007) questioned this conclusion, performing simulations showing that fast-moving jets do not excite significant supersonic motions in a smooth ambient medium. However, *Cunningham et al.* (2009) showed that jets running into a turbulent ambient medium are more efficient in driving turbulent motions. Indeed, *Carroll et al.* (2009) were able to demonstrate explicitly that fully developed turbulence can be driven and maintained by a collection of collimated outflows, even in the absence of any magnetic field. Magnetic fields tend to couple different parts of the clump material together, which enables the outflows to deposit their energy and momentum in the ambient medium more efficiently (*Nakamura and Li*, 2007; *Wang et al.*, 2010).

The effects of magnetic fields and outflow feedback on cluster formation are illustrated in Fig. 2, which shows the rates of star formation in a parsec-scale clump with mass on the order of 10^3 M$_\odot$ for four simulations of increasing complexity. In the simplest case of no turbulence, magnetic field, or outflow feedback (the top left line in the figure), the clump collapses rapidly, forming stars at a rate approaching the characteristic free-fall rate (the dashed line). This rate is reduced progressively with the inclusion of initial turbulence (second solid line from top), turbulence and magnetic field (third line), and turbulence, magnetic field, and outflow feedback (bottom line). In the case with all three ingredients, the star-formation rate is kept at ~10% of the free-fall rate (the dotted line). One implication of these and other simulations is that the majority of the cluster members may be formed in a relatively leisurely manner in an outflow-driven, magnetically mediated, turbulent state, rather than rapidly in a free-fall time. Simple analytic estimates by *Nakamura and Li* (2011) suggest that outflows should be able to maintain such low star-formation rates in most of the observed clumps in the solar neighborhood.

Another implication is that outflows from low-mass stars can influence the formation of the massive stars that form in the same cluster. For example, the same simulations that show a reduction in the star-formation rate due to outflow feedback also show that outflows prevent rapid mass infall toward the massive stars that tend to reside at the bottom of the gravitational potential of protoclusters. Outflows can also have important interactions with other forms of feedback from both low- and high-mass stars, a topic we defer to section 5.1.

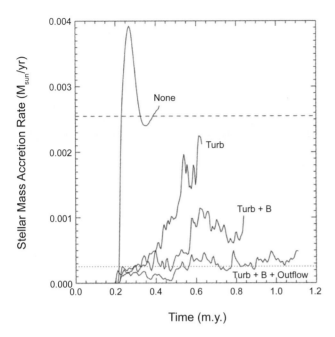

Fig. 2. The rates of star formation as a function of time for four simulations of cluster formation (adopted from *Wang et al., 2010*). The curves from top left to bottom right are for models that include, respectively, neither turbulence nor magnetic field nor outflow feedback (line labeled "None"), turbulence only ("Turb"), turbulence and magnetic field ("Turb + B"), and all three ingredients ("Turb + B + outflow"; see text for discussion). The dashed horizontal line indicates a star-formation rate such that the depletion time t_{dep} is equal to the free-fall time t_{ff}, while the dotted line indicates a star-formation rate corresponding to $t_{dep} = 10\ t_{ff}$.

2.1.2. Observations. Several recent studies have searched for observational signatures of the outflow feedback effects described in the previous section. Even though not all studies use the same procedure to estimate the molecular outflow energetics and different studies use different methods to assess the combined impact from all outflows on the cluster gas, there are some points of consensus. One is that, in the vast majority of regions, the combined action of all outflows seems to be sufficient to drive the observed level of turbulence. The simplest method used in observational studies to determine this has been to compare the total mechanical energy of molecular outflows (i.e., the total kinetic energy of the molecular gas that has been entrained by the protostellar wind) with the turbulent energy of the cluster gas. Studies show that for a number of clusters the current total molecular outflow energy is ~30% or more of the total turbulent energy (e.g., *Arce et al., 2010; Curtis et al., 2010; Graves et al., 2010; Duarte-Cabral et al., 2012*), yet for other regions the total outflow energy is just 1–20% of the total turbulent energy (e.g., *Arce et al., 2010; Narayanan et al., 2012*). It is unclear to what extent these differences in outflow energy are correlated with other cloud properties. *Sun et al.* (2006) study CO(3–2) emission from different regions of Perseus, and find that those with active star forma-

tion, such as NGC 1333, show steeper velocity power spectral indices than more quiescent regions. However, it is not clear if differences in the turbulent energy injection are the cause, and the observations themselves are at relatively low (~80″) resolution. Further study of this question is needed.

Although such simple comparisons are useful to gauge the relative importance of outflows, they do not necessarily indicate whether outflows can maintain the observed turbulence. A better way to address this is by comparing the total outflow power (L_{flow}) or mechanical luminosity (i.e., the rate at which the outflows inject energy into their surroundings through entrainment of molecular gas by the protostellar wind) with the turbulent energy dissipation rate (L_{turb}). Although there are many uncertainties associated with the estimation of both L_{flow} and L_{turb} from observations (e.g., *Williams et al., 2003; Arce et al., 2010*), it is clear that for most protostellar clusters observed thus far $L_{flow} \sim L_{turb}$ (*Williams et al., 2003; Stanke and Williams, 2007; Swift and Welch, 2008; Maury et al., 2009; Arce et al., 2010; Nakamura et al., 2011a,b*). The usual interpretation has been that outflows have sufficient power to sustain (or at least provide a major source of power for maintaining) the turbulence in the region.

The physical assumption behind these observational comparisons is that the gas that has been put in motion through the interaction of the protostellar wind and the ambient medium (i.e., the gas that makes up the bipolar molecular outflow) will eventually slow down and feed the turbulent motions of the cloud through some (not well understood) mechanism. However, it is unclear how efficiently outflow motions convert into cloud turbulence, and observational studies typically do not address this issue. Among the few exceptions are the studies by *Swift and Welch* (2008) and *Duarte-Cabral et al.* (2012), which use observations of ^{13}CO and C^{18}O (which are much better at tracing cloud structure than the more commonly observed ^{12}CO) to investigate how outflows create turbulence. These studies show direct evidence of outflow-induced turbulence, but both conclude that only a fraction of the outflow mechanical luminosity is used to sustain the turbulence in the cloud while a significant amount is deposited outside the cloud. They also suggest that the typical outflow energy injection scale, the scale at which the outflow momentum is most efficiently injected, is around a few tenths of a parsec, which agrees with the theoretical and numerical prediction (*Matzner, 2007; Nakamura and Li, 2007*). However, the clouds in these studies are relatively small and host large outflows. Similar observations of larger and denser clouds are needed to further investigate if the "outflow-to-turbulence" efficiency depends on cloud environment.

We note that even though most studies show that outflows have the potential to have significant impact on the cluster environment, recent cloud-wide surveys have shown that outflows lack the power needed to sustain the observed turbulence on the scale of a molecular cloud complex (*Walawender et al., 2005; Arce et al. 2010; Narayanan et al., 2012*) or a giant molecular cloud (*Dent et al.*, 2009; *Ginsburg et al.*, 2011), with sizes of more than 10 pc. Outflows in cloud complexes and giant molecular clouds (GMCs) are mostly clustered in regions with sizes of 1–4 pc, and there are large extents inside clouds with few or no outflows. This implies that an additional energy source is responsible for turbulence on a global cloud scale. For more discussion of this topic, see the chapter by Dobbs et al. in this volume.

Several investigators have also performed observations to study the role of outflows in gas dispersal (e.g., *Knee and Sandell*, 2000; *Swift and Welch*, 2008; *Arce et al.*, 2010; *Curtis et al.*, 2010; *Graves et al.*, 2010; *Narayanan et al.*, 2012; *Plunkett et al.*, 2013). As with the work on turbulent driving, different studies use different methods to attack this problem. One common practice has been to compare the total kinetic energy of all molecular outflows with the cluster-forming clump's gravitational binding energy, with the typical result being that total outflow kinetic energy is less than 20% of the binding energy (e.g., *Arce et al.*, 2010; *Narayanan et al.*, 2012). This simple analysis seems to indicate that in most regions outflows do not have enough energy to significantly disrupt their host clouds. An equivalent way of phrasing this conclusion is that, if all the detected (current) outflow momentum were used to accelerate gas to the region's escape velocity, at most 5–10% of the clump's mass could potentially be dispersed (*Arce et al.*, 2010). This is in stark contrast with the theoretical work of *Matzner and McKee* (2000), which suggests ejection fractions of 50–70%. One possible explanation for the discrepancy is that a significant fraction of the outflow energy is deposited outside the cloud and is not detected by the observations. Another possibility is that the momentum in currently visible outflows might be only a fraction of the total momentum from all outflows throughout the life of the cloud. However, the difference between current and total momentum would need to be a factor of ~10 to reconcile observations with theory, and a more likely scenario is that outflows disperse some of the gas, while other mechanisms, such as stellar winds and UV radiation (e.g., *Arce et al.*, 2011) remove the rest. Certainly, further observations are needed in order to better understand the role of outflows in cluster gas dispersal.

Finally, we caution that most detailed observations of outflows have concentrated on relatively nearby (d < 500 pc) clusters, which, for the most part, are only forming low- to intermediate-mass stars. These regions do not accurately represent the galactic cluster population, introducing a bias. For example, the nearby regions do not reach the stellar densities or total stellar masses found in more distant regions. There have been a number of outflow observations of high-mass star-forming regions (e.g., *Beuther et al.*, 2002; *Gibb et al.*, 2004; *Wu et al.*, 2005; *Zhang et al.*, 2005; *López-Sepulcre et al.*, 2009; *Sánchez-Monge et al.*, 2013). However, such regions are typically more than 1–2 kpc away, so only the largest (and usually most powerful) outflows in each region are resolved. High angular resolution is essential to untangle the emission from the numerous outflows present in high-mass star-forming regions (e.g., *Beuther et al.*, 2003; *Leurini et al.*, 2009; *Varricatt et al.*, 2010). Consequently, a complete census of outflows (especially those from low-mass stars) in these far-away clusters has been difficult with current instruments, and more detailed observations of high-mass star-forming regions are needed in order to better understand the role of outflows in clusters. Such observations will be complicated by the fact that more massive regions will contain massive stars that produce other forms of feedback, as discussed below. This will require disentangling outflows from these other effects. The best approach may be to focus on regions that are dense and massive, but are also relatively young and thus have formed few or no massive stars yet. The Atacama Large Millimeter/sub-millimeter Array (ALMA) will certainly be instrumental in conducting these studies.

2.2. Radiation Pressure

2.2.1. Theory and radiation trapping. In clusters containing massive stars, a second form of momentum feedback comes into play: radiation pressure. Radiation feedback in general consists of the transfer of both energy and momentum from the radiation field generated by stars to the surrounding gas, but in this section we shall focus on the transfer of momentum. Except for photons above 13.6 eV, this transfer is mostly mediated by dust grains. In comparison to protostellar outflows, which deliver a momentum per unit mass of stars formed $V_{out} \sim 20$–40 km s^{-1} (over a time comparable to the accretion time, ~0.1 m.y.), a zero-age stellar population drawn from a fully sampled IMF produces a radiation field of 1140 $L_\odot$ per $M_\odot$ of stars (*Murray and Rahman*, 2010), which carries a momentum per unit time per unit stellar mass $\dot{V}_{rad} \sim 24$ km s^{-1} m.y.$^{-1}$. Thus, over the time $t_{form} \sim 1$–3 m.y. that it takes a star cluster to form, the total momentum per unit stellar mass $V_{rad} = \dot{V}_{rad} t_{form}$ injected by the radiation field can be competitive with or even exceed that of the outflows.

In the immediate vicinity of forming massive stars, on scales too small to fully sample the IMF, $\dot{V}_{rad}$ can be a factor of ~10–100 larger — an individual massive star can have a light-to-mass ratio in excess of 10^4 $L_\odot/M_\odot$, a factor of 10 higher than the IMF average. Conversely, the very steep mass-luminosity relation of stars (roughly $L \propto M^{3.5}$ near 1 $M_\odot$, although flattening at much higher masses) ensures that radiation pressure feedback is dominated by extremely massive stars. As a result, in clusters smaller than ~10^4 $M_\odot$ that do not fully sample the IMF, the light-to-mass ratio is typically much smaller than the mean of a fully sampled IMF (*Cerviño and Luridiana*, 2004; *da Silva et al.*, 2012). Radiation pressure is therefore less important compared to outflows, which to first order simply follow the mass. Even the Orion Nebula cluster, the region of massive star formation nearest to the Sun, has a light-to-mass ratio well below the expected value for a fully sampled IMF (*Kennicutt and Evans*, 2012). Thus radiation feedback from stars is likely to play a crucial role in the formation of individual high-mass stars, in the formation of massive clusters, and possibly even on galactic scales, but is likely to be unimportant in comparison to outflows in low-mass star clusters such as those closest to the Sun.

Radiation pressure begins to become significant for a population of stars once the light-to-mass ratio exceeds ~ 1000 $L_\odot/M_\odot$, which corresponds to a mass of ~ 20 $M_\odot$ for a single star (*Krumholz et al., 2009*), and $\sim 10^{3.5}$ $M_\odot$ for a star cluster that samples the IMF (*Cerviño and Luridiana, 2004; Krumholz and Thompson, 2012, 2013*). For such stars and star clusters, much of the luminosity will come in the form of ionizing photons, and thus radiation pressure and photoionization feedback will act together; we defer a more general discussion of the latter process to section 3.2. While most classical treatments of HII regions have ignored the effects of radiation pressure, recent analytic models by *Krumholz and Matzner* (2009), *Krumholz and Dekel* (2010), *Fall et al.* (2010), and *Murray et al.* (2010, 2011) have begun to include it, as did earlier models of starburst galaxies (*Thompson et al., 2005*). Their general approach is to solve a simple ordinary differential equation for the rate of momentum change of the thin shell bounding an evolving HII region due to both gas and radiation pressure.

In this treatment the authors introduce a factor f_{trap} [called τ_{IR} in the *Murray et al.* (2010, 2011) models] to account for the boosting of direct radiation pressure force by radiation energy trapped in the expanding shell. The momentum per unit time per unit stellar mass delivered by the radiation field to the gas is $f_{trap}\dot{V}_{rad}$; thus if $f_{trap} \gg 1$, then radiation pressure can be the dominant feedback mechanism almost anywhere massive stars are present. As discussed by *Krumholz and Thompson* (2012, 2013), this factor crudely interpolates between a flow driven purely by the momentum of the radiation field and one that is partly driven by a buildup of radiation energy, in analogy with the "explosive" case introduced in section 1.2. Values of $f_{trap} \gg 1$ occur if each photon undergoes many interactions before escaping, while $f_{trap} \sim 1$ corresponds to each stellar photon being absorbed only once, depositing its momentum, and then escaping. In spherical symmetry, it is straightforward to calculate f_{trap} by solving the one-dimensional equation of radiative transfer or some approximation to it (e.g., the diffusion approximation). Several authors have done this over the years, both analytically and numerically, and found $f_{trap} \gg 1$ (e.g., *Kahn*, 1974; *Yorke and Kruegel*, 1977; *Wolfire and Cassinelli*, 1986, 1987). Once one drops the assumption of spherical symmetry, however, the problem becomes vastly more complicated. As a result, the actual value of f_{trap} has been subject to considerable debate both observationally and theoretically, as we discuss below.

2.2.2. Observations of radiation pressure effects. Only a few observations to date have investigated the importance of radiation pressure feedback. *Scoville et al.* (2001) studied the central regions of M51 and found that radiation pressure from young clusters forming there exceeds their self-gravity. They proposed that this sets an upper limit on cluster masses of ~ 1000 $M_\odot$. More recent work has studied the giant HII

region 30 Doradus in the Large Magellanic Clouds (LMC) (*Lopez et al., 2011; Pellegrini et al., 2011*) (Fig. 1), as well as a larger sample of HII regions in the Magellanic Clouds (*Lopez et al., 2014*). The *Lopez et al.* (2011) study finds that f_{trap} is generally small, but that nonetheless radiation pressure dominates within 75 pc of the R136 cluster at the center of 30 Doradus. In contrast, *Pellegrini et al.* (2011) argue that radiation pressure is nowhere important in 30 Doradus. (They do not consider the pressure associated with any trapped IR radiation field, and thus do not address the question of f_{trap}.) This discrepancy is more a matter of definitions than of physics. *Lopez et al.* (2011) adopt the formal definition of radiation pressure as simply the trace of the radiation pressure tensor, while *Pellegrini et al.* (2011) attempt to compute the actual force exerted on matter by radiation. These two definitions produce very different results in the optically thin interior of 30 Doradus, since in a transparent medium the force experienced by the matter can be small even if the radiation pressure exceeds the gas pressure by orders of magnitude. Regardless of this difference in definition, both sets of authors agree that, in its present configuration, warm ionized gas pressure exceeds radiation pressure at the edge of the swept-up shell of material that bounds 30 Doradus. The two studies differ, however, on how this compares to the pressure of shock-heated gas, a topic we defer to section 3.1.

2.2.3. Simulations of radiation pressure feedback. There has been much more work on simulations of the effects of radiation pressure feedback. On the scales of the formation of individual stars (see the chapter by Tan et al. in this volume for more details), *Yorke and Sonnhalter* (2002) performed two-dimensional radiation-hydrodynamic simulations, and *Krumholz et al.* (2005b, 2007b, 2009, 2010) performed three-dimensional ones. The general picture established by these simulations, illustrated in Fig. 3, is that, despite radiation force formally being stronger than gravity on the small scales studied, radiation pressure nevertheless fails to halt accretion. Gravitational and Rayleigh-Taylor (RT) instabilities that develop in the surrounding gas channel the gas onto the star system through non-axisymmetric disks and filaments that self-shield against radiation while allowing radiation to escape through optically thin bubbles in the RT-unstable flow. The radiation-RT instability has been formally analyzed, and linear growth rates calculated, by *Jacquet and Krumholz* (2011) and *Jiang et al.* (2013).

Some details of this picture have recently been challenged by *Kuiper et al.* (2011, 2012), who use a more sophisticated radiative transfer method than *Krumholz et al.* (2009). Their improved treatment of the direct stellar radiation field increases the rate at which matter is driven away from the star, and as a result the radiation-RT instability does not have time to set in before matter is expelled. While this claim seems very likely to be true as applied to the idealized simulations performed by *Kuiper et al.* (2011, 2012), it is unclear whether it would apply in the more realistic case of a turbulent or magnetized protostellar core (*Myers et al., 2013*), or one with a disk that is gravitationally unstable as found in the

Krumholz et al. (2009) simulations. In these situations the initial "seeds" for the instability will be much larger, and the growth to nonlinear scale presumably faster. Unfortunately

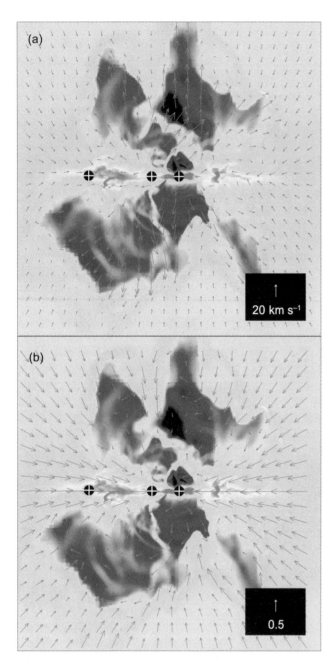

Fig. 3. Slices through a simulation of the formation of a 70-$M_\odot$ binary system, taken from *Krumholz et al.* (2009). Both panels show a region 6000 AU on a side; shades of gray show volume density on a scale from 10^{-20} to 10^{-14} g cm^{-3}, and plus signs show stars. **(a)** Arrows show the velocity; **(b)** arrows show the net radiation plus gravity force $f_{rad} + f_{grav}$, with the arrow direction indicating the force direction, and the arrow length scales by $|f_{rad} + f_{grav}|/|f_{grav}|$. These slices show the simulation at a time of 1.0 mean-density free-fall times, at which point the total stellar mass is ≈ 60 $M_\odot$ and the mass of the primary is ≈ 36 $M_\odot$.

Kuiper et al.'s (2011, 2012) numerical method is limited to treating the case of a single star fixed at the origin of the computational grid, and thus cannot simulate turbulent flows or provide a realistic treatment of the gravitational instabilities expected in massive star disks, which involve displacement of the star from the center of mass (*Kratter et al.*, 2010), or fragmentation of the disk into multiple stars (*Kratter and Matzner*, 2006; *Kratter et al.*, 2008; *Krumholz et al.*, 2007b; *Peters et al.*, 2010a,b, 2011). In reality the issue is likely moot in any event, as *Krumholz et al.* (2005a) show analytically and *Cunningham et al.* (2011) numerically that protostellar outflows should punch holes through which radiation can escape independent of whether radiation-RT instability occurs or not.

Much less work has been done on larger scales. A number of authors have introduced subgrid models for radiation pressure feedback, along with other forms of feedback (*Hopkins et al.*, 2011, 2012a,b; *Genel et al.*, 2012; *Agertz et al.*, 2013; *Aumer et al.*, 2013; *Stinson et al.*, 2013; *Bournaud et al.*, 2014). Others explicitly solve the equation of radiative transfer along rays emanating from stellar sources, but do not make any attempt to account for radiation that is absorbed and then reemitted (*Wise et al.*, 2012; *Kim et al.*, 2013a,b). However, none of these simulations include a fully self-consistent treatment of the interaction of the radiation field with the ISM, and as result they are forced to adopt a value of f_{trap}, either explicitly or implicitly. The outcome of the simulations depends strongly on this choice. At one extreme, some authors adopt values of $f_{trap} \gg 1$, in some cases $f_{trap} \sim 50$ (e.g., *Hopkins et al.*, 2011; *Aumer et al.*, 2013), and find that radiation pressure is a dominant regulator of star formation in rapidly-star-forming galaxies. At the other extreme, models with more modest values of f_{trap} find correspondingly modest effects (e.g., *Kim et al.*, 2013b; *Agertz et al.*, 2013). The only large-scale fully radiation-hydrodynamic simulations published thus far are those of *Krumholz and Thompson* (2012, 2013), who find that real galaxies on large scales likely have $f_{trap} \sim 1$, because radiation-RT instability reduces the efficiency of radiation-matter coupling far below the value for a laminar radiation distribution. [*Socrates and Sironi* (2013) have also argued against values of $f_{trap} \gg 1$, for somewhat different reasons.] There is some potential worry about the treatment of the direct radiation field in the *Krumholz and Thompson* (2012, 2013) models, following the points made by *Kuiper et al.* (2012). *Krumholz and Thompson* (2012, 2013) argue that if $f_{trap} \gg 1$ then the direct radiation pressure force is by definition unimportant, and thus that their treatment of radiation is adequate for investigating whether f_{trap} is indeed large. The simulations of *Jiang et al.* (2013), which use a more sophisticated treatment of radiation pressure than either *Krumholz and Thompson* (2012, 2013) or *Kuiper et al.* (2012), are qualitatively consistent with this conclusion. However, there is clearly a need for further numerical investigations with more accurate radiation-hydrodynamic methods to fully pin down the correct value of f_{trap} for use in subgrid models.

3. EXPLOSIVE FEEDBACK

3.1. Main-Sequence Winds from Hot Stars

3.1.1. Budget and relative importance. While all stars that accrete from disks appear to produce protostellar outflows, only those with surface temperatures above ~2.5 × 10^4 K produce strong winds (*Vink et al.,* 2000). Main-sequence stars reach this temperature at a mass of ~40 $M_\odot$, and stars this massive have such short Kelvin-Helmholtz times that, even for very high accretion rates, they reach their main-sequence surface temperatures while still forming (*Hosokawa and Omukai,* 2009). As a result, hot stellar winds will begin to appear very early in the star-formation process. These winds carry slightly less momentum than the stellar radiation field (*Kudritzki et al.,* 1999); a calculation using Starburst99 (*Leitherer et al.,* 1999; *Vázquez and Leitherer,* 2005) gives $\dot{V}_{msw}$ = 9 km s^{-1} m.y.$^{-1}$ per unit mass of star formed for a zero-age population.

This estimate is based on nonrotating stars of solar metallicity, using the wind prescriptions of *Leitherer et al.* (1992) and *Vink et al.* (2000). Stars of lower metallicity will have significantly lower wind momentum fluxes, due to the reduced efficiency of line-driving in a stellar atmosphere containing fewer heavy elements (*Vink et al.,* 2001). Conversely, stellar rotation can increase the instantaneous momentum flux by a factor of a few, and the integrated momentum over the stellar lifetime by a factor of ~10, with significant uncertainty arising from the poorly known distribution of birth rotation rates (*Maeder and Meynet,* 2000, 2010; *Ekström et al.,* 2012). There is also significant uncertainty on the momentum budget at ages greater than a few million years, stemming from our poor knowledge of how exactly massive stars evolve into luminous blue variables, red and yellow supergiants, and Wolf-Rayet stars.

Despite these uncertainties, even the highest plausible stellar wind momentum estimates yield injection rates at most comparable to the stellar radiation field. Thus, if stellar winds represent a momentum-driven form of feedback, they should only provide a mild enhancement of radiation pressure. However, hot star winds can have terminal velocities of several times 10^3 km s^{-1} (e.g., *Castor et al.,* 1975b; *Leitherer et al.,* 1992), so when they shock against one another or the surrounding ISM, the post-shock temperature can exceed ~10^7 K. At this temperature radiative cooling times are long (*Castor et al.,* 1975a; *Weaver et al.,* 1977), so shocked stellar winds might build up an energy-driven, adiabatic flow that would make them far more effective than radiation. On the other hand, they might also leak out of their confining shells of dense interstellar matter, which would greatly reduce the pressure build-up and lead to something closer to the momentum-limited case.

Whether shocked stellar wind gas does actually build up an energy-driven flow and thereby become an important feedback mechanism has been subject to considerable debate, and we discuss the available observational and theoretical evidence in the next section. To frame the discussion, consider an H$_{II}$ region with a volume V and a pressure at its outer edge P, within which some volume V_w is occupied by X-ray emitting post-shock wind gas at pressure P_w. *Yeh and Matzner* (2012) introduce the wind parameter

$$\Omega \equiv \frac{P_w V_w}{PV - P_w V_w} \qquad (2)$$

as a measure of the relative strength of winds. The virial theorem implies that PV is what controls the large-scale dynamics, so $\Omega \gg 1$ [as expected for models such as those of *Castor et al.* (1975a) and *Weaver et al.* (1977)] indicates that the large-scale dynamics are determined primarily by winds, while $\Omega \ll 1$ [as expected in models where the post-shock wind gas undergoes free expansion (e.g., *Chevalier and Clegg,* 1985)] indicates they are unimportant. Note that P and V include any form of feedback that contributes pressure to and occupies volume within the H$_{II}$ region, not just the pressure and volume associated with the ~10^4 K photoionized gas (see below). Thus, Ω should be thought of as measuring the contribution of stellar winds to the total dynamical budget. As we discuss in the next section, the true value of Ω remains an open question in both observations and theory.

3.1.2. Observations and theory. Observations of stellar wind feedback were revolutionized by the launch of the Chandra X-Ray Observatory, which for the first time made it possible to detect X-ray emission from the hot post-shock wind gas in H$_{II}$ regions (*Townsley et al.,* 2003). The sample of H$_{II}$ regions with X-ray measurements includes M17 and the Rosette Nebula (*Townsley et al.,* 2003), the Carina Nebula (*Townsley et al.,* 2011), the Tarantula Nebula/30 Doradus (*Townsley et al.,* 2006; *Lopez et al.* 2011; *Pellegrini et al.,* 2011), and a few tens of smaller H$_{II}$ regions in the Magellanic Clouds (*Lopez et al.,* 2014). A measurement of the X-ray luminosity and spectrum can be used to infer P_w, at least up to an unknown volume filling factor. The observations conducted to date strongly rule out the largest predicted values of Ω, but the exact value is still debated. In 30 Doradus, probably the best-studied case, *Lopez et al.* (2011) and *Pellegrini et al.* (2011) report similar estimates for the pressures of 10^4-K photoionized gas and radiation, but *Pellegrini et al.*'s (2011) estimate of the X-ray-emitting gas pressure is ~2 orders of magnitude larger. Most of this discrepancy is due to differing assumptions about the volume-filling factor of the emitting gas, with *Lopez et al.* (2014) assuming it is on the order of unity and *Pellegrini et al.* (2011) arguing for a much smaller value, which would imply higher P_w but also lower V_w. *Lopez et al.*'s (2014) reported values give $\Omega \ll 1$, while the value of Ω based on *Pellegrini et al.*'s (2011) modeling is unclear because they do not report values for V_w. However, they likely obtain $\Omega \ll 1$ too, since, all other things being equal, a reduction in the filling factor tends to lower V_w more than it raises P_w.

While X-rays are the most direct way of constraining Ω, some optical and IR line ratios are sensitive to it as well (*Yeh and Matzner,* 2012; *Yeh et al.,* 2013; *Verdolini et al.,* 2013). There are significant modeling uncertainties associated with the assumed geometry, but these are quite different from the filling factor issues that hamper X-ray measurements. *Yeh and Matzner* (2012) find that available data favor $\Omega < 1$, but this is a preliminary analysis.

Several authors have also made theoretical models of wind feedback. Much of this work has treated the ISM surrounding the star as a simple uniform-density medium, and has focused on instead on the circumstellar medium, within which there can be complex interactions between components of the wind launched at different stages of stellar evolution (*Garcia-Segura et al.,* 1996a,b; *Freyer et al.,* 2003, 2006; *Toalá and Arthur,* 2011). This work, while clearly important for the study of circumstellar bubbles, offers limited insight into how wind feedback affects the process of star formation. Similarly, building on the classical *Castor et al.* (1975a) and *Weaver et al.* (1977) models, a number of authors have made increasingly sophisticated spherically symmetric models of stellar wind bubbles, including better treatments of conduction and radiative cooling (*Capriotti and Kozminski,* 2001; *Tenorio-Tagle et al.,* 2007; *Arthur,* 2012; *Silich and Tenorio-Tagle,* 2013). They find that stellar wind feedback can be dominant, but since these models necessarily exclude leakage, it is unclear how much weight to give to this conclusion.

Studies that include multi-dimensional effects in a complex, star-forming ISM are significantly fewer. *Harper-Clark and Murray* (2009) present analytic models for cold shells driven by a combination of ionized gas pressure, radiation pressure, and wind pressure, including parameterized treatments of wind leakage. These models are able to fit the observations by adopting fairly strong leakage of hot gas, and the required values of the leakage parameter suggest that hot gas is subdominant compared to other forms of feedback. Two-dimensional simulations by *Tenorio-Tagle et al.* (2007) and three-dimensional ones by *Dale and Bonnell* (2008) and *Rogers and Pittard* (2013) generally find that leakage is a very significant effect, as illustrated in Fig. 4, with a majority of the injected wind energy escaping rather than being used to do work on the cold ISM. However, even with these losses, the multi-dimensional grid-based simulations do show that a wind of hot gas is eventually able to entrain the cold ISM via Kelvin-Helmholtz instabilities and eventually remove all the cold gas from a forming star cluster. [Simulations using older formulations of smoothed particle hydrodynamics (SPH) cannot capture this effect due to their difficulties in modeling the Kelvin-Helmholtz instability (see *Agertz et al.,* 2007). Newer SPH methods can overcome this limitation (e.g., *Price,* 2008; *Read and Hayfield,* 2012; *Hopkins,* 2013), but all SPH simulations of stellar wind feedback published to date use the older methods.] These simulations, however, do not include radiation pressure or other forms of feedback, and it is unclear if winds would be dominant in competition with other mechanisms.

In addition, all the multidimensional simulations performed to date lack the resolution and the sophisticated microphysics required to handle a number of other poten-

tially important effects. For example, the development of a turbulent interface between the cold and hot gas might greatly enhance the rate of conduction, thus lowering the temperature in the hot gas to $\sim 10^5$ K so that radiative losses via far-UV metal lines become rapid (*McKee et al.,* 1984). Another possibly important effect is mixing of dust grains into the hot gas, where, until they are destroyed by sputtering, they can remove energy via collisional heating followed by thermal radiation. In light of the continuing controversy over the importance of stellar winds, reinvestigation of these topics using modern hydrodynamic techniques is urgently needed.

3.2. Photoionization Feedback

Stars with masses ≥ 10 $M_\odot$ emit very large quantities of ionizing photons, creating ionized bubbles — HII regions. Equilibrium between heating and cooling processes inside HII regions results in them having remarkably constant temperatures of $\approx 10^4$ K and internal sound speeds of ≈ 10 km s^{-1} (*Osterbrock and Ferland,* 2006). The overpressure in the bubble causes it to expand at velocities on the order of the sound speed. In a uniform medium, this leads to the well-known Spitzer solution (*Spitzer,* 1978). As is the case with stellar winds, the interaction between the cold molecular gas and the overpressured, expanding hot gas is physically complex, and thus it is not trivial to assign a momentum

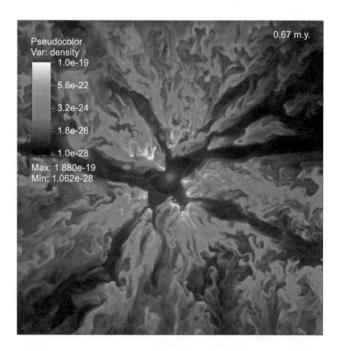

Fig. 4. Slice through a simulation of a molecular clump with hot stellar winds launched by a central star cluster. The region shown is 32 pc on a side at a time 0.67 m.y. after wind launching begins. Shades of gray show the density; the low-density channels shown in black are filled with hot, escaping wind material. Adapted from Fig. 3 of *Rogers and Pittard* (2013).

budget to it and compare it to other sources of feedback. We will return to this topic below.

Observationally, H II regions are extremely bright at radio wavelengths (due to radio recombination lines and bremsstrahlung) and in the IR (due to reprocessing of stellar radiation by dust), making it possible to study them over large distances. H II regions are often divided into (ultra- or hyper-) compact and diffuse types, and this was originally thought to be an evolutionary sequence resulting from expansion. However, observations by, e.g., *Wood and Churchwell* (1989) and *Kurtz et al.* (1994) revealed that UC H II regions rarely resemble classical Strömgren spheres. Common morphologies are cometary, core-halo or shell-like, and irregular. Simulations suggest that these morphologies result from variations in the mass distribution in the immediate vicinity of the ionizing stars, and are likely to be variable over approximately a thousand years or even shorter timescales (*Peters et al.,* 2010a). *Kurtz et al.* (1999) and *Kim and Koo* (2001) found that many compact H II regions are embedded in larger diffuse ionized regions, leading *Kim and Koo* (2003) to propose that UC H II regions are dense cores embedded in champagne flows. More recent observational work has concentrated on the interaction of H II regions with IR dust bubbles (e.g., *Watson et al.,* 2008; *Deharveng et al.,* 2010; *Anderson et al.,* 2011), molecular gas (e.g., *Anderson et al.,* 2011), and stellar winds (e.g., *Townsley et al.,* 2003).

On small scales, photoionization is sometimes suggested to limit the growth of OB stars. However, *Walmsley* (1995) showed that the expansion of an H II region could be stalled or even reversed by an accretion flow. *Keto* (2003) generalized this result by showing that accretion onto an ionizing star can proceed through a gravitationally trapped H II region; observations of ionized accretion flows support this picture (*Keto and Wood,* 2006; *Klaassen and Wilson,* 2007). Simulations by *Peters et al.* (2010a,b) also found that ionizing sources were unable to disrupt accretion flows onto them until material was drained from the flows by other, lower-mass accretors. Moreover, *Hosokawa et al.* (2010, 2012) point out that accretion onto massive stars at high but not unreasonable rates causes the stars to expand, cooling their photospheres and reducing their ionizing fluxes, further easing accretion. *Klassen et al.* (2012) model the consequences of accretion-induced expansion and conclude that the shrinking of the H II region due to the drop in ionizing flux may be observable with facilities such as the Extended Very Large Array (EVLA) or ALMA. Statistical analysis of the correlation between the bolometric luminosities of massive young stellar objects and the ionizing photon fluxes required to drive the H II regions around them provides direct evidence for this effect (*Mottram et al.,* 2011; *Davies et al.,* 2011).

At larger scales, there are three major outstanding questions regarding H II regions. The first — whether they are able to trigger star formation — is discussed in section 5.2. The second is whether H II region expansion is able to drive GMC turbulence. There have been several simulations of H II region expansion in turbulent clouds (e.g., *Mellema et al.,* 2006; *Mac Low et al.,* 2007; *Tremblin et al.,* 2012b;

Dale et al., 2012b, 2013b), but few authors have addressed this issue in detail. *Mellema et al.* (2006) simulated the expansion of an H II region in a turbulent cloud and found that substantial kinetic energy was deposited in the neutral gas, although they did not show explicitly that this actually kept the cold gas turbulent in the sense of maintaining a self-similar velocity field over some range of scales. In their simulations of an ionizing source inside a fractal cloud, *Walch et al.* (2012) also showed that the kinetic energy of the cold gas was strongly influenced by ionization — more so than by gravity — and that a large fraction of this energy resided in random motions, which they identified as turbulence.

Gritschneder et al. (2009) simulated plane-parallel photoionization of a turbulent box and analyzed the power spectra of the velocity field both with and without the influence of feedback. They found that ionization was an efficient driver of turbulence, although with a substantially flatter power spectrum than the Kolmogorov velocity field with which the box was seeded. However, because they were irradiating the whole of one side of their simulation domain, they were driving turbulence on the largest scale available and it is not clear that this result applies more generally. In the case of point-like ionizing sources inside a large cloud, the H II regions must grow to fill a large fraction of the system volume in order for turbulent driving to be effective on the largest scales and for the turbulent cascade to operate. Semi-analytic models of GMCs including H II region feedback by *Krumholz et al.* (2006) and *Goldbaum et al.* (2011), and simulations of H II regions evolving in isolated clouds by *Walch et al.* (2012), *Dale et al.* (2012b), and *Dale et al.* (2013b) suggest that is often likely to be the case, consistent with observations that many H II regions are champagne flows.

The final question is whether H II regions can disrupt protoclusters and terminate star formation at low efficiencies. *Whitworth* (1979), *Franco et al.* (1990), *Franco et al.* (1994), *Matzner* (2002), *Krumholz et al.* (2006), and *Goldbaum et al.* (2011) found that photoionization should be effective in destroying clouds. However, these authors considered the effects of ionization on smooth clouds. The picture from recent numerical simulations of H II regions expanding in highly structured clouds is less clear. While O stars located on the edges of clouds can drive very destructive champagne flows (e.g., *Whitworth,* 1979), massive stars are usually to be found embedded deep inside clouds. In addition, molecular clouds usually possess complex density fields, and the massive stars are often located inside the densest regions. *Dale et al.* (2005) found that the influence of a photoionizing source could be strongly limited by dense large-scale structures and accretion flows. *Walch et al.* (2012) found that ionization was very destructive to $\sim 10^4$ M$_\odot$ fractal clouds on timescales of 1 m.y., but *Dale and Bonnell* (2011) and *Dale et al.* (2012b, 2013b) simulated expanding H II regions in turbulent clouds with a range of radii (~ 1–100 pc) and masses (10^4–10^6 M$_\odot$) and found that the influence of ionization depends critically on cloud escape velocities. It is very effective in clouds with escape speeds well below ~10 km s⁻¹, but becomes ineffective once the escape speed reaches this value. Figure 5 shows an example. In high-escape-speed clouds, radiation pressure may dominate instead (see section 2.2).

A final caution is that, with a few exceptions (*Krumholz et al.*, 2007a; *Arthur et al.*, 2011; *Peters et al.*, 2011; *Gendelev and Krumholz*, 2012), the simulations of Hɪɪ regions performed to date ignore magnetic fields, and the coupling between magnetic fields and ionizing radiation can generate unexpected effects. We discuss this further in section 5.1.

3.3. Supernovae

Supernovae represent a final source of explosive feedback. For every 100 $M_\odot$ of stars formed, ~1 star will end its life in a type II supernova; the supernovae are distributed roughly uniformly in time from ≈4 to 40 m.y. after the onset of star formation, with a slight peak during the first million years after the explosions begin (e.g., *Matzner,* 2002). Each supernova yields ~10^{51} erg, much of which ends up as thermal energy in a hot phase with a long cooling time. However, as pointed out by *Krumholz and Matzner* (2009) and *Fall et al.* (2010), supernovae are of quite limited importance on the scales of individual star clusters simply due to timescale issues. The first supernovae do not occur until roughly 4 m.y. after the onset of star formation. In comparison, the crossing time of a protocluster is $t_{cr} = 2R/\sigma$, where R and σ are the radius and velocity dispersion, respectively. Since protoclusters have virial ratios $\alpha_{vir} \approx 1$, and form a sequence of roughly constant surface density with $\Sigma \sim$ 1 g cm⁻² (*Fall et al.,* 2010), the crossing time varies with protocluster mass as $t_{cr} \approx 0.25(M/10^4\ M_\odot)^{1/4}$ m.y. (*Tan et*

al., 2006). Thus protoclusters have crossing times smaller than the time required for supernovae to start occurring unless their masses are ≥10^6 $M_\odot$. In the absence of feedback, protocluster gas clouds convert the great majority of their gas to stars in roughly a crossing time, so supernovae cannot be the dominant form of feedback in clusters smaller than ~10^6 $M_\odot$ — such clusters would convert all of their mass to stars before the first supernova occurred, unless some other mechanism were able to delay star formation for several crossing times and allow time for supernovae to begin.

This conclusion is consistent with observations of the most massive clusters that host stars that will end their lives as supernovae. In 30 Doradus there is only one detectable supernova remnant (visible as a bright spot in the lower right corner of Fig. 1) (*Lopez et al.,* 2011), and its radius is far smaller than that of the evacuated bubble. Similarly, in Westerlund 1, which has also ejected its central gas, there has been a supernova (*Muno et al.,* 2006a), but no corresponding supernova remnant has been detected (*Muno et al.,* 2006b). This is likely because the gas had already been expelled before the supernova occurred, and thus the ejecta have yet to encounter material dense enough to produce on observable shock. However, we emphasize that the conclusion that supernovae are unimportant for clusters does not apply on the larger scales of diffuse giant molecular clouds or galaxies, which have crossing times that are comparable to or significantly larger than the lifetime of a massive star.

4. THERMAL FEEDBACK

Although radiative transfer was included in some of the very earliest calculations of star formation (e.g., *Larson,* 1969), the importance of thermal feedback for star forma-

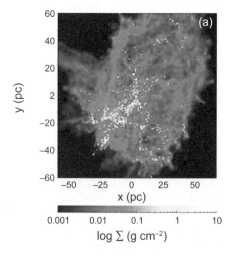 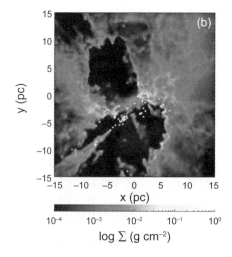

Fig. 5. Column density maps from two simulations of photoionization feedback. **(a)** Cloud with an escape speed >10 km s⁻¹, within which photoionization has done little to inhibit star formation or remove mass. **(b)** Cloud with an escape speed <10 km s⁻¹, where a majority of the mass has been ejected by photoionization feedback. Adapted from *Dale et al.* (2012b).

tion has only been recognized recently. This is true even for the case of massive star formation, where for a long time in the literature "radiative feedback" meant radiation pressure (see section 2.2), not radiative heating. This began to change in the 1980s, when three-dimensional calculations of low-mass star formation began to include radiative transfer using the Eddington approximation (*Boss*, 1983, 1984, 1986). Such calculations showed that perturbed molecular cloud cores containing several Jeans masses initially readily fragmented to form binary systems, and that dynamical collapse and fragmentation is terminated by the thermal heating after the clouds become opaque [the so-called opacity limit for fragmentation (*Low and Lynden-Bell*, 1976; *Rees*, 1976)]. However, over the past few years it has been realized that thermal feedback could play a much greater role in star formation than simply setting the minimum fragment mass.

4.1. Origin of Thermal Feedback

Thermal feedback is inevitable in the star-formation process as gravitational potential energy is converted to kinetic and thermal energy during collapse. Initially, the rate of compression of the gas is low, and the additional thermal energy is quickly radiated, resulting in an almost isothermal collapse (*Larson*, 1969). However, as the rate of collapse increases, compressional heating eventually exceeds the radiative losses (*Masunaga and Inutsuka*, 1999) and the collapse at the center of the cloud transitions to an almost adiabatic phase leading to the formation of a pressure-supported object with a mass of a few Jupiter masses and a radius of ~5 AU [the so-called first hydrostatic core (*Larson*, 1969)]. The first core accretes through a supercritical shock from which most of the accretion luminosity is radiated away (*Tomida et al.*, 2010a,b; *Commerçon et al.*, 2011b; *Schönke and Tscharnuter*, 2011). This results in heating of the surrounding gas, which, although modest, may affect fragmentation (*Boss et al.*, 2000; *Whitehouse and Bate*, 2006).

Once the center of the first hydrostatic core reaches ≈2000 K, the dissociation of molecular hydrogen initiates a second dynamical collapse, resulting in the formation of the second, or stellar, core (*Larson*, 1969; *Masunaga and Inutsuka*, 2000). The stellar core forms with a few Jupiter masses of gas and a radius of ~2 $R_{\odot}$. Because gravitational potential energy scales inversely with radius, the formation of the stellar core is associated with a dramatic increase in the luminosity of the protostar and significant heating to distances of hundreds of astronomical units from the stellar core (e.g., *Whitehouse and Bate*, 2006). Recent radiation hydrodynamical simulations of stellar core formation have shown that this burst of thermal feedback, due to accretion rates of ~10^{-3} $M_{\odot}$ yr^{-1}, which last for a few years, can be great enough to launch pressure-driven bipolar outflows (in the absence of magnetic fields) as gas and dust heated by the accretion luminosity expands and bursts out of the first hydrostatic core in which the stellar core is embedded (*Bate*, 2010, 2011; *Schönke and Tscharnuter*, 2011).

Once a stellar core forms, there are three sources of thermal feedback: (1) radiation originating from the core itself, (2) luminosity from accretion onto the star, and (3) luminosity from continued collapse of the cloud and disk accretion. For low-mass protostars (≲3 $M_{\odot}$) accreting at rates ≳10^{-6} $M_{\odot}$ yr^{-1}, accretion luminosity dominates both the intrinsic stellar luminosity (e.g., *Palla and Stahler*, 1991, 1992; *Hosokawa and Omukai*, 2009) and the luminosity of the larger-scale collapse (*Offner et al.*, 2009; *Bate*, 2012). For example, the accretion luminosity

$$L_{acc} \approx \frac{GM_* \dot{M}_*}{R_*} \qquad (3)$$

of a star of mass $M_{\odot} = 1$ $M_{\odot}$ with a radius of $R_* = 2$ $R_{\odot}$ (e.g., *Hosokawa and Omukai*, 2009) accreting at $\dot{M}_* = 1 \times 10^{-6}$ $M_{\odot}$ yr^{-1} is ≈15 $L_{\odot}$, whereas the luminosity of the stellar object itself is ≈1 $L_{\odot}$. For intermediate-mass protostars ($M_* > 3$–9 $M_{\odot}$), whether the accretion luminosity or the intrinsic luminosity dominates depends on accretion rate, while for masses greater than ≈9 $M_{\odot}$ the intrinsic stellar luminosity dominates for all reasonable accretion rates (≲10^{-3} $M_{\odot}$ yr^{-1}).

Several authors have considered the impact of accretion luminosity on the temperature distribution in protostellar cores of a variety of masses, both analytically and numerically (*Chakrabarti and McKee*, 2005; *Krumholz*, 2006; *Robitaille et al.*, 2006, 2007). These models show that even sub-solar-mass protostars could heat the interior of cores to temperatures in excess of 100 K to distances ~100 AU or 30 K to distances ~1000 AU. *Krumholz* (2006) points out that this could significantly inhibit fragmentation of massive cores to form stellar groups and multiple star formation in low-mass cores.

Due to the inverse radial dependence of equation (3), the luminosity from accretion onto a star will generally dominate that produced by either the accretion disk or continued collapse on larger scales. However, there are a large number of uncertainties that make accurate determination of the luminosity difficult. The evolution of the stellar core depends both on its initial structure at formation and on how much energy is advected into the star as it accretes (*Hartmann et al.*, 1997; *Tout et al.*, 1999; *Baraffe et al.*, 2009; *Hosokawa et al.*, 2011). Different assumptions give different intrinsic luminosities and stellar radii. The latter uncertainty translates into an uncertainty in the accretion luminosity. Furthermore, an unknown fraction of the energy will drive jets and outflows rather than being emitted as accretion luminosity. Furthermore, protostars may accrete much of their mass in bursts (see the chapter by Audard et al. in this volume). If this is the case, protostars may spend the majority of their time in a low-luminosity state with only brief periods of high luminosity. Thus, *Stamatellos et al.* (2011, 2012) recently argued that numerical calculations assuming continuous radiative feedback may overestimate its effects. For a detailed discussion of these issues and numerical issues related to modeling protostars with sink particles, see *Bate* (2012).

4.2. Observations of Thermal Feedback

Observations provide strong evidence for thermal feedback. The first indirect hints came from observations of a narrow CO(6–5) component around low-mass young stellar objects (YSOs), which models suggested was produced by a ~1000-AU scale region heated to ~100 K by UV photons interacting with the walls of an outflow cavity (*Spaans et al.*, 1995). More recently, several new telescopes and instruments have allowed us to obtain much larger samples to study the effects of thermal feedback on subparsec scales. Combining information on the thermal structure, thermodynamic properties, and fragmentation of cluster-forming clumps can potentially constrain the importance of thermal feedback on cluster formation (*Zhang et al.*, 2009; *Zhang and Wang*, 2011; *Longmore et al.*, 2011; *Wang et al.*, 2011).

In the vicinity of low-mass stars, *van Kempen et al.* (2009a,b) used APEX-CHAMP$^+$ to obtain spatially resolved maps of high-J CO lines that trace warm gas around ~30 nearby sources; *van Kempen et al.* (2010) complement this with Herschel/Photodetector Array Camera and Spectrometer (PACS) spectroscopy to study the spectra in detail. *Visser et al.* (2012) and *Yıldız et al.* (2012) model the data and confirm that they are consistent with heating by a combination of stellar photons and UV produced by shocks when the jet interacts with the circumstellar medium. All these observations point to the conclusion that ~1000-AU scale, ~100-K regions are ubiquitous around the outflow cavities produced by embedded low-mass protostars.

In more massive regions, *Longmore et al.* (2011) analyzed the density and temperature structure of the massive protocluster G8.68-0.37 with an estimated mass of ≈1500 M$_\odot$. Combining Australia Telescope Compact Array (ATCA) and Submillimeter Array (SMA) observations with radiative transfer modeling, they found radial temperature profiles T $\propto$ r$^{-0.35}$ with temperatures of ≈40 K at distances of 0.3 pc from the cluster center. *Zhang et al.* (2009) and *Wang et al.* (2012a) used Very Large Array (VLA) ammonia observations of another massive protocluster, G28.34+0.06, and found that warmer gas seemed to be associated with outflows, but that the protostars themselves did not seem to provide significant thermal feedback on scales of 0.06 pc. The chemical species observed with Herschel also provide evidence for thermal feedback associated with outflows (*Bruderer et al.*, 2010).

By using multi-wavelength imaging, temperature maps of star-forming regions can be constructed. *Hatchell et al.* (2013) used James Clerk Maxwell Telescope (JCMT) Submillimetre Common-User Bolometer Array-2 (SCUBA-2) observations to map the temperature structure in NGC 1333, detecting heating from a nearby B star, other young IR/optical stars in the cluster, and embedded protostars. Temperatures ranged from 40 K at distances of a few thousand astronomical units from some of the more luminous stars to

20–30 K on scales of ≈0.2 pc in the northern portion of the star-forming region. They argued that heating from existing stars may lead to increased masses to the next generation of stars to be formed in the region. *Sicilia-Aguilar et al.* (2013) used multi-wavelength Herschel observations to create temperature maps of the Corona Australis region, also detecting heating from protostars on scales of thousands of astronomical units.

Therefore, from both theory and observation it is now clear that even low-mass protostars produce substantial thermal feedback on the gas and dust surrounding them. As we will discuss in the next two sections, thermal feedback may be a crucial ingredient in producing the stellar IMF.

4.3. Influence on the Initial Mass Funtion: Low-Mass End

About 15 years ago it became computationally feasible to perform hydrodynamic simulations of the gravitational collapse of molecular clouds to produce groups of protostars (e.g., *Bonnell et al.*, 1997; *Klessen et al.*, 1998; *Bate et al.*, 2003). Early simulations treated the gas either isothermally or using simple barotropic equations of state. These calculations were able to produce IMF-like stellar mass distributions, but with two major problems. First, simulations systematically over-produced brown dwarfs compared with observed Galactic star-forming regions (*Bate et al.*, 2003; *Bate and Bonnell*, 2005; *Bate*, 2009a), particularly in the case of decaying turbulence (*Offner et al.*, 2008). Second, the characteristic mass of stars formed in the simulations was proportional to the initial Jeans mass (*Klessen and Burkert*, 2000, 2001; *Bate and Bonnell*, 2005; *Jappsen et al.*, 2005; *Bonnell et al.*, 2006), while there is no firm evidence for such environmental dependence in reality (*Bastian et al.*, 2010).

To explain why the characteristic stellar mass does not vary strongly with environment, several authors have suggested that it might be set by microphysical processes that cause the equation of state to deviate subtly from isothermal, e.g., a changeover from cooling being dominated by line emission to being dominated by dust emission at some characteristic density (*Larson*, 1985, 2005). Simulations using such non-isothermal equations of state show that they are capable of producing a characteristic stellar mass that does not depend on the mean density or similar properties of the initial cloud (*Jappsen et al.*, 2005). However, these models neglected the effects of stellar radiative feedback, which, as discussed above, heats the gas near existing protostars and inhibits fragmentation. Indeed, radiation-hydrodynamic simulations show that, once radiative feedback is included, the proposed non-isothermal equations of state do not provide a good description of the actual temperature structure (*Krumholz et al.*, 2007b; *Urban et al.*, 2009).

The first cluster-scale calculations to include radiative transfer (*Bate*, 2009b; *Offner et al.*, 2009; *Urban et al.*, 2010) showed that this drastically reduces the amount of fragmentation even in regions that produce only low-mass stars. As a result, the typical stellar mass is greater than without thermal feedback, greatly reducing the ratio of brown dwarfs to stars and bringing it into good agreement with the observed Galactic IMF. However, potentially of even more importance, *Bate* (2009b) also showed that radiative feedback apparently removed the dependence of the IMF on the initial Jeans mass of the cloud, and therefore could be a crucial ingredient for producing an invariant IMF. *Krumholz* (2011) took Bate's argument even further and proposed that the characteristic mass of the IMF may be linked, through thermal feedback, to a combination of fundamental constants.

More recent radiation-hydrodynamic simulations of larger clouds that produce hundreds of protostars have yielded populations of protostars whose mass distributions are statistically indistinguishable from the observed IMF (*Bate,* 2012; *Krumholz et al.,* 2012a). Figure 6 shows some example results. This led a number of authors to conclude that gravity, hydrodynamics, and thermal feedback may be the primary ingredients for producing the statistical properties of low-mass stars. However, none of the above simulations included magnetic fields.

4.4. Influence on the Initial Mass Function: High-Mass End

Thermal feedback from protostars is fundamentally a local effect, and is stronger for higher mass protostars and/or greater accretion rates (equation (3)). Therefore, it becomes even more important if the molecular cloud core has a high density and/or produces a massive protostar. In the absence of thermal feedback, a massive dense molecular cloud core is prone to fragment into many protostars since the Jeans mass scales inversely with the square root of the density. Such fragmentation typically leads to a dense cluster of protostars

that evolve according to competitive accretion, resulting in a cluster with a wide range of stellar masses (e.g., *Bonnell et al.,* 2004). However, the inclusion of thermal feedback can substantially alter this picture since the heating may be strong enough to exclude the vast majority of fragmentation, with the result that only a few massive stars are produced rather than a populous cluster (*Krumholz et al.,* 2007b). This implies that massive stars may be preferentially produced in regions with high densities (*Krumholz and McKee,* 2008; *Krumholz et al.,* 2010).

While thermal feedback can be important for reducing the level of fragmentation and producing massive stars, in some calculations it can be too dominant. *Krumholz et al.* (2011) found that as the star formation proceeds in a dense cluster-forming cloud and the thermal feedback becomes more intense, the rate of production of new protostars can decrease. Since the protostars in the cloud continue to accrete more and more mass, this can lead to a situation in which the characteristic mass of the stellar population increases with time. This means that the stellar mass distribution evolves with time, rather than always being consistent with the observed IMF as in calculations of stellar clusters forming in lower-density clouds (*Bate,* 2012). In the long term, this would result in a top-heavy IMF. Combining the effects of radiative feedback and protostellar outflows may provide a way to reduce this effect, however (see section 5.1).

Peters et al. (2010a,b, 2011) obtained a somewhat different result in their simulations, finding that radiative feedback only modestly reduced the degree of fragmentation. They term the phenomenon they observe "fragmentation-induced starvation," a process by which, if a secondary protostar manages to form in orbit around a massive protostar, it may reduce the growth rate of the massive protostar by accreting material that it would otherwise accrete. *Girichidis et al.* (2012a,b) show that fragmentation-induced starvation

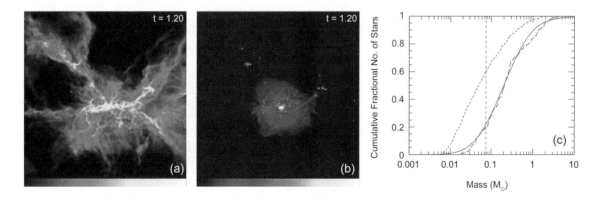

Fig. 6. Results from a radiation-hydrodynamic simulation of the formation of a star cluster in a 500-M$_\odot$ gas cloud by *Bate* (2012). **(a)** Gas column density Σ at the end of the simulation on a logarithmic scale from $-1.4 < \log \Sigma < 1.0$ with Σ in g cm^{-2}. **(b)** Density-weighted temperature on a linear scale from 9 to 50 K. **(c)** Cumulative mass distribution of the stars at the end of the simulation (dashed line), and for comparison the observed IMF [solid line (*Chabrier,* 2005)] and the mass distribution from a simulation with identical initial conditions and evolved to the same time using a barotropic equation of state [dotted line (*Bate,* 2009a)]. Figures adapted from *Bate* (2012).

also occurs in more general geometries, albeit in simulations that do not include radiative feedback. The relatively modest effects of radiative feedback in the *Peters et al.* (2010a,b, 2011) simulations stands in contrast to the much stronger effects identified by *Krumholz et al.* (2011, 2012b) and *Bate* (2012).

Some of this difference may originate in the numerical method for treating radiation, with *Peters et al.* (2010a,b, 2011) using a ray-tracing method that only follows photons directly emitted by the star, while *Krumholz et al.* (2011, 2012b) and *Bate* (2012) use a diffusion method that follows the dust-reprocessed radiation but only indirectly treats direct stellar photons. However, a more likely explanation is a difference of initial conditions. *Krumholz and McKee* (2008) and *Krumholz et al.* (2010) argue that the surface density of a region is the key parameter that determines how effective radiative feedback will be, since it determines how effectively stellar radiation is trapped. In their simulations, *Krumholz et al.* (2011, 2012b) consider a region similar to the center of the Orion Nebula cluster, with a surface density $\Sigma \approx 1$ g cm^{-2} (≈ 5000 M$_\odot$ pc^{-2}), while *Bate* (2012) simulates a region with $\Sigma \approx 0.2$ g cm^{-2} (≈ 1000 M$_\odot$ pc^{-2}), similar to nearby low-mass star-forming regions such as Serpens or ρ Ophiuchus. In contrast, *Peters et al.* (2010a,b, 2011) use an initial condition with $\Sigma \approx 0.03$ g cm^{-2} (≈ 100 M$_\odot$ pc^{-2}), comparable to the surface density in giant molecular clouds averaged over >10-pc scales (see the chapter by Dobbs et al. in this volume). *Peters et al.*'s (2010a,b, 2011) surface density is low enough that their cloud is optically thin in the near-IR, and it is not surprising that radiative heating has minimal effects in such an environment, since any stellar radiation absorbed by the dust escapes as soon as it is reemitted, rather than having to diffuse outward and heat the cloud in the process.

Simultaneous observations of fragmentation and temperature distribution in cluster-forming clumps can provide a key observational test of the numerical results. Recent high-angular-resolution observations of dense, massive IR dark clouds, precursors to cluster-forming regions, have begun to detect massive cores at the early phases of cluster formation (*Rathborne et al.*, 2008; *Zhang et al.*, 2009; *Bontemps et al.*, 2010; *Zhang and Wang*, 2011; *Wang et al.*, 2011, 2012b; *Longmore et al.*, 2011). Indeed, these cores contain masses at least a factor of 10 larger than the Jeans mass (*Zhang et al.*, 2009). However, they do not appear to lie in the high-temperature sections of the clump (*Wang et al.*, 2012b), which appears to conflict with numerical simulations. Future observations of larger samples of massive cluster-forming clumps will provide a statistical trend that further constrains the role of thermal feedback to the formation of massive stars (see section 6.1).

5. PUTTING IT ALL TOGETHER

5.1. Interactions Between Feedback Mechanisms

5.1.1. Combined effects of multiple mechanisms. In the preceding sections, we have discussed the effects of various different types of feedback individually. However, in reality, feedback mechanisms frequently act simultaneously. For low-mass star formation, the dominant mechanisms are thought to be protostellar outflows and thermal feedback. Radiation pressure is negligible and supernovae do not occur. Photoionization will only affect the immediate vicinities of protostars, although it may be crucial for the erosion of protoplanetary disks (*Hollenbach et al.*, 1994; *Yorke and Welz*, 1996; *Richling and Yorke*, 1997; *Clarke et al.*, 2001). In contrast, the feedback from high-mass protostars involves all the above mechanisms.

Both analytic and numerical investigations of multiple mechanisms are few. *McKee et al.* (1984) considered the interaction of a stellar wind and photoionization with a clumpy medium, and concluded that the photoevaporation of clumps would control the dynamics, either by creating an ionized medium that would pressure-confine the wind or by providing a mass load that would limit its expansion. *Krumholz et al.* (2005a) showed that protostellar outflows would significantly weaken the effects of radiation pressure feedback by creating escape routes for photons. *Cunningham et al.* (2011) confirmed this prediction with simulations, and also showed that focusing of the radiation by the outflow cavity prevents the formation of radiation-pressure-driven bubbles and the associated Rayleigh-Taylor instabilities seen in earlier calculations (*Krumholz et al.*, 2009).

Hansen et al. (2012) investigated the combined effects of radiative transfer and protostellar outflows on low-mass star formation. As the outflows reduce the accretion rates of the protostars, they also reduce their masses and luminosities and hence the level of thermal feedback. They found that the outflows did not have a significant impact on the kinematics of the star-forming cloud, but the calculations did not include magnetic fields. In section 4.4, we mentioned that *Krumholz et al.* (2011) found that in massive dense star-forming clouds, thermal feedback could be so effective at inhibiting fragmentation that it led to a top-heavy IMF. However, *Krumholz et al.* (2012a) showed that this "overheating problem" could be reduced by including the effects of large-scale turbulent driving and protostellar outflows, since both of these processes lower protostellar accretion rates and thus the effects of thermal feedback. This is one case where the details of, and uncertainties in, accurately determining the luminosities of protostars (see section 4.1) can play a crucial role in the outcome of star formation.

A number of authors have also simulated the interaction of stellar winds with photoionized regions (e.g., *Garcia-Segura et al.*, 1996a,b; *Freyer et al.*, 2003, 2006; *Toalá and Arthur*, 2011). However, these simulations generally begin with a single star placed in a uniform, non-self-gravitating medium. While this setup is useful for studying the internal dynamics of wind bubbles, it limits the conclusions that can be drawn about how the feedback affects the formation of star clusters, where the surrounding medium is highly structured and strongly affected by gravity.

5.1.2. Feedback, turbulence, and magnetic fields. Some feedback effects are enhanced by the presence of turbulence and/or magnetic fields, while others are reduced. For example, *Banerjee et al.* (2007) examined jets propagating into a quiescent medium and concluded that they do not drive supersonic turbulence, while *Cunningham et al.* (2009) shows that jets propagating into a preexisting turbulent medium could inject energy into the turbulence, thus potentially allowing outflows to sustain existing turbulence in a star-forming region. It has been noted, however, that the velocity power spectrum of the turbulence generated by outflows in magnetized clouds may be slightly steeper than that generated by isotropic forcing (*Carroll et al.*, 2009).

By themselves, stronger magnetic fields have been shown to reduce the star-formation rate in molecular clouds (*Nakamura and Li*, 2007; *Price and Bate*, 2008, 2009). However, magnetic fields can also interact with feedback effects in subtle ways. We already discussed (section 2.1.1) how magnetic fields enhance the effects of protostellar outflows by raising the efficiency with which they deposit their energy in clouds. *Gendelev and Krumholz* (2012) demonstrated a similar magnetic enhancement in the effects of photoionization feedback. On the other hand, *Peters et al.* (2011) investigate the interaction of magnetic fields with ionizing radiation on smaller scales, and find that ionizing radiation tends to make it harder for massive stars to drive collimated magnetized outflows, because the pressure of the photoionized gas disrupts the magnetic tower configuration that can drive outflows from lower mass stars.

In a similar vein, while it has been known for some time that magnetic fields alone reduce fragmentation in collapsing gas (e.g., *Hennebelle et al.*, 2011), they also appear to greatly enhance the effects of thermal feedback. *Commerçon et al.* (2010, 2011a) and *Myers et al.* (2013) find that magnetic fields provide an efficient mechanism for angular momentum transport and thus tend to increase accretion rates onto forming stars. This in turn raises their accretion luminosities, and thus the strength of thermal feedback. Moreover, thermal feedback and magnetic fields work well in combination to reduce fragmentation in both low- and high-mass star-formation calculations. For low-mass clusters, *Price and Bate* (2009) found that thermal feedback inhibits fragmentation on small scales, while magnetic fields provide extra support on large scales. The result is that the combination of magnetic fields and thermal feedback is much more effective than one would naively guess. For massive cores, *Commerçon et al.* (2010, 2011a) find a similar effect operating at early times, up to the formation of a Larson's first core. However, they are unable to address the question of fragmentation at later times.

Myers et al. (2013) use a subgrid stellar evolution model that allows them to run for much longer times than *Commerçon et al.* (2010, 2011a), and find that thermal feedback inhibits fragmentation in the dense central regions, while magnetic fields inhibit it in the diffuse outer regions. They find that strong magnetic fields and thermal feedback in massive dense cores make it very difficult to form anything other than a single massive star or, perhaps, a binary. Figure 7 illustrates this effect. It shows the results of three simulations by *Myers et al.* (2013) using identical resolution and initial conditions, one with magnetic fields but no radiation, one with radiation but no magnetic fields,

and one with both radiation and magnetic fields. The run with both forms many fewer stars than one might naively have guessed based on the results with radiation or magnetic fields alone.

These results are in contrast to those of *Peters et al.* (2011), who also include both radiation and magnetic fields, but find only a modest reduction in fragmentation. As discussed in section 4.4, this is likely a matter of initial conditions: *Commerçon et al.* (2010, 2011a) and *Myers et al.* (2013) consider dense prestellar cores with surface density $\Sigma \approx 1$ g cm^{-2} [chosen to match observed IR dark cloud cores (*Swift*, 2009)], while *Peters et al.* (2011) simulate much more diffuse regions with $\Sigma \approx 0.03$ g cm^{-2}, and is not clear if their calculations ever evolve to produce the sorts of structures from which *Commerçon et al.* (2010, 2011a) and *Myers et al.* (2013) begin.

The combined effects of magnetic fields and thermal feedback can even modify star-formation rates. Thermal feedback by itself reduces star-formation rates by at most tens of percents (*Bate*, 2009b, 2012; *Krumholz et al.*, 2010), but adding magnetic fields can reduce the star-formation rate in low-mass clusters by almost an order of magnitude over purely hydrodynamic collapse (*Price and Bate*, 2009) and significantly more than magnetic fields alone.

5.2. Triggering

Thus far we have primarily focused on negative feedback, in the sense of restraining or terminating star formation. However, it is also possible for feedback to be positive, in the sense of promoting or accelerating feedback. The statistical arguments outlined in section 1.1 would tend to suggest that negative feedback must predominate, but this does not necessarily imply that positive feedback never occurs or cannot be important in some circumstances.

Positive feedback is usually referred to as triggered or induced star formation. This phrase can mean increasing the star-formation rate, increasing the star-formation efficiency, or increasing the total number of stars formed. These definitions can all be applied locally or globally. *Dale et al.* (2007) draw a distinction between weak triggering — temporarily increasing the star-formation rate by inducing stars to form earlier — and strong triggering — increasing the star-formation efficiency by causing the birth of stars that would not otherwise form. They note that it may be very difficult for observations to distinguish these possibilities.

Analytic studies by *Whitworth et al.* (1994a) suggest that the gravitational instability operating in swept-up shells driven into uniform gas by expanding Hɪɪ regions or wind bubbles should be an efficient triggering process. *Whitworth et al.* (1994b) extended this work to show that this process should result in a top-heavy IMF, a result also found in simulations of fragmenting shells by *Wünsch et al.* (2010) and *Dale et al.* (2011). This raises the intriguing prospect of star formation as a self-propagating process (e.g., *Shore*, 1981, 1983). However, simulations of ionizing feedback in fractal (*Walch et al.*, 2013) and turbulent clouds (*Dale et al.*, 2007, 2012a, 2013a; *Dale and Bonnell*, 2012) suggest that

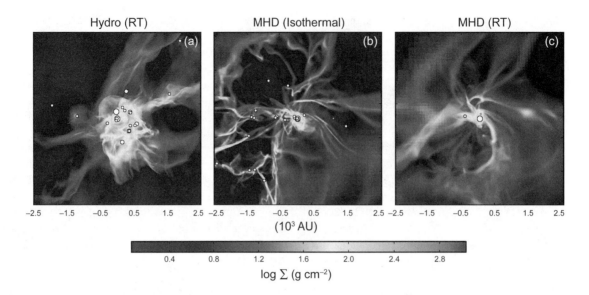

Fig. 7. Results from three simulations by *Myers et al.* (2013). The gray color scale shows the column density, and white circles show stars, with the size of the circle indicating mass: <1 M$_\odot$ (small circles), 1–8 M$_\odot$ (medium circles), and >8 M$_\odot$ (large circles). **(a)** Simulation including radiative transfer (RT) but no magnetic fields; **(b)** simulation with magnetic fields but no radiative transfer; **(c)** simulation with both radiative transfer and magnetic fields. All runs began from identical initial conditions, and have been run to 60% of the free-fall time at the initial mean density.

this is not the case. They find that, while ionization feedback can modestly change the rate, efficiency, and number of stars, it does not significantly alter the IMF.

Pillars or "Elephant Trunks" are a widespread and distinctive feature of star-forming regions and have often been invoked as signposts of triggering (e.g., *Smith et al.*, 2000, 2005; *Billot et al.*, 2010). *Williams et al.* (2001), *Miao et al.* (2006), *Gritschneder et al.* (2010), *Mackey and Lim* (2011), *Tremblin et al.* (2012a,b), and *Walch et al.* (2013) have simulated pillar formation in a wide variety of initial conditions and provide several plausible mechanisms for their origins. However, it is not clear to what extent these morphological features are actually indicative of triggered star formation.

There is also large body of literature on the induced collapse of initially stable density configurations, such as Bonner-Ebert spheres, by winds or by Hɪɪ regions, known as radiation-driven implosion (e.g., *Sandford et al.*, 1982, 1984; *Klein et al.*, 1983; *Bertoldi*, 1989; *Bertoldi and McKee*, 1990; *Kessel-Deynet and Burkert*, 2003; *Bisbas et al.*, 2011). This process is able to produce not only single stars, but small groups, and since the initial conditions are stable by construction, this is a good example of strong triggering.

Establishing the occurrence of triggered star formation analytically or numerically is relatively straightforward. One can use either initial conditions that are stable in the absence of feedback, as in the radiation-driven implosion simulations, or, for more complex initial conditions, control simulations without feedback. Detecting triggering in

observed systems is much more difficult, since neither of these paths are open to the observer. Instead, observers must search for circumstantial evidence for triggering. Examples of such attempts in the literature include surveys of young stars near bubbles, ionization fronts, or bright-rimmed clouds (e.g., *Urquhart et al.*, 2007; *Deharveng et al.*, 2008; *Snider et al.*, 2009; *Smith et al.*, 2010) or pillars, campaigns to find clusters elongated toward feedback sources or showing strong age gradients in young stars (e.g., *Sugitani et al.*, 1995; *Chauhan et al.*, 2009; *Getman et al.*, 2012), and searches for regions of unusually high star-formation rate or efficiency (e.g., *Bisbas et al.*, 2011). In these observational campaigns, the standard practice has been to claim evidence for triggered star formation whenever there are coherent structures within which the stellar age differences are smaller than the crossing time of the cloud structure. Generally such structures are compressed shells of molecular gas with tens of parsec radii, such as the CepOB2 bubble (*Patel et al.*, 1998) or the MonR2 GMC (*Xie and Goldsmith*, 1994).

This approach must be treated with caution, because star formation correlates with the presence of dense gas on parsec scales even in the absence of obvious feedback or triggering (*Heiderman et al.*, 2010; *Gutermuth et al.*, 2011). Thus one expects to find correlations between YSOs and nebulosity features even in the absence of triggering. For evidence of triggering to be convincing, one must show that the correlation between star formation and feedback-driven features is in excess of what one would expect simply from the baseline correlation of YSOs and dense gas. For example, in some cases the ratio of YSO mass to molecular gas mass is far higher than one normally finds in active star-forming regions (e.g., *Getman et al.,* 2009). However, even in such cases it is unclear whether the enhancement is a matter of positive feedback (enhanced star formation raising the stellar mass) or negative feedback (ablation of gas reducing the gas mass without creating any additional stars).

The contribution of triggered star formation to the global star-formation rate is unclear. Most GMCs have uniformly low star-formation efficiencies (*Evans et al.,* 2009), showing no obvious correlation with numbers of OB stars or signs of feedback such as bubbles. *Kendrew et al.* (2012) and *Thompson et al.* (2012) use statistical correlations between large catalogs of bubbles and young stellar objects to infer that triggered star formation may contribute at most tens of percents to the Galactic population of massive stars. Such a small variation is well within the cloud-to-cloud scatter in star-formation efficiency seen by *Evans et al.* (2009); indeed, it is smaller than the typical observational uncertainty on the star-formation rate, particularly within a single cloud. Similarly, simulations by *Dale et al.* (2012a, 2013a) imply that, while triggering certainly does occur, it is overwhelmed by negative feedback on the scale of GMCs.

Ultimately, quantifying the sizes and timescales for the destructive influences of feedback is essential; outside these zones of negative influence, large-scale effects from feedback may indeed turn positive, especially in aggregate with other moderately nearby feedback sources. Are these aggregate effects truly triggering, or are they simply the source of the large-scale turbulence that drives the creation of the next generation of star-forming molecular material?

6. FUTURE PROSPECTS

6.1. Observations

In order to place more stringent constraints on how feedback affects the star- and cluster-formation process, new observations must tackle the problem on two fronts. Naturally, we must continue to push to observe statistically significant samples of ever more extreme (and more distant) environments as they come within reach of new facilities across the electromagnetic (EM) spectrum. In addition, we must endeavor to improve our understanding of star formation within nearby, low-feedback environments in order to set an interpretive framework that will be essential for determining the net effect of more extreme feedback. In both cases, characterization of the properties and kinematics of star-forming gas and the forming stars themselves will be essential.

High-angular-resolution observations of protocluster-forming regions have revealed fragmentation of dense molecular gas at spatial scales of $<10^3$ AU (*Rathborne et al.,* 2008; *Zhang et al.,* 2009; *Bontemps et al.,* 2010; *Zhang and Wang,* 2011; *Wang et al.,* 2011, 2012a; *Longmore et al.,* 2011; *Palau et al.,* 2013). Despite limited sensitivity

and dynamic range, these observations provided valuable snapshots of density and temperature distributions of the star-forming gas. The next generation of interferometers such as ALMA will provide even more detailed images of density distribution and linewidth of star-forming cores in protocluster-forming molecular clumps. At the same time, the temperature distribution within the molecular clump can be readily obtained from observations with the Karl Jansky VLA (e.g., *Wang et al.,* 2008, 2012a). The combined information will constrain the thermal dynamic properties and fragmentation of cluster-forming clumps, providing direct information on the initial physical properties of molecular gas that gives rise to a stellar cluster.

A survey of nearby clusters, within ~500 pc, using existing millimeter interferometers [e.g., Combined Array for Research in Millimeter-wave Astronomy (CARMA) (see *Plunkett et al.,* 2013)] or heterodyne receiver arrays in (sub)-millimeter (single-dish) telescopes [e.g., SEQUOIA on Large Millimeter Telescope (LMT) or the 64-pixel array Supercam on Submillimeter Telescope (SMT)] allow for efficient mapping of outflows using different CO transition lines. Even though in many cases the angular resolution will not be enough to resolve individual outflows in regions with a high density of protostars, these observations will be necessary to obtain total molecular outflow energetics and compare them with cloud energy and gas distribution for a large sample of clusters. Observations of these regions should also include maps of higher (column) density tracers, such as ^{13}CO and $C^{18}O$, in order to investigate the impact of outflows and winds in the cloud structure and turbulence.

In the near future, when ALMA is completed and on-the-fly (OTF) mapping becomes available, it will be feasible to conduct studies similar to the ones described above for a (larger) sample of clusters (within about 10 kpc) that is representative of the galactic cluster population. High-resolution ALMA observations of massive (far-away) clusters will also help in studying the impact of compact HII regions on the surrounding molecular cloud. In addition, multi-epoch VLA continuum observations of compact HII regions will allow measurement of the expansion velocity of the ionized bubble, and in concert with multi-wavelength data, will place constraints on the impact of winds and radiation pressure on the surrounding cluster environment. However, these observations may be complicated by the fact that, at least during the hypercompact stage, some HII regions are observed to shrink rather than expand (*Galván-Madrid et al.,* 2008), likely as a result of the motions of dense material near the ionizing source causing part of the HII region to be shadowed (*Galván-Madrid et al.,* 2011).

At larger scales, OTF observations using heterodyne receiver arrays in millimeter single-dish telescopes will allow fast mapping of molecular gas that has been swept-up by bubbles and signal-to-noise (SNR) in high-mass star-forming regions. In lower-density regions, where there is little or no molecular gas, large-scale galactic HI surveys with Square Kilometre Array (SKA) precursor telescopes, like the Galactic Australian Square Kilometre Array Pathfinder (GASKAP) (*Dickey et al., 2013*), will provide useful information on the effects of stellar feedback on the low-density outskirts of clusters.

Confidently obtaining the census of low-mass young stars forming in more diverse feedback-affected environments is essential to establishing correlations between feedback sources (e.g., their nature, position, and intensity) and changes in star-gas column density correlation or other aspects of the star-formation process. Given the simultaneous needs of moderate to high extinction penetration and membership isolation from field stars, IR and X-ray imaging capabilities remain the central means for identifying YSOs. Recent IR YSO membership surveys with Spitzer have considerably expanded our knowledge of YSOs in the nearest kiloparsec. However, they are limited in their ability to discern members projected on bright nebulosity, thus only a relatively narrow range of galactic environments have been thoroughly searched for forming stellar content throughout the stellar mass range (*Allen et al., 2007; Evans et al., 2009; Gutermuth et al., 2009; Megeath et al., 2012*).

X-ray observations, particularly deep and high-resolution imaging with Chandra, have proven effective at bypassing the nebulosity limitations of IR surveys, detecting substantial numbers of YSOs in particularly IR-bright high-mass star-forming regions. X-ray emission from YSOs is generally considered to be a product of magnetic field activity, and thus not strongly correlated with the presence of a disk (*Feigelson et al., 2007*). The resulting YSO census derived from X-ray imaging therefore trades the disk bias of IR surveys for a broad completeness decay as a function of luminosity, as stochastic flaring events of gradually increasing strength, and therefore rarity, are required to detect lower-mass sources with smaller quiescent luminosities.

In the near term, improving capabilities in adaptive optics on ever-larger-aperture groundbased optical and near-IR telescopes will facilitate other means of young stellar membership isolation, both via facilitating high throughput spectroscopy as well as yielding sufficient astrometric precision for proper motion characterization (e.g., *Lu et al., 2013*). Looking further ahead, the considerable improvement in angular resolution and sensitivity at near- and mid-IR wavelengths afforded by the James Webb Space Telescope (JWST) will dramatically improve the contrast between nebulosity and point sources in Milky Way star-forming regions, enabling Spitzer-like mid-IR YSO surveys out to much greater distances and in regions influenced by much more significant feedback sources. The resulting censuses of young stars with IR excess will reach well down the stellar mass function in regions found both within the Molecular Ring as well as toward the outskirts of the Milky Way. Unfortunately, next-generation X-ray space telescopes of similar angular resolution and better sensitivity relative to Chandra [e.g., the Astrophysics Experiment for Grating and Imaging Spectroscopy (AEGIS), Advanced X-ray Spectroscopic Imaging Observatory (AXSIO), and Square Meter Arcsecond Resolution X-ray Telescope (SMART-X)] remain in the planning phase for launch on a time frame outside the scope of this review.

6.2. Simulations and Theory

In the area of simulations and theory, in the next few years we can expect improvements in several areas. The first need that should be apparent from the preceding discussion is for simulations that include a number of different feedback mechanisms, and that can assess their relative importance. As of now, there are simulations and analytic models addressing almost every potentially important form of feedback: pre-main-sequence outflows, main-sequence winds, ionizing radiation, non-ionizing radiation, radiation pressure, and supernovae. However, no simulation or model includes all of them, and only a few include more than one. Moreover, many simulations including feedback do not include magnetic fields. As discussed in section 5.1, interactions between different feedback mechanisms, and between feedback and preexisting turbulence and magnetic fields, are potentially important, but remain largely unexplored.

This limitation is mostly one of code development. Designing and implementing the numerical methods required to treat even one form of feedback in the context of an adaptive mesh refinement, smoothed particle hydrodynamics, or other code capable of the high dynamic range required to study star formation requires an effort lasting several years. As a result, no one code includes treatments of all the potentially important mechanisms. However, that situation is improving as code development progresses, and the pace of development is increasing as at least some of the remaining work involves porting existing techniques from one code to another, rather than developing entirely new ones. By the time of the next Protostars and Planets conference, it seems likely that there will be more than a few published simulations that include magnetohydrodynamics (MHD), protostellar outflows, main-sequence winds, and multiple radiation effects, including pressure and heating by both ionizing and non-ionizing photons. There are also likely to be improvements in the numerical techniques used for many of these processes, particularly radiative transfer, where it seems likely that in the next few years many codes will be upgraded to use variable Eddington tensor methods (e.g., *Davis et al., 2012; Jiang et al., 2012*) or other high-order methods such as S_n transport.

A second major area in need of progress is the initial conditions used in simulations of star-cluster formation. At present, most simulations begin with highly idealized initial conditions: either spherical regions that may or may not be centrally concentrated, or turbulent periodic boxes. In simulations without feedback, *Girichidis et al.* (2011, 2012a,b) show that the results can depend strongly on which of these setups is used. For simulations with feedback, those that run long enough and contain forms of feedback such that they are able to reach a statistical steady state are probably fairly insensitive to the initial conditions. For the vast majority of simulations, though, particularly those where gas expulsion is rapid and occurs at most ~1 dynamical time after the onset of star formation, the initial conditions likely matter a great deal. In reality, the dense regions that form clusters are embedded within larger giant molecular clouds, which are themselves embedded in a galactic disk. They likely begin forming stars while they are still accreting mass, and the larger environments that are missing in most simulations can provide substantial inputs of mass, kinetic energy, and confining pressure. Simulations that include both the formation of a cloud and feedback have begun to appear in the context of studies of GMCs on larger scales (e.g., *Vázquez-Semadeni et al.,* 2010), and there is clearly a need to extend this approach to the smaller, denser scales required to study the formation of star clusters.

A third, closely related problem is that the current generation of simulations usually explore a very limited range of parameters — for example, only a single cloud mass and density, or a single magnetic field strength or orientation. As a result, it is difficult to draw general conclusions, particularly when results differ between groups. For example, *Wang et al.* (2010) find that including protostellar outflows dramatically reduces the star-formation rate in their simulations of cluster formation, while *Krumholz et al.* (2012b), using essentially the same prescription to model outflows, find that the effects on the star-formation rate are much more modest. Is this because *Wang et al.*'s (2010) simulations include magnetic fields and those of *Krumholz et al.* (2012b) do not? Because Krumholz et al.'s simulations include radiation and *Wang et al.*'s (2010) do not? Or because Wang et al. simulate a region modeled after a relatively low-density region like NGC 2264, while *Krumholz et al.* (2012b) choose parameters appropriate for a much-higher-density region like the core of the Orion Nebula cluster? Since each paper simulated only a single environment, the answers remain unknown. While there are analytic models that provide some guidance as to which feedback mechanisms might be important under what conditions (e.g., *Fall et al.,* 2010), there are precious few parameter studies. [See *Krumholz et al.* (2010), *Myers et al.* (2011), and *Dale et al.* (2013b) for some of the few exceptions.] This will need to change in the coming years.

All these advances are likely to require fundamental changes in the algorithms and code architecture used for numerical simulations of star-cluster formation. Using present algorithms, parallel simulations of star-cluster formation that go to resolutions high enough to (for example) resolve fragmentation to the IMF, and that include even one or two feedback mechanisms, often require many months of run time. Adding more physical processes, or more accurate treatments of the ones already included, will only exacerbate the problem. Part of the problem is that the techniques currently in use do not scale particularly well on modern massively parallel architectures. This is partly a matter of physics: The problem of star formation is inherently computationally difficult due to the wide range of time and spatial scales that must be treated. However, it is also partly a matter of code design: Few modern multi-physics codes have been ported to hybrid threaded/message-passing architectures, and even fewer have been optimized to run on GPUs or similar special-purpose hardware. In addition to improving the physics in our codes, a great deal of software engineering will be required to meet the goals laid out above in time for the next Protostars and Planets conference.

Acknowledgments. This work was supported by the following sources: the Alfred P. Sloan Foundation (M.R.K.); National Science Foundation grants AST-0955300 (M.R.K.), AST-0845619 (H.G.A.), AST-1313083 (Z.-Y.L.), AST-0908553 (R.I.K.), and NSF12-11729 (R.I.K.); NASA ATP grants NNX13AB84G (M.R.K. and R.I.K.), and NNX10AH30G (Z.-Y.L.); NASA ADAP grants NNX11AD14G (R.A.G.) and NNX13AF08G (R.A.G.); NASA through Chandra Award Number GO2-13162A (M.R.K.) issued by the Chandra X-ray Observatory Center, which is operated by the Smithsonian Astrophysical Observatory for and on behalf of the National Aeronautics Space Administration under contract NAS8-03060; NASA through Hubble Award Number 13256 (M.R.K.) issued by the Space Telescope Science Institute, which is operated by the Association of Universities for Research in Astronomy, Inc., under NASA contract NAS 5-26555; NASA by Caltech/JPL awards 1373081 (R.A.G.), 1424329 (R.A.G.), and 1440160 (R.A.G.) in support of Spitzer Space Telescope observing programs; the U.S. Department of Energy at the Lawrence Livermore National Laboratory under contract DE-AC52-07NA2734 (R.I.K.); the DFG cluster of excellence "Origin and Structure of the Universe" (J.E.D.); the hospitality of the Aspen Center for Physics, which is supported by the National Science Foundation Grant PHY-1066293 (M.R.K., R.I.K., and R.A.G.).

REFERENCES

Agertz O. et al. (2007) *Mon. Not. R. Astron. Soc., 380,* 963.
Agertz O. et al. (2013) *Astrophys. J., 770,* 25.
Allen L. et al. (2007) In *Protostars and Planets V* (B. Reipurth et al., eds.), pp. 361–376. Univ. of Arizona, Tucson.
Anderson L. D. et al. (2011) *Astrophys. J. Suppl., 194,* 32.
Arce H. G. et al. (2010) *Astrophys. J., 715,* 1170.
Arce H. G. et al. (2011) *Astrophys. J., 742,* 105.
Arthur S. J. (2012) *Mon. Not. R. Astron. Soc., 421,* 1283.
Arthur S. J. et al. (2011) *Mon. Not. R. Astron. Soc., 414,* 1747.
Aumer M. et al. (2013) *Mon. Not. R. Astron. Soc., 434,* 3142.
Banerjee R. et al. (2007) *Astrophys. J., 668,* 1028.
Baraffe I. et al. (2009) *Astrophys. J. Lett., 702,* L27.
Bastian N. et al. (2010) *Annu. Rev. Astron. Astrophys., 48,* 339.
Bate M. R. (2009a) *Mon. Not. R. Astron. Soc., 392,* 590.
Bate M. R. (2009b) *Mon. Not. R. Astron. Soc., 392,* 1363.
Bate M. R. (2010) *Mon. Not. R. Astron. Soc., 404,* L79.
Bate M. R. (2011) *Mon. Not. R. Astron. Soc., 417,* 2036.
Bate M. R. (2012) *Mon. Not. R. Astron. Soc., 419,* 3115.
Bate M. R. and Bonnell I. A. (2005) *Mon. Not. R. Astron. Soc., 356,* 1201.
Bate M. R. et al. (2003) *Mon. Not. R. Astron. Soc., 339,* 577.
Bertoldi F. (1989) *Astrophys. J., 346,* 735.
Bertoldi F. and McKee C. F. (1990) *Astrophys. J., 354,* 529.
Beuther H. et al. (2002) *Astron. Astrophys., 383,* 892.
Beuther H. et al. (2003) *Astron. Astrophys., 408,* 601.
Billot N. et al. (2010) *Astrophys. J., 712,* 797.

Binney J. and Tremaine S. (1987) In *Galactic Dynamics* (J. Ostriker, ed.), Princeton Univ., Princeton. 755 pp.

Bisbas T. G. et al. (2011) *Astrophys. J., 736*, 142.

Bonnell I. A. et al. (1997) *Mon. Not. R. Astron. Soc., 285*, 201.

Bonnell I. A. et al. (2003) *Mon. Not. R. Astron. Soc., 343*, 413.

Bonnell I. A. et al. (2004) *Mon. Not. R. Astron. Soc., 349*, 735.

Bonnell I. A. et al. (2006) *Mon. Not. R. Astron. Soc., 368*, 1296.

Bontemps S. et al. (2010) *Astron. Astrophys., 524*, A18.

Boss A. P. (1983) *Icarus, 55*, 181.

Boss A. P. (1984) *Astrophys. J., 277*, 768.

Boss A. P. (1986) *Astrophys. J. Suppl., 62*, 519.

Boss A. P. et al. (2000) *Astrophys. J., 528*, 325.

Bournaud F. et al. (2014) *Astrophys. J., 780*, 57.

Bruderer S. et al. (2010) *Astrophys. J., 720*, 1432.

Capriotti E. R. and Kozminski J. F. (2001) *Publ. Astron. Soc. Pac., 113*, 677.

Carroll J. J. et al. (2009) *Astrophys. J., 695*, 1376.

Castor J. et al. (1975a) *Astrophys. J. Lett., 200*, L107.

Castor J. I. et al. (1975b) *Astrophys. J., 195*, 157.

Cerviño M. and Luridiana V. (2004) *Astron. Astrophys., 413*, 145.

Chabrier G. (2005) In *The Initial Mass Function 50 Years Later* (E. Corbelli et al., eds.), p. 41. Astrophys. Space Sci. Library, Vol. 327, Kluwer, Dordrecht.

Chakrabarti S. and McKee C. F. (2005) *Astrophys. J., 631*, 792.

Chauhan N. et al. (2009) *Mon. Not. R. Astron. Soc., 396*, 964.

Chevalier R. A. and Clegg A. W. (1985) *Nature, 317*, 44.

Chu Y.-H. and Kennicutt R. C. Jr. (1994) *Astrophys. J., 425*, 720.

Clark P. C. et al. (2005) *Mon. Not. R. Astron. Soc., 359*, 809.

Clarke C. J. et al. (2001) *Mon. Not. R. Astron. Soc., 328*, 485.

Commerçon B. et al. (2010) *Astron. Astrophys., 510*, L3.

Commerçon B. et al. (2011a) *Astrophys. J. Lett., 742*, L9.

Commerçon B. et al. (2011b) *Astron. Astrophys., 530*, A13.

Crutcher R. M. (2012) *Annu. Rev. Astron. Astrophys., 50*, 29.

Cunningham A. J. et al. (2009) *Astrophys. J., 692*, 816.

Cunningham A. J. et al. (2011) *Astrophys. J., 740*, 107.

Curtis E. I. et al. (2010) *Mon. Not. R. Astron. Soc., 408*, 1516.

Dale J. E. and Bonnell I. (2011) *Mon. Not. R. Astron. Soc., 414*, 321.

Dale J. E. and Bonnell I. A. (2008) *Mon. Not. R. Astron. Soc., 391*, 2.

Dale J. E. and Bonnell I. A. (2012) *Mon. Not. R. Astron. Soc., 422*, 1352.

Dale J. E. et al. (2005) *Mon. Not. R. Astron. Soc., 358*, 291.

Dale J. E. et al. (2007) *Mon. Not. R. Astron. Soc., 377*, 535.

Dale J. E. et al. (2011) *Mon. Not. R. Astron. Soc., 411*, 2230.

Dale J. E. et al. (2012a) *Mon. Not. R. Astron. Soc., 427*, 2852.

Dale J. E. et al. (2012b) *Mon. Not. R. Astron. Soc., 424*, 377.

Dale J. E. et al. (2013a) *Mon. Not. R. Astron. Soc., 431*, 1062.

Dale J. E. et al. (2013b) *Mon. Not. R. Astron. Soc., 430*, 234.

da Silva R. L. et al. (2012) *Astrophys. J., 745*, 145.

Davies B. et al. (2011) *Mon. Not. R. Astron. Soc., 416*, 972.

Davis S. W. et al. (2012) *Astrophys. J. Suppl., 199*, 9.

Deharveng L. et al. (2008) *Astron. Astrophys., 482*, 585.

Deharveng L. et al. (2010) *Astron. Astrophys., 523*, A6.

Dent W. R. F. et al. (2009) *Mon. Not. R. Astron. Soc., 395*, 1805.

Dickey J. M. et al. (2013) *Publ. Astron. Soc. Australia, 30*, e003.

Duarte-Cabral A. et al. (2012) *Astron. Astrophys., 543*, A140.

Ekström S. et al. (2012) *Astron. Astrophys., 537*, A146.

Evans N. J. et al. (2009) *Astrophys. J. Suppl., 181*, 321.

Fall S. M. and Chandar R. (2012) *Astrophys. J., 752*, 96.

Fall S. M. et al. (2010) *Astrophys. J. Lett., 710*, L142.

Federrath C. (2013) *Mon. Not. R. Astron. Soc., 436*, 3167.

Federrath C. and Klessen R. S. (2012) *Astrophys. J., 761*, 156.

Feigelson E. et al. (2007) In *Protostars and Planets V* (B. Reipurth et al., eds.), pp. 313–328. Univ. of Arizona, Tucson.

Franco J. et al. (1990) *Astrophys. J., 349*, 126.

Franco J. et al. (1994) *Astrophys. J., 436*, 795.

Freyer T. et al. (2003) *Astrophys. J., 594*, 888.

Freyer T. et al. (2006) *Astrophys. J., 638*, 262.

Galván-Madrid R. et al. (2008) *Astrophys. J. Lett., 674*, L33.

Galván-Madrid R. et al. (2011) *Mon. Not. R. Astron. Soc., 416*, 1033.

Garcia-Segura G. et al. (1996a) *Astron. Astrophys., 305*, 229.

Garcia-Segura G. et al. (1996b) *Astron. Astrophys., 316*, 133.

Gendelev L. and Krumholz M. R. (2012) *Astrophys. J., 745*, 158.

Genel S. et al. (2012) *Astrophys. J., 745*, 11.

Getman K. V. et al. (2009) *Astrophys. J., 699*, 1454.

Getman K. V. et al. (2012) *Mon. Not. R. Astron. Soc., 426*, 2917.

Gibb A. G. et al. (2004) *Astrophys. J., 616*, 301.

Ginsburg A. et al. (2011) *Mon. Not. R. Astron. Soc., 418*, 2121.

Girichidis P. et al. (2011) *Mon. Not. R. Astron. Soc., 413*, 2741.

Girichidis P. et al. (2012a) *Mon. Not. R. Astron. Soc., 420*, 613.

Girichidis P. et al. (2012b) *Mon. Not. R. Astron. Soc., 420*, 3264.

Goldbaum N. J. et al. (2011) *Astrophys. J., 738*, 101.

Graves S. F. et al. (2010) *Mon. Not. R. Astron. Soc., 409*, 1412.

Gritschneder M. et al. (2009) *Astrophys. J. Lett., 694*, L26.

Gritschneder M. et al. (2010) *Astrophys. J., 723*, 971.

Gutermuth R. A. et al. (2009) *Astrophys. J. Suppl., 184*, 18.

Gutermuth R. A. et al. (2011) *Astrophys. J., 739*, 84.

Hansen C. E. et al. (2012) *Astrophys. J., 747*, 22.

Harper-Clark E. and Murray N. (2009) *Astrophys. J., 693*, 1696.

Hartmann L. et al. (1997) *Astrophys. J., 475*, 770.

Hatchell J. et al. (2013) *Mon. Not. R. Astron. Soc., 429*, L10.

Heiderman A. et al. (2010) *Astrophys. J., 723*, 1019.

Hennebelle P. et al. (2011) *Astron. Astrophys., 528*, A72.

Hollenbach D. et al. (1994) *Astrophys. J., 428*, 654.

Holmberg J. and Flynn C. (2000) *Mon. Not. R. Astron. Soc., 313*, 209.

Hopkins P. F. (2013) *Mon. Not. R. Astron. Soc., 428*, 2840.

Hopkins P. F. et al. (2011) *Mon. Not. R. Astron. Soc., 417*, 950.

Hopkins P. F. et al. (2012a) *Mon. Not. R. Astron. Soc., 421*, 3522.

Hopkins P. F. et al. (2012b) *Mon. Not. R. Astron. Soc., 421*, 3488.

Hosokawa T. and Omukai K. (2009) *Astrophys. J., 691*, 823.

Hosokawa T. et al. (2010) *Astrophys. J., 721*, 478.

Hosokawa T. et al. (2011) *Astrophys. J., 738*, 140.

Hosokawa T. et al. (2012) *Astrophys. J. Lett., 760*, L37.

Jacquet E. and Krumholz M. R. (2011) *Astrophys. J., 730*, 116.

Jappsen A.-K. et al. (2005) *Astron. Astrophys., 435*, 611.

Jiang Y.-F. et al. (2012) *Astrophys. J. Suppl., 199*, 14.

Jiang Y.-F. et al. (2013) *Astrophys. J., 763*, 102.

Juneau S. et al. (2009) *Astrophys. J., 707*, 1217.

Kahn F. D. (1974) *Astron. Astrophys., 37*, 149.

Kendrew S. et al. (2012) *Astrophys. J., 755*, 71.

Kennicutt R. C. and Evans N. J. (2012) *Annu. Rev. Astron. Astrophys., 50*, 531.

Kessel-Deynet O. and Burkert A. (2003) *Mon. Not. R. Astron. Soc., 338*, 545.

Keto E. (2003) *Astrophys. J., 599*, 1196.

Keto E. and Wood K. (2006) *Astrophys. J., 637*, 850.

Kim J.-h. et al. (2013a) *Astrophys. J., 775*, 109.

Kim J.-h. et al. (2013b) *Astrophys. J., 779*, 8.

Kim K.-T. and Koo B.-C. (2001) *Astrophys. J., 549*, 979.

Kim K.-T. and Koo B.-C. (2003) *Astrophys. J., 596*, 362.

Klaassen P. D. and Wilson C. D. (2007) *Astrophys. J., 663*, 1092.

Klassen M. et al. (2012) *Astrophys. J., 758*, 137.

Klein R. I. et al. (1983) *Astrophys. J. Lett., 271*, L69.

Klessen R. S. and Burkert A. (2000) *Astrophys. J. Suppl., 128*, 287.

Klessen R. S. and Burkert A. (2001) *Astrophys. J., 549*, 386.

Klessen R. S. et al. (1998) *Astrophys. J., 501*, L205.

Knee L. B. G. and Sandell G. (2000) *Astron. Astrophys., 361*, 671.

Kratter K. M. and Matzner C. D. (2006) *Mon. Not. R. Astron. Soc., 373*, 1563.

Kratter K. M. et al. (2008) *Astrophys. J., 681*, 375.

Kratter K. M. et al. (2010) *Astrophys. J., 708*, 1585.

Krumholz M. R. (2006) *Astrophys. J. Lett., 641*, L45.

Krumholz M. R. (2011) *Astrophys. J., 743*, 110.

Krumholz M. R. and Dekel A. (2010) *Mon. Not. R. Astron. Soc., 406*, 112.

Krumholz M. R. and Matzner C. D. (2009) *Astrophys. J., 703*, 1352.

Krumholz M. R. and McKee C. F. (2008) *Nature, 451*, 1082.

Krumholz M. R. and Tan J. C. (2007) *Astrophys. J., 654*, 304.

Krumholz M. R. and Thompson T. A. (2012) *Astrophys. J., 760*, 155.

Krumholz M. R. and Thompson T. A. (2013) *Mon. Not. R. Astron. Soc., 434*, 2329.

Krumholz M. R. et al. (2005a) *Astrophys. J. Lett., 618,* L33.

Krumholz M. R. et al. (2005b) In *Massive Star Birth: A Crossroads of Astrophysics* (R. Cesaroni et al., eds.), pp. 231–236. IAU Symp. 227, Cambridge Univ., Cambridge.

Krumholz M. R. et al. (2006) *Astrophys. J., 653,* 361.

Krumholz M. R. et al. (2007a) *Astrophys. J., 671,* 518.

Krumholz M. R. et al. (2007b) *Astrophys. J., 656,* 959.

Krumholz M. R. et al. (2009) *Science, 323,* 754.

Krumholz M. R. et al. (2010) *Astrophys. J., 713,* 1120.

Krumholz M. R. et al. (2011) *Astrophys. J., 740,* 74.

Krumholz M. R. et al. (2012a) *Astrophys. J., 745,* 69.

Krumholz M. R. et al. (2012b) *Astrophys. J., 754,* 71.

Kudritzki R. P. et al. (1999) *Astron. Astrophys., 350,* 970.

Kuiper R. et al. (2011) *Astrophys. J., 732,* 20.

Kuiper R. et al. (2012) *Astron. Astrophys., 537,* A122.

Kurtz S. et al. (1994) *Astrophys. J. Suppl., 91,* 659.

Kurtz S. E. et al. (1999) *Astrophys. J., 514,* 232.

Lada C. J. and Lada E. A. (2003) *Annu. Rev. Astron. Astrophys., 41,* 57.

Larson R. B. (1969) *Mon. Not. R. Astron. Soc., 145,* 271.

Larson R. B. (1985) *Mon. Not. R. Astron. Soc., 214,* 379.

Larson R. B. (2005) *Mon. Not. R. Astron. Soc., 359,* 211.

Leitherer C. et al. (1992) *Astrophys. J., 401,* 596.

Leitherer C. et al. (1999) *Astrophys. J. Suppl., 123,* 3.

Leurini S. et al. (2009) *Astron. Astrophys., 507,* 1443.

Li Z.-Y. and Nakamura F. (2006) *Astrophys. J. Lett., 640,* L187.

Longmore S. N. et al. (2011) *Astrophys. J., 726,* 97.

Lopez L. A. et al. (2011) *Astrophys. J., 731,* 91.

Lopez L. A. et al. (2014) *Astrophys. J.,* in press, arXiv:1309.5421.

López-Sepulcre A. et al. (2009) *Astron. Astrophys., 499,* 811.

Low C. and Lynden-Bell D. (1976) *Mon. Not. R. Astron. Soc., 176,* 367.

Lu J. R. et al. (2013) *Astrophys. J., 764,* 155.

Mackey J. and Lim A. J. (2011) *Mon. Not. R. Astron. Soc., 412,* 2079.

Mac Low M.-M. et al. (2007) *Astrophys. J., 668,* 980.

Maeder A. and Meynet G. (2000) *Astron. Astrophys., 361,* 159.

Maeder A. and Meynet G. (2010) *New Astron. Rev., 54,* 32.

Masunaga H. and Inutsuka S.-I. (1999) *Astrophys. J., 510,* 822.

Masunaga H. and Inutsuka S.-I. (2000) *Astrophys. J., 531,* 350.

Matzner C. D. (2002) *Astrophys. J., 566,* 302.

Matzner C. D. (2007) *Astrophys. J., 659,* 1394.

Matzner C. D. and McKee C. F. (2000) *Astrophys. J., 545,* 364.

Maury A. J. et al. (2009) *Astron. Astrophys., 499,* 175.

McKee C. F. (1989) *Astrophys. J., 345,* 782.

McKee C. F. et al. (1984) *Astrophys. J. Lett., 278,* L115.

Megeath S. T. et al. (2012) *Astron. J., 144,* 192.

Mellema G. et al. (2006) *Astrophys. J., 647,* 397.

Miao J. et al. (2006) *Mon. Not. R. Astron. Soc., 369,* 143.

Mottram J. C. et al. (2011) *Astrophys. J. Lett., 730,* L33.

Muno M. P. et al. (2006a) *Astrophys. J. Lett., 636,* L41.

Muno M. P. et al. (2006b) *Astrophys. J., 650,* 203.

Murray N. and Rahman M. (2010) *Astrophys. J., 709,* 424.

Murray N. et al. (2010) *Astrophys. J., 709,* 191.

Murray N. et al. (2011) *Astrophys. J., 735,* 66.

Myers A. T. et al. (2011) *Astrophys. J., 735,* 49.

Myers A. T. et al. (2013) *Astrophys. J., 766,* 97.

Nakamura F. and Li Z.-Y. (2007) *Astrophys. J., 662,* 395.

Nakamura F. and Li Z.-Y. (2011) *Astrophys. J., 740,* 36.

Nakamura F. et al. (2011a) *Astrophys. J., 737,* 56.

Nakamura F. et al. (2011b) *Astrophys. J., 726,* 46.

Narayanan G. et al. (2012) *Mon. Not. R. Astron. Soc., 425,* 2641.

Norman C. and Silk J. (1980) *Astrophys. J., 238,* 158.

Offner S. S. R. et al. (2008) *Astrophys. J., 686,* 1174.

Offner S. S. R. et al. (2009) *Astrophys. J., 703,* 131.

Osterbrock D. E. and Ferland G. J. (2006) In *Astrophysics of Gaseous Nebulae and Active Galactic Nuclei* (J. Murdzek, ed.), University Science Books, Mill Valley, California. 461 pp.

Padoan P. and Nordlund Å. (2011) *Astrophys. J., 730,* 40.

Palau A. et al. (2013) *Astrophys. J., 762,* 120.

Palla F. and Stahler S. W. (1991) *Astrophys. J., 375,* 288.

Palla F. and Stahler S. W. (1992) *Astrophys. J., 392,* 667.

Parker J. W. (1993) *Astron. J., 106,* 560.

Patel N. A. et al. (1998) *Astrophys. J., 507,* 241.

Pellegrini E. W. et al. (2011) *Astrophys. J., 738,* 34.

Peters T. et al. (2010a) *Astrophys. J., 711,* 1017.

Peters T. et al. (2010b) *Astrophys. J., 719,* 831.

Peters T. et al. (2011) *Astrophys. J., 729,* 72.

Plunkett A. L. et al. (2013) *Astrophys. J., 774,* 22.

Portegies Zwart S. F. et al. (2010) *Annu. Rev. Astron. Astrophys., 48,* 431.

Price D. J. (2008) *J. Comp. Phys., 227,* 10040.

Price D. J. and Bate M. R. (2008) *Mon. Not. R. Astron. Soc., 385,* 1820.

Price D. J. and Bate M. R. (2009) *Mon. Not. R. Astron. Soc., 398,* 33.

Rathborne J. M. et al. (2008) *Astrophys. J., 689,* 1141.

Read J. I. and Hayfield T. (2012) *Mon. Not. R. Astron. Soc., 422,* 3037.

Rees M. J. (1976) *Mon. Not. R. Astron. Soc., 176,* 483.

Richling S. and Yorke H. W. (1997) *Astron. Astrophys., 327,* 317.

Robitaille T. P. et al. (2006) *Astrophys. J. Suppl., 167,* 256.

Robitaille T. P. et al. (2007) *Astrophys. J. Suppl., 169,* 328.

Rogers H. and Pittard J. M. (2013) *Mon. Not. R. Astron. Soc., 431,* 1337.

Sánchez-Monge Á. et al. (2013) *Astron. Astrophys., 557,* A94.

Sandford M. T. II et al. (1982) *Astrophys. J., 260,* 183.

Sandford M. T. II et al. (1984) *Astrophys. J., 282,* 178.

Schönke J. and Tscharnuter W. M. (2011) *Astron. Astrophys., 526,* A139.

Scoville N. Z. et al. (2001) *Astron. J., 122,* 3017.

Shore S. N. (1981) *Astrophys. J., 249,* 93.

Shore S. N. (1983) *Astrophys. J., 265,* 202.

Shu F. H. et al. (1999) In *The Origin of Stars and Planetary Systems* (C. J. Lada and N. D. Kylafis, eds.), p. 193. NATO ASIC Proc. 540, Crete, Greece.

Sicilia-Aguilar A. et al. (2013) *Astron. Astrophys., 551,* A34.

Silich S. and Tenorio-Tagle G. (2013) *Astrophys. J., 765,* 43.

Silva-Villa E. and Larsen S. S. (2011) *Astron. Astrophys., 529,* A25.

Smith N. et al. (2000) *Astrophys. J. Lett., 532,* L145.

Smith N. et al. (2005) *Astron. J., 129,* 888.

Smith N. et al. (2010) *Mon. Not. R. Astron. Soc., 406,* 952.

Snider K. D. et al. (2009) *Astrophys. J., 700,* 506.

Socrates A. and Sironi L. (2013) *Astrophys. J. Lett., 772,* L21.

Spaans M. et al. (1995) *Astrophys. J. Lett., 455,* L167.

Spitzer L. (1978) In *Physical Processes in the Interstellar Medium* (M. F. Bode and A. Evans, eds.), Wiley-VCH, New York. 335 pp.

Stamatellos D. et al. (2011) *Astrophys. J., 730,* 32.

Stamatellos D. et al. (2012) *Mon. Not. R. Astron. Soc., 427,* 1182.

Stanke T. and Williams J. P. (2007) *Astron. J., 133,* 1307.

Stinson G. S. et al. (2013) *Mon. Not. R. Astron. Soc., 428,* 129.

Sugitani K. et al. (1995) *Astrophys. J. Lett., 455,* L39.

Sun K. et al. (2006) *Astron. Astrophys., 451,* 539.

Swift J. J. (2009) *Astrophys. J., 705,* 1456.

Swift J. J. and Welch W. J. (2008) *Astrophys. J. Suppl., 174,* 202.

Tan J. C. et al. (2006) *Astrophys. J. Lett., 641,* L121.

Tenorio-Tagle G. et al. (2007) *Astrophys. J., 658,* 1196.

Thompson M. A. et al. (2012) *Mon. Not. R. Astron. Soc., 421,* 408.

Thompson T. A. et al. (2005) *Astrophys. J., 630,* 167.

Toalá J. A. and Arthur S. J. (2011) *Astrophys. J., 737,* 100.

Tomida K. et al. (2010a) *Astrophys. J. Lett., 725,* L239.

Tomida K. et al. (2010b) *Astrophys. J. Lett., 714,* L58.

Tout C. A. et al. (1999) *Mon. Not. R. Astron. Soc., 310,* 360.

Townsley L. K. et al. (2003) *Astrophys. J., 593,* 874.

Townsley L. K. et al. (2006) *Astron. J., 131,* 2140.

Townsley L. K. et al. (2011) *Astrophys. J. Suppl., 194,* 15.

Tremblin P. et al. (2012a) *Astron. Astrophys., 538,* A31.

Tremblin P. et al. (2012b) *Astron. Astrophys., 546,* A33.

Urban A. et al. (2009) *Astrophys. J., 698,* 1341.

Urban A. et al. (2010) *Astrophys. J., 710,* 1343.

Urquhart J. S. et al. (2007) *Astron. Astrophys., 467,* 1125.

van Kempen T. A. et al. (2009a) *Astron. Astrophys., 501,* 633.

van Kempen T. A. et al. (2009b) *Astron. Astrophys., 507,* 1425.

van Kempen T. A. et al. (2010) *Astron. Astrophys., 518,* L121.

Varricatt W. P. et al. (2010) *Mon. Not. R. Astron. Soc., 404,* 661.

Vázquez G. A. and Leitherer C. (2005) *Astrophys. J., 621,* 695.

Vázquez-Semadeni E. et al. (2010) *Astrophys. J., 715*, 1302.

Verdolini S. et al. (2013) *Astrophys. J., 769*, 12.

Vink J. S. et al. (2000) *Astron. Astrophys., 362*, 295.

Vink J. S. et al. (2001) *Astron. Astrophys., 369*, 574.

Visser R. et al. (2012) *Astron. Astrophys., 537*, A55.

Walawender J. et al. (2005) *Astron. J., 129*, 2308.

Walawender J. et al. (2008) In *Handbook of Star Forming Regions, Volume I* (B. Reipurth, ed.), p. 346. ASP Monograph, Astronomical Society of the Pacific, San Francisco.

Walch S. et al. (2013) *Mon. Not. R. Astron. Soc., 435*, 917.

Walch S. K. et al. (2012) *Mon. Not. R. Astron. Soc., 427*, 625.

Walmsley M. (1995) *Rev. Mex. Astron. Astrofis. Ser. Conf., 1*, 137.

Wang K. et al. (2011) *Astrophys. J., 735*, 64.

Wang K. et al. (2012a) *Astrophys. J. Lett., 745*, L30.

Wang K. et al. (2012b) *Astrophys. J. Lett., 745*, L30.

Wang P. et al. (2010) *Astrophys. J., 709*, 27.

Wang Y. et al. (2008) *Astrophys. J. Lett., 672*, L33.

Watson C. et al. (2008) *Astrophys. J., 681*, 1341.

Weaver R. et al. (1977) *Astrophys. J., 218*, 377.

Whitehouse S. C. and Bate M. R. (2006) *Mon. Not. R. Astron. Soc., 367*, 32.

Whitworth A. (1979) *Mon. Not. R. Astron. Soc., 186*, 59.

Whitworth A. P. et al. (1994a) *Astron. Astrophys., 290*, 421.

Whitworth A. P. et al. (1994b) *Mon. Not. R. Astron. Soc., 268*, 291.

Williams J. P. et al. (2003) *Astrophys. J., 591*, 1025.

Williams R. J. R. et al. (2001) *Mon. Not. R. Astron. Soc., 327*, 788.

Wise J. H. et al. (2012) *Mon. Not. R. Astron. Soc., 427*, 311.

Wolfire M. G. and Cassinelli J. P. (1986) *Astrophys. J., 310*, 207.

Wolfire M. G. and Cassinelli J. P. (1987) *Astrophys. J., 319*, 850.

Wood D. O. S. and Churchwell E. (1989) *Astrophys. J. Suppl., 69*, 831.

Wu Y. et al. (2005) *Astron. J., 129*, 330.

Wünsch R. et al. (2010) *Mon. Not. R. Astron. Soc., 407*, 1963.

Xie T. and Goldsmith P. F. (1994) *Astrophys. J., 430*, 252.

Yeh S. C. C. and Matzner C. D. (2012) *Astrophys. J., 757*, 108.

Yeh S. C. C. et al. (2013) *Astrophys. J., 769*, 11.

Yildiz U. A. et al. (2012) *Astron. Astrophys., 542*, A86.

Yorke H. W. and Kruegel E. (1977) *Astron. Astrophys., 54*, 183.

Yorke H. W. and Sonnhalter C. (2002) *Astrophys. J., 569*, 846.

Yorke H. W. and Welz A. (1996) *Astron. Astrophys., 315*, 555.

Zhang Q. and Wang K. (2011) *Astrophys. J., 733*, 26.

Zhang Q. et al. (2005) *Astrophys. J., 625*, 864.

Zhang Q. et al. (2009) *Astrophys. J., 696*, 268.

Zuckerman B. and Evans N. J. (1974) *Astrophys. J. Lett., 192*, L149.

Reipurth B., Clarke C. J., Boss A. P., Goodwin S. P., Rodríguez L. F., Stassun K. G., Tokovinin A., and Zinnecker H. (2014) Multiplicity in early stellar evolution. In *Protostars and Planets VI* (H. Beuther et al., eds.), pp. 267–290. Univ. of Arizona, Tucson, DOI: 10.2458/azu_uapress_9780816531240-ch012.

Multiplicity in Early Stellar Evolution

Bo Reipurth
University of Hawaii

Cathie J. Clarke
Institute of Astronomy, Cambridge

Alan P. Boss
Carnegie Institution of Washington

Simon P. Goodwin
University of Sheffield

Luis Felipe Rodríguez
Universidad Nacional Autónoma de Mexico

Keivan G. Stassun
Vanderbilt University

Andrei Tokovinin
Cerro Tololo Interamerican Observatory

Hans Zinnecker
NASA Ames Research Center

Observations from optical to centimeter wavelengths have demonstrated that multiple systems of two or more bodies is the norm at all stellar evolutionary stages. Multiple systems are widely agreed to result from the collapse and fragmentation of cloud cores, despite the inhibiting influence of magnetic fields. Surveys of class 0 protostars with millimeter interferometers have revealed a very high multiplicity frequency of about 2/3, even though there are observational difficulties in resolving close protobinaries, thus supporting the possibility that all stars could be born in multiple systems. Near-infrared adaptive optics observations of class I protostars show a lower binary frequency relative to the class 0 phase, a declining trend that continues through the class II/III stages to the field population. This loss of companions is a natural consequence of dynamical interplay in small multiple systems, leading to ejection of members. We discuss observational consequences of this dynamical evolution, and its influence on circumstellar disks, and we review the evolution of circumbinary disks and their role in defining binary mass ratios. Special attention is paid to eclipsing PMS binaries, which allow for observational tests of evolutionary models of early stellar evolution. Many stars are born in clusters and small groups, and we discuss how interactions in dense stellar environments can significantly alter the distribution of binary separations through dissolution of wider binaries. The binaries and multiples we find in the field are the survivors of these internal and external destructive processes, and we provide a detailed overview of the multiplicity statistics of the field, which form a boundary condition for all models of binary evolution. Finally, we discuss various formation mechanisms for massive binaries, and the properties of massive trapezia.

1. INTRODUCTION

Many reviews have been written on pre-main-sequence binaries over the past 25 years, e.g., *Reipurth* (1988), *Zinnecker* (1989), *Mathieu* (1994), and *Goodwin* (2010), and particular mention should be made of IAU Symposium No. 200 (*Zinnecker and Mathieu*, 2001), which is still today a useful reference. Most recently, *Duchêne and Kraus* (2013) review the binarity for stars of all masses and ages.

Stimulated by the growing discoveries of multiple systems among young stars, there is increasing interest in the idea, first formulated by *Larson* (1972), that all stars may be born in small multiple systems, and that the mixture of single, binary, and higher-order multiples we observe at different ages and in different environments may result from the dynamical evolution, driven either internally or externally, of a primordial population of multiple systems. While more work needs to be done to determine the multiplicity of newborn protostars, at least — as has been widely accepted for some time — binarity and multiplicity is clearly established as the principal channel of star formation. The inevitable implication is that dynamical evolution is an essential part of early stellar evolution. In the following we explore the processes and phenomena associated with the early evolution of multiple systems, with a particular emphasis on triple systems.

2. PHYSICS OF MULTIPLE STAR FORMATION

The collapse and fragmentation of molecular cloud cores (*Boss and Bodenheimer,* 1979) is generally agreed to be the mechanism most likely to account for the formation of the majority of binary and multiple star systems. Major advances in our physical understanding of the fragmentation process have occurred in the last decade as a result of the availability of adaptive mesh refinement (AMR) hydrodynamics (HD) codes, which allow the computational effort to be concentrated where it is needed, in regions with large gradients in the physical variables. Many of these AMR codes, as well as smoothed particle hydrodynamics (SPH) codes with variable smoothing lengths, have been extended to include such effects as radiative transfer (RHD) and magnetic fields (MHD), allowing increasingly realistic three-dimensional numerical models to be developed. We concentrate here on the theoretical progress made on three-dimensional models of the fragmentation process since *Protostars and Planets V* (*PPV*) appeared in 2007.

In *PPV*, the focus was on purely HD models of the collapse of turbulent clouds initially containing many Jeans masses, leading to abundant fragmentation and the formation of multiple protostar systems and protostellar clusters (*Bonnell et al.,* 2007; *Goodwin et al.,* 2007; *Whitworth et al.,* 2007). Three-dimensional HD modeling work has continued on initially turbulent, massive clouds, with an eye toward determining cluster properties such as the initial mass function (e.g., *Clark et al.,* 2008; *Offner et al.,* 2009) and the number of brown dwarfs formed (e.g., *Bonnell et al.,* 2008; *Bate,* 2009a,b; *Attwood et al.,* 2009). Three-dimensional HD SPH calculations by *Bate* (2009a) made predictions of the frequency of single, binary, triple, and quadruple star systems formed during the collapse of a highly unstable cloud with an initial mass of 500 $M_{\odot}$, a Jeans mass of 1 $M_{\odot}$, and a turbulent, high-Mach-number (13.7) velocity field. This simulation involved a sufficiently large population of stars and brown dwarfs (1250) so as to provide an excellent basis for comparison with observed multiple systems. It is remarkable that this simulation — which clearly omits important physical ingredients such as magnetic fields and radiative feedback — nevertheless results in a reasonable

match to a wide range of observed binary parameters. In parallel with this study of binarity within the context of cluster formation, other groups have instead pursued high-resolution core scale simulations of HD collapse of much lower mass, initially Bonnor-Ebert-like clouds, delineating how factors such as the initial rotation rate, metallicity, turbulence, and density determine whether the cloud forms a single or multiple protostar system [see *Arreaga-García et al.* (2010) and *Walch et al.* (2010) for SPH and *Machida* (2008) for AMR calculations of this type].

Despite this striking agreement between the outcomes of the simplest barotropic models and observations, it is nevertheless essential to conduct simulations that incorporate a more realistic set of physical processes. *Offner et al.* (2009) found that radiative feedback in three-dimensional RHD AMR calculations could indeed have an important effect on stellar multiplicity, primarily by reducing the number of stars formed. They also emphasized (*Offner et al.,* 2010) that the inclusion of radiative feedback changes the dominant mode of fragmentation: With a barotropic equation of state, fragmentation normally occurs at the point when the flow is centrifugally supported — i.e., when it collapses into a disk at radii <100 AU. This mode is relatively suppressed when radiative feedback is included and the fragments mainly form from turbulent fluctuations within the natal core, at separations ~1000 AU. Such initially wide pairs, however, spiral in to smaller separations, an effect also found in the simulations of *Bate* (2012), which are the radiative counterparts of the previous (*Bate,* 2009a) calculations (see also *Bate,* 2009b). The resulting binary statistics are scarcely distinguishable from those in the earlier barotropic calculations and again in good agreement with observations (see Fig. 1).

Observations of molecular clouds have shown that magnetic fields are generally more dynamically important than turbulence, but are only one source of cloud support against gravitational collapse for cloud densities in the range of 10^{3}–10^{4} cm^{-3} (*Crutcher,* 2012). While it has long been believed that magnetic field support is lost through ambipolar diffusion, leading to gravitational collapse, current observations do not support this picture (*Crutcher,* 2012), but rather one where magnetic reconnection eliminates the magnetic flux that would otherwise hinder star formation (*Lazarian et al.,* 2012). Three-dimensional MHD calculations of collapse and fragmentation have become increasingly commonplace, although usually assuming ideal magneto-hydrodynamics (i.e., frozen-in fields) rather than processes such as ambipolar diffusion or magnetic reconnection.

Machida et al. (2008) found that fragmentation into a wide binary could occur provided that the initial magnetic cloud core rotated fast enough, while close binaries resulted when the initial magnetic energy was larger than the rotational energy. *Hennebelle and Fromang* (2008) and *Hennebelle and Teyssier* (2008) found that initially uniform density and rotation magnetic clouds could fragment if a density perturbation was large enough (50% amplitude), as in the standard isothermal test case of *Boss and Boden-*

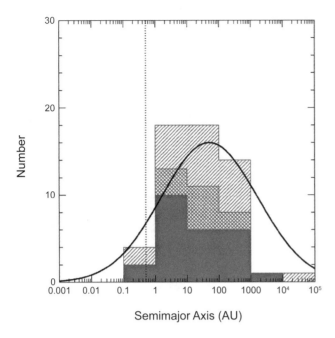

Fig. 1. Distributions of semimajor axes for primaries with masses greater than 0.1 $M_\odot$ (histogram) from *Bate* (2012), compared to observations [solid line (*Raghavan et al.,* 2010)]. Solid, double-hatched, and single-hatched histograms are for binaries, triples, and quadruples, respectively. The vertical line is the resolution limit of the SPH calculation.

heimer (1979). *Price and Bate* (2007), *Bürzle et al.* (2011), and *Boss and Keiser* (2013) all studied the collapse of initially spherical, 1-$M_\odot$ magnetic cloud cores, with uniform density, rotation, and magnetic fields, the MHD version of *Boss and Bodenheimer* (1979). They found that clouds could collapse to form single, binary, or multiple protostar systems, depending on such factors as the initial magnetic field strength and its orientation with respect to the rotation axis. When fragmentation did occur, binary star systems were the typical outcome, along with a few higher-order systems. *Joos et al.* (2012) found that the initial direction of the magnetic field with respect to the rotation axis had an important effect on whether the collapse produced a protostellar disk that might later fragment into a multiple system.

Radiative transfer effects were included in the models of *Commerçon et al.* (2010), who studied the collapse of 1-$M_\odot$ clouds with an AMR RMHD code, finding that frozen-in fields always inhibited cloud fragmentation. *Boss* (2009) used a three-dimensional pseudo-MHD code with radiative transfer in the Eddington approximation to study the collapse and fragmentation of prolate and oblate magnetic clouds, including the effects of ambipolar diffusion, finding that the oblate clouds collapsed to form rings, susceptible to subsequent fragmentation, while prolate clouds collapsed to form either single, binary, or quadruple protostar systems. *Kudoh and Basu* (2008, 2011) also included ambipolar diffusion in their true MHD models, finding that collapse could be accelerated by supersonic turbulence.

There has also been progress in adding magnetic fields and feedback from radiation and outflows into simulations of more massive clouds, although many such simulations do not resolve fragmentation on scales less than ~100 AU (e.g., *Krumholz et al.,* 2012; *Hansen et al.,* 2012) and are thus not the simulations of choice for following binary formation. *Hennebelle et al.* (2011) followed the collapse of a 100-$M_\odot$ cloud with their AMR MHD code, and found that the magnetic field could reduce the degree of fragmentation, compared to a nonmagnetic cloud collapse, by as much as a factor of 2. *Commerçon et al.* (2011) extended their previous work on 1-$M_\odot$ clouds to include 100-$M_\odot$ clouds, but again found that magnetic fields and radiative transfer combined to inhibit fragmentation. *Seifried et al.* (2012) found that their 100-$M_\odot$ turbulent, magnetic clouds collapsed to form just a relatively small number of protostars. Likewise, the high-resolution simulations of *Myers et al.* (2013) (which combine the inclusion of radiative transfer and magnetic fields with a resolution of 10 AU within a 1000-$M_\odot$ cloud) find that these effects in combination strongly suppress binary formation within the cloud.

While powerful theoretical tools now exist, along with widespread access to large computational clusters, the huge volume of parameter space that needs to be explored has to date prevented a comprehensive theoretical picture from emerging. Nevertheless, it is clear that in spite of the various magnetic field effects, MHD collapse and fragmentation remains as a possibility in at least some portions of the parameter space of initial conditions. When fragmentation does occur in the collapse of massive, magnetic clouds, relatively small numbers of fragments are produced, compared to the results of three-dimensional models of nonmagnetic, often turbulent collapse, where much larger numbers of fragments tend to form (e.g., *Bonnell et al.,* 2007; *Whitworth et al.,* 2007). While such massive clouds might form small clusters of stars, low-mass magnetized clouds are more likely to form single or binary star systems.

In summary, then, it is premature to draw definitive conclusions about the conditions required to produce a realistic population of binary systems. Those simulations that can offer a statistical ensemble of binary star systems for comparison with observations are able to match the data very well, regardless of whether thermal feedback is employed (*Bate,* 2009a, 2012). The thermal feedback in these latter simulations is, however, underestimated somewhat (*Offner et al.,* 2010), and therefore represents an interim case between the full-feedback and no-feedback case. It remains to be seen whether simulations with magnetic fields and full feedback do an equally good job at matching the binary statistics, despite the indications from the studies listed above that these effects tend to suppress binary fragmentation.

3. DEFINITION OF MULTIPLICITY

In order to discuss observational results and compare the multiplicity for different evolutionary stages and/or in different regions, we need simple and precise terminology.

Following *Batten* (1973), the fractions of systems containing exactly n stars are denoted as f_n. The multiplicity frequency or multiplicity fraction MF = $1-f_1 = f_2 + f_3 + f_4 + \ldots$ gives the fraction of nonsingle systems in a given sample. This is more commonly written

$$MF = \frac{B + T + Q}{S + B + T + Q} \qquad (1)$$

where S, B, T, and Q are the number of single, binary, triple, and quadruple, etc., systems (*Reipurth and Zinnecker,* 1993).

Another common characteristic of multiplicity, the companion star fraction CSF = $f_2 + 2f_3 + 3f_4 + \ldots$, quantifies the average number of stellar companions per system; it is commonly written

$$CSF = \frac{B + 2T + 3Q}{S + B + T + Q} \qquad (2)$$

which is the average number of companions in a population, and in principle can be larger than 1 (e.g., *Ghez et al.,* 1997). Measurements of MF are less sensitive to the discovery of all subsystems than CSF, explaining why MF is used more frequently in comparing theory with observations. The fraction of higher-order multiples is simply HF = $1-f_1-f_2 = f_3 + f_4 + \ldots$

The vast majority of observed multiple systems are *hierarchical*: The ratio of separations between their inner and outer pairs is large, ensuring long-term dynamical stability. Stellar motions in stable hierarchical systems are represented approximately by Keplerian orbits. Hierarchies can be described by binary graphs or trees (Fig. 2). The position of each subsystem in this graph can be coded by its level. The outermost (widest) pair is at the root of the tree (level 1). Inner pairs associated with primary and secondary components of the outer pair are called levels 11 and 12,

respectively, and this notation continues to deeper levels. Triple systems can have inner pairs at level 11 or 12. When both subsystems are present, we get the so-called 2 + 2 *quadruple*. Alternatively, a *planetary* quadruple system consists of levels 1, 11, and 111; it has two companions associated with the same primary star.

In principle, the most precise description of multiplicity statistics would be the joint distribution of the main orbital parameters (period or semimajor axis, mass ratio, and eccentricity) at all hierarchical levels. But even for simple binaries such a three-dimensional distribution is poorly known, and the number of variables and complexity increases quickly when dealing with triples, quadruples, etc. To first order, the multiplicity is characterized by the fractions f_n or by their combinations such as MF (which equals the fraction of level-1 systems), CSF, and HF.

4. OBSERVATIONS OF PROTOSTELLAR BINARIES AND MULTIPLES

Studies of binaries and multiples during the protostellar stage are important, since they offer the best chance of seeing the results of fragmentation of molecular clouds, as discussed in section 2. However, most protostars are still deeply embedded, so such observations are hampered by extinctions that can exceed $A_V \sim 100$ mag. Hence infrared, submillimeter, or radio continuum observations are required.

4.1. Infrared Observations

Class I protostars are often detectable at near-infrared wavelengths, although for disk orientations near edge-on, one sees them only in scattered light. In contrast, the massive circumstellar environment of class 0 sources make them detectable only at longer wavelengths. *Haisch et al.* (2004) performed a near-infrared imaging survey of 76 class I sources and found a companion star fraction of 18 ± 4% in the separation range ~300–2000 AU. In a similar study, *Duchêne et al.* (2004) obtained a companion star fraction of 27 ± 6% in the range 110–1400 AU. To detect closer companions, *Duchêne et al.* (2007) used adaptive optics to survey 45 protostars, and found a companion star fraction of 47 ± 8% in the range 14–1400 AU; comparison of the two numbers indicates the prevalence of close protostellar companions. In a major survey of 189 class I sources, *Connelley et al.* (2008a,b) detected 89 companions, and the separation distribution function is shown in Fig. 3a. For the closer separations, it is seen to be very similar to that of T Tauri binaries. But for larger separations, a clear excess of wide companions (with separations up to 4500 AU) becomes evident, which is not seen for the more evolved T Tauri stars. When plotting the binary fraction as a function of spectral index, which measures the amount of circumstellar material and is used as a proxy for stellar age, they find a dramatic decline in these wide companions (Fig. 3b), from ~50% to <5%. In other words, powerful dynamical processes must occur during the class I phase,

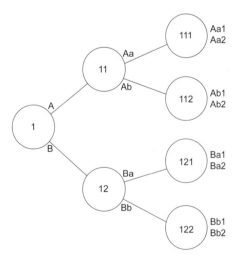

Fig. 2. The structure of hierarchical multiple systems can be represented by a binary graph; the figure describes all possible multiples up to an octuple system. The position of each subsystem is coded by levels indicated in the circles. The nomenclature follows the IAU recommendation.

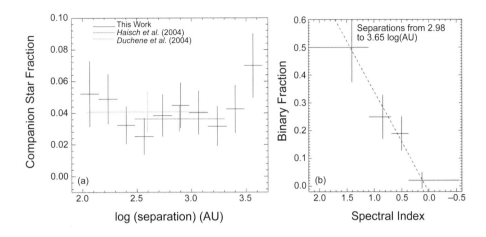

Fig. 3. (a) The separation distribution function of embedded protostellar binaries. There is a strong excess of widely separated companions with separations larger than 1000 AU. **(b)** The population of wide companions is found to disappear with decreasing spectral index, which is a proxy for age. From *Connelley et al.* (2008a,b).

leading to the dispersal of a significant population of wide companions. These observations can be understood in the context of the dynamical evolution of newborn multiple systems, which in most cases break up, leading to the ejection of one of the components (see section 5). Such ejected components should be observable for a while, and *Connelley et al.* (2009) found that of 47 protostars observed with adaptive optics, every target with a close companion has another young star within a projected separation of 25,000 AU.

The study of even closer companions to protostars is still in its infancy. In a pilot program, *Viana Almeida et al.* (2012) found large radial velocity variations in three out of seven embedded sources in Ophiuchus, and speculated that they could be evidence for spectroscopic protobinaries. *Muzerolle et al.* (2013) used the Spitzer Space Telescope to monitor IC 348 and found a protostar showing major luminosity changes on a period of 25.34 days; the most likely explanation is that a companion in an eccentric orbit drives pulsed accretion around periastron.

4.2. Submillimeter Observations

The study of binarity of the youngest protostars, the class 0 sources, requires longer wavelength observations, and millimeter interferometry has become a powerful tool to study binarity of protostars. Pioneering work was done by *Looney et al.* (2000), who observed seven class 0 and I sources; further small samples of embedded sources were observed by, e.g., *Chen et al.* (2008, 2009), *Maury et al.* (2010), *Enoch et al.* (2011), and *Tobin et al.* (2013). All these studies suffer from very small samples, and hence yield uncertain statistics. This problem has been alleviated by *Chen et al.* (2013), who presented high-angular-resolution 1.3-mm and 850-µm dust continuum data from the Submillimeter Array for 33 class 0 sources. No less than 21 of the sources show evidence for companions in the projected separation

range from 50 to 5000 AU. This leads to a multiplicity frequency MF = 0.64 ± 0.08 and a companion star fraction CSF = 0.91 ± 0.05 for class 0 protostars. As noted by *Chen et al.* (2013), their survey is complete only for systems larger than ~1800 AU, and hence these values must be regarded as lower limits. Given that numerous class 0 binaries may have much closer companions, these results are consistent with the possibility that virtually all stars are born as binaries or multiples, an idea that dates back to *Larson* (1972). Figure 4 shows in graphical form the observed decrease in binarity as a function of evolutionary stage, a result that strongly supports a view of early stellar evolution in which small multiple systems evolve dynamically and then break up, and the decay products eventually evolve into the distribution of singles, binaries, and higher-order multiples we observe in the field.

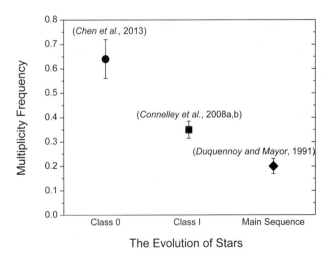

Fig. 4. The multiplicity frequency declines through the protostellar phase because of the breakup of small multiple systems. From *Chen et al.* (2013).

4.3. Radio Continuum Observations

As is clear from the discussion above, it is critically important to study protostars with much higher resolution in order to determine the multiplicity at small separations. Radio observations are the only technique available at present that allows the study with high angular resolution of the earliest stages of star formation. These studies can be performed with an angular resolution on the order of 0.1 arcsec with radio interferometers such as the Jansky Very Large Array (VLA) and the expanded Multi-Element Radio-Linked Interferometer Network (eMERLIN). What is detected here is the free-free emission from the base of the ionized outflows that are frequently present at early evolutionary stages. These structures trace the star with high precision, and favor the detection of very young class 0, I, and II objects.

A series of VLA studies (e.g., *Rodríguez et al.,* 2003, 2010; *Reipurth et al.,* 2002, 2004) show binary and multiple sources clustered on scales of a few hundred astronomical units. A binary frequency on the order of ~33% is found in these studies. Since not all sources show free-free emission, and those that do are often found to be variable, such statistics provide only lower limits.

If the star has strong magnetospheric activity, the resulting gyrosynchrotron emission is compact and intense enough to be observed with the technique of Very Long Baseline Interferometry (VLBI), which can reach angular resolutions on the order of 1 milliarcsecond and better, and which allows the study of stellar motions with great detail. This technique favors the detection of the more evolved class III stars. It should be noted, however, that at least one class I protostar, IRS 5b in Corona Australis, has been detected with VLBI techniques (*Deller et al.,* 2013). In a series of studies to determine the parallax of young stars in Gould's belt (*Loinard,* 2013), it has been found that several are binary and it has been possible to follow their orbital motions (e.g., *Torres et al.,* 2012) and study the radio emission as a function of separation, finding evidence of interaction between the individual magnetospheres. Radio emission of nonthermal origin has been detected all the way down to the ultracool dwarfs (late M, L, and T types), in some sources in the form of periodic bursts of extremely bright, 100% circularly polarized, coherent radio emission (e.g., *Hallinan et al.,* 2007).

With the new generation of centimeter and millimeter interferometers, especially the Atacama Large Millimeter Array (ALMA), the field of radio emission from binary and multiple young stellar systems faces a new era of opportunity that should result in much better statistics, especially in the protostellar stage.

5. DYNAMICS OF MULTIPLE STARS

If three bodies are randomly placed within a volume, then more than 98% of the systems will be in a nonhierarchical configuration, i.e., the third body is closer than ~10× the separation of the other two bodies. It is well known that such configurations are inherently unstable, and will on a timescale of around 100 crossing times decay into a hierarchical configuration, in a process where the third body is ejected, either into a distant orbit or into an escape (see Fig. 5) (e.g., *Anosova,* 1986; *Sterzik and Durisen,* 1998; *Umbreit et al.,* 2005). The energy to do this comes from the binding energy of the remaining binary, which as a result shrinks and at the same time frequently gets a highly eccentric orbit. For such an ejection to take place, the three bodies must first meet in a close triple approach, during which energy and momentum can be exchanged. A detailed analysis of the dynamics of triple systems can be found in *Valtonen and Karttunen* (2006).

N-body simulations that include the gravitational potential of a cloud core reveal that many systems break up shortly after formation, sending the third body into an escape, but the majority goes through several or many ejections that are too weak to escape the potential well, and the third body thus falls back (*Reipurth et al.,* 2010). As the cloud core gradually shrinks through accretion, outflows, and irradiation, the third body eventually manages to escape. In some cases the triple remains bound until after the core has disappeared (see Fig. 6), but only about ~10% of triples are stable enough to survive on long timescales. The body that is ejected is most often the lowest-mass member, but complex dynamics can lead to many other configurations and outcomes. Stochastic events play an essential but unpredictable role in the early stages of triple systems, and so their evolution can only be understood statistically.

A stability analysis of hierarchical bound triple systems formed in N-body simulations shows that they divide into stable and unstable systems. Any time that a distant third component passes through a periastron passage and comes close to the inner binary, there is the possibility of an in-

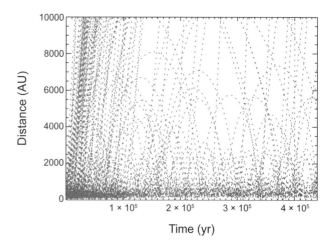

Fig. 5. One hundred simulations showing the dynamical evolution of a triple system of three 0.5-M$_\odot$ stars with initial mean separations of 100 AU embedded in a 3-M$_\odot$ cloud core. Many of the initial ejections are escapes, but the majority fall back, to be ejected sometimes again and again. From *Reipurth et al.* (2010).

stability of the system, depending on the configuration of the inner binary. Stable triples remain bound for hundreds of millions or billions of years, but unstable systems can break apart at any time. Figure 7 shows the fraction of triples that after 100 million years (m.y.) are stable, unstable, or already disrupted, as function of their projected separation, from a major N-body simulation. For separations less than 10,000 AU (vertical line) the majority is stable, but for wider separations unstable systems dominate. For young systems in star-forming regions, however, unstable systems significantly dominate at all separations. These unstable systems will soon break apart. For young ages, one therefore observes many more triple systems than at older ages (*Reipurth and Mikkola*, 2012).

Triple systems can be classified in the triple diagnostic diagram (Fig. 8a), where the mass ratio of the binary is plotted as a function of the mass of the third body relative to the total system mass (Reipurth and Mikkola, in preparation). In the righthand side of the diagram reside the systems that are dominated by a massive single star (S-type), to the left are those where a massive binary dominates (B-type), and in the middle are the systems where the mass is about equally distributed in the binary and the single (E-type). Subdivisions can additionally be made depending on the mass ratio of the binary (high, medium, low). Note that since the axes represent ratios, i.e., dimensionless numbers, then the absolute mass of the system is not involved. This simple classification system encompasses all categories of triple systems. As the name indicates, the distribution of systems in the diagram harbors important diagnostics for understanding the early evolution of triple systems. Figure 8b shows the result of N-body calculations that include

accretion as the three bodies move around each other inside the cloud core. All systems in the diagram are long-term stable. To better isolate the interplay between dynamics and accretion, all three components started out with equal masses, i.e., they were initially placed at (0.333,1.000). As is evident, the interplay between dynamics and accretion can lead to very different outcomes, with some areas of the diagram populated much more densely than others. Comparison with complete, unbiased samples of triples will provide much insight into the formation processes of triple systems.

5.1. Origin of Brown Dwarf/Very-Low-Mass Binaries

The formation of very-low-mass ($M \leq 0.1\ M_\odot$) objects has been debated for a long time, and three basic ideas have emerged: A very-low-mass (VLM) object can form if the nascent core has too little mass (*Padoan and Nordlund*, 2004), or it can form if the stellar seed is removed from the infall zone through dynamical ejection (*Reipurth and Clarke*, 2001; *Stamatellos et al.*, 2007; *Basu and Vorobyov*, 2012), or the cloud core can photoevaporate if a nearby OB star is formed (*Whitworth and Zinnecker*, 2004). The emerging consensus is that all three mechanisms are likely to operate under different circumstances, and that the relevant question is not which mechanism is correct, but how big their relative contributions are to the production of VLM objects (*Whitworth et al.*, 2007). Similarly, brown dwarf/VLM binaries are likely to have several formation mechanisms.

Extensive numerical studies combining N-body simulations with accretion have shown that the large majority of brown dwarf ejections are not violent events, but rather the

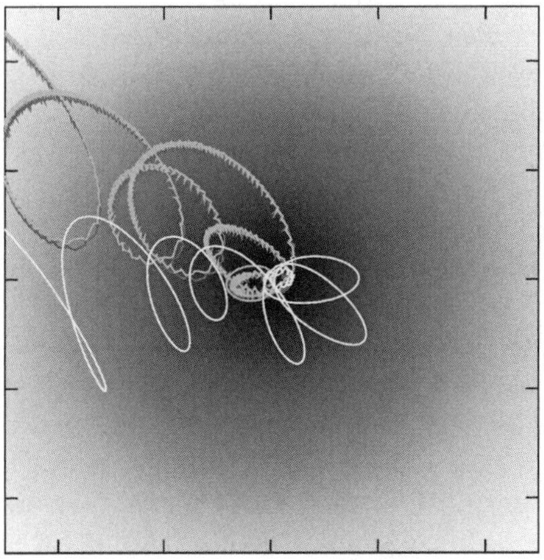

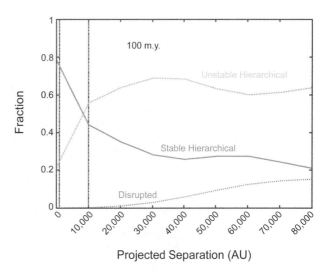

Fig. 6. An example of the chaotic orbits of three bodies born in and accreting from a cloud core. The triple system in this simulation remains bound as it drifts away from the core, but is unstable and will eventually break apart. The figure is 10,000 AU across. From Reipurth and Mikkola (in preparation).

Fig. 7. The relative numbers of bound stable, bound unstable, and unbound triple systems as function of the projected separation of the outer pairs, from a major N-body simulation after 100 m.y. The majority of very wide binaries is unstable at this age. At much younger ages, say 1 m.y., unstable systems dominate at all separations. From *Reipurth and Mikkola* (2012).

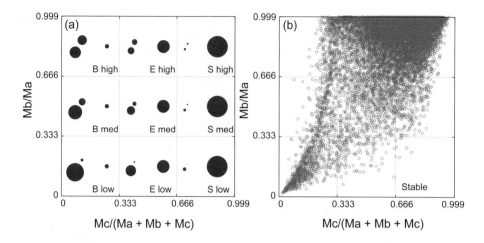

Fig. 8. **(a)** Location and definition of nine different types of triple systems in the triple diagnostic diagram. **(b)** The location of 15,524 stable triple systems in the triple diagnostic diagram at an age of 1 m.y. Since all triple systems in these simulations were started out with three identical bodies, the original systems were all located at the point (0.333,1.000). Their final location is determined by their dynamical evolution and resulting accretion. From Reipurth and Mikkola (in preparation).

result of unstable triple systems that eventually drift apart at very small velocities, typically within the first 100 m.y. (Reipurth and Mikkola, in preparation). When brown dwarfs are released from triple systems, by far the majority of the remaining binaries are VLM objects. These binaries gently recoil and become isolated VLM binaries. These VLM binaries have a semimajor axis distribution that peaks at around 10–15 AU, but with a tail stretching out to ~250 AU. At shorter separations, the simulations show a steep decline in number of systems, although the simulations underestimate the number of close binaries because they do not take viscous orbital evolution into account. Brown dwarf and VLM binaries formed through dynamical interactions can in principle have much larger separations, of many hundreds or thousands of astronomical units, but in that case they must be bound triple systems, where one component is a close, often unresolved, binary. More than 90% of bound triple systems at 1 m.y. have dispersed by 100 m.y., and all VLM triple systems with outer semimajor axes less than a few hundred astronomical units have broken up. In this context, it is interesting that *Biller et al.* (2011) found an excess of 10–50 AU young brown dwarf binaries in the 5-m.y.-old Upper Scorpius association compared to the field.

5.2. Origin of Spectroscopic Binaries

Spectroscopic binaries is a generic term for all binaries that have separations so close that their orbital motion is measurable with radial velocity techniques. In practical terms, the large majority of known spectroscopic binaries has a period less than ~4000 days. Surveys of metal-poor field stars (for which statistics are particularly good) find that $18 \pm 4\%$ are spectroscopic binaries (*Carney et al.,* 2003).

Radial velocity studies of young stars are complicated by the sometimes wide and/or complex line profiles, but an increasing number of pre-main-sequence spectroscopic

binaries are now known (e.g., *Mathieu,* 1994; *Melo et al.,* 2001; *Prato,* 2007; *Joergens,* 2008; *Nguyen et al.,* 2012). In the Orion Nebula Cluster (ONC), *Tobin et al.* (2009) have found so far that 11.5% of the observed members are spectroscopic binaries, but the survey is still ongoing.

Spectroscopic binaries have semimajor axes that are often measured in units of stellar radii, and rarely exceed a few astronomical units. These binaries cannot form with such close separations, and they must therefore result from processes that cause a spiral-in of an initially wider binary system. Given that newborn binaries and multiples are surrounded by a viscous medium during the protostellar phase, components can naturally spiral in during the star-forming process because of dynamical friction with the surrounding medium (e.g., *Gorti and Bhatt,* 1996; *Stahler,* 2010; *Korntreff et al.,* 2012) (see section 9).

The evolution of embedded triple systems can enhance or initiate this process (e.g., *Bate et al.,* 2002). When a newborn triple system is transformed from a nonhierarchical configuration to a hierarchical one, the newly bound binary shrinks in order for the third body to be ejected into a more distant orbit or into an escape. The binary also gets a highly eccentric orbit as a result of this process. Given that the components are surrounded by abundant circumstellar material at early evolutionary stages, the shrinkage and the eccentric orbits force regular dissipative interactions, leading to orbital decay. As discussed by *Stahler* (2010), the ultimate result can be a merger. But if the infall of new material from an envelope ceases before that, then the orbital decay is halted, and the binary ends up with the orbital parameters it happens to have when the viscous interactions cease [except for very close binaries, where orbital circularization will occur (*Zahn and Bouchet,* 1989)]. Spectroscopic binaries originating in a triple system therefore have an important stochastic element in their evolution, depending on when the triple system broke up and when circumstellar material became exhausted.

5.3. Origin of Single Stars

The strong increase in number of stars (single and nonsingle) for decreasing mass combined with the strong decrease in number of binaries also for decreasing mass (see section 12) led *Lada* (2006) to conclude that the majority of stars in our Galaxy are single. And for solar-type stars in the solar neighborhood, *Raghavan et al.* (2010) found that 56% are single. This high preponderance of single stars is not consistent with the very high multiplicity frequency determined among protostars (see section 2), and leads to interesting questions about the origin of single stars.

When we observe a single star, it may have one of three origins: It can be born in isolation; it may have been born in a multiple system that decayed and ejected one component; or it may even be the product of two stars in a binary that spiraled in and merged during the protostellar phase. Historically, mergers have been considered almost exclusively in the contexts of late stellar evolutionary stages or massive stars. Intriguingly, new N-body simulations of low-mass small-N multiple systems and studies of orbital decay in a viscous medium indicate that mergers may occur in a nonnegligible number of cases during early stellar evolution (*Rawiraswattana et al.*, 2012; *Stahler*, 2010; *Leigh and Geller*, 2012; *Korntreff et al.*, 2012).

Small triple systems, whether formed in isolation or in a cluster, will evolve dynamically, and ~90% break up, each producing a single star, which drifts away with a velocity around 1 km s^{-1}. This corresponds to ~100,000 AU in half a million years, so very soon such ejecta will disperse and any trace of their origin will be lost. Because of dynamical processing, it is the lowest-mass components that tend to escape. Newborn higher-order multiples such as quadruples, pentuples, sextuples, etc., may produce more than one single star per star-forming event.

The formation of a single star from a collapse event is — not surprisingly — the standard view of the origin of single stars. However, the very high multiplicity of protostars (see section 4.2) has by now made it clear that single-star collapse is not the principal channel of star formation. And it should by no means be automatically assumed that young single stars found in a low-mass star-forming region represent cases of single, isolated star formation.

6. OBSERVATIONAL CONSEQUENCES

The dynamical evolution discussed above has observational consequences, as discussed in the sections below.

6.1. FUor Eruptions

The close triple encounters in triple systems, which are prerequisites for the ejection of one of the components, are statistically most likely to occur during the protostellar stage (*Reipurth*, 2000). At this stage the three bodies are surrounded by significant amounts of circumstellar material, which will interact and cause a major brightening, from accretion and shock-heating. These events we here call "encounters of type 1." After the hierarchical configuration has been achieved, the shrinking of the binary orbit and its high eccentricity will lead to a series of disk-disk interactions at each periastron passage (Fig. 9). The disks will be seriously disturbed, causing eruptions, but much of the mass will fall back and reassemble in the disk again (*Clarke and Pringle*, 1993; *Hall et al.*, 1996; *Umbreit et al.*, 2011) (see section 9.6). As a result of this viscous evolution, the binary shrinks until the point when the stars are so close that the circum*stellar* material shifts from being in two circumstellar disks to instead assemble in one circum*binary* disk (*Reipurth and Aspin*, 2004). This sequence of eruptions is called "encounters of type 2." Finally, if the triple evolution occurs so early that abundant gas is present, then the inspiral phase of the binary can result in the coalescence of the two stars (e.g., *Stahler*, 2010; *Rawiraswattana et al.*, 2012; *Leigh and Geller*, 2012); such events are called "encounters of type 3." Observations have revealed various types of outbursts among young stars, the main one being the FUor eruptions (*Herbig*, 1977) (see the chapter by Audard et al. in this volume). Once enough detailed observations have become available, it may be possible to identify those that result from triple evolution, since each of the above types of encounters are likely to have characteristic energy releases and timescales, which may make them identifiable. It will be challenging to disentangle the various types of eruptions observed, since disks obviously can be disturbed also internally through instabilities, and disks have limited ways to react to perturbations, whether internal or external.

6.2. Herbig-Haro Flows

Accretion and outflow is generally coupled, and so the above-mentioned encounters will give rise to different outflow characteristics, at young ages manifested as Herbig-Haro flows (*Reipurth and Bally*, 2001). Encounters of type 1 from close triple approaches will result in one or a few giant bow shocks, while a sequence of type 2 encounters will produce

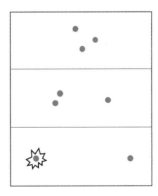

Fig. 9. A schematic plot of the evolution of a triple system, from nonhierarchical to hierarchical (top two panels) followed by the binary viscous in-spiral phase leading to disk-disk interactions and in some cases stellar mergers.

closely spaced knots, driven by cyclic accretion modulated on an orbital timescale, as seen in the finely collimated Herbig-Haro jets. Once the binary components have spiraled in so close that disk truncation rips up the magnetic field anchoring that supports the jet launch platform, the collimated outflow phase is terminated, and subsequent mass loss will appear as massive but uncollimated winds, like those seen in the spectra of FUor eruptions. Seen in this perspective, giant Herbig-Haro flows represent a fossil record of the accretion history primarily dictated by the orbital evolution of their driving sources, which are expected to be multiple, as frequently observed (*Reipurth*, 2000). Other disk instabilities can also form Herbig-Haro flows, but on a smaller scale.

6.3. Orphaned Protostars

The many dynamical ejections in which the third body fails to escape the potential well of the core plus remaining binary instead lead to large excursions, where the third body for long periods is tenuously bound in the outskirts or outside the cloud core. If such ejections occur during the protostellar stage, as many do, then these orphaned protostars open the possibility to study naked protostars still high up on their Hayashi tracks at near-intrared and even at optical wavelengths (*Reipurth et al.,* 2010). The triple system T Tauri may be such a case.

6.4. Formation of Wide Binaries

Binaries with semimajor axes as large as 10,000 AU are now frequently found thanks to increasing astrometric precision. They challenge our understanding of star formation because their separations exceed the typical size of a collapsing cloud core. The dynamical evolution of multiple systems offers a simple way to form such wide binaries: Although born compact, a triple system can dynamically scatter a component to very large distances, thus unfolding the triple system into an extreme hierarchical architecture. Many very wide binaries are therefore likely to be triples or higher-order multiples, although true wide binaries can also form when a merger has taken place in an encounter of type 3 (*Reipurth and Mikkola,* 2012). Another independent mechanism that forms wide binaries in clusters is discussed in section 10.2.

7.   PRE-MAIN-SEQUENCE BINARIES/MULTIPLES

It has been known since the early studies of T Tauri stars that some are binaries (*Joy and van Biesbroeck,* 1944; *Herbig,* 1962). Further interest was spurred by the discovery of an infrared companion to T Tauri by *Dyck et al.* (1982). In 1993, three major surveys appeared that established that T Tauri stars have about twice the binary frequency compared to field stars, at least for the wider pairs (*Reipurth and Zinnecker,* 1993; *Leinert et al.,* 1993; *Ghez et al.,* 1993). In the following, we examine the status 20 years later.

7.1.   Statistics and Environment

When comparing the multiplicity of young stars to the field, the key reference for solar-type field stars has been *Duquennoy and Mayor* (1991). Since that study, observational techniques have improved, and *Raghavan et al.* (2010) have studied 454 F6-K3 dwarf and subdwarf stars within 25 pc using many different techniques. Their observed fractions of single, binary, triple, and quadruple stars are 56 ± 2%: 33 ± 2%: 8 ± 2%: 3 ± 1%, yielding a completion corrected multiplicity frequency of 46%, and implying that among *solar-type* stars, the majority are single. They also found that 25% of nonsingle stars are higher-order multiples, and that the percentage of triple and quadruple systems is roughly twice that estimated by *Duquennoy and Mayor* (1991). Systems with larger cross sections, i.e., those with more than two components or with long orbital periods, tend to be younger, indicating the loss of components with time. *De Rosa et al.* (2014) have studied 435 A-type stars and, within the errors, find the precise same fractions of singles, binaries, etc., as *Raghavan et al.* (2010) did for later-type stars.

Among more massive stars, the radial velocity study of *Chini et al.* (2012) examined 250 O-type stars and 540 B-type stars and found that more than 82% of stars with masses above 16 M$_\odot$ form *close* binaries, but that this high frequency drops monotonically to less than 20% for stars of 3 M$_\odot$ (see section 11). For late-type stars, *Fischer and Marcy* (1992) found a binary frequency of 42 ± 9% among nearby M dwarfs, while *Bergfors et al.* (2010) for M0–M6 dwarfs measured 32 ± 6% in the range 3–180 AU. For very late-type stars (M6 and later), *Allen* (2007) determined a binary frequency of 20–22%, consistent with the ~24% binary frequency found for L dwarfs by *Reid et al.* (2006). And *Kraus and Hillenbrand* (2012) found a smooth decline in binary frequency from 0.5 M$_\odot$ to 0.02 M$_\odot$. Altogether, these results confirm the trends seen in various other investigations (e.g., *Raghavan et al.,* 2010), namely that binarity is a strongly decreasing function with decreasing stellar mass.

For young stars, getting good statistics is obviously more difficult. The more massive young stars, the Herbig Ae/Be stars, have long been known to have a high binarity. *Leinert et al.* (1997) used speckle interferometry to find a binary frequency of 31% to 42%, while *Baines et al.* (2006) used spectroastrometry to determine a binary frequency of 68 ± 11%, with a hint that the binarity of Herbig Be stars is higher than for the Herbig Ae stars. To this should be added the spectroscopic binaries, which *Corporon and Lagrange* (1999) found to be around 10%. *Kouwenhoven et al.* (2007) analyzed several datasets on the Upper Sco association, and found that intermediate-mass stars have a binary frequency >70% at a 3σ confidence level.

The most thoroughly examined low-mass star-forming region is Taurus-Auriga, and in a detailed study *Kraus et al.* (2011) found an observed multiplicity frequency of ~60% for separations in the range 3–5000 AU. When corrections are done to account for missing very close and very wide companions, the multiplicity frequency rises to ~67–75%.

Taurus-Auriga, however, appears to be different from other low-mass star-forming regions (e.g., *Correia et al.,* 2006) (see section 10.3). Chamaeleon I was studied by *Lafrenière et al.* (2008), who found a multiplicity frequency of 30 ± 6% over the interval ~16–1000 AU. In Ophiuchus, *Ratzka et al.* (2005) determined a multiplicity frequency of 29 ± 4% in the range 18–900 AU, while in the Upper Scorpius region of the Scorpius-Centaurus OB association, *Kraus et al.* (2008) found a binary frequency of 35 ± 5% in the 6–435-AU range. When properly scaled and compared, these values are consistent within their errors, suggesting that Taurus is atypical.

Other observations indicate that multiplicity differs among some regions. For example, *Reipurth and Zinnecker* (1993) found that young stars in clouds with 10 or fewer stars were twice as likely to have a visual companion as clouds with more stars. *Brandeker et al.* (2006) found a deficit of wide binaries in the η Chamaeleontis cluster. *Kraus and Hillenbrand* (2007) noted that the wide binary frequencies in four star-forming regions are dependent on both the mass of the primary star and the environment, but did not find a relation with stellar density. *Connelley et al.* (2008b) found that the binary separation distribution of class I sources in a distributed population in Orion (not near the ONC) is significantly different from other nearby, low-mass star-forming regions.

Naturally, these results raise the question of whether the environment plays a role for the population of binaries and multiples. It is conceivable that different physical conditions can affect the frequency and properties of newborn binaries (*Durisen and Sterzik,* 1994; *Sterzik et al.,* 2003). And longer-term dynamical interactions between binaries and single stars will depend on the stellar density in the birth environment (e.g., *Kroupa,* 1998; *Kroupa and Bouvier,* 2003; *Kroupa and Petr-Gotzens,* 2011). Assuming all stars are formed as binaries in groups and clusters of different densities, *Marks and Kroupa* (2012) show that — using an inverse dynamical population synthesis — the above-mentioned binary properties in different star-forming regions can be reproduced. This is further discussed in section 10.

7.2. The Separation Distribution Function

Binaries have separations spanning an enormous range, from contact binaries to tenuously bound ultrawide binaries and proper motion pairs with separations up to a parsec (and possibly even more). The way binaries are distributed along this vast range in separations carries information on both the mechanisms of formation and subsequent dynamical (and sometimes viscous) evolution. We note that almost all authors for practical reasons use projected separations. Because most binaries are eccentric and therefore spend more time near apastron than at periastron, one can show that — for reasonable assumptions about eccentricity — the mean instantaneous projected separations and mean semimajor axes differ by only ~5% (e.g., *van Albada,* 1968).

Öpik (1924) suggested that the distribution of separations for field binaries follows a log-flat distribution $f(a) \propto 1/a$, whereas *Kuiper* (1942) found a log-normal distribution; the latter has been supported by both *Duquennoy and Mayor* (1991) and *Raghavan et al.* (2010), who found the peak of the distribution of solar-type binaries to be at ~30 AU and ~50 AU, respectively, and Öpik's Law is no longer considered for closer binaries. But the distribution of the widest binaries can be fitted with a power law, although with an exponent between –1.5 and –1.6, decreasing somewhat faster than Öpik's Law (*Lépine and Bongiorno,* 2007; *Tokovinin and Lépine,* 2012).

For young low-mass stars the separation distribution function is less well known. For clusters, the absence of wide binaries has been noted in the ONC (*Scally et al.,* 1999; *Reipurth et al.,* 2007) (see section 10.3). Among less densely populated low-mass star-forming regions, the most detailed study is of Taurus by *Kraus et al.* (2011). They find that the separation distribution function for stars in the mass range from 0.7 to 2.5 $M_\odot$ is nearly log-flat over the wide separation range 3–5000 AU, i.e., there are relatively more wide binaries among young stars than in the field.

For VLM objects, it has been known for some time that the mean separation and separation range of binaries, both young and old, shrink with decreasing mass (see *Burgasser et al.,* 2007, and references therein). Where *Fischer and Marcy* (1992) found that M-star binary separations peak around 4–30 AU, *Burgasser et al.* (2007) estimated that VLM objects peak around ~3–10 AU. *Kraus and Hillenbrand* (2012) studied low-mass (0.02–0.5 $M_\odot$) young stars and brown dwarfs in nearby associations and found that the mean separation and separation range of binaries decline smoothly with mass; a degeneracy between total binary frequency and mean binary separation, however, precludes a more precise description of this decline.

7.3. Mass Ratios

The mass ratios ($q = M_2/M_1$) we observe for young stars are dominated by processes during the protobinary accretion phase, and subsequent circumbinary disk accretion will have only a limited effect on the mass ratios (see section 9.2). Spectroscopic determinations of YSO binary component masses are still rare (e.g., *Daemgen et al.,* 2012, 2013; *Correia et al.,* 2013), and estimates of mass ratios for young stars are mostly based on component photometry, with the significant caveats that come from accretion luminosity, differences in extinction of the components, and biases toward detecting brighter companions. For young intermediate-mass stars in the Sco OB2 association, *Kouwenhoven et al.* (2005) could fit the mass ratio distribution with a declining function for rising mass ratios [$f(q) \sim q^{-0.33}$], revealing a clear preference for low-q systems. In contrast, low-mass YSOs have a gently rising distribution for rising mass ratios, which becomes increasingly steep for VLM objects, showing a clear preference for q ~ 1 binaries (*Kraus et al.,* 2011; *Kraus and Hillenbrand,* 2012), as do VLM binaries in the field (e.g., *Burgasser et al., 2007*). It is noteworthy that this naturally results from dynamical interactions in VLM triple systems (Reipurth and Mikkola, in preparation).

7.4. Eccentricities

The eccentricity of binaries in the solar neighborhood has been studied by *Raghavan et al.* (2010), who finds an essentially flat distribution from circular out to e ~ 0.6 for binaries with periods longer than ~12 days (to avoid the effects of circularization). Higher eccentricities are less common, but this may be due to observational bias. For VLM binaries, *Dupuy and Liu* (2011) found eccentricities with a distribution very similar to the solar neighborhood.

Little is known about eccentricities of young binaries, except for the ~50 mostly short-period spectroscopic PMS binaries that have been analyzed to date; *Melo et al.* (2001) found that binaries with periods less than 7.5 days have already circularized during pre-main-sequence evolution, in agreement with theory (*Zahn and Bouchet, 1989*).

8. PRE-MAIN-SEQUENCE ECLIPSING BINARIES

Accurate measurements of the basic physical properties — masses, radii, temperatures, metallicities — of PMS stars and brown dwarfs are essential to our understanding of the physics of star formation. Dynamical masses and radii from eclipsing binaries (EBs) remain the gold standard, and represent the fundamental test bed with which to assess the performance of theoretical PMS evolution models. In turn, these models are the basis for determining the basic properties of all other young stars generally — individual stellar masses and ages, mass accretion rates — and thus help to constrain key aspects of star-forming regions, such as cluster star-formation histories and initial mass functions.

8.1. Performance of Pre-Main-Sequence Evolutionary Models

The *PPV* volume included a summary of the properties of the 10 PMS stars that are components of EBs known at that time (*Mathieu et al., 2007*), and summarized the performance of four different sets of PMS evolutionary tracks (*D'Antona and Mazzitelli, 1997; Baraffe et al., 1998; Palla and Stahler, 1999; Siess et al., 2000*) in predicting the dynamically measured masses of these stars from their Hertzsprung-Russell diagram positions. To summarize briefly the current status: (1) All the above models correctly predict the measured masses to ~10% above 1 $M_\odot$; (2) the models overall perform poorly below 1 $M_\odot$, generally predicting masses larger than the observed masses by up to 100%; and (3) the models of *Palla and Stahler* (1999) and *Siess et al.* (2000) are performing the best, predicting the observed masses to 5% on average although with a large scatter of 25%.

There are now as of this writing 23 PMS stars that are components of 13 EBs, including two brown dwarfs in one

EB (*Stassun et al., 2006, 2007*). An important development is the emergence of new models — the first in more than a decade — with physics attuned to PMS stars, namely the Pisa models (*Tognelli et al., 2011*) and the Dartmouth models (e.g., *Dotter et al., 2008; Feiden and Chaboyer, 2012*). A full assessment of the latter models against the sample of PMS EBs is underway (Stassun et al., in preparation), but preliminary results are promising. For example, the dynamically measured masses are correctly predicted by the Dartmouth models to ~15% over the range of masses 0.2–1.8 $M_\odot$.

The major review by *Torres et al.* (2010), while focused on main-sequence EBs, highlights the importance of reliable metallicities, temperatures, and (when possible) apsidal motions. Among PMS EBs, metallicity determinations are not commonly reported, but should in principle be determinable from the spectra used for the radial-velocity measurements. Temperatures remain vexing because of uncertainties over the spectral type to temperature scale for PMS stars, especially at very low masses. Only recently was the first apsidal motion for a very young EB reported [V578 Mon (*Garcia et al., 2011, 2013a*)]. As demonstrated by *Feiden and Dotter* (2013), such apsidal motion measurements can provide particularly stringent constraints on the models, specifically on the interior structure evolution with age, a critically important physical ingredient.

Importantly, *Torres et al.* (2013) have used the quadruple PMS system LkCa3 to perform a stringent test of various PMS evolutionary models. They find clearly that the Dartmouth models perform best, and moreover find that these models can fit another benchmark quadruple system, GG Tau, whereas previous generation models cannot (Fig. 10).

8.2. Impact of Activity on Temperatures, Radii, and Estimated Masses of Young Stars

Stars in short-period binaries are often chromospherically active, and thus may suffer from activity-reduced temperatures and/or -inflated radii, causing them to appear discrepant relative to standard model isochrones (e.g., *Torres, 2013*). In particular, such activity-reduced temperatures can cause the derived stellar masses to be underestimated by up to a factor of ~2.

A particularly important case in point is 2M0535-05, the only known EB comprising two brown dwarfs (*Stassun et al., 2006, 2007*), which proved enigmatic from the start. The system is a member of the very young ONC, and thus the expectation is that both components of the EB should have an age of ~1 m.y. However, a very peculiar feature of the system is a reversal of temperatures with mass — the higher-mass brown dwarf is cooler than its lower-mass brown dwarf companion — making the higher-mass brown dwarf appear younger than the lower-mass companion and a factor of 2 lower in mass than its true mass. *Reiners et al.* (2007) showed that the higher-mass brown dwarf is highly chromospherically active as measured by the luminosity of its Hα emission, whereas the lower-mass brown dwarf is

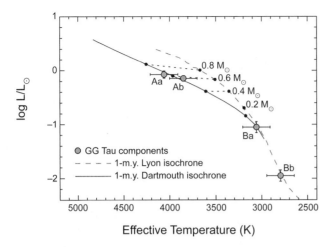

Fig. 10. Application of Dartmouth models to the quadruple PMS system GG Tau (*Torres et al.,* 2013). Previous-generation models (here compared to the Lyon models), which have been used to calibrate the PMS temperature scale, do not perform as well.

a factor of 10 less active and appears "normal" relative to the evolutionary tracks.

Motivated by this peculiar but important system, *Stassun et al.* (2012) have used a sample of low-mass EBs to determine empirical corrections to stellar temperatures and radii as a function of chromospheric Hα activity (*Morales et al.,* 2010). Notably, these corrections indicate that the nature of the temperature reduction and radius inflation is such that the bolometric luminosity is roughly conserved. The *Stassun et al.* (2012) relations are able to fully explain the anomalous temperature reversal found in the 2M0535-05 brown-dwarf EB.

However, there is not as yet consensus on the underlying physical cause of this effect. *Chabrier et al.* (2007) suggest that surface spots and convection inhibited at the surface are the driver, whereas *MacDonald and Mullan* (2009) suggest a global inhibition of convection through strong fields threading the interiors of the stars. *Mohanty and Stassun* (2012) performed detailed spectrocopic analysis of the eclipsing brown-dwarf EB 2M0535-05, the results of which appear to disfavor the *Chabrier et al.* (2007) hypothesis. However, questions remain as to the physical plausibility of the magnitude of interior fields required in the *MacDonald and Mullan* (2009) hypothesis.

At the same time, the Dartmouth models also now incorporate the effects of internal magnetic fields, which successfully accounts for the effects of temperature reduction and radius inflation in a physically self-consistent fashion (*Feiden and Chaboyer,* 2012).

8.3. Impact of Triplicity on Properties of Pre-Main-Sequence Stars

There is increasing evidence that the presence of third bodies in young binaries can significantly alter the proper-

ties of the component stars, either directly through tidal heating effects and/or indirectly by impacting the accretion history of the system. As an exemplar case, *Stassun et al.* (2008) identified Par 1802 to be an unusual PMS EB whose component stars have identical masses (q = 0.99) yet radii that differ by 7%, temperatures that differ by 9%, and luminosities that differ by 60%. Thus the pair cannot be fit by any standard PMS stellar evolution models under the usual assumption of coevality for the component stars because the stars' highly unequal luminosities cause them to appear highly noncoeval. *Gómez Maqueo Chew et al.* (2012) used 15 years of eclipse timing measurements to reveal the presence of a wide tertiary component in the system. Modeling the tidal heating on the EB pair arising from the previous orbital evolution can explain the overluminosity of the primary eclipsing component, and moreover suggests a close three-body (perhaps exchange) interaction in the past.

Relatedly, recent theoretical work has suggested that accretion history (e.g., FU Ori outbursts, differential accretion in protobinaries) can alter the PMS mass-radius relationship (e.g., *Simon and Obbie,* 2009; *Baraffe and Chabrier,* 2010). Consequently, new-generation PMS evolution models are seeking to simulate these effects. For example, the new Pisa models are being further developed to include thin-disk accretion episodes during the early PMS phase.

As suggested by the example of Par 1802 above, PMS EBs provide a unique opportunity to assess the frequency of higher-order multiples among close binaries, because of the high quality and multi-faceted ways in which these benchmark systems are studied. Among the sample of 11 PMS EB systems that have detailed EB solutions published as of this writing [i.e., excluding PMS EBs with preliminary reports such as the six systems announced in *Morales-Calderon et al.* (2012)] and that have stellar mass components (i.e., excluding the double brown-dwarf EB 2M0535-05), six are now known to include a third body. This preliminary census implies a very high ratio of triples to binaries, consistent with the view that tertiaries may be critical to the formation of tight pairs.

9. GAS IN BINARIES AND MULTIPLE SYSTEMS

9.1. Observations of Circumbinary Structures

Circumbinary disks play an important role in shaping binary orbital properties: Mass flow from the disk affects the ultimate binary mass ratio while the flow of angular momentum from binary to disk drives changes in the binary period and eccentricity. The observational study of circumbinary accretion flows is, however, challenging: Massive circumbinary disks are rare among binaries with separations in the range of a few to 100 AU (*Jensen et al.,* 1996; *Harris et al.,* 2012), which constitute the bulk of the pre-main-sequence binary population. To date, only a handful of circumbinary disks have been imaged directly (see *Hioki et al.,* 2009), and here the limitations of coronagraphic imaging do not allow the study of the structures — critical to the binary's evolution — that link the disk to the binary. Circumbinary disks are considerably more abundant around the closest binaries (a few astronomical units or less); on the main sequence such binaries are — unlike wider pairs — preferentially associated with circumbinary debris disks (*Trilling et al.,* 2007) and are in the regime where Kepler has recently revealed a number of circumbinary planets (e.g., *Doyle et al.,* 2011; *Welsh et al.,* 2012). The reason for the higher incidence of massive circumbinary disks in close pairs is unclear — i.e., whether it reflects the initial configuration at formation or whether such disks drive binaries to small orbital separations. Alternatively, this association may be a matter of disk survival: *Alexander* (2012) has argued that disks around close binaries should be long lived, since viscous draining is impeded by the binary's tidal barrier while the gas is too tightly bound to be readily photoevaporated.

Interferometric studies are just beginning to probe the dust morphology in these systems (*Boden et al.,* 2009; *Garcia et al.,* 2013b), and so the bulk of our knowledge derives from time-domain studies. For example, in eccentric binaries, periodic optical and X-ray variations have been attributed to a dynamically modulated accretion flow [e.g., *Mathieu et al.* (1997) in DQ Tau; *Gomez de Castro et al.* (2013) in AK Sco], although optical variability also accompanies synchotron flares at millimeter wavelengths, which can be understood as reconnection events when the two stellar magnetospheres interact at periastron (e.g., *Salter et al.,* 2010). *Muzerolle et al.* (2013) have recently interpreted large-scale periodic variations in a protostellar source as deriving from a binary-modulated pulsed accretion flow. Variations in the observer's viewing angle also modulate line emission in low eccentricity binaries: For example, in V4046 Sgr, HD modeling (*de Val-Borro et al.,* 2011) reproduces the periodic changes in the wings of the Balmer lines observed by *Stempels and Gahm* (2004). In CS Cha, the binary's variable illumination of dusty accretion streams has been invoked to explain its periodic infrared variability (*Nagel et al.,* 2012); it is, however, notable that the spectral energy distribution of CS Cha implies that the inner edge of the optically thick circumbinary disk is at about 10× the binary orbital separation (*Espaillat et al.,* 2011), which is several times larger than what is expected from dynamical truncation by the binary. This finding exemplifies the difficulty of connecting models and observations, since the dust emission is apparently not merely being shaped by the response of the gas to the binary potential.

9.2. Simulations of Circumbinary Disks

While observed circumbinary disks are generally low in mass during the classical T Tauri phase, this was almost certainly not the case at earlier evolutionary phases. In hydrodynamic collapse simulations, protobinaries are surrounded by circumbinary disks formed from higher-angular-momentum material in the natal core, and the interaction between the disk and the binary is key to shaping the system's ultimate orbital elements (a, e, q). A complete theory of binary formation should require not only the creation of the protobinary fragments but should contain a clear prescription for the evolution of these quantities as a function of the properties of the circumbinary disk.

Unfortunately, this goal remains elusive, despite comprehensive (SPH) studies devoted to this problem (e.g., *Bate and Bonnell,* 1997; *Bate,* 2000). Qualitative features of these studies (especially the preferential accretion of gas onto the secondary and hence the increase of the binary mass ratio) were challenged by *Ochi et al.* (2005) and *Hanawa et al.* (2010), whose AMR simulations were morphologically distinct and involved a preferential accretion of gas onto the primary (thus driving $q = M_2/M_1$ downward). It now seems likely that these differences arose from the different parameters of the latter studies (i.e., warm, two-dimensional flows) rather than from a code difference; nevertheless, there are no fully converged simulations of circumbinary accretion that have been run to a steady state, and this probably explains the variety of results reported in the literature with regard to the sign and magnitude of effects associated with circumbinary accretion (see also *Fateeva et al.,* 2011; *de Val-Borro et al.,* 2011). This raises a cautionary note with regard to the fidelity of cluster-scale simulations in modeling this process, since disks in such simulations are always relatively poorly resolved.

If there is still no clear consensus in the purely HD case, the situation becomes still more complicated when magnetic fields are involved. This is illustrated by two recent studies. *Zhao and Li* (2013) modeled accretion onto a "seed binary" placed within a moderately magnetized core and found that severe magnetic braking of the accreting gas has two notable effects: The binary shrinks to small separations, while the low angular momentum of the braked gas ensures that the flow is predominantly to the primary, thus lowering the binary mass ratio. In another study, *Shi et al.* (2012) conducted the first simulation of binary/circumbinary disk interaction, which — rather than adopting a parameterized "α-type" viscosity in the disk (*Shakura and Sunyaev,* 1973) as in most previous works (e.g., *Artymowicz and Lubow,* 1994; *MacFadyen and Milosavljevic,* 2008; *Cuadra et al.,* 2009; *Hanawa et al.,* 2010) — instead simulated the self-consistent angular momentum transfer associated with the development of magneto-hydrodynamic turbulence in the disk. The simulation considered the limit of ideal MHD and is therefore not applicable to "dead zones" of low ionization (*Bai,* 2011; *Mohanty et al.,* 2013): In practice this limits it to radii within ~0.5 AU or beyond ~10 AU.

It is found that the effective efficiency of angular momentum transport (e.g., as parameterized by the Shakura and Sunyaev α parameter) is about an order of magnitude higher in the accretion streams that link the binary to the disk than in the body of the disk, and this results in a much more vigorous flow through the accretion streams than in previous simulations that do not treat the development of magneto-turbulent stresses self-consistently (see *MacFadyen and Milosavljevic,* 2008). Indeed, in the *Shi et al.* (2012)

simulations, the flow through the accretion streams is ~30% of the flow through the outer disk. In such a situation the net evolution of the binary is governed by two nearly canceling terms (the spin-up effect of accretion and the spin-down torque associated with the non-axisymmetric disk/accretion streams) and is thus very sensitive to numerical inaccuracies/uncertainties in the disk thermodynamics. So although this simulation is undoubtedly more realistic than previous calculations, it raises awkward issues: Apparently the derivation of a simple relationship between circumbinary disk properties and associated orbital evolution may be more elusive than ever.

9.3. Disk Lifetimes in Binaries

Since the review of this subject by *Monin et al.* (2007) in *PPV*, a number of studies have charted the relative lifetimes of disks in binaries compared with single stars, and studied the relative lifetimes of the primaries' and secondaries' disks. Early studies in this area (e.g., *Prato and Simon,* 1997) had argued that circumstellar disks must be replenished from a circumbinary reservoir during the classical T Tauri phase, a requirement that was puzzling given the observed lack of circumbinary material in all but the closest binaries. This conclusion was based on (1) the fact that disks in binaries were not apparently shorter lived than disks in single stars and (2) the scarcity of "mixed pairs" (i.e., those with only one disk). Resupply of circumstellar disks would extend their lifetimes and coordinate the disappearance of the disks, since otherwise — in isolation — the secondary's disk would drain first on account of its smaller tidal truncation radius and shorter viscous timescale (*Armitage et al.,* 1999).

However, recent studies have undermined the observational basis for these arguments. *Cieza et al.* (2009) and *Kraus and Hillenbrand* (2009) demonstrated that the lifetime of disks in close binaries is indeed reduced compared with single stars or wide pairs, concluding that incompleteness of the census of close binaries, the use of unresolved disk indicators, and projection effects had all previously masked this correlation in smaller samples (see also *Kraus et al.,* 2012; *Daemgen et al.,* 2013). Moreover, the census of binaries for which spectral diagnostics have been measured for each component has been augmented by *Daemgen et al.* (2012, 2013) in the ONC and Cha I. These new results have reinforced the suggestion of *Monin et al.* (2007) that the early conclusions about the absence of mixed pairs was skewed by results from Taurus that are not borne out in other regions. Kneller and Clarke (in preparation) argue that the observed incidence of disks in binaries as a function of q and separation is compatible with clearing by combined viscous draining and X-ray photoevaporation. Such models predict a strong tendency for the secondary's disk to disappear before the primary's for binaries closer than 100 AU while predicting that in wider mixed pairs, disks are equally likely to exist around the secondary and primary components: This latter prediction needs to be tested observationally in larger samples.

9.4. Eclipses by Disks

Disks may cause eclipses of their central stars, and these events present rare but valuable opportunities to study the detailed structure of disks during the planet-forming era. The best studied example is KH15D (e.g., *Herbst et al.,* 2010). KH15D is a binary that is occulted by a circumbinary screen of material that moves slowly across the binary components, occulting them in turn. Modeling of the screen suggests its origin to be a precessing, warped circumbinary ring of material several astronomical units from the tight binary. The obscuring ring has very sharp edges — it is well modeled as a knife edge — indicating a high degree of coherence to the material despite the dynamics of the system.

RW Aur is a newly discovered exemplar of this class (*Rodriguez et al.,* 2013). In this case, light curve observations by the Kilodegree Extremely Little Telescope (KELT) exoplanet transit survey (*Pepper et al.,* 2007) witnessed a sudden dramatic eclipse of the star with a depth of ~2 mag and lasting approximately 180 days. Archival photometric observations can rule out a similarly long and deep eclipse over the past 100 years. This singular event is interpreted and modeled as an occultation of the primary star (RW Aur A) by the long tidal arm observed by *Cabrit et al.* (2006) resulting from tidal disruption of its circumstellar disk by the recent flyby of RW Aur B (see Fig. 11). RW Aur B may itself be a tight binary, making the RW Aur system a triple (*Ghez et al.,* 1993). The eclipse observations indicate a knife-edge structure to the occulting feature, consistent with the dynamical simulations of *Clarke and Pringle* (1993), which demonstrate a high degree of coherence in the tidal arm persisting long after the flyby (see section 9.6).

9.5. Alignment of Orbital Planes in Young Binaries

A number of systems show evidence that they have undergone dynamical events that have perturbed the orientation of the binary and/or its circumstellar disks. For example, *Bisikalo et al.* (2012) present evidence that the disk around RW Aur A (discussed above) is counterrotating with respect to the binary orbital motion. Similarly, *Albrecht et al.* (2009) used Rossiter-McLaughlin measurements during the eclipse of the massive binary DI Her to show that the projected spin of one of its B-type components was highly misaligned (72°) with respect to the binary orbital plane. There is no unique explanation for such systems. One idea is that, whereas the spin of the stars reflects that of the local gas reservoir (material that collapsed first), the spin direction of circumbinary structures (or the orbital plane of binary systems) may inherit a different direction from a larger region within the turbulent medium, because of chaotic changes in the mean angular momentum vector of accreting material [this effect is significant in the whole-cluster simulations of *Bate* (2009a), where misaligned systems are common]. Alternatively, dynamical interactions (for example, an exchange interaction within a nonhierarchical system) can play a similar role, although this again requires that the natal gas contains a range of spin

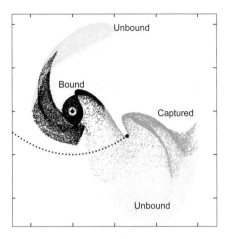

Fig. 11. Severe disk disturbances occur during close periastron passages. This simulation shows an encounter between two solar-type stars, one with and one without a disk; black dots show material that remains bound, gray dots show material captured by the intruder, and light gray dots show unbound material. The box is 500 × 500 AU. Figure courtesy of S. Pfalzner and M. Steinhausen.

directions. On the other hand, the Kozai-Lidov (*Kozai,* 1962; *Lidov,* 1962) mechanism within triple systems can induce spin-orbit misalignment even in the absence of exchange interactions: *Fabrycky and Tremaine* (2007) suggested that while Kozai-Lidov induced oscillations in eccentricity and inclination can deliver companions to small pericenter distances, tidal dissipation could allow such systems to free themselves from the Kozai-Lidov regime, trapping them in a state where their spins are decoupled from their orbital inclination. Triple companions, however, have not been detected to date in either of these systems. Other recent measurements of misalignment of orbital planes within pre-main-sequence binaries include KH15D (*Capelo et al.,* 2012) and FS Tau (*Hioki et al.,* 2011), while circumbinary debris disks present a mixed picture with respect to the alignment of orbital and circumbinary disk planes (*Kennedy et al.,* 2012a,b).

Facchini and Lodato (2013) and *Foucart and Lai* (2013) have recently presented analytic and numerical calculations of the evolution of the warp and twist of a disk that is initially misaligned with the binary orbit. Foucart and Lai showed that the back-reaction on the binary orbit realigns the system on a timescale that is short compared with that required for the binary to accrete significant mass from the circumbinary disk. They therefore argued that close binaries (which gain significant mass from the circumbinary disk) should become aligned with their disks during the pre-main-sequence period, thus explaining the surprising abundance of (necessarily aligned) planets in circumbinary orbit in the Kepler sample (e.g., *Doyle et al.,* 2011; *Welsh et al.,* 2012).

9.6. Retention of Disks in Dynamical Encounters

Dynamical interactions within multiple systems result in the pruning of circumstellar disks, leading *Reipurth and*

Clarke (2001) to argue that disk *size* may provide a diagnostic of an object's previous history of close encounters in a few-body system. The influence of stellar flybys on disk structure was first examined by *Clarke and Pringle* (1993), while *Hall* (1997) reconstructed disk surface profiles post-encounter from ballistic calculations through the assumption that bound particles should recircularize while retaining their individual angular momenta. The SPH calculations of *Pfalzner et al.* (2005) (see Fig. 11) showed that this is a reasonable approximation, and more recently *Umbreit et al.* (2011) have applied the same procedure to stars undergoing close encounters within triple systems. This study (which started from nonhierarchical co-planar triples with co-planar disks) showed that the reconstructed density profiles show a boosted power law profile in the inner disk and an exponential cut-off at a radius of a few tenths of the minimum encounter distance. It is found that disk stripping during triple decays is qualitatively very similar and only slightly stronger than that occurring during two-body flybys.

9.7. Planetary Systems in Multiple Stellar Systems

Around 7% of currently detected planets are in binary systems, most of which are located in circumstellar orbits in wide systems. Two categories of systems have attracted considerable recent interest, i.e., the circumbinary (P-type) planets around close (sub-astronomical-unit) separation binaries discovered by Kepler (*Doyle et al.,* 2011; *Welsh et al.,* 2012), and the circumstellar (S-type) planets discovered in relatively close (a < 20 AU) pairs (*Chauvin et al.,* 2011; *Dumusque et al.,* 2012). Planetesimal accumulation in binary environments faces a well-known problem (*Thebault et al.,* 2008) on account of the pumping of the planetesimal velocity dispersion by gravitational perturbations from the binary; a high-velocity dispersion implies destructive collisions (*Leinhardt and Stewart,* 2012) and thus limits the possibilities for planetesimal growth. Some suggested solutions to this problem fall into the category of simply forming planets in a more benign dynamical regime (i.e., further from the perturber) and then invoking migration — of either the planet or the binary companion — to achieve the observed planet/binary architecture (see *Payne et al.,* 2009; *Thebault et al.,* 2009). Alternatively, *Xie et al.* (2010) have explored the effect of a modest inclination between the binary and disk plane. In terms of *in situ* formation models, gas drag has been invoked as a mechanism for enforcing apsidal alignment of perturbed planetesimal orbits: This produces local velocity coherence in objects of a given size. However, since gas drag is a size-dependent phenomenon, this does not prevent destructive collisions in a planetesimal population with a realistic size distribution and is thus unlikely to solve the problem (*Thebault et al.,* 2006, 2008).

Recent works on this topic concentrate on the effect of the disk's gravitational field upon the growth of planetesimal velocity dispersion. *Rafikov* (2013) argued that if the disk is approximately axisymmetric, then its gravitational influence induces size-independent apsidal precession, which acts to reduce the eccentricity excitation by the companion. However, the simulations of *Marzari et al.* (2013) show that the planet induces a strong eccentricity in the disk, and that gravitational coupling with the eccentric disk actually amplifies the stirring of the planetesimal population. In effect, therefore, these studies come to qualitatively opposite conclusions as to whether binarity is a major obstacle to planet formation. These divergent conclusions essentially hinge on the axisymmetry of the gas disk in the region of interest; further hybrid HD/N-body modeling is required in order to delineate the areas of parameter space in which planetesimal growth is possible.

Finally, although planets are known in stellar triple systems (e.g., *Bechter et al.,* 2014), the issue of planet formation in higher-order multiples is unexplored [although several works have examined the stability boundaries of particle disks within triple systems (e.g., *Verrier and Evans,* 2007; *Domingos et al.,* 2012)]. While the presence of three bodies in general restricts the stability regions available, there are certain configurations where a third body can stabilize particle orbits. In particular, whereas particles in a circumstellar disk with an inclined companion can be subject to Kozai-Lidov instability, this can be suppressed if the central object within the disk is itself a binary, since in this case the binary induces nodal libration, which stabilizes the particles (*Verrier and Evans,* 2009). Such studies will be important in interpreting the statistics of debris disks within multiple star systems.

10. BINARIES IN CLUSTERS

We have seen (see section 5) that multiple systems can change due to internal (secular) interactions. But dynamical interactions with other systems can play an extremely important role in altering multiple systems (changing their orbital parameters, or even destroying them). In the relatively dense environments of star-forming regions or star clusters, the initial multiple population can be very significantly altered. This means that any multiple population we observe is almost certainly not what formed, and different populations can evolve in (very) different ways depending on the environment. Therefore, to extrapolate back to formation from any observations, we must fold in the (possibly very complex) dynamical evolution [called "inverse population synthesis" by *Kroupa* (1995)].

10.1. External Dynamical Interactions

The dynamical destruction of binaries was first studied in detail by *Heggie* (1975) and *Hills* (1975) (see also *Hills,* 1990). They placed binaries into two broad categories: "hard" and "soft." Hard binaries are those that are so tightly bound that encounters are extremely unlikely to alter their properties (to destroy them, or even to change their orbital parameters much). Soft binaries are those that are so weakly bound that they are almost certain to be destroyed by any encounter. Generally, if the kinetic energy of a perturber

is greater than the binding energy of the binary, then the binary is destroyed. An alternative way of looking at this is that a binary is hard if its orbital speed is greater than the speed of an encounter (*Hills,* 1990). Investigation of the dynamics of encounters leads to the Heggie-Hills law, that states that hard binaries get harder (encounters typically take energy from the binary), and soft binaries get softer (encounters typically give energy to the binary).

The hard-soft boundary is basically set by the velocity dispersion of the perturbing stars and the mass of the binary. The faster the perturbers are moving, the closer the hard-soft boundary, and the more massive the binary is, the more difficult it is to destroy. It might be expected that binary destruction will depend on the mass ratio; however, simulations show that destructive encounter energies are almost always significantly greater than the binding energy, and so destruction does not depend on the mass ratio (*Parker and Reggiani,* 2013).

However, the "hardness" of a binary is not the only thing that decides if a binary will survive. To destroy even a soft binary an encounter is required, therefore the encounter rate is crucial. In the field, many formally soft binaries survive for significant amounts of time, because the encounter rate is very low.

Therefore, the survival of a binary depends on (1) the energy of the binary, (2) the energy of encounters, and (3) the frequency of encounters. The more massive and dense a star-forming region or star cluster is, the more frequent and energetic encounters will be, and so binary destruction/alteration should be more efficient.

In any given environment, hard binaries should survive, and soft binaries will almost certainly be destroyed. The most interesting binaries, however, are often "intermediates" between hard and soft, which may or may not survive depending on the exact details of their dynamical histories (e.g., *Parker and Goodwin,* 2012).

Let us take an "average" binary system of component masses m, where m ~ 0.4 $M_\odot$ is the average mass of a star. Let us put this binary in a virialized cluster of N stars of total mass M = Nm, and radius R. The hard-soft boundary, a_{hs}, will be at approximately

$$a_{hs} \sim 10^5 \left(\frac{R}{pc} \right) \left(\frac{1}{N} \right) AU \qquad (1)$$

(see *Parker and Goodwin,* 2012). Numerical experiments show that a safe value for a hard binary that will almost certainly survive is about $a_{hs}/4$. For clusters like the ONC (N ~ 10^3, R ~ 1 pc) a_{hs} ~ 100 AU. For relatively nearby regions, the ONC is very massive and dense, and so in local regions we tend not to expect processing of binaries with a < 100 AU. This means that the a < 100 AU population of binaries is "pristine" (i.e., unprocessed), while a > 100 AU binaries may (or may not) have been processed (*Goodwin,* 2010; *King et al.,* 2012b).

It is important to remember, however, that it is not necessarily the *current* density of a region that is important in assessing the possible impact of binary processing. Rather, it is the (usually unknown) density history of the region.

The values of N ~ 10³ and R ~ 1 pc used above for the ONC are the present-day values, and the calculated hard-soft boundary of a ~ 100 AU is the current safe hard-soft boundary. If the ONC was much denser in the past [as has been argued (see *Scally et al.*, 2005; *Parker et al.*, 2009)], then the hard-soft boundary in the past could have been much smaller. If a region spends at least a crossing time in a dense state, then it is that dense state that imposes itself in binary destruction (*Parker et al.*, 2009). This could be very important if regions undergo expansion due to gas expulsion (e.g., *Marks and Kroupa*, 2011, 2012), or process binaries in short-lived substructures (*Kroupa et al.*, 2003; *Parker et al.*, 2011).

10.2. Binary Formation Through Encounters

Dynamics are usually associated with binary destruction rather than formation. But hard binaries can be formed by three-body encounters (with the third body carrying away energy). The rate of binary creation per unit volume, $\dot{N}_b$, depends on the stellar masses (m), velocity dispersion (σ), and number density of stars (n)

$$\dot{N}_b = 0.75 \frac{G^5 m^5 n^3}{\sigma^9} \qquad (2)$$

(*Goodman and Hut*, 1993). In the Galactic field this number is essentially zero [~10⁻²¹ pc⁻³ gigayears (G.y.)⁻¹]. However, in dense star-forming regions and clusters, the rate may be significant, especially for higher-mass stars. Simulations show that initially single massive stars can pair-up in hard binaries, and can form complex higher-order systems similar to the Trapezium (*Allison and Goodwin*, 2011). This is due to the very strong dependency of $\dot{N}_b$ on the higher m and n, and the lower σ in clusters (which can make 30 orders of magnitude difference).

Kouwenhoven et al. (2010) and *Moeckel and Bate* (2010) independently found that dissolving dense regions can also form very wide binaries by "chance," when two stars leaving the region find themselves bound once outside the region. Similarly, *Moeckel and Clarke* (2011) find that dense regions constantly form soft binaries. While the region remains dense, these binaries are destroyed as fast as they are made. However, when the region dissolves into the field they can be "frozen in" at lower densities and survive. On average, one region produces one wide binary with a median separation of about 10⁴ AU, almost independently of the number of stars in that region (*Kouwenhoven et al.*, 2010). Since the stars are paired randomly, it is quite possible for the wide binaries to be made of one or two hard binaries (making triple or quadruple systems). The mass ratio distribution of wide binaries would be expected to be randomly paired from the IMF. This process acts independently of the wide binaries that form through the unfolding of triple systems, as discussed in section 5.

10.3. Observations of Young Multiples in Clusters

It is only in nearby star-forming regions that we can examine in any detail the (especially low-mass) binary properties. Locally, young star-forming regions cover a wide range of densities from a few stars pc⁻³ (e.g., Taurus) to a few thousand stars pc⁻³ [e.g., the ONC (see *King et al.*, 2012a,b)]. These are often — rather arbitrarily — divided by density into low-density "associations" and high-density "clusters." More formally, "clusters" are often thought of as bound objects, or objects at least a few crossing times old (e.g., *Gieles and Portegies Zwart*, 2011). We will take the Gieles and Portegies Zwart definition of a cluster as dynamically old systems, since these are systems that we might expect to have significantly processed their multiple populations. Locally, this probably safely includes the ONC and IC 348 as "clusters" for which we have some detailed information on the stellar multiplicity.

It is worth noting that the level of processing of binaries will not simply depend on mass, but rather on the dynamical age of a system. For example, *Becker et al.* (2013) suggest that the binary properties (and unusual IMF) of the low-mass "cluster" η Cha could be explained by an initially very high density and rapid dynamical evolution.

Observations of the ONC and IC 348 show a lower-binary frequency than associations [e.g., *Köhler et al.* (2006) and *Reipurth et al.* (2007) for the ONC; *Duchêne et al.* (1999) for IC 348]. The ONC is also found to have an almost complete lack of wide (>1000 AU) systems (*Scally et al.*, 1999).

King et al. (2012a,b) collated binary statistics for seven young regions and attempted to correct for the different selection effects and produce directly comparable samples. Only in the range 62–620 AU is it possible to compare regions as diverse as Taurus (with an average density of <10 stars pc⁻³) to the ONC (around 5000 stars pc⁻³). In this separation range, the binary fraction of Taurus is around 21 ± 5%, compared to around 10% in regions with densities greater than a few hundred stars pc⁻³ (Cha I, Ophiuchus, IC 348, and the ONC). The solar field values in the same range are roughly 10%.

Given the densities of the ONC, it is almost impossible to imagine a scenario in which we are observing the birth population. The binaries we observe have separations of 62–620 AU, almost all above the hard-soft boundary in the ONC. Taking a size for the ONC of 1 pc, a density of 5000 stars pc⁻³, and a velocity dispersion of 2 km s⁻¹, the typical encounter timescale at 1000 AU is about 1 m.y. — roughly the age of the ONC.

It is often stated that clusters have a field-like binary distribution; however, this is somewhat misleading. Binary studies of the ONC and IC 348, the only dense clusters analyzed so far, are of a limited range of around 50–700 AU and in this range they have a similar binary fraction to the field. However, we have no information on smaller separations, and they are certainly not field-like at large separations, where there is an almost complete lack of systems.

Reipurth et al. (2007) find a significant (factor of 2–3) difference between the ratio of wide (200–620 AU) to close (62–200 AU) binaries between the inner parsec of the ONC and outside this region. This could suggest a difference in dynamical age, and hence the degree of processing, between the inner and outer regions of the ONC (*Parker et al.,* 2009).

Interestingly, *King et al.* (2012b) find that while the binary frequency in the ONC is significantly lower than in associations, the binary separation distribution looks remarkably similar. Such distributions in the 62–620-AU range are always approximately log-flat in all regions and show no statistically significant differences. Taurus has twice as many binaries as the ONC in the same separation range, but the distribution of binary separations is the same.

This is worth remarking on, because it is very unexpected. A reasonable assumption would be that associations and clusters form the same primordial population, but clusters are much more efficient at processing that population. The field is then the sum of relatively unprocessed binaries from associations, and relatively highly processed binaries from clusters (e.g., *Kroupa,* 1995; *Marks and Kroupa,* 2011, 2012). But processing is separation dependent, and wider binaries should be processed more efficiently than closer binaries. Therefore, if we take initially the same binary frequency and separation distribution in the 62–620-AU range in both associations and clusters, we would expect (1) a lower final binary frequency in clusters (which we see), and (2) fewer wider binaries in clusters than associations (which we do not see). Note that by wide, we do not mean >1000 AU, which are missing in the ONC (*Scally et al.,* 1999), but rather fewer, say, 200–620-AU binaries than 62–200-AU binaries.

That the separation distributions in low-density and high-density environments is the same could suggest that high-density regions somehow overproduce slightly wider systems, which are then preferentially destroyed to fortuitously produce the same final separation distribution. This would seem rather odd (*King et al.,* 2012b).

The 62–620-AU range of binaries for which we have observations in the ONC are mostly (rather frustratingly) intermediate binaries whose processing depends on the details of their dynamical histories. *Parker and Goodwin* (2012) show that in ONC-like systems the tendency is to preferentially destroy wider systems, but small-N statistics means that some clusters can produce separation distributions in the observed range that sometimes retain the initial shape. So it is possible that the separation distribution in the ONC is statistically slightly unusual. However, the difference between the inner and outer ratio of wide (200–620 AU) to close (62–200 AU) binaries observed by *Reipurth et al.* (2007) suggests that the inner regions of the ONC have been efficient at processing the wider binaries.

In summary, in clusters we expect significant binary destruction. However, interpreting observations of binaries in clusters is difficult. This is due to the lack of nearby clusters and the limited range of binary separations that are observable. But in the binary populations of clusters should be clues to the formation and assembly of clusters, and differences between star formation in different environments.

11. THE MULTIPLICITY OF MASSIVE STARS

We here define massive stars to be OB stars on the main sequence, above ~10 M$_\odot$ (about B2V) capable of ionizing atomic hydrogen, with the dividing line between O and B stars around 16 M$_\odot$ (about B0V) (*Martins et al.,* 2005). Massive stars occur mostly in young clusters and associations, but to a small degree also in the field and as runaway stars. There are some 370 O stars known in the Galactic O Star Catalog (*Maiz-Apellaniz et al.,* 2004; *Sota et al.,* 2008), with 272 located in young clusters and associations, 56 in the field, and 42 classified as runaway stars.

11.1. Recent Observational Progress

A comprehensive review of the multiplicity of massive stars was given by *Zinnecker and Yorke* (2007), emphasizing the difference in multiplicity between high- and low-mass stars and its implication for their different origins. In the meantime, *Mason et al.* (2009) in a statistical analysis summarized the multiplicity of massive stars based on the Galactic O Star Catalog (see above), both for visual and for spectroscopic multiple systems. *Chini et al.* (2012), in a vast spectroscopic study, presented evidence for a nearly 100% binary frequency among the most massive stars, dropping substantially for later-type B stars, thus confirming the mass dependence of the multiplicity. At the same time, *Sana et al.* (2012) for the first time derived the distributions of orbital periods and mass ratios for an unbiased sample of some 70 O stars based on a multi-epoch, spectroscopic monitoring effort. Three important results emerged: (1) The mass-ratio distribution is nearly flat with no statistically significant peak at q = 1 (identical twins); (2) the distribution of orbital periods peaks at very short periods (3–5 days) and declines toward longer periods; and (3) a large fraction (>70%) of massive binaries are so close that the components will be interacting in the course of their lifetime, thus affecting the statistics of WR stars, X-ray binaries, and supernovae, and of these one-third will actually merge (*Sana et al.,* 2012).

In yet another recent study, based on the Very Large Telescope (VLT)-Fibre Large Array Multi Element Spectrograph (FLAMES) Tarantula Survey, *Sana et al.* (2013) probed the spectroscopic binary fraction of 360 massive stars in the 30 Doradus starburst region in the Large Magellanic Cloud. They discovered that at least 40% of the massive stars in the region are spectroscopic binaries (both single and double lined). The unmistakable conclusion of all these studies is that the processes that form massive stars strongly favor the production of (mostly tight) binary and multiple systems.

Detailed studies of the multiplicity and orbital parameters of massive stars in young clusters [NGC 6231, NGC 6611 (*Sana et al.,* 2008, 2009)] and OB associations [Cyg OB2 (*Kiminki and Kobulnicky,* 2012)] have also been published, in an effort to find correlations with cluster properties and statistical differences between cluster and "field" stars. None were found (see review by *Sana and Evans, 2011*). A contentious issue is the multiplicity among bona fide runaway O stars, which was believed to be low (*Gies and Bolton, 1986; Mason et al., 2009*), but following the new results of *Chini et al.* (2012), it seems to be very high (75%). In the case of runaway O stars, it may eventually be useful to discriminate between high-velocity runaway stars (>40 km s^{-1}), presumably originating from supernovae explosions in binary systems (*Blaauw, 1961*), and slow runaways ("walkaways," <10 km s^{-1}, which are harder to identify) whose origin is likely due to dynamical ejection from dense young clusters (*Poveda et al., 1967; Clarke and Pringle, 1992; Kroupa, 2000*). The multiplicity of truly isolated field O stars [if they do exist (cf. *de Wit et al., 2005; Bressert et al., 2012; Oey et al., 2013*)] still needs to be investigated.

11.2. Origin of Short-Period Massive Binary Systems

In recent years it has become evident that at least 44% of all O stars are close spectroscopic binaries (see review by *Sana and Evans, 2011*). There are several — at least five — ideas to explain the origin of such close massive spectroscopic binaries; these are briefly discussed below. In addition, we need to explain the origin of hierarchical triple systems among massive stars; such systems could either result from inner and outer disk fragmentation or from a more chaotic dynamical N-body interaction.

Massive tight binaries cannot originate from the simple gravitational fragmentation of massive cloud cores and filaments into two Jeans masses. The Jeans radius (10,000 AU) is far too large compared with the separations of the two binary components (1–10 AU). More sophisticated physical processes must be at play, such as:

1. *Inner disk fragmentation* (*Kratter and Matzner, 2006*) followed by circumbinary accretion, to make the components grow in mass (*Artymowicz and Lubow, 1996*).

2. *Roche lobe overflow* of a close rapidly accreting bloated protobinary (*Krumholz and Thompson, 2007*).

In both cases, the authors argue that the accretion flow would drive the component masses to near equality (massive twins). These theories, however, do not explain how to get the initially lower-mass close binaries in the first place.

3. *Accretion onto a low-mass initially wide binary system* (*Bonnell and Bate, 2005*). While growing in mass by accretion, the orbital separation of the binary system keeps shrinking. In this case, one can show analytically that — depending on the angular momentum of the accreting gas — the wide binary, while growing in mass, will shrink its orbital separation substantially (for example, two 1-M$_\odot$ protostars at 30-AU separation can easily end up as two 30-M$_\odot$ components at about 1 AU separation if the specific angular momentum of the accreted gas is constant; i.e., if accreting gas angular momentum scales linearly with the accreted gas mass).

4. *Magnetic effects on fragmentation.* As noted in section 2, three-dimensional MHD calculations are just now becoming commonplace, and effects such as magnetic torques on rotating clouds might well lead to the formation of closer binary star systems than those found to date by three-dimensional HD and RHD models of the fragmentation process (*Price and Bate, 2007*).

5. *Viscous evolution and orbital decay.* When a triple system breaks up and ejects a component, the orbit of the remaining binary tightens and becomes highly eccentric (see section 5). When this occurs at early evolutionary stages while the binary components are still surrounded by dense circumstellar material, the components interact viscously during periastron passages and their orbits decay (e.g., *Stahler, 2010; Korntreff et al., 2012*). At the same time, the UV radiation field photoevaporates the circumstellar material, leaving many binaries stranded in close orbits.

11.3. Trapezia

The famous Orion Trapezium (e.g., *Herbig and Terndrup, 1986; Close et al., 2012*) is the prototype of nonhierarchical compact groups of OB stars. The concept was first introduced by *Ambartsumian* (1954), who recognized that such systems are inherently unstable. Kinematic studies of trapezia show the internal motions expected for bound, virialized small clusters, but occasionally having components with velocities exceeding the escape speed (*Allen et al., 2004*).

High-precision astrometry from radio interferometry has demonstrated that three of the sources in the Becklin-Neugebauer/Kleinman-Low (BN/KL) region in Orion have large motions and are receding from a point in between them, suggesting that they were all part of a small stellar group that disintegrated ~500 years ago (*Rodríguez et al., 2005; Gómez et al., 2006*) (but see also *Tan, 2004*). Just like the disintegration of small low-mass, very young stellar systems can lead to giant Herbig-Haro bow shocks (*Reipurth, 2000*), so will the breakup of a trapezium of massive protostars with abundant gas lead to an energetic, explosive event, as observed around the BN/KL region (*Bally and Zinnecker, 2005; Zapata et al., 2009; Bally et al., 2011*).

Trapezia are common in regions of massive star formation (e.g., *Salukvadze and Javakhishvili, 1999; Abt and Corbally, 2000*). Of particular interest are studies of the earliest stages of formation of a trapezium at centimeter and millimeter wavelengths (e.g., *Rodón et al., 2008*). N-body simulations of massive trapezia in clusters demonstrate that these systems are highly dynamical entities, interacting and exchanging members with the surrounding cluster before eventually breaking apart (*Allison and Goodwin, 2011*).

12.   STATISTICAL PROPERTIES OF MULTIPLE STARS

Multiple systems with three or more components (hereafter *multiples*) are a natural and rather frequent outcome of star formation. Compared to binaries, they have more parameters (periods, mass ratios, etc.), so their statistics bring additional insights on the formation mechanisms.

We focus here on stars with primary components of about one solar mass, as their multiplicity statistics are known best. *Raghavan et al.* (2010) estimated a multiplicity fraction MF = 0.46 and a higher-order fraction (triples and up) HF ≈ 0.12 in a sample of 454 solar-mass dwarfs within 25 pc of the Sun. A much larger sample is needed, however, for a meaningful statistical study of hierarchical systems. Here we present preliminary results on F and G dwarfs within 67 pc selected from Hipparcos, the FG-67pc sample (*Tokovinin, 2014*). It contains a few hundred hierarchical systems among ~5000 stars.

12.1. Period Ratio and Dynamical Stability

Figure 12 compares the inner, *short* periods P_S at levels 11 and 12 to the outer, *long* periods P_L at level 1 for the FG-67pc sample (for a definition of levels, see section 3). Note that orbital periods of wide pairs are estimated statistically by assuming that projected separation equals orbital semimajor axis. Such estimates P^* are unbiased and differ from the true periods P by less than 3×, in most cases.

The points in Fig. 12 fill the space above the dashed line, reflecting the fact that all combinations of inner and outer periods allowed dynamically are actually possible. The minimum period (or separation) ratio allowed by dynamical stability has been studied by several authors. The stability criterion of *Mardling and Aarseth* (2001), for example, can be written as

$$P_L/P_S > 4.7 \left(1 - e_L\right)^{-1.8} \left(1 + e_L\right)^{0.6} \left(1 + q_{out}\right)^{0.1} \quad (3)$$

where e_L is the eccentricity of the outer orbit, while the ratio of the distant-companion mass to the combined mass of the inner binary q_{out} plays only a minor role. The dashed line in Fig. 12 corresponds to P_L/P_S = 4.7; all points are above it (with one exception caused by the uncertainty of P^*). Although orbits of outer systems tend to have moderate e_L (*Shatsky*, 2001), its variation spreads the value of the P_L/P_S threshold over at least 1 order of magnitude.

Outer systems with P < 10^3 d do not exist or are rare (see the empty lower-left corner in Fig. 12). Such triples can be readily discovered by radial-velocity variations superposed on the short (inner) orbit, so their absence is not an observational bias. However, tight triples are found among massive stars.

12.2. Distribution of Periods and Mass Ratios, Fraction of Hierarchies

Accounting for the observational selection is critical. In the FG-67pc sample, the probability of detecting a companion to the main target over the full range of periods and mass ratios has been determined to be about 78%. This means that only $0.78^2 \approx 0.61$ fraction of level-11 triples is actually discovered (assuming that detections of inner and outer companions are uncorrelated). The observed fractions

of S:B:T:Q systems are 64:29:6:1%. The selection corrected fractions are 54:31:6:7. The difference between observed (raw) and corrected fractions increases with increasing multiplicity. Some systems known presently as binaries are in fact triples, some triples are quadruples, etc.

The joint distribution of period P and mass ratios q = M_2/M_1 is frequently approximated by the Gaussian distribution of x = $\log_{10}(P/1d)$ and by the power-law distribution of q (see, e.g., *Duchêne and Kraus*, 2013)

$$f(x, q) = C \, \varepsilon \, q^{\beta} \exp\left[-(x - x_0)^2 \big/ (2\sigma^2)\right] \quad (4)$$

where ε is the fraction of systems and C is the normalization constant. It is likely that the mass-ratio distribution depends on period, but this is still being debated.

The parameters of equation (4) for the FG-67pc sample are found by maximum likelihood, accounting for the incomplete detections and missing data. When all stellar pairs are considered regardless of their hierarchical levels, the result is ε = CSF = 0.57 ± 0.02, while the median period is x_0 = 4.53 ± 0.09. If, on the other hand, we count only the outer level-1 systems, the result is different: ε = MF = 0.47 ± 0.01 and x_0 = 4.97 ± 0.06. Binary periods at the outer hierarchical level are thus almost 3× longer than the periods of all binaries. Similarly, for the inner pairs at levels 11 and 12 we derive much shorter median periods: x_0 = 3.12 and x_0 = 2.45, respectively. Note that the formal errors quoted above are only lower statistical limits; the results are influenced by several assumptions and approximations made in the analysis, making the real uncertainty larger. The exponent of the mass-ratio distribution turns out to be small, β ≈ 0.2, meaning that the distribution of q is almost uniform.

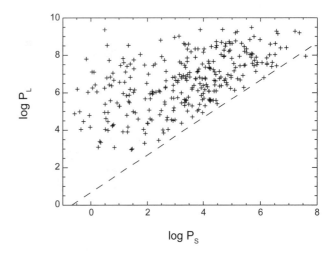

Fig. 12. Orbital periods P_S at inner hierarchical levels 11 and 12 are compared to the periods of outer systems P_L from the FG-67pc sample. The periods are expressed in days and plotted on the logarithmic scale. The dashed line marks the dynamical stability limit P_L/P_S = 4.7.

We derive the selection-corrected fractions of subsystems of level 11 and level 12 as 10% and 8%, respectively. Discovery of subsystems in the *secondary* companions (level 12) is more problematic than at level 11. Usually researchers concentrate on discovering companions to their primary targets and forget that some of those companions may, in turn, be close pairs. The estimated detection rate of level-12 subsystems in the FG-67pc sample is only ~0.2, so their true frequency depends on the large, hence uncertain, correction. However, there is a strong evidence that the occurrence of subsystems in the secondary components is nearly as frequent as in the main (primary) targets.

Among the 88 subsystems of level 12, about half also have subsystems of level 11. There is hence a correlation between those levels: The frequency of 2 + 2 quadruples is larger than could be inferred from the frequency of levels 11 and 12 if they were independent (uncorrelated). Among the 8% of systems containing secondary pairs of level 12, half also contain level-11 pairs; they are 2 + 2 quadruples. Considering this, the fraction of systems with at least three companions is HF ≈ 0.10 + 0.04 = 0.14, not 0.10 + 0.08 = 0.18 as one might naively assume by summing up the frequencies of levels 11 and 12.

12.3. Statistical Model of Hierarchical Multiplicity

It is remarkable that inner pairs in hierarchical multiples are statistically similar to simple binaries. The mass ratios in spectroscopic binaries with and without distant tertiary companions are distributed in the same way (*Tokovinin et al.,* 2006). The frequency of spectroscopic subsystems in visual binaries is similar to the frequency of spectroscopic binaries in the open-cluster and field populations (*Tokovinin and Smekhov,* 2002). The frequency of resolved subsystems in wide binaries is again comparable to binaries in the field (*Tokovinin et al.,* 2010). To first order, we can construct a hierarchical triple by selecting two binaries randomly from the same generating distribution of periods and keeping only stable (hierarchical) combinations. This recipe is applied recursively to simulate higher-order multiples.

To test this idea, *Tokovinin* (2014) simulated multiples, filtered them by the average detection probability, and compared to the real sample, following the strategy of *Eggleton* (2009). The parameters of the generating distribution (equation (4)) were taken from the maximum-likelihood analysis and could be further adjusted to improve the agreement between the simulated and real samples. If the multiplicity fraction ε is kept constant, the HF in the simulated sample is too low. So, to reach an agreement between simulations and reality, we had to increase ε at inner hierarchical levels and introduce a correlation between levels 11 and 12. Alternatively, the agreement can be obtained by assuming a variable (stochastic) binary frequency ε. Cases with a high ε produce many hierarchies, while the cases of small ε generate mostly single and binary stars. This finding suggests that the field population is a mixture coming from binary-rich and binary-poor environments. Differences of

the multiplicity fraction among star-forming regions are well documented (see section 10.3).

Interestingly, the simulated quadruples outnumber triples, resembling in this respect the HD simulations of *Bate* (2012). The 2 + 2 quadruples are much more frequent (~4–5% of all stars) than the 3 + 1 quadruples. The large number of 2 + 2 quadruples in the FG-67pc sample predicted by this model can be verified observationally.

A loose correlation between orientations of the angular momentum vectors in the inner and outer subsystems of triples was found in early works and confirmed by *Sterzik and Tokovinin* (2002). This correlation becomes stronger at moderate P_L/P_S ratios, i.e., in triples with weak hierarchy. These authors tried to match the observational result with simulations of dynamically decaying N-body systems. Agreement could be achieved for certain initial conditions (rotating and/or flattened clusters). However, multiple systems produced by the pure N-body decay without gas drag and accretion are statistically very different from the real multiples in their eccentricities and period ratios (*Tokovinin,* 2008), pointing to the importance of viscous interactions and accretion during the earliest phases of multiple evolution.

13. CONCLUDING REMARKS

In summary, it appears that the large majority — and potentially all — of stars are born in small multiple systems. A picture is emerging where the field population of single, binary, and multiple stars derives from a birth population that has been transformed by both internal and external dynamical processes. These processes sculpt the (still unknown) separation distribution function at birth into the log-normal distribution (with a power-law tail for the wider binaries) observed in the evolved field population.

Acknowledgments. We thank the referee for a helpful report. B.R. acknowledges support through the NASA Astrobiology Institute under Cooperative Agreement No. NNA09DA77A issued through the Office of Space Science. The work of A.P.B. was partially supported by the NSF under grant AST-1006305.

REFERENCES

Abt H. A. and Corbally C. J. (2000) *Astrophys. J., 541,* 841.
Albrecht S. et al. (2009) *Nature, 461,* 373.
Alexander R. (2012) *Astrophys. J. Lett., 757,* L29.
Allen C. et al. (2004) In *The Environments and Evolution of Double and Multiple Stars* (C. Allen and C. Scarfe, eds.), p. 195. Rev. Mex. Astron. Astrofis. Ser. Conf. 21.
Allen P. R. (2007) *Astrophys. J., 668,* 492.
Allison R. J. and Goodwin S. P. (2011) *Mon. Not. R. Astron. Soc., 415,* 1967.
Ambartsumian V. A. (1954) *Contrib. Byurakan Obs., 15,* 3.
Anosova J. P. (1986) *Astrophys. Space Sci., 124,* 217.
Armitage P. J. et al. (1999) *Mon. Not. R. Astron. Soc., 304,* 425.
Arreaga-García G. et al. (2010) *Astron. Astrophys., 509,* A96.
Artymowicz P. and Lubow S. H. (1994) *Astrophys. J., 421,* 651.
Artymowicz P. and Lubow S. H. (1996) *Astrophys. J. Lett., 467,* L77.
Attwood R. E. et al. (2009) *Astron. Astrophys., 495,* 201.
Bai X.-N. (2011) *Astrophys. J., 739,* A50.
Baines D. et al. (2006) *Mon. Not. R. Astron. Soc., 367,* 737.
Bally J. and Zinnecker H. (2005) *Astron. J., 129,* 2281.
Bally J. et al. (2011) *Astrophys. J., 727,* A113.

Baraffe I. and Chabrier G. (2010) *Astron. Astrophys., 521,* A44.
Baraffe I. et al. (1998) *Astron. Astrophys., 337,* 403.
Basu S. and Vorobyov E. I. (2012) *Astrophys. J., 750,* A30.
Bate M. R. (2000) *Mon. Not. R. Astron. Soc., 314,* 33.
Bate M. R. (2009a) *Mon. Not. R. Astron. Soc., 392,* 590.
Bate M. R. (2009b) *Mon. Not. R. Astron. Soc., 397,* 232.
Bate M. R. (2012) *Mon. Not. R. Astron. Soc., 419,* 3115.
Bate M. R. and Bonnell I. A. (1997) *Mon. Not. R. Astron. Soc., 285,* 33.
Bate M. R. et al. (2002) *Mon. Not. R. Astron. Soc., 336,* 705.
Batten A. H. (1973) *Binary and Multiple Star Systems.*
    Pergamon, Oxford.
Bechter E. B. et al. (2014) *Astrophys J., 788,* A2.
Becker C. et al. (2013) *Astron. Astrophys., 552,* A46.
Bergfors C. et al. (2010) *Astron. Astrophys., 520,* A54.
Biller B. et al. (2011) *Astrophys. J., 730,* A39.
Bisikalo D. V. et al. (2012) *Astron. Rept., 56,* 686.
Blaauw A. (1961) *Bull. Astron. Inst. Netherlands, 15,* 265.
Boden A. F. et al. (2009) *Astrophys. J. Lett., 696,* L111.
Bonnell I. A. and Bate M. R. (2005) *Mon. Not. R. Astron. Soc.,
    362,* 915.
Bonnell I. A. et al. (2007) In *Protostars and Planets V* (B. Reipurth
    et al., eds.), p. 149. Univ. of Arizona, Tucson.
Bonnell I. A. et al. (2008) *Mon. Not. R. Astron. Soc., 389,* 1556.
Boss A. P. (2009) *Astrophys. J., 697,* 1940.
Boss A. P. and Bodenheimer P. (1979) *Astrophys. J., 234,* 289.
Boss A. P. and Keiser S. A. (2013) *Astrophys. J., 763,* A1.
Bürzle F. et al. (2011) *Mon. Not. R. Astron. Soc., 412,* 171.
Brandeker A. et al. (2006) *Astrophys. J., 652,* 1572.
Bressert E. et al. (2012) *Astron. Astrophys., 542,* A49.
Burgasser A. J. et al. (2007) In *Protostars and Planets V* (B. Reipurth
    et al., eds.), p. 427. Univ. of Arizona, Tucson.
Cabrit S. et al. (2006) *Astron. Astrophys., 452,* 897.
Capelo H. et al. (2012) *Astrophys. J. Lett., 757,* L18.
Carney B. W. et al. (2003) *Astron. J., 125,* 293.
Chabrier G. et al. (2007) *Astron. Astrophys., 472,* L17.
Chauvin G. et al. (2011) *Astron. and Astrophys., 528,* A8.
Chen X. et al. (2008) *Astrophys. J., 683,* 862.
Chen X. et al. (2009) *Astrophys. J., 691,* 1729.
Chen X. et al. (2013) *Astrophys. J., 768,* A110.
Chini R. et al. (2012) *Mon. Not. R. Astron. Soc., 424,* 1925.
Cieza L. et al. (2009) *Astrophys. J. Lett., 696,* L84.
Clark P. C. et al. (2008) *Mon. Not. R. Astron. Soc., 386,* 3.
Clarke C. J. and Pringle J. E. (1992) *Mon. Not. R. Astron. Soc.,
    255,* 423.
Clarke C. J. and Pringle J. E. (1993) *Mon. Not. R. Astron. Soc.,
    261,* 190.
Close L. M. et al. (2012) *Astrophys. J., 749,* A180.
Commerçon B. et al. (2010) *Astron. Astrophys., 510,* L3.
Commerçon B. et al. (2011) *Astrophys. J Lett., 742,* L9.
Connelley M. S. et al. (2008a) *Astron. J., 135,* 2496.
Connelley M. S. et al. (2008b) *Astron. J., 135,* 2526.
Connelley M. S. et al. (2009) *Astron. J., 138,* 1193.
Corporon P. and Lagrange A.-M. (1999) *Astron. Astrophys. Suppl.,
    136,* 429.
Correia S. et al. (2006) *Astron. Astrophys., 459,* 909.
Correia S. et al. (2013) *Astron. Astrophys., 557,* A63.
Crutcher R. M. (2012) *Ann. Rev. Astron. Astrophys., 50,* 29.
Cuadra J. et al. (2009) *Mon. Not. R. Astron. Soc., 393,* 1423.
Daemgen S. et al. (2012) *Astron. Astrophys., 540,* A46.
Daemgen S. et al. (2013) *Astron. Astrophys., 554,* A43.
D'Antona F. and Mazzitelli I. (1997) *Mem. Soc. Astron. Ital., 68,* 807.
Deller A. T. et al. (2013) *Astron. Astrophys., 552,* A51.
De Rosa R. J. et al. (2014) *Mon. Not. R. Astron. Soc., 437,* 1216.
de Val-Borro M. et al. (2011) *Mon. Not. R. Astron. Soc., 413,* 2679.
de Wit W. J. et al. (2005) *Astron. Astrophys., 437,* 247.
Domingos R. et al. (2012) *Astron. Astrophys., 544,* A63.
Dotter A. et al. (2008) *Astrophys. J. Suppl., 178,* 89.
Doyle L. R. et al. (2011) *Science, 333,* 1602.
Duchêne G. and Kraus A. (2013) *Annu. Rev. Astron. Astrophys., 51,* 269.
Duchêne G. et al. (1999) *Astron. Astrophys., 343,* 831.
Duchêne G. et al. (2004) *Astron. Astrophys., 427,* 651.
Duchêne G. et al. (2007) *Astron. Astrophys., 476,* 229.
Dumusque X. et al (2012) *Nature, 491,* 207.
Dupuy T. J. and Liu M. C. (2011) *Astrophys. J., 733,* A122.
Duquennoy A. and Mayor M. (1991) *Astron. Astrophys., 248,* 485.
Durisen R. H. and Sterzik M. F. (1994) *Astron. Astrophys., 286,* 84.
Dyck H. M. et al. (1982) *Astrophys. J. Lett., 255,* L103.
Eggleton P. (2009) *Mon. Not. R. Astron. Soc., 399,* 1471.
Enoch M. L. et al. (2011) *Astrophys. J. Suppl., 195,* A21.
Espaillat C. et al. (2011) *Astrophys. J., 728,* A49.

Fabrycky D. and Tremaine S. (2007) *Astrophys. J., 669,* 1298.
Facchini S. and Lodato G. (2013) *Mon. Not. R. Astron. Soc., 433,* 2142.
Fateeva A. M. et al. (2011) *Astrophys. Space Sci., 335,* 125.
Feiden G. and Chaboyer B. (2012) *Astrophys. J., 761,* A30.
Feiden G. A. and Dotter A. (2013) *Astrophys. J., 765,* A86.
Fischer D. A. and Marcy G. W. (1992) *Astrophys. J., 396,* 178.
Foucart F. and Lai D. (2013) *Astrophys. J., 764,* A106.
Garcia V. et al. (2011) *Astron. J., 142,* A27.
Garcia V. et al. (2013a) *Astrophys. J., 769,* A114.
Garcia P. J. V. et al. (2013b) *Mon. Not. R. Astron. Soc., 430,* 1839.
Ghez A. M. et al. (1993) *Astron. J., 106,* 2005.
Ghez A. M. et al. (1997) *Astrophys. J., 481,* 378.
Gieles M. and Portegies Zwart S. F. (2011) *Mon. Not. R. Astron. Soc.,
    410,* L6.
Gies D. R. and Bolton C. T. (1986) *Astrophys. J. Suppl., 61,* 419.
Gómez L. et al. (2006) *Astrophys. J., 635,* 1166.
Gomez de Castro et al. (2013) *Astrophys. J., 766,* 62.
Gómez Maqueo Chew Y. et al. (2012) *Astrophys. J., 745,* A58.
Goodman J. and Hut P. (1993) *Astrophys. J., 403,* 271.
Goodwin S. P. (2010) *Philos. Trans. R. Soc. Ser. A, 368,* 851.
Goodwin S. P. et al. (2007) In *Protostars and Planets V* (B. Reipurth
    et al., eds.), p. 133. Univ. of Arizona, Tucson.
Gorti U. and Bhatt H. C. (1996) *Mon. Not. R. Astron. Soc., 283,* 566.
Haisch K. E. et al. (2004) *Astron. J., 127,* 1747.
Hall S. (1997) *Mon. Not. R. Astron. Soc., 287,* 148.
Hall S. et al. (1996) *Mon. Not. R. Astron. Soc., 278,* 303.
Hallinan G. et al. (2007) *Astrophys. J. Lett., 663,* L25.
Hanawa T. et al. (2010) *Astrophys. J., 708,* 485.
Hansen C. et al. (2012) *Astrophys. J., 747,* A22.
Harris R. J. et al. (2012) *Astrophys. J., 751,* A115.
Heggie D. C. (1975) *Mon. Not. R. Astron. Soc., 173,* 729.
Hennebelle P. and Fromang S. (2008) *Astron. Astrophys., 477,* 9.
Hennebelle P. and Teyssier R. (2008) *Astron. Astrophys., 477,* 25.
Hennebelle P. et al. (2011) *Astron. Astrophys., 528,* A72.
Herbig G. H. (1962) *Adv. Astron. Astrophys., 1,* 47.
Herbig G. H. (1977) *Astrophys. J., 217,* 693.
Herbig G. H. and Terndrup D. M. (1986) *Astrophys. J., 307,* 609.
Herbst W. et al. (2010) *Astron. J., 140,* 2025.
Hills J. G. (1975) *Astron. J., 80,* 809.
Hills J. G. (1990) *Astron. J., 99,* 979.
Hioki T. et al. (2009) *Pub. Astron. Soc. Japan, 61,* 1271.
Hioki T. et al. (2011) *Pub. Astron. Soc. Japan, 63,* 543.
Jensen E. L. et al. (1996) *Astrophys. J., 458,* 312.
Joergens V. (2008) *Astron. Astrophys., 492,* 545.
Joos M. et al. (2012) *Astron. Astrophys., 543,* A128.
Joy A. H. and van Biesbroeck G. (1944) *Publ. Astron. Soc. Pac.,
    56,* 123.
Kennedy G. M. et al. (2012a) *Mon. Not. R. Astron. Soc., 421,* 2264.
Kennedy G. M. et al. (2012b) *Mon. Not. R. Astron. Soc., 426,* 2115.
Kiminki D. C. and Kobulnicky H. A. (2012) *Astrophys. J., 751,* A4.
King R. R. et al. (2012a) *Mon. Not. R. Astron. Soc., 421,* 2025.
King R. R. et al. (2012b) *Mon. Not. R. Astron. Soc., 427,* 2636.
Köhler R. et al. (2006) *Astron. Astrophys., 458,* 461.
Korntreff C. et al. (2012) *Astron. Astrophys., 543,* A126.
Kouwenhoven M. B. N. et al. (2005) *Astron. Astrophys., 430,* 137.
Kouwenhoven M. B. N. et al. (2007) *Astron. Astrophys., 474,* 77.
Kouwenhoven M. B. N. et al. (2010) *Mon. Not. R. Astron. Soc.,
    404,* 1835.
Kozai Y. (1962) *Astron. J., 67,* 591.
Kratter K.M. and Matzner C. D. (2006) *Mon. Not. R. Astron. Soc.,
    373,* 1563.
Kraus A. L. and Hillenbrand L. A. (2007) *Astrophys. J., 662,* 413.
Kraus A. L. and Hillenbrand L. A. (2009) *Astrophys. J., 784,* 531.
Kraus A. L. and Hillenbrand L. A. (2012) *Astrophys. J., 757,* A141.
Kraus A. L. et al. (2008) *Astrophys. J., 679,* 762.
Kraus A. L. et al. (2011) *Astrophys. J., 731,* A8.
Kraus A. L. et al. (2012) *Astrophys. J., 745,* A19.
Kroupa P. (1995) *Mon. Not. R. Astron. Soc., 277,* 1491.
Kroupa P. (1998) *Mon. Not. R. Astron. Soc., 298,* 231.
Kroupa P. (2000) In *Massive Stellar Clusters* (A. Lancon and C. Boily,
    ed.), p. 233. ASP Conf. Ser. 211, Astronomical Society of the
    Pacific, San Francisco.
Kroupa P. and Bouvier J. (2003) *Mon. Not. R. Astron. Soc., 346,* 343.
Kroupa P. and Petr-Gotzens M. G. (2011) *Astron. Astrophys., 529,* A92.
Kroupa P. et al. (2003) *Mon. Not. R. Astron. Soc., 346,* 354.
Krumholz M. R. and Thompson T. A. (2007) *Astrophys. J., 661,* 1034.
Krumholz M. et al. (2012) *Astrophys. J., 754,* A71.
Kudoh T. and Basu S. (2008) *Astrophys. J. Lett., 679,* L97.
Kudoh T. and Basu S. (2011) *Astrophys. J., 728,* A123.
Kuiper G. P. (1942) *Astrophys. J., 95,* 201.

Lada C. J. (2006) *Astrophys. J. Lett., 640,* L63.
Lafrenière D. et al. (2008) *Astrophys. J., 683,* 844.
Larson R. B. (1972) *Mon. Not. R. Astron. Soc., 156,* 437.
Lazarian A. et al. (2012) *Astrophys. J., 757,* A154.
Leigh N. and Geller A. M. (2012) *Mon. Not. R. Astron. Soc., 425,* 2369.
Leinert Ch. et al. (1993) *Astron. Astrophys., 278,* 129.
Leinert Ch. et al. (1997) *Astron. Astrophys., 318,* 472.
Leinhardt Z. and Stewart S. (2012) *Astrophys. J., 745,* A79.
Lépine S. and Bongiorno B. (2007) *Astron. J., 133,* 889.
Lidov M. (1962) *Planet. Space Sci., 9,* 719.
Loinard L. (2013) In *Advancing the Physics of Cosmic Distances* (R. de Grijs, ed.), p. 36. IAU Symp. 289, Cambridge Univ., Cambridge.
Looney L. G. et al. (2000) *Astrophys. J., 529,* 477.
MacDonald J. and Mullan D. J. (2009) *Astrophys. J., 700,* 387.
MacFadyen A. I. and Milosavljevic M. (2008) *Astrophys. J., 672,* 83.
Machida M. N. (2008) *Astrophys. J. Lett., 682,* L1.
Machida M. N. et al. (2008) *Astrophys. J., 677,* 327.
Maiz-Appellaniz J. et al. (2004) *Astrophys. J. Suppl., 151,* 103.
Mardling R. A. and Aarseth S. J. (2001) *Mon. Not. R. Astron. Soc., 321,* 398.
Marks M. and Kroupa P. (2011) *Mon. Not. R. Astron. Soc., 417,* 1702.
Marks M. and Kroupa P. (2012) *Astron. Astrophys., 543,* A8.
Martins F. et al. (2005) *Astron. Astrophys., 436,* 1049.
Marzari F. et al. (2013) *Astron. Astrophys., 553,* A71.
Mason B. D. et al. (2009) *Astron. J., 137,* 3358.
Mathieu R. D. (1994) *Annu. Rev. Astron. Astrophys., 32,* 465.
Mathieu R. D. et al. (1997) *Astron. J., 113,* 1841.
Mathieu R. D. et al. (2007) In *Protostars and Planets V* (B. Reipurth et al., eds.), p. 411. Univ. of Arizona, Tucson.
Maury A. J. et al. (2010) *Astron. Astrophys., 512,* A40.
Melo C. H. F. et al. (2001) *Astron. Astrophys., 378,* 898.
Moeckel N. and Bate M. R. (2010) *Mon. Not. R. Astron. Soc., 404,* 721.
Moeckel N. and Clarke C. J. (2011) *Mon. Not. R. Astron. Soc., 415,* 1179.
Mohanty S. and Stassun K. G. (2012) *Astrophys. J., 758,* A12.
Mohanty S. et al. (2013) *Astrophys. J., 764,* A65.
Monin J.-L. et al. (2007) In *Protostars and Planets V* (B. Reipurth et al., eds.), p. 395. Univ. of Arizona, Tucson.
Morales J. C. et al. (2010) *Astrophys. J., 718,* 502.
Morales-Calderon M. et al. (2012) *Astrophys. J., 753,* A149.
Muzerolle J. et al. (2013) *Nature, 493,* 378.
Myers A. et al. (2013) *Astrophys. J., 766,* A97.
Nagel E. et al. (2012) *Astrophys. J., 747,* A139.
Nguyen D. C. et al. (2012) *Astrophys. J., 745,* A119.
Ochi Y. et al. (2005) *Astrophys. J., 623,* 922.
Oey M. S. et al. (2013) *Astrophys. J., 768,* A66.
Offner S. S. R. et al. (2009) *Astrophys. J., 703,* 131.
Offner S. S. R. et al. (2010) *Astrophys. J., 725,* 1485.
Öpik E. (1924) *Publ. Tartu Obs., 25(6).*
Padoan P. and Nordlund Å. (2004) *Astrophys. J., 617,* 559.
Palla F. and Stahler S. (1999) *Astrophys. J., 525,* 772.
Parker R. J. and Goodwin S. P. (2012) *Mon. Not. R. Astron. Soc., 424,* 272.
Parker R. J. and Reggiani M. M. (2013) *Mon. Not. R. Astron. Soc., 432,* 2378.
Parker R. J. et al. (2009) *Mon. Not. R. Astron. Soc., 397,* 1577.
Parker R. J. et al. (2011) *Mon. Not. R. Astron. Soc., 418,* 2565.
Payne M. et al. (2009) *Mon. Not. R. Astron. Soc., 336,* 973.
Pepper J. et al. (2007) *Publ. Astron. Soc. Pac., 119,* 923.
Pfalzner S. et al. (2005) *Astrophys. J., 629,* 526.
Poveda A. et al. (1967) *Bol. Observ. Ton. Tac., 4,* 86.
Prato L. (2007) *Astrophys. J., 657,* 338.
Prato L. and Simon M. (1997) *Astrophys. J., 474,* 455.
Price D. J. and Bate M. R. (2007) *Mon. Not. R. Astron. Soc., 377,* 77.
Rafikov R. (2013) *Astrophys. J. Lett., 765,* L8.
Raghavan D. et al. (2010) *Astrophys. J. Suppl., 190,* 1.
Ratzka T. et al. (2005) *Astron. Astrophys., 437,* 611.
Rawiraswattana K. et al. (2012) *Mon. Not. R. Astron. Soc., 419,* 2025.
Reid I. N. et al. (2006) *Astron. J., 132,* 891.
Reiners A. et al. (2007) *Astrophys. J. Lett., 671,* L149.
Reipurth B. (1988) In *Formation and Evolution of Low-Mass Stars* (A. K. Dupree and M. T. V. T. Lago, eds.), p. 305. Kluwer, Dordrecht.
Reipurth B. (2000) *Astron. J., 120,* 3177.
Reipurth B. and Aspin C. (2004) *Astrophys. J. Lett., 608,* L65.
Reipurth B. and Bally J. (2001) *Annu. Rev. Astron. Astrophys., 39,* 403.
Reipurth B. and Clarke C. (2001) *Astron. J., 122,* 432.
Reipurth B. and Mikkola S. (2012) *Nature, 492,* 221.
Reipurth B. and Zinnecker H. (1993) *Astron. Astrophys., 278,* 81.
Reipurth B. et al. (2002) *Astron. J., 124,* 1045.
Reipurth B. et al. (2004) *Astron. J., 127,* 1736.

Reipurth B. et al. (2007) *Astron. J., 134,* 2272.
Reipurth B. et al. (2010) *Astrophys. J. Lett., 725,* L56.
Rodón J. A. et al. (2008) *Astron. Astrophys., 490,* 213.
Rodriguez J. et al. (2013) *Astron. J., 146,* A112.
Rodríguez L. F. et al. (2003) *Astrophys. J., 598,* 1100.
Rodríguez L. F. et al. (2005) *Astrophys. J. Lett., 627,* L65.
Rodríguez L. F. et al. (2010) *Astron. J., 140,* 968.
Salter D. M. et al. (2010) *Astron. Astrophys., 521,* A32.
Salukvadze G. N. and Javakhishvili G. Sh. (1999) *Astrophysics, 42,* 431.
Sana H. and Evans C. J. (2011) In *Active OB Stars: Structure, Evolution, Mass Loss, and Critical Limits* (C. Neiner et al., eds.), p. 474. IAU Symp 272, Cambridge Univ., Cambridge.
Sana H. et al. (2008) *Mon. Not. R. Astron. Soc., 386,* 447.
Sana H. et al. (2009) *Mon. Not. R. Astron. Soc., 400,* 1479.
Sana H. et al. (2012) *Science, 337,* 444.
Sana H. et al. (2013) *Astron. Astrophys., 550,* A107.
Scally A. et al. (1999) *Mon. Not. R. Astron. Soc., 306,* 253.
Scally A. et al. (2005) *Mon. Not. R. Astron. Soc., 358,* 742.
Seifried D. et al. (2012) *Mon. Not. R. Astron. Soc., 423,* L40.
Shakura N. and Sunyaev R. A. (1973) *Astron. Astrophys., 24,* 337.
Shatsky N. (2001) *Astron. Astrophys., 380,* 238.
Shi J.-M. et al. (2012) *Astrophys. J., 747,* A118.
Siess I. et al. (2000) *Astron. Astrophys., 358,* 593.
Simon M. and Obbie R. C. (2009) *Astron. J., 137,* 3442.
Sota A. et al. (2008) *Rev. Mex. Astron. Astrofis. Ser. Conf., 33,* 56.
Stahler S. W. (2010) *Mon. Not. R. Astron. Soc., 402,* 1758.
Stamatellos D. et al. (2007) *Mon. Not. R. Astron. Soc., 382,* L30.
Stassun K. G. et al. (2006) *Nature, 440,* 311.
Stassun K. G. et al. (2007) *Astrophys. J., 664,* 1154.
Stassun K. G. et al. (2008) *Nature, 453,* 1079.
Stassun K. G. et al. (2012) *Astrophys. J., 756,* A47.
Stempels H. and Gahm G. (2004) *Astron. Astrophys., 421,* 1159.
Sterzik M. and Durisen R. (1998) *Astron. Astrophys., 339,* 95.
Sterzik M. and Tokovinin A. (2002) *Astron. Astrophys., 384,* 1030.
Sterzik M. et al. (2003) *Astron. Astrophys., 411,* 91.
Tan J. (2004) *Astrophys. J. Lett., 607,* L47.
Thebault P. et al. (2006) *Icarus, 183,* 193.
Thebault P. et al. (2008) *Mon. Not. R. Astron. Soc., 388,* 1528.
Thebault P. et al. (2009) *Mon. Not. R. Astron. Soc., 393,* L21.
Tobin J. J. et al. (2009) *Astrophys. J., 697,* 1103.
Tobin J. J. et al. (2013) *Astrophys. J., 779,* A93.
Tognelli E. et al. (2011) *Astron. Astrophys., 533,* A109.
Tokovinin A. (2008) *Mon. Not. R. Astron. Soc., 389,* 925.
Tokovinin A. and Lépine S. (2012) *Astron. J., 144,* A102.
Tokovinin A. and Smekhov M. (2002) *Astron. Astrophys., 382,* 118.
Tokovinin A. et al. (2006) *Astron. Astrophys., 450,* 681.
Tokovinin A. et al. (2010) *Astron. J., 140,* 510.
Tokovinin A. (2014) *Astron J., 147,* A87.
Torres G. (2013) *Astron. Nachr., 334,* 4.
Torres G. et al. (2010) *Astron. Astrophys. Rev., 18,* 67.
Torres G. et al. (2013) *Astrophys. J., 773,* A40.
Torres R. M. et al. (2012) *Astrophys. J., 747,* A18.
Trilling D. et al. (2007) *Astrophys. J., 658,* 1264.
Umbreit S. et al. (2005) *Astrophys. J., 623,* 940.
Umbreit S. et al. (2011) *Astrophys. J., 743,* A106.
Valtonen M. and Karttunen H. (2006) *The Three-Body Problem,* Cambridge Univ., Cambridge.
van Albada T. S. (1968) *Bull. Astron. Inst. Netherlands, 20,* 57.
Verrier P. and Evans N. (2007) *Mon. Not. R. Astron. Soc., 382,* 1432.
Verrier P. and Evans N. (2009) *Mon. Not. R. Astron. Soc., 394,* 1721.
Viana Almeida P. et al. (2012) *Astron. Astrophys., 539,* A62.
Walch S. et al. (2010) *Mon. Not. R. Astron. Soc., 402,* 2253.
Welsh W. F. et al. (2012) *Nature, 481,* 475.
Whitworth A. P. and Zinnecker H. (2004) *Astron. Astrophys., 427,* 299.
Whitworth A. P. et al. (2007) In *Protostars and Planets V* (B. Reipurth et al., eds.), p. 459. Univ. of Arizona, Tucson.
Xie J.-W. et al. (2010) *Astrophys. J., 708,* 1566.
Zahn J.-P. and Bouchet L. (1989) *Astron. Astrophys., 223,* 112.
Zapata L. A. et al. (2009) *Astrophys. J. Lett., 704,* L45.
Zhao B. and Li Z.-Y. (2013) *Astrophys. J., 763,* A7.
Zinnecker H. (1989) In *Low Mass Star Formation and Pre-Main Sequence Objects* (B. Reipurth, ed.), p. 447. ESO Conf. Workshop Proc. No. 33.
Zinnecker H. and Mathieu R., eds. (2001) *The Formation of Binary Stars,* IAU Symp. 200, Astronomical Society of the Pacific, San Francisco.
Zinnecker H. and Yorke H.W. (2007) *Annu. Rev. Astron. Astrophys., 45,* 481.

Longmore S. N., Kruijssen J. M. D., Bastian N., Bally J., Rathborne J., Testi L., Stolte A., Dale J., Bressert E., and Alves J. (2014) The formation and early evolution of young massive clusters. In *Protostars and Planets VI* (H. Beuther et al., eds.), pp. 291–314. Univ. of Arizona, Tucson, DOI: 10.2458/azu_uapress_9780816531240-ch013.

The Formation and Early Evolution of Young Massive Clusters

Steven N. Longmore
Liverpool John Moores University

J. M. Diederik Kruijssen
Max-Planck Institut für Astrophysik

Nate Bastian
Liverpool John Moores University

John Bally
University of Boulder, Colorado

Jill Rathborne
Commonwealth Scientific and Industrial Research Organisation (CSIRO) Astronomy and Space Science

Leonardo Testi
European Southern Observatory

Andrea Stolte
Argelander Institut für Astronomie, Bonn

James Dale
Excellence Cluster Universe

Eli Bressert
Commonwealth Scientific and Industrial Research Organisation (CSIRO) Astronomy and Space Science

Joao Alves
University of Vienna

We review the formation and early evolution of the most massive (> few 10^4 $M_\odot$) and dense (radius of a few parsecs) young stellar clusters, focusing on the role that studies of these objects in our Galaxy can play in our understanding of star and planet formation as a whole. Comparing the demographics of young massive cluster (YMC) progenitor clouds and YMCs across the Galaxy shows that gas in the Galactic Center can accumulate to a high enough density that molecular clouds already satisfy the criteria used to define YMCs, without forming stars. In this case formation can proceed *in situ* — i.e., the stars form at protostellar densities close to the final stellar density. Conversely, in the disk, the gas either begins forming stars while it is being accumulated to high density, in a "conveyor belt" mode, or the timescale to accumulate the gas to such high densities must be much shorter than the star-formation timescale. The distinction between the formation regimes in the two environments is consistent with the predictions of environmentally dependent density thresholds for star formation. This implies that stars in YMCs of similar total mass and radius can have formed at widely different initial protostellar densities. The fact that no strong, systematic variations in fundamental properties (such as the IMF) are observed between YMCs in the disk and Galactic Center suggests that, statistically speaking, stellar mass assembly is not affected by the initial protostellar density. We then review recent theoretical advances and summarize the debate on three key open questions: the initial (proto)stellar distribution, infant (im)mortality, and age spreads within YMCs. We conclude that (1) the initial protostellar distribution is likely hierarchical, (2) YMCs likely experienced a formation history that was dominated by gas exhaustion rather than gas expulsion, (3) YMCs are dynamically stable from a young age, and (4) YMCs have age spreads much smaller than their mean age. Finally, we show that it is plausible that metal-rich globular clusters may have formed in a similar way to YMCs in nearby galaxies. In summary, the study of YMC formation bridges star/planet formation in the solar neighborhood to the oldest structures in the local universe.

1. OVERVIEW OF YOUNG MASSIVE CLUSTERS AND THEIR ROLE IN A GLOBAL UNDERSTANDING OF STAR AND PLANET FORMATION

1.1. Motivation

As exemplified by the chapters in this volume, the basic theoretical framework describing the formation of isolated, low-mass stars is now well established. This framework is underpinned by detailed observational studies of the closest star-forming regions. But how typical is the star and planet formation in Taurus, Perseus, or even Orion compared to the formation environment of most stars across cosmological timescales?

The fact that half the star formation in the Galaxy is currently taking place in the 24 most massive giant molecular clouds (GMCs) (*Lee et al., 2012*) suggests that even in the Milky Way at the present day, star-formation regions in the local neighborhood are not typical. Environmental conditions were likely even more different at earlier epochs of the universe. The epoch of peak star-formation-rate density is thought to lie between redshifts of 2 and 3 (*Madau et al., 1998; Hopkins and Beacom, 2006; Bouwens et al., 2011; Moster et al., 2013*), when the average gas surface density (and hence inferred protostellar densities) were significantly higher (see, e.g., *Carilli and Walter, 2013*). Given that most stars in the solar neighborhood are at a similar age to the Sun [~5 gigayears (G.y.)] (*Nordström et al., 2004*), does it then make sense to compare the planetary populations observed around these stars to the protoplanetary disks in nearby star-forming regions?

The fundamental question underlying this line of reasoning is, "Does the process of stellar and planetary mass assembly care about the environment in which the stars form?" If the answer is "no," then studying the nearest star-formation regions will tell us all there is to know about star and planet formation. If the answer is "yes," it is crucial to understand how and why the environment matters. The potential implications of these answers provide a strong motivation for comparing the process of star and planet formation in extreme environments with that in nearby, well-studied, less-extreme star-formation regions.

1.2. Young Massive Clusters: Ideal Probes of Star and Planet Formation in Extreme Environments

Stars are observed to form in a continuous range of environments and densities, from isolated molecular gas clouds expected to form single low-mass stellar systems [e.g., B68 (*Alves et al., 2001*)], through GMC complexes expected to form hundreds of thousands of stars across the full stellar mass range (see the chapter by Molinari et al. in this volume). Similarly, young stellar systems are found at a continuous range of mass and stellar densities (*Bressert et al., 2010*). Given this apparently continuous distribution in mass and density of both gas and stars, what motivates

a definition for a distinct class of stellar systems beyond mere phenomenology?

The answer lies in the fact that most stellar systems dissolve shortly after forming, thereby feeding the field star populations of galaxies (*Lada and Lada, 2003*). Only a small fraction are simultaneously both massive and dense enough to remain gravitationally bound long after their formation and subsequent removal of the remaining natal molecular gas cloud. Being able to study an ensemble single stellar population long after formation offers many advantages, not least of which is the ability to retrospectively investigate the conditions under which the stars may have formed.

The most extreme examples of such stellar systems are globular clusters, which formed at the earliest epochs of the universe and survive to the present day (*Brodie and Strader, 2006*). One of the groundbreaking discoveries of the Hubble Space Telescope (HST) was that massive stellar clusters, with properties that rival those found in globular clusters in terms of mass and stellar density, are still forming in the universe today (*Holtzman et al., 1992*). These young massive clusters (YMCs) have stellar masses and densities orders of magnitude larger than typical open clusters and comparable to those in globular clusters. Crucially, they are also gravitationally bound and likely to be long lived. As such, these stellar systems are potentially local-universe analogs of the progenitors of globular clusters.

At the same time, the apparent continuum of young cluster properties (e.g., *Bressert et al., 2010*) suggests that YMCs merely represent extreme examples of their less massive and dense counterparts — open clusters. As such, characterizing and understanding how YMCs form is critical to help make the connection between the range of physical conditions for star and planet formation between galactic and extragalactic cluster-formation environments.

1.2.1. Young massive clusters: Definition and general properties. In their recent review, *Portegies Zwart et al.* (2010) defined YMCs as stellar systems with mass $\geq 10^4$ M$_\odot$ and ages less than 100 Ma [i.e., formed less than 100 million years (m.y.) ago], but substantially exceeding the current dynamical time (the orbital time of a typical star). While the ultimate longevity of a stellar system will depend on the environment it experiences over time (*Spitzer, 1958*), this last criterion effectively distinguishes between presently bound clusters and unbound associations (see section 3.2.1 for further details).

Given previous confusion in the literature caused by loose and varied definitions of what constitutes a stellar cluster (see section 3.2.1), it is important to point out the implications of the above criteria. First, there are many, well-known, massive associations of stars that do not pass these criteria (e.g., the Cygnus OB association and the Orion Nebular Cluster). Second, YMCs that do pass these criteria (e.g., Westerlund 1, NGC 3603, Trumpler 14) may lie within a much larger stellar association, which as a whole does not pass these criteria. This is a direct consequence of the "continuum" of stellar properties discussed above. We emphasize that the focus here is on the YMCs and not the more distributed stellar populations. This will impact the discussion in section 3 on gas expulsion, longevity, and the presence of age spreads within clusters.

To date, nearly 100 YMCs have been discovered in the local group and out to distances of a few megaparsecs. The properties of many of these are cataloged in *Portegies Zwart et al.* (2010). For convenience, we summarize their characteristic properties below. Young massive clusters typically have radii of ~1 pc and core stellar densities $\geq 10^3$ M$_\odot$ pc^{-3}. They are generally spherical, centrally concentrated, and often mass segregated (i.e., more massive stars are preferentially found toward the center of the cluster). The initial cluster mass distribution is not trivial to measure, but over many orders of magnitude in mass appears to be reasonably well approximated by a power law, dN/dM $\propto$ M^{-2}, across all environments. Young massive clusters are found predominantly in starburst galaxies and mergers — a couple of thousand are known to exist in the Antennae and NGC 3256, for example. These YMCs are typically more massive than those found in the local group and Milky Way. In the local universe (i.e., not starbursts/mergers), YMCs are typically found in the disks of galaxies. Globular clusters are predominantly found in galactic halos. Rotation has been observed in one YMC [R136 (*Hénault-Brunet et al.,* 2012a)] as well as two intermediate-aged massive, dense clusters [GLIMPSE-C01, NGC 1846 (*Davies et al.,* 2011; *Mackey et al.,* 2013)]. Given the difficulty in measuring rotation, it is currently unknown how common this property is among YMCs.

1.2.2. The role of young massive clusters in the broader context of planet, star, and cluster formation. The properties of YMCs make them ideal probes of star and planet formation in extreme environments. Stars forming at such high (proto)stellar densities suffer the maximal effects of feedback from surrounding stars. Also, the very short dynamical time increases the likelihood of interactions with nearby stars at all stages of the formation process. Therefore, studying the formation of stars within a YMC compared to low stellar density systems, offers an opportunity to quantify how dynamical encounters and stellar feedback affect the process of stellar mass assembly.

Young massive clusters contain a very large number of stars of a similar age (age spreads $\lesssim 1$ m.y.; see section 3.2.3). These stars likely formed from the same gas cloud, and therefore were born in the same global environmental conditions and have the same chemical composition. This makes YMC precursor gas clouds the perfect test beds to study the origin of the stellar initial mass function (IMF). For example, by studying YMC progenitor clouds before the onset of star formation, it should be possible to determine if the final stellar mass is set by the initial mass distribution of gas fragments, or alternatively, by these initial fragments subsequently accreting unbound gas from the surrounding environment (see the chapter by Offner et al. in this volume for a more detailed discussion of the origin of the IMF).

Young massive cluster precursor clouds are also, statistically speaking, the best place to search for the progenitors of the most massive stars. While the progenitors of many late-O and early-B stars have been identified, precursors to stars destined to be hundreds of solar masses still prove elusive. Identifying such precursors will help in the theoretical challenge to understand how the most massive stars form (see the chapter by Tan et al. in this volume for a review on high-mass star formation).

Young massive clusters occupy a unique position in understanding cluster formation. As a bridge in the apparent continuum of cluster mass and density distributions between open and globular clusters, studying their global and environmental properties can provide insight into what conditions are required in order for bound clusters to form. Is there a single, scalable formation mechanism applicable to all clusters? Or are additional mechanisms required to accumulate such a large gas mass in a small volume for the most massive clusters? Young massive clusters may be used as a direct probe to understand the conditions required for globular cluster formation.

1.3. Scope of the Review

Several fundamental, unanswered questions about the formation and early evolution of YMCs currently limit their utility as probes of star and planet formation in extreme environments. For example, while the spatial distribution of stars in YMCs older than a few million years is relatively well known (*King,* 1966; *Elson et al.,* 1987), it is not clear how this relates to the initial protostellar or gas distribution (e.g., *Rolffs et al.,* 2011). Any initial substructure that existed in the gas and protostars is erased quickly (*McMillan et al.,* 2007). Therefore, if the stars actually formed at a much lower density — and hence in a much less extreme environment than assumed from the present-day stellar density — and then grew into a massive, dense cluster over time, there would be little evidence of this in the final stellar surface density distribution as the structure would have been erased by violent relaxation. A potential, new method of deriving the initial conditions of cluster formation *a posteriori* would be to consider quantities that are conserved during violent relaxation, such as the degree of mass segregation, and to combine these with a measure of the remaining substructure. Collapsing, virialized, and unbound stellar structures may follow distinct evolutionary histories in the plane defined by these quantities (*Parker et al.,* 2014).

More generally, it is not clear if all clusters of the same mass and radius form from gas with similar properties. Are there different ways to form bound clusters of similar final stellar properties? If so, and if stellar mass assembly depends on the protostellar environment, it is important to understand how and when these different mechanisms operate.

Understanding these questions requires making the link between the evolution of the initial progenitor gas clouds toward the final, gas-free stellar populations. However, the properties of YMCs have been derived almost exclusively from optical/infrared observations. This has strongly biased YMC detection toward clusters with relatively low extinctions ($A_v \leq 30$), preferentially selecting clusters that are already gas-free — i.e., older than a (few) million years. This bias means that very little is known about YMCs younger than this, or their progenitor gas clouds.

In this review we focus on (1) the initial conditions of proto-YMCs; (2) the gas-rich, first (few) approximately million years in the life of YMCs as they are forming stars; and (3) the evolution shortly thereafter. This is intended to complement the *Portegies Zwart et al.* (2010) review, which focused on the aspects of YMCs older than a few million years.

2. MOLECULAR CLOUD PROGENITORS OF YOUNG MASSIVE CLUSTERS — THE INITIAL CONDITIONS

Understanding the formation of YMCs requires first finding samples of YMC progenitor clouds that represent the initial conditions (i.e., before star formation commences), which can be directly compared to their more evolved stellar counterparts. However, very few pre-star-forming YMC progenitor clouds have been identified. In an attempt to understand the plausible range of properties for the initial molecular cloud progenitors of YMCs, we consider some simplified formation scenarios below.

2.1. Simplified Young Massive Cluster Formation Scenarios

The most basic initial condition for YMC formation is a gas reservoir with enough mass, M_{gas}^{init}, to form a stellar cluster of mass, $M_* \geq 10^4\ M_\odot$. These two quantities are trivially related via the star-formation efficiency, ε, through $M_{gas}^{init} = M_*/\varepsilon$. To span the expected range of molecular cloud progenitor properties, we investigate two extremes in the initial spatial distribution of the gas (i.e., before the onset of any star formation) relative to that of the final stellar cluster. First, we consider the case where the initial extent of the gas, R_{gas}^{init}, equals that of the resulting cluster, R_*. Then we consider the case where R_{gas}^{init} is substantially (factors of several or more) larger than R_*.

2.1.1. $R_{gas}^{init} = R_*$: "In situ formation." In this scenario, all the required gas is gathered into the final star cluster volume *before* star formation commences (i.e., *in situ* star formation). In principle, a direct observational prediction of this model would be that one would expect to find gas clouds with mass M_{gas} and radius R_* with no signs of active star formation. However, the probability of finding such a cloud under this scenario depends on the ratio of the time taken to accumulate the gas within the final cluster volume to the time taken for star formation to get underway there. The very high densities required to form a YMC implies that the gas inside the final cluster volume will have a correspondingly short free-fall time in this scenario. If star formation happens on a dynamical timescale, this implies that either the time taken to accumulate the gas reservoir there must also be very short, or that star formation inside

the final cluster volume is somehow delayed or suppressed while the gas accumulates.

In the former case, which we term "*in situ* fast formation," the accumulation time is very short and star formation is almost instantaneous once the gas is accumulated. It is therefore very unlikely that a YMC progenitor cloud with mass M_{gas} and radius R_* with no signs of active star formation would be observed, but significant numbers of such clouds exhibiting ongoing star formation should be observed. In the latter case, dubbed "*in situ* slow formation," the accumulation time is long and star formation is delayed until most of the mass required to build the YMC has accumulated inside R_*. Significant numbers of clouds with mass close to M_{gas} and radius R_* but with no active star formation would therefore be observed in this case.

2.1.2. $R_{gas}^{init} > R_*$: "Conveyor belt formation." In this scenario, the gas that eventually ends up in the YMC is initially much more (factors of several or greater) extended than that of the final cluster. The initial, quiescent gas clouds begin forming stars at a much lower global surface/volume density than in the previous *in situ* scenario. In order for the protocluster to reach the required final stellar densities, the gas and forming stars must converge into a bound stellar system. The most likely agents to enhance gas density are the convergence of two initially independent gas flows, or the gravitational collapse of a single cloud. In this scenario, one would never expect to see clouds of M_{gas}^{init} and R_* with no signs of active star formation.

As outlined in section 3.1, the long-term survival of the cluster is strongly influenced by the mechanism and timescale for gas removal. The time for gas dispersal, t_{disp}, therefore places a strong upper limit on the time during which it is possible to form a cluster through convergence/collapse. Given a convergence velocity, V_{conv}, this sets an upper limit to the initial radius of the gas to be $R_{gas}^{init} = R_* + V_{conv}t_{disp}$.

The timescale for star formation and the observed age spreads are key diagnostics for distinguishing between these scenarios. We look at the observational evidence for variation in these properties in YMCs in section 3.1.

2.2. Comparing Young Massive Cluster and Progenitor Cloud Demographics

We now demonstrate how one can use the observed demographics of molecular cloud populations, compared with those of the stellar cluster populations in the same region, to test these different YMC-formation scenarios.

First, we assume that in a region with a large enough volume to sample all stages of the star/cluster-formation process, the present-day molecular cloud population will create similar clusters as those observed at the present day. In practice this implies that the star-formation rate, cluster-formation rate, and distribution of stars into clusters of a given mass and density should have been constant over several star-formation life cycles. This seems a reasonable assumption for disks in nearby galaxies, but may not hold in mergers, starburst systems, or dwarf galaxies (see *Kruijssen and Longmore, 2014*).

The most massive gas clouds (M_{gas}^{max}) seem the obvious birth sites for the most massive clusters (M_*^{max}). If no existing gas clouds have enough mass to form the observed most massive clusters (i.e., $M_{gas}^{max} \ll M_*^{max}/\varepsilon$), these clouds must gain additional mass from elsewhere (e.g., through merging gas flows or accreting lower density gas from outside the present-day boundary) — i.e., "conveyor belt" formation.

On the other hand, if there are gas clouds of sufficient mass (i.e., $M_{gas}^{max} \geq M_*^{max}/\varepsilon$), then the spatial/kinematic substructure of this gas and the distribution of star-formation activity within these clouds can provide a key to the formation mechanism. If concentrations of gas with M_*/ε within $\sim R_*$ exist, then finding a sizable fraction with no star-formation activity would indicate YMCs are forming *in situ*. If the gas in the most massive clouds is spatially distributed such that no subregion of any cloud contains a mass concentration of M_*/ε within $\sim R_*$, then *in situ* formation seems highly unlikely, in which case the stars forming in the gas must converge and become gravitationally bound before the star formation can disrupt the cloud. Evidence for such convergence should be imprinted in the gas kinematics, e.g., velocity dispersions on the order of $V_{conv} = (R_{gas}^{init} - R_*)/t_{disp}$. Inverse P-Cygni profiles and red/blue-shifted line profile asymmetries may also be observed, but care must be taken interpreting such spectral line diagnostics (*Smith et al., 2012, 2013*).

2.2.1. Observational tracers and diagnostics. We now investigate the feasibility of directly comparing YMC and progenitor cloud demographics given current observational facilities. A fundamental limitation is the distance to which it is possible to detect a precursor cloud of a given mass. The Atacama Large Millimeter/submillimeter Array's (ALMA) factor >10 improvement in sensitivity compared to existing (sub)millimeter interferometers makes it the optimal facility for detecting gas clouds out to large distances. At a frequency of 230 GHz (wavelength of 1.26 mm), ALMA will achieve a 10σ continuum sensitivity limit for a 1-hr integration of approximately 0.1 mJy (assuming 8-GHz bandwidth). Assuming gas and dust properties similar to those in massive star-forming regions in the Milky Way [gas temperature of 20 K, gas:dust ratio of 100:1, *Ossenkopf and Henning* (1994) dust opacities], this sensitivity limit corresponds very roughly to a mass limit of $\{10^5, 10^7\}$ $M_\odot$ at a distance of $\{0.5, 5\}$ Mpc. This simplistic calculation neglects several subtleties (e.g., the effects of beam dilution, higher gas temperatures in vigorously star-forming systems, and metallicity variations on the gas-to-dust ratio and dust opacity). However, it illustrates that the gas cloud populations previously accessible within the Large Magellanic Cloud (LMC)/Small Magellanic Cloud (SMC) can now be probed out to M31/M33 distances, and similar studies currently being done on M31/M33 GMC populations will be possible out to more extremely star-forming galaxies like M82 and NGC 253.

Emission from the CO molecule is another standard tracer of GMC populations. A combination of the excitation conditions and abundance means for a gas cloud of a given mass, low-J transitions of CO are usually brighter than the dust emission in extragalactic observations. This means that the mass estimates above provide a lower mass limit to the detectability of gas clouds in CO.

However, the expected high volume and column densities of YMC progenitor clouds means that CO may not be the ideal molecular line tracer for identification purposes. To illustrate this point, we note that a fiducial YMC progenitor cloud of 10^5 $M_\odot$ with a radius of 1 pc (e.g., as would be expected to form a 3×10^4 $M_\odot$ cluster through *in situ* formation, assuming a 30% star-formation efficiency) would have an average volume and column density of 2×10^4 $M_\odot$ pc^{-3} (4×10^5 cm^{-3}) and 3×10^4 $M_\odot$ pc^{-2} (2×10^{24} cm^{-2}), respectively. This column density corresponds to a visual extinction of ~ 2000 mag. At such high densities, even if observations can resolve the gas emission down to parsec scales, the CO emission will be optically thick. Therefore such observations can only probe a $\tau = 1$ surface, not the bulk of the gas mass. Similar resolution observations of molecular transitions with a higher critical density (comparable to that of the average volume density in the YMC progenitor cloud) are required to pinpoint these clouds. As an interesting aside, such high-column densities render Hα — the traditional star-forming (SF) indicator in extragalactic observations — completely unusable. Probing gas clouds with and without prodigious embedded star-formation activity will therefore rely on complementary observations to measure star-formation tracers less affected by extinction [e.g., centimeter-continuum emission to get the free-free luminosity, or far-infrared (IR) observations, to derive the bolometric luminosity].

The gas mass inferred from observations is a beam averaged quantity. In other words, if a gas cloud is much smaller than the observational resolution and sits within a lower-density environment, the measured beam-averaged column/volume density will be lower than the true value, leading to incomplete YMC progenitor samples. However, even when not operating at its best resolution, ALMA should routinely resolve the approximately parsec-scale YMC progenitor cloud sizes out to several tens of megaparsecs.

To measure what influence the high protostellar density environment has on forming protostars and their planetary systems, it is necessary to resolve individual stellar systems. In practice the projected protostellar separation will vary, both from source to source, and as a function of radius within an individual region. However, relying on the fact that the average core mass is ~ 1 $M_\odot$, the characteristic projected separation of protostars within a protocluster of mass M_* and radius R_* is proportional to $R_*(M_*/M_\odot)^{-1/2}$. The typical projected angular separation of protostars within a protocluster as a function of distance to the protocluster, D, is (very roughly) $4(R_*/pc)(M_*/10^4 \, M_\odot)^{-1/2}(D/kpc)^{-1}$ arcsec. Even at the maximum resolution of ALMA of 0.01″ [i.e., using the most extended 10-km baselines at the highest frequency (Band 9)], it will not be possible to resolve individual stellar systems in YMC progenitor clouds beyond about 100 kpc (i.e., LMC and SMC distances). The maximum angular resolution limit for ALMA is comparable to that expected from future 30–40-m aperture optical/IR telescopes. For *at least* the next several decades, observations probing the physics shaping the IMF in dense stellar systems must be limited to star-forming regions in the LMC/SMC and closer.

Assuming it is possible to resolve individual protostellar systems, the observational limit then becomes one of mass sensitivity. Even with ALMA and choosing the closest possible targets in the Galaxy, deep integrations will be required to probe the gas expected to form stars across the full stellar mass range.

Understanding the gas kinematics across a range of densities and spatial scales is necessary to distinguish between the different formation scenarios of YMCs. The "conveyor belt" model, for example, suggests that large amounts of low- or moderate-density gas should be rapidly infalling. Given the new frontier in sensitivity being opened up by ALMA, it is not clear at this stage what the best transitions for this purpose might be. However, studies of (less extreme) massive and dense high-mass star-forming regions are paving the way (e.g., *Peretto et al.*, 2013). Deriving the spatial and kinematic distribution of mass as a function of size scale will likely require simultaneously observing many different transitions to solve for opacity, excitation, and chemistry variations. Extreme environments, like the Galactic Center, will prove especially challenging in this regard.

2.3. Young Massive Clusters and Progenitor Clouds in the Milky Way

Extragalactic observations will be crucial to probe the formation of the most massive YMCs in a wide range of environments (e.g., galaxy mergers). However, the discussion in section 2.2.1 shows that for the foreseeable future the Milky Way, and to a lesser extent the LMC and SMC, are the only places in the universe where it will be possible to resolve sites of individual forming protostellar/planetary systems in regions that have protostellar densities $>10^4$ $M_\odot$ pc^{-3}. This means they are also the only places where it will be possible to directly test the effect of extreme environments on individual protostellar systems. This provides a strong motivation to identify a complete sample of YMCs and their progenitor clouds in the Galaxy. Such a catalog does not yet exist due to the difficulty in finding clouds at certain stages of the formation process.

On the one hand, it is straightforward to find all the clouds in the Galaxy with embedded stellar populations $>10^4$ $M_\odot$. Their high-bolometric luminosity ($>10^6$ $L_\odot$) and ionizing flux ($Q > 10^{51}$/s) make them very bright at far-IR wavelengths (where the spectral energy distribution peaks) and centimeter wavelengths (which traces the free-free emission from the ionized gas at wavelengths where the clouds and the rest of the Galaxy are optically thin). As a result, these sources with prodigious embedded star formation have been known since the early Galactic plane

surveys at these wavelengths (e.g., *Westerhout* 1958), and many such objects have been studied in detail (e.g., *Plume et al.*, 1997; *Sridharan et al.*, 2002; *Beuther et al.*, 2002; *Lumsden et al.*, 2013).

The difficulty in generating a complete catalog of YMC progenitor clouds has been finding those before star formation has begun. At this early stage there is no ionizing radiation and the luminosity is low. Therefore, these regions do not stand out in centimeter or far-IR wavelength surveys. However, as discussed above, in all three scenarios they must have a large gas mass in a small volume. As such, they should be easy to pick out as bright, compact objects at millimeter and submillimeter wavelengths. However, observational limitations have meant that Galactic plane surveys at these wavelengths have only been possible over the last few years. Previous targeted surveys for young massive protoclusters have not found any starless gas clouds with $>10^5$ $M_\odot$ at parsec-sized scales (e.g., *Faúndez et al.*, 2004; *Hill et al.*, 2005; *Simon et al.*, 2006; *Rathborne et al.*, 2006; *Purcell et al.*, 2006; *Peretto and Fuller*, 2009).

However, thanks to a concerted effort from the observational Galactic star-formation community over the last few years (see the chapter by Molinari et al. in this volume), the data will soon be available to compile a complete list of YMC progenitor clouds in the Milky Way needed to make definitive statements about the relative populations of YMC progenitor clouds with and without prodigious star-formation activity. To date, systematic, blind, large-area searches for YMC progenitor clouds at all stages of the cluster-formation process have been published for two regions of the Galaxy: the first quadrant (*Ginsburg et al.*, 2012) and the inner 200 pc (*Longmore et al.*, 2013a). In the near future, results from continuum surveys like the APEX Telescope Large Area Survey of the Galaxy (ATLASGAL) (*Schuller et al.*, 2009; *Contreras et al.*, 2013), combined with spectral line studies [e.g., Millimetre Astronomy Legacy Team 90 GHz (MALT90) survey, Census of High and Medium Mass Protostars (CHaMP), Three-Millimeter Ultimate Mopra Milky Way Survey (ThrUMMS), Mopra Southern Galactic Place CO survey (*Foster et al.*, 2011; *Jackson et al.*, 2013; *Barnes et al.*, 2011; *Burton et al.*, 2013)], will extend the search to the fourth quadrant. For example, *Urquhart et al.* (2013) have already identified a sample of YMC candidates with signs of active star formation, and Contreras et al. (in preparation) have identified YMC candidates at all evolutionary stages through the MALT90 survey. In the longer term, Herschel Infrared Galactic Plane Survey (HiGAL) (*Molinari et al.*, 2010) will provide a sensitive, uniform survey across the whole Galaxy. However, our analysis relies on having complete samples at all stages of the cluster-formation process, so we focus on the extant surveys of the first quadrant and inner 200 pc of the Galaxy below.

2.3.1. The first quadrant of the Galaxy. Ginsburg et al. (2012) used Bolocam Galactic Plane Survey (BGPS) data (*Aguirre et al.* 2011) to carry out a systematic search for YMC progenitor clouds in the first Galactic quadrant, $l = 6°-90°$ |b| < 0.5°. This region is equivalent to ~30% of the total Galactic surface area, assuming an outer Galactic radius of 15 kpc. In this region *Ginsburg et al.* (2012) identified 18 clouds with mass $M_{gas} > 10^4\ M_\odot$ and radius $r \leq 2.5$ pc. All these clouds have gravitational escape speeds comparable to or larger than the sound speed in photoionized gas, and therefore pass the *Bressert et al.* (2012a) criteria for YMC progenitor clouds. Crucially, all 18 of these clouds are prodigiously forming stars. None of them are starless. *Ginsburg et al.* (2012) use this to place an upper limit of 0.5 m.y. to the starless phase for the clouds in their sample. This is similar to the upper limit on the lifetimes of clouds forming high-mass stars by *Tackenberg et al.* (2012). Assuming a star-formation efficiency of 30%, only 3 of the 18 identified clouds are massive enough to form bound stellar clusters of $10^4\ M_\odot$.

2.3.2. The inner 200 pc of the Galaxy. Longmore et al. (2013a) conducted a systematic search for likely YMC progenitor clouds in the inner 200 pc of the Galaxy by combining dust continuum maps with spectral line maps tracing molecular gas at high volume density. Based on maps of the projected enclosed mass as a function of radius, they identified six clouds as potential YMC progenitors. Intriguingly, despite having extremely high column densities (up to $\sim 10^{24}$ cm^{-2}; $2 \times 10^4\ M_\odot$ pc^{-2}) and being opaque up to 70 μm, four of the six potential YMC progenitor candidates show almost no signs of star formation. The upper limit to the free-free emission from sensitive centimeter continuum observations shows that there are, at most, a small number of early B stars in these four clouds (*Immer et al.,* 2012; *Rodríguez and Zapata,* 2013). This is in stark contrast to the clouds of similar mass and density seen in the disk of the Galaxy, which are all prodigiously forming stars (see section 2.3.1).

2.3.3. Comparison of the first quadrant and inner 200 pc. Following the arguments outlined in section 2.2, if the molecular cloud population in a given region can be expected to produce the stellar populations in the same region, the cloud and stellar demographics can be used to infer something about the underlying formation mechanism. We now attempt this for the first quadrant and inner 200 pc of the Galaxy.

The first step is testing whether the assumption of the observed gas clouds producing the observed stellar populations holds for these regions. The region observed by *Ginsburg et al.* (2012) covers 30% of the surface area of the Galaxy (assuming a Galactic radius of 15 kpc). The inner few hundred parsecs of the Galaxy contains roughly 10% of the molecular gas in the Galaxy (for mass estimates, see *Pierce-Price et al.,* 2000; *Ferrière et al.,* 2007; *Kalberla and Kerp,* 2009; *Molinari et al.,* 2011). If the star-formation rate in these regions has remained constant over several star-formation cycles, it seems reasonable to assume such large gas reservoirs will produce statistically similar stellar populations as observed at the present day. However, once a stellar system has formed, the environment in the Galactic Center is potentially a lot more disruptive than in the disk. Indeed, even dense clusters like YMCs are not

expected to live longer than (or be detectable after) a few million years in the Galactic Center (e.g., *Portegies Zwart et al.,* 2001, 2002; *Kim and Morris,* 2003). For that reason, it is crucial to only compare the demographics of stellar clusters in the first quadrant and Galactic Center that are younger than this age.

We conclude that a reliable metric to investigate the different formation mechanisms is to compare the number of YMCs younger than a few million years (N_{YMC}), to the number of YMC progenitor clouds with prodigious star-formation activity (N_{cloud}^{active}), to the number of YMC progenitor clouds with no discernible star-formation activity ($N_{cloud}^{no\ SF}$). In other words, the ratio of $N_{YMC}:N_{cloud}^{active}:N_{cloud}^{no\ SF}$ contains information about the relative lifetime of these three stages.

The inner 200 pc of the Galaxy contains two YMCs — the Arches and Quintuplet clusters [we exclude the nuclear cluster as this most likely has a different formation mode; see *Genzel et al.* (2010b) for a review] — and two SF active clouds, Sgr B2 and Sgr C. Combined with the four quiescent clouds from section 2.3.2, the $N_{YMC}:N_{cloud}^{active}:N_{cloud}^{no\ SF}$ ratio in the inner 200 pc of the Galaxy is then 2:2:4.

Turning to the first quadrant, there is presently one known YMC in W49 (*Alves and Homeier,* 2003). Given the observational difficulties in finding unembedded YMCs at large distances through the Galactic disk, others may well exist. Completeness is not an issue for the two earlier stages (see section 2.2.1). Combined with the number of SF active and quiescent clouds from section 2.3.1, the $N_{YMC}:N_{cloud}^{active}:N_{cloud}^{no\ SF}$ ratio is then 1:3:0.

Comparing the $N_{YMC}:N_{cloud}^{active}:N_{cloud}^{no\ SF}$ ratios between the inner 200 pc and first quadrant shows that both regions are producing a similar number of YMCs with ages less than a few million years. However, there is a large disparity between the number of progenitor clouds with/without star formation in the two regions: $N_{cloud}^{no\ SF} = 0$ for the first quadrant but $N_{cloud}^{no\ SF} = 4$ for the inner 200 pc. Comparing to the predictions of the scenarios in section 2.2, the Galactic Center appears to be forming YMCs in an "*in situ*, slow formation" mode, whereas the disk appears to be consistent with a "conveyor belt" or "fast *in situ*" mode of formation.

In summary, studying the currently available data in the Galaxy suggests that YMCs in different regions accumulate their mass differently. The two regions studied contain a sizeable fraction of the gas in the Milky Way, so it seems reasonable to conclude that this is representative of YMC formation as a whole in the Galaxy. Of course, when similar data becomes available for the rest of the Milky Way — in particular the fourth quadrant, which contains a large fraction of the gas in the Galactic disk — it is important to test this result.

However, these Galactic regions only represent a small fraction of all the environments in the universe known to be forming YMCs. Clearly it would be foolhardy at this stage to draw any general claims about YMC formation from a dataset sampling such a small fraction of the total number of regions forming YMCs. Future observational studies comparing the gas and stellar demographics across the full range of environments are required to make any such general, empirically based statements about YMC formation. In the upcoming ALMA, James Webb Space Telescope (JWST), and Extremely Large Telescopes (ELT) era, the datasets needed to solve this problem should become available.

In the meantime, we can still make progress in a general understanding of the YMC formation process from what we learn in the Galaxy if we can understand two key aspects: (1) if/how the underlying physical mechanism for YMC formation in the Galaxy depends on the environment, and (2) how those environmental conditions compare to other YMC-forming environments across cosmological timescales.

2.4. The Role of the Environment for Young Massive Cluster Formation

We now investigate how differences in the environmental conditions may be playing a role in YMC formation. Following from the previous discussion, we start by comparing the general properties of the gas in the Galactic Center and the disk, before focusing on the properties of individual YMC progenitor clouds in the two regions.

2.4.1. Comparison of gas properties across the Milky Way. The general properties of the gas in the disk and the center of the Galaxy are both well characterized, and are known to vary substantially from each other (for reviews see *Morris and Serabyn,* 1996; *Ferrière et al.,* 2007) (also see the chapter by Molinari et al. in this volume). In summary, the gas in the Galactic Center lies at a much higher column and volume density (*Longmore et al.,* 2013b), has a much larger velocity dispersion at a given physical size (*Shetty et al.,* 2012), and has a higher gas kinetic temperature (*Ao et al.,* 2013; *Mills and Morris,* 2013). The offset between the gas and dust temperature (*Molinari et al.,* 2011) in the Galactic Center is thought to be either due to a cosmic-ray flux that is orders of magnitude larger than that found in the disk, or due to the widespread shocks that are observed throughout the gas (*Ao et al.,* 2013; *Yusef-Zadeh et al.,* 2013). The disk has a well-known metallicity gradient with galactocentric radius of −0.03 to −0.07 dex kpc^{-1} (*Balser et al.,* 2011). The metallicity in the Galactic Center is measured to vary within a factor of 2 of the solar value (*Shields and Ferland,* 1994; *Najarro et al.,* 2009).

There is evidence that a combination of the environmental factors and the global properties of the gas leads to differences in how the star formation proceeds between the two regimes. Given the large reservoir of dense gas in the Galactic Center, the present-day star-formation rate is at least an order of magnitude lower than that predicted by star-formation relations where the star formation scales with the gas density (e.g., *Lada et al.,* 2012; *Krumholz et al.,* 2012a; *Beuther et al.,* 2012; *Longmore et al.,* 2013b; *Kauffmann et al.,* 2013). *Kruijssen et al.* (2013) find that the currently low star-formation rate (SFR) in the Galactic Center is consistent with an elevated critical density for star

formation due to the high turbulent pressure. They propose a self-consistent cycle of star formation in the Galactic Center, in which the plausible star-formation inhibitors are combined. However, the fact that (1) there is a nonzero star-formation rate in the Galactic Center (albeit an order of magnitude lower than predicted given the amount of dense gas), and (2) at least two YMCs are found in the Galactic Center, means that some mechanism must be able to overcome any potential suppression in star formation in a small fraction of the gas. As the details of this mechanism are of potential interest in understanding why the YMC-formation mode in the Galactic Center may be different from the disk, in section 2.4.2 we examine this further before turning in section 2.4.3 to YMC formation in the disk.

2.4.2. Young massive cluster formation in the Galactic Center. A global understanding of star formation in the Galactic Center is hampered by the difficulty in determining the three-dimensional distribution of the gas and stars. Building on earlier efforts (e.g., *Binney et al.,* 1991), *Molinari et al.* (2011) put forward a model that the "twisted ring" of dense molecular gas of projected radius ~100 pc that they identified as very bright submillimeter continuum emission in the HiGAL data was on elliptical X2 orbits (i.e., orbits perpendicular to the long axis of the stellar bar). In this scenario, the two prominent sites of star formation in the ring — Sgr B2 and Sgr C — lie at the location where the X2 orbits intersect with the X1 orbits (i.e., gas streams funneled along the leading edge of the stellar bar from the disk to the Galactic Center). In this picture, the collision of dense gas clouds may lead to YMC formation (see, e.g., *Stolte et al.,* 2008).

Based on the observed mass distribution and kinematics, *Longmore et al.* (2013a) postulated that the gas in this ring may be affected by the varying gravitational potential it experiences. They hypothesized that the net effect of the interaction is a compression of the gas, aided by the gas dissipating the tidally injected energy through shocks. If the gas was previously sitting close to gravitational stability, the additional net compression of the gas might be enough for it to begin collapsing to form stars. If this hypothesis proves correct, one can use the known time since pericenter passage to effectively follow the physics shaping the formation of the most massive stellar clusters in the Galaxy, and by inference the next generation of the most massive stars in the Galaxy, as a function of absolute time. Numerical simulations by several different groups show that this scenario is plausible and can accurately reproduce the observed gas properties (Lucas and Bonnell, in preparation; Kruijssen et al., in preparation). The fact that the gas in this region has already assembled itself into clouds of ~10^5 M$_\odot$ and radius of a few parsecs before any star formation has begun suggests that once the gas becomes gravitationally bound, it will form a young massive cluster.

The extreme infrared-dark cloud, G0.253+0.016 [M0.25, the "Lima Bean," the "Brick (*Lis et al.,* 1994; *Lis and Menten,* 1998; *Bally et al.,* 2010; *Longmore et al.,* 2012)] is the best-studied example of such a cloud. Despite containing ~10^5 M$_\odot$ of gas in a radius of ~3 pc, the only signs of potential star-formation activity are one 22-GHz H$_2$O maser (*Lis et al.,* 1994) and several compact radio sources at its periphery (*Rodríguez and Zapata,* 2013), indicating that at most a few early B stars have formed. As expected, the average gas column density is very high [>10^{23} cm^{-3} (*Molinari et al.,* 2011)]. *Kauffmann et al.* (2013) showed that at 0.1-pc scales, there are very few subregions with high column density contrast (corresponding to densities 2×10^5 cm^{-3}) compared to the ambient cloud. In the scenario proposed by *Longmore et al.* (2013a), this cloud has recently passed pericenter with the supermassive black hole at the center of the Galaxy and is being tidally compressed perpendicular to the orbit and stretched along the orbit. Preliminary numerical modeling results suggest the diffuse outer layers of the cloud may be removed in the process, leading to the large observed velocity dispersions and explaining the observed cloud morphology (Kruijssen et al., in preparation.). The bulk of the cloud mass can remain bound, even though standard virial analysis would suggest the cloud is globally unbound (e.g., *Kauffmann et al.,* 2013). The prediction is that the tidally injected energy is presently supporting the cloud against collapse, but as the cloud continues on its orbit this energy will be dissipated through shocks, and the cloud will eventually collapse to form a YMC. Given the large difference between the observed dust temperature (*Molinari et al.,* 2011) and gas temperature (*Guesten et al.,* 1981; *Mills and Morris,* 2013), complicated chemistry, extreme excitation conditions, evidence for widespread shocks and a high cosmic-ray rate (*Ao et al.,* 2013; *Yusef-Zadeh et al.,* 2013), detailed observational studies (*Kendrew et al.,* 2013; *Johnston et al.,* 2013; Rathborne et al., in preparation) and numerical modeling (e.g., *Clark et al.,* 2013; Kruijssen et al., in preparation; Lucas and Bonnell, in preparation) are required to test this hypothesis. Understanding the origin and star-formation potential of these extreme Galactic Center clouds promises to be an exciting avenue for study in the next few years.

2.4.3. Young massive cluster formation in the Galactic disk. To date, no clouds of gas mass ~10^5 M$_\odot$ and radius ~1 pc with no signs of star formation have been found outside the Galactic Center. So what were the initial conditions for the YMCs that are known to have formed in the disk of the Milky Way? Clues to their origin can be gleaned from the properties of the present-day molecular cloud population in the disk. The mass distribution follows a power law, $dN/dM \propto M^{-\gamma}$, with $1.5 < \gamma < 1.8$ (*Elmegreen and Falgarone,* 1996; *Rosolowsky,* 2005). In order to produce 10^4 M$_\odot$ of stars, mass conservation means that YMCs must have formed within progenitor clouds of gas mass at least 10^4 M$_\odot/\varepsilon$ ~ 10^5 M$_\odot$. Clouds more massive than this have virial ratios close to unity (albeit with significant scatter) (*Rosolowsky,* 2007; *Dobbs et al.,* 2011b). While there are many difficulties in using virial ratios to unambiguously determine if an individual cloud is gravitationally bound, it seems reasonable to assume that many of the most massive clouds are likely to be close to being gravitationally bound.

This places an interesting constraint — the fundamental gas reservoir limitation means that the clouds in the disk where one expects to find YMC progenitors are also clouds that are possibly undergoing global gravitational collapse.

The clouds of gas mass >10^5 M$_\odot$ in the disk are typically many tens to hundreds of parsecs in size. The average volume and column density is therefore low (e.g., a few 10^2 cm^{-3}, a few 10^{21} cm^{-2}), especially compared to similar mass clouds in the Galactic Center. However, these global size scales are much larger than the parsec scales of interest for YMC formation. The immediate conclusion is that the YMCs embedded within these clouds can therefore only make up a small volume-filling factor of the whole cloud.

The fact that no 10^5-M$_\odot$-parsec-scale, starless subregions have been found within GMCs in the disk suggests that the GMCs do not begin life with such dense subregions. However, we know of at least 18 dense, parsec-scale subregions of ≥10^4 M$_\odot$ that have prodigious star formation (*Ginsburg et al.,* 2012). By learning how these couple to the larger (10–100-pc scale) cloud it may be possible to understand how YMCs assemble their mass in the disk. As mentioned earlier, each of the regions in the first quadrant containing a candidate progenitor cloud has been well studied and much is known about the gas properties and (embedded) star-formation activity. Therefore, such a study is feasible. However, an indepth review of these detailed studies is beyond the scope of this work. Instead, we focus on two regions: W49 and W43. The former is the most luminous star-forming region in the Galaxy and contains the most massive and dense progenitor cloud in the *Ginsburg et al.* (2012) sample. As such, this is the most likely site of future YMC formation in the first quadrant. In terms of large-scale Galactic structure, W49 and most other YMC progenitor cloud candidates are not found at any "special" place in the Galaxy (other than potentially lying within spiral arms). W43 is the possible exception to the rule, and is postulated to lie at the interface between the Scutum-Centaurus (or Scutum-Crux) arm and the stellar bar.

2.4.3.1. *W49*: Lying at a distance of $11.11^{+0.79}_{-0.69}$ kpc from Earth (*Zhang et al.,* 2013), W49 is the most luminous star-forming region in the Galaxy [$10^{7.2}$ L$_\odot$ (*Sievers et al.,* 1991, scaled to the more accurate distance of *Zhang et al.,* 2013)], embedded within one of the most massive molecular clouds [~1.1×10^6 M$_\odot$ within a radius of 60 pc (*Galván-Madrid et al.,* 2013, and references therein)]. The spatial extent of the entire cloud is 120 pc, but the prominent star-formation region, W49A, is confined to an inner radius of ~10 pc. W49A comprises three subregions — W49N, W49S, and W49SW — each with radii of a few parsecs and separated from each other by less than 10 pc. Approximately 20% of the mass (and practically all the dense gas) lies within 0.1% of the volume [~2×10^5 M$_\odot$ of gas within a radius of 6 pc (*Nagy et al.,* 2012; *Galván-Madrid et al.,* 2013)]. W49N is the most actively star forming of these, containing both a cluster of stars >4×10^4 M$_\odot$ and a well-known ring of H II regions, all within a radius of a few parsecs (*Welch et al.,* 1987; *Alves and Homeier,* 2003; *Homeier and Alves,* 2005).

Ginsburg et al. (2012) identified both W49N and W49SW as likely YMC progenitor clouds. Historically, several scenarios have been put forward to explain the interaction of these dense clumps with the larger-scale cloud, from global gravitational collapse to cloud collisions and triggering (e.g., *Welch et al.,* 1987; *Buckley and Ward-Thompson,* 1996). In the most recent, multi-scale dense gas survey, *Galván-Madrid et al.* (2013) show the larger-scale gas cloud is constructed of a hierarchical network of filaments that radially converge on to the densest YMC progenitor clouds, which act as a hub for the filaments [reminiscent of the "hub-filament" formation scenario described in *Myers* (2009) and observed toward the very luminous massive star-formation region G10.6 by *Liu et al.* (2012)]. Based on kinematic evidence, they conclude the region as a whole is undergoing gravitational collapse with localized collapse onto the YMC progenitor clouds.

2.4.3.2. *W43:* Located within the so-called "molecular ring," the region lying between Galactic longitudes of 29.5° and 31.5° contains a particularly high concentration of molecular clouds, several well-known star-formation complexes such as the mini-starburst W43-Main, and four YMC precursors (*Motte et al.,* 2003; *Ginsburg et al.,* 2012). The ^{13}CO emission in the region shows complicated velocity structure over 60 km s^{-1} along the line of sight. *Nguyen Luong et al.* (2011b) and *Carlhoff et al.* (2013) conclude that almost all this ^{13}CO gas is associated with W43 and that the gas within ~20 km s^{-1} of the 96 km s^{-1} velocity component is part of a single W43 molecular cloud complex. This complex has an equivalent diameter of ~140 pc, total mass of ~7×10^6 M$_\odot$, and many subregions of high gas density. The measured distance is consistent with the complex lying at the meeting point of the Scutum-Crux arm at the end of bar. *Nguyen Luong et al.* (2011b) and Motte et al (in preparation) argue that the large-velocity dispersion and complicated kinematic structure indicate the convergence point of high-velocity gas streams. Three YMC progenitor clouds lie within this longitude-velocity range, making it a particularly fertile place for YMC formation. Two of the three progenitor candidates — W43-MM1 and W43-MM2 — have projected separations of less than a few parsecs, and are very likely to be associated. *Nguyen-Lu'o'ng et al.* (2013) conclude that these have formed via colliding flows driven by gravity. Given that several red supergiant (RSG) clusters are found at a similar location and distance (*Figer et al.,* 2006; *Davies et al.,* 2007), this region of the Galaxy appears to have been forming dense clusters of >10^4 M$_\odot$ for at least 20 m.y. It also flags the other side of the bar as an interesting place to search for YMCs and YMC progenitor clouds [indeed, *Davies et al.* (2012) have already identified one YMC candidate there].

2.4.4. *Summary: Environment matters.* In summary, the disk and Galactic Center are assembling gas into YMCs in different ways. In the Galactic Center, the mechanism is "*in situ* slow formation," where the gas is able to reach very high densities without forming stars. Something, possibly

cloud-cloud collisions or tidal forces and the gas dissipating energy through shocks, allows some parts of the gas reservoir to collapse under its own gravity to form a YMC.

Conversely, in the disk, the lack of starless 10^5 M$_\odot$, r < 1-pc gas clouds suggest YMCs either form in a "conveyor belt" mode, where stars begin forming as the mass is being accumulated to high density, or the timescale to accumulate the gas to such high densities must be much shorter than the star-formation timescale. In two of the most fertile YMC-forming regions in the first quadrant (W49 and W43), recent studies have shown evidence of large-scale gas flows and gravitational collapse feeding the YMC progenitor clouds.

After YMCs have formed, the remains of their natal clouds can also provide clues to the formation mechanism. Observations of the remaining gas associated with the formation of Westerlund 2 and NGC 3603 suggest these YMCs formed at the interaction zones of cloud-cloud collisions (*Furukawa et al.,* 2009; *Ohama et al.,* 2010; *Fukui et al.* 2014; see also *Dib et al.,* 2013).

From the above evidence alone, it would be premature to claim large-scale gas flows as a necessary condition to form YMCs in the disk. However, combined with the fact that the most massive gas clouds have virial ratios closest to unity (*Rosolowsky,* 2007; *Dobbs et al.,* 2011b), and numerous numerical/observational studies of other (generally less massive/dense) cluster-forming regions show evidence for large-scale gas flows feeding gas to protocluster scales [e.g., W3(OH), G34.3+0.2, G10.6-0.4, SDC335.579-0.292, DR21, K3-50A, Serpens South, GG035.39-00.33, G286.21+0.17, G20.08-0.14 N (*Keto et al.,* 1987a,b; *Liu et al.,* 2012; *Peretto et al.,* 2013; *Csengeri et al.,* 2011; *Hennemann et al.,* 2012; *Klaassen et al.,* 2013; *Kirk et al.,* 2013; *Henshaw et al.,* 2013; *Nguyen Luong et al.,* 2011a; *Barnes et al.,* 2010; *Smith et al.,* 2009; *Galván-Madrid et al.,* 2009)], often with "hub-filament" morphology (*Myers,* 2009; *Liu et al.,* 2012), suggests this is a fruitful area for further investigation.

If YMCs in the disk of the Milky Way form predominantly as a result of large-scale gas flows, one would not expect to see parsec-scale regions of ~10^5 M$_\odot$ with no star formation. Rather, the initial conditions of the next YMC generation must be massive GMCs with little signs of current star formation, and kinematic signatures of either large-scale infall or converging flows. Searching for these clouds is another interesting avenue for further investigation.

So what can we learn about YMC formation more generally from this analysis? One potential interpretation of the difference between the YMC progenitor cloud demographics between the first quadrant and inner 200 pc is evidence for two "modes" of YMC formation. However, there are several reasons to suspect this may be misleading. First, the physical properties in the interstellar medium are observed to vary continuously, as are the range of environmental conditions in which star formation occurs. Second, gas is scale-free. Therefore, despite the fixed mass/radius criterion for YMC progenitor clouds being physically well motivated in terms of forming bound stellar systems, imposing *any* discrete scale in a scale-free system is arbitrary by definition. The distinction between the simplistic scenarios in section 2.1 is therefore sensitive to the criteria used to define likely progenitor clouds, and essentially boils down to whether or not stars have begun forming at the imposed density threshold.

In theories of star formation where turbulence sets the molecular cloud structure (e.g., *Krumholz and McKee*, 2005; *Padoan and Nordlund*, 2011; *Federrath et al.*, 2010), in the absence of magnetic fields, the critical overdensity for star formation to begin, x_{crit}, is given by $x_{crit} = \rho_c/\rho_0 = \alpha_{vir}\mathcal{M}^2$, where ρ_c is the critical density for star formation, ρ_0 is the mean density, α_{vir} is the virial parameter, and $\mathcal{M}$ is the turbulent Mach number of the gas. The critical density is then given by $\rho_c = \alpha_{vir}\rho_0\mathcal{M}^2 = \alpha_{vir}\rho_0 v_{turb}^2/c_s^2 = \alpha_{vir}P_{turb}/c_s^2$, where v_{turb} is the turbulent velocity dispersion, c_s is the sound speed, and P_{turb} is the turbulent pressure. Therefore, given an imposed mass and radius threshold, M_{th} and R_{th}, a molecular cloud in a particular environment will have begun forming stars if it passes the following YMC density threshold, ρ_{YMC}, criterion

$$\rho_{YMC} \propto M_{th}/R_{th}^3 > (4\pi/3)\alpha_{vir}P_{turb}\mu m_H/k_B T \qquad (1)$$

where μ, m_H, k_B, and T are the molecular weight, hydrogen mass, Boltzmann constant, and gas temperature. Figure 1 shows the implications of this critical density for gas clouds in different Galactic environments. Clouds with densities below the critical value, but high enough to satisfy the YMC progenitor candidate criteria, form according to the "*in situ*" scenario. Clouds that satisfy the YMC progenitor candidate criteria but with densities exceeding the critical value should already be forming stars, and hence forming YMCs in the "conveyor belt" mode, or the timescale for gas accumulation is much shorter than the star-formation timescale. The YMC candidates in the disk/Galactic Center have average volume densities above/below the critical density for star formation in that environment, respectively. This provides a potential explanation as to why the candidate YMC progenitor clouds in the disk and Galactic Center have similar global properties, but all those in the disk have prodigious star-formation activity and many of those in the galactic center do not.

Building on this progress toward a general understanding of YMC formation requires knowing (1) how the dense gas reservoirs got there in the first place (e.g., looking at the kinematics of H I halos surrounding GMCs), and (2) how similar the Galactic conditions are to other YMC-forming environments across cosmological timescales.

Numerical simulations and extragalactic observations offer alternative ways to tackle point (1). The chapter by Dobbs et al. in this volume shows there are many potential molecular cloud formation mechanisms that become dominant in different environments. All these mechanisms involve flows with some degree of convergence, and it seems plausible that the processes forming YMC progenitor clouds involve those with the largest mass flux to the size scales of interest. Extragalactic observations in starburst systems like the Antennae and M82 show that gas clouds thought to be the sites of future YMC formation lie at locations of colliding gas flows and regions of very high gas compression (*Keto et al.*, 2005; *Wei et al.*, 2012).

While there appear to be qualitative similarities across a wide range of environments (e.g., converging gas flows and gravitational collapse in the Milky Way disk and star-bursting environments), a more quantitative approach is needed to rigorously compare the environmental conditions and gas properties. Ultimately, the extent to which we can apply what we learn about star and planet formation in our Galaxy to other locations in the universe depends on how similar the Galactic conditions are to other environments across cosmological timescales. Observational resolution and sensitivity limitations make a direct comparison of regions across such a wide range of distances challenging. However, by taking due care when comparing widely heterogenous datasets, it is possible. In a recent study, *Kruijssen and Longmore* (2013) show that in terms of their baryonic composition, kinematics, and densities, the clouds in the solar neighborhood are similar to those in nearby galaxies. At the current level of observational precision, the clouds and regions in the Galactic Center are indistinguishable from high-redshift clouds and galaxies. The Milky Way therefore contains large reservoirs of gas with properties directly comparable to most of the known range of star-formation environments and is therefore an excellent template for studying star and planet formation across cosmological timescales.

Returning to the focus of the review, how does what we are learning about YMC formation affect the role of YMCs as a probe of star and planet formation in extreme environments? Irrespective of the details, the above analysis suggests that clusters of the same final mass and radius can have very different formation histories, in which case, if investigating the role of formation environment on the resulting stellar/planetary population in a YMC, it is crucial to understand the cluster's formation history. Both in terms of the initial (proto)stellar densities and gas conditions, the YMCs forming in the Galactic Center represent the most extreme conditions for star/planet formation in the Galaxy. This strongly motivates comparing the detailed star-formation process in the Galactic Center with that in the rest of the Galaxy. The fact that to date, key parameters (like the IMF) of YMCs in the Galactic Center clusters and those in the disk are consistent to the uncertainty level of current observations [e.g., *Bastian et al.*, 2010; *Habibi et al.*, 2013; *Hussmann et al.*, 2012, although the nuclear cluster may be an exception (*Lu et al.*, 2013)] suggests that the extreme environment and high protostellar densities do not affect how stars assemble their mass, at least in a statistical sense. Finally, the fact that the range of star-forming environments in the Milky Way are so similar to the range of known star-forming environments across cosmological timescales suggests one can directly apply what we learn about star and planet formation in different parts of the Galaxy to much of the universe.

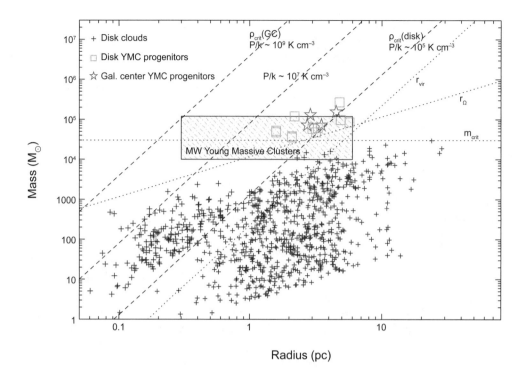

Fig. 1. Mass vs. radius for gas clouds and YMCs in the Milky Way. Observations of gas clouds are shown as different symbols and the hatched rectangle shows the location of Galactic YMCs (from *Portegies Zwart et al.*, 2010). The plus symbols are infrared-dark clouds and dense ammonia clouds in the disk (*Rathborne et al.*, 2006; *Walsh et al.*, 2011; *Purcell et al.*, 2012), the squares are YMC progenitor candidates in the disk (*Ginsburg et al.*, 2012; *Urquhart et al.*, 2013), and the stars are YMC progenitor cloud candidates in the Galactic Center (*Immer et al.*, 2012; *Longmore et al.*, 2012, 2013a). The dotted lines show the criteria for YMC progenitor clouds put forward by *Bressert et al.* (2012b): m_{crit} is the minimum mass reservoir requirement, r_{Ω} is the criterium that the escape speed of the cloud is greater than the sound speed in ionized gas (~10 km s^{-1}), and r_{vir} is the virial mass given a crossing time of 1 m.y. Based on these criteria, clouds lying above all three dotted lines are candidate YMC progenitor clouds. The dashed lines show the critical density for star formation expressed in equation (1) for three different environments. From left to right, the gas properties used to calculate the critical densities are v_{turb} = (15, 7, 2) km s^{-1}, ρ_0 = (2 × 10^4, 10^3, 10^2) cm^{-3}, T = (75, 40, 20) K, with α_{vir} = 1 in all cases. The corresponding turbulent pressure, expressed as P/k, are given next to each line. The gas properties for the left and right dashed lines are chosen to resemble those for clouds in the Galactic Center and disk, respectively.

3. THE EARLY EVOLUTION OF YOUNG MASSIVE CLUSTERS

This section is devoted to the first ~10 m.y. of a YMC's lifetime. Over the past few years we have seen substantial progress in our understanding of this evolutionary phase, both observationally and theoretically. In two subsections, we first review the current theoretical framework (section 3.1) before building on this to summarize the intense recent debate on two key open questions in the field (section 3.2).

3.1. Theoretical Framework

The formation and early evolution of YMCs is dominated by a plethora of complex and interacting physical processes. A theoretical understanding of this phase requires insight in the conversion of gas into stars (*Elmegreen*, 2000; *Clarke et al.*, 2000; *Krumholz and Tan*, 2007), in

the feedback effects of individual stars on the cloud as a whole and locally (*Dale et al.*, 2005, 2012; *Murray et al.*, 2010; *Krumholz et al.*, 2012b), as well as the (possibly differing) phase-space distribution of the gas and the stars — both before and after feedback effects become important (*Offner et al.*, 2009a; *Moeckel and Bate*, 2010; *Kruijssen et al.*, 2012b; *Girichidis et al.*, 2012). The interplay of these processes determines the star-formation efficiencies (SFEs) in the proto-YMC clouds, the timescale on which gas not involved in star formation is expelled, and therefore, ultimately, what fraction of the stars reside in bound clusters once star formation is complete (*Kruijssen*, 2012). Once all the gas in such clouds has been converted to stars or expelled by feedback, the modeling of a YMC reduces to solving the gravitational N-body problem. Even in the absence of external forces, the evolution of such a dense stellar system is very complex, involving processes such as mass segregation, core collapse, stellar collisions, and the dynamical evaporation of the YMC (*Portegies Zwart et al.*, 2010). In addition, YMCs can be affected by close interactions with molecular clouds or other density peaks in their galactic environment (*Spitzer*, 1987; *Gieles et al.*, 2006b; *Kruijssen et al.*, 2011). In this section, we review the new generation of hydrodynamical and N-body simulations that are able to model these processes and their effects on star formation in massive, dense molecular clouds.

Despite the recent advances in the numerical modeling of cluster formation (for a review, see *Kruijssen*, 2013), there are very few simulations of YMC formation, mainly due to the large dynamic range that needs to be covered in terms of mass, size, and time. Current state-of-the-art simulations can model the formation of a cluster with a stellar mass spectrum that is complete down to the hydrogen-burning limit (~0.08 $M_\odot$) up to mass scales of a few 10^3 $M_\odot$ (*Tilley and Pudritz*, 2007; *Bonnell et al.*, 2008; *Krumholz et al.*, 2012b), still well below the lower mass limit of the YMC regime. As a result, any current simulation aiming to model the formation and early evolution of YMCs necessarily has to neglect possibly important physics, such as resolving the assembly of the full stellar mass spectrum. A second problem is that simulating feedback processes (e.g., ionization, stellar winds, supernovae) is computationally expensive for several reasons. For instance, a full, three-dimensional modeling of radiative transfer is intrinsically very expensive. Additionally, feedback effects exacerbate the problem of the dynamic range that needs to be covered, in terms of mass, time, and spatial scales (gas ejected by stellar winds or supernovae can reach velocities of several 1000 km s^{-1}, while cold gas has a typical velocity dispersion of 1–10 km s^{-1}).

Because of the above limitations, it is necessary that current and future studies addressing different parts of the YMC formation problem are well connected, self-consistently linking the different physics. We therefore review the theoretical framework for the formation and early evolution of YMCs with a view to setting out the initial conditions from which such simulations should begin. This framework is divided in three steps that are often probed in different numerical experiments. Linking these steps will be one of the major challenges of the coming years.

3.1.1. Young massive cluster formation from the hierarchical interstellar medium. A theory of YMC formation (or cluster formation in general) first needs to acknowledge the hierarchical nature of the interstellar medium (ISM) (*Elmegreen and Falgarone*, 1996; *Klessen et al.*, 1998; *Bate et al.*, 1998; *Padoan and Nordlund*, 2002), which persists in the condensation and assembly of GMCs and protostellar cores, reaching all the way from the galactic gas disk scale height (~100 pc) to the ambipolar diffusion length (~0.01 pc). The term *hierarchy* is often used very loosely to indicate scale-free or fractal-like substructure — for clarity, we adopt the definition that hierarchicality indicates the growth of structure from the merging of smaller structures. Hierarchical growth is crucial for understanding YMC formation, as even the most extreme and densest protocluster clouds like the Brick (*Longmore et al.*, 2012;

Rathborne et al., 2014) show substructure. If star formation is driven by self-gravity, then the hierarchical structure of gas clouds implies that the collapse to protostellar cores occurs more rapidly locally than on a global scale, and hence the conversion to the near-spherical symmetry that characterizes YMCs has to take place when protostellar cores and/or stars have already formed. The hierarchical growth of proto-YMC clouds therefore continues during star formation and the assembly of the actual YMC, during which the gravitational pull of the global gas structure may provide a "conveyor belt" for transporting stellar aggregates to the center of mass.

Numerical simulations that include self-gravity, hydrodynamics, and sink particle formation automatically capture the hierarchical growth of YMCs, provided that they start from turbulent initial conditions (see, e.g., *Mac Low and Klessen*, 2004) or convergent flows. The turbulence generates scale-free structure in which perturbations on the smallest length-scales naturally collapse first, leading to star formation dispersed throughout much of the cloud volume. The work by *Maschberger et al.* (2010) (see their Fig. 3) illustrates how the resulting merger trees of the stellar structure in numerical simulations of cluster formation can be used to trace and quantify hierarchical growth — much like their use in galaxy-formation studies.

3.1.2. The early disruption of stellar structure. While the global dynamics of gas and stars are initially coupled, there are several reasons why their co-evolution cannot persist for longer than a dynamical time. First, gas is dissipational, whereas the stars are ballistic, which inevitably leads to differing kinematics. Second, gas is being used to form stars — because gravitational collapse and star formation proceed most rapidly near the density peaks, these attain much higher local SFEs than their surroundings, possibly appearing as relatively gas-free, gravitationally bound stellar groups within a cocoon of gas (*Peters et al.*, 2010; *Kruijssen et al.*, 2012b; *Girichidis et al.*, 2012). In numerical simulations, these stellar groups are virialized even when omitting the gravitational influence of the gas. The local gas exhaustion is driven by accretion as well as the accretion-induced shrinkage of the stellar structure (while gas expulsion leads to the expansion of an initially virialized system, gas accretion has the opposite effect), implying that the gas and stars are indeed decoupled [as is also indicated by the velocity dispersions of the gas and stars (see *Offner et al.*, 2009a)].

If the gas density is not high enough to deplete the gas through secular star-formation processes, the co-evolution of gas and stars ceases when stellar feedback becomes important. Protostellar outflows (*Nakamura and Li*, 2007; *Wang et al.*, 2010; *Hansen et al.*, 2012; *Krumholz et al.*, 2012b), photoionizing radiation (*Dale et al.*, 2012), radiative feedback (*Offner et al.*, 2009b; *Murray et al.* 2010), stellar winds, and supernovae (*Pelupessy and Portegies Zwart*, 2012) can potentially blow out large mass fractions. Whether or not the expulsion of residual gas by feedback affects the boundedness of stellar structure depends on the division between gas exhaustion and gas expulsion. If the density was high enough to lead to gas exhaustion, then gas expulsion cannot affect the dynamical state of the stars, whereas a low SFE (and hence density) implies that the stars themselves are only held together by the gravitational potential of the gas, and therefore will disperse when the gas is expelled. The obvious question to ask is whether such low-density stellar associations were gravitationally bound at any point during the star-formation process. In part, this depends on the dynamical state of their nascent clouds. Current evidence suggests that at least some substantial fraction of clouds and star-forming regions is initially globally unbound (e.g., *Heyer et al.,* 2009; *Dobbs et al.,* 2011b). Simulations of globally unbound clouds by *Clark et al.* (2005) show that the clusters formed in such an environment resemble OB-associations containing several locally bound but mutually unbound subgroups.

It is clear that the importance of feedback and gas expulsion crucially depends on the outcome of the star-formation process. Dense, virialized YMCs likely experienced a formation history that was dominated by gas *exhaustion*, whereas unbound associations suffered from the effects of gas *expulsion*. The balance between both mechanisms can vary among different parts of the same region, or when considering different spatial scales. It seems logical that there exists a critical density, irrespective of the spatial scale, above which gas exhaustion is more important than gas expulsion. A second, independent mechanism for the early disruption of stellar structure is formed by the tidal perturbations coming from the dense, star-forming environment [the "cruel cradle effect" (*Kruijssen et al.,* 2011; also see *Elmegreen and Hunter,* 2010)], which is thought to suppress the formation of bound clusters at galactic gas surface densities of $\Sigma > 10^3$ $M_\odot$ pc^{-2} (*Kruijssen,* 2012) and primarily affects low-density clusters. The key conclusion is that, irrespective of the particular mechanism, the early disruption of stellar structure does not depend on mass or spatial scale, but exclusively on the volume density.

For a long time, our picture of the early evolution of stellar clusters was based on numerical simulations lacking a live gas component, i.e., N-body simulations of stars existing in virial equilibrium with a background potential representing the gas — the latter being removed to probe the effects of gas expulsion (*Lada et al.,* 1984; *Goodwin,* 1997; *Adams,* 2000; *Geyer and Burkert,* 2001; *Boily and Kroupa,* 2003; *Goodwin and Bastian,* 2006; *Baumgardt and Kroupa,* 2007). These simulations naturally showed that if a large part of the gravitational potential is removed, a corresponding part of the structure is unbound [as can be shown analytically (see *Hills,* 1980)]. Together with the observation that after some 10 m.y. only 10% of all stars are situated in bound clusters (*Lada and Lada,* 2003), this has led to the classical picture of early cluster evolution, in which 90% of all clusters are destroyed on a short timescale by gas expulsion. By contrast, the modern view of cluster formation discussed above connects gas expulsion to unbound associations, standing separately from the early

evolution of the gravitationally bound, dense clusters (of which YMCs are the most extreme examples) that represent the high-density end of a continuous-density spectrum of star formation. This separation implies a prediction that the fraction of star formation occurring in bound clusters varies with the galactic environment (*Kruijssen,* 2012), and it crucially depends on the spatial variation of the SFE as a function of the local volume density and the resulting relative spatial distributions of gas and stars. It is therefore key to develop physically motivated initial conditions for future numerical simulations of early cluster evolution that are designed to accurately represent these distributions.

3.1.3. When do stellar dynamics take over? In dynamically young systems, the hierarchical growth of gaseous structure translates to a hierarchy in the young stellar structure. However, in gravitationally bound systems [i.e., those with ages older than a dynamical timescale (see *Gieles and Portegies Zwart,* 2011)], this hierarchy is short-lived — once the stellar mass dominates self-gravity on a certain spatial scale (starting at the smallest scales), the hierarchy is erased by N-body dynamics on a dynamical timescale. The step toward stellar-dynamics-dominated evolution may occur due to gas exhaustion at high densities, gas expulsion at low densities, or a combination of both on large spatial scales covering a wide range of local environments.

Once the gravitational support from the gas has vanished, the hierarchical growth of stellar structure may continue if the region is globally subvirial, but only for another dynamical timescale (*Allison et al.,* 2010). The merging of subclusters leads to violent relaxation and the growth of a centrally concentrated stellar system, during which the degree of mass segregation in the progenitor systems is preserved (*Allison et al.,* 2009). As a result, mass segregation can be achieved on a dynamical timescale, even in massive YMCs. The further structural evolution of a cluster or YMC is governed by two-body relaxation and possible external tidal perturbations.

3.1.4. Toward physically motivated initial conditions for numerical simulations. We have divided the cluster-formation process into three different stages: (1) the condensation of stars and stellar structure from the hierarchically structured ISM; (2) the early disruption of stellar structure by gas expulsion or tidal perturbations (by contrast, violent relaxation erases substructure but does enable the formation of YMCs and long-lived clusters); and (3) the transition to the phase where YMC evolution is dominated by stellar dynamics. Each of these stages has been studied separately in numerical simulations of cluster formation, using a variety of numerical methods. It is possible to base the initial conditions for each of these steps on the outcomes of the previous ones. For instance, the initial conditions for YMC formation from the turbulent, hierarchical ISM requires one to resimulate (i.e., zoom in on) GMCs identified in galactic-scale simulations (e.g., *Tasker and Tan,* 2009; *Dobbs et al.,* 2011a). The next step demands the inclusion of realistic feedback and star-formation models — as reviewed in *Kruijssen* (2013), such simulations should ideally include radiative feedback, protostellar outflows, and magnetic fields. Using the hierarchically structured initial conditions from the preceding step, and ensuring these conditions accurately reflect the observed properties of such regions, will ensure a physically accurate setup.

The transition from an embedded cluster to a gas-poor system is nontrivial and benefits from a fully self-consistent treatment of both the collisional stellar dynamics as well as the hydrodynamics. This has been enabled by a new generation of numerical methods such as the Astronomical Multipurpose Software Environment (AMUSE) (*Portegies Zwart et al.,* 2009) and SEREN (*Hubber et al.,* 2013). If the classical approach of integrating the N-body dynamics in a background potential representing the gas is to remain viable, it is key that the gravitational potential is taken from hydrodynamical simulations (e.g., *Moeckel and Bate,* 2010). These should either be seeded with stars at the density peaks, or the sink particle data from the prior simulation could be used for the initial conditions (e.g., *Parker and Dale,* 2013). While this method is intrinsically flawed with respect to more advanced numerical approaches that self-consistently solve for sink-particle formation, feedback, and the co-evolution of gas and stars, it can still lead to new insights if it is applied with care.

In the near future, simulations of YMC formation are facing two main challenges. The first lies in establishing a consensus on scale-free (or generalized) initial conditions for any of the above three stages. The above discussion provides a physically motivated outline for how this can be done. The second challenge is one of physical scale — a YMC by definition has a stellar mass $M > 10^4 M_\odot$, which has not successfully been reproduced in numerical simulations without sacrificing the accuracy of low-mass star formation and the resulting stellar dynamics (*Tilley and Pudritz,* 2007; *Krumholz et al.,* 2012b; *Dale et al.,* 2012). As long as computational facilities cannot yet support the fully self-consistent modeling of YMC formation, such physical compromises may even be desirable. After all, answering problems such as the appearance of multiple generations of stars in massive clusters (*Gratton et al.,* 2012) or the formation of massive black holes (*Portegies Zwart et al.,* 2004) likely require the numerical modeling of YMC formation. Especially in the case of massive YMCs, the omission of accurate stellar dynamics will not necessarily be insurmountable — the relaxation times of YMCs exceed the timescales of interest when modeling their formation, and the first large-scale numerical simulations of spatially resolved YMC formation in a galactic context (*Hopkins et al.,* 2013) provide an excellent benchmark for future, more detailed efforts.

3.2. Key Open Questions

3.2.1. The distribution of young stars. How long do clusters live and what causes their destruction? This appears to be a straightforward question — surely one can simply look at a population of young clusters and look for a charac-

teristic timescale at which they begin to disappear. In reality, this is a nuanced problem. On the one hand, exposed (i.e., non-embedded), dynamically evolved clusters are distinct objects. However, young clusters often contain smaller substructures with no distinct boundaries within much larger ensembles (cf. *Allen et al.,* 2007). This is presumably due to the scale-free nature of the gas out of which clusters form (e.g., *Hopkins,* 2013).

In their seminal work, *Lada and Lada* (2003) studied a collection of embedded young regions, and defined clusters to be groups with >35 members with a stellar mass density of >1 $M_\odot$ pc^{-3}, which translates to a stellar surface density of 3 YSOs pc^{-2}. With this definition, they determined that ~90% of stars formed in clusters. When comparing this value to the number of open clusters (relative to the number expected from embedded clusters), they concluded that the majority of clusters disrupted when passing from the embedded to exposed phase. One caveat to this work is that the open cluster catalog used by *Lada and Lada* (2003) was only complete to a few hundred parsecs, whereas they assumed that it was complete to 2 kpc. This does not strongly affect their results for the early evolution, but it does explain why their expected open cluster numbers continue to diverge when compared to observations of clusters older than 5–10 Ma.

Bressert et al. (2010) used a comprehensive Spitzer Space Telescope Survey of nearly all YSOs in the solar neighborhood (within a distance of 500 pc) to measure their surface density distribution. They found a continuous distribution (i.e., without a characteristic density), with a peak at ~20 YSOs pc^{-2}, log-normally distributed (potentially with an extended tail at high densities). The authors concluded that the fraction of stars forming in clusters is a sensitive function of the criteria used to define a cluster. By adopting different criteria proposed in the literature, one could conclude that between 20% and 90% of stars form in clusters. The ambiguity in defining a cluster makes it difficult to quantify a single mechanism or efficiency that can explain the disruption of clusters across the full range of mass and density.

The initial spatial distribution of stars plays an important role in the subsequent evolution of the cluster. Both a centrally concentrated cluster profile (e.g., *Bate et al.,* 1998; *Parker and Meyer,* 2012; *Pfalzner and Kaczmarek,* 2013) and a rapid expansion of initially dense low-N clusters due to dynamical relaxation (e.g., *Gieles et al.,* 2012) can reproduce the continuous surface density distribution observed by *Bressert et al.* (2010). However, the observed spatial correlation of many YSOs with filamentary gas (cf. *Allen et al.,* 2007; *Gutermuth et al.,* 2011), combined with the lack of strong deviations between YSOs in various phases, argues against this set of initial conditions.

The continuous distribution of structures as a function of spatial scale makes it difficult to define what constitutes a (bound) cluster or an (unbound) association. While it is clear that stars rarely form in isolation, it does not appear to be the case that all stars form in gravitationally bound clusters. The terminology used for these young regions (e.g., clusters, groupings, "clusterings," associations) has led to much confusion and misunderstanding within the field.

In order to better characterize the different stellar structures found in SF regions, *Gieles and Portegies Zwart* (2011) introduced a dynamical definition of clusters/associations. The authors focus on massive stellar populations, $>10^4$ M$_\odot$, and found multiple examples where the same region was referred to as a cluster and an association by different studies. To address this problem, they introduced the parameter Π — the ratio between the stellar age and the crossing time of a region. Since unbound regions expand, the crossing time continually increases, so Π stays roughly constant or decreases with time. By comparison, the crossing time for bound structures remains fairly constant, so Π values increase with age. With this definition, *Gieles and Portegies Zwart* (2011) looked at a collection of regions discussed in the literature. They found that at young ages (<3 Ma) there was a continuous distribution of Π values. After 10 Ma, the distribution became discontinuous with a break at Π = 1, clearly separating expanding associations (Π < 1) and bound clusters (Π > 1).

To summarize, since the gas distribution from which stars form is scale-free, it is not expected that all stars form in dense or gravitationally bound clusters. Rather, young stellar structures are expected to have a continuous distribution of densities (*Hopkins* 2013). Hence, at young ages it is nontrivial to define which stars will end up in bound clusters. It is only once a structure has dynamically evolved (i.e., is older than a few crossing times) that structures become distinct from their surroundings.

While it may not be possible to define the fraction of stars that are born in clusters, we can measure the fraction of stars that are in dynamically evolved (>10 Ma) clusters at a given age. A variety of studies have found fractions of ~10% for the Milky Way (e.g., *Miller and Scalo,* 1978; *Lamers and Gieles,* 2006), for ages between a few million years and a few hundred million years. While limited by resolution, extragalactic studies have found fractions between 2–30% (*Adamo et al.,* 2011). In addition, evidence has been found for a potential relationship between the fraction of stars forming within clusters and the star-formation rate surface density [i.e., clusters are more prevalent in starbursts (e.g., *Adamo et al.,* 2011; *Silva-Villa and Larsen,* 2011)]. The techniques used for extragalactic studies are still developing, providing a rich potential avenue for future works in this area.

3.2.2. Infant (im)mortality. In the simplistic view of cluster formation discussed in section 3.1, whether or not a cluster survives the transition from the embedded to exposed phase is related to the effective star-formation efficiency [eSFE — a measure of how far out of virial equilibrium the cluster is after gas expulsion (see, e.g., *Goodwin,* 2009)]. In this picture, if the eSFE is higher than a critical value (~30%) a cluster is expected to remain bound. The destruction of clusters based on the removal of their natal gas (gas expulsion) is generally referred to as "infant mortality." This is often thought to be mass independent

due to the similarity between the embedded and exposed cluster mass functions.

For massive clusters, the rapid removal of gas and the subsequent expansion of the cluster (due to the resulting shallower gravitational potential) is predicted to have a measurable effect on the clusters' dynamical stability. For example, *Goodwin and Bastian* (2006) ran a series of N$_{body}$ simulations of massive clusters undergoing rapid gas removal. They found that the light-to-mass ratio (L/M) varied by more than a factor of 10 for eSFEs of 10–60%. Measurements of the L/M ratio of extragalactic YMCs appeared to confirm that many were expanding. However, *Gieles et al.* (2010) showed that massive binaries are likely affecting the results.

In order to overcome the problem of unresolved binaries, a number of works have focused on YMCs in the galaxy and LMC. In these systems it is possible to determine velocities of individual stars either through proper motions or radial velocities, and multiple-epoch velocity measurements can remove binaries from the sample. So far, this has been done for NGC 3603 (*Rochau et al.,* 2010), Westerlund 1 (*Mengel and Tacconi-Garman,* 2007; *Cottaar et al.,* 2012), the Arches (*Clarkson et al.,* 2012), and R136 in the LMC (*Hénault-Brunet et al.,* 2012b). All clusters studied to date are consistent with being in virial equilibrium, and do not show evidence for being affected by gas expulsion. Young massive clusters are expected to revirialize within 20–50 crossing times, which would happen quickly if YMCs begin life at very high densities (*Banerjee and Kroupa,* 2013). However, for typical observed initial sizes of ~1 pc and velocity dispersions of ~5 km s^{-1} (*Portegies Zwart et al.,* 2010), the revirialization time is >4 m.y. (*Baumgardt and Kroupa,* 2007; *Goodwin and Bastian,* 2006). We would expect to see evidence of this in the above studies. Therefore, it appears that YMCs are dynamically stable from a very young age [the Arches is 2.5–4 m.y. old (*Martins et al.,* 2008)]. Section 3.1 discussed possible interpretations of this result.

The apparent long-term survival of clusters (after the gas expulsion process) is also seen in the age distribution of clusters in the Milky Way (e.g., *Wielen,* 1971). For example, using complete open cluster catalogs, *Lamers et al.* (2005) and *Piskunov et al.* (2006) found a flat age distribution (number of clusters per million year) until >300 m.y. in the solar neighborhood (600 and 800 pc, respectively). However, the estimated disruption timescale for these samples are significantly less than that found for the LMC and SMC (e.g., *Hodge,* 1987; *Portegies Zwart et al.,* 2010).

The cause of this cluster disruption is likely to be the combined effects of tidal fields, dynamical relaxation, and most importantly for young clusters, interactions between clusters and GMCs (*Gieles et al.,* 2006b; *Kruijssen et al.,* 2011). Hence, it is the gas surface density that likely sets the average lifetimes of YMCs in galaxies (cf. *Portegies Zwart et al.,* 2010).

3.2.3. Age spreads within young massive clusters. It is often assumed that all stars within a massive cluster are coeval, but how close is this approximation to reality? If star formation is hierarchical (e.g., *Efremov and Elmegreen,* 1998; *Hopkins,* 2013), then we would expect the age spread within any region to be proportional to the size of the region studied. Hence, regions of 20–50 pc across may have spreads of ~10 m.y., whereas dense compact regions of cluster sizes (1–3 pc) are expected to have spreads <1 m.y. (e.g., *Efremov and Elmegreen,* 1998). Here we focus on cluster scales, i.e., ≲3 pc. However, we note that young clusters are generally not found in isolation, and that when one looks at the surrounding distribution, age spreads are often found [again, proportional to the size scale probed (see, e.g., *Melena et al.,* 2008; *Román-Zúñiga et al.,* 2008; *Da Rio et al.,* 2010a; *Preibisch et al.,* 2011)].

As a general rule, the current observational and theoretical limitations mean the detectable spread in stellar ages within a cluster is proportional to the mean age of that system. Observations of the youngest clusters therefore provide the best targets to quantify absolute age spreads. There are two complimentary ways to determine the ages of young stellar systems (see the chapter by Soderblom et al. in this volume for a review of this topic). The first is through spectroscopy of the high-mass stars, whose evolutionary timescale is quite short, to place them on the Hertzsprung-Russell diagram. The second method is through photometry of many low-mass pre-main-sequence (PMS) stars in order to place them in a color magnitude diagram (CMD) and compare the mean age and scatter relative to theoretical models. As an aside, it is worth noting that in addition to being a fundamental question in cluster formation, potential age spreads must be taken into account for accurate derivations of the stellar initial mass function (see the chapter by Offner et al. in this volume).

One potential caveat to studies of young clusters is that clusters tend not to form in isolation but rather within larger star-forming complexes, often containing multiple clusters and a distributed network of stars. These complexes often have age spreads on the order of 10 m.y., which means that simply due to projection effects, some relatively old (10-Ma) stars may appear to be part of a cluster. Hence, any measured spread is always an upper limit. An example of such a region is the Carina Nebula complex, which extends for ~30 pc and contains compact clusters with ages between ~1 and ~8 Ma, along with a large (~50% of the total) distributed stellar population (e.g., *Preibisch et al.,* 2011; *Ascenso et al.,* 2007).

The nearest and best-studied cluster, with a mass in excess of 1000 $M_\odot$, is the ONC at a distance of 417 ± 7 pc (*Menten et al.,* 2007). There has been a long-standing debate on the absolute age and relative age spread within the cluster, with initial reports of an accelerating star-formation history (SFH) over the past ~10 m.y., peaking 1–2 m.y. ago (*Palla and Stahler,* 1999). *Da Rio et al.* (2010b) found an age spread of ~2–3 m.y. in the cluster by studying the PMS color distributions with a collection of archival HST images. However, *Alves and Bouy* (2012) have shown that the cluster/association NGC 1980, which has an older age

(4–5 Ma), overlaps significantly in projection, explaining much of the previously reported spread. Additionally, rotation in young stars can significantly add the dispersions in CMDs of PMSs, which will lead to an overestimation of any age spread present (*Jeffries* 2007).

The most massive YMC known in the galaxy, Westerlund 1, has a mean age of ~5 Ma. *Kudryavtseva et al.* (2012) have studied the PMS color distribution and reported an upper limit to the relative age spread of 0.4 m.y. It appears that the statistic used did not show the age spread, but rather this is the potential error on the best fitting age. However, it is clear that the age spread is much smaller than the age of the cluster. A similar conclusion was reached for the massive cluster R136 in the LMC, based on spectroscopy of the high-mass stars (e.g., *Massey and Hunter,* 1998). There is also evidence of a 1–2-m.y. age spread within R136, potentially due to the merger of two clusters with slightly different ages (*Sabbi et al.,* 2012).

The Galactic YMC NGC 3603 has been subject to significant debate on the potential age spread within clusters. Using the early release science data from Wide Field Camera 3 (WFC3) on HST, *Beccari et al.* (2010) reported that approximately two-thirds of the PMS stars have ages ~3 Ma, while approximately one-third have ages >10 Ma (from 0.2 to 2 pc from the cluster center). However, while the spatial distribution of the young stars was centrally concentrated on the cluster, the older stars appear to have a uniform distribution across the field of view. NGC 3603 is part of a large star-forming region, with ongoing star formation. The presence of the evolved massive star Sher 25 has been used as evidence for a significant duration of star-formation within the region (>10 m.y.), not necessarily confined to the cluster itself. However, *Melena et al.* (2008) find the age of Sher 25 to be 4 Ma and subsequently revise the upper age limit in NGC 3603 from 10 to 4 Ma. *Kudryavtseva et al.* (2012) also studied the inner 0.5 pc of NGC 3603 and found an upper limit to the age spread of 0.1 m.y. (although see note above). Hence, this cluster is consistent with a near coeval population, with an older population of stars in the region around it.

Contrary to the above studies, which are consistent with age spreads much less than the mean stellar age, the young Galactic cluster W3-main (~4 × 10³ $M_\odot$) appears to have an age spread of ~2–3 m.y., similar to the age of the cluster. This is based on spectroscopy of a massive star that appears to have already left the zero-age main sequence along with the presence of multiple ultracompact HII regions, all within a projected radius of less than 2 pc (*Bik et al.,* 2012). Given the proximity of other nearby star forming complexes, projection effects may explain the observations. However, *Bik et al.* (2012) propose that the spatial distribution of the high-mass stars and ultracompact HII regions argues against this. The cluster also appears to have a significant amount of dense gas within the cluster region, potentially explaining why star formation has been able to continue.

There has been increasing evidence that high-mass and intermediate-mass stars cannot be explained by evolutionary models with the same age. Such an age discrepancy was discussed for Westerlund 1 in *Negueruela et al.* (2010), where red supergiants are expected to be of apparently older age (>6 Ma), while the rich WR population should be younger than 5 Ma (see also *Crowther et al.*, 2006), which is consistent with the derived age of blue supergiants. A similar trend is found in the Arches and Quintuplet Galactic Center clusters, where the apparent age of intermediate-mass, less-evolved supergiants appears to exceed the age of the WN population (*Martins et al.*, 2008; *Liermann et al.*, 2010, 2012). In all these clusters, particularly young ages are derived from single-star evolution models for the hydrogen-rich WN population. While these discrepancies have led to claims of age spreads in young star clusters, there is increasing evidence that the superluminous WNs (and LBVs) are rejuvenated binary products [see especially the discussion in *Liermann et al.* (2012)]. Hence, while age spreads in the range of ±2 m.y. [e.g.,W3 Main (*Bik et al.*, 2012)] are present in young associations, there is little evidence for extreme age spreads in the spatially confined population of YMCs (see above references), especially when the expected consequences of binary evolution are taken into account.

3.2.4. Timescale of the removal of gas from young massive clusters. Presumably, the lack of significant age spreads within most YMCs is due to the fact that the dense gas, which is required for star formation, is removed on short timescales. For example, YMCs such as the Arches (2.5–4 Ma), NGC 3603 (1–2 Ma), Trumpler 14 (~1 Ma), and Westerlund 1 (5–7 Ma) are all devoid of dense gas. It is often assumed that supernovae play a role in the removal of the gas, but as the above examples show, the gas is removed well before the first supernovae can occur (~3 m.y.). Even for relatively low-mass clusters ($\leq 10^3$ M$_\odot$), the timescale of gas removal is <1 m.y. (*Seale et al.* 2012). Potential causes of the rapid gas removal are discussed in section 3.1.

This rapid removal naturally explains the lack of significant age spreads in clusters, but places strong constraints on the dynamical stability of young clusters.

3.3. Age Spreads Within Older Clusters

The ability to discern age spreads within clusters is linked to the (mean) age of the cluster itself, with an uncertainty of approximately 10–20% of the cluster age. While the constraints from older clusters are significantly less in an absolute sense, they are still worth looking into as some surprising features have been found.

The relatively nearby double cluster, h and χ Persei, have masses of 4.7 and 3.7×10^3 M$_\odot$, respectively, within the clusters, and a total mass of 2×10^4 M$_\odot$, including the common halo around the clusters (*Currie et al.*, 2010). Their common age is derived to be 14 Ma (*Currie et al.*, 2010). An upper limit of 2 Ma for the possible age spread has been put on this population using an extensive spectroscopic survey of cluster members (*Currie et al.*, 2010).

There have been claims of extended star-formation epi-sodes (200–500 m.y.) within 1–2 Ga clusters in the LMC/SMC (e.g., *Mackey and Broby Nielsen*, 2007; *Goudfrooij et al.*, 2011) based primarily on the presence of extended main-sequence turnoffs in these clusters. *Goudfrooij et al.* (2011) and *Conroy and Spergel* (2011) suggest that such age spreads can happen in massive clusters, which have high enough escape velocities to retain stellar ejecta and accreted material from the surroundings. *Bastian and Silva-Villa* (2013) have tested this claim by studying two massive (10^5 M$_\odot$) clusters in the LMC with ages of 180 and 280 Ma, with resolved HST photometry. The authors do not find any evidence for an age spread within these clusters, and put an upper limit of 35 m.y. on any potential age spread.

Bastian et al. (2013a) have compiled a list of all known Galactic and extragalactic massive clusters with ages between 10–1000 Ma and masses between 10^4–10^8 M$_\odot$ with published integrated spectroscopy and/or resolved photometry of individual stars. The authors looked for emission lines associated with the clusters, in order to find evidence for ongoing star formation within the existing clusters. Of the 129 clusters in their sample, no clusters were found with ongoing star formation. This effectively rules out theories and interpretations that invoke extended (tens to hundreds of million years) star-formation periods within massive clusters (e.g., *Conroy and Spergel*, 2011).

We will return to the possibility of multiple star-formation events within massive clusters in section 5, when we discuss globular clusters and how they relate to YMCs.

4. THE EFFECT OF ENVIRONMENT ON PLANET FORMATION AND THE ORIGIN OF THE FIELD

Young stars in YMCs show infrared excess emission and evidence for accretion [e.g., *Stolte et al.* (2004) and *Beccari et al.* (2010) for the Galactic YMC NGC 3603] suggesting the presence of circumstellar disks. Whether these disks are similar to those in nearby environments (see *Stolte et al.*, 2010) or can survive and produce — possibly even habitable — planetary systems in YMCs remain interesting questions. Considering the disruptive effects of stellar clustering on the survival of protoplanetary disks and planetary systems in small-N systems (*Scally and Clarke*, 2001; *Olczak et al.*, 2010; *Parker and Quanz*, 2012; *de Juan Ovelar et al.*, 2012, *Rosotti et al.*, 2014), the formation of (habitable) planets in YMCs may still be possible, but their chances of long-term (>10^8 yr) survival are most likely very limited. This is underlined by the absence of planets in the 10-G.y.-old globular clusters NGC 104 and NGC 6397 (*Gilliland et al.*, 2000; *Nascimbeni et al.*, 2012), whereas Neptune-like planets have been detected in the intermediate-aged (τ ~ 1 G.y.) open cluster NGC 6811 (*Meibom et al.*, 2013), which has a much lower stellar density (Σ ~ 10^3 pc^{-2}) than globular clusters or YMCs (Σ > 10^4 pc^{-2}).

Provided that long-term habitable planets cannot form in YMCs, we can now ask the question, what fraction of all stars may be born in such a hostile environment? It was proposed by *de Juan Ovelar et al.* (2012) that a stellar surface density of at most $\Sigma < 10^3$ M$_\odot$ pc^{-2} is needed to keep the habitable zone of G dwarfs unperturbed over billion-year timescales, whereas *Thompson* (2013) showed that protoplanetary systems with densities $\Sigma > 6 \times 10^3$ M$_\odot$ pc^{-2} may not have an ice line, inhibiting the formation of gas and ice giants (if they form by core accretion) as well as potentially habitable planets. In the following, we adopt a maximum density for the long-term existence of habitable planets of $\Sigma_{crit} = 10^3$ M$_\odot$ pc^{-2}. Given the very weak mass-radius relation of young clusters R = 3.75 pc (M/10^4 M$_\odot$)$^{0.1}$ from *Larsen* (2004), this can be converted to a critical cluster mass of M$_{crit} \sim 10^5$ M$_\odot$.

We point out here that a seemingly opposite conclusion was reached by *Dukes and Krumholz* (2012) and *Craig and Krumholz* (2013), who find that the dense cluster environment does not affect the properties of disks or planetary systems. However, this apparent dichotomy arises from the adopted definition of "cluster." In this review we use the term "cluster" to refer to bound systems, whereas in the above two papers the term refers to any overdensity of stars, reflected by the short lifetimes (of a few crossing times) and constant surface densities (as appropriate for GMCs and unbound associations). This results in a decreasing volume density with mass and hence a decreasing importance of environmental effects. The low-volume density systems considered by *Dukes and Krumholz* (2012) and *Craig and Krumholz* (2013) constitute the majority of star-forming environments.

The initial cluster mass function is a power law dN/dM $\propto$ M^{-2} for masses M > 10^2 M$_\odot$ up to some maximum mass (*Portegies Zwart et al.*, 2010), which in the nearby universe is M $\sim 2 \times 10^5$ M$_\odot$ (*Larsen*, 2009), and in the high-redshift universe may have been M $\sim 10^7$ M$_\odot$ (see section 5). Because the slope of the initial cluster mass function is −2, equal numbers of stars are formed in each dex of cluster mass. Given the above mass limits, in the nearby universe therefore about 10% of all gravitationally bound clusters are hostile to habitable planets, whereas in the high-redshift universe this may have been 40%. Note that this assumes that clusters in the early universe followed the same mass-radius relation as clusters do today. If they were denser, the 40% quoted here is a lower limit. Locally, only 10% of all stars are born in bound clusters (*Lada and Lada*, 2003), whereas at the high densities of the young universe, this may have been up to 50% (*Kruijssen*, 2012).

Combining all the above numbers, we find that in the present-day universe, only 1% of all stars are born in an environment of too high density to allow habitable planet formation, whereas at high redshift (z = 2–3) up to 20% of all stars did. A rough interpolation suggests that in the formation environment of the solar system (z ~ 0.5) at least 90% of all stars resided in sufficiently low-density environments to potentially host habitable planetary systems. This number is representative for the universe as a whole — making the highly simplifying assumption that about half the stars in the universe formed in present-day galactic environments, whereas the other half formed in a

high-density environment. We find that some *10% of all stars in the universe may not have planets in their habitable zone due to environmental effects;* the remaining 90% of stars formed in a setting that in terms of planet formation was as benign as the field. We conclude that while very interesting in its own right, the environmental inhibition of habitable-planet formation is likely not important for the cosmic-planet population.

5. LINKING YOUNG MASSIVE CLUSTER FORMATION TO GLOBULAR CLUSTERS AND OTHER MAJOR STAR-FORMATION EVENTS IN THE UNIVERSE

Globular clusters (GCs) (extremely dense clusters of old stars) are found in galaxies of all masses, but the most prominent (N$_{GC}$ > 100) populations exist in galaxies of stellar masses exceeding a few 10^{10} M$_\odot$ (*Peng et al.*, 2008). The Milky Way contains over 150 old globular clusters, which have ages of 5–13 Ga (i.e., formed 5–13 gigayears ago) (*Forbes and Bridges*, 2010) and masses of roughly 10^4–10^6 M$_\odot$ (*Harris*, 1996; *Brodie and Strader*, 2006). The detection of YMCs with masses comparable to or greater than GCs in nearby merging and starburst galaxies (*Schweizer*, 1982; *Holtzman et al.*, 1992; *Schweizer et al.*, 1996; *Whitmore et al.*, 1999) has reinvigorated discussion of their origins. Did the most massive YMCs form in a way similar to GCs? If true, this would provide a unique window to understanding extreme cluster formation at or before the peak of cosmic star formation [z ~ 2–3 (*Madau et al.*, 1996; *Hopkins and Beacom*, 2006; *Bouwens et al.*, 2011; *Moster et al.*, 2013)].

The mass spectrum of GCs is fundamentally different than that of young clusters. While the latter follows a power law dN/dM $\propto$ M^{-2} over most of its mass range M = 10^2–10^8 M$_\odot$ [*Zhang and Fall* (1999) and *Larsen* (2009), except for possibly an exponential truncation at the high-mass end (see, e.g., *Gieles et al.*, 2006a)], GCs have a characteristic mass scale of M ~ 10^5 M$_\odot$ (*Harris*, 1996; *Brodie and Strader*, 2006). The modern interpretation is that GCs initially followed the same mass spectrum as young clusters in the local universe (e.g., *Elmegreen and Efremov*, 1997; *Kravtsov and Gnedin*, 2005; *Kruijssen and Cooper*, 2012), but that it subsequently evolved into its present, peaked form. Clusters much more massive than the characteristic GC mass likely existed, but were rare due to the steep slope of the mass spectrum. In addition, the maximum mass of GCs has been influenced by their formation environment (see below) and/or dynamical friction (e.g., *Tremaine et al.*, 1975). On the other side of the mass spectrum, the numerous low-mass GCs were disrupted and did not survive until the present day (*Fall and Zhang*, 2001). However, this explanation is not undisputed. First, the characteristic mass scale of GCs is remarkably universal (*Jordán et al.*, 2007), whereas the disruption of stellar clusters is sensitive to the galactic environment (e.g., *Kruijssen et al.*, 2011). Second, in the Fornax dwarf spheroidal galaxy the metal-poor GCs constitute 25% of the field stellar mass at similar metallicities (*Larsen et al.,* 2012), which is 2 orders of magnitude higher than the "universal" baryonic mass fraction of GCs in giant elliptical galaxies found by *McLaughlin* (1999). This suggests that very specific conditions are required for a full cluster mass spectrum to have been present initially — in particular, all coeval stars need to have formed in bound clusters [at odds with theory and observations (see, e.g., *Kruijssen,* 2012)], of which the surviving GCs did not lose any mass at all, whereas the lower-mass clusters were all completely destroyed. Several adjustments to the simple disruption model have therefore been proposed (see below). For the comparison to YMCs, the key point is that the conditions of GC formation must have been such that they could survive a Hubble time of dynamical evolution, irrespective of their particular formation mechanism(s). This can be used to constrain their formation environment.

Early GC formation theories attributed the characteristic mass scale of GCs to specific conditions at the time of GC formation, which historically was mainly driven by the fact that no open clusters were known to have masses comparable to GCs. Many of these theories addressed the problem of producing a sufficient mass concentration of dense gas needed to form a GC. Since the Jeans mass following recombination is $M_{Jeans} = 10^5 - 10^6$ $M_\odot$, *Peebles and Dicke* (1968) suggested that GCs were the first bound structures to form. Similarly, *Fall and Rees* (1985) suggested that proto-GC clouds formed though thermal instabilities in galactic halos from metal-poor gas with a Jeans mass above 10^6 $M_\odot$. *Ashman and Zepf* (1992) put forward the idea that the shocks in gas-rich galaxy mergers lead to the formation of extremely massive GMCs and GCs, in line with observations of ongoing mergers in the local universe (e.g., *Whitmore et al.,* 1999). Alternatively, it has been proposed that GCs formed in low-mass dark matter halos, which were subsequently lost due to a Hubble time of tidal stripping and/or two-body relaxation (e.g., *Bekki et al.,* 2008; *Griffen et al.,* 2010). Finally then, GCs may represent the former nuclear clusters of cannibalized dwarf galaxies (*Lee et al.,* 2009) — note that in general, the hierarchical growth of galaxies implies that some fraction of the GC population in a galaxy previously belonged to smaller systems (e.g., *Muratov and Gnedin,* 2010).

All the above models require that the distribution of GCs should follow the spatial structure of the galactic stellar halo — because either (1) they were initially directly associated with dark matter; (2) they formed at very early times, before the main bodies of present-day galaxies were in place; or (3) their production/accretion took place during hierarchical galaxy growth, which predominantly populates galaxy halos (e.g., *Sales et al.,* 2007). It has been shown that metal-rich ([Fe/H] > −1) GCs are associated with the stellar mass in galaxies [in the Milky Way, the metal-rich GC population even exhibits a net rotation of 60 km s⁻¹ (see *Dinescu et al.,* 1999)], whereas the metal-poor ([Fe/H] < −1) part of the GC population extends further and is associated with the stellar halo (*Brodie and Strader,* 2006; *Strader et al.,* 2011). The above GC-formation scenarios may therefore possibly apply to metal-poor GCs, but are unlikely to hold for the metal-rich ones. Overall though, a reexamination of these models has mainly been prompted by the HST discovery that YMCs with masses exceeding those of GCs are still forming in the universe today (e.g., *Whitmore et al.,* 1999).

The elevated gas accretion rates in high-redshift galaxies lead to extremely turbulent (with Mach numbers $\mathcal{M} \sim 100$), clumpy galaxies, which have high Toomre (or Jeans) masses, allowing the clumps to reach masses up to 10^9 $M_\odot$ (*Elmegreen and Elmegreen,* 2005; *Förster Schreiber et al.,* 2009; *Krumholz and Burkert,* 2010; *Forbes et al.,* 2012). In these conditions, it seems plausible that the high stellar masses required for GC progenitors can easily be assembled (*Shapiro et al.* 2010), and considering the high gas densities this is likely accompanied by locally highly efficient star formation and hence a negligible effect of any disruption due to gas expulsion (see section 3.1). The formation of gravitationally bound, extremely massive clusters seems inevitable in the high-redshift universe – examples of similar conditions in the nearby universe exist as well (*Schweizer,* 1982; *Holtzman et al.,* 1992; *Miller et al.* 1997; *Bastian et al.,* 2006). While these examples mainly refer to galaxy mergers, which are not thought to be the main source of star formation in the early universe (e.g., *Genzel et al.,* 2010a), the conditions in the interstellar medium are similar (*Kruijssen and Longmore,* 2013). Importantly though, these nearby examples are all accompanied by a population of low-mass clusters (see above and, e.g., *Zhang and Fall,* 1999).

Despite the fact that GCs presently reside in widely differing galactic environments, their common ancestry may provide an explanation for their characteristic mass-scale, even if they formed in the same way as clusters in nearby galaxies. The high-mass end of the cluster mass function is likely truncated by the Toomre mass, which is the largest mass-scale in a galaxy over which gas clumps can become self-gravitating, and ranges from several 10^6 $M_\odot$ in Milky Way-like spirals to several 10^9 $M_\odot$ in extreme, high-redshift environments. For a global star-formation efficiency of 10% (appropriate for the ~100-pc scales considered here), this gives maximum cluster masses between several 10^5 $M_\odot$ and several 10^8 $M_\odot$. Evidence for such a truncation exists in observations of nearby galaxies (*Gieles et al.,* 2006a; *Larsen,* 2009) and it has also been derived indirectly from the present-day GC mass spectrum (*Jordán et al.,* 2007; *Kruijssen and Portegies Zwart,* 2009). As discussed above, the high initial masses required for GCs greatly constrain their possible formation environment to the vigorously star-forming, highly turbulent environments seen at high redshift, where the Toomre mass is sufficiently high to form proto-GCs ($M_T \sim 10^8$ $M_\odot$; see below). These galaxies are characterized by high accretion rates and densities, and are forming large numbers of clusters per unit time, leading to the formation of extremely massive clusters (see *Bastian,* 2008, for local examples). However, these high ambient densities also lead to very efficient cluster disruption — predominantly due to tidal shocking (*Spitzer*, 1958; *Gieles et al.*, 2006b). This suggests a picture in which the vast majority of GC disruption occurred at high redshift, removing the low-mass clusters from the GC population and leaving only the most massive ones intact (*Elmegreen*, 2010; *Kruijssen et al.*, 2012a).

Given a high-mass truncation of the initial cluster mass function, the destruction of low-mass clusters eventually leads to a saturation of the characteristic mass-scale at ~10% of the truncation mass (*Gieles*, 2009). For a Toomre mass of $M_T \sim 10^8 M_\odot$, a global star-formation efficiency of 10% (as assumed above) gives an initial truncation mass of $M_{max,i} \sim 10^7 M_\odot$. A key requirement is that the resulting GCs survived until the present day and hence at some point decoupled from their high-density natal environment. In the most extreme environments, cluster disruption may typically remove up to 90% of the mass from massive clusters before the disruption rate starts decreasing due to cluster migration (*Kruijssen et al.*, 2011), and hence the truncation mass would have decreased to a few $10^6 M_\odot$ by the time the most massive GCs escaped to a more quiescent environment (such as the galaxy halo). At 10% of the truncation mass, this implies a *universal* characteristic mass-scale of GCs of a few $10^5 M_\odot$, which is indeed what is seen in the universe today. In this interpretation, the mass-scale of GCs indirectly reflects the Toomre mass at the time of their formation — much like the masses of YMCs do in nearby galaxies.

A fundamental difficulty in trying to reconstruct the conditions of GC formation from the present-day GC populations that exist in nearby galaxies is that the evolution of dense stellar systems with two-body relaxation times shorter than a Hubble time is convergent (*Gieles et al.*, 2011). In other words, the characteristics imprinted by their formation process will be washed out by stellar dynamics. It is therefore necessary to address the formation of GCs from more direct avenues.

During the last decade, the discovery of multiple main sequences and chemical abundance variations in GCs (for a review, see *Gratton et al.*, 2012) has triggered a reevaluation of GC formation mechanisms, suggesting that they formed during multiple star-formation episodes (spaced by 10–100 m.y.) and therefore underwent chemical self-enrichment (e.g., *Conroy and Spergel*, 2011). These new results have even led to the idea that GCs formed by a fundamentally different process than YMCs in the nearby universe. However, without any direct observations to support such a far-reaching interpretation, it remains conceivable that "normal" YMC formation at the low metallicities and high densities that characterize the high-redshift universe naturally lead to the observed properties of GCs.

There exist no nearby YMC-forming regions of adequately low metallicity ([Fe/H] = −1.7 to −0.7) *and* high mass ($M_{YMC} > 10^6 M_\odot$) to probe the formation of GC-like YMCs. Hence, the question is to what extent nearby YMCs are representative of young GCs. No indication of ongoing star formation has been detected in massive ($M = 10^4–10^8 M_\odot$) and young to intermediate-aged ($\tau =$

10–1000 m.y.) clusters in nearby galaxies (*Bastian et al.*, 2013a), suggesting that these all formed in a single burst of star formation. A recently proposed model by *Bastian et al.* (2013b) argues that the high stellar densities of young GCs imply that stellar winds and binary ejecta would have been efficiently swept up by the long-lived protostellar disks around low-mass stars. Because low-mass stars are still fully convective in the PMS phase, they are then uniformly polluted. In this "early disk accretion" model, only the stars that passed through the core of the cluster during the right time interval exhibit abundance variations, naturally leading to relatively bimodal abundance variations and the (incorrect) impression of multiple episodes of star formation. The model will need to be verified in future work, but for now it illustrates that GC formation may very well have proceeded in a way that is consistent with the known physics of YMC formation.

Given the available evidence, it seems plausible that at least the metal-rich part of the GC population formed in a way similar to YMCs in nearby galaxies. The study of YMC formation can therefore provide key insights in the formation of the oldest structures in the local universe. However, the extended spatial distribution of metal-poor clusters with respect to the galaxy light (*Brodie and Strader*, 2006) suggests that a different formation mechanism may have been at play. Tentative support is provided by the lack of field stars with metallicities similar to the (metal-poor) GCs in the dwarf spheroidal galaxy Fornax (*Larsen et al.*, 2012). If these GCs formed like YMCs as part of a full mass spectrum of lower-mass clusters that were subsequently disrupted, a larger population of coeval field stars would be expected. Fornax is an extreme case though — for more massive galaxies the metal-poor and metal-rich GC subpopulations may emerge naturally in the context of hierarchical galaxy formation (*Muratov and Gnedin*, 2010; *Tonini*, 2013). Studies of the co-formation of galaxies and their cluster populations will constrain GC formation further in the coming years.

6. SUMMARY AND OUTLOOK

We conclude that the study of YMC formation will play a key role in our future understanding of star and planet formation across cosmological timescales. Young massive clusters act as a natural bridge linking what we learn from star-forming regions in the solar neighborhood to starburst systems in the local universe, globular-cluster formation, and star/planet formation at the earliest epochs of the universe.

In the same way that the discovery and characterization of YMCs with HST and 8-m-class optical/infrared telescopes revolutionized our understanding of the relationship between open and globular clusters, the advent of new survey data and new/upcoming facilities [ALMA, Very Large Array (VLA), JWST, Gaia, ELTs] are set to revolutionize our understanding of the YMC-formation process. In 5 to 10 years we should hopefully have a complete sample of

YMC progenitor clouds in the Galaxy and demographics of YMC progenitors and YMCs in a large number of external galaxies, with a wide range of environments (mergers/starbursts, low-metallicity systems, centers of galaxies, etc.). With this data it will be possible to determine (1) the global environmental conditions required to assemble gas to such high density, (2) whether YMCs in different environments assemble their mass differently, (3) how this is related to the environmental variation of the critical density for star formation, and (4) whether the properties of stars and planets in YMCs are affected by differences in the initial protostellar density.

The fact that for the foreseeable future the Milky Way is the only place it will be possible to resolve individual (proto)stars means Galactic studies will be crucial. Follow-up studies connecting the initial gas conditions to the subsequent stellar populations will be able to probe the assembly of stellar and planetary mass as a function of environment. Gaia and complementary, follow-up spectroscopic surveys will nail down the six-dimensional stellar structure of the nearest clusters, unambiguously measuring any primordial rotation or expansion as well as any dynamical evidence for subclusters. This will directly answer several of the key questions outlined in this review.

We conclude that the next few years are set to be an exciting and productive time for YMC studies and will likely lead to major breakthroughs in our understanding of YMC formation.

Acknowledgments. We would like to thank G. Beccari, H. Beuther, Y. Contreras, A. Font, R. Galvan-Madrid, J. Kauffmann, M. Krumholz, B. Mills, F. Motte, C. Lang, Q. Nguyen Luong, T. Pillai, J. Urquhart, Q. Zhang, and the anonymous referee for helpful comments on the manuscript. We acknowledge the hospitality of the Aspen Center for Physics, which is supported by National Science Foundation Grant No. PHY-1066293.

REFERENCES

Adamo A. et al. (2011) *Mon. Not. R. Astron. Soc., 417,* 1904.
Adams F. C. (2000) *Astrophys. J., 542,* 964.
Aguirre J. E. et al. (2011) *Astrophys. J. Suppl., 192,* 4.
Allen L. et al. (2007) In *Protostars and Planets V* (B. Reipurth et al., eds.), pp. 361–376. Univ. of Arizona, Tucson.
Allison R. J. et al. (2009) *Astrophys. J. Lett., 700,* L99.
Allison R. J. et al. (2010) *Mon. Not. R. Astron. Soc., 407,* 1098.
Alves J. and Bouy H. (2012) *Astron. Astrophys., 547,* A97.
Alves J. and Homeier N. (2003) *Astrophys. J. Lett., 589,* L45.
Alves J. F. et al. (2001) *Nature, 409,* 159.
Ao Y. et al. (2013) *Astron. Astrophys., 550,* A135.
Ascenso J. et al. (2007) *Astron. Astrophys., 476,* 199.
Ashman K. M. and Zepf S. E. (1992) *Astrophys. J., 384,* 50.
Bally J. et al. (2010) *Astrophys. J., 721,* 137.
Balser D. S. et al. (2011) *Astrophys. J., 738,* 27.
Banerjee S. and Kroupa P. (2013) *Astrophys. J., 764,* 29.
Barnes P. J. et al. (2010) *Mon. Not. R. Astron. Soc., 402,* 73.
Barnes P. J. et al. (2011) *Astrophys. J. Suppl., 196,* 12.
Bastian N. (2008) *Mon. Not. R. Astron. Soc., 390,* 759.
Bastian N. and Silva-Villa E. (2013) *Mon. Not. R. Astron. Soc., 431,* L122.
Bastian N. et al. (2006) *Astron. Astrophys., 448,* 881.
Bastian N. et al. (2010) *Annu. Rev. Astron. Astrophys., 48,* 339.
Bastian N. et al. (2013a) *Mon. Not. R. Astron. Soc., 436,* 2852.

Bastian N. et al. (2013b) *Mon. Not. R. Astron. Soc., 436,* 2398.
Bate M. R. et al. (1998) *Mon. Not. R. Astron. Soc., 297,* 1163.
Baumgardt H. and Kroupa P. (2007) *Mon. Not. R. Astron. Soc., 380,* 1589.
Beccari G. et al. (2010) *Astrophys. J., 720,* 1108.
Bekki K. et al. (2008) *Mon. Not. R. Astron. Soc., 387,* 1131.
Beuther H. et al. (2002) *Astrophys. J., 566,* 945.
Beuther H. et al. (2012) *Astrophys. J., 747,* 43.
Bik A. et al. (2012) *Astrophys. J., 744,* 87.
Binney J. et al. (1991) *Mon. Not. R. Astron. Soc., 252,* 210.
Boily C. M. and Kroupa P. (2003) *Mon. Not. R. Astron. Soc., 338,* 665.
Bonnell I. A. et al. (2008) *Mon. Not. R. Astron. Soc., 389,* 1556.
Bouwens R. J. et al. (2011) *Astrophys. J., 737,* 90.
Bressert E. et al. (2010) *Mon. Not. R. Astron. Soc., 409,* L54.
Bressert E. et al. (2012a) *Astrophys. J. Lett., 758,* L28.
Bressert E. et al. (2012b) *Astron. Astrophys., 542,* A49.
Brodie J. P. and Strader J. (2006) *Annu. Rev. Astron. Astrophys., 44,* 193.
Buckley H. D. and Ward-Thompson D. (1996) *Mon. Not. R. Astron. Soc., 281,* 294.
Burton M. G. et al. (2013) *Publ. Astron. Soc. Australia, 30,* e044.
Carilli C. L. and Walter F. (2013) *Annu. Rev. Astron. Astrophys., 51,* 105.
Carlhoff P. et al. (2013) *Astron. Astrophys., 560,* A24.
Clark P. C. et al. (2005) *Mon. Not. R. Astron. Soc., 359,* 809.
Clark P. C. et al. (2013) *Astrophys. J. Lett., 768,* L34.
Clarke C. J. et al. (2000) In *Protostars and Planets IV* (V. Mannings et al., eds.), p. 151. Univ. of Arizona, Tucson.
Clarkson W. I. et al. (2012) *Astrophys. J., 751,* 132.
Conroy C. and Spergel D. N. (2011) *Astrophys. J., 726,* 36.
Contreras Y. et al. (2013) *Astron. Astrophys., 549,* A45.
Cottaar M. et al. (2012) *Astron. Astrophys., 539,* A5.
Craig J. and Krumholz M. R. (2013) *Astrophys. J., 769,* 150.
Crowther P. A. et al. (2006) *Mon. Not. R. Astron. Soc., 372,* 1407.
Csengeri T. et al. (2011) *Astrophys. J. Lett., 740,* L5.
Currie T. et al. (2010) *Astrophys. J. Suppl., 186,* 191.
Dale J. E. et al. (2005) *Mon. Not. R. Astron. Soc., 358,* 291.
Dale J. E. et al. (2012) *Mon. Not. R. Astron. Soc., 424,* 377.
Da Rio N. et al. (2010a) *Astrophys. J., 722,* 1092.
Da Rio N. et al. (2010b) *Astrophys. J., 723,* 166.
Davies B. et al. (2011) *Mon. Not. R. Astron. Soc., 411,* 1386.
Davies B. et al. (2012) *Mon. Not. R. Astron. Soc., 419,* 1860.
Davies R. I. et al. (2007) *Astrophys. J., 671,* 1388.
de Juan Ovelar M. et al. (2012) *Astron. Astrophys., 546,* L1.
Dib S. et al. (2013) *Mon. Not. R. Astron. Soc., 436,* 3727.
Dinescu D. I. et al. (1999) *Astron. J., 117,* 1792.
Dobbs C. L. et al. (2011a) *Mon. Not. R. Astron. Soc., 417,* 1318.
Dobbs C. L. et al. (2011b) *Mon. Not. R. Astron. Soc., 413,* 2935.
Dukes D. and Krumholz M. R. (2012) *Astrophys. J., 754,* 56.
Efremov Y. N. and Elmegreen B. G. (1998) *Mon. Not. R. Astron. Soc., 299,* 588.
Elmegreen B. G. (2000) *Astrophys. J., 530,* 277.
Elmegreen B. G. (2010) *Astrophys. J. Lett., 712,* L184.
Elmegreen B. G. and Efremov Y. N. (1997) *Astrophys. J., 480,* 235.
Elmegreen B. G. and Elmegreen D. M. (2005) *Astrophys. J., 627,* 632.
Elmegreen B. G. and Falgarone E. (1996) *Astrophys. J., 471,* 816.
Elmegreen B. G. and Hunter D. A. (2010) *Astrophys. J., 712,* 604.
Elson R. A. W. et al. (1987) *Astrophys. J., 323,* 54.
Fall S. M. and Rees M. J. (1985) *Astrophys. J., 298,* 18.
Fall S. M. and Zhang Q. (2001) *Astrophys. J., 561,* 751.
Fáundez S. et al. (2004) *Astron. Astrophys., 426,* 97.
Federrath C. et al. (2010) *Astron. Astrophys., 512,* A81.
Ferrière K. et al. (2007) *Astron. Astrophys., 467,* 611.
Figer D. F. et al. (2006) *Astrophys. J., 643,* 1166.
Forbes D. A. and Bridges T. (2010) *Mon. Not. R. Astron. Soc., 404,* 1203.
Forbes J. et al. (2012) *Astrophys. J., 754,* 48.
Förster Schreiber N. M. et al. (2009) *Astrophys. J., 706,* 1364.
Foster J. B. et al. (2011) *Astrophys. J. Suppl., 197,* 25.
Fukui Y. et al. (2014) *Astrophys. J., 780,* 36.
Furukawa N. et al. (2009) *Astrophys. J. Lett., 696,* L115.
Galván-Madrid R. et al. (2009) *Astrophys. J., 706,* 1036.
Galván-Madrid R. et al. (2013) *Astrophys. J., 779,* 121.
Genzel R. et al. (2010a) *Mon. Not. R. Astron. Soc., 407,* 2091.
Genzel R. et al. (2010b) *Reviews of Modern Physics, 82,* 3121.
Geyer M. P. and Burkert A. (2001) *Mon. Not. R. Astron. Soc., 323,* 988.

Gieles M. (2009) *Mon. Not. R. Astron. Soc., 394,* 2113.

Gieles M. and Portegies Zwart S. F. (2011) *Mon. Not. R. Astron. Soc., 410,* L6.

Gieles M. et al. (2006a) *Astron. Astrophys., 446,* L9.

Gieles M. et al. (2006b) *Mon. Not. R. Astron. Soc., 371,* 793.

Gieles M. et al. (2010) *Mon. Not. R. Astron. Soc., 402,* 1750.

Gieles M. et al. (2011) *Mon. Not. R. Astron. Soc., 413,* 2509.

Gieles M. et al. (2012) *Mon. Not. R. Astron. Soc., 426,* L11.

Gilliland R. L. et al. (2000) *Astrophys. J. Lett., 545,* L47.

Ginsburg A. et al. (2012) *Astrophys. J. Lett., 758,* L29.

Girichidis P. et al. (2012) *Mon. Not. R. Astron. Soc., 420,* 613.

Goodwin S. P. (1997) *Mon. Not. R. Astron. Soc., 284,* 785.

Goodwin S. P. (2009) *Astrophys. Space Sci., 324,* 259.

Goodwin S. P. and Bastian N. (2006) *Mon. Not. R. Astron. Soc., 373,* 752.

Goudfrooij P. et al. (2011) *Astrophys. J., 737,* 3.

Gratton R. G. et al. (2012) *Astron. Astrophys. Rev., 20,* 50.

Griffen B. F. et al. (2010) *Mon. Not. R. Astron. Soc., 405,* 375.

Guesten R. et al. (1981) *Astron. Astrophys., 103,* 197.

Gutermuth R. A. et al. (2011) *Astrophys. J., 739,* 84.

Habibi M. et al. (2013) *Astron. Astrophys., 556,* A26.

Hansen C. E. et al. (2012) *Astrophys. J., 747,* 22.

Harris W. E. (1996) *Astron. J., 112,* 1487.

Hénault-Brunet V. et al. (2012a) *Astron. Astrophys., 545,* L1.

Hénault-Brunet V. et al. (2012b) *Astron. Astrophys., 546,* A73.

Hennemann M. et al. (2012) *Astron. Astrophys., 543,* L3.

Henshaw J. D. et al. (2013) *Mon. Not. R. Astron. Soc., 428,* 3425.

Heyer M. et al. (2009) *Astrophys. J., 699,* 1092.

Hill T. et al. (2005) *Mon. Not. R. Astron. Soc., 363,* 405.

Hills J. G. (1980) *Astrophys. J., 235,* 986.

Hodge P. (1987) *Publ. Astron. Soc. Pac., 99,* 724.

Holtzman J. A. et al. (1992) *Astron. J., 103,* 691.

Homeier N. L. and Alves J. (2005) *Astron. Astrophys., 430,* 481.

Hopkins A. M. and Beacom J. F. (2006) *Astrophys. J., 651,* 142.

Hopkins P. F. (2013) *Mon. Not. R. Astron. Soc., 428,* 1950.

Hopkins P. F. et al. (2013) *Mon. Not. R. Astron. Soc., 430,* 1901.

Hubber D. A. et al. (2013) *Mon. Not. R. Astron. Soc., 430,* 3261.

Hussmann B. et al. (2012) *Astron. Astrophys., 540,* A57.

Immer K. et al. (2012) *Astron. Astrophys., 537,* A121.

Jackson J. M. et al. (2013) *Publ. Astron. Soc. Australia, 30,* e057.

Jeffries R. D. (2007) *Mon. Not. R. Astron. Soc., 381,* 1169.

Johnston K. et al. (2013) In *Protostars and Planets VI,* Heidelberg, July 15–20, 2013. Poster #1S018, p. 18.

Jordán A. et al. (2007) *Astrophys. J. Suppl., 171,* 101.

Kalberla P. M. W. and Kerp J. (2009) *Annu. Rev. Astron. Astrophys., 47,* 27.

Kauffmann J. et al. (2013) *Astrophys. J. Lett., 765,* L35.

Kendrew S. et al. (2013) *Astrophys. J. Lett., 775,* L50.

Keto E. et al. (2005) *Astrophys. J., 635,* 1062.

Keto E. R. et al. (1987a) *Astrophys. J. Lett., 323,* L117.

Keto E. R. et al. (1987b) *Astrophys. J., 318,* 712.

Kim S. S. and Morris M. (2003) *Astrophys. J., 597,* 312.

King I. R. (1966) *Astron. J., 71,* 64.

Kirk H. et al. (2013) *Astrophys. J., 766,* 115.

Klaassen P. D. et al. (2013) *Astron. Astrophys., 556,* A107.

Klessen R. S. et al. (1998) *Astrophys. J. Lett., 501,* L205.

Kravtsov A. V. and Gnedin O. Y. (2005) *Astrophys. J., 623,* 650.

Kruijssen J. M. D. (2012) *Mon. Not. R. Astron. Soc., 426,* 3008.

Kruijssen J. M. D. (2013) *ArXiv e-prints,* arXiv:1304.4600.

Kruijssen J. M. D. and Cooper A. P. (2012) *Mon. Not. R. Astron. Soc., 420,* 340.

Kruijssen J. M. D. and Longmore S. N. (2013) *Mon. Not. R. Astron. Soc., 435,* 2598.

Kruijssen J. M. D. and Longmore S. N. (2014) *Mon. Not. R. Astron. Soc., 439,* 3239.

Kruijssen J. M. D. and Portegies Zwart S. F. (2009) *Astrophys. J. Lett., 698,* L158.

Kruijssen J. M. D. et al. (2011) *Mon. Not. R. Astron. Soc., 414,* 1339.

Kruijssen J. M. D. et al. (2012a) *Mon. Not. R. Astron. Soc., 421,* 1927.

Kruijssen J. M. D. et al. (2012b) *Mon. Not. R. Astron. Soc., 419,* 841.

Kruijssen J. M. D. et al. (2013) *Mon. Not. R. Astron. Soc., 440,* 3370–3391.

Krumholz M. and Burkert A. (2010) *Astrophys. J., 724,* 895.

Krumholz M. R. and McKee C. F. (2005) *Astrophys. J., 630,* 250.

Krumholz M. R. and Tan J. C. (2007) *Astrophys. J., 654,* 304.

Krumholz M. R. et al. (2012a) *Astrophys. J., 745,* 69.

Krumholz M. R. et al. (2012b) *Astrophys. J., 754,* 71.

Kudryavtseva N. et al. (2012) *Astrophys. J. Lett., 750,* L44.

Lada C. J. and Lada E. A. (2003) *Annu. Rev. Astron. Astrophys., 41,* 57.

Lada C. J. et al. (1984) *Astrophys. J., 285,* 141.

Lada C. J. et al. (2012) *Astrophys. J., 745,* 190.

Lamers H. J. G. L. M. and Gieles M. (2006) *Astron. Astrophys., 455,* L17.

Lamers H. J. G. L. M. et al. (2005) *Astron. Astrophys., 441,* 117.

Larsen S. S. (2004) *Astron. Astrophys., 416,* 537.

Larsen S. S. (2009) *Astron. Astrophys., 494,* 539.

Larsen S. S. et al. (2012) *Astron. Astrophys., 544,* L14.

Lee E. J. et al. (2012) *Astrophys. J., 752,* 146.

Lee J.-W. et al. (2009) *Nature, 462,* 480.

Liermann A. et al. (2010) *Astron. Astrophys., 524,* A82.

Liermann A. et al. (2012) *Astron. Astrophys., 540,* A14.

Lis D. C. and Menten K. M. (1998) *Astrophys. J., 507,* 794.

Lis D. C. et al. (1994) *Astrophys. J. Lett., 423,* L39.

Liu H. B. et al. (2012) *Astrophys. J., 745,* 61.

Longmore S. N. et al. (2012) *Astrophys. J., 746,* 117.

Longmore S. N. et al. (2013a) *Mon. Not. R. Astron. Soc., 433,* L15.

Longmore S. N. et al. (2013b) *Mon. Not. R. Astron. Soc., 429,* 987.

Lu J. R. et al. (2013) *Astrophys. J., 764,* 155.

Lumsden S. L. et al. (2013) *Astrophys. J. Suppl., 208,* 11.

Mackey A. D. and Broby Nielsen P. (2007) *Mon. Not. R. Astron. Soc., 379,* 151.

Mackey A. D. et al. (2013) *Astrophys. J., 762,* 65.

Mac Low M.-M. and Klessen R. S. (2004) *Rev. Mod. Phys., 76,* 125.

Madau P. et al. (1996) *Mon. Not. R. Astron. Soc., 283,* 1388.

Madau P. et al. (1998) *Astrophys. J., 498,* 106.

Martins F. et al. (2008) *Astron. Astrophys., 478,* 219.

Maschberger T. et al. (2010) *Mon. Not. R. Astron. Soc., 404,* 1061.

Massey P. and Hunter D. A. (1998) *Astrophys. J., 493,* 180.

McLaughlin D. E. (1999) *Astron. J., 117,* 2398.

McMillan S. L. W. et al. (2007) *Astrophys. J. Lett., 655,* L45.

Meibom S. et al. (2013) *Nature, 499,* 55.

Melena N. W. et al. (2008) *Astron. J., 135,* 878.

Mengel S. and Tacconi-Garman L. E. (2007) *Astron. Astrophys., 466,* 151.

Menten K. M. et al. (2007) *Astron. Astrophys., 474,* 515.

Miller B. W. et al. (1997) *Astron. J., 114,* 2381.

Miller G. E. and Scalo J. M. (1978) *Publ. Astron. Soc. Pac., 90,* 506.

Mills E. A. C. and Morris M. R. (2013) *Astrophys. J., 772,* 105.

Moeckel N. and Bate M. R. (2010) *Mon. Not. R. Astron. Soc., 404,* 721.

Molinari S. et al. (2010) *Astron. Astrophys., 518,* L100.

Molinari S. et al. (2011) *Astrophys. J. Lett., 735,* L33.

Morris M. and Serabyn E. (1996) *Annu. Rev. Astron. Astrophys., 34,* 645.

Moster B. P. et al. (2013) *Mon. Not. R. Astron. Soc., 428,* 3121.

Motte F. et al. (2003) *Astrophys. J., 582,* 277.

Muratov A. L. and Gnedin O. Y. (2010) *Astrophys. J., 718,* 1266.

Murray N. et al. (2010) *Astrophys. J., 709,* 191.

Myers P. C. (2009) *Astrophys. J., 700,* 1609.

Nagy Z. et al. (2012) *Astron. Astrophys., 542,* A6.

Najarro F. et al. (2009) *Astrophys. J., 691,* 1816.

Nakamura F. and Li Z.-Y. (2007) *Astrophys. J., 662,* 395.

Nascimbeni V. et al. (2012) *Astron. Astrophys., 541,* A144.

Negueruela I. et al. (2010) *Astron. Astrophys., 516,* A78.

Nguyen Luong Q. et al. (2011a) *Astron. Astrophys., 535,* A76.

Nguyen Luong Q. et al. (2011b) *Astron. Astrophys., 529,* A41.

Nguyen-Lu'o'ng Q. et al. (2013) *Astrophys. J., 775,* 88.

Nordström B. et al. (2004) *Astron. Astrophys., 418,* 989.

Offner S. S. R. et al. (2009a) *Astrophys. J. Lett., 704,* L124.

Offner S. S. R. et al. (2009b) *Astrophys. J., 703,* 131.

Ohama A. et al. (2010) *Astrophys. J., 709,* 975.

Olczak C. et al. (2010) *Astron. Astrophys., 509,* A63.

Ossenkopf V. and Henning T. (1994) *Astron. Astrophys., 291,* 943.

Padoan P. and Nordlund A. (2002) *Astrophys. J., 576,* 870.

Padoan P. and Nordlund A. (2011) *Astrophys. J., 730,* 40.

Palla F. and Stahler S. W. (1999) *Astrophys. J., 525,* 772.

Parker R. J. and Dale J. E. (2013) *Mon. Not. R. Astron. Soc..*

Parker R. J. and Meyer M. R. (2012) *Mon. Not. R. Astron. Soc., 427,* 637.

Parker R. J. and Quanz S. P. (2012) *Mon. Not. R. Astron. Soc., 419,* 2448.

Parker R. J. et al. (2014) *Mon. Not. R. Astron. Soc., 438,* 620.

Peebles P. J. E. and Dicke R. H. (1968) *Astrophys. J., 154,* 891.
Pelupessy F. I. and Portegies Zwart S. (2012) *Mon. Not. R. Astron. Soc., 420,* 1503.
Peng E. W. et al. (2008) *Astrophys. J., 681,* 197.
Peretto N. and Fuller G. A. (2009) *Astron. Astrophys., 505,* 405.
Peretto N. et al. (2013) *Astron. Astrophys., 555,* A112.
Peters T. et al. (2010) *Astrophys. J., 725,* 134.
Pfalzner S. and Kaczmarek T. (2013) *Astron. Astrophys., 555,* A135.
Pierce-Price D. et al. (2000) *Astrophys. J. Lett., 545,* L121.
Piskunov A. E. et al. (2006) *Astron. Astrophys., 445,* 545.
Plume R. et al. (1997) *Astrophys. J., 476,* 730.
Portegies Zwart S. et al. (2009) *New Astron., 14,* 369.
Portegies Zwart S. F. et al. (2001) *Astrophys. J. Lett., 546,* L101.
Portegies Zwart S. F. et al. (2002) *Astrophys. J., 565,* 265.
Portegies Zwart S. F. et al. (2004) *Nature, 428,* 724.
Portegies Zwart S. F. et al. (2010) *Annu. Rev. Astron. Astrophys., 48,* 431.
Preibisch T. et al. (2011) *Astron. Astrophys., 530,* A34.
Purcell C. R. et al. (2006) *Mon. Not. R. Astron. Soc., 367,* 553.
Purcell C. R. et al. (2012) *Mon. Not. R. Astron. Soc., 426,* 1972.
Rathborne J. M. et al. (2006) *Astrophys. J., 641,* 389.
Rathborne J. M. et al. (2014) *ArXiv e-prints,* arXiv:1403.0996.
Rochau B. et al. (2010) *Astrophys. J. Lett., 716,* L90.
Rodríguez L. F. and Zapata L. A. (2013) *Astrophys. J. Lett., 767,* L13.
Rolffs R. et al. (2011) *Astron. Astrophys., 527,* A68.
Román-Zúñiga C. G. et al. (2008) *Astrophys. J., 672,* 861.
Rosolowsky E. (2005) *Publ. Astron. Soc. Pac., 117,* 1403.
Rosolowsky E. (2007) *Astrophys. J., 654,* 240.
Rosotti G. P. et al. (2014) *Mon. Not. R. Astron. Soc., 441,* 2094–2110.
Sabbi E. et al. (2012) *Astrophys. J. Lett., 754,* L37.
Sales L. V. et al. (2007) *Mon. Not. R. Astron. Soc., 379,* 1464.
Scally A. and Clarke C. (2001) *Mon. Not. R. Astron. Soc., 325,* 449.
Schuller F. et al. (2009) *Astron. Astrophys., 504,* 415.
Schweizer F. (1982) *Astrophys. J., 252,* 455.
Schweizer F. et al. (1996) *Astron. J., 112,* 1839.
Seale J. P. et al. (2012) *Astrophys. J., 751,* 42.

Shapiro K. L. et al. (2010) *Mon. Not. R. Astron. Soc., 403,* L36.
Shetty R. et al. (2012) *Mon. Not. R. Astron. Soc., 425,* 720.
Shields J. C. and Ferland G. J. (1994) *Astrophys. J., 430,* 236.
Sievers A. W. et al. (1991) *Astron. Astrophys., 251,* 231.
Silva-Villa E. and Larsen S. S. (2011) *Astron. Astrophys., 529,* A25.
Simon R. et al. (2006) *Astrophys. J., 639,* 227.
Smith R. J. et al. (2009) *Mon. Not. R. Astron. Soc., 400,* 1775.
Smith R. J. et al. (2012) *Astrophys. J., 750,* 64.
Smith R. J. et al. (2013) *Astrophys. J., 771,* 24.
Spitzer L. (1987) *Dynamical Evolution of Globular Clusters.* Princeton Univ., Princeton. 191 pp.
Spitzer L. Jr. (1958) *Astrophys. J., 127,* 17.
Sridharan T. K. et al. (2002) *Astrophys. J., 566,* 931.
Stolte A. et al. (2004) *Astron. J., 128,* 765.
Stolte A. et al. (2008) *Astrophys. J., 675,* 1278.
Stolte A. et al. (2010) *Astrophys. J., 718,* 810.
Strader J. et al. (2011) *Astrophys. J. Suppl., 197,* 33.
Tackenberg J. et al. (2012) *Astron. Astrophys., 540,* A113.
Tasker E. J. and Tan J. C. (2009) *Astrophys. J., 700,* 358.
Thompson T. A. (2013) *Mon. Not. R. Astron. Soc., 431,* 63.
Tilley D. A. and Pudritz R. E. (2007) *Mon. Not. R. Astron. Soc., 382,* 73.
Tonini C. (2013) *Astrophys. J., 762,* 39.
Tremaine S. D. et al. (1975) *Astrophys. J., 196,* 407.
Urquhart J. S. et al. (2013) *Mon. Not. R. Astron. Soc., 435,* 400.
Walsh A. J. et al. (2011) *Mon. Not. R. Astron. Soc., 416,* 1764.
Wang P. et al. (2010) *Astrophys. J., 709,* 27.
Wei L. H. et al. (2012) *Astrophys. J., 750,* 136.
Welch W. J. et al. (1987) *Science, 238,* 1550.
Westerhout G. (1958) *Bull. Astron. Inst. Netherlands, 14,* 215.
Whitmore B. C. et al. (1999) *Astron. J., 118,* 1551.
Wielen R. (1971) *Astron. Astrophys., 13,* 309.
Yusef-Zadeh F. et al. (2013) *Astrophys. J. Lett., 764,* L19.
Zhang B. et al. (2013) *Astrophys. J., 775,* 79.
Zhang Q. and Fall S. M. (1999) *Astrophys. J. Lett., 527,* L81.

Part II:
Disk Formation and Evolution

Dutrey A., Semenov D., Chapillon E., Gorti U., Guilloteau S., Hersant F., Hogerheijde M., Hughes M., Meeus G., Nomura H., Piétu V., Qi C., and Wakelam V. (2014) Physical and chemical structure of planet-forming disks probed by millimeter observations and modeling. In *Protostars and Planets VI* (H. Beuther et al., eds.), pp. 317–338. Univ. of Arizona, Tucson, DOI: 10.2458/azu_uapress_9780816531240-ch014.

Physical and Chemical Structure of Planet-Forming Disks Probed by Millimeter Observations and Modeling

Anne Dutrey
University of Bordeaux, France

Dmitry Semenov
Max-Planck-Institut für Astronomie, Königstuhl

Edwige Chapillon
Institute of Astronomy and Astrophysics, Academia Sinica

Uma Gorti
SETI Institute

Stéphane Guilloteau and Franck Hersant
University of Bordeaux, France

Michiel Hogerheijde
Leiden University

Meredith Hughes
Wesleyan University

Gwendolyn Meeus
Universidad Autónoma de Madrid

Hideko Nomura
Kyoto University

Vincent Piétu
Domaine Universitaire

Chunhua Qi
Harvard-Smithsonian Center for Astrophysics

Valentine Wakelam
University of Bordeaux, France

Protoplanetary disks composed of dust and gas are ubiquitous around young stars and are commonly recognized as nurseries of planetary systems. Their lifetime, appearance, and structure are determined by an interplay between stellar radiation, gravity, thermal pressure, magnetic field, gas viscosity, turbulence, and rotation. Molecules and dust serve as major heating and cooling agents in disks. Dust grains dominate the disk opacities, reprocess most of the stellar radiation, and shield molecules from ionizing ultraviolet (UV)/X-ray photons. Disks also dynamically evolve by building up planetary systems, which drastically change their gas and dust density structures. Over the past decade, significant progress has been achieved in our understanding of disk chemical composition thanks to the upgrade or advent of new millimeter/infrared (IR) facilities [Submillimeter Array (SMA), Plateau de Bure Interferometer (PdBI), Combined Array for Research in Millimeter-wave Astronomy (CARMA), Herschel, Expanded Very Large Array (e-VLA), Atacama Large Millimeter/submillimeter Array (ALMA)]. Some major breakthroughs in our comprehension of the disk physics and chemistry have been done since *Protostars and Planets V* (*Reipurth et al.,* 2007). This review will present and discuss the impact of such improvements on our understanding of the disk physical structure and chemical composition.

317

1. INTRODUCTION

The evolution of the gas and dust in protoplanetary disks is a key element that regulates the efficiency, diversity, and timescale of planet formation. The situation is complicated by the fact that the dust and gas physically and chemically interact atop the disk structure, which evolves with time. Initially, small dust grains are dynamically well-coupled to the gas and are later assembled by grain growth in bigger centimeter-sized particles that settle toward the disk mid-plane (see the chapter by Testi et al. in this volume). After large grains become dynamically decoupled from the gas, they become subject to head wind from the gas orbiting at slightly lower velocities and spiral rapidly inward or experience mutual destructive collisions. Collisionally generated small grains are either swept out by larger grains or stirred by turbulence into the disk atmosphere. As a result, the dust-to-gas ratio and average dust sizes vary throughout the disk. All these processes affect the disk thermal and density structures and thus disk chemical composition.

In the dense disk mid-plane, thermal equilibrium between gas and dust is achieved, with dust transferring heat to gas by rapid gas-grain collisions. The disk mid-plane is well-shielded from high-energy stellar radiation and thus is "dark," being heated indirectly via infrared (IR) emission from the upper layers, and has a low-ionization degree and low-turbulence velocities. The ratio of ions to neutral molecules determines the level of turbulence and regulates the redistribution of the angular momentum. The outer disk mid-plane is so cold that many gaseous molecules are frozen out onto dust grains, leading to the formation of icy mantles that are steadily processed by cosmic rays.

Above the mid-plane a warmer, less-dense region is located, where gas-grain collisional coupling can no longer be efficient, and dust and gas temperatures start to depart from each other, with the gas temperature usually being higher. The intermediate disk layer is only partly shielded from the ionizing radiation by the dust and thus is more ionized and dynamically active than the mid-plane. Rich gas-phase and gas-grain chemistries enable synthesis of many molecules in this so-called molecular layer.

Finally, in the heavily irradiated, hot and dilute atmosphere, only simple atoms, ions, photostable radicals, and polycyclic aromatic hydrocarbons (PAHs) are able to survive. The global chemical evolution is dominated by a limited set of gas-phase reactions in this disk region.

Observations of protoplanetary disks of various ages and sizes surrounding Sun-like and intermediate-mass stars help to resolve some of the related ambiguities. Detailed studies of protoplanetary disks remain an observationally challenging task, however, as disks are compact objects and have relatively low masses. At visual and IR wavelengths, disks are typically opaque, and (sub)millimeter imaging with single-dish telescopes and antenna arrays is used to peer through their structure. Since observations of the most dominant species in disks, H_2, are impossible [except for the hot upper layers via their weak quadrupole IR transitions or the inner disk via the fluorescent far-ultraviolet (FUV) lines], other molecules are employed to trace disk kinematics, temperature, density, and chemical structure. The poorly known properties, such as (sub)millimeter dust emissivities and dust-to-gas ratio, make it hard to derive a total disk mass from dust emission alone. Apart from a handful of simple molecules, like CO, HCO^+, H_2CO, CS, CN, HCN, and HNC, the molecular content of protoplanetary disks remains largely unknown.

In the last 10 years, upgraded and new millimeter or submillimeter facilities [Institut de Radioastronomie Millimétrique (IRAM), CARMA, SMA, Herschel, and ALMA] have permitted the detection of a few other molecular species (DCN, N_2H^+, H_2O, HC_3N, C_3H_2, and HD) at better spatial and frequency resolution. In the meantime, disk models have benefited from improvements in astrochemistry databases [such as the Kinetic Database for Astrochemistry (KIDA) and the UMIST Database for Astrochemistry 2012 (UDFA'12)], as well as the development of coupled thermochemical disk physical models, line radiative transfer codes, and better analysis tools. The spatial distribution of molecular abundances in disks is still poorly determined, hampering a detailed comparison with existing chemical models. Due to the complexity of the molecular line excitation, unambiguous interpretation of the observational results necessitates advanced modeling of the disk physical structure and evolution, chemical history, and radiative transfer.

Figure 1 illustrates the state of the art prior to the Protostars and Planets VI conference held in Heidelberg in June 2013. The disk structure (density and temperature) has been calculated for a T Tauri star with the mass of $1 M_\odot$. With the exception of the very inner disk (R ≤ 10 AU), where dissipation of accretion energy can be a source of mechanical heat, disks are mostly heated by stellar radiation, including UV and X-rays, the interstellar UV field, and cosmic rays. Both the gas and dust have vertical and radial temperature gradients, with the upper disk layers being superheated by the central star(s). As grains grow and rapidly settle toward the disk mid-plane (see the chapter by Testi et al. in this volume), the vertical temperature gradient changes and the mid-plane cools off further. This may have an impact on the location and size of the freeze-out zones of various molecules, where they mainly remain bound to the dust grains.

In summary, there are three different chemical zones in disks: (1) the disk upper layer, similar to a dense photon-dominated region (PDR), consisting of simple ions, neutral species, and small dust grains; (2) the warm intermediate molecular layer, with many molecules (including some complex ones) in the gas phase and rapid gas-dust interactions; and (3) the cold mid-plane, devoid of many gaseous molecules but icy-rich at radii ≥20–200 AU, similar to dense prestellar cores.

This review will present the recent improvements on our understanding of disk physical structure and chemical composition and their evolution with time, both from observational and theoretical perspectives. Other recent

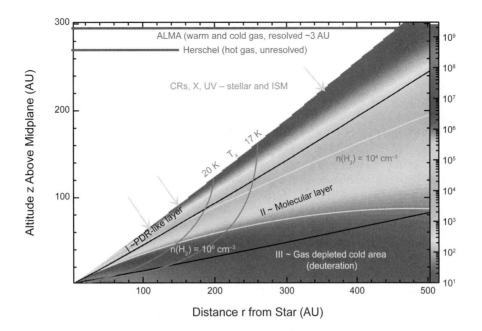

Fig. 1. See Plate 12 for color version. Molecular disk: state of the art after *Protostars and Planets V* (*Reipurth et al.,* 2007). The model is done for a disk orbiting a young star of 1 $M_\odot$. The volume density as a function of radius r and altitude z above the mid plane is given in color. The location of the isotherms at 17 and 20 K are also shown. I, II, and III correspond to the location of the three important layers of gas: the PDR-like atmosphere irradiated by the star, the UV interstellar field and the cosmic rays (I), the molecular layer typically located between one and three scale heights (II), and the mid-plane where only molecular hydrogen and deuterated isotopologs of simple molecules are abundant in the gas phase (III). The line at the very top (blue in the color version) corresponds to the area sampled by ALMA (with the best resolution for a source at 150 pc); the line immediately underneath that (red in color version) gives the area sampled by Herschel observations (unresolved).

comprehensive reviews on the subject of disk physics and chemistry are those by *Bergin* (2011), *Semenov* (2010), and *Henning and Semenov* (2013).

2. DATA ANALYSIS AND BIAS

The physical structure of protoplanetary disks is so unique because it has both strong density and temperature gradients on relatively short spatial scales. For example, at a radius of 100 AU, the mid-plane temperature is expected to be around 7–12 K, while at the 1–3 scale heights (H) above (at about 15–50 AU), the temperature is warm enough to facilitate the presence of a molecular layer. In the meantime, the gas density drops by about one to four orders of magnitude. As a consequence, the chemical conditions rapidly change along the line of sight, as well as the excitation conditions, leading in some cases to non-long term evolution (LTE) effects. Last but not least, a proper analysis of any line detection in disks requires an adequate handling of the line formation process, including possible non-LTE excitation, but also gas kinematics, in particular, the disk Keplerian shear (*Guilloteau et al.,* 2006; *Dutrey et al.,* 2007a).

Line analysis falls into two main categories: inversion methods, which attempt to retrieve the physical parameters (e.g., the excitation temperature, the molecular surface density distributions, etc.) and their confidence intervals, and forward modeling, which evaluate whether a given model can represent the observations. In the radiative transfer codes used to analyze observations or predict millimeter/submillimeter line emissions, the physical parameters (density, gas temperature, scale height, or turbulence) are usually specified as explained below:

Temperature. Disk temperature is treated in three main ways: (1) the gas temperature is assumed to be isothermal in the vertical direction and to follow a power law in the radial direction, $T_k(r) \propto r^{-q}$ (*Dutrey et al.,* 1994); (2) a vertical gradient is also included (*Dartois et al.,* 2003; *Qi et al.,* 2011; *Rosenfeld et al.,* 2012), using simple parameterization of dust disk models possibly with the use of a simplified approach to the radiative transfer problem (*Chiang and Goldreich,* 1997; *D'Alessio et al.,* 1998); or (3) the temperature is directly calculated by the dust radiative equilibrium modeling (*Pinte et al.,* 2008; *Isella et al.,* 2009). This last category of models has a number of hidden and often not well-constrained parameters. The surface density distribution must be specified, as well as the dust properties (minimum and maximum radii, size distribution index, and composition) to calculate the dust absorption coefficients. These models can accommodate grain growth and even dust settling (e.g., *Dullemond et al.,* 2007; *Hasegawa and Pudritz,* 2010b; *Williams and*

Cieza, 2011). The dust disk properties are usually adjusted to fit the spectral energy distribution (SED), sometimes simultaneously with resolved dust maps in the millimeter/submillimeter (*Isella et al.*, 2009; *Gräfe et al.*, 2013). A last refinement is to also evaluate the gas heating and cooling and to account for the differences between the gas and dust temperature as a function of disk location (this will be discussed in section 4.1).

Density. Two types of surface density must be considered: the total (mass) surface density and individual molecule distributions. Some parametric models allow adjustment of the molecular distributions, either as power-law surface densities (*Piétu et al.*, 2007), or more sophisticated variations, such as piece-wise power-law-integrated abundance profiles (*Qi et al.*, 2008; *Guilloteau et al.*, 2012) atop a prescribed H_2 surface density. In forward modeling, models computing the thermal balance (dust and/or gas) in general assume a power-law (mass) surface density profile. Exponentially tapered profiles have also been invoked to explain the different radii observed in CO isotopologs (*Hughes et al.* 2009), but are not widely used in forward modeling.

Scale height. The gas is usually assumed to be in hydrostatic equilibrium. The scale height is prescribed as a power law or self-consistently calculated from the temperature profile, but the feedback of the temperature profile on the vertical density profile is neglected in most cases. The scale height can be expressed as (e.g., *Dartois et al.*, 2003; *Dutrey et al.*, 2007b)

$$H(r) = \sqrt{\frac{2kr^3T(r)}{GM_*m_0}} = \sqrt{\frac{2k}{m_0}}\frac{r}{V_K(r)}\sqrt{T(r)} = \sqrt{2}c_s/\Omega \quad (1)$$

where $T(r)$ is the kinetic temperature, M_* is the stellar mass, m_0 is the mean molecular weight of the gas, $V_K(r)$ is the Keplerian velocity, c_s is the sound speed, and Ω is the angular velocity. Self-consistent calculation of the hydrostatic equilibrium with vertical temperature gradients is, however, taken into account in some models (e.g., *D'Alessio et al.*, 1998; *Gorti and Hollenbach*, 2004).

Turbulence. Turbulence is usually mimicked by setting the local line width $\Delta V = \sqrt{v_{th}^2 + v_{turb}^2}$ where v_{th} is the thermal broadening and v_{turb} the turbulent term (*Dartois et al.*, 2003; *Piétu et al.*, 2007; *Hughes et al.*, 2011).

Excitation conditions. The first rotational levels of most of the simple molecules observed so far with current arrays such as CO, HCO^+, CN, and HCN should be thermalized in the molecular layer, given the expected H_2 densities. This strongly simplifies the analysis of the data. This is, however, less true for higher levels (above J=3), as noted by *Pavlyuchenkov et al.* (2007), and a correct treatment of the non-LTE excitation conditions is needed. These non-LTE effects will become more common with submillimeter ALMA observations sensitive to weaker lines and to more warmer, inner disk regions with strong dust continuum background emission.

Thermochemical models. The most sophisticated approach is to couple the calculation of the thermal and density structures to the derivation of the chemistry to make molecular-line predictions (*Woitke et al.*, 2009b; *Kamp et al.*, 2010; *Akimkin et al.*, 2013). However, the computing requirements limit this kind of approach to forward modeling only. Despite the interest of such models for understanding the role of various processes, it is important to remember that they also have their own limitations, such as the use of equilibrium chemistry, the choice of dust properties, or an underlying surface density profile. We review these models in section 4.1.

2.1. Unresolved Data

For most unresolved data, the SED at near-, mid-, and far-IR and (sub)millimeter wavelengths is commonly used to derive the disk structure from radii <1 AU all the way to the outer radius (*Dullemond et al.*, 2007), as the SED traces the distribution of the dust in the disk. The dust is supported by the gas, so information on the gas radial distribution and scale heights can also be derived. However, this assumes that the dust is dynamically coupled to the gas, which is only true for small particles (≤ 100 μm). Complicating details are the presence of multiple dust populations, vertical settling and radial drift of dust constituents, generally poorly constrained dust opacities (see also the chapter by Testi et al. in this volume), and the gas and dust thermal decoupling.

Moreover, the IR part of the SED is in general insensitive to the total dust content, which is only constrained by the (sub)millimeter range. Neither of them is sensitive to the disk outer radius. Spatially integrated, spectrally resolved line profiles can be the tools used to sample disk properties, because the line formation in a Keplerian disk links velocities to radial distances. This can be used to obtain disk radii (*Guilloteau et al.*, 2013) or as evidence for central holes (*Dutrey et al.*, 2008) (see also section 5).

2.2. Resolved Interferometric Data

Resolved interferometric molecular maps on nearby protoplanetary disks are routinely obtained from most millimeter/submillimeter arrays such as the SMA, CARMA, and IRAM PdBI.

The *uv* coverage of these facilities is still a limiting factor and data are generally compared to disk models by χ^2 minimizations performed in the Fourier plane in order to avoid nonlinear effects due to the deconvolution. This later step should no longer be necessary for most ALMA configurations.

Other important limitations are due to the assumed density and temperature laws. Current arrays do not have enough sensitivity and angular resolution to allow a fine determination of the radial and vertical structure. Depending on the way the temperature is calculated (from the dust) or determined (from thermalized CO lines such the 1–0 or the 2–1 transitions), the biases are different. In the first case, the gas temperature is directly dependent on the poorly known dust properties, and there is no direct gas temperature measurement. In the second case, an excitation temperature is determined, but its interpretation as the kinetic temperature depends on the robustness of the LTE hypothesis and the region sampled by the measurement, as it depends on the chemical behavior of the observed molecule.

3. MOLECULAR OBSERVATIONS

We discuss in this section how molecular observations obtained at millimeter/submillimeter wavelengths can constrain the disk structures. Some far-IR results obtained mostly with Herschel are also discussed, but the gas properties of the warm surface and inner disk derived from IR observations (Spitzer, Herschel) are discussed in detail in the chapter by Pontoppidan et al. in this volume.

3.1. Detected Species

Since H_2 cannot be used as a tracer of the bulk of gas mass, the study of the more abundant molecules after H_2 is mandatory to improve our knowledge of the gas disk density and mass distribution. After the detection of a few simple molecules in the T Tauri disks surrounding the 0.5 $M_\odot$ DM Tau and the binary system GG Tau (1.2 $M_\odot$) by *Dutrey et al.* (1997), millimeter/submmilliter facilities have observed several protoplanetary disks around young stars with mass ranging between ~0.3 and 2.5 $M_\odot$ (e.g., *Kastner et al.,* 1997; *Thi et al.,* 2004). The main result of millimeter/submillimeter molecular studies is that detections are limited to the most abundant molecules found in cold molecular clouds: CO (with ^{13}CO and $C^{18}O$), HCO^+ (with $H^{13}CO^+$ and DCO^+), CS, HCN (with HNC and DCN), CN, H_2CO, N_2H^+, C_2H, and, very recently, HC_3N (*Chapillon et al.,* 2012b), followed by the detection of cyclic C_3H_2 in the disk surrounding HD 163296 using ALMA (see Fig. 3) (*Qi et al.,* 2013a). H_2O has also been detected by the Herschel satellite in TW Hya and DM Tau (marginal detection) (*Bergin et al.,* 2010; *Hogerheijde et al.,* 2011), while the main reservoir of elemental deuterium, HD (Fig. 2), has been detected in TW Hya by *Bergin et al.* (2013). The millimeter/submillimeter molecular detections are summarized in Table 1.

3.2. Outer Disk Structure (R > 30 AU)

3.2.1. Carbon monoxide as a tracer of the disk structure. Resolved spectroimaging observations of CO isotopologs have so far been the most powerful method for constraining the geometry (outer radius, orientation, and inclination) and velocity pattern of protoplanetary disks (*Koerner et al.,* 1993; *Dutrey et al.,* 1994; *Guilloteau and Dutrey,* 1998).

As the first lowest rotational levels of CO are thermalized, the observations of the J=1–0 and J=2–1 transitions permit a direct measurement of the gas temperature and surface densities. Furthermore, because of their different

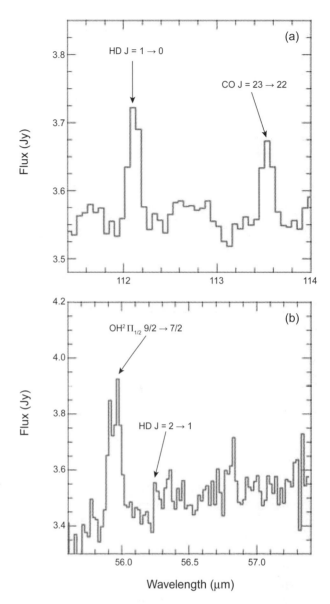

Fig. 2. Hydrogen deuteride around TW Hydra observed with Herschel. (a) The HD J=1–0 line at 112 μm is detected at 5σ level; (b) the HD J=2–1 line has only an upper limit. From *Bergin et al.* (2013).

opacities, the ^{12}CO, ^{13}CO J=1–0 and J=2–1 lines sample different disk layers, probing the temperature gradient as a function of height (see *Dutrey et al.,* 2007b, for a detailed discussion). With the optically thicker lines (e.g., ^{12}CO J=2–1), the temperature can be measured in a significant disk fraction, while the estimate of the molecular surface density is only possible in the outer optically thin region (*Dartois et al.,* 2003; *Piétu et al.,* 2007).

3.2.2. Temperature structure. Closer to the mid-plane, several studies on various molecular lines such as CCH 1–0 and 2–1 (*Henning et al.,* 2010), HCN 1–0 and CN 2–1 (*Chapillon et al.,* 2012a), and CS 3–2 (*Guilloteau et al.,* 2012) suggest that the gas is cold, with apparent temperatures on the order of ~10–15 K at 100 AU for T Tauri

TABLE 1. Detected species in T Tauri (cold) and Herbig Ae (warm)
outer disks using millimeter/submillimeter facilities.

Species	T Tauri Cold Disk T < 15 K at R > 50 AU	Herbig Ae Warm Disk T < 15 K at R > 200 AU	Telescope
CO, ^{13}CO, C^{18}O	Many	Many	Many
CN	Many	Many	Many
HCN, HNC	Several, DM Tau		IRAM, SMA
DCN	TW Hydra	—	SMA, ALMA
CS	Several	Several	IRAM, SMA
C$_2$H	Several	Several	IRAM, SMA
H$_2$CO	Several	A few	IRAM, SMA
HCO$^+$	Several	Several	IRAM, SMA
DCO$^+$	TW Hya, DM Tau	HD 163296	IRAM, SMA
N$_2$H$^+$	3–4	HD 163296	IRAM, SMA, ALMA
HC$_3$N	3–4		IRAM
c-C$_3$H$_2$	—	HD 163296	ALMA
H$_2$O	TW Hya, DM Tau		Herschel
HD	TW Hya		Herschel

disks. As this is below the CO freeze-out temperature (17–9 K), CO and most other molecules should be severely depleted from the gas phase (apart from H$_2$, H$_3^+$, and their deuterated isotopologs). Disks around Herbig Ae (HAe) stars appear warmer (*Piétu et al.*, 2007). To explain the low-molecular temperatures observed in the T Tauri disks, several possibilities can be invoked. With the physical structure predicted by standard disk models, the observed molecular transitions of CN, CCH, HCN, and CS should be thermalized in a large area of these disks. Hence, a subthermal excitation for these is unlikely, especially as the derived low temperatures are similar to those obtained from (thermalized) CO 1–0 and 2–1 transitions (*Piétu et al.*, 2007). *Chapillon et al.* (2012a) have also investigated the possibility of having a lower gas-to-dust ratio (by about a factor of ~6). In that case subthermal excitation becomes possible, but the predicted column densities for HCN and CN are low compared to the observed ones. Turbulence may play a role by transporting gaseous molecules into the cold mid-plane regions on a shorter timescale than the freeze-out timescale (*Semenov et al.*, 2006; *Aikawa*, 2007), but may be insufficient (*Hersant et al.*, 2009; *Semenov and Wiebe*, 2011) (see also section 4.2.2.1). Finally, another possibility would be that "cold" chemistry for molecules such as CN or CCH, and in particular photodesorption rates, are poorly known (*Öberg et al.*, 2009a; *Hersant et al.*, 2009; *Walsh et al.*, 2012). Tracing the very cold mid-plane is challenging since most molecules should be frozen onto grains. The best candidates to trace this zone are the H$_2$D$^+$ or D$_2$H$^+$ ions, because of their smaller molecular masses, but they have yet to be detected in protoplanetary disks (see section 5.3 and the chapter by Ceccarelli in this volume).

Higher in the disk, higher transitions of CO permit to constrain the vertical structure. Using CO 3–2 and 6–5 maps obtained with the SMA, *Qi et al.* (2004, 2006) have investigated the temperature and heating processes in the T Tauri

disk surrounding TW Hya. They found that the intensity of CO 6–5 emission can only be explained in the presence of additional heating by stellar UV and X ray photons.

While a significant fraction of T Tauri disks exhibit low-gas temperatures, several studies of HAe disks found that these disks are warmer, as predicted by the thermochemical disk models. *Piétu et al.* (2007) and *Chapillon et al.* (2012a) found from the CO, HCN, and CN studies with the PdBI that the molecular disk surrounding MWC 480 has a typical temperature of about 30 K at radius of 100 AU around the mid-plane. In the case of HD 163296, *Qi et al.* (2011) used a multi-transition and multi-isotopolog study of CO (from SMA and CARMA arrays) to determine the location of the CO snowline, where the CO column density changes by a factor of 100. They found a radius of ~155 AU for this snowline. This result, consistent with the SMA observation of H$_2$CO (*Qi et al.*, 2013b), is confirmed by the ALMA observation of DCO$^+$ (*Mathews et al.*, 2013). More recently, *Qi et al.* (2013c) found that the CO snowline in TW Hya is likely located at a radius of ~30 AU, where the inner edge of the N$_2$H$^+$ ring is detected. This N$_2$H$^+$ ring is the result of an increase in the N$_2$H$^+$ column density in the region where CO abundances are low, because CO efficiently destroys N$_2$H$^+$ by the proton transfer reaction N$_2$H$^+$ + CO → N$_2$ + HCO$^+$.

3.2.3. Molecular density structure. The chemical behavior and the associated abundance variations of the molecules used to characterize the gas preclude an absolute determination of the gas mass. Recently, the far-IR fundamental rotational line of HD has been detected in the TW Hya disk with Herschel (*Bergin et al.*, 2013). The abundance distribution of HD closely follows that of H$_2$. Therefore, HD, which has a weak permanent dipole moment, can be used as a direct probe of the disk gas mass. The inferred mass of the TW Hya disk is ≥0.05 M$_\odot$, which is surprising for a 3–10-million-year (m.y.)-old system. Other detections of HD using the Stratospheric Observatory for Infrared Astronomy (SOFIA) could be a very powerful way to directly constrain the gas mass, assuming the thermal structure is reasonably known.

Whereas the absolute determination of the gas mass is not yet possible, the radial and vertical distributions of the molecular gas are better known. Several complementary studies of the gas and dust distributions, assuming either power law or exponential decay for the surface density distribution, have been recently performed (*Isella et al.*, 2009; *Hughes et al.*, 2009, 2011; *Guilloteau et al.*, 2011). Most of these studies are based on the analysis of dust maps (see also the chapter by Testi et al. in this volume), but these results are still limited in sensitivity and angular resolution.

The vertical location of the molecular layer was investigated in several studies, but in an indirect way. *Piétu et al.* (2007) found that the apparent scale heights of CO are larger than the expected hydrostatic value in DM Tau, LkCa 15, and MWC 480. *Guilloteau et al.* (2012) showed that the CS 3–2 and 5–4 PdBI maps are best explained if the CS layer is located about one scale height above the mid-plane, in agreement with model predictions. However, the first *direct* measurement of the location of the molecular layer has only been recently obtained with ALMA observations of HD 163296, where the CO emission clearly originates from a layer at ~15° above the disk plane (*Rosenfeld et al.*, 2013; *de Gregorio-Monsalvo et al.*, 2013), at a few hundred astronomical units.

3.2.4. Gas-grain coupling. Several observational studies indicate the importance of the grain surface chemistry for disks. Using the IRAM 30-m telescope, *Dutrey et al.* (2011) failed to detect SO and H_2S in DM Tau, GO Tau, LkCa 15, and MWC 480 disks. They compared the molecular column densities derived from the observations with chemical predictions made using Nautilus (*Hersant et al.*, 2009) and the density and temperature profiles taken from previous analysis (e.g., *Piétu et al.*, 2007). They reproduced the SO upper limits and CS column densities reasonably well, but failed to match the upper limits obtained on H_2S by at least 1 order of magnitude. This suggests that at the high densities and low temperatures encountered around disk mid-planes, H_2S remains locked onto the grain surfaces, where it also gets destroyed to form other species. Indeed, some recent experiments by *Garozzo et al.* (2010) have shown that H_2S on grains is easily destroyed by cosmic rays and leads to the formation of C_2S, SO_2, and OCS on grains. These studies also suggest that most of the sulfur may be in the form of a sulfur-rich residuum, which could be polymers of sulfur or amorphous aggregates of sulfur (*Wakelam et al.*, 2004). The associated grain surface reactions are not yet incorporated in chemical models.

The low apparent CO-to-dust ratio observed in all disks has in general been attributed to the freeze-out of CO onto the dust grains in the outer disk regions at $r \geq 200$ AU with $T \leq 20$ K. The importance of this mechanism is demonstrated by the observations of CO isotopologs in the HD 163296 disk by *Mathews et al.* (2013). However, this mechanism, although unavoidable, may not be sufficient to explain the low-CO content. Strong apparent CO depletion (factor of ~100) has

also been observed in warm disks, with temperatures >30 K, where thermal desorption should occur: the disks around the Herbig stars CQ Tau and MWC 758 (*Chapillon et al.*, 2008) and BP Tau (*Dutrey et al.*, 2003). *Chapillon et al.* (2008) suggested that CO may have been removed from the gas phase during the cold prestellar phase by adsorption on small grains. During the warmer protostellar phase, grain growth occurs, and CO may stay locked in the ice mantles of large grains that may remain sufficiently colder than small grains during the reheating phase. Also, from a complete modeling of CO isotopologs in TW Hya, *Favre et al.* (2013) concluded that depletion alone could not account for the low CO-to-H_2 ratio, the H_2 content being derived from detection of HD (*Bergin et al.*, 2013). They invoke CO conversion to carbon chains, or perhaps CO_2, that can remain locked in ice mantles at a higher temperature than CO.

3.2.5. Molecular complexity. Prior to ALMA, with the exception of HC_3N (*Chapillon et al.*, 2012b), the molecules that have been detected are the simplest, lighter molecules. *Qi et al.* (2013a) recently report the detection with ALMA of c-C_3H_2 in the warm disk surrounding the HAe star HD 163296 (see Fig. 3).

3.2.6. Perspectives. A major challenge of ALMA will be to refine our knowledge of the vertical and radial structure by studying the excitation conditions through multi-isotopolog, multi-transition studies of several molecular tracers such as CN, CS, HCO^+, or HCN at a high angular resolution. Particularly, the CO isotopologs will remain robust tracers of the disk structure. Defining the physical conditions in the mid-plane will likely remain difficult and should require a long integration time to detect species such as H_2D^+, DCO^+, or N_2D^+.

3.3. Inner Disk Structure (<30 AU)

Unlike the outer disk, the inner (<30 AU) disk has remained unresolved with (sub)millimeter interferometers (see, e.g., *Dullemond and Monnier*, 2010). In the near future, the longest baselines of ALMA will start to image these regions, but up to now all information about molecular gas has been obtained from *spatially* unresolved observations. However, *spectrally* resolved observations can be used to establish the region from which line emission originates if a Keplerian velocity curve is assumed. Infrared interferometry provides information on the innermost regions at radii of <1 AU (e.g., *Kraus et al.*, 2009), but has been mostly limited to continuum observations, with the exception of several solid-state features (e.g., *van Boekel et al.*, 2004).

Spectral energy distribution modeling also provides at first order a good description of the density and temperature structure of the dust throughout the disk. However, especially in the inner disk and also at several scale heights, gas temperatures may exceed dust temperatures because of high-energy radiation (*Kamp and Dullemond*, 2004). The (local) gas-to-dust ratio, dust-size distribution, and effects of dust settling are all factors in determining the relative dust and gas temperatures, and complicate the interpretation of

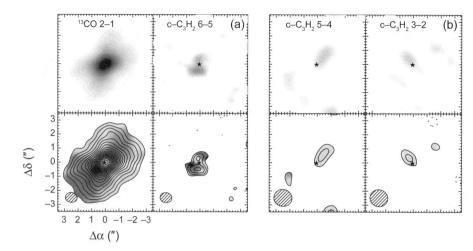

Fig. 3. See Plate 13 for color version. The integrated intensity maps and intensity-weighted mean velocity fields of ^{13}CO 2–1 and c–C_3H_2 6-5 lines (left panels), as well as c–C_3H_2 5–4 and 3–2 lines (right panels) toward HD 163296. The resolved velocity field of the c–C_3H_2 6-5 line agrees with the CO kinematics. In the c–C_3H_2 maps, the first contour marks 3σ followed by the 1σ contour increases. The rms varies between 6 and 9 mJy km s^{-1} per beam. Synthesized beams are presented in the lower left corners. The star symbol denotes the continuum (stellar) position. The axes are offsets from the pointing center in arcseconds. From *Qi et al.* (2013a).

molecular line observations of the inner disk. Of particular interest for the inner disk are constraints on any gaps present in this region (e.g., *Maaskant et al.*, 2013) and the shape of the gap walls (*McClure et al.*, 2013), which depend on the gap-opening mechanism and accretion (*Mulders et al.*, 2013). The disk gaps and inner holes provide a directly irradiated, warm surface visible in excited molecular lines, and the question is whether any gas remains in these regions (see chapters by Espaillat et al. and Pontoppidan et al. in this volume).

In recent years, the high-fidelity (HIFI) and Photodetector Array Camera and Spectrometer (PACS) instruments on the Herschel Space Observatory have probed the inner disks through spatially unresolved, but, in the case of HIFI, spectrally resolved, emission lines of gas species and several solid-state features from the dust. Several authors (*Meeus et al.*, 2012; *Riviere-Marichalar et al.*, 2012, 2013; *Fedele et al.*, 2012) present Herschel observations of H_2O, CO, [OI], OH, CH$^+$, and [CII] lines from T Tauri and HAe/Be disks. Atomic oxygen is firmly detected in most disks and correlates with the far-IR dust continuum, CO is observed in ~50% of disks and is stronger in flaring disks, while [CII] emission is often not confined to the disk and difficult to separate from the surrounding environment (*Fedele et al.*, 2012, 2013). H_2O and OH lines are only observed in a few Herbig stars (*Meeus et al.*, 2012; *Fedele et al.* 2012), while they are strong in T Tauri stars with outflows (*Podio et al.*, 2012). Furthermore, the emission of hot water located around 2–3 AU is observed in 24% of gas-rich T Tauri disks (*Riviere-Marichalar et al.*, 2012). Lastly, CH$^+$, tracing hot gas was discovered in two Herbig stars with high-UV flux: HD 97048 and HD 100546 (*Thi et al.*, 2011a; *Meeus et al.*, 2012).

Using thermochemical models with a disk structure derived from continuum observations, it was shown that these lines probe the inner disk as well as the upper disk layers directly illuminated, and heated, by UV radiation, where PAH heating can play an important role (e.g., *Woitke et al.*, 2010; *Bruderer et al.*, 2012). Herbig stars and the T Tauri stars have an important difference in this context: While in Herbig stars the UV radiation is stellar, in T Tauri stars it is mainly due to accretion shocks. Almost universally, these lines show that the gas in the upper disk regions is warmer than the dust (e.g., *Kamp and Dullemond*, 2004; *Bruderer et al.*, 2012), in accordance with models including photon heating (PDRs). Considering multiple transitions of CO in the HD 100546 disk, *Fedele et al.* (2013) found a steeper temperature gradient for the gas compared to the dust, providing further direct proof of thermal decoupling between gas and dust. *Aresu et al.* (2012) showed that the [OI] line flux increases with UV flux when $L_X < 10^{30}$ erg s^{-1}, and with increasing LX (when it is higher that the UV luminosity). *Thi et al.* (2011a) detected CH$^+$ in HD 100546, and concluded that this species is most abundant at the disk rim at 10–13 AU. In HD 163296, *Tilling et al.* (2012) studied the effect of dust settling, and found that in settled models the line fluxes of species formed deeper in the disk are increased. *Thi et al.* (2013) found that the line observations of the disk around 51 Oph can only be explained if it is compact (<10–15 AU), although an outer, cold disk devoid of gas may be extended up to 400 AU. For all these lines observed with Herschel, it is good to remember that, while they dominate the far-IR spectrum, they only trace a small fraction of the disk, and that the dominant disk mass reservoir near the mid-plane at these radii is much cooler and will require high-resolution millimeter (ALMA) observations.

Solid-state features present in the Herschel wavelength range provided information about the dust composition of the inner disk. Emission of the 69-μm forsterite feature in the disk of HD 100546 (*Sturm et al.*, 2010, 2013) was shown by *Mulders et al.* (2011) to be dominated by emission from the inner disk wall located between 13–20 AU. Detailed modeling of the line shape of the feature yielded a crystalline mass fraction of 40–60% in this region, with a low-iron content of <0.3%. A tentative detection of crystalline water ice using PACS was also reported by *McClure et al.* (2012) in the disk surrounding the T Tauri star GQ Lup.

The chapter by Pontoppidan et al. in this volume extensively discusses the volatile content of disks. Here we only mention that significant decrease of the H_2O abundances across the expected snowline has been observed in several disks using unresolved Herschel and Spitzer observations (e.g., *Meijerink et al.*, 2009; *Zhang et al.*, 2013). *Najita et al.* (2013) and *Qi et al.* (2013b) concluded that species such as HCN, N_2H^+, and H_2CO reveal the presence of H_2O and CO snowlines through chemical signatures.

Intriguing conclusions can be drawn from spatially unresolved observations using spectrally resolved emission lines, when an underlying Keplerian velocity profile is assumed. Using science verification data from ALMA and the high signal-to-noise CO line data, *Rosenfeld et al.* (2012) found that the inner disk of TW Hya is likely warped by a few degrees on scales of 5 AU. Such a warp is consistent with what was earlier deduced by optical scattered light (*Roberge et al.*, 2005), and may be explained by a (planetary?) companion inside the disk.

With the rollout of ALMA, high-sensitivity imaging of regions as small as a few astronomical units will become feasible, providing direct information on the region where planets form.

4. THERMOCHEMICAL PROCESSES

Protoplanetary disks are characterized by strong vertical and radial temperature and density gradients and varying UV/X-ray radiation intensities. These conditions favors diverse chemical processes, including photochemistry, molecular-ion reactions, neutral-neutral reactions, gas-grain surface interactions, and grain surface reactions. Local density, temperature, and high-energy irradiation determine which of these processes will be dominating.

Disk chemistry is driven by high-energy radiation and cosmic rays (*Aikawa and Herbst*, 1999; *Aikawa et al.*, 2002; *van Zadelhoff et al.*, 2003; *Gorti and Hollenbach*, 2004; *van Dishoeck*, 2006; *van Dishoeck et al.*, 2006; *Fogel et al.*, 2011; *Vasyunin et al.*, 2011; *Walsh et al.*, 2012; *Akimkin et al.*, 2013). T Tauri stars possess intense nonthermal UV radiation from the accretion shocks, while the hotter HAe/Be stars produce intense thermal UV emission. The overall intensity of the stellar UV radiation at 100 AU can be higher by a factor of 100–1000 for a T Tauri disk (*Bergin et al.*, 2003) and 10^5 for a HAe disk (*Semenov et al.*, 2005; *Jonkheid et al.*, 2006), respectively, compared to

the interstellar radiation field (*Draine*, 1978). Photodissociation of molecules will depend sensitively on the shape and strength of the UV radiation field. For example, Lyman-α photons will selectively dissociate HCN and H_2O, while other molecules that dissociate between 91.2 and 110 nm such as CO and H_2 remain unaffected (*van Dishoeck and Black*, 1988; *van Dishoeck et al.*, 2006). Selective photodissociation of CO by the interstellar radiation field can also play an important role in the outer disk region (*Dartois et al.*, 2003).

The ~1–10-keV X-ray radiation is another important energy source for disk chemistry. The median value for the X-ray luminosity of the T Tauri stars is $\log(L_X/L_{bol}) \approx -3.5$ or $L_X \approx 3 \times 10^{29}$ erg s^{-1} [with an uncertainty of an order of magnitude (e.g., *Preibisch et al.*, 2005; *Getman et al.*, 2009)]. This radiation is generated by coronal activity, driven by magnetic fields generated by an αω dynamo mechanism in convective stellar interiors. Herbig Ae stars have weak surface magnetic fields due to their nonconvective interiors. Their typical X-ray luminosities are ≥10× lower than those of T Tauri stars (*Güdel and Nazé*, 2009). The key role of X-rays in disk chemistry is their ability to penetrate through high-gas columns (~0.1–1 g cm^{-2}) and ionize He, which destroys tightly bound molecules like CO and replenishes gas with elemental carbon and oxygen. The similar effect for disk chemistry is provided by cosmic-ray particles (CRPs), which can penetrate even through very high-gas columns, ~100 g cm^{-2} (*Umebayashi and Nakano*, 1980; *Gammie*, 1996; *Cleeves et al.*, 2013a).

In the terrestrial planet-forming region, ~1–10 AU, temperatures are ≥100 K and densities may exceed 10^{12} cm^{-3}. Despite intense gas-grain interactions, surface processes do not play a role and disk chemistry is determined by fast gas-phase reactions, which quickly reaches a steady state. In the absence of intense sources of ionizing radiation and high temperatures, neutral-neutral reactions with barriers become important (*Harada et al.*, 2010). At densities ≥10^{12} cm^{-3}, three-body reactions also become important (*Aikawa et al.*, 1999).

The outer disk regions (≥10 AU) can be further divided into three chemically different regimes (*Aikawa and Herbst*, 1999): (1) cold chemistry in the mid-plane dominated by surface processes, (2) rich gas-grain chemistry in the warm intermediate layer, and (3) restricted PDR-like gas-phase chemistry in the atmosphere.

In the disk atmosphere stellar UV radiation and the interstellar radiation field ionize and dissociate molecules and drive gas-phase PDR-like chemistry. Adjacent to the disk surface a warm, ≈30–300-K molecular layer is located. This region is partly shielded from stellar and interstellar UV/X-ray radiation, and chemistry is active in the gas and on the dust surfaces. The ionization produces ions, like H_3^+, driving rapid proton transfer and other ion-molecule processes (*Herbst and Klemperer*, 1973; *Watson*, 1974; *Woodall et al.*, 2007; *Wakelam et al.*, 2012). If water is frozen onto dust grains, a relatively high C/O ratio close to or even >1 can be reached in the gas phase, leading to a carbon-based chemistry. Ices are released from grains by thermal or photodesorption. In the outer mid-plane, the cosmic-ray particles and locally produced UV photons are the only ionizing sources. The temperature drops below ~20–50 K and freeze-out of molecules and hydrogenation reactions on grain surfaces are dominating the chemistry. The most important desorption process for volatiles such as CO, N_2, and CH_4 is thermal desorption. In addition, cosmic-ray and X-ray spot heating may release mantle material back to the gas phase (*Leger et al.,* 1985; *Hasegawa and Herbst,* 1993; *Najita et al.,* 2001), as well as reactive desorption (*Garrod et al.,* 2007).

In the outer cold disk regions, the addition of nuclear-spin-dependent chemical reactions involving ortho- and parastates of key species is required.

4.1. Line Emission and Thermochemical Models

Increasing detections of molecular, atomic, and ionic emission lines from groundbased IR and (sub)millimeter facilities and Spitzer and Herschel space observatories have led to a recent shift in emphasis on modeling the *gas* structure in disks.

Gas line emission is sensitive to both the abundances of trace-emitting species (and hence chemistry) and the excitation conditions (density, temperature, irradiation) at the surface. Meanwhile, the dominant cooling process that controls the gas temperature in the disk surface is radiative cooling by transition lines (e.g., OI, CII, Lyα). Therefore, we need to treat chemistry and thermal processes self-consistently in order to obtain gas temperature structure and predict line fluxes. Thermochemical models are also important in order to understand the photoevaporation process as a gas dispersal mechanism from disks (*Gorti and Hollenbach,* 2009; *Gorti et al.,* 2009; see also the chapter by Espaillat et al. in this volume).

4.1.1. Thermochemical models. Early disk models assumed that the gas temperature was equal to that of the dust. In the surface layers, however, low densities and/or heating of gas by stellar X-ray/UV radiation result in weaker thermal coupling of gas to dust, and gas and dust temperatures can differ significantly. This is more evident in disks where the dust has evolved and the collisional cross-sections are lower (see section 4.2.1). Importantly, the vertical density distribution is determined by the gas thermal pressure gradient. These considerations have given rise to the development of thermochemical disk models, where the gas and dust temperatures are determined separately and self-consistently solved with chemistry to allow more accurate interpretations of observed line emission (see also *Dullemond et al.,* 2007).

The first thermochemical models consistently solved for the gas temperature structure coupled with chemistry, using a density distribution derived from the dust radiative transfer and temperature solutions. The temperature of the gas deviated significantly from that of the dust in optically thin dust-rich disks (*Kamp and van Zadelhoff,* 2001). These

models were extended to optically thick disks (*Kamp and Dullemond,* 2004), and it was determined that gas/dust temperatures were well-coupled in the mid-plane but not in the higher optically thin dust surface. Ultraviolet heating was found to be important in the thermally decoupled regions, probed by molecules such as H_2O, CO, and OH. *Jonkheid et al.* (2004, 2007) considered the effects of dust evolution to find that dust settling and the resulting increased decoupling allowed gas temperatures to rise; this often increased the intensities of atomic and molecular lines. While these studies focused on UV heating of gas, Glassgold and collaborators (e.g., *Glassgold et al.,* 2004, 2007, 2009, 2012; *Meijerink et al.,* 2008; *Najita et al.,* 2011) have been examining the effect of X-rays on gas heating and chemistry with increasing levels of detail. X-ray heating dominates surface gas heating in these models and leads to ionization deep in the disk, driving ion-molecule chemistry and the formation of water and other organics. *Glassgold et al.* (2007) predicted an important strongly emitting tracer of ionized gas, the mid-IR line of [NeII] at 12.8 µm. The primary conclusion of these investigations is that the disk structure closely resembles photodissociation regions [(PDRs and X-ray dominated regions (XDRs)], with a cold, shielded molecular interior and warmer, atomic/ionic surface regions.

A further refinement to thermochemical models was to self-consistently solve for vertical hydrostatic equilibrium in the disk as set by the gas temperature structure (*Gorti and Hollenbach,* 2004, 2008; *Nomura and Millar,* 2005; *Aikawa and Nomura,* 2006; *Nomura et al.,* 2007) and to consider irradiation of the disk surface by both UV and X-rays (*Gorti and Hollenbach,* 2004, 2008; *Nomura et al.,* 2007; *Akimkin et al.,* 2013). Gas in the surface layer is hotter than the dust grains, which results in a more vertically extended disk atmosphere.

The increased UV/X-ray attenuation in the inner disk lowers the surface gas temperature at intermediate and larger radii compared to the solutions obtained without a gas-determined equilibrium density structure. X-ray heating is found to dominate in the surface layers of disks, often heating gas to ≥1000–5000 K in the inner disk, while FUV heating and both X-ray and UV photodissociation are important in the intermediate (A_V ~ 0.1–1) regions. Dust thermal balance dominates in the dense, shielded interior layers, although ionization by hard X-ray photons can drive ion-molecule chemistry even in the denser regions of the disk. The change in temperature obtained with a self-consistent density determination is inferred to affect the calculated intensities of molecular and atomic line emission from disks.

More recently, *Akimkin et al.* (2013) have introduced the state-of-the-art thermochemical ANDES model, which treats time-dependent chemical reactions self-consistently with gas and dust thermal processes and calculations of dust coagulation, fragmentation, and settling (see section 4.2.1).

4.1.2. Modeling line emission. Recent development of thermochemical models focused more on modeling observable line emissions from disks. *Gorti and Hollenbach*

(2008) found that [NeII] and [ArII] lines are tracers of X-ray ionized regions, and that X-rays and FUV are both significant contributors to gas heating. They concluded that the [OI] 63-μm line was a strong, luminous coolant in disks. *Nomura et al.* (2007) investigated the strength of various H_2 lines, including dust coagulation calculations, to suggest that IR line ratios of rovibrational transitions of H_2 are sensitive to gas temperature and the FUV radiation field, and could be a useful tracer of grain evolution in the surface layers of the inner disks.

In order to statistically model and predict line fluxes from disks for observations by Herschel, ALMA, and forthcoming facilities, a tool called the DENT grid was developed to calculate line fluxes with a large number of parameters, such as stellar mass, age, and UV excess, disk masses, scale heights, inner and outer radii, dust properties, and inclination angles, as a part of the Herschel open-time key program of Gas in Protoplanetary Systems (GASPS) (*Woitke et al.,* 2010; *Kamp et al.,* 2011). The DENT grid used the three-dimensional Monte Carlo radiative transfer code, MCFOST (*Pinte et al.,* 2006, 2009), for calculating dust temperature and line radiative transfer, and the thermochemical model, ProDiMo, for calculating gas temperature and abundances of species. ProDiMo is a sophisticated model that takes into consideration all the processes mentioned in the previous subsection. *Woitke et al.* (2009b) utilized full two-dimensional dust continuum transfer, a modest chemical network, and computed the gas temperature and density structure (in one + one-dimension), but initially only including UV heating and simplified photochemistry. These were updated to improve upon the calculations of the rates of photoprocesses and line radiative transfer (*Kamp et al.,* 2010). X-ray irradiation was added in *Aresu et al.* (2011) and *Meijerink et al.* (2012). *Thi et al.* (2011c) studied the effects of the inner rim structure on the disk. These models have been successfully applied to observational data of Herschel GASPS objects of protoplanetary disks and debris disks to infer the important disk properties, such as dust evolution, radial extent, and disk masses (*Meeus et al.,* 2010; *Thi et al.,* 2011b; *Tilling et al.,* 2012; *Podio et al.,* 2013). *Bruderer et al.* (2009) developed another thermochemical disk model.

Bruderer et al. (2012) adapted it to a disk around a Herbig Be star, HD 100546, and used it to explain the high-J CO lines and the [OI] 63-μm line observed by Herschel, together with an upper limit on the [CI] 370-μm line [obtained by the Atacama Pathfinder Experiment (APEX)]. They discussed the variability of the gas/dust ratio and the amount of volatile carbon in the disk atmosphere. *Chapillon et al.* (2008) adopted the Meudon PDR code (*Le Bourlot et al.,* 1993; *Le Petit et al.,* 2006) with a number of variable parameters such as UV field, grain size, and gas-to-dust ratio to explain millimeter CO-line observations toward HAe disks.

While thermochemical models have been developed, the non-LTE line radiative transfer methods for calculating transition lines from disks also have been studied. In *Pavlyuchenkov et al.* (2007), the comparison between various approximate line radiative transfer approaches and a well-tested accelerated Monte Carlo code was performed. Using a T Tauri-like flared disk model and various distributions of molecular abundances (layered and homogeneous), the excitation temperatures, synthetic spectra, and channel maps for a number of rotational lines of CO, $C^{18}O$, HCO^+, DCO^+, HCN, CS, and H_2CO were simulated. It was found that the LTE approach widely assumed by observers is accurate enough for modeling the low molecular rotational lines, whereas it may significantly overestimate the high-J line intensities in disks. The full escape probability (FEP) approach works better for these high-J rotational lines but fails sometimes for the low-J transitions. *Semenov et al.* (2008) adopted the code to simulate the ALMA observations of maps of molecular line emission from disks, using the resulting molecular abundance profiles of their calculations of chemical reactions.

Brinch and Hogerheijde (2010) developed a line radiative transfer code, LIME, based on the RATRAN code (*Hogerheijde and van der Tak,* 2000), in which grids are laid out adaptive to the opacity. Therefore, the code can deal with objects having inhomogeneous three-dimensional structure with large density contrast as well as overlapping lines of multiple species. The code is available online at *http://www.nbi.dk/sbrinch/lime.php*. The LIME code is used in the ARTIST software package, which is designed to model two-dimensional/three-dimensional lines and continuum radiative transfer and synthesize interferometric images (*Padovani et al.,* 2012) (*http://youngstars.nbi.dk/artist/Welcome.html*).

Non-long-term evolution effects are more significant for higher transition lines whose critical densities are high. *Meijerink et al.* (2009) modeled mid-IR water lines observed by Spitzer (*Rothman et al.,* 2005) and showed that the differences in line fluxes obtained by LTE and non-LTE calculations are within 1 order of magnitude for most of the lines. *Woitke et al.* (2009a) also studied water line fluxes, including the submillimeter lines and using their full thermochemical model. They showed that the differences are smaller for low-excitation water lines.

4.2. Chemistry Versus Physical Processes

Dynamical processes that lead to planet formation, such as grain evolution and gas dispersal, affect chemical structure in protoplanetary disks. Thus these processes, as well as some environmental effects, could appear in molecular line emission from the disks.

4.2.1. Effect of grain evolution. Grain size distributions in protoplanetary disks are thought to be very different from that of the interstellar grains because of coagulation, settling toward the disk mid-plane and transferring toward the central star under the influence of the gravity of the star (see the chapter by Testi et al. in this volume).

One of the recent advancements in studies of disk chemistry is the treatment of grain evolution (Fig. 4). Grain growth depletes the upper disk layers of small grains and hence reduces the opacity of disk matter, allowing the FUV radiation to penetrate more efficiently into the disk and to heat and dissociate molecules deeper in the disk. Also, larger grains populating the disk mid-plane will delay the depletion of gaseous species because of the reduced total surface area (*Aikawa and Nomura*, 2006; *Fogel et al.*, 2011;

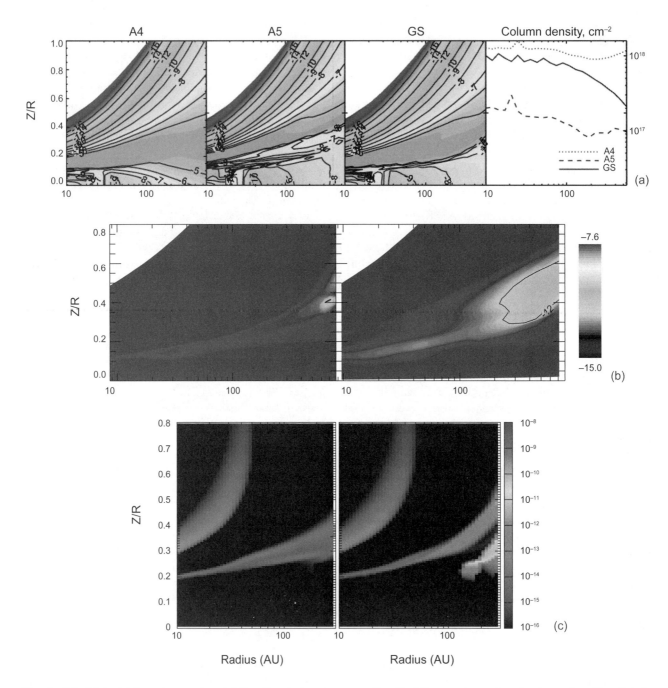

Fig. 4. See Plate 14 for color version. **(a)** Fractional abundance profiles (first three columns) and column densities (fourth column) of the gas-phase carbon monoxide in a T Tauri disk calculated with three distinct dust grain models: A4, the uniform grains with radius of 1 μm; A5, the uniform grains with radius of 0.1 μm; and GS, the model with grain growth and settling. Note how grain evolution affects the depletion of molecules onto grains near the mid-plane of the cold outer disk (*Vasyunin et al.*, 2011). **(b)** Gas-phase methanol abundances in a T Tauri disk calculated with a laminar gas-grain chemical model (left) and a two-dimensional turbulent-mixing chemical model (right). The transport of methanol ice from dark mid-plane to upper, more UV-irradiated disk layer enables more methanol ice to be desorbed in the gas phase (*Semenov and Wiebe*, 2011). **(c)** Gas-phase methanol abundances in a T Tauri disk calculated with two gas-grain chemical models: with no surface chemistry taken into account (left) and with surface chemistry (right) (*Walsh et al.*, 2010). The surface hydrogenation of CO leads to efficient production of methanol ice, which can be partly UV-desorbed in disk layers above mid-plane.

Vasyunin et al., 2011). Grain coagulation, fragmentation, sedimentation, turbulent stirring, and radial transport are all important processes to control the grain size distribution.

Nomura et al. (2007) investigated the effect of dust evolution only on gas temperature profile and molecular hydrogen excitation by using the result of calculations of coagulation equation with settling of dust particles. *Vasyunin et al.* (2011) studied the effect on chemical reactions, taking into account the fragmentation and cratering of dust particles as well.

Akimkin et al. (2013) have developed the ANDES model, which is based on combined description of the one + one-dimensional continuum radiative transfer, detailed two-dimensional dust evolution model, and time-dependent laminar chemistry. This model was applied to investigate the impact of grain evolution on disk chemistry. It was found that due to grain growth and dust removal from the disk atmosphere, the molecular layer shifts closer toward the disk mid-plane, which remains shielded by dust from the ionizing radiation. Larger grains that settled to the mid-plane lower the frequency of gas-grain collisions, thus making depletion of gaseous molecules less severe and hindering the growth of ices on grains.

4.2.2. Chemodynamical models. Gas in protoplanetary disks is thought to disperse in 1–10 m.y. Magneto-rotational instability driven turbulence will cause angular momentum transfer and the gas accretes toward the central star as a result, while photoevaporation disperses the gas in the region where the thermal energy is high enough for the gas to escape from the gravity of the central star (see chapters by Turner et al., Alexander et al., and Espaillat et al. in this volume). The gas motion associated with this gas dispersal process is expected to affect the chemical structure, especially near the boundaries between the three layers: (1) photon-dominated surface layer, (2) warm molecular-rich middle layer, and (3) cold molecule-depleted layer near the mid-plane. The effect appears significant when the timescale of the gas motion is shorter than the timescales of the dominant chemical reactions. So far, in addition to laminar disk models, a number of chemodynamical models of protoplanetary disks have been developed.

On the other hand, chemodynamical models have been also developed to treat chemistry in the dynamically evolving early phase of star and disk formation where circumstellar disks are still embedded in a dense envelope. Some information about the chemistry in protoplanetary disks could remain in this early phase after the surrounding envelope clears up, especially regarding ice molecules on grains.

In the coming years, we expect that sophisticated multidimensional magneto-hydrodynamical (MHD) models will be coupled with time-dependent chemistry. The first steps in this direction have been made (*Turner et al.,* 2007), where a simple chemical model has been coupled to a local three-dimensional MHD simulation.

4.2.2.1. Turbulent mixing and gas accretion: The chemical evolution in protoplanetary disks with one-dimensional radial advective mass transport was studied by *Woods and*

Willacy (2007, 2009a), *Willacy and Woods* (2009), and *Nomura et al.* (2009). Another class of chemodynamical disk models is based on turbulent diffusive mixing, which uniformizes chemical gradients, and is modeled in one dimension, two dimensions, or even full three dimensions (*Ilgner and Nelson,* 2006a,b,c, 2008; *Semenov et al.,* 2006; *Aikawa,* 2007; *Semenov and Wiebe,* 2011; *Willacy et al.,* 2006; *Hersant et al.,* 2009; *Turner et al.,* 2006, 2007). In several disk models, both advective and turbulent transport were considered (*Heinzeller et al.,* 2011), while *Gorti et al.* (2009) studied photoevaporation of disks and loss of the gas due to the stellar FUV and X-ray radiation.

The turbulence in disks is a three-dimensional-phenomenon driven by a magneto-rotational instability (*Balbus and Hawley,* 1991; see also the chapter by Turner et al. in this volume). Global MHD simulations show that advection has no specified direction in various disk regions, and in each location goes both inward and outward. The corresponding turbulent velocity of the gas V_{turb} depends on the viscosity parameter α (*Shakura and Sunyaev,* 1973) and scales with it somewhere between linear and square-root dependence: $\alpha < V_{turb}/c_s < \sqrt{\alpha}$; here c_s is the sound speed. The calculated α-viscosity stresses have values in a range of 10^{-4}–10^{-2} in the mid-plane and the molecular layer, respectively, and rise steeply to transonic values of ~0.5 in the disk atmosphere.

Magneto-rotational instability (MRI) requires a modest gas ionization to be operational. The ionization structure of protoplanetary disks is largely controlled by a variety of chemical processes. Recent studies of the chemistry coupled to the dynamics in protoplanetary disks are briefly summarized below (see also section 5.1). *Ilgner and Nelson* (2006c) have studied the ionization structure of inner disks (r < 10 AU), considering vertical mixing and other effects like X-ray flares, various elemental compositions, etc. They found that mixing has no profound effect on electron concentration if metals are absent in the gas since recombination timescales are faster than dynamical timescales. However, when T ≥ 200 K and alkali atoms are present in the gas, chemistry of ionization becomes sensitive to transport, such that diffusion may reduce the size of the turbulent-inactive disk "dead" zone (see also section 5.2).

Willacy et al. (2006) have attempted to study systematically the impact of disk diffusivity on the evolution of various chemical families. They used a steady-state α-disk model similar to that of *Ilgner et al.* (2004) and considered one-dimensional vertical mixing in the outer disk region with r > 100 AU. They found that vertical transport can increase column densities (vertically integrated concentrations) by up to 2 orders of magnitude for some complex species. Still, the layered disk structure was largely preserved even in the presence of vertical mixing. *Semenov et al.* (2006) and *Aikawa* (2007) have found that turbulence allows the transport of gaseous CO from the molecular layer down toward the cold mid-plane where it otherwise remains frozen out, which may explain the large amount of cold (≤5 K) CO gas detected in the disk of DM Tau (*Dartois et al.,* 2003). *Hersant et al.* (2009) have studied various mechanisms to retain gas-phase CO in very cold disk regions, including vertical mixing. They found that photodesorption in the upper, less-obscured molecular layer greatly increases the gas-phase CO concentration, whereas the role of vertical mixing is less important.

Later, in *Woods and Willacy* (2007), the formation and destruction of benzene in turbulent protoplanetary disks at $r \leq 35$ AU has been investigated. These authors found that radial transport allows efficient synthesis of benzene at ≤ 3 AU, mostly due to ion-molecule reactions between C_3H_3 and $C_3H_4^+$ followed by grain dissociative recombination. The resulting concentration of C_6H_6 at larger radii of 10–30 AU is increased by turbulent diffusion up to 2 orders of magnitude.

In a similar study, *Nomura et al.* (2009) have considered an inner disk model with radial advection (≤ 12 AU). They found that the molecular concentrations are sensitive to the transport speed, such that in some cases gaseous molecules are able to reach the outer, cooler disk regions where they should be depleted. This increases the production of many complex or surface-produced species such as methanol, ammonia, hydrogen sulfide, acetylene, etc.

Tscharnuter and Gail (2007) have considered a two-dimensional disk chemo-hydrodynamical model in which global circulation flow patterns exist, transporting disk matter outward in the disk mid-plane and inward in elevated disk layers. They found that gas-phase species produced by warm neutral-neutral chemistry in the inner region can be transported into the cold outer region and freeze out onto dust grain surfaces. The presence of such large-scale meridional flows in protoplanetary accretion disks was called into question in later MHD studies (*Fromang et al.,* 2011; *Flock et al.,* 2011).

Heinzeller et al. (2011) have studied the chemical evolution of a protoplanetary disk with radial viscous accretion, vertical mixing, and a vertical disk wind (in the atmosphere). They used a steady-state disk model with $\alpha = 0.01$ and $\dot{M} = 10^{-8}$ $M_\odot$ yr^{-1}. They found that mixing lowers concentration gradients, enhancing abundances of NH_3, CH_3OH, C_2H_2, and sulfur-containing species. They concluded that such a disk wind has an effect similar to turbulent mixing on chemistry, while the radial accretion changes molecular abundances in the mid-plane, and the vertical turbulent mixing enhances abundances in the intermediate molecular layer.

A detailed study of the effect of two-dimensional radial-vertical mixing on gas-grain chemistry in a protoplanetary disk has been performed in *Semenov and Wiebe* (2011). These authors used the α-model of a ~5-m.y.-old DM Tau-like disk coupled to the large-scale gas-grain chemical code "ALCHEMIC" (*Semenov et al.,* 2010). To account for production of complex molecules, an extended set of surface processes was added. A constant value of the viscosity parameter $\alpha = 0.01$ was assumed.

In this study it was shown that turbulent transport changes the abundances of many gas-phase species and particularly ices. Vertical mixing is more important than radial mixing, mainly because radial temperature and density gradients in disks are weaker than vertical ones. The simple molecules not responding to dynamical transport include C_2H, C^+, CH_4, CN, CO, HCN, HNC, H_2CO, and OH, as well as water and ammonia ices. The species sensitive to transport are carbon chains and other heavy species, in particular sulfur-bearing and complex organic molecules (COMs) frozen onto the dust grains. Mixing transports ice-coated grains into warmer disk regions where efficient surface recombination reactions proceed. In a warm molecular layer these complex ices evaporate and return to the gas phase.

4.2.2.2. *Dynamics in early phase:* In the protostellar phase, materials in envelopes surrounding circumstellar disks dynamically fall onto the disks, and then accrete toward the central star in the disks. Chemical evolution, including gas-grain interaction, along two-dimensional advection flow from envelopes to disks was studied by *Visser et al.* (2009, 2011). Using axisymmetric two-dimensional semi-analytical dynamical models, they discussed the origin of organic molecules and other species in comets in our solar system.

Assumption of axisymmetry will be adoptable for quiescent disks where mass is transported locally with low accretion rates, but may be inappropriate for disks in the early phase, which are massive enough to trigger gravitational instabilities (*Vorobyov and Basu,* 2006). Gravitational instabilities produce transient non-axisymmetric structures in the form of spiral waves, density clumps, etc. (*Boley et al.,* 2010; *Vorobyov and Basu,* 2010). The results of Vorobyov and collaborators have been challenged by *Zhu et al.* (2010a,b, 2012), who found that ongoing accretion of an envelope surrounding a forming protoplanetary disks can alone drive nonstationary accretion and hence outbursts similar to the FU Ori phenomenon. More importantly, gravitational instabilities lead to efficient mass transport and angular momentum redistribution, which could be characterized by a relatively high-viscosity parameter $\alpha \sim 1$. *Ilee et al.* (2011) performed a first study of time-dependent chemistry along paths of fluid elements in three-dimensional radiative hydrodynamical simulations of a gravitationally unstable massive disk (0.39 $M_\odot$ within a disk radius of 50 AU) by *Boley et al.* (2007) and showed the importance of continuous stirring by rotating spiral density waves with weak shocks on chemical properties in the disk.

4.2.3. Environmental effects. Since chemical processes are sensitive to local density, temperature, and radiation fields, we expect to see different chemical structure and molecular line emission if a protoplanetary disk is in a specific environment.

Cleeves et al. (2011) investigated the chemical structure of a transition disk with a large central hole (with a radius of 45 AU). Transition disks are interesting objects to study dispersal mechanisms of disks, for example, by giant planets and/or photoevaporation, and so on (see the chapters by Espaillat et al. and Alexander et al. in this volume, as well as section 5.4). Cleeves et al. showed that the inner edge of the disk is heated by direct irradiation from the central star so that the even molecules with high-desorption energies can survive in the gas phase near the mid-plane. Line radiative transfer calculations predicted that this actively evolving truncation region in transition disks will be probed by observations of high-transition lines of such molecules by ALMA with high-spatial resolution.

Walsh et al. (2013) studied the chemical structure of a protoplanetary disk irradiated externally by a nearby massive star. Although it is observationally suggested that stars are often formed in young star clusters (e.g., *Lada and Lada,* 2003), star and planet formation in isolated systems have been mainly studied because detailed observations have been only available so far in isolated star-forming regions close to our Sun. However, ALMA is expected to provide detailed (sub)millimeter observations of protoplanetary disks in young clusters, such as the Trapezium cluster in the Orion nebula, allowing us to study star and planet formation in varied environments. *Walsh et al.* (2013) showed that external irradiation from nearby OB stars can heat gas and dust even near the disk mid-planes, especially in the outer regions, affecting snowlines of molecules with a desorption temperature of ~30 K. Also, due to the externally irradiated FUV photons, the ionization degree becomes higher in the disk surface. Predictions of molecular line emission showed that observations of line ratios will be useful to probe physical and chemical properties of the disk, such as ionization degree and photoevaporation condition.

Cleeves et al. (2013a) investigated the ionization rate in protoplanetary disks excluded by cosmic rays. It has been suggested that strong winds from the central young star with strong magnetic activity may decrease the cosmic-ray flux penetrating into the disks by analogy with the solar wind modulating cosmic-ray photons in our solar system. Cleeves et al. showed that the ionization rate by cosmic rays can be reduced by stellar winds to a value lower than that due to short-lived radionuclides of $\zeta_{RN} \sim 10^{-18}$ s^{-1} (*Umebayashi and Nakano,* 2009), with which magneto-rotational instability will be locally stabilized in the inner disk.

4.3. New Advances in Treatment of Chemical Processes

4.3.1. Photochemistry. Under the strong UV and X-ray irradiation from the central star, photochemistry is dominant in the surface layers of protoplanetary disks. A detailed two-dimensional/three-dimensional treatment of the X-ray and UV radiation transfer with anisotropic scattering is an essential ingredient for realistic disk chemical models (e.g., *Igea and Glassgold,* 1999; *van Zadelhoff et al.,* 2003). The photochemistry dominant region expands in the disks with grain evolution as dust opacities become lower at UV wavelengths. In heavily irradiated disk atmospheres, many species will exist in excited (ro)vibrational states, which may then react differently with other species and require the addition of state-to-state processes in the models (*Pierce and A'Hearn,* 2010).

An accurately calculated UV spectrum, including Lyα resonant scattering, is required to calculate photodissociation and photoionization rates and shielding factors for CO, H$_2$, and H$_2$O (*Bethell and Bergin,* 2009, 2011). *Fogel et al.* (2011) adopt their Lyα and UV-continuum radiation field to time-dependent chemistry in disks with grain settling to show that the Lyα radiation impacts the column densities of HCN, NH$_3$, and CH$_4$, especially when grain settling is significant.

Walsh et al. (2012) studied the impact of photochemistry and X-ray ionization on the molecular composition in a disk to show that detailed treatment of X-ray ionization mainly affects N$_2$H$^+$ distribution, while the shape of the UV spectrum affects the distribution of many molecules in the outer disk.

Photodesorption of molecules from icy mantles on grains is also an important mechanism in the surface and middle layer of protoplanetary disks. The photodesorption rates were recently derived experimentally for CO, H$_2$O, CH$_4$, and NH$_3$ (*Öberg et al.,* 2007, 2009a,b; *Fayolle et al.,* 2011). *Walsh et al.* (2010) showed using a chemical model in a protoplanetary disk that photodesorption significantly affects molecular column densities of HCN, CN, CS, H$_2$O, and CO$_2$ (see also *Willacy,* 2007; *Semenov and Wiebe,* 2011).

4.3.2. Chemistry of organic molecules. Although various complex organic molecules, such as glycoaldehyde and cyanomethanimine (e.g., *Zaleski et al.,* 2013), have been detected toward luminous hot cores and hot corinos, H$_2$CO, HC$_3$N (*Chapillon et al.,* 2012b), and c-C3H$_2$ (*Qi et al.,* 2013a) are the largest molecules detected toward less luminous protoplanetary disks so far. On the other hand, more complex organic molecules have been found in our solar system by IR and radio observations toward comets (e.g., *Mumma and Charnley,* 2011), in a sample return mission (Stardust) from Comet 81P/Wild2 (*Elsila et al.,* 2009) and in meteorites. Investigating how these complex molecules were formed in the early solar system is an interesting topic.

Since the detection of mid-IR lines of small organic molecules and water by the Spitzer Space Telescope (e.g., *Carr and Najita,* 2008), they have been treated in many chemical models of protoplanetary disks (e.g., *Woods and Willacy,* 2009b; *Nomura et al.,* 2009; *Woitke et al.,* 2009a). *Agúndez et al.* (2008) pointed out an importance of gas-phase synthesis with molecular hydrogen to form carbon-bearing organic molecules, while *Glassgold et al.* (2009) indicated that water abundance is significantly affected by the molecular hydrogen formation rate from which water is mainly formed through a gas-phase reaction in a warm disk atmosphere. *Heinzeller et al.* (2011) showed that transport of molecular hydrogen from a UV-photon-shielded molecular-rich layer to a hot disk surface by turbulent mixing enhances abundances of water and small organic molecules.

Meanwhile, more complex organic molecules, which have not been detected toward protoplanetary disks yet, are believed to form on grain surfaces (e.g., *Herbst and van Dishoeck,* 2009). *Walsh et al.* (2010) showed that grain surface chemistry, together with photodesorption, enhances gas-phase complex organic molecules, such as HCOOH, CH_3OH, and $HCOOCH_3$ in the middle layer of the outer disk. Also, *Semenov and Wiebe* (2011) showed that grain surface chemistry and then organic molecule formation are affected by turbulent mixing. *Walsh et al.* (2014) further treated complex grain surface chemistry [following *Garrod and Herbst* (2006) and *Garrod et al.* (2008)], which includes heavy-radical recombination reactions efficient on warm grains together with photodesorption, reactive desorption, and photodissociation of ice on grains. They showed that complex organic molecules are formed by the surface reactions on warm grains near the mid-plane at the disk radius of ≥20 AU. The resulting grain-surface fractional abundances are consistent with those observed toward comets. Also, based on model calculations of transition lines, they predict that strong methanol lines will be excellent candidates for ALMA observations toward disks.

4.4. Sensitivity Analysis and Intrinsic Uncertainties of Chemical Models

To improve our knowledge of the physics and chemistry of protoplanetary disks, it is necessary to understand the robustness of the chemical models to the uncertainties in the reaction rate coefficients that define the individual reaction efficiency. Sensitivity methods have been developed for such purposes and applied to several types of sources (*Vasyunin et al.*, 2004; *Wakelam et al.*, 2005, 2010). Those methods follow the propagation of the uncertainties during the calculation of the abundance species. Error bars on the computed abundances can be determined and "key" reactions can be identified. *Vasyunin et al.* (2008) applied such methods for protoplanetary disk chemistry, showing that the dispersion in the molecular column densities is within an order of magnitude at 10^6 yr when chemical reactions are in a quasi-equilibrium state. Abundances of some molecules could be very sensitive to rate uncertainties, especially those of X-ray ionization rates in specific regions.

Results of model calculations of chemical reactions also depend on the applied chemical network in some degree. Chemical networks used for disk chemistry are usually derived from the public networks: UMIST (*http://www.udfa.net/*) (*McElroy et al.*, 2013) and OSU (*http://www.physics.ohio-state.edu/~eric/research.html*). Those two networks have been developed and updated over the years since the 1990s. The latest version of the OSU network has recently been published under the name kida.uva.2011 (*Wakelam et al.*, 2012), updated from the online database KIDA (*http://kida.obs.u-bordeaux1.fr/*). The UMIST network contains a number of chemical reactions with activation barriers, and even three-body assisted reactions that can become important in the inner regions of the disks. The OSU/kida.uva network is primarily designed for the outer parts of the disk where the temperatures stay <300 K. Additions have been made to the OSU network by *Harada et al.* (2010, 2012) to use it up to temperatures of about 800 K. In *Wakelam et al.* (2012), differences in the resulting molecular abundances

computed by different chemical networks were studied in a dense cloud model to show that errors are typically within an order of magnitude. It is worth noticing that these networks contain approximate values for the gas-phase molecular photodissociation rates, which are valid for only the UV interstellar radiation field. Photorates for different star spectra have been published by *van Dishoeck et al.* (2006) (see also *Visser et al.,* 2009).

Contrary to the gas phase, as yet there has been no detailed database of surface reactions for the interstellar medium. Only a few networks have been put online by authors, and some of these can be found on the KIDA website (*http://kida.obs.u-bordeaux1.fr/models*).

5. TOWARD THE ATACAMA LARGE MILLIMETER ARRAY ERA

5.1. Measuring the Turbulence

Turbulence is a key ingredient for the evolution and composition of disks. On one hand, the turbulence strength, more precisely the α-viscosity parameter, can be derived indirectly through its impact on disk spreading (*Guilloteau et al.,* 2011) and dust settling (*Mulders and Dominik,* 2012). From the (dust) disk shapes and sizes at millimeter wavelengths (which sample the disk mid-plane), *Guilloteau et al.* (2011) found $\alpha \simeq 10^{-3}$–10^{-2} at a characteristic radius of 100 AU, possibly decreasing with time. Infrared SED analyses, which sample the top layers and are more sensitive to 10 AU scales, provided similar values, although the derived α depends on the assumed grain sizes and gas-to-dust ratio (*Mulders and Dominik,* 2012). On another hand, high-resolution spectroscopy can directly probe the magnitude of turbulence by separating the bulk motions (Keplerian rotation) from local broadening and then isolating the thermal and turbulent contributions to this broadening. For spatially unresolved spectra, turbulent broadening can smear out the double-peaked profiles expected for the Keplerian disk. This property was used by *Carr et al.* (2004), who found supersonic turbulence from CO overtone bands in SVS 13 but lower values from H_2O lines, suggesting a highly turbulent disk surface in the inner few astronomical units.

With spatially resolved spectroimaging, the Keplerian motions lead to isovelocity regions that follow an arc shape, given by

$$r(\theta) = \left(GM_* / V_{obs}^2\right) \sin^2 i \, \cos^2 \theta \qquad (2)$$

The dynamical mass and disk inclination control the curvature, while local broadening controls the width of the emission across the arc. Thus, local broadening is easily separated from the Keplerian shear and can be recovered usually through a global fit of a (semi)parametric disk model. Using the PdBI, local linewidths on the order of 0.05–0.15 km s⁻¹ were found in the early work of *Guilloteau and Dutrey* (1998) using CO observations of DM Tau, and subsequently for LkCa 15 and MWC 480 by *Piétu et al.* (2007). Work by *Qi et al.* (2004) found that a smaller linewidth in the range 0.05–0.1 km s⁻¹ was required to reproduce CO J=3–2 observations of the TW Hya disk. Using CO isotopologs, *Dartois et al.* (2003) suggested the turbulence could be smaller in the upper layers of the disk, a result not confirmed by *Piétu et al.* (2007) for LkCa 15 and MWC 480 [for GM Aur, see also *Dutrey et al.* (2008)]. Similar values were reported from other molecular tracers such as CN, HCN (*Chapillon et al.*, 2012a), or HCO⁺ (*Piétu et al.*, 2007). The inferred turbulent velocities are subsonic, Mach ≃ 0.1–0.5, which corresponds to viscosity α values of ≃0.01–0.1 (*Cuzzi et al.*, 2001).

These initial measurements suffered from several limitations: (1) sensitivity, resulting in only CO being observed at high S/N; (2) spectral resolution; and (3) limited knowledge of the thermal linewidth. Several authors, including *Piétu et al.* (2007), *Isella et al.* (2007), and *Qi et al.* (2008), have particularly noted the difficulty of meaningfully constraining turbulent linewidths whose magnitudes are comparable to the spectral resolution of the data. *Hughes et al.* (2011) utilized the higher spectral resolution provided by the SMA correlator to measure the nonthermal broadening of the CO lines in TW Hya and HD 163296. They used two independent thermal structures to derive consistent estimates of turbulent broadening that achieve consistent results: one derived from modeling the SED of the stars to estimate the temperature in the CO emitting layer, and one freely variable parametric model. Their results indicate a very small level of turbulence in TW Hya, <0.04 km s⁻¹, but a substantially higher value in HD 163296 (0.3 km s⁻¹).

Because of its high optical depth, CO unfortunately only samples a very thin layer high up in the disk atmosphere. Measurements with other, optically thinner, tracers are essential to probe the turbulence across the disk. However, the distribution of molecules like CN is insufficiently well understood to remove the thermal component. For CN (and also C₂H), the apparent excitation temperatures are found to be low, 10–15 K (*Henning et al.*, 2010; *Chapillon et al.*, 2012a), while the temperature in the molecular-rich layer is expected to be around 20–30 K, high enough to contribute significantly to the local linewidth. To minimize the contribution of thermal motions to the local linewidth such that the turbulence component can be measured, *Guilloteau et al.* (2012) used CS, a relatively "heavy" molecule (μ = 44, compared to 25–28 for CN, CCH, or CO). Observing the J=3–2 transition with the IRAM PdBI, and with accurate modeling of the correlator response, they derive a turbulent width of 0.12 km s⁻¹ and show this value to be robust against the unknowns related to the CS spatial distribution (kinetic temperature, location above disk plane). It corresponds to a Mach number on the order of 0.5, a value that is only reached at several scale heights above the disk plane in MHD simulations (e.g., *Fromang and Nelson*, 2006; *Flock et al.*, 2011), while chemical models predict CS to be around 1–2 scale heights.

All measurements so far provide disk-averaged values. The radial variations are as yet too poorly constrained:

Assuming a power law radial distribution for the turbulent width, *Guilloteau et al.* (2012) inferred an exponent 0.38 ± 0.45, which can equally accommodate a constant turbulent velocity or a constant Mach number. It is also worth emphasizing that the "turbulent" width is an adjustable parameter to catch (to first order) all deviations around the mean Keplerian motion, as could occur in the case of, e.g., spiral waves or disk warps [see, for example, the complex case of AB Aur in the papers by *Piétu et al.* (2005) and *Tang et al.* (2012)].

The extremely limited number of sources studied so far precludes any general conclusion to be drawn from the relation between nonthermal linewidth and, e.g., disk size (which would be expected due to viscous evolution), stellar mass (which determines the spectrum of ionizing radiation), or evolutionary state (transitional or nontransitional). Substantial progress is expected with ALMA, which will provide much higher sensitivity, higher spectral resolution, and higher angular resolution, but also the possibility of using transitions with different optical depths to probe different altitudes above the disk plane. The CS molecule is well suited for this, along with a number of other rotational lines falling in the ALMA bands.

5.2. Estimating the Ionization and Dead Zone

One of the most important properties of protoplanetary disks is their ionization structure, which has been poorly constrained by the observations until now. Weakly ionized plasma in a rotating configuration is subject to MRI (e.g., *Balbus and Hawley*, 1991), which is thought to drive turbulence in disks. Turbulence causes anomalous viscosity of the gas and thus enables angular momentum transport and regulates the global disk evolution. It is operational even at very low-ionization degree values of ≤10⁻¹⁰ in the inner disk mid-plane regions completely shielded from the external UV/X-ray radiation (but rarely from cosmic-ray particles) (see also *Cleeves et al.*, 2013b). In disks that are sufficiently massive the inner region can be shielded even from the CRPs, either due to magneto-spherical activity of the central star or by a high obscuring gas column of ≥100 g cm⁻³ (*Umebayashi and Nakano*, 1980; *Gammie*, 1996; *Cleeves et al.*, 2013a). This leads to the formation of a "dead zone" with damped turbulence, through which transport of matter is severely reduced (see, e.g., *Sano et al.*, 2000; *Semenov et al.*, 2004; *Flock et al.*, 2012b; *Martin et al.*, 2012; *Mohanty et al.*, 2013; *Dzyurkevich et al.*, 2013). This has implications on the efficiency of grain growth and their radial migration (*Hasegawa and Pudritz*, 2010a; *Meheut et al.*, 2012a), planet formation (*Oishi et al.*, 2007), disk chemistry, physics (*Suzuki et al.*, 2010; *Lesniak and Desch*, 2011; *Flock et al.*, 2012a), and development of spiral waves, gaps, and other asymmetric structures (*Meheut et al.*, 2010, 2012b; *Morishima*, 2012; *Gressel et al.*, 2012).

Even with modern computational facilities, the complex interplay between weakly charged plasma and magnetic fields in disks is very difficult to simulate in full detail.

Ilgner and Nelson (2006b,c) investigated in detail ionization chemistry in disks and its sensitivity to various physical and chemical effects at r < 10 AU: (1) the X-ray flares from the young T Tauri star (*Ilgner and Nelson*, 2006c) and (2) the vertical turbulent mixing transport and the amount of gas-phase metals (*Ilgner and Nelson*, 2006b). Using an α-disk model with stellar X-ray-irradiation, they demonstrated that the simple network from *Oppenheimer and Dalgarno* (1974), using only five species, tends to overestimate the ionization degree since metals exchange charges with polyatomic species, and that magnetically decoupled "dead" regions may exist in disks as long as small grains and metals are removed from gas. The vertical diffusion has no effect on the electron abundances when the metals are absent in the gas because recombination timescales for polyatomic ions are rapid. When the dominant ions are metals, the characteristic ionization timescale becomes so long that the ionization chemistry becomes sensitive to transport, which drastically reduces the size of the "dead" zone. In the disk model with sporadic X-ray flares (by up to a factor of 100 in the luminosity) the outer part of the "dead" zone disappears, whereas the inner part of the "dead" zone evolves along with variations of the X-ray flux.

The first attempts to self-consistently model disk chemical, physical, and turbulent structures in full three dimension have been performed by *Turner et al.* (2006) and *Ilgner and Nelson* (2008). Both studies have employed a shearing-box approximation to calculate a patch of a three-dimensional MHD disk at radii of ~1–5 AU, treated the development of the MRI-driven turbulence, and focused on the multi-fluid evolution of the disk ionization state.

Ilgner and Nelson (2008) have confirmed their earlier findings (*Ilgner and Nelson*, 2006a,b) that turbulent mixing has no effect on the disk ionization structure in the absence of the gas-phase metals. The presence of metals, however, prolongs the recombination timescale, and the mixing is thus able to enliven the "dead" zone at r $\geq$ 5 AU (with the resulting $\alpha = 1$–5×10^{-3}).

The first three-dimensional MHD-disk model coupled with a limited time-dependent ionization chemistry was made by *Turner et al.* (2007). They have again confirmed that large-scale turbulent transport brings charged particles and radial magnetic fields toward the mid-plane, resulting in accretion stresses there that are only a few times lower than in the active surface layers. Later, *Flaig et al.* (2012) performed local three-dimensional radiative MHD simulations of different radii of a protosolar-like disk, including a simplified chemical network with recombination of charged particles on dust grains in the presence of the stellar X-rays, cosmic rays, and the decay of radionuclides. They have found that at the distance between 2 and 4 AU, a "dead zone" appears.

A detailed study of the global gas-grain ionization chemistry in the presence of turbulent mixing for a DM Tau-like disk model has been performed by *Semenov and Wiebe* (2011). It was found that turbulent diffusion does not affect abundances of simple metal ions and molecular ions

such as C^+, Mg^+, Fe^+, He^+, H^+_3, CH^+_3, and NH^+_4. On the other hand, charged species sensitive to the mixing include hydrocarbon ions, electrons, H^+, N_2H^+, HCO^+, N^+, OH^+, H_2O^+, etc. They have reconfirmed that the global ionization structure has a layered structure even in the presence of transport processes: (1) a heavily irradiated and ionized, hot atmosphere, where the dominant ions are C^+ and H^+; (2) a partly UV-shielded, warm molecular layer where carbon is locked in CO and the major ions are H^+, HCO^+, and H^+_3; and (3) a dark, dense, and cold mid-plane, where most of the molecules are frozen out onto dust grains, and the most abundant charged particles are dust grains and H^+_3.

Walsh et al. (2012) have investigated the impact of photochemistry and X-ray ionization on the molecular composition and ionization fraction in a T Tauri disk. The photoreaction rates were calculated using the local UV-wavelength spectrum and the wavelength-dependent photo cross sections. The same approach was utilized to model the transport of the stellar X-ray radiation. They have found that photochemistry is more important for global disk chemistry than the X-ray radiation. More accurate photochemistry affects the location of the H/H_2 and C^+/C dissociation fronts in the upper disk layer and increases abundances of neutral molecules in the outer disk region. *Walsh et al.* (2012) concluded that there is a potential "dead zone" with suppressed accretion located within the inner ~200 AU of the mid-plane region.

Several observational campaigns have begun to test some of these theoretical predictions. *Dutrey et al.* (2007a) have performed high-sensitivity interferometric observations with the IRAM PdBI array of N_2H^+ 1–0 in the disks around DM Tau, LkCa 15, and MWC 480. These data were used to derive the N_2H^+ column densities and, together with the HCO^+ measurements, were compared with a steady-state disk model with a vertical temperature gradient coupled to a gas-grain chemistry network. The derived N_2H+/HCO^+ ratio is on the order of 0.02–0.03, similar to the value observed in colder and darker prestellar cores. The observed values qualitatively agree with the modeled column densities at a disk age of a few million years, but the radial distribution of the molecules were not reproduced. The estimated ionization degree values from the HCO^+ and N_2H^+ data are ~2–7×10^{-9} (with respect to the total hydrogen density), which are typical for the warm molecular layers in the outer disk regions.

Öberg et al. (2011b) have presented interferometric observations with SMA of CO 3–2 in the DM Tau disk, and upper limits on H_2D^+ $1_{1,0}$–$0_{1,1}$ and N_2H^+ 4–3. With these data, IRAM 30-m observations of $H^{13}CO^+$ 3–2, and previous SMA observations of N_2H^+ 3–2, HCO^+ 3–2, and DCO^+ 3–2 (see *Öberg et al.*, 2010), they constrained ionization fraction using a parametric physical disk model. They have found that (1) in a warm molecular layer (T > 20 K), $HCO+$ is the dominant ion and the ionization degree is ~4×10^{-10}; (2) in a cooler layer around mid-plane where CO is depleted ($\leq$15–20 K), N_2H^+ and DCO^+ are the dominant ions, and the fractional ionization is ~3×10^{-11}; and (3) in the cold, dense mid-plane (T < 15 K), the isotopologs of H^+_3 are the main charge carriers, and the fractional ionization drops below $\sim 3 \times 10^{-10}$. These observations confirm that the disk ionization degree decreases toward deeper, denser, and better-shielded disk regions. Unfortunately, the best tracer of the ionization in the disk mid-plane, ortho-H_2D^+, still remains to be firmly detected in protoplanetary disks (see section 5.3 and the chapter by Ceccarelli et al. in this volume).

5.3. Deuteration

Deuterated molecules are important tracers of the thermal and chemical history in protoplanetary disks and the ISM (see also the chapter by Ceccarelli et al. in this volume). The cosmic D/H ratio in the local ISM is D/H $\approx 1.5 \times 10^{-5}$ (*Linsky et al.*, 2006), with a major reservoir of D locked in HD in dense regions. The deuterium fractionation processes redistribute the elemental deuterium from HD to other species at low temperatures, enhancing abundances and D/H ratios of many polyatomic species by orders of magnitude, as observed in the cold ISM (e.g., *Bacmann et al.*, 2003; *Bergin and Tafalla*, 2007; *Henning and Semenov*, 2013).

This is due to the difference in zero point vibrational energy between the H-bearing and D-bearing species, which implies barriers for backward reactions and thus favors production of deuterated species. The key process for gas-phase fractionation at $T \lesssim 10$–30 K is deuterium enrichment of H_3^+: $H_3^+ + HD \leftrightarrows H_2D^+ + H_2 + 232$ K (*Millar et al.*, 1989; *Gerlich et al.*, 2002). Upon fractionation of H_3^+, proton exchange reactions transfer the D enhancement to more complex gaseous species. One of the key reactions of this kind in protoplanetary disks is the formation of DCO^+: $H_2D^+ + CO \leftrightarrows DCO^+ + H_2$, which produces an enhanced DCO^+/HCO^+ ratios in outer cold disk mid-planes (at ~ 10–30 K). A similar reaction produces N_2D^+ in the coldest regions where gaseous CO is frozen-out: $H_2D^+ + N_2 \leftrightarrows N_2D^+ + H_2$.

In the outer mid-plane regions where CO and other molecules are severely depleted, multiply deuterated H_3^+ isotopologs attain large abundances (*Albertsson et al.*, 2013). Upon dissociative recombination large amounts of atomic H and D are produced. Atomic hydrogen and deuterium can then stick to grain surfaces and hydrogenate or deuterate the first generation of relatively simple ices. After that, CRP-driven/UV-driven photo and thermal processing of ice mantles allow more complex (organic) deuterated ices to be synthesized (e.g., multiply deuterated H_2O, H_2CO, CH_3OH, HCOOH, etc.).

Another pathway to redistribute the elemental D from HD via gas-phase fractionation is effective at higher temperatures, $\lesssim 70$–80 K: $CH_3^+ + HD \leftrightarrows CH_2D^+ + H_2 + 390$ K (*Asvany et al.*, 2004) and $C_2H_2^+ + HD \leftrightarrows C_2HD^+ + H_2 + 550$ K (*Herbst et al.*, 1987). After that, DCN molecules can be produced: $N + CH_2D^+ \rightarrow DCN^+ + H_2$, followed by protonation of DCN^+ by H_2 or HD and dissociative recombination of $DCNH^+$ (*Roueff et al.*, 2007).

Only three deuterated species have been detected in disks: DCO^+, DCN, and HD. The detection of the ground

transition $1_{0,1}$–$0_{0,0}$ of HDO in absorption in the disk of DM Tau by *Ceccarelli et al.* (2005) was later refuted by *Guilloteau et al.* (2006), and remains unproven. DCO^+ was first detected in the TW Hya disk (*van Dishoeck et al.*, 2003; *Qi et al.*, 2008), then in DM Tau by *Guilloteau et al.* (2006), and LkCa 15 by *Öberg et al.* (2010). The interpretation of the apparent $[DCO^+]/[HCO^+]$ ratio, ~ 0.02, is not simple, as HCO^+ is in general optically thick (e.g., *Öberg et al.*, 2011a), and both molecules have different spatial distributions (*Öberg et al.*, 2012).

DCN has been detected so far in two disks: in TW Hya by *Qi et al.* (2008) and in LkCa 15 by *Öberg et al.* (2010). The SMA and ALMA science verification observations show different spatial distributions of DCN and DCO^+, which imply different formation pathways (see *Qi et al.*, 2008; *Öberg et al.*, 2012). Deuteration for DCN proceeds more efficiently through proton exchange with deuterated light hydrocarbons like CH_3^+, whereas for DCO^+ low-temperature fractionation via H_3^+ isotopologs is important (see above).

Recently, the far-IR fundamental rotational line of HD has been detected in the TW Hya disk with Herschel (*Bergin et al.* 2013). The abundance distribution of HD closely follows that of H_2. Therefore, HD, which has a weak permanent dipole moment, can be used for probing the disk gas mass. The inferred mass of the TW Hya disk is $\gtrsim 0.05$ $M_\odot$, which is surprising for a ~ 3–10-m.y.-old system.

Among deuterated molecular species, H_2D^+ is potentially a critical probe of disk mid-planes, where it can be a dominant charged species (*Ceccarelli and Dominik*, 2005). It is rapidly destroyed by the ion-molecule reactions with CO and other molecules, and abundant enough only in the high-density cold mid-plane regions, where most of the molecules are frozen onto grains. Its lowest transition is unfortunately at 372 GHz, where the atmospheric transparency is limited. Current searches failed to detect it, the initial tentative reports of *Ceccarelli et al.* (2004) being superseded by $3\times$ more sensitive upper limits of *Chapillon et al.* (2011) using the James Clark Maxwell Telescope (JCMT) and Atacama Pathfinder Experiment (APEX) in DM Tau and TW Hya. *Chapillon et al.* (2011) compared the upper limit with the prediction from several disk models, varying the density, temperature, and outer radius), grain sizes (0.1, 1, and 10 μm), CO abundances (10^{-4}, 10^{-5}, and 10^{-6}), and rate of cosmic-ray ionization (10^{-17}, 3×10^{-17}, and 10^{-16} s^{-1}). The data only firmly exclude cases with both a low-CO abundance and small grains. This study also indicates that H_2D^+ is more difficult to detect than expected, and that even with the full ALMA, significant integration times will be needed to probe the physics and chemistry of the cold depleted mid-plane of T Tauri disks through H_2D^+ imaging.

5.4. Gas in Inner Cavities

It is worth mentioning that (sub)millimeter observations are in some cases sensitive to the gaseous content of the inner disks. The simple technique that comes to mind is the use of observations spatially resolving the inner disk (whatever arbitrary size an "inner disk" refers to). Unless the inner region under consideration is large, this requires a very fine angular resolution, which limits its applicability to the bright disks (and bright lines) and large holes. However, it is not necessary to spatially resolve the inner parts of the disk to have access to its content. Since disks are geometrically thin with a well-ordered velocity field, there is a mapping between velocity axis and position within the disks. Specifically, in a Keplerian disk, the line wings originate from the inner parts, so for example, the absence of high-velocity wings indicates an inner truncation. However, the line wings are also fainter than the core, which makes such analysis challenging from a sensitivity point of view. Another complication arises from the fact that the flux density from these wings is comparable if not weaker than the continuum flux density, calling for both a very good bandpass calibration and continuum handling.

The archetype of a large inner hole lies at the center of GG Tau, the famous multiple system, with a circumbinary ring surrounding GG Tau A. *Dutrey et al.* (1994) and *Guilloteau et al.* (1999) found that the dust and the gas circumbinary ring plus disk had the same inner radius. *Guilloteau and Dutrey* (2001) later found ^{12}CO 2–1 emission within the ring that possibly traces streaming material from the circumbinary ring feeding the circumprimary disks that would otherwise disappear through accretion of their material onto the host stars.

A handful of other systems have been imaged at high-angular resolution in molecular lines, providing interesting complementary information to that of the continuum (see the chapter by Espaillat et al. in this volume). For example, *Piétu et al.* (2007) found from CO 2–1 observations that CO is present in the inner part of the LkCa 15 disk, where dust emission presents a drop in emissivity, a situation similar to that in the AB Aur system (*Piétu et al.,* 2006; *Tang et al.,* 2012).

Using a spectroscopic approach, *Dutrey et al.* (2008) analyzed the J=2–1 line emission of CO isotopologs in the GM Aur disk (PdBI data) and used a global fit to the data to overcome the sensitivity issue. They showed that the inner radius in CO was comparable to that of the dust (*Hughes et al.,* 2009). A similar method was used to infer the presence of an inner gas cavity in the 40-m.y.-old 49 Ceti system (*Hughes et al.,* 2008). More recently, *Rosenfeld et al.* (2012) performed a similar analysis on TW Hya, taking advantage of the brightness of the relatively close system and enhanced sensitivity of the ALMA array. This allowed them to not only trace CO down to about 2 AU from the central star, but also to point out that a simple model fails to reproduce the observed line wings, calling for alternative models such as, e.g., a warped inner disk.

ALMA is already providing examples of the kind of structures that may be present in the inner disk, by the analogous examples at larger radii. Using ALMA, *Casassus et al.* (2013) were able to image multiple molecular gas

species within the dust inner hole of HD 142527 and utilize the varying optical depths of the different species to detect a flow of gas crossing the gap opened by a planet, while *van der Marel et al.* (2013) recently imaged a dust trap in the disk surrounding Oph IRS 48. Over the next several years, ALMA will likely uncover similar structures, and many more unexpected ones, at radii between a few and 30 AU.

6. SUMMARY AND OUTLOOK

Since *Protostars and Planets V* (*Reipurth et al.,* 2007), thermochemical models of disks, as chemical models, have been significantly improved but the detection rate of new molecules in disks has not significantly increased. However, there are some trends that appear from the recent observational results. Among them, it is important to mention the following:

1. So far, only simple molecules have been detected, the most complex being HC_3N and c-C_3H_2, showing that progress in understanding the molecular complexity in disks will require substantial observational efforts, even with ALMA. Similarly, the upper limits obtained on H_2D^+ indicate that probing the disk mid-plane will be difficult.

2. Multi-line, multi-isotopolog CO interferometric studies allow observers and modelers to retrieve the basic density and temperature structures of disks. This method will become more and more accurate with ALMA.

3. The CO, CN, HCN, CS, and CCH molecular studies apparently reveal low temperatures in the gas phase in several T Tauri disks. These observations suggest that the molecular layer is at least partly colder than predicted and that there are some ingredients still missing in our understanding of the disk physics and/or chemistry. More accurate ALMA data will refine these studies.

4. Dust is one of the major agents shaping the disk physics and chemistry. In particular, it plays a fundamental role in controlling the UV disk illumination and hence the temperature. It also strongly influences the chemistry since a significant part of the disk is at a temperature that is below the freeze-out temperature of most molecules. Although substantial progress has been made in chemical modeling, a more comprehensive treatment of chemical reactions on grain surfaces remains to be incorporated.

5. Deriving the total amount of gas mass is the most difficult task because the bulk of the gas, H_2, cannot be directly traced. The recent detection of HD with Herschel is therefore very promising. In the near future, the German Receiver for Astronomy at Terahertz Frequencies (GREAT) on SOFIA may allow detection of HD lines in other disks. Probing the gas-to-dust ratio will be then an even more complex step, as recent dust observations reveal that the dust properties also change radially. ALMA will fortunately provide better accuracy on the observable aspect of dust properties, the emissivity index.

6. With adequate angular resolution and sensitivity, ALMA will provide the first images of inner dust and gas disks ($R \leq 30$ AU), revealing the physics and chemistry of

the inner disk. In particular, these new observations will provide the first quantitative constraint on the ionization status, the turbulence, and the kinematics in the planet-forming area, as a departure to two-dimensional symmetry.

Acknowledgments. A.D. would like to acknowledge all CID members for a very fruitful collaboration, in particular Th. Henning, R. Launhardt, F. Gueth, and K. Schreyer. We also thank all the KIDA team (*http://kida.obs.u-bordeaux1.fr*) for providing chemical reaction rates for astrophysics. A.D. and S.G. thank the French Program PCMI for providing financial support. D.S. acknowledges support by the Deutsche Forschungsgemeinschaft through SPP 1385: The First Ten Million Years of the Solar System — A Planetary Materials Approach (SE 1962/1-2 and 1962/1-3).

REFERENCES

Agúndez M. et al. (2008) *Astron. Astrophys., 483,* 831.
Aikawa Y. (2007) *Astrophys. J. Lett., 656,* L93.
Aikawa Y. and Herbst E. (1999) *Astron. Astrophys., 351,* 233.
Aikawa Y. and Nomura H. (2006) *Astrophys. J., 642,* 1152.
Aikawa Y. et al. (1999) *Astrophys. J., 519,* 705.
Aikawa Y. et al. (2002) *Astron. Astrophys., 386,* 622.
Akimkin V. et al. (2013) *Astrophys. J., 766,* 8.
Albertsson T. et al. (2013) *Astrophys. J. Suppl., 207,* 27.
Aresu G. et al. (2011) *Astron. Astrophys., 526,* 163.
Aresu G. et al. (2012) *Astron. Astrophys., 547,* A69.
Asvany O. et al. (2004) *Astrophys. J., 617,* 685.
Bacmann A. et al. (2003) *Astrophys. J. Lett., 585,* L55.
Balbus S. A. and Hawley J. F. (1991) *Astrophys. J., 376,* 214.
Bergin E. A. (2011) In *Physical Processes in Circumstellar Disks Around Young Stars* (P. J. V. Garcia, ed.), pp. 55–113. Univ. of Chicago, Chicago.
Bergin E. A. and Tafalla M. (2007) *Annu. Rev. Astron. Astrophys., 45,* 339.
Bergin E. et al. (2003) *Astrophys. J. Lett., 591,* L159.
Bergin E. A. et al. (2010) *Astron. Astrophys., 521,* L33.
Bergin E. A. et al. (2013) *Nature, 493,* 644.
Bethell T. and Bergin E. (2009) *Science, 326,* 1675.
Bethell T. J. and Bergin E. A. (2011) *Astrophys. J., 739,* 78.
Boley A. C. et al. (2007) *Astrophys. J., 665,* 1254.
Boley A. C. et al. (2010) *Icarus, 207,* 509.
Brinch C. and Hogerheijde M. R. (2010) *Astron. Astrophys., 523,* A25.
Bruderer S. et al. (2009) *Astrophys. J., 700,* 872.
Bruderer S. et al. (2012) *Astron. Astrophys., 541,* A91.
Carr J. S. and Najita J. R. (2008) *Science, 319,* 1504.
Carr J. S. et al. (2004) *Astrophys. J., 603,* 213.
Casassus S. et al. (2013) *Nature, 493,* 191.
Ceccarelli C. and Dominik C. (2005) *Astron. Astrophys., 440,* 583.
Ceccarelli C. et al. (2004) *Astrophys. J. Lett., 607,* L51.
Ceccarelli C. et al. (2005) *Astrophys. J. Lett., 631,* L81.
Chapillon E. et al. (2008) *Astron. Astrophys., 488,* 565.
Chapillon E. et al. (2011) *Astron. Astrophys., 533,* A143.
Chapillon E. et al. (2012a) *Astron. Astrophys., 537,* A60.
Chapillon E. et al. (2012b) *Astrophys. J., 756,* 58.
Chiang E. I. and Goldreich P. (1997) *Astrophys. J., 490,* 368.
Cleeves L. I. et al. (2011) *Astrophys. J. Lett., 743,* L2.
Cleeves L. I. et al. (2013a) *Astrophys. J., 772,* 5.
Cleeves L. I. et al. (2013b) *Astrophys. J., 777,* 28.
Cuzzi J. N. et al. (2001) *Astrophys. J., 546,* 496.
D'Alessio P. et al. (1998) *Astrophys. J., 500,* 411.
Dartois E. et al. (2003) *Astron. Astrophys., 399,* 773.
de Gregorio-Monsalvo I. et al. (2013) *Astron. Astrophys., 557,* A133.
Draine B. T. (1978) *Astrophys. J., 36,* 595.
Dullemond C. P. and Monnier J. D. (2010) *Annu. Rev. Astron. Astrophys., 48,* 205.
Dullemond C. P. et al. (2007) In *Protostars and Planets V* (B. Reipurth et al., eds.), pp. 555–572. Univ. of Arizona, Tucson.
Dutrey A. et al. (1994) *Astron. Astrophys., 286,* 149.
Dutrey A. et al. (1997) *Astron. Astrophys., 317,* L55.

Dutrey A. et al. (2003) *Astron. Astrophys., 402,* 1003.
Dutrey A. et al. (2007a) *Astron. Astrophy., 464,* 615.
Dutrey A. et al. (2007b) In *Protostars and Planets V* (B. Reipurth et al.), pp. 495–506. Univ. of Arizona, Tucson.
Dutrey A. et al. (2008) *Astron. Astrophys., 490,* L15.
Dutrey A. et al. (2011) *Astron. Astrophys., 535,* A104.
Dzyurkevich N. et al. (2013) *ArXiv e-prints,* arXiv:1301.1487.
Elsila J. E. et al. (2009) *Meteoritics & Planet. Sci., 44,* 1323.
Favre C. et al. (2013) *Astrophys. J. Lett., 776,* L38.
Fayolle E. C. et al. (2011) *Astrophys. J. Lett., 739,* L36.
Fedele D. et al. (2012) *Astron. Astrophys., 544,* L9.
Fedele D. et al. (2013) *ArXiv e-prints,* arXiv:1308.1578.
Flaig M. et al. (2012) *Mon. Not. R. Astron. Soc., 420,* 2419.
Flock M. et al. (2011) *Astrophys. J., 735,* 122.
Flock M. et al. (2012a) *Astrophys. J., 744,* 144.
Flock M. et al. (2012b) *Astrophys. J., 761,* 95.
Fogel J. K. J. et al. (2011) *Astrophys. J., 726,* 29.
Fromang S. and Nelson R. P. (2006) *Astron. Astrophys., 457,* 343.
Fromang S. et al. (2011) *Astron. Astrophys., 534,* A107.
Gammie C. F. (1996) *Astrophys. J., 457,* 355.
Garozzo M. et al. (2010) *Astron. Astrophys., 509,* A67.
Garrod R. T. and Herbst E. (2006) *Astron. Astrophys., 457,* 927.
Garrod R. T. et al. (2007) *Astron. Astrophys., 467,* 1103.
Garrod R. T. et al. (2008) *Astrophys. J., 682,* 283.
Gerlich D. et al. (2002) *Planet. Space Sci., 50,* 1275.
Getman K. V. et al. (2009) *Astrophys. J., 699,* 1454.
Glassgold A. E. et al. (2004) *Astrophys. J., 615,* 972.
Glassgold A. E. et al. (2007) *Astrophys. J., 656,* 515.
Glassgold A. E. et al. (2009) *Astrophys. J., 701,* 142.
Glassgold A. E. et al. (2012) *Astrophys. J., 756,* 157.
Gorti U. and Hollenbach D. (2004) *Astrophys. J., 613,* 424.
Gorti U. and Hollenbach D. (2008) *Astrophys. J., 683,* 287.
Gorti U. and Hollenbach D. (2009) *Astrophys. J., 690,* 1539.
Gorti U. et al. (2009) *Astrophys. J., 705,* 1237.
Gräfe C. et al. (2013) *Astron. Astrophys., 553,* A69.
Gressel O. et al. (2012) *Mon. Not. R. Astron. Soc., 422,* 1140.
Güdel M. and Nazé Y. (2009) *Astron. Astrophys., 17,* 309.
Guilloteau S. and Dutrey A. (1998) *Astron. Astrophys., 339,* 467.
Guilloteau S. and Dutrey A. (2001) In *The Formation of Binary Stars* (H. Zinnecker and R. Mathieu, eds.), p. 229. IAU Symp. 200, Cambridge Univ., Cambridge.
Guilloteau S. et al. (1999) *Astron. Astrophys., 348,* 570.
Guilloteau S. et al. (2006) *Astron. Astrophys., 448,* L5.
Guilloteau S. et al. (2011) *Astron. Astrophys., 529,* A105.
Guilloteau S. et al. (2012) *Astron. Astrophys., 548,* A70.
Guilloteau S. et al. (2013) *Astron. Astrophys., 549,* A92.
Harada N. et al. (2010) *Astrophys. J., 721,* 1570.
Harada N. et al. (2012) *Astrophys. J., 756,* 104.
Hasegawa T. I. and Herbst E. (1993) *Mon. Not. R. Astron. Soc., 263,* 589.
Hasegawa Y. and Pudritz R. E. (2010a) *Astrophys. J. Lett., 710,* L167.
Hasegawa Y. and Pudritz R. E. (2010b) *Mon. Not. R. Astron. Soc., 401,* 143.
Heinzeller D. et al. (2011) *Astrophys. J., 731,* 115.
Henning T. and Semenov D. (2013) *ArXiv e-prints,* arXiv:1310.3151.
Henning T. et al. (2010) *Astrophys. J., 714,* 1511.
Herbst E. and Klemperer W. (1973) *Astrophys. J., 185,* 505.
Herbst E. and van Dishoeck E. F. (2009) *Annu. Rev. Astron. Astrophys., 47,* 427.
Herbst E. et al. (1987) *Astrophys. J., 312,* 351.
Hersant F. et al. (2009) *Astron. Astrophys., 493,* L49.
Hogerheijde M. R. and van der Tak F. F. S. (2000) *Astron. Astrophys., 362,* 697.
Hogerheijde M. R. et al. (2011) *Science, 334,* 338.
Hughes A. M. et al. (2008) *Astrophys. J., 681,* 626.
Hughes A. M. et al. (2009) *Astrophys. J., 698,* 131.
Hughes A. M. et al. (2011) *Astrophys. J., 727,* 85.
Igea J. and Glassgold A. E. (1999) *Astrophys. J., 518,* 848.
Ilee J. D. et al. (2011) *Mon. Not. R. Astron. Soc., 417,* 2950.
Ilgner M. and Nelson R. P. (2006a) *Astron. Astrophys., 445,* 205.
Ilgner M. and Nelson R. P. (2006b) *Astron. Astrophys., 445,* 223.
Ilgner M. and Nelson R. P. (2006c) *Astron. Astrophys., 455,* 731.
Ilgner M. and Nelson R. P. (2008) *Astron. Astrophys., 483,* 815.
Ilgner M. et al. (2004) *Astron. Astrophys., 415,* 643.
Isella A. et al. (2007) *Astron. Astrophys., 469,* 213.
Isella A. et al. (2009) *Astrophys. J., 701,* 260.

Jonkheid B. et al. (2004) *Astron. Astrophys., 428,* 511.
Jonkheid B. et al. (2006) *Astron. Astrophys., 453,* 163.
Jonkheid B. et al. (2007) *Astron. Astrophys., 463,* 203.
Kamp I. and Dullemond C. P. (2004) *Astrophys. J., 615,* 991.
Kamp I. and van Zadelhoff G.-J. (2001) *Astron. Astrophys., 373,* 641.
Kamp I. et al. (2010) *Astron. Astrophys., 510,* A18.
Kamp I. et al. (2011) *Astron. Astrophys., 532,* A85.
Kastner J. H. et al. (1997) *Science, 277,* 67.
Koerner D. W. et al. (1993) *Icarus, 106,* 2.
Kraus S. et al. (2009) *Astron. Astrophys., 508,* 787.
Lada C. J. and Lada E. A. (2003) *Annu. Rev. Astron. Astrophys., 41,* 57.
Le Bourlot J. et al. (1993) *Astron. Astrophys., 267,* 233.
Le Petit F. et al. (2006) *Astrophys. J. Suppl., 164,* 506.
Leger A. et al. (1985) *Astron. Astrophys., 144,* 147.
Lesniak M. V. and Desch S. J. (2011) *Astrophys. J,* 118.
Linsky J. L. et al. (2006) *Astrophys. J., 647,* 1106.
Maaskant K. M. et al. (2013) *Astron. Astrophys, 555,* A64.
Martin R. G. et al. (2012) *Mon. Not. R. Astron. Soc., 420,* 3139.
Mathews G. S. et al. (2013) *Astron. Astrophys., 557,* A132.
McClure M. K. et al. (2012) *Astrophys. J. Lett., 759,* L10.
McClure M. K. et al. (2013) *Astrophys. J., 775,* 114.
McElroy D. et al. (2013) *Astron. Astrophys., 550,* A36.
Meeus G. et al. (2010) *Astron. Astrophys., 518,* L124.
Meeus G. et al. (2012) *Astron. Astrophys., 544,* A78.
Meheut H. et al. (2010) *Astron. Astrophys., 516,* A31.
Meheut H. et al. (2012a) *Astron. Astrophys., 545,* A134.
Meheut H. et al. (2012b) *Mon. Not. R. Astron. Soc., 422,* 2399.
Meijerink R. et al. (2008) *Astrophys. J., 676,* 518.
Meijerink R. et al. (2009) *Astrophys. J., 704,* 1471.
Meijerink R. et al. (2012) *Astron. Astrophys., 547,* A68.
Millar T.-J. et al. (1989) *Astrophys. J., 340,* 906.
Mohanty S. et al. (2013) *Astrophys. J., 764,* 65.
Morishima R. (2012) *Mon. Not. R. Astron. Soc., 420,* 2851.
Mulders G. D. and Dominik C. (2012) *Astron. Astrophys., 539,* A9.
Mulders G. D. et al. (2011) *Astron. Astrophys., 531,* A93.
Mulders G. D. et al. (2013) *ArXiv e-prints,* arXiv:1306.4264.
Mumma M. J. and Charnley S. B. (2011) *Annu. Rev. Astron. Astrophys., 49,* 471.
Najita J. et al. (2001) *Astrophys. J., 561,* 880.
Najita J. R. et al. (2011) *Astrophys. J., 743,* 147.
Najita J. R. et al. (2013) *Astrophys. J., 766,* 134.
Nomura H. and Millar T. J. (2005) *Astron. Astrophys., 438,* 923.
Nomura H. et al. (2007) *Astrophys. J., 661,* 334.
Nomura H. et al. (2009) *Astron. Astrophys., 495,* 183.
Öberg K. I. et al. (2007) *Astrophys. J. Lett., 662,* L23.
Öberg K. I. et al. (2009a) *Astron. Astrophys., 496,* 281.
Öberg K. I. et al. (2009b) *Astrophys. J., 693,* 1209.
Öberg K. I. et al. (2010) *Astrophys. J., 720,* 480.
Öberg K. I. et al. (2011a) *Astrophys. J., 734,* 98.
Öberg K. I. et al. (2011b) *Astrophys. J., 743,* 152.
Öberg K. I. et al. (2012) *Astrophys. J., 749,* 162.
Oishi J. S. et al. (2007) *Astrophys. J., 670,* 805.
Oppenheimer M. and Dalgarno A. (1974) *Astrophys. J., 192,* 29.
Padovani M. et al. (2012) *Astron. Astrophys., 543,* A16.
Pavlyuchenkov Y. et al. (2007) *Astrophys. J., 669,* 1262.
Pierce D. M. and A'Hearn M. F. (2010) *Astrophys. J., 718,* 340.
Piétu V. et al. (2005) *Astron. Astrophys., 443,* 945.
Piétu V. et al. (2006) *Astron. Astrophys., 460,* L43.
Piétu V. et al. (2007) *Astron. Astrophys., 467,* 163.
Pinte C. et al. (2006) *Astron. Astrophys., 459,* 797.
Pinte C. et al. (2008) *Astron. Astrophys., 489,* 633.
Pinte C. et al. (2009) *Astron. Astrophys., 498,* 967.
Podio L. et al. (2012) *Astron. Astrophys., 545,* A44.
Podio L. et al. (2013) *Astrophys. J. Lett., 766,* L5.
Preibisch T. et al. (2005) *Astrophys. J., 160,* 401.
Qi C. et al. (2004) *Astrophys. J. Lett., 616,* L11.
Qi C. et al. (2006) *Astrophys. J. Lett., 636,* L157.
Qi C. et al. (2008) *Astrophys. J., 681,* 1396.
Qi C. et al. (2011) *Astrophys. J., 740,* 84.
Qi C. et al. (2013a) *Astrophys. J. Lett., 765,* L14.
Qi C. et al. (2013b) *Astrophys. J., 765,* 34.
Qi C. et al. (2013c) *Science, 341,* 630.
Reipurth B. et al., eds. (2007) *Protostars and Planets V.* Univ. of Arizona, Tucson.

Riviere-Marichalar P. et al. (2012) *Astron. Astrophys., 538,* L3.
Riviere-Marichalar P. et al. (2013) *Astron. Astrophys., 555,* A67.
Roberge A. et al. (2005) *Astrophys. J., 622,* 1171.
Rosenfeld K. A. et al. (2012) *Astrophys. J., 757,* 129.
Rosenfeld K. A. et al. (2013) *Astrophys. J., 774,* 16.
Rothman L. S. et al. (2005) *J. Quant. Spec. Radiat. Transf., 96,* 139.
Roueff E. et al. (2007) *Astron. Astrophys., 464,* 245.
Sano T. et al. (2000) *Astrophys. J., 543,* 486.
Semenov D. (2010) *ArXiv e-prints,* arXiv:1011.4770.
Semenov D. and Wiebe D. (2011) *Astrophys. J., 196,* 25.
Semenov D. et al. (2004) *Astron. Astrophys., 417,* 93.
Semenov D. et al. (2005) *Astrophys. J., 621,* 853.
Semenov D. et al. (2006) *Astrophys. J. Lett., 647,* L57.
Semenov D. et al. (2008) *Astrophys. J. Lett., 673,* L195.
Semenov D. et al. (2010) *Astron. Astrophys., 522,* A42.
Shakura N. I. and Sunyaev R. A. (1973) *Astron. Astrophys., 24,* 337.
Sturm B. et al. (2010) *Astron. Astrophys., 518,* L129.
Sturm B. et al. (2013) *Astron. Astrophys., 553,* A5.
Suzuki T. K. et al. (2010) *Astrophys. J., 718,* 1289.
Tang Y.-W. et al. (2012) *Astron. Astrophys., 547,* A84.
Thi W.-F. et al. (2004) *Astron. Astrophys., 425,* 955.
Thi W.-F. et al. (2011a) *Astron. Astrophys., 530,* L2.
Thi W.-F. et al. (2011b) *Astron. Astrophys., 530,* L2.
Thi W.-F. et al. (2011c) *Mon. Not. R. Astron. Soc., 412,* 711.
Thi W.-F. et al. (2013) *Astron. Astrophys., 557,* A111.
Tilling I. et al. (2012) *Astron. Astrophys., 538,* A20.
Tscharnuter W. M. and Gail H.-P. (2007) *Astron. Astrophys., 463,* 369.
Turner N. J. et al. (2006) *Astrophys. J., 639,* 1218.
Turner N. J. et al. (2007) *Astrophys. J., 659,* 729.
Umebayashi T. and Nakano T. (1980) *Publ. Astron. Soc. Japan, 32,* 405.
Umebayashi T. and Nakano T. (2009) *Astrophys. J., 690,* 69.
van Boekel R. et al. (2004) *Nature, 432,* 479.
van der Marel N. et al. (2013) *Science, 340,* 1199.
van Dishoeck E. F. (2006) *Proc. Natl. Acad. Sci., 103,* 12249.
van Dishoeck E. F. and Black J. H. (1988) *Astrophys. J., 334,* 771.
van Dishoeck E. F. et al. (2003) *Astron. Astrophys., 400,* L1.
van Dishoeck E. F. et al. (2006) In *Photoprocesses in Protoplanetary Disks,* p. 231. Faraday Discussions Vol. 133, Royal Society of Chemistry, Cambridge.
van Zadelhoff G.-J. et al. (2003) *Astron. Astrophys., 397,* 789.
Vasyunin A. I. et al. (2004) *Astron. Lett., 30,* 566.
Vasyunin A. I. et al. (2008) *Astrophys. J., 672,* 629.
Vasyunin A. I. et al. (2011) *Astrophys. J., 727,* 76.
Visser R. et al. (2009) *Astron. Astrophys., 503,* 323.
Visser R. et al. (2011) *Astron. Astrophys., 534,* A132.
Vorobyov E. I. and Basu S. (2006) *Astrophys. J., 650,* 956.
Vorobyov E. I. and Basu S. (2010) *Astrophys. J., 719,* 1896.
Wakelam V. et al. (2004) *Astron. Astrophys., 422,* 159.
Wakelam V. et al. (2005) *Astron. Astrophys., 444,* 883.
Wakelam V. et al. (2010) *Astron. Astrophys., 517,* A21.
Wakelam V. et al. (2012) *Astrophys. J. Suppl., 199,* 21.
Walsh C. et al. (2010) *Astrophys. J., 722,* 1607.
Walsh C. et al. (2012) *Astrophys. J., 747,* 114.
Walsh C. et al. (2013) *Astrophys. J. Lett., 766,* L23.
Walsh C. et al. (2014) *Astron. Astrophys., 563,* A33.
Watson W. D. (1974) *Astrophys. J., 188,* 35.
Willacy K. (2007) *Astrophys. J., 660,* 441.
Willacy K. and Woods P. M. (2009) *Astrophys. J., 703,* 479.
Willacy K. et al. (2006) *Astrophys. J., 644,* 1202.
Williams J. P. and Cieza L. A. (2011) *Annu. Rev. Astron. Astrophys., 49,* 67.
Woitke P. et al. (2009a) *Astron. Astrophys., 501,* L5.
Woitke P. et al. (2009b) *Astron. Astrophys., 501,* 383.
Woitke P. et al. (2010) *Mon. Not. R. Astron. Soc., 405,* L26.
Woodall J. et al. (2007) *Astron. Astrophys., 466,* 1197.
Woods P. M. and Willacy K. (2007) *Astrophys. J. Lett., 655,* L49.
Woods P. M. and Willacy K. (2009a) *Astrophys. J., 693,* 1360.
Woods P. M. and Willacy K. (2009b) *Astrophys. J., 693,* 1360.
Zaleski D. P. et al. (2013) *Astrophys. J. Lett., 765,* L10.
Zhang K. et al. (2013) *Astrophys. J., 766,* 82.
Zhu Z. et al. (2010a) *Astrophys. J., 713,* 1134.
Zhu Z. et al. (2010b) *Astrophys. J., 713,* 1143.
Zhu Z. et al. (2012) *Astrophys. J., 746,* 110.

Testi L., Birnstiel T., Ricci L., Andrews S., Blum J., Carpenter J., Dominik C., Isella A., Natta A., Williams J. P., and Wilner D. J. (2014)
Dust evolution in protoplanetary disks. In *Protostars and Planets VI* (H. Beuther et al., eds.), pp. 339–361. Univ. of Arizona, Tucson,
DOI: 10.2458/azu_uapress_9780816531240-ch015.

Dust Evolution in Protoplanetary Disks

Leonardo Testi
European Southern Observatory, Istituto Nazionale di Astrofisica, and Excellence Cluster Universe

Tilman Birnstiel
Harvard-Smithsonian Center for Astrophysics

Luca Ricci
California Institute of Technology

Sean Andrews
Harvard-Smithsonian Center for Astrophysics

Jürgen Blum
Technische Universität Braunschweig

John Carpenter
California Institute of Technology

Carsten Dominik
University of Amsterdam

Andrea Isella
California Institute of Technology

Antonella Natta
Dublin Institute for Advanced Studies and Istituto Nazionale di Astrofisica

Jonathan P. Williams
University of Hawaii

David J. Wilner
Harvard-Smithsonian Center for Astrophysics

In the core-accretion scenario for the formation of planetary rocky cores, the first step toward planet formation is the growth of dust grains into larger and larger aggregates and eventually planetesimals. Although dust grains are thought to grow up to micrometer-sized particles in the dense regions of molecular clouds, the growth to pebbles and kilometer-sized bodies must occur at the high densities within protoplanetary disks. This critical step is the last stage of solids evolution that can be observed directly in extrasolar systems before the appearance of large planetary-sized bodies. In this chapter we review the constraints on the physics of grain-grain collisions as they have emerged from laboratory experiments and numerical computations. We then review the current theoretical understanding of the global processes governing the evolution of solids in protoplanetary disks, including dust settling, growth, and radial transport. The predicted observational signatures of these processes are summarized. We briefly discuss grain growth in molecular cloud cores and in collapsing envelopes of protostars, as these likely provide the initial conditions for the dust in protoplanetary disks. We then review the observational constraints on grain growth in disks from millimeter surveys, as well as the very recent evidence for radial variations of the dust properties in disks. We also include a brief discussion on the small end of the grain size distribution and dust settling as derived from optical, near-, and mid-infrared observations. Results are discussed in the context of global dust-evolution models; in particular, we focus on the emerging evidence for a very efficient early growth of grains and the radial distribution of maximum grain sizes as the result of growth barriers. We also highlight the limits of the current models of dust evolution in disks, including the need to slow the radial drift of grains to overcome the migration/fragmentation barrier.

1. INTRODUCTION

In this chapter we will discuss the evolution of dust in protoplanetary disks, focusing on the processes of grain growth and the observational consequences of this process. In the standard scenario for planet formation, this is the phase in which the solids grow from micrometer-sized particles, which are present in the molecular cloud cores out of which stars and protoplanetary disks are formed, to centimeter-sized and beyond on the path to become planetesimals. This is the last stage of solid growth that is directly observable before the formation of large, planetary-sized bodies that can be individually observed. As this phase of growth is directly observable, it has the potential of setting strong constraints on the initial stages of the planet-formation process. In this review, we focus on the growth of particles on the disk mid-plane, as this is where most of the solid mass of the disk is concentrated and where planets are expected to form. Grain growth in the disk inner regions and atmosphere can be effectively investigated in the IR and have been extensively reviewed in *Natta et al.* (2007), although at the time of that review the observational evidence and theoretical understanding of the global dust evolution processes in the disk were limited.

This phase of growth up to centimeter-sized particles in protoplanetary disks can also be connected to the study of the most pristine solids in our own solar system. We can directly follow grain growth in the solar nebula through the study of primitive rocky meteorites known as chondrites. Chondrites are composed mainly of millimeter-sized, spherical chondrules with an admixture of smaller, irregularly shaped calcium-aluminum-rich inclusions (CAIs) embedded in a fine-grained matrix (e.g., *Scott and Krot*, 2005). Some chondrules may be the splash from colliding planetesimals, but most have properties that are consistent with being flash-heated agglomerates of the micrometer-sized silicate dust grains in the matrix. Calcium-aluminum-rich inclusions are the first solids to condense from a hot, low-pressure gas of solar composition. Their ages provide the zero point for cosmochemical timescales and the astronomical age of the Sun, 4.567 Ga (i.e., formed 4.567 billion years ago). Relative ages can be tracked by the decay of short-lived radionuclides and allow the formation timescales of dust agglomerates and planetesimals to be placed in context with astronomical observations (*Dauphas and Chaussidon*, 2011). Recent measurements of absolute ages of the chondrules found that these were formed over a period that span from CAI formation to a few million years beyond (*Connelly et al.*, 2012). As chondrule dating refers to the time when the dust agglomerate was melted, this new result tells us either that dust grains rapidly grew up to millimeter sizes in the early solar nebula and were melted progressively over time while agglomeration into larger bodies continued over the lifetime of the protoplanetary disk, or that the growth of these millimeter- and centimeter-sized agglomerates from smaller particles continued over a period of several million years. In any case, the process of assembling CAIs and chondrules into

larger bodies had to occur throughout the disk lifetime (see also the chapter by Johansen et al. in this volume).

In this review we will thus concentrate on the most recent developments of the theoretical models for grain evolution in disks and the constraints on the grain-grain collisions from laboratory experiments as well as on the new wealth of data at (sub)millimeter and centimeter wavelengths that is becoming available. The goal will be to give an overview of the current astronomical constraints on the process and timescale of grain growth in disks.

The processes and observations that we discuss in this chapter are expected to occur in protoplanetary disks, and both the theoretical models and the interpretation of the observations rely on assumptions on the disk structure and its evolution. The structure and evolution of protoplanetary disks during the pre-main-sequence phase of stellar evolution has been discussed extensively in recent reviews and books (e.g., *Dullemond et al.*, 2007; *Hartmann*, 2009; *Armitage*, 2010; *Williams and Cieza*, 2011). Throughout this chapter, our discussion will assume a flared, irradiated disk structure in hydrostatic equilibrium with a constant gas to dust ratio of 100 by mass, unless explicitly stated otherwise. While the detailed disk structure and its evolution under the effects of viscous accretion, chemical evolution, photoevaporation, and planet-disk interaction are actively investigated in detail (see, e.g., the chapters by Alexander et al., Audard et al., Dutrey et al., and Turner et al. in this volume), this assumed disk structure is an adequate representation of the early phases of disk evolution and a good reference for the processes of dust evolution.

Figure 1 shows a sketch of a protoplanetary disk where we illustrate pictorially the physical processes involved in grain growth on the lefthand side. On the righthand side we illustrate the regions probed by the various observational techniques and the angular resolutions offered by the forthcoming generation of facilities and instruments in the IR and submillimeter regions.

In section 2, we introduce the main concepts at the basis of the transport, dynamics, and evolution of solid particles in disks. First we introduce the interactions between the solids and gas and the complex dynamics of solids in disks, then we describe the basic processes of the growth of solids toward larger aggregates. In section 3 we discuss constraints on the outcomes of grain-grain collisions from laboratory experiments and numerical computations. These provide the basic constraints to the global-evolution models of solids in disks, which we describe in section 4. In section 5 we discuss the effects of growth on the dust opacities and the observational consequences. The possible recent evidence for grain evolution at and before the disk-formation epoch is discussed in section 6. The observational constraints for grain growth at IR wavelengths are described in section 7. We discuss grain properties in the disk mid-plane from (sub) millimeter observations in section 8, beginning with the methodology in section 8.1, then the results of low resolution, multi-wavelength continuum surveys in section 8.2, ending with the most recent resolved studies in section 8.3.

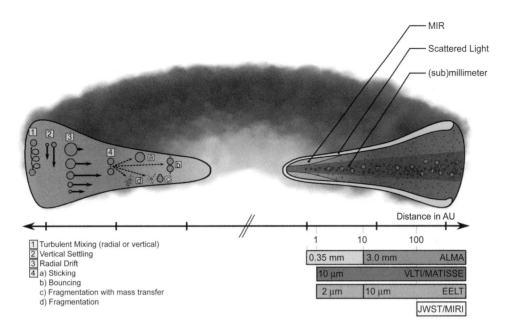

Fig. 1. Illustration of the structure, grain evolution processes, and observational constraints for protoplanetary disks. On the lefthand side we show the main grain transport and collision mechanism properties. The different lengths of the arrows illustrate the different velocities of the different grains. On the righthand side, we show the areas of the disk that can be probed by the various techniques. The axis shows the logarithmic radial distance from the central star. The horizontal bars show the highest angular resolutions (left edge of the bars) that can be achieved with a set of upcoming facilities and instruments for the typical distance of the nearest star-forming regions.

2. DUST TRANSPORT AND GROWTH PROCESSES IN DISKS

2.1. Dust Transport Processes in Disks

2.1.1. Drag forces. Dust particles embedded in gaseous protoplanetary disks are not orbiting freely, but feel friction when moving with respect to the gas. The force exerted on them depends not only on the relative motion between gas and dust, but also on the particle size: Small particles that are observable at up to centimeter wavelength can quite safely be assumed to be smaller than the mean free path of the gas molecules and are thus in the Epstein regime. If the particles are larger than about the mean free path of the gas molecules, a flow structure develops around the dust particle and the drag force is said to be in the Stokes drag regime (*Whipple,* 1972; *Weidenschilling,* 1977a). Large particles in the inner few astronomical units of the disk could be in this regime, and the transition into the Stokes drag regime might be important for trapping of dust particles and the formation of planetesimals (e.g., *Birnstiel et al.,* 2010a; *Laibe et al.,* 2012; *Okuzumi et al.,* 2012). An often used quantity is the stopping time, or friction time, which is the characteristic timescale for the acceleration or deceleration of the dust particles $\tau_s = mv/F$, where m and v are the particle mass and velocity, and F is the drag force. Even more useful is the concept of the Stokes number, which in this context is defined as

$$St = \Omega_K \tau_s \qquad (1)$$

a dimensionless number, which relates the stopping time to the orbital period Ω_K. The concept of the Stokes number is useful because particles of different shapes, sizes, or composition, or in a different environment, have identical aerodynamical behavior if they have the same Stokes number.

2.1.2. Radial drift. The simple concept of drag force leads to important implications, the first of which — radial drift — was realized by *Whipple* (1972), *Adachi et al.* (1976), and *Weidenschilling* (1977a): An orbiting parcel of gas is in a force balance between gravitational, centrifugal, and pressure forces. The pressure gradient is generally pointing outward because densities and temperatures are higher in the inner disk. This additional pressure support results in a slightly sub-Keplerian orbital velocity for the gas. In contrast, a freely orbiting dust particle feels only centrifugal forces and gravity, and should therefore be in a Keplerian orbit. This slight velocity difference between gas and a free-floating dust particle thus causes an efficient deceleration of the dust particle, once embedded in the gaseous disk. Consequently, the particle loses angular momentum and spirals toward smaller radii. This inward drift velocity is only a small fraction of the total orbital velocity (a few permil), but, for St ~ 1 particles, it still leads to an inward drift speed on the order of 50 m s⁻¹. It also means that particles of different sizes acquire very different radial velocities and also that at any given radius, dust of the right size may be quickly moving toward the central star.

Nakagawa et al. (1986) investigated the equations of motion for arbitrary gas-to-dust ratios. For a value of 100 by mass, gas is dynamically dominating and the classical results of *Weidenschilling* (1977a) are recovered; however for a decreasing ratio, the drag that the dust exerts on the gas becomes more important, and eventually may be reversed: Dust would not drift inward and gas would be pushed outward instead. However, much lower gas-to-dust ratios, approaching unity, also lead to other effects such as the streaming instability (*Youdin and Goodman,* 2005) or self-induced stirring (for details, see the chapter by Johansen et al. in this volume).

The radial drift process does not necessarily mean that all particles are falling into the star, as dust may pile up at some specific locations in the innermost regions of the disk where the gas-to-dust ratio will become very low [this process has actually been suggested as an explanation for the abundance of rocky exoplanets close to the host star (*Chatterjee and Tan,* 2014)]; however, a significant fraction of large grains need to be kept in the outer disk for long timescales, otherwise it would result in a stark contrast to the observed population of millimeter-sized grains in the outer disk (see section 8). In the next paragraph we will discuss some processes that may oppose and slow down the radial drift.

2.1.3. Dust trapping. Many works have addressed details of radial drift (e.g., *Whipple,* 1972; *Weidenschilling,* 1977a; *Youdin and Shu,* 2002; *Takeuchi and Lin,* 2002; *Brauer et al.,* 2007); however, the main conclusions remain: Unless the gas to dust ratio is very low, or disks are much more massive than seems reasonable, observable particles should drift to the inner regions within only a small fraction of the lifetimes of protoplanetary disks. Observable here means detectable with current (sub)millimeter observatories, i.e., mainly dust in the outer regions (beyond ~20 AU) of the disk at current resolutions. The only other way to stop particles from spiraling inward is to locally reverse the pressure gradient, as can be seen from the drift velocity (*Nakagawa et al.,* 1986)

$$u_{drift} = \frac{1}{St + St^{-1}(1+\epsilon)^2} \frac{c_s^2}{V_K} \frac{\partial \ln P}{\partial \ln r} \quad (2)$$

where P is the gas pressure, $c_s = \sqrt{k_B T / \mu m_p}$ the isothermal sound speed, V_K the Keplerian velocity, and ϵ the dust-to-gas ratio. If the pressure gradient is zero or positive, there is no radial drift or particles drift outward.

This basic mechanism of dust drifting to regions of higher pressure has been explored in different settings: *Barge and Sommeria* (1995) showed that anticyclonic vortices represent a high-pressure region that can accumulate dust particles (see also *Klahr and Henning,* 1997; *Fromang and Nelson,* 2005). The dust drift mechanism also works efficiently in the azimuthal direction if there exist regions of azimuthal overpressure (e.g., *Birnstiel et al.,* 2013); it should be stressed, however, that the mechanism relies on relative motion between gas and dust. As an example of this exception, an overdensity caused by an eccentric disk does not trap dust particles, as shown by *Hsieh and Gu*

(2012) and *Ataiee et al.* (2013). For an eccentric disk, the overdensity, and thus the pressure maximum, arises due to the nonconstant azimuthal velocity along the elliptic orbits, but the same holds for the dust, i.e., the velocity difference between dust and the partially pressure supported gas remain, and consequently the radial drift mechanism is still active but the dust is not concentrated azimuthally with respect to the gas.

2.1.4. Radial mixing and meridional flows. The drift motion toward higher pressure is not the only mode of transport of dust; there are two additional ones: mixing and advection. These are interesting as they may oppose radial drift under certain circumstances and may provide an explanation for processed dust in the outer disk. Outward mixing processes, including winds (e.g., *Shu et al.,* 1994, 2001), have been invoked to explain the presence of crystalline solids in disks (*Keller and Gail,* 2004; *Ciesla,* 2009) and comets in our own solar system (*Bockelée-Morvan et al.,* 2002). Evidence for dust processing in the inner regions of disks and subsequent radial transport has been recently observed directly by Spitzer (*Ábrahám et al.,* 2009; *Juhász et al.,* 2012). In this review we will not discuss those topics that have been extensively covered elsewhere (e.g., *Natta et al.,* 2007; *Dullemond and Monnier,* 2010; *Williams and Cieza,* 2011); instead, we provide below a brief account of the radial transport processes.

Advection is a result of the viscous evolution of the gas, which essentially drags dust particles along with the radial gas velocity, as long as they are efficiently coupled to the gas. This drag velocity was derived, e.g., in *Takeuchi and Lin* (2002) as

$$u_{r,drag} = \frac{u_{r,gas}}{1 + St^2} \quad (3)$$

which states that only small particles (St < 1) are effectively coupled with the gas. It was first shown by *Urpin* (1984) that the gas velocity of a region with net inward motion can be positive at the mid-plane. This flow pattern, called meridional flow, allows for outward transport of dust grains. This process is typically much weaker than radial drift and thus unable to explain millimeter-sized particles in the outer disk. In addition, large-scale circulation patterns of meridional flows, while present in viscous disk simulations (*Kley and Lin,* 1992; *Rozyczka et al.,* 1994), are not reproduced in magnetorotational instability (MRI) turbulent simulations (*Fromang et al.,* 2011).

In addition to this systematic motion, the gas is also thought to be turbulent (see also the chapters by Dutrey et al. and Turner et al. in this volume), and the dust is therefore mixed by this turbulent stirring as well. The fact that the dust is not necessarily perfectly coupled to the gas motion leads to some complications (see *Youdin and Lithwick,* 2007). The ratio between gas viscosity ν_{gas} and gas diffusivity D_{gas} is called the Schmidt number. It is usually assumed that the gas diffusivity equals the gas viscosity, in which case the Schmidt number for the dust is defined as

$$Sc = \frac{D_{gas}}{D_{dust}} \qquad (4)$$

Youdin and Lithwick (2007) included effects of orbital dynamics, which were neglected in previous studies of *Cuzzi et al.* (1993) and *Schräpler and Henning* (2004), and showed that the Schmidt number can be approximated by $Sc \simeq 1 + St^2$. It is currently believed that none of these radial transport mechanisms can overcome the radial drift induced by the pressure gradient, because the vertically integrated net transport velocity is dominated by the much stronger radial drift velocity (e.g., *Jacquet, 2013*).

2.1.5. Vertical mixing and settling. A dust particle elevated above the gas mid-plane, orbiting at the local Keplerian velocity, would vertically oscillate due to its orbital motion if it were not for the gas drag force, which damps motion relative to the gas flow. Following the concepts discussed in the previous sections, particles with Stokes number smaller than unity are thus damped effectively within one orbit. Larger particles will experience damped vertical oscillations. The terminal velocity of small particles can be calculated by equating the vertical component of the gravitational acceleration and the deceleration by drag forces. The resulting settling velocity for $St < 1$, $v_{sett} = St\,\Omega_K\,z$, increases with particle size and height above the mid-plane.

Vertical settling was already the focus of the earliest models of planetesimal formation (e.g., *Safronov*, 1969; *Goldreich and Ward*, 1973; *Weidenschilling*, 1980; *Nakagawa et al.*, 1986), and has remained an active topic of debate (e.g., *Cuzzi et al.*, 1993; *Johansen and Klahr*, 2005; *Carballido et al.*, 2006; *Bai and Stone*, 2010). The main question, however, is usually not the effectiveness of the settling motion itself, but the effectiveness of the opposing mechanism: turbulent mixing. *Dubrulle et al.* (1995) calculated the vertical structure of the dust disk by solving for an equilibrium between settling and mixing effects. The resulting scale height of the dust concentration in an isothermal disk with scale height $H_p = c_s/\Omega_K$ then becomes $H_{dust} = H_p/\sqrt{1 + St/\alpha_t}$, where we used the canonical turbulence prescription with parameter α_t (*Shakura and Sunyaev*, 1973). The detailed magneto-hydrodynamic (MHD) models of MRI turbulent disks of *Fromang and Nelson* (2009) show that the simple result from *Dubrulle et al.* (1995) is a good approximation only near the mid-plane. Above roughly one scale height, the vertical variations in the strength of the turbulence and other quantities cause strong variations from the result of *Dubrulle et al.* (1995).

The effects that dust settling has on the observational appearance of disks have been derived, e.g., by *Chiang et al.* (2001), *D'Alessio et al.* (2001), and *Dullemond and Dominik* (2004). While the settling process could lead to the rapid growth of particles and the formation of a thin mid-plane layer of large pebbles, it is expected that small dust is replenished in the disk upper layers, e.g., by small fragments from shattering collisions between dust grains (e.g., *Dullemond and Dominik*, 2005; *Birnstiel et al.*, 2009;

Zsom et al., 2011) or by continuous infall (e.g., *Mizuno et al.*, 1988; *Dominik and Dullemond*, 2008).

2.2. Grain Growth Processes

The transport mechanisms discussed in the previous sections all lead to large differential vertical and radial motion of dust particles. These in turn imply frequent grain-grain collisions, potentially leading to growth. The two main ingredients for a model of dust growth are the collision frequency and the collision outcome. The growth process is modeled as the result of primordial dust particles, referred to as monomers, that can stick together to form larger aggregates. The latter depends on many different parameters, such as the composition (e.g., fraction of icy, silicate, or carbonaceous particles), monomer size distribution, structure (i.e., compact, porous, fractal grains, layers, etc.), impact parameter, and impact velocity. We will first discuss here the expected ranges of impact velocities and then, in section 3, some of the recent laboratory constraints on the collision outcomes.

The final ingredient, the collision frequency, depends on the one hand on the cross section of the particles and their number density, which are results of the modeling itself, and on the other hand on the relative velocity of grains. We will describe these in the context of the global dust-evolution models in section 4.

2.2.1. Relative velocities. The relative velocities between particles in disks under the combined effects of settling and radial drift can directly be derived from the terminal dust velocities of *Nakagawa et al.* (1986) for two different-sized particles (e.g., *Weidenschilling and Cuzzi*, 1993; *Brauer et al.*, 2008a), as well as many others. As an example, Fig. 2a shows the expected relative velocities between grains of different sizes as computed for 1 AU by *Weidenschilling and Cuzzi* (1993). Both radial drift and vertical settling velocities peak at a Stokes number of unity. Relative motions increase with the size difference of the particles, because particles with the same Stokes number have the same systematic velocities. The maximum relative velocity via radial drift is

$$\Delta v_{drift} = \frac{c_s^2}{V_K}\frac{\partial \ln P}{\partial \ln r} = \frac{1}{2} \qquad (5)$$

Relative azimuthal velocities are small for particles of Stokes number less than unity; approach a constant, high value for $St > 1$; and also increase with the Stokes number difference.

In addition to these systematic motions, random motions also induce relative velocities, even between particles of the same Stokes number. Brownian motion of the particles is negligible for large particles, but it is the dominating source of relative motion for small particles, roughly submicrometer in size.

A topic of current research that is much more complicated than the relative motion discussed above is turbulent motion. The most frequently used formalism of this problem

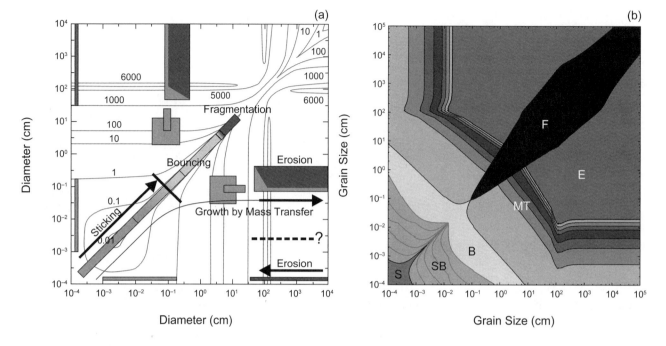

Fig. 2. Schematic representation of the outcomes of dust collisions in protoplanetary disks. **(a)** The background plot shows the collision velocities (in units of centimeter per second) between two nonfractal dust agglomerates with sizes indicated on the axes in a minimum-mass solar nebula (MMSN) model at 1 AU (*Weidenschilling and Cuzzi,* 1993). The shaded boxes denote the explored parameter space and results of laboratory experiments. Here, medium gray represents sticking or mass transfer, light gray bouncing, and dark gray (and black) fragmentation or erosion. The arrow denoted "Sticking" indicates the direct growth of millimeter-sized dust aggregates as found in the simulations by *Zsom et al.* (2010). Further growth is prevented by bouncing. A possible direct path to the formation of planetesimals is indicated by the arrow "Growth by mass transfer." **(b)** The collision outcomes parameter space as used by numerical models of dust evolution in protoplanetary disks by *Windmark et al.* (2012b), as derived interpolating the results of the laboratory experiments across the entire parameter space. The labels illustrate the dominant outcomes in the various regimes: S for sticking, B for bouncing, MT for mass transfer, E for erosion, and F for fragmentation. Note that these collisional outcomes refer only to collisions between bare silicate grains.

was introduced by *Voelk et al.* (1980) and *Markiewicz et al.* (1991), and recently, closed-form expressions were derived by *Cuzzi and Hogan* (2003) and *Ormel and Cuzzi* (2007). Their results show that, similar to radial drift and vertical settling, turbulent relative velocities increase with the Stokes number difference between the colliding particles and generally increase with the Stokes number until a St = 1. Beyond, it drops off again, but slower than, e.g., relative velocities induced by radial drift or vertical settling.

The maximum turbulent velocity according to *Ormel and Cuzzi* (2007) is

$$\Delta v_{turb} = c_s \left(9\alpha_t / 2 \right)^{1/2} \qquad (6)$$

and a factor of $\sqrt{3}$ smaller for collisions between equal-sized particles. For typical parameter choices ($\partial \ln P / \partial \ln r = -2.75$), equations (5) and (6) imply that the turbulence parameter α_t has to be only larger than about $2(H_p/r)^2$ to be the dominant source of relative velocity between grains. The dominant contributions to the relative velocities are also shown in Fig. 3a. For planetesimals, the gravitational torques of the turbulent gas also play a role (not shown in Fig. 3, but

see the chapter by Johansen et al. in this volume).

The works mentioned above generally treat only the root mean square (rms) velocity between dust grains, but the distribution of the collision velocity can also be important. Recently, some numerical works have tested the analytically derived collision velocities and started to derive distributions of collision velocities (e.g., *Carballido et al.,* 2010; *Pan and Padoan,* 2010, 2013; *Hubbard,* 2012). We will come back to the treatment of grain velocities in the context of global-disk models in section 4.2.

2.2.2. Condensation. A different physical process for growing (destroying) dust grains without grain collisions is condensation (evaporation) of material from the gas phase or the sublimation of mantles and solids. Dust growth via this mechanism was mentioned in previous works (*Goldreich and Ward,* 1973). It is commonly discussed in the context of dust formation and evolution in the interstellar medium (e.g., *Zhukovska et al.,* 2008; *Hirashita and Kuo,* 2011), but also in the context of the "condensation sequence" (e.g., *Lodders,* 2003). The problem with growing large grains via condensation in a protoplanetary disk is twofold: First, there is usually not enough material in the gas phase to grow a macroscopic dust/ ice mantle on every microscopic grain; second, accretion is a

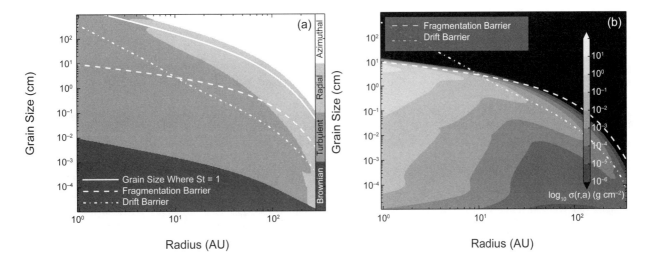

Fig. 3. (a) Important particle sizes and regimes of relative velocities in a fiducial disk model. The solid white line denotes the particle size corresponding to a Stokes number of unity, i.e., the fastest-drifting dust particles. The dashed and dash-dotted lines are the size barriers due to fragmentation and radial drift, respectively (see equations (8) and (9)). The shaded areas denote the dominating source of relative velocities at a given particle size and radius. This plot used a disk mass of 1% of the stellar mass, a stellar mass of 1 $M_\odot$, a turbulence parameter α_t of 10^{-3}, a fragmentation velocity of 3 m s^{-1}, a solid density of 1.6 g cm^{-3}, and a self-similar gas surface density (see *Lynden-Bell and Pringle*, 1974) with viscosity index p = 1 and characteristic radius of 60 AU, and a temperature profile of T = 50 K(r/10 AU)$^{-0.5}$. **(b)** Grain size distribution as a function of radius in the disk after 3 Ma for the same parameters as above. The initial dust-to-gas ratio is 10^{-2}, but the resulting dust-to-gas ratio at 3 Ma was used in **(a)**.

surface effect and the dust surface area is strongly dominated by the smallest grains. Condensation will therefore happen preferentially on the smallest grains, until all condensable material is consumed. Growth of large grains by condensation in protoplanetary disks can therefore only happen if there is some continuous source of condensable material (e.g., near an evaporation zone) and if some mechanism is able to preferentially transport the condensing material onto the largest particles. Some recent work by *Ros and Johansen* (2013) simulated growth of decimeter-sized particles consisting entirely of ice. However, this work relies on the absence of both dust and radial drift; i.e., dust nuclei are not left over after evaporation and vapor thus recondenses on another ice particle instead. Whether large particles can be formed by condensation under realistic disk conditions remains a topic for future research.

3. CONSTRAINTS ON THE PHYSICAL PROCESSES FROM LABORATORY EXPERIMENTS

The constraints on the collision outcomes as a function of particle size, composition, shape, and relative velocities are an essential input to the models of grain evolution in disks. It is not easy to explore the full parameter space for the conditions present in protoplanetary disks; here we will cover the emerging trends from laboratory experiments and numerical computations of grain-grain (or aggregate-aggregate) low-velocity collisions (for recent reviews, see also *Dominik et al.*, 2007; *Blum and Wurm*, 2008). Dust grains in protoplanetary disks are subject to relatively low-velocity

mutual collisions, which can have a very wide spectrum of results ranging from the complete fragmentation of the two particles leading to a swarm of smaller fragments, to the formation of a single larger particle containing the total mass of the system. The experimental techniques of investigating dust-particle and dust-aggregate collisions have considerably improved over the past decades so that reliable conclusions on the dust evolution in disks can be drawn.

The main questions that we will address are the following: (1) Why do dust particles or dust aggregates grow? (2) What are the structures of growing dust particles? (3) How fast do dust aggregates grow? (4) What is the maximum size that dust aggregates can reach? (5) Is the formation of larger aggregates by direct sticking collisions possible?

We will review the status of research on the above points one by one. In Fig. 2, we provide a schematic representation of the various processes discussed below.

3.1. Why Do Dust Particles or Dust Aggregates Grow?

The cause of dust growth is probably the easiest to answer. It is generally assumed that in the dense regions of protoplanetary disks not too close to the central star, electrostatic charges and magnetic materials do not play a dominant role in the interaction between grains; however, this is currently debated and several studies have investigated their effects on grain evolution in disks (*Dominik and Nübold*, 2002; *Matthews et al.*, 2012; *Okuzumi et al.*, 2011b). In the absence of these effects, dust grains in contact still possess an adhesive (van der Waals) force (*Heim et al.,* 1999; *Gundlach et al.,* 2011). If the dissipation of kinetic energy during the collision is strong enough, hysteresis effects at the contact point can lead to the sticking of the dust grains (*Chokshi et al.,* 1993; *Dominik and Tielens,* 1997). This direct hit-and-stick coagulation process has been experimentally investigated by *Poppe et al.* (2000). They found that ~1-μm-sized silica grains can stick to one another if they collide at less than ~1 m s⁻¹. Smaller grains can stick at higher velocities, whereas larger grains possess a lower threshold for sticking. This threshold velocity, below which sticking is dominant, is possibly strongly affected by the grain material. Recent investigations have shown that micrometer-sized water-ice particles seem to stick up to about 30 m s⁻¹ collision velocity, in agreement with the much higher surface energy of water ice (*Gundlach et al.,* 2011). Such high sticking thresholds have been used before in numerical simulations of ice-aggregate growth (see, e.g., *Wada et al.,* 2009, and the chapter by Johansen et al. in this volume). That larger particles have a lower sticking threshold is in principle still true for dust aggregates, although the details of the hit-and-stick process for soft-matter (or granular) particles are somewhat different. Based on laboratory experiments on collisions among (sub) millimeter-sized high-porosity dust aggregates (consisting of ~1-μm-sized silica grains), *Kothe et al.* (2013) derived a mass-dependent threshold velocity of the form $v \propto m^{-3/4}$. However, the threshold velocity for sticking of dust aggregates depends on many parameters, e.g., grain size and material (submicrometer vs. micrometer, silicates vs. ice), aggregate morphology (fractal, fluffy, compact, hierarchical), or mass ratio (through reduced mass).

3.2. What Are the Structures of Growing Dust Particles?

As long as the collision velocity is well below the threshold for sticking (see above), a mutual collision is in the so-called hit-and-stick regime, in which the projectile particle sticks to the target aggregate at the point of first contact. Numerical simulations as well as experiments have shown that in this regime, the dust aggregates grow to extremely fluffy structures, which have a fractal dimension D of well below 2. This is particularly true as long as Brownian motion dominates the collisions, which lead to fractal clusters with $D \approx 1.5$ (*Blum et al.,* 2000; *Krause and Blum,* 2004; *Paszun and Dominik,* 2006). This is typically the case during the very first stages of growth, when the dust aggregates are not much larger than tens of micrometers. With increasing dust-aggregate mass, the hit-and-stick regime is being replaced by the compaction stage (*Blum and Wurm,* 2000; *Dominik and Tielens,* 1997), so that the fractal dimension increases (whether to values of D = 3 or slightly below is still under debate) and dust aggregates are better described by their density or porosity. Typical volume filling factors of 0.05–0.50 are being expected for most of the mass range (*Zsom et al.,* 2010), although

Wada et al. (2008) derived in their numerical simulations growing aggregates with D = 2.5 (and correspondingly extremely small volume filling factors) when 0.1-μm-sized ice/dust monomers were considered. For monomer-sized distributions, according to the one derived by *Mathis et al.* (1977) (known as MRN-type distribution) for interstellar dust particles, the bulk of the mass of an aggregate is dominated by the largest monomers, whereas the number of particles, and thus the contacts between monomer grains, are determined by the smallest particles. *Ormel et al.* (2009) show that such an aggregate has the same strength as an aggregates of single-sized monomers of 0.056 μm. These results may also hold for aggregates consisting of ice and silicate particles. Here, the weaker-bound silicate grains might determine the strength and growth properties of the ice-dust bodies.

3.3. How Fast Do Dust Aggregates Grow?

Both experiments and numerical computations show that the first stage of growth is characterized by sticking. In this early phase their growth rate is simply given by the product of the number density of available collision partners, their mutual collision cross section, and their relative velocity. Under the assumption of nonfractal growth, the latter two are easily determined, while the first is a result of the growth process (see *Blum,* 2006, for a detailed description of dust-growth processes). The initial growth is also connected with a compaction of the aggregates as larger and larger grains collide and stick, absorbing part of the kinetic energy and rearranging their internal structure. This phase is followed by the so-called "bouncing barrier," when compact grains will not easily grow through the hit-and-stick mechanism (*Zsom et al.,* 2010). The detailed outcome of the full process of grain growth will be discussed in the context of the global models in section 4; numerical experiments show that due to compaction processes, the dust aggregates start very fluffy but are ultimately relatively compact. It is thus realistic to constrain the collision of particles beyond the "bouncing barrier" with the results of laboratory experiments based on compact aggregates.

3.4. What is the Maximum Size that Dust Aggregates Can Reach?

Although many uncertainties about the collision properties of growing protoplanetary dust aggregates exist (see, e.g., the discussion in *Kothe et al.,* 2013), it seems to be evident from laboratory experiments that direct sticking between dust aggregates in mutual collisions is limited (a possible mass-velocity threshold relation has been mentioned above). Thus, there is a maximum dust-aggregate size that can be formed in direct sticking collisions. The limiting factor for further growth is the bouncing and compaction of the dust aggregates, which has been observed in many laboratory experiments (see Fig. 2). However, another physical process that has been found in the laboratory is a possible loophole out of this "bouncing barrier." Above a certain velocity threshold, which is typically close to the fragmentation barrier (~1 m s^{-1}), collisions between small and large dust aggregates can lead to a substantial mass transfer from the smaller to the larger object (*Kothe et al.,* 2010; *Teiser et al.,* 2011; *Teiser and Wurm,* 2009a,b; *Wurm et al.,* 2005) (see also Fig. 2). Recent numerical simulations indicate that the gap between the onset of bouncing and the onset of the mass-transfer process can be bridged (*Windmark et al.,* 2012a,b; *Garaud et al.,* 2013) so that growth to planetesimal sizes seems to be possible. As most of the laboratory experiments have been performed with silicates, the above statements might not be applicable to icy particles, i.e., for distances to the central star beyond the snow line. *Okuzumi et al.* (2012) studied the mass evolution of aggregates consisting of 0.1-μm ice aggregates and found that there is no bouncing barrier, so icy planetesimals with extremely high porosity (density below a few 10^{-4} g cm^{-3}) may be formed by direct collisional growth.

3.5. Is Planetesimal Formation by Direct Sticking Collisions Possible?

As indicated above, mass transfer from projectile to target agglomerate has been identified in the laboratory and confirmed in numerical simulations as a potential process by which in principle arbitrarily large dust aggregates, i.e., planetesimals, can be formed. This mass-transfer process in collisions at velocities at which the smaller of the dust aggregates fragment can continually add mass to the growing larger dust aggregate as long as the absolute size and impact velocity of the (smaller) projectile aggregate is below a threshold curve derived by *Teiser and Wurm* (2009b). However, *Schräpler and Blum* (2011) showed in laboratory experiments that there is also a lower threshold mass for the mass-transfer process to be active (see Fig. 2). When growing dust aggregates are continuously bombarded by micrometer-sized monomer dust grains, they can be eroded quite efficiently. Thus, the formation of planetesimals by mass transfer can only be possible if the mass distribution of dust aggregates does not favor erosion (too many monomer dust grains) or fragmentation (too many similar-sized large dust aggregates). Much more work, both in the laboratory and with numerical simulations, is required before the question of whether planetesimals can form by collisional sticking can be finally answered.

4. MODELS OF DUST EVOLUTION IN DISKS

4.1. Introduction and Early Work

The previous sections have laid the basis for dust evolution, such as transport processes, collision velocities, and collisional outcomes. In this section, we will now bring all this together and review how grains in protoplanetary disks grow, how they are transported, and how growth and transport effects work with or against each other.

Some of the earliest works on dust particle growth in the context of planet formation were done by *Safronov* (1969), who considered both a toy model (growth of a single particle sweeping up other nongrowing particles) as well as analytical solutions to the equation that governs the time evolution of a particle size distribution, often called the Smoluchowski equation (*Smoluchowski,* 1916). A rather general way of writing this equation is

$$\dot{n}(x) =$$
$$\iint_{x_1, x_2}^{\infty} n(x_1)n(x_2)K(x_1, x_2)L(x_1, x_2, x)dx_1dx_2 \qquad (7)$$

Here x denotes a vector of properties such as composition, porosity, charge, or others, but to keep it simple, we can think of it just as particle mass m. Equation (7) means that the particle number density n(m) changes if particles with masses m_1 and m_2 collide at a rate of $K(m_1, m_2)$, where $L(m_1, m_2, m)$ denotes the mass distribution of the collisional outcome. For example, for perfect sticking, $L(m_1, m_2, m)$ is positive for all combinations of m_1 and $m_2 = m - m_1$, creating a particle of mass m, negative for all cases where $m = m_1$ or $m = m_2$, and 0 for all other combinations. Care has to be taken that the double integral does not count each collision twice. In principle, the collision rate K and outcome L can be a function of other properties as well, but in most applications the collision rate is simply a product of collisional cross section and relative velocity.

The early works of *Safronov* (1969), *Weidenschilling* (1980), and *Nakagawa et al.* (1981) considered dust grains settling toward the mid-plane and sweeping up other dust grains on the way. It was found that within a few 10^3 orbits, centimeter-sized dust grains can form near the mid-plane and the gravitational instability of this dust layer was investigated. *Weidenschilling* (1984) showed that turbulence could cause high collision velocities and lead to break-up of aggregates instead of continuing growth.

Numerical modeling of the full coagulation equation was found to be very difficult, as the dynamical range from submicrometer to decimeter sizes spans 18 orders of magnitude in mass, and it was shown that a rather high number of mass sampling points is needed in order to reach agreement between numerical and analytical results (e.g., *Ohtsuki et al.,* 1990; *Lee,* 2000). Many works therefore considered simplified models that use averaged values and only a single size.

In the following, we will focus on more recent works, first, local models of collisional dust evolution, followed by simplified and full global models. For more historical reviews on the subject, we refer to the previous reviews of *Weidenschilling and Cuzzi* (1993), *Beckwith et al.* (2000), and *Dominik et al.* (2007).

4.2. Recent Growth Models and Growth Barriers

Due to the complications mentioned above, modeling dust growth, even only at a single point in space, can be challenging, depending on which effects and parameters are taken into account. One of the most detailed methods of dust modeling is the N-body-like evolution of monomers, as done by *Kempf et al.* (1999); however, it is impossible to use this method for following the growth processes even only to millimeter-sized particles, due to the sheer number of involved monomers.

Monte Carlo methods make it computationally much more feasible to include several additional dust properties, such as porosity, and were first applied in the context of protoplanetary disks by *Ormel et al.* (2007, 2008). Similar Monte Carlo methods, as well as the experiment-based, detailed collision model of *Güttler et al.* (2010), were used in *Zsom et al.* (2010), which followed the size and porosity evolution of a swarm of particles. In agreement with previous studies, the initial growth occurs in the hit-and-stick phase, which leads to the formation of highly porous particles. As particle sizes and impact velocities increase, the main collision outcome shifts to bouncing with compaction, in which two colliding particles do not stick to each other, but are only compacted due to the collision. The final outcome in these models was a deadlock where all particles are of similar sizes and the resulting impact velocities are such that only bouncing collisions occurred, i.e., further growth was found to be impossible and this situation was called the bouncing barrier.

Even if bouncing is not considered, and particles are assumed to transition from sticking directly to a fragmenting/eroding collision outcome (as in the statistical Smoluchowski models of *Brauer et al.*, 2008a), a similar problem was found: As particles become larger, they tend to collide at higher impact velocities (see section 2.2.1). At some point the impact velocity exceeds the threshold velocity for shattering. Equating this threshold velocity above which particles tend to experience disruptive collisions with an approximation of the size-dependent relative velocity yields the according size threshold of (see *Birnstiel et al.*, 2012, and Fig. 3)

$$a_{\text{frag}} \simeq \frac{2}{3\pi} \frac{\Sigma_{\text{gas}}}{\rho_s \alpha_t} \frac{u_{\text{frag}}^2}{c_s^2} \qquad (8)$$

where Σ_{gas} is the gas surface density, ρ_s the material density of the dust grains, and u_{frag} the fragmentation threshold velocity derived from laboratory measurements. Earlier models used energy arguments to decide whether fragmentation happens (e.g., *Weidenschilling*, 1997), or did not include the effects of erosion (e.g., *Dullemond and Dominik*, 2005). The more recent models of *Windmark et al.* (2012a), which use an experiment-based collisional outcome model, indicate that most of the dust mass is still expected to remain in small, observationally accessible particles (≤ 1 cm), while allowing the formation of a small number of large bodies.

If fragmentation is limiting further growth of particles, a steady-state distribution is quickly established in which grains undergo numerous cycles of growth and subsequent

shattering, i.e., growth and fragmentation balance each other. The resulting grain size distributions tend to be power laws or broken power laws, which do not necessarily follow the MRN-type distribution (see *Birnstiel et al.*, 2011). A common feature of these distributions is the fact that most of the dust mass is concentrated in the largest grains. However, the total dust surface area is typically dominated by small grains of about a few micrometers in radius. Below a few micrometers, the main source of collision velocities switches from turbulence to Brownian motion (see Fig. 3), which leads to a kink in the size distribution, such that particles much smaller than a micrometer do not contribute much to the total grain surface area (see *Birnstiel et al.*, 2011; *Ormel and Okuzumi*, 2013).

Another recent development in dust modeling was the introduction by *Okuzumi et al.* (2009) of a method to self-consistently evolve the porosity in addition to the size distribution with the conventional Smoluchowski method. Including another particle property, such as porosity, increases the dimensionality of the problem from one dimension (mass) to two dimensions (mass and porosity), which makes conventional grid-based methods prohibitively slow. By including an additional property in the Smoluchowski equation and assuming it to be narrowly distributed, Okuzumi et al. derived the moment equations, which describe the time evolution of the size distribution and the time evolution of the mean porosity of each particle size. These moment equations are one-dimensional equations of the particle mass, which can be solved individually with the conventional Smoluchowski solvers. The results of this method showed good agreement with Monte Carlo methods at significantly increased computational speeds.

This method was also used for including grain charge in *Okuzumi et al.* (2011a) to confirm a scenario of *Okuzumi* (2009), in which dust grain charging and the subsequent repelling force could halt grain growth. This was termed the "charging barrier." They found general agreement with their previous scenario. In particular, they found that considering a distribution of relative velocities (for Brownian motion and turbulence), in contrast to assuming that all grains collide with the rms velocity, does not allow growth beyond the charging barrier. If turbulent mixing is considered, however, the charging barrier could be overcome (*Okuzumi et al.*, 2011b).

The effects of velocity distributions were taken into account in *Windmark et al.* (2012a) and *Garaud et al.* (2013) to also show that the bouncing barrier can be overcome: Not all particles collide with the rms velocity. Instead, collisions at higher velocities (possibly causing fragmentation or erosion) or lower velocities (possibly causing sticking) also occur. These additional outcomes open up a channel of growth, as some particles might always be in the "lucky" regime of velocity space and continue to stick, while all particles can also sweep up the fragments produced by high-velocity impacts. In this sense, when a distribution of collision speeds and/or turbulent mixing is taken into account, both the charging as well as the bouncing barrier become "porous barriers." They can slow down particle growth, but they cannot prevent the formation of larger bodies.

Even if particles beyond the usual growth barriers are formed, i.e., particles with sizes of around centimeters or meters, depending on the position in the disk, it still seems unrealistic to assume that these boulders stick to each other at any velocity. The laboratory results of *Wurm et al.* (2005), however, might provide a solution to this problem: It was found that impacts of small grains onto larger targets at velocities on the order of 25–50 m s^{-1} still lead to net growth of the target, while some of the projectile mass is redistributed to smaller fragments (see also *Teiser and Wurm,* 2009a,b; *Kothe et al.,* 2010). This mechanism of fragmentation with mass transfer was shown to be a way in which larger particles, embedded in a large number of small dust grains, can continue to grow by sweeping up the small grains at high impact velocities (*Windmark et al.,* 2012b).

4.3. Simplified Global Models

Modeling the evolution of a full particle size distribution faces many conceptual challenges and comes at significant computational costs. These complications have led to several models of the global evolution of dust in protoplanetary disks that focus on the transport side of the evolution and use only a simplified size evolution.

Some earlier works focused only on the very early evolution or on a limited radial range (e.g., *Morfill and Voelk,* 1984; *Sterzik and Morfill,* 1994). The models of *Stepinski and Valageas* (1996, 1997), and *Hughes and Armitage* (2012) also followed the monodisperse growth of dust particles; however, they only considered pure sticking, i.e., particles could grow without limits. They showed that particles can be quickly depleted from the disk if they grow and drift. While in *Youdin and Shu* (2002), particles of a fixed size accumulate because the drift velocity becomes slower in the inner regions of the disk, in the case of *Stepinski and Valageas* (1997) and *Kornet et al.* (2001), particles grow as they drift inward, which counteracts the decrease in velocity. In the case of a massive disk, the entire dust population is lost toward the inner boundary. Similar results were found by *Laibe et al.* (2008), who implemented the growth prescription of *Stepinski and Valageas* (1997) into a smoothed particle hydrodynamics (SPH) code.

Garaud (2007) expounded upon this idea by assuming the particle size distribution to be a MRN-like power law (see *Mathis et al.,* 1977) up to a certain maximum size. The maximum size was assumed to evolve as it sweeps up other grains, but the small grains were not evolved accordingly, i.e., they were assumed to be produced by fragmentation, even though fragmentation was not taken into account. This way, similar to previous works, particles could grow unhindered to planetesimals. One important finding of *Garaud* (2007) was that the dust mass in the outer disk represents a reservoir for the inner disk. As the small grains in the outer disk grow, they start to spiral inward due to radial drift and feed the inner disk.

Another approach was taken by *Birnstiel et al.* (2012), where results of full numerical simulations (see section 4.4) were taken as a template for a simplified model. In this model, the dust distribution is divided into small grains, which are basically following the gas motion, and large grains, which contain most of the mass and are subject to significant radial drift. The full numerical results showed that the shape and upper limit of the size distributions mostly represent two limiting cases, where either radial drift or grain fragmentation is the size-limiting factor. The resulting grain size yields an effective transport velocity that can be used to calculate the evolution of the surface density, which agrees well with the full numerical solutions.

4.4. Detailed Global Dust Evolution Models

The first models treating both the dust collisional evolution and radial transport in a protoplanetary disk were done by *Schmitt et al.* (1997) and *Suttner and Yorke* (2001); however, these works could only simulate the evolution for 100 and 10^4 yr, respectively. Still, they showed that especially in the innermost regions, grain growth occurs very quickly, and the initial growth is driven by Brownian motion until turbulence-induced relative velocities dominate at larger sizes.

The models of *Dullemond and Dominik* (2005) and *Tanaka et al.* (2005) were global in the sense that they simulated dust evolution and vertical transport at several radial positions of the disk, thus being able to calculate spectral energy distributions of the disk, but they did not include the radial transport.

The models of *Brauer et al.* (2008a) and *Birnstiel et al.* (2010a) were among the first to simulate dust evolution and radial transport for several millions of years, i.e., covering the typical lifetime of protoplanetary disks. This was possible because treating dust evolution was treated in a vertically integrated way, which reduces the problem to one spatial and the mass dimension. They showed that both radial drift and grain shattering collisions pose a serious obstacle to the collisional formation of planetesimals. *Brauer et al.* (2008b) used such a model to show that in the scenario of *Kretke and Lin* (2007), dust particles can accumulate in pressure bumps and form larger bodies, as long as turbulence is weak enough to avoid particle fragmentation.

A common feature of basically all simplified and full global models is the radial drift barrier, which limits the maximum size that particles can reach. Technically speaking, it is not a growth barrier, as it is not limiting the particle growth itself, but it still enforces an upper size cutoff by quickly removing particles larger than a particular size. Setting aside the other possible barriers and assuming that particles perfectly stick upon collision, we can define a growth timescale $t_{grow} = a/\dot{a}$, which is the timescale on which the size is doubled. For a monodisperse size distribution of spherical grains, the growth rate can be written as $\dot{a} = \Delta u \rho_{dust}/\rho_s$, where ρ_{dust} is the dust density and Δu the collision velocity. $\dot{a}$ can be thought of as the velocity of motion along the vertical axis in Fig. 3, while the radial drift velocity is the motion along the horizontal axis. The size limit enforced by drift is therefore approximately where the growth timescale exceeds the drift timescale $t_{drift} = r/u_{drift}$, which is approximately given by (see *Birnstiel et al., 2012*, and Fig. 3)

$$a_{drift} \simeq 0.35 \frac{\Sigma_{dust}}{\rho_s} \frac{V_K^2}{c_s^2} \left| \frac{d\ln P}{d\ln r} \right|^{-1} \qquad (9)$$

Having most of the dust mass contained in the largest grains of a given size $a_{max}(r)$, the transport velocity of the dust surface density can roughly be approximated as $u_{drift}(a_{max}(r))$. In a quasistationary situation, where the dust is flowing inward at a rate $\dot{M}_{dust}$, the surface density profile follows directly from the mass conservation

$$\Sigma_{dust} = \frac{\dot{M}_{dust}}{2\pi r u_{drift}} \qquad (10)$$

If the largest grain size is set by fragmentation, which is mostly the case in the inner disk (cf. Fig. 3), then the dust surface density was shown to follow a profile of $\Sigma_{dust} \propto r^{-1.3}$, which is in agreement with both the estimates for the solar system (*Weidenschilling, 1977b; Hayashi, 1981*) and the estimates from the extrasolar planets of *Chiang and Laughlin* (2013).

If radial drift is at play, and the dust surface density drops, then the outer disk in particular will at some point become dominated by the radial drift barrier, in which case the resulting dust surface density profile was shown to be $\Sigma_{dust} \propto r^{-(2p+1)/4}$ (where $\Sigma_{gas} \propto r^{-p}$), in agreement with the current observations of *Andrews et al.* (2012).

A possible way of overcoming the drift barrier for a limited spatial range was, however, discussed in *Okuzumi et al.* (2012). They used the method of *Okuzumi et al.* (2009) to simulate the porosity evolution in addition to the global evolution of nonfragmenting dust particles, based on a collisional model of *Suyama et al.* (2012). They found that outside the snow line, but inside ~10 AU, extremely porous particles, as formed in this collisional model (internal densities on the order of 10^{-5} g cm^{-3}), have a strongly decreased growth time due to their increased collisional cross section, which allows them to cross the drift barrier. Sintering (e.g., *Sirono, 2011a,b*) is expected to prevent this mechanism inside ~7 AU (*Okuzumi et al., 2012*). However, the precise impact that sintering — or fragmentation — has on this mechanism remains to be investigated.

4.5. Summary: Modeling

Current modeling efforts are standing on the proverbial shoulders of giants: Laboratory work, numerical modeling of individual particle collisions, and theoretical studies on disk structure, turbulence, collision velocities, and other physical effects represent the foundation on which global models of dust evolution are built.

Models predict grains to grow to different sizes, migrate, and mix radially and vertically throughout the disk. Large grains are expected to settle on the mid-plane and be transported radially to the inner disk regions, while small grains are mixed vertically to the disk surface by turbulence and transported outward as the gas spreads out as part of the viscous evolution of the disk. The growth timescale is also longer in the outer regions of the disk, and as a consequence, this region acts as a reservoir for the inner disk. Dust evolution in the planet-forming regions of the disk is strongly influenced by the rate at which large solids are transported inward from the outer disk. Understanding how to slow down the radial transport is one of the important goals for future research in this field.

Current models of nonfractal grain growth can explain the formation of larger bodies by incremental growth, but most of the dust mass tends to be trapped either by fragmentation or by radial drift. Size limits associated with these effects lead to a characteristic dust surface density distribution that can be compared directly with observations (see section 8). Fractal dust aggregates with internal densities on the order of 10^{-5}–10^{-3} g cm^{-3} experience accelerated growth and could break through the radial drift barrier. A detailed analysis of the evolution of these aggregates and their observational properties are an active area of research in which we also expect rapid progress in the coming years.

5. EFFECTS OF GRAIN GROWTH ON DUST OPACITY

The wavelength-dependent absorption and scattering properties of dust grains change as they grow in size. This behavior provides a tool to study the grain growth through panchromatic observations of circumstellar disks from optical to centimeter wavelengths.

A significant fraction of the disk emission at optical and near-IR wavelengths consists of stellar light scattered by the dust grains in the superficial layer of the disk (see Fig. 1). The morphology and color of the scattered light is most sensitive to grains with sizes between 0.01 and 10 μm, which correspond to the transition between the Rayleigh and the Mie scattering. As a general trend, as grains grow, the scattering phase function becomes less isotropic and the color of the scattered light becomes redder (*Bohren and Huffman*, 1983). In addition, the strength of mineral spectroscopical features, e.g., the silicate resonance feature at 10 μm, decreases (see *Kessler-Silacci et al.*, 2006, and references therein). As discussed in section 7, spatially resolved and spectroscopic observations of disks therefore allow measurements of the size of the dust grains in the surface layer of the disk.

Observations at longer wavelengths probe the dust properties in the disk interior where most of the mass is located. From far-IR to millimeter wavelengths, the disk emission is dominated by the thermal emission from cold dust, which is controlled by the dust opacity κ_ν. For a distribution of dust grains with minimum and maximum sizes a_{min} and a_{max}, a good approximation is a power law, $\kappa_\nu \propto \nu^\beta$, where the spectral index β is sensitive to a_{max} but does not depend on a_{min} for $a_{min} < 1$ μm (*Miyake and Nakagawa*, 1993; *Draine*, 2006).

The dust-opacity spectral index β depends not only on the grain sizes of the emitting dust, but also on other factors such as dust chemical composition, porosity, and geometry, as well as on the the grain size distribution, normally assumed as a power law of the form $dN = n(a)^{-q}da$ (e.g., *Natta et al.*, 2007). The main result is that, regardless of all the uncertainties of the dust model, dust grains with sizes on the order of 1 mm or larger are required to explain β values lower than about 1 [Fig. 4; see also *Natta and Testi* (2004)].

The inferred range of β values in young disks also has an important consequence for the derivation of dust masses through submillimeter/millimeter photometry, which implies knowledge of the dust-opacity coefficient. Solids with sizes much larger than the wavelength of the observations do not efficiently emit/absorb radiation at that wavelength and are therefore characterized by low values of the dust-opacity coefficient. Physical models of dust emission have been used to quantify the effect of changing β on the millimeter dust-opacity coefficient. For example, for the dust models of *D'Alessio et al.* (2006) with a slope of 2.5 for the grain-size distribution, a dust population with $\beta \approx 0.2$ has a dust-opacity coefficient, 1 mm, lower by nearly 2 orders of magnitude than a dust population with $\beta \approx 1.0$–1.5. This difference can be understood by the fact that, for a given slope of the grain-size distribution, a value of β much lower than 1 requires extending the distribution to very large grains, much larger than the observing wavelength. Similar results are obtained with the dust models considered in *Ricci et al.* (2010b). Despite the fact that the absolute value of the dust opacity at any wavelength strongly depends on the adopted dust model, this shows how inferring relations that involve disk dust masses without accounting for possible variations of the β parameter throughout the sample can lead to potentially large biases and errors.

6. DUST EVOLUTION BEFORE THE DISK FORMATION

This review focuses on the dust evolution in protoplanetary disks as the first step of the formation of planetesimals and larger bodies in planetary systems. Nevertheless, many of the physical processes described in the first two sections are also expected to occur in molecular cloud cores and in the envelopes of protostars before the disk-formation stage.

Grain coagulation and growth in cores and protostellar envelopes have been modeled by several authors (e.g., *Ossenkopf*, 1993; *Weidenschilling and Ruzmaikina*, 1994; *Suttner et al.*, 1999; *Ormel et al.*, 2009; *Hirashita and Li*, 2013). These models show that grains can form fluffy aggregates very efficiently as a function of core density and time. The growth is favored by the presence of ice mantles, which are expected to form in the denser and cooler interior of the clouds. Taking the most recent calculations by *Ormel et al.* (2009), on a timescale of ~1 Ma, grains can grow to several micrometer-sized particles at densities of ~10^5 cm^{-3}, and reach several hundred micrometer-sized aggregates at higher densities. The effects of these changes on the dust scattering and absorption properties have also been computed, and it has been shown that the growth can be traced by combining absorption, scattering, and emission at different wavelengths (*Kruegel and Siebenmorgen*, 1994; *Ossenkopf and Henning*, 1994; *Ormel et al.*, 2011).

Detailed modeling of the IR scattered light from dark cores in molecular clouds has shown widespread evidence for the presence of dust grains that have grown to several micrometer-sized particles, even in relatively low-density regions (*Pagani et al.*, 2010; *Steinacker et al.*, 2010). This is consistent with the models described above if the cores survive for timescales on the order of ≥1 Ma. These findings are in agreement with studies of the optical and IR extinction law in nearby molecular clouds (e.g., *Foster et al.*, 2013). Evidence of growth to larger sizes in the densest regions of molecular cloud cores has been investigated by measuring the dust emission at far-IR and submillimeter wavelengths and comparing the emission properties with extinction in the optical and IR (e.g., *Stepnik et al.*, 2003;

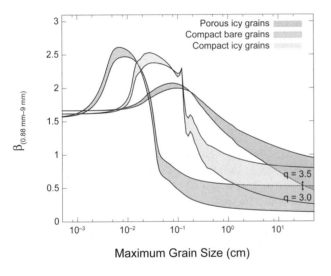

Fig. 4. Spectral index β of the dust opacity $\kappa_\nu \propto \nu^\beta$ calculated between the wavelengths of 0.88 and 9 mm as a function of the maximum grain size, for a grain size distribution $n(a) \propto a^{-q}$ characterized by a minimum grain size of 0.01 μm. Different colors corresponds to grains with different chemical composition and porosity. Dark gray: grains composed of astronomical silicates, carbonaceous material, and water ice, with relative abundances as in *Pollack et al.* (1994) and a porosity of 50%. Light gray: compact grains with the same composition as above. Medium gray: compact grains composed only of astronomical silicates and carbonaceous material. For each composition, the shaded region shows the values of β in the range $q = 3.0$ to $q = 3.5$. Despite the dependence of β on the grain composition and the value of q, maximum grain sizes larger than about 1 mm are required to explain values of β less than unity. These opacities have been computed following the prescription as in *Natta and Testi* (2004).

Roy et al., 2013; *Suutarinen et al.*, 2013). Solid results in this context have proven elusive so far; the main limitations of the current studies are the sensitivity of wide-area IR surveys that do not see through the densest regions of cores. The use of the submillimeter spectral index as a probe of the grain size distribution has been questioned by several studies that showed a possible dependence of the dust opacity with temperature (*Paradis et al.*, 2010; *Veneziani et al.*, 2010). However, recent detailed studies of dense and cold clouds, which include a broader range of (sub)millimeter wavelengths, suggest that, in some cases, the observed anticorrelation between the dust-opacity spectral index β and temperature may be the result of the uncertainties resulting from using simplified single-temperature modified blackbody fits to observations covering a limited range of wavelengths (*Shetty et al.*, 2009; *Kelly et al.*, 2012; *Sadavoy et al.*, 2013; *Juvela et al.*, 2013). Indirect evidence supporting grain growth in the dense regions of prestellar cores has been obtained by *Keto and Caselli* (2008), invoking significant grain growth at densities exceeding 10^5 cm^{-3} to explain the inferred change in dust opacity in the inner regions of the studied prestellar cores.

A recent development has been the study of dust properties in protostellar envelopes and young disks. The dust-opacity index in the youngest protostars indicate possible evidence for large grains in the collapsing envelopes (*Jørgensen et al.*, 2009; *Kwon et al.*, 2009). These initial results have been recently followed up with better data and more detailed modeling, which confirmed the initial suggestion that β ≤ 1 in the inner envelopes of the youngest protostars (*Chiang et al.*, 2012; *Tobin et al.*, 2013; *Miotello*, 2014.). These findings suggest that large dust aggregates, perhaps up to millimeter size, can form during the disk-formation stage in the infalling envelopes and are broadly in agreement with a very fast formation of CAIs in carbonaceous chondritic meteorites (*Connelly et al.*, 2012) (see also the chapter by Johansen et al. in this volume).

7. CONSTRAINTS ON DUST GROWTH IN DISKS FROM NEAR-INFRARED AND MID-INFRARED OBSERVATIONS

Because of the impact of density on timescales for growth, it is clear that most of the dust growth has to happen in the mid-plane of the disk, where densities are highest. The outer surfaces of the disk — specifically the upper disk surface at the inner rim — are mainly acting as display cases where products of this dust growth, more or less modified by transport processes from the mid-plane to the surface, are displayed to the observer by their interaction with optical, near-IR, and mid-IR radiation. The upper disk surface is relevant because it is directly illuminated (if the disk is flaring), and it also allows radiation at wavelengths well below the submillimeter regime to escape. As vertical mixing happens on timescales that are short compared to radial transport, grains in the disk surface are related to the population on the mid-plane at the same distance from the star. The inner rim

of the disk is important because dust gets exposed, often at temperatures close to its evaporation temperature. The inner rim also in principle allows study of the mid-plane dust by constraining the evaporation surface; initial results show that grains larger than micrometer-sized particles are required, but this avenue of research has so far been only marginally explored (e.g., *Isella et al.*, 2006; *Dullemond and Monnier*, 2010; *McClure et al.*, 2013). The tracers of dust size at these surfaces are (1) angular and wavelength dependence of scattered light, both intensity and polarization, and (2) the shape and strength of dust emission features (*Dullemond and Monnier*, 2010).

7.1. Scattered Light

Scattered light images can resolve the disks of nearby young stars down to distances of about 20 AU (e.g., *Ardila et al.*, 2007) from the star, in some cases even down to about 10 AU (*Quanz et al.*, 2012). Scattered light carries information about grain size in the intensity of the scattered light, and in particular in the angular dependence of the intensity, i.e., the phase function. The brightness of scattered light is often low, indicating grains with low albedo (*Ardila et al.*, 2007; *Fukagawa et al.*, 2010). The color of disks can be redder than that of the star (*Ardila et al.*, 2007; *Clampin et al.*, 2003; *Wisniewski et al.*, 2008), an effect that can be explained by the presence of large grains with a wavelength-dependent effective albedo caused by a narrowing of the forward-scattering peak at shorter wavelengths (*Mulders and Dominik*, 2012). Observed disks often differ in brightness between the front- and backside (e.g., *Kudo et al.*, 2008), and a brighter frontside is often taken as a sign for large grains whose scattering phase function is dominated by forward scattering (*Quanz et al.*, 2011). However, *Mulders et al.* (2013) warn that large grains, if present in the disk surface, may have phase functions with a very narrow forward-scattering peak of only a few degrees that can never be observed except in edge-on disks. The behavior of the phase function at intermediate angles (the angles actually seen in an inclined disk) may depend on details of the grain composition and structure.

Model fitting of scattered light images taken at multiple wavelengths indicate the presence of stratification, with smaller grains higher up in the disk atmosphere and larger grains settled deeper (e.g., *Pinte et al.*, 2007, 2008; *Duchêne et al.*, 2010), in accordance with the predictions of models (e.g., *Dullemond and Dominik*, 2004).

7.2. Dust Features

The usefulness of mid-IR features as tracers of dust growth is limited because the spatial resolution currently available at these wavelengths makes it hard to study the shape of features as a function of distance from the star, so only an average profile is observed. Interferometric observations allow us to extract the spectrum emitted by the innermost few astronomical units (*van Boekel et al.*, 2004), showing that the inner regions are much more crystallized than the disk-integrated spectrum indicated. Pioneering work on the integrated spectrum with the Infrared Space Observatory (ISO) on Herbig stars (*Bouwman et al.*, 2001; *van Boekel et al.*, 2005) has in recent years been extended to larger samples of Herbig stars and to T Tauri stars, making use of the better sensitivity provided by Spitzer. The result of these studies is that grains of sizes around 1 to a few micrometers must form quickly as the resulting broadening and weakening of the 10- and 20-μm features are present in most disks, essentially independent of age and other stellar parameters (*Kessler-Silacci et al.*, 2006; *Furlan et al.*, 2006; *Juhász et al.*, 2010; *Oliveira et al.*, 2011). *Juhasz et al.* (2010) find that larger grains are more prominent in sources that show a deficit of far-IR emission, interpreted as an indication of a flatter disk structure. These findings are in apparent contradiction with the results of *Sicilia-Aguilar et al.* (2007), who find an apparent trend for smaller grains in the atmospheres of older disks in Tr37. The interpretation for these observations is that larger grains settle more rapidly onto the mid-plane as compared to small grains, and hence, at later ages, are not accessible to IR observations. As IR observations probe a thin surface layer of the disk at a radius that is a strong function of the central star parameters (e.g., *Kessler-Silacci et al.*, 2007; *Natta et al.*, 2007), it is very difficult, and possibly misleading, to use spatially unresolved spectroscopy as a probe of global dust growth in disks. Nevertheless, an important result shown by all these studies is that the dust producing the 10- and 20-μm features contains a significant fraction of crystalline material, indicating that this dust was heated to temperatures of around 1000 K during processing in the disk. *Ábrahám et al.* (2009) and *Juhász et al.* (2012) clearly detected the increased production and subsequent transport to the outer disk of crystalline dust during the 2008 outburst of the star EX Lup. These findings suggest that eruptive phenomena in young disk-star systems play an important role in the processing and mixing of dust in protoplanetary disks.

7.3. Constraints for Mid-Plane Processes

It is now generally believed that dust in the inner parts of the disk quickly develops into a steady-state situation with ongoing coagulation and reproduction of small grains by erosion and fragmentation (see section 4.2). The dust present in the disk surface therefore does not seem to be a good tracer of the evolution of the mean and maximum particles size in the mid-plane of the disk, but rather reports on the presence of this steady state. While the suggestion of a possible correlation between mid-plane and surface dust processing was made by *Lommen et al.* (2009), many previous and subsequent attempts based on larger surveys have so far shown a distinct lack of correlation (*Natta et al.*, 2007; *Ricci et al.*, 2010a; *Juhász et al.*, 2010), probably caused by the fact that mid-IR observations are tracing the atmosphere of the inner disk (0.1–10 AU depending on the stellar properties), while mid-plane dust has so far

been probed only in the outer disk (beyond ~20 AU; see section 8).

8. OBSERVATIONS OF GRAIN GROWTH IN THE DISK MID-PLANE

The denser regions of the disk mid-plane can be investigated at (sub)millimeter and centimeter wavelengths, where dust emission is more optically thin. Long wavelength observations are thus the unique tool that allows us to probe grain evolution on the disk mid-plane, where most of the mass of the solids is confined and where planetesimal and planet formation is thought to occur. In this section we will focus on the observational constraints on grain growth on the disk mid-plane from long-wavelength observations, describing the methodology and limitations of this technique (section 8.1) as well as the results of the relatively extensive photometric surveys (section 8.2) and of the more limited — but very promising for the future — spatially resolved multiwavelength observations (section 8.3).

8.1. Methodology

As described in section 5, as grains grow to sizes comparable with the observing wavelength, the dust opacity changes significantly, imprinting the signature of growth in the disk emission. In particular, at millimeter and centimeter wavelengths, the spectral index of the emission of optically thin dust at a given temperature can be directly related to the spectral dependence of the dust-opacity coefficient, which in turn is related to the maximum grain size.

Values of β for dust in protoplanetary disks can be derived by measuring the slope α_{mm} of the submillimeter spectral energy distribution (SED) ($F_\nu \propto \nu^{\alpha_{mm}}$). Under the simple assumptions of optically thin dust emission in the Rayleigh-Jeans tail of the spectrum, $\beta = \alpha_{mm}-2$. The first single-dish and interferometric observations that measured the submillimeter slope of the SED of young disks came in the late 1980s and early 1990s (*Weintraub et al.*, 1989; *Woody et al.*, 1989; *Adams et al.*, 1990; *Beckwith and Sargent*, 1991). These pioneering works already revealed a wide interval of β values ranging from ≈ 0 up to the values typical of the interstellar medium (ISM) (≈ 1.7). The presence of dust grains with sizes larger than about 0.1 mm was soon recognized as a possible explanation for the disks showing $\beta < 1$ (*Weintraub et al.*, 1989; *Beckwith and Sargent*, 1991).

While this simplified approach has been useful in the initial studies, it is obvious that it has serious limitations and should not be used now that quick and efficient programs to self-consistently compute the dust emission from a protoplanetary disk with an arbitrary dust opacity are fast and widely available. All the results that we present in this chapter are derived using disk models that include an approximate, but proper, treatment of the density and temperature profile in the disk and use opacities computed from physical dust models (e.g., *Chiang and Goldreich*, 1997; *Dullemond et al.*, 2001, 2007; *Natta and Testi*, 2004, and successive improvements). These models include an approximate treatment of the radiation transfer in the regions where the disk becomes optically thick even at millimeter wavelengths under the assumption of a smooth disk structure. The effect of including optically thick regions due to local overdensities at large distances from the star has been investigated and shown by *Ricci et al.* (2012b) to be unlikely to play a major role in protoplanetary disks. Another serious source of uncertainty, if proper disk-emission models are not considered, comes from the assumption that the whole disk emission is in the Rayleigh-Jeans regime, as the dust temperatures in the outer disk mid-plane easily reach values below 15–20 K. Neglecting the effects of optical depth and low dust temperatures can result in a significant underestimate of β, leading to incorrect results regarding grain growth (see, e.g., *Weintraub et al.*, 1989; *Testi et al.*, 2001, 2003; *Wilner et al.*, 2005).

Another aspect to be considered is whether other sources of emission might be contaminating the signal from the large grains at the observing wavelengths. The major source of uncertainties comes from the contamination of the emission at long wavelengths from gas in the stellar chromosphere or in the wind/jets. The typical approach used to resolve this uncertainty is to combine the millimeter observations with long-wavelength observations that probe the gas emission (*Testi et al.*, 2001, 2003; *Wilner et al.*, 2005; *Rodmann et al.*, 2006). All these studies show that for T Tauri stars not affected by intense external photoionization, the contribution of the gas emission at wavelength shorter than 3 mm is below the 30% level. It is important to note that this is an estimate and, especially for the fainter disks and the higher-mass central stars, the contribution may be significantly higher and needs to be accounted for properly. In regions where there is a strong external ionizing radiation, the contribution of the gas to the emission may be dominant at 3 mm and still be significant at 1 mm. An example of this condition is the inner regions of the Trapezium cluster in Orion (e.g., *Williams et al.*, 2005; *Eisner et al.*, 2008; *Mann and Williams*, 2010, and references therein). Since nonthermal emission and thermal free-free emission from an ionized wind vary on various timescales (e.g., *Salter et al.*, 2010; *Ubach et al.*, 2012, and references therein), nearly simultaneous observations at (sub)millimeter and centimeter wavelengths should be used to quantify the gas spectrum and subtract it from the measured fluxes to infer the emission from dust only.

8.2. Results from Multi-Wavelength Submillimeter Photometry

Natta et al. (2007) covered the main results obtained in this field until the first half of the last decade. The main conclusion at the time was that, while evidence for large grains had been found in several bright disks, no clear trend with the properties of the system or age was evident. Since then, many photometric surveys have targeted young disks in several nearby star-forming regions (distances <500 pc

from the Sun) at long wavelengths, with the goal of constraining dust evolution in less-biased samples.

The most extensive studies aimed at a derivation of the grain growth properties have been performed for the Taurus-Auriga (*Andrews and Williams*, 2005; *Rodmann et al.*, 2006; *Ricci et al.*, 2010b) and Ophiuchus (*Andrews and Williams*, 2007; *Ricci et al.*, 2010a) star-forming regions. Less-extensive studies have also been carried out in southern star-forming regions (*Lommen et al.*, 2009; *Ubach et al.*, 2012) and the more distant Orion Nebula Cluster (*Mann and Williams*, 2010; *Ricci et al.*, 2011). It is important to note that all the surveys of disks conducted to date with the aim of characterizing the spectral index β of the dust absorption coefficient are far from being complete in any star-forming region, and the level of completeness decreases with decreasing disk mass. In many cases the studies included a detailed analysis of the possible contribution of the gas emission at long wavelengths and enough resolution to confirm that the disks are mostly optically thin and, for the large part, followed the methodology described in section 8.1 to derive the level of grain growth.

Following the method laid out by *Ricci et al.* (2010b), we have selected a subsample of all the published measurements of the 1.1–3-mm spectral index for disks surrounding single stars (or wide separation binaries) of spectral type from K and early M. The measured spectral indices are plotted in Fig. 5 against the flux measured at 1.1 mm (scaled at a common distance of 140 pc). The flux is roughly proportional to the total dust mass in the disk (assuming similar dust properties in all disks).

The main results of these surveys is that the dust in the outer disk regions appears to have grown to sizes of at least ~1 mm for the vast majority of the disks. Within the relatively small samples investigated so far, the distribution of spectral indices is consistent with being the same for nearly all the regions probed so far. The general picture that is emerging from this comparison is that dust appears to quickly grow to large sizes, but then it needs to be retained in the disk for a relatively long time, comparable to the disk lifetime. The only region where there may be a hint for possibly different distribution of spectral index values is Chamaeleon, where *Ubach et al.* (2012) derived a range of β values between 0.9 and 1.8 for 8 disks. Chamaeleon is among the oldest regions in the sample [albeit still young, with an estimated median age of ~2 Ma (*Luhman*, 2007)]. It is possible that the different values of α in Chamaeleon could be an indication for a time evolution of the grain-size distribution, with a loss of millimeter/centimeter-sized pebbles relative to smaller grains. This suggestion will be tested when statistically significant samples in younger and older star-forming regions are observed with ALMA.

Following *Birnstiel et al.* (2010b) and *Pinilla et al.* (2012b), we show in Fig. 5 the prediction of global grain-evolution models in disks. *Birnstiel et al.* (2010b) found that the measured 1.1–3-mm spectral indices can be well reproduced by models with reasonable values for parameters regulating grain fragmentation, gas turbulence, and disk

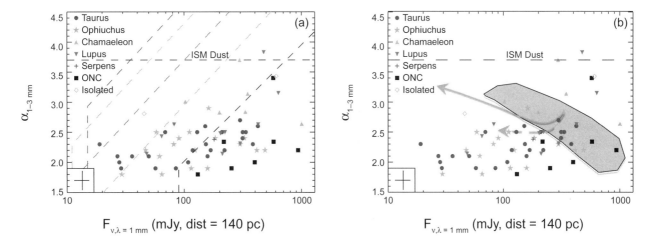

Fig. 5. (a) Spectral index between 1.1 and 3 mm plotted against the flux at 1.1 mm (scaled for a common distance of 140 pc) for disks around single stars (or wide binaries) with spectral types early M to K in nearby star-forming regions. The dashed lines mark the typical sensitivity limits of the surveys in Taurus, Ophiuchus, Lupus, Chamaeleon, and the Orion Nebula Cluster. **(b)** The gray area illustrates the range of predictions for global dust evolution models without radial drift (*Birnstiel et al.,* 2010b); the two arrows illustrate the evolutionary trajectories in the first few million years as predicted by the global models including the effect of radial drift (solid line) and including pressure traps in the gas distribution to slow the rate of drift (dashed line) (*Pinilla et al.,* 2012b).

structure. However, these models are able to explain only the upper envelope of the measured fluxes (i.e., the most massive disks in the sample). There is a large population of disks that are difficult to reconcile with the model predictions: those with low millimeter flux and low spectral index (low-mass disks containing a substantial amount of large grains). This discrepancy cannot be solved by simply reducing the mass of the disk models, as disks with lower surface densities would hardly grow grains (*Birnstiel et al.,* 2010b), as can be seen by the fragmentation and drift-limited growth in equations (8) and (9), which show that the maximum grain size depends on the gas and dust surface density, respectively.

Pinilla et al. (2012b) investigated the effect of time evolution on these modeling results, finding that while the radial drift process would progressively reduce the disk mass, evolving the models over a few million years to lower 1-mm fluxes, the drift and fragmentation processes will more efficiently remove the large grains from the disk, resulting in a steep increase of α that would not be consistent with the observations. *Pinilla et al.* (2012b) showed that the drift of large particles needs to be slowed down, but not halted completely, in order to explain the observed distribution in a framework of disk evolution. In this context, it is important to point out that no correlation has so far been found between individual stellar ages and α (e.g., *Ricci et al.,* 2010a).

Several mechanisms have been proposed to slow down the radial drift of large grains. For example, in MHD simulations of disks with zonal flows (*Johansen and Klahr,* 2005; *Johansen et al.,* 2009; *Uribe et al.,* 2011), the pressure field in the disk would be modified, and this may be a viable mechanism to create local pressure maxima that could efficiently trap large grains. Another possibility that

has been explored by some authors is that very large grains are injected very early in the outer disk, and then their migration is impaired as they are decoupled from the gas (e.g., *Laibe et al.,* 2012). This scenario would require the formation of very large (centimeter-sized) grains in cores and protostellar envelopes before disk formation, or an extremely efficient growth in very massive (young) disks. These hypotheses have not yet been modeled in detail to explore their feasibility.

An important prediction of models of dust evolution in gaseous disks is that grains are expected to grow to a maximum size that depends on the local density of gas and dust (e.g., *Birnstiel et al.,* 2009). Therefore larger values for the millimeter spectral indices should be expected for disks with lower flux at 1 mm, i.e., lower mass in dust and presumably lower densities as well. An extreme example of such systems could be disks around young brown dwarfs, if they are sufficiently large that the surface density is low. Initial measurements of these systems have confirmed the presence of large grains and relatively large radii (*Ricci et al.,* 2012a, 2013). In spite of the very-low-millimeter fluxes and estimated low dust mass, the millimeter spectral indices measured for the disks of ρ Oph 102 and 2M0444+2512 are as low as for the bulk of the T Tauri stars, taken together with the information on the disk spatial extent from the interferometric observations. *Ricci et al.* (2012a, 2013) constrain a value of $\beta \approx 0.5$ for both these brown dwarf disks.

Pinilla et al. (2013) and *Meru et al.* (2013) have investigated in detail the grain growth process in brown dwarf disks, showing that it is indeed possible to have grain growth and explain the values of millimeter flux and spectral index, at least in the conditions derived for the disks that have been observed so far. A more serious problem is represented by the radial drift. To slow down the radial motion of the grains, *Pinilla et al.* (2013) had to assume rather extreme gas pressure inhomogeneities. More observational constraints on larger samples of brown dwarf disks will soon be offered by ALMA, allowing a more thorough test of the grain-growth models.

8.2.1. Evolution of submillimeter fluxes for disks in different star-forming regions. Additional evidence for dust grain growth comes from (sub)millimeter continuum surveys of star-forming regions with different ages. The key finding is that the disk millimeter luminosity distribution declines rapidly with time, much quicker than the IR fraction, such that very few disks are detected at all in regions older than a few million years.

The Taurus and Ophiuchus clouds host two of the best-studied, nearby, low-mass star-forming regions and provide a benchmark for comparisons with other regions. Each contains about 200 class II sources that have been well-characterized at IR wavelengths through Spitzer surveys (*Evans et al.,* 2009; *Luhman et al.,* 2010). Almost all these sources have been observed at (sub)millimeter wavelengths in Taurus (*Andrews et al.,* 2013, and references therein), and a large survey was carried out in Ophiuchus by *Andrews and Williams* (2007). Both regions are relatively young, in the sense of having an IR disk fraction $f_{disk} = N_{disk}/N_{tot} \geq$ 60%, and the disk millimeter distributions are broadly similar, lognormal with a mean flux density mean F(1.3 mm) = 4 mJy and standard deviation 0.9 dex (*Andrews et al.,* 2013).

However, only a handful of disks are detected at millimeter wavelengths in more evolved regions such as Upper Sco [f_{disk} = 19% (*Mathews et al.,* 2012)] and σ Ori [f_{disk} = 27% (*Williams et al.,* 2013)]. Statistical comparisons must of course take into account not only the varying survey depths, but also the stellar properties of the samples, as disk masses depend on both stellar binarity and mass (*Andrews et al.,* 2013). The results of a Monte Carlo sampling technique to allow for these effects demonstrates that IR class II disk millimeter luminosities decrease significantly as regions age and f_{disk} decreases (*Williams et al.,* 2013).

The precise ages (and age spreads) of the compared regions are not well known but they are all young enough, ≪10 Ma, that planet formation should be ongoing. Therefore, the decrease in millimeter luminosity is attributable more to a decrease in the emitting surface area per unit mass, i.e., grain growth, than a decrease in the solid mass in disks (*Greaves and Rice,* 2010). The relative age differences of Upper Sco and σ Ori with respect to Taurus are better constrained and are ~3–5 Ma. This is therefore an upper limit to the typical timescale on which most of the solid mass in disks is locked up into particles greater than about a millimeter in size.

8.3. Resolved Images at Millimeter/Radio Wavelengths

The theoretical models for the evolution of solid particles in protoplanetary disks that were described in sec-

tion 4 make several physical predictions that should have direct observational consequences: (1) Settling, growth, and inward drift conspire to produce a size-sorted vertical and radial distribution of solids, such that larger particles are preferentially more concentrated in the disk mid-plane and near the host star; (2) drift alone should substantially increase the gas-to dust mass ratio at large disk radii (especially for millimeter/centimeter-sized particles); and (3) dust transport and fragmentation processes imply that the growth of solids to planetesimals has to happen in spatially confined regions of the disks. With their high sensitivity to cool gas and dust emission at a wide range of angular scales, observations with millimeter/radio-wavelength interferometers are uniquely suited for an empirical investigation of these physical effects. Ultimately, such data can be used to benchmark the theoretical models and then provide observational constraints on their key input parameters (e.g., turbulence, particle properties, growth timescales, etc.; see sections 3 and 4).

8.3.1. Constraints on large-scale radial variation of dust properties in disks. In the context of the dust continuum emission, we already highlighted in section 8.2 how the disk-averaged millimeter/radio "color" — typically parameterized in terms of the spectral index of the dust opacity, β — provides a global view of the overall level of particle growth in the disk. While this is a useful and efficient approach to study the demographics of large surveys, it cannot tell us about the expected strong spatial variations of the dust properties in individual disks (e.g., *Birnstiel et al.,* 2010b). A more useful technique would be to map out the spatial dependence of the millimeter/radio colors, β(r) or its equivalent, by resolving the continuum emission at a range of observing wavelengths. The underlying principle behind this technique rests on the (well justified) assumption that the continuum emission at sufficiently long wavelengths is optically thin, so that the surface brightness profile scales like

$$I_\nu(r) \propto \kappa_\nu B_\nu(T)\zeta\Sigma_g \qquad (11)$$

where κ_ν is the opacity spectrum, $B_\nu(T)$ is the Planck function at the local temperature, ζ is the inverse of the gas-to-dust mass ratio, and Σ_g is the gas surface density profile — each of which is thought to vary spatially in a given disk.

The spectral behavior in equation (11) has been exploited to interpret multi-wavelength-resolved continuum images in two basic approaches. The first approach acknowledges that the forward-modeling problem is quite difficult, since we do not really have an *a priori* parametric model for the spatial variations of κ_ν, T, or Σ_d (where the dust surface density profile, $\Sigma_d = \zeta \Sigma_g$). *Isella et al.* (2010) reasoned that, for a suitable assumption or model of T(r), the spectral gradient of the surface brightness profile itself should provide an empirical measurement of the resolved millimeter/radio color regardless of Σ_d. In essence, parametric fits for the optical depth profiles at each individual wavelength, $\tau_\nu(r) \approx \kappa_\nu \Sigma_d$, can be converted into an opacity index profile

$$\beta(r) = \frac{\partial \log \kappa_\nu(r)}{\partial \log \nu} \approx \frac{\partial \log \tau_\nu(r)}{\partial \log \nu} \qquad (12)$$

Although the initial studies that adopted this approach had insufficient data to conclusively argue for a nonconstant $\beta(r)$, they did establish an empirically motivated technique that provides a straightforward means of mapping millimeter/radio colors (e.g., *Isella et al.*, 2010; *Banzatti et al.*, 2011). Put simply, this approach allows one to reconstruct the $\beta(r)$ required to reconcile seemingly discrepant continuum emission structures at different observing wavelengths.

A slight variation on this approach is to adopt a more typical forward-modeling technique, where an assumption is made for a parametric formulation for both κ_ν and Σ_d. For example, *Guilloteau et al.* (2011) explored their dual-wavelength observations of Taurus disks with power-law and stepfunction $\beta(r)$ profiles, and argued that a spectral index that increases with disk radius provides a substantially improved fit quality compared to a global, constant index. In a recent refinement of the *Banzatti et al.* (2011) work, *Trotta et al.* (2013) have developed a more physically motivated prescription for $\kappa_\nu(r)$ (parameterized as a function of the local grain size distribution) that clearly calls for an increasing $\beta(r)$ in the disk around CQ Tau.

In these initial studies, the fundamental technical obstacle was really the limited wavelength range over which sensitive, resolved continuum measurements were available (typically $\lambda = 1.3$–2.7 mm). An extension of this work to centimeter wavelengths offers a substantially increased leverage on determining $\beta(r)$ and uniquely probes the largest detectable solid particles, while also overcoming the systematic uncertainties related to the absolute amplitude calibration for individual datasets. In two recent studies of the disks around AS 209 (*Pérez et al.*, 2012) and UZ Tau E (Harris, personal communication), resolved continuum emission at $\lambda \approx 9$ mm from the upgraded Jansky Very Large Array (JVLA) was folded into the analysis to provide robust evidence for significant increases in their $\beta(r)$ profiles (as shown together in Fig. 6). The analysis of the $\beta(r)$ profiles, as inferred from the optical depth profiles, for these two objects suggest at least an order of magnitude increase in particle sizes when moving from $r \approx 100$-AU to ~ 10-AU scales.

Taken together, these multi-wavelength interferometric dust continuum measurements reveal a fundamental, and apparently general, observational feature in young protoplanetary disks: The size of the dust emission region is anti-correlated with the observing wavelength. The sense of that relationship is in excellent agreement with the predictions of theoretical models for the evolution of disk solids, where the larger particles that emit more efficiently at longer wavelengths are concentrated at small disk radii due to the combined effects of growth and drift. The quan-

titative characterization of this feature in individual disks is just getting started, but the promise of a new opportunity to leverage current observing facilities to constrain planetesimal formation in action is exciting. However, there is a downside: It is now clear that resolved observations at a single millimeter/radio wavelength are not sufficient to constrain fundamental parameters related to the dust density structure. Given the lingering uncertainty on an appropriate general parameterization for Σ_d, it is worthwhile to point out that the optimized mechanics of the modeling approach for these resolved multi-wavelength continuum data are still in a stage of active development. Yet, as more data become available, rapid advances are expected from both the theoretical and observational communities.

The behavior of the multi-wavelength dust continuum emission makes a strong case for the spatial variation of κ_ν induced by the growth and migration of disk solids. But, as mentioned at the start of this section, those same physical processes should also produce a complementary discrepancy between the spatial distribution of the gas and dust phases in a disk: The dust should be preferentially more concentrated toward the stellar host, and ζ should decrease dramatically with radius as the gas-to-dust ratio increases in the outer disk. The fundamental problem is that the derived dust densities are very uncertain; even more critical is the fact that we do not yet really understand how to measure gas densities in these disks, so we cannot infer $\zeta(r)$ with sufficient quantitative reliability.

However, an approach that relies on a comparison between the spatial extents of molecular line and dust continuum emission can provide an indirect constraint on the spatial variation of $\zeta(r)$, even if its normalization remains uncertain. It has been recognized for some time that the sizes of the gas disks traced by optically thick CO line emission appear systematically larger than their optically thin dust continuum (e.g., *Piétu et al.*, 2007; *Isella et al.*, 2007). Although previous work suggested that this may be an artificial feature caused by optical depth effects or a misleading density model (*Hughes et al.*, 2008), the CO-dust size discrepancies have persisted with more sophisticated models and improved sensitivity. The evidence for a decreasing $\zeta(r)$ profile is derived through a process of logical negation. First, the dust structure of an individual disk is modeled, based on radiative transfer calculations that match the broadband SED and one or more resolved continuum images. Then, a gas structure model is calculated assuming a spatially invariant ζ. A comparison of the corresponding synthetic CO-emission model and the data invariably demonstrates that the model produces line emission that is much too compact. Small grains are found to be well mixed with the gas out to very large radii, e.g., from scattered light observations (see section 7). These findings imply that the large grains are confined to a smaller region of the disk, as compared to the gas and small dust grains. It is reasonable to deduce that a combination of grain-sized radial segregation and a decreasing $\zeta(r)$ would reconcile the line and continuum data.

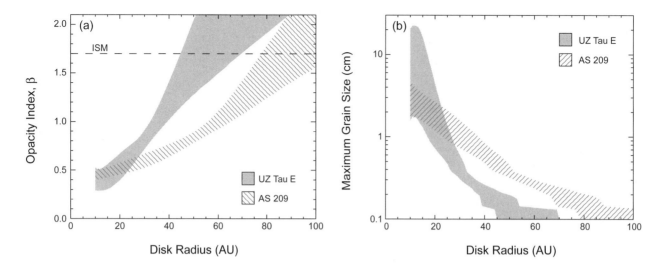

Fig. 6. **(a)** Confidence ranges for the profiles of β as a function of radius for the two disks surrounding the young stars AS 209 and UZ Tau E, as derived from submillimeter- through centimeter-wave high-signal-to-noise and high-angular-resolution interferometric observations (*Pérez et al.,* 2012; Harris, personal communication). **(b)** Values of a_{max} as a function of radius for the AS 209 and UZ Tau E disks derived with our reference dust model. The dashed and continuous lines show the fragmentation and radial drift limit, respectively, for a representative global model of dust evolution in disks (section 4).

As of this writing, only five disks have been modeled in this manner to indirectly infer a decreasing $\zeta(r)$: around IM Lup (*Panić et al.,* 2008), TW Hya (*Andrews et al.,* 2012), LkCa 15 (*Isella et al.,* 2012), V4046 Sgr (*Rosenfeld et al.,* 2013), and HD 163296 (*de Gregorio-Monsalvo et al.,* 2013). Ultimately, an empirical measurement of the CO-dust size discrepancy could be combined with the $\beta(r)$ profile inferences to help better constrain drift rates and how the migration of solids depends on the local disk conditions.

8.3.2. Constraints on vertical stratification of dust properties in disks. In section 7 we discussed the important constraints that IR and scattered-light observations provide on the grain populations in the disk atmosphere and on the dust settling (an extensive discussion of these can also be found in *Natta et al.,* 2007). The initial attempts to relate the grain properties in the disk atmosphere, as derived from mid-IR spectroscopy, with the properties on the mid-plane, as derived from (sub)millimeter photometry, have met with very limited success, even when considering large samples (e.g., *Lommen et al.,* 2009; *Ricci et al.,* 2010a; *Juhász et al.,* 2010; *Ubach et al.,* 2012). While models do predict a relationship between the growth level on the mid-plane and the atmosphere of the disks, the atmosphere grains should reach a steady-state balance between growth and fragmentation while mid-plane dust is still growing; in addition, another likely cause of the difficulty in observationally recovering a correlation between the dust properties is most likely linked to the vastly different regions of the disks that are currently sampled in the two wavelength regimes (e.g., *Natta et al.,* 2007). A meaningful comparison will only be possible with data sampling the same regions of the disk, which will require substantial improvement of the mid-IR observations, both in terms of sensitivity and angular resolu-

tion; this will become possible with the advent of the 40-m class Extremely Large Telescopes (ELT).

Better results are being obtained by comparing the dust properties as derived from high-angular-resolution optical IR images and millimeter dust emission maps. Constraints on the dust settling in the upper layers of the disk atmosphere have been discussed extensively in the last decade (e.g., *Wolf et al.,* 2003, 2008, 2012; *Liu et al.,* 2012) (see section 7); so far, the main limitation to combined multiwavelength studies that could constrain dust properties across the whole disk scale height has been the limited sensitivity and angular resolution of (sub)millimeter observations. Some initial studies in this direction are promising, although these are currently limited to very few objects and with very favorable viewing geometries (e.g., *Gräfe et al.,* 2013). It is expected that ALMA will provide the angular resolution and sensitivity for significant progress in this area, although obviously the best constraints will continue to be possible only for disks with very favorable geometries (*Boehler et al.,* 2013).

8.3.3. Constraints on small-scale variation of dust properties and dust trapping. One key development that is expected in the near future is the direct observation of the expected small-scale structures within the disk, where solids will be able to overcome the many growth barriers discussed in sections 3 and 4 and reach the planetesimal stage (see also the chapter by Johansen et al. in this volume). All the proposed solutions to the long-standing meter-sized barrier problem in planet formation (*Weidenschilling,* 1977a and all subsequent evolutions) involve the confinement of the largest particles in localized areas of the disk where they can locally overcome the barriers on their way to become planetesimals. The typical dimensions of these large grain "traps," as they are normally referred to, is on the order of the local scale height of the disk or less. Several studies have simulated in detail the observability of these features in young protoplanetary disks with ALMA, showing that there are good prospects to reveal them both in the gas and/or dust emission (e.g., *Cossins et al.,* 2010; *Pinilla et al.,* 2012b, 2013; *Douglas et al.,* 2013).

A very recent and extremely encouraging development in this direction is the detection with ALMA of a large dust trap in a transition disk by *van der Marel et al.* (2013). These authors revealed clearly the segregation of large dust particles in the outer regions of the disk caused by the radial and azimuthal inhomogeneity in the gas density induced by the likely presence of a planetary companion within the disk inner hole. These data provide a direct observational support to the simulations of dust trapping in transitional disks, as discussed by *Pinilla et al.* (2012a). While the results of *van der Marel et al.* (2013) provide direct proof of the dust-trapping concept, the characteristics of the observed system are such that the trap may support the formation of an analog to the solar system Kuiper belt, but cannot obviously explain the formation of planets in that system, as the presence of a planet is a requirement for the formation of the trap. Nevertheless, the result illustrates how we can expect to constrain dust trapping in disks with ALMA in the future.

Sensitive and high-angular-resolution observations with ALMA will also very soon allow us to understand the role of snow lines in the evolution of solids in protoplanetary disks. Grains that are migrating inward across a snow line will lose part of their icy mantles, and models predict that the cycle of sublimation and condensation will allow efficient growth and trapping across the snow line (e.g., *Ros and Johansen,* 2013). *Mathews et al.* (2013) and *Qi et al.* (2013) spatially resolved the CO snow line in HD 163296 and TW Hya, paving the road for an investigation of the role of this particular snow line in the evolution of solids.

9. SUMMARY AND OUTLOOK

Global models of grain evolution in disks, constrained by the results of laboratory and numerical calculations of grain and aggregate collisions, predict that in the conditions of protoplanetary disks, grains should very rapidly grow to centimeter sizes. The growth process is not without difficulties, and there are numerous "barriers" to overcome. At this time, the most critical problem seems to be the drift-fragmentation "barrier" caused by the large differential radial speeds of grains of different sizes induced by the aerodynamical drag. While the level of growth predicted by models is consistent with observational data, the time evolution is too fast unless the radial drift motion of large grains is slowed down in pressure "traps." Several ideas exit for plausible mechanisms to produce these traps: from disk instabilities to snow lines. The predictions should now be tested observationally, and ALMA will offer a unique opportunity to do this in the coming years thanks to its superb angular resolution and sensitivity.

Observations show that the majority of disks around isolated pre-main-sequence stars in the nearby star-forming regions contain a significant amount of grains grown to at least millimeter sizes. No clear evolutionary trends have emerged so far, although some indirect evidence of disk aging is also starting to emerge from millimeter observations. This finding strongly supports the notion that grain growth is a common and very fast process in disks, possibly occurring in the earliest phases of disk formation. The other implication is that the large grains are either kept in the disk mid-plane or they are continuously reformed for several million years. This is required to explain the presence of these dust aggregates in disks around pre-main-sequence stars, and is also consistent with cosmochemical evidence in our own solar system.

Multi-wavelength radially resolved observations of grain properties, which were just barely becoming possible at the time of the last Protostars and Planets V conference, are now providing a wealth of new constraints to the models. The overall radial segregation of dust aggregates of different sizes predicted by models is in general agreement with observations; however, the fast draining of solids toward the inner disk predicted by models cannot reproduce the observations. A mechanism to slow down the radial drift is definitely required.

While the interpretation of the millimeter observations in terms of grain growth appears to be solid, the values of the dust opacities for the dust aggregates are still very uncertain. New computations or laboratory measurements for a range of grain composition and structure would represent a major step toward putting the study of dust evolution in disks on very solid ground. Similarly, a major step in the understanding of the physical processes of grain growth in disks will be the extension of the grain-grain laboratory experiments to icy dust particles. These experiments are now being carried out and initial results are expected in the coming years.

Armed with better constraints on the grain-grain collision outcomes, the global models of dust evolution in disks will also have to be advanced to the next stage. This will have to include a proper account of the gas and dust coevolution and the extension of the models to two and three dimensions in order to account for vertical transport and azimuthal inhomogeneities, which are now routinely observed with ALMA.

On the observational side, new and upgraded observing facilities at millimeter and centimeter wavelengths are offering an unprecedented opportunity to expand the detailed studies to faint objects and to resolve the detailed radial and vertical structure of the grain properties. In the coming years we expect that it will be possible to place strong constraints on the evolutionary timescale for the dust on the disk mid-plane, and therefore explore this process as a function of the central host star parameters and as a function of environment. The study of the dust and gas small-scale structures in the disk, as well as their global distribution, will most likely allow us to solve the long-standing problem of radial drift.

Acknowledgments. We thank M. Bizzarro, P. Caselli, C. Chandler, C. Dullemond, A. Dutrey, Th. Henning, A. Johansen, C. Ormel, R. Waters, and especially the referee, S. Okuzumi, for valuable comments and discussions on the content of this chapter. We thank F. Windmark for providing data for Fig. 2. The work reported here has been partly supported over the years by grants to INAF-Osservatorio Astrofisico di Arcetri from the MIUR, PRIN-INAF, and ASI, and the grant Progetti Premiali 2012-iALMA (CUP C52I12000140001). Extended support from the Munich-IMPRS as well as ESO Studentship/Internship, DGDF and Scientific Visitor programs over the period 2009–2013 is gratefully acknowledged. S.A. and T.B. acknowledge support from NASA Origins of Solar Systems grant NNX12AJ04G. A.I. and J.M.C. acknowledge support from NSF award AST-1109334. J.P.W. acknowledges support from NSF award AST-1208911.

REFERENCES

Ábrahám P. et al. (2009) *Nature, 459,* 224.
Adachi I. et al. (1976) *Prog. Theor. Phys., 56,* 1756.
Adams F. C. et al. (1990) *Astrophys. J., 357,* 606.
Andrews S. M. and Williams J. P. (2005) *Astrophys. J., 631,* 1134.
Andrews S. M. and Williams J. P. (2007) *Astrophys. J., 671,* 1800.
Andrews S. M. et al. (2012) *Astrophys. J., 744,* 162.
Andrews S. M. et al. (2013) *Astrophys. J., 771,* 129.
Ardila D. R. et al. (2007) *Astrophys. J., 665,* 512.
Armitage P. J. (2010) *Astrophysics of Planet Formation,* Cambridge Univ., Cambridge.
Ataiee S. et al. (2013) *Astron. Astrophys., 553,* L3.
Bai X.-N. and Stone J. M. (2010) *Astrophys. J., 722,* 1437.
Banzatti A. et al. (2011) *Astron. Astrophys., 525,* A12.
Barge P. and Sommeria J. (1995) *Astron. Astrophys., 295,* L1.
Beckwith S. V. W. and Sargent A. I. (1991) *Astrophys. J., 381,* 250.
Beckwith S. V. W. et al. (2000) In *Protostars and Planets IV* (V. Mannings et al., eds.), p. 533. Univ. of Arizona, Tucson.
Birnstiel T. et al. (2009) *Astron. Astrophys., 503,* L5.
Birnstiel T. et al. (2010a) *Astron. Astrophys., 513,* A79.
Birnstiel T. et al. (2010b) *Astron. Astrophys., 516,* L14.
Birnstiel T. et al. (2011) *Astron. Astrophys., 525,* A11.
Birnstiel T. et al. (2012) *Astron. Astrophys., 539,* A148.
Birnstiel T. et al. (2013) *Astron. Astrophys., 550,* L8.
Blum J. (2006) *Adv. Phys., 55,* 881.
Blum J. and Wurm G. (2000) *Icarus, 143,* 138.
Blum J. and Wurm G. (2008) *Annu. Rev. Astron. Astrophys., 46,* 21.
Blum J. et al. (2000) *Phys. Rev. Lett., 85,* 2426.
Bockelée-Morvan D. et al. (2002) *Astron. Astrophys., 384,* 1107.
Boehler Y. et al. (2013) *Mon. Not. R. Astron. Soc., 431,* 1573.
Bohren C. F. and Huffman D. R. (1983) *Absorption and Scattering of Light by Small Particles,* Wiley, New York.
Bouwman J. et al. (2001) *Astron. Astrophys., 375,* 950.
Brauer F. et al. (2007) *Astron. Astrophys., 469,* 1169.
Brauer F. et al. (2008a) *Astron. Astrophys., 480,* 859.
Brauer F. et al. (2008b) *Astron. Astrophys., 487,* L1.
Carballido A. et al. (2006) *Mem. Natl. R. Astron. Soc., 373,* 1633.
Carballido A. et al. (2010) *Mon. Not. R. Astron. Soc., 405,* 2339.
Chatterjee S. and Tan J. C. (2014) *Astrophys. J., 780,* 53.
Chiang E. I. and Goldreich P. (1997) *Astrophys. J., 490,* 368.
Chiang E. and Laughlin G. (2013) *Mon. Not. R. Astron. Soc., 431,* 3444.
Chiang E. I. et al. (2001) *Astrophys. J., 547,* 1077.
Chiang H.-F. et al. (2012) *Astrophys. J., 756,* 168.
Chokshi A. et al. (1993) *Astrophys. J., 407,* 806.
Ciesla F. J. (2009) *Icarus, 200,* 655.
Clampin M. et al. (2003) *Astron. J., 126,* 385.
Connelly J. N. et al. (2012) *Science, 338,* 651.
Cossins P. et al. (2010) *Mon. Not. R. Astron. Soc., 407,* 181.
Cuzzi J. N. and Hogan R. C. (2003) *Icarus, 164,* 127.
Cuzzi J. N. et al. (1993) *Icarus, 106,* 102.
D'Alessio P. et al. (2001) *Astrophys. J., 553,* 321.
D'Alessio P. et al. (2006) *Astrophys. J., 638,* 314.
Dauphas N. and Chaussidon M. (2011) *Annu. Rev. Earth Planet. Sci., 39,* 351.

de Gregorio-Monsalvo I. et al. (2013) *Astron. Astrophys., 557,* A133.
Dominik C. and Dullemond C. P. (2008) *Astron. Astrophys., 491,* 663.
Dominik C. and Nübold H. (2002) *Icarus, 157,* 173.
Dominik C. and Tielens A. G. G. M. (1997) *Astrophys. J., 480,* 647.
Dominik C. et al. (2007) In *Protostars and Planets V* (B. Reipurth et al., eds.), pp. 783. Univ. of Arizona, Tucson.
Douglas T. A. et al. (2013) *Mon. Not. R. Astron. Soc., 433,* 2064.
Draine B. T. (2006) *Astrophys. J., 636,* 1114.
Dubrulle B. et al. (1995) *Icarus, 114,* 237.
Duchêne G. et al. (2010) *Astrophys. J., 712,* 112.
Dullemond C. P. and Dominik C. (2004) *Astron. Astrophys., 421,* 1075.
Dullemond C. P. and Dominik C. (2005) *Astron. Astrophys., 434,* 971.
Dullemond C. P. and Monnier J. D. (2010) *Annu. Rev. Astron. Astrophys., 48,* 205.
Dullemond C. P. et al. (2001) *Astrophys. J., 560,* 957.
Dullemond C. P. et al. (2007) In *Protostars and Planets V* (B. Reipurth et al., eds.), pp. 555–572. Univ. of Arizona, Tucson.
Eisner J. A. et al. (2008) *Astrophys. J., 683,* 304.
Evans N. J. II et al. (2009) *Astrophys. J. Suppl., 181,* 321.
Foster J. B. et al. (2013) *Mon. Not. R. Astron. Soc., 428,* 1606.
Fromang S. and Nelson R. P. (2005) *Mon. Not. R. Astron. Soc., 364,* L81.
Fromang S. and Nelson R. P. (2009) *Astron. Astrophys., 496,* 597.
Fromang S. et al. (2011) *Astron. Astrophys., 534,* A107.
Fukagawa M. et al. (2010) *Publ. Astron. Soc. Japan, 62,* 347.
Furlan E. et al. (2006) *Astrophys. J. Suppl., 165,* 568.
Garaud P. (2007) *Astrophys. J., 671,* 2091.
Garaud P. et al. (2013) *Astrophys. J., 764,* 146.
Goldreich P. and Ward W. R. (1973) *Astrophys. J., 183,* 1051.
Gräfe C. et al. (2013) *Astron. Astrophys., 553,* A69.
Greaves J. S. and Rice W. K. M. (2010) *Mon. Not. R. Astron. Soc., 407,* 1981.
Guilloteau S. et al. (2011) *Astron. Astrophys., 529,* A105.
Gundlach B. et al. (2011) *Icarus, 214,* 717.
Güttler C. et al. (2010) *Astron. Astrophys., 513,* A56.
Hartmann L. (2009) *Accretion Processes in Star Formation, 2nd edition,* Cambridge Univ., Cambridge.
Hayashi C. (1981) *Prog. Theor. Phys. Suppl., 70,* 35.
Heim L.-O. et al. (1999) *Phys. Rev. Lett., 83,* 3328.
Hirashita H. and Kuo T.-M. (2011) *Mon. Not. R. Astron. Soc., 416,* 1340.
Hirashita H. and Li Z.-Y. (2013) *Mon. Not. R. Astron. Soc., 434,* L70.
Hsieh H.-F. and Gu P.-G. (2012) *Astrophys. J., 760,* 119.
Hubbard A. (2012) *Mon. Not. R. Astron. Soc., 426,* 784.
Hughes A. L. H. and Armitage P. J. (2012) *Mon. Not. R. Astron. Soc., 423,* 389.
Hughes A. M. et al. (2008) *Astrophys. J., 678,* 1119.
Isella A. et al. (2006) *Astron. Astrophys., 451,* 951.
Isella A. et al. (2007) *Astron. Astrophys., 469,* 213.
Isella A. et al. (2010) *Astrophys. J., 714,* 1746.
Isella A. et al. (2012) *Astrophys. J., 747,* 136.
Jacquet E. (2013) *Astron. Astrophys., 551,* 75.
Johansen A. and Klahr H. (2005) *Astrophys. J., 634,* 1353.
Johansen A. et al. (2009) *Astrophys. J., 697,* 1269.
Jørgensen J. K. et al. (2009) *Astron. Astrophys., 507,* 861.
Juhász A. et al. (2010) *Astrophys. J., 721,* 431.
Juhász A. et al. (2012) *Astrophys. J., 744,* 118.
Juvela M. et al. (2013) *Astron. Astrophys., 556,* A63.
Keller C. and Gail H.-P. (2004) *Astron. Astrophys., 415,* 1177.
Kelly B. C. et al. (2012) *Astrophys. J., 752,* 55.
Kempf S. et al. (1999) *Icarus, 141,* 388.
Kessler-Silacci J. et al. (2006) *Astrophys. J., 639,* 275.
Kessler-Silacci J. E. et al. (2007) *Astrophys. J., 659,* 680.
Keto E. and Caselli P. (2008) *Astrophys. J., 683,* 238.
Klahr H. H. and Henning T. (1997) *Icarus, 128,* 213.
Kley W. and Lin D. N. C. (1992) *Astrophys. J., 397,* 600.
Kornet K. et al. (2001) *Astron. Astrophys., 378,* 180.
Kothe S. et al. (2010) *Astrophys. J., 725,* 1242.
Kothe S. et al. (2013) *Icarus, 225,* 75.
Krause M. and Blum J. (2004) *Phys. Rev. Lett., 93(2),* 021103.
Kretke K. A. and Lin D. N. C. (2007) *Astrophys. J. Lett., 664,* L55.
Kruegel E. and Siebenmorgen R. (1994) *Astron. Astrophys., 288,* 929.
Kudo T. et al. (2008) *Astrophys. J. Lett., 673,* L67.
Kwon W. et al. (2009) *Astrophys. J., 696,* 841.
Laibe G. et al. (2008) *Astron. Astrophys., 487,* 265.
Laibe G. et al. (2012) *Astron. Astrophys, 537,* 61.
Lee M. H. (2000) *Icarus, 143,* 74.

Liu Y. et al. (2012) *Astron. Astrophys., 546,* A7.
Lodders K. (2003) *Astrophys. J., 591,* 1220.
Lommen D. et al. (2009) *Astron. Astrophys., 495,* 869.
Luhman K. L. (2007) *Astrophys. J. Suppl., 173,* 104.
Luhman K. L. et al. (2010) *Astrophys. J. Suppl., 186,* 111.
Lynden-Bell D. and Pringle J. E. (1974) *Mon. Not. R. Astron. Soc., 168,* 603.
Mann R. K. and Williams J. P. (2010) *Astrophys. J., 725,* 430.
Markiewicz W. J. et al. (1991) *Astron. Astrophys., 242,* 286.
Mathews G. S. et al. (2012) *Astrophys. J., 745,* 23.
Mathews G. S. et al. (2013) *Astron. Astrophys., 557,* A132.
Mathis J. S. et al. (1977) *Astrophys. J., 217,* 425.
Matthews L. S. et al. (2012) *Astrophys. J., 744,* 8.
McClure M. K. et al. (2013) *Astrophys. J., 775,* 114.
Meru F. et al. (2013) *Astrophys. J. Lett., 774,* L4.
Miotello A. et al. (2014) *Astron. Astrophys., 567,* 32.
Miyake K. and Nakagawa Y. (1993) *Icarus, 106,* 20.
Mizuno H. et al. (1988) *Astron. Astrophys., 195,* 183.
Morfill G. E. and Voelk H. J. (1984) *Astrophys. J., 287,* 371.
Mulders G. D. and Dominik C. (2012) *Astron. Astrophys., 539,* A9.
Mulders G. D. et al. (2013) *Astron. Astrophys., 549,* A112.
Nakagawa Y. et al. (1981) *Icarus, 45,* 517.
Nakagawa Y. et al. (1986) *Icarus, 67,* 375.
Natta A. and Testi L. (2004) In *Star Formation in the Interstellar Medium: In Honor of David Hollenbach* (D. Johnstone et. al., eds.), p. 279. ASP Conf. Ser. 323, Astronomical Society of the Pacific, San Francisco.
Natta A. et al. (2007) In *Protostars and Planets V* (B. Reipurth et al., eds.), pp. 767–781. Univ. of Arizona, Tucson.
Ohtsuki K. et al. (1990) *Icarus, 83,* 205.
Okuzumi S. (2009) *Astrophys. J., 698,* 1122.
Okuzumi S. et al. (2009) *Astrophys. J., 707,* 1247.
Okuzumi S. et al. (2011a) *Astrophys. J., 731,* 95.
Okuzumi S. et al. (2011b) *Astrophys. J., 731,* 96.
Okuzumi S. et al. (2012) *Astrophys. J., 752,* 106.
Oliveira I. et al. (2011) *Astrophys. J., 734,* 51.
Ormel C. W. and Cuzzi J. N. (2007) *Astron. Astrophys., 466,* 413.
Ormel C. W. and Okuzumi S. (2013) *Astrophys. J., 771,* 44.
Ormel C. W. et al. (2007) *Astron. Astrophys., 461,* 215.
Ormel C. W. et al. (2008) *Astrophys. J., 679,* 1588.
Ormel C. W. et al. (2009) *Astron. Astrophys., 502,* 845.
Ormel C. W. et al. (2011) *Astron. Astrophys., 532,* A43.
Ossenkopf V. (1993) *Astron. Astrophys., 280,* 617.
Ossenkopf V. and Henning T. (1994) *Astron. Astrophys., 291,* 943.
Pagani L. et al. (2010) *Science, 329,* 1622.
Pan L. and Padoan P. (2010) *J. Fluid Mech., 661,* 73.
Pan L. and Padoan P. (2013) *Astrophys. J., 776,* 12.
Panić O. et al. (2008) *Astron. Astrophys., 491,* 219.
Paradis D. et al. (2010) *Astron. Astrophys., 520,* L8.
Paszun D. and Dominik C. (2006) *Icarus, 182,* 274.
Pérez L. M. et al. (2012) *Astrophys. J. Lett., 760,* L17.
Piétu V. et al. (2007) *Astron. Astrophys., 467,* 163.
Pinilla P. et al. (2012a) *Astron. Astrophys., 545,* A81.
Pinilla P. et al. (2012b) *Astron. Astrophys., 538,* A114.
Pinilla P. et al. (2013) *Astron. Astrophys., 554,* A95.
Pinte C. et al. (2007) *Astron. Astrophys., 469,* 963.
Pinte C. et al. (2008) *Astron. Astrophys., 489,* 633.
Pollack J. B. et al. (1994) *Astrophys. J., 421,* 615.
Poppe T. et al. (2000) *Astrophys. J., 533,* 472.
Qi C. et al. (2013) *Science, 341,* 630.
Quanz S. P. et al. (2011) *Astrophys. J., 738,* 23.
Quanz S. P. et al. (2012) *Astron. Astrophys., 538,* A92.
Ricci L. et al. (2010a) *Astron. Astrophys., 521,* A66.
Ricci L. et al. (2010b) *Astron. Astrophys., 512,* A15.
Ricci L. et al. (2011) *Astron. Astrophys., 525,* A81.
Ricci L. et al. (2012a) *Astrophys. J. Lett., 761,* L20.
Ricci L. et al. (2012b) *Astron. Astrophys., 540,* A6.
Ricci L. et al. (2013) *Astrophys. J. Lett., 764,* L27.
Rodmann J. et al. (2006) *Astron. Astrophys., 446,* 211.
Ros K. and Johansen A. (2013) *Astron. Astrophys., 552,* 137.
Rosenfeld K. A. et al. (2013) *Astrophys. J., 775,* 136.
Roy A. et al. (2013) *Astrophys. J., 763,* 55.
Rozyczka M. et al. (1994) *Astrophys. J., 423,* 736.
Sadavoy S. I. et al. (2013) *Astrophys. J., 767,* 126.
Safronov V. S. (1969) *Evolution of the Protoplanetary Cloud and Formation of the Earth and Planets.* Nauka, Moscow (in Russian); translated in English (1972) as NASA TT F-677.
Salter D. M. et al. (2010) *Astron. Astrophys., 521,* A32.
Schmitt W. et al. (1997) *Astron. Astrophys., 325,* 569.
Schräpler R. and Blum J. (2011) *Astrophys. J., 734,* 108.
Schräpler R. and Henning T. (2004) *Astrophys. J., 614,* 960.
Scott E. R. D. and Krot A. N. (2005) *Astrophys. J., 623,* 571.
Shakura N. I. and Sunyaev R. A. (1973) *Astron. Astrophys., 24,* 337.
Shetty R. et al. (2009) *Astrophys. J., 696,* 2234.
Shu F. et al. (1994) *Astrophys. J., 429,* 781.
Shu F. H. et al. (2001) *Astrophys. J., 548,* 1029.
Sicilia-Aguilar A. et al. (2007) *Astrophys. J., 659,* 1637.
Sirono S.-I. (2011a) *Astrophys. J. Lett., 733,* L41.
Sirono S.-I. (2011b) *Astrophys. J., 735,* 131.
Smoluchowski M. V. (1916) *Phys. Zeit, 17,* 557.
Steinacker J. et al. (2010) *Astron. Astrophys., 511,* A9.
Stepinski T. F. and Valageas P. (1996) *Astron. Astrophys., 309,* 301.
Stepinski T. F. and Valageas P. (1997) *Astron. Astrophys., 319,* 1007.
Stepnik B. et al. (2003) *Astron. Astrophys., 398,* 551.
Sterzik M. F. and Morfill G. E. (1994) *Icarus, 111,* 536.
Suttner G. and Yorke H. W. (2001) *Astrophys. J., 551,* 461.
Suttner G. et al. (1999) *Astrophys. J., 524,* 857.
Suutarinen A. et al. (2013) *Astron. Astrophys., 555,* A140.
Suyama T. et al. (2012) *Astrophys. J., 753,* 115.
Takeuchi T. and Lin D. N. C. (2002) *Astrophys. J., 581,* 1344.
Tanaka H. et al. (2005) *Astrophys. J., 625,* 414.
Teiser J. and Wurm G. (2009a) *Astron. Astrophys., 505,* 351.
Teiser J. and Wurm G. (2009b) *Mon. Not. R. Astron. Soc., 393,* 1584.
Teiser J. et al. (2011) *Icarus, 215,* 596.
Testi L. et al. (2001) *Astrophys. J., 554,* 1087.
Testi L. et al. (2003) *Astron. Astrophys., 403,* 323.
Tobin J. J. et al. (2013) *Astrophys. J., 771,* 48.
Trotta F. et al. (2013) *Astron. Astrophys., 558,* A64.
Ubach C. et al. (2012) *Mon. Not. R. Astron. Soc., 425,* 3137.
Uribe A. L. et al. (2011) *Astrophys. J., 736,* 85.
Urpin V. A. (1984) *Soviet Astron., 28,* 50.
van Boekel R. et al. (2004) *Nature, 432,* 479.
van Boekel R. et al. (2005) *Astron. Astrophys., 437,* 189.
van der Marel N. et al. (2013) *Science, 340,* 1199.
Veneziani M. et al. (2010) *Astrophys. J., 713,* 959.
Voelk H. J. et al. (1980) *Astron. Astrophys., 85,* 316.
Wada K. et al. (2008) *Astrophys. J., 677,* 1296.
Wada K. et al. (2009) *Astrophys. J., 702,* 1490.
Weidenschilling S. J. (1977a) *Mon. Not. R. Astron. Soc., 180,* 57.
Weidenschilling S. J. (1977b) *Astrophys. J. Suppl., 51,* 153.
Weidenschilling S. J. (1980) *Icarus, 44,* 172.
Weidenschilling S. J. (1984) *Icarus, 60,* 553.
Weidenschilling S. J. (1997) *Icarus, 127,* 290.
Weidenschilling S. J. and Cuzzi J. N. (1993) In *Protostars and Planets III* (E. H. Levy and J. I. Lunine, eds.), pp. 1031–1060. Univ. of Arizona, Tucson.
Weidenschilling S. J. and Ruzmaikina T. V. (1994) *Astrophys. J., 430,* 713.
Weintraub D. A. et al. (1989) *Astrophys. J. Lett., 340,* L69.
Whipple F. L. (1972) In *From Plasma to Planet* (A. Elvius, ed.), pp. 211–232. Wiley, New York.
Williams J. P. and Cieza L. A. (2011) *Annu. Rev. Astron. Astrophys., 49,* 67.
Williams J. P. et al. (2005) *Astrophys. J., 634,* 495.
Williams J. P. et al. (2013) *Mon. Not. R. Astron. Soc., 435,* 1671.
Wilner D. J. et al. (2005) *Astrophys. J. Lett., 626,* L109.
Windmark F. et al. (2012a) *Astron. Astrophys., 544,* L16.
Windmark F. et al. (2012b) *Astron. Astrophys., 540,* A73.
Wisniewski J. P. et al. (2008) *Astrophys. J., 682,* 548.
Wolf S. et al. (2003) *Astrophys. J., 588,* 373.
Wolf S. et al. (2008) *Astrophys. J. Lett., 674,* L101.
Wolf S. et al. (2012) *Astron. Astrophys. Rev., 20,* 52.
Woody D. P. et al. (1989) *Astrophys. J. Lett., 337,* L41.
Wurm G. et al. (2005) *Icarus, 178,* 253.
Youdin A. N. and Goodman J. (2005) *Astrophys. J., 620,* 459.
Youdin A. N. and Lithwick Y. (2007) *Icarus, 192,* 588.
Youdin A. N. and Shu F. H. (2002) *Astrophys. J., 580,* 494.
Zhukovska S. et al. (2008) *Astron. Astrophys., 479,* 453.
Zsom A. et al. (2010) *Astron. Astrophys., 513,* A57.
Zsom A. et al. (2011) *Astron. Astrophys., 534,* 73.

Pontoppidan K. M., Salyk C., Bergin E. A., Brittain S., Marty B., Mousis O., and Öberg K. I. (2014) Volatiles in protoplanetary disks. In *Protostars and Planets VI* (H. Beuther et al., eds.), pp. 363–385. Univ. of Arizona, Tucson, DOI: 10.2458/azu_uapress_9780816531240-ch016.

Volatiles in Protoplanetary Disks

Klaus M. Pontoppidan
Space Telescope Science Institute

Colette Salyk
National Optical Astronomy Observatory

Edwin A. Bergin
University of Michigan

Sean Brittain
Clemson University

Bernard Marty
Université de Lorraine

Olivier Mousis
Université de Franche-Comté

Karin I. Öberg
Harvard University

Volatiles are compounds with low sublimation temperatures, and they make up most of the condensible mass in typical planet-forming environments. They consist of relatively small, often hydrogenated, molecules based on the abundant elements carbon, nitrogen, and oxygen. Volatiles are central to the process of planet formation, forming the backbone of a rich chemistry that sets the initial conditions for the formation of planetary atmospheres, and act as a solid mass reservoir catalyzing the formation of planets and planetesimals. Since *Protostars and Planets V* (*Reipurth et al.,* 2007), our understanding of the evolution of volatiles in protoplanetary environments has grown tremendously. This growth has been driven by rapid advances in observations and models of protoplanetary disks, and by a deepening understanding of the cosmochemistry of the solar system. Indeed, it is only in the past few years that representative samples of molecules have been discovered in great abundance throughout protoplanetary disks (CO, H_2O, HCN, C_2H_2, CO_2, HCO^+) — enough to begin building a complete budget for the most abundant elements after hydrogen and helium. The spatial distributions of key volatiles are being mapped, snow lines are directly seen and quantified, and distinct chemical regions within protoplanetary disks are being identified, characterized, and modeled. Theoretical processes invoked to explain the solar system record are now being observationally constrained in protoplanetary disks, including transport of icy bodies and concentration of bulk condensibles. The balance between chemical *reset* — processing of inner disk material strong enough to destroy its memory of past chemistry — and *inheritance* —— the chemically gentle accretion of pristine material from the interstellar medium (ISM) in the outer disk — ultimately determines the final composition of preplanetary matter. This chapter focuses on making the first steps toward understanding whether the planet-formation processes that led to our solar system are universal.

1. INTRODUCTION

The study of the role of ices and volatile compounds in the formation and evolution of planetary systems has a long and venerable history. Up until the late twentieth century, due to the lack of knowledge of exoplanetary systems, the solar system was the only case study. Consequently, some of the earliest pieces of firm evidence for abundant ices in planetary systems came from cometary spectroscopy, showing the photodissociation products of what could only be common ices, such as water and ammonia (*Whipple,* 1950). Following early suggestions by, e.g., *Kuiper* (1953), the presence of water ice in the rings of Saturn (*McCord et al.,* 1971) and in the jovian satellite system (*Pilcher et al.,* 1972) was confirmed using infrared spectroscopy. The characterization of ices in the outer solar system continues to this day (*Mumma and Charnley,* 2011; *Brown et al.,* 2012). Similarly, the presence of ices in the dense ISM — the material out of which all planetary systems form — has been recognized since the 1970s and conjectured even earlier. With the advent of powerful infrared satellite observatories, such as the Infrared Space Observatory (ISO) and the Spitzer Space Telescope, we know that interstellar ices are commonplace and carry a large fraction of the solid mass in protostellar environments (*Gibb et al.,* 2004; *Öberg et al.,* 2011b).

The next logical investigative steps include the tracking of the volatiles as they take part in the formation and evolution of protoplanetary disks — the intermediate evolutionary stage between the ISM and evolved planetary systems during which the planets actually form. This subject, however, has seen little progress until recently. A decade ago, volatiles in protoplanetary disks were essentially beyond our observational capabilities, and the paths the volatiles take from the initial conditions of a protoplanetary disk until they are incorporated into planets and planetesimals were not well understood. What happened over the past decade, and in particular since the conclusion of *Protostars and Planets V* (*PPV*) (*Reipurth et al.,* 2007), is that our observational knowledge of volatiles in protoplanetary disks has been greatly expanded, opening up a new era of comparative cosmochemistry. That is, the solar nebula is no longer an isolated case study, but a data point — albeit an important one — among hundreds. The emerging complementary study of volatiles in protoplanetary disks thus feeds on comparisons between the properties of current-day solar system material and planet-forming gas and dust during the critical first few million years of the development of exoplanetary systems. This is an area of intense contemporary study, and is the subject of this chapter.

1.1. Emerging Questions

Perhaps one of the most central questions in astronomy today is whether our solar system, or any of its characteristics, is common or an oddity. We already know that most planetary systems have orbital architectures that do not resemble the solar system, but is the chemistry of the solar system also uncommon? Another matter is the degree to which volatiles are *inherited* from the parent molecular cloud, or whether their chemistry is *reset* as part of typical disk evolution. That is, can we recover evidence for an interstellar origin in protoplanetary and planetary material, or is that early history lost in the proverbial furnace of planet formation? There is currently an apparent disconnect between the cosmochemical idea of an ideal condensation sequence from a fully vaporized and hot early phase in the solar nebula, supported by a wealth of data from meteoritic

material, and the astrophysical idea of a relatively quiescent path of minimally processed material during the formation of protoplanetary disks. The answer is likely a compromise, in which parts of the disk are violently reset, while others are inherited and preserved over the lifetime of the central star, with most regions showing some evidence for both, due to a variety of mixing processes. In this chapter, we will consider the solar nebula as a protoplanetary disk among many others — and indeed call it the solar protoplanetary disk. We will discuss the issue of inheritance vs. reset, and suggest a division of disks into regions characterized by these very different chemical circumstances.

1.2. Defining Protoplanetary Disks and Volatiles

The term protoplanetary disk generally refers to the rotationally supported, gas-rich accretion disk surrounding a young pre-main-sequence star. The gas-rich disk persists during planetesimal and giant planet formation, but not necessarily during the final assembly of terrestrial planets. During the lifetime of a protoplanetary disk, both solid and gas-phase chemistry is active, shaping the initial composition of planets, asteroids, comets, and Kuiper belt objects. Many of the chemical properties and architecture of the current-day solar system were set during the nebular/protoplanetary disk phase, and are indeed preserved to this day, although somewhat obscured by dynamical mixing processes during the later debris disk phase.

Most of the mass in protoplanetary disks is in the form of molecular hydrogen and helium gas. Some of the solid mass is carried as dust grains, mostly composed of silicates and with some contribution of carbon-dominated material. This material is generally *refractory*, i.e., very high temperatures (≥ 1000 K) are needed to sublimate it. Only in a very limited region, in a limited period of time or under unusual circumstances, are refractory grains returned to the gas phase. The opposite of refractory is *volatile*. In the context of this chapter, disk volatiles are molecular or atomic species with relatively low sublimation temperatures (less than a few 100 K) that are found in the gas phase throughout a significant portion of a typical disk under typical, quiescent disk conditions. In cosmochemistry the use of the words "refractory" and "volatile" is somewhat different; while they are still related to condensation temperature, their use is usually reserved for atomic elements, rather than molecules, for interpreting elemental abundances in meteorites. It is also possible to define a subclass of volatile material that includes all *condensible* species — i.e., the class of volatiles that are found in both their solid and gaseous forms in significant parts of the disk, but that excludes species that never condense in bulk. It is the ability of condensible volatiles to relatively easily undergo dramatic phase changes that lead to them having a special role in the evolution of protoplanetary disks and the formation of planets. An example of a volatile that is not considered condensible is the dominant mass compound, H_2, while one of the most important condensible species is water.

2. SOLAR NEBULA AS A VOLATILE-RICH PROTOPLANETARY DISK

The solar system was formed from a protoplanetary disk during a time frame of 2–3 million years (m.y.), spanning ages of at least 4567–4564 Ma — a number known to a high degree of precision, thanks to accurate radiometric dating of primitive meteoritic material formed by gas-condensation processes (*Scott,* 2007). The composition of solar system bodies, including primitive material in meteorites, comets, and planets, provide a detailed window into their formation. The primordial elemental composition of the solar protoplanetary disk has been systematically inferred from measurements of the solar atmosphere, generally under assumptions of efficient convective mixing. Any compositional or isotopic differences measured in primitive solar system materials, such as comets and meteorites, can therefore be ascribed to processes taking place during the formation and evolution of the solar system, generally after the formation of the Sun.

One of the initial models to account for the volatility patterns of meteorites is one where the solar disk started hot (>1400 K), such that all material was vaporized. As the gas cooled, the elements formed minerals governed by the condensation sequence in thermodynamic equilibrium (e.g., *Grossman,* 1972; *Wood and Hashimoto,* 1993; *Ebel and Grossman,* 2000), with some level of incompletion due to grain growth and disk gas dispersal (*Wasson and Chou,* 1974; *Cassen,* 1996). Condensation models predict that calcium-aluminum compounds would be the first elements to condense (*Lord,* 1965; *Grossman,* 1972), a proposition that is consistent with calcium-aluminum inclusions (CAIs) being the oldest material found in the solar system (*Amelin et al.,* 2002).

The widespread presence of presolar grains, albeit in extremely dilute quantities (*Zinner,* 1998), the high levels of deuterium enrichments seen in Oort cloud comets and meteorites (*Mumma and Charnley,* 2011; *Alexander et al.,* 2012), and the presence of highly volatile CO in comets present the contrary perspective that at least part of the nebula remained cold or represent material provided to the disk at later cooler stages. It is with these fundamental contradictions in mind that we strive to understand the solar system and general protoplanetary disks under a common theoretical umbrella.

2.1. Bulk Abundances of Carbon, Oxygen, and Nitrogen

For instance, consider the bulk abundances of common elements in Earth, and in particular carbon, oxygen and nitrogen, which are the main building blocks, along with hydrogen, of volatile chemistry. With a surface dominated by carbon-based life, coated by water, with a nitrogen-rich atmosphere, our planet is actually a carbon-, nitrogen- and water-poor world, compared to solar abundances and, presumably, to the abundance ratios of the molecular cloud out of which the solar system formed. This observation is illustrated in Fig. 1, where Earth's relative elemental abundances are compared to those of the Sun. While the refractory elements are essentially of solar abundances, others are depleted by orders of magnitude, most prominently carbon and nitrogen, and to a lesser degree, oxygen. Going beyond Earth, carbon, nitrogen, and oxygen abundances in various solar system bodies are compared (Earth, meteorites, comets, and the Sun) in Fig. 2.

A great amount of information about the relative distribution of volatiles in the solar nebula is contained in these plots. Specifically, they indicate the volatility of the chemical compounds carrying the bulk of each element: If an element is highly depleted, most of that element must either have been in a form too volatile to condense at the location and time that a specific solar system body formed, or the material was subjected to subsequent, possibly transient, heating at a later time while still in a form where outgassing and loss could be efficient. In other words, while the plot will not directly tell us what the main elemental carriers were, it will provide us with some of their basic physical and chemical properties. In the case of Earth, we can see that many elements, and not only carbon, nitrogen, and oxygen, are depleted to some extent, giving rise to the idea that Earth formed from material exposed to temperatures well above 150 K, and going as high as 1000 K (*Allègre et al.,* 2001).

The ISM and the Sun represent the total amount of material available. Beyond the Sun, most bodies in the solar system are hydrogen-poor. This reflects the fact that molecular hydrogen is the most volatile molecular compound present in a protoplanetary disk and is never efficiently incorporated into solids anywhere in the solar system. Comets formed in the coldest regions of the nebula and incorporate the largest amount of hydrogen, mostly trapped in the form of water. In contrast, the rocky planetesimals that formed close to the Sun contain successively less water and therefore less hydrogen.

2.1.1. Oxygen. The oxygen content in the solar system is also, to a large degree, a story of water. Oxygen is generally the least-depleted element among the carbon-nitrogen-oxygen (CNO) trifecta across the solar system, indicating that a relatively more refractory compound carried much of the oxygen. A likely candidate for this carrier is silicates (mostly oxides built with the anions SiO_4^{4-} or SiO_3^{2-}). Indeed, comets have the highest O/Si ratio, above that even of the Sun, although there is some uncertainty in the calculation of silicon abundances in comets as these depend on an assumed mass opacity (*Min et al.,* 2005). However, one value comes from *in situ* measurements of Comet Halley by the Soviet Vega 1 spacecraft (*Jessberger et al.,* 1988), and supports this general picture. Thus, comets incorporated oxygen in the form of silicates, as well as CNO ices. Rocky bodies also contain significant amounts of oxygen, but below the total atomic nebular reservoir, by factors of 2–10. This allows a rough estimate of the fraction of oxygen in volatiles vs. refractory compounds — a few volatile oxygen atoms for each refractory oxygen atom. Furthermore, it demonstrates

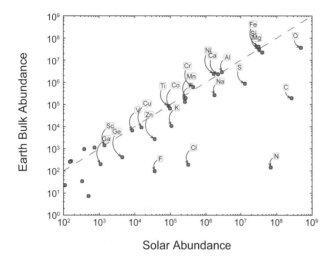

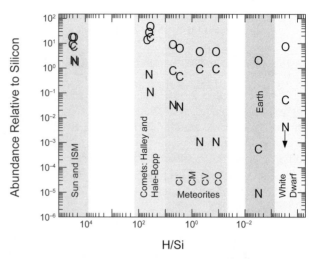

Fig. 1. Solid Earth bulk elemental abundances (*Allègre et al., 2001*) compared to solar abundances (*Grevesse et al., 2010*). The solar abundances are normalized to an H abundance of 10^{12}. For Earth, the silicon abundance is scaled to that of the Sun. All abundances are in units of relative number density of nuclei. The dashed line guides the eye to the 1:1 ratio. For instance, the solar abundances of C, O, and N are $10^{8.43}$, $10^{8.69}$, and $10^{7.83}$, respectively.

Fig. 2. The relative CNO abundances in the solar system. The abundances are computed relative to Si, the primary refractory element in the solar system and interstellar space. The abundances are shown as a function of the H/Si ratio, which separates the various bodies. The figure is taken and updated from *Lee et al.* (2010) and *Geiss* (1987). The white dwarf values are for GD 40 (*Jura et al., 2012*) and are provided with an arbitrary H/Si ratio (since this is unknown).

that in the terrestrial-planet-forming regions of the nebula (inside the snow line), most oxygen was in a gaseous form. Overall, the variation in bulk oxygen abundance throughout the solar system suggests that the main oxygen carrier was a condensible species — frozen out in the outer disk, yet in the gas phase in the inner disk. From these considerations, likely candidates are H_2O, CO_2, and CO.

2.1.2. Carbon. The bulk carbon abundance shows a stronger signal than that of oxygen. As we traverse the solar system, from comets over the asteroids to Earth, we discover a gradient in the relative amounts of condensed carbon, culminating in a carbon depletion of 3 orders of magnitude in Earth. The caveat is that Earth's C/Si ratio is estimated from the mantle (*Allègre et al., 2001*), but Earth's total carbon content is highly uncertain as there might be deep reservoirs of carbon in the lower mantle and/or core (*Wood, 1993; Dasgupta and Hirschmann, 2010*). However, meteorites have been posited as tracers of Earth's starting materials and even unprocessed carbonaceous chondrites have nearly an order of magnitude carbon deficiency relative to the total amount that was available in the nebula. The likely inference is that within the inner nebula volatile carbon must have been predominantly in a volatile, gaseous form, in contrast with the ISM, where much, perhaps most, of the carbon is sequestered as graphite or other highly refractory forms. This so-called carbon deficit problem will be discussed further in section 2.2. The carbon deficit in planetesimals is contrasted with a carbon enhancement in the giant planet atmospheres; the C/H in the jovian atmosphere is enhanced by a factor of 3.5 over solar (*Niemann et al., 1996; Wong et al., 2004*), while the saturnian ratio

is 7× solar (*Flasar et al., 2005*). This is consistent with the giant planets forming outside the condensation front of a primary volatile carbon carrier.

The carbon that made it to Earth was likely delivered after the protoplanetary disk phase, but during the bulk Earth formation, and there is significant evidence that it is mostly of chondritic origin (*Marty, 2012; Albarède et al., 2013*). Carbonaceous chondrites are composed primarily of silicates, but contain up to 10–20% equivalent water in the form of hydrated silicates, and a few percent of carbon, by mass (*Scott, 2007*). While strongly depleted in bulk carbon relative to solar abundances, the remaining chondritic carbon (and nitrogen) is mostly carried by refractory organics. The origin of chondritic organics is often traced by the isotopic ratios of noble gases (Ar, Ne, Xe, . . .), trapped in an ill-defined organic phase [called "Q" (*Ott et al., 1981; Marrocchi et al., 2005*)]. Present in very small fractions (<1% by mass), chondritic noble gases generally have a recurring isotopic signature, strongly fractionated relative to that of the Sun and the solar protoplanetary disk (chondritic noble gases are heavy). The isotopic signature in noble gases in Earth's atmosphere is consistent with a mixture that is 90% chondritic and at most 10% solar (*Marty, 2012*). The chondritic noble gas fractionation has been reproduced in the laboratory during gas-solid exchanges under ionizing conditions (*Frick et al., 1979*), consistent with entrapment during photochemical formation of the organics from ices in the solar disk (*Jenniskens et al., 1993; Gerakines et al., 1996; Öberg et al., 2009*). Another way of correctly fractionating noble gases is by selective trapping in forming ices. *Notesco et al.* (1999) found experimentally that noble gases are enriched according to their isotopic mass ratios ($\sqrt{m_1/m_2}$) during the formation of amorphous water ice. That is, a similar effect can be generated as a purely thermal effect. In summary, there is significant evidence that Earth's carbon is delivered with a volatile reservoir. That said, it is not clear whether the delivery vector was from a chondritic reservoir, as opposed to a cometary reservoir. There is support for chondritic delivery, but an unambiguous identification was for a long time not possible due to a lack of accurate measurements of noble gases in comets. Recently, samples returned by the Stardust mission from Comet 81P/Wild 2 showed similar noble gas signatures as those of chondrites, indicating that the noble gas hosts may be of similar origin in comets and chondrites (*Marty et al.*, 2008).

2.1.3. Nitrogen. Nitrogen is the most depleted element in our selection. In Earth, it is depleted by 5 orders of magnitude. While Earth's atmosphere is dominated by N_2, this represents about half of Earth's nitrogen (*Marty*, 1995). In comparison, the total mass of the atmosphere is a bit less than 1 ppm of the total mass of Earth. While the absolute abundance of nitrogen is uncertain in comets, it is clear that it is highly depleted relative to the solar value. The best estimates come from the Halley *in situ* measurements (*Wyckoff et al.*, 1991), as this accounts for nitrogen incorporated in the coma as evaporated ices and in the refractory component. The Hale-Bopp estimate [taken from the summary of *Mumma and Charnley* (2011)], which has the lower N/Si ratio in Fig. 2, only includes the ices and is therefore a lower limit. One interesting aspect is that *Wyckoff et al.* (1991) estimates that Comet Halley carried most of its nitrogen (90%) in a carbonaceous (CHON) refractory component. As in the other heavy element pools, there is a gradient in the N/Si ratio as we approach the terrestrial-planet-forming region with both meteorites and Earth exhibiting large nitrogen depletions, relative to the total amount available. Clearly, the vast majority of the nitrogen resided in the gas in the solar nebula in a highly volatile form, which could be NH_3, N_2, and/or N. In this context, laboratory measurements suggest that both CO and N_2 have similar binding strengths to the grain surface (*Öberg et al.*, 2005). Thus the difference between the presence of CO in comets and the relative lack of N_2 stands out. Note, however, that in the jovian atmosphere, nitrogen in the form of NH_3 is enhanced relative to solar by a factor of 3 (*Wong et al.*, 2004), similar to that of carbon, and similarly indicating that the jovian core formed outside the primary nitrogen condensation front.

2.2. Origin of the Carbon Deficit

As we have seen in Fig. 2, solids in the terrestrial region of the solar system, as represented by Earth and meteorites, have far less elemental carbon than what was available, on average, in the solar protoplanetary disk. The implication is that the carbon was either entirely missing in the terrestrial region, or it was in a volatile form, not available for con-

densation. In stark contrast, the dense ISM sequesters ~60% of the carbon in some unknown combination of amorphous carbon grains, large organics, or, perhaps, as polycyclic aromatic hydrocarbons (PAHs) (*Savage and Sembach*, 1996; *Draine*, 2003). If terrestrial worlds formed directly from solid grains that are supplied to the young disk via collapse, then they would have 2× more carbon than silicon by mass. This is strikingly different from the observed ratio of 1–2% of C/Si by mass (*Allègre et al.*, 2001), and essentially all the primordial, highly refractory, carbon carriers must be destroyed prior to terrestrial planet formation to account for the composition of Earth and carbonaceous chondrites.

A similar signature is known to exist in exoplanetary systems from measurements of asteroidal material accreted onto white dwarfs (*Jura*, 2006). A small fraction of white dwarfs show lines from a wide range of heavy elements. Since any heavy elements are rapidly depleted from white dwarf photospheres by settling, their presence indicates that they must have been recently accreted (*Paquette et al.*, 1986). In such polluted white dwarfs, the C/Fe and C/Si abundance ratios are well below solar and consistent with accretion of asteroidal material deficient in carbon (*Jura*, 2008; *Gänsicke et al.*, 2012; *Farihi et al.*, 2013). This suggests that Earth's missing carbon and the carbon deficit in the inner solar system are a natural outcome of star and planet formation, i.e., the process leading to a strong carbon deficit is a universal property of protoplanetary disks.

Theoretical models suggest the carbon grains can be destroyed via oxidation either by OH at the inner edge of the accretion disk (*Gail*, 2002) or by oxygen atoms created by photodissociation of oxygen-carrying volatiles, such as water, on disk surfaces exposed to ultraviolet (UV) radiation from the central star (or other nearby stars) (*Lee et al.*, 2010). Although the details vary depending upon the uncertain carrier of carbon-grain absorption in the ISM, the destruction of the grains could produce CO or CO_2 (graphite/amorphous carbon) or, to a lesser degree, hydrocarbons with oxygen from PAHs (*Gail*, 2002; *Lee et al.*, 2010; *Kress et al.*, 2010). The theoretical implication is that CO or CO_2 is the likely dominant carrier of carbon in the inner disk, rather than either solid-phase carbon grains or gas-phase molecules like CH_4, and that the sum of these species should be close to the solar carbon abundance.

2.3. Solar System Snow Line

We know that the distribution of planets in the solar system is skewed, with low-mass terrestrial planets residing inside 3 AU and massive gas and ice giant planets orbiting beyond this radius, and it has long been thought that this is related to the presence of a water ice condensation front (the so-called "snow line") being located at a few astronomical units (*Hayashi*, 1981). However, the present-day distribution of water in the asteroid belt is somewhat ambiguous, and the location of the end-stage snow line in the solar disk may be preserved only as a fossil. The lifetime of water ice on the surfaces of all but the largest asteroids is short-lived inside 5 AU (*Lebofsky*, 1980; *Levison and Duncan*, 1997). Exceptions are 1 Ceres, which has some amount of surface water ice (*Lebofsky et al.*, 1981), and 24 Themis (*Campins et al.*, 2010; *Rivkin and Emery*, 2010). In both cases, this may be evidence for subsurface ice revealed by recent activity. For most asteroids, the water content has been modified by heating within a few million years of disk formation. This heating led to the melting of interior ice, followed by aqueous alteration and the formation of hydrated silicates that are stable up to the current age of the solar system.

The direct methods of searching for asteroid surface water in the form of hydrated silicates have encountered some interpretational difficulties. Hydrated silicates on the surfaces of outer belt asteroids with semimajor axes in the range 2.5–5.2 AU do not show clear evidence for a snow line (*Jones et al.*, 1990), and their presence may therefore be more a tracer of local thermal evolution than the primordial water ice distribution (*Rivkin et al.*, 2002). That is, they may be a tracer of planetesimals that were exposed to transient heating strong enough to melt water and cause the formation of the hydrated silicates, rather than the presence of water per se.

The localization of the solar system snow line relies on the measured water content of meteorites, coupled with spectroscopic links to taxonomic asteroid classes (*Chapman et al.*, 1975; *Chapman*, 1996). Carbonaceous chondrites (CI and CM) contain up to 5–10% of water by mass (*Kerridge*, 1985), and have been associated with C-type asteroids, which have semimajor axes predominantly beyond 2.5 AU (*Bus and Binzel*, 2002). Conversely, ordinary chondrites are relatively dry (0.1% of water by mass) (*McNaughton et al.*, 1981) and are associated with S-type asteroids, which are found within 2.5 AU (*Bus and Binzel*, 2002). The different asteroid types, however, are radially mixed, and any present-day water distribution is not as sharp as the snow line once was. This is likely, at least in part, related to dynamical mixing of planetesimals as described by the Nice and Grand Tack models (*Walsh et al.*, 2011).

In summary, the study of the solar nebula snow line is associated with significant uncertainties due to a lack of *in situ* measurements of asteroids and smearing out by dynamical mixing that astronomical measurements of snow lines in protoplanetary disks may provide important new constraints on this critical feature of a forming planetary system.

2.4. Isotope Fractionation in Solar System Volatiles

One of the classical methods for investigating the origin of solar system materials is the measurement and comparison of accurate isotopic ratios. This is in particular true for carbon-nitrogen-oxygen-hydrogen (CNOH) as the main volatile building blocks, and strong fractionation patterns are indeed observed in the bulk carriers of these elements in the solar system. The most important systems include the deuterium/hydrogen (D/H) fraction, the relative amounts of the three stable isotopes of oxygen — ^{16}O, ^{17}O, and ^{18}O — as well as the ratio of the two stable nitrogen isotopes,

$^{14}N/^{15}N$. Since these are abundant species — in particular relative to the many rare refractory systems traditionally used in cosmochemistry — they are also accessible, albeit with difficulty, to astronomical observations of the ISM and protoplanetary disks. These elements show great isotopic heterogeneity in various solar system reservoirs, in sharp contrast to more refractory elements, of which most are isotopically homogenized at the 0.1% level — including elements with somewhat lower condensation temperatures, such as Cu, K, and Zn (*Zhu et al.*, 2001; *Luck et al.*, 2001). Of particular interest is meteoritic oxygen, which scatters along the famous mass-independent fractionation line in a three-isotope plot (*Clayton et al.*, 1973) (see Fig. 3). The $^{14}N/^{15}N$ ratio exhibits large variations of up to a factor of 5 among different solar system bodies (*Briani et al.*, 2009). The ubiquitous strong fractionation of the CNOH systems suggests that these elements were carried by molecular compounds accessible to gas-phase fractionation processes.

We now know the location of the primordial material in fractionation diagrams. Recently, the Genesis mission accurately measured the elemental composition of the Sun and, by extension, that of the solar protoplanetary disk (*Burnett et al.*, 2003). Over a period of three years, the solar wind was sampled by passive implantation of ions into pure target material. The spacecraft was located at Lagrangian point L1 to prevent any alteration of the solar wind composition by the terrestrial magnetic field. A result from Genesis of particular interest to this chapter was the measurement of the

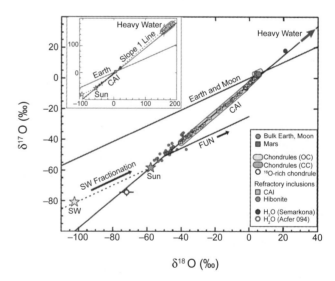

Fig. 3. The three-isotope plot of oxygen in the solar system, showing the fractionation line of meteorites along a mass-independent slope-1 line. The solar point is a recent addition from Genesis samples of the solar wind. This point, labeled "Sun," is deduced from the sample measurements, labeled "SW," by assuming a correction for mass-dependent fractionation in the solar wind. The inset figure shows an extended view highlighting the very-^{16}O-depleted samples of heavy water in the carbonaceous chondrite Acfer 094. Figure originally from *McKeegan et al.* (2011).

isotopic composition of solar oxygen, nitrogen, and noble gases. Because the young Sun burned deuterium, the D/H ratio of the solar disk has been measured independently using the jovian atmosphere. Genesis established that all solar system reservoirs (except for the atmosphere of Jupiter and probably of the other giant planets) are enriched, relative to the Sun and solar protoplanetary disk, in the rare and heavy isotopes of hydrogen (D), nitrogen (^{15}N), and oxygen (^{17}O and ^{18}O). There are also indications of an enrichment in ^{13}C in solar system solids and gases relative to the Sun (*Hashizume et al.*, 2004).

The origin of the fractionation patterns in CNOH in the solar system is still a matter of vigorous debate, and, while data have improved in the past decade, few questions have been conclusively settled; we still do not have a clear idea of whether isotopic ratios in volatiles have an interstellar origin or whether they evolved as a consequence of protoplanetary disk evolution. In meteorites, the organic matter tends to show the strong isotopic anomalies of hydrogen and nitrogen (*Epstein et al.*, 1987). In the ISM the $^{14}N/^{15}N$ ratio is typically just above 400 with relatively little variation (*Aléon*, 2010, and references therein). This is matched by the ratio in the jovian atmosphere measured by the Galileo probe [435 ± 57 (*Owen et al.*, 2001)], and in the Sun [442 ± 66 (*Marty et al.*, 2010)], but contrasts with cometary and meteoritic values of <150 (*Aléon*, 2010, and references therein). The D/H ratio in organics is similarly enhanced by factors of up to 5 in highly localized and heterogenous "hot spots" in meteorites, which are likely explained by the presence of organic radicals with D/H as high as ~0.02 (*Remusat et al.*, 2009). The D/H ratio is generally correlated with the $^{14}N/^{15}N$ fractionation in organics in meteorites and comets, relative to planetary atmospheres and the Sun (see Fig. 4). Given the high levels of deuteration observed in the ISM (*Ceccarelli et al.*, 2007) and in protoplanetary disks (*Qi et al.*, 2008), it is tempting to use the meteoritic patterns to argue for an interstellar origin. However, the deuteration in dense clouds and disks is typically higher than in meteorites, leading to disk models that use a highly fractionated initial condition that evolves toward a more equilibrated (less fractionated) system at the high temperatures and densities in a disk (e.g., *Drouart et al.*, 1999). In this way, the protoplanetary disk phase may represent a modified "intermediate" fractionation state, relatively close to the solar D/H ratio, compared to the extreme primordial values. This is discussed in greater detail in the context of comets in section 2.5. We also refer the reader to the chapter by Ceccarelli et al. in this volume, which focuses specifically on deuterium fractionation.

Oxygen isotopes in the solar system uniquely display a puzzling mass-independent fractionation pattern (*Clayton et al.*, 1973), as shown in detail in Fig. 3. The location of nearly all inner solar system materials on a single mass-independent fractionation line led to a search for physical processes capable of producing the observed fractionation slope, and affecting a large portion of the inner solar nebula. A leading candidate is isotope-selective photodissocia-

tion of CO (*Kitamura and Shimizu*, 1983; *Thiemens and Heidenreich*, 1983; *Clayton*, 2002; *Yurimoto and Kuramoto*, 2004; *Lyons and Young*, 2005), a process that can now be investigated in protoplanetary disks, as we will discuss in section 3.7.

2.5. Origin of Cometary Volatiles in the Solar System

In the recent review of cometary chemistry by *Mumma and Charnley* (2011), the observed composition of comets is compared to that of the ISM — due to a lack of appropriate observations of protoplanetary disks. Using recent disk data, we can now begin to make a more direct comparison. The motivation for moving beyond the solar system is clear: Comets sample some of the oldest and most primitive material in the solar system, and are likely essentially unaltered since the protoplanetary disk stage. They are thus our best window into the volatile composition of the solar protoplanetary disk. A first, essential question to ask in preparation for a comparison to contemporary protoplanetary disks is this: Is the volatile reservoir sampled by comets formed in the solar protoplanetary disk, or is it interstellar?

The analysis of grains from the Stardust mission suggests that the answer is both, with captured samples of Comet Wild 2 representing instead a wide range of origins. The comet has incorporated some presolar material (*Brownlee et al.*, 2006), as evidenced by deuterium and ^{15}N excesses in some grains (*Sandford et al.*, 2006), consistent with unprocessed materials in the outer solar system, where the comet is believed to have formed. However, the cometary samples seem to most resemble samples of primitive meteorites, believed to have formed much closer to the young Sun. The oxygen-, deuterium-, and nitrogen-isotopic compositions of most Wild 2 grains are strikingly similar to carbonaceous chondrite meteorites (*McKeegan et al.*, 2006), and neon-isotope ratios of noble gases are similar to those trapped

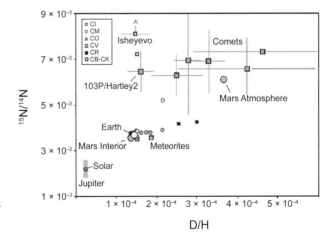

Fig. 4. Relation between the nitrogen ($^{15}N/^{14}N$) and deuterium (D/H) fractionation in the solar system. Figure from *Marty* (2012).

in chondritic organics (*Marty et al.*, 2008). In addition, the grains are primarily crystalline, rather than amorphous, with compositions inconsistent with being formed via low-temperature condensation of interstellar grains, and therefore requiring a high-temperature origin (*Brownlee et al.*, 2006). Thus, Stardust suggests that cometary matter is a mixture of outer and inner solar system materials.

An alternative tracer of formation location may be the spin temperature of water. In molecules with two or more hydrogen atoms, transitions that change the total nuclear spin are strictly forbidden. For water, therefore, as long as the chemical bonds are not broken, it is expected that the ratio between the ortho (s = 0) and para (s = 1) states persists at the value set at the formation temperature (spin temperature) of the molecule. Thus, the ortho/para ratio in comets could potentially measure the formation location of the comet, and determine the relative contributions of *in situ* ice formation vs. inheritance of ice from the molecular cloud. In comets, the water spin temperature has been measured to span from 26 K (*Bockelée-Morvan et al.*, 2009) to more than 50 K (*Mumma et al.*, 1988). These temperatures are significantly higher than those of the water-forming regions in molecular clouds (10–15 K), and perhaps represent a wide range of formation radii — 5–50 AU for typical protoplanetary disks (*D'Alessio et al.*, 1997; *Chiang and Goldreich*, 1997). However, recent laboratory results complicate this picture considerably; two experiments show that the spin temperature of water produced at very low temperature and subsequently sublimated retains no history of its formation condition (*Sliter et al.*, 2011; *Hama et al.*, 2011).

Deuterium fractionation is another tracer of the cometary formation environment. The water ice in Oort cloud comets are typically enriched by a factor of 2 in deuterium (normalized to ^1H) with respect to the terrestrial value and most primitive meteorites. However, the recent analysis of Jupiter-family Comet 103P/Hartley 2 (*Hartogh et al.*, 2011) and Oort cloud comet Garradd (*Bockelée-Morvan et al.*, 2012) have shown that lower, even terrestrial-like, D/H ratios exist in comets. These observations were used to support the idea that Earth's water was delivered by comets — a hypothesis that was otherwise falling out of favor based on the high D/H ratios in Oort comets. However, it also suggests that there is heterogeneity in the formation environments of comets, and that there may even be a gradient in the water D/H ratios in the solar disk, as predicted by models of the solar nebula (*Horner et al.*, 2007). This suggests a partial disk origin for cometary volatiles, or at least that a uniform interstellar D/H ratio evolved during the protoplanetary disk phase.

The basic idea underlying this model is that the solar disk starts with an initial condition of highly deuterated water, inherited from the protosolar molecular cloud. The water then evolves, via deuterium exchange with HD, toward less deuteration during the disk phase — a process that happens faster at high temperatures in the inner disk, but not fast

enough in the outer disk to account for the observed D/H ratios throughout the solar system [in chondrites as well as comets (*Drouart et al.*, 1999)]. Consequently, models have invoked various mixing processes, such as turbulent diffusion or viscous expansion, to reproduce the observed deuteration from *in situ* disk chemistry (*Mousis et al.*, 2000; *Hersant et al.*, 2001; *Willacy and Woods*, 2009; *Jacquet and Robert*, 2013). In particular, *Yang et al.* (2012) find that Jupiter-family comets can obtain a terrestrial D/H ratio in the outermost part of a viscously evolving disk. The final solar system D/H ratios are therefore likely dependent on both inheritance and *in situ* chemical evolution.

The cometary $^{14}N/^{15}N$ ratios have been measured in both CN and HCN. In these species, the nitrogen-isotopic ratio is a factor of 2 lower than that of Earth and most meteorites, and a factor of 3 lower than that of the Sun. Enrichment of the heavy ^{15}N isotope has been proposed to be evidence for interstellar fractionation, although chemical models of dense clouds do not readily produce fractionation in any nitrogen carrier as strong as that seen in the solar system (*Terzieva and Herbst*, 2000), and the usability of the nitrogen-isotopic ratio as an origin tracer remains unclear.

A final complication in the story of comets is that it is now understood that the orbital parameters of a given comet are a poor tracer of its formation radius in the solar disk. The Nice model (*Gomes et al.*, 2005) suggests that comets forming anywhere from 5 to >30 AU may have contributed to both the Kuiper belt comets as well as the Oort cloud comets. Cometary origins may be even more interesting if, as suggested by *Levison et al.* (2010), ~50% of comets are inherited from other stellar systems.

2.6. Total Mass of Condensible Volatile Molecules

The magnitude of the local solid surface density is linked to the efficiency of planetesimal formation, and by extension the formation of planets via accretionary processes. *Hayashi* (1981) originally used an ice/rock ratio of 4 in his formulation of the classical minimum-mass solar nebula (MMSN). Some investigations in later years have revised this value slightly downward, but it is still high. *Stevenson* (1985) found a value of 2–3, based on comparisons to the icy moons of the solar system giant planets. *Lodders* (2003) found ice/rock = 2, based on equilibrium chemistry. The recent update to the MMSN by *Desch* (2007) also uses ice/rock = 2, while *Dodson-Robinson et al.* (2009) used a full chemical model coupled with a dynamical disk model to predict an ice/rock ratio close to the original ice/rock = 4. This variation is in part due to differences in assumptions about the elemental oxygen abundance, and about the chemistry. The higher values for the ice/rock ratio require that both the bulk carbon and nitrogen are readily condensible with sublimation temperatures higher than those of, e.g., CO and N_2. Typically, this requires carriers such as CH_4 and NH_3. A high value for the ice/rock ratio therefore predicts large amount of gaseous CH_4 and NH_3 just inside their respective snow lines. The theoretical expectation for protoplanetary disks is therefore that the value most likely lies somewhere between ice/rock ~ 2–4, and precise observations of all the major carriers of carbon, oxygen, and nitrogen are needed to determine the true ice/rock ratios, as well as their potential variation from disk to disk.

In the dense ISM and in protostellar envelopes, a nearly complete inventory of ice and dust has been observed and quantified (e.g., *Gibb et al.*, 2004; *Whittet et al.*, 2009; *Öberg et al.*, 2011b) using infrared absorption spectroscopy toward young stellar objects and background stars. *Whittet* (2010) finds that as much as half of the oxygen is unaccounted for in dense clouds when taking into account refractories — such as silicates, as well as volatiles in the form of water, CO_2 ices, and CO gas. This potentially lowers the ice/rock ratio if the missing oxygen is in the form of a chondritic-like refractory organic phase (as suggested by Whittet) or as extremely volatile O_2. In the densest regions of dark clouds and protostellar envelopes, it has been suggested that the ice abundances increase, in particular those of H_2O, CO_2, and CH_3OH, and some observations find that as much as 90% of the oxygen may be accounted for in one region (*Pontoppidan et al.*, 2004) (see also Fig. 5). In the case where the ice abundances are the highest, this yields an observed ice/rock ratio of at least ~1.5. Strictly speaking, this is a lower limit, since some ice species are as yet unidentified (*Boogert et al.*, 2008), and some species, such as N_2 (*Pontoppidan et al.*, 2003), may not be detectable. Importantly, in the ISM, NH_3 is not very abundant [~5% relative to water (*Boogert et al.*, 2008)]. The remaining uncertain components are the main reservoirs of carbon and nitrogen, with detected reservoirs of the former only accounting for 35% of the solar abundance, and detected reservoirs of the latter accounting for less than 10% of the solar abundance. Likely reservoirs in the ISM are refractory carbon and N_2, neither of which is directly identifiable using infrared spectroscopy. The difference up to the putative solar value of ice/rock = 4 is essentially recovered if the missing oxygen, carbon, and nitrogen are fully sequestered into ices; in *Dodson-Robinson et al.* (2009), for example, 50% of the carbon is in the form of CH_4, and 80% of nitrogen in the form of NH_3. Thus, whether the bulk disk chemistry resembles the ISM or not is of fundamental importance for computing the disk's solid surface density. Recent observations have begun to reveal some of the chemical demographics of protoplanetary disks, as we shall see in the following section.

3. DISTRIBUTION AND EVOLUTION OF VOLATILES IN PROTOPLANETARY DISKS

Protoplanetary disks are structures characterized by a wide range of physical environments. They contain regions with densities in excess of 10^{15} cm^{-3} and ISM-like conditions with densities $<10^4$ cm^{-3}, they have temperatures in the range 10 K < T < 10,000 K, and parts of them are deeply hidden beneath optically thick layers of dust with depths of

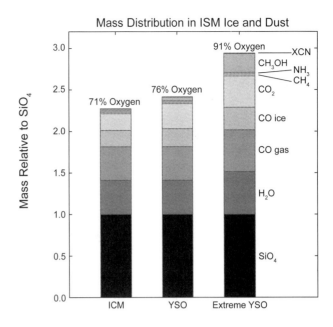

Fig. 5. The mass distribution of molecules in the interstellar medium derived from ice observations (*Whittet et al.*, 2007; *Boogert et al.*, 2008; *Öberg et al.*, 2011b). The distribution assumes that all available silicon is bound in the form of tetrahedral silicates, such as forsterite (taking up four oxygen atoms per silicon atom), although other common silicate species contain slightly less oxygen (three oxygen atoms per silicon atom), including enstatite and forsterite. The plot displays only observed carriers (except for silicates) for three environments: The intracloud medium (ICM), young stellar objects (YSOs), and "extreme" YSO with the highest observed ice abundances are from *Pontoppidan et al.* (2004). Above each column is shown the fraction of the solar elemental oxygen abundance that is included in each budget. The observed oxygen fractions are slightly higher than in *Whittet* (2010) due to our use of the lower *Grevesse et al.* (2010) value for O/H = 4.89 × 10^{-4}, instead of the O/H = 5.75 × 10^{-4} value of *Przybilla et al.* (2008).

more than 1000 mag. From an astrophysical perspective, disk conditions are often characterized in radial or vertical terms, depending on circumstances, although of course any disk has a full three-dimensional structure. Radially, disks are often divided into vaguely defined regions that are roughly separated by an assumed mode of planet formation: terrestrial (0–5 AU), giant planet (5–20 AU), and Kuiper/cometary belt (>20 AU). The first two are characterized by a series of condensation fronts, and are often bundled into an "inner disk" nomenclature. The latter region is universally known as the "outer disk." Vertically, disks are thought of as consisting of three regions: a low-density, molecule-poor region near the disk surface; an intermediate region, with sufficient radiation shielding and temperatures to sustain and excite gas-phase molecules (often called the "warm molecular layer"); and a colder region near the disk mid-plane that is often shielded by optically thick dust and in which volatile molecules may condense beyond a certain radius. A more general review of protoplanetary disk structure as it relates to chemistry can be found in *Henning and Semenov* (2013).

The methods used to infer the presence and relative abundances of disk volatiles draw from a heterogenous set of observational facilities, spanning the UV to radio regimes. Advances in observational capabilities across the electromagnetic spectrum have allowed more comprehensive descriptions of the bulk chemical composition of protoplanetary disks than previously possible. In place of a picture in which disks are composed simplistically of molecular hydrogen and helium with trace amounts of refractory dust, they are now seen as complex environments with distinct regions characterized by fundamentally different chemical compositions. Some of these regions are more accessible to observations than others, and the more readily observed regions are often used to infer the properties of unseen or hidden regions. With a few notable exceptions of disks with very high accretion rates, protoplanetary disks are generally characterized by having a surface that is hotter than the interior. This property leads to observables (line as well as continuum emission) that are dominated by the surface of the molecular layer. This tends to be true even at radii and at observing wavelengths where the disks are optically thin in the vertical direction (*Semenov et al.*, 2008). Thus, most of our direct knowledge of volatiles in disks pertains to surface conditions and to a fraction of the vertical column density.

3.1. History of Observations of Volatiles in Disks

Until a few years ago, direct measurements of the amount and distribution of bulk volatiles in protoplanetary disks were limited. Part of the reason was that the strong transitions of the most common volatile species were not visible from the ground, and spacebased observatories were still too insensitive to obtain spectroscopy of typical protoplanetary disks. Millimeter observations of a small number of large, massive disks revealed the presence of trace species ($\leq 1\%$ of the total volatile mass) in the cold, outer regions of disks (beyond 100 AU), such as CS, HCO^+, and HCN (*Dutrey et al.*, 1997; *Thi et al.*, 2004; *Öberg et al.*, 2010). An exception to these millimeter-detected trace species is CO, which is one of the most important bulk volatiles. A key conclusion of the early millimeter work was that efficient freeze-out leads to strong depletions in the outer disk (e.g., *Qi et al.*, 2004). This was supported by a few detections of ices in disks. *Malfait et al.* (1998) and *Chiang et al.* (2001) discovered emission from crystalline water ice in two disks in the far-infrared. At shorter wavelengths, ices do not produce emission features, since the temperatures required to excite existing resonances below ~40 μm will effectively desorb all bulk volatile species, including water. Scattered near-infrared light has since revealed water ice in at least one disk (*Honda et al.*, 2009), and attempts have been made to measure ices in absorption toward edge-on disks (*Thi et al.*, 2002), although difficulties in locating the absorbing material along the line of sight has led to

ambiguities between ice located in the disk itself and ice located in foreground clouds (*Pontoppidan et al.*, 2005). More recently, sensitive searches for ices in edge-on disks have made progress in detecting ice absorption in edge-on disks with no potential contribution from cold foreground cloud material (*Aikawa et al.*, 2012). In addition, *McClure et al.* (2012) find a tentative detection of water ice emission from a disk observed with the Herschel Space Observatory, and recent upgrades to PACS data reduction routines may enable additional detections in the near future.

Exploration of the molecular content of warm inner disks (≤ 10 AU) has greatly expanded since *PPV*. Inner disks have been readily detected in CO rovibrational lines (*Najita et al.*, 2003; *Blake and Boogert*, 2004), where it is essentially ubiquitous in protoplanetary disks (*Brown et al.*, 2013). *Carr et al.* (2004) detected lines from hot water near 2.5 μm toward one young star and argued that the origin was in a Keplerian disk. However, it was with the launch of the Spitzer Space Telescope in August 2003 that this field underwent a major advance. C_2H_2, HCN, and CO_2 were first detected by Spitzer in absorption toward a young low-mass star (*Lahuis et al.*, 2006), with C_2H_2, HCN absorption also being detected from the ground in a second source (*Gibb et al.*, 2007). But Spitzer also initiated the very first comprehensive surveys of the molecular content in the inner regions of protoplanetary disks. *Carr and Najita* (2008) realized that the mid-infrared spectrum from 10 to 20 μm of the classical T Tauri star AA Tau showed a low-level complex forest of molecular emission, mostly from pure rotational lines of water, along with lines and bands from OH, HCN, C_2H_2, and CO_2. The emission was characteristic of abundant water with temperatures of 500–1000 K. *Salyk et al.* (2008) discovered an additional two disks with strong mid-infrared molecular emission, and showed that these disks also have emission from hot water as traced by rovibrational transitions near 3 μm. A survey of more than 60 protoplanetary disks around low-mass (≤ 1 $M_\odot$) stars revealed the presence of mid-infrared emission bands from organics (HCN in up to 30% of the disks, and up to 10% for C_2H_2) (*Pascucci et al.*, 2009). Advances in the data reduction procedure for high-resolution Spitzer spectra subsequently revealed water emission, along with emission from the organics, OH, and CO_2, in about half the sources in a sample of ~50 disks around low-mass stars (*Pontoppidan et al.*, 2010a). Following the initial flurry of results based on Spitzer data, the later years have focused on modeling and interpreting these data (*Salyk et al.*, 2011; *Carr and Najita*, 2011), as well as obtaining follow-up observations in the near- and mid-infrared of, e.g., water and OH, by groundbased facilities at high spectral resolution (*Pontoppidan et al.*, 2010b; *Mandell et al.*, 2012).

3.2. Retrieval of Volatile Abundances from Protoplanetary Disks

Before proceeding to the implications of the observations of volatiles in protoplanetary disks, it is worthwhile to briefly consider the technical difficulties in accurately measuring the basic physical and chemical parameters of disks from observations of molecular emission lines. The assumptions that are imposed on the data analysis essentially always generate model dependencies that can cloud interpretation. For this reason intense work has been underway for the past decade to minimize the model dependencies in this process and to better understand the physics that forms emission lines from disks, as well as other analogous nebular objects. Retrieval refers to the modeling process of inverting line fluxes and profiles to fundamental physical parameters such as the fractional abundance [X(species) = n(species)/n(H) (R, z)], the total gas mass, or the kinetic gas temperature.

Because molecular emission line strengths are dependent on abundance as well as excitation, retrieval is necessarily a model-dependent, usually iterative, process, even for disks that are spatially resolved. A common procedure uses broadband multi-wavelength observations (1–1000 μm) to measure the emission from dust in the disk. Structural models for the distribution of dust in the disk are fit to the data, typically under some assumption of dust properties. Assuming that the dust traces the bulk molecular gas, the spatial fractional abundance structure of molecular gas-phase species can be retrieved by modeling the line emission implied by the physical structure fixed by the dust (*Zhang et al.*, 2013; *Bergin et al.*, 2013). Another analysis approach does not attempt to retrieve molecular abundances directly, but attempts to constrain the global disk physical structure by fixing the input elemental abundances and using a thermochemical model to calculate line fluxes. Such models simultaneously solve the chemistry and detailed radiative transfer, but rely on having incorporated all important physical and chemical processes (e.g., *Aikawa et al.*, 2002; *Thi et al.*, 2010; *Woitke et al.*, 2011; *Tilling et al.*, 2012). Both approaches carry with them significant uncertainties and degeneracies, so molecular abundances from disks are probably still inaccurate to an order of magnitude, although progress is continuously being made to reduce this uncertainty (*Kamp et al.*, 2013).

3.3. Demographics of Volatiles in Disks

It is now possible to summarize some general observed properties of volatiles in protoplanetary disks. Their composition is dominated by the classical CNO elemental trifecta as shown in Fig. 6. In addition, there are likely trace species including other nucleosynthetic elements (e.g., sulfur and fluorine), although the only firm detections to date are of sulfur-bearing species, and those appear to indicate that the bulk of the sulfur is sequestered in a refractory carrier (*Dutrey et al.*, 2011).

Comparing the observed column densities of CNO-bearing volatiles in the inner disks, *Salyk et al.* (2011) find that, while the data are consistent with most of the oxygen being carried by water and CO, carbon seems to be slightly underabundant by a factor of 2–3, while nitrogen (only observed in the form of HCN) is still largely unaccounted

for by about 2 orders of magnitude. *Carr and Najita* (2011) used a marginally different analysis method, but found slightly higher HCN and C_2H_2 abundances by factors of a few in similar, albeit not identical, disks. While underlining that molecular abundances retrieved from low-resolution spectra are still uncertain, this does not change the basic demographic picture. The (slight) carbon deficit may be explained by some sequestration in a carbonaceous refractory component, similar to the situation in the ISM. This is, however, not consistent with the carbon deficit in the inner solar system, as discussed in section 2.2. The nitrogen

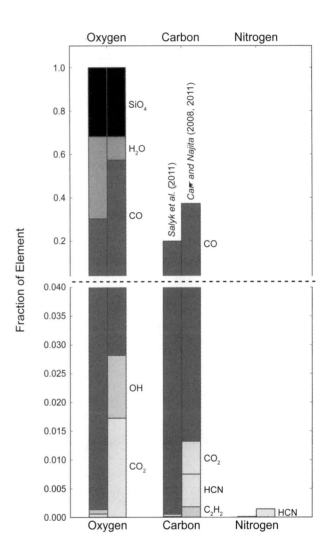

Fig. 6. Observed elemental abundance budget at ~1 AU in protoplanetary disks. The values retrieved by *Salyk et al.* (2011) and *Carr and Najita* (2008, 2011) are compared side-by-side to provide a rough indication of current uncertainties. All budgets are normalized to the assumption that all oxygen is accounted for in the warm molecular layer (fraction of O = 1). With this assumption, about one-third of the carbon is accounted for, and almost none of the nitrogen. The implication is that any missing contribution is located in an as yet unobserved carrier. Note that the plot has been split into two different y-axis scales to simultaneously show major as well as minor carriers.

deficit is potentially interesting, as NH_3 can be ruled out as a significant carrier in the inner disk by stringent upper limits on the presence of emission bands at 10 μm (*Salyk et al., 2011*) and 3 μm (*Mandell et al., 2012*). This has led to speculation that most of the nitrogen in the terrestrial region is stored in N_2, and is largely unavailable for planetesimal formation. Beyond the terrestrial region toward the outer disk, we would still expect to find nitrogen in the form of NH_3 to explain the cometary and giant planet abundances. The inner disk elemental abundances are summarized in Table 1 and illustrated in Fig. 6.

As a comparison to the elemental budgets determined from molecular lines, a number of optical integral-field spectroscopic measurements of the abundances of many elements in the photoevaporative flows from protoplanetary disks in Orion are available (*Tsamis et al.,* 2011, 2013; *Mesa-Delgado et al.,* 2012). The Orion disks are being eroded by the harsh interstellar radiation field from young massive stars, providing a unique opportunity to measure their bulk composition. They show almost no depletion. In the cases of carbon and oxygen, the abundances are within a factor 1.5 relative to solar, while the nitrogen abundance is a little more than a factor of 2 lower than solar. This suggests that most of the unseen molecular carriers are being entrained in the photoevaporative flows.

In the outer disk, the CNO budget is observationally still incomplete due to strong depletions from the gas phase by freeze-out. In general, most molecules are depleted relative to cloud abundances (*Dutrey et al.,* 1997) and CO is underabundant by about an order of magnitude (*van Zadelhoff et al.,* 2001; *Qi et al.,* 2004). Water is depleted by up to 6–7 orders of magnitude (*Bergin et al.,* 2010; *Hogerheijde et al.,* 2011) — possibly even more than can be explained by pure freeze-out.

3.4. Chemical Dependencies on Stellar Type

There are large differences in the observational signatures of volatiles as a function of the mass and evolutionary stage of the central star. It is generally difficult to detect molecular emission from disks around Herbig Ae stars (see Fig. 7), and the implications of this strong observable are currently being debated. The effect is strongest in the inner disk, where very few disks around stars hotter than ~7000 K show any molecular emission at all (*Pontoppidan et al.,* 2010a), with the exception of CO (*Blake and Boogert,* 2004) and OH (*Mandell et al.,* 2008; *Fedele et al.,* 2011). The apparent increases of the OH/H_2O ratio in Herbig stars have led to speculation that the strong dependence of molecular line detections on spectral type is due to the photodestruction of water in the photolayers of Herbig disks, rather than an actual deficit of bulk mid-plane water. However, it is possible that an enhanced abundance of the photoproducts of water, and radicals in general, in combination with efficient vertical mixing, can lead to large changes in bulk chemistry dependent on spectral type. It should be noted, however, that it is still debated how much of a depletion

TABLE 1. Distribution of C, N, and O in inner disks.[*]

	Oxygen (%)	Carbon (%)	Nitrogen (%)
H_2O	38	—	—
CO	30	57	—
HCN	—	0.1–0.6[†]	0.4–2.5
NH_3	—	—	<1.1
C_2H_2	—	0.03–0.25	—
CO_2	(0.04)	(0.05)	—
OH	(0.04)	—	—
Silicates	32	—	—
Total	100[‡]	57–58	0.4–2.5

[*]Values are relevant for disks around stars of roughly 0.5–1.5 $M_\odot$.
[†]Ranges indicate differences in median derived values from *Salyk et al.* (2011) and *Carr and Najita* (2011).
[‡]It is assumed that all the oxygen is accounted for, and the other fractions are scaled using the solar elemental abundances.

the nondetections actually correspond to, as the dynamic range of the infrared Spitzer spectra is relatively low. A detection of water in the disk around HD 163296 suggests that the depletion in the inner disk in some cases may be as little as 1–2 orders of magnitude, although in this case, the water emission originates from somewhat larger radii (15–20 AU) than traced by the multitude of mid-infrared molecular lines [~1 AU (*Fedele et al.,* 2012)].

Toward the lower end of the stellar mass spectrum, *Pascucci et al.* (2009) find that disks around young stars with spectral types M5–M9 have significantly less HCN relative to C_2H_2, possibly requiring a greater fraction of the elemental nitrogen being sequestered as highly volatile N_2. Further emphasizing the dependence of stellar type on inner disk chemistry, *Pascucci et al.* (2013) recently found strongly enhanced carbon chemistry in brown dwarf disks, as measured by HCN and C_2H_2, relative to oxygen chemistry, as measured by H_2O. Observationally, solar-mass young stars therefore seem to be the most water-rich, and research in the near future is likely to explore how this may affect the structure and composition of planetary systems around different stars.

3.5. Origin of Disk Volatiles: Inheritance and Reset

How much of the protostellar volatile material survives incorporation into the disk? That is, are initial disk conditions predominantly protostellar (*Visser et al.,* 2009), and to what degree do the initial chemical conditions survive the physical evolution of protoplanetary disks? Which regions of the disk are inherited, giving rise to planetesimals of circum- or interstellar origin, and which regions are chemically reset, leading to disk-like planetesimals? Chemical models indicate that essentially all disk regions are chemically active on timescales shorter than the lifetime of the disk. The dominant chemical pathways and the timescales for equilibration, however, are expected to vary substantially from region to region, depending on the physical proper-

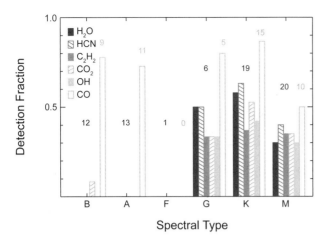

Fig. 7. The relation between Spitzer detection rates of the infrared molecular forest and stellar spectral type (*Pontoppidan et al., 2010a*).

ties of the local environment (*Semenov and Wiebe,* 2011). Partial answers to these questions may be found in the observed volatile demographics and their comparison to protostellar chemistry. Different scenarios should result in significant differences in volatile composition and isotope fractionation. Searching for such differences or similarities is the subject of broad ongoing observational investigations.

Prestellar and protostellar volatile compositions have been systematically investigated through gas and ice observations (*van Dishoeck,* 2004; *Öberg et al.,* 2011b; *Caselli and Ceccarelli,* 2012). There is some familial resemblance between the protostellar and comet compositions in our solar system, which has been taken as evidence of a direct inheritance of the comet-forming volatile reservoir (*Mumma and Charnley,* 2011). The possibility of such a connection is supported by recent two-dimensional hydrodynamical simulations of the chemical evolution on trajectories from protostellar envelopes up to incorporation into the disks (*Aikawa et al.,* 1999; *Brinch et al.,* 2008; *van Weeren et al.,* 2009; *Visser et al.,* 2009, 2011; *Hincelin et al.,* 2013). While different trajectories are exposed to different levels of heat and radiation, most ices that are deposited into the comet-forming region survive incorporation, and only some of the most volatile ices sublimate during disk formation (e.g., N_2 and CO).

Some ice evolution during disk formation is likely, however, since the grains spend a significant time at 20–40 K — temperatures known to be favorable for complex molecule formation (*Garrod and Herbst,* 2006; *Garrod et al.,* 2008; *Öberg et al.,* 2009). Once the disk is formed, the initial volatile abundances may evolve through gas and grain surface reactions. The importance of this *in situ* processing compared to the chemistry preceding the disk stage will depend on the exposure of disk material to conditions that favor chemical reactions, i.e., UV and X-ray radiation, high densities, and heat. Disk models that begin with mostly atomic structures implicitly assume that the disk chemistry

is reset. This assumption is plausible for regions inside the snow line because of the short chemical timescales (10^4 yr or less) for regions dominated by gas-phase chemistry (*Woitke et al.,* 2009). This is consistent with the meteoritic evidence that has always favored a complete chemical reset; most of the collected meteorites originated from feeding zones within the solar snow line (see also section 2 for a discussion of the evidence for a reset from a cosmochemical perspective).

The addition of accretion flows and advection, which transport volatiles from the outer to the inner disk, as well as vertical and radial mixing driven by local turbulence, result in mixing between a chemically active surface and a relatively dormant mid-plane. *Semenov and Wiebe* (2011) explored the different turbulence and chemistry timescales in disks, deriving characteristic timescales to determine which chemical processes can be efficiently erased by mixing (with necessary assumptions about disk parameters, including ionization rates, structure, etc.). They found that vertical mixing is faster than the disk lifetime ($<10^6$ yr) interior of 200 AU in a typical disk. Ion-molecule reaction rates depend on the abundances of ions involved in the reaction, but can be fast; as a representative example, they estimate rates leading to the production of HCO^+ to be $<10^4$ yr throughout the disk. In such a case, the gas-phase chemistry will not retain any memory of the original composition, and, inside the snow line, it is very difficult to retain any primordial signature of volatiles over the disk life time. Photochemistry is fast outside the mid-plane. Outside the snow line, the mid-plane is dominated by fast freeze-out and a slower desorption by cosmic rays that still has a timescale shorter than the disk lifetime, implying that some gas-grain circulation will take place. Grain surface chemistry can in some cases be very slow, however, and thus icy grains that do not get dredged up into the disk atmosphere or into the inner disk may retain an inherited chemistry. Gas-phase emission from disks should mainly reflect disk *in situ* chemistry, especially from regions inside the main condensation fronts, including, e.g., isotopic fractionation patterns, while comet compositions and probes of disk ices from beyond a few tens of astronomical units should exhibit protostellar-like abundances and fractionation patterns.

In situ deuterium fractionation in disks provides the most obvious explanation for the recent observations of different disk distributions of two deuterated molecules, DCN and DCO^+, with formation pathways that present different temperature dependencies (*Öberg et al.,* 2012). As discussed in section 2.4, cometary H_2O deuterium fractionation may or may not be consistent with protostellar values, while nitrogen fractionation as well as NH_3 spin temperatures both point to single cold origin (*Mumma and Charnley,* 2011), which would require that part of the disk (most likely the disk mid-plane) retain a history of the initial (cloud) conditions.

So how do the chemical abundances of solar-type inner disks compare to those of protostellar clouds and envelopes? Figure 8 shows a comparison of the abundances of HCN, CO_2, H_2O, and NH_3 (relative to CO) derived from Spitzer

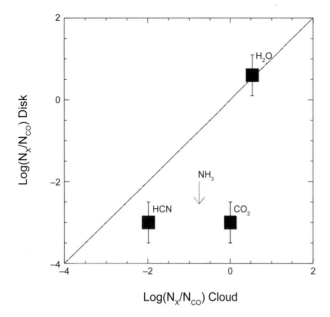

Fig. 8. Comparison between the observed abundances of gas-phase inner disk volatiles derived from Spitzer-IRS spectra (*Salyk et al.*, 2011) relative to those in ices in protostellar clouds (*Öberg et al.*, 2011b; *Lahuis and van Dishoeck*, 2000). Disk abundances are appropriate for the inner disk, as the Spitzer-IRS emission lines originate primarily in the few-astronomical-unit region.

Infrared Spectrograph (IRS) observations of disks (*Salyk et al.*, 2011) with those derived from protostellar clouds (*Öberg et al.*, 2011b). While the abundance of H_2O is similar for both clouds and disks, inner disk abundances of other observed species are lower than in clouds. In summary, evidence for significant changes in inner disk chemistry during disk formation and evolution is emerging, emphasizing that protoplanetary disks have inner regions strongly affected by a reset-type chemistry. To what degree this is consistent with the cosmochemical record will be an active research subject in the near future.

3.6. Volatiles During Transient Accretion Events

Chemical models of disk evolution often assume a relatively quiescent disk, experiencing processes no more dynamic than viscous accretion. However, evidence is mounting that accretion of material from the natal cloud to the disk must occur in short, intense bursts that may drive the chemical reset over a relatively wide range of disk radii (e.g., *Dunham et al.*, 2010; see also the chapter on young eruptive variables elsewhere in this book). The outbursts of FU Ori and EXor stars reveal violent and energetic accretion episodes (e.g., *Herbig*, 1966, 1977; *Hartmann and Kenyon*, 1996) that must have profound effects on the volatile contents of their disks. Not surprisingly, the molecular emission lines from eruptive variables show remarkable variability throughout outbursting events. For example,

observations of rovibrational CO in the EXor V1647 Ori revealed fundamental and overtone CO emission at the beginning of its outburst (*Vacca et al.*, 2004; *Rettig et al.*, 2005), cooler emission and a blue-shifted absorption after return to preoutburst luminosity, and finally a ceasing of all emission after one year (*Brittain et al.*, 2007). Variable rovibrational lines of CO have also been observed towards EX Lupi, show evidence for at least two gas components at different temperatures, with one tightly correlated with the system luminosity, and displaying asymmetries due to a possible disk hotspot, and evidence for a warm molecular outflow (*Goto et al.*, 2011). Recently, a number of additional molecular species have been observed in EX Lupi both before and during an outburst taking place during 2008, viz. H_2O, OH, C_2H_2, HCN, and CO_2 (*Banzatti et al.*, 2012). When the star underwent the outburst, the organic lines went away while the water and OH lines brightened, suggestive of the simultaneous evaporation of large amounts of water locked in ice pre-burst, and active photochemistry in the disk surface.

3.7. Was the Solar Nebula Chemically Typical?

Although destined to grow dramatically in the future, comparisons of protoplanetary disk chemistry with aspects of the cosmochemical evidence from the solar nebula are emerging. It indeed appears that critical structures in the solar nebula with respect to volatiles are reproduced in protoplanetary disks. The astrophysical perspective shows a clear distribution (in the gas) of a volatile-rich inner nebula and volatile-poor outer nebula, with different snow lines between water and more volatile CO.

For instance, do the D/H fractionation patterns in protoplanetary disks match those of the solar system? While still elusive in disks, HDO and $H_2^{18}O$ have been detected in the warm envelopes in the immediate surroundings of young solar-mass stars. The measurements have a wider degree of diversity than observed in solar system materials, with ratios of warm HDO to H_2O ranging from $<6 \times 10^{-4}$ (*Jørgensen and van Dishoeck*, 2010) to as high as $6–8 \times 10^{-2}$ [*Coutens et al.* (2012) and *Taquet et al.* (2013); corresponding to D/H ratios a factor of 2 lower]. Part of the explanation for this heterogeneity could be that the derivation of the D/H ratio depends sensitively on whether the observations resolve the innermost (presumably disk-like) regions (*Persson et al.*, 2013), as well as on the details of the retrieval method, with recent models deriving lower values of D/H (*Visser et al.*, 2013; *Persson et al.*, 2013). Nevertheless, the lower values of D/H for many sources are similar to those of cometary ices. We direct the reader to the chapter by van Dishoeck et al. in this volume for more details about the analysis and interpretation of water vapor observations in young, embedded disks.

Protoplanetary disks also show tentative evidence in support of proposed solutions to the carbon deficit problem described in section 2.2. [C I] has not yet been detected in disks, with strong upper limits suggesting a general depletion of gas-phase carbon relative to the observed dust mass, even after accounting for CO freeze-out (*Panić et al.*, 2010; *Chapillon et al.*, 2010; *Casassus et al.*, 2013). While a possible explanation for this depletion is that the overall gas/dust ratio of the disk is depleted, *Bruderer et al.* (2012) show that, for at least one disk — HD 100546 — a low gas/dust ratio is inconsistent with observed CO line fluxes and instead suggests that the disk gas is depleted in carbon. A natural way to preferentially deplete carbon in the disk gas is to convert it into a less-volatile form. This idea is also supported by the first attempt to directly measure the CO/H_2 ratio in a disk; using observations of HD and CO in the TW Hya disk, *Favre et al.* (2013) derive a CO/H_2 ratio a few to 100× lower than the canonical value, suggesting the conversion of much of the carbon in CO into more refractory compounds, such as organics.

Protoplanetary disk observations have also been used to search for evidence for isotope-selective photodissocation of CO, predicted to be the cause of the solar system's mass-independent oxygen anomalies (see section 2.4). Using high-resolution spectrosopic observations of protoplanetary disks, *Smith et al.* (2009) directly measured the abundances of $C^{16}O$, $C^{17}O$, and $C^{18}O$. They found evidence for mass-independent fractionation, consistent with a scenario in which the fractionation is caused by isotope selective photodissocation in the disk surface due to CO self-shielding (*Lyons and Young*, 2005; *Visser et al.*, 2009).

4. DISK VOLATILES AND EXOPLANETS

In section 2, we discussed how the presence of a water snow line shaped the composition of the different planets and other bodies in the solar system, depending on their formation location and, perhaps, dynamical history. We also demonstrated that molecular species other than water must have had similar strong radial gas-to-solid gradients, as preserved in elemental and isotopic abundances. Now that we have an increasingly detailed, if not yet fully comprehensive, picture of protoplanetary disk chemistry and thermodynamics, supported by astronomical data, it is natural to ask how volatiles impact the architecture and composition of exoplanets. Planetary properties that may be influenced by the natal volatile composition include system characteristics, such as the distribution of masses and orbital radii, their bulk chemistry (the fractions of rock, ice, and gas), and their atmospheric and surface chemistry.

While the combination of inner, dry terrestrial planets and outer icy giants and planetesimals in the solar system strongly suggests that the water snow line is irreversibly imprinted on the large-scale architecture of a planetary system, the discovery of hot-Jupiter planets on short-period orbits showed that there is, in fact, no such universal relationship (*Mayor and Queloz*, 1995; *Marcy and Butler*, 1996). The solution does not appear to be to allow giant planets to form close to the parent star (well within 1 AU) — the consensus is that this is not possible because sufficient amounts of mass cannot be maintained within such small radii. Rather,

the planets migrate radially through some combination of gas interactions during the protoplanetary disk phase (*Lin et al.*, 1996), dynamical interactions with a planetesimal cloud (*Hahn and Malhotra*, 1999), and three-body interactions with a stellar companion (*Fabrycky and Tremaine*, 2007). While a discussion of the dynamical evolution of planetary systems is beyond the scope of this chapter, the important point, when relating exoplanets to the volatile system in protoplanetary disks, is that observed planetary orbits cannot be assumed to reflect their birth radii and feeding zones. The chemical composition of planets may have been set by disk conditions that were never present at their ultimate mean orbital radius. In the solar system, it was difficult to deliver Earth's water, while exoplanetary systems may have common water worlds that formed beyond the snow line, but heated up either due to orbital or snow line migration (*Tarter et al.*, 2007).

Planetary migration may be seen as a complication when interpreting observations, but perhaps it is also an opportunity. We review two different aspects relating disks to exoplanets. One is the question of whether exoplanet architecture (bulk mass and orbital structure) reflects the properties of the volatile distribution in their parent disks, even though some planetary migration takes place. Another is whether the chemical structure of exoplanet atmospheres constrain their birth radii, and therefore provides an independent measure of the migration process by allowing a comparison between the initial and final orbital radii.

4.1. Importance of Solid Volatiles in Planet Formation

Do condensible volatiles play a central role in the basic ability of protoplanetary disks to form planets of all types, or are they merely passengers as the rocky material and H_2/He gas drive planet formation? The answer to this question generally seems to be that condensible volatiles do play a central role. As discussed in section 2.6, the total mass of condensible volatiles available for planetesimal formation is at least twice that of rock, and maybe as much as 3–4× as much, at radii beyond the main condensation fronts. The large increase in solid mass offered by condensing volatiles affects the growth of planets on all scales — in grain-grain interactions, in the formation of planetesimals, and in the formation of gas giants.

At the microscopic level, icy grains may stick more efficiently at higher collision velocities than bare rocky grains. Simulations suggest that the breakup velocity for icy grains may be 10× higher than that of silicate grains (*Wada et al.*, 2009). Other mechanisms that aid in the growth of icy grains include the potential to melt and rapidly refreeze upon high-velocity collisions, leading to sticking even in energetic collisions (*Wettlaufer*, 2010).

An increase in solid mass may also catalyze the formation of planetesimals — a critical step without which no planet formation could take place in any but the most massive disks. At sufficiently high-solid-surface densities, the so-called streaming instability can lead to the rapid formation of planetesimals by gravitational collapse. This is due to a combination of pressure gradient flows and drag forces exerted by solid particles back onto the gas (e.g., *Youdin and Shu*, 2002; *Johansen et al.*, 2007).

In the context of giant planet formation by core accretion, *Pollack et al.* (1996) demonstrated that a factor of 2 in solid surface density can lead to a reduction in giant planet formation times by a factor of 30. The timescale is determined by the time it takes for a planet to reach the crossover mass — the mass at which runaway gas accretion occurs. Since the lifetime of the gas in protoplanetary disks is comparable to the timescale for effective core accretion, factors of a few in solid mass can make a large difference in whether a system forms giant planets or not. Planet-formation models that include more complexity, such as orbital migration (*Ida and Lin*, 2004a) and evolving disks across a wide parameter space (*Mordasini et al.*, 2009), confirm this basic conclusion and emphasize the need for solid-state volatiles around solar-metallicity stars to form gas giants.

The dependence of planet formation on the availability of solid mass is famously reflected in the positive relation between stellar metallicity and the frequency of gas giants with masses $\geq M_{Jup}$ (*Santos et al.*, 2004; *Fischer and Valenti*, 2005), coupled with a lack of such a relation with Neptune-mass planets (*Sousa et al.*, 2008). The interpretation of these observations is that low-metallicity disks cannot form giant planets within the disk lifetime because the solid cores never grow sufficiently massive to accrete large gas envelopes, while the Neptune-mass planets can form over a wider range of stellar metallicity (*Ida and Lin*, 2004b).

4.2. Condensation Fronts: A Generalized Concept of Snow Lines

When searching for links between protoplanetary disk chemistry and exoplanets, it becomes necessary to generalize the concept of the snow line. Any molecular, or even atomic, species with condensation temperatures between 10 K and 2000 K will have a curve demarcating the border beyond which the partial vapor pressure drops dramatically (the species freezes out). Furthermore, the border is rarely a straight line, but a surface that roughly follows an isotherm, with some departures due to the pressure dependency on the condensation temperature. In a disk, a condensation front is typically strongly curved and becomes nearly horizontal for disk radii dominated by external irradiation heating (*Meijerink et al.*, 2009), or as a pressure-gradient effect in regions that are mostly isothermal (*Ros and Johansen*, 2013). Use of the term "condensation front" to describe this surface for a given molecular or atomic species allows the snow line nomenclature to be retained as a specific reference to water freeze-out at the disk mid-plane, a distinction we will utilize in this text.

Condensation fronts evolve strongly with time as the disk temperature changes in response to evolving stellar insolation and internal disk heating (e.g., *Lissauer*, 1987),

and with stellar luminosity (*Kennedy and Kenyon*, 2008). At the earliest stages, high accretion rates can push the snow line out to 5 AU or more around a young solar-mass star. At intermediate stages (1–2 m.y.) during which giant planets may accumulate most of their mass, disk accretion rates drop toward 10^{-9} $M_\odot$ yr^{-1}, decreasing viscous heating to the point where the snow line moves within 1 AU. At the latest stages of evolution, the disk clears, and the snow line radius is again pushed outward as the disk becomes optically thin in the radial direction exposing the disk mid-plane to direct stellar irradiation (*Garaud and Lin*, 2007; *Zhang et al.*, 2013). We show the predicted evolution of mid-plane snow line distances for a range of stellar masses in Fig. 9 [based on the models of *Kennedy and Kenyon* (2008)]. Importantly, it is seen that, while the end-stage snow line is located well outside the habitable zone of a solar-mass star [where liquid water may exist on a planetary surface (*Kasting et al.*, 1993)], it was actually inside the habitable zone during the critical period in the planet-formation process that sees the formation of giant planets and the planetesimal system. Consequently, there is a possibility that habitable water worlds are more common among exoplanetary systems than the solar system would initially suggest.

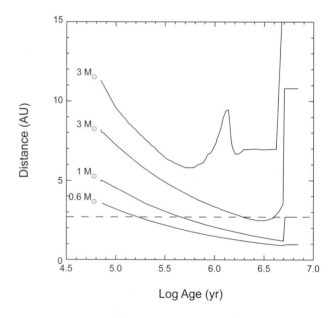

Fig. 9. The evolution of the mid-plane snow line as a function of stellar mass. The data are taken from *Kennedy and Kenyon* (2008) and modified with an optically thin end stage due to disk dissipation. Aside from the final optically thin stage, the shapes of the curves are determined by the stellar luminosity and accretion history of the disk. Note that the 3-$M_\odot$ star undergoes large changes in luminosity as it approaches the main sequence, while the lower-mass stars never reach the main sequence during the disk lifetime. The dashed line represents the predicted snowline for an optically thin disk around a Sun-like star, originally derived by *Hayashi* (1981).

We can also consider the dependence of stellar mass on this basic form of the snow line evolutionary curve. While stars with $M_* \geq 1\ M_\odot$ have disks that during the optically thick phase experience a snow line radius significantly closer to the star than the optically thin end stage configuration, disks around lower-mass stars follow a different path. Around these stars, the optically thin end-stage snow line is the closest it gets. This is due to the relative dominance of accretion heating vs. direct stellar irradiation around the low-luminosity low-mass stars, suggesting that planetary migration is more important for forming water worlds in low-mass systems.

4.3. Transport of Volatiles

Protoplanetary disks are evolving systems that transport a variety of quantities, angular momentum, mass, and energy [see *Armitage* (2011) for a general review]. The transport and mixing of solids in protoplanetary disks is a large, general area of study, but one that has important consequences for disk volatiles. Like refractory material, condensed-phase volatiles can move, in bulk, relative to each other and to the primary H/He gas reservoir. However, as opposed to the refractories, a property of volatiles that make them a special case is their ability to coexist in solid and gaseous forms and to go through rapid phase changes across relatively sharp condensation fronts. Two different transport mechanisms may act in conjunction to concentrate volatiles, and they are therefore often discussed together: the movement of gaseous volatiles by diffusion across the steep partial pressure gradients of a condensation front and the migration of boulder-sized icy bodies against a shallow bulk (H/He) gas pressure gradient.

Radial pressure gradients in protoplanetary disks are well-known catalysts of solid transport through advection. In the absence of other effects, they are sufficiently efficient to move large fractions of the entire solid mass in the disk. Such pressure gradients support the gas by a radial force component, allowing it to orbit at slightly sub- or supra-Keplerian velocities, depending on the direction of the gradient. Gas that is not in a Keplerian orbit in turn produces a dissipative force on solid particles, causing them to lose or gain angular momentum and migrate radially (*Weidenschilling*, 1977; *Stepinski and Valageas*, 1997). The rate of advection is a sensitive function of particle size, with the highest rates seen for sizes of 1–100 cm. For viscously evolving disks, the pressure gradient is monotonically decreasing with radius, leading to rapid inward advection. This mechanism has the power to dramatically change the surface density of, in particular, volatiles, as icy bodies are transported from the outer disk toward, and across, the snow line (*Ciesla and Cuzzi*, 2006).

As the ices cross over their respective condensation fronts and subsequently sublimate, the partial pressure of each volatile species sharply increases. Such steep partial pressure gradients are not stable, and the volatile gas will diffuse out across the condensation front again where it refreezes (mostly due to turbulent mixing, although other types of mixing may also operate). In the absence of inward advection, this can lead to the complete depletion of inner disk volatiles. The process of diffusion against a partial pressure gradient is known as the cold finger effect (*Stevenson and Lunine*, 1988). Together, these effects — inward solid advection and outward gas diffusion — may, for certain disk parameters, increase the solid volatile surface density just beyond the condensation fronts by an order of magnitude (*Cuzzi and Zahnle*, 2004; *Ciesla and Cuzzi*, 2006).

An important consequence of the transport of volatile species toward condensation fronts located at different disk radii may be to impose a strong radial dependence on the elemental C/O ratio, as suggested by models of meteorite formation (*Hutson and Ruzicka*, 2000; *Pasek et al.*, 2005). In direct analog to other astrochemical environments where the elemental C/O ratio dictates a water- or carbon-dominated chemistry, disks are expected to be similarly sensitive. It may therefore be possible to measure the disk C/O ratios at various radii by measuring the relative abundances of water- and carbon-bearing volatile species. A range of HCN/H_2O ratios measures using Spitzer spectra (*Salyk et al.*, 2011; *Carr and Najita*, 2011), and an observed correlation of HCN/H_2O line strengths with disk mass is at least consistent with the possibility that planetesimal growth could sequester O at or beyond the snow line and increase the inner disk C/O ratio (*Najita et al.*, 2013).

Vertical transport of volatiles, while involving less bulk mass than radial transport, may act to significantly alter the observables of disk volatiles. In the range of radii in the disk where the condensation fronts are nearly horizontal (and very close in height) due to direct stellar irradiation of the disk, the cold finger effect may act in the vertical direction. That is, volatile vapor turbulently mixes across the condensation front, below which it freezes out. Some of the resulting solids will then settle toward the mid-plane, leading to a net loss of volatiles from the disk surface. This was used as a proposed, but qualitative, explanation for the observed water vapor distribution in the AS 205N disk, in which Spitzer spectroscopy suggested a snow line near 1 AU, well within the ~150-K region normally demarcating the snow line (*Meijerink et al.*, 2009). The effect was subsequently supported by the numerical models of *Ros and Johansen*, (2013).

Beyond the condensation front, grain settling may also deplete the surface of all solid volatiles. This was suggested to explain the extremely low abundances of water vapor observed with the Heterodyne Instrument for the Far Infrared (HIFI) onboard the Herschel Space Observatory in the outer TW Hya disk ($\geq$50 AU) (*Hogerheijde et al.*, 2011). While it may seem counterintuitive to constrain the abundance of water ice using water vapor, this works in the disk surface because UV and X-ray radiation will desorb water ice into the gas phase at a constant rate at a given incident intensity (*Ceccarelli et al.*, 2005; *Dominik et al.*, 2005). In the case of TW Hya, the water vapor abundance was found to be 1–2 orders of magnitude lower than expected from photodesorption, leading to a suggestion that the water ice was depleted by a similar factor through a process of preferential settling. Similarly low amounts of water vapor are consistent with upper limits measured by *Bergin et al.* (2010) around DM Tau.

4.4. Observations of Condensation Fronts

Based on the development of high-fidelity mid-infrared spectroscopy targeting warm volatiles, including water, as well as some of the first Atacama Large Millimeter/sub-millimeter Array (ALMA) line imaging results, it has now become feasible to measure, and even image, the location of condensation fronts. The observations address specific technical challenges in measuring a condensation front.

One challenge is related to the small angular size of the gas-rich region (tens of milli-arcseconds for water and ~0.5 arcsec for CO in nearby star-forming regions); to measure the snow line with a direct line image, the spatial resolution of the observation must be significantly better than these sizes. To characterize the shape and sharpness of the condensation front, the resolution must be even higher. As an alternative to direct imaging, the many transitions of water covering virtually all excitation temperatures provide an opportunity to spectrally map its radial distribution in a disk, using temperature as a proxy for radius. The first qualitative attempt was made by *Meijerink et al.* (2009) for the disk around AS 205N, and the method was refined and quantified for the transitional disk TW Hya by *Zhang et al.* (2013).

In TW Hya, the inner few astronomical units are depleted of small dust grains (*Calvet et al.*, 2002), and the snow line is located further from the star than would be expected for a classical optically thick disk. Instead, the decreased radial optical depth of grains allows radiation to penetrate to larger radii in the disk, and moves the snow line to just beyond the transition radius where the disk becomes optically thick. This observation is evidence that the snow line location moves outward near the end of the gas-rich, optically thick phase of protoplanetary disk evolution.

One challenge to the temperature-mapping method is that the disk mid-plane is shielded, either by highly optically thick dust or by interfering line emission from a superheated surface layer. For condensation fronts at small radii and relatively high temperatures, such as for water, the relevant mid- to far-infrared lines are therefore formed in the disk surface above one scale height (*Woitke et al.*, 2009), as opposed to in the mid-plane. In practice, this may skew measured condensation fronts to larger radii.

While water has the most famous condensation front, others, characterized by lower temperatures, are large enough to image directly with facilities like ALMA. Evidence for CO freeze-out has been previously suggested due to a measured overall depletion of CO in disks relative to dark cloud values (e.g., *Qi et al.*, 2004). However, attempts to image the CO condensation front directly in disks using rotational CO lines have proven to be more challenging than initially expected (*Qi et al.*, 2011). The CO condensation

front tends to get washed out in line images due to the high opacity of low-J CO lines and filling in by warm surface gas. Effectively, the disk atmosphere hides the depletion in the disk mid-plane. The solution to this problem has been to use chemical tracers, i.e., molecules that only exist in abundance when the CO is absent from the gas phase and/or abundant in the solid phase, thus avoiding interference by emission from CO-rich regions. Four molecules have been suggested: N_2H^+, which is efficiently destroyed by gas-phase CO (*Qi et al.*, 2013a); H_2CO and CH_3OH, which form from CO ice hydrogenation; and DCO^+, which is enhanced as CO depletion allows the formation of one of its parent molecules, H_2D^+ (*Mathews et al.*, 2013). Examples of CO condensation fronts measured using DCO^+ (HD 163296) and N_2H^+ (TW Hya) are shown in Fig. 10.

N_2H^+, H_2CO, and CH_3OH should all be present exclusively outside the CO snow line, with an inner edge that traces its position. The case of DCO^+ is complicated because CO is also a parent molecule, and its abundance is a result of a competition between CO and H_2D^+, leading to efficient formation in a narrow temperature range just around the CO condensation front. Recent images of DCO^+ in the disk around HD 163296 show a ring-like structure, with a radius consistent with that expected for CO condensation, suggesting that at least for this disk, DCO^+ is a good CO snow line tracer (*Mathews et al.*, 2013) (see Fig. 10). However, TW Hya does not show the same ring-like structure in DCO^+ (*Qi et al.*, 2008), while a more recent ALMA image of N_2H^+ in this disk confirms a clear ring-like structure, with a location consistent with the expected radius of the CO snow line (*Qi et al.*, 2013b).

4.5. Influence of Disk Volatiles on Planetary Composition

It is a basic expectation that the bulk elemental composition of a planet reflects that of the protoplanetary disk material that formed it. Furthermore, as planets never fully mix and equilibrate, interiors and atmospheres may also reflect differences in the composition of material accreted during different stages of formation. This dependence is supported by the wide diversity of planets (and likely moons) in exoplanetary systems. Clearly, not all planets are the same; to what degree is this due to disk volatiles and the processes that sculpt their distribution in a protoplanetary disk? We are just beginning to be able to study the compositions of exoplanets as well, and diversity appears to be the rule, rather than the exception. Thanks to the success of the Kepler mission (*Borucki et al.*, 2010), other exoplanet transit studies (e.g., *Barge et al.*, 2008), and radial velocity surveys (*Mayor et al.*, 2003; *Cumming et al.*, 2008) combined with transit observations, we can now determine the average density of some exoplanets and study their bulk chemistry (e.g., *Sasselov*, 2003; *Burrows et al.*, 2007).

Furthermore, the first observations of the atmospheric compositions of exoplanets are being made possible with studies of primary and secondary planetary transits (*Knutson*

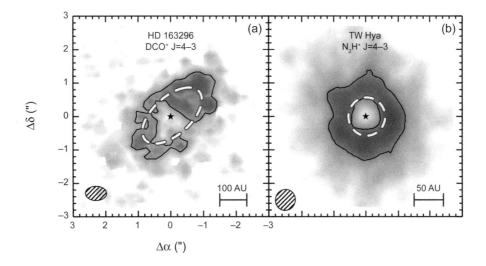

Fig. 10. The CO snowline as observed with ALMA. The CO snowline is observed using chemical tracers of the absence of CO from the gas-phase [left, DCO+ (*Mathews et al.,* 2013) and right, N_2H^+ (*Qi et al.,* 2013b)]. The dashed curves delineate the 17-K isotherms, where CO freezes out.

et al., 2007) as well as direct imaging (*Konopacky et al.,* 2013). However, the relatively small number of planets that have been studied have atmospheric signatures that are not trivially reproduced by existing atmospheric models (e.g., *Barman et al.,* 2011; *Skemer et al.,* 2012).

One important parameter, among others, in modeling exoplanet atmospheres is the elemental C/O ratio. The possible detection of a high C/O ratio in the exoplanet WASP-12b prompted a study suggesting a simple way in which the disk architecture might affect the composition of planetary atmospheres (*Madhusudhan et al.,* 2011). *Öberg et al.* (2011a) suggested that if a giant planet atmosphere does not mix, or mixes incompletely with volatiles sequestered in its core, it is possible to create an atmosphere with a high C/O ratio near unity. This is possible in regions of the disk where oxygen preferentially condenses relative to carbon (for instance, outside the water condensation front, but inside the CH_4, CO_2, and CO fronts). If planets form via core accretion from solids followed by the accretion of a gaseous envelope, their atmospheres will inherit the high C/O ratio, driving an atmospheric carbon-dominated chemistry. This predicts different planetary atmospheric compositions as a function of initial formation radii, but independently of the final planetary orbits.

5. PHYSICAL STRUCTURE AND KINEMATICS OF PROTOPLANETARY DISKS AS TRACED BY VOLATILES

Our knowledge of the physical structure and kinematics of protoplanetary disks is fundamentally dependent on how well we understand the behavior of their volatiles. By their nature, volatiles can be found in the gas phase throughout the disk. Indeed, the total mass of protoplanetary disks is entirely dominated by a volatile molecule — H_2. As we have seen in the previous sections, many of the most abundant heavy elements are also often found in the form of gas-phase volatiles. This is fortunate, because in their absence, our current knowledge of the physical structure and kinematics of protoplanetary disks would be far less advanced; we rely on molecular line tracers to measure velocities, temperatures, and mass of the gas. The problem with using volatiles as physical tracers arises because the disks are far from well-mixed, chemically simple systems. Thus, kinematic, thermal, and abundance measurements derived from molecular lines depend on a model of the volatile system: Are most of the tracer molecules frozen out? Are we tracing a specific layer where chemical formation happens to be more effective? Has the total gas + solid volatile reservoir been depleted due to transport?

In this section, we discuss the links between the physichemical behavior of disk volatiles and their use as physical tracers, and in particular point out areas where our improved understanding of volatiles improve the physical measurements. A more general discussion of measurements of physical properties of protoplanetary disks can be found elsewhere in this book.

5.1. Impact of Volatiles on Disk Mass Measurements

The bulk gas masses and the gas-to-dust ratios of protoplanetary disks are some of the most fundamental observational parameters. They are often used to place the disks into a global evolutionary scheme, or to compare their properties to those inferred for the earliest stages of solar system formation.

The bulk mass of disks has most commonly been measured by observing the dust mass and converting to a gas mass. More recently, attempts have been made to measure the gas mass directly, via various gas-phase tracers. However, no matter the method of bulk mass measurement, an understanding of the volatile content of the disk is required, as the volatiles represent most of the disk mass. Therefore, many of the intrinsic properties of volatiles discussed elsewhere in this chapter also create difficulties for attempts to measure bulk disk masses. Nevertheless, there has been significant progress on this front, which we describe here.

The use of the dust continuum emission is the oldest method for estimating the total disk mass, and is still the most frequently used (*Beckwith et al.*, 1990; *Andrews and Williams*, 2005; *Mann and Williams*, 2010). In converting from dust mass to total disk mass, the gas-to-dust ratio and dust opacities are typically assumed. However, both of these assumed values are traditionally based on dust models that often do not include any of the CNOH volatiles. Are these assumptions actually reasonable, given the knowledge that the disk dust mass is dominated by ices? For disk material in the limiting (but well-mixed) case that everything, except for hydrogen and helium, is in the solid state, the solar abundances of *Grevesse et al.* (2010) imply a gas-to-dust mass ratio of 72. The consequence is that disk masses are probably slightly lower than if the gas-to-dust ratio is assumed to have the canonical value of 100, but only by a relatively small fraction. In terms of the dust opacity, the addition of ices has a complex behavior due to several competing effects. Ice itself generally has lower mass opacity than refractory dust, so the opacity of bulk disk material will decrease significantly with the addition of ice. Secondary effects, such as a potentially increased sticking efficiency, may change the opacity of icy grains by catalyzing grain growth. All together, the net effect seems to be to slightly decrease the mass opacity of ice by 20–30% (*Ossenkopf and Henning*, 1994; *Semenov et al.*, 2003). Thus, volatiles actually have a relatively minor effect on determination of bulk dust masses in disks. The one potential caveat is if ices convert from simple species to complex CNOH organics, as suggested by solar system meteoritics. The organics have much higher opacities than, say, water, CO_2, or NH_3 ice, and may dominate the total dust opacity at high abundances (*Pollack et al.*, 1994).

One approach to estimate disk masses with gas-phase tracers has been to observe multiple rotational transitions of multiple isotopologues of CO with millimeter interferometers. The more optically thin isotopologues are sensitive to the disk mass, while the optically thick transitions (especially from higher excitation states) are sensitive to the disk temperature structure. Using this method, *Panić et al.* (2008) simultaneously measure the gas and dust masses in the disk around HD 169142 and find a gas/dust ratio in the range of 25–100.

A new method for estimating disk mass was explored by *Bergin et al.* (2013), who used observations of HD J=1–0 obtained with Herschel-PACS to measure the disk mass around TW Hya. HD represents an attractive mass tracer for two main reasons: It directly probes the bulk molecu-

lar component (H_2) with a conversion factor from HD to H_2 expected to be near the interstellar value for nearly all regions of the disk, and HD, unlike CO, is not expected to freeze out in disk conditions. However, as for all gas tracers, the result is still model-dependent through the assumed gas temperature. *Bergin et al.* (2013) used observations of CO-millimeter transitions to independently constrain the gas temperature distribution in the disk. They found a disk mass consistent with a gas/dust ratio near 100, suggesting that even this evolved disk has a canonical gas/dust ratio and sufficient mass to form giant planets. Measurements of the gas mass in TW Hya were also attempted by *Thi et al.* (2010) and *Gorti et al.* (2011), both of whom used a large suite of emission lines, including millimeter-wave CO lines, to constrain the disk properties. However, the model dependencies in this approach were exposed by the discrepant results produced by the two groups, which allowed for TW Hya to have either a very subcanonical or nearly canonical gas/dust ratio. Use of the method presented by *Thi et al.* (2010) also gives gas-to-dust ratios significantly below 100 for other disks (e.g., *Tilling et al.*, 2012), demonstrating a strong need to reach a consensus using different methods for disk mass measurements.

5.2. Kinematics of Disk Volatiles

The bulk mass of protoplanetary disks resides close to the mid-plane, and can be described by a Keplerian flow; known departures from this, for instance, due to radial pressure support (*Weidenschilling*, 1977), are very slight ($\lesssim 0.01 \, V_{Kepler}$), and still beyond our observational capabilities. The surface layers of the disk, however, may be subject to much greater dynamical effects in the form of enhanced ionization-dependent turbulence and winds. Some chemical models now begin to consider the efficiency of processing volatiles as they are mixed into a turbulent surface layer and exposed to high temperatures and direct irradiation before being returned to the mid-plane (*Semenov and Wiebe*, 2011; *Furuya et al.*, 2013). As all chemical timescales in the mid-plane are much longer than the photochemical timescale on the disk surface, such mixing can lead to significant changes to the bulk chemistry, if efficient.

Volatile emission line shapes also hint at the tidal influence of planets. Models of forming gas giant planets embedded in circumstellar disks indicate that their interactions cause the inner edge of the cleared region exterior to their orbit to grow eccentric (*Papaloizou et al.*, 2001; *Kley and Dirksen*, 2006; *Lubow*, 1991). When the companion mass grows to more than 0.3% of the mass of the central star, simulations show that azimuthally averaged eccentricities can grow as large as e = 0.25 at the inner edge of the disk, yet fall off as fast as r^{-2} (*Kley and Dirksen*, 2006). Gas arising from material in a Keplerian orbit with a nonzero eccentricity reveals a distinct shape (*Liskowsky et al.*, 2012) due to the uneven illumination of the inner edge of the disk and a slight Doppler shift relative to a circular orbit due to the noncircular velocities. *Regály et al.* (2010) describe the rovibrational

CO-line profiles that could arise from a disk harboring a suprajovian mass companion. However, the CO emission originates from a large radial extent, diluting the signal from the asymmetric inner rim (*Liskowsky et al.,* 2012). A better tracer of the inner rim is rovibrational OH, which may be excited preferentially in a ring at the inner edge of the outer disk. Such asymmetric rovibrational OH emission has been observed toward V380 Ori (*Fedele et al.,* 2011) and HD 100546 (*Liskowsky et al.,* 2012). It should be noted that other scenarios could produce asymmetries; for example, fluctuations of turbulence (e.g., *Fromang and Nelson,* 2006) could give rise to non-axisymmetric density structures that result in asymmetric emission from OH, which originates in a region near the critical density for thermalizing its vibrational levels. Further temporal studies of this effect are in progress to distinguish between the possible scenarios.

6. FUTURE DIRECTIONS

Our understanding of protoplanetary disk evolution and planet formation is in a transitional stage. Where low-resolution observations once led to models of disks as monolithic objects, characterized by a few scalar parameters (mass, size, dust grain size, etc.), disks are now seen as well-resolved composite entities with different, highly physically and chemically distinct regions. We are no longer averaging properties over entire disks, but are taking the first steps toward deconstructing them by zooming in on specific subregions to derive their causal interrelations. In this chapter, we have shown how volatiles play a key role in sculpting a changing environment as we move from the inner terrestrial region of protoplanetary disks, across the giant planet region, and toward the Kuiper belt and cometary regions. We expect that during the next 5–10 years, we will see dramatic progress in our understanding of disk chemistry, the transport of bulk volatiles, and their links to exoplanetary systems:

1. *Elemental budgets in disks:* One of the most direct, and observationally feasible, comparisons of protoplanetary disks to the conditions in the solar nebula is the distribution of elements in volatile and refractory molecules. We expect that significant progress in detecting all the bulk elemental volatile carriers throughout protoplanetary disks will be made in the next years. However, the success of this task will depend not only on new data, but also on improving the accuracy with which radiative transfer calculations can retrieve the abundances. An important question is whether our inability to observe N_2 will result in the nitrogen budget remaining elusive for the foreseeable future.

2. *Inheritance or reset:* A central question is that of inheritance. To what degree is the volatile composition transferred into disks from the dense ISM through the protostellar phase, and to what degree are the volatiles formed *in situ* within chemically active regions in the disk, resetting previous chemical signatures and losing memory of the interstellar phase? Likely, the answer will be that both types of evolution play a role, with the latter being more impor-

tant for the inner disk. Improvements in the accuracy with which chemical abundances can be retrieved as a function of location in the disk, with spatial fidelities as high as a few astronomical units, are required. This will be provided by a synergistic combination of submillimeter interferometric line imaging with ALMA with mid- to far-infrared spectroscopy with James Webb Space Telescope (JWST) and a range of groundbased optical/IR telescopes. Importantly, no single observatory will be sufficient in isolation because of the need to cover a wide range in gas temperatures and corresponding wavelength ranges.

3. *Transport of volatiles:* Models have long predicted that the local concentration of volatiles may vary by orders of magnitude in different disk regions due to a combination of gas diffusion and solid particle advection. Collectively, these transport processes may alter the observational signature of the disk surface, dramatically change the local mode of chemistry by changing the C/O ratio and increasing the local solid surface density to aid in the formation of planetesimals and planets. The near future is likely to bring observational searches for bulk transport of volatiles that test dynamochemical models of protoplanetary disks.

4. *Planet composition as a function of volatile content:* The composition of individual planets must necessarily reflect that of the disk material used in their formation. Thus there is an expectation that the chemical demographics of disks will match the compositional demographics of exoplanetary systems. The available data are still far too insufficient to make such connections, but the rapid advances in both disk and planet observations are likely to soon make such comparisons common.

Acknowledgments. This work was supported by NASA Origins of the Solar System Grant No. OSS 11-OSS11-0120, the NASA Planetary Geology and Geophysics Program under grant NAG 5-10201, and the National Science Foundation under grant AST-99-83530. The authors are grateful to E. van Dishoeck and T. Henning for comments that greatly improved this chapter.

REFERENCES

Aikawa Y. et al. (1999) *Astrophys. J., 519,* 705.
Aikawa Y. et al. (2002) *Astron. Astrophys., 386,* 622.
Aikawa Y. et al. (2012) *Astron. Astrophys., 538,* A57.
Albarède F. et al. (2013) *Icarus, 222,* 44.
Aléon J. (2010) *Astrophys. J., 722,* 1342.
Alexander C. M. O. et al. (2012) *Science, 337,* 721.
Allègre C. et al. (2001) *Earth Planet. Sci. Lett., 185,* 49.
Amelin Y. et al. (2002) *Science, 297,* 1678.
Andrews S. M. and Williams J. P. (2005) *Astrophys. J., 631,* 1134.
Armitage P. J. (2011) *Annu. Rev. Astron. Astrophys., 49,* 195.
Banzatti A. et al. (2012) *Astrophys. J., 745,* 90.
Barge P. et al. (2008) *Astron. Astrophys., 482,* L17.
Barman T. S. et al. (2011) *Astrophys. J., 733,* 65.
Beckwith S. V. W. et al. (1990) *Astron. J., 99,* 924.
Bergin E. A. et al. (2010) *Astron. Astrophys., 521,* L33.
Bergin E. A. et al. (2013) *Nature, 493,* 644.
Blake G. A. and Boogert A. C. A. (2004) *Astrophys. J. Lett., 606,* L73.
Bockelée-Morvan D. et al. (2009) *Astrophys. J., 696,* 1075.
Bockelée-Morvan D. et al. (2012) *Astron. Astrophys., 544,* L15.
Boogert A. C. A. et al. (2008) *Astrophys. J., 678,* 985.
Borucki W. J. et al. (2010) *Science, 327,* 977.
Briani G. et al. (2009) In *Lunar and Planetary Science XL,*

Abstract #1642. Lunar and Planetary Institute, Houston.

Brinch C. et al. (2008) *Astron. Astrophys., 489,* 617.

Brittain S. et al. (2007) *Astrophys. J. Lett., 670,* L29.

Brown J. M. et al. (2013) *Astrophys. J., 770,* 94.

Brown M. E. et al. (2012) *Astron. J., 143,* 146.

Brownlee D. et al. (2006) *Science, 314,* 1711.

Bruderer S. et al. (2012) *Astron. Astrophys., 541,* A91.

Burnett D. S. et al. (2003) *Space Sci. Rev., 105,* 509.

Burrows A. et al. (2007) *Astrophys. J., 661,* 502.

Bus S. J. and Binzel R. P. (2002) *Icarus, 158(1),* 146.

Calvet N. et al. (2002) *Astrophys. J., 568,* 1008.

Campins H. et al. (2010) *Nature, 464,* 1320.

Carr J. S. and Najita J. R. (2008) *Science, 319,* 1504.

Carr J. S. and Najita J. R. (2011) *Astrophys. J., 733,* 102.

Carr J. S. et al. (2004) *Astrophys. J., 603,* 213.

Casassus S. et al. (2013) *Astron. Astrophys., 553,* A64.

Caselli P. and Ceccarelli C. (2012) *Astron. Astrophys. Rev., 20,* 56.

Cassen P. (1996) *Meteoritics & Planet. Sci., 31,* 793.

Ceccarelli C. et al. (2005) *Astrophys. J. Lett., 631,* L81.

Ceccarelli C. et al. (2007) In *Protostars and Planets V* (B. Reipurth et al., eds.), pp. 47–62. Univ. of Arizona, Tucson.

Chapillon E. et al. (2010) *Astron. Astrophys., 520,* A61.

Chapman C. R. (1996) *Meteoritics & Planet. Sci., 31,* 699.

Chapman C. R. et al. (1975) *Icarus, 25,* 104.

Chiang E. I. and Goldreich P. (1997) *Astrophys. J., 490,* 368.

Chiang E. I. et al. (2001) *Astrophys. J., 547,* 1077.

Ciesla F. J. and Cuzzi J. N. (2006) *Icarus, 181,* 178.

Clayton R. N. (2002) *Nature, 415,* 860.

Clayton R. N. et al. (1973) *Science, 182,* 485.

Coutens A. et al. (2012) *Astron. Astrophys., 539,* A132.

Cumming A. et al. (2008) *Publ. Astron. Soc. Pac., 120,* 531.

Cuzzi J. N. and Zahnle K. J. (2004) *Astrophys. J., 614,* 490.

D'Alessio P. et al. (1997) *Astrophys. J., 474,* 397.

Dasgupta R. and Hirschmann M. M. (2010) *Earth Planet. Sci. Lett., 298,* 1.

Desch S. J. (2007) *Astrophys. J., 671,* 878.

Dodson-Robinson S. E. et al. (2009) *Icarus, 200,* 672.

Dominik C. et al. (2005) *Astrophys. J. Lett., 635,* L85.

Draine B. T. (2003) *Annu. Rev. Astron. Astrophys., 41,* 241.

Drouart A. et al. (1999) *Icarus, 140,* 129.

Dunham M. M. et al. (2010) *Astrophys. J., 710,* 470.

Dutrey A. et al. (1997) *Astron. Astrophys., 317,* L55.

Dutrey A. et al. (2011) *Astron. Astrophys., 535,* A104.

Ebel D. S. and Grossman L. (2000) *Geochim. Cosmochim. Acta, 64,* 339.

Epstein S. et al. (1987) *Nature, 326,* 477.

Fabrycky D. and Tremaine S. (2007) *Astrophys. J., 669,* 1298.

Farihi J. et al. (2013) *Science, 342,* 218.

Favre C. et al. (2013) *Astrophys. J. Lett., 776,* L38.

Fedele D. et al. (2011) *Astrophys. J., 732,* 106.

Fedele D. et al. (2012) *Astron. Astrophys., 544,* L9.

Fischer D. A. and Valenti J. (2005) *Astrophys. J., 622,* 1102.

Flasar F. M. et al. (2005) *Science, 307,* 1247.

Frick U. et al. (1979) In *Proc. Lunar Planet. Sci. 10th,* pp. 1961–1972.

Fromang S. and Nelson R. P. (2006) *Astron. Astrophys., 457,* 343.

Furuya K. et al. (2013) *Astrophys. J., 779,* 11.

Gail H.-P. (2002) *Astron. Astrophys., 390,* 253.

Gänsicke B. T. et al. (2012) *Mon. Not. R. Astron. Soc., 424,* 333.

Garaud P. and Lin D. N. C. (2007) *Astrophys. J., 654,* 606.

Garrod R. T. and Herbst E. (2006) *Astron. Astrophys., 457,* 927.

Garrod R. T. et al. (2008) *Astrophys. J., 682,* 283.

Geiss J. (1987) *Astron. Astrophys., 187,* 859.

Gerakines P. A. et al. (1996) *Astron. Astrophys., 312,* 289.

Gibb E. L. et al. (2004) *Astrophys. J. Suppl., 151,* 35.

Gibb E. L. et al. (2007) *Astrophys. J., 660,* 1572.

Gomes R. et al. (2005) *Nature, 435,* 466.

Gorti U. et al. (2011) *Astrophys. J., 735,* 90.

Goto M. et al. (2011) *Astrophys. J., 728,* 5.

Grevesse N. et al. (2010) *Astrophys. Space Sci., 328,* 179.

Grossman L. (1972) *Geochim. Cosmochim. Acta, 36,* 597.

Hahn J. M. and Malhotra R. (1999) *Astron. J., 117,* 3041.

Hama T. et al. (2011) *Astrophys. J. Lett., 738,* L15.

Hartmann L. and Kenyon S. J. (1996) *Annu. Rev. Astron. Astrophys., 34,* 207.

Hartogh P. et al. (2011) *Nature, 478,* 218.

Hashizume K. et al. (2004) *Astrophys. J., 600,* 480.

Hayashi C. (1981) *Progr. Theor. Phys. Suppl., 70,* 35.

Henning T. and Semenov D. (2013) *Chem. Rev., 113,* 9016.

Herbig G. H. (1966) *Vistas Astron., 8,* 109.

Herbig G. H. (1977) *Astrophys. J., 217,* 693.

Hersant F. et al. (2001) *Astrophys. J., 554,* 391.

Hincelin U. et al. (2013) *Astrophys. J., 775,* 44.

Hogerheijde M. R. et al. (2011) *Science, 334,* 338.

Honda M. et al. (2009) *Astrophys. J. Lett., 690,* L110.

Horner J. et al. (2007) *Earth Moon Planets, 100,* 43.

Hutson M. and Ruzicka A. (2000) *Meteoritics & Planet. Sci., 35,* 601.

Ida S. and Lin D. N. C. (2004a) *Astrophys. J., 604,* 388.

Ida S. and Lin D. N. C. (2004b) *Astrophys. J., 616,* 567.

Jacquet E. and Robert F. (2013) *Icarus, 223,* 722.

Jenniskens P. et al. (1993) *Astron. Astrophys., 273,* 583.

Jessberger E. K. et al. (1988) *Nature, 332,* 691.

Johansen A. et al. (2007) *Nature, 448,* 1022.

Jones T. D. et al. (1990) *Icarus, 88,* 172.

Jørgensen J. K. and van Dishoeck E. F. (2010) *Astrophys. J. Lett., 725,* L172.

Jura M. (2006) *Astrophys. J., 653,* 613.

Jura M. (2008) *Astron. J., 135,* 1785.

Jura M. et al. (2012) *Astrophys. J., 750,* 69.

Kamp I. et al. (2013) *Astron. Astrophys., 559,* A24.

Kasting J. F. et al. (1993) *Icarus, 101,* 108.

Kennedy G. M. and Kenyon S. J. (2008) *Astrophys. J., 673,* 502.

Kerridge J. F. (1985) *Geochim. Cosmochim. Acta, 49(8),* 1707 .

Kitamura Y. and Shimizu M. (1983) *Moon Planets, 29,* 199.

Kley W. and Dirksen G. (2006) *Astron. Astrophys., 447,* 369.

Knutson H. A. et al. (2007) *Nature, 447,* 183.

Konopacky Q. M. et al. (2013) *Science, 339,* 1398.

Kress M. E. et al. (2010) *Adv. Space Res., 46,* 44.

Kuiper G. P. (1953) *Proc. Natl. Acad. Sci., 39,* 1153.

Lahuis F. and van Dishoeck E. F. (2000) *Astron. Astrophys., 355,* 699.

Lahuis F. et al. (2006) *Astrophys. J. Lett., 636,* L145.

Lebofsky L. A. (1980) *Astron. J., 85,* 573.

Lebofsky L. A. et al. (1981) *Icarus, 48,* 453.

Lee J.-E. et al. (2010) *Astrophys. J. Lett., 710,* L21.

Levison H. F. and Duncan M. J. (1997) *Icarus, 127(1),* 13.

Levison H. F. et al. (2010) *Science, 329,* 187.

Lin D. N. C. et al. (1996) *Nature, 380,* 606.

Liskowsky J. P. et al. (2012) *Astrophys. J., 760,* 153.

Lissauer J. J. (1987) *Icarus, 69,* 249.

Lodders K. (2003) *Astrophys. J., 591,* 1220.

Lord H. C. III (1965) *Icarus, 4,* 279.

Lubow S. H. (1991) *Astrophys. J., 381,* 259.

Luck J. M. et al. (2001) *Meteoritics & Planet. Sci. Suppl., 36,* 118.

Lyons J. R. and Young E. D. (2005) *Nature, 435,* 317.

Madhusudhan N. et al. (2011) *Nature, 469,* 64.

Malfait K. et al. (1998) *Astron. Astrophys., 332,* L25.

Mandell A. M. et al. (2008) *Astrophys. J. Lett., 681,* L25.

Mandell A. M. et al. (2012) *Astrophys. J., 747,* 92.

Mann R. K. and Williams J. P. (2010) *Astrophys. J., 725,* 430.

Marcy G.W. and Butler R. P. (1996) *Astrophys. J. Lett., 464,* L147.

Marrocchi Y. et al. (2005) *Earth Planet. Sci. Lett., 236,* 569.

Marty B. (1995) *Nature, 377,* 326.

Marty B. (2012) *Earth Planet. Sci. Lett., 313,* 56.

Marty B. et al. (2008) *Science, 319,* 75.

Marty B. et al. (2010) *Geochim. Cosmochim. Acta, 74,* 340.

Mathews G. S. et al. (2013) *Astron. Astrophys., 558,* A66.

Mayor M. and Queloz D. (1995) *Nature, 378,* 355.

Mayor M. et al. (2003) *The Messenger, 114,* 20.

McClure M. K. et al. (2012) *Astrophys. J. Lett., 759,* L10.

McCord T. B. et al. (1971) *Astrophys. J., 165,* 413.

McKeegan K. D. et al. (2006) *Science, 314,* 1724.

McKeegan K. D. et al. (2011) *Science, 332,* 1528.

McNaughton N. J. et al. (1981) *Nature, 294,* 639.

Meijerink R. et al. (2009) *Astrophys. J., 704,* 1471.

Mesa-Delgado A. et al. (2012) *Mon. Not. R. Astron. Soc., 426,* 614.

Min M. et al. (2005) *Icarus, 179,* 158.

Mordasini C. et al. (2009) *Astron. Astrophys., 501,* 1139.

Mousis O. et al. (2000) *Icarus, 148,* 513.

Mumma M. J. and Charnley S. B. (2011) *Annu. Rev. Astron. Astrophys., 49,* 471.

Mumma M. J. et al. (1988) *Bull. Am. Astron. Soc., 20,* 826.

Najita J. et al. (2003) *Astrophys. J., 589,* 931.

Najita J. R. et al. (2013) *Astrophys. J., 766,* 134.

Niemann H. B. et al. (1996) *Science, 272,* 846.

Notesco G. et al. (1999) *Icarus, 142,* 298.

Öberg K. I. et al. (2005) *Astrophys. J. Lett., 621,* L33.

Öberg K. I. et al. (2009) *Astron. Astrophys., 504,* 891.

Öberg K. I. et al. (2010) *Astrophys. J., 720,* 480.

Öberg K. I. et al. (2011a) *Astrophys. J. Lett., 743,* L16.

Öberg K. I. et al. (2011b) *Astrophys. J., 740,* 109.

Öberg K. I. et al. (2012) *Astrophys. J., 749,* 162.

Ossenkopf V. and Henning T. (1994) *Astron. Astrophys., 291,* 943.

Ott U. et al. (1981) *Geochim. Cosmochim. Acta, 45,* 1751.

Owen T. et al. (2001) *Astrophys. J. Lett., 553,* L77.

Panić O. et al. (2008) *Astron. Astrophys., 491,* 219.

Panić O. et al. (2010) *Astron. Astrophys., 519,* A110.

Papaloizou J. C. B. et al. (2001) *Astron. Astrophys., 366,* 263.

Paquette C. et al. (1986) *Astrophys. J. Suppl., 61,* 197.

Pascucci I. et al. (2009) *Astrophys. J., 696,* 143.

Pascucci I. et al. (2013) *Astrophys. J., 779,* 178.

Pasek M. A. et al. (2005) *Icarus, 175,* 1.

Persson M. V. et al. (2013) *Astron. Astrophys., 549,* L3.

Pilcher C. B. et al. (1972) *Science, 178,* 1087.

Pollack J. B. et al. (1994) *Astrophys. J., 421,* 615.

Pollack J. B. et al. (1996) *Icarus, 124,* 62.

Pontoppidan K. M. et al. (2003) *Astron. Astrophys., 408,* 981.

Pontoppidan K. M. et al. (2004) *Astron. Astrophys., 426,* 925.

Pontoppidan K. M. et al. (2005) *Astrophys. J., 622,* 463.

Pontoppidan K. M. et al. (2010a) *Astrophys. J., 720,* 887.

Pontoppidan K. M. et al. (2010b) *Astrophys. J. Lett., 722,* L173.

Przybilla N. et al. (2008) *Astrophys. J. Lett., 688,* L103.

Qi C. et al. (2004) *Astrophys. J. Lett., 616,* L11.

Qi C. et al. (2008) *Astrophys. J., 681,* 1396.

Qi C. et al. (2011) *Astrophys. J., 740,* 84.

Qi C. et al. (2013a) *Astrophys. J., 765,* 34.

Qi C. et al. (2013b) *Science, 341,* 630.

Regály Z. et al. (2010) *Astron. Astrophys., 523,* A69.

Reipurth B. et al., eds. (2007) *Protostars and Planets V.* Univ. of Arizona, Tucson.

Remusat L. et al. (2009) *Astrophys. J., 698,* 2087.

Rettig T. W. et al. (2005) *Astrophys. J., 626,* 245.

Rivkin A. S. and Emery J. P. (2010) *Nature, 464,* 1322.

Rivkin A. S. et al. (2002) In *Asteroids III* (W. F. Bottke Jr. et al., eds.), pp. 235–253. Univ. of Arizona, Tucson.

Ros K. and Johansen A. (2013) *Astron. Astrophys., 552,* A137.

Salyk C. et al. (2008) *Astrophys. J. Lett., 676,* L49.

Salyk C. et al. (2011) *Astrophys. J., 731,* 130.

Sandford S. A. et al. (2006) *Science, 314,* 1720.

Santos N. C. et al. (2004) *Astron. Astrophys., 415,* 1153.

Sasselov D. D. (2003) *Astrophys. J., 596,* 1327.

Savage B. D. and Sembach K. R. (1996) *Annu. Rev. Astron. Astrophys., 34,* 279.

Scott E. R. D. (2007) *Annu. Rev. Earth Planet. Sci., 35,* 577.

Semenov D. and Wiebe D. (2011) *Astrophys. J. Suppl., 196,* 25.

Semenov D. et al. (2003) *Astron. Astrophys., 410,* 611.

Semenov D. et al. (2008) *Astrophys. J. Lett., 673,* L195.

Skemer A. J. et al. (2012) *Astrophys. J., 753,* 14.

Sliter R. et al. (2011) *J. Phys. Chem. A, 115,* 9682.

Smith R. L. et al. (2009) *Astrophys. J., 701,* 163.

Sousa S. G. et al. (2008) *Astron. Astrophys., 487,* 373.

Stepinski T. F. and Valageas P. (1997) *Astron. Astrophys., 319,* 1007.

Stevenson D. J. (1985) *Icarus, 62,* 4.

Stevenson D. J. and Lunine J. I. (1988) *Icarus, 75,* 146.

Taquet V. et al. (2013) *Astrophys. J. Lett., 768,* L29.

Tarter J. C. et al. (2007) *Astrobiology, 7,* 30.

Terzieva R. and Herbst E. (2000) *Mon. Not. R. Astron. Soc., 317,* 563.

Thi W. F. et al. (2002) *Astron. Astrophys., 394,* L27.

Thi W.-F. et al. (2004) *Astron. Astrophys., 425,* 955.

Thi W.-F. et al. (2010) *Astron. Astrophys., 518,* L125.

Thiemens M. H. and Heidenreich J. E. III (1983) *Science, 219,* 1073.

Tilling I. et al. (2012) *Astron. Astrophys., 538,* A20.

Tsamis Y. G. et al. (2011) *Mon. Not. R. Astron. Soc., 412,* 1367.

Tsamis Y. G. et al. (2013) *Mon. Not. R. Astron. Soc., 430,* 3406.

Vacca W. D. et al. (2004) *Astrophys. J. Lett., 609,* L29.

van Dishoeck E. F. (2004) *Annu. Rev. Astron. Astrophys., 42,* 119.

van Weeren R. J. et al. (2009) *Astron. Astrophys., 497,* 773.

van Zadelhoff G.-J. et al. (2001) *Astron. Astrophys., 377,* 566.

Visser R. et al. (2009) *Astron. Astrophys., 495,* 881.

Visser R. et al. (2011) *Astron. Astrophys., 534,* A132.

Visser R. et al. (2013) *Astrophys. J., 769,* 19.

Wada K. et al. (2009) *Astrophys. J., 702,* 1490.

Walsh K. J. et al. (2011) *Nature, 475,* 206.

Wasson J. T. and Chou C.-L. (1974) *Meteoritics, 9,* 69.

Weidenschilling S. J. (1977) *Mon. Not. R. Astron. Soc., 180,* 57.

Wettlaufer J. S. (2010) *Astrophys. J., 719,* 540.

Whipple F. L. (1950) *Astrophys. J., 111,* 375.

Whittet D. C. B. (2010) *Astrophys. J., 710,* 1009.

Whittet D. C. B. et al. (2007) *Astrophys. J., 655,* 332.

Whittet D. C. B. et al. (2009) *Astrophys. J., 695,* 94.

Willacy K. and Woods P. M. (2009) *Astrophys. J., 703,* 479.

Woitke P. et al. (2009) *Astron. Astrophys., 501,* 383.

Woitke P. et al. (2011) *Astron. Astrophys., 534,* A44.

Wong M. H. et al. (2004) *Icarus, 171,* 153.

Wood B. J. (1993) *Earth Planet. Sci. Lett., 117,* 593.

Wood J. A. and Hashimoto A. (1993) *Geochim. Cosmochim. Acta, 57,* 2377.

Wyckoff S. et al. (1991) *Astrophys. J., 367,* 641.

Yang L. et al. (2012) In *Lunar and Planetary Science XLIII,* Abstract #2023. Lunar and Planetary Institute, Houston.

Youdin A. N. and Shu F. H. (2002) *Astrophys. J., 580,* 494.

Yurimoto H. and Kuramoto K. (2004) *Science, 305,* 1763.

Zhang K. et al. (2013) *Astrophys. J., 766,* 82.

Zhu X. K. et al. (2001) *Nature, 412,* 311.

Zinner E. (1998) *Annu. Rev. Earth Planet. Sci., 26,* 147.

Audard M., Ábrahám P., Dunham M. M., Green J. D., Grosso N., Hamaguchi K., Kastner J. H., Kóspál Á., Lodato G., Romanova M. M., Skinner S. L., Vorobyov E. I., and Zhu Z. (2014) Episodic accretion in young stars. In *Protostars and Planets VI* (H. Beuther et al., eds.), pp. 387–410. Univ. of Arizona, Tucson, DOI: 10.2458/azu_uapress_9780816531240-ch017.

Episodic Accretion in Young Stars

Marc Audard
University of Geneva

Péter Ábrahám
Konkoly Observatory

Michael M. Dunham
Yale University

Joel D. Green
University of Texas at Austin

Nicolas Grosso
Observatoire Astronomique de Strasbourg

Kenji Hamaguchi
National Aeronautics and Space Administration and University of Maryland, Baltimore County

Joel H. Kastner
Rochester Institute of Technology

Ágnes Kóspál
European Space Agency

Giuseppe Lodato
Università Degli Studi di Milano

Marina M. Romanova
Cornell University

Stephen L. Skinner
University of Colorado at Boulder

Eduard I. Vorobyov
University of Vienna and Southern Federal University

Zhaohuan Zhu
Princeton University

In the last 20 years, the topic of episodic accretion has gained significant interest in the star-formation community. It is now viewed as a common, although still poorly understood, phenomenon in low-mass star formation. The FU Orionis objects (FUors) are long-studied examples of this phenomenon. FU Orionis objects are believed to undergo accretion outbursts during which the accretion rate rapidly increases from typically 10^{-7} to a few 10^{-4} M$_\odot$ yr^{-1}, and remains elevated over several decades or more. EXors, a loosely defined class of pre-main-sequence stars, exhibit shorter and repetitive outbursts, associated with lower accretion rates. The relationship between the two classes, and their connection to the standard pre-main-sequence evolutionary sequence, is an open question: Do they represent two distinct classes, are they triggered by the same physical mechanism, and do they occur in the same evolutionary phases? Over the past couple of decades, many theoretical and numerical models have been developed to explain the origin of FUor and EXor outbursts. In parallel, such accretion bursts have been detected at an increasing rate, and as observing techniques improve, each individual outburst is studied in increasing detail. We summarize key observations of pre-main-sequence star outbursts, and review the latest thinking on outburst triggering mechanisms, the propagation of outbursts from star/disk to disk/jet systems, the relation between classical EXors and FUors, and newly discovered outbursting sources — all of which shed new light on episodic accretion. We finally highlight some of the most promising directions for this field in the near- and long-term.

1. INTRODUCTION

Episodic accretion has become a recent focus of attention in the star- and planet-formation community, turning into a central topic to understand the evolution of protostars and accreting young stars. Initially identified as young stellar objects (YSOs) with strong, long-lived optical outbursts (*Herbig,* 1966, 1977), FU Orionis objects (hereafter FUors) have triggered many investigations to understand the eruptive phenomenon. Several reviews have been published (*Herbig,* 1977; *Reipurth,* 1990; *Hartmann et al.,* 1993; *Hartmann,* 1991; *Kenyon,* 1995a,b; *Bell et al.,* 2000; *Hartmann and Kenyon,* 1996) [see also the specific chapter on the FU Ori phenomenon in *Hartmann* (2008) and the recent review by *Reipurth and Aspin* (2010)]. The field has exploded in the last 15 years thanks to new ground- and spacebased facilities.

Observationally FUor candidates — and their possible short timescale counterparts, EX Lupi objects, dubbed EXors by *Herbig* (1989) — have been studied across the electromagnetic spectrum, while theoretical studies have further explored the origin of the outburst mechanism. Erupting young stars are no longer oddities but are now placed prominently on the grand scheme of star formation and time evolution of mass-accretion rates, from embedded protostars to classical T Tauri stars (CTTS), and eventually weak-line T Tauri stars (WTTS). In parallel, recent studies have led to doubt as to the need for separate observational classification of FUors and EXors, as discoveries of new outbursting sources have resulted in a less definitive separation. Episodic accretion has also possibly resolved, among other issues, the so-called luminosity problem in low-mass embedded protostars.

In this review, we provide a summary of the literature published on the "historical" FUor and EXor classes and on the theoretical and numerical studies relevant to episodic accretion. We start with the review by *Hartmann and Kenyon* (1996), although we refer to older studies whenever needed. Finally, we emphasize that this review focuses on episodic accretion, i.e., strong variability due to accretion events: We do not address small-scale variability or variability caused by geometrical effects, clearing of dust, etc., although some of these aspects will be mentioned when observed in parallel with episodic accretion.

2. OBSERVATIONS

2.1. Episodic Accretion During Star Formation

The general picture of the evolution of pre-main-sequence accretion (*Hartmann and Kenyon,* 1996) suggests that much of the material added to the central star, and the material available for planet-forming disks, is influenced by the frequency and intensity of eruptive bursts followed by long periods of relative quiescence. It is suspected that this process occurs at *all* early stages of star formation after the prestellar core, but becomes observable only as

the circumstellar envelope thins. In one picture, FUors and EXors are part of this continuum. In this picture, the FUor bursts are longer and stronger compared to the bursts of EXors. The bursts would occur in repeated cycles and are fueled by additional material falling from the circumstellar envelope to the disk in between bursts, halted by some mechanism, and released in a dramatic flood quasiperiodically. In an alternative picture, EXors would be a separate phenomenon associated with instabilities in the disks of T Tauri stars (TTS), while FUors span the divide between protostars with disks and envelopes and TTS with disks. Observations reveal a more complicated picture in which strong, long outbursts can also occur in previously identified CTTS and EXor-type short outbursts in relatively embedded young stars with envelopes.

In the next sections, we have kept this historical, observational separation between FUors and EXors with the aim to draw commonalities within the classes. We aim to build on the observational and theoretical results to address the validity of the separation and to propose future steps to help determine how and if FUors and EXors are related.

2.2. Characteristics of FU Orionis Objects

The initial class of FUors was comprised of FU Ori, V1057 Cyg, and V1515 Cyg, all showing strong outbursts with amplitudes of several magnitudes, albeit over significantly different timescales and durations (*Herbig,* 1977). V1735 Cyg was added to the list shortly thereafter (*Elias,* 1978). Although FU Ori has been slowly fading since its 1936 outburst (*Kenyon et al.,* 2000), it is still in a high state at present. Notice that V1331 Cyg has often been included among lists of FUors [following *Welin* (1976) who considered it a pre-FUor], but there is no support for such a classification. Since then, several candidates have been added (e.g., *Graham and Frogel,* 1985; *Eislöffel et al.,* 1990; *Staude and Neckel,* 1991, 1992; *Strom and Strom,* 1993; *McMuldroch et al.,* 1995; *Shevchenko et al.,* 1997; *Sandell and Aspin,* 1998; *Aspin and Sandell,* 2001; *Aspin and Reipurth,* 2003; *Movsessian et al.,* 2003, 2006; *Quanz et al.,* 2007a,b; *Tapia et al.,* 2006; *Kóspál et al.,* 2008; *Magakian et al.,* 2010, 2013; *Reipurth et al.,* 2012). Many objects that are spectroscopically similar to classical FUors but have never been seen to erupt are instead classified as FUor-like objects [analogous to nova-like objects; see *Reipurth et al.* (2002)]. Many FUor candidates are significantly more extinguished, with redder/cooler spectral energy distributions (SEDs) than classical FUors, which exhibit relatively evolved/blue SEDs reminiscent of TTS. In Table 1 we provide a non-exhaustive list of eruptive objects, including FUor-like objects.

In the optical, classical FUors exhibit similar spectra, with F/G supergiant spectral types but broad lines compared to TTS, and little indication of magnetospheric accretion. In contrast, they show K/M supergiant spectral types in the near-infrared. A supergiant spectrum can also be seen in the ultraviolet (UV) (*Kravtsova et al.,* 2007). They are associated with reflection nebulae and are associated with a moderate level of extinction ($A_V \sim$ 1.8–3.5 mag). In the near-infrared, FUors show first-overtone CO absorption at 2.2 µm. The Fe I, Li I, and Ca I optical lines, as well as the near-infrared CO lines, are typically double-peaked and show line broadening that is kinematically consistent with a rotating disk origin (*Hartmann and Kenyon*, 1996), although a wind may also be required to explain these profiles (*Eisner and Hillenbrand*, 2011). A different origin for the optical line splitting was put forward by *Petrov and Herbig* (1992) and *Petrov et al.* (1998), who argued that the profile can be explained simply by the presence of central emission cores in the broad absorption lines (see also *Herbig et al.*, 2003). Similarly, a model invoking a large, dark polar spot was put forward to explain the line profiles in the optical for FU Ori (*Petrov and Herbig*, 2008). Despite the controversy at optical wavelengths [which may in fact relate to different aspects of the same underlying phenomenon; see *Kravtsova et al.* (2007)], the accretion disk model generally works well (e.g., *Hartmann et al.*, 2004).

In the infrared, the similarity between classical FUors begins to waver, as FU Ori exhibits pristine silicate emission features and a relatively blue SED consistent with that of a mildly flared disk (*Green et al.*, 2006; *Quanz et al.*, 2007c), while V1515 Cyg and V1057 Cyg are closer to flat-spectrum sources with weaker silicate emission. They are further distinguished in the far-infrared/submillimeter where envelopes often dominate SEDs: FU Ori and V1515 Cyg have weak continuum beyond 100 µm and very little CO-line emission, while V1057 Cyg shows warm CO gas and substantially stronger continuum (*Green et al.*, 2013b). Nevertheless, near-infrared interferometry shows that classical FUors all show significant contributions from envelopes (*Millan-Gabet et al.*, 2006). Such envelopes have masses of a few tenths of a solar mass and are comparable to class I protostars rather than class II stars (*Sandell and Weintraub*, 2001; *Pérez et al.*, 2010).

During their outbursts, classical FUors have bolometric luminosities of 100–300 $L_\odot$, with accretion rates between 10^{-6} and 10^{-4} $M_\odot$ yr^{-1} (section 4.4). One of the challenges is identifying candidate FUors before they occur: V1057 Cyg is one of only two identified FUors (with HBC 722, discussed below) with a preoutburst optical spectrum (*Herbig*, 1977). It was found to have optical emission lines typical of a TTS. Modern large spectral surveys of active star-forming regions should increase the likelihood of serendipitous preoutburst observations (section 5).

Since few FUor outbursts have been detected, the question arises as to whether some existing YSOs may have previously undergone undetected FUor eruptions. For instance, some driving sources of Herbig-Haro (HH) objects might be FUor-like, because their near-infrared spectra are substantially similar to those of FUors (*Reipurth and Aspin*, 1997; *Greene et al.*, 2008). These authors noted that young stars with FUor-like characteristics might be more common than projected from the relatively few known classical FUors.

2.3. FU Orionis Temporal Behavior

2.3.1. Long-term evolution. In the optical, the initial brightening of ~5 mag is followed by a longer plateau phase, during which classical FUors display a relatively long decay timescale: FU Ori has faded by about 1 mag since its 1936 outburst, at a rate of 14 mmag yr^{-1} (*Kenyon et al.*, 2000). BBW 76 (also known as Bran 76) showed a decay of about 23 mmag yr^{-1} in V between 1983 and 1994, with a slow-down in the infrared around 1989, although a historical search indicated that BBW 76 has remained at a similar brightness level since at least 1900 (*Reipurth et al.*, 2002). To document the long-term flux evolution of FUors, *Ábrahám et al.* (2004b) compared 1–100-µm photometric observations, obtained by the Infrared Space Observatory (ISO) in 1995–1998, with earlier data taken from the Infrared Astronomical Satellite (IRAS) catalog (from 1983). Both satellite datasets were supplemented by contemporaneous groundbased infrared observations from the literature. In two cases (Z CMa, Parsamian 21) no significant difference between the two epochs was seen. V1057 Cyg, V1515 Cyg, and V1735 Cyg became fainter at near-infrared wavelengths, while V346 Nor had become slightly brighter. At $\lambda \geq 60$ µm most of the sources remained constant; only V346 Nor seemed to fade. A detailed case study of V1057 Cyg revealed that the long-term evolution of the system at near and mid-infrared wavelengths was consistent with model predictions of *Kenyon and Hartmann* (1991) and *Turner et al.* (1997), except at wavelengths longward of 25 µm.

2.3.2. Variability and episodic accretion. In addition to broad light curve evolution, FUors exhibit a variety of short-timescale (hours to days) variability, both semi-periodic and stochastic. *Kenyon et al.* (2000) present rapid cadence photometry of classical FUors. They find evidence for low-amplitude "flickering" (~0.03–0.2 mag) over a timescale of ≤ 1 d that they attribute to the inner accretion disk. Similar results were found by *Clarke et al.* (2005). From high-spectral-resolution optical observations of FU Ori, *Powell et al.* (2012) confirmed the modulation of wind lines (P ~ 14 d) and photospheric lines (P ~ 3.5 d) discovered by *Herbig et al.* (2003), while *Siwak et al.* (2013) used Microvariability and Oscillations of Stars (MOST) satellite and found the possible existence of 2–9-d quasiperiodic features in FU Ori, which they interpreted as plasma condensations or localized disk heterogeneities. The flickering and presence of periodicity was also found in other erupting stars, such as V1647 Ori (*Bastien et al.*, 2011). *Green et al.* (2013a) found a rotational period for HBC 722 of 5.8 d, and a variety of subperiods between 0.9 d and 1.3 d, with a peak in the periodogram at 1.28 d. This shorter period was attributed to instability near the inner disk edge (see also section 3.4.3).

In any case, episodic accretion can be distinguished from "normal" TTS variability that results from either small-scale accretion events, geometrical effects, or magnetic activity (*Costigan et al.*, 2012; *Chou et al.*, 2013; *Scholz et al.*, 2013). Indeed, accreting TTS tend to vary in mass-

TABLE 1. Nonexhaustive list of eruptive young stars.

Name	Type	Distance (pc)	Onset (yr)	Duration (yr)	A_V (mag)	L_{bol} ($L_\odot$)	$\dot{M}_{acc}$ ($M_\odot$ yr^{-1})	Companion	References
RNO 1B	FUor-like	850	—	>12	9.2	—	—	Y (RNO 1C, 4")	[44,72,78]
RNO 1C	FUor-like	850	—	12.0	—	—	—	Y (RNO 1B, 4")	[44,72]
V1180 Cas	EXor?	600	2000, 2004	2.5, 7	4.3	0.07 (L)	>1.6e-7(L)	Y? (6.2")	[51]
V512 Per	EXor	300	>1988, <1990	>4	~40	66 (L)	—	Y (0.3")	[5,12,22]
PP13S	FUor-like	350	—	—	~40	30	—	N?	[19,73]
XZ Tau	EXor?	140	1998	>3	1.4	0.5	1e-7	Y(0.3")	[18,23,31,33]
UZ Tau E	EXor	140	1921	0.5?	1.5	1.7	1-3e-7	Y(SB+4")	[23,36,43,56]
VY Tau	EXor	140	many	0.5-2.0	0.85	0.75	—	Y(0.66")	[23,36,56]
LDN 1415 IRS	EXor?	170	>2002, <2006	—	—	>0.13(L)	—	—	[80]
V582 Aur	FUor	—	>1984, <1986	>26	—	—	—	—	[74]
V1118 Ori	EXor	414	2004, many	~1.2	0-2	1.4(L),7-25(H)	2.5e-7(L), 1e-6(H)	Y(0.18")	[14,36,39, 55,59,68]
Haro 5a IRS	FUor-like	450	—	—	22	50	—	—	[69]
NY Ori	EXor	414	many	>0.3	0.3	—	—	N	[14,36,39,45,59]
V1143 Ori	EXor	500	many	~1	—	—	—	—	[39,64]
V883 Ori	FUor-like	460	—	—	—	400	—	—	[72,81]
Reipurth 50 N IRS 1	FUor-like	460	—	—	—	300	—	—	[16,81]
V2775 Ori	FUor	420	>2005, <2007	>5	8-12	2-4.5(L),22-28(H)	2e-6(L),1e-5(H)	Y?(11")	[24]
FU Ori	FUor	450	1936	~100	1.5-2.6	340-500	—	Y(0.5")	[1,34,77]
V1647 Ori	EXor?	400	1966, 2003, 2008	0.4-1.7, 2.5,>4.3	8-19	3.5-5.6, 34-44	6e-7,4e-6-1e-5	—	[3,4,7,8,9,10,15, 25,63,83,84]
AR 6A	FUor-like	800	—	>13	18	450	—	Y(AR6B,2.8")	[13]
AR 6B	FUor-like	800	—	—	>18	—	—	Y(AR6A,2.8")	[13]
V900 Mon	FUor-like	1100	>1953,<2010	>16	13	106(H)	—	N	[70]
Z CMa	FUor	930-1100	many	5-10	1.8-3.5	400-600	1e-3	Y(0.1")	[34,35,46, 53,72,85]
BBW 76	FUor-like	1700	<1900	~40	2.2	287	7.2e-5	N	[1,27,3]
V723 Car	EXor	—	—	—	—	—	—	N	[82]
GM Cha	EXor?	160	many	>1.9	≥13	>1.5	1e-7	Y(10")	[66]
EX Lup	EXor	155	2008, many	<1	0	0.7,2	4e-10,2e-7	Y?(BD)	[2,6,30,36,37, 38,54,76]
V346 Nor	FUor	700	~1980	>5	>12	135	—	N	[1,26,27,34,67]
OO Ser	FUor	311	1995	>16	42	4.5(L),26-36(H)	—	N	[32,40,41,42,48]
Parsamian 21	FUor-like	400	—	—	8?	3,4,10	—	N?	[20,47,79]
V1515 Cyg	FUor	1000,1050	~1950	~30	2.8-3.2	200	3.5e-5	—	[1,27,34,77]
PV Cep	EXor?	325	repetitive	~2	12.0	41(L),100(H)	2e-7-2.6e-6(L),5.2e-6(H)	—	[52]
V2492 Cyg	EXor?	600	>2009,<2010	>3	6-12,10-20	14(L),43(H)	2.5e-7(H)	—	[21,49,50]
HBC 722	FUor	600	2009	>4	3.4,3.1	0.7-0.85(L),8.7-12(H)	1e-6(H)	—	[29,49,60]
V2494 Cyg	FUor-like	700-800	>1952, <1989	>20	5.8	14-18	—	—	[57,69]
V1057 Cyg	FUor	600,700	1970	~10	3.0-4.2	250-800	4.5e-5	N	[1,27,34,77]

TABLE 1 (continued).

Name	Type	Distance (pc)	Onset (yr)	Duration (yr)	A_V (mag)	L_{bol} ($L_\odot$)	$\dot{M}_{acc}$ ($M_\odot$ yr^{-1})	Companion	References
V2495 Cyg	FUor	800	1999	>8	—	—	—	—	[11,62]
RNO 127	FUor	800	1999	>6	—	—	—	—	[61,62]
V1735 Cyg	FUor	900	>1957, <1965	>20	8.0–10.8	235	—	Y? (20–24")	[1,34,77]
HH 354 IRS	FUor-like	750	—	—	—	73.0	—	—	[20,69]
V733 Cep	FUor-like	800	>1953, <1984	>38	8	135(H)	—	—	[65,71]

L: quiescent or low state, H: outburst or high state, Companion: The parentheses show the angular separation of the companion from FUor/EXor.

References: [1] Ábrahám et al. (2004b); [2] Ábrahám et al. (2009); [3] Ábrahám et al. (2004a); [4] Andrews et al. (2004); [5] Anglada et al. (2004); [6] Aspin et al. (2010); [7] Aspin et al. (2009b); [8] Aspin et al. (2006); [9] Aspin (2011b); [10] Aspin et al. (2008); [11] Aspin et al. (2009c); [12] Aspin and Sandell (1994); [13] Aspin and Reipurth (2003); [14] Audard et al. (2010); [15] Briceño et al. (2004); [16] Casali (1991); [17] Chavarría (1981); [18] Coffey et al. (2004); [19] Cohen et al. (1983); [20] Connelley and Greene (2010); [21] Covey et al. (2011); [22] Eislöffel et al. (1991); [23] Elias (1978); [24] Fischer et al. (2012); [25] Gibb (2008); [26] Graham and Frogel (1985); [27] Green et al. (2006); [28] Green et al. (2013b); [29] Green et al. (2011); [30] Grosso et al. (2010); [31] Haas et al. (1990); [32] Haisch et al. (2004); [33] Hartigan and Kenyon (2003); [34] Hartmann and Kenyon (1996); [35] Hartmann and Kenyon (1985); [36] Herbig (1989); [37] Herbig (1977); [38] Herbig (2007); [39] Herbig (2008); [40] Hodapp (1996); [41] Hodapp (1999); [42] Hodapp et al. (2012); [43] Jensen et al. (2007); [44] Kenyon et al. (1993); [45] Köhler et al. (2006); [46] Koresko et al. (1991); [47] Kóspál et al. (2008); [48] Kóspál et al. (2007); [49] Kóspál et al. (2011a); [50] Kóspál et al. (2013); [51] Kun et al. (2011a); [52] Kun et al. (2011b); [53] Leinert and Haas (1987); [54] Lombardi et al. (2008); [55] Lorenzetti et al. (2006); [56] Lorenzetti et al. (2007); [57] Magakian et al. (2013); [58] McMuldroch et al. (1993); [59] Menten et al. (2007); [60] Miller et al. (2011); [61] Movsessian et al. (2003); [62] Movsessian et al. (2006); [63] Muzerolle et al. (2005); [64] Parsamian et al. (1987); [65] Peneva et al. (2010); [66] Persi et al. (2007); [67] Prusti et al. (1993); [68] Reipurth et al. (2007b); [69] Reipurth and Aspin (1997); [70] Reipurth et al. (2012); [71] Reipurth et al. (2007a); [72] Sandell and Weintraub (2001); [73] Sandell and Aspin (1998); [74] Semkov et al. (2011); [75] Shevchenko et al. (1991); [76] Sipos et al. (2009); [77] Skinner et al. (2009); [78] Staude and Neckel (1991); [79] Staude and Neckel (1992); [80] Stecklum et al. (2007); [81] Strom et al. (1993); [82] Tapia et al. (2006); [83] Teets et al. (2011); [84] Venkata Raman et al. (2013); [85] Whelan et al. (2010).

accretion rates by 0.37–0.83 dex or less, depending on the accretion diagnostic. Such variability occurs over timescales of ≤15 months, with a major portion of the variability occurring over shorter timescales, 8–25 d, i.e., comparable to stellar rotational periods. In contrast, episodic accretion is characterized by stronger variations in mass-accretion rate and occurs on longer timescales (i.e., months to years).

2.4. Cold Environments of FU Orionis Objects

(Sub-)millimeter observations originally suggested that classical FUors are extremely young, more similar to class I protostars than to class II stars (*Sandell and Weintraub,* 2001). These single-dish continuum maps showed resolved emission and typically elongated disk-like morphology, with sizes of several thousand astronomical units (e.g., *Dent et al.,* 1998; *Henning et al.,* 1998). The mass in these structures was estimated at a few tenths of solar mass. Such reservoirs are sufficient to replenish circumstellar disks after episodes of rapid accretion. Molecular outflows were generally detected, except for the most evolved FUors (*Evans et al.,* 1994; *Moriarty-Schieven et al.,* 2008), with mass loss rates between a few times 10^{-8} $M_\odot$ yr^{-1} and 10^{-6} $M_\odot$ yr^{-1}, without correcting for optical depth; mass loss rates likely peak at 10^{-5} $M_\odot$ yr^{-1}. Dense gas traced by HCO$^+$ was also detected (*McMuldroch et al.,* 1993, 1995; *Evans et al.,* 1994). Newly obtained line profiles with Herschel (*Green et al.,* 2013b) are consistent with the above results. However, with single-dish observations, it is often difficult to determine whether the progenitor of the outflow is the FUor or in fact nearby young protostars, such as in the case of HBC 722, that remained undetected in the continuum, with an upper limit of 0.02 $M_\odot$ for the mass of the disk (*Dunham et al.,* 2012), low enough to rule out gravitational instabilities in the disk as the driving mechanism for the outburst.

Further evidence of gas emission associated with FUors was obtained with ISO by *Lorenzetti et al.* (2000) who found [OI], [CII], and [NII], and some cold molecular lines, both on and off-source. All FUors observed with Herschel by *Green et al.* (2013b) exhibited strong [OI] 63-μm emission, indicative of high-mass loss rates or UV excitation. On the other hand, the warm CO emission lines, generally observed in class 0/I protostars, were not detected, with only V1057 Cyg showing compact high-J CO emission.

Interferometric observations are, however, required to distinguish between the circumstellar envelope and cloud emission. *Kóspál* (2011) detected strong, narrow ^{13}CO J=1–0 line emission in Cygnus FUors. The emission was spatially resolved in all cases, with deconvolved sizes of a few thousand astronomical units. For V1057 Cyg, the emitting area was rather compact, suggesting that the origin of the emission is a circumstellar envelope surrounding the central star. For V1735 Cyg, the ^{13}CO emission was offset from the stellar position, indicating that the source of this emission might be a small foreground cloud, also responsible for the high reddening of the central star. The

[13]CO emission toward V1515 Cyg was the most extended in the sample, and it apparently coincided with the ring-like optical reflection nebula associated with V1515 Cyg.

Future interferometric observations in the (sub-)millimeter will better disentangle the envelope and disk emissions, and constrain the mass reservoir around FUors.

2.5. Observational Classification Schema for FU Orionis Objects

There are a number of methods used to classify young stars, such as the infrared spectral index α (*Greene et al.*, 1994) or the bolometric temperature T_{bol} (*Chen et al.*, 1995; *Robitaille et al.*, 2006). Some FUors exhibit class II SEDs, while others exhibit flat SEDs [characterized as having 350 K < T_{bol} < 950 K (*Evans et al.*, 2009; *Fischer et al.*, 2012)], near the class I/II boundary (*Green et al.*, 2013b). The envelope mass derived from (sub-)millimeter dust continuum can also be used to discriminate between evolutionary stages. A thorough discussion on the distinction between stages and the observationally defined infrared class is provided in the chapter by Dunham et al. in this volume.

Several studies have focused on the silicate feature around 10 μm and on ice properties (*Polomski et al.*, 2005; *Schütz et al.*, 2005; *Green et al.*, 2006; *Quanz et al.*, 2007c). The silicate feature can be found either in absorption or in emission. The FUors with silicate absorption generally show an amorphous silicate structure, similar to the interstellar medium, although some extra emission or large silicates can be found. In addition, H_2O, methanol, and CO_2 ice absorption is detected. In contrast, FUors with silicate in emission can show evidence for absorption in the disk photosphere from blended H_2O vapor absorption [5–8 μm (*Green et al.*, 2006)]. The silicate feature in emission does not show evidence of crystals, with some FUors showing weak, pristine silicate features (e.g., V1515 Cyg and V1057 Cyg). This led *Quanz et al.* (2007c) to propose that FUors can be classified in two categories, defined by silicate absorption vs. emission. The emission features arise from the disk surface layers, and represent evidence for grain growth but no processing (Fig. 1). Objects with silicate in absorption are likely young FUors, still embedded in an envelope, whereas objects with silicates in emission are likely more evolved FUors with (partially) depleted envelopes.

Green et al. (2013b) found that the mid-infrared sequence proposed by *Quanz et al.* (2007c) broadly described the continuum of FUors for which there was sufficient (sub-)millimeter data. However, the line observations provide a different picture from the continuum, as FUors may not be well characterized by the stage I/II sequence. While FU Ori and V1515 Cyg, two of the most evolved FUors, were found to have little warm (~100 K) gas, V1057 Cyg exhibited CO rotational lines typical of class 0/I protostars. If this CO originates in the shocked layer of the envelope surrounding the outflow, FU Ori's lack of emission can be explained by a tenuous envelope; however, the mid-infrared

similarity between V1515 Cyg and V1057 Cyg (in contrast with FU Ori) would not predict the difference in their sub-millimeter emission lines. The foregoing makes apparent the difficulty of classifying FUors in the stage I/II sequence.

2.6. Characteristics of EXors

Classical EXors (*Herbig*, 1989, 2008) were initially defined as a list of variable stars showing large-scale outbursts and TTS spectra. The prototype, EX Lup, largely influenced the definition of the class, with its repetitive, short-lived outbursts, M-type dwarf spectrum in quiescence, and absence of features typical of FUors (see section 2.2). The original list of EXors (*Herbig*, 1989) has changed little since that time. New potential EXors were found, although their classifications as EXor or FUor are sometimes unclear (e.g., *Eislöffel et al.*, 1991; *Aspin and Sandell*, 1994; *Stecklum et al.*, 2007; *Persi et al.*, 2007; *Kun et al.*, 2011a). Abundant photometric observations during and after outbursts indicated brightness increases of several magnitudes over durations of several months (e.g., *Coffey et al.*, 2004; *Audard et al.*, 2010). Bolometric luminosities are about 1–2 $L_{\odot}$ in quiescence (including about 0.3–0.5 $L_{\odot}$ of stellar photospheric luminosity) and peak around several lunar masses to a few tens of lunar masses (*Lorenzetti et al.*, 2006; *Audard et al.*, 2010; *Sipos et al.*, 2009; *Aspin et al.*, 2010). Coverage tends to become less frequent after outburst, preventing precise determinations of EXor outburst durations. Nevertheless, eruptions occur frequently in the same objects, with separations of a few years between outbursts, and durations of about 1–2 yr (e.g., *Herbig*, 2008). Oscillations (≈30–60 d), longer than the rotation period, can be detected during outbursts and can be explained by magnetospheric accretion models of trapped disks near corotation (e.g., *Acosta-Pulido et al.*, 2007; *D'Angelo and Spruit*, 2012).

EXors show absorption spectra typical of K or M dwarfs with T Tauri-like emission spectral features during their minimum light (*Parsamian and Mujica*, 2004; *Herbig*, 2008). Optical and near-infrared SEDs during outbursts can be well fitted with an extra thermal component, e.g., a single blackbody component with temperatures varying from 1000 K to 4500 K and emitting radii of 0.01 AU to 0.1 AU, typical of inner disk regions (*Lorenzetti et al.*, 2012). Alternatively, the outburst SED can be fitted with a hot-spot model whose emission dominates in the optical (*Audard et al.*, 2010; *Lorenzetti et al.*, 2012). The coverage factor can be large during the outburst, although *Lorenzetti et al.* (2012) argue that it does not exceed about 10% of the stellar surface. In any case, color variations observed during outbursts cannot be explained by extinction effects alone. Signatures of infall and outflow are detected in Na I $D_{1,2}$, in a similar fashion as in CTTS. Near-infrared spectra show line emission from hydrogen recombination lines, with CO overtone features and weaker atomic features commonly detected both in emission and absorption and variable on short timescales (*Lorenzetti et al.*, 2009).

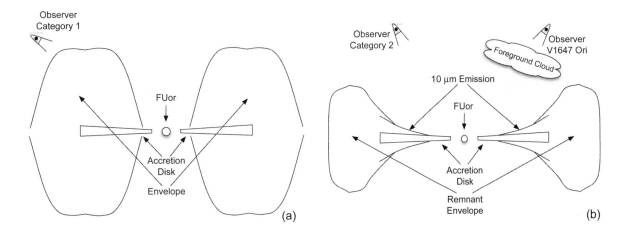

Fig. 1. Sketch of the two categories of FUors (*Quanz et al.,* 2007c). The first category is related to embedded sources with silicate features in absorption and younger ages than FUors of the second category, which show silicate features in emission. Reproduced by permission of the AAS.

The SEDs of EXors span the class I/II divide symmetrically, with a peak in the "flat spectrum" category (*Giannini et al.,* 2009). In the mid-infrared, EXors show the 10-μm silicate feature in emission (*Audard et al.,* 2010; *Kóspál et al.,* 2012). Some of them show wavelength-independent flux changes, probably due to varying accretion. Others are more variable in the 10-μm silicate feature than in the adjacent continuum, which can be interpreted as possible structural changes in the inner disk.

2.7. New Eruptive Stars

Several newly discovered eruptive young stars have been found, many in the last decade. Some of them often show spectral characteristics typical of FUors, but smaller luminosities (see Table 1). HBC 722 (also known as V2493 Cyg) started with a 0.6–0.85-$L_\odot$ luminosity and a SED typical of CTTS, then its luminosity rose to 4–12 $L_\odot$ in outburst (*Semkov et al.,* 2010, 2013; *Miller et al.,* 2011; *Kóspál et al.,* 2011a). Similarly, V2775 Ori changed its bolometric luminosity from 4.5 $L_\odot$ to ≈51 $L_\odot$ (*Caratti o Garatti et al.,* 2011; *Fischer et al.,* 2012). LDN 1415 IRS could also be a low-luminosity erupting star [0.13 $L_\odot$ in low state (*Stecklum et al.,* 2007)], although its status as FUor or EXor is yet unclear. Moderate luminosities are also observed in OO Ser (*Hodapp et al.,* 1996, 2012), with 26–36 $L_\odot$ in outburst, whereas V733 Cep displayed stronger luminosity (135 $L_\odot$) and has slowly faded since 2007 (*Peneva et al.,* 010). The luminosity of V2494 Cyg (=HH 381 IRS) is difficult to constrain (14–45 $L_\odot$) in light of the uncertainty in its distance (*Magakian et al.,* 2013). The recently discovered V1647 Ori also showed moderate luminosity at peak [about 20–40 $L_\odot$ (*Andrews et al.,* 2004; *Muzerolle et al.,* 2005; *Aspin,* 2011b)], although it may more closely resemble an EXor than a FUor. Some new sources are young (typically class I) and deeply embedded, hidden behind thick envelopes [e.g., OO Ser (*Kóspál et al.,* 2007), V900 Mon

(*Reipurth at al.,* 2012)]. Others have no detectable envelopes, suggesting a relatively evolved state [e.g., V733 Cep (*Reipurth et al.,* 2007a), HBC 722 (*Green et al.,* 2011, 2013b; *Dunham et al.,* 2012)], despite sometimes displaying heavy extinction due to surrounding clouds.

Even more interestingly, timescales for outbursts were found *in between* those of classical EXors and FUors (Fig. 2). OO Ser was in outburst for about 8 yr (*Hodapp et al.,* 1996, 2012). Together with its moderate luminosity, it bears resemblance to V1647 Ori, which has been twice in outburst since 2004 (see section 2.9.3). HBC 722 also displayed quite peculiar behavior: It started fading after peak brightness with a rate much faster than typical for FUors, but the fading stopped, and the object started brightening again, with no clear signs that its outburst will end any time soon (*Semkov et al.,* 2012) (Fig. 2). These discoveries demonstrate that eruptive phenomena span a considerable range in evolutionary state, envelope mass, stellar mass, time behavior, and accretion rates.

2.8. Rearrangements in the Circumstellar Matter

Several unusual eruptive stars were recently identified: sources where the brightening may be due to the combination of two effects (*Hillenbrand et al.,* 2013). One effect is an intrinsic brightening related to enhanced accretion, as in all eruptive stars (e.g., *Sicilia-Aguilar et al.,* 2008). The other effect is a dust-clearing event that reduces the extinction along the line of sight, such as in UX Ori-type stars (e.g., *Xiao et al.,* 2010; *Chen et al.,* 2012; *Semkov and Peneva,* 2012). Table 1 does not include sources for which extinction effects dominate the time variability (e.g., GM Cep, V1184 Tau). Accretion and extinction changes often happen synchronously, suggesting that they are physically linked. The extinction can reach high values (e.g., *Covey et al.,* 2011; *Kóspál et al.,* 2011a). The variations are likely due to obscuring structures lying close to the stars

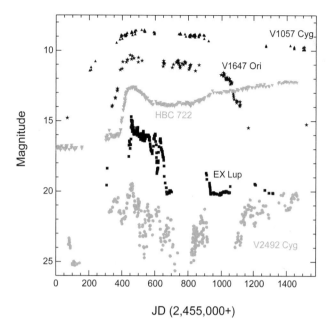

Fig. 2. Comparison light curves for the FUor V1057 Cyg, the intermediate case V1647 Ori, the new sources HBC 722 and VSX J2025120.1+440523 (also known as PTF 10nvg or V2492 Cyg), and the classical EXor EX Lup, showing the continuum of outburst durations. Adapted from *Kóspál et al.* (2011a).

[i.e., within a few tenths of an astronomical unit (*Kóspál et al., 2013*)]. Interestingly, the objects in question can also be highly embedded class I objects, such as V2492 Cyg, and they can display rich and variable emission-line spectra like EXors (*Aspin, 2011a*). The intermediate-mass star PV Cep, originally classified as an EXor by *Herbig* (1989) on the basis of its large outburst in 1977–1979 (*Cohen et al., 1981*) — despite its mass, embeddedness, and higher accretion rate than typical EXors — also shows variability resulting from an interplay between variable accretion and circumstellar extinction, hinting at a rapidly changing dust distribution close to the star (*Kun et al., 2011b; Lorenzetti et al., 2011; see also Caratti o Garatti et al., 2013*).

These new observations indicate that structural changes often happen in the innermost part of the disk or circumstellar matter, typically within 1 AU, supporting the conclusion that these changes are related to the outburst. It is remarkable that all the objects described above show EXor-like, repetitive outbursts. This implies that whatever structural change happens in the disk, it should be reversible on a relatively short timescale, setting strong constraints on the physical processes involved.

2.9. Multi-Wavelength Studies of Outbursts

Recent outbursts in classical EXors and in new eruptive young stars triggered strong interest in obtaining detailed follow-up observations at all possible wavelengths. We focus on three well-studied objects — the classical EXors EX Lup

and V1118 Ori, and the recently discovered V1647 Ori — that reflect the diversity of outburst properties.

2.9.1. The prototype EX Lupi. EX Lupi is a young low-mass (0.6 $M_\odot$) M0V star (*Gras-Velázquez and Ray, 2005*), with a quiescent $L_{bol} = 0.7$ $L_\odot$. Its disk has a mass of 0.025 $M_\odot$, with an inner hole within 0.2–0.3 AU, and an outer radius of 150 AU (*Sipos et al., 2009*). There is no hint of any envelope. This prototype of the EXor class shows small-amplitude variations in quiescence, but has displayed several outbursts (*Lehmann et al., 1995; Herbig et al., 2001; Herbig, 2007*) during which the photospheric spectrum is veiled by a hot continuum, the equivalent widths of the optical emission lines decrease, inverse P Cygni absorption components appear at the higher Balmer lines, the emission-line structures undergo striking variations, and many emission lines exhibit both a narrow and a broad line profile component. All these signatures indicate mass infall in a magnetospheric accretion event. In quiescence, permitted emission lines and numerous metallic lines are detected (*Kóspál et al., 2011b; Sicilia-Aguilar et al., 2012*), likely originating from a hot (6500 K), dense, non-axisymmetric, and non-uniform accretion column that displays velocity variations along the line of sight on timescales of days. Further evidence supporting a magnetospheric accretion model is given by the spectroastrometry of near-infrared hydrogen lines (*Kóspál et al., 2011b*), which suggest a funnel flow or disk wind origin rather than an equatorial boundary layer.

A strong outburst ($\Delta V \sim 5$ mag) was intensively monitored in 2008. The source reached peak brightness in three to four weeks, remained in a high state for six months with a slight fading, and returned to its initial state within a few more weeks. The rapid recovery of EX Lup after the outburst and the similarity between the pre- and postoutburst states suggest that the geometry of the accretion channels did not change between quiescence and outburst, and only the accretion rate varied (e.g., *Sicilia-Aguilar et al., 2012*).

During the outburst, the SED indicates a hot single-temperature blackbody component that emitted 80–100% of the total accretion luminosity (*Juhász et al., 2012*). Strong correlation is also found between the decreasing optical and X-ray fluxes, while UV and soft X-rays are associated with accretion shocks (see section 4.5). From CO fundamental band emission lines, *Goto et al.* (2011) identified a broad-line component that was highly excited, and decreased as the source faded. This gas component likely orbited the star at 0.04–0.4 AU, implying that it is the inner 0.4 AU of the disk that became involved in the outburst, a region coinciding with the optically thin, but gas-rich, inner hole. Furthermore, *Juhász et al.* (2012) concluded, mainly on the basis of accretion timescales, that thermal instability was likely not the triggering mechanism of the 2008 outburst. It remains an open question whether the inner disk hole, whose radius is larger than the dust sublimation radius, is related to the eruptive phenomenon.

Spitzer spectra obtained at peak light displayed features of crystalline forsterite, while such crystals were not detected in quiescence (Fig. 3) (*Ábrahám et al.*, 2009). The crystals were likely produced through thermal annealing in the surface layer of the inner disk by heat from the outburst, a scenario that could produce raw material for primitive comets. *Juhász et al.* (2012) presented additional multi-epoch Spitzer spectra, and showed that the strength of the crystalline bands increased after the end of the outburst, but six months later the crystallinity decreased. Although vertical mixing in the disk would be a potential explanation, fast radial transport of crystals (e.g., by stellar/disk winds) was preferred. The outburst also influenced the gas chemistry of the disk: *Banzatti et al.* (2012a,b) found that the H_2O and OH-line fluxes increased; new OH, H_2, and HI transitions were detected; and organics were no longer seen.

In summary, the EX Lup outburst indicated that its enhanced accretion probably proceeded through magnetospheric accretion channels that were present also in quiescence but delivered less material onto the star.

2.9.2. The classical EXor V1118 Ori.

The classical EXor V1118 Ori has displayed several outbursts (*Herbig et al.*, 2008, and references therein). *Parsamian et al.* (2002) showed that optical spectra taken during the 1989 and 1992–1994 outbursts were similar to those of TTS with strong H and CaII lines. The star is a close binary (0."18) with similar fluxes in Hα (*Reipurth et al.*, 2007b). V1118 Ori was followed during its 2005–2006 outburst in the optical, infrared, and X-ray (Fig. 4) (*Audard et al.*, 2005, 2010; *Lorenzetti et al.*, 2006, 2007) until it returned to quiescence in 2008. The X-ray results are described in section 2.10. The optical and near-infrared emission at the peak of the outburst was dominated by a hotspot (*Audard et al.*, 2010; see also *Lorenzetti et al.*, 2012), whereas disk emission dominated in the mid-infrared. Star + disk + hot-spot models suggested that the mass-accretion rate increased from quiescence to peak of outburst from 2.5×10^{-7} to 1.0×10^{-6} $M_\odot$ yr^{-1} (*Audard et al.*, 2010), together

with a significant increase in fractional area of the hot spot. *Lorenzetti et al.* (2012) used a different approach and fitted the difference of the outburst and quiescent SEDs with a blackbody, obtaining a temperature of ~4500 K with a radius for the emitting region (assumed to be a uniform disk) of 0.01 AU. I-band polarimetry indicated that V1118 Ori is polarized at the level of 2.5%, with higher and more variable values observed in quiescence, suggesting that the spotted and magnetized photosphere can be seen once the envelope partially disappears (*Lorenzetti et al.*, 2007). Color-color diagrams showed variations but no signature of reddening caused by circumstellar matter (*Audard et al.*, 2010), consistent with the low extinction ($A_V \leq 2$) found during all activity phases by *Lorenzetti et al.* (2006). Spitzer spectra showed a slight increase in flux of the crystalline feature at the peak of the outburst compared to postoutburst, but no variation in shape that would indicate a change in grain population. From the CO bandhead emission observed during the declining phase, *Lorenzetti et al.* (2006) derived a mass loss rate of $(3-8) \times 10^{-8}$ $M_\odot$ yr^{-1}. A spectrum taken after the outburst detected no emission lines and 2.3-μm CO in absorption. *Herbig* (2008) also reported a switch from emission during outburst to absorption in quiescence for LiI and KI lines, and wind emission in Hα in the decay phase but a symmetric profile after the outburst. Strong wind losses have, thus, likely occurred only transiently.

2.9.3. A new outbursting source: V1647 Ori.

The outburst of V1647 Ori was discovered in January 2004, illuminating a surrounding nebula. Archival data showed that the source was previously in outburst for 5–20 months in 1966–1967 (*Aspin et al.*, 2006). The start of the twenty-first century outburst occurred in late 2003, leading to an increase of ~5 mag in the I-band in 4 months (*Briceño et al.*, 2004). Preoutburst data indicated a flat-spectrum source with an estimated preoutburst bolometric luminosity of ≈3.5–5.6 $L_\odot$ and age of 0.4 million years (m.y.), typical

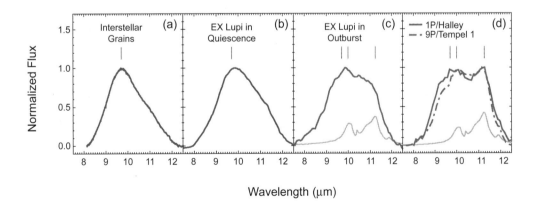

Fig. 3. Silicate emission spectra for **(a)** the interstellar medium, **(b)** EX Lup in quiescence, **(c)** Ex Lup during its 2008 outburst, and **(d)** two comets. The bottom curves in **(c)** and **(d)** show the emissivity curve of pure forsterite, for grain temperatures of 1250 K and 300 K, respectively. The outburst spectrum of EX Lup shows evidence of forsterite, not observed during quiescence, and produced through thermal annealing in the surface layer of the inner disk by heat from the outburst. Adapted from *Ábrahám et al.* (2009).

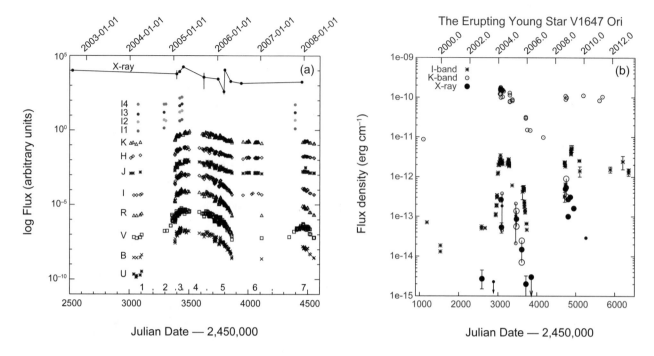

Fig. 4. Optical, infrared, and X-ray light curves of **(a)** V1118 Ori [from *Audard et al.* (2010), reproduced with permission from *Astronomy and Astrophysics,* copyright ESO] and **(b)** V1647 Ori (from M. Richmond, personal communication; adapted from *Teets et al.* (2011)]. The light curves show that X-rays follow the accretion event, albeit with a weak flux increase in V1118 Ori but a strong increase in V1647 Ori.

of CTTS (*Ábrahám et al.,* 2004a; see also *Andrews et al.,* 2004). The luminosity increased during the outburst by a factor of 10–15 (*Andrews et al.,* 2004; *Muzerolle et al.,* 2005; *Aspin,* 2011b). The peculiarity of V1647 Ori is that its SED differs from the classical EXors, and more closely resembles those of FUors, with strong extinction [A_V ~ 6 mag (*Ojha et al.,* 2005)]. The circumstellar mass, however, is typical of TTS (*Andrews et al.,* 2004; *Ábrahám et al.,* 2004a; *Tsukagoshi et al.,* 2005).

During the outburst, the near-infrared color indices of V1647 Ori shifted to bluer colors along the reddening vector, mainly due to intrinsic brightening and partly to decreasing extinction (*Reipurth and Aspin,* 2004a; *McGehee et al.,* 2004; *Acosta-Pulido et al.,* 2007). The mass-accretion rate during the outburst was estimated at a few 10^{-6} to 10^{-5} M$_\odot$ yr^{-1} (*Muzerolle et al.,* 2005; *Aspin,* 2011b; *Mosoni et al.,* 2013). CO bandhead emission was detected along with strong H and He emission lines with P Cyg profiles, indicating a strong wind with velocities up to 600 km s^{-1}, together with ice absorption from H_2O and CO and in Na I D (*Reipurth and Aspin,* 2004a; *Vacca et al.,* 2004; *Walter et al.,* 2004). *Rettig et al.* (2005) showed that the CO emission was hot (2500 K) and attributed the emission to the accretion zone of the inner disk. They also detected narrow CO absorption lines superimposed on the low-J emission lines that originated in a foreground cold (18 K) cloud. The ices are unprocessed with a temperature of ~20 K, also indicating a cloud origin. In later spectra taken in 2004–2005, evidence for a slow decline in bright-

ness was measured, including a decline of the hot, fast wind (measured from the He I 1.083-μm absorption strength), and of the temperature of the inner disk (*Gibb et al.,* 2006; *Ojha et al.,* 2006). From radio data taken in early 2005, *Vig et al.* (2006) proposed a homogeneous H II region to explain the radio, Hα, and X-ray measurements. Further X-ray results are described in section 2.10.

In late 2005, V1647 Ori suddenly returned to quiescence over a period of a few months (*Chochol et al.,* 2006; *Acosta-Pulido et al.,* 2007; *Aspin et al.,* 2008). The He I 1.083-μm line was then observed in emission, in contrast to the outburst phase (*Acosta-Pulido et al.,* 2007). Blueshifted CO absorption lines (30 km s^{-1}, 700 K) appeared in February 2006 superimposed on the previously observed CO emission lines, but were not reobserved in the spectra obtained in December 2006 and February 2007 (*Brittain et al.,* 2007; *Aspin et al.,* 2009a). This transient postoutburst outflow was possibly launched by the outward motion of the magnetospheric radius during the rapid decrease of the accretion rate. In January 2006, *Fedele et al.* (2007) also detected forbidden emission lines indicative of shocked gas likely produced by a HH-like object driven by V1647 Ori.

Mid-infrared interferometric data obtained during outburst revealed that the emitting region was extended (7 AU at 10 μm), no close companion could be seen, and the 8–13-μm spectrum exhibited no strong spectral features (*Ábrahám et al.,* 2006). Using a disk + envelope geometry and varying the mass-accretion rate from (1.6–7) × 10^{-6} M$_\odot$ yr^{-1}, *Mosoni et al.* (2013) reproduced the SEDs taken at different epochs during the outburst. The models suggested an *increase* in the inner radius of the disk/envelope at the beginning of the eruption. However, the system was more resolved at the later epoch when the outbursting region was already shrinking, indicating either that the envelope inner radius suddenly increased at late stages of the outburst, or that the fading of the central source was not immediately followed by the fading of the outer regions.

V1647 Ori returned to the spotlight when a second outburst was reported in August–September 2008 (*Aspin et al.,* 2009b). The source photometry, accretion rate, luminosity, and morphology were similar to those seen after the onset of the previous outburst. However, CO overtone emission was not detected, despite being seen shortly after the onset of the 2003 outburst. *Aspin et al.* (2009b) concluded that the quiescent period between these two outbursts was due to a partial decline in the accretion rate (factor of 2–3) and reformation of dust in the immediate circumstellar environment. They argued that the 2008 outburst was caused by the combination of enhanced accretion and sublimation of dust due to this energetic event. In the high state of the second outburst, the Hα and Ca II lines did not change remarkably (*Aspin,* 2011b). The CO overtone bandhead was still not detected, while the water vapor absorption remained strong. *Brittain et al.* (2010) further found that the temperature of the CO emission had varied with the accretion rate, and showed a direct relation between the Brγ luminosity and line full width at half maximum and the source brightness, indicating that the accretion rate had driven the variability.

In view of the duration of the outburst, its recurrence, and the various spectroscopic features observed before, during, and after the outburst, V1647 Ori does not fit well either with classical EXors or FUors; it may instead be an intermediate case. In fact, *Aspin et al.* (2009a) presented a spectrum of V1647 Ori, taken in quiescence, that does not resemble those of late-type standards, class I protostars, or EXors observed in quiescence, but does show considerable correspondence with several classical FUors observed *during* their outbursts. Given that V1647 Ori shares some characteristics of both FUors and EXors, they proposed the existence of a "continuum" of eruption characteristics rather than two distinct classes.

2.10. High-Energy Processes

Strong X-ray emission is characteristic of young stars (*Feigelson and Montmerle,* 1999), but our knowledge of the high-energy behavior of eruptive young stars is limited to just a few examples observed over the past decade or so. X-rays probe high-energy processes, such as coronal activity and accretion shocks, and they are an important source of ionization and heating of accretion disk atmospheres, and thus may influence disk chemistry and strengthen the coupling of disk gas to the stellar magnetic field.

2.10.1. X-rays from classical FUors. The first X-ray spectrum of FU Ori obtained with the X-Ray Multi-Mirror Mission (XMM)-Newton appeared quite unusual. The emission consisted of a hot plasma viewed through very high absorption — much higher than anticipated based on A_V — and a cooler component seen through 10× lower absorption (*Skinner et al.,* 2006). Subarcsecond X-ray imaging by Chandra subsequently provided an explanation for the sharply contrasting absorbing columns (*Skinner et al.,* 2010): The high-temperature component was positionally coincident with FU Ori, while the centroid of the cooler component was offset from FU Ori and displaced toward the infrared companion FU Ori S. The "excess" absorption toward FU Ori is likely a result of accreting gas, FU Ori's powerful wind, or both. Time variability in the hot component — implying a magnetic origin — was detected, but no such variability was seen in the fainter, less-absorbed, cooler component. The classical FUor V1735 Cyg was bright (log $L_X \approx 31.0$ ergs s⁻¹) and displayed very hot (T > 70 MK), heavily absorbed plasma but no cool plasma, possibly because the latter is more susceptible to absorption by intervening cool gas. In contrast, V1057 Cyg and V1515 Cyg both escaped detection (*Skinner et al.,* 2009). Interestingly, a faint soft X-ray jet was detected with Chandra in Z CMa after the outburst (*Stelzer et al.,* 2009), with a position angle consistent with that of the microjet launched by the FUor component of this close binary (*Whelan et al.,* 2010). The jet was not detected in a preoutburst observation, suggesting that mass ejection occurred. Clearly more observations are needed to characterize the X-ray properties of the FUor class as a whole.

2.10.2. X-rays from classical EXors. The eruptions of V1118 Ori in 2005–2006 and EX Lup in 2008 were both monitored in X-rays. *Teets et al.* (2012) followed the EX Lup outburst beginning two months after the outburst onset until just after its conclusion, while *Audard et al.* (2005, 2010) and *Lorenzetti et al.* (2006) followed V1118 Ori until it returned in quiescence in 2008 (Fig. 4). In both outbursts, there were strong correlations between the decreasing optical and X-ray fluxes following the peak of the optical outburst, suggesting a relation with the accretion rate. However, in contrast to V1647 Ori (see below), the X-ray flux increased only mildly in both cases, with a decrease in flux after outburst, and relatively cool plasma temperatures (5−8 MK) were observed. The temperature in V1118 Ori contrasted with a serendipitous preoutburst observation in 2002 that showed a dominant 25-MK plasma (*Audard et al.,* 2005). The plasma temperature then gradually returned to higher values in later phases of the outburst (*Audard et al.,* 2010), in similar fashion to EX Lup (*Teets et al.,* 2012) — although the cool plasma was still detected before the end of the 2008 EX Lup outburst (*Grosso et al.,* 2010).

One possible explanation for the anticorrelation between optical/infrared flux and X-ray temperature observed for both V1118 Ori and EX Lup is that their soft X-ray outputs were enhanced during eruption, due to emission arising in accretion shocks (along with possible changes in magnetospheric configurations). In the case of EX Lup, this hypothesis is consistent with the strong variability in

UV emission detected by XMM-Newton, which was likely due to accretion spots; in addition, the X-ray emission at energies above 1.5 keV showed far stronger photoelectric absorption than the cool plasma, suggesting the coronal emission was subject to absorption by overlying, high-density gas in accretion streams (*Grosso et al.*, 2010).

2.10.3. X-rays from V1647 Ori. *Kastner et al.* (2004, 2006) followed the entire 2003–2005 duration of V1647 Ori's outburst with Chandra, including a serendipitous preoutburst observation. The X-ray flux from V1647 Ori closely tracked the rise and fall of its optical-infrared flux (Fig. 4), thereby providing definitive evidence for the production of high-energy emission via star-disk interactions. The monitoring data also indicated hardening of the X-ray emission during outburst. Further monitoring of V1647 Ori in 2008–2009 — i.e., soon after the onset of its second outburst — established the striking similarity between the two outbursts in terms of X-ray/near-infrared flux ratio and X-ray spectral hardness (*Teets et al.*, 2011).

Deep XMM-Newton and Suzaku exposures were obtained during V1647 Ori's first and second outbursts, respectively (*Grosso et al.*, 2005; *Hamaguchi et al.*, 2010). The observations showed warm (10 MK) and hot (50 MK) plasmas, the former indicative of high-density plasma associated with small magnetic loops around coronally active stars (*Preibisch et al.*, 2005), the latter consistent with magnetic reconnection events. Strong fluorescent Fe Kα emission was detected, providing evidence for irradiation of neutral material residing either in the circumstellar disk or at the stellar surface. The comparison of A_V ($\approx$10 mag) with the hydrogen column density ($N_H \sim 4 \times 10^{22}$ cm^{-2}) points out a significant excess of X-ray absorption from relatively dust-free material.

Via time-series analysis, *Hamaguchi et al.* (2012) established that the puzzling short-term (hours-timescale) X-ray variability of V1647 Ori was due to rotational modulation. The ~1-d X-ray period corresponds to rotation at near-breakup rotation speed. *Hamaguchi et al.* (2012) further demonstrated that a model consisting of two pancake-shaped hot spots of high plasma density ($\geq 5 \times 10^{10}$ cm^{-3}), located at or near the stellar surface, reproduced well the X-ray rotational modulation signature. Under the assumption that the hot spots are generated via magnetic reconnection activity, the stability of the X-ray light curve during the course of two accretion-related outbursts further suggests that the star-disk magnetic configuration of V1647 Ori has remained relatively unchanged over timescales of years.

3. THEORY

3.1. Spectral Energy Distribution Fits and Accretion Disk Origin

Spectral energy distribution fitting based on theoretical models is a powerful tool in understanding the origin of these outbursting objects. The most successful model to explain the peculiarities of FUors was proposed by *Hartmann*

and Kenyon (1985, 1987a,b) and *Kenyon et al.* (1988), who suggested that the SED was dominated by a protostellar disk accreting at a high rate ($\dot{M} \approx 10^{-5}$–10^{-4} M$_\odot$ yr^{-1}). Within this picture, it is easy to account for the main peculiarities observed in FUor spectra, such as the changing supergiant spectral types from optical to infrared, and the double-peaked line profiles, with peak separation decreasing with increasing wavelength (*Hartmann and Kenyon*, 1996).

Self-consistent disk atmospheric models including the accretion process and the radiative transfer in the disk atmosphere are needed to constrain the SED and, in particular, disk properties (e.g., *Calvet et al.*, 1991a,b; *Popham et al.*, 1996). *Zhu et al.* (2007, 2008, 2009c) extended the first models with a more complete opacity database (*Kurucz et al.*, 1974). This simple steady accretion disk model could fit FU Ori's SED (Fig. 5) (*Zhu et al.*, 2007, 2008), and confirmed the Keplerian rotation of FU Ori's disk (*Zhu et al.*, 2009a). The derived size of the high-accretion-rate inner disk was from 5 R$_\odot$ to ~0.5–1 AU, beyond which is a passively heated outer disk. The outer value is consistent with the hot disk size found by *Eisner and Hillenbrand* (2011). Incidentally, such a size leads to a viscosity parameter α ~ 0.02–0.2 in outburst (*Zhu et al.*, 2007).

Different from this traditional steady viscous disk approach, *Lodato and Bertin* (2001) argued that FUor disks might be gravitationally unstable. The radial structure of such a gravitationally unstable disk should be significantly different from a non-self-gravitating one. They contended that the observed flat SED in the infrared might be related to a combination of extra heating induced by nonlocal dissipation of large-scale spiral structures and by the possible

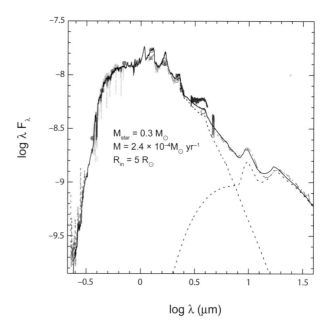

$M_{star} = 0.3$ M$_\odot$
$M = 2.4 \times 10^{-4}$ M$_\odot$ yr^{-1}
$R_{in} = 5$ R$_\odot$

Fig. 5. Observed SED of FU Ori (solid line) fitted with a steady accretion disk model with contributions from the inner hot disk (light dotted curve) and the flared outer disk (light dashed curve). Adapted from *Zhu et al.* (2008).

departure from a purely Keplerian rotation, if the disk mass is a sizeable fraction of the central object (see also *Adams et al., 1988; Bertin and Lodato, 1999*). *Lodato and Bertin* (2003) then investigated how such effects would modify the shape of the double-peaked line profiles.

3.2. Origin of the Outburst

A closely related question for the disk-accretion model pertains to the origin of the outburst. Two main schools of thought have been proposed to explain the triggering of FUor outbursts: (1) disk instability and (2) perturbation of the disk by an external body.

Thermal instability is due to the thermal runaway process when the disk temperature reaches the hydrogen ionization temperature. In detail, a hydrostatic viscous disk can be thermally stable only if the opacity changes slowly with the temperature. When the disk temperature reaches the hydrogen ionization temperature, the opacity increases dramatically with the temperature ($\kappa \sim T^{10}$). A slight temperature increase will cause a significant amount of energy trapping in the disk and the disk temperature continues to rise, which leads to the thermal runaway. The latter ends when the disk is fully ionized at $\sim 10^5$ K and the opacity changes slowly with temperature again. After the thermal runaway, a high disk temperature can lead to a high disk-accretion rate since the viscosity ν is proportional to the disk temperature ($\nu = \alpha c_s^2/\Omega \propto \alpha T/\Omega$, with c_s the sound speed, α the viscosity parameter, and Ω the angular velocity).

The thermal instability model can naturally explain the fast rise time of FU Ori. However, since thermal instability needs to be triggered at $T \sim 5000$ K, the outbursting disk is small with a size of ~ 20 R$_\odot$. In order to maintain the outburst for hundreds of years, the α parameter needs to be as low as 10^{-3} during the outburst. Furthermore, in order to produce enough accretion rate contrast before and during the outburst, α needs to be 10^{-4} in the quiescent state (*Bell and Lin, 1994; Bell et al., 1995*). Two-dimensional radiation hydrodynamic simulations have been carried out to study the triggering of thermal instability in disks and the thermal structure of the boundary layer (*Kley and Lin, 1996, 1999; Okuda et al., 1997*). The SED fitting for FU Ori does not require an additional hot boundary layer (Fig. 5), which may suggest that the boundary layer can be different between FUors and TTS (*Popham, 1996*).

Several models have been put forward to cope with the limitations of the standard thermal instability model. These include variations on the thermal instability model itself, or an altogether new trigger mechanism, for example, associated with the onset of the magneto-rotational or gravitational instability. They are discussed in detail below. A comparison of the light curves produced by these different models can be found in Figs. 6 and 7, which show their long-term evolution (over a period of thousands of years) and the detailed evolution of a single outburst, respectively.

3.2.1. Thermal instabilities induced by a planet. In thermal instability models, it has been noted that, in the

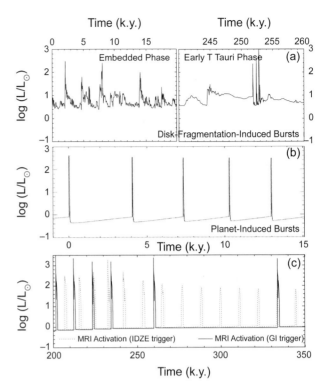

Fig. 6. **(a)** Total (accretion plus photospheric) luminosity in the disk fragmentation model showing typical outbursts in the embedded phase of star formation (left) and in the early T Tauri phase (right). **(b)** Bolometric light curves for the planet-induced thermal instability model. Within this model the recurrence time between outbursts is reduced as time increases. **(c)** Total luminosity for the MRI thermally activation models. The solid curves are from models where MRI is triggered by GI, while the dotted curves are from models where MRI is activated at the inner dead zone edge (IDZE) due to a nonzero dead zone viscosity.

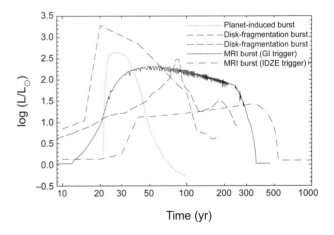

Fig. 7. Time evolution of individual luminosity outbursts in different burst-triggering models. The zero-time is chosen arbitrarily to highlight distinct models. The two distinct outburst curves in the disk fragmentation model stem from the state of the fragment when accreted onto the star. Tidally smeared fragments give rise to a slowly rising curve (predominantly, in the embedded phase), while a sharply rising curve is produced by weakly perturbed fragments in the early T Tauri phase.

absence of a trigger, the instability would first occur in the innermost parts of the disk and then propagate inside-out, in a "snowplow" fashion (*Lin et al.,* 1985), decelerating as it propagates out and thus leading to a slow rise in the light curve (as observed in the case of V1515 Cyg). A fast rise (as observed in FU Ori and V1057 Cyg) is produced only if the outburst is "triggered" somehow far from the disk inner edge, so that the instability can propagate outside-in, in an "avalanche" fashion. *Clarke et al.* (1990) have confirmed this behavior by including an *ad hoc* density perturbation to produce triggered outbursts. The latter approach has been developed initially by *Clarke and Syer* (1996) and then by *Lodato and Clarke* (2004), who considered outbursts triggered by a massive planet. During quiescence the planet opens up a gap in the disk. For sufficiently massive planets, the resulting type II migration is slow and the inner disk is rapidly emptied out (in what would resemble a transitional disk), leading to a steepening of the density profile in the outer disk, that can trigger the thermal-viscous instability at the outer gap edge. Observationally, there are a number of clues that indicate the presence of a planet in FUor disks. *Clarke and Armitage* (2003) have suggested that a planet embedded in the disk of a FUor would lead to a clear spectroscopic signature in the form of a periodic modulation of the double-peaked line profiles on the orbital period of the planet. Such periodic modulations of the line profiles have been observed only for the fast-rise FUors (FU Ori and V1057 Cyg), with a period of ~3 d (*Herbig et al.,* 2003), corresponding to a semimajor axis of 7–10 $R_\odot$ (see also *Powell et al.,* 2012).

The results of *Lodato and Clarke* (2004) confirm that a planet with a mass of a few Jupiter masses can produce a fast-rise outburst (with a rise time on the order of one year, as observed for FU Ori and V1057 Cyg). The issue with this class of models is the duration of the outburst, predicted to be on the order of ~50 yr by such models, with a relatively shallow dependence on the planet mass. This timescale is set by the viscous time on the ionized branch of the stability curve at the outermost location reached by the instability front that is generally confined within ~50 $R_\odot$, thus making this timescale too small, unless α is assumed to be unrealistically low in the upper branch (α ~ 10^{-4}).

Clarke et al. (2005) compared planet-triggered outburst models to a long-term monitoring campaign of FU Ori and V1057 Cyg optical light curves. The time-dependent models were able to match the color evolution of the light curves much better than a simple series of steady-state models with varying $\dot{M}$. *Clarke et al.* (2005) also discussed the sudden luminosity dips observed for V1057 Cyg at the end of the outburst and for V1515 Cyg, while FU Ori does not appear to show any such behavior. This nonperiodic variability was interpreted as a consequence of the interaction of a disk wind with the infalling envelope. Numerical simulations show that powerful winds are able to push away the infalling envelope to large radii clearing up our view to the inner disk, while for less powerful ones the envelope "crushes down" the wind occulting the inner disk. If the

wind strength is proportional to the accretion rate in the inner disk, such models can explain the observed behavior.

The interaction between the disk and a planet can be much more complex if there is mass transfer between the planet and the star. *Nayakshin and Lodato* (2013) considered the case of a few-Jupiter-mass inflated planet, migrating within a disk. When the planet opens up a gap, mass transfer between the planet and the star leads to the planet losing mass but retaining angular momentum, thus migrating out and switching off the mass transfer. Conversely, if the gap is only partially open, a runaway configuration can occur where the planet migrates further in, enhancing the mass loss rate and giving rise to a powerful flare. This mechanism gives rise to variability on several timescales and of different amplitudes depending on parameters.

3.2.2. A combination of gravitational instability and magneto-rotational instability. Both magneto-rotational instability (MRI) (*Balbus and Hawley,* 1998, and references therein) (see also the accompanying chapter by Turner et al. in this volume) and gravitational instability (GI) (see *Durisen et al.,* 2007, and references therein) are promising mechanisms to transfer angular momentum in disks. However, MRI only operates when the disk reaches enough ionization (i.e., in the hotter parts of the disk), and GI only operates when the disk is sufficiently cold to become gravitationally unstable, according to the classical criterion (*Toomre,* 1964)

$$Q = \frac{c_s \kappa}{\pi G \Sigma} \approx \frac{c_s \Omega}{\pi G \Sigma} \lesssim 1 \tag{1}$$

where Q is the Toomre stability criterion, Σ is the surface density, and the epicyclic frequency κ can be approximated with the angular frequency Ω for a quasi-Keplerian disk. A realistic protoplanetary disk is likely to have a complicated accretion structure: The inner disk is MRI-active due to the thermal ionization, the region between ~1 AU to tens of astronomical units has a layered accretion structure with a MRI stable dead zone, and the outer region can be MRI-active due to cosmic/X-ray ionization. The outer disk can also be gravitationally unstable when the disk is at the infall phase with a significant amount of mass loading.

This complicated structure is unlikely to maintain a steady accretion (*Gammie,* 1996; *Armitage et al.,* 2001; *Zhu et al.,* 2009b, 2010a; *Martin et al.,* 2012a,b). The outer disk transfers mass to the inner disk either due to GI or due to the layered accretion. With more and more mass shoveled to the inner disk, the inner disk becomes gravitationally unstable and continues to transfer mass to small radii. Gravitational instability can also heat up the disk. Since the electron fraction in protoplanetary disks depends nearly exponentially on temperature due to the collisional ionization of potassium (*Gammie,* 1996; *Umebayashi,* 1983), the gaseous disk will be well coupled to the magnetic field when T ~ 1000 K and MRI starts to operate. This sudden activation of MRI leads to disk outbursts. During the outburst, a thermal instability should also be triggered at the inner disk as a byproduct. Two-dimensional radiation hydrodynamic simulations have been carried out to verify such mechanisms (*Zhu et al.*, 2009b).

This MRI + GI mechanism shares similarities with the traditional thermal instability mechanism. For a given disk surface density, the disk has two possible structures: (1) MRI-inactive, and (2) MRI-active. The switch from one to another occurs at $T \sim 1000$ K and leads to outbursts (*Martin and Lubow*, 2011, 2013). In the MRI + GI picture, GI triggers the switch. However, if there are other mechanisms to heat up the disk and trigger the switch, the disk can also experience outbursts. For example, *Bae et al.* (2013) found that the radial energy diffusion from the inner MRI-active disk to the dead zone can trigger the switch and lead to outbursts, although these outbursts are weaker and shorter than the MRI + GI case. This may have implications for weaker FUors or EXors.

After the envelope infall stage, the layered accretion can still pile up mass in the dead zone and leads to disk outbursts (*Zhu et al.*, 2010b), which is consistent with the fact that some FUors are in the T Tauri phase. However, this massive dead zone would persist throughout the whole T Tauri phase and should be easily observed by the Atacama Large Millimeter/submillimeter Array (ALMA).

3.2.3. Disk fragmentation. Another accretion burst mechanism is related to the phenomenon of disk gravitational fragmentation and subsequent inward migration of the fragments onto the protostar. Observations of jets and outflows suggest that the formation of protostellar disks often occurs in the very early stage of star formation, when a protostar is deeply embedded in a parental cloud core.

The forming disk becomes gravitationally unstable, if the Q parameter drops below unity. A higher initial mass and angular momentum of the parental core both favor the formation of disks with stronger gravitational instability (*Kratter et al.*, 2008; *Vorobyov*, 2011). For relatively long cooling timescales (above a few dynamical timescales), the outcome of the instability is to produce a persistent spiral structure (*Gammie*, 2001; *Lodato and Rice*, 2004, 2005; *Mejía et al.*, 2005; *Boley et al.*, 2006), able to transport angular momentum efficiently through the disk (*Cossins et al.*, 2009, 2010) and trigger the MRI burst (section 3.2.2).

In the opposite regime, where the cooling timescale is comparable to or shorter than the dynamical timescale, these disks can experience gravitational fragmentation and form bound fragments with mass ranging from giant planets to very-low-mass stars (*Gammie*, 2001; *Johnson and Gammie*, 2003; *Rice et al.*, 2005; *Stamatellos and Whitworth*, 2009a,b; *Vorobyov and Basu*, 2010; *Vorobyov*, 2013).

The likelihood of survival of the fragments is, however, low. Gravitational instability in the embedded phase is strong, fueled with a continuing infall of gas from a parent cloud core, and resultant gravitational and tidal torques are rampant. As a consequence, the fragments tend to be driven into the inner disk and likely onto the protostar due to the loss of angular momentum caused by the gravitational interaction with the trailing spiral arms (*Vorobyov and*

Basu, 2006, 2010; *Cha and Nayakshin*, 2011; *Machida et al.*, 2011). The infall of the fragments can trigger mass accretion bursts and associated luminosity outbursts that are similar in magnitude to FUor or EXor events (*Vorobyov and Basu*, 2005, 2006, 2010), during which a notable fraction of the protostellar mass can be accumulated (*Dunham and Vorobyov*, 2012). The protostellar accretion pattern in the disk fragmentation models is highly variable (*Vorobyov*, 2009) and is characterized by short-duration ($\lesssim 100-200$ yr) bursts with $\dot{M} \gtrsim$ a few $\times 10^{-5}$ $M_\odot$ yr^{-1} alternated with longer (10^3-10^4 yr) quiescent periods with $\dot{M} \lesssim 10^{-6}$ $M_\odot$ yr^{-1}. The most intense accretion bursts can reach 10^{-3} $M_\odot$ yr^{-1}.

This burst mechanism is most effective in the embedded stage of disk evolution and diminishes once the parental core has accreted most of its mass reservoir onto the protostar + disk system. Only occasional bursts associated with the fragments that happened to survive through the embedded stage can occur in the T Tauri stage. Moreover, the initial conditions in the parental core set the likelihood of disk fragmentation and the burst occurrence; the relevant rotational energy vs. initial core mass diagram is provided in *Basu and Vorobyov* (2012) and *Vorobyov* (2013). Magnetic fields and strong background irradiation both act to suppress disk fragmentation and associated accretion bursts, but are unlikely to quench this phenomenon completely (*Vorobyov and Basu*, 2006, 2010).

3.2.4. External environment triggers. The accretion and luminosity burst mechanisms considered in the previous sections are induced by "internal" triggers (*Tassis and Mouschovias*, 2005). These triggers depend on the physical conditions in the system and if they are suppressed, the protostar is likely to accrete in a quiescent manner. Nevertheless, accretion bursts can still be induced in quiescent systems by the so-called "external" triggers, of which a close encounter in a binary system was the first proposed candidate (*Bonnell and Bastien*, 1992). *Reipurth and Aspin* (2004b) considered the case where dynamical interactions in small-N systems might lead to close encounters. Starting from the fact that a couple of FUors are found to be in a binary system where both stars are FUor, they argued that whatever triggered the FUor eruption in one component is likely to also have triggered the eruption in the other component, and identified the breakup of a small multiple system as a natural common precursor event. Indeed, dynamical interaction in small-N systems can easily result in the formation of a close binary. *Reipurth and Aspin* (2004b) then argued that the decay of the binary due to interaction with a circumbinary disk may lead to triggering of instabilities in the individual circumstellar disks. However, a quantitative modeling of such processes is still lacking.

The idea of external triggering received further development by *Pfalzner et al.* (2008) and *Pfalzner* (2008), who considered accretion processes in young and dense stellar clusters, choosing the Orion nebular cluster as representative. They combined cluster simulations performed with the N-body code and particle simulations that described the effect of a close passage on the disk around a young star to determine the induced mass accretion. They concluded that the close encounters can reproduce the accretion bursts typical for the FUor events, albeit with certain limitations.

Close encounters with nearby stars have also been considered as a way to induce fragmentation in a disk that is gravitationally stable, but not fragmenting in isolation (*Mayer et al.,* 2005), although the effectiveness of such process is probably limited (*Lodato et al.,* 2007). However, the short rise times are difficult to achieve unless the matter is stored somewhere close to the protostar and the inferred duration of FU Ori events requires a rather high mass ratio between the intruder and the target.

3.3. Episodic Accretion and Its Implications

Episodic accretion has several key implications for star and planet formation and evolution, and these are described in detail below.

3.3.1. Resolving the luminosity problem in embedded sources. In one of the first statistically significant studies of the luminosities of embedded protostars, *Kenyon et al.* (1990, 1994) and *Kenyon and Hartmann* (1995) found that the observed luminosities of 23 protostars in Taurus were much lower than expected from simple theoretical predictions. Current samples of protostars with accurately determined luminosities measure in the hundreds (e.g., *Evans et al.,* 2009; *Enoch et al.,* 2009; *Kryukova et al.,* 2012; *Dunham et al.,* 2013) (see also the chapter by Dunham et al. in this volume). While the details of these studies differ, they all confirm the existence of the luminosity problem and aggravate it by showing that the luminosity distribution of protostars extends to even lower luminosities than found by the *Kenyon et al.* (1990, 1994) surveys.

As originally proposed by *Kenyon and Hartmann* (1995), one possible resolution to the "luminosity problem" is episodic accretion. If a significant fraction of the total accretion onto a star occurs in short-lived bursts, most protostars will be observed during periods of accretion below the mean accretion rate, thus most protostars will emit less accretion luminosity than expected assuming constant accretion at the mean rate. To test the ability and necessity of episodic accretion for resolving the luminosity problem, *Dunham et al.* (2010) modified existing evolutionary models of collapsing cores first published by *Young and Evans* (2005) and showed that a very simple implementation of episodic accretion into the accretion model was capable of resolving the luminosity problem and matching the observed protostellar luminosity distribution. *Offner and McKee* (2011) also found that episodic accretion can contribute a significant fraction of the stellar mass. *Dunham and Vorobyov* (2012) revisited this topic with more sophisticated evolutionary models incorporating the exact time evolution of the accretion process predicted by hydrodynamical simulations, and confirmed these findings. However, these results only demonstrate that episodic accretion is a *possible* solution to the luminosity problem.

3.3.2. Impact on disk fragmentation. Stellar irradiation is known to have a profound impact on the gravitational

stability of protostellar disks (*Stamatellos and Whitworth,* 2009a,b; *Offner et al.,* 2009, *Rice et al.,* 2011). Flared disks can intercept a notable fraction of the stellar UV and X-ray flux, which first heats dust and then gas via collisions with dust grains. The net effect is an increase in the gas/dust temperature leading to disk stabilization. X-rays can also ionize gas. Smoothed particle hydrodynamics (SPH) simulations by *Stamatellos et al.* (2011, 2012) demonstrate that quiescent periods between the luminosity bursts can be sufficiently long for the disk to cool and fragment. Thus, episodic accretion can enable the formation of low-mass stars, brown dwarfs, and planetary mass objects by disk fragmentation, as confirmed by the recent numerical hydrodynamical simulations presented by *Vorobyov* (2013).

3.3.3. Luminosity spread in young clusters. The luminosity spread observed in Hertzsprung-Russell (HR) diagrams of young star-forming regions is a well-known feature (*Hillenbrand,* 2009), but its origin is uncertain. It can be attributed to observational uncertainties, significant age spread, or yet unknown physical processes. As *Baraffe et al.* (2009) demonstrated, protostellar evolutionary models assuming episodic accretion are able to reproduce the observed luminosity spread for objects with final ages of a few million years. In contrast, non-accreting models require an age spread of at least 10 m.y. *Baraffe et al.* (2009) thus concluded that the observed HR luminosity spread does not stem entirely from an age spread, but rather from the impact of episodic accretion on the physical properties of the protostar. This idea was recently called into question by *Hosokawa et al.* (2011), who argued that accretion variability had little effect on the evolution of low-mass protostars with effective temperature below 4000 K.

Baraffe et al. (2012) noted that *Hosokawa et al.* (2011) considered models with the initial protostellar "seed" mass of 10 M_{Jup}. Having provided justification for a smaller initial mass of protostars on the order of 1.0 M_{Jup}, *Baraffe et al.* (2012) showed that the luminosity spread in the HR diagram and the inferred properties of FU Ori events (stellar radius, accretion rate) can both be explained by the "hybrid" accretion scenario with variable accretion histories derived from disk-fragmentation models (section 3.2.2). In this accretion scenario, a protostar absorbs no accretion energy below a threshold accretion rate of 10^{-5} $M_{\odot}$ yr^{-1} and 20% of the accretion energy above this value.

Several important implications of protostellar evolutionary models with variable accretion were emphasized by *Baraffe et al.* (2012). First, each protostar/brown dwarf experiences its own accretion history and ends up randomly in the HR diagram at the end of the accretion phase. It is likely that the concept of a birthline does not apply to low-mass (<1.0 $M_{\odot}$) objects. Moreover, age determinations from "standard" non-accreting isochrones may overestimate the age of young protostars by a factor of several or more. Finally, inferring masses from the HR diagram using isochrones of non-accreting protostars can yield severely incorrect determinations, possibly overestimating the mass by as much as 40% or more.

3.3.4. Impact on chemistry. Large changes in the accretion luminosity of young stars due to episodic accretion can drive significant chemical changes in the surrounding core and disk. Several authors have published chemical models exploring these effects and have shown that the CO ice evaporates into the gas phase in the surrounding envelope during episodes of increased luminosity, affecting the abundances of many other species through chemical reactions (e.g., *Lee, 2007*; *Visser and Bergin, 2012*; *Vorobyov et al., 2013*). Many of these effects can endure long after an accretion burst has subsided, leading to non-equilibrium chemistry compared to that expected from the currently observed protostellar luminosity. In particular, the abundance of gas-phase CO in the envelope can be used as an indicator of past accretion bursts and perhaps even a means of measuring the time since the last burst (Fig. 8) (*Vorobyov et al., 2013*). Episodic accretion bursts can also affect the abundances and chemical compositions of various molecular ices frozen onto dust grains. *Kim et al.* (2012) showed that a chemical evolutionary model including episodic accretion could provide the necessary thermal processing and match the 15.2-μm CO_2 ice absorption features of low-luminosity protostars.

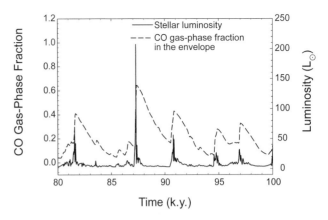

Fig. 8. Predicted CO gas-phase fraction ξ_{CO} (dashed lines) and total stellar luminosity L_* (solid lines) vs. time elapsed since the formation of the protostar. The correlation between ξ_{CO} and L_* is evident. In particular, ξ_{CO} steeply rises during the burst to a maximum value and gradually declines to a minimum value after the burst. The relaxation time to the preburst stage is notably longer than the burst duration. Adapted from *Vorobyov et al.* (2013).

3.4. Disk-Magnetosphere Interactions

If a protostar has a strong magnetic field, then the disk magnetosphere interaction can lead to episodic accretion, variabilities at different timescales, and outflows. In most cases, the magnetic fields of protostars are not known. However, the measurements of the field in several CTTS indicate the presence of a magnetic field of a few kilogauss (e.g., *Johns-Krull, 2007*; *Donati et. al., 2008*). The possible presence of a 1-kG magnetic field in 0.05 AU of FU Ori was reported by *Donati et al.* (2005), and *Green et al.* (2013a) estimated the magnetic field of HBC 722 to be 2.2–2.7 kG. There is also indirect evidence of a magnetic field from the X-ray emissions of FUors and EXors. If a star has a magnetic field of a few kilogauss, then the disk will be truncated by the magnetosphere at distance r_m, where the magnetic stress is equal to the matter stress in the disk (*Königl, 1991*). In EXors, this radius may be as large as a few stellar radii, while in FUors it can be smaller than one stellar radius, or the field may be buried. Numerical simulations predict that even in the case of a tiny magnetosphere, the magnetic flux of the star is not buried, but rather partially inflated into the corona, and may drive strong outflows (*Lii et al., 2012*).

3.4.1. Cyclic accretion in a weak "propeller" regime. A model of cyclic accretion was proposed for stars accreting in a weak propeller regime. In the propeller regime, the magnetosphere rotates more rapidly than the inner disk, and matter of the disk can be ejected to outflows by a rapidly rotating magnetosphere (*Illarionov and Sunyaev, 1975*; *Lovelace et al., 1999*). However, in a weak propeller regime, the magnetospheric radius, r_m, is only slightly larger than the corotation radius, r_c (where the angular velocity of the Keplerian disk is equal to the angular velocity of the star),

and such a weak propeller cannot drive outflows. Instead, the star transfers its excess angular momentum to the disk, matter accumulates in the disk for a long period of time, and a "dead disk" is formed; then, part of this disk matter accretes to the star, the magnetosphere expands, and the process repeats in a cyclic fashion (*Baan, 1977*; *Sunyaev and Shakura, 1977*; *Spruit and Taam, 1993*). The timescale of accretion episodes is determined by the accretion rate in the disk and other parameters. *D'Angelo and Spruit* (2010, 2012) investigated this model for a wide range of parameters at which accretion is either cyclic or steady and showed that the model can explain EXor outbursts.

3.4.2. Outflows from the disk-magnetosphere boundary. Different models have been proposed to explain high-velocity winds of FUors and EXors (see section 4.4), including the disk-winds model, where matter is driven by centrifugal force along the inclined field lines of the disk (e.g., *Blandford and Payne, 1982*; *Zanni et al., 2007*); accretion-powered stellar winds (e.g., *Matt and Pudritz, 2005*); and X-winds, which are launched centrifugally from the disk-magnetosphere boundary (*Shu et al., 1994*). The reader is encouraged to read the chapter by Frank et al. in this volume.

3.4.2.1. Conical winds: Recent numerical simulations show a new type of wind that can be important in the cases of EXors and FUors. These winds form at the disk magnetosphere boundary during the episodes of high accretion rate. The newly incoming matter compresses the magnetosphere of the star, the field lines inflate due to differential rotation between the disk and the star, and conically shaped winds flow from the inner disk (*Romanova et al., 2009*; *Kurosawa and Romanova, 2012*). These winds are driven by the magnetic force, $F_M \sim -\nabla(rB_\phi)^2$, which arises due to the wrapping of the field lines above the disk (*Lovelace et al.,* 1991). They are also gradually collimated by the magnetic hoop-stress, and can be strongly collimated for high accretion rates (*Lii et al.,* 2012). Moreover, the star can rotate much more slowly than the inner disk (at $r_m \ll r_c$). This is different from the X-winds, which require the condition of $r_m \approx r_c$. Conical winds appear during a burst of accretion and continue for the entire duration of the burst. A magnetic field of a few kilogauss is required for FUors, while in EXors the field can be weaker. The conical wind model was compared with the empirical model based on the spectral analysis of the winds in FU Ori (*Calvet et al.,* 1993; *Hartmann and Calvet,* 1995). A reasonably good agreement was achieved between these models (*Königl et al.,* 2011).

3.4.2.2. *Propeller-driven winds:* If a protostar rotates much more rapidly than the inner disk (strong propeller regime) and the accretion rate in the disk is relatively high, then a significant part of the disk can be redirected to the outflows by the rapidly rotating magnetosphere (*Romanova et al.,* 2005, 2009; *Ustyugova et al.,* 2006). In this regime, accretion and outflows also occur in cycles, where matter accumulates in the inner disk, diffuses through the field lines of the rapidly rotating magnetosphere, and is ejected to the outflows; the magnetosphere then expands and the cycle repeats (*Goodson et al.,* 1997; *Lii et al.,* 2013). The timescale of the cycle varies from a few weeks to a few months, and depends on a number of parameters, such as the accretion rate in the disk and the diffusivity at the disk-magnetosphere boundary.

In the case of a rapidly rotating star, outflows have a second component: a magnetically dominated, low-density, high-velocity jet, where matter is accelerated rapidly by the magnetic force that appears due to the winding of the stellar field lines (a "magnetic tower"). The jet carries significant energy and angular momentum from the star to the corona, causing the star to spin down rapidly (e.g., *Romanova et al.,* 2005). If protostars accrete most of the mass during episodes of enhanced accretion, then powerful outflows observed can be associated with the episodes of strongest accretion (e.g., *Reipurth,* 1989). These outflows carry matter and angular momentum into the cloud and may influence the overall dynamics of star formation.

3.4.3. *Variability during bursts of accretion.* During a burst of accretion, different processes are expected to occur at the disk-magnetosphere boundary. Matter may flow to the star above the magnetosphere in two ordered funnel streams and form two hot spots on the stellar surface (*Bertout et al.,* 1988; *Romanova et al.,* 2004). Alternatively, matter may accrete through the magnetic Rayleigh-Taylor instability, where several unstable "tongues" penetrate the field lines and form spots of chaotic shape and position (*Kulkarni and Romanova,* 2008; *Romanova et al.,* 2008). In these cases, the light curve and spectral changes appear chaotic, with a few accretion events per period of the inner disk (*Kurosawa and Romanova,* 2013). Observations of young stars often show variability at this timescale (e.g., *Alencar et al.,* 2010; *Cody et al.,* 2013). In the unstable regime, the tongues rotate with the angular velocity of the inner

disk, and the frequency of the inner disk may be present in the Fourier spectrum of the light curves. Variations of the accretion rate will lead to variations of the inner disk radius and this frequency will vary in time. In cases of small magnetospheres, $r_m \lesssim (1-2)R_\star$, one or two regular tongues rotate orderly with the frequency of the inner disk (*Romanova and Kulkarni,* 2009; *Bachetti et al.,* 2010). This phenomenon can potentially explain the quasiperiodic variability of 2–9 d, which has been recently observed in FU Ori and Z CMa by *Siwak et al.* (2013). Alternatively, it can be connected with the waves excited in the disk by the tilted dipole (e.g., *Bouvier et al.,* 1999; *Romanova et al.,* 2013). The modulation of the blue-shifted spectral lines in FU Ori with a period of 14 d (*Powell et al.,* 2012) may be a sign of modulation of the wind by the waves in the disk. Different longer-period variabilities (e.g., *Hillenbrand et al.,* 2013) can also be connected with the waves in the disk, excited at different distances from the star.

Another type of variability may be connected with *episodic inflation* and reconnection of the field lines connecting a star with the inner disk (*Aly and Kuijpers,* 1990). The signs of such variability on a timescale of a few stellar rotations have been observed by *Bouvier et al.* (2003) in AA Tau and in young bursting stars by *Findeisen et al.* (2013). This process may also lead to the phenomenon of episodic X-ray flares during the burst.

4.   CURRENT VIEW OF EPISODIC ACCRETION

Despite great progress both observationally and theoretically, there is still a great deal of uncertainty concerning the origin of the outbursts, whether the mechanism is different for FUors and EXors, and whether both classes of outbursting sources can be various examples of the same phenomenon in the time history of a forming star. In this section, we aim to highlight specific characteristics — such as outburst timescales, object evolutionary stages as inferred from infrared/submillimeter-spectral features and SEDs, the presence/absence of outflows, accretion rates, binarity, and high-energy activity levels — in which classical FUors and EXors, along with analogous newly discovered eruptive objects, converge as well as diverge. However, we caution that many critical behaviors are not easily disentangled, and the quest to understand the origin of pre-main-sequence outbursts will still require sustained efforts in the coming years.

4.1.   Outburst Timescales and Repetition

The recently discovered examples of outbursts have begun to fill in the notional gap between long-duration (classical FUor) and short-duration (classical EXor) eruptions (Fig. 2). Several objects (e.g., OO Ser, V1647 Ori) appear to display outburst decay times of a few years, i.e., intermediate between the two classes. Such a conceptual progression from a bimodal distribution to a continuum of outburst timescales is hence perhaps merely the natural consequence of the discovery of additional outbursting YSOs and pre-main-sequence stars, combined with the longer baseline and more comprehensive arsenal of observational data available to measure outburst decay times for both previously and newly identified eruptive objects.

On the other hand, the question of a clear distinction between FUor and EXor classes in terms of outburst repetition and duty cycle remains open. The repetitive nature of EXor outbursts might be considered a defining characteristic of such objects — one that potentially links strong, short-timescale EXor eruptions to the lesser variability of TTS more generally. In contrast, we simply have not had time to observe the cessation of any of the classical FUor outbursts, and hence we are unable to assess the potential for repetition of these long-duration eruptions. But essentially all theoretical models predict that FUor outbursts should also occur more than once during early stellar evolution, with an average time span of thousands of years between outbursts (section 3.2; Fig. 6); *Scholz et al.* (2013) estimated time intervals between outbursts of 5–50 k.y.

FU Orionis and EXor systems are often surrounded by ring- or cometary-shaped nebulae that evidently reside in the outflow and/or the parent cloud (e.g., *Goodrich*, 1987). These bright optical/infrared reflection nebulae become illuminated by the central stars as their activity levels increase [sometimes revealing light-travel-time effects that facilitate distance and luminosity determinations (e.g., *Briceño et al.*, 2004)]. Furthermore, HH objects and jets tend to be associated with sources that have recently entered elevated activity states (e.g., *Reipurth and Aspin*, 1997; *Takami et al.*, 2006). Thus, outflow structures might be used to infer protostellar and pre-main-sequence outburst duty cycles. However, any direct connection between outflow and outburst activity is difficult to infer; e.g., the time intervals between the appearance of large working surfaces in HH flows, typically 500–1000 yr, imply that HH driving sources are more likely to be outside of a higher accretion FUor or EXor state (*Reipurth and Bally*, 2001).

4.2. Evolutionary Stages

Just as the outburst duration gap between FUors and EXors has closed, the long-held view that longer duration eruptions involve higher protostellar accretion rates and occur at earlier stages of pre-main-sequence stellar evolution (e.g., *Hartmann and Kenyon*, 1996; *Sandell and Weintraub*, 2001) has become subject to question. This increasing ambiguity is in large part the result of the recent scrutiny of FUors via higher-resolution infrared and submillimeter imaging and spectroscopy. It is now apparent that classical FUors present a highly heterogeneous set of near- to mid-infrared spectral features and SEDs that could possibly relate to an evolutionary sequence within the FUor class (e.g., *Quanz et al.*, 2007c). In this scheme, the presence or (partial) depletion of an envelope may significantly modify the observational properties (e.g., silicate in absorption or emission, strong far-infrared excess). However, new outbursting sources have

demonstrated that the case may not be this simple: There are often more deeply embedded protostars lying in close proximity to optically identified eruptive objects that show FUor characteristics (e.g., *Green et al.*, 2013b; *Dunham et al.*, 2012). While some FUors appear relatively devoid of significant circumstellar envelopes, it also appears that some FUors do possess massive, molecule-rich circumstellar envelopes (e.g., *Kóspál et al.*, 2011a). The behavior and strength of outbursts may also be influenced by the presence of a wind/outflow and its interaction with a surrounding envelope (*Clarke et al.*, 2005).

Theoretically, it is difficult to provide a single evolutionary framework to explain both EXors and FUors. The difficulty with providing such a unified scenario is probably due to different causes. First, it is extremely difficult to self-consistently model the evolution of the entire disk, starting from sub-astronomical-unit to hundred-astronomical-unit scales, and as a result most theoretical models focus on either the inner or outer disk. Second, few attempts have been made up to now to directly compare theoretical models to actual observations, e.g., in terms of detailed light curve or color evolution modeling for specific events. As the models become more sophisticated and observations richer, this is certainly a direction to pursue in the future. Third, very little has been done from the theoretical standpoint in order to describe EXors. The models of *D'Angelo and Spruit* (2010, 2012) appear to be promising, but still lack a detailed time-dependent calculation. The explanation of the origin of EXor outbursts as related to the innermost parts of the disk might establish them as a separate class with respect to FUors.

When focusing on individual theoretical models, those presented in *Nayakshin and Lodato* (2012) (presented in section 3.2.1) do show variability on a variety of timescales and amplitude that might reproduce, within a single scenario, both EXors and FUors. The models of *Vorobyov and Basu* (2005, 2006, 2010) (presented in section 3.2.3) produce luminosity outbursts with amplitudes typical for both FUors and EXors but fail to reproduce the short rising times of EXors, possibly due to limitations of the numerical code. Certainly, theoretical models do require significant development before a proper comparison with data can be made.

Finally, radiative transfer models should attempt to model the behavior of disk interiors and their atmospheres during outbursts, e.g., as a function of mass-accretion rate or disk and envelope masses. Such models should determine whether such parameters can explain the CO bandhead in absorption in FUors but in emission in EXors, and pursue whether such CO features can change depending on the conditions, possibly explaining, e.g., the reversal of CO observed early in the outburst of V1647 Ori.

4.3. Binarity

The search for tight binaries in FUors and EXors is linked with the quest to identify the origin of erupting events. *Bonnell and Bastien* (1992) proposed that FUor outbursts may be due to a perturbation induced by a companion at periastron passage. Noting that there is already a list of known FUor binaries, *Reipurth and Aspin* (2004b) proposed that FUors may be newborn binaries that have become bound when a small nonhierarchical multiple system breaks up and the two components spiral in toward each other, perturbing their disks. This model derives particular motivation from the observation that FU Ori is the northern component in a close (0.″5), pre-main-sequence binary system (*Reipurth and Aspin*, 2004b; *Wang et al.*, 2004) whose non-outbursting component, FU Ori S, is likely more massive than the FUor namesake (*Beck and Aspin*, 2012; *Pueyo et al.*, 2012). The scenario advanced by *Reipurth and Aspin* (2004b) implies that FU Ori must be a close binary (<10 AU) (see also *Malbet et al.*, 1998, 2005; *Quanz et al.*, 2006) and, if so, the newly discovered companion is the outlying member in a triple system. Z CMa, a 0.″1 FUor/Herbig Be binary surrounded by a circumbinary disk (*Alonso-Albi et al.*, 2009) — and with jets emanating from both components (*Whelan et al.*, 2010; *Benisty et al.*, 2010; *Canovas et al.*, 2012) — is another potential example of a system that has undergone binary-disk interactions, although the observed outbursts in this binary do not always stem from FUor variability (*Teodorani et al.*, 1997; *van den Ancker et al.*, 2004; *Szeifert et al.*, 2010; *Hinkley et al.*, 2013).

If such binary-disk interaction is the dominant mechanism to trigger FUor outbursts, then FUor eruptions should preferentially occur in close binaries, i.e., in about 20% of all stars. However, V1057 Cyg and V1515 Cyg, both "classical" FUors, are not known to harbor close companions, although it is difficult to probe the 1–10-AU separation range, the most relevant for the binarity model, due to their distances. In the case of EXors, some stars are known visual or spectroscopic binaries [e.g., V1118 Ori at 72-AU separation (*Reipurth et al.*, 2007b); UZ Tau E with an ~0.15 AU (*Prato et al.*, 2002]; EX Lup might also harbor a brown dwarf companion (Á. Kóspál, personal communication), while others show no evidence of binarity (e.g., *Melo*, 2003; *Herbig*, 2007, 2008).

In summary, the jury is still out on whether binarity plays any role in triggering eruption events in EXors or FUors; further studies aiming at discovering faint companions or possibly planets in disks of erupting stars are clearly needed in the coming years. The reader is encouraged to read the chapter by Reipurth et al. in this volume.

4.4. Accretion, Infall, Winds, and Outflows

Pre-main-sequence accretion rates are difficult to determine and, for a given class of object, are determined via a variety of means and therefore highly inhomogeneous. Eruptive stars are no exception to this rule. For instance, the photometric data and flux-calibrated spectra of the outburst phase can be compared with SED models to constrain the accretion disk parameters. Correlations between mass-accretion rates and the emission line fluxes, obtained in the

framework of magnetospheric accretion, can also be used.

Nevertheless, some general statements can be made about accretion rates. The most luminous FUors have mass-accretion rates that can reach 10^{-4}–10^{-3} $M_\odot$ yr^{-1} (see Table 1). However, eruptive FUor-like objects with lower luminosities imply lower mass-accretion rates, as low as 10^{-6} $M_\odot$ yr^{-1}. Hence, there is a clear overlap between the ranges of mass-accretion rates observed during outburst in FUors and classical EXors (10^{-8}–10^{-6} $M_\odot$ yr^{-1}) or intermediate objects (~10^{-5} $M_\odot$ yr^{-1}). Quiescent mass-accretion rates can start as low 10^{-10} $M_\odot$ yr^{-1} (e.g., *Sipos et al.*, 2009) for EXors with little or no envelope, i.e., probing episodic accretion in later evolutionary stages. In summary, the peak mass-accretion rate is likely not the only physical parameter that determines the nature of the eruption.

In optical/infrared spectra, EXors ubiquitously show inverse P Cyg profiles due to infall (e.g., *Herbig*, 2007). In addition, blue-shifted absorption and/or P Cyg profiles are observed in spectral lines such as Hα and NaI D, indicating strong winds. The wind velocities are typically −50 km s^{-1} to −300 km s^{-1}, with maximum values up to −600 km s^{-1} in FUors (e.g., *Croswell et al.*, 1987; *Vacca et al.*, 2004; *Reipurth and Aspin*, 2004a). The maximum velocities appear generally lower in EXors, although some fast wind can be detected as well [e.g., V1647 Ori (*Aspin and Reipurth*, 2009)]. In addition, blue-shifted absorption is stronger in FUors than in EXors (e.g., *Herbig*, 2007, 2008, 2009). Many FUors also show millimeter signatures of CO outflows. The typical velocity and mass loss rate of the outflows are 10–40 km s^{-1} and 10^{-8}–10^{-6} $M_\odot$ yr^{-1}, respectively (*Evans et al.*, 1994). However, some objects do not show significant CO emission associated with outflows. As discussed in section 4.1, the connection between HH objects and outbursts is also difficult to infer. However, the strength of the disk wind increases during EXor outbursts [e.g., EX Lup (*Sicilia-Aguilar et al.*, 2012)] and decays as the outburst decays (e.g., *Aspin et al.*, 2010). This relation indicates that the wind strength is indeed related to the mass-accretion rate, as generally observed in CTTS (*Calvet*, 1997).

4.5. High-Energy Processes

The scant existing high-energy data make it apparent that episodes of high accretion rate, whether of sustained or more transient nature, are usually accompanied by enhanced X-ray emission (see section 2.10). In the case of sustained outbursts (i.e., for "classical" FUors and V1647 Ori), the emission (when detected) is overall relatively hard, betraying an origin in magnetic activity. For shorter-duration outbursts (as in the "classical" EXors), at least some of the enhanced X-ray flux appears to arise in accretion shocks. V1647 Ori appears to represent something of a hybrid case, in that its emission during high-accretion states is dominated by plasma that is too hot to be due to accretion shocks, yet is confined to "hot spots" very near the stellar surface (*Hamaguchi et al.*, 2012). Further X-ray observations are needed to more firmly establish whether and how the X-ray flux levels and plasma temperatures of eruptive young stars correlate with both long- and short-term variations in optical/infrared fluxes and other (e.g., emission-line-based) accretion and outflow signatures.

5. FUTURE DIRECTIONS

We identify here a number of potentially interesting directions that will or should be explored in the field of episodic accretion:

1. Continuum and molecular line images with ALMA will provide new opportunities to firmly establish the envelope vs. disk masses of FUors and EXors, so as to compare them with each other and with those of deeply embedded protostars and CTTS, and to study the chemistry taking place in disks and being modified due to episodic accretion events.

2. The output of present and forthcoming generations of large-field optical monitoring facilities (e.g., the Digitized Sky Survey, Palomar Transient Factory, and Large Synoptic Survey Telescope) will continue to enlarge the sample of eruptive pre-main-sequence objects. We can potentially take advantage of these data to deduce the frequency of eruptive objects, and determine accretion burst duty cycles, as functions of mass and class. However, these identifications will not include protostars at very early (cloud- and/or envelope-embedded) pre-main-sequence evolutionary stages. Nor will they allow continuous time monitoring to study time variability over long timescales.

3. Our present understanding of episodic accretion is potentially heavily biased, due to our "traditional" reliance on optically detected eruptions in identifying FUors and EXors. The identification of eruptions in the near-infrared has begun to mitigate this bias somewhat, mainly thanks to the Two Micron All Sky (2MASS) survey. The Wide-field Infrared Survey Explorer (WISE) has now provided an all-sky mid-infrared snapshot against which future wide-field mid-infrared imaging surveys can be compared, so as to identify eruptions associated with much more deeply embedded protostars (see *Antoniucci et al.*, 2013; *Johnstone et al.*, 2013; *Scholz et al.*, 2013). Armed with such identifications, we can begin to more accurately pinpoint the epoch of onset of episodic accretion during protostellar evolution, and obtain follow-up observations from the ground or in space.

4. It is worth considering the extent to which dramatic accretion-driven outbursts effectively cause young stars to "revert" to earlier stages of protostellar evolution. In other words, if we observe a class II object or flat-spectrum source enter a FUor outburst (e.g., HBC 722), are we in effect seeing a born-again class I protostar?

5. The understanding of outburst feedback on the inner disk structure (crystallization, chemistry) would profit from further investigation, especially in the region of the formation of Earth-like planets. Such outbursts may indeed have taken place in the history of our solar system. High angular resolution observations will thus help discern structure and physical conditions in the inner disk and search for very close companions.

6. Further modeling of the effect of episodic accretion on the disk structure should be considered. The CO spectrum in absorption observed continuously in FUors but rarely or transiently in EXors has traditionally been explained by heating of the disk interior during the accretion event, assuming built-up material falling from the envelope. However, some FUors, including FU Ori, do not show evidence of massive envelopes. Thus, it remains unclear why they exhibit these typical FUor spectral characteristics.

7. Finally, it could be worthwhile to investigate numerically the link between EXors and FUors by treating the inner and outer disk simultaneously, although this may be out of reach in the near (and mid) term. These and other future efforts should continue to focus on the fundamental physical processes underlying outbursts, such as narrowing down the possible mechanisms that can lead to accretion bursts, identifying the key system parameters that control burst energetics (amplitude, duration, repetition), and constraining the range of key system parameters such as accretion rates, outflow rates, and star/disk/outflow geometries.

Acknowledgments. We dedicate this review to the late G. H. Herbig, who passed away on October 12, 2013, for his pioneering and long-lasting work on eruptive young stars. We acknowledge the fantastic works of a long list of authors mentioned in the reference list that have contributed to our knowledge of episodic accretion in star and planet formation, among them B. Reipurth, whom we wish to thank for carefully reading the manuscript and providing detailed comments as referee of this chapter. H. Beuther is also thanked for providing further editorial comments to improve the manuscript. We are grateful to M. Richmond for providing the V1647 Ori light curves in Fig. 4 and M. Kun for reading the manuscript and providing corrections. Finally, we thank several participants at the Protostars and Planets VI conference for coming forward and providing useful comments and feedback to improve this review, among them W. Fischer, C. McKee, and S. Offner. We have attempted to include all refereed publications from 1996 until 2013 that are relevant to the study of episodic accretion. We apologize if we have unintentionally missed publications. M.M.R. was supported by NSF grant AST-1211318; G.L. by PRIN MIUR 2010-2011, project 2010LY5N2T; and P.Á. and Á.K. partly by OTKA 101393. E. I. V. by RFBR grant 14-02-00719.

REFERENCES

Ábrahám P. et al. (2004a) *Astron. Astrophys., 419,* L39.
Ábrahám P. et al. (2004b) *Astron. Astrophys., 428,* 89.
Ábrahám P. et al. (2006) *Astron. Astrophys., 449,* L13.
Ábrahám P. et al. (2009) *Nature, 459,* 224.
Acosta-Pulido J. A. et al. (2007) *Astron. J., 133,* 2020.
Adams F. C. et al. (1988) *Astrophys. J., 326,* 865.
Alencar S. H. P. et al. (2010) *Astron. Astrophys., 519,* A88.
Alonso-Albi T. et al. (2009) *Astron. Astrophys., 497,* 117.
Aly J. J. and Kuijpers J. (1990) *Astron. Astrophys., 227,* 473.
Andrews S. M. and Williams J. P. (2005) *Astrophys. J., 631,* 1134.
Andrews S. M. et al. (2004) *Astrophys. J. Lett., 610,* L45.
Antoniucci S. et al. (2013) *New Astron., 23,* 98.
Armitage P. J. et al. (2001) *Mon. Not. R. Astron. Soc., 324,* 705.
Aspin C. (2011a) *Astron. J., 141,* 196.

Aspin C. (2011b) *Astron. J., 142,* 135.
Aspin C. and Reipurth B. (2003) *Astron. J., 126,* 2936.
Aspin C. and Reipurth B. (2009) *Astron. J., 138,* 1137.
Aspin C. and Sandell G. (1994) *Astron. Astrophys., 288,* 803.
Aspin C. and Sandell G. (2001) *Mon. Not. R. Astron. Soc., 328,* 751.
Aspin C. et al. (2006) *Astron. J., 132,* 1298.
Aspin C. et al. (2008) *Astron. J., 135,* 423.
Aspin C. et al. (2009a) *Astron. J., 137,* 2968.
Aspin C. et al. (2009b) *Astrophys. J. Lett., 692,* L67.
Aspin C. et al. (2010) *Astrophys. J. Lett., 719,* L50.
Audard M. et al. (2005) *Astrophys. J. Lett., 635,* L81.
Audard M. et al. (2010) *Astron. Astrophys., 511,* A63.
Baan W. (1977) *Astrophys. J., 214,* 245.
Bachetti M. et al. (2010) *Mon. Not. R. Astron. Soc., 403,* 1193.
Bae J. et al. (2013) *Astrophys. J., 764,* 141.
Balbus S. A. and Hawley J. F. (1998) *Rev. Modern Phys., 70,* 1.
Banzatti A. et al. (2012a) *Astrophys. J., 745,* 90.
Banzatti A. et al. (2012b) *Astrophys. J., 751,* 160.
Baraffe I. et al. (2009) *Astrophys. J. Lett., 702,* L27.
Baraffe I. et al. (2012) *Astrophys. J., 756,* 118.
Bastien F. A. et al. (2011) *Astron. J., 142,* 141.
Basu S. and Vorobyov E. I. (2012) *Astrophys. J., 750,* 30.
Beck T. L. and Aspin C. (2012) *Astron. J., 143,* 55.
Bell K. R. and Lin D. N. C. (1994) *Astrophys. J., 427,* 987.
Bell K. R. et al. (1995) *Astrophys. J., 444,* 376.
Bell K. R. et al. (2000) In *Protostars and Planets IV* (V. Mannings
    et al., eds.), pp. 897–926. Univ. of Arizona, Tucson.
Benisty M. et al. (2010) *Astron. Astrophys., 517,* L3.
Bertin G. and Lodato G. (1999) *Astron. Astrophys., 350,* 694.
Bertout C. et al. (1988) *Astrophys. J., 330,* 350.
Blandford R. D. and Payne D. G. (1982) *Mon. Not. R. Astron. Soc.,
    199,* 883.
Boley A.C. et al. (2006) *Astrophys. J., 651,* 517.
Bonnell I. and Bastien P. (1992) *Astrophys. J. Lett., 401,* L31.
Bouvier J. et al. (1999) *Astron. Astrophys., 349,* 619.
Bouvier J. et al. (2003) *Astron. Astrophys., 409,* 169.
Briceño C. et al. (2004) *Astrophys. J. Lett., 606,* L123.
Brittain S. D. et al. (2007) *Astrophys. J. Lett., 670,* L29.
Brittain S. D. et al. (2010) *Astrophys. J., 708,* 109.
Calvet N. (1997) In *Herbig-Haro Flows and the Birth of Stars*
    (B. Reipurth and C. Bertout, eds.), pp. 417–432. IAU Symp. 182,
    Kluwer, Dordrecht.
Calvet N. et al. (1991a) *Astrophys. J., 383,* 752.
Calvet N. et al. (1991b) *Astrophys. J., 380,* 617.
Calvet N. et al. (1993) *Astrophys. J., 402,* 623.
Canovas H. et al. (2012) *Astron. Astrophys., 543,* A70.
Caratti o Garatti A. et al. (2011) *Astron. Astrophys., 526,* L1.
Caratti o Garatti A. et al. (2013) *Astron. Astrophys., 554,* A66.
Casali M. M. (1991) *Mon. Not. R. Astron. Soc., 248,* 229.
Cha S.-H. and Nayakshin S. (2011) *Mon. Not. R. Astron. Soc.,
    415,* 3319.
Chavarria C. (1981) *Astron. Astrophys., 101,* 105.
Chen H. et al. (1995) *Astrophys. J., 445,* 377.
Chen W. P. et al. (2012) *Astrophys. J., 751,* 118.
Chochol D. et al. (2006) *Contrib. Astron. Obs. Skalnate Pleso, 36,* 149.
Chou M.-Y. et al. (2013) *Astron. J., 145,* 108.
Clarke C. J. and Armitage P. J. (2003) *Mon. Not. R. Astron. Soc.,
    345,* 691.
Clarke C. J. and Syer D. (1996) *Mon. Not. R. Astron. Soc., 278,* L23.
Clarke C. J. et al. (1990) *Mon. Not. R. Astron. Soc., 242,* 439.
Clarke C. J. et al. (2005) *Mon. Not. R. Astron. Soc., 361,* 942.
Cody A.-M. et al. (2013) *Astron. J., 145,* 79.
Coffey D. et al. (2004) *Astron. Astrophys., 419,* 593.
Cohen M. et al. (1981) *Astrophys. J., 245,* 920.
Cohen M. et al. (1983) *Astrophys. J., 273,* 624.
Connelley M. S. and Greene T. P. (2010) *Astron. J., 140,* 1214.
Cossins P. et al. (2009) *Mon. Not. R. Astron. Soc., 393,* 1157.
Cossins P. et al. (2010) *Mon. Not. R. Astron. Soc., 401,* 2587.
Costigan G. et al. (2012) *Mon. Not. R. Astron. Soc., 427,* 1344.
Covey K. R. et al. (2011) *Astron. J., 141,* 40.
Croswell K. et al. (1987) *Astrophys. J., 312,* 227.
D'Angelo C. R. and Spruit H. C. (2010) *Mon. Not. R. Astron. Soc.,
    406,* 1208.
D'Angelo C. R. and Spruit H. C. (2012) *Mon. Not. R. Astron. Soc.,
    420,* 416.

Dent W. R. F. et al. (1998) *Mon. Not. R. Astron. Soc., 301,* 1049.
Donati J. F. et al. (2005) *Nature, 438,* 466.
Donati J.-F. et al. (2008) *Mon. Not. R. Astron. Soc., 386,* 1234.
Dunham M. M. and Vorobyov E. I. (2012) *Astrophys. J., 747,* 52.
Dunham M. M. et al. (2010) *Astrophys. J., 710,* 470.
Dunham M. M. et al. (2012) *Astrophys. J., 755,* 157.
Dunham M. M. et al. (2013) *Astron. J., 145,* 94.
Durisen R. H. et al. (2007) In *Protostars and Planets V* (B. Reipurth
    et al., eds.), pp. 607–622. Univ. of Arizona, Tucson.
Eislöffel J. et al. (1990) *Astron. Astrophys., 232,* 70.
Eislöffel J. et al. (1991) *Astrophys. J. Lett., 383,* L19.
Eisner J. A. and Hillenbrand L. A. (2011) *Astrophys. J., 738,* 9.
Elias J. H. (1978) *Astrophys. J., 224,* 857.
Enoch M. L. et al. (2009) *Astrophys. J., 692,* 973.
Evans N. J. II et al. (1994) *Astrophys. J., 424,* 793.
Evans N. J. II et al. (2009) *Astrophys. J. Suppl., 181,* 321.
Fedele D. et al. (2007) *Astron. Astrophys., 472,* 207.
Feigelson E. D. and Montmerle T. (1999) *Annu. Rev. Astron. Astrophys.,
    37,* 363.
Findeisen K. et al. (2013) *Astrophys. J., 768,* 93.
Fischer W. J. et al. (2012) *Astrophys. J., 756,* 99.
Gammie C. (1996) *Astrophys. J., 457,* 355.
Gammie C. (2001) *Astrophys. J., 553,* 174.
Giannini T. et al. (2009) *Astrophys. J., 704,* 606.
Gibb A. G. (2008) In *Handbook of Star Forming Regions, Volume I:
    The Northern Sky* (B. Reipurth, ed.), pp. 693–731. ASP Monograph
    Publications, Astronomical Society of the Pacific, San Francisco.
Gibb E. L. et al. (2006) *Astrophys. J., 641,* 383.
Goodrich R. W. (1987) *Publ. Astron. Soc. Pacific, 99,* 116.
Goodson A. P. et al. (1997), *Astrophys. J., 489,* 199.
Goto M. et al. (2011) *Astrophys. J., 728,* 5.
Graham J. A. and Frogel J. A. (1985) *Astrophys. J., 289,* 331.
Gras-Velázques À. and Ray T. P. (2005) *Astron. Astrophys., 443,* 541.
Green J. D. et al. (2006) *Astrophys. J., 648,* 1099.
Green J. D. et al. (2011) *Astrophys. J. Lett., 731,* L25.
Green J. D. et al. (2013a) *Astrophys. J., 764,* 22.
Green J. D. et al. (2013b) *Astrophys. J., 772,* 117.
Greene T. P. et al. (1994) *Astrophys. J., 434,* 614.
Greene T. P. et al. (2008) *Astron. J., 135,* 1421.
Grosso N. et al. (2005) *Astron. Astrophys., 438,* 159.
Grosso N. et al. (2010) *Astron. Astrophys., 522,* A56.
Haas M. et al. (1990) *Astron. Astrophys., 230,* L1.
Haisch K. E. Jr. et al. (2004) *Astron. J., 127,* 1747.
Hamaguchi K. et al. (2010) *Astrophys. J. Lett., 714,* L16.
Hamaguchi K. et al. (2012) *Astrophys. J., 741,* 32.
Hartigan P. and Kenyon S. J. (2003) *Astrophys. J., 583,* 334.
Hartmann L. (1991) In *Physics of Star Formation and Early Stellar
    Evolution* (C. J. Lada and N. D. Kylafis, eds.), pp. 623–648.
    Kluwer, Dordrecht.
Hartmann L. (2008) In *Accretion Processes in Star Formation* (A. King
    and D. Lin, eds.), Cambridge Univ., Cambridge.
Hartmann L. and Calvet N. (1995) *Astron. J., 109,* 1846.
Hartmann L. and Kenyon S. J. (1985) *Astrophys. J., 299,* 462.
Hartmann L. and Kenyon S. J. (1987a) *Astrophys. J., 322,* 393.
Hartmann L. and Kenyon S. J. (1987b) *Astrophys. J., 312,* 243.
Hartmann L. and Kenyon S. J. (1996) *Annu. Rev. Astron. Astrophys.,
    34,* 207.
Hartmann L. et al. (1989) *Astrophys. J., 338,* 1001.
Hartmann L. et al. (1993) In *Protostars and Planets III* (E. H. Levy and
    J. I. Lunine, eds.), pp. 497–520. Univ. of Arizona, Tucson.
Hartmann L. et al. (2004) *Astrophys. J., 609,* 906.
Henning Th. et al. (1998) *Astron. Astrophys., 336,* 565.
Herbig G. H. (1966) *Vistas Astron., 8,* 109.
Herbig G. H. (1977) *Astrophys. J., 217,* 693.
Herbig G. H. (1989) In *ESO Workshop on Low Mass Star Formation
    and Pre-Main Sequence Objects* (B. Reipurth, ed.), pp. 233–246.
    European Southern Observatory, Garching.
Herbig G. H. (2007) *Astron. J., 133,* 2679.
Herbig G. H. (2008) *Astron. J., 135,* 637.
Herbig G. H. (2009) *Astron. J., 138,* 448.
Herbig G. H. et al. (2001) *Publ. Astron. Soc. Pac., 133,* 1547.
Herbig G. H. et al. (2003) *Astrophys. J., 595,* 384.
Hillenbrand L. (2009) In *The Age of Stars* (E. Mamajek et al., eds.),
    pp. 81–88. IAU Symp. 258, Cambridge Univ., Cambridge.
Hillenbrand L. A. et al. (2013) *Astron. J., 145,* 59.

Hinkley S. et al. (2013) *Astrophys. J. Lett., 763*, L9.
Hodapp K. W. (1999) *Astron. J., 118*, 1338.
Hodapp K. W. et al. (1996) *Astrophys. J., 468*, 861.
Hodapp K. W. et al. (2012) *Astrophys. J., 744*, 56.
Hosokawa T. et al. (2011) *Astrophys. J. 738*, 140.
Illarionov A. F. and Sunyaev R. A. (1975) *Astron. Astrophys., 39*, 185.
Jensen E. L. et al. (2007) *Astron. J., 134*, 241.
Johns-Krull C. M. (2007) *Astrophys. J., 664*, 975.
Johnson B. M. and Gammie C. F. (2003) *Astrophys. J., 597*, 131.
Johnstone D. et al. (2013) *Astrophys. J., 765*, 133.
Juhász A. et al. (2012) *Astrophys. J., 744*, 118.
Kastner J. H. et al. (2004) *Nature, 430*, 429.
Kastner J. H. et al. (2006) *Astrophys. J., 648*, 43.
Kenyon S. J. (1995a) *Astrophys. Space Sci., 233*, 3.
Kenyon S. J. (1995b) *Rev. Mex. Astron. Astrophys. Ser. Conf., 1*, 237.
Kenyon S. J. and Hartmann L. W. (1991) *Astrophys. J., 383*, 664.
Kenyon S. J. and Hartmann L.W. (1995) *Astrophys. J. Suppl., 101*, 117.
Kenyon S. J. et al. (1988) *Astrophys. J., 325*, 231.
Kenyon S. J. et al. (1990) *Astron. J., 99*, 869.
Kenyon S. J. et al. (1993) *Astron. J., 105*, 1505.
Kenyon S. J. et al. (1994) *Astron. J., 108*, 251.
Kenyon S. J. et al. (2000) *Astrophys. J., 531*, 1028.
Kim H. J. et al. (2012) *Astrophys. J., 758*, 38.
Kley W. and Lin D. N. C. (1996) *Astrophys. J., 461*, 933–949.
Kley W. and Lin D. N. C. (1999) *Astrophys. J., 518*, 833.
Köhler R. et al. (2006) *Astron. Astrophys., 458*, 461.
Königl A. (1991) *Astrophys. J. Lett., 370*, L39.
Königl A. et al. (2011) *Mon. Not. R. Astron. Soc., 416*, 757.
Koresko C. D. et al. (1991) *Astron. J., 102*, 2073.
Kóspál Á. (2011) *Astron. Astrophys., 535*, 125.
Kóspál Á. et al. (2007) *Astron. Astrophys., 470*, 211.
Kóspál Á. et al. (2008) *Mon. Not. R. Astron. Soc., 383*, 1015.
Kóspál Á. et al. (2011a) *Astron. Astrophys., 527*, 133.
Kóspál Á. et al. (2011b) *Astrophys. J., 736*, 72.
Kóspál Á. et al. (2012) *Astrophys. J. Suppl., 201*, 11.
Kóspál Á. et al. (2013) *Astron. Astrophys., 551*, 62.
Kratter K. M. et al. (2008) *Astrophys J., 681*, 375.
Kravtsova A. S. et al. (2007) *Astron. Letters, 33*, 755.
Kryukova E. et al. (2012) *Astron. J., 144*, 31.
Kulkarni A. K. and Romanova M. M. (2008) *Astrophys. J., 386*, 673.
Kun M. et al. (2011a) *Astrophys. J. Lett., 733*, L8.
Kun M. et al. (2011b) *Mon. Not. R. Astron. Soc., 413*, 2689.
Kurosawa R. and Romanova M. M. (2012) *Mon. Not. R. Astron. Soc., 426*, 2901.
Kurosawa R. and Romanova M. M. (2013) *Mon. Not. R. Astron. Soc., 431*, 2673.
Kurucz R. L. et al. (1974) *Blanketed Model Atmospheres for Early-Type Stars.* Smithsonian Institution, Washington.
Lee J.-E. (2007) *J. Korean Astron. Soc., 40*, 83.
Lehmann T. et al. (1995) *Astron. Astrophys., 300*, L9.
Leinert Ch. and Haas M. (1987) *Astron. Astrophys., 182*, L47.
Lii P. et al. (2012) *Mon. Not. R. Astron. Soc., 420*, 2020.
Lii P. S et al. (2013) *Mon. Not. R. Astron. Soc.*, arXiv:1304.2703.
Lin D. N. C. et al. (1985) *Mon. Not. R. Astron. Soc, 212*, 105.
Lodato G. and Bertin G. (2001) *Astron. Astrophys., 375*, 455.
Lodato G. and Bertin G. (2003) *Astron. Astrophys., 408*, 1015.
Lodato G. and Clarke C. J. (2004) *Mon. Not. R. Astron. Soc., 353*, 841.
Lodato G. and Rice W. K. M. (2004) *Mon. Not. R. Astron. Soc., 351*, 630.
Lodato G. and Rice W. K. M. (2005) *Mon. Not. R. Astron. Soc., 358*, 1489.
Lodato G. et al. (2007) *Mon. Not. R. Astron. Soc., 374*, 590.
Lombardi M. et al. (2008) *Astron. Astrophys., 480*, 785.
Lorenzetti D. et al. (2000) *Astron. Astrophys., 357*, 1035.
Lorenzetti D. et al. (2006) *Astron. Astrophys., 453*, 579.
Lorenzetti D. et al. (2007) *Astrophys. J., 665*, 1182.
Lorenzetti D. et al. (2009) *Astrophys. J., 693*, 1056.
Lorenzetti D. et al. (2011) *Astrophys. J., 732*, 69.
Lorenzetti D. et al. (2012) *Astrophys. J., 749*, 188.
Lovelace R. V. E. et al. (1991) *Astrophys. J., 379*, 696.
Lovelace R. V. E. et al. (1999) *Astrophys. J., 514*, 368.
Machida M. N. et al. (2011) *Astrophys. J., 729*, 42.
Magakian T. Y. et al. (2013) *Mon. Not. R. Astron. Soc., 432*, 2685.
Magakian T. Y. et al. (2010) *Astron. J., 139*, 969.
Malbet F. et al. (1998) *Astrophys. J. Lett., 507*, L149.

Malbet F. et al. (2005) *Astron. Astrophys., 437*, 627.
Martin R. G. and Lubow S. H. (2011) *Astrophys. J. Lett., 740*, L6.
Martin R. G. and Lubow S. H. (2013) *Mon. Not. R. Astron. Soc., 432*, 1616.
Martin R. G. et al. (2012a) *Mon. Not. R. Astron. Soc., 420*, 3139.
Martin R. G. et al. (2012b) *Mon. Not. R. Astron. Soc., 423*, 2718.
Matt S. and Pudritz R. (2005) *Astrophys. J. Lett., 632*, L135.
Mayer L. et al. (2005) *Mon. Not. R. Astron. Soc., 363*, 641.
McGehee P. M. et al. (2004) *Astrophys. J., 616*, 1058.
McMuldroch S. et al. (1993) *Astron. J., 106*, 2477.
McMuldroch S. et al. (1995) *Astron. J., 110*, 354.
Mejía A. C. et al. (2005) *Astrophys. J., 619*, 1098.
Melo C. H. F. (2003), *Astron. Astrophys., 410*, 269.
Menten K. M. et al. (2007) *Astron. Astrophys., 474*, 515.
Millan-Gabet R. et al. (2006) *Astrophys. J., 641*, 547.
Miller A. A. et al. (2011) *Astrophys. J., 730*, 80.
Moriarty-Schieven G. H. et al. (2008) *Astron. J., 136*, 1658.
Mosoni L. et al. (2013) *Astron. Astrophys., 552*, A62.
Movsessian T. A. et al. (2003) *Astron. Astrophys., 412*, 147.
Movsessian T. A. et al. (2006) *Astron. Astrophys., 455*, 1001.
Muzerolle J. et al. (2005) *Astrophys. J. Lett., 620*, L107.
Nayakshin S. and Lodato G. (2012) *Mon. Not. R. Astron. Soc., 426*, 70.
Offner S. S. R. and McKee C. F. (2011) *Astrophys. J., 736*, 53.
Offner S. S. R. et al. (2009) *Astrophys. J., 703*, 131.
Ojha D. et al. (2005) *Publ. Astron. Soc. Jap., 57*, 203.
Ojha D. K. et al. (2006) *Mon. Not. R. Astron. Soc., 368*, 825.
Okuda T. et al. (1997) *Publ. Astron. Soc. Jap., 49*, 679.
Parsamian E. S. and Mujica R. (2004) *Astrophysics, 47*, 433.
Parsamian E. S. et al. (2002) *Astrophysics, 45*, 393.
Peneva S. P. et al.. (2010) *Astron. Astrophys. 515*, 24.
Pérez L. M. et al. (2010) *Astrophys. J., 724*, 493.
Persi P. et al. (2007) *Astron. J., 133*, 1690.
Petrov P. P. and Herbig G. H. (1992) *Astrophys. J., 392*, 209.
Petrov P. P. and Herbig G. H. (2008) *Astron. J., 136*, 676.
Petrov P. et al. (1998) *Astron. Astrophys., 331*, L53.
Pfalzner S. (2008) *Astron. Astrophys., 492*, 735.
Pfalzner S. et al. (2008) *Astron. Astrophys., 487*, L45.
Polomski E. F. et al. (2005) *Astron. J., 129*, 1035.
Popham R. (1996) *Astrophys. J., 467*, 749.
Popham R. et al. (1996) *Astrophys. J., 473*, 422.
Powell S. L. et al. (2012) *Mon. Not. R. Astron. Soc., 426*, 3315.
Prato L. et al. (2002) *Astrophys. J. Lett., 579*, L99.
Preibisch T. et al. (2005) *Astrophys. J. Suppl., 160*, 401.
Prusti T. et al. (1993) *Astron. Astrophys., 279*, 163.
Pueyo L. et al. (2012) *Astrophys. J., 757*, 57.
Quanz S. P. et al. (2006) *Astrophys. J., 648*, 472.
Quanz S. P. et al. (2007a) *Astrophys. J., 656*, 287.
Quanz S. P. et al. (2007b) *Astrophys. J., 658*, 487.
Quanz S. P. et al. (2007c) *Astrophys. J., 668*, 359.
Reipurth B. (1989) *Nature, 340*, 42.
Reipurth B. (1990) In *Flare Stars in Star Clusters, Associations and the Solar Vicinity* (L. V. Mirzoyan et al., eds.), pp. 229–251. IAU Symp. 137, Kluwer, Dordrecht.
Reipurth B. and Aspin C. (1997) *Astron. J., 114*, 2700.
Reipurth B. and Aspin C. (2004a) *Astrophys. J. Lett., 606*, L119.
Reipurth B. and Aspin C. (2004b) *Astrophys. J. Lett., 608*, L65.
Reipurth B. and Aspin C. (2010) In *Victor Ambartsumian Centennial Volume, Evolution of Cosmic Objects Through Their Physical Activity* (H. Harutyunyan et al., eds.), pp. 19–38. Gitutyun, Yerevan.
Reipurth B. and Bally J. (2001) *Annu. Rev. Astron. Astrophys., 39*, 403.
Reipurth B. et al. (2002) *Astron. J., 124*, 2194.
Reipurth B. et al. (2007a) *Astron. J., 133*, 1000.
Reipurth B. et al. (2007b) *Astron. J., 134*, 2272.
Reipurth B. et al. (2012) *Astrophys. J. Lett., 748*, L5.
Rettig T. W. et al. (2005) *Astrophys. J., 626*, 245.
Rice W. K. M et al. (2005) *Mon. Not. R. Astron. Soc., 364*, L56.
Rice W. K. M. et al. (2011) *Mon. Not. R. Astron. Soc., 418*, 1356.
Romanova M. M. and Kulkarni A. K. (2009) *Mon. Not. R. Astron. Soc., 398*, 1105.
Romanova M. M. et al. (2004) *Astrophys. J., 610*, 920.
Romanova M. M. et al. (2005) *Astrophys. J Lett., 635*, L165.
Romanova M. M. et al. (2008) *Astrophys. J. Lett., 673*, L171.
Romanova M. M. et al. (2009) *Mon. Not. R. Astron. Soc., 399*, 1802.
Romanova M. M. et al. (2013) *Mon. Not. R. Astron. Soc., 430*, 699.
Sandell G. and Aspin C. (1998) *Astron. Astrophys., 333*, 1016.

Sandell G. and Weintraub D. A. (2001) *Astrophys. J. Suppl., 134,* 115.
Scholz A. et al. (2013) *Mon. Not. R. Astron. Soc., 430,* 2910.
Schütz O. et al. (2005) *Astron. Astrophys., 431,* 165.
Semkov E. H. and Peneva S. P. (2012) *Astron. Sp. Sci., 338,* 95.
Semkov E. H. et al. (2010) *Astron. Astrophys., 523,* L3.
Semkov E. H. et al. (2011) *Bulg. Astron. J., 15,* 49.
Semkov E. H. et al. (2012) *Astron. Astrophys., 542,* 43.
Semkov, E. H. et al. (2013) *Astron. Astrophys., 556,* A60.
Shevchenko V. S. et al. (1991) *Sov. Astron., 35,* 135.
Shevchenko V. S. et al. (1997) *Astron. Astrophys. Suppl., 124,* 33.
Shu F. et al. (1994) *Astrophys. J., 429,* 781.
Sicilia-Aguilar A. et al. (2008) *Astrophys. J., 673,* 382.
Sicilia-Aguilar A. et al. (2012) *Astron. Astrophys. 544,* 93.
Sipos N. et al. (2009) *Astron. Astrophys., 507,* 881.
Siwak M. et al. (2013) *Mon. Not. R. Astron. Soc., 432,* 194.
Skinner S. L. et al. (2006) *Astrophys. J., 643,* 995.
Skinner S. L. et al. (2009) *Astrophys. J., 696,* 766.
Skinner S. L. et al. (2010) *Astrophys. J., 722,* 1654.
Spruit H. C. and Taam R. E. (1993) *Astrophys. J., 402,* 593.
Stamatellos D. and Whitworth A. P. (2009a) *Mon. Not. R. Astron. Soc., 392,* 413.
Stamatellos D. and Whitworth A. P. (2009b) *Mon. Not. R. Astron. Soc., 400,* 1563.
Stamatellos D. et al. (2011) *Astrophys. J., 730,* 32.
Stamatellos D. et al. (2012) *Mon. Not. R. Astron. Soc., 427,* 1182.
Staude H. J. and Neckel Th. (1991) *Astron. Astrophys., 244,* L13.
Staude H. J. and Neckel Th. (1992) *Astrophys. J., 400,* 556.
Stecklum B. et al. (2007) *Astron. Astrophys., 463,* 621.
Stelzer B. et al. (2009) *Astron. Astrophys., 499,* 529.
Strom K. M. and Strom S. E. (1993) *Astrophys. J. Lett., 412,* L63.
Sunyaev R. A. and Shakura N. I. (1977) *Sov. Astron. Lett., 3,* 138.
Szeifert T. et al. (2010) *Astron. Astrophys., 509,* L7.
Takami M. et al. (2006) *Astrophys. J., 641,* 357.
Tapia M. et al. (2006) *Mon. Not. R. Astron. Soc., 367,* 513.
Tassis K. and Mouschovias T. C. (2005) *Astrophys. J., 618,* 783.
Teets W. K. et al. (2011) *Astrophys. J., 741,* 83.

Teets W. K. et al. (2012) *Astrophys. J., 760,* 89.
Teodorani M. et al. (1997) *Astron. Astrophys. Suppl., 126,* 91.
Toomre A. (1964) *Astrophys. J.,* 139, 1217.
Tsukagoshi T. et al. (2005) *Publ. Astron. Soc. Jap., 57,* L21.
Turner N. J. J. et al. (1997) *Astrophys. J., 480,* 754.
Umebayashi T. (1983) *Prog. Theor. Phys., 69,* 480.
Ustyugova G. V. et al. (2006) *Astrophys. J., 646,* 304.
Vacca W. D. et al. (2004) *Astrophys. J. Lett., 609,* L29.
van den Ancker M. E. et al. (2004) *Mon. Not. R. Astron. Soc., 349,* 1516.
Venkata Raman V. et al. (2013) *Research in Astron. Astrophys., 13,* 1107.
Vig S. et al. (2006) *Astron. Astrophys., 446,* 1021.
Visser R. and Bergin E. A. (2012) *Astrophys. J., 754,* 18.
Vorobyov E. I. (2009) *Astrophys. J., 704,* 715.
Vorobyov E. I. (2011) *Astrophys. J., 729,* 146.
Vorobyov E. I. (2013) *Astron. Astrophys., 552,* 129.
Vorobyov E. I. and Basu S. (2005) *Astrophys. J. Lett., 633,* L137.
Vorobyov E. I. and Basu S. (2006) *Astrophys. J., 650,* 956.
Vorobyov E. I. and Basu S. (2010) *Astrophys. J., 719,* 1896.
Vorobyov E. I. et al. (2013) *Astron. Astrophys., 557,* A35.
Walter F. M. et al. (2004) *Astron. J., 128,* 1872.
Wang H. et al. (2004) *Astrophys. J. Lett., 601,* L83.
Welin G. (1976) *Astron. Astrophys., 49,* 145.
Whelan E. T. et al. (2010) *Astrophys. J. Lett., 720,* L119.
Xiao L. et al. (2010) *Astron. J., 139,* 1527.
Young C. H. and Evans N. J. II (2005) *Astrophys. J., 627,* 293.
Zanni C. et al. (2007) *Astron. Astrophys., 469,* 811–828.
Zhu Z. et al. (2007) *Astrophys. J., 669,* 483.
Zhu Z. et al. (2008) *Astrophys. J., 684,* 1281.
Zhu Z. et al. (2009a) *Astrophys. J. Lett., 694,* L64.
Zhu Z. et al. (2009b) *Astrophys. J., 694,* 1045.
Zhu Z. et al. (2009c) *Astrophys. J., 701,* 620.
Zhu Z. et al. (2010a) *Astrophys. J., 713,* 1134.
Zhu Z. et al. (2010b) *Astrophys. J., 713,* 1143.

Turner N. J., Fromang S., Gammie C., Klahr H., Lesur G., Wardle M., and Bai X.-N. (2014) Transport and accretion in planet-forming disks. In *Protostars and Planets VI* (H. Beuther et al., eds.), pp. 411–432. Univ. of Arizona, Tucson, DOI: 10.2458/azu_uapress_9780816531240-ch018.

Transport and Accretion in Planet-Forming Disks

Neal J. Turner
Jet Propulsion Laboratory, California Institute of Technology

Sébastien Fromang
Service d'Astrophysique de Saclay

Charles Gammie
University of Illinois at Urbana-Champaign

Hubert Klahr
Max Planck Institute for Astronomy

Geoffroy Lesur
Institute for Planetology and Astrophysics of Grenoble

Mark Wardle
Macquarie University

Xuc-Ning Bai
Harvard-Smithsonian Center for Astrophysics

Planets appear to form in environments shaped by the gas flowing through protostellar disks to the central young stars. The flows in turn are governed by orbital angular momentum transfer. In this chapter we summarize current understanding of the transfer processes best able to account for the flows, including magneto-rotational turbulence, magnetically launched winds, self-gravitational instability, and vortices driven by hydrodynamical instabilities. For each, in turn, we outline the major achievements of the past few years and the outstanding questions. We underscore the requirements for operation, especially ionization for the magnetic processes and heating and cooling for the others. We describe the distribution and strength of the resulting flows and compare them with the long-used phenomenological α-picture, highlighting issues where the fuller physical picture yields substantially different answers. We also discuss the links between magnetized turbulence and magnetically launched outflows, and between magnetized turbulence and hydrodynamical vortices. We end with a summary of the status of efforts to detect specific signatures of the flows.

1. INTRODUCTION

A key to the evolution of the planet-forming material in protostellar disks is the angular momentum. The angular momentum per unit mass of gas orbiting just above the star's surface is only 1% that of gas near the disk's outer edge. Sustaining the accretion on the star therefore requires taking almost all the angular momentum out of the accreting matter. Since the angular momentum cannot be destroyed, it must be handed off to other material. The gas receiving the angular momentum could move radially outward, or could join outflows escaping above and below the disk. The gas losing angular momentum and spiraling inward releases gravitational potential energy into turbulence, vortices, or magnetic dissipation, which can stir and heat the flow.

Angular momentum transport processes that could be important under the right circumstances include magneto-

hydrodynamical (MHD) turbulence initiated by the magneto-rotational instability, outflows, hydrodynamical processes, and gravitational instability. In several of these the large-scale transport is built up from smaller-scale gas motions such as turbulent eddies, spiral waves, or long-lived vortices. In addition to governing the flow of gas to the star, these small-scale motions can control the mixing of gas molecules and dust grains along the radial and vertical gradients in temperature, density, and radiation intensity (*Fromang and Papaloizou,* 2006; *Ciesla and Cuzzi,* 2006; *Turner et al.,* 2010; *Semenov and Wiebe,* 2011). The small-scale motions can also alter the growth of dust grains into bigger bodies (*Brauer et al.,* 2008; *Okuzumi,* 2009; *Zsom et al.,* 2010), the collisional evolution of planetesimals (*Gressel et al.,* 2011), and the orbital migration of protoplanets (*Nelson and Papaloizou,* 2004).

In this chapter we review what is known about the basic operation of these transport processes and when and in what regions they work. We restrict ourselves to processes originating in the disk. For example, we do not consider either the influence of a companion star (*Lin and Papaloizou,* 1993), the shear between the disk surface and infalling material (*Ruzmaikina,* 1982), or the interaction of the disk with the young star's magnetosphere (*Romanova et al.,* 2013). Furthermore, we devote the most attention to work published since the previous volume in this series (*Reipurth et al.,* 2007).

We examine the magneto-rotational instability's (MRI) linear growth (section 2) and development into saturated turbulence (section 3), disk-driven winds (section 4), and self-gravitational (section 5) and hydrodynamical instabilities (section 6). Finally, we discuss the prospects for observing the signatures of each transport process (section 7) and the outlook for the next few years (section 8).

2. THE LINEAR MAGNETO-ROTATIONAL INSTABILITY

Study of the linear MRI is warranted by the desire to understand the mechanism driving MHD turbulence and the conditions under which driving occurs, so we can map the MRI-active regions in protoplanetary disks. The disks' weak ionization can lead to magnetic fluctuations being damped. While linear analysis does not address the nonlinear saturated state, it provides a good check on numerical codes during the initial departure from equilibrium. Local linear analyses have also proven surprisingly good at pinpointing the boundaries of the MHD turbulent regions in stratified simulations.

We first review the behavior of the MRI in ideal MHD and then describe the different behaviors in the presence of Ohmic and ambipolar diffusion and Hall drift. We conclude with a discussion of the ionization equilibrium and the implications for magnetic activity.

2.1. Magneto-Rotational Instability Under Ideal Magnetohydrodynamics

Magneto-rotational instability in ideal MHD Couette flows was discovered in the late 1950s (*Chandrasekhar,* 1960; *Velikhov,* 1959), but its astrophysical applications were appreciated only with rediscovery in the context of accretion disks (*Balbus and Hawley,* 1991). The basic mechanism relies on angular momentum transfer by magnetic tension coupling fluid elements at neighboring radii. As the fluid elements drift apart under the orbital shear, magnetic tension transfers angular momentum from the inner fluid element to the outer one. The inner element moves inward and speeds up, and the outer moves outward and slows down, increasing the rate at which the elements drift apart and leading to a runaway. The key is that the magnetic field is weak enough so magnetic tension cannot overcome the tendency of the fluid elements to separate due to the shear.

The result is a robust instability, relying only on a magnetic field embedded in a rotating shear flow in which angular momentum decreases outward. For a simple initially vertical magnetic field, the most unstable mode is axisymmetric with growth rate $\frac{3}{4}\Omega$ (*Balbus and Hawley,* 1991) and vertical wavelength near $2\pi v_{Az}/\Omega$. This rate has proven to be a general maximum in collisional plasmas under Keplerian rotation. The growth rate is reduced if the plasma beta parameter $\beta = P_{gas}/P_{magnetic} < 1$, because the fastest growing mode no longer fits within the disk thickness. Adding a toroidal magnetic component such that $\beta < 1$ reduces the fastest linear growth rate without changing the wavelength much. The growth rate approaches zero as $\beta \to 0$ if the field is within 30° of toroidal (*Kim and Ostriker,* 2000). For strictly toroidal fields, the fastest modes' azimuthal wavelength again is near $2\pi v_{A\phi}/\Omega$, but the vertical wavelength is vanishingly small (*Terquem and Papaloizou,* 1996). As with inclined fields, growth is slower if $\beta \leq 1$ (*Kim and Ostriker,* 2000). When the toroidal fields' $\beta \leq 10$, the Parker magnetic buoyancy instability grows faster than the MRI (*Terquem and Papaloizou,* 1996) and expels some of the magnetic energy into the disk atmosphere (*Hirose and Turner,* 2011).

2.2. Magneto-Rotational Instability Under Ohmic Magnetohydrodynamics

The effect of Ohmic resistivity η on the MRI is straightforward: It tends to erase magnetic fluctuations on length scales shorter than η/v_A. The important length scale for the MRI is the fastest-growing wavelength $2\pi v_A/\Omega$, so the dimensionless number that determines its effect on the MRI is the Elsasser number $\Lambda = v_A^2/(\eta\Omega)$. Local and global linear instabilities are damped when the time to diffuse across the fastest-growing wavelength is less than the growth time, corresponding to $\Lambda < 1$ (*Jin,* 1996; *Sano and Miyama,* 1999). This linear criterion is also a good estimator for the onset of turbulence (*Sano and Inutsuka,* 2001; *Turner et al.,* 2007).

2.3. Magneto-Rotational Instability Under Ambipolar Magnetohydrodynamics

Magneto-rotational instability under ambipolar diffusion behaves similarly to the Ohmic case, but there are some subtle differences because ambipolar diffusion does not dissipate field-parallel currents. For strictly vertical initial fields, there are no field-parallel currents in the initial and perturbed states and the linear MRI behaves as in the Ohmic case, but with Λ replaced by $Am = v_A^2/(\eta_A\Omega)$, where η_A is the diffusivity due to ambipolar drift (*Blaes and Balbus,* 1994; *Wardle,* 1999).

This degeneracy is broken if the initial field has toroidal and vertical components. Then perturbations with a restricted range of wave-vector orientations may be destabilized by ambipolar diffusion. This only partly compensates for the lower growth rate associated with the tilt in the field or with Ohmic resistivity if this is also present (*Desch,* 2004; *Kunz and Balbus,* 2004), and so appears to be of minor consequence.

2.4. Magneto-Rotational Instability Under Hall Magnetohydrodynamics

Hall drift, characterized by transport coefficient η_H and dimensionless number $\Lambda_H = v_A^2/(\eta_H\Omega)$, differs fundamentally from resistivity and ambipolar diffusion. It makes magnetic field lines drift through the gas in the direction of the current density, giving a tendency to twist; no dissipation is involved. Hall drift therefore permits the magnetic field to restructure in strange ways with no energy penalty. For example, Hall drift tends to rotate the MRI-induced buckling of an initially vertical field clockwise about the initial direction (*Wardle*, 1999). If this is parallel to the rotation axis, Hall drift acts in concert with the Keplerian shear in generating toroidal from radial field. The maximum growth rate remains at $\frac{3}{4}\Omega$ but the most unstable wavelength becomes longer. If the initial field and rotation are antiparallel, Hall drift acts against the Keplerian shear, and completely suppresses the MRI if $\Lambda_H < 0.5$.

Because Hall drift acts directly with or against the Keplerian shear in generating B_ϕ from B_R, it strongly modifies the linear MRI when $\Lambda_H < 1$, irrespective of the magnitudes of Λ and Am. For the "favorable" case $B_z > 0$, if $\Lambda_H < \min(\Lambda, Am)$ the maximum growth rate is $\frac{3}{4}\Omega$ even if Λ and/or Am are small. Even if Λ and/or Am $< \Lambda_H$, Hall drift mitigates the damping effects of Ohmic or ambipolar diffusion, increasing the effective Elsasser number by the factor $\Lambda_H^{-1/2}$ (*Wardle*, 1999; *Wardle and Salmeron*, 2012). A toroidal component in the initial field introduces trigonometric corrections to the MRI growth rate and wavelength, and additional mitigation of Ohmic damping by ambipolar diffusion (*Pandey and Wardle*, 2012).

Hall drift is able to restructure the field with no associated dissipation. Its ability to mitigate or exacerbate damping of the MRI, and the fact that Ohmic and ambipolar diffusion and Hall drift all scale differently with the field strength B, may give rise to interesting behavior in the nonlinear state.

2.5. Ionization and Recombination

The transport coeficients η, η_A, and η_H depend on the abundances of the charged particles in the weakly ionized gas in protoplanetary disks. The ionization sources in protoplanetary disks are thermal collisions, stellar ultraviolet (UV) and X-ray photons, interstellar cosmic rays, and the decay of radionuclides. Collisional ionization is important at $\gtrsim 1000$ K, temperatures found only within 0.1–1 AU of the star depending on the mass flow rate, and also high in the disk atmosphere. Only the elements with the lowest ionization potentials are ionized, so the ionization fraction is capped at the gas-phase abundance of the most common alkali metals.

Far-ultraviolet continuum photons are absorbed in the disk's uppermost 0.001–0.1 g cm^{-2} (*Bergin et al.*, 2007; *Perez-Becker and Chiang*, 2011a) unless absorbed first by an intervening disk wind emerging from smaller radii (*Ferro-Fontán and Gómez de Castro*, 2003; *Panoglou et al.*, 2012). Lyman-α photons propagate by resonant scatter-

ing and therefore may enhance ionization rates somewhat deeper in the disk (*Bethell and Bergin*, 2011).

Stellar X-rays are absorbed in the uppermost 10 g cm^{-2} (*Igea and Glassgold*, 1999; *Mohanty et al.*, 2013), yielding unattenuated ionization rates around 10^{-10} s^{-1} at 1 AU for the median X-ray luminosity of 10^{30} erg s^{-1} measured among young solar-mass stars in Orion (*Garmire et al.*, 2000).

Interstellar cosmic rays are largely screened out by the young star's wind, according to an extrapolation of the solar relation between spot coverage and cosmic-ray modulation to young stars' greater spot-covering fractions (*Cleeves et al.*, 2013). On the other hand, if the stellar wind is restricted to the poles and the incoming cosmic rays interact mostly with a disk wind threaded by poloidal magnetic fields, the energetic particles are both focused and mirrored by the fields' pinching toward the equatorial plane. The first increases and the second decreases the ionization rate, on balance yielding only small changes compared with ambient conditions (*Desch et al.*, 2004; *Cleeves et al.*, 2013). If cosmic rays do reach the disk, then they will penetrate much deeper than X-rays, to 100 g cm^{-2} (*Umebayashi and Nakano*, 1981, 2009), but yield a lower unattenuated ionization rate of about 10^{-17} s^{-1}.

The radionuclides ionize at lower rates, and the decay products are absorbed inside the solid material if contained in particles bigger than ~1 mm (*Umebayashi et al.*, 2013). Most of the radionuclide ionization comes from ^{26}Al, if present. Its half-life of 0.7 million years (m.y.) and initial abundance inferred from decay products in primitive meteorites yield an ionization rate $(7–10) \times 10^{-19}$ s^{-1} (*Umebayashi and Nakano*, 2009). Clearly there is a great deal of variation in the overall ionization rate with distance from the star and depth in the disk.

The electrons and molecular ions created by ionizations may recombine directly, or by charge transfer to metal ions followed by radiative recombination (e.g., *Umebayashi and Nakano*, 1990; *Ilgner and Nelson*, 2006a). The latter rate is relatively low, making metal ions the dominant positive species in the gas phase except at the highest ionization levels. At low temperatures the metal atoms become adsorbed on grain surfaces, removing them from the picture. Dust grains play a major role through the competitive sticking of electrons and ions from the gas phase and subsequent recombination on the grains (*Sano et al.*, 2000).

For a given grain population, the important parameter is ζ/n_H, the ratio of the ionization rate per H nucleus to the number of nuclei per unit volume. At high ionization levels most charge resides in the gas phase. Grains acquire a net negative charge because of the higher thermal speed of the electrons, but the fraction of negative charge held by grains is small. As an electron is more likely to encounter an ion in the gas phase than a grain, the majority of recombinations occur in the gas phase. At intermediate ionization levels, electrons are more likely to stick to a grain surface before encountering an ion in the gas phase. In this regime, most recombinations occur via the sticking of ions to negatively charged grains. At low ionization levels, most grains are uncharged, and ions and electrons tend to stick to neutral grains before recombining. Recombinations occur primarily via the collision of oppositely charged grains.

2.6. Conductivity

Ionization fractions are highest in the disk atmosphere, where the UV and X-ray photons are absorbed (in the upper 0.01 and 10 g cm^{-2}, respectively), and low densities reduce the recombination rates. Deeper in the disk, the ionization levels plummet due to shielding by the overlying layers and the increase in recombination rates with density.

The overall ionization level and the relative abundances of ions, electrons, and charged grains therefore vary strongly with depth, with concomitant changes to the Ohmic and ambipolar diffusivities and Hall drift. The relative importance of the non-ideal terms is determined by the ratio n_H/B, which controls the degree to which neutral collisions decouple charged species from the magnetic field. At successively higher densities, first grains, then ions, and finally electrons are decoupled. Roughly speaking, ambipolar diffusion dominates at low densities on strong magnetic fields, when ions and electrons remain coupled. The Ohmic term typically dominates at the high densities near the mid-plane and on weak magnetic fields. The Hall term dominates over a broad range of intermediate conditions, when ions are decoupled from the fields by collisions with neutrals, but electrons are not (*Wardle*, 2007; *Salmeron and Wardle*, 2008).

Ohmic diffusivity is always determined by the electron fraction even when electrons are not the most abundant charged particle. Ambipolar diffusivity is typically determined by the ion density, but can be set by the charged grains if these are small and abundant (*Bai*, 2011b); at low ionization levels, charged grains again play a role. The Hall drift is more complex, as it is controlled by the mismatch in the degree to which neutral collisions decouple positive and negative species from the magnetic field. At low densities negatively charged grains are decoupled and ions and electrons are coupled, implying $\eta_H < 0$. At intermediate densities ions decouple too and $\eta_H > 0$.

2.7. Active Layers and Dead Zone

A simple local criterion for MRI turbulence is that the linear instability has an unstable mode with wavelength shorter than the disk scale height — i.e., Ohmic and ambipolar diffusion are not too severe. Hall drift enters by modifying the twisting of the field in response to the Keplerian shear, and may aid or hinder the growth. By way of example, for vertical fields this criterion may be written

$$\left(\frac{1}{\Lambda} + \frac{1}{Am}\right)^2 < \left(2 + \frac{s}{\Lambda_H}\right)\left(\frac{3\beta}{16\pi^2} - \frac{1}{2} - \frac{s}{\Lambda_H}\right) \qquad (1)$$

where $s = \text{sign}(B_z)$ [*Wardle and Salmeron* (2012) used $1/k$ rather than λ in the criterion $kh = 1$, so in their equivalent expression β has a coefficient 3/4]. This criterion describes

reasonably well whether turbulence develops in simulations in unstratified boxes in resistive MHD (*Sano and Inutsuka*, 2001), ambipolar diffusion (*Hawley and Stone*, 1998; *Bai and Stone*, 2011), and Hall drift (*Sano and Stone*, 2002a,b; *Kunz and Lesur*, 2013). The criterion also adequately describes the height of the boundary in stratified local simulations treating resistivity (*Turner and Sano*, 2008; *Okuzumi and Hirose*, 2011).

Most assessments of the dead zone extent in protoplanetary disks included only Ohmic resistivity. However, ambipolar diffusion is more effective in the surface layers. The MRI turbulence's upper and outer edges tend to be set by ambipolar diffusion (*Perez-Becker and Chiang*, 2011b; *Bai*, 2011a) or by the fields' stiffness at high magnetic pressure (*Miller and Stone*, 2000; *Turner et al.*, 2010). Polycyclic aromatic hydrocarbon grains (PAHs), if found at abundances similar to those in Herbig Ae/Be disks, would limit the turbulence to a thin layer ionized by the stellar far-UV (FUV) in the disk near the star (*Perez-Becker and Chiang*, 2011a,b), but at the lower densities found further out, PAHs enhance magnetic coupling (*Bai*, 2011b).

Hall drift plays a critical role throughout much of the disk (Fig. 1), increasing or decreasing the estimated active column by orders of magnitude (*Wardle and Salmeron*, 2012). However, for $\Lambda_H \leq 5 \, v_A/c_s$, the Hall drift leads to most of the magnetic flux being confined into zonal regions in unstratified calculations, shutting down turbulent transport (*Kunz and Lesur*, 2013). This has the potential to greatly reduce the extent of the MRI turbulent layer. The following section deals with the saturation of the MRI and the properties of the resulting flows.

3. SATURATION OF MAGNETOHYDRODYNAMICAL TURBULENCE

While the last section covered the linear MRI, this one focuses on the nonlinear evolution. By the mid-1990s it was established that the MRI robustly leads to MHD turbulence, transporting angular momentum outward (*Hawley and Balbus*, 1992; *Hawley et al.*, 1995, 1996; *Brandenburg et al.*, 1995). However, in recent years, our view of disk turbulence has been challenged by subtle effects both numerical and physical, thanks to faster computers and a better understanding of the microphysics. The purpose of this section is to review these findings and discuss the saturation amplitude of the turbulence. The amplitude is commonly measured by α, the ratio of accretion stress to gas pressure. The stress $-B_R B_\phi/4\pi + \rho v_R \delta v_\phi$ depends on the density ρ and the radial and azimuthal components of the magnetic field and velocity, the latter measured relative to the background orbital rotation (*Hawley et al.*, 1995).

3.1. Numerical Approaches

Many of the codes now in use employ variations of the conservative Godunov scheme. This is the case for example in Athena (*Stone et al.*, 2008), Pluto (*Mignone et al.*,

2007) and Nirvana-III (*Ziegler,* 2005, 2008). Alternatives include the pseudo-spectral code Snoopy (*http://ipag.osug. fr/~glesur/snoopy.html*) and the sixth-order finite difference Pencil Code (*http://pencil-code.nordita.org*), while Zeus (*Stone and Norman,* 1992; *Hawley and Stone,* 1995), which dominated the early days, is still used by some teams. With this wealth of methods, a large variety of setups has been developed for studying the MRI's nonlinear evolution. Here we have space for only a sampling.

The most popular of all the approaches is the local shearing box model. The differential rotation is linearized in a small Cartesian volume rotating around the central object. Compressible and incompressible formulations are used and the vertical stratification can be included (Fig. 2). The shearing box is useful as a simplified framework for the dynamics. The problem is physically well posed and boundary conditions are straightforward. High numerical resolutions can be reached and the consequences of dissipative processes investigated. We detail recent findings in sections 3.2–3.5. Of course, the shearing box's advantages come at a cost: Curvature terms are ignored and large-scale gradients are neglected. These limitations have motivated the development of global simulations treating the radial structure. Spatial resolution is sacrificed, and boundary conditions require care, but in other respects the realism is better and the results can be compared directly with observations. We review recent global findings in section 3.6.

A key aspect of all simulations is the magnetic field topology. In the shearing box this is particularly important, because the magnetic flux threading the domain is conserved over time due to the boundary conditions: For example, any net (domain-averaged) vertical field present initially remains for all time. Different configurations have been investigated: Vertical and/or toroidal fields yield "net flux simulations" in which the ever-present magnetic flux is a permanent seed for turbulence. Configurations with vanishing mean magnetic field have also been investigated. In this peculiar situation, the flow must constantly regenerate its own magnetic field through a dynamo process.

3.2. Ideal Magnetohydrodynamical Simulations

Here we describe the lessons learned in recent years using local numerical calculations neglecting all explicit non-ideal magnetic effects. To further simplify the problem, the equation of state is most often taken to be isothermal. The flow then depends *a priori* on the magnetic field topology, box size, grid resolution, and whether the vertical stratification is included.

The special zero-net-flux or dynamo case has proven to be peculiar. In the absence of vertical stratification, MRI-driven turbulent activity decreases as resolution is increased (*Fromang and Papaloizou,* 2007; *Guan et al.,* 2009; *Simon et al.,* 2009; *Bodo et al.,* 2011), with no sign of converging to a well-defined value even at resolutions up to 256 cells per density scale height. This is a surprise because when extrapolated to infinite resolution, it suggests the flow remains laminar in the absence of a net magnetic flux. It is still not clear whether this result is a numerical artifact, as claimed early on (*Fromang and Papaloizou,* 2007), or is due to limitations of the shearing box (*Kitchatinov and Rüdiger,* 2010; *Bodo et al.,* 2011).

Nonconvergence, however, appears to be specific to the homogeneous shearing box without net magnetic field. When the density stratification is treated, numerical convergence is reached (*Davis et al.,* 2010; *Shi et al.,* 2010). It is unclear why this differs from the homogeneous case. A suggestion is that the episodic escape of toroidal magnetic fields revealed in the so-called butterfly diagram (*Miller and Stone,* 2000; *Shi et al.,* 2010; *Hirose and Turner,* 2011) mediates a mean-field dynamo (*Gressel,* 2010). Care is needed because the grid resolutions in stratified simulations are limited, owing to the high Alfvén speeds in the disk corona that necessitate timesteps shorter than in the homogeneous case. The lack of convergence of turbulent transport also disappears when a net magnetic flux, toroidal (*Guan et al.,* 2009) or vertical (*Simon et al.,* 2009), threads the computational domain. Again, the reason for the difference compared with zero-net-flux boxes has not been identified, even if it seems natural

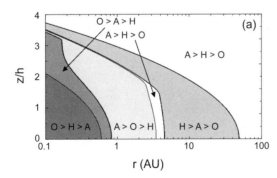
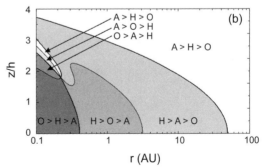

Fig. 1. Effect of grains on magnetic diffusion and drift in the minimum-mass solar nebula where $\Sigma = 1700 \, (r/AU)^{-1.5}$ g cm^{-2}, $T = 280 \, (r/AU)^{-0.5}$ K, and B is constant with height and chosen so $v_A = 0.1 \, c_s$ at the mid-plane. The axes are the radius in astronomical units and height in scale-heights. Shading indicates the ordering of the Ohmic, Hall, and ambipolar diffusivities for a population of 0.1-μm grains with **(a)** 1% or **(b)** 0.01% of the gas mass, under ionization by X-rays, cosmic rays, and radioactivity.

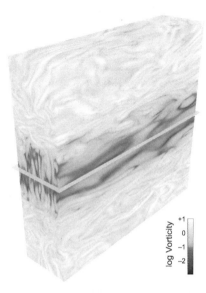

log Vorticity

+1
0
−1
−2

Fig. 2. Stratified shearing-box calculation of a small patch of protostellar disk. The Ohmic resistivity is low enough for magneto-rotational turbulence only in the X-ray ionized layers on the disk's top and bottom surfaces. Colors indicate the vorticity $|\nabla \times \mathbf{v}|$ of the velocity $\mathbf{v}$ relative to the Keplerian shear. The ionized, turbulent layers show folded eddy structures, while the flow near the mid-plane is dominated by shearing waves (*Gressel et al.,* 2011).

to relate it to the linear MRI modes in that case. With a net vertical magnetic field, the turbulent activity depends on the box's radial extent (*Bodo et al.,* 2008), and this comes from the linear MRI modes and how they are destabilized by parasitic instabilities.

Clarifying these issues will be tremendously valuable. However, regardless of its origins, the convergence issue highlighted in this section teaches us that the grid resolution in current numerical simulations affects the outcome in some cases, and thus dissipation should be treated explicitly. This is the topic of the following sections.

3.3. Magneto-Rotational Instability with Ohmic Resistivity

To address the convergence issue, several investigations have included a constant Ohmic resistivity, η, often along with a constant kinematic viscosity ν. The result that has emerged is that both dissipation coefficients are important: α is an increasing function of their ratio, the magnetic Prandtl number $Pm = \nu/\eta$ (Fig. 3). This result is robust, having appeared in homogeneous shearing boxes with net vertical (*Lesur and Longaretti,* 2007; *Longaretti and Lesur,* 2010) and toroidal (*Simon and Hawley,* 2009) magnetic fluxes, with no net flux (*Fromang et al.,* 2007), and in stratified boxes without net flux, although it appears weaker in that case (*Simon et al.,* 2011a). As an example, in homogeneous boxes with a vertical magnetic field having $\beta = 10^3$, α ranges from 2×10^{-2} to 10^{-1} when Pm varies from 0.0625 to 1, all

else being equal. Recall that in the ideal MHD limit (section 3.2) this case shows numerical convergence. Clearly, numerical and physical convergence are two different things, and small-scale dissipation matters even in situations that appear numerically converged, at least over the limited range of Reynolds numbers that is numerically accessible.

These results raise questions about the flow structure in the inner parts of protoplanetary disks where the resistivity is much larger than the viscosity (*Balbus and Henri,* 2008) (Pm $\ll$ 1). The problem is that this regime is numerically challenging to investigate as it requires extremely large spatial resolution: The smallest published value of Pm amounts to 1/16. Does α approach an asymptote at lower Pm? If an asymptote exists, how small is α there, especially when stratification is included? The broad implications of these questions make them a focus of ongoing research.

While these results highlight the properties of the MRI in the inner parts of protoplanetary disks, they are not directly applicable to the planet-forming region of protoplanetary disks, where magnetic coupling is good only in the surface layers. In the scenario suggested early on by *Gammie* (1996), the disk is divided into a laminar equatorial "dead zone" and MRI-turbulent surface "active layers." Modeling this scenario, *Fleming and Stone* (2003) used an Ohmic diffusivity that is a fixed function of time and height above the equatorial plane. The MRI is stabilized in the dead zone while the active layers are turbulent. The dead zone shows hydrodynamical stresses excited by sound waves propagating from the active layers, confirmed by *Oishi and Mac Low* (2009), among others. Another modification to the original scenario is that large-scale mean magnetic fields can diffuse downward to transport angular momentum within the dead zone itself (*Turner and Sano,* 2008). The α values reported are around 10^{-4}. Semi-analytic recipes are now available describing the vertical structure of such layered disks (*Okuzumi and Hirose,* 2011).

Two further aspects of the layered disk picture have been investigated intensively since the previous volume in this series. The first involves the changes over time in the Ohmic diffusivity profile as a function the local flow properties (*Ilgner and Nelson,* 2006b; *Turner et al.,* 2007; *Ilgner and Nelson,* 2008). The second involves better treating the disk thermodynamics (*Flaig et al.,* 2010; *Hirose and Turner,* 2011). The vertical distribution of heating differs significantly from that assumed in α-models.

3.4. Magneto-Rotational Instability with Hall Effect

The first investigation of the MRI with Hall effect was performed by *Sano and Stone* (2002a,b), who found that MRI turbulence with both Hall and Ohmic terms was rather similar to MRI turbulence with the Ohmic term alone. The stress depended weakly on the Hall parameter, α increasing with η_H. However, *Wardle and Salmeron* (2012) pointed out that there was room for stronger effects in the Hall-dominated regime, where the Hall term is the largest in the induction equation.

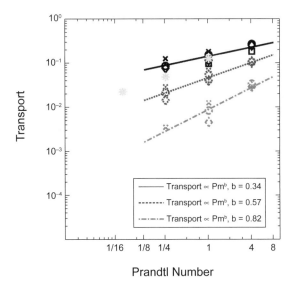

Fig. 3. Turbulent angular momentum transport coefficient α vs. Prandtl number Pm and plasma β for various Reynolds numbers. Solid, dotted, and dot-dashed indicate β = 10^2, 10^3, and 10^4. Squares, pluses, diamonds, circles, and x's mark Re = 400, 800, 1600, 3200, and 6400. A power-law fit is shown for each value of β. Stars mark three higher-resolution runs at Re = 20,000 and β = 10^3 (*Longaretti and Lesur,* 2010).

The problem was revisited by *Kunz and Lesur* (2013), who showed that under a Hall term 10–100× stronger than in *Sano and Stone* (2002b), the MRI develops into self-sustaining large-scale zonal magnetic field structures that produce only very weak angular momentum transport (α ~ 10^{-5}). The transition to this regime happens where $\Lambda_H \lesssim 5\, v_A/c_s$. Therefore, even if the linear analysis shows fast growth at large scales, the instability does not necessarily lead to MHD turbulence and enhanced angular momentum transport. Note that these results are from unstratified shearing boxes. It is thus too soon to rule out Hall-dominated MRI as a source of turbulent transport in disks, since stratification is likely to modify these results.

3.5. Magneto-Rotational Instability with Ambipolar Diffusion

Numerical implementation of ambipolar diffusion comes in two flavors. In the two-fluid approach, ions and neutrals are treated as separate fluids coupled by collisional drag, but typically not by chemistry, with the magnetic field carried by the ions. In the single-fluid approach the neutrals are the fluid, with the ions, whose density is determined by equilibrium chemistry, providing dissipation through ion-neutral drag. Generally speaking, the two-fluid approach applies to well-ionized systems with slow recombination, while the single-fluid approach applies in the opposite ("strong-coupling") limit. Chemical modeling indicates the strong-coupling limit is almost always valid in protostellar disks (*Bai,* 2011a). Early on, *Mac Low et al.* (1995) incorporated ambipolar diffusion in their single-fluid simulations, but a systematic study was

lacking. *Hawley and Stone* (1998) studied the MRI with ambipolar diffusion using the two-fluid approach, so the results do not apply directly to protostellar disks.

Bai and Stone (2011) made a series of unstratified shearing-box simulations of the MRI with ambipolar diffusion in the strong-coupling limit. They systematically explored the ambipolar diffusivity, magnetic field strength, and field geometry, finding that the MRI operates at any value of Am as long as the field is weak enough that the most unstable mode fits within the disk scale height, which is consistent with linear analysis (section 2.3). The saturation level of the turbulence depends on field strength and configuration, and when ambipolar diffusion is strong (Am < 10), fields with comparable net vertical and toroidal components are best, while a pure toroidal field suppresses the MRI. Summarizing their results, the saturation level of the MRI turbulence with the most favorable field geometry is

$$\alpha_{max} \approx \frac{1}{2}\left[\left(\frac{50}{Am^{1.2}}\right)^2 + \left(\frac{8}{Am^{0.3}}+1\right)^2\right]^{-1/2} \quad (2)$$

The next step has been to treat ambipolar diffusion in stratified simulations. Ambipolar diffusion is the dominant non-ideal MHD effect in low-density regions, including the surface layer of the inner disk together with the entire outer disk (section 2). For the inner disk at 1–10 AU, *Bai and Stone* (2013b) and *Bai* (2013) performed stratified shearing-box simulations that include both Ohmic resistivity (dominant in the mid-plane) and ambipolar diffusion (dominant at the disk surface) with diffusion coeficients based on the equilibrium chemistry calculations of *Bai* (2011a), as well as the FUV ionization model of *Perez-Becker and Chiang* (2011a). They found that without net vertical magnetic flux, the system evolves into extremely weak turbulence in the FUV layer, far too weak to drive appreciable accretion. When a net vertical magnetic flux is included, even if the initial field configuration is subject to the MRI, the instability is suppressed and the disk's whole vertical extent relaxes into a laminar state, while a wind is launched and efficiently extracts angular momentum from the disk (section 4). In both cases the MRI is suppressed by ambipolar diffusion. For the outer disk at ≥30 AU, *Simon et al.* (2013a,b) performed similar calculations and found first that net vertical magnetic flux is essential, otherwise the resulting MRI turbulence is too weak. In the presence of net vertical flux, the turbulence is stronger and takes place mainly in the surface FUV layer. Both results qualitatively agree with expectations from unstratified simulations (*Bai and Stone,* 2011), except that with increasing net vertical magnetic flux, large-scale field comes to dominate over turbulent field as the driver of angular momentum transport.

3.6. Global Calculations

Local simulations are invaluable for studying turbulence, but their reach is limited. Thus several teams have embraced the challenges of global simulations. Resolutions do not yet approach those achieved in the shearing box, where >100 cells per scale height is commonplace. Consequently it is difficult at present to include small-scale dissipation in global work. The only successful example is *Dzyurkevich et al.* (2010), where the Ohmic diffusion is treated. Given all the subtleties uncovered when the non-ideal effects were taken into account in shearing boxes, doubts remain about our ability to characterize the saturation of the turbulence through global calculations. The global results to date should therefore be interpreted with care. Nevertheless, progress has been made thanks to growing computer power and there have been claims that the best-resolved models are numerically converged (*Sorathia et al.*, 2012). Of course the lesson of the shearing box is that numerically converged simulations do not necessarily yield the correct stresses, as these can depend on effects not treated (*Hawley et al.*, 2013).

The basic result of global simulations is boring and reassuring at the same time: The transport is consistent with shearing box simulations under similar conditions (*Fromang and Nelson*, 2006). The turbulent velocity dispersion increases with height (*Fromang and Nelson*, 2009; *Flock et al.*, 2011) as in shearing boxes (*Miller and Stone*, 2000). There is growing evidence that large-scale flow features are similar in wide shearing boxes (*Simon et al.*, 2012) and global calculations (*Beckwith et al.*, 2011; *Flock et al.*, 2012). Likewise, the zonal flows, axisymmetric variations in the orbital speed that persist for many orbits, seem to be a natural outcome of MHD turbulence in both setups (*Johansen et al.*, 2009; *Uribe et al.*, 2011; *Dittrich et al.*, 2013).

Some issues can only be addressed using global calculations. For example, *Sorathia et al.* (2010) reported that small patches of disk behave as if threaded by a vertical net field, even if the total vertical flux threading the disk as a whole vanishes. Such fields connect the disk atmosphere with its equatorial plane, suggesting that coronal magnetic fields are central in determining the accretion flow (*Beckwith et al.*, 2011). Large-scale flow in protoplanetary disks, important for our understanding of particle transport, remains poorly known. The prediction from traditional α-models of a meridional circulation, with matter flowing outward near the mid-plane while accretion proceeds through the upper layers, fails in fully turbulent (and thus unrealistic) disks (*Fromang et al.*, 2011; *Flock et al.*, 2011). It is likely that mixing within disks made up of both active and dead zones differs from the models published so far.

Besides the resolution issue already mentioned, global simulations suffer from two limitations of special importance. First, the equation of state is typically locally isothermal with no explicit heating and cooling. These are potentially important for the flow in the disk corona, the structure of the disk interior, and the dead zone nearest the star. Second, no published global simulation of turbulent protoplanetary disks yet includes a net vertical magnetic flux. Given the differences between shearing-box results with and without vertical fluxes (sections 3.5 and 4), such simulations would be extremely valuable and should be a focus in coming years.

3.7. Outstanding Issues

While small-scale dissipation has been (and still is!) studied in detail, the results have mainly taught us to be humble: More than 20 years after the discovery of the MRI, we still do not know how the turbulence saturates as a function of parameters such as the surface density, temperature, and magnetic flux.

The question of how disk dynamos work is still debated. The results of the last few years suggest that brute force is of limited interest due to the enormous resolutions required. An interesting alternative is provided by the discovery of limit cycles near marginal stability (*Herault et al.*, 2011), which might bear some significance for fully developed turbulence.

Perhaps most importantly for the transport, all the results based on the MRI linear properties agree in their predictions that protoplanetary disks are MRI-inefficient at distances from the star between about 1 AU and 10 AU (*Turner and Drake*, 2009; *Bai*, 2011a,b; *Perez-Becker and Chiang*, 2011a,b; *Mohanty et al.*, 2013; *Dzyurkevich et al.*, 2013). In this section, we have seen that many predictions of the linear stability analyses carry over to the nonlinear regime, with the notable exception of the Hall effect. Together these results suggest that something is missing in our understanding of the basic processes driving the accretion flows in the planet-forming region. This has revived interest in an old alternative transport mechanism, namely disk winds, the focus of the next section.

4. DISK-DRIVEN WINDS AND MAGNETOHYDRODYNAMICAL TURBULENCE

In previous sections we explored the possibility of driving accretion by transporting angular momentum inside the disk through the MRI. However, this is not the only way accretion can occur. An alternative is to extract the angular momentum vertically by applying a torque at the disk surface. Such a torque usually comes from a large-scale wind driven by magnetic processes. This section is dedicated to the launching of these winds and their impact on accretion in protoplanetary disks. Jets formed as disk winds propagate to large distances are not discussed in this section. The interested reader may refer to the chapter by Frank et al. in this volume.

To understand how a cold wind can drive accretion, consider angular momentum conservation in a thin disk

$$\overline{\rho v_R} \frac{\partial \Omega R^2}{\partial R} = \frac{1}{R} \frac{\partial}{\partial R} R^2 \left[\overline{B_\phi B_R} - \overline{\rho v_\phi v_R} \right] + R \left[B_\phi B_z - \rho v_\phi v_z \right]_{z=-h}^{+h} \quad (3)$$

where Ω is the Keplerian frequency, h is the disk scale-height, and overbars mean $\int_{-h}^{h} dz$. Equation (3) shows that accretion (negative $\overline{\rho v_R}$) can be driven either by angular momentum transport inside the disk (first term on the righthand side) or by a torque exerted at the disk surface (second term on the righthand side). It is the second term that connects winds to accretion in the bulk of the disk. Equation (3) also shows that a magnetically driven wind can easily dominate accretion in thin disks. Neglecting the kinetic term, one easily deduces that the turbulent torque $\sim hB_\phi B_R$, while the wind torque $\sim RB_\phi B_z$. Assuming $B_\phi B_R \sim B_\phi B_z$ within the disk, one finds that the torque due to the wind is $\sim R/h \times$ larger than the torque due to turbulence.

4.1. Launching Mechanism

Protoplanetary disks are thin ($h/R \lesssim 0.1$) and dynamically cold, so winds cannot be launched exclusively by thermal processes at the disk surface. Driving a wind involves accelerating the flow to escape speed, which requires some extra power source. The power can come either from heating the gas as it is ejected (see the chapter by Alexander et al. in this volume) or from rotational energy extracted using a large-scale magnetic field, a process known as magneto-centrifugal acceleration (*Blandford and Payne,* 1982). The latter is extensively described in the literature [for an introduction, see *Spruit* (1996)], so we here give only a brief qualitative picture.

Consider a frame co-rotating with the disk at radius R_0 from the central star. In this frame, fluid parcels feel an effective potential combining the gravitational and centrifugal potentials. If poloidal (R,z) magnetic field lines act as rigid wires for fluid parcels, then a parcel initially at rest at ($R = R_0$, $z = 0$) can undergo a runaway if the field line to which it is attached is more inclined than a critical angle. Along such a field line, the effective potential decreases with distance, leading to an acceleration of magneto-centrifugal origin. This yields the inclination angle criterion $\theta > 30°$ for the disk-surface poloidal field with respect to the vertical. Here fluid parcels rotate at constant angular velocity and so increase their specific angular momentum.

Magnetic field lines act as rigid wires so long as the poloidal flow is slower than the local Alfvén speed — i.e., most of the energy is contained in the magnetic fields. The accelerating flow eventually reaches a poloidal speed exceeding the Alfvén speed, at a location known as the Alfvén point. Above here, magneto-centrifugal acceleration ceases and the flow winds up the field into toroidal coils. From this point on, fluid parcels rotate at constant specific angular momentum.

Angular momentum is thus extracted from the disk first by the magnetic torque at the disk surface ($B_\phi B_z$ in equation (3)) and stored in the magnetic field's toroidal component. As acceleration proceeds, the angular momentum is slowly transferred to the ejected material until the Alfvén point is reached. Winds therefore efficiently drive accretion by purely and simply removing the angular momentum from the disk. Moreover, one can easily see that the efficiency of this process is directly connected to the disk's magnetic

field strength, with a stronger field leading to a bigger torque and therefore faster accretion.

We have seen that a wind implies an accretion flow inside the disk. The poloidal field lines threading the flow are accreted as well, if the ideal MHD is valid. Such a flow must soon stop once a large magnetic flux accumulates near the disk's center. A long-lived wind requires the flow to slip past the field lines. This is usually modeled assuming the disk is either turbulent, producing a turbulent diffusivity (*Ferreira and Pelletier,* 1993, 1995; *Ferreira,* 1997), or subject to strong ambipolar diffusion (*Wardle and Königl,* 1993; *Königl et al.,* 2010; *Salmeron et al.,* 2011), a reasonable assumption at large distances ($\gtrsim$10 AU).

These very simple arguments show three basic requirements for a steady magneto-centrifugal wind: (1) The field is strong enough for the poloidal component to act as a rigid wire. This means that at the base of the wind the plasma $\beta \lesssim 1$, where the magnetic pressure is that in the mean poloidal field. Fluctuations due to turbulence can exceed the mean field, but are not directly relevant for the existence of a steady wind. (2) The field is sufficiently inclined, with $\theta > 30°$. (3) Accreting gas slips through magnetic field lines, via either turbulent diffusion, or Ohmic or ambipolar diffusion.

4.2. Local Disk Wind Solutions

A natural question is whether the wind can coexist with turbulence inside the disk. However, numerically spanning both the small turbulent eddy scales and the large global scales associated with the wind propagation is a challenge. Two complementary approaches are used: (1) Local methods such as the shearing box involve treating the disk up to a few scale heights while neglecting the global geometry. (2) The global approach means approximating turbulence by some crude mean-field method. We focus here on the first approach since our topic is disk transport, but we also discuss the connection to global solutions to demonstrate the limitations of local solutions.

The shearing box outlined in section 3 is shear-periodic in the radial or x-direction and periodic in the toroidal or y-direction. The vertical or z-boundaries have outflow conditions to cope with winds. This approach has several limitations:

1. *Symmetries:* The shearing box does not specify whether the central object lies at $x \to +\infty$ or $x \to -\infty$, leaving the two equivalent. This symmetry is apparent in shearing box results, which can exhibit winds bent toward either large or small x.

2. *Magnetic flux transport:* Because curvature terms are neglected in the shearing box, magnetic flux tubes can cross the box again and again without the fields piling up anywhere, since the flux escaping matches that entering on the box's farside. In other words, shearing boxes allow spurious solutions with constant $E_\phi \neq 0$. This unrealistic situation should be kept in mind when one compares local and global wind solutions.

3. *Boundary conditions:* The third and most stringent limitation comes from the effective potential (gravitational

plus centrifugal) in the shearing box being a local expansion of the real potential around R_0 of the form

$$\psi \propto -\frac{3}{2}x^2 + \frac{1}{2}z^2 \qquad (4)$$

where x and z are respectively local equivalents of $R–R_0$ and z. Clearly this potential does not allow a particle to escape to $z \rightarrow \infty$ if x is kept finite. Because magneto-centrifugal wind solutions cannot propagate radially past the Alfvén radius, disk winds cannot be gravitationally unbound in a finite shearing box. The winds that nevertheless appear owe their existence to the outflow vertical boundary conditions.

Below we show how these limitations affect the numerical results.

All the local numerical wind solutions obtained in recent years can be understood in the framework of the magneto-centrifugal acceleration mechanism described above. Here we distinguish "strong field" solutions, for which magnetic and thermal pressure are near equipartition $\beta_{mid} \sim 1$ in the mid-plane, from "weak field" solutions, where thermal pressure dominates at the mid-plane $\beta_{mid} \gg 1$ (but not necessarily in the disk corona). We furthermore separate the solutions by the non-ideal effects included.

4.2.1. Ideal and weakly resistive winds. Ideal MHD solutions are the simplest obtainable, but of course neglect the non-ideal effects, which can be important in protoplanetary disks. Moreover, under ideal MHD, the numerical convergence of MRI turbulence can be an issue (section 3). To address these limitations, some solutions have been computed in the resistive MHD framework. A weak resistivity and viscosity seem not to affect the wind dynamics (*Fromang et al., 2013*).

The first ideal-MHD solutions exhibiting both a disk wind and the MRI on weak fields ($\beta_{mid} \gg 1$) were obtained by *Suzuki and Inutsuka* (2009). These involve weak poloidal fields $\beta_{mid} = 10^6–10^4$. Standard MRI turbulence appears and a short-lived magneto-centrifugal wind slowly empties the domain, since the shearing-box boundaries allow no replenishment by accretion flow. Because the mid-plane fields are weak, the wind is launched about two scale-heights up, so that $\beta \sim 1$ at the launching surface as expected from the phenomenology presented above. The wind mass loss rate $\dot{M}_W$ increases steeply as β decreases, almost emptying the box after 75 rotations for $\beta = 10^4$. *Suzuki et al.* (2010) explored the consequences of such a wind for global disk models, demonstrating that a wind could empty the inner disk and create an inner hole. *Bai and Stone* (2013a) explored stronger fields up to $\beta = 100$. They confirmed some of the *Suzuki and Inutsuka* (2009) results, showing that $\dot{M}_W \propto 1/\beta$ holds for the β values they explored. However, *Fromang et al.* (2013) and *Bai and Stone* (2013a) demonstrated that $\dot{M}_W$ was highly sensitive to the domain height: $\dot{M}_W \propto 1/L_z$ where L_z is the box vertical size. This is a clear signature of the third shearing-box

limitation detailed above. Box simulations yield unreliable results for the mass flow rate extracted by disk winds and the efficiency of disk dispersal.

The shearing box's extra symmetries also allow the wind to switch erratically between ejection toward larger and smaller radii (*Fromang et al.,* 2013; *Bai and Stone,* 2013b). The configuration switchings happen independently above and below the disk. For example, the upper disk can drive a wind toward larger radii, while the lower disk drives a wind toward smaller radii. Such artificial outcomes are unsatisfactory for a global overview of the wind structure.

The strong-field regime $\beta_{mid} \sim 1$ is well studied for jet formation because it is the only one where self-similar global wind solutions exist (*Ferreira,* 1997). This regime has also been explored in the local approximation. *Ogilvie* (2012) obtained one-dimensional quasisteady strong-field wind solutions for which $\dot{M}_W$ *decreases* sharply when β decreases. Wind solutions possibly even cease to exist for sufficiently strong fields ($\beta_{mid} \simeq 0.3$). This study was revisited in three dimensions by *Lesur et al.* (2013) in a regime where wind solutions of *Ogilvie* (2012) coexist with large-scale MRI modes (*Latter et al.,* 2010). They showed that MRI modes naturally evolve into quasisteady wind solutions when the field is strong enough ($\beta \sim 10$), making a natural connection between the MRI and strong winds. However, similarly to the weak field case, $\dot{M}_W$ depends on the box vertical extent, casting doubt on the conclusions. *Moll* (2012) and *Lesur et al.* (2013) showed that the quasisteady wind solutions of *Ogilvie* (2012) were radially unstable, producing time-dependent wind solutions. Several mechanisms potentially lead to unstable winds (*Lubow et al.,* 1994b; *Lubow and Spruit,* 1995; *Cao and Spruit,* 2002). However, none seems entirely compatible with numerical results (*Moll,* 2012; *Lesur et al.,* 2013).

In all these numerical solutions, the toroidal electromotive force $E_\phi \neq 0$ and magnetic flux tubes accrete. This is to be expected since the solutions are computed in ideal or weakly resistive regimes.

4.2.2. Winds with ambipolar diffusion and Hall effect. The Hall effect and ambipolar diffusion are believed to dominate in many protoplanetary disks (section 2). Ambipolar diffusion is especially important since it lets the gas cross magnetic field lines rather than sweeping them along. In principle, this avoids the flux accumulation problem faced by the ideal-MHD and weakly resistive models described in section 4.2.

Local solutions including both Hall effect and ambipolar diffusion were initially presented by *Wardle and Königl* (1993), who treated the ions, electrons, and neutrals as independent fluids coupled together by a drag force. With this approach, the neutrals (which make up most of the mass) can accrete without advecting field lines (which are attached to the charged species). *Wardle and Königl* (1993) demonstrated steady wind solutions in the $\beta_{mid} \simeq 1$ regime, both with and without accreting magnetic field lines. Surprisingly, solutions with and without magnetic accretion are similar, indicating the magnetically accreting solution presented in section 4.2 might be qualitatively correct.

Comparable winds can be launched when the Hall and Ohmic diffusivities are large (*Königl et al.*, 2010; *Salmeron et al.*, 2011). Interestingly, the mass-loss rate and torque depend on the poloidal field polarity (i.e. the sign of **B** × Ω). In particular, some wind solutions are obtained only for positive polarities.

The problem of outflows in the presence of ambipolar diffusion was revisited by *Bai and Stone* (2013b). Using a complex chemical network to compute ambipolar and Ohmic diffusivities, they found mid-plane ambipolar diffusion several orders of magnitude stronger than anticipated by *Wardle and Königl* (1993) and concluded that non-ideal effects suppress MRI turbulence at 1 AU. However, even poloidal fields as weak as $\beta_{mid} = 10^5$ trigger a quasisteady wind starting about 4 scale-heights above the mid-plane, where ambipolar diffusion is significantly reduced thanks to FUV ionization. While the shearing box is ambiguous regarding the direction to the star, global magneto-centrifugal wind solutions generally have the horizontal components of the magnetic field changing sign across the disk. In the local weak-field ambipolar-dominated calculations, the horizontal field flips in a thin layer offset from the mid-plane by several scale-heights. This layer, a fraction of a scale-height thick, receives the wind torque and carries the entire accretion flow. Based on these results, *Bai* (2013) proposed that the accretion rates in the planet-forming region at ~1–10 AU can be explained by such a steady wind alone, the bulk of the disk being essentially quiet. The ambipolar-diffusion-dominated outer disk is likely MRI turbulent but can also launch an outflow (*Simon et al.*, 2013b).

4.2.3. Connecting local and global solutions. In one of the first attempts at connecting local and global solutions, *Wardle and Königl* (1993) matched their local results to a global *Blandford and Payne* (1982) wind, ruling out some of the local solutions. Similarly, *Lesur et al.* (2013) compared their local solution against global solutions computed *à la Ferreira and Pelletier* (1995). There was qualitative agreement but significant differences remained, e.g., in mass loss rate and β_{mid}. Interestingly, they found no global solutions for fields as weak as they considered in the local solutions, suggesting that if a global wind exists for given parameters, then the local approach will catch the right solution. However, the local approach might also catch solutions that simply do not exist in global geometry. These spurious winds are impossible to distinguish from "real" winds solely by a local approach. Future work on this topic will therefore have to extend the recent findings on local solutions to global geometries.

4.3. Global Disk Wind Solutions

To overcome the limitations of local solutions, one can perform global calculations. Here, however, a scale separation problem occurs. On one hand, any turbulence must be resolved. On the other hand, the solution must be computed far from the disk in the vertical direction (typically to tens or even hundreds of disk radii). For this reason, all global disk wind solutions have been obtained in a 2.5-dimensional approximation with turbulent motions either ignored or treated as viscosities and resistivities. This approximation is brought into question by three-dimensional results on torus accretion by a black hole (*Beckwith et al.*, 2009). The main difficulties with the global models are two-fold:

1. They need a strong magnetic flux in the disk mid-plane ($\beta_{mid} \sim 1$), which can either build up as flux accretes from the outer disk [a scenario with several problems; see *Lubow et al.* (1994a)] or be put in place when the disk forms (in which case the entire question of jet formation is linked to the disk-formation scenario).

2. They rely on strong resistivities. A resistivity η_t, supposed to mimic the local effects of three-dimensional turbulence, is quantified by the parameter α_m in $\eta_t = \alpha_m c_s H$. To avoid magnetic flux accumulating, these models require $\alpha_m \sim 1$. However, no physical mechanism justifies such a high turbulent diffusivity with $\beta_{mid} \sim 1$. In particular, MRI turbulence is believed to produce at best $\alpha_m \sim 0.1$ (*Lesur and Longaretti*, 2009).

The behavior of global models as a function of the turbulent resistivity was investigated by *Zanni et al.* (2007) in 2.5-dimensional simulations. For $\alpha_m = 1$ and $\beta_{mid} = 1$, they found quasisteady ejection resembling the self-similar models of *Ferreira* (1997). Lower values of α_m invariably lead to unsteady ejection, probably due to episodes when magnetic flux accretes followed by sudden releases of the flux in numerical reconnection or ejection events. This indicates that ejection is possible for weak diffusivities, even though the resulting winds and jets are likely to be highly variable.

Issues of disk magnetization were studied by *Tzeferacos et al.* (2009) and *Murphy et al.* (2010). They demonstrated the possibility of launching unsteady magnetically driven winds up to $\beta_{mid} \sim 500$. However, these models rely on large diffusivities with $\alpha_m \simeq 1$, which are probably too high to be realistic if turbulence is driven solely by the MRI. They also pointed out the sensitivity of the mass-loading rate (the fraction of the mass that is ejected in the wind) to the resolution. The high mass-loading rate achieved in these simulations is likely due to numerical diffusion of gas at the top of the disk, where the launching occurs. Such resolution issues do not arise in the global semi-analytic disk-wind model developed by *Teitler* (2011), which furthermore treats both mass and magnetic flux transport under the full tensor conductivity.

All the models considered above are "cold," with the turbulent heating canceled by radiative cooling. *Tzeferacos et al.* (2013) showed explicitly that if weak turbulent heating is not balanced by radiative cooling at the disk surface, the mass-loading rate and ejection efficiency are significantly affected. This demonstrates the sensitivity of winds to thermodynamics near the launch point, which must be modeled accurately if we are to predict the large-scale properties of winds and match jet observations (see the chapter by Frank et al. in this volume).

4.4. Outstanding Issues

The new disk-wind scenario differs significantly from the $\beta_{mid} \simeq 1$ picture investigated in the 1990s. The magnetization is weak, with $\beta_{mid} \sim 10^4$, and the ambipolar diffusivity varies with height. This scenario is still developing and global modeling may soon clarify the picture. Outflowing material could shield the disk from ionizing stellar photons (*Ferro-Fontán and Gómez de Castro*, 2003; *Panoglou et al.*, 2012), preventing wind launching. The sensitivity of winds to radiative transfer and chemistry must be investigated further. Finally, we note that this scenario is unlikely to explain the protostellar jets launched within 1 AU, where non-ideal MHD effects are weak thanks to thermal ionization.

5. SELF-GRAVITATING TURBULENCE

Self-gravitating, non-axisymmetric structures transport angular momentum with a flux density

$$W_{r\phi} = \int dz \left(\frac{g_R g_\phi}{4\pi G} + \rho v_R \delta v_\phi \right) \quad (5)$$

where g_R and g_ϕ are the radial and azimuthal components of the gravitational acceleration. This transport is nonlocal in the sense that it can couple together parts of the disk that are widely separated in radius (*Balbus and Papaloizou*, 1999). Gravitational torques may, however, be described by an effective α, if the characteristic scale λ_{char} is small compared to the local radius (*Lodato and Rice*, 2004, 2005; *Vorobyov and Basu*, 2009; *Cossins et al.*, 2009; *Forgan et al.*, 2011; *Michael et al.*, 2012), or equivalently the disk mass is small compared to the star (*Vorobyov and Basu*, 2010).

Self-gravitating disk structures are detectable in the millimeter continuum and lines in nearby (d ≤ 150 pc) systems with the Atacama Large Millimeter/submillimeter Array (ALMA) (e.g., *Cossins et al.*, 2010; *Douglas et al.*, 2013), although self-gravitating disks are likely to be found in young, short-lived systems, and are therefore expected to be rare. If they are found, torques and a disk-evolution timescale can be measured directly from a surface density map (e.g., *Gnedin et al.*, 1995), and molecular line observations may allow us to measure the ratio of vertical to in-plane velocity dispersion well enough to separate the values expected for gravitationally driven and MRI-driven turbulence (*Forgan et al.*, 2012).

When is the disk self-gravitating? Self-gravity becomes important when the Toomre parameter

$$Q \equiv \frac{c_s \kappa}{\pi G \Sigma} \sim 1 \quad (6)$$

where c_s is the sound speed; κ is the epicyclic frequency, which $\simeq \Omega$ in a Keplerian disk; and Σ is the surface density.

If we define the scale height $H \simeq c_s/\Omega$ and the disk mass $M_d \equiv \pi R^2 \Sigma$, then we can rewrite Q to obtain the useful relation $Q = (R/H)M_d/M_*$.

The Q parameter describes the balance between the stabilizing influence of pressure (c_s) and rotation (κ) and the destablizing influence of self-gravity ($G\Sigma$). Pressure and rotation are weakest in comparison to self-gravity at a characteristic wavelength

$$\lambda_c \equiv \frac{2c_s}{G\Sigma} = 2\pi Q H \quad (7)$$

Measuring this characteristic scale in a face-on disk may then be a direct probe of the surface density, through Q; numerical experiments show that even in the nonlinear regime much of the power is concentrated near λ_c (*Cossins et al.*, 2009).

The formal analysis leading to the conclusion that Q > 1 implies stability has formal limitations: It relies on an axisymmetric analysis of a zero-thickness disk, and requires that $\lambda_c \ll r$ [it is a local, or lowest-order Wentzel-Kramers-Brillouin (WKB), analysis]. But a self-gravitating disk is most responsive to non-axisymmetric perturbations; finite thickness is stabilizing (*Goldreich and Lynden-Bell*, 1965; *Mamatsashvili and Rice*, 2010); and long-wavelength modes can be stabilized or destabilized by global features in the disk (e.g., *Meschiari and Laughlin*, 2008). In the end, a good rule of thumb is that Q ≤ 2 disks are strongly self-gravitating.

The outcome of gravitational instability depends on how the disk is driven unstable. In one commonly analyzed model the disk is driven unstable by cooling that is characterized entirely by the timescale τ_{cool} (*Gammie*, 2001), or equivalently by the dimensionless $\beta_c = \tau_{cool}\Omega$. Then there is a β_{crit} (*Shlosman and Begelman*, 1987) dividing models that fragment immediately ($\beta_c < \beta_{crit} \simeq 3$) from those that enter a quasisteady, or gravito-turbulent, state at $\beta_c > \beta_{crit}$. Since cooling times typically decrease outward in protoplanetary disks, this implies that there is a critical radius at a few tens of astronomical units beyond which a massive disk can fragment (*Matzner and Levin*, 2005; *Rafikov*, 2005). The disk can also be driven unstable by mass loading via infall. In this case, whether the disk fragments depends on the ratio of infall to accretion timescales (e.g., *Boley*, 2009; *Rafikov*, 2009; *Kratter et al.*, 2010; *Harsono et al.*, 2011).

In a gravito-turbulent state the disk may settle into a condition of marginal stability, with Q slightly greater than the marginally stable value (*Paczynski*, 1978; *Vorobyov and Basu*, 2007). Transport in the gravito-turbulent state can be described by an effective α, and is predominantly local, if $\lambda_c/R \leq 1$. If angular momentum transport can be described by an effective α, its value is $(4/9)\beta_c^{-1}/(\gamma(\gamma-1))$ (*Gammie*, 2001). Then a minimum β_c implies a maximum α in the gravito-turbulent state; at larger values, the disk fragments. The $\alpha(\tau_{cool}, \Omega)$ relationship also applies to more complicated cooling laws, but if there are variations in the surface density on the order of unity, then it may be difficult to estimate τ_{cool}.

Angular momentum transport will not cease if the disk fragments. Even low-mass bound objects will generate wakes in a self-gravitating disk due to tidal interactions (*Julian and Toomre,* 1966), and superposition of these wakes transports angular momentum (*Johnson et al.,* 2006). Massive, embedded objects can lead to rapid disk evolution (e.g., *Krumholz et al.,* 2007; *Vorobyov and Basu,* 2010) or a complicated interplay between disk and planetary orbital evolution (*Boley,* 2009; *Michael et al.,* 2011).

The critical value of β_c remains unclear (*Meru and Bate,* 2011). Numerical outcomes are sensitive to numerical diffusion (*Paardekooper et al.,* 2011; *Paardekooper,* 2012; *Meru and Bate,* 2012), the use of two-dimensional vs. three-dimensional models, gravitational softening (*Müller et al.,* 2012), cooling with realistic opacities (e.g., *Johnson and Gammie,* 2003), and perhaps most important, irradiation (e.g., *Matzner and Levin,* 2005; *Clarke et al.,* 2007; *Cai et al.,* 2008; *Rice et al.,* 2011; *Forgan and Rice,* 2013), which at the least increases the temperature and therefore stabilizes the disk.

The most surprising new discovery about fragmentation is the finding by *Paardekooper* (2012) (see also *Hopkins and Christiansen,* 2013), stimulated by work of *Meru and Bate* (2011), that a two-dimensional local disk model can still fragment for $\beta_c \gg \beta_{crit}$, if one waits long enough! This is illustrated in Fig. 4, showing that an apparently steady gravito-turbulent state later fragments at sufficiently high resolution. The result is best described as stochastic fragmentation: For $\beta_c > \beta_{crit}$, the disk has a finite probability of producing a bound object per unit area per unit time. It is not yet known whether this extends to models in three dimensions, or with more realistic cooling. Stochastic fragmentation may revive direct planet formation by gravitational instability for disks with $\beta_c > \beta_{crit}$ (e.g., *Boss,* 1997), or it may imply a $\beta_{crit,eff} > \beta_{crit}$ where the fragmentation time is the disk lifetime.

Evidently there are many open questions about angular momentum transport by gravitational instability. We think two in particular merit further exploration: (1) Magnetized, self-gravitating disks: Does magnetically driven turbulence (in surface layers, or throughout the disk) enhance or limit the development of bound structures in protoplanetary disks? Pioneering simulations (*Fromang,* 2005) need to be revisited at higher numerical resolution. (2) Thermodynamically self-consistent disks with realistic opacities and external irradiation: While much work has already been done (*Boss,* 2007b; *Cai et al.,* 2008; *Steiman-Cameron et al.,* 2013), tracking cooling inside protofragments requires very high spatial resolution.

6. HYDRODYNAMICAL PROCESSES OR DISK WEATHER

Our understanding of the flows triggered solely by gas pressure and viscous forces has advanced greatly since the previous book in this series (*Reipurth et al.,* 2007). Giant vortices, which may accelerate the planet-formation

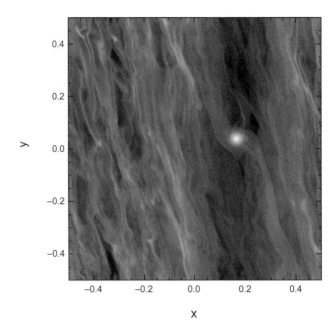

Fig. 4. A two-dimensional (razor-thin) numerical model of a cooling, gravitationally unstable disk with $\beta_c = 5$ is in an apparently steady gravito-turbulent state at $\Omega t = 100$ but by $\Omega t = 400$ produces a bound object as shown. Shading indicates surface density in units of the mean. From *Paardekooper* (2012).

process (*Barge and Sommeria,* 1995; *Tanga et al.,* 1996; *Klahr and Bodenheimer,* 2006), have gained observational support through ALMA observations of emission perhaps from millimeter-sized particles concentrated to one side of a young star (*van der Marel et al.,* 2013). The feature resembles predictions of the vortices' appearance by *Wolf and Klahr* (2002) but has higher contrast, since the big particles concentrate more strongly than the submicrometer grains in the model (*Lyra and Lin,* 2013).

Hydrodynamical turbulence is less robust than the MRI and self-gravity — otherwise it would have appeared in numerical models years ago. When magnetic fields couple to the gas, the MRI governs the flow (*Lyra and Klahr,* 2011; *Nelson et al.,* 2013). Thus the most promising sites for nonmagnetic instabilities are the magnetically decoupled dead zones. It seems likely that the solar nebula at 1–10 AU from the young Sun was a suitable location.

Since protostellar disks have enormous Reynolds numbers Re > 10^9, it might seem that their orbital shear would easily lapse into turbulence. However, as rotating fluids, the disks are a special case of the Taylor-Couette flow, which is linearly stable when the angular momentum increases with radius, as it does on the Keplerian rotation curve. This is embodied in the Rayleigh stability criterion $\partial_R(\Omega^2 R^4) > 0$ (*Rayleigh,* 1917) at radius R and rotation frequency Ω. Based on contradictory Taylor-Couette experiments spanning almost a century (*Taylor,* 1923, 1936; *Wendt,* 1959; *Schultz-Grunow,* 1959; *Lathrop et al.,* 1992; *Richard and*

Zahn, 1999; Ji et al., 2006; Paoletti et al., 2012; Schartman et al., 2012), disks could be nonlinearly either stable or unstable. Yet the experiments depart significantly from astrophysical disks, especially at the vertical boundaries. The containers' upper and lower lids are argued to drive any turbulence observed in the experiments (*Avila, 2012*). Also, even if accretion disks become turbulent at some critical Reynolds number above the maximum Re ~ 2 × 10^4 achieved in numerical simulations, the resulting turbulent viscosity must still be relatively low, in fact, inversely proportional to the critical Reynolds number (*Lesur and Longaretti, 2005*). Thus we are compelled to investigate instabilities that occur under flow conditions relevant to protostellar disks and can be studied numerically.

Thermal convection perpendicular to the disk's midplane was once a favorite to cause turbulence. Yet it suffers a self-consistency problem, producing too little heating to sustain the vertical entropy gradient that drives it (as reviewed by *Klahr, 2007*). Convection pumped artificially by heating the mid-plane can, however, transport angular momentum outward (*Klahr and Bodenheimer, 2003; Lesur and Ogilvie, 2010*).

More promising are three classes of instabilities driven by radial stratification. We consider in turn the gradients in pressure, temperature, and entropy: (1) Radial pressure gradients shift the rotation profile away from Keplerian. (2) Radial temperature gradients with Coriolis forces yield vertical shear, or orbital speed gradients in height, called "thermal winds" in geophysical settings. (3) Radial entropy gradients make the disk gas radially buoyant, but not so buoyant as to directly overcome the angular momentum barrier. In the next three sections we briefly describe what is known about the instabilities arising in these settings.

6.1. Rossby-Wave Instability

A bump in the radial pressure profile yields a local extremum in the rotation speed, or more precisely in the vortensity (vorticity divided by surface density). The extremum is linearly unstable to Rossby-wave instability (RWI), which develops into vortices in a few orbits (*Lovelace et al., 1999; Li et al., 2000, 2001*). Since the orbital speed's radial gradient is key, the RWI is a special kind of Kelvin-Helmholtz instability found in rotating systems. While the RWI develops under conditions readily studied in numerical simulations (*Meheut et al., 2010, 2012b,c*) the unstable initial state has to be explained in the first place. Suitable conditions can occur near the outer edge of a gap opened in the disk by a planet (*Balmforth and Korycansky, 2001; de Val-Borro et al., 2007*). In another scenario, a pressure bump forms at a dead zone's inner edge (*Varnière and Tagger, 2006; Lyra et al., 2009*). *Varnière and Tagger* (2006) used a radially varying α value for this experiment, while *Lyra and Mac Low* (2012) performed a global non-ideal MHD simulation with radially increasing resistivity and also found a pressure bump, and subsequently a large vortex. In conclusion, the RWI does not make vortices in an unperturbed laminar disk, but requires either planets or neighboring MHD active zones.

6.2. Goldreich-Schubert-Fricke Instability

In protostellar disks the temperature and density vary independently. Temperatures are set by starlight heating outside a central accretion-dominated region (*Dullemond et al., 2007*), while densities are determined by the accretion flow. Surfaces of constant pressure therefore tilt relative to surfaces of constant density, making protostellar disks at least in part baroclinic. The baroclinic regions reach a balance between the gas pressure, gravity, and rotational forces by forming a "thermal wind" in which the orbital frequency increases with the distance from the mid-plane (*Tassoul, 2007; Fromang et al., 2011*). The increase is quadratic where the temperature is constant with height. The resulting shear is locally stable against the Kelvin-Helmholtz instability (KHI) flowing to the vertical density stratification (*Rüdiger et al., 2002*). KHI can nevertheless occur if the embedded solid particles sediment with respect to the gas, forming a thin mid-plane layer rotating at Keplerian speed. The shear relative to the slower-rotating gas can yield Richardson numbers low enough for KHI to produce mild local turbulence (*Cuzzi et al., 1993; Johansen et al., 2006a; Barranco, 2009; Lee et al., 2010*).

Accretion disks' radial shear also permits an instability first studied for rotating stars' interiors, the Goldreich-Schubert-Fricke (GSF) instability (*Goldreich and Schubert, 1967; Fricke, 1968*). This can be triggered when the radial shear is Rayleigh stable but receives a boost from vertical shear. Under the combined shear, there exist paths inclined slightly with respect to the vertical, along which the specific angular momentum is constant. Exchanging gas parcels along these paths can be unstable if the disk's vertical stratification is neutrally buoyant, with Brunt-Väisälä frequency zero. Any real, nonzero buoyancy frequency stabilizes the exchange of gas parcels. The instability thus occurs only if the disk either is adiabatically stratified, or relaxes thermally on very short timescales. Quick-enough heating and cooling can overcome the stable vertical density stratification. Indeed, in the two limiting cases of instantaneous cooling (*Urpin and Brandenburg, 1998*) and isentropic or uniform-entropy stratification (*Rüdiger et al., 2002*), the buoyancy frequency and thus the Richardson number are effectively zero and the Rayleigh stability criterion generalizes to $\partial_R \Omega^2(R,z)R^4 - (k_R/k_z)R^4 \partial_z \Omega^2(R,z) > 0$ (*Urpin, 2003; Nelson et al., 2013*). Hence perturbing the vertical rotation profile can lead to weak, transient turbulence until the shear decays, even in an isothermal flow with an unsheared background state (*Arlt and Urpin, 2004*). When the vertical shear is part of the background profile, the disk undergoes narrow near-vertical overturning motions, leading to sustained accretion stresses at levels α ~ 10^{-3} (*Nelson et al., 2013*). The instability is suppressed when thermal relaxation enforces an isothermal atmosphere slower than 0.01 Ω^{-1}. The threshold is more easily reached when the vertical structure is driven toward isentropy: Instability occurs for thermal relaxation times up to 1 Ω^{-1}. Candidates for regions with thermodynamics suitable for the GSF are disks with moderate optical depths, and in particular the outer reaches of T Tauri disks well above the mid-plane.

6.3. Baroclinic Vortex Formation

Vortices can also arise in disks having a radial entropy gradient (*Klahr and Bodenheimer*, 2003). Unlike the baroclinic instability of stars and planetary atmospheres, which relies on the vertical shear and seems to be suppressed in disks by the radial shear, the baroclinic instability in disks depends on the radial entropy gradient through the Brunt-Väisälä frequency $N^2 = -\gamma^{-1}\beta_P\beta_S(H/R)^2\Omega^2$. This is proportional to the product of the logarithmic gradients in the pressure P and entropy $S = \log(P\rho^{-\gamma})$, the latter written as a potential temperature. Here the local pressure scale height $H = c_s/\Omega$ and the gradients are $\beta_P = d\log P/d\log R$ and $\beta_S = \beta_T + (1-\gamma)\beta_\rho$, with β_ρ the logarithmic radial density and β_T the logarithmic radial temperature gradient, all for the vertically unstratified case. Positive N^2 corresponds to convective stability and negative to unstable stratification if the fluid is nonrotating. In a rotating system the radial motions are affected if the modified Rayleigh criterion $(1/R^3)\partial_R[\Omega^2(R)R^4] + N^2 > 0$ holds [a special case of the Solberg-Høiland criterion with rotation (*Tassoul*, 2007; *Chandrasekhar*, 1961)]. Since typically $-N^2 \ll \Omega^2$, all radial convective modes are linearly stable (*Klahr*, 2004; *Johnson and Gammie*, 2005, 2006). Yet numerical models indicate vortices preexisting in the disk are amplified if thermal relaxation occurs over approximately the vortices' internal rotation periods (*Petersen et al.*, 2007a,b). Because growth requires perturbations of finite amplitude, *Lesur and Papaloizou* (2010) coined the name subcritical baroclinic instability (SBI). The vortex growth rate is proportional to $-\tau N^2$, with τ the thermal relaxation time, provided $\Omega\tau < 1$ (*Lesur and Papaloizou*, 2010; *Raettig et al.*, 2013). The thermal relaxation times needed for SBI are generally 100 longer than the maximum for the GSF in an isothermal background, and therefore SBI seems likely to occur in regions of greater optical depth.

Vortices in three dimensions are susceptible to the elliptic instability, a parasitic process (*Lesur and Papaloizou*, 2009). The instability drastically disturbs the flow in vortices rotating fast enough. Rotation is faster in vortices with smaller azimuthal-to-radial aspect ratios, and elliptic instability typically is important at aspect ratios <4. More azimuthally extended vortices may be safe from destruction by the elliptic instability but still internally unstable, and thus turbulent at some level. At least in local three-dimensional calculations, the parasitic growth does not completely destroy vortices, but weakens them to levels where the elliptic instability is suppressed allowing for new baroclinic growth. The saturated vortices' turbulent transport coefficient α is

0.001–0.01 depending on the cooling time and entropy gradient (Fig. 5), a range compatible with vortices providing the angular momentum transport behind disks' observed surface density profiles (*Andrews et al.*, 2009, 2010).

6.4. General Properties of Vortices

Independently of how they form, vortices extend up to two pressure scale heights in radius and eight or more in azimuth, in both two- and three-dimensional calculations. Wider vortices experience a supersonic difference in orbital speed between their inner and outer edges, so a shock truncates the flow pattern. Because the structures are large and can trap solid particles, they might be observable and indeed may already have been observed (*Brown et al.*, 2009; *Regály et al.*, 2012; *van der Marel et al.*, 2013).

With magnetic fields well coupled, vortices are quickly disrupted by a vortex version of the MRI coined the magneto-elliptic instability (*Lyra and Klahr*, 2011) and occurring only in three-dimensional calculations. Yet if the disk is magnetically dead this effect should be suppressed.

Like their cousins in Earth's atmosphere, vortices in the body of the disk are not fixed at a radial or azimuthal location, but tend to migrate on reaching a certain amplitude as they exchange angular momentum with their surroundings by emitting spiral waves (*Paardekooper et al.*, 2010). Consequently vortices are amplified by the radial buoyancy, then migrate inward and may disappear into the star or magnetically active part of the disk. Thus a reliable mechanism is needed for initiating vortices and this is yet to be identified. Possibilities are a RWI relying on planets or other disk turbulence, the spiral density waves launched by the previous generation of vortices, and the pattern generated by the GSF instability. Or, perhaps a further hydrodynamic instability

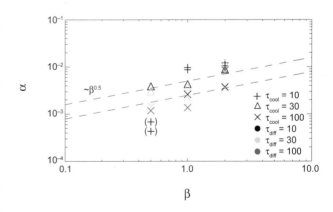

Fig. 5. Reynolds stress parameter α in saturated two-dimensional baroclinic turbulence vs. the logarithmic radial gradient in both entropy and pressure, $\beta_S = \beta_P$. Each calculation is shown by a symbol whose color indicates the timescale for thermal diffusion within the disk plane, and shape indicates the timescale for heating and cooling via the disk surface (*Raettig et al.*, 2013).

triggers vortex growth. Vortices could create copies of themselves through a nonlinear "critical layer" instability (*Marcus et al., 2013*). This occurs in disks that are stably stratified in the vertical direction with little or no thermal relaxation — diametrically opposed to the conditions favoring the GSF. How strict these requirements are and how well they are realized in protostellar disks is still unclear.

In summary, we emphasize that no hydrodynamical instability is known to act on a strictly barotropic, vertically unstratified disk without a local extremum in its rotation profile, nor does any known mechanism sustain long-lived vortices within (*Shen et al., 2006*). However, at least two instabilities can operate globally in a baroclinic disk under suitable thermal conditions: the GSF and baroclinic vortex driving instabilities.

Motivated by the terrestrial atmospheric weather patterns arising from individual instabilities driven by solar heating, we have named the total effect of the hydrodynamical instabilities the "disk weather." A key issue is when and where in protostellar disks the heating and cooling timescales support such weather. Calculations to date span only a small fraction of the parameter space. For instance, there exists no global, vertically stratified calculation of baroclinic vortices with realistic entropy gradients and radiation transport, flowing to the high spatial resolution required. Such calculations will eventually be needed if we are to study the origin, stability, and ultimate fate of vortices as weather patterns in protostellar disks.

7. OBSERVATIONAL SIGNATURES AND CONSTRAINTS

To understand the evolution of protostellar disks, we must know which transport processes are at work. The many possible mechanisms mean that a reliable answer requires an empirical approach. Observations probe the disks' kinematics, density and temperature structure, ionization state, magnetic fields, and composition, each of which carry signatures of the transport processes. While few definitive conclusions can yet be reached, some predictions are now falsifiable. It is worth remembering that transport process may differ from one disk annulus to the next and from one star to another. For example, gravitational instability is more likely at the higher surface densities found in class I young stellar objects (YSOs), while magnetic coupling is easier to achieve at the lower surface densities found in T Tauri disks.

7.1. Kinematics

Each transport process has a distinct flow pattern that would serve as a fingerprint if we could only observe with enough spatial and spectral resolution. Disk weather makes vortices, gravitational instability produces spiral shocks, disk winds mean helical outflow, and MRI turbulence is transonic near the disk surface.

Nonthermal velocity dispersions in the outer reaches of a few disks have been measured using millimeter and submil-limeter lines. TW Hya yielded an upper limit $<0.1\ c_s$, while velocities $\sim 0.5\ c_s$ were found in the outer disks of HD 163296 (*Hughes et al., 2011*) and DM Tau (*Guilloteau et al., 2012*).

Velocities nearer the star are also detected using infrared (IR) lines formed in the disk atmosphere. The evidence again is for turbulent speeds comparable to the sound speed (*Carr et al., 2004; Hartmann et al., 2004; Najita et al., 2009*). In Fig. 6, the disk's rotation and microturbulence are jointly constrained by the widths of the isolated water and CO lines, the CO lines' beating immediately redward of the bandhead, and their strengths far from the bandhead compared to the bandhead itself.

Both MRI turbulence (*Hawley et al., 1995; Miller and Stone, 2000; Simon et al., 2011b*) and gravito-turbulence (*Forgan et al., 2012*) have anisotropies that could distinguish them from other transport processes. The velocity dispersion ellipsoid is best accessed using saturated emission lines, especially those of heavy species for which the thermal speed is less than the turbulent speed (*Horne, 1995*).

Warm disk outflows inferred from atomic oxygen emission line profiles in the optical (*Rigliaco et al., 2013*) could arise either in a photoevaporative wind (see the chapter by Alexander et al. in this volume) or in a magneto-centrifugal wind. If a cold outflow were observed, it would uniquely indicate magnetic acceleration.

7.2. Structure in the Gas and Dust

Mapping surface densities at millimeter and submillimeter wavelengths (*Andrews et al., 2009, 2010; Isella et al., 2009*) offers the potential to determine both whether transport mechanisms can operate and whether they are operating. Questions of the first kind include whether the mass column is low enough for ionizing radiation to reach the mid-plane and whether the Toomre Q parameter is low enough for gravitational instability. A question of the second kind, now accessible with ALMA, is whether any spiral

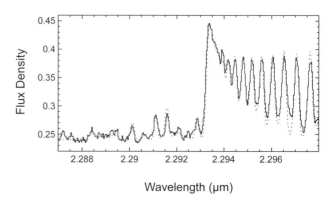

Fig. 6. Observed 2.3-μm spectrum of young star V1331 Cyg (histogram) overlaid with a model (dotted). The microturbulent broadening used, 4 km s⁻¹, exceeds the 0.9 km s⁻¹ CO thermal dispersion and is near the sound speed at 2500 K (*Najita et al., 2009*).

features more closely resemble the density waves raised by self-gravity (*Cossins et al.,* 2009, 2010) or those excited in MRI turbulence (*Heinemann and Papaloizou,* 2009, 2012).

Millimeter mapping also enables detection of vortices. Modeling indicates vortices trap solid particles (*Inaba and Barge,* 2006; *Lyra et al.,* 2008, 2009; *Meheut et al.,* 2012a), and the continuum emission from millimeter-sized grains shows big asymmetries in several cases (*Brown et al.,* 2009). Among the most spectacular is a crescent feature observed at wavelength 0.44 mm near the inner rim of the transitional disk around Oph IRS 48 (*van der Marel et al.,* 2013). The feature is >100× brighter than emission on the opposite side of the star, while no such asymmetry appears either in the mid-IR emission from micrometer-sized dust, or in the line emission from CO gas. Developing such a strong feature in the particles appears to require a long-lived asymmetry in the gas (*Birnstiel et al.,* 2013; *Ataiee et al.,* 2013; *Lyra and Lin,* 2013). Along similar lines, zonal flows will be detectable by axisymmetric rings and gaps (*Ruge et al.,* 2013). Care is needed to distinguish gaps made by different processes.

Young stars' IR spectral energy distributions (SEDs) and scattered light imaging sample the distribution of sub-micrometer grains. Turbulence of some kind appears to be needed to keep enough of the dust suspended in the disk atmospheres to account for the reprocessed and scattered starlight (Fig. 7) (*Watson et al.,* 2007; *Dullemond and Dominik,* 2005). Typical SEDs of disk-bearing young stars can be explained using dust stirred by turbulence at 1% of the sound speed, almost independent of stellar mass over the range from 0.08 $M_\odot$ to 2 $M_\odot$ (*Mulders and Dominik,* 2012). Higher turbulent speeds are allowed if balanced by reduced dust abundances or reduced gas masses.

Dust evolution is intimately coupled to turbulence driven by magnetic forces, because recombination on grains can reduce the gas conductivity below the threshold for MRI turbulence (*Sano et al.,* 2000; *Ilgner and Nelson,* 2006a; *Wardle,* 2007). In the absence of turbulence, grains grow and settle within a fraction of the disk lifetime. On the other hand, turbulent stirring enhances the collision rates of solid particles, speeding growth if the collisions are slow and causing fragmentation if the collisions are fast (*Brauer et al.,* 2008; *Zsom et al.,* 2010; *Birnstiel et al.,* 2010). It remains to be seen whether the mutual coupling can account for the correlation observed between grain size and settling (*Sargent et al.,* 2009) and the lack of correlation of grain size with other parameters.

Further information on transport processes is contained in the disk thickness. MRI turbulence leads to the formation of a magnetically supported disk atmosphere that is optically thin to its own thermal emission, because good magnetic coupling requires recombination that is not too fast, and the opacity, like the recombination, depends on the dust cross-section (*Bai and Goodman,* 2009). At the same time, the magnetized atmosphere is optically thick to the starlight entering at a grazing angle (*Hirose and Turner,* 2011). The disk thickness then reveals the magnetic field strengths rather than the internal temperatures. A contrary

effect comes from magnetic fields generated in a wind with mid-plane plasma beta near unity. These can compress the disk so that its thickness is less than the hydrostatic value (*Königl et al.,* 2010).

Variability in the IR excess is widespread among young stars with disks, having amplitudes 0.1–0.5 mag and time-scales as short as a week (*Luhman et al.,* 2008, 2010; *Muzerolle et al.,* 2009; *Flaherty et al.,* 2011). In some cases the variations correlate with markers of the star's accretion and magnetic activity (*Flaherty et al.,* 2013), while in others they do not (*Morales-Calderón et al.,* 2011). The widespread occurrence of the variability suggests a universal process, such as interaction with the stellar magnetosphere in the first situation and turbulence-driven changes in the size or shape of the starlight reprocessing disk material in the second.

7.3. Temperature Profiles

While accretion flows by their very nature involve the release of gravitational potential energy, the transport processes described above differ in where the power is deposited. A variety of temperature profiles may result. Since spectral lines sample the temperature gradients at the depths where they form, spectroscopy could be used to constrain which mechanisms are at work, even where it is not feasible to reach spectral resolutions high enough to measure kinematics.

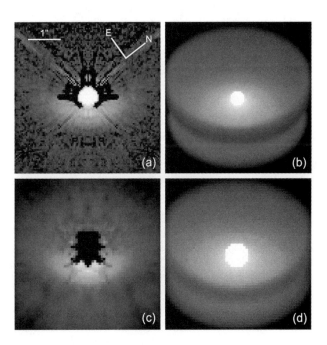

Fig. 7. Images of the disk around IM Lupi at wavelengths **(a,b)** 0.6 and **(c,d)** 1.6 μm from **(a,c)** Hubble, alongside **(b,d)** models fitted jointly to the images and optical-to-millimeter-wavelength SED. While millimeter data show substantial grain growth and settling, the brightness of these scattered light images indicates that submicrometer grains remain suspended in the disk atmosphere, requiring stirring (*Pinte et al.,* 2008).

The accretion power per unit disk area falls off with distance from the star as $\sim r^{-3}$, a steeper dependence than the illuminating starlight, so that with increasing mass flow rate the accretion power becomes dominant first near the star. The accretion power determines the mid-plane temperature within 6 AU and the surface temperature within 2 AU of an 0.9-$M_\odot$ T Tauri star accreting at 10^{-7} $M_\odot$ yr^{-1} with the heating distributed in proportion to the mass (*Dullemond et al.*, 2007). Thus the signatures of accretion heating are most likely to appear in spectral lines formed within a few astronomical units of the star.

A well-established example is the CO rovibrational first overtone band, which arises in the disk atmosphere within 1 AU of the star and can be either in emission, when the atmosphere is hotter than the interior due to stellar illumination, or in absorption, when the internal heating is strong enough to dominate (*Calvet et al.*, 1991; *Najita et al.*, 2007; *Connelley and Greene*, 2010). The CO rovibrational bands' power source can be the accretion flow if the associated heating corresponds to a stress approaching the pressure in the disk atmosphere (*Glassgold et al.*, 2004). Stellar X-ray heating as the power source is insufficient.

Magneto-rotational instability turbulence yields dissipation distributed throughout the disk thickness when the gas is fully ionized (*Flaig et al.*, 2010, 2012). The dissipation occurs higher on average than the accretion flow, flowing to the magnetic fields' buoyancy, and is of course further concentrated toward the surface if the interior is too weakly ionized to couple to the magnetic fields. Consequently, the mid-plane is cooler than if the dissipation were distributed uniformly in the column (*Hirose and Turner*, 2011). The MRI turbulent layer emits much of its power in spectral lines. Line surface brightnesses might one day reveal the mass flow rate as a function of distance from the star (*Bai and Goodman*, 2009).

7.4. Magnetic Field Measurements

To distinguish the two magnetic transport mechanisms — MRI turbulence and magneto-centrifugal winds — it would be valuable to know how the fields are oriented and whether they are straight or tangled. Grains aligned by toroidal magnetic fields in the outer parts of T Tauri disks ought to detectably polarize the thermal emission (*Cho and Lazarian*, 2007), but the submillimeter flux from scales >100 AU is weakly polarized in the five bright protostellar disks examined so far, with P < 1% (*Hughes et al.*, 2009, 2013). The lack of polarization could be due to magnetic fields that are either tangled or, for systems viewed face-on, poloidal rather than toroidal. Magnetized turbulence and winds thus cannot yet be separated. Better submillimeter spatial resolution and far-IR polarimeters will enable similar tests to be applied to material nearer the star.

A record of the magnetic fields in the protosolar disk was once thought to be preserved in the remnant magnetization of chondritic meteorites last melted during the planet-forming epoch. The remnants now appear more likely to sample the magnetic fields in the unmelted chondritic surface layer of a partially differentiated planetesimal that sustained dynamo action in its convecting metallic core (*Weiss et al.*, 2010).

7.5. Detecting Transport Through Its Effects on Composition

Departures from local chemical equilibrium can indicate material being transported along gradients, especially gradients in temperature and radiation intensity. A classic case is the presence of crystalline silicate grains in comets, which because of their icy makeup surely formed beyond the solar nebula's snow line. Grains in the interstellar medium are almost entirely amorphous (*Kemper et al.*, 2005), and crystalline structure is destroyed by cosmic rays within 70 m.y. (*Bringa et al.*, 2007). The comet grains' crystallinity thus most likely arose by heating to 1000 K or more within the solar nebula. Crystalline silicate grains are detected in comets through IR spectroscopy (*Min et al.*, 2005; *Lisse et al.*, 2006) and found in the sample brought back from Comet 81P/Wild 2 by the Stardust mission. The returned sample shows a wide range of olivine and pyroxene compositions. Among the grains is a refractory, calcium-aluminum-rich-inclusion (CAI)-like particle processed at temperatures above 2000 K (*Zolensky et al.*, 2006). Taken together, these properties appear to require large-scale radial transport in the protosolar disk.

Crystalline silicates also occur in the disks around other young stars. The silicate grains' crystalline fraction is similar across the range of annuli in T Tauri disks producing the 10–33-μm IR bands (*Sargent et al.*, 2009), indicating either a dispersed source or efficient transport. Variations in the 10- and 20-μm signatures of crystalline forsterite during an outburst of EX Lupi suggest supersonic transport, compatible with grains dragged along by a wind (*Juhász et al.*, 2012).

A new kind of probe of the transport is the mid-IR emission from water and organic molecules discovered in the atmospheres of many T Tauri disks (*Salyk et al.*, 2008; *Carr and Najita*, 2008, 2011; *Pontoppidan et al.*, 2010; *Salyk et al.*, 2011). In particular, there is a trend between the HCN-to-water ratio at astronomical unit distances and the submillimeter continuum flux outside 20 AU, suggested to arise from water ice accumulating into solid bodies (*Najita et al.*, 2013).

The compositions of disks can be easier to measure than their kinematics, but carry fewer clear signatures of specific transport processes since the details of the flow often matter less than whether the material has been exposed to high temperatures or energetic radiation. Nevertheless, the transport processes discussed in this chapter have some distinctive features with the potential to yield differing chemical signatures:

1. Magneto-rotational instability turbulence yields mixing and angular momentum transfer coefficients roughly equal (*Johansen and Klahr*, 2005; *Turner et al.*, 2006; *Fro-*

mang and Papaloizou, 2006), with the angular momentum transfer increasing faster than the mixing as the background vertical magnetic field is made stronger (*Johansen et al.,* 2006b). As discussed in section 2, the magnetic activity at planet-forming distances is strongest in the disk surface layers. The global meridional circulation predicted in simple viscous models has not so far been seen in MHD calculations (*Fromang et al.,* 2011; *Flock et al.,* 2011).

2. Magneto-centrifugal winds drive the gas radially inward near the mid-plane and outward above the height where the inclined field lines' azimuthal speed reaches Keplerian, in radially localized stationary solutions (*Salmeron et al.,* 2011). Rapid inward and outward radial transport near the surface of a wind-driving disk could also come from episodic channel flows (*Suzuki and Inutsuka,* 2009) whose repeated breakup would cause some vertical mixing.

3. Gravito-turbulence can redistribute material in radius as rapidly as can MRI turbulence (*Michael et al.,* 2012). Even so, a contaminant may remain inhomogeneous during the global mixing timescale if the unstable features are bigger than the disk thickness (*Boss,* 2007a, 2013) or if shock passages affect desorption from grains or destruction in chemical reactions (*Ilee et al.,* 2011). The shock heating associated with gravitational instability appears to drive little convection in the vertical direction when an accurate boundary condition is applied at the disk photosphere (*Cai et al.,* 2010).

Several attempts have been made to bring these ideas about transport into contact with composition measurements. Crystalline grains can be moved through the disk to the comet-forming region with either a sustained radially outward mid-plane flow, or a diffusivity for turbulent mixing exceeding that for angular momentum transfer (*Hughes and Armitage,* 2010). Vertical mixing with a coefficient in the range inferred for the radial angular momentum transfer enhances the abundances of some carbon- and sulfur-bearing molecules in T Tauri disks by orders of magnitude, by bringing reactive atoms and ions from the overlying photodissociation layers into more shielded molecular regions (*Willacy et al.,* 2006; *Semenov and Wiebe,* 2011). The molecular abundances are affected more by turbulent mixing than by a steady-state disk wind (*Heinzeller et al.,* 2011). Comparing these models with the molecular column densities obtained from millimeter line measurements, mostly near and outside 100 AU, yields mixed results, suggesting further complexities in the chemistry, dynamics, or both.

8. SUMMARY AND OUTLOOK

Magneto-rotational turbulence is suppressed by ambipolar diffusion in local stratified models of T Tauri disks' planet-forming regions, even when the Elsasser number and plasma beta criteria indicate turbulence in a surface layer about one scale height thick (*Bai and Stone,* 2013b). Furthermore, the turbulence is suppressed by a strong Hall effect in unstratified calculations (*Kunz and Lesur,* 2013). These results raise serious doubts about whether MRI turbulence is effective at 1–10 AU even in surface layers.

Further progress might involve (1) bringing the temperatures in MHD calculations up to the level of sophistication found in thermochemical models, where stellar X-ray and UV photons heat the disk atmosphere to >2000 K (*Glassgold et al.,* 2004; *Ercolano et al.,* 2008; *Owen et al.,* 2010; *Aresu et al.,* 2011; *Akimkin et al.,* 2013), surely affecting the conductivity; and (2) allowing the Ohmic, Hall, and ambipolar terms all to vary with height. No nonlinear MHD calculation yet treats this full problem. Magneto-rotational turbulence likely remains important in the thermally ionized gas inside 1 AU and the less-dense material beyond 10 AU. A key unresolved issue here is the saturation amplitudes at realistic Prandtl numbers.

Magneto-centrifugal disk winds appear to be favored in disks with the interior dominated by Ohmic diffusion and the surface dominated by ambipolar diffusion, threaded by magnetic fields that are not too weak. Disk winds and MRI turbulence coexist if the magnetic fields are weak enough that the MRI wavelength fits in the disk thickness. The dependences of both MRI turbulence and magneto-centrifugal winds on the net field point to the large-scale magnetic flux transport as a major issue that must be adequately dealt with if we are to understand the long-term evolution of protostellar disks.

Gravitational instability occurs in disks whose mass ratios with their central stars exceed their aspect ratios. It is most easily triggered in the outer reaches, since $c_s\Omega$ falls off faster with radius than the typical surface density profile. Gravitational instability can be important during the protostellar phase when the disk is fed rapidly with material falling in from the surrounding envelope. The outcome of the instability depends on how quickly the compressed gas cools, and room remains for dynamical modeling with more complete treatment of the radiative losses.

All the disk weather processes discussed here, including Rossby and baroclinic vortices and the GSF instability, depend on radiative forcing for their energy supply. There is an urgent need for calculations with better heating and cooling. The instabilities grow slower than the MRI and therefore can reach significant amplitudes only in regions lacking MRI turbulence. Whether they survive in the presence of a magnetically launched wind is unclear.

While all the transport processes we discuss are commonly measured and compared using the Shakura-Sunyaev α-parameter, none behaves like the simple *Shakura and Sunyaev* (1973) model. Magneto-rotational turbulent stresses fall off with height slower than the gas pressure, even when the magnetic coupling is good throughout. Magneto-centrifugal winds expel the angular momentum vertically rather than radially. The energy escapes in a Poynting flux that dissipates far away, leaving the disk unheated. Gravitational instability leads to intermittent shock dissipation, and disk weather yields transport localized around vortices. Any treatment of these processes using an α-model should therefore be evaluated carefully.

Observations offer tantalizing clues to the transport processes. Nonthermal motions are detected, but the spatial resolution is too coarse to unequivocally distinguish turbulence from winds or vortices. In order of decreasing distance from the central star:

1. Millimeter lines formed outside ~100 AU in two disks show transonic superthermal widths (*Hughes et al.,* 2011; *Guilloteau et al.,* 2012). These are explained most easily using turbulence but could also arise from outflows.

2. Keeping enough dust suspended in the atmosphere to explain both the scattered starlight at ~50 AU and the SEDs requires some kind of stirring. Magneto-rotational turbulence can do the job where magnetic coupling is good. Outflows and vortices could potentially also reproduce the dust distributions, but better models are called for.

3. Asymmetric submillimeter surface brightness maps in a few disks (*van der Marel et al.,* 2013) have so far been reproduced only using long-lived nonaxisymmetric gas density disturbances. Vortices are the most viable explanation.

4. Applying angular momentum constraints to jets yields launch points within a few astronomical units of the stars (*Anderson et al.,* 2003; *Coffey et al.,* 2007). The inner regions of disks clearly launch outflows.

5. Superthermal CO overtone linewidths indicate transonic turbulence in the atmosphere inside 1 AU, where the gas is thermally ionized, consistent with MRI (*Carr et al.,* 2004; *Hartmann et al.,* 2004; *Najita et al.,* 2009). On the other hand, the blue-shifted central peaks in many young stars' CO fundamental bands suggest outflows (*Brown et al.,* 2013). The two kinds of flows likely occur together in some disks.

A key to further progress in understanding protostellar disk evolution is to bring the models closer to observations like these, seeking clearer signatures of the proposed transport processes.

Acknowledgments. N.T. was supported at the Jet Propulsion Laboratory, California Institute of Technology, by NASA grant 11-OSS11-0074; S.F. by the European Research Council under the E. U. Seventh Framework Programme (FP7/2007-2013)/ERC Grant agreement no. 258729; H.K. by the Deutsche Forschungsgemeinschaft Schwerpunktprogramm (DFG SPP) 1385 "The First Ten Million Years of the Solar System"; G.L. by the European Community via contract PCIG09-GA-2011-294110; and M.W. by Australian Research Council grant DP120101792.

REFERENCES

Akimkin V. et al. (2013) *Astrophys. J., 766,* 8.
Anderson J. M. et al. (2003) *Astrophys. J. Lett., 590,* L107.
Andrews S. M. et al. (2009) *Astrophys. J., 700,* 1502.
Andrews S. M. et al. (2010) *Astrophys. J., 723,* 1241.
Aresu G. et al. (2011) *Astron. Astrophys., 526,* A163.
Arlt R. and Urpin V. (2004) *Astron. Astrophys., 426,* 755.
Ataiee S. et al. (2013) *Astron. Astrophys., 553,* L3.
Avila M. (2012) *Phys. Rev. Lett., 108,* 12, 124501.
Bai X.-N. (2011a) *Astrophys. J., 739,* 50.
Bai X.-N. (2011b) *Astrophys. J., 739,* 51.
Bai X.-N. (2013) *Astrophys. J., 772,* 96.
Bai X.-N. and Goodman J. (2009) *Astrophys. J., 701,* 737.
Bai X.-N. and Stone J. M. (2011) *Astrophys. J., 736,* 144.
Bai X.-N. and Stone J. M. (2013a) *Astrophys. J., 767,* 30.
Bai X.-N. and Stone J. M. (2013b) *Astrophys. J., 769,* 76.

Balbus S. A. and Hawley J. F. (1991) *Astrophys. J., 376,* 214.
Balbus S. A. and Henri P. (2008) *Astrophys. J., 674,* 408.
Balbus S. A. and Papaloizou J. C. B. (1999) *Astrophys. J., 521,* 650.
Balmforth N. J. and Korycansky D. G. (2001) *Mon. Not. R. Astron. Soc., 326,* 833.
Barge P. and Sommeria J. (1995) *Astron. Astrophys., 295,* L1.
Barranco J. A. (2009) *Astrophys. J., 691,* 907.
Beckwith K. et al. (2009) *Astrophys. J., 707,* 428.
Beckwith K. et al. (2011) *Mon. Not. R. Astron. Soc., 416,* 361.
Bergin E. A. et al. (2007) In *Protostars and Planets V* (B. Reipurth et al., eds.), pp. 751–766. Univ. of Arizona, Tucson.
Bethell T. J. and Bergin E. A. (2011) *Astrophys. J., 739,* 78.
Birnstiel T. et al. (2010) *Astron. Astrophys., 513,* A79.
Birnstiel T. et al. (2013) *Astron. Astrophys., 550,* L8.
Blaes O. M. and Balbus S. A. (1994) *Astrophys. J., 421,* 163.
Blandford R. D. and Payne D. G. (1982) *Mon. Not. R. Astron. Soc., 199,* 883.
Bodo G. et al. (2008) *Astron. Astrophys., 487,* 1.
Bodo G. et al. (2011) *Astrophys. J., 739,* 82.
Boley A. C. (2009) *Astrophys. J. Lett., 695,* L53.
Boss A. P. (1997) *Science, 276,* 1836.
Boss A. P. (2007a) *Astrophys. J., 660,* 1707.
Boss A. P. (2007b) *Astrophys. J. Lett., 661,* L73.
Boss A. P. (2013) *Astrophys. J., 773,* 5.
Brandenburg A. et al. (1995) *Astrophys. J., 446,* 741.
Brauer F. et al. (2008) *Astron. Astrophys., 480,* 859.
Bringa E. M. et al. (2007) *Astrophys. J., 662,* 372.
Brown J. M. et al. (2009) *Astrophys. J., 704,* 496.
Brown J. M. et al. (2013) *Astrophys. J., 770,* 94.
Cai K. et al. (2008) *Astrophys. J., 673,* 1138.
Cai K. et al. (2010) *Astrophys. J. Lett., 716,* L176.
Calvet N. et al. (1991) *Astrophys. J., 380,* 617.
Cao X. and Spruit H. C. (2002) *Astron. Astrophys., 385,* 289.
Carr J. S. and Najita J. R. (2008) *Science, 319,* 1504.
Carr J. S. and Najita J. R. (2011) *Astrophys. J., 733,* 102.
Carr J. S. et al. (2004) *Astrophys. J., 603,* 213.
Chandrasekhar S. (1960) *Proc. Natl. Acad. Sci., 46,* 253.
Chandrasekhar S., ed. (1961) *Hydrodynamic and Hydromagnetic Stability,* Dover, Mineola, New York. 704 pp.
Cho J. and Lazarian A. (2007) *Astrophys. J., 669,* 1085.
Ciesla F. J. and Cuzzi J. N. (2006) *Icarus, 181,* 178.
Clarke C. J. et al. (2007) *Mon. Not. R. Astron. Soc., 381,* 1543.
Cleeves L. I. et al. (2013) *Astrophys. J., 772,* 5.
Coffey D. et al. (2007) *Astrophys. J., 663,* 350.
Connelley M. S. and Greene T. P. (2010) *Astron. J., 140,* 1214.
Cossins P. et al. (2009) *Mon. Not. R. Astron. Soc., 393,* 1157.
Cossins P. et al. (2010) *Mon. Not. R. Astron. Soc., 407,* 181.
Cuzzi J. N. et al. (1993) *Icarus, 106,* 102.
Davis S. W. et al. (2010) *Astrophys. J., 713,* 52.
Desch S. J. (2004) *Astrophys. J., 608,* 509.
Desch S. J. et al. (2004) *Astrophys. J., 602,* 528.
de Val-Borro M. et al. (2007) *Astron. Astrophys., 471,* 1043.
Dittrich K. et al. (2013) *Astrophys. J., 763,* 117.
Douglas T. A. et al. (2013) *Mon. Not. R. Astron. Soc., 433,* 2064.
Dullemond C. P. and Dominik C. (2005) *Astron. Astrophys., 434,* 971.
Dullemond C. P. et al. (2007) In *Protostars and Planets V* (B. Reipurth et al., eds.), pp. 555–572. Univ. of Arizona, Tucson.
Dzyurkevich N. et al. (2010) *Astron. Astrophys., 515,* A70.
Dzyurkevich N. et al. (2013) *Astrophys. J., 765,* 114.
Ercolano B. et al. (2008) *Astrophys. J., 688,* 398.
Ferreira J. (1997) *Astron. Astrophys., 319,* 340.
Ferreira J. and Pelletier G. (1993) *Astron. Astrophys., 276,* 625.
Ferreira J. and Pelletier G. (1995) *Astron. Astrophys., 295,* 807.
Ferro-Fontán C. and Gómez de Castro A. I. (2003) *Mon. Not. R. Astron. Soc., 342,* 427.
Flaherty K. M. et al. (2011) *Astrophys. J., 732,* 83.
Flaherty K. M. et al. (2013) *Astron. J., 145,* 66.
Flaig M. et al. (2010) *Mon. Not. R. Astron. Soc., 409,* 1297.
Flaig M. et al. (2012) *Mon. Not. R. Astron. Soc., 420,* 2419.
Fleming T. and Stone J. M. (2003) *Astrophys. J., 585,* 908.
Flock M. et al. (2011) *Astrophys. J., 735,* 122.
Flock M. et al. (2012) *Astrophys. J., 744,* 144.
Forgan D. and Rice K. (2013) *Mon. Not. R. Astron. Soc., 430,* 2082.
Forgan D. et al. (2011) *Mon. Not. R. Astron. Soc., 410,* 994.
Forgan D. et al. (2012) *Mon. Not. R. Astron. Soc., 426,* 2419.
Fricke K. (1968) *Z. Astrophys., 68,* 317.

Fromang S. (2005) *Astron. Astrophys., 441,* 1.
Fromang S. and Nelson R. P. (2006) *Astron. Astrophys.,457,* 343.
Fromang S. and Nelson R. P. (2009) *Astron. Astrophys., 496,* 597.
Fromang S. and Papaloizou J. (2006) *Astron. Astrophys., 452,* 751.
Fromang S. and Papaloizou J. (2007) *Astron. Astrophys., 476,* 1113.
Fromang S. et al. (2007) *Astron. Astrophys., 476,* 1123.
Fromang S. et al. (2011) *Astron. Astrophys., 534,* A107.
Fromang S. et al. (2013) *Astron. Astrophys., 552,* A71.
Gammie C. F. (1996) *Astrophys. J., 457,* 355.
Gammie C. F. (2001) *Astrophys. J., 553,* 174.
Garmire G. et al. (2000) *Astron. J., 120,* 1426.
Glassgold A. E. et al. (2004) *Astrophys. J., 615,* 972.
Gnedin O. Y. et al. (1995) *Astron. J., 110,* 1105.
Goldreich P. and Lynden-Bell D. (1965) *Mon. Not. R. Astron. Soc., 130,* 125.
Goldreich P. and Schubert G. (1967) *Astrophys. J., 150,* 571.
Gressel O. (2010) *Mon. Not. R. Astron. Soc., 405,* 41.
Gressel O. et al. (2011) *Mon. Not. R. Astron. Soc., 415,* 3291.
Guan X. et al. (2009) *Astrophys. J., 694,* 1010.
Guilloteau S. et al. (2012) *Astron. Astrophys., 548,* A70.
Harsono D. et al. (2011) *Mon. Not. R. Astron. Soc., 413,* 423.
Hartmann L. et al. (2004) *Astrophys. J., 609,* 906.
Hawley J. F. and Balbus S. A. (1992) *Astrophys. J., 400,* 595.
Hawley J. F. and Stone J. M. (1995) *Comp. Phys. Commun., 89,* 127.
Hawley J. F. and Stone J. M. (1998) *Astrophys. J., 501,* 758.
Hawley J. F. et al. (1995) *Astrophys. J., 440,* 742.
Hawley J. F. et al. (1996) *Astrophys. J., 464,* 690.
Hawley J. F. et al. (2013) *Astrophys. J., 772,* 102.
Heinemann T. and Papaloizou J. C. B. (2009) *Mon. Not. R. Astron. Soc., 397,* 64.
Heinemann T. and Papaloizou J. C. B. (2012) *Mon. Not. R. Astron. Soc., 419,* 1085.
Heinzeller D. et al. (2011) *Astrophys. J., 731,* 115.
Herault J. et al. (2011) *Phys. Rev. E, 84(3),* 36321.
Hirose S. and Turner N. J. (2011) *Astrophys. J. Lett., 732,* L30.
Hopkins P. F. and Christiansen J. L. (2013) *Astrophys. J., 776,* 48.
Horne K. (1995) *Astron. Astrophys., 297,* 273.
Hughes A. L. H. and Armitage P. J. (2010) *Astrophys. J., 719,* 1633.
Hughes A. M. et al. (2009) *Astrophys. J., 704,* 1204.
Hughes A. M. et al. (2011) *Astrophys. J., 727,* 85.
Hughes A. M. et al. (2013) *Astron. J., 145,* 115.
Igea J. and Glassgold A. E. (1999) *Astrophys. J., 518,* 848.
Ilee J. D. et al. (2011) *Mon. Not. R. Astron. Soc., 417,* 2950.
Ilgner M. and Nelson R. P. (2006a) *Astron. Astrophys., 445,* 205.
Ilgner M. and Nelson R. P. (2006b) *Astron. Astrophys., 445,* 223.
Ilgner M. and Nelson R. P. (2008) *Astron. Astrophys., 483,* 815.
Inaba S. and Barge P. (2006) *Astrophys. J., 649,* 415.
Isella A. et al. (2009) *Astrophys. J., 701,* 260.
Ji H. et al. (2006) *Nature, 444,* 343.
Jin L. (1996) *Astrophys. J., 457,* 798.
Johansen A. and Klahr H. (2005) *Astrophys. J., 634,* 1353.
Johansen A. et al. (2006a) *Astrophys. J., 643,* 1219.
Johansen A. et al. (2006b) *Mon. Not. R. Astron. Soc., 370,* L71.
Johansen A. et al. (2009) *Astrophys. J., 697,* 1269.
Johnson B. M. and Gammie C. F. (2003) *Astrophys. J., 597,* 131.
Johnson B. M. and Gammie C. F. (2005) *Astrophys. J., 626,* 978.
Johnson B. M. and Gammie C. F. (2006) *Astrophys. J., 636,* 63.
Johnson E. T. et al. (2006) *Astrophys. J., 647,* 1413.
Juhász A. et al. (2012) *Astrophys. J., 744,* 118.
Julian W. H. and Toomre A. (1966) *Astrophys. J., 146,* 810.
Kemper F. et al. (2005) *Astrophys. J., 633,* 534.
Kim W.-T. and Ostriker E. C. (2000) *Astrophys. J., 540,* 372.
Kitchatinov L. L. and Rüdiger G. (2010) *Astron. Astrophys., 513,* L1.
Klahr H. (2004) *Astrophys. J., 606,* 1070.
Klahr H. (2007) In *Convection in Astrophysics* (F. Kupka et al., eds.), pp. 405–416. IAU Symp. 239, Cambridge Univ., Cambridge.
Klahr H. H. and Bodenheimer P. (2003) *Astrophys. J., 582,* 869.
Klahr H. and Bodenheimer P. (2006) *Astrophys. J., 639,* 432.
Königl A. et al. (2010) *Mon. Not. R. Astron. Soc., 401,* 479.
Kratter K. M. et al. (2010) *Astrophys. J., 708,* 1585.
Krumholz M. R. et al. (2007) *Astrophys. J., 656,* 959.
Kunz M. W. and Balbus S. A. (2004) *Mon. Not. R. Astron. Soc., 348,* 355.
Kunz M. W. and Lesur G. (2013) *Mon. Not. R. Astron. Soc., 434,* 2295.
Lathrop D. P. et al. (1992) *Phys. Rev. Lett., 68,* 1515.
Latter H. N. et al. (2010) *Mon. Not. R. Astron. Soc., 406,* 848.

Lee A. T. et al. (2010) *Astrophys. J., 718,* 1367.
Lesur G. and Longaretti P.-Y. (2005) *Astron. Astrophys., 444,* 25.
Lesur G. and Longaretti P.-Y. (2007) *Mon. Not. R. Astron. Soc., 378,* 1471.
Lesur G. and Longaretti P.-Y. (2009) *Astron. Astrophys., 504,* 309.
Lesur G. and Ogilvie G. I. (2010) *Mon. Not. R. Astron. Soc., 404,* L64.
Lesur G. and Papaloizou J. C. B. (2009) *Astron. Astrophys., 498,* 1.
Lesur G. and Papaloizou J. C. B. (2010) *Astron. Astrophys., 513,* A60+.
Lesur G. et al. (2013) *Astron. Astrophys., 550,* A61.
Li H. et al. (2000) *Astrophys. J., 533,* 1023.
Li H. et al. (2001) *Astrophys. J., 551,* 874.
Lin D. N. C. and Papaloizou J. C. B. (1993) In *Protostars and Planets III* (E. H. Levy and J. I. Lunine, eds.), pp. 749–835. Univ. of Arizona, Tucson.
Lisse C. M. et al. (2006) *Science, 313,* 635.
Lodato G. and Rice W. K. M. (2004) *Mon. Not. R. Astron. Soc., 351,* 630.
Lodato G. and Rice W. K. M. (2005) *Mon. Not. R. Astron. Soc., 358,* 1489.
Longaretti P.-Y. and Lesur G. (2010) *Astron. Astrophys., 516,* A51.
Lovelace R. V. E. et al. (1999) *Astrophys. J., 513,* 805.
Lubow S. H. and Spruit H. C. (1995) *Astrophys. J., 445,* 337.
Lubow S. H. et al. (1994a) *Mon. Not. R. Astron. Soc., 267,* 235.
Lubow S. H. et al. (1994b) *Mon. Not. R. Astron. Soc., 268,* 1010.
Luhman K. L. et al. (2008) *Astrophys. J., 675,* 1375.
Luhman K. L. et al. (2010) *Astrophys. J. Suppl., 186,* 111.
Lyra W. and Klahr H. (2011) *Astron. Astrophys., 527,* A138.
Lyra W. and Lin M.-K. (2013) *Astrophys. J., 775,* 17.
Lyra W. and Mac Low M.-M. (2012) *Astrophys. J., 756,* 62.
Lyra W. et al. (2008) *Astron. Astrophys., 491,* L41.
Lyra W. et al. (2009) *Astron. Astrophys., 497,* 869.
Mac Low M.-M. et al. (1995) *Astrophys. J., 442,* 726.
Mamatsashvili G. R. and Rice W. K. M. (2010) *Mon. Not. R. Astron. Soc., 406,* 2050.
Marcus P. S. et al. (2013) *Phys. Rev. Lett., 111(8),* 084501.
Matzner C. D. and Levin Y. (2005) *Astrophys. J., 628,* 817.
Meheut H. et al. (2010) *Astron. Astrophys., 516,* A31.
Meheut H. et al. (2012a) *Astron. Astrophys., 545,* A134.
Meheut H. et al. (2012b) *Astron. Astrophys., 542,* A9.
Meheut H. et al. (2012c) *Mon. Not. R. Astron. Soc., 422,* 2399.
Meru F. and Bate M. R. (2011) *Mon. Not. R. Astron. Soc., 411,* L1.
Meru F. and Bate M. R. (2012) *Mon. Not. R. Astron. Soc., 427,* 2022.
Meschiari S. and Laughlin G. (2008) *Astrophys. J. Lett., 679,* L135.
Michael S. et al. (2011) *Astrophys. J. Lett., 737,* L42.
Michael S. et al. (2012) *Astrophys. J., 746,* 98.
Mignone A. et al. (2007) *Astrophys. J. Suppl., 170,* 228.
Miller K. A. and Stone J. M. (2000) *Astrophys. J., 534,* 398.
Min M. et al. (2005) *Icarus, 179,* 158.
Mohanty S. et al. (2013) *Astrophys. J., 764,* 65.
Moll R. (2012) *Astron. Astrophys., 548,* A76.
Morales-Calderón M. et al. (2011) *Astrophys. J., 733,* 50.
Mulders G. D. and Dominik C. (2012) *Astron. Astrophys., 539,* A9.
Müller T. W. A. et al. (2012) *Astron. Astrophys., 541,* A123.
Murphy G. C. et al. (2010) *Astron. Astrophys., 512,* A82.
Muzerolle J. et al. (2009) *Astrophys. J. Lett., 704,* L15.
Najita J. R. et al. (2007) In *Protostars and Planets V* (B. Reipurth et al., eds.), pp. 507–522. Univ. of Arizona, Tucson.
Najita J. R. et al. (2009) *Astrophys. J., 691,* 738.
Najita J. R. et al. (2013) *Astrophys. J., 766,* 134.
Nelson R. P. and Papaloizou J. C. B. (2004) *Mon. Not. R. Astron. Soc., 350,* 849.
Nelson R. P. et al. (2013) *Mon. Not. R. Astron. Soc., 435,* 2610.
Ogilvie G. I. (2012) *Mon. Not. R. Astron. Soc., 423,* 1318.
Oishi J. S. and Mac Low M.-M. (2009) *Astrophys. J., 704,* 1239.
Okuzumi S. (2009) *Astrophys. J., 698,* 1122.
Okuzumi S. and Hirose S. (2011) *Astrophys. J., 742,* 65.
Owen J. E. et al. (2010) *Mon. Not. R. Astron. Soc., 401,* 1415.
Paardekooper S.-J. (2012) *Mon. Not. R. Astron. Soc., 421,* 3286.
Paardekooper S.-J. et al. (2010) *Astrophys. J., 725,* 146.
Paardekooper S.-J. et al. (2011) *Mon. Not. R. Astron. Soc., 416,* L65.
Paczynski B. (1978) *Acta Astron., 28,* 91.
Pandey B. P. and Wardle M. (2012) *Mon. Not. R. Astron. Soc., 423,* 222.
Panoglou D. et al. (2012) *Astron. Astrophys., 538,* A2.
Paoletti M. S. et al. (2012) *Astron. Astrophys., 547,* A64.
Perez-Becker D. and Chiang E. (2011a) *Astrophys. J., 735,* 8.

Perez-Becker D. and Chiang E. (2011b) *Astrophys. J., 727*, 2.

Petersen M. R. et al. (2007a) *Astrophys. J., 658*, 1236.

Petersen M. R. et al. (2007b) *Astrophys. J., 658*, 1252.

Pinte C. et al. (2008) *Astron. Astrophys., 489*, 633.

Pontoppidan K. M. et al. (2010) *Astrophys. J., 720*, 887.

Raettig N. et al. (2013) *Astrophys. J., 765*, 115.

Rafikov R. R. (2005) *Astrophys. J. Lett., 621*, L69.

Rafikov R. R. (2009) *Astrophys. J., 704*, 281.

Rayleigh L. (1917) *Philos. Trans. R. Soc. London Ser. A., 93*, 148.

Regály Z. et al. (2012) *Mon. Not. R. Astron. Soc., 419*, 1701.

Reipurth B. et al., eds. (2007) *Protostars and Planets V*, Univ. of Arizona, Tucson.

Rice W. K. M. et al. (2011) *Mon. Not. R. Astron. Soc., 418*, 1356.

Richard D. and Zahn J.-P. (1999) *Astron. Astrophys., 347*, 734.

Rigliaco E. et al. (2013) *Astrophys. J., 772*, 60.

Romanova M. M. et al. (2013) *Mon. Not. R. Astron. Soc., 430*, 699.

Rüdiger G. et al. (2002) *Astron. Astrophys., 391*, 781.

Ruge J. P. et al. (2013) *Astron. Astrophys., 549*, A97.

Ruzmaikina T. V. (1982) *Mitt. Astron. Ges. Hamburg, 57*, 45.

Salmeron R. and Wardle M. (2008) *Mon. Not. R. Astron. Soc., 388*, 1223.

Salmeron R. et al. (2011) *Mon. Not. R. Astron. Soc., 412(2)*, 1162.

Salyk C. et al. (2008) *Astrophys. J. Lett., 676*, L49.

Salyk C. et al. (2011) *Astrophys. J., 731*, 130.

Sano T. and Inutsuka S.-i. (2001) *Astrophys. J. Lett., 561*, L179.

Sano T. and Miyama S. M. (1999) *Astrophys. J., 515*, 776.

Sano T. and Stone J. M. (2002a) *Astrophys. J., 570*, 314.

Sano T. and Stone J. M. (2002b) *Astrophys. J., 577*, 534.

Sano T. et al. (2000) *Astrophys. J., 543*, 486.

Sargent B. A. et al. (2009) *Astrophys. J. Suppl., 182*, 477.

Schartman E. et al. (2012) *Astron. Astrophys., 543*, A94.

Schultz-Grunow F. (1959) *Z. Ang. Math. Mech., 39*, 101.

Semenov D. and Wiebe D. (2011) *Astrophys. J. Suppl., 196*, 25.

Shakura N. I. and Sunyaev R. A. (1973) *Astron. Astrophys., 24*, 337.

Shen Y. et al. (2006) *Astrophys. J., 653*, 513.

Shi J. et al. (2010) *Astrophys. J., 708*, 1716.

Shlosman I. and Begelman M. C. (1987) *Nature, 329*, 810.

Simon J. B. and Hawley J. F. (2009) *Astrophys. J., 707*, 833.

Simon J. B. et al. (2009) *Astrophys. J., 690*, 974.

Simon J. B. et al. (2011a) *Astrophys. J., 730*, 94.

Simon J. B. et al. (2011b) *Astrophys. J., 743*, 17.

Simon J. B. et al. (2012) *Mon. Not. R. Astron. Soc., 422*, 2685.

Simon J. B. et al. (2013a) *Astrophys. J., 764*, 66.

Simon J. B. et al. (2013b) *Astrophys. J., 775*, 73.

Sorathia K. A. et al. (2010) *Astrophys. J., 712*, 1241.

Sorathia K. A. et al. (2012) *Astrophys. J., 749*, 189.

Spruit H. C. (1996) In *Evolutionary Processes in Binary Stars* (R. A. M. J. Wijers et al., eds.), pp. 249–286. NATO ASIC Proc. 477.

Steiman-Cameron T. Y. et al. (2013) *Astrophys. J., 768*, 192.

Stone J. M. and Norman M. L. (1992) *Astrophys. J. Suppl., 80*, 753.

Stone J. M. et al. (2008) *Astrophys. J. Suppl., 178*, 137.

Suzuki T. K. and Inutsuka S.-i. (2009) *Astrophys. J. Lett., 691*, L49.

Suzuki T. K. et al. (2010) *Astrophys. J., 718, 2*, 1289.

Tanga P. et al. (1996) *Icarus, 121*, 158.

Tassoul J.-L. (2007) *Stellar Rotation*, Cambridge Univ., Cambridge.

Taylor G. I. (1923) *Philos. Trans. R. Soc. London Ser. A, 223*, 289.

Taylor G. I. (1936) *Philos. Trans. R. Soc. London Ser. A, 157*, 546.

Teitler S. (2011) *Astrophys. J., 733*, 57.

Terquem C. and Papaloizou J. C. B. (1996) *Mon. Not. R. Astron. Soc., 279*, 767.

Turner N. J. and Drake J. F. (2009) *Astrophys. J., 703*, 2152.

Turner N. J. and Sano T. (2008) *Astrophys. J. Lett., 679*, L131.

Turner N. J. et al. (2006) *Astrophys. J., 639*, 1218.

Turner N. J. et al. (2007) *Astrophys. J., 659*, 729.

Turner N. J. et al. (2010) *Astrophys. J., 708*, 188.

Tzeferacos P. et al. (2009) *Mon. Not. R. Astron. Soc., 400*, 820.

Tzeferacos P. et al. (2013) *Mon. Not. R. Astron. Soc., 428*, 3151.

Umebayashi T. and Nakano T. (1981) *Publ. Astron. Soc. Japan, 33*, 617.

Umebayashi T. and Nakano T. (1990) *Mon. Not. R. Astron. Soc., 243*, 103.

Umebayashi T. and Nakano T. (2009) *Astrophys. J., 690*, 69.

Umebayashi T. et al. (2013) *Astrophys. J., 764*, 104.

Uribe A. L. et al. (2011) *Astrophys. J., 736*, 85.

Urpin V. (2003) *Astron. Astrophys., 404*, 397.

Urpin V. and Brandenburg A. (1998) *Mon. Not. R. Astron. Soc., 294*, 399.

van der Marel N. et al. (2013) *Science, 340*, 1199.

Varnière P. and Tagger M. (2006) *Astron. Astrophys., 446*, L13.

Velikhov E. (1959) *J. Exp. Theor. Phys. (USSR), 36*, 1398.

Vorobyov E. I. and Basu S. (2007) *Mon. Not. R. Astron. Soc., 381*, 1009.

Vorobyov E. I. and Basu S. (2009) *Mon. Not. R. Astron. Soc., 393*, 822.

Vorobyov E. I. and Basu S. (2010) *Astrophys. J., 719*, 1896.

Wardle M. (1999) *Mon. Not. R. Astron. Soc., 307*, 849.

Wardle M. (2007) *Astrophys. Space Sci., 311*, 35.

Wardle M. and Königl A. (1993) *Astrophys. J., 410*, 218.

Wardle M. and Salmeron R. (2012) *Mon. Not. R. Astron. Soc., 422*, 2737.

Watson A. M. et al. (2007) In *Protostars and Planets V* (B. Reipurth et al., eds.), pp. 523–538. Univ. of Arizona, Tucson.

Weiss B. P. et al. (2010) *Space Sci. Rev., 152*, 341.

Wendt F. (1959) *Ingenieur-Archiv, 4*, 557.

Willacy K. et al. (2006) *Astrophys. J., 644*, 1202.

Wolf S. and Klahr H. (2002) *Astrophys. J. Lett., 578*, L79.

Zanni C. et al. (2007) *Astron. Astrophys., 469*, 811.

Ziegler U. (2005) *Astron. Astrophys., 435*, 385.

Ziegler U. (2008) *Comp. Phys. Commun., 179*, 227.

Zolensky M. E. et al. (2006) *Science, 314*, 1735.

Zsom A. et al. (2010) *Astron. Astrophys., 513*, A57.

Color Section

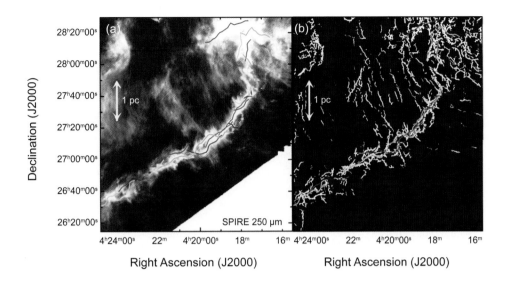

Plate 1. (a) Herschel/SPIRE 250-μm dust continuum image of the B211/B213/L1495 region in Taurus (*Palmeirim et al.,* 2013). The colored curves display the velocity-coherent "fibers" identified within the B213/B211 filament by *Hacar et al.* (2013) using $C^{18}O(1-0)$ observations. **(b)** Fine structure of the Herschel/SPIRE 250-μm dust continuum emission from the B211/B213 filament obtained by applying the multi-scale algorithm *getfilaments* (*Men'shchikov,* 2013) to the 250-μm image shown in **(a)**. Note the faint striations perpendicular to the main filament and the excellent correspondence between the small-scale structure of the dust continuum filament and the bundle of velocity-coherent fibers traced by *Hacar et al.* (2013) in $C^{18}O$ [same colored curves as in **(a)**].

Accompanies chapter by André et al. (pp. 27–51).

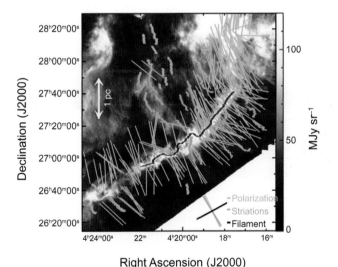

Plate 2. Display of optical and infrared polarization vectors from, e.g., *Heyer et al.* (2008) and *Chapman et al.* (2011) tracing the magnetic field orientation, overlaid on the Herschel/SPIRE 250-μm image of the B211/B213/L1495 region in Taurus (*Palmeirim et al.,* 2013) (see Fig. 2a). The plane-of-the-sky projection of the magnetic field appears to be oriented perpendicular to the B211/B213 filament and roughly aligned with the general direction of the striations overlaid in blue. A very similar pattern is observed in the Musca cloud (Cox et al., in preparation).

Accompanies chapter by André et al. (pp. 27–51).

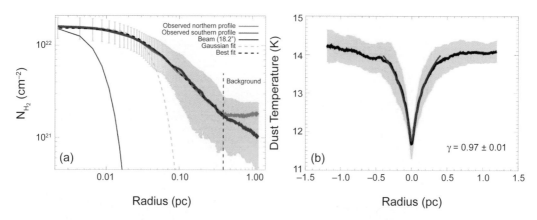

Plate 3. (a) Mean radial column density profile observed perpendicular to the B213/B211 filament in Taurus (*Palmeirim et al., 2013*), for both the northern (blue curve) and southern part (red curve) of the filament. The yellow area shows the ($\pm 1\sigma$) dispersion of the distribution of radial profiles along the filament. The inner solid purple curve shows the effective 18" HPBW resolution (0.012 pc at 140 pc) of the Herschel column density map used to construct the profile. The dashed black curve shows the best-fit Plummer model (convolved with the 18" beam) described in equation (1) with $p = 2.0 \pm 0.4$ and a diameter $2 \times R_{flat} = 0.07 \pm 0.01$ pc, which matches the data very well for $r \leq 0.4$ pc. **(b)** Mean dust temperature profile measured perpendicular to the B213/B211 filament in Taurus. The solid red curve shows the best polytropic model temperature profile obtained by assuming $T_{gas} = T_{dust}$ and that the filament has a density profile given by the Plummer model shown in **(a)** (with $p = 2$) and obeys a polytropic equation of state, $P \propto \rho^\gamma$ [and thus $T(r) \propto \rho(r)^{(\gamma-1)}$]. This best fit corresponds to a polytropic index $\gamma = 0.97 \pm 0.01$ (see *Palmeirim et al., 2013*, for further details).

Accompanies chapter by André et al. (pp. 27–51).

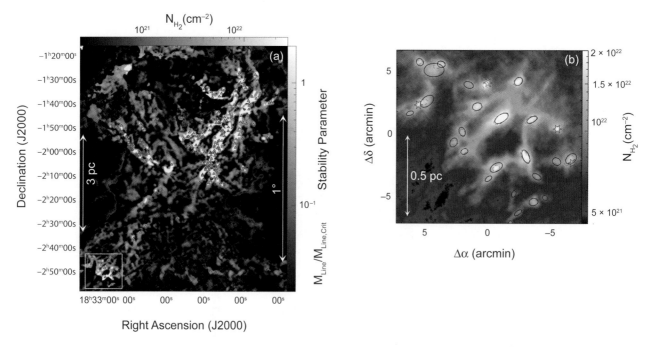

Plate 4. (a) Column density map of a subfield of the Aquila star-forming region derived from Herschel data (*André et al., 2010*). The contrast of the filaments has been enhanced using a curvelet transform (cf. *Starck et al., 2003*). Given the typical width ~0.1 pc of the filaments (*Arzoumanian et al., 2011*) (see Fig. 5), this map is equivalent to a map of the mass per unit length along the filaments. The areas where the filaments have a mass per unit length larger than half the critical value $2c_s^2/G$ (cf. *Inutsuka and Miyama, 1997*) (section 5.1) and are thus likely gravitationally unstable are highlighted in white. The bound prestellar cores identified by *Könyves et al.* (2010) are shown as small blue triangles. **(b)** Close-up column density map of the area shown as a yellow box on the left. The black ellipses mark the major and minor FWHM sizes of the prestellar cores found with the source extraction algorithm *getsources* (*Men'shchikov et al., 2012*); four protostellar cores are also shown by red stars. The dashed purple ellipses mark the FHWM sizes of the sources independently identified with the *csar* algorithm (*J. Kirk et al., 2013*). The effective resolution of the image is ~18" or ~0.02 pc at d ~ 260 pc.

Accompanies chapter by André et al. (pp. 27–51).

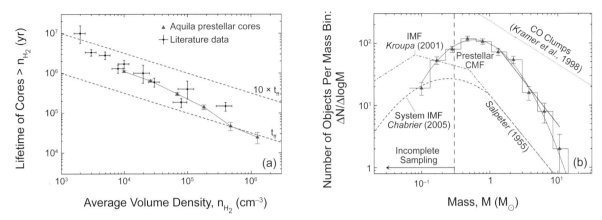

Plate 5. (a) Plot of estimated lifetime against minimum average core density (after *Jessop and Ward-Thompson,* 2000) for the population of prestellar and starless cores identified with Herschel in the Aquila cloud complex (blue triangles; Könyves et al., in preparation) and literature data [black crosses; see *Ward-Thompson et al.* (2007) and references therein]. Note the trend of decreasing core lifetime with increasing average core density. For comparison, the dashed lines correspond to the free-fall timescale (t_{ff}) and a rough approximation of the ambipolar diffusion timescale ($10 \times t_{ff}$). (b) Core mass function (blue histogram with error bars) of the ~500 candidate prestellar cores identified with Herschel in Aquila (*Könyves et al.,* 2010; *André et al.,* 2010). The IMF of single stars [corrected for binaries; e.g., *Kroupa* (2001)], the IMF of multiple systems (e.g., *Chabrier,* 2005), and the typical mass spectrum of CO clumps (e.g., *Kramer et al.,*1998) are shown for comparison. A lognormal fit to the observed CMF is superimposed (red curve); it peaks at ~0.6 $M_\odot$, close to the Jeans mass within marginally critical filaments at T ~ 10 K (see section 6.3).

Accompanies chapter by André et al. (pp. 27–51).

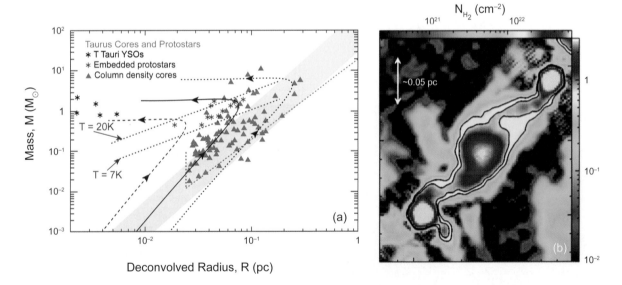

Plate 6. (a) Mass vs. size diagram for the cores identified with Herschel in the Taurus S1/S2 region (*J. Kirk et al.,* 2013). The two red solid lines mark the loci of critically self-gravitating isothermal Bonnor-Ebert spheres at T = 7 K and T = 20 K, respectively, while the yellow band shows the region of the mass-size plane where low-density CO clumps are observed to lie (cf. *Elmegreen and Falgarone,* 1996). The blue dotted curve indicates both the mass sensitivity and the resolution limit of the *J. Kirk et al.* (2013) study. Model evolutionary tracks from *Simpson et al.* (2011) are superposed as black curves (dashed, solid, and dotted for ambient pressures $P_{ext}/k_B = 10^6$ K cm^{-3}, 10^5 K cm^{-3}, and 10^4 K cm^{-3}, respectively), in which starless cores (blue triangles) first grow in mass by accretion of background material before becoming gravitationally unstable and collapsing to form protostars (red stars) and T Tauri stars (black stars). (b) Blow-up Herschel column density map of a small portion of the Taurus B211/B213 filament (cf. Fig. 2a) showing two protostellar cores on either side of a starless core (Palmeirim et al., in preparation). The contrast of filamentary structures has been enhanced using a curvelet transform (see *Starck et al.,* 2003). Note the bridges of material connecting the cores along the B211/B213 filament, suggesting that the central starless core may still be growing in mass through filamentary accretion, a mechanism seen in some numerical simulations (cf. *Balsara et al.,* 2001) (section 5).

Accompanies chapter by André et al. (pp. 27–51).

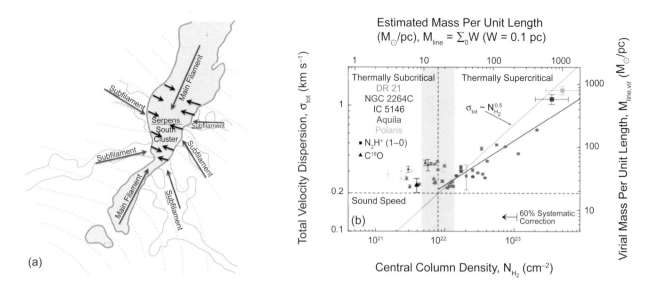

Plate 7. (a) Sketch of the typical velocity field inferred from line observations (e.g., *H. Kirk et al.,* 2013) within and around a supercritical filament [here Serpens-South in the Aquila complex; adapted from *Sugitani et al.* (2011)]: Red arrows mark longitudinal infall along the main filament; black arrows indicate radial contraction motions; and blue arrows mark accretion of background cloud material through striations or subfilaments, along magnetic field lines (green dotted curves from Sugitani et al.). **(b)** Total (thermal + nonthermal) velocity dispersion vs. central column density for a sample of 46 filaments in nearby interstellar clouds (*Arzoumanian et al.,* 2013). The horizontal dashed line shows the value of the thermal sound speed ~0.2 km s⁻¹ for T = 10 K. The vertical gray band marks the border zone between thermally subcritical and thermally supercritical filaments where the observed mass per unit length M_{line} is close to the critical value $M_{line,crit}$ ~ 16 M⊙/pc for T = 10 K. The blue solid line shows the best power-law fit $\sigma_{tot} \propto N_{H2}^{0.36\pm0.14}$ to the data points corresponding to supercritical filaments.

Accompanies chapter by André et al. (pp. 27–51).

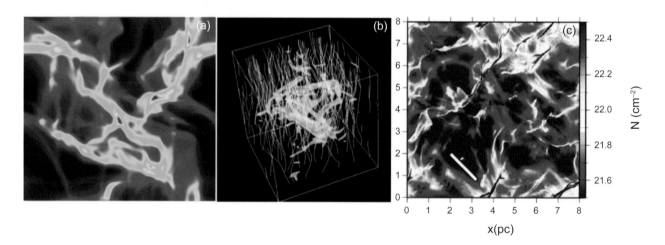

Plate 8. (a) Formation of filaments and cores from shock compression in numerical simulations of isothermal supersonic turbulence with rms Mach number ~10 (*Padoan et al.,* 2001). **(b)** Three-dimensional view of flattened filaments resulting from mass accumulation along magnetic field lines in the MHD model of *Li et al.* (2010). The magnetic field favors the formation of both low- and high-mass stars compared to intermediate-mass stars, due to fragmentation of high-density filaments and global clump collapse, respectively. **(c)** Face-on column density view of a shock-compressed dense layer in numerical MHD simulations by Inutsuka et al. (in preparation). The color scale (in cm⁻²) is shown on the right. The mean magnetic field is in the plane of the layer and its direction is shown by a white bar. Note the formation of dense, magnetized filaments whose axes are almost perpendicular to the mean magnetic field. Fainter "striation"-like filaments can also be seen that are almost perpendicular to the dense filaments.

Accompanies chapter by André et al. (pp. 27–51).

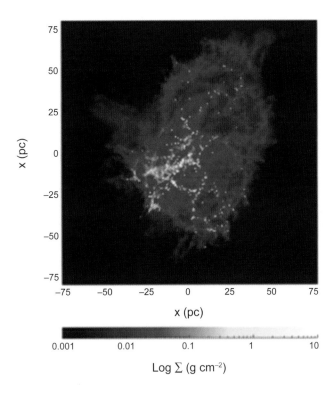

Plate 9. Distribution of stars, cold gas filaments (in red), voids, and the low-density ionized gas (in purple) that fills the cavities in the SPH numerical simulations of *Dale and Bonnell* (2011), in which a massive stellar cluster forms at the junction of a converging network of filaments, from the gravitational collapse of a turbulent molecular cloud.

Accompanies chapter by André et al. (pp. 27–51).

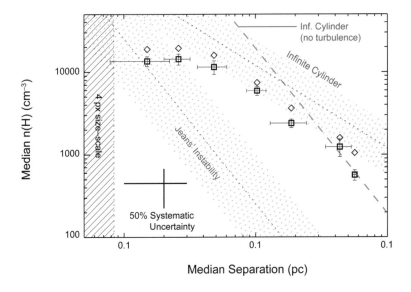

Plate 10. Median density of significant substructures as a function of separation in the high-(2" or 0.035 pc)-resolution column density map of the massive, filamentary IRDC G11.11-0.12 by *Kainulainen et al.* (2013). The red dotted line shows the relation expected from standard Jeans' instability arguments. The blue dotted line shows the relation predicted by the linear stability analysis of self-gravitating isothermal cylinders. Note how the structure of G11 is consistent with global cylinder fragmentation on scales ≥0.5 pc and approaches local Jeans' fragmentation on scales ≥0.2 pc. From *Kainulainen et al.* (2013.)

Accompanies chapter by André et al. (pp. 27–51).

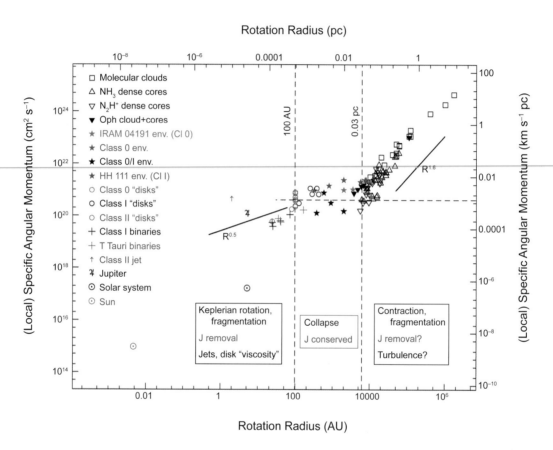

Plate 11. Progression of specific angular momentum as function of scale and/or evolutionary stage of YSOs. From *Belloche* (2013).

Accompanies chapter by Li et al. (pp. 173–194).

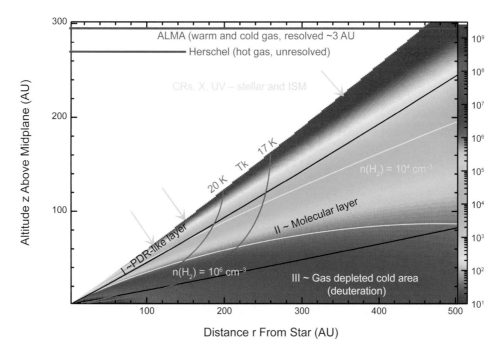

Plate 12. Molecular disk: state of the art after *Protostars and Planets V* (*Reipurth et al., 2007*). The model is done for a disk orbiting a young star of 1 M$_\odot$. The volume density as a function of radius r and altitude z above the mid-plane is given in color. The location of the isotherms at 17 and 20 K are also shown. I, II, and III correspond to the location of the three important layers of gas: the PDR-like atmosphere irradiated by the star, the UV interstellar field and the cosmic rays (I), the molecular layer typically located between one and three scale heights (II), and the mid-plane where only molecular hydrogen and deuterated isotopologs of simple molecules are abundant in the gas phase (III). The line at the very top (blue) corresponds to the area sampled by ALMA (with the best resolution for a source at 150 pc); the line immediately underneath that (red) gives the area sampled by Herschel observations (unresolved).

Accompanies chapter by Dutrey et al. (pp. 317–338).

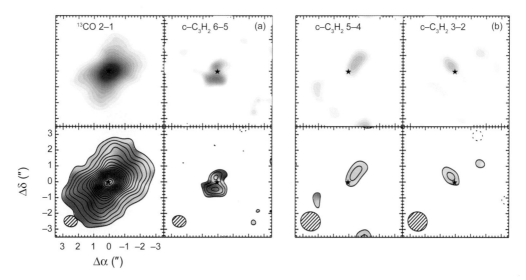

Plate 13. The integrated intensity maps and intensity-weighted mean velocity fields of ^{13}CO 2–1 and c–C$_3$H$_2$ 6–5 lines (left panels), as well as c–C$_3$H$_2$ 5–4 and 3–2 lines (right panels) toward HD 163296. The resolved velocity field of the c–C$_3$H$_2$ 6–5 line agrees with the CO kinematics. In the c–C$_3$H$_2$ maps, the first contour marks 3σ followed by the 1σ contour increases. The rms varies between 6 and 9 mJy km s^{-1} per beam. Synthesized beams are presented in the lower left corners. The star symbol denotes the continuum (stellar) position. The axes are offsets from the pointing center in arcseconds. From *Qi et al.* (2013a).

Accompanies chapter by Dutrey et al. (pp. 317–338).

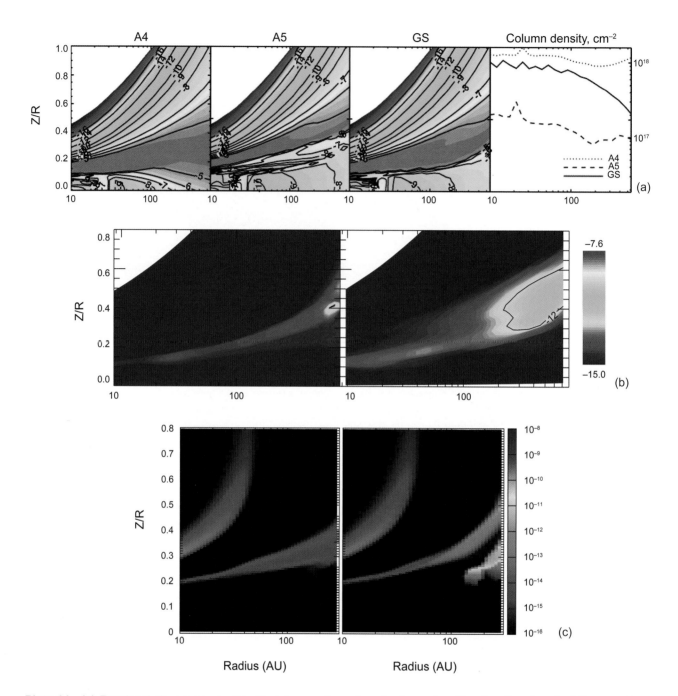

Plate 14. **(a)** Fractional abundance profiles (first three columns) and column densities (fourth column) of the gas-phase carbon monoxide in a T Tauri disk calculated with three distinct dust grain models: A4, the uniform grains with radius of 1 µm; A5, the uniform grains with radius of 0.1 µm; and GS, the model with grain growth and settling. Note how grain evolution affects the depletion of molecules onto grains near the mid-plane of the cold outer disk (*Vasyunin et al.*, 2011). **(b)** Gas-phase methanol abundances in a T Tauri disk calculated with a laminar gas-grain chemical model (left) and a two-dimensional turbulent-mixing chemical model (right). The transport of methanol ice from dark mid-plane to upper, more UV-irradiated disk layer enables more methanol ice to be desorbed in the gas phase (*Semenov and Wiebe*, 2011). **(c)** Gas-phase methanol abundances in a T Tauri disk calculated with two gas-grain chemical models: with no surface chemistry taken into account (left) and with surface chemistry (right) (*Walsh et al.*, 2010). The surface hydrogenation of CO leads to efficient production of methanol ice, which can be partly UV-desorbed in disk layers above mid-plane.

Accompanies chapter by Dutrey et al. (pp. 317–338).

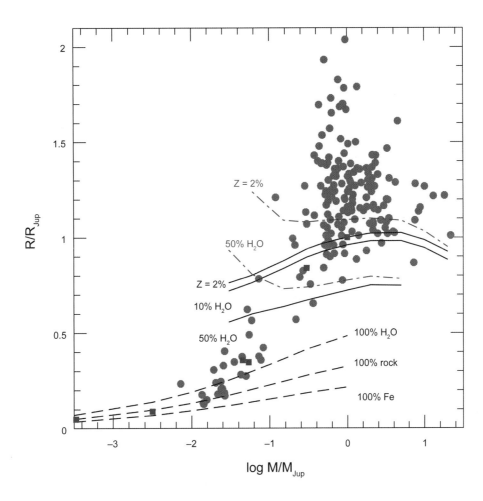

Plate 15. Mass vs. radius of known planets, including solar system planets (blue squares: Jupiter, Saturn, Neptune, Uranus, Earth, and Mars) and transiting exoplanets (magenta dots, from the *Extrasolar Planet Encyclopedia* at *http://exoplanet.eu*). Note that the sample shown in this figure has been cleaned up and only planets where the parameters are reasonably precise are included. The curves correspond to models from $0.1\ M_\oplus$ to $20\ M_{Jup}$ at 4.5 G.y. with various internal chemical compositions. The solid curves correspond to a mixture of H, He, and heavy elements (models from *Baraffe et al., 2008*). The long dashed lines correspond to models composed of pure water, rock, or iron from *Fortney et al.* (2007). The "rock" composition here is olivine (forsterite Mg_2SiO_4) or dunite. Solid and long-dashed lines (in black) are for non-irradiated models. Dash-dotted (red) curves correspond to irradiated models at 0.045 AU from a Sun.

Accompanies chapter by Baraffe et al. (pp. 763–786).

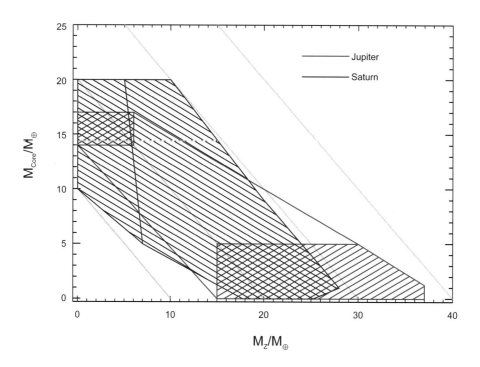

Plate 16. Inferred constraints on the interior of Jupiter and Saturn from fully convective models with a distinct core. The y-axis shows metals in the core, while the x-axis shows metals within the H-He envelope. Gray lines are constant metal mass. See text (section 4.3) for explanations of hatched regions and lines.

Accompanies chapter by Baraffe et al. (pp. 763–786).

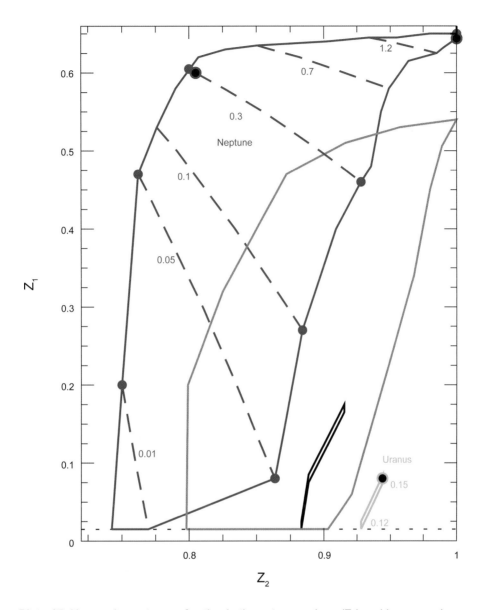

Plate 17. Heavy element mass fraction in the outer envelope (Z_1) and inner envelope (Z_2) of Uranus models (black) and Neptune models (gray) as labeled with the Voyager shape and rotation data. The solid lines frame the full set of solutions for each planet. Dashed lines within the box of Neptune models indicate solutions of same transition pressure between inner and outer envelopes (in M_{bar} as labeled). The dotted line is a guide to the eye for the solar metallicity $Z = 0.015$. The colored areas use the modified shape and rotation data for Uranus (green) and Neptune (blue). Adapted from *Nettelmann et al.* (2013a).

Accompanies chapter by Baraffe et al. (pp. 763–786).

Bouvier J., Matt S. P., Mohanty S., Scholz A., Stassun K. G., and Zanni C. (2014) Angular momentum evolution of young low-mass stars
and brown dwarfs: Observations and theory. In *Protostars and Planets VI* (H. Beuther et al., eds.), pp. 433–450. Univ. of Arizona, Tucson,
DOI: 10.2458/azu_uapress_9780816531240-ch019.

Angular Momentum Evolution of Young Low-Mass Stars and Brown Dwarfs: Observations and Theory

Jérôme Bouvier
Observatoire de Grenoble

Sean P. Matt
Exeter University

Subhanjoy Mohanty
Imperial College London

Aleks Scholz
University of St Andrews

Keivan G. Stassun
Vanderbilt University

Claudio Zanni
Osservatorio Astrofisico di Torino

This chapter aims at providing the most complete review of both the emerging concepts and the latest observational results regarding the angular momentum evolution of young low-mass stars and brown dwarfs. In the time since *Protostars and Planets V* (*Reipurth et al.,* 2007) (PPV), there have been major developments in the availability of rotation-period measurements at multiple ages and in different star-forming environments that are essential for testing theory. In parallel, substantial theoretical developments have been carried out in the last few years, including the physics of the star-disk interaction, numerical simulations of stellar winds, and the investigation of angular momentum transport processes in stellar interiors. This chapter reviews both the recent observational and theoretical advances that prompted the development of renewed angular momentum evolution models for cool stars and brown dwarfs. While the main observational trends of the rotational history of low-mass objects seem to be accounted for by these new models, a number of critical open issues remain that are outlined in this review.

1. INTRODUCTION

The angular momentum content of a newly born star is one of the fundamental quantities, like mass and metallicity, that durably impacts the star's properties and evolution. Rotation influences the star's internal structure, energy transport, and the mixing processes in the stellar interior, which are reflected in surface elemental abundances. It is also the main driver for magnetic activity, from X-ray luminosity to ultraviolet (UV) flux and surface spots, which are the ultimate source of stellar winds. Studying the initial angular momentum content of stars and its evolution throughout the star's lifetime brings unique clues to the star-formation process, to the accretion/ejection phenomenon in young stellar objects, to the history and future of stellar activity and its impact on surrounding planets, and

to physical processes that redistribute angular momentum in stellar interiors.

Spectacular progress has been made, both on the observational and theoretical sides, on the issue of the angular momentum evolution of young stellar objects since *Protostars and Planets V* (*Reipurth et al.,* 2007) (PPV). On the observational side, thousands of new rotational periods have been derived for stars over the entire mass range from solar-type stars down to brown dwarfs (BDs) at nearly all stages of evolution between birth and maturity. The picture we have of the rotational evolution of low-mass and very-low-mass stars and brown dwarfs has never been as well documented as today. On the theoretical side, recent years have seen a renaissance in numerical simulations of magnetized winds that are the prime agent of angular momentum loss, new attempts have been made to understand how young stars exchange angular momentum with their disks via magnetic interactions, and new insights have been gained on the way angular momentum is transported in stellar interiors.

In the following sections, we review the latest developments that shed new light on the processes governing the angular momentum evolution of young stars and brown dwarfs, and also provide important context for other chapters in this volume to explore possible connections between the rotational history of stars and the formation, migration, and evolution of planetary systems (star-disk interaction, inner disk warps and cavities, planet engulfment, irradiation of young planets, etc.). In section 2, we review the latest advances in the derivation of rotation rates for low-mass stars and brown dwarfs from birth to the early main sequence. In section 3, we provide an account of the physical mechanisms thought to dictate the evolution of stellar rotation during the pre-main-sequence (PMS) and early main-sequence (MS), including star-disk interaction, stellar winds, and angular momentum transport in stellar interiors. In section 4, we discuss various classes of angular momentum evolution models that implement the latest theoretical developments to account for the observed evolution of stellar rotation in cool stars and brown dwarfs.

2. OBSERVATIONAL STUDIES OF STELLAR ROTATION

The measurement of rotational periods for thousands of stars in molecular clouds and young open clusters provide the best way to trace their angular momentum evolution from about 1 million years ago (Ma) to 1 billion years ago (Ga). This section discusses the observational studies of stellar rotation performed since the PPV review by *Herbst al.* (2007), for solar-type stars and lower-mass stars, down to the brown dwarf regime.

2.1. Solar-Mass and Low-Mass Stars

In the last 7 yr, more than 5000 new rotational periods have been measured for cool stars in star-forming regions and young open clusters, over an age range from 1 Ma to 1 Ga. In parallel, dedicated photometric monitoring of nearby M dwarfs, aimed at planetary transit searches, have reported tens of periods for the field very-low-mass population over the age range 1 Ga to 13 Ga. These recent studies are listed in Table 1, while Fig. 1 provides a graphical summary of the results. A compilation of prior results was published by *Irwin and Bouvier* (2009).

In addition, long-term monitoring by Kepler and the Convection, Rotation and planetary Transits (CoRoT) has provided rotation periods for more than 10,000 GKM field dwarfs (e.g., *Nielsen et al., 2013; McQuillan et al., 2013; Harrison et al., 2012; Affer et al., 2012*). These results offer a global view of stellar rotation as a function of mass on the main sequence, exhibiting a large dispersion at each spectral type, which possibly reflects the age distribution of the stellar samples.

The evolution of rotational distributions from 1 Ma to the old disk population shown in Fig. 1 reveals a number of features:

1. The initial distribution of spin rates at an age of about 2 Ma is quite wide over the whole mass range from 0.2 to 1.0 $M_\odot$, with the bulk of rotational periods ranging from 1 to 10 d. The lower envelope of the period distribution is located at about 0.7 d, which corresponds to about 40–50% of the breakup limit over the mass range 0.2–1.0 $M_\odot$. The origin of the initial scatter of stellar angular momentum for low-mass stars remains an open issue and probably reflects physical processes taking place during the embedded protostellar stage. *Gallet and Bouvier* (2013) suggested that the dispersion of initial angular momenta may be linked to the protostellar disk mass.

2. From 1 to 5 Ma, i.e., during the early PMS evolution, the rotation rates of solar-type stars hardly evolve. In contrast, the lowest-mass stars significantly spin up. *Henderson and Stassun* (2012) suggested that the increasing period-mass slope for lower-mass stars can be used as an age proxy for very young clusters. *Littlefairet al.* (2010), however, reported that the similarly aged (5 Ma) NGC 2362 and Cep OB3b clusters exhibit quite a different rotational period distribution at low masses, which may point to the impact of environmental effects on rotation properties.

3. Past the end of the PMS accretion phase, the rotational distribution of the 13-Ma h Per cluster members is remarkably flat over the 0.4–1.2 $M_\odot$ range. The lower envelope of the period distribution, now located at about 0.2–0.3 d, bears strong evidence for PMS spinup, as the freely evolving stars contract toward the zero age main sequence (ZAMS). In contrast, the slow rotators still retain periods close to 8–10 d, a result interpreted as evidence for core-envelope decoupling in these stars (*Moraux et al., 2013*). Similar results are seen in the 40-Ma clusters IC 4665 and NGC 2547, with the addition of very-low-mass stars that are faster rotators and exhibit a steep period-mass relationship.

4. Once on the early MS (0.1–0.6 Ga), a well-defined sequence of slow rotators starts to appear over the mass range 0.6–1.1 $M_\odot$, while the lower-mass stars still retain fast rotation. This suggests a spindown timescale on the order of a few 0.1 gigayears (G.y.) for solar-type stars, as angular momentum is carried away by magnetized winds. The development of a slow rotator sequence and its gradual evolution toward longer periods indeed serves as a basis for main-sequence gyrochronology (*Barnes, 2007*).

5. By an age of 0.5–0.6 Ga, all solar-type stars down to a mass of 0.6 $M_\odot$ have spun down, thus yielding a tight period-mass relationship, with the rotation rate decreasing toward lower masses (*Delorme et al., 2011*). In contrast, the very-low-mass stars still exhibit a large scatter in spin rates at that age (*Agüeros et al., 2011*). It is only in the old disk population, by about 10 Ga, that the majority of lowest-mass stars join the slow rotator sequence (*Irwin et al., 2011*). Clearly, the spindown timescale is a strong function of stellar mass, being much longer for the lowest-mass stars than for solar-type ones (e.g., *McQuillan et al., 2013*).

TABLE 1. Post-PPV rotational period distributions for young (≤1 Ga) stars.

Reference	Target	Age (Ma)	Mass range ($M_\odot$)	$N_\star$
Grankin (2013)	Taurus	1–3	0.4–1.6	61
Xiao et al. (2012)	Taurus	1–3	0.3–1.2	18
Artemenko et al. (2012)	Various SFRs	1–5	0.3–3.0	52
Henderson and Stassun (2012)	NGC 6530	2	0.2–2.0	244
Rodríguez-Ledesma et al. (2009)	ONC	2	0.015–0.5	487
Affer et al. (2013)	NGC 2264	3	0.2–3.0	209
Cody and Hillenbrand (2010)	σ Ori	3	0.02–1.0	64
Littlefair et al. (2010)	Cep OB3b	4–5	0.1–1.3	475
Irwin et al. (2008a)	NGC 2362	5	0.1–1.2	271
Messina et al. (2011)	Young assoc.	6–40	0.2–1.0	80
Messina et al. (2010)	Young assoc.	8–110	0.2–1.0	165
Moraux et al. (2013)	h Per	13	0.4–1.4	586
Irwin et al. (2008b)	NGC 2547	40	0.1–0.9	176
Scholz et al. (2009)	IC 4665	40	0.05–0.5	20
Hartman et al. (2010)	Pleiades	125	0.4–1.3	383
Irwin et al. (2009)	M 50	130	0.2–1.1	812
Irwin et al. (2007)	NGC 2516	150	0.15–0.7	362
Meibom et al. (2009)	M35	150	0.6–1.6	310
Sukhbold and Howell (2009)	NGC 2301	210	0.5–1.0	133
Meibom et al. (2011b)	M34	220	0.6–1.2	83
Hartman et al. (2009)	M 37	550	0.2–1.3	371
Scholz and Eislöffel (2007)	Praesepe	578	0.1–0.5	5
Agüeros et al. (2011)	Praesepe	578	0.27–0.74	40
Scholz et al. (2011)	Praesepe	578	0.16–0.42	26
Collier Cameron et al. (2009)	Coma Ber	591	FGK	46
Delorme et al. (2011)	Praesepe/Hyades	578/625	FGK	52/70
Meibom et al. (2011a)	NGC 6811	1000	FGK	71
Irwin et al. (2011)	Field M dwarfs	500–13000	0.1–0.3	41
Kiraga and Stepien (2007)	Field M dwarfs	3000–10000	0.1–0.7	31

A long-standing and somewhat controversial issue remains the so-called "disk-locking" process, i.e., the observational evidence that stars magnetically interacting with their accretion disk during the first few million years of PMS evolution are prevented from spinning up in spite of contracting toward the ZAMS (e.g., *Rebull et al.,* 2004). A number of post-PPV studies tend to support the view that, at a given age, disk-bearing PMS stars are, on average, slower rotators than diskless ones, with periods typically in the range from 3 to 10 d for the former, and between 1 and 7 d for the latter (e.g., *Affer et al.,* 2013; *Xiao et al.,* 2012; *Henderson and Stassun,* 2012; *Littlefair et al.,* 2010; *Rodríguez-Ledesma et al.,* 2009; *Irwin et al.,* 2008a; *Cieza and Baliber,* 2007). A significant overlap between the rotational distributions of classical and weak-line T Tauri stars is apparent in all star-forming regions investigated thus far. Furthermore, *Cody and Hillenbrand* (2010) failed to find any evidence for a disk-rotation connection among the very-low-mass members of the 3-Ma σ Ori cluster, which suggests it may only be valid over a restricted mass range.

The lack of a clear relationship between rotation and disk accretion may have various causes. Observationally,

the determination of rotational period distributions relies on the assumption that the photometric periods derived from monitoring studies arise from surface spot modulation and therefore accurately reflect the star's rotational period. Recently, *Artemenko et al.* (2012) questioned the validity of this assumption for classical T Tauri stars. Based on the comparison of photometric periods and v sin i measurements, they claimed that in a fraction of classical T Tauri stars, the measured periods correspond to the Keplerian motion of obscuring circumstellar dust in the disk and are significantly longer than the stellar rotational periods (see also *Percy et al.,* 2010). *Alencar et al.* (2010), however, found that the photometric periods of classical T Tauri stars undergoing cyclical disk obscuration were statistically similar to those of classical T Tauri stars dominated by surface spots, thus suggesting that the obscuring dust is located close to the co-rotation radius in the disk.

On the theoretical side, the star-disk interaction may impact the star's rotation rate in various ways, depending in particular on the ratio between the disk truncation and co-rotation radii. *Le Blanc et al.* (2011) have modeled the spectral energy distribution of young stars in IC 348 in

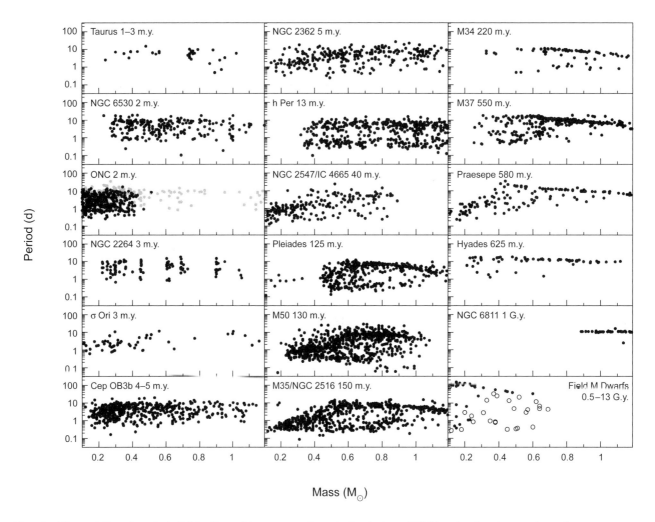

Fig. 1. The rotational period distribution of low-mass stars derived since PPV in star-forming regions, in young open clusters, and in the field. The panels are ordered by increasing age, from top to bottom and left to right. The ONC panel includes previous measurements by *Herbst et al.* (2002) shown as gray dots. In the lower right panel, young disk M dwarfs are shown as open circles, old disk M dwarfs are marked with filled circles. References are listed in Table 1.

an attempt to derive the disk inner radius and evaluate its relationship with the star's rotational period. No clear trend emerges from the ratio of inner disk radius to co-rotation radius when comparing slow and fast rotators. It should be cautioned, however, that spectral energy distribution (SED) modeling actually measures the inner dusty disk radius, while the gaseous disk may extend further in (e.g., *Carr* 2007). Also, scattered light in the near-infrared (NIR) may substantially alter the measurement of the inner dust disk radius in T Tauri stars (*Pinte et al.*, 2008).

2.2. Very-Low-Mass Stars and Brown Dwarfs

Significant progress has recently been made in evaluating the rotational properties of very-low-mass objects (VLM) (masses below ~0.3 $M_\odot$), including brown dwarfs, i.e., objects with masses too low to sustain stable hydrogen burning (M < 0.08 $M_\odot$). In the last Protostars and Planets review on this subject (*Herbst et al.*, 2007), about 200

periods for VLM objects in the ONC and NGC 2264, two 1–3 million-year (m.y.)-old star-forming regions, were discussed. In addition, smaller samples in other clusters were already available at that time. For brown dwarfs the total sample was limited to about 30 periods, only a handful for ages >10 m.y., complemented by v sin i measurements. We summarize in this section the most recent advances regarding the measurement of spin rates for very-low-mass stars and brown dwarfs, and recapitulate the emerging picture for the angular momentum evolution in the VLM domain.

2.2.1. Star-forming regions. For the Orion Nebula Cluster, at an age of 1–2 Ma, *Rodríguez-Ledesma et al.* (2009) have published several hundred new VLM periods. This includes more than 100 periods for brown dwarf candidates, the largest brown dwarf period sample in any region studied so far. A new period sample across the stellar/substellar regime in the slightly older σ Ori cluster has been presented by *Cody and Hillenbrand* (2010), including about 40 periods for VLM objects. In addition, the period sample from the Monitor project in the 4–5-m.y.-old cluster NGC 2362 contains about 20–30 periods in the VLM regime (*Irwin et al., 2008a*) and the new period sample in the 4–5-m.y.-old Cep OB3b region published by *Littlefair et al.* (2010) extends into the VLM regime. Taken together with the previously reported samples, there are now more than hundred VLM periods available at ages of 3–5 Ma.

From the period distributions in these very young regions, the following features are noteworthy:

1. Very-low-mass objects at young ages show a wide range of periods, similar to more massive stars, from a few hours up to at least 2 weeks.

2. In all these samples there is a consistent trend of faster rotation toward lower masses. In the ONC, the median period drops from 5 d for M > 0.4 $M_\odot$ to 2.6 d for VLM stars and to 2 d for brown dwarfs (*Rodríguez-Ledesma et al.*, 2009). As noted by *Cody and Hillenbrand* (2010) and earlier by *Herbst et al.* (2001), this period-mass trend is consistent with specific angular momentum being only weakly dependent on mass below about 1 $M_\odot$. An intriguing case in the context of the period-mass relation is the Cep OB3b region (*Littlefair et al.*, 2010). While the same trend is observed, it is much weaker than in the other regions. The VLM stars in Cep OB3b rotate more slowly than in other clusters with similar age. This may be a sign that rotational properties are linked to environmental factors, a possibility that needs further investigation.

3. A controversial aspect of the periods in star-forming regions is their lower limit. The breakup limit, where centrifugal forces balance gravity, lies between 3 and 5 hr at these young ages. *Zapatero Osorio et al.* (2003), *Caballero et al.* (2004), and *Scholz and Eislöffel* (2004a, 2005) report brown dwarf periods that are very close to that limit. On the other hand, the *Cody and Hillenbrand* (2010) sample contains only one period shorter than 14 hr, although their sensitivity increases toward shorter periods. Thus, it remains to be confirmed whether some young brown dwarfs indeed rotate close to breakup speed.

4. Whether the disk-rotation connection observed for solar-mass and low-mass young stars extends down to the VLM and brown dwarf domains remains unclear. *Cody and Hillenbrand* (2010) found the same period distribution for disk-bearing and diskless VLM stars in the 3-m.y.-old σ Ori cluster while in the 2-m.y.-old ONC, *Rodríguez-Ledesma et al.* (2010) find that objects with NIR excess tend to rotate slower than objects without NIR excess in the mass regime between 0.075 and 0.4 $M_\odot$. No such signature is seen in the substellar regime, with the possible caveat that many brown dwarf disks show little or no excess emission in the NIR and require MIR data to be clearly detected. Finally, *Mohanty et al.* (2005), *Nguyen et al.* (2009), *Biazzo et al.* (2009), and *Dahm et al.* (2012) report somewhat conflicting results regarding the existence of a disk-rotation connection in the very-low-mass regime based on v sin i measurements of members of 1–5-m.y.-old clusters.

2.2.2. Pre-main-sequence clusters. For the pre-main-sequence age range between 5 and 200 Ma, about 200

VLM periods are now available, an increased factor of 20 compared with 2007. About 80 of them have been measured in IC 4665 and NGC 2547, two clusters with ages around 40 Ma (*Scholz et al.*, 2009; *Irwin et al.*, 2008b). Approximately 100 periods are available for VLM objects in NGC 2516 (≈150 Ma) from *Irwin et al.* (2007). In addition, a few more VLM periods are contained in the samples for M34 (*Irwin et al.*, 2006) and M50 (*Irwin et al.*, 2009), although the latter sample might be affected by substantial contamination. Note that in these clusters (as well as in the main-sequence Praesepe cluster discussed below) the number of measured brown dwarf periods is very low (probably <5).

The most significant feature in the period distributions in this mass and age regime is the distinctive lack of slow rotators. Essentially all VLM periods measured thus far in these clusters are shorter than 2 d, with a clear preference for periods less than 1 d. The median period is 0.5–0.7 d. The lowest period limit is around 3 hr. This preference for fast rotators cannot be attributed to a bias in the period data for two reasons. First, most of the studies cited above are sensitive to longer periods. Second, *Scholz and Eislöffel* (2004b) demonstrate that the lower envelope of the v sin i for Pleiades VLM stars (*Terndrup et al.*, 2000) translates into an upper period limit of only 1–2 d, consistent with the period data.

Compared with the star-forming regions, both the upper period limit and the median period drop significantly. As discussed by *Scholz et al.* (2009), this evolution is consistent with angular momentum conservation plus weak rotational braking, but cannot be accommodated by a Skumanich-type wind-braking law.

2.2.3. Main-sequence clusters. For the Praesepe cluster, with an age of 580 Ma, a cornerstone for tracing the main-sequence evolution of stars, around 30 rotation periods have been measured for VLM stars (*Scholz and Eislöffel*, 2007; *Scholz et al.*, 2011). In combination with the samples for more massive stars by *Delorme et al.* (2011) and *Agüeros et al.* (2011), this cluster has now a well-defined period sample for FGKM dwarfs (cf. Fig. 1). With one exception, all VLM stars in the *Scholz et al.* (2011) sample have periods less than 2.5 d, with a median around 1 d and a lower limit of 5 hr. *Scholz et al.* compared the Praesepe sample with the pre-main-sequence clusters. While the evolution of the lower period limit is consistent with zero or little angular momentum loss between 100 and 600 Ma, the evolution of the upper period limit implies significant rotational braking. In this paper, this is discussed in terms of a mass-dependent spindown on the main sequence. With an exponential spindown law, the braking timescale τ is ~0.5 G.y. for 0.3 $M_\odot$, but >1 G.y. for 0.1 $M_\odot$. Thus, wind braking becomes less efficient toward lower masses. Similar to the pre-main-sequence clusters, the rotation of brown dwarfs is unexplored in this age regime.

2.2.4. Field populations. The largest (in fact, the only large) sample of periods for VLM stars in the field has been published recently by *Irwin et al.* (2011), with 41 periods for stars with masses between the hydrogen-burning limit and 0.35 $M_\odot$. A few more periods in this mass domain are available from *Kiraga and Stepien* (2007). Interestingly, the *Irwin et al.* (2011) sample shows a wide spread of periods, from 0.28 d up to 154 d, in stark contrast to the uniformly fast rotation in younger groups of objects. Based on kinematical age estimates, *Irwin et al.* (2011) find that the majority of the oldest objects in the sample (thick disk, halo) are slow rotators, with a median period of 92 d. For comparison, the younger thin disk objects have a median period of 0.7 d. This provides a firm constraint on the spin-down timescale of VLM stars that should be comparable with the thick disk age, i.e., 8–10 G.y. Similar conclusions were reached by *Delfosse et al.* (1998) and *Mohanty and Basri* (2003) based on v sin i measurements.

Brown dwarfs in the field have spectral types of L, T, and Y, and effective temperatures below 2500 K. At these temperatures, magnetic activity as it is known for M dwarfs is no longer observed, and thus periodic flux modulations from magnetically induced spots as in VLM stars are not expected. Some L and T dwarfs, however, do exhibit persistent periodic variability, which is usually attributed to the presence of a non-uniform distribution of atmospheric clouds (see *Radigan et al.,* 2012), which again allows for a measurement of the rotation period. About a dozen periods for field L and T dwarfs are reported in the literature and are most likely the rotation periods (*Bailer-Jones and Mundt* 1999, 2001; *Clarke et al.,* 2002; *Koen,* 2006; *Lane et al.,* 2007; *Artigau et al.,* 2009; *Radigan et al.,* 2012; *Heinze et al.,* 2013). All these periods are shorter than 10 hr.

Again it is useful to compare these findings with v sin i data. Rotational velocities have been measured for about 100 brown dwarfs by *Mohanty and Basri* (2003), *Zapatero Osorio et al.* (2006), *Reiners and Basri* (2008, 2010), *Blake et al.* (2010), and *Konopacky et al.* (2012). The composite figure by *Konopacky et al.* (2012, their Fig. 3) contains about 90 values for L dwarfs and 8 for T dwarfs. From this combined dataset it is clear that essentially all field brown dwarfs are fast rotators with v sin i > 7 km s⁻¹, corresponding to periods shorter than 17 hr, thus confirming the evidence from the smaller sample of periods. The lower limit in v sin i increases toward later spectral types, from 7 km s⁻¹ for early L dwarfs to more than 20 km s⁻¹ for late L dwarfs and beyond. Because brown dwarfs cool down as they age and thus spectral type is a function of age and mass, these values are difficult to compare with models. However, they strongly suggest that rotational braking becomes extremely inefficient in the substellar domain. Extrapolating from the trend seen in the M dwarfs, the braking timescales for brown dwarfs are expected to be longer than the age of the universe.

The results of rotational studies of the VLM stars in Praesepe and in the field clearly indicate that the spindown timescale increases toward lower masses in the VLM regime. There is no clear mass threshold at which the long-term rotational braking ceases to be efficient, as might be expected in a scenario where the dynamo mode switches

due to a change in interior structure. The observational data rather suggests that the rotational braking becomes gradually less efficient toward lower masses, until it essentially shuts down for brown dwarfs.

2.3. Gyrochronology

By the time low-mass stars reach the ZAMS (100–200 Ma), the observations show clear evidence for two distinct sequences of fast and slow rotators in the mass vs. period plane, presumably tracing the lower and upper envelopes of stellar rotation periods at the ZAMS. Observations in yet older open clusters show a clear convergence in the angular momentum evolution for all FGK dwarfs toward a single, well-defined, and mass-dependent rotation period by the age of the Hyades (~600 Ma, cf. Fig.1).

These patterns — and in particular the observed sequence of slow rotators with stellar mass — have been used as the basis for "gyrochronology" as an empirical tool with which to measure the ages of main-sequence stars (e.g., *Barnes,* 2003, 2007; *Mamajek and Hillenbrand,* 2008; *Meibom et al.,* 2009; *Collier Cameron et al.,* 2009; *Delorme et al.,* 2011). The method has so far been demonstrated and calibrated for solar-type stars with convective envelopes (i.e., mid-F to early-M dwarfs), with ages from the ZAMS to the old field population.

In the gyrochronology paradigm, the principal assumptions are that a given star begins its main-sequence rotational evolution with a certain "initial" rotation period on the ZAMS, on either a rapid-rotator sequence (so-called "C" sequence) or a slow-rotator sequence (so-called "I" sequence). All stars on the rapid-rotator sequence evolve onto the slow-rotator sequence on a timescale governed by the convective turnover time of the convective envelope (*Barnes and Kim,* 2010). Once on the slow-rotator sequence, the star then spins down in a predictable way with time, thus allowing its age to be inferred from its rotation period.

As discussed by *Epstein and Pinsonneault* (2012), there are limitations to the technique, particularly in the context of very young stars. First, to convert a given star's observed rotation period into a gyro-age requires assuming the star's initial rotation period. At older ages, this is not too problematic because the convergence of the gyrochrones makes the star eventually forget its own initial ZAMS rotation period. However, at ages near the ZAMS, the effect of the (generally unknown) initial period is more important.

More importantly, the technique has not yet been demonstrated or calibrated at ages earlier than the ZAMS. Presumably, the rotational scatter observed at the ZAMS and the relationships between stellar mass and surface rotation period must develop at some stage during the PMS and should be understood in the context of the early angular momentum evolution of individual stars (see section 4). Interestingly, there is now observational evidence that specific patterns may be emerging during the PMS in the period-mass diagrams, which encode the age of the youngest clusters (*Henderson and Stassun,* 2012).

3. THE PHYSICAL PROCESSES GOVERNING ANGULAR MOMENTUM EVOLUTION

The evolution of the rotation period and the angular momentum of a forming protostar is determined both by internal and external processes. External processes include all the mechanisms of angular momentum exchange with the surrounding ambient medium, with the accretion disk and the circumstellar outflows in particular. Internal processes determine the angular momentum redistribution throughout the stellar interior. These various processes are discussed in this section.

Figure 2a shows the evolution of the surface rotation rate of a 1-M$_\odot$ star, as predicted by *Gallet and Bouvier*'s (2013) models. The blue and red tracks correspond to the upper and lower envelopes of the observed spin distributions, and represent the range of spin evolutions exhibited by the majority of solar-mass stars (cf. Fig. 6). The dotted lines show the evolution of the rotation rate, assuming that the star conserves angular momentum [assuming solid body rotation and structural evolution from *Baraffe et al.* (1998)]. These angular momentum conserved tracks are shown for a few arbitrary "starting points," at the earliest time shown in the plot and at the time where the rotation rate is no longer constant in time. Assuming angular momentum conservation, the stars are expected to spin up, due to their contraction as they evolve toward the zero-age main sequence (at ~40 Ma) and then to have a nearly constant rotation rate on the main sequence (since then the structure changes much more slowly). It is clear from a comparison between the dotted lines and solid lines that the observed evolution is completely inconsistent with angular momentum conservation and instead requires substantial angular momentum loss.

Figure 2b shows the external torque on the star that is necessary to produce the red and blue tracks of the upper panel, assuming solid-body rotation. If the star is also accreting from a Keplerian disk, this should in principle result in an additional spinup torque on the star, given approximately by $\tau_a \gtrsim \dot{M}_a (GM_* R_*)^{1/2}$, where $\dot{M}_a$ is the accretion rate (e.g., *Ghosh and Lamb*, 1978; *Matt and Pudritz*, 2005b). As a simple quantitative example, the dotted lines in the lower panel show the torque required to produce the gyrotracks of the upper panel, assuming the stars are accreting at a constant rate of $\dot{M}_a = 10^{-7}$ M$_\odot$ yr^{-1}, during the time when the rotation rate is constant.

It is clear from the figure that the observed evolution of the spin rates of solar-mass stars requires substantial angular momentum loss at nearly all stages, and the required torque is largest at the youngest ages. During the first few million years of the T Tauri phase, the observed spin-rate distributions do not appear to evolve substantially. The torques required simply to prevent these stars from spinning up, due to contraction, are ~10^6× larger than the torque on the present-day Sun. Accreting stars require even larger torques to counteract the additional spinup effects of accretion, which depends primarily on the accretion rate.

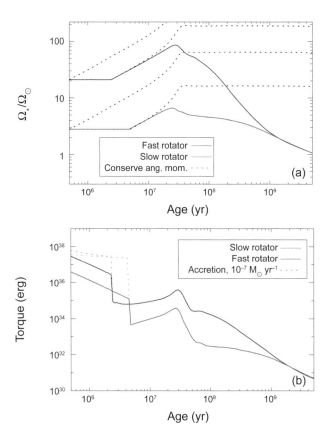

Fig. 2. (a) Angular rotation rate of a 1-M$_\odot$ star as a function of age. The lower and upper solid lines show the evolution of the surface rotation rate from the models by *Gallet and Bouvier* (2013), which best reproduce the lower and upper (respectively) range of the observed spin distributions (cf. Fig. 6). The dotted lines show the expected evolution of rotation rate, if the star were to conserve angular momentum, shown at a few different "starting points." **(b)** The external torque on the star that is required to produce the spin evolution tracks of the upper panel. The two solid lines show the torque required, assuming only the stellar structural evolution from a model of *Baraffe et al.* (1998) and the (solid-body) spin evolution of the corresponding lines in the upper panel. The dotted lines show the torque that would be required if the stars were also accreting at a rate of 10^{-7} M$_\odot$ yr^{-1}, during the time when the rotation rate is constant on the upper panel tracks.

After an age of ~5 Ma, and until the stars reach the ZAMS (at ~40 Ma), the stars spin up, due to their contraction. Although the surface rotation rates suggest a substantial fraction of their angular momentum is lost during this spinup phase, the torque is apparently much less than during the first few million years. This apparent sudden change in the torque suggests a change in the mechanism responsible for angular momentum loss. The fact that a substantial fraction of stars younger than ~5 m.y. are still accreting suggests that the star-disk interaction may in some way be responsible for the largest torques (*Koenigl*, 1991). In this case, the eventual dissipation of the disk (i.e., the cessation of accretion) provides a natural explanation (at least in principle) for the transition to much weaker torques.

3.1. Star-Disk Interaction

Our understanding of the various processes that are involved in the magnetic interaction between the star and its surrounding accretion disk has improved significantly in the last few years (e.g., *Bouvier et al.,* 2007). A number of recent results have spurred the development of new models and ideas for angular momentum transport, as well as further development and modification to existing models. In particular, it has been clear for some time that, due to the differential twisting of the magnetic field lines, the stellar magnetic field cannot connect to a very large region in the disk (e.g., *Shu et al.,* 1994; *Lovelace et al.,* 1995; *Uzdensky et al.,* 2002; *Matt and Pudritz,* 2005b). In addition, the competition between accretion and diffusion is likely to reduce the magnetic field intensity in the region beyond the co-rotation radius (*Bardou and Heyvaerts,* 1996; *Agapitou and Papaloizou,* 2000; *Zanni and Ferreira,* 2009). As a consequence, the widespread "disk-locking" paradigm, as proposed in the classical *Ghosh and Lamb* (1979) picture, has been critically reexamined. Also, it has been realized that the strength of the dipolar components of magnetic fields are generally significantly weaker than the average surface fields, which indicates that the latter is dominated by higher-order multi-poles (*Gregory et al.,* 2012). As the large-scale dipolar field is thought to be the key component for angular momentum loss, mechanisms for extracting angular momentum from the central star are thus required to be even more efficient. These issues have prompted different groups to reconsider and improve various scenarios based on the presence of outflows that could efficiently extract angular momentum from the star-disk system, as schematically illustrated in Fig. 3. The key new developments have been (1) the generalization of the X-wind model to multi-polar fields, (2) a renewed exploration of stellar winds as a significant angular momentum loss mechanism, and (3) the recognition that magnetospheric ejections that naturally arise from the star-disk interaction may actually have a significant contribution to the angular momentum transport.

Note that in the following sections we deal with only one specific class of disk winds, namely the X-wind, and neglect other popular models, the "extended disk wind" in particular, presented in the chapter by Frank et al. in this volume. First, in its "standard" formulation (e.g., *Blandford and Payne,* 1982; *Ferreira,* 1997), an extended disk wind exerts a torque onto the disk without modifying its Keplerian angular momentum distribution. In this respect, such a solution simply allows the disk to accrete and it does not affect the stellar angular momentum evolution. Second, we only discuss models that exploit the stellar magnetic flux. It has been shown (e.g., *Zanni and Ferreira,* 2013) that the stellar magnetic field cannot provide enough open flux to the disk to produce a relevant extended disk wind: The

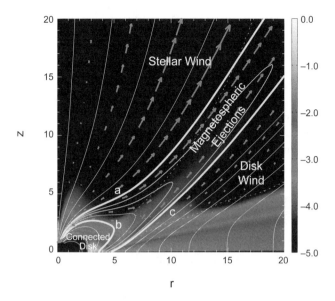

Fig. 3. Schematic view of the outflows that can be found in a star-disk interacting system. (1) Stellar winds accelerated along the open magnetic flux anchored onto the star; (2) magnetospheric ejections associated with the expansion and reconnection processes of closed magnetic field lines connecting the star and the disk; and (3) disk winds (including X-winds) launched along the open stellar magnetic surfaces threading the disk. From *Zanni and Ferreira* (2013).

latter needs a proper disk field distribution to be powered. Scenarios in which a disk field interacts with the stellar magnetic flux have been proposed (see *Ferreira et al.,* 2000) and they deserve future investigation.

3.1.1. Accretion-driven stellar winds. The idea that stellar winds may be important for removing angular momentum from accreting stars has been around since the first measurements of the rotational properties of young stars (e.g., *Shu et al.,* 1988; *Hartmann and Stauffer,* 1989). For stars that are actively accreting from a disk, the amount of angular momentum carried onto the star by the disk is proportional to the accretion rate. Stellar winds could be important for counteracting the spinup effect due to accretion, if the mass outflow rate is a significant fraction (~10%) of the accretion rate (*Hartmann and Stauffer,* 1989; *Matt and Pudritz,* 2005a). Due to the substantial energy requirements for driving such a wind, *Matt and Pudritz* (2005a) suggested that a fraction of the gravitational potential energy released by accretion (the "accretion power") may power a stellar wind with sufficiently enhanced mass outflow rates.

The torque from a stellar wind (discussed further below) depends upon many factors, and generally increases with magnetic field strength and also with mass loss rate. Thus, the problem discussed above of having weak dipolar fields can in principle be made up for by having a larger mass loss rate. However, as the mass loss rate approaches a substantial fraction of the mass accretion rate, there will not be enough accretion power to drive the wind. *Matt and Pudritz* (2008b) derived a "hard" upper limit of $\dot{M}_{wind}/\dot{M}_{acc} \lesssim 60\%$.

By considering that the accretion power must be shared between (at least) the stellar wind and the observed accretion diagnostics (e.g., the UV excess luminosity), *Zanni and Ferreira* (2011) made a quantitative comparison between the predictions of accretion-powered stellar winds and an observed sample of accreting stars, to test whether there was enough accretion power available to drive a wind capable of removing the accreted angular momentum. The range of observational uncertainties in quantities such as the UV excess and dipolar magnetic field strength was sufficient to span the range of possibilities from having enough accretion power to not having enough accretion power. However, this work demonstrated again that accretion-powered stellar winds need substantial large-scale magnetic fields in order to be efficient, and the strength of the observed fields are (within uncertainties) near the minimum required strengths (*Gregory et al.*, 2012).

Even if there is enough accretion power to drive a wind, there is still a question of whether or how accretion power may transfer to the open field regions of a star and drive a wind. Based on calculations of the emission properties and cooling times of the gas, *Matt and Pudritz* (2007) ruled out solar-like, hot coronal winds for mass loss rates greater than ~10–11 $M_\odot$/yr for "typical" T Tauri stars of ~0.5 $M_\odot$. They concluded that more massive winds would be colder (colder than ~10,000 K) and thus must be driven by something other than thermal pressure, such as Alfvén waves (*Decampli*, 1981). Taking a "first principles" approach and adopting a simplified, one-dimensional approach, *Cranmer* (2008, 2009) developed models whereby the energy associated with variable accretion drove MHD waves that traveled from the region of accretion on the star to the polar regions where it led to enhanced MHD wave activity and drove stellar wind. Those models typically reached mass loss rates that were equal or less than a few percent of the accretion rate. These values appear to be on the low end of what is needed for significant angular momentum loss for most observed systems. Further theoretical work is needed to explore how accretion power may transfer to a stellar wind.

3.1.2. Magnetospheric ejections. Magnetospheric ejections (MEs) are expected to arise because of the expansion and subsequent reconnection of the closed magnetospheric field lines connecting the star to the disk (*Hayashi et al.*, 1996; *Goodson et al.*, 1999; *Zanni and Ferreira*, 2013). The inflation process is the result of the star-disk differential rotation and the consequent buildup of toroidal magnetic field pressure. This is the same phenomenon that bounds the size of the closed magnetosphere and limits the efficiency of the *Ghosh and Lamb* (1979) mechanism. Initially, MEs are launched along magnetic field lines that still connect the star with the disk so that they can exchange mass, energy, and angular momentum both with the star and the disk. On a larger spatial scale, the MEs disconnect from the central region of the disk-star system in a magnetic reconnection event and propagate ballistically as magnetized plasmoids. Because of magnetic reconnection, the inner magnetic surfaces close again and the process repeats periodically

(see Fig. 4). This phenomenon is likely to be related to the "conical winds" simulated by *Romanova et al.* (2009).

Magnetospheric ejections contribute to control the stellar rotation period in two ways (*Zanni and Ferreira*, 2013). First, they extract angular momentum from the disk close to the truncation region so that the spinup accretion torque is sensibly reduced. This effect closely resembles the action of an X-wind (see section 3.1.3), which represents the limit solution capable of extracting all the angular momentum carried by the accretion flow. Second, if the ejected plasma rotates slower than the star, the MEs can extract angular momentum directly from the star thanks to a differential rotation effect. In such a situation, MEs are powered by both the stellar and the disk rotation, as in a huge magnetic slingshot. In agreement with other popular scenarios (*Koenigl*, 1991; *Ostriker and Shu*, 1995), the spindown torque exerted by the MEs can balance the accretion torque when the disk is truncated close to the co-rotation radius. In a propeller phase (see, e.g., *Ustyugova et al.*, 2006; *D'Angelo and Spruit*, 2011), when the truncation radius gets even closer to co-rotation, the spindown torque can even balance the spinup due to contraction.

Despite these first encouraging results, various issues remain. The MEs phenomenon is based on a charge and discharge process whose periodicity and efficiency depend on magnetic reconnection events that are controlled by numerical diffusion only in the solutions proposed by *Zanni and Ferreira* (2013). In order to produce an efficient spindown torque, this scenario requires a rather strong kilogauss dipolar field component, which has been only occasionally observed in classical T Tauri stars (e.g., *Donati et al.*, 2008, 2010b). In the propeller regime, which provides the most efficient spindown torque, the accretion rate becomes intermittent on a dynamical timescale, corresponding to a few rotation periods of the star. Even though this effect is enhanced by the axial symmetry of the models, there is as yet no observational evidence for such a behavior.

3.1.3. X-winds. The X-wind model invokes the interaction of the stellar magnetosphere with the surrounding disk to explain the slow spin rates of accreting T Tauri stars, well below breakup, within a single theoretical framework, via the central concept of trapped flux. (Other types of outflows are considered in the chapter by Frank et al. in this volume.) In steady-state, the basic picture is as follows (see Fig. 5): All the stellar magnetic flux initially threading the entire disk is trapped within a narrow annulus (the X-region) at the disk inner edge. The X-region straddles the co-rotation radius R_X (where the disk Keplerian angular velocity, $\Omega_X = \sqrt{GM_*/R_X^3}$, equals the stellar angular velocity, Ω_*; R_X lies near, but exterior to, the inner edge), a feature known as disk-locking. The resulting dominance of the magnetic pressure over gas within the X-region makes the entire annulus rotate as a solid body at the co-rotation angular velocity $\Omega_X = \Omega_*$. Consequently, disk material slightly interior to R_X rotates at sub-Keplerian velocities, allowing it to climb efficiently onto field lines that bow sufficiently inward and accrete onto the star; conversely,

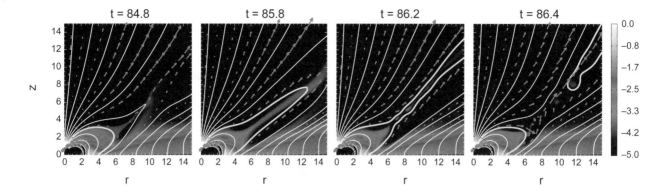

Fig. 4. Temporal evolution of the periodic inflation/reconnection process associated with magnetospheric ejections. Speed vectors (arrows) and magnetic field lines (solid lines) are superimposed to logarithmic density maps. The thicker solid lines in the second, third, and fourth panels follow the evolution of a single magnetic surface. Time is given in units of the stellar rotation period. From *Zanni and Ferreira* (2013).

material within the X-region but slightly exterior to R_X rotates at super-Keplerian velocities, enabling it to ascend field lines that bend sufficiently outward and escape in a wind. The magnetic torques associated with the accretion funnels transfer excess specific angular momentum (excess relative to the amount already residing on the star) from the infalling gas to the disk material at the foot points of the funnel flow field lines in the inner parts of the X-region, which tends to push this material outward. Conversely, the magnetic torques in the wind cause the outflowing gas to gain angular momentum at the expense of the disk material connected to it by field lines rooted in the outer parts of the X-region, pushing this material inward. The pinch due to this outward push on the inside and inward push on the outside of the X-region is what keeps the flux trapped within it, and truncates the disk at the inner edge in the first place. The net result is a transfer of angular momentum from the accreting gas to the wind, allowing the star to remain slowly rotating.

The X-wind accretion model was originally formulated assuming a dipole stellar field (*Ostriker and Shu, 1995*). However, detailed spectropolarimetric reconstructions of the stellar surface field point to more complex field configurations (e.g., *Donati et al., 2010a, 2011*). In view of this, and noting that the basic idea of flux trapping, as outlined above, does not depend on the precise field geometry, *Mohanty and Shu* (2008) generalized the X-wind accretion model to arbitrary multiple fields. The fundamental relationship in this case, for a star of mass M_*, radius R_*, and angular velocity Ω_*, is

$$F_h R_*^2 \bar{B}_h = \bar{\beta} f^{1/2} \left(GM_* \dot{M}_D / \Omega_*\right)^{1/2} \qquad (1)$$

Here F_h is the fraction of the surface area $2\pi R_*^2$ of one hemisphere of the stellar surface (either above or below the equatorial plane) covered by accretion hot spots with mean field strength $\bar{B}_h$; $\bar{\beta}$ is a dimensionless, inverse mass-loading

parameter that measures the ratio of magnetic field to mass flux; and f is the fraction of the total disk accretion rate $\dot{M}_D$ that flows into the wind (so 1–f is the fraction that accretes onto the star). Equation (1) encapsulates the concept of flux trapping; it relates the amount of observed flux in hot spots on the lefthand side (which equals the amount of trapped flux in the X-region that is loaded with infalling gas) to independently observable quantities on the righthand side, without any assumptions about the specific geometry of the stellar field that ultimately drives the funnel flow.

There is now some significant evidence for the generalized X-wind model. First, surveys of classical T Tauri stars in Taurus and NGC 2264 strongly support the correlation $F_h R_*^2 \propto (GM_* \dot{M}_D / \Omega_*)^{1/2}$, predicted by equation (1) if $\bar{B}_h$ can be assumed constant, an assumption admittedly open to debate (*Johns-Krull and Gafford, 2002; Cauley et al., 2012*). Second, in two stars with directly measured $\bar{B}_h$ from spectropolarimetric data, as well as estimates of F_h, $\dot{M}$, and stellar parameters [V2129 Oph and BP Tau (*Donati et al., 2007, 2008*, respectively)], *Mohanty and Shu* (2008) find that equation (1) (with some simplistic assumptions about β and f; see also below) produces excellent agreement with the observed $\bar{B}_h F_h$; they also find that numerical models incorporating the mix of multiple components observed on these stars give F_h, $\bar{B}_h$, and hot-spot latitudes consistent with the data. More generally, there is substantial evidence that disks are involved in regulating the angular momentum of the central star, but evidence for disk-locking (disk truncation close to the co-rotation radius R_X), as specifically proposed by X-wind theory, is as yet inconclusive; more detailed studies of statistical significant samples are required (see section 2.1).

Support for the X-wind model from numerical simulations has so far been mixed. Using dipole stellar fields, *Romanova et al.* (2007) have obtained winds and funnel flows consistent with the theory over extended durations, but many others (e.g., *Long et al., 2007, 2008*) have failed. It is noteworthy in this regard that in the presence of finite

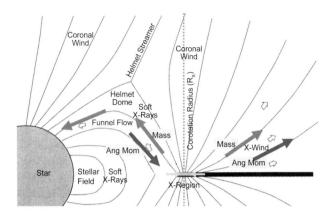

Fig. 5. Schematic of steady-state X-wind model. Black thick line in the equatorial plane is the truncated disk; black solid curves show the magnetic field; vertical dotted line shows the co-rotation radius R_X; thick gray horizontal line shows the X-region straddling R_X. Arrows along the field lines show the direction of mass and angular momentum transport: Interior to R_X, material flows from the X-region onto the star in a funnel flow along field lines that bow sufficiently inward, and the excess angular momentum in this gas flows back into the X-region via magnetic stresses; exterior to R_X, material flows out of the X-region in a wind, along field lines that bow sufficiently outward, and carries away with it angular momentum from the X-region. Horizontal arrows in the equatorial plane show the pinching of gas in the X-region due to the angular momentum transport, which truncates the disk at the inner edge and keeps magnetic flux trapped in the X-region.

resistivity η, the flux trapping that is key to the X-wind model requires that field diffusion out of the X-region be offset by fluid advection of field into it. This in turn demands that η/ν ≪ 1, where ν is the disk viscosity (*Shu et al.,* 2007). *Romanova et al.* (2007) explicitly show that this condition is critical for attaining an X-type magnetic configuration, while other simulations typically have ν ~ η instead.

Finally, as an ideal-MHD, steady-state, axisymmetric semi-analytic model with a stellar field aligned with the rotation axis, the X-wind theory clearly has limitations. For instance, with nonzero resistivity, reconnection events can lead to episodic outbursts; similarly, changes in the stellar field or disk accretion rate, and tilted and/or nonaxisymmetric fields, can all lead to temporally varying phenomena, as indeed observed in classical T Tauri stars (e.g., *Alencar et al.,* 2012). X-wind theory can at best represent only the time-averaged behavior of an intrinsically time-variable system; numerical simulations are essential for quantitatively exploring these issues in detail (e.g., *Romanova et al.,* 2011; *Long et al.,* 2011). Similarly, in the idealized model, the X-region shrinks to a mathematical point, and how the infinitesimal ν and η then load field lines is not answerable within the theory, leading to somewhat *ad hoc* estimates for f and β (see *Mohanty and Shu,* 2008; *Cai et*

al., 2008). An associated question is what the appropriate ratio η/ν is for realistic disks. Global numerical simulations of the magneto-rotational instability (MRI), with nonzero magnetic flux, are vital to shed light on these issues (see *Shu et al.,* 2007).

3.2. Stellar Winds

Most stars spend the majority of their lives in isolation, in a sense that after the age of ~10 Ma, they are no longer accreting material and most do not possess companions that are within reach of their magnetospheres nor close enough for significant tidal interactions. For isolated stars, the only available way of losing substantial angular momentum is by losing mass. It has been known for a long time that low-mass stars (those with substantial convective envelopes) are magnetically active and spin down via stellar winds (e.g., *Parker,* 1958; *Kraft,* 1967; *Skumanich,* 1972; *Soderblom,* 1983). The coupling of the magnetic field with the wind can make the angular momentum loss very efficient, in a sense that the fractional loss of angular momentum can be a few orders of magnitude larger than the fractional amount of mass lost (e.g., *Schatzman,* 1962; *Weber and Davis,* 1967; *Mestel,* 1968; *Reiners et al.,* 2009b).

In order to calculate the amount of angular momentum loss due to magnetized stellar winds, the theory generally assumes the conditions of ideal MHD and a steady-state flow. These assumptions are acceptable for characterizing the average, global wind properties, as needed to understand the long-timescale evolution of stellar rotation. In this case, the torque on the star can generally be expressed as (e.g., *Matt et al.,* 2012a)

$$T_w = K \, (2GM_*)^{-m} \, R_*^{5m+2} \, \dot{M}_w^{1-2m} \, B_*^{4m} \, \Omega_* \qquad (2)$$

where M_* and R_* are the stellar mass and radius, $\dot{M}_w$ the mass loss rate, B_* the stellar magnetic field, and Ω_* the stellar angular velocity. K and m are dimensionless numbers that depend upon the interaction between the accelerating flow and rotating magnetic field.

Until recently, most models for computing the angular momentum evolution of stars (discussed below; see section 4) have used the stellar wind torque formulation of *Kawaler* (1988) (based on *Mestel,* 1984), which is equivalent to equation (2) with a value of m = 0.5 and fitting the constant K in order to match the present-day solar-rotation rate. This formulation is convenient because, for m = 0.5, the stellar wind torque is independent of the mass loss rate (see equation (2)). For given stellar parameters, one only needs to specify how the surface magnetic field strength depends upon rotation rate, discussed further below.

However, Kawaler's formulation relies upon a one-dimensional approach and adopts simple power-law relationships for how the magnetic field strength and the wind velocity vary with distance from the star. *Matt and Pudritz* (2008a) pointed out that these assumptions are not generally valid in multi-dimensional winds and that numerical simulations are needed to accurately and self-consistently determine the values of K and m. Using two-dimensional (axisymmetric) numerical MHD simulations, *Matt and Pudritz* (2008a) and *Matt et al.* (2012a) computed steady-state solutions for coronal (thermally driven) winds in the case of stars with a dipolar magnetic field aligned with the rotation axis. Using parameter studies to determine how the torque varies, *Matt et al.* (2012a) found

$$m \approx 0.2177 \quad K \approx 6.20 \left[1 + \left(f/0.0716 \right)^2 \right]^{-m} \quad (3)$$

where f is the stellar rotation rate expressed as a fraction of breakup (Keplerian rate at the stellar surface), $f^2 \equiv \Omega_*^2 R_*^3 (GM_*)^{-1}$. Note that the dimensionless factor K now contains a dependence on the stellar spin rate (in addition to that appearing in equation (2)), and it is nearly constant when the star rotates slower than a few percent of breakup speed.

The formulation of equations (2) and (3) is derived from simulations with fixed assumptions about the wind driving (e.g., coronal temperature) and a particular (dipolar) field geometry on the stellar surface. Thus, further work is needed to determine how the torque depends upon the wind-driving properties and varies for more complex magnetic geometries. However, this is the most dynamically self-consistent stellar wind torque formulation to date. [*Ud-Doula et al.* (2009) reported m ≈ 0.25, derived from their simulations of massive star winds, driven by radiation, rather than thermal pressure. This suggests that the value of m does not strongly vary with the wind driving properties.]

After adopting equation (3), it is clear that the torque in equation (2) depends upon both the surface magnetic field strength and the mass loss rate. To specify the magnetic field strength, most spin evolution models adopt the relationship suggested by *Kawaler* (1988), $B_* \propto \Omega_*^a R_*^{-2}$. The value of a is usually taken to be unity for slow rotators, but above some critical rotation rate of approximately $10\,\Omega_\odot$ for solar-type stars, the magnetic field is taken to be independent of rotation rate by setting a = 0. The so-called "dynamo saturation" above some critical velocity is observationally supported both by direct magnetic field measurements (e.g., *Reiners et al.*, 2009) and activity proxies (e.g., *Wright et al.*, 2011).

Recently, *Reiners and Mohanty* (2012) pointed out that both observations of magnetic fields (e.g., *Saar*, 1996; *Reiners et al.*, 2009a) and dynamo theory (e.g., *Durney and Stenflo*, 1972; *Chabrier and Küker*, 2006) were more consistent with a dynamo relationship following $B_* \propto \Omega_*^a$ — i.e., the average magnetic field strength goes as some power of Ω, instead of the magnetic flux (as in the Kawaler formulation). This has important implications for the dependence of the torque on the mass (and radius) of the star. Furthermore, *Reiners and Mohanty* (2012) derived a new torque formulation, based on similar assumptions to Kawaler's, arriving at the equivalent of equation (2), with m = 2/3. Although this value of m is inconsistent with the

MHD simulation results discussed above, *Reiners and Mohanty* (2012) demonstrated that, in order to simultaneously explain the observed spin evolution of both solar mass and very-low-mass stars, the stellar wind torque must depend much more strongly on the stellar mass (and radius) than the Kawaler formulation.

To calculate the torque, we also need to know how the mass loss rate depends on stellar properties. Due to the relatively low loss rates of non-accreting Sun-like and low-mass stars, observational detections and measurements are difficult. So far, most of what we know is based on approximately a dozen measurements by *Wood et al.* (2002, 2005), which suggest that the mass loss rates vary nearly linearly with the X-ray luminosity, up to a threshold X-ray flux, above which the mass loss rates saturate. More recently, *Cranmer and Saar* (2011) have developed a theoretical framework for predicting the mass loss rates of low-mass stars, based upon the propagation and dissipation of Alfvén waves (see also *Suzuki et al.*, 2012). The models of *Cranmer and Saar* (2011) and *Suzuki et al.* (2012) compute the mass loss rate as a function of stellar parameters in a way that is self-consistent with the scaling of magnetic field strength with stellar rotation rate (including both saturated and nonsaturated regimes). These models can now be used, in conjunction with equations (2) and (3), to compute the stellar wind torque during most of the lifetime of an isolated, low-mass star.

As another means of probing the mass loss rates, *Aarnio et al.* (2012) used an observed correlation between coronal mass ejections and X-ray flares, together with the observed flare rate distributions derived from T Tauri stars in Orion (*Albacete Colombo et al.*, 2007), to infer mass-loss rates due to coronal mass ejections alone. *Drake et al.* (2013) presented a similar analysis for main-sequence stars. Furthermore, *Aarnio et al.* (2012) explored how coronal mass ejections may influence the angular momentum evolution of pre-main-sequence stars. They concluded that they are not likely to be important during the accretion phase or during early contraction, but they could potentially be important after ~10 Ma. Although there are a number of uncertainties associated with estimating CME mass-loss rates from observed X-ray properties, this is an interesting area for further study.

3.3. Internal Processes

As angular momentum is removed from the stellar surface by external processes, such as star-disk interaction and stellar winds, the evolution of the surface rotation rate depends in part on how angular momentum is transported in the stellar interior. Two limiting cases are (1) solid-body rotation, where it is assumed that angular momentum loss at the stellar surface is instantaneously redistributed throughout the whole stellar interior, i.e., the star has uniform rotation from the center to the surface; and (2) complete core-envelope decoupling, where only the outer convective zone is spun down while the inner radiative core accelerates as it develops during the PMS, thus yielding a large velocity gradient at the core-envelope interface. Presumably, the actual rotational profile of solar-type and low-mass stars lies between these two extremes. Until recently, only the internal rotation profile of the Sun was known (*Schou et al.,* 1998). Thanks to Kepler data, internal rotation has now been measured for a number of giant stars evolving off the main sequence from the seismic analysis of mixed gravity and pressure modes (e.g., *Deheuvels et al.,* 2012). The internal rotation profile of PMS stars and of field main-sequence stars is, however, still largely unconstrained by the observations.

A number of physical processes act to redistribute angular momentum throughout the stellar interior. These include various classes of hydrodynamical instabilites (e.g., *Decressin et al.,* 2009; *Pinsonneault* 2010; *Lagarde et al.,* 2012; *Eggenberger et al.,* 2012), magnetic fields (e.g., *Denissenkov and Pinsonneault,* 2007; *Spada et al.,* 2010; *Strugarek et al.,* 2011), and gravity waves (*Charbonnel et al.,* 2013; *Mathis* 2013), all of which may be at work during PMS evolution. At the time of this writing, only a few of the current models describing the angular momentum evolution of young stars include these processes from first principles (e.g., *Denissenkov et al.,* 2010; *Turck-Chièze et al.,* 2010; *Charbonnel et al.,* 2013; *Marques et al.,* 2013; *Mathis,* 2013), highlighting the need for additional physics to account for the observations. Pending fully consistent physical models, most current modeling efforts adopt empirical prescriptions for core-envelope angular momentum exchange as discussed below.

4. ANGULAR MOMENTUM EVOLUTION MODELS

In an attempt to account for the observational results described in section 2, most recent models of angular momentum evolution rest on the three main physical processes described in section 3: star-disk interaction, wind braking, and angular momentum redistribution in the stellar interior. Each of these processes is included in the models in a variety of ways, as described below:

Star-disk interaction: Only a few recent models attempt to provide a physical description of the angular momentum exchange taking place between the star and its accretion disk. For instance, *Matt et al.* (2012b) computed the evolution of the torque exerted by accretion-powered stellar winds onto the central star during the early accreting PMS phase. Another example is the work of Gallet and Zanni (in preparation), who combined the action of accretion-driven winds and magnetospheric ejections to account for the nearly constant angular velocity of young PMS stars in spite of accretion and contraction. Both models require dipolar magnetic field components of about 1–2 kG, i.e., on the high side of the observed range of magnetic field strength in young stars (*Donati and Landstreet,* 2009; *Donati et al.,* 2013; *Gregory et al.,* 2012). Most other models, however, merely assume constant angular velocity for the central star

as long as it accretes from its disk [as originally proposed by *Koenigl* (1991)], with the disk lifetime being a free parameter in these models (e.g., *Irwin et al.,* 2007, 2008a; *Bouvier,* 2008; *Irwin and Bouvier,* 2009; *Denissenkov,* 2010; *Reiners and Mohanty,* 2012; *Gallet and Bouvier,* 2013).

Wind braking: Until a few years ago, most models used *Kawaler*'s (1988) semi-empirical prescription, with the addition of saturation at high velocities [as originally suggested by *Stauffer and Hartmann* (1987)] to estimate the angular momentum loss rate due to magnetized winds (see, e.g., *Bouvier et al.,* 1997; *Krishnamurthi et al.,* 1997; *Sills et al.,* 2000). Recently, more physically sounded braking laws have been proposed. *Reiners and Mohanty* (2012) revised *Kawaler*'s (1988) prescription on the basis of a better understanding of dynamo-generated magnetic fields, while *Matt et al.* (2012a) used two-dimensional MHD simulations to derive a semi-analytical formulation of the external torque exerted on the stellar surface by stellar winds. The latter result has been used in the angular momentum evolution models developed for solar-type stars by *Gallet and Bouvier* (2013), who also provide a detailed comparison between the various braking laws.

Internal angular momentum transport: While some models do include various types of angular momentum transport processes (e.g., *Denissenkov et al.,* 2010; *Turck-Chièze et al.,* 2010; *Charbonnel et al.,* 2013; *Marques et al.,* 2013), the most popular class of models so far rely on the simplifying assumption that the star consists of a radiative core and a convective envelope that are both in solid-body rotation but at different rates. In these so-called double-zone models, angular momentum is exchanged between the core and the envelope at a rate set by the core-envelope coupling timescale, a free parameter of this class of models (e.g., *Irwin et al.,* 2007, 2009; *Bouvier,* 2008; *Denissenkov,* 2010; *Spada et al.,* 2011). When dealing with fully convective interiors, whether PMS stars on their Hayashi track or very-low-mass stars, models usually assume solid-body rotation throughout the star.

We illustrate below how these classes of models account for the observed spin-rate evolution of solar type stars, low-mass and very low-mass stars, and brown dwarfs.

4.1. Solar-Type Stars

Figure 6 (from *Gallet and Bouvier,* 2013) shows the observed and modeled angular momentum evolution of solar-type stars in the mass range 0.9–1.1 M$_\odot$, from the start of the PMS at 1 Ma to the age of the Sun. The rotational distributions of solar-type stars are shown at various time steps corresponding to the age of the star-forming regions and young open clusters to which they belong (cf. Fig 1). Three models are shown, which start with initial periods of 10, 7, and 1.4 d, corresponding to slow, median, and fast rotators, respectively. The models assume constant angular velocity during the star-disk interaction phase in the early PMS, and implement the *Matt et al.* (2012a) wind-braking prescription, as well as core-envelope decoupling. The free

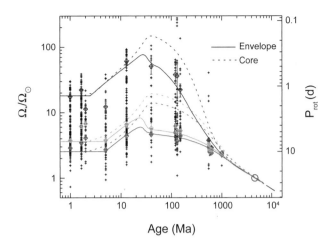

Fig. 6. The rotational angular velocity of solar-type stars is plotted as a function of age. The left y-axis is labeled with angular velocity scaled to the angular velocity of the present Sun while the right y-axis is labeled with rotational period in days. On the x-axis the age is given in million years. *Observations:* The black crosses shown at various age steps are the rotational periods measured for solar-type stars in star-forming regions and young open clusters over the age range of ~1 Ma to ~1 Ga. At every age, the lower, middle, and upper diamonds represent the 25, 50, and 90th percentiles of the observed rotational distributions, respectively. The open circle at ~4.56 Ga is the angular velocity of the present Sun. *Models:* The angular velocity of the convective envelope (solid line) and of the radiative core (dashed lines) is shown as a function of time for slow (lower), median (middle), and fast (upper) rotator models, with initial periods of 10, 7, and 1.4 d, respectively. The dashed black line at the age of the Sun illustrates the asymptotic Skumanich relationship, $\Omega \propto t^{-1/2}$. From *Gallet and Bouvier* (2013).

parameters of the models are the initial periods, chosen to fit the rotational distributions of the earliest clusters; the star-disk interaction timescale τ_d during which the angular velocity is held constant at its initial value; the core-envelope coupling timescale τ_{ce}; and the calibration constant KW for wind-driven angular momentum losses. The latter is fixed by the requirement to fit the Sun's angular velocity at the Sun's age. These parameters are varied until a reasonable agreement with observations is obtained. In this case, the slow-, median-, and fast-rotator models aim at reproducing the 25th, 50th, and 90th percentiles of the observed rotational distributions and their evolution from the early PMS to the age of the Sun.

This class of models provides a number of insights into the physical processes at work. The star-disk interaction lasts for a few million years in the early PMS, and possibly longer for slow rotators ($\tau_d \simeq 5$ m.y.) than for fast ones ($\tau_d \simeq 2.5$ m.y.). As the disk dissipates, the star begins to spin up as it contracts toward the ZAMS. The models then suggest much longer core-envelope coupling timescales for slow rotators ($\tau_{ce} \simeq 30$ m.y.) than for fast ones ($\tau_{ce} \simeq$

12 m.y.). Hence, on their approach to the ZAMS, only the outer convective envelope of slow rotators is spun down while their radiative core remains in rapid rotation. They consequently develop large angular velocity gradients at the interface between the radiative core and the convective envelope on the ZAMS and, indeed, most of their initial angular momentum is then hidden in their radiative core (cf. *Gallet and Bouvier,* 2013). (Note that the effect of hiding some angular momentum in the radiative core would "smooth out" the torques shown in Fig. 2, where solid-body rotation was assumed. Namely, the ZAMS torques will be slightly less and the post-ZAMS will be slightly larger to an age of ~1 Ga, due to the effects of differential rotation and core-envelope decoupling.) As stars evolve on the early MS, wind braking eventually leads to the convergence of rotation rates for all models by an age of $\simeq$1 Ga. This is due to the strong dependency of the braking rate onto the angular velocity: Faster rotators brake more efficiently than slow ones. Also, the early-MS spin evolution of slow rotators is flatter than that of fast rotators, in part because the angular momentum hidden in the radiative core at the ZAMS resurfaces on a timescale of a few 0.1 G.y. on the early MS. These models illustrate the strikingly different rotational histories solar-type stars may experience prior to about 1 Ga, depending mostly on their initial period and disk lifetime. In turn, the specific rotational history a young star undergoes may have a long-lasting impact on its properties, such as lithium content, even long after rotational convergence is completed (e.g., *Bouvier,* 2008; *Randich,* 2010).

These models describe the spin evolution of isolated stars while many cool stars belong to multiple stellar systems (cf. *Duchêne and Kraus,* 2013). For short-period binaries ($P_{orb} \leq 12$ d), tidal interaction enforces synchronization between the orbital and rotational periods (*Zahn,* 1977), and the spin evolution of the components of such systems will clearly differ from that of single stars (*Zahn and Bouchet,* 1989). However, the fraction of such tight, synchronized systems among solar-type stars is low, on the order of 3% (*Raghavan et al.,* 2010), so that tidal effects are unlikely to play a major role in the angular momentum evolution of most cool stars. Another potentially important factor is the occurrence of planetary systems (e.g., *Mayor et al.,* 2011; *Bonfils et al.,* 2013). The frequency of hot Jupiters, i.e., massive planets close enough to their host star to have a significant tidal or magnetospheric influence (cf. *Dobbs-Dixon et al.,* 2004; *Lanza,* 2010; *Cohen et al.,* 2010), is quite low, amounting to a mere 1% around FGK stars (e.g., *Wright et al.,* 2012). However, there is mounting evidence that the formation of planetary systems is quite a dynamic process, with gravitational interactions taking place between forming and/or migrating planets (*Albrecht et al.,* 2012) (see also the chapters by Davies et al. and Baruteau et al. in this volume), which may lead to planet scattering and even planet engulfment by the host star. The impact of such catastrophic events onto the angular momentum evolution of planet-bearing stars has been investigated by *Bolmont et al.* (2012), who showed it could significantly modify the instantaneous spin rate of planet host stars both during the PMS and on the main sequence.

4.2. Very-Low-Mass Stars

Models similar to those described above for solar-type stars have been shown to apply to lower-mass stars, at least down to the fully convective boundary ($\approx$0.3 $M_\odot$), with the core-envelope coupling timescale apparently lengthening as the convective envelope thickens (e.g., *Irwin et al.*, 2008b). In the fully convective regime, i.e., below 0.3 $M_\odot$, models ought to be simpler as the core-envelope decoupling assumption becomes irrelevant and uniform rotation is usually assumed throughout the star. Yet, the rotational evolution of very-low-mass stars actually appears more complex than that of their more massive counterparts and still challenges current models. Figure 7 (from *Irwin et al.*, 2011) shows that disk locking still seems to be required for VLM stars in order to account for their slowly evolving rotational period distributions during the first few million years of PMS evolution. Yet, as discussed above (see section 2.2), the evidence for a disk rotation connection in young VLM stars is, at best, controversial. Equally problematic, the rotational period distribution of field M dwarfs appears to

be bimodal, with pronounced peaks at fast (0.2–10 d) and slow (30–150 d) rotation (*Irwin et al.*, 2011). Most of the slow rotators appear to be thick disk members, i.e., they are on average older than the fast ones that are kinematically associated to the thin disk, and the apparent bimodality could thus simply result from a longer spindown timescale for VLM stars, on the order of a few billion years, as advocated by *Reiners and Mohanty* (2012) and *McQuillan et al.* (2013).

However, as shown in Fig. 7, this bimodality may not be easily explained for field stars at an age of several billion years. Indeed, contrary to solar-type stars whose rotational scatter decreases from the ZAMS to the late-MS (cf. Fig. 6), the distribution of spin rates of VLM stars widens from the ZAMS to later ages. The large dispersion of rotation rates observed at late ages for VLM stars thus requires drastically different model assumptions. Specifically, for a given model mass (0.25 $M_\odot$ in Fig. 7), the calibration of the wind-driven angular momentum loss rate has to differ by 1 order of magnitude between slow and fast rotators (*Irwin et al.*, 2011). Why a fraction of VLM stars remain as fast rotators over nearly 10 G.y. while another fraction is slowed down on a timescale of only a few billion years is currently unclear. A promising direction to better understand the rotational evolution of VLM stars is the recently reported evidence

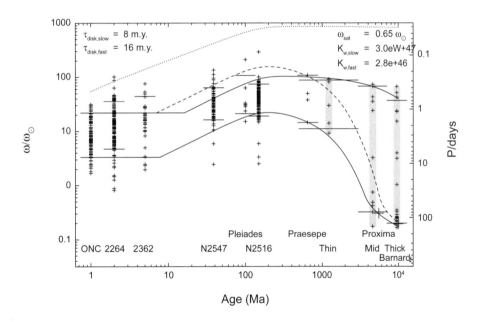

Fig. 7. The rotational angular velocity of very-low-mass stars (0.1–0.35 $M_\odot$) is plotted as a function of age. The left y-axis is labeled with angular velocity scaled to the angular velocity of the present Sun, while the right y-axis is labeled with rotational period in days. On the x-axis the age is given in million years. *Observations:* The black crosses shown at various age steps are the rotational periods measured for very-low-mass stars in star-forming regions, in young open clusters, and in the field over the age range of 1 Ma–10 Ga. Short horizontal lines show the 10th and 90th percentiles of the angular velocity distributions at a given age, used to characterize the slow and fast rotators, respectively. *Models:* The solid curves show rotational evolution models for 0.25-$M_\odot$ stars, fit to the percentiles, with the upper curve for the rapid rotators (with parameters $\tau_{d,fast}$ and $K_{W,fast}$) and the lower curve for the slow rotators (with parameters $\tau_{d,slow}$ and $K_{W,slow}$). Note the factor of 10 difference between $K_{W,fast}$ and $K_{W,slow}$. The dashed curve shows the result for the rapid rotators if the wind parameter $K_{W,fast}$ is assumed to be the same as for the slow rotators rather than allowing it to vary. The dotted curve shows the breakup limit. From *Irwin et al.* (2011).

for a bimodality in their magnetic properties. Based on spectropolarimetric measurements of the magnetic topology of late M dwarfs obtained by *Morin et al.* (2010), *Gastine et al.* (2013) have suggested that a bistable dynamo operates in fully convective stars, which results in two contrasting magnetic topologies: strong axisymmetric dipolar fields, or weak multipolar fields. Whether the different magnetic topologies encountered among M dwarfs is at the origin of their rotational dispersion at late ages remains to be assessed.

4.3. Brown Dwarfs

Figure 8 illustrates the current rotational data and models for brown dwarfs from 1 Ma to the field substellar population. As discussed in section 2.2 above, substellar objects are characterized by fast rotation from their young age throughout their whole evolution, with a median period of about 2 d at 1 Ma and 3–4 hr at 1 Ga. Somewhat controversial evidence for disk locking has been reported among young brown dwarfs (see section 2.2 above), although sensitive mid-infrared surveys are still needed for large samples in order to better characterize the disk frequency. Figure 8 shows models computed with and without angular momentum losses from (sub)stellar winds. The evolution of substellar rotational distributions in the first few million years is consistent with either no or moderate disk locking, as previously advocated by *Lamm et al.* (2005). At an age of a few million years, the observed rotation rates suggest that substellar objects experience little angular momentum loss on this timescale. By an age of a few billion years, however, some angular momentum loss has occurred. The best fit to the observational constraints is obtained with models featuring an angular momentum loss rate for brown dwarfs that is about 10,000× weaker than that assumed for solar-type stars (cf. Fig. 6). Whether the unefficient rotational braking of a brown dwarf results from a peculiar magnetic topology, their predominantly neutral atmospheres, or some other cause is currently unclear.

4.4. Summary

Current models of the spin evolution of low-mass stars appear to converge toward the following consensus:

1. At all masses, the initial distribution of angular momentum exhibits a large dispersion that must reflect some process operating during the core collapse and/or the embedded protostellar stage. Current models do not solve for this initial rotational scatter but adopt it as initial conditions.

2. Some disk-related process is required at least for solar-type and low-mass stars during the first few million years in order to account for their hardly evolving rotational period distributions during the early PMS. Whether this process is still instrumental in the VLM and substellar regimes remains to be assessed. The disk lifetimes required by angular momentum evolution models are consistent with those empirically derived from the evolution of infrared excess in young stars (e.g., *Bell et al.*, 2013).

3. Rotational braking due to magnetized winds is strongly mass dependent, being much less efficient at very low masses. At a given mass, angular momentum loss must also scale with the spin rate in order to account for the rotational convergence of solar-type and low-mass stars on a timescale of a few 0.1 G.y. The spindown timescale from the ZAMS increases toward lower-mass stars (from ~0.1 G.y. at 1 $M_\odot$ to ~1 G.y. at 0.3 $M_\odot$, and ≥10 G.y. at ≤0.1 $M_\odot$), but once completed, the rotational convergence usually occurs at a lower spin rate for lower-mass stars (*McQuillan et al.*, 2013).

4. Some form of core-envelope decoupling must be introduced in the models in order to simultaneously account for the specific spin evolution of initially slow and fast rotators. The empirically derived core-envelope coupling timescale is found to be longer in slow rotators than

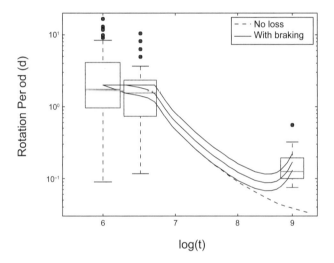

Fig. 8. Boxplot showing rotation periods for brown dwarfs as a function of age. The plot contains periods for ~1 m.y. [ONC (*Rodríguez-Ledesma et al.*, 2009)], for ~3 m.y. (*Cody and Hillenbrand*, 2010; *Scholz and Eislöffel*, 2004a, 2005; *Bailer-Jones and Mundt*, 2001; *Zapatero Osorio et al.*, 2003; *Caballero et al.*, 2004), and for the field population, for convenience plotted at an age of 1 Ga although individual ages may vary (*Bailer-Jones and Mundt*, 2001; *Clarke et al.*, 2002; *Koen*, 2006; *Lane et al.*, 2007; *Artigau et al.*, 2009; *Radigan et al.*, 2012; *Girardin et al.*, 2013; *Gillon et al.*, 2013; *Heinze et al.*, 2013). Horizontal line in the boxes: median; box: lower and upper quartile; "whiskers": range of datapoints within 1.5× (75–25%) range. Outliers outside that range are plotted as individual datapoints. Note that a few more brown dwarf periods have been measured at other ages, not shown here due to the small sample sizes. The dashed line illustrates evolution models without angular momentum loss. The three solid lines correspond to models including saturated angular momentum losses and disk-locking phases lasting for 1, 2, and 5 m.y., respectively, with object radii taken from the 0.05-$M_\odot$ BT-Settl evolutionary models of *Allard et al.* (2011). The best fit to the current observational constraints is obtained by assuming an angular momentum loss rate for brown dwarfs that is ~10,000× weaker than that used for solar-type stars shown in Fig. 6.

in fast ones at a given mass, thus providing some hints at the underlying physical process responsible for angular momentum transport in stellar interiors.

5. CONCLUSION

In the last few years, we have reached a stage where the rotational evolution of cool stars and brown dwarfs is relatively well constrained by the observations. Additional rotational period measurements for homogeneous and coeval populations are still required to fill a few age and mass gaps, e.g., old (≥ 1 Ga) field dwarfs and ZAMS cluster (0.1–0.5 Ga) brown dwarfs, so as to provide a complete picture of the spin evolution of stars and substellar objects. The physical description of the mechanisms that dictate the spin evolution of cool stars has also tremendously progressed over the last years, with the exploration of new processes and the refinement of prior ones. Yet the slow rotation rates of young stars still very much remain a challenge to these models. A better characterization of the critical quantities involved in the star-disk interaction and in stellar winds, such as the stellar magnetic field intensity and topology, the mass accretion rate onto the star, and the amount of mass loss a star experiences during its lifetime, is sorely needed in order to progress on these issues. In spite of these limitations, the semi-empirical angular momentum evolution models developed to date appear to grasp some of the major trends of the observed spin evolution of cool stars and brown dwarfs. Undoubtedly, the main area of progress to be expected in the next few years lies in the improved physical modeling of these processes.

Acknowledgments. The authors would like to dedicate this contribution to the memory of Jean Heyvaerts, who passed away the week before the Protostars and Planets VI conference. J.B. acknowledges the grant ANR 2011 Blanc SIMI5-6 020 01 Toupies: Towards Understanding the Spin Evolution of Stars (*http://ipag. osug.fr-/Anr Toupies/*). K.G.S. and S.P.M. acknowledge the U.S. National Science Foundation (NSF) grant AST-0808072. K.G.S. also acknowledges NSF grant AST-0849736. S.M. acknowledges the support of the UK STFC grant ST/H00307X/1, and many invaluable discussions with F. Shu and A. Reiners. We thank F. Gallet and J. Irwin for providing Fig. 6 and Fig. 7 respectively, and we are indebted to the many authors who have provided us with the data used to build Fig. 1.

REFERENCES

Aarnio A. N. et al. (2012) *Astrophys. J., 760,* 9.
Affer L. et al. (2012) *Mon. Not. R. Astron. Soc., 424,* 11.
Affer L. et al. (2013) *Mon. Not. R. Astron. Soc., 430,* 1433.
Agapitou V. and Papaloizou J. C. B. (2000) *Mon. Not. R. Astron. Soc., 317,* 273.
Agüeros M. A. et al. (2011) *Astrophys. J., 740,* 110.
Albacete Colombo J. F. et al. (2007) *Astron. Astrophys., 474,* 495.
Albrecht S. et al. (2012) *Astrophys. J., 757,* 18.
Alencar S. H. P. et al. (2012) *Astron. Astrophys., 541,* A116.
Alencar S. H. P. et al. (2010) *Astron. Astrophys., 519,* A88.
Allard F. et al. (2011) In *Cool Stars, Stellar Systems, and the Sun* (C. Johns-Krull et al., eds.), p. 91. ASP Conf. Ser. 448, Astronomical Society of the Pacific, San Francisco.

Artemenko S. A. (2012) *Astron. Lett., 38,* 783.
Artigau É. et al. (2009) *Astrophys. J., 701,* 1534.
Bailer-Jones C. A. L. and Mundt R. (1999) *Astron. Astrophys., 348,* 800.
Bailer-Jones C. A. L. and Mundt R. (2001) *Astron. Astrophys., 367,* 218.
Baraffe I. et al. (1998) *Astron. Astrophys., 337,* 403.
Bardou A. and Heyvaerts J. (1996) *Astron. Astrophys., 307,* 1009.
Barnes S. A. (2003) *Astrophys. J., 586,* 464.
Barnes S. A. (2007) *Astrophys. J., 669,* 1167.
Barnes S. A. and Kim Y.-C. (2010) *Astrophys. J., 721,* 675.
Bell C. P. M. et al. (2013) *Mon. Not. R. Astron. Soc., 434,* 806.
Biazzo K. et al. (2009) *Astron. Astrophys., 508,* 1301.
Blake C. H. et al. (2010) *Astrophys. J., 723,* 684.
Blandford R. D. and Payne D. G. (1982) *Mon. Not. R. Astron. Soc., 199,* 883.
Bolmont E. et al. (2012) *Astron. Astrophys., 544,* A124.
Bonfils X. et al. (2013) *Astron. Astrophys., 549,* A109.
Bouvier J. (2008) *Astron. Astrophys., 489,* L53.
Bouvier J. et al. (1997) *Astron. Astrophys., 326,* 1023.
Bouvier J. et al. (2007) In *Protostars and Planets V* (B. Reipurth et al., eds.), pp. 479–494. Univ. of Arizona, Tucson.
Caballero J. A. et al. (2004) *Astron. Astrophys., 424,* 857.
Cai M. J. et al. (2008) *Astrophys. J., 672,* 489.
Carr J. S. (2007) In *Star-Disk Interaction in Young Stars* (J. Bouvier and I. Appenzeller, eds.), pp. 135–146. IAU Symp. 243, Cambridge Univ., Cambridge.
Cauley P. W. et al. (2012) *Astrophys. J., 756,* 68.
Chabrier G. and Küker M. (2006) *Astron. Astrophys., 446,* 1027.
Charbonnel C. et al. (2013) *Astron. Astrophys., 554,* A40.
Cieza L. and Baliber N. (2007) *Astrophys. J., 671,* 605.
Clarke F. J. et al. (2002) *Mon. Not. R. Astron. Soc., 332,* 361.
Cody A. M. and Hillenbrand L. A. (2010) *Astrophys. J. Suppl., 191,* 389.
Cohen O. et al. (2010) *Astrophys. J. Lett., 723,* L64.
Collier Cameron A. et al. (2009) *Mon. Not. R. Astron. Soc., 400,* 451.
Cranmer S. R. (2008) *Astrophys. J., 689,* 316.
Cranmer S. R. (2009) *Astrophys. J., 706,* 824.
Cranmer S. R. and Saar S. H. (2011) *Astrophys J., 741,* 54.
Dahm S. E. et al. (2012) *Astrophys. J., 745,* 56.
D'Angelo C. R. and Spruit H. C. (2011) *Mon. Not. R. Astron. Soc., 416,* 893.
Decampli W. M. (1981) *Astrophys. J., 244,* 124.
Decressin T. et al. (2009) *Astron. Astrophys., 495,* 271.
Deheuvels S. et al. (2012) *Astrophys. J., 756,* 19.
Delfosse X. et al. (1998) *Astron. Astrophys., 331,* 581.
Delorme P. et al. (2011) *Mon. Not. R. Astron. Soc., 413,* 2218.
Denissenkov P. A. (2010) *Astrophys. J., 719,* 28.
Denissenkov P. A. and Pinsonneault M. (2007) *Astrophys. J., 655,* 1157.
Denissenkov P. A. et al. (2010) *Astrophys. J., 716,* 1269.
Dobbs-Dixon I. et al. (2004) *Astrophys. J., 610,* 464.
Donati J.-F. and Landstreet J. D. (2009) *Annu. Rev. Astron. Astrophys., 47,* 333.
Donati J.-F. et al. (2007) *Mon. Not. R. Astron. Soc., 380,* 1297.
Donati J.-F. et al. (2008) *Mon. Not. R. Astron. Soc., 386,* 1234.
Donati J.-F. et al. (2010a) *Mon. Not. R. Astron. Soc., 402,* 1426.
Donati J.-F. et al. (2010b) *Mon. Not. R. Astron. Soc., 409,* 1347.
Donati J.-F. et al. (2011) *Mon. Not. R. Astron. Soc., 417,* 1747.
Donati J. et al. (2013) *ArXiv e-prints,* arXiv:1308.5143.
Drake J. J. et al. (2013) *ArXiv e-prints,* arXiv:1302.1136.
Duchêne G. and Kraus A. (2013) *Annu. Rev. Astron. Astrophys., 51,* 269.
Durney B. R. and Stenflo J. O. (1972) *Astrophys. Space Sci., 15,* 307.
Eggenberger P. et al. (2012) *Astron. Astrophys., 544,* L4.
Epstein C. R. and Pinsonneault M. H. (2012) *ArXiv e-prints,* arXiv:1203.1618.
Ferreira J. (1997) *Astron. Astrophys., 319,* 340.
Ferreira J. et al. (2000) *Mon. Not. R. Astron. Soc., 312,* 387.
Gallet F. and Bouvier J. (2013) *Astron. Astrophys., 556,* A36.
Gastine T. et al. (2013) *Astron. Astrophys., 549,* L5.
Ghosh P. and Lamb F. K. (1978) *Astrophys. J. Lett., 223,* L83.
Ghosh P. and Lamb F. K. (1979) *Astrophys. J., 234,* 296.
Gillon M. et al. (2013) *Astron. Astrophys., 555,* L5.
Girardin F. et al. (2013) *Astrophys. J., 767,* 61.
Goodson A. P. et al. (1999) *Astrophys. J., 524,* 142.
Grankin K. N. (2013) *Astron. Lett., 39,* 251.
Gregory S. G. et al. (2012) *Astrophys. J., 755,* 97.
Harrison T. E. et al. (2012) *Astron. J., 143,* 4.
Hartman J. D. et al. (2010) *Mon. Not. R. Astron. Soc., 408,* 475.
Hartman J. D. et al. (2009) *Astrophys. J., 691,* 342.

Hartmann L. and Stauffer J. R. (1989) *Astron. J., 97,* 873.
Hayashi M. R. et al. (1996) *Astrophys. J. Lett., 468,* L37.
Heinze A. N. et al. (2013) *Astrophys. J., 767,* 173.
Henderson C. B. and Stassun K. G. (2012) *Astrophys. J., 747,* 51.
Herbst W. et al. (2001) *Astrophys. J. Lett., 554,* L197.
Herbst W. et al. (2002) *Astron. Astrophys., 396,* 513.
Herbst W. et al. (2007) In *Protostars and Planets V* (B. Reipurth et al., eds.), pp. 297–311. Univ. of Arizona, Tucson.
Irwin J. and Bouvier J. (2009) In *The Ages of Stars* (E. E. Mamajek et al., eds.), pp. 363–374. IAU Symp. 258, Cambridge Univ., Cambridge.
Irwin J. et al. (2006) *Mon. Not. R. Astron. Soc., 370,* 954.
Irwin J. et al. (2007) *Mon. Not. R. Astron. Soc., 377,* 741.
Irwin J. et al. (2008a) *Mon. Not. R. Astron. Soc., 384,* 675.
Irwin J. et al. (2008b) *Mon. Not. R. Astron. Soc., 383,* 1588.
Irwin J. et al. (2009) *Mon. Not. R. Astron. Soc., 392,* 1456.
Irwin J. et al. (2011) *Astrophys. J., 727,* 56.
Johns-Krull C. M. and Gafford A. D. (2002) *Astrophys. J., 573,* 685.
Kawaler S. D. (1988) *Astrophys. J., 333,* 236.
Kiraga M. and Stepien K. (2007) *Acta Astron., 57,* 149.
Koen C. (2006) *Mon. Not. R. Astron. Soc., 367,* 1735.
Koenigl A. (1991) *Astrophys. J. Lett., 370,* L39.
Konopacky Q. M. et al. (2012) *Astrophys. J., 750,* 79.
Kraft R. P. (1967) *Astrophys. J., 150,* 551.
Krishnamurthi A. et al. (1997) *Astrophys. J., 480,* 303.
Lagarde N. et al. (2012) *Astron. Astrophys., 543,* A108.
Lamm M. H. et al. (2005) *Astron. Astrophys., 430,* 1005.
Lane C. et al. (2007) *Astrophys. J. Lett., 668,* L163.
Lanza A. F. (2010) *Astron. Astrophys., 512,* A77.
Le Blanc T. S. (2011) *Astron. J., 142,* 55.
Littlefair S. P. et al. (2010) *Mon. Not. R. Astron. Soc., 403,* 545.
Long M. et al. (2007) *Mon. Not. R. Astron. Soc., 374,* 436.
Long M. et al. (2008) *Mon. Not. R. Astron. Soc., 386,* 1274.
Long M. et al. (2011) *Mon. Not. R. Astron. Soc., 413,* 1061.
Lovelace R. V. E. et al. (1995) *Mon. Not. R. Astron. Soc., 275,* 244.
Mamajek E. E. and Hillenbrand L. A. (2008) *Astrophys. J., 687,* 1264.
Marques J. P. et al. (2013) *Astron. Astrophys., 549,* A74.
Mathis S. (2013) In *Studying Stellar Rotation and Convection: Theoretical Background and Seismic Diagnostics* (M. Goupil et al., eds.), p. 23. Lecture Notes in Physics, Vol. 865, Springer-Verlag, Berlin.
Matt S. and Pudritz R. E. (2005a) *Astrophys. J. Lett., 632,* L135.
Matt S. and Pudritz R. E. (2005b) *Mon. Not. R. Astron. Soc., 356,* 167.
Matt S. and Pudritz R. E. (2007) In *Star-Disk Interaction in Young Stars* (J. Bouvier and I. Appenzeller, eds.), pp. 299–306. IAU Symp. 243, Cambridge Univ., Cambridge.
Matt S. and Pudritz R. E. (2008a) *Astrophys. J., 678,* 1109.
Matt S. and Pudritz R. E. (2008b) *Astrophys. J., 681,* 391.
Matt S. P. et al. (2012a) *Astrophys. J. Lett., 754,* L26.
Matt S. P. et al. (2012b) *Astrophys. J., 745,* 101.
Mayor M. et al. (2011) *ArXiv e-prints,* arXiv:1109.2497.
McQuillan A. et al. (2013) *Mon. Not. R. Astron. Soc., 432,* 1203.
Meibom S. (2009) *Astrophys. J., 695,* 679.
Meibom S. et al. (2011a) *Astrophys. J. Lett., 733,* L9.
Meibom S. et al. (2011b) *Astrophys. J., 733,* 115.
Messina S. et al. (2010) *Astron. Astrophys., 520,* A15.
Messina S. et al. (2011) *Astron. Astrophys., 532,* A10.
Mestel L. (1968) *Mon. Not. R. Astron. Soc., 138,* 359.
Mestel L. (1984) In *Cool Stars, Stellar Systems, and the Sun* (S. L. Baliunas and L. Hartmann, eds.), p. 49. Lecture Notes in Physics, Vol. 193, Springer-Verlag, Berlin.
Mohanty S. and Basri G. (2003) *Astrophys. J., 583,* 451.
Mohanty S. and Shu F. H. (2008) *Astrophys. J., 687,* 1323.
Mohanty S. et al. (2005) *Mem. Soc. Astron. Ital., 76,* 303.
Moraux E. et al. (2013) *ArXiv e-prints,* arXiv:1306.6351.
Morin J. et al. (2010) *Mon. Not. R. Astron. Soc., 407,* 2269.
Nguyen D. C. et al. (2009) *Astrophys. J., 695,* 1648.
Nielsen M. B. et al. (2013) *Astron. Astrophys., 557,* L10.
Ostriker E. C. and Shu F. H. (1995) *Astrophys. J., 447,* 813.

Parker E. N. (1958) *Astrophys. J., 128,* 664.
Percy J. R. et al. (2010) *Publ. Astron. Soc. Pac., 122,* 753.
Pinsonneault M. H. (2010) In *Light Elements in the Universe* (C. Charbonnel et al., eds.), pp. 375–380. IAU Symp. 268, Cambridge Univ., Cambridge.
Pinte C. et al. (2008) *Astrophys. J. Lett., 673,* L63.
Radigan J. et al. (2012) *Astrophys. J., 750,* 105.
Raghavan D. et al. (2010) *Astrophys. J. Suppl., 190,* 1.
Randich S. (2010) In *Light Elements in the Universe* (C. Charbonnel et al., eds.), pp. 275–283. IAU Symp. 268, Cambridge Univ., Cambridge.
Rebull L. M. et al. (2004) *Astron. J., 127,* 1029.
Reiners A. and Basri G. (2008) *Astrophys. J., 684,* 1390.
Reiners A. and Basri G. (2010) *Astrophys. J., 710,* 924.
Reiners A. and Mohanty S. (2012) *Astrophys. J., 746,* 43.
Reiners A. et al. (2009a) *Astrophys. J., 692,* 538.
Reiners A. et al. (2009b) In *Cool Stars, Stellar Systems, and the Sun* (E. Stempels, ed.), pp. 250–257. AIP Conf. Ser. 1094, American Institute of Physics, Melville, New York.
Reipurth B. et al., eds. (2007) *Protostars and Planets V.* Univ. of Arizona, Tucson.
Rodríguez-Ledesma M. V. et al. (2009) *Astron. Astrophys., 502,* 883.
Rodríguez-Ledesma M. V. et al. (2010) *Astron. Astrophys., 515,* A13.
Romanova M. M. et al. (2007) In *Star-Disk Interaction in Young Stars* (J. Bouvier and I. Appenzeller, eds.), pp. 277–290. IAU Symp. 243, Cambridge Univ., Cambridge.
Romanova M. M. et al. (2009) *Mon. Not. R. Astron. Soc., 399,* 1802.
Romanova M. M. et al. (2011) *Mon. Not. R. Astron. Soc., 411,* 915.
Saar S. H. (1996) In *Stellar Surface Structure* (K. G. Strassmeier and J. L. Linsky, eds.), p. 237. IAU Symp. 176, Cambridge Univ., Cambridge.
Schatzman E. (1962) *Annal. Astrophys., 25,* 18.
Scholz A. and Eislöffel J. (2004a) *Astron. Astrophys., 419,* 249.
Scholz A. and Eislöffel J. (2004b) *Astron. Astrophys., 421,* 259.
Scholz A. and Eislöffel J. (2005) *Astron. Astrophys., 429,* 1007.
Scholz A. and Eislöffel J. (2007) *Mon. Not. R. Astron. Soc., 381,* 1638.
Scholz A et al. (2009) *Mon. Not. R. Astron. Soc., 400,* 1548.
Scholz A. et al. (2011) *Mon. Not. R. Astron. Soc., 413,* 2595.
Schou J. et al. (1998) *Astrophys. J., 505,* 390.
Shu F. H. et al. (1988) *Astrophys. J. Lett., 328,* L19.
Shu F. et al. (1994) *Astrophys. J., 429,* 781.
Shu F. H. et al. (2007) *Astrophys. J., 665,* 535.
Sills A. et al. (2000) *Astrophys. J., 534,* 335.
Skumanich A. (1972) *Astrophys. J., 171,* 565.
Soderblom D. R. (1983) *Astrophys. J. Suppl., 53,* 1.
Spada F. et al. (2010) *Mon. Not. R. Astron. Soc., 404,* 641.
Spada F. et al. (2011) *Mon. Not. R. Astron. Soc., 416,* 447.
Stauffer J. R. and Hartmann L. W. (1987) *Astrophys. J., 318,* 337.
Strugarek A. et al. (2011) *Astron. Astrophys., 532,* A34.
Sukhbold T. and Howell S. B. (2009) *Publ. Astron. Soc. Pac., 121,* 1188.
Suzuki T. K. et al. (2012) *ArXiv e-prints,* arXiv:1212.6713.
Terndrup D. M. et al. (2000) *Astron. J., 119,* 1303.
Turck-Chièze S. et al. (2010) *Astrophys. J., 715,* 1539.
Ud-Doula A. et al. (2009) *Mon. Not. R. Astron. Soc., 392,* 1022.
Ustyugova G. V. et al. (2006) *Astrophys. J., 646,* 304.
Uzdensky D. A. et al. (2002) *Astrophys. J., 565,* 1191.
Weber E. J. and Davis L. (1967) *Astrophys. J., 148,* 217.
Wood B. E. et al. (2002) *Astrophys. J., 574,* 412.
Wood B. E. et al. (2005) *Astrophys. J., 628,* L143.
Wright J. T. et al. (2012) *Astrophys. J., 753,* 160.
Wright N. J. et al. (2011) *Astrophys. J., 743,* 48.
Xiao H. Y. et al. (2012) *Astrophys. J. Suppl., 202,* 7.
Zahn J.-P. (1977) *Astron. Astrophys., 57,* 383.
Zahn J.-P. and Bouchet L. (1989) *Astron. Astrophys., 223,* 112.
Zanni C. and Ferreira J. (2009) *Astron. Astrophys., 508,* 1117.
Zanni C. and Ferreira J. (2011) *Astrophys. J. Lett., 727,* L22.
Zanni C. and Ferreira J. (2013) *Astron. Astrophys., 550,* A99.
Zapatero Osorio M. R. et al. (2003) *Astron. Astrophys., 408,* 663.
Zapatero Osorio M. R. et al. (2006) *Astrophys. J., 647,* 1405.

Frank A., Ray T. P., Cabrit S., Hartigan P., Arce H. G., Bacciotti F., Bally J., Benisty M., Eislöffel J., Güdel M., Lebedev S., Nisini B., and Raga A. (2014) Jets and outflows from star to cloud: Observations confront theory. In *Protostars and Planets VI* (H. Beuther et al., eds.), pp. 451–474. Univ. of Arizona, Tucson, DOI: 10.2458/azu_uapress_9780816531240-ch020.

Jets and Outflows from Star to Cloud: Observations Confront Theory

A. Frank
University of Rochester

T. P. Ray
Dublin Institute for Advanced Studies

S. Cabrit
Observatoire de Paris

P. Hartigan
Rice University

H. G. Arce
Yale University

F. Bacciotti
Osservatorio Astrofisico di Arcetri, Florence

J. Bally
University of Colorado at Boulder

M. Benisty
University of Grenoble

J. Eislöffel
Thüringer Landessternwarte

M. Güdel
University of Vienna

S. Lebedev
Imperial College London

B. Nisini
Osservatorio Astronomico di Roma

A. Raga
Universidad Nacional Autónoma de México

In this review we focus on the role jets and outflows play in the star- and planet-formation process. Our essential question can be posed as follows: Are jets/outflows merely an epiphenomenon associated with star formation, or do they play an important role in mediating the physics of assembling stars both individually and globally? We address this question by reviewing the current state of observations and their key points of contact with theory. Our review of jet/outflow phenomena is organized into three length-scale domains: (1) *source and disk scales* (0.1–10^2 AU) where the connection with protostellar and disk evolution theories is paramount; (2) *envelope scales* (10^2–10^5 AU) where the chemistry and propagation shed further light on the jet launching process, its variability and its impact on the infalling envelope; and (3) *parent cloud scales* (10^5–10^6 AU) where global momentum injection into cluster/cloud environments become relevant. Issues of feedback are of particular importance on the smallest scales, where planet formation regions in a disk may be impacted by the presence of disk winds, irradiation by jet shocks, or shielding by the winds. Feedback on envelope scales may determine the final stellar mass (core-to-star efficiency) and envelope dissipation. Feedback also plays an important role on the larger scales with outflows contributing to turbulent support within clusters, including alteration of cluster star-formation efficiencies (SFEs) (feedback on larger scales currently appears unlikely). In describing these observations we also look to the future and consider the questions that new facilities such as the Atacama Large Millimeter/submillimeter Array (ALMA) and the Jansky Array can address. A particularly novel dimension of our review is that we consider results on jet dynamics from the emerging field of high-energy-density laboratory astrophysics (HEDLA), which is now providing direct insights into the three-dimesional dynamics of fully magnetized, hypersonic, radiative outflows.

1. INTRODUCTION

In many ways the discovery that star formation involves *outflow* as well as *inflow* from gravitational collapse marked the beginning of modern studies of the assembly of stars. Jets and outflows were the first and most easily observed recognition that the narrative of star formation would include many players and processes beyond the spherical collapse of clouds. The extraordinary progress made in the study of protostellar jets and outflows since the first discovery of Herbig-Haro (HH) objects (1950s), HH jets (1980s), and molecular outflows (1980s) also reflects the growing power and sophistication of star-formation science. The combination of ever-higher-resolution observational and computational methods, combined with innovative laboratory experiments, have allowed many aspects of the protostellar outflow problem to be clarified, although as we shall see, crucial issues such as the launching process(es) remain debated.

Hypersonic collimated protostellar mass loss appears to be a ubiquitous aspect of the star-formation process. The observations currently indicate that most, if not all, low- and high-mass stars produce accretion-powered collimated ejections during their formation. These ejections are traced in two ways [see, e.g., the excellent *Protostars and Planets V* (PPV) observational reviews by *Ray et al.* (2007), *Arce et al.* (2007), *Bally et al.* (2007)].

First there are the narrow, highly collimated "jets" of atomic and/or molecular gas with velocities on the order of v ~ 100–1000 km s^{-1} (v increasing with central source mass). These jets are believed to arise through magnetohydrodynamic (MHD) processes in the rotating star-disk system. The other tracer is the less-collimated, more-massive "molecular outflows" with velocities on the order of v ~ 1–30 km s^{-1} that are believed to consist of shells of ambient gas swept up by the jet bow shock and a surrounding slower wider-angle component. The fast and dense jet quickly escapes from the protostellar envelope (the still-infalling remains of the original "core" from which the star formed) and propagates into the surrounding environment to become a "parsec-scale outflow." The less-dense wide-angle wind and the swept-up outflow expand more slowly, carving out a cavity that widens over time into the envelope and the surrounding cloud. Most stars are born in clustered environments where the stellar separation is < 1 pc. Thus, these large-scale outflows affect the interstellar medium (ISM) within a cluster and, perhaps, the cloud as a whole. The important elements and processes on each scale from star to cloud are illustrated in Fig. 1.

Given their ubiquity and broad range of scales, a central question is whether jets and outflows constitute a mere epiphenomenon of star formation, or whether they are an essential component in the regulation of that process. In particular, winds/outflows are currently invoked to solve several major outstanding issues in star formation: (1) the low-SFE in turbulent clouds (see, e.g., the chapters by Padoan et al. and Krumholtz et al. in this volume); (2) the systematic shift between the core mass function and the stellar initial mass function, suggesting a core-to-star efficiency of only 30% (see, e.g., the chapters by Offner et al. and Padoan et al. in this volume); and (3) the need to efficiently remove angular momentum from the young star and its disk. The former is important to avoid excessive spin-up by accretion and contraction (cf. the chapter by Bouvier et al. in this volume), while the latter is required to maintain accretion at observed rates, in particular across the "deadzone" where MHD turbulence is inefficient (cf. the chapter by Turner et al. in this volume). Last but not least, protostellar winds may also affect disk evolution and planet formation through disk irradiation or shielding, and enhanced radial mixing of both gas and solids.

In this chapter, we address this central question of outflow feedback on star and planet formation while also reviewing the current state of jet/outflow science. From the description above, it is clear that the degree of feedback will differ according to outflow properties on different spatial scales. The impact on the star and disk will depend on the physics of jet launching and angular momentum extraction (small scales). The impact on core-to-star efficiencies will depend both on the intrinsic jet structure and the jet propagation/interaction with surrounding gas. Finally, the impact on global star-formation efficiency will depend on the overall momentum injection and on the efficacy of its coupling to cloud turbulence.

Thus in what follows we review the current state of understanding of protostellar jets and outflows by breaking the chapter into three sections through the following division of scales: (1) *star and disk* (1–10^2 AU); (2) *envelope and parent clump* (10^2 AU–0.5 pc); and (3) *clusters and molecular clouds* (0.5–10^2 pc). In each section we review the field *and* present new results obtained since PPV. Where appropriate we also address how new results speak to issues of feedback on star and planet formation. We also attempt to point to ways in which new observing platforms such as the Atacama Large Millimeter/submillimeter Array (ALMA) can be expected to influence the field in the near future. We note that we will focus on outflows from nearby low-mass stars (< 500 pc), which offer the best resolution into the relevant processes. An excellent review of outflows from high-mass sources was presented in *Arce et al.* (2007), which showed that some (but not all) appear as scaled-up versions of the low-mass case (see also, e.g., *Codella et al.*, 2013; *Zhang et al.*, 2013). ALMA will revolutionize our view of these distant and tightly clustered objects so much that our current understanding will greatly evolve in the next few years. Finally, we note that our review includes, for the first time, results on jet dynamics from the emerging field of high-energy-density laboratory astrophysics (HEDLA). These experiments and their theoretical interpretation provide direct insights into the three-dimensional dynamics of fully magnetized, hypersonic, radiative outflows.

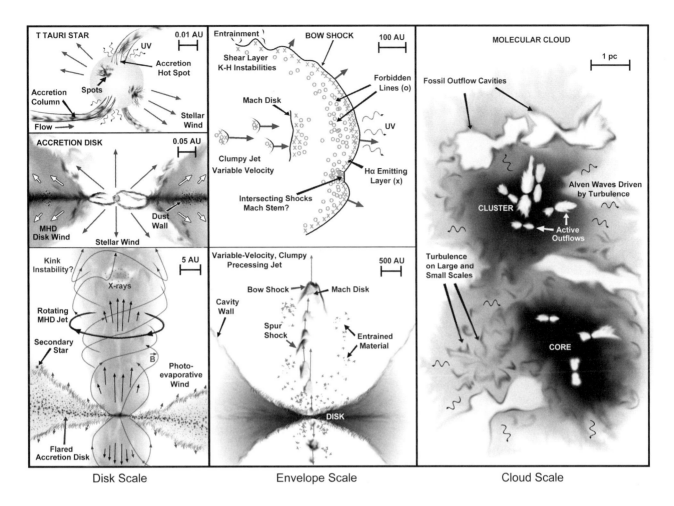

Fig. 1. A schematic view of jets and outflows across 7 orders of magnitude in scale. Note the presence of the scale bar in each figure as one moves from the physics of launching near the star out to the physics of feedback on cluster and cloud scales. See text for reference to specific processes and classes of objects.

2. SOURCE AND DISK SCALES (1–10² AU)

2.1. The Accretion-Ejection Link

While the precise origin of jets from young stars is still hotly debated, there is general consensus that the launching process involves the dynamical interaction of accreted matter with the stellar and/or disk magnetic field. However, launch distances depend on the model: a few solar radii for stellar winds, ≈0.05 AU for those launched at the stellar magnetosphere-disk interface (see the chapter by Bouvier et al. in this volume) (*Shang et al.,* 2007), and possibly as far as several astronomical units for magneto-centrifugal disk (D)-winds (see the chapter by Turner et al. in this volume) (*Pudritz et al.,* 2007). Unfortunately, the projected dimensions on the sky for even the largest proposed launch regions are tens of milliarcseconds. We do not, however, have to spatially resolve the launch zone to at least begin to test various models. Jet properties such as the ejection to accretion ratio, collimation, and velocity structure can be measured on larger scales, which then al-

low mechanisms working on smaller scales to be inferred. With that being said, however, we note that interferometric studies are beginning to resolve the smaller scales directly (see section 2.5).

We first consider the ratio of jet mass flux to accretion rate. Measuring the mass outflow rate in an atomic jet can be achieved since in principle we know all the necessary quantities from observation. For example, through spectroscopy and multi-epoch imaging, we can determine both the radial and tangential velocity of a jet and hence its true velocity. In addition, the jet radius, ionization fraction, electron density, and hence total density can be found from a combination of imaging and consideration of various line ratios (*Podio et al.,* 2011). One can then calculate $\dot{M}_{jet} \approx \pi r_{jet}^2 \rho_{jet} V_{jet}$. Typical outflow rates are found to be 10^{-7}–10^{-9} $M_\odot$ yr^{-1} for jets from low-mass classical T Tauri stars (CTTS). As one might suspect, higher rates are found for more embedded sources of comparable mass (*Caratti o Garatti et al.,* 2012). We caution that what we see as a jet in fact consists of a string of shocks with a wide range of conditions. The measurements described above represent a bulk average over the shocked gas, unless the knots are well-resolved spatially. Moreover, there are a number of methods of measuring the mass-loss rate that give somewhat different values (within a factor of 3–10). For a more detailed examination of this problem, the reader is referred to *Dougados et al.* (2010).

Measuring the accretion rate, $\dot{M}_{acc}$, is also challenging. Assuming material is accreted onto the star through magnetospheric accretion (*Bouvier et al.*, 2007) from the vicinity of the disk's inner radius R_{in} yields the accretion luminosity $L_{acc} \approx GM_*\dot{M}_{acc}(1-R_*/R_{in})/R_*$. Note that the disk inner radius is often considered to be its co-rotation radius with the star. In the case of CTTS, this energy is mainly observed in the ultraviolet (UV) band (*Gullbring et al.,* 2000), but direct observation of this UV excess can be difficult as it may be highly extincted, particularly in more embedded sources. Fortunately, the strength of the UV excess has been found to be related to the luminosity of a number of optical and infrared (IR) emission lines such as Hα, CaII, Paβ, and Brγ (e.g., *Natta et al.,* 2006), which are thought to be mainly produced in the (magnetospheric) accretion funnel flow. The relationships between the various line luminosities and the UV excess has been tested for objects from young brown dwarfs up to intermediate-mass young stars and has been found to be robust (e.g., *Rigliaco et al.,* 2012).

These emission line "proxies" can be used to determine the accretion luminosity, and hence accretion rates, with a good degree of certainty. The large instantaneous spectral coverage made possible by new instruments such as XSHOOTER on the Very Large Telescope (VLT) is particularly well suited to simultaneously cover both accretion and jet-line indicators and thus constrain the ejection/accretion ratio (*Ellerbroek et al.,* 2013).

However, a number of important caveats must be raised in considering these methods. First, spectroastrometric or interferometric studies of certain lines show that some portion of their emission must arise from the outflow, i.e., not all the line's luminosity can be from magnetospheric accretion close to the star (*Whelan et al.,* 2009a) (section 2.5). In such cases, the good correlation with UV excess would trace in part the underlying ejection-accretion connection.

Moreover, as the accretion does not seem to be uniform, i.e., there may be an unevenly spaced number of accretion columns (see Fig. 1), individual line strengths can vary over periods of days with the rotation phase of the star (*Costigan et al.,* 2012). Accretion can also be intrinsically time variable on shorter timescales than those probed by forbidden lines in jets (several years). Thus time-averaged accretion values should be used when comparisons are made with mass-flux rates derived from such jet tracers.

Studies of accretion onto young stellar objects (YSOs) suggest a number of findings that are directly relevant to outflow studies. In particular. it is found that:

1. Once the dependence on stellar mass ($\propto M_*^2$) is taken into account, the accretion rate seems to fall off with time t with an approximate t^{-1} law (*Caratti o Garatti et al.,* 2012). This also seems to be reflected in outflow proxies,

with similar ejection/accretion ratios in class I and class II sources (e.g., *Antoniucci et al.,* 2008).

2. Many embedded sources appear to be accreting at instantaneous rates that are far too low to acquire final masses consistent with the initial mass function (*Evans et al.,* 2009; *Caratti o Garatti et al.,* 2012). This suggests accretion and associated outflows may be episodic.

3. Typical ratios of jet mass flux to accretion rate for low-mass CTTS are ≈10% (e.g., *Cabrit,* 2007). Similar ratios are obtained for jets from intermediate-mass T Tauri stars (*Agra-Amboage et al.,* 2009) and Herbig Ae/Be stars (e.g., *Ellerbroek et al.,* 2013). In contrast, the lowest-mass class II objects (e.g., young brown dwarfs) show larger ratios (*Whelan et al.,* 2009b). This begs the obvious question, do very-low-mass objects have difficulty accreting because of their magnetic ejection configuration?

If the jet is responsible for extracting excess angular momentum from the accretion disk, then the angular momentum flux in the wind/jet, $\dot{J}_W$, should equal that to be removed from the accreting flow, $\dot{J}_{acc}$. Since $\dot{J}_W \simeq \dot{M}_W \Omega r_A^2$ and $\dot{J}_{acc} \simeq \Omega r_l^2 \dot{M}_{acc}$ (where Ω is the angular velocity at the launch radius r_l, and r_A is the Alfvén radius), it follows that $\dot{M}_W/\dot{M}_{acc} \simeq (r_l/r_A)^2 = 1/\lambda$, with λ defined as the magnetic lever arm parameter of the disk wind (*Blandford and Payne,* 1982). The observed ejection/accretion ratio of 10% in CTTS is then consistent with a moderate $\lambda \simeq 10$, while the higher ratio in brown dwarfs would indicate that the Alfvén radius is much closer to the launch point.

2.2. The Collimation Zone

Since PPV, great strides have been made in our understanding of how jets are collimated from both theoretical and observational perspectives. In particular, recent images of jets within 100 AU of their source give us clues as to how the flows are focused. These observations involve both high-spatial-resolution instrumentation in space, e.g., the Hubble Space Telescope (HST), as well as ground-based studies, [e.g., various optical/IR adaptive optics (AO) facilities and millimeter/radio interferometers]. At PPV it was already known that optically visible jets from classical T Tauri stars (i.e., class II sources) begin with wide (10°–30°) opening angles close to the source and are rapidly collimated to within a few degrees in the innermost 50–100 AU (*Ray et al.,* 2007).

Perhaps the most interesting finding since PPV is that rapid focusing of jets occurs not only in the case of class II sources, but also in *embedded protostars as well*, i.e., class 0 and class I sources. Jets from these early phases are difficult to observe optically, due to the large amount of dust present, and certainly one cannot trace them optically back to their source as in the case of classical T Tauri stars. Nevertheless, their inner regions can be probed through molecular tracers such as SiO in the millimeter range, or [FeII] and H_2 lines in the near- and mid-IR (see section 3.1). Using, e.g., the Institut de Radioastronomie Millimétrique (IRAM) Plateau de Bure Interferometer (PdBI) with 0.″3 resolution, *Cabrit et al.* (2007) has shown that the SiO jet from the class 0 source HH 212 is collimated on scales similar to jets from class II sources. This suggests that the infalling envelope does not play a major role in focusing the jet. Thus a more universal collimation mechanism must be at work at all stages of star formation to produce a directed beam of radius about 15 AU on 50-AU scales (see Fig. 2). The same applies to class I jets, where both the ionized and molecular jet components show similar opening angles at their base as class II jets (*Davis et al.,* 2011). Observations of this type rule out collimation by the ambient thermal pressure (*Cabrit,* 2009), and instead favor the idea, first proposed by *Kwan and Tademaru* (1988), that magnetic fields *anchored in the disk* force the jet to converge. The required poloidal disk field would be $B_D \simeq$ 10 mG$(\dot{M}_w V_w/10^{-6}\ M_\odot$ km s^{-1} yr$^{-1})^{0.5}$, where the scaling is for typical CTTS jet parameters.

Thanks to the additional toroidal field that develops in the centrifugal launch process, MHD disk winds could provide the required collimation with an even smaller poloidal disk field (*Meliani et al.,* 2006). In this case, we expect distortion of the magnetic field, from a largely poloidal to a largely toroidal geometry, to begin in the vicinity of the Alfvén radius ($r_A = \sqrt{\lambda}r_l$). The jet, however, may have to traverse many Alfvén radii before being effectively focused, since its collimation depends not simply on the magnetic lever arm but also on the poloidal field strength at the disk surface. Several models of truncated MHD disk winds reproduce the point spread function (PSF)-convolved widths of *atomic* class II jets with launch radii in the range 0.1–1 AU, despite widely differing magnetic lever arms (*Stute et al.,* 2010; *Shang et al.,* 2010).

In the future, ALMA and then the James Webb Space Telescope (JWST) will be able to carry out even more detailed

high-spatial-resolution studies of outflows that should better distinguish between these models. In the interim, the new class of high-sensitivity radio interferometers, such as the Jansky Very Large Array (JVLA) and Multi-Element Radio Linked Interferometer Network (MERLIN), are already online and show potential for measuring the collimation of jets within 10 AU of the source (*Ainsworth et al.,* 2013; *Lynch et al.,* 2013).

2.3. Angular Momentum Transport in Jets

Irrespective of the precise nature of the central engine, a basic requirement of any complete model is that angular momentum must be removed from the accreted material before it can find its final "resting place" on the star. As matter rains down on the disk from the surrounding envelope before being accreted, this process must involve the disk. As reviewed in the chapter by Turner et al. in this volume, ordinary particle viscosity is too small to make the *horizontal transport* of angular momentum from inner to outer regions of the disk efficient, and additional mechanisms have to be considered. A promising mechanism appeared to be the generation of a "turbulent viscosity" by magneto-rotational instabilities, the so-called magneto-rotational instability (MRI) (*Balbus and Hawley,* 2006). Recent studies, however, demonstrate that in an initially MRI-unstable disk, the inclusion of a significant vertical magnetic flux, and of ambipolar diffusion coupled with Ohmic dissipation, suppresses MRI turbulence, and instead a powerful magneto-centrifugal wind is generated (see the chapter by Turner et al. in this volume) (*Bai and Stone,* 2013; *Lesur et al.,* 2013). Indeed, MHD centrifugal models for jet launching (*Blandford and Payne,* 1982; *Pudritz and Norman,* 1983) indicate that protostellar jets can provide a valid solution to the angular momentum problem via *vertical transport* along the ordered component of the strong magnetic field threading the disk. In D-wind models this occurs in an extended region where the footpoints of the flow are located (see, e.g., *Ferreira* 1997; *Pudritz et al.,* 2007), while in the X-wind model it is assumed that the wind only extracts the angular momentum from the inner boundary of the disk (*Camenzind,* 1990; *Shu et al.,* 1994; *Fendt,* 2009; *Čemeljić et al.,* 2013). Note that in this case the angular momentum "problem" is only partially resolved, as material still has to be transported within the disk to its inner truncation radius. Finally, MHD stellar winds flowing along open field lines attached to the star's surface (e.g., *Sauty et al.,* 1999; *Matt and Pudritz,* 2005), and episodic plasmoid ejections by magnetospheric field lines linking the star and the disk (*Ferreira et al.,* 2000; *Zanni and Ferreira,* 2013), contribute to the braking of the star (see the chapter by Bouvier et al. in this volume). In fact, such stellar and/ or magnetospheric winds must be active, at some level, to explain the low observed spin rates of young stars. Thus several MHD ejection sites probably coexist in young stars, and the difficulty is to determine the relative contribution of each to the observed jets.

A key observational diagnostic to discriminate between these theories is the detection of possible signatures of

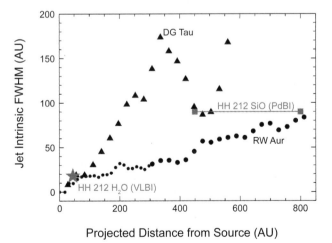

Fig. 2. A plot showing the width of the jet in HH 212 as observed in SiO with the PdB Interferometer vs. distance from the source in astronomical units. A comparison with jets from class II sources (e.g., DG Tau and RW Aur as illustrated) shows that this outflow, from an embedded class 0 source, is collimated on similar scales. From *Cabrit et al.* (2007).

rotation in protostellar jets. The review by *Ray et al.* (2007) in PPV describes the detection in five objects of asymmetric Doppler shifts in emission lines from opposite borders of the flow (*Davis et al.*, 2000; *Bacciotti et al.*, 2002; *Woitas et al.*, 2005; *Coffey et al.*, 2004, 2007). Since PPV, Doppler shifts have been searched for in many other outflows, in atomic *and* molecular lines. These studies are very demanding, as they require both high angular and spectral resolution, pushing the instrumentation, even on HST, to its limits. Possible signatures of rotation, with toroidal and poloidal velocities v_ϕ, v_p consistent with magneto-centrifugal acceleration, appear in many of the cases studied [HL Tau: *Movsessian et al.* (2007); HH 26, HH 72: *Chrysostomou et al.* (2008); HH 211: *Lee et al.* (2007, 2009); HH 212: *Lee et al.* (2008), *Coffey et al.* (2011); CB 26: *Launhardt et al.* (2009); Ori-S6: *Zapata et al.* (2010); NGC 1333 IRAS 4A2: *Choi et al.* (2011)].

The collection of the rotation data allowed for the determination of the specific angular momentum rv_ϕ. Assuming an axisymmetric, stationary magneto-centrifugal wind, the ratio rv_ϕ/v_p^2 gives the location r_l of the footpoint in the disk of the sampled streamline, while the product $rv_\phi v_p$ gives the magnetic lever arm parameter λ (e.g., *Anderson et al.*, 2003; *Ferreira et al.*, 2006). Hence this is a powerful tool to discriminate between proposed jet spatial origins and launch models. As shown in Fig. 5 of *Cabrit* (2009), the observed signatures when interpreted as steady jet rotation are only consistent with an origin in an extended, warm D-wind, launched from between 0.1 and 3–5 AU. *The significant implication is that jets and the associated magnetic fields may strongly affect the disk structure in the region where terrestrial planets form.* The inferred magnetic lever arm parameter is moderate, $\lambda \le 10$, in line with the mean observed ejection to accretion ratio (see section 2.1).

Note, however, that due to limited angular resolution, only the external streamlines of the flow are sampled (*Pesenti et al.*, 2004), and the current measurements cannot exclude the existence of inner stellar or X-winds. In addition, all the measurements are based on emission lines produced in shocks that can also self-generate rotational motions (*Fendt*, 2011). Finally, if the jet is observed far from the star, the interaction with the environment can hide and confuse rotation signatures. The primary hypothesis to be tested, however, is the veracity or otherwise of the rotation interpretation. Simulations including an imposed rotation motion were successful in reproducing the observed spectra (e.g., *Staff et al.*, 2010). In contrast, other studies claim that rotation can be mimicked by, e.g., asymmetric shocking against a warped disk (*Soker*, 2005), jet precession (*Cerqueira et al.*, 2006), and internal shocks (*Fendt*, 2011). Although it is unlikely that these processes apply in all cases, they may contribute to the Doppler shift, confusing the real rotation signature.

From an observational perspective it was found that out of the examined disks associated with rotating jets, one, the RW Aur disk, clearly appeared to counterrotate with respect to the jet (*Cabrit et al.*, 2006). Since this result potentially undermined the rotation hypothesis, the bipolar jet from RW Aur was observed again after the SM4 repair, twice with the Space Telescope Imaging Spectrograph (STIS) in UV lines, at an interval of six months (*Coffey et al.*, 2012). The result was again puzzling: The rotation sense for one lobe was in agreement with the disk, and hence opposite to that measured in the optical years before. Moreover, no signature was detected at that time from the other lobe and, after six months, it had disappeared from both lobes. Despite these findings, *Sauty et al.* (2012) has recently demonstrated that disagreement with the disk rotation can be accommodated within the classical magnetocentrifugal theory, as toroidal velocity reversals can occur occasionally without violating the total (kinetic plus magnetic) angular momentum conservation. Their simulations also show that the rotation sense can change in time, thereby accounting for the detected variability. Thus it appears that observations are still compatible with the jets being a robust mechanism for the extraction of angular momentum from the inner disk. The gain in resolution offered by ALMA and JWST will be crucial to test and confirm this interpretation.

2.4. Wide-Angle Structure and Blue/Red Asymmetries

Other constraints for jet-launching models come from the overall kinematics in the inner few 100 AU (where internal shocks and interaction with ambient gas are still limited). Obtaining such information is very demanding as it requires spectroimaging at subarcsecond resolution, either with HST or with powerful AO systems from the ground, coupled with a long-slit or integral field unit (IFU). It is therefore only available for a handful of bright jets. A useful result of such studies is that while jet-acceleration scales and terminal velocities seem equally compatible with stellar wind, X-wind, or D-wind models (*Cabrit*, 2009), there is a *clear drop in velocity toward the jet edges* (as illustrated, e.g., in Fig. 3). This "onion-like" velocity structure, first discovered in the DG Tau jet, has been seen whenever the jet base is resolved laterally and thus may be quite general (*Beck et al.*, 2007; *Coffey et al.*, 2008; *Pyo et al.*, 2009; *Agra-Amboage et al.*, 2011). It argues against the "classical" X-wind model where the ejection speed is similar at all angles (*Shang et al.*, 2007), and instead requires that the optically bright jet beam is closely surrounded by a slower wide-angle "wind." A natural explanation for such transverse velocity decrease is a range of launch radii in an MHD disk wind (*Agra-Amboage et al.*, 2011), or a magnetospheric wind surrounded by a disk wind (*Pyo et al.*, 2009). Turbulent mixing layers and material ejected sideways from internal working surfaces may also contribute to this low-velocity "sheath" (see Fig. 1) (*Garcia Lopez et al.*, 2008). Studies combining high spectral and spatial resolution will be essential to shed further light on this issue.

Asymmetries in jet velocity, density, and opening angle between the blue and red lobes are seen in many jets (*Hirth et al.*, 1994; *Podio et al.*, 2011) (see also Fig. 3). They

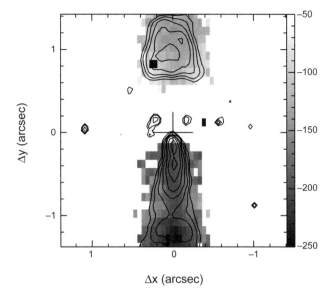

Fig. 3. Map of centroid velocities in the DG Tau jet, as determined from the [FeII]1.64-µm line observed with the SINFONI IFU on the VLT. Note the fast drop in velocity away from the jet axis, and the velocity asymmetry between blue and red lobes (velocities in the redshifted lobe (on top) were given a minus sign to ease comparison). One arcsecond is 140 pc. From *Agra-Amboage et al.* (2011).

hold another fundamental clue to the jet-launch process, because they remove ill-known variables like stellar mass, disk truncation radius, etc., which are the same for both sides of the flow. Recent studies of RW Aur show that the velocity asymmetry varies over time, while the velocity dispersion remains the same fraction of jet speed in both lobes (*Melnikov et al.,* 2009; *Hartigan and Hillenbrand,* 2009). Such asymmetries could be modeled by MHD disk winds where the launch radii or magnetic lever arms differ on either side (*Ferreira et al.,* 2006; *Shang et al.,* 2010). Possible physical reasons for this are, e.g., different ionization or magnetic diffusivities on the two faces of the disk (*Bai and Stone,* 2013; *Fendt and Sheikhnezami,* 2013). Investigating if and how strong blue/red asymmetries can be produced in magnetospheric or stellar winds, along with testable differences compared with D-winds, should be a priority for future theoretical studies.

2.5. Resolving the Central Engine

At the time of PPV, near-IR (NIR) interferometric measurements in young stars were only possible in the dust continuum, revealing sizes and fluxes compatible with puffed-up rims at the dust sublimation radius [see Fig. 1 and *Millan-Gabet et al.* (2007)]. Magneto-hydrodynamic disk winds capable of lifting dust particles have recently been suggested as an alternative means for producing the interferometric sizes and NIR excess in Herbig Ae/Be stars (*Bans and König* 2012). The ability to spatially/spectrally

resolve hydrogen lines has also recently been achieved by the Very Large Telescope Interferometer (VLTI), the Keck Interferometer, and the Center for High Angular Resolution Astronomy (CHARA) array. These studies now enable in-depth studies of the spatial distribution and kinematics of the *gas* on subastronomical-unit scales and bring new constraints on the connection between accretion and ejection.

Strong hydrogen emission lines are among the most prominent manifestation of an actively accreting young star. In T Tauri and Herbig Ae stars, they are considered as good proxies of mass accretion onto the star, as their luminosity correlates with the accretion rate measured from the UV excess (see *Calvet et al.,* 2004) (section 2.1). Yet their precise origin is still unclear. Interferometric observations at low spectral resolution (R ~ 1500–2000) with the Keck Interferometer and the Astronomical Multi BEam Recombiner (AMBER) instrument at VLTI provided the first average size measurements in Brγ in about 20 young stars. Gaseous emission is generally more compact than K-band dust continuum (normally located at 0.2–0.5 AU). *Kraus et al.* (2008) fitted typical ring radii ~0.15 to 2.22 AU for five intermediate-mass young stars. *Eisner et al.* (2010) retrieved smaller extents from 0.04 to 0.28 AU for 11 solar-mass and intermediate-mass young stars. The common interpretation is that the smallest sizes are dominated by magnetospheric accretion, while sizes larger than ~0.1 AU trace compact outflows. These results firmly establish the contribution of ejection processes to hydrogen line formation.

The connection between accretion and ejection processes on astronomical unit scales has recently been specifically addressed in young spectroscopic binaries, where numerical models predict enhanced accretion near periastron. In the close Herbig Ae binary HD 104237 (separation ~0.22 AU), more than 90% of the Brγ line emission is unresolved and explained by magnetospheric emission that increases at periastron. The large-scale jet should be fed/collimated by the circumbinary disk (*Garcia et al.,* 2013). The wider, massive Herbig Be binary HD 200775 (separation ~5 AU) was studied in Hα with the Visible spEctroGraph and polArimeter (VEGA) instrument at the CHARA array. The large size increase near periastron (from 0.2 to ~0.6 AU) indicates simultaneously enhanced ejection in a nonspherical wind (*Benisty et al.,* 2013). Centroid shifts with 0.1 mas precision across the Brγ line profile have also been achieved. They reveal a bipolar outflow in the binary Herbig Be star Z CMa, with a clear connection between its accretion outburst and episodic ejection (see Fig. 4) (*Benisty et al.,* 2010). These findings suggest that the accretion-ejection connection seen in T Tauri stars extends well into the Herbig Ae/Be mass regime.

Finally, spectrally resolved interferometric observations of the Herbig Ae star AB Aur in Hα (*Rousselet-Perraut et al.,* 2010) and of the Herbig Be star MWC 297 in Brγ (*Weigelt et al.,* 2011) have been modeled using radiative transfer codes simulating stellar and/or disk winds. These studies show that HI line emission is enhanced toward the equator, lending support to the scenario of magneto-centrifugal launching of jets through disk winds, rather than through

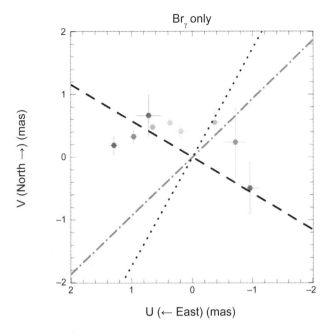

Fig. 4. The Brγ two-dimensional photocenter position in Z CMa as a function of velocity across the Brγ profile, using AMBER/VLTI. The velocity goes from −350 (SW) to +350 (NE) km s⁻¹. The position angle of the binary (dashed-dotted line), known large-scale outflow (dashed line), and direction perpendicular to the jet (dotted line) are overplotted. Note that information is being recovered on 0.1-mas scales. From *Benisty et al.* (2010).

stellar winds. In the near future, spatially resolved multiwavelength observations of lines emitted at different optical depths (combining, e.g., Hα and Brγ) will bring additional constraints. The next generation VLTI imaging instrument Gravity is expected to bring the first model-independent images of the central engine in NIR, and enable statistical studies of solar-mass young stars.

2.6. Jets on Small Scales: A High-Energy Perspective

At the time of PPV, observations of jets and outflows from young stars were largely confined to optical or longer wavelength regimes with the occasional foray into the UV. Since then, however, jets from protostars and T Tauri stars have been found to contain plasma at temperatures of several million kelvin. While this discovery came as a surprise, it was not completely unpredictable. Indeed, jet flow velocities of 300–400 km s⁻¹ can, when flowing against a stationary obstacle, easily shock-heat gas up to ≈1 MK. The ensuing X-ray emission would then serve as a valuable jet diagnostic (*Raga et al.*, 2002c). However, optical observations of jets within 100 AU of their source typically indicate low shock velocities 30–100 km s⁻¹ (e.g., *Lavalley-Fouquet et al.*, 2000; *Hartigan and Morse*, 2007). The detection of strong X-rays in jets on small scales was therefore still not fully anticipated.

The Taurus jets of L1551 IRS-5, DG Tau, and RY Tau show luminous ($L_X \approx 10^{28}$–10^{29} erg s⁻¹) X-ray sources at distances corresponding to 30–140 AU from the driving star (*Favata et al.*, 2002; *Bally et al.*, 2003; *Güdel et al.*, 2008; *Schneider et al.*, 2011; *Skinner et al.*, 2011). A compact X-ray jet was also detected in the eruptive variable Z CMa (*Stelzer et al.*, 2009). The X-ray spectra of these jet sources are soft, but still require electron temperatures of ≈3−7 MK. Further spatially unresolved jets have been discovered based on Two-Absorber X-ray (TAX) spectra. These composite spectra reveal a hard, strongly absorbed spectral component (the star) on top of a soft, little absorbed component (the X-ray jet) (*Güdel et al.*, 2007), an identification explicitly demonstrated for DG Tau (see Fig. 5). The strongly differing absorption column densities between the two components (by a factor of ~10) indicate that the jet X-ray source is located far outside the immediate stellar environment, where hard coronal X-rays are subject to strong absorption. So far, objects like GV Tau, CW Tau, HN Tau, DP Tau, and Sz 102 belong to this class (*Güdel et al.*, 2009); the Beehive proplyd in the Orion Nebula cluster is another obvious example (*Kastner et al.*, 2005).

The best-studied bright, central X-ray jet of DG Tau has been found to be stationary on timescales of several years on spatial scales of about 30 AU from the central star (*Schneider and Schmitt*, 2008, *Güdel et al.*, 2013; Güdel et al., in preparation). Its extent along the jet axis seems to be solely determined by plasma cooling. An assessment of the relevant cooling mechanisms (*Güdel et al.*, 2008; *Schneider and Schmitt*, 2008) suggests that radiative cooling dominates for $n_e > 10^4$–10^5 cm⁻³, which may be appropriate for these central sources. For example, an electron density of $n_e \approx 10^6$ cm⁻³ and a flow speed of 300 km s⁻¹ are in agreement with the observed extension of DG Tau's inner X-ray source of ≈0″.3–0″.5 as a result of radiative cooling (*Schneider and Schmitt*, 2008). The location of the X-ray source relative to emission sources at lower temperatures may also be revealing. *Schneider et al.* (2013a) obtained high-resolution HST observations in optical and UV lines and found a high-velocity (≈200 km s⁻¹) CIV-emitting cloud slightly downwind from the X-ray source, but its luminosity is too high to be explained by cooling plasma previously emitting in X-rays. This observation suggests local heating even beyond the X-ray source.

How is the plasma heated to several megakelvin within tens of astronomical units of the central star? Shocks that produce X-rays require very high jet velocities v_s = 370−525 km s⁻¹ (for T = 4 MK) depending on the ionization degree of the preshock material (*Bally et al.*, 2003). The fixed nature of the inner sources suggests they are associated with a stationary heating process in the launching or collimation region (*Güdel et al.*, 2008; *Schneider and Schmitt*, 2008). Viable models include the following: (1) Shocks form when an initially wide wind is deflected and collimated into a jet, perhaps by magnetic fields that act as a nozzle for the heated plasma. X-ray luminosity and plasma cooling indicate preshock densities on the order of

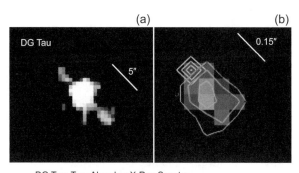

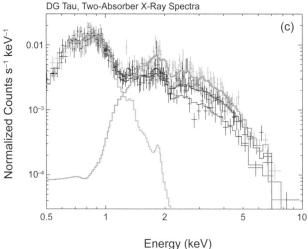

Fig. 5. The X-ray jet of DG Tau. **(a)** Larger-scale structure of the jet (SW) and counterjet (NE) observed by Chandra in the 0.6–1.7-keV range (*Güdel et al.*, 2008). **(b)** Innermost region of the soft forward jet (extended gray contours) peaking 30 AU to the SW from DG Tau itself (compact hard source) (from *Güdel et al.*, in preparation). **(c)** Three two-absorber X-ray (TAX) spectra of DG Tau observed a few days apart. The thick gray histograms show model fits to the soft (below 1 keV) and hard (above 1.5 keV) spectral domains, respectively. The strongly absorbed, variable hard spectrum originates in the stellar corona; the constant, little-absorbed soft spectrum comes from the jet base in **(b)** (*Güdel et al.*, 2013; Güdel et al., in preparation).

$10^3–10^4$ cm^{-3} (*Bally et al.*, 2003). Specifically, a diamond shock forming at the opening of a (magnetic?) nozzle and producing a hot, standing shock was modeled by *Bonito et al.* (2011). (2) Randomly accelerated and ejected clouds of gas at different velocities produce, through collisions, chains of moving but also stationary knots along the jet with X-ray emission characteristics similar to what is observed (*Bonito et al.*, 2010). One potential drawback of this model is that very high initial velocities are required to reproduce moderate-velocity X-ray knots.

A major problem with all these shock models is that the high velocities required to reach the observed temperatures are not observed in any jet spectral lines so far. However, the X-ray-emitting plasma component contributes only a minor fraction to the total mass loss rate of the associated atomic jet: $\approx 10^{-3}$ in DG Tau (*Schneider and Schmitt*, 2008).

It is therefore conceivable that the X-rays are produced in a superfast but rather minor jet component not detected at other wavelengths, e.g., a stellar or magnetospheric wind (*Günther et al.*, 2009). In this context, it may be relevant that X-ray jet models based on radiative cooling times indicate very small filling factors f on the order of 10^{-6} but high electron densities n_e, e.g., $n_e > 10^5$ cm^{-3} (*Güdel et al.*, 2008); the resulting pressure would then far exceed that in the cooler 10^4-K atomic jet, and might contribute to transverse jet expansion.

It is also conceivable that the standing X-ray structures are not actually marking the location of a stationary heating process, but only the exit points from denser gaseous environments within which X-rays are absorbed and which obscure our view to the initial high-energy source. This is an attractive explanation for L1551 IRS-5 with its deeply embedded protostellar binary (*Bally et al.*, 2003); it could also hold for the soft emission in DG Tau, which is seen to be produced near the base of a converging cone of H_2 emitting material that may block the view to the source closer than 0.15 of the star (*Schneider et al.*, 2013b, *Güdel et al.*, 2013; Güdel et al., in preparation). With these ideas in mind, an alternative model could involve the production of hot plasma in the immediate stellar environment through magnetic reconnection of star-disk magnetic fields, ejecting high-velocity plasma clouds analogous to solar coronal mass ejections (*Hayashi et al.*, 1996). If these cool, they may eventually collide with the (slower) jet gas and therefore shock-heat gas further out (*Skinner et al.*, 2011).

2.7. Connection with Laboratory Experiments: Magnetic Tower Jets

A key development since PPV, and of direct relevance to the launching mechanism, has been the first successful production of laboratory jets driven by the pressure gradient of a toroidal magnetic field (*Lebedev et al.*, 2005a,b; *Ciardi et al.*, 2007), in a topology similar to the "magnetic tower" model of astrophysical jets (*Lynden-Bell*, 1996, 2006). The generated outflow consists of a current-carrying central jet, collimated by strong toroidal fields in a surrounding magnetically dominated expanding cavity, which in turn is confined by the pressure of the ambient medium. The most recent configurations even allow for the generation of several eruptions within one experimental run (*Ciardi et al.*, 2009; *Suzuki-Vidal et al.*, 2010) (see Fig. 6). The experiments are scalable to astrophysical flows because critical dimensionless numbers such as the plasma collisionality, radiative cooling parameter ($\chi \approx 0.1$), and the ratio of thermal to magnetic pressure ($\beta \approx 1$), are all in the appropriate ranges. Furthermore, the viscous Reynolds number (Re $\approx 10^6$) and magnetic Reynolds number (Re$_M \approx 200–500$) are much greater than unity, ensuring that transport arises predominantly by advection with the flow.

The main findings from these magnetic tower experiments are the following: an efficient conversion of magnetic energy into flow kinetic energy; a high degree of jet

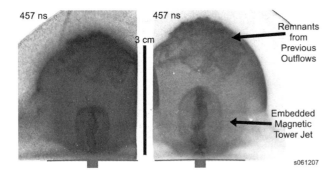

Fig. 6. Episodic "magnetic tower" jet produced in laboratory experiments with the MAGPIE pulsed-power facility. The self-emission XUV images show the growth of the latest magnetic jet and cavity inside the broader cavity created by previous episodes. Adapted from *Ciardi et al.* (2009).

collimation (<10°) for sufficiently strong radiative cooling; an enhanced collimation for episodic jets, as magnetic fields trapped in the previously ejected plasma add to the collimation of the later episodes; the generation of an X-ray pulse at each new eruption, as the central jet is compressed on-axis by the magnetic field; and the development of current-driven MHD instabilities leading to variability in density (≈100%) and velocity (≈30%). In particular, these experiments show that kink-mode instabilities *disrupt but do not destroy the MHD jet*, despite a dominant toroidal field. Instead, the nonlinear saturation of the unstable modes fragments the beam into chains of dense knots that propagate at a range of velocities around the average beam speed. Compelling similarities of the episodic jet behavior in laboratory experiments with observations of transient bubble-like structures in the XZ Tau and DG Tau jets are discussed by *Ciardi et al.* (2009) and *Agra-Amboage et al.* (2011). The stability and possible observational signatures from different configurations of magnetic tower jets were recently studied using numerical simulations with the AstroBEAR code (*Huarte-Espinosa et al.,* 2012).

2.8. Impact on Planet Formation

About 25% of the CTTS in Taurus with known jets show detectable X-ray emission from the jet base (M. Güdel, personal communication). This could be relevant for the processing of circumstellar material in their protoplanetary disks. Apart from driving chemical processing, direct heating of the disk surface by X-ray jets may induce photoevaporation (see Fig. 1) that competes with that induced by stellar X-rays and UV photons, because of the more favorable illumination geometry. Simple estimates for DG Tau suggest that photoevaporation outside about 20 AU would be dominated by X-rays from the jet (*Güdel et al.,* 2013; Güdel et al., in preparation).

At the same time, dusty MHD disk winds, if present, could effectively screen the disk against the *stellar* FUV and X-ray photons. For an accretion rate ≈10^{-7} M$_\odot$ yr^{-1},

an MHD disk wind launched out to 1 AU would attenuate stellar photons reaching the disk surface by $A_V \simeq 10$ mag and a factor of ≈500 in coronal X-rays, while the star would remain visible to an outside observer with $A_V \leq 1$ mag for inclinations up to 70° from pole-on (*Panoglou et al.,* 2012).

Another important dynamical feedback of MHD disk winds on the planet formation zone would be to induce fast radial accretion at sonic speeds due to efficient angular momentum removal by the wind torque (see the chapter by Turner et al. in this volume and references therein), and to modify planet migration through the associated strong magnetic disk fields (see the chapter by Baruteau et al. in this volume and references therein). The thermal processing, coagulation, and fall-back of dust grains ejected in an MHD disk wind from 1 to 3 AU was also recently invoked as a means to form and radially redistribute chondrules in our solar system (*Salmeron and Ireland,* 2012).

3. ENVELOPE AND PROPAGATION SCALES (10^2 AU–0.5 PC)

3.1. Jet Physical Conditions Across Star-Formation Phases

Since PPV, the Spitzer, Herschel, and Chandra missions along with improved groundbased facilities have allowed the study of jets on intermediate scales in younger sources, and in temperature/chemical regimes unexplored in the past. Such studies reveal a frequent coexistence of molecular gas at 500–2000 K with atomic gas at 10^4 K, and in a few cases with hot plasma at several million kelvin. This broad range of conditions was unanticipated and may arise from several factors: (1) the intrinsic spread of physical and chemical conditions present in the cooling zones behind radiative shocks, (2) the interaction of the jet with its environment (e.g., entrainment of molecules along the jet beam), and (3) the simultaneous contributions of (possibly molecular) disk winds, stellar winds, and magnetospheric ejections. Disentangling these three factors is essential to obtain accurate jet properties and understand its interaction with the natal core. The evolution of jet composition as the source evolves from class 0 to class II holds important clues to this issue.

Molecular jets from the youngest protostars (class 0 sources) have benefited most from recent progress in the sub/millimeter and IR ranges. Interferometric maps in CO and SiO show that they reach deprojected velocities of several hundred kilometers per second, as expected for the "primary wind" from the central source (*Arce et al.,* 2007; *Codella et al.,* 2007; *Hirano et al.,* 2010). This is further supported by chemical abundances that are clearly distinct from those in swept-up ambient gas (*Tafalla et al.,* 2010). Multi-line SiO observations show that they are warm (T$_{kin}$ in the range 100–500 K) and dense [n(H$_2$) 10^5–10^6 cm^{-3}] (*Nisini et al.,* 2007; *Cabrit et al.,* 2007). This result was confirmed via H$_2$ mid-IR observations of the class 0 jets L1448 and HH 211 (*Dionatos et al.,* 2009, 2010) and

Herschel observations of water lines in L1448 (*Kristensen et al.*, 2011; *Nisini et al.*, 2013). ALMA observations will soon provide an unprecedented view of these warm, dense molecular jets, as already illustrated by first results in the CO(6–5) line (e.g., *Kristensen et al.*, 2013a; *Loinard et al.*, 2013). In particular, they should clarify the corresponding ejection/accretion ratio [currently subject to significant uncertainties (e.g., *Lee*, 2010)].

Spitzer also revealed for the first time an embedded *atomic* component associated with these molecular class 0 jets, via mid-IR lines of [FeII], [SI], and [SiII] (*Dionatos et al.*, 2009, 2010). It is characterized by a low electron density (~100–400 cm^{-3}), moderate ionization fractions of <10^{-3}, and T<3000 K. However, its contribution to the overall jet dynamics and its relationship to the molecular jet are still very uncertain. This issue will be likely revised thanks to Herschel Photodetector Array Camera and Spectrometer (PACS) observations that resolve strong collimated [OI] 63-μm emission in several class 0 jets (see Fig. 7), this line being a better tracer of mass flux.

On larger scales, the shocks caused by the interaction of class 0 jets with the ambient medium have been probed with much better resolution and sensitivity than possible in the 1990s with the Infrared Space Observatory (ISO) satellite. Spitzer line maps demonstrate a smooth H$_2$ temperature stratification between 100 K and 4000 K (*Neufeld et al.*, 2009), where H$_2$ pure rotational lines are the main cooling agent (*Nisini et al.*, 2010b). The other two main molecular shock coolants, CO and H$_2$O, were studied in detail with Herschel. The H$_2$O and far-IR CO lines (at J ≥ 13) are strictly correlated with H$_2$ v = 0 and trace high-pressure postshock gas (*Santangelo et al.*, 2012; *Tafalla et al.*, 2013). Contrary to simple expectations, the contribution of water to total cooling is never larger than ~20–30% (*Nisini et al.*, 2010a) (see the chapter by van Dishoek et al. in this volume for a more general discussion of water). Detailed analysis of CO and H$_2$O line profiles and maps reveals multiple shock components within 30″, with different temperatures, sizes, and water abundances that further complicate the analysis (e.g., *Lefloch et al.*, 2012; *Santangelo et al.*, 2013). As for more complex organic molecules, their abundances and deuteration levels in outflow shocks have proven to offer a useful "fossil" record of ice mantles formed in the cold preshock ambient cloud (*Arce et al.*, 2008; *Codella et al.*, 2012).

The dissociative "reverse shock" (or "Mach disk") where the jet currently impacts on the leading bowshock (see Fig. 1) was also unambiguously identified for the first time in class 0 jets, in [NeII], [SiII], and [FeII] lines with Spitzer (*Neufeld et al.*, 2006; *Tappe et al.*, 2008, 2012), and in OH and [OI] with Herschel/PACS (*Benedettini et al.*, 2012). In the latter case, the momentum flux in the reverse shock seems sufficient to drive the whole swept-up CO cavity.

The abundance, excitation, and collimation of molecules in jets clearly evolve in time. In contrast to class 0 sources, older class I jets are undetectable (or barely so) in low-J CO and SiO emission. Hot molecular gas at ≈1000–2000 K is still seen, in the form of rovibrational H$_2$ emission and

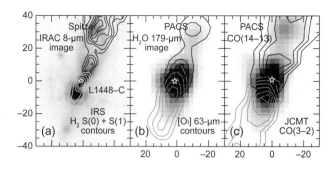

Fig. 7. Spitzer and Herschel spectral images of the jet from the class 0 object L1448-C, revealing the spatial distribution of warm molecular gas at 300–2000 K and the presence of an embedded bipolar atomic jet. **(a)** Contours of H$_2$ S(0) and S(1) line emission from Spitzer IRS superimposed on the IRAC 8-μm image in grayscale. Adapted from *Giannini et al.* (2011). **(b)** Blue-shifted (black contours) and red-shifted (gray contours) emission of [OI] 63 μm superimposed on a grayscale image of the H$_2$O 179-μm line, both obtained with Herschel/PACS. Adapted from *Nisini et al.* (2013) and Nisini et al. 2013 (in preparation). **(c)** Grayscale PACS image of the CO(14–13) emission with superposed contours of CO(3–2) from JCMT. Adapted from *Nisini et al.* (2013) and Nisini et al. 2013 (in preparation).

more rarely v = 0–1 CO absorption (see *Davis et al.*, 2011; *Herczeg et al.*, 2011, and references therein). While some H$_2$ may be associated with the fast atomic jet, it mainly traces a slower "intermediate velocity component" (IVC) ≈10–50 km s^{-1} near the jet base, and in all cases carries a 10–1000× smaller mass flux than the atomic jet (*Nisini et al.*, 2005; *Podio et al.*, 2006; *Garcia Lopez et al.*, 2008; *Davis et al.*, 2011). In the later class II stage, hot H$_2$ generally peaks at even smaller velocities ≤15 km s^{-1} and traces a wider flow around the atomic jet (see, e.g., *Herczeg et al.*, 2006; *Takami et al.*, 2007; *Beck et al.*, 2008; *Schneider et al.*, 2013b).

Concerning the atomic jet component in class I jets, optical and NIR line ratios indicate similar temperatures ≈10^4 K and ionization fraction x_e ~ 0.05–0.9 as in class II jets, indicating moderate shock speeds ~30–70 km s^{-1}, but with higher electron and total density (*Nisini et al.*, 2005; *Podio et al.*, 2006; *Antoniucci et al.*, 2008; *Garcia Lopez et al.*, 2008). This implies a higher mass flux rate (although the ejection to accretion ratio remains similar; see section 2.1). A more complete view of the different excitation components present in jet beams can be obtained by combining emission lines in a wide wavelength range of 0.3–2 μm. The first such studies in class II jets reveal a broader range of ionization states (including [SiII] and [OIII]) than seen in optical lines, probing faster shocks ≥100 km s^{-1} that must be accounted for in mass-flux-rate derivations (*Bacciotti et al.*, 2011).

Finally, extended X-ray emission has been resolved with Chandra along the L1551-IRS5 (class I) and DG Tau (class II) jets out to distances of 1000 AU, revealing hot plasma at several million kelvin that was totally unanticipated from optical data on similar scales. X-rays have been detected as far as 0.1 pc to 2.5 pc from the driving source, associated with high-excitation HH objects. The relationship between X-ray and optical emission, however, is not always clear. HH 80/81 shows X-rays, radio continuum, and optical lines all coinciding at arcsecond resolution. The X-rays, however, point to a factor of 10 lower density and 2–5× lower speed than the jet (*Pravdo et al., 2004*). An inverse situation is encountered in the Cepheus A east/west region (*Pravdo and Tsuboi, 2005*). Here, the required shock speeds are comparable to flow speeds, but the head of the expanding region is detected in Hα and not in X-rays. Hence the hot plasma appears to be heated significantly upstream of the leading working surface, possibly in a reverse shock (*Schneider et al., 2009*) or in a collision with another jet (*Cunningham et al., 2009b*). Clearly, further work is needed to fully understand the link between optical and X-ray emission from jets and HH objects on intermediate scales.

3.2. Magnetic and Chemical Diagnostics on Intermediate Scales

Modern models of jet launching all invoke magnetic fields to achieve the desired terminal velocities and narrow collimation angles of ~5°. However, measuring field strengths within bright optical jets has proved very difficult because Zeeman splitting is undetectable in optical lines. *Carrasco-González et al.* (2010) were able to measure polarized synchrotron radio emission in the shocked jet of HH 80/81 and inferred an average field of ~200 μG at 0.5 pc, with a helical structure about the jet axis. But this jet, driven by a massive protostar, is quite exceptional in regard to its speed (1000 km s^{-1}) and brightness in the radio range.

In optical jets, it is still possible to estimate B fields through the effect they have on postshock compression and the resulting emission line ratios. *Morse et al.* (1992, 1993) inferred a preshock field of ~20–30 μG in distant bowshocks of two class I jets with a density of 100–200 cm^{-3}. More recently, *Teşileanu et al.* (2009, 2012) estimated B ≃ 500 μG at n_H ≃ 1–5 × 10^4 cm^{-3} in two class II microjets within 500 AU of the source. These all yield transverse Alfvén speeds $V_{A,\phi}$ ≃ 4 km s^{-1}, typically one-fiftieth of the jet speed. This value is a lower limit, as the B field could have been partly dissipated by reconnection or ambipolar diffusion between the point where the jet is launched to where shock waves are observed, and further lowered by "velocity stretching" between internal working surfaces (*Hartigan et al., 2007*). Hence these results provide interesting constraints for MHD jet-launching models. Resolved spatial maps of the ionization fraction, temperature, and density obtained with HST for the HH 30 jet in two epochs (*Hartigan and Morse, 2007*) also reveal an unexplained new phenomenon where the highest ionization lies upstream from the emission knots and does not show a correlated density increase [such behavior was also observed by HST in RW Aur (*Melnikov et al., 2009*)]. Models of line emis-

sion from magnetized jet shocks have yet to fully confront these observational constraints.

The magnetic field in *molecular* jets from class 0 sources could also, in principle, be constrained by shock modeling. Such efforts are complicated by the fact that two kinds of shocks may exist: the sudden "J-type" shock fronts and the broader C-type shocks, where the magnetic field is strong enough to decouple ions and neutrals and energy is dissipated by ambipolar diffusion. Given uncertainties in the beam-filling factor, H$_2$ data alone are often insufficient to constrain the B value, unless it is large enough to make C-type shock-cooling regions spatially resolvable [e.g, as in Orion BN-KL (*Kristensen et al., 2008*; *Gustafsson et al., 2010*)] or when the transition from C- to J-type shock can be located along a large bowshock surface (e.g., *Giannini et al., 2008*). However, B values in bowshocks may be more relevant to the external medium than to the jet itself. Shock chemistry offers additional clues [see the excellent review of this topic in *Arce et al.* (2007)] but requires complex modeling. For example, SiO was long believed to offer an unambiguous tracer of C-type shocks, but recent models now also predict substantial SiO in dense J-type shocks, from grain-grain shattering (*Guillet et al., 2009*). One must also account for the fact that young C-type shocks in jets will contain an embedded J-type front. Herschel observations bring additional C-/J-type shock diagnostics such as [O I], OH, and NH$_3$ lines (*Flower and Pineau des Forêts, 2013*) so that our understanding of the B field in class 0 jets should greatly progress in the coming years.

Another indirect clue to the launch region is whether the jet is depleted in refractory elements, as one would expect if it originates from beyond the dust sublimation radius (≈0.1–1 AU) where these elements would be mainly locked up in grains. Refractory gas-phase abundances have been measured in bright HH objects for decades (e.g., *Brugel et al., 1981*). Only recently, however, have such studies been extended to the much fainter jet beams. Assuming solar abundances, measurements indicate significant gas-phase depletions of Fe, Si, and/or Ca (*Nisini et al., 2005*; *Podio et al., 2006, 2009, 2011*; *Dionatos et al., 2009, 2010*; *Agra-Amboage et al., 2011*). While the data are not extensive and the measurements difficult, the general consensus indicates that dust exists in *atomic* jets at all evolutionary stages (from class 0 to class II), in larger amounts at lower velocities, and gets progressively destroyed along the jet in strong shocks. The dust could be entrained from the surrounding cloud, or may be carried along with gas ejected from the circumstellar disk (see Fig. 1). In any case, a large fraction of the dust should survive the acceleration process. Such measurements also argue against the lower-velocity gas-tracing sideways ejections from internal jet-working surfaces; if this were the case, it should be more shock-processed and less depleted in refractories than high-velocity gas, whereas the opposite is observed (*Agra-Amboage et al., 2011*).

The dust content is more difficult to constrain in the *molecular* component of class 0 jets. Silicon monoxide is the only detected molecule involving a refractory species, and unfortunately it is optically thick in the inner 500 AU of class 0 jets (*Cabrit et al.*, 2007). The lower limit on SiO gas-phase abundance is ≃10% of elemental silicon, still compatible with an initially dusty jet (*Cabrit et al.*, 2012). One possible indirect indication that molecular jets might arise from *dusty* MHD disk winds are the predicted chemical and temperature structures (*Panoglou et al.*, 2012). When ionization by coronal X-rays is included, ion-neutral coupling is sufficient to lift molecules from the disk without destroying them, while efficient dust shielding enables high abundances of H_2, CO, and H_2O. As the wind density drops in the class I and II phases, dust shielding is less efficient. The molecular region moves to larger launch radii ≃0.5–1 AU, while heating by ion-neutral drag increases. This trend would agree with observations of decreasing speed, mass flux, and collimation of H_2 and increasing temperatures in class 0 to class II jets (see section 3.1). The broad H_2O line wings recently discovered toward class 0 and class I sources with Herschel's Heterodyne Instrument for the Far Infrared (HIFI) (*Kristensen et al.*, 2012) can also be reproduced by this model as well as the correlation with envelope density (Yvart et al., in preparation). ALMA and IR IFUs with laser guide stars will bring key constraints on this scenario and on the origin of molecular jets, in particular through more detailed characterization of their peculiar chemical abundances (cf. *Tafalla et al.*, 2010) and the confrontation with model predictions.

3.3. Ejection Variability and Implications for Source and Disk Properties

Since jets are accretion driven, outflow properties that change with distance from the source provide important constraints on past temporal variations in the ejection/accretion system, over a huge range of timescales from <5 to 10^5 yr that cannot be probed by any other means.

Optical, IR, and millimeter observations show that both atomic and molecular jet beams exhibit a series of closely spaced inner "knots" within 0.1 pc of the source (see Fig. 1), together with more distant, well-separated larger bows or "bullets" that in most cases have a clear correspondence in the opposite lobe, HH 212 being the most spectacular example to date (*Zinnecker et al.*, 1998). Atomic jet knots and bows have line ratios characteristic of internal shock waves. Therefore, they cannot just trace episodes of enhanced jet density, which alone *would not produce shocks*. Significant variations in speed or ejection angles are also required. Several lines of evidence imply that these shocks are caused by supersonic velocity jumps where fast material catches up with slower ejecta (e.g., *Raga et al.*, 2002b; *Hartigan et al.*, 2005). The same was recently demonstrated for their CO "bullet" counterparts (*Santiago-García et al.*, 2009; *Hirano et al.*, 2010).

The most natural origin for such velocity jumps is initial variability in the ejection speed (*Raga et al.*, 1990). This is supported, for example, by numerical simulations of the resulting jet structure, and by HST proper motions at the base of the HH 34 jet clearly showing a velocity increase of 50 km s^{-1} over the last 400 yr (*Raga et al.*, 2012). The knot/bow spacing and velocity patterns in class I atomic jets then suggest that up to three modes of velocity variability are present in parallel, with typical periods of a few 10 yr, a few 100 yr, and a few 1000 yr respectively and velocity amplitudes of 20–140 km s^{-1} (*Raga et al.*, 2002a). A strikingly similar hierarchy of knot/bullet spacings is seen in class 0 jets, suggesting a similar variability behavior (see, e.g., *Cabrit*, 2002). Time series of Taurus class II jets at 0".15 resolution show new knots that emerge from within 50–100 AU of the source with an even shorter interval of 2.5–5 yr (*Hartigan and Morse*, 2007; *Agra-Amboage et al.*, 2011). Spitzer observations further reveal that the 27-yr period knots in HH 34 are synchronized to within 5 yr between the two jet lobes, implying that the initial perturbation is less than 3 AU across at the jet base (*Raga et al.*, 2011).

These results set interesting constraints for jet launching and variable accretion models. Proposed physical origins for quasiperiodic jet variability include stellar magnetic cycles or global magnetospheric relaxations of the star-disk system (3–30 yr), perturbations by unresolved (possibly excentric) binary companions, and EXOr-FUOr outbursts (see the chapter by Audard et al. in this volume). Dedicated monitoring of at least a few prototypical sources should be a priority to clarify the link between these phenomena and jet variability. We note that care must be taken in interpreting such results, however, as jet are likely to be inherently clumpy on subradial scales. The internal dynamics of clumps of different size and velocity represents an essentially different form of dynamics than pure velocity pulsing across the jet cross-section (*Yirak et al.*, 2012). In particular, as clumps collide and potentially merge, they can mimic the appearance of periodic pulsing (*Yirak et al.*, 2009).

Internal shocks may also be produced without velocity pulsing if the jet axis wanders sufficiently that dense packets of gas can shock against ambient or slow cocoon material (e.g., *Lim and Steffen*, 2001). Jet axis wandering with time is indeed a common characteristic among sources of various masses and evolutionary stages. Mass ejected from a young stellar object should follow approximately a linear trajectory once it leaves the star-disk system, unless it is deflected by a dense clump or a sidewind. And indeed, most knot proper motions are radial to within the errors (e.g., *Hartigan et al.*, 2005). Hence, jet wiggles or misaligned sections, commonly seen in the optical, IR, and millimeter, most likely indicate a variation in ejection angle. Jet precession produces point-symmetric (S-shaped) wiggles between the jet and counterjet, while orbital motion of the jet source in a binary system will produce mirror-symmetric wiggles. Precession by a few degrees has long been known in class 0/I jets (see, e.g., Figs. 8a,b), with typical periods ranging from 400 to 50,000 yr (e.g., *Eislöffel et al.*, 1996; *Gueth et al.*, 1996; *Devine et al.*, 1997), and larger axis changes of up to 45° in a few sources (e.g., *Cunningham et al.*, 2009b). Mirror-symmetric signatures of jet orbital motion have been identified more recently, e.g., in the HH 211 class 0 jet (see Fig. 8c), the HH 111 class I jet, and the HH 30 class II jet, with orbital periods of 43 yr, 1800 yr, and 114 yr respectively (*Lee et al.,* 2010; *Noriega-Crespo et al.,* 2011; *Estallela et al.,* 2012). It is noteworthy that secular disk precession driven by tidal interaction with the orbiting companion (assumed non-co-planar) could explain the longer precession timescale observed on larger scales in HH 111 (*Terquem et al.,* 1999; *Noriega-Crespo et al.,* 2011). Such a coincidence suggests that jet axis precession is due to precession of the disk axis, rather than of the stellar spin axis. Although more examples are needed to confirm this hypothesis, it supports independent conclusions that jet collimation (and possibly ejection) is controlled by *the disk B field* (see section 2.2). Observations of jet orbital motions also provide unique constraints on the mass and separation of close companions that would be otherwise difficult to resolve. Interestingly, the inferred binary separation of 18 AU in HH 30 is consistent with the size of its inner disk hole (*Estallela et al.,* 2012). The jet from the Herbig Be member of ZCMa shows wiggles with a 4–8-yr period similar to the timescale of its EXOr outbursts, suggesting that such outbursts may be driven by a yet-undetected companion (*Whelan et al.,* 2010).

3.4. Jet Propagation and Shock Structure: Connection with Laboratory Experiments

Another major development in the time-domain has been the acquisition of multiple-epoch emission-line images from HST, which now span enough time (10 yr) to reveal not only proper motions of individual knots, but also to begin to show how the shock waves evolve and interact. Images of the classic large-scale bow shocks in HH 1, HH 2, HH 34, and HH 47 (*Hartigan et al.,* 2011) show evidence for a variety of phenomena related to jet propagation (see Fig. 1), including standing deflection shocks where the precessing jet beam encounters the edges of a cavity, and where a strong bow shock encounters a dense obstacle on one side. Knots along the jet may brighten suddenly as denser material flows into the shock front, and fade on the cooling timescale of decades. Multiple bow shocks along working surfaces sometimes overlap to generate bright spots where they intersect, and the morphologies of Mach disks range from well-defined almost planar shocks to small reverse bow shocks as the jet wraps around a denser clump. The bow shocks themselves exhibit strong shear on the side where they encounter slower material, and show evidence for Kelvin-Helmholtz instabilities along the wings of the bows.

In support of the interpretation of observations, innovative laboratory experiments on jet propagation have been carried out, where magnetic fields are not present or not dynamically significant. Experiments on pulsed-power facilities investigated the gradual deflection (bending) of supersonic jets by the ram pressure of a sidewind (*Lebedev et al.,* 2004, 2005b). The experimental results were used to benchmark numerical simulations and the same computer code was used to simulate astrophysical systems with scaled-up initial conditions (*Ciardi et al.,* 2008). Both the experiments and the astrophysical simulations show that the

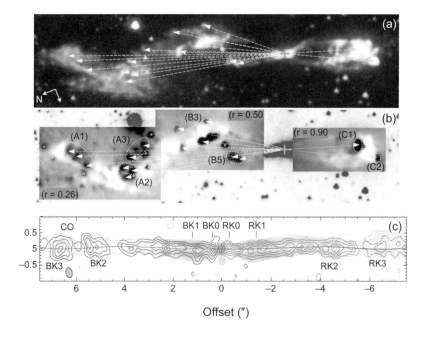

Fig. 8. Examples of S-shaped (precession) and mirror-symmetric (orbital) wiggling in protostellar jets. **(a)** Three-color Spitzer IRAC image of the L1157 outflow (3.6, 4.5, and 8.0 μm, respectively). **(b)** Difference image of L1157 where warm/dense H₂ knots show up in black and are connected by arrows to the central protostar (cross). **(c)** Jet orbital motion model (curve) superposed onto the CO and SiO knots in the HH 211 jet (contours). Adapted from *Takami et al.* (2011) and *Lee et al.* (2010).

jet can be deflected by a significant angle ($\approx 30^{\circ}$) without being destroyed. The interaction between the jet and the sidewind also leads to variability in the initially laminar flow, driven by the onset of Kelvin-Helmholtz instabilities.

Experiments with laser-driven jets (*Foster*, 2010; *Hansen et al.*, 2011) have been primarily devoted to studies of hydrodynamical instabilities in a jet interacting with, and deflected by, localized dense obstacles (*Hartigan et al.*, 2009), in a geometry similar to the HH 110 jet. The experimental results have been compared in detail with numerical simulations and show good agreement. Another recent experiment (*Yirak et al.*, 2012) investigated the formation of Mach stems in collisions between bowshocks, which is relevant to observations of similar structures in HH objects with HST. Such collisions are expected in the interaction of clumpy jets with ambient gas, and could lead to the formation of shocks normal to the flow and a localized increase in emission (*Hartigan et al.*, 2011).

Finally, several experiments (*Nicolaï et al.*, 2008; *Suzuki-Vidal et al.*, 2012) investigated the jet-ambient interaction in conditions where radiative cooling is very strong ($\chi \ll 1$). It was found that small-scale clumps, attributed to cooling instabilities, rapidly develop in the bowshock region and that the clump size decreases for increasing radiative cooling (*Suzuki-Vidal et al.*, 2013). These experiments represent the first investigation of cooling instabilities evolving into their highly nonlinear stages, which may have observable consequences, e.g., online ratios of high- vs. low-ionization stages.

3.5. Core-to-Star Efficiency and Envelope Dissipation

A comparison of the prestellar core mass function with the initial mass function suggests that only one-third of the core mass ends up into the star (see, e.g., the chapters by Offner et al. and Padoan et al. in this volume). Since circumstellar disks are seen to contain only a small fraction of the final stellar mass, protostellar jets/winds are prime candidates to explain this low "core-to-star" efficiency (e.g., *Myers*, 2008).

An attractive possibility is that a substantial fraction of infalling core gas is reejected during early class 0 collapse via magnetically driven outflows. Three-dimensional MHD simulations of rotating collapse over 3×10^4 yr suggest that MHD ejection results in a final accreted mass of only $\approx 20\%/\cos\alpha$ of the initial core mass, where α is the initial angle between the B field and the core-rotation axis (*Ciardi and Hennebelle*, 2010). Longer simulations extending to $\approx 10^5$ yr with $\alpha = 0$ suggest that mass accretion during the class I phase brings the final core-to-star efficiency closer to 50% (*Machida and Hosokawa*, 2013). Although slightly larger than the observed 30%, this result indicates that early protostellar MHD ejections could play a key role in determining the core-to-star efficiencies. The ejected mass in this early phase is at relatively low velocity and may constitute part of the low-velocity V-shaped cavities later observed around class 0 jets (see the chapter by H. Li et al.

in this volume). Slow outflows recently attributed to very young first or second hydrostatic cores should provide a test of this scenario.

Another complementary scenario is that swept-up outflow cavities driven by wide-angle winds halt infall by dispersing the infalling envelope (*Myers*, 2008); early evidence suggestive of envelope dispersion by outflows was presented in PPV (*Arce et al.*, 2007). Here we discuss new elements relevant to this issue, and their resulting implications. Interferometric CO observations in a sample of nearby protostars have demonstrated an increasing opening angle of the outflow cavity with age (*Arce and Sargent* 2006). Class 0 cavities show opening angles of 20°–50°, class I outflows show 80°–120°, and class II outflows show cavities of about 100° to 160°. A similar trend is seen in scattered light (*Seale and Looney*, 2008). These results may be understood if protostellar winds are wide-angled with a denser inner part along the outflow axis. At early times, only the fastest and densest axial component of the wind punctures the circumstellar environment. As the protostar evolves, entrainment by the outflow decreases the density in the envelope, allowing material at larger angles from the outflow axis to be swept up, widening the outflow cavity (*Arce et al.*, 2007). Three-dimensional simulations and synthetic CO observations of protostellar outflows in a turbulent core do show a gradual increase of the cavity opening angle, up to 50° at 5×10^4 yr, even though the angular distribution of injected momentum remains constant over time (*Offner et al.*, 2011).

A caveat to this interpretation comes from recent studies claiming mass-flux rates in CO outflows are too low to alone disperse the surrounding envelopes within their disappearance timescale of ≈ 2–3×10^5 yr (*Hatchell et al.*, 2007; *Curtis et al.*, 2010). If this were indeed the case, the observed broadening of outflow cavities with age would then be a *consequence*, rather than the *cause*, of envelope dispersal. Efforts are needed to reduce uncertainties in line opacity, gas temperature, and hidden gas at low velocity or in atomic form (e.g., *Downes and Cabrit*, 2007) to obtain accurate estimates of outflow rates on the relevant scales. It is also noteworthy that the envelope mass typically drops by an order of magnitude between the class 0 and class I phases (*Bontemps et al.*, 1996), while the cavity full opening angle $\theta \le 100^{\circ}$ encompasses a fraction $1 - \cos(\theta/2) \le 36\%$ of the total envelope solid angle, and an even smaller fraction of the envelope *mass* (concentrated near the equator by rotational and magnetic flattening). This seems to indicate that outflow cavities will be too narrow during the early phase where stellar mass is assembled to affect the core-to-star efficiency, although they could still be essential for dissipating the residual envelope at later stages.

Another open question that bears more on the issue of the launching mechanism is the nature of the wide-angle component responsible for the observed opening of outflow cavities. While a *fast* ≈ 100 km s^{-1} wide-angle wind has often been invoked (*Shang et al.*, 2006; *Arce et al.*, 2007), this now appears to be ruled out by recent observations indicating a fast drop in velocity away from the jet axis (see section 3.2). On the other hand, a *slow* wide-angle wind may still be present. In particular, an MHD disk wind launched out to several astronomical units naturally produces a slow wide-angle flow around a much faster and denser axial jet (see, e.g., *Pudritz et al.*, 2007; *Panoglou et al.*, 2012), with an angular distribution of momentum similar to that used in *Offner et al.* (2011). Sideways splashing by major working surfaces (see section 3.3) could also contribute to gradually broaden the cavity base. Herschel studies of warm >300 K molecular gas have started to reveal the *current* shock interaction between the jet/wind and the envelope (*Kristensen et al.*, 2013b). ALMA maps of outflow cavities promise to shed new light on this issue, thanks to their superb dynamic range and sensitivity to faint features (*Arce et al.*, 2013).

4. PARENT CLOUD SCALES (0.5–10² PC)

4.1. Parsec-Scale Jets, Outflows, and Large-Scale Shells

Optical and NIR wide-field camera surveys in the late 1990s and early 2000s revealed that atomic and H₂ jets with projected extensions on the plane of the sky larger than 1 pc (so-called *parsec-scale jets*) are a common phenomenon (e.g., *Eislöffel and Mundt*, 1997; *Reipurth et al.*, 1997; *Mader et al.*, 1999; *Eislöffel*, 2000; *Stanke et al.*, 2000; *McGroarty et al.*, 2004; *McGroarty and Ray*, 2004). More recent cloud-wide surveys continue to find new giant jets, indicating that young stars of all masses can power flows that interact with their surroundings at parsec-scale distances (e.g., *Davis et al.*, 2008, 2009; *Bally et al.*, 2012; *Ioannidis and Froebrich*, 2012). In many cases these wide-field observations reveal that jets originally thought to extend less than about 0.5 pc in reality extend 2 to 3 pc (or even more) on the sky.

The fact that parsec-scale protostellar jets are a common phenomenon should not have come as a surprise since, assuming (constant) jet velocities of 100 to 300 km s⁻¹ and timescales of at least 2×10^5 yr (the approximate lifetime of the class I stage), their expected size would be (at least) 20 to 60 pc. Even when deceleration of the ejecta is considered, they are expected to reach sizes of a few parsec at an age of $\sim 10^4$–10^5 yr (*Cabrit and Raga*, 2000; *Goodman and Arce*, 2004). Hence, parsec-scale protostellar flows should be a common phenomenon (if not the norm).

In many cases, however, these giant jets have been hard to detect in the optical and NIR as very wide-field images are needed to cover their entire extent. Moreover, jet time variability leads to gas accumulation in knots (working surfaces) with low density between them. Thus giant HH flows do not show a continuous bright emission (unlike, for example, microjets or HH jets within a few 0.1 pc of their powering source, which exhibit closely spaced bright knots arising from shorter modes of variability; see section 3.3). Instead, giant jets appear as a sparse chain of diffuse and fragmented HH or H₂ knots separated by distances of 0.1 pc to 2 pc. Without proper-motion studies, it is sometimes

hard to distinguish between knots from different jets and properly identify their source.

Millimeter CO observations have shown that giant jets can entrain the ambient molecular gas and produce large (>1 pc), massive (a few solar masses or more) bipolar shells of swept-up gas at medium velocity (≈10 km s⁻¹), often referred to as "(giant) molecular outflows" (e.g., *Tafalla and Myers*, 1997; *Arce and Goodman*, 2001, 2002; *Stojimirović et al.*, 2006). These observations show that even when they are too narrow to fully disperse the dense envelope around their source, jets can impact the density and kinematic distribution of their (less dense) parent clump and cloud, out to distances greater than a parsec away from the source.

Recent cloud-wide CO maps, like optical and IR surveys, have shown that molecular outflows can be much larger than previously thought (see Fig. 9), and have helped increase the number of known giant flows (*Stojimirović et al.*, 2007; *Arce et al.*, 2010; *Narayanan et al.*, 2012). For example, in an unbiased search using a cloud-wide CO map of Taurus, *Narayanan et al.* (2012) found that 40% of the 20 detected outflows have sizes larger than 1 pc. Given the difficulty in detecting the entirety of giant outflows (see above), it would not be surprising if most outflows from late class 0 sources (and older) have scales of 1 pc or more.

We note that many giant HH jets extend beyond the confines of their parent molecular cloud. Thus an observed molecular outflow only traces the swept-up gas lying within

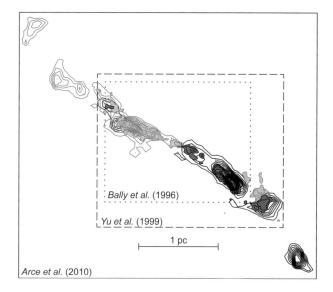

Fig. 9. Our changing view of the size of the giant molecular outflow from the class I source B5-IRS1 (black cross symbol at the center shows the position of the source). The dotted square shows the original extent from *Bally et al.* (1996). The dashed square shows the region mapped by *Yu et al.* (1999). The light (dark) gray lobe represents the blue (red) lobe of the molecular outflow presented by *Yu et al.* (1999). The thick gray (black) contours show the map of the blue (red) lobes from the cloud-scale CO outflow survey of *Arce et al.* (2010).

the molecular cloud [see, e.g., HH 111 (*Cernicharo and Reipurth,* 1996; *Lefloch et al.,* 2007), HH 300 (*Arce and Goodman,* 2001), and HH 46/47 (*van Kempen et al.,* 2009; *Arce et al.,* 2013)]. This implies that giant HH jets most likely drive *atomic hydrogen outflows* in the intercloud medium as well, and may also be a source of turbulence in the low-density (atomic) ISM. Future galactic H I interferometric surveys should help assess the impact of giant outflows on the atomic medium.

4.2. Fundamental Issues in Jet/Outflow Feedback on Clouds

Outflow feedback touches on two critical issues facing modern theories of star formation: the relative inefficiency of star formation, and the origin of turbulence in clouds (*McKee and Ostriker,* 2007; *Elmegreen and Scalo,* 2004). The first issue relates to the fact that observations find surprisingly low values of the SFE in clouds, with typical values ranging from 0.01 to 0.1. Theoretical accounts for the low values of SFE rely on some form of support such as supersonic turbulence within the cloud to keep it from collapsing. But while turbulence can provide an isotropic pressure support, both hydrodynamic and MHD turbulence decay quickly (*Mac Low,* 1999; *Stone et al.,* 1998). Thus turbulent motions must be continually driven, either internally via gravitational contraction and stellar feedback, or externally via turbulence in the general ISM if clouds are long-lived. How this "driving" takes place and (self) regulates the SFE is the second critical issue.

Thus a fundamental question facing studies of both turbulence and SFE is the role of stellar feedback, via both radiation and outflows. The various feedback mechanisms in star formation are reviewed in the chapter by Krumholtz et al. in this volume and in *Vázquez-Semadeni* (2011). Here we focus on *outflow-driven* feedback, and the circumstances under which it could *substantially* change conditions in a star-forming cloud. With respect to turbulence, the question becomes: (1) Do protostellar outflows inject enough momentum to counteract turbulence decay in clouds, and (2) can outflows couple to cloud gas on the correct scales to drive turbulent (rather than organized) motions?

Answering question 1 requires that a steady state can be established between the dissipation rate of the turbulent momentum in a cloud (dP_{turb}/dt) and the momentum injection rate by protostellar outflows (dP_{out}/dt). When such a steady state is achieved the cloud is close to virial equilibrium (Fig. 1). Note that the dissipation rate of turbulent momentum can be written as a function of the cloud mass M_{cl}, its turbulent velocity dispersion V_{vir}, and the dissipation time t_{diss} as (*Nakamura and Li,* 2011a)

$$\frac{dP_{turb}}{dt} = \alpha \frac{M_{cl} V_{vir}}{t_{diss}} \quad (1)$$

where α is a factor close to unity. The outflow momentum injection rate can be written as

$$\frac{dP_{out}}{dt} = \varepsilon_{SFR} \times f_w V_w \quad (2)$$

where ε_{SFR} is the star-formation rate in solar masses per year, f_w is the fraction of stellar mass injected as wind, and V_w is the wind velocity. As we will see, both observations and simulations suggest that on *cluster scales* a balance between these terms can be achieved. Thus the implication is that outflow momentum deposition is sufficient to lead to the observed values of ε_{SFR}.

It is worth noting that essential elements of the problem can be captured via dimensional analysis (*Matzner,* 2007). By considering a cloud of mean density ρ_0, with outflows occurring at a rate per volume S and with momentum I, one can define characteristic outflow scales of mass, length, and time

$$\mathcal{M} = \frac{\rho_0^{4/7} I^{3/7}}{S^{3/7}}, \mathcal{L} = \frac{I^{1/7}}{\rho_0^{1/7} S^{1/7}}, \mathcal{T} = \frac{\rho_0^{3/7}}{I^{3/7} S^{4/7}} \quad (3)$$

Combining these gives other characteristic quantities. Of particular interest is the characteristic velocity

$$\mathcal{V} = \frac{\mathcal{L}}{\mathcal{T}} = \frac{I^{4/7} S^{3/7}}{\rho_0^{4/7}} = \frac{I}{\rho_0 \mathcal{L}^3} \quad (4)$$

Assuming typical values for ρ_0, I, and S in cluster environments yields a supersonic characteristic mach number of $M = \mathcal{V}/c > 1$. This suggests that outflows contain enough momentum to drive supersonic turbulence. Note, however, that these relations are for spherical outflows and an open question relates to how more narrow bipolar outflows will couple to the cluster/cloud gas. Jet wandering may play a key role here. One must also be careful when measuring the typical momentum in outflows I to account for non-emitting molecular gas (due to dissociation) and to very-low-velocity gas representing decelerating fossil cavities just before they are subsummed by background turbulence. We will discuss this issue in section 4.4.

The question of outflow feedback altering the star-forming properties of a cluster/cloud is a more complex issue as there are a number of ways to characterize the problem. Outflows can directly alter SFE by providing turbulent support against gravity as discussed above, or they may help unbind gas from either individual cores or the cluster environment as a whole. In addition, outflows could change the global properties of star formation by starving still-forming higher-mass stars of their reservoirs of gas and therefore shifting the mean stellar mass of a cluster to lower values.

Note that we have been careful to distinguish between feedback on clump/cluster scales and that on scales of the larger parent clouds. Outflows likely represent the lowest rung of a "feedback-ladder," a sequence of ever-more-powerful momentum and energy injection mechanisms that operate as more massive stars form. *Thus collimated outflows are likely to be most effective in driving feedback on cluster rather than full giant molecular cloud (GMC) scales*. We note, however, that magnetic fields may provide more effective coupling between outflows and larger cloud scales (*De Colle and Raga, 2005*).

4.3. Observations of Jet-Cloud Interactions: Momentum Budget and Turbulence-Driving Scale

Numerous attempts have been made to investigate observationally how much feedback protostellar outflows are providing to their surrounding cloud (Fig. 10). We note that in the line of virial analyses, many such studies have focused on comparing the kinetic *energy* in outflows to that in cloud turbulence or gravitational binding; however, since energy is not conserved (because of strong radiative losses) as the outflow sweeps up mass and eventually slows down to merge with the background cloud, it is very important to put more emphasis on the measurement of the outflow *momentum*, a conserved quantity, to reach meaningful conclusions. We will still quote the energy budgets here for completeness, but will focus on the relevant momentum estimates to reach our conclusion.

Graves et al. (2010) present various CO-line maps of the Serpens molecular cloud obtained in the course of the James Clerk Maxwell Telescope (JCMT) Gould Belt Legacy Survey. Because of the complexity of spatially overlapping outflows in this crowded star-forming region, the analysis of outflow properties is based on the blue-/red-shift deviation of the ^{12}CO velocity with respect to C^{18}O. The latter is optically thin across the cloud, does not trace outflows, and thus defines a kind of local rest velocity. After correction for a random inclination distribution of the flows, this study finds the total outflow energy to be approximately 70% of the total turbulent energy of the region. Similar conclusions have come from studies of other regions such as ρ Ophiuchi (*Nakamura et al.,* 2011a), Serpens South (*Nakamura et al.,* 2011b), L1641-N (*Nakamura et al.,* 2012), and NGC 2264C (*Maury et al.,* 2009).

Arce et al. (2010, 2011) analyzed the COordinated Molecular Probe Line Extinction Thermal Emission (COMPLETE) survey CO datasets of the entire Perseus star-forming complex to find new outflows, and wind-driven shells around more evolved class II stars resulting from the interaction of wide-angle winds with the cloud material. This study more than doubled the amount of outflowing mass, momentum, and kinetic energy of protostellar outflows in Perseus. They calculate that the total outflow kinetic energy in the various star-forming regions within the Perseus cloud complex (e.g., B1, B5, IC 348, L1448, NGC 1333) amounts to about 14–80% of the local total turbulent energy, and 4–40% of the total gravitational binding energy in these regions. In the same regions, the total outflow *momentum* is typically 10% of the cloud turbulent momentum (up to 35% in B5). If one takes into account that these outflows most likely have ages of only 0.2 million years (m.y.) or less, while molecular clouds have lifetimes of about 3–6 m.y. (e.g., *Evans et al.,* 2009), it becomes clear that a few generations of outflows will suffice to provide a very significant source of momentum input to each cloud. With the flow timescales and turbulence dissipation timescales estimated by *Arce et al.* (2010), the mean rate of momentum injection by outflows is only a factor of 2.5 less than the rate of turbulent momentum dissipation. We note that a comparable fraction of outflow momentum could be hidden in the form of atomic gas, swept-up and dissociated in shocks faster than 25 km s^{-1} (*Downes and Cabrit,* 2007). Taken together, these observations indicate that directed momentum injection by outflows could significantly contribute to sustaining observed levels of turbulence. Similar detailed studies of other regions are certainly necessary in the future to quantify the impact of jets and outflows on their surrounding clouds.

This raises the question of whether outflows are also able to ultimately disrupt their cloud by dispersing and unbinding cloud material. *Arce et al.* (2010) find that outflows in Perseus currently carry momentum enough to accelerate only 4–23% of the mass of their respective clouds to the local escape velocity. But multiple generations of outflows will again increase their impact. Hence, it is clear that outflows are within range of unbinding some fraction of their parent clusters. A plausible scenario proposed by *Arce et al.* (2010) would be that outflows help disperse a fraction of their surrounding gas, and other mechanisms, such as dispersion by stellar winds and erosion by radiation, help dissipate the rest of the gas that does not end up forming stars.

Observational studies have also been used to estimate the role played by outflows in the injection of cloud turbulence.

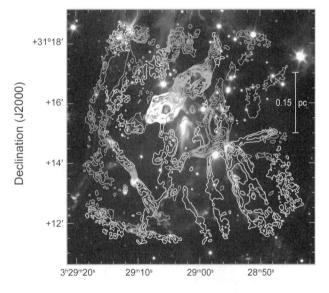

Fig. 10. Map of molecular outflows in the central region of the protostellar cluster NGC 1333, overlaid on a map of the IRAC 4.5-μm emission. White and gray contours show the integrated intensity of the CO(1-0) blue-shifted and red-shifted outflow emission, from the CARMA interferometric observations by *Plunkett et al.* (2013).

These efforts consist of attempts to constrain the scale at which turbulence is driven into molecular clouds. *Brunt et al.* (2009) and *Padoan et al.* (2009) conducted such studies on the NGC 1333 star-forming region using principal component analysis (PCA) in the former case and velocity component analysis (VCS) in the latter. In both cases, analysis of CO line maps from the COMPLETE survey (*Ridge et al., 2006*) were compared with a corresponding analysis on synthesized maps from numerical simulations of clouds with turbulence driven at various scales (in Fourier space). Both find that the observations are only consistent with simulated turbulence driven at large scales on the size of the entire NGC 1333 region. Thus, they come to the conclusion that turbulence should mostly be driven externally, and that outflows — as small-scale driving sources within the molecular cloud — should not play a major role.

This appears to contradict the above observations indicating that outflows are a major source of momentum input, at least at the cluster levels. However, *Arce et al.* (2010) comment that simulations of turbulence in Fourier space, with a necessarily limited range of wave numbers, may cause a difference between flows as they appear in these simulations and turbulence in nature. Simulations of outflow-driven feedback do lend support to this interpretation (*Carroll et al., 2010*), and we come back to this crucial issue in section 4.6.

4.4. Physical Processes in Jet-Cloud Feedback

Propagating jets can, in principle, entrain environmental material through two (not totally unrelated) processes. The first of these is *prompt entrainment,* which is the incorporation of material in shocks such as the working surface at the leading head of a jet, or internal working surfaces produced by a time-dependent ejection (see *Masson and Chernin,* 1993; *Raga and Cabrit,* 1993). A second mechanism is *side entrainment,* which is the incorporation of material through a turbulent mixing layer at the outer edge of the jet beam (*Raga et al.,* 1993). While both processes are likely to shape the interaction of jets with their environments, prompt entrainment is likely to be more important for feedback on cluster and cloud scales since it will often be the fossil swept-up shells (bounded by shocks) that couple outflow momenta to the cloud.

The jet-to-cloud momentum transfer efficiency varies inversely with the jet-to-cloud density ratio (*Masson and Chernin,* 1993). An overdense jet will "punch" fast through the cloud without depositing much momentum into the swept-up shell. The efficiency will increase when a jet impacts a *denser* region of the molecular cloud (e.g., a molecular cloud core). In such an interaction, the jet will initially be deflected along the surface of the dense core, but at later times the jet will slowly burrow a hole into the core (*Raga and Murdin,* 2002). During this burrowing process, most of the momentum of the jet is transferred to the cloud core material.

Efficient momentum deposition also occurs if the jet ejection direction is time-dependent (due to precession of

the jet axis or orbital motion of the source; see section 3) (*Raga et al.,* 2009). This will be particularly true when a variable jet direction is combined with a variable ejection velocity modulus (Fig. 11). These effects break the jet into a series of "bullets" traveling in different directions (*Raga and Biro,* 1993). Alternatively, the ejection itself might be in the form of discrete "plasmoids" ejected along different paths (*Yirak et al.,* 2008). These bullets differ from the leading head of a well-aligned jet in that they are not resupplied by material ejected at later times. Therefore, they slow down as they move through the molecular cloud due to ram pressure braking, as seen in the giant HH 34 jet complex (*Cabrit and Raga,* 2000; *Masciadri et al.,* 2002). Most of the jet momentum could then be deposited within the molecular cloud, instead of escaping into the atomic ISM.

When such "bullets" eventually become sub-Alfvénic and/or subsonic (depending on the cloud magnetization), their momentum can be efficiently converted into MHD wave-like motions of the molecular cloud (*De Colle and Raga,* 2005). This is true for any form of decaying jet flow (mass loss decreasing over time). Thus "fossil" swept-up shells that expand and slow down after the brief class 0 phase should be the main agents coupling outflow momentum to cloud turbulence. Numerous such "fossil cavities" have been discovered in regions such as NGC 1333 (*Quillen et al.,* 2005) and across the Perseus cloud (*Arce et al.,* 2010; *Arce,* 2011). Observationally derived scaling properties for momentum injection in such flows (*Quillen et al.,* 2005) have been recovered in simulations of shells driven by "decaying" outflows (*Cunningham et al.,* 2009a).

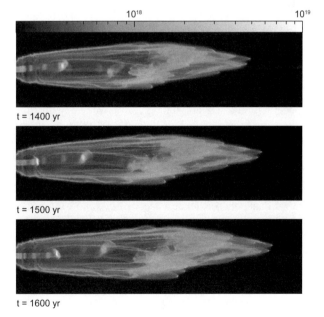

Fig. 11. Column-density time-sequence computed from a model of a jet with a variable ejection velocity and a precession of the outflow axis. The initial jet radius is resolved with 10 grid points at the highest resolution of a 5-level adaptive grid.

The direct turbulent driving and/or coupling of *individual* outflows to the cloud has been investigated numerically by a number of authors. While *Banerjee et al.* (2007) showed that a single active outflow (i.e., a class 0 source) in a quiescent medium would not drive turbulent motions, *Cunningham et al.* (2009a) demonstrated that a fossil outflow in an already turbulent cloud will fragment and reenergize those turbulent motions. This speaks to the complex issue of "detrainment" (i.e., the eventual merging of material from the outflow into the surrounding environment). While it is clear that jets can reenergize turbulence, the end states of detrainment remain an important issue needing resolution in order to calculate the full "feedback" of the momentum provided by jets into the turbulent motions of the placental molecular cloud.

Most importantly, large-scale simulations (discussed in the next section) show that interactions (collisions) between *multiple* outflows on scale $\mathcal{L}$ (from equation (3)) may be the principle mechanism for converting directed outflow momentum into random turbulent motions. Conversely, the role of cloud turbulence in altering outflow properties was explored in *Offner et al.* (2011). In that study, turbulent motions associated with collapse produced asymmetries between the red and blue swept-up outflow lobes. This study also showed that some caution must be used in converting observations of outflows into measurements of injected momentum, as low-velocity outflow material can be misidentified as belonging to turbulent cores.

4.5. Large-Scale Simulations of Outflow Feedback

Analytic models such as those of *Matzner and McKee* (2000) and *Matzner* (2007) have articulated basic features of outflow-driven feedback such as the scaling laws discussed in section 4.3. The inherently three-dimensional and time-dependent nature of outflow feedback, however, requires study through detailed numerical simulations (see also the review by *Vázquez-Semadeni,* 2011).

The ability of multiple outflows to generate turbulent support within a self-gravitating clump was first addressed in the simulations of *Li and Nakamura* (2006) and *Nakamura and Li* (2007), which started with a centrally condensed turbulent clump containing many Jeans masses. The initial turbulence generated overdense regions that quickly became Jeans unstable, initiating local regions of gravitational collapse. Once a density threshold was crossed, these collapsing regions were identified as protostars. Mass and momentum were then driven back into the grid in the form of outflows.

The most important conclusion of these studies was that once star formation and its outflows commenced, the young cluster achieved a dynamic equilibrium between momentum input and turbulent dissipation. It is noteworthy that these studies found bipolar outflows more effective than spherical winds for turbulent support, as the former could propagate across longer distances. Star-formation efficiencies of just a few percent were achieved in simulations with outflow feedback.

Another key point to emerge from simulations of outflow feedback is the nature of the turbulence it produces. In *Nakamura and Li* (2007), a break in the velocity power spectra $E(k)$ was identified, below which (i.e., longer scale-lengths) the spectrum flattened. This issue was addressed again in *Carroll et al.* (2009a), who ran simulations of the interaction of randomly oriented interacting bipolar outflows. In this work, the outflow momentum injection rate was made time-dependent to explore the role of fossil shells in coupling to the cloud turbulence. *Carroll et al.* (2009b) also found a well-defined "knee" in the spectrum at $\mathcal{K} \propto 1/\mathcal{L}$, the interaction scale defined via dimensional analysis (see equation (3)). Thus, the collision of fossil outflow cavities and the subsequent randomization of directed momenta were responsible for generating the observed turbulence.

Carroll et al. (2009b) also found that outflow-driven turbulence produced a power spectrum that steepened above the knee as $E(k) \propto k^{-3}$ (see Fig. 12). In contrast, standard turbulence simulations using forcing in Fourier space typically find "Burger's" values of k^{-2}. The steeper slope was caused by outflow shells sweeping up eddies with wavenumbers higher than $\mathcal{K}$. The presence of both a knee and a steeper slope in the spectrum of outflow-driven turbulence offers the possibility for observation of these, and perhaps other, signatures of outflow feedback. Note that changes to the turbulent spectra via outflows remained even in the presence of driving at scales larger than $1/\mathcal{K}$ (*Carroll et*

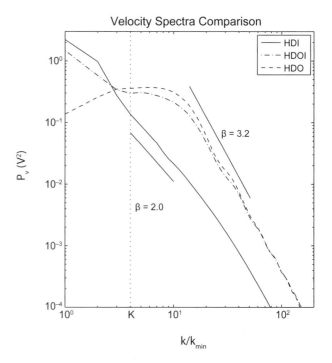

Fig. 12. Velocity power spectrum for runs with pure Fourier driving (HDI, solid line), Fourier + outflow driving (HDOI, dash-dotted line), and pure outflow driving (HDO, dashed line) from simulations by *Carroll et al.* (2010). The vertical dashed line corresponds to the outflow interaction wave number $\mathcal{K} = 1/\mathcal{L}$. Dotted lines show $k^{-\beta}$ for $\beta = 2$ and $\beta = 3.2$.

al., 2010). Modifications of density probability distribution functions (PDFs) of the ambient medium via outflow-driven turbulence were also reported in *Moraghan et al.* (2013).

Principal component analysis methods applied to data cubes from simulations of outflow-driven turbulence demonstrate that the discreet, small-scale sources can artificially appear overwhelmed by larger-scale flows, even if those flows have far less power (*Carroll et al.*, 2010). From these results it is likely that the issue of observational determination of the correct driving scale(s) of turbulence remains an open question. Note that the issue is not just the largest scales at which driving occurs, but which process dominates on the scales where star formation occurs. Thus even if turbulence cascades down from GMC scales, outflow feedback on cluster scales may still be important in determining local SFEs and related properties.

Because magnetic fields are closely tied to the origin of protostellar outflows, exploring the combined role of magnetic field and outflows feedback has been an important issue. Using adaptive mesh refinement (AMR) methods, *Wang et al.* (2010) began with a turbulent, moderately condensed clump of ~1600 $M_\odot$ and found that in the absence of regulation by magnetic fields and outflow feedback, massive stars would readily form within a cluster of hundreds of lower-mass stars. These simulations showed that the massive stars were fed by material infalling from large scales (i.e., clump-fed rather than core-fed accretion). The importance of large-scale accretion modes made high-mass star formation particularly susceptible to disruption by outflows. Once mass loss was initiated by lower-mass stars, their outflows eroded the dense filaments feeding massive star formation. In addition, at later times the induced turbulent motions of interacting outflows slowed down the global collapse modes that had continued to fuel the young massive stars. Thus *Wang et al.* (2010) found global accretion rates were reduced, leading to fewer high-mass stars by the end of the simulations. *Wang et al.* (2010) and *Nakamura and Li* (2011a) also found strong links between outflow feedback and magnetic fields. Even an initially weak field could retard star formation as the field was amplified to equipartition strength by the outflow-driven turbulence, with the "turbulent" field component dominating the uniform one.

Using AMR methods, *Hansen et al.* (2012) studied low-mass star formation in the presence of outflow *and* radiation feedback. These simulations found that outflows reduce protostellar masses and accretion rates by a factor of 3 each. In this way, outflows also led to a reduction in protostellar luminosities by an order of magnitude. This reduced the radiation feedback and enhanced fragmentation. In contrast with previous results, *Hansen et al.* (2012) found that the outflows did not change the global dynamics of the cloud because they were narrow and did not couple well to the dense gas. *Krumholz et al.* (2012) studied the role of outflow (and radiation) feedback in high-mass star-forming regions. Their results also indicated a smaller impact from outflows. Note that both these simulations did not include magnetic fields.

Finally, we note that almost all simulations of outflow feedback rely on common parameterizations of the individual outflows. In particular, the total outflow momentum is expressed as $P_0 = f_w V_w M_*$, making the combination $f_w V_w$ the outflow momentum per unit stellar mass. In their analytic description of outflow feedback and star formation, *Matzner and McKee* (2000) assumed a value of $f_w V_w = 40$ km s^{-1}. A review of the literature yields an observational range for this parameter of $10 < f_w V_w < 25$ km s^{-1} (*Henriksen et al.*, 1997; *Bontemps et al.*, 1996; *Richer et al.*, 2000; *Plunkett et al.*, 2013). Assuming $V_w = 100$ km s^{-1} yields $0.01 < f_w < 0.25$. The presence of a wide-angle wind relative to a collimated flow component is another key parameter used in simulations, sometimes expressed as the ratio of momentum in a fully collimated component to a spherical one, $\varepsilon = P_c/P_s$. To the extent that observations provide a guide for this parameter, it would appear that $\varepsilon \gg 1$ is favored, since wide-angle winds do not appear to carry as much momentum as the jets (at least in the class 0 phase, where most of the momentum is injected). Finally, we note that care should be taken in how "outflow" momentum is added to the grid in feedback simulations. Given that the large speeds associated with the winds can slow down simulations [via Courant-Friedrich-Levy (CFL) conditions], momentum is sometimes added via lower-speed higher-mass flows, or given directly to ambient material in the vicinity of the source. Further work should be done to test the effect these assumptions have on turbulence injection and outflow feedback on star formation.

5. EXPLOSIVE OUTFLOWS

Finally, we note another class of mass loss may play an important role in delivering momentum back into the parent cloud, i.e., explosive though nonterminal (i.e., non-supernova) outflows. An archetype of this phenomena is the Becklin-Neugebauer/Kleinmann-Low (BN/KL) region in Orion, which produced a powerful (~10^{47-48} erg) wide-angle explosion approximately 500–1000 yr ago (*Allen et al.*, 1993; *Doi et al.*, 2002; *Bally et al.*, 2011; *Goddi et al.*, 2011). The origin of the outflow appears to lay in a nonhierarchical multiple star system that experienced a dynamical interaction, leading to the ejection of two members and the formation of a tight binary or possibly a merger (*Rodríguez et al.*, 2005; *Bally and Zinnecker*, 2005; *Gómez et al.*, 2005, 2008; *Zapata et al.*, 2009; *Bally*, 2011). Proposed scenarios for powering the outflow involve the release of energy from envelope orbital motions, gravitational binding of the tight pair, or magnetic shear. The rapid release of energy leads to the fastest ejecta emerging from deep within the gravitational potential of the decaying cluster. Rayleigh-Taylor instabilities are then triggered as this material plows through a slower-moving, previously ejected envelope. The fragmented ejecta that are created will be effective at driving turbulence in their surroundings, like the "bullets" in precessing jets.

If such a mechanism operates in other massive star-forming regions, it may be an important source of outflow feedback. Spitzer detected at 4.5 μm a wide-angle outflow similar to BN/KL in the 10^6 $L_\odot$ hot core G34.25+0.16 [located at 5 kpc in the inner Galaxy (*Cyganowski et al.,* 2008)]. Source G in W49, the most luminous water-maser outflow in the Milky Way, may be yet another example (*Smith et al.,* 2009). Finally, *Sahai et al.* (2008) found evidence for interstellar bullets having a similar structure to the BN/KL "fingers" in the outflow from the massive young protostar IRAS 05506+2414.

6. CONCLUSIONS AND FUTURE DIRECTIONS

In this chapter we have attempted to demonstrate that protostellar jets and outflows are not only visually beautiful and important on their own as examples of astrophysical magnetofluid dynamical processes, but they are also an essential player in the assembly of stars across a remarkable range of size scales. We find that issues of feedback from jets/outflows back to the star-formation process are apparent on three scales: those associated with planet formation, those associated with the natal core, and those associated with clustered star formation.

On scales associated with planet-forming disks, jets can impact planet assembly through disk irradiation/shielding and MHD effects associated with outflow launching. These processes can alter disk properties in those regions where planets will be forming. Future work should focus on articulating the feedback between jet driving and the disk mechanisms associated with creating planets.

On the scales of the natal cores, jets and outflows seem capable of explaining the low observed 30% core-to-star efficiency, through a combination of powerful MHD ejection during the earliest collapse phase, and envelope clearing by wide-angle winds during the later phases.

On scales associated with clusters (or perhaps clouds), multiple jets/outflows can drive turbulence, alter SFEs, and affect the mean stellar mass. It is also possible that outflows may help unbind cluster gas. Future work should focus on providing simulations with the best parameterizations of jet properties (such as momentum injection history and distributions), characterization of physical mechanisms for feedback, and exploration of feedback observational signatures. The full range of scales, in which both collimated and uncollimated outflows from young stars impact the star-formation process, must also be articulated. Observational signatures of such feedback must also be explored.

With regard to the physics of jets themselves, new results make it clear that they are collimated magnetically on inner disk scales, and include multiple thermal and chemical components surrounded by a *slower,* wide-angle wind. The emerging picture is that of a powerful MHD disk wind, collimating the inner stellar wind and magnetospheric ejections responsible for braking the star. Detailed analysis and modeling is needed to confirm this picture and articulate the properties of these different components as a function of source age.

We have also shown how HEDLA experiments have already contributed new and fundamental insights into the hydrodynamic and MHD evolution of jets. We expect HEDLA studies to grow beyond jet research in the future as they hold the promise of touching on many issues relevant to star and planet formation (i.e., cometary globules and hot Jupiters).

Observationally we expect new platforms to hold great promise for jet and outflow studies. In particular, ALMA and NIR IFUs should prove crucial to resolving jet-rotation profiles, shocks, and chemical stratification in statistically relevant jet samples, and to better understand their interaction with the surrounding envelope. Such data will provide definitive tests of disk-wind models. Near-infrared interferometry of CTTS (e.g., with GRAVITY on VLTI) promises to be a powerful test of atomic jet models. Synchrotron studies with the Expanded Very Large Array (EVLA) and the Low Frequency Array (LOFAR) should allow jet magnetic fields to finally come into view. Finally, long baseline monitoring of the short quasiperiodic knot modulation in jets (~3–15 yr) should allow clarification of the origin of these features and their link with stellar and disk physics (magnetic cycles, accretion outbursts) and source binarity.

Acknowledgments. A.F. acknowledges support from the National Science Foundation (NSF), Department of Energy (DOE), and National Aeronautics and Space Administration (NASA). T.P.R. acknowledges support from Science Foundation Ireland under grant 11/RFP/AST3331. S.C. acknowledges support from Centre National de la Recherche Scientifique (CNRS) and Centre National d'Etudes Spatiales (CNES) under the PCMI program. H.G.A. acknowledges support from NSF CAREER award AST-0845619.

REFERENCES

Agra-Amboage V. et al. (2009) *Astron. Astrophys., 493,* 1029–1041.
Agra-Amboage V. et al. (2011) *Astron. Astrophys., 532,* A59.
Ainsworth R. E. et al. (2013) *Mon. Not. R. Astron. Soc., 436,* L64–L68.
Allen D. A. et al. (1993) *Publ. Astron. Soc. Australia, 10,* 298.
Anderson J. M. et al. (2003) *Astrophys. J. Lett., 590,* L107–L110.
Antoniucci S. et al. (2008) *Astron. Astrophys., 479,* 503–514.
Arce H. G. (2011) *Comput. Star Form., 270,* 287–290.
Arce H. G. and Goodman A. A. (2001) *Astrophys. J., 554,* 132–151.
Arce H. G. and Goodman A. A. (2002) *Astrophys. J., 575,* 911–927.
Arce H. G. and Sargent A. I. (2006) *Astrophys. J., 646,* 1070–1085.
Arce H. G. et al. (2007) In *Protostars and Planets V* (B. Reipurth et al., eds.), pp. 245–260. Univ. of Arizona, Tucson.
Arce H. G. et al. (2008) *Astrophys. J. Lett., 681,* L21.
Arce H. G. et al. (2010) *Astrophys. J., 715,* 1170–1190.
Arce H. G. et al. (2011) *Astrophys. J., 742,* 105.
Arce H. G. et al. (2013) *Astrophys. J., 774,* A39.
Bacciotti F. et al. (2002) *Astrophys. J., 576,* 222–231.
Bacciotti F. et al. (2011) *Astrophys. J. Lett., 737,* L26.
Bai X.-N. and Stone J. M. (2013) *Astrophys. J., 769,* 76.
Balbus S. A. and Hawley J. F. (2006) *Astrophys. J., 652,* 1020–1027.
Bally J. (2011) *Comput. Star Form., 270,* 247–254.
Bally J. and Zinnecker H. (2005) *Astron. J., 129,* 2281–2293.
Bally J. et al. (1996) *Astrophys. J., 473,* 921.
Bally J. et al. (2003) *Astrophys. J., 584,* 843–852.
Bally J. et al. (2007) In *Protostars and Planets V* (B. Reipurth et al., eds.), pp. 215–230. Univ. of Arizona, Tucson.
Bally J. et al. (2011) *Astrophys. J., 727,* 113.
Bally J. et al. (2012) *Astron. J., 144,* 143.
Banerjee R. et al. (2007) *Astrophys. J., 668,* 1028–1041.
Bans A. and Königl A. (2012) *Astrophys. J., 758,* 100.
Beck T. L. et al. (2007) *Astron. J., 133,* 1221.
Beck T. L. et al. (2008) *Astrophys. J., 676,* 472–489.
Benedettini M. et al. (2012) *Astron. Astrophys., 539,* L3.

Benisty M. et al. (2010) *Astron. Astrophys., 517*, L3.

Benisty M. et al. (2013) *Astron. Astrophys., 555*, A113.

Blandford R. D. and Payne D. G. (1982) *Mon. Not. R. Astron. Soc., 199*, 883–903.

Bonito R. et al. (2010) *Astron. Astrophys., 517*, A68.

Bonito R. et al. (2011) *Astrophys. J., 737*, 54.

Bontemps S. et al. (1996) *Astron. Astrophys., 311*, 858–872.

Bouvier J. et al. (2007) In *Protostars and Planets V* (B. Reipurth et al., eds.), pp. 479–494. Univ. of Arizona, Tucson.

Brugel E. W. et al. (1981) *Astrophys. J. Suppl., 47*, 117–138.

Brunt C. M. et al. (2009) *Astron. Astrophys., 504*, 883–890.

Cabrit S. (2002) In *Gaia: A European Space Project* (O. Bienaymé et al., eds.), pp. 147–182. EAS Publications Series, Vol. 3, Cambridge Univ., Cambridge.

Cabrit S. (2007) In *Star-Disk Interaction in Young Stars* (J. Bouvier et al., eds.), pp. 203–214. IAU Symp. 243, Cambridge Univ., Cambridge.

Cabrit S. (2009) In *Protostellar Jets in Context* (K. Tsinganos et al., eds.), pp. 247–257. Springer-Verlag, Berlin.

Cabrit S. and Raga A. (2000) *Astron. Astrophys., 354*, 667–673.

Cabrit S. et al. (2006) *Astron. Astrophys., 452*, 897–906.

Cabrit S. et al. (2007) *Astron. Astrophys., 468*, L29–L32.

Cabrit S. et al. (2012) *Astron. Astrophys., 548*, L2.

Calvet N. et al. (2004) *Astron. J., 128*, 1294–1318.

Camenzind M. (1990) *Rev. Mod. Astron., 3*, 234–265.

Caratti o Garatti A. et al. (2012) *Astron. Astrophys., 538*, A64.

Carrasco-González C. et al. (2010) *Science, 330*, 1209.

Carroll J. J. et al. (2009a) *Astrophys. J., 695*, 1376–1381.

Carroll J. et al. (2009b) *Am. Astron. Soc. Meeting Abstracts #214*, #425.06.

Carroll J. J. et al. (2010) *Astrophys. J., 722*, 145–157.

Čemeljić M. et al. (2013) *Astrophys. J., 768*, 5.

Cernicharo J. and Reipurth B. (1996) *Astrophys. J. Lett., 460*, L57.

Cerqueira A. H. et al. (2006) *Astron. Astrophys., 448*, 231–241.

Choi M. et al. (2011) *Astrophys. J. Lett., 728*, L34.

Chrysostomou A. et al. (2008) *Astron. Astrophys., 482*, 575–583.

Ciardi A. and Hennebelle P. (2010) *Mon. Not. R. Astron. Soc., 409*, L39–L43.

Ciardi A. et al. (2007) *Phys. Plasmas, 14*, 056501.

Ciardi A. et al. (2008) *Astrophys. J., 678*, 968–973.

Ciardi A. et al. (2009) *Astrophys. J. Lett., 691*, L147–L150.

Codella C. et al. (2007) *Astron. Astrophys., 462*, L53–L56.

Codella C. et al. (2012) *Astrophys. J., 757*, L9.

Codella C. et al. (2013) *Astron. Astrophys., 550*, A81.

Coffey D. et al. (2004) *Astron. J., 604*, 758–765.

Coffey D. et al. (2007) *Astrophys. J., 663*, 350–364.

Coffey D. et al. (2008) *Astrophys. J., 689*, 1112–1126.

Coffey D. et al. (2011) *Astron. Astrophys., 526*, A40.

Coffey D. et al. (2012) *Astrophys. J., 749*, 139.

Costigan G. et al. (2012) *Mon. Not. R. Astron. Soc., 427*, 1344–1362.

Cunningham A. J. et al. (2009a) *Astrophys. J., 692*, 816–826.

Cunningham N. J. et al. (2009b) *Astrophys. J., 692*, 943–954.

Curtis E. I. et al. (2010) *Mon. Not. R. Astron. Soc., 408*, 1516–1539.

Cyganowski C. J. et al. 2008, *Astron. J., 136*, 2391.

Davis C. J. et al. (2000) *Mon. Not. R. Astron. Soc., 314*, 241–255.

Davis C. J. et al. (2008) *Mon. Not. R. Astron. Soc., 387*, 954–968.

Davis C. J. et al. (2009) *Astron. Astrophys., 496*, 153–176.

Davis C. J. et al. (2011) *Astron. Astrophys., 528*, A3.

De Colle F. and Raga A. C. (2005) *Mon. Not. R. Astron. Soc., 359*, 164–170.

Devine D. et al. (1997) *Astron. J., 114*, 2095.

Dionatos O. et al. (2010) *Astron. Astrophys., 521*, A7.

Dionatos O. et al. (2009) *Astrophys. J., 692*, 1–11.

Doi T. et al. (2002) *Astron. J., 124*, 445–463.

Dougados C. et al. (2010) In *Jets from Young Stars IV: From Models to Observations and Experiments* (P. J. V. Garcia and J. M. Ferreira, eds.), p. 213. Lecture Notes in Physics Vol. 793, Springer-Verlag, Berlin.

Downes T. P. and Cabrit S. (2007) *Astron. Astrophys., 471*, 873–884.

Eislöffel J. (2000) *Astron. Astrophys., 354*, 236–246.

Eislöffel J. and Mundt R. (1997) *Astron. J., 114*, 280–287.

Eislöffel J. et al. (1996) *Astron. J., 112*, 2086.

Eisner J. A. et al. (2010) *Astrophys. J., 718*, 774–794.

Ellerbroek L. E. et al. (2013) *Astron. Astrophys., 551*, A5.

Elmegreen B. G. and Scalo J. (2004) *Annu. Rev. Astron. Astrophys., 42*, 211–273.

Estallela R. et al. (2012) *Astron. J., 144*, 61.

Evans N. J. II et al. (2009) *Astrophys. J. Suppl., 181*, 321–350.

Favata F. et al. (2002) *Astron. Astrophys., 386*, 204–210.

Fendt C. (2009) *Astrophys. J., 692*, 346–363.

Fendt C. (2011) *Astrophys. J., 737*, 43.

Fendt C. and Sheikhnezami S. (2013) *Astrophys. J., 774*, 12.

Ferreira J. (1997) *Astron. Astrophys., 319*, 340–359.

Ferreira J. et al. (2000) *Mon. Not. R. Astron. Soc., 312*, 387–397.

Ferreira J. et al. (2006) *Astron. Astrophys., 453*, 785–796.

Flower D. R. and Pineau des Forêts G. (2013) *Mon. Not. R. Astron. Soc., 436(3)*, 2143–2150.

Foster J. M. (2010) *Am. Astron. Soc. Meeting Abstracts #216*, #205.02.

Garcia P. J. V. et al. (2013) *Mon. Not. R. Astron. Soc., 430*, 1839–1853.

Garcia Lopez R. et al. (2008) *Astron. Astrophys., 487*, 1019–1031.

Giannini T. et al. (2008) *Astron. Astrophys., 481*, 123–139.

Giannini T. et al. (2011) *Astrophys. J., 738*, 80.

Goddi C. et al. (2011) *Astrophys. J. Lett., 739*, L13.

Gómez L. et al. (2005) *Astrophys. J., 635*, 1166–1172.

Gómez L. et al. (2008) *Astrophys. J., 685*, 333–343.

Goodman A. A. and Arce H. G. (2004) *Astrophys. J., 608*, 831–845.

Graves S. F. et al. (2010) *Mon. Not. R. Astron. Soc., 409*, 1412–1428.

Güdel M. et al. (2007) *Astron. Astrophys., 468*, 515–528.

Güdel M. et al. (2008) *Astron. Astrophys., 478*, 797–807.

Güdel M. et al. (2009) In *Protostellar Jets in Context* (K. Tsinganos et al., eds.), pp. 347–352. Springer-Verlag, Berlin.

Gueth F. et al. (1996) *Astron. Astrophys., 307*, 891.

Guillet V. et al. (2009) *Astron. Astrophys., 497*, 145.

Gullbring E. et al. (2000) *Astrophys. J., 544*, 927–932.

Günther H. M. et al. (2009) *Astron. Astrophys., 493*, 579–584.

Gustafsson M. et al. (2010) *Astron. Astrophys., 513*, A5.

Hansen J. F. et al. (2011) *Phys. Plasmas, 18*, 082702.

Hansen C. E. et al. (2012) *Astrophys. J., 747*, 22.

Hartigan P. and Hillenbrand L. (2009) *Astrophys. J., 705*, 1388–1394.

Hartigan P. and Morse J. (2007) *Astrophys. J., 660*, 426–440.

Hartigan P. et al. (2005) *Astron. J., 130*, 2197–2205.

Hartigan P. et al. (2007) *Astrophys. J., 661*, 910–918.

Hartigan P. et al. (2009) *Astrophys. J., 705*, 1073–1094.

Hartigan P. et al. (2011) *Astrophys. J., 736*, 29.

Hatchell J. et al. (2007) *Astron. Astrophys., 472*, 187–198.

Hayashi M. et al. (1996) *Astrophys. J., 468*, L37.

Henriksen R. et al. (1997) *Astron. Astrophys., 323*, 549.

Herczeg G. J. et al. (2006) *Astrophys. J. Suppl., 165*, 256.

Herczeg G. J. et al. (2011) *Astron. Astrophys., 533*, A112.

Hirano N. et al. (2010) *Astrophys. J., 717*, 58–73.

Hirth G. A. et al. (1994) *Astrophys. J. Lett., 427*, L99–L102.

Huarte-Espinosa M. et al. (2012) *Astrophys. J., 757*, 66.

Ioannidis G. and Froebrich D. (2012) *Mon. Not. R. Astron. Soc., 425*, 1380–1393.

Kastner J. H. et al. (2005) *Astrophys. J. Suppl., 160*, 511–529.

Kraus S. et al. (2008) *Astron. Astrophys., 489*, 1157–1173.

Kristensen L. E. et al. (2008) *Astron. Astrophys., 477*, 203–211.

Kristensen L. E. et al. (2011) *Astron. Astrophys., 531*, L1.

Kristensen L. E. et al. (2012) *Astron. Astrophys., 542*, A8.

Kristensen L. E. et al. (2013a) *Astron. Astrophys., 549*, L6.

Kristensen L. E. et al. (2013b) *Astron. Astrophys., 557*, A23.

Krumholz M. R. et al. (2012) *Astrophys. J., 754*, 71.

Kwan J. and Tademaru E. (1988) *Astrophys. J., 332*, L41.

Launhardt R. et al. (2009) *Astron. Astrophys., 494*, 147–156.

Lavalley-Fouquet C. et al. (2000) *Astron. Astrophys., 356*, L41–L44.

Lebedev S. V. et al. (2004) *Astrophys. J., 616*, 988–997.

Lebedev S. V. et al. (2005a) *Mon. Not. R. Astron. Soc., 361*, 97–108.

Lebedev S. V. et al. (2005b) *Plasma Phys. Controlled Fusion, 47*, 465–479.

Lee C.-F. et al. (2007) *Astrophys. J., 670*, 1188–1197.

Lee C.-F. et al. (2008) *Astrophys. J., 685*, 1026–1032.

Lee C.-F. et al. (2009) *Astrophys. J., 699*, 1584–1594.

Lee C.-F. et al. (2010) *Astrophys. J., 713*, 731–737.

Lefloch B. et al. (2007) *Astrophys. J., 658*, 498–508.

Lefloch B. et al. (2012) *Astrophys. J. Lett., 757*, L25.

Lesur G. et al. (2013) *Astron. Astrophys., 550*, A61.

Li Z.-Y. and Nakamura F. (2006) *Astrophys. J. Lett., 640*, L187–L190.

Lim A. J. and Steffen W. (2001) *Mon. Not. R. Astron. Soc., 322*, 166–176.

Loinard L. et al. (2013) *Mon. Not. R. Astron. Soc., 430*, L10–L14.

Lynch C. et al. (2013) *Astrophys. J., 766*, 53.

Lynden-Bell D. (1996) *Mon. Not. R. Astron. Soc., 279*, 389–401.

Lynden-Bell D. (2006)*Mon. Not. R. Astron. Soc., 369,* 1167–1188.
Machida M. N. and Hosokawa T. (2013) *Mon. Not. R. Astron. Soc., 431,* 1719–1744.
Mac Low M.-M. (1999) *Astrophys. J., 524,* 169–178.
Mader S. L. et al. (1999) *Mon. Not. R. Astron. Soc., 310,* 331–354.
Masciadri E. et al. (2002) *Astrophys. J., 580,* 950–958.
Masson C. R. and Chernin L. M. (1993) *Astrophys. J., 414,* 230–241.
Matt S. and Pudritz R. E. (2005) *Astrophys. J. Lett., 632,* L135–L138.
Matzner C. D. (2007) *Astrophys. J., 659,* 1394–1403.
Matzner C. D. and McKee C. F. (2000) *Astrophys. J., 545,* 364.
Maury A. J. et al. (2009) *Astron. Astrophys., 499,* 175–189.
McGroarty F. and Ray T. P. (2004) *Astron. Astrophys., 420,* 975–986.
McGroarty F. et al. (2004) *Astron. Astrophys., 415,* 189–201.
McKee C. F. and Ostriker E. C. (2007) *Annu. Rev. Astron. Astrophys., 45,* 565–687.
Meliani Z. et al. (2006) *Astron. Astrophys., 460,* 1.
Melnikov S. Y. et al. (2009) *Astron. Astrophys., 506,* 763–777.
Millan-Gabet R. et al. (2007) In *Protostars and Planets V* (B. Reipurth et al., eds.), pp. 539–554. Univ. of Arizona, Tucson.
Moraghan A. et al. (2013) *Mon. Not. R. Astron. Soc., 432,* L80.
Morse J. A. et al. (1992) *Astrophys. J., 399,* 231–245.
Morse J. A. et al. (1993) *Astrophys. J., 410,* 764.
Movsessian T. A. et al. (2007) *Astron. Astrophys., 470,* 605–614.
Myers P. C. (2008) *Astrophys. J., 687,* 340.
Nakamura F. and Li Z.-Y. (2007) *Astrophys. J., 662,* 395–412.
Nakamura F. and Li Z.-Y. (2011a) *Astrophys. J., 740,* 36.
Nakamura F. and Li Z.-Y. (2011b) *Comput. Star Form., 270,* 115–122.
Nakamura F. et al. (2011a) *Astrophys. J., 726,* 46.
Nakamura F. et al. (2011b) *Astrophys. J., 737,* 56.
Nakamura F. et al. (2012) *Astrophys. J., 746,* 25.
Narayanan G. et al. (2012)*Mon. Not. R. Astron. Soc., 425,* 2641–2667.
Natta A. et al. (2006) *Astron. Astrophys., 452,* 245–252.
Neufeld D. A. et al. (2006) *Astrophys. J., 649,* 816.
Neufeld D. A. et al. (2009) *Astrophys. J., 706,* 170–183.
Nicolaï P. et al. (2008) *Phys. Plasmas, 15,* 082701.
Nisini B. et al. (2005) *Astron. Astrophys., 441,* 159–170.
Nisini B. et al. (2007) *Astron. Astrophys., 462,* 163–172.
Nisini B. et al. (2010a) *Astron. Astrophys., 518,* L120.
Nisini B. et al. (2010b) *Astrophys. J., 724,* 69–79.
Nisini B. et al. (2013) *Astron. Astrophys., 549,* A16.
Noriega-Crespo A. et al. (2011) *Astrophys. J. Lett., 732,* L16.
Offner S. S. R. et al. (2011) *Astrophys. J., 743,* 91.
Padoan P. et al. (2009) *Astrophys. J. Lett., 707,* L153–L157.
Panoglou D. et al. (2012) *Astron. Astrophys., 538,* A2.
Pesenti N. et al. (2004) *Astron. Astrophys., 416,* L9.
Plunkett A. L. et al. (2013) *Astrophys. J., 774,* 22.
Podio L. et al. (2006) *Astron. Astrophys., 456,* 189–204.
Podio L. et al. (2009) *Astron. Astrophys., 506,* 779–788.
Podio L. et al. (2011) *Astron. Astrophys., 527,* A13.
Pravdo S. H. and Tsuboi Y. (2005) *Astrophys. J., 626,* 272–282.
Pravdo S. H. et al. (2004) *Astrophys. J., 605,* 259–271.
Pudritz R. E. and Norman C. A. (1983) *Astrophys. J., 274,* 677–697.
Pudritz R. E. et al. (2007) In *Protostars and Planets V* (B. Reipurth et al., eds.), pp. 277–294. Univ. of Arizona, Tucson.
Pyo T.-S. et al. (2009) *Astrophys. J., 694,* 654.
Quillen A. C. et al. (2005) *Astrophys. J., 632,* 941–955.
Raga A. C. and Biro S. (1993) *Mon. Not. R. Astron. Soc., 264,* 758.
Raga A. and Cabrit S. (1993) *Astron. Astrophys., 278,* 267–278.
Raga A. and Murdin P. (2002) *Encyclopedia of Astronomy and Astrophysics,* Cambridge Univ., Cambridge.
Raga A. C. et al. (1990) *Astrophys. J., 364,* 601–610.
Raga A. C. et al. (1993) *Astron. Astrophys., 276,* 539.
Raga A. C. et al. (2002a) *Astrophys. J., 395,* 647–656.
Raga A. C. et al. (2002b) *Astrophys. J. Lett., 565,* L29.
Raga A. C. et al. (2002c) *Astrophys. J. Lett., 576,* L149.
Raga A. C. et al. (2009) *Astrophys. J. Lett., 707,* L6–L11.
Raga A. C. et al. (2011) *Astrophys. J. Lett., 730,* L17.
Raga A. C. et al. (2012) *Astrophys. J. Lett., 744,* L12.
Ray T. et al. (2007) In *Protostars and Planets V* (B. Reipurth et al., eds.), pp. 231–244. Univ. of Arizona, Tucson.

Reipurth B. et al. (1997) *Astron. J., 114,* 2708.
Richer J. et al. (2000) In *Protostars and Planets IV* (V. Mannings et al., eds.), pp. 867–894. Univ. of Arizona, Tucson.
Ridge N. A. et al. (2006) *Astron. J., 131,* 2921–2933.
Rigliaco E. et al. (2012) *Astron. Astrophys., 548,* A56.
Rodríguez L. F. et al. (2005) *Astrophys. J. Lett., 627,* L65–L68.
Rousselet-Perraut K. et al. (2010) *Astron. Astrophys., 516,* L1.
Sahai R. et al. (2008) *Astrophys. J., 680,* 483–494.
Salmeron R. and Ireland T. R. (2012) *Earth Planet. Sci. Lett., 327,* 61–67.
Santangelo G. et al. (2012) *Astron. Astrophys., 538,* A45.
Santangelo G. et al. (2013) *Astron. Astrophys., 557,* A22.
Santiago-García J. et al. (2009) *Astron. Astrophys., 495,* 169–181.
Sauty C. et al. (1999) *Astron. Astrophys., 348,* 327–349.
Sauty C. et al. (2012) *Astrophys. J. Lett., 759,* L1.
Schneider P. C. and Schmitt J. H. M. M. (2008) *Astron. Astrophys., 488,* L13–L16.
Schneider P. C. et al. (2009) *Astron. Astrophys., 508,* 717–724.
Schneider P. C. et al. (2011) *Astron. Astrophys., 530,* A123.
Schneider P. C. et al. (2013a) *Astron. Astrophys., 550,* L1.
Schneider P. C. et al. (2013b) *Astron. Astrophys., 557,* A110.
Seale J. P. and Looney L. W. (2008) *Astrophys. J., 675,* 427–442.
Shang H. et al. (2006) *Astrophys. J., 649,* 845.
Shang H. et al. (2007) In *Protostars and Planets V* (B. Reipurth et al., eds.), pp. 261–276. Univ. of Arizona, Tucson.
Shang H. et al. (2010) *Astrophys. J., 714,* 1733–1739.
Shu F. et al. (1994) *Astrophys. J., 429,* 781–796.
Skinner S. L. et al. (2011) *Astrophys. J., 737,* 19.
Smith N. et al. (2009) *Mon. Not. R. Astron. Soc., 399,* 952–965.
Soker N. (2005) *Astron. Astrophys., 435,* 125–129.
Staff J. E. et al. (2010) *Astrophys. J., 722,* 1325–1332.
Stanke T. et al. (2000) *Astron. Astrophys., 355,* 639–650.
Stelzer B. et al. (2009) *Astron. Astrophys., 499,* 529–533.
Stojimirović I. et al. (2007) *Astrophys. J., 660,* 418–425.
Stojimirović I. et al. (2006) *Astrophys. J., 649,* 280–298.
Stone J. M.et al. (1998) *Astrophys. J. Lett., 508,* L99–L102.
Stute M. et al. (2010) *Astron. Astrophys., 516,* A6.
Suzuki-Vidal F. et al. (2010) *Phys. Plasmas, 17,* 112708.
Suzuki-Vidal F. et al. (2012) *Phys. Plasmas, 19,* 022708.
Suzuki-Vidal F. et al. (2013) *High Energy Density Phys., 9,* 141–147.
Tafalla M. and Myers P. C. (1997) *Astrophys. J., 491,* 653.
Tafalla M. et al (2010) *Astron. Astrophys., 522,* A91.
Tafalla M. et al. (2013) *Astron. Astrophys., 551,* A116.
Takami M. et al. (2007) *Astrophys. J., 670,* 33.
Takami M. et al. (2011) *Astrophys. J., 743,* 193.
Tappe A. et al. (2008) *Astrophys. J., 680,* 117.
Tappe A. et al. (2012) *Astrophys. J., 751,* 9.
Teşileanu O. et al. (2009) *Astron. Astrophys., 507,* 581–588.
Teşileanu O. et al. (2012) *Astrophys. J., 746,* 96.
Terquem C. et al. (1999) *Astrophys. J., 512,* L131.
van Kempen T. A. et al. (2009) *Astron. Astrophys., 501,* 633–646.
Vázquez-Semadeni E. (2011) *Comput. Star Form., 270,* 275–282.
Wang P. et al. (2010) *Astrophys. J., 709,* 27–41.
Weigelt G. et al. (2011) *Astron. Astrophys., 527,* A103.
Whelan E. T. et al. (2009a) *Astrophys. J. Lett., 691,* L106–L110.
Whelan E. T. et al. (2009b) *Astrophys. J., 706,* 1054–1068.
Whelan E. T. et al. (2010) *Astrophys. J. Lett., 720,* L119.
Woitas J. et al. (2005) *Astron. Astrophys., 432,* 149–160.
Yirak K. et al. (2008) *Astrophys. J., 672,* 996–1005.
Yirak K. et al. (2009) *Astrophys. J., 695,* 999–1005.
Yirak K. et al. (2012) *Astrophys. J., 746,* 133.
Yu K. C. et al. (1999), *Astron. J., 118,* 2940.
Zanni C. and Ferreira J. (2013) *Astron. Astrophys., 550,* A99.
Zhang Y. et al. (2013) *Astrophys. J., 767,* 58.
Zapata L. A. et al. (2009) *Astrophys. J. Lett., 704,* L45–L48.
Zapata L. A. et al. (2010) *Astron. Astrophys., 510,* A2.
Zinnecker H. et al. (1998) *Nature, 394,* 862.

Alexander R., Pascucci I., Andrews S., Armitage P., and Cieza L. (2014) The dispersal of protoplanetary disks. In *Protostars and Planets VI* (H. Beuther et al., eds.), pp. 475–496. Univ. of Arizona, Tucson, DOI: 10.2458/azu_uapress_9780816531240-ch021.

The Dispersal of Protoplanetary Disks

Richard Alexander
University of Leicester

Ilaria Pascucci
University of Arizona

Sean Andrews
Harvard-Smithsonian Center for Astrophysics

Philip Armitage
University of Colorado

Lucas Cieza
Universidad Diego Portales

Protoplanetary disks are the sites of planet formation, and the evolution and eventual dispersal of these disks strongly influences the formation of planetary systems. Disk evolution during the planet-forming epoch is driven by accretion and mass loss due to winds, and in typical environments photoevaporation by high-energy radiation from the central star is likely to dominate final gas disk dispersal. We present a critical review of current theoretical models, and discuss the observations that are used to test these models and inform our understanding of the underlying physics. We also discuss the role disk dispersal plays in shaping planetary systems, considering its influence on both the process(es) of planet formation and the architectures of planetary systems. We conclude by presenting a schematic picture of protoplanetary disk evolution and dispersal, and discussing prospects for future work.

1. INTRODUCTION

The evolution and eventual dispersal of protoplanetary disks play crucial roles in planet formation. Protoplanetary disks are a natural consequence of star formation, spun up by angular momentum conservation during gravitational collapse. The simple fact that these disks are observed to accrete tells us that they evolve, and observations of diskless stars show that final gas disk dispersal is very efficient. Disk dispersal therefore sets a strict limit on the timescale for gas-giant planet formation. Removal of disk gas can also alter the disk's chemical composition, which has important implications for planet formation, and as disk clearing halts planet migration it also influences the initial architectures of planetary systems. In this chapter we review the physics of protoplanetary disk dispersal and its implications for the formation of planetary systems.

1.1. Observational Constraints on Disk Dispersal

Gas-rich protoplanetary disks were discovered more than 25 years ago (e.g., *Sargent and Beckwith,* 1987), and are now commonly observed. The chapters in this volume by Dutrey et al., Espaillat et al., Pontoppidan et al., and Testi et al. present a comprehensive summary of disk observations (see also *Pascucci and Tachibana,* 2010; *Williams and Cieza,* 2011); here we merely highlight the key observations that motivate and constrain theoretical models of disk dispersal. Note also that, for reasons of length, our discussion focuses on stars of approximately solar mass, which are generally better studied than higher- or lower-mass stars.

Young, solar-like stars are traditionally classified either by the slope of their infrared (IR) spectral energy distribution (SED), or by the strength of emission lines in their spectrum. As circumstellar material absorbs stellar radiation and reemits it at longer wavelengths, a redder SED is broadly associated with more circumstellar dust. Objects with class II SEDs are therefore inferred to be stars with disks, while near-stellar class III SEDs are indicative of young stars that have shed their disks. [Class 0 and I sources are embedded objects, at an earlier evolutionary stage (*Lada,* 1987; *Andre et al.,* 1993).] The major source of optical emission lines from young stars is accretion: Objects with bright emission lines (such as Hα) are referred to as "classical T Tauri stars" (CTTs), while similar stars that lack accretion signatures are designated "weak-lined T Tauri stars" (WTTs). Classical T Tauri stars generally have class II SEDs and WTTs class III SEDs, and although there is not a perfect correspondence between the different classifications we use these terms interchangably.

The dust (solid) component of the disk represents only a small fraction of the disk mass but dominates the opacity, and dust in the disk absorbs stellar radiation and reemits it at longer wavelengths. Continuum emission at different wavelengths probes dust at different temperatures, and therefore different radii, in the disk: Warm dust in the inner few astronomical units is observed in the near-IR, while millimeter emission traces cold dust in the outer disk. Observations of gas are more challenging, primarily because the majority of the disk mass is cold molecular hydrogen that emits only through weak quadrupole transitions. Line emission from H_2, as well as from CO and other trace species (both atomic and molecular), is detected (see, e.g., *Najita et al.*, 2007b; *Williams and Cieza*, 2011), but gas in protoplanetary disks is most readily observed through the signatures of accretion onto the stellar surface.

This wide variety of observational tracers allows us to build up a broad picture of protoplanetary disk evolution. In the youngest clusters [≤1 million years (m.y.)] the IR excess fraction for single stars is close to 100%, but this declines dramatically with age and is typically 10% or less for ages ≥5 m.y. (e.g., *Haisch et al.*, 2001; *Mamajek*, 2009; *Kraus et al.*, 2012). A similar decline in disk fraction is seen in accretion signatures (*Fedele et al.*, 2010), and the mass of cold dust and gas in outer disks is also substantially depleted in older clusters (e.g., *Mathews et al.*, 2012). Protoplanetary disk lifetimes are therefore inferred to be a few million years, with order-of-magnitude scatter.

These observations also allow us to make quantitative measurements of disk properties. Resolved observations of CO emission lines show Keplerian rotation profiles on scales of tens to hundreds of astronomical units (e.g., *Simon et al.*, 2000). Disk surface densities typically decline with radius (at least at radii ≥10 AU), with power-law indices ($\Sigma \propto R^{-p}$) measured to be $p \approx 0.5$–1 (*Andrews et al.*, 2009). Stellar accretion rates range from $\dot{M} \geq 10^{-7}\ M_\odot\ yr^{-1}$ to $\leq 10^{-10}\ M_\odot\ yr^{-1}$ (e.g., *Muzerolle et al.*, 2000), while disk masses estimated from (sub)millimeter continuum emission range from $M_d \sim 0.1\ M_\odot$ to $\leq 0.001\ M_\odot$ (*Andrews and Williams*, 2005). The accretion timescales inferred from these measurements (i.e., $t \sim M_d/\dot{M}$) are therefore also approximately millions of years, implying that protoplanetary disks evolve substantially during their lifetimes.

Observations of young disk-less stars show that disk dispersal is extremely efficient. Searches for gas around WTTs yield only upper limits: Nondetections of H_2 rovibrational transitions and other mid-IR gas emission lines imply warm gas surface densities $\Sigma < 1\ g\ cm^{-2}$ at approximately astronomical-unit radii (*Pascucci et al.*, 2006), while nondetections of H_2 fluorescent electronic transitions suggest $\Sigma < 10^{-5}\ g\ cm^{-2}$ (*Ingleby et al.*, 2009). This latter limit is

~10^{-7} of the surface densities inferred for accreting CTTs, and suggests that gas disk dispersal is almost total. We also see that the various disk and accretion signatures are very strongly correlated and usually vanish together, implying that clearing occurs nearly simultaneously across the entire radial extent of the disk (e.g., *Andrews and Williams*, 2005; *Cieza et al.*, 2008).

Finally, relatively few objects show evidence of partial disk clearing (the so-called "transitional" disks; see section 3.4 and the chapter by Espaillat et al. in this volume), and there is a striking dearth of objects with properties intermediate between CTTs and WTTs (e.g., *Kenyon and Hartmann*, 1995; *Duvert et al.*, 2000; *Padgett et al.*, 2006). This suggests that the transition from disk-bearing to disk-less is rapid, as few objects are "caught in the act" of disk clearing (see Fig. 1). Statistical estimates find the that dispersal timescale is ~10× shorter than the typical disk lifetime (*Simon and Prato*, 1995; *Wolk and Walter*, 1996; *Andrews and Williams*, 2005; *Luhman et al.*, 2010; *Koepferl et al.*, 2013). The mechanism(s) that drive final disk dispersal must therefore efficiently remove both gas and dust, from <0.1 AU to >100 AU, on a timescale ~10^5 yr, after a disk lifetime of a few million years.

1.2. Disk Dispersal Mechanisms

Observed disk lifetimes require disks to survive for at least thousands of orbital periods even at large radii. Disks are therefore dynamically long-lived, and evolve relatively

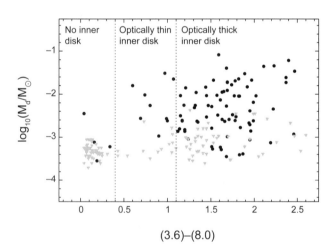

Fig. 1. Compilation of IR and (sub)millimeter data for disks in Taurus-Auriga. The horizontal axis shows the Spitzer IR color (in magnitudes), while the vertical axis shows the disk mass derived from millimeter continuum emission (circles denote detections; triangles upper limits). Infrared excesses, which trace inner dust disks, correlate strongly with disk mass, and there is a striking lack of objects with properties between disk-bearing CTTs and disk-less WTTs. Figure adapted from *Cieza et al.* (2008), using data from *Andrews and Williams* (2005), *Luhman et al.* (2010) and *Andrews et al.* (2013).

slowly. A comprehensive review of protoplanetary disk physics is given by *Armitage* (2011), while *Hollenbach et al.* (2000) discussed disk dispersal mechanisms in detail; here we summarize the key physical processes.

Disk accretion is a major driver of disk evolution, and is generally thought to dominate at early times. The observed evolution of CTT disks on million-year timescales is broadly consistent with viscous accretion disk theory (e.g., *Shakura and Sunyaev*, 1973; *Lynden-Bell and Pringle*, 1974; *Hartmann et al.*, 1998). In this picture disks evolve due to the exchange (transport) of angular momentum between neighboring annuli, which is traditionally attributed to viscous stresses. In reality these stresses are due to turbulence (and possibly also laminar magnetic torques) in the disk (e.g., *Balbus*, 2011) (see also the chapter by Turner et al. in this volume), and in the modern interpretation the Shakura-Sunyaev α-parameter represents the efficiency of turbulent transport. The characteristic viscous timescale is

$$t_\nu = \frac{1}{\alpha\Omega}\left(\frac{H}{R}\right)^{-2} \qquad (1)$$

where Ω is the (Keplerian) orbital frequency, and the disk aspect ratio is typically $H/R \sim 0.1$. Observations suggest that $\alpha \sim 0.01$ (e.g., *Hartmann et al.*, 1998), so the local viscous timescale is at least thousands of orbital periods, and $t_\nu \sim$ m.y. at radii $\gtrsim 100$ AU. Protoplanetary disks are therefore observed to live for at most a few viscous timescales in their outer regions, implying that some other mechanism drives disk dispersal.

A variety of other mechanisms can remove mass and/or angular momentum from disks. The presence of a binary companion strongly affects disk formation and evolution, particularly for close binaries (e.g., *Harris et al.*, 2012). In star clusters disks undergo tidal stripping during close stellar encounters, and can be evaporated or ablated by radiation and winds from massive stars. However, these processes dominate in only a small fraction of disks (e.g., *Scally and Clarke*, 2001; *Adams et al.*, 2006), and the majority of disks in massive clusters have similar properties to those in less hostile environments (e.g., *Mann and Williams*, 2010). Thus, while environment clearly plays a major role in some cases (see section 2.2.4), disk evolution around single stars must primarily be due to "internal" processes.

Magnetically launched jets and winds extract both mass and angular momentum, and may drive accretion in protoplanetary disks (e.g., *Königl and Salmeron*, 2011). Magnetocentrifugal winds are perhaps the most plausible explanation for protostellar jets and outflows, and may well play a major role in disk evolution at early times. High-density winds from the star or inner disk may also strip the disk of gas at greater than approximately astronomical-unit radii (*Lovelace et al.*, 2008; *Matsuyama et al.*, 2009). We defer detailed discussion of magnetic winds to the chapter by Frank et al. in this volume, but consider their role in disk dispersal in section 2.5.

We also note in passing that the processing of disk material into planets appears not to be a major driver of disk evolution or dispersal. It is now clear that planet formation is ubiquitous, but observations of both the solar system and exoplanets show that planets account for ≤1% of the initial disk mass in most systems (e.g., *Wright et al.*, 2011; *Mayor et al.*, 2013). Thus, while planet formation may involve a significant fraction of the total mass of heavy elements in the disk, planets represent only a small fraction of the disk mass budget. In addition, planet formation (by core accretion) is a rather slow process [typically requiring approximately million-year timescales (e.g., *Pollack et al.*, 1996)], so it seems unlikely that planet formation plays a major role in driving the rapid disk dispersal required by observations.

The final commonly considered mechanism for disk dispersal is photoevaporation. High-energy radiation [ultraviolet (UV) and/or X-rays] heats the disk surface to high temperatures ($\sim 10^3$–10^4 K), and beyond a few astronomical units this heated layer is unbound and flows from the disk surface as a pressure-driven wind. *Hollenbach et al.* (2000) and *Dullemond et al.* (2007) considered various disk dispersal processes in detail in previous Protostars and Planets volumes, and concluded that photoevaporation is the dominant mechanism for removing disk gas at large radii (see also the more recent review by *Clarke*, 2011). Considerable progress has been made in this field in recent years, and we devote much of this review to the theory and observations of photoevaporative winds. In section 2 we review the theory of disk dispersal, while section 3 discusses the observations that constrain our theoretical models. In section 4 we discuss the implications of these results for the formation and evolution of planetary systems, and we conclude by presenting a schematic picture of the evolution and dispersal of protoplanetary disks.

2. MODELS OF DISK DISPERSAL

A variety of processes influence the evolution and dispersal of protoplanetary disks but, for the reasons discussed in section 1.2, we focus primarily on accretion and photoevaporation. We first review the basic physics of disk photoevaporation, then discuss the current state-of-the-art in theoretical modeling and the observational predictions of these models. We also review recent work suggesting that magneto-hydrodynamic (MHD) turbulence can drive winds; mass-loss rates from MHD winds are highly uncertain, but may be high enough to contribute to disk dispersal.

2.1. Disk Photoevaporation: Theoretical Basics

The basic principles of disk photoevaporation are readily understood. When high-energy radiation is incident on a disk, its upper layers are heated to well above the mid-plane temperature. At sufficiently large radius (i.e., high enough in the potential well) the thermal energy of the heated layer exceeds its gravitational binding energy, and the heated gas escapes. The result is a centrifugally launched, pressure-driven flow, which is referred to as a photoevaporative wind.

Photoevaporation was applied to disks around young stars as long ago as *Bally and Scoville* (1982), and the first detailed models were presented by *Shu et al.* (1993) and *Hollenbach et al.* (1994). We follow their approach in defining the characteristic length-scale (the "gravitational radius") as the (cylindrical) radius where the Keplerian orbital speed is equal to the sound speed of the hot gas

$$R_g = \frac{GM_*}{c_s^2} \qquad (2)$$

Here M_* is the stellar mass and c_s is the (isothermal) sound speed of the heated disk surface layer. The simplest case we can consider is an isothermal wind launched from a thin disk in Keplerian rotation, which is analogous to the problem of Compton-heated winds around active galactic nuclei (e.g., *Begelman et al.*, 1983). The flow is launched subsonically from the base of the heated atmosphere, and passes through a sonic transition (typically $\sim R_g/2$ along each streamline) (see Fig. 2) before becoming supersonic at large radii.

A purely pressure-driven wind exerts no torque, and depletes the disk without altering its specific angular momentum. Photoevaporation therefore has two distinct timescales: the flow timescale (which by definition is approximately the dynamical timescale), and the much longer mass-loss timescale. If the wind structure does not vary rapidly, these timescales can be considered independently. We can therefore construct dynamical models of photoevaporation in order to determine the mass-loss profile $\dot{\Sigma}(R,t)$, which can then be incorporated into secular disk evolution models as a sink term.

In the case of an isothermal wind, simple arguments show that the rate of mass-loss per unit area $\dot{\Sigma}(R)$ peaks at $R_c \simeq R_g/5$ (*Liffman*, 2003; *Font et al.*, 2004; *Dullemond et al.*, 2007), and this is now commonly referred to in the literature as the "critical radius" (see also *Adams et al.*, 2004). However, even in the isothermal case the flow solution is analytically intractable [as pointed out by *Begelman et al.* (1983)], as the bulk properties of the flow depend on the local divergence of the streamlines (which is not known *a priori*). We can compute solutions numerically if the base density profile $\rho_0(R)$ is known, but in general this is not the case. The base density profile is determined by the balance between radiative heating and cooling, and the heating rate in turn depends on how radiation is transported through the disk atmosphere. The irradiation can be either "external" (i.e., from nearby massive stars) or "internal" (i.e., from the central star): The former is dominant in the central regions of massive star clusters (see section 2.2.4), but for the reasons discussed in section 1.2 we focus on the latter. Central star-driven photoevaporation has two limiting cases: (1) an optically thick disk, where the atmosphere is irradiated obliquely; and (2) a disk with an optically thin inner hole (expected during disk dispersal at late times), where the base of the flow is heated normally (i.e., face-on). Disk photoevaporation is a coupled problem in radiative transfer, thermodynamics, and hydrodynamics, but unfortunately full

radiation hydrodynamic simulations remain prohibitively expensive. Instead we must use simplified models, with the choice of simplifications depending primarily on the dominant heating mechanism. Three wavelength regimes are particularly relevant to protoplanetary disks: ionizing, extreme UV (EUV) radiation (13.6–100 eV); far-ultraviolet (FUV) radiation (6–13.6 eV), which is capable of dissociating H_2 and other molecules; and X-rays (0.1–10 keV). In the following sections we discuss models of these processes in turn, highlighting the main results and addressing the shortcomings of the different approaches.

2.2. Models of Photoevaporative Winds

2.2.1. Extreme ultraviolet heating. The simplest of these three cases is EUV heating, where the incident photons are sufficiently energetic to ionize hydrogen atoms. The absorption cross section at the Lyman limit ($h\nu = 13.6$ eV; $\lambda = 912$ Å) is very large ($\sigma = 6.3 \times 10^{-18}$ cm^2) and decreases approximately as ν^{-3} (*Osterbrock and Ferland*, 2006), so the dominant contribution to the ionization rate comes from photons at or close to the threshold energy. The EUV heating rate is therefore not very sensitive to the incident spectrum, and to a good approximation depends only on the ionizing photon luminosity Φ. The resulting radiative transfer problem is analogous to an ionization-bounded H II region, and results in a near-isothermal ionized atmosphere (with $T \simeq 10^4$ K and $c_s \simeq 10$ km s^{-1}), separated from the neutral underlying disk by an ionization front (see discussion in *Clarke*, 2011). The typical critical radius is therefore

$$R_{c,\text{EUV}} \simeq 1.8 \left(\frac{M_*}{1\,M_\odot} \right) \text{AU} \qquad (3)$$

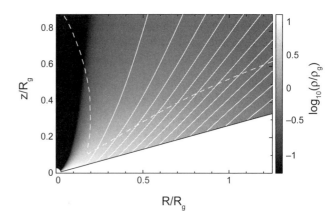

Fig. 2. Typical structure of a photoevaporative wind. The grayscale shows the gas density (normalized to the base density at $R = R_g$), with streamlines plotted at 5% intervals of the total mass flux. The dashed line shows the location of the sonic surface. This example shows the structure of an isothermal, EUV-driven wind, from the models used in *Pascucci et al.* (2011). For a 1-$M_\odot$ star with ionizing luminosity $\Phi = 10^{41}$ photon s^{-1}, the units correspond to $R_g = 8.9$ AU and $n_g = \rho_g/\mu m_H = 2.8 \times 10^4$ cm^{-3}.

Close to the star (i.e., $R \ll R_c$) the ionized disk atmosphere is bound, but at larger radii the ionization front represents the launching surface for the photoevaporative wind. The radiative transfer problem therefore reduces to one of ionization balance, as specifying the position and density of the ionization front uniquely determines the flow solution.

The first quantitative models of this process (for the case of an optically thick disk) were performed by *Hollenbach et al.* (1994, hereafter *HJLS94*), using "one + one-dimensional" numerical radiative transfer calculations. Approximately one-third of radiative recombinations of hydrogen produce another ionizing photon, and *HJLS94* found that this diffuse (recombination) field dominates at the ionization front for all radii of interest. They divided the atmosphere into "static" and "flow" regions, and showed that the base density profile, $\rho_0(R)$, can be estimated from a Strömgren-like condition (see also *Alexander*, 2008a; *Clarke*, 2011). The heating process is recombination-limited, and consequently the base density scales as $\Phi^{1/2}$. *HJLS94*'s models involved no hydrodynamics, and instead simply estimated the photoevaporation rate by assuming $\dot{\Sigma}(R) = 2\rho_0(R)c_s$ beyond R_g.

Richling and Yorke (1997) subsequently introduced numerical hydrodynamics, and studied the effects of dust opacity, in the massive star regime. The first hydrodynamic models of central-star-driven photoevaporation around solar-mass stars were presented by *Font et al.* (2004). These models used two-dimensional numerical hydrodynamics, assuming an isothermal equation of state and adopting the base density profile $\rho_0(R)$ derived by *HJLS94* as a (input) boundary condition. This allows numerical solution of a steady-state wind profile, which in turn quantifies several of the estimates and assumptions made above; the wind structure resulting from such a calculation is shown in Fig. 2. The launch velocity at the base of the flow is typically 0.3–0.4 c_s, and the mass-loss profile $\dot{\Sigma}(R)$ peaks at $\simeq 0.14\ R_g$ (see Fig. 3, solid line). The total mass loss per logarithmic interval in radius, $2\pi R^2\dot{\Sigma}$, peaks at approximately 9 AU. The integrated mass-loss rate over the entire disk is

$$\dot{M}_{w,EUV} \simeq 1.6 \times 10^{-10} \left(\frac{\Phi}{10^{41}\,\mathrm{s}^{-1}}\right)^{1/2} \times \left(\frac{M_*}{1\,M_\odot}\right)^{1/2} M_\odot\ \mathrm{yr}^{-1} \quad (4)$$

The behavior of the EUV wind changes significantly in the case of a disk with an optically thin inner hole, as the heating is dominated by direct irradiation from the central star. *Alexander et al.* (2006a) studied this problem using both analytic arguments and two-dimensional numerical hydrodynamics. As the direct field dominates, they were able to compute the location of the ionization front "on the fly" in their hydrodynamic calculations, and find a self-consistent flow solution. In this case *Alexander et al.* (2006a) found that

$$\dot{M}_w \simeq 1.4 \times 10^{-9} \left(\frac{\Phi}{10^{41}\,\mathrm{s}^{-1}}\right)^{1/2} \left(\frac{R_{in}}{3\,\mathrm{AU}}\right)^{1/2} M_\odot\ \mathrm{yr}^{-1} \quad (5)$$

where R_{in} is the radius of the inner disk edge. Direct irradiation therefore increases the wind rate by approximately an order of magnitude, and the mass-loss rate increases for larger inner holes.

From a theoretical perspective, EUV photoevaporation is now well understood, and there is good agreement between analytic and numerical models. The remaining weakness in this approach is that (in the optically thick case) the hydrodynamic models still rely on the radiative transfer solution of *HJLS94*, and are not strictly self-consistent. This is only a minor issue, however, and is dwarfed by the much larger uncertainty in the input parameters: The ionizing luminosities of T Tauri stars (TTs) are very poorly constrained by observations (see section 3.1). Moreover, both the accretion columns and any jets or winds close to the star are extremely optically thick to ionizing photons (*Alexander et al.*, 2004a; *Gorti et al.*, 2009; *Owen et al.*, 2012), so estimating the ionizing flux that reaches the disk at approximately astronomical-unit radii is by far the dominant uncertainty in calculations of EUV photoevaporation.

2.2.2. X-ray heating. We next consider photoevaporation by stellar X-rays. T Tauri stars are known to be bright (although highly variable) X-ray sources, with median luminosity $L_X \simeq 10^{30}$ erg s^{-1} and a spectrum that typically peaks around 1 keV (e.g., *Feigelson et al.*, 2007). X-rays have long been known to play an important role in sustaining the level of disk ionization required to drive MHD turbulence (*Glassgold et al.*, 1997, 2000), but recently attention has also turned to the thermal effects of X-ray irradiation. Approximately kiloelectron-volt X-rays are absorbed by K-shell ionization of heavy elements (primarily O, but also C and Fe), and the resulting photoelectrons then collisionally ionize and/or heat the hydrogen atoms/molecules in the gas. In general, the gas and dust temperatures are decoupled, and the dominant cooling channels are (1) metal line emission

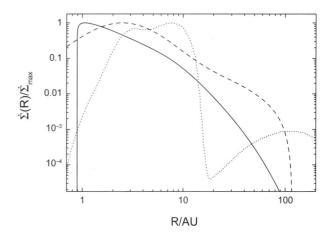

Fig. 3. Normalized mass-loss profiles for different photoevaporation models around a 1-$M_\odot$ star: EUV [solid line (*Font et al.*, 2004)]; X-ray [dashed line (*Owen et al.*, 2012)], and FUV-dominated [dotted line (*Gorti et al.*, 2009)]. Note that for fiducial parameters the absolute mass-loss rates differ by 1–2 orders of magnitude.

and (2) gas cooling via collisions with grains (e.g., *Ercolano et al.*, 2008). For solar abundances the absorption cross section is $\sigma \simeq 2 \times 10^{-22}$ (hν/1 keV)$^{-2.485}$ cm^2 (*Glassgold et al.*, 1997), and 0.3–1-keV X-rays therefore provide significant heating up to a (neutral hydrogen) column depth $N_H \sim 10^{21}$–10^{22} cm^{-2}.

The first models of X-ray photoevaporation were similar in spirit to *HJLS94*, and used only hydrostatic calculations. *Alexander et al.* (2004b) used a simple heating model, and *Ercolano et al.* (2008, 2009) improved on this approach by using two-dimensional Monte Carlo radiative transfer (see also *Gorti and Hollenbach*, 2009). The resulting disk structure consists of a very tenuous hot ($\sim 10^6$ K) corona, above a partially ionized atmosphere at T $\sim 10^3$–10^4 K. There is a smooth transition from this atmosphere to the (cold) underlying disk, unlike in the EUV case, and the varying temperature means that the flow is not well characterized by a single critical radius. The vertical density gradient is steep, so a small uncertainty in the vertical location of the launch point leads to a large uncertainty in the mass flux. Estimating photoevaporation rates from static calculations is thus fraught with difficulty, and despite deriving similar structures, the studies of *Alexander et al.* (2004b), *Ercolano et al.* (2008, 2009), and *Gorti and Hollenbach* (2009) came to qualitatively different conclusions about the importance of X-ray heating.

This problem was essentially solved by *Owen et al.* (2010), who coupled the radiative transfer models of *Ercolano et al.* (2009) to numerical hydrodynamics. These authors used radiative transfer calculations to establish a monotonic relationship between the X-ray ionization parameter ($\xi = L_X/nd^2$) and the gas temperature, which can be used in lieu of an energy equation in hydrodynamic calculations. This allows the wind structure to be computed numerically, and the steady-state solution can be verified *a posteriori* against the radiative transfer code. The photoevaporative flow is launched from the atomic layer, at temperatures $\simeq 3000$–5000 K, and the flow structure is largely determined by the temperature and density at the sonic point (*Owen et al.*, 2012). The resulting mass-loss profile $\dot{\Sigma}(R)$ is broader than in the EUV case, and peaks at $\simeq 3$ AU (see Fig. 3, dashed line); the mass-loss rate $2\pi R^2 \dot{\Sigma}$ peaks at 40–60 AU. Heating at the base of the flow is dominated by X-rays with h$\nu \geq 0.3$–0.4 keV, and the integrated wind rate scales almost linearly with L_X (and is largely insensitive to stellar mass). Assuming a fixed input spectrum with variable luminosity, *Owen et al.* (2011, 2012) fit the following scaling relation

$$\dot{M}_{w,X} \simeq 6.3 \times 10^{-9} \left(\frac{L_X}{10^{30} \text{ergs}^{-1}} \right)^{1.14} \times$$
$$\left(\frac{M_*}{1 M_\odot} \right)^{-0.068} M_\odot \text{ yr}^{-1} \tag{6}$$

Strikingly, this is $\sim 40\times$ larger than the fiducial EUV-driven wind rate (equation (4)). As X-ray heating is always local

to the initial absorption, the integrated wind rate increases only modestly (by a factor of ~ 2) in the presence of an optically thin disk inner hole. However, *Owen et al.* (2012, 2013b) find that X-ray photoevaporation of an inner hole has a dramatic effect if the disk surface density is sufficiently small (typically $\lesssim 0.1$–1 g cm^{-2}). In this case the physical depth of the X-ray-heated column is comparable to the disk scale-height, rendering the inner edge of the disk dynamically unstable. This instability leads to very rapid (dynamical) dispersal of the disk, and *Owen et al.* (2012, 2013b) dub this process "thermal sweeping."

The use of the ionization parameter-temperature relation restricts this method to the regime where X-ray heating is dominant, but the resulting wind solutions are robust. As in the EUV case, however, the input radiation field remains a significant source of uncertainty. Although TTs are readily observed in X-rays, observations lack the spectral resolution to be used as inputs for these models. Instead, *Ercolano et al.* (2008, 2009) and *Owen et al.* (2010) created synthetic input spectra, using a coronal emission measure distribution derived primarily from studies of RS CVn-type binaries. It is notable, however, that *Gorti and Hollenbach* (2009) found very different results using a somewhat harder X-ray spectrum. Some discrepancies between competing radiative transfer codes also remain unresolved, particularly with regard to the temperatures in the X-ray-heated region: The models of Gorti and Hollenbach consistently predict lower temperatures this region than the models of Ercolano, Owen, and collaborators (see, e.g., the discussion in *Ercolano et al.*, 2009). In addition, as X-rays are primarily absorbed by heavy elements the heating rates are sensitive to assumptions about disk chemistry, and effects such as dust settling have not yet been considered in detail. X-ray photoevaporation models are thus subject to potentially significant systematic uncertainties, and although the individual calculations are mature and self-consistent, further exploration of these issues is desirable.

2.2.3. Far-ultraviolet heating. Finally we consider photoevaporation by non-ionizing, H$_2$-dissociating, FUV radiation. This is by far the most complex case, and is still not fully understood (see *Clarke*, 2011, for a more detailed review). The radiative transfer problem is similar to that in photodissociation regions (PDRs) (e.g., *Tielens and Hollenbach*, 1985), but is complicated substantially by the disk geometry. The heating/cooling balance in PDRs depends strongly on the ratio of the incident flux to the gas density G$_0$/n. For FUV photoevaporation we are generally in the high G$_0$/n limit, so dust attenuation of the incident flux dominates and the heated column has a roughly constant depth $N_H \sim 10^{21}$–10^{23} cm^{-3} that depends primarily on the dust properties (and is comparable to the depth of the X-ray-heated region). Most of the FUV flux is absorbed by dust grains and reradiated as (IR) continuum, although absorption and reemission by polycyclic aromatic hydrocarbons (PAHs) is also significant. Gas heating is due to collisions with photoelectrons from grains and PAHs, or FUV-pumping of H$_2$ (followed by fluorescence and collisional deexcitation).

The gas and dust temperatures are decoupled at low column density, with gas temperatures ranging from a few hundred to several thousand Kelvin (e.g., *Adams et al.*, 2004).

Far-ultraviolet heating usually dominates in the case of photoevaporation by nearby massive stars (e.g., *Johnstone et al.*, 1998) (see also section 2.2.4), but understanding FUV irradiation by the central star remain a work in progress. *Gorti and Hollenbach* (2004, 2008, 2009) have constructed a series of models of this process, using "one + one-dimensional" radiative transfer and a thermochemical network to compute the structure of the disk atmosphere. The computational expense of these calculations precludes the addition of numerical hydrodynamics; mass fluxes are instead estimated analytically, using a method similar to that of *Adams et al.* (2004). These models use an input spectrum that spans the EUV, FUV, and X-rays, and includes contributions from both coronal emission and the stellar accretion shock (as well as attenuation near the star). The wind is again mainly launched from the atomic layer, but the temperature in the launching region varies substantially, ranging from >1000 K at a few astronomical units to <100 K at ~100 AU. The resulting mass-loss profile has a peak at 5–10 AU, and the FUV irradiation also drives significant mass loss at large radii ($\gtrsim$100 AU; see Fig. 3, dotted line). The integrated wind rate depends primarily on the total FUV flux, and for fiducial parameters ($L_{FUV} = 5 \times 10^{31}$ erg s^{-1}, $M_* = 1$ $M_\odot$) is 3×10^{-8} $M_\odot$ yr^{-1} (*Gorti and Hollenbach*, 2009). This is again 2 orders of magnitude larger than the fiducial EUV wind rate, and comparable to the X-ray-dominated wind rates found by *Owen et al.* (2010, 2012). As much of the mass loss originates at large radii, FUV photoevaporation can play a major role in depleting the disk's mass reservoir, and can also potentially truncate disks at radii $\gtrsim$100 AU.

However, due to the complexity of this problem, these models still suffer from significant uncertainties. In particular, the thermal physics can be very sensitive to the abundance of PAHs in the disk atmosphere. The abundance and depletion of PAHs is highly uncertain (e.g., *Geers et al.*, 2009), and recent calculations suggest that changes to the PAH and dust abundances may alter the heating/cooling rates significantly (Gorti et al., in preparation). In addition, these models still lack detailed hydrodynamics. Estimating mass fluxes from hydrostatic calculations is problematic (as discussed in section 2.2.2), and introduces an additional uncertainty in the derived wind rates. FUV-dominated photoevaporation rates are therefore still uncertain at the order-of-magnitude level, and further work is needed to understand this complex process fully.

2.2.4. External irradiation. Our discussion has focused on photoevaporation of disks by their central stars, but it has long been recognized that external irradiation dominates in some cases. The geometry of the radiative transfer problem (essentially plane-parallel irradiation) is much simpler than in the central-star case, and consequently the flow structure is amenable to a semi-analytic solution. *Johnstone et al.* (1998) (see also *Richling and Yorke*, 2000) constructed detailed models of disk evaporation by radiation from nearby O-type stars, as expected in the cores of massive clusters, and found that the heating is dominated by the photospheric EUV and FUV flux. Again the basic picture is that the wind is launched from a PDR on the disk surface, with the thickness of the PDR determined by the incident flux (and hence the distance d from the ionizing source). For disks in the immediate vicinity of an O-type star (d $\leq$ 0.03 pc), the ionization front is roughly coincident with the disk surface, and the mass-loss rate is determined by the incident EUV flux. At larger distances, where the ionizing flux is weaker, the PDR thickens and the wind is launched from the neutral, FUV-heated layer. The wind is optically thick to EUV photons, so the ionization front is offset from the base of the flow. For spatially extended disks (with size $R_d \geq R_c$) the wind is launched almost vertically from the irradiated disk surface, while for more compact ("subcritical") disks the wind is launched radially from the disk outer edge (*Adams et al.*, 2004). The flow subsequently interacts with ionizing photons from the irradiating star, leading to an ionization front with a characteristic cometary shape. Typical mass-loss rates are $\dot{M}_w \sim 10^{-7}$ $M_\odot$ yr^{-1}, with a strong dependence on the disk size R_d.

External photoevaporation is best seen in the Orion Nebula Cluster (ONC), where a small number (~100) of so-called proplyds ("PROtoPLanetarY DiskS") are observed in silhouette against the background nebula. These objects show bright emission lines, offset ionization fronts, and cometary shapes, and when discovered they were quickly recognized as disks undergoing external photoevaporation (e.g., *O'Dell et al.*, 1993; *McCaughrean and O'Dell*, 1996). The study of proplyds is now relatively mature, and there is excellent agreement between models and observations of their photoevaporative flows (e.g., *Störzer and Hollenbach*, 1999; *Henney and O'Dell*, 1999; *Mesa-Delgado et al.*, 2012). The principal factor controlling the long-term evolution of disks subject to external photoevaporation is the initial disk mass (*Clarke*, 2007), but most disks around solar mass stars do not experience such harsh environments. Dynamical models of clusters find that external photoevaporation, and other "environmental" factors such as stellar encounters, play a significant role in the evolution of only a small fraction of disks [$\leq$10% (*Scally and Clarke*, 2001; *Adams et al.*, 2006)]. Observations of disk masses in the ONC suggest that the disks within d $\leq$ 0.3 pc of the cluster core are significantly depleted compared to those at larger distances (*Eisner et al.*, 2006; *Mann and Williams*, 2010), consistent with this scenario. Thus, although external photoevaporation drives the evolution of proplyds in the centers of massive clusters, it is not thought to play a major role in the evolution and dispersal of the majority of disks.

2.3. Coupling to Models of Disk Evolution

The mass-loss rates discussed above exceed the observed (stellar) accretion rates of many CTTs [typically $\dot{M} \sim 10^{-9}$–10^{-8} $M_\odot$ yr^{-1} (e.g., *Hartmann et al.*, 1998; *Muzerolle et al.*, 2000)], which suggests that photoevaporation plays a major role in the evolution and dispersal of protoplanetary disks. However, the local mass-loss timescale (t ~ $\Sigma/\dot{\Sigma}$) exceeds the viscous timescale in much of the disk, so understanding how photoevaporation influences disk evolution requires us to consider these competing processes simultaneously. *Clarke et al.* (2001) presented the first such models, combining a one-dimensional viscous accretion disk model with the photoevaporation prescription of *HJLS94*. In this scenario the low (EUV) wind rate is initially negligible, but the disk accretion rate declines with time and eventually becomes comparable to the mass-loss rate due to photoevaporation. After this time the outer disk is no longer able to resupply the inner disk, as the mass loss is concentrated at a particular radius (≃R_c; see Fig. 3), and all the accreting material is "lost" to the wind. The wind first opens a gap in the disk at ≃R_c, and the interior gas then accretes onto the star on its (short) viscous timescale, removing the inner disk in ~10^5 yr (see Fig. 4, black solid line). *Alexander et al.* (2006a,b) subsequently extended this model to consider direct photoevaporation of the outer disk, and found that photoevaporation efficiently clears the entire disk from the inside out on a timescale of a few × 10^5 yr (after a disk lifetime of a few million years). This is consistent with the two-timescale behavior described in section 1.1. Moreover, *Alexander and Armitage* (2007) found that dust grains are accreted from the inner disk even more rapidly than the gas, confirming that this process efficiently clears both the gas and dust disks. Note, however, that in order for disk clearing to operate in this manner, we require that photoevaporation be powered by something *other* than the accretion luminosity, as otherwise the wind fails to overcome the accretion flow (e.g., *Matsuyama et al.*, 2003b). The source of high-energy photons is usually assumed to be the stellar chromosphere or corona and, due to the role of photoevaporation in precipitating this rapid disk clearing, models of this type are generally referred to as "UV-switch" models (after *Clarke et al.*, 2001).

These models considered only EUV photoevaporation, but the evolution is qualitatively similar regardless of the photoevaporation mechanism: Once the disk accretion rate drops below the wind rate, photoevaporation takes over and rapidly clears the disk from the inside out (e.g., *Gorti et al.*, 2009; *Owen et al.*, 2010). However, the much higher photoevaporation rates predicted by models of X-ray and FUV photoevaporation qualitatively change how the disk evolves. Mass loss due to EUV photoevaporation is much lower than typical disk accretion rates, and therefore only influences the disk at late times, after it has undergone substantial viscous evolution. By contrast, the X-ray/FUV models of *Owen et al.* (2010, 2012) and *Gorti et al.* (2009) predict photoevaporation rates that are comparable to the median accretion rate for CTTs (~10^{-9}–10^{-8} M$_\odot$ yr^{-1}). In these models photoevaporation therefore represents a significant mass sink even in the early stages of disk evolution (t ≲ 1 m.y.), and the total mass lost to photoevaporation on million-year timescales may exceed that accreted onto the star (see Fig. 4). Such high wind rates challenge the

conventional paradigm of viscous accretion in protoplanetary disks, and are arguably the most controversial recent development in the theory of disk evolution and dispersal.

However, despite these important differences between competing models, the major uncertainty in our theory of disk evolution remains our ignorance of how angular momentum is transported. All the models discussed above adopt simplified viscosity laws, typically by assuming a constant α. A constant α-parameter is arguably valid as a global average in space and time (e.g., *Hartmann et al.*, 1998; *King et al.*, 2007), but is clearly not an accurate description of the disk microphysics. To date, the only time-dependent models that have attempted to incorporate both photoevaporation and a more physical treatment of angular momentum transport are those of *Morishima* (2012) and *Bae et al.* (2013), which considered the evolution of layered disks (e.g., *Gammie*, 1996; *Armitage et al.*, 2001) subject to X-ray photoevaporation. In this scenario the presence of a dead zone at the disk mid-plane acts as a bottleneck for the accretion flow, preventing steady accretion through the inner disk at ≳10^{-8} M$_\odot$ yr^{-1}. If the photoevaporation rate exceeds the bottleneck accretion rate, then the wind can open a gap in the disk while a substantial dead zone is still present. The gap generally opens beyond the dead zone, and consequently the inner disk drains much more slowly than in the canonical picture of *Clarke et al.* (2001); *Morishima* (2012) argues that this may be the origin of the (small) population

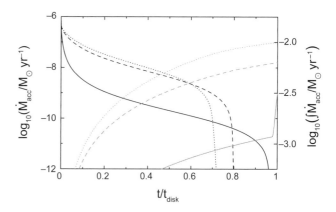

Fig. 4. Stellar accretion rate (black lines, left axis) and total mass lost to photoevaporation (gray lines, right axis) as a function of time for a viscous disk model with different photoevaporation models. The disk model is the median disk from *Alexander and Armitage* (2009), which has α = 0.01 and an initial disk mass of $10^{-1.5}$ M$_\odot$. The various curves show how this disk model evolves when subject to EUV (solid lines), X-ray (dashed), or FUV photoevaporation (dotted), using the fiducial wind profiles described in section 2.2. In all cases the accretion rate drops rapidly to zero once photoevaporation triggers inner disk clearing. For clarity, the horizontal axis shows relative time, as the absolute timescales are primarily a result of the adopted initial conditions and viscosity law. Note, however, that for the same disk parameters the absolute lifetimes for these models differ by factors of several.

of transitional disks known to have large inner dust holes and ongoing gas accretion. Real disks may therefore behave rather differently than the simple models described above, but a better understanding of angular momentum transport is required to make significant further progress.

The disk evolution models described here also make a number of other simplifications. In particular, the model initial conditions are highly idealized, and the stellar irradiation that drives photoevaporation is usually assumed to be constant over the disk lifetime. The latter assumption seems plausible (see section 3.1), but a more realistic treatment of the early stages of disk evolution, incorporating infall and gravitational instabilities, is clearly desirable. If photoevaporation only influences the disk significantly at late times then disk dispersal is largely insensitive to the choice of initial conditions, but this is not necessarily true if the mass-loss rate is high. Moreover, if we are to use these models to further our understanding of planet formation and migration (e.g., *Alexander and Armitage, 2009*), then care must be taken to ensure that our treatment of disk formation and early evolution does not unduly influence our results.

2.4. Disk Photoevaporation: Observational Predictions

These models are now relatively mature, and make a series of explicit observational predictions. Many of these predictions are testable with current facilities, offering the opportunity both to identify the mechanisms driving disk dispersal, and to discriminate between competing models. We discuss the theory behind the most important predictions here, and then review the observational evidence in detail in section 3.

As with the models, predictions from disk dispersal models generally take one of two forms: direct diagnostics of the photoevaporative wind, and predictions for the properties of evolving disks (which are usually statistical in nature). The primary diagnostic of disk photoevaporation is line emission from the wind, as the low gas density ($\lesssim 10^6$ cm^{-3}) gives rise to numerous forbidden emission lines. The first detailed calculations of line emission from (EUV) photoevaporative winds driven by the central star were performed by *Font et al.* (2004), who used numerical hydrodynamics and a simplified radiative transfer scheme to compute synthetic line profiles for various optical forbidden lines ([NII] 6583 Å, [SII] 6716/6731 Å, and [OI] 6300 Å). They found that the high flow velocities result in broad line-widths (~30 km s^{-1}) from the ionized species, and showed that for low disk inclinations these lines are blue-shifted by 5–10 km s^{-1}. Their predicted line fluxes (~10^{-5} L$_\odot$) were consistent with observations of TTs (e.g., *Hartigan et al., 1995*), but they noted that an EUV wind cannot be the origin of the observed [OI] emission, as little or no neutral oxygen exists in the ionized flow. *Alexander* (2008b) subsequently made similar calculations for the [NeII] 12.81-μm line. The high ionization potential of Ne (21.6 eV) means that [NeII] emission can only arise in photoionized gas (*Glassgold*

et al., 2007), and the high critical density (5 × 10^5 cm^{-3}) means that most of the emission comes from close to the base of the photoevaporative flow. The predicted line luminosity is a few × 10^{-6} L$_\odot$, and scales linearly with the ionizing luminosity Φ. Moreover, the predicted [NeII] line profile varies strongly with disk inclination: For edge-on disks the (Keplerian) rotation dominates, leading to broad, double-peaked lines, while for face-on disks the line is narrower (≈10 km s^{-1}) and blue-shifted by 5–7 km s^{-1} (see also Fig. 5). *Alexander* (2008b) noted that detection of this blue-shift is possible with current mid-IR spectrographs (i.e., at λ/Δλ ≈ 30,000), and would represent a clear signature of a low-velocity, ionized wind.

Glassgold et al. (2007) showed that disk irradiation by stellar X-rays can result in strong [NeII] (and [NeIII]) emission from the disk surface layers. *Hollenbach and Gorti* (2009) subsequently presented analytic calculations of a number of fine-structure and hydrogen recombination lines from the EUV and X-ray heated layers. They found that the IR fine-structure lines scale linearly with the EUV and X-ray luminosities, and suggested that the ratios of the [NeII] 12.81-μm, [NeIII] 15.55-μm, and [ArII] 6.99-μm lines can be used as diagnostics of the incident spectrum, potentially distinguishing between EUV and X-ray irradiation. By comparing with the available data (from Spitzer) they concluded that internal shocks (in jets) or X-ray excitation dominates the production of the [NeII] line, unless the incident EUV spectrum is very soft. *Ercolano and Owen* (2010)

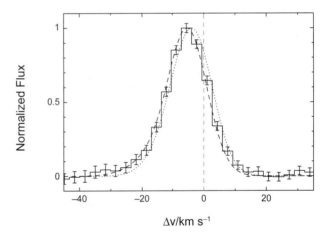

Fig. 5. [NeII] 12.81-μm line profile from the near face-on disk around TW Hya. The solid line (and error bars) show the observed profile from *Pascucci et al.* (2011), obtained by co-adding the spectra from the different position angles at which they observed. The dashed curve shows the theoretical prediction for an EUV-driven wind (*Alexander, 2008b*), while the dotted curve shows the corresponding prediction for X-ray photoevaporation (*Ercolano and Owen, 2010*); both profiles have been normalized by their peak flux. The observed blue-shift of 5.4 ± 0.6 km s^{-1} represents an unambiguous detection of a slow, ionized wind, but [NeII] observations alone do not distinguish between the different models.

then combined Monte Carlo radiative transfer calculations with the hydrodynamic wind solution of *Owen et al.* (2010) to construct a detailed atlas of emission lines from X-ray-heated winds. They computed fluxes and profiles for almost 100 lines, looking at both collisionally excited lines from metals and recombination lines of H/He, and investigated how the line emission varied with inclination and L_X. *Ercolano and Owen* (2010) again found blue-shifted [Ne II] emission [detected by *Pascucci and Sterzik* (2009)] to be the "smoking gun" of disk photoevaporation (see Fig. 5), but the relatively low ionization fraction ($\chi_e \simeq 0.01$) of the X-ray-heated wind means that, despite the much larger wind rate, the predicted [Ne II] luminosity and line profile are both very similar to those predicted for an EUV-driven wind (which has $\chi_e \simeq 1$). *Ercolano and Owen* (2010) were also able to reproduce the observed low-velocity component of the [O I] 6300-Å line, which they found to be excited by collisions with neutral hydrogen in the X-ray-heated wind. *Gorti et al.* (2011) also find that the [Ne II] emission is likely to originate in a predominantly neutral, X-ray-heated region but, for the specific case of TW Hya (where the [O I] 6300 Å line is not blue-shifted; see section 3.5), *Gorti et al.* (2011) instead argue that the observed [O I] emission is primarily nonthermal, and find that OH photodissociation (by stellar FUV photons) in a bound disk layer at ~10 AU naturally reproduces the luminosity and profile of the [O I] line (see also *Rigliaco et al.*, 2013). Taken together, these results suggest that observations of line emission should allow us to distinguish between different photoevaporation models.

The simplest statistical predictions of disk dispersal theory are the evolutionary timescales. The predictions of photoevaporative clearing models are generic in this respect: All predict rapid inside-out disk dispersal after approximately million-year disk lifetimes, satisfying the two-timescale condition discussed in section 1.1. The absolute timescales primarily reflect the choice of disk model (particularly the viscosity and initial mass distribution), but the relative timescales (e.g., the ratio of the clearing time to the lifetime) are only weakly dependent on these parameters (*Clarke et al.*, 2001; *Alexander et al.*, 2006b; *Alexander*, 2008a). The main difference between the models is that the clearing time is longer in the X-ray and FUV cases than the EUV (e.g., *Gorti et al.*, 2009; *Owen et al.*, 2010) (see also Fig. 4). The larger critical radius in these models results in a longer viscous timescale at the gap-opening radius, and the higher wind rates mean that disk clearing begins at a higher disk accretion rate. It therefore takes somewhat longer for the inner disk to drain onto the star [this phase was termed "photoevaporation-starved accretion" by *Drake et al.* (2009) and *Owen et al.* (2011)] and, because the disk mass is higher when the gap opens, photoevaporation also takes longer to remove the outer disk.

The most robust statistical prediction these models make is the distribution of disk accretion rates. In viscous disk models the accretion rate $\dot{M}(t)$ declines as a power-law (e.g., *Lynden-Bell and Pringle*, 1974; *Hartmann et al.*, 1998), and the photoevaporative wind essentially sets a lower cut-off

to this power-law (as seen in Fig. 4). We therefore expect a power-law distribution of accretion rates, truncated below the wind rate $\dot{M}_w$. *Alexander and Armitage* (2009) showed that, for EUV photoevaporation, a modest spread in disk parameters broadly reproduces both the magnitude and scatter of observed accretion rates. X-ray photoevaporation rates scale with the X-ray luminosity (equation (6)), and consequently the statistical models of *Owen et al.* (2011) found a negative correlation between accretion rate $\dot{M}$ and L_X (in a coeval population). We can make similar predictions for the distribution of disk masses, but the model disk masses are inevitably degenerate with the magnitude of the disk viscosity (as $\dot{M} = 3\pi\nu\Sigma$ in a steady-state accretion disk). For $\alpha \sim 0.01$ and EUV photoevaporation, the disk mass at the start of the clearing phase is ~0.001 $M_\odot$ (*Clarke et al.*, 2001; *Alexander et al.*, 2006b). The higher wind rates in X-ray and FUV models result in larger disk masses before disk clearing begins (*Gorti et al.*, 2009; *Owen et al.*, 2010), and these models generally favor lower disk viscosities in order to reproduce observed disk lifetimes and masses.

Moving beyond global disk properties, *Alexander et al.* (2006b) used a simple prescription to model the SEDs of their evolving disks. They showed that their models were consistent with the observed spread of CTT fluxes across a wide range of wavelengths (from near-IR to submillimeter) and, crucially, showed that the rapid clearing phase of the evolution corresponds to the poorly populated region between the observed loci of CTTs and WTTs (see Fig. 1). *Alexander and Armitage* (2009) found that a modest spread in disk parameters results in a significant scatter in disk lifetimes, and that the disk fraction of the population declines in a manner consistent with observations (e.g., *Mamajek*, 2009). The X-ray models of *Owen et al.* (2011) also successfully reproduce the observed decline in disk fraction with age. However, in this case the clearing timescale is significantly longer than in the EUV models, as discussed above, and consequently *Owen et al.* (2011) also predicted a significant population of non-accreting disks with large inner holes, which should only be detected at ≥50 μm (although thermal sweeping may rapidly disperse these disks).

Finally, we consider the predicted properties of so-called transitional disks. Broadly speaking, these are objects that are observed to have properties between those typical of CTTs and WTTs (see section 3.4 and the chapter by Espaillat et al. in this volume for more details), and most observational definitions of "transitional" require some degree of inner dust disk depletion (e.g., *Strom et al.*, 1989). *Alexander et al.* (2006b) noted that the inside-out clearing characteristic of UV-switch models invariably gives rise to a short "inner hole" phase, and suggested that some subset of the known transitional disks may be undergoing photoevaporative clearing. The predicted properties of such objects are again fairly generic: an inner disk cavity with size $\geq R_c$; little or no ongoing accretion ($\dot{M} \leq \dot{M}_w$); and a small outer disk mass. However, a number of other mechanisms have been also proposed to explain the observed transitional disks, ranging from the dynamical influence of planets to the evolution and growth of small dust grains (e.g., *Rice et al.,* 2003; *Quillen et al.,* 2004; *Dullemond and Dominik,* 2005; *Chiang and Murray-Clay,* 2007; *Krauss et al.,* 2007). *Alexander and Armitage* (2007) noted that the properties of planet-cleared inner disks are distinct from those cleared by photoevaporation, and suggested that the distribution of transitional disks in the $\dot{M}$-M_d plane can be used to distinguish between different mechanisms for producing disk inner holes [see also *Najita et al.* (2007a) and the reviews by *Najita et al.* (2007b) and *Alexander* (2008a)]. *Alexander and Armitage* (2009) showed that both of these mechanisms can operate simultaneously, and noted that the observational definition of "transitional" has a strong influence on how samples of such objects are interpreted (see also *Alexander,* 2008a). *Owen et al.* (2011) found that many of the observed transitional disks are consistent with models of X-ray photoevaporation, and suggested that the distribution of transitional disks in the $\dot{M}$-R_{hole} plane also represents an important diagnostic. *Owen et al.* (2012) also find that if the process of "thermal sweeping" operates as predicted, then outer disk clearing is extremely rapid and the population of non-accreting transitional disks should be relatively small. Interpreting observations of transitional disks remains challenging, but these results show that the properties of transitional disks can offer important insights into the process(es) of disk dispersal.

2.5. Magneto-Hydrodynamic Disk Winds

Blandford and Payne (1982) showed that an accretion disk threaded by a poloidal magnetic field can drive a MHD outflow. Unlike thermally driven winds, MHD outflows remove mass while also exerting a torque on the disk surface, and hence have a qualitatively different impact on secular disk evolution (see the chapters by Frank et al. and Turner et al. in this volume). Magneto-hydrodynamic winds can result in rapid disk evolution, and could even preclude disk formation entirely (*Li et al.,* 2011) (see also the chapter by Z.-Y. Li et al. in this volume). Stars with million-year-old disks evidently avoided that fate, perhaps as a consequence of having formed from gas with a relatively weak field (*Krumholz et al.,* 2013), but the remnant flux could still be dynamically important over longer timescales. The physics of "wind-driven" disks has been considered in detail by *Salmeron et al.* (2011), and MHD effects could contribute to disk dispersal if the typical magnetic flux threading protostellar disks is sufficiently large.

An organized magnetic field that supports a wind is more effective at driving disk evolution [by a factor $\sim(H/R)^{-1}$] than a turbulent field of the same strength. In general, however, there is no reason why winds and turbulent transport cannot coexist (*Balbus and Hawley,* 1998; *Shu et al.,* 2007). *Suzuki and Inutsuka* (2009) simulated the evolution of the magneto-rotational instability (MRI) in local stratified domains, threaded by a vertical field with a mid-plane plasma $\beta = 8\pi\rho c_s^2/B_z^2$ that varied between 10^4 and 10^7. In their fiducial case, with $\beta = 10^6$ (physically, a vertical

field of $\simeq 10^{-2}$ G at 1 AU), a few percent of the disk mass was lost within a hundred orbits, and these mass-loss rates are easily large enough to be important for disk dispersal (*Suzuki et al.,* 2010).

Outflows have been observed to form robustly in local disk simulations whenever a vertical field is present. *Fromang et al.* (2013) greatly extended the work of *Suzuki and Inutsuka* (2009), and studied how the derived outflows depended on critical numerical parameters (the vertical domain size, boundary conditions, and resolution). Mass loss was observed, but angular momentum transport (for $\beta = 10^4$) was dominated by turbulent rather than wind stresses. *Bai and Stone* (2013a,b) studied both the ideal-MHD limit and the specific case of protoplanetary disks, simulating the MRI in a local domain with vertical profiles of Ohmic and ambipolar diffusion appropriate to conditions at 1 AU. Mass loss was again observed, but in this dead zone regime (*Gammie,* 1996) the wind also dominated the evolution of angular momentum.

The existing simulations exhibit an unphysical dependence of the outlow properties on the boundary conditions, which is likely to be associated with the inherent approximation of a local geometry (*Lesur et al.,* 2013). Estimated mass-loss rates are as high as $\sim 10^{-8}$ $M_\odot$ yr^{-1} (*Bai,* 2013; *Simon et al.,* 2013), but global simulations are required to quantify the true mass-loss rate and to make testable observational predictions. Nonetheless, it seems clear that there is a continuum between the classical limits of viscous and wind-driven disks, and that as mass is accreted the dynamical importance of any nonzero vertical flux must rise, potentially becoming important during the dispersal phase (*Armitage et al.,* 2013).

3. OBSERVATIONS OF DISK DISPERSAL

Having outlined the theory of disk dispersal, we now consider the observational evidence, focusing in particular on observations that inform and constrain our theoretical models. We consider the high-energy radiation fields of TTs, direct diagnostics of disk photoevaporation, and indirect, statistical studies of disk evolution and dispersal. We also review what we can learn from observations of transitional disks, before presenting a detailed case study of TW Hya (our nearest and best-studied protoplanetary disk).

3.1. High-Energy Emission from T Tauri Stars

As discussed in section 2.1, the input radiation fields remain a major uncertainty in models of disk photoevaporation. This problem is particularly acute in the EUV, where interstellar absorption prohibits direct observation of ionizing photons. *Kamp and Sammar* (2004) used a scaling argument (based on solar observations) to estimate the high-energy spectrum of a young, active G-type star, and their spectrum has an ionizing luminosity $\Phi \simeq 2.5 \times 10^{41}$ s^{-1}. *Alexander et al.* (2005) then used previously derived emission measures to estimate the ionizing emission from several massive, luminous CTTs, and derived values $\Phi \geq 10^{42}$ s^{-1}. *Alexander et al.* (2005) also suggested that the UV HeII/CIV line ratio may be used as a diagnostic of the chromospheric emission from TTs; however, recent high-resolution spectra show that for CTTs these lines in fact originate primarily in the accretion flow, and do not trace the chromospheric emission well (*Ardila et al., 2013*). *Herczeg* (2007) estimated $\Phi \simeq 5 \times 10^{41}$ s^{-1} for TW Hya, and *Espaillat et al.* (2013) estimated $\Phi \simeq 10^{42}$–10^{43} s^{-1} for SZ Cha, but accurate measurements of the EUV luminosity remain scarce (although free-free emission offers a promising alternative diagnostic, as discussed below).

By contrast, X-rays from TTs have been observed for more than 30 years (e.g., *Feigelson and Decampli*, 1981), and are now well-characterized (*Feigelson et al., 2007; Güdel and Nazé, 2009*). X-ray luminosities range from $L_X \lesssim 10^{28}$ erg s^{-1} to $L_X \gtrsim 10^{32}$ erg s^{-1}, with a spectrum that peaks around 1 keV and is broadly consistent with emission from a ~10^7-K plasma. Some fraction of the observed X-ray emission (particularly at low energies) may originate in the accretion flow (*Kastner et al., 2002; Dupree et al., 2012*), and this poorly characterized "soft excess" (at 0.3–0.4 keV) dominates the X-ray luminosity of a small number of CTTs (*Güdel and Nazé, 2009*). However, magnetic reconnection events in the stellar chromosphere and/or corona are thought to dominate the X-ray emission from the majority of TTs (e.g., *Feigelson and Montmerle, 1999*). X-ray emission from TTs is highly variable, and shows a weak anticorrelation with measured accretion rates (*Feigelson et al., 2007*), but young stars show only a modest decline in their X-ray emission on timescales ~100 m.y. (*Ingleby et al., 2011a; Stelzer et al., 2013*).

Far-ultraviolet observations of TTs are more difficult to interpret, as the bulk of the FUV emission from CTTs originates in the accretion shock and to first order $L_{FUV} \propto \dot{M}$ (e.g., *Calvet and Gullbring, 1998; Gullbring et al., 1998; Yang et al., 2012*). However, for high $\dot{M}$, the accretion columns and any magnetically driven jet or outflow strongly shield the disk from UV photons produced in the accretion shock, and in the models of *Gorti et al.* (2009) the photoevaporative mass-loss rate is in fact only weakly dependent on the stellar accretion rate. As discussed in section 2.3, however, one cannot self-consistently use the accretion luminosity to shut off disk accretion, so the chromospheric FUV emission is most critical for disk dispersal. Recent observations by *Ingleby et al.* (2012) find that the chromospheric FUV in the range 1230–1800 Å (h$\nu \simeq$ 7–10 eV) saturates at $\simeq 10^{-4}$ L$_*$. However, *France et al.* (2012) and *Schindhelm et al.* (2012) (see also *Herczeg et al., 2002, 2004*) were recently able to reconstruct the FUV radiation fields of several CTTs from spectra of fluorescent H$_2$ emission, and found that the FUV luminosity is dominated by line emission, with $\geq$90% of the total FUV flux being emitted in Lyα. The integrated FUV luminosity is therefore $L_{FUV} \sim 10^{-3}$ L$_*$. *Gorti et al.* (2009) adopt a constant stellar/chromospheric luminosity of $L_{FUV} \simeq 5 \times 10^{-4}$ L$_*$, consistent with these observations, but the models

do not yet include the large contribution from Lyα. This is unlikely to alter the heating rates dramatically, but should be taken into account in future studies.

HJLS94 and *Lugo et al.* (2004) computed the free-free emission from photoevaporative winds around massive stars, and *Pascucci et al.* (2012) have recently calculated the free-free continuum emission and H radio recombination lines arising from a fully (EUV) or partially ionized (X-ray) protoplanetary disk surface. They show that the free-free continuum produces excess emission on top of the dust continuum at centimeter wavelengths, and is detectable with current radio instruments. Such excess emission at 3.5 cm is detected from the photoevaporating disk around TW Hya (*Wilner et al., 2005*, and references therein). *Pascucci et al.* (2012) show that if the stellar X-ray luminosity is known, one can estimate the X-ray contribution to the free-free emission and thus find the EUV contribution; in other words, it is possible to measure the stellar EUV flux that the disk receives. In the case of TW Hya, *Pascucci et al.* (2012) find that EUV photons dominate the observed free-free emission and estimate $\Phi \sim 5 \times 10^{40}$ s^{-1} at the disk surface. *Owen et al.* (2013a) have recently extended this analysis with detailed numerical calculations of free-free emission from EUV- and X-ray-irradiated disks, and find that the free-free emission scales approximately linearly with either Φ or L_X, in agreement with *Pascucci et al.* (2012). *Owen et al.* (2013a) also argue that if disks can be observed close to the end of their lifetimes (i.e., where photoevaporation starts to overcome disk accretion), then the free-free flux should scale $\propto \dot{M}^2$ in the EUV-driven case, but $\propto \dot{M}$ in the X-ray driven case.

3.2. Direct Observations

As discussed in section 2.4, directly probing photoevaporative flows requires us to identify the gas lines that trace the heated disk surface layers. Spitzer Infrared Spectrograph (IRS) observations were the first to discover such possible tracers, via the [NeII] emission line at 12.81 μm (*Pascucci et al., 2007; Lahuis et al., 2007*). Due to the low spatial and spectral resolution of the Spitzer IRS, these data cannot prove that the [NeII] line is indeed a disk diagnostic for individual sources. However, studies of over 100 TTs show that sources with known jets/outflows have systematically higher [NeII] luminosities (by 1–2 orders of magnitude) than sources with no jets, and also find a weak correlation between $L_{[NeII]}$ and L_X (*Güdel et al., 2010; Baldovin-Saavedra et al., 2011*). These results point to shock-induced emission in circumstellar gas dominating the Spitzer [NeII] fluxes of jet sources, but lend support to the disk origin for evolved and transitional disks with no jets. *Szulágyi et al.* (2012) subsequently considered the detection statistics of the [NeII] 12.81-μm, [NeIII] 15.55-μm, and [ArII] 6.98-μm lines in a large sample of transitional disks and measured the line flux ratios. Although the number of detections is small, the [NeII] line is typically 10× brighter than the [NeIII] line, and similar in flux to the [ArII] line. Charge exchange between Ne^{++} and H in partially ionized gas naturally leads to a [NeII]/[NeIII] ratio ~10 (*Glassgold et al., 2007; Hollenbach and Gorti, 2009*), which suggests that X-ray irradiation dominates the heating and ionization of the disk surface traced by these forbidden lines [although [NeII]/[NeIII] ratios less than unity have recently been reported for some class I and II sources (*Espaillat et al., 2013; Kruger et al., 2013*)].

As the line falls in an atmospheric window, bright [NeII] emission can be followed up with high-resolution (~10 km s^{-1}) groundbased spectrographs, allowing us to disentangle the disk from the jet/outflow contribution. The first such observations hinted at a flux enhancement on the blue side of the [NeII] line from the transitional disk TW Hya (*Herczeg et al., 2007*). Higher signal-to-noise spectra then found unequivocal evidence of central star-driven photoevaporation in three transitional disks (TW Hya, T Cha, and CS Cha): modest linewidths (FWHM ≃ 15–40 km s^{-1}) accompanied by small (≃3–6 km s^{-1}) blue-shifts in the peak centroid (*Pascucci and Sterzik, 2009*, see also Fig. 5 and section 3.5). As of this writing, 55 Spitzer [NeII] detections have been followed up with groundbased high-resolution spectrographs (*van Boekel et al., 2009; Najita et al., 2009; Pascucci and Sterzik, 2009; Sacco et al., 2012; Baldovin-Saavedra et al., 2012*). Twenty-four of these resulted in detections, with all the detected lines also being spectrally resolved. Eight of these detections are class I/II sources where most of the unresolved Spitzer [NeII] emission clearly arises in jets/outflows: The [NeII] emission is broad (≥40 km s^{-1}) and blue-shifted by ≥50 km s^{-1} [and in the case of T Tau is also spatially resolved (*van Boekel et al., 2009*)]. The [NeII] lines from three sources (AA Tau, CoKu Tau/1, and GM Aur) were interpreted as tracing bound gas in a disk, heated by stellar X-rays (*Najita et al., 2009*). However, AA Tau and CoKu Tau/1 are seen close to edge-on (i > 70°) and are known to power jets (e.g., *Hartigan et al., 1995; Baldovin-Saavedra et al., 2012*), which may contaminate the observed [NeII] lines. Finally, 13 sources have narrow [NeII] lines (~15–50 km s^{-1}) and small blue-shifts (~2–18 km s^{-1}), consistent with photoevaporative winds. Among these objects the transitional disks typically show smaller linewidths and blue-shifts than the class I/II sources, but the interpretation of these wind sources is not straightforward. [NeII] linewidths of 15–25 km s^{-1} can be produced in photoevaporative winds (*Alexander, 2008b; Ercolano and Owen, 2010*), but broader lines are difficult to reconcile with photoevaporation models unless the disks are viewed close to edge-on. In addition, the measured blue-shifts cluster around ≃10 km s^{-1} (*Sacco et al., 2012*), somewhat larger than predicted for X-ray winds and more in line with EUV-driven wind models (*Alexander, 2008b*). Thus, while small blue-shifts in the [NeII] emission unambiguously point to ongoing photoevaporation, [NeII] lines alone cannot be used to measure photoevaporation rates (*Pascucci et al., 2011*). This is due to the degeneracy discussed in section 2.4: A low-density wind with a high ionization fraction (as predicted for EUV photoevaporation) and a higher-density wind with a lower ionization fraction (as predicted for X-ray photoevaporation) both result in

very similar [NeII] emission. Further diagnostics are therefore necessary to determine the primary heating/ionization mechanism, and to measure photoevaporative wind rates.

These results, and the predictions of photoevaporative wind models, have recently motivated a reanalysis of the optical forbidden lines detected toward TTs. In particular, oxygen forbidden lines have long been known to display two components: a high-velocity component (HVC), blue-shifted by hundreds of kilometers per second with respect to the stellar velocity, and a low-velocity component (LVC), blue-shifted by a few to several kilometers per second (e.g., *Hartigan et al., 1995*). While the HVC unambiguously traces accretion-driven jets, as with the [NeII] HVC, the origin of the LVC has remained a mystery. As discussed in section 2.4, reproducing the large [OI] line luminosities via thermal excitation in a wind requires a mostly neutral layer at high temperatures (~8000 K), as predicted for soft X-ray heating (*Font et al., 2004; Hollenbach and Gorti, 2009; Ercolano and Owen, 2010*). Alternatively, the [OI] LVC could trace a cooler (<1000 K) disk layer where neutral oxygen is produced by OH photodissociation, as proposed for TW Hya by *Gorti et al.* (2011). In this case the observed [OI] emission is not blue-shifted (somewhat unusually), hinting at a disk rather than wind origin (*Pascucci et al., 2011*). Comparison of [NeII] and [OI] line profiles in this manner is still limited to a small sample. However, in the five sources where contamination from jet emission can be excluded, the [NeII] line shows a larger peak blue-shift than the [OI] line, and the line profiles are sufficiently different to suggest that the two lines originate in physically distinct regions (*Pascucci et al., 2011; Rigliaco et al., 2013*). *Rigliaco et al.* (2013) also reanalyzed the Taurus TT sample of *Hartigan et al.* (1995) and found (1) a tight correlation between the luminosity of the [OI] LVC and the stellar accretion rate (and therefore the stellar FUV flux); and (2) a relatively small range of [OI] 6300/5577-Å line ratios over a very large range in luminosity, which they argue is difficult to reproduce in thermally heated gas (see also *Gorti et al., 2011*). These results suggest that the [OI] LVC traces the region where stellar FUV photons dissociate OH molecules, and the typical blue-shifts (~5 km s^{-1}) point to the emission arising in unbound gas. Whether this wind is FUV-, X-ray-, or magnetically driven remains unclear. However, if the [OI] LVCs do trace photoevaporative winds, then photoevaporation must be ubiquitous in class II disks, with mass-loss rates $\dot{M}_w$ ≥ 10^{-9} M$_\odot$yr^{-1}.

Finally, large disk surveys of rovibrational CO-line emission at 4.7 μm have recently identified an interesting subclass of single-peaked CO-line sources (e.g., *Brown et al., 2013*) (see also the chapter by Pontoppidan et al. in this volume). The spectroastrometric signal of the highest S/N examples is consistent with a combination of gas in Keplerian rotation plus a slow (few kilometers per second) disk wind, at approximately astronomical-unit radii (*Pontoppidan et al., 2011*). *Brown et al.* (2013) also find that the majority of the CO profiles have excess emission on the blue side of the line, which further supports the wind hypothesis and suggests that molecular winds are common in class I/II sources. Whether these winds are thermally or magnetically driven is not clear, but several CO wind sources are also known to have strong [O I] HVC emission (e.g., *Hartigan et al., 1995; Rigliaco et al., 2013*). This hints at a magnetic origin, and cold molecular outflows at approximately astronomical-unit radii are difficult to reconcile with a purely thermal wind scenario, but detailed comparisons with models have not yet been possible. Future studies of spectrally resolved line profiles should allow us to understand the relationship between these different wind diagnostics, and to measure mass-loss rates empirically.

3.3. Indirect Observations

Much of our knowledge about the evolution and eventual dispersal of circumstellar disk material comes from indirect, demographic studies of the fundamental observational tracers of disk gas and dust. In section 1 we discussed the basic constraints imposed by these studies. Here we review the observations behind these constraints in more detail, and discuss the extent to which demographic surveys can be used to test theoretical models. The most common approach is to characterize a sample of tracer measurements in a young star cluster of a given age, and then compare that ensemble to similar results obtained for star clusters with different ages: in essence, a relatively straightforward statistical comparison of how the disk tracer varies with time. Alternatively, specific disk dispersal mechanisms can be constrained by studying how these disk tracers vary with respect to environment, stellar host properties, or some other evolutionary proxy. In principle, the relationship between these tracers and age (or its proxy) can then be directly compared with the predictions of disk evolution models.

Arguably the most robust and testable of those predictions is the decay of accretion rates with time demanded by viscous evolution models (e.g., *Hartmann et al., 1998*) (see also section 2.3). Accretion rates are usually derived from ultraviolet continuum excesses (*Calvet and Gullbring, 1998; Gullbring et al., 1998*) or H recombination lines (*Muzerolle et al., 1998, 2001*), benchmarked against magnetospheric flow and accretion shock calculations. Combining these accretion rates with stellar ages, estimated via grids of pre-main-sequence stellar evolution models, there is significant observational evidence to support the standard viscous disk paradigm: Both the frequency of accretors and their typical $\dot{M}$ values decrease substantially from ~1 to 10 m.y. (e.g., *Muzerolle et al., 2000; Fang et al., 2009; Sicilia-Aguilar et al., 2010; Fedele et al., 2010*), although the inferred stellar ages remain subject to significant uncertainties (see the chapter by Soderblom et al. in this volume). However, current surveys of TT accretion rates do not yet allow detailed comparisons with models of disk dispersal; in particular, the lack of useful upper limits for weakly- and non-accreting disks severely limits the statistical power of these data (see, e.g., the discussion in *Clarke and Pringle, 2006*).

Accretion signatures are definitive evidence for the presence of gas in the inner disk, but the converse is not necessarily true and measured accretion rates do not quantify the available gas mass. However, when grounded on a set of detailed physico-chemical models (*Gorti and Hollenbach, 2004; Woitke et al., 2009, 2010; Kamp et al., 2010, 2011*), observations of mid-IR (e.g., [S I], H_2) and far-IR (e.g., [O I], [C II]) cooling lines can be sensitive and direct tracers of even small amounts of gas at radii <50 AU. Observations of these lines with Spitzer (e.g., *Hollenbach et al., 2005; Pascucci et al., 2006; Chen et al., 2006*) and Herschel (e.g., *Mathews et al., 2010; Woitke et al., 2011; Lebreton et al., 2012*) indicate that these gas reservoirs are depleted within ≲10 m.y. Millimeter-wave spectroscopic searches for rotational transitions of the abundant CO molecule in older (debris) disks suggest that this depletion timescale applies at much larger radii as well (*Zuckerman et al., 1995; Dent et al., 1995, 2005; Najita and Williams, 2005*). Additional constraints on the gas mass in the inner disk come from far-UV spectra of H_2 electronic transitions (e.g., *Lecavelier des Etangs et al., 2001; Roberge et al., 2005*), which indicate that stars older than ~10 m.y. or that exhibit no signatures of accretion have virtually no gas at radii <1 AU (*Ingleby et al., 2009, 2012; France et al., 2012*).

These statistical signatures of gas dispersal in the inner disk are also seen in analogous trends for disk solids. Even a small amount of dust emits a substantial IR continuum luminosity, so determining the presence or absence of warm dust grains in the inner disk (at ≲1 AU) is relatively straightforward, even for large samples. Infrared photometric surveys have been conducted for ~20 nearby young stellar associations, providing estimates of the fraction of young stars with excess emission from warm dust at ages ranging from <1 to ~30 m.y. (e.g., *Haisch et al., 2001; Luhman, 2004; Lada et al., 2006; Hernández et al., 2007*). Summarizing those results, *Mamajek* (2009) showed that the inner disk fraction decays exponentially with a characteristic timescale of ≈2.5 m.y. As with the gas tracers, there is a substantial population of young stars (up to 40% at ~2 m.y.) that exhibit no near-IR excess from warm dust in an inner disk, and a small (but not negligible, ≲10%) sample of older (~10–20 m.y.) pre-main-sequence stars that retain their dust (and gas) signatures (e.g., TW Hya; see section 3.5). However, recent high-resolution imaging surveys revealed that a significant fraction of the youngest "disk-less" stars and WTTs are in fact close (<40 AU) binaries (*Ireland and Kraus, 2008; Kraus et al., 2008, 2011*). In this case dynamical clearing is expected to suppress most inner disk tracers, and the "corrected" disk fraction for single stars in young clusters (with ages ≲1–2 m.y.) is close to 100% (*Kraus et al., 2012*). Finally, although in most cases there is a correspondence between accretion and dust signatures (e.g., *Hartigan et al., 1990; Fedele et al., 2010*), recent studies have identified a small but significant population of young stars that show weak (or very red) dust emission but no hints of accretion (e.g., *Lada et al., 2006; Cieza et* al., 2007, 2013); this may be evidence for substantial radial evolution in the disk at late times (see section 3.4 and the chapter by Espaillat et al. in this volume).

The luminosity of the optically thin millimeter-wave emission from a disk is the best available quantitative diagnostic of its dust mass (*Beckwith et al.,* 1990). However, until recently, such observations were difficult to obtain for large samples, so demographic studies are not yet mature. In nearby low-mass clusters with ages of ~1–3 m.y., millimeter-wave emission consistent with a dust mass of 1–1000 $M_\oplus$ is found for essentially all stars with accretion signatures and near-IR excess emission (*Andrews and Williams,* 2005, 2007). Comparisons with samples in older associations suggest that the typical dust mass has declined substantially by ~5 m.y. (e.g., *Carpenter et al.,* 2005; *Mathews et al.,* 2012).

These global properties of protoplanetary disks suggest that the accreting stage is mostly driven by viscous evolution, while the non-accreting phase is dominated by a different dispersal process such as photoevaporation (e.g., *Williams and Cieza,* 2011). However, demographic surveys do not currently provide a strong means of discriminating between the theoretical models of disk dispersal discussed in section 2. The broad conclusions of early demographic work provided much of the original motivation for these models, but the statistical power of these studies has not advanced significantly in the intervening period. There is a large dispersion in the properties of individual systems at any given stellar age, and most demographic tracers also show a systematic dependence on stellar mass (e.g., *Muzerolle et al.,* 2005; *Andrews et al.,* 2013; *Mohanty et al.,* 2013). Moreover, most surveys are biased against disk-less stars and WTTs, and uniform samples of nondetections or upper limits are rare. Environmental factors such as external photoevaporation (e.g., *Johnstone et al.,* 1998; *Mann and Williams,* 2010) (see section 2.2.4) or tidal stripping by binary companions (e.g., *Artymowicz and Lubow,* 1994; *Harris et al.,* 2012) also "contaminate" demographic data, further complicating comparisons with models. Thus, while demographic surveys provide important clues to our understanding of disk evolution and dispersal, their ability to discriminate between models is currently limited by a combination of selection biases and poor number statistics. This approach is potentially very powerful, however, and we urge that future demographic surveys strive toward uniform sensitivity in unbiased samples.

3.4. Transitional Disks

Broadly speaking, transitional disks are protoplanetary disks with a significant deficit of near-IR and/or mid-IR flux with respect to the median SED of CTTs (e.g., *Strom et al.,* 1989). This definition includes both objects with IR SEDs that are smoothly falling with wavelength, and systems with clear "dips" in their SEDs. While the former group may contain objects with continuous disks extending inward to the dust destruction radius, disks in the latter group show clear evidence for inner holes and gaps and are the focus of our discussion. (The chapter by Espaillat et al. in this volume discusses observations of transitional disks in much greater detail.) As discussed in section 2.4, a variety of different mechanisms have been proposed to explain the holes and gaps in transitional disks, including photoevaporation (*Alexander et al.,* 2006b), grain growth (*Dullemond and Dominik,* 2005), and dynamical clearing by giant planets or (sub)stellar companions (*Artymowicz and Lubow,* 1994; *Lubow and D'Angelo,* 2006). These processes are not mutually exclusive, and are likely to operate simultaneously (e.g., *Williams and Cieza,* 2011). For our purposes, the key question regarding transitional disks is whether or not their gaps and cavities are mainly due to photoevaporation, or to the other processes listed above; essentially, we would like to know what fraction (if any) of the observed transitional disks are undergoing disk dispersal.

Near-IR interferometry (*Pott et al.,* 2010), adaptive optics imaging (*Cieza et al.,* 2012), and aperture masking observations (*Kraus et al.,* 2011) have shown that the observed inner holes and gaps in transitional disks are rarely due to close stellar companions or brown dwarfs. Similarly, grain growth models have difficulties explaining the large (sub)millimeter cavities observed in resolved images of transitional disks (*Birnstiel et al.,* 2012). Moreover, *Najita et al.* (2007a) and *Espaillat et al.* (2012) find that transitional disks tend to have lower accretion rates than CTTs with similar disk masses, which suggests that the radial distribution of gas in transitional disks is different from that in CTTs. Photoevaporaton and dynamical clearing by forming planets thus remain the leading explanations for most inner holes and gaps. In particular, photoevaporative clearing nicely explains the incidence and properties of transitional disks around WTTs. Such systems represent ~10% of the pre-main-sequence population in nearby molecular clouds (*Cieza et al.,* 2007), and tend to be very faint at millimeter wavelengths. Moreover, their SEDs appear to trace the inside-out dispersal of protoplanetary disks, and are consistent with objects seen during passage from the primordial to the debris disk stage (*Wahhaj et al.,* 2010; *Cieza et al.,* 2013).

The importance of photoevaporation in accreting transitional disks, however, is more controversial. As discussed in section 2.4, whether or not photoevaporation can produce inner holes and gaps in accreting transitional disks depends on the mass-loss rate, and for sufficiently high rates a gap opens while the disk is still relatively massive. Under these circumstances the inner disk accretes onto the star at detectable levels ($\gtrsim 10^{-10}$ $M_\odot$ yr^{-1}) for a significant period of time, and during this phase appears as an accreting transitional disk. The low mass-loss rates predicted for EUV photoevaporation cannot therefore explain observed accreting transitional disks, but the much higher wind rates predicted for X-ray and FUV photoevaporation can account for some accreting transitional objects. However, even these models cannot account for the transitional disks with strong ongoing accretion that show very large cavities (R ≥ 20–80 AU) in resolved (sub)millimeter images (*Owen et al.,* 2011; *Morishima,* 2012). The masses of the known large-cavity disks are higher than the median disk population but, as (sub)millimeter imaging surveys have so far focused on the brightest, most massive disks, the significance of this result is not yet clear (*Andrews et al.,* 2011). Dynamical clearing (by giant planets) seems to be the most likely explanation for these systems, but even this scenario has difficulty accounting for all the observed properties of these unusual objects (e.g., *Zhu et al.,* 2012; *Clarke and Owen,* 2013).

Transitional disks also provide observational tests for X-ray photoevaporation models, as the mass-loss rates scale with L_X (equation (6)). This implies that (1) accreting transitional disks should have, on average, higher L_X than coeval CTTs with "full" disks; and (2) there should be a weak anticorrelation between $\dot{M}$ and the size of the cavity, and few objects with large (≥20 AU) cavities and detectable accretion rates (*Owen et al.,* 2011, 2012). These theoretical predictions have not been verified in the largest samples of transitional disks studied to date (*Kim et al.,* 2013). Similarly, observations of protoplanetary disks show no difference between the FUV emission levels of systems with transitional disks and those with full disks (*Ingleby et al.,* 2011a). These results suggest that either the observational uncertainties and/or selection biases are large enough to mask the predicted correlations, or that the inner holes and gaps of accreting transitional disks are mostly not the result of X-ray- or FUV-driven photoevaporation. Higher photoevaporation rates throughout disk lifetimes are also difficult to reconcile with systems that have stringent upper limits for their accretion rates (<10^{-10} M$_\odot$ yr^{-1}), yet show no evidence for holes or gaps in their SEDs (*Ingleby et al.,* 2011b), and the increasing evidence for non-axisymmetric structures and dynamical clearing in a number of accreting transitional disks (e.g., *Kraus and Ireland,* 2012; *Casassus et al.,* 2013; *van der Marel et al.,* 2013) also argues against a photoevaporative origin. Moreover, the (sub)millimeter fluxes and accretion rates of transitional disks suggests that there may in fact be two distinct populations of transitional disks (*Owen and Clarke,* 2012): those with low disk masses and modest to nondetectable accretion (whose inner holes are primarily due to photoevaporation); and those with large disks masses and high accretion rates (whose inner holes could be caused by giant planet formation). Overall, the properties of transitional disks favor models where photoevaporative mass-loss rates are typically low (≤10^{-10}–10^{-9} M$_\odot$ yr^{-1}), and only overcome accretion when disks masses and accretion rates are also low.

3.5. TW Hya: A Case Study of Late-Stage Disk Evolution

The transitional disk TW Hya is a unique benchmark for protoplanetary disk physics. In addition to being our nearest [54 ± 6 pc (*van Leeuwen,* 2007)] and best-studied protoplanetary disk, it is one of the oldest known gas-rich systems [~10 m.y. (*Torres et al.,* 2008)]. The original identification of TW Hya as a transitional object was made by *Calvet et al.* (2002), who found that the observed flux deficit at λ ≤ 10 μm can be modeled with a disk that is depleted of small (≤micrometer) dust grains within ≤4 AU, leaving an optically thin inner cavity. The cavity is not completely empty, however; near- and mid-IR interferometric observations have spatially resolved the warm dust emission, and suggest that an optically thick dust component is present within 4 AU (*Eisner et al.,* 2006; *Ratzka et al.,* 2007; *Akeson et al.,* 2011). The exact location and nature of this component remain unclear, and both a dust ring (*Akeson et al.,* 2011) and a self-luminous companion (*Arnold et al.,* 2012) have been suggested. Interferometric observations at 7 mm find a lack of emission at small radii, which is consistent with the presence of a 4-AU dust cavity (*Hughes et al.,* 2007). The millimeter-sized dust disk is found to extend out to 60 AU, where it appears to be sharply truncated, while the gas component extends to >200 AU (*Andrews et al.,* 2012).

TW Hya's average accretion rate, obtained from eight different optical diagnostics, is ≈7 × 10^{-10} M$_\odot$ yr^{-1} (*Curran et al.,* 2011). This is an order of magnitude below the median accretion rate of CTTs in Taurus (e.g., *Gullbring et al.,* 1998), but still clearly places TW Hya in the group of accreting transitional disks (see section 3.4). Variability in some of these optical diagnostics points to variable mass accretion [hinting at rates as high as ~10^{-8} M$_\odot$ yr^{-1} at times (*Alencar and Batalha,* 2002)], but contamination by a stellar wind may contribute to the observed variations (*Dupree et al.,* 2012). Reproducing all the observed gas emission lines also requires significant depletion of gas within the 4-AU dust cavity, by 1–2 orders of magnitude compared to the gas surface density in the outer disk (*Gorti et al.,* 2011). Estimates for the outer disk mass range from 5 × 10^{-4} M$_\odot$ (*Thi et al.,* 2010), suggesting substantial depletion, to 0.06 M$_\odot$ (*Gorti et al.,* 2011), suggesting little or no depletion with respect to "normal" disks. The recent *Herschel* detection of HD in the TW Hya disk also favors a large disk mass (*Bergin et al.,* 2013).

While the presence of a giant planet is the leading hypothesis to explain the 4-AU cavity (e.g., *Calvet et al.,* 2002), the moderate depletion of dust and gas and the relatively low stellar accretion rate are also consistent with some of the star-driven photoevaporation models discussed in section 2. Moreover, the detection of a small blue-shift (≈5 km s^{-1}) in the [NeII] 12.81-μm line demonstrates that the TW Hya disk is indeed losing mass via photoevaporation (*Pascucci and Sterzik,* 2009; *Pascucci et al.,* 2011) (see also Fig. 5). In addition, these data show that more than 80% of the [NeII] emission comes from beyond the dust cavity and is confined within ~10 AU (*Pascucci et al.,* 2011), in agreement with the predictions of EUV- and X-ray-driven photoevaporation models. However, the [OI] 6300-Å line from TW Hya has only a moderate width (≈10 km s^{-1}) and is centered on the stellar velocity (*Alencar and Batalha,*

2002; *Pascucci et al.,* 2011), which suggests that the [OI] emission originates in a bound disk layer rather than a wind (*Gorti et al.,* 2011). We also note that if TW Hya's excess flux at 3.5 cm is due to free-free emission, this implies an EUV luminosity of $\sim 5 \times 10^{40}$ s^{-1} incident on the disk surface (*Pascucci et al.,* 2012). This is close to the fiducial luminosity assumed in models (equation (4)), and is similar to the value derived by assuming that the [NeII] emission is entirely due to EUV photoevaporation [7.5×10^{40} s^{-1} (*Alexander,* 2008b; *Pascucci et al.,* 2011)].

While it is clear that the disk of TW Hya is currently undergoing photoevaporation, current data are not sufficient to conclude that photoevaporation is responsible for the 4-AU cavity. Empirical measurements of the mass-loss rate are needed, and only rates higher than the current accretion rate would be consistent with a cavity carved by photoevaporation. A further drawback of the photoevaporation hypothesis is that there is a relatively short window ($\sim 2 \times 10^5$ yr) during which we could observe a photoevaporation-induced gap and still detect accretion onto the star (e.g., *Owen et al.,* 2011). This, coupled with the apparently large disk mass, favors the giant planet hypothesis for clearing the inner disk (e.g., *Gorti et al.,* 2011). However, regardless of which mechanism is ultimately responsible for the formation of the cavity, the disk of TW Hya offers a unique insight into how multiple disk dispersal mechanisms can operate concurrently, and may even couple to one another in driving final disk clearing. As our observational capabilities improve in the coming years we should be able to study many more objects in this level of detail, and build up a comprehensive picture of how disk dispersal operates in a large number of systems.

4. IMPLICATIONS AND CONSEQUENCES OF DISK DISPERSAL

Having reviewed both the theory and observations of disk evolution and dispersal, we now move on to consider how these processes affect the formation and evolution of planetary systems. Changes in disk properties have the potential to alter the microphysics of planet formation, and the effects of disk dispersal have important consequences for the dynamics of forming planetary systems. Here we consider each of these in turn, before summarizing our conclusions and discussing prospects for future work.

4.1. Chemical Effects and Impact on Planet Formation

We have seen that photoevaporative mass loss, whether driven by the central star or by nearby massive stars, removes material from the disk surface over much of the planet-forming epoch. However, for most of the disk lifetime the column directly affected by photoevaporation is only a small fraction of the disk surface density. For typical disk parameters, an X-ray heated surface layer with column density (along the line of sight to the star) $N_H \sim 10^{22}$ cm^{-2}

reaches down to only 3–4 H above the disk mid-plane at approximately astronomical-unit radii. In such low-density gas the vertical settling timescale for dust is very short (*Dullemond and Dominik,* 2005), and turbulence will lift only the smallest (submicrometer) particles (*Dubrulle et al.,* 1995) into the launching zone of the wind. Photoevaporation therefore preferentially removes dust-poor material from the disk, increasing the dust-to-gas ratio as the disk evolves. This latter quantity is known to be crucial in driving collective mechanisms for planetesimal formation, including the streaming and gravitational instabilities (e.g., *Chiang and Youdin,* 2010), and consequently photoevaporation may play a significant role in the formation of planets.

Throop and Bally (2005) studied the effect of photoevaporation on planetesimal formation in the context of externally irradiated protoplanetary disks. They considered disks around low-mass stars, and modeled the dust distribution under the limiting assumption of a nearly laminar disk, where the only source of turbulence is the Kelvin-Helmoltz instability generated when the dust layer becomes too dense (*Sekiya,* 1998). The disks were then subjected to "external" FUV/EUV photoevaporation, as expected close to massive stars in the center of massive star-forming regions such as Orion (see section 2.2.4). *Throop and Bally* (2005) adopted the mass-loss rates of *Johnstone et al.* (1998), and assumed that small dust grains were entrained in the wind so long as their volume density did not exceed the gas density in the wind. Under these conditons, *Throop and Bally* (2005) found significant photoevaporative enhancement in the dust-to-gas ratio, reaching values high enough to meet the gravitational instability threshold of *Youdin and Shu* (2002) between 5 and 50 AU.

It is, however, highly unlikely that photoevaporation is a prerequisite for all planetesimal formation. As *Throop and Bally* (2005) observed, if planetesimal formation does not begin until the disk-dispersal epoch, then there is insufficient time to form giant planet cores before all the gas is gone. The most plausible role for photoevaporation in planetesimal formation is instead as a possible mechanism for forming a second generation of planetesimals, at later times or at larger radii than would otherwise be possible. The predicted evolution of the gas surface density during photoevaporation by the central star favors such a scenario. Using a one-dimensional model of EUV-driven photoevaporation, *Alexander and Armitage* (2007) found that radial pressure gradients can lead to the formation of a ring of enhanced dust-to-gas ratio as the gas disk is dispersed from the inside out. Again, too little gas remains at this stage for the results to be important for gas-giant formation, but planetesimal formation at a late epoch could still play a role in the formation of terrestrial planets or debris disks (*Wyatt,* 2008). The key uncertainty is how much solid material remains at relatively large radii late in the disk evolution. Absent an efficient particle trapping mechanism (e.g., *Pinilla et al.,* 2012), the outer region of an evolving disk will be depleted of solids under the action of radial drift long before photoevaporation becomes dominant (*Takeuchi and Lin,* 2005; *Takeuchi et al.,* 2005; *Hughes and Armitage,* 2012).

Photoevaporation may also affect the chemical evolution of protoplanetary disks. As disks evolve, cool refractory elements condense onto dust grains, while volatiles (H and He) primarily remain in the gas phase. As photoevaporation removes mass from the disk surface, the mid-plane gradually evolves and becomes enriched in refractory elements. This process was invoked by *Guillot and Hueso* (2006) as part of an explanation for the Ar, Kr, and Xe enrichment with respect to H measured in Jupiter's atmosphere by the Galileo probe (*Owen et al.,* 1999). *Guillot and Hueso* (2006) considered a disk around a solar-mass star, undergoing viscous evolution and subject to central-star-EUV and external-FUV photoevaporation. In their model hydrogen is lost in the photovaporative wind, while noble gases condense onto grains in the cold outer disk and have a smaller escape rate. The noble gases are then vaporized again in the warmer disk region where Jupiter forms, and delivered to the envelope in the gas phase. The significance of this enrichment process should be reevaluated in light of the potentially higher X-ray- or FUV-driven photoevaporation rates (see section 2.1), and current thinking as to the efficiency of vertical mixing processes within the disk [*Guillot and Hueso* (2006) assumed that vertical mixing was dominated by convection].

4.2. Dynamical Effects

The manner in which protoplanetary disks are dispersed influences the mass and final orbital properties of planets, through its effects on planetary growth, migration, and orbital stability. The rapid decrease in surface density as the disk is dispersed can starve late-forming cores of gas, preventing them growing into fully formed gas giants. This mechanism was proposed by *Shu et al.* (1993) to explain why Saturn, Uranus, and Neptune are gas-poor, with smaller envelopes than Jupiter. The same effect halts inward migration, stranding planets at radii between their formation radius (which may be beyond the snow line) and the locations of hot Jupiters (with semimajor axis a < 0.1 AU). If several planets form in close proximity, the removal of gas will stop disk damping of eccentricity and inclination, with further evolution of the system occurring via purely N-body perturbations. These effects are all generic to any disk-dispersal mechanism. However, whether they impart identifiable features on the properties of observed planetary systems depends on how quickly, and from which radii, gas is lost during disk dispersal.

Angular momentum exchange between massive planets (≥0.5 M_{Jup}) and the protoplanetary gas disk results in type II migration, in which the planet's orbital evolution within a gap is coupled to the evolution of the disk (e.g., *Kley and Nelson,* 2012). Migration in this regime is typically inward, although outward migration is possible if planets form in a region where the viscous flow of the gas is away from the star, and mass loss from the outer regions of the disk also promotes outward migration (*Veras and Armitage,* 2004; *Martin et al.,* 2007). The rate of migration is generally a nonlinear function of the local gas disk conditions, and can be estimated in one-dimensional viscous disk models given knowledge of how angular momentum is transported in the disk (*Ivanov et al.,* 1999). Disk dispersal inevitably marks the endpoint of type II migration, and hydrodynamic simulations show that migrating giant planets are indeed stranded by final disk dispersal (*Rosotti et al.,* 2013).

The influence of photoevaporation on the orbital distribution of extrasolar gas giants was included in early population synthesis calculations by *Armitage et al.* (2002), using a simple analytic prescription for external FUV photoevaporation (see also *Matsuyama et al.,* 2003a). These models were extended by *Alexander and Armitage* (2009), who studied giant planet migration and the formation of transition disks using a one-dimensional disk model that included both viscous transport of angular momentum and EUV photoevaporation. *Alexander and Armitage* (2009) assumed that the time at which gas giants form is uniformly distributed toward the end of the disk lifetime, and found that the observed distribution of exoplanet semimajor axes (within a few astronomical units) is consistent with planet formation further out (≥5 AU), followed by type II migration and stranding when the disk is dispersed. Integrated over all (giant) planet masses, the distribution depends primarily upon the nature of angular momentum transport in the disk [and hence is modified in the presence of a dead zone (e.g., *Armitage,* 2007; *Matsumura et al.,* 2009)], but is also affected by uncertainties in the rate of mass and angular momentum accretion across gaps in the disk (*Lubow and D'Angelo,* 2006). The integrated distribution is essentially independent of the details of the photoevaporation model, but sensitivity to the disk dispersal mechanism becomes apparent when the distribution of planet semimajor axes is broken down into different mass bins. *Alexander and Pascucci* (2012) found that variations in the migration rate near the radius where photoevaporation first opens a gap lead to mass-dependent deserts and pileups in the planetary distribution. These effects may be observable in the case of EUV photoevaporation, because the photoevaporative gap in this case falls at small radii (≈1–2 AU) where most of the observed planets are likely to have migrated, rather than formed *in situ*.

State-of-the-art population-synthesis models incorporate a much broader range of physical processes than just disk evolution and planet migration, including simplified treatments of core formation, type I migration, and envelope evolution (see chapter by Benz et al. in this volume). *Mordasini et al.* (2012) include simple prescriptions for both external (FUV) and internal (EUV) photoevaporation, and their models were able to reproduce the observed distribution of planets [f(a,M_p,R_p)] reasonably well for planets ≥2 $R_\oplus$. However, in these models uncertainties other than those associated with disk dispersal are dominant. The models of *Hasegawa and Pudritz* (2012) similarly invoke internal FUV photoevaporation to drive disk dispersal, but find that the strongest features in the resulting planet population are due to changes in the migration rate at specific locations in the disk [so-called "planet traps" (e.g., *Masset et al.,* 2006)]. It may therefore be difficult to distinguish the effects of photoevaporation from other physical processes.

Additional dynamical effects arise when multiple planets interact with a dispersing gas disk. In systems with well-separated planets, the changing gravitational potential of the disk during dispersal alters the precession rates of planets and leads to a radial sweeping of secular resonances (*Nagasawa et al.,* 2005). By contrast, in closely packed planetary systems gravitational torques between planets and the gas damp eccentricity and inclination (*Kominami and Ida,* 2002; *Agnor and Ward,* 2002), and the presence of a gas disk therefore suppress planet-planet interactions and scattering. *Moeckel and Armitage* (2012) studied the development of dynamical instabilities during the final phase of X-ray-driven disk clearing, using two-dimensional hydrodynamics to simulate the evolution of a gas disk with three embedded planets. They found that the outcome of dynamical instabilities in the presence of a dispersing disk was similar to gas-free simulations, although a significant number of stable resonant systems formed due to gas-driven orbital migration. However, it remains computationally challenging to model the formation and growth of multiple planets in two-dimensional simulations. As a result, it is unclear whether multiple massive planets typically evolve to resonant, packed, and rapidly unstable configurations when the gas disk is dispersed, or if stable configurations are preferred (*Marzari et al.,* 2010; *Lega et al.,* 2013). Moreover, dynamical simulations to date have generally focused on gas-giant planets, which migrate in the type II regime. The effects of disk dispersal on the migration and dynamics of lower-mass planets are potentially more significant, but remain largely unexplored by current models.

Finally, we note that disks in binary systems may also represent an interesting test of disk dispersal theory. Disk masses are systematically lower in binary systems than for single stars, and disk formation is apparently strongly suppressed around binaries with small (≤40 AU) separations (*Harris et al.,* 2012; *Kraus et al.,* 2012). However, long-lived circumbinary disks do exist (e.g., *Ireland and Kraus,* 2008; *Rosenfeld et al.,* 2012), and the recent discovery that circumbinary planets are relatively common (*Doyle et al.,* 2011; *Welsh et al.,* 2012) has revived interest in the evolution and dispersal of disks around young binary stars. *Alexander* (2012) constructed one-dimensional models of circumbinary disk evolution, and found that the suppression of disk accretion by the tidal torque from the binary greatly enhances the role played by photoevaporation. These results suggest that circumbinary disks may provide a useful laboratory for studying disk dispersal.

4.3. Schematic Picture of Disk Evolution and Dispersal

We have reviewed the theory and observations underpinning our understanding of how protoplanetary disks are dispersed. There have been significant advances in this field since *Protostars and Planets V* (*Reipurth et al.,* 2007), and we now have a robust theoretical framework that we are beginning to test directly with observations. Although a number of details remain unresolved, the results discussed above allow us to draw several interesting conclusions:

1. Protoplanetary disk evolution on approximately million-year timescales is primarily driven by accretion, but winds (both magnetically launched and photoevaporative) may drive significant mass loss throughout the lifetimes of many disks.

2. Disk photoevaporation, driven by high-energy radiation from the central star, is now directly detected in a number of systems, and is the most plausible mechanism for gas disk dispersal.

3. Theoretical models predict that photoevaporative mass-loss rates range from ~10^{-10} to 10^{-8} M$_\odot$ yr^{-1}. Current models suggest that X-ray or FUV heating dominates in typical systems, yielding mass-loss rates toward the upper end of this range.

4. Inferred mass-loss rates, from both direct and indirect observational tracers, are broadly consistent with these predictions (although the lower end of this range is weakly favored by current demographic data).

5. High-resolution spectroscopy of emission lines offers a direct test of disk wind models, and may allow mass-loss rates to be measured empirically.

6. Photoevaporation can explain the properties of some, but not all, transitional disks, and it seems likely that multiple disk-clearing mechanisms operate concurrently in these systems.

7. Disk dispersal ends the epoch of giant planet formation, and can have a strong influence on the architectures of forming planetary systems.

8. Mass loss throughout the disk lifetime may also influence (or even trigger) planet formation, by depleting the disk of gas and enhancing the fractional abundances of both dust and heavy elements.

Based on these conclusions, we are able to construct a tentative schematic picture of protoplanetary disk evolution and dispersal, which is illustrated in Fig. 6. The earliest stages of disk evolution (broadly described as the class I phase) are dominated by infall onto the disk, and accretion is driven primarily by gravitational instabilities. This phase is also characterized by strong jets and outflows, launched from close to the star by magnetic effects, although their significance in terms of the global evolution of the disk is unclear. The disk then evolves toward a more quiescent evolutionary phase, of which class II sources and CTTs are typical. Here the evolution is primarily driven by disk accretion, but mass loss in low-velocity winds is also significant in many, perhaps most, systems. At this stage these winds are primarily neutral, and driven by some combination of X-ray and/or FUV photoevaporation and magnetic fields; the dominant mechanism, and the extent to which such winds deplete or truncate the disk, may well vary from disk to disk. Final disk dispersal begins when mass loss begins to dominate over accretion. The winds

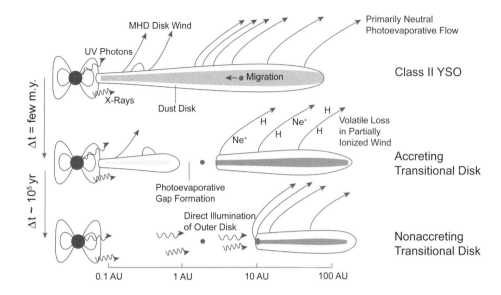

Fig. 6. Schematic representation of the disk evolution and dispersal sequence outlined in section 4.3. At early times accretion dominates the evolution, while mass loss is dominated by some combination of X-ray and/or FUV photoevaporation and magnetic fields. At later times the photoevaporative flow is at least partially ionized, driven by EUV and/ or X-ray irradiation, and once this wind overcomes the accretion flow the disk is rapidly cleared from the inside out.

are now at least partially ionized (with ionization fraction $\chi_e \geq 10^{-2}$), and significant mass loss is apparently driven by some combination of X-ray and EUV photoevaporation (which again may vary between disks). This evolutionary phase is broadly associated with the transition from class II to class III SEDs, but we stress that a variety of different physical processes contribute to the observed properties of individual "transitional" disks. Once photoevaporation becomes the major driver of disk evolution, the effect is dramatic. The disk gas is rapidly cleared from the inside out, stranding any migrating planets at their present locations and profoundly altering the spatial distribution of the remaining disk solids. From this point onward the nascent planetary system is dominated by gravitational and collisional dynamics, and gradually evolves through the debris disk phase to stability as a mature planetary system.

This picture is obviously somewhat idealized, but is now supported by mature theoretical models and observational evidence. Several important uncertainties remain, however, and we conclude by highlighting the most important areas for future progress. In the short term, emission line studies are perhaps the most promising diagnostic, with the potential to provide empirical measurements of disk mass-loss rates in addition to offering precise tests of theoretical models. The evolution of disk dispersal theory also continues apace, and recent developments in our understanding of magnetically driven winds have the potential to alter this field significantly in the coming years. On longer timescales we expect new, high-resolution observational facilities and techniques [especially the Atacama Large Millimeter/ submillimeter Array (ALMA)] to revolutionize our under-

standing of protoplanetary disk physics; recent studies of young binaries and tentative detections of forming planets represent only the tip of this coming iceberg. We also continue to extend our understanding of how disk evolution and dispersal influences planet formation and the architectures of planetary systems, and to build links between protoplanetary disks and our rapidly expanding knowledge of exoplanets. The future of this field is therefore bright, and we look forward to discussing a plethora of exciting new developments at the Protostars and Planets VII conference.

Acknowledgments. We thank S. Cabrit, C. Clarke, A. Dunhill, S. Edwards, B. Ercolano, C. Espaillat, U. Gorti, G. Herczeg, D. Hollenbach, J. Owen, K. Pontoppidan, and E. Rigliaco for a number of insightful discussions. We also thank U. Gorti and J. Owen for providing some of the data used in Figs. 3, 4, and 5. We are grateful to the referee, D. Hollenbach, and the editor, K. Dullemond, for their thoughtful and detailed comments. R.A. acknowledges support from STFC through an Advanced Fellowship (ST/G00711X/1) and Consolidated Grant ST/K001000/1. I.P. acknowledges support from an NSF Astronomy and Astrophysics research grant (AST0908479). P.A. acknowledges support from NASA's Origins of Solar Systems Program (NNX13AI58G). L.C. was supported by NASA through the Sagan Fellowship Program, under an award from Caltech.

REFERENCES

Adams F. C. et al. (2004) *Astrophys. J., 611,* 360.
Adams F. C. et al. (2006) *Astrophys. J., 641,* 504.
Agnor C. B. and Ward W. R. (2002) *Astrophys. J., 567,* 579.
Akeson R. L. et al. (2011) *Astrophys. J., 728,* 96.
Alencar S. H. P. and Batalha C. (2002) *Astrophys. J., 571,* 378.

Alexander R. (2008a) *New Astron. Rev., 52,* 60.

Alexander R. D. (2008b) *Mon. Not. R. Astron. Soc., 391,* L64.

Alexander R. (2012) *Astrophys. J. Lett., 757,* L29.

Alexander R. D. and Armitage P. J. (2007) *Mon. Not. R. Astron. Soc., 375,* 500.

Alexander R. D. and Armitage P. J. (2009) *Astrophys. J., 704,* 989.

Alexander R. D. and Pascucci I. (2012) *Mon. Not. R. Astron. Soc., 422,* L82.

Alexander R. D. et al. (2004a) *Mon. Not. R. Astron. Soc., 348,* 879.

Alexander R. D. et al. (2004b) *Mon. Not. R. Astron. Soc., 354,* 71.

Alexander R. D. et al. (2005) *Mon. Not. R. Astron. Soc., 358,* 283.

Alexander R. D. et al. (2006a) *Mon. Not. R. Astron. Soc., 369,* 216.

Alexander R. D. et al. (2006b) *Mon. Not. R. Astron. Soc., 369,* 229.

Andre P. et al. (1993) *Astrophys. J., 406,* 122.

Andrews S. M. and Williams J. P. (2005) *Astrophys. J., 631,* 1134.

Andrews S. M. and Williams J. P. (2007) *Astrophys. J., 671,* 1800.

Andrews S. M. et al. (2009) *Astrophys. J., 700,* 1502.

Andrews S. M. et al. (2011) *Astrophys. J., 732,* 42.

Andrews S. M. et al. (2012) *Astrophys. J., 744,* 162.

Andrews S. M. et al. (2013) *Astrophys. J., 771,* 129.

Ardila D. R. et al. (2013) *Astrophys. J. Suppl., 207,* 1.

Armitage P. J. (2007) *Astrophys. J., 665,* 1381.

Armitage P. J. (2011) *Annu. Rev. Astron. Astrophys., 49,* 195.

Armitage P. J. et al. (2001) *Mon. Not. R. Astron. Soc., 324,* 705.

Armitage P. J. et al. (2002) *Mon. Not. R. Astron. Soc., 334,* 248.

Armitage P. J. et al. (2013) *Astrophys. J. Lett., 778,* L14.

Arnold T. J. et al. (2012) *Astrophys. J., 750,* 119.

Artymowicz P. and Lubow S. H. (1994) *Astrophys. J., 421,* 651.

Bae J. et al. (2013) *Astrophys. J., 774,* 57.

Bai X.-N. (2013) *Astrophys. J., 772,* 96.

Bai X.-N. and Stone J. M. (2013a) *Astrophys. J., 767,* 30.

Bai X.-N. and Stone J. M. (2013b) *Astrophys. J., 769,* 76.

Balbus S. A. (2011) In *Physical Processes in Circumstellar Disks Around Young Stars* (P. J. V. Garcia, ed.), pp. 237–282. Univ. of Chicago, Chicago.

Balbus S. A. and Hawley J. F. (1998) *Rev. Mod. Phys., 70,* 1.

Baldovin-Saavedra C. et al. (2011) *Astron. Astrophys., 528,* A22.

Baldovin-Saavedra C. et al. (2012) *Astron. Astrophys., 543,* A30.

Bally J. and Scoville N. Z. (1982) *Astrophys. J., 255,* 497.

Beckwith S. V. W. et al. (1990) *Astron. J., 99,* 924.

Begelman M. C. et al. (1983) *Astrophys. J., 271,* 70.

Bergin E. A. et al. (2013) *Nature, 493,* 644.

Birnstiel T. et al. (2012) *Astron. Astrophys., 544,* A79.

Blandford R. D. and Payne D. G. (1982) *Mon. Not. R. Astron. Soc., 199,* 883.

Brown J. M. et al. (2013) *Astrophys. J., 770,* 94.

Calvet N. and Gullbring E. (1998) *Astrophys. J., 509,* 802.

Calvet N. et al. (2002) *Astrophys. J., 568,* 1008.

Carpenter J. M. et al. (2005) *Astron. J., 129,* 1049.

Casassus S. et al. (2013) *Nature, 493,* 191.

Chen C. H. et al. (2006) *Astrophys. J. Suppl., 166,* 351.

Chiang E. and Murray-Clay R. (2007) *Nature Phys., 3,* 604.

Chiang E. and Youdin A. N. (2010) *Annu. Rev. Earth Planet. Sci., 38,* 493.

Cieza L. et al. (2007) *Astrophys. J., 667,* 308.

Cieza L. A. et al. (2008) *Astrophys. J. Lett., 686,* L115.

Cieza L. A. et al. (2012) *Astrophys. J., 750,* 157.

Cieza L. A. et al. (2013) *Astrophys. J., 762,* 100.

Clarke C. J. (2007) *Mon. Not. R. Astron. Soc., 376,* 1350.

Clarke C. (2011) In *Physical Processes in Circumstellar Disks Around Young Stars* (P. J. V. Garcia, ed.), pp. 355–418. Univ. of Chicago, Chicago.

Clarke C. J. and Owen J. E. (2013) *Mon. Not. R. Astron. Soc., 433,* L69.

Clarke C. J. and Pringle J. E. (2006) *Mon. Not. R. Astron. Soc., 370,* L10.

Clarke C. J. et al. (2001) *Mon. Not. R. Astron. Soc., 328,* 485.

Curran R. L. et al. (2011) *Astron. Astrophys., 526,* A104.

Dent W. R. F. et al. (1995) *Mon. Not. R. Astron. Soc., 277,* L25.

Dent W. R. F. et al. (2005) *Mon. Not. R. Astron. Soc., 359,* 663.

Doyle L. R. et al. (2011) *Science, 333,* 1602.

Drake J. J. et al. (2009) *Astrophys. J. Lett., 699,* L35.

Dubrulle B. et al. (1995) *Icarus, 114,* 237.

Dullemond C. P. and Dominik C. (2005) *Astron. Astrophys., 434,* 971.

Dullemond C. P. et al. (2007) In *Protostars and Planets V* (B. Reipurth et al., eds.), pp. 555–572. Univ. of Arizona, Tucson.

Dupree A. K. et al. (2012) *Astrophys. J., 750,* 73.

Duvert G. et al. (2000) *Astron. Astrophys., 355,* 165.

Eisner J. A. et al. (2006) *Astrophys. J. Lett., 637,* L133.

Ercolano B. and Owen J. E. (2010) *Mon. Not. R. Astron. Soc., 406,* 1553.

Ercolano B. et al. (2008) *Astrophys. J., 688,* 398.

Ercolano B. et al. (2009) *Astrophys. J., 699,* 1639.

Espaillat C. et al. (2012) *Astrophys. J., 747,* 103.

Espaillat C. et al. (2013) *Astrophys. J., 762,* 62.

Fang M. et al. (2009) *Astron. Astrophys., 504,* 461.

Fedele D. et al. (2010) *Astron. Astrophys., 510,* A72.

Feigelson E. D. and Decampli W. M. (1981) *Astrophys. J. Lett., 243,* L89.

Feigelson E. D. and Montmerle T. (1999) *Annu. Rev. Astron. Astrophys., 37,* 363.

Feigelson E. et al. (2007) In *Protostars and Planets V* (B. Reipurth et al., eds.), pp. 313–328. Univ. of Arizona, Tucson.

Font A. S. et al. (2004) *Astrophys. J., 607,* 890.

France K. et al. (2012) *Astrophys. J., 756,* 171.

Fromang S. et al. (2013) *Astron. Astrophys., 552,* A71.

Gammie C. F. (1996) *Astrophys. J., 457,* 355.

Geers V. C. et al. (2009) *Astron. Astrophys., 495,* 837.

Glassgold A. E. et al. (1997) *Astrophys. J., 480,* 344.

Glassgold A. E. et al. (2000) In *Protostars and Planets IV* (V. Mannings et al., eds.), p. 429. Univ. of Arizona, Tucson.

Glassgold A. E. et al. (2007) *Astrophys. J., 656,* 515.

Gorti U. and Hollenbach D. (2004) *Astrophys. J., 613,* 424.

Gorti U. and Hollenbach D. (2008) *Astrophys. J., 683,* 287.

Gorti U. and Hollenbach D. (2009) *Astrophys. J., 690,* 1539.

Gorti U. et al. (2009) *Astrophys. J., 705,* 1237.

Gorti U. et al. (2011) *Astrophys. J., 735,* 90.

Güdel M. and Nazé Y. (2009) *Astron. Astrophys. Rev., 17,* 309.

Güdel M. et al. (2010) *Astron. Astrophys., 519,* A113.

Guillot T. and Hueso R. (2006) *Mon. Not. R. Astron. Soc., 367,* L47.

Gullbring E. et al. (1998) *Astrophys. J., 492,* 323.

Haisch K. E. Jr. et al. (2001) *Astrophys. J. Lett, 553,* L153.

Harris R. J. et al. (2012) *Astrophys. J., 751,* 115.

Hartigan P. et al. (1990) *Astrophys. J. Lett., 354,* L25.

Hartigan P. et al. (1995) *Astrophys. J., 452,* 736.

Hartmann L. et al. (1998) *Astrophys. J., 495,* 385.

Hasegawa Y. and Pudritz R. E. (2012) *Astrophys. J., 760,* 117.

Henney W. J. and O'Dell C. R. (1999) *Astron. J., 118,* 2350.

Herczeg G. J. (2007) In *Star-Disk Interaction in Young Stars* (J. Bouvier and I. Appenzeller, eds.), pp. 147–154. IAU Symp. 243, Cambridge Univ., Cambridge.

Herczeg G. J. et al. (2002) *Astrophys. J., 572,* 310.

Herczeg G. J. et al. (2004) *Astrophys. J., 607,* 369.

Herczeg G. J. et al. (2007) *Astrophys. J., 670,* 509.

Hernández J. et al. (2007) *Astrophys. J., 671,* 1784.

Hollenbach D. and Gorti U. (2009) *Astrophys. J., 703,* 1203.

Hollenbach D. et al. (1994) *Astrophys. J., 428,* 654.

Hollenbach D. J. et al. (2000) In *Protostars and Planets IV* (V. Mannings et al., eds.), p. 401. Univ. of Arizona, Tucson.

Hollenbach D. et al. (2005) *Astrophys. J., 631,* 1180.

Hughes A. L. H. and Armitage P. J. (2012) *Mon. Not. R. Astron. Soc., 423,* 389.

Hughes A. M. et al. (2007) *Astrophys. J., 664,* 536.

Ingleby L. et al. (2009) *Astrophys. J. Lett., 703,* L137.

Ingleby L. et al. (2011a) *Astron. J., 141,* 127.

Ingleby L. et al. (2011b) *Astrophys. J., 743,* 105.

Ingleby L. et al. (2012) *Astrophys. J. Lett., 752,* L20.

Ireland M. J. and Kraus A. L. (2008) *Astrophys. J. Lett., 678,* L59.

Ivanov P. B. et al. (1999) *Mon. Not. R. Astron. Soc., 307,* 79.

Johnstone D. et al. (1998) *Astrophys. J., 499,* 758.

Kamp I. and Sammar F. (2004) *Astron. Astrophys., 427,* 561.

Kamp I. et al. (2010) *Astron. Astrophys., 510,* A18.

Kamp I. et al. (2011) *Astron. Astrophys., 532,* A85.

Kastner J. H. et al. (2002) *Astrophys. J., 567,* 434.

Kenyon S. J. and Hartmann L. (1995) *Astrophys. J. Suppl., 101,* 117.

Kim K. H. et al. (2013) *Astrophys. J., 769,* 149.

King A. R. et al. (2007) *Mon. Not. R. Astron. Soc., 376,* 1740.

Kley W. and Nelson R. P. (2012) *Annu. Rev. Astron. Astrophys., 50,* 211.

Koepferl C. M. et al. (2013) *Mon. Not. R. Astron. Soc., 428,* 3327.

Kominami J. and Ida S. (2002) *Icarus, 157,* 43.

Königl A. and Salmeron R. (2011) In *Physical Processes in*

Circumstellar Disks Around Young Stars (P. J. V. Garcia, ed.), pp. 283–352. Univ. of Chicago, Chicago.

Kraus A. L. and Ireland M. J. (2012) *Astrophys. J., 745,* 5.

Kraus A. L. et al. (2008) *Astrophys. J., 679,* 762.

Kraus A. L. et al. (2011) *Astrophys. J., 731,* 8.

Kraus A. L. et al. (2012) *Astrophys. J., 745,* 19.

Krauss O. et al. (2007) *Astron. Astrophys., 462,* 977.

Kruger A. J. et al. (2013) *Astrophys. J., 764,* 127.

Krumholz M. R. et al. (2013) *Astrophys. J. Lett., 767,* L11.

Lada C. J. (1987) In *Star Forming Regions* (M. Peimbert and J. Jugaku, eds.), pp. 1–17. IAU Symp. 115, Cambridge Univ., Cambridge.

Lada C. J. et al. (2006) *Astron. J., 131,* 1574.

Lahuis F. et al. (2007) *Astrophys. J., 665,* 492.

Lebreton J. et al. (2012) *Astron. Astrophys., 539,* A17.

Lecavelier des Etangs A. et al. (2001) *Nature, 412,* 706.

Lega E. et al. (2013) *Mon. Not. R. Astron. Soc., 431,* 3494.

Lesur G. et al. (2013) *Astron. Astrophys., 550,* A61.

Li Z.-Y. et al. (2011) *Astrophys. J., 738,* 180.

Liffman K. (2003) *Publ. Astron. Soc. Australia, 20,* 337.

Lovelace R. V. E. et al. (2008) *Mon. Not. R. Astron. Soc., 389,* 1233.

Lubow S. H. and D'Angelo G. (2006) *Astrophys. J., 641,* 526.

Lugo J. et al. (2004) *Astrophys. J., 614,* 807.

Luhman K. L. (2004) *Astrophys. J., 617,* 1216.

Luhman K. L. et al. (2010) *Astrophys. J. Suppl., 186,* 111.

Lynden-Bell D. and Pringle J. E. (1974) *Mon. Not. R. Astron. Soc., 168,* 603.

Mamajek E. E. (2009) In *Exoplanets and Disks: Their Formation and Diversity* (T. Usuda et al., eds.), pp. 3–10. AIP Conf. Ser. 1158, American Institute of Physics, Melville, New York.

Mann R. K. and Williams J. P. (2010) *Astrophys. J., 725,* 430.

Martin R. G. et al. (2007) *Mon. Not. R. Astron. Soc., 378,* 1589.

Marzari F. et al. (2010) *Astron. Astrophys., 514,* L4.

Masset F. S. et al. (2006) *Astrophys. J., 642,* 478.

Mathews G. S. et al. (2010) *Astron. Astrophys., 518,* L127.

Mathews G. S. et al. (2012) *Astrophys. J., 745,* 23.

Matsumura S. et al. (2009) *Astrophys. J., 691,* 1764.

Matsuyama I. et al. (2003a) *Astrophys. J. Lett., 585,* L143.

Matsuyama I. et al. (2003b) *Astrophys. J., 582,* 893.

Matsuyama I. et al. (2009) *Astrophys. J., 700,* 10.

Mayor M. et al. (2013) *Astron. Astrophys.,* in press, arXiv:1109.2497.

McCaughrean M. J. and O'Dell C. R. (1996) *Astron. J., 111,* 1977.

Mesa-Delgado A. et al. (2012) *Mon. Not. R. Astron. Soc., 426,* 614.

Moeckel N. and Armitage P. J. (2012) *Mon. Not. R. Astron. Soc., 419,* 366.

Mohanty S. et al. (2013) *Astrophys. J., 773,* 168.

Mordasini C. et al. (2012) *Astron. Astrophys., 547,* A112.

Morishima R. (2012) *Mon. Not. R. Astron. Soc., 420,* 2851.

Muzerolle J. et al. (1998) *Astron. J., 116,* 2965.

Muzerolle J. et al. (2000) *Astrophys. J. Lett., 535,* L47.

Muzerolle J. et al. (2001) *Astrophys. J., 550,* 944.

Muzerolle J. et al. (2005) *Astrophys. J., 625,* 906.

Nagasawa M. et al. (2005) *Astrophys. J., 635,* 578.

Najita J. and Williams J. P. (2005) *Astrophys. J., 635,* 625.

Najita J. R. et al. (2007a) *Mon. Not. R. Astron. Soc., 378,* 369.

Najita J. R. et al. (2007b) In *Protostars and Planets V* (B. Reipurth et al., eds.), pp. 507–522. Univ. of Arizona, Tucson.

Najita J. R. et al. (2009) *Astrophys. J., 697,* 957.

O'Dell C. R. et al. (1993) *Astrophys. J., 410,* 696.

Osterbrock D. E. and Ferland G. J. (2006) *Astrophysics of Gaseous Nebulae and Active Galactic Nuclei.* University Science Books, Herndon, Virginia. 496 pp.

Owen J. E. and Clarke C. J. (2012) *Mon. Not. R. Astron. Soc., 426,* L96.

Owen J. E. et al. (2010) *Mon. Not. R. Astron. Soc., 401,* 1415.

Owen J. E. et al. (2011) *Mon. Not. R. Astron. Soc., 412,* 13.

Owen J. E. et al. (2012) *Mon. Not. R. Astron. Soc., 422,* 1880.

Owen J. E. et al. (2013a) *Mon. Not. R. Astron. Soc., 434,* 3378.

Owen J. E. et al. (2013b) *Mon. Not. R. Astron. Soc., 436,* 1430.

Owen T. et al. (1999) *Nature, 402,* 269.

Padgett D. L. et al. (2006) *Astrophys. J., 645,* 1283.

Pascucci I. and Sterzik M. (2009) *Astrophys. J., 702,* 724.

Pascucci I. and Tachibana S. (2010) In *Protoplanetary Dust: Astrophysical and Cosmochemical Perspectives* (D. A. Apai and D. S. Lauretta, eds.), pp. 263–298. Cambridge Univ., Cambridge.

Pascucci I. et al. (2006) *Astrophys. J., 651,* 1177.

Pascucci I. et al. (2007) *Astrophys. J., 663,* 383.

Pascucci I. et al. (2011) *Astrophys. J., 736,* 13.

Pascucci I. et al. (2012) *Astrophys. J. Lett., 751,* L42.

Pinilla P. et al. (2012) *Astron. Astrophys., 538,* A114.

Pollack J. B. et al. (1996) *Icarus, 124,* 62.

Pontoppidan K. M. et al. (2011) *Astrophys. J., 733,* 84.

Pott J.-U. et al. (2010) *Astrophys. J., 710,* 265.

Quillen A. C. et al. (2004) *Astrophys. J. Lett., 612,* L137.

Ratzka T. et al. (2007) *Astron. Astrophys., 471,* 173.

Reipurth B. et al., eds. (2007) *Protostars and Planets V,* Univ. of Arizona, Tucson.

Rice W. K. M. et al. (2003) *Mon. Not. R. Astron. Soc., 342,* 79.

Richling S. and Yorke H. W. (1997) *Astron. Astrophys., 327,* 317.

Richling S. and Yorke H. W. (2000) *Astrophys. J., 539,* 258.

Rigliaco E. et al. (2013) *Astrophys. J., 772,* 60.

Roberge A. et al. (2005) *Astrophys. J. Lett., 626,* L105.

Rosenfeld K. A. et al. (2012) *Astrophys. J., 759,* 119.

Rosotti G. P. et al. (2013) *Mon. Not. R. Astron. Soc., 430,* 1392.

Sacco G. G. et al. (2012) *Astrophys. J., 747,* 142.

Salmeron R. et al. (2011) *Mon. Not. R. Astron. Soc., 412,* 1162.

Sargent A. I. and Beckwith S. (1987) *Astrophys. J., 323,* 294.

Scally A. and Clarke C. (2001) *Mon. Not. R. Astron. Soc., 325,* 449.

Schindhelm E. et al. (2012) *Astrophys. J. Lett., 756,* L23.

Sekiya M. (1998) *Icarus, 133,* 298.

Shakura N. I. and Sunyaev R. A. (1973) *Astron. Astrophys., 24,* 337.

Shu F. H. et al. (1993) *Icarus, 106,* 92.

Shu F. H. et al. (2007) *Astrophys. J., 665,* 535.

Sicilia-Aguilar A. et al. (2010) *Astrophys. J., 710,* 597.

Simon J. B. et al. (2013) *Astrophys. J., 775,* 73.

Simon M. and Prato L. (1995) *Astrophys. J., 450,* 824.

Simon M. et al. (2000) *Astrophys. J., 545,* 1034.

Stelzer B. et al. (2013) *Mon. Not. R. Astron. Soc., 431,* 2063.

Störzer H. and Hollenbach D. (1999) *Astrophys. J., 515,* 669.

Strom K. M. et al. (1989) *Astron. J., 97,* 1451.

Suzuki T. K. and Inutsuka S.-i. (2009) *Astrophys. J. Lett., 691,* L49.

Suzuki T. K. et al. (2010) *Astrophys. J., 718,* 1289.

Szulágyi J. et al. (2012) *Astrophys. J., 759,* 47.

Takeuchi T. and Lin D. N. C. (2005) *Astrophys. J., 623,* 482.

Takeuchi T. et al. (2005) *Astrophys. J., 627,* 286.

Thi W.-F. et al. (2010) *Astron. Astrophys., 518,* L125.

Throop H. B. and Bally J. (2005) *Astrophys. J. Lett., 623,* L149.

Tielens A. G. G. M. and Hollenbach D. (1985) *Astrophys. J., 291,* 722.

Torres C. A. O. et al. (2008) In *Handbook of Star Forming Regions, Volume II* (B. Reipurth, ed.), p. 757. Astronomical Society of the Pacific, San Francisco.

van Boekel R. et al. (2009) *Astron. Astrophys., 497,* 137.

van der Marel N. et al. (2013) *Science, 340,* 1199.

van Leeuwen F. (2007) *Astron. Astrophys., 474,* 653.

Veras D. and Armitage P. J. (2004) *Mon. Not. R. Astron. Soc., 347,* 613.

Wahhaj Z. et al. (2010) *Astrophys. J., 724,* 835.

Welsh W. F. et al. (2012) *Nature, 481,* 475.

Williams J. P. and Cieza L. A. (2011) *Annu. Rev. Astron. Astrophys., 49,* 67.

Wilner D. J. et al. (2005) *Astrophys. J. Lett., 626,* L109.

Woitke P. et al. (2009) *Astron. Astrophys., 501,* 383.

Woitke P. et al. (2010) *Mon. Not. R. Astron. Soc., 405,* L26.

Woitke P. et al. (2011) *Astron. Astrophys., 534,* A44.

Wolk S. J. and Walter F. M. (1996) *Astron. J., 111,* 2066.

Wright J. T. et al. (2011) *Publ. Astron. Soc. Pacific, 123,* 412.

Wyatt M. C. (2008) *Annu. Rev. Astron. Astrophys., 46,* 339.

Yang H. et al. (2012) *Astrophys. J., 744,* 121.

Youdin A. N. and Shu F. H. (2002) *Astrophys. J., 580,* 494.

Zhu Z. et al. (2012) *Astrophys. J., 755,* 6.

Zuckerman B. et al. (1995) *Nature, 373,* 494.

Espaillat C., Muzerolle J., Najita J., Andrews S., Zhu Z., Calvet N., Kraus S., Hashimoto J., Kraus A., and D'Alessio P. (2014) An observational perspective of transitional disks. In *Protostars and Planets VI* (H. Beuther et al., eds.), pp. 497–520. Univ. of Arizona, Tucson, DOI: 10.2458/azu_uapress_9780816531240-ch022.

An Observational Perspective of Transitional Disks

Catherine Espaillat
Boston University

James Muzerolle
Space Telescope Science Institute

Joan Najita
National Optical Astronomy Observatory

Sean Andrews
Harvard-Smithsonian Center for Astrophysics

Zhaohuan Zhu
Princeton University

Nuria Calvet
University of Michigan

Stefan Kraus
University of Exeter

Jun Hashimoto
The University of Oklahoma

Adam Kraus
University of Texas

Paola D'Alessio
Universidad Nacional Autónoma de México

Transitional disks are objects whose inner disk regions have undergone substantial clearing. The Spitzer Space Telescope produced detailed spectral energy distributions (SEDs) of transitional disks that allowed us to infer their radial dust disk structure in some detail, revealing the diversity of this class of disks. The growing sample of transitional disks also opened up the possibility of demographic studies, which provided unique insights. There now exist (sub)millimeter and infrared images that confirm the presence of large clearings of dust in transitional disks. In addition, protoplanet candidates have been detected within some of these clearings. Transitional disks are thought to be a strong link to planet formation around young stars and are a key area to study if further progress is to be made on understanding the initial stages of planet formation. Here we provide a review and synthesis of transitional disk observations to date with the aim of providing timely direction to the field, which is about to undergo its next burst of growth as the Atacama Large Millimeter/submillimeter Array (ALMA) reaches its full potential. We discuss what we have learned about transitional disks from SEDs, color-color diagrams, and imaging in the (sub)millimeter and infrared. We note the limitations of these techniques, particularly with respect to the sizes of the clearings currently detectable, and highlight the need for pairing broadband SEDs with multi-wavelength images to paint a more detailed picture of transitional disk structure. We review the gas in transitional disks, keeping in mind that future observations with ALMA will give us unprecedented access to gas in disks, and also observed infrared variability pointing to variable transitional disk structure, which may have implications for disks in general. We then distill the observations into constraints for the main disk-clearing mechanisms proposed to date (i.e., photoevaporation, grain growth, and companions) and explore how the expected observational signatures from these mechanisms, particularly planet-induced disk clearing, compare to actual observations. Finally, we discuss future avenues of inquiry to be pursued with ALMA, the James Webb Space Telescope (JWST), and the next generation of groundbased telescopes.

1. INTRODUCTION

Disks around young stars are thought to be the sites of planet formation. However, many questions exist concerning how the gas and dust in the disk evolve into a planetary system. Observations of T Tauri stars (TTS) may provide insights into these questions, and a subset of TTS, the "transitional disks," have gained increasing attention in this regard. The unusual spectral energy distributions (SEDs) of transitional disks (which feature infrared excess deficits) may indicate that they have developed significant radial structure.

Transitional disk SEDs were first identified by *Strom et al.* (1989) and *Skrutskie et al.* (1990) from near-infrared (NIR) groundbased photometry and Infrared Astronomical Satellite (IRAS) mid-infrared (MIR) photometry. These systems exhibited small NIR and/or MIR excesses, but significant MIR and far-infrared (FIR) excesses indicating that the *dust* distribution of these disks had an inner hole (i.e., a region that is mostly devoid of small dust grains from a radius R_{hole} down to the central star). They proposed that these disks were in transition from objects with optically thick disks that extend inward to the stellar surface (i.e., class II objects) to objects where the disk has dissipated (i.e., class III objects), possibly as a result of some phase of planet formation. A few years later, *Marsh and Mahoney* (1992, 1993) proposed that such transitional disk SEDs were consistent with the expectations for a disk subject to tidal effects exerted by companions, either stars or planets.

More detailed studies of transitional disks became possible as increasingly sophisticated instruments became available. The spectrographs onboard the Infrared Space Observatory (ISO) were able to study the brightest stars, and *Bouwman et al.* (2003) inferred a large hole in the disk of the Herbig Ae/Be star HD 100546 based on its SED. Usage of the term "transitional disk" gained substantial momentum in the literature after the Spitzer Space Telescope's (*Werner et al.,* 2004) Infrared Spectrograph (IRS) (*Houck et al.,* 2004) was used to study disks with inner holes (e.g., *D'Alessio et al.,* 2005; *Calvet et al.,* 2005). Spitzer also detected disks with an annular "gap" within the disk as opposed to holes (e.g., *Brown et al.,* 2007; *Espaillat et al.,* 2007). In this review, we use the term *transitional disk* to refer to an object with an inner disk hole and *pre-transitional disk* to refer to a disk with a gap. For many (pre-)transitional disks, the inward truncation of the outer dust disk has been confirmed, predominantly through (sub)millimeter interferometric imaging (e.g., *Hughes et al.,* 2007, 2009; *Brown et al.,* 2008, 2009; *Andrews et al.,* 2009, 2011b; *Isella et al.,* 2010b). We note that (sub)millimeter imaging is not currently capable of distinguishing between a hole or gap in the disk (i.e., it can only detect a generic region of clearing or a "cavity" in the disk). Also, it has not yet been confirmed if these clearings detected in the dust disk are present in the gas disk as well. The combination of these dust cavities with the presence of continuing gas accretion onto the central star is a challenge to theories of disk clearing.

The distinct SEDs of (pre-)transitional disks have lead many researchers to conclude that these objects are being caught in an important phase in disk evolution. One possibility is that these disks are forming planets given that cleared disk regions are predicted by theoretical planet-formation models (e.g., *Paardekooper and Mellema,* 2004; *Zhu et al.,* 2011; *Dodson-Robinson and Salyk,* 2011). Potentially supporting this, there exist observational reports of protoplanet candidates in (pre-)transitional disks [e.g., LkCa 15, T Cha (*Kraus et al.,* 2011; *Huélamo et al.,* 2011)]. Stellar companions can also clear the inner disk (*Artymowicz and Lubow,* 1994), but many stars harboring (pre-)transitional disks are single stars (*Kraus et al.,* 2011). Even if companions are not responsible for the clearings seen in all (pre-)transitional disks, these objects still have the potential to inform our understanding of how disks dissipate, primarily by providing constraints for disk-clearing models involving photoevaporation and grain growth.

In this chapter, we will review the key observational constraints on the dust and gas properties of (pre-)transitional disks and examine these in the context of theoretical disk-clearing mechanisms. In section 2, we look at SEDs (section 2.1) as well as (sub)millimeter (section 2.2) and infrared (IR) (section 2.3) imaging. We also review IR variability in (pre-)transitional disks (section 2.4) and gas observations (section 2.5). In section 3, we turn the observations from section 2 into constraints for the main disk-clearing mechanisms proposed to date (i.e., photoevaporation, grain growth, and companions) and discuss these mechanisms in light of these constraints. In section 4, we examine the demographics of (pre-)transitional disks (i.e., frequencies, timescales, disk masses, accretion rates, stellar properties) in the context of disk clearing, and in section 5 we conclude with possibilities for future work in this field.

2. OVERVIEW OF OBSERVATIONS

In the two decades following *Strom et al.*'s (1989) identification of the first transitional disks using NIR and MIR photometry, modeling of the SEDs of these disks, enabled largely by the Spitzer IRS, inferred the presence of holes and gaps that span several astronomical units (section 2.1). Many of these cavities were confirmed by (sub)millimeter interferometric imaging (section 2.2) and IR polarimetric and interferometric (section 2.3) images. Later on, MIR variability was detected, which pointed to structural changes in these disks (section 2.4). While there are not currently as many constraints on the gas in the disk as there are for the dust, it is apparent that the nature of gas in the inner regions of (pre-)transitional disks differs from other disks (section 2.5). These observational results have significant implications for our understanding of planet formation and we review them in the following sections.

2.1. Spectral Energy Distributions

Spectral energy distributions are a powerful tool in disk studies as they provide information over a wide range of wavelengths, tracing different emission mechanisms and material at different stellocentric radii. In a SED, one can see the signatures of gas accretion [in the ultraviolet (UV); see the *Protostars and Planets IV* (PPIV) review by *Calvet et al.* (2000)], the stellar photosphere (typically ~1 μm in TTS), and the dust in the disk (in the IR and longer wavelengths). However, SEDs are not spatially resolved and this information must be supplemented by imaging, ideally at many wavelengths (see sections 2.2–2.3). Here we review what has been learned from studying the SEDs of (pre-)transitional disks, particularly using the Spitzer IRS, the Infrared Array Camera (IRAC), and the Multiband Imaging Photometer for Spitzer (MIPS).

2.1.1. Spectral energy distribution classification. A popular method of identifying transitional disks is to compare individual SEDs to the median SED of disks in the Taurus star-forming region (Figs. 1a–c, dashed line in panels). The median Taurus SED is typically taken as representative of an optically thick full disk (i.e., a disk with no significant radial discontinuities in its dust distribution). The NIR emission (1–5 μm) seen in the SEDs of full disks is dominated by the "wall" or inner edge of the dust disk. This wall is located where there is a sharp change at the radius at which the dust destruction temperature is reached and dust sublimates. T Tauri stars in Taurus have NIR excess emission that can be fit by blackbodies with temperatures within the observed range of dust sublimation temperatures [1000–2000 K (*Monnier and Millan-Gabet*, 2002)], indicating that there is optically thick material located at the dust destruction radius in full disks (*Muzerolle et al.*, 2003). Roughly, the MIR emission in the SED traces the inner tens of astronomical units in disks and emission at longer wavelengths comes from outer radii of the disk.

The SEDs of transitional disks are characterized by NIR (1–5 μm) and MIR emission (5–20 μm) similar to that of a stellar photosphere, while having excesses at wavelengths ~20 μm and beyond comparable to Taurus median (Fig. 1c) (*Calvet et al.*, 2005). From this we can infer that small, hot dust that typically emits at these wavelengths in full disks has been removed and that there is a large hole in the inner disk, larger than can be explained by dust sublimation (Fig. 2c). Large clearings of dust in the submillimeter regime have been identified in disks characterized by this

Fig. 1. Spectral energy distributions of **(a)** an evolved disk [RECX 11 (*Ingleby et al.*, 2011)], **(b)** a pre-transitional disk [LkCa 15 (*Espaillat et al.*, 2007)], and **(c)** a transitional disk [GM Aur (*Calvet et al.*, 2005)]. The stars are all K3–K5 and the fluxes have been corrected for reddening and scaled to the stellar photosphere (dot-long-dashed line) for comparison. Relative to the Taurus median [short-dashed line (*D'Alessio et al.*, 1999)], an evolved disk has less emission at all wavelengths; a pre-transitional disk has a MIR deficit (5–20 μm, ignoring the 10-μm silicate emission feature), but comparable emission in the NIR (1–5 μm) and at longer wavelengths; and a transitional disk has a deficit of emission in the NIR and MIR with comparable emission at longer wavelengths.

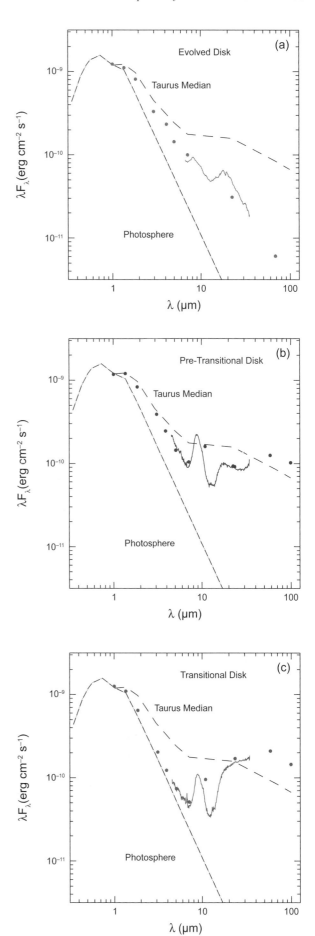

type of SED (section 2.2) (e.g., *Hughes et al., 2009; Brown et al., 2009; Andrews et al., 2011b*), confirming the SED interpretation. We note that disks with holes have also been referred to as cold disks (*Brown et al., 2007*) or weak excess transitional disks (*Muzerolle et al., 2010*), but here we use term "transitional disks."

A subset of disks with evidence of clearing in the submillimeter show significant NIR excesses relative to their stellar photospheres, in some cases comparable to the median Taurus SED, but still exhibit MIR dips and substantial excesses beyond ~20 μm (Fig. 1b). This NIR excess is blackbody-like with temperatures expected for the sublimation of silicates (*Espaillat et al., 2008b*), similar to the NIR excesses in full disks discussed earlier. This similarity indicates that these disks still have optically thick material close to the star, possibly a remnant of the original inner disk, and these gapped disks have been dubbed pre-transitional disks (*Espaillat et al., 2007*), cold disks (*Brown et al., 2007*), or warm transitional disks (*Muzerolle et al., 2010*). Here we adopt the term "pre-transitional disks" for these objects. In Table 1 we summarize some of properties of many of the well-known (pre-)transitional disks.

We note that some SEDs have emission that decreases steadily at all wavelengths (Fig. 1a). These disks have been called a variety of names: anemic (*Lada et al., 2006*), homologously depleted (*Currie and Sicilia-Aguilar, 2011*), evolved (*Hernández et al., 2007b, 2008, 2010*), or weak excess transition (*Muzerolle et al., 2010*). Here we adopt the terminology "evolved disk" for this type of object. Some researchers include these objects in the transitional disk class. However, these likely comprise a heterogenous class of disks, including cleared inner disks, debris disks (see the chapter by Matthews et al. in this volume), and disks with significant dust grain growth and settling. This has been an issue in defining this subset of objects. We include evolved disks in this review for completeness, but focus on disks with more robust evidence for disk holes and gaps.

2.1.2. Model fitting. Detailed modeling of many of the above-mentioned SEDs has been performed in order to infer the structure of these disks. Spectral energy distributions of transitional disks (i.e., objects with little or no NIR and MIR emission) have been fit with models of inwardly truncated optically thick disks (e.g., *Rice et al., 2003; Calvet et al., 2002*). The inner edge or "wall" of the outer disk is frontally illuminated by the star, dominating most of the emission seen in the IRS spectrum, particularly from ~20 to 30 μm. Some of the holes in transitional disks are relatively dust-free (e.g., DM Tau), while SED model fitting indicates that others with strong 10-μm silicate emission have a small amount of optically thin dust within their disk holes to explain this feature [e.g., GM Aur (*Calvet et al., 2005*)]. Beyond ~40 μm, transitional disks have a contribution to their SEDs from the outer disk. In pre-transitional disks, the observed SED can be fit with an optically thick inner disk separated by an optically thin gap from an optically thick outer disk (e.g., *Brown et al., 2007; Espaillat et al., 2007; Mulders et al., 2010; Dong et al., 2012*). There is

an inner wall located at the dust sublimation radius that dominates the NIR (2–5 μm) emission and should cast a shadow on the outer disk (see section 2.4). In a few cases, the optically thick inner disk of pre-transitional disks has been confirmed using NIR spectra (*Espaillat et al., 2010*) following the methods of *Muzerolle et al.* (2003). Like the transitional disks, there is evidence for relatively dust-free gaps (e.g., UX Tau A) as well as gaps with some small, optically thin dust to explain strong 10-μm silicate emission features [e.g., LkCa 15 (*Espaillat et al., 2007*)]. The SEDs of evolved disks can be fit with full disk models (*Sicilia-Aguilar et al., 2011*), particularly in which the dust is very settled toward the mid-plane (e.g., *Espaillat et al., 2012*).

There are many degeneracies to keep in mind when interpreting SED-based results. First, there is a limit to the

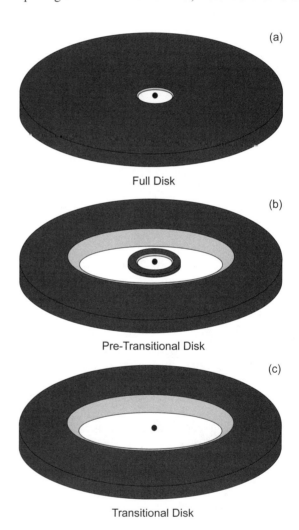

Fig. 2. Schematic of **(a)** full, **(b)** pre-transitional, and **(c)** transitional disk structure. For the full disk, progressing outward from the star (black) is the inner disk wall (light gray) and outer disk (dark gray). Pre-transitional disks have an inner disk wall (light gray) and inner disk (dark gray) followed by a disk gap (white), then the outer disk wall (light gray) and outer disk (dark gray). The transitional disk has an inner disk hole (white) followed by an outer disk wall (light gray) and outer disk (dark gray).

TABLE 1. Overview of selected (pre-)transitional disks.

Object	Class*	Submillimeter Cavity Radius (AU)	NIR Cavity Detected?	0.1-$M_\odot$ Companion Detection Limit	Accretion Rate ($M_\odot$ yr^{-1})	References[†]
AB Aur	PTD	70	—	6 AU	1.3×10^{-7}	[1,25,38]
CoKu Tau/4	TD	—	—	binary[§]	$<10^{-10}$	[26,39]
DM Tau	TD	19	—	6 AU	2.9×10^{-9}	[2,25,40]
GM Aur	TD	28	yes	6 AU	9.6×10^{-9}	[2,11,25,40]
HD 100546	PTD	—	yes	—	5.9×10^{-8}	[12,41]
HD 141569	PTD	—	yes	—	7.4×10^{-9}	[13,38]
HD 142527	PTD	140	yes	binary[¶]	9.5×10^{-8}	[3,14,27,38]
HD 169142	PTD	—	yes	—	9.1×10^{-9}	[15,38]
IRS 48	TD[‡]	60	—	8 AU	4.0×10^{-9}	[4,28,38]
LkCa 15	PTD	50	yes	6 AU	3.1×10^{-9}	[2,16,25,40]
MWC 758	PTD	73	—	28 AU	4.5×10^{-8}	[2,29,38]
PDS 70	PTD	—	yes	6 AU	$<10^{-10}$	[17,30,42]
RX J1604-2130	TD	70	yes	6 AU	$<10^{-10}$	[5,18,31]
RX J1615-3255	TD	30	no	8 AU	4×10^{-10}	[2,19,33,43]
RX J1633-2442	TD	25	—	6 AU	1.3×10^{-10}	[6,25,44]
RY Tau	PTD	14	—	6 AU	$6.4–9.1 \times 10^{-8}$	[7,25,45]
SAO 206462	PTD	46	—	25 AU	4.5×10^{-9}	[2,32,38]
SR 21	TD	36	no	8 AU	$<1.4 \times 10^{-9}$	[2,20,33,46]
SR 24 S	PTD	29	—	8 AU	7.1×10^{-8}	[2,33,46]
Sz 91	TD	65	yes	25 AU	1.4×10^{-9}	[8,21,34]
TW Hya	TD	4	no	3 AU	1.8×10^{-9}	[9,22,35,40]
UX Tau A	PTD	25	—	6 AU	1.1×10^{-9}	[2,25,40,47]
V4046 Sgr	TD	29	—	binary[**]	5.0×10^{-9}	[10,36,48]
DoAr 44	PTD	30	no	8 AU	3.7×10^{-9}	[2,23,33,38]
LkHα 330	PTD	68	no	8 AU	2.2×10^{-8}	[2,24,33,38]
WSB 60	PTD	15	—	25 AU	3.7×10^{-9}	[2,37,46]

*We classify objects as either a pre-transitional disk (PTD) or transitional disk (TD).

[†]Submillimeter cavity radii are from [1] *Piétu et al.* (2005); [2] *Andrews et al.* (2011b); [3] *Casassus et al.* (2012); [4] *Bruderer et al.* (2014); [5] *Mathews et al.* (2012); [6] *Cieza et al.* (2012); [7] *Isella et al.* (2010a); [8] *Tsukagoshi et al.* (2013); [9] *Hughes et al.* (2007); [10] *Rosenfeld et al.* (2013). Near-infrared cavities are from [11] Hashimoto et al. (in preparation); [12] *Tatulli et al.* (2011); [13] *Weinberger et al.* (1999); [14] *Avenhaus et al.* (2014); [15] *Quanz et al.* (2013b); [16] *Thalmann et al.* (2010); [17] *Hashimoto et al.* (2012); [18] *Mayama et al.* (2012); [19] Kooistra et al. (in preparation); [20] *Follette et al.* (2013); [21] *Tsukagoshi et al.* (2013); [22] Akiyama et al. (in preparation); [23] Kuzuhara et al. (in preparation); [24] Bonnefoy et al. (in preparation). Companion detection limits are from [6]; [25] *Kraus et al.* (2011); [26] *Ireland and Kraus* (2008); [27] *Biller et al.* (2012); [28] Lacour et al. (in preparation); [29] *Grady et al.* (2013); [30] Kenworthy et al. (in preparation); [31] *Kraus et al.* (2008); [32] *Vicente et al.* (2011); [33] Ireland et al. (in preparation); [34] *Romero et al.* (2012); [35] *Evans et al.* (2012); [36] *Stempels and Gahm* (2004); [37] *Ratzka et al.* (2005). Accretion rates are from [5]; [38] *Salyk et al.* (2013); [39] *Cohen and Kuhi* (1979); [40] *Ingleby et al.* (2013); [41] *Pogodin et al.* (2012); [42] *Dong et al.* (2012); [43] *Krautter et al.* (1997); [44] *Cieza et al.* (2010); [45] *Calvet et al.* (2004); [46] *Natta et al.* (2006); [47] *Alcalá et al.* (2014); [48] *Donati et al.* (2011).

[‡]The classification of IRS 48 as a TD or PTD is uncertain due to the presence of strong PAH emission in this object.

[§]The CoKu Tau/4 binary has a separation of 8 AU, which is within the range of semimajor axes required by tidal interaction theory to explain the 14-AU-radius SED-inferred inner disk hole of this object (*Nagel et al.,* 2010). A binary system with an eccentricity of 0.8 can clear out a region 3.5× the semimajor axis (e.g., *Artymowicz and Lubow,* 1994).

[¶]The HD 142527 binary is separated by 13 AU. This is not large enough to explain the disk gap. Note that *Casassus et al.* (2013) have disputed the presence of a companion, but recent results from *Close et al.* (2014) may confirm it.

[**]The V4046 Sgr binary has a separation of 0.045 AU, too small to clear out the disk hole. *Rosenfeld et al.* (2013) provide an aperture-masking companion limit that suggests there are no 0.1 $M_\odot$ companions down to 3 AU.

gap sizes that can be detected with Spitzer IRS. Over 80% of the emission at 10 μm comes from within 1 AU in the disk (e.g., *D'Alessio et al.,* 2006). Therefore, the Spitzer IRS is most sensitive to clearings in which a significant amount of dust located at radii <1 AU has been removed, and so it will be easier to detect disks with holes (i.e., transitional disks) as opposed to disks with gaps (i.e., pre-transitional

disks). The smallest gap in the innermost disk that will cause a noticeable "dip" in the Spitzer spectrum would span ~0.3–4 AU. It would be very difficult to detect gaps whose inner boundary is outside 1 AU [e.g., a gap spanning 5–10 AU in the disk (*Espaillat et al.,* 2010)]. Therefore, with current data we cannot exclude that any disk currently thought to be a full disk contains a small gap, nor can we exclude that currently known (pre-)transitional disks have additional clearings at larger radii (e.g., *Debes et al.*, 2013). It will be largely up to ALMA and the next generation of IR interferometers to detect such small disk gaps (e.g., *de Juan Ovelar et al.*, 2013). One should also keep in mind that millimeter data are necessary to break the degeneracy between dust settling and disk mass (see *Espaillat et al.*, 2012). Also, the opacity of the disk is controlled by dust and in any sophisticated disk model the largest uncertainty lies in the adopted dust opacities. We will return to disk-model limitations in section 2.2.

2.1.3. Implications for color-color diagrams. Another method of identifying transitional disks is through color-color diagrams (Fig. 3). This method grew in usage as more Spitzer IRAC and MIPS data became available and (pre-)transitional disks well characterized by Spitzer IRS spectra could be used to define the parameter space populated by these objects. In these diagrams, transitional disks are distinct from other disks since they have NIR colors or slopes (generally taken between two IRAC bands, or K and an IRAC band) significantly closer to stellar photospheres than other disks in Taurus, but MIR colors (generally taken between K or one IRAC band and MIPS [24]) comparable or higher than other disks in Taurus (e.g., *Hernández et al.*, 2007a, 2008, 2010; *Merín et al.*, 2010; *Muzerolle et al.*, 2010; *Luhman et al.*, 2010; *Williams and Cieza*, 2011; *Luhman and Mamajek*, 2012). Color-color diagrams are limited in their ability to identify pre-transitional disks because their fluxes in the NIR are comparable to many other disks in Taurus. Spitzer IRS data can do a better job of identifying pre-transitional disks using the equivalent width of the 10-μm feature or the NIR spectral index (e.g., n_{2-6}) vs. the MIR spectral index (e.g., n_{13-31}) (*Furlan et al.*, 2011; *McClure et al.*, 2010; *Manoj et al.*, 2011). Evolved disks are easier to identify in color-color diagrams since they show excesses over their stellar photospheres that are consistently lower than most disks in Taurus, both in the NIR *and* MIR.

In the future, JWST's sensitivity will allow us to expand Spitzer's SED and color-color work to many more disks, particularly to fainter objects in older and farther star-forming regions, greatly increasing the known number of transitional, pre-transitional, and evolved disks. Upcoming high-resolution imaging surveys with (sub)millimeter facilities in the near-future (i.e., ALMA) and IR interferometers further in the future [i.e., Very Large Telescope (VLT)/ Multi AperTure mid-Infrared SpectroScopic Experiment (MATISSE)] will give us a better understanding of the small-scale spatial structures in disks that SEDs cannot access.

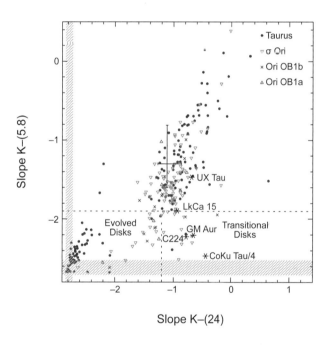

Fig. 3. Spitzer color-color diagram of objects in different associations and clusters. Observations are shown for populations of different ages: Taurus, 1–2 m.y. (*Luhman et al.*, 2010); σ Ori, 3 m.y. (*Hernández et al.*, 2007a); Ori OB1b, 7 m.y.; Ori OB1a, 10 m.y. (*Hernández et al.*, 2007b). The hatched region corresponds to stellar photospheric colors. The error bars represent the median and quartiles of Taurus objects (i.e., where most full disks are expected to lie). Well-characterized transitional disks [GM Aur, CoKu Tau/4, CVSO 224 (*Espaillat et al.*, 2008a)] and pre-transitional disks (LkCa 15, UX Tau) are indicated with asterisks. The dotted lines correspond to the lower quartile of disk emission in σ Ori, and roughly separate the evolved disks (lower left) from the transitional disks (lower right). Note that the pre-transitional disks do not lie below the dotted line, highlighting that it is harder to identify disk gaps based on colors alone. Figure adapted from *Hernández et al.* (2007b).

2.2. Submillimeter/Radio Continuum Imaging

The dust continuum at (sub)millimeter/radio wavelengths is an ideal probe of cool material in disks. At these wavelengths, dust emission dominates over the contribution from the stellar photosphere, ensuring that contrast limitations are not an issue. Moreover, interferometers give access to the emission structure on a wide range of spatial scales, and will soon provide angular resolution that regularly exceeds 100 mas. The continuum emission at these long wavelengths is also thought to have relatively low optical depths, meaning the emission morphology is sensitive to the density distribution of millimeter- and centimeter-sized grains (*Beckwith et al.*, 1990). These features are especially useful for observing the dust-depleted inner regions of (pre-)transitional disks, as will be illustrated in the following subsections.

2.2.1. (Sub)millimeter disk cavities. With sufficient resolution, the (sub)millimeter dust emission from disks with cavities exhibits a "ring"-like morphology, with limb-brightened ansae along the major axis for projected viewing geometries. In terms of the actual measured quantity, the interferometric visibilities, there is a distinctive oscillation pattern (effectively a Bessel function) where the first "null" is a direct measure of the cavity dimensions (see *Hughes et al., 2007*).

As of this writing, roughly two dozen disk cavities have been directly resolved at (sub)millimeter wavelengths. A gallery of representative continuum images, primarily from observations with the Submillimeter Array (SMA), is shown in Fig. 4. For the most part, these discoveries have been haphazard: Some disks were specifically targeted based on their infrared SEDs (section 2.1) (e.g., *Brown et al., 2009*), while others were found serendipitously in high-resolution imaging surveys aimed at constraining the radial distributions of dust densities (e.g., *Andrews et al., 2009*).

Perhaps the most remarkable aspect of these searches is the frequency of dust-depleted disk cavities at (sub)millimeter wavelengths, especially considering that the imaging census of all disks has so far been severely restricted by both sensitivity and resolution limitations. In the nearest star-forming regions accessible to the northern hemisphere (Taurus and Ophiuchus), only about half the disks in the bright half of the millimeter luminosity (approximately disk mass) distribution have been imaged with sufficient angular resolution (~0.3″) to find large (>20 AU in radius) disk cavities (*Andrews et al., 2009, 2010; Isella et al., 2009; Guilloteau et al., 2011*). Even with these strong selection biases, the incidence of resolved cavities is surprisingly high compared to expectations from IR surveys (see section 4). *Andrews et al.* (2011b) estimated that at least one in three of these millimeter-bright (massive) disks exhibit large cavities.

2.2.2. Model fitting. The basic structures of disk cavities can be quantified through radiative transfer modeling of their SEDs (see section 2.1) simultaneously with resolved millimeter data. These models often assume that the cavity can be described as a region of sharply reduced dust

surface densities (e.g., *Piétu et al., 2006; Brown et al., 2008; Hughes et al., 2009; Andrews et al., 2011b; Cieza et al., 2012; Mathews et al., 2012*). Such work finds cavity radii of ~15–75 AU, depletion factors in the inner disk of ~10^2–10^5 relative to optically thick full disks, and outer regions with sizes and masses similar to those found for full disks. However, there are subtleties in this simple modeling prescription. First, the depletion levels are usually set by the infrared SED, not the millimeter data: The resolved images have a limited dynamic range and can only constrain an intensity drop by a factor of <100. Second, and related, is that the "sharpness" of the cavity edge is unclear. The most popular model prescription implicitly imposes a discontinuity, but the data only directly indicate that the densities substantially decrease over a narrow radial range (a fraction of the still-coarse spatial resolution; ~10 AU). Alternative models with a smoother taper at the cavity edge can explain the data equally well in many cases (e.g., *Isella et al., 2010b, 2012; Andrews et al., 2011a*), and might alleviate some of the tension with IR scattered-light measurements (see section 2.3).

Some additional problematic issues with these simple models have been illuminated, thanks to a new focus on the details of the resolved millimeter data. For example, in some disks the dust-ring morphology is found to be remarkably narrow — with nearly all the emission coming from a belt 10–20 AU across (or less) — even as we trace gas with molecular line emission extending hundreds of astronomical units beyond it (e.g., *Rosenfeld et al., 2013*). This hints at the presence of a particle "trap" near the cavity edge, as might be expected from local dynamical interactions between a planet and the gas disk (see section 3.2) (*Zhu et al., 2012; Pinilla et al., 2012*). In a perhaps related phenomenon (e.g., *Regály et al., 2012; Birnstiel et*

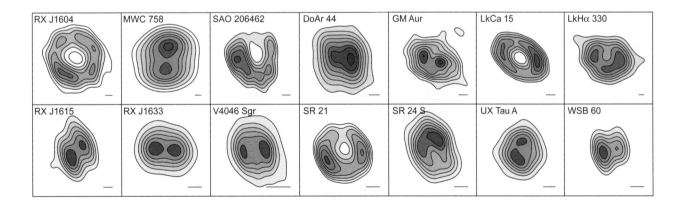

Fig. 4. A gallery of 880-μm dust continuum images from the Submillimeter Array for (pre-)transitional disks in nearby star-forming regions. From left to right, the panels show the disks around RX J1604.3-2130 (*Mathews et al., 2012*), MWC 758 (*Isella et al., 2010b*), SAO 206462 (*Brown et al., 2009*), DoAr 44 (*Andrews et al., 2009*), GM Aur (*Hughes et al., 2009*), LkCa 15 (*Andrews et al., 2011b*), LkHα 330 (*Brown et al., 2008*), RX J1615.3-3255 (*Andrews et al., 2011b*), RX J1633.9-2442 (*Cieza et al., 2012*), V4046 Sgr (*Rosenfeld et al., 2013*), SR 21 (*Brown et al., 2009*), SR 24 S (*Andrews et al., 2010*), UX Tau A (*Andrews et al., 2011b*), and WSB 60 (*Andrews et al., 2009*). A projected 30-AU scale bar is shown at the lower right corner of each panel.

al., 2013), new high-fidelity images of (pre-)transitional disks are uncovering evidence that strong azimuthal asymmetries are common features of the millimeter emission rings (*Brown et al.*, 2009; *Tang et al.*, 2012; *Casassus et al.*, 2013; *van der Marel et al.*, 2013; *Isella et al.*, 2013).

A number of pressing issues will soon be addressed by the ALMA project. Regarding the incidence of the disk holes and gaps, an expanded high-resolution imaging census should determine the origin of the anomalously high occurrence of dust cavities in millimeter-wave images. If the detection rate estimated by *Andrews et al.* (2011b) is found to be valid at all luminosities, it would confirm that even the small amount of dust inside the disk cavities sometimes produces enough IR emission to hide the standard (pre-)transitional disk signature at short wavelengths, rendering IR selection inherently incomplete — something already hinted at in the current data. Perhaps more interesting would be evidence that the (pre-)transitional disk frequency depends on environmental factors, like disk mass (i.e., a selection bias) or stellar host properties (e.g., *Owen and Clarke*, 2012). More detailed analyses of the disk structures are also necessary, both to develop a more appropriate modeling prescription and to better characterize the physical processes involved in clearing the disk cavities. Specific efforts toward resolving the depletion zone at the cavity boundary, searching for material in the inner disk, determining ring widths and measuring their millimeter/radio colors to infer the signatures of particle evolution and trapping (e.g., *Pinilla et al.*, 2012; *Birnstiel et al.*, 2013), and quantifying the ring substructures should all impart substantial benefits on our understanding of disks.

2.3. Infrared Imaging

Infrared imaging has been used successfully to observe disks around bright stars. Spacebased observations free from atmospheric turbulence [e.g., Hubble Space Telescope (HST)] have detected fine disk structure such as spiral features in the disk of HD 100546 (*Grady et al.*, 2001) and a ring-like gap in HD 141569 (*Weinberger et al.*, 1999). Mid-infrared images of (pre-)transitional disks are able to trace the irradiated outer wall, which effectively emits thermal radiation (e.g., *Geers et al.*, 2007; *Maaskant et al.*, 2013). More recently, high-resolution NIR polarimetric imaging and IR interferometry has become available, allowing us to probe much further down into the inner few tens of astronomical units in disks in nearby star-forming regions. Here we focus on NIR polarimetric and IR interferometric imaging of (pre-)transitional disks and also the results of IR imaging searches for companions in these objects.

2.3.1. Near-infrared polarimetric imaging. Near-infrared polarimetric imaging is capable of tracing the spatial distribution of submicrometer-sized dust grains located in the uppermost, surface layer of disks. One of the largest NIR polarimetry surveys of disks to date was conducted as part of the Strategic Explorations of Exoplanets and Disks with Subaru (SEEDS) (*Tamura*, 2009) project [e.g.,

Taurus at 140 pc (*Thalmann et al.*, 2010; *Tanii et al.*, 2012; *Hashimoto et al.*, 2012; *Mayama et al.*, 2012; *Follette et al.*, 2013; *Takami et al.*, 2013, *Tsukagoshi et al.*, 2014)]. There is also imaging work being done by VLT's Nasmyth Adaptive Optics System (NAOS) and COude Near Infrared CAmera (CONICA) (collectively known as NACO), predominantly focusing on disks around Herbig Ae/Be stars (*Quanz et al.*, 2011, 2012, 2013a,b; *Rameau et al.*, 2012). These surveys access the inner tens of astronomical units in disks, reaching a spatial resolution of 0.06″ (8 AU) in nearby star-forming regions. Such observations have the potential to reveal fine structures such as spirals, warps, offsets, gaps, and dips in the disk.

Most of the (pre-)transitional disks around TTS observed by SEEDS have been resolved. This is because the stellar radiation can reach the outer disk more easily given that the innermost regions of (pre-)transitional disks are less dense than full disks. Many of these disks can be sorted into the three following categories based on their observed scattered light emission (i.e., their "polarized intensity" appearance) at 1.6 μm: category A — no cavity in the NIR with a smooth radial surface brightness profile at the outer wall (e.g., SR 21, DoAr 44, RX J1615-3255; Figs. 5a,b); category B — similar to category A, but with a broken radial brightness profile (e.g., TW Hya, RX J1852.3-3700, LkHα 330); these disks display a slight slope in the radial brightness profile in the inner portion of the disk, but a steep slope in the outer regions (Figs. 5c,d); category C — a clear cavity in the NIR polarized light (e.g., GM Aur, Sz 91, PDS 70, RX J1604-2130; Figs. 5e,f).

The above categories demonstrate that the spatial distribution of small and large dust grains in the disk are not necessarily similar. Based on previous submillimeter images (see section 2.2), the large, millimeter-sized dust grains in the inner regions of each of the above disks is significantly depleted. However, in categories A and B there is evidence that a significant amount of small, submicrometer-sized dust grains remains in the inner disk, well within the cavity seen in the submillimeter images. In category C, the small dust grains appear to more closely trace the large dust distribution, as both are significantly depleted in the inner disk. One possible mechanism that could explain the differences between the three categories presented above is dust filtration (e.g., *Rice et al.*, 2006; *Zhu et al.*, 2012), which we will return to in more detail in section 3.2. More high-resolution imaging observations of disks at different wavelengths is necessary to develop a fuller picture of their structure given that the disk's appearance at a certain wavelength depends on the dust opacity.

2.3.2. Infrared interferometric imaging. Infrared interferometers, such as the Very Large Telescope Interferometer (VLTI), the Keck Interferometer (KI), and the Center for High Angular Resolution Astronomy (CHARA) array, provide milliarcsecond angular resolution in the NIR and MIR regime (1–13 μm), enabling new constraints on the structure of (pre-)transitional disks. Such spatially resolved studies are important to reveal complex structure in the small dust

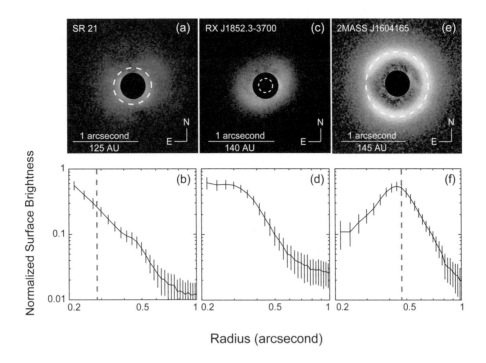

Fig. 5. Near-infrared polarimetric images of transitional disks in the H band from the Subaru Telescope (top; 1.6 μm) along with their averaged radial surface brightness profiles along the major disk axis (bottom). Broken lines in the top and bottom panels correspond to the radius of the outer wall as measured with submillimeter imaging and/or SEDs. We note that the regions of the inner disk that cannot be resolved are masked out in the panels. From left to right the objects are SR 21 (*Follette et al.,* 2013), RX J1852-3700 (Kudo et al., in preparation), and RX J1604-2130 (*Mayama et al.,* 2012).

distribution within the innermost region of the disk, testing the basic constructs of models that have been derived based on spatially unresolved data (e.g., SEDs).

The visibility amplitudes measured with interferometry permit direct constraints on the brightness profile, and, through radiative transfer modeling (see sections 2.1 and 2.2 for discussion of limitations), on the distribution and physical conditions of the circumstellar material. The NIR emission (H and K band, 1.4–2.5 μm) in the pre-transitional disks studied most extensively with IR interferometry (i.e., HD 100546, T Cha, and V1247 Ori) is dominated by hot optically thick dust, with smaller contributions from scattered light (*Olofsson et al.,* 2013) and optically thin dust emission (*Kraus et al.,* 2013). The measured inner disk radii are in general consistent with the expected location of the dust sublimation radius, while the radial extent of this inner emission component varies significantly for different sources [HD 100546: 0.24–4 AU (*Benisty et al.,* 2010; *Tatulli et al.,* 2011); T Cha: 0.07–0.11 AU (*Olofsson et al.,* 2011, 2013); V1247Ori: 0.18–0.27 AU (*Kraus et al.,* 2013)].

The MIR regime (N band, 8–13 μm) is sensitive to a wider range of dust temperatures and stellocentric radii. In the transitional disk of TW Hya, the region inside 0.3–0.52 AU is found to contain only optically thin dust (*Eisner et al.,* 2006; *Ratzka et al.,* 2007; *Akeson et al.,* 2011), followed by an optically thick outer disk (*Arnold et al.,* 2012b), in agreement with SED modeling (*Calvet et al.,* 2002) and (sub)millimeter imaging (*Hughes et al.,* 2007).

The gaps in the pre-transitional disks of the Herbig Ae/Be star HD 100546 (*Mulders et al.,* 2013) and the TTS T Cha (*Olofsson et al.,* 2011, 2013) were found to be highly depleted of (sub)micrometer-sized dust grains, with no significant NIR or MIR emission, consistent with SED-based expectations (i.e., no substantial 10-μm silicate emission). The disk around the Herbig Ae/Be star V1247 Ori, on the other hand, exhibits a gap filled with optically thin dust. The presence of such optically thin material within the gap is not evident from the SED, while the interferometric observations indicate that this gap material is the dominant contributor at MIR wavelengths. This illustrates the importance of IR interferometry for unraveling the physical conditions in disk gaps and holes.

We note that besides the dust continuum emission, some (pre-)transitional disks exhibit polycyclic aromatic hydrocarbon (PAH) spectral features, e.g., at 7.7 μm, 8.6 μm, and 11.3 μm. For a few objects, it was possible to locate the spatial origin of these features using adaptive optics (AO) imaging (*Habart et al.,* 2006) or long-baseline interferometry from the VLT's Mid-Infrared Instrument (MIDI) (*Kraus,* 2013). These observations showed that these molecular bands originate from a significantly more extended region than the NIR/MIR continuum emission, including the gap region and the outer disk. This is consistent with the scenario that these particles are transiently heated by UV photons and can be observed over a wide range of stellocentric radii.

One of the most intriguing findings obtained with IR interferometry is the detection of nonzero phase signals, which indicate the presence of significant asymmetries in the inner, astronomical-unit-scale disk regions. Keck's Near InfraRed Camera 2 (NIRC2) aperture masking observations of V1247 Ori (*Kraus et al., 2013*) revealed asymmetries whose direction is not aligned with the disk minor axis and also changes with wavelength. Therefore, these asymmetries are neither consistent with a companion detection, nor with disk features. Instead, these observations suggest the presence of complex, radially extended disk structures, located within the gap region. It is possible that these structures are related to the spiral-like inhomogeneities that have been detected with coronagraphic imaging on scales ~10× larger (e.g., *Hashimoto et al., 2011; Grady et al., 2013*) and that they reflect the dynamical interaction of the gap-opening body/bodies with the disk material. Studying these complex density structures and relating the asymmetries to the known spectrophotometric variability of these objects (section 2.4) will be a major objective of future interferometric imaging studies.

The major limitations from the existing studies arise from sparse UV coverage, which has so far prevented the reconstruction of direct interferometric images for these objects. Different strategies have been employed in order to relax the UV-coverage restrictions, including the combination of long-baseline interferometric data with single-aperture interferometry techniques [e.g., speckle and aperture masking interferometry (*Arnold et al., 2012b; Olofsson et al., 2013; Kraus et al., 2013*)] and the combination of data from different facilities (*Akeson et al., 2011; Olofsson et al., 2013; Kraus et al., 2013*). Truly transformational results can be expected from the upcoming generation of imaging-optimized long-baseline interferometric instruments, such as the four-telescope MIR beam combiner MATISSE, which will enable efficient long-baseline interferometric imaging on scales of several astronomical units.

2.3.3. Companion detections. Near-infrared imaging observations can directly reveal companions within the cleared regions of disks. Both theory and observations have long shown that stellar binary companions can open gaps (e.g., *Artymowicz and Lubow* 1994; *Jensen and Mathieu* 1997; *White and Ghez,* 2001), based on numerous moderate-contrast companions (~0–3 mag contrast, or companion masses >0.1 $M_\odot$) that have been identified with radial velocity (RV) monitoring, HST imaging, AO imaging, and speckle interferometry. For example, NIR imaging of CoKu Tau/4, a star surrounded by a transitional disk, revealed a previously unknown stellar-mass companion that is likely responsible for the inner clearing in this disk, demonstrating that it is very important to survey stars with (pre-)transitional disks for binarity in addition to exploring other possible clearing mechanisms (see section 3.2) (*Ireland and Kraus, 2008*).

The detection of substellar or planetary companions has been more challenging, due to the need for high contrast ($\Delta L' = 5$ to achieve $M_{lim} = 30$ M_{Jup} for a 1-$M_\odot$ primary star) near or inside the formal diffraction limit of large telescopes. Most of the high-contrast candidate companions identified to date have been observed with interferometric techniques such as nonredundant mask (NRM) interferometry (*Tuthill et al., 2000; Ireland* 2012) that measure more stable observable quantities (such as closure phase) to achieve limits of $\Delta L' = 7$–8 at λ/D (3–5 M_{Jup} at 8 AU). The discoveries of NRM include a candidate planetary mass companion to LkCa 15 (*Kraus and Ireland,* 2012) and a candidate low-mass stellar companion to T Cha (*Huélamo et al.,* 2011). A possible candidate companion was reported around FL Cha, although these asymmetries could be associated with disk emission instead (*Cieza et al., 2013*).

Advanced imaging techniques are also beginning to reveal candidate companions at intermediate orbital radii that correspond to the optically thick outer regions of (pre-) transitional disks (e.g., *Quanz et al.,* 2013a), beyond the outer edge of the hole or gap region. The flux contributions of companions can be difficult to distinguish from scattered light due to disk features (*Olofsson et al., 2013; Cieza et al.,* 2013; *Kraus et al.,* 2013). However, the case of LkCa 15 shows that the planetary hypothesis can be tested using multi-epoch, multi-wavelength data (to confirm colors and Keplerian orbital motion) and by direct comparison to resolved submillimeter maps (to localize the candidate companion with respect to the inner disk edge).

Even with the enhanced resolution and contrast of techniques like NRM, current surveys are only able to probe super-Jupiter masses in outer solar systems. For bright stars (I ≤ 9), upcoming extreme AO systems like the Gemini Planet Imager (GPI) and Spectro-Polarimetric High-contrast Exoplanet REsearch (SPHERE) will pave the way for higher contrasts with both imaging and NRM, achieving contrasts of $\Delta K \geq 10$ at λ/D (~1–2 M_{Jup} at 10 AU). However, most young solar-type and low-mass stars fall below the optical flux limits of extreme AO. Further advances for those targets will require observations with JWST that probe the sub-Jupiter regime for outer solar systems (ΔM ~10 at λ/D, or <1 M_{Jup} at >15 AU) or with future groundbased telescopes that probe the Jupiter regime near the snow line (achieving ΔK ~10 at λ/D or ~1–2 M_{Jup} at >2–3 AU).

2.4. Time Domain Studies

Infrared variability in TTS is ubiquitous and several groundbased studies have been undertaken to ascertain the nature of this variability (e.g., *Joy,* 1945; *Rydgren et al.,* 1976; *Carpenter et al.,* 2001). With the simultaneous MIR wavelength coverage provided by Spitzer IRS, striking variability in (pre-)transitional disks was discovered, suggestive of structural changes in these disks with time. We review this variability along with the mechanisms that have been proposed to be responsible for it in the following subsections.

2.4.1. "Seesaw" variability. The flux in many pre-transitional disks observed for variability to date "seesaws," i.e., as the emission decreases at shorter wavelengths in the IRS data, the emission increases at longer wavelengths (Fig. 6)

(*Muzerolle et al.,* 2009; *Espaillat et al.,* 2011; *Flaherty et al.,* 2012). Mid-infrared variability with IRS was also seen in some transitional disks [e.g., GM Aur and LRLL 67 (*Espaillat et al.,* 2011; *Flaherty et al.,* 2012)], although in these objects the variability was predominantly around the region of the silicate emission feature. Typically, the flux in the pre-transitional disks and transitional disks observed changed by about 10% between epochs, but in some objects the change was as high as 50%.

This variability may point to structural changes in disks. Spectral energy distribution modeling can explain the seesaw behavior seen in pre-transitional disks by changing the height of the inner wall of these disks (Fig. 7) (*Espaillat et al.,* 2011). When the inner wall is taller, the emission at the shorter wavelengths is higher since the inner wall dominates the emission at 2–8 μm. The taller inner wall casts a larger shadow on the outer disk wall, leading to less emission at wavelengths beyond 20 μm where the outer wall dominates. When the inner wall is shorter, the emission at the shorter wavelengths is lower and the shorter inner wall casts a smaller shadow on the outer disk wall, leading to more emission at longer wavelengths. This "seesaw" variability confirms the presence of optically thick material in the inner disk of pre-transitional disks. The variability seen in transitional disks may suggest that while the disk is vertically optically thin, there is a radially optically thick structure in the inner disk, perhaps composed of large grains

and/or limited in spatial extent so that it does not contribute substantially to the emission between 1 and 5 μm while still leading to shadowing of the outer disk. One intriguing possibility involves accretion streams connecting multiple planets, as predicted in the models of *Zhu et al.* (2011) and *Dodson-Robinson and Salyk* (2011) and claimed to be seen by ALMA (*Casassus et al.,* 2013).

2.4.2. Possible underlying mechanisms. When comparing the nature of the observed variability to currently known physical mechanisms, it seems unlikely that star spots, winds, and stellar magnetic fields are the underlying cause. The star spots proposed to explain variability at shorter wavelengths in other works (e.g., *Carpenter et al.,* 2001) could change the irradiation heating, but this would cause an overall increase or decrease of the flux, not seesaw variability. A disk wind that carries dust may shadow the outer disk. However, *Flaherty et al.* (2011) do not find evidence for strong winds in their sample, which displays seesaw variability. Stellar magnetic fields that interact with dust beyond the dust sublimation radius may lead to changes if the field expands and contracts or is tilted with respect to plane of the disk (*Goodson and Winglee,* 1999; *Lai and Zhang,* 2008). However, it is thought that the stellar magnetic field truncates the disk within the co-rotation radius and for many objects. The co-rotation radius is within the dust sublimation radius, making it unlikely that the stellar magnetic field is interacting with the dust in the disk (*Flaherty et al.,* 2011).

It is unclear what role accretion or X-ray flares may play in disk variability. Accretion rates are known to be variable in young objects, but *Flaherty et al.* (2012) do not find that the observed variations in accretion rate are large enough to reproduce the magnitude of the MIR variability observed. Strong X-ray flares can increase the ionization of dust and lead to a change in scale height (*Ke et al.,* 2012). However, while TTS are known to have strong X-ray flares (*Feigelson*

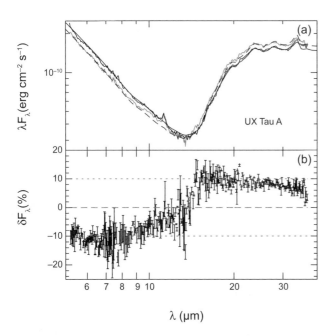

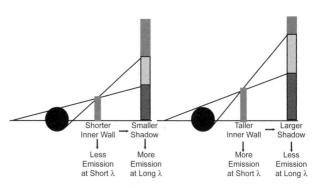

Fig. 6. **(a)** Observed Spitzer IRS spectra (gray) and models (solid and broken black lines) for the pre-transitional disk of UX Tau A. The variability between the two spectra can be reproduced by changing the height of the inner disk wall in the model by 17%. **(b)** Percentage change in flux between the two IRS spectra above. The observed variability cannot be explained by the observational uncertainties of IRS (error bars). Figure adapted from *Espaillat et al.* (2011).

Fig. 7. Schematic depicting the proposed link between the observed variability and disk structure based on the SED modeling from Fig. 6. Medium gray corresponds to visible areas of the disk wall, while light gray and dark gray areas are in the penumbra and umbra, respectively. The MIR variability observed in pre-transitional disks can be explained by changes in the height of the inner disk wall, which results in variable shadowing of the outer wall.

et al., 2007), it is unlikely that all the MIR disk variability observed overlapped with strong X-ray flares.

The MIR variability seen is most likely due to perturbations in the disk, possibly by planets or turbulence. Planets are thought to create spiral density waves in the disk [see the *Protostars and Planets V* (PPV) review by *Durisen et al.* (2007)], which may have already been detected in disks (e.g., *Hashimoto et al.,* 2011; *Grady et al.,* 2013). Such spiral-density waves may affect the innermost disk, causing the height of the inner disk wall to change with time and creating the seesaw variability observed. The timescales of the variability discussed here span ~1–3 yr down to 1 week or less. If the variability is related to an orbital timescale, this corresponds to ~1–2 AU and <0.07 AU in the disk, plausible locations for planetary companions given our own solar system and detections of hot Jupiters (*Marcy et al.,* 2005). Turbulence is also a viable solution. Magnetic fields in a turbulent disk may lift dust and gas off the disk (*Turner et al.,* 2010; *Hirose and Turner,* 2011). The predicted magnitude of such changes in the disk scale height are consistent with the observations.

Observations with JWST can explore the range of timescales and variability present in (pre-)transitional disks to test all the scenarios explored above. It is also likely that a diverse range of variability is present in most disks around young stars [e.g., Spitzer's Young Stellar Object Variability (YSOVAR) program (*Morales-Calderón et al.,* 2011)] and JWST observations of a wide range of disks will help us more fully categorize disk variability.

2.5. Gaseous Emission

The structure of the gas in (pre-)transitional disks is a valuable probe, because the mechanisms proposed to account for the properties of these objects (section 3; e.g., grain growth, photoevaporation, companions) may impact the gas in different ways from the dust. The chapters in this volume by Pontoppidan et al. and Alexander et al., as well as the earlier PPV review by *Najita et al.* (2007b), describe some of the available atomic and molecular diagnostics and how they are used to study gas disks. In selecting among these specifically for the purpose of probing radial disk structure, it is important to consider how much gas is expected to remain in a hole or gap. Although theoretical studies suggest the gas column density is significantly reduced in the cleared dust region [e.g., by ~1000 (*Regály et al.,* 2010)], given a typical TTS's disk column density of 100 g cm^{-2} at 1 AU, a fairly hefty gas column density could remain. Many stars hosting (pre-)transitional disks also show signs of significant gas accretion (section 4.2). These considerations suggest that molecular diagnostics, which probe larger disk column densities ($N_H > 1 \times 10^{21}$ cm^{-2}), are more likely to be successful in detecting a hole or gap in the gas disk. In the following, we review what is known to date about the radial structure of gas in disks.

2.5.1. Mid-infrared and (sub)millimeter spectral lines. As there are now gas diagnostics that probe disks over a wide range of radii (~0.1–100 AU), the presence or absence of these diagnostics in emission can give a rough idea of whether gas disks are radially continuous or not. At large radii, the outer disks of many (pre-)transitional disks (TW Hya, GM Aur, DM Tau, LkCa 15, etc.) are well studied at millimeter wavelengths (e.g., *Koerner and Sargent,* 1995; *Guilloteau and Dutrey,* 1994; *Dutrey et al.,* 2008) (see also section 2.2). For some, the millimeter observations indicate that rotationally excited CO exists within the cavity in the dust distribution [e.g., TW Hya, LkCa 15 (*Rosenfeld et al.,* 2012; *Piétu et al.,* 2007)], although these observations cannot currently constrain how that gas is distributed. Many (pre-)transitional disks also show multiple signatures of gas close to the star: ongoing accretion (section 4.2), UV H$_2$ emission (e.g., *Ingleby et al.,* 2009; *France et al.,* 2012), and rovibrational CO emission (*Salyk et al.,* 2009, 2011; *Najita et al.,* 2008, 2009).

In contrast, the 10–20-µm Spitzer spectra of these objects conspicuously lack the rich molecular emission (e.g., H$_2$O, C$_2$H$_2$, HCN) that characterizes the MIR spectra of full disks around classical TTS (CTTS) (i.e., stars that are accreting) (*Najita et al.,* 2010; *Pontoppidan et al.,* 2010). As these MIR molecular diagnostics probe radii within a few astronomical units of the star, their absence is suggestive of a missing molecular disk at these radii, i.e., a gap between an inner disk (traced by CO and UV H$_2$) and the outer disk (probed in the millimeter). Alternatively, the disk might be too cool to emit at these radii, or the gas may be abundant but in atomic form. Further work, theoretical and/or observational, is needed to evaluate these possibilities.

2.5.2. Velocity resolved spectroscopy. Several approaches can be used to probe the distribution of the gas in greater detail. In the absence of spatially resolved imaging, which is the most robust approach, velocity-resolved spectroscopy coupled with the assumption of Keplerian rotation can probe the radial structure of gaseous disks (see PPIV review by *Najita et al.,* 2000). The addition of spectroastrometric information (i.e., the spatial centroid of spectrally resolved line emission as a function of velocity) can reduce ambiguities in the disk properties inferred with this approach (*Pontoppidan et al.,* 2008). These techniques have been used to search for sharp changes in gaseous emission as a function of radius as evidence of disk cavities, and to identify departures from azimuthal symmetry such as those created by orbiting companions.

Velocity-resolved spectroscopy of CO rovibrational emission provides tentative evidence for a truncated inner gas disk in the evolved disk V836 Tau (*Strom et al.,* 1989). An optically thin gap, if there is one, would be found at small radii (~1 AU) and plausibly overlap the range of disk radii probed by the CO emission. Indeed, the CO emission from V836 Tau is unusual in showing a distinct double-peaked profile consistent with the truncation of the CO emission beyond ~0.4 AU (Fig. 8) (*Najita et al.,* 2008). In comparison, other disks show much more centrally peaked rovibrational CO line profiles (*Salyk et al.,* 2007, 2009; *Najita et al.,* 2008, 2009).

Spectroscopy also indicates possible differences in the radial extent of the gas and dust in the inner disk. The CO emission profile from the pre-transitional disk of LkCa 15 (Fig. 8) (*Najita et al.,* 2008) spans a broad range of velocities, indicating that the inner gas disk extends over a much larger range of radii (from ~0.08 AU out to several astronomical units) than in V836 Tau. This result might be surprising given that SED modeling suggests that the inner optically thick dust disk of LkCa 15 extends over a narrow annular region [0.15–0.19 AU (*Espaillat et al.,* 2010)]. The origin of possible differences in the radial extent of the gas and dust in the inner disk region is an interesting topic for future work.

Some of the best evidence to date for the truncation of the outer gas disks of (pre-)transitional disks comes from studies of Herbig Ae/Be stars. For nearby systems with large dust cavities, groundbased observations can spatially resolve the inner edge of the CO rovibrational emission from the outer disks around these bright stars [e.g., HD 141569 (*Goto et al.,* 2006)]. Line profile shapes and constraints from UV fluorescence modeling have also been used to show that rovibrational CO and OH emission is truncated at the same radius as the disk continuum in some systems (*Brittain et al.,* 2007, 2009; *van der Plas et al.,* 2009; *Liskowsky et al.,* 2012), although the radial distribution of the gas and dust appear to differ in other systems [e.g., IRS 48 (*Brown et al.,* 2012)]. The gas disk may sometimes be truncated significantly inward of the outer dust disk and far from the star. For example, the CO emission from SR 21 extends much further in (to ~7 AU) than the inner hole size of ~18 AU (*Pontoppidan et al.,* 2008).

Several unusual aspects of the gaseous emission from HD 100546 point to the possibility that the cavity in the molecular emission is created by an orbiting high-mass giant planet. *Acke and van den Ancker* (2006) inferred the presence of a gap in the gas disk based on a local minimum in the [O I] 6300-Å emission from the disk. The rovibrational OH emission from HD 100546 is found to show a strong line asymmetry that is consistent with emission from an eccentric inner rim [e ≥ 0.18 (*Liskowsky et al.,* 2012)]. Eccentricities of that magnitude are predicted to result from planet-disk interactions (e.g., *Papaloizou et al.,* 2001; *Kley and Dirksen,* 2006). In addition, the CO rovibrational emission varies in its spectroastrometric signal. The variations can be explained by a CO emission component that orbits the star just within the 13-AU inner rim (Fig. 9) (*Brittain et al.,* 2013). The required emitting area (0.1 AU²) is similar to that expected for a circumplanetary disk surrounding a 5-M_{Jup} planet at that distance (e.g., *Quillen and Trilling,* 1998; *Martin and Lubow,* 2011). Further studies with ALMA and Extremely Large Telescopes (ELTs) will give us new opportunities to explore the nature of disk holes and gaps in the gas disk (e.g., *van der Marel et al.,* 2013; *Casassus et al.,* 2013).

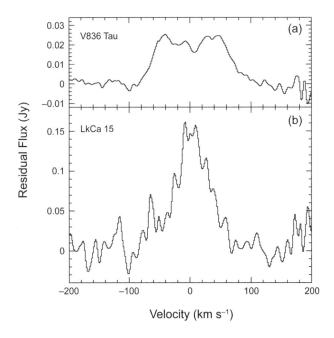

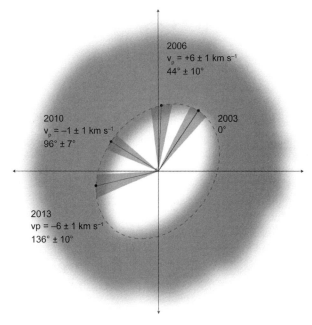

Fig. 8. Rovibrational line profiles of CO v = 1–0 emission from **(a)** V836 Tau and **(b)** LkCa 15 after correction for stellar CO absorption (*Najita et al.,* 2008). The double-peaked profile of V836 Tau indicates that the gas emission is truncated beyond ~0.4 AU. The centrally peaked LkCa 15 profile indicates gas emission extending to much larger radii.

Fig. 9. Spectroastrometry of the CO rovibrational emission from HD 100546, followed over 10 yr, reveals evidence for a compact source of CO emission that orbits the star near the inner disk edge (*Brittain et al.,* 2013, 2014). Possible interpretations include emission from a circumplanetary disk or a hot spot heated by an orbiting companion.

3. COMPARISON TO THEORETICAL MECHANISMS

There are three main observational constraints every theory must confront when attempting to explain (pre-)transitional disk observations. First, the cleared regions in (pre-)transitional disks studied to date are generally large (sections 2.1 and 2.2). For transitional disks, the optically thin region extends from tens of astronomical units all the way down to the central star. For pre-transitional disks, SED modeling suggests that the optically thick region may only extend up to ~1 AU, followed by a gap that is significantly depleted of dust up to tens of astronomical units, as seen in the transitional disks. At the same time, submillimeter imaging has revealed the existence of some disks that have inner regions significantly depleted of large dust grains without exhibiting MIR SED deficits (section 2.2), suggesting that a large amount of small dust grains still remain in the inner disk, as also shown by NIR polarimetric images (section 2.3). Second, in order to be discernible in the SED, the gaps/holes need to be optically thin, which implies that the mass of small dust grains (on the order of a micrometer or less) must be extremely low. Third, while most (pre-)transitional disks accrete onto the central star at a rate that is lower than the median accretion rate for TTS [~10^{-8} $M_\odot$ yr^{-1} (*Hartmann et al.*, 1998)] (section 4.2), their rates are still substantial (section 4.2). This indicates that considerable gas is located within the inner disk, although we currently do not have many constraints on how this gas is spatially distributed. In the following sections, we review the clearing mechanisms that have been applied to explain (pre-)transitional disk observations in light of the above constraints.

3.1. Nondynamical Clearing Mechanisms

A diverse set of physical mechanisms has been invoked to explain (pre-)transitional disk observations, with varying levels of success. Much of the attention is focused on dynamical interactions with one or more companion objects, although that subject will be addressed separately in section 3.2. There are a number of alternative scenarios that merit review here, including the effects of viscous evolution, particle growth and migration, and dispersal by (photoevaporative) winds.

3.1.1. Viscous evolution. The interactions of gravitational and viscous torques comprise the dominant mechanism for the structural evolution of disks over most of their lifetimes (*Lynden-Bell and Pringle*, 1974; *Hartmann et al.*, 1998). An anomalous kinematic viscosity, presumably generated by magneto-hydrodynamic (MHD) turbulence, drives a persistent inward flow of gas toward the central star (e.g., *Hartmann et al.*, 2006). Angular momentum is transported outward in that process, resulting in some disk material being simultaneously spread out to large radii. As time progresses, the disk mass and accretion rate steadily decline. To first order, this evolution is self-similar, so there is no preferential scale for the depletion of disk material.

The nominal viscous evolution (approximately depletion) timescales at tens of astronomical units are long, comparable to the stellar host ages.

If the gas and dust were perfectly coupled, we would expect viscous evolution acting alone to first produce a slight enhancement in the FIR/millimeter SED (due to the spreading of material out to larger radii) and then settle into a slow, steady dimming across the SED (as densities decay). Coupled with the sedimentation of disk solids, these effects are a reasonable explanation for the evolved disks (section 2.1). That said, there is no reason to expect that viscous effects alone are capable of the preferential depletion of dust at small radii needed to produce the large cleared regions typical of (pre-)transitional disks. Even in the case of enhanced viscosity in the outer wall due to the magneto-rotational instability (MRI) (*Chiang and Murray-Clay*, 2007), this needs a preexisting inner hole to have been formed via another mechanism in order to be effective.

3.1.2. Grain growth. The natural evolution of dust in a gas-rich environment offers two complementary avenues for producing the observable signatures of holes and gaps in disks. First is the actual removal of material due to the inward migration and subsequent accretion of dust particles. This "radial drift" occurs because thermal pressure causes the gas to orbit at sub-Keplerian rates, creating a drag force on the particles that saps their orbital energy and sends them spiraling in to the central star (*Whipple*, 1972; *Weidenschilling*, 1977; *Brauer et al.*, 2008). Swept up in this particle flow, the reservoir of emitting grains in the inner disk can be sufficiently depleted to produce a telltale dip in the IR SED (e.g., *Birnstiel et al.*, 2012). A second process is related to the actual growth of dust grains. Instead of a decrease in the dust densities, the inner disk only appears to be cleared due to a decrease in the grain emissivities: Larger particles emit less efficiently (e.g., *D'Alessio et al.*, 2001; *Draine*, 2006). Because growth timescales are short in the inner disk, the IR emission that traces those regions could be suppressed enough to again produce a dip in the SED.

The initial models of grain growth predicted a substantial reduction of small grains in the inner disk on short timescales, and therefore a disk-clearing signature in the IR SED (*Dullemond and Dominik*, 2005; *Tanaka et al.*, 2005). More detailed models bear out those early predictions, even when processes that decrease the efficiency (e.g., fragmentation) are taken into account. *Birnstiel et al.* (2012) demonstrated that such models can account for the IR SED deficit of transitional disks by tuning the local conditions so that small (approximately micrometer-sized) particles in the inner disk grow efficiently to millimeter/centimeter sizes. However, those particles cannot grow much larger before their collisions become destructive: The resulting population of fragments would then produce sufficient emission to wash out the infrared SED dip. The fundamental problem is that those large particles emit efficiently at millimeter/centimeter wavelengths, so these models do not simultaneously account for the ring-like emission morphologies observed with interferometers (Fig. 4). *Birnstiel et al.* (2012) argued that the dilemma of this conflicting relationship between growth/fragmentation and the IR/millimeter emission diagnostics means that particle evolution *alone* is not the underlying cause of cavities in disks. A more in-depth discussion of these and other issues related to the observational signatures of grain growth and migration are addressed in the chapter by Testi et al. in this volume.

3.1.3. Photoevaporation. Another mechanism for sculpting (pre-)transitional disk structures relies on the complementary interactions of viscous evolution, dust migration, and disk dispersal via photoevaporative winds (*Hollenbach et al.*, 1994; *Clarke et al.*, 2001; *Alexander et al.*, 2006; *Alexander and Armitage*, 2007; *Gorti and Hollenbach*, 2009; *Gorti et al.*, 2009; *Owen et al.*, 2010, 2011; *Rosotti et al.*, 2013). Here, the basic idea is that the high-energy irradiation of the disk surface by the central star can drive mass loss in a wind that will eventually limit the resupply of inner disk material from accretion flows. Once that occurs, the inner disk can rapidly accrete onto the star, leaving behind a large (and potentially growing) hole at the disk center. The detailed physics of this process can be quite complicated, and depend intimately on how the disk is irradiated. The chapter by Alexander et al. in this volume provides a more nuanced perspective on this process, as well as on the key observational features that support its presumably important role in disk evolution in general, and the (pre-)transitional disk phenomenon in particular.

However, for the subsample of (pre-)transitional disks that have been studied in detail, the combination of large sizes and tenuous (but nonnegligible) contents of the inner disks (sections 2.1, 2.2, and 2.3), substantial accretion rates (section 4.2), and relatively low X-ray luminosities (section 4.3) indicate that photoevaporation does not seem to be a viable mechanism for the depletion of their inner disks (e.g., *Alexander and Armitage*, 2009; *Owen et al.*, 2011; *Bae et al.*, 2013a). In addition, there exist disks with very low accretion rates onto the star that do not show evidence of holes or gaps in the inner disk (*Ingleby et al.*, 2012).

3.2. Dynamical Clearing by Companions

Much of the theoretical work conducted to explain the clearings seen in disks has focused on dynamical interactions with companions. When a second gravitational source is present in the disk, it can open a gap (*Papaloizou et al.*, 2007; *Crida et al.*, 2006; *Kley and Nelson*, 2012) (see also the chapter by Baruteau et al. in this volume). The salient issue is whether this companion is a star or a planet. It has been shown both theoretically and observationally that a stellar-mass companion can open a gap in a disk. For example, CoKu Tau/4 is surrounded by a transitional disk (*D'Alessio et al.*, 2005) and it has a nearly equal-mass companion with a separation of 8 AU (section 2.3) (*Ireland and Kraus*, 2008). Such a binary system is expected to open a cavity at 16–24 AU (*Artymowicz and Lubow*, 1994), which is consistent with the observations (*D'Alessio et al.*, 2005; *Nagel et al.*, 2010). Here we focus our efforts on discuss-

ing dynamical clearing by planets. A confirmed detection of a planet in a disk around a young star does not yet exist. Therefore, it is less clear if dynamical clearing by planets is at work in some or most (pre-)transitional disks. Given the challenges in detecting young planets in disks, the best we can do at present is theoretically explore what observational signatures would be present if there were indeed planets in disks and to test if this is consistent with what has been observed to date, as we will do in the following subsections.

3.2.1. Maintaining gas accretion across holes and gaps. A serious challenge for almost all theoretical disk-clearing models posed to date is the fact that some (pre-)transitional disks exhibit large dust clearings while still maintaining significant gas accretion rates onto the star. Compared to other disk-clearing mechanisms (see section 3.1), gap opening by planets can more easily maintain gas accretion across the gap since the gravitational force of a planet can "pull" the gas from the outer disk into the inner disk.

The gap's depth and the gas accretion rate across the gap are closely related. The disk accretion rate at any radius R is defined as $\dot{M} = 2\pi R \langle \Sigma v_r \rangle$, where Σ and v_r are the gas surface density and radial velocity at R. If we further assume $\langle \Sigma v_r \rangle = \langle \Sigma \rangle \langle v_r \rangle$ and the accretion rate across the gap is a constant, the flow velocity is accelerated by about a factor of 100 in a gap that is a factor of 100 in depth.

In a slightly more realistic picture, $\langle \Sigma v_R \rangle = \langle \Sigma \rangle \langle v_R \rangle$ breaks down within the gap since the planet-disk interaction is highly asymmetric in the R-ϕ two-dimensional plane. Inside the gap, the flow only moves radially at the turnover of the horseshoe orbit, which is very close to the planet. This high-velocity flow can interact with the circumplanetary material, shock and accrete onto the circumplanetary disk, and eventually onto the protoplanet (*Lubow et al.*, 1999). Due to the great complexity of this process, the ratio between the accretion onto the planet and the disk's accretion rate across the gap is unclear. Thus, we will parameterize this accretion efficiency onto the planet as ξ. After passing the planet, the accretion rate onto the star is only $1-\xi$ of the accretion rate of the case where no planet is present. Note that this parameterization assumes that the planet mass is larger than the local disk mass so that the planet migration is slower than the typical type II rate. For a disk with $\alpha = 0.01$ and $\sim 10^{-8}$ $M_\odot$ yr^{-1}, the local disk mass is 1.5 M_{Jup} at 20 AU. *Lubow and D'Angelo* (2006) carried out three-dimensional viscous hydrodynamic simulations with a sink particle as the planet and found a ξ of 0.9. *Zhu et al.* (2011) carried out two-dimensional viscous hydrosimulations, but depleted the circumplanetary material at different timescales, and found that ξ can range between 0.1 and 0.9 depending on the circumplanetary disk-accretion timescale. The accretion efficiency onto the planet plays an essential role in the accretion rate onto the star.

3.2.2. Explaining infrared spectral energy distribution deficits. Regardless of the accretion efficiency, there is an intrinsic tension between significant gas accretion rates onto the star and the optically thin inner disk region in (pre-)transitional disks. This is because the planet's influence on the accretion flow of the disk is limited to the gap region (*Crida et al., 2006*), whose outer and inner edge hardly differ by a factor of more than 2. After passing through the gap, the inner disk surface density is again controlled only by the accretion rate and viscosity (e.g., MHD turbulence), similar to a full disk. Since a full disk's inner disk produces strong NIR emission, it follows that transitional disks should also produce strong NIR emission, but they do not.

With this simple picture it is very diffcult to explain transitional disks, which have strong NIR deficits, compared to pre-transitional disks. For example, the transitional disk GM Aur has very weak NIR emission. It has an optically thin inner disk at 10 μm with an optical depth of ~0.01 and an accretion rate of ~10^{-8} $M_\odot$ yr^{-1}. Using a viscous disk model with $\alpha = 0.01$, Σ_g is derived to be 10–100 g cm^{-2} at 0.1 AU. Considering that the nominal opacity of ISM dust at 10 μm is 10 cm^2 g^{-1}, the optical depth at 10 μm for the inner disk is 100–1000 (*Zhu et al., 2011*), which is 4–5 orders of magnitude larger than the optical depth (~0.01) derived from observations.

In order to resolve the conflict between maintaining gas accretion across large holes and gaps while explaining weak NIR emission, several approaches are possible. In the following sections, we will outline two of these, namely multiple giant planets and dust filtration along with their observational signatures.

3.2.3. A possible solution: Multiple giant planets. One possibility is that multiple planets are present in (pre-)transitional disks. If multiple planets from 0.1 AU to tens of astronomical units can open a mutual gap, the gas flow can be continuously accelerated and passes from one planet to another so that a low disk-surface density can sustain a substantial disk accretion rate onto the star (*Zhu et al., 2011; Dodson-Robinson and Salyk, 2011*). However, hydrodynamical simulations have shown that each planet pair in a multiple planet system will move into 2:1 mean-motion resonance (*Pierens and Nelson, 2008*). Even in a case with four giant planets, with the outermost one located at 20 AU, the mutual gap is from 2 to 20 AU (*Zhu et al., 2011*). Therefore, to affect the gas flow at 0.1 AU, we need to invoke an even higher number of giant planets.

If there are multiple planets present in (pre-)transitional disks, the planet accretion efficiency parameter (ξ) cannot be large in order to maintain a moderate accretion rate onto the star. With N planets in the disk, the accretion rate onto the star will be $(1-\xi)^N \dot{M}_{full}$. If $\xi = 0.9$ and the full disk has a nominal accretion rate $\dot{M}_{full} = 10^{-8}$ $M_\odot$ yr^{-1}, after passing two planets the accretion rate is 0.01×10^{-8} $M_\odot$ $yr^{-1} = 10^{-10}$ $M_\odot$ yr^{-1}, which is already below the observed accretion rates in (pre-)transitional disks (see section 4.2). On the other hand, if $\xi = 0.1$, even with four planets, the disk can still accrete at $(1-0.1)^4 \dot{M}_{full} = 0.66 \dot{M}_{full}$. Thus, ξ is one key parameter in the multi-planet scenario that demands further study.

3.2.4. Another possible solution: Dust filtration. Another possibility is that the small dust grains in the inner disk are highly depleted by physical removal or grain growth, to the point where the dust opacity in the NIR is far smaller than the interstellar medium (ISM) opacity. Generally, this dust opacity depletion factor is ~10^3–10^5 (*Zhu et al., 2011*). Dust filtration is a promising mechanism to deplete dust to such a degree.

Dust filtration relies on the fact that dust and gas in disks are not perfectly coupled. Due to gas drag, dust particles will drift toward a pressure maximum in a disk (see the chapter by Johansen et al. in this volume) (*Weidenschilling, 1977*). The drift speed depends on the particle size, and particles with the dimensionless stopping time $T_s = t_s\Omega \sim 1$ drift fastest in disks. If the particle is in the Epstein regime (when a particle is smaller than molecule mean-free-path) and has a density of 1 g cm^{-3}, T_s is related to the particle size, s, and the disk surface density, Σ_g, as $T_s = 1.55 \times 10^{-3}(s/1$ mm$)$ $(100$ g $cm^{-2}/\Sigma_g)$ (*Weidenschilling, 1977*). In the outer part of the disk (e.g., ~50 AU) where $\Sigma_g \sim 1$ or 10 g cm^{-2}, 1-cm particles have $T_s \sim 1$ or 0.1. Thus, submillimeter and centimeter observations are ideal to reveal the effects of particle drift in disks. At the outer edge of a planet-induced gap, where the pressure reaches a maximum, dust particles drift outward, possibly overcoming their coupling to the inward-accreting gas. Dust particles will then remain at the gap's outer edge while the gas flows through the gap. This process is called "dust filtration" (*Rice et al., 2006*), and it depletes the dust interior to the semimajor axis of the planet-induced gap, forming a dust-depleted inner disk.

Dust trapping at the gap edge was first simulated by *Paardekooper and Mellema* (2006) and *Fouchet et al.* (2007). However, without considering particle diffusion due to disk turbulence, millimeter- and centimeter-sized particles will drift from tens of astronomical units into the star within 200 orbits. *Zhu et al.* (2012) included particle diffusion due to disk turbulence in two-dimensional simulations evolving over viscous timescales, where a quasi-steady-state for both gas and dust has been achieved, and found that micrometer-sized particles are difficult to filter by a Jupiter-mass planet in a $\dot{M} = 10^{-8}$ $M_\odot$ yr^{-1} disk. *Pinilla et al.* (2012) has done one-dimensional calculations considering dust growth and dust fragmentation at the gap edge and suggested that micrometer-sized particles may also be filtered. Considering the flow pattern is highly asymmetric within the gap, two-dimensional simulations including both dust growth and dust fragmentation may be needed to better understand the dust-filtration process.

Dust filtration suggests that a deeper gap opened by a more massive planet can lead to the depletion of smaller dust particles. Thus transitional disks may have a higher-mass planet(s) than pre-transitional disks. Since a higher-mass planet exerts a stronger torque on the outer disk, it may slow down the accretion flow passing the planet and lead to a lower disk-accretion rate onto the star. This is consistent with observations that transitional disks have lower accretion rates than pre-transitional disks (*Espaillat et al., 2012; Kim et al., 2013*). Furthermore, dust filtration can explain the observed differences in the relative distributions of micrometer-sized dust grains and millimeter-sized dust grains. Disks where the small dust grains appear to closely trace the large dust distribution (i.e., category C disks from section 2.3) could be in an advanced stage of dust filtration, after the dust grains in the inner disk have grown to larger sizes and can no longer be detected in NIR scattered light. Disks with NIR scattered light imaging evidence for a significant amount of small, submicrometer-sized dust grains within the cavities in the large, millimeter-sized dust grains seen in the submillimeter (i.e., the category A and C disks from section 2.3) could be in an earlier stage of disk clearing via dust filtration.

Future observations will further our understanding of the many physical processes that can occur when a giant planet is in a disk, such as dust growth in the inner disk, dust growth at the gap edge (*Pinilla et al.,* 2012), tidal stripping of the planets (*Nayakshin,* 2013), and planets in the dead zone (*Morishima,* 2012; *Bae et al.,* 2013b). In addition, here we have explored how to explain large holes and gaps in disks, yet there may be smaller gaps present in disks (section 2.1) that may not be observable with current techniques. The gap-opening process is tied to the viscosity parameter assumed. If the disk is inviscid, a very-low-mass perturber can also open a small gap [e.g., 10 $M_\oplus$ (*Li et al.,* 2009; *Muto et al.,* 2010; *Duffell and MacFadyen,* 2012; *Zhu et al.,* 2013; *Fung et al.,* 2014)]. Disk ionization structure calculations have indeed suggested the existence of such low-turbulence regions in protoplanetary disks (i.e., the dead zone; see the chapter by Turner et al. in this volume), which may indicate that gaps in disks are common. Interestingly, in a low viscosity disk, vortices can also be produced at the gap edge (*Li et al.,* 2005), which can efficiently trap dust particles (Fig. 10) (*Lyra et al.,* 2009; *Ataiee et al.,* 2013; *Lyra and Lin,* 2013, *Zhu et al.,* 2014) and is observable with ALMA (*van der Marel et al.,* 2013). We note that other physical processes can also induce vortices (*Klahr and Henning,* 1997; *Wolf and Klahr,* 2002; see also the chapter by Turner et al. in this volume), thus the presence of vortices cannot be viewed as certain evidence for the presence of a planet within a disk gap. In the near future, new observations with ALMA will provide further constraints on the distributions of dust and gas in the disk, which will help illuminate our understanding of the above points.

4. DEMOGRAPHICS

Basic unresolved questions underlie and motivate the study of disk demographics around TTS. Do all disks go through a (pre-)transitional disk phase as stars evolve from disk-bearing stars to diskless stars? Assuming an affirmative answer, we can measure a *transition timescale* by multiplying the fraction of disk sources with (pre-)transitional disk SEDs with the TTS disk lifetime (*Skrutskie et al.,* 1990). Are (pre-)transitional disks created by a single process or multiple processes? Theory predicts that several mechanisms can generate (pre-)transitional disk SEDs, all of which are important for disk evolution. Table 2 summarizes the generic properties expected for (pre-)transitional disks produced by different mechanisms, as previously described in section 3 and the literature (e.g., *Najita et al.,* 2007a;

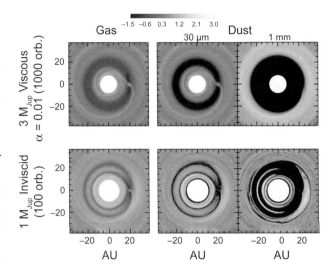

Fig. 10. The disk surface density for the gas (left) and dust (right) in viscous (α = 0.01, top) and inviscid (bottom) simulations. In the viscously accreting disk, the accretion flow carries small particles from the outer disk to the inner disk, while big particles are deposited at the gap edge. In the inviscid disk, particles are trapped in the vortex at the gap edge and the planet co-orbital region. Figure adapted from *Zhu et al.* (2012) and *Zhu et al.* (2014).

Alexander and Armitage, 2007). Which of these occur commonly (or efficiently) and on what timescale(s)? These questions can, in principle, be addressed demographically because different mechanisms are predicted to occur more readily in different kinds of systems (e.g., those of different disk masses and ages) and are expected to impact system parameters beyond the SED (e.g., the stellar accretion rate). We highlight developments along these lines below, taking into account points raised in sections 2 and 3.

4.1. Frequency and the Transition Timescale

Because different clearing mechanisms likely operate on different timescales, demographic studies aim to identify the processes that produce (pre-)transitional disks in populations of different ages. One area that has been investigated thoroughly is the overall frequency of transitional disks. Spitzer surveys have had tremendous success in cataloging young stellar object (YSO) populations, identifying dust emission from disks around stars within 500 pc across the full stellar mass range and into the brown dwarf regime at 3.6–8 μm, and for most of the stellar mass range at 24 μm. Based on the photometric SEDs, a number of studies have identified transitional disk candidates, allowing demographic comparisons with stellar and disk properties. We note that reported transitional disk frequencies to date should be taken as a lower limit on the frequency of holes and gaps in disks given that, as noted in section 2.1, it is harder to identify pre-transitional disks based on photometry alone as well as smaller gaps in disks with current data.

TABLE 2. Observable characteristics of proposed disk-clearing mechanisms.

Mechanism	Dust Distribution	Gas Distribution	Accretion Rate	Disk Mass	L_X
Viscous evolution	No hole/gap	No hole/gap	Low accretion	Low mass	No dependence
Grain growth	No hole/gap	No hole/gap	Unchanged	All masses	No dependence
Photoevaporation	R_h-radius hole	No/little gas within R_h	No/low accretion	Low mass	Correlated
~0.1-M_{Jup} planet	Gap	No hole/gap	Unchanged	All masses	No dependence
~1-M_{Jup} planet	Gap	Gap	~0.1–0.9 CTTS	Higher masses*	No dependence
Multiple giant planets	Gap/R_h-radius hole	No/little gas within R_h	No/low accretion	Higher masses*	No dependence

*Higher-mass disks may form planets easier according to core accretion theory (see the chapter by Helled et al. in this volume). Observations are needed to test this.

Note that here we refer to the dust and gas distribution of the inner disk. Relative terms are in comparison to the properties of otherwise comparable disks around CTTS (e.g., objects of similar age, mass).

There has been some controversy regarding the transition timescale, which as noted above can be estimated from the frequency of transitional disks relative to the typical disk lifetime. Early pre-Spitzer studies (e.g., *Skrutskie et al.,* 1990; *Wolk and Walter,* 1996) suggested that the transition timescale was short, on the order of 10% of the total disk lifetime, a few 10^5 yr. The values estimated from Spitzer surveys of individual clusters/star-forming regions have a wider range, from as small as these initial estimates (*Cieza et al.,* 2007; *Hernández et al.,* 2007b; *Luhman et al.,* 2010; *Koepferl et al.,* 2013) to as long as a time comparable to the disk lifetime itself (*Currie et al.,* 2009; *Sicilia-Aguilar et al.,* 2009; *Currie and Sicilia-Aguilar,* 2011).

The discrepancies among these studies are largely the result of differing definitions of what constitutes a disk in transition, as well as how to estimate the total disk lifetime. The shorter timescales are typically derived using more restrictive selection criteria. In particular, the evolved disks, in which the IR excess is small at all observed wavelengths, tend to be more common around older stars [≥3 million years (m.y.)] and low-mass stars and brown dwarfs; the inclusion of this SED type will lead to a larger transitional disk frequency and hence timescale for samples that are heavily weighted to these stellar types. Moreover, the status of the evolved disks as bona fide (pre-)transitional disks (i.e., a disk with inner clearing) is somewhat in dispute (see section 2.1). As pointed out by *Najita et al.* (2007a), the different SED types may indicate multiple evolutionary mechanisms for disks. More complete measurements of disk masses, accretion rates, and resolved observations of these objects are needed to better define their properties and determine their evolutionary status.

Combining data from multiple regions and restricting the selection to transitional disk SEDs showing evidence for inner holes, *Muzerolle et al.* (2010) found evidence for an increase in the transition frequency as a function of mean cluster age. However, there was considerable uncertainty because of the small number statistics involved, especially at older ages where the total disk frequency is typically 10% or less. Figure 11 shows the fraction of transitional disks relative to the total disk population in a given region

as a function of the mean age of each region. Here we supplement the statistics from the IRAC/MIPS flux-limited photometric surveys listed in *Muzerolle et al.* (2010) with new results from IRS studies of several star-forming regions (*McClure et al.,* 2010; *Oliveira et al.,* 2010; *Furlan et al.,* 2011; *Manoj et al.,* 2011; *Arnold et al.,* 2012a), and a photometric survey of the 11-m.y.-old Upper Scorpius association (*Luhman and Mamajek,* 2012). A weak age dependence remains, with frequencies of a few percent at t ≤ 2 m.y. and ~10% at older ages. Some of this correlation can be explained as a consequence of the cessation of star formation in most regions after ~3 m.y.; as noted by *Luhman et al.* (2010), the frequency of transitional disks relative to full disks in a region no longer producing new disks should increase as the number of full disks decreases with time. Note that the sample selection does not include pre-transitional disks, which cannot be reliably identified without MIR spectroscopy. Interestingly, for the regions with reasonably complete IRS observations (the four youngest regions in Fig. 11), the transition fraction would increase to ~10–15% if the pre-transitional disks were included. Note also that the IRS surveys represent the known membership of each region, and are reasonably complete down to spectral types of about M5. We refer the reader to the cited articles for a full assessment of sample completeness. All transitional disks in this combined sample are spectroscopically confirmed pre-main-sequence stars.

An important caveat to the above statistics is that SED morphology is an imperfect tracer of disk structure and is most sensitive to large structures that span many astronomical units (see section 2.1). Smaller gaps can be masked by emission from even tiny amounts of dust. As mentioned in section 2.2, resolved submillimeter imaging has found a greater frequency of disks with inner dust cavities among the 1–2-m.y.-old stars in Taurus and Ophiuchus [~30% (*Andrews et al.,* 2011b)] than has been estimated from NIR flux deficits alone. The statistics for older regions are incomplete, limited by weaker (sub)millimeter emission that may be a result of significant grain growth compared to younger disks (*Lee et al.,* 2011; *Mathews et al.,* 2012). Improved statistics await more sensitive observations such as with ALMA.

4.2. Disk Masses and Accretion Rates

Another powerful tool has been to compare stellar accretion rates and total disk mass. Looking at full disks around CTTS and (pre-)transitional disks in Taurus, *Najita et al.* (2007a) found that compared to single CTTS, (pre-)transitional disks have stellar accretion rates ~10× lower at the same disk mass and median disk masses ~4× larger. These properties are roughly consistent with the expectations for Jupiter-mass giant planet formation. Some of the low-accretion-rate, low-disk-mass sources could plausibly be produced by UV photoevaporation, although none of the (pre-)transitional disks (GM Aur, DM Tau, LkCa 15, UX Tau A) are in this group. Notably, none of these disks were found to overlap the region of the $\dot{M}_*$-M_d plane occupied by most disks around CTTS.

With the benefit of work in the literature, a preliminary $\dot{M}_*$-M_d plot can also be made for disks around CTTS and (pre-)transitional disks in Ophiuchus (Fig. 12). Most of the Oph sources are co-located with the Taurus CTTS, while several fall below this group. The latter includes DoAr 25 and the transitional disk SR 21. The other two sources have high extinction. Such sources ($A_V > 14$; boxes in Fig. 12) are found to have lower $\dot{M}_*$, on average, for their disk masses.

This is as might be expected, if scattered light leads to an underestimate of A_V and therefore the line luminosity from which $\dot{M}_*$ is derived. A source like DoAr 25, which has a very low accretion rate for its disk mass and also lacks an obvious hole based on the SED or submillimeter continuum imaging (*Andrews and Williams, 2008; Andrews et al., 2009, 2010*), is interesting. Does it have a smaller gap than can be measured with current techniques? Higher-angular-resolution observations of such sources could explore this possibility.

Like SR 21, both SR 24 S and DoAr 44 have also been identified as having cavities in their submillimeter continuum (*Andrews et al., 2011b*), although their accretion rates place them in the CTTS region of the plot. DoAr 44 has been previously identified as a pre-transitional disk based on its SED (*Espaillat et al., 2010*). Whether SR 24S has a strong MIR deficit in its SED is unknown, as its SED has not been studied in as much detail.

What accounts for the CTTS-like accretion rates of these systems? One possibility is that these systems may be undergoing low-mass planet formation. Several studies have described how low-mass planets can potentially clear gaps in the dust disk while having little impact on the gas, with the effect more pronounced for the larger grains probed in

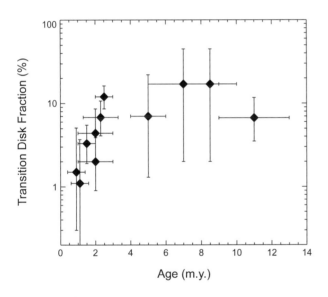

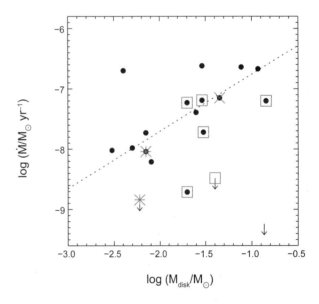

Fig. 11. The transitional disk fraction for 10 star-forming regions and young clusters as a function of mean stellar age. Fractions were estimated by taking the ratio of the number of objects that exhibit a transitional disk SED with the total number of disks in each region. Evolved disks and pre-transitional disks are not included here. The data are taken from (in order of mean age) *Arnold et al.* (2012a), NGC 1333; *McClure et al.* (2010), Ophiuchus; *Furlan et al.* (2011), Taurus; *Manoj et al.* (2011), Chamaeleon I; *Flaherty and Muzerolle* (2008), NGC 2068/2071; *Oliveira et al.* (2010), Serpens; *Muench et al.* (2007), IC 348; *Hernández et al.* (2007a), Orion OB1a/b; *Megeath et al.* (2005), η Chamaeleontis association; *Luhman and Mamajek* (2012), Upper Scorpius association.

Fig. 12. Preliminary $\dot{M}_*$-M_d for disks in Ophiuchus. The dotted line indicates the locus occupied by CTTS in Taurus (*Najita et al., 2007a*). Disks with submillimeter cavities and (pre-)transitional disk SEDs are overlaid with a star (SR 21, bottom left; DoAr 44, middle); the disk with a submillimeter cavity but no obvious SED MIR deficit (SR 24S) is overlaid with an "X." Sources with $A_V > 14$ (*McClure et al., 2010*) are indicated with a box. Objects where only upper limits were available for the accretion rates are indicated with an arrow (including SR 21, GSS 26, DoAr 25). Accretion rates are from *Natta et al.* (2006) if available, otherwise from *Eisner et al.* (2005) and *Valenti et al.* (1993) (scaled). Disk masses are from *Andrews and Williams* (2007) and *Andrews et al.* (2010).

the submillimeter (section 3.2) (see also *Paardekooper and Mellema*, 2006; *Fouchet et al.*, 2007).

In related work, recent studies find that (pre-)transitional disks have lower stellar-accretion rates on average than other disks in the same star-forming region and that the accretion rates of pre-transitional disks are closer to that of CTTS than the transitional disks (*Espaillat et al.*, 2012; *Kim et al.*, 2013). However, other studies find little difference between the accretion rates of (pre-)transitional and other disks (*Keane et al.*, 2014; *Fang et al.*, 2013). This could be due to different sample selection (i.e., colors vs. IRS spectra) and/or different methods for calculating accretion rates (i.e., H$_\alpha$ vs. NIR emission lines). More work has to be done to better understand disk accretion rates, ideally with a large IR and submillimeter selected sample and consistent accretion-rate measurement methods.

4.3. Stellar Host Properties

One would naively expect an age progression from full disks to (pre-)transitional disks to diskless stars in a given region if these classes represent an evolutionary sequence. Observational evidence for age differences, however, is decidedly mixed. From improved parallax measurements in Taurus, *Bertout et al.* (2007) found that the four stars surrounded by (pre-)transitional disks with age measurements (DM Tau, GM Aur, LkCa 15, UX Tau A) were intermediate between the mean CTTS and weak TTS (i.e., stars that are not accreting) ages. However, the quoted age uncertainties are roughly equal to the ages of individual objects, and the small sample is not statistically robust. Moreover, studies of other regions have found no statistical difference in ages between stars with and without disks [NGC 2264, Orion (*Dahm and Simon*, 2005; *Jeffries et al.*, 2011)]. In general, the uncertainties associated with determining stellar ages remain too large, and the samples of (pre-)transitional disks too small, to allow any definitive conclusion of systematic age differences for (pre-)transitional disks.

The frequency of transitional disks as a function of stellar mass also provides some clues to their origins. *Muzerolle et al.* (2010) showed that transitional disks appeared to be underrepresented among mid- to late-M-type stars compared with the typical initial mass function of young stellar clusters. Using the expanded sample adopted for Fig. 11, this initial finding appears to be robust (Fig. 13). The deficit remains even after adding the known pre-transitional disks. However, the evolved disks (as opposed to the *transitional* disks) do appear to be much more common around lower-mass stars (e.g., *Sicilia-Aguilar et al.*, 2008). As *Ercolano et al.* (2009) pointed out, there may be a strong bias here as optically thick inner disks around mid- to late-M-type stars produce less short-wavelength excess and can be confused with true inner disk clearing. Nevertheless, this discrepancy may provide another indication of different clearing mechanisms operating in different transitional disks. For example, the relative dearth of (pre-)transitional disks around lower-mass stars may reflect a decreased rate

of giant planet formation, as would be expected given their typically smaller initial disk masses.

Among other stellar properties, the X-ray luminosity can provide further useful constraints. L$_X$ has been invoked as a major contributor to theoretical photoevaporation rates (e.g., *Gorti and Hollenbach* 2009; *Owen et al.*, 2012), as well as accretion rates related to MRI at the inner disk edge (*Chiang and Murray-Clay*, 2007). Examining the (pre-)transitional disks in Taurus and Chamaleon I, *Kim et al.* (2009) found correlations between the size of the inner cleared regions and several stellar/disk properties including stellar mass, X-ray luminosity, and mass-accretion rate. These possibly point to the MRI or photoevaporation mechanisms. However, known independent correlations between stellar mass, L$_X$, and mass-accretion rate may also be responsible. In a follow-up study with a large sample of 62 (pre-)transitional disks in Orion, *Kim et al.* (2013) attempted to correct for the stellar-mass dependencies and found no residual correlation between accretion rate or L$_X$ and the size of the inner hole. The measured properties of most (pre-)transitional disks do not overlap with the ranges of parameter space predicted by these models. *Kim et al.* (2013) concluded that the demographics of their sample are most consistent with giant planet formation being the dominant process responsible for creating (pre-)transitional disk SEDs.

To better understand the connection between underlying physical mechanisms and observed disk demographics, more theoretical and observational work needs to be done. Improved theoretical estimates on stellar ages, accretion

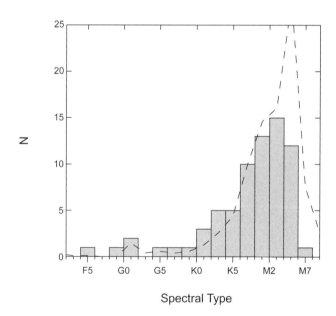

Fig. 13. The distribution of spectral types for the known transitional disks with spectral type F5 or later. The dashed line indicates the combined parent sample of all disks in the 1–3-m.y.-old regions, scaled down by a factor of 10, highlighting the apparent deficit of transitional disks around stars later than ~M4.

efficiencies, and disk masses for different clearing mechanisms will aid in interpretation of the observations. At the same time, larger groundbased surveys are essential to confidently constrain basic properties such as stellar accretion rates, ages, and disk masses. In the near future, the higher resolution of ALMA will reveal the extent of gaps in disks, leading to better statistics to measure the (pre-)transitional disk frequency and the transition timescale. The JWST will be key in addressing how the (pre-)transitional frequency depends on age. The best constraints on disk demographics clearly require a multi-wavelength approach.

5. CONCLUDING REMARKS AND FUTURE PROSPECTS

It is thought that practically all stars have planets (see the chapter by Fischer et al. in this volume). If we start with the reasonable assumption that these planets formed out of disks, then the question is not if there are planets in disks around young stars, but what are their observable signatures. To date we have detected large, optically thin holes and gaps in the dust distribution of disks, and theoretical planet-clearing mechanisms can account for some of their observed properties. These (pre-)transitional disks have captured the interest of many scientists since they may be a key piece of evidence in answering one of the fundamental questions in astronomy: How do planets form?

While other disk-clearing mechanisms (e.g., grain growth, photoevaporation; see section 3.1) should certainly be at play in most, if not all, disks [and may even work more effectively in concert with planets (*Rosotti et al.*, 2013)], here we speculate as to what kinds of planets could be clearing (pre-)transitional disks, beginning with massive giant planets. If a massive giant planet (or multiple planets) is forming in a disk, it will cause a large cavity in the inner disk since it will be very efficient at cutting off the inner disk from the outer disk. This will lead to significant depletion of small and large dust grains in the inner disk and lower accretion onto the central star (section 3.2). This is consistent with (pre-)transitional disks, which have submillimeter and NIR scattered light cavities and MIR deficits in the SED (e.g., LkCa 15, GM Aur, Sz 91, RX J1604-2130; sections 2.1–2.3). This is also consistent with the lower accretion rates measured for these objects (section 4.2). One caveat is that massive planets would be the easiest to detect, but there has not yet been a robust detection of a protoplanet in a (pre-)transitional disk.

A less-massive planet (or fewer planets) would still lead to substantial clearing of the inner disk, but would be less efficient at cutting off the inward flow of material from the outer disk (section 3.2). This is consistent with those (pre-) transitional disks that have a submillimeter cavity but no MIR deficit in the SED, such as WSB 60, indicating that small dust still makes it through into the inner disk (sections 2.1–2.3). Interestingly, WSB 60 is in Oph, and overall (pre-)transitional disks SEDs are rare in this region. Given that Oph is quite young (as it formed approximately 1 m.y.

ago), this may be indicative of the effects of dust evolution and/or planet growth. Disks in Oph could be in the initial stages of gap opening by planets, whereas in older systems, the dust grains in the inner disk have grown to larger sizes and/or the planets have grown large enough to more efficiently disrupt the flow of material into the inner disk. The above suggests that most disks with planets go through a pre-transitional disk phase and that there are many disks with small gaps that have escaped detection with currently available techniques. More theoretical work is needed to explore if all pre-transitional disks eventually enter a transitional disk phase as planets grow more massive and the inner disk is accreted onto the star. Note also that there are some disks that have submillimeter cavities and MIR SED deficits, but no evidence of clearing in NIR scattered light (e.g., SR 21, DoAr 44, RX J1615-3255). Apparently, there is some important disk physics regarding accretion efficiencies that we are missing. It is necessary to study disks in other star-forming regions with ALMA, both to probe a range of ages and to increase the statistical sample size, in order to address questions of this kind. Detections of protoplanets would be ideal to test the link between planet mass and disk structure.

While much progress has been made in understanding the nature of the disks described above, future work may focus more closely on disks with smaller optically thin gaps that could plausibly be created by low-mass planets. The presence of low-mass orbiting companions (~0.1 M_{Jup}) (*Paardekooper and Mellema*, 2006) is expected to alter the dust disk significantly with little impact on the gas disk, whereas high-mass companions (≥1 M_{Jup}) can create gaps or holes in the gas disk and possibly alter its dynamics. Kepler finds that super-Earth-mass objects are very common in mature planetary systems (see the chapter by Fischer et al. in this volume), so it may be that smaller-mass objects form fairly commonly in disks and open holes and gaps that are not cleared of dust. These systems are more difficult to identify with SED studies and current interferometers, so potentially there are many more disks with gaps than currently known. High-spatial-resolution images from ALMA will be able to detect these small gaps, in some cases even those with sizes down to about 3 AU (*Wolf and D'Angelo*, 2005; *Gonzalez et al.*, 2012).

There are many remaining avenues that researchers can take to fully characterize clearing in disks around TTS. Theoretically, we can simulate the influence of various-mass planets on different-sized dust particles, which can be compared with observations at various wavelengths to constrain a potential planet's mass. We can use ALMA, VLTI, and JWST to test the full extent of disk holes and gaps as well as their frequency with respect to age. We can also make substantial progress studying the gas distributions in disks in the near future with ALMA, particularly to determine if the structure inferred for the disk from studies of the dust is the same as that in the gas. Gas tracers may also reveal the presence of an orbiting planet via emission from a circumplanetary disk. Finally, and perhaps most importantly, we need robust detections of protoplanets in disks around young stars. These future advances will help us understand how gas/ice giant planets and terrestrial planets form out of disks and hopefully antiquate this review by the time of Protostars and Planets VII.

Acknowledgments. The authors thank the anonymous referee, C. Dullemond, L. Hartmann, and L. Ingleby for insightful comments that helped improve this review. Work by C.E. was performed in part under contract with Caltech funded by NASA through the Sagan Fellowship Program executed by the NASA Exoplanet Science Institute. Z.Z. acknowledges support by NASA through Hubble Fellowship grant HST-HF-51333.01-A awarded by the Space Telescope Science Institute, which is operated by the Association of Universities for Research in Astronomy, Inc., for NASA under contract NAS 5-26555.

Finally, we offer a special recognition of the contribution of Paola D'Alessio, who passed away in November 2013. She is greatly missed as a scientist, colleague, and friend. Among Paola's most important papers are those that addressed the issue of dust growth and mixing in T Tauri disks. In *D'Alessio et al.* (1999), she showed that disk models with well-mixed dust with properties like that of the diffuse interstellar medium produced disks that were too vertically thick and produced too much infrared excess, while lacking sufficient millimeter-wave fluxes. In the second paper of the series (*D'Alessio et al., 2001*), Paola and collaborators showed that models with power-law size distributions of dust with maximum sizes around 1 mm produced much better agreement with the millimeter and infrared emission of most T Tauri disks, but failed to exhibit the 10-μm silicate emission feature usually seen, leading to the conclusion that the large grains must have settled closer to the mid-plane while leaving a population of small dust suspended in the upper disk atmosphere. This paper provided the first clear empirical evidence for the expected evolution of dust as a step in growing larger bodies in protoplanetary disks. Another important result was the demonstration that once grain growth proceeds past sizes comparable to the wavelengths of observation, the spectral index of the disk emission is determined only by the size distribution of the dust, not its maximum. Along with quantitative calculations of dust properties, the results imply that the typical opacities used to estimate dust masses will generally lead to underestimates of the total solid mass present, a point that is frequently forgotten or ignored. In the third paper of this series (*D'Alessio et al., 2006*), Paola and her collaborators developed models that incorporated a thin central layer of large dust along with depleted upper disk layers containing small dust. The code that developed these models has been used in over 30 papers to compare with observations, especially those from the IRS (*Furlan et al., 2005; McClure et al., 2010*) as well as the IRAC camera (*Allen et al., 2004*) onboard the Spitzer Space Telescope, and more recently from the Physics and Astronomy Classification Scheme (PACS) onboard the Herschel Space Telescope (*McClure et al., 2012*), and from the SMA (*Qi et al., 2011*). Paola's models also played a crucial role in the recognition of transitional and pre-transitional T Tauri disks (*Calvet et al., 2002; D'Alessio et al., 2005; Espaillat et al., 2007, 2010*). In many cases, particularly those of the pre-transitional disks, finding evidence for an inner hole or gap from the spectral energy distribution depends upon careful and detailed modeling. The inference of gaps and holes from the SEDs is now being increasingly confirmed directly by millimeter and submillimeter imaging (*Hughes et al., 2009*),

showing reasonable agreement in most cases with the hole sizes predicted by the models. Combining imaging with SED modeling in the future will place additional constraints on the properties of dust in protoplanetary disks. Paola's influence in the community extended well beyond her direct contributions to the literature in over 100 refereed papers. She provided disk models for many other researchers as well as detailed dust opacities. The insight provided by her calculations informed many other investigations, as studies of X-ray heating of protoplanetary disk atmospheres (*Glassgold et al., 2004, 2007, 2009*), of the chemical structure of protoplanetary disks and propagation of high-energy radiation (*Fogel et al., 2011; Bethell and Bergin, 2011*), and on the photo-evaporation of protoplanetary disks (*Ercolano et al., 2008*). Paola's passing is a great loss to the star-formation community and we remain grateful for her contributions to our field.

REFERENCES

Acke B. and van den Ancker M. E. (2006) *Astron. Astrophys., 449,* 267.
Akeson R. L. et al. (2011) *Astrophys. J., 728,* 96.
Alcalá J. M. et al. (2014) *Astron. Astrophys., 561,* A2.
Alexander R. D. and Armitage P. J. (2007) *Mon. Not. R. Astron. Soc., 375,* 500.
Alexander R. D. and Armitage P. J. (2009) *Astrophys. J., 704,* 989.
Alexander R. D. et al. (2006) *Mon. Not. R. Astron. Soc., 369,* 229.
Allen L. E. et al. (2004) *Astrophys. J. Suppl., 154,* 363.
Andrews S. M. and Williams J. P. (2007) *Astrophys. J., 659,* 705.
Andrews S. M. and Williams J. P. (2008) *Astrophys. Space Sci., 313,* 119.
Andrews S. M. et al. (2009) *Astrophys. J., 700,* 1502.
Andrews S. M. et al. (2010) *Astrophys. J., 723,* 1241.
Andrews S. M. et al. (2011a) *Astrophys. J. Lett., 742,* L5.
Andrews S. M. et al. (2011b) *Astrophys. J., 732,* 42.
Arnold L. A. et al. (2012a) *Astrophys. J. Suppl., 201,* 12.
Arnold T. J. et al. (2012b) *Astrophys. J., 750,* 119.
Artymowicz P. and Lubow S. H. (1994) *Astrophys. J., 421,* 651.
Ataiee S. et al. (2013) *Astron. Astrophys., 553,* L3.
Avenhaus H. et al. (2014) *Astrophys. J., 781,* 87.
Bae J. et al. (2013a) *Astrophys. J., 774,* 57.
Bae J. et al. (2013b) *Astrophys. J., 764,* 141.
Beckwith S. V. W. et al. (1990) *Astron. J., 99,* 924.
Benisty M. et al. (2010) *Astron. Astrophys., 511,* A75.
Bertout C. et al. (2007) *Astron. Astrophys., 473,* L21.
Bethell T. J. and Bergin E. A. (2011) *Astrophys. J., 740,* 7.
Biller B. et al. (2012) *Astrophys. J. Lett., 753,* L38.
Birnstiel T. et al. (2012) *Astron. Astrophys., 544,* A79.
Birnstiel T. et al. (2013) *Astron. Astrophys., 550,* L8.
Bouwman J. et al. (2003) *Astron. Astrophys., 401,* 577.
Brauer F. et al. (2008) *Astron. Astrophys., 480,* 859.
Brittain S. D. et al. (2007) *Astrophys. J., 659,* 685.
Brittain S. D. et al. (2009) *Astrophys. J., 702,* 85.
Brittain S. D. et al. (2013) *Astrophys. J., 767,* 159.
Brittain S. D. et al. (2014) *Astrophys. J., 791,* 136.
Brown J. M. et al. (2007) *Astrophys. J. Lett., 664,* L107.
Brown J. M. et al. (2008) *Astrophys. J. Lett., 675,* L109.
Brown J. M. et al. (2009) *Astrophys. J., 704,* 496.
Brown J. M. et al. (2012) *Astrophys. J. Lett., 758,* L30.
Bruderer S. et al. (2014) *Astron. Astrophys., 562,* A26.
Calvet N. et al. (2000) In *Protostars and Planets IV* (V. Mannings et al., eds.), p. 377. Univ. of Arizona, Tucson.
Calvet N. et al. (2002) *Astrophys. J., 568,* 1008.
Calvet N. et al. (2004) *Astron. J., 128,* 1294.
Calvet N. et al. (2005) *Astrophys. J. Lett., 630,* L185.
Carpenter J. M. et al. (2001) *Astron. J., 121,* 3160.
Casassus S. et al. (2012) *Astrophys. J. Lett., 754,* L31.
Casassus S. et al. (2013) *Nature, 493,* 191.
Chiang E. and Murray-Clay R. (2007) *Nature Phys., 3,* 604.
Cieza L. et al. (2007) *Astrophys. J., 667,* 308.
Cieza L. A. et al. (2010) *Astrophys. J., 712,* 925.
Cieza L. A. et al. (2012) *Astrophys. J., 752,* 75.
Cieza L. A. et al. (2013) *Astrophys. J. Lett., 762,* L12.

Clarke C. J. et al. (2001) *Mon. Not. R. Astron. Soc., 328,* 485.
Close L. M. et al. (2014) *Astrophys. J. Lett., 781,* L30.
Cohen M. and Kuhi L. V. (1979) *Astrophys. J. Suppl., 41,* 743.
Crida A. et al. (2006) *Icarus, 181,* 587.
Currie T. and Sicilia-Aguilar A. (2011) *Astrophys. J., 732,* 24.
Currie T. et al. (2009) *Astrophys. J., 698,* 1.
Dahm S. E. and Simon T. (2005) *Astron. J., 129,* 829.
D'Alessio P. et al. (1999) *Astrophys. J., 527,* 893.
D'Alessio P. et al. (2001) *Astrophys. J., 553,* 321.
D'Alessio P. et al. (2005) *Astrophys. J., 621,* 461.
D'Alessio P. et al. (2006) *Astrophys. J., 638,* 314.
Debes J. H. et al. (2013) *Astrophys. J., 771,* 45.
de Juan Ovelar M. et al. (2013) *Astron. Astrophys., 560,* A111.
Dodson-Robinson S. E. and Salyk C. (2011) *Astrophys. J., 738,* 131.
Donati J.-F. et al. (2011) *Mon. Not. R. Astron. Soc., 417,* 1747.
Dong R. et al. (2012) *Astrophys. J., 760,* 111.
Draine B. T. (2006) *Astrophys. J., 636,* 1114.
Duffell P. C. and MacFadyen A. I. (2012) *Astrophys. J., 755,* 7.
Dullemond C. P. and Dominik C. (2005) *Astron. Astrophys., 434,* 971.
Durisen R. H. et al. (2007) In *Protostars and Planets V* (B. Reipurth et al., eds.), pp. 607–622. Univ. of Arizona, Tucson.
Dutrey A. et al. (2008) *Astron. Astrophys., 490,* L15.
Eisner J. A. et al. (2005) *Astrophys. J., 623,* 952.
Eisner J. A. et al. (2006) *Astrophys. J. Lett., 637,* L133.
Ercolano B. et al. (2008) *Astrophys. J., 688,* 398.
Ercolano B. et al. (2009) *Astrophys. J., 699,* 1639.
Espaillat C. et al. (2007) *Astrophys. J. Lett., 670,* L135.
Espaillat C. et al. (2008a) *Astrophys. J. Lett., 689,* L145.
Espaillat C. et al. (2008b) *Astrophys. J. Lett., 682,* L125.
Espaillat C. et al. (2010) *Astrophys. J., 717,* 441.
Espaillat C. et al. (2011) *Astrophys. J., 728,* 49.
Espaillat C. et al. (2012) *Astrophys. J., 747,* 103.
Evans T. M. et al. (2012) *Astrophys. J., 744,* 120.
Fang M. et al. (2013) *Astrophys. J. Suppl., 207,* 5.
Feigelson E. et al. (2007) In *Protostars and Planets V* (B. Reipurth et al., eds.), pp. 313–328. Univ. of Arizona, Tucson.
Flaherty K. M. and Muzerolle J. (2008) *Astron. J., 135,* 966.
Flaherty K. M. ct al. (2011) *Astrophys. J., 732,* 83.
Flaherty K. M. et al. (2012) *Astrophys. J., 748,* 71.
Fogel J. K. J. et al. (2011) *Astrophys. J., 726,* 29.
Follette K. B. et al. (2013) *Astrophys. J., 767,* 10.
Fouchet L. et al. (2007) *Astron. Astrophys., 474,* 1037.
France K. et al. (2012) *Astrophys. J., 756,* 171.
Fung J. et al. (2014) *Astrophys. J., 782,* 88.
Furlan E. et al. (2005) *Astrophys. J. Lett., 628,* L65.
Furlan E. et al. (2011) *Astrophys. J. Suppl., 195,* 3.
Geers V. C. et al. (2007) *Astron. Astrophys., 469,* L35.
Glassgold A. E. et al. (2004) *Astrophys. J., 615,* 972.
Glassgold A. E. et al. (2007) *Astrophys. J., 656,* 515.
Glassgold A. E. et al. (2009) *Astrophys. J., 701,* 142.
Gonzalez J.-F. et al. (2012) *Astron. Astrophys., 547,* A58.
Goodson A. P. and Winglee R. M. (1999) *Astrophys. J., 524,* 159.
Gorti U. and Hollenbach D. (2009) *Astrophys. J., 690,* 1539.
Gorti U. et al. (2009) *Astrophys. J., 705,* 1237.
Goto M. et al. (2006) *Astrophys. J., 652,* 758.
Grady C. A. et al. (2001) *Astron. J., 122,* 3396.
Grady C. A. et al. (2013) *Astrophys. J., 762,* 48.
Guilloteau S. and Dutrey A. (1994) *Astron. Astrophys., 291,* L23.
Guilloteau S. et al. (2011) *Astron. Astrophys., 529,* A105.
Habart E. et al. (2006) *Astron. Astrophys., 449,* 1067.
Hartmann L. et al. (1998) *Astrophys. J., 495,* 385.
Hartmann L. et al. (2006) *Astrophys. J., 648,* 484.
Hashimoto J. et al. (2011) *Astrophys. J. Lett., 729,* L17.
Hashimoto J. et al. (2012) *Astrophys. J. Lett., 758,* L19.
Hernández J. et al. (2007a) *Astrophys. J., 662,* 1067.
Hernández J. et al. (2007b) *Astrophys. J., 671,* 1784.
Hernández J. et al. (2008) *Astrophys. J., 686,* 1195.
Hernández J. et al. (2010) *Astrophys. J., 722,* 1226.
Hirose S. and Turner N. J. (2011) *Astrophys. J. Lett., 732,* L30.
Hollenbach D. et al. (1994) *Astrophys. J., 428,* 654.
Houck J. R. et al. (2004) *Astrophys. J. Suppl., 154,* 18.
Huélamo N. et al. (2011) *Astron. Astrophys., 528,* L7.
Hughes A. M. et al. (2007) *Astrophys. J., 664,* 536.
Hughes A. M. et al. (2009) *Astrophys. J., 698,* 131.
Ingleby L. et al. (2009) *Astrophys. J. Lett., 703,* L137.

Ingleby L. et al. (2011) *Astrophys. J., 743,* 105.
Ingleby L. et al. (2012) *Astrophys. J. Lett., 752,* L20.
Ingleby L. et al. (2013) *Astrophys. J., 767,* 112.
Ireland M. J. (2012) In *Adaptive Optics Systems III* (B. L. Ellerbroek et al., eds.), SPIE Conf. Ser. 8447, Bellingham, Washington.
Ireland M. J. and Kraus A. L. (2008) *Astrophys. J. Lett., 678,* L59.
Isella A. et al. (2009) *Astrophys. J., 701,* 260.
Isella A. et al. (2010a) *Astrophys. J., 714,* 1746.
Isella A. et al. (2010b) *Astrophys. J., 725,* 1735.
Isella A. et al. (2012) *Astrophys. J., 747,* 136.
Isella A. et al. (2013) *Astrophys. J., 775,* 30.
Jeffries R. D. et al. (2011) *Mon. Not. R. Astron. Soc., 418,* 1948.
Jensen E. L. N. and Mathieu R. D. (1997) *Astron. J., 114,* 301.
Joy A. H. (1945) *Astrophys. J., 102,* 168.
Ke T. T. et al. (2012) *Astrophys. J., 745,* 60.
Keane J. T. et al. (2014) *Astrophys. J., 787,* 153.
Kim K. H. et al. (2009) *Astrophys. J., 700,* 1017.
Kim K. H. et al. (2013) *Astrophys. J., 769,* 149.
Klahr H. H. and Henning T. (1997) *Icarus, 128,* 213.
Kley W. and Dirksen G. (2006) *Astron. Astrophys., 447,* 369.
Kley W. and Nelson R. P. (2012) *Annu. Rev. Astron. Astrophys., 50,* 211.
Koepferl C. M. et al. (2013) *Mon. Not. R. Astron. Soc., 428,* 3327.
Koerner D. W. and Sargent A. I. (1995) *Astron. J., 109,* 2138.
Kraus A. L. and Ireland M. J. (2012) *Astrophys. J., 745,* 5.
Kraus A. L. et al. (2008) *Astrophys. J., 679,* 762.
Kraus A. L. et al. (2011) *Astrophys. J., 731,* 8.
Kraus S. (2013) *Astrophys. J., 768,* 80.
Kraus S. et al. (2013) *Astrophys. J., 768,* 80.
Krautter J. et al. (1997) *Astron. Astrophys. Suppl., 123,* 329.
Lada C. J. et al. (2006) *Astron. J., 131,* 1574.
Lai D. and Zhang H. (2008) *Astrophys. J., 683,* 949.
Lee N. et al. (2011) *Astrophys. J., 736,* 135.
Li H. et al. (2005) *Astrophys. J., 624,* 1003.
Li H. et al. (2009) *Astrophys. J. Lett., 690,* L52.
Liskowsky J. P. et al. (2012) *Astrophys. J., 760,* 153.
Lubow S. H. and D'Angelo G. (2006) *Astrophys. J., 641,* 526.
Lubow S. H. et al. (1999) *Astrophys. J., 526,* 1001.
Luhman K. L. and Mamajek E. E. (2012) *Astrophys. J., 758,* 31.
Luhman K. L. et al. (2010) *Astrophys. J. Suppl., 186,* 111.
Lynden-Bell D. and Pringle J. E. (1974) *Mon. Not. R. Astron. Soc., 168,* 603.
Lyra W. and Lin M.-K. (2013) *Astrophys. J., 775,* 17.
Lyra W. et al. (2009) *Astron. Astrophys., 493,* 1125.
Maaskant K. M. et al. (2013) *Astron. Astrophys., 555,* A64.
Manoj P. et al. (2011) *Astrophys. J. Suppl., 193,* 11.
Marcy G. et al. (2005) *Progr. Theor. Phys. Suppl., 158,* 24.
Marsh K. A. and Mahoney M. J. (1992) *Astrophys. J. Lett., 395,* L115.
Marsh K. A. and Mahoney M. J. (1993) *Astrophys. J. Lett., 405,* L71.
Martin R. G. and Lubow S. H. (2011) *Mon. Not. R. Astron. Soc., 413,* 1447.
Mathews G. S. et al. (2012) *Astrophys. J., 753,* 59.
Mayama S. et al. (2012) *Astrophys. J. Lett., 760,* L26.
McClure M. K. et al. (2010) *Astrophys. J. Suppl., 188,* 75.
McClure M. K. et al. (2012) *Astrophys. J. Lett., 759,* L10.
Megeath S. T. et al. (2005) *Astrophys. J. Lett., 634,* L113.
Merín B. et al. (2010) *Astrophys. J., 718,* 1200.
Monnier J. D. and Millan-Gabet R. (2002) *Astrophys. J., 579,* 694.
Morales-Calderón M. et al. (2011) *Astrophys. J., 733,* 50.
Morishima R. (2012) *Mon. Not. R. Astron. Soc., 420,* 2851.
Muench A. A. et al. (2007) *Astron. J., 134,* 411.
Mulders G. D. et al. (2010) *Astron. Astrophys., 512,* A11.
Mulders G. D. et al. (2013) *ArXiv e-prints,* arXiv:1306.4264.
Muto T. et al. (2010) *Astrophys. J., 724,* 448.
Muzerolle J. et al. (2003) *Astrophys. J. Lett., 597,* L149.
Muzerolle J. et al. (2009) *Astrophys. J. Lett., 704,* L15.
Muzerolle J. et al. (2010) *Astrophys. J., 708,* 1107.
Nagel E. et al. (2010) *Astrophys. J., 708,* 38.
Najita J. R. et al. (2000) In *Protostars and Planets IV* (V. Mannings et al., eds.), p. 457. Univ. of Arizona, Tucson.
Najita J. R. et al. (2007a) *Mon. Not. R. Astron. Soc., 378,* 369.
Najita J. R. et al. (2007b) In *Protostars and Planets V* (B. Reipurth et al., eds.), pp. 507–522. Univ. of Arizona, Tucson.
Najita J. R. et al. (2008) *Astrophys. J., 687,* 1168.
Najita J. R. et al. (2009) *Astrophys. J., 697,* 957.
Najita J. R. et al. (2010) *Astrophys. J., 712,* 274.

Natta A. et al. (2006) *Astron. Astrophys., 452,* 245.
Nayakshin S. (2013) *Mon. Not. R. Astron. Soc., 431,* 1432.
Oliveira I. et al. (2010) *Astrophys. J., 714,* 778.
Olofsson J. et al. (2011) *Astron. Astrophys., 528,* L6.
Olofsson J. et al. (2013) *Astron. Astrophys., 552,* A4.
Owen J. E. and Clarke C. J. (2012) *Mon. Not. R. Astron. Soc., 426,* L96.
Owen J. E. et al. (2010) *Mon. Not. R. Astron. Soc., 401,* 1415.
Owen J. E. et al. (2011) *Mon. Not. R. Astron. Soc., 412,* 13.
Owen J. E. et al. (2012) *Mon. Not. R. Astron. Soc., 422,* 1880.
Paardekooper S.-J. and Mellema G. (2004) *Astron. Astrophys., 425,* L9.
Paardekooper S.-J. and Mellema G. (2006) *Astron. Astrophys., 453,* 1129.
Papaloizou J. C. B. et al. (2001) *Astron. Astrophys., 366,* 263.
Papaloizou J. C. B. et al. (2007) In *Protostars and Planets V* (B. Reipurth et. al., eds.), pp. 655–668. Univ. of Arizona, Tucson.
Pierens A. and Nelson R. P. (2008) *Astron. Astrophys., 482,* 333.
Piétu V. et al. (2005) *Astron. Astrophys., 443,* 945.
Piétu V. et al. (2006) *Astron. Astrophys., 460,* L43.
Piétu V. et al. (2007) *Astron. Astrophys., 467,* 163.
Pinilla P. et al. (2012) *Astron. Astrophys., 538,* A114.
Pogodin M. A. et al. (2012) *Astron. Nachr., 333,* 594.
Pontoppidan K. M. et al. (2008) *Astrophys. J., 684,* 1323.
Pontoppidan K. M. et al. (2010) *Astrophys. J., 720,* 887.
Qi C. et al. (2011) *Astrophys. J., 740,* 84.
Quanz S. P. et al. (2011) *Astrophys. J., 738,* 23.
Quanz S. P. et al. (2012) *Astron. Astrophys., 538,* A92.
Quanz S. P. et al. (2013a) *Astrophys. J. Lett., 766,* L1.
Quanz S. P. et al. (2013b) *Astrophys. J. Lett., 766,* L2.
Quillen A. C. and Trilling D. E. (1998) *Astrophys. J., 508,* 707.
Rameau J. et al. (2012) *Astron. Astrophys., 546,* A24.
Ratzka T. et al. (2005) *Astron. Astrophys., 437,* 611.
Ratzka T. et al. (2007) *Astron. Astrophys., 471,* 173.
Regály Z. et al. (2010) *Astron. Astrophys., 523,* A69.
Regály Z. et al. (2012) *Mon. Not. R. Astron. Soc., 419,* 1701.
Rice W. K. M. et al. (2003) *Mon. Not. R. Astron. Soc., 342,* 79.
Rice W. K. M. et al. (2006) *Mon. Not. R. Astron. Soc., 373,* 1619.
Romero G. A. et al. (2012) *Astrophys. J., 749,* 79.
Rosenfeld K. A. et al. (2012) *Astrophys. J., 757,* 129.
Rosenfeld K. A. et al. (2013) *Astrophys. J., 775,* 136.
Rosotti G. P. et al. (2013) *Mon. Not. R. Astron. Soc., 430,* 1392.
Rydgren A. E. et al. (1976) *Astrophys. J. Suppl., 30,* 307.
Salyk C. et al. (2007) *Astrophys. J. Lett., 655,* L105.
Salyk C. et al. (2009) *Astrophys. J., 699,* 330.

Salyk C. et al. (2011) *Astrophys. J., 743,* 112.
Salyk C. et al. (2013) *Astrophys. J., 769,* 21.
Sicilia-Aguilar A. et al. (2008) *Astrophys. J., 687,* 1145.
Sicilia-Aguilar A. et al. (2009) *Astrophys. J., 701,* 1188.
Sicilia-Aguilar A. et al. (2011) *Astrophys. J., 742,* 39.
Skrutskie M. F. et al. (1990) *Astron. J., 99,* 1187.
Stempels H. C. and Gahm G. F. (2004) *Astron. Astrophys., 421,* 1159.
Strom K. M. et al. (1989) *Astron. J., 97,* 1451.
Takami M. et al. (2013) *Astrophys. J., 772,* 145.
Tamura M. (2009) In *Exoplanets and Disks: Their Formation and Diversity* (T. Usuda et al., eds.), pp. 11–16. AIP Conf. Ser. 1158, American Institute of Physics, Melville, New York.
Tanaka H. et al. (2005) *Astrophys. J., 625,* 414.
Tang Y.-W. et al. (2012) *Astron. Astrophys., 547,* A84.
Tanii R. et al. (2012) *Publ. Astron. Soc. Japan, 64,* 124.
Tatulli E. et al. (2011) *Astron. Astrophys., 531,* A1.
Thalmann C. et al. (2010) *Astrophys. J. Lett., 718,* L87.
Tsukagoshi T. et al. (2013) In *New Trends in Radio Astronomy in the Alma Era* (R. Kawabe et al., eds.), p. 391. ASP Conf. Ser. 476, Astronomical Society of the Pacific, San Francisco.
Tsukagoshi T. et al. (2014) *Astrophys. J., 783,* 90.
Turner N. J. et al. (2010) *Astrophys. J., 708,* 188.
Tuthill P. G. et al. (2000) *Publ. Astron. Soc. Pac., 112,* 555.
Valenti J. A. et al. (1993) *Astron. J., 106,* 2024.
van der Marel N. et al. (2013) *Science, 340,* 1199.
van der Plas G. et al. (2009) *Astron. Astrophys., 500,* 1137.
Vicente S. et al. (2011) *Astron. Astrophys., 533,* A135.
Weidenschilling S. J. (1977) *Mon. Not. R. Astron. Soc., 180,* 57.
Weinberger A. J. et al. (1999) *Astrophys. J. Lett., 525,* L53.
Werner M. W. et al. (2004) *Astrophys. J. Suppl., 154,* 1.
Whipple F. L. (1972) In *From Plasma to Planet* (A. Elvius, ed.), p. 211. Wiley, New York.
White R. J. and Ghez A. M. (2001) *Astrophys. J., 556,* 265.
Williams J. P. and Cieza L. A. (2011) *Annu. Rev. Astron. Astrophys., 49,* 67.
Wolf S. and D'Angelo G. (2005) *Astrophys. J., 619,* 1114.
Wolf S. and Klahr H. (2002) *Astrophys. J. Lett., 578,* L79.
Wolk S. J. and Walter F. M. (1996) *Astron. J., 111,* 2066.
Zhu Z. et al. (2011) *Astrophys. J., 729,* 47.
Zhu Z. et al. (2012) *Astrophys. J., 755,* 6.
Zhu Z. et al. (2013) *Astrophys. J., 768,* 143.
Zhu Z. et al. (2014) *Astrophys. J., 785,* 122.

Matthews B. C., Krivov A. V., Wyatt M. C., Bryden G., and Eiroa C. (2014) Observations, modeling, and theory of debris disks. In *Protostars and Planets VI* (H. Beuther et al., eds.), pp. 521–544. Univ. of Arizona, Tucson, DOI: 10.2458/azu_uapress_9780816531240-ch023.

Observations, Modeling, and Theory of Debris Disks

Brenda C. Matthews
National Research Council of Canada — Herzberg Astronomy and Astrophysics

Alexander V. Krivov
Friedrich-Schiller-Universität Jena

Mark C. Wyatt
University of Cambridge

Geoffrey Bryden
Jet Propulsion Laboratory

Carlos Eiroa
Universidad Autónoma de Madrid

Main-sequence stars, like the Sun, are often found to be orbited by circumstellar material that can be categorized into two groups: planets and debris. The latter is made up of asteroids and comets, as well as the dust and gas derived from them, which makes debris disks observable in thermal emission or scattered light. These disks may persist over billions of years through steady-state evolution and/or may also experience sporadic stirring and major collisional breakups, rendering them atypically bright for brief periods of time. Most interestingly, they provide direct evidence that the physical processes (whatever they may be) that act to build large oligarchs from micrometer-sized dust grains in protoplanetary disks have been successful in a given system, at least to the extent of building up a significant planetesimal population comparable to that seen in the solar system's asteroid and Kuiper belts. Such systems are prime candidates to host even larger planetary bodies as well. The recent growth in interest in debris disks has been driven by observational work that has provided statistics, resolved images, detection of gas in debris disks, and discoveries of new classes of objects. The interpretation of this vast and expanding dataset has necessitated significant advances in debris disk theory, notably in the physics of dust produced in collisional cascades and in the interaction of debris with planets. Application of this theory has led to the realization that such observations provide a powerful diagnostic that can be used not only to refine our understanding of debris disk physics, but also to challenge our understanding of how planetary systems form and evolve.

1. INTRODUCTION

Our evolving understanding of debris disks through the Protostars and Planets (PP) series was succintly summarized by *Meyer et al.* (2007), emphasizing the important role they play in studies of planetary systems and stressing the need to resolve disks to break the degeneracies inherent in spectral energy distribution (SED) modeling, setting well the stage for the near-decade of debris disk science that has come and gone since. The Spitzer Space Telescope's debris disk surveys are complete and now in the literature, and these are augmented significantly by those of the Herschel Space Observatory, which will have a lasting legacy owing to its resolving power and wavelength coverage. In parallel, there have been similar or even superior resolving power breakthroughs at wavelengths from the near-infrared

(IR) [e.g., Center for High Angular Resolution Astronomy (CHARA), Very Large Telescope Interferometer (VLTI), Large Binocular Telescope Interferometer (LBTI)] to (sub) millimeter [i.e., Atacama Large Millimeter/submillimeter Array (ALMA), Submillimeter Array (SMA), and Combined Array for Research in Millimeter-wave Astronomy (CARMA)] on the ground.

In this chapter we focus exclusively on debris disks, circumstellar dust (and potentially gas) disks created through destructive processes acting on larger, unseen planetesimals within the disk systems. A key diagnostic of a debris disk is the low fractional luminosity of the dust; even the brightest debris disks are ~100× fainter than protoplanetary disks, and most are ~100× fainter still. Our goal for this chapter remains the same as previous PP proceedings authors: To understand the evolution of planetary systems through observations of the circumstellar dust and gas that surrounds many of these systems throughout their lifetime using physical models.

Our chapter is divided as follows. In section 2, we discuss how disks are detected and characterized, highlighting results from recent surveys followed by section 3 on the origin of dust in collisional cascades. We then discuss the birth and evolution of debris disks in section 4. In section 5 we present the significant advances in debris disk imaging, followed by section 6 on the origin of structure through perturbations. Finally, we highlight observations and origin scenarios of hot dust (section 7) and gas (section 8) in debris disks. We summarize our chapter and provide an outlook to the future in section 9.

2. DETECTION AND DISTRIBUTION OF DUST

Observations of debris disks help not only to study the properties of individual disks, but also to ascertain their incidence and their correlation with stellar properties — e.g., age, spectral type, metallicity, stellar and/or planetary companions. Therefore, their study is imperative to understand the diversity of planetary system architectures and to thereby place the solar system's debris disk, primarily composed of warm dust in the terrestrial planet zone, the asteroid belt (0.5–3 AU) and cold dust and planetesimals in the Edgeworth-Kuiper belt (EKB), in context.

While at optical and near-IR wavelengths, scattered light from dust grains highlights regions of disks where small grains dominate; dust is most effectively traced by its thermal emission. Typically, evidence of a disk comes from "excess" IR emission detected above the level of the stellar photosphere. Figure 1 highlights the dust temperatures that are probed by different observing wavelength regimes. If a disk is composed of multiple components at a range of distances from the star, then observations at different wavelengths can probe the different components, with shorter wavelengths probing closer in material, as illustrated for the Infrared Astronomical Satellite (IRAS), the Multi-Band Imaging Photometer for Spitzer (MIPS), and the Herschel Photodetector Array Camera and Spectrometer (PACS) observing wavelengths by Fig. 2.

The emerging picture is that disks commonly are composed of multiple components as Fig. 1 illustrates, but that does not mean that all components are necessarily present or detectable around all stars. For many disks, just one component, typically (but not always) that in the Kuiper belt zone, dominates the detected emission, and in such cases, observations over a range of wavelengths provide a probe of different dust grain sizes, with longer wavelengths probing larger grains. In addition, the comparison and complementarity of the different statistical studies are modulated by the different observational strategies and target properties, e.g., spectral type (see Fig. 2), and on distance according to the nature of the survey (sensitivity vs. calibration-limited strategies) (*Wyatt*, 2008). Therefore, the flux contrast between the stellar photosphere and a potentially existing debris disk is mainly determined by the stellar temperature, modulated by the distance to the star (e.g., *Eiroa et al.,* 2013). Both Spitzer and Herschel are well suited to carry out detailed statistical studies relating debris disk properties with stellar ones, i.e., to determine the frequency and characterize the nature of disks. Good summaries of the Spitzer results and statistics are given by *Wyatt* (2008) and *Krivov* (2010).

As sensitivity toward faint debris disks improves, so too does sensitivity to extragalactic contamination. This is important because it impacts disk incidences, interpretation of structures, and the identification of new cold disk populations. The impact of background galaxy contamination has typically been a problem for surveys in the far-IR and submillimeter, but it has recently been identified as a major problem for the Wide-field Infrared Survey Explorer (WISE) as well (*Kennedy and Wyatt,* 2012).

In the following sections, we consider in turn our understanding of the different disk components as traced predominantly by observations in different wavelength regimes, but we stress that this does not represent a one-to-one correspondence with disk temperatures. For example, Fig. 2 shows that MIPS/24 and PACS/100 are equally sensitive to dust at ~3 AU around G-type stars (akin to our asteroid belt's warm dust). As we discuss below, ALMA at (sub) millimeter wavelengths now has the potential to detect even the warm dust components of disks, so it is sensitivity as much as wavelength that is key to detecting dust over a range of temperature (or emitting radii).

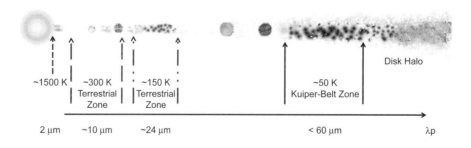

Fig. 1. Locations of dust emission observed in debris disks and the typical temperatures observed at those locations, and the primary observing wavelengths at which the dust is detected. Figure courtesy of K. Su.

2.1. Far-Infrared to Millimeter Observations

Many debris disks are detected only at long wavelengths (≥60 μm), corresponding to cold (≤100 K) dust orbiting at tens to hundreds of astronomical units. The debris disks detected with IRAS were Kuiper belt analogs seen around ~15% of main-sequence stars (*Aumann, 1985*). *Rhee et al.* (2007) have reanalyzed IRAS data and found the detection rate around nearby A stars to be 20%. In the past decade, several surveys by Spitzer, i.e., A stars (*Rieke et al., 2005*) and the Formation and Evolution of Planetary Systems (FEPS) (*Meyer et al., 2006*) and by Herschel, i.e., Disk Emission via a Bias-free Reconnaissance in the Infrared/Submillimetre (DEBRIS) (Matthews et al., in preparation) and DUst around NEarby Stars (DUNES) (*Eiroa et al., 2013*) have measured the incidence of debris disks for spectral types A through M at 70–160 μm.

Keeping in mind the relative sensitivities of the various surveys (Fig. 2), the highest detection rates are measured for A stars: 33% and 25% at 70 μm (*Su et al., 2006*) and 100 μm (Thureau et al., in preparation), respectively. For solar-type stars (FGK), the FEPS survey detected disks around 10% of stars (*Hillenbrand et al., 2008*), while *Trilling et al.* (2008) measured a rate of 16% in the field, comparable to the rate detected by the DEBRIS survey [17% (Sibthorpe et al., in preparation)]. The DUNES data yield an increase in the debris disk detection rate up to ~20.2% [in contrast to the Spitzer detection rate of ~12.1% for the same sample of FGK stars (*Eiroa et al., 2013*)]. Although there is an apparent decrease of excess rates from A to K spectral types, this appears to be largely an age effect (*Su*

et al., 2006; Trilling et al., 2008). In this respect, *Trilling et al.* (2008) note that there is no trend among FGK stars of similar age in the FEPS survey and that the rates for AFGK stars are statistically indistinguishable, a result supported by DUNES data (*Eiroa et al., 2013*).

The amount of dust in debris disks is frequently quantified in terms of the fractional luminosity of the dust, $f_d \equiv L_d/L_\star$, which is usually estimated assuming that the dust behaves as a pure or modified blackbody emitter (see section 3.5). f_d values are found in the range of ~10^{-3}–10^{-6} with a clear decrease toward older systems, although with a large dispersion at any given age (e.g., *Siegler et al., 2007; Hillenbrand et al., 2008; Moór et al., 2011b*). (The evolution of debris disks is discussed in more detail in section 4.4 below.) There is, however, no clear dependence of fractional luminosity on spectral type from late B to M, as shown in Fig. 3, although the scatter in f_d is high for all spectral types.

In spite of the remarkable contribution of Spitzer, its moderate sensitivity in terms of the dust fractional luminosity [$f_d \sim 10^{-5}$ (*Trilling et al., 2008*)] makes elusive the detection of debris disks with levels similar to the EKB [~10^{-7} (*Vitense et al., 2012*)]. This constraint is somewhat mitigated by the Herschel surveys, which observed closer to the peak of the SEDs and therefore could detect fainter disks, e.g., DUNES observations of 133 FGK stars located at distances less than 20 pc achieves an average sensitivity of fd ~ 10^{-6}, although with a dependence on stellar flux (*Eiroa et al., 2013*). The DEBRIS survey, with flux-limited observations of 446 A through M nearby stars (Matthews et al., in preparation) has a median sensitivity of f_d of just 2 × 10^{-5} but a faintest detected disk of 7 × 10^{-7}, approaching the level of the EKB.

Based on Figs. 2 and 3, increased sensitivity to fainter luminosity disks around A stars is expected in Herschel observations compared to Spitzer. Interestingly, DEBRIS

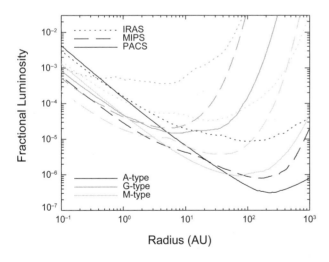

Fig. 2. This plot shows, based on the DEBRIS survey data (Matthews et al., in preparation), the lines of fractional luminosity vs. radius above which 25% of that sample could have been detected by IRAS, MIPS (24 and 70 μm), and PACS (100 and 160 μm), assuming each star to have a single temperature belt dominating its emission spectrum. The 25% level is comparable to the detection levels for many surveys given that DEBRIS is a survey of very nearby stars. Figure courtesy of G. Kennedy.

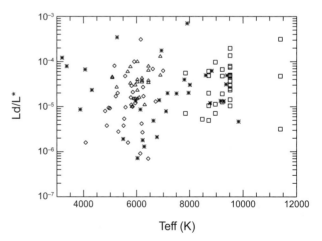

Fig. 3. Fractional luminosity of dust vs. effective stellar temperatures for main-sequence stars. Squares: A stars (*Su et al., 2006*); triangles: FGK stars (*Trilling et al., 2008*); diamonds: FGK stars (*Eiroa et al., 2013*); asterisks: AFGKM stars (Matthews et al., in preparation).

data indicate there is not a significant number of detections of A-star disks with fractional luminosities below the levels detected by Spitzer (Thureau et al., in preparation), implying that there is not a significant fraction of undetected, colder (i.e., larger) disks around A stars awaiting discovery. In comparison, Fig. 3 shows that, for the FGK stars, many disks of lower fractional luminosities have been detected by Herschel relative to Spitzer.

Detection of M-star-hosted debris disks at rates comparable to those around more massive stars has proven difficult. There are several factors at play around M stars, e.g., wind-related grain-removal processes that are more efficient than around more massive stars (e.g., *Plavchan et al., 2005*), which effectively removes small grains from the system, necessitating surveys at longer wavelengths to find disks comparable to those seen around earlier type stars (*Matthews et al., 2007*). That said, given their prevalence in the galaxy, the detection of protoplanetary disks at earlier phases of evolution (e.g., *Andrews and Williams, 2005*), and the detection of extrasolar planets around a number of M dwarfs (e.g., *Endl et al., 2008*, and references therein), including multiple systems around GJ 876 (*Rivera et al., 2005*), GJ 581(*Udry et al., 2007*), and GJ 676A (*Anglada-Escudé and Tuomi, 2012*), we expect M stars to host planetesimal populations.

Figure 2 shows that existing observations of M stars are not sensitive to the same levels of fractional luminosity detectable toward earlier type stars. As might be expected, low detection rates around M stars are reported by many Spitzer studies, generally excepting only the very young systems (e.g., *Low et al., 2005; Gautier et al., 2007; Plavchan et al., 2009; Koerner et al., 2010*). One particular exception is the study of *Forbrich et al.* (2008), which detected excesses around 11 M stars in the 30–40-Ma (i.e., formed 30–40 million years ago) NGC 2547 cluster at 24 μm, tracing warm material around 1 AU from the parent stars (close to the snow line). The M star disk incidence rate appears, in that cluster, to exceed that of G and K stars of the same age. The highest detection rate is quoted by *Lestrade et al.* (2006) using combined submillimeter studies to calculate an excess fraction of 13^{+6}_{-8}% from 20 to 200-Ma targets, which is not markedly different from the field FGK incidence rates discussed above. *Avenhaus et al.* (2012) report no detections toward M stars in the WISE sample, which *Heng and Malik* (2013) attribute to the sensitivity of the WISE data to the ~1-AU region of the disk, where the planetesimals cannot persist for periods beyond 300 million years (m.y.); however, this prediction is at odds with the WISE disks reported around older, solar-type stars. Some disks are seen around older M star systems, however, such as the GJ 842.2 disk (*Lestrade et al., 2006*) and the resolved disk around the old [2–8 Ga (i.e., formed 2–8 gigayears ago)] multiplanet host GJ 581 (*Lestrade et al., 2012*).

(Sub)millimeter observations of debris disks typically trace the Rayleigh-Jeans tail of the outer cold dust [e.g., TWA 7 (*Matthews et al., 2007*)]. They provide relevant constraints on the debris dust, e.g., measurement of the power-law exponent β of the opacity law and determination of the dust mass for sizes up to ~1 mm since emission at these wavelengths is optically thin. While most observations have concentrated on disks previously identified at other wavelengths, a few statistical surveys have been carried out, e.g., the JCMT survey by *Najita and Williams* (2005) and the CSO/IRAM survey of *Roccatagliata et al.* (2009). *Najita and Williams* (2005) detected 3 sources out of 13 nearby solar-mass stars, while 5 out of 27 FEPS sources older than 10 Ma were detected in the statistical analysis at 350 μm and/or 1.2 mm by *Roccatagliata et al.* (2009). An incidence rate of ~60% among a sample of 22 sources with previously known far-IR excesses was found in the APEX survey of *Nilsson et al.* (2010). We note, however, that this sample includes very young objects whose disks are still in a protostellar or transitional phase (for more information on these disk phases, see the chapters by Z. Li et al., Dunham et al., and Espaillat et al. in this volume). The SCUBA-2 Observations of Nearby Stars (SONS) Survey, currently underway, has achieved a detection rate of 40% for 850-μm detections of cold dust toward known disk hosts, including many not previously detected at submillimeter wavelengths (*Panić et al., 2013*).

2.2. Mid-Infrared Observations

Warm dust in the habitable zones of solar-type stars is of particular interest, both as a potential indicator of ongoing terrestrial planet formation and as a source of obscuration for future imaging of such planets. Mid-IR excess is difficult to detect, since the dust emission tends to be overwhelmed by the star. The zodiacal light in the solar system, for example, would be less than 10^{-4} as bright as the Sun at 10 μm, if viewed from afar (*Kelsall et al., 1998*). Due to this limitation, photometric surveys for warm dust in the mid-IR are generally limited by their systematic floor, e.g., uncertainty in the instrument calibration and/or in the estimate for the stellar photosphere. Nevertheless, considerable progress has been made in detecting potential analogs to the solar system's warm dust around other stars, so-called exozodiacal dust.

Spitzer photometry and spectroscopy in the mid-IR yields detection rates that are a strong function of wavelength, spectral type, and age. Only a few percent of mature solar-type stars exhibit mid-IR excess. For example, *Hillenbrand et al.* (2008) and *Dodson-Robinson et al.* (2011) find rates of ~4% at 24 μm in contrast to ~16% at 70 μm (*Trilling et al., 2008*), while *Lawler et al.* (2009) find a rate of ~1% in the range 8.5–12 μm vs. ~11.8% in the 30–34-μm wavelength range.

Detection rates are higher for younger stars [e.g., in the Pleiades (*Gorlova et al., 2006*)]. In the 10–20-Ma Sco-Cen association, the rates at 24 μm with Spitzer are comparable to those seen in the far-IR, i.e., 25% and 33% for B + A stars (*Chen et al., 2012*) and F + G stars (*Chen et al., 2011*), respectively. The rate for A stars of all ages is measured to be 32% at 24 μm (*Su et al., 2006*), comparable to the rate measured at 70 μm for the same stars. *Morales et al.* (2009, 2011) compiled a sample of 69 young (<1 Ga) nearby stars with 24 μm excess, most of which (50) occur around A-type stars. Within the FEPS sample, the rate of 24 μm excess is markedly higher (15%) for stars with ages <300 Ma, compared to 2.7% at >200 Ma (*Carpenter et al., 2009*).

Spectra or longer wavelength measurements not only help in constraining dust composition (section 3.5), they also help in measuring the dust temperature, which can otherwise be ambiguous from just one or two photometric measurements (see Fig. 2). So while a Spitzer/IRS spectral survey of nearby solar-type stars found a relatively high rate of excess at 32 μm (*Lawler et al., 2009*), most of these detections were identified as the short-wavelength side of previously detected cold dust, not new detections of warm emission. Only two systems (1% of those observed) were detected at 10 μm — HD 69830 and η Crv — both identified as having dust at ~1 AU (*Beichman et al., 2005a; Wyatt et al., 2005*).

Additional systems with warm excess have been detected within the IRAS, AKARI, and WISE datasets, which, as all-sky surveys, are well suited for identifying bright, rare objects. Examples include BD+20 307 (*Stencel and Backman, 1991; Song et al., 2005*), HD 106797 (*Fujiwara et al., 2009*), and HD 15407 (*Melis et al., 2010*). The spectra of these systems generally have strong silicate features at 10 μm, which provide useful constraints on the dust size and composition (see section 3.5). These dust constraints, combined with the large amount of dust present and its short collisional lifetime, suggest that its origin lies in unusual collisional events (see section 7). Most recently, *Kennedy and Wyatt* (2013) created a uniform catalog of bright warm-dust systems based on WISE 12-μm excess. They find that such systems occur around ~1% of young stars (<120 m.y. old), but are relatively rare for ~1-G.y.-old stars like BD+20 307 (10^{-4} frequency). While their survey only probes the bright end of the luminosity function, they extend their results down to fainter disks assuming a gradual collisional evolution, predicting that at least 10% of nearby stars will have sufficient dust in their habitable zones to pose problems for future efforts at direct imaging of exo-Earths. On the other hand, *Melis et al.* (2012) report a rapid loss of IR excess for the young solar-type star TYC 8241 2652 1 — a factor of 30 reduction in less than two years. The dramatic change of this system remains unexplained.

Measurement of the typical amount of exozodiacal dust around nearby stars is an important step toward planning future imaging of Earth-like planets (*Roberge et al., 2012*). While it is possible to extrapolate from the brightest disks (e.g., *Lawler and Gladman*, 2012; *Kennedy and Wyatt*, 2013), the typical exozodi is still quite uncertain, where a zodi refers to the equivalent luminosity of a disk identical to the solar system's zodiacal cloud (*Roberge et al., 2012*). Future advancement in the detection of faint warm dust requires improved control over measurement systematics, as well as an understanding of the potential sources of confusion (e.g., *Kennedy and Wyatt,* 2012). While spectroscopy is

an improvement over photometry (calibration against ancillary photometry is not necessary; one can instead look for changes in slope within the spectrum itself), interferometry provides the best opportunity for robust detection of dust at levels much fainter than the star itself. A mid-IR survey by the Keck interferometer nuller (KIN) detected 10-μm emission from two known disks — η Crv and γ Oph — and placed a 3-σ upper limit on the mean exozodi level at <150 zodis (*Millan-Gabet et al.,* 2011). Over the next few years, the LBTI will conduct a similar survey with much better sensitivity, extending our knowledge of the exozodiacal luminosity function an order of magnitude lower (*Hinz et al.,* 2008).

2.3. Near-Infrared Observations

Interior to the warm habitable-zone dust, a new category of systems with very hot dust (~1000 K) has been discovered via near-IR interferometry. The Vega system has been studied in the most detail, based on its detection with the CHARA interferometer (*Absil et al.,* 2006). The best fit to its 1.3% excess at K band, combined with photometric upper limits at mid-IR wavelengths, is a disk with ~1-μm-sized grains piled up at an inner radius of 0.2 AU, roughly consistent with a sublimation temperature of 1700 K. Examples of other prominent debris disks with detected near-IR excess include τ Ceti (*di Folco et al.,* 2007) and β Leo (*Akeson et al.,* 2009). More broadly, *Absil et al.* (2013) have performed a near-IR survey of 42 nearby stars with CHARA, detecting excess for 28 ± 7% of the targets. They find that near-IR excess is detected more frequently around A stars (50 ± 13%) than FGK stars (18 ± 7%). Interestingly, K-band excess is found more frequently around solar-type stars with detected far-IR excess, although not at a statistically significant level, and no such correlation is seen for A stars. Somewhat lower detection rates were obtained with a Very Large Telescope (VLT)/Precision Integrated-Optics Near-infrared Imaging ExpeRiment (PIONIER) survey in the H-band for a sample of 89 stars in the southern hemisphere, and the frequencies appear more uniform across the spectral types (Ertel et al., in preparation). We discuss the origin of hot dust in section 7.

2.4. Impact of Stellar Metallicity and Multiplicity

In contrast to the well-known positive correlation between giant planets and stellar metallicity, *Greaves et al.* (2006) and *Beichman et al.* (2006) found that the incidence of debris disks is uncorrelated with metallicity. These results are corroborated by subsequent works (*Maldonado et al.,* 2012, and references therein). The only subtle difference might be a "deficit" of stars with disks at very low metallicities (−0.50 < [Fe/H] < −0.20) with respect to stars without detected disks (*Maldonado et al.,* 2012), although this requires confirmation.

Debris disks are common around binary/multiple stars, giving information on the disks of multiple components that can be compared with those of protoplanetary disks (*Monin et al., 2007*). Statistically, *Trilling et al.* (2007) found an incidence rate of ~50% for binary systems (A3–F8) with small (<3 AU) and large (>50 AU) separations. They find a marginally higher rate of excess for old (age > 600 Ma) stars in binaries than for single old AFGK stars. In contrast, *Rodriguez and Zuckerman* (2012) found that IRAS-detected disks are less common around multiple stars [a result consistent with that of the DEBRIS survey (Rodriguez et al., in preparation)], and that, at any given age, the fractional luminosity of the dust in multiple systems is lower than in single stars, which would imply that disks around multiple systems are cleared out more efficiently.

A few multiple systems also host multiple disks, namely the δ Sculptoris quadruple system (Phillips et al., in preparation) and the Fomalhaut tertiary system (*Kennedy et al., 2014*).

2.5. Correlation with Exoplanet Populations

Many theories predict statistical relationships between planet and debris populations. Based on models of disk evolution as a function of disk mass/metallicity, *Wyatt et al.* (2007a) predict a correlation between planets and debris disk brightness. *Raymond et al.* (2012), on the other hand, predict that orbital instability among neighboring giant planets will produce an anti-correlation between eccentric Jupiters and debris disks, but a positive correlation between terrestrial planets and debris.

As a test of such theories, many attempts have been made to identify statistical correlations between debris and planet properties. While the long-period planets detected by direct imaging generally reside in prominent debris disks [e.g., β Pic and HR 8799 (*Lagrange et al.,* 2009; *Marois et al.,* 2008)], unbiased surveys with larger sample size are required to quantify any possible correlation. For this reason, debris disk surveys generally focus on planetary systems discovered via their induced radial velocities, such that the planets tend to orbit much closer to their central stars than outer debris detected in the far-IR. While initial Spitzer results found a relatively high detection rate of IR excess for known planet-bearing stars (*Beichman et al.,* 2005b), analysis of the full set of Spitzer observations did not find a statistically significant correlation between planets and debris (*Bryden et al.,* 2009; *Kóspál et al.,* 2009). More recent analysis of Spitzer data combined with ongoing discoveries of lower mass planets, however, finds within the nearest 60 G stars a correlation between debris and the mass of a planet, where 4 out of 6 stars with only sub-Saturn-mass planets also have Spitzer-detected IR excess (*Wyatt et al.,* 2012).

Within the Herschel key program surveys of nearby stars, 12 of 37 known planet-bearing stars are identified as having IR excess (*Marshall et al.,* 2014). While this detection rate is not significantly higher than for other solar-type stars [20% within DUNES (*Eiroa et al.,* 2013)], debris disks are again found to be more common around systems with low-mass planets (Moro-Martín et al., in preparation).

Furthermore, combining the unbiased legacy surveys with Herschel observations of an additional 69 solar-type stars known to have planets reveals a statistically significant (>3-σ) correlation between the presence of planets and the brightness of orbiting debris (Fig. 4) (Bryden et al., in preparation). A planet-debris trend is also seen among Herschel-observed M stars, albeit with extremely small number statistics; in contrast to the 1% detection rate of IR excess for M stars not known to have planets (Matthews et al., in preparation), one of the three observed planet-bearing M stars has detected debris (*Lestrade et al.,* 2012).

3. DUST ORIGIN IN COLLISIONAL CASCADES

The baseline theoretical model for debris disks is that planet formation processes leave one or more belts of planetesimals orbiting the star, similar to the asteroid and Kuiper belts in the solar system, that collide at sufficiently high velocities for those bodies to be destroyed, replenishing the dust that is observed. The process of grinding of larger bodies, planetesimals, into ever-smaller bodies, down to dust, is called a collisional cascade. This section presents analytic laws and reviews numerical modeling of such collisional cascades, showing how the complex interaction between different small body forces, and the composition and internal structure of those bodies, affect the structure and evolution of the disks and their observable properties.

3.1. Modeling Methods

Many properties of collisional cascades can be obtained analytically. However, the derivation of dust distributions with an accuracy sufficient to interpret the observations often necessitates numerical modeling of the cascade, together with radiation pressure and drag forces. Suitable numerical methods can be classified into three major groups.

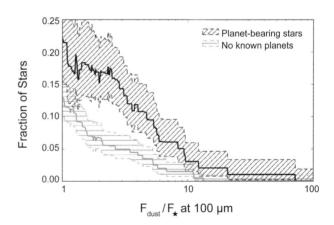

Fig. 4. Stars that are known to have planets tend to have brighter debris disks. Herschel far-IR observations of planet-bearing stars (top curve) and stars not known to have planets (bottom curve) find that the disks' fractional brightnesses ($F_{dust}/F_\star$ at 100 μm) are significantly larger for systems with planets. Figure courtesy of G. Bryden.

N-body codes (e.g., *Lecavelier des Etangs et al.,* 1996b; *Wyatt,* 2006; *Krivov et al.,* 2009; *Stark and Kuchner,* 2009; *Thébault,* 2009, 2012; *Kuchner and Stark,* 2010; *Thébault et al.,* 2012) follow trajectories of individual disk objects by numerically integrating their equations of motion and store their instantaneous positions and velocities. Assuming that the objects are produced and lost at constant rates, this yields steady-state density maps of different-sized particles in the disk. Then, collisional velocities and rates can be computed and collisions can be applied to modify the collisionless distributions. This collisional "post-processing" can also be done iteratively. As the N-body methods are able to handle an arbitrary large array of forces, they are superior to the others in studies of structures in debris disks arising from interactions with planets, ambient gas, or interstellar medium (hereafter ISM) (section 6). However, they cannot treat a large number of objects sufficiently to cover a broad range of particle masses and often use rough prescriptions of collisional outcomes. As a result, an accurate characterization of the size distribution with N-body codes is not possible.

Statistical methods effectively replace particles with their distribution in an appropriate phase space. In applications to planetesimal and debris disks, one introduces a mesh of several variables comprising, e.g., mass, distance, and velocity (e.g., *Kenyon and Bromley,* 2008, 2010; *Thébault et al.,* 2003; *Thébault and Augereau,* 2007; *Thébault and Wu,* 2008) or mass and orbital elements (*Krivov et al.,* 2005, 2006, 2008; *Löhne et al.,* 2012). The number of particles in each bin of the mesh at successive time instants is computed by solving equations that describe gain and loss of objects by collisions and other physical processes. Statistical codes are much more accurate in handling collisions than N-body ones, but treat dynamics in a simplified way, e.g., by averaging over angular orbital elements. For this reason, they are less suitable for simulation of structures in debris disks.

In recent years, *hybrid codes* have been developed that combine accurate treatments of both dynamics and collisions (*Kral et al.,* 2013; *Nesvold et al.,* 2013). The basic construct in these codes is the superparticle (SP) (*Grigorieva et al.,* 2007). An SP represents a swarm of like-sized particles sharing the same orbit that can be obtained through N-body integrations to account for all perturbations. Collisions between SPs in intersecting orbits are modeled by generating a set of new SPs that represent the modified remnants and newly born collisional fragments. The algorithm is complemented by procedures of merging similar SPs and deleting redundant or unnecessary ones to keep the total number of SPs at a computationally affordable level.

3.2. Steady-State Collisional Cascade

To model collisional cascades with statistical methods, one usually considers how mass flows between bins associated with objects of a given mass. Collisions lead to redistribution of mass among the bins. An important "reference case" is the steady-state collisional cascade, in which all bins gain mass from the breakup of larger bodies

at the same rate at which it is lost in collisional destruction or erosion. Under the assumption that the fragment redistribution function (i.e., the size distribution of fragments from each collision) is scale-independent, the mass loss (or production) rate of a steady-state collisional cascade is the same in all logarithmic bins (*Wyatt et al.,* 2011), with well-characterized implications for the size distribution. In the absence of major collisional breakup events, a steady-state cascade represents a reasonable approximation to reality for all sizes with collisional timescales that are shorter than the age over which the cascade has been evolving.

3.3. Size Distribution

A steady-state cascade gives a size distribution of solids that is usually approximated by a power law, $n(D) \propto D^{-\alpha}$, or by a combination of such power laws with different indices in different size ranges. The index α does not usually have a strong dependence on the redistribution function, but is sensitive to the critical energy for fragmentation and dispersal, $Q_D^\star$. If $Q_D^\star$ is independent of size, and neglecting the cratering collisions, then $\alpha = 3.5$ regardless of the redistribution function (*Dohnanyi,* 1969), as long as it is scale-independent. For $Q_D^\star \propto D^b$, this generalizes to $\alpha = (7 + b/3)/(2 + b/3)$ (*Durda and Dermott,* 1997; *O'Brien and Greenberg,* 2003; *Wyatt et al.,* 2011). With the typical values, $b \sim -0.3$ for objects in the strength regime ($D \lesssim 0.1$ km) and $b \sim 1.5$ for objects in the gravity regime ($D \gtrsim 0.1$ km), this leads to α between 3 and 4. Including further effects, such as the size-dependent velocity evolution of objects in the cascade, can somewhat flatten or steepen the slope α, but usually does not drive α outside that range (e.g., *Belyaev and Rafikov,* 2011; *Pan and Schlichting,* 2012). For $3 < \alpha < 4$, the cross section, and thus the amount of observed thermal emission or scattered light, are dominated by small grains, while most of the mass resides in the biggest solids.

For dust sizes below 1 mm nongravitational forces modify the size distribution. Radiation pressure effectively reduces the mass of the central star by a factor $1-\beta$, where $\beta = 1.15Q_{pr}(D)(L_\star/L_\odot)(M_\odot/M_\star)(1\text{ g cm}^{-3}/\rho)(1\text{ }\mu\text{m}/D)$, with Q_{pr} being the radiation pressure efficiency averaged over the stellar spectrum, and ρ standing for the grains' bulk density. As a result, dust released from parent bodies in nearly circular orbits immediately acquire eccentricities $e = \beta/(1-\beta)$. Thus if $\beta \geq 0.5$, the grains will be blown away on hyperbolic orbits. The corresponding size D_{bl}, referred to as the blowout size (~ 1 μm for Sun-like stars), is a natural lowest cutoff to the size distribution, since the amount of grains with $D < D_{bl}$ is by several orders of magnitude (the ratio of the collisional lifetime of bound grains to the disk-crossing time of unbound ones) smaller than that of the bound grains. Note that for solar-type stars, *very* small grains ($D \ll D_{bl}$) may also stay in bound orbits, as their β becomes smaller than 0.5 at least for some material compositions. Furthermore, for later-type stars $\beta < 0.5$ at all sizes; the fate of submicrometer-sized motes in disks of these stars is unknown.

Thus a reasonable approximation to the size distribution is to use $\alpha = 3.5$ (or the appropriate slopes in the strength and gravity regimes) from the largest planetesimals involved in the cascade (i.e., those whose collisional lifetime is shorter than the time elapsed after the ignition of the cascade) down to the blowout size (see, e.g., the simulations for $\tau > 10^{-6}$ and $e = 0.1$ of Fig. 5a). This can particularly be helpful in estimating the total disk mass from the dust mass that can be inferred from submillimeter observations (e.g., *Wyatt and Dent*, 2002; *Krivov et al.*, 2008). However, the nongravitational forces acting on micrometer-sized grains further modify the size distribution at these sizes.

A natural consequence of mass loss rate being independent of size in logarithmic bins when a cut-off below some minimum particle size is imposed is that the size distribution becomes wavy (*Campo Bagatin et al.*, 1994; *Wyatt et al.*, 2011). The shape of the size distribution close to the blowout limit would be complicated by a natural spread in mechanical strength, optical properties, and even the densities of individual dust grains in realistic disks. The waviness also depends on the treatment of cratering collisions, since it does not arise in the simulations of *Müller et al.* (2010) (see also Fig. 5a).

Whether the waviness exists or not, caution is required when using a single power-law down to D_{bl}. The blowout limit is not sharp, and the maximum in the size distribution can be shifted to larger sizes, both in disks of low dynamical excitation (section 4.2) and in transport-dominated disks (e.g., *Krivov et al.*, 2000), either because of the disks' low optical depth or because transport mechanisms are particularly strong [e.g., owing to strong stellar winds that amplify the Poynting-Robertson (P-R) effect (*Plavchan et al.*, 2005)]. Figure 5a shows that this effect is evident both for the $\tau < 10^{-8}$ and $e = 0.001$ simulations.

Even in a transport-dominated disk, the largest planetesimals are dominated by collisions. This is because the lifetime against catastrophic collisions in a disk with a size distribution slope $\alpha < 4$ increases more slowly with particle size, $T_{coll} \propto D^{\alpha-3}$ (e.g., *Wyatt et al.*, 1999), than the transport timescale, $T_{drag} \propto D$. Defining the size D_c to be that at which the two timescales are equal (i.e., $T_{coll} = T_{drag}$), for a disk to have particles of any size in the transport dominated regime requires $D_c > D_{bl}$. The rough criterion for this (assuming transport to be caused by the P-R force) is $\tau < v/c$, where τ is normal geometrical optical depth, v is the circular Keplerian velocity at the disk radius, and c is the speed of light (*Kuchner and Stark*, 2010). At sizes $D < D_c$, the slope $\alpha = \alpha_c - 1$, where α_c is the slope of the redistribution function (*Wyatt et al.*, 2011), meaning that it is grains of size $\sim D_c$ that dominate the cross section.

3.4. Radial Distribution

Collisions, together with radiation pressure and drag forces, set not only the size distribution, but also the radial distribution of dust. As long as transport is negligible (i.e., $T_{drag} \gg T_{coll}$), a narrow parent belt of planetesimals should be surrounded by a halo of small grains in eccentric orbits, with pericenters residing within a "birth ring" and the apocenters located outside it (*Lecavelier des Etangs et al.*, 1996a). The smaller the grains, the more extended their halo. Assuming $\alpha \approx 3.5$, the radial profile of dust surface density (or the normal optical depth) in such a halo should be close to $\tau \propto r^{-1.5}$ (*Strubbe and Chiang*, 2006; *Krivov et al.*, 2006); see, e.g., the $\tau > 10^{-4}$, $e = 0.1$ simulation of Fig. 5b. For comparison, the halo formed by unbound grains on hyperbolic orbits would exhibit a flatter falloff, $\tau \propto r^{-1}$ (e.g., *Lecavelier des Etangs et al.*, 1998).

In the opposite limiting case of a transport-dominated disk (i.e., $T_{drag} \ll T_{coll}$), drag forces steepen the profile of the halo to $\tau \propto r^{-2.5}$ (*Strubbe and Chiang*, 2006; *Krivov et al.*, 2006) and the inner gap of the birth ring (see the lower τ simulations of Fig. 5b) is filled by dust with a nearly uniform density profile, $\tau \propto r^0$ (e.g., *Briggs*, 1962) — which can, of course, be altered by planets if any orbit in the gap (section 6). Since a disk's fractional luminosity is $f_d \approx (\tau/2)(dr/r)$, where dr is the characteristic disk width, many of the debris disks observed by Herschel at $f_d \sim 10^{-6}$ (see section 2.1) have optical depths such that $T_{coll} \sim T_{drag}$. For such low-density disks, one expects both a "leak" of dust inward from the parent ring location and a halo of small grains spreading outward.

The slopes, and the extent, of both the halo and inward leak components depend not only on T_{coll}/T_{drag} (see *Wyatt*, 2005b), but also on many other factors. Major — and essentially unknown — ones are again the level of stirring of dust-producing planetesimals (characterized by their typical eccentricities e and inclinations I) and their mechanical strength (i.e., $Q_D^\star$). The stirring level and $Q_D^\star$ both determine the degree of "destructiveness" of the collisions and, through that, strongly affect both the size and radial distribution of dust. As an example, lowering the stirring level would largely suppress the outer halos.

3.5. Observational Constraints on Dust Properties

The models above describe the size and spatial distribution of dust. These already rely on a vast number of parameters and assumptions, most of which are poorly known or not known at all. These are mostly related to the largest planetesimals (e.g., their bulk composition, mechanical strength, primordial size distribution, and dynamical excitation, as well as location, width, and total mass of the belts), possible planetary perturbers that would affect the disks, and even the central stars (ages, stellar mass loss rates, etc.). However, there is another important unknown, namely the properties of the observable dust (e.g., shape, porosity, and chemical composition).

In the discussion above we tacitly assumed that the observations are sensitive to the cross-sectional area of the dust. This would be the case if the dust acts like a blackbody, but the situation is more complicated in ways that need to be understood if observations are to be correctly interpreted. For a start, the small particles that dominate the cross-section are inefficient emitters, which means that

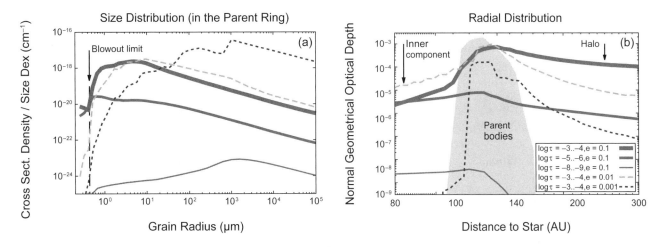

Fig. 5. (a) Typical size distribution and **(b)** radial profile for fiducial disks with a 130-AU radius around a Sun-like star. All disks were generated numerically with a collisional code ACE (*Krivov et al.,* 2005, 2006, 2008; *Müller et al.,* 2010; *Löhne et al.,* 2012) and evolved until a quasi-steady state at dust sizes were reached. Dust (and planetesimals) were assumed to be a mixture of astrosilicate and water ice in equal mass fractions. Disks have different peak optical depths (thick lines: $\tau \sim 10^{-3} \ldots 10^{-4}$, medium lines: $\tau \sim 10^{-5} \ldots 10^{-6}$, thin lines: $\tau \sim 10^{-8} \ldots 10^{-9}$), and dynamical excitations (solid: $e \sim 2I \sim 0.1$, dashed: 0.01, dotted: 0.001). The figure demonstrates that both in weakly-stirred and transport-dominated disks the dominant grain sizes get larger [**(a)**], and the outer halos weaker [**(b)**]. Figure courtesy of A. V. Krivov.

they emit at a temperature above blackbody (see also section 5.1). This fact means that the SED, which constrains the disk temperature, can be used in concert with a disk radius measured directly from imaging to constrain the particle size distribution [e.g., *Wyatt and Dent* (2002), who found a size distribution slope $\alpha \approx 3.5$ for Fomalhaut] and composition [e.g., porous grains constituting a heterogeneous mixure of silicates, carbonaceous compounds, and ices have been favored for HD 181327 and HD 32297 (*Lebreton et al.,* 2012; *Donaldson et al.,* 2013)].

Additional information can be gained from scattered light and polarimetric data. *Hinkley et al.* (2009) use groundbased polarimetry to image the HR 4796A disk, measuring its scale height and determining that the disk is dominated by micrometer-sized grains, while *Maness et al.* (2009) find that submicrometer-sized grains dominate the asymmetric disk of HD 61005. *Graham et al.* (2007) use polarized imaging with HST of the AU Mic disk to reveal that the disk is significantly (>300×) depleted of scattering material inside the 40–50-AU birth ring (*Strubbe and Chiang,* 2006). The source of the emission is found to be highly porous (91–94%), micrometer-sized grains (with an upper size of decimeters for the parent bodies) in an optically thin disk. *Shen et al.* (2009) use random aggregates to model the AU Mic data and find these require a lower porosity of just 60% in 0.2–0.5-µm-sized grains.

The chemical composition of the debris dust can be directly probed by the presence of emission features in the spectrum. This composition should trace that of planetesimals, and thus its knowledge could give cosmochemical and mineralogical hints to the processes and conditions from the protoplanetary phase (see chapters by Dutrey et al. and Pontoppidan et al. in this volume), including possible radial

mixing and thermal history. The chemical composition of dust, together with — equally unknown — grain shape and porosity, determines radiation pressure strength and D_{bl}, absorption efficiency in both the optical range and in the IR, and therefore the dust temperature and thermal emission efficiency. It also affects the mechanical strength of the particles as, e.g., silicate particles show different collisional outcomes than icy grains (Kilias et al., in preparation).

Direct probes of composition through solid state features are rare, however. Apart from the β Pic (*de Vries et al.,* 2012) and possibly HD 181327 (*Chen et al.,* 2008) disks, this method is confined to systems with hot dust emitting in the mid-IR (see section 7), for which the 10-µm silicate feature is an important diagnostic of the dust composition and size. For instance, the presence of the silicate feature requires grains with radii ≤10/2π µm, and so relatively few A stars with 10-µm excess exhibit strong silicate features (*Chen et al.,* 2006; *Morales et al.,* 2009). Less-luminous stars, on the other hand, can more easily maintain small grains; later-type stars with 10-µm excess generally have silicate features. Yet, in some systems the discovery of sharp features uncovers a problem, i.e., that grains producing them must be smaller than the blowout limit (e.g., *Fujiwara et al.,* 2012), contrary to the theoretical expectation that such grains should be underabundant.

The shapes of these features allow distinct mineral species to be identified and enables comparative astrogeology between systems and a consideration of their evolutionary state. Spitzer spectra of dust around HD 69830, for example, were originally explained by cometary parent bodies (*Beichman et al.,* 2005a), but more detailed analysis subsequently found evidence of olivines, pyroxenes, and sulfides but no water ice, consistent with break-up of a C-type asteroid

(*Lisse et al.,* 2008; *Beichman et al.,* 2011). The spectrum of HD 172555, on the other hand, is dominated by silica with a size distribution heavily weighted toward fine grains (*Lisse et al.,* 2009). The impact velocities required to produce this composition also imply that the dust was created by a large (>150 km radius) asteroid impact [see *Lisse et al.* (2012) for a compilation of mid-IR spectra].

For disks without the benefit of direct imaging or spectroscopy, the degeneracy between particle size and disk radius, as well as the dust optical properties, means that none of these properties can be well constrained. Yet it is possible to estimate the sizes of emitting grains from the shape of the SEDs. Submillimeter data can be fit using a modified blackbody approximation, in which the flux longward of wavelength λ_0 is divided by a factor $(\lambda/\lambda_0)^{-\beta}$ where β is the emissivity power index. The measured β values are typically lower than the value of 2 observed for ISM grains (e.g., *Andrews and Williams,* 2005), suggesting debris disk grains are larger than those populating the ISM. The λ_0 value can be thought of as representative of the grain sizes, and the values vary significantly. *Booth et al.* (2013) find $\lambda_0 \sim 70$–170 μm and $\beta \sim 0.7$–1.6, as is consistent with many measured values for debris disks. In HR 8799, the best fit yields $\lambda_0 \sim 40$ μm, likely indicative of the small grain-dominated halo's contribution to the SED's cold component (*Matthews et al.,* 2014).

4. BIRTH AND EVOLUTION OF DEBRIS DISKS

4.1. Observations of the Transition Phase

At the young end, it is appropriate to pose the question: When is a disk a debris disk as opposed to a late-stage protoplanetary disk (or "transition disk"; see chapter by Espaillat et al. in this volume)? Several studies have measured the disk frequency in young clusters and these convincingly show that 50% of protoplanetary disks (as identified by a large near-IR excess from dust close to the star) are lost by 3 Ma, and that the fraction remaining is neglible by 10 Ma (*Haisch et al.,* 2001; *Hernández et al.,* 2007). Measurements of disk masses from submillimeter data (which probe the outer disk regions) also indicate that the 10-Ma age is significant (see Fig. 3) (*Panić et al.,* 2013). Indeed, the TW Hydra Association [~8 Ma (*Zuckerman and Song,* 2004)], hosts stars in the protoplanetary phase, such as TW Hydrae itself (*Hughes et al.,* 2008b; *Rosenfeld et al.,* 2012; *Andrews et al.,* 2012) and in the debris phase, such as HR 4796A and TWA 7 (*Schneider et al.,* 1999, 2005; *Matthews et al.,* 2007; *Thalmann et al.,* 2011) as could be expected within this transition period (*Riviere-Marichalar et al.,* 2013).

4.2. Stirring

To produce dust by collisions, planetesimals in debris disks must have relative velocities sufficient for their fragmentation, i.e., the disks must be stirred. They may be prestirred by pro-

cesses acting during the protoplanetary phase (*Wyatt,* 2008). In typical protoplanetary disk models, relative velocities of solids are set by turbulence, Brownian motion, and differential settling and radial drift. This means that for millimeter-sized grains at ~100 AU, they are expected to lie in the m s^{-1} range (*Brauer et al.,* 2008). This translates to eccentricities and inclinations $e \sim I \sim 0.001$. While there may be additional sources of velocity excitation that are not considered in such models, e.g., additional excitation from binary companions (*Marzari et al.,* 2012; *Thébault et al.,* 2010), the prestirring level is not expected to be much higher than $e \sim I \sim 0.01$–0.001 if planetesimals are to form by coagulation (e.g., *Brauer et al.,* 2008; *Zsom et al.,* 2010, 2011), unless collective dust phenomena are invoked (see *Chiang and Youdin,* 2010, for a review).

The activation of disks some time after gas dispersal is usually referred to as delayed stirring (*Dominik and Decin,* 2003). One viable mechanism could be stellar flybys (*Kenyon and Bromley,* 2002). Since the timescale of cluster dissipation is ≲100 million years (*Kroupa,* 1995, 1998), the ignition of the cascade by flybys can be expected to occur early in the systems' history. At that time, stars may experience encounters at various distances R, with several hundreds of astronomical units probably being typical for a low- to medium-density cluster. Each encounter truncates the disk at r ~ 0.3 R, and the zone between ~0.2 R and ~0.3 R becomes the most excited, and thus the brightest during that "encounter era." However, because the collisional depletion time at that region is also the shortest, that part of the disk is short-lived. What remains afterwards is a weakly excited disk inside ~0.2 R, where the dynamical excitation scales as $e \propto r^{5/2}$ (*Kobayashi and Ida,* 2001).

At later times, assuming a standard picture of collisional growth following planetesimal formation, it is expected that eventually the planetesimals will grow large enough to stir the debris disk, so-called self-stirring (*Kenyon and Bromley,* 2010; *Kennedy and Wyatt,* 2010). Formation of 1000-km-sized planetesimals is normally considered sufficient to trigger the cascade. In the models of *Kenyon and Bromley* (2008), assuming a standard minimum-mass solar nebula with a solid surface density of ~1 $M_\oplus$ AU^{-2} r$^{-3/2}$ around a solar-mass star, such objects would form on a timescale ~400 (r/80 AU)^3 million years. This suggests that Pluto-sized objects can grow quickly enough to explain the dust production in Kuiper-belt-sized disks starting from ages of hundreds of millions of years. These self-stirring models also predict an increase in disk radius with age, which has been suggested by some observations (see section 4.4), although *Wyatt et al.* (2007b) noted that the rapid decay of close-in disks (see section 4.4) means that the mean radius of detectable disks should increase with age, even if the disks themselves are not changing in radius.

Alternatively, or in addition to self-stirring, stirring by planets in the system is possible (*Mustill and Wyatt,* 2009). The secular perturbations from a planet with mass m_{pl}, semimajor axis a_{pl}, and eccentricity e_{pl} stir the disk (see Fig. 7) on a timescale $\propto r^{9/2}/(m_{pl}e_{pl}a_{pl}^3)$ to $e \sim 2(a_{pl}/r)e_{pl}$. This timescale may be shorter than the self-stirring timescale for systems with a moderately eccentric planet orbiting close to the inner edge of the disk (like Fomalhaut) and even for systems with a close-in radial velocity planet in a more eccentric orbit (like ε Eri).

4.3. Disks at Different Stirring Levels

In the debris disk of our solar system, the EKB, both self-stirring and planetary stirring mechanisms are at work, and the stirring level is high. The resonant and scattered populations acquired their e ~I > 0.1 from Neptune, while cold classical EKB objects have e ~ I ≤ 0.1 consistent with self-stirring by their largest embedded members (D ≤ 200 km). It is also possible that the Sun was born in a massive cluster and that flybys played a significant role in the early history of the EKB (e.g., *Dukes and Krumholz,* 2012). In contrast, the origin of stirring in extrasolar debris disks remains completely unknown. Case studies (e.g., *Müller et al.,* 2010; *Wyatt et al.,* 2012) and interpretations of debris disk statistics (e.g., *Wyatt et al.,* 2007c; *Löhne et al.,* 2008; *Kennedy and Wyatt,* 2010) have shown that both self-stirring and planet-stirring models are consistent with the available data. The true level of stirring is not known either. Most of the "pre-Herschel" models assumed values at the EKB level, e ~ I ~ 0.1. However, recent discoveries of disks with sharp outer and smooth inner edges, and those of thermally cold disks, have challenged this assumption. It is therefore important to review how the level (and origin) of stirring affect the disk observables.

If the dust-producing planetesimals have a low dynamical excitation that, however, is still high enough for collisions to be mostly destructive, then the low collision velocities between large grains, which are not susceptible to radiation pressure, would decrease the rate at which small grains are produced. However, the destruction rate of these small grains is set by eccentricities induced by radiation pressure and remains the same (*Thébault and Wu,* 2008). This should result in a dearth of small dust. The maximum of the size distribution shifts to larger values (Fig. 5a), and the outer edge of the disks, formed by small "halo" grains, gets sharper (Fig. 5b). This is the scenario favored in the collisional simulations of *Löhne et al.* (2012), who modeled the Herschel disk around the Sun-like star HD 207129 (*Marshall et al.,* 2011). Their best match to the data assumed a dynamical excitation at the e ~ I ~ 0.03 level, for which the dominant grain size lies at ≈3–4 μm.

At a stirring level below e ~ I ~ 0.01, i.e., at random velocities below a few tens of meters per second, collisions are not necessarily destructive — with the caveat that the collisional outcome, and thus the fragmentation threshold, depend on many factors, which include not only the impact velocity, but also the impact angle, masses, materials, porosities, morphology, and hardnesses of projectile and target (*Blum and Wurm,* 2008). At e ~ I ~ 0.001, i.e., at the level consistent with the natural prestirring from the protoplanetary phase, impact experiments in the laboratory reveal a rather complex mixture of outcomes, including

disruption, cratering and bouncing with mass transfer, and agglomeration in comparable fractions (*Güttler et al.,* 2010). Using these results, *Krivov et al.* (2013) considered belts of primordial grains that could have grown on the periphery of a protoplanetary disk (where stirring from close-in planets, if any, is negligible) to sizes larger than ~1 mm (to avoid the gas or radiative drag losses), but smaller than ~1 km (to keep self-stirring at a low level). Their collisional simulations demonstrate that such a belt can largely preserve the primordial size distribution over billions of years, with only a moderate accretional growth of the largest solids and a moderate production of collisional fragments in the submillimeter range (Fig. 5b, dotted line). The existence of such belts was previously predicted by *Heng and Tremaine* (2010), who considered a regime in which collisions are low enough in velocity for the outcome of collisions to be bouncing rather than fragmentation (or accretion). This scenario may provide an explanation for Herschel observations that identified several candidate "cold disks" where the emitting dust material is nearly as cold as blackbody and thus should be dominated by large grains (*Eiroa et al.,* 2011, 2013).

4.4. Evolution on the Main Sequence

The large samples of main-sequence stars observed for IR emission have built up a picture of how debris disks evolve over time. It has been conclusively shown that excess rates decrease with the age of the stars independently of spectral type (e.g., *Su et al.,* 2006; *Wyatt et al.,* 2007b; *Trilling et al.,* 2008; *Carpenter et al.,* 2009; *Chen et al.,* 2011, 2012; *Urban et al.,* 2012). The decline of the incidence rate, which was already noticed in ISO surveys (e.g., *Habing et al.,* 2001), is significant up to ages <1 Ga, but is not apparent for older ages (*Trilling et al.,* 2008). Such a nonobvious decay is also supported by the Herschel DUNES survey (*Eiroa et al.,* 2013).

The decline in excess emission is observed to be much faster at 24 μm than at 70 μm for A stars (*Su et al.,* 2006). Thureau et al. (in preparation) also note an apparent decline in the excess at 100 and 160 μm. The decay timescales increase with wavelength. The upper envelope to the ratio of observed flux to photospheric flux at each wavelengths can be approximated by decay curves of $1 + t_0/t$, where t_0 is 150, 400, and 800 Ma for 24, 70, and 100/160 μm, respectively. Among A stars in the DEBRIS survey, the incidence rate for stars <450 Ma is double that of stars >450 Ma, but substantial levels of excess are still observed for the oldest stars in the sample (~800 Ma). For FGK stars, *Carpenter et al.* (2009) and *Siegler et al.* (2007) both find that 24-μm excesses decline with age for the FEPS targets and a compilation of FGK stars observed with Spitzer respectively, whereas *Hillenbrand et al.* (2008) do not see such an apparent trend in the FEPS Sun-like stellar sample at 70 μm.

Many studies do not find any trend in disk properties such as temperature with the stellar temperature or age (e.g., *Hillenbrand et al.,* 2008; *Trilling et al.,* 2008; *Lawler et al., 2009; Chen et al., 2011, 2012; Dodson-Robinson et al., 2011; Moór et al., 2011b*), regardless of the nature of the dust (i.e., cold or warm). However, *Rhee et al.* (2007) and *Moór et al.* (2011b) claim that for early and late-type stars, respectively, there is an increase of the radial location of the dust with the stellar age. In addition, Herschel DUNES results suggest trends might exist between the mean blackbody radius for each F, G, and K spectral types, and a correlation of disk sizes and an anticorrelation of disk temperatures with the age (*Eiroa et al.,* 2013). At submillimeter wavelengths, *Nilsson et al.* (2010) find an evolution of disk radii with age (for <300-Ma stars), in contrast with the results from *Najita and Williams* (2005).

The observed decay of debris disks with time is expected, because planetesimal families undergo collisional depletion over long time spans and are not replenished. Useful analytic scaling laws that describe timescales of the quasi-steady-state evolution of collision-dominated disks have been worked out (*Wyatt et al.,* 2007c; *Löhne et al.,* 2008; *Krivov et al.,* 2008). For example, for a narrow parent ring with initial mass M_0 and radius r, at a time t_{age} from the onset of the cascade

$$F\left(xM_0, r, t_{age}\right) = xF\left(M_0, r, xt_{age}\right) \qquad (1)$$

$$F\left(M_0, xr, t_{age}\right) = F\left(M_0, r, x^{-13/3} t_{age}\right) \qquad (2)$$

where x > 0 is an arbitrary factor, and $F(M_0, r, t)$ stands for any quantity directly proportional to the amount of disk material, such as the total disk mass, mass of dust, and total cross section. Equation (2) explains why the fastest declines occur in warm dust that is located closer to the star, while the colder outer dust remains in much later times.

One consequence of these scalings is that, once the largest planetesimals have come to collisional equilibrium, the dust fractional luminosity should decay as $f_d \propto t_{age}^{-1}$, and that there is a maximum possible fractional luminosity at any t_{age} (*Wyatt et al.,* 2007c)

$$f_{max} = 2.4 \times 10^{-8} \left(\frac{r}{AU}\right)^{7/3} \left(\frac{dr}{r}\right) \left(\frac{D_{max}}{60\,km}\right)^{0.5}$$
$$\times \left(\frac{Q_D^\star}{300\,J/kg}\right)^{5/6} \left(\frac{e}{0.1}\right)^{-5/3} \left(\frac{t_{age}}{Gyr}\right)^{-1} \qquad (3)$$

for a central star with solar mass and luminosity. These results are valid under certain assumptions, including independence of $Q_D^\star$ on size, as well as the requirement that the collisional lifetime of the largest planetesimals $T_{coll}(D_{max})$ is shorter than t_{age}. More detailed models that lift these assumptions predict shallower decay laws, $f_d \propto t_{age}^{-0.3...08}$, and allow f_{max} to be, by about an order of magnitude, larger than equation (3) suggests (*Löhne et al.,* 2008; *Gáspár et al.,* 2013). Yet these results imply that the hot dust in some systems, having $f_d \gg f_{max}$, cannot stem from a steady-state cascade in "asteroid belts" close to the stars (see section 7).

These models for the long-term collisional evolution of planetesimal belts have been particularly useful for the interpretation of statistics on the incidence of debris as a function of age. They have been taken as the basis of population synthesis models (analogous to population synthesis models used to explain exoplanet populations; see the chapter by Benz et al. in this volume) that can reproduce the observed evolution, and in so doing provide information on the distribution of planetesimal belt radii and largest planetesimal sizes (*Wyatt et al.,* 2007b; *Löhne et al.,* 2008; *Kains et al.,* 2011; *Gáspár et al.,* 2013).

4.5. Debris Disks around Post-Main-Sequence Stars

Observational evidence of disk evolution after the main sequence phase has advanced significantly since Protostars and Planets V, when there was one detected disk around a white dwarf (*Zuckerman and Becklin,* 1987). Debris disks now have been detected around many white dwarfs (*Dufour et al.,* 2012; *Kilic et al.,* 2005, 2006; *Kilic and Redfield,* 2007), including those at the center of planetary nebulae (*Su et al.,* 2007; *Bilíková et al.,* 2012), primarily in the near- and mid-IR. In fact, the growth in the body of work on this subject is such that it could encompass a chapter unto itself, and we can only scratch the surface here.

Typical incidence rates for warm debris disks around cool white dwarfs are on the order of 1% (*Girven et al.,* 2011; *Kilic et al.,* 2009). *Barber et al.* (2012) estimate a disk incidence rate of $4.3^{+2.7}_{-1.2}$% in a metallicity unbiased sample of 117 cool, hydrogen-atmosphere white dwarfs. Many detected disks have gaseous as well as solid components (*Brinkworth et al.,* 2012). *Gänsicke et al.* (2008) find that the hydrogen-to-metal abundance in a white dwarf gas disk is more than 1000× below solar, supporting the idea that these disks of dust and gas are created by the disruption of rocky planetesimals.

In addition to dust in orbital configurations, an increasing number of studies demonstrate that tidally stripped asteroids or planetary material has been deposited on the surface of white dwarfs themselves (*Zuckerman et al.,* 2003; *Graham et al.,* 1990). This type of "pollution" provides important compositional information about rocky bodies in planetary systems. Significant numbers of metal-contaminated white dwarfs are now known, and many of these are now known to have IR excesses, detected from groundbased facilities or Spitzer. Based on K-band data, *Kilic and Redfield* (2007) suggest the incidence rate of debris disks around DAZ (hydrogen-dominated atmosphere) white dwarfs is ~14%.

Several authors have modeled the evolution of such disks. The debris present should be a descendent of the main-sequence debris population, but the physics are not yet fully understood; *Bonsor and Wyatt* (2010) looked at the evolution of the population of planetesimals known around main-sequence A stars and their detectability due to collisions and the changing radiation and wind forces throughout the post-main-sequence, while *Dong et al.* (2010) modeled the dynamical evolution of a debris/giant planet system through to the white dwarf phase, predicting final debris ring sizes of 30–50 AU. Models for the transport of material onto the star explain transport rates consistent with what can be provided through PR drag (*Rafikov*, 2011a), but a number of stars are observed to have higher accretion rates that are not so readily explained (*Rafikov*, 2011b). *Metzger et al.* (2012) suggest that sublimation of solids enhances the gas component of the disk, leading to runaway accretion; coupled with a long metal settling phase on the star, this could explain the polluted white dwarfs that do not show IR excess.

5. RESOLVED DISKS AND DISK STRUCTURE

In 2005, the year of the Protostars and Planets V meeting, the submillimeter camera SCUBA (*Holland et al.*, 1998) was retired from the James Clark Maxwell Telescope (JCMT). SCUBA had been the most effective imaging instrument for debris disks, resolving more than half of the 14 systems imaged at that time. Excepting β Pic and Fomalhaut, each disk was resolved at a single wavelength only. Nearly a decade later, significant strides have been made in resolving disk structures, owing to ongoing campaigns on the Hubble Space Telescope (HST) (e.g., *Hines et al.*, 2007; *Maness et al.*, 2009) and groundbased instruments in the optical [e.g., VLT (*Buenzli et al.*, 2010)], near-IR (e.g., *Janson et al.*, 2013; *Thalmann et al.*, 2011), mid-IR (e.g., *Smith et al.*, 2009, 2012; *Moerchen et al.*, 2010) and (sub)millimeter (e.g., *Maness et al.*, 2008; *Wilner et al.*, 2012; *MacGregor et al.*, 2013). Spitzer also contributed to the resolved images library, resolving several disks, many with multiple components [Vega, HR 8799, Fomalhaut (*Su et al.*, 2005, 2009; *Stapelfeldt et al.*, 2004, respectively)]. Most notably, the launch of Herschel in 2009 has yielded a vast gallery of resolved disks; both the DUNES and DEBRIS key programs have resolved half the detected disks (i.e., *Matthews et al.*, 2010; *Eiroa et al.*, 2010, 2011, 2013; *Marshall et al.*, 2011; *Churcher et al.*, 2011; *Wyatt et al.*, 2012; *Löhne et al.*, 2012; *Lestrade et al.*, 2012; *Kennedy et al.*, 2012a,b; *Broekhoven-Fiene et al.*, 2013; *Booth et al.*, 2013). Herschel has also produced the first resolved debris disk around a subgiant, κ CrB (*Bonsor et al.*, 2013a), and a resolved image of the gas-rich debris disk 49 Ceti (*Roberge et al.*, 2013).

Such emphasis is placed on resolved disk images because of the immense amount of information that can be gleaned from them compared to an SED alone. Resolving disks even marginally at one wavelength places very meaningful constraints on the disk structure, e.g., giving a direct measurement of the radial location of the dust, but resolving disks at multiple wavelengths at high resolution allows significant observational constraints on the radial and temperature structure of disks, as well as their component grain sizes and compositions. Azimuthal variations, warps, and offsets between the disk and star can also be evidence of unseen planets in such systems (see sections 5.5 and 6).

5.1. The Physical Extent of Disks

Measured sizes of debris disks from resolved images are consistently larger than those inferred from SED analyses in which temperature (typically found in the range ~40–200 K) is used as a proxy for radial separation from the star and the dust grains are assumed to be blackbodies. Grains are typically graybodies, meaning they exhibit a warmer temperature than a blackbody given their distance from the parent star (see section 3.5). It is therefore not surprising that the ratio, Γ, of the radius measured from the image to that inferred from the disk temperature, assuming blackbody grains, is often greater than unity. *Booth et al.* (2013) show Γ to range from 1 to 2.5 for nine resolved A stars in the DEBRIS survey. *Rodriguez and Zuckerman* (2012) find ratios as high as 5 in their sample of IRAS binary star diskhosts. These results are consistent with typical grain size being on the order of micrometers, roughly comparable with the radiation pressure blowout limit discussed in section 3.

The resolved physical extent of disk emission typically ranges from tens to hundreds of astronomical units from the parent star, but it is important to differentiate this emission from the underlying distribution of planetesimals, which is best traced by longer-wavelength observations of larger (and typically cooler) grains, as we discuss below.

5.2. Planetesimal Rings, Belts, and Gaps

Many disks show evidence for a narrow birth ring, harboring planetesimals on nearly circular orbits, as the underlying sources of the debris disk material (e.g., *Strubbe and Chiang*, 2006). Collisional modeling for the disks of Vega (*Müller et al.*, 2010) and HD 207129 (*Löhne et al.*, 2012) demonstrates that these are compatible with a steady-state collisional cascade in a narrow planetesimal annulus, even though the dust distribution appears broader. An appreciable spread of the dust sheet should be naturally caused by radiation pressure and drag forces acting on collisional fragments, and may be either outward (Vega) or inward (HD 207129) from the birth ring, as discussed in section 3 (see also Fig. 5b).

In systems without detected planets, one of the key measurable quantities is the sharpness of the inner edge, which sets constraints on the mass of any planet if it sits close to the inner edge of the belt (*Chiang et al.*, 2009), as discussed further in section 6. Figure 6 shows one of the first resolved images from ALMA, which promises to be a key instrument in the resolution of planetesimal belt locations around nearby stars. The combination of long observing wavelengths and high resolution places significant constraint on the belt width (13–19 AU) and reveals the sharpness of both the inner and outer boundary (*Boley et al.*, 2012). In addition, ALMA imaging of the AU Mic debris disk has firmly placed the planetesimal belt at 40 AU for that edge-on system, and a second, unresolved component centered on the star could be a warm asteroid belt (*MacGregor et al.*, 2013).

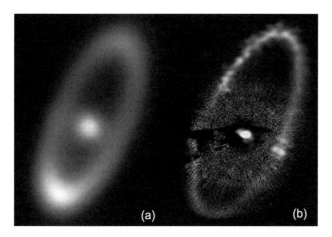

Fig. 6. The **(a)** Herschel 70-μm (*Acke et al.,* 2012) and **(b)** composite HST/ALMA 850-μm (*Kalas et al.,* 2005; *Boley et al.,* 2012) images of Fomalhaut reveal the impact of observing wavelength and resolution on the detected structure. Both Herschel and ALMA have sufficient sensitivity to detect an unresolved central component [associated with the warm dust seen at 24 μm by *Stapelfeldt et al.* (2004)] and show sharp inner edges and the offset of the ring from the central star first detected by HST. It is this offset and the obvious brightness asymmetry of the disk seen by Herschel that strongly support the case for an unseen planet in the Fomalhaut system. Figures courtesy of ESA/Herschel/PACS/B. Acke/KU Leuven (Herschel); ESO/NOAJ/NRAO (ALMA); and NASA/ESA (HST).

For many of the disks resolved by Herschel, however, the cold dust emission is not consistent with distribution in a narrow ring — even when taking into account that the distribution of dust is usually broader than that of the parent planetesimals. The full sample of resolved disks around A stars between 20 and 40 pc (from the DEBRIS sample) reveal that only three of the nine systems are well fit by narrow rings, while four definitely require multiple rings to explain the observations (*Booth et al.,* 2013).

Such systems can instead be explained as a broad belt. Such a disk can be modeled with a set of non-interacting narrow radial annuli, and the same models of section 3 can be applied, allowing one to predict the radial distribution of dust (e.g., *Kennedy and Wyatt,* 2010; *Wyatt et al.,* 2012). Sufficiently close to the star, where $T_{coll}(D_{max}) < t_{age}$, equation (3) suggests that the radial profile of the dust optical depth should be rising outward as $\tau \propto r^{7/3}$. Farther out — at the distances that have already been reached by the stirring front (section 4.2), but the largest planetesimals have not yet started to deplete collisionally — a profile that is flat or declining with radius is expected for $\tau(r)$. This shows a possible way of explaining inner gaps in debris disks by collisional erosion, i.e., without the assistance of planets (see section 6).

However, within the limits of low-resolution imaging, broad disks could alternatively be interpreted as multiple belts, and there is evidence for this in the SEDs of some

disks (e.g., *Hillenbrand et al.,* 2008). *Morales et al.* (2011) studied a sample of B8–K0 stars, many of which contain both warm and cold dust, reminiscent of the two-belt debris architecture of the solar system. Fitting the SEDs of these two-belt systems, *Morales et al.* (2011) find a bimodal distribution to the measured dust temperatures (their Fig. 2), where ~50-K cold belts are distinct from ~200-K warm belts, although *Ballering et al.* (2013) find less distinction in temperature. The break between the two populations occurs at the same temperature for both solar-type and A-type stars, whereas the dust location varies considerably with spectral type if other grain properties, e.g., optical constants or size distribution, are ignored. *Morales et al.* (2011) suggest that the break in temperatures is related to the ice line at ~150 K, either by setting the location for dust-releasing comet sublimation or by creating a favorable location for giant planets to form and remove neighboring debris.

Evidence for multiple, narrow components in well-resolved images of disks are becoming abundant and provide more definitive evidence of the presence of gaps or holes from which dust is excluded (i.e., *Moro-Martín et al.,* 2010). For example, in addition to cold dust components, both Vega and Fomalhaut exhibit warm (170 K) dust, identified by Spitzer (*Su et al.,* 2013; *Stapelfeldt et al.,* 2004), and hot dust, revealed by 2-μm excess (see section 2.3) (*Absil et al.,* 2006; *di Folco et al.,* 2004). *Su et al.* (2013) show that the three components are spatially separated with orbital ratios of ~10. Even richer, and more reminiscent of the solar system, is the architecture of the HR 8799 system, with warm and cold disk components (*Su et al.,* 2009; *Reidemeister et al.,* 2009; *Matthews et al.,* 2014) and four planets between them (*Marois et al.,* 2008, 2010).

5.3. Halos

Halos refer to disk components detected at very large distances from a star, typically far beyond the expected location of a planetesimal belt. The origin of halos is readily explained by the collisional processes that create grains close to the blowout size. These grains are pushed outward by radiation pressure to extended elliptic or even hyperbolic orbits, to distances far outside the birth ring (see section 3). The first image of a debris disk around β Pic was due to its extended halo seen in scattered light (*Smith and Terrile,* 1984), which typically extends further in radius than thermal emission. Halos have now also been detected in emission, however [e.g., Vega (*Sibthorpe et al.,* 2010)]. The resolved disk thermal emission toward HR 8799 was identified with Spitzer to extend to 1000 AU (*Su et al.,* 2009); Herschel data resolve a very extended halo with an outer bound of 2000 AU (*Matthews et al.,* 2014), more extensive than the scattered light halo around β Pic. Interestingly, the halo is not distinct from the planetesimal belt in temperature, and HR 8799's SED is well fit by two temperature components, one of warm dust close to the star (*Su et al.,* 2009), and the other a cold outer component. Radial profiles reveal the distinction in the distribution around 300 AU between the shallower profile of the planetesimal belt and the steep profile of the halo (*Matthews et al., 2014*).

Not all disks have extended halos. The absence of a detectable halo can equally be attributed to a low stirring level of dust-producing planetesimals (e.g., *Löhne et al., 2012*), to grains that are mechnically "harder" than assumed, to the dearth of high-β grains caused by their peculiar composition, or even to strong stellar winds that might enhance the inward drift of grains by the P-R effect in some systems (e.g., *Augereau and Beust, 2006; Reidemeister et al., 2011*).

5.4. Disk Orientation

One of the fundamental unknowns is the inclination of the disk system relative to us, which can be estimated straightforwardly from a well-resolved image with an underlying assumption of a circular disk. There is growing evidence that most systems in which inclinations have been measured for disks and stars independently show no evidence of misalignment (*Watson et al., 2011; Greaves et al., 2014*). There are also examples of systems in which the star, planets, and disk are all aligned (*Kennedy et al., 2013*). Thus for systems with radial-velocity planets, resolving a disk in the system provides an inclination that (if shared) can better constrain the masses of the planets, which are lower limits when it is unknown. However, misaligned disks are occasionally detected. For example, *Kennedy et al.* (2012a) present a steady-state circumbinary polar-ring model for the disk around 99 Herculis based on Herschel observations, in contrast to the coplanar systems around two other DEBRIS-resolved disks in binary systems (*Kennedy et al., 2012b*).

5.5. Asymmetric Structures

Resolved imaging can highlight areas of dust concentration and avoidance. These locations frequently exhibit strong asymmetries in emission, such as eccentric offsets, inclined warps, and dense clumps. It is worth noting that the presence of planets in both the β Pic (*Heap et al., 2000*) and Fomalhaut (*Kalas et al., 2005; Quillen, 2006*) systems was predicted based on asymmetries in their disk structures, and companions were subsequently found in both disks (*Lagrange et al., 2010; Kalas et al., 2008*).

The best known case for an eccentric ring is Fomalhaut (see Fig. 6), but other systems have now been observed to show such offsets, i.e., HR 4796A (*Thalmann et al., 2011*), HD 202628 (*Krist et al., 2012*) and ζ² Ret (*Eiroa et al., 2010*). The classic example of a warped disk is β Pic (*Heap et al., 2000*), which was later revealed to have two distinct disk components, one inclined to the main disk (*Golimowski et al., 2006*) but aligned with the orbit of the detected planet (*Lagrange et al., 2012*).

In the nearby disk around ε Eridani, significant clumpy structure has been observed at multiple wavelengths, most strikingly in the submillimeter (*Greaves et al., 1998*). Due to the close proximity of this star (3.3 pc), its high proper motion allows the confirmation that some of the clumps are co-moving with the star (*Greaves et al., 2005*). It is also possible to search for evidence of orbital rotation of these features, which is currently detected at 2–3σ significance both for ε Eridani (*Poulton et al., 2006*) and for a clump in the β Pic disk (*Li et al., 2012*).

Generally, the most easily observed asymmetry is a difference in brightness of one side of the disk over another. As discussed in section 6, particles on eccentric orbits have an asymmetric distribution around the star and glow near the pericenter due to their higher temperature there. Very little eccentricity need be imposed on the dust to create this effect [e.g., β Leo (*Churcher et al., 2011*)]. Giant planets in young disks may even induce spiral structures, e.g., HD 141569 (*Clampin et al., 2003*).

Not all asymmetries need arise from planetary influence (see section 6.4). Several resolved disks have been inferred to show structure resulting from interaction with the ISM, i.e., HD 15115 (*Kalas et al., 2007; Rodigas et al., 2012*), HD 32297 (*Debes et al., 2009*), and HD 61005 (*Maness et al., 2009*). Such ISM interactions do not preclude the presence of an underlying debris disk, however; *Buenzli et al.* (2010) discovered an off-center ring in the HD 61005 system with high-resolution imaging with the VLT, which could point to an underlying planetary companion.

6. PLANETARY AND STELLAR PERTURBATIONS

Most particles in a debris disk simply orbit the star on Keplerian orbits until they collide with another particle, at which point there is some redistribution of mass into particles that then follow new Keplerian orbits (see section 3), usually only slightly modified from the original orbit unless the particles are small enough for radiation pressure to be significant. The dominant perturbation to this scenario in the solar system comes from the gravitational perturbations of the planets. If a star has any planets in orbit around it, it is inevitable that its debris disk will be affected by such perturbations, since the debris orbits in a gravitational potential that is modified by the planets. Such perturbations can be split into three different components, both mathematically and physically (*Murray and Dermott, 1999*), and each of these can be linked to specific types of structure that would be imposed on any gravitationally perturbed disk.

6.1. Secular Perturbations

Secular perturbations are the long-term effect of a planet's gravity, and are equivalent to the perturbations that would arise from spreading the mass of the planet along a wire that follows its orbit. For moderately circular and co-planar orbits, the effect of a planet's eccentricity and inclination are decoupled, but both play a similar role. Planetary eccentricities impose an eccentricity and pericenter orientation on all disk material, while planetary inclinations affect the orbital plane of that material. In the case of a single planet system, the disk tends to align with the planet. The alignment takes place on long timescales that also depend on distance from the planet. This means that the effect of a planet can cause disk evolution over tens of millions of years before it reaches steady state.

If a planet is introduced into a system on an orbital plane that is misaligned with the disk midplane, a warp will propagate through the disk (see Fig. 7), with more distant material yet to notice the planet and so retaining the original plane, and closer material already having been aligned with the planet. A warp at 80 AU was used to infer the presence of an inner planet in the 12-Ma β Pic disk at ~9 AU and 9 M_{Jup} (*Mouillet et al.,* 1997; *Augereau et al.,* 2001) that was later identified in direct imaging (*Lagrange et al.,* 2010). Such a warp can also be a steady-state feature in a system of multiple (misaligned) planets, although such a scenario also needs to acknowledge that the planets secularly perturb each other, providing additional constraints (e.g., *Dawson et al.,* 2012).

A planet that is introduced on an eccentric orbit causes a tightly wound spiral to propagate through a disk of planetesimals that were initially on circular orbits (see Fig. 7) (*Wyatt,* 2005a). The spiral is caused by differential precession between neighboring planetesimal orbits, which eventually also causes these orbits to cross, potentially resulting in catastrophic collisions that can ignite a collisional cascade (see section 4.2) (*Mustill and Wyatt,* 2009). In other words, secular perturbations allow a planet's gravitational reach to extend far beyond its own orbit.

On long timescales an eccentric planet will cause a disk to become eccentric. This may be observed as an offset center of symmetry, or as the consequent brightness asymmetry caused by one side being closer to the star (*Wyatt et al.,* 1999). Although such eccentricities were initially discussed from low significance brightness asymmetries in HR 4796 (*Telesco et al.,* 2000), the predicted offset in this system has now been confirmed (*Thalmann et al.,* 2011), and the offset is quite striking in some disks (*Kalas et al.,* 2005; *Krist et al.,* 2012), and becoming more ubiquitous as disks can be imaged at higher resolution (see section 5.5).

Secular perturbations from binary companions would play a similar role to that of a planet, although in this case the companion may be expected to have a relatively high eccentricity or inclination, resulting in the evolution of the eccentricities and inclinations of disk material being coupled. At high enough mutual inclinations, the character of the evolution changes dramatically as disk material undergoes Kozai oscillations, in which high inclinations can be converted into high eccentricities (*Kozai,* 1962). It is notable that such large oscillations do not necessarily imply long-term instability, although this might not be compatible with a disk that is narrow both radially and vertically. However, there are solutions for circumbinary orbits for which narrow disks are possible. For example, there is a stationary (i.e., non-evolving) orbit that is orthogonal to the binary orbital plane, leading to the possibility that stable circumpolar rings may exist (*Kennedy et al.,* 2012b) (see section 5.4).

The above discussion on structures from secular per-

turbations only considered the distribution of planetesimal orbits. In general, the orbits of dust grains trace that of the planetesimals. However, the effect of radiation pressure does need to be taken into consideration, particularly where it is evident that observed structure originates from a halo component [see section 5.3 and Nesvold *et al.* (2013)].

6.2. Resonant Perturbations

At locations where debris orbits the star, there arise an integer (p) number of times for every integer (p + q) number of planet orbits. As this definition suggests, there are an infinite number of resonances, but the strongest are usually the first-order resonances (i.e., q = 1). As the example of the Plutinos in the EKB and the Kirkwood gaps in the asteroid belt tell us, specific resonances can be either over- or underpopulated. One location where resonances are always underpopulated is close to a planet where, because resonances have finite width, the first-order resonances overlap (*Wisdom,* 1980). This overlap region is chaotic so debris does not remain there for long. The shape of this chaotic region has been the subject of much discussion, since the slope of the edge of the debris, and its offset

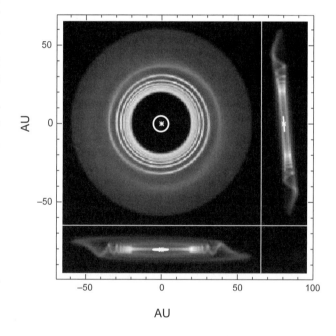

Fig. 7. Structure of an initially coplanar low eccentricity 20–60-AU debris disk 100 m.y. after the introduction of a 1-M_{Jup} planet on an orbit (shown in white) at 5 AU with eccentricity e = 0.1 and inclined 5° to the disk mid-plane. The planetesimals exhibit a tightly wound spiral structure far from the planet (*Wyatt,* 2005), which is wound so tightly at the inner edge that the planetesimals are on crossing orbits, ensuring destructive collisions (*Mustill and Wyatt,* 2009). Coincident with the spiral is a warp (*Augereau et al.,* 2001) with an appearance that depends on the viewer's orientation to the line-of-nodes (see bottom and right panels for two different edge-on views). Figure courtesy of M. Wyatt.

from the planet, can be used to determine the mass of the planet — e.g., sharper edges require lower-mass planets (*Quillen*, 2006; *Chiang et al.*, 2009), although there are degeneracies with the eccentricities of the planetesimals' orbits (*Mustill and Wyatt*, 2012).

It is generally suspected that planets shape the inner edge of debris disks, but it has yet to be demonstrated, and a debris disk that was born with a sharp inner edge would likely maintain it for several billions of years. Similarly, the fact that the debris in the solar system traces the only regions of the solar system that are stable over its 4.5-Ga age (*Lecar et al.*, 2001) is suggestive that extrasolar debris also traces regions of dynamical stability, and therefore there are planets hiding in the gaps in the debris (*Faber and Quillen*, 2007), causing instabilities due to both overlapping mean-motion resonances and secular resonances (*Moro-Martín et al.*, 2010). This is not necessarily the case, however, since debris may be absent from regions in between planets for reasons associated with the formation of the system.

A more specific observable caused by resonances is a clumpy structure. This is because of the special geometry of resonant orbits, which have closed patterns in the frame rotating with the planet, with each resonance having its own geometry (i.e., its own clumpy pattern). Resonances occupy relatively narrow regions of parameter space, and to form a detectable clumpy structure these would have to be filled by resonance sweeping. For example, if a planet migrates through a disk, its resonances will sweep through the disk; some planetesimals can be trapped into resonance and migrate out with the planet (*Malhotra*, 1993; *Wyatt*, 2003). Because trapping probabilities into different resonances depend on the planet's mass and migration rate (*Wyatt*, 2003), as well as the particle eccentricity (*Reche et al.*, 2008; *Mustill and Wyatt*, 2011), these parameters can be constrained from observations of a clumpy structure; the planet's location can also be pinpointed, since it will lie at a radial distance just inside the clumps, at an azimuthal angle that is evident from the clumpy structure, but generally lies far from clumps. An understanding of dust physics is also important for this comparison, since radiation pressure eventually causes the dust to fall out of resonance, so that observations at short enough wavelengths to probe small bound grains have an axisymmetric structure (*Wyatt*, 2006; *Krivov et al.*, 2007), while those that probe small unbound grains have spiral structure emanating from the clumps (*Wyatt*, 2006). This can explain why disk structure is so wavelength dependent, and underscores the need for multiwavelength imaging to constrain the models.

Another mechanism for resonance sweeping is dust migration through a disk by P-R drag; the dust crosses the resonances and becomes trapped, thus halting the migration (*Weidenschilling and Jackson*, 1993). Various papers have studied the structures expected in this case (*Ozernoy et al.*, 2000; *Quillen and Thorndike*, 2002; *Kuchner and Holman*, 2003; *Deller and Maddison*, 2005). Early papers ignored the importance of collisions, however, which tend to destroy

particles in mutual collisions before they migrate far from their source in disks that are dense enough to detect (*Wyatt*, 2005b). Fully consistent N-body simulations of this process that incorporate collisions are now being performed (*Stark and Kuchner*, 2009; *Kuchner and Stark*, 2010), confirming that this is an important factor, and concluding that clumpy structures formed by this mechanism kick in at optical depths below around 10^{-5} (see also *Krivov et al.*, 2007).

There is controversy over some claims of clumps (*Hughes et al.*, 2012), so it is important that the models make testable predictions. One of these is that the clumps should orbit the star with the planet.

6.3. Scattering

Scattering processes are those that are best visualized as the hyperbolic encounter of a planetesimal with a planet. This results in a large impulsive change in the planetesimal's orbit, and a minor change in that of the planet. In the solar system, scattering is the dominant process governing the evolution of comets that are passed in from the EKB by interactions with the giant planets (*Levison and Duncan*, 1997), which may also be true in extrasolar systems (see section 7 and the chapter by Davies et al. in this volume). Another solar system population dominated by scattering is the scattered disk of the EKB, which extends out toward the Oort cloud through interactions with Neptune (*Morbidelli et al.*, 2004). Planet formation may result in all systems having a scattered disk at some level, and indeed there is evidence in some planet formation models for mini-Oort clouds (*Raymond and Armitage*, 2013). The collisional evolution of such high eccentricity populations has its own challenges, as discussed in *Wyatt et al.* (2010).

Scattering is also inevitable if a planet is placed in the middle of a debris disk, and can be invoked to place constraints on the presence of embedded planets. For example, a disk would be naturally broadened by such scattering, so a narrow disk requires the most massive object to be between the masses of Pluto and Earth. The recent discovery that Fomalhaut b crosses, and may even go through, its debris disk (*Kalas et al.*, 2013) emphasizes that scattering processes may be a significant evolutionary mechanism.

If the debris disk is sufficiently massive, then multiple scatterings can significantly alter the orbit of an embedded planet. Typically the planet's eccentricity would be damped by dynamical friction and its semimajor axis would evolve, causing the planet to migrate (*Fernandez and Ip*, 1984; *Levison et al.*, 2007). The direction of the migration depends on various factors (*Kirsh et al.*, 2009; *Capobianco et al.*, 2011), and it is possible for a planet to migrate all the way to the edge of a sufficiently massive debris disk (*Gomes et al.*, 2004). The resulting migration could be invoked to cause resonance trapping (section 6.2), although it is challenging to model this process due to the large number of planetesimals required to make the migration smooth enough for resonance trapping to occur.

The planet need not migrate far for this process to have a significant effect on the debris disk. For example, if there are multiple planets in the system, then this migration can push them toward an unstable configuration (*Moore and Quillen*, 2013). In the Nice model for the solar system, the ensuing instability resulted in the depletion of most of the mass of the EKB (*Gomes et al.*, 2005). With the right configuration this instability could have been delayed for several hundreds of millions of years, explaining the late heavy bombardment (LHB). If such long delays are common in other systems there could be a signature of this in statistics, although this appears not to be the case because the primary consequence would be a sharp drop in the number of disks detected (*Booth et al.*, 2009).

Scattering can also have a consequence on the distribution of dust, even if it no longer has an effect on the planetesimal belt; e.g., *Thébault et al.* (2010) considered how the action of a planet exterior to a planetesimal belt would truncate the halo of small dust created in collisions in the belt and put onto high-eccentricity orbits by radiation pressure (see also *Lagrange et al.*, 2012).

6.4. Nongravitational Perturbations

It is worth pointing out that although the above discussion describes ways in which planets will inevitably impose structure on a debris disk, the existence of such features in a disk is not necessarily a signpost for planets, as there may be a nonplanetary explanation. For example, clumps in debris disks can also arise from recent collisions (*Wyatt and Dent*, 2002), although these are usually expected to be hard to detect except in the inner few astronomical units (*Kenyon and Bromley*, 2005). Confirmation of more systems in which structures can be linked to planets that are known from other means, such as the case of β Pic b, is needed for greater confidence in a planetary interpretation.

Nongravitational perturbations can often be ruled out, but there are some that have already been shown to dominate the structure of some disks, most notably interaction with the ISM (*Artymowicz and Clampin*, 1997). Bow shocklike structure can be evidence of interstellar dust streaming past the star that is deflected by radiation pressure (*Gáspár et al.*, 2008). Forcing from the ISM can also have a secular effect on particle orbits (*Maness et al.*, 2009; *Debes et al.*, 2009) and is most important for low-density disks (*Marzari and Thébault*, 2011). Of key importance to such analyses is the direction and speed of motion through the ISM, which is known for most nearby stars.

7. ORIGIN OF HOT DUST

The origin of hot dust around nearby stars remains a mystery, partly because its rarity and proximity to the star make it difficult to study. Regardless, its very existence causes some theoretical conundrums, and several models have been proposed to circumvent these.

7.1. Asteroid Belt

While it is tempting to interpret hot exozodiacal dust found at a few astronomical units (section 2.2) as asteroid belt analogs, such belts would rapidly deplete due to mutual collisions and so become undetectable within a few tens of millions of years (see equation (3)), and so it is not possible to explain hot dust around stars that are older than a few hundred million years in this way (*Wyatt et al.*, 2007c; *Heng and Tremaine*, 2010). This is, however, a perfectly reasonable explanation for the hot dust found around young stars. The age up to which asteroid belt emission can be detected is a strong function of the radial location of the belt, but otherwise the main constraint is that sufficient mass remains in the belt at the end of planet formation. The ubiquity or absence of such belts could thus indicate whether planetary systems are commonly fully packed (leaving no regions of long-term dynamical stability in which an asteroid belt could reside) or are otherwise depleted by processes such as planet migration or secular resonance sweeping.

7.2. Terrestrial Planet Formation

Terrestrial planet formation (see the chapter by Raymond et al. in this volume) is another model that can be invoked to explain the existence of hot dust around young stars, since the process of building up terrestrial planets by the coalescence of planetary embryos is thought to take 10–100 million years. Many models of this process include a significant planetesimal population that co-exists with the embryos throughout this time, with a consequently high level of mid-IR emission from dust produced in collisions among this planetesimal population (e.g., *Kenyon and Bromley*, 2004). Even if such planetesimals are not retained from the protoplanetary disk phase, collisions between the embryos are thought to have occurred as late as 50 Ma in the solar system, the date of the Moon-forming collision. The debris produced in that collision would not be dynamically removed for tens of millions of years, and so unless the escaping debris is all placed in small objects that can then collisionally deplete very rapidly, this debris would be detectable for ~15 million years (*Jackson and Wyatt*, 2012). This is an exciting possibility, since it means that we can witness the aftermath of such massive collisions, and this has been proposed as the explanation of some hot excesses on compositional grounds (*Lisse et al.*, 2008, 2009). However, this can also cause problems for planet formation models, because hot dust is found around just ~1% of young Sun-like stars (*Kennedy and Wyatt*, 2013). This could mean that late giant collisions are rare, and that terrestrial planet formation is largely complete by the time the protoplanetary disk disperses (although see *Rhee et al.*, 2008; *Melis et al.*, 2010).

7.3. Cometary Populations

The fraction of dust in the solar system's zodiacal cloud that comes from asteroids and from comets is a matter for debate, but it seems that both contribute (*Nesvorný et al.,* 2010). The main obstacle to resolving that question is how to model the production of cometary dust. In the context of extrasolar systems, a bigger obstacle is the lack of knowledge about the planetary system, since that determines the rate at which cometary material is scattered into the inner system where it may be seen as a hot dust excess. However, this also introduces the exciting possibility that hot cometary dust, and how it relates to an outer Kuiper belt, can be used as a probe of the intervening planetary dynamics. Some progress has been made on understanding these dynamics recently, by using analytical considerations to assess how far in planetary systems can scatter comets, which depends on their spacing (*Bonsor and Wyatt,* 2012), and N-body simulations to show that tightly packed systems of low-mass planets would create high levels of hot dust (*Bonsor et al.,* 2012). Nevertheless, further work is still needed on this process, including how to model the level of dust production by comets, which arises not just from sublimation, but also from disintegration, and (if the comet population is sufficiently massive) collisions.

7.4. Extreme Eccentricity Populations

A variant of the cometary explanation is the possibility that the comets are not continually being replenished from an outer belt, rather that they were implanted on highly eccentric orbits at the end of planet formation and have remained there ever since. *Wyatt et al.* (2010) showed that a population that has pericenters at ~1 AU, and apocenters at hundreds of astronomical units, could survive for billions of years of collisional evolution, with properties similar to those observed. The viability of this explanation thus rests on the likelihood that planet formation processes result in such a highly eccentric population. Such extreme eccentricities were not found in models that invoked planet migration (*Payne et al.,* 2009), but planet-planet scattering may be a plausible formation mechanism (*Raymond and Armitage,* 2013).

7.5. Dynamical Instability

The remaining explanations for hot dust around old stars envisage this as a transient population. One explanation is that there was a large influx of comets in a single event, possibly related to a dynamical instability in the planetary system, similar to the LHB experienced by the solar system ~600 Ma after formation. Modeling of both the solar system (*Booth et al.,* 2009) and instabilities in a larger set of planet formation models (*Bonsor et al.,* 2013b) find that the resulting hot dust enhancement is relatively short-lived. Thus even if late-stage instabilities were common [which we know is not the case (*Booth et al.,* 2009)], there would be very few nearby stars currently undergoing this phenomenon. However, that does not exclude rare systems such as η Corvi, which has hot dust within a massive Kuiper belt (*Wyatt et al.,* 2005), from being in such a state.

7.6. Recent Collisions

While collisions would have depleted an asteroid belt to below detectable levels, the proximity of the asteroid belt to the star means that it only takes the breakup of one relatively small asteroid (e.g., 10–100 km) to create an observable amount of dust [if it is broken into small enough fragments (e.g., *Kenyon and Bromley,* 2005)]. Indeed, the evolution of the dust content of the asteroid belt is known to have been punctuated by many such events, visible today in the large-body population as the Hirayama asteroid families (*Nesvorný et al.,* 2003), and in the dust population as the dust bands (*Low et al.,* 1984). Thus it is tempting to explain hot dust in this way (*Weinberger et al.,* 2011), especially in systems like BD+20 307 and HD 69830 toward which we see no cold reservoir of planetesimals, as would be required by most other scenarios [although such a reservoir can have evaded detection (*Wyatt et al.,* 2007c)].

7.7. Temporal Evolution

One way to distinguish between the models will be to consider how the disks evolve with time (e.g., *Meng et al.,* 2012). For example, as asteroid belts deplete, the frequency of the massive collisions considered in section 7.6 also goes down. This means that for every 1–3-Ga collision there would be 10× more in the 0.3–1-Ga age range. The age dependence of this phenomenon is not well characterized due to small number statistics, but the existence of several >1-Ga hot dust systems argues against such a dramatic decline. However, this age dependence does not hold if the parent bodies are large enough, and few enough, to be stranded from collisional equilibrium constraints (e.g., *Kennedy and Wyatt,* 2011). Thus an origin in collisions between planetary embryos, rather than between the largest members of an asteroid belt, remains a plausible explanation, if planet formation models can be shown to retain such embryos for billions of years.

The discovery of a hot disk undergoing rapid decay is a significant advance that remains without adequate theoretical explanation (*Melis et al.,* 2012). If such rapid evolution is the norm, this would require revision to the models that attempt to explain hot dust. *Kennedy and Wyatt* (2013) also pointed out that another way to characterize the way hot disks evolve is to look at the frequency of fainter disks; those that are detectable with current technology are the outliers, but if they are transient they must evolve through lower levels of excess that can be detectable.

7.8. Extremely Hot Dust

The majority of the discussion above focused on dust that is at ~1 AU. However, a more ubiquitous phenomenon seems to be the excess inferred to be an order of magnitude closer to the star (see section 2.3) (*Absil et al.,* 2013). The explanations for this excess would be broadly similar to those outlined above, with typically more stringent constraints. However, in this case it might be possible to invoke processes related to the star itself, rather than a debris disk (e.g., *Cranmer et al., 2013*).

8. GAS IN YOUNG DEBRIS DISKS

8.1. Observations

The gas-to-dust ratio of most debris disks is not well constrained. While the disks are identified based on their dust emission, there are relatively few detections of accompanying gas. Indeed, debris disks are sometimes defined as being gas-poor. The best example of a gas-bearing debris disk is the young β Pic system [~12 Ma (*Zuckerman et al., 2001*)], where small amounts of gas have been identified via ultraviolet (UV)/optical absorption lines for a range of species [e.g., CaII, NaI, CII, CIII, and OI (*Hobbs et al., 1985; Roberge et al., 2006*)]. The velocity structure of the disk has recently been mapped in CO with ALMA (*Dent et al., 2014*); however, absorption by molecular hydrogen is not detected (*Lecavelier des Etangs et al., 2001*). While observations of absorption lines can provide a very sensitive tracer of orbiting gas [indeed, β Pic's gas was detected before its IR excess (*Slettebak, 1975*)], they require alignment of the disk with the line of sight; emission line measurements are more effective at ruling out gas for systems not viewed edge-on. These longer wavelength emission measurements, however, are limited to specific temperatures/regions of the disk. Mid-IR spectra, for example, are generally sensitive to warm (hundreds of Kelvins) gas, while millimeter observations probe colder material. For this reason, *Pascucci et al.* (2006) used a comprehensive approach — combining Spitzer spectra (H_2 at 17 μm, [FeII] at 26 μm, and [SI] at 25.23 μm) with millimeter observations of ^{12}CO transitions — to place gas limits on a sample of young stars. For their ~10–100-Ma disks, they rule out gas masses more than a few Earth masses for the outer disk (10–40 AU) and constrain the surface density at 1 AU to be <1 g cm^{-2}, i.e., <10^{-4} of the minimum mass solar nebula (MMSN).

While most debris disk observations are consistent with the early removal of most gas (at ages ≤10 Ma), 49 Ceti appears to be an exceptional case. One of only two debris disks with detected CO emission [the other is HD 21997 (*Moór et al., 2011a*)], 49 Ceti is the oldest known disk with significant amounts of gas [~40 Ma (*Zuckerman and Song, 2012*)]. Its resolved CO emission is modeled as a gas disk extending out to 200 AU, with an empty region within 40 AU (*Hughes et al., 2008a*). The presence of a significant amount of gas at such an age has implications for the formation of giant planets. If the gas is primordial, i.e., there is a large amount of hydrogen alongside the CO, then the total gas mass is estimated at ~10 $M_\oplus$ (*Hughes et al., 2008a*). While this is not enough to form a Jupiter-mass planet, it could easy supply the envelope for Neptune/Uranus analogs (~1 $M_\oplus$ of hydrogen). On the other hand, Herschel recently made the first detection of ionized carbon emission from the disk, at 158 μm (*Roberge et al., 2013*).

The [OI] 63-μm line, however, was unexpectedly not detected, suggesting an enhanced C/O ratio. Thermal modeling of the overall dataset (lines plus the resolved dust emission) suggests that the source of the gas may be cometary rather than primordial (*Roberge et al., 2013*).

8.2. Origin of Gas

It cannot be completely ruled out that what is observed is unusually long-lived remnants of the protoplanetary disks. Indeed, the fact that the gas and dust in HD 21997 are not co-spatial argues for a primordial origin for some of its gas (*Kóspál et al., 2013*). Also, about 10 $M_\oplus$ of gas, if not more, could still remain in many young debris disks where gas was searched for and not found, without violating observations (*Hillenbrand, 2008*). However, there are more arguments in favor of secondary, rather than primordial, origin of the observed gas (*Fernández et al., 2006*). For instance, CO observed around β Pic (*Vidal-Madjar et al., 1994; Jolly et al., 1998; Roberge et al., 2000*), 49 Cet (*Hughes et al., 2008a*), and HD 21997 (*Moór et al., 2011a*) should be photodissociated on timescales of hundreds of years in the absence of shielding (*van Dishoeck and Black, 1988; Roberge et al., 2000; Moór et al., 2011a*), and thus likely needs continuous replenishment.

Evaporation of short-period and especially infalling comets, as inferred from observed variable absorption lines of β Pic (e.g., *Ferlet et al., 1987; Beust and Valiron, 2007*), can certainly contribute to the observed gas, but — as long as the process only works close to the star — is unable to explain why gas is observed so far from the star, as inferred from the double-peaked line profiles of β Pic (*Olofsson et al., 2001*) and HD 21997 (*Moór et al., 2011a*).

To account for similarity between the gas and dust distributions, *Czechowski and Mann* (2007) proposed vaporization of solids in dust-dust collisions. However, the typical threshold speed for vaporization by pressurizing shocks is a few kilometers per second for ice and in excess of 10 km s^{-1} for silicate and carbon (*Tielens et al., 1994*). Thus this mechanism can only work very close to the stars where the Keplerian speeds are high — or in exceptionally dusty disks, such as β Pic's, where abundant high-speed β-meteoroids (grains put on hyperbolic trajectories by radiation pressure; see section 3) are predicted to sweep through the disk and vaporize bound grains.

One more possibility is photon-stimulated desorption of dust, which can release various species such as sodium, observed around β Pic (*Chen et al., 2007*). This process requires strong UV emission from the star, which would be consistent with the fact that gas has been detected predominantly in debris disks of A-type stars.

Collisions of volatile-rich dust grains can also release gas through sublimation, rather than collisional vaporization. In contrast to protoplanetary disks, the gas pressure in debris disks is very low. As a result, sublimation occurs at lower temperatures. For instance, ice sublimates at ~100–110 K instead of 145–170 K as in protoplanetary disks (*Lecar et* *al.*, 2006). The sublimation distance for "dirty" (organics-coated) ice bodies ranges from 20 to 35 AU for $L_* < 30 L_\odot$ and 65 $(L_*/100 L_\odot)^{1/2}$ AU for $L_* \geq 30 L_\odot$ (*Kobayashi et al.*, 2008). The sublimation distances are similar for ammonia ice and is even larger for other ices such as CO (*Dodson-Robinson et al.*, 2009). CO ice, in particular, sublimates at temperatures as low as 18–35 K and thus would vaporize in EKB-sized debris disks even around solar-type stars following collisions between the dust grains. However, the gas production rate needed to account for the amount of CO observed around 49 Cet (*Hughes et al.*, 2008a) and HD 21997 (*Moór et al.*, 2011a) is by at least an order of magnitude larger than the estimated dust production rate. To mitigate the problem of a relatively rapid rate of CO production (compared to dust production rate), *Zuckerman and Song* (2012) suggested that most of the CO is produced in collisions of larger, cometary bodies that sequester CO in their deeper, subsurface layers. Indeed, the mass loss rate is the same in all logarithmic bins of a steady-state collisional cascade (see section 3.2) (*Wyatt et al.*, 2011). Assuming that the gas production rate is a fixed fraction of the mass loss rate, the gas production rate in collisions of all bodies from, say, 1 μm to 10 km in size can easily be an order of magnitude larger than in collisions of dust grains in the 1–10-μm range.

All the processes outlined above should occur in every disk at some level. The only question is how efficient they are and how that efficiency depends on the system's parameters (e.g., system age, stellar luminosity, disk radius, and disk mass). For example, the mechanisms differ in whether they result from interaction of a single solid object (a dust grain or a comet) with the radiation field or require pairwise collisions between such solids. These two types of processes would imply that the gas production efficiency scales with the amount of solids linearly and quadratically, respectively. Unfortunately, with only a few "gaseous" debris disks discovered so far, the statistics are too scarce to identify the dominant process based on this argument. Nor is it clear whether the same or different mechanisms are responsible for the gas production in these systems. Finally, it is debated whether gas production at observable levels occurs in a steady-state regime or if it is only observed in those systems where a major "stochastic" event recently occurred, such as an individual collision between two gas-rich comets or even planetary embryos (*Moór et al.*, 2011a; *Zuckerman and Song*, 2012).

8.3. Dynamical Effect on Dust

Assuming gas is not influenced by dust, dust "aerodynamics" was laid down by *Weidenschilling* (1977), mostly for protoplanetary disks. This treatment was generalized to optically thin disks, where dust is subject to radiation pressure, by *Takeuchi and Artymowicz* (2001). They pointed out that while the motion of gas is sub-Keplerian (usually due to the pressure gradient), that of dust grains is sub-Keplerian as well because of the radiation pressure.

Sufficiently small grains should be slower than the gas. As a result, while larger grains drift inward, grains smaller than several times the blowout size should spread outward. Using these results, *Thébault and Augereau* (2005) described the radial profile of dust distribution in a debris disk with gas, attempting to put an upper limit on the gas density from the observed scattered-light brightness profiles in resolved disks. *Krivov et al.* (2009) revisited the issue by considering a wide range of gas densities, temperatures, and compositions. They showed that, although individual trajectories of grains are sensitive to these parameters, the overall dust distribution remains nearly the same. They concluded that the observed brightness profiles of debris disks do not pose stringent constraints on the gas component in the systems.

Once the gas-to-dust ratio becomes comparable to unity, it is vital to include the back reaction of gas from dust. Gas drag is known to concentrate dust at pressure maxima (*Takeuchi and Artymowicz*, 2001). If the disk is optically thin to the stellar light, and the star emits enough in UV, gas is mainly heated by dust through photoelectric heating. In this case, the larger the dust concentration, the hotter the gas, and the higher the pressure, which causes the dust to concentrate more, creating an instability (*Klahr and Lin*, 2005; *Besla and Wu*, 2007; *Lyra and Kuchner*, 2013). This has been suggested as a "planetless" explanation for rings and spirals seen in systems like HR 4796A, HD 141569, and β Pic.

The key problem with all the models of gas-dust interaction is that, for a standard primordial composition, most of the gas mass should be contained in hydrogen. Its amount is hard to measure directly, and it is difficult to deduce it from the observed species, since H_2/CO and other ratios in debris disks are not reliably known.

9. SUMMARY/OUTLOOK

The past decade has brought a wealth of new data and modeling advancement for the study of debris disks as components of extrasolar planetary systems. Much of this has been achieved with data from the spacebased IR surveys of Spitzer, Herschel, and WISE, windows that are now closed for the foreseeable future. While *Meyer et al.* (2007) tabulated 14 resolved debris disks around other stars, disks have now been resolved around nearly 100 nearby stars, many at multiple wavelengths. These resolved images effectively break the degeneracies inherent in modeling the disks for dust location and grain sizes. Many resolved cold disks are being found to have inner, warm disk components as well. The HR 8799 system, with warm and cold debris components with four gas giant planets between them, is very similar to the solar system, but exhibits an extended halo seen around several, predominantly A-type, stars.

Now coming to the forefront, however, are ground-based facilities that promise even better sensitivity and the resolution of disk structures that are key to any indepth study of individual disks and the detection of signatures associated with planets. Advances made possible by interferometers over a large range of wavelengths (i.e., LBTI,

VLTI, CHARA, along with ALMA and the Jansky Very Large Array) will be particularly illuminating. At short wavelengths, interferometers probe the presence of hot dust, and its origins are clearly a key question to be addressed in the coming decade. At long wavelengths, ALMA will provide unprecedented resolution of cold disk components, including potentially the position of the central star through detection of warm inner components in the same observation (*Boley et al.,* 2012; *MacGregor et al.,* 2013). High-resolution imaging of disks will also be possible through facilities such as the James Webb Space Telescope, Cerro Chajnantor Atacama Telescope, and Gemini Planet Imager.

There are several key questions that are likely to be the focus of study in the coming decade. One of the most challenging will be, What is the main mechanism that stirs debris disks: planets, embedded big planetesimals, or both? And what range of stirring levels is present in disks? The challenge for theoretical models will be to balance detailed studies of the dynamical and collisional evolution of specific systems, with the need to explain observations of the broad ensemble of disks in such a way that meaningful constraints can be extracted on the underlying physics. For example, disks are seen to become fainter, and potentially larger, with age, as expected for material undergoing collisional evolution in a steady state. However, as the disks are being characterized in ever-increasing detail, it is clear that we do not yet have an explanation for the diverse properties of their halos, and an explanation for the range of disk widths and the relation between multiple components (and the implications of such components on our interpretation of the statistics) are still to be determined.

Cold and hot disk components appear to be detected toward 20–30% of stars, with rates higher toward A stars, while warm dust components are detected ~10× less frequently. Measured incidence rates are suggestively lower around solar-type stars, but not statistically so, and current surveys have failed to detect significant numbers of disks around M stars, most exceptions being very young, although they lack the sensitivity to fractional luminosities around M stars comparable to earlier spectral types. Thus, the true incidences of cold and warm debris dust remain not entirely clear, and the potential differences of disk frequencies among stars of different spectral types are still awaiting explanation.

Recent work suggests a positive correlation between debris disks and planets where the most massive planet in the system is a Saturn mass or less (*Wyatt et al.,* 2012; Moro-Martin et al., in preparation), and Herschel data show a statistically significant correlation between planets and the brightness of debris (Bryden et al., in preparation), as predicted by *Wyatt et al.* (2007c). However, whether or not the presence of dusty debris around a star implies the existence of planets remains an open question.

A related, more focused issue for modelers and observers is whether inner cavities or gaps in debris disks are indeed populated by planets. Direct imaging of planets will not be possible for all systems, but higher-resolution disk images can solidify the link between disk structures and underlying planet populations. Eccentricities of disks can place significant constraints on the mass and orbits of inner planets, as can the sharpness of the inner disk edges when resolved. In addition, the warps and offsets that were the fingerprints of planets in the β Pic and Fomalhaut systems can be discerned in other systems when higher-resolution images are available.

A traditional hallmark of debris disks has been the absence of detected gas in most systems. Herschel and now ALMA have made significant progress toward the detection of various species in the debris disks β Pic and 49 Ceti, both previously known gas hosts. Both are relatively young, which may be a prerequisite for detected gas in debris systems.

In addition to the disks detected around white dwarfs, metal-rich white dwarfs themselves may be a class related to the debris disk phenomenon. Such stars may be "polluted" by the deposition of material from planetesimals onto the stellar surface. These objects therefore provide another means of probing the composition of planetesimals in extrasolar debris disk systems.

Acknowledgments. C.E. is partly supported by the Spanish grant AYA 2011-26202. A.V.K. acknowledges support from DFG grant Kr 2164/10-1. M.C.W. is grateful for support from the EU through ERC grant number 279973.

REFERENCES

Absil O. et al. (2006) *Astron. Astrophys., 452,* 237.
Absil O. et al. (2013) *Astron. Astrophys., 555,* A104.
Acke B. et al. (2012) *Astron. Astrophys., 540,* A125.
Akeson R. L. et al. (2009) *Astrophys. J., 691,* 1896.
Andrews S. M. and Williams J. P. (2005) *Astrophys. J., 631,* 1134.
Andrews S. M. et al. (2012) *Astrophys. J., 744,* 162.
Anglada-Escudé G. and Tuomi M. (2012) *Astron. Astrophys., 548,* A58.
Artymowicz P. and Clampin M. (1997) *Astrophys. J., 490,* 863.
Augereau J. and Beust H. (2006) *Astron. Astrophys., 455,* 987.
Augereau J.-C. et al. (2001) *Astron. Astrophys., 370,* 447.
Aumann H. H. (1985) *Publ. Astron. Soc. Pac., 97,* 885.
Avenhaus H. et al. (2012) *Astron. Astrophys., 548,* A105.
Ballering N. P. et al. (2013) *Astrophys. J., 775,* 55.
Barber S. D. et al. (2012) *Astrophys. J., 760,* 26.
Beichman C. A. et al. (2005a) *Astrophys. J., 626,* 1061.
Beichman C. A. et al. (2005b) *Astrophys. J., 622,* 1160.
Beichman C. A. et al. (2006) *Astrophys. J., 652,* 1674.
Beichman C. A. et al. (2011) *Astrophys. J., 743,* 85.
Belyaev M. A. and Rafikov R. R. (2011) *Icarus, 214,* 179.
Besla G. and Wu Y. (2007) *Astrophys. J., 655,* 528.
Beust H. and Valiron P. (2007) *Astron. Astrophys., 466,* 201.
Bilíková J. et al. (2012) *Astrophys. J. Suppl., 200,* 3.
Blum J. and Wurm G. (2008) *Annu. Rev. Astron. Astrophys., 46,* 21.
Boley A. C. et al. (2012) *Astrophys. J. Lett., 750,* L21.
Bonsor A. and Wyatt M. (2010) *Mon. Not. R. Astron. Soc., 409,* 1631.
Bonsor A. and Wyatt M. C. (2012) *Mon. Not. R. Astron. Soc., 420,* 2990.
Bonsor A. et al. (2012) *Astron. Astrophys., 548,* A104.
Bonsor A. et al. (2013a) *Mon. Not. R. Astron. Soc., 431,* 3025.
Bonsor A. et al. (2013b) *Mon. Not. R. Astron. Soc., 433,* 2938.
Booth M. et al. (2009) *Mon. Not. R. Astron. Soc., 399,* 385.
Booth M. et al. (2013) *Mon. Not. R. Astron. Soc., 428,* 1263.
Brauer F. et al. (2008) *Astron. Astrophys., 480,* 859.
Briggs R. E. (1962) *Astron. J., 67,* 710.
Brinkworth C. S. et al. (2012) *Astrophys. J., 750,* 86.
Broekhoven-Fiene H. et al. (2013) *Astrophys. J., 762,* 52.
Bryden G. et al. (2009) *Astrophys. J., 705,* 1226.

Buenzli E. et al. (2010) *Astron. Astrophys., 524*, L1.
Campo Bagatin A. et al. (1994) *Planet. Space Sci., 42*, 1079.
Capobianco C. C. et al. (2011) *Icarus, 211*, 819.
Carpenter J. M. et al. (2009) *Astrophys. J. Suppl., 181*, 197.
Chen C. H. et al. (2006) *Astrophys. J. Suppl., 166*, 351.
Chen C. H. et al. (2007) *Astrophys. J., 666*, 466.
Chen C. H. et al. (2008) *Astrophys. J., 689*, 539.
Chen C. H. et al. (2011) *Astrophys. J., 738*, 122.
Chen C. H. et al. (2012) *Astrophys. J., 756*, 133.
Chiang E. and Youdin A. N. (2010) *Annu. Rev. Earth Planet. Sci., 38*, 493.
Chiang E. et al. (2009) *Astrophys. J., 693*, 734.
Churcher L. J. et al. (2011) *Mon. Not. R. Astron. Soc., 417*, 1715.
Clampin M. et al. (2003) *Astron. J., 126*, 385.
Cranmer S. R. et al. (2013) *Astrophys. J., 772*, 149.
Czechowski A. and Mann I. (2007) *Astrophys. J., 660*, 1541.
Dawson R. I. et al. (2012) *Astrophys. J., 761*, 163.
Debes J. H. et al. (2009) *Astrophys. J., 702*, 318.
Deller A. T. and Maddison S. T. (2005) *Astrophys. J., 625*, 398.
Dent W. et al. (2014) *Science, 343*, 1490.
de Vries B. L. et al. (2012) *Nature, 490*, 74.
di Folco E. et al. (2004) *Astron. Astrophys., 426*, 601.
di Folco E. et al. (2007) *Astron. Astrophys., 475*, 243.
Dodson-Robinson S. E. et al. (2009) *Icarus, 200*, 672.
Dodson-Robinson S. E. et al. (2011) *Astron. J., 141*, 11.
Dohnanyi J. S. (1969) *J. Geophys. Res., 74*, 2531.
Dominik C. and Decin G. (2003) *Astrophys. J., 598*, 626.
Donaldson J. K. et al. (2013) *Astrophys. J., 772*, 17.
Dong R. et al. (2010) *Astrophys. J., 715*, 1036.
Dufour P. et al. (2012) *Astrophys. J., 749*, 6.
Dukes D. and Krumholz M. R. (2012) *Astrophys. J., 754*, 56.
Durda D. D. and Dermott S. F. (1997) *Icarus, 130*, 140.
Eiroa C. et al. (2010) *Astron. Astrophys., 518*, L131.
Eiroa C. et al. (2011) *Astron. Astrophys., 536*, L4.
Eiroa C. et al. (2013) *Astron. Astrophys., 555*, A11.
Endl M. et al. (2008) *Astrophys. J., 673*, 1165.
Faber P. and Quillen A. C. (2007) *Mon. Not. R. Astron. Soc., 382*, 1823.
Ferlet R. et al. (1987) *Astron. Astrophys., 185*, 267.
Fernandez J. A. and Ip W.-H. (1984) *Icarus, 58*, 109.
Fernández R. et al. (2006) *Astrophys. J., 643*, 509.
Forbrich J. et al. (2008) *Astrophys. J., 687*, 1107.
Fujiwara H. et al. (2009) *Astrophys. J. Lett., 695*, L88.
Fujiwara H. et al. (2012) *Astrophys. J. Lett., 749*, L29.
Gänsicke B. T. et al. (2008) *Mon. Not. R. Astron. Soc., 391*, L103.
Gáspár A. et al. (2008) *Astrophys. J., 672*, 974.
Gáspár A. et al. (2013) *Astrophy. J., 768*, 25.
Gautier T. N. III et al. (2007) *Astrophys. J., 667*, 527.
Girven J. et al. (2011) *Mon. Not. R. Astron. Soc., 417*, 1210.
Golimowski D. A. et al. (2006) *Astron. J., 131*, 3109.
Gomes R. S. et al. (2004) *Icarus, 170*, 492.
Gomes R. et al. (2005) *Nature, 435*, 466.
Gorlova N. et al. (2006) *Astrophys. J., 649*, 1028.
Graham J. R. et al. (1990) *Astrophys. J., 357*, 216.
Graham J. R. et al. (2007) *Astrophys. J., 654*, 595.
Greaves J. S. et al. (1998) *Astrophys. J. Lett., 506*, L133.
Greaves J. S. et al. (2005) *Astrophys. J. Lett., 619*, L187.
Greaves J. S. et al. (2006) *Mon. Not. R. Astron. Soc., 366*, 283.
Greaves J. S. et al. (2014) *Mon. Not. R. Astron. Soc., 438*, L31–L35.
Grigorieva A. et al. (2007) *Astron. Astrophys., 461*, 537.
Güttler C. et al. (2010) *Astron. Astrophys., 513*, A56.
Habing H. J. et al. (2001) *Astron. Astrophys., 365*, 545.
Haisch K. E. et al. (2001) *Astrophys. J. Lett., 553*, L153.
Heap S. R. et al. (2000) *Astrophys. J., 539*, 435.
Heng K. and Malik M. (2013) *Mon. Not. R. Astron. Soc., 432*, 2562.
Heng K. and Tremaine S. (2010) *Mon. Not. R. Astron. Soc., 401*, 867.
Hernández J. et al. (2007) *Astrophys. J., 662*, 1067.
Hillenbrand L. A. (2008) *Phys. Scr. T, 130(1)*, 014024.
Hillenbrand L. A. et al. (2008) *Astrophys. J., 677*, 630.
Hines D. C. et al. (2007) *Astrophys. J. Lett., 671*, L165.
Hinkley S. et al. (2009) *Astrophys. J., 701*, 804.
Hinz P. M. et al. (2008) In *Optical and Infrared Astronomy* (M. Schöller et al., eds.), SPIE Conf. Series 7013.
Hobbs L. M. et al. (1985) *Astrophys. J. Lett., 293*, L29.
Holland W. S. et al. (1998) In *Advanced Technology MMW, Radio, and Terahertz Telescopes* (T. G. Phillips, eds.), pp. 305–318. SPIE Conf. Series 3357.
Hughes A. M. et al. (2008a) *Astrophys. J., 681*, 626.

Hughes A. M. et al. (2008b) *Astrophys. J., 678*, 1119.
Hughes A. M. et al. (2012) *Astrophys. J., 750*, 82.
Jackson A. P. and Wyatt M. C. (2012) *Mon. Not. R. Astron. Soc., 425*, 657.
Janson M. et al. (2013) *Astrophys. J., 773*, 73.
Jolly A. et al. (1998) *Astron. Astrophys., 329*, 1028.
Kains N. et al. (2011) *Mon. Not. R. Astron. Soc., 414*, 2486.
Kalas P. et al. (2005) *Nature, 435*, 1067.
Kalas P. et al. (2007) *Astrophys. J. Lett., 661*, L85.
Kalas P. et al. (2008) *Science, 322*, 1345.
Kalas P. et al. (2013) *Astrophys. J., 775*, 56.
Kelsall T. et al. (1998) *Astrophys. J., 508*, 44.
Kennedy G. M. and Wyatt M. C. (2010) *Mon. Not. R. Astron. Soc., 405*, 1253.
Kennedy G. M. and Wyatt M. C. (2011) *Mon. Not. R. Astron. Soc., 412*, 2137.
Kennedy G. M. and Wyatt M. C. (2012) *Mon. Not. R. Astron. Soc., 426*, 91.
Kennedy G. M. and Wyatt M. C. (2013) *Mon. Not. R. Astron. Soc., 433*, 2334.
Kennedy G. M. et al. (2012a) *Mon. Not. R. Astron. Soc., 421*, 2264.
Kennedy G. M. et al. (2012b) *Mon. Not. R. Astron. Soc., 426*, 2115.
Kennedy G. M. et al. (2013) *Mon. Not. R. Astron. Soc., 436*, 898.
Kennedy G. M. et al. (2014) *Mon. Not. R. Astron. Soc., 438*, L96–L100.
Kenyon S. J. and Bromley B. C. (2002) *Astron. J., 123*, 1757.
Kenyon S. J. and Bromley B. C. (2004) *Astrophys. J. Lett., 602*, L133.
Kenyon S. J. and Bromley B. C. (2005) *Astron. J., 130*, 269.
Kenyon S. J. and Bromley B. C. (2008) *Astrophys. J. Suppl., 179*, 451.
Kenyon S. J. and Bromley B. C. (2010) *Astrophys. J. Suppl., 188*, 242.
Kilic M. and Redfield S. (2007) *Astrophys. J., 660*, 641.
Kilic M. et al. (2005) *Astrophys. J. Lett., 632*, L115.
Kilic M. et al. (2006) *Astrophys. J., 646*, 474.
Kilic M. et al. (2009) *Astrophys. J., 696*, 2094.
Kirsh D. R. et al. (2009) *Icarus, 199*, 197.
Klahr H. and Lin D. N. C. (2005) *Astrophys. J., 632*, 1113.
Kobayashi H. and Ida S. (2001) *Icarus, 153*, 416.
Kobayashi H. et al. (2008) *Icarus, 195*, 871.
Koerner D. W. et al. (2010) *Astrophys. J. Lett., 710*, L26.
Kóspál Á. et al. (2009) *Astrophys. J. Lett., 700*, L73.
Kóspál Á. et al. (2013) *Astrophys. J., 776*, 77.
Kozai Y. (1962) *Astron. J., 67*, 591.
Kral Q. et al. (2013) *Astron. Astrophys., 558*, A121.
Krist J. E. et al. (2012) *Astron. J., 144*, 45.
Krivov A. V. (2010) *Rev. Astron. Astrophys., 10*, 383.
Krivov A. V. et al. (2000) *Astron. Astrophys., 362*, 1127.
Krivov A. V. et al. (2005) *Icarus, 174*, 105.
Krivov A. V. et al. (2006) *Astron. Astrophys., 455*, 509.
Krivov A. V. et al. (2007) *Astron. Astrophys., 462*, 199.
Krivov A. V. et al. (2008) *Astrophys. J., 687*, 608.
Krivov A. V. et al. (2009) *Astron. Astrophys., 507*, 1503.
Krivov A. V. et al. (2013) *Astrophys. J., 772*, 32.
Kroupa P. (1995) *Mon. Not. R. Astron. Soc., 277*, 1507.
Kroupa P. (1998) *Mon. Not. R. Astron. Soc., 298*, 231.
Kuchner M. J. and Holman M. J. (2003) *Astrophys. J., 588*, 1110.
Kuchner M. J. and Stark C. C. (2010) *Astron. J., 140*, 1007.
Lagrange A.-M. et al. (2009) *Astron. Astrophys., 493*, L21.
Lagrange A. et al. (2010) *Science, 329*, 57.
Lagrange A.-M. et al. (2012) *Astron. Astrophys., 542*, A40.
Lawler S. M. and Gladman B. (2012) *Astrophys. J., 752*, 53.
Lawler S. M. et al. (2009) *Astrophys. J., 705*, 89.
Lebreton J. et al. (2012) *Astron. Astrophys., 539*, A17.
Lecar M. et al. (2001) *Annu. Rev. Astron. Astrophys., 39*, 581.
Lecar M. et al. (2006) *Astrophys. J., 640*, 1115.
Lecavelier des Etangs A. et al. (1996a) *Astron. Astrophys., 307*, 542.
Lecavelier des Etangs A. et al. (1996b) *Icarus, 123*, 168.
Lecavelier des Etangs A. et al. (1998) *Astron. Astrophys., 339*, 477.
Lecavelier des Etangs A. et al. (2001) *Nature, 412*, 706.
Lestrade J.-F. et al. (2006) *Astron. Astrophys., 460*, 733.
Lestrade J.-F. et al. (2012) *Astron. Astrophys., 548*, A86.
Levison H. F. and Duncan M. J. (1997) *Icarus, 127*, 13.
Levison H. F. et al. (2007) In *Protostars and Planets V* (B. Reipurth et al., eds.), pp. 669–684. Univ. of Arizona, Tucson.
Li D. et al. (2012) *Astrophys. J., 759*, 81.
Lisse C. M. et al. (2008) *Astrophys. J., 673*, 1106.
Lisse C. M. et al. (2009) *Astrophys. J., 701*, 2019.
Lisse C. M. et al. (2012) *Astrophys. J., 747*, 93.

Löhne T. et al. (2008) *Astrophys. J., 673,* 1123.
Löhne T. et al. (2012) *Astron. Astrophys., 537,* A110.
Low F. J. et al. (1984) *Astrophys. J. Lett., 278,* L19.
Low F. J. et al. (2005) *Astrophys. J., 631,* 1170.
Lyra W. and Kuchner M. J. (2013) *Nature, 499,* 184.
MacGregor M. A. et al. (2013) *Astrophys. J. Lett., 762,* L21.
Maldonado J. et al. (2012) *Astron. Astrophys., 541,* A40.
Malhotra R. (1993) *Nature, 365,* 819.
Maness H. L. et al. (2008) *Astrophys. J. Lett., 686,* L25.
Maness H. L. et al. (2009) *Astrophys. J., 707,* 1098.
Marois C. et al. (2008) *Science, 322,* 1348.
Marois C. et al. (2010) *Nature, 468,* 1080.
Marshall J. P. et al. (2011) *Astron. Astrophys., 529,* A117.
Marshall J. P. et al. (2014) *Astron. Astrophys., 565,* 15.
Marzari F. and Thébault P. (2011) *Mon. Not. R. Astron. Soc., 416,* 1890.
Marzari F. et al. (2012) *Astron. Astrophys., 539,* A98.
Matthews B. C. et al. (2007) *Astrophys. J., 663,* 1103.
Matthews B. C. et al. (2010) *Astron. Astrophys., 518,* L135.
Matthews B. C. et al. (2014) *Astrophys. J. 780,* 97.
Melis C. et al. (2010) *Astrophys. J. Lett., 717,* L57.
Melis C. et al. (2012) *Nature, 487,* 74.
Meng H. Y. A. et al. (2012) *Astrophys. J. Lett., 751,* L17.
Metzger B. D. et al. (2012) *Mon. Not. R. Astron. Soc., 423,* 505.
Meyer M. R. et al. (2006) *Publ. Astron. Soc. Pac., 118,* 1690.
Meyer M. R. et al. (2007) In *Protostars and Planets V* (B. Reipurth
    et al., eds.), pp. 573–588. Univ. of Arizona, Tucson.
Millan-Gabet R. et al. (2011) *Astrophys. J., 734,* 67.
Moerchen M. M. et al. (2010) *Astrophys. J., 723,* 1418.
Monin J.-L. et al. (2007) In Protostars and Planets V (B. Reipurth
    et al., eds.), pp. 395–409. Univ. of Arizona, Tucson.
Moór A. et al. (2011a) *Astrophys. J. Lett., 740,* L7.
Moór A. et al. (2011b) *Astrophys. J. Suppl., 193,* 4.
Moore A. and Quillen A. C. (2013) *Mon. Not. R. Astron. Soc., 430,* 320.
Morales F. Y. et al. (2009) *Astrophys. J., 699,* 1067.
Morales F. Y. et al. (2011) *Astrophys. J. Lett., 730,* L29.
Morbidelli A. et al. (2004) *Mon. Not. R. Astron. Soc., 355,* 935.
Moro-Martín A. et al. (2010) *Astrophys. J., 717,* 1123.
Mouillet D. et al. (1997) *Mon. Not. R. Astron. Soc., 292,* 896.
Müller S. et al. (2010) *Astrophys. J., 708,* 1728.
Murray C. D. and Dermott S. F. (1999) *Solar System Dynamics.*
    Cambridge Univ. Press.
Mustill A. J. and Wyatt M. C. (2009) *Mon. Not. R. Astron. Soc.,
    399,* 1403.
Mustill A. J. and Wyatt M. C. (2011) *Mon. Not. R. Astron. Soc.,
    413,* 554.
Mustill A. J. and Wyatt M. C. (2012) *Mon. Not. R. Astron. Soc.,
    419,* 3074.
Najita J. and Williams J. P. (2005) *Astrophys. J., 635,* 625.
Nesvold E. R. et al. (2013) *Astrophys. J., 777,* 144.
Nesvorný D. et al. (2003) *Astrophys. J., 591,* 486.
Nesvorný D. et al. (2010) *Astrophys. J., 713,* 816.
Nilsson R. et al. (2010) *Astron. Astrophys., 518,* A40.
O'Brien D. P. and Greenberg R. (2003) *Icarus, 164,* 334.
Olofsson G. et al. (2001) *Astrophys. J., 563,* L77.
Ozernoy L. M. et al. (2000) *Astrophys. J. Lett., 537,* L147.
Pan M. and Schlichting H. E. (2012) *Astrophys. J., 747,* 113.
Panić O. et al. (2013) *Mon. Not. R. Astron. Soc., 435,* 1037.
Pascucci I. et al. (2006) *Astrophys. J., 651,* 1177.
Payne M. J. et al. (2009) *Mon. Not. R. Astron. Soc., 393,* 1219.
Plavchan P. et al. (2005) *Astrophys. J., 631,* 1161.
Plavchan P. et al. (2009) *Astrophys. J., 698,* 1068.
Poulton C. J. et al. (2006) *Mon. Not. R. Astron. Soc., 372,* 53.
Quillen A. C. (2006) *Mon. Not. R. Astron. Soc., 372,* L14.
Quillen A. C. and Thorndike S. (2002) *Astrophys. J., 578,* L149.
Rafikov R. R. (2011a) *Astrophys. J. Lett., 732,* L3.
Rafikov R. R. (2011b) *Mon. Not. R. Astron. Soc., 416,* L55.
Raymond S. N. and Armitage P. J. (2013) *Mon. Not. R. Astron. Soc.,
    429,* L99.
Raymond S. N. et al. (2012) *Astron. Astrophys., 541,* A11.
Reche R. et al. (2008) *Astron. Astrophys., 480,* 551.
Reidemeister M. et al. (2009) *Astron. Astrophys., 503,* 247.
Reidemeister M. et al. (2011) *Astron. Astrophys., 527,* A57.
Rhee J. H. et al. (2007) *Astrophys. J., 660,* 1556.
Rhee J. H. et al. (2008) *Astrophys. J., 675,* 777.
Rieke G. H. et al. (2005) *Astrophys. J., 620,* 1010.
Rivera E. J. et al. (2005) *Astrophys. J., 634,* 625.

Riviere-Marichalar P. et al. (2013) *Astron. Astrophys., 555,* A67.
Roberge A. et al. (2000) *Astrophys. J., 538,* 904.
Roberge A. et al. (2006) *Nature, 441,* 724.
Roberge A. et al. (2012) *Publ. Astron. Soc. Pac., 124,* 799.
Roberge A. et al. (2013) *Astrophys. J., 771,* 69.
Roccatagliata V. et al. (2009) *Astron. Astrophys., 497,* 409.
Rodigas T. J. et al. (2012) *Astrophys. J., 752,* 57.
Rodriguez D. R. and Zuckerman B. (2012) *Astrophys. J., 745,* 147.
Rosenfeld K. A. et al. (2012) *Astrophys. J., 757,* 129.
Schneider G. et al. (1999) *Astrophys. J. Lett., 513,* L127.
Schneider G. et al. (2005) *Astrophys. J. Lett., 629,* L117.
Shen Y. et al. (2009) *Astrophys. J., 696,* 2126.
Sibthorpe B. et al. (2010) *Astron. Astrophys., 518,* L130.
Siegler N. et al. (2007) *Astrophys. J., 654,* 580.
Slettebak A. (1975) *Astrophys. J., 197,* 137.
Smith B. A. and Terrile R. J. (1984) *Science, 226,* 1421.
Smith R. et al. (2009) *Astron. Astrophys., 503,* 265.
Smith R. et al. (2012) *Mon. Not. R. Astron. Soc., 422,* 2560.
Song I. et al. (2005) *Nature, 436,* 363.
Stapelfeldt K. R. et al. (2004) *Astrophys. J., 154,* 458.
Stark C. C. and Kuchner M. J. (2009) *Astrophys. J., 707,* 543.
Stencel R. E. and Backman D. E. (1991) *Astrophys. J. Suppl., 75,* 905.
Strubbe L. E. and Chiang E. I. (2006) *Astrophys. J., 648,* 652.
Su K. Y. L. et al. (2005) *Astrophys. J., 628,* 487.
Su K. Y. L. et al. (2006) *Astrophys. J., 653,* 675.
Su K. Y. L. et al. (2007) *Astrophys. J. Lett., 657,* L41.
Su K. Y. L. et al. (2009) *Astrophys. J., 705,* 314.
Su K. Y. L. et al. (2013) *Astrophys. J., 763,* 118.
Takeuchi T. and Artymowicz P. (2001) *Astrophys. J., 557,* 990.
Telesco C. M. et al. (2000) *Astrophys. J., 530,* 329.
Thalmann C. et al. (2011) *Astrophys. J. Lett., 743,* L6.
Thébault P. (2009) *Astron. Astrophys., 505,* 1269.
Thébault P. (2012) *Astron. Astrophys., 537,* A65.
Thébault P. and Augereau J.-C. (2005) *Astron. Astrophys., 437,* 141.
Thébault P. and Augereau J.-C. (2007) *Astron. Astrophys., 472,* 169.
Thébault P. and Wu Y. (2008) *Astron. Astrophys., 481,* 713.
Thébault P. et al. (2003) *Astron. Astrophys., 408,* 775.
Thébault P. et al. (2010) *Astron. Astrophys., 524,* A13.
Thébault P. et al. (2012) *Astron. Astrophys., 547,* A92.
Tielens A. G. G. M. et al. (1994) *Astrophys. J., 431,* 321.
Trilling D. E. et al. (2007) *Astrophys. J., 658,* 1289.
Trilling D. E. et al. (2008) *Astrophys. J., 674,* 1086.
Udry S. et al. (2007) *Astron. Astrophys., 469,* L43.
Urban L. E. et al. (2012) *Astrophys. J., 750,* 98.
van Dishoeck E. F. and Black J. H. (1988) *Astrophys. J., 334,* 771.
Vidal-Madjar A. et al. (1994) *Astron. Astrophys., 290,* 245.
Vitense A. et al. (2012) *Astron. Astrophys., 540,* A30.
Watson C. A. et al. (2011) *Mon. Not. R. Astron. Soc., 413,* L71.
Weidenschilling S. J. (1977) *Mon. Not. R. Astron. Soc., 180,* 57.
Weidenschilling S. J. and Jackson A. A. (1993) *Icarus, 104,* 244.
Weinberger A. J. et al. (2011) *Astrophys. J., 726,* 72.
Wilner D. J. et al. (2012) *Astrophys. J. Lett., 749,* L27.
Wisdom J. (1980) *Astron. J., 85,* 1122.
Wyatt M. C. (2003) *Astrophys. J., 598,* 1321.
Wyatt M. C. (2005a) *Astron. Astrophys., 440,* 937.
Wyatt M. C. (2005b) *Astron. Astrophys., 433,* 1007.
Wyatt M. C. (2006) *Astrophys. J., 639,* 1153.
Wyatt M. C. (2008) *Annu. Rev. Astron. Astrophys., 46,* 339.
Wyatt M. C. and Dent W. R. F. (2002) *Mon. Not. R. Astron. Soc.,
    334,* 589.
Wyatt M. C. et al. (1999) *Astrophys. J., 527,* 918.
Wyatt M. C. et al. (2005) *Astrophys. J., 620,* 492.
Wyatt M. C. et al. (2007a) *Mon. Not. R. Astron. Soc., 380,* 1737.
Wyatt M. C. et al. (2007b) *Astrophys. J., 663,* 365.
Wyatt M. C. et al. (2007c) *Astrophys. J., 658,* 569.
Wyatt M. C. et al. (2010) *Mon. Not. R. Astron. Soc., 402,* 657.
Wyatt M. C. et al. (2011) *Cel. Mech. Dyn. Astron., 111,* 1.
Wyatt M. C. et al. (2012) *Mon. Not. R. Astron. Soc., 424,* 1206.
Zsom A. et al. (2010) *Astron. Astrophys., 513,* A57.
Zsom A. et al. (2011) *Astron. Astrophys., 534,* A73.
Zuckerman B. and Becklin E. E. (1987) *Nature, 330,* 138.
Zuckerman B. and Song I. (2004) *Annu. Rev. Astron. Astrophys., 42,* 685.
Zuckerman B. and Song I. (2012) *Astrophys. J., 758,* 77.
Zuckerman B. et al. (2001) *Astrophys. J. Lett., 562,* L87.
Zuckerman B. et al. (2003) *Astrophys. J., 596,* 477.

Part III:

Planetary Systems — Search, Formation, and Evolution

Johansen A., Blum J., Tanaka H., Ormel C., Bizzarro M., and Rickman H. (2014) The multifaceted planetesimal formation process. In *Protostars and Planets VI* (H. Beuther et al., eds.), pp. 547–570. Univ. of Arizona, Tucson, DOI: 10.2458/azu_uapress_9780816531240-ch024.

The Multifaceted Planetesimal Formation Process

Anders Johansen
Lund University

Jürgen Blum
Technische Universität Braunschweig

Hidekazu Tanaka
Hokkaido University

Chris Ormel
University of California, Berkeley

Martin Bizzarro
Copenhagen University

Hans Rickman
Uppsala University and Polish Academy of Sciences Space Research Center

Accumulation of dust and ice particles into planetesimals is an important step in the planet formation process. Planetesimals are the seeds of both terrestrial planets and the solid cores of gas and ice giants forming by core accretion. Leftover planetesimals in the form of asteroids, transneptunian objects, and comets provide a unique record of the physical conditions in the solar nebula. Debris from planetesimal collisions around other stars signposts that the planetesimal formation process, and hence planet formation, is ubiquitous in the Galaxy. The planetesimal formation stage extends from micrometer-sized dust and ice to bodies that can undergo runaway accretion. The latter ranges in size from 1 km to 1000 km, dependent on the planetesimal eccentricity excited by turbulent gas density fluctuations. Particles face many barriers during this growth, arising mainly from inefficient sticking, fragmentation, and radial drift. Two promising growth pathways are mass transfer, where small aggregates transfer up to 50% of their mass in high-speed collisions with much larger targets, and fluffy growth, where aggregate cross sections and sticking probabilities are enhanced by a low internal density. A wide range of particle sizes, from 1 mm to 10 m, concentrate in the turbulent gas flow. Overdense filaments fragment gravitationally into bound particle clumps, with most mass entering planetesimals of contracted radii from 100 km to 500 km, depending on local disk properties. We propose a hybrid model for planetesimal formation where particle growth starts unaided by self-gravity but later proceeds inside gravitationally collapsing pebble clumps to form planetesimals with a wide range of sizes.

1. INTRODUCTION

Most stars are born surrounded by a thin protoplanetary disk with a characteristic mass between 0.01% and 10% of the mass of the central star (*Andrews and Williams*, 2005). Planetesimal formation takes place as the embedded dust and ice particles collide and grow to ever larger bodies. Tiny particles collide gently due to Brownian motion, while larger aggregates achieve higher and higher collision speeds as they gradually decouple from the smallest eddies of the turbulent gas and thus no longer inherit the incompressibility of the gas flow (*Voelk et al.*, 1980). The gas disk is partially pressure-supported in the radial direction, causing particles of sizes from centimeters to meters to drift toward the star (*Whipple*, 1972; *Weidenschilling*, 1977a). Drift speeds depend on the particle size and hence different-sized particles experience high-speed collisions.

The growth from dust and ice grains to planetesimals — the seeds of both terrestrial planets as well as the cores of gas giants and ice giants — is an important step toward the assembly of a planetary system. A planetesimal can be defined as a body that is held together by self-gravity rather than material strength, corresponding to minimum sizes on the order of 100–1000 m (*Benz*, 2000). The planetesimal formation stage, on the other hand, must extend to sizes where the escape speed of the largest bodies exceeds the random motion of the planetesimals in order to enter the next stage of gravity-driven collisions. The random speed of planetesimals and preplanetesimals (the latter can be roughly defined as bodies larger than 10 m in size) is excited mainly by the stochastic gravitational pull from density fluctuations in the turbulent gas; under nominal turbulence conditions the largest planetesimals need to reach sizes as large as 1000 km to start runaway growth (*Ida et al.*, 2008). In disk regions with less vigorous turbulence, planetesimal sizes between 1 km and 10 km may suffice to achieve significant gravitational focusing (*Gressel et al.*, 2012). Planetesimals must grow to runaway sizes despite bouncing and disruptive collisions (*Blum and Wurm*, 2000).

Planetesimals must also grow rapidly — radial drift timescales of centimeter- to meter-sized particles are as short as a few hundred orbits. Despite these difficulties there is ample evidence from cosmochemistry that large planetesimals formed in the solar nebula within a few million years (*Kleine et al.*, 2004; *Baker et al.*, 2005), early enough to melt and differentiate by decay of short-lived radionuclides.

The time since the *Protostars and Planets V* (*Reipurth et al.*, 2007), review on planetesimal formation (*Dominik et al.*, 2007) has seen a large expansion in the complexity and realism of planetesimal formation studies. In this review we focus on three areas in which major progress has been obtained, namely (1) the identification of a bouncing barrier at millimeter sizes and the possibility of growth by mass transfer in high-speed collisions (*Wurm et al.*, 2005; *Johansen et al.*, 2008; *Güttler et al.*, 2010; *Zsom et al.*, 2010; *Windmark et al.*, 2012b), (2) numerical simulations of collisions between highly porous ice aggregates that can grow past the bouncing and radial drift barriers due to efficient sticking and cross sections that are greatly enhanced by a low internal density (*Wada et al.*, 2008, 2009; *Seizinger and Kley*, 2013), and (3) hydrodynamical and magneto-hydrodynamical simulations that have identified a number of mechanisms for concentrating particles in the turbulent gas flow of protoplanetary disks, followed by a gravitational fragmentation of the overdense filaments to form planetesimals with characteristic radii larger than 100 km (*Johansen et al.*, 2007, 2009a; *Lyra et al.*, 2008a, 2009; *Bai and Stone*, 2010a; *Raettig et al.*, 2013).

The review is laid out as follows. In section 2 we give an overview of the observed properties of planetesimal belts around the Sun and other stars. In section 3 we review the physics of planetesimal formation and derive the planetesimal sizes necessary to progress toward planetary systems. Section 4 concerns particle concentration and the density environment in which planetesimals form, while section 5 describes laboratory experiments to determine the outcome of collisions and methods for solving the coagulation equation. In section 6 we discuss the bouncing barrier for silicate dust and the possibility of mass transfer in high-speed collisions. Section 7 discusses computer simulations of highly porous ice aggregates whose low internal density can lead

to high growth rates. Finally, in section 8 we integrate the knowledge gained over recent years to propose a unified model for planetesimal formation involving both coagulation, particle concentration, and self-gravity.

2. PLANETESIMAL BELTS

The properties of observed planetesimal belts provide important constraints on the planetesimal formation process. In this section we review the main properties of the asteroid belt and transneptunian population as well as debris disks around other stars.

2.1. Asteroids

The asteroid belt is a collection of mainly rocky bodies residing between the orbits of Mars and Jupiter. Asteroid orbits have high eccentricities (e ~ 0.1) and inclinations, excited by resonances with Jupiter and also a speculated population of embedded super-Ceres-sized planetary embryos that were later dynamically depleted from the asteroid belt (*Wetherill*, 1992). The high relative speeds imply that the asteroid belt is in a highly erosive regime where collisions lead to fragmentation rather than to growth.

Asteroids range in diameters from D ≈ 1000 km (Ceres) down to the detection limit at subkilometer sizes (*Gladman et al.*, 2009). The asteroid size distribution can be parameterized in terms of the cumulative size distribution $N_>(D)$ or the differential size distribution $dN_>/dD$. The cumulative size distribution of the largest asteroids resembles a broken power law $N_> \propto D^{-p}$, with a steep power law index p ≈ 3.5 for asteroids above 100 km diameter and a shallower power-law index p ≈ 1.8 below this knee (*Bottke et al.*, 2005).

Asteroids are divided into a number of classes based on their spectral reflectivity. The main classes are the moderate-albedo S-types, which dominate the region from 2.1 to 2.5 AU, and the low-albedo C-types, which dominate regions from 2.5 to 3.2 AU (*Gradie and Tedesco*, 1982; *Bus and Binzel*, 2002). If these classes represent distinct formation events, then their spatial separation can be used to constrain the degree of radial mixing by torques from turbulent density fluctuations in the solar nebula (*Nelson and Gressel*, 2010). Another important class is M-type asteroids, of which some are believed to be the metallic cores remaining from differentiated asteroids (*Rivkin et al.*, 2000). Two of the largest asteroids in the asteroid belt, Ceres and Vesta, are known to be differentiated [from their shape and measured gravitational moments, respectively (*Thomas et al.*, 2005; *Russell et al.*, 2012)].

Asteroids are remnants of solar system planetesimals that have undergone substantial dynamical depletion and collisional erosion. Dynamical evolution models can be used to link their current size distribution to the primordial birth size distribution. *Bottke et al.* (2005) suggested, based on the observed knee in the size distribution at sizes around 100 km, that asteroids with diameters above D ≈ 120 km are primordial and that their steep size distribution reflects their formation sizes, while smaller asteroids are fragments of collisions between their larger counterparts. *Morbidelli et al.* (2009) tested a number of birth-size distributions for the asteroid belt, based on either classical coagulation models starting with kilometer-sized planetesimals or turbulent concentration models predicting birth sizes in the range between 100 and 1000 km (*Johansen et al.,* 2007; *Cuzzi et al.,* 2008), and confirmed that the best match to the current size distribution is that the asteroids formed big, in that the asteroids above 120 km in diameter are depleted dynamically but have maintained the shape of the primordial size distribution. On the other hand, *Weidenschilling* (2011) found that an initial population of 100-meter-sized planetesimals can reproduce the current observed size distribution of the asteroids, including the knee at 100 km. However, gap formation around large planetesimals and trapping of small planetesimals in resonances is not included in this particle-in-a-box approach (*Levison et al.,* 2010).

2.2. Meteorites

Direct information about the earliest stages of planet formation in the solar nebula can be obtained from meteorites that record the interior structure of planetesimals in the asteroid belt. The reader is referred to the chapter by Davis et al. in this volume for more details on the connection between cosmochemistry and planet formation. Meteorites are broadly characterized as either primitive or differentiated (*Krot et al.,* 2003). Primitive meteorites (chondrites) are fragments of parent bodies that did not undergo melting and differentiation, and therefore contain pristine samples of the early solar system. In contrast, differentiated meteorites are fragments of parent bodies that underwent planetesimal-scale melting and differentiation to form a core, mantle, and crust.

The oldest material to crystallize in the solar nebula is represented in chondrites as millimeter- to centimeter-sized calcium-aluminum-rich inclusions (CAIs) and ferromagnesian silicate spherules (chondrules) of typical sizes from 0.1 mm to 1 mm (Fig. 1). The chondrules originated in an unidentified thermal processing mechanism (or several mechanisms) that melted preexisting nebula solids (e.g., *Desch and Connolly,* 2002; *Ciesla et al.,* 2004); alternatively, chondrules can arise from the crystallization of splash ejecta from planetesimal collisions (*Sanders and Taylor,* 2005; *Krot et al.,* 2005).

The majority of CAIs formed as fine-grained condensates from a ^{16}O-rich gas of approximately solar composition in a region with high-ambient temperature (>1300 K) and low total pressures (~10^{-4} bar). This environment may have existed in the innermost part of the solar nebula during the early stage of its evolution, characterized by high mass accretion rates (~10^{-5} M$_\odot$ yr^{-1}) to the proto-Sun (*D'Alessio et al.,* 2005). Formation of CAIs near the proto-Sun is also indicated by the presence in these objects of short-lived radionuclide ^{10}Be formed by solar energetic particle irradiation (*McKeegan et al.,* 2000). In contrast, most chondrules represent coalesced

^{16}O-poor dust aggregates that were subsequently rapidly melted and cooled in lower-temperature regions (<1000 K) and higher ambient vapor pressures (~10^{-3} bar) than CAIs, resulting in igneous porphyritic textures we observe today (*Scott,* 2007). Judging by their sheer abundance in chondrite meteorites, the formation of chondrules must reflect a common process that operated in the early solar system. Using the assumption-free uranium-corrected Pb-Pb dating method, *Connelly et al.* (2012) recently showed that CAIs define a brief formation interval corresponding to an age of 4567.30 ± 0.16 Ma, whereas chondrule ages range from 4567.32 ± 0.42 to 4564.71 ± 0.30 Ma. These data indicate that chondrule formation started contemporaneously with CAIs and lasted ~3 million years (m.y.).

A consequence of accretion of planetesimals on million-year timescales or less is the incorporation of heat-generating short-lived radioisotopes such as ^{26}Al, resulting in widescale melting, differentiation, and extensive volcanic activity. Both long-lived and short-lived radioisotope chronometers have been applied to study the timescale of asteroidal differentiation. Of particular interest are the ^{26}Al-^{26}Mg and ^{182}Hf-^{182}W decay systems, with half-lives of 0.73 m.y. and 9 m.y., respectively. Aluminum and Mg are refractory and lithophile elements and thus remain together in the mantle after core formation. In contrast, the different geochemical behavior of Hf and W during melting results in W being preferentially partitioned in the core and Hf being partitioned into the silicate mantle and crust. Therefore, the short-lived ^{182}Hf-^{182}W system is useful to study the timescales of core formation in asteroids, as well as in planets, which can be used to constrain planetesimal formation models.

Eucrite and angrite meteorites are the two most common groups of basaltic meteorites, believed to be derived from the mantles of differentiated parent bodies. The howardite-

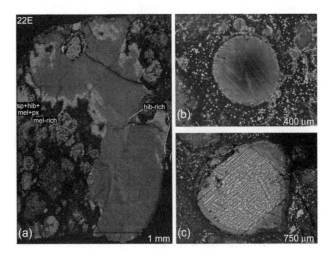

Fig. 1. X-ray elemental maps composed of Mg, Ca, and Al of a fined-grained CAI from the **(a)** Efremovka CV3 chondrite and **(b,c)** two barred-olivine chondrules from the NWA 3118 CV3 chondrite.

eucrite-diogenite (HED) meteorite clan provides our best samples of any differentiated asteroid and could come from the 500-km-diameter asteroid 4 Vesta (*Binzel and Xu,* 1993) [although see *Schiller et al.* (2011) for a different view]. The interior structure of Vesta was recently determined by the Dawn mission (*Russell et al.,* 2012), indicating an iron core of approximately 110 km in radius (see Fig. 2). The ^{26}Al-^{26}Mg systematics of the angrite and HED meteorites indicate that silicate differentiation on these parent bodies occurred within the first few million years of solar system formation (*Schiller et al.,* 2010, 2011; *Bizzarro et al.,* 2005; *Spivak-Birndorf et al.,* 2009). Similarly, the Mg-isotope composition of olivines within pallasite meteorites — a type of stony-iron meteorites composed of centimeter-sized olivine crystals set in an iron-nickel matrix — suggests silicate differentiation within 1.5 m.y. of solar system formation (*Baker et al.,* 2012). Chemical and isotopic diversity of iron meteorites show that these have sampled approximately 75 distinct parent bodies (*Goldstein et al.,* 2009). The ^{182}Hf-^{182}W systematics of some magmatic iron meteorites require the accretion and differentiation of their parent bodies to have occurred within less than 1 m.y. of solar system formation (*Kleine et al.,* 2009).

An important implication of the revised absolute chronology of chondrule formation of *Connelly et al.* (2012), which is not based on short-lived radionuclides for which homogeneity in the solar nebula has to be assumed, is that the accretion and differentiation of planetesimals occurred during the epoch of chondrule formation. This suggests that, similarly to chondrite meteorites, the main original constituents of early-accreted asteroids may have been chondrules. The timing of accretion of chondritic parent bodies can be constrained by the ages of their youngest chondrules, which requires that most chondrite parent bodies accreted

>2 m.y. after solar system formation. Therefore, chondrites may represent samples of asteroidal bodies that formed after the accretion of differentiated asteroids, at a time when the levels of ^{26}Al were low enough to prevent significant heating and melting.

2.3. Transneptunian Objects

The transneptunian objects are a collection of rocky/icy objects beyond the orbit of Neptune (*Luu and Jewitt,* 2002). Transneptunian objects are categorized into several dynamical classes, the most important for planetesimal formation being the classical Kuiper belt objects, the scattered disk objects, and the related centaurs, which have orbits that cross the orbits of one or more of the giant planets. The classical Kuiper belt objects do not approach Neptune at any point in their orbits and could represent the pristine population of icy planetesimals in the outer solar nebula. The so-called cold component of the classical Kuiper belt has a large fraction (at least 30%) of binaries of similar size (*Noll et al.,* 2008), which strongly limits the amount of collisional grinding that these bodies can have undergone (*Nesvorný et al.,* 2011). The scattered disk objects, on the other hand, can have large semimajor axes, but they all have perihelia close to Neptune's orbit. The centaurs move between the giant planets and are believed to be the source of short-period comets (see below).

The largest transneptunian objects are much larger than the largest asteroids, with Pluto, Haumea, Makemake, and Eris defined as dwarf planets of 1.5–3× the diameter of Ceres (*Brown,* 2008). Nevertheless, the size distribution of transneptunian objects shows similarities with the asteroid belt, with a steep power law above a knee at around D ~ 100 km (*Fuentes and Holman,* 2008). The turnover at the knee implies that there are fewer intermediate-mass planetesimals than expected from an extrapolation from larger sizes — this has been dubbed the "missing intermediate-sized planetesimals" problem (*Sheppard and Trujillo,* 2010; *Shankman et al.,* 2013), and suggests that the characteristic planetesimal birth size was ~100 km. The accretion ages of Kuiper belt objects are not known, in contrast to the differentiated asteroids, where the inclusion of large amounts of ^{26}Al requires early accretion. While the largest Kuiper belt objects are likely differentiated into a rocky core and an icy mantle — this is clear, e.g., for Haumea, which is the dense remnant of a differentiated body (*Ragozzine and Brown,* 2009) — differentiation could be due to long-lived radionuclides such as U, Th, and ^{40}K and hence happen over much longer timescales (~100 m.y.) than the ~million-year timescale characteristic of asteroid differentiation by ^{26}Al decay (*McKinnon et al.,* 2008).

2.4. Comets

Comets are typically kilometer-sized volatile-rich bodies that enter the inner solar system. They bring with them a wealth of information about the conditions during the

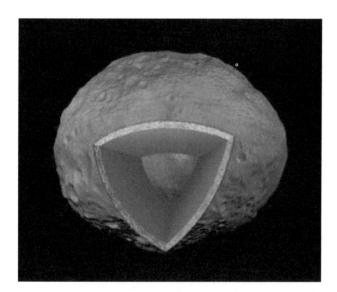

Fig. 2. Surface and interior structure of the 500-km-diameter asteroid 4 Vesta, obtained by the NASA Dawn satellite. The iron core is 110 km in radius, surrounded by a silicate mantle and a basaltic crust. Credit: NASA/JPL-Caltech.

planetesimal formation epoch in the outer solar system. Outgassing provides knowledge about the compositions and heating histories of icy planetesimals, and the volumes and masses of comet nuclei can be used to derive their density, which can be compared to the density expected from planetesimal formation models. Comet nuclei are (or are related to) icy planetesimals, i.e., planetesimals that formed beyond the snow line in a formation zone extending from roughly 15 to 30 AU from the Sun — at least in the framework of the so-called Nice model, where the giant planets form in a compact configuration between 5 and 12 AU and later migrate by planetesimal scattering to their current positions (*Levison et al.*, 2011). This zone is wide enough to allow for the appearance of chemical zoning, but such a zoning would likely not be reflected in any separation between the short- and long-period comet source regions.

Comets have to be considered together with the Centaurs and transneptunians, because these are dynamically related to the Jupiter-family comets, and most of the data we have on comets come from observations and modeling of Jupiter-family comets. The emergent picture is a very wide size spectrum going from subkilometer to 10^3 km in diameter. However, the slope of the size distribution is difficult to establish, since small objects are unobservable in the outer populations, and large ones are lacking in the Jupiter family. Generally, the measurements of masses and sizes of comet nuclei are consistent with a rather narrow range of densities at about 0.5 g cm^{-3} (*Weissman et al.*, 2004; *Davidsson et al.*, 2007). Most of the determinations use the nongravitational force due to asymmetric outgassing. Assuming a composition that is roughly a 50-50 mix of ice and refractories (*Greenberg and Hage,* 1990), the porosity comes out as roughly two-thirds. But the mass determinations only concern the bulk mass, so one cannot distinguish between mesoscale porosity (*rubble piles*) and small-scale porosity (*pebble piles*). Estimates of comet tensile strengths have generally been extremely low, as expected for porous bodies. For Comet Shoemaker-Levy 9, modeling of its tidal breakup led to (nonzero) values so low that the object was described as a strengthless rubble pile (*Asphaug and Benz,* 1996). The nontidal splittings often observed for other comets appear to be so gentle that, again, an essentially zero strength has been inferred (*Sekanina,* 1997).

The current results on volatile composition of comets have not indicated any difference between the Jupiter-family comets — thought to probe the scattered disk — and the Halley-type and long-period comets, which should come from the Oort cloud (*A'Hearn et al.*, 2012). From a dynamical point of view, this result is rather expected, since both these source populations have likely emerged from the same ultimate source, a disk of icy planetesimals extending beyond the giant planet zone in the early solar system (*Brasser and Morbidelli,* 2013).

The origin of comets is closely related to the issue of the chemical pristineness of comets. If the total mass of the planetesimal disk was dominated by the largest planetesimals, several hundreds of kilometers in size, then the

kilometer-sized comet nuclei that we are familiar with could arise from collisional grinding of the large planetesimals. Nevertheless, there are severe problems with this idea. One is that comet nuclei contain extremely volatile species, like the S$_2$ molecule (*Bockelée-Morvan et al.*, 2004), which would hardly survive the heating caused by a disruptive impact on the parent body. The alternative — that tidal disruptions of large planetesimals at mutual close encounters may lead to large numbers of smaller objects — seems more viable, but this too suffers from the second pristineness problem, namely, the geologic evolution expected within a large planetesimal due to short-lived radio nuclei (*Prialnik et al.*, 2004). If the large planetesimal thus becomes chemically differentiated, there seems to be no way to form the observed comet nuclei by breaking it up — no matter which mechanism we invoke.

The pristine nature of comets is thus consistent with the formation of small, kilometer-scale cometesimals in the outer part of the solar nebula, avoiding (because of their small size) melting and differentiation due to release of short-lived radionuclides. Another possibility is that comets accreted from material comprising an early-formed ^{26}Al free component (*Olsen et al.*, 2013), which would have prevented differentiation of their parent bodies. The recent chronology of ^{26}Al-poor inclusions found in primitive meteorites indicates that this class of objects formed coevally with CAIs [which record the canonical ^{26}Al/^{27}Al value of 5×10^{-5}; (*Holst et al.*, 2013)]. This provides evidence for the existence of ^{26}Al-free material during the earliest stages of the protoplanetary disk. The discovery of refractory material akin to CAIs in the Stardust samples collected from Comet 81P/Wild 2 shows efficient outward transport of material from the hot inner disk regions to cooler environments far from the Sun during the epoch of CAI formation (*Brownlee et al.*, 2006). Efficient outward transport during the earliest stages of solar system evolution would have resulted in the delivery of a significant fraction of the ^{26}Al-poor material to the accretion region of cometary bodies. Analysis of the Coci refractory particle returned from Comet 81P/Wild 2 did not show detectable ^{26}Al at the time of its crystallization (*Matzel et al.*, 2010), which, although speculative, is consistent with the presence of an early-formed ^{26}Al-free component in comets.

2.5. Debris Disks

Planetesimal belts around other stars show their presence through the infrared emission of the dust produced in collisions (*Wyatt,* 2008). The spectral energy distribution of the dust emission reveals the orbital distance of the planetesimal belt. Warm debris disks resembling the asteroid belt in the solar system are common around young stars of ~50-m.y. age (at least 30%), but their occurrence falls to a few percent within 100–1000 Ma (*Siegler et al.*, 2007). Thus planetesimal formation appears to be ubiquitous around the Sun as well as around other stars. This is in agreement with results from the Kepler mission that flat planetary systems exist around a high fraction of solar-type stars (*Lissauer et al.,* 2011; *Tremaine and Dong,* 2012; *Johansen et al.,* 2012b; *Fang and Margot,* 2012).

3. THE PLANETESIMAL FORMATION STAGE

The dust and ice particles embedded in the gas in protoplanetary disks collide and merge, first by contact forces and later by gravity. This process leads eventually to the formation of the terrestrial planets and the cores of gas giants and ice giants forming by core accretion. The planetesimal formation stage can broadly be defined as the growth from dust grains to particle sizes where gravity contributes significantly to the collision cross section of two colliding bodies.

3.1. Drag Force

Small particles are coupled to the gas via drag force. The acceleration by the drag force can be written as

$$\dot{\mathbf{v}} = -\frac{1}{\tau_f}(\mathbf{v} - \mathbf{u}) \tag{1}$$

where $\mathbf{v}$ is the particle velocity, $\mathbf{u}$ is the gas velocity at the position of the particle, and τ_f is the friction time, which contains all the physics of the interaction of the particle with the gas flow (*Whipple,* 1972; *Weidenschilling,* 1977a). The friction time can be divided into different regimes, depending on the mean free path of the gas molecules, λ, and the speed of the particle relative to the gas, $\delta v = |\mathbf{v} - \mathbf{u}|$. The Epstein regime is valid when the particle size is smaller than the mean free path. The flux of impinging molecules is set in this regime by their thermal motion and the friction time is independent of the relative speed

$$\tau_f = \frac{R\rho_\bullet}{c_s\rho_g} \tag{2}$$

Here R is the radius of the particle, assumed to be spherical. The other parameters are the material density $\rho_\bullet$, the gas sound speed c_s, and the gas density ρ_g. Particles with sizes above 9/4× the mean free path of the molecules enter the Stokes regime (see *Whipple,* 1972, and references therein), with

$$\tau_f = \frac{R\rho_\bullet}{c_s\rho_g}\frac{4}{9}\frac{R}{\lambda} \tag{3}$$

Here the friction time is proportional to the squared radius and independent of gas density, since λ is inversely proportional to the gas density. The flow Reynolds number past the particle, Re = $(2R\delta v)/\nu$, determines the further transition to drag regimes that are nonlinear in the relative speed, with the kinematic viscosity given by $\nu = (1/2)c_s\lambda$. At unity flow Reynolds number the drag transitions to an intermediate regime with the friction time proportional to

$(\delta v)^{-0.4}$. Above Re = 800 the drag force finally becomes quadratic in the relative velocity, with friction time

$$\tau_f = \frac{6R\rho_\bullet}{(\delta v)\rho_g} \tag{4}$$

Following the descriptions above, the stepwise transition from Epstein drag to fully quadratic drag happens in the optically thin minimum mass solar nebula (MMSN) (*Hayashi,* 1981), with a power-law index −1.5 for the surface density (*Weidenschilling,* 1977b) and −0.5 for the temperature, at particle sizes

$$R_1 = \frac{9\lambda}{4} = 3.2 \text{ cm}\left(\frac{r}{\text{AU}}\right)^{2.75} \tag{5}$$

$$R_2 = \frac{\nu}{2(\delta v)} \approx 6.6 \text{ cm}\left(\frac{r}{\text{AU}}\right)^{2.5} \tag{6}$$

$$R_3 = \frac{800\nu}{2(\delta v)} \approx 52.8 \text{ m}\left(\frac{r}{\text{AU}}\right)^{2.5} \tag{7}$$

Here R_1 denotes the Epstein-to-Stokes transition, R_2 the Stokes-to-nonlinear transition, and R_3 the nonlinear-to-quadratic transition. The latter two equations are only approximate because of the dependence of the friction time on the relative speed. Here we used the sub-Keplerian speed Δv as the relative speed between particle and gas, an approximation that is only valid for large Stokes numbers (see equation (8) below).

A natural dimensionless parameter to construct from the friction time is the Stokes number St = $\Omega\tau_f$, with Ω denoting the Keplerian frequency at the given orbital distance. The inverse Keplerian frequency is the natural reference timescale for a range of range of physical effects in protoplanetary disks, hence the Stokes number determines (1) turbulent collision speeds, (2) sedimentation, (3) radial and azimuthal particle drift, (4) concentration in pressure bumps and vortices, and (5) concentration by streaming instabilities. Figure 3 shows the particle size corresponding to a range of Stokes numbers, for both the nominal density of 1 g cm^{-3} and extremely fluffy particles with an internal density of 10^{-5} g cm^{-3} (which could be reached when ice aggregates collide; see section 7).

3.2. Radial Drift

Protoplanetary disks are slightly pressure-supported in the radial direction, due to gradients in both mid-plane density and temperature. This leads to sub-Keplerian motion of the gas, $v_{gas} = v_K - \Delta v$, with v_K denoting the Keplerian speed and the sub-Keplerian velocity difference Δv defined as (*Nakagawa et al.,* 1986)

$$\Delta v \equiv \eta v_K = -\frac{1}{2}\left(\frac{H}{r}\right)^2\frac{\partial \ln P}{\partial \ln r}v_K \tag{8}$$

In the MMSN the aspect ratio H/r rises proportional to $r^{1/4}$ and the logarithmic pressure gradient is $\partial \ln P / \partial \ln r = -3.25$ in the mid-plane. This gives a sub-Keplerian speed that is constant $\Delta v = 53$ m s^{-1}, independent of orbital distance. Radiative transfer models with temperature and density dependent dust opacities yield disk aspect ratios H/r with complicated dependency on r and thus a sub-Keplerian motion that depends on r (e.g., *Bell et al.*, 1997). Nevertheless, a sub-Keplerian speed of ~50 m s^{-1} can be used as the nominal value for a wide range of protoplanetary disk models.

The drag force on the embedded particles leads to particle drift in the radial and azimuthal directions (*Whipple*, 1972; *Weidenschilling*, 1977a)

$$v_r = -\frac{2\Delta v}{St + St^{-1}} \qquad (9)$$

$$v_\phi = v_K - \frac{\Delta v}{1 + St^2} \qquad (10)$$

These equations give the drift speed directly in the Epstein and Stokes regimes. In the nonlinear and quadratic drag regimes, where the Stokes number depends on the relative speed, the equations can be solved using an iterative method to find consistent v_r, v_ϕ, and St.

The azimuthal drift peaks at $v_\phi = v_K - \Delta v$ for the smallest Stokes numbers where the particles are carried passively with the sub-Keplerian gas. The radial drift peaks at unity Stokes number where particles spiral in toward the star at $v_r = -\Delta v$. The radial drift of smaller particles is slowed down by friction with the gas, while particles larger than Stokes number unity react to the perturbing gas drag by entering mildly eccentric orbits with low radial drift.

Radial drift puts requirements on the particle growth at Stokes number around unity (generally from St = 0.1 to St – 10) to occur within the timescale $t_{drift} \sim r/\Delta v \sim 100$–1000 orbits (*Brauer et al.*, 2007, 2008a), depending on the location in the protoplanetary disk. However, three considerations soften the — at first glance — very negative impact of the radial drift. First, the ultimate fate of drifting particles is not to fall into the star, but rather to sublimate at evaporation fronts (or snow lines). This can lead to pile up of material around evaporation fronts (*Cuzzi and Zahnle*, 2004) and to particle growth by condensation of water vapor onto existing ice particles (*Stevenson and Lunine*, 1988; *Ros and Johansen*, 2013). Second, the radial drift flow of particles is linearly unstable to streaming instabilities (*Youdin and Goodman*, 2005), which can lead to particle concentration in dense filaments and planetesimal formation by gravitational fragmentation of the filaments (*Johansen et al.*, 2009a; *Bai and Stone*, 2010a). Third, very fluffy particles with low internal density reach unity Stokes number, where the radial drift is highest, in the Stokes drag regime (*Okuzumi et al.*, 2012). In this regime the Stokes number, which determines radial drift, increases as the square of the particle size and hence growth to "safe" Stokes numbers with low radial drift is much faster than for compact particles. These possibilities are discussed more in the following sections.

3.3. Collision Speeds

Drag from the turbulent gas excites both large-scale random motion of particles as well as collisions (small-scale random motion). The two are distinct because particles may have very different velocity vectors at large separations,

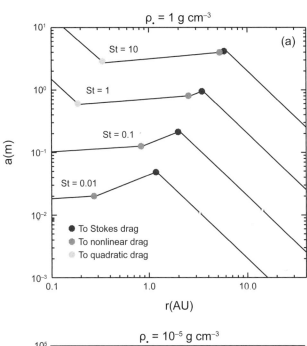

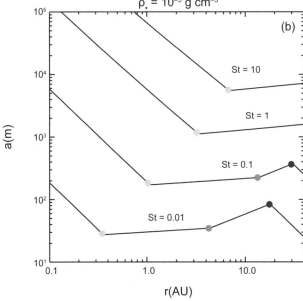

Fig. 3. The particle size corresponding to Stokes numbers from 0.01 to 10: **(a)** for compact particles with material density $\rho_\bullet = 1$ g cm^{-3} and **(b)** for extremely porous aggregates with $\rho_\bullet = 10^{-5}$ g cm^{-3}. Such fluffy particles may be the result of collisions between ice aggregates (discussed in section 7). Transitions between drag force regimes are indicated with large dots.

but these will become increasingly aligned as the particles approach each other due to the incompressible nature of the gas flow. Particles decouple from turbulent eddies with turnover times shorter than the friction time, and the gradual decoupling from the smallest eddies results in crossing particle trajectories. This decoupling can be interpreted as singularities in the particle dynamics [so-called "caustics" (*Gustavsson and Mehlig,* 2011)]. Caustics in turn give rise to collisions as small particles enter regions of intense clustering (clustering is discussed further in section 4.1.1). The random motion is similar to the turbulent speed of the gas for small particles, but particles experience decreased random motion as they grow to Stokes numbers above unity.

The problem when calculating the turbulent collision speed is that at close separations (when the particles are about to collide) they interact with the same eddies, which causes their motions to become highly correlated. The framework set out by *Voelk et al.* (1980), which employs a Langevin approach, is still widely used, and *Ormel and Cuzzi* (2007) provided closed-form analytical approximations to their results [but see *Pan and Padoan* (2010) for a criticism of the simplifications made in the Völk model]. The closed-form expressions of *Ormel and Cuzzi* (2007) require numerical solution of a single algebraic equation for each colliding particle pair (defined by their friction times). With knowledge of the properties of the turbulence, particularly the turbulent root mean square (rms) speed and the frequency of the smallest and the largest eddies, the collision speeds can then be calculated at all locations in the disk.

Another important contribution to turbulent collision speeds is the gravitational pull from turbulent gas density fluctuations. The eccentricity of a preplanetesimal increases as a random walk due to uncorrelated gravitational kicks from the turbulent density field (*Laughlin et al.,* 2004; *Nelson and Papaloizou,* 2004). The eccentricity would grow unbounded with time as $e \propto t^{1/2}$ in the absence of dissipation. Equating the eccentricity excitation timescale with the timescale for damping by tidal interaction with the gas disk (from *Tanaka and Ward,* 2004), aerodynamic gas drag, and inelastic collisions with other particles, *Ida et al.* (2008) provide parameterizations for the equilibrium eccentricity as a function of particle mass and protoplanetary disk properties. The resulting collision speeds dominate over the contributions from the direct drag from the turbulent gas at sizes above approximately 10 m.

Ida et al. (2008) adopt the nomenclature of *Ogihara et al.* (2007) for the eccentricity evolution, where a dimensionless parameter γ determines the proportionality between the eccentricity and $t^{1/2}$. The parameter γ is expected to scale with the density fluctuations $\delta\rho/\rho$ but can be directly calibrated with turbulence simulations. The shearing box simulations by *Yang et al.* (2009) of turbulence caused by the magneto-rotational instability (*Balbus and Hawley,* 1991) suggest that $\delta\rho/\rho \propto \sqrt{\alpha}$ where α is the dimensionless measure of the turbulent viscosity (*Shakura and Sunyaev,* 1973). In their nominal ideal-MHD turbulence model

with $\alpha \approx 0.01$, *Yang et al.* (2012) find $\gamma \approx 6 \times 10^{-4}$. This leads to an approximate expression for γ as a function of the strength of the turbulence, $\gamma \approx 0.006\sqrt{\alpha}$. The resulting eccentricities are in broad agreement with the resistive magneto-hydrodynamic simulations of *Gressel et al.* (2012), where the turbulent viscosity from Reynolds stresses fall below $\alpha = 10^{-4}$ in the mid-plane.

In Fig. 4 we show the collision speeds of two particles of sizes from 1 μm to 100 km, in a figure similar to Fig. 3 of the classic review by *Weidenschilling and Cuzzi* (1993). We take into account the Brownian motion, the differential drift, and the gas turbulence [from the closed-form expressions of *Ormel and Cuzzi* (2007)]. We consider an MMSN model at r = 1 AU and $\alpha = 10^{-2}$ in Figs. 4a,b and $\alpha = 10^{-4}$ in Figs. 4c,d. The gravitational pull from turbulent density fluctuations driven by the magneto-rotational instability is included in Figs. 4b,d. The collision speed approaches 100 m s^{-1} for meter-sized boulders when $\alpha = 10^{-2}$ and 30 m s^{-1} when $\alpha = 10^{-4}$, due to the combined effect of the drag from the turbulent gas and differential drift. The collision speed of large, equal-sized particles (with St > 1) would drop as

$$\delta v = c_s \sqrt{\frac{\alpha}{St}} \qquad (11)$$

in the absence of turbulent density fluctuations [for a discussion of this high-St regime, see *Ormel and Cuzzi* (2007)]. This region has been considered a safe haven for preplanetesimals after crossing the ridges around unity Stokes number (*Weidenschilling and Cuzzi,* 1993). However, the oasis vanishes when including the gravitational pull from turbulent gas density fluctuations, and instead the collision speeds continue to rise toward larger bodies, as they are damped less and less by gas drag (Figs. 4b,d).

The collision speed of equal-sized particles is explored further in Fig. 5. Here we show results of both a model with fully developed MRI turbulence and a more advanced turbulent stirring model that includes the effect of a dead zone in the mid-plane, where the ionization degree is too low for the gas to couple with the magnetic field (*Fleming and Stone,* 2003; *Oishi et al.,* 2007), with weak stirring by density waves traveling from the active surface layers (*Okuzumi and Ormel,* 2013; *Ormel and Okuzumi,* 2013).

The collision speeds rise with size until peaking at unity Stokes number with collision speed approximately $\sqrt{\alpha}c_s$. The subsequent decoupling from the gas leads to a decline by a factor of 10, followed by an increase due to the turbulent density fluctuations starting at sizes of approximately 10–100 m. The planetesimal formation stage can be defined as the growth until the gravitational cross section of the largest bodies is significantly larger than their geometric cross section, a transition that occurs when the escape speed of the largest bodies approach their random speed. In Fig. 5 we indicate also the escape speed as a function of the size of the preplanetesimal. The transition to runaway accretion can only start at 1000-km bodies in a disk with nominal turbulence $\alpha = 10^{-2}$ (*Ida et al.,* 2008).

Lower values of α, e.g., in regions of the disk where the ionization degree is too low for the magneto-rotational instability, lead to smaller values for the planetesimal size needed for runaway accretion, namely 1–10 km.

3.4. Sedimentation

While small micrometer-sized grains follow the gas density tightly, larger particles gradually decouple from the gas flow and sediment toward the mid-plane. An equilibrium mid-plane layer is formed when the turbulent diffusion of the particles balances the sedimentation. The Stokes number introduced in section 3.1 also controls sedimentation (because the gravity toward the mid-plane is proportional to Ω^2). Balance between sedimentation and turbulent diffusion (*Dubrulle et al.,* 1995) yields a mid-plane layer thickness H_p, relative to the gas scale-height H, of

$$\frac{H_p}{H} = \sqrt{\frac{\delta}{St + \delta}} \qquad (12)$$

Here δ is a measure of the diffusion coefficient $D = \delta c_s H$, similar to the standard definition of α for turbulent viscosity (*Shakura and Sunyaev,* 1973). In general $\delta \approx \alpha$ in turbulence driven by the magneto-rotational instability (*Johansen and Klahr,* 2005; *Turner et al.,* 2006), but this equality may be invalid if accretion is driven, for example, by disk winds (*Blandford and Payne,* 1982; *Bai and Stone,* 2013).

Equation (12) was derived assuming particles to fall toward the mid-plane at their terminal velocity $v_z = -\tau_f \Omega^2 z$. Particles with Stokes number larger than unity do not reach terminal velocity before arriving at the midplane and hence undergo oscillations that are damped by drag and excited by turbulence. Nevertheless, *Carballido et al.* (2006) showed that equation (12) is in fact valid for all values of the Stokes

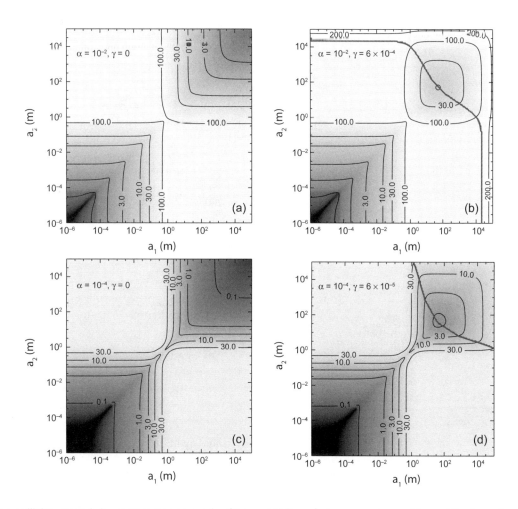

Fig. 4. The collision speed, in meters per second, of two particles of size a_1 and a_2, with contributions from Brownian motion, differential radial and azimuthal drift, and gas turbulence. **(a),(b)** Collision speeds for $\alpha = 10^{-2}$; **(c),(d)** collision speeds for $\alpha = 10^{-4}$. The gravitational pull from turbulent gas density fluctuations is included in **(b)** and **(d)**. The thick black line marks the transition from dominant excitation by direct drag to dominant excitation by turbulent density fluctuations. The "oasis" of low collision speeds for particles above 10 m vanishes when including eccentricity pumping by turbulent density fluctuations.

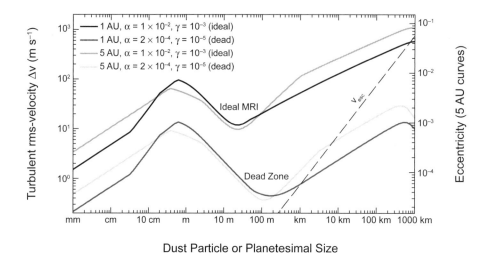

Fig. 5. The collision speed of equal-sized particles as a function of their size, from dust to planetesimals, based on the stirring model of *Ormel and Okuzumi* (2013). The collision speeds of small particles (below ~10 m) are excited mainly by direct gas drag, while the gravitational pull from turbulent gas density fluctuations dominates for larger particles. The transition from relative-speed-dominated to escape-speed-dominated can be used to define the end of the planetesimal formation stage (dashed line). The planetesimal sizes that must be reached for runaway accretion range from near 1000 km in highly turbulent disks ($\alpha = 10^{-2}$) to around 1–10 km in disk regions with an extended dead zone that is stirred by sheared density waves from the active regions ($\alpha = 2 \times 10^{-4}$).

number. *Youdin and Lithwick* (2007) interpreted this as a cancellation between the increased sedimentation time of oscillating particles and their decreased reaction to the turbulent motion of the gas.

The particle density in the equilibrium mid-plane layer is

$$\rho_p = Z\rho_g \frac{H}{H_p} = Z\rho_g \sqrt{\frac{St + \delta}{\delta}} \qquad (13)$$

Here Z is the ratio of the particle column density to the gas column density. In the limit of large particle sizes the mid-plane layer thickness is well approximated by $H_p = c_p/\Omega$, where c_p is the random particle motion. This expression is obtained by associating the collision speed of large particles in equation (11) with their random speed (with $\alpha = \delta$) and inserting this into equation (12) in the limit $St \gg \delta$. The mass flux density of particles can then be written as

$$\mathcal{F} = c_p\rho_p = Z\rho_g H\Omega \qquad (14)$$

independent of the collision speed as well as the degree of sedimentation, since the increased mid-plane density of larger particles is canceled by the decrease in collision speeds. Hence the transition from Stokes numbers above unity to planetesimal sizes with significant gravitational cross sections is characterized by high collision speeds and slow, ordered growth ($\dot{R}$ is independent of size when $\mathcal{F}$ is constant), which does not benefit from increased particle sizes nor from decreased turbulent diffusion.

Mid-plane solid-to-gas ratios above unity are reached for St = 1 when $\delta < 10^{-4}$. Weaker stirring allows successively smaller particles to reach unity solid-to-gas ratios in the mid-plane. This marks an important transition to where particles exert a significant drag on the gas and become concentrated by streaming instabilities (*Youdin and Goodman*, 2005; *Johansen et al.*, 2009a; *Bai and Stone*, 2010a,c). This effect will be discussed further in the next section.

4. PARTICLE CONCENTRATION

High local particle densities can lead to the formation of planetesimals by gravitational instability in the sedimented mid-plane layer. Particle densities above the Roche density

$$\rho_R = \frac{9\Omega^2}{4\pi G} = 4.3 \times 10^{-7} \text{ g cm}^{-3} \left(\frac{r}{AU}\right)^{-3}$$
$$= 315\rho_g \left(\frac{r}{AU}\right)^{-1/4} \qquad (15)$$

are bound against the tidal force from the central star and can contract toward solid densities. The scaling with gas density assumes a distribution of gas according to the MMSN. Sedimentation increases the particle density in the mid-plane and can trigger bulk gravitational instabilities in the mid-plane layer (*Goldreich and Ward*, 1973). Sedimentation is nevertheless counteracted by turbulent diffusion (see section 3.4 and equation (12)) and particle densities are prevented from reaching the Roche density by global turbulence or by mid-plane turbulence induced by the friction of the particle on the gas (*Weidenschilling, 1980; Johansen et al., 2009a; Bai and Stone, 2010a*). High enough densities for gravitational collapse can nevertheless be reached when particles concentrate in the gas turbulence to reach the Roche density in local regions.

4.1. Passive Concentration

The turbulent motion of gas in protoplanetary disks can lead to trapping of solid particles in the flow. The size of an optimally trapped particle depends on the length-scale and turnover timescale of the turbulent eddies. We describe here particle trapping progressively from the smallest scales unaffected by disk rotation (*eddies*) to the largest scales in near-perfect geostrophic balance between the pressure gradient and Coriolis accelerations (*vortices* and *pressure bumps*).

We consider rotating turbulent structures with length scale ℓ, rotation speed v_e, and turnover time $t_e = \ell/v_e$. These quantities are approximate and all factors on the order of unity are ignored in the following. Dust is considered to be passive particles accelerating toward the gas velocity in the friction time τ_f, independent of the relative velocity. Particle trapping by streaming instabilities, where the particles play an active role in the concentration, is described in section 4.2. The three main mechanisms for concentrating particles in the turbulent gas flow in protoplanetary disks are sketched in Fig. 6.

4.1.1. Isotropic turbulence. On the smallest scales of the gas flow, where the Coriolis force is negligible over the turnover timescale of the eddies, the equation governing the structure of a rotating eddy is

$$\frac{dv_r}{dt} = -\frac{1}{\rho}\frac{\partial P}{\partial r} \equiv f_P \qquad (16)$$

Here f_P is the gas acceleration caused by the radial pressure gradient of the eddy. We use r as the radial coordinate

in a frame centered on the eddy. The pressure must rise outward, $\partial P/\partial r > 0$, to work as a centripetal force. In such low-pressure eddies the rotation speed is set by

$$f_P = -\frac{v_e^2}{\ell} \qquad (17)$$

Very small particles with $\tau_f \ll t_e$ reach their terminal velocity

$$v_p = -\tau_f f_P \qquad (18)$$

on a timescale much shorter than the eddy turnover timescale. This gives

$$v_p = -\tau_f f_P = \tau_f \frac{v_e^2}{\ell} = \frac{\tau_f}{t_e} v_e \qquad (19)$$

The largest particles to reach their terminal velocity in the eddy turnover timescale have $\tau_f \sim t_e$. This is the optimal particle size to be expelled from small-scale eddies and cluster in regions of high pressure between the eddies. Larger particles do not reach their terminal velocity before the eddy structure breaks down and reforms with a new phase, and thus their concentration is weaker.

Numerical simulations and laboratory experiments have shown that particles coupling at the turnover timescale of eddies at the Kolmogorov scale of isotropic turbulence experience the strongest concentrations (*Squires and Eaton, 1991; Fessler et al., 1994*). In an astrophysics context, such turbulent concentration of submillimeter-sized particles between small-scale eddies has been put forward to explain the narrow size ranges of chondrules found in primitive meteorites (*Cuzzi et al., 2001*), as well as the formation of asteroids by gravitational contraction of rare, extreme concentration events of such particles (*Cuzzi et al., 2008*). This model was nevertheless criticized by *Pan et al.* (2011), who found that efficiently concentrated particles have a

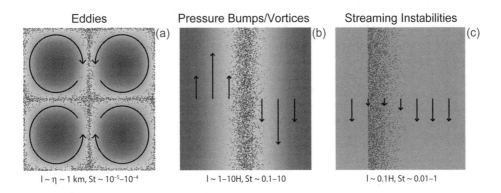

| Eddies | Pressure Bumps/Vortices | Streaming Instabilities |

Fig. 6. The three main ways to concentrate particles in protoplanetary disks. **(a)** Turbulent eddies near the smallest scales of the turbulence, η, expel tiny particles to high-pressure regions between the eddies. **(b)** The zonal flow associated with large-scale pressure bumps and vortices, of sizes from one scale height up to the global scale of the disk, trap particles of Stokes number from 0.1 to 10. **(c)** Streaming instabilities on intermediate scales trap particles of Stokes number from 0.01 to 1 by accelerating the pressure-supported gas to near the Keplerian speed, which slows down the radial drift of particles in the concentration region.

narrow size range and that concentration of masses sufficiently large to form the primordial population of asteroids is hard to achieve.

4.1.2. Turbulence in rigid rotation. On larger scales of protoplanetary disks, gas and particle motion is dominated by Coriolis forces and shear. We first expand our particle-trapping framework to flows dominated by Coriolis forces and then generalize the expression to include shear.

In a gas rotating rigidly at a frequency Ω, the equilibrium of the eddies is now given by

$$2\Omega v_e - \frac{1}{\rho}\frac{\partial P}{\partial r} = -\frac{v_e^2}{\ell} \tag{20}$$

For slowly rotating eddies with $v_e/\ell \ll \Omega$ we can ignore the centripetal term and get

$$v_e = -\frac{f_p}{2\Omega} \tag{21}$$

High-pressure regions have $v_e < 0$ (clockwise rotation), while low-pressure regions have $v_e > 0$ (counterclockwise rotation).

The terminal velocity of inertial particles can be found by solving the equation system

$$\frac{dv_x}{dt} = +2\Omega v_y - \frac{1}{\tau_f}v_x \tag{22}$$

$$\frac{dv_y}{dt} = -2\Omega v_x - \frac{1}{\tau_f}(v_y - v_e) \tag{23}$$

for $v_x \equiv v_p$. Here we have fixed a coordinate system in the center of an eddy with x pointing along the radial direction and y along the rotation direction at $x = \ell$. The terminal velocity is

$$v_p = \frac{v_e}{\left(2\Omega\tau_f\right)^{-1} + 2\Omega\tau_f} \tag{24}$$

Thus high-pressure regions, with $v_e < 0$, trap particles, while low-pressure regions, with $v_e > 0$, expel particles. The optimally trapped particle has $2\Omega\tau_f = 1$. Since we are now on scales with $t_e \gg \Omega^{-1}$, the optimally trapped particle has $\tau_f \ll t_e$ and thus these particles have ample time to reach their terminal velocity before the eddies turn over. An important feature of rotating turbulence is that the optimally trapped particle has a friction time that is *independent* of the eddy turnover timescale, and thus all eddies trap particles of $\tau_f \sim 1/(2\Omega)$ most efficiently.

4.1.3. Turbulence with rotation and shear. Including both Keplerian shear and rotation the terminal velocity changes only slightly compared to equation (24)

$$v_p = \frac{2v_e}{\left(\Omega\tau_f\right)^{-1} + \Omega\tau_f} \tag{25}$$

Equilibrium structures in a rotating and shearing frame are axisymmetric pressure bumps surrounded by super-Keplerian/sub-Keplerian zonal flows (Fig. 6). The optimally trapped particle now has friction time $\Omega\tau_f = 1$. This is in stark contrast to isotropic turbulence, where each scale of the turbulence traps a small range of particle sizes with friction times similar to the turnover timescale of the eddy. The eddy speed is related to the pressure gradient through

$$v_e = -\frac{1}{2\Omega}\frac{1}{\rho}\frac{\partial P}{\partial r} \tag{26}$$

Associating v_e with $-\Delta v$ defined in equation (8), we recover the radial drift speed

$$v_r = -\frac{2\Delta v}{St^{-1} + St} \tag{27}$$

However, this expression now has the meaning that particles drift radially proportional to the *local* value of Δv. Particles pile up where Δv vanishes, i.e., in a pressure bump.

4.1.4. Origin of pressure bumps. Trapping of $St \sim 1$ particles (corresponding to centimeter-sized pebbles to meter-sized rocks and boulders, depending on the orbital distance; see Fig. 3) in pressure bumps in protoplanetary disks has been put forward as a possible way to cross the meter-barrier of planetesimal formation (*Whipple,* 1972; *Haghighipour and Boss,* 2003). *Rice et al.* (2004) identified this mechanism as the cause of particle concentration in spiral arms in simulations of self-gravitating protoplanetary disks. Shearing box simulations of turbulence caused by the magneto-rotational instability (*Balbus and Hawley,* 1991) show the emergence of long-lived pressure bumps surrounded by a super-Keplerian/sub-Keplerian zonal flow (*Johansen et al.,* 2009b; *Simon et al.,* 2012). Similarly strong pressure bumps have been observed in global simulations (*Fromang and Nelson,* 2005; *Lyra et al.,* 2008b). High-pressure anti-cyclonic vortices concentrate particles in the same way as pressure bumps (*Barge and Sommeria,* 1995). Vortices may arise naturally through the baroclinic instability thriving in the global entropy gradient (*Klahr and Bodenheimer,* 2003), although the expression of the baroclinic instability for realistic cooling times and in the presence of other sources of turbulence is not yet clear (*Lesur and Papaloizou,* 2010; *Lyra and Klahr,* 2011; *Raettig et al.,* 2013).

Pressure bumps can also be excited by a sudden jump in the turbulent viscosity or by the tidal force of an embedded planet or star. *Lyra et al.* (2008a) showed that the inner and outer edges of the dead zone, where the ionization degree is too low for coupling the gas and the magnetic field, extending broadly from 0.5 to 30 AU in nominal disk models (*Dzyurkevich et al.,* 2013; see also the chapter by Turner et al. in this volume), develop steep pressure gradients as the gas piles up in the low-viscosity region. The inner edge is associated with a pressure maximum and can directly trap particles. Across the outer edge of the dead zone the pressure transitions from high to low, hence there is no local maximum that can trap particles. The hydrostatic equilibrium nevertheless breaks into large-scale particle-trapping vortices through the Rossby wave instability (*Li et al.,* 2001) in the variable-α simulations of *Lyra et al.* (2008a). *Dzyurkevich et al.* (2010) identified the inner pressure bump in resistive simulations of turbulence driven by the magneto-rotational instability. The jump in particle density, and hence in ionization degree, at the water snow line has been proposed to cause a jump in the surface density and act as a particle trap (*Kretke and Lin,* 2007; *Brauer et al.,* 2008b; *Drążkowska et al.,* 2013). However, the snow line pressure bump remains to be validated in magneto-hydrodynamical simulations.

Lyra et al. (2009) found that the edge of the gap carved by a Jupiter-mass planet develops a pressure bump that undergoes Rossby wave instability. The vortices concentrate particles very efficiently. Observational evidence for a large-scale vortex structure overdense in millimeter-sized pebbles was found by *van der Marel et al.* (2013). These particles have a Stokes number approximately unity at the radial distance of the vortex, in the transitional disk IRS 48. This discovery marks the first confirmation that dust traps exist in nature. The approximate locations of the dust-trapping mechanisms that have been identified in the literature are shown in Fig. 7.

4.2. Streaming Instability

The above considerations of passive concentration of particles in pressure bumps ignore the back-reaction friction force exerted by the particles onto the gas. The radial drift of particles leads to outward motion of the gas in the mid-plane, because of the azimuthal frictional pull of the particles on the gas.

Youdin and Goodman (2005) showed that the equilibrium streaming motion of gas and particles is linearly unstable to small perturbations, a result also seen in the simplified mid-plane layer model of *Goodman and Pindor* (2000). The eight dynamical equations (six for the gas and particle velocity fields and two for the density fields) yield eight linear modes, one of which is unstable and grows exponentially with time. The growth rate depends on both the friction time and the particle mass-loading. Generally the growth rate increases proportional to the friction time (up to St ~ 1), as the particles enjoy increasing freedom to move relative to the gas. The dependence on the particle mass-loading is more complicated; below a mass-loading of unity the growth rate increases slowly (much more slowly than linearly) with mass-loading, but after reaching unity in dust-to-gas ratio the growth rate jumps by one or more orders of magnitude. The e-folding timescale of the unstable mode is as low as a few orbits in the regime of high mass-loading.

The linear mode of the streaming instability is an exact solution to the coupled equations of motion of gas and particles, valid for very small amplitudes. This property can be exploited to test a numerical algorithm for solving the full nonlinear equations. *Youdin and Johansen* (2007) tested their numerical algorithm for two-way drag forces against two modes of the streaming instability, the modes having different wavenumbers and friction times. These

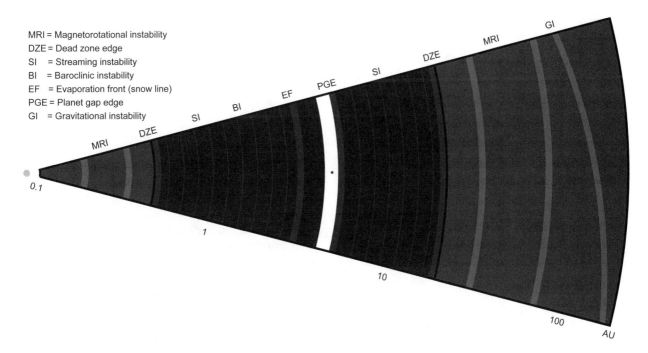

Fig. 7. Sketch of the particle concentration regions in a wedge of a protoplanetary disk seen from above. The left and right outer regions show areas where the magnetorotational instability is expected to operate, while the two inner regions show the extent of the dead zone in a nominal protoplanetary disk model. The particle trapping mechanisms are described in the main text.

modes have subsequently been used in other papers (e.g., *Bai and Stone,* 2010b) to test the robustness and convergence of numerical algorithms for the coupled dynamics of gas and solid particles.

The nonlinear evolution of the streaming instability can be studied either with or without particle stratification. The case without particle stratification is closest to the linear stability analysis. In this case the mean particle mass-loading in the simulation domain must be specified, as well as the friction time of the particles. The initial gas and particle velocities are set according to drag force equilibrium (*Nakagawa et al.,* 1986), with particles drifting in toward the star and gas drifting out.

In simulations not including the component of the stellar gravity toward the mid-plane, i.e., nonstratified simulations, high particle densities are reached mainly for particles with $St = 1$ (*Johansen and Youdin,* 2007; *Bai and Stone,* 2010b). At a particle mass-loading of 3 ($\times$ the mean gas density), the particle density reaches almost 1000$\times$ the gas density. The overdense regions appear nearly axisymmetric in both local and semiglobal simulations (*Kowalik et al.,* 2013). Smaller particles with $St = 0.1$ reach only between 20 and 60$\times$ the gas density when the particle mass-loading is unity or higher. Little or no concentration is found at a dust-to-gas ratio of 0.2 for $St = 0.1$ particles. *Bai and Stone* (2010b) presented convergence tests of nonstratified simulations in two-dimensions and found convergence in the particle density distribution functions at 1024^2 grid cells.

While nonstratified simulations are excellent for testing the robustness of a numerical algorithm and comparing to results obtained with different codes, stratified simulations including the component of the stellar gravity toward the mid-plane are necessary to explore the role of the streaming instability for planetesimal formation. In the stratified case the mid-plane particle mass-loading is no longer a parameter that can be set by hand. Rather its value is determined self-consistently in a competition between sedimentation and turbulent diffusion. The global metallicity $Z = \Sigma_p/\Sigma_g$ is now a free parameter and the mass-loading in the mid-plane depends on the scale height of the layer and thus on the degree of turbulent stirring.

Stratified simulations of the streaming instability display a binary behavior determined by the heavy element abundance. At Z around the solar value or lower, the mid-plane layer is puffed up by strong turbulence and shows little particle concentration. *Bai and Stone* (2010a) showed that the particle mid-plane layer is stable to Kelvin-Helmholtz instabilities thriving in the vertical shear of the gas velocity, so the mid-plane turbulence is likely a manifestation of the streaming instability which is not associated with particle concentration. Above solar metallicity very strong particle concentrations occur in thin and dominantly axisymmetric filaments (*Johansen et al.,* 2009a; *Bai and Stone,* 2010a). The metallicity threshold for triggering strong clumping may be related to reaching unity particle mass-loading in the mid-plane, necessary for particle pileups by the streaming instability. The particle density can reach several thousand times the local gas density

even for relatively small particles of $St = 0.3$ (*Johansen et al.,* 2012a). Measurements of the maximum particle density as a function of time are shown in Fig. 8, together with a column density plot of the overdense filaments.

Particles of $St = 0.3$ reach only modest concentration in nonstratified simulations. The explanation for the higher concentration seen in stratified simulations may be that the mid-plane layer is very thin, on the order of 1% of a gas scale height, so that slowly drifting clumps are more likely to merge in the stratified simulations. The maximum particle concentration increases when the resolution is increased, by a factor of ~4 each time the number of grid cells is doubled in each direction. This scaling may arise from thin filamentary structures that resolve into thinner and thinner filaments at higher resolution. The overall statistical properties of the turbulence nevertheless remain unchanged as the resolution increases (*Johansen et al.,* 2012a).

The ability of the streaming instability to concentrate particles depends on both the particle size and the local metallicity in the disk. Carrera et al. (in preparation) show that particles down to $St = 0.01$ are concentrated at a metallicity slightly above the solar value, while smaller particles down to $St = 0.001$ require significantly increased metallicity, e.g., by photoevaporation of gas or pileup of particles from the outer disk. Whether the streaming instability can explain the presence of millimeter-sized chondrules in primitive meteorites is still not known. Alternative models based on small-scale particle concentration (*Cuzzi et al.,* 2008) and particle sedimentation (*Youdin and Shu,* 2002; *Chiang and Youdin,* 2010; *Lee et al.,* 2010) may be necessary.

4.3. Gravitational Collapse

Planetesimals forming by gravitational contraction and collapse of the overdense filaments are generally found to be massive, corresponding to contracted radii between 100 km and 1000 km, depending on the adopted disk model (*Johansen et al.,* 2007, 2009a, 2011, 2012a; *Kato et al.,* 2012). Figure 9 shows the results of a high-resolution simulation of planetesimal formation through streaming instabilities, with planetesimals from 50- to 200-km radius forming when the model is applied to $r = 3$ AU. The formation of large planetesimals was predicted previously by *Youdin and Goodman* (2005) based on the available mass in the linear modes. The planetesimals that form increase in size with increasing disk mass (*Johansen et al.,* 2012a), with super-Ceres-sized planetesimals arising in massive disks approaching Toomre instability in the gas (*Johansen et al.,* 2011). Increasing the numerical resolution maintains the size of the largest planetesimals, but allows smaller-mass clumps to condense out of the turbulent flow alongside their massive counterparts (*Johansen et al.,* 2012a). Although most of the particle mass enters planetesimals with characteristic sizes above 100 km, it is an open question, which can only be answered with very-high-resolution computer simulations, whether there is a minimum size to the smallest planetesimals that can form in particle concentration models.

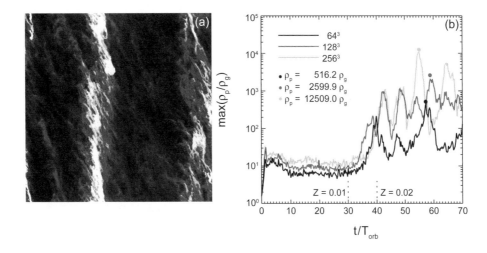

Fig. 8. (a) Particle concentration by the streaming instability is mainly elongated along the Keplerian flow direction showing the column density of particles with St = 0.3 in a local frame where the horizontal axis represents the radial direction and the vertical axis the orbital direction. The substructure of the overdense region appears fractal with filamentary structure on many scales. **(b)** Shown here is the measured maximum particle density as a function of time for three different grid resolutions. The metallicity is gradually increased from the initial Z = 0.01 to Z = 0.02 between 30 and 40 orbits, triggering strong particle concentration. Higher resolution resolves smaller embedded substructures and hence higher densities.

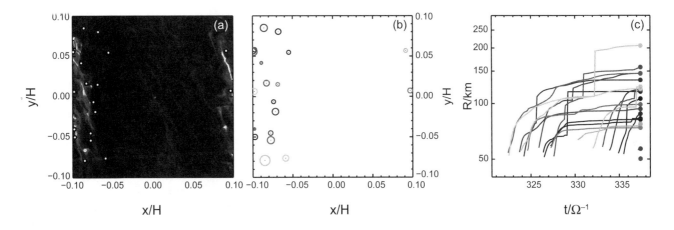

Fig. 9. Planetesimal formation in a particle filament formed by the streaming instability. **(a)** The column density of particles with St = 0.3, with dots marking the newly formed planetesimals. **(b)** The positions and Hill spheres of the planetesimals. **(c)** Contracted radii when the scale-free simulation is applied to the asteroid belt.

Strictly speaking, simulations of particle concentration and self-gravity only find the mass spectrum of gravitationally bound clumps. Whether these clumps will contract to form one or more planetesimals is currently not known in detail. *Nesvorný et al.* (2010) took bound particle clumps similar to those arising in numerical simulations and evolved in a separate N-body code with collisions and merging. The rotating clumps typically collapse to a binary planetesimal, with orbital properties in good agreement with the observed high binary fraction in the classical cold Kuiper belt (*Noll et al.,* 2008). The limited mass resolution that is allowed in direct N-body simulations nevertheless highlights the necessity for further studies of the fate of collapsing pebble clumps. Information from laboratory experiments

on coagulation, bouncing, and fragmentation of particles is directly applicable to self-gravitating particle clumps as well (*Güttler et al.,* 2010), and including realistic particle interaction in collapse simulations should be a high priority for future studies.

5. PARTICLE GROWTH

The size distribution of solid particles in protoplanetary disks evolves due to processes of coagulation, fragmentation, sublimation, and condensation. The particle concentration mechanisms discussed in the previous section are only relevant for particle sizes from millimeters to meters, much larger than the canonical submicrometer-sized dust and ice grains that enter the protoplanetary disk. Hence growth to macroscopic sizes must happen in an environment of relatively uniform particle densities.

Primitive meteorites show evidence of thermal processing of protoplanetary disk solids in the form of CAIs, which likely condensed directly from the cooling gas near the proto-Sun, and chondrules, which formed after rapid heating and melting of preexisting dust aggregates (section 2.2). Thermal processing must also have taken place near the water snow line, where the continued process of sublimation and condensation can lead to the formation of large hail-like ice particles (*Ros and Johansen,* 2013), or more distantly around the CO snow line identified observationally at 30 AU around the young star TW Hya (*Qi et al.,* 2013).

Direct sticking is nevertheless the most general mechanism for dust growth, as the densities in protoplanetary disks are high enough for such coagulation to be important over the lifetime of the gaseous disk. Physically, the collisional evolution is determined by the following factors:

1. The spatial density of particles. This will depend on both the location in the disk and on the degree of particle concentration (see section 4).

2. The collision cross section among particles, σ_{col}. An appropriate assumption is to take spherical particles, for which $\sigma_{col,ij} = \pi(R_i + R_j)^2$, where R_i and R_j are simply the radii of two particles in the absence of gravitational focusing.

3. The relative velocity among the particles, δv_{ij}. The relative velocity affects both the collision rate and the collision outcome.

4. The collision outcome, which is determined by the collision energy, porosity, and other collision parameters like the impact parameter.

The product of δv_{ij} and σ_{ij}, which together with the particle density determine the collision rate, is most often referred to as the collision kernel K_{ij}.

5.1. Numerical Approaches to the Coagulation Equation

The evolution of the dust size distribution is most commonly described by the Smoluchowski equation (*Smoluchowski,* 1916), which reads

$$\frac{dn(m)}{dt} = \frac{1}{2} \int dm' \, K(m', m-m') n(m') n(m-m') - n(m) \int dm' K(m, m') n(m') \tag{28}$$

where n(m) is the particle number density *distribution* [i.e., n(m)dm gives the number density of particles between mass m and m + dm]. The terms on the righthand side of equation (28) simply account for the particles of mass m that are removed due to collisions with any other particle and those gained by a collision between two particles of mass m' and m−m' (the factor of 1/2 prevents double counting).

In numerical applications one most often subdivides the mass grid in bins (often logarithmically spaced) and then counts the interactions among these bins.

However, equation (28), an already complex integro-differential equation, ignores many aspects of the coagulation process. For example, collisional fragmentation is not included and the dust spatial distribution is assumed to be homogeneous (i.e., there is no local concentration). Equation (28) furthermore assumes that each particle pair (m, m') collides at a single velocity and ignores the effects of a velocity distribution two particles may have. Indeed, perhaps the most important limitation of equation (28) is that the time-evolution of the system is assumed to be only a function of the particles' masses. For the early coagulation phases this is certainly inappropriate, as the dust internal structure is expected to evolve and influence the collision outcome. The collision outcomes depend very strongly on the internal structure (porosity, fractal exponent) of dust aggregates as well on their composition [ices or silicates (*Wada et al.,* 2009; *Güttler et al.,* 2010)].

In principle, the Smoluchowski coagulation equation can be extended to include these effects (*Ossenkopf,* 1993). However, binning in additional dimensions besides mass may render this approach impractical. To circumvent the multi-dimensional binning, *Okuzumi et al.* (2009) outlined an extension to Smoluchowski's equation, which treats the mean value of the additional parameters at every mass m. For example, the dust at a mass bin m is in addition characterized by a filling factor ϕ, which is allowed to change with time. If the distribution in ϕ at a certain mass m is expected to be narrow, such an approach is advantageous, as one may not have to deviate from Smoluchoski's one-dimensional framework.

A more radical approach is to use a direct-simulation Monte Carlo (MC) method, which drops the concept of the distribution function altogether (*Ormel et al.,* 2007). Instead, the MC method computes the collision probability between each particle pair of the distribution (essentially the kernel function K_{ij}), which depends on the properties of the particles. Random numbers then determine which particle pair will collide and the time-step that is involved. The outcome of the collision between these two particles must be summarized in a set of physically motivated recipes (the analogy with equation (28) is simply $m_1 + m_2 \rightarrow m$). A new set of collision rates is computed, after which the procedure repeats itself.

Since there are many more dust grains in a disk than any computer can handle, the computational particles should be chosen such to accurately *represent* the physical size distribution. A natural choice is to sample the mass of the distribution (*Zsom and Dullemond,* 2008). This method also allows fragmentation to be naturally incorporated. The drawback, however, is that the tails of the size distribution are not well resolved. The limited dynamic range is one of the key drawbacks of MC methods. *Ormel and Spaans* (2008) have described a solution, but at the expense of a more complex algorithm.

5.2. Laboratory Experiments

Numerical solutions to the coagulation equation rely crucially on input from laboratory experiments on the outcome of collisions between dust particles. Collision experiments in the laboratory as well as under microgravity conditions over the past 20 yr have proven invaluable for the modeling of the dust evolution in protoplanetary disks. Here we briefly review the state-of-the-art of these experiments (more details are given in the chapter by Testi et al. in this volume). A detailed physical model for the collision behavior of silicate aggregates of all masses, mass ratios, and porosities can be found in *Güttler et al.* (2010), and recent modifications are published by *Kothe et al.* (2013).

The three basic types of collisional interactions between dust aggregates are:

1. Direct collisional sticking. When two dust aggregates collide gently enough, contact forces (e.g., van der Waals forces) are sufficiently strong to bind the aggregates together. Thus, a more massive dust aggregate has formed. Laboratory experiments summarized by *Kothe et al.* (2013) suggest that there is a mass-velocity threshold $v \propto m^{-3/4}$ for spherical medium-porosity dust aggregates such that only dust aggregates less massive and slower than this threshold can stick. However, hierarchical dust aggregates still stick at higher velocities/masses than given by the threshold for homogeneous dust aggregates (*Kothe et al.*, 2013).

2. Fragmentation. When the collision energy is sufficiently high, similar-sized dust aggregates break up so that the maximum remaining mass decreases under the initial masses of the aggregates and a power-law-type tail of smaller-mass fragments is produced. The typical fragmentation velocity for dust aggregates consisting of micrometer-sized silicate monomer grains is 1 m s^{-1}.

3. Bouncing. Above the sticking threshold and below the fragmentation limit, dust aggregates bounce off one another. Although the conditions under which bouncing occurs are still under debate (see sections 6 and 7), some aspects of bouncing are clear. Bouncing does not directly lead to further mass gain so that the growth of dust aggregates is stopped ("bouncing barrier"). Furthermore, bouncing leads to a steady compression of the dust aggregates so that their porosity decreases to values of typically 60% (*Weidling et al.*, 2009).

Although the details of the three regimes are complex — regime boundaries are not sharp, and other more subtle processes exist — this simplified physical picture shows that in protoplanetary disks, in which the mean collision velocity increases with increasing dust-aggregate size (see Fig. 4), a *maximum aggregate mass* exists. Detailed numerical simulations using the dust-aggregate collision model by *Güttler et al.* (2010) find such a maximum mass (*Zsom et al.*, 2010; *Windmark et al.*, 2012b). For a MMSN model at 1 AU, the maximum dust-aggregate size is in the range of millimeters to centimeters (e.g., *Zsom et al.*, 2010).

Even if the bouncing barrier can be overcome, another perhaps even more formidable obstacle will present itself at the size scale where collision velocities reach ~1 m s^{-1} (Fig. 4). At this size, collisions among two silicate, similar-sized, dust particles are seen to fragment, rather than accrete. Consequently, growth will stall and the size distribution will, as with the bouncing case, settle into a steady-state (*Birnstiel et al.*, 2010, 2011, 2012). Highly porous icy aggregates may nevertheless still stick at high collision speeds (see section 7).

6. GROWTH BY MASS TRANSFER

Over the last two decades much effort has been invested to study experimentally the collisions among dust particles and to assess the role of collisions in planetesimal formation. As was seen in section 5, a plethora of outcomes — sticking, compaction, bouncing, fragmentation — is possible depending mainly on the collision velocity and the particle size ratio. Using the most updated laboratory knowledge of collision outcomes, *Zsom et al.* (2010, 2011) solved the coagulation equation with a MC method and found that in the ice-free, inner disk regions, the dust size distribution settles into a state dominated by millimeter-sized particles. Further growth is impeded because collisions among two such millimeter-sized particles will mainly lead to bouncing and compactification.

The real problem behind this "bouncing barrier" may, counterintuitively, not be the bouncing but rather the absence of erosion and fragmentation events. Based upon laboratory experiments (*Wurm et al.*, 2005; *Teiser and Wurm*, 2009a,b; *Kothe et al.*, 2010; *Teiser et al.*, 2011; *Meisner et al.*, 2013), it has become clear that above the fragmentation limit, the collision between two dust aggregates can lead to a mass gain of the larger (target) aggregate if the smaller (projectile) aggregate is below a certain threshold. For impact velocities in the correct range and relatively small projectile aggregates, up to 50% of the mass of the projectile can be firmly transferred to the target. This process has been shown to continue to work after multiple collisions with the target and under a wide range of impact angles (see Fig. 10).

Growth by mass transfer requires a high fraction of the dust mass present in small dust aggregates, which can be swept up by the larger particles. The number of small dust aggregates can be maintained by catastrophic collisions between larger bodies. A simple two-component model for such coagulation-fragmentation growth was developed by *Johansen et al.* (2008). They divided the size distribution into small particles and large particles and assumed that (1) a small particle will stick to a large particle, (2) a small particle will bounce off another small particle, and (3) a large particle will shatter another large particle. Small particles collide with large particles at speed v_{12}, while large particles collide with each other at speed v_{22}. Under these conditions an equilibrium in the surface density ratio of large and small particles, Σ_1/Σ_2, can be reached, with equal mass flux from large to small particles by fragmentation as from small to large particles by mass transfer in collisions.

In the equilibrium state where the mass flux from large to small particles balances the mass flux from small to large particles, the growth rate of the large particles is

$$\dot{R} = \frac{\Sigma_p / \left(\sqrt{2\pi} H_1 \right)}{\rho_\bullet} \times \frac{v_{22}}{\left[2 / \left(1 + H_1^2 / H_2^2 \right) \right]^{1/2} + 4 v_{22} / v_{12}} \quad (29)$$

Here H_1 and H_2 are the scale heights of the small particles and the large particles, respectively, and $\Sigma_p = \Sigma_1 + \Sigma_2$ is the total column density of the dust particles. The complexity of the equation arises from the assumption that the small grains are continuously created in collisions between the large particles. Setting $H_1 = H$ and $H_2 \ll H_1$, the equations are valid for a very general sweep-up problem where a few large particles sweep up a static population of small particles, resulting in growth of the large particles at the rate

$$\dot{R} = \frac{\varepsilon Z \rho_g}{4 \rho_\bullet} v_{12} = 2.7 \text{ mm yr}^{-1} \left(\frac{\varepsilon}{0.5} \right) \left(\frac{r}{AU} \right)^{-2.75} \times$$
$$\left(\frac{Z}{0.01} \right) \left(\frac{\rho_\bullet}{1 \text{ g cm}^{-3}} \right)^{-1} \left(\frac{v_{12}}{50 \text{ m s}^{-1}} \right) \quad (30)$$

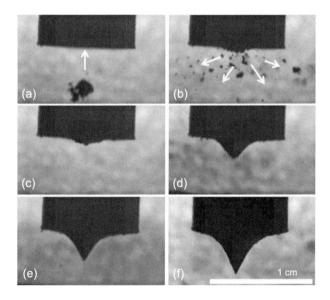

Fig. 10. Experimental example of mass transfer in fragmenting collisions. All experiments were performed in vacuum. **(a)** A millimeter-sized fluffy dust aggregate is ballistically approaching the centimeter-sized dusty target at a velocity of 4.2 m s⁻¹. Projectile and target consist of monodisperse SiO₂ spheres 1.5 μm in diameter. **(b)** Shortly after impact, most of the projectile's mass flies off the target in the form of small fragments (as indicated by the white arrows); part of the projectile sticks to the target. **(c)–(e)** The same target after **(c)** 3, **(d)** 24, **(e)** 74, and **(f)** 196 consecutive impacts on the same spot. Credit: Stefan Kothe, TU Braunschweig.

We introduced here a sticking coefficient ε, which measures either the sticking probability or the fraction of mass that is transferred from the projectile to the target. The growth rate is constant and depends only on the particle size through v_{12}. If the small particles are smaller than Stokes number unity and the large particles are larger, then we can assume that the flux of small particles is carried with the sub-Keplerian wind, $v_{12} \approx \Delta v$, and solve equation (30) for the timescale to cross the radial drift barrier. The end of the radial drift barrier can be set at reaching roughly St = 10. This Stokes number corresponds in the Epstein regime to the particle size

$$R_{10}^{(Ep)} = \frac{10 H \rho_g}{\rho_\bullet} \quad (31)$$

In the Stokes regime the particle size is

$$R_{10}^{(St)} = \sqrt{\frac{90 H \rho_g \lambda}{4 \rho_\bullet}} \quad (32)$$

Figure 3 shows that compact ice particles grow to Stokes number 10 in the Epstein regime outside r = 5 AU and in the Stokes or nonlinear regimes inside this radius. Fluffy ice particles with a very low internal density of $\rho_\bullet = 10^{-5}$ g cm⁻³ grow to Stokes number 10 in the nonlinear or quadratic regimes in the entire extent of the disk. In the Epstein regime the time to grow to Stokes number 0, $\tau_{10} = R_{10}/\dot{R}$, is independent of both material density and gas density

$$\tau_{10}^{(Ep)} = \frac{40 H}{\varepsilon Z v_{12}} \approx 25000 \text{ yr} \left(\frac{\varepsilon}{0.5} \right)^{-1} \left(\frac{r}{AU} \right)^{5/4} \quad (33)$$

The growth-time is much shorter in the Stokes regime where the Stokes number is proportional to the squared particle radius, giving

$$\tau_{10}^{(St)} = \frac{\sqrt{360 H \rho_\bullet (m/\sigma)}}{\varepsilon Z \rho_g v_{12}} \approx 500 \text{ yr} \times$$
$$\left(\frac{\varepsilon}{0.5} \right)^{-1} \left(\frac{\rho_\bullet}{1 \text{ g cm}^{-3}} \right)^{1/2} \left(\frac{r}{AU} \right)^{27/8} \quad (34)$$

with a direct dependence on both material density and gas density. Here m and σ are the mass and collisional cross section of a hydrogen molecule. Regarding the sticking coefficient, *Wurm et al.* (2005) find mass transfer efficiencies of $\varepsilon \approx 0.5$ for collision speeds up to $v_{12} = 25$ m s⁻¹, hence we have used $\varepsilon = 0.5$ as a reference value in the above equations. It is clearly advantageous to have a low internal density and cross the radial drift barrier in the Stokes regime (*Okuzumi et al.,* 2012). The nonlinear and quadratic regimes are similarly beneficial for the growth, with the timescale for crossing the meter barrier in the quadratic drag force regime approximately a factor $6c_s/\Delta v$ faster than in the Epstein regime (equation (33)).

The two-component model presented in *Johansen et al.* (2008) is analytically solvable, but very simplified in the collision physics. *Windmark et al.* (2012b) presented solutions to the full coagulation equation including a velocity-dependent mass transfer rate. They showed that artificially injected seeds of centimeter sizes can grow by sweeping up smaller millimeter-sized particles that are stuck at the bouncing barrier (*Zsom et al.*, 2010), achieving at 3-AU growth to 3 m in 100,000 yr and to 100 m in 1 m.y. The timescale to grow across the radial drift barrier is broadly in agreement with the simplified model presented above in equation (34). This time is much longer than the radial drift timescale, which may be as short as 100–1000 yr. It is therefore necessary to invoke pressure bumps to protect the particles from radial drift while they grow slowly by mass transfer. However, in that case the azimuthal drift of small particles vanishes and a relatively high turbulent viscosity of $\alpha = 10^{-2}$ must be used to regain the flux onto the large particles. The pressure bumps must additionally be very long-lived, on the order of the lifetime of the protoplanetary disk.

While the discovery of a pathway for growth by mass transfer at high-speed collisions is a major experimental breakthrough, it seems that this effect cannot in itself lead to widespread planetesimal formation, unless perhaps in the innermost part of the protoplanetary disk where dynamical timescales are short (equation (34)). The planetesimal sizes that are reached within a million years are also quite small, on the order of 100 m. These sizes are too small to undergo gravitational focusing even in weakly turbulent disks (see Fig. 5). An additional concern is the erosion of the preplanetesimal by tiny dust grains carried with the sub-Keplerian wind (*Schräpler and Blum*, 2011; *Seizinger et al.*, 2013).

Particles stuck at the bouncing barrier are at the lower end of the sizes that can undergo particle concentration and gravitational collapse. However, if a subset of "lucky" particles experience only low-speed, sticking collisions and manage to grow past the bouncing barrier, this can eventually lead to the crossing of the bouncing barrier by a very small fraction of the particles (*Windmark et al.*, 2012a, *Garaud et al.*, 2013), although this still requires a pressure bump to stop radial drift.

7. FLUFFY GROWTH

The last years have seen major improvements in N-body molecular-dynamics simulations of dust aggregate collisions by a number of groups (e.g., *Wada et al.*, 2008, 2009, 2011; *Paszun and Dominik*, 2008, 2009; *Seizinger and Kley*, 2013). These N-body simulations show that fluffy aggregates have the potential to overcome barriers in dust growth. In N-body simulations of dust aggregates, all surface interactions between monomers in contact in the aggregates are calculated by using a particle interaction model.

N-body simulations of dust aggregates have some merits compared to laboratory experiments. Precise information such as the channels through which the collisional energy is dissipated can be readily obtained and subsequent data analysis is straightforward. Another advantage of N-body simulations is that one can study highly fluffy aggregates, which are otherwise crushed under Earth's gravity. On the other hand, an accurate interaction model of constituent particles is required in N-body simulations of dust aggregates. In the interaction model used in most N-body simulations, the constituent particles are considered as adhesive elastic spheres. The adhesion force between them is described by the Johnson-Kendall-Roberts (JKR) theory (*Johnson et al.*, 1971). As for tangential resistive forces against sliding, rolling, or twisting motions, the model of Dominik and Tielens (DT) is used (*Dominik and Tielens*, 1995, 1996, 1997; *Wada et al.*, 2007). In order to better reproduce the results of laboratory experiments, the interaction model must be calibrated against laboratory experiments (*Paszun and Dominik*, 2008; *Seizinger et al.*, 2012; *Tanaka et al.*, 2012).

In protoplanetary disks, dust aggregates are expected to have fluffy structures with low bulk densities if impact compactification is negligible (e.g., *Okuzumi et al.*, 2009, 2012; *Zsom et al.*, 2010, 2011). Especially in the early stage of dust growth, low-speed impacts result in hit-and-stick of dust aggregates, with little or no compression. This growth mode makes fluffy aggregates with a low fractal dimension of ~2. Thus it is necessary to examine the outcome of collisions between fluffy dust aggregates and the resulting compression. The results of N-body simulations can then be used in numerical models of the coagulation equation, including the evolution of the dust porosity.

7.1. Critical Impact Velocity for Dust Growth

Dominik and Tielens (1997) first carried out N-body simulation of aggregate collisions, using the particle interaction model they constructed. They also derived a simple recipe for outcomes of aggregate collisions from their numerical results, although they only examined head-on collisions of two-dimensional small aggregates containing as few as 40 particles. The collision outcomes are classified into hit-and-stick, sticking with compression, and catastrophic disruption, depending on the impact energy. Bouncing of two aggregates was not observed in their simulations.

According to the DT recipe (*Dominik and Tielens*, 1997), growth is possible at collisions with $E_{imp} < An_kE_{break}$, where $A \sim 10$ and n_k is the total number of contacts in two colliding aggregates. For relatively fluffy aggregates, n_k is approximately equal to the total number of constituent particles in the two aggregates, N. The energy for breaking one contact between two identical particles, E_{break}, is given by (e.g., *Chokshi et al.*, 1993; *Wada et al.*, 2007)

$$E_{break} = 23\left[\gamma^5 r^4 \left(1 - v^2\right)^2 \Big/ \mathcal{E}^2\right]^{1/3} \qquad (35)$$

where r, γ, $\mathcal{E}$, and v are the radius, surface energy, Young's modulus, and Poisson's ratio of constituent particles, respectively. For icy particles the breaking energy is obtained as $5.2 \times 10^{-10}(r/0.1 \ \mu m)^{4/3}$ erg, while the value for silicate particles is $1.4 \times 10^{-11}(r/0.1 \ \mu m)^{4/3}$ erg, so icy aggregates are much more sticky than silicate aggregates.

Using the impact velocity, v_{imp}, the impact energy is expressed as $(Nm/4)v_{imp}^2/2$, where m is the mass of a constituent particle and Nm/4 is the reduced mass of two equal-sized aggregates. Substituting this expression into the above energy criterion for dust growth and noting that $n_k \simeq N$, we obtain the velocity criterion as

$$v_{imp} < \sqrt{8AE_{break}/m} \qquad (36)$$

Here A is a dimensionless parameter that needs to be calibrated with experiments. Note that this velocity criterion is dependent only on properties of constituent particles and, thus independent of the mass of the aggregates.

To evaluate the critical impact velocity for growth more accurately, *Wada et al.* (2009) performed N-body simulations of large aggregates made of up to $\sim 10^4$ submicrometer icy particles, including offset collisions (see Fig. 11). They consider two kinds of aggregate structure, the so-called cluster-cluster agglomeration (CCA) and particle-cluster

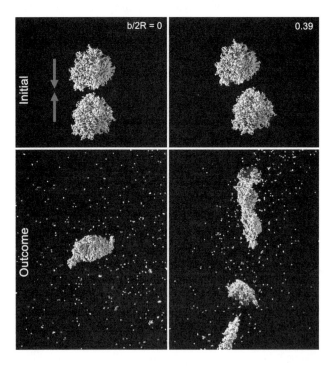

Fig. 11. Examples of collision outcomes of icy PCA clusters consisting of 8000 particles for two values of the impact parameter b (lower panels). The upper panels represent initial aggregates. The collision velocity is 70 m s⁻¹ in both cases and R is the radius of the initial aggregates. The head-on collision (left panel) results in sticking with minor fragmentation while aggregates tend to pass by each other in off-set collisions (right panels). Adapted from Fig. 2 of *Wada et al.* (2009), reproduced by permission of the AAS.

agglomeration (PCA) clusters. The CCA clusters have an open structure with a fractal dimension of 2, while the PCA clusters have a fractal dimension of 3 and a volume filling factor of 0.15. The PCA clusters are rather compact, compared to the CCA. Since dust aggregates are expected to be much more compact than CCA clusters due to compression (see section 7.2), the growth and disruption process of PCA clusters is of particular importance for elucidating the planetesimal formation.

For PCA clusters (composed of 0.1-μm-sized icy particles), the critical impact velocity is obtained as 60 m s⁻¹ from their simulations, independent of the aggregate mass within the mass range examined in the simulations of *Wada et al.* (2009). This indicates that icy dust aggregates can circumvent the fragmentation barrier and grow toward planetesimal sizes via collisional sticking. Note that the critical velocity actually increases with the aggregate mass when only head-on collisions are considered, as seen in Fig. 11. For an accurate evaluation of the critical velocity for growth, offset collisions should be considered as well, as in *Wada et al.* (2009).

The above result also fixes the corresponding constant A in the critical energy at 30. By using E_{break} of silicate particles in equation (36) with $A = 30$, the critical velocity for growth of silicate aggregates is obtained as $v_{imp} = 1.3(r/0.6 \ \mu m)^{-5/6}$ m s⁻¹. It agrees well with the laboratory experiments and numerical simulations for small silicate aggregates (e.g., *Blum and Wurm*, 2000; *Güttler et al.*, 2010; *Paszun and Dominik*, 2009; *Seizinger and Kley*, 2013).

In the case of CCA clusters, *Wada et al.* (2009) found that the constant A is the same as in the DT recipe ($\simeq 10$). Considering much smaller volume filling factors of CCA clusters than PCA, it indicates that the critical velocity is only weakly dependent on the volume filling factor.

7.2. Compression of Dust Aggregates

The DT recipe does not describe the amount of changes in the porosity (or the volume filling factor) at aggregate collisions. The first attempt to model porosity changes was done by *Ormel et al.* (2007), using simple prescriptions for the collision outcome of porosity.

Wada et al. (2008) examined compression at head-on collisions of two equal-sized CCA clusters, using a high number of N-body simulations. Compression (or restructuring) of an aggregate occurs through rolling motions between constituent particles. Thus it is governed by the rolling energy E_{roll} (i.e., the energy for rolling of a particle over a quarter of the circumference of another particle in contact). At low-energy impacts with $E_{imp} \lesssim E_{roll}$, aggregates just stick to each other, as indicated by the DT recipe. For higher-energy impacts, the resultant aggregates are compressed, depending on the impact energy. The radius of resulting aggregates, R, is fitted well with the power-law function

$$R \simeq 0.8 \left[E_{imp}/(N E_{roll}) \right]^{-0.1} N^{1/2.5} r \qquad (37)$$

where r is the radius of a monomer and N the number of constituent particles. Analyzing the structure of the compressed aggregates, it can be shown that the compressed aggregates have a fractal dimension of 2.5. This fractal dimension is consistent with equation (37) since equation (37) gives the relation of $N \propto R^{2.5}$ for maximally compressed aggregates (with $E_{imp} \sim NE_{roll}$). The compression observed in the N-body simulation is less much extreme than in the simple porosity model used by *Ormel et al.* (2007) because of the low fractal dimension of 2.5.

While *Wada et al.* (2008) examined the compression in a single collision, dust aggregates will be gradually compressed by successive collisions in realistic systems. In order to examine such a gradual compression process during growth, *Suyama et al.* (2008) performed N-body simulations of sequential collisions of aggregates. Even after multiple collisions, the compressed aggregates maintain a fractal dimension of 2.5. *Suyama et al.* (2012) further extend their porosity model to unequal-mass collisions.

Okuzumi et al. (2012) applied the porosity evolution model of *Suyama et al.* (2012) to dust growth outside the snow line in a protoplanetary disk. As a first step, collisional fragmentation is neglected because a relatively high impact velocity ($\sim$60 m s^{-1}) is required for significant disruption of icy dust aggregates. They found that dust particles evolve into highly porous aggregates (with bulk densities much less than 0.1 g cm^{-3}) even if collisional compression is taken into account. This is due to the ineffective compression at aggregate collisions in the porosity model of *Suyama et al.* (2012). Another important aspect of fluffy aggregates is that they cross the radial drift barrier in the Stokes or nonlinear drag force regimes (see Fig. 3 and discussion in section 6), where it is easier to grow to sizes where radial drift is unimportant. This mechanism accelerates dust growth at the radial drift barrier effectively and enables fluffy aggregates to overcome this barrier inside 10 AU (*Okuzumi et al.*, 2012; *Kataoka et al.*, 2013).

7.3. Bouncing Condition in N-Body Simulations

Until 2011, bouncing events were not reported in N-body simulations of aggregate collisions, while bouncing events have been frequently observed in laboratory experiments (e.g., *Blum and Münch*, 1993; *Langkowski et al.*, 2008; *Weidling et al.*, 2009, 2012). *Wada et al.* (2011) found bouncing events in their N-body simulations of both icy and silicate cases for rather compact aggregates with filling factor $\phi \geq$ 0.35. Such compact aggregates have a relatively large coordination number (i.e., the mean number of particles in contact with a particle), which inhibits the energy dissipation through the rolling deformation and helps bouncing. *Seizinger and Kley* (2013) further investigated the bouncing condition and proposed that the more realistic condition is $\phi > 0.5$.

This critical volume filling factor for bouncing is a few times as large as that in the laboratory experiments (e.g., *Langkowski et al.*, 2008). The origin of this discrepancy between these two approaches is not clear yet. From a qualitative point of view, however, laboratory experiments show that fluffy silicate aggregates with $\phi < 0.1$ tend to stick to each other (*Blum and Wurm*, 2000; *Langkowski et al.*, 2008; *Kothe et al.*, 2013). This qualitative trend is consistent with the N-body simulations. In protoplanetary disks, dust aggregates are expected to be highly porous. In the growth of such fluffy aggregates, the bouncing barrier would not be a strong handicap.

8. TOWARD A UNIFIED MODEL

Three main scenarios for the formation of planetesimals have emerged in the last years: (1) formation via coagulation-fragmentation cycles and mass transfer from small to large aggregates (*Wurm et al.*, 2005; *Windmark et al.*, 2012b), (2) growth of fluffy particles and subsequent compactification by self-gravity (*Wada et al.*, 2008, 2009), and (3) concentration of pebbles in the turbulent gas and gravitational fragmentation of overdense filaments (*Johansen et al.*, 2007, 2009a; *Kato et al.*, 2012).

The collision speeds at which planetesimal-sized bodies form are typically 50 m s^{-1} in models (1) and (2), as small dust aggregates are carried onto the growing planetesimal with the sub-Keplerian wind, or even higher if the turbulent density fluctuations are strong. During the gravitational collapse of pebble clouds, formation mechanism (3) above, the collision speeds reach a maximum of the escape speed of the forming body. In the cometesimal-formation zone any subsequent impacts by high-speed particles, brought in with the sub-Keplerian flow, will at most add a few meters of compact debris to the pebble pile over the life-time of the solar nebula (equation (30) applied to r = 10 AU).

This difference in impact velocity leads to substantial differences in the tensile strengths of the planetesimals/cometesimals. Due to the relatively high impact speeds in models (1) and (2), the growing bodies are compacted and possess tensile strengths on the order of 1–10 kPa (*Blum et al.*, 2006). *Skorov and Blum* (2012) pointed out that these tensile-strength values are too high to explain the continuous dust emission of comets approaching the Sun and favor model (3) above, for which the tensile strengths of loosely packed millimeter- to centimeter-sized pebbles should be much smaller.

Skorov and Blum (2012) base their comet-nucleus model on a few assumptions: (1) dust aggregates in the inner parts of the protoplanetary disk can only grow to sizes of millimeter to centimeter; (2) turbulent diffusion can transport dust aggregates to the outer disk; (3) in the outer disk, dust and ice aggregates become intermixed with a dust-to-ice ratio found in comets; and (4) cometesimals form via gravitational instability of overdense regions of millimeter- to centimeter-sized particles, which will form planetesimals with a wide spectrum of masses. While kilometer-scale planetesimals have not yet been observed to form in hydrodynamical simulations of planetesimal formation, this may be an artifact of the limited numerical resolution (see discussion in section 4.3).

If the comet nuclei as we find them today have been dormant in the outer reaches of the solar system since their formation and have not been subjected to intense bombardment and aqueous or thermal alteration, then they today represent the cometesimals of the formation era of the solar system. *Skorov and Blum* (2012) derive for their model tensile strengths of the ice-free dusty surfaces of comet nuclei

$$T = T_0 \left(\frac{r}{1\,\mathrm{mm}} \right)^{-2/3} \qquad (38)$$

with $T_0 = 0.5$ Pa and r denoting the radius of the dust and ice aggregates composing the cometary surface.

These extremely low tensile-strength values are, according to the thermophysical model of *Skorov and Blum* (2012), just sufficiently low to explain a continuous outgassing and dust emission of comet nuclei inside a critical distance to the Sun. This model provides a strong indication that kilometer-sized bodies in the outer solar system were formed by the gravitational contraction of smaller subunits (pebbles) at relatively low (on the order of meters per second) velocities. Such pebble-pile planetesimals are in broad agreement with the presence of large quantities of relatively intact millimeter-sized chondrules and CAIs in chondrites from the asteroid belt.

The cometesimals could thus represent the smallest bodies that formed by gravitational instability of millimeter- to centimeter-sized icy/rocky pebbles. These pebbles will continue to collide inside the gravitationally collapsing clump. Low-mass clumps experience low collision speeds and contract as kinetic energy is dissipated in inelastic collisions, forming small planetesimals that are made primarily of pristine pebbles. Clumps of larger mass have higher collision speeds between the constituent pebbles, and hence growth by coagulation-fragmentation cycles inside massive clumps determines the further growth toward one or more solid bodies that may go on to differentiate by decay of short-lived radionuclides. Therefore coagulation is not only important for forming pebbles that can participate in particle concentration, but continues inside the collapsing clumps to determine the birth size distribution of planetesimals.

There is thus good evidence that pebbles are the primary building blocks of planetesimals both inside the ice line (today's asteroids) and outside the ice line (today's comets and Kuiper belt objects). This highlights the need to better understand pebble formation and dynamics. The bouncing barrier is a useful way to maintain a high number of relatively small pebbles in the protoplanetary disk (*Zsom et al.,* 2010). As pebbles form throughout the protoplanetary disk, radial drift starts typically when reaching millimeter to centimeter sizes. Such small pebbles drift relatively fast in the outer disk but slow down significantly (to speeds on the order of 10 cm s^{-1}) in the inner few astronomical units. Pebbles with a high ice fraction fall apart at the ice line around 3 AU, releasing refractory grains, which can go on to form new (chondrule-like) pebbles inside the ice line (*Sirono,* 2011), as well as water vapor, which boosts

formation of large, icy pebbles outside the ice line (*Ros and Johansen,* 2013). Similar release of refractories and rapid pebble growth will occur at ice lines of more volatile species like CO (*Qi et al.,* 2013). In the optically thin very inner parts of the protoplanetary disk, illumination by the central star leads to photophoresis, which causes chondrules to migrate outward (*Loesche et al.,* 2013). Radial drift of pebbles will consequently not lead to a widespread depletion of planetesimal building blocks from the protoplanetary disk. Instead, pebble formation can be thought of as a continuous cycle of formation, radial drift, destruction, and reformation.

We propose therefore that particle growth to planetesimal sizes starts unaided by self-gravity, but after reaching Stokes numbers roughly between 0.01 and 1, proceeds inside self-gravitating clumps of pebbles (or extremely fluffy ice balls with similar aerodynamic stopping times). In this picture coagulation and self-gravity are not mutually exclusive alternatives, but rather two absolutely necessary ingredients in the multifaceted planetesimal formation process.

Acknowledgments. We thank the referee for useful comments that helped improve the manuscript. A.J. was supported by the Swedish Research Council (grant 2010-3710), the European Research Council under ERC Starting Grant agreement 278675-PEBBLE2PLANET, and the Knut and Alice Wallenberg Foundation. M.B. acknowledges funding from the Danish National Research Foundation (grant number DNRF97) and the European Research Council under ERC Consolidator grant agreement 616027-STARDUST2ASTEROIDS. C.W.O. acknowledges support for this work by NASA through Hubble Fellowship grant no. HST-HF-51294.01-A awarded by the Space Telescope Science Institute, which is operated by the Association of Universities for Research in Astronomy, Inc., for NASA, under contract NAS 5-26555. H.R. was supported by the Swedish National Space Board (grant 74/10:2) and the Polish National Science Center (grant 2011/01/B/ST9/05442). This work was granted access to the HPC resources of Ju-RoPA/FZJ, Stokes/ICHEC, and RZG P6/RZG, made available within the Distributed European Computing Initiative by the PRACE-2IP, receiving funding from the European Community's Seventh Framework Programme (FP7/2007-2013) under grant agreement no. RI-283493.

REFERENCES

A'Hearn M. F. et al. (2012) *Astrophys. J., 758,* 29.
Andrews S. M. and Williams J. P. (2005) *Astrophys. J., 631,* 1134.
Asphaug E. and Benz W. (1996) *Icarus, 121,* 225.
Bai X.-N. and Stone J. M. (2010a) *Astrophys. J., 722,* 1437.
Bai X.-N. and Stone J. M. (2010b) *Astrophys. J. Suppl., 190,* 297.
Bai X.-N. and Stone J. M. (2010c) *Astrophys. J. Lett., 722,* L220.
Bai X.-N. and Stone J. M. (2013) *Astrophys. J., 769,* 76.
Baker J. et al. (2005) *Nature, 436,* 1127.
Baker J. A. et al. (2012) *Geochim. Cosmochim. Acta, 77,* 415.
Balbus S. A. and Hawley J. F. (1991) *Astrophys. J., 376,* 214.
Barge P. and Sommeria J. (1995) *Astron. Astrophys., 295,* L1.
Bell K. R. et al. (1997) *Astrophys. J., 486,* 372.
Benz W. (2000) *Space Sci. Rev., 92,* 279.
Binzel R. P. and Xu S. (1993) *Science, 260,* 186.
Birnstiel T. et al. (2010) *Astron. Astrophys., 513,* A79.
Birnstiel T. et al. (2011) *Astron. Astrophys., 525,* A11.
Birnstiel T. et al. (2012) *Astron. Astrophys., 544,* A79.
Bizzarro M. et al. (2005) *Astrophys. J. Lett., 632,* L41.

Blandford R. D. and Payne D. G. (1982) *Mon. Not. R. Astron. Soc.,* *199,* 883.

Blum J. and Münch M. (1993) *Icarus, 106,* 151.

Blum J. and Wurm G. (2000) *Icarus, 143,* 138.

Blum J. et al. (2006) *Astrophys. J., 652,* 1768.

Bockelée-Morvan D. et al. (2004) *Icarus, 167,* 113.

Bottke W. F. et al. (2005) *Icarus, 175,* 111.

Brasser R. and Morbidelli A. (2013) *Icarus, 225,* 40.

Brauer F. et al. (2007) *Astron. Astrophys., 469,* 1169.

Brauer F. et al. (2008a) *Astron. Astrophys., 480,* 859.

Brauer F. et al. (2008b) *Astron. Astrophys., 487,* L1.

Brown M. E. (2008) In *The Solar System Beyond Neptune* (M. A. Barucci et al., eds.), pp. 335–344. Univ. of Arizona, Tucson.

Brownlee D. et al. (2006) *Science, 314,* 1711.

Bus S. J. and Binzel R. P. (2002) *Icarus, 158,* 146.

Carballido A. et al. (2006) *Mon. Not. R. Astron. Soc., 373,* 1633.

Chiang E. and Youdin A. N. (2010) *Annu. Rev. Earth Planet. Sci., 38,* 493.

Chokshi A. et al. (1993) *Astrophys. J., 407,* 806.

Ciesla F. J. et al. (2004) *Meteoritics & Planet. Sci., 39,* 1809.

Connelly J. N. et al. (2012) *Science, 338,* 651.

Cuzzi J. N. and Zahnle K. J. (2004) *Astrophys. J., 614,* 490.

Cuzzi J. N. et al. (2001) *Astrophys. J., 546,* 496.

Cuzzi J. N. et al. (2008) *Astrophys. J., 687,* 1432.

D'Alessio P. et al. (2005) In *Chondrites and the Protoplanetary Disk* (A. N. Krot et al., eds.), p. 353. ASP Conf. Ser. 341, Astronomical Society of the Pacific, San Francisco.

Davidsson B. J. R. et al. (2007) *Icarus, 187,* 306.

Desch S. J. and Connolly H. C. Jr. (2002) *Meteoritics & Planet. Sci., 37,* 183.

Dominik C. and Tielens A. G. G. M. (1995) *Philos. Mag. A, 72,* 783.

Dominik C. and Tielens A. G. G. M. (1996) *Philos. Mag. A, 73,* 1279.

Dominik C. and Tielens A. G. G. M. (1997) *Astrophys. J., 480,* 647.

Dominik C. et al. (2007) In *Protostars and Planets V* (B. Reipurth et al., eds.), pp. 783–800. Univ. of Arizona, Tucson.

Drążkowska J. et al. (2013) *Astron. Astrophys., 556,* A37.

Dubrulle B. et al. (1995) *Icarus, 114,* 237.

Dzyurkevich N. et al. (2010) *Astron. Astrophys., 515,* A70.

Dzyurkevich N. et al. (2013) *Astrophys. J., 765,* 114.

Fang J. and Margot J.-L. (2012) *Astrophys. J., 761,* 92.

Fessler J. R. et al. (1994) *Phys. Fluids, 6,* 3742.

Fleming T. and Stone J. M. (2003) *Astrophys. J., 585,* 908.

Fromang S. and Nelson R. P. (2005) *Mon. Not. R. Astron. Soc., 364,* L81.

Fuentes C. I. and Holman M. J. (2008) *Astron. J., 136,* 83.

Garaud P. et al. (2013) *Astrophys. J., 764,* 146.

Gladman B. J. et al. (2009) *Icarus, 202,* 104.

Goldreich P. and Ward W. R. (1973) *Astrophys. J., 183,* 1051.

Goldstein J. I. et al. (2009) *Chem. Erde–Geochem., 69,* 293.

Goodman J. and Pindor B. (2000) *Icarus, 148,* 537.

Gradie J. and Tedesco E. (1982) *Science, 216,* 1405.

Greenberg J. M. and Hage J. I. (1990) *Astrophys. J., 361,* 260.

Gressel O. et al. (2012) *Mon. Not. R. Astron. Soc., 422,* 1140.

Gustavsson K. and Mehlig B. (2011) *Phys. Rev. E, 84(4),* 045304.

Güttler C. et al. (2010) *Astron. Astrophys., 513,* A56.

Haghighipour N. and Boss A. P. (2003) *Astrophys. J., 598,* 1301.

Hayashi C. (1981) *Progr. Theor. Phys. Suppl., 70,* 35.

Holst J. C. et al. (2013) *Proc. Natl. Acad. Sci., 110,* 8819.

Ida S. et al. (2008) *Astrophys. J., 686,* 1292.

Johansen A. and Klahr H. (2005) *Astrophys. J., 634,* 1353.

Johansen A. and Youdin A. (2007) *Astrophys. J., 662,* 627.

Johansen A. et al. (2007) *Nature, 448,* 1022.

Johansen A. et al. (2008) *Astron. Astrophys., 486,* 597.

Johansen A. et al. (2009a) *Astrophys. J. Lett., 704,* L75.

Johansen A. et al. (2009b) *Astrophys. J., 697,* 1269.

Johansen A. et al. (2011) *Astron. Astrophys., 529,* A62.

Johansen A. et al. (2012a) *Astron. Astrophys., 537,* A125.

Johansen A. et al. (2012b) *Astrophys. J., 758,* 39.

Johnson K. L. et al. (1971) *Proc. R. Soc. London Ser. A, 324,* 301.

Kataoka A. et al. (2013) *Astron. Astrophys., 554,* A4.

Kato M. T. et al. (2012) *Astrophys. J., 747,* 11.

Klahr H. H. and Bodenheimer P. (2003) *Astrophys. J., 582,* 869.

Kleine T. et al. (2004) *Geochim. Cosmochim. Acta, 68,* 2935.

Kleine T. et al. (2009) *Geochim. Cosmochim. Acta, 73,* 5150.

Kothe S. et al. (2010) *Astrophys. J., 725,* 1242.

Kothe S. et al. (2013) *ArXiv e-prints,* arXiv:1302.5532K.

Kowalik K. et al. (2013) *ArXiv e-prints,* arXiv:1306.3937K.

Kretke K. A. and Lin D. N. C. (2007) *Astrophys. J. Lett., 664,* L55.

Krot A. N. et al. (2003) In *Treatise on Geochemistry, Vol. 1: Meteorites, Comets, and Planets* (A. M. Davis, ed.), p. 83. Elsevier-Pergamon, Oxford.

Krot A. N. et al. (2005) *Nature, 436,* 989.

Langkowski D. et al. (2008) *Astrophys. J., 675,* 764.

Laughlin G. et al. (2004) *Astrophys. J., 608,* 489.

Lee A. T. et al. (2010) *Astrophys. J., 718,* 1367.

Lesur G. and Papaloizou J. C. B. (2010) *Astron. Astrophys., 513,* A60.

Levison H. F. et al. (2010) *Astron. J., 139,* 1297.

Levison H. F. et al. (2011) *Astron. J., 142,* 152.

Li H. et al. (2001) *Astrophys. J., 551,* 874.

Lissauer J. J. et al. (2011) *Astrophys. J. Suppl., 197,* 8.

Loesche C. et al. (2013) *Astrophys. J., 778,* 101.

Luu J. X. and Jewitt D. C. (2002) *Annu. Rev. Astron. Astrophys., 40,* 63.

Lyra W. and Klahr H. (2011) *Astron. Astrophys., 527,* A138.

Lyra W. et al. (2008a) *Astron. Astrophys., 491,* L41.

Lyra W. et al. (2008b) *Astron. Astrophys., 479,* 883.

Lyra W. et al. (2009) *Astron. Astrophys., 493,* 1125.

Matzel J. E. P. et al. (2010) *Science, 328,* 483.

McKeegan K. D. et al. (2000) *Science, 289,* 1334.

McKinnon W. B. et al. (2008) In *The Solar System Beyond Neptune* (M. A. Barucci et al., eds.), pp. 213–241. Univ. of Arizona, Tucson.

Meisner T. et al. (2013) *Astron. Astrophys., 559,* A123.

Morbidelli A. et al. (2009) *Icarus, 204,* 558.

Nakagawa Y. et al. (1986) *Icarus, 67,* 375.

Nelson R. P. and Gressel O. (2010) *Mon. Not. R. Astron. Soc., 409,* 639.

Nelson R. P. and Papaloizou J. C. B. (2004) *Mon. Not. R. Astron. Soc., 350,* 849.

Nesvorný D. et al. (2010) *Astron. J., 140,* 785.

Nesvorný D. et al. (2011) *Astron. J., 141,* 159.

Noll K. S. et al. (2008) *Icarus, 194,* 758.

Ogihara M. et al. (2007) *Icarus, 188,* 522.

Oishi J. S. et al. (2007) *Astrophys. J., 670,* 805.

Okuzumi S. and Ormel C. W. (2013) *Astrophys. J., 771,* 43.

Okuzumi S. et al. (2009) *Astrophys. J., 707,* 1247.

Okuzumi S. et al. (2012) *Astrophys. J., 752,* 106.

Olsen M. B. et al. (2013) *Astrophys. J. Lett., 776,* L1.

Ormel C. W. and Cuzzi J. N. (2007) *Astron. Astrophys., 466,* 413.

Ormel C. W. and Okuzumi S. (2013) *Astrophys. J., 771,* 44.

Ormel C. W. and Spaans M. (2008) *Astrophys. J., 684,* 1291.

Ormel C. W. et al. (2007) *Astron. Astrophys., 461,* 215.

Ossenkopf V. (1993) *Astron. Astrophys., 280,* 617.

Pan L. and Padoan P. (2010) *J. Fluid Mech., 661,* 73.

Pan L. et al. (2011) *Astrophys. J., 740,* 6.

Paszun D. and Dominik C. (2008) *Astron. Astrophys., 484,* 859.

Paszun D. and Dominik C. (2009) *Astron. Astrophys., 507,* 1023.

Prialnik D. et al. (2004) In *Comets II* (M. C. Festou et al., eds.), pp. 359–387. Univ. of Arizona, Tucson.

Qi C. et al. (2013) *Science, 341,* 630.

Raettig N. et al. (2013) *Astrophys. J., 765,* 115.

Ragozzine D. and Brown M. E. (2009) *Astron. J., 137,* 4766.

Reipurth B. et al., eds. (2007) *Protostars and Planets V.* Univ. of Arizona, Tucson.

Rice W. K. M. et al. (2004) *Mon. Not. R. Astron. Soc., 355,* 543.

Rivkin A. S. et al. (2000) *Icarus, 145,* 351.

Ros K. and Johansen A. (2013) *Astron. Astrophys., 552,* A137.

Russell C. T. et al. (2012) *Science, 336,* 684.

Sanders I. S. and Taylor G. J. (2005) In *Chondrites and the Protoplanetary Disk* (A. N. Krot et al., eds.), p. 915. ASP Conf. Ser. 341, Astronomical Society of the Pacific, San Francisco.

Schiller M. et al. (2010) *Geochim. Cosmochim. Acta, 74,* 4844.

Schiller M. et al. (2011) *Astrophys. J. Lett., 740,* L22.

Schräpler R. and Blum J. (2011) *Astrophys. J., 734,* 108.

Scott E. R. D. (2007) *Annu. Rev. Earth Planet. Sci., 35(1),* 577–620.

Seizinger A. and Kley W. (2013) *Astron. Astrophys., 551,* A65.

Seizinger A. et al. (2012) *Astron. Astrophys., 541,* A59.

Seizinger A. et al. (2013) *Astron. Astrophys., 560,* A45.

Sekanina Z. (1997) *Astron. Astrophys., 318,* L5.

Shakura N. I. and Sunyaev R. A. (1973) *Astron. Astrophys., 24,* 337.

Shankman C. et al. (2013) *Astrophys. J. Lett., 764,* L2.

Sheppard S. S. and Trujillo C. A. (2010) *Astrophys. J. Lett., 723,* L233.
Siegler N. et al. (2007) *Astrophys. J., 654,* 580.
Simon J. B. et al. (2012) *Mon. Not. R. Astron. Soc., 422,* 2685.
Sirono S.-i. (2011) *Astrophys. J. Lett., 733,* L41.
Skorov Y. and Blum J. (2012) *Icarus, 221,* 1.
Smoluchowski M. V. (1916) *Z. Phys., 17,* 557.
Spivak-Birndorf L. et al. (2009) *Geochim. Cosmochim. Acta, 73,* 5202.
Squires K. D. and Eaton J. K. (1991) *Phys. Fluids, 3,* 1169.
Stevenson D. J. and Lunine J. I. (1988) *Icarus, 75,* 146.
Suyama T. et al. (2008) *Astrophys. J., 684,* 1310.
Suyama T. et al. (2012) *Astrophys. J., 753,* 115.
Tanaka H. and Ward W. R. (2004) *Astrophys. J., 602,* 388.
Tanaka H. et al. (2012) *Progr. Theor. Phys. Suppl., 195,* 101.
Teiser J. and Wurm G. (2009a) *Astron. Astrophys., 505,* 351.
Teiser J. and Wurm G. (2009b) *Mon. Not. R. Astron. Soc., 393,* 1584.
Teiser J. et al. (2011) *Icarus, 215,* 596.
Thomas P. C. et al. (2005) *Nature, 437,* 224.
Tremaine S. and Dong S. (2012) *Astron. J., 143,* 94.
Turner N. J. et al. (2006) *Astrophys. J., 639,* 1218.
van der Marel N. et al. (2013) *Science, 340,* 1199.
Voelk H. J. et al. (1980) *Astron. Astrophys., 85,* 316.
Wada K. et al. (2007) *Astrophys. J., 661,* 320.
Wada K. et al. (2008) *Astrophys. J., 677,* 1296.
Wada K. et al. (2009) *Astrophys. J., 702,* 1490.
Wada K. et al. (2011) *Astrophys. J., 737,* 36.
Weidenschilling S. J. (1977a) *Mon. Not. R. Astron. Soc., 180,* 57.

Weidenschilling S. J. (1977b) *Astrophys. Space Sci., 51,* 153.
Weidenschilling S. J. (1980) *Icarus, 44,* 172.
Weidenschilling S. J. (2011) *Icarus, 214,* 671.
Weidenschilling S. J. and Cuzzi J. N. (1993) In *Protostars and Planets III* (E. H. Levy and J. I. Lunine, eds.), pp. 1031–1060. Univ. of Arizona, Tucson.
Weidling R. et al. (2009) *Astrophys. J., 696,* 2036.
Weidling R. et al. (2012) *Icarus, 218,* 688.
Weissman P. R. et al. (2004) In *Comets II* (M. C. Festou et al., eds.), pp. 337–357. Univ. of Arizona, Tucson.
Wetherill G. W. (1992) *Icarus, 100,* 307.
Whipple F. L. (1972) In *From Plasma to Planet* (A. Elvius, ed.), p. 211. Wiley, New York.
Windmark F. et al. (2012a) *Astron. Astrophys., 544,* L16.
Windmark F. et al. (2012b) *Astron. Astrophys., 540,* A73.
Wurm G. et al. (2005) *Icarus, 178,* 253.
Wyatt M. C. (2008) *Annu. Rev. Astron. Astrophys., 46,* 339.
Yang C.-C. et al. (2009) *Astrophys. J., 707,* 1233.
Yang C.-C. et al. (2012) *Astrophys. J., 748,* 79.
Youdin A. and Johansen A. (2007) *Astrophys. J., 662,* 613.
Youdin A. N. and Goodman J. (2005) *Astrophys. J., 620,* 459.
Youdin A. N. and Lithwick Y. (2007) *Icarus, 192,* 588.
Youdin A. N. and Shu F. H. (2002) *Astrophys. J., 580,* 494.
Zsom A. and Dullemond C. P. (2008) *Astron. Astrophys., 489,* 931.
Zsom A. et al. (2010) *Astron. Astrophys., 513,* A57.
Zsom A. et al. (2011) *Astron. Astrophys., 534,* A73.

Gail H.-P., Trieloff M., Breuer D., and Spohn T. (2014) Early thermal evolution of planetesimals and its impact on processing and dating of meteoritic material. In *Protostars and Planets VI* (H. Beuther et al., eds.), pp. 571–593. Univ. of Arizona, Tucson, DOI: 10.2458/azu_uapress_9780816531240-ch025.

Early Thermal Evolution of Planetesimals and Its Impact on Processing and Dating of Meteoritic Material

Hans-Peter Gail and Mario Trieloff
Universität Heidelberg

Doris Breuer and Tilman Spohn
Deutsches Zentrum für Luft- und Raumfahrt, Berlin

Radioisotopic ages for meteorites and their components provide constraints on the evolution of small bodies: timescales of accretion, thermal and aqueous metamorphism, differentiation, cooling, and impact metamorphism. Realizing that the decay heat of short-lived nuclides (e.g., ^{26}Al, ^{60}Fe) was the main heat source driving differentiation and metamorphism, thermal modeling of small bodies is of the utmost importance to set individual meteorite age data into the general context of the thermal evolution of their parent bodies, and to derive general conclusions about the nature of planetary building blocks in the early solar system. As a general result, modeling easily explains that iron meteorites are older than chondrites, as early formed planetesimals experienced a higher concentration of short-lived nuclides and more severe heating. However, core formation processes may also extend to 10 million years (m.y.) after the formation of calcium-aluminum-rich inclusions (CAIs). A general effect of the porous nature of the starting material is that relatively small bodies (less than a few kilometers) will also differentiate if they form within 2 m.y. after CAIs. A particular interesting feature to be explored is the possibility that some chondrites may derive from the outer undifferentiated layers of asteroids that are differentiated in their interiors. This could explain the presence of remnant magnetization in some chondrites due to a planetary magnetic field.

1. INTRODUCTION

Radioisotopic ages for meteorites and their components provide constraints on the evolution of small bodies: timescales of accretion, thermal and aqueous metamorphism, differentiation, cooling, and impact metamorphism. There have been many debates about the nature of the heating source of small bodies, realizing that the decay heat of long-lived nuclides (e.g., ^{40}K, U, and Th) is insufficient. Most studies prefer the decay heat of ^{26}Al as a main heat source driving differentiation and metamorphism, and possibly ^{60}Fe, although recent studies (*Tang and Dauphas,* 2012; *Telus et al.,* 2012) indicate that its abundance is probably too low to induce major effects. Thermal modeling of small bodies is of the utmost importance to set individual meteorite age data into the general context of thermal evolution of their parent bodies, and to derive general conclusions about the nature of planetary building blocks in the early solar system.

For undifferentiated chondritic planetesimals, a number of thermal evolution models were constructed that yielded onion-shell-type parent body structures with strongly metamorphosed and slowly cooling rocks in deep planetesimal interiors, and weakly metamorphosed fast-cooled rocks in shallow outer layers. Models ranged from mere analytical

(e.g., *Miyamoto et al.,* 1982; *Trieloff et al.,* 2003; *Kleine et al.,* 2008) to numerical models considering initial porosity, sintering, or regolith development (e.g., *Bennett and McSween,* 1996; *Akridge et al.,* 1998; *Harrison and Grimm,* 2010; *Henke et al.,* 2012a,b, 2013; *Monnereau et al.,* 2013). There is no final agreement, however, if specific asteroids cooled in an undisturbed onion-shell-type structure, i.e., some authors argue that this structure was preserved — at least for a part of the H-chondrite asteroid — as long as 100 million years (m.y.), while others argue that this structure was disrupted and reassembled by a large impact event after reaching peak metamorphic temperatures (*Taylor et al.,* 1987; *Scott and Rajan,* 1981; *Ganguly et al.,* 2013; *Scott et al.,* 2013, 2014). However, the modeling of onion-shell-type bodies nevertheless allows insights into thermal evolution without additional complexities involving major collision events.

For differentiated planetesimals, numerical models were developed to describe, e.g., the differentiation of Vesta (*Ghosh and McSween,* 1998). Important processes that need to be considered are formation time, but also duration of accretion (*Merk et al.,* 2002), convection, and melt migration (*Hevey and Sanders,* 2006; *Sahijpal et al.,* 2007; *Moskovitz and Gaidos,* 2011; *Neumann et al.,* 2012). Only few models take into account initial porosity, sintering, the accretion process, and redistribution of heat sources during core and mantle/crust formation. Chronological constraints are derived from iron meteorites (e.g., *Kleine et al.,* 2005, 2009) and basaltic achondrites (e.g., *Bizzarro et al.,* 2005).

1.1. Meteorites and Their Components

Much of our knowledge about the formation of first solids and planetary bodies in the solar system is derived from studies of extraterrestrial matter. Meteorites were available for laboratory investigations long before extraterrestrial samples returned by space probes. Most meteorites are from asteroids, which in turn are the remnants of a swarm of planetesimals leftover from the planet-formation process 4.6 gigayears (G.y.) ago. The diversity of meteorite classes available in worldwide collections implies that they are derived from about 100 different parent bodies. Only a minority of meteorites are from asteroids that were differentiated like the terrestrial planets, with metallic Fe-Ni cores, silicate mantles, and crusts. These meteorites comprise various classes of iron meteorites and meteorites of mainly silicate composition (e.g., basaltic, lherzolitic, etc.). The largest group of differentiated silicate meteorites belong to the howardites, eucrites, and diogenites group (HEDs) and are commonly assumed to stem from Vesta, based on dynamical reasons and spectroscopy (see section 6). Another important group of basaltic meteorites are the angrites, for which no parent body has been conclusively assigned. Iron meteorites and basaltic achondrites yield surprisingly young radiometric ages (*Trieloff,* 2009), consistent with the picture that they formed and differentiated early in solar system history. This provides evidence for a heat source providing differentiation energy even for small bodies in the early solar system.

On the other hand, the majority of meteorites are undifferentiated. Such meteorites have to a first order solar abundances of Fe, where Fe is partly present as metal, and partly oxidized residing in silicates (see Table 1). The term "undifferentiated" also refers to the simultaneous existence of phases formed at low temperatures — represented by fine-grained matrix rich in volatile elements — and phases formed at high temperature. Prominent examples are rare, centimeter-sized calcium-aluminum-rich inclusions (CAIs) and abundant submillimeter- to millimeter-sized droplets, crystallized from liquid silicates, that are called chondrules — after which undifferentiated meteorites are named chondrites. These tiny objects formed by yet-unknown flash heating processes in the solar nebula, were rapidly cooled (on the order of 1 K hr^{-1} or K day^{-1}) and predate accumulation of planetesimals.

The two major classes of undifferentiated meteorites are ordinary chondrites — the most abundant class of meteorites delivered to Earth — and carbonaceous chondrites. Carbonaceous chondrites can have significant amounts of water and carbon and contain more matrix and fewer chondrules. Matrix often contains hydrous minerals resulting from ancient interaction of liquid water and primary minerals, indicating mild thermal heating of the parent body initiating fluid flow.

In general, chondrites represent undifferentiated, chemically "primitive" material with approximately solar element abundances of nonvolatile elements. However, the various chondrite groups have characteristic compositional differences, e.g., in major-element ratios like Mg/Si and Al/Si, or oxygen-isotopic composition. A further well-known classification criterion is the strong variation in the abundance of Fe in reduced (metallic) and oxidized form, i.e., the Fe content of silicates.

The most reduced groups are EH and EL enstatite chondrites, named after the Fe-free Mg-end member enstatite of the mineral group pyroxene. Here, Fe is not present in oxidized form in silicates; it occurs mainly in metals and sulfides. Ordinary chondrites are classified according to the total abundance of Fe, and the ratio of reduced (metallic) Fe to oxidized Fe: H (high total Fe), L (low total Fe), and LL (low total Fe, low metal). Carbonaceous chondrites are classified according to their contents of volatile elements, but also other parameters. They are named after a prominent member of the respective group: CI (Ivuna), CM (Mighei), CO (Ornans), CK (Karoonda), CV (Vigarano), and CR (Renazzo). Some of the characteristic properties of these meteorite groups are shown in Table 1.

Based on several lines of reasoning, each chondrite group and its unique composition is assigned to a specific independent parent body. Although they remained undifferentiated, chondritic rocks were heated to variable peak metamorphic temperatures described by the so called petrologic type. In general, petrologic type 3 represents the most pristine, unaltered material. In the case of (essentially dry) ordinary chondrites, petrologic types 4–6 describe a sequence of increasing metamorphism, at temperatures up to 1200–1300 K. In the case of water-bearing carbonaceous chondrites, the sequence of type 3-type 2-type 1 is a sequence of increasing intensity of aqueous metamorphism and increasing alteration by low-temperature fluids [see *Weisberg et al.* (2006) for details of classification].

The existence of different petrologic types among individual chondrite groups demonstrates that different heating effects and/or cooling conditions prevailed on individual parent bodies. Thermal modeling should explain these variable heating and cooling scenarios reflected by chondritic rocks.

1.2. Radioactives in the Early Solar System

A number of now-extinct short-lived and long-lived radioactive nuclei were present in the early solar system that play a key role for heating of planetesimals and for dating key processes during formation of the solar system (for a review, see, e.g., *McKeegan and Davis,* 2003; *Krot et al.,* 2009; *Dauphas and Chaussidon,* 2001). Table 2 gives an overview of some important radioisotopes present in the early solar system along with their daughter nuclides, decay modes, and initial abundances at the time of CAI formation. Radioactive isotopes are important for early solar system evolution in two respects:

TABLE 1. Properties of important chondrite groups.

Chondrite group	EH	EL	H	L	LL	CI	CM	CR[*]	CO	CK	CV
Chondrules[†] (vol%)	60–80	60–80	60–80	60–80	60–80	≪1	20	50–60	48	15	45
Avg. diameter[†] (mm)	0.2	0.6	0.3	0.7	0.9	–	0.3	0.7	0.15	0.7	1.0
Molten chondr. [‡] (%)	17	17	16	16	16	–	≈5	<1	4	<1	6
CAIs[†] (vol%)	0.1–1?	0.1–1?	0.1–1?	0.1–1?	0.1–1?	≪1	5	0.5	13	4	10
Matrix[†] (vol%)	<2–15?	<2–15?	10–15	10–15	10–15	>99	70	30–50	34	75	40
FeNi metal[†,§] (vol%)	8?	15?	10	5	2	0	0.1	5–8	1–5	<0.01	0–5
Fe_{metal}/Fe_{total}[**]	0.76	0.83	0.58	0.29	0.11	0	0	–	0–0.2	–	0–0.3
Fayalite in Olv[¶] (mol%)	–	–	16–20	23–26	27–32	–	–	–	–	–	–
Mg/Si[**]	0.75	0.85	0.95	0.93	0.94	1.05	1.05	1.06	1.06	1.08	1.05
Al/Si[**]	0.051	0.055	0.065	0.065	0.065	0.085	0.093	0.079	0.092	0.097	0.111
Ca/Si[**]	0.036	0.038	0.050	0.050	0.049	0.061	0.071	0.060	0.700	0.075	0.082
Fe/Si[**]	0.92	0.66	0.80	0.59	0.53	0.86	0.84	0.80	0.80	0.73	0.75
$(Refract./Mg)_{CI}$[††]	0.87	0.83	0.93	0.94	0.90	1.00	1.15	1.03	1.13	1.21	1.35
Petrol. types[¶]	3–5	3–6	3–6	3–6	3–6	1	2	2	3	3–6	3
Max. temp.[¶] (K)	1020	1220	1220	1220	1220	430	670	670	870	1220	870

Adapted from *Trieloff and Palme* (2006), with modifications.

[*]CR group without CH chondrites.
[†]From *Brearley and Jones* (1998).
[‡]From *Rubin and Wasson* (1995).
[§]In matrix.
[¶]From *Sears and Dodd* (1988).
[**]From *Hoppe* (2009).
[††]From *Scott et al.* (1996).

TABLE 2. Radioactives important for thermochronology and heating of planetary bodies.

Parent	Daughter	Half-Life[*] (yr)	Abundance[†]	Reference	Heating Rate (W Kg⁻¹)	Main Decay Modes	Percentage
^{26}Al	^{26}Mg	7.17×10^5	5.2×10^{-5}	^{27}Al	1.67×10^{-7}	$^{26}Al \rightarrow ^{26}Mg + \beta^+ + \nu$	82
						$^{26}Al + e \rightarrow ^{26}Mg + \bar{\nu}$	18
^{60}Fe	^{60}Ni	2.6×10^6	5.8×10^{-9}	^{56}Fe	2.74×10^{-8}	$^{60}Fe \rightarrow ^{60}Co + \beta^- + \nu$	100
		5.27×100				$^{60}Co \rightarrow ^{60}Ni + \beta^- + \nu$	100
^{53}Mn	^{53}Cr	3.7×10^6	6.3×10^{-6}	^{55}Mn		$^{53}Mn + e \rightarrow ^{53}Cr + \bar{\nu}$	100
^{182}Hf	^{182}W	8.9×10^6	9.7×10^{-5}	^{180}Hf		$^{182}Hf \rightarrow ^{182}Ta + \beta^- + \bar{\nu}$	100
		114.43 d				$^{182}Ta \rightarrow ^{182}W + \beta^- + \bar{\nu}$	100
^{129}I	^{129}Xe	1.6×10^7	$1.2 \times 10+$	^{127}I		$^{129}I \rightarrow ^{129}Xe + \beta^- + \bar{\nu}$	100
^{244}Pu	$^{131-136}Xe$	8.2×10^7	6.6×10^{-3}	^{238}U		$^{244}Pu \rightarrow$ spont. fission	
^{40}K	^{40}Ca	1.26×10^9	1.58×10^{-3}	^{39}K	2.26×10^{-11}	$^{40}K \rightarrow$ -Ca $+ \beta^- + \bar{\nu}$	89
	^{40}Ar					$^{40}K + e \rightarrow ^{40}Ar + \bar{\nu}$	10.7
^{232}Th	^{208}Pb	1.401×10^{10}	4.40×10^{-8}	Si	1.30×10^{-12}	$^{232}Th \rightarrow ^{208}Pb + 6\alpha + 4\beta$	
^{235}U	^{207}Pb	7.038×10^8	5.80×10^{-9}	Si	3.66×10^{-12}	$^{235}U \rightarrow ^{207}Pb + 7\alpha + 4\beta$	
^{238}U	^{206}Pb	4.468×10^9	1.80×10^{-8}	Si	1.92×10^{-12}	$^{238}U \rightarrow ^{206}Pb + 8\alpha + 7\beta$	

[*]Half-lives are from *Rugel et al.* (2009) for ^{60}Fe, from *Matson et al.* (2009) for ^{26}Al.
[†]Abundances for alive nuclei at formation time of the solar system are from *Lodders et al.* (2009), for extinct nuclei for ^{26}Al from *Jacobsen et al.* (2008), ^{60}Fe from *Tang and Dauphas* (2012), ^{53}Mn from *Trinquier et al.* (2008), ^{182}Hf from *Burkhardt et al.* (2008), ^{129}I from *Brazzle* (1999), ^{244}Pu from *Hudson et al.* (1989).

1. Radioisotopes can be used to trace the cooling histories of meteorites and their parent bodies via isotopic-dating methods (see section 3). While the use of long-lived nuclides like ^{238}U, ^{235}U, and ^{40}K are widespread in geochronology (U-Pb-Pb and Ar-Ar dating), short-lived nuclides with half-lives of <100 m.y., such as ^{26}Al, ^{53}Mn, ^{182}Hf, ^{129}I, or ^{244}Pu, are mainly used for meteorites that formed within the first tens of millions of years of the solar system. Ages based on one specific short-lived nuclide define only relative timescales that need to be calibrated to an absolute timescale (typically by fast-cooled precisely dated rocks called tie points). However, the chronologic interpretation of initial isotope abundances as "ages" requires a homogeneous distribution of the parent nuclide in the solar nebula, which is for some cases still a matter of debate.

2. Radioisotopes were furthermore an important heat source for early formed planetesimals and caused aqueous and thermal metamorphism, or even melting and differentiation. Particularly, ^{26}Al was present in an abundance that should have severe heating effects on early formed bodies larger than a few kilometers in size (see section 4.5).

2. INTERNAL CONSTITUTION AND THERMAL EVOLUTION OF PLANETESIMALS

The evolution of small planetary bodies such as asteroids begins with its accretion from the protoplanetary dust and ice as a porous aggregate (e.g., *Morbidelli et al.,* 2012) (see also the chapters by Johansen et al. and Raymond et al. in this volume). Heating by radioactive decay combined with self-gravity leads to substantial compaction at subsolidus temperatures and, in the case of icy bodies, to crystallization of amorphous ice and phase transitions of volatile species. Provided strong radiogenic heating by ^{26}Al and ^{60}Fe, the iron and silicate phases start to melt upon reaching the respective solidus temperature — heating by impacts has played a minor role in melting on bodies smaller than 1000 km (*Keil et al.,* 1997; *Srámek et al.,* 2012). If melting is sufficiently widespread, the planetary body can differentiate (at least partially) into an iron core and a silicate mantle. Differentiation is initiated by the iron melt sinking downward and silicate melt percolating toward the surface. In the case of icy bodies, melting of the ice starts before iron and silicates and thus we have a third component that is possibly separated. Whether a small planetary body remains undifferentiated or starts to melt and differentiate depends mainly on the composition, the accretion time associated with the available amount of short-lived radioactive elements, and the size of the body. The earlier the onset of accretion and the larger the body, the higher the interior temperatures and the more likely the differentiation (see Fig. 1). In the following we discuss in more detail the evolution and interior structure of the different types of small planetary bodies, i.e., undifferentiated, differentiated, and icy bodies.

2.1. Undifferentiated Bodies

In the initial state, the planetesimals are assumed to be highly porous as they initially grew as aggregates of dusty and unconsolidated material of the protoplanetary disks. As most meteorites are well consolidated (*Britt and Consolmagno,* 2003), the planetesimals have experienced some form of compaction from the unconsolidated and highly porous to consolidated state. Two subsequent stages of compaction are generally assumed: cold compaction simply due to self-gravitation (e.g., *Güttler et al.,* 2009), and hot pressing, i.e., so-called sintering, which is associated with the deformation of the crystals due to temperature and pressure (*Rao and Chaklader,* 1972; *Yomogida and Matsui,* 1984). The surface porosity is further influenced by impact processes that also result in a compaction of the dusty material (*Weidling et al.,*

2009). For planetary bodies bigger than 10 km most of the interior is already compacted by cold isostatic pressing to about a density close to that of the densest random packing of equal-sized spheres. The porosity ϕ (i.e., the fraction of voids per unit volume) of the closest random packing is about 0.36–0.4 (*Song et al.,* 2008; *Henke et al.,* 2012a). In the subsequent evolution, when the planetary body is heated by the radioactive decay, sintering starts in the deep interior where pressure and temperature are highest. The threshold temperature, at which sintering becomes active and the porosity rapidly approaches zero, is about 700–750 K assuming the simple treatment of sintering from *Yomogida and Matsui* (1984). With further heating, the compacted region moves toward the surface and the planetary body shrinks in size as the porosity (volume) decreases. Figure 2 shows a typical evolution of the radial distribution of the filling factor D (defined as D = 1−ϕ; note that in some of the literature the filling factor D is denoted as ϕ). Assuming an initial porosity of the planetary body between 40% and 60%, the radius shrinks by about 15–25% of its initial value in about 0.6 decay timescales of ^{26}Al.

A porous layer remains at the surface, as the threshold temperature of sintering cannot be reached there. The thickness of this porous layer varies with the onset time of formation and the size of the planetary body — the later the onset time and the smaller the body, the less heat that is provided by radioactive decay and the larger the relative thickness of the outer porous regolith layer. An important effect of sintering is the associated change in the thermal conductivity of the material. The thermal conductivity is about 2–3 orders of magnitude higher for compacted material in comparison to highly porous material (e.g., *Krause et al.,* 2011a,b) (see also section 4.3). Various thermal evolution models for planetesimals (*Yomogida and Matsui,* 1984; *Sahijpal et al.,* 2007; *Gupta and Sahijpal,* 2010; *Henke et al.,* 2012a; *Neumann et al.,* 2014a,b) have demonstrated the importance of sintering and the associated change in thermal conductivity: The porous layer thermally insulates the interior, resulting in temperatures in the outer layers that increase more rapidly than in the models assuming a homogeneous body without porosity changes (*Miyamoto et al.,* 1982). This observation in particular has consequences on the burial depths of meteorites and predicts shallower outer layers for the petrologic types (see section 5). It should be noted that sintering as well as the rate of change of the interior temperatures are dampened by any kind of non-instantaneous accretion (*Merk et al.,* 2002; *Neumann et al.,* 2012). Hence the thickness of the porous regolith layer increases with the duration of accretion.

The typical cooling history shows that the planetary body is first heated by decay of ^{26}Al and possibly ^{60}Fe and then cools down over an extended period. The temperature distribution in an undifferentiated but sintered planetesimal shows a typical onion-like thermal structure for which the temperature decreases from the center toward the surface (*Bennett and McSween,* 1996; *Harrison and Grimm,* 2010; *Henke et al.,* 2012a,b; *Monnereau et al.,* 2013). The

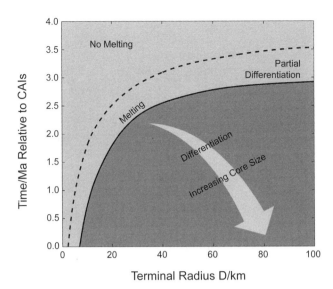

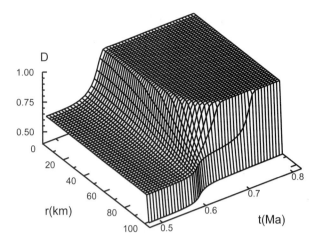

Fig. 1. Formation time (assuming instantaneous accretion) as a function of the terminal radius of a planetesimal indicating the parameter space for which a planetesimal forms a core (dark gray); melting occurs but no core differentiation (light gray area in between solid and dashed line) and no melting occurs (light gray area above dashed line). From *Neumann et al.* (2012).

Fig. 2. Evolution of the filling factor D by sintering of a body with initially 100 km radius and an inital filling factor of D = 0.64 (= random densest packing of equal-sized spheres). At about one half-life ($\tau_{1/2}$) of ^{26}Al after (instantaneous) formation of the body there is a sudden shrinking of the body due to onset of sintering. From *Henke et al.* (2012a).

coupling of accretion and sintering, however, leads to an episodic evolution of the central peak temperature with at least two heating phases during the first few million years (*Neumann et al.*, 2012). The heating by the long-lived isotopes of U, Th, and K becomes significant only after ≈8 m.y. after CAIs. These nuclides do not provide an effective heat source because in asteroid-sized bodies their heat generation rate is of the same magnitude as the rate of heat loss through the surface (*Yomogida and Matsui*, 1984). The temperature distribution is consistent with an onion-shell structure in which the degree of thermal metamorphism is a function of depth within the parent body. The timescale of the heating and cooling phase depends on the size of the body and its formation time — thus results from thermal evolution models can be used to fit the chronological data of meteorites to obtain the onset time of formation, the size of the parent body, and the burial depths of the meteorites (see sections 3 and 5).

2.2. Differentiated Bodies

In contrast to undifferentiated bodies, melting has occurred in the interior of differentiated bodies, and processes like heat transport via magma segregation or convection and redistribution of the radioactive isotopes play an important role in their thermochemical evolution — effects that also tend to dampen the interior temperature increase. As in the case of the undifferentiated bodies, in the initial stage the differentiated bodies accrete as porous material and compact due to cold compaction and hot pressing. Melting starts

when the solidus temperature of the iron and silicate phase is reached. Typically the iron-rich phase starts to melt at a lower temperature than the silicates for a volatile-poor body.

It has been shown that the heat production by the short-lived radioactive nuclei coupled with an initially porous state suffices to achieve melting even in a small planetesimal with a radius of only a few kilometers in case they form simultaneously with the CAIs (*Hevey and Sanders*, 2006; *Moskovitz and Gaidos*, 2011; *Henke et al.*, 2012a; *Neumann et al.*, 2012) (Fig. 1). With increasing formation time a larger planetesimal is required for melting to occur. After a formation time of about 2.8–3 m.y. melting and differentiation is unlikely, assuming a planetesimal with a composition consistent to ordinary chondrites (e.g., *Ghosh and McSween*, 1998; *Sahijpal et al.*, 2007; *Moskovitz and Gaidos*, 2011; *Neumann et al.*, 2012). This threshold may shift to later formation times if the material consists of volatiles reducing the melting temperature.

The occurrence of melt does not necessarily result in differentiation (core-mantle or mantle-crust differentiation) of the planetary body. Whether partial melt can separate from the solid matrix depends on the density contrast between the molten and the solid material, the amount of partial melt, and the melt network. In particular, an interconnected melt network is required as melt will not separate from the residual solid matrix if trapped in disconnected pockets. These pockets occur when the so-called dihedral angle exceeds a value of 60° and the melt fraction stays below a certain critical melt fraction (*von Bargen and Waff*, 1986). For a dihedral angle smaller than 60°, a stable interconnected melt channel occurs for all melt fractions. In most partial melts, including basic Si melts, the dihedral angle is less than 60° (*Beere*, 1975; *Waff and Bulau*, 1979). Thus, even if the silicate melt fraction is as small as a few percent, an interconnected network forms and is able to segregate (*Taylor et al., 1993; Maaloe, 2003*). For iron melt the situation is more controversial: Experimental studies — most at higher pressures than expected for melting in small planetary bodies — have shown that partial melting does not lead to metal melt migration unless a substantial fraction of partial melt is present (e.g., *Takahashi, 1983; Agee and Walker, 1988; Gaetani and Grove, 1999*). Only recent experiments suggest an interconnected melt network for pressures below 2–3 GPa (*Terasaki et al., 2008*). Furthermore, Fe,Ni-FeS veins have been identified in the acapulcoites and lodranites, suggesting melt migration for a low degree of melting (*McCoy et al., 1996*). However, these veins represent only short-distance migration, suggesting that a sulfide-rich early partial melt was not separated from the metal-rich residue (*McCoy et al., 1997*).

Different scenarios and relative timing between the core mantle and mantle-crust differentiation have been proposed, of which the following three main scenarios can be found in the literature: Iron segregation and core formation already occurs for a small melt fraction and even before silicate starts to melt (*Hewins and Newsom, 1988; Sahijpal et al., 2007*). Thus, mantle-crust differentiation follows core formation. In contrast to the first scenario, it is suggested that core formation requires a substantially larger melt fraction, including melt of the silicates. Assuming a melt threshold of about 40%, mantle-crust differentiation starts before core formation with an onset time of melt extrusions between 0.15 and 6 m.y. after the CAIs (*Gupta and Sahijpal, 2010*). In the third scenario, a threshold of more than 50% melt is assumed for metal to efficiently segregate into the center (e.g., *Taylor, 1992; Taylor et al., 1993; Righter and Drake, 1997; Drake, 2001; Hevey and Sanders, 2006; Gupta and Sahijpal, 2010; Moskovitz and Gaidos, 2011*), and therefore the presence of an early magma ocean is required to form a core. These models further imply that core formation is almost instantaneous and is followed — as in the first scenario — by mantle-crust differentiation. The existence of an early magma ocean on planetesimals that form with or shortly after the CAIs has been also suggested to explain the observed remnant magnetization in meteorites (*Weiss et al., 2008, 2010; Carporzen et al., 2010; Fu et al., 2012*). It should be noted, however, that in most models studying the influence of differentiation on the thermal evolution of a planetesimal, a certain scenario is prescribed, i.e., segregation of either silicates or iron occurs at a fixed melt fraction and is not calculated self-consistently.

A first attempt to model the differentiation process self-consistently via porous flow by *Neumann et al.* (2012) indicates that the timing of core formation with respect to the occurrence of silicate melting also depends on the amount of light elements, such as sulfur, in the iron phase, which lower the melting temperature: For a sulfur-rich composition, core formation is prior to silicate melting, which is similar to the findings by *Hewins and Newsom* (1988), but for a low sulfur content, as in fact suggested for H chondrites (*Keil, 1962*),

core formation starts simultaneously with silicate melting but is completed earlier than the mantle-crust differentiation. This difference in timing is caused by the variation in the migration velocity through porous media that depends on the melt fraction (among others). In the case of low sulfur content the temperature difference between solidus and liquidus of the Fe-FeS mixture is a few hundred degrees, whereas it is only a few tens of degrees for a high sulfur content. Thus, assuming the same temperature increase by radioactive heat sources and the same mass of the entire iron-rich phase, a smaller melt fraction is available for low sulfur content, which then migrates on timescales longer than the mean life of the short-lived radioactive isotopes — the temperature and the amount of melting increase before melt segregates. As a further consequence, core formation takes more than 2 m.y. and up to 10 m.y. for planetesimals that experience less than 50% degree of melting.

As mentioned above, for degrees of melting higher than about 50%, a magma ocean is present in a planetesimal. At this melt threshold a change in the matrix structure can be observed that is associated with a strong decrease of the viscosity to values representing the liquid material (≈0.1–1 Pas). In this region of high degree of melting, the melt and heat transport is not via porous flow, but vigorous convection can set in to efficiently cool the interior. A first numerical study to model the convective cooling of a magma ocean is by *Hevey and Sanders* (2006). They incorporate increased heat transfer by convection once the melt fraction exceeds a value of about 50%. Due to efficient cooling, further heating by ^{26}Al and ^{60}Fe does not increase the magma ocean temperature and thus the degree of partial melting, but increases the extent of the magma ocean — the planetesimal becomes a partly molten sphere undergoing vigorous convection below a thin rigid shell. A similar process has been suggested by *Weiss et al.* (2008) and *Elkins-Tanton et al.* (2011), who show that the efficient cooling of the magma ocean might be responsible for the generation of a magnetic field in the iron-rich core of parent bodies lasting more than 10 m.y. These studies, however, neglected the partitioning of ^{26}Al into the silicate melt. Considering that migration of ^{26}Al enriched melt toward the surface strongly dampens the internal temperatures even before a large-scale internal magma ocean can be generated (*Hevey and Sanders, 2006; Moskovitz and Gaidos, 2011; Neumann et al., 2014a*). Precluding a global magma ocean is supported by *Wilson and Keil* (2012), who suggest that volatile-rich magmas in the mantles would be removed very efficiently and only small amounts of melt in the interior would be present at any time.

Although details about the timing of the differentiation processes differ between the various models, most thermochemical models suggest that both chondritic meteorites and differentiated meteorites may originate from one and the same parent body (*Sahijpal and Soni, 2005; Bizzarro et al., 2005; Hevey and Sanders, 2006; Sahijpal et al., 2007; Weiss et al., 2010; Elkins-Tanton et al., 2011; Neumann et al., 2012; Formisano et al., 2013a*). These models show that a typical structure consists of a differentiated interior with an iron core and a silicate mantle that is covered by an undifferentiated upper layer with its lower part being sintered and compacted and its upper part still porous. An interior structure that may also explain the remnant magnetization of the CV meteorite Allende (*Weiss et al.,* 2010) — although the origin of the magnetization is controversially discussed (*Bland et al.,* 2011) — can develop as the basaltic magmas rising by porous flow cool and solidify while rising toward the surface, leaving the upper layer unaffected (*Elkins-Tanton et al.,* 2011; *Neumann et al.,* 2012). Furthermore, Lutetia, with its high bulk density of 3400 ± 300 kg m^{-3} (*Pätzold et al.,* 2011) and primitive surface, could also be a potential candidate (*Weiss et al.,* 2012; *Formisano et al.,* 2013b; *Neumann et al.,* 2013). An undisturbed primordial surface layer is, however, more difficult to preserve for a high degree of partial melting (e.g., early formation relative to CAIs) and further considering the enrichment of ^{26}Al into the melt (*Neumann et al.,* 2014a). This also applies in the case of an increase in melt buoyancy due to exsolution of volatiles (*Wilson and Keil,* 1991; *Keil and Wilson,* 1993) and/or a decrease in the resistance to flow due to the coalescence of melt channels into larger veins and dikes (*Wilson et al.,* 2008). These processes also tend to preclude a deep magma ocean (see above). Instead, a much more efficient transport of melt toward the surface and a "destruction" of the upper primordial layer is likely. In the work by *Wilson and Keil* (1991), the authors further argue that basaltic melt that contains more than a few hundred parts per million of volatiles is lost to space by explosive volcanism for parent bodies <100 km in radius. This process might explain the lack of observed asteroids with a basaltic crust. For larger bodies such as Vesta, which is one of the few asteroids showing a basaltic crust at its surface (*Binzel et al.,* 1997; *De Sanctis et al.,* 2012), explosive silicate eruptions cannot escape the gravity field and remain at the surface. In conclusion, an early formation time relative to the CAIs and/or a volatile-rich composition favors the existence of a basaltic crust (unless it is lost to space), whereas a late formation time and a dry composition is more in favor of an undifferentiated layer rather than a differentiated interior.

2.3. Icy Bodies

The thermal and structural evolution of a planetesimal can vary substantially depending on the concentration of water. The presence of water ice may moderate the temperature increase during heating, as substantial amounts of heat may be absorbed during melting. Moreover, the resulting warm melt water may redistribute heat within the body by hydrothermal convection and porous flow (*Grimm and McSween,* 1989). In addition, other processes that are not common in dry silicate bodies may influence the thermal evolution of icy bodies. Among these are exothermic processes such as the crystallization of amorphous ice [one of the major possible morphological transformations inside small icy planetary bodies, according to *Prialnik*

and Bar-Nun (1992) and *Prialnik and Podolak* (1999)], the formation of hydrous silicates (e.g., serpentine), and endothermic dehydration reactions. The release of latent heat through hydration reactions, in particular, may cause runaway melting of ice, as has been demonstrated by *Cohen and Coker* (2000). The initiation of the above processes depends mostly on temperature and occurs at times determined by the thermal evolution of the planetesimal, and they are usually associated with density changes in the interiors. Expansion or shrinkage of the planetesimals due to density changes may even shape the surfaces of these bodies (see Fig. 3). The following list details the processes in order of occurrence during heating of an icy planetesimal:

1. The exothermic transition from amorphous to crystalline ice occurs at about 100 K (*Merk and Prialnik,* 2003, 2006), thereby the density of ice (940 kg m^{-3} for amorphous ice) decreases, causing expansion and an increase in thermal conductivity of the ice by a factor of ≈20.

2. Water-ice melting is an endothermic reaction, and during melting of ice the density increases to 1000 kg m^{-3}, causing shrinkage of the body. However, with further increasing temperature up to the evaporation temperature, the density again decreases and results in expansion. Ice melting is also related to hydrothermal convection and differentiation of ice, water, and silicates and the associated heat transport.

3. Hydration of silicate minerals (e.g., the formation of serpentine from forsterite and H$_2$O) can occur simultaneously with ice melting but across a larger temperature interval. Hydration reduces the density of the rock and is an exothermic reaction.

4. The dehydration of silicates occurring at higher temperatures is endothermic and consumes more energy than is produced by hydration. Release of the water bonded in minerals increases the density of the rock phase and causes shrinking of the body.

5. Silicate and iron melting are both endothermic processes that require energy and are associated with expansion of the body. The temperature interval for which melting occurs depends on the composition of the planetesimals and in particular on the amount of water in the anhydrous mineral phase. Iron and silicate melting are also related to porous flow, differentiation of silicates and iron, as well as the associated heat transport.

Wakita and Sekiya (2011) investigated the occurrence of liquid water in bodies with radii of up to 1000 km that are heated by ^{26}Al. They considered differentiation of ice and rock via Stokes flow and hydration reactions, and varied the time of the (instantaneous) formation and the initial temperature. According to their results, icy planetesimals that formed 2.4 m.y. after CAIs will not experience melting of water ice and thus differentiation, regardless of their size. It should be noted, however, that this conclusion must likely be revised toward later formation times if long-lived nuclides and ^{60}Fe are taken into account (e.g., *McCord and Sotin,* 2005). The thermal history of an icy body (mostly depending on the formation time relative to the CAIs), de-

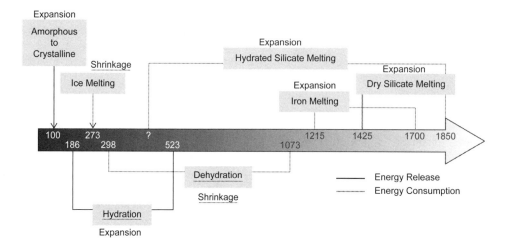

Fig. 3. Sketch showing processes that occur at certain temperatures or temperature ranges in icy planetesimals. Numbers are temperature in Kelvin; note that absolute values will vary with the assumed composition. Considering that a body heats up with time during and after its accretion, these processes change subsequently during the thermal evolution and influence the temperature and the structural evolution of a planetesimal. Note that when the body cools after the initial heating phase, energy will be released by crystallization of melt and the volume changes are reversed.

pending on which of the above processes actually occurred, may result in a variety of interior structures. Figure 4 shows several possibilities ranging from a homogeneous body made of a mixture of H₂O and high-density silicates to a fully differentiated structure with even an inner iron core. Whether the latter structure is possible, however, is questionable for small icy bodies because large amounts of ²⁶Al or early accretion may instead result in a dehydrated planetesimal, as water may vaporize and escape the accreting body (*McCord and Sotin*, 2005).

Among the icy bodies, the asteroid Ceres, the second target of the Dawn mission, is of particular interest and its thermal evolution has been repeatedly modeled. Ceres is the largest body in the asteroid belt with a radius of 470 km, and its low bulk density of 2077 ± 36 kg m⁻³ (*Thomas et al.*, 2005) suggests a global ice mass fraction of 17–27 wt.% if the average porosity is negligible (*McCord and Sotin*, 2005). The surface material displays spectral properties of carbonates, brucite, and carbon-rich clays (*Rayner et al.*, 2003; *Cohen et al.*, 1998; *Rivkin et al.*, 2006; *Milliken and Rivkin*, 2009). Based on the inferred surface composition, the primordial material of Ceres has a composition like that of carbonaceous chondrites and was formed during the first 10 m.y. of the solar system. Considering the results of both spectral data and mapping, Ceres is most likely an ice-rich world whose surface has substantially been changed by aqueous alteration, which could be a result of the eruptions driven by interior aqueous activity and subsequent sublimation of ice from the surface. Assuming an ice-silicate composition, thermal models by *McCord and Sotin* (2005), which considered short- and long-lived radioactive nuclide heating, suggest that Ceres is likely differentiated into a

rocky core, an icy mantle, and possibly a shallow still-liquid layer above the core (Figs. 4b,c). Even if only long-lived radionuclide heating is assumed, the water ice in Ceres would melt quickly and a water mantle would form, except for an upper crust that would not melt. Whether this structure actually formed depends critically on the amount of water present at the time of its accretion, as well as the amount of ²⁶Al present in the pre-Ceres objects. The likely presence of a liquid layer above the silicate core has been supported by *Castillo-Rogez and McCord* (2010), who emphasize that the surface temperatures of 180 K would increase the likelihood of a deep ocean in today's Ceres, provided a sufficient concentration of ammonium and salts (Fig. 4c) are present.

3. THERMOCHRONOMETRY OF METEORITES

The physical properties (e.g., size, formation time, short-lived nuclide concentration, initial porosity, and heat conduction) of meteorite parent bodies significantly influence their cooling history. Isotopic dating of meteorites can define points in time when cooling caused temperatures to fall below a certain critical value called "closure temperature."

The age of a mineral or rock, usually expressed in million years (Ma) or billion years (Ga), can be understood as the accumulation time of a daughter isotope by radioactive decay from a parent nuclide. The system to be dated must meet certain conditions to obtain geologically meaningful ages:

1. "Closed system" condition: During the accumulation time no loss, gain (except radioactive decay), or diffusion processes affecting concentration profiles of both parent and daughter nuclides should occur, unless such changes can be corrected for.

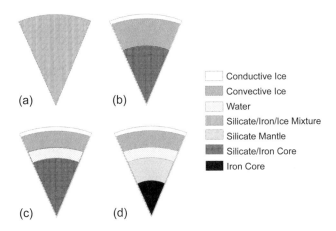

Fig. 4. Structural end members of small icy bodies based on different evolution scenarios: **(a)** homogeneous body made of a mixture of H_2O and high-density silicates; **(b)** differentiated body with high-density silicate core or low-density serpentine and outer ice layer; **(c)** same as **(b)**, but with the presence of anti-freeze material (ammonia) maintaining a liquid layer; and **(d)** fully differentiated model with an inner iron core. Adapted from *McCord and Sotin* (2005).

2. During formation of the system, parent and daughter elements were fractionated, i.e., partitioned differently into different phases due to their different geochemical behavior. On the other hand, daughter element isotopes were equilibrated, i.e., all cogenetic phases have identical daughter isotope ratios. This allows the determination of the initial abundance of daughter atoms in order to quantify the daughter isotope excess due to *in situ* radioactive decay.

If both of these conditions are fulfilled, radioisotope ages date the event of fractionation of parent from daughter element. This may be caused by the crystallization of cogenetic minerals, yielding the crystallization age of a rock. Strictly speaking, these ages are cooling ages, as the radiometric clock only starts below the so-called closure temperature when diffusion of daughter element isotopes becomes ineffective. Closure temperatures depend on the dating system and the diffusion properties of isotopes in the mineral(s) dated. Particularly low closure temperatures are inherent to isotopic dating systems involving high diffusion coefficients such as noble gases. Cooling ages may significantly postdate the crystallization of a rock, if cooling required long intervals of time and closure temperatures are low.

The following radioisotopic dating systems can be used to infer constraints on a meteorite's cooling history:

1. ^{182}Hf-^{182}W: This system is based on the decay of the short-lived nuclide ^{182}Hf and dates the separation of metal and silicate during parent-body metamorphism or differentiation. For chondrite metamorphic processes, the closure temperature of 1150 ± 75 K for H6 chondrites (*Kleine et al.,* 2008) was inferred by lattice strain models.

2. U-Pb-Pb: This system is based on the simultaneous decay of ^{238}U to ^{206}Pb and ^{235}U to ^{207}Pb. High precision ages are usually obtained by phases rich in U, e.g., phosphates (*Göpel et al.,* 1994; *Blinova et al.,* 2007). Phosphates date cooling through the closure temperature of 720 K as inferred from diffusion experiments (*Cherniak and Watson,* 2001). Silicates hamper Pb diffusion at higher temperatures; *Amelin et al.* (2005) estimated 1050 ± 100 K for pyroxenes of chondrules from chondritic meteorites. Silicate dates (*Amelin et al.,* 2005; *Blinova et al.,* 2007; *Bouvier et al.,* 2007) are usually less precise, and are used only if the maximum temperature to which the meteorite was subjected was comparable or exceeded the closure temperature of 1050 ± 100 K. For mildly metamorphosed meteorites, this is frequently not the case.

3. ^{26}Al-^{26}Mg: The ^{26}Al decays with a half-life of 0.72 m.y. There are many Al-Mg dates for CAIs and chondrules, and very few for parent-body processes. For example, two dates exist for the formation and cooling of secondary feldspar for the H chondrites Forest Vale and Ste. Marguerite (*Zinner and Göpel,* 2002). Diffusion data exist for Ca-rich feldspar (*LaTourette and Wasserburg,* 1998), from which a closure temperature of 750 ± 130 K can be calculated for a cooling rate of 1000 K m.y.$^{-1}$ and 2-μm-sized feldspar grains, using the *Dodson* (1973) formula.

4. ^{40}Ar-^{39}Ar: This is a classical technique based on the dual decay of ^{40}K to ^{40}Ar and ^{40}Ca; numerous whole-rock Ar-Ar data exist and few data exist for mineral separates, e.g., feldspar that usually dominates whole-rock ages of equilibrated ordinary chondrites. Argon diffusion data were used to infer a closure temperature of 550 ± 20 K (*Trieloff et al.,* 2003). Currently there is a debate about the revision of the K decay constant (*Schwarz et al.,* 2011; *Renne et al.,* 2010). However, most recent recommendations (*Schwarz et al.,* 2012; *Renne et al.,* 2011) agree that the *Steiger and Jäger* (1977) convention should be revised upward by 22–28 Ma at an absolute age of 4500 Ma.

5. ^{244}Pu fission tracks: This method uses tracks from the now-extinct nuclide ^{244}Pu registered by phosphates and adjacent silicates (e.g., orthopyroxene). Due to different retention temperatures [550 K for orthopyroxene, 390 K for merrillite obtained by annealing experiments (*Pellas et al.* 1997)], minerals can record different fission track densities in the case of slow cooling, whereas fast cooling results in indistinguishable densities. Due to complex correction procedures — for irradiation geometry, cosmic-ray tracks, and cosmic-ray spallation recoil tracks – this method is very time consuming, which explains why only few groups worldwide have applied this technique (*Pellas et al.,* 1997; *Trieloff et al.,* 2003).

6. ^{53}Mn-^{53}Cr: For this system, isotopic ages can be obtained for certain mineral phases as well as acid leachates. The closure temperature for orthopyroxene and olivine for typically 100-μm-sized grains is around 800–900 K (*Ito and Ganguly,* 2006; *Ganguly et al.,* 2007).

7. ^{129}I-^{129}Xe: The ^{129}I-^{129}Xe system can be applied to phosphate or feldspar in ordinary chondrites as well as hydrous phases in carbonacous chondrites (*Brazzle,* 1999). The closure temperature of the I-Xe system is largely

Legend:
- ☐ Conductive Ice
- Convective Ice
- Water
- Silicate/Iron/Ice Mixture
- Silicate Mantle
- Silicate/Iron Core
- ■ Iron Core

unconstrained; however, the agreement between some Pb-Pb and I-Xe phosphate data for slowly cooled ordinary chondrites suggest a similar closure temperature of 720 K for the I-Xe system in chondritic phosphates.

In order to derive meaningful constraints on the parent-body thermal evolution, it is necessary to ascertain that the thermochronological data are due to the primordial cooling history of the parent body, and not due to secondary impact-induced reheating or other shock effects. As meteorites are derived from asteroids that were fragmented by collisions occurring throughout the 4.6-G.y. history of the solar system, meteorite samples that display shock metamorphism and implicitly experienced disturbances of radioisotopic clocks should be avoided. In practice, this means that only meteorites displaying a low degree of shock metamorphism should be selected for thermochronological evaluation.

Apart from that, reasonable thermochronological constraints can only be obtained for meteorites for which several points on a cooling curve exist, i.e., different radioisotopic-dating techniques should have been applied that define the points of time at which temperatures fell below different — system-specific — closure temperatures. The biggest such dataset exists for H chondrites, and is shown in Table 3.

4. MODEL CALCULATIONS FOR CHONDRITIC PLANETESIMALS

4.1. Structure of Chondritic Material

Detailed studies of the thermal evolution of planetesimals with particular emphasis on comparison with the meteoritic record are presently only available for parent bodies of ordinary chondrites, hence we concentrate on such objects. The main constituents present in chondritic meteorites are chondrules and a matrix of fine dust grains. The typical relative abundances of these constituents are shown in Table 1; they have typical sizes of micrometers in the relatively unprocessed matrix (*Scott and Krot,* 2005), and typical sizes of millimeters in the chondrules, with variations specific to each chondrite group (*Scott,* 2007; *Weisberg et al.,* 2006). Although chondrules may also contain fine-grained mineral assemblages, they entered the meteorite parent body as solidified melt droplets after they formed by unspecified flash-heating events in the solar nebula.

The compacted solid material has a density ρ_b. This material is a complex mixture of minerals, dominated by olivine and pyroxenes. The porous material has a density

TABLE 3.   Closure ages determined for selected H chondrites for a number of thermochronometers.

Meteorite	Type	Hf-W[*]	Pb-Pb[†]	Al-Mg[‡]	U-Pb-Pb[§]	Ar-Ar[¶]	Fission Tracks[**]
				Thermochronometer			
Closure Temperature		1150 ± 75	1050 ± 100	750 ± 130	720 ± 50	550 ±20	390 ± 25
				Closure time			
Estacado	H6	4557.3 ± 1.6	4561 ± 7		4501.6 ± 2.2	4465 ± 5	4401 ± 10
Guareña	H6				4504.4 ± 0.5	4458 ± 10	4402 ± 14
Kernouvé	H6	4557.9 ± 1.0	4537.0 ± 1.1		4522.5 ± 2.0	4499 ± 6	4438 ± 10
Mt. Browne	H6		4554.8 ± 6.3		4543 ± 27	4516 ± 5	4471 ± 13
Richardton	H5	4561.6 ± 0.8	4562.7 ± 1.7		4551.4 ± 0.6	4525 ± 11	4469 ± 14
Allehan	H5				4550.2 ± 0.7	4541 ± 11	4490 ± 14
Nadiabondi	H5		4558.9 ± 2.3		4555.6 ± 3.4	4535 ±10	4543 ± 20
Forest Vale	H4			4562.5 ± 0.2	4560.9 ± 0.7	4551 ± 8	4544 ± 14
Ste. Marguerite	H4			4563.1 ± 0.2	4562.7 ± 0.6	4562 ± 16	4550 ± 17

Only meteorites for which at least three data points are available are listed. All temperatures are in Kelvin, all ages in million years.

[*]From *Kleine et al.* (2008), recalculated relative to the $^{182}Hf/^{180}Hf$ of the angrite D'Orbigny, which has a Pb-Pb age of t = 4563.4 ± 0.3 Ma (*Kleine et al.,* 2012). Closure temperature calculated using lattice strain models.

[†]Closure temperature for Pb diffusion in chondrule pyroxenes estimated by *Amelin et al.* (2005). Age data by *Blinova et al.* (2007) for Estacado and Mt. Browne, *Amelin et al.* (2005) for Richardton, and *Bouvier et al.* (2007) for Nadiabondi and Kernouvé.

[‡]Data from *Zinner and Göpel* (2002). Activation energy and frequency factor by *LaTourette and Wasserburg* (1998): E = 274 kJ mol^{-1}, D_0 = 1.2 × 10^6. Closure temperature calculated according to *Dodson* (1973) at 1000-K m.y.$^{-1}$ cooling rate and 2-μm feldspar grain size.

[§]Closure temperature by *Cherniak et al.* (1991), phosphate U-Pb-Pb age data by *Göpel et al.* (1994), and *Blinova et al.* (2007) for Estacado and Mt. Browne.

[¶]Ar-Ar ages by *Trieloff et al.* (2003) and *Schwarz et al.* (2006) for Mt. Browne, recalculated for miscalibration of K decay constant (*Renne et al.,* 2011; *Schwarz et al.,* 2011, 2012) (see text). Closure temperature by *Trieloff et al.* (2003) and *Pellas et al.* (1997).

[**]Calculated age at 390 K from time interval between Pu-fission-track retention in merrillite at 390 K and Pu-fission track retention by pyroxene at 550 K (corresponds to Ar-Ar feldspar age at 550 K). Data by *Trieloff et al.* (2003), closure temperature by *Pellas et al.* (1997).

$\varrho = \varrho_b(1-\phi) + \varrho_p\phi$, where ϱ_p is the density of the material in the pores. For ordinary chondrites, the pores are filled with a low-pressure gas, and this may be neglected; for carbonaceous chondrites the pores could be filled with water. Besides porosity ϕ, one also considers the filling factor $D = 1-\phi$.

The porous material may approximately be described as a packing of spheres. For a packing of equal-sized spheres, there are two critical filling factors. One is the random close packing with a filling factor of $D = 0.64$, i.e., a porosity of $\phi = 0.36$ (*Scott, 1962; Jaeger and Nagel, 1992; Song et al., 2008*). The other one is the loosest close packing that is just stable under the application of external forces (in the limit of vanishing force), which has $D = 0.56$ or $\phi = 0.44$ (*Onoda and Liniger, 1990; Jaeger and Nagel, 1992*).

At the basis of the planetesimal-formation process is the agglomeration of fine dust grains, like those found in interplanetary dust particles (IDPs), into dust aggregates of increasing size. Such an agglomerated material from very fine grains that was not subject to any pressure has a porosity as high as $\phi \approx 0.8 \ldots 0.9$ (e.g., *Yang et al., 2000; Krause et al., 2011b*). Such high-porosity material seems to be preserved in some comets (*Blum et al., 2006*).

Collisions of dust aggregates during the growth process of planetesimals leads to compaction of the material. The experiments of *Weidling et al.* (2009) have shown that the porosity can be reduced to $\phi \approx 0.64$ [or even lower; see *Kothe et al.* (2010)] by repeated impacts. This porosity is still higher than the random loose packing; lower porosities of $\phi \leq 0.4$ were obtained by applying static pressures of more than 10 bar (*Güttler et al., 2009*). Since planetesimals form by repeated collisions with other planetesimals and impact velocities can be rather high (see the chapter by Johansen et al. in this volume), one can assume in model calculations that the initial porosity of the material from which the parent bodies of the asteroids formed is already compacted to $\phi \approx 0.4$.

4.2. Equations

Essentially all the models for the internal constitution and the thermal evolution of the parent bodies of meteorites considered so far are one-dimensional spherically symmetric models. The basic equations to be considered in this case are the heat conduction equation

$$\frac{\partial T}{\partial t} = \frac{1}{\varrho c_p r^2} \frac{\partial}{\partial r} r^2 K(T,\phi) \frac{\partial T}{\partial r} + \frac{H}{c_p} \quad (1)$$

where T is the temperature, c_p the specific heat per unit mass of the material, K the heat conductivity, H the heating rate per unit mass, and ϱ the mass-density, and the hydrostatic pressure equation

$$\frac{\partial P}{\partial r} = \frac{GM_r}{r^2} \cdot \varrho \quad \text{with} \quad M_r = \int_0^r 4\pi\varrho(s)s^2 ds \quad (2)$$

where G is the gravitational constant and M_r is the mass inside a sphere of radius r. These equations must be combined with an equation of state $\varrho = \varrho$ (P,T,D) and an equation for the compaction of the chondritic material that describes how the filling factor D of the granular material develops over time under the action of pressure and temperature. The equations have to solved subject to the appropriate initial and boundary conditions.

If compaction and melting of the material is not considered, one can assume ϱ to be constant. The pressure equation and the equation of state usually need not be considered in that case. This is an approach that is frequently followed if only the effects of heating and cooling on the chondrite material are of interest (see section 4.5).

For the initial conditions the question of how the large planetesimals that could be the parent bodies of meteorites are formed becomes important. Although the details of this are not precisely known, the general assumption is that they assemble from smaller planetesimals that in turn somehow assembled from the dust in the protoplanetary accretion disk. Essential for the problem of thermal evolution of the bodies is how the time during which the parent body acquired most of its material is related to the duration of the main heating processes acting in the body. If decay of short-lived radioactive nuclei is the main heat source, the body is supplied with the whole latent heat content of the radioactive nuclei, if the essential growth period is short compared to the decay constant λ of the dominating heat source, while in the opposite case much of the latent heat present at the onset of growth is liberated and radiated away before the material is added to the growing body.

The case in which the body forms rapidly compared to the timescale λ is called *instantaneous formation*. In case of slow formation, one has to add to the above equations a prescription how the mass M of the body increases during the growth process. In principle, the mass of the large planetesimals increases stepwise during runaway growth by the rare event of impact with the most massive bodies from the background planetesimal swarm. This growth mode was studied, e.g., by *Ghosh et al.* (2003). Due to the difficulties in modeling unsteady growth, most of the few studies considering growth assume a continuous growth with some prescribed rate $\dot{M}(t)$ of mass increase. In such calculations one typically starts with a small seed body and follows its growth as matter is deposited at the surface.

The temperature at the surface, T_{srf}, is determined by the gain and loss of energy at the surface in form of a boundary condition at radius R

$$-K\frac{\partial T}{\partial r}\bigg|_{r=R} = H_{srf} - L \quad (3)$$

where the lefthand side is the heat flux from the interior to the surface, H_{srf} is the energy gain from the outside, and L is the energy lost to exterior space. The most important processes to be considered are heat gain by absorption of radiative energy from the outside and heat loss by radiation into empty space ($\propto T_{srf}^4$). For growing bodies the heat content of the infalling material and impact heating also have to be considered, but for bodies <500 km in size, impact heating can be neglected. [More accurately stated, impact heating is important as a local effect, but not for the global energy budget (e.g., *Keil et al., 1997; Srámek et al., 2012*).]

Since the heating by irradiation by the Sun (or by the accretion disk during the earliest phases) varies with distance, it is not practically feasible to determine T_{srf} from this boundary condition, because the location of a particular parent body of meteorites at its formation time is not known. Therefore one has to introduce some *ad hoc* prescription for T_{srf}; typically one prescribes some representative value for the surface temperature of bodies in the asteroid belt and assumes that this temperature is constant over the first few hundred million years of thermal evolution.

4.3. Heat Capacity and Heat Conductivity

The heat conduction equation depends on the specific heat c_p per unit mass. This can be calculated from the values of c_p for the components of the mineral mixture. For H- and L-chondrite material, *Yomogida and Matsui* (1984) give an analytic fit for the temperature variation of c_p that can be used in model calculations. In many model calculations a fixed value for c_p corresponding to room temperature is used, but this significantly underestimates the real c_p at higher temperatures and results in temperatures that are too high. The importance of using realistic temperature-dependent specific heats for modeling the thermal evolution of planetesimals has been stressed by *Ghosh and McSween* (1999).

Knowledge of the heat conduction coefficient K is particular important for calculating reliable thermal models. Unfortunately, heat conduction strongly depends on the structure of the material, and no simple method is known to calculate this for the highly complex structured material of chondrites. *Yomogida and Matsui* (1983) measured the heat conductivity of chondritic material for a sample of H and L chondrites in the temperature range 200–500 K. The result shows a moderate temperature variation and a strong dependence on porosity. They developed a fit formula (*Yomogida and Matsui*, 1984) that describes the temperature-porosity variation of an average heat conductivity K for this material. A more recent experimental investigation of K for meteoritic material (*Opeil et al.*, 2010) found similar results.

There is, however, the problem that the measurements are done for material that is already strongly compacted, with porosities generally less than 15% for unshocked material (*Britt and Consolmagno, 2003; Consolmagno et al., 2008*), while the initial porosity of the material must have been much higher than ≈40%. The heat conductivity of highly porous material relevant for the initial dust material from which planetesimals formed has been measured by *Krause et al.* (2011b) in the range of porosities from 44% to 82%. *Krause et al.* (2011a) and *Henke et al.* (2012a) have matched this with the results of *Yomogida and Matsui*

(1983) into a unified analytic approximation for the whole range of porosities of the form

$$K = K_b \mathcal{K}(\phi) \tag{4}$$

where K_b is the (possibly temperature dependent) heat conductivity of the bulk material, and $\mathcal{K}$ describes the variation with porosity ($\mathcal{K} = 1$ for $\phi = 0$)

$$\mathcal{K}(\phi) = \left(e^{-4\phi/\phi_1} + e^{4(a-\phi/\phi_2)} \right)^{1/4} \tag{5}$$

with constants a, ϕ_1, ϕ_2 given in its latest form in *Henke et al.* (2013). An essentially equivalent approach was developed by *Warren* (2011). Figure 5 illustrates how strongly the heat conductivity changes with porosity, because in a granular material, heat flows only through the small contact areas of the particles. The heat conductivity changes by 2 orders of magnitude if the porosity varies from $\phi \approx 40\%$ from the loosely packed initial stage to $\phi \approx 0$ for consolidated material. By this calculation, a porous surface layer that escapes compaction completely changes the internal thermal structure by its insulating effect, as is demonstrated in *Akridge et al.* (1998) (see also Fig. 8).

The bulk heat conductivity K_b of the material is that of a complex mixture of minerals with iron and FeS. There is no exact method to calculate this for composite media, but a number of recipes have been developed for approximate calculations (see *Berryman*, 1995, for a review).

4.4. Heating

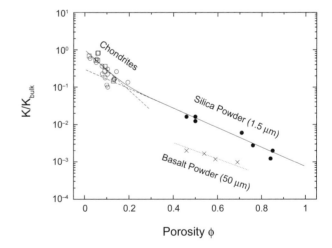

Fig. 5. Variation of heat conductivity K with porosity ϕ. Results for fine-grained silica powder (filled circles) from experiments of *Krause et al.* (2011a,b) and for particulate basalt (crosses) from *Fountain and West* (1970). Typical grain size is indicated for both cases. Open squares and open circles are experimental results for heat conductivity and porosity for ordinary H and L chondrites, respectively (from *Yomogida and Matsui*, 1983), and the solid line is the fit according to equation (5) (from *Henke et al.*, 2012a).

The parent bodies of ordinary chondrites were subject to substantial heating because the mineral content of high-petrologic-type meteorites was equilibrated at temperatures up to 1000 K and higher (e.g., *Slater-Reynolds and McSween,* 2005). The possible heat sources responsible for heating have been the subject of controversial debate (see *McSween et al.,* 2003, for a review). Presently the most efficient sources are believed to be heating by decay of short-lived radioactive nuclei and, for bodies with diameters in excess of about 500 km, the liberation of gravitational energy during growth of the bodies. The most efficient sources of heat are shown in Table 2 (see also *Cohen and Coker,* 2000). The dominating heat source is ²⁶Al decay; ⁶⁰Fe was also considered, but now its low abundance (*Tang and Dauphas,* 2012) seems to render its contribution insignificant.

The rate of a radioactive heat source varies with time as

$$H(t) = H_0 e^{-\lambda/t} \qquad (6)$$

where

$$\lambda = \ln 2 / \tau_{1/2} \qquad (7)$$

is the decay constant of the isotope that is responsible for heating and H_0 the heating rate by radioactive decay at some initial instant $t = 0$. The values for H_0 at the time of CAI formation and $\tau_{1/2}$ for the main heat sources are given in Table 2.

4.5. Analytic Model

The simplest case that could be considered is if the heat capacity c_p, heat conductivity K, and matter density ρ are all assumed to be constant and if the body is formed instantaneously at some point in time taken as the origin of time axis. Only the heat transport equation (1) has to be solved in this case. The initial-boundary value problem with constant temperature T_0 at initial time $t = 0$, and at the fixed outer boundary at $r = R$ and a heat source varying with time as in equation (6), allows for an analytic solution (*Carslaw and Jaeger,* 1959)

$$T(r,t) = T_0 + \frac{H_0}{c_p \lambda} e^{-\lambda t} \left[\frac{R \sin r \left(\frac{\lambda}{\kappa}\right)^{\frac{1}{2}}}{r \sin R \left(\frac{\lambda}{\kappa}\right)^{\frac{1}{2}}} - 1 \right]$$

$$+ \frac{2R^3 H_0 \rho}{r \pi^3 K} \sum_{n=1}^{\infty} \frac{(-1)^n \sin \frac{n\pi r}{R}}{n \left(n^2 - \frac{\lambda R^2}{\kappa \pi^2} \right)} e^{-\kappa n^2 \pi^2 t / R} \qquad (8)$$

where $\kappa = K / \rho c_p$ is the heat diffusivity. If more than one radioactive heat source is present, the second and third term on the righthand side of the equation have to be repeated for each decay system. Although this solution neglects some important variables (significant spatial and temporal variations of K, c_p, ρ, finite formation time, etc.), it is still used in many studies of thermal evolution of asteroids because

it enables the qualitative study of some basic features of the problem in a simple way.

Figure 6 shows an example of the thermal evolution of a body 50 km in size calculated with the parameters used by *Hevey and Sanders* (2006), who studied the evolution of such a body through the stages of sintering and melting. The figure shows the critical temperatures of T ≈ 750 K for the onset of sintering of the granular material (see section 4.6), T ≈ 1200 K for the onset of the melting of the Fe-FeS eutectic [see *Fei et al.* (1997) for a discussion of the melting behavior of the Fe-FeS system], and T ≈ 1700 K, where the last silicate components of the chondritic material melt [see *Agee et al.* (1995) for a study of the melting behavior of the Allende chondrite].

4.6. Compaction

The compaction of material in planetesimals is a two-step process. The initially very loosely packed dust material

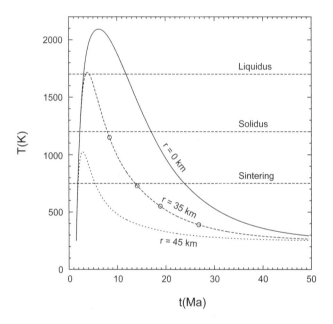

Fig. 6. Thermal evolution of a model asteroid satisfying the prerequisites for applying the analytic solution in equation (8). The radius R is 50 km; the formation time is t_{form} = 1.5 × 10⁶ yr after CAI formation. Shown are the temperature evolution at the center (solid line), at a depth below the surface of 15 km (thick dashed line), and at a depth of 5 km (dotted line). The thin horizontal dashed lines show three characteristic temperatures important for the thermal evolution of ordinary chondrite parent bodies: onset of compaction by sintering, onset of melting of the Fe-FeS eutectic (solidus), and final melting of the silicates (liquidus). The four circles denote closing temperatures of 1150, 730, 550, and 390 K for different thermochronological systems (see Table 3). The model parameters are chosen as in *Hevey and Sanders* (2006): K = 2.1 W mK⁻¹, c_p = 837 J kgK⁻¹, ρ = 3300 kg m⁻³, T_0 = 250 K, heating by decay of ²⁶Al with λ = 9.5 × 10⁻⁷ yr⁻¹ and H_0 = 1.9 × 10⁻⁷ W kg⁻¹.

in the planetesimals comes under increasing pressure by the increasing self-gravity of the growing bodies. The granular material can adjust by mutual gliding and rolling of the granular components to the exerted force and evolves into configurations with closer packing. The ongoing collisions with other bodies during the growth process enhances this compaction of the material. This mode of compaction, referred to as "cold pressing," by its very nature does not depend on temperature as it already operates at low temperature. This is discussed in *Güttler et al.* (2009) and applied to planetesimals in *Henke et al.* (2012a).

A second mode of compaction commences if radioactive decays heat the planetesimal material to such an extent that creep processes are thermally activated in the lattice of the solid material. The granular components are then plastically deformed under pressure and voids are gradually closed. This kind of compaction, called "hot pressing" or "sintering," is what obviously operated in ordinary chondrite material, and the different chondrite petrologic types 4–6 are obviously different stages of compaction that is occurring via hot pressing. *Yomogida and Matsui* (1984) were the first to perform a quantitative study of this process by applying the theory of *Rao and Chaklader* (1972). The basic assumption in this and all modern theories is that the strain rate $\dot{\varepsilon}$ in the material is related to the applied stress σ and the size d of the granular units by a relation of the form

$$\dot{\varepsilon} = A \cdot \frac{\sigma_1^n}{d^m} e^{E_{act}/RT} \qquad (9)$$

with some constants A, n, m and activation energy E_{act} [see *Karato* (2013) for a review of the deformation behavior of minerals], and $\dot{\varepsilon}$ is given in terms of the rate of change of the filling factor as

$$\dot{\varepsilon} = -\dot{D}/D \qquad (10)$$

The stress σ_1 is the pressure acting at the contact faces of the granular units. The quantities A . . . have to be determined experimentally for each material [see *Karato* (2013) for a compilation of data].

The stress σ_1 is given by the pressure acting at the contact areas between the granular units. It is assumed that this is given in terms of the applied pressure P and the areas of contact faces, πa^2, as well as the average cross-section of the cell occupied by one granular unit, C_{av}, as $\pi a^2 \sigma_1 = C_{av}P$. The second basic assumption in the theory refers to the relation between the filling factor D, the radius a of the contact areas, the average number of contact points Z, and the average cross-section C_{av}. Complete sets of equations for calculating this from the geometry of packing are given in *Kakar and Chaklader* (1967), *Rao and Chaklader* (1972), *Arzt et al.* (1983), *Yomogida and Matsui* (1984), and *Helle et al.* (1985). In conjunction with these equations, equation (10) forms a differential equation for calculating the time evolution of D that has to be solved together with the heat conduction and pressure equation.

Calculations for sintering of the material in planetesimals on this basis have been performed by *Yomogida and Matsui* (1984), *Akridge et al.* (1997), *Senshu* (2004), and *Henke et al.* (2012a). The results suggest sintering in 100-km-sized planetesimals to occur around 750 K if coefficients in equation (9) are taken for olivine from *Schwenn and Goetze* (1978). However, the use of these coefficients at a much lower temperatures than those for which they have been measured (typically in the range 1300 to 1600 K) bears the risk for a loss of reliability of the model results, but no better information is presently available.

5. RECONSTRUCTION OF PARENT-BODY PROPERTIES

5.1. Onion-Shell Model

In the onion-shell model it is assumed that the different meteorite petrologic types 3–6 result from different evolution histories of the material under the different temperature and pressure conditions encountered at the burial depth within their parent bodies (see Fig. 7). If the onion-shell model is accepted, it is reasonable to assume that meteorites of different petrologic types — if they can be assumed for some reason to originate from the same parent body — carry information about the properties and thermal history of their parent bodies, as well as the metamorphic processes within the body in general and the processes acting at the initial burial depths of each of the meteorites in particular. This allows one to retrieve the major properties of the corresponding parent body and its early evolutionary history by empirically comparing

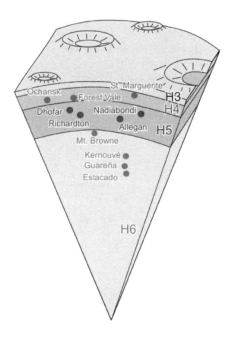

Fig. 7. Schematic representation of the onion shell model and the burial depths of the set of H chondrites used for reconstructing the parent body from a thermal history model of the body and the individual meteorites.

determined thermal evolution histories of the individual meteorites with thermal evolution models of asteroids.

The thermal history of a meteorite is derived from the closure ages and corresponding closure temperatures of a number of radioactive decay systems, which allow us to determine the instant at which the temperature of a meteorite cooled below the closure temperature for diffusion of the decay products of an isotope out of its host mineral (section 3). If such data are available for a meteorite for more than one decay system, one can try to reconstruct the properties of the parent body and the burial depth of the meteorite within its parent body because the details of the temperature evolution at some location within the body shows characteristic differences for different locations and different bodies. Because of the significant errors in the empirical data, however, this cannot be applied to a single meteorite (although it is not completely impossible if at least four data points are available). If, on the other hand, there is a group of meteorites of different petrologic types with likely origin from one parent body, it is possible to determine the characteristic parameters of the parent body (e.g., size, formation time) and the individual burial depths of the meteorites by minimizing the differences between observed closure ages and temperatures and individual temperature histories for the meteorites at their burial depths within the unknown parent body.

A good example of a case study describing early heating and cooling of an asteroid could be performed with the H-chondrite parent body. There exist a substantial set of H chondrites for which a number of relatively precise radioisotopic ages is available with different closure temperatures. This dataset comprises nine H chondrites of petrologic types 4–6. Table 3 shows a collection of data available for such studies. Many of these data have only become available during about the last decade, and therefore it is only recently that a sufficient database exists for such work.

Unfortunately, other than H chondrites, there is currently no other meteorite class for which there exists a dataset of comparable size and quality to use for promising attempts to determine its parent body, although there have been a few thermochronological studies for other classes. In the case of L chondrites, in fact, there are many thermochronological data available, but the parent body probably suffered a catastrophic collision at 500 Ma (*Korochantseva et al.,* 2007) and all thermochronometers are likely to be shifted in this case. For this reason we restrict our discussions to the case of H chondrites.

Although well-defined complete age datasets are rare, recent studies yielded important progress for Mn-Cr dating of aqueous alteration or metamorphism of the carbonaceous chondrites (*Fujiya et al.,* 2012; *Doyle et al.,* 2013). *Fujiya et al.* (2012) found indistinguishable Mn-Cr ages, implying CM carbonate formation 4.8 ± 0.5 m.y. after CAIs; their parent-body modeling yielded the determination of parent-body accretion 3.5 m.y. after CAIs. This study also solved a critical problem in carbonate Mn-Cr dating, i.e., evaluation of precise dates by establishing correct sensitivity coefficients of Mn and Cr in carbonates, invalidating some previously published Mn-Cr age data.

The onion-shell model is criticized heavily by some authors (e.g., *Taylor et al.,* 1987; *Scott and Rajan,* 1981; *Ganguly et al.,* 2013; *Scott et al.,* 2013, 2014), because metallographic cooling data are argued to suggest an impact-dominated thermal evolution. Although such problems demonstrate that the thermal evolution does not proceed fully in such a simple way as assumed in the onion-shell model, the models to be discussed later show, on the other hand, that it is possible to obtain a fully consistent picture for the thermal history of the H-chondrite parent body from the onion-shell model. The solution for this contradictory situation has yet to be determined.

5.2. Reconstructions of the H-Chondrite Parent Body

There were a number of attempts to reconstruct the properties of the H-chondrite parent body using empirically determined cooling histories of H chondrites from different initial burial depths within the parent body. Table 4 gives an overview of the results, values of key parameters of these models, and assumptions and methods used. The table lists all the models that attempted to obtain in some way a "best fit" between properties and cooling histories derived from laboratory studies and a thermal evolution model of the parent body. The results are radius R and formation time t_{form} of the putative parent body, and the burial depths of the meteorites. All studies showed that the observed properties of the investigated H chondrites are compatible with a parent body with a radius on the order of magnitude of 100 km and an "onion-shell" model. In this model the body is accreted cold, then heats up within a period of ~3 m.y. by decay of radioactive nuclei, mainly ^{26}Al, and finally cools down over an extended period of ~100 m.y., without suffering catastrophic collisions that would have disrupted the body and thereby disturbing its thermal evolution record. All models published so far are based on the instantaneous formation hypothesis; in all cases the body is assumed to have a spherically symmetric structure.

The first such model was constructed by *Minster and Allegre* (1979). The small number of data available at that time allowed only some very crude estimations of parent-body radius and formation time. This model led to further studies based on better data and methods. Many such models are based on the analytic solution of the heat conduction equation (1) with constant coefficients and fixed boundary temperature, and with ^{26}Al as heat source (see section 4.5) (see *Miyamoto et al.,* 1982; *Bennett and McSween,* 1996; *Trieloff et al.,* 2003; *Kleine et al.,* 2008). The basic characteristics of these models are listed in Table 4. The models depend on the formation time, t_{form}, and on the radius R of the parent body and the surface temperature T_{srf} as free parameters. The size and formation time of the body is estimated by comparing such models to meteoritic properties. For the surface temperature a value is assumed (different

TABLE 4. Properties of the parent body of H chondrites derived by different thermal evolution models
and some key data used in the modeling.

Quantity	Miy81	Ben96	Akr98	Tri03	Kle08	Har10	Hen12	Mon13	Unit
				Models					
t_{form}	2.5	2.2	2.3	2.5	2.7	2.2	1.84	1.98	Ma
R_{core}	85	99.0	97.5	100	100	99.2	110.8	126	km
d_{srf}	—	1.0	2.5	—	—	0.8	0.73	0.0	km
Φ_{srf}	—			—	—		22.0		%
T_{srf}	200	300	200	300	300	170	178	292	K
M	0.82	na	1.4	1.3	1.3	1.4	2.2	3.2	10^{19} kg
K	1	*	*	1	1	*	1.75†	‡	W/mK
c_p	625	§	¶	625	625	§	§	¶	J/kgK
ρ_{bulk}	3200	3780	3450	3200	3200	3250	3780	3800	kg/m³
H	1.82	na		1.82	1.82	1.82	1.87	1.93	10^{-7} W/kg
Heating	^{26}Al	^{26}Al	^{26}Al, ^{60}Fe l.l.	^{26}Al	^{26}Al	^{26}Al	^{26}Al, ^{60}Fe l.l.	^{26}Al	
Solution method	analytic	analytic	numeric	analytic	analytic	numeric	numeric	numeric	
Sintering	—	yes	no	—	—	no	yes	no	
Regolith	—	yes	yes	—	—	yes	no	yes	

Models are [Miy81] *Miyamoto et al.* (1982); [Be96] *Bennett and McSween* (1996); [Ak98] *Akridge et al.* (1998); [Tr03] *Trieloff et al.* (2003); [Kl08] *Kleine et al.* (2008); [Ha10] *Harrison and Grimm* (2010); [He12] *Henke et al.* (2012b), Table 6: [Mon13] *Monnereau et al.* (2013). na = no data given; l.l. = Contribution of ^{40}K, Th, U considered.

*Fit of K to meteoritic data (*Yomogida and Matsui,* 1984). At T = 500 K the value of the consolidated material is 3.66 W mK^{-1}. It varies only slightly with temperature.

†Variation of K with porosity according to equation (5).

‡Temperature dependence $K = K_0(T_{srf}/T)^a$ with K = 4 W/mK and a = 0, . . . 2, prefered value a = 1/2.

§Fit of c_p to meteoritic data (*Yomogida and Matsui,* 1984). At T = 500 K the value of the consolidated material is 865 J kgK^{-1}.

¶Akr98: Data for forsterite, Mon13: data for olivine with admixture of iron.

authors assumed different values) that is considered as reasonable for bodies at some location in the asteroid belt.

Miyamoto et al. (1982) constructed models for the H- and L-chondrite parent bodies and estimated size and formation time by comparing maximum metamorphic temperatures of the different petrologic types and their relative abundances in observed falls, along with a 100-m.y. difference in time between cooling through the closure temperature of the Rb-Sr system. The frequency of fall of the different petrologic types is now considered as unsuitable for determining parent-body properties because of the highly random nature of their delivery to Earth (see *Bennett and McSween,* 1996). Bennett and McSween calculated models for the H- and L-chondrite parent bodies, in which the properties of the parent bodies and their formation times were estimated by comparing maximum metamorphic temperatures of the different petrologic types and using a 60-m.y. difference in time between cooling through the closure temperature of the Pb-Pb system. *Trieloff et al.* (2003) calculated models for the H-chondrite parent body. Model parameters were estimated by comparison with a large set of observed cooling ages for four thermochronometers: Ar-Ar, Pb-Pb, and Pu-fission-track retention from nine meteorites, shown in Table 3. The model parameters were fitted such that the resulting cooling curves for all meteorites pass as close as

possible to the closing-temperature/closing-age data points. *Kleine et al.* (2008) calculated models for the H-chondrite parent body and fitted to meteoritic data using the dataset of *Trieloff et al.* (2003), augmented by new data on the Hf-W thermochronometry of four meteorites (see Table 3).

The weak point of the analytic models following the method of *Miyamoto et al.* (1982) is their inability to account for the nonconstancy of the coefficients in the heat conduction equation or the addition of further important processes. To allow for more realistic input, the basic equations have to be solved numerically, either based on the methods of finite elements (used by *Harrison and Grimm,* 2010) or finite differences (used by *Akridge et al.,* 1998; *Henke et al.,* 2012a,b; *Monnereau et al.,* 2013). [There are many more studies on thermal evolution of asteroids, but without attempting to reconstruct the parent bodies of specific meteoritic classes (see review by *Ghosh et al.,* 2006).]

Akridge et al. (1998) calculated models for the H-chondrite parent body that considered temperature- and porosity-dependent heat conductivity (from *Yomogida and Matsui,* 1983) and specific heat (from *Robie et al.,* 1979), and in particular the insulating effect of a highly porous outer layer is considered by assuming an outer layer consisting of what they call a megaregolith layer with 30% porosity and an outer regolith layer with ≈50% porosity.

Model parameters for the parent body were estimated by comparison with observed Pb-Pb closure ages, with metallographic cooling rates, and with maximum metamorphic temperatures of the different petrologic types. The results show the importance of considering the insulating effect of an outer regolith layer that results in a flat inner temperature distribution and a steep descent to the surface temperature within a rather thin outer layer (see Figs. 8b,c). This is in contrast to models assuming constant properties of the material across the body. The porous mantle results in rather shallow burial depths of H3 to H5 chondrites while most of the interior is of type H6 (see Fig. 7).

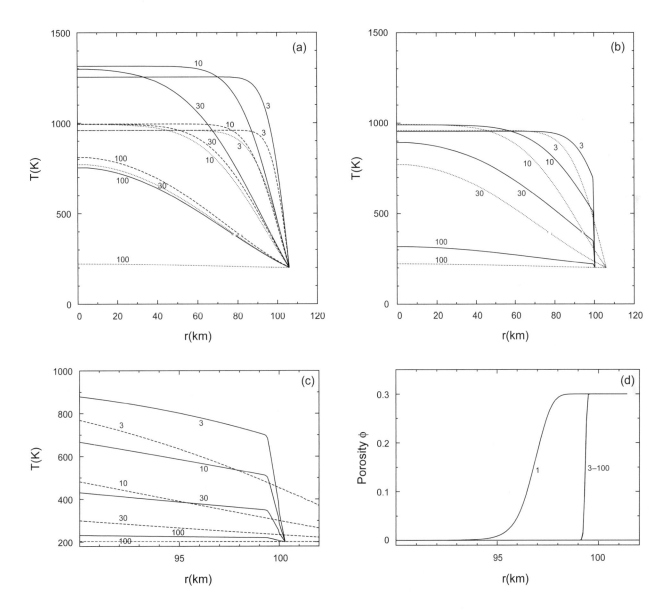

Fig. 8. Comparison of models for the thermal evolution of planetesimals based on different approximations. The formation time is assumed to be $t_{form} = 2.2 \times 10^6$ yr after CAI formation, instantaneous formation assumed. The surface temperature is fixed at $T_{srf} = 200$ K. The heating is assumed to be due to ^{26}Al only, the rate is h = 1.82×10^{-7} W kg^{-1} at time of CAI formation. The density of the bulk material is ρ_b = 3800 kg m^{-3}; the initial density ρ_0 is lower in some models, corresponding to a certain initial porosity Φ of the material. The initial radius is determined such that all models have the same mass of 1.6×10^{19} kg, which corresponds to a body of R = 100 km radius if the matter is compacted to ρ_b. Shown are the radial temperature variations at the indicated instances (in Ma after t_{form}). **(a)** Models without sintering or outer regolith layer. Full line: Model like *Miyamoto et al.* (1982) with ρ_0 = 3200 kg m^{-3}, fixed specific heat capacity c_p = 625 J kgK^{-1}, fixed heat conductivity K = 1 W mK^{-1}. Dashed line: Same model, but temperature-dependent heat capacity in the analytic approximation given by *Yomogida and Matsui* (1983). Dotted line: Same model, but K = 3.7 W mK^{-1}, corresponding to almost compacted material. **(b)** Comparison of models with and without insulating outer layer. Dotted line: Same model as in **(a)**. Full line: Sintering included (as in *Henke et al.*, 2012a) with an assumed initial porosity of Φ = 0.3. **(c)** Outermost layers of the models in **(b)**. **(d)** Distribution of porosity in the model with sintering.

Harrison and Grimm (2010) calculated models for the H-chondrite parent body based on essentially the same approximations as in *Akridge et al.* (1998). The model parameters, including thicknesses of regolith and megaregolith layers, are determined so as to obtain an optimum fit to cooling age data for 8 meteorites (see Table 3, without Mt. Browne), metallographic cooling rates for 20 meteorites, and cooling rates determined from Pu-fission tracks for 7 meteorites. Also compared is the maximum temperature at the burial depths with (assumed) maximum metamorphic temperatures of the petrologic type.

Henke et al. (2012a,b) calculated models for the H-chondrite parent body, including modeling of the evolution of porosity, by solving the equation for compaction by cold pressing (*Güttler et al.,* 2009) and the equations for sintering of granular material by hot pressing at elevated temperatures similar to *Yomogida and Matsui* (1984). In this model the body develops a compacted core, but retains a surface layer of significant porosity (see Fig. 2), analogous to *Akridge et al.* (1998) and *Harrison and Grimm* (2010), but now determined self-consistently from the sintering process. The model parameters were determined by a least-squares method that minimizes the distances between a cooling curve and the corresponding data points and an automated search strategy for the model parameters based on a variant of the genetic algorithm (*Charbonneau,* 1995).

In *Henke et al.* (2013) the model was somewhat improved. It was found that the best fit between models and meteoritic data is obtained with a rather low ^{60}Fe abundance of the order of 10^{-8} for the ^{60}Fe/^{56}Fe abundance ratio, i.e., close to the latest value of $5.9 \pm 1.3 \times 10^{-9}$ for the ^{60}Fe/^{56}Fe ratio found by *Tang and Dauphas* (2012). In some models (see Table 4), ^{60}Fe is considered as a heat source, but at the low abundance now assumed, ^{60}Fe does not substantially contribute to the heating of planetesimals and therefore can be neglected. The model was also extended to include a finite growth time of the parent body of the H chondrites by considering mass influx over some finite period and introducing a moving outer boundary. It turned out that only short accretion times of up to 0.3 m.y. are compatible with the empirical data on the cooling history of H chondrites. The instantaneous accretion hypothesis widely used in model calculations is supported by this finding.

Monnereau et al. (2013) calculated models for the H-chondrite parent body based on similar approximations as in *Akridge et al.* (1998). A power-law temperature dependence of heat conductivity was assumed. A growth of the body was allowed for by the method of a moving grid as in *Merk et al.* (2002). The same set of thermochronological data as in *Henke et al.* (2012b) was used for comparison. A large number of models were calculated for widely varying parameter values to determine the region in parameter space inside of which a close fit of the meteoritic cooling histories is possible. The consequences of an extended accretion period were studied, and it turned out in this model calculation as well that good fits to the meteoritic record can only be obtained for growth periods less than about 0.2 m.y.

Figure 8 shows for comparison the thermal structure of some "standard" body of a fixed mass that corresponds to a fully consolidated body of 100 km radius with the composition of the H-chondrite parent body. The heating rate is also assumed to be the same in all models in order that the total amount of heat released is transferred to the same mass in all cases considered (for details, see the caption to Fig. 8). The thermal structure is shown for models based on the different kinds of approximations used so far for modeling the parent body of the H chondrites. Two important points can be recognized: (1) models with a constant heat capacity result in temperatures that are much too high, and (2) models without insulating outer layers result in temperatures in the shallow outer layers that are much too low. A correct treatment of the heat capacity and inclusion of an outer regolith layer are indispensable for realistic models.

However, despite the different assumptions made in the model calculations — from the rather simplistic model of a homogeneous body with constant material properties that can be treated analytically (*Miyamoto et al.,* 1982), to models with rather complex implemented physics like that of *Harrison and Grimm* (2010) and *Henke et al.* (2012b) — the derived properties of the parent body are rather similar (see Table 4). These do not seem to critically depend on the particular assumptions. The more-strongly-deviating results for the burial depths of the meteorites found in the different models cannot be used to discriminate between models, because no independent information on them is presently available. This would require that methods are developed to determine the maximum pressure experienced by the material at the rather low pressures in meteoritic parent bodies.

There are also a number of other model calculations (e.g., *Bennett and McSween,* 1996; *Hevey and Sanders,* 2006; *Sahijpal et al.,* 2007; *Warren,* 2011) that studied the thermal evolution of asteroids and gave important insight into their properties and the processes operating in them, but did not attempt to obtain definite results for the properties of the parent body of H chondrites from some kind of "best fit" to meteoritic data. Only the range of possible values is studied in these papers. This does not enable a direct comparison, but the range of possible values for radii and formation times given are in accord with the results shown in Table 4.

5.3. Optimized Fit to Meteoritic Data

A systematic method for fitting thermochronological data with results of model calculations for the thermal evolution of planetesimals was applied by *Henke et al.* (2012b, 2013). This method allows the use of a large number of input data of meteorites and can deal with a large number of parameters. For this purpose, the code for modeling the thermal evolution of planetesimals is coupled with a code for minimum search in order to obtain optimized fits of thermal evolution models to a set of empirical data for the cooling history of meteorites. The quality function to be minimized corresponds to the χ^2 method; details are described in *Henke et al.* (2012b). The so-called genetic algorithm in a variation of the method described by *Charbonneau* (1995) was chosen as the method for the minimum search. This mimics the basic principles that are thought to rule the biological evolution of species, and is believed to find optimal values even under extremely unfavorable conditions. This automated search for a minimum allows much more complex comparisons with meteoritic data (larger numbers of meteorites and parameters, inclusion of meteorites with incomplete or lower-quality data) than is possible if the optimization is done manually, as in previous work.

The method was applied to the case of H chondrites because there are nine chondrites (see Table 3) satisfying the requirements that their characteristics suggest they could have a common origin from one parent body, at least three data points are available for each of them, and they are of shock stage S0, such that no reset of the thermochronologic clocks is expected. For these meteorites a total of 37 data points are available, while the parent-body model depends on four parameters (radius R, formation time t_{form}, surface temperature T_{srf}, initial porosity Φ_{srf}). Because the measured bulk heat conductivity K_b of chondrites shows large scatter (*Yomogida and Matsui,* 1983) this quantity was also considered as a free parameter that is included in the optimization process. Additionally, one has the nine unknown burial depths of the meteorites as parameters. In total, the problem requires fitting the 37 data points with a model depending on 14 parameters. It turns out that a very good fit is possible, resulting in a value of $\chi^2 < 7$. The quality of the fit can therefore be considered as excellent, because an acceptable fit only requires a value of $\chi^2 < 23$ [= number of data points (37) less number of parameters (14)].

Figure 9 shows the evolution of the temperature at the burial depth for the optimized model for the meteorites and the corresponding data points for closure time and closure temperature. Figure 7 shows a schematic graphical representation of the burial depths of all the meteorites used for the optimization process. The results demonstrate that it is possible to find a solution for the properties of the putative common parent body of these meteorites and their burial depths that reproduces all experimental findings in a satisfactory manner.

This finding lends some credit to the onion-shell model to explain the petrologic types of the meteorites. It would be desirable to extend such calculations to other meteoritic classes, but this requires that more thermochronological data are acquired.

6. VESTA, THE PARENT BODY OF THE HOWARDITE-EUCRITE-DIOGENITE METEORITES

The HED meteorites are a subgroup of the achondrites and originate from a differentiated parent body that experienced extensive igneous processing. The spectral similarity between these meteorites and the asteroid 4 Vesta indicated their genetic link (e.g., *McCord et al.,* 1970), and recently the Dawn mission, which orbited Vesta between July 2011 and September 2012, has supported the assumption of Vesta as the parent body. HEDs consist mainly of noncumulative eucrites (pigeonite-plagioclase basalts) and diogenites that are composed of orthopyroxene and olivine of different mixtures. Howardites are impact breccias, composed predominantly of eucrite and diogenite clasts. The igneous lithology is completed with more rare rock types, including cumulative eucrites. A eucritic upper crust and orthopyroxene-rich lower crustal layer have been identified by Dawn in the Rheasilvia basin (*De Sanctis et al.,* 2012; *Prettyman et al.,* 2012). This huge basin allows the identification of material that was formed down to a depth of 30–45 km and was excavated or exposed by an impact (*Jutzi and Asphaug,* 2011; *Ivanov and Melosh,* 2013). Olivine-rich and thus mantle material has not been found in a significant amount, i.e., not above the detection limit of <25%, suggesting a crustal thickness of at least 30–45 km (*McSween et al.,* 2013). This finding may also explain the lack of olivine-rich samples in the HED collection.

The lithology of the HEDs described above has been used to constrain the interior structure of the parent body. Vesta is believed to have a layered structure consisting of at least a metallic core, a rocky olivine-rich mantle, and a crust consisting mainly of an upper basaltic (eucritic) and a lower orthopyroxene-rich layer (e.g., *Ruzicka et al.,* 1997; *Righter and Drake,* 1997; *Mandler and Elkins-Tanton,* 2013). The details in the interior structure (i.e., core size, mantle, and crust thickness), however, vary between the models and depend (among other things) on the assumption of the unknown bulk composition of Vesta. However, the recent gravity measurements from the Dawn mission have provided a further constraint on the interior structure, i.e., the core radius with a size of 107–113 km assuming core densities between 7100 and 7800 kg m^{-3} (*Russell et al.,* 2012). Furthermore, some HEDs show a remnant magnetization, suggesting a formerly active dynamo in a liquid metallic core (*Fu et al.,* 2012).

Various radioisotopic dating systems have been used to infer the differentiation events of Vesta. From these data it has been concluded that the core-mantle differentiation likely precedes the mantle-crust differentiation. Eucrites exhibit siderophile depletions that indicate the formation of an iron-rich core within ≈1 to 4 m.y. of the beginning of the solar system and prior to their crystallization (*Palme and Rammensee,* 1981; *Righter and Drake,* 1997; *Kleine et al.,* 2009). The Hf-W ages suggest a crystallization time for eucrite zircons with up to ≈7 m.y. after CAI formation, whereas for eucrite metals it is ≈20 Ma (*Srinivasan et al.,* 2007). The latter, however, may reflect impact-triggered thermal metamorphism in the crust of Vesta (*Srinivasan et al.,* 2007). In contrast to these findings, the ^{26}Mg composition of the most primitive diogenites requires the onset of crystallization within 0.6 ± 0.4 m.y. after CAIs, and the variation in ^{26}Mg in diogenites and eucrites indicates complete solidification, i.e., rapid cooling of the body, even within the first 2–3 m.y. after CAIs (*Schiller et al.,* 2011).

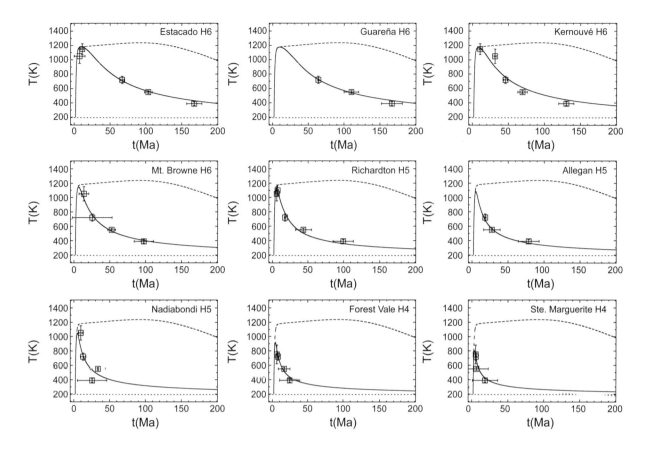

Fig. 9. Thermal evolution history of individual meteorites in an optimized evolution model for the parent body of H chondrites. The values of the free parameters found for the optimized model of *Henke et al.* (2012a, Table 6) are given in Table 4, model "Hen12." The solid lines show the temperature evolution at the burial depths of the meteorites, the dashed lines show the temperature evolution at the center of the asteroid, and the dotted lines correspond to the surface temperature. The square boxes with error bars indicate the individual data points and the estimated uncertainty of closing temperature and age determinations.

Two possible differentiation scenarios have been associated with the formation of the basaltic achondrites (i.e., eucrites and diogenites). The first scenario suggests that eucrites and diogenites originated from the partial melting of the silicates (e.g., *Stolper,* 1975, 1977; *Jones,* 1984) with the extraction of basaltic (euritic) magma leaving behind a harzburgite, orthopyroxenite, or dunite residual depending on the degree of partial melting. The other scenario favored by geochemical arguments suggests that achondrites were cumulates formed by magma fractionation. In that scenario, the diogenites could have either crystallized in a magma ocean (e.g., *Ikeda and Takeda,* 1985; *Righter and Drake,* 1997; *Ruzicka et al.,* 1997; *Takeda,* 1997; *Warren,* 1997; *Drake,* 2001; *Greenwood et al.,* 2005; *Schiller et al.,* 2011) or in multiple, smaller magma chambers (e.g., *Shearer et al.,* 1997; *Barrat et al.,* 2008; *Beck and McSween,* 2010; *Mandler and Elkins-Tanton,* 2013). Eucrites are therefore products from the magmas that had earlier crystallized diogenites.

Several numerical and experimental studies (e.g., *Ghosh and McSween,* 1998; *Righter and Drake,* 1997; *Drake,* 2001; *Gupta and Sahijpal,* 2010; *Elkins-Tanton et al.,* 2011;

Formisano et al., 2013a; *Neumann et al.,* 2014a) have been performed in order to model Vesta's thermal and geological evolution. As a general conclusion from the thermal models, one can state that Vesta should have finished its accretion by $\approx$1.5–3 m.y. after the CAIs to efficiently differentiate due to the decrease of the radiogenic heating by ^{26}Al and ^{60}Fe with time. The timing and duration of crust and core formation in particular then vary between the models, as in most cases differentiation processes have not been modeled self-consistently; rather, some specific scenario has been assumed and its consequences for the thermal evolution have been studied. For instance, *Righter and Drake* (1997) considered core formation and crystallization of a cooling magma ocean using a numerical physicochemical model. The core is assumed to have separated during a global magma ocean episode at a melt fraction of 65–77%. Then, after cooling to a crystal fraction of 80%, residual melt percolated from the former extensive magma ocean to form eucrites at the surface and diogenites in shallow layers by further crystallization. In the model of Righter and Drake, complete crystallization occurred within 20 m.y. after the formation of Vesta. *Gupta and Sahijpal* (2010) performed a thermal model and considered convective cooling in a global magma ocean. They investigated two evolution paths, i.e., the formation of basaltic achondrites via partial melting of silicates or as residual melts after crystallization of a convecting magma ocean. For the partial melting model, they concluded that, depending on the formation time, melt extraction is possible between 0.15 and 6 m.y. after the CAIs and that differentiation proceeds rapidly within $O(10^4)$ yr. For the scenarios where accretion was completed within 2 m.y. after the CAIs, a magma ocean formed that does not crystallize completely for at least 6–10 m.y. *Ghosh and McSween* (1998) investigated the differentiation of Vesta by assuming instantaneous core formation in the temperature interval of 1213–1223 K, and assuming that HED meteorites are the product of 25% partial melting. To obtain such a scenario, they concluded that Vesta must have accreted at 2.85 m.y., differentiated at 4.58 m.y., and formed a basaltic crust at 6.58 m.y. relative to the time of formation of the CAIs. Furthermore, they suggested that the mantle partially remained for 100 m.y. after its formation and that some near-surface layers may have remained undifferentiated. A recent study by *Formisano et al.* (2013a) considers core formation by porous flow of iron melt. Core differentiation takes place in all scenarios in which Vesta completes its accretion in less than 1.4 m.y. after CAIs and reaches 100% of silicate melting throughout the whole asteroid if Vesta formed before 1 m.y. after CAIs. The model, however, neglects volcanic heat transport, convection in the magma ocean. Taking those factors into consideration as well results in a different evolution path: Efficient cooling by melt migration and redistribution of ^{26}Al toward the surface inhibits the formation of a global magma ocean, yet a shallow magma ocean forms close to the surface (*Neumann et al.*, 2014a).

Thus, whether the basaltic crust formed after the silicate-iron separation due to the crystallization of a global magma ocean, or prior to the silicate-iron separation due to the extrusion of the first partial melts (partial melting origin or multiple, smaller magma chambers), remains an open issue. Which scenario is more likely depends on the competition between heating by short-lived isotopes and the depletion of these heat sources due to their partitioning into the melt and the migration velocity of this melt (see the discussion in section 2.2 about the global magma ocean). Only if the extrusion of melt containing ^{26}Al is comparatively slow would an asteroid-scale magma ocean have formed, otherwise HEDs must have formed by the extrusion of early partial melts that are either collected in smaller magma chambers or produced by a shallow rapidly [within $O(10^4–10^6 \text{ m.y.})$] freezing magma ocean close to the surface (*Neumann et al.*, 2014a). The latter models also suggest that mantle-crust differentiation precedes core formation, which seems to contradict the chronological records of HEDs, indicating a reversed sequence of differentiation (*Kleine et al.*, 2009).

On the other hand, a global magma ocean is difficult to reconcile with the chronological records of the basalts, which suggest their early origin as a global magma ocean, producing residual basaltic melts much later than during

the first 10 m.y. (*Gupta and Sahijpal*, 2010; *Schiller et al.*, 2011). The shallow magma ocean model (*Neumann et al.*, 2014) is consistent with an early formation time of $t_0 < 1$ m.y. of diogenites and rapid cooling of the body, even within the first 2–3 m.y. after CAIs (*Schiller et al.*, 2011). It further implies that the first noncumulate eucrites form before diogenites, consistent with internal ^{26}Al–^{26}Mg isochrons recently obtained by *Hublet et al.* (2013) that infer younger ages for the diogenites compared to those of the eucrites, and younger ages for the cumulative eucrites compared to the noncumulative ones.

Acknowledgments. This work was supported by "Schwerpunktprogramm1385" of the "Deutsche Forschungsgemeinschaft (DFG)." M. T. acknowledges support by Klaus Tschira foundation gGmbH. We also acknowledge W. H. Schwarz and S. Henke for providing some of the figures.

REFERENCES

Agee C. B. and Walker D. (1988) *Earth Planet. Sci. Lett., 90,* 144.
Agee C. B. et al. (1995) *J. Geophys. Res., 100,* 17725.
Akridge G. et al. (1997) In *Lunar and Planetary Science XXVIII,* Abstract #1178. Lunar and Planetary Institute, Houston.
Akridge G. et al. (1998) *Icarus, 132,* 185.
Amelin Y. et al. (2005) *Geochim. Cosmochim. Acta, 69,* 505.
Arzt E. et al. (1983) *Metall. Trans. A, 14A,* 211.
Barrat J. A. et al. (2008) *Meteoritics & Planet. Sci., 43,* 1759.
Beck A. W. and McSween H. Y. Jr. (2010) *Meteoritics & Planet. Sci., 45,* 850.
Beere W. (1975) *Acta Metall. Sin., 23,* 131.
Bennett M. and McSween H. Y. (1996) *Meteoritics & Planet. Sci., 31,* 783.
Berryman J. G. (1995) In *Rock Physics and Phase Relations: A Handbook of Physical Constants* (T. J. Ahrens, ed.), pp. 205–228. AGU Reference Shelf Vol. 3, American Geophysical Union, Washington, DC.
Binzel R. P. et al. (1997) *Icarus, 128,* 95.
Bizzarro M. et al. (2005) *Astrophys. J. Lett., 632,* L41.
Bland P. A. et al. (2011) *Meteoritics & Planet. Sci., 74,* 5275.
Blinova A. and Amelin Y. and Samson C. (2007) *Meteoritics & Planet. Sci., 7/8,* 1337.
Blum J. et al. (2006) *Astrophys. J., 652,* 1768.
Bouvier A. et al. (2007) *Geochim. Cosmochim. Acta, 71,* 1583.
Brazzle R. H. (1999) *Geochim. Cosmochim. Acta, 63,* 739.
Brearley A. J. and Jones R. H. (1998) In *Reviews in Mineralogy, Vol. 36: Planetary Materials* (J. J. Papike, ed.), pp. 3:1–3:398. Mineralogical Society of America, Chantilly, Virginia.
Britt D. T. and Consolmagno G. J. (2003) *Meteoritics & Planet. Sci., 38,* 1161.
Burkhardt C. et al. (2008) *Geochim. Cosmochim. Acta, 72,* 6177.
Carporzen L. et al. (2010) In *EGU General Assembly Conference Abstracts, 12,* 13971.
Carslaw H. S. and Jaeger J. C. (1959) *Conduction of Heat in Solids.* Oxford Univ., Oxford. 520 pp.
Castillo-Rogez J. C. and McCord T. B. (2010) *Icarus, 205,* 443.
Charbonneau P. (1995) *Astrophys. J. Suppl., 101,* 309.
Cherniak D. J. and Watson E. B. (2001) In *Eleventh Annual V. M. Goldschmidt Conference,* p. 3260.
Cherniak D. et al. (1991) *Geochim. Cosmochim. Acta, 55,* 1663.
Cohen B. A. and Coker R. F. (2000) *Icarus, 145,* 369.
Cohen M. et al. (1998) *Astron. J., 115,* 1671.
Consolmagno G. et al. (2008) *Chem. Erde–Geochem., 68,* 1.
Dauphas N. and Chaussidon M. (2011) *Annu. Rev. Astron. Astrophys., 39,* 351.
De Sanctis M. C. et al. (2012) *Science, 336,* 697.
Dodson M. H. (1973) *Contrib. Mineral. Petrol., 40,* 259.
Doyle P. M. et al. (2013) In *Lunar and Planetary Science XLIV,* Abstract #1793. Lunar and Planetary Institute, Houston.
Drake M. J. (2001) *Meteoritics & Planet. Sci., 36,* 501.

Elkins-Tanton L. et al. (2011) *Earth Planet. Sci. Lett., 305,* 1.
Fei Y. et al. (1997) *Science, 275,* 1621.
Formisano M. et al. (2013a) *Meteoritics & Planet. Sci., 48,* 2316.
Formisano M. et al. (2013b) *Astrophys. J., 770,* 50.
Fountain J. A. and West E. A. (1970) *J. Geophys. Res., 75,* 4063.
Fu R. R. et al. (2012) *Science, 338,* 238.
Fujiya W. et al. (2012) *Nature Commun., 3,* 627.
Gaetani G. A. and Grove T. L. (1999) *Earth Planet. Sci. Lett., 169,* 147.
Ganguly J. et al. (2007) *Geochim. Cosmochim. Acta, 71,* 3915.
Ganguly J. et al. (2013) *Geochim. Cosmochim. Acta, 105,* 206.
Ghosh A. and McSween H. Y. (1998) *Icarus, 134,* 187.
Ghosh A. and McSween H. Y. (1999) *Meteoritics & Planet. Sci., 34,* 121.
Ghosh A. et al. (2003) *Meteoritics & Planet. Sci., 38,* 711.
Ghosh A. et al. (2006) In *Meteorites and the Early Solar System II* (D. S. Lauretta and H. Y. McSween Jr., eds.), pp. 555–566. Univ. of Arizona, Tucson.
Göpel C. et al. (1994) *Earth Planet. Sci. Lett., 121,* 153.
Greenwood R. C. et al. (2005) *Nature, 435,* 916.
Grimm R. E. and McSween H. Y. (1989) *Icarus, 82,* 244.
Gupta G. and Sahijpal S. (2010) *J. Geophys. Res., 115,* E08001.
Güttler C. et al. (2009) *Astrophys. J., 701,* 130.
Harrison K. P. and Grimm R. E. (2010) *Geochim. Cosmochim. Acta, 74,* 5410.
Helle A. S. et al. (1985) *Acta Metall., 33,* 2163.
Henke S. et al. (2012a) *Astron. Astrophys., 537,* A45.
Henke S. et al. (2012b) *Astron. Astrophys., 545,* A135.
Henke S. et al. (2013) *Icarus, 226,* 212.
Hevey P. J. and Sanders I. S. (2006) *Meteoritics & Planet. Sci., 41,* 95.
Hewins R. H. and Newsom H. E. (1988) In *Meteorites and the Early Solar System* (J. F. Kerridge and M. S. Matthews, eds.), pp. 73–101. Univ. of Arizona, Tucson.
Hoppe P. (2009) In *Landolt-Börnstein Group VI: Astronomy, Astrophysics and Cosmology, Vol. 4B* (J. E. Trümper, ed.), pp. 450–466. Springer, Berlin.
Hublet G. et al. (2013) *Mineral. Mag., 77(5),* 1343.
Hudson G. B. et al. (1989) *Proc. Lunar Planet. Sci. Conf. 19th,* p. 547.
Ikeda Y. and Takeda H. (1985) *J. Geophys. Res.–Solid Earth, 90(S02),* C649.
Ito M. and Ganguly J. (2006) *Geochim. Cosmochim. Acta, 70,* 799.
Ivanov B. A. and Melosh H. J. (2013) *J. Geophys. Res.–Planets, 118(7),* 1545.
Jacobsen B. et al. (2008) *Earth Planet. Sci. Lett., 272,* 353.
Jaeger H. M. and Nagel S. R. (1992) *Science, 255,* 1523.
Jones J. H. (1984) *Geochim. Cosmochim. Acta, 48, 4,* 641.
Jutzi M. and Asphaug E. (2011) *Geophys. Res. Lett., 38,* L01102.
Kakar A. K. and Chaklader A. C. D. (1967) *J. Appl. Phys., 38,* 3223.
Karato S. (2013) In *Physics and Chemistry of the Deep Earth* (S. Karato, ed.), pp. 94–144. Wiley-Blackwell, Oxford.
Keil K. (1962) *J. Geophys. Res., 67,* 4055.
Keil K. and Wilson L. (1993) *Earth Planet. Sci. Lett., 117,* 111.
Keil K. et al. (1997) *Meteoritics & Planet. Sci., 32,* 349.
Kleine T. et al. (2005) *Geochim. Cosmochim. Acta, 69,* 5805.
Kleine T. et al. (2008) *Earth Planet. Sci. Lett., 270,* 106.
Kleine T. et al. (2009) *Geochim. Cosmochim. Acta, 73,* 5150.
Kleine T. et al. (2012) *Geochim. Cosmochim. Acta, 84,* 186.
Korochantseva E. V. et al. (2007) *Meteoritics & Planet. Sci., 42,* 113.
Kothe S. et al. (2010) *Astrophys. J., 725,* 1242.
Krause M. et al. (2011a) In *Lunar and Planetary Science XLII,* Abstract #2696. Lunar and Planetary Institute, Houston.
Krause M. et al. (2011b) *Icarus, 214,* 286.
Krot A. N. et al. (2009) *Geochim. Cosmochim. Acta, 73,* 4963.
LaTourette T. and Wasserburg G. (1998) *Earth Planet. Sci. Lett., 158,* 91.
Lodders K. et al. (2009) In *Landolt-Börnstein Group VI: Astronomy, Astrophysics and Cosmology, Vol. 4B* (J. E. Trümper, ed.), pp. 560–599. Springer, Berlin.
Maaloe S. (2003) *J. Petrol., 44,* 223.
Mandler B. E. and Elkins-Tanton L. T. (2013) *Meteoritics & Planet. Sci., 48,* 2333.
Matson D. L. et al. (2009) In *Lunar and Planetary Science XL,* Abstract #2191. Lunar and Planetary Institute, Houston.
McCord T. B. and Sotin C. (2005) *J. Geophys. Res.–Planets, 110,* E05009.
McCord T. B. et al. (1970) *Science, 168,* 1445.
McCoy T. J. et al. (1996) *Geochim. Cosmochim. Acta, 60,* 2681.

McCoy T. J. et al. (1997) *Geochim. Cosmochim. Acta, 61,* 639.
McKeegan K. D. and Davis A. M. (2003) In *Treatise on Geochemistry, Vol. 1: Meteorites, Comets, and Planets* (A. M. Davis, ed.), pp. 431–460, Elsevier.
McSween H. et al. (2003) In *Asteroids III* (W. F. Bottke Jr. et al., eds.), pp. 559–571. Univ. of Arizona, Tucson.
McSween H. Y. et al. (2013) In *Lunar and Planetary Science XLIV,* Abstract #1529. Lunar and Planetary Institute, Houston.
Merk R. and Prialnik D. (2003) *Earth Moon Planets, 92,* 359.
Merk R. and Prialnik D. (2006) *Icarus, 183,* 283.
Merk R. et al. (2002) *Icarus, 159,* 183.
Milliken R. E. and Rivkin A. S. (2009) *Nature Geosci., 2,* 258.
Minster J. F. and Allegre C. J. (1979) *Earth Planet. Sci. Lett., 42,* 333.
Miyamoto M. et al. (1982) *Proc. Lunar Planet. Sci. 12B,* p. 1145 .
Monnereau M. et al. (2013) *Geochim. Cosmochim. Acta, 119,* 302.
Morbidelli A. et al. (2012) *Annu. Rev. Astron. Astrophys., 40,* 251.
Moskovitz N. and Gaidos E. (2011) *Meteoritics & Planet. Sci., 46,* 903.
Neumann W. et al. (2012) *Astron. Astrophys., 543,* A141.
Neumann W. et al. (2013) *Icarus, 224,* 126.
Neumann W. et al. (2014a) *Earth Planet. Sci. Lett., 395,* 267.
Neumann W. et al. (2014b) *Astron. Astrophys., 567,* A120.
Onoda G. Y. and Liniger E. G. (1990) *Phys. Rev. Lett., 64,* 2727.
Opeil C. P. et al. (2010) *Icarus, 208,* 449.
Palme H. and Rammensee W. (1981) *Earth Planet. Sci. Lett., 55, 3,* 356.
Pätzold M. et al. (2011) *Science, 334,* 491.
Pellas P. et al. (1997) *Geochim. Cosmochim. Acta, 61,* 3477.
Prettyman T. H. et al. (2012) *Science, 338,* 242.
Prialnik D. and Bar-Nun A. (1992) *Astron. Astrophys., 258,* L9.
Prialnik D. and Podolak M. (1999) *Space Science Rev., 90,* 169.
Rao A. S. and Chaklader A. C. D, (1972) *J. Am. Ceramic Soc., 55,* 596.
Rayner J. T. et al. (2003) *Publ. Astron. Soc. Pac., 115,* 362.
Renne P. R. et al. (2010) *Geochim. Cosmochim. Acta, 74,* 5349.
Renne P. R. et al. (2011) *Geochim. Cosmochim. Acta, 75,* 5097.
Righter K. and Drake M. J. (1997) *Meteoritics & Planet. Sci., 32,* 929.
Rivkin A. S. et al. (2006) *Icarus, 185,* 563.
Robie R. A. et al. (1979) *U.S. Geol. Surv. Bull. 1452,* U.S. Govt. Printing Office, Washington, DC.
Rubin A. E. and Wasson J. T. (1995) *Meteoritics, 30,* 569.
Rugel G. et al. (2009) *Phys. Rev. Lett., 103,* 072502.
Russell C. T. et al. (2012) *Science, 336,* 684.
Ruzicka A. et al. (1997) In *Lunar and Planetary Science XXVIII,* Abstract #1215. Lunar and Planetary Institute, Houston.
Sahijpal S. and Soni P. (2005) *Lunar Planet. Sci. Conf. Lett., 36,* 1296.
Sahijpal S. et al. (2007) *Meteoritics & Planet. Sci., 42,* 1529.
Schiller M. et al. (2011) *Astrophys. J. Lett., 740,* L22.
Schwarz W. H. et al. (2006) *Meteoritics & Planet. Sci., Supplement, 41,* 5132.
Schwarz W. H. et al. (2011) *Geochim. Cosmochim. Acta, 75,* 5094.
Schwarz W. H. et al. (2012) In *34th Intl. Geol. Congress, Brisbane, Australia,* p. 3139.
Schwenn M. B. and Goetze C. (1978) *Tectonophysics, 48,* 41.
Scott E. R. D. (2007) *Annu. Rev. Earth Planet. Sci., 35,* 577.
Scott E. R. D. and Krot A. N. (2005) *Astrophys. J., 623,* 571.
Scott E. R. D. and Rajan R. S. (1981) *Geochim. Cosmochim. Acta, 45,* 53.
Scott E. R. D. et al. (1996) In *Chondrules and the Protoplanetary Disk* (R. H. Hewins et al., eds.), pp. 87–96. Cambridge Univ., Cambridge.
Scott E. R. D. et al. (2013) In *Lunar and Planetary Science XLIV,* Abstract #1826. Lunar and Planetary Institute, Houston.
Scott E. et al. (2014) *Geochim. Cosmochim. Acta., 136,* 13.
Scott G. D. (1962) *Nature, 194,* 956.
Sears D.W. G. and Dodd R. T. (1988) In *Meteorites and the Early Solar System* (J. F. Kerridge and M. S. Matthews, eds.), pp. 3–31. Univ. of Arizona, Tucson.
Senshu H. (2004) In *Lunar and Planetary Science XXXV,* Abstract #1557. Lunar and Planetary Institute, Houston.
Shearer C. K. et al. (1997) *Meteoritics & Planet. Sci., 32,* 877.
Slater-Reynolds V. and McSween J. (2005) *Meteoritics & Planet. Sci., 40,* 745.
Song C. et al. (2008) *Nature, 453,* 629.
Srámek O. et al. (2012) *Icarus, 217,* 339.
Srinivasan G. et al. (2007) *Science, 317,* 345.
Steiger R. H. and Jäger E. (1977) *Earth Planet. Sci. Lett., 36,* 359.
Stolper E. (1975) *Nature, 258,* 220.
Stolper E. (1977) *Geochim. Cosmochim. Acta, 41,* 587.

Takahashi E. (1983) *Mem. Natl. Inst. Polar Res., 30,* 168.

Takeda H. (1997) *Meteoritics & Planet. Sci., 32,* 841.

Tang H. and Dauphas N. (2012) In *Lunar and Planetary Science XLIII,* Abstract #1703. Lunar and Planetary Institute, Houston.

Taylor G. J. (1992) *J. Geophys. Res., 97,* 14717.

Taylor G. J. et al. (1987) *Icarus, 69,* 1.

Taylor G. J. et al. (1993) *Meteoritics, 28,* 34.

Telus M. et al. (2012) *Meteoritics & Planet. Sci., 47,* 2013.

Terasaki H. et al. (2008) *Earth Planet. Sci. Lett., 273,* 132.

Thomas P. C. et al. (2005) *Nature, 437,* 224.

Trieloff M. (2009) In *Landolt-Börnstein Group VI, Astronomy and Astrophysics, Vol. 4B* (J. Trümper, ed.), pp. 599–612. Springer, Berlin-Heidelberg.

Trieloff M. and Palme H. (2006) In *Planet Formation* (H. Klahr and W. Brandner, eds.), p. 64. Cambridge Univ., Cambridge.

Trieloff M. et al. (2003) *Nature, 422,* 502.

Trinquier A. et al. (2008) *Geochim. Cosmochim. Acta, 72,* 5146.

von Bargen N. and Waff H. S. (1986) *J. Geophys. Res., 91,* 9261.

Waff H. S. and Bulau J. R. (1979) *J. Geophys. Res., 84,* 6109.

Wakita S. and Sekiya M. (2011) *Earth Planets Space, 63,* 1193.

Warren P. H. (1997) *Meteoritics & Planet. Sci., 32,* 945.

Warren P. H. (2011) *Meteoritics & Planet. Sci., 46,* 53.

Weidling R. et al. (2009) *Astrophys. J., 696,* 2036.

Weisberg M. K. et al. (2006) In *Meteorites and the Early Solar System II* (D. S. Lauretta and H. Y. McSween, eds.), pp. 19–52. Univ. of Arizona, Tucson.

Weiss B. P. et al. (2008) *Science, 322,* 713.

Weiss B. P. et al. (2010) *Space Sci. Rev., 152,* 341.

Weiss B. P. et al. (2012) *Planet. Space Sci., 66,* 137.

Wilson L. and Keil K. (1991) *Earth Planet. Sci. Lett., 104,* 505.

Wilson L. and Keil K. (2012) *Chem. Erde–Geochem., 72,* 289.

Wilson L. et al. (2008) *Geochim. Cosmochim. Acta, 72,* 6154.

Yang R. Y. et al. (2000) *Phys. Rev. E, 62,* 3900.

Yomogida K. and Matsui T. (1983) *J. Geophys. Res., 88,* 9513.

Yomogida K. and Matsui T. (1984) *Earth Planet. Sci. Lett., 68,* 34.

Zinner E. and Göpel C. (2002) *Meteoritics & Planet. Sci., 37,* 1001.

Raymond S. N., Kokubo E., Morbidelli A., Morishima R., and Walsh K. J. (2014) Terrestrial planet formation at home and abroad. In *Protostars and Planets VI* (H. Beuther et al., eds.), pp. 595–618. Univ. of Arizona, Tucson, DOI: 10.2458/azu_uapress_9780816531240-ch026.

Terrestrial Planet Formation at Home and Abroad

Sean N. Raymond
Laboratoire d'Astrophysique de Bordeaux

Eiichiro Kokubo
National Astronomical Observatory of Japan

Alessandro Morbidelli
Observatoire de la Cote d'Azur

Ryuji Morishima
University of California, Los Angeles

Kevin J. Walsh
Southwest Research Institute

We review the state of the field of terrestrial planet formation with the goal of understanding the formation of the inner solar system and low-mass exoplanets. We then review the dynamics and timescales of accretion from planetesimals to planetary embryos and from embryos to terrestrial planets. We discuss radial mixing and water delivery, planetary spins, and the importance of parameters regarding the disk and embryo properties. Next, we connect accretion models to exoplanets. We first explain why the observed hot super-Earths probably formed by *in situ* accretion or inward migration. We show how terrestrial planet formation is altered in systems with gas giants by the mechanisms of giant planet migration and dynamical instabilities. Standard models of terrestrial accretion fail to reproduce the inner solar system. The "Grand Tack" model solves this problem using ideas first developed to explain the giant exoplanets. Finally, we discuss whether most terrestrial planet systems form in the same way as ours, and highlight the key ingredients missing in the current generation of simulations.

1. INTRODUCTION

The term "terrestrial planet" evokes landscapes of a rocky planet like Earth or Mars but given recent discoveries it has become somewhat ambiguous. Does a 5-$M_\oplus$ super-Earth count as a terrestrial planet? What about the Mars-sized moon of a giant planet? These objects are terrestrial planet-sized, but their compositions and corresponding landscapes probably differ significantly from those of our terrestrial planets. In addition, while Earth is thought to have formed via successive collisions of planetesimals and planetary embryos, the other objects may have formed via different mechanisms. For instance, under some conditions, a 10-$M_\oplus$ or larger body can form by accreting only planetesimals, or even only centimeter-sized pebbles. In the context of the classical stages of accretion this might be considered a "giant embryo" rather than a planet (see section 7.1).

What criteria should be used to classify a planet as terrestrial? A bulk density higher than a few grams per cubic meters probably indicates a rock-dominated planet, but densities of low-mass exoplanets are extremely challenging to pin down (see *Marcy et al., 2014*). A planet with a bulk density of 0.5–2 g cm^{-3} could either be rocky with a small H-rich envelope or an ocean planet (*Fortney et al., 2007; Valencia et al., 2007; Adams et al., 2008*). Bulk densities larger than 3 g cm^{-3} have been measured for planets as massive as 10–20 $M_\oplus$, although higher-density planets are generally smaller (*Weiss et al., 2013*). Planets with radii $R \lesssim 1.5$–2 $R_\oplus$ or masses $M \lesssim 5$–10 $M_\oplus$ are likely to preferentially have densities of 3 g cm^{-3} or larger and thus be rocky (*Weiss and Marcy, 2014; Lopez and Fortney, 2014*).

In this review we address the formation of planets in orbit around stars that are between roughly a lunar mass (~0.01 $M_\oplus$) and 10 $M_\oplus$. Although the compositions of planets in this mass range certainly vary substantially, these planets are capable of having solid surfaces, whether they are covered by thick atmospheres or not. These planets are also below the expected threshold for giant planet cores (e.g., *Lissauer and Stevenson,* 2007). We refer to these as terrestrial planets. We start our discussion of terrestrial planet formation when planetesimals have already formed; for a discussion of planetesimal formation please see the chapter by Johansen et al. in this volume.

Our understanding of terrestrial planet formation has undergone a dramatic improvement in recent years. This was driven mainly by two factors: increased computational power and observations of extrasolar planets. Computing power is the currency of numerical simulations, which continually increase in resolution and have become more and more complex and realistic. At the same time, dramatic advances in exoplanetary science have encouraged many talented young scientists to join the ranks of the planet formation community. This manpower and computing power provided a timely kick in the proverbial butt.

Despite the encouraging prognosis, planet formation models lag behind observations. Half of all Sun-like stars are orbited by close-in "super-Earths," yet we do not know how they form. There exist ideas as to why Mercury is so much smaller than Earth and Venus, but they remain speculative and narrow. Only recently was a cohesive theory presented to explain why Mars is smaller than Earth, and more work is needed to confirm or refute it.

We first present the observational constraints in the solar system and extrasolar planetary systems in section 2. Next, we review the dynamics of accretion of planetary embryos from planetesimals in section 3, and of terrestrial planets from embryos in section 4, including a discussion of the importance of a range of parameters. In section 5 we apply accretion models to extrasolar planets and in section 6 to the solar system. We discuss different modes of accretion and current limitations in section 7 and summarize in section 8.

2. OBSERVATIONAL CONSTRAINTS

Given the explosion of new discoveries in extrasolar planets and our detailed knowledge of the solar system, there are ample observations with which to constrain accretion models. Given the relatively low resolution of numerical simulations, accretion models generally attempt to reproduce large-scale constraints such as planetary mass- and orbital distributions rather than smaller-scale ones like the exact characteristics of each planet. We now summarize the key constraints for the solar system (see Table 1) and exoplanets.

2.1. The Solar System

2.1.1. The masses and orbits of the terrestrial planets. There exist metrics to quantify different aspects of a planetary system and to compare it with the solar system. The angular momentum deficit (AMD) (*Laskar,* 1997) measures the difference in orbital angular momentum between the planets' orbits and the same planets on circular, coplanar orbits. The AMD is generally used in its normalized form

$$AMD = \frac{\sum_j m_j \sqrt{a_j} \left(1 - \cos\left(i_j\right)\sqrt{1 - e_j^2}\right)}{\sum_j m_j \sqrt{a_j}} \quad (1)$$

where a_j, e_j, i_j, and m_j are planet j's semimajor axis, eccentricity, inclination, and mass. The AMD of the solar system's terrestrial planets is 0.0018.

The radial mass concentration (RMC) [defined as S_c by *Chambers* (2001)] measures the degree to which a system's mass is concentrated in a small region

$$RMC = \max\left(\frac{\sum m_j}{\sum m_j \left[\log_{10}\left(a/a_j\right)\right]^2}\right) \quad (2)$$

Here, the function in brackets is calculated for a across the planetary system, and the RMC is the maximum of that function. For a single-planet system the RMC is infinite. The RMC is higher for systems in which the total mass is packed in smaller and smaller radial zones. The RMC is thus smaller for a system with equal-mass planets than a system in which a subset of planets dominate the mass. The RMC of the solar system's terrestrial planets is 89.9.

2.1.2. The geochemically-determined accretion histories of Earth and Mars. Radiogenic elements with half-lives of a few to 100 m.y. can offer concrete constraints on the accretion of the terrestrial planets. Of particular interest is the ^{182}Hf-^{182}W system, which has a half-life of 9 million years (m.y.) Hafnium is lithophile ("rock-loving") and W is siderophile ("iron-loving"). The amount of W in a planet's mantle relative to Hf depends on the timing of core formation (*Nimmo and Agnor,* 2006). Early core formation (also called "core closure") would strand still-active Hf and later its product W in the mantle, while late core formation would cause all W to be sequestered in the core and leave behind a W-poor mantle. Studies of the Hf-W system have concluded that the last core formation event on Earth happened roughly 30–100 m.y. after the start of planet formation (*Kleine et al.,* 2002, 2009; *Yin et al.,* 2002; *König et al.,* 2011). Similar studies on martian meteorites

TABLE 1.   Key inner solar system constraints.

Angular momentum deficit (AMD)	0.0018
Radial mass concentration (RMC)	89.9
Mars' accretion timescale[*]	3–5 m.y.
Earth's accretion timescale[†]	~50 m.y.
Earth's late veneer[‡]	$(2.5–10) \times 10^{-3}$ M$_\oplus$
Total mass in asteroid belt	5×10^{-4} M$_\oplus$
Earth's water content by mass[§]	5×10^{-4}–3×10^{-3}

[*]*Dauphas and Pourmand* (2011).
[†]*Kleine et al.* (2009), *König et al.* (2011).
[‡]*Day et al.* (2007), *Walker* (2009); see also *Bottke et al.* (2010), *Schlichting et al.* (2012), and *Raymond et al.* (2013).
[§]*Lécuyer et al.* (1998), *Marty* (2012).

show that Mars' accretion finished far earlier, within 5 m.y. (*Nimmo and Kleine*, 2007; *Dauphas and Pourmand*, 2011).

The highly-siderophile element (HSE) contents of the terrestrial planets' mantles also provide constraints on the total amount of mass accreted by a planet after core closure (*Drake and Righter*, 2002). This phase of accretion is called the *late veneer* (*Kimura et al.*, 1974). Several unsolved problems exist regarding the late veneer, notably the very high Earth/Moon HSE abundance ratio (*Day et al.*, 2007; *Walker*, 2009), which has been proposed to be the result of either a top-heavy (*Bottke et al.*, 2010; *Raymond et al.*, 2013) or bottom-heavy (*Schlichting et al.*, 2012) distribution of planetesimal masses.

2.1.3. The large-scale structure of the asteroid belt. Reproducing the asteroid belt is not the main objective of formation models. But any successful accretion model must be consistent with the asteroid belt's observed structure, and that structure can offer valuable information about planet formation. Populations of small bodies can be thought of as the "blood spatter on the wall" that helps detectives solve the "crime," figuratively speaking of course.

The asteroid belt's total mass is just 5×10^{-4} $M_\oplus$, about 4% percent of a lunar mass. This is 3–4 orders of magnitude smaller than the mass contained within the belt for any disk surface density profile with a smooth radial slope. In addition, the inner belt is dominated by more volatile-poor bodies such as E-types and S-types, whereas the outer belt contains more volatile-rich bodies such as C-types and D-types (*Gradie and Tedesco*, 1982; *DeMeo and Carry*, 2013). There are no large gaps in the distribution of asteroids — apart from the Kirkwood gaps associated with strong resonances with Jupiter — and this indicates that no large (≥ 0.05 $M_\oplus$) embryos were stranded in the belt after accretion, even if the embryos could have been removed during the late heavy bombardment (*Raymond et al.*, 2009).

2.1.4. The existence and abundance of volatile species — especially water — on Earth. Although it contains just 0.05–0.1% water by mass (*Lécuyer et al.*, 1998; *Marty*, 2012), Earth is the wettest terrestrial planet. It is as wet as ordinary chondrite meteorites, thought to represent the S-type asteroids that dominate the inner main belt, and wetter than enstatite chondrites that represent E-types interior to the main belt [see, e.g., Fig. 5 of *Morbidelli et al.* (2012)]. We think that this means that the rocky building blocks in the inner solar system were dry. In addition, heating mechanisms such as collisional heating and radiogenic heating from ^{26}Al may have dehydrated fast-forming planetesimals (e.g., *Grimm and McSween*, 1993). The source of Earth's water therefore requires an explanation.

The isotopic composition of Earth's water constrains its origins. The D/H ratio of water on Earth is a good match to carbonaceous chondrite meteorites thought to originate in the outer asteroid belt (*Marty and Yokochi*, 2006). The D/H of most observed comets is $2\times$ higher — although one comet was recently measured to have the same D/H as Earth (*Hartogh et al.*, 2011) — and that of the Sun (and presumably the gaseous component of the protoplanetary disk) is

$6\times$ smaller (*Geiss and Gloeckler*, 1998). It is interesting to note that, while the D/H of Earth's water can be matched with a weighted mixture of material with solar and cometary D/H, that same combination does not match the ^{15}N/^{14}N isotopic ratio (*Marty and Yokochi*, 2006). Carbonaceous chondrites, on the other hand, match both measured ratios.

The bulk compositions of the planets are another constraint. For example, the core/mantle (iron/silicate) mass ratio of the terrestrial planets ranges from 0.4 (Mars) to 2.1 (Mercury). The bulk compositions of the terrestrial planets depend on several factors in addition to orbital dynamics and accretion: the initial compositional gradients of embryos and planetesimals, evolving condensation fronts, and the compositional evolution of bodies due to collisions and evaporation. Current models for the bulk composition of terrestrial planets piggyback on dynamical simulations such as the ones discussed in sections 4–6 below (e.g., *Bond et al.*, 2010; *Carter-Bond et al.*, 2012; *Elser et al.*, 2012). These represent a promising avenue for future work.

2.2. Extrasolar Planetary Systems

2.2.1. The abundance and large-scale characteristics of "hot super-Earths." These are the terrestrial exoplanets whose origin we want to understand. Radial velocity and transit surveys have shown that roughly 30–50% of main-sequence stars host at least one planet with $M_p \lesssim 10$ $M_\oplus$ with orbital period $P \lesssim 85$–100 d (*Mayor et al.*, 2011; *Howard et al.*, 2010, 2012; *Fressin et al.*, 2013). Hot super-Earths are preferentially found in multiple systems (e.g., *Udry et al.*, 2007; *Lissauer et al.*, 2011). These systems are in compact orbital configurations that are similar to those of the solar system's terrestrial planets as measured by the orbital period ratios of adjacent planets. The orbital spacing of adjacent Kepler planet candidates is also consistent with that of the solar system's planets when measured in mutual Hill radii (*Fang and Margot*, 2013).

Figure 1 shows eight systems each containing 4–5 presumably terrestrial exoplanets discovered by the Kepler mission. The largest planet in each system is less than 1.5 $R_\oplus$, and in one system the largest planet is actually smaller than Earth (KOI-2169). The solar system is included for scale, with the orbit of each terrestrial planet shrunk by a factor of 10 (but with their actual masses). Given that the x axis is on a log scale, the spacing between planets is representative of the ratio between their orbital periods (for scale, the Earth/Venus period ratio is about 1.6).

Given the uncertainties in the orbits of extrasolar planets and observational biases that hamper the detection of low-mass, long-period planets, we do not generally apply the AMD and RMC metrics to these systems. Rather, the main constraints come from the systems' orbital spacing, masses, and mass ratios.

2.2.2. The existence of giant planets on exotic orbits. Simulations have shown in planetary systems with giant planets that the giants play a key role in shaping the accretion of terrestrial planets (e.g., *Chambers and Cassen*,

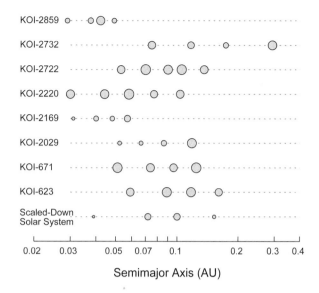

Fig. 1. Systems of (presumably) terrestrial planets. The top eight systems are candidate Kepler systems containing four or five planets that do not contain any planets larger than 1.5 R_⊕ (from *Batalha et al.*, 2013). The bottom system is the solar system's terrestrial planets with semimajor axes scaled down by a factor of 10. The size of each planet is scaled to its actual measured size (the Kepler planet candidates do not have measured masses).

2002; *Levison and Agnor,* 2003; *Raymond et al.,* 2004). Giant exoplanets have been discovered on diverse orbits that indicate rich dynamical histories. Gas giants exist on orbits with eccentricities as high as 0.9. It is thought that these planets formed in systems with multiple gas giants that underwent strong dynamical instabilities that ejected one or more planets and left behind surviving planets on eccentric orbits (*Chatterjee et al.,* 2008; *Jurić and Tremaine,* 2008; *Raymond et al.,* 2010). Hot Jupiters — gas giants very close to their host stars — are thought to have either undergone extensive inward gas-driven migration (*Lin et al.,* 1996) or been recircularized by star-planet tidal interactions from very eccentric orbits produced by planet-planet scattering (*Nagasawa et al.,* 2008; *Beaugé and Nesvorný,* 2012) or other mechanisms (e.g., *Fabrycky and Tremaine,* 2007; *Naoz et al.,* 2011) (see also the chapter by Davies et al. in this volume). There also exist gas giants on nearly-circular Jupiter-like orbits (e.g., *Wright et al.,* 2008). However, from the current discoveries, systems of gas giants like the solar system's — with giant planets confined beyond 5 AU on low-eccentricity orbits — appear to be the exception rather than the rule.

Of course, many planetary systems do not host currently detected giant planets. Radial velocity surveys show that at least 14% of Sun-like stars have gas giants with orbits shorter than 1000 d (*Mayor et al.,* 2009), and projections to somewhat larger radii predict that ~20% have gas giants within 10 AU (*Cumming et al.,* 2008). Although they are limited by small number statistics, the statistics of high-

magnification (planetary) microlensing events suggest that 50% or more of stars have gas giants on wide orbits (*Gould et al.,* 2010). In addition, the statistics of short-duration microlensing events suggests that there exists a very abundant population of gas giants on orbits that are separated from their stars; these could either be gas giants on orbits larger than ~10 AU or free-floating planets (*Sumi et al.,* 2011).

2.2.3. The planet-metallicity correlation. Gas giants — at least those easily detectable with current techniques — are observed to be far more abundant around stars with high metallicities (*Gonzalez,* 1997; *Santos et al.,* 2001; *Laws et al.,* 2003; *Fischer and Valenti,* 2005). However, this correlation does not hold for low-mass planets, which appear to be able to form around stars with a wide range of metallicities (*Ghezzi et al.,* 2010; *Buchhave et al.,* 2012; *Mann et al.,* 2012). It is interesting to note that there is no observed trend between stellar metallicity and the presence of debris disks (*Greaves et al.,* 2006; *Moro-Martín et al.,* 2007), although disks do appear to dissipate faster in low-metallicity environments (*Yasui et al.,* 2009). The planet-metallicity correlation in itself does strongly constrain the planet formation models we discuss here. What is important is that the formation of systems of hot super-Earths does not appear to depend on the stellar metallicity, i.e., the solids-to-gas ratio in the disk.

Additional constraints on the initial conditions of planet formation come from observations of protoplanetary disks around other stars (*Williams and Cieza,* 2011). These observations measure the approximate masses and radial surface densities of planet-forming disks, mainly in their outer parts. They show that protoplanetary disks tend to have masses on the order of 10^{-3}–$10^{-1}\times$ the stellar mass (e.g., *Scholz et al.,* 2006; *Andrews and Williams,* 2007a; *Eisner et al.,* 2008; *Eisner,* 2012), with typical radial surface density slopes of $\Sigma \propto r^{-(0.5-1)}$ in their outer parts (*Mundy et al.,* 2000; *Looney et al.,* 2003; *Andrews and Williams,* 2007b). In addition, statistics of the disk fraction in clusters with different ages show that the gaseous component of disks dissipate within a few million years (*Haisch et al.,* 2001; *Hillenbrand et al.,* 2008; *Fedele et al.,* 2010). It is also interesting to note that disks appear to dissipate more slowly around low-mass stars than solar-mass stars (*Pascucci et al.,* 2009).

3. FROM PLANETESIMALS TO PLANETARY EMBRYOS

In this section we summarize the dynamics of accretion of planetary embryos. We first present the standard model of runaway and oligarchic growth from planetesimals (section 3.1). We next present a newer model based on the accretion of small pebbles (section 3.2).

3.1. Runaway and Oligarchic

3.1.1. Growth modes. There are two growth modes: "orderly" and "runaway." In orderly growth, all planetesimals grow at the same rate, so the mass ratios between planetesimals tend to unity. During runaway growth, on the other hand, larger planetesimals grow faster than smaller ones and mass ratios increase monotonically. Consider the evolution of the mass ratio between two planetesimals with masses M_1 and M_2, assuming $M_1 > M_2$. The time derivative of the mass ratio is given by

$$\frac{d}{dt}\left(\frac{M_1}{M_2}\right) = \frac{M_1}{M_2}\left(\frac{1}{M_1}\frac{dM_1}{dt} - \frac{1}{M_2}\frac{dM_2}{dt}\right) \qquad (3)$$

It is the relative growth rate $(1/M)dM/dt$ that determines the growth mode. If the relative growth rate decreases with M, $d(M_1/M_2)/dt$ is negative, and the mass ratio tends to be unity. This corresponds to orderly growth. If the relative growth rate increases with M, $d(M_1/M_2)/dt$ is positive and the mass ratio increases, leading to runaway growth.

The growth rate of a planetesimal with mass M and radius R that is accreting field planetesimals with mass m (M > m) can be written as

$$\frac{dM}{dt} \simeq n_m \pi R^2 \left(1 + \frac{v_{esc}^2}{v_{rel}^2}\right) v_{rel} m \qquad (4)$$

where n_m is the number density of field planetesimals, and v_{rel} and v_{esc} are the relative velocity between the test and the field planetesimals and the escape velocity from the surface of the test planetesimal, respectively (e.g., *Kokubo and Ida*, 1996). The term v_{esc}^2/v_{rel}^2 indicates the enhancement of collisional cross-section by gravitational focusing.

3.1.2. Runaway growth of planetesimals. The first dramatic stage of accretion through which a population of planetesimals passes is runaway growth (*Greenberg et al.*, 1978; *Wetherill and Stewart*, 1989; *Kokubo and Ida*, 1996). During planetesimal accretion gravitational focusing is efficient because the velocity dispersion of planetesimals is kept smaller than the escape velocity due to gas drag. In this case equation (4) reduces to

$$\frac{dM}{dt} \propto \Sigma_{dust} M^{4/3} v^{-2} \qquad (5)$$

where Σ_{dust} and v are the surface density and velocity dispersion of planetesimals and we used $n_m \propto \Sigma_{dust} v^{-1}$, $v_{esc} \propto M^{1/3}$, $R \propto M^{1/3}$, and $v_{rel} \simeq v$. During the early stages of accretion, Σ_{dust} and v barely depend on M, i.e., the reaction of growth on Σ_{dust} and v can be neglected since the mass in small planetesimals dominate the system. In this case we have

$$\frac{1}{M}\frac{dM}{dt} \propto M^{1/3} \qquad (6)$$

which leads to runaway growth.

During runaway growth, the eccentricities and inclinations of the largest bodies are kept small by dynamical friction from smaller bodies (*Wetherill and Stewart*, 1989; *Ida*

and Makino, 1992). Dynamical friction is an equipartitioning of energy that maintains lower random velocities — and therefore lower-eccentricity and lower-inclination orbits — for the largest bodies. The mass distribution relaxes to a distribution that is well approximated by a power-law distribution. Among the large bodies that form in simulations of runaway growth, the mass follows a distribution $dn_c/dm \propto m^y$, where $y \simeq -2.5$. This index can be derived analytically as a stationary distribution (*Makino et al.*, 1998). The power index smaller than –2 is characteristic of runaway growth, as most of the system mass is contained in small bodies. We also note that runaway growth does not necessarily mean that the growth time decreases with mass, but rather that the mass ratio of any two bodies increases with time.

3.1.3. Oligarchic growth of planetary embryos. During the late stages of runaway growth, embryos grow while interacting with one another. The dynamics of the system become dominated by a relatively small number — a few tens to a few hundred — of oligarchs (*Kokubo and Ida*, 1998, 2000; *Thommes et al.*, 2003).

Oligarchic growth is the result of the self-limiting nature of runaway growth and orbital repulsion of planetary embryos. The formation of similar-sized planetary embryos is due to a slow-down of runaway growth (*Lissauer*, 1987; *Ida and Makino*, 1993; *Ormel et al.*, 2010). When the mass of a planetary embryo M exceeds about 100× that of the average planetesimal, the embryo increases the random velocity of neighboring planetesimals to be $v \propto M^{1/3}$ [but note that this depends on the planetesimal size (*Ida and Makino*, 1993; *Rafikov*, 2004; *Chambers*, 2006)]. The relative growth rate (from equation (5)) becomes

$$\frac{1}{M}\frac{dM}{dt} \propto \Sigma_{dust} M^{-1/3} \qquad (7)$$

Σ_{dust} decreases through accretion of planetesimals by the embryo as M increases (*Lissauer*, 1987). The relative growth rate is a decreasing function of M, which changes the growth mode to orderly. Neighboring embryos grow while maintaining similar masses. During this stage, the mass ratio of an embryo to its neighboring planetesimals increases because for the planetesimals with mass m, $(1/m)dm/dt \propto \Sigma_{dust}m^{1/3}M^{-2/3}$, such that

$$\frac{(1/M)dM/dt}{(1/m)dm/dt} \propto \left(\frac{M}{m}\right)^{1/3} \qquad (8)$$

The relative growth rate of the embryo is larger than the planetesimals' by a factor of $(M/m)^{1/3}$. A bimodal embryo-planetesimal system is formed. While the planetary embryos grow, a process called orbital repulsion keeps their orbital separations at roughly 10 mutual Hill radii $R_{H,m}$, where $R_{H,m} = 1/2 (a_1 + a_2) [(M_1 + M_2)/(3M_\star)]^{1/3}$; here subscripts 1 and 2 refer to adjacent embryos. Orbital repulsion is a coupling effect of gravitational scattering between planetary embryos that increases their orbital separation and eccentricities and dynamical friction from small planetesimals that decreases the eccentricities (*Kokubo and Ida*, 1995). Essentially, if two embryos come too close to each other, their eccentricities are increased by gravitational perturbations. Dynamical friction from the planetesimals recircularizes their orbits at a wider separation.

An example of oligarchic growth is shown in Fig. 2 (*Kokubo and Ida*, 2002). About 10 embryos form with masses comparable to Mars' ($M \approx 0.1\ M_\oplus$) on nearly circular non-inclined orbits with characteristic orbital separations of 10 $R_{H,m}$. At large a the planetary embryos are still growing at the end of the simulation.

Although oligarchic growth describes the accretion of embryos from planetesimals, it implies giant collisions between embryos that happen relatively early and are followed by a phase of planetesimal accretion. Consider the last pairwise accretion of a system of oligarchs on their way to becoming planetary embryos. The oligarchs have masses M_{olig} and are spaced by N mutual Hill radii $R_{H,m}$, where $N \approx 10$ is the rough stability limit for such a system. The final system of embryos will likewise be separated by $N\ R_{H,m}$, but with larger masses M_{emb}. The embryos grow by accreting material within an annulus defined by the interembryo separation. Assuming pairwise collisions between equal-mass oligarchs to form a system of equal-mass embryos, the following simple relation should hold: $N\ R_{H,m}(M) = 2N\ R_{H,m}(M_{emb})$. Given that $R_{H,m}(M) \sim (2M)^{1/3}$, this implies that $M_{emb} = 8\ M_{olig}$. After the collision between a pair of oligarchs, each embryo must therefore accrete the remaining three-quarters of its mass from planetesimals.

We can estimate the dynamical properties of a system of embryos formed by oligarchic growth. We introduce a protoplanetary disk with surface density of dust and gas Σ_{dust} and Σ_{gas} defined as

$$\Sigma_{dust} = f_{ice}\ \Sigma_1 \left(\frac{a}{1\ AU} \right)^{-x} g\ cm^{-2}$$

$$\Sigma_{gas} = f_{gas}\ \Sigma_1 \left(\frac{a}{1\ AU} \right)^{-x} g\ cm^{-2}$$
(9)

where Σ_1 is simply a reference surface density in solids at 1 AU and x is the radial exponent. f_{ice} and f_{gas} are factors that enhance the surface density of ice and gas with respect to dust. In practice, f_{ice} is generally taken to be 2–4 (see *Kokubo and Ida*, 2002; *Lodders*, 2003), and $f_{gas} \approx 100$. Given an orbital separation b of embryos, the isolation (final) mass of a planetary embryo at orbital radius a is estimated as (*Kokubo and Ida*, 2002)

$$M_{iso} \simeq 2\pi a b \Sigma_{dust} = 0.16 \left(\frac{b}{10 r_H} \right)^{3/2} \left(\frac{f_{ice}\ \Sigma_1}{10} \right)^{3/2}$$
$$\left(\frac{a}{1\ AU} \right)^{(3/2)(2-x)} \left(\frac{M_\star}{M_\odot} \right)^{-1/2} M_\oplus$$
(10)

where $M_\star$ is the stellar mass. The time evolution of an oligarchic body is (*Thommes et al.*, 2003; *Chambers*, 2006)

$$m(t) = M_{iso} \tanh^3 \left(\frac{t}{\tau_{grow}} \right)$$
(11)

The growth timescale τ_{grow} is estimated as

$$\tau_{grow} = 1.1 \times 10^6 f_{ice}^{-1/2} \left(\frac{f_{gas}}{240} \right)^{-2/5} \left(\frac{\Sigma_1}{10} \right)^{-9/10}$$
$$\left(\frac{b}{10\ r_H} \right)^{1/10} \left(\frac{a}{1\ AU} \right)^{8/5 + 9x/10} \left(\frac{M_\star}{M_\odot} \right)^{-8/15}$$
$$\left(\frac{\rho_p}{2\ g\ cm^{-3}} \right)^{11/15} \left(\frac{r_p}{100\ km} \right)^{2/5} yr$$
(12)

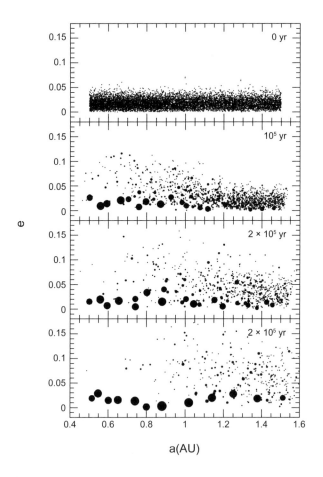

Fig. 2. Oligarchic growth of planetary embryos. Snapshots of the planetesimal system on the a-e plane are shown for t = 0, 10^5, 2×10^5, and 4×10^5 yr. The circles represent planetesimals with radii proportional to their true values. The initial planetesimal system consists of 10,000 equal-mass (m = $2.5 \times 10^{-4}\ M_\oplus$) bodies. In this simulation, a six-fold increase in the planetesimal radius was used to accelerate accretion. In 4×10^5 yr, the number of bodies decreases to 333. From *Kokubo and Ida* (2002).

where r_p and ρ_p are the physical radius and internal density of planetesimals. Equation (11) indicates that the embryo gains 44%, 90%, and 99% of its final mass during $1\tau_{grow}$, $2\tau_{grow}$, and $3\tau_{grow}$.

For the standard disk model defined above, $M_{iso} \sim 0.1 \, M_\oplus$ in the terrestrial planet region. This suggests that if they formed by oligarchic growth, Mercury and Mars may simply represent leftover planetary embryos. A short growth timescale ($\tau_{grow} < 2$ m.y.) of Mars estimated by the Hf-W chronology (*Dauphas and Pourmand*, 2011) would suggest that Mars accreted from a massive disk of small planetesimals (*Kobayashi and Dauphas*, 2013; *Morishima et al.*, 2013). Alternately, accretion of larger planetesimals might have been truncated as proposed by the Grand Tack model (see section 6.3). Unlike Mars and Mercury, further accretion of planetary embryos is necessary to complete Venus and Earth. This next, final stage is called late-stage accretion (see section 4).

3.2. Embryo Formation by Pebble Accretion

Lambrechts and Johansen (2012) (hereafter LJ12) proposed a new model of growth for planetary embryos and giant planet cores. They argued that if the disk's mass is dominated by pebbles of a few decimeters in size, the largest planetesimals accrete pebbles very efficiently and can rapidly grow to several Earth masses (see also *Johansen and Lacerda*, 2010; *Ormel and Klahr*, 2010; *Murray-Clay et al.*, 2011). This model builds on a recent planetesimal formation model in which large planetesimals (with sizes from ~100 up to ~1000 km) form by the collapse of a self-gravitating clump of pebbles, concentrated to high densities by disk turbulence and the streaming instability (*Youdin and Goodman*, 2005; *Johansen et al.*, 2006, 2007, 2009) (see also the chapter by Johansen et al. in this volume). The pebble accretion scenario essentially describes how large planetesimals continue to accrete. There is observational evidence for the existence of pebble-sized objects in protoplanetary disks (*Wilner et al.*, 2005; *Rodmann et al.*, 2006; *Lommen et al.*, 2007; *Pérez et al.*, 2012), although their abundance relative to larger objects (planetesimals) is unconstrained.

Pebbles are strongly coupled with the gas so they encounter the already-formed planetesimals with a velocity Δv that is equal to the difference between the Keplerian velocity and the orbital velocity of the gas, which is slightly sub-Keplerian due to the outward pressure gradient. LJ12 define the planetesimal Bondi radius as the distance at which the planetesimal exerts a deflection of one radian on a particle approaching with a velocity Δv

$$R_B = \frac{GM}{\Delta v^2} \quad (13)$$

where G is the gravitational constant and M is the planetesimal mass (the deflection is larger if the particle passes closer than R_B). LJ12 showed that all pebbles with a stopping time t_f smaller than the Bondi time $t_B = R_B/\Delta v$ that

pass within a distance $R = (t_f/t_B)^{1/2}R_B$ spiral down toward the planetesimal and are accreted by it (see Fig. 3). Thus, the growth rate of the planetesimal is

$$dM/dt = \pi \rho R^2 \Delta v \quad (14)$$

where ρ is the volume density of the pebbles in the disk. Because $R \propto M$, the accretion rate $dM/dt \propto M^2$. Thus, pebble accretion is at the start a super-runaway process that is faster than the runaway accretion scenario (see section 3.1) in which $dM/dt \propto M^{4/3}$. According to LJ12, this implies that in practice, only planetesimals more massive than $\sim 10^{-4} \, M_\oplus$ (comparable to Ceres' mass) undergo significant pebble accretion and can become embryos/cores. The super-runaway phase cannot last very long. When the Bondi radius exceeds the scale height of the pebble layer, the accretion rate becomes

$$dM/dt = 2R \, \Sigma \, \Delta v \quad (15)$$

where Σ is the surface density of the pebbles. This rate is proportional to M at the boundary between runaway and orderly (oligarchic) growth.

Moreover, when the Bondi radius exceeds the Hill radius $R_H = a \, [M/(3M_\star)]^{1/3}$, the accretion rate becomes

$$dM/dt = 2R_H \, \Sigma \, v_H \quad (16)$$

where v_H is the Hill velocity (i.e., the difference in Keplerian velocities between two circular orbits separated by R_H). Here $dM/dt \propto M^{2/3}$ and pebble accretion enters an oligarchic regime.

For a given surface density of solids Σ the growth of an embryo is much faster if the solids are pebble-sized than planetesimal-sized. This is the main advantage of the pebble-accretion model. However, pebble accretion ends when the gas disappears from the protoplanetary disk, whereas runaway/oligarchic accretion of planetesimals can continue. Also, the ratio between $\Sigma_{planetesimals}/\Sigma_{pebbles}$ remains to be quantified, and ultimately it is this ratio that determines which accretion mechanism is dominant.

An important problem in solar system formation is that the planetary embryos in the inner solar system are thought to have grown only up to at most a Mars mass, whereas in the outer solar system some of them reached many Earth masses, enough to capture a primitive atmosphere and become giant planets. The difference between these masses can probably be better understood in the framework of the pebble-accretion model than in the planetesimal-accretion model.

The dichotomy in embryo mass in the inner/outer solar system might have been caused by radial drift of pebbles. We consider a disk with a "pressure bump" (*Johansen et al.*, 2009) at a given radius R_{bump}. At this location the gas' azimuthal velocity v_θ is larger than the Kepler velocity v_K. Pebbles cannot drift from beyond R_{bump} to within R_{bump}

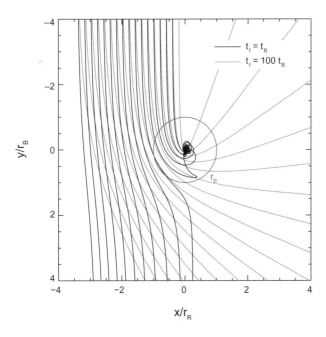

Fig. 3. Trajectories of particles in the vicinity of a growing embryo. The black curves represent particles strongly coupled to the gas and the gray curves particles that are weakly coupled, as measured by the ratio of the stopping time t_f to the Bondi time t_B. The orbits of weakly coupled particles are deflected by the embryo's gravity, but the strongly coupled particles spiral inward and are quickly accreted onto the embryo. From *Lambrechts and Johansen* (2012).

because they are too strongly coupled to the gas. Embryos growing interior to R_{bump} are thus "starved" in the sense that they can only accrete pebbles within R_{bump}, and are not in contact with the presumably much larger pebble reservoir beyond R_{bump}. Of course, embryos growing exterior to R_{bump} would not be starved and could grow much faster and achieve much larger masses within the gaseous disk's lifetime. On the contrary, the planetesimal accretion model does not seem to present a sharp radial boundary for slow/fast accretion and so it is harder to understand the dichotomy of embryo masses in that framework.

Ida and Lin (2008) argued that a pressure bump could be located at the snow line. If this is true, then we can speculate that giant planet cores should form in the icy part of the disk and sub-Mars-mass planetary embryos in the rocky part of the disk. This seems to be consistent with the structure of the solar system.

4. FROM PLANETARY EMBRYOS TO TERRESTRIAL PLANETS

The final accumulation of terrestrial planets — sometimes called late-stage accretion — is a chaotic phase characterized by giant embryo-embryo collisions. It is during this phase that the main characteristics of the planetary system are set: the planets' masses and orbital architecture, feeding zones and thus bulk compositions, and spin

rates and obliquities [although their spins may be altered by other processes on long timescales; see, e.g., *Correia and Laskar* (2009)].

Whether embryos form by accreting planetesimals or pebbles, the late evolution of a system of embryos is likely in the oligarchic regime. The transition from oligarchic growth to late-stage accretion happens when there is insufficient damping of random velocities by gas drag and dynamical friction from planetesimals (*Kenyon and Bromley*, 2006). The timescale of the orbital instability of an embryo system has been numerically calculated by N-body simulations to be

$$\log t_{inst} \simeq c_1 \left(\frac{b_{ini}}{r_H} \right) + c_2 \tag{17}$$

where b_{ini} is the initial orbital separation of adjacent embryos and c_1 and c_2 are functions of the initial $\langle e^2 \rangle^{1/2}$ and $\langle i^2 \rangle^{1/2}$ of the system (*Chambers et al.*, 1996; *Yoshinaga et al.*, 1999).

The most important quantity in determining the outcome of accretion is the level of eccentricity excitation of the embryos. This is determined by a number of parameters including forcing from any giant planets that exist in the system (*Chambers and Cassen*, 2002; *Levison and Agnor*, 2003; *Raymond et al.*, 2004). Although giant planets are far larger than terrestrials, they are thought to form far faster and strongly influence late-stage terrestrial accretion. The lifetimes of gaseous protoplanetary disks are just a few million years (*Haisch et al.*, 2001), whereas geochemical constraints indicate that Earth took 50–100 m.y. to complete its formation. (*Touboul et al.*, 2007; *Kleine et al.*, 2009; *König et al.*, 2011). The dynamics described in this section are assumed to occur in a gas-free environment (we consider the effects of gas in other sections). We first describe the dynamics of accretion and radial mixing (section 4.1), then the effect of accretion on the final planets' spins (section 4.2), and then the effect of embryo and disk parameters on accretion (section 4.3). We explain the consequences of taking into account imperfect accretion (section 4.4) and the effect of giant planets on terrestrial accretion (section 4.5).

4.1. Timescales and Radial Mixing

Figure 4 shows the evolution of a simulation of late-stage accretion from *Raymond et al.* (2006b) that included a single Jupiter-mass giant planet on a circular orbit at 5.5 AU. The population of embryos is excited from the inside out by mutual scattering among bodies and from the outside in by secular and resonant excitation by the giant planet. Accretion is faster closer in and proceeds as a wave sweeping outward in time. At 10 m.y. the disk inside 1 AU is dominated by four large embryos with masses close to Earth's. The population of close-in (red) planetesimals has been strongly depleted, mainly by accretion but also by some scattering to larger orbital radii. Over the rest of the simulation the wave of accretion sweeps outward across the entire system. Small bodies are scattered onto highly-eccentric orbits and either collide with growing embryos or venture too close to the giant planet and are ejected from the system. Embryos maintain modest eccentricities by dynamical friction from the planetesimals. Nonetheless, strong embryo-embryo gravitational scattering events spread out the planets and lead to giant impacts such as the one thought to be responsible for creating Earth's Moon (*Ćuk and Stewart,* 2012; *Canup,* 2012).

After 200 m.y. three terrestrial planets remain in the system with masses of 1.54, 2.04, and 0.95 $M_\oplus$ (inner to outer). Although modestly more massive, the orbits of the two inner planets are decent analogs for Earth and Venus. The outer planet does a poor job of reproducing Mars: It is nine times too massive and too far from the star. This underscores the "small Mars problem": Simulations that do not invoke strong external excitation of the embryo swarm systematically produce Mars analogs that are far too massive (*Wetherill,* 1991; *Raymond et al.,* 2009). We will return to this problem in section 6.

A large reservoir of water-rich material is delivered to the terrestrial planets in the simulation shown in Fig. 4. By 10 m.y. four large embryos have formed inside 1 AU, but they remain dry because to this point their feeding zones have been restricted to the inner planetary system. Over the following 20 m.y. planetesimals and embryos from the outer planetary system are scattered inward by repeated gravitational encounters with growing embryos. These bodies sometimes collide with the growing terrestrial planets. This effectively widens the feeding zones of the terrestrial planets to include objects that condensed at different temperatures and therefore have different initial compositions (see also *Bond et al.,* 2010; *Carter-Bond et al.,* 2012; *Elser et al.,* 2012). The compositions of the terrestrial planets become mixtures of the compositions of their constituent embryos and planetesimals. The planets' feeding zones represent those constituents. When objects from past 2.5 AU are accreted, water-rich material is delivered to the planet in the form of hydrated embryos and planetesimals. In the simulations, from 30 to 200 m.y. the terrestrial planets accrete objects from a wide range of initial locations and more water is delivered to them.

Given that the water delivered to the planets in this simulation originated in the region between 2.5 and 4 AU, its composition should be represented by carbonaceous chondrites that provide a very good match to Earth's water (*Morbidelli et al.,* 2000; *Marty and Yokochi,* 2006). The planets are delivered a volume of water that may be too large. For example, the Earth analog's final water content by mass was 8×10^{-3}, roughly 8–20 times the actual value. However, water loss during giant impacts was not taken into account in the simulation (see, e.g., *Genda and Abe,* 2005).

4.2. Planetary Spins

Giant impacts impart large amounts of spin angular momentum on the terrestrial planets (e.g., *Safronov,* 1969;

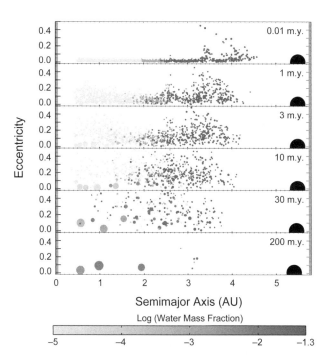

Fig. 4. Six snapshots of a simulation of terrestrial planet formation (adapted from *Raymond et al.,* 2006b). The simulation started from 1885 self-gravitating sub-lunar-mass bodies spread from 0.5 to 5 AU following an $r^{-3/2}$ surface density profile, comprising a total of 9.9 $M_\oplus$. The large black circle represents a Jupiter-mass planet. The size of each body is proportional to its mass$^{1/3}$. The grayscale represents each body's water content (see color bar).

Lissauer and Kary, 1991; *Dones and Tremaine,* 1993). The last few giant impacts tend to dominate the spin angular momentum (*Agnor et al.,* 1999; *Kokubo and Ida,* 2007; *Kokubo and Genda,* 2010). Using a "realistic" accretion condition of planetary embryos [*Genda et al.* (2012); see also section 4.4], *Kokubo and Genda* (2010) found that the spin angular velocity of accreted terrestrial planets follows a Gaussian distribution with a nearly mass-independent average value of about 70% of the critical angular velocity for rotational breakup

$$\omega_{cr} = \left(\frac{GM}{R^3}\right)^{1/2} \tag{18}$$

where M and R are the mass and radius of a planet. This appears to be a natural outcome of embryo-embryo impacts at speeds slightly larger than escape velocity. At later times, during the late veneer phase, the terrestrial planets' spins are further affected by impacts with planetesimals (*Raymond et al.,* 2013).

The obliquity of accreted planets ranges from 0° to 180° and follows an isotropic distribution (*Agnor et al.,* 1999; *Kokubo and Ida,* 2007; *Kokubo and Genda,* 2010). Both prograde and retrograde spins are equally probable. The isotropic distribution of ε is a natural outcome of giant impacts. During the giant impact stage, the thickness of a planetary embryo system is $\sim a\langle i^2\rangle^{1/2} \sim 10\ r_H$, far larger than the radius R of planetary embryos $R \sim 10^{-2}r_H$, where a, i, and r_H are the semimajor axis, inclination, and Hill radius of planetary embryos. Thus, collisions are fully three-dimensional and isotropic, which leads to isotropic spin angular momentum. This result clearly shows that prograde spin with small obliquity, which is common to the terrestrial planets in the solar system except for Venus, is not a common feature for planets assembled by giant impacts. Note that the initial obliquity of a planet determined by giant impacts can be modified substantially by stellar tide if the planet is close to the star and by satellite tide if the planet has a large satellite.

4.3. Effect of Disk and Embryo Parameters

The properties of a system of terrestrial planets are shaped in large part by the total mass and mass distribution within the disk, and the physical and orbital properties of planetary embryos and planetesimals within the disk. However, while certain parameters have a strong impact on the outcome, others have little to no effect.

Kokubo et al. (2006) performed a suite of simulations of accretion of populations of planetary embryos to test the importance of the embryo density, mass, spacing, and number. They found that the bulk density of the embryos had little to no effect on the accretion within the range that they tested, $\rho = 3.0$–5.5 g cm^{-3}. One can imagine that the dynamics could be affected for extremely high values of ρ, if the escape speed from embryos were to approach a significant fraction of the escape speed from the planetary system (*Goldreich et al.*, 2004). In practice this is unlikely to occur in the terrestrial planet-forming region because it would require unphysically large densities. The initial spacing likewise had no meaningful impact on the outcome, at least when planetary embryos were spaced by 6–12 mutual Hill radii (*Kokubo et al.*, 2006). Likewise, for a fixed total mass in embryos, the embryo mass was not important.

The total mass in embryos does affect the outcome. A more massive disk of embryos and planetesimals produces fewer, more-massive planets than a less-massive disk (*Kokubo et al.*, 2006; *Raymond et al.*, 2007a). Embryos' eccentricities are excited more strongly in massive disks by encounters with massive embryos. With larger mean eccentricities, the planets' feeding zones are wider than if the embryos' eccentricities were small, simply because any given embryos crosses a wider range of orbital radii. The scaling between the mean accreted planet mass and the disk mass is therefore slightly steeper than linear: the mean planet mass M_p scales with the local surface density Σ_0 as $M_p \propto \Sigma_0^{1.1}$ (*Kokubo et al.*, 2006). It is interesting to note that this scaling is somewhat shallower than the $\Sigma_0^{1.5}$ scaling of embryo mass with the disk mass (*Kokubo and Ida*, 2000). Accretion also proceeds faster in high-mass disks, as the timescale for interaction drops.

Terrestrial planets that grow from a disk of planetesimals and planetary embryos retain a memory of the surface density profile of their parent disk. In addition, the dynamics is influenced by which part of the disk contains the most mass. In disks with steep density profiles — i.e., if the surface density scales with orbital radius as $\Sigma \propto r^{-x}$, disks with large values of x — more mass is concentrated in the inner parts of the disk, where the accretion times are faster and protoplanets are dry. Compared with disks with shallower density profiles (with small x), in disks with steep profiles the terrestrial planets tend to be more massive, form more quickly, form closer in, and contain less water (*Raymond et al.*, 2005; *Kokubo et al.*, 2006).

4.4. Effect of Imperfect Accretion

As planetesimals' eccentricities are excited by growing embryos, they undergo considerable collisional grinding. Collisional disruption can be divided into two types: catastrophic disruption due to high-energy impacts and cratering due to low-energy impacts. *Kobayashi and Tanaka* (2010a) found that cratering collisions are much more effective in collisional grinding than collisions causing catastrophic disruption, simply because the former impacts occur much more frequently than the latter ones. Small fragments are easily accreted by embryos in the presence of nebular gas (*Wetherill and Stewart*, 1993), although they rapidly drift inward due to strong gas drag, leading to small embryo masses (*Chambers*, 2008; *Kobayashi and Tanaka*, 2010b).

Giant impacts between planetary embryos often do not result in net accretion. Rather, there exists a diversity of collisional outcomes. These include near-perfect merging at low impact speeds and near head-on configurations, partial accretion at somewhat higher impact speeds and angles, "hit-and-run" collisions at near-grazing angles, and even net erosion for high-speed, near head-on collisions (*Agnor and Asphaug*, 2004; *Asphaug et al.*, 2006; *Asphaug*, 2010). Two recent studies used large suites of smoothed-particle hydrodynamics (SPH) simulations to map out the conditions required for accretion in the parameter space of large impacts (*Genda et al.*, 2012; *Leinhardt and Stewart*, 2012). However, most N-body simulations of terrestrial planet formation to date have assumed perfect accretion in which all collisions lead to accretion.

About half of the embryo-embryo impacts in a typical simulation of late-stage accretion do not lead to net growth (*Agnor and Asphaug*, 2004; *Kokubo and Genda*, 2010). Rather, the outcomes are dominated by partially accreting collision, hit-and-run impacts, and graze-and-merge events in which two embryos dissipate sufficient energy during a grazing impact to become gravitationally bound and collide (*Leinhardt and Stewart*, 2012).

Taking into account only the accretion condition for embryo-embryo impacts, the final number, mass, orbital elements, and even growth timescale of planets are barely affected (*Kokubo and Genda*, 2010; *Alexander and Agnor*, 1998). This is because even though collisions do not lead to accretion, the colliding bodies stay on the colliding orbits after the collision and thus the system is unstable and the next collision occurs shortly.

However, by allowing nonaccretionary impacts to both erode the target embryo and produce debris particles, *Chambers* (2013) found that fragmentation does have a noted effect on accretion. The final stages of accretion are lengthened by the sweep-up of collisional fragments. The planets that formed in simulations with fragmentation had smaller masses and smaller eccentricities than their counterparts in simulations without fragmentation.

Imperfect accretion also affects the planets' spin rates. *Kokubo and Genda* (2010) found that the spin angular momentum of accreted planets was 30% smaller than in simulations with perfect accretion. This is because grazing collisions that have high angular momentum are likely to result in a hit and run, while nearly head-on collisions that have small angular momentum lead to accretion. The production of unbound collisional fragments with high angular momentum could further reduce the spin angular velocity. The effect of nonaccretionary impacts on the planetary spins has yet to be carefully studied.

A final consequence of fragmentation is on the core mass fraction. Giant impacts lead to an increase in the core mass fraction because the mantle is preferentially lost during imperfect merging events (*Benz et al.,* 2007; *Stewart and Leinhardt,* 2012; *Genda et al.,* 2012). However, the sweep-up of these collisional fragments on 100-m.y. timescales rebalances the composition of planets to roughly the initial embryo composition (*Chambers,* 2013). We speculate that a net increase in core mass fraction should be retained if the rocky fragments are allowed to collisionally evolve and lose mass.

4.5. Effect of Outer Giant Planets

We now consider the effect of giant planets on terrestrial accretion. We restrict ourselves to systems with giant planets similar to our own Jupiter and Saturn, i.e., systems with nonmigrating giant planets on stable orbits exterior to the terrestrial planet-forming region. In section 5.2 we will consider the effects of giant planet migration and planet-planet scattering.

The most important effect of giant planets on terrestrial accretion is the excitation of the eccentricities of planetary embryos. This generally occurs by the giant planet-embryo gravitational eccentricity forcing followed by the transmission of that forcing by embryo-embryo or embryo-planetesimal forcing. The giant planet forcing typically occurs via mean motion or secular resonances, or secular dynamical forcing. Giant planet-embryo excitation is particularly sensitive to the giant planets' orbital architecture (*Chambers and Cassen,* 2002; *Levison and Agnor,* 2003; *Raymond,* 2006). Figure 5 shows the eccentricities of test particles excited for 1 m.y. by two different configurations of Jupiter and Saturn (*Raymond et al.,* 2009), both of which are consistent with the present-day solar system (see section 6). The spikes in eccentricity seen in Fig. 5 come from specific resonances:

in the "Jupiter and Saturn in RESonance" (JSRES) configuration, the ν_5 secular resonance at 1.3 AU and the 2:1 mean-motion resonance with Jupiter at 3.4 AU; and in the "Extra-Eccentric Jupiter and Saturn" (EEJS) configuration, the ν_5 and ν_6 secular resonances at 0.7 and 2.1 AU, and a hint of the 2:1 mean-motion resonance with Jupiter at 3.3 AU. The "background" level of excitation seen in Fig. 5 comes from secular forcing, following a smooth function of the orbital radius.

The eccentricity excitation of terrestrial embryos is significant even for modest values of the giant planets' eccentricity. In Fig. 5, Jupiter and Saturn have eccentricities of 0.01–0.02 in the JSRES configuration and 0.1 in the EEJS configuration. The test particles in the JSRES system are barely excited by the giant planets interior to 3 AU; the magnitude of the spike at 1.3 AU is far smaller than the secular forcing anywhere in the EEJS simulation. Note also that this figure represents just the first link in the chain. The eccentricities imparted to embryos are systematically transmitted to the entire embryo swarm, and it is the mean eccentricity of the embryo swarm that dictates the outcome of accretion.

In a population of embryos with near-circular orbits, the communication zone — the radial distance across which a given embryo has gravitational contact with its neighbors — is very narrow. Embryos grow by collisions with their immediate neighbors. The planets that form are thus limited in mass by the mass in their immediate vicinity. In contrast, in a population of embryos with significant eccentricities, the communication zone of embryos is wider. Each embryo's orbit crosses the orbits of multiple other bodies and, by secular forcing, gravitationally affects the orbits of even more. This of course does not imply any imminent collisions, but it does mean that the planets that form will sample a wider radial range of the disk than in the case of very low embryo eccentricities. This naturally produces a smaller number of more-massive planets. Given that collisions preferentially occur at pericenter, the terrestrial planets that form tend to also be located closer in when the mean embryo eccentricity is larger (*Levison and Agnor,* 2003).

In systems with one or more giant planets on orbits exterior to the terrestrial planet-forming region, the amplitude of excitation of the eccentricities of terrestrial embryos is larger when the giant planets' orbits are eccentric or closer in. The timescale for excitation is shorter when the giant planets are more massive. Thus, the strongest perturbations come from massive eccentric gas giants.

Simulations have indeed shown that systems with massive or eccentric outer gas giants systematically produce fewer, more-massive terrestrial planets (*Chambers and Cassen,* 2002; *Levison and Agnor,* 2003; *Raymond et al.,* 2004). However, the efficiency of terrestrial accretion is smaller in the presence of a massive or eccentric gas giant because a fraction of embryos and planetesimals are excited onto orbits that are unstable and are thus removed from the system. The most common mechanism for the removal of such bodies is by having their eccentricities

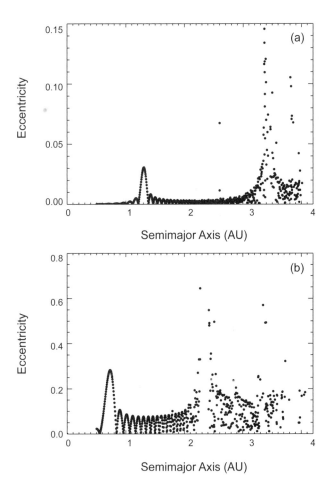

Fig. 5. Excitation of test particles by two configurations of Jupiter and Saturn. Both panels show the eccentricities of massless test particles after 1 m.y. (giant planets not shown). Note the difference in the y-axis scale between the two panels. Each scenario is consistent with the present-day solar system (see discussion in section 6). **(a)** Jupiter and Saturn are in 3:2 mean-motion resonance with semimajor axes of 5.4 and 7.2 AU and low eccentricities in the JSRES configuration. **(b)** The gas giants are at their current semimajor axes of 5.2 and 9.5 AU with eccentricities of 0.1 in the EEJS configuration. From *Raymond et al.* (2009).

increased to the point where their orbits cross those of a giant planet, then being ejected entirely from the system into interstellar space.

The strong outside-in perturbations produced by massive or eccentric outer gas giants also act to accelerate terrestrial planet formation. This happens for two reasons. First, when embryos have significant mean eccentricities, the typical time between encounters decreases, as long as eccentricities are more strongly perturbed than inclinations. Second, accretion is slower in the outer parts of planetary systems because of the longer orbital and encounter timescales, and it is these slow-growing regions that are most efficiently cleared by the giant planets' perturbations.

Given their outside-in influence, outer gas giants also

play a key role in water delivery to terrestrial planets. It should be noted up front that the gas giants' role in water delivery is purely detrimental, at least in the context of outer giant planets on static orbits. Stimulating the eccentricities of water-rich embryos at a few astronomical units can in theory cause some embryos to be scattered inward and deliver water to the terrestrial planets. In practice, a much larger fraction of bodies is scattered outward, encounters the giant planets, and is ejected from the system than is scattered inward to deliver water (*Raymond et al.*, 2006b).

Finally, simulations with setups similar to the one from Fig. 4 confirm that the presence of one or more giant planets strongly anticorrelates with the water content of the terrestrial planets in those systems (*Chambers and Cassen*, 2002; *Raymond et al.*, 2004, 2006b, 2007b, 2009; *O'Brien et al.*, 2006). There is a critical orbital radius beyond which a giant planet must lie for terrestrial planets to accrete and survive in a star's liquid water habitable zone (*Raymond*, 2006). This limit is eccentricity dependent: A zero-eccentricity (single) giant planet must lie beyond 2.5 AU to allow a terrestrial planet to form between 0.8 and 1.5 AU, whereas a giant planet with an eccentricity of 0.3 must lie beyond 4.2 AU. For water to be delivered to the terrestrial planets from a presumed source at 2–4 AU (as in Fig. 4), the giant planet must be farther still (*Raymond*, 2006).

5. TERRESTRIAL ACCRETION IN EXTRASOLAR PLANETARY SYSTEMS

Extrasolar planetary systems do not typically look like the solar system. To extrapolate to extrasolar planetary systems is therefore not trivial. Additional mechanisms must be taken into account, in particular orbital migration both of planetary embryos (type I migration) and of gas giant planets (type II migration) and dynamical instabilities in systems of multiple gas giant planets.

There exists ample evidence that accretion does indeed occur around other stars. Not only has an abundance of low-mass planets been detected (*Mayor et al.*, 2011; *Batalha et al.*, 2013), but the dust produced during terrestrial planet formation (*Kenyon and Bromley*, 2004) has also been detected (e.g., *Meyer et al.*, 2008; *Lisse et al.*, 2008), including the potential debris from giant embryo-embryo impacts (*Lisse et al.*, 2009). In this section we first address the issue of the formation of hot super-Earths. Then we discuss how the dynamics shaping the known systems of giant planets may have sculpted unseen terrestrial planets in those systems.

5.1. Hot Super-Earths

Hot super-Earths are extremely common. Roughly one-third to one-half of Sun-like (FGK) stars host at least one planet with a mass less than 10 $M_\oplus$ and a period of less than 50–100 d (*Howard et al.*, 2010; *Mayor et al.*, 2011). The frequency of hot super-Earths is at least as high around M stars as around FGK stars and possibly higher (*Howard et al.,* 2012; *Bonfils et al.,* 2013; *Fressin et al.,* 2013). Hot super-Earths are typically found in systems of many planets on compact but nonresonant orbits (e.g., *Udry et al.,* 2007; *Lovis et al.,* 2011; *Lissauer et al.,* 2011).

Several mechanisms have been proposed to explain the origin of hot super-Earths (see *Raymond et al.,* 2008): (1) *in situ* accretion from massive disks of planetary embryos and planetesimals; (2) accretion during inward type I migration of planetary embryos; (3) shepherding in interior mean-motion resonances with inward-migrating gas giant planets; (4) shepherding by inward-migrating secular resonances driven by dissipation of the gaseous disk; (5) circularization of planets on highly eccentric orbits by star-planet tidal interactions; and (6) photoevaporation of close-in gas giant planets.

Theoretical and observational constraints effectively rule out mechanisms 3–6. The shepherding of embryos by migrating resonances (mechanisms 3 and 4) can robustly transport material inward (*Zhou et al.,* 2005; *Fogg and Nelson,* 2005, 2007; *Raymond et al.,* 2006a; *Mandell et al.,* 2007; *Gaidos et al.,* 2007). An embryo that finds itself in resonance with a migrating giant planet will have its eccentricity simultaneously excited by the giant planet and damped by tidal interactions with the gaseous disk (*Tanaka and Ward,* 2004; *Cresswell et al.,* 2007). As the tidal damping process is nonconservative, the embryo's orbit loses energy and shrinks, removing the embryo from the resonance. The migrating resonance catches up to the embryo and the process repeats itself, moving the embryo inward, potentially across large distances. This mechanism is powered by the migration of a strong resonance. This requires a connection between hot super-Earths and giant planets. If a giant planet migrated inward, and the shepherd was a mean-motion resonance (likely the 3:2, 2:1, or 3:1 resonance) then hot super-Earths should be found just interior to close-in giant planets, which is not observed. If a strong secular resonance migrated inward then at least one giant planet on an eccentric orbit must exist exterior to the hot super-Earth, and there should only be a small number of hot super-Earths. This is also not observed.

Tidal circularization of highly-eccentric hot super-Earths (mechanism 5) is physically possible but requires extreme conditions (*Raymond et al.,* 2008). Star-planet tidal friction of planets on short-pericenter orbits can rapidly dissipate energy, thereby shrinking and recircularizing the planets' orbits. This process has been proposed to explain the origin of hot Jupiters (*Ford and Rasio,* 2006; *Fabrycky and Tremaine,* 2007; *Beaugé and Nesvorný,* 2012), and the same mechanism could operate for low-mass planets. Very close pericenter passages — within 0.02 AU — are required for significant radial migration (*Raymond et al.,* 2008). Although such orbits are plausible, another implication of the model is that, given their large prior eccentricities, hot super-Earths should be found in single systems with no other planets nearby. This is not observed.

The atmospheres of very close-in giant planets can be removed by photoevaporation from the host star (mechanism 6) (*Lammer et al.,* 2003; *Baraffe et al.,* 2004, 2006; *Yelle,* 2004; *Erkaev et al.,* 2007; *Hubbard et al.,* 2007a; *Raymond et al.,* 2008; *Murray-Clay et al.,* 2009; *Lopez and Fortney,* 2014). The process is driven by ultraviolet (UV) heating from the central star. Mass loss is most efficient for planets with low surface gravities extremely close to UV-bright stars. Within ~0.02 AU, planets as large as Saturn can be photoevaporated down to their cores on billion-year timescales. Since both the photoevaporation rate and the rate of tidal evolution depend on the planet mass, a very close-in rocky planet like Corot-7b (*Léger et al.,* 2009) could have started as a Saturn-mass planet on a much wider orbit (*Jackson et al.,* 2010). Although photoevaporation may cause mass loss in some very close-in planets, it cannot explain the systems of hot super-Earths. *Hubbard et al.* (2007b) showed that the mass distributions of very highly-irradiated planets within 0.07 AU was statistically indistinguishable from the mass distribution of planets at larger distances. In addition, given the very strong radial dependence of photoevaporative mass loss, the mechanism is likely to produce systems with a single hot super-Earth as the closest-in planet rather than multiple systems of hot super-Earths.

Given the current constraints from Kepler and radial velocity observations, mechanisms 1 and 2 — *in situ* accretion and type I migration — are the leading candidates to explain the formation of the observed hot super-Earths. Of course, we cannot rule out additional mechanisms that have yet to come to light.

For systems of hot super-Earths to have accreted *in situ* from massive populations of planetesimals and planetary embryos, their protoplanetary disks must have been very massive (*Raymond et al.,* 2008; *Hansen and Murray,* 2012, 2013; *Chiang and Laughlin,* 2013; *Raymond and Cossou,* 2014). The observed systems of hot super-Earths often contain 20–40 $M_\oplus$ in planets within a fraction of an astronomical unit of the star (*Batalha et al.,* 2013). Let us put this in the context of simplified power-law disks

$$\Sigma = \Sigma_0 \left(\frac{r}{1\,\text{AU}} \right)^{-x} \tag{19}$$

The minimum-mass solar nebula (MMSN) model (*Weidenschilling,* 1977; *Hayashi et al.,* 1985) has x = 3/2, although modified versions have x = 0.5 (*Davis,* 2005) and x ≈ 2 (*Desch,* 2007). *Chiang and Laughlin* (2013) created a minimum-mass *extrasolar* nebula using the Kepler sample of hot super-Earths and found a best fit for x = 1.6–1.7 with a mass normalization roughly 10× higher than the MMSN. However, *Raymond and Cossou* (2014) showed that minimum-mass disks based on Kepler multiple-planet systems actually cover a broad range in surface density slopes and are inconsistent with a universal underlying disk profile.

Only steep power-law disks allow for a significant amount of mass inside 1 AU. Consider a disk with a mass of 0.05 $M_\odot$ extending from 0 to 50 AU with an assumed dust-to-gas ratio of 1%. This disk contains a total of 150 $M_\oplus$ in solids. If the disk follows an $r^{-1/2}$ profile (i.e., with x = 1/2), then it only contains 0.4 $M_\oplus$ in solids inside 1 AU. If the disk has x = 1, then it contains 3 $M_\oplus$ inside 1 AU. If the disk has x = 1.5–1.7, then it contains 21–46 $M_\oplus$ inside 1 AU. Submillimeter observations of cold dust in the outer parts of nearby protoplanetary disks generally find values of x between 0.5 and 1 (*Mundy et al., 2000; Looney et al., 2003; Andrews and Williams, 2007b*). However, the inner parts of disks have yet to be adequately measured.

The dynamics of *in situ* accretion of hot super-Earths would presumably be similar to the well-studied dynamics of accretion presented in sections 3 and 4. Accretion would proceed faster than at 1 AU due to the shorter relevant timescales, but would consist of embryo-embryo and embryo-planetesimal impacts (*Raymond et al., 2008*). However, even if super-Earths accrete modest gaseous envelopes from the disk, these envelopes are expected to be lost during the dispersal of the protoplanetary disk under most conditions (*Ikoma and Hori, 2012*). This loss process is most efficient at high temperatures, making it hard to explain the large radii of some detected super-Earths. Nonetheless, super-Earths that form by *in situ* accretion appear to match several other features of the observed population, including their low mutual inclination orbits and the distributions of eccentricity and orbital spacing (*Hansen and Murray, 2013*).

Alternately, the formation of hot super-Earths may involve long-range orbital migration (*Terquem and Papaloizou, 2007*). Once they reach ~0.1 $M_\oplus$, embryos are susceptible to type I migration (*Goldreich and Tremaine, 1980; Ward, 1986*). Type I migration may be directed inward or outward depending on the local disk properties and the planet mass (*Paardekooper et al., 2010; Masset and Casoli, 2010; Kretke and Lin, 2012*). In most disks outward migration is only possible for embryos larger than a few Earth masses. All embryos therefore migrate inward when they are small. If they grow quickly enough during the migration, then in some regions they can activate their corotation torque and migrate outward.

A population of inward-migrating embryos naturally forms a resonant chain. Migration is stopped at the inner edge of the disk (*Masset et al., 2006*) and the resonant chain piles up against the edge (*Ogihara and Ida, 2009*). If the resonant chain gets too long, cumulative perturbations from the embryos act to destabilize the chain, leading to accretionary collisions and a new shorter resonant chain (*Morbidelli et al., 2008; Cresswell and Nelson, 2008*). This process can continue throughout the lifetime of the gaseous disk and include multiple generations of inward-migrating embryos or populations of embryos.

Figure 6 shows the formation of a system of hot super-Earths by type I migration from *Cossou et al.* (2014). In this simulation 60 $M_\oplus$ in embryos with masses of 0.1–2 $M_\oplus$ started from 2–15 AU. The embryos accreted as they migrated inward in successive waves. One embryo (the thick black curve in Fig. 6) grew large enough to trigger outward migration and stabilized at a zero-torque zone in

the outer disk, presumably to become a giant planet core. The system of hot super-Earths that formed is similar in mass and spacing to the Kepler-11 system (*Lissauer et al., 2011*). The four outer super-Earths are in a resonant chain, but the inner one was pushed interior to the inner edge of the gas disk and removed from resonance.

It was proposed by *Raymond et al.* (2008) that transit measurements of hot super-Earths could differentiate between the *in situ* accretion and type I migration models. They argued that planets formed *in situ* should be naked high-density rocks, whereas migrated planets are more likely to be dominated by low-density material such as ice. It has been claimed that planets that accrete *in situ* can have thick gaseous envelopes and thus inflated radii (*Hansen and Murray, 2012; Chiang and Laughlin, 2013*). However, detailed atmospheric calculations by *Ikoma and Hori* (2012) suggest that it is likely that low-mass planets generally lose their atmospheres during disk dispersal. This is a key point. If these planets can indeed retain thick atmospheres, then simple measurements of the bulk density of super-Earths would not provide a mechanism for differentiation between the models. However, if hot super-Earths cannot retain thick atmospheres after forming *in situ*, then low-density

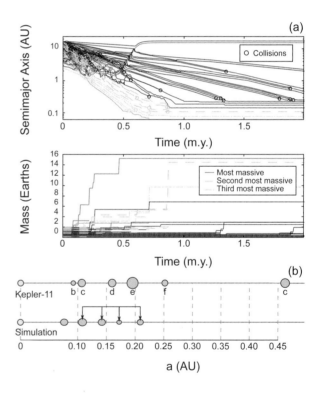

Fig. 6. Formation of a system of hot super-Earths by type I migration. **(a)** Evolution of the embryos' orbital radii; **(b)** mass growth. The thick black, gray dashed, and gray dash-dotted curves represent embryos that coagulated into the three most massive planets. Only the most massive (thick black) planet grew large enough to trigger outward migration before crossing into a zone of pure inward migration. From *Cossou et al.* (2013).

planets must have formed at larger orbital distances and migrated inward.

It is possible that migration and *in situ* accretion both operate to reproduce the observed hot super-Earths. The main shortcoming of *in situ* accretion models is that the requisite inner disk masses are extremely large and do not fit the surface density profiles measured in the outskirts of protoplanetary disks. Type I migration of planetary embryos provides a natural way to concentrate solids in the inner parts of protoplanetary disks. One can envision a scenario that proceeds as follows: Embryos start to accrete locally throughout the disk. Any embryo that grows larger than roughly a Mars-mass type I migrates inward. Most embryos migrate all the way to the inner edge of the disk, or at least to the pileup of embryos bordering on the inner edge. There are frequent close encounters and impacts between embryos. The embryos form long resonant chains that are successively broken by perturbations from other embryos or by stochastic forcing from disk turbulence (*Terquem and Papaloizou*, 2007; *Pierens and Raymond*, 2011). As the disk dissipates the resonant chain can be broken, leading to a last phase of collisions that effectively mimics the *in situ* accretion model. There remains sufficient gas and collisional debris to damp the inclinations of the surviving super-Earths to values small enough to be consistent with observations. However, it is possible that many super-Earths actually remain in resonant orbits but with period ratios altered by tidal dissipation (*Batygin and Morbidelli*, 2013).

5.2. Sculpting by Giant Planets: Type II Migration and Dynamical Instabilities

The orbital distribution of giant exoplanets is thought to have been sculpted by two dynamical processes: type II migration and planet-planet scattering (*Moorhead and Adams*, 2005; *Armitage*, 2007). These processes each involve long-range radial shifts in giant planets' orbits and have strong consequences for terrestrial planet formation in those systems. In fact, each of these processes has been proposed to explain the origin of hot Jupiters (*Lin et al.*, 1996; *Nagasawa et al.*, 2008), so differences in the populations of terrestrial planets, once observed, could help resolve the question of the origin of hot Jupiters.

Only a fraction of planetary systems contain giant planets. About 14% of Sun-like stars host a gas giant with period shorter than 1000 d (*Mayor et al.*, 2011), although the fraction of stars with more distant giant planets could be significantly higher (*Gould et al.*, 2010).

When a giant planet becomes massive enough to open a gap in the protoplanetary disk, its orbital evolution becomes linked to the radial viscous evolution of the gas. This is called type II migration (*Lin and Papaloizou*, 1986; *Ward*, 1997). As a giant planet migrates inward, it encounters other small bodies in various stages of accretion. Given the strong damping of eccentricities by the gaseous disk, a significant fraction of the material interior to the giant planet's initial orbit is shepherded inward by strong resonances as

explained in section 5.1 (*Zhou et al.*, 2005; *Fogg and Nelson*, 2005, 2007, 2009; *Raymond et al.*, 2006a; *Mandell et al.*, 2007). Indeed, the simulation from Fig. 7a formed two hot super-Earth planets, one just interior to the 2:1 and 3:1 resonance. The orbits of the two planets became destabilized after several million years, collided, and fused into a single 4-$M_\oplus$ hot super-Earth. There also exists a population of very close-in planetesimals in the simulation from Fig. 7; these were produced by the same shepherding mechanism as the hot super-Earths but, because the dissipative forces from gas drag were so much stronger for these objects than the damping due to disk-planet tidal interactions felt by the embryos (*Adachi et al.*, 1976; *Ida et al.*, 2008), they were shepherded by a much higher-order resonance, here the 8:1.

Planetesimals or embryos that come too close to the migrating giant are scattered outward onto eccentric orbits. These orbits are slowly recircularized by gas drag and dynamical friction. On 10–100-m.y. or longer timescales, a second generation of terrestrial planets can form from this scattered material (*Raymond et al.*, 2006a; *Mandell et al.*, 2007). The building blocks of this new generation of planets are significantly different than the original ones. This new distribution is comprised of two components: bodies that originated across the inner planetary system that were scattered outward by the migrating gas giant, and bodies that originated exterior to the gas giant. When taking into account the original location of these protoplanets, the effective feeding zone of the new terrestrial planets essentially spans the entire planetary system. This new generation of terrestrial planets therefore inevitably contains material that condensed at a wide range of orbital distances. Their volatile contents are huge. Indeed, the water content of the 3-$M_\oplus$ planet that formed at 0.9 AU (in the shaded habitable zone) in Fig. 7 is roughly 10% by mass. Even if 90% of the water were lost during accretion, that still corresponds to 10× Earth's water content (by mass), meaning that this planet is likely to be covered by global oceans.

The simulation from Fig. 7 showed the simple case of a single giant planet on a low eccentricity ($e \approx 0.05$) migrating through a disk of growing planetesimals and embryos. Migration would be more destructive to planet formation under certain circumstances. For example, if migration occurs very late in the evolution of the disk, then less gas remains to damp the eccentricities of scattered bodies. This is probably more of an issue for the formation of hot super-Earths than for scattered embryos: Since the viscous timescale is shorter closer in, much of the inner disk may in fact drain onto the star during type II migration (*Thommes et al.*, 2008) and reduce the efficiency of the shepherding mechanism. In addition, multiple giant planets may often migrate inward together. In that case the giant planets' eccentricities would likely be excited to modest values, and any embryo scattered outward would likely encounter another giant planet, increasing the probability of very strong scattering events onto unbound orbits.

Although type II migration certainly does not provide a comfortable environment for terrestrial accretion, planet-

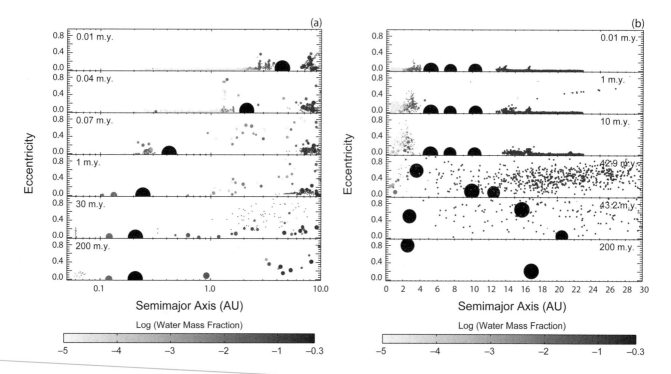

Fig. 7. The effect of **(a)** giant planet migration and **(b)** dynamical instabilities on terrestrial planet formation. In each panel the large black circles represent roughly Jupiter-mass gas giant planets and the smaller circles each represent a planetary embryo or planetesimal. The grayscale corresponds to each body's water content (see color bars), and the relative size of each particle (giant planets excepted) refers to their mass$^{1/3}$. Adapted from simulations by **(a)** *Raymond et al.* (2006a) and **(b)** *Raymond et al.* (2012).

planet scattering is far more disruptive. The broad eccentricity distribution of observed giant exoplanets is naturally reproduced if at least 75% of the observed planets are the survivors of violent dynamical instabilities (*Chatterjee et al.*, 2008; *Jurić and Tremaine*, 2008; *Raymond et al.*, 2010). It is thought that giant planets form in multiple systems on near-circular orbits but in time, perturbations destabilize these systems and lead to a phase of close gravitational encounters. Repeated planet-planet scattering usually leads to the ejection of one or more giant planets (*Rasio and Ford*, 1996; *Weidenschilling and Marzari*, 1996) (see also the chapter by Davies et al. in this volume). The large eccentricities of the observed planets are essentially the scars of past instabilities.

Instabilities are also destructive for terrestrial planets or their building blocks. The timing of instabilities is poorly constrained, although it is thought that many instabilities may be triggered by either migration in systems of multiple gas giants (*Adams and Laughlin*, 2003; *Moorhead and Adams*, 2005) or by the removal of damping during the dissipation of the gaseous disk (*Moeckel et al.*, 2008; *Matsumura et al.*, 2010; *Moeckel and Armitage*, 2012). On the other hand, systems of planets on more widely spaced orbits or systems with wide binary companions may naturally experience instabilities on billion-year timescales (*Marzari and Weidenschilling*, 2002; *Kaib et al.*, 2013). Although early instabilities may allow for additional sources

of damping via gas drag from remaining gas and dynamical friction from abundant planetesimals, in practice the timing of the instability makes little difference for the survival of terrestrial bodies (*Raymond et al.*, 2012).

Instabilities between Jupiter-sized planets typically only last for ~10^5 yr. When a giant planet is scattered onto a highly eccentric orbit, even if it only lasts for a relatively short time, very strong secular forcing can drive the orbits of inner terrestrial bodies to very high eccentricities. The outcome of the perturbation is extremely sensitive to the proximity of the giant planet to the terrestrial planet zone: Giant planets whose pericenter distances come within a given separation act so strongly that terrestrial planets or embryos are driven entirely into the central star (*Veras and Armitage*, 2005, 2006; *Raymond et al.*, 2011, 2012). The giant planet instabilities that are the least disruptive to the terrestrial planets are those that are very short in duration, that are confined to the outer parts of the planetary system, or that result in a collision between giant planets.

Figure 7b shows a simulation in which all terrestrial bodies were removed from the system by an instability between three ~Jupiter-mass giant planets that occurred after 42 m.y. During the first 42 m.y. of the simulation, accretion in the inner disk proceeded in the same manner as in Fig. 4. Once the instability was triggered after 42.8 m.y., the inner disk of planets — including two planets that had grown to nearly an Earth mass — were driven into the central star. The entire outer disk of planetesimals was ejected by repeated giant planet-planetesimal scattering over the next few million years (*Raymond et al.*, 2012).

Instabilities systematically perturb both the terrestrial planet-forming region and outer disks of planetesimals. The dynamics of gas giant planets thus creates a natural correlation between terrestrial planets and outer planetesimal disks. On billion-year timescales planetesimal disks collisionally grind down and produce cold dust that is observable at wavelengths as debris disks (*Wyatt*, 2008; *Krivov*, 2010). On dynamical grounds, *Raymond et al.* (2011, 2012) predicted a correlation between debris disks and systems of low-mass planets, as each of these forms naturally in dynamically calm environments, i.e., in systems with giant planets on stable orbits or in systems with no gas giants.

6. FORMATION OF THE SOLAR SYSTEM'S TERRESTRIAL PLANETS

A longstanding goal of planet formation studies has been to reproduce the solar system using numerical simulations. Although that goal has not yet been achieved, substantial progress has been made.

Jupiter and Saturn are key players in this story. Their large masses help shape the final stages of terrestrial accretion (section 4.5). However, there exist few constraints on their orbits during late-stage terrestrial accretion, and these are model-dependent.

The Nice model (e.g., *Tsiganis et al.*, 2005; *Morbidelli et al.*, 2007) proposes that the late heavy bombardment (LHB) — a spike in the impact rate on multiple solar system bodies that lasted from roughly 400 until 700 m.y. after the start of planet formation (*Tera et al.*, 1974; *Cohen et al.*, 2000; *Chapman et al.*, 2007) — was triggered by an instability in the giant planets' orbits. The instability was triggered by gravitational interactions between the giant planets and a disk of planetesimals exterior to the planets' orbits comprising perhaps 30–50 $M_\oplus$. Before the Nice model instability, the giant planets' orbits would have been in a more compact configuration, with Jupiter and Saturn interior to the 2:1 resonance and perhaps lodged in 3:2 resonance. Although there is no direct constraint, hydrodynamical simulations indicate that the gas giants' eccentricities were likely lower than their current values, probably around 0.01-0.02 (*Morbidelli et al.*, 2007).

An alternate but still self-consistent assumption is that the gas giants were already at their current orbital radii during terrestrial accretion. In that case, Jupiter and Saturn must have had slightly higher eccentricities than their current ones because scattering of embryos during accretion tends to modestly decrease eccentricities (e.g., *Chambers and Cassen*, 2002). In this scenario, an alternate explanation for the LHB is needed.

In this section we first consider "classical" models that assume that the orbits of the giant planets were stationary (section 6.1). Based on the above arguments we consider two reasonable cases. In the first case, Jupiter and Saturn were trapped in 3:2 mean-motion resonance at 5.4 and 7.2 AU with low eccentricities ($e_{giants} \approx 0.01$–0.02). In the second, Jupiter and Saturn were at their current orbital radii but with higher eccentricities ($e_{giants} = 0.07$–0.1).

Of course, Jupiter's and Saturn's orbits need not have been stationary at this time. It is well known that giant planets' orbits can migrate long distances, inward or outward, driven by exchanges with the gaseous protoplanetary disk (e.g., *Lin and Papaloizou*, 1986; *Veras and Armitage*, 2004) or a disk of planetesimals (e.g., *Fernandez and Ip*, 1984; *Murray et al.*, 1998). Although the last phases of accretion are constrained by Hf-W measurements of Earth samples to occur after the dissipation of the typical gas disk, giant planet migration at early times can sculpt the population of embryos and thus affect the "initial conditions" for late-stage growth.

While the Nice model relies on a delayed planetesimal-driven instability, earlier planetesimal-driven migration of the giant planets has recently been invoked (*Agnor and Lin*, 2012). In section 6.2 we consider the effect of this migration, which must occur on a timescale shorter than Mars' measured accretion time of a few million years (*Dauphas and Pourmand*, 2011) to have an effect. Finally, in section 6.3 we describe a new model called the Grand Tack (*Walsh et al.*, 2011) that invokes early gas-driven migration of Jupiter and Saturn. It is possible that disks are not radially smooth, or at least that planetesimals do not form in a radially uniform way (e.g., *Johansen et al.*, 2007; *Chambers*, 2010). *Jin et al.* (2008) proposed that a discontinuity in viscosity regimes at ~2 AU could decrease the local surface density and thus form a small Mars. However, the dip produced is too narrow to cut off Mars' accretion (*Raymond et al.*, 2009). It has also been known for decades that an embryo distribution with an abrupt radial edge naturally forms large planets within the disk but small planets beyond the edge (*Wetherill*, 1978). This "edge effect" can explain the large Earth/Mars mass ratio (see below).

Table 2 summarizes the ability of various models to reproduce the observational constraints discussed in section 2.

6.1. Classical Models with Stationary Gas Giants

Figure 5 shows how the giant planets excite the eccentricities of test particles for each assumption (*Raymond et al.*, 2009). In Fig. 5a (labeled JSRES), the giant planets are in a low-eccentricity compact configuration consistent with the Nice model, whereas in Fig. 5b (labeled EEJS) the giant planets have significant eccentricities and are located at their current orbital radii. The much stronger eccentricity excitation imparted by eccentric gas giants and the presence of strong resonances such as the ν_6 resonance seen at 2.1 AU in Fig. 5b have a direct influence on terrestrial planet formation.

Simulations with Jupiter and Saturn on circular orbits reproduce several aspects of the terrestrial planets (*Wetherill*, 1978, 1996, 1985; *Chambers and Wetherill*, 1998; *Morbidelli et al.*, 2000; *Chambers*, 2001; *Raymond et al.*, 2004, 2006b, 2007b, 2009; *O'Brien et al.*, 2006; *Morishima*

TABLE 2. Success of different models in matching inner solar system constraints.

Model	AMD	RMC	M_{Mars}	T_{form}	Ast. Belt	$WMF_{\oplus}$	Comments
Resonant Jup, Sat	✓	×	×	✓	×	✓	Consistent with Nice model
Eccentric Jup, Sat	~	~	✓	✓	✓	×	Not consistent with Nice model
Grand Tack	✓	✓	✓	~	✓	✓	Requires tack at 1.5 AU
Planetesimal-driven migration	✓	×	×	✓	×	✓	Requires other source of LHB

A check ("✓") represents success in reproducing a given constraint, a cross ("×") represents a failure to reproduce the constraint, and an approximate sign ("~") represents a "maybe," meaning success in reproducing the constraints in a fraction of cases. The constraints are, in order, the terrestrial planets' angular momentum deficit (AMD) and radial mass concentration (RMC) (see also Fig. 8), Mars' mass, Earth's formation timescale, the large-scale structure of the asteroid belt, and the delivery of water to Earth (represented by Earth's water mass fraction $WMF_{\oplus}$).

et al., 2010). Simulations typically form about the right number (3–5) of terrestrial planets with masses comparable to their actual masses. Earth analogs tend to complete their accretion on 50–100-m.y. timescales, consistent with geochemical constraints. Simulations include late giant impacts between embryos with similar characteristics to the one that is thought to have formed the Moon (*Ćuk and Stewart,* 2012; *Canup,* 2012). Embryos originating at 2.5–4 AU, presumed to be represented by carbonaceous chondrites and therefore to be volatile-rich, naturally deliver water to Earth during accretion (see Fig. 4).

There are three problems. First and most importantly, simulations with Jupiter and Saturn on circular orbits are unable to form good Mars analogs. Rather, planets at Mars' orbital distance are an order of magnitude too massive, a situation called the "small Mars problem" (*Wetherill,* 1991; *Raymond et al.,* 2009). Second, the terrestrial planet systems that form tend to be far too spread out radially. Their radial mass concentration (RMC) (see equation (2)) are far smaller than the solar system's value of 89.9 (see Table 1). Third, large (~Mars-sized) embryos are often stranded in the asteroid belt. All three of these problems are related: The large RMC in these systems is a consequence of too much mass existing beyond 1 AU. This mass is in the form of large Mars analogs and embryos in the asteroid belt.

Simulations starting with Jupiter and Saturn at their current orbital radii but with larger initial eccentricities (e = 0.07–0.1) reproduce many of the same terrestrial planet constraints (*Raymond et al.,* 2009; *Morishima et al.,* 2010). Simulations tend to again form the same number of terrestrial planets with masses comparable to the actual planets'. Moon-forming impacts also occur. Beyond this the accreted planets contrast with those that accrete in simulations with circular gas giants. With eccentric Jupiter and Saturn, the terrestrial planets accrete faster, in modest agreement with Earth's geochemical constraints. The delivery of water to Earth is much less efficient. But Mars analogs form with about the right mass!

In these simulations, a strong secular resonance with Saturn — the ν_6 at 2.1 AU — acts to clear out the material in the inner asteroid belt and in Mars' vicinity. The resonance is so strong that bodies that are injected into it are driven to very high eccentricities and collide with the

Sun within a few million years (*Gladman et al.,* 1997). Any embryo from the inner planetary system that is scattered out near the ν_6 is quickly removed from the system. The Mars region is quickly drained and a small Mars forms. The ν_6 acts as a firm outer edge such that the terrestrial planet systems form in more compact configurations, with RMC values that approach the solar system's (but still remain roughly a factor of 2 too small; see Fig. 8). The AMD of the terrestrial planets are systematically higher than the solar system value because the planetesimals that could provide damping at late times are too efficiently depleted. The terrestrial planet-forming region is effectively cut off from the asteroid belt by the resonance, and water delivery is inefficient. If the gravitational potential from the dissipating gas disk is accounted for, the ν_5 and ν_6 resonances sweep inward and can perhaps shepherd water-rich embryos into Earth's feeding zone by the same mechanism presented in section 5.2 (*Thommes et al.,* 2008; *Morishima et al.,* 2010). However, hydrodynamical simulations suggest that Jupiter's and Saturn's eccentricities are unlikely to remain high enough during the gaseous disk phase for this to occur (e.g., *Morbidelli et al.,* 2007; *Pierens and Raymond,* 2011).

The early orbits of Jupiter and Saturn sculpt dramatically different terrestrial planet systems. Systems with gas giants on circular orbits form Mars analogs that are far too large and strand embryos in the asteroid belt. Systems with gas giants on eccentric orbits do not deliver water to Earth and have eccentricities that are too large. To date, no other configuration of Jupiter and Saturn with static orbits has been shown to satisfy all constraints simultaneously.

To quantify the failings of the classical model, Fig. 8 shows the AMD and RMC statistics for simulated terrestrial planets under the two assumptions considered here. The accreted planets are far too radially spread out (have small RMC values). In many cases their orbits are also too excited, with larger AMD values than the actual terrestrial planets'.

6.2. Accretion with Planetesimal-Driven Migration of Jupiter and Saturn

If Jupiter and Saturn formed in a more compact orbital configuration, then the migration to their current configuration may have perturbed the terrestrial planets, or even the building blocks of the terrestrial planets if their formation was not complete. *Brasser et al.* (2009, 2013) and *Agnor and Lin* (2012) simulated the influence of planetesimal-driven migration of the giant planets on the terrestrial planets assuming that the migration occurred late, after the terrestrial planets were fully formed. They found that if Jupiter and Saturn migrated with eccentricities comparable to their present-day values, a smooth migration with an exponential timescale characteristic of planetesimal-driven migration ($\tau \sim 5$–10 m.y.) would have perturbed the eccentricities of the terrestrial planets to values far in excess of the observed ones. To resolve this issue, *Brasser et al.* (2009, 2013) suggested a jumping Jupiter in which encounters between an ice giant and Jupiter caused Jupiter's and Saturn's orbits to spread much faster than if migration were driven solely by encounters with planetesimals (see also *Morbidelli et al.*, 2010). On the other hand, *Agnor and Lin* (2012) suggested that the bulk of any giant planet migration occurred during accretion of terrestrial planets.

Whenever the migration occurred, the degree of eccentricity excitation of Jupiter and Saturn is constrained by the dynamics of resonance crossing. Jupiter and Saturn are naturally excited to $e_{giants} \sim 0.05$ but cannot reach the higher eccentricities invoked by the eccentric Jupiter and Saturn model described above (*Tsiganis et al.*, 2005). Given that the eccentricity excitation is the key difference between this model and those with stationary giant planets

discussed above, the only free parameter is the timing of the eccentricity excitation.

Two recent papers simulated the effect of planetesimal-driven migration of Jupiter's and Saturn's orbits on terrestrial planet formation (*Walsh and Morbidelli*, 2011; *Lykawka and Ito*, 2013). In both studies terrestrial planets accrete from a disk of material that stretches from ~0.5 AU to 4.0 AU. In *Walsh and Morbidelli* (2011), Jupiter and Saturn are initially at 5.4 and 8.7 AU respectively (slightly outside the 2:1 mean-motion resonance), with eccentricities comparable to the current ones, and migrate to 5.2 and 9.4 AU with an e-folding time of 5 m.y. In their simulations Mars is typically far too massive and the distribution of surviving planetesimals in the asteroid belt is inconsistent with the observed distribution. *Lykawka and Ito* (2013) performed similar simulations but included the 2:1 resonance crossing of Jupiter and Saturn, which provides a sharp increase in the giant planets' eccentricities and thus in the perturbations felt by the terrestrial planets. They tested the timing of the giant planets' 2:1 resonance crossing between 1 and 50 m.y. They found the expected strong excitation in the asteroid belt once the giant planets' eccentricities increased, but the perturbations were too small to produce a small Mars. Although they produced four Mars analogs in their simulations, they remained significantly more massive than the real Mars, accreted on far longer timescales than the geochemically constrained one, and stranded large embryos in the asteroid belt. Their AMD and RMC values remain incompatible with the real solar system (Fig. 8).

If another mechanism is invoked to explain the LHB, planetesimal-driven migration of Jupiter and Saturn is plausible. However, it does not appear likely to have occurred, as it is incapable of solving the Mars problem.

6.3. The Grand Tack Model

Prior to 2009, several studies of terrestrial accretion had demonstrated an edge effect in terrestrial accretion. A distribution of embryos with an abrupt edge naturally produces a large mass gradient between the massive planets that formed within the disk and the smaller planets that were scattered beyond the disk's edge (*Wetherill*, 1978, 1991; *Chambers and Wetherill*, 1998; *Agnor et al.*, 1999; *Chambers*, 2001; *Kominami and Ida*, 2004). These studies had outer edges at 1.5–2 AU and generally considered their initial conditions a deficiency imposed by limited computational resources.

Hansen (2009) turned the tables by proposing that, rather than a deficiency, initial conditions with edges might actually represent the true initial state of the disk. Indeed, *Morishima et al.* (2008) and *Hansen* (2009) showed that most observed constraints could be reproduced by a disk of embryos spanning only from 0.7 to 1 AU. Earth and Venus are massive because they formed within the annulus, whereas Mars' and Mercury's small masses are explained as edge effects, embryos that were scattered exterior and interior respectively to the annulus at early times, stranding and starving them. Mars analogs consistently accrete on the

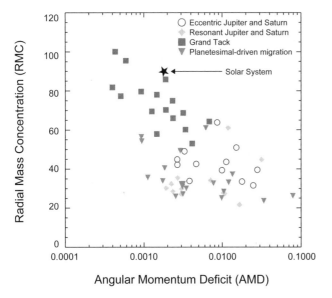

Fig. 8. Orbital statistics of the terrestrial planet systems formed in different models. The configuration of each system is represented by its angular momentum deficit and radial mass concentration values; see section 2.1 for the definition of these terms. The simulations with eccentric and resonant gas giants are from *Raymond et al.* (2009), those including planetesimal-driven migration of the gas giants are from *Lykawka and Ito* (2013), and the Grand Tack simulations are from *O'Brien et al.* (2014).

short observed timescale. The main unanswered question in these studies was the origin of the edges of the annulus.

Walsh et al. (2011) presented a mechanism to produce the outer edge of the disk by invoking migration of the giant planets to dramatically sculpt the distribution of solid material in the inner solar system. Given that gas giant planets must form in the presence of gaseous disks and that these disks invariably drive radial migration (*Ward,* 1997), it is natural to presume that Jupiter and Saturn must have migrated to some extent. A Jupiter-mass planet naturally carves an annular gap in the gaseous disk and migrates inward on the local viscous timescale (*Lin and Papaloizou,* 1986). In contrast, a Saturn-mass planet migrates much more quickly because of a strong gravitational feedback during disk clearing (*Masset and Papaloizou,* 2003). Assuming that Jupiter underwent rapid gas accretion before Saturn, hydrodynamical simulations show that Jupiter would have migrated inward relatively slowly. When Saturn underwent rapid gas accreted it migrated inward quickly, caught up to Jupiter, and became trapped in 2:3 resonance. At this point the direction of migration was reversed and Jupiter "tacked," i.e., it changed its direction of migration (*Masset and Snellgrove,* 2001; *Morbidelli et al.,* 2007; *Pierens and Nelson,* 2008; *Pierens and Raymond,* 2011; *D'Angelo and Marzari,* 2012). The outward migration of the two gas giants slowed as the gaseous disk dissipated, stranding Jupiter and Saturn on resonant orbits. This naturally produces the initial conditions for a recently revised version of the Nice model (*Morbidelli et al.,* 2007; *Levison et al.,* 2011), with Jupiter at 5.4 AU and Saturn at 7.2 AU.

This model is called the Grand Tack. One cannot know the precise migration history of the gas giants *a priori* given uncertainties in disk properties and evolution. *Walsh et al.* (2011) anchored Jupiter's migration reversal point at 1.5 AU because this truncates the inner disk of embryos and planetesimals at 1.0 AU, creating an outer edge at the same location as invoked by *Hansen* (2009). Jupiter's formation zone was assumed to be ~3–5 AU [although a range of values was tested by *Walsh et al.* (2011)], in the vicinity of the snow line (e.g., *Sasselov and Lecar,* 2000; *Kornet et al.,* 2004; *Martin and Livio,* 2012), presumably a favorable location for giant planet formation. The Grand Tack model also proposes that the compositional gradient seen in the asteroid belt can be explained by the planetesimals' formation zones. Volatile-poor bodies ("S-class") are primarily located in the inner belt and volatile-rich bodies ("C-class") primarily in the outer belt (*Gradie and Tedesco,* 1982; *DeMeo and Carry,* 2013). The Grand Tack scenario presumes that S-class bodies formed interior to Jupiter's initial orbit and that C-class bodies formed exterior.

The evolution of the Grand Tack is illustrated in Fig. 9 (*Walsh et al.,* 2011). Jupiter's and Saturn's inward migration scattered S-class planetesimals from the inner disk, with ~10% ending on eccentric orbits beyond the giant planets. Meanwhile, a large fraction of planetesimals and embryos were shepherded inward by the same mechanism discussed in section 5.2 onto orbits inside 1 AU. Follow-

ing Jupiter's "tack" the outward-migrating gas giants first encountered the scattered S-class planetesimals, about 1% of which were scattered inward onto stable orbits in the asteroid belt. The giant planets then encountered the disk of C-class planetesimals that originated beyond Jupiter's orbit. Again, a small fraction (~1%) were scattered inward and trapped in the asteroid belt. The final position of a scattered body depends on the orbital radius of the scattering body, in this case Jupiter. Jupiter was closer in when it scattered the S-class planetesimals and farther out when it scattered the C-class planetesimals. The S-class bodies were therefore preferentially implanted in the inner part of the asteroid belt and the C-class bodies preferentially in the outer part of the belt, as in the present-day belt (*Gradie and Tedesco,* 1982; *DeMeo and Carry,* 2013). The total mass of the asteroid population is set by the need to have ~2 $M_\oplus$ of material remaining in the inner truncated disk of embryos and planetesimals (to form the planets). This requirement for the planets sets the total mass in S-class bodies implanted into the asteroid belt as they originate from the same inner disk. The current ratio of S-class to C-class asteroids sets the mass in outer disk planetesimals.

The Grand Tack model reproduces many aspects of the terrestrial planets. Planets that accrete from a truncated disk have similar properties to those in *Hansen* (2009) and *Morishima et al.* (2008). Earth/Mars mass ratios are close matches to the actual planets, and Mars' accretion timescale

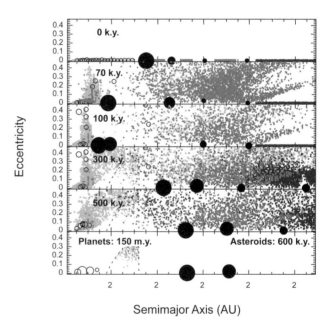

Fig. 9. Evolution of the Grand Tack model (*Walsh et al.,* 2011). The large black dots represent the four giant planets, with sizes that correspond to their approximate masses. Light gray symbols indicate S-class bodies and darker ones C-class bodies. There exist two categories of C-class objects (dark gray and black) that originate between and beyond the giant planets' orbits. Open circles indicate planetary embryos. The evolution of the particles includes drag forces imparted by an evolving gaseous disk.

is a good match to Hf/W constraints. Figure 8 shows that the AMD is systematically lower than in simulations of the classical model (section 6.1) and the RMC is systematically higher (*Walsh et al.,* 2011; *O'Brien et al.,* 2014). In contrast with other models, the Grand Tack simulations provide a reasonable match to the inner solar system.

The Grand Tack delivers water-rich material to the terrestrial planets by a novel mechanism. As Jupiter and Saturn migrate outward, they scatter about 1% of the C-class asteroids that they encountered onto stable orbits in the asteroid belt. And for every implanted C-type asteroid, 10–20 C-class bodies are scattered onto *unstable* orbits that cross the orbits of the terrestrial planets. These scattered C-class planetesimals accrete with the growing terrestrial planets and naturally deliver water. The amount of water-rich material accreted by Earth is less than in classical simulations with stationary giant planets like the one presented in Fig. 4, but is still significantly larger than Earth's current water budget (*O'Brien et al.,* 2014). The chemical signature of the delivered water is the same as C-type asteroids (and therefore carbonaceous chondrites), and thus provides a match to the signature of Earth's water (*Marty and Yokochi,* 2006). Thus, in the Grand Tack model, water was delivered to Earth not by C-type asteroids but by the same parent population as for C-type asteroids.

There remain some issues with the Grand Tack model. The accretion timescales are much faster for all the planets than what was typically found in previous models. This is a consequence of the removal of embryos beyond 1 AU, where growth timescales are long. In simulations, Mars analogs typically form in less than 10 m.y. (*O'Brien et al.,* 2014). Earth analogs form in 10–20 m.y., with giant embryo-embryo impacts occurring after 20 m.y. in only a modest fraction (~20%) of simulations. This is roughly a factor of 2 faster than the Hf-W constraints (*Touboul et al.,* 2007; *Kleine et al.,* 2009; *König et al.,* 2011). However, new simulations show that the accretion timescale of the terrestrial planets can be lengthened to match observations simply by increasing the total embryo-to-planetesimal mass ratio in the annulus, which is itself an unconstrained parameter (*Jacobson et al.,* 2013).

An open question related to the origin of Mercury's small mass is the origin of the *inner* edge of the annulus proposed by *Hansen* (2009). One possibility is that, as embryos grow larger from the inside out, they also become subject to type I migration from the inside out (*McNeil et al.,* 2005; *Daisaka et al.,* 2006; *Ida and Lin,* 2008). For embryo-mass objects, migration is directed inward (*Paardekooper et al.,* 2010), so as each embryo forms it migrates inward. If, by some process, inward-migrating planets are removed from the system (presumably by colliding with the star), then an inner edge in the distribution of *surviving* embryos could correspond to the outermost orbital radius at which an embryo formed and was destroyed. Another possibility is that planetesimals could only form in a narrow annulus. If a pressure bump (*Johansen et al.,* 2009) were located in that region it could act to concentrate small particles (*Haghighipour and Boss,* 2003; *Youdin and Chiang,* 2004)

and efficiently form planetesimals (see the chapter by Johansen et al. in this volume).

7. DISCUSSION

7.1. Terrestrial Planets Versus Giant Embryos

We think that Earth formed via successive collisions between planetesimals and planetary embryos, including a protracted stage of giant impacts between embryos. But does the formation of most terrestrial planets follow the same blueprint as Earth?

The alternative is that terrestrial exoplanets are essentially giant planetary embryos. They form from planetesimals or pebbles and do not undergo a phase of giant impacts after the dissipation of the gaseous disk. This is a wholly reasonable possibility. Imagine a disk that only forms planetesimals in a few preferred locations, perhaps at pressure bumps. The planetesimals in each location could be efficiently swept up into a single large embryo, perhaps by the largest planetesimal undergoing a rapid burst of super-runaway pebble accretion. Isolated giant embryos would evolve with no direct contact with other embryos. Only if several embryos formed and migrated toward a common location would embryo-embryo interactions become important, and collisions would only occur if a critical number of embryos was present [the critical number is about 5 (*Morbidelli et al.,* 2008; *Pierens et al.,* 2013)].

Terrestrial planets and giant embryos should differ in terms of their accretion timescales, atmospheres, and perhaps geological evolution. The timescale for the completion of Earth's accretion is at least 10× longer than the typical gas disk lifetime (see section 2). Giant embryos must form within the lifetime of the gaseous disk, while the mechanisms to efficiently concentrate are active. How would Earth be different if it had accreted 10× faster? The additional heat of formation and from trapped short-lived radionuclides could act to rapidly devolatilize the giant embryo's interior. However, giant embryos may be able to gravitationally capture thick envelopes of gas from the disk, at least at cooler locations within the disk (*Ikoma and Hori,* 2012). The fate of giant embryos' volatiles remain unstudied. Nonetheless, given that only a very small amount of H and He are needed to significantly inflate a planet's radius (*Fortney et al.,* 2007), giant embryos would likely have low bulk densities. Many low-density planets have indeed been discovered (*Marcy et al.,* 2014; *Weiss et al.,* 2013), although we stress that this does not indicate that these are giant embryos.

How could we tell observationally whether late phases of giant impacts are common? Perhaps the simplest approach would be to search for signatures of such impacts around stars that no longer harbor gaseous disks. The evolution of warm dust, detected as excess emission at mid-infrared wavelengths, has recently been measured to decline on 100-m.y. timescales (*Meyer et al.,* 2008; *Carpenter et al.,* 2009; *Melis et al.,* 2010). This dust is thought to trace the terrestrial planet-forming region (*Kenyon and Bromley*, 2004) and indicates the presence of planetesimals or other large dust-producing bodies in that region. In some cases the signature of specific minerals in the dust can indicate that it originated in a larger body. In fact, the signature of a giant impact was reported by *Lisse et al.* (2009) around the ~12-m.y.-old A star HD 172555. Given the 1–10-m.y. interval between giant impacts in accretion simulations and the short lifetime of dust produced (*Kenyon and Bromley*, 2005; *Melis et al.*, 2012), a direct measure of the frequency of systems in which giant impacts occur will require a large sample of young stars surveyed at mid-infrared wavelengths (e.g., *Kennedy and Wyatt*, 2012).

7.2. Limitations of the Simulations

Despite marked advances in the last few years, simulations of terrestrial planet formation remain both computationally and physically limited. Even the best numerical integrators (*Chambers*, 1999; *Duncan et al.*, 1998; *Stadel*, 2001) can follow the orbits of at most a few thousand particles at ~1 AU for the >100-m.y. timescales of terrestrial planet formation. There is a 3–4 order-of-magnitude uncertainty about the sizes of initial planetesimals, and a corresponding 9–12 order-of-magnitude uncertainty about the initial number of planetesimals. It is clear that current simulations cannot fully simulate the conditions of planet formation except in very constrained settings (e.g., *Barnes et al.*, 2009). Simulations thus resort to including planetesimals that are far more massive than they should be.

There exist several processes thought to be important in planet formation that have yet to be adequately modeled. For example, the full evolution of a planetesimal swarm including growth, dynamical excitation, and collisional grinding has yet to be fully simulated (but see *Bromley and Kenyon*, 2011; *Levison et al.*, 2012). In addition to the numerical and computational challenges, this task is complicated by the fact that the initial distribution and sizes of planetesimals, pebbles, and dust remain at best modestly constrained by models and observations (see chapters by Johansen et al. and Testi et al. in this volume). Likewise, the masses, structure, and evolution of the dominant, gaseous components of protoplanetary disks is an issue of ongoing study.

8. SUMMARY

This chapter has flown over a broad swath of the landscape of terrestrial planet formation. We now summarize the key points:

1. The term "terrestrial planet" is well-defined in the confines of our solar system but not in extrasolar planetary systems (section 1).

2. There exist ample observed and measured constraints on terrestrial planet formation models in the solar system and in extrasolar planetary systems (section 2).

3. There exist two models for the growth of planetary embryos (section 3). Oligarchic growth proposes that embryos grow from swarms of planetesimals. Pebble accretion proposes that they grow directly from centimeter-sized pebbles.

4. Starting from systems of embryos and planetesimals, the main factors determining the outcome of terrestrial accretion have been determined by simulations (section 4). The most important one is the level of eccentricity excitation of the embryo swarm — by both gravitational self-stirring and perturbations from giant planets — as this determines the number, masses, and feeding zones of terrestrial planets.

5. The observed systems of hot super-Earths probably formed by either *in situ* accretion from massive disks or inward migration of embryos driven by interactions with the gaseous disk (section 5.1). The key debate in differentiating between the models is whether rocky planets that accrete *in situ* could retain thick gaseous envelopes.

6. The dynamical histories of giant exoplanets are thought to include gas-driven migration and planet-planet scattering. An inward-migrating gas giant forms low-mass planets interior to strong resonances and stimulates the formation of outer volatile-rich planets. Dynamical instabilities among giant planets can destroy terrestrial planets or their building blocks, and this naturally produces a correlation between debris disks and terrestrial planets (section 5.2).

7. Historical simulations of terrestrial planet formation could not reproduce Mars' small mass (section 6.1). This is called the "small Mars problem." Simulations can reproduce Mars' small mass by invoking large initial eccentricities of Jupiter and Saturn at their current orbital radii. Invoking early planetesimal-driven migration of Jupiter and Saturn does not produce a small Mars (section 6.2).

8. The Grand Tack model proposes that Jupiter migrated inward to 1.5 AU then back outward due to disk torques before and after Saturn's formation (section 6.3). The inner disk was truncated at 1 AU, producing a large Earth/Mars mass ratio. Water was delivered to the terrestrial planets in the form of C-class bodies scattered inward during the gas giants' outward migration.

9. Finally, it remains unclear whether most systems of terrestrial planets undergo phases of giant collisions between embryos during their formation (section 7.1).

Acknowledgments. We are grateful to a long list of colleagues whose work contributed to this chapter. S.N.R. thanks NASA's Astrobiology Institute for their funding via the University of Washington, University of Colorado, and Virtual Planetary Laboratory lead teams. S.N.R. and A.M. thank the CNRS' Programme Nationale de Planetologie. K.J.W. acknowledges support from NLSI CLOE.

REFERENCES

Adachi I. et al. (1976) *Prog. Theor. Phys.*, *56*, 1756.
Adams E. R. et al. (2008) *Astrophys. J.*, *673*, 1160.
Adams F. C. and Laughlin G. (2003) *Icarus*, *163*, 290.
Agnor C. and Asphaug E. (2004) *Astrophys. J. Lett.*, *613*, L157.
Agnor C. B. and Lin D. N. C. (2012) *Astrophys. J.*, *745*, 143.
Agnor C. B. et al. (1999) *Icarus*, *142*, 219.

Alexander S. G. and Agnor C. B. (1998) *Icarus, 132,* 113.
Andrews S.M. and Williams J. P. (2007a) *Astrophys. J., 671,* 1800.
Andrews S. M. and Williams J. P. (2007b) *Astrophys. J., 659,* 705.
Armitage P. J. (2007) *Astrophys. J., 665,* 1381.
Asphaug E. (2010) *Chemie der Erde/Geochemistry, 70,* 199.
Asphaug E. et al. (2006) *Nature, 439,* 155.
Baraffe I. et al. (2004) *Astron. Astrophys., 419,* L13.
Baraffe I. et al. (2006) *Astron. Astrophys., 450,* 1221.
Barnes R. et al. (2009) *Icarus, 203,* 626.
Batalha N. M. et al. (2013) *Astrophys. J. Suppl., 204,* 24.
Batygin K. and Morbidelli A. (2013) *Astron. J., 145,* 1.
Beaugé C. and Nesvorný D. (2012) *Astrophys. J., 751,* 119.
Benz W. et al. (2007) *Space Sci. Rev., 132,* 189.
Bond J. C. et al. (2010) *Icarus, 205,* 321.
Bonfils X. et al. (2013) *Astron. Astrophys., 549,* A109.
Bottke W. F. et al. (2010) *Science, 330,* 1527.
Brasser R. et al. (2009) *Astron. Astrophys., 507,* 1053.
Brasser R. et al. (2013) *Mon. Not. R. Astron. Soc..*
Bromley B. C. and Kenyon S. J. (2011) *Astrophys. J., 731,* 101.
Buchhave L. A. et al. (2012) *Nature, 486,* 375.
Canup R. M. (2012) *Science, 338,* 1052.
Carpenter J. M. et al. (2009) *Astrophys. J. Suppl., 181,* 197.
Carter-Bond J. C. et al. (2012) *Astrophys. J., 760,* 44.
Chambers J. E. (1999) *Mon. Not. R. Astron. Soc., 304,* 793.
Chambers J. E. (2001) *Icarus, 152,* 205.
Chambers J. (2006) *Icarus, 180,* 496.
Chambers J. (2008) *Icarus, 198,* 256.
Chambers J. E. (2010) *Icarus, 208,* 505.
Chambers J. E. (2013) *Icarus, 224,* 43.
Chambers J. E. and Cassen P. (2002) *Meteoritics & Planet. Sci., 37,* 1523.
Chambers J. E. and Wetherill G. W. (1998) *Icarus, 136,* 304.
Chambers J. E. et al. (1996) *Icarus, 119,* 261.
Chapman C. R. et al. (2007) *Icarus, 189,* 233.
Chatterjee S. et al. (2008) *Astrophys. J., 686,* 580.
Chiang E. and Laughlin G. (2013) *Mon. Not. R. Astron. Soc., 431,*3444.
Cohen B. A. et al. (2000) *Science, 290,* 1754.
Correia A. C. M. and Laskar J. (2009) *Icarus, 201,* 1.
Cossou C. et al. (2014) *ArXiv e-prints,* arXiv:1407.6011C.
Cresswell P. and Nelson R. P. (2008) *Astron. Astrophys., 482,* 677.
Cresswell P. et al. (2007) *Astron. Astrophys., 473,* 329.
Ćuk M. and Stewart S. T. (2012) *Science, 338,* 1047.
Cumming A. et al. (2008) *Publ. Astron. Soc. Pac., 120,* 531.
Daisaka J. K. et al. (2006) *Icarus, 185,* 492.
D'Angelo G. and Marzari F. (2012) *Astrophys. J., 757,* 50.
Dauphas N. and Pourmand A. (2011) *Nature, 473,* 489.
Davis S. S. (2005) *Astrophys. J. Lett., 627,* L153.
Day J. M. D. et al. (2007) *Science, 315,* 217.
DeMeo F. E. and Carry B. (2013) *Icarus, 226,* 723.
Desch S. J. (2007) *Astrophys. J., 671,* 878.
Dones L. and Tremaine S. (1993) *Icarus, 103,* 67.
Drake M. J. and Righter K. (2002) *Nature, 416,* 39.
Duncan M. J. et al. (1998) *Astron. J., 116,* 2067.
Eisner J. A. (2012) *Astrophys. J., 755,* 23.
Eisner J. A. et al. (2008) *Astrophys. J., 683,* 304.
Elser S. et al. (2012) *Icarus, 221,* 859.
Erkaev N. V. et al. (2007) *Astron. Astrophys., 472,* 329.
Fabrycky D. and Tremaine S. (2007) *Astrophys. J., 669,* 1298.
Fang J. and Margot J.-L. (2013) *Astrophys. J., 767,* 115.
Fedele D. et al. (2010) *Astron. Astrophys., 510,* A72.
Fernandez J. A. and Ip W. (1984) *Icarus, 58,* 109.
Fischer D. A. and Valenti J. (2005) *Astrophys. J., 622,* 1102.
Fogg M. J. and Nelson R. P. (2005) *Astron. Astrophys., 441,* 791.
Fogg M. J. and Nelson R. P. (2007) *Astron. Astrophys., 461,* 1195.
Fogg M. J. and Nelson R. P. (2009) *Astron. Astrophys., 498,* 575.
Ford E. B. and Rasio F. A. (2006) *Astrophys. J. Lett., 638,* L45.
Fortney J. J. et al. (2007) *Astrophys. J., 659,* 1661.
Fressin F. et al. (2013) *Astrophys. J., 766,* 81.
Gaidos E. et al. (2007) *Science, 318,* 210.
Geiss J. and Gloeckler G. (1998) *Space Sci. Rev., 84,* 239.
Genda H. and Abe Y. (2005) *Nature, 433,* 842.
Genda H. et al. (2012) *Astrophys. J., 744,* 137.
Ghezzi L. et al. (2010) *Astrophys. J., 720,* 1290.
Gladman B. J. et al. (1997) *Science, 277,* 197.
Goldreich P. and Tremaine S. (1980) *Astrophys. J., 241,* 425.
Goldreich P. et al. (2004) *Astrophys. J., 614,* 497.
Gonzalez G. (1997) *Mon. Not. R. Astron. Soc., 285,* 403.

Gould A. et al. (2010) *Astrophys. J., 720,* 1073.
Gradie J. and Tedesco E. (1982) *Science, 216,* 1405.
Greaves J. S. et al. (2006) *Mon. Not. R. Astron. Soc., 366,* 283.
Greenberg R. et al. (1978) *Icarus, 35,* 1.
Grimm R. E. and McSween H. Y. (1993) *Science, 259,* 653.
Haghighipour N. and Boss A. P. (2003) *Astrophys. J., 583,* 996.
Haisch K. E. Jr. et al. (2001) *Astrophys. J. Lett., 553,* L153.
Hansen B. M. S. (2009) *Astrophys. J., 703,* 1131.
Hansen B. M. S. and Murray N. (2012) *Astrophys. J., 751,* 158.
Hansen B. M. S. and Murray N. (2013) *Astrophys. J., 775,* 53.
Hartogh P. et al. (2011) *Nature, 478,* 218.
Hayashi C. et al. (1985) In *Protostars and Planets II* (D. C. Black and
    M. S. Matthews, eds.), pp. 1100–1153. Univ. of Arizona, Tucson.
Hillenbrand L. A. et al. (2008) *Astrophys. J., 677,* 630.
Howard A. W. et al. (2010) *Science, 330,* 653.
Howard A. W. et al. (2012) *Astrophys. J. Suppl., 201,* 15.
Hubbard W. B. et al. (2007a) *Astrophys. J. Lett., 658,* L59.
Hubbard W. B. et al. (2007b) *Icarus, 187,* 358.
Ida S. and Lin D. N. C. (2008) *Astrophys. J., 673,* 487.
Ida S. and Makino J. (1992) *Icarus, 96,* 107.
Ida S. and Makino J. (1993) *Icarus, 106,* 210.
Ida S. et al. (2008) *Astrophys. J., 686,* 1292.
Ikoma M. and Hori Y. (2012) *Astrophys. J., 753,* 66.
Jackson B. et al. (2010) *Mon. Not. R. Astron. Soc., 407,* 910.
Jacobson S. et al. (2013) In *Protostars and Planets VI,* Heidelberg,
    July 15–20, 2013, Poster #2H024, p. 24.
Jin L. et al. (2008) *Astrophys. J. Lett., 674,* L105.
Johansen A. and Lacerda P. (2010) *Mon. Not. R. Astron. Soc., 404,* 475.
Johansen A. et al. (2006) *Astrophys. J., 636,* 1121.
Johansen A. et al. (2007) *Nature, 448,* 1022.
Johansen A. et al. (2009) *Astrophys. J., 697,* 1269.
Jurić M. and Tremaine S. (2008) *Astrophys. J., 686,* 603.
Kaib N. A. et al. (2013) *Nature, 493,* 381.
Kennedy G. M. and Wyatt M. C. (2012) *Mon. Not. R. Astron. Soc., 426,* 91.
Kenyon S. J. and Bromley B. C. (2004) *Astrophys. J. Lett., 602,*L133.
Kenyon S. J. and Bromley B. C. (2005) *Astron. J., 130,* 269.
Kenyon S. J. and Bromley B. C. (2006) *Astron. J., 131,* 1837.
Kimura K. et al. (1974) *Geochim. Cosmochim. Acta, 38,* 683.
Kleine T. et al. (2002) *Nature, 418,* 952.
Kleine T. et al. (2009) *Geochim. Cosmochim. Acta, 73,* 5150.
Kobayashi H. and Dauphas N. (2013) *Icarus, 225,* 122.
Kobayashi H. and Tanaka H. (2010a) *Icarus, 206,* 735.
Kobayashi H. and Tanaka H. (2010b) *Icarus, 206,* 735.
Kokubo E. and Genda H. (2010) *Astrophys. J. Lett., 714,* L21.
Kokubo E. and Ida S. (1995) *Icarus, 114,* 247.
Kokubo E. and Ida S. (1996) *Icarus, 123,* 180.
Kokubo E. and Ida S. (1998) *Icarus, 131,* 171.
Kokubo E. and Ida S. (2000) *Icarus, 143,* 15.
Kokubo E. and Ida S. (2002) *Astrophys. J., 581,* 666.
Kokubo E. and Ida S. (2007) *Astrophys. J., 671,* 2082.
Kokubo E. et al. (2006) *Astrophys. J., 642,* 1131.
Kominami J. and Ida S. (2004) *Icarus, 167,* 231.
König S. et al. (2011) *Geochim. Cosmochim. Acta, 75,* 2119.
Kornet K. et al. (2004) *Astron. Astrophys., 417,* 151.
Kretke K. A. and Lin D. N. C. (2012) *Astrophys. J., 755,* 74.
Krivov A. V. (2010) *Res. Astron. Astrophys., 10,* 383.
Lambrechts M. and Johansen A. (2012) *Astron. Astrophys., 544,* A32.
Lammer H. et al. (2003) *Astrophys. J. Lett., 598,* L121.
Laskar J. (1997) *Astron. Astrophys., 317,* L75.
Laws C. et al. (2003) *Astron. J., 125,* 2664.
Lécuyer C. et al. (1998) *Chem. Geol., 145,* 249.
Léger A. et al. (2009) *Astron. Astrophys., 506,* 287.
Leinhardt Z. M. and Stewart S. T. (2012) *Astrophys. J., 745,* 79.
Levison H. F. and Agnor C. (2003) *Astron. J., 125,* 2692.
Levison H. F. et al. (2011) *Astron. J., 142,* 152.
Levison H. F. et al. (2012) *Astron. J., 144,* 119.
Lin D. N. C. and Papaloizou J. (1986) *Astrophys. J., 309,* 846.
Lin D. N. C. et al. (1996) *Nature, 380,* 606.
Lissauer J. J. (1987) *Icarus, 69,* 249.
Lissauer J. J. and Kary D. M. (1991) *Icarus, 94,* 126.
Lissauer J. J. and Stevenson D. J. (2007) *Protostars and Planets V*
    (B. Reipurth et al., eds.), pp. 591–606. Univ. of Arizona, Tucson.
Lissauer J. J. et al. (2011) *Nature, 470,* 53.
Lisse C. M. et al. (2008) *Astrophys. J., 673,* 1106.
Lisse C. M. et al. (2009) *Astrophys. J., 701,* 2019.
Lodders K. (2003) *Astrophys. J., 591,* 1220.

Lommen D. et al. (2007) *Astron. Astrophys., 462,* 211.
Looney L. W. et al. (2003) *Astrophys. J., 592,* 255.
Lopez E. and Fortney J. (2014) *Astrophys. J., 792,* 1.
Lovis C. et al. (2011) *Astron. Astrophys., 528,* A112.
Lykawka P. S. and Ito T. (2013) *Astrophys. J., 773,* 65.
Makino J. et al. (1998) *New Astron., 3,* 411.
Mandell A. M. et al. (2007) *Astrophys. J., 660,* 823.
Mann A. W. et al. (2012) *Astrophys. J., 753,* 90.
Marcy G. W. et al. (2014) *Astrophys. J. Suppl., 210,* article ID 20.
Martin R. G. and Livio M. (2012) *Mon. Not. R. Astron. Soc., 425,* L6.
Marty B. (2012) *Earth Planet. Sci. Lett., 313,* 56.
Marty B. and Yokochi R. (2006) *Rev. Mineral Geophys., 62,* 421.
Marzari F. and Weidenschilling S. J. (2002) *Icarus, 156,* 570.
Masset F. S. and Casoli J. (2010) *Astrophys. J., 723,* 1393.
Masset F. S. and Papaloizou J. C. B. (2003) *Astrophys. J., 588,* 494.
Masset F. and Snellgrove M. (2001) *Mon. Not. R. Astron. Soc., 320,* L55.
Masset F. S. et al. (2006) *Astrophys. J., 642,* 478.
Matsumura S. et al. (2010) *Astrophys. J., 714,* 194.
Mayor M. et al. (2009) *Astron. Astrophys., 507,* 487.
Mayor M. et al. (2011) *ArXiv e-prints,* arXiv:1109.2497.
McNeil D. et al. (2005) *Astron. J., 130,* 2884.
Melis C. et al. (2010) *Astrophys. J. Lett., 717,* L57.
Melis C. et al. (2012) *Nature, 487,* 74.
Meyer M. R. et al. (2008) *Astrophys. J. Lett., 673,* L181.
Moeckel N. and Armitage P. J. (2012) *Mon. Not. R. Astron. Soc.,419,* 366.
Moeckel N. et al. (2008) *Astrophys. J., 688,* 1361.
Moorhead A. V. and Adams F. C. (2005) *Icarus, 178,* 517.
Morbidelli A. et al. (2000) *Meteoritics & Planet. Sci., 35,* 1309.
Morbidelli A. et al. (2007) *Astron. J., 134,* 1790.
Morbidelli A. et al. (2008) *Astron. Astrophys., 478,* 929.
Morbidelli A. et al. (2010) *Astron. J., 140,* 1391.
Morbidelli A. et al. (2012) *Annu. Rev. Earth Planet. Sci., 40,* 251.
Morishima R. et al. (2008) *Astrophys. J., 685,* 1247.
Morishima R. et al. (2010) *Icarus, 207,* 517.
Morishima R. et al. (2013) *Earth Planet. Sci. Lett., 366,* 6.
Moro-Martín A. et al. (2007) *Astrophys. J., 658,* 1312.
Mundy L. G. et al. (2000) In *Protostars and Planets IV* (V. Mannings et al., eds.), pp. 355–376. Univ. of Arizona, Tucson.
Murray N. et al. (1998) *Science, 279,* 69.
Murray-Clay R. A. et al. (2009) *Astrophys. J., 693,* 23.
Murray-Clay R. et al. (2011) In *Extreme Solar Systems II,* #8.04.
Nagasawa M. et al. (2008) *Astrophys. J., 678,* 498.
Naoz S. et al. (2011) *Nature, 473,* 187.
Nimmo F. and Agnor C. B. (2006) *Earth Planet. Sci. Lett., 243,* 26.
Nimmo F. and Kleine T. (2007) *Icarus, 191,* 497.
O'Brien D. P. et al. (2006) *Icarus, 184,* 39.
O'Brien D. P. et al. (2014) *Icarus, 239,* 74.
Ogihara M. and Ida S. (2009) *Astrophys. J., 699,* 824.
Ormel C.W. and Klahr H. H. (2010) *Astron. Astrophys., 520,* A43.
Ormel C. W. et al. (2010) *Astrophys. J. Lett., 714,* L103.
Paardekooper S.-J. et al. (2010) *Mon. Not. R. Astron. Soc., 401,* 1950.
Pascucci I. et al. (2009) *Astrophys. J., 696,* 143.
Pérez L. M. et al. (2012) *Astrophys. J. Lett., 760,* L17.
Pierens A. and Nelson R. P. (2008) *Astron. Astrophys., 482,* 333.
Pierens A. and Raymond S. N. (2011) *Astron. Astrophys., 533,* A131.
Pierens A. et al. (2013) *Astron. Astrophys., 558,* A105.
Rafikov R. R. (2004) *Astron. J., 128,* 1348.
Rasio F. A. and Ford E. B. (1996) *Science, 274,* 954.
Raymond S. N. (2006) *Astrophys. J. Lett., 643,* L131.
Raymond S. N. and Cossou C. (2014) *Mon. Not. R. Astron. Soc., 440,* L11.
Raymond S. N. et al. (2004) *Icarus, 168,* 1.

Raymond S. N. et al. (2005) *Astrophys. J., 632,* 670.
Raymond S. N. et al. (2006a) *Science, 313,* 1413.
Raymond S. N. et al. (2006b) *Icarus, 183,* 265.
Raymond S. N. et al. (2007a) *Astrophys. J., 669,* 606.
Raymond S. N. et al. (2007b) *Astrobiology, 7,* 66.
Raymond S. N. et al. (2008) *Mon. Not. R. Astron. Soc., 384,* 663.
Raymond S. N. et al. (2009) *Icarus, 203,* 644.
Raymond S. N. et al. (2010) *Astrophys. J., 711,* 772.
Raymond S. N. et al. (2011) *Astron. Astrophys., 530,* A62.
Raymond S. N. et al. (2012) *Astron. Astrophys., 541,* A11.
Raymond S. N. et al. (2013) *Icarus, 226,* 671.
Rodmann J. et al. (2006) *Astron. Astrophys., 446,* 211.
Safronov V. S. (1969) *Evoliutsiia Doplanetnogo Oblaka.*
Santos N. C. et al. (2001) *Astron. Astrophys., 373,* 1019.
Sasselov D. D. and Lecar M. (2000) *Astrophys. J., 528,* 995.
Schlichting H. E. et al. (2012) *Astrophys. J., 752,* 8.
Scholz A. et al. (2006) *Astrophys. J., 645,* 1498.
Stadel J. G. (2001) Cosmological N-body simulations and their analysis, Ph.D. thesis, University of Washington.
Stewart S. T. and Leinhardt Z. M. (2012) *Astrophys. J., 751,* 32.
Sumi T. et al. (2011) *Nature, 473,* 349.
Tanaka H. and Ward W. R. (2004) *Astrophys. J., 602,* 388.
Tera F. et al. (1974) *Earth Planet. Sci. Lett., 22,* 1.
Terquem C. and Papaloizou J. C. B. (2007) *Astrophys. J., 654,* 1110.
Thommes E. W. et al. (2003) *Icarus, 161,* 431.
Thommes E. W. et al. (2008) *Science, 321,* 814.
Touboul M. et al. (2007) *Nature, 450,* 1206.
Tsiganis K. et al. (2005) *Nature, 435,* 459.
Udry S. et al. (2007) *Astron. Astrophys., 469,* L43.
Valencia D. et al. (2007) *Astrophys. J., 665,* 1413.
Veras D. and Armitage P. J. (2004) *Mon. Not. R. Astron. Soc., 347,* 613.
Veras D. and Armitage P. J. (2005) *Astrophys. J. Lett., 620,* L111.
Veras D. and Armitage P. J. (2006) *Astrophys. J., 645,* 1509.
Walker R. J. (2009) *Chemie der Erde/Geochemistry, 69,* 101.
Walsh K. J. and Morbidelli A. (2011) *Astron. Astrophys., 526,* A126.
Walsh K. J. et al. (2011) *Nature, 475,* 206.
Ward W. R. (1986) *Icarus, 67,* 164.
Ward W. R. (1997) *Icarus, 126,* 261.
Weidenschilling S. J. (1977) *Astrophys. Space Sci., 51,* 153.
Weidenschilling S. J. and Marzari F. (1996) *Nature, 384,* 619.
Weiss L. and Marcy G. W. (2014) *Astrophys. J. Lett., 783,* L6.
Weiss L. M. et al. (2013) *Astrophys. J., 768,* 14.
Wetherill G. W. (1978) In *Protostars and Planets* (T. Gehrels, ed.), pp. 565–598. Univ. of Arizona, Tucson.
Wetherill G. W. (1985) *Science, 228,* 877.
Wetherill G. W. (1991) In *Lunar and Planetary Science XXII,* p. 1495. Lunar and Planetary Institute, Houston.
Wetherill G. W. (1996) *Icarus, 119,* 219.
Wetherill G. W. and Stewart G. R. (1989) *Icarus, 77,* 330.
Wetherill G. W. and Stewart G. R. (1993) *Icarus, 106,* 190.
Williams J. P. and Cieza L. A. (2011) *Annu. Rev. Astron. Astrophys., 49,* 67.
Wilner D. J. et al. (2005) *Astrophys. J. Lett., 626,* L109.
Wright J. T. et al. (2008) *Astrophys. J. Lett., 683,* L63.
Wyatt M. C. (2008) *Annu. Rev. Astron. Astrophys., 46,* 339.
Yasui C. et al. (2009) *Astrophys. J., 705,* 54.
Yelle R. V. (2004) *Icarus, 170,* 167.
Yin Q. et al. (2002) *Nature, 418,* 949.
Yoshinaga K. et al. (1999) *Icarus, 139,* 328.
Youdin A. N. and Chiang E. I. (2004) *Astrophys. J., 601,* 1109.
Youdin A. N. and Goodman J. (2005) *Astrophys. J., 620,* 459.
Zhou J.-L. et al. (2005) *Astrophys. J. Lett., 631,* L85.

Chabrier G., Johansen A., Janson M., and Rafikov R. (2014) Giant planet and brown dwarf formation. In *Protostars and Planets VI* (H. Beuther et al., eds.), pp. 619–642. Univ. of Arizona, Tucson, DOI: 10.2458/azu_uapress_9780816531240-ch027.

Giant Planet and Brown Dwarf Formation

Gilles Chabrier
Ecole Normale Supérieure de Lyon and University of Exeter

Anders Johansen
Lund University

Markus Janson
Princeton University and Queen's University Belfast

Roman Rafikov
Princeton University

Understanding the dominant brown dwarf and giant planet formation processes, and finding out whether these processes rely on completely different mechanisms or share common channels, represents one of the major challenges of astronomy and remains the subject of heated debate. It is the aim of this review to summarize the latest developments in this field and to address the issue of origin by comparing different brown dwarf and giant planet formation scenarios with presently available observational constraints. As examined in the review, if objects are classified as "brown dwarfs" or "giant planets" on the basis of their formation mechanism, it has now become clear that their mass domains overlap and that there is no mass limit between these two distinct populations. Furthermore, while there is increasing observational evidence for the existence of non-deuterium (D)-burning brown dwarfs, some giant planets, characterized by a significantly metal-enriched composition, might be massive enough to ignite D burning in their core. Deuterium burning (or lack thereof) thus plays no role in either brown dwarf or giant planet formation. Consequently, we argue that the International Astronomical Union (IAU) definition for distinguishing these two populations has no physical justification and results in scientific confusion. In contrast, brown dwarfs and giant planets might bear some imprints of their formation mechanism, notably in their mean density and the physical properties of their atmosphere. Future direct imaging surveys will undoubtedly provide crucial information and perhaps provide some clear observational diagnostics to unambiguously distinguish these different astrophysical objects.

1. INTRODUCTION

The past decade has seen a wealth of substellar object (SSO) discoveries. Substellar objects are defined as objects not massive enough to sustain hydrogen burning and thus encompass brown dwarfs (BD) and planets, more specifically giant planets (GP), which we will focus on in the present review. For the sake of clarity, we will first specify which objects we refer to as "brown dwarfs" and "giant planets," respectively. For reasons that will be detailed in this review, we do not use the International Astronomical Union (IAU) definition to distinguish between these two populations. Instead, we adopt a classification that, as argued along the lines below, (1) appears to be more consistent with observations; (2) is based on a physically motivated justification, invoking two drastically different dominant formation mechanisms; and (3) allows a mass overlap between these

two populations, as indeed supported by observations but excluded by the present official definition. This classification is the following. We will denominate "brown dwarfs" as (1) all free-floating objects below the hydrogen-burning minimum mass, ~0.075 $M_\odot$ for solar composition (*Chabrier et al.*, 2000b), whether or not they are massive enough to ignite D burning; and (2) objects that are companions to a parent star or another BD, but exhibit compositional and mechanical (mass radius) properties consistent with those of a gaseous sphere of global chemical composition similar to that of the parent star/BD. In contrast, "giant planets" are defined as companions of a central, significantly more massive object, a necessary condition, whose bulk properties strongly depart from those just mentioned. We are well aware of the fact that these definitions are ambiguous in some cases (e.g., some planets might be ejected onto a hyperbolic orbit and become genuine "free-floating planets"), but, as argued in the review, these cases are likely to be statistically rare. We will come back to this issue in the conclusion of the review.

According to the above classification, the detection of BDs now extends down to objects nearly as cool as our jovian planets, both in the field and in cluster associations, closely approaching the bottom of the stellar initial mass function (IMF). Extrasolar planets, on the other hand, are discovered by radial velocity techniques and transit surveys at an amazing pace. The wealth of discoveries now extends from gaseous giants of several Jupiter masses to Earth-mass objects.

Both types of objects raise their own fundamental questions and generate their own debates. Do BDs form like stars, as a natural extension of the stellar-formation process in the substellar regime? Or are BDs issued from a different formation scenario and represent a distinctive class of astrophysical bodies? Similarly, can GPs form in a fashion similar to stars as the result of a gravitational instability (GI) in a protoplanetary disk, or do they emerge from a completely different process, involving the early accretion of planetesimals? What are the distinctive observational signatures of these different processes? From a broader point of view, major questions concerning the very identification of BDs and gaseous planets remain to be answered. Among these questions are, non-exhaustively: Do the mass domains of BDs and GPs overlap or is there a clear mass cut-off between these two populations? Does D burning (or lack thereof) play a determinant role in BD (or planet) formation, providing a key diagnostic to distinguish these two populations, or is this process inconsequential? There is also a related, crucial question: What is the minimum mass for star/BD formation and the maximum mass for planet formation?

It is the aim of the present review to address these questions and to find out whether or not we have at least some reasonably robust answers about BD and GP dominant formation mechanisms by confronting predictions of the various suggested formation scenarios with observations. The general outline of the review is as follows: In sections 2 and 3, we first summarize the most recent observational determinations of BD and GP properties. This includes both field and star-forming regions for BDs as well as transit and radial velocity surveys for planets. We will particularly focus on direct imaging observations of Jupiter-mass bodies as wide companions of a central star, which may encompass both types of objects. In section 4, we briefly summarize the observable properties of protostellar and protoplanetary disks and examine the required conditions for such disks to be gravitational unstable and fragment. Section 5 is devoted to a brief summary of the different existing formation scenarios for BDs and planets. We will identify the distinctive concepts and physical mechanisms at the heart of these scenarios and examine their viability by confronting them with the previous observational constraints. In section 6, we will examine whether BDs and GPs can be distinguished according to various signatures that have been suggested in

the literature, namely early evolution, D burning, or mass domain. Finally, as the outcome of these challenges, we will try in the conclusion to determine what is or are the dominant formation mechanisms for BDs and planets and whether or not they may share some common channels.

2. OBSERVATIONAL PROPERTIES OF BROWN DWARFS

An excellent summary of BD observational properties can be found in *Luhman et al.* (2007) and *Luhman* (2012). We refer the reader to these reviews, and all appropriate references therein, for a detailed examination of BD observational properties. We will only briefly summarize some results here.

2.1. Brown Dwarf Census and Mass Function

It is now widely admitted that the stellar IMF, defined hereafter as dN/d log M (*Salpeter*, 1955), flattens below a mass ~ 0.6 $M_\odot$ compared to a Salpeter IMF (*Kroupa*, 2001; *Chabrier*, 2003). Although the IMF below this mass is often parameterized by a series of multiple power-law segments, such a form does not have physical foundation. In contrast, the IMF can also be parameterized by a power-law + log normal form, a behavior that has been shown to emerge from the gravoturbulent picture of star formation (*Hennebelle and Chabrier*, 2008; *Hopkins*, 2012). The behavior of such a stellar IMF in the BD regime can now be confronted directly with observations both in the field and in young clusters and star-forming regions. Figure 1 compares the predictions of the number density distributions of BDs as a function of T_{eff} and J-absolute magnitude for an age of ~ 1 gigayear (G.y.), assuming a Galactic constant star-formation rate, a double-exponential Galactic disk density distribution, and a *Chabrier* (2005) (hereafter *C05*) IMF [see *Chabrier* (2002, 2005) for details of the calculations], with the most recent observational determinations. The *C05* IMF is based on a slightly revised version of the *Chabrier* (2003) IMF (*Chabrier*, 2003) (see *C05*). The observations include in particular the recent results from the Wide-field Infrared Survey Explorer (WISE) survey, which reaches nearly the bottom end of the BD mass distribution ($T_{eff} <$ 500 K) in the field (*Kirkpatrick et al.*, 2011). It is striking to see how the WISE discoveries, as well as those from previous surveys, are consistent with a *C05* IMF extending from the stellar into to the BD regime. As shown in Table 1 of *C05*, the BD-to-star ratio predicted with this IMF was ~ 1:4, while the ratio revealed so far by the (still partly incomplete at the $\sim 20\%$ level) WISE survey is ~ 1:5. The determination of the age of the WISE objects is an extremely delicate task. Evolutionary models (mass, age) and synthetic spectra (T_{eff}, log g) in this temperature range remain too uncertain so far to provide a reliable answer. The WISE survey, however, probes the local thin-disk population, so the discovered BDs should be on average less than about 3 G.y. old. According to BD theoretical evolution models (*Baraffe et al.*,

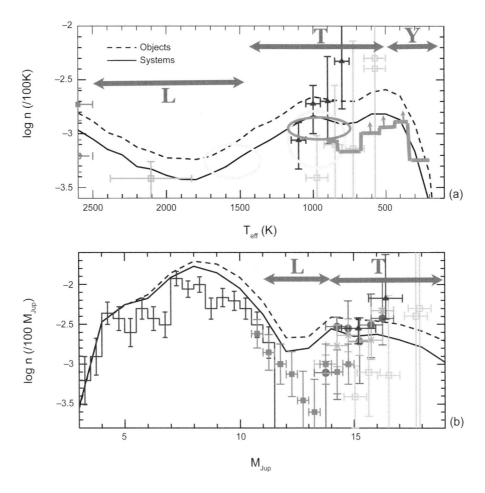

Fig. 1. Brown dwarf number density as a function of **(a)** T_{eff} and **(b)** absolute J-magnitude predicted with a *Chabrier* (2005) IMF for resolved objects (dash) and unresolved binary systems (solid). The letters L, T, and Y correspond to the various BD spectral types. The solid triangles, squares, and circles indicate various observational determinations (*Burgasser et al.*, 2004; *Cruz et al.*, 2007; *Metchev et al.*, 2008; *Burningham et al.*, 2010; *Reylé et al.*, 2010), while the histogram with the arrows portrays the WISE (uncomplete) observations (*Kirkpatrick et al.*, 2011).

2003; *Burrows et al.*, 2003; *Saumon and Marley*, 2008), for an age of 3 G.y. (respectively 1 G.y.), BDs with effective temperatures below 400 K (respectively ~550–600 K) must be less massive than 0.012 $M_\odot$, i.e., below the deuterium (D)-burning limit for solar composition (*Chabrier et al.*, 2000a). So far, the WISE survey has even discovered one field BD with $T_{eff} \lesssim 350$ K, reaching the temperature domain of Jupiter-like objects. As just mentioned, the determination of the effective temperature in this regime still retains significant uncertainties and these values should be taken with due caution, but these results suggest that some of the field objects discovered by WISE lie below the D-burning limit. At any rate, the WISE survey, as both statistics and models improve, holds the perspective to nail down the minimum mass for star/BD formation and the IMF, a result of major importance.

Notice that a *C05* IMF would globally correspond to a negative α value, in terms of dN/dM $\propto$ M$^{-\alpha}$, below about

$T_{eff} \lesssim 1300$ K, indicating a decreasing number of BDs with decreasing mass below ~30 M_{Jup} for an age of 1 G.y. This is in stark contrast with the planetary mass function, which reveals *an increasing number of objects* below about the same mass (*Mayor et al.*, 2011; *Howard et al.*, 2010). Similarly, the BD mass distribution has now been determined in more than a dozen clusters or star-forming regions down to ~0.02 $M_\odot$ or less, and these distributions are found to agree quite well, with some expected scatter, with the very same *C05* IMF for unresolved systems (see, e.g., Figs. 8 and 9 of *Jeffries*, 2012).

Brown dwarfs are now routinely discovered with masses below the D-burning limit (*Lucas et al.*, 2006; *Caballero et al.*, 2007; *Luhman and Muench*, 2008; *Marsh et al.*, 2010; *Scholz et al.*, 2012b,a; *Lodieu et al.*, 2011; *Delorme et al.*, 2012a). Although these objects are sometimes denominated free-floating "planetary mass objects" based on the IAU definition, such a distinct appelation is not justified, as these objects are regular genuine BDs not massive enough to ignite D-burning, in the same way that stars not massive enough to ignite hydrogen (H)-burning via the carbon-nitrogen-oxygen (CNO) cycles are still called "stars" and have never been named differently.

All these BD discoveries are consistent with the very same underlying "universal" IMF extending continuously from the stellar into the BD regime down to about the opacity fragmentation limit, with no noticeable variations between various regions (under Milky Way-like conditions) nor evidence whatsoever for a discontinuity near the H-burning limit. This strongly argues in favor of a scenario in which BDs and stars share dominantly the same formation mechanism.

There have been some claims based on microlensing observations (*Sumi et al.,* 2011) for an excess of a very-low-mass-object population (approximately a few Jupiter masses) in the field, formed in protoplanetary disks and subsequently scattered into unbound orbits. The validity of these results, however, can be questioned for several reasons. First, the excess number is based on five events — which depart from the expected distribution only at the ~1σ limit — and thus has no statistical significance. Second, when observing a short duration *isolated* event with low S/N and noncomplete coverage, the symmetry of the event, a characteristic property of lensing events, is very difficult to assess properly, in contrast to the case when the planetary event occurs as a caustic during a stellar-lensing event. Free-floating planet candidate events are thus much more uncertain than star-planet candidate events and should be considered with extreme caution. Third, planet-planet scattering usually produces objects with high velocity dispersions compared with the parent star. As examined below (section 2.2), no such population of high-velocity objects in the BD or planet mass range has been discovered so far, at least down to present observational limits. Most importantly, the excess of low-mass objects claimed by the microlensing results would correspond to a population of free-floating Jupiter-mass objects in the field almost twice as common as stars, a census clearly excluded so far by observations (e.g., *Scholz et al.,* 2012b).

2.2. Brown Dwarf Velocity and Spatial Distributions

Available radial velocity measurements of BDs in young clusters and star-forming regions clearly indicate that young BDs and young stars have similar velocity dispersions (e.g., *Joergens,* 2006; *White and Basri,* 2003; *Kurosawa et al.,* 2006). Similar observations show that BDs have the same spatial distributions as stars [for recent determinations, see, e.g., *Parker et al.* (2011), *Bayo et al.* (2011), *Scholz et al.* (2012b)]. These observations clearly contradict at least the early versions of the accretion-ejection model for BD formation (section 5.2.2) (*Reipurth and Clarke,* 2001; *Kroupa and Bouvier,* 2003), which predicted that young BDs should have higher velocities than stars and thus should be more widely distributed.

2.3. Brown Dwarfs or Giant Planets as Star or Brown Dwarf Companions

The statistics of BD and GP companions to stars or BDs potentialy bears important information on the formation process of these objects. Various studies have indeed compared the multiplicity distributions of prestellar core fragmentation simulations with the presently observed values in young clusters or star-forming environments. Such comparisons, however, are of limited utility to infer the specific formation process. Indeed, while simulations address the precise prestellar cloud/core collapse stages, i.e., the initial epoch (~10^5 yr) of star formation, observations probe present conditions, i.e., typical ages of one to several million years. Dynamical interactions and interactions with the disk have continued to occur between these two epochs and have strongly influenced the multiplicity properties. It is well established and intuitively expected, for instance, that the most weakly bound systems, which include preferentially those with BD or very-low-mass star companions, are more prone to disruption than those that are more massive, so that low-mass binary properties are expected to differ substantially from their primordial distribution (e.g., *Duchêne and Kraus,* 2013). This naturally explains the decreasing binarity frequency with mass (see below). Therefore, presently observed values are of limited relevance to infer primordial multiplicity properties and their dependence on formation conditions.

Keeping that in mind, observations today point to star and BD binary properties that appear to be continuous from the stellar to the BD regime, with binary fractions and average separations decreasing continuously with primary mass, and mass ratios increasing with it (e.g., *Faherty et al.,* 2011; *Kraus and Hillenbrand,* 2012; *Luhman,* 2012). The continuity of the multiplicity properties thus mirrors that found for the IMF.

An important observational constraint for BD formation theories also stems from the observations of wide binaries. A particular example is the wide binary FU Tau a,b, with no identified stellar companion in its vicinity (*Luhman et al.,* 2009). This is the best example that BDs can form in isolation, with no nearby stellar companion, in low-density environments (stellar density $n_\star \sim 1$–10 pc^{-3}). Several wide ($\approx$400–4000 AU) binary substellar companions, down to ~5 M_{Jup} (*Luhman et al.,* 2011; *Aller et al.,* 2013), have also been observed in star-forming regions. One of these objects has an extreme mass ratio $q = M_2/M_1 \approx 0.005$. Worth mentioning are also the isolated and rather fragile BD wide binary Two Micron All-Sky Survey (2MASS) J12583501/J12583798 (6700 AU) (*Radigan et al.,* 2009) or the quadrupole system containing 2MASS 04414489+23015113 (1700 AU) (*Todorov et al.,* 2010). All these systems seem rather difficult to explain with dynamical processes or disk fragmentation, clearly indicating that these mechanisms are not necessary for BD formation. As discussed in section 5.2.4, such wide binaries are most likely the indication that core fragmentation into binaries might occur at the very early stages of star formation (see section 5.2.4).

Other relevant observational inputs for understanding BD and GP formation include the BD desert (e.g., *Grether and Lineweaver,* 2006), which is a lack of companions in the mass range 10–100 M_{Jup} (i.e., 1–10% of the primary mass) in the separation range covered with radial velocity (up to a few astronomical units) around Sun-like stars. The frequency of companions increases with smaller mass below the desert (e.g., *Howard et al.,* 2010), and increases with larger mass above the desert, clearly indicating separate formation mechanisms between planetary and stellar companions. Moreover, GPs appear to become less frequent around lower-mass stars and more frequent around higher-mass stars (e.g., *Johnson et al.,* 2010); meanwhile, BD companions become more frequent around low-mass stars and BDs in the same separation range (e.g., *Joergens,* 2008). This further implies that the desert is primarily determined by system mass ratios, and is not specific to BD secondaries. As mentioned in section 2.1, the fact that the mass function of planetary companions to stars (*Mayor et al.,* 2011; *Howard et al.,* 2010) differs drastically — qualitatively and quantitatively — from the BD mass function clearly points to a different dominant formation mechanism for these two populations. More quantitative results about the statistics of wide orbit SSOs around stars and BDs in the BD/GP mass range will be given in section 3.2.

2.4. Observational Identifications of Isolated Proto- and Pre-Brown-Dwarf Cores

The recent results of the Herschel survey confirm that the mass function of prestellar cores [the so-called core mass function (CMF)] in molecular clouds resembles the stellar IMF down to the present observational sensitivity limit, which lies within the BD domain (*André et al.,* 2010; *Könyves et al.,* 2010) (see also the chapter by Offner et al. in this volume). Although caution is necessary at this stage before definitive answers can be claimed, these results strongly argue in favor of the stellar + BD IMF being determined at the very early stages of star formation rather than at the final gas-to-star conversion stage, an issue we will come back to in section 5.2.

Several class 0 objects have been identified with Spitzer surveys with luminosities L < 0.1 $L_{\odot}$, indicative of young proto-BDs (e.g., *Huard et al.,* 2006; *Dunham et al.,* 2008; *Kauffmann et al.,* 2011) (see further references in *Luhman,* 2012). Although some of these objects may eventually accrete enough material to become stars, some have such a small accretion reservoir that they will definitely remain in the BD domain (*Lee and Kim,* 2009; *Kauffmann et al.,* 2011; *Lee et al.,* 2013). Most importantly, there are now emerging discoveries of isolated proto-BDs (*Lee et al.,* 2009a,b, 2013; *Palau et al.,* 2012). The existence of such isolated proto-BDs again indicates that dynamical interactions, disk fragmentation, or photoionizing radiation are not essential for BD formation.

Also of notable importance is the discovery of the isolated, very-low-mass BD OTS 44 (spectral type M9.5, M ~ 12 M_{Jup}), with significant accretion and a substantial disk (*Luhman et al.,* 2004, 2005; *Joergens et al.,* 2013). This

demonstrates that the processes that characterize the canonical star-like mode of formation apply to isolated BDs down to few Jupiter masses.

But the most striking result concerning BD formation is the recent observation of an isolated pre-BD core (*André et al.,* 2012). The mass and size of this core have been determined with the Plateau de Bure Interferometer (PdBI), yielding a core mass ~30 M_{Jup} with a radius <460 AU, i.e., a mean density n ~ 10^7–10^8 cm^{-3}, consistent with theoretical predictions of the gravoturbulent scenario (section 5.2.3). It should be stressed that this core lies outside dense filaments of Oph and thus is not expected to accrete a significant amount of mass (see supplementary material in *André et al.,* 2012). Indeed, the surrounding gas has been detected in follow-up observations with the Institut de Radioastronomie Millimétrique (IRAM) telescope and the most conservative estimates suggest mass growth by at most a factor of 2, leaving the final object in the BD regime. This is the unambiguous observation of the possibility for BD cores to form in isolation by the collapse of a surrounding mass reservoir, as expected from gravoturbulent cloud fragmentation. Given the nature of the CMF/IMF, however, the number of low-mass cores is expected to decrease exponentially, making their discoveries very challenging.

2.5. Summary

In summary [again, see *Luhman* (2012) for a more thorough review], it seems rather reasonable to affirm that young BDs and young stars (1) exhibit similar radial velocity dispersions; (2) have similar spatial distribution in young clusters; (3) have mass distributions consistent with the same underlying IMF; (4) both have wide binary companions; (5) exhibit similar accretion + disk signatures such as large blue/ultraviolet (UV) excess, with large asymmetric emission lines (UV, H_α, Ca II, etc...), with accretion rates scaling roughly with the square of the mass, $\dot{M} \propto M^2$, from 2 down to about ~0.01 $M_{\odot}$, signatures all consistent with a natural extension of the CTT phase in the BD domain; (6) have similar disk fractions; (7) both exhibit outflows; and (8) have both been identified at the early prestellar/BD stage of core formation in isolated environments. It seems that the most natural conclusion we can draw at this stage from such observational results is that star and BD formations seem to share many observational signatures, pointing to a common dominant underlying mechanism, and that dynamical interactions, disk fragmentation, or photoionizing radiation are not essential for BD formation.

3. OBSERVATIONAL PROPERTIES OF GIANT PLANETS

3.1. Observed Correlations

During the two decades over which exoplanetary observation has been an active area in astronomy, the field has expanded rapidly, both in terms of the number of detected planets and in terms of the parameter space of planets that has opened up for study. As a result, GPs can now be studied comprehensively over a rather wide domain of parameters. By combining radial velocity and transits, both radius and mass as well as a range of other properties can be studied, and additional constraints can be acquired from astrometry and microlensing as well as direct imaging, which is described separately below. Furthermore, the host star properties can be studied and correlated with planet occurrence, in order to gain additional clues about planet formation.

One of the most well-known and important relations between host star properties and planet occurrence is the planet-metallicity correlation (e.g., *Santos et al.,* 2004; *Fischer and Valenti,* 2005), which refers to the rapid increase in GP frequency as a function of stellar host metallicity. While only a few percent of solar abundance stars have GPs with orbital periods of less than 4 yr, ~25% of stars with with [Fe/H] > 0.3 have such planets. This is strong evidence in favor of core accretion (hereafter CA) as a formation scenario, since a framework in which planets grow bottom-up from solids in a protoplanetary disk favors a high concentration of dust and ices. Equally interesting in this regard is the fact that the correlation appears to weaken for lower planetary masses (e.g., *Sousa et al.,* 2008; *Mayor et al.,* 2011; *Buchhave et al.,* 2012). This again fits well into the CA scenario, since it would be natural to assume that the gas GPs are most dependent on high concentrations of solids, as they need to rapidly reach the critical core mass for gas accretion before the gas disk dissipates. It also appears that planet frequency and/or mass scales with stellar mass (e.g., *Lovis and Mayor,* 2007; *Hekker et al.,* 2008; *Johnson et al.,* 2010), although there is not as clear a convergence on this issue in the literature as for the metallicity correlation (see, e.g., *Lloyd,* 2011; *Mortier et al.,* 2013). In any case, this is once again easy to understand in the CA paradigm, as more mass in the star means more mass in the disk, and thus a more generous reservoir from which to grow planets rapidly and to large sizes.

An aspect in the study of GPs that has received considerable attention in recent years is the study of alignment (or misalignment) between the orbital plane of the planet and the rotational plane of the star. A projection of the differential angle between these planes can be measured in transiting systems using the Rossiter-McLaughlin (RM) effect (*Rossiter,* 1924; *McLaughlin,* 1924). The RM effect is a sequence of apparent red-/blue-shift of the light from a rotating star, as a planet blocks approaching/receding parts of the stellar surface during transit. This is the most commonly used method for measuring spin-orbit alignment (e.g., *Queloz et al.,* 2000; *Winn et al.,* 2005), but other methods have recently been implemented as well. These include spot crossings during transit (e.g., *Sanchis-Ojeda et al.,* 2012) and independent determinations of the inclination of the stellar plane; the latter can be accomplished through relating the projected rotational velocity to the rotational period and radius of the star (e.g., *Hirano et al.,* 2012), or

through observing mode splitting in asteroseismology (e.g., *Chaplin et al.,* 2013).

In almost all the above cases, the measurement only gives a projection of the relative orientation. As a consequence, a planet can be misaligned, but still appear aligned through the projection, so it is typically not possible to draw conclusions from single cases. However, firm statistical conclusions can be drawn from studying populations of objects. The studies that have been performed show that GPs are sometimes well aligned with the stellar spin, but sometimes strongly misaligned or even retrograde (e.g., *Triaud et al.,* 2010). Since there are indications for a correlation between orbital misalignment and expected tidal timescale, where systems with short timescales have low obliquities (*Albrecht et al.,* 2012), it has been hypothesized that all systems with a close-in GP may feature a misalignment between the stellar spin and planetary orbit, but that alignment occurs over tidal timescales. There is a possible contrast here with lower-mass planets, which often occur in multiply transiting configurations (e.g., *Lissauer et al.,* 2011b; *Johansen et al.,* 2012b; *Fabrycky et al.,* 2012). This implies a high degree of co-planarity, at least between the planes of the different planets. Furthermore, the few stellar spin measurements that have been performed for small-planet systems imply co-planarity also in a spin-orbit sense (e.g., *Albrecht et al.,* 2013). If these trends hold up with increasing data, it could indicate that while gas giants and smaller planets probably pass through a similar stage of evolution (core formation), there may be differences in the evolutionary paths that bring them into small orbits.

Since transiting GPs can be relatively easily followed up with radial velocity, both mass and radius can be determined in general for such planets, which allows one to make inferences about their composition. Due to nearly equal compensation between electron degeneracy and interionic electrostatic interactions (*Chabrier and Baraffe,* 2000), an isolated gas GP without a substantial core or other heavy element enrichment will have a radius that only very weakly depends on mass. Indeed, the radius will consistently be close to that of Jupiter, apart from during the earliest evolutionary phases, when the planet is still contracting after formation (e.g., *Baraffe et al.,* 2003; *Burrows et al.,* 2003). However, with a larger fraction of heavy elements, the radius decreases, thus providing a hint of this compositional difference (*Fortney et al.,* 2007; *Baraffe et al.,* 2008; *Leconte et al.,* 2009). In practice, when studying transiting planets for this purpose, there is a primary complicating factor: The planets under study are far from being isolated, but rather are at a short distance from their parent star. The influence of the star may substantially inflate the planet, sometimes causing a radius as high as 2 R_{Jup} (e.g., *Hartman et al.,* 2011; *Anderson et al.,* 2011; *Leconte et al.,* 2009). Many different mechanisms have been proposed to account for precisely how this happens, but so far it remains an unsolved and interesting problem. For the purpose of our discussion, however, we merely establish that it is a complicating factor when evaluating the composition of the planet. Nonetheless, some transiting gas giants do feature subjovian radii [e.g., HD 149026 b (*Sato et al.,* 2005) and HAT-P-2b (*Bakos et al.,* 2007)], which has led to the inference of very large cores with tens or even hundreds of Earth masses [e.g., >200 $M_\oplus$ for HAT-P-2b (*Baraffe et al.,* 2008; *Leconte et al.,* 2009, 2011)]. Although it is unclear how exactly these large cores could form, such a heavy material enrichment can hardly be explained by accretion of solids on a gaseous planet embryo, as produced by GI (see section 4.2). This may thus imply that the formation of these objects occurred through the process of CA, possibly involving collisions between Jupiter-mass planets (*Baraffe et al.,* 2008).

Other relevant observational inputs for understanding GP formation include the BD desert and the BD mass function, an issue we have already addressed in section 2.3. The observations show that the frequency of companions increases with smaller mass below the desert and increases with larger mass above the desert (*Mayor et al.,* 2011; *Howard et al.,* 2010), clearly indicating separate formation mechanisms between planetary and stellar/BD companions.

Finally, both microlensing data (e.g., *Gould et al.,* 2010) as well as the asymptotic trends in radial velocity data (e.g., *Cumming et al.,* 2008) indicate that GPs are more common beyond the ice line than they are at smaller separations. This may again provide a clue regarding formation, although it should probably be regarded primarily as providing constraints on migration rather than formation.

3.2. Direct Imaging Constraints

Thanks to developments in both adaptive-optics-based instrumentation and high-contrast imaging techniques, direct imaging of wide massive planets has become possible over the past decade. This has led to a number of surveys being performed, ranging from large-scale surveys, covering a broad range of targets (e.g., *Lowrance et al.,* 2005; *Chauvin et al.,* 2010), to more targeted surveys of specific categories, such as low-mass stars (e.g., *Masciadri et al.,* 2005; *Delorme et al.,* 2012b), intermediate-mass stars (e.g., *Lafrenière et al.,* 2007; *Biller et al.,* 2007; *Kasper et al.,* 2007), high-mass stars (e.g., *Janson et al.,* 2011; *Vigan et al.,* 2012; *Nielsen et al.,* 2013), or stars hosting debris disks (e.g., *Apai et al.,* 2008; *Janson et al.,* 2013; *Wahhaj et al.,* 2013). A common trait of all such surveys is that they generally aim for targets that are quite young (a few million years to at most a few billion years). This is due to the fact that distant GPs (which are not strongly irradiated by the primary star) continuously cool over time, and hence become continuously fainter with age, which increases the brightness contrast between the planet and star and makes the planet more difficult to detect at old stellar ages.

The number of giant exoplanets that have been imaged so far is problematic to precisely quantify. This is partly due to the lack of a clear consensus concerning the definition of what a planet is in terms of mass vs. formation criteria (for example), and partly due to uncertainties in evaluating

a given object against a given criterion. For instance, the errors in the mass estimate often lead to an adopted mass range that encompasses both D-burning and non-D-burning masses. Likewise, it is often difficult to assess the formation scenario for individual objects, given the relatively few observables that are typically available. As a result, the directly imaged planet count could range between ~5 and ~30 planets in a maximally conservative and maximally generous sense, respectively. Two systems that would arguably count as planetary systems under any circumstance are HR 8799 (*Marois et al.,* 2008a, 2010) and β Pic (*Lagrange et al.,* 2009, 2010). All four planets around HR 8799 and the single detected planet around β Pic have best-fit masses well into the non-D-burning regime (~5–10 M_{Jup}), ratios of mass to primaries of less than 1%, and indications of sharing a common orbital geometry, both with each other (for the HR 8799 system) and with the debris disks that exist in both systems. Recently, another planet has been imaged around the Sun-like star GJ 504 (*Kuzuhara et al.,* 2013), which has a similarly small mass ratio and a mass that remains in the non-D-burning regime independently of the choice of initial conditions in existing theoretical mass luminosity relationships, and thus a similar reliability in its classification. The most puzzling properties of the HR 8799 planets in a traditional planet-forming framework are their large orbital separations (~10–70 AU for HR 8799b/c/d/e), as will be discussed further in section 5.3.

Besides the above relatively clear-cut cases, it is informative to probe some of the more ambiguous cases, which may or may not be counted as planets, depending on definitions and uncertainties. The B9-type star κ And hosts a companion with a mass of ~10–20 M_{Jup} (*Carson et al.,* 2013), straddling the D-burning limit. However, due to the very massive primary, the mass ratio of the κ And system is only 0.5–1.0%. This is very similar to the HR 8799 and β Pic planets, and the companion resides at a comparable orbital separation of ~40 AU. This might imply a similar formation mechanism, which would support the fact that planet formation can produce companions that approach or exceed the D-burning limit, as has been suggested by theoretical models (see the chapter by Baraffe et al. in this volume).

A candidate planet presently in formation has been reported in the LkCa15 system (*Kraus and Ireland,* 2012). With the potential for constraining the timescale of formation and initial conditions of a forming planet, such systems would have a clear impact on our understanding of planet formation, but more data will be needed to better establish the nature of this candidate before any conclusion can be drawn in that regard. The young solar analog 1RXS 1609 has a companion with a mass of ~6–11 M_{Jup} (*Lafrenière et al.,* 2008) and a system mass ratio of 1%, but the projected separation is 330 AU, which may be larger than the size of the primordial disk. It is therefore unclear if it could have formed in such a disk, unless it was scattered outward in three-body interactions, for example. One of the clearest hints of incompatibility between a D-burning-based and a formation-based definition is 2M 1207 b (*Chauvin et al.,* 2005), which is a ~4 M_{Jup} companion to a ~20 M_{Jup} BD (e.g., *Mamajek,* 2005). Thus the companion is firmly in the non-D-burning mass regime, but with a system mass ratio of 20%, the couple appears to be best described as an extension of the BD binary population, rather than a planetary system.

Despite these detections, the general conclusion from the direct imaging surveys performed so far is that very wide and massive objects in the GP mass domain are rare, with many of the surveys yielding null results. These results can be used to place formal constraints on the underlying planet population. For example, the Gemini Deep Planet Survey (GDPS) survey (*Lafrenière et al.,* 2007) has indicated that the fraction of FGKM-type stars harboring >2-M_{Jup} objects in the 25–420 AU range is less than 23%, at 95% confidence. For >5-M_{Jup} objects, the corresponding upper limit is 9%. In order to examine how this translates into total GP frequencies, *Lafrenière et al.* (2007) assume a mass distribution of $dN/dM \propto M^{-1.2}$ and a semimajor axis distribution of $dN/da \propto a^{-1}$. This leads to the conclusion (at 95% confidence) that less than 28% of the stars host at least one 0.5–13-M_{Jup} planet in the semimajor axis range of 10–25 AU, less than 13% host a corresponding planet at 25–50 AU, and less than 9% host such a planet at 50–250 AU. *Nielsen and Close* (2010) combine the GDPS data with spectral differential imaging (SDI) data from the Very Large Telescope (VLT) and Multiple Mirror Telescope (MMT). They find that if power-law distributions in mass and semimajor axis that fit the radial velocity distribution (*Cumming et al.,* 2008) are adopted and extrapolated to larger distances, planets following these particular distributions cannot exist beyond 65 AU at 95% confidence. This limit is moved out to 82 AU if the stellar host mass is taken into consideration, since the most stringent detection limits in imaging are acquired around the lowest-mass primaries, which is also where the GP frequency is expected to be the lowest from RV data (e.g., *Johnson et al.,* 2010).

The above studies assume a continuous distribution from small separation planets to large separation planets, and thus essentially assume the same formation mechanism (primarily CA) for all planets. However, at wide separations, disk instability becomes increasingly efficient (and CA less efficient), such that one might imagine that two separate populations of planets exist, formed by separate mechanisms. Hence, an alternative approach to statistically examine the direct imaging data is to interpret them in the context of predictions from disk instability formation. This was done in *Janson et al.* (2011), where the detection limits from direct imaging were compared to the calculated parameter range in the primordial disk in which a forming planet or BD could simultaneously fulfill the Toomre criterion and the cooling criterion for fragmentation (see section 4.2 for a discussion of these criteria) for a sample of BA-type stars. It was concluded that less than 30% of such stars can host disk instability-formed companions at 99% confidence. This work was later extended to FGKM-type stars in *Janson et al.* (2012) using the GDPS sample, where additional effects such as disk migration were taken into account, assuming a wide range of migration rates from virtually no migration to very rapid migration. There, it was found that <10% of the stars could host disk-instability-formed companions at 99% confidence. Since a much larger fraction than this are known to host planets believed to have formed by CA (e.g., *Mayor et al.,* 2011; *Batalha et al.,* 2013), it strongly suggests that disk instability cannot be a dominant mechanism of planet formation. This upper limit is tightened further by the discovery of rapid branches of the CA scenario that can lead to formation of gas giants in wide orbits, as discussed in section 5.3.

It is important to note that several assumptions are included in the various analyses described above, some of which are quite uncertain. In particular, all the above studies translate brightness limits into mass limits using evolutionary models that are based on so-called "hot-start" conditions (e.g., *Baraffe et al.,* 2003; *Burrows et al.,* 2003) (see section 6.1). It has been suggested that the initial conditions may be substantially colder (e.g., *Fortney et al.,* 2008a) or have a range of values in between (*Spiegel and Burrows,* 2012), which could make planets of a given mass and age significantly fainter than under hot-start conditions. This would lead to overly restrictive upper limits on planet frequency in the statistical analyses. However, the measured temperatures of most of the planets that have been imaged so far are consistent with hot-start predictions, and robustly exclude at least the coldest range of initial conditions (e.g., *Marleau and Cumming,* 2013; *Bowler et al.,* 2013), an issue we will address further in section 6.1. As will be shown in this section, cold vs. hot/warm-start conditions are not necessarily equivalent to formation by CA vs. disk instability. Nonetheless, the fact remains that the detection limits are based on mass-luminosity models that have not been fully calibrated in this mass and age range, which is an uncertainty that should be kept in mind. In any case, direct imaging surveys have already started to put meaningful constraints on the distributions of distant GP-mass objects, and near-future instruments such as the Spectro-Polarimetric High-contrast Exoplanet REsearch (SPHERE) (*Beuzit et al.,* 2008), Gemini Planet Imager (GPI) (*Macintosh et al.,* 2008), and Coronagraphic High Angular Resolution Imaging Spectrograph (CHARIS) (*Peters et al.,* 2012) will substantially enhance the capacity of this technique, thus leading to tighter constraints and more detections of closer-in and less-massive objects.

Aside from providing information about the distributions of planet-mass objects, the future instruments mentioned above will also be able to provide substantial information about atmospheric properties, through their capacity to provide planetary spectra. This has already been achieved in the HR 8799 system with present-day instrumentation (e.g., *Janson et al.,* 2010; *Bowler et al.,* 2010; *Oppenheimer et al.,* 2013; *Konopacky et al.,* 2013), but can be delivered at much higher quality and for a broader range of targets with improved instruments. Spectroscopy enables studies of chemical abundances in the planetary atmospheres, which may in turn provide clues to their formation. For instance, CA might result in a general metallicity enhancement, similar to how the GPs in the solar system are enriched in heavy elements with respect to the Sun. This may be possible to probe with metal-dominated compounds such as CO_2 (e.g., *Lodders and Fegley,* 2002). Another possibility is that the formation may leave traces in specific abundance ratios, such as the ratio of carbon to oxygen (e.g., *Madhusudhan et al.,* 2011; *Öberg et al.,* 2011). The C/O ratio of planets should remain close to stellar if they form through GI. However, if they form by CA, the different condensation temperatures of water- and carbon-bearing compounds may lead to a nonstellar abundance ratio in the gas atmosphere that is accreted onto the planet during formation, unless it gets reset through subsequent planetesimal accretion. The most robust observational study addressing this issue to date is that of *Konopacky et al.* (2013), but while the results do indicate a modestly nonstellar C/O ratio, they are not yet unambiguous. Again, future studies have great promise for clarifying these issues.

4. PROTOSTELLAR AND PROTOPLANETARY DISKS

4.1. Observations of Protostellar and Protoplanetary Disks

Circumstellar disks around young objects are generally studied via their thermal and line emission at infrared and millimeter wavelengths, where they dominate the stellar radiation. Unfortunately, the determination of the radii of most disks cannot be done with current technology, although that will change with The Atacama Large Millimeter/submillimeter Array (ALMA) and Submillimeter Array (SMA). Noticeable exceptions are nearly edge-on disks, which can be detected in scattered light while occulting the star. Using Spitzer spectroscopy and high-resolution imaging from the Hubble Space Telescope (HST), *Luhman et al.* (2007) were able to determine the disk radius of a young BD in Taurus, and the latter was found to be small in size, with $R \approx 20$–40 AU.

This is also what seems to be found at the early class 0 stage with observations of isolated, massive, compact ($R \sim 20$ AU) disks (*Rodríguez et al.,* 2005), while large and massive disks prone to fragmentation so far appear to be very rare at this stage of evolution (*Maury et al.,* 2010). Two noticeable exceptions are the $R \sim 70$–120 AU, $M \sim 0.007$ $M_\odot$ disk around the ~0.2-$M_\odot$ protostar L1527 (*Tobin et al.,* 2013a) and the $R \sim 150$ AU, $M \sim 0.02$ $M_\odot$ disk around VLA1623 (*Murillo et al.,* 2013). Moreover, *Tobin et al.* (2013b) observed density structures around the class 0/I protostars CB230 IRS1 and L1165-SMM1 that they identify as companion protostars, because of the strong ionizing radiation, along with an extended (~100 AU), possibly circumbinary-disk-like structure for L1165-SMM1. Whether these companions indeed formed in an originally

massive disk or during the very collapse of the prestellar core itself, however, remains an open question. In contrast, the extended emission around the class 0 L1157-mm is suggestive of a rather small (≤ 15 AU) disk. It should be noticed, however, that L1527, CB230 IRS1, and L1165-SMM1 are more evolved than genuine class 0 objects like L1157-mm, and protostellar disks are expected to expand significantly with time due to internal angular momentum redistribution (e.g., *Basu,* 1998; *Machida et al.,* 2010; *Dapp et al.,* 2012). Large disks are indeed observed at later class 0/I or class I/II stages (see the chapter by Z.-Y. Li et al. in this volume). At such epochs, however, they are unlikely to be massive enough relative to the central star to be prone to fragmentation (see next section). Indeed, observed class I disks generally have masses less than 10% of the star mass.

Clearly, this issue of massive disks at the early stages of star formation is not quite settled for now and will strongly benefit from the ALMA. In any case, as discussed in detail in section 5.1, compact disk observations are in agreement with magneto-hydrodynamical (MHD) simulations and thus suggest that magnetic field is a key element in shaping up protostellar disks.

4.2. Disk Instability

The idea that massive gaseous disks may fragment under the action of their own gravity into bound, self-gravitating objects is rather old (*Kuiper,* 1951). However, the quantitative analysis of its operation was not around until the works of *Safronov* (1960) and *Goldreich and Lynden-Bell* (1965). They were the first to show that a two-dimensional (i.e., zero thickness) disk of rotating fluid becomes unstable due to its own gravity provided that the so-called Toomre Q satisfies

$$Q \equiv \frac{\Omega c_s}{\pi G \Sigma} < 1 \tag{1}$$

where c_s and Σ are the sound speed and surface density of the disk. This particular form assumes a Keplerian rotation, when the so-called epicyclic frequency coincides with the angular frequency Ω.

The physics behind this criterion is relatively simple and lies in the competition between the self-gravity of gas, which tends to destabilize the disk, and the stabilizing influence of both pressure and Coriolis force. If we consider a parcel of gas with characteristic dimension L and surface density Σ in a zero-thickness differentially rotating disk, one can evaluate the attractive force it feels due to its own self-gravity as $F_G \sim G \Sigma^2 L^2$. However, at the same time it feels a repulsive pressure force $F_P \sim c_s^2 \Sigma L$, where c_s is the sound speed. In addition, differential rotation of the disk causes the fluid element to spin around itself at the rate $\sim \Omega$, resulting in a stabilizing centrifugal force $F_C \sim \Omega^2 \Sigma L^3$. Gravity overcomes pressure ($F_G > F_P$) when the gas element is sufficiently large, $L \geq c_s^2/(G\Sigma)$, which is in line with the conventional Jeans criterion for instability in a three-dimensional homogeneous self-gravitating gas.

However, the case of a disk, self-gravity also has to exceed the Coriolis force ($F_G > F_C$), which results in an upper limit on the fluid parcel size, $L \lesssim G\Sigma/\Omega^2$. Gravity wins, and fluid element collapses when both of these inequalities are fulfilled, which is only possible when the condition in equation (1) is fulfilled. This condition is often recast in an alternative form, as a lower limit on the local volumetric gas density $\rho \gtrsim M_\star/a^3$ at semimajor axis a.

Applying this criterion to a disk around a solar-type star, one finds that disk gravitational instability (hereafter GI) can occur only if the disk surface density is very high

$$\Sigma > \frac{\Omega c_s}{\pi G} \approx 10^5 \, \mathrm{g \, cm}^{-2} \left(\frac{AU}{r} \right)^{7/4} \qquad (2)$$

where we used temperature profile $T(r) = 300 \, \mathrm{K}(r = AU)^{-1/2}$. This considerably (by more than an order of magnitude) exceeds surface density in the commonly adopted minimum mass solar nebula (MMSN). Such a disk extending to 50 AU would have a mass close to $M_\odot$. For that reason, with the development of the CA paradigm in the late 1970s the interest in planet formation by GI decreased (but see *Cameron, 1978*).

This situation began to change in the 1990s, when the problems with the CA scenario for planet formation at large separations mentioned in section 2 started emerging. Motivated by this, *Boss* (1997) resurrected the GI scenario in a series of papers that were founded on grid-based numerical simulations of three-dimensional disks. These calculations adopted an isothermal equation of state (EOS) (the validity of this assumption will be discussed below) and found that massive disks with Σ exceeding the limit in equation (2) indeed readily fragment, which would be expected if disk fragmentation was governed by equation (1) alone.

However, the logic behind linking the outcome of GI unambiguously to the value of Toomre Q is not without caveats. Indeed, both the simple qualitative derivation presented above and the more rigorous analyses of the dispersion relation explicitly assume that the pressure of the collapsing fragment is the unperturbed pressure. This is a good approximation in the initial, linear phase of the instability, meaning that Toomre Q is a good predictor of whether the instability can be triggered. However, in the nonlinear collapse phase, gas density grows and pressure increases, eventually stopping contraction provided that F_P grows faster than F_G as the gas element contracts. In real disks this is possible, e.g., if contraction is isentropic and gas has the adiabatic index $\gamma > 4 = 3$, which is the case for molecular gas in protoplanetary disks. As a result, in the absence of radiative losses, GI cannot result in disk fragmentation, which agrees with numerical calculations.

Cooling of collapsing fluid elements lowers their entropy, thus reducing pressure support and making fragmentation possible. Using two-dimensional hydrosimulations in the shearing-sheet approximation, *Gammie* (2001) has shown that fragmentation of a gravitationally unstable disk patch is possible only if the local cooling time of the disk t_c satisfies

$$t_c \Omega \lesssim \beta \qquad (3)$$

where $\beta \approx 3$. In the opposite limit of a long cooling time, the disk settles into a quasistationary gravitoturbulent state, in which it is marginally gravitationally unstable.

A simple way to understand this result is to note that the fluid element contracts and rebounds on a dynamical timescale $\sim \Omega^{-1}$, so unless cooling proceeds faster than that, pressure forces would at some point become strong enough to prevent collapse. This result has been generally confirmed in different settings, e.g., in global three-dimensional geometry and with both grid-based and smoothed particle hydrodynamic (SPH) numerical schemes (*Rice et al., 2005*). The only difference found by different groups is the precise value of β at which fragmentation becomes possible; e.g., *Rice et al.* (2005) find $\beta \approx 5$ in their three-dimensional SPH calculations. The value of β is also a function of the gas EOS, with lower γ resulting in higher β (*Rice et al., 2005*). This is natural as gas with softer EOS needs less cooling for the role of pressure to be suppressed. It should nevertheless be noted that some authors found the cooling criterion in equation (3) to be non-universal, with β depending on the resolution of simulations, their duration, and the size of the simulation domain (*Meru and Bate, 2011; Paardekooper, 2012*). It is not clear at the moment whether these claimed deviations from simple cooling criterion have a physical or numerical origin (*Paardekooper et al., 2011; Meru and Bate, 2012*).

Cooling time in a disk can be estimated using a simple formula

$$t_c \sim \frac{\Sigma c_s^2}{\sigma T_{\mathrm{eff}}^4} = \frac{\Sigma (k_B/\mu)}{\sigma T^3 f(\tau)} \qquad (4)$$

where T_{eff} is the effective temperature at the disk photosphere, related to the mid-plane temperature T via $T^4 = T_{\mathrm{eff}}^4 f(\tau)$. The explicit form of the function $f(\tau)$ depends on a particular mode of vertical energy transport in the disk, but is expected to scale with the optical depth of the disk τ. As disk cooling is inefficient both in the optically thick ($\tau \gg 1$) and thin ($\tau \ll 1$) regimes, $f(\tau) \gg 1$ in both limits. In particular, if energy is transported by radiation, then $f(\tau) \approx \tau + \tau^{-1}$, and a similar expression has been derived by *Rafikov* (2007) if energy transport is accomplished by convection. Interestingly, the cooling constraint is easier to satisfy at higher disk temperatures since for a fixed $f(\tau)$, the radiative flux from the disk surface increases with temperature as T^4. This is a much faster scaling than that of the thermal content of the disk, which is linear in T. Thus, to satisfy equation (4), one would need to increase T, forcing Σ to grow to keep $Q \lesssim 1$. As a result, at a given radius, fragmentation of a gravitationally unstable disk is possible only if the disk is hot enough (short cooling time) and very dense (*Rafikov, 2005, 2007*).

Even under the most optimistic assumptions about the efficiency of cooling ($f(\tau) \sim 1$), a gravitationally unstable disk with a surface density profile (equation (2)) has

$$t_c \Omega \sim \frac{\Omega^2}{\pi G \sigma} \left(\frac{k_B}{\mu} \right)^{3/2} T^{-5/2} \approx 600 \left(\frac{AU}{r} \right)^{7/4} \qquad (5)$$

for the previously adopted temperature profile. The fragmentation criterion (equation (3)) is clearly not fulfilled at separations of several astronomical units in such a disk, making planet or BD formation via GI problematic in the inner regions of the protoplanetary disk (*Rafikov*, 2005). *Boss* (2002) resorted to convective cooling as a possible explanation for rapid heat transport between the disk midplane and photosphere, but *Rafikov* (2007) has demonstrated that this is an unlikely possibility. Subsequently, *Boley and Durisen* (2006) have interpreted the vertical motion seen in the *Boss* (2002) simulations as being the shock bores rather than true convective motions.

Combining GI with the cooling criteria in equations (1) and (4), one can make more rigorous statements and show that disk fragmentation is very difficult to achieve close to the star, within several tens of astronomical units (*Rafikov*, 2005, 2007). In particular, for the disk opacity scaling as $\kappa \propto T^2$ thought to be appropriate for outer, cool disk regions, fragmentation of the optically thick disk is inevitable outside 60–100 AU around a solar-type star (*Matzner and Levin*, 2005; *Rafikov*, 2009), but cannot occur interior to this zone. This result may be relevant for understanding the origin of the outermost planets in the HR 8799 system, although, as discussed in section 5.3, these planets might also have formed by CA.

In optically thin disks this boundary is pushed considerably further from the star (*Matzner and Levin*, 2005; *Rafikov*, 2009), but if the disk is very massive or, equivalently, has a high mass-accretion rate of $\dot{M} \geq 10^{-5}$ $M_\odot$ yr^{-1}, then it must be optically thick at these separations. Such high mass-accretion rates may be typical during the early, embedded phase of the protostellar evolution, but are unlikely during the classical T Tauri phase. As a result, objects formed by disk fragmentation must be born early and then survive in a disk for several million years, which may not be trivial. Indeed, gravitational coupling to a massive protoplanetary disk in which clumps form should naturally drive their rapid radial migration, potentially resulting in their accretion by the star (*Vorobyov and Basu*, 2006a; *Machida et al.*, 2011; *Baruteau et al.*, 2011). Moreover, as equation (1) for disk instability is fulfilled, spiral modes tend to develop in the disk and carry out angular momentum. As a result, the outer parts of the disk will disperse without fragmenting while the inner parts will accrete onto the central star (*Laughlin and Bodenheimer*, 1994). Therefore, in order for a bound object to be able to form by GI and survive, a fragment must both (1) cool (i.e., radiate away the energy induced by the PdV compression work) and (2) lose angular momentum fast enough to condense out before it is either dispersed out, sheared apart, or accreted. Some other processes affecting the survival of collapsing objects have been reviewed in *Kratter and Murray-Clay* (2011).

All these mechanisms — radiative cooling, angular momentum redistribution, and migration — are by nature strongly nonlinear and cannot be captured by linear stability analysis such as that leading to the condition in equation (1). Instead, one must rely on numerical simulations to explore

this issue. Numerical studies of planet or BD formation by GI, however, have not yet converged to a definitive conclusion regarding the viability of this mechanism. Part of the reason for this lies in the markedly different numerical schemes employed in these studies. Smoothed particle hydrodynamics (SPH) is well suited for modeling self-gravitating systems and following the collapse of forming fragments to very high densities. However, this method has known problems with treating gas dynamics, which plays a key role in disk fragmentation. Grid-based methods have their own problems and so far many of the numerical studies employing this numerical scheme have reported conflicting results. In contrast, simulations including a detailed thermal evolution of the disk, including stellar irradiation, show that these effects severely quench fragmentation (*Matzner and Levin*, 2005).

A very important source of this ambiguity lies in how the disk cooling is treated in simulations. We saw that the rate at which the disk cools may be a key factor controlling the disk's ability to fragment. A number of early studies of disk fragmentation adopted an isothermal EOS for the gas, which is equivalent to assuming that $t_{cool} \rightarrow 0$ (*Armitage and Hansen*, 1999; *Mayer et al.*, 2002). Others assumed a less-compressible EOS $P = K\rho^\gamma$ with $\gamma > 1$, but this still neglects gas heating in shocks and makes the disk easier to fragment (*Mayer et al.*, 2004). Not surprisingly, fragmentation has been found to be a natural outcome in such simulations (*Boss*, 2007).

To definitively settle the question of disk fragmentation one must be able to follow disk thermodynamics as accurately as possible, and be able to calculate transport of radiation from the mid-plane of the disk to its photosphere. Treatment of disk properties at the photosphere is a key issue, since the cooling rate is a very sensitive function of the photospheric temperature T_{eff} (see equation (4)). All existing studies of disk fragmentation incorporating radiation transfer rely on the use of the flux-limited diffusion approximation. This approximation works well in the deep optically thick regions of the disk, as well as in the outer, rarefied optically thin regions. Unfortunately, it is not so accurate precisely at the transition between the two limits, i.e., at the disk photosphere, where one would like to capture disk properties most accurately.

Treatment of radiation transfer in SPH simulations suffers from similar issues, exacerbated by the fact that in SPH even the very definition of the disk photosphere is highly ambiguous and subjective (*Mayer et al.*, 2007). There are also other technical issues that might have affected disk fragmentation in some studies, such as the treatment of the EOS at low temperatures (*Boley et al.*, 2007; *Boss*, 2007), artificial constraints on the disk temperature (*Cai et al.*, 2010), and so on. A number of these numerical issues, technical but nevertheless quite crucial, were discussed in detail in *Durisen et al.* (2007) and *Dullemond et al.* (2009).

Despite these ambiguities, the general consensus that seems to be emerging from numerical and analytical studies of GI is that disk fragmentation is extremely unlikely in the innermost parts of the disk, where the cooling time is very long compared to the local dynamical time. At the same time, formation of self-gravitating objects may be possible in the outer parts of the disk, beyond 50–100 AU from the star, even though some three-dimensional radiative, gravitational hydrodynamical models claim that even at such distances disk instability is unlikely (*Boss,* 2006). In any case, as mentioned above, it is still not clear if objects formed by GI can avoid disruption or migration into the star during their residence in a dense protoplanetary disk out of which they formed (e.g., *Machida et al.,* 2011; *Baruteau et al.,* 2011; *Vorobyov,* 2013).

5. FORMATION SCENARIOS FOR BROWN DWARFS AND GIANT PLANETS

5.1. Common Formation Scenario: Disk Fragmentation

Disk fragmentation has been invoked both for the formation of BDs and GPs and thus its pertinence must be examined as a general mechanism for the formation of SSOs.

As described in the previous section, for a disk to become gravitationally unstable and lead eventually to the formation of GP or BD companions, the disk must be massive enough and fulfill the appropriate cooling conditions. Some simulations (*Stamatellos et al.,* 2007; *Vorobyov and Basu,* 2006b) do predict BD formation at large orbital distances under such conditions, provided the disk is massive and extended enough [typically $M_d \geq 0.3\ M_\star$, $R_d \geq 100\ (M_\star/M_\odot)^{1/3}$ AU (*Stamatellos and Whitworth,* 2009)]. These simulations, however, lack a fundamental physical ingredient, namely the magnetic field. Indeed, it has been shown by several studies that magnetic fields prevent significant mass growth and stabilize the disk, severely hampering fragmentation (e.g., *Hennebelle and Teyssier,* 2008; *Price and Bate,* 2009; *Commerçon et al.,* 2010; *Machida et al.,* 2010). Magnetic fields drive outflows and produce magnetic braking, carrying out most of the angular momentum. These simulations, however, were conducted with ideal MHD. Subsequent calculations have shown that non-ideal MHD effects (e.g., *Dapp et al.,* 2012; *Machida and Matsumoto,* 2011), misalignment of the field with the rotation axis (*Hennebelle and Ciardi,* 2009; *Joos et al.,* 2012; *Li et al.,* 2013), and the turbulent nature of the velocity field (*Joos et al.,* 2013; *Seifried et al.,* 2012, 2013) all decrease the efficiency of magnetic braking. However, although large disks can form in these cases, they still remain smaller and less massive than those produced in pure hydrodynamical simulations. Three-dimensional calculations, including radiation hydrodynamics, turbulence, and complete MHD equations, suggest that, for typical observed values of cloud magnetizations and rotation rates, large Keplerian disks do form during the initial core collapse phase but do not seem to be massive enough to lead to fragmentation (Masson et al., in preparation). This importance of the magnetic field has received strong support from the confrontation of observations carried out at the PdBI with synthetic images derived from simulations of both pure hydrodynamical and MHD simulations (*Maury et al.,* 2010). It was shown that the disk synthetic images derived from the MHD simulations were similar to those observed, with rather compact disks of typical FWHM ~ 0.2″–0.6″, while the hydrodynamical simulations were producing both disks that were too extended and (multiple) structures too fragmented compared with observations. Moreover, as mentioned in section 4.1, observations of isolated disks at the early class 0 stage have revealed compact (R ~ 20 AU) disks (*Rodríguez et al.,* 2005), while so far extended massive disks prone to fragmentation appear to be rare. However, as mentioned in section 4.1, better statistics are certainly needed to reach more robust conclusions. Therefore, although disk fragmentation probably occurs in some disks at the early stages of star formation, the fragmentation process is largely overestimated in purely hydrodynamical simulations. Furthermore, as mentioned in section 4.2, it is not clear that fragments, if they form in the disk, will be able to cool quickly enough to form bound objects and to survive turbulent motions or rotational shear, nor that they will not migrate quickly inward and end up being accreted by the star, leading to episodic accretion events.

Other observational constraints seem to contradict GI in a disk as the main route for BD or GP formation. If indeed this mechanism was dominant, most class 0 objects should have massive disks, a conclusion that so far does not seem to be supported by observations. An argument sometimes raised by some proponents of this scenario (e.g., *Stamatellos and Whitworth,* 2009) is that the lifetime of the massive and extended disks produced in the simulations is too short (~a few 10^3 yr) for the disks to be observed. This argument, however, does not hold since class 0 objects last over a significantly longer period, about ~10^5 yr. In addition, these simulations do not include an accreting envelope and thus end up too early compared with realistic situations, yielding an inaccurate diagnostic. As mentioned above, more accurate calculations, including an accreting envelope, magnetic field, turbulence, and feedback (e.g., *Seifried et al.,* 2013; Masson et al., in preparation), show the emergence of large centrifugal disks but, for typical observed magnetic flux conditions, the disks do not appear to be massive enough to lead to fragmentation — at least for low-mass protostars, the bulk of the distribution. It is fair to say, however, that at the time of this writing this important issue is far from being settled and further exploration of this topic is needed. As mentioned above, however, it is mandatory to include all the proper physics in these studies to reach plausible conclusions.

Most importantly, the observational constraints arising from the existence of wide BD or GP companions (see sections 2.3 and 3.2) strongly argue against GI as an efficient formation mechanism. Besides the objects mentioned in section 2.3, an interesting example is the triple system LHS6343 (*Johnson et al.,* 2011), with the presence of a BD companion with mass $M_C = 0.063\ M_\odot$ to a low-mass star with $M_A = 0.37 \, M_\odot$. Assuming disk-to-star standard mass fractions around ~10%, the disk around M_A should thus have a mass $M_d \approx 0.04 \, M_\odot$. Although admittedly crude, this estimate shows that there is not enough mass in the disk to have formed the BD companion, even if assuming 100% disk-to-BD mass conversion efficiency. Finally, the statistics arising from direct imaging in section 3.2 that reveal, at least so far, the scarcity of wide and massive SSOs in the BD or GP mass range around FGKM-type stars strongly argues against GI as the main formation mechanism for BDs or GPs. One might invoke the possibility that objects form by GI at large orbits and then migrate inward closer to the star. This argument, however, does not hold. First, it must be stressed that the possibility of migration is already included in the general analysis mentioned above. Second, if there was a "planetary" population forming by GI and migrating inward, there should also be a "brown dwarf" population undergoing the same process. But this can be firmly observationally excluded. Finally, if we interpret the existing close-in planet population as objects that formed by GI and migrated inward, this would not explain the aforementioned planet bottom-heavy mass distribution and planet-metallicity correlation. Gravitational instability, however, might occur in some particular situations, for instance, in very massive disks around binary systems (*Delorme et al., 2013*), even though other formation mechanisms are also possible in such situations. As mentoned in section 3.2, the combination of ALMA and future direct imaging projects will definitely help nail down this issue.

5.2. Formation Scenarios for Brown Dwarfs

In this section, we address formation scenarios other than GI that have been specifically suggested for BDs. Note that a more specific examination of the star and BD formation theories can be found in the short review by *Hennebelle and Chabrier* (2011).

5.2.1. Photoionization. The fact that the BD mass function is about the same regardless of the presence of O stars shows that halting of accretion by photoionizing radiation (*Whitworth and Zinnecker, 2004*) is clearly not a major mechanism for BD formation. Furthermore, as mentioned in section 2, BDs have been observed in isolated environments, excluding a necessary connection between the presence of photoionizing radiation and BD formation.

5.2.2. Accretion-ejection. In the accretion-ejection scenario (*Reipurth and Clarke, 2001*), BDs are the result of accreting ~1-M_{Jup} stellar embryos, formed by the efficient dynamical fragmentation of a molecular clump, that get ejected from the surrounding gas reservoir and thus end up remaining in the substellar domain. The characteristic conditions of this scenario are (1) dynamical interactions are responsible for the formation of both high-mass (by merging) and low-mass (by ejection) objects, implying that star formation occurs essentially, if not only, in dense cluster environments; and (2) the IMF is determined essentially at the latest stages of the collapse, namely the ultimate gas-

to-star conversion, with no correlation whatsoever with the initial core mass function. In fact, prestellar cores do not really exist in this scenario.

The most achieved simulations exploring this scenario are those conducted by *Bate* (2012), which, although still lacking a magnetic field, include a treatment of gas heating and cooling. One of the most striking results of these simulations is that the resulting IMF reproduces quite well the *C05* IMF (see section 2.1). These simulations, however, remain of questionable relevance for most Milky Way molecular cloud conditions. Indeed, it has been established observationally that star-forming molecular clouds in the MW follow the so-called Larson's relations (*Larson, 1981*), although with significant scatter, in terms of cloud size vs. mean density and velocity dispersion (e.g., *Hennebelle and Falgarone, 2012*)

$$\bar{n} \simeq 3 \times 10^3 \left(\frac{L}{1 \, pc} \right)^{-\eta_d} cm^{-3} \qquad (6)$$

$$\sigma_{rms} \simeq 0.8 \left(\frac{L}{1 \, pc} \right)^{\eta} km \, s^{-1} \qquad (7)$$

with $\eta_d \sim 0.7$–1.0 and $\eta_d \sim 0.4$. *Bate*'s (2012) initial conditions, however, correspond to a 500-$M_\odot$ cloud at 10 K with size L = 0.4 pc, thus mean density $\bar{n} \simeq 3.2 \times 10^4$ cm^{-3} and surface density $\Sigma \approx 10^3 \, M_\odot$ pc^{-2} [while typical values for MW molecular clouds are around ~60–100 $M_\odot$ pc^{-2} (*Heyer et al., 2009*)], and Mach number $\mathcal{M} = 14$. These values correspond to rather extreme cloud conditions, ~4–5× denser and more turbulent than the aforementioned typical observed ones. Such conditions will strongly favor fragmentation and dynamical interactions and it is thus not surprising that they produce a significant number of ejected BD embryos. Interestingly enough, these simulations can be directly confronted to observations. Indeed, the simulated cloud mass and size are very similar to that of the young cluster NGC 1333 (*Scholz et al., 2012a*). The simulations, however, produce a significantly larger number of stars + BDs than those observed, with a total stellar mass ~191 $M_\odot$ against ~50 $M_\odot$ in NGC 1333 (*Scholz et al., 2012a*); interestingly, this echos the aforementioned factors of ~4 in density and velocity. Other inputs in the simulations — for example, the assumption of a uniform initial density profile, the lack of a magnetic field, and the underestimated radiative feedback — all favor fragmentation and thus probably overestimate the number of protostellar or proto-BD embryos formed in a collapsing core, again favoring dynamical interactions/ejections. Observations in fact tend to suggest that fragmentation within prestellar cores is rather limited, with most of the mass of the core ending up in one or just a few smaller cores (*Bontemps et al., 2010; Tachihara et al., 2002*). This again casts doubts on the relevance of such initial conditions to explore star/BD formation under typical Milky Way molecular cloud conditions. At the very least, if indeed BD formation by dynamical ejections might occur under some circumstances, existing simulations severely overestimate the efficiency of this process. In fact, one is entitled to suspect that the similarity between the IMF produced by the simulations and that representative of the Galactic field reflects in reality the result of the initial gravoturbulent collapse of the cloud rather than the results of dynamical interactions, in which case the IMF in the simulations should already be largely determined at the beginning of the simulation. If this is the case, these simulations in fact provide support to star/BD formation by gravoturbulent collapse. At any rate, until simulations with more realistic initial conditions are conducted, those performed by *Bate* (2012) cannot be considered as a reliable demonstration of dominant BD formation by accretion-ejection.

The accretion-ejection scenario faces other important issues, some of which have already been mentioned in section 2.3. Without being exhaustive, one can add a few more: (1) How can BDs form, according to this scenario, in low-density environments? The Taurus cloud for instance, even though having a stellar density about 3 orders of magnitude smaller than other common clusters, has a comparable abundance of BDs (*Luhman,* 2012). As noted by *Kraus et al.* (2011), the low-stellar density (≤5 stars/pc^2) and low-velocity dispersion of Taurus members (σ ~ 0.2 km s^{-1}) indicate that there are no small N-body clusters from which stars or BDs could have been ejected, as advocated in the accretion-ejection scenario. The fact that BDs form as efficiently in such low-density environments strongly argues against dynamical interactions. (2) Various observations show average dispersion velocities of prestellar cores of about ⟨σ⟩ ~ 0.4 km s^{-1} (*André et al.,* 2009; *Walsh et al.,* 2007; *Muench et al.,* 2007; *Gutermuth et al.,* 2009; *Bressert et al.,* 2010). This indicates a typical collision timescale between prestellar cores significantly larger than their dynamical timescale, suggesting little dynamical evolution during prestellar core formation. (3) How can this scenario explain the observed similarity between the CMF and the IMF, if indeed such a similarity is confirmed?

All these observational constraints (notably the observed low-velocity dispersion between protostar/BDs) severely argue against the accretion-ejection scenario for BD or star formation as a dominant mechanism, except possibly for the most massive stars, which represent only a very small fraction of the stellar population.

5.2.3. Gravoturbulent fragmentation. In this scenario, large-scale turbulence injected at the cloud scale by various sources cascades to smaller scales by shocks and generates a field of density fluctuations down to the dissipative scale (e.g., *Mac Low and Klessen,* 2004). Overdense regions inside which gravity overcomes all other sources of support collapse and form self-gravitating cores that isolate themselves from the surrounding medium. In this scenario, the CMF/IMF is set up by the spectrum of turbulence, at the very early stages of the star-formation process. The first theory combining turbulence and gravity was proposed by *Padoan and Nordlund* (2002) but has been shown to suffer from various caveats (e.g., *McKee and Ostriker,* 2007; *Hen-*

nebelle and Chabrier, 2011). A different theory was derived more recently by *Hennebelle and Chabrier* (2008, 2009, 2013) and *Hopkins* (2012) (see the chapter by Offner et al. in this volume). These latter theories nicely reproduce the *C05* IMF down to the least-massive BDs, for appropriate (Larson-like) conditions. A common and important feature of these theories is that it is inappropriate to use the average thermal Jeans mass as an estimate of the characteristic mass for fragmentation. This would, of course, preclude significant BD formation. Indeed, in both the Hennebelle-Chabrier and Hopkins theories, the spectrum of collapsing prestellar cores strongly depends on the Mach number, shifting the low-mass tail of the IMF to much smaller scales than naively expected from a purely gravitational Jeans argument. The theories indeed yield a reasonably accurate number of pre-BD cores for adequate molecular-cloud-like conditions and Mach values ($\mathcal{M}$ ~ 3–8). Densities required for the collapse of BD-mass cores, ~10^7–10^8 g cm^{-3} (see section 2.4), are indeed produced by turbulence-induced shock compression (remembering that shock conditions imply density enhancements ∝$\mathcal{M}^2$). Note that a similar scenario, taking into account the filamentary nature of the star-forming dense regions, was developed by *Inutsuka* (2001) (see the chapter by André et al. in this volume).

Support for the gravoturbulent scenario arises from several observational facts: (1) It explains, within the framework of the same theory, the observed mass spectra of both *unbound* CO-clumps and *bound* prerestellar cores (*Hennebelle and Chabrier,* 2008); (2) it implies that the IMF is already imprinted in the cloud conditions (mean density, temperature, and Mach number), naturally explaining the resemblance of the CMF with the IMF; and (3) it relies on one single "universal" parameter, namely the velocity power spectrum index of turbulence, which explains as well the Larson relations for molecular clouds.

A major problem against the gravoturbulent scenario to explain the final IMF would arise if the cores were fragmenting significantly into smaller pieces during their collapse. As mentioned in the previous section, however, observations tend to show that such fragmentation is rather limited. Numerical simulations of collapsing dense cores indeed show that radiative feedback and magnetic fields drastically reduce the fragmentation process (e.g., *Krumholz et al.,* 2007; *Offner et al.,* 2009; *Commerçon et al.,* 2010, 2011; *Hennebelle et al.,* 2011; *Seifried et al.,* 2013). The analysis of simulations aimed at exploring the CMF-to-IMF conversion (*Smith et al.,* 2009) also show a clear correlation between the initial core masses and the final sink masses up to a few local free-fall times (*Chabrier and Hennebelle,* 2010), suggesting that, at least for the bulk of the stellar mass spectrum, the initial prestellar cores do not fragment into many objects, as indeed suggested by the CMF/IMF similarity. But the most conclusive support for BD formation by gravoturbulent fragmentation comes from the emerging observations of isolated proto-BDs and of the pre-BD core Oph B-11, with densities in agreement with the aforementioned value (see section 2.4).

5.2.4. Formation of binaries. Newly formed stars must disperse a tremendous amount of angular momentum in condensing through more than 6 orders of magnitude in radius (*Bodenheimer,* 1995). Besides magnetic braking, binary formation offers a convenient way to redistribute at least part of this excess angular momentum.

As mentioned in section 2.3, several binaries have been observed at large projected separations (>500 AU) with masses down to ~5 M_{Jup}. *Looney et al.* (2000) showed that multiple systems in the class 0 and I phases are prevalent on large spatial scales (≥1000 AU), while binary systems at smaller scales (≤500 AU) seem to be quite sparse (*Maury et al.,* 2010). Whether this lack of small scale binaries at the class 0/I stages is real or stems from a lack of observations with sufficient resolution or sensitivity, however, is still unclear so far. In any case, prestellar and protoplanetary disks do not extend out to thousands of astronomical units, so formation of such wide BD binaries by disk GI seems to be clearly excluded. In contrast, such distances can be compared with the characteristic sizes of starless prestellar core envelopes, ~10^4 AU (*Men'shchikov et al.,* 2010). The existence of these wide systems thus suggests that prestellar core fragmentation into binaries might occur at the very early stages (typically at the class 0 stage or before) of the collapse. Indeed, there are some observational evidences for binary fragmentation at this stage (*Looney et al.,* 2000; *Duchêne et al.,* 2003). Moreover, the similarity, at least in a statistical sense, between some of the star and BD multiplicity properties (see section 2.3) suggests that most BD companions form similarly as stellar companions. This hypothesis has received theoretical and numerical support from the recent work of *Jumper and Fisher* (2013). These authors show that BD properties, including the BD desert, wide BD binaries, and the tendency for low-mass (in particular BD) binaries to be more tightly bound than stellar binaries — arguments often used as an evidence for distinct formation mechanisms for BDs — can be adequately reproduced by angular momentum scaling during the collapse of turbulent clouds. Indeed, the early stages of cloud collapse/fragmentation and core formation are characterized by the formation of puffy disk-like structures that keep accreting material from the surrounding core envelope. These structures, however, are not relaxed and differ from structures purely supported by rotation, characteristic of relaxed, equilibrium disks. Such "pseudo-disks" may fragment (as the result of global nonlinear GI) during the (first or second) collapse of the prestellar core and end up forming (wide or tight) binaries, possibly of BD masses (*Bonnell and Bate,* 1994; *Machida,* 2008). The occurence of such fragmentation at the cloud collapse stage has been found in radiation hydrodynamic simulations (*Commerçon et al.,* 2010; *Peters et al.,* 2011; *Bate,* 2012). They remain to be explored in the context of resistive MHD before more definitive conclusions can be reached concerning this important issue. As just mentioned, such fragmentation into binaries occurs at the very early stages of the core collapse, during the main accretion phase. Whether it occurs dominantly by redistribution of angular momentum during the collapse itself or by global instability in the mass-loaded growing pseudodisk remains an open question so far, and a clear-cut distinction between the two processes is rather blurry at this stage. It must be kept in mind, however, that the conditions for the formation *and* survival of bound fragments in a disk are subject to very restrictive constraints, as examined in section 4.2.

5.3. Formation Scenarios for Giant Planets: Core Accretion

5.3.1. Core accretion mechanism. In the CA scenario gas GPs form in two steps. First, a solid core is accumulated by accretion of planetesimals. The growing core attracts a hydrostatic envelope of protoplanetary disk gas, extending from the surface out to the Bondi radius where the sound speed equals the escape speed of the core. Since accreting planetesimals not only increase the core mass but also release gravitational energy, the energy released at the core surface must be transported through the envelope, making the latter nonisothermal. The mass of the envelope increases with the core mass faster than linearly, so at some point the envelope starts to dominate the gravitational potential. Hydrostatic solutions cease to exist beyond a critical core mass that depends mainly on the accretion rate of solids and on the gas opacity. As a result, the core starts accreting mass on its own thermal timescale, turning itself into a GP.

Perri and Cameron (1974) assumed the hydrostatic envelope to be fully convective and found critical core masses in excess of 100 $M_{\oplus}$; they used this result to support the alternative GI scenario for GP formation. *Mizuno* (1980) allowed for both convective and radiative regions in the envelope and constructed the equilibrium model using gas opacities depending on the density and temperature, with a stepwise constant dust opacity below the dust sublimation temperature. The radiative solution drastically reduces the critical core mass to approximately 10 $M_{\oplus}$, in agreement with constraints from the gravitational potentials of Jupiter and Saturn (*Guillot,* 2005), for an assumed mass-accretion rate of 10^{-6} $M_{\oplus}$ per year. Planetary envelopes can still be fully convective at very high planetesimal-accretion rates (*Wuchterl,* 1993; *Ikoma et al.,* 2001; *Rafikov,* 2006), typical for cores at approximately astronomical units from the star, but this is unlikely to be an issue at large separations (*Rafikov,* 2006).

A generic property of the hydrostatic envelope is that its mass is inversely proportional to the luminosity (and hence the mass-accretion rate) and the opacity (*Stevenson,* 1982). The resulting critical core mass roughly scales as $M_c \propto \dot{M}^{1/4}$ (*Ikoma et al.,* 2000). Hence the growth rate of the core partially determines the critical core mass. Beyond the critical core mass the envelope will emit more energy than provided by the gravitational potential release of the accreted solids and undergo runaway contraction on the Kelvin-Helmholtz timescale (*Bodenheimer and Pollack,* 1986). Interestingly, the concept of a critical core mass seems to be supported observationally by the mass-radius relationship of the recently detected Kepler low-mass transit planets *(Lissauer et al.,* 2011a,b; *Carter et al.,* 2012; *Ofir et al.,* 2013), keeping in mind the uncertainties in the mass determinations of Kepler objects. Indeed, while planets above about ~6 $M_\oplus$ seem to have a radius requiring a substantial (≥10% by mass) gaseous H/He envelope, so far planets below this mass have a radius consistent with a much lower, or even negligible, gas mass fraction. Even the rather extended gaseous envelope of the lowest transiting planet discovered so far, KOI-314c ($1.0^{+0.4}_{-0.3}$ $M_\oplus$) *(Kipping et al.,* 2014), with a radius R_{env} ~ 7^{+12}_{-13}% of the planet's radius, represents a negligible fraction (≤3%) of its mass.

5.3.2. Timescale for accumulation of the core. In the classical CA scenario, the core grows by accreting planetesimals. The Hill radius

$$R_H = \left[GM_p / \left(3\Omega^2 \right) \right]^{1/3} = \frac{1}{p} R_p \tag{8}$$

denotes the maximal distance over which the core can deflect planetesimals that pass by with (linearized) Keplerian shear flow. Here $p = [9 M_\star/(4\pi\rho a^3)]^{1/3} \ll 1$ is a parameter that depends on semimajor axis a and material density ρ for a given stellar mass $M_\star$ *(Goldreich et al.,* 2004). An additional random component to the particle motion will reduce the scattering cross section of the core to below ~R_H^2.

The gravitational radius of the core denotes the maximum impact parameter at which a planetesimal approaching with velocity v gets accreted by a core with radius R at closest approach. This distance is given by

$$R_G = R\sqrt{1 + \frac{v_e^2}{v^2}} \tag{9}$$

where R is the radius of the core, v_e the escape speed at the surface, and v the relative approach speed. In dynamically cold disks, planetesimals enter the Hill sphere of the core with the Hill speed $v_H = \Omega R_H$. In that case the gravitational radius can be simplified as

$$R_G \approx \sqrt{p} R_H \tag{10}$$

A small fraction of the planetesimals that enter the Hill sphere of the core actually collide with it. The mass-accretion rate is ultimately set by the gravitational radius and the scale height of the planetesimals. The random motion of planetesimals enter both these quantities and should ideally be calculated self-consistently using an N-body approach (e.g., *Levison et al.,* 2010).

Assuming that the random particle motion is similar to ΩR_H — the Keplerian velocity shear across the Hill radius — one can find that the core grows at the rate *(Dones and Tremaine,* 1993)

$$dR/dt \sim p^{-1}\Sigma_s\Omega/\rho \approx 50\,m\,yr^{-1}\left(r/AU\right)^{-2} \tag{11}$$

assuming surface density of solids typical for the MMSN (~30 g cm^{-2} at 1 AU). This corresponds to a mass growth rate on the order of 10^{-6} $M_\oplus$ yr^{-1} at 5 AU, where Jupiter presumably formed *(Dodson-Robinson et al.,* 2009). This leads to core formation in 10^7 yr, in marginal agreement with the observed lifetimes of protoplanetary disks *(Haisch et al.,* 2001), but the timescale can be lessened by assuming a planetesimal column density that is 6–10× the value in the MMSN *(Pollack et al.,* 1996). Note that equation (11) assumes uniform planetesimal surface density near the embryo, an assumption that may be violated if the embryo is capable of clearing a gap in the planetesimal population around its orbit *(Tanaka and Ida,* 1997; *Rafikov,* 2001, 2003a), which would slow down its growth *(Rafikov,* 2003b).

While the agreement between the core mass (or more precisely, the total amount of heavy material) of Jupiter and the mass-accretion rate necessary to form the core is a success for the CA model, the mass-accretion rate scales roughly as the semimajor axis to the inverse second power in accordance with equation (11). Hence core formation beyond 5 AU is impossible in the million-year timescale of the gaseous protoplanetary disk. This conflicts both with the high gas contents of Saturn, the small H/He envelope of Uranus and Neptune, and the observations of gas giant exoplanets in wide orbits beyond 10 AU *(Marois et al.,* 2008b).

This discussion makes it clear that gas giants can form by CA at large separations only if the core-formation timescale can be reduced dramatically. Ironically, acceleration of the core growth makes it harder to reach the threshold for CA, because the critical core mass is inversely proportional to the core luminosity, which is predominantly derived from the heat released in planetesimal accretion. Nevertheless, one can show *(Rafikov,* 2011) that the higher planetesimal-accretion rate $\dot{M}$ still reduces the total time until CA sets in, even though the critical core mass is larger and is more difficult to achieve.

5.3.3. Fragment accretion. One pathway to rapid core formation is to accrete planetesimal fragments damped by the gas to a scale height lower than the gravitational accretion radius *(Rafikov,* 2004).

Numerical and analytical studies of planetesimal dynamics generically find that soon after a dominant core emerges it takes over dynamical stirring of the surrounding planetesimals, increasing their random velocities to high values. Physical collisions between planetesimals then lead to their destruction rather than accretion, giving rise to a fragmentation cascade that extends to small sizes. *Rafikov* (2004) demonstrated that random velocities of particles in a size range 1–10 m are effectively damped by gas drag against excitation by the core gravity. Gas drag also does not allow such small fragments to be captured in mean-motion resonances with the embryo (e.g., *Levison et al.,* 2010). This velocity damping is especially effective in the vertical direction, allowing small debris to settle into a geometrically thin disk with thickness below the maximum impact parameter, still resulting in planetesimal accretion (~$p^{1/2}R_H$). From the accretion point of view the disk of such planetesimal debris is essentially two-dimensional, resulting in the highest mass-accretion rate possible in the absence of any external agents such as described next (see section 5.3.4 below)

$$dR/dt \sim p^{-3/2} \Sigma_s \Omega / \rho \approx 1 \, km \, yr^{-1} \left(r/AU \right)^{-3/2} \qquad (12)$$

This formula assumes that most of the surface density of solids Σ_s ends up in small (~10 m) fragments dynamically cooled by gas drag. Because of the dramatically reduced vertical scale height of the fragment population, this rate is higher than the previously quoted equation (11) for the velocity dispersion $\sim \Omega R_H$ and decreases somewhat more slowly with distance (as $r^{-3/2}$). As a result, a Neptune-sized core can be formed at 30 AU within several million years.

A key ingredient of this scenario is the collisional grinding of planetesimals, which allows them to efficiently couple to gas and become dynamically cold, accelerating core growth. Fragmentation plays the role of an agent that transfers mass from a dynamically hot mass reservoir (population of large planetesimals, accreted at low rate because of relatively weak focusing) to dynamically cold debris. Even though at a given moment the latter population may have lower surface density than the former, it is accreted at much higher efficiency and easily dominates core growth. The importance of fragment accretion for accelerating core growth has been confirmed in recent coagulation simulations by *Kenyon and Bromley* (2009) (see also *Levison et al.*, 2010).

In the framework of this scenario, *Rafikov* (2011) sets a constraint on the distance from the star at which gas giants can still form by CA in a given amount of time (several million years). By adopting the maximally efficient accretion of small debris confined to a two-dimensional disk, he showed that GP formation can be extended out to 40–50 AU in the MMSN and possibly even further in a more massive planetesimal disk (out to 200 AU in the marginally gravitationally unstable gaseous protoplanetary disk). This estimate makes some rather optimistic assumptions, such as that most of the planetesimal mass is in small (1–100 m) objects (e.g., as a result of efficient fragmentation), and that seed protoplanetary embryos with sizes of at least several hundred kilometers can be somehow produced far from the star within several million years. Thus, it becomes very difficult to explain the origin of planets beyond 100 AU by CA. This issue is examined below.

5.3.4. Pebble accretion. The arguably most fundamental particles to accrete are those forming directly by coagulation and condensation as part of the planetesimal-formation process. Several particle growth mechanisms predict that particles will stop growing efficiently when they reach pebbles of millimeter and centimeter sizes.

While small dust grains grow as fluffy porous aggregates, they are eventually compacted by collisions around millimeter sizes and enter a regime of bouncing rather than sticking (*Zsom et al.*, 2010). This bouncing barrier maintains a component of small pebbles that can feed a growing core. In the alternative mechanism, where particles

grow mainly by ice condensation near evaporation fronts, particles sediment out of the gas when reaching pebble sizes. This reduces their growth rate drastically as they must collectively compete for the water vapor in the thin midplane layer (*Ros and Johansen*, 2013). Icy particles may be able to retain their fluffy structure against compaction and hence avoid the bouncing barrier (*Wada et al.*, 2009), but even under perfect sticking conditions, particles grow more slowly with increasing mass and spend significant time as fluffy snowballs whose dynamics are similar to compact pebbles (*Okuzumi et al.*, 2012).

The dynamics of pebbles in the vicinity of a growing core are fundamentally different from the dynamics of a planetesimal or a planetesimal fragment. Pebbles are influenced by drag forces, which act on a timescale comparable to or shorter than the orbital period, which is also the characteristic timescale for the gravitational deflection of particles passing the planet with the Keplerian shear. The gravity of the core pulls the pebbles out of their Keplerian orbits. The resulting motion across gas streamlines leads to frictional dissipation of the kinetic energy of the particles and subsequent accretion by the core. While planetesimals are only accreted from within the gravitational radius, which is a small fraction of the Hill radius, pebbles are in fact accreted from the entire Hill sphere.

The potential of drag-assisted accretion was explored by *Weidenschilling and Davis* (1985), who showed that particles below 1–10 m in size experience sufficient friction to avoid getting trapped in mean-motion resonances with the growing core. *Kary et al.* (1993) confirmed these results but warned that only a small fraction (10–40%) of the drifting pebbles are accreted by the core due to their radial drift. This concern is based on the assumption that pebbles form once and are subsequently lost from the system by radial drift. This loss of solids would be in disagreement with observations of dust in disks of million-year age (*Brauer et al.*, 2007). The solution to the drift problem may be that pebbles are continuously formed and destroyed by coagulation/ fragmentation and condensation/sublimation processes (*Birnstiel et al.*, 2010; *Ros and Johansen*, 2013). Hence any material that is not immediately accreted by one or more growing cores will get recycled into new pebbles that can then be accreted.

Johansen and Lacerda (2010) simulated the accretion of small pebbles onto large planetesimals or protoplanets. They found that pebbles with friction times less than the inverse orbital frequency are accreted from the entire Hill sphere, in agreement with the picture of drag-assisted accretion. *Ormel and Klahr* (2010) interpreted the rapid accretion as a sedimentation of particles toward the core. *Johansen and Lacerda* (2010) also observed that the pebbles arrive at the protoplanet surface with positive angular momentum, measured relative to the angular momentum of the disk, which leads to prograde rotation in agreement with the dominant rotation direction of the largest asteroids.

The dependence of the pebble-accretion rate on the size of the core was parameterized in *Lambrechts and Johansen*

(2012). Below a transition core mass of around 0.01 $M_\oplus$, pebbles are blown past the core with the sub-Keplerian gas. Pebbles are captured from within the Bondi radius of the sub-Keplerian flow, which is much smaller than the Hill radius. With increasing core mass the Bondi radius grows larger than the Hill radius, starting the regime of efficient pebble accretion. Growth rates in this regime are 1000× higher than for accretion of large planetesimals with velocity dispersion on the order of ΩR_H (see equation (11)), at 5 AU, and 10,000× faster, at 50 AU, depending on the degree of sedimentation of pebbles. Thus it is possible to create cores of 10 $M_\oplus$ even in the outer parts of protoplanetary disks, which may explain observed gas giant exoplanets in wide orbits (*Marois et al.,* 2008b).

The accretion rate of pebbles depends on the (unknown) fraction of the solid mass in the disk, which has grown to pebble sizes. There is observational evidence for large pebble masses in some protoplanetary disks — the disk around TW Hya should, for example, contain 0.001 $M_\odot$ of approximately centimeter-sized pebbles to match the emission at centimeter wavelengths (*Wilner et al.,* 2005). Such high pebble masses are surprising, given the efficiency for turning pebbles into planetesimals via various mechanisms for particle concentration (discussed in the next section). However, if planetesimals form through streaming instabilities, then the simulations show that only around 50% of the available pebble mass is incorporated into planetesimals (*Johansen et al.,* 2012b). The remaining pebbles are unable to concentrate in the gas flow, due to their low mass loading in the gas, and are available for a subsequent stage of pebble accretion onto the largest planetesimals.

The range of optimal particle sizes for pebble accretion falls significantly below the planetesimal fragments described above. Direct accretion is most efficient for particles that couple to the gas on timescales from the inverse Keplerian frequency down to 1% of that value. That gives typical particle sizes in the range of centimeter to meter in the formation zone of Jupiter and millimeter to centimeter for planet formation in wide orbits, due to the lower gas density in the outer parts of protoplanetary disks. However, *Morbidelli and Nesvorny* (2012) showed that even 10-m-sized boulders can be accreted with high efficiency, following a complex interaction with the core.

Although pebbles are not trapped in mean-motion resonances, their motion is sensitive to the pressure profile of the gas component. The global pressure gradient drives radial drift of pebbles toward the star, but any local pressure maximum will trap pebbles and prevent their drift to the core. *Paardekooper and Mellema* (2006) showed that planets above approximately 15 $M_\oplus$ make a gap in the pebble distribution and shut off pebble accretion. The shut off is explained in the analytical framework of *Muto and Inutsuka* (2009) as a competition between radial drift of particles and the formation of a local gas pressure maximum at the outside of the planetary orbit. The core mass is already comparable to the inferred core masses of Jupiter and Saturn when the accretion of pebbles shuts down. *Morbidelli and Nesvorny*

(2012) suggested that the termination of pebble accretion will in fact be the trigger of gravitational collapse of gas.

5.3.5. Implications of rapid core accretion models. In the pebble-accretion scenario, cores of 10 $M_\oplus$ or more can form at any location in the protoplanetary disk, provided that a sufficient amount of material is available. This can now be confronted with observational constraints from exoplanet surveys. Direct imaging surveys now have good upper limits for the occurrence rate of a Jupiter-mass planet in wide orbits (section 3.2). While many aspects of the observations and the luminosity of young planetary objects are still unknown, it seems that gas GPs are rare beyond 25 AU (with <23% of FGKM stars hosting gas giants in such orbits). This can be interpreted in three ways: (1) most of the solids (pebbles, planetesimals) in protoplanetary disks reside in orbits within 20 AU, (2) wide orbits are mostly populated by ice giants as in the solar system, and (3) gas GPs do form often in wide orbits but migrate quickly to the inner planetary system or are parts of systems that are unstable to planet-planet interactions shortly after their formation. Regarding interpretation (1), observations of protoplanetary disks in millimeter and centimeter wavelengths often find opacities that are consistent with large populations of pebbles (e.g., *Wilner et al.,* 2005; *Rodmann et al.,* 2006; *Lommen et al.,* 2007), The fact that these pebbles are only partially aerodynamically coupled to the gas flow, an important requirement for pebble accretion, was confirmed in the observations of a pebble-filled vortex structure in the transitional disk orbiting the star Oph IRS 48 (*van der Marel et al.,* 2013). However, *Pérez et al.* (2012) showed that particles are generally larger closer to the star, in agreement with a picture where the outer regions of protoplanetary disks are drained of pebbles by radial drift. Regarding interpretation (2), namely that wide orbits may be dominated by ice giants, it is an intriguing possibility, but its validation requires a better understanding of why ice giants in our own solar system did not accrete massive gaseous envelopes. Finally, regarding interpretation (3), it is important to keep in mind that massive planets in wide orbits have a large gravitational influence, which can lead to a disruption of the system. The HR 8799 system may owe its stability to a 4:2:1 resonance (*Fabrycky and Murray-Clay,* 2010), which indicates that a significant fraction of similar systems, less fortunately protected by resonances, may have been disrupted early on.

5.3.6. Formation of the seeds. Efficient accretion of pebbles from the entire Hill sphere requires cores that are already more massive than 0.1–1% of an Earth mass. Massive seed embryos are also needed in the rapid-fragment-accretion scenario (*Rafikov,* 2011) (see section 5.3.3). Below this transition mass pebbles are swept past the core with the sub-Keplerian wind. One can envision three ways to make such large seeds: (1) direct formation of very large planetesimals, (2) runaway accretion of planetesimals, and (3) inefficient pebble accretion from large Ceres-sized planetesimals to the transition mass.

The growth from dust to planetesimals is reviewed in the chapters by Testi et al. and Johansen et al. in this volume.

Dust and ice grains initially grow by colliding and sticking to particle sizes that react to the surrounding gas flow on an orbital timescale. These pebbles and rocks, ranging from millimeter to meter in size, experience strong concentrations in the turbulent gas flow. Particles get trapped in vortices (*Barge and Sommeria*, 1995; *Klahr and Bodenheimer*, 2003) and in large-scale pressure bumps, which arise spontaneously in the turbulent gas flow (*Johansen et al.*, 2009b). Particles can also undergo concentration through streaming instabilities, which occur in mid-plane layers of dust-to-gas ratios on the order of unity, as overdense filaments of particles catch more and more particles drifting in from further out (*Youdin and Goodman*, 2005; *Johansen and Youdin*, 2007; *Johansen et al.*, 2009a; *Bai and Stone*, 2010). *Youdin and Goodman* (2005) performed a linear stability analysis of the equilibrium flow of gas and particles in the presence of a radial pressure gradient. The free energy in the streaming motion of the two components — particles orbiting faster than the gas and drifting inward, pushing the gas outward — forms the base of an unstable mode with growth rate depending on the particle friction time and on the local dust-to-gas ratio. Higher friction time and higher dust-to-gas ratio lead to shorter growth time. The growth time decreases by 1 to 2 orders of magnitude around a dust-to-gas ratio on the order of unity, which shows the importance of reaching this threshold mass-loading by sedimentation.

The streaming instability can concentrate particles up to several thousand times the local gas density in filamentary structures (*Johansen et al.*, 2012a). When particle densities reach the Roche density, the overdense filaments fragment gravitationally into a number of bound clumps that contract to form planetesimals (*Johansen et al.*, 2007, 2009a; *Nesvorný et al.*, 2010). The characteristic sizes of planetesimals formed by a GI of overdense regions seems to be similar to or smaller than the dwarf planet Ceres, the largest asteroid in the asteroid belt. Smaller planetesimals form as well, particularly in high-resolution simulations (*Johansen et al.*, 2012a). However, for pebble accretion the largest planetesimal is the most interesting, since the accretion efficiency increases strongly with increasing size. Nevertheless, a planetesimal with the mass of Ceres is still a factor of 10 to 100 too light to undergo pebble accretion from the full Hill sphere. Hence suggestion (1) above, the direct formation of very massive planetesimals, is not supported by the simulations. A more likely scenario is a combination of suggestions (2) and (3), i.e., that a large Ceres-mass planetesimal grows in mass by a factor of 100 by accreting other planetesimals through its gravitational cross section and by accreting pebbles from the Bondi radius. This intermediate step between planetesimals and the seeds of the cores could act as a bottleneck mechanism that prevents too many planetesimals from reaching the pebble-accretion stage. Hence only a small number of cores are formed from the largest and luckiest of the planetesimal population, in agreement with the low number of GPs in the solar system. The growth from 100-km-scale planetesimals to 1000-km-scale planetary seeds, capable of rapid pebble

accretion from the entire Hill sphere, has nevertheless not been yet modeled in detail and represents an important priority for the future. The formation of too many competing cores could drastically reduce the efficiency of this growth path. *Nesvorný and Morbidelli* (2012) nevertheless argued for an initial number of no more than 10 growing cores that were subsequently reduced in number by collisions and ejections.

6. BROWN DWARF VERSUS GIANT PLANET: QUESTIONING THE INTERNATIONAL ASTRONOMICAL UNION DEFINITION

6.1. Hot Start Versus Cold Start

Marley et al. (2007) suggested that young GPs formed by CA in a protoplanetary disk should have a much lower entropy content at young ages than objects of the same mass and age formed by gravitational collapse, either from a prestellar core or by disk instability. Consequently, the first type of these objects should be significantly smaller, cooler, and (about 100×) fainter at young ages than the second type. This gave rise to the so-called "cold-start" observational signature, characteristic of young GPs formed by CA, vs. the "hot-start" one, typical of young objects, GPs or BDs, formed by collapse to distinguish these objects from their distinct formation mechanisms. This suggestion, however, directly relies on two assumptions. First, the assumption that GPs formed by CA have a low-entropy content implies that all the energy of the accretion shock through which most of the planetary mass is processed is radiated away, leaving the internal energy content of the nascent planet unaffected. This is characteristic of a so-called supercritical shock. In that case, the radiative losses at the accretion shock act as a sink of entropy. This assumption, however, has never been verified. In fact, a proper treatment of the accretion shock at the onset of prestellar core formation, the so-called second Larson's core, shows that the shock in that case is subcritical, with essentially all the energy from the infalling material being absorbed by the stellar embryo (*Vaytet et al.*, 2013; *Tomida et al.*, 2013; *Bate et al.*, 2014). Although the two formation conditions differ in several ways, they share enough common processes to at least question the assumption of a supercritical shock for planet formation. The second underlying assumption about this scenario is the assumption that BDs form with a high-entropy content. Again, such an assumption is not necessarily correct. Initial conditions for BD formation are rather uncertain. Figure 2 (see also *Mordasini et al.*, 2012a; *Mordasini*, 2013; *Spiegel and Burrows*, 2012) compares the early evolution of the luminosity for a 5-M_{Jup} object under several assumptions. The blue solid and dot-dashed lines portray GP early evolution tracks kindly provided by C. Mordasini assuming that either the accretion energy is entirely converted to radiation, i.e., a supercritical shock condition as in *Marley et al.* (2007), or that this energy is absorbed by the planet, i.e., a subcritical shock with no radiative loss at the shock. The former case yields a low-entropy content and therefore a low luminosity at the end of the accretion shock, while in the second case the luminosity slowly decreases from its value at the end of the shock, 2 orders of magnitude brighter, for several million years. The red long-dashed and short-dashed lines correspond to the early evolution of a BD (*Baraffe et al., 2003*) assuming an initial radius $R_i \sim 8\ R_{Jup}(\simeq 0.8\ R_\odot)$ (short dash) or $R_i \sim 1.6\ R_{Jup}$ (long dash). This corresponds to specific entropies $\widetilde{S} \simeq 1.1 \times 10^9$ and $\simeq 8.0 \times 10^8$ erg g^{-1} K^{-1}, respectively.

As seen in Fig. 2, depending on the outcome of the accretion shock episode in one case and on the initial radius (thus entropy content) in the second case, and given the long Kelvin-Helmholtz timescale for such low-mass objects (several million years), young planets formed by CA can be as bright and even brighter than young BDs for several million years. (Magneto)radiation-hydrodynamics calculations of the collapse of prestellar cores (*Tomida et al., 2013; Vaytet et al., 2013*) seem to exclude the above rather extreme "cold-start" initial condition for a proto-BD, even though calculations have not been performed yet for such low masses. They suggest initial entropy contents and radii closer to the "hot-start" case, the outer region of the protostellar core having been heated by the shock, attaining a higher entropy. Hot-start conditions for core-accretion GPs, probably between the two extreme cases above, however, are presently not excluded. Interestingly enough, measured temperatures and luminosities of some directly imaged exoplanets, notably β Pic b and κ And b, are consistent with

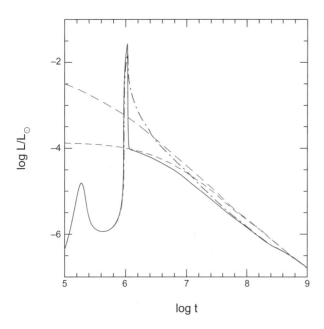

Fig. 2. Early evolution of a 5-M$_{Jup}$ object according to different formation scenarios. Blue: Object formed by CA assuming either a supercritical shock (solid line) or a subcritical shock (dash-dot line) at the end of the accretion process, after *Mordasini et al.* (2012a). Red: Object formed by gravitational collapse for two arbitrary initial conditions (see text), after *Baraffe et al.* (2003).

hot-start-like conditions and seem, so far, to exclude the coldest range of initial conditions for these objects (*Currie et al., 2013; Marleau and Cumming, 2013*). Therefore, at least in the absence of a better knowledge of the accretion shock condition at the end of the CA process, the cold-start vs. hot-start argument does not provide a reliable diagnostic to distinguish CA- from GI-formed objects. Certainly there is no one-to-one correspondence between cold-start vs. hot-start conditions and CA vs. disk or core collapse.

6.2. Deuterium Burning

The distinction between BDs and GPs has become a topic of intense debate these days. In 2003, the IAU has adopted the D-burning minimum mass, ~10 M$_{Jup}$, as the official distinction between the two types of objects. We have discussed this limit in previous reviews (*Chabrier, 2003; Chabrier et al., 2007*) and shown that it does not rely on any robust physical justification and is a purely semantic definition. Deuterium burning has no impact on star formation and a negligible impact on stellar/BD evolution (*Chabrier and Baraffe, 2000*). This is in stark contrast with the lifetime impact of H-burning, making H-burning a genuine physical mechanism for distinguishing objects in nuclear equilibrium for most of their lifetime, defined as stars, from objects that lack significant support against gravitational contraction and continue to contract throughout their lifetimes, defined as BDs.

As mentioned in section 3.2, one of the strongest arguments against D-burning to distinguish planets from BDs is 2M 1207 b (*Chauvin et al., 2005*), which is a ~4-M$_{Jup}$ companion to a ~20-M$_{Jup}$ BD. Thus, the companion is firmly in the non-D-burning mass regime, but with a system mass ratio of 20%, the couple appears to be best described as an extension of the BD binary population rather than a planetary system. In contrast, it is presently not excluded that genuine planets formed by CA, characterized by a significant heavy element enrichment, reach masses above the D-burning limit and thus ignite D-burning in their core (*Baraffe et al., 2008; Mollière and Mordasini, 2012; Bodenheimer et al., 2013*) (see the chapter by Baraffe et al. in this volume).

6.3. Brown Dwarf/Planet Overlapping Mass Regime

As mentioned in section 2.1, there is now ample evidence for the existence of free-floating BDs with masses on the order of a few Jupiter masses in (low-extinction) young clusters and in the field (see, e.g., *Caballero et al., 2007*), with a mass distribution consistent with the extension of the stellar IMF into the BD regime. The BD and planet mass domains thus clearly overlap, arguing against a clear mass separation. The fundamentally different mass distribution of exoplanets detected by radial velocity surveys, with the mass function rising below ~30 M$_{Jup}$ (*Mayor et al., 2011*), in stark contrast with the BD mass distribution (see section 2.1), clearly suggests two distinct populations with different origins.

Of particularly note at this stage are the transiting objects HAT-P-2b, with a mass of 9 M_{Jup} (*Bakos et al.,* 2007), and HAT-P-20b, with a mass of 7.2 M_{Jup} (*Bakos et al.,* 2011). Both objects are too dense to be BDs. Assuming that the observational error bars on the radius are reliable, and given the age inferred for these systems, the observed mass-radius determinations imply significant enrichment in heavy material, revealing their planetary nature (*Leconte et al.,* 2009, 2011). This shows that planets at least 9× more massive than Jupiter, close to the D-burning limit, can form according to the core-accretion scenario, possibly from the merging of lower-mass planet embryos. Note that while massive objects like HAT-P-20b approach the upper limit of the mass distribution predicted by the CA scenario (*Mordasini et al.,* 2012b), its large metal enrichment [$M_Z \sim 340\ M_\oplus$ (*Leconte et al.,* 2011)] certainly excludes formation by gravitational collapse.

According to the arguments developed in this and previous sections, the present IAU definition, based on a clear-cut mass limit between BDs and planets, is clearly incorrect and confusing and should be abandonned. We come back to this point below.

7. CONCLUSION

Even though it is probably still premature to reach definitive conclusions about BD and GP formation and we must remain open to all possibilities, the confrontation of the various theories with the observational constraints described in this review suggests some reasonably sound conclusions: the ability of BDs to form in isolation and in wide binaries; the similarity of BD number-density in low- (e.g., Taurus) and high-density environments, indicating no significant dependence of BD abundances upon stellar density; the emerging observations of isolated proto-BDs and pre-BD cores; the many observational properties shared by young BDs and young stars; the observed close similarity between the prestellar/BD CMF and the final stellar/BD IMF; and the consistency between the BD IMF and the natural extension of the same stellar IMF down to nearly the bottom of the BD domain. All these points provide evidence that dynamical interactions and dense cluster environments, disk fragmentation, or photoionizing radiation are not required for BD formation. In contrast, all these properties are consistent with BD formation being a natural scaled-down version of star formation by the turbulence-induced fragmentation of molecular clumps, leading to the formation of prestellar and pre-BD cores. It is not excluded, however, that, under some specific circumstances (very dense environment, very massive disks), the other aforementioned mechanisms might play some role, but in the absence of clear observational evidence for this so far, they seem unlikely to be the dominant mechanisms for star and BD formation.

Conversely, the overwhelming majority of planet discoveries are consistent with planet formation by CA. As examined in section 5.3, this scenario might also explain planet formation at large orbital distances, not to mention the possibility of such objects forming by planet scattering or outward migration. Here again, alternative scenarios such as disk fragmentation might occur in some places, notably in massive circumbinary disks, and thus explain some fraction of the planet population [possibly ~10% or so (e.g., *Vorobyov,* 2013)], but they can hardly be considered as dominant scenarios for planet formation. Hybrid scenarios invoking both gravitational fragmentation at large orbital distances followed by inward migration, invoking even in some cases evaporation (see the chapter by Helled et al. in this volume), seem to raise even more problems than they bring solutions, as migration can only exacerbate problems with the GI. Not only does one need to form planets by GI, but one also needs to prevent them from migrating rapidly all the way into the star (*Vorobyov and Basu,* 2006b; *Machida et al.,* 2011; *Baruteau et al.,* 2011; *Vorobyov,* 2013). Migration, however, is very likely and fast given that, in that case, the disk mass must be very high, a requirement for GI to occur. If planets indeed form this way, this requires very fine tuning conditions, making the branching ratio for this route very small.

Finally, as discussed in section 6, there is ample evidence that the planet and BD domains overlap and that D-burning plays no particular role in the formation process. There is now growing evidence, possibly including the WISE survey for the field population (see section 2.1), for the existence of non-D-burning free-floating BDs and, conversely, no physical arguments against the possibility for genuine planets to ignite D-burning in their core. This again shows that the IAU definition has no scientific justification and only brings about confusion for not only scientists but also the media. It also argues against the use of specific appelations for free-floating objects below the D-burning limit, these latter being simply non D-burning BDs.

Given the arguments examined along this review, it seems rather safe to argue that BDs and GPs represent two distinct populations of astrophysical bodies that arise *dominantly* from two different formation mechanisms. While BDs appear to form preferentially like stars, from the gravoturbulent fragmentation of a parent (possibly filamentary) molecular clump, GPs arguably form essentially by CA in a protoplanetary disk, i.e., from the growth of solids (planetesimals, pebbles) that eventually yield the accretion of a surrounding gas-rich H/He envelope. So the very definition of a BD or a GP is intrinsically, tightly linked to its formation mechanism. As briefly discussed below and in section 3.2, the latter should leave imprints that might be observationally detectable. As examined in section 6.1, the luminosity or effective temperature at early ages, however, cannot be used as a diagnostic to distinguish between these two populations. On the other hand, we argue in this review that planets are necessarily companions of a central, significantly more massive object. Consequently, free-floating objects down to a few Jupiter masses can rather unambiguously be identified as genuine (non-D-burning) BDs, and one should stop giving them different names, which simply adds to the confusion. This is the most direct observational distinction between the two types of objects. Ejected planets probably exist and weaken this statement, but they are unlikely to represent a significant fraction of the population. The diagnostic is less clear for wide companions to stars, except if the mass ratio can safely exclude one of the two possibilities. According to the distinct formation scenarios, planets should have a substantial enrichment in heavy elements compared with their parent star, as observed for our own solar GPs, whereas BDs of the same mass should have the same composition as their parent cloud. As suggested in *Chabrier et al.* (2007) and *Fortney et al.* (2008b) and examined in section 3.2, GPs should bear the signature of this enhanced metallicity in their atmosphere. Spectroscopy or even photometry of atmospheric chemical abundances, notably metal-dominated compounds such as CO or CO_2, for example, may provide clues about the formation mechanism and therefore help identify the nature of the object. In the absence (so far) of clear observational diagnostics (mean density, atmospheric abundances, oblateness, etc.) to get clues about these formation conditions, the very nature of some of these objects might remain uncertain. As frustrating as this may sound, we will have to admit that such uncertainties presently remain, as ignorance is sometimes part of science.

Acknowledgments. The authors are grateful to the referee, A. Morbidelli, for a careful reading of the manuscript and constructive comments. The research leading to these results has received funding from the European Research Council under the European Community 7th Framework Programme (FP7/2007-2013 Grant Agreement no. 247060).

REFERENCES

Albrecht S. et al. (2012) *Astrophys. J., 757*, 18.
Albrecht S. et al. (2013) *Astrophys. J., 771*, 11.
Aller K. M. et al. (2013) *Astrophys. J., 773*, 63.
Anderson D. R. et al. (2011) *Mon. Not. R. Astron. Soc., 416*, 2108.
André P. et al. (2009) In *Structure Formation in Astrophysics* (G. Chabrier, ed.), p. 254. Cambridge Univ., Cambridge.
André P. et al. (2010) *Astron. Astrophys., 518*, L102.
André P. et al. (2012) *Science, 337*, 1196.
Apai D. et al. (2008) *Astrophys. J., 672*, 1196.
Armitage P. J. and Hansen B. M. S. (1999) *Nature, 402*, 633.
Bai X.-N. and Stone J. M. (2010) *Astrophys. J., 722*, 1437.
Bakos G. Á. et al. (2007) *Astrophys. J., 670*, 826.
Bakos G. Á. et al. (2011) *Astrophys. J., 742*, 116.
Baraffe I. et al. (2003) *Astron. Astrophys., 402*, 701.
Baraffe I. et al. (2008) *Astron. Astrophys., 482*, 315.
Barge P. and Sommeria J. (1995) *Astron. Astrophys., 295*, L1.
Baruteau C. et al. (2011) *Mon. Not. R. Astron. Soc., 416*, 1971.
Basu S. (1998) *Astrophys. J., 509*, 229.
Batalha N. M. et al. (2013) *Astrophys. J. Suppl., 204*, 24.
Bate M. R. (2012) *Mon. Not. R. Astron. Soc., 419*, 3115.
Bate M. R. et al. (2014) *Mon. Not. R. Astron. Soc., 437*, 77.
Bayo A. et al. (2011) *Astron. Astrophys., 536*, A63.
Beuzit J.-L. et al. (2008) In *Ground-Based and Airborne Instrumentation for Astronomy II* (I. McLean and M. Casali, eds.) SPIE Conf. Series 7014, Bellingham, Washington.
Biller B. A. et al. (2007) *Astrophys. J. Suppl., 173*, 143.
Birnstiel T. et al. (2010) *Astron. Astrophys., 513*, A79.
Bodenheimer P. (1995) *Annu. Rev. Astron. Astrophys., 33*, 199.
Bodenheimer P. and Pollack J. B. (1986) *Icarus, 67*, 391.
Bodenheimer P. et al. (2013) *Astrophys. J., 770*, 120.
Boley A. C. and Durisen R. H. (2006) *Astrophys. J., 641*, 534.

Boley A. C. et al. (2007) *Astrophys. J. Lett., 656*, L89.
Bonnell I. A. and Bate M. R. (1994) *Mon. Not. R. Astron. Soc., 271*, 999.
Bontemps S. et al. (2010) *Astron. Astrophys., 524*, A18.
Boss A. P. (1997) *Science, 276*, 1836.
Boss A. P. (2002) *Astrophys. J., 576*, 462.
Boss A. P. (2006) *Astrophys. J. Lett., 637*, L137.
Boss A. P. (2007) *Astrophys. J. Lett., 661*, L73.
Bowler B. P. et al. (2010) *Astrophys. J., 723*, 850.
Bowler B. P. et al. (2013) *Astrophys. J., 774*, 55.
Brauer F. et al. (2007) *Astron. Astrophys., 469*, 1169.
Bressert E. et al. (2010) *Mon. Not. R. Astron. Soc., 409*, L54.
Buchhave L. A. et al. (2012) *Nature, 486*, 375.
Burgasser A. J. et al. (2004) *Astron. J., 127*, 2856.
Burningham B. et al. (2010) *Mon. Not. R. Astron. Soc., 406*, 1885.
Burrows A. et al. (2003) *Astrophys. J., 596*, 587.
Caballero J. A. et al. (2007) *Astron. Astrophys., 470*, 903.
Cai K. et al. (2010) *Astrophys. J. Lett., 716*, L176.
Cameron A. G. W. (1978) *Moon Planets, 18*, 5.
Carson J. et al. (2013) *Astrophys. J. Lett., 763*, L32.
Carter J. A. et al. (2012) *Science, 337*, 556.
Chabrier G. (2002) *Astrophys. J., 567*, 304.
Chabrier G. (2003) *Publ. Astron. Soc. Pac., 115*, 763.
Chabrier G. (2005) In *The Initial Mass Function 50 Years Later* (E. Corbelli et al., eds.), p. 41. Astrophys. Space Sci. Library, Vol. 327, Kluwer, Dordrecht.
Chabrier G. and Baraffe I. (2000) *Annu. Rev. Astron. Astrophys., 38*, 337.
Chabrier G. and Hennebelle P. (2010) *Astrophys. J. Lett., 725*, L79.
Chabrier G. et al. (2000a) *Astrophys. J. Lett., 542*, L119.
Chabrier G. et al. (2000b) *Astrophys. J., 542*, 464.
Chabrier G. et al. (2007) In *Protostars and Planets V* (B. Reipurth et al., eds.), pp. 623–638. Univ. of Arizona, Tucson.
Chaplin W. J. et al. (2013) *Astrophys. J., 766*, 101.
Chauvin G. et al. (2005) *Astron. Astrophys., 438*, L25.
Chauvin G. et al. (2010) *Astron. Astrophys., 509*, A52.
Commerçon B. et al. (2010) *Astron. Astrophys., 510*, L3.
Commerçon B. et al. (2011) *Astrophys. J. Lett., 742*, L9.
Cruz K. L. et al. (2007) *Astron. J., 133*, 439.
Cumming A. et al. (2008) *Publ. Astron. Soc. Pac., 120*, 531.
Currie T. et al. (2013) *Astrophys. J., 776*, 15.
Dapp W. B. et al. (2012) *Astron. Astrophys., 541*, A35.
Delorme P. et al. (2012a) *Astron. Astrophys., 548*, A26.
Delorme P. et al. (2012b) *Astron. Astrophys., 539*, A72.
Delorme P. et al. (2013) *Astron. Astrophys., 553*, L5.
Dodson-Robinson S. E. et al. (2009) *Astrophys. J., 707*, 79.
Dones L. and Tremaine S. (1993) *Icarus, 103*, 67.
Duchêne G. and Kraus A. (2013) *Annu. Rev. Astron. Astrophys., 51*, 269.
Duchêne G. et al. (2003) *Astrophys. J., 592*, 288.
Dullemond C. P. et al. (2009) In *Structure Formation in Astrophysics* (G. Chabrier, ed.), p. 350. Cambridge Univ., Cambridge.
Dunham M. M. et al. (2008) *Astrophys. J. Suppl., 179*, 249.
Durisen R. H. et al. (2007) In *Protostars and Planets V* (B. Reipurth et al., eds.), pp. 607–622. Univ. of Arizona, Tucson.
Fabrycky D. C. and Murray-Clay R. A. (2010) *Astrophys. J., 710*, 1408.
Fabrycky D. C. et al. (2012) *ArXiv e-prints*, arXiv:1202.6328.
Faherty J. K. et al. (2011) *Astron. J., 141*, 71.
Fischer D. A. and Valenti J. (2005) *Astrophys. J., 622*, 1102.
Fortney J. J. et al. (2007) *Astrophys. J., 659*, 1661.
Fortney J. J. et al. (2008a) *Astrophys. J., 683*, 1104.
Fortney J. J. et al. (2008b) *Astrophys. J., 683*, 1104.
Gammie C. F. (2001) *Astrophys. J., 553*, 174.
Goldreich P. and Lynden-Bell D. (1965) *Mon. Not. R. Astron. Soc., 130*, 97.
Goldreich P. et al. (2004) *Annu. Rev. Astron. Astrophys., 42*, 549.
Gould et al. (2010) *Astrophys. J., 720*, 1073.
Grether D. and Lineweaver C. H. (2006) *Astrophys. J., 640*, 1051.
Guillot T. (2005) *Annu. Rev. Earth Planet. Sci., 33*, 493.
Gutermuth R. A. et al. (2009) *Astrophys. J. Suppl., 184*, 18.
Haisch K. E. Jr. et al. (2001) *Astrophys. J. Lett., 553*, L153.
Hartman J. D. et al. (2011) *Astrophys. J., 742*, 59.
Hekker S. et al. (2008) *Astron. Astrophys., 480*, 215.
Hennebelle P. and Chabrier G. (2008) *Astrophys. J., 684*, 395.
Hennebelle P. and Chabrier G. (2009) *Astrophys. J., 702*, 1428.
Hennebelle P. and Chabrier G. (2011) In *Computational Star Formation* (J. Alves et al., eds.), pp. 159–168. IAU Symp. 270, Cambridge Univ., Cambridge.
Hennebelle P. and Chabrier G. (2013) *Astrophys. J., 770*, 150.

Hennebelle P. and Ciardi A. (2009) *Astron. Astrophys., 506,* L29.
Hennebelle P. and Falgarone E. (2012) *Astron. Astrophys. Rev., 20,* 55.
Hennebelle P. and Teyssier R. (2008) *Astron. Astrophys., 477,* 25.
Hennebelle P. et al. (2011) *Astron. Astrophys., 528,* A72.
Heyer M. et al. (2009) *Astrophys. J., 699,* 1092.
Hirano T. et al. (2012) *Astrophys. J., 756,* 66.
Hopkins P. F. (2012) *Mon. Not. R. Astron. Soc., 423,* 2037.
Howard A. W. et al. (2010) *Science, 330,* 653.
Huard T. L. et al. (2006) *Astrophys. J., 640,* 391.
Ikoma M. et al. (2000) *Astrophys. J., 537,* 1013.
Ikoma M. et al. (2001) *Astrophys. J., 553,* 999.
Inutsuka S.-I. (2001) *Astrophys. J. Lett., 559,* L149.
Janson M. et al. (2010) *Astrophys. J. Lett., 710,* L35.
Janson M. et al. (2011) *Astrophys. J., 736,* 89.
Janson M. et al. (2012) *Astrophys. J., 745,* 4.
Janson M. et al. (2013) *Astrophys. J., 773,* 73.
Jeffries R. D. (2012) In *Low Mass Stars and the Transition Stars/ Brown Dwarfs — Evry Schatzman School on Stellar Physics XXIII* (C. Reylé et al., eds.), pp. 45–89. EAS Publ. Series, Vol. 57, Cambridge Univ., Cambridge.
Joergens V. (2006) *Astron. Astrophys., 448,* 655.
Joergens V. (2008) *Astron. Astrophys., 492,* 545.
Joergens V. et al. (2013) *Astron. Astrophys., 558,* L7.
Johansen A. and Lacerda P. (2010) *Mon. Not. R. Astron. Soc., 404,* 475.
Johansen A. and Youdin A. (2007) *Astrophys. J., 662,* 627.
Johansen A. et al. (2007) *Nature, 448,* 1022.
Johansen A. et al. (2009a) *Astrophys. J. Lett., 704,* L75.
Johansen A. et al. (2009b) *Astrophys. J., 697,* 1269.
Johansen A. et al. (2012a) *Astron. Astrophys., 537,* A125.
Johansen A. et al. (2012b) *Astrophys. J., 758,* 39.
Johnson J. A. et al. (2010) *Publ. Astron. Soc. Pac., 122,* 905.
Johnson J. A. et al. (2011) *Astrophys. J., 730,* 79.
Joos M. et al. (2012) *Astron. Astrophys., 543,* A128.
Joos M. et al. (2013) *Astron. Astrophys., 554,* A17.
Jumper P. H. and Fisher R. T. (2013) *Astrophys. J., 769,* 9.
Kary D. M. et al. (1993) *Icarus, 106,* 288.
Kasper M. et al. (2007) *Astron. Astrophys., 472,* 321.
Kauffmann J. et al. (2011) *Mon. Not. R. Astron. Soc., 416,* 2341.
Kenyon S. J. and Bromley B. C. (2009) *Astrophys. J. Lett., 690,* L140.
Kipping D. M. et al. (2014) *Astrophys. J., 784,* 28.
Kirkpatrick J. D. et al. (2011) *Astrophys. J. Suppl., 197,* 19.
Klahr H. H. and Bodenheimer P. (2003) *Astrophys. J., 582,* 869.
Konopacky Q. M. et al. (2013) *Science, 339,* 1398.
Könyves et al. (2010) *Astron. Astrophys., 518,* L106.
Kratter K. M. and Murray-Clay R. A. (2011) *Astrophys. J., 740,* 1.
Kraus A. L. and Hillenbrand L. A. (2012) *Astrophys. J., 757,* 141.
Kraus A. L. and Ireland M. J. (2012) *Astrophys. J., 745,* 5.
Kraus A. L. et al. (2011) *Astrophys. J., 731,* 8.
Kroupa P. (2001) *Mon. Not. R. Astron. Soc., 322,* 231.
Kroupa P. and Bouvier J. (2003) *Mon. Not. R. Astron. Soc., 346,* 369.
Krumholz M. R. et al. (2007) *Astrophys. J., 656,* 959.
Kuiper G. P. (1951) *Proc. Natl. Acad. Sci., 37,* 1.
Kurosawa R. et al. (2006) *Mon. Not. R. Astron. Soc., 372,* 1879.
Kuzuhara M. et al. (2013) *Astrophys. J., 774,* 11.
Lafrenière D. et al. (2007) *Astrophys. J., 670,* 1367.
Lafrenière D. et al. (2008) *Astrophys. J. Lett., 689,* L153.
Lagrange A.-M. et al. (2009) *Astron. Astrophys., 493,* L21.
Lagrange A.-M. et al. (2010) *Science, 329,* 57.
Lambrechts M. and Johansen A. (2012) *Astron. Astrophys., 544,* A32.
Larson R. B. (1981) *Mon. Not. R. Astron. Soc., 194,* 809.
Laughlin G. and Bodenheimer P. (1994) *Astrophys. J., 436,* 335.
Leconte J. et al. (2009) *Astron. Astrophys., 506,* 385.
Leconte J. et al. (2011) *EPJ Web Conf., 11,* 3004.
Lee C.-F. et al. (2009a) *Astrophys. J., 699,* 1584.
Lee C. W. et al. (2009b) *Astrophys. J., 693,* 1290.
Lee C. W. et al. (2013) *Astrophys. J., 777,* 50.
Lee J.-E. and Kim J. (2009) *Astrophys. J. Lett., 699,* L108.
Levison H. F. et al. (2010) *Astron. J., 139,* 1297.
Li Z.-Y. et al. (2013) *Astrophys. J., 774,* 82.
Lissauer J. J. et al. (2011a) *Nature, 470,* 53.
Lissauer J. J. et al. (2011b) *Astrophys. J. Suppl., 197,* 8.
Lloyd J. P. (2011) *Astrophys. J. Lett., 739,* L49.
Lodders K. and Fegley B. (2002) *Icarus, 155,* 393.
Lodieu N. et al. (2011) *Astron. Astrophys., 527,* A24.
Lommen D. et al. (2007) *Astron. Astrophys., 462,* 211.
Looney L. W. et al. (2000) *Astrophys. J., 529,* 477.

Lovis C. and Mayor M. (2007) *Astron. Astrophys., 472,* 657.
Lowrance P. J. et al. (2005) *Astron. J., 130,* 1845.
Lucas P. W. et al. (2006) *Mon. Not. R. Astron. Soc., 373,* L60.
Luhman K. L. (2012) *Annu. Rev. Astron. Astrophys., 50,* 65.
Luhman K. L. and Muench A. A. (2008) *Astrophys. J., 684,* 654.
Luhman K. et al. (2004) *Astrophys. J., 617,* 565.
Luhman K. L. et al. (2005) *Astrophys. J. Lett., 635,* L93.
Luhman K. L. et al. (2007) In *Protostars and Planets V* (B. Reipurth et al., eds.), pp. 443–457. Univ. of Arizona, Tucson.
Luhman K. L. et al. (2009) *Astrophys. J., 691,* 1265.
Luhman K. L. et al. (2011) *Astrophys. J. Lett., 730,* L9.
Mac Low M.-M. and Klessen R. S. (2004) *Rev. Mod. Phys., 76,* 125.
Machida M. N. (2008) *Astrophys. J. Lett., 682,* L1.
Machida M. N. and Matsumoto T. (2011) *Mon. Not. R. Astron. Soc., 413,* 2767.
Machida M. N. et al. (2010) *Astrophys. J., 724,* 1006.
Machida M. N. et al. (2011) *Astrophys. J., 729,* 42.
Macintosh B. A. et al. (2008) In *Adaptive Optics Systems* (N. Hubin et al., eds.), p. 701518. SPIE Conf. Series 7015, Bellingham, Washington.
Madhusudhan N. et al. (2011) *Astrophys. J., 743,* 191.
Mamajek E. E. (2005) *Astrophys. J., 634,* 1385.
Marleau G.-D. and Cumming A. (2013) *ArXiv e-prints,* arXiv:1302.1517.
Marley M. S. et al. (2007) *Astrophys. J., 655,* 541.
Marois C. et al. (2008a) *Science, 322,* 1348.
Marois C. et al. (2008b) *Science, 322,* 1348.
Marois C. et al. (2010) *Nature, 468,* 1080.
Marsh K. A. et al. (2010) *Astrophys. J. Lett., 709,* L158.
Masciadri E. et al. (2005) *Astrophys. J., 625,* 1004.
Matzner C. D. and Levin Y. (2005) *Astrophys. J., 628,* 817.
Maury A. J. et al. (2010) *Astron. Astrophys., 512,* A40.
Mayer L. et al. (2002) *Science, 298,* 1756.
Mayer L. et al. (2004) *Astrophys. J., 609,* 1045.
Mayer L. et al. (2007) *Astrophys. J. Lett., 661,* L77.
Mayor M. et al. (2011) *ArXiv e-prints,* arXiv:1109.2497.
McKee C. F. and Ostriker E. C. (2007) *Annu. Rev. Astron. Astrophys., 45,* 565.
McLaughlin D. B. (1924) *Astrophys. J., 60,* 22.
Men'shchikov A. et al. (2010) *Astron. Astrophys., 518,* L103.
Meru F. and Bate M. R. (2011) *Mon. Not. R. Astron. Soc., 411,* L1.
Meru F. and Bate M. R. (2012) *Mon. Not. R. Astron. Soc., 427,* 2022.
Metchev S. A. et al. (2008) *Astrophys. J., 676,* 1281.
Mizuno H. (1980) *Progr. Theor. Phys., 64,* 544.
Mollière P. and Mordasini C. (2012) *Astron. Astrophys., 547,* A105.
Morbidelli A. and Nesvorny D. (2012) *Astron. Astrophys., 546,* A18.
Mordasini C. (2013) *Astron. Astrophys., 558,* A113.
Mordasini C. et al. (2012a) *Astron. Astrophys., 547,* A111.
Mordasini C. et al. (2012b) *Astron. Astrophys., 541,* A97.
Mortier A. et al. (2013) *Astron. Astrophys., 551,* A112.
Muench A. A. et al. (2007) *Astrophys. J., 671,* 1820.
Murillo N. M. et al. (2013) *Astron. Astrophys., 560,* A103.
Muto T. and Inutsuka S.-I. (2009) *Astrophys. J., 695,* 1132.
Nesvorný D. and Morbidelli A. (2012) *Astron. J., 144,* 117.
Nesvorný D. et al. (2010) *Astron. J., 140,* 785.
Nielsen E. L. and Close L. M. (2010) *Astrophys. J., 717,* 878.
Nielsen E. L. et al. (2013) *Astrophys. J., 776,* 4.
Öberg K. I. et al. (2011) *Astrophys. J. Lett., 743,* L16.
Offner S. S. R. et al. (2009) *Astrophys. J., 703,* 131.
Ofir A. et al. (2013) *ArXiv e-prints,* arXiv:1310.2064.
Okuzumi S. et al. (2012) *Astrophys. J., 752,* 106.
Oppenheimer B. R. et al. (2013) *Astrophys. J., 768,* 24.
Ormel C.W. and Klahr H. H. (2010) *Astron. Astrophys., 520,* A43.
Paardekooper S.-J. (2012) *Mon. Not. R. Astron. Soc., 421,* 3286.
Paardekooper S.-J. and Mellema G. (2006) *Astron. Astrophys., 453,* 1129.
Paardekooper S.-J. et al. (2011) *Mon. Not. R. Astron. Soc., 416,* L65.
Padoan P. and Nordlund Å. (2002) *Astrophys. J., 576,* 870.
Palau A. et al. (2012) *Mon. Not. R. Astron. Soc., 424,* 2778.
Parker R. J. et al. (2011) *Mon. Not. R. Astron. Soc., 412,* 2489.
Pérez L. M. et al. (2012) *Astrophys. J. Lett., 760,* L17.
Perri F. and Cameron A. G. W. (1974) *Icarus, 22,* 416.
Peters M. A. et al. (2012) In *Ground-Based and Airborne Instrumentation for Astronomy IV* (I. S. McLean et al., eds.), p. 84467U. SPIE Conf. Series 8446, Bellingham, Washington.
Peters T. et al. (2011) *Astrophys. J., 729,* 72.

Pollack J. B. et al. (1996) *Icarus, 124,* 62.
Price D. J. and Bate M. R. (2009) *Mon. Not. R. Astron. Soc., 398,* 33.
Queloz D. et al. (2000) *Astron. Astrophys., 359,* L13.
Radigan J. et al. (2009) *Astrophys. J., 698,* 405.
Rafikov R. R. (2001) *Astron. J., 122,* 2713.
Rafikov R. R. (2003a) *Astron. J., 125,* 922.
Rafikov R. R. (2003b) *Astron. J., 125,* 942.
Rafikov R. R. (2004) *Astron. J., 128,* 1348.
Rafikov R. R. (2005) *Astrophys. J. Lett., 621,* L69.
Rafikov R. R. (2006) *Astrophys. J., 648,* 666.
Rafikov R. R. (2007) *Astrophys. J., 662,* 642.
Rafikov R. R. (2009) *Astrophys. J., 704,* 281.
Rafikov R. R. (2011) *Astrophys. J., 727,* 86.
Reipurth B. and Clarke C. (2001) *Astron. J., 122,* 432.
Reylé C. et al. (2010) *Astron. Astrophys., 522,* A112.
Rice W. K. M. et al. (2005) *Mon. Not. R. Astron. Soc., 364,* L56.
Rodmann J. et al. (2006) *Astron. Astrophys., 446,* 211.
Rodríguez L. F. et al. (2005) *Astrophys. J. Lett., 621,* L133.
Ros K. and Johansen A. (2013) *Astron. Astrophys., 552,* A137.
Rossiter R. A. (1924) *Astrophys. J., 60,* 15.
Safronov V. S. (1960) *Ann. Astrophys., 23,* 979.
Salpeter E. E. (1955) *Astrophys. J., 121,* 161.
Sanchis-Ojeda R. et al. (2012) *Nature, 487,* 449.
Santos N. C. et al. (2004) *Astron. Astrophys., 415,* 1153.
Sato B. et al. (2005) *Astrophys. J., 633,* 465.
Saumon D. and Marley M. S. (2008) *Astrophys. J., 689,* 1327.
Scholz A. et al. (2012a) *Astrophys. J., 744,* 6.
Scholz A. et al. (2012b) *Astrophys. J., 756,* 24.
Seifried D. et al. (2012) *Mon. Not. R. Astron. Soc., 423,* L40.
Seifried D. et al. (2013) *Mon. Not. R. Astron. Soc., 432,* 3320.
Smith R. J. et al. (2009) *Mon. Not. R. Astron. Soc., 396,* 830.

Sousa S. G. et al. (2008) *Astron. Astrophys., 487,* 373.
Spiegel D. S. and Burrows A. (2012) *Astrophys. J., 745,* 174.
Stamatellos D. and Whitworth A. P. (2009) *Mon. Not. R. Astron. Soc., 392,* 413.
Stamatellos D. et al. (2007) *Mon. Not. R. Astron. Soc., 382,* L30.
Stevenson D. J. (1982) *Planet. Space Sci., 30,* 755.
Sumi et al. (2011) *Nature, 473,* 349.
Tachihara K. et al. (2002) *Astron. Astrophys., 385,* 909.
Tanaka H. and Ida S. (1997) *Icarus, 125,* 302.
Tobin J. J. et al. (2013a) *Astrophys. J., 771,* 48.
Tobin J. J. et al. (2013b) *Astrophys. J., 779,* 93.
Todorov K. et al. (2010) *Astrophys. J. Lett., 714,* L84.
Tomida K. et al. (2013) *Astrophys. J., 763,* 6.
Triaud A. H. M. J. et al. (2010) *Astron. Astrophys., 524,* A25.
van der Marel N. et al. (2013) *Science, 340,* 1199.
Vaytet N. et al. (2013) *Astron. Astrophys., 557,* A90.
Vigan A. et al. (2012) *Astron. Astrophys., 544,* A9.
Vorobyov E. I. (2013) *Astron. Astrophys., 552,* A129.
Vorobyov E. I. and Basu S. (2006a) *Astrophys. J., 650,* 956.
Vorobyov E. I. and Basu S. (2006b) *Astrophys. J., 650,* 956.
Wada K. et al. (2009) *Astrophys. J., 702,* 1490.
Wahhaj Z. et al. (2013) *Astrophys. J., 773,* 179.
Walsh A. J. et al. (2007) *Astrophys. J., 655,* 958.
Weidenschilling S. J. and Davis D. R. (1985) *Icarus, 62,* 16.
White R. J. and Basri G. (2003) *Astrophys. J., 582,* 1109.
Whitworth A. P. and Zinnecker H. (2004) *Astron. Astrophys., 427,* 299.
Wilner D. J. et al. (2005) *Astrophys. J. Lett., 626,* L109.
Winn J. N. et al. (2005) *Astrophys. J., 631,* 1215.
Wuchterl G. (1993) *Icarus, 106,* 323.
Youdin A. N. and Goodman J. (2005) *Astrophys. J., 620,* 459.
Zsom A. et al. (2010) *Astron. Astrophys., 513,* A57.

Helled R., Bodenheimer P., Podolak M., Boley A., Meru F., Nayakshin S., Fortney J. J., Mayer L., Alibert Y., and Boss A. P. (2014)
Giant planet formation, evolution, and internal structure. In *Protostars and Planets VI* (H. Beuther et al., eds.), pp. 643–665. Univ. of Arizona,
Tucson, DOI: 10.2458/azu_uapress_9780816531240-ch028.

Giant Planet Formation, Evolution, and Internal Structure

Ravit Helled
Tel-Aviv University

Peter Bodenheimer
University of California, Santa Cruz

Morris Podolak
Tel-Aviv University

Aaron Boley
University of Florida and The University of British Columbia

Farzana Meru
Eidgenössische Technische Hochschule (ETH) Zürich

Sergei Nayakshin
University of Leicester

Jonathan J. Fortney
University of California, Santa Cruz

Lucio Mayer
University of Zürich

Yann Alibert
University of Bern

Alan P. Boss
Carnegie Institution

The large number of detected giant exoplanets offers the opportunity to improve our understanding of the formation mechanism, evolution, and interior structure of gas giant planets. The two main models for giant planet formation are core accretion and disk instability. There are substantial differences between these formation models, including formation timescale, favorable formation location, ideal disk properties for planetary formation, early evolution, planetary composition, etc. First, we summarize the two models including their substantial differences, advantages, and disadvantages, and suggest how theoretical models should be connected to available (and future) data. We next summarize current knowledge of the internal structures of solar — and extrasolar — giant planets. Finally, we suggest the next steps to be taken in giant planet exploration.

1. INTRODUCTION

Giant planets play a critical role in shaping the architectures of planetary systems. Their large masses, orbital angular momentum, and fast formation make them prime candidates for driving rich dynamics among nascent planets, including exciting the orbits of small bodies and possibly delivering volatiles to terrestrial planets. Their bulk properties are also key for exploring the physical and chemical conditions of protoplanetary disks in which planets form. Furthermore, the diversity in properties (e.g., mass, radius, semimajor axis, and density) of exoplanets calls for multiple formation and evolution mechanisms to be explored. As a result, detailed investigations of the formation mechanism, evolution, and interior structure of giant planets have been conducted for decades, although for most of that time, our understanding of giant planets was limited to the four outer planets in the solar system. Naturally, theory was driven to

643

explain the properties of those planets and their particular characteristics. The detections of giant planets outside the solar system have led to the discovery of a rich and diverse population of giant exoplanets, many of which have measured radii and masses.

Clues regarding the nature of giant planet formation might be revealed from the two giant planet correlations with stellar metallicity [Fe/H] of main-sequence stars. The first is the correlation of the frequency of giant planets with stellar metallicity (*Gonzalez, 1997; Santos et al., 2004; Fischer and Valenti, 2005; Mortier et al., 2013*), which has now been determined for stellar masses from ~0.3 to 2 $M_\odot$ and metallicities [Fe/H] between –1.0 and 0.5 (*Johnson et al., 2010*). Nevertheless, the robustness and generality of the correlation between giant planet occurrence and stellar metallicity is still under investigation (*Maldonado et al., 2013*). The second correlation suggests that the heavy-element mass in giant planets is a function of stellar metallicity. Metal-rich stars tend to have metal-rich planets. There is a large scatter, however, and planetary mass is a better predictor of metal enrichment than stellar metallicity (see section 5.2). This explosion of new information demonstrates the need to understand planet formation in general, and presents an opportunity to test planet-formation theories. While the basic ideas of how giant planets are formed have not changed much, substantial progress has been made recently in highlighting the complexity of the models, in which effects that were perceived as second order may actually be fundamental in shaping the distribution of giant planets.

Accurate measurements of the giant planets in the solar system can provide valuable information on the internal structures of the planets, which are extremely important to test formation models. We anticipate major progress in that direction when data from the Juno and Cassini Solstice missions are available. In addition, the available information on the mean densities of giant exoplanets teaches us about the composition of giant planets in general. This chapter summarizes the recent progress in giant planet studies.

2. GIANT PLANET FORMATION MODELS

In the following section we summarize the two formation models: core accretion (CA), the standard model, and disk instability (DI). In the CA model the formation of a giant planet begins with planetesimal coagulation and core formation, similar to terrestrial planets, which is then followed by accretion of a gaseous envelope. In the DI model, gas giant planets form as a result of gravitational fragmentation in the disk surrounding the young star.

2.1. Core Accretion

The initial step in the formation of giant planets by CA is the same as that for terrestrial planets, namely the buildup of planetesimals with sizes from a few tens of meters to a few hundred kilometers. Planetesimal formation is itself

undergoing current research and is addressed in the chapter by Johansen et al. in this volume. Here we focus on the later stages, starting at the point where runaway growth of the planetesimals has produced a planetary embryo having a fraction of an Earth mass ($M_\oplus$), still surrounded by a swarm of planetesimals. The planetesimals accrete onto the embryo to form the heavy-element core of the giant planet. Once the core reaches a fraction of $M_\oplus$ and its escape velocity exceeds the local thermal speed of the gas in the disk, the core can begin to capture gas from the surrounding disk. Solids and gas then accrete concurrently. When the mass of gas (M_{env}) approaches that of the core (M_{core}), gas accretion can become very rapid and can build the planet up to the Jupiter-mass (M_{Jup}) range. Gas accretion is cut off at some point, either because of the dissipation of the disk and/or by the gravitational interaction of the planet with the disk, which tends to open up a gap in the planet's vicinity as described below.

2.1.1. Physics of the core-accretion model.

2.1.1.1. Core-accretion rate: Core accretion calculations generally start with an embryo of ~0.01–0.1 $M_\oplus$ surrounded by solid planetesimals, which would be expected to have a range of sizes. Safronov's equation (*Safronov, 1969*) is a useful approximation for the planetesimal accretion rate onto an embryo

$$\frac{dM_{solid}}{dt} = \dot{M}_{core} = \pi R_{capt}^2 \sigma_s \Omega F_g \tag{1}$$

where πR_{capt}^2 is the geometrical capture cross section, Ω is the orbital frequency, σ_s is the disk's solid surface density, and F_g is the gravitational enhancement factor. If no gas is present, then $R_{capt} = R_{core}$. In the presence of an envelope, gas drag and ablation can result in planetesimal deposition outside the core, with $R_{capt} > R_{core}$.

In the two-body approximation in which the tidal effects of the central star are not included, F_g is given by

$$F_g = \left[1 + \left(\frac{v_e}{v} \right)^2 \right] \tag{2}$$

where v_e is the escape velocity from the embryo, and v is the relative velocity of the embryo and the accreting planetesimal. This approximation is valid only when the effect of the star is negligible, otherwise three-body effects must be taken into account, and the determination of F_g is much more complicated (*Greenzweig and Lissauer, 1992*).

2.1.1.2. The disk of planetesimals: The gravitational focusing factor depends strongly on the excitation state of planetesimals (eccentricity and inclination), which itself is the result of excitation by the forming planet (and other planetesimals) and damping by gas drag. Therefore, the value of F_g must be determined (numerically) by computing the planet's growth, the structure of the gas disk, and the characteristics of planetesimals (dynamics and mass function) self-consistently. Gravitational interactions among planetesimals and, more importantly, the influence of the planetary embryo tend to stir up the velocities of the planetesimals, reducing F_g. *Fortier et al.* (2013) show that giant planet formation is suppressed when this effect is included.

The large number of planetesimals in the feeding zone ($>10^{10}$) requires that the evolution, growth, and accretion rate of the planetesimals onto the embryo be handled in a statistical manner (*Inaba et al.*, 2003; *Weidenschilling*, 2008, 2011; *Bromley and Kenyon*, 2011; *Kobayashi et al.*, 2010). Planetesimals are distributed into a finite number of mass bins, with masses ranging from tens of meters to tens of kilometers. The planetesimals are also distributed into radial distance zones. The codes calculate the evolution of the size distribution and the orbital element distribution, as well as the accretion rate of each size onto the planetary embryo. The size distribution evolves through collisions, and, if relative velocities between planetesimals become large enough, through fragmentation. The velocity distributions are affected by viscous stirring, dynamical friction, gravitational perturbations between non-overlapping orbits, and damping by gas drag from the disk. The outcome of fragmentation depends on specific impact energies, material strengths, and gravitational binding energies of the objects. Also to be included are migration of planetesimals through the disk, and interactions between the protoplanetary envelope and the various-sized planetesimals (see below). These simulations give $\dot{M}_{core}$ as time progresses.

Including all these physical processes in planet-formation models is numerically challenging, and all models involve some degree of approximation. The full problem of the self-consistent buildup to Jupiter mass, including the evolution of the planet, the gas disk, and the planetesimal distribution, has not been completed to date.

2.1.1.3. *Envelope accretion rate:* The outer radius of the planet cannot be larger than the smaller of the Bondi radius R_B or the Hill radius R_H. Here $R_B = GM_p/c_s^2$ and $R_H = a (M_p/3M_\star)^{1/3}$, where M_p and $M_\star$ are the masses of the planet and star, respectively, a is the distance of the planet from the star, and c_s is the sound speed in the disk. However, three-dimensional numerical simulations of the gas flow around a planet embedded in a disk (*Lissauer et al.*, 2009) show that not all the gas flowing through the Hill volume is actually accreted by the planet. Much of it just flows through the Hill volume and back out. A modified Hill radius, within which gas does get accreted onto the planet, is found to be ~0.25 R_H. Thus one can define an effective outer radius for the planet

$$R_{eff} = \frac{GM_p}{c_s^2 + \frac{GM_p}{0.25R_H}} \qquad (3)$$

During the earlier phases of the evolution, when M_{env} is less than or comparable to M_{core}, the gas accretion rate is determined by the requirement that the outer planet radius $R_p = R_{eff}$. As the envelope contracts and R_{eff} expands, gas is added at the outer boundary to maintain this condition.

Once the planet has entered the phase of rapid gas accretion, the envelope is relatively massive, and it can contract rapidly. At some point the mass addition rate required by

the above condition exceeds the rate at which matter can be supplied by the disk. The disk-limited rates are calculated from three-dimensional simulations of the flow around a planet embedded in a disk (*Lubow and D'Angelo*, 2006; *Lissauer et al.*, 2009; *D'Angelo et al.*, 2011; *Bodenheimer et al.*, 2013; *Uribe et al.*, 2013). The rates are eventually limited as the mass of the planet grows, exerts tidal torques on the disk, and opens up a gap. Although gas can still accrete through the gap, the rate becomes slower as the gap width increases. Planet-disk interactions, however, can change the eccentricities and increase $\dot{M}_{env}$ substantially even after gap formation (*Kley and Dirksen*, 2006).

2.1.1.4. *Structure and evolution of the envelope:* The envelope of the forming planets is typically considered to consist mainly of hydrogen and helium (H, He), with a small fraction of heavy elements (high-Z material). The Z-component of the planetary envelope can be stellar, substellar, or superstellar. If all the planetesimals reach the core without depositing mass in the envelope, the planetary envelope is likely to have a composition similar to the gas in the protoplanetary disk. However, enrichment of a gas giant through CA is not necessarily automatic, since the accreted gas could be depleted in solids as a result of planetesimal formation. As a result, the accreted gas can be substellar. On the other hand, if planetesimals suffer a strong mass ablation during their journey toward the core, the planetary envelope will be substantially enriched with heavy elements compared to the disk gas, in particular during the first phases of planetary growth, when the solid accretion rate can be much larger than the gas accretion rate.

Typically, CA models assume that the planetary envelope is spherically symmetric, a good assumption, up to the phase of rapid gas accretion. The stellar structure equations of hydrostatic equilibrium, energy transport by radiation or convection, and energy conservation are used to determine its structure. During phases when $R_p = R_{eff}$, the density and temperature at the outer boundary are set to the disk values ρ_{neb} and T_{neb}, respectively.

The differential equations of stellar structure are then supplemented with the equation of state (EOS) and the opacity. The grains in the envelope are typically assumed to have an interstellar size distribution. However, *Podolak* (2003) shows that when grain coagulation and sedimentation in the envelope are considered, the actual grain opacities, while near interstellar values near the surface of the protoplanet, can drop by orders of magnitude, relative to interstellar, deeper in the envelope.

2.1.1.5. *Interaction of planetesimals with the envelope:* The accreted planetesimals are affected by gas drag, which can result in a considerable enhancement of the effective cross section for their capture, larger than πR_{core}^2. The deposition of planetesimal material also has significant effects on the composition, mean molecular weight, opacity, and EOS of the envelope. Individual orbits of planetesimals that pass through the envelope are calculated (*Podolak et al.*, 1988). A variety of initial impact parameters are chosen, and the critical case is found inside of which the planetesimal is captured. Then R_{capt} is the pericenter of the critical orbit. Even if $M_{env} < 0.01$ $M_\oplus$, this effect can be significant, especially for smaller planetesimals that are easily captured. The value of R_{capt} can be up to $10\times$ R_{core}, depending on M_{env} and the planetesimal size.

The determination of the planetesimal mass ablation is in turn a complex process, and depends strongly on the (poorly known) characteristics of planetesimals, among them their mass and mechanical properties. Low-mass planetesimals and/or low-tensile-strength planetesimals are likely to suffer strong mass ablation. These properties depend on the (yet unknown) planetesimal formation process (see the chapter by Johansen et al. in this volume) and their interaction with forming planets and gas in the disk. Interestingly enough, the enrichment of planetary envelopes by incoming planetesimals has a strong influence on the planetary growth by gas accretion itself. In particular, as shown by *Hori and Ikoma* (2011), envelope pollution can strongly reduce the critical mass — i.e., the core mass required to initiate rapid gas accretion. This can speed up the formation of gas giant planets, leading to objects with small cores and enriched envelopes at the end of the formation process.

On the other hand, it is often assumed that the ablated planetesimal material eventually sinks to the core, releasing gravitational energy in the process. Detailed calculations (*Iaroslavitz and Podolak,* 2007) show that for particular envelope models the rocky or organic components do sink, while the ices remain dissolved in the envelope.

2.1.2. Phases of planetary formation. In classical CA models (i.e., without migration and disk evolution) the important parameters for defining the formation of a giant planet are σ_s, $M_\star$, and a. These three quantities determine the isolation mass, $M_{iso} = (8/\sqrt{3})(\pi C)^{3/2}M_\star^{-1/2}\sigma_s^{3/2}a^3$, which gives the maximum mass to which the embryo can grow, at which point it has accreted all the planetesimals in the feeding zone. C is the number of Hill-sphere radii on each side of the planet that is included in the feeding zone (*Pollack et al.,* 1996). M_{iso} must roughly exceed 3 $M_\oplus$ in order to get mass accretion up to the Jupiter-mass range during the lifetime of the disk. Although it is commonly stated that a core of ≈ 10 $M_\oplus$ is necessary to initiate rapid gas accretion, the critical core mass can be both larger and smaller than this standard value, depending on the disk's properties, the physics included in the planetary formation process, and the core accretion rate. Additional parameters include the gas-to-solid ratio in the disk, which for solar composition is on the order of 100, but which decreases as the metallicity of the star increases. The gas surface density determines ρ_{neb} and a disk model determines T_{neb}. The size distribution and composition of the planetesimals in the disk are important for the calculation of $\dot{M}_{core}$. Many calculations are simplified in that a single-sized planetesimal is used, with a radius subkilometer-sized and up to 100 km.

The main phases associated with the buildup of a giant planet in the CA model are:

Phase 1: Primary core-accretion phase. The core accretes planetesimals until it has obtained most of the solid mass within its gravitational reach, i.e., its mass approaches the isolation mass. M_{env} grows as well, but it remains only a tiny fraction of M_{core}.

Phase 2: Slow-envelope accretion. The $\dot{M}_{core}$ drops considerably, and $\dot{M}_{env}$ increases until it exceeds $\dot{M}_{core}$. As M_{env} increases, the expansion of the zone of gravitational influence allows more planetesimals to be accreted, but at a slow, nearly constant rate. $\dot{M}_{env}$ is limited by the ability of the envelope to radiate energy and contract, so this rate is also slow, about 2–3$\times$ that of the core. Thus the timescale depends on the opacity in the envelope, and if grain settling is incorporated into the calculations, it falls in the range 0.5 to a few million years.

Phase 3: Rapid-gas accretion. After crossover ($M_{core} = M_{env} \approx \sqrt{2} M_{iso}$), the rate of gas accretion increases continuously, although the structure remains in quasihydrostatic equilibrium. $\dot{M}_{env}$ greatly exceeds that of the core. Eventually the point is reached where the protoplanetary disk cannot supply gas fast enough to keep up with the contraction of the envelope. $\dot{M}_{env}$, instead of being determined by the radiation of energy by the planet, is then determined by the properties of the disk in the vicinity of the planet. Beyond this point, the accretion flow of the gas is hydrodynamic, nearly in free-fall, onto a protoplanet that is in hydrostatic equilibrium at a radius much smaller than R_{eff}. The core mass remains nearly constant on the timescale of the growth of the envelope. The timescale to build up to ≈ 1 M_{Jup} is 10^4–10^6 yr, depending on the assumed disk viscosity.

During disk-limited accretion, the gas envelope may be considered to consist of a hydrostatic contracting inner region, containing almost all the mass, plus a hydrodynamic accretion flow of gas onto it from the disk. The flow will be supersonic, so the boundary condition at the outer edge of the hydrostatic region will be determined by the properties of the shock that forms at that interface (*Bodenheimer et al.,* 2000). Interestingly enough, the properties of the shock have strong implications for the entropy of the gas that is finally accreted by the forming planet. Depending on the fraction of the incoming energy that is radiated away through the shock, the entropy of accreted gas could differ substantially. This in turn governs the evolution of the post-formation luminosity (*Mordasini et al.,* 2012b), a quantity that could be accessible by direct imaging observations of planets.

Phase 4: Contraction and cooling. The planet evolves at constant mass. The main energy source is slow contraction in hydrostatic equilibrium, while an energy sink arises from cooling by radiation. Stellar radiation heats up the surface. Once accretion terminates, there is no longer an external energy source from planetesimal accretion, but for a planet very close to its star, tidal dissipation and magnetic dissipation may provide additional energy. If the planet accretes a mass of ≈ 13 M_{Jup}, deuterium burning can set in, providing a significant energy supply.

An example of a calculation of the first three phases is shown in Fig. 1. The first luminosity peak represents the maximum of $\dot{M}_{core}$ during phase 1, while the second peak corresponds to the maximum disk-limited accretion rate of gas during phase 3. These three phases can change when the exchange of angular momentum between the forming planet and the disk is considered (see the chapter by Baruteau et al. in this volume). When the orbital parameters of a forming planet change with time, its formation is clearly affected. First, the planet can form more rapidly (all other parameters being equal), and second, its content in heavy elements is expected to be higher. It should be noted that the resulting higher content in heavy elements does not imply a higher core mass, and does not require a starting point in a massive and/or very metal-rich disk.

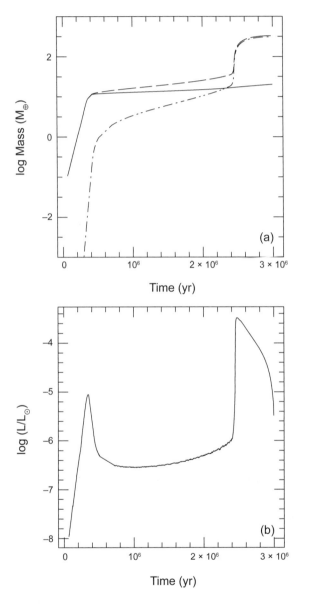

Fig. 1. **(a)** An example of the growth of a protoplanet (at 5.2 AU with σ_s = 10 g cm^{-2}) by core accretion. The planet mass is shown vs. time. Solid line: core mass; dash-dot line: envelope mass; dashed line: total mass. Grain opacities are reduced by a factor of 50 from interstellar values. Crossover mass is reached at 2.3 m.y. and 1 M_{Jup} at 3 m.y. **(b)** The radiated luminosity vs. time. Adapted from *Lissauer et al.* (2009).

Alibert et al. (2004) have investigated how orbital migration of a forming planet, whatever its origin, prevents the planet from reaching isolation. The same effect can also appear as a result of the orbital drift of planetesimals by gas drag. As a consequence, phase 2 as presented above is suppressed, and a forming planet would transit directly from phase 1 to phase 3. In that case the total formation timescale is substantially reduced. This process is illustrated in Fig. 2, which shows the formation of a Jupiter-mass planet for two models that are identical except that one assumes *in situ* formation (dashed lines), while the second includes planetary migration (solid lines). In the migrating case, the initial planetary location is 8 AU with σ_s = 3.2 g cm^{-2}, and the planet migrates down to 5 AU, where the *in situ* model is computed with σ_s = 7.5 g cm^{-2}. For the *in situ* model, the crossover mass is reached after ~50 million years (m.y.), while in the migrating model, crossover mass is reached in less than 1 m.y. One should also note that when migration is included the total amount of heavy elements is increased by a factor of ~2. Both models are computed assuming the same disk mass, which is around 3× the minimum mass solar nebula (MMSN), compatible with disk-mass estimations from observations (e.g., *Andrews et al.*, 2011) (see section 2.1.4).

2.1.3. Termination of accretion. Two mechanisms have been discussed for the termination of the rapid gas accretion onto a protoplanet: disk dissipation and gap opening. The general picture regarding the evolution of a disk involves a gradual reduction in disk mass and disk gas density over

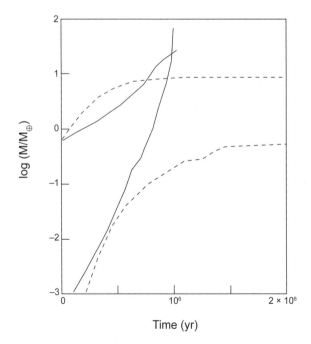

Fig. 2. The total mass of heavy elements (core + envelope) and gaseous envelope (H/He) mass vs. time until crossover mass is reached. Solid lines: a migrating planet starting at 8 AU, without gap formation; dashed lines: *in situ* formation. Adapted from *Alibert et al.* (2005a).

a period of 2–4 m.y., after which a relatively rapid (a few 10^5 yr) clearing phase occurs. If the giant planet fully forms before the clearing phase occurs, then the disk dissipation mechanism is not relevant. However, this mechanism could work to explain the structure of Uranus and Neptune (*Pollack et al.,* 1996). The core-accretion time for these planets at their present orbital positions is so long that it is very unlikely that they formed there. It has been suggested (e.g., *Tsiganis et al.,* 2005) that these planets formed at closer distances, between 5 and 20 AU, and were later scattered outward. Even at these distances the formation time for the core could be long enough so that substantial disk dissipation occurs during phase 2, and the planet never reaches the crossover mass.

For planets that reach the phase of rapid gas accretion it seems as though gap opening is the critical factor. Once the forming planet reaches a sufficient mass, its gravitational influence on the disk results in a torque on disk material interior to the planet's orbit; the torque removes angular momentum from the material and allows it to move closer to the star. At the same time, a torque is also exerted on the material external to the planet; this material gains angular momentum and is forced to move to larger distances. In this way a gap, or at least a disk region of reduced density, tends to form near the planet's orbit (*Lin and Papaloizou,* 1979). However, there is an opposing effect: The viscous frictional torque in the differentially rotating disk tends to transfer angular momentum outward and mass inward; this effect tends to fill in the gap.

The condition for a gap to open then requires that the gravitational effect of the planet dominates over the viscous and pressure effects (*Crida et al.,* 2006). If $R_H > H$ (the disk scale height), which is the normal situation for the appropriate mass range, then the minimum planet mass M_p for gap opening is (*D'Angelo et al.,* 2011)

$$\left(\frac{M_p}{M_\star}\right)^2 \approx 3\pi\alpha f \left(\frac{H}{a}\right)^2 \left(\frac{R_H}{a}\right)^3 \qquad (4)$$

where f is a parameter on the order of unity, and α is the disk viscosity parameter. For $H/a = 0.05$ and $\alpha = 10^{-2}$ the minimum estimated mass is about 0.2 M_{Jup} around a solar-mass star. Numerical simulations of disk evolution with an embedded planet of various masses show that by the time this mass is reached, the gas density near the orbit has been reduced to about 60% of the unperturbed value; at 1.0 M_{Jup} the reduction is over 95% (*D'Angelo et al.,* 2011).

Even after gap opening has been initiated, gas can still flow through it and accrete onto the planet, although the rate of accretion decreases as the gap becomes deeper and wider. Figure 3 shows that, depending on the viscosity, the peak in $\dot{M}_{env}$ occurs for a planet of mass 0.1 to 1 M_{Jup} around a solar-mass star. However, beyond that point, gas continues to be accreted through the gap until its density is reduced by a factor of 1000, leading to a final planet mass of up to 5–10 M_{Jup} for $\alpha \approx 4 \times 10^{-3}$. Further accretion is possible if the protoplanet has an eccentric orbit. Thus the maximum

mass of a planet appears to be close to the upper limit of the observed mass range for exoplanets. Planets that end up in the Jupiter-Saturn mass range would therefore have to be explained by formation in a disk region with cold temperatures (low H/a) and low viscosity, or at least in a disk that has evolved to such conditions at the time of rapid gas accretion. Angular momentum is expected to be transferred from the disk to the planet during the gap-opening process, leading to a subdisk around the planetary core (*Ayliffe and Bate,* 2012). One might think that the presence of this subdisk would slow down the accretion process onto the planet; this effect must still be characterized, taking into account accretion through the planet's polar regions. In any case, the limiting gas accretion rate depends, in particular, on the disk viscosity and gas surface density. The latter quantity is known to evolve on timescales comparable to that of buildup of planetary cores (a few million years). As a consequence, at a time when the critical mass — i.e., the core mass needed to initiate rapid gas accretion (e.g. *Mizuno,* 1980) — is reached, the limiting accretion rate has decreased by up to 2 orders of magnitude, compared to a classical young disk, reaching values ~10^{-4} $M_\oplus$ yr^{-1} or lower. Under these circumstances, the accretion of 1 M_{Jup} takes $\approx$3 m.y., comparable with disk lifetimes. Such calculations have been used in order to explain the heavy-element mass in Jupiter and Saturn (*Alibert et al.,* 2005b), as well as

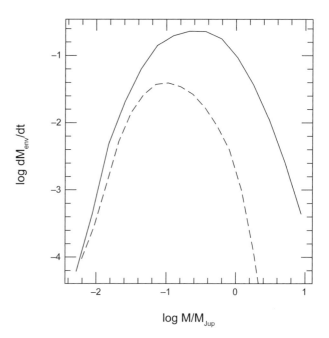

Fig. 3. Disk-limited gas accretion rates ($M_\oplus$/yr^{-1}) vs. total planetary mass for a planet at 5.2 AU around a 1-$M_\odot$ star in a disk with initial gas density $\sigma_g = 700$ g cm^{-2}. Solid line: disk viscosity $\alpha = 4 \times 10^{-3}$; dashed line: disk viscosity $\alpha = 4 \times 10^{-4}$. At masses below ~0.25 M_{Jup} the actual rates fall below these curves because they are limited by the rate at which the gaseous envelope contracts. These rates scale as $\sigma_g a^2 \Omega$, where a is the orbital distance, and Ω is the orbital frequency. Adapted from *Lissauer et al.* (2009).

the observed mass function of exoplanets (*Mordasini et al.,* 2009a,b). An example is the model shown in Fig. 1, which reached a final mass of 1 M_{Jup} without any assumed cutoff, in a disk with a viscosity parameter $\alpha = 4 \times 10^{-4}$ and a disk lifetime of 3 m.y. (*Lissauer et al.,* 2009).

2.1.4. Core accretion as a function of parameters.
2.1.4.1. Effect of position in disk: The presence of exoplanets near their stars has led to the suggestion (*Lin et al.,* 1996) that giant planets form in the 5–10-AU region according to the picture just discussed, but during the formation process they migrate inward. However, one must also consider the possibility that these planets could actually form at or near their present locations. In the very inner regions of disks, ~0.05 AU where the hot Jupiters are found, there are several difficulties. First, the temperatures are high, so there is very little condensable material. Second, the typical mass distribution in the inner disk does not provide sufficient material to form 5–10-$M_{\oplus}$ cores, necessary for giant planet formation. Third, even if a planet did form, it would very likely migrate into the star [although outward migration may be feasible for low-mass planets (*Kley et al.,* 2009; *Hellary and Nelson,* 2012)]. A possible way of circumventing these problems is to build up the core by migration of planetesimals or small solid objects inward through the disk and then collecting them in the inner regions of the disk (e.g., *Ward,* 1997; *Bodenheimer et al.,* 2000). Also, planet formation is more likely in a metal-rich or massive inner disk. The third difficulty can be averted if one assumes that the inner disk has been cleared, for example, by the stellar magnetic field, so that the torques on the planet would be negligible.

The possibility of migration introduces an additional complexity. In this case, it is difficult to define the formation location of a planet, since they can migrate over very large distances, experiencing very diverse conditions in different parts of the disk. Nonetheless, the following facts are still valid: (1) The solid accretion rate decreases as a function of distance, and (2) the amount of available solid material decreases as one moves inward toward the star. If there is no significant migration, the optimum location for core accretion formation of a giant planet is between 5 and 10 AU for a solar-mass star. The maximum distance at which a giant planet can form in a MMSN is 40–50 AU (*Rafikov,* 2011), based on the assumption of a maximum rate of planetesimal accretion onto the core.

2.1.4.2. The effect of the mass of the star: Typical observed disk masses decrease roughly linearly with the mass of the central star from $M_{\star}$ of a few solar masses down to less than 0.1 $M_{\odot}$ with a typical $M_{disk}/M_{\star} = 0.002–0.006$ (*Andrews et al.,* 2013). However, observed disk masses show a large spread at a given stellar mass, from 0.001 $M_{\odot}$ to 0.1 $M_{\odot}$ for a 1-$M_{\odot}$ star. Furthermore, disk masses are uncertain because they are determined from dust continuum observations that do not measure the gas mass and are affected by uncertainty in the dust opacity. Characteristic disk lifetimes extend from a median of 3 m.y. to a maximum of 10 m.y. (*Hillenbrand,* 2008). Separating out the disk dispersal times as a function of mass has proved to be difficult (*Kennedy and Kenyon,* 2009), with no strong dependence observed. As a useful first approximation, one can say that disk lifetimes are independent of stellar mass for $M_{\star} < 1.5\ M_{\odot}$, but there is a decrease in lifetimes for $M_{\star} > 1.5\ M_{\odot}$, possibly because of more effective dissipation by ultraviolet and X-ray irradiation (see the chapter by Alexander et al. in this volume). Again, for a given stellar mass, there is a range of disk lifetimes, from about 1 to 10 m.y.

If the disk mass in solids scales with stellar mass, then the solid surface density in disks around stars less massive than the Sun is expected to be low, leading to slower formation. In addition, the dynamical time Ω^{-1} is longer, resulting in longer accretion times. Thus it should be more difficult to form a Jupiter-like planet at a given radial distance around a low-mass star than around a solar-mass star (*Laughlin et al.,* 2004), on a timescale comparable to the characteristic 2.5-m.y. lifetime of disks around low-mass stars (*Mamajek,* 2009). While the existence of a disk with high enough surface density to form a giant planet is possible, the probability of forming the planet around the low-mass star is clearly smaller than for a 1-$M_{\odot}$ star. A similar argument, in the opposite sense, applies to stars that are more massive than the Sun. These qualitative considerations are consistent with the observed correlation that the probability of a star hosting a planet of ~0.5 M_{Jup} or greater is proportional to $M_{\star}$ (*Johnson et al.,* 2010). In addition, more detailed theoretical calculations going beyond these order of magnitude estimates predict that giant planets are less frequent around low-mass stars (*Ida and Lin,* 2005; *Alibert et al.,* 2011).

2.1.4.3. Effect of the metallicity of the star: Clearly if the metallicity of the star [Fe/H] increases, the metallicity of its disk also increases. If the abundances in the grain-forming material — namely ices, organics, and silicates — increase in proportion to the iron abundance, one gets higher σ_s at a given distance around a star of higher metallicity. Indeed, observations show that the planet occurrence probability at a given iron abundance increases with the stellar silicon abundance (*Robinson et al.,* 2006), consistent with this picture. The higher σ_s gives a shorter formation time for the core at a given distance, with time inversely proportional to $\sqrt{\sigma_s}$. However, there is an opposing effect: The higher dust opacity in the protoplanet's envelope would tend to lengthen the time for gas accretion. On the other hand, the higher dust density speeds up the process of dust coagulation and settling, which reduces the opacity, so the envelope effect may not be very significant. In fact, the observed probability of giant planet formation increases faster than linearly with the metallicity. A possible explanation is that a higher M_{iso} gives a higher luminosity during phase 2 and a significantly reduced accretion time for the gas (*Ida and Lin,* 2004). Thus there is an increased probability that the giant planet can form before the disk dissipates.

Clearly there are other parameters that affect the CA process, including the disk mass (at a given stellar mass), disk lifetime, disk viscosity, and dust opacity in the protoplanetary envelope. The first two of these effects, along with disk metallicity, have been studied by *Mordasini et al.* (2012a), while a detailed model of the last effect was presented by *Movshovitz et al.* (2010).

2.2. DISK INSTABILITY

Since *Protostars and Planets V* (*Reipurth et al.,* 2007) (PPV), DI studies have made large advancements in isolating conditions that are likely to lead to disk fragmentation, i.e., the breakup of a protoplanetary disk into one or more self-gravitating clumps. Recent studies show that while the period of clump formation may be short-lived, fragments, whether permanent or transient, can have a fundamental influence on the subsequent evolution of a nascent stellar and/or planetary system. In this section, we summarize the DI model for giant planet formation.

2.2.1. Disk fragmentation. The formation of giant planets in the DI model is conditioned by disk fragmentation. Disk perturbations will grow and form density enhancements throughout regions of a disk wherever the destabilizing effects of self-gravity become as important as, or dominate over the stabilizing effects of pressure and shear. The threshold for the growth of axisymmetric density perturbations in a thin gaseous disk is given by the *Toomre* (1964) criterion $Q = c_s \kappa / \pi G \sigma_g \sim 1$, where c_s is the sound speed, κ is the epicyclic frequency, and σ_g is the gas surface density. The Q criterion is strictly derived for an axisymmetric, infinitesimally thin disk, and describes the disk's response at a given radius to small (linear) axisymmetric perturbations. However, it is routinely applied to global, vertically stratified disks subject to non-axisymmetric perturbations with considerable degree of success.

Numerous numerical studies in two and three dimensions have demonstrated that a disk will develop spiral structure for $Q \leq 1.7$, well before $Q = 1$ is reached (*Durisen et al.,* 2007). These non-axisymmetric perturbations are the physical outcome of gravitational instabilities in realistic disks. They produce disk torques and shocks that redistribute mass and angular momentum, and provide a source of heating throughout gravitationally unstable regions. Thus, spiral arms can act to stabilize the disk by increasing the local sound speed and spreading the disk mass out. In contrast, radiative cooling (with or without convection present) will decrease the sound speed and destabilize the disk.

When the heating and mass transport from spiral arms can balance cooling and/or infall of gas from the molecular cloud core when the disk is still young, persistent spiral structure can exist in the disk. In other cases, cooling and/ or mass infall can lead to a second instability, i.e., the collapse of regions of spiral arms into bound, self-gravitating clumps. Determining whether or not clumps formed by gravitational instability can evolve into gas giant planets requires a combination of simulations that span several orders of magnitude in densities and spatial scales.

The exact conditions that lead to a disk breaking up into self-gravitating clumps are still under investigation. The principal and firmly established condition is that the disk must be strongly self-gravitating, where $Q < 1.4$ is necessary for an isothermal disk to fragment (e.g., *Nelson et al.,* 1998; *Mayer et al.,* 2004). This value is higher than the often assumed $Q < 1$. For non-isothermal disks, the Q condition alone is insufficient, and a cooling timescales is usually used to identify regions that are expected to form clumps. *Gammie* (2001) found that for two-dimensional disks with an adiabatic index $\Gamma = 2$, disks fragmented when $\beta < 3$ where $\beta = t_{cool}\Omega$ and t_{cool} is the cooling timescale. Here, $\Omega \approx \kappa$ is the local orbital frequency. *Rice et al.* (2005) studied this cooling criterion in more detail and found that fragmentation occurred in three-dimensional disks when $\beta \leq 6$ for $\gamma = 5/3$ and $\beta \leq 12$ for $\gamma = 7/5$. While there have been suggestions that the critical cooling timescales might be affected by the disk's thermal history (*Clarke et al.,* 2007), the temperature dependence of the cooling law (*Cossins et al.,* 2010), and the star and disk properties (*Meru and Bate,* 2011b), the β condition for fragmentation was recently questioned anew by *Meru and Bate* (2011a), who found that the cooling time constraint was not converged with numerical resolution in smoothed particle hydrodynamics (SPH) studies. Additional investigations of the topic were presented by *Lodato and Clarke* (2011) and *Paardekooper et al.* (2011).

Meru and Bate (2012) showed that the previous nonconvergent results were due to the effects of artificial viscosity in SPH codes and suggested that the critical cooling timescale may be at least as large as $\beta = 20$ and possibly even as high as $\beta = 30$. However, convergence between different codes has not yet been seen. Regardless, while the exact value might change the radial location beyond which fragmentation can occur, there is a growing consensus that fragmentation is more common at large than at small disk radii. Moreover, if the disk is still accreting gas from the molecular core, a high mass accretion rate onto the disk may become larger than mass transport through the disk by spiral waves, leading to disk fragmentation in disks that are cooling relatively slowly.

The mass of the star can also affect disk fragmentation. On the one hand, an increase in stellar mass will increase the Toomre parameter via the angular frequency. On the other hand, more-massive stars are thought to harbor more-massive disks, thus decreasing the Toomre parameter. In addition, the temperature of different mass stars will also affect the disk temperatures and hence the corresponding Toomre parameter. The combination of these effects suggests that disk fragmentation around different mass stars is not straightforward and has to be investigated carefully. *Boss* (2011) explored this effect and found that increasing the stellar mass resulted in an increase in the number of fragments. *Boss* (2006b) also showed that gravitational instability can offer a mechanism for giant planet formation around M dwarfs where CA often fails due to the long formation timescales in disks around low-mass stars. However, there is not yet a consensus on the conditions that lead to disk fragmentation. For example, *Cai et al.* (2008) found that clumps are not likely to form around low-mass stars.

It is clear that follow-up studies are needed to determine how the efficiency of disk fragmentation varies with stellar mass. Moreover, clump evolution itself may be dependent on the host star's mass, further leading to observational differences.

2.2.2. Differences between numerical models of disk instability. One of the major difficulties in understanding the conditions that lead to clump formation is that fragmentation is inherently a highly nonlinear process. While analytical models have been presented (e.g., *Rafikov,* 2007; *Clarke,* 2009), theoretical efforts to model DI must ultimately be done numerically. In this section we restrict attention to numerical models, and in particular to models focused on whether DI can lead to the formation of self-gravitating clumps, as opposed to other processes in DI models. Further details on the numerical methods for DI models can be found in *Durisen et al.* (2007). Here we concentrate on DI numerical models published since PPV.

Tables 1 and 2 list some of the primary differences between global numerical DI models, with the former table listing Lagrangian, three-dimensional SPH codes, and the latter grid-based, finite-differences (FD) codes. The SPH codes include GADGET2 (*Alexander et al.,* 2008), GASOLINE (*Mayer et al.,* 2007), GADGET3 (*Cha and Nayakshin,* 2011), SPH/MPI (message-passing interface) (*Meru and Bate,* 2010, 2012), and DRAGON (*Stamatellos and Whitworth,* 2009). The FD codes include FARGO (*Meru and Bate,* 2012; *Paardekooper et al.,* 2011), EDTONS (*Boss,* 2012), CHYMERA/BDNL (*Boley et al.,* 2007), Indiana University Hydrodynamics Group (IUHG) (*Michael and Durisen,* 2010), and ORION (*Kratter et al.,* 2010a), which is an adaptive mesh refinement (AMR) code.

Artificial viscosity (e.g., *Boss,* 2006a; *Pickett and Durisen,* 2007) and spatial resolution (e.g., *Nelson,* 2006; *Meru and Bate,* 2011a) have been shown repeatedly to be important factors in the accuracy of DI numerical calculations and are also not mutually exclusive (*Meru and Bate,* 2012). For both FD and SPH codes, it is important to have sufficient spatial resolution to resolve the Jeans and Toomre length scales (*Nelson,* 2006), otherwise numerically derived fragmentation might result. Finite-differences codes avoid this outcome by monitoring these criteria throughout the grid, while SPH codes enforce a minimum number of nearest-neighbor particles (typically 32 or 64) in order to minimize the loss of spatial resolution during an evolution. For SPH codes, some of the other key parameters are the

TABLE 1. Comparison of numerical techniques for smoothed particle hydrodynamics (SPH) codes.

Code	α_{SPH}	β_{SPH}	N_{max}	Thermodynamics	Ref.
GADGET2	1.0	2.0	1.3×10^7	β cooling	[1]
GASOLINE	1.0	2.0	1.0×10^6	flux-limited RT	[2]
GADGET3	0.0	0.0	1.5×10^6	β cooling	[3]
SPH/MPI	0.1	0.2	2.5×10^5	flux-limited RT	[4]
SPH/MPI	0.01–10	0.2–2.0	1.6×10^7	β cooling	[5]
DRAGON	0.1	0.2	2.5×10^5	diffusion RT	[6]

α_{SPH} and β_{SPH} are coefficients of artificial viscosity, N_{max} is the maximum total number of SPH particles used to represent the disk, and the disk thermodynamics are classified as either a simple β cooling prescription, or more detailed, flux-limited radiative transfer (RT).

References: [1] *Alexander et al.* (2008); [2] *Mayer et al.* (2007); [3] *Cha and Nayakshin* (2011); [4] *Meru and Bate* (2010); [5] *Meru and Bate* (2012); [6] *Stamatellos and Whitworth* (2009).

TABLE 2. Comparison of numerical techniques for finite differences (FD) codes.

Code	Dimensionality	C_q	N_{max}	Thermodynamics	Ref.
FARGO	two-dimensional	0–2.5	5×10^7	β cooling	[1]
IUHG	three-dimensional	3.0	2×10^6	β cooling	[2]
CHYMERA/BDNL	three-dimensional	3.0	8×10^6	flux-limited RT	[3]
EDTONS	three-dimensional	0.0	8×10^6	flux-limited RT	[4]
ORION(AMR)	three-dimensional	$\neq 0$	3×10^{14}	isothermal	[5]

Dimensionality is the number of dimensions in the solution (two-dimensional or three-dimensional), C_q is a coefficient of artificial viscosity, N_{max} is the maximum total effective number of grid points (including mid-plane symmetry), and disk thermodynamics is classified as isothermal, simple β cooling, or more detailed, flux-limited radiative transfer (RT).

References: [1] *Meru and Bate* (2012); [2] *Michael and Durisen* (2010); [3] *Boley et al.* (2007); [4] *Boss* (2008; 2012); [5] *Kratter et al.* (2010a).

smoothing length, h, which is variable in modern implementations, and the softening parameter used to derive the disk's self-gravity from that of the ensemble of particles, the latter also having an analogous importance in grid codes (*Müller et al.,* 2012). Differences in the handling of radiative transfer can also lead to spurious results, prompting the testing of numerical schemes against analytical results (*Boley et al.,* 2007; *Boss,* 2009). Even indirect factors such as whether the central protostar is allowed to wobble to preserve the center of mass of the star-disk system may have an effect (cf. *Michael and Durisen,* 2010; *Boss,* 2012).

While SPH codes with spatially adaptive smoothing lengths are able to follow arbitrarily high-density clumps, most of the FD codes used to date have either fixed or locally refined grids, which limits their abilities to follow clumps in evolving disks. Pioneering DI calculations with the FLASH AMR code have been shown to achieve very good convergence with SPH simulations conducted with the SPH code GASOLINE, although only simple thermodynamics with a prescribed equation of state was employed (*Mayer and Gawryszczak,* 2008). A challenge for future work is to develop AMR simulations that retain the crucial physics of the DI mechanism.

2.2.3. Efficiency of disk instability with stellar/disk metallicity. While more giant planets have been discovered around metal-rich stars (*Fischer and Valenti,* 2005; *Mayor et al.,* 2011), some planets have been discovered around metal-poor hosts (e.g., *Santos et al.,* 2011). Furthermore, disks around low-metallicity stars are dispersed more rapidly [in ≤1 m.y. (*Yasui et al.,* 2009)], compared to classical T Tauri disks, which may survive for up to 10 m.y. If planets form in such systems the process must be rapid.

Fragmentation is heavily dependent on disk thermodynamics, which is affected by the disk opacity and mean molecular weight. Both of these scale directly with the disk metallicity. With recent developments of complex radiative transfer models, one can try to understand the effects of metallicity via these parameters. A decreased opacity is equivalent to simulating a lower metallicity environment or substantial grain growth. *Cai et al.* (2006) showed that gravitational instability becomes stronger when the opacity is reduced. This is because the cooling is more rapid in an optically thin environment, as the energy is able to stream out of the disk at a faster rate. Earlier studies investigating the effect of opacity on disk fragmentation suggested that varying it by a factor of 10 or even 50 had no effect on the fragmentation outcome (*Boss,* 2002; *Mayer et al.,* 2007). These were simulations carried out in the inner disk (≤30 AU), where it is optically thick, and so changes of this order did not significantly affect the cooling. On the other hand, decreasing the opacity by 2 orders of magnitude allowed the inner disks to fragment when otherwise identical disks with higher Rosseland mean opacities did not (*Meru and Bate,* 2010).

At large radii big opacity changes are not required for fragmentation to occur: Decreasing the opacity by a factor of a few to 1 order of magnitude is sufficient to increase the cooling rate to allow such disks to fragment when otherwise they would not have done so with interstellar opacity values (*Meru and Bate,* 2010; *Rogers and Wadsley,* 2012). Nevertheless, in general, when considering the effects of radiative transfer and stellar irradiation, opacity values that are lower than interstellar Rosseland mean values appear to be required. These results suggest that in a lower-metallicity disk, the critical radius beyond which the disk can fragment will move to smaller radii since more of the disk will be able to cool at the fast rate needed for fragmentation. Alternatively, sudden opacity changes, occurring on extremely short timescales, could be induced by the response of realistic composite dust grains to spiral shocks. Using post-processing analysis of three-dimensional disk simulations, *Podolak et al.* (2011) have found that the ice shells of grains with refractory cores are rapidly lost as they pass through the shock. This alters the grain-size distribution as ice migrates between grains with different sizes, which results in a change in the opacity to several times lower than standard values. Since the opacity change would occur on timescales much shorter than the orbital time, cooling would be locally enhanced, possibly promoting fragmentation even in the inner disk.

Furthermore, earlier studies showed that a larger mean molecular weight (>2.4, i.e., solar or supersolar metallicity) may be required for disk fragmentation to occur, as this reduces the compressional heating (*Mayer et al.,* 2007). This study was carried out for the inner disk, where small opacity changes are not as important (big changes are required for any effect to be seen, as illustrated above). Therefore the effect of a reduced compressional heating was more important. However, varying both parameters for the outer disk, which may be the most interesting region for disk fragmentation, has not yet been conducted. Variations of molecular weight from the nominal solar value (~2.4) to larger values (2.6–2.7) might occur as a result of the sublimation of ice coatings on grains as those grains pass through spiral shocks, if the dust-to-gas ratio is increased by at least an order of magnitude relative to mean nebular values (*Podolak et al.,* 2011). This increase could occur, e.g., through dust migration into arms via gas drag. Further investigation of the topic by coupling chemistry and hydrodynamical simulations will be a major task of future research on DI.

Furthermore, the effects of opacity in the spiral arms of self-gravitating disks also play an important role. Once fragments form, their survivability needs to be considered; it is affected by metallicity via a variety of processes that can determine changes in the opacity inside the collapsing clump and therefore affect its ability to cool and contract.

In summary, from a disk lifetime perspective, planet formation in low-metallicity systems has to occur faster than those in high-metallicity systems — making gravitational instability a more viable formation mechanism in such environments than core accretion. However, a clear trend between the efficiency of planet formation via gravitational instability and metallicity has not yet been established, since there are competing effects whose interplay at large radii (where gravitational instability is more likely to operate) is yet to be determined. For the outer disk, opacity changes are clearly important and suggest lower metallicity favors fragmentation. However, its importance relative to the effects of the mean molecular weight and that of line cooling is unclear. From the observational perspective the statistics are simply too low to rule in or rule out the possibility that gravitational instability is important at the low-metallicity end (*Mortier et al.*, 2013).

2.2.4. Clump evolution. The initial mass of the protoplanets that are formed via gravitational instability is not well constrained, but estimates suggest a mass range from ~1 M_{Jup} to ~10 M_{Jup} (*Boley et al.*, 2010; *Forgan and Rice*, 2011; *Rogers and Wadsley*, 2012). This uncertainty also leads to different predictions for the destiny of such clumps. For example, while *Boley et al.* (2010) and *Cha and Nayakshin* (2011) find that most clumps migrate inward and get tidally destroyed at several tens of astronomical units from the star, *Stamatellos and Whitworth* (2009) and *Kratter et al.* (2010b) find that the formation of much more massive objects is possible (brown dwarfs of tens of Jupiter masses) at radial distances of ~100 AU from the parent star. Global disk simulations are therefore not yet able to predict the outcome of clump formation in a robust fashion.

After a gaseous clump has formed in the disk, its subsequent evolution — namely whether or not a protoplanet with size, mass, density, and angular momentum comparable to a giant planet arises — depends on the initial conditions of the clump and on the nature of the formation process. Clumps indeed start with a fairly well-constrained density and temperature profile, as well as with a distribution of angular momentum inherited from the disk material (*Boley et al.*, 2010). Their initial composition can vary from subnebular to highly enriched compared to the star, depending on the birth environment in which the protoplanet has formed (see section 2.2.5 for details). The composition of the clump also affects the evolution since it determines the internal opacity (see *Helled and Bodenheimer*, 2010). The clump should also become convective, fully or partially, depending on its temperature structure and composition.

Once formed, the evolution of clumps consists of three main stages. During the first stage, the newly formed planetary object is cold and extended with a radius of a few thousand to 10,000× Jupiter's present radius (R_{Jup}), with hydrogen being in molecular form (H_2). The clump contracts quasistatically on timescales of 10^4–10^6 yr, depending on mass, and as it contracts, its internal temperature increases. When a central temperature of ~2000 K is reached, the molecular hydrogen dissociates, and a dynamical collapse of the entire protoplanet occurs (the second stage). The extended phase is known as the pre-collapse stage (*DeCampli and Cameron*, 1979), and during that phase the object is at most risk of being destroyed by tidal disruption and disk interactions. After the dynamical collapse, the planet becomes compact and dense, with the radius being a few times Jupiter's radius. During this third stage it is therefore

less likely to be disrupted, although the planetary object still has the danger of falling into its parent star due to inward migration, as in the CA model. The protoplanet then continues to cool and contract on a much longer timescale (10^9 yr). At this later stage, the evolution of the planet is similar for both CA and DI.

Helled and colleagues (*Helled et al.*, 2006, 2008; *Helled and Bodenheimer*, 2010; 2011) have studied the one-dimensional evolution of spherically symmetric, nonrotating clumps assuming various masses and compositions. Essentially, similarly to CA models, a system of differential equations analogous to those of stellar evolution is solved. These models were used to investigate the possibility of core formation and planetary enrichment, as we discuss below, and in addition, to explore the change in the pre-collapse evolution under different assumptions and formation environments. The pre-collapse timescale of protoplanets depends on the planetary mass, composition, and distance from the star, and it can vary from a few times 10^3 to nearly 10^6 yr. These timescales are long enough that the clump might be exposed to several harmful mechanisms while it is still a relatively low-density object, such as tidal mass loss during inward migration or photoevaporation by the central star and/or by neighboring massive stars, as we discuss below. However, during this stage, physical processes such as grain settling and planetesimal capture, which can significantly change the planetary composition, can occur. Typically, massive planets evolve faster, and therefore have a higher chance to survive as they migrate toward the central star. For massive objects (>5 M_{Jup}), the pre-collapse timescales can be reduced to a few thousand years if the protoplanets have low opacity (and therefore more efficient cooling), due to either low metallicities or due to opacity reduction via grain growth and settling (*Helled and Bodenheimer*, 2011). Likewise, the pre-collapse timescale can be increased to more than 10^5 yr if the metallicity is as large as 3× solar and the protoplanet has only a few Jupiter masses.

The pre-collapse evolution of protoplanets with masses of 3 and 10 M_{Jup} is presented in Fig. 4. As can be seen from the figure, more-massive protoplanets have shorter pre-collapse timescales. The initial radius increases with increasing planetary mass, as more-massive objects can be more extended and still be gravitationally bound due to stronger gravity. The central temperature, however, is higher for more-massive bodies.

Three-dimensional hydrodynamical collapse simulations of clumps formed by DI are in their early stages. In this case clumps are identified in simulations of global, self-gravitating protoplanetary disks, extracted from the disk with a numerically careful procedure, and then resampled at much higher resolution in order to follow the collapse to much higher densities (*Galvagni et al.*, 2012). Such clumps possess appreciable amounts of angular momentum with the highest values being above the expected threshold for the bar instability (*Durisen et al.*, 1986). Indeed, a bar-like mode is observed to arise in the clumps with very fast rotation, while in the others only strong spiral arms develop.

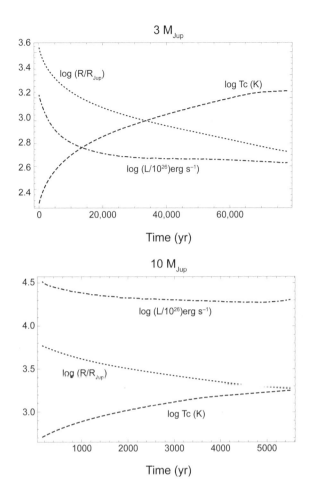

Fig. 4. Pre-collapse evolution of 3 and 10 M_{Jup} assuming solar composition and interstellar dust opacity. The luminosity, radius, and central temperature vs. time are presented. Adapted from *Helled and Bodenheimer* (2010).

The current three-dimensional simulations are purely hydrodynamical, with no account for solid-gas interactions such as planetesimal accretion and dust sedimentation, at variance with what can be done with the aforementioned one-dimensional models. Only the solar metallicity case has been considered so far. On the other hand, they follow, in the most general way possible, the hydrodynamics and self-gravity of a collapsing clump, and include also radiative cooling and compressional/shock heating. Strong compression and shocks are not included by construction in one-dimensional models. Radiative cooling is employed with different degrees of sophistication. The simulations adopt an EOS that takes into account the rotational, vibrational, and dissociation modes of H_2, and employ a simple cooling function (*Boley et al.,* 2010) that interpolates between optically thin and optically thick regimes. In the study by *Galvagni et al.* (2012), the pre-collapse timescales were found to be ~1000 yr, comparable to the shortest found in one-dimensional calculations.

2.2.4.1. *Opacity/metallicity and the contraction of protoplanets:* The opacity plays an important role as well

in terms of cooling during the post-formation contraction. Thus, metallicity has a direct impact on the pre-collapse evolution of the newly formed planets, and possibly on their survival. *Helled and Bodenheimer* (2011) have shown that when the planetary opacity is scaled with the stellar metallicity, the pre-collapse timescale is proportional to planetary metallicity. Shorter pre-collapse stages of metal-poor protoplanets can support their survival. Metal-rich protoplanets are more vulnerable to destruction; however, if they do manage to survive, they have a better opportunity to accrete solid planetesimals and form heavy-element cores.

When grain coagulation and sedimentation are included, it is found that the pre-collapse stages are significantly shorter for all the planetary masses and metallicities considered. The timescale is found to be ~10^3 yr for masses between 3 and 7 M_{Jup}, and is relatively insensitive to planetary composition (*Helled and Bodenheimer,* 2011). It is found that the pre-collapse evolution of a metal-rich protoplanet can actually be shorter than that of a metal-poor planet, a result of very efficient opacity reduction caused by the larger amounts of grains initially present in the atmosphere, leading to rapid grain growth and settling. The shorter timescales lead to two consequences: (1) a reduction in the risk of clump disruption, and (2) smaller final masses, because clumps accrete gas (and solids) most efficiently during their early evolution. On the other hand, the short times would lead to less-significant enrichment with heavy elements, at least via planetesimal capture, and would suppress the formation of cores in these objects. Whether DI or CA, the outcome of planet formation seems to be driven by solids.

2.2.4.2. *The effect of the disk on the planetary evolution:* Recently, *Vazan and Helled* (2012) modeled the pre-collapse evolution of protoplanets coupled directly to disk models and explicitly included accretion onto the protoplanet. The change in the pre-collapse timescale of a Jupiter-mass protoplanet for various radial distances was presented, and it was found that a 1-M_{Jup} protoplanet cannot evolve and contract at radial distances smaller than ~11 AU due to the influence of the disk. This is because of the sufficiently high pressure and temperature of the disk, which prevent the protoplanet from contracting to reach a dynamical collapse and to become a gravitationally bound object. While the exact radial distance for which contraction is no longer possible depends on the assumed disk model, the results of *Vazan and Helled* (2012) suggest that the presence of the disk has a nonnegligible effect on the planetary evolution, and that the pre-collapse timescale of protoplanets can significantly lengthen for protoplanets as they get closer to the star. The pre-collapse evolution timescale of a 1-M_{Jup} protoplanet, with solar composition and interstellar grain opacity, was found to range between 10^5 and 10^6 yr for radial distances between 50 and 11 AU, respectively. A Saturn-mass protoplanet was found to dissipate at a radial distance of ~12 AU, while 3- and 5-M_{Jup} protoplanets are found to dissipate only at 9 and 7 AU, respectively. Clearly, the influence/presence of the protoplanetary disk is less significant for more-massive protoplanets.

Since protoplanets are embedded in the gaseous disk, gas can be accreted as they evolve; therefore, the effect of gas accretion on the planetary evolution was investigated as well. It was shown that an increase of mass due to efficient gas accretion results in a faster contraction. It is also found that the planetary location has no significant influence on the planetary evolution when high accretion rates are considered. High accretion rates lead to significantly higher luminosity, which leads to shorter pre-collapse timescales (for details, see *Vazan and Helled*, 2012).

The conclusion of Vazan and Helled that the formation of planets by DI is limited to relatively large radial distances is in good agreement with previous research (*Rafikov*, 2007; *Boley*, 2009), but for the first time, it was based on planetary evolution considerations. However, if protoplanets are formed at large radial distances and migrate inward after the dynamical collapse, when the protoplanets are denser and more compact, the planets could survive at small radial distances. Alternatively, protoplanets could form and survive at small radial distances if their initial masses are sufficiently large ($\gg 1 \, M_{Jup}$).

2.2.4.3. *Clump migration in gravitationally unstable disks:* The angular momentum exchange and resulting migration of planets in gravitationally unstable disks varies noticeably from migration in laminar disks. Both *Baruteau et al.* (2011) and *Michael et al.* (2011) found, using independent methods, that planets in gravitationally unstable disks can migrate rapidly inward on roughly type I migration timescales (a few orbital times). In addition to the usual torques that are exerted on planets, stochastic torques are present due to disk turbulence, which may kick a planet inward or outward (the effects of such stochastic kicks are greater for lower-mass planets). Furthermore, there are some indications that planets may stall (*Michael et al.*, 2011) or even open gaps in the disk (*Zhu et al.*, 2012; *Vorobyov*, 2013), resulting in slower migration; however, the opening of clear gaps is not efficient for planets even as massive as 5 M_{Jup} due to strong turbulence in the disk resulting from gravitational instabilities and the stochastic orbital evolution of the planet (e.g., *Baruteau et al.*, 2011). Moreover, the evolution of the planet itself can also affect the migration history. For example, mass-losing planets may suffer from an outward torque, resulting in outward migration (*Nayakshin and Lodato*, 2012). Clearly, giant planet formation models that include planetary evolution and migration self-consistently are required to provide a more complete picture on the formation and evolution of DI planets.

2.2.5. *Composition of protoplanets.* Planets formed by DI are often thought to have stellar abundances. Recent studies, however, show that the composition of DI planets can range from substellar to superstellar.

2.2.5.1. *Metal enrichment from birth:* Spiral structure creates local gas pressure maxima that can collect solids through aerodynamic forces (e.g., *Weidenschilling*, 1977; *Rice et al.*, 2006; *Clarke and Lodato*, 2009), increasing the local solids-to-gas ratio. Spiral arms also act as local potential minima, which, when coupled with gas dissipa-

tion, can also help to maintain a high solid concentration. Because clumps are born from the fragmentation of spiral arms, they have the potential to form in regions of high solid concentrations. Hydrodynamics simulations with gas-solid coupling show that fragments can indeed be born with superstellar metallicities (*Boley and Durisen*, 2010; *Boley et al.*, 2011). However, the degree of enrichment strongly depends on the size distribution of solids. Very small solids will be well entrained with the gas and will not show a high degree of concentration, while very large solids have long coupling timescales, and are not expected to be captured by transient non-axisymmetric structure, such as spiral arms. The stopping time of a solid in a gaseous disk is $t_s \sim \rho_s s / (\rho_g c_s)$ for a solid of size s with internal density ρ_s embedded in a gas with density ρ_g and sound speed c_s. Solids that are expected to have the greatest degree of trapping in spiral structure have stopping times that are within a factor of a few of their orbital period. In a self-gravitating disk, maximum trapping corresponds to rock-sized objects (tens of centimeters) and boulders (~100 m).

If most of the solids are in the 10-cm- to 100-m size range, clumps can be enhanced at birth by factors of ~1.5–2 (*Boley and Durisen*, 2010). The overall enhancement is nontrivial but fairly modest, because some solid-depleted gas becomes mixed with the spiral arm during the formation of the clump. On the other hand, if most of the solids are in kilometer-sized objects or larger, then depletion at birth may actually be a possibility, as the solids do not necessarily follow the contraction of gas. The type of object that forms initially depends, in part, on the evolution of solids.

2.2.5.2. *Metal enrichment via planetesimal capture:* Another mechanism in which DI planets can be enriched with heavy elements is planetesimal accretion. Clumps that form in the disk are surrounded by planetesimals, which are affected by the gravitational field of the protoplanet and can therefore capture a large portion of solid material. *Helled et al.* (2006) and *Helled and Schubert* (2009) have shown that the final composition of the protoplanet (clump) can vary substantially when planetesimal capture is considered. The available solid mass for capture depends on the solid surface density σ_s at the planetary location. This density changes with the disk mass, its radial density profile, and stellar metallicity. The total available mass of solids in the planet's feeding zone depends strongly on the radial distance a, on the radially decreasing $\sigma_s(a)$, and weakly on the planetary mass. The actual planetary enrichment depends on the ability of the protoplanet to capture these solids during its pre-collapse contraction. The final mass that can be captured also depends on the physical properties of the planetesimals such as their sizes, compositions, and relative velocities.

The planetesimal accretion rate is given by equation (1), but here the capture radius is significantly larger. Enrichment by planetesimal capture is efficient as long as the protoplanet is extended (a few tenths of an astronomical unit). It then fills most of its feeding zone, so planetesimals are slowed down by gas drag, and are absorbed by the protoplanet. Therefore the longer the clump remains in the pre-collapse stage, the longer the clump will be able to accrete. *Helled et al.* (2006) have shown that a 1-M_{Jup} protoplanet at 5.2 AU with σ_s = 10 g cm^{-2} could accrete up to 40 $M_\oplus$ of heavy elements during its pre-collapse evolution. *Helled and Schubert* (2009) have shown that a 1-M_{Jup} protoplanet formed between 5 and 30 AU can accrete 1–110 $M_\oplus$ of heavy elements, depending on disk properties, and concluded that the final composition of a giant planet formed by DI is strongly determined by its formation environment.

Enrichment of massive protoplanets at large radial distances such as those discovered around HR 8799 (*Marois et al.,* 2008), where planets are likely to form by DI, has been investigated by *Helled and Bodenheimer* (2010). Since the timescale of the pre-collapse stage is inversely proportional to the square of the planetary mass, massive protoplanets would have less time to accrete solids.

The final composition of protoplanets can change considerably depending on the "birth environment" in which the planet forms. The large variations in heavy-element enrichment that are derived under different assumed physical conditions could, in principle, explain the diversity in the derived compositions of gas giant exoplanets. Since higher surface density leads to larger heavy-element enrichments, the DI scenario is also consistent with the correlation between heavy-element enrichment and stellar metallicity (*Guillot et al.,* 2006; *Burrows et al.,* 2007), highlighting the danger in assuming that metallicity correlations are prima facie evidence in support of the CA model. Note, however, that the correlation between the probability of detection of a giant planet and the metallicity of the star is a separate issue.

2.2.5.3. *Enrichment through tidal stripping:* Another mechanism that can lead to planetary enrichment in the DI scenario is enrichment through tidal stripping, which is one outcome of the tidal disruption/ downsizing hypothesis (*Boley et al.,* 2010; 2011; *Nayakshin,* 2010). Clumps create significant potential wells that allow solids to settle toward their center. This settling can lead to the formation of heavy-element cores (see section 2.2.6) that depletes the outer envelope of the clump. If the clump can be stripped of this metal-poor gas, the final object can become significantly enhanced in metals, producing objects that have extremely massive cores for their bulk mass (e.g, HAT-P-22b). These Brobdingnags may be a natural outcome of the DI model for the following reasons: (1) Clumps that form through fragmentation roughly fill their Hill sphere, to a factor of a few. At distances of ~100 AU, this implies initial clump sizes ~5 AU. (2) Clumps are highly susceptible to migration through the disk via clump-clump scattering (*Boley et al.,* 2010) and clump-disk torques (*Baruteau et al.,* 2011; *Michael et al.,* 2011). If the inward migration rate is faster than the contraction rate, the clump can overfill its Hill sphere, resulting in significant mass loss. If solid settling has occurred by this time, this overflow will preferentially remove metal-poor gas.

The various mechanisms for planetary enrichment in the DI model are much more efficient when the planets are formed in a metal-rich environment, i.e., around metal-rich stars.

2.2.6. *Core formation.* Protoplanets formed by DI can still have cores in their centers. However, unlike in the CA model, the existence of a core is not directly linked to the formation mechanism, and the core can also form after the formation of the planet. One way in which cores can form is via enrichment from birth. *Boley and Durisen* (2010) and *Boley et al.* (2011) have shown [building on the work of *Rice et al.* (2004, 2006)] that the solids that are collected by the forming planets tend to concentrate near the center, and can therefore end up as heavy-element cores.

Another way in which cores can be formed is via grain coagulation and settling. During the pre-collapse evolution the internal densities and temperature are low enough to allow grains to coagulate and sediment to the center (*De-Campli and Cameron,* 1979; *Boss,* 1998). Both of these studies, however, assumed that the planetary interior is radiative, while protoplanets are now found to be convective (*Wuchterl et al.,* 2000; *Helled et al.,* 2006). A detailed analysis of coagulation and sedimentation of grains of various sizes and compositions in evolving protoplanets including the presence of convection has been presented by *Helled et al.* (2008) and *Helled and Schubert* (2008). It was confirmed that silicate grains can grow and sediment to form a core both for convective and nonconvective envelopes, although the sedimentation times of grains with sizes smaller than about 1 cm are substantially longer if the envelope is convective. The reason is that grains must grow and decouple from the convective flux in convective regions in order to sediment to the center. Small icy/organic grains were found to dissolve in the planetary envelope, not contributing to core formation. Settling of volatiles into the planetary center is possible only if their sizes are sufficiently large (meters and larger).

A third mechanism for core formation is settling of larger bodies, i.e., planetesimals that are captured by the protoplanet. Core formation is favorable in low-mass protoplanets due to their longer contraction timescales, which allows more planetesimal accretion. In addition, they have lower internal temperatures and lower convective velocities, which support core formation (*Helled and Schubert,* 2008). The final core mass changes with the planetary mass, the accretion rate of solids (planetesimals), and radial distance. If planetesimals are captured during the pre-collapse stage, they are likely to settle to the center, and therefore increase the core mass significantly. As a result, there is no simple prediction for the core masses of giant planets in the DI model; however, simulations suggest a possible range of 0 to ~100 $M_\oplus$. In addition, it seems that the cores are primarily composed of refractory materials with small (or no) fractions of volatile materials (ices/organics). If grain settling works together with, e.g., enrichment at birth, then it may be possible to form much larger cores than what one would expect from pure settling.

2.2.7. *Mass loss.* Protoplanets formed by DI could lose substantial envelope mass, which can lead not just to enrichment through removing dust-depleted gas, but might also lead to formation of rocky cores/planets (*Nayakshin, 2010; Boley et al.,* 2010). The cooling and migration time-scales determine the degree of mass loss that can occur and where in the disk mass loss can commence.

The planet-star separation at which the mass loss commences depends on the mass and the age of the protoplanet, as well as the conditions in the protoplanetary disk itself. If H is still predominantly molecular in the protoplanet (very young and/or relatively less massive, e.g., $M_p \lesssim 5\text{-}M_{Jup}$ planet), then disruption is likely to occur at distances of ~1 AU, although the survival of clumps at this evolutionary stage and this distance has been shown to be unlikely (*Vazan and Helled,* 2012). If the planet is additionally puffed up by the internal energy release such as a massive core formation (*Nayakshin and Cha,* 2012) or by external irradiation (*Boss et al.,* 2002), then tidal disruption may occur at much greater distances, e.g., tens of astronomical units, where clumps are more likely to exist.

On the other hand, if the protoplanet is far along in its contraction sequence (more-massive $M_p \gtrsim 10\text{-}M_{Jup}$ and/or older protoplanets), then it may be in a "second stage," where H is atomic and partially ionized. In this case the protoplanet is at least an order of magnitude more compact and cannot be tidally disrupted unless it migrates as close as ~0.1 AU to its star (*Nayakshin,* 2011). It has been suggested that such tidal disruption events may potentially explain the population of close-in planets, both terrestrial and giant (see *Nayakshin,* 2011, and references therein).

Because fragments become convective as they contract, their internal structure can be approximated by a polytrope with n = 2.3, which is appropriate for a mixture of hydrogen and helium. In this regime, molecular protoplanets should have runaway disruptions, leaving behind only the much-denser parts, e.g., the cores. On the other hand, tidal disruptions of protoplanets in the immediate vicinity of the protostar, e.g., at ≤0.1 AU, are much more likely to be a steady-state process. *Nayakshin and Lodato* (2012) suggest that the protoplanets actually migrate outward during this process. The accretion rates onto the protostars during a tidal disruption process are as large as ~10^{-4} $M_{\odot}$ yr^{-1}, sufficiently large to present a natural model to explain the FU Ori outbursts of young protostars (*Hartmann and Kenyon,* 1996; *Vorobyov and Basu,* 2006; *Boley et al.,* 2010).

3. POST-FORMATION EVOLUTION

The formation phase of a giant planet lasts for ~10^4–10^6 yr, depending on the formation model, while the following phase of slow contraction and cooling takes place over billions of years. Thus it is the latter phase in which an exoplanet is most likely to be detected. The salient features are the gradual decline in luminosity with time, and the mass-luminosity relation, which gives luminosity of a planet roughly proportional to the square of the mass, so the prime targets for direct detection are relatively massive planets around the younger stars.

After accretion is terminated, the protoplanet evolves at constant mass on a timescale of several billion years with energy sources that include gravitational contraction, cooling of the warm interior, and surface heating from the star. The latter source is important for giant planets close to their stars because the heating of the surface layers delays the release of energy from the interior and results in a somewhat larger radius at late times than for an unheated planet. For giant planets close to their stars, tidal dissipation in the planet, caused by circularization of its orbit and synchronization of its rotation with its orbital motion, can provide a small additional energy source (*Bodenheimer et al.,* 2001). In addition, magnetic field generation in close-in planets can result in Ohmic dissipation in the interior leading to inflation (*Batygin and Stevenson,* 2010).

The initial conditions for the final phase depend only weakly on the formation process. In the DI model the evolution goes through a phase of gravitational collapse induced by molecular dissociation, and equilibrium is regained, 10^5 to 10^6 yr after formation, at a radius of only 1.3 R_{Jup} for the case of 1.0 M_{Jup} (*Bodenheimer et al.,* 1980). In the case of CA models, calculations (*Bodenheimer and Pollack,* 1986; *Bodenheimer et al.,* 2000) show that after the formation phase, which lasts a few million years, the radius declines to ≈10^{10} cm or 1.4 R_{Jup}, on a timescale of 10^5 yr for 1 M_{Jup}. This comparison has not been made, however, for planets in the 10-M_{Jup} range.

An important point is that in both cases there is an accretion shock on the surface of the planet. In the case of DI it occurs during the collapse induced by molecular dissociation. The outer layers accrete supersonically onto an inner hydrostatic core. In the case of CA it results from accretion of disk material onto the protoplanet during the phase of rapid gas accretion when the planetary radius is well inside the effective radius. The heat generated just behind the shock escapes the object by radiation through the infalling envelope. Thus much of the gravitational energy liberated by the infalling gas does not end up as internal energy of the planet; it is lost. The planet starts the final phase of evolution in a stage of relatively low entropy, sometimes referred to as a cold start. The actual entropy of the newly formed planet is somewhat uncertain, as it depends on the details of how much energy is actually radiated from behind the shock and how much is retained in the interior of the planet.

However, many numerical simulations do not take into account the formation phase in the calculation of the final contraction phase. They simply assume that the evolution starts at a radius 2–3× that of the present Jupiter, and assume the model is fully convective at that radius. The shock dissipation and radiation of energy is not taken into account, and the result is that the initial models are warmer and have higher entropy than the "cold start" models. They are known as "hot start" models. After a certain amount of time, both types of models converge to the same track of radius or luminosity vs. time, but at young ages the results under the two assumptions can be quite different, depending on the mass of the planet. Jupiter-mass planets converge to the same track after 20 m.y.; however, 10-M_{Jup} planets take over 10^9 yr to converge (*Marley et al., 2007*). Thus young, relatively massive planets calculated according to the "cold start" are considerably fainter than those, at the same age, calculated with the "hot start."

Note that it is often assumed that, as a result of gas accretion through a shock, planets formed by CA finish their formation with lower luminosities than planets formed via DI. This conclusion, however, depends on (1) the efficiency of energy radiation through the shock, and (2) the core mass of the planet at the end of accretion. *Mordasini et al.* (2012b) show that a CA calculation in which no energy is radiated from the shock produces a model very close to a "hot start." Furthermore, "cold start" calculations show that the entropy and luminosity of a planet after formation depend on its core mass (*Bodenheimer et al.*, 2013). For example, a 12-M_{Jup} planet (which is in the suspected mass range for directly imaged exoplanets), just after formation, could have a luminosity ranging from log $L/L_\odot = -4.8$ to -6.6, corresponding to core masses ranging from 31 to 4.8 $M_\oplus$. For lower final masses or higher core masses, the luminosities can be even higher (*Mordasini, 2013*). In fact, both formation models could result in either "cold" or "hot" starts. In order to better understand the evolutions of giant planets when accretion and shock calculations are included, the development of three-dimensional-hydrodynamical models is required.

Future planetary evolution models should include physical processes such as core erosion, rotation, settling, etc., which are typically not included in evolution models due to their complexity. Such processes could change the contraction histories of the planets as well as the final internal structure. While we are aware of the challenge in including such processes in planetary evolution models, we hope that by the time of Protostars and Protoplanets VII (PPVII), progress in that direction will be reported.

4. SUMMARY OF GIANT PLANET FORMATION MODELS

4.1. Core Accretion

A giant planet forming according to the CA model goes through the following steps:

1. Accretion of dust particles and planetesimals results in a solid core of a few Earth masses, accompanied by a very-low-mass gaseous envelope.

2. Further accretion of gas and solids results in the mass of the envelope increasing faster than that of the core until a crossover mass is reached.

3. Runaway gas accretion occurs with relatively little accretion of solids.

4. Accretion is terminated by tidal truncation (gap formation) or dissipation of the nebula.

5. The planet contracts and cools at constant mass to its present state.

The *strengths* of this scenario include:

1. The model can lead to both the formation of giant and icy planets.

2. The model can explain the correlation between stellar metallicity and giant planet occurrence and the correlation between the stellar and planetary metallicity.

3. The model is consistent with the enhancement of the heavy-element abundances in Jupiter, Saturn, Uranus, and Neptune.

4. The model predicts that giant planets should be rare around stars with lower mass than the Sun. There are in fact some giant planets around low-mass stars, but the frequency (for masses above 0.5 M_{Jup}) is only about 3% for stars of about 0.5 $M_\odot$ (*Johnson et al., 2010*), at solar metallicity.

5. The long formation time, caused by the slow gas accretion during phase 2, can be considerably reduced when the effect of grain settling and coagulation on the dust opacity is taken into account (*Movshovitz et al., 2010*). Similarly, including migration resulting from disk-planet angular momentum exchange can suppress phase 2, therefore substantially decreasing the formation time.

A number of *problems* also emerge:

1. The model depends on the processes of planetesimal formation and early embryo growth, which are poorly understood.

2. The timescale for orbital decay of the planet due to type I migration, although uncertain, may still be shorter than the time required to build up the core to several Earth masses. Further details on these processes are described in the chapter by Baruteau et al. in this volume.

3. The model faces extreme difficulties in explaining planets around stars of very low heavy-element abundance or massive giant planets at radial distances greater than about 20 AU.

4. In the absence of migration, the formation timescale is close to the limits imposed by observations of protoplanetary disks.

5. Final envelope abundances and gas giant formation timescales are dependent on the dynamics of planetesimals, which is still very uncertain.

6. The opacity of the envelope plays a critical role in gas giant planet formation timescales. The actual opacity of accreted gas onto a growing gas giant and the role grain growth plays in the evolution of solids in the envelope still have significant uncertainties.

4.2. Disk Instability

While as old as the CA paradigm, if not older, DI is taking considerably longer to mature as a theory. In part this is due to the complexities of understanding the conditions under which disks can fragment, but is also in part due to the realization that additional branches of the theory are required, including recognizing possible paths for forming rocky objects.

Generally speaking, the main weakness of the DI model is the uncertainty regarding whether clumps can indeed form in realistic disks, and even if they do form, it is not clear if they can actually survive and evolve to become gravitationally bound planets. Giant planet formation by DI is conditioned by:

1. Disks must at some point become strongly self-gravitating. This can be achieved by mass loading sections of a disk through accretion of the protostellar envelope or through variations of the accretion rate in the disk itself.

2. There must be significant cooling in the disk to increase the likelihood of fragmentation of spiral structure, although the exact limits are still being explored.

The main *advantages* to this mechanism are:

1. Fragmentation can in principle lead to a variety of outcomes, including gas giants with and without cores, metal-rich and metal-poor gas giants, brown dwarfs, and possibly terrestrial planets. These various outcomes depend on a complex competition between the cooling, dust coagulation and settling, accretion, and disk-migration timescales.

2. Planet formation can begin during the earliest stages of disk evolution, well within the embedded phase.

3. Disk instability may lead to many failed attempts at planet formation through fragmentation followed by tidal disruption, which could be linked with outburst activities of young protoplanetary disks, as well as to the processing of some of the first disk solids.

4. Disk instability can take place at large disk radii and in low metal environments, consistent with the HR 8799 (*Marois et al.*, 2008) planetary system.

The main *disadvantages* of this mechanism are:

1. The exact conditions that can lead to disk fragmentation and the frequency of real disks that become self-gravitating are not well understood.

2. Even if protoplanets can be formed by DI, whether they can survive (both tidal disruption and rapid inward migration) to become gravitationally bound planets is still questionable.

3. The model cannot easily explain the formation of intermediate-mass planets.

4. The model does not provide a natural explanation for the correlation of giant planet occurrence and stellar metallicity.

5. Grain evolution is a principal branching mechanism for the different types of objects that form by DI. Thus, a predictive model for DI cannot be developed without a more complete model of grain physics.

6. The detailed effects of mass accretion onto clumps and mass removal by tides must still be explored.

7. As summarized above for CA, the detailed interactions between clumps and planetesimals, if present, must be further developed.

4.3. Core Accretion and Disk Instability: Complementarity

Core accretion and DI are not necessarily competing processes. Disk instability is likely most common during the early embedded phases of disk evolution (~few 10^5 yr), while CA occurs at later stages (~few million years). In this view, DI could represent the first trials of planet formation, which may or may not be successful. If successful, it does not preclude formation by CA at later stages.

Core accretion may be the dominant mechanism for forming ice giants and low-mass gas giants, while DI may become significant for the high end of the mass distribution of giant planets. Nevertheless, it seems possible that both scenarios can lead to the formation of an object on either end of the mass spectrum.

Overall, CA explains many of the properties of solar system planets and exoplanets. Nevertheless, a number of exoplanets cannot naturally be explained by the CA model, such as giant planets at very large radii and planets around metal-poor stars, both of which have been observed. Models of both CA and DI are continuously rejuvenating, and we hope that by PPVII, we will have a better understanding of the types and relative frequency of planets under those two paradigms.

5. GIANT PLANET INTERIORS

5.1. Giant Planets in the Solar System

Although there are now hundreds of known giant exoplanets, we can usually measure only the most basic parameters: their mass and/or radius. While the mean density is a powerful diagnostic, it is insufficient to give a detailed picture of the planetary interior. At present, only the planets in our own solar system can be studied in sufficient detail to allow us to form such a picture. Details of these modeling efforts are presented in the chapter by Baraffe et al. in this volume, but because the results of such models are of great importance in helping us constrain formation scenarios, we devote this section to looking at some of the difficulties and limitations of interior models.

Models of the planetary interior have three main components, which can be characterized as simple physics, complex physics, and philosophy. In addition, the issues that need to be considered in each of these components will be different for the class of bodies that are mostly H/He, like Jupiter and Saturn, than for the class comprising Uranus and Neptune. The latter contain substantial amounts of H/He as well, but the bulk of their mass is high-Z material.

5.1.1. Gas giants: Jupiter and Saturn.

5.1.1.1. Simple physics: The basic components of any model are the conservation of mass and of momentum. The former is written as

$$\frac{dM}{dr} = 4\pi r^2 \rho(r) \tag{5}$$

where M(r) is the mass contained in a sphere of radius r and $\rho(r)$ is the density at r. Conservation of momentum is simply an expression of force balance, and is given by

$$\frac{dP}{dr} = -\frac{GM(r)\rho(r)}{r^2} + \frac{2}{3}\omega^2 r \rho(r) \tag{6}$$

where P is the pressure, and ω is the rotation rate. The second term on the righthand side is an average centrifugal term that is added to allow for the effect of rotation. For Jupiter this amounts to a correction of ~2% in the interior, rising to ~5% near the surface. For Saturn it is about twice that. For Uranus and Neptune it is less than 1% except near the surface. As the inputs to the models become more precise, equations (5) and (6) must be replaced by their two-dimensional versions.

5.1.1.2. *Complex physics:* A description of how the temperature changes within the planet must be added to the framework described above, and this, in turn, depends on the mechanism of energy transport. *Hubbard* (1968) showed that in order to support the high thermal emission observed for Jupiter, most of its interior must be convective, and therefore this region can be well represented by an adiabat. It is generally assumed that the same is true for Saturn. Although *Guillot et al.* (1994) have suggested the possibility that Jupiter and Saturn might have radiative windows at temperatures of ~2000 K, and *Leconte and Chabrier* (2012) have studied how composition gradients can modify the usual adiabatic gradient through double diffusive convection, the assumption of a strictly adiabatic temperature gradient is still quite standard.

Even with the assumption of an adiabatic temperature gradient, there is still the nontrivial problem of computing the density profile. Equations of state have been measured experimentally for H, He, H_2O, and some additional materials of interest, but there are still large uncertainties in the experiments themselves and in their theoretical interpretation (e.g., *Saumon et al.,* 1995; *Nettelmann et al.,* 2008; *Militzer et al.,* 2008). This leads to significant uncertainties in, for example, the estimated core mass of Jupiter and Saturn. The lack of knowledge on the depth of differential rotation in Jupiter and Saturn also introduces a major source of uncertainty to structure models.

Another feature that is difficult to model is the effect of the magnetic field. The effects in Jupiter and Saturn are expected to be small, and have mostly been ignored, but as measurements improve, including magnetic effects will become necessary. These will include pressure effects that must be added to equation (6), as well as contributions to the viscosity that will affect the convective motions, and, as a result, the temperature gradient. These motions may also contribute to differential rotation in these planets, and may themselves be driven by differential rotation. In return, magnetic field modeling may eventually allow us to set useful limits on location of conductive regions of the planet's interior (e.g., *Stanley and Bloxham,* 2006; *Redmer et al.,* 2011), and on the extent of the differential rotation.

5.1.1.3. *Philosophy:* Perhaps the most difficult aspect of modeling is the choice of overall structure and composition. Traditional models of Jupiter and Saturn assumed a core of heavy material surrounded by a H/He envelope with some additional high-Z component (*Podolak and Cameron,* 1974; *Saumon and Guillot,* 2004), but more recent models have divided the envelope into a heavy-element-rich layer under

a layer with more nearly solar composition (*Chabrier et al.,* 1992; *Guillot,* 1999; *Nettelmann et al.,* 2012). The motivation for dividing the envelope is based on the idea that at high pressures not only does H change from the molecular to the metallic phase, but He may become immiscible in H and rain out to the deeper interior (*Stevenson,* 1975; *Wilson and Militzer,* 2010). The depletion of He in Saturn's upper atmosphere has been observed (*Gautier et al.,* 2006) and He rain has been advanced as an explanation for Saturn's high thermal emission (*Fortney and Nettelmann,* 2010).

The assumption of a core was originally based on the need to match the low gravitational quadrupole moments of these planets (*Demarcus,* 1958). This assumption was later strengthened by the CA model, which requires a large heavy-element core to initiate the capture of a H/He envelope. However, for both Jupiter and Saturn, interior models without cores can be found. This highlights the fact that the model results are due in part to the underlying picture used by the modelers in deriving their results. This has led to a second general approach to interior modeling where some function is assumed for the density distribution as a function of radius, and the free parameters of this function are chosen to fit the observables such as the average density, and the moments of the gravitational field. One then tries to interpret this density distribution in terms of composition using equations of state.

5.1.1.4. *Ambiguities in the models:* The reason the core mass varies so widely can be understood by considering a simplified toy model in which the density distribution is given by

$$\rho(r) = \rho_c \qquad r \le R_c$$
$$\rho(r) = \rho_0 \left[1 - \left(\frac{r}{R}\right)^\alpha \right] \qquad r > R_c \qquad (7)$$

where R_c is the core radius, R is the planet's radius, and ρ_c, ρ_0, and α are constants. α = 2 gives a fair representation of Jupiter's density distribution as derived from detailed interior models. Once the core density ρ_c is chosen, R_c and ρ_0 are found by requiring that the distribution reproduce Jupiter's mass and inertia factor, which are determined using the Radau approximation (*Podolak and Helled,* 2012).

As can be seen from Fig. 5, $\rho_c = 10$ g cm^{-3} (dashed) or $\rho_c = 100$ g cm^{-3} (dotted) give nearly identical values for the envelope's density, and this density distribution is very similar to one derived using realistic equations of state and matching the observed gravitational moments (solid). In other words, not only are the mass and moment of inertia insufficient to distinguish between a core density of 10 g cm^{-3} and one of 100 g cm^{-3} and core masses differing by a factor of 1.5, but probing the density distribution in the outer parts of the planet is not enough either.

We cannot measure the internal density distribution of a planet directly, but we can measure some of the moments of the distribution. Because of the cylindrical hemispheric symmetries of a rotating planet the moments are even functions of only the radius r and co-latitude θ and are given by

$$J_{2\ell} = -\frac{1}{Ma^{2\ell}} \int r'^{2\ell} P_{2\ell}(\cos\theta') \rho(r',\theta') d^3r' \qquad (8)$$

where M is the mass, a is the average radius, and P_n is the nth-order Legendre polynomial. Interior models are constructed to fit the mass (essentially J_0) and as many of the $J_{2\ell}$'s as have been measured. Although each higher-order J gives additional information on $\rho(r)$, this information is more and more strongly weighted by high orders of r. In fact, only J_0 (i.e., the mass) probes the region of the core. The higher J's are only sensitive to the density distribution in the envelope. Since this is nearly independent of the core density, small changes in the envelope density caused by differences in the equation of state or in the assumed composition can cause large changes in the calculated mass of the core, which ranges between 0 and up to ~15 $M_\oplus$.

The composition of the envelope must also be treated with caution. The density of a mixture of materials is given, to an excellent approximation, by the additive volume assumption. If ρ_i is the density of the ith component and X_i is its mass fraction, then assuming the volume of the mixture is the sum of the volumes of the individual components gives $1/\rho = \Sigma_i X_i/\rho_i$. For Jupiter, H and He account for ~90% of the envelope by mass, while the high-Z component makes up the remainder. On the other hand, the density of the high-Z component is much larger than that of H. Thus the major contribution to ρ is due to H. As a result, if we try to match the density at some point in the envelope with a particular mixture, a small error in the equation of state of H may cause a much larger error in the estimated abundance of high-Z material (*Podolak and Hubbard*, 1998).

Finally, there is the ambiguity due to the uncertainty in the composition itself. The same pressure-density relation can be well reproduced by different mixtures, even over a

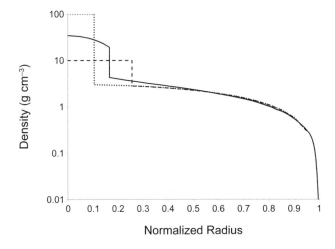

Fig. 5. Density vs. radius for a realistic Jupiter model (solid curve), a toy model with ρ_c = 10 g cm^{-3} (dashed curve), and a toy model with ρ_c = 100 g cm^{-3} (dotted curve). Although the core densities differ greatly, the densities for r ≥ 0.5 are nearly identical.

range of pressures. Measurements of the magnetic field, which gives information about the conductivity as a function of depth, may help break this degeneracy, but such measurements require detailed dynamo models for their interpretation, and the construction of such models is still in the early stages. In addition, a more detailed knowledge of the planet's differential rotation (if any) is required.

5.1.2. Icy giants: Uranus and Neptune. Typically, giant planet formation models concentrate on Jupiter-like planets, while lower-mass planets, such as Uranus and Neptune, are simply considered to be "failed giant planets." These planets are more naturally explained by CA, in which the forming planets remain gas-poor due to their slow growth at farther distances where runaway gas accretion cannot be reached before the gas dissipates from the young planetary system. It should be noted, however, as was discussed earlier, that icy/rock low-mass planets could form by DI if there is gaseous mass loss due to tidal stripping or photoevaporation. The efficiency of forming planets like Uranus and Neptune in such a scenario is still unknown, and further investigation of this topic is required.

Modeling the interior structures of Uranus and Neptune is even more difficult. In the first place, the high-Z component is a much larger fraction of the total mass and must therefore be treated more carefully. In the second place, the thermal structure is not well known. Neptune has a strong internal heat source, so an adiabatic interior is a reasonable assumption, but Uranus is in equilibrium with solar insulation (e.g., *Pearl and Conrath*, 1991). In addition, thermal evolution models of Uranus give too long a cooling time for the planet (*Fortney et al.*, 2011). It seems likely that some process like layered diffusion is inhibiting the heat flow in the planet, and temperatures in the interior might be much higher than adiabatic.

In addition, the observational data for Uranus and Neptune have much larger uncertainties than for Jupiter and Saturn. The observed gravitational moment J_2, flattening f, and angular velocity ω are related to first order by

$$J_2 - \frac{2f}{3} + \frac{q}{3} \approx 0 \qquad (9)$$

where $q \equiv \omega^2 a^3/GM$ is the ratio of centrifugal to gravitational forces. The rotation periods determined by the Voyager flybys together with the measured values of J_2 and f do not satisfy equation (9) (*Podolak and Helled*, 2012). Indeed, *Helled et al.* (2010) suggested that Uranus' rotation period is significantly shorter, and Neptune's significantly longer than the values determined by Voyager.

Uranus and Neptune have usually been treated using three-shell models, where the structure was divided into a core of rock, a shell of "ice," and a H/He envelope in the solar ratio together with an admixture of high-Z material. More recent models have maintained the three-shell structure but have replaced the ice shell and envelope with an inner envelope containing H/He together with a large fraction of high-Z material and an outer envelope composed of H/He and a small mass fraction of high-Z material.

As is the case for Jupiter and Saturn, there is no compelling reason to limit the number of shells to three. *Helled et al.* (2010) have assumed that ρ(r) for these planets could be fit by a sixth-order polynomial and have found the coefficients required to fit all the observed parameters. The "empirical" equation of state generated by these models could be interpreted as requiring a continuous increase in the H/He mass fraction with increasing radius. This would be the equivalent of infinitely many shells. Without a clearer understanding of the temperature structure inside these bodies, it is difficult to generate more detailed models. In addition, *Helled et al.* (2010) have suggested that Uranus and Neptune could be rather "dry." It was shown that the gravity data can be fit as well with silicates as with water. This also raises the question as to whether Uranus and Neptune are truly icy planets or, alternatively, silicate-dominated planets, similar to Pluto (*Simonelli and Reynolds,* 1989).

A puzzling problem to face modelers is the fact that although Uranus and Neptune have similar masses and radii, and are both found in the outer regions of the solar system, they appear to be quite different internally. In addition to the differences in heat flow mentioned above, Uranus' radius is larger than Neptune's but it is smaller in mass. This means that Neptune is more dense than Uranus by 30%. This difference is likely to be the result of differences in the formation history of these bodies. *Stevenson* (1986) and *Podolak and Helled* (2012) have discussed the possibility that giant impacts have significantly affected the internal structure of these planets, which led to the observed dichotomy between them. If this is so, then details of this formation history may eventually be extracted from interior models. This exciting possibility is still not within our grasp, and should also be considered when modeling interiors of exoplanets with similar masses.

5.2. Giant Exoplanets

5.2.1. Structure and composition of exoplanets. 5.2.1.1. Basic concepts: There are two main ways that one might try to constrain the composition of a giant exoplanet. Since gravity field or moment of inertia information about distant planets will be rarely available, if at all, observers must utilize more indirect methods. One avenue is from spectroscopy. Since our solar system's giant planets are enhanced in heavy elements (at least in carbon, which can be observed as methane), spectra of exoplanet atmospheres could potentially also be used to determine if the H/He envelopes of such planets are enriched in heavy elements (*Chabrier et al.,* 2007; *Fortney et al.,* 2008). In warm planets, H_2O, CO, CH_4, and NH_3 could all be detectable in the infrared. However, high S/N observations will be needed across a wide wavelength range for strong constraints, either for hot Jupiters observed during transit or occultation, or young planets that are directly imaged. There are no firm

constraints on giant planet envelope metallicity at this time.

A simpler avenue, at least conceptually, is direct observations of a transiting planet's mass and radius, which yield its mean density. Models for a solar composition giant planet at a given measured mass and age can be calculated. If the observed planet has a smaller radius (has a higher mean density) than the model planet, this is strong evidence that the planet is enhanced in heavy elements compared to the solar composition model. It is readily apparent that Jupiter and Saturn are enhanced in heavy elements compared to the Sun, from a measurement of their mass and radius alone, without knowledge of their gravity field. Of course for an exoplanet, we will generally remain ignorant of whether these heavy elements are predominantly within a core or are mixed throughout the H/He envelope.

5.2.1.2. Recent work: The use of mass and radius measurements of transiting planets as a structure probe for giant planets has been strongly hindered in practice. The most strongly irradiated planets ($T_{eq} > 1000$ K) have radii larger than expected from standard models of giant planet cooling and contraction (*Bodenheimer et al.,* 2001; *Guillot and Showman,* 2002; *Baraffe et al.,* 2003; *Fortney et al.,* 2007; *Burrows et al.,* 2007). Some unmodeled physical process yields planetary radii (and hence interior temperatures) that are inflated compared to the predictions of simple contraction models. Since for any given planet the level of additional interior energy (or other process) that causes radius inflation is unknown, the amount of heavy elements within a planet (which would cause a smaller radius) is therefore unconstrained. In cases where the planet is still smaller than a pure H/He object, a lower limit on its heavy-element mass can be obtained.

Even with the uncertainty inherent in the radius inflation mechanism, progress has been made on constraining exoplanet composition with transiting planets. With a sample size of only nine planets, *Guillot et al.* (2006) were able to show that there is a strong correlation between stellar metallicity and the inferred mass of heavy elements within a planet. The most iron-rich stars possessed the most heavy-element-rich planets. However, *Guillot et al.* (2006) made the *ad hoc* assumption that the radius inflation mechanism (additional power in the convective interior) scaled as the stellar flux incident upon each planet. With a somewhat larger sample size, *Burrows et al.* (2007) came to a similar relation between stellar and planetary metal enrichment. *Burrows et al.* (2007) made the assumption that *ad hoc* high atmospheric opacity slowed the contraction of all planets, leading to larger-than-expected radii at ages of 1 Ga or greater.

Miller and Fortney (2011) recognized that anomalously large radii disappear at incident fluxes less than 2 × 10^8 erg s^{-1} cm^{-2} ($T_{eq} \sim 1000$ K). By treating only these relatively cool planets, and assuming no additional internal energy source for them, the heavy-element mass of each planet can be derived, as all these cooler planets are smaller than solar-composition H/He objects. With their sample of 14 planets, they find that all need at least ~10–15 $M_\oplus$

of heavy elements. The enrichments of Saturn and Jupiter were found to fit in well within their exoplanetary "cousin" planets. When the heavy-element mass is plotted vs. metallicity, there is a clear correlation between the mass of heavy elements in a giant planet and the metallicity of the host star, as already noted by *Guillot et al.* (2006) and *Burrows et al.* (2007). However, the enrichment (i.e., Z_{planet}/Z_{star}) is not very sensitive to the metallicity of the star. The data suggest that it is the planetary mass that determines the relative enrichment and not the metallicity of the star. This tendency is consistent with the finding that the occurrence rate of low-mass exoplanets is not increasing with increasing stellar metallicity (*Buchhave et al.,* 2012). The sample implies that for low metallicity ([Fe/H]< 0), the planets that exist are the ones with the large enrichments, i.e., the more-terrestrial planets. For higher metallicities ([Fe/H]> 0), a large range of enrichments is possible. Clearly, a new area of modeling work will be the comparison of population synthesis models (see chapter by Benz et al. in this volume) to observed giant planet composition vs. planetary mass, stellar mass, stellar metallicity, orbital separation, eccentricity, orbital alignment, and other parameters. As the sample size of cooler transiting gas giants expands, such avenues of research are likely to bear fruit.

One important finding from the past decade is that it is clearly incorrect to uniformly assume that all giant planets possess ~15 $M_\oplus$ of heavy elements, an estimate that would work fairly well for Jupiter, Saturn, Uranus, and Neptune. The detection of transits of the relatively small-radius Saturn-mass planet HD 149026 by *Sato et al.* (2005) showed that even relatively low mass "gas giants" can possess 60–90 $M_\oplus$ of heavy elements. Relatively small radii for massive giant planets from ~2 to 10 M_{Jup} show that hundreds of Earth masses of heavy elements must be incorporated into some planets (*Baraffe et al.,* 2008; *Miller and Fortney,* 2011). It seems likely that for massive giant planets with large enrichment in heavy elements, the bulk of the heavy-element mass is in the H/He envelope, rather than in a core. This is an area that should be investigated in future formation models.

5.2.1.3. *Degeneracy in composition:* While it is certainly essential to derive the heavy-element mass fractions for giant planets, it would be even more interesting to derive additional details, like the ice/rock ratio. This could in principle be easier for the lower-mass (Neptune-like) giant planets, where the bulk of their mass is in heavy elements, not H/He. However, there is significant compositional degeneracy, as rock-H/He mixtures yield similar mass-radius relations as three-component models that include rock, ices, and H/He. This is not a new phenomenon, and as discussed earlier, it has long been appreciated that Neptune and Uranus can be modeled with a wide array of mixtures of rock, ices, and H/He (e.g., *Podolak et al.,* 1991). Therefore there is no observational evidence that exoplanets in the ice giant mass range are actually composed mostly of fluid ices. There is at least one possible way around this degeneracy that could yield some information on heavy-element com-

position. One could obtain correlations for a large number of planets for planetary heavy-element enrichment as a function of stellar [O/H] and [Si/H], in addition to [Fe/H].

6. THE FUTURE

Observations of hundreds of exoplanets, as well as increasingly detailed information about the giant planets in our own solar system, have inspired modelers to explore planet-formation scenarios in more detail. To date we have explored the effects of such processes as migration, planetesimal capture, disk metallicity, disk magnetic fields, etc. It is perhaps not surprising that these processes often work in opposite directions. We are entering a new stage in our understanding of giant planets, one in which we realize that in order to untangle the complex interplay of the relevant physical processes it is necessary to build models that self-consistently combine various physical processes. In particular, we suggest the following specific investigations:

1. Work toward a self-consistent model for the evolution of the growing planet, the gaseous disk, and the planetesimal distribution in the CA paradigm.

2. Develop a better model for planetesimal properties: velocities, inclinations, sizes, compositions, morphology, and the time dependence of these properties.

3. Track the deposition of heavy elements that are accreted in the envelope of a forming planet and account for their EOS and opacity in CA models.

4. Determine the final masses and identify the primary physical effects that determine the range of final masses in the CA, DI, and hybrid paradigms.

5. Develop opacity calculations that account for the initial size distribution of interstellar grains, grain fragmentation, various possible morphologies, and grain composition for both CA and DI.

6. Develop self-consistent models of the disk-planet interaction and co-evolution in the DI model.

7. Determine whether *in situ* formation of short-period massive and intermediate-mass planets is plausible.

8. Explore the earliest stages of disk evolution to search for observational signatures of planet-formation mechanisms. For example, if DI is a viable formation pathway, then spiral structure in very young disks should be detectable. The Atacama Large Millimeter/sub-millimeter Array (ALMA) and future instruments may make this possible.

9. Determine what physical factors lead to fragmentation, how metallicity affects planet formation by DI, and the observational consequences for the DI model.

10. Perform high-resolution two-dimensional/three-dimensional simulations of clumps through their contraction phases, including the effects of rotation, tidal perturbations, and background pressures and temperatures.

11. Develop predictions of the two formation models that can be tested by future observations.

12. Perform two-dimensional structure model calculations for the solar system planets accounting for the static

and dynamical contributions and determine improved equations of state of H, He, high-Z, and their interactions, as well as possibly magnetic fields.

13. Include additional constraints for the planetary internal structure of giant exoplanets such as the Love number/flattening and atmospheric composition.

This work is now beginning, and it is hoped that by the time of PPVII we will understand planetary systems better — both our own and those around other stars.

Acknowledgments. Support for A.C.B. was provided under contract with the California Institute of Technology funded by NASA via the Sagan Fellowship Program. The work of A.P.B. was partially supported by the NASA Origins of Solar Systems Program (NNX09AF62G). F.M. is supported by the ETH Zurich Post-Doctoral Fellowship Program and by the Marie Curie Actions for People COFUND program. P.B. was supported in part by a grant from the NASA Origins program. M.P. acknowledges support from ISF grant 1231/10. Y.A. is supported by the European Research Council through grant 239605 and thanks the International Space Science Institute, in the framework of an ISSI Team.

REFERENCES

Alexander R. D. et al. (2008) *Mon. Not. R. Astron. Soc., 389,* 1655–1664.
Alibert Y. et al. (2004) *Astron. Astrophys., 417,* L25–L28.
Alibert Y. et al. (2005a) *Astron. Astrophys., 434,* 343–353.
Alibert Y. et al. (2005b) *Astrophys. J. Lett., 626,* L57–L60.
Alibert Y. et al. (2011) *Astron. Astrophys., 526,* A63.
Andrews S. M. et al. (2011) *Astrophys. J., 732,* 42.
Andrews S. M. et al. (2013) *Astrophys. J., 771,* 129.
Ayliffe B. and Bate M. R. (2012) *Mon. Not. R. Astron. Soc., 427,* 2597–2612.
Baraffe I. et al. (2003) *Astron. Astrophys., 402,* 701–712.
Baraffe I. et al. (2008) *Astron. Astrophys., 482,* 315–332.
Baruteau C. et al. (2011) *Mon. Not. R. Astron. Soc., 416,* 1971–1982.
Batygin K. and Stevenson D. J. (2010) *Astrophys. J. Lett., 714,* L238–L243.
Bodenheimer P. and Pollack J. B. (1986) *Icarus, 67,* 391–408.
Bodenheimer P. et al. (1980) *Icarus, 41,* 293–308.
Bodenheimer P. et al. (2000) *Icarus, 143,* 2–14.
Bodenheimer P. et al. (2001) *Astrophys. J., 548,* 466–472.
Bodenheimer P. et al. (2013) *Astrophys. J., 770,* 120.
Boley A. C. (2009) *Astrophys. J. Lett., 695,* L53.
Boley A. C. and Durisen R. H. (2010) *Astrophys. J., 724,* 618–639.
Boley A. C. et al. (2007) *Astrophys. J., 665,* 1254–1267.
Boley A. C. et al. (2010) *Icarus, 207,* 509–516.
Boley A. C. et al. (2011) *Astrophys. J., 735,* 30.
Boss A. P. (1998) *Astrophys. J., 503,* 923.
Boss A. P. (2002) *Astrophys. J. Lett., 567,* L149–L153.
Boss A. P. (2006a) *Astrophys. J., 641,* 1148–1161.
Boss A. P. (2006b) *Astrophys. J. Lett., 644,* L79–L82.
Boss A. P. (2008) *Astrophys. J., 677,* 607.
Boss A. P. (2009) *Astrophys. J., 694,* 107–114.
Boss A. P. (2011) *Astrophys. J., 731,* 74–87.
Boss A. P. (2012) *Mon. Not. R. Astron. Soc., 419,* 1930–1936.
Boss A. P. et al. (2002) *Icarus, 156,* 291–295.
Bromley B. C. and Kenyon S. J. (2011) *Astrophys. J., 731,* 101–119.
Buchhave L. A. et al. (2012) *Nature, 486,* 375–377.
Burrows A. et al. (2007) *Astrophys. J., 661,* 502–514.
Cai K. et al. (2006) *Astrophys. J. Lett., 636,* L149–L152.
Cai K. et al. (2008) *Astrophys. J., 673,* 1138–1153.
Cha S. H. and Nayakshin S. (2011) *Mon. Not. R. Astron. Soc., 415,* 3319–3334.
Chabrier G. et al. (1992) *Astrophys. J., 391,* 817–826.
Chabrier G. et al. (2000) *Astrophys. J., 542,* 464–472.
Chabrier G. et al. (2007) In *Protostars and Planets V* (B. Reipurth et al., eds.), pp. 623–638. Univ. of Arizona, Tucson.

Clarke C. J. (2009) *Mon. Not. R. Astron. Soc., 396,* 1066.
Clarke C. J. and Lodato G. (2009) *Mon. Not. R. Astron. Soc., 398,* L6–L10.
Clarke C. J. et al. (2007) *Mon. Not. R. Astron. Soc., 381,* 1543–1547.
Cossins P. et al. (2010) *Mon. Not. R. Astron. Soc., 401,* 2587–2598.
Crida A. et al. (2006) *Icarus, 181,* 587–604.
D'Angelo G. et al. (2011) In *Exoplanets* (S. Seager, ed.), pp. 319–346. Univ. of Arizona, Tucson.
DeCampli W. M. and Cameron A. G. W. (1979) *Icarus, 38,* 367–391.
Demarcus W. C. (1958) *Astron. J., 63,* 2–28.
Durisen R. H. et al. (1986) *Astrophys. J., 305,* 281–308.
Durisen R. H. et al. (2007) In *Protostars and Planets V* (B. Reipurth et al., eds.), pp. 607–621. Univ. of Arizona, Tucson.
Fischer D. A. and Valenti J. (2005) *Astrophys. J., 622,* 1102–1117.
Forgan D. and Rice K. (2011) *Mon. Not. R. Astron. Soc., 417,* 1928–1937.
Fortier A. et al. (2013) *Astron. Astrophys., 549,* A44.
Fortney J. J. and Nettelmann N. (2010) *Space Sci. Rev., 152,* 423–447.
Fortney J. J. et al. (2007) *Astrophys. J., 659,* 1661–1672.
Fortney J. J. et al. (2008) *Astrophys. J., 683,* 1104–1116.
Fortney J. J. et al. (2011) *Astrophys. J., 729,* 32–46.
Galvagni M. et al. (2012) *Mon. Not. R. Astron. Soc., 427,* 1725–1740.
Gammie C. F. (2001) *Astrophys. J., 553,* 174–183.
Gautier D. et al. (2006) In *36th COSPAR Scientific Assembly,* Abstract #867.
Gonzalez G. (1997) *Mon. Not. R. Astron. Soc., 285,* 403.
Greenzweig Y. and Lissauer J. (1992) *Icarus, 100,* 440–463.
Guillot T. (1999) *Planet. Space Sci., 47,* 1183–1200.
Guillot T. and Showman A. P. (2002) *Astron. Astrophys., 385,* 156–165.
Guillot T. et al. (1994) *Icarus, 112,* 337–353.
Guillot T. et al. (2006) *Astron. Astrophys., 453,* L21–L24.
Hartmann L. and Kenyon S. J. (1996) *Annu. Rev. Astron. Astrophys., 34,* 207–240.
Hellary P. and Nelson R. P. (2012) *Mon. Not. R. Astron. Soc., 419,* 2737.
Helled R. and Bodenheimer P. (2010) *Icarus, 207,* 503–508.
Helled R. and Bodenheimer P. (2011) *Icarus, 211,* 939–947.
Helled R. and Schubert G. (2008) *Icarus, 198,* 156–162.
Helled R. and Schubert G. (2009) *Astrophys. J., 697,* 1256–1262.
Helled R. et al. (2006) *Icarus, 185,* 64–71.
Helled R. et al. (2008) *Icarus, 195,* 863–870.
Helled R. et al. (2010) *Icarus, 210,* 446–454.
Hillenbrand L. (2008) *Phys. Script., 130,* 014024.
Hori Y. and Ikoma M. (2011) *Mon. Not. R. Astron. Soc., 416,* 1419–1429.
Hubbard W. B. (1968) *Astrophys. J., 152,* 745–754.
Iaroslavitz E. and Podolak M. (2007) *Icarus, 187,* 600–610.
Ida S. and Lin D. N. C. (2004) *Astrophys. J., 616,* 567–572.
Ida S. and Lin D. N. C. (2005) *Astrophys. J., 626,* 1045–1060.
Inaba S. et al. (2003) *Icarus, 166,* 46–62.
Johnson J. A. et al. (2010) *Publ. Astron. Soc. Pac., 122,* 905–915.
Kennedy G. M. and Kenyon S. J. (2009) *Astrophys. J., 695,* 1210–1226.
Kley W. and Dirksen G. (2006) *Astron. Astrophys., 447,* 369–377.
Kley W. et al. (2009) *Astron. Astrophys., 506,* 971–987.
Kobayashi H. et al. (2010) *Icarus, 209,* 836–847.
Kratter K. M. et al. (2010a) *Astrophys. J., 708,* 1585–1597.
Kratter K. M. et al. (2010b) *Astrophys. J., 710,* 1375–1386.
Laughlin G. et al. (2004) *Astrophys. J. Lett., 612,* L73–L76.
Leconte J. and Chabrier G. (2012) *Astron. Astrophys., 540,* A20.
Lin D. N. C. and Papaloizou J. C. B. (1979) *Mon. Not. R. Astron. Soc., 186,* 799–812.
Lin D. N. C. et al. (1996) *Nature, 380,* 606–607.
Lissauer J. J. et al. (2009) *Icarus,199,* 338–350.
Lodato G. and Clarke C. J. (2011) *Mon. Not. R. Astron. Soc., 413,* 2735–2740.
Lubow S. H. and D'Angelo G. (2006) *Astrophys. J., 641,* 526–533.
Maldonado J. et al. (2013) *Astron. Astrophys., 544,* A84.
Mamajek E. E. (2009) In *Exoplanets and Disks: Their Formation and Diversity* (T. Usuda et al., eds.) pp. 3–10. AIP Conf. Ser. 1158, American Institute of Physics, Melville, New York.
Marley M. et al. (2007) *Astrophys. J., 655,* 541–549.
Marois C. et al. (2008) *Science, 322,* 1348–1352.
Mayer L. and Gawryszczak A J. (2008) In *Extreme Solar Systems* (D. Fischer et al., eds.), p. 243. ASP Conf. Ser. 398, Astronomical Society of the Pacific, San Francisco.
Mayer L. et al. (2004) *Astrophys. J, 609,* 1045.
Mayer L. et al. (2007) *Astrophys. J. Lett., 661,* L77–L80.

Mayor M. et al. (2011) *ArXiv e-prints,* arXiv:1109.2497.

Meru F. and Bate M. R. (2010) *Mon. Not. R. Astron. Soc., 406,* 2279–2288.

Meru F. and Bate M. R. (2011a) *Mon. Not. R. Astron. Soc., 411,* L1–L5.

Meru F. and Bate M. R. (2011b) *Mon. Not. R. Astron. Soc., 410,* 559–572.

Meru F. and Bate M. R. (2012) *Mon. Not. R. Astron. Soc., 427,* 2022–2046.

Michael S. and Durisen R. H. (2010) *Mon. Not. R. Astron. Soc., 406,* 279–289.

Michael S. et al. (2011) *Astrophys. J., 737,* 42–48.

Militzer B. et al. (2008) *Astrophys. J. Lett., 688,* L45–L48.

Miller N. and Fortney J. J. (2011) *Astrophys. J. Lett., 736,* L29–L34.

Mizuno H. (1980) *Prog. Theor. Phys., 64,* 544–557.

Mordasini C. (2013) *Astron. Astrophys., 558,* A113.

Mordasini C. et al. (2009a) *Astron. Astrophys., 501,* 1139–1160.

Mordasini C. et al. (2009b) *Astron. Astrophys., 501,* 1161–1184.

Mordasini C. et al. (2012a) *Astron. Astrophys., 541,* A97.

Mordasini C. et al. (2012b) *Astron. Astrophys., 547,* A111.

Mortier A. et al. (2013) *Astron. Astrophys., 551,* A112.

Movshovitz N. et al. (2010) *Icarus, 209,* 616–624.

Müller T. W. A. et al. (2012) *Astron. Astrophys., 541,* A123.

Nayakshin S. (2010) *Mon. Not. R. Astron. Soc., 408,* 2381–2396.

Nayakshin S. (2011) *Mon. Not. R. Astron. Soc., 416,* 2974–2980.

Nayakshin S. and Cha S. H. (2012) *Mon. Not. R. Astron. Soc., 423,* 2104–2119.

Nayakshin S. and Lodato G. (2012) *Mon. Not. R. Astron. Soc., 426,* 70–90.

Nelson A. F. et al. (1998) *Astrophys. J., 502,* 342–371.

Nelson A. F. (2006) *Mon. Not. R. Astron. Soc., 373,* 1039–1070.

Nettelmann N. et al. (2008) *Astrophys. J., 683,* 1217–1228.

Nettelmann N. et al. (2012) *Astrophys. J., 750,* 52.

Paardekooper S. et al. (2011) *Mon. Not. R. Astron. Soc., 416,* L65–L69.

Pearl J. C. and Conrath B. J. (1991) *J. Geophys. Res., 96,* 18921–18930.

Pickett M. K. and Durisen R. H. (2007) *Astrophys. J. Lett., 654,* L155–L158.

Podolak M. (2003) *Icarus, 165,* 428–437.

Podolak M. and Cameron A. G. W. (1974) *Icarus, 22,* 123–148.

Podolak M. and Helled R. (2012) *Astrophys. J. Lett., 759,* L32.

Podolak M. and Hubbard W. B. (1998) In *Solar System Ices* (M. F. B. Schmitt and C. de Bergh, eds.), pp. 735–748. Kluwer, Dordrecht.

Podolak M. et al. (1988) *Icarus, 73,* 163–179.

Podolak M. et al. (1991) In *Uranus* (J. T. Bergstralh et al., eds.), pp. 29–61. Univ. of Arizona, Tucson.

Podolak M. et al. (2011) *Astrophys. J., 734,* 56.

Pollack J. B. et al. (1996) *Icarus, 124,* 62–85.

Rafikov R. R. (2007) *Astrophys. J., 662,* 642–650.

Rafikov R. R. (2011) *Astrophys. J., 727,* 86.

Redmer R. et al. (2011) *Icarus, 211,* 798–803.

Rice W. K. M. et al. (2004) *Mon. Not. R. Astron. Soc., 355,* 543–552.

Rice W. K. et al. (2005) *Mon. Not. R. Astron. Soc., 364,* L56–L60.

Rice W. K. M. et al. (2006) *Mon. Not. R. Astron. Soc., 372,* L9–L13.

Robinson S. E. et al. (2006) *Astrophys. J., 643,* 484–500.

Rogers P. D. and Wadsley J. (2012) *Mon. Not. R. Astron. Soc., 423,* 1896–1908.

Safronov V. S. (1969) *Evolution of the Protoplanetary Cloud and Formation of the Earth and Planets.* Nauka, Moscow (in Russian).

Santos N. C. et al. (2004) *Astron. Astrophys., 415,* 1153–1166.

Santos N. C. et al. (2011) *Astron. Astrophys., 526,* A112.

Sato B. et al. (2005) *Astrophys. J., 633,* 465–473.

Saumon D. and Guillot T. (2004) *Astrophys. J., 609,* 1170–1180.

Saumon D. et al. (1995) *Astrophys. J. Suppl., 99,* 713–741.

Simonelli D. P. and Reynolds R. T. (1989) *Space Sci. Rev., 16,* 1209–1212.

Stamatellos D. and Whitworth A. P. (2009) *Mon. Not. R. Astron. Soc., 400,* 1563.

Stanley S. and Bloxham J. (2006) *Icarus, 184,* 556–572.

Stevenson D. J. (1975) *Phys. Rev. B, 12,* 3999–4007.

Stevenson D. J. (1986) In *Lunar Planet. Sci. XVII,* p. 1011. Lunar and Planetary Institute, Houston.

Toomre A. (1964) *Astrophys. J., 139,* 1217–1238.

Tsiganis K. et al. (2005) *Nature, 435,* 459–461.

Uribe A. L. et al. (2013) *Astrophys. J., 769,* 97.

Vazan A. and Helled R. (2012) *Astrophys. J., 756,* 90.

Vorobyov E. I. (2013) *Astron. Astrophys., 552,* 15.

Vorobyov E. I. and Basu S. (2006) *Astrophys. J., 650,* 956–969.

Ward W. R. (1997) *Astrophys. J. Lett., 482,* L211–L214.

Weidenschilling S. J. (1977) *Astrophys. Space Sci., 51,* 153–158.

Weidenschilling S. J. (2008) *Phys. Script., 130,* 014021.

Weidenschilling S. J. (2011) *Icarus, 214,* 671–684.

Wilson H. F. and Militzer B. (2010) *Phys. Rev. Lett., 104,* 121101.

Wuchterl G. et al. (2000) In *Protostars and Planets IV* (V. Mannings et al., eds.), pp. 1081–1109. Univ. of Arizona, Tucson.

Yasui C. et al. (2009) *Astrophys. J., 705,* 54–63.

Zhu Z. et al. (2012) *Astrophys. J., 746,* 110.

Baruteau C., Crida A., Paardekooper S.-J., Masset F., Guilet J., Bitsch B., Nelson R., Kley W., and Papaloizou J. (2014) Planet-disk interactions and early evolution of planetary systems. In *Protostars and Planets VI* (H. Beuther et al., eds.), pp. 667–689. Univ. of Arizona, Tucson, DOI: 10.2458/azu_uapress_9780816531240-ch029.

Planet-Disk Interactions and Early Evolution of Planetary Systems

Clément Baruteau
University of Cambridge

Aurélien Crida
Universtité Nice Sophia Antipolis and Observatoire de la Côte d'Azur

Sijme-Jan Paardekooper
University of Cambridge

Frédéric Masset
Universidad Nacional Autónoma de México

Jérôme Guilet
University of Cambridge

Bertram Bitsch
Universtité Nice Sophia Antipolis and Observatoire de la Côte d'Azur

Richard Nelson
Queen Mary, University of London

Wilhelm Kley
Universität Tübingen

John Papaloizou
University of Cambridge

The great diversity of extrasolar planetary systems has challenged our understanding of how planets form and how their orbits evolve as they form. Among the various processes that may account for this diversity, the gravitational interaction between planets and their parent protoplanetary disk plays a prominent role in shaping young planetary systems. Planet-disk forces are large, and the characteristic times for the evolution of planets' orbital elements are much shorter than the lifetime of protoplanetary disks. The determination of such forces is challenging, because it involves many physical mechanisms and requires detailed knowledge of the disk structure. Yet intense research over the past few years, with the exploration of many new avenues, represents a significant improvement on the state of the discipline. This chapter reviews the current understanding of planet-disk interactions and highlights their role in setting the properties and architecture of observed planetary systems.

1. INTRODUCTION

Since the fifth edition of Protostars and Planets in 2005 (PPV) (see *Reipurth et al.,* 2007), the number of extrasolar planets has increased from about 200 to nearly 1000, with several thousand transiting planet candidates awaiting confirmation. These prolific discoveries have emphasized the amazing diversity of exoplanetary systems. They have brought crucial constraints on models of planet formation and evolution, which need to account for the many flavors in which exoplanets come. Some giant planets, widely known as the hot Jupiters, orbit their star in just a couple of days, like 51 Pegasus b (*Mayor and Queloz,* 1995). Some others orbit their star in a few tens to hundreds of years, like the four planets known to date in the HR 8799 planetary system (*Marois et al.,* 2010). At the time of PPV, it was already established that exoplanets have a broad distribution of eccentricity, with a median value ≈0.3 (*Takeda and*

Rasio, 2005). Since then, measurements of the Rossiter McLaughlin effect in about 50 planetary systems have revealed several hot Jupiters on orbits highly misaligned with the equatorial plane of their star (e.g., *Albrecht et al.,* 2012), suggesting that exoplanets could also have a broad distribution of inclination. Not only should models of planet formation and evolution explain the most exotic flavors in which exoplanets come, they should also reproduce their statistical properties. This is challenging, because the predictions of such models depend sensitively on the many key processes that come into play. One of these key processes is the interaction of forming planets with their parent protoplanetary disk, which is the scope of this chapter.

Planet-disk interactions play a prominent role in the orbital evolution of young forming planets, leading to potentially large variations not only in their semimajor axis (a process known as planet migration), but also in their eccentricity and inclination. Observations of (1) hot Jupiters on orbits aligned with the equatorial plane of their star, (2) systems with several co-planar low-mass planets with short and intermediate orbital periods (like those discovered by the Kepler mission), and (3) many near-resonant multi-planet systems are evidence that planet disk interactions are one major ingredient in shaping the architecture of observed planetary systems. But it is not the only ingredient; planet-planet and star-planet interactions are also needed to account for the diversity of exoplanetary systems. The long-term evolution of planetary systems after the dispersal of their protoplanetary disk is reviewed in the chapter by Davies et al. in this volume.

This chapter commences with a general description of planet-disk interactions in section 2. Basic processes and recent progress are presented with the aim of letting non-experts pick up a physical intuition and a sense of the effects in planet-disk interactions. Section 3 continues with a discussion on the role played by planet-disk interactions in the properties and architecture of observed planetary systems. Summary points follow in section 4.

2. THEORY OF PLANET-DISK INTERACTIONS

2.1. Orbital Evolution of Low-Mass Planets: Type I Migration

Embedded planets interact with the surrounding disk mainly through gravity. Disk material, in orbit around the central star, feels a gravitational perturbation caused by the planet that can lead to an exchange of energy and angular momentum between planet and disk. In this section, we assume that the perturbations due to the planet are small, so that the disk structure does not change significantly due to the planet, and that migration is slow, so that any effects of the radial movement of the planet can be neglected. We will return to these issues in sections 2.2 and 2.3. If these assumptions are valid, we are in the regime of type I migration, and we are dealing with low-mass planets (typically up to Neptune's mass).

The perturbations in the disk induced by the planet are traditionally split into two parts: (1) a wave part, where the disk response comes in the form of spiral density waves propagating throughout the disk from the planet location, and (2) a part confined in a narrow region around the planet's orbital radius, called the planet's horseshoe region, where disk material executes horseshoe turns relative to the planet. An illustration of both perturbations is given in Fig. 1. We will deal with each of them separately below. For simplicity, we will focus on a two-dimensional disk, characterized by vertically averaged quantities such as the surface density Σ. We make use of cylindrical polar coordinates (r, φ) centered on the star.

2.1.1. Wave torque. It has been known for a long time that a planet exerting a gravitational force on its parent disk can launch waves in the disk at Lindblad resonances (*Goldreich and Tremaine,* 1979, 1980). These correspond to locations where the gas azimuthal velocity relative to the planet matches the phase velocity of acoustic waves in the azimuthal direction. This phase velocity depends on the azimuthal wavenumber, the sound speed, and the epicyclic frequency, i.e., the oscillation frequency of a particle in the disk subject to a small radial displacement (e.g., *Ward,* 1997). At large azimuthal wavenumber, the phase velocity tends to the sound speed, and Lindblad resonances therefore pile up at r = a_p ± 2H/3, where a_p is the semimajor axis of the planet and H ≪ r is the pressure scale-height of the disk (*Artymowicz,* 1993b). The superposition of the waves launched at Lindblad resonances gives rise to a one-armed spiral density wave (*Ogilvie and Lubow,* 2002), called the wake (see Fig. 1).

It is possible to solve the wave excitation problem in the linear approximation and calculate analytically the resulting torque exerted on the disk using the Wentzel-Kramers-Brillouin (WKB) approximation (*Goldreich and Tremaine,* 1979; *Lin and Papaloizou,* 1979). Progress beyond analytical calculations for planets on circular orbits has been made by solving the linear perturbation equations numerically, as done in two dimensions in *Korycansky and Pollack* (1993). The three-dimensional case was tackled in *Tanaka et al.* (2002), which resulted in a widely used torque formula valid for isothermal disks only. It is important to note that two-dimensional calculations only give comparable results to more realistic three-dimensional calculations if the gravitational potential of the planet is softened appropriately. Typically, the softening length has to be a sizeable fraction of the disk scale-height (*Müller et al.,* 2012). Using this two-dimensional softened gravity approach, *Paardekooper et al.* (2010) found that the dependence of the wave torque (Γ_L) (the wave torque exerted on the planet is also commonly referred to as the Lindblad torque) on disk gradients in a non-isothermal, adiabatic disk is

$$\gamma \Gamma_L / \Gamma_0 = -2.5 - 1.7\beta + 0.1\alpha \qquad (1)$$

where γ is the ratio of specific heats, α and β are negatives of the (local) power-law exponents of the surface density

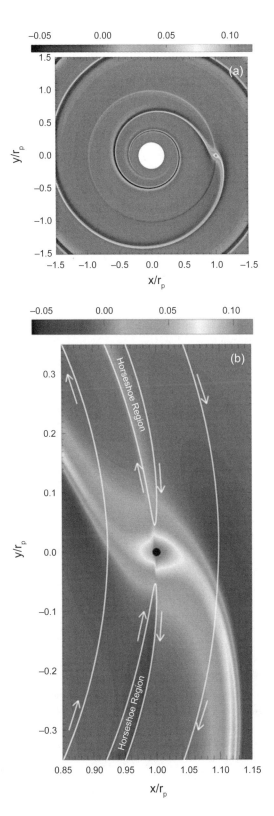

Fig. 1. Relative perturbation of the surface density of a gaseous protoplanetary disk perturbed by a 5-$M_{\oplus}$ planet located at x = r_p and y = 0. The planet induces a one-armed spiral density wave — the wake — that propagates throughout the disk, and density perturbations confined in the planet's horseshoe region. Typical gas trajectories relative to the planet are shown with thick curves and arrows in the bottom panel.

Σ and temperature T($\Sigma \propto r^{-\alpha}$, T $\propto r^{-\beta}$) and the torque is normalized by

$$\Gamma_0 = \frac{q^2}{h^2} \Sigma_p \, r_p^4 \Omega_p^2$$

(2)

where q is the planet-to-star mass ratio, h = H/r is the aspect ratio and quantities with subscript p refer to the location of the planet. Note that in general we expect α, $\beta > 0$, i.e. both surface density and temperature decrease outward. For reasonable values of α, the wave torque on the planet is negative: It decreases the orbital angular momentum of the planet, and thus its semimajor axis (the planet being on a circular orbit), leading to inward migration. The linear approximation remains valid as long as q $\ll$ h^3 (*Korycansky and Papaloizou*, 1996). For a disk around a solar mass star with h = 0.05, this means that the planet mass needs to be much smaller than 40 $M_{\oplus}$.

The factor γ in equation (1) is due to the difference in sound speed between isothermal and adiabatic disks (*Baruteau and Masset*, 2008a). For disks that can cool efficiently, we expect the isothermal result to be valid ($\gamma \to$ 1), while for disks that cannot cool efficiently, the adiabatic result should hold. It is possible to define an effective γ that depends on the thermal diffusion coefficient so that the isothermal regime can be connected smoothly to the adiabatic regime (*Paardekooper et al.*, 2011).

A generalized expression for the Lindblad torque has been derived by *Masset* (2011) for two-dimensional disks where the density and temperature profiles near the planet are not power laws, like at opacity transitions or near cavities. This generalized expression agrees well with equation (1) for power-law disks. We stress that there is to date no general expression for the wave torque in three-dimensional non-isothermal disks. Analytics are involved (*Tanaka et al.*, 2002; *D'Angelo and Lubow*, 2010) and it is difficult to measure the wave torque independently from the co-rotation torque in three-dimensional numerical simulations of planet-disk interactions.

The above discussion neglected possible effects of self-gravity. *Pierens and Huré* (2005) showed that in a self-gravitating disk, Lindblad resonances get shifted toward the planet, thereby making the wave torque stronger. This was confirmed numerically by *Baruteau and Masset* (2008b). The impact of a magnetic field in the disk and of possibly related magneto-hydrodynamic (MHD) turbulence will be considered in section 2.1.4.

The normalization factor Γ_0 sets a timescale for type I migration of planets on circular orbits:

$$\tau_0 = \frac{r_p}{|dr_p/dt|} = \frac{1}{2} \frac{h^2}{q} \frac{M_{\star}}{\Sigma_p r_p^2} \Omega_p^{-1}$$

(3)

where $M_{\star}$ denotes the mass of the central star. Assuming a typical gas surface density of 2000 (r_p/1 AU)$^{-3/2}$ g cm^{-2}, $M_{\star} = M_{\odot}$, and h = 0.05, the migration timescale in years at 1 AU is given approximately by 1/q. This means that an Earth-mass planet at 1 AU would migrate inward on a timescale of ~3×10^5 yr, while the timescale for Neptune would only be ~2×10^4 yr. These timescales are shorter than the expected disk lifetime of 10^6–10^7 yr, making this type of migration far too efficient for planets to survive on orbits of several astronomical units. A lot of work has been done recently on how to stop type I migration or make it less efficient (see sections 2.1.2 and 2.1.3).

2.1.2. Co-rotation torque. Most progress since PPV (*Papaloizou et al., 2007*) has been made in understanding the torque due to disk material that on average co-rotates with the planet, the co-rotation torque. By solving the linearized disk equations in the same way as for the wave torque, it is possible to obtain a numerical value for the co-rotation torque. One can show that in the case of an isothermal disk, this torque scales with the local gradient in specific vorticity or vortensity (*Goldreich and Tremaine, 1979*). (Note that vorticity is defined here as the vertical component of the curl of the gas velocity. In a two-dimensional model, specific vorticity, or vortensity, refers to vorticity divided by surface density.) It therefore has a stronger dependence on background surface density gradients than the wave torque, with shallower profiles giving rise to a more positive torque. It was nevertheless found in *Tanaka et al.* (2002) that, except for extreme surface density profiles, the wave torque always dominates over the linear co-rotation torque ($\Gamma_{c,lin}$). Using the two-dimensional softened gravity approach, *Paardekooper et al.* (2010) obtained for a non-isothermal disk

$$\gamma \Gamma_{c,lin}/\Gamma_0 = 0.7\left(\frac{3}{2} - \alpha - \frac{2\xi}{\gamma}\right) + 2.2\xi \qquad (4)$$

where $\xi = \beta - (\gamma-1)\alpha$ is the negative of the (local) power-law exponent of the specific entropy. For an isothermal disk, $\xi = 0$ and the co-rotation torque is proportional to the vortensity gradient.

An alternative expression for the co-rotation torque was derived by *Ward* (1991) by considering disk material on horseshoe orbits relative to the planet. This disk material defines the planet's horseshoe region (see bottom panel in Fig. 1). The torque on the planet due to disk material executing horseshoe turns, the horseshoe drag, scales with the vortensity gradient in an isothermal disk, just like the linear co-rotation torque. It comes about because material in an isothermal inviscid fluid conserves its vortensity. When executing a horseshoe turn, which takes a fluid element to a region of different vorticity because of the Keplerian shear, conservation of vortensity dictates that the surface density should change (*Ward, 1991*). In a gas disk, this change in surface density is smoothed out by evanescent pressure waves (*Casoli and Masset, 2009*). Nevertheless, this change in surface density results in a torque being applied on the planet.

For low-mass planets, for which $q \ll h^3$, the half-width of the horseshoe region, x_s, is (*Masset et al., 2006; Paardekooper and Papaloizou, 2009b*)

$$x_S \approx 1.2 r_p \sqrt{q/h} \qquad (5)$$

Thus it is only a fraction of the local disk thickness. The horseshoe drag, which scales as x_s^4 (*Ward, 1991*), therefore has the same dependence on q and h as the wave torque and the linear co-rotation torque. The numerical coefficient in front is generally much larger than for the linear co-rotation torque, however (*Paardekooper and Papaloizou, 2009a; Paardekooper et al., 2010*). Since both approaches aim at describing the same thing, the co-rotation torque, it was long unclear as to which result should be used. It was shown in *Paardekooper and Papaloizou* (2009a) that whenever horseshoe turns occur, the linear co-rotation torque gets replaced by the horseshoe drag, unless a sufficiently strong viscosity is applied. It should be noted that horseshoe turns do not exist within linear theory, making linear theory essentially invalid for low-mass planets as far as the co-rotation torque is concerned.

The co-rotation torque in the form of horseshoe drag is very sensitive to the disk's viscosity and thermal properties near the planet. *Paardekooper and Mellema* (2006) found that in three-dimensional radiation hydrodynamical simulations, migration can in fact be directed outward due to a strong positive co-rotation torque counterbalancing the negative wave torque over the short duration (15 planet orbits) of their calculations. This was subsequently interpreted as being due to a radial gradient in disk specific entropy, which gives rise to a new horseshoe drag component due to conservation of entropy in an adiabatic disk (*Baruteau and Masset, 2008a; Paardekooper and Mellema, 2008; Paardekooper and Papaloizou, 2008*). However, the situation is not as simple as for the isothermal case. It turns out that the entropy-related horseshoe drag arises from the production of vorticity along the downstream separatrices of the planet's horseshoe region (*Masset and Casoli, 2009*). Conservation of entropy during a horseshoe turn leads to a jump in entropy along the separatrices whenever there is a radial gradient of entropy in the disk. This jump in entropy acts as a source of vorticity, which in turn leads to a torque on the planet. Crucially, the amount of vorticity produced depends on the streamline topology, in particular the distance of the stagnation point to the planet. An analytical model for an adiabatic disk where the background temperature is constant was developed in *Masset and Casoli* (2009), while *Paardekooper et al.* (2010) used a combination of physical arguments and numerical results to obtain the following expression for the horseshoe drag

$$\gamma \Gamma_{c,HS}/\Gamma_0 = 1.1\left(\frac{3}{2} - \alpha\right) + 7.9\frac{\xi}{\gamma} \qquad (6)$$

where the first term on the righthand side is the vortensity-related part of the horseshoe drag, and the second term is the entropy-related part. The model derived in *Masset and Casoli* (2009), under the same assumptions as for the stagnation point, yields a numerical coefficient for the entropy-related part of the horseshoe drag of 7.0 instead of 7.9.

Comparing the linear co-rotation torque to the nonlinear horseshoe drag (see equations (4) and (6)) we see that both depend on surface density and entropy gradients, but also that the horseshoe drag is always stronger. In the inner regions of disks primarily heated by viscous heating, the entropy profile should decrease outward ($\xi > 0$). The co-rotation torque should then be positive, promoting outward migration.

The results presented above were for adiabatic disks, while the isothermal result can be recovered by setting $\gamma = 1$ and $\beta = 0$ (which makes $\xi = 0$ as well). Real disks are neither isothermal nor adiabatic. When the disk can cool efficiently, which happens in the optically thin outer parts, the isothermal result is expected to be valid [or rather the locally isothermal result — a disk with a fixed temperature profile, which behaves slightly differently from a truly isothermal disk (see *Casoli and Masset*, 2009)]. In the optically thick inner parts of the disk, cooling is not efficient and the adiabatic result should hold. An interpolation between the two regimes was presented in *Paardekooper et al.* (2011) and *Masset and Casoli* (2010).

2.1.3. Saturation of the co-rotation torque. While density waves transport angular momentum away from the planet, the horseshoe region only has a finite amount of angular momentum to play with since there are no waves involved. Sustaining the co-rotation torque therefore requires a flow of angular momentum into the horseshoe region: Unlike the wave torque, the co-rotation torque is prone to saturation. In simulations of disk-planet interactions, sustaining (or unsaturating) the co-rotation torque is usually established by including a Navier-Stokes viscosity. In a real disk, angular momentum transport is likely due to turbulence arising from the magneto-rotational instability (MRI) (see the chapter by Turner et al. in this volume), and simulations of turbulent disks give comparable results to viscous disks (*Baruteau and Lin*, 2010; *Baruteau et al.*, 2011a; *Pierens et al.*, 2012) as far as saturation is concerned (see also section 2.1.4). The main result for viscous disks is that the viscous diffusion timescale across the horseshoe region has to be shorter than the libration timescale in order for the horseshoe drag to be unsaturated (*Masset*, 2001, 2002). This also holds for non-isothermal disks, and expressions for the co-rotation torque have been derived in two-dimensional disk models for general levels of viscosity and thermal diffusion or cooling (*Masset and Casoli*, 2010; *Paardekooper et al.*, 2011). Results from two-dimensional radiation hydrodynamic simulations, where the disk is self-consistently heated by viscous dissipation and cooled by radiative losses, confirm this picture (*Kley and Crida*, 2008).

The torque expressions of *Masset and Casoli* (2010) and *Paardekooper et al.* (2011) were derived using two-dimensional disk models, and three-dimensional disk models are still required to get definitive, accurate predictions for the wave and the co-rotation torques. Still, the predictions of the aforementioned torque expressions are in decent agreement with the results of three-dimensional simulations of planet-disk interactions. The simulations of *D'Angelo and Lubow* (2010) explored the dependence of the total torque with density and temperature gradients in three-dimensional locally isothermal disks. They found the total normalized torque $\Gamma_{tot}/\Gamma_0 \approx -1.4 - 0.4\beta - 0.6\alpha$. For the planet mass and disk viscosity of these authors, the co-rotation torque reduces to the linear co-rotation torque. Summing equations (1) and (4) with $\gamma = 1$ and $\xi = \beta$ yields $\Gamma_{tot}/\Gamma_0 = -1.4 - 0.9\beta - 0.6\alpha$, which is in good agreement with the results of *D'Angelo and Lubow* (2010), except for the coefficient in front of the temperature gradient. The simulations of *Kley et al.* (2009), *Ayliffe and Bate* (2011), and *Bitsch and Kley* (2011b) were for non-isothermal radiative disks. *Ayliffe and Bate* (2011) considered various temperature power-law profiles and showed that the total torque does not always exhibit a linear dependence with temperature gradient. *Bitsch and Kley* (2011b) highlighted substantial discrepancies between the torque predictions of *Masset and Casoli* (2010) and *Paardekooper et al.* (2011). These discrepancies originate from a larger half-width (x_s) of the planet's horseshoe region adopted in *Masset and Casoli* (2010), which was suggested as a proxy to assess the co-rotation torque in three dimensions. Adopting the same, standard value for x_s given by equation (5), one can show that the total torques of *Masset and Casoli* (2010) and *Paardekooper et al.* (2011) are actually in very good agreement. Both are less positive than in the numerical results of *Bitsch and Kley* (2011b). Discrepancies originate from inherent torque differences between two-dimensional and three-dimensional disk models (see, e.g., *Kley et al.*, 2009), and possibly from a nonlinear boost of the positive co-rotation torque in the simulations of *Bitsch and Kley* (2011b), for which $q \geq h^3$ [for locally isothermal disks (see *Masset et al.*, 2006)].

The thermal structure of protoplanetary disks is determined not only by viscous heating and radiative cooling, but also by stellar irradiation. The effects of stellar irradiation on the disk structure have been widely investigated using a one + one-dimensional numerical approach (e.g., *Bell et al.*, 1997; *Dullemond et al.*, 2001), with the goal of fitting the spectral energy distributions of observed disks. Stellar heating dominates in the outer regions of disks, viscous heating in the inner regions. The disk's aspect ratio should then slowly increase with increasing distance from the star (*Chiang and Goldreich*, 1997). This increase may have important implications for the direction and speed of type I migration, which, as we have seen above, are quite sensitive to the aspect ratio (local value and radial profile). This is illustrated in Fig. 2, which displays the torque acting on type I migrating planets sitting in the mid-plane of their disk, with and without inclusion of stellar irradiation. The disk structure was calculated in *Bitsch et al.* (2013b) for standard opacities and a constant disk viscosity (thereby fixing the density profile). Two regions of outward migration originate from opacity transitions at the silicate condensation line (near 0.8 AU) and at the water ice line (near 5 AU). In this model, heating by stellar irradiation becomes prominent beyond ≈8 AU. The resulting increase

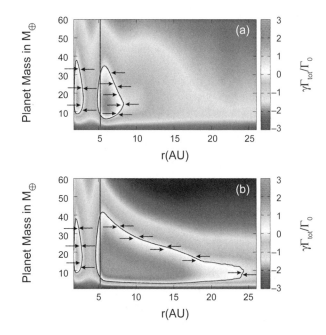

Fig. 2. Type I migration torque on planets of different masses in disk models **(a)** with stellar irradiation and **(b)** without. The torque expressions in *Paardekooper et al.* (2011) are used and expressed in units of Γ_0, given by equation (2). Black lines delimit areas where migration is directed outward. Black arrows show the direction of migration, and the vertical line the ice line location at 170 K. Adapted from *Bitsch et al.* (2013b).

in the disk's aspect ratio profile gives a shallower entropy profile (β and thus ξ take smaller values) and therefore a smaller (although positive) entropy-related horseshoe drag (see equation (6)). Inclusion of stellar irradiation therefore reduces the range of orbital separations at which outward migration may occur (see also *Kretke and Lin*, 2012). The outer edge of regions of outward migration are locations where type I planetary cores converge, which may lead to resonant captures (*Cossou et al.*, 2013) and could enhance planet growth if enough cores are present to overcome the trapping process. Note in Fig. 2 that planets up to ~5 $M_\oplus$ do not migrate outward. This is because for such planet masses, and for the disk viscosity in this model ($\alpha_v \sim$ a few $\times 10^{-3}$), the co-rotation torque is replaced by the (smaller) linear co-rotation torque. Figure 2 provides a good example of how sensitive predictions of planet-migration models can be to the structure of protoplanetary disks . Future observations of disks, e.g., with the Atacama Large Millimeter/submillimeter Array (ALMA), will give precious information that will help constrain migration models.

2.1.4. Effect of disk magnetic field and turbulence. So far we have considered the migration of a planet in a purely hydrodynamical laminar disk where turbulence is modeled by an effective viscosity. As stressed in section 2.1.2, a turbulent viscosity is essential to unsaturate the co-rotation torque. The most likely (and best studied) source of turbulence in protoplanetary disks is the MRI

(*Balbus and Hawley*, 1991) which can amplify the magnetic field and drive MHD turbulence by tapping energy from the Keplerian shear. Furthermore, the powerful jets observed to be launched from protoplanetary disks are thought to arise from a strong magnetic field [likely through the magneto-centrifugal acceleration (see *Ferreira et al.*, 2006)]. Magnetic field and turbulence thus play a crucial role in the dynamics and evolution of protoplanetary disks, and need to be taken into account in theories of planet-disk interactions.

2.1.4.1. Stochastic torque driven by turbulence: Magneto-hydrodynamic turbulence excites non-axisymmetric density waves (see Fig. 3a), which cause a fluctuating component of the torque on a planet in a turbulent disk (*Nelson and Papaloizou*, 2004; *Nelson*, 2005; *Laughlin et al.*, 2004). This torque changes sign stochastically with a typical correlation time of a fraction of an orbit. Because the density perturbations are driven by turbulence rather than the planet itself, the specific torque due to turbulence is independent of planet mass, while the (time-averaged) specific torque driving type I migration is proportional to the planet mass. Type I migration should therefore outweigh the stochastic torque for sufficiently massive planets and on a long-term evolution, whereas stochastic migration arising from turbulence should dominate the evolution of planetesimals and possibly small-mass planetary cores (*Baruteau and Lin*, 2010; *Nelson and Gressel*, 2010). The stochastic torque adds a random walk component to planet migration, which can be represented in a statistical sense by a diffusion process acting on the probability distribution of planets (*Johnson et al.*, 2006; *Adams and Bloch*, 2009). A consequence of this is that a small fraction of planets may migrate to the outer parts of their disk even if the laminar type I migration is directed inward. Note that the presence of a dead zone around the disk's mid-plane, where MHD turbulence is quenched due to the low ionization, reduces the amplitude of the stochastic torque (*Oishi et al.*, 2007; *Gressel et al.*, 2011, 2012).

2.1.4.2. Mean migration in a turbulent disk: Two-dimensional hydrodynamical simulations of disks with stochastically forced waves have been carried out to mimic MHD turbulence and to study its effects on planet migration. Using the model by *Laughlin et al.* (2004), *Baruteau and Lin* (2010) and *Pierens et al.* (2012) showed that, when averaged over a sufficiently long time, the torque converges toward a well-defined average value, and that the effects of turbulence on this average torque are well described by an effective turbulent viscosity and heat diffusion. In particular, the wave torque is affected little by turbulence, while the co-rotation torque can be unsaturated by this wake-like turbulence. *Baruteau et al.* (2011a) and *Uribe et al.* (2011) have shown that a similar conclusion holds in simulations with fully developed MHD turbulence arising from the MRI. *Baruteau et al.* (2011a), however, discovered the presence of an additional component of the co-rotation torque, which has been attributed to the effect of the magnetic field (*Guilet et al.*, 2013) (see below). Note

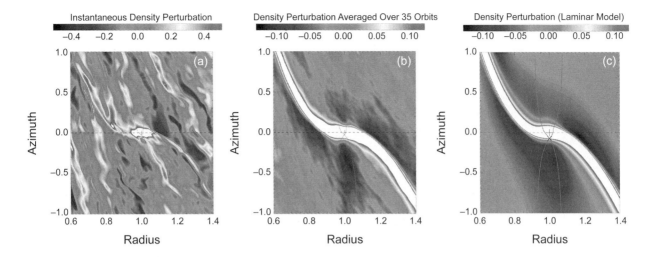

Fig. 3. Surface density perturbation of a three-dimensional MRI-turbulent disk with a planet embedded at r = 1 and φ = 0 [**(a)** and **(b)** adapted from *Baruteau et al., 2011a*)] and **(c)** in an equivalent two-dimensional laminar disk with an azimuthal magnetic field (adapted from *Guilet et al., 2013*). In the turbulent simulation, the planet wake is barely visible in the instantaneous density perturbation [**(a)**], because of the large turbulent density fluctuations. When averaged over 35 orbits, density perturbations agree very well with those of the equivalent laminar disk model [compare **(b)** and **(c)**], revealing not only the planet wake, but also an asymmetric underdensity confined in the planet's horseshoe region and arising from the azimuthal magnetic field. The separatrices of the planet's horseshoe region are shown by thin gray curves.

that the mean migration in a turbulent disk has been conclusively studied only for fairly massive planets (typically $q/h^3 \sim 0.3$), as less-massive planets need a better resolution and a longer averaging time. It is an open question whether the migration of smaller planets is affected by turbulence in a similar way as by a diffusion process. One may wonder indeed whether the diffusion approximation remains valid when the width of the planet's horseshoe region is a small fraction of the disk scale height and therefore of the typical correlation length of turbulence.

2.1.4.3. *Wave torque with a magnetic field:* In addition to driving turbulence, the magnetic field has a direct effect on planet migration by modifying the response of the gas to the planet's potential. In particular, wave propagation is modified by the magnetic field, and three types of waves exist: fast and slow magneto-sonic waves as well as Alfvén waves. *Terquem (2003)* showed that for a strong azimuthal magnetic field, slow MHD waves are launched at the so-called magnetic resonances, which are located where the gas azimuthal velocity relative to the planet matches the phase velocity of slow MHD waves. The angular momentum carried by the slow MHD waves gives rise to a new component of the torque. If the magnetic field strength is steeply decreasing outward, this new torque is positive and may lead to outward migration (*Terquem, 2003; Fromang et al., 2005*). A vertical magnetic field also impacts the resonances (*Muto et al., 2008*) but its effect on the total torque remains to be established. In the inner parts of protoplanetary disks, the presence of a strong vertical magnetic field is needed to explain the launching of observed jets. A better understanding of the strength and evolution of such a vertical field (*Guilet and Ogilvie, 2012, 2013*) and its effect

on planet migration will improve the description of planet migration near the central star.

2.1.4.4. *Co-rotation torque with a magnetic field:* A strong azimuthal magnetic field can prevent horseshoe motions so that the co-rotation torque is replaced by the torque arising from magnetic resonances, as discussed in the previous paragraph. *Guilet et al. (2013)* showed that horseshoe motions take place and suppress magnetic resonances for weak enough magnetic fields, when the Alfvén speed is less than the shear velocity at the separatrices of the planet's horseshoe region. Using two-dimensional laminar simulations with effective viscosity and resistivity, these authors showed that advection of the azimuthal magnetic field along horseshoe trajectories leads to an accumulation of magnetic field near the downstream separatrices of the horseshoe region. This accumulation in turns leads to an underdensity at the same location to ensure approximate pressure balance (see Fig. 3c). The results of these laminar simulations agree very well with those of the MHD turbulent simulations of *Baruteau et al. (2011a)*. A rear-front asymmetry in the magnetic field accumulation inside the horseshoe region gives rise to a new component of the co-rotation torque, which may cause outward migration even if the magnetic pressure is less than 1% of the thermal pressure (*Guilet et al., 2013*). This new magnetic co-rotation torque could take over the entropy-related co-rotation torque to sustain outward migration in the radiatively efficient outer parts of protoplanetary disks . Future studies should address the behavior of the co-rotation torque in the dead zone, and in regions of the disk threaded by a vertical magnetic field. Also, other non-ideal MHD effects, such as the Hall effect and ambipolar diffusion, can have a significant impact on the MRI turbulence and on the disk structure (e.g., *Bai and Stone*, 2011; *Kunz and Lesur*, 2013; *Simon et al.*, 2013). These non-ideal MHD effects still need to be explored in the context of planet-disk interactions.

2.1.5. Evolution of eccentric or inclined low-mass planets. We have so far considered planet-disk interactions for low-mass planets on circular orbits. Interaction between two or more planets migrating in their parent disk may increase eccentricities (e) and inclinations (i) (see section 2.4). We examine below the orbital evolution of protoplanets on eccentric or inclined orbits due to planet-disk interactions.

In the limit of small eccentricities, it can be shown that the effect of the disk is to damp the eccentricity of type I migrating planets (*Goldreich and Tremaine*, 1980; *Artymowicz*, 1993a; *Masset*, 2008). Similar arguments can be made for inclined low-mass planets, for which planet-disk interactions damp the inclination in time (*Tanaka and Ward*, 2004). *Papaloizou and Larwood* (2000) have provided an analytic expression for the eccentricity damping timescale in two dimensions, while *Tanaka and Ward* (2004) derived expressions for the eccentricity and inclination damping timescales in three dimensions

$$\tau_e = \frac{e}{|de/dt|} \approx 2.6h^2\tau_0, \ \tau_i = \frac{i}{|di/dt|} \approx 3.7h^2\tau_0 \qquad (7)$$

where τ_0 is given by equation (3). Note that, because h^2 is very small, damping of e and i is much faster than migration. A single low-mass planet should therefore migrate on a circular and co-planar orbit. The above results were confirmed by hydrodynamical simulations (*Cresswell et al.*, 2007; *Bitsch and Kley*, 2010, 2011a). Eccentricity and inclination damping rates can be alternatively derived using a dynamical friction formalism (*Muto et al.*, 2011; *Rein*, 2012b).

We have stressed above that the migration of low-mass planets can be directed outward if the disk material in the horseshoe region exerts a strong positive co-rotation torque on the planet. Numerical simulations by *Bitsch and Kley* (2010) have shown that for small eccentricities, the magnitude of the co-rotation torque decreases with increasing eccentricity, which restricts the possibility of outward migration to planets with eccentricities below a few percent. A consequence of this restriction is to shift regions of convergent migration to smaller radii for mildly eccentric low-mass planets [see last two paragraphs in section 2.1.2 and *Cossou et al.* (2013)]. In a very recent study, *Fendyke and Nelson* (2014) have explored in detail the influence of orbital eccentricity on the co-rotation torque for a range of disk and planet parameters. Their study indicates that the reason why the co-rotation torque decreases with increasing e is because the width of the horseshoe region narrows as e increases. Furthermore, by fitting the results from their suite of simulations with an analytic function, they find that the co-rotation torque scales with eccentricity according to the expression $\Gamma_c(e) = \Gamma_c(0)$ exp $(-e/e_f)$, where $\Gamma_c(e)$ is the co-rotation torque at eccentricity e, $\Gamma_c(0)$ that at zero eccentricity, and e_f is an e-folding eccentricity that scales linearly with the disk aspect ratio at

the planet's orbital radius (the expression $e_f = h/2 + 0.01$ provides a good overall fit to the simulations). Furthermore, the Lindblad torque becomes more positive with increasing e (*Papaloizou and Larwood*, 2000). This sign reversal does not necessarily lead to outward migration as the torque on an eccentric planet changes both the planet's semimajor axis and eccentricity (see, e.g., *Masset*, 2008).

Orbital migration also changes when a low-mass planet acquires some inclination. The larger the inclination, the less time the planet interacts with the dense gas near the disk mid-plane, and the smaller the co-rotation torque and the migration rate. Thus, inclined low-mass planets can only undergo outward migration if their inclination remains below a few degrees (*Bitsch and Kley*, 2011a).

2.2. Orbital Evolution of Massive, Gap-Opening Planets: Type II Migration

2.2.1. Gap opening. The wave torque described in section 2.1.1 is the sum of a positive torque exerted on the planet by its inner wake, and a negative torque exerted by its outer wake. Equivalently, the planet gives angular momentum to the outer disk (the disk beyond the planet's orbital radius), and it takes some from the inner disk. If the torque exerted by the planet on the disk is larger in absolute value than the viscous torque responsible for disk spreading, an annular gap is carved around the planet's orbit (*Lin and Papaloizou*, 1986a). In this simple one-dimensional picture, the gap width is the distance from the planet where the planet torque and the viscous torque balance each other (*Varnière et al.*, 2004). However, *Crida et al.* (2006) showed that the disk material near the planet also feels a pressure torque that comes about because of the non-axisymmetric density perturbations induced by the planet. In a steady state, the torques due to pressure, viscosity, and the planet all balance together, and such conditions determine the gap profile. The half-width of a planetary gap barely exceeds about twice the size of the planet's Hill radius, defined as $r_H = r_p(q/3)^{1/3}$. A gap should therefore be understood as a narrow depleted annulus between an inner disk and an outer disk. The width of the gap carved by a Jupiter-mass planet does not exceed about 30% of the star-planet orbital separation.

Based on a semi-analytic study of the above torque balance, *Crida et al.* (2006) showed that a planet opens a gap with bottom density less than 10% of the background density if the dimensionless quantity $\mathcal{P}$ defined as

$$\mathcal{P} = \frac{3}{4}\frac{H}{r_H} + \frac{50}{q\mathcal{R}} \qquad (8)$$

is ≤ 1. In equation (8), $\mathcal{R} = r_p^2\Omega_p/\nu$ is the Reynolds number, ν the disk's kinematic viscosity. Adopting the widely used alpha prescription for the disk viscosity, $\nu = a_\nu H^2\Omega$, the above gap-opening criterion becomes

$$\frac{h}{q^{1/3}} + \frac{50\alpha_\nu h^2}{q} \leq 1 \qquad (9)$$

This criterion is essentially confirmed by simulations of MRI-turbulent disks, although the width and depth of the gap can be somewhat different from an equivalent viscous disk model (*Papaloizou et al., 2004; Zhu et al., 2013*).

We point out that equation (8) with $\mathcal{P} = 1$ can be solved analytically; the minimum planet-to-star mass ratio for opening a deep gap as defined above is given by

$$q_{min} = \frac{100}{\mathcal{R}}\left[(X+1)^{1/3} - (X-1)^{1/3}\right]^{-3} \qquad (10)$$

with $X = \sqrt{1 + 3\mathcal{R}h^3/800}$, and where above quantities are to be evaluated at the planet's orbital radius. Taking a Sun-like star, equation (10) shows that in the inner regions of protoplanetary disks, where typically h = 0.05 and $\alpha_v \sim$ a few 10^{-3}, planets with a mass on the order of that of Jupiter, or larger, will open a deep gap. In the dead zone of a protoplanetary disk, where α_v can be 1 or 2 orders of magnitude smaller, planet masses $\geq 50\ M_\oplus$ will open a gap. At larger radii, where planets could form by gravitational instability, h is probably near 0.1, $\alpha_v \sim 10^{-2}$, and only planets above 10 M_{Jup} could open a gap (but see section 3.2).

Equations (9) and (10) give an estimate of the minimum planet-to-star mass ratio for which a gap with density contrast $\geq 90\%$ is carved. Recent simulations by *Duffell and MacFadyen* (2013) indicate that similarly deep gaps could be opened for mass ratios smaller than given by these equations in disks with very low viscosities, such as what is expected in dead zones. Note also that planets such that $\mathcal{P}$ is a few can open a gap with a density contrast of few tens of percent. This may concern planets of few Earth to Neptune masses in dead zones (e.g., *Rafikov, 2002; Muto et al., 2010; Dong et al., 2011*).

2.2.2. Type II migration. The formation of an annular gap around a planet splits the protoplanetary disk into an inner disk and an outer disk, which both repel the planet toward the center of the gap. While the planet is locked in its gap, it continues to migrate as the disk accretes onto the star. In other words, the planet follows the migration trajectory imposed by the disk (*Lin and Papaloizou, 1986b*), and the migration timescale is then the viscous accretion time, $\tau_v = r_p^2/\nu$. However, if the planet is much more massive than the gas outside the gap, the planet will slow down the disk viscous accretion. This occurs if $M_p > 4\pi\Sigma_o r_p^2$, where Σ_o is the surface density of the outer disk just outside the gap. When this occurs, the inner disk still accretes onto the star, while the outer disk is held by the planet. This leads to the partial (or total) depletion of the inner disk (see below). The migration timescale, which is then set by the balance between the viscous torque and the planet's inertia, is given by $\tau_v \times (M_p/4\pi\Sigma_o r_p^2)$.

The above considerations show that the timescale for type II migration (τ_{II}) is given by

$$\tau_{II} = \tau_v \times \max\left(1, \frac{M_p}{4\pi\Sigma_o r_p^2}\right) \qquad (11)$$

Equation (11) applies when a planet carves a deep gap, i.e., when $\mathcal{P} < 1$. However, when $\mathcal{P} \geq 1.5$ and the density inside the gap exceeds about 20% of the background density, the planet and the gas are no longer decoupled. The gas in the gap exerts a co-rotation torque on the planet, which is usually positive. Therefore, the migration of planets that marginally satisfy the gap-opening criterion can be slower than in the standard type II migration. In particular, if the gap density is large enough, and depending on the local density and temperature gradients, the co-rotation torque can overcome the viscous torque and lead to outward migration (*Crida and Morbidelli, 2007*). Note that the drift of the planet relative to the gas may lead to a positive feedback on migration (see section 2.3).

2.2.3. Link with observations. Recently, gap structures similar to what are predicted by planet-disk interactions have been observed in the disks around HD 169142 (*Quanz et al., 2013*) and TW Hya (*Debes et al., 2013*). The gap around HD 169142 is located 50 AU from the star, and seems to be quite deep (the surface brightness at the gap location is decreased by 10). The gap around TW Hya is located 80 AU from the star, and is much shallower (the decrease in surface brightness is only 30%). If confirmed, these would be the first observations of gaps in protoplanetary disks that could be carved by a planet.

Cavities have been observed in several circumstellar disks in the past few years. Contrary to a gap, a cavity is characterized by the absence of an inner disk. In each of these transition disks, observations indicate a lack of dust below some threshold radius that extends from a few astronomical units to a few tens of astronomical units. This lack of dust is sometimes considered to track a lack of gas, but observational evidence for accretion onto the star in some cases shows that the cavities may be void of dust but not of gas. A narrow ring of hot dust is sometimes detected in the central astronomical units of these disks (e.g., *Olofsson et al., 2012*), and this structure is often claimed to be the signpost of a giant planet carving a big gap in the disk. It should be kept in mind, however, that the gap opened by a planet is usually much narrower than these observed depletions (see section 2.2.1), and rarely (completely) gas proof. There is here a missing ingredient between the numerical simulations of gas disks and observations. The outer edge of the gap opened by a giant planet corresponds to a pressure maximum. Dust decoupled from the gas tends to accumulate there, and does not drift through the gap. For typical disk densities and temperatures between 1 and 10 AU, decoupling is most efficient for dust particles of a few centimeters to a meter (e.g., *Ormel and Klahr, 2010*). Consequently, gaps appear wider in the dust component than in the gas component, and the inner disk could be void of dust even if not of gas (*Fouchet et al., 2010; Zhu et al., 2012b*). Note that these authors find that the smallest dust grains, which are well coupled to the gas, should have a distribution identical to that of the gas and be present inside the cavity. Therefore, the observation of a cavity should depend on the size of the tracked dust, i.e., on the wavelength. Interpreting observations thus requires to decouple the dynamics of the gas and dust components. The coming years should see exciting, high-resolution observations of protoplanetary disks with, e.g., ALMA and the Multi Aperture mid-Infrared SpectroScopic Experiment (MATISSE).

2.2.4. Formation of a circumplanetary disk. The formation of a circumplanetary disk accompanies the formation of a gap. The structure of a circumplanetary disk and the gas accretion rate onto the planet have been investigated through two-dimensional hydrodynamical simulations (e.g., *Rivier et al.*, 2012), three-dimensional hydrodynamical simulations (*D'Angelo et al.*, 2003; *Ayliffe and Bate*, 2009; *Machida et al.*, 2010; *Tanigawa et al.*, 2012), and more recently through three-dimensional MHD simulations (*Uribe et al.*, 2013; *Gressel et al.*, 2013). *Gressel et al.* (2013) find accretion rates $\sim$0.01 $M_\oplus$ yr^{-1}, in good agreement with previous three-dimensional hydrodynamical calculations of viscous laminar disks. Also, in agreement with previous assessment in nonmagnetic disk environments, they find that the accretion flow in the planet's Hill sphere is intrinsically three-dimensional, and that the flow of gas toward the planet moves mainly from high latitudes, rather than along the mid-plane of the circumplanetary disk.

Another issue is to address how a circumplanetary disk impacts migration. Being bound to the planet, the circumplanetary disk migrates at the same drift rate as the rate of the planet. Issues arise in hydrodynamical simulations that discard the self-gravity of the disk, as in this case the wave torque can only apply to the planet, and not to its circumplanetary material. This causes the circumplanetary disk to artificially slow down migration, akin to a ball and chain. This issue can be particularly important for type III migration (see section 2.3). A simple workaround to this problem in simulations of non-self-gravitating disks is to exclude the circumplanetary disk in the calculation of the torque exerted on the planet (see *Crida et al.*, 2009b). Another solution suggested by these authors and also adopted by *Peplínski et al.* (2008a) is to imprint to the circumplanetary disk the acceleration felt by the planet.

2.2.5. Evolution of the eccentricity and inclination of gap-opening planets. The early evolution of the solar system in the primordial gas solar nebula should have led the four giant planets to be in a compact resonant configuration, on quasicircular and co-planar orbits (*Morbidelli et al.*, 2007; *Crida*, 2009). Their small but nonzero eccentricities and relative inclinations are supposed to have been acquired after dispersal of the nebula, during a late global instability in which Jupiter and Saturn crossed their 2:1 mean-motion resonance (*Tsiganis et al.*, 2005). This is the so-called Nice model.

Many massive exoplanets have much higher eccentricities than the planets in the solar system. Also, recent measurements of the Rossiter-McLaughlin effect have reported several hot Jupiters with large obliquities, which indicates that massive planets could also acquire a large inclination during their evolution.

Planet-disk interactions usually tend to damp the ec-

centricity and inclination of massive gap-opening planets (e.g., *Bitsch et al.*, 2013a; *Xiang-Gruess and Papaloizou*, 2013). Expressions for the damping timescales of eccentricity and inclination can be found in *Bitsch et al.* (2013a). In particular, this would indicate that type II migration should only produce hot Jupiters on circular and non-inclined orbits. There are, however, circumstances in which planets may acquire fairly large eccentricities and obliquities while embedded in their disk, which we summarize below.

2.2.5.1. *Eccentricity:* The 3:1 mean-motion resonance between a planet and the disk excites eccentricity (*Lubow*, 1991). Thus, if a planet carves a gap that is wide enough for the eccentricity pumping effect of the 3:1 resonance to overcome the damping effect of all closer resonances, the planet eccentricity will grow [e.g., *Papaloizou et al.* (2001); the disk eccentricity will grow as well]. Hydrodynamical simulations show that planet-disk interactions can efficiently increase the eccentricity of planets over $\sim$5–10 M_{Jup} (*Papaloizou et al.*, 2001; *Bitsch et al.*, 2013a; *Dunhill et al.*, 2013). Eccentricity values up to $\approx$0.25 have been obtained in *Papaloizou et al.* (2001).

2.2.5.2. *Obliquity:* Planets formed in a disk could have nonzero obliquities if the rotation axes of the star and the disk are not aligned. Several mechanisms causing misalignment have been proposed. One possibility is that the protoplanetary disk had material with differing angular momentum directions added to it at different stages of its life (e.g., *Bate et al.*, 2010). Alternatively, in dense stellar clusters, the interaction with a temporary stellar companion could tilt the disk's rotation axis (*Batygin*, 2012). However, both mechanisms should be extended by including the interaction between the disk and the magnetic field of the central star. This interaction might tilt the star's rotation axis, and lead to misalignments even in disks that are initially aligned with their star (*Lai et al.*, 2011).

2.3. Feedback of the Co-Orbital Dynamics on Migration

Under most circumstances, such as those presented in the previous sections, the migration rate of a planet is provided by the value of the disk torque, which depends on the local properties of the underlying disk, but not on the migration rate itself. There are some circumstances, however, in which the torque also depends on the drift rate, in which case one has the constituting elements of a feedback loop, with potentially important implications for migration. This, in particular, is the case of giant or subgiant planets embedded in massive disks, which deplete (at least partially) their horseshoe region.

2.3.1. Criterion for migration to run away. The co-rotation torque comes from material that executes horseshoe U-turns relative to the planet. Most of this material is trapped in the planet's horseshoe region. However, if there is a net drift of the planet with respect to the disk, material outside the horseshoe region will execute a unique horseshoe U-turn relative to the planet, and by doing so will exchange angular momentum with the planet. This drift may come about because of migration, and/or because the disk has a radial viscous drift. The torque arising from orbit crossing material naturally scales with the mass flow rate across the orbit, which depends on the relative drift of the planet and the disk. It thus depends on the migration rate.

For the sake of definiteness, we consider hereafter a planet moving inward relative to the disk, but it should be kept in mind that the processes at work here are essentially reversible, so that they can also be applied to an outward-moving planet. The picture above shows that the co-rotation torque on a planet migrating inward has three contributions:

1. The contribution of the inner disk material flowing across the orbit. As this material gains angular momentum, it exerts a negative torque on the planet that scales with the drift rate. It therefore tends to increase the migration rate, and yields a positive feedback on the migration.

2. The contribution of the co-orbital material in the planet's horseshoe region. It is two-fold. A first component arises from the material that exerts a horseshoe drag on the planet, which corresponds to the same horseshoe drag as if there was no drift between the disk and the planet (see section 2.1.2).

3. Furthermore, as the material in the horseshoe region can be regarded as trapped in the vicinity of the planetary orbit, it has to move inward at the same rate as the planet. The planet then exerts on this material a negative torque, which scales with the drift rate. By the law of action-reaction, this trapped material yields a second, positive component of the horseshoe drag on the planet that scales with the drift rate. Thus, the contribution of the drifting trapped horseshoe material also yields a negative feedback on the migration.

If the surface density profile of the disk is unaltered by the planet, i.e., if the angular momentum profile of the disk is unaltered, then contributions (1) and (3) cancel each other out exactly (*Masset and Papaloizou*, 2003). In that case, the net co-rotation torque reduces to contribution (2), and the co-rotation torque expressions presented in section 2.1.2, which have been derived assuming that the planet is on a fixed circular orbit, are valid regardless of the migration rate. Conversely, if the planet depletes, at least partly, its horseshoe region (when $\mathcal{P} \lesssim$ a few), contributions (1) and (3) do not cancel out, and the net co-rotation torque depends on the migration rate.

The above description shows that the co-orbital dynamics cause a feedback on migration when planets open a gap around their orbit. There are two key quantities to assess in order to determine when the feedback causes the migration to run away. The first quantity is called the co-orbital mass deficit (δM). It represents the mass that should be added to the planet's horseshoe region so that it has the average surface density of the orbit-crossing flow. The second quantity is the sum of the planet mass (M_p) and of the circumplanetary disk mass (M_{CPD}), i.e., the quantity $\tilde{M}_p = M_p + M_{CPD}$. As shown in Masset and Papaloizou, two regimes may occur. If $\tilde{M}_p > \delta M$, the coorbital dynam-

ics accelerate the migration, but there is no runaway. On the contrary, if $\tilde{M}_p < \delta M$, migration runs away. A more rigorous derivation performed by Masset and Papaloizou shows that the co-orbital mass deficit actually features the inverse vortensity in place of the surface density, but the same qualitative picture holds.

2.3.2. Properties of type III migration. When migration runs away, the drift rate has an exponentially growing behavior until the so-called fast regime is reached, in which the planet migrates a sizable fraction of the horseshoe width in less than a libration time. When that occurs, the drift rate settles to a finite, large value, which defines the regime of type III migration. As stressed earlier, this drift can either be outward or inward. The occurrence of type III migration with variations in the planet mass, disk mass, aspect ratio, and viscosity was discussed in detail in *Masset* (2008). The typical migration timescale associated with type III migration, which depends on the disk mass (*Lin and Papaloizou*, 2010), is on the order of a few horseshoe libration times. For the large planetary masses prone to type III migration, which have wide horseshoe regions hence short libration times, this typically amounts to a few tens of orbits.

Type III migration can in principle be outward. For this to happen, an initial seed of outward drift needs to be applied to the planet (*Peplínski et al.*, 2008b; *Masset and Papaloizou*, 2003; *Masset*, 2008). Nevertheless, all outward episodes of type III migration reported so far have been found to stall and revert back to inward migration. Interestingly, gravitationally unstable outer gap edges may provide a seed of outward type III migration and can bring massive planets to large orbital distances (*Lin and Papaloizou*, 2012; *Cloutier and Lin*, 2013). An alternative launching mechanism for outward type III migration is the outward migration of a resonantly locked pair of giant planets (*Masset and Snellgrove*, 2001) (see also section 2.4.5), which is found to trigger outward runaways at longer times (*Masset*, 2008).

The migration regime depends on how the coorbital mass deficit compares with the mass of the planet and its circumplanetary disk. It is thus important to describe correctly the buildup of the circumplanetary material and the effects of this mass on the dynamics of the gas and planet. *D'Angelo et al.* (2005) find, using a nested grid code, that the mass reached by the circumplanetary disk mass (CPD) depends heavily on the resolution for an isothermal equation of state. *Peplínski et al.* (2008a) circumvent this problem by adopting an equation of state that not only depends on the distance to the star, but also on the distance to the planet, in order to prevent an artificial flood of the CPD and to obtain numerical convergence at high resolution. Furthermore, as we have seen in section 2.2.4, in simulations that discard self-gravity, the torque exerted on the planet should exclude the circumplanetary disk. *D'Angelo et al.* (2005) find indeed that taking that torque into account may inhibit type III migration.

Type III migration has allowed ruling out a recent model of the solar nebula (*Desch*, 2007), more compact than the standard model of *Hayashi* (1981). Indeed, *Crida* (2009) has shown that Jupiter would be subject to type III migration in Desch's model and would not survive over the disk's lifetime. The occurrence of type III migration may thus provide an upper limit to the surface density of the disk models in systems known to harbor giant planets at sizable distances from their host stars.

It has been pointed out above that the exact expression of the co-orbital mass deficit involves the inverse vortensity rather than the surface density across the horseshoe region. This has some importance in low-viscosity disks: Vortensity can be regarded as materially conserved except during the passage through the shocks triggered by a giant planet, where vortensity is gained or destroyed. The corresponding vortensity perturbation can be evaluated analytically (*Lin and Papaloizou*, 2010). Eventually, the radial vortensity distribution around a giant planet exhibits a characteristic two-ring structure on the edges of the gap, which is unstable (*Lovelace et al.*, 1999; *Li et al.*, 2000) and prone to the formation of vortices (*Lin and Papaloizou*, 2010). The resulting vortensity profile determines the occurrence of type III migration. If vortices form at the gap edges, the planet can undergo nonsmooth migration with episodes of type III migration that bring the planet inward over a few Hill radii (a distance that is independent of the disk mass), followed by a stalling and a rebuild of the vortensity rings (*Lin and Papaloizou*, 2010).

The above results have been obtained for fixed-mass planets. However, for the high gas densities required by the type III migration regime, rapid growth may be expected. Using three-dimensional hydrodynamical simulations with simple prescriptions for gas accretion, *D'Angelo and Lubow* (2008) find that a planetary core undergoing rapid runaway gas accretion does not experience type III migration, but goes instead from the type I to the type II migration regime. Future progress will be made using more realistic accretion rates, like those obtained with three-dimensional radiation-hydrodynamics calculations (e.g., *D'Angelo and Bodenheimer*, 2013).

2.3.3. Feedback of coorbital dynamics on type I migration. It has been shown that type I migration in adiabatic disks could feature a kind of feedback reminiscent of type III migration. The reason is that in adiabatic disks, the co-rotation torque depends on the position of the stagnation point relative to the planet (see section 2.1.2). This position, in turn, depends on the migration rate, so that is here as well as feedback of the co-orbital dynamics on migration. *Masset and Casoli* (2009) found that this feedback on type I migration is negative, and that it has only a marginal impact on the drift rate for typical disk masses.

2.4. Orbital Evolution of Multi-Planet Systems

So far, we have examined the orbital evolution of a single planet in a protoplanetary disk, while about one-third of confirmed exoplanets reside in multi-planetary systems (see *exoplanets.org*). In such systems, the gravitational interaction between planets can significantly influence the planet orbits, leading, in particular, to important resonant processes that we describe below. A fair number of multi-planetary systems are known to have at least two planets in mean-motion resonance. *Szuszkiewicz and Podlewska-Gaca* (2012) list, for example, 32 resonant or near-resonant systems, with many additional Kepler candidate systems. The mere existence of these resonant systems is strong evidence that dissipative mechanisms changing planet semimajor axes must have operated. The probability of forming resonant configurations *in situ* is likely small (e.g., *Beaugé et al.*, 2012).

2.4.1. Capture in mean-motion resonance. We consider a system of two planets that undergo migration in their disk. If the migration drift rates are such that the mutual separation between the planets increases, i.e., when divergent migration occurs, the effects of planet-planet interactions are small and no resonant capture occurs. Conversely, resonant capture occurs for convergent migration under quite general conditions, which we discuss below.

Planets can approach each other from widely separated orbits if they have fairly different migration rates, or if they form in close proximity and are sufficiently massive to carve a common gap (see Fig. 4). In the latter case, the outer disk pushes the outer planet inward, and the inner disk pushes the inner planet outward, causing convergence. If the planets approach a commensurability, where the orbital periods are the ratio of two integers, orbital eccentricities

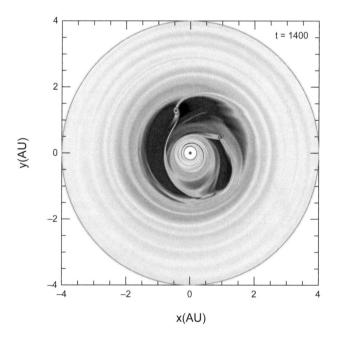

Fig. 4. Surface density of a disk with two Jupiter-mass planets located in a common gap, and engaged in a 2:1 mean-motion resonance. The inner planet mainly interacts with the inner disk, and the outer planet with the outer disk, which helps to maintain the resonant configuration. From *Kley and Nelson* (2012).

or not resonant capture occurs hinges primarily on the time the planets take to cross the resonance. Capture requires the convergent migration timescale to be longer than the libration timescale associated with the resonance width (*Snellgrove et al.*, 2001). Otherwise, the two-planet system does not have enough time for the resonance to be excited: The two planets will pass through the resonance and no capture will occur (e.g., *Quillen*, 2006; *Mustill and Wyatt*, 2011). Due to the sensitivity of the resonant capture to the migration process, the interpretation of observed resonant planetary systems provides important clues about the efficiency of disk-planet interactions.

A mean-motion resonance between two planets occurs when their orbital frequencies satisfy

$$(p + q)\Omega_2 - p\Omega_1 = 0 \tag{12}$$

where Ω_i is the angular velocity of the two planets, and where subscripts 1 and 2 refer to quantities of the inner and outer planets, respectively. In equation (12), q and p are positive integers, and q denotes the order of the resonance. The above condition for a mean-motion resonance can be recast in terms of the planets' semimajor axes, a_i, as

$$\frac{a_2}{a_1} = \left(\frac{p+q}{p}\right)^{2/3} \tag{13}$$

Formally, a system is said to be in a p + q : p mean-motion resonance if at least one of the resonant angles is librating, i.e., has a dynamical range smaller than 2π. The resonant angles ($\phi_{1,2}$) are defined as

$$\phi_{1,2} = (p + q)\lambda_2 - p\lambda_1 - q\varpi_{1,2} \tag{14}$$

where λ_i denotes the planets' longitude, and ϖ_i the longitude of their pericenter. The difference in pericenter longitudes is often used to characterize resonant behavior. For instance, when that quantity librates, the system is said to be in apsidal co-rotation. This means that the two apsidal lines of the resonant planets are always nearly aligned, or maintain a constant angle between them. This is the configuration that the planets end up with when they continue their inward migration after capture in resonance (e.g., *Kley et al.*, 2004).

Several bodies in our solar system are in mean-motion resonance. For example, the jovian satellites Io, Europa, and Ganymede are engaged in a so-called 1:2:4 Laplace resonance, while Neptune and Pluto (as well as the plutinos) are in a 3:2 mean-motion resonance. However, out of the eight planets in the solar system, not a single pair is presently in a mean-motion resonance. According to the Nice model for the early solar system, this might have been different in the past (see section 2.2.5).

The question of which resonance the system may end up in depends on the mass, the relative migration speed, and the initial separation of the planets (*Nelson and Papaloizou*, 2002). Because the 2:1 resonance (p = 1, q = 1) is the first first-order resonance that two initially well-separated planets encounter during convergent migration, it is common for planets to become locked in that resonance, provided convergent migration is not too rapid.

After a resonant capture, the eccentricities increase and the planets generally migrate as a joint pair, maintaining a constant orbital period ratio (see, however, the last two paragraphs in section 3.1.2). Continued migration in resonance drives the eccentricities up, and they would increase to very large values in the absence of damping agents, possibly rendering the system unstable. The eccentricity damping rate by the disk, $\dot{e}$, is often parameterized in terms of the migration rate $\dot{a}$ as

$$\left|\frac{\dot{e}}{e}\right| = K\left|\frac{\dot{a}}{a}\right| \tag{15}$$

with K a dimensionless constant. For low-mass planets, typically below 10 to 20 $M_\oplus$, eccentricity damping occurs much more rapidly than migration: $K \sim O[h^{-2}] \sim$ a few hundred in locally isothermal disks (see section 2.1.5) and may take even larger values in radiative disks. High-mass planets create gaps in their disk (see section 2.2) and the eccentricity damping is then strongly reduced. For the massive planets in the GJ 876 planetary system, the three-body integrations of *Lee and Peale* (2002), in which migration is applied to the outer planet only, showed that $K \sim 100$ can reproduce the observed eccentricities. If massive planets orbit in a common gap, as in Fig. 4, the disk parts on each side of the gap may act as damping agents, and it was shown by the two-dimensional hydrodynamical simulations of *Kley et al.* (2004) that such a configuration gives $K \sim 5$–10.

Disk turbulence adds a stochastic component to convergent migration, which may prevent resonant capture or maintenance of a resonant configuration. This has been examined in N-body simulations with prescribed models of disk-planet interactions and disk turbulence (e.g., *Ketchum et al.*, 2011; *Rein*, 2012a), and in hydrodynamical simulations of planet-disk interactions with simplified turbulence models (*Pierens et al.*, 2011; *Paardekooper et al.*, 2013).

2.4.2. Application to specific multi-planet systems. Application of the above considerations leads to excellent agreement between theoretical evolution models of resonant capture and the best observed systems, in particular GJ 876 (*Lee and Peale*, 2002; *Kley et al.*, 2005; *Crida et al.*, 2008). Because resonant systems most probably echo an early formation via disk-planet interactions, the present dynamical properties of observed systems can give an indicator of evolutionary history. This has been noticed recently in the system HD 45364, where two planets in 3:2 resonance have been discovered by *Correia et al.* (2009). Fits to the data give semimajor axes $a_1 = 0.681$ AU and $a_2 = 0.897$ AU, and eccentricities $e_1 = 0.168$ and $e_2 = 0.097$, respectively. Nonlinear hydrodynamic simulations of disk-planet interactions have been carried out for this system by *Rein et al.*

(2010). For suitable disk parameters, the planets enter the 3:2 mean-motion resonance through convergent migration. After the planets reached their observed semimajor axis, a theoretical radial velocity (RV) curve was calculated. Surprisingly, even though the simulated eccentricities (e_1 = 0.036, e_2 = 0.017) differ significantly from the data fits, the theoretical model fits the observed data points as well as the published best fit solution (*Rein et al., 2010*). The pronounced dynamical differences between the two orbital fits, which both match the existing data, can only be resolved with more observations. Hence, HD 45364 serves as an excellent example of a system in which a greater quantity and quality of data will constrain theoretical models of this interacting multi-planetary system.

2.4.3. Eccentricity and inclination excitations. Another interesting observational aspect where convergent migration due to disk-planet interactions may have played a prominent role is the high mean eccentricity of extrasolar planets (~0.3). As discussed above, for single planets, disk-planet interactions nearly always lead to eccentricity damping, or, at best, to modest growth for planets of a few Jupiter masses (see section 2.2.5). Strong eccentricity excitation may occur, however, during convergent migration and resonant capture of two planets. Convergent migration of three massive planets in a disk may lead to close encounters that significantly enhance planet eccentricities (*Marzari et al., 2010*) and inclinations. In the latter case, an inclination at least ≥20°–40° between the planetary orbit and the disk may drive Kozai cycles. Under disk-driven Kozai cycles, the eccentricity increases to large values and undergoes damped oscillations with time in antiphase with the inclination (*Teyssandier et al., 2013; Bitsch et al., 2013a; Xiang-Gruess and Papaloizou, 2013*).

As the disk slowly dissipates, damping will be strongly reduced. This may leave a resonant system in an unstable configuration, triggering dynamical instabilities (*Adams and Laughlin, 2003*). Planet-planet scattering may then pump eccentricities to much higher values. This scenario has been proposed to explain the observed broad distribution of exoplanet eccentricities (*Chatterjee et al., 2008; Jurić and Tremaine, 2008; Matsumura et al., 2010*). However, note that the initial conditions taken in these studies are unlikely to result from the evolution of planets in a protoplanetary disk (*Lega et al., 2013*).

Using N-body simulations with two or three giant planets and prescribed convergent migration, but no damping, *Libert and Tsiganis* (2009, 2011) found that, as the eccentricities raise to about 0.4 due to resonant interactions, planet inclinations could also be pumped under some conditions. The robustness of this mechanism needs to be checked by including eccentricity and inclination damping by the disk.

2.4.4. Low-mass planets. Disk-planet interactions of several protoplanets (5–20 $M_\oplus$) undergoing type I migration leads to crowded systems (*Cresswell and Nelson, 2008*). These authors find that protoplanets often form resonant groups with first-order mean-motion resonances having commensurabilities between 3:2 and 8:7 (see also *McNeil*

et al., 2005; Papaloizou and Szuszkiewicz, 2005). Strong eccentricity damping allows these systems to remain stable during their migration. In general terms, these simulated systems are reminiscent of the low-mass planet systems discovered by the Kepler mission, such as Kepler-11 (*Lissauer et al., 2011a*). The proximity of the planets to the star in that system, and their near co-planarity, hints strongly toward a scenario of planet formation and migration in a gaseous protoplanetary disk.

2.4.5. Possible reversal of the migration. *Masset and Snellgrove* (2001) have shown that a pair of close giant planets can migrate outward. In this scenario, the inner planet is massive enough to open a deep gap around its orbit and undergoes type II migration. The outer, less-massive planet opens a partial gap and migrates inward faster than the inner planet. If convergent migration is rapid enough for the planets to cross the 2:1 mean-motion resonance, and to lock in the 3:2 resonance, the planets will merge their gap and start migrating outward together. As the inner planet is more massive than the outer one, the (positive) torque exerted by the inner disk is larger than the (negative) torque exerted by the outer disk, which results in the planet pair moving outward. Note that in order to maintain joint outward migration for the long term, gas in the outer disk has to be funneled to the inner disk upon embarking onto horseshoe trajectories relative to the planet pair. Otherwise, gas would pile up at the outer edge of the common gap, much like a snow plow, and the torque balance as well as the direction of migration would ultimately reverse.

The above mechanism of joint outward migration of a resonant planet pair relies on an asymmetric density profile across the common gap around the two planets. This mechanism is therefore sensitive to the disk's aspect ratio, its viscosity, and the mass ratio between the two planets, since they all impact the density profile within and near the common gap. If, for instance, the outer planet is too small, the disk density beyond the inner planet's orbit will be too large to reverse the migration of the inner planet (and thus of the planet pair). Conversely, if the outer planet is too big, the torque imbalance on each side of the common gap will favor joint inward migration. Numerical simulations by *D'Angelo and Marzari* (2012) showed that joint outward migration works best when the mass ratio between the inner and outer planets is comparable to that of Jupiter and Saturn (see also *Pierens and Nelson, 2008*). In particular, *Morbidelli and Crida* (2007) found that the Jupiter-Saturn pair could avoid inward migration and stay beyond the ice line during the gas disk phase. Their migration rate depends on the disk properties, but could be close to stationary for standard values.

More recently, *Walsh et al.* (2011) have proposed that Jupiter first migrated inward in the primordial solar nebula down to ~1.5 AU, where Saturn caught it up. Near that location, after Jupiter and Saturn have merged their gap and locked into the 3:2 mean-motion resonance, both planets would have initiated joint outward migration until the primordial nebula dispersed. This scenario is known as the

Grand Tack, and seems to explain the small mass of Mars and the distribution of the main asteroid belt.

3. PLANET FORMATION AND MIGRATION: COMPARISON WITH OBSERVATIONS

The previous section has reviewed the basics of, and recent progress on, planet-disk interactions. We continue in this section with a discussion of the role played by planet-disk interactions in the properties and architecture of observed planetary systems. Section 3.1 starts with planets on short-period orbits. Emphasis is put on hot Jupiters, including those with large spin-orbit misalignments (section 3.1.1), and on the many low-mass candidate systems uncovered by the Kepler mission (section 3.1.2). Section 3.2 then examines how planet-disk interactions could account for the massive planets recently observed at large orbital separations by direct imaging techniques. Finally, section 3.3 addresses how well global models of planet formation and migration can reproduce the statistical properties of exoplanets.

3.1. Planets on Short-Period Orbits

3.1.1. Giant planets. The discovery of 51 Pegasi b (*Marcy and Butler*, 1995) on a close-in orbit of 4.2 d as the first example of a hot Jupiter led to the general view that giant planets, which are believed to have formed beyond the ice line at ≥1 AU, must have migrated inward to their present locations. Possible mechanisms for this include type II migration (see section 2.2) and either planet-planet scattering, or Kozai oscillations induced by a distant companion leading to a highly eccentric orbit that is then circularized as a result of tidal interaction with the central star (see, e.g., *Papaloizou and Terquem*, 2006; *Kley and Nelson*, 2012; *Baruteau and Masset*, 2013, and references therein). The relative importance of these mechanisms is a matter of continuing debate.

The relationship between planet mass and orbital period for confirmed exoplanets on short-period orbits is shown in Fig. 5. The hot Jupiters are seen to be clustered in circular orbits at periods in the range 3–5 d. For planet masses in the range 0.01–0.1 M_{Jup}, there is no corresponding clustering of orbital periods, indicating that this is indeed a feature associated with hot Jupiters.

Measurements of the Rossiter-McLaughlin effect (e.g., *Triaud et al.*, 2010) indicate that around one-third of hot Jupiters orbit in planes that are significantly misaligned with the equatorial plane of the central star. This is not expected from disk-planet interactions leading to type II migration, and so has led to alternative mechanisms being favored. Thus *Albrecht et al.* (2012) propose that hot Jupiters are placed in an isotropic distribution through dynamical interactions and are then circularized by tides, with disk-driven migration playing a negligible role. They account for the large fraction of misaligned hot Jupiters around stars with effective temperatures ≥6200 K, as well as the large fraction

of aligned hot Jupiters around cooler stars, being due to a very large increase in the effectiveness of the tidal processes causing alignment for the cooler stars.

However, there are a number of indications that the process of hot-Jupiter formation does not work in this way, and that a more gentle process such as disk-driven migration has operated on the distribution. We begin by remarking the presence of significant eccentricities for periods ≥6 d in Fig. 5. The period range over which tidal effects can circularize the orbits is thus very limited. Giant planets on circular orbits with periods greater than 10 d, which are interior to the ice line, must have been placed there by a different mechanism. For example, 55 Cnc b, a 0.8-M_{Jup} planet on a 14-d, near-circular orbit exterior to hot super-Earth 55 Cnc e (*Dawson and Fabrycky*, 2010), is a good case for type II migration having operated on that system. There is no reason to suppose that a smooth delivery of hot Jupiters through type II migration would not function at shorter periods.

There are also issues with the effectiveness of the tidal process (see *Rogers and Lin*, 2013). It has to align inclined circular orbits without producing inspiral into the central star. To avoid the latter, the components of the tidal potential that act with nonzero forcing frequency in an inertial frame when the star does not rotate have to be ineffective. Instead, one has to rely on components that appear to be stationary in this limit. These have a frequency that is a multiple of the stellar rotation frequency, expected to be significantly less than the orbital frequency, as viewed from the star when it rotates. As such components depend only on the time-averaged orbit, they are insensitive to the

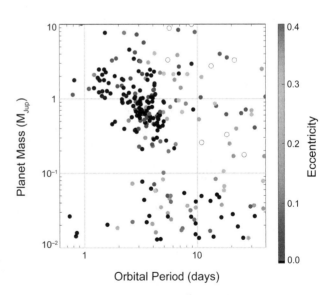

Fig. 5. Mass as a function of orbital period for confirmed exoplanets on short-period orbits. The grayscale of the dots represents the magnitude of the orbital eccentricity as indicated in the grayscale bar (white dots with a gray circle correspond to eccentricities larger than 0.4). Data were extracted from *exoplanets.org*.

sense of rotation in the orbit. Accordingly, there is a symmetry between prograde- and retrograde-aligning orbits with respect to the stellar equatorial plane. This is a strict symmetry when the angular momentum content of the star is negligible compared to that of the orbit, otherwise there is a small asymmetry (see *Rogers and Lin*, 2013). Notably, a significant population of retrograde orbits with aligned orbital planes, which is expected in this scenario, is not observed.

Dawson and Murray-Clay (2013) have recently examined the dependence of the relationship between mass and orbital period on the metallicity of the central star. They find that the pileup for orbital periods in the range 3–5 d characteristic of hot Jupiters is only seen at high metallicity. In addition, high eccentricities, possibly indicative of dynamical interactions, are also predominantly seen at high metallicity. This indicates multi-planet systems in which dynamical interactions leading to close orbiters occur at high metallicity, and that disk-driven migration is favored at low metallicity.

Finally, misalignments between stellar equators and orbital planes may not require strong dynamical interactions. Several mechanisms may produce misalignments between the protoplanetary disk and the equatorial plane of its host star (see section 2.2.5). Another possibility is that internal processes within the star, such as the propagation of gravity waves in hot stars, lead to different directions of the angular momentum vector in the central and surface regions (e.g., *Rogers et al.*, 2012).

3.1.2. Low-mass planets. The Kepler mission has discovered tightly packed planetary systems orbiting close to their star. Several have been determined to be accurately co-planar, which is a signature of having formed in a gaseous disk. These include KOI 94 and KOI 25 (*Albrecht et al.*, 2013), KOI 30 (*Sanchis-Ojeda et al.*, 2012), and Kepler 50 and Kepler 55 (*Chaplin et al.*, 2013). We remark that if one supposed that formation through *in situ* gas-free accumulation had taken place, for planets of a fixed type, the formation timescale would be proportional to the product of the local orbital period and the reciprocal of the surface density of the material making up the planets (e.g., *Papaloizou and Terquem*, 2006). For a fixed mass this is proportional to $r^{3.5}$. Scaling from the inner solar system, where this timescale is taken to be ~3 × 10^8 yr (*Chambers and Wetherill*, 1998), it becomes ≤10^5 yr for r < 0.1 AU. Note that this timescale is even shorter for more massive and more compact systems. This points to a possible formation during the disk lifetime, although the formation process should be slower and quieter in a disk, as protoplanets are constrained to be in nonoverlapping near-circular orbits. Under this circumstance, disk-planet interactions cannot be ignored.

Notably, *Lissauer et al.* (2011b) found that a significant number of Kepler multiplanet candidate systems contain pairs that are close to first-order resonances. They also found a few multi-resonant configurations. An example is the four-planet system KOI-730, which exhibits the mean-

motion ratios 8:6:4:3. More recently, *Steffen et al.* (2013) confirmed the Kepler 60 system, which has three planets with the inner pair in a 5:4 commensurability and the outer pair in a 4:3 commensurability. However, most of the tightly packed candidate systems are nonresonant.

The period ratio histogram for all pairs of confirmed exoplanet systems is shown in Fig. 6a. This shows prominent spikes near the main first-order resonances. However, this trend is biased because many of the Kepler systems were validated through observing transit timing variations, which are prominent for systems near resonances. Figure 6b shows the same histogram for Kepler candidate systems announced at the time of writing (quarters 1–8). In this case, although there is some clustering in the neighborhood of the 3:2 commensurability and an absence of systems at exact 2:1 commensurability, because there is an overall tendency for systems to have period ratios slightly larger than exact resonant values, there are many nonresonant systems.

At first sight, this appears to be inconsistent with results from the simplest theories of disk-planet interactions, for which convergent migration is predicted to form either resonant pairs (*Papaloizou and Szuszkiewicz*, 2005) or resonant chains (*Cresswell and Nelson*, 2006). However, there are features not envisaged in that modeling that could modify these results, which we briefly discuss below. These fall into two categories: (1) those operating while the disk is still present, and (2) those operating after the disk has

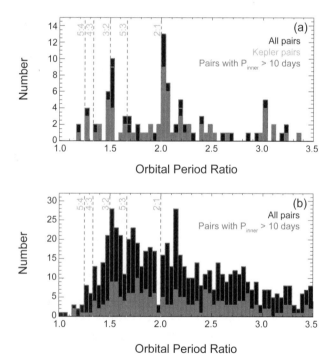

Fig. 6. (a) The period ratio histogram for confirmed exoplanets. (b) The same histogram for Kepler candidates from quarters 1 to 8. Vertical dashed lines show the period ratio of a few mean-motion resonances. Data were extracted from *exoplanets.org*.

dispersed. An example of the latter type is the operation of tidal interactions with the central star, which can cause two-planet systems to increase their period ratios away from resonant values (*Papaloizou, 2011; Lithwick and Wu, 2012; Batygin and Morbidelli, 2013*). However, this cannot be the only process operating, as Fig. 6 shows that the same period ratio structure is obtained for periods both less than and greater than 10 d. We also note that compact systems with a large number of planets can be close to dynamical instability through the operation of Arnold diffusion [see, e.g., the analysis of Kepler 11 by *Migaszewski et al.* (2012)]. Thus it is possible that memory of the early history of multiplanet systems is absent from their current configurations.

When the disk is present, stochastic forcing due to the presence of turbulence could ultimately cause systems to diffuse out of resonance (e.g., *Pierens et al., 2011; Rein, 2012a*). When this operates, resonances may be broken and period ratios may both increase and decrease away from resonant values. *Paardekooper et al.* (2013) employed stochastic fluctuations in order to enable lower-order resonances to be diffused through slow convergent migration. In this way, they could form a 7:6 commensurability in their modeling of the Kepler 36 system.

Another mechanism that could potentially prevent the formation of resonances in a disk involves the influence of each planet's wakes on other planets. The dissipation induced by the wake of a planet in the co-orbital region of another planet causes the latter to be effectively repelled. This repulsion might either prevent the formation of resonances, or result in an increase in the period ratio from a resonant value. *Podlewska-Gaca et al.* (2012) considered a 5.5-$M_\oplus$ super-Earth migrating in a disk toward a giant planet. Figure 7a shows the time evolution of the super-Earth's semimajor axis when orbiting exterior to a planet of 1 or 2 M_{Jup}. We see that the super-Earth's semimajor axis attains a minimum and ultimately increases, and thus a resonance is not maintained. Figure 7b shows the disk's surface density at one illustrative time: The super-Earth (gray arrow) feels a headwind from the outer wake of the hot Jupiter, which leads to the super-Earth being progressively repelled. This mechanism could account for the observed scarcity of super-Earths on near-resonant orbits exterior to hot Jupiters.

A similar effect was found in disk-planet simulations with two partial gap-opening planets by *Baruteau and Papaloizou* (2013). This is illustrated in Fig. 8. The orbital period ratio between the planets initially decreases until the planets get locked in the 3:2 mean-motion resonance. The period ratio then increases away from the resonant ratio as a result of wake-planet interactions. The inset shows a density contour plot where the interaction of the planets with each other's wakes is clearly seen. Divergent evolution of a planet pair through wake-planet interactions requires some nonlinearity, and so will not work with pure type I migration ($\mathcal{P} \gg 1$), but not so much that the gaps become totally cleared (*Baruteau and Papaloizou, 2013*). This mechanism works best for partial gap-opening planets ($\mathcal{P} \sim$ a few, q ~

h³), which concern super-Earth to Neptune-mass planets in disks with aspect ratio h ≤ 3% (expected in inner disk regions), or Saturn-mass planets if h ~ 5%. These results show circumstances where convergent migration followed by attainment of stable strict commensurability may not be an automatic consequence of disk-planet interactions. Wake-planet interactions could explain why near-resonant planet pairs among Kepler's multiple candidate systems tend to have period ratios slightly greater than resonant.

3.2. Planets on Long-Period Orbits

In the past decade, spectacular progress in direct imaging techniques have uncovered more than 20 giant planets with orbital separations ranging from about ten to a few hundred astronomical units. The four planets in the HR 8799 system (*Marois et al., 2010*) or β Pictoris b (*Lagrange et al., 2010*) are remarkable examples. The discoveries of these cold Jupiters have challenged theories of planet formation and evolution. We review below the mechanisms that have been proposed to account for the cold Jupiters.

3.2.1. Outward migration of planets formed further in? In the core-accretion scenario for planet formation, it

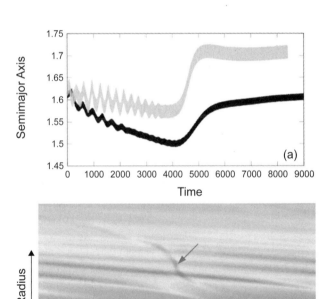

Fig. 7. (a) Semimajor axis evolution of a 5.5-$M_\oplus$ super-Earth orbiting exterior to an inwardly migrating giant planet of 1 M_{Jup} (dark line) or 2 M_{Jup} (light gray line) that is initially located at a = 1. The time shown in the x axis is 2π× the initial orbital period of the giant planet. (b) Disk's surface density for the 1-M_{Jup} case. The super-Earth's location is spotted by a gray arrow. Adapted from *Podlewska-Gaca et al.* (2012).

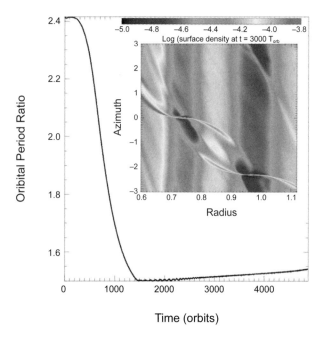

Fig. 8. Evolution of the orbital period ratio between a 13-$M_\oplus$ outer planet and a 15-$M_\oplus$ inner planet. The period ratio increases from the 3:2 resonant ratio due to wake-planet interactions. These interactions are visible in the disk's surface density shown in the inset image. Adapted from *Baruteau and Papaloizou (2013)*.

is difficult to form Jupiter-like planets in isolation beyond 10 AU from a Sun-like star (*Pollack et al., 1996; Ida and Lin, 2004*) (see the chapter by Helled et al. in this volume). Could forming Jupiters move out to large orbital separations in their disk? Outward type I migration followed by rapid gas accretion is possible, but the maximum orbital separation attainable through type I migration is uncertain (see sections 2.1.2 and 2.1.4). Planets in the Jupiter-mass range are expected to open an annular gap around their orbit (see section 2.2.1). If a deep gap is carved, inward type II migration is expected. If a partial gap is opened, outward type III migration could occur under some circumstances (see section 2.3), but numerical simulations have shown that it is difficult to sustain this type of outward migration over long timescales (*Masset and Papaloizou, 2003; Peplínski et al., 2008b*). It is therefore unlikely that a single massive planet formed through the core-accretion scenario could migrate to several tens or hundreds of astronomical units.

It is possible, however, that a pair of close giant planets may migrate outward according to the mechanism described in section 2.4.5. For non-accreting planets, this mechanism could deliver two near-resonant giant planets at orbital separations comparable to those of the cold Jupiters (*Crida et al., 2009a*). However, this mechanism relies on the outer planet to be somewhat less massive than the inner one. Joint outward migration may stall and eventually reverse if the outer planet grows faster than the inner one (*D'Angelo and Marzari, 2012*). Numerical simulations by these authors

showed that it is difficult to reach orbital separations typical of the cold Jupiters.

3.2.2. Planets scattered outward? The fraction of confirmed planets known in multi-planetary systems is about one-third (see, e.g., *exoplanets.org*), from which nearly two-thirds have an estimated minimum mass [the remaining one-third comprises Kepler multiple planets confirmed by transit timing variation (TTV), and for which an upper mass estimate has been obtained based on dynamical stability; see, e.g., *Steffen et al.* (2013)]. More than half the confirmed multiple planets having a lower mass estimate are more massive than Saturn, which indicates that the formation of several giant planets in a protoplanetary disk should be quite common. Smooth convergent migration of two giant planets in their parent disk should lead to resonant capture followed by joint migration of the planet pair (e.g., *Kley et al., 2004*). Dispersal of the gas disk may trigger the onset of dynamical instability, with close encounters causing one of the two planets to be scattered to large orbital separations (*Chatterjee et al., 2008*). A system of three giant planets is more prone to dynamical instability, and disk-driven convergent migration of three giant planets may induce planet scattering even in quite massive protoplanetary disks (*Marzari et al., 2010; Moeckel and Armitage, 2012*). Planet scattering before or after disk dispersal could thus be a relevant channel for delivering one or several massive planets to orbital separations comparable to the cold Jupiters'. It could also account for the observed free-floating planets.

3.2.3. Formation and evolution of planets by gravitational instability? Giant planets could also form after the fragmentation of massive protoplanetary disks into clumps through gravitational instability (GI). The GI is triggered as the well-known Toomre-Q parameter is ~1 and the disk's cooling timescale approaches the dynamical timescale (*Gammie, 2001; Rafikov, 2005*). The later criterion is prone to some uncertainty due to the stochastic nature of fragmentation (*Paardekooper, 2012*). The GI could trigger planet formation typically beyond 30 AU from a Sun-like star.

How do planets evolve once fragmentation is initiated? First, GI-formed planets are unlikely to stay in place in their gravito-turbulent disk. Since these massive planets form in about a dynamical timescale, they rapidly migrate to the inner parts of their disk, having initially no time to carve a gap around their orbit (*Baruteau et al., 2011b; Zhu et al., 2012a; Vorobyov, 2013*) (see Fig. 9). These inner regions should be too hot to be gravitationally unstable, and other sources of turbulence, like the MRI, will set the background disk profiles and the level of turbulence. The rapid inward migration of GI-formed planets could then slow down or even stall, possibly accompanied by the formation of a gap around the planet's orbit. Gap-opening may also occur if significant gas accretion occurs during the initial stage of rapid migration (*Zhu et al., 2012a*), which may promote the survival of inwardly migrating clumps. Planet-planet interactions, which may result in scattering events, mergers, or resonant captures in a disk, should also play a prominent role in shaping planetary systems formed by GI. The near-

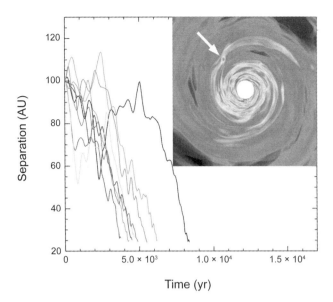

Fig. 9. Semimajor axis evolution of a Jupiter-mass planet formed in a self-gravitating turbulent disk at 100 AU from a solar-mass star. Different curves are for different simulations with varying initial conditions. The inset image shows the disk's surface density in one of these simulations, and the arrow spots the planet's location. Adapted from *Baruteau et al.* (2011b).

resonant architecture of the HR 8799 planet system could point to resonant captures after convergent migration in a gravito-turbulent disk.

Furthermore, gas clumps progressively contract as they cool down. As clumps initially migrate inward, they may experience some tidal disruption, a process known as tidal downsizing. This process could deliver a variety of planet masses in the inner parts of protoplanetary disks (*Boley et al.*, 2010; *Nayakshin*, 2010).

3.3. Planet Formation with Migration: Population Synthesis and N-Body Simulations

All the disk-driven migration scenarios discussed in this review have some dependence on the planet mass, so it is necessary to consider the combined effects of mass growth and migration when assessing the influence of migration on the formation of planetary systems. Two approaches that have been used extensively for this purpose are planetary population synthesis and N-body simulations, both of which incorporate prescriptions for migration.

Population synthesis studies use Monte Carlo techniques to construct synthetic planetary populations, with the aim of determining which combinations of model ingredients lead to statistically good fits to the observational data [orbital elements and masses in particular, but in more recent developments planetary radii and luminosities are also calculated as observables (see, e.g., *Mordasini et al.*, 2012)]. In principle, this allows the mass-period and mass-radius relation for gaseous exoplanets to be computed. Input

variables that form the basis of the Monte Carlo approach include initial gas disk masses, gas-to-dust ratios, and disk photoevaporation rates, constrained by observational data. The advantages of population synthesis studies lie in their computational speed and ability to include a broad range of physical processes. This allows the models to treat elements of the physics (such as gas envelope accretion, or ablation of planetesimals as they pass through the planet envelope, for example) much more accurately than is possible in N-body simulations. A single realization of a Monte Carlo simulation consists of drawing a disk model from the predefined distribution of possibilities and introducing a single low-mass planetary embryo in the disk at a random location within a predefined interval in radius. Accretion of planetesimals then proceeds, followed by gas envelope settling as the core mass grows. Runaway gas accretion to form a gas giant may occur if the core mass reaches the critical value. Further implementation details are provided in the chapter by Benz et al. in this volume.

The main advantages of the N-body approach are that they automatically include an accurate treatment of planet-planet interactions that is normally missing from the "single-planet-in-a-disk" Monte Carlo models, they capture the competitive accretion that is inherently present in the oligarchic picture of early planet formation, and they incorporate giant impacts between embryos that are believed to provide the crucial last step in terrestrial planet formation. At present, however, gas accretion has been ignored or treated in a crude manner in N-body models. As such, population-synthesis models can provide an accurate description of the formation of a gas giant planet, whereas N-body models are well suited to examining the formation of systems of lower-mass terrestrial, super-Earth, and core-dominated Neptune-like bodies.

As indicated above, the basis of almost all published population-synthesis models has been the core-accretion scenario of planetary formation, combined with simple prescriptions for type I and type II migration and viscous disk evolution (*Armitage et al.*, 2002; *Ida and Lin*, 2004; *Alibert et al.*, 2005; *Mordasini et al.*, 2009a). A notable exception is the recent population synthesis study based on the disk fragmentation model (*Forgan and Rice*, 2013). Almost all studies up to the present time have adopted type I migration rates similar to those arising from equation (1), supplemented with an arbitrary reduction factor that slows the migration. The influence of the vortensity and entropy-related horseshoe drag discussed in section 2.1.2 has not yet been explored in detail, although a couple of recent preliminary explorations that we describe below have appeared in the literature.

Ida and Lin (2008), *Mordasini et al.* (2009a), and *Mordasini et al.* (2009b) consider the effects of type I and type II migration in their population-synthesis models. Although differences exist in the modeling procedures, these studies all conclude that unattenuated type I migration leads to planet populations that do not match the observed distributions of planet mass and semimajor axis. Models presented in *Ida and Lin* (2008), for example, fail to produce giant planets at all if full-strength type I migration operates. Statistically acceptable giant-planet populations are reported for reductions in the efficiency of type I migration by factors of 0.01 to 0.03, with type II migration being required to form "hot Jupiters." With the type II timescale of ~10^5 yr being significantly shorter than disk lifetimes, numerous giant planets migrate into the central star in these models. The survivors are planets that form late as the disk is being dispersed (through viscous evolution and photoevaporation), but just early enough to accrete appreciable gaseous envelopes. *Mordasini et al.* (2009a,b) present models with full-strength type I migration that are able to form a sparse population of gas giants. Cores that accrete very late in the disk lifetime are able to grow to large masses as they migrate because they do not exhaust their feeding zones. Type I migration of the forming planetary cores in this case, however, strongly biases the orbital radii of planetary cores to small values, leading to too many short-period massive gas giants that are in contradiction of the exoplanet data.

The above studies focused primarily on forming gas-giant planets, but numerous super-Earth and Neptune-mass planets have been discovered by both groundbased surveys and the Kepler mission (e.g., *Mayor et al.*, 2011; *Borucki et al.*, 2011; *Fressin et al.*, 2013). Based on 8 yr of High Accuracy Radial velocity Planet Searcher (HARPS) data, the former publication in this list suggests that at least 50% of solar-type stars hosts at least one planet with a period of 100 d or less. Based on an analysis of the false-positive rate in Kepler detections, *Fressin et al.* (2013) suggest that 16.5% of FGK stars have at least one planet between 0.8 and 1.25 $R_\oplus$ with orbital periods up to 85 d. These results appear consistent with the larger numbers of super-Earth and Neptune-like planets discovered by Kepler. In a recent study, *Howard et al.* (2010) performed a direct comparison between the predictions of population-synthesis models with RV observations of extrasolar planets orbiting within 0.25 AU around 166 nearby G-, K-, and M-type stars (the η_{Earth} survey). The data indicate a high density of planets with $M_p = 4$–10 $M_\oplus$ with periods <10 d, in clear accord with the discoveries made by Kepler. This population is not present in the Monte Carlo models because of rapid migration and mass growth. *Ida and Lin* (2010) recently considered specifically the formation of super-Earths using population synthesis, incorporating for the first time a treatment of planet-planet dynamical interactions. In the absence of an inner disk cavity (assumed to form by interaction with the stellar magnetic field) the simulations failed to form systems of short-period super-Earths because of type I migration into the central star. This requirement for an inner disk cavity to halt inward migration, in order to explain the existence of the observed short-period planet population, appears to be a common feature in planetary-formation models that include migration. Given that planets are found to have a wide-range of orbital radii, however, it seems unlikely that this migration-stopping mechanism can apply to all systems. Given the large numbers of planets that migrate

into the central star in the population-synthesis models, it would appear that such a stopping mechanism when applied to all planet-forming disks would predict the existence of a significantly larger population of short-period planets than is observed. This point is illustrated by Fig. 7, which shows the mass-period relation for planets with masses $0.1 \leq M_p \leq 1$ M_{Jup} in Fig. 7a and $10^{-3} \leq M_p \leq 10^{-1}$ M_{Jup} in Fig. 7b. Although a clustering of giant planets between orbital periods 3–5 d is observed, there is no evidence of such a pile-up for the lower-mass planets. This suggests that an inner cavity capable of stopping the migration of planets of all masses may not be a prevalent feature of planet-forming disks.

N-body simulations with prescriptions for migration have been used to examine the interplay between planet growth and migration. We primarily concern ourselves here with simulations that include the early phase of oligarchic growth when a swarm of Mars-mass embryos embedded in a disk of planetesimals undergo competitive accretion. A number of studies have considered dynamical interaction between much more massive bodies in the presence of migration, but we will not consider these here. Early work included examination of the early phase of terrestrial planet formation in the presence of gas (*McNeil et al.*, 2005), which showed that even unattenuated type I migration was not inconsistent with terrestrial planet formation in disks with a moderately enhanced solids abundance. N-body simulations that explore short-period super-Earth formation and demonstrate the importance of tidal interaction with the central star for disk models containing inner cavities have been presented by *Terquem and Papaloizou* (2007). *McNeil and Nelson* (2009, 2010) examined the formation of hot super-Earth and Neptune-mass planets using N-body simulations combined with type I migration (full strength and with various attenuation factors). The motivation here was to examine whether or not the standard oligarchic growth picture of planet formation combined with type I migration could produce systems such as Gliese 581 and HD 69830 that contain multiple short-period super-Earth and Neptune-mass planets. These hot and warm super-Earth and Neptune systems probably contain up to 30–40 $M_\oplus$ of rocky or icy material orbiting within 1 AU. The models incorporated a purpose-built multiple time-step scheme to allow planet-formation scenarios in global disk models extending out to 15 AU to be explored. The aim was to examine whether or not hot super-Earths and Neptunes could be explained by a model of formation at large radius followed by inward migration, or whether instead smaller building blocks of terrestrial mass could migrate in and form a massive disk of embryos that accretes *in situ* to form short-period bodies. As such, this was a precursor study to the recent *in situ* models that neglect the migration described in *Hansen and Murray* (2013). The suite of some 50 simulations led to the formation of a few individual super-Earth and Neptune-mass planets, but failed to produce any systems with more than 12 $M_\oplus$ of solids interior to 1 AU.

Thommes et al. (2008) presented a suite of simulations of giant-planet formation using a hybrid code in which emerging embryos were evolved using an N-body integrator combined with a one-dimensional viscous disk model. Although unattenuated type I and type II migration were included, a number of models led to successful formation of systems of surviving gas giant planets. These models considered an initial population of planetary embryos undergoing oligarchic growth extending out to 30 AU from the star, and indicate that the right combination of planetary growth times, disk masses, and lifetimes can form surviving giant planets through the core-accretion model, provided embryos can form and grow at rather large orbital distances before migrating inward.

The role of the combined vorticity- and entropy-related co-rotation torque, and its ability to slow or reverse type I migration of forming planets, has not yet been explored in detail. The survival of protoplanets with masses in the range $1 \leq M_p \leq 10\ M_\oplus$ in global one-dimensional disk models has been studied by *Lyra et al.* (2010). These models demonstrate the existence of locations in the disk where planets of a given mass experience zero migration due to the cancellation of Lindblad and co-rotation torques (zero-migration radii or planetary-migration traps). Planets have a tendency to migrate toward these positions, where they then sit and drift inward slowly as the gas disk disperses. Preliminary results of population synthesis calculations have been presented by *Mordasini et al.* (2011), and N-body simulations that examine the oligarchic growth scenario under the influence of strong co-rotation torques have been presented by *Hellary and Nelson* (2012). These studies indicate that the convergent migration that arises as planets move toward their zero-migration radii can allow a substantial increase in the rate of planetary accretion. Under conditions where the disk hosts a strongly decreasing temperature gradient, *Hellary and Nelson* (2012) computed models that led to outward migration of planetary embryos to radii ~50 AU, followed by gas accretion that formed gas giants at these large distances from the star. The temperature profiles required for this were substantially steeper than those that arise from calculations of passively heated disks; however, it remains to be determined whether these conditions can ever be realized in a protoplanetary disk. Following on from the study of co-rotation torques experienced by planets on eccentric orbits by *Bitsch and Kley* (2010), *Hellary and Nelson* (2012) incorporated a prescription for this effect and found that planet-planet scattering causes eccentricity growth to values that effectively quench the horseshoe drag, such that crowded planetary systems during the formation epoch may continue to experience rapid inward migration. Further work is clearly required to fully assess the influence of the co-rotation torque on planet formation in the presence of significant planet-planet interactions.

Looking to the future, it is clear that progress in making accurate theoretical predictions that apply across the full range of observed exoplanet masses will be best achieved by bringing together the best elements of the population synthesis and N-body approaches. Some key issues that require particular attention include the structure of the disk close to the star, given its influence in shaping the short-period planet population (see section 3.1). This will require developments in both observation and theory to constrain the nature of the magnetospheric cavity and its influence on the migration of planets of all masses. Significant improvements in underlying global disk models are also required, given the sensitivity of migration processes to the detailed disk physics. Particular issues at play are the roles of magnetic fields, the thermal evolution, and the nature of the turbulent flow in disks that sets the level of the effective viscous stress. These are all active areas of research at the present time and promise to improve our understanding of planet formation processes in the coming years.

4. SUMMARY

The main points to take away from this chapter are summarized below:

1. Disk-planet interactions are a natural process that inevitably operates during the early evolution of planetary systems, when planets are still embedded in their protoplanetary disk. They modify all orbital elements of a planet. While eccentricity and inclination are usually damped quickly, the semimajor axis may increase or decrease more or less rapidly depending on the planet-to-star mass ratio (q) and the disk's physical properties (including its aspect ratio h).

2. Planet migration comes in three main flavors: (1) Type I migration applies to low-mass planets ($\mathcal{P} \gg 1$, which is the case if $q \ll h^3$) that do not open a gap around their orbit. Its direction (inward or outward) and speed are very sensitive to the structure of the disk and its radiative and turbulent properties in a narrow region around the planet. While major progress in understanding the physics of type I migration has been made since PPV, robust predictions of its direction and pace will require more knowledge of protoplanetary disks in regions of planet formation. ALMA should bring precious constraints in that sense. (2) Type II migration is for massive planets ($\mathcal{P} \leq 1$, or $q \geq q_{min}$ given by equation (10)) that carve a deep gap around their orbit. Type II migrating planets drift inward in a timescale comparable to or longer than the disk's viscous timescale at their orbital separation. (3) Type III migration concerns intermediate-mass planets ($q \sim h^3$) that open a partial gap around their orbit. This very rapid, preferentially inward migration regime operates in massive disks.

3. Planet-disk interaction is one major ingredient for shaping the architecture of planetary systems. The diversity of migration paths predicted for low-mass planets probably contributes to the diversity in the mass/semimajor-axis diagram of observed exoplanets. Convergent migration of several planets in a disk could provide the conditions for exciting planets' eccentricity and inclination.

4. The distribution of spin-orbit misalignments among hot Jupiters is very unlikely to have an explanation based on a single scenario for the large-scale inward migration required to bring them to their current orbital separations. Hot Jupiters on orbits aligned with their central star point preferentially to a smooth disk delivery, via type II

migration, rather than to dynamical interactions with a planetary or a stellar companion, followed by star-planet tidal realignment.

5. Convergent migration of two planets in a disk does not necessarily result in the planets being in mean-motion resonance. Turbulence in the disk, the interaction between a planet and the wake of its companion, or late star-planet tidal interactions could explain why many multi-planet candidate systems discovered by the Kepler mission are near- or nonresonant. Wake-planet interactions could account for the observed scarcity of super-Earths on near-resonant orbits exterior to hot Jupiters.

6. Recent observations of circumstellar disks have reported the existence of cavities and large-scale vortices in millimeter-sized grains. These features do not necessarily track the presence of a giant planet in the disk. It should be kept in mind that the gaps carved by planets of ~1 M_{Jup} or less are narrow annuli, not cavities.

7. Improving theories of planet-disk interactions in models of planet-population synthesis is essential for progress in understanding the statistical properties of exoplanets. Current discrepancies between theory and observations point to uncertainties in planet-migration models as much as to uncertainties in planet mass growth, physical properties of protoplanetary disks, or the expected significant impact of planet-planet interactions.

Acknowledgments. We thank C. Dullemond and the anonymous referee for their constructive reports. C.B. was supported by a Herchel Smith Postdoctoral Fellowship of the University of Cambridge, S.J.P. by a Royal Society University Research Fellowship, J.G. by the Science and Technology Facilities Council, and B.B. by the Helmholtz Alliance Planetary Evolution and Life.

REFERENCES

Adams F. C. and Bloch A. M. (2009) *Astrophys. J., 701,* 1381.
Adams F. C. and Laughlin G. (2003) *Icarus, 163,* 290.
Albrecht S. et al. (2012) *Astrophys. J., 757,* 18.
Albrecht S. et al. (2013) *Astrophys. J., 771,* 11.
Alibert Y. et al. (2005) *Astron. Astrophys., 434,* 343.
Armitage P. J. et al. (2002) *Mon. Not. R. Astron. Soc., 334,* 248.
Artymowicz P. (1993a) *Astrophys. J., 419,* 166.
Artymowicz P. (1993b) *Astrophys. J., 419,* 155.
Ayliffe B. A. and Bate M. R. (2009) *Mon. Not. R. Astron. Soc., 397,* 657.
Ayliffe B. A. and Bate M. R. (2011) *Mon. Not. R. Astron. Soc., 415,* 576.
Bai X.-N. and Stone J. M. (2011) *Astrophys. J., 736,* 144.
Balbus S. A. and Hawley J. F. (1991) *Astrophys. J., 376,* 214.
Baruteau C. and Lin D. N. C. (2010) *Astrophys. J., 709,* 759.
Baruteau C. and Masset F. (2008a) *Astrophys. J., 672,* 1054.
Baruteau C. and Masset F. (2008b) *Astrophys. J., 678,* 483.
Baruteau C. and Masset F. (2013) In *Tides in Astronomy and Astrophysics* (J. Souchay et al., eds.), p. 201. Lecture Notes in Physics Vol. 861, Springer-Verlag, Berlin.
Baruteau C. and Papaloizou J. C. B. (2013) *Astrophys. J., 778,* 7.
Baruteau C. et al. (2011a) *Astron. Astrophys., 533,* A84.
Baruteau C. et al. (2011b) *Mon. Not. R. Astron. Soc., 416,* 1971.
Bate M. R. et al. (2010) *Mon. Not. R. Astron. Soc., 401,* 1505.
Batygin K. (2012) *Nature, 491,* 418.
Batygin K. and Morbidelli A. (2013) *Astron. J., 145,* 1.
Beaugé C. et al. (2012) *Res. Astron. Astrophys., 12,* 1044.
Bell K. R. et al. (1997) *Astrophys. J., 486,* 372.

Bitsch B. and Kley W. (2010) *Astron. Astrophys., 523,* A30.
Bitsch B. and Kley W. (2011a) *Astron. Astrophys., 530,* A41.
Bitsch B. and Kley W. (2011b) *Astron. Astrophys., 536,* A77.
Bitsch B. et al. (2013a) *Astron. Astrophys., 555,* A124.
Bitsch B. et al. (2013b) *Astron. Astrophys., 549,* A124.
Boley A. C. et al. (2010) *Icarus, 207,* 509.
Borucki W. J. et al. (2011) *Astrophys. J., 736,* 19.
Casoli J. and Masset F. S. (2009) *Astrophys. J., 703,* 845.
Chambers J. E. and Wetherill G. W. (1998) *Icarus, 136,* 304.
Chaplin W. J. et al. (2013) *Astrophys. J., 766,* 101.
Chatterjee S. et al. (2008) *Astrophys. J., 686,* 580.
Chiang E. I. and Goldreich P. (1997) *Astrophys. J., 490,* 368.
Cloutier R. and Lin M.-K. (2013) *Mon. Not. R. Astron. Soc., 434,* 621.
Correia A. C. M. et al. (2009) *Astron. Astrophys., 496,* 521.
Cossou C. et al. (2013) *Astron. Astrophys., 553,* L2.
Cresswell P. and Nelson R. P. (2006) *Astron. Astrophys., 450,* 833.
Cresswell P. and Nelson R. P. (2008) *Astron. Astrophys., 482,* 677.
Cresswell P. et al. (2007) *Astron. Astrophys., 473,* 329.
Crida A. (2009) *Astrophys. J., 698,* 606.
Crida A. and Morbidelli A. (2007) *Mon. Not. R. Astron. Soc., 377,* 1324.
Crida A. et al. (2006) *Icarus, 181,* 587.
Crida A. et al. (2008) *Astron. Astrophys., 483,* 325.
Crida A. et al. (2009a) *Astrophys. J. Lett., 705,* L148.
Crida A. et al. (2009b) *Astron. Astrophys., 502,* 679.
D'Angelo G. and Bodenheimer P. (2013) *Astrophys. J., 778,* 77.
D'Angelo G. and Lubow S. H. (2008) *Astrophys. J., 685,* 560.
D'Angelo G. and Lubow S. H. (2010) *Astrophys. J., 724,* 730.
D'Angelo G. and Marzari F. (2012) *Astrophys. J., 757,* 50.
D'Angelo G. et al. (2003) *Astrophys. J., 586,* 540.
D'Angelo G. et al. (2005) *Mon. Not. R. Astron. Soc., 358,* 316.
Dawson R. I. and Fabrycky D. C. (2010) *Astrophys. J., 722,* 937.
Dawson R. I. and Murray-Clay R. A. (2013) *Astrophys. J. Lett., 767,* L24.
Debes J. H. et al. (2013) *Astrophys. J., 771,* 45.
Desch S. J. (2007) *Astrophys. J., 671,* 878.
Dong R. et al. (2011) *Astrophys. J., 741,* 57.
Duffell P. C. and MacFadyen A. I. (2013) *Astrophys. J., 769,* 41.
Dullemond C. P. et al. (2001) *Astrophys. J., 560,* 957.
Dunhill A. C. et al. (2013) *Mon. Not. R. Astron. Soc., 428,* 3072.
Fendyke S. M. and Nelson R. P. (2014) *Mon. Not. R. Astron. Soc., 437,* 96.
Ferreira J. et al. (2006) *Astron. Astrophys., 453,* 785.
Forgan D. and Rice K. (2013) *Mon. Not. R. Astron. Soc., 432,* 3168.
Fouchet L. et al. (2010) *Astron. Astrophys., 518,* A16.
Fressin F. et al. (2013) *Astrophys. J., 766,* 81.
Fromang S. et al. (2005) *Mon. Not. R. Astron. Soc., 363,* 943.
Gammie C. F. (2001) *Astrophys. J., 553,* 174.
Goldreich P. and Tremaine S. (1979) *Astrophys. J., 233,* 857.
Goldreich P. and Tremaine S. (1980) *Astrophys. J., 241,* 425.
Gressel O. et al. (2011) *Mon. Not. R. Astron. Soc., 415,* 3291.
Gressel O. et al. (2012) *Mon. Not. R. Astron. Soc., 422,* 1140.
Gressel O. et al. (2013) *Astrophys. J., 779,* 59.
Guilet J. and Ogilvie G. I. (2012) *Mon. Not. R. Astron. Soc., 424,* 2097.
Guilet J. and Ogilvie G. I. (2013) *Mon. Not. R. Astron. Soc., 430,* 822.
Guilet J. et al. (2013) *Mon. Not. R. Astron. Soc., 430,* 1764.
Hansen B. M. S. and Murray N. (2013) *Astrophys. J., 775,* 53.
Hayashi C. (1981) *Progr. Theor. Phys. Suppl., 70,* 35.
Hellary P. and Nelson R. P. (2012) *Mon. Not. R. Astron. Soc., 419,* 2737.
Howard A. W. et al. (2010) *Science, 330,* 653.
Ida S. and Lin D. N. C. (2004) *Astrophys. J., 604,* 388.
Ida S. and Lin D. N. C. (2008) *Astrophys. J., 673,* 487.
Ida S. and Lin D. N. C. (2010) *Astrophys. J., 719,* 810.
Johnson E. T. et al. (2006) *Astrophys. J., 647,* 1413.
Jurić M. and Tremaine S. (2008) *Astrophys. J., 686,* 603.
Ketchum J. A. et al. (2011) *Astrophys. J., 726,* 53.
Kley W. and Crida A. (2008) *Astron. Astrophys., 487,* L9.
Kley W. and Nelson R. P. (2012) *Annu. Rev. Astron. Astrophys., 50,* 211.
Kley W. et al. (2004) *Astron. Astrophys., 414,* 735.
Kley W. et al. (2005) *Astron. Astrophys., 437,* 727.
Kley W. et al. (2009) *Astron. Astrophys., 506,* 971.
Korycansky D. G. and Papaloizou J. C. B. (1996) *Astrophys. J. Suppl., 105,* 181.
Korycansky D. G. and Pollack J. B. (1993) *Icarus, 102,* 150.
Kretke K. A. and Lin D. N. C. (2012) *Astrophys. J., 755,* 74.
Kunz M. W. and Lesur G. (2013) *Mon. Not. R. Astron. Soc., 434,* 2295.
Lagrange A. et al. (2010) *Science, 329,* 57.

Lai D. et al. (2011) *Mon. Not. R. Astron. Soc., 412,* 2790.
Laughlin G. et al. (2004) *Astrophys. J., 608,* 489.
Lee M. H. and Peale S. J. (2002) *Astrophys. J., 567,* 596.
Lega E. et al. (2013) *Mon. Not. R. Astron. Soc., 431,* 3494.
Li H. et al. (2000) *Astrophys. J., 533,* 1023.
Libert A.-S. and Tsiganis K. (2009) *Mon. Not. R. Astron. Soc., 400,* 1373.
Libert A.-S. and Tsiganis K. (2011) *Cel. Mech. Dyn. Astron., 111,* 201.
Lin D. N. C. and Papaloizou J. (1979) *Mon. Not. R. Astron. Soc., 186,* 799.
Lin D. N. C. and Papaloizou J. (1986a) *Astrophys. J., 307,* 395.
Lin D. N. C. and Papaloizou J. (1986b) *Astrophys. J., 309,* 846.
Lin M.-K. and Papaloizou J. C. B. (2010) *Mon. Not. R. Astron. Soc., 405,* 1473.
Lin M.-K. and Papaloizou J. C. B. (2012) *Mon. Not. R. Astron. Soc., 421,* 780.
Lissauer J. J. et al. (2011a*) Nature, 470,* 53.
Lissauer J. J. et al. (2011b) *Astrophys. J. Suppl., 197,* 8.
Lithwick Y. and Wu Y. (2012) *Astrophys. J. Lett., 756,* L11.
Lovelace R. V. E. et al. (1999) *Astrophys. J., 513,* 805.
Lubow S. H. (1991) *Astrophys. J., 381,* 259.
Lyra W. et al. (2010) *Astrophys. J. Lett., 715,* L68.
Machida M. N. et al. (2010) *Mon. Not. R. Astron. Soc., 405,* 1227.
Marcy G. W. and Butler R. P. (1995) *Bull. Am. Astron. Soc., 27,* 1379.
Marois C. et al. (2010*) Nature, 468,* 1080.
Marzari F. et al. (2010) *Astron. Astrophys., 514,* L4.
Masset F. S. (2001) *Astrophys. J., 558,* 453.
Masset F. S. (2002) *Astron. Astrophys., 387,* 605.
Masset F. S. (2008) In *Tidal Effects in Stars, Planets and Disks* (M.-J. Goupin and J.-P. Zahn, eds.), pp. 165–244. EAS Publ. Series, Vol. 29, Cambridge Univ., Cambridge.
Masset F. S. (2011*) Cel. Mech. Dyn. Astron., 111,* 131.
Masset F. S. and Casoli J. (2009) *Astrophys. J., 703,* 857.
Masset F. S. and Casoli J. (2010) *Astrophys. J., 723,* 1393.
Masset F. S. and Papaloizou J. C. B. (2003) *Astrophys. J., 588,* 494.
Masset F. and Snellgrove M. (2001) *Mon. Not. R. Astron. Soc., 320,* L55.
Masset F. S. et al. (2006) *Astrophys. J., 652,* 730.
Matsumura S. et al. (2010) *Astrophys. J., 714,* 194.
Mayor M. and Queloz D. (1995) *Nature, 378,* 355.
Mayor M. et al. (2011) *ArXiv e-prints,* arXiv:1109.2497.
McNeil D. S. and Nelson R. P. (2009) *Mon. Not. R. Astron. Soc., 392,* 537.
McNeil D. S. and Nelson R. P. (2010) *Mon. Not. R. Astron. Soc., 401,* 1691.
McNeil D. et al. (2005) *Astron. J., 130,* 2884.
Migaszewski C. et al. (2012) *Mon. Not. R. Astron. Soc., 427,* 770.
Moeckel N. and Armitage P. J. (2012) *Mon. Not. R. Astron. Soc., 419,* 366.
Morbidelli A. and Crida A. (2007) *Icarus, 191,* 158.
Morbidelli A. et al. (2007) *Astron. J., 134,* 1790.
Mordasini C. et al. (2009a) *Astron. Astrophys., 501,* 1139.
Mordasini C. et al. (2009b) *Astron. Astrophys., 501,* 1161.
Mordasini C. et al. (2011) In *The Astrophysics of Planetary Systems: Formation, Structure, and Dynamical Evolution* (A. Sozzetti et al., eds.), pp. 72–75. IAU Symp. 276, Cambridge Univ., Cambridge.
Mordasini C. et al. (2012) *Astron. Astrophys., 547,* A112.
Müller T. W. A. et al. (2012) *Astron. Astrophys., 541,* A123.
Mustill A. J. and Wyatt M. C. (2011) *Mon. Not. R. Astron. Soc., 413,* 554.
Muto T. et al. (2008) *Astrophys. J., 679,* 813.
Muto T. et al. (2010) *Astrophys. J., 724,* 448.
Muto T. et al. (2011*) Astrophys. J., 737,* 37.
Nayakshin S. (2010) *Mon. Not. R. Astron. Soc., 408,* L36.
Nelson R. P. (2005) *Astron. Astrophys., 443,* 1067.
Nelson R. P. and Gressel O. (2010) *Mon. Not. R. Astron. Soc., 409,* 639.
Nelson R. P. and Papaloizou J. C. B. (2002) *Mon. Not. R. Astron. Soc., 333,* L26.
Nelson R. P. and Papaloizou J. C. B. (2004) *Mon. Not. R. Astron. Soc., 350,* 849.
Ogilvie G. I. and Lubow S. H. (2002) *Mon. Not. R. Astron. Soc., 330,* 950.
Oishi J. S. et al. (2007) *Astrophys. J., 670,* 805.
Olofsson J. et al. (2012) *Astron. Astrophys., 542,* A90.

Ormel C.W. and Klahr H. H. (2010) *Astron. Astrophys., 520,* A43.
Paardekooper S.-J. (2012) *Mon. Not. R. Astron. Soc., 421,* 3286.
Paardekooper S.-J. and Mellema G. (2006) *Astron. Astrophys., 459,* L17.
Paardekooper S.-J. and Mellema G. (2008) *Astron. Astrophys., 478,* 245.
Paardekooper S.-J. and Papaloizou J. C. B. (2008) *Astron. Astrophys., 485,* 877.
Paardekooper S.-J. and Papaloizou J. C. B. (2009a) *Mon. Not. R. Astron. Soc., 394,* 2283.
Paardekooper S.-J. and Papaloizou J. C. B. (2009b) *Mon. Not. R. Astron. Soc., 394,* 2297.
Paardekooper S.-J. et al. (2010) *Mon. Not. R. Astron. Soc., 401,* 1950.
Paardekooper S.-J. et al. (2011) *Mon. Not. R. Astron. Soc., 410,* 293.
Paardekooper S.-J. et al. (2013) *Mon. Not. R. Astron. Soc., 434,* 3018.
Papaloizou J. C. B. (2011) *Cel. Mech. Dyn. Astron., 111,* 83.
Papaloizou J. C. B. and Larwood J. D. (2000) *Mon. Not. R. Astron. Soc., 315,* 823.
Papaloizou J. C. B. and Szuszkiewicz E. (2005) *Mon. Not. R. Astron. Soc., 363,* 153.
Papaloizou J. C. B. and Terquem C. (2006) *Rept. Prog. Phys., 69,* 119.
Papaloizou J. C. B. et al. (2001) *Astron. Astrophys., 366,* 263.
Papaloizou J. C. B. et al. (2004) *Mon. Not. R. Astron. Soc., 350,* 829.
Papaloizou J. C. B. et al. (2007) In *Protostars and Planets V* (B. Reipurth et al., eds.), pp. 655–668. Univ. of Arizona, Tucson.
Pepliński A. et al. (2008a) *Mon. Not. R. Astron. Soc., 386,* 164.
Pepliński A. et al. (2008b) *Mon. Not. R. Astron. Soc., 387,* 1063.
Pierens A. and Huré J.-M. (2005) *Astron. Astrophys., 433,* L37.
Pierens A. and Nelson R. P. (2008) *Astron. Astrophys., 482,* 333.
Pierens A. et al. (2011) *Astron. Astrophys., 531,* A5.
Pierens A. et al. (2012) *Mon. Not. R. Astron. Soc., 427,* 1562.
Podlewska-Gaca E. et al. (2012) *Mon. Not. R. Astron. Soc., 421,* 1736.
Pollack J. B. et al. (1996) *Icarus, 124,* 62.
Quanz S. P. et al. (2013) *Astrophys. J. Lett., 766,* L2.
Quillen A. C. (2006) *Mon. Not. R. Astron. Soc., 365,* 1367.
Rafikov R. R. (2002) *Astrophys. J., 572,* 566.
Rafikov R. R. (2005) *Astrophys. J. Lett., 621,* L69.
Rein H. (2012a) *Mon. Not. R. Astron. Soc., 427,* L21.
Rein H. (2012b) *Mon. Not. R. Astron. Soc., 422,* 3611.
Rein H. et al. (2010) *Astron. Astrophys., 510,* A4.
Reipurth B. et al., eds. (2007) *Protostars and Planets V.* Univ. of Arizona, Tucson.
Rivier G. et al. (2012) *Astron. Astrophys., 548,* A116.
Rogers T. M. and Lin D. N. C. (2013) *Astrophys. J. Lett., 769,* L10.
Rogers T. M. et al. (2012) *Astrophys. J. Lett., 758,* L6.
Sanchis-Ojeda R. et al. (2012) *Nature, 487,* 449.
Simon J. B. et al. (2013) *Astrophys. J., 775,* 73.
Snellgrove M. D. et al. (2001) *Astron. Astrophys., 374,* 1092.
Steffen J. H. et al. (2013*) Mon. Not. R. Astron. Soc., 428,* 1077.
Szuszkiewicz E. and Podlewska-Gaca E. (2012*) Origins Life Evol. Biosph., 42,* 113.
Takeda G. and Rasio F. A. (2005) *Astrophys. J., 627,* 1001.
Tanaka H. and Ward W. R. (2004) *Astrophys. J., 602,* 388.
Tanaka H. et al. (2002) *Astrophys. J., 565,* 1257.
Tanigawa T. et al. (2012) *Astrophys. J., 747,* 47.
Terquem C. and Papaloizou J. C. B. (2007) *Astrophys. J., 654,* 1110.
Terquem C. E. J. M. L. J. (2003) *Mon. Not. R. Astron. Soc., 341,* 1157.
Teyssandier J. et al. (2013) *Mon. Not. R. Astron. Soc., 428,* 658.
Thommes E. W. et al. (2008) *Science, 321,* 814.
Triaud A. H. M. J. et al. (2010) *Astron. Astrophys., 524,* A25.
Tsiganis K. et al. (2005) *Nature, 435,* 459.
Uribe A. L. et al. (2011) *Astrophys. J., 736,* 85.
Uribe A. L. et al. (2013) *Astrophys. J., 769,* 97.
Varnière P. et al. (2004) *Astrophys. J., 612,* 1152.
Vorobyov E. I. (2013) *Astron. Astrophys., 552,* A129.
Walsh K. J. et al. (2011*) Nature, 475,* 206.
Ward W. R. (1991) In *Lunar Planet. Sci. XXII,* p. 1463. Lunar and Planetary Institute, Houston.
Ward W. R. (1997) *Icarus, 126,* 261.
Xiang-Gruess M. and Papaloizou J. C. B. (2013) *Mon. Not. R. Astron. Soc., 431,* 1320.
Zhu Z. et al. (2012a) *Astrophys. J., 746,* 110.
Zhu Z. et al. (2012b) *Astrophys. J., 755,* 6.
Zhu Z. et al. (2013) *Astrophys. J., 768,* 143.

Benz W., Ida S., Alibert Y., Lin D., and Mordasini C. (2014) Planet population synthesis. In *Protostars and Planets VI* (H. Beuther et al., eds.), pp. 691–713. Univ. of Arizona, Tucson, DOI: 10.2458/azu_uapress_9780816531240-ch030.

Planet Population Synthesis

Willy Benz
University of Bern

Shigeru Ida
Tokyo Institute of Technology

Yann Alibert
University of Bern

Douglas Lin
University of California at Santa Cruz

Christoph Mordasini
Max Planck Institute for Astronomy

With the increasing number of exoplanets discovered, statistical properties of the population as a whole become unique constraints on planet-formation models, provided a link between the description of the detailed processes playing a role in this formation and the observed population can be established. Planet population synthesis provides such a link. This approach allows us to study how different physical models of individual processes (e.g., protoplanetary disk structure and evolution, planetesimal formation, gas accretion, migration, etc.) affect the overall properties of the population of emerging planets. By necessity, planet population synthesis relies on simplified descriptions of complex processes. These descriptions can be obtained from more detailed specialized simulations of these processes. The objective of this chapter is twofold: (1) provide an overview of the physics entering in the two main approaches to planet population synthesis, and (2) present some of the results achieved as well as illustrate how it can be used to extract constraints on the models and to help interpret observations.

1. INTRODUCTION

The number of known exoplanets has increased dramatically in recent years (see *www.exoplanet.eu*). As of this writing, over 1000 confirmed exoplanets were known, mostly found through precise radial velocity (RV) surveys. Additionally, there are more than 3600 transiting candidate planets found by the Kepler satellite (see, e.g., *www.kepler.nasa.gov*). All these detections have revealed that planets are quite common and that the diversity of existing systems is much larger than was expected from studies of our own solar system. Finally, with the increasing number of planets, the search for correlations and structures in the properties of planets and planetary systems becomes increasingly meaningful. The correlations and structures have pinpointed the importance of complex interaction processes taking place during the formation stages of the planets (e.g., planetary migration).

These insights were essentially gained by the fact that, for the first time, a large set of planets was available to study, and statistical analysis became possible. The analysis of the characteristics of an ensemble of objects, as well as the differences between individual objects, is a standard approach in astrophysics and has been applied successfully in a number of areas (e.g., galactic evolution).

Planet population synthesis in the context of the core accretion scenario has been pioneered by *Ida and Lin* (2004a) in an effort to develop a deterministic model of planetary formation allowing a direct comparison with the observed population of exoplanets. They presented formation models for planets orbiting solar-type stars but neglected the effect of type I migration on the basis that its efficiency was poorly determined. They simulated the mass-semimajor axis distribution of planets for stars of different metallicity and masses (*Ida and Lin*, 2004b, 2005). *Burkert and Ida* (2007) applied this model to discuss a potential period gap in the observed gas giant distribution orbiting stars more massive than the Sun. *Currie* (2009) discussed the same issue with a similar model. *Payne and Lodato* (2007) discussed planets orbiting brown dwarfs using the model by *Ida and Lin* (2004b).

In *Ida and Lin* (2008a), type I migration was incorporated, using a conventional isothermal formula (e.g., *Tanaka et al.,* 2002), with an efficiency factor that uniformly decreases the migration speed, because the predicted formation efficiency of gas giants with the full strength of the migration is too low to be consistent with observations and to allow for uncertainties in the theoretically derived migration speed. *Ida and Lin* (2008b) considered the effect of a potential migration trap due to a snow line. With a similar model, *Miguel et al.* (2011a,b) studied a dependence on initial disk models. *Mordasini et al.* (2009a) developed a model based on more detailed calculation of gas envelope contraction, disk evolution, and planetesimal dynamics. Type I migration was treated in a similar way to *Ida and Lin* (2008a). They applied their model (*Mordasini et al.,* 2009b) for statistical comparisons with the population of giant extrasolar planets known at that time. *Alibert et al.* (2011) studied the impact of the stellar mass on planetary populations, while *Mordasini et al.* (2012c) investigated how important properties of the protoplanetary disk (mass, metallicity, lifetime) translate into planetary properties. In an attempt to further couple formation models to the major observable physical characteristics of a planet (besides mass and semimajor axis, also radius, luminosity, and bulk composition), *Mordasini et al.* (2012a) added to the model the long-term evolution of the planets after formation (cooling and contraction). This enabled statistical comparisons with results of transit and (in future) direct imaging surveys, and in particular the result of the Kepler satellite (*Mordasini et al.,* 2012b).

Since planet-planet scattering is a chaotic process, including such effects in an otherwise deterministic calculation was not easy. *Alibert et al.* (2013) incorporated a full N-body integrator with collision detections in order to simulate planet-planet interactions. While this approach adds a significant computational burden, it has the advantage of handling all dynamical aspects (including resonances) correctly. [Note that *Bromley and Kenyon* (2006, 2011), *Thommes et al.* (2008), and *Hellary and Nelson* (2012) also developed hybrid N-body simulations, although they did not present much statistical discussion of predicted planet distributions.] *Ida and Lin* (2010) and *Ida et al.* (2013) took another approach in which the planet-planet interactions (scatterings, ejections, collisions) are treated in a Monte Carlo fashion, calibrated by numerical simulations. Although relatively complicated multiple steps are needed for the Monte Carlo method to reproduce predictions accurate enough for the statistical purposes of population synthesis, it is much faster than direct N-body simulations.

2. THE PHYSICS OF POPULATION SYNTHESIS

As its name indicates, the goal of planet population synthesis is to allow the computing of a full planet population given a suitable set of initial conditions. Practically, this requires a full planet-formation model that computes the final characteristics of planets from specific initial conditions. The physics behind the formation model will be discussed in this section, while matters related to initial conditions are presented in section 3.

By nature, an end-to-end simulation of the formation of even a single planet is probably impossible to carry out in a complete and detailed manner. Hence, assumptions have to be made in order to make the problem tractable. In this approach, the difficulty is to identify how far the problem can be simplified while still conserving the overall properties of the emerging planet population, including their mean values and dispersions. Guidance must come as much as possible from observations and from detailed modeling of all the individual processes entering in the computations of a planet population.

Figure 1 provides an overview of all the elements that enter in a self-consistent planet-formation model. As can be seen from this figure, a large number of processes enter in the physical computation of the formation of a planet. By necessity, each process has to be described in rather simplified physical terms. We stress that most of these descriptions are not specific to population synthesis but are commonly used throughout the literature to discuss these individual processes. The essence of population synthesis consists therefore of coupling these processes in a physically meaningful and consistent manner. This is especially important as many of the processes have comparable timescales.

In what follows we describe the essential ingredients of the planet-formation model as they have been worked out in a series of papers by Ida and Lin (*Ida and Lin,* 2004a,b, 2005, 2008a,b, 2010; *Ida et al.,* 2013, hereafter referred to as "*IL*") and by a series of papers by Alibert, Mordasini, and Benz (*Mordasini et al.,* 2009a,b; *Alibert et al.,* 2011; *Mordasini et al.,* 2012b,c; *Alibert et al.,* 2013, hereafter referred to as "*AMB*") with significant contributions by various collaborators at the University of Bern (A. Fortier in particular) and at the Max Planck Institute for Astronomy. We opted for this relatively detailed approach as we believe that the extent of the physical description of the processes entering in these models is not always fully appreciated as it is disseminated throughout a number of papers.

2.1. Structure and Evolution of the Protoplanetary Disk

Capturing the structure and evolution of the protoplanetary disk is important since the migration rate of planets, as well as their internal structure (through the planetary surface conditions), and the amount of gas they can accrete, is determined, at least partially, by the disk (modules 1 and 2 in Fig. 1). Different levels of complexity can be used to describe the disk. Existing state-of-the-art, fully three-dimensional, self-consistent magneto-hydrodynamical models are not applicable for population synthesis as the computational resources involved are such that only relatively short timespans can be modeled. Given both observational incompleteness and theoretical uncertainties in the detailed structure of disks, one simple approach consists of assuming

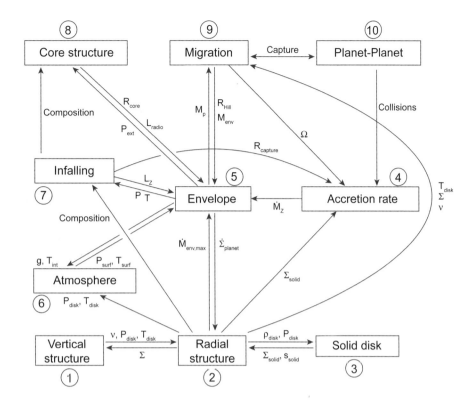

Fig. 1. Schematic of the coupling between the different processes entering in the computation of a self-consistent planet-formation model. Quantities exchanged by the different modules are indicated along arrows.

that the gas surface density follows exponential decay with a characteristic disk evolution timescale (comparable to observed disk lifetimes, i.e., of a few million years) (e.g., *Ida and Lin,* 2004a). This approach has the advantage of being extremely fast and allows exploration of a large number of models. However, this approach does not self-consistently provide a relation between the disk surface density and the local pressure and temperature that enter in the calculation of the structure of the growing planets, as well as their migration rates. An intermediate approach consists of using a model of viscously evolving disks (*Shakura and Sunyaev,* 1973) for which the local vertical structure can be computed (*Papaloizou and Terquem,* 1999; *Alibert et al.,* 2005a). Such an intermediate approach has the advantage of providing both the full structure of the disk and the framework to self-consistently include additional physical processes such as photoevaporation from the central star and/or nearby stars, the existence of dead zones, and irradiation from the central star. Both approaches are briefly outlined below.

IL adopt the minimum mass solar nebula (MMSN) model (*Hayashi,* 1981) as a fiducial set of initial conditions and introduce multiplicative factors (f_d and f_g) to scale the MMSN disk surface densities of gas (Σ) and planetesimals (Σ_s). *IL* set

$$\Sigma = \Sigma_{10} f_g \left(r/10\text{AU} \right)^{-q} \tag{1}$$

where a normalization factor $\Sigma_{10} = 75\text{g cm}^{-2}$ corresponds to $1.4 \times \Sigma$ at 10 AU of the MMSN model. The inner disk

boundary where Σ vanishes is set at ~0.04 AU. *IL* often use a power-law exponent q = 1 that corresponds to the self-similar steady accretion disk model with a constant α viscosity, rather than the original MMSN model for which q = 1.5. However, they found that this does not considerably affect the results.

Neglecting the detailed energy balance in the disk, *IL* adopt the equilibrium temperature distribution of optically thin disks prescribed by *Hayashi* (1981)

$$T = 280 \left(\frac{r}{1\text{AU}} \right)^{-1/2} \left(\frac{L_*}{L_\odot} \right)^{1/4} \text{K} \tag{2}$$

where L_* and $L_\odot$ are respectively stellar and solar luminosity. *IL* determine the position of the ice line (a_{ice}) as the location at which T = 170 K, which translates for an optically thin disk into (equation (2))

$$a_{ice} = 2.7 \left(L_*/L_\odot \right)^{1/2} \text{AU} \tag{3}$$

Due to viscous diffusion and photoevaporation, f_g decreases with time. For simplicity, *IL* adopt

$$f_g = f_{g,0} \exp\left(-t/\tau_{dep} \right) \tag{4}$$

where τ_{dep} is the disk lifetime (for detailed discussion, see *Ida and Lin* 2008a). *IL* use τ_{dep} as a free parameter ranging from 10^6 yr to 10^7 yr. The self-similar solution with

$\Sigma \propto r^{-1}$ has an asymptotic exponential cut-off at radius r_m of the maximum viscous couple. In the region at $r < r_m$, Σ decreases uniformly independent of r as the exponential decay does, although the time dependence is slightly different. Note that this treatment is relevant in the regime where the disk-mass depletion rate by photoevaporation is so low that it does not affect disk evolution by viscous diffusion until the last phase of disk depletion.

The disk structure and evolution in the *AMB* model does not assume any power-law structure but is calculated from assuming local hydrostatic equilibrium. In this case, the vertical structure of the disk can be computed from

$$\frac{1}{\rho}\frac{\partial P}{\partial z} = -\Omega^2 z \tag{5}$$

where z is the vertical coordinate, ρ the density, P the pressure, and Ω is the angular frequency of the disk. The disk is assumed to be Keplerian, therefore $\Omega^2 = GM_*/r^3$, G being the gravitational constant and M_* the mass of the central star. This equation is solved together with the energy equation, which states that the energy produced by viscosity is removed by the radiative flux

$$\frac{\partial F}{\partial z} = \frac{9}{4}\rho v \Omega^2 \tag{6}$$

where F is the radiative flux (see also *Ruden and Lin, 1986*). Assuming an optically thick medium, the radiative flux is written

$$F = -\frac{16\pi\sigma T^3}{3\kappa\rho}\frac{\partial T}{\partial z} \tag{7}$$

where T is the temperature, κ is the opacity, and σ is the Stefan-Boltzmann constant. The viscosity is calculated using the standard α-parameterization $v = \alpha c_s^2/\Omega$ where the speed of sound c_s^2 is determined from the equation of state. This set of three differential equations can be solved with the addition of suitable boundary conditions (*Papaloizou and Terquem, 1999*)

$$P_s = \frac{\Omega^2 H \tau_{ab}}{\kappa_s} \tag{8}$$

$$F_s = \frac{3}{8\pi}\dot{M}_*\Omega^2 \tag{9}$$

$$2\sigma\left(T_s^4 - T_b^4\right) - \frac{\alpha k T_s \Omega}{8\mu m_H \kappa_s} - \frac{3}{8\pi}\dot{M}_*\Omega^2 = 0 \tag{10}$$

$$F(z = 0) = 0 \tag{11}$$

where the subscript s refers to values taken at z = H, τ_{ab} is the optical depth between the surface of the disk (z = H) and infinity, T_b is the background temperature, k is the Boltzmann constant, μ is the mean molecular mass of the

gas, and m_H is the mass of the hydrogen atom. In the above equations, $\dot{M}_*$ is the equilibrium accretion rate defined by $\dot{M} \equiv 3\pi\tilde{v}\Sigma$ where $\Sigma \equiv \int_{-H}^{H}\rho dz$ is the surface density, and $\tilde{v}$ the effective viscosity defined by

$$\tilde{v} \equiv \frac{\int_{-H}^{H}v\rho dz}{\Sigma} \tag{12}$$

The effect of the irradiation of the central star can, in its simplest form, be included by suitably modifying the surface temperature of the disk (*Fouchet et al., 2012*) in the form

$$T_s^4 = T_{s,noirr}^4 + T_{s,irr}^4 \tag{13}$$

where $T_{s,noirr}$ is the temperature due to viscous heating given in the above and the irradiation temperature is given by *Hueso and Guillot (2005)*

$$T_{s,irr} = T_*\left[\frac{2}{3\pi}\left(\frac{R_*}{r}\right)^3 + \frac{1}{2}\left(\frac{R_*}{r}\right)^2\left(\frac{H_P}{r}\right)\right.$$
$$\left.\left(\frac{d\ln H_p}{d\ln r} - 1\right)\right]^{1/4} \tag{14}$$

where R_* is the stellar radius, r is the distance to the star, and H_P is the pressure scale height defined as $\rho(z = H_P) = e^{-1/2}\rho(z = 0)$ (see also *Chiang and Goldreich, 1997; Garaud and Lin, 2007*).

The radial evolution of the disk (module 2 in Fig. 1) is provided by the standard viscous disk evolution equation (*Lynden-Bell and Pringle, 1974*) complemented by appropriate terms describing the gas accreted by the planets and the one lost by photoevaporation. It is customary to write this equation in terms of the evolution of the surface density $\Sigma(r)$

$$\frac{\partial \Sigma}{\partial t} = \frac{3}{r}\frac{\partial}{\partial r}\left[r^{1/2}\frac{\partial}{\partial r}\left(\tilde{v}\Sigma r^{1/2}\right)\right] + \dot{\Sigma}_w(r) + \dot{Q}_{planet} \tag{15}$$

where $\dot{\Sigma}_w(r)$ describes the sink term associated with photoevaporation caused by the host star itself (internal photoevaporation), as well as close-by massive stars (external photoevaporation) (for a detailed discussion, see *Mordasini et al., 2012b*), and $\dot{Q}_{planet}$ represents the rate at which gas is being accreted by the growing planets.

Internal photoevaporation due to the stellar extreme ultraviolet (EUV) radiation is modeled according to *Clarke et al. (2001)* based on the "weak stellar wind" scenario of *Hollenbach et al. (1994)*. It leads to mass loss concentrated on an annulus around $\beta_{II}R_{g,II}$, where $R_{g,II} \approx 7$ AU for a 1 $M_\odot$ star is the gravitational radius for ionized hydrogen (mass m_H) and a speed of sound $c_{s,II}$ associated with a temperature of approximately 10^4 K, and β_{II} is a parameter reflecting that some mass loss occurs already inside $R_{g,II}$. The decay rate of disk surface density due to the mass loss is

$$\dot{\Sigma}_{w,int} = \begin{cases} 0 & \text{for } r < \beta_{II}R_{g,II} \\ 2c_{s,II}n_0(r)m_H & \text{otherwise} \end{cases} \quad (16)$$

The density of ions at the base of the wind $n_0(r)$ as a function of distance is approximately

$$n_0(r) = n_0\left(R_{14}\right)\left(\frac{r}{\beta_{II}R_{g,II}}\right)^{-5/2} \quad (17)$$

The density n_0 at a normalization radius R_{14} is given by radiation-hydrodynamic simulations (*Hollenbach et al.,* 1994) and depends on the ionizing photon luminosity of the central star.

For far-ultraviolet (FUV)-driven external photoevaporation, the mass loss outside a critical radius $\beta_I R_{g,I}$ can be written as (*Matsuyama et al.,* 2003)

$$\dot{\Sigma}_{w,ext} = \begin{cases} 0, & r \le \beta_I R_{g,I} \\ \dfrac{\dot{M}_{wind,ext}}{\pi\left(R_{max}^2 - \beta_I^2 R_{g,I}^2\right)}, & r > \beta_I R_{g,I} \end{cases} \quad (18)$$

where $R_{g,I}$ is the gravitational radius for neutral hydrogen at a temperature of approximately 1000 K ($\approx$140 AU for a 1-$M_\odot$ star), while R_{max} is a fixed outer radius. The total mass lost through photoevaporation is a free parameter and is set by $\dot{M}_{wind,ext}$ in such a way that, together with the viscous evolution, the distribution of disk lifetimes is in agreement with observations.

The term $\dot{Q}_{planet}$ is obtained from the computation of the amount of gas accreted by the planet(s) as described in section 2.4. The amount accreted is removed from disk over an annulus centered on the planet(s), with a width equal to the corresponding Hill radius

$$R_H = a\left(\frac{M}{3M_*}\right)^{1/3} \quad (19)$$

where M is the mass of the planet, a its semimajor axis, and M_* the mass of the central star.

2.2. Structure and Evolution of the Planetesimal Disk

The structure and dynamical evolution of the planetesimal disk (module 3 in Fig. 1) is essential in order to capture the essence of the growth of the planets. Both *IL* and *AMB* have considered so far only two types of planetesimals: rocky and icy. A planetesimal is declared rocky or icy according to its initial position in the disk and does not subsequently change its nature. Planetesimals located inside the ice line are rocky while those locate outside are icy. The ice line (a_{ice}) is defined as the distance from the star where the temperature of the gas drops below the ice condensation temperature of approximately 170 K (*IL*) and 160 K (*AMB*) for the typical pressure range encountered

(see section 2.1 for how this temperature is computed). Note that while "icy" planetesimals means planetesimals composed mainly of ice, they also have a rocky component. Following *Hayashi* (1981), *IL* and *AMB* assume the mass fraction of the rocky component in icy planetesimals to be 1/4.2 and 1/4, respectively.

As far as the radial distribution of solids is concerned, *IL* set the surface density of planetesimals (Σ_s)

$$\Sigma_s = \Sigma_{s,10}\eta_{ice}f_d\left(r/10AU\right)^{-q_s} \quad (20)$$

where a normalization factor $\Sigma_{s,10} = 0.32$ g cm^{-2} corresponds to $1.4\times \Sigma_s$ at 10 AU of the MMSN model, and the step function $\eta_{ice} = 1$ inside the ice line at a_{ice} (equation (3)) and 4.2 for $r > a_{ice}$. *IL* usually adopt $q_s = 1.5$ according to the MMSN, because dust grains suffer inward migration due to gas drag and can be more concentrated in the inner regions than gas components. However, q_s can be similar to the exponent of the radial distribution of gas ($\sim$1). The initial disk metallicity [Fe/H] is used to relate $f_{d,0}$ and $f_{g,0}$ as $f_{d,0} = f_{g,0}10^{[Fe/H]}$, where $f_{d,0}$ and $f_{g,0}$ are initial values of f_d and f_g, respectively.

AMB set the initial surface density of planetesimals proportional to the gas surface density

$$\Sigma_s\left(r, t = 0\right) = f_{D/G}f_{R/I}(r)\Sigma\left(r, t = 0\right) \quad (21)$$

where $f_{D/G}$ is the dust-to-gas ratio, which is equal to $f_{d,0}/f_{g,0}$ in the *IL* expression, and $f_{R/I}$ is the rock-to-ice ratio (i.e., a factor taking into account the degree of condensation of the volatiles), which has the same role as η_{ice} in the *IL* expression.

The above equation provides the spatial distribution in terms of surface density of the planetesimals. In addition, the dynamics of these planetesimals needs to be specified, as the motion of the planetesimals will ultimately determine their collision rate and the growth rate of planets. The dynamics of the planetesimals is determined by three different processes: (1) nebular gas drag, (2) gravitational stirring by growing protoplanets also known as viscous stirring, and (3) mutual gravitational interactions between the planetesimals.

IL directly give a protoplanet's growth timescale ($\tau_{c,acc}$) as a function of its mass (M_c) and semimajor axis and disk surface density (f_d and f_g). The expression for the growth timescale used by *IL* is provided in section 2.3. Because runaway growth is quickly transformed into oligarchic growth (*Ida and Makino,* 1993), the system is reduced to a bimodal population of protoplanets and small planetesimals (*Kokubo and Ida,* 1998, 2002). Since the stirring of planetesimal velocity dispersion by the protoplanets is dominating over the mutual stirring between the planetesimals, *IL* neglected the latter effect. In this approach, the growth rate of a protoplanet is determined by spatial mass density of the planetesimals and relative velocity between a protoplanet and the planetesimals. Dynamical friction from the small planetesimals damps velocity dispersion (eccentricity and inclination) of the protoplanets below those of the planetesimals (*Ida and Makino, 1992*), so that the relative velocity is dominated by the velocity dispersion of the planetesimals. The velocity dispersion of the planetesimals is determined by a balance between gravitational stirring by a nearby protoplanet and damping due to gas drag.

In contrast, *AMB* solve much more detailed equations of evolution of eccentricity e and inclination i of the planetesimals instead of using a final formula derived from simulations. In this approach, the evolution of these quantities is determined by computing explicitly the contribution from all three processes mentioned above (*Fortier et al., 2013*).

$$\frac{de^2}{dt} = \left(\frac{de^2}{dt}\right)_{drag} + \left(\frac{de^2}{dt}\right)_{VS,M} + \left(\frac{de^2}{dt}\right)_{VS,m} \quad (22)$$

$$\frac{di^2}{dt} = \left(\frac{di^2}{dt}\right)_{drag} + \left(\frac{di^2}{dt}\right)_{VS,M} + \left(\frac{di^2}{dt}\right)_{VS,m} \quad (23)$$

where the first terms are damping due to gas drag, the second terms are excitations due to scattering by protoplanets, and the third terms are those by the other planetesimals. No hypothesis is made as far as an equilibrium between any two terms is concerned in solving the above equations.

2.3. Accretion of Solids

A protoplanet grows in mass by accreting planetesimals (module 4 in Fig. 1) and nebular gas. Since protoplanets are seeded by very small masses (typically with a mass of 10^{20} g $\sim 10^{-8}$ $M_\oplus$ for *IL* and 10^{-2} $M_\oplus$ for *AMB*; note, however, that the initial mass does not affect the results since planetesimal accretion is slower in the later phase), they are growing initially essentially through the accretion of planetesimals. Later, as the protoplanets reach larger masses, they will be able to grow a gaseous envelope by gravitationally binding nebular gas.

In the *IL* approach, the growth of the protoplanet is specified by a growth timescale that is defined by its mass (M_c) and semimajor axis (a) and the local surface density. This timescale can be computed (*Ida and Lin, 2004a, 2010*) from the analytically evaluated relative velocity based on planetesimal dynamics revealed by N-body simulations (*Kokubo and Ida, 1998, 2002*) and Monte Carlo three-body simulations (*Ohtsuki et al., 2002*) and gas drag laws derived by *Adachi et al.* (1976)

$$\tau_{c,acc} = 3.5 \times 10^5 \eta_{ice}^{-1} f_d^{-1} f_g^{-2/5} \left(\frac{a}{1AU}\right)^{5/2} \times$$
$$\left(\frac{M_c}{M_\oplus}\right)^{1/3} \left(\frac{M_*}{M_\odot}\right)^{-1/6} yr \quad (24)$$

where the mass of the typical field planetesimals is set to be m = 10^{20} g.

In the *AMB* approach, the accretion rate of solid by a core of mass M_{core} is explicitly computed (*Chambers, 2006; Fortier et al., 2013*)

$$\frac{dM_{core}}{dt} = \left(\frac{2\pi\Sigma_s R_H^2}{P_{orb}}\right) P_{coll} \quad (25)$$

where Σ_s is the surface density of planetesimals at the planet location and P_{orb} the orbital period of the planet. The collision probability P_{coll} for a planet to accrete planetesimals depends upon two important parameters: (1) the relative velocities between planets and planetesimals, and (2) the presence of an atmosphere large enough to dissipate enough kinetic energy of the planetesimals to result in merging. The relative velocities between planetesimals and protoplanets as a function of time can be derived from the knowledge of the time evolution of their respective eccentricities and inclinations (see section 2.2).

When the planetary core reaches a mass larger than ~1 $M_\oplus$, its gas envelope becomes massive enough to affect the dynamics of planetesimals that penetrate it. As a result of gas drag, the effective cross section of the planet is increased. Such an effect must be computed (module 7 in Fig. 1) by numerically solving for the motion of planetesimals in the planetary envelope under the effects of gravity, gas drag, thermal ablation, and mechanical disruption (*Podolak et al., 1988; Alibert et al., 2005a*). Alternatively, it is possible to use fits of similar calculations that directly provide the planet cross section (e.g., *Inaba et al., 2001*). The same module also calculates the mass (and energy) deposition of the planetesimals in the protoplanetary envelope (*Mordasini et al., 2006*), yielding the heavy element enrichment of the envelope. Observationally, this is of interest for transit and spectroscopic studies of extrasolar planets (e.g., *Fortney et al., 2013*).

The relations used so far by both *IL* and *AMB* are derived from theoretical considerations and comparisons with N-body simulations assuming a single isolated protoplanet embedded in a large number of smaller planetesimals. However, during the formation of planetary systems, several protoplanets grow concurrently and sometimes sufficiently close to each other for the feeding zone to overlap. In this case, two or more protoplanets compete for the available planetesimals. N-body simulations of such a situation (*Alibert et al., 2013*) have shown that the protoplanets are so efficient in scattering the planetesimals that the latter are homogenized over the sum of the feeding zone of the neighboring growing protoplanets. This in turn changes the mass reservoir of solids available to accrete from, and therefore changes the growth of the protoplanets. Finally, gas drag combined with the tidal perturbation of protoplanets may lead to the clearing of planetesimal gaps, which may also reduce dM_{core}/dt (*Zhou and Lin, 2007*).

2.4. Envelope Structure and Accretion of Gas

The knowledge of the planetary interior structure is essential to compute the accretion rate of gas (modules 5 and 6 in Fig. 1). Indeed, gas accretion depends crucially on the ability of a planet to cool and radiate away the energy gained by the accreting of planetesimals. This is nicely exemplified by the dependence of planetary internal structure on the typical opacity in the envelope (see section 5).

IL use a fitting formula for the critical core mass ($M_{c,hydro}$) beyond which atmospheric pressure no longer supports a gas envelope against the planetary gravity (i.e., no hydrostatic equilibrium exists), as well as to describe the quasistatic envelope contraction afterward. These formulas were obtained by detailed one-dimensional calculations of envelope structure and radiative/convective heat transfer as described below.

Ikoma et al. (2000) carried out a one-dimensional calculation of an envelope structure similar to *Bodenheimer and Pollack* (1986) (also see below) with a broad range of parameters and derived the critical core mass as

$$M_{c,hydro} \simeq 10 \left(\frac{\dot{M}_c}{10^{-6} M_\oplus / yr} \right)^{0.25} M_\oplus \qquad (26)$$

where the dependence on the opacity in the envelope (e.g., *Ida and Lin*, 2004a; *Hori and Ikoma*, 2010) is neglected, because opacity in the envelope is highly uncertain. Note that $M_{c,hydro}$ depends on the planetesimal accretion rate $\dot{M}_c$. Since $M_{c,hydro}$ can be comparable to an Earth mass ($M_\oplus$) after the core accretes most of planetesimals in its feeding zone, whether the core becomes a gas giant planet is actually regulated by a timescale of the subsequent quasistatic envelope contraction [Kelvin-Helmholtz (KH) contraction time] rather than the value of $M_{c,hydro}$.

Because the contraction of the gas envelope also releases energy to produce pressure to support the gas envelope itself, the contraction is quasistatic. Its rate is still regulated by the efficiency of radiative/convective transfer in the envelope such that

$$\frac{dM_{planet}}{dt} \simeq \frac{M_{planet}}{\tau_{KH}} \qquad (27)$$

where M_{planet} is the planet mass including the gas envelope. Based on the results by one-dimensional calculations (*Ikoma et al.*, 2001), *IL* approximate the KH contraction timescale τ_{KH} of the envelope with

$$\tau_{KH} \simeq \tau_{KH1} \left(\frac{M_{planet}}{M_\oplus} \right)^{-k2} \qquad (28)$$

where τ_{KH1} is the contraction timescale for $M_{planet} = M_\oplus$. Since there are uncertainties associated with dust sedimentation and opacity in the envelope (*Pollack et al.*, 1996;

Helled et al., 2008; *Hori and Ikoma*, 2011), *IL* adopt a range of values $\tau_{KH1} = 10^8$–10^{10} yr and k2 = 3–4 with nominal parameters of k2 = 3 and $\tau_{KH1} = 10^9$ yr. Equation (27) shows that dM_{planet}/dt rapidly increases as M_{planet} grows. However, it is limited by the global gas accretion rate throughout the disk and by the process of gap formation near the protoplanets' orbits, as discussed later.

The *AMB* approach consists of solving the standard internal structure equations (*Bodenheimer and Pollack*, 1986)

$$\frac{dr}{dM_r} = \frac{1}{4\pi\rho r^2} \qquad (29)$$

$$\frac{dP}{dM_r} = -\frac{GM_r}{4\pi r^4} \qquad (30)$$

$$\frac{dT}{dP} = \nabla_{ad} \text{ or } \nabla_{rad} \qquad (31)$$

where r, P, and T are the radius, pressure, and temperature that are specified as a function of the mass M_r, which represents the mass inside a sphere of radius r (including the mass of the core M_{core}). Stability against convection is checked using the Schwarzschild criterion (e.g., *Kippenhahn and Weigert*, 1994). Depending on whether convection is present or not, the adiabatic gradient (∇_{ad}) or the radiative gradient (∇_{rad}) is used. These equations are solved together with the equation of state (EOS) by *Saumon et al.* (1995). The opacity is taken from *Bell and Lin* (1994). *Podolak* (2003) and *Movshovitz and Podolak* (2008) have argued that grain opacities are significantly reduced in planetary envelopes as compared to the interstellar medium. Reducing the grain opacity allows runaway accretion to occur at smaller core masses and therefore speeds up the giant formation timescale (*Pollack et al.*, 1996; *Hubickyj et al.*, 2005).

In order to gain computing time and to avoid numerical convergence difficulties, the energy equation is not solved. Instead, the procedure outlined by *Mordasini et al.* (2012a), based on total energy conservation, is adopted, with a small improvement that allows taking the energy of the core into account, as well as described in *Fortier et al.* (2013). The total luminosity $L = L_{cont} + L_{acc}$ is the sum of the energy gained through the contraction of the envelope L_{cont} and by the accretion of planetesimals L_{acc}. The contraction luminosity, assumed to be constant throughout the envelope, is computed from the change of energy of the planet between the time t and t + dt

$$L_{cont} = -\frac{E_{tot}(t + dt) - E_{tot}(t) - E_{gas,acc}}{dt} \qquad (32)$$

where E_{tot} is the total planetary energy and $E_{gas,acc} = dt\dot{M}_{gas}u_{int}$ is the energy gained by the accretion of nebular gas with a specific internal energy u_{int} at a rate $\dot{M}_{gas}$. The luminosity associated with the accretion of planetesimals, which are assumed to deposit their energy onto the core, can be written

$$L_{acc} = G \frac{\dot{M}_{core} M_{core}}{R_{core}} \qquad (33)$$

where $\dot{M}_{core}$ is the mass accretion rate of the planetesimals, which results in an increase in the core mass M_{core} and radius R_{core}. It has to be noted that L_{cont} cannot be computed in a straightforward manner since in order to compute $E_{tot}(t + dt)$, the structure of the envelope at $t + dt$ needs to be known. This difficulty can be circumvented with the help of an iterative scheme (*Fortier et al.*, 2013).

The internal structure equations are solved with four boundary conditions: (1) the radius of the core R_{core}, (2) the total radius of the planet R_M, (3) the surface temperature of the planet T_{surf}, and (4) the surface pressure P_{surf}. With these boundary conditions the structure equations provide a unique solution for a given planet mass.

The core radius can be calculated (module 8 in Fig. 1) for a given core mass, composition (rocky or icy), and pressure at its surface (relevant for planets with a massive H/He envelope). *AMB* solve the internal structure equations for a differentiated core using a simple modified polytropic equation of state for the density ρ as a function of pressure P (*Seager et al.*, 2007)

$$\rho(P) = \rho_0 + cP^n \qquad (34)$$

where ρ_0, c, and n are material parameters. This EOS neglects the relatively small temperature dependency of ρ for solids. Therefore it is sufficient to consider only the equations of mass conservation and hydrostatic equilibrium (equations (29) and (30)) to calculate the core's internal structure and radius (for details, see *Mordasini et al.*, 2012b). Regarding the composition, for rocky material, a silicate-iron ratio of 2:1 in mass is assumed, as for Earth, and the ice fraction is self-consistently given by the formation model, as it is known whether the planet accretes rocky or icy planetesimals.

While the planet is embedded in the nebula, and the core is subcritical ($M_{core} \lesssim 10\ M_\oplus$) for gas accretion, the gaseous envelope of the protoplanet smoothly transitions into the background nebula. During this so-called attached phase, the total radius of the planet R_M is given by (*Lissauer et al.*, 2009)

$$R_M = \frac{GM}{\dfrac{c_s^2}{k_1} + \dfrac{GM}{k_2 R_H}} \qquad (35)$$

where c_s^2 is the square of the sound speed in the mid-plane of the gaseous nebula at the location of the planet, k_1 and k_2, with values of 1 and 1/4 respectively. The temperature and the pressure at the surface of the planet (module 6 in Fig. 1) are specified by a matching condition with the local properties of the disk

$$T_{surf} = \left(T_{disk}^4 + \frac{3\tau L}{16\pi\sigma R_M^2} \right)^{1/4} \qquad (36)$$

$$P_{surf} = P_{disk}(a) \qquad (37)$$

where $\tau = \kappa(T_{disk}, \rho_{disk})\rho_{disk} R_M$ (*Papaloizou and Terquem*, 1999), L is the luminosity of the planet, and T_{disk}, ρ_{disk}, P_{disk} are respectively the temperature, density, and pressure in the mid-plane of the disk at the planet's location. By solving these equations and requiring that pressure and temperature match the values of the nebula R_M not only provides the internal structure but also the mass of the gaseous envelope. Hence, as long as the planetary envelope matches continuously the nebula at R_M, this procedure also determines the rate of gas accretion $\dot{M}_{gas}$ by comparison of the envelope mass at t and t + dt.

As the core and/or envelope mass grows so that the planet becomes supercritical (its mass being larger than the critical mass), the gas accretion rate accelerates and eventually reaches a point when the disk can no longer sustain this rate. At this moment, the envelope detaches from the nebula. The planet's outer radius R_P is no longer equal to the Roche limit but must be calculated. In fact, during this phase, the radius rapidly contracts from its original value down to $R_P \approx 2$ to 5 jovian radii (R_{Jup}), depending upon the planet's entropy. In this detached phase, the planetary growth rate by gas accretion no longer depends on the planet's internal structure, but rather on the structure and evolution of the disk. For a one + one-dimensional viscous disk, at a given time t, the radial mass flux at r is given by

$$F(r) = 3\pi\nu(r)\Sigma(r) + 6\pi r \frac{\partial(\nu\Sigma)}{\partial r} \qquad (38)$$

Hence, the maximum mass delivery rate by the disk to the planet $\dot{M}_{gas,max}$ is given by the net mass flux entering and leaving the gas feeding zone of the planet $a_p \pm R_H$. This can be written as

$$\dot{M}_{gas,max} = \max\left[F\left(a_p + R_H\right), 0 \right] + \min\left[F\left(a_p - R_H\right), 0 \right] \qquad (39)$$

During a time dt, the maximum gas-mass that the protoplanet can accrete is given by $\dot{M}_{gas,max} \times dt$. For simplicity, it is assumed that a fixed fraction [0.75 to 0.9 (*Lubow and D'Angelo*, 2006)] of the disk's mass flux is accreted onto the planet.

Being detached from the disk, the continuity in pressure between the envelope and the disk is no longer a suitable boundary condition. In fact, both the ram pressure of the gas falling in from the boundary of the gas feeding zone to the planetary surface, where a standing shock is formed, and the photospheric pressure needs to be accounted for. This provides the new pressure and temperature at the surface of the planet (e.g., *Bodenheimer et al., 2000*; *Papaloizou and Nelson, 2005*)

$$P_{surf} = P_{disk}(a_p) + \frac{\dot{M}_{gas}}{4\pi R_P^2} v_{ff} + \frac{2g}{3\kappa} \quad (40)$$

where g is the gravitational acceleration at R_P, and

$$T_{surf}^4 = (1-A)T_{neb}^4 + T_{int}^4 \quad (41)$$

with A the albedo and v_{ff} the free-fall velocity from the limit of the boundary of the feeding zone to the surface given by

$$v_{ff} = \sqrt{\frac{2GM}{R_P} - \frac{2GM}{R_H}} \quad (42)$$

and

$$T_{int}^4 = \frac{3\tau L_{int}}{8\pi\sigma R_P^2} \quad (43)$$

where $\tau = \max [\kappa(T_{disk},\rho_{disk})\rho_{disk}R_P, 2/3]$.

These boundary conditions basically envision a miniature version of spherical accretion onto a stellar core as in *Stahler et al.* (1980). The actual geometry, and therefore the boundary conditions as well, are more complex and require in principle three-dimensional radiation-hydrodynamic simulations (e.g., *Klahr and Kley*, 2006). For the calculation of L_{int}, assumptions have to be made regarding the structure of the accretion shock: If the shock is radiatively (in)efficient, the potential energy of the gas liberated at the shock is (is not) radiated away, so that material gets incorporated into the planet at low (high) entropy, resulting in a low (high) luminosity (e.g., *Marley et al.*, 2007; *Spiegel and Burrows*, 2012; *Mordasini et al.*, 2012a). Typically, the limiting case of completely cold/hot accretion are considered. Together with D-burning in more massive objects (*Spiegel et al.*, 2011; *Mollière and Mordasini*, 2012; *Bodenheimer et al.*, 2013), this yields the post-formation luminosity (and radius) of giant planets (hot/cold start), which is crucial for the interpretation of directly imaged planets.

The final evolutionary (or isolated) phase occurs after the protoplanetary nebula has dissipated so that the planet cools and contracts, conserving the total mass (neglecting further accretion or mass loss, e.g., through atmospheric escape for close-in planets). The simplest possibility to model this phase is with a gray atmosphere so that

$$P_{surf} = \frac{2g}{3\kappa} \quad (44)$$

where the opacity κ is now given by the grain-free opacities of *Freedman et al.* (2008), and

$$T_{surf}^4 = (1-A)T_{eq}^4 + T_{int}^4 \quad (45)$$

with T_{eq} the equilibrium temperature due to stellar irradiation (equation (2)). As noted by *Bodenheimer et al.* (2000), these simple models lead to luminosities and radii as a function of time that agree relatively well with full nongray models (e.g., *Burrows et al.*, 1997; *Baraffe et al.*, 2003). This enables us to compare the radii (and luminosities) calculated by synthetic populations with results of transit (and direct imaging) surveys.

While the approach used by *IL* is computationally extremely rapid and therefore allows the tests of many models, it includes uncertainties in the mass and structure of planets, especially Earth-like or super-Earth planets with small mass atmospheres. On the other hand, it should also be pointed out that full computations by *AMB* are only accurate as long as the ingredients used are well justified, which are not always the case. For example, the EOS, the opacity, the structure of the accretion shock (e.g., *Marley et al.*, 2007), or the models of convection (e.g., *Baraffe et al.*, 2012; *Leconte and Chabrier*, 2012) are uncertain.

2.5. Interactions Between Growing Planets

Interactions between planets growing within a given protoplanetary disk are key ingredients in a model of the formation of planetary systems (module 10 in Fig. 1). Because eccentricity damping due to dynamical friction with the gaseous disk (e.g., *Tanaka and Ward*, 2004) is very strong, protoplanets formed through oligarchic growth are usually isolated from each other and in nearly circular orbits. Once the disk of gas is sufficiently depleted, secular perturbations pump up eccentricities (*Chambers et al.*, 1996), leading to orbit crossing and hence to collisions. Because a single collision can double the mass of bodies involved, planetary masses can eventually increase by orders of magnitude. On the other hand, scattering among gas giant planets often results in ejection of one or two planets, leaving other planets in highly eccentric isolated orbits (e.g., *Marzari and Weidenschilling*, 2000; *Nagasawa et al.*, 2007). The perturbations from gas giants on such eccentric orbits can alter orbital configurations of a whole planetary system. Even rocky planets in the inner regions that are far from the interacting gas giants in outer regions can be severely affected (e.g., *Matsumura et al.*, 2013).

In early population-synthesis models, interactions between planets have been neglected, because they are chaotic and highly nonlinear and hence difficult to model. However, recent models (*Ida and Lin*, 2010; *Ida et al.*, 2013; *Alibert et al.*, 2013) have now incorporated planet-planet gravitational interactions.

Because explicit N-body simulations are computationally very expensive, *IL* developed semi-analytical Monte Carlo models to compute planet-planet collisions and scatterings. Their approach is based on detailed planetesimal dynamic studies revealed by prior detailed N-body simulations and statistical formulations (e.g., *Ida and Nakazawa*, 1989; *Ida*, 1990; *Ida and Makino*, 1993; *Palmer et al.*, 1993; *Aarseth et al.*, 1993; *Kokubo and Ida*, 1998, 2002; *Ohtsuki et al.*,

2002). As a result, their models reproduce quantitatively statistical distributions of N-body simulation outcomes. Because the prescriptions of planet-planet collisions and scatterings are rather complicated multi-step schemes, we omit the descriptions of their prescriptions and refer for details to *Ida and Lin* (2010)and *Ida et al.* (2013).

In contrast, *AMB* combines direct N-body simulations with population-synthesis calculations, in order to accurately take into account the effects of planet-planet interactions. These authors use a standard integration scheme (e.g., Bulirsch-Stoer). The equation of motions for the protoplanets in a heliocentric reference frame are written as

$$\ddot{r}_i = -G\left(m_0 + m_i\right)\frac{r_i}{r_i^3}$$

$$-G\sum_{j=1,j\neq i}^{n} m_j\left\{\frac{r_i - r_j}{\left|r_i - r_j\right|^3} + \frac{r_j}{r_j^3}\right\} \qquad (46)$$

with i = 1, 2, 3 . . . N, m_i the mass of the planets and M_* the mass of the central star. The integration uses an adaptive timestep that ensures reaching a desired precision. We note that since the integration is carried out over the protoplanetary disk lifetime [≤10 million years (m.y.)] and that dissipative forces exist, a simplectic integrator is not necessary. While this approach is relatively straightforward it has the disadvantage of being relatively expensive in integration time for a large number of protoplanets and/or for small timesteps. It has, however, the great merit of capturing all possible effects (e.g., resonances, collisions).

Collisions between the planets are detected by checking whether two planets come closer to each other than $d_{col} = R_1 + R_2$ (where R_1 and R_2 are the radii of the two bodies). This is achieved by searching among all possible pairs of protoplanets for those that, during the time t and $t + \Delta t$, will come closer than d_{col} to each other, Δt being the timestep of the N-body integrator. In practice, this is done by extrapolating the positions of all protoplanets using Taylor expansions, and searching for a time τ such that for $t \leq \tau \leq t + \Delta t$ the distance between any possible pair of protoplanets is less than d_{col}. This approach has been proposed by *Richardson et al.* (2000), who used a first-order Taylor expansion, and has more recently been improved by (*Alibert et al.*, 2013) by using second-order extrapolations.

2.6. Planet-Disk Interactions: Migration

Planet-disk interactions lead to planet migration and to the damping of eccentricity and inclination. The migration results from planet-disk angular momentum exchange and the determination of the migration rate requires the computation of the two-dimensional or three-dimensional structure (including the thermodynamics) of the protoplanetary disk. The computer time required for carrying out such detailed multi-dimensional simulations of these processes over a timescale covering planet formation is prohibitively high.

Hence, in the population synthesis approach, planetary migration is computed using fits to migration rates resulting from hydrodynamical calculations (see the chapter by Baruteau et al. in this volume and references therein).

Planetary migration occurs in different regimes depending upon the mass of the planet. For low-mass planets, i.e., planets not massive enough to open a gap in the protoplanetary disk, migration occurs due to the imbalance between the Lindblad and co-rotation torques exerted on the planet by the inner and outer regions of the disk. This regime is called "type I migration" and the corresponding migration rate has been derived by linear and numerical calculations (e.g., *Ward*, 1997; *Tanaka et al.*, 2002; *Paardekooper and Papaloizou*, 2009; *Paardekooper et al.*, 2011). For highermass planets, i.e., planets massive enough to open a gap in the protoplanetary disk, the planet is confined in the gap by Lindblad torques and thus follows the global disk accretion. This regime is called "type II migration." Type II migration is itself subdivided in two modes: disk-dominated type II migration, the case in which the local disk mass exceeds the planetary mass, and planet-dominated type II migration, the opposite case (see also *Lin and Papaloizou*, 1986; *Ida and Lin*, 2004a; *Mordasini et al.*, 2009a). In the former case, the migration rate is simply given by the local viscous evolution of the protoplanetary disk, while the migration is decelerated by the inertia of the planet in the latter case.

Initial population-synthesis models by both *IL* and *AMB* made use of the conventional formula of type I migration derived for locally isothermal disks (*Tanaka et al.*, 2002). In order to investigate how sensitive the results are on the magnitude of this migration, a scaling factor C_1 was introduced

$$\tau_{mig1} = \frac{a}{\dot{a}}$$

$$= \frac{1}{C_1}\frac{1}{3.81}\left(\frac{c_s}{a\Omega_K}\right)^2\frac{M_*}{M_{planet}}\frac{M_*}{a^2\Sigma_g}\Omega_K^{-1}$$

$$\simeq 1.5\times10^5\frac{1}{C_1 f_g}\left(\frac{M_c}{M_\oplus}\right)^{-1}\left(\frac{a}{1AU}\right) \qquad (47)$$

$$\times\left(\frac{M_*}{M_\odot}\right)^{3/2} yr$$

The expression of *Tanaka et al.* (2002) corresponds to $C_1 = 1$, while $C_1 < 1$ implies slower migration rates. *IL* assume type I migration ceases inside the inner boundary of the disk.

Since the publication of *Tanaka et al.* (2002), radiative effects on the type I migration rate have been investigated (e.g., *Paardekooper and Mellema*, 2006; *Masset and Casoli*, 2010; *Paardekooper et al.*, 2011). It was shown that the migration velocity, as well as its direction, depend sensitively upon the detailed dynamical and thermal structure of disks, leading to a number of subregimes of type I migration [locally isothermal, adiabatic, (un)saturated]. Recently, a new semi-analytic description of type I migration, which reproduces the results of *Paardekooper et al.* (2011), has been derived (*Mordasini et al.,* 2011a; *Kretke and Lin,* 2012). It includes the effect of co-rotation torques that can lead to outward migration in non-isothermal disks. This new formalism has been implemented in recent simulations by *AMB,* who have shown that the scaling factor determining the migration speed introduced in earlier models (equation (47)) becomes much less important (*Alibert et al.,* 2013).

The transition mass between type I and type II migration is $M_{g,vis}$ (equation (52)) in *IL*'s prescription and $M_{g,th}$ (equation (53)) (more precisely, the condition derived by *Crida et al.,* 2006) in *AMB*'s prescription. The comparison between gap-opening criteria is treated in more detail in section 2.7.

Initially, population-synthesis models have assumed an isothermal migration rate reduced by $C_1 \sim 10^{-2}–10^{-1}$ in *IL*'s simulations and $C_1 \sim 10^{-3}–10^{-2}$ in *AMB*'s simulations. The values of C_1 less than unity were needed to prevent cores of growing giant planets to fall into the host star. These findings by population syntheses were an important motivation to develop physically more realistic non-isothermal migration models. This is one out of several examples in which population synthesis can be used to test in a statistical sense detailed modeling of individual processes. Population-synthesis models did not provide a better understanding of the migration itself but pointed out that the current prescription did not result in planet populations with the observed characteristics. It is worth pointing out that with this new formalism for type I migration, which has been implemented in recent simulations by *AMB,* an arbitrary scaling factor (equation (47)) slowing down migration is no longer an absolute necessity (*Alibert et al.,* 2013). While this represents a definitive progress, difficulties remain. They are linked to the sensitivity of the migration rate to the saturation of the corotation torque, to a partial gap formed by relatively large migrating planet, and to orbital eccentricity. A further difficulty is due to the fact that the onset of efficient gas accretion onto the core, and the saturation of the co-rotation torque, occur at a similar mass (on the order of 10 $M_\oplus$), so that a self-consistent coupled approach of the two processes is necessary.

For type II migration, as long as the mass of the planet remains smaller than the local disk mass (on the order of $\pi\Sigma r^2$), the migration timescale ($\tau_{mig2}(a) = a/|v_r|$) is given by the local viscous diffusion time, $t_{vis}(a) \sim (2/3)(a^2/\nu)$. For a steady accretion disk with $F \sim 3\pi\Sigma\nu$ and $\Sigma \propto 1/a$, $M_{disk}(a) = \int^a 2\pi a\Sigma da = 2\pi\Sigma a^2$. Then, $\tau_{mig2}(a) \sim t_{vis}(a) \sim M_{disk}(a)/F$ (*Hasegawa and Ida,* 2013). For a planet more massive than the inner disk mass $M_{disk}(a)$, the viscous torque from the outer disk pushes the planet (mass M_{planet}) rather than the inner disk. Then, replacing M_{disk} with M_{planet}, $\tau_{mig}(a) \sim M_{planet}/F$ (*Hasegawa and Ida,* 2013). In summary, the type II migration rate ($v_r \sim a/\tau_{mig2}$) is roughly given by

$$\frac{da}{dt} = -\frac{3\nu}{2a} \times \min\left[1, \frac{M_{disk}(a)}{M_{planet}}\right] \quad (48)$$

Migrations for $M_{planet} < M_{disk}(a)$ and $M_{planet} > M_{disk}(a)$ are called "disk-dominated" and "planet-dominated" type II migrations, respectively.

IL and *AMB* use more detailed prescriptions for a planet-dominated regime. *IL* adopt (*Ida and Lin,* 2008a)

$$\frac{da}{dt} = -\frac{3\nu}{2a} \times \min\left[1, 2C_2\left(\frac{30AU}{a}\right)^{1/2}\frac{M_{disk}(a)}{M_{planet}}\right] \quad (49)$$

Because they use $C_2 = 0.1$ and usually $a \sim O(1)$, the extra factor $2C_2(30\,AU/a)^{1/2}$ is ~1 and *IL*'s prescription is the same as the simple formula given by equation (48).

AMB use

$$\frac{da}{dt} = -\frac{3\nu}{2a} \times \min\left[1, \frac{2\Sigma a^2}{M_{planet}}\right] \quad (50)$$

In the region of $\Sigma \propto 1/a$, this formula becomes

$$\frac{da}{dt} = -\frac{3\nu}{2a} \times \min\left[1, \frac{1}{\pi}\frac{M_{disk}(a)}{M_{planet}}\right] \quad (51)$$

Thus, *AMB*'s type II migration rate is slower than *IL*'s by a factor of π in the planet-dominated regime, while the rates are identical in the disk-dominated regime. More detailed hydrodynamical simulations are required to determine which migration rate is more appropriate in the planet-dominated regime. However, as discussed below, different treatments of gas accretion onto planets after gap formation more significantly affect the efficiency of type II migration in population-synthesis simulations.

2.7. Planet-Disk Interactions: Gap Formation

Gap formation in a gas disk by the perturbations exerted by a planet has two important consequences: (1) It reduces or even terminates the gas flow onto the planet; and (2) the planet switches its migration from type I to type II. While gravitational torques exerted by the planet on the disk work toward the opening of a gap, two physical processes tend to prevent this opening: viscous diffusion and pressure gradients. The planet's tidal torque exceeds the disk's intrinsic viscous stress at the mass (*Lin and Papaloizou,* 1986)

$$
\begin{aligned}
M_{planet} > M_{g,vis} &\simeq \frac{40M_\star}{Re} \\
&\simeq 40\alpha\left(\frac{H_{disk}}{a}\right)^2 M_\star \\
&\simeq 30\left(\frac{\alpha}{10^{-3}}\right)\left(\frac{a}{1AU}\right)^{1/2} \\
&\quad \times\left(\frac{L_\star}{L_\odot}\right)^{1/4} M_\oplus
\end{aligned}
\quad (52)
$$

where H_{disk} is the disk scale-height at the location of the planet, and $Re = a^2\Omega/\nu$ is the Reynolds number at the location of the planet (a). This is called the "viscous condition" for gap formation. If the gap half-width is less than $\sim H_{disk}$, the pressure gradient inhibits gap formation. The gap half-width may be $\sim R_H$ where R_H is the planet's Hill radius. Then the critical mass is given by $R_H \gtrsim \beta H_{disk}$, where $\beta \sim O(1)$, i.e.

$$M_{planet} > \beta^3 M_{g,th} \simeq 3\left(\frac{\beta H_{disk}}{r}\right)^3 M_\star$$
$$\simeq 120\beta^3 \left(\frac{a}{1AU}\right)^{3/4}\left(\frac{L_*}{L_\odot}\right)^{3/8}$$
$$\times \left(\frac{M_*}{M_\odot}\right)^{-1/2} M_\oplus \qquad (53)$$

where $M_{g,th}$ is defined by $3(H_{rmdisk}/r)^3 M_\odot$ (note that the definition is different from that in *IL*'s papers by a factor of β^3). This is called the "thermal condition."

Lin and Papaloizou (1993) and *Crida et al.* (2006) found through numerical calculations that the gap-opening condition is

$$\frac{3}{4}\frac{H_{disk}}{R_H} + \frac{50 M_\star}{M_{planet} Re} < 1 \qquad (54)$$

which is equivalent to a combination of viscous condition and thermal condition (with $\beta = 3/4$).

IL adopt the following prescription:

1. For $M_{planet} > M_{g,vis}$ ($\beta = 2$), a gap is formed and type I migration is switched to type II migration. Here, a gap is assumed to be partial (low-density region along the planet's orbit), so that the gas disk still crosses the gap.

2. For $M_{planet} > 8M_{g,th}$, gas accretion onto the planet is completely terminated, because hydrodynamical simulations show that gas flow across the gap rapidly decays as M_p increases beyond Jupiter mass (e.g., *D'Angelo et al.*, 2002; *Lubow and D'Angelo*, 2006), and *Dobbs-Dixon et al.* (2007) suggests that gas accretion onto the planet is terminated if R_H well exceeds H_{disk}.

3. As a result, the accretion rate onto the planet is given by [F is the mass flux in the disk (equation (38))]

$$\frac{dM_{planet}}{dt} = f_{gap}F \qquad (55)$$

where f_{gap} is a reduction factor due to gap opening

$$f_{gap} = \begin{cases} 1\left[M_p < M_{g,vis}\right] \\ 0\left[M_p > 8M_{g,th}\right] \end{cases} \qquad (56)$$

and for $M_{g,vis} < M_p < 8M_{g,th}$

$$f_{gap} = \frac{\log M_{planet} - \log M_{g,vis}}{\log 8M_{g,th} - \log M_{g,vis}} \qquad (57)$$

This formula is constructed to avoid any abrupt truncation.

4. The gas accretion also decays according to global disk depletion. This effect is automatically included through F, which is proportional to Σ.

AMB adopt a different prescription:

1. When condition (54) is satisfied, a gap is formed and the migration mode switches from type I to type II. Note that for $\beta = 3/4$, the thermal condition $M_{g,th} > (3/4)^3 M_{g,th}$ is comparable to the viscous condition $M_{planet} > M_{g,vis}$.

2. After gap opening and provided the planet is in the detached phase, the gas accretion onto the planet remains given by F× a fixed factor <1 (see section 2.4) without any further reduction. The planet keeps growing until the disk gas becomes globally depleted, which manifests itself by a gradual decrease of F to zero. This limiting assumption of no reduction due to gap formation is motivated by the results of isothermal hydrodynamic simulations of *Kley and Dirksen* (2006). They found that for planetary masses above a certain minimum mass (3–5 M_{Jup}, depending upon viscosity), the disk makes a transition from a circular state into an eccentric state. In this state, the mass accretion rate onto the planet is greatly enhanced relative to the case of a circular, clean gap because the edge of the gap periodically approaches the protoplanet, which can even become (re) engulfed in the disk gas for large eccentricities.

Because *AMB* assume that the planet continues accreting gas at the unimpeded disk accretion rate (equation (38)) even well after gap opening, and they also consider relatively efficient photoevaporation (which is not considered by *IL*), the migration soon enters the planet-dominated regime and becomes slower and slower as the planet keeps growing. With *AMB*'s prescription for a planet-dominated regime, it can be shown that d $\log M_p$/d log a = $(a/M_p)(dM_p/dt)/(da/dt) \sim -\pi$. This is clearly in contrast to *IL*'s approach described above. As a consequence, even though the type II formalism is very similar, the predicted distributions of semimajor axis of gas giants between the two approaches are different. This is essentially due to the increased role of planet inertia in the *AMB* formalism, which slows down the migration of massive planets. This leads *IL* to predict a much larger frequency of hot Jupiters (section 5.2).

3. INITIAL CONDITIONS

The single most important ingredient of the population-synthesis method is a global planet-formation model that "translates" properties of a protoplanetary disk (which are the initial conditions for planet formation) into properties of the emerging planetary system. This formation model has been described in the previous section. The other most important ingredients are the probabilities of occurrences (distributions) of these initial conditions.

The initial conditions of population-synthesis calculations are of two different types. The first one is related to the properties of the protoplanetary disk, while the second is related to the properties of the protoplanets themselves. As far as the first type is concerned, the basic assumption of planetary population synthesis is that the (observed) diversity of planetary systems is a consequence of the (observed) diversity of the properties of protoplanetary disks. This assumption is verified *a posteriori* by the large diversity of planets resulting from this assumption. The initial conditions characterizing a disk are therefore treated as Monte Carlo variables that can be drawn from probability distributions.

Ideally, the probability distributions of the properties of protoplanetary disks should be taken directly from observations (see the chapters by Dutrey et al. and Testi et al. in this volume). Unfortunately, this is not a straightforward task, as present-day observations do not constrain the innermost parts of disks (where planets actually form) very well, and the number of well-characterized disks is presently small. Both *IL* and *AMB* consider three fundamental disk properties as Monte Carlo variables:

1. *The (initial) surface density of gas in the protoplanetary disk.* In the *IL* models (equation (1)), it is given by the scaling factor $f_{g,0}$ [a scaling factor for $\Sigma_g(r = 10$ AU, $t = 0)$]. According to the distributions of total disk masses inferred by radio observations of T Tauri disks, *IL* assume a Gaussian distribution of $\log_{10} f_{g,0}$ with a mean and standard deviation of 0 and 1 respectively. In the *AMB* models, the different disk masses are represented by different Σ ($r = 5.2$ AU, $t = 0$) in equation (15). For the distribution of disk masses, *AMB* fit the disk mass distribution observed by *Andrews et al.* (2010) with a Gaussian distribution with a mean and standard deviation in $\log_{10}(M_{disk}/M_\odot)$ of -1.66 and 0.56, respectively, or alternatively directly bootstrap from the observed distribution (*Fortier et al.,* 2013). Note that the mean value, $\log_{10}(M_{disk} = M_\odot) = -1.66$, is comparable to the disk mass corresponding to $\log_{10} f_{g,0} \sim 0$ with the disk size ~100 AU.

2. *The lifetime of the protoplanetary disk.* In the *IL* models (equation (4)), it is represented by τ_{dep}. According to infrared (IR) and radio observations, *IL* assume a Gaussian distribution of $\log_{10} \tau_{dep}$ with a mean and standard deviation 6.5 and 0.5. In the *AMB* models, the distribution of disk lifetimes is obtained by specifying a distribution of external photoevaporation rates (equation (18)). This distribution is adjusted in a way (for details, see *Mordasini et al.,* 2009a) that the distribution of the resulting lifetimes of the synthetic disks agrees with the observed distribution as derived from the fraction of stars with an IR excess (*Haisch et al.,* 2001) (see also the chapter by Dutrey et al. in this volume).

3. *The surface density of solids.* In the *IL* models (equation (20)), it is represented by the scaling factor f_d. The initial value, $f_{d,0}$, is given by $f_{g,0} 10^{[Fe/H]}$, where [Fe/H] is the stellar metallicity and the same dust-to-gas ratio is assumed between the stellar surface and the disk. Consequently, the initial dust-to-gas ratio $f_{D/G}$ is given by $f_{D/G,0}(f_{d,0}/f_{g,0})$, where $f_{D/G,0}$ is the ratio associated with the solar composi-

tion (*IL* adopt $f_{D/G,0} = 1/240$). In the *AMB* models, $f_{D/G} = f_{D/G,0} 10^{[Fe/H]}$ is first specified and then the surface density of solids is given by equation (21). Thus, the *IL* and *AMB* models are equivalent as far as the description of the surface density of solids is concerned. For the probability distribution, *IL* use a Gaussian distribution of [Fe/H] with a mean and standard deviation of 0 and 0.2 dex, respectively, while *AMB* use that of -0.02 and 0.22 dex, corresponding to the CORALIE planet search sample (*Udry et al.,* 2000).

Besides the total disk mass, it is also necessary to specify the radial profile of the gas (and solid) surface density. One approach is to assume some theoretically inspired surface density profile, mass, and composition. First models of population synthesis have indeed assumed disk profiles similar to the minimum solar nebula, with a surface density slope of typically $-3/2$.

Recently, as the number of well-characterized disks has grown, new models started to consider disk profiles that come directly from fits of observations (e.g., *Andrews et al.,* 2010). Typical disk profiles are given by

$$\Sigma(r) = (2 - \gamma)\frac{M_{disk}}{2\pi a_{core}^{2-\gamma} r_0^\gamma}\left(\frac{r}{r_0}\right)^{-\gamma} \times \exp\left[-\left(\frac{r}{a_{core}}\right)^{2-\gamma}\right] \tag{58}$$

where r_0 is equal to 5.2 AU, and M_{disk}, a_{core}, γ are derived from observations. The observations (e.g., *Andrews et al.,* 2010) suggest $\gamma \sim 1$, which is consistent with the self-similar solution with constant α. Accordingly, *IL* adopt $\Sigma(r) \propto 1/r$ in recent models, as already mentioned. *AMB* initially adopted $\Sigma(r) \propto r^{-1/5}$. In more recent models, the gas surface density is taken to follow the distribution of equation (58), the disk parameters being directly the ones derived in *Andrews et al.* (2010).

To start a population-synthesis calculation it is necessary to distribute planetary seeds within the disk. The growth of these seeds is then followed in time. The initial location and mass of these seeds are not constrained from observations and are derived from theoretical arguments. Seeds are often assumed to have an initial location distribution that is uniform in log, following N-body calculations of the early stage of planetary growth (e.g., *Kokubo and Ida,* 1998, 2002). *IL* use a somewhat different approach. The masses of protoplanets formed by oligarchic growth are predicted by Σ_s distribution as "isolation" masses in inner regions or final masses predicted by simple formula in outer regions (where planetesimal accretion is so slow that the protoplanets' mass does not reach their isolation mass), so that seeds are set up with orbital separations comparable to the feeding zone width of the protoplanets.

One should note that recent models of planetesimals and protoplanet formation (see the chapters by Johansen et al., Raymond et al., and Helled et al. in this volume) predict that protoplanets might form under precise circumstances (e.g., close to the ice line or near a pressure maximum). Under these circumstances, the distribution of initial locations of protoplanets would be far from uniform in log-scale but rather concentrated at specific locations in the disk. A better understanding of these issues is necessary for future progress in population-synthesis models.

The initial mass of the seeds is similarly observationally undetermined. Ideally, the result of population-synthesis models should be independent of the assumed initial mass of seeds, provided the initial protoplanetary disk model is consistent with the time required to grow these seeds. More massive seeds taking longer to grow should be implanted in already more evolved disks. This obviously raises the question of how to define time zero. In all cases, the initial mass of seeds should be small enough so that any processes such as migration or gas accretion remains negligible. In practice, it is often a good choice to assume that the initial mass of seeds corresponds to the one obtained at the end of the local runaway growth phase, as the latter is believed to be very rapid.

4. OBSERVATIONAL BIASES

One of the key objectives of population-synthesis models is to compare models with observations, even in a quantitative way if possible. From this comparison, constraints on some of the key processes acting during planet formation should be gained. However, for this comparison to be meaningful, it is essential that observational selection biases associated with the exoplanet detection methods are well understood. Large, homogeneous surveys with a well characterized detection bias like the High Accuracy Radial velocity Planet Searcher (HARPS) or Eta-Earth surveys for the RV technique (*Mayor et al., 2011; Howard et al., 2010*), or the Kepler mission for the transit technique (*Borucki et al., 2011*), are, in this respect, of particular interest. Since the majority of exoplanets have been discovered by these two techniques, their selection biases are also best known. For RVs, the simplest approach is to consider a detection criterion based on the induced RV amplitude and a maximal period. A more sophisticated approach uses a tabulated detection probability that is a function of the planet's mass and orbital period, and takes the instrumental characteristics as well as the actual measurement schedule into account (*Mordasini et al., 2009b*). In addition, recent "controlled experiment" microlensing surveys (*Gould et al., 2010*) and forthcoming large direct imaging [with the Gemini Planet Imager (GPI) at the Gemini Observatory, and the Spectro-Polarimetric High-contrast Exoplanet REsearch (SPHERE) at the Very Large Telescope (VLT)] and astrometric surveys (e.g., Gaia) are equally important, as different techniques typically probe different subpopulation of planets, yielding complementary constraints for the modeling. It would be particular informative if the systematic surveys can provide well-determined upper limits on both the presence and absence of planets in domains of parameter space. Such data can be used to verify or falsify predictions on different planetary characteristics made with population-synthesis models.

5. HIGHLIGHTS OF POPULATION-SYNTHESIS MODELS

5.1. Individual Systems

Population-synthesis predictions are made by the superposition of individual systems with different initial conditions (section 5.2). Before we start statistical discussions on the distributions predicted by population synthesis in comparison with the observed ones, we first show examples of the evolution of individual systems.

Figure 2 shows an example of evolution of a system using the *IL* approach with $C_1 = 0.1$, $f_g = f_d = 2.0$ ([Fe/H] = 0), $\tau_{KH1} = 10^9$ yr, $\tau_{dep} = 3 \times 10^6$ yr, and orbiting a solar-mass star ($M_* = 1\ M_\odot$). Figures 2a,b show time and mass evolution. The light, dark and middle gray lines represent rocky, icy, and gas giant planets with their main component being rock, ice, and gas, respectively. The bulk composition of the planets can change over time through gas or planetesimal accretion or planet-planet collisions. At 0.1–1 AU small protoplanets grow *in situ* until their masses reach ~0.1–1 $M_\oplus$ and then they undergo type I migration and accumulate near the inner boundary of the disk, which is set at 0.04 AU. Many resonant protoplanets actually accumulate at the vicinity of this boundary (Fig. 2a) and are preserved until the gas disk decays enough to allow orbit crossing and merging starts. Just outside the ice line at a ~ 3 AU, a core reaches $M_p \sim 5\ M_\oplus$ and starts runaway gas accretion without any significant type I migration. After it has evolved into a gas giant with a surrounding gap, it undergoes type II migration. The emerging gas giant scatters and ejects nearby protoplanets. Finally, a system with closely packed close-in super-Earths, a gas giant at an intermediate distance, and outer icy planets in nearly circular orbits is formed (Figs. 2c,d). Since the super-Earths are formed by scattering and merging, they have been kicked out from resonances. Such closely packed, nonresonant, close-in super-Earths are found to be common by Kepler observations.

Note that in our solar system, no planet exists inside Mercury's orbit at a = 0.39 AU, while RV and Kepler observations suggest that more than 50% of solar-type stars have close-in planets. Hence, a planetary system like our own solar system may only form if either we lose these innermost planets to the Sun (no inner cavity exists) or outward type I migration took place. Furthermore, for giant planets such as Jupiter and Saturn to remain at large distances, they must have formed late, at a time when the disk was severely depleted, in order to avoid extensive type II migration. Since a significant time lag may exist between the formation of two giant planets, it is not easy for the current model to explain the presence of two gas giants in the outer regions. A mechanism to form two gas giants almost simultaneously might be required, such as in the induced-formation model by *Kobayashi et al.* (2012). (Such an effect has not yet been incorporated into population-synthesis simulations.)

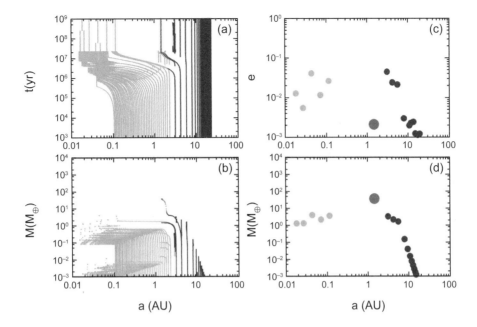

Fig. 2. Growth and migration of planets in a system with $C_1 = 0.1$ and $f_{d,0} = 2.0$. The light, dark, and middle gray lines represent planets with their main component as rock, ice, and gas, respectively. **(a)**,**(b)** Time and mass evolution of planets with different initial semimajor axes. **(c)**,**(d)** Orbital eccentricity and **(d)** planetary mass as a function of semimajor axis in the final state. Adapted from *Ida et al.* (2013).

5.2. Comparisons with Observations

In this section we present some aspects of the statistical comparisons between synthetic populations and observations. To this effect, a population of planets is built by running the planet-formation model using a large number of different initial conditions. As explained above, the initial conditions are drawn at random following a Monte Carlo procedure in which observations and theoretical arguments are used to determine the probability of occurrence of a given initial condition.

One of the key results of population-synthesis models is the computation of the mass vs. semimajor axis diagram of planets. Such a diagram might be of similar importance for planetary physics as the Hertzsprung-Russell diagram is for stellar astrophysics. First models by *Ida and Lin* (2004a), and later by *Mordasini et al.* (2009b), provided a consistent global picture, with some interesting differences, however. Indeed, both models concluded that type I migration had to be highly reduced compared to the theoretical estimates available at the time (e.g., *Tanaka et al.,* 2002). Moreover, it appeared naturally from the very concept of the core-accretion model that there should be a lack of planets of masses between Neptune and Saturn. This results simply from the fact that the accretion of gas is low for subcritical planets (with a core mass less than $\sim$10 $M_\oplus$), whereas it is quite rapid for supercritical planets (*Pollack et al.,* 1996), up to the point when additional processes hinder gas accretion (for masses larger than $\sim$100 $M_\oplus$). Since it is unlikely that the protoplanetary disk, and therefore the gas supply, disappears exactly during the short timescale of gas run-

away accretion, fewer planets with intermediate masses are expected. *Ida and Lin* (2004a) called this potential deficit of intermediate-mass planets the "planetary desert."

How much of a *desert* the "planetary desert" actually is depends upon the details of the computation of the gas accretion rate (for a dedicated discussion, see *Mordasini et al.,* 2011b). In particular, the *AMB* models and *Ida et al.* (2013) limit the gas accretion rate onto the planet by the rate at which the disk can actually provide gas to the planet. This modification reduces the gas accretion rate, and the "planetary desert" becomes less pronounced than in the earlier models by *Ida and Lin* (2004a). This difference shows that the comparison of the actual and synthetic a–M diagram helps to better understand the mechanism of gas accretion.

This is illustrated in Fig. 3 from *Alibert et al.* (2013). One certainly notes that there are fewer planets in the 20–100-$M_\oplus$ mass range than less- and/or more-massive planets. However, this part of the diagram is far from being totally empty. *Ida and Lin* (2010) and *Ida et al.* (2013) showed that super-Earths can be formed by *in situ* collisional coalescence of two or more protoplanets after disk-gas depletion. Some of these planets have migrated to the vicinity of their host stars before disk-gas depletion, which is illustrated in Fig. 2. This effect, in addition to the limitation of gas accretion onto the planet, also supplies a population of intermediate-mass close-in planets, making the "planetary desert" less conspicuous. Nevertheless, this mass range remains an underpopulated region in the M_p–a distribution (see Fig. 4).

A certain depletion of intermediate-mass planets is also evident in Fig. 5. It shows the planetary initial mass function

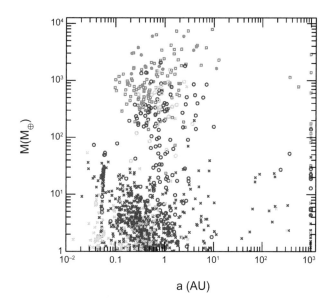

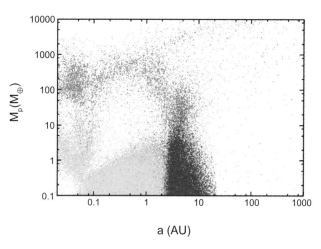

a (AU)

Fig. 3. Synthetic mass-distance diagram at the time the protoplanetary nebula vanishes. The light gray (black) crosses are rocky (icy) planets with a gaseous envelope less massive than the core. The open light gray (black) circles are rocky (icy) planets with an envelope 1–10× more massive than the core. The dark filled circles (empty squares) are giant planets with a rocky (icy) core. The envelope is at least 10× more massive than the core. The model assumes 10 protoplanets concurrently forming per disk.

Fig. 4. Synthetic mass-distance diagram at the time the protoplanetary nebula vanishes obtained by *Ida et al.* (2013). The shades of gray have the same meaning as in as Fig. 3. The population of 0.3–30 $M_\oplus$ at a $\leq$ 0.1 AU is formed by *in situ* merging of protoplanets that have migrated to the vicinity of the host stars.

(P-IMF), i.e., the distribution of planetary masses during the phase when the protoplanetary disks disappear, as found by *AMB*. The P-IMF is one of the most important results of planetary population synthesis. Two peaks are clearly present, a smaller one at a few hundred Earth masses representing gaseous giant planets, and a much larger second one for low-mass, solid-dominated planets. This is a simple consequence of the fact that typically, the conditions in the protoplanetary disks are such that only low-mass planets can form. Note that the abundant population of low-mass planets was predicted by population synthesis long before its existence was confirmed by RV surveys and the Kepler mission. The mass function presented in Fig. 5 is derived from models considering the formation of systems, seeded by 10 protoplanets per disk (*Alibert et al.*, 2013). Interestingly enough, the mass function obtained in this case is quite similar to the one obtained with only one protoplanet present in each disk (*Mordasini et al.*, 2009b).

If we plot all the exoplanets discovered by RV surveys (e.g., *exoplanets.org*), a deficit of planets in 30–100 $M_\oplus$ is suggested at a $\leq$ 0.1 AU (see the top panel of Fig. 6c), even though it is not very pronounced. Results from RV surveys for controlled samples (e.g., *Howard et al.*, 2010; *Mayor et al.*, 2011) do not show this deficit. However, their statistical significance is limited by the size of the samples. The Kepler data does not show a clear deficit either, although the distribution function of physical radii found by Kepler can only be translated into a mass distribution if the planetary

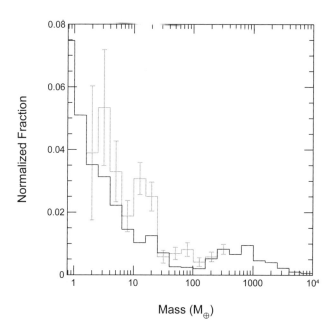

Fig. 5. Synthetic planetary initial mass function P-IMF (thick line). The strong increase toward small masses and the transition between solid- and gas-dominated planets at about 40 $M_\oplus$ is visible. The thin gray line is the bias-corrected observed mass distribution from *Mayor et al.* (2011). It has been normalized to the value of the synthetic distribution in the bin at 1 M_{Jup}.

mass-radius relationship is known, as discussed below [for a comparison of the synthetic planetary radius distribution and the Kepler results, see *Mordasini et al.* (2012b)]. The issue of the "planetary desert" must be further investigated from both observational and theoretical sides.

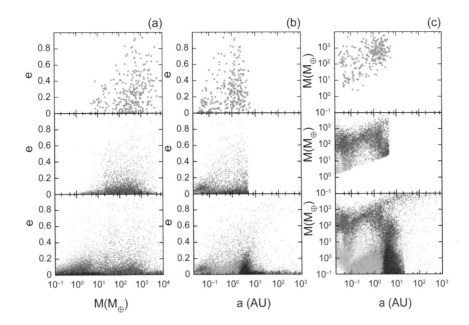

Fig. 6. Comparison of theoretical results with RV observational data: **(a)** e–M_p, **(b)** e–a, and **(c)** M_p–a distributions. The top row shows data obtained from RV surveys. The bottom row shows distributions of planets in 10^4-systems simulated by *Ida et al.* (2013) [for detailed simulation parameters, see *Ida et al.* (2013)]. The middle row only represents simulated observable planets (RV greater than 1 m s^{-1} and periods less than 10 yr). Adapted from *Ida et al.* (2013).

Interestingly, even if the bias-corrected observed mass distribution of *Mayor et al.* (2011) does not show a deficit at intermediate masses, it nevertheless exhibits a very clear change in the slope of the mass function at about 20–30 M$_\oplus$ (Fig. 5). In the synthetic mass distribution, a similar, even though less-abrupt, change can be seen. It arises from the fact that when the total mass of the planet is approximately 30 M$_\oplus$, the accretion of gas becomes very rapid. This is because at this mass, the core already exeeds significantly the critical/crossover core mass, which is typically about 15 M$_\oplus$ (*Pollack et al., 1996*). The change in slope at about 30–40 M$_\oplus$ therefore potentially represents the transition from solid- to gas-dominated planets. In other words, it is evidence of the existence of a critical core mass, a key concept within the core-accretion paradigm. If this imprint of core accretion into the planetary mass function is confirmed, it would represent a key statistical finding for both theory and observation.

The observed M$_p$–a distribution (the top panel of Fig. 6c) also shows a pile-up of gas giants at ≤0.05 AU (hot Jupiters) and at ≥1 AU (cool Jupiters), the latter being particularly well pronounced. The theoretical prediction by *Ida et al.* (2013) at the bottom panel of Fig. 6c shows too many hot Jupiters and no clear pile-up of cool Jupiters. The inconsistency may be due to type II migration that was too efficient, a condition that *IL* adopted in their prescriptions as discussed in section 2.7. In comparison, the M$_p$–a distribution obtained by *Alibert et al.* (2013) (Fig. 3) shows too few hot Jupiters, which may be due to the underefficient prescription for type II migration they adopted (see section 2.7). Their result does not show a clear pile-up of cool Jupiters, either. Therefore, the ob-

served M$_p$–a distribution shows that the prescription as currently implemented does not completely capture all aspects of the problem.

One of the most important aspects discovered by observations is that many gas giants have large orbital eccentricity, sometimes up to e ~ 0.9. This is in contrast to Jupiter and Saturn in our solar system, which have eccentricities on the order of ~0.05. Gravitational scattering between gas giants is one of the most plausible mechanisms that can produce high eccentricities. Many N-body simulations of scattering of two or three gas giants have been carried out and they show a functional form of the eccentricity distribution that is consistent with observations (e.g., *Marzari and Weidenschilling*, 2000; *Ford et al.*, 2000; *Zhou et al.*, 2007; *Jurić and Tremaine*, 2008; *Chatterjee et al.*, 2008; *Ford and Rasio*, 2008). However, initial conditions for these simulations have been artificially specified and did not result from system formation calculations. Therefore, it remains to be seen if systems of giant planets compact enough to result in such scattering can be formed. Such information can only be obtained from population-synthesis simulations that follow the formation of truly interacting systems of planets (*Alibert et al.*, 2013; *Ida et al.*, 2013).

In Figs. 6a,c, e–M$_p$ and e–a distributions predicted by *Ida et al.* (2013) are compared with observed data. In the observed e–a distribution, close-in planets with a ≤ 0.1 AU generally have less-eccentric orbits than those with a ~ 1 AU. This correlation has been attributed to the orbital circularization of close-in planets (e.g., *Rasio and Ford*, 1996; *Dobbs-Dixon et al.*, 2004; *Jackson et al.*, 2008; *Matsumura et al.*, 2010; *Nagasawa and Ida*, 2011).

Although tidal effects have not been implemented, the e–a correlation is well established in the results obtained. The maximum eccentricity excited by close scattering between gas giants is $\sim v_{esc}/v_K$ (*Safronov and Zvjagina*, 1969). Since surface escape velocity of the giants v_{esc} is independent of a, whereas the Kepler velocity (v_K) is $\propto a^{-1/2}$, e of giants resulted by scattering should be $\propto a^{1/2}$. Thus, we obtain the maximum eccentricity at a ~ 0.1 AU. This is $3\times$ smaller than that at a ~ 1 AU in the simulated models, which well reproduces the observed e–a correlation (*Ida et al.*, 2013).

The actually observed e–M_p distribution shows that e increases with mass M_p. This correlation, which is totally counterintuitive, is also reproduced by the simulations. These show that multiple massive giants are preferentially formed in relatively massive disks and these systems are more prone to dynamical instabilities, orbit crossing, and excitation of high eccentricities. This trend is actually responsible for this correlation between e and M_p (*Ida et al.*, 2013). This correlation was also suggested by *Thommes et al.* (2008) and N-body simulations of giants with various masses (*Raymond et al.*, 2010), although their initial conditions were somewhat artificial.

Population-synthesis models also generate a population of massive gas giants ($M_p \gtrsim 10$ M_{Jup}) with large semimajor axes (a $\gtrsim 30$ AU) (Figs. 3, 4, and the bottom panels of Figs. 6b,c), which could correspond to planets discovered by direct imaging. In these results, the fraction of stars that host such planets is limited to a few percent and most of the planets have low eccentricity (e ≤ 0.1). If they are formed by scattering between gas giants, they should have large e. Close inspection of the results shows an alternative path for the formation of distant gas giant planets. In systems that contain a gas giant(s), the rapid gas accretion of the first generation of gas giant(s) destabilizes the orbits of nearby residual protoplanets and some protoplanets are scattered to large distances. Since planetesimal accretion rate is low at the large distance, some of the protoplanets start to accrete gas efficiently. The scattered protoplanets initially have high eccentric orbits and they take longer time to pass through their apoastrons, so that they tend to accrete gas from that region that has relatively large specific angular momentum. As a result, their orbits become circularized with a radius comparable to their apoastron radius. This path was already found by E. Thommes (personal communication, 2010) through a hybrid N-body and two-dimensional hydrodynamical simulation. Because even a single gas giant can scatter multiple protoplanets outward, we also found systems with multiple distant gas giants in nearly circular orbits.

Another important result of population-synthesis models has been the quantitative explanation of the so-called "metallicity effect" for gas giant planets. It has been well known for more than a decade that the probability of observing a giant planet orbiting a given star increases with the metallicity of the latter (e.g., *Gonzalez*, 1997; *Santos et al.*, 2001; *Fischer and Valenti*, 2005). Models by *Ida and Lin* (2005) and later by *Mordasini et al.* (2009b, 2012c) have quantitatively demonstrated that this metallicity effect

is a natural outcome of the core-accretion model. Indeed, in this model, the formation of a gas giant follows the initial building of a planetary core of ~ 10 $M_\oplus$ (the critical mass; see equation (26)). Although the critical core mass becomes smaller if the core has accreted most of planetesimals in its feeding zone, leading to a smaller rate of planetesimal accretion, envelope contraction occurs within a timescale of 1 m.y. only if the core mass is larger than several Earth masses (equation (28)). Such relatively massive cores are more easily formed in metal-rich disks, as explained below.

Around a metal-rich star, the total solid mass is larger for the same total disk mass as long as metallicity in star and disk are proportional. Then, the building of a planetary core by accretion of planetesimals is faster and the core's isolation mass is larger (e.g., *Kokubo and Ida*, 1998), assuming that the amount of planetesimals increases with the mass fraction of heavy elements, as indicated by planetesimal-formation models (e.g., *Brauer et al.*, 2008). Planets growing in such environments have a larger likelihood of reaching the critical mass before the gas disk has vanished, and are therefore more prone to becoming observable giant planets. Figure 7 presents the distribution of the metallicity of stars harboring a synthetic planet in a certain mass range, as found by *AMB* in the population presented in *Alibert et al.* (2013). As the initial condition, the total population of stars is assumed to have a metallicity distribution that follows the observed distribution in the solar neighborhood as described in section 3 (solid curve). It is clear that the [Fe/H] distribution of stars around which at least one synthetic giant planet forms (dashed curve) is shifted toward higher metallicities, which is the manifestation of the aforementioned "metallicity effect." On the other hand, the distribution of stars with Earth to super-Earth planets (dotted line) is very similar to that of all stars, showing that there is no metallicity effect for this kind of planet as observed by RV surveys (*Sousa et al.*, 2011; *Mayor et al.*, 2011).

Matsumura et al. (2013) showed that in systems with multiple gas giants, their secular perturbations often destabilize the orbits of Earths/super-Earths even if the Earths/super-Earths are in innermost regions far from the gas giants. *Ida et al.* (2013) showed that in disks with larger amounts of solids, gas giants often become dynamically active and only a few gas giants survive in the systems. These results suggest that survival of Earths/super-Earths is inhibited in metal-rich disks, which is consistent with the observation (*Mayor et al.*, 2011). In other words, coexistence of Earths/super-Earths and gas giants may be relatively rare.

5.3. Population Synthesis as an Exploratory Tool

Even though the current models of population synthesis do not account for all observed characteristics of the discovered exoplanet population, they can nevertheless be used as an exploratory tool for (1) inferring the combined effects of different processes, and identifying which processes are dominant in shaping the population of extrasolar planets;

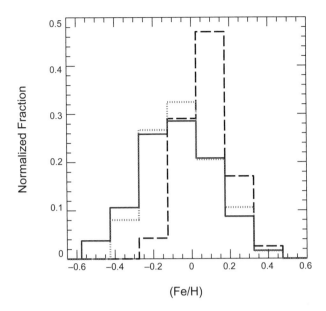

Fig. 7. Distribution of stellar metallicities in the synthetic population of *Alibert et al.* (2013). The solid line shows the hosts of all synthetic planetary systems. The dashed line are stars with at least one giant planet (M ≥ 100 M$_\oplus$) inside 1 AU, while the dotted line are stars with at least one low-mass planet with $1 \leq$ M/M$_\oplus \leq 10$. This plot can be compared to Fig. 16 in *Mayor et al.* (2011).

and (2) inferring the differential effects of different processes for populations of planets and planetary systems. To carry out this exploration, two sets of similar models are run by changing only one parameter (or by including/neglecting one process) at a time. We present here a few examples that illustrate the unique power of such an approach.

5.3.1. Type I migration. Since the publication of the first models of migration (*Goldreich and Tremaine*, 1980), it has been recognized that the very high migration rate derived from linear studies were hardly compatible with the existence of planets. In fact, the first models of type I migration predicted a very short migration timescale of ~10^4 yr for a core of ~10 M$_\oplus$ at a ~ 5 AU, a time actually much shorter than the growth timescale of the planetary core itself. More detailed later models predicted smaller migration rates (e.g., *Tanaka et al.*, 2002), on the order of the accretion timescale of planetesimals. However, the predicted migration timescale still remained much shorter than the observationally inferred typical disk lifetime of a few million years. The question arose as to the possibility for planets to form and survive given such short migration rates.

Integrated population-synthesis models, including the effects of type I migration, have shown that the formation and survival of planets was indeed possible even with these short migration rates (e.g., *Ida and Lin*, 2004a; *Alibert et al.*, 2005a). However, the statistics of the planet population obtained with these migration rates (*Tanaka et al.*, 2002) was quite different from that of the known planets. These studies also showed that the migration rate had to

be reduced by a factor of ~10^{-1-3} in order to obtain a good agreement between models and observations (*Ida and Lin*, 2008a; *Mordasini et al.*, 2009b; *Ida et al.*, 2013). Interestingly enough, the same migration rates [from *Tanaka et al.* (2002), reduced by a factor of 10–1000] was also required to match the internal structure and composition of Jupiter and Saturn, using integrated planet-formation models (*Alibert et al.*, 2005b).

Population-synthesis models have therefore demonstrated that there were some missing physical effects in the original prescription of type I migration rates, and that these effects should lead to a global reduction of the overall extent of inward migration of low-mass planets. These findings have helped motivate many investigations on planet-disk interactions. The missing physical effects were subsequently found to be related to the co-rotation torque in non-isothermal disks (see the chapter by Baruteau et al. in this volume).

The co-rotation torque can lead to outward migration provided non-isothermal effects are included. Then, the boundary between outward and inward migration regions, the so-called "convergence zones," can act as a migration trap in which high-mass planet formation is enhanced (e.g., *Lyra et al.*, 2010; *Mordasini et al.*, 2011a; *Kretke and Lin*, 2012). Other possible "migration traps" are the inner edge of the disk or dead zone (e.g., *Masset et al.*, 2006; *Ogihara et al.*, 2010), the ice line (e.g., *Kretke and Lin*, 2007), and the outer edge of the dead zone (e.g., *Matsumura et al.*, 2009; *Hasegawa and Pudritz*, 2012; *Regály et al.*, 2013). In order to understand how these migration traps are reflected by distributions of final planets, tests using population-synthesis simulations can be useful.

5.3.2. Type II migration. As discussed in section 2.6, the description for type II migration rate should be somewhat less uncertain than that for type I migration. More detailed prescriptions for the transition from type I to type II migration and from disk- to planet-dominated migration are nevertheless needed. However, while there is close agreement on these aspects, different assumptions regarding gas accretion across the gap can lead to significant differences in the end distribution of gas giants.

After the gap is formed, the disk gas accretion rate F can be divided in this region into three components: f_{mig}F, the component that pushes the planet (and the inner disk), f_pF, the fraction that is accreted by the planet, and f_{cross}F, the fraction that crosses the gap without being accreted by the planet. In an equilibrium, $f_{mig} + f_p + f_{cross} = 1$ ($0 < f_{mig}, f_p, f_{cross} < 1$), and the type II migration timescale in the planet-dominated regime is given by ~M$_p$ = (f_{mig}F) (*Hasegawa and Ida*, 2013). The current version of both *IL*'s and *AMB*'s prescriptions may be oversimplified because they do not take into account the mass flow across the gap.

For further progress, a prescription for f_{mig}, f_p, and f_{cross} as functions of the planet mass is needed. Several hydrodynamical simulations and analysis have studied these components (e.g., *D'Angelo et al.*, 2002; *Veras and Armitage*, 2004; *Lubow and D'Angelo*, 2006; *Dobbs-Dixon et al.,* 2007; *Alexander and Armitage,* 2007; *Alexander and Pascucci,* 2012). However, the three components have not been consistently given as functions of planet mass and disk parameters.

5.3.3. The effect of multiplicity. By computing two populations based on identical models but with one parameter changed at any one time, it is possible to assess the differential effect of some particular processes. To illustrate this, we consider the effect of the number of planets that grow within a given disk. Ultimately, this allows us to compare the formation of a single planet to the formation of a planetary system.

Models by *Ida and Lin* (2010) and *Ida et al.* (2013) have been the first to consider the emergence of multiple-planet systems in their population-synthesis studies. Using a novel approach that avoids the use of an N-body integrator to compute the orbital evolution of the system, they modeled the formation of systems starting with a very large number of growing seeds that grow, migrate, and potentially collide with each other. Their approach allows very efficient exploration of the parameter space (since the computation of a population is very rapid), at the expense of neglecting some subtle effects (e.g., low-order mean-motion resonances).

Models by *Alibert et al.* (2013) are based on a different approach, as they rely on standard N-body simulations. While this approach captures virtually all the potential dynamical effects, it comes at a significant computing cost. It is interesting to note that the two approaches are highly complementary; the first approximation allows us to explore extensively the parameter space, selecting the most interesting cases, whereas the second method can be used to study these interesting cases in more detail.

Both studies have shown that (1) the effect of multiplicity is important for low-mass planets, and (2) the characteristics of gas giant planets are less affected by the presence of multiple planets in a system (with an exception in the case of multiple long-period distant gas giants and dynamical interactions in more closely packed systems).

The effects on low-mass planets are: (1) close scattering (eccentricity excitation or ejection) and resonant trapping by large planets, and (2) merger events between low-mass planets. The above discussions indicate that close scattering by giant planets is very important for orbital evolution and survival of low-mass planets in systems with gas giants, in particular, in dynamically active systems.

The oligarchic growth model (*Kokubo and Ida,* 1998, 2002) predicts that in MMSN, the isolation mass of protoplanets is ~0.1–0.2 $M_\oplus$ at ~1 AU. While disk gas is present, planet-disk interactions suppress eccentricity of these protoplanets sufficiently to avoid orbit crossing. After disk gas depletion, they begin to undergo orbit crossing and collisional coalescence among themselves. N-body simulations showed that Earth-sized bodies are formed after multiple collisions (e.g., *Chambers and Wetherill,* 1998; *Agnor et al.,* 1999; *Kokubo et al.,* 2006). This implies that merging of low-mass planets after disk gas depletion can increase

planetary masses by an order of magnitude. In the earlier calculations of *IL,* this effect was partially included by the artificial expansion of feeding zones after disk gas depletion (*Ida and Lin,* 2004a). But this merging tendency is consistently included in their recent calculations, which directly take into account planet-planet interactions. As stated above, closely packed multiple super-Earths can be formed near the disk inner edge in a similar way (*Ida and Lin,* 2010). Even in the presence of disk gas, merging between low-mass planets occurs when feeding zones overlap due to growth of protoplanets (*Chambers,* 2006) or convergent migration between protoplanets occurs (*Ida et al.,* 2013). These effects enhance the formation of sufficiently massive cores to initiate runaway gas accretion.

5.3.4. Dependence on the mass of the host stars. Models by *Ida and Lin* (2005) and *Alibert et al.* (2011) have shown that the population of planets depends upon the mass of the central star. Because disk mass is generally smaller around less-massive stars, the frequency of cores reaching a sufficiently large mass to evolve into gas giants is lower in these systems. This is a natural consequence of the core-accretion model. In contrast, the disk-instability model does not necessarily imply such a trend. Thus, the correlation between the mass of the host stars and the fraction of stars with gas giants is a good test for discriminating between core-accretion and disk-instability models (for detailed discussions, see *Ida and Lin,* 2005).

Laughlin et al. (2004) and *Ida and Lin* (2005) predicted that gas giants are rare around M_{stars}, while Neptunes are rather abundant. This trend was confirmed later by RV surveys, although microlensing surveys suggest the possibility of an abundant population of gas giants around M dwarfs (*Gould et al.,* 2010). Around intermediate-mass stars, the fraction of stars with gas giants also increases with mass because there are more heavy elements in their disks to provide building blocks for cores of gas giant planets. Radial velocity surveys suggest that η_J increases with the stellar mass for M_* up to ~1.8 $M_\odot$ (*Johnson et al.,* 2010). These planets and their progenitor cores form preferentially beyond the ice line. Around massive stars with high stellar luminosity, the gas depletion process is rapid and the ice line is located at large disk radii, where the growth time for sufficiently massive cores is long. It has been suggested that the frequency of gas giants (η_J) may reach a maximum for some critical stellar mass and then decreases above this value (*Ida and Lin,* 2005; *Kennedy et al.,* 2007). Further observational determination of the η_J–M_* correlation in the high stellar mass limit will provide an important constraint for the theoretical models.

Recent RV surveys for GK clump giant stars (stars that were A dwarfs during the main-sequence phase) show that there may be a lack of giant planets inside 0.6 AU for stars with mass ≥1.5 $M_\odot$ (*Sato et al.,* 2010, and references therein). *Kunitomo et al.* (2011) suggested that tidal decay may not be responsible for the depletion for $M_* \geq 2 M_\odot$. A possible deficit of intermediate-period gas giants around

F dwarfs were addressed by population synthesis, taking into account the possible dependence of disk structure (*Kretke et al.*, 2009) or lifetime on stellar mass (*Burkert and Ida*, 2007; *Currie*, 2009; *Alibert et al.*, 2011). The period distribution of gas giants around A and F dwarfs is another important issue that should be addressed by population-synthesis simulation.

The characteristics of planetary systems are sensitively affected by the dependence of disk dynamical/thermal structure and its evolution on stellar mass. These disk properties will be revealed by observations by the Atacama Large Millimeter/submillimeter Array (ALMA). Population synthesis is an ideal tool for testing the effects of these disk properties on planet formation [see *Ida and Lin* (2005) and *Mordasini et al.* (2012c) for systematic studies of how disk properties translate into planetary properties].

5.3.5. The planetary mass-radius relationship. In the last few years, the study of exoplanets has gone beyond the discovery of data points in the mass-distance diagram. Thanks to various observational techniques complementary to RV surveys, namely transit, direct imaging, and spectroscopic observations, it has become possible to derive a characterization of exoplanets in terms of their basic physical properties such as mean density, intrinsic luminosity, and atmospheric composition. In particular, the planetary mass-radius relationship has emerged as a new observational constraint for formation theory, since it allows the fundamental geophysical classification of planets (rocky, ice-dominated, gas-dominated).

The population syntheses of *IL* and *AMB* predict not only the total mass of synthetic planets, but also their bulk composition (see Figs. 3 and 4). This bulk composition acquired during formation determines (for a given orbital distance and entropy for gas-dominated planets) the planetary radius (neglecting special evolutionary effects such as envelope heating and inflation for close-in planets). The planets' mass-radius distribution adds new constraints for population-synthesis models.

As an illustration, we first consider the bulk composition of close-in, low-mass planets. These planets have been found in large numbers both by RV surveys and by the Kepler mission. If these planets form inside the ice line, either approximately *in situ* (e.g., *Chiang and Laughlin*, 2013) or with migration only inside a_{ice} (as in Fig. 2), they would have rocky cores. Alternatively, if these planets (or their building blocks) form outside the ice line, and then migrated inward over large distances, they would mostly contain ices. If it is observationally possible to (statistically) distinguish these compositions, it would add strong constraints on type I migration models. Some complications may arise from the (partial) degeneracy of the mass-radius relationship (e.g., *Valencia et al.*, 2007). A second example is the abundant low-mass, low-density planets [e.g., around Kepler-11 (*Lissauer et al.*, 2013)]. These low-mass planets seem to have accreted significant amounts of H/He. The efficiency with which a core can accrete gas during the nebular phase depends on the opacity due to grains in the

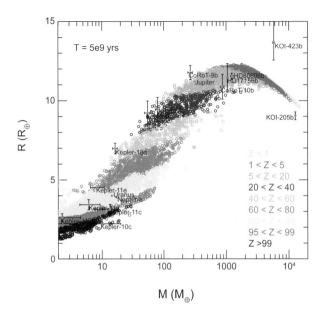

Fig. 8. Synthetic mass-radius relationship of planets with primordial H/He envelopes at an age of 5 G.y. compared to solar system planets and exoplanets with a semimajor axis of at least 0.1 AU. The mass fraction of heavy elements Z (in percent) is given by the scale shown in the plot. Adapted from *Mordasini et al.* (2012b).

protoplanetary atmosphere. Theoretical grain-evolution models (*Podolak*, 2003; *Movshovitz and Podolak*, 2008) predict low-grain opacities, allowing low-mass cores to accrete considerable H/He, resulting in large radii.

In order to take advantage of this new information, *AMB* expanded their original formation model (*Alibert et al.*, 2005a) into a self-consistently coupled formation and evolution model (for details, see *Mordasini et al.*, 2012a) that predicts on a population level planetary radii (and luminosities). Figure 8 (from *Mordasini et al.*, 2012b) shows the mass-radius relationship for synthetic planets around solar-like stars at an age of 5 G.y. The mass-radius distribution has a typical S-shape, with large (in term of radius) planets absent for small masses, and small planets absent in the high-mass domain. It is also clear that there are correlations between the composition and radius and between composition and mass (gas-dominated planets do not exist in the low-mass domain). These correlations are a natural consequence of the core-accretion mechanism (and of the EOS of different materials): Low-mass cores have long KH timescales for envelope accretion (equation (28)), and therefore remain solid-dominated and small, leaving the upper left part of the figure empty. On the other hand, the lower right part of the figure remains empty because massive, supercritical cores must accreted massive H/He envelopes (at least if they form during the presence of the nebula), so that their radius is large. As can be seen in Fig. 8, all the observations can be relatively well reproduced by theoretical models. However, this good match depends on the assumed grain opacity in planetary envelopes during their formation. By comparing the observed mass-radius relationship with the synthetic models obtained with different grain opacities, it becomes possible to observationally constrain this important quantity.

6. CONCLUSIONS AND OUTLOOK

As demonstrated in this chapter, planet population synthesis is based on a physical description of a large number of processes playing a role in the formation of planets. By necessity, this description is simplified in order to keep the problem tractable and to allow the simulation of the formation of planets for a large set of initial conditions. We have presented the two different approaches that have been most used in the literature. These two sets of prescriptions differ essentially by the degree of simplification being adopted and/or the use of fitting formulas stemming from more complete and detailed calculations of individual processes. While differing in these aspects, both approaches aim at developing a self-consistent model in which the relevant processes operate on the proper timescale. This is necessary to account for the numerous feedbacks occurring between the various processes (see Fig. 1).

A detailed physical understanding of the many aspects of planet formation, from the early condensation of solids and the formation of planetesimals to the runaway accretion of gas in the late stages of giant planet formation taking place within a time-evolving protoplanetary disk, is essential. As illustrated in the relevant chapters of this volume, some aspects can be studied observationally and some only theoretically, essentially by means of large-scale numerical simulations of growing physical accuracy. While all these efforts are essential to our understanding, they are all restricted in either spatial or temporal dimensions.

The goal of population synthesis is to put these space and/or time snapshots together in order to obtain a full picture and a complete history of planet formation from the initial protoplanetary disk to the observed planetary systems. This is essential for at least two reasons: (1) planet formation is not directly observed, only initial conditions (protoplanetary disks) and end products (planetary systems) are — the link between the two must be provided by theory at least for now; and (2) the diversity of characteristics of the ensemble of exoplanets sets statistical constraints on formation models.

In the end, population synthesis is similar to idealized models developed by theorists in an attempt to understand the basic behavior of complex physical systems. The success of the approach does not lie in having all aspects described in detail but to include the key physical processes and their respective feedback. Hence, population synthesis relies on detailed numerical simulations to provide sufficient understanding to allow the derivation of simplified descriptions that capture the essence of the phenomenon. In return, the approach allows visualizing the effect on the ensemble population of planets of a given physical description of individual processes (e.g., disk structure, migration, opacity), which when compared to observation, allows setting constraints on the models.

We have shown in this chapter that this approach has been useful in the past to identify key problems in the theoretical descriptions of key processes active in planet formation (e.g., planetary migration). We also have pointed out a number of areas where further detailed modeling is needed. Among those, we highlight (1) disk structure and evolution and the associated transport of gas and solids; (2) the formation of planetesimals and the resulting chemical composition and size distribution as a function of distance to the star; (3) the fact that migration is still too fast, even taking into account the non-isothermal effects associated with the co-rotation torques; and (4) the fact that the gas flow through the gap is still not clearly established for the accurate determination of the asymptotic masses of gas giant planets. The predictive power of population synthesis will rest on the progress that will be achieved in the future in the physical understanding of these processes.

Acknowledgments. W.B. would like to acknowledge partial support from the Swiss National Science Foundation. S.I. acknowledges support from the JSPS grant. C.M. acknowledges support from the Max Planck Society through the Reimar-Lüst Fellowship. Y.A. acknowledges the support of the European Research Council under grant 239605. The authors have been supported by the International Space Science Institute (ISSI), in the framework of an ISSI Team. D.L. acknowledges support from NASA, the National Space Foundation (NSF), and University of California (UC)/Lab fee grants.

REFERENCES

Aarseth S. J. et al. (1993) *Astrophys. J., 403,* 351.
Adachi I. et al. (1976) *Progr. Theor. Phys., 56,* 1756.
Agnor C. B. et al. (1999) *Icarus, 142,* 219.
Alexander R. D. and Armitage P. J. (2007) *Mon. Not. R. Astron. Soc., 375,* 500.
Alexander R. D. and Pascucci I. (2012) *Mon. Not. R. Astron. Soc., 422,* L82.
Alibert Y. et al. (2005a) *Astron. Astrophys., 434,* 343.
Alibert Y. et al. (2005b) *Astrophys. J. Lett., 626,* L57.
Alibert Y. et al. (2011) *Astron. Astrophys., 526,* A63.
Alibert Y. et al. (2013) *Astron. Astrophys., 558,* A109.
Andrews S. M. et al. (2010) *Astrophys. J., 723,* 1241.
Baraffe I. et al. (2003) *Astron. Astrophys., 402,* 701.
Baraffe I. et al. (2012) In *New Horizons in Time Domain Astronomy* (E. Griffin et al., eds.), pp. 138–138. IAU Symp. 285, Cambridge Univ., Cambridge.
Bell K. R. and Lin D. N. C. (1994) *Astrophys. J., 427,* 987.
Bodenheimer P. and Pollack J. B. (1986) *Icarus, 67,* 391.
Bodenheimer P. et al. (2000) *Icarus, 143,* 2.
Bodenheimer P. et al. (2013) *Astrophys. J., 770,* 120.
Borucki W. J. et al. (2011) *Astrophys. J., 736,* 19.
Brauer F. et al. (2008) *Astron. Astrophys., 480,* 859.
Bromley B. C. and Kenyon S. J. (2006) *Astron. J., 131,* 2737.
Bromley B. C. and Kenyon S. J. (2011) *Astrophys. J., 731,* 101.
Burkert A. and Ida S. (2007) *Astrophys. J., 660,* 845.
Burrows A. et al. (1997) *Astrophys. J., 491,* 856.
Chambers J. (2006) *Icarus, 180,* 496.
Chambers J. E. and Wetherill G. W. (1998) *Icarus, 136,* 304.
Chambers J. E. et al. (1996) *Icarus, 119,* 261.
Chatterjee S. et al. (2008) *Astrophys. J., 686,* 580.
Chiang E. and Laughlin G. (2013) *Mon. Not. R. Astron. Soc., 431,* 3444.

Chiang E. I. and Goldreich P. (1997) *Astrophys. J., 490*, 368.
Clarke C. J. et al. (2001) *Mon. Not. R. Astron. Soc., 328*, 485.
Crida A. et al. (2006) *Icarus, 181*, 587.
Currie T. (2009) *Astrophys. J. Lett., 694*, L171.
D'Angelo G. et al. (2002) *Astron. Astrophys., 385*, 647.
Dobbs-Dixon I. et al. (2004) *Astrophys. J., 610*, 464.
Dobbs-Dixon I. et al. (2007) *Astrophys. J., 660*, 791.
Fischer D. A. and Valenti J. (2005) *Astrophys. J., 622*, 1102.
Ford E. B. and Rasio F. A. (2008) *Astrophys. J., 686*, 621.
Ford E. B. et al. (2000) *Astrophys. J., 535*, 385.
Fortier A. et al. (2013) *Astron. Astrophys., 549*, A44.
Fortney J. J. et al. (2013) *Astrophys. J., 775*, 80.
Fouchet L. et al. (2012) *Astron. Astrophys., 540*, A107.
Freedman R. S. et al. (2008) *Astrophys. J. Suppl., 174*, 504.
Garaud P. and Lin D. N. C. (2007) *Astrophys. J., 654*, 606.
Goldreich P. and Tremaine S. (1980) *Astrophys. J., 241*, 425.
Gonzalez G. (1997) *Mon. Not. R. Astron. Soc., 285*, 403.
Gould A. et al. (2010) *Astrophys. J., 720*, 1073.
Haisch K. E. Jr. et al. (2001) *Astrophys. J. Lett., 553*, L153.
Hasegawa Y. and Ida S. (2013) *Astrophys. J., 774*, 146.
Hasegawa Y. and Pudritz R. E. (2012) *Astrophys. J., 760*, 117.
Hayashi C. (1981) *Progr. Theor. Phys. Suppl., 70*, 35.
Hellary P. and Nelson R. P. (2012) *Mon. Not. R. Astron. Soc., 419*, 2737.
Helled R. et al. (2008) *Icarus, 195*, 863.
Hollenbach D. et al. (1994) *Astrophys. J., 428*, 654.
Hori Y. and Ikoma M. (2010) *Astrophys. J., 714*, 1343.
Hori Y. and Ikoma M. (2011) *Mon. Not. R. Astron. Soc., 416*, 1419.
Howard A. W. et al. (2010) *Science, 330*, 653.
Hubickyj O. et al. (2005) *Icarus, 179*, 415.
Hueso R. and Guillot T. (2005) *Astron. Astrophys., 442*, 703.
Ida S. (1990) *Icarus, 88*, 129.
Ida S. and Lin D. N. C. (2004a) *Astrophys. J., 604*, 388.
Ida S. and Lin D. N. C. (2004b) *Astrophys. J., 616*, 567.
Ida S. and Lin D. N. C. (2005) *Astrophys. J., 626*, 1045.
Ida S. and Lin D. N. C. (2008a) *Astrophys. J., 673*, 487.
Ida S. and Lin D. N. C. (2008b) *Astrophys. J., 685*, 584.
Ida S. and Lin D. N. C. (2010) *Astrophys. J., 719*, 810.
Ida S. and Makino J. (1992) *Icarus, 96*, 107.
Ida S. and Makino J. (1993) *Icarus, 106*, 210.
Ida S. and Nakazawa K. (1989) *Astron. Astrophys., 224*, 303.
Ida S. et al. (2013) *Astrophys. J., 775*, 42.
Ikoma M. et al. (2000) *Astrophys. J., 537*, 1013.
Ikoma M. et al. (2001) *Astrophys. J., 553*, 999.
Inaba S. et al. (2001) *Icarus, 149*, 235.
Jackson B. et al. (2008) *Mon. Not. R. Astron. Soc., 391*, 237.
Johnson J. A. et al. (2010) *Publ. Astron. Soc. Pac., 122*, 905.
Jurić M. and Tremaine S. (2008) *Astrophys. J., 686*, 603.
Kennedy G. M. et al. (2007) *Astrophys. Space Sci., 311*, 9.
Kippenhahn R. and Weigert A. (1994) *Stellar Structure and Evolution.*
    Astronomy and Astrophysics Library, Springer-Verlag, Berlin. 484 pp.
Klahr H. and Kley W. (2006) *Astron. Astrophys., 445*, 747.
Kley W. and Dirksen G. (2006) *Astron. Astrophys., 447*, 369.
Kobayashi H. et al. (2012) *Astrophys. J., 756*, 70.
Kokubo E. and Ida S. (1998) *Icarus, 131*, 171.
Kokubo E. and Ida S. (2002) *Astrophys. J., 581*, 666.
Kokubo E. et al. (2006) *Astrophys. J., 642*, 1131.
Kretke K. A. and Lin D. N. C. (2007) *Astrophys. J. Lett., 664*, L55.
Kretke K. A. and Lin D. N. C. (2012) *Astrophys. J., 755*, 74.
Kretke K. A. et al. (2009) *Astrophys. J., 690*, 407.
Kunitomo M. et al. (2011) *Astrophys. J., 737*, 66.
Laughlin G. et al. (2004) *Astrophys. J. Lett., 612*, L73.
Leconte J. and Chabrier G. (2012) *Astron. Astrophys., 540*, A20.
Lin D. N. C. and Papaloizou J. (1986) *Astrophys. J., 309*, 846.
Lin D. N. C. and Papaloizou J. C. B. (1993) In *Protostars and Planets III*
    (E. H. Levy et al., eds.), pp. 749–835. Univ. of Arizona, Tucson.
Lissauer J. J. et al. (2009) *Icarus, 199*, 338.
Lissauer J. J. et al. (2013) *Astrophys. J., 770*, 131.
Lubow S. H. and D'Angelo G. (2006) *Astrophys. J., 641*, 526.
Lynden-Bell D. and Pringle J. E. (1974) *Mon. Not. R. Astron. Soc., 168*, 603.

Lyra W. et al. (2010) *Astrophys. J. Lett., 715*, L68.
Marley M. S. et al. (2007) *Astrophys. J., 655*, 541.
Marzari F. and Weidenschilling S. J. (2000) *Bull. Am. Astron. Soc., 32*, 1099.
Masset F. S. and Casoli J. (2010) *Astrophys. J., 723*, 1393.
Masset F. S. et al. (2006) *Astrophys. J., 642*, 478.
Matsumura S. et al. (2009) *Astrophys. J., 691*, 1764.
Matsumura S. et al. (2010) *Astrophys. J., 725*, 1995.
Matsumura S. et al. (2013) *Astrophys. J., 767*, 129.
Matsuyama I. et al. (2003) *Astrophys. J., 582*, 893.
Mayor M. et al. (2011) *ArXiv e-prints*, arXiv:1109.2497.
Miguel Y. et al. (2011a) *Mon. Not. R. Astron. Soc., 417*, 314.
Miguel Y. et al. (2011b) *Mon. Not. R. Astron. Soc., 412*, 2113.
Mollière P. and Mordasini C. (2012) *Astron. Astrophys., 547*, A105.
Mordasini C. et al. (2006) In *Tenth Anniversary of 51 Peg-b: Status of and Prospects for Hot Jupiter Studies* (L. Arnold et al., eds.), pp. 84–86. Frontier Group, Paris.
Mordasini C. et al. (2009a) *Astron. Astrophys., 501*, 1139.
Mordasini C. et al. (2009b) *Astron. Astrophys., 501*, 1161.
Mordasini C. et al. (2011a) In *The Astrophysics of Planetary Systems: Formation, Structure, and Dynamical Evolution* (A. Sozzetti et al., eds.), pp. 72–75. IAU Symp. 276, Cambridge Univ., Cambridge.
Mordasini C. et al. (2011b) *Astron. Astrophys., 526*, A111.
Mordasini C. et al. (2012a) *Astron. Astrophys., 547*, A111.
Mordasini C. et al. (2012b) *Astron. Astrophys., 547*, A112.
Mordasini C. et al. (2012c) *Astron. Astrophys., 541*, A97.
Movshovitz N. and Podolak M. (2008) *Icarus, 194*, 368.
Nagasawa M. and Ida S. (2011) *Astrophys. J., 742*, 72.
Nagasawa M. et al. (2007) In *Protostars and Planets V* (B. Reipurth et al., eds.), pp. 639–654. Univ. of Arizona, Tucson.
Ogihara M. et al. (2010) *Astrophys. J., 721*, 1184.
Ohtsuki K. et al. (2002) *Icarus, 155*, 436.
Paardekooper S.-J. and Mellema G. (2006) *Astron. Astrophys., 459*, L17.
Paardekooper S.-J. and Papaloizou J. C. B. (2009) *Mon. Not. R. Astron. Soc., 394*, 2283.
Paardekooper S.-J. et al. (2011) *Mon. Not. R. Astron. Soc., 410*, 293.
Palmer P. L. et al. (1993) *Astrophys. J., 403*, 336.
Papaloizou J. C. B. and Nelson R. P. (2005) *Astron. Astrophys., 433*, 247.
Papaloizou J. C. B. and Terquem C. (1999) *Astrophys. J., 521*, 823.
Payne M. J. and Lodato G. (2007) *Mon. Not. R. Astron. Soc., 381*, 1597.
Podolak M. (2003) *Icarus, 165*, 428.
Podolak M. et al. (1988) *Icarus, 73*, 163.
Pollack J. B. et al. (1996) *Icarus, 124*, 62.
Rasio F. A. and Ford E. B. (1996) *Science, 274*, 954.
Raymond S. N. et al. (2010) *Astrophys. J., 711*, 772.
Regály Z. et al. (2013) *Mon. Not. R. Astron. Soc., 433*, 2626.
Richardson D. C. et al. (2000) *Icarus, 143*, 45.
Ruden S. P. and Lin D. N. C. (1986) *Astrophys. J., 308*, 883.
Safronov V. S. and Zvjagina E. V. (1969) *Icarus, 10*, 109.
Santos N. C. et al. (2001) *Astron. Astrophys., 373*, 1019.
Sato B. et al. (2010) *Publ. Astron. Soc. Japan, 62*, 1063.
Saumon D. et al. (1995) *Astrophys. J. Suppl., 99*, 713.
Seager S. et al. (2007) *Astrophys. J., 669*, 1279.
Shakura N. I. and Sunyaev R. A. (1973) *Astron. Astrophys., 24*, 337.
Sousa S. G. et al. (2011) *Astron. Astrophys., 533*, A141.
Spiegel D. S. and Burrows A. (2012) *Astrophys. J., 745*, 174.
Spiegel D. S. et al. (2011) *Astrophys. J., 727*, 57.
Stahler S. W. et al. (1980) *Astrophys. J., 241*, 637.
Tanaka H. and Ward W. R. (2004) *Astrophys. J., 602*, 388.
Tanaka H. et al. (2002) *Astrophys. J., 565*, 1257.
Thommes E. et al. (2008) *Astrophys. J., 676*, 728.
Udry S. et al. (2000) *Astron. Astrophys., 356*, 590.
Valencia D. et al. (2007) *Astrophys. J., 665*, 1413.
Veras D. and Armitage P. J. (2004) *Mon. Not. R. Astron. Soc., 347*, 613.
Ward W. R. (1997) *Astrophys. J. Lett., 482*, L211.
Zhou J.-L. and Lin D. N. C. (2007) *Astrophys. J., 666*, 447.
Zhou J.-L. et al. (2007) *Astrophys. J., 666*, 423.

Fischer D. A., Howard A. W., Laughlin G. P., Macintosh B., Mahadevan S., Sahlmann J., and Yee J. C. (2014) Exoplanet detection techniques. In *Protostars and Planets VI* (H. Beuther et al., eds.), pp. 715–737. Univ. of Arizona, Tucson, DOI: 10.2458/azu_uapress_9780816531240-ch031.

Exoplanet Detection Techniques

Debra A. Fischer
Yale University

Andrew W. Howard
University of Hawai'i at Manoa

Greg P. Laughlin
University of California at Santa Cruz

Bruce Macintosh
Stanford University

Suvrath Mahadevan
The Pennsylvania State University

Johannes Sahlmann
Université de Genève

Jennifer C. Yee
Harvard-Smithsonian Center for Astrophysics

We are still in the early days of exoplanet discovery. Astronomers are beginning to model the atmospheres and interiors of exoplanets and have developed a deeper understanding of processes of planet formation and evolution. However, we have yet to map out the full complexity of multi-planet architectures or to detect Earth analogs around nearby stars. Reaching these ambitious goals will require further improvements in instrumentation and new analysis tools. In this chapter, we provide an overview of five observational techniques that are currently employed in the detection of exoplanets: optical and infrared (IR) Doppler measurements, transit photometry, direct imaging, microlensing, and astrometry. We provide a basic description of how each of these techniques works and discuss forefront developments that will result in new discoveries. We also highlight the observational limitations and synergies of each method and their connections to future space missions.

1. INTRODUCTION

Humans have long wondered whether other solar systems exist around the billions of stars in our galaxy. In the past two decades, we have progressed from a sample of one to a collection of hundreds of exoplanetary systems. Instead of an orderly solar nebula model, we now realize that chaos rules the formation of planetary systems. Gas giant planets can migrate close to their stars. Small rocky planets are abundant and dynamically pack the inner orbits. Planets circle outside the orbits of binary star systems. The diversity is astonishing.

Several methods for detecting exoplanets have been developed: Doppler measurements, transit observations, microlensing, astrometry, and direct imaging. Clever in-novations have advanced the precision for each of these techniques; however, each of the methods have inherent observational incompleteness. The lens through which we detect exoplanetary systems biases the parameter space that we can see. For example, Doppler and transit techniques preferentially detect planets that orbit closer to their host stars and are larger in mass or size, while microlensing, astrometry, and direct imaging are more sensitive to planets in wider orbits. In principle, the techniques are complementary; in practice, they are not generally applied to the same sample of stars, so our detection of exoplanet architectures has been piecemeal. The explored parameter space of exoplanet systems is a patchwork quilt that still has several missing squares.

2. THE DOPPLER TECHNIQUE

2.1. Historical Perspective

The first Doppler-detected planets were met with skepticism. *Campbell et al.* (1988) identified variations in the residual velocities of γ Ceph, a component of a binary star system, but attributed them to stellar activity signals until additional data confirmed it as a planet 15 years later (*Hatzes et al., 2003*). *Latham et al.* (1989) detected a Doppler signal around HD 114762 with an orbital period of 84 d and a mass $M_P \sin i = 11\ M_{Jup}$. Since the orbital inclination was unknown, they expected that the mass could be significantly larger and interpreted their data as a probable brown dwarf. When *Mayor and Queloz* (1995) modeled a Doppler signal in their data for the Sun-like star, 51 Pegasi, as a Jupiter-mass planet in a 4.23-d orbit, astronomers wondered if this could be a previously unknown mode of stellar oscillations (*Gray,* 1997) or nonradial pulsations (*Hatzes et al., 1997*). The unexpected detection of significant eccentricity in exoplanet candidates further raised doubts among astronomers, who argued that although stars existed in eccentric orbits, planets should reside in circular orbits (*Black,* 1997). It was not until the first transiting planet (*Henry et al.,* 2000; *Charbonneau et al.,* 2000) and the first multi-planet system (*Butler et al.,* 1999) were detected (almost back-to-back) that the planet interpretation of the Doppler velocity data was almost unanimously accepted.

The Doppler precision improved from about 10 m s⁻¹ in 1995 to 3 m s⁻¹ in 1998, and then to about 1 m s⁻¹ in 2005 when the High Accuracy Radial velocity Planet Searcher (HARPS) was commissioned (*Mayor et al.,* 2003). A Doppler precision of 1 m s⁻¹ corresponds to shifts of stellar lines across 1/1000th of a CCD pixel. This is a challenging measurement that requires high signal-to-noise (S/N), high resolution, and large spectral coverage. Echelle spectrometers typically provide these attributes and have served as the workhorse instruments for Doppler planet searches.

Figure 1 shows the detection history for planets identified with Doppler surveys (planets that also are observed to transit their host star are shown in light gray). The first planets were similar in mass to Jupiter and there has been a striking decline in the lower envelope of detected planet mass with time as instrumentation improved.

2.2. Radial Velocity Measurements

The Doppler technique measures the reflex velocity that an orbiting planet induces on a star. Because the star-planet interaction is mediated by gravity, more massive planets result in larger and more easily detected stellar velocity amplitudes. It is also easier to detect close-in planets, both because the gravitational force increases with the square of the distance and because the orbital periods are shorter and therefore more quickly detected. *Lovis and Fischer* (2011) provide a detailed discussion of the technical aspects of Doppler analysis with both an iodine cell and a thorium-argon simultaneous reference source.

The radial velocity semi-amplitude, K_1, of the star can be expressed in units of cm s⁻¹ with the planet mass in units of $M_\oplus$

$$K_* = \frac{8.95\ \text{cm s}^{-1}}{\sqrt{1-e^2}} \frac{M_P \sin i}{M_\oplus} \left(\frac{M_* + M_P}{M_\odot}\right)^{-2/3} \left(\frac{P}{\text{yr}}\right)^{-1/3} \quad (1)$$

The observed parameters (velocity semi-amplitude K_*, orbital period P, eccentricity e, and orientation angle ω) are used to calculate a minimum mass of the planet $M_P \sin i$ if the mass of the star M_* is known. The true mass of the planet is unknown because it is modulated by the unknown inclination. For example, if the orbital inclination is 30°, the true mass is a factor of 2× the Doppler-derived $M_P \sin i$. The statistical probability that the orbit inclination is within an arbitrary range $i_1 < I < i_2$ is given by

$$\mathcal{P}_{incl} = |\cos(i_2) - \cos(i_1)| \quad (2)$$

Thus, there is a roughly 87% probability that random orbital inclinations are between 30° and 90°, or equivalently, an 87% probability that the true mass is within a factor of 2 of the minimum mass $M_P \sin i$.

Radial velocity observations must cover one complete orbit in order to robustly measure the orbital period. As a result, the first detected exoplanets resided in short-period orbits. Doppler surveys that have continued for a decade or more (*Fischer et al.,* 2014; *Marmier et al.,* 2013) have been able to detect gas giant planets in Jupiter-like orbits.

2.3. The Floor of the Doppler Precision

An important question is whether the Doppler technique can be further improved to detect smaller planets at wider orbital radii. The number of exoplanets detected each year rose steadily until 2011 and has dropped precipitously after that year. This is due in part to the fact that significant telescope time has been dedicated to transit follow-up and also because observers are working to extract the smallest possible planets, requiring more Doppler measurement points given current precision. Further gains in Doppler precision and productivity will require new instruments with greater stability as well as analytical techniques for decorrelating stellar noise.

Figure 2, reproduced from *Pepe et al.* (2011), shows an example of one of the lowest-amplitude exoplanets, detected with HARPS. The velocity semi-amplitude for this planet is K = 0.769 m s⁻¹ and the orbital period is 58.43 d. The data comprised 185 observations spanning 7.5 yr. The residual velocity scatter after fitting for the planet was reported to be 0.77 m s⁻¹, showing that high precision can be achieved with many data points to beat down the single measurement precision.

One promising result suggests that it may be possible for stable spectrometers to average over stellar noise signals

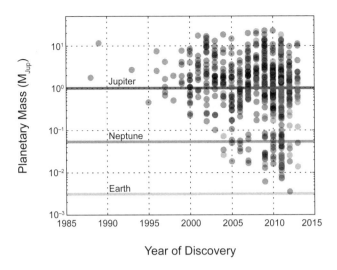

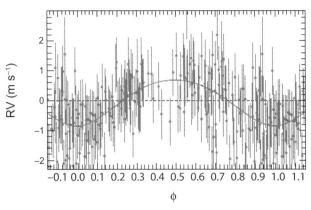

Fig. 2. The phase-folded data for the detection of a planet orbiting HD 85512. From Fig. 13 of *Pepe et al.* (2011).

Fig. 1. Planet mass is plotted as a function of the year of discovery. The color coding is black for planets with no known transit, whereas light gray is planets that do transit.

and reach precisions below 0.5 m s^{-1}, at least for some stars. After fitting for three planets in HD 20794, *Pepe et al.* (2011) found that the RMS of the residual velocities decreased from 0.8 m s^{-1} to 0.2 m s^{-1} as they binned the data in intervals from 1 to 40 nights. Indeed, a year later, the High Accuracy Radial velocity Planet Searcher (HARPS) team published the smallest-velocity signal ever detected: a planet candidate that orbits α Centauri B (*Dumusque et al.,* 2012) with a velocity amplitude K = 0.51 m s^{-1}, planet mass M sin i = 1.13 M$_{\oplus}$, and an orbital period of 3.24 d. This detection required 469 Doppler measurements obtained over seven years and fit for several time-variable stellar noise signals. Thus, the number of observations required to solve for the five-parameter Keplerian model increases exponentially with decreasing velocity amplitude.

2.4. The Future of Doppler Detections

It is worth pondering whether improved instruments with higher resolution, higher sampling, greater stability, and more precise wavelength calibration will ultimately be able to detect analogs of Earth with 0.1-m s^{-1} velocity amplitudes. An extreme precision spectrometer will have stringent environmental requirements to control temperature, pressure, and vibrations. The dual requirements of high resolution and high S/N lead to the need for moderate- to large-aperture telescopes (*Strassmeier et al.,* 2008; *Spanò et al.,* 2012). The coupling of light into the instrument must be exquisitely stable. This can be achieved with a double fiber scrambler (*Hunter and Ramsey,* 1992) where the near field of the input fiber is mapped to the far field of the output fiber, providing a high level of scrambling in both the radial and azimuthal directions. At some cost to throughput, the double fiber scrambler stabilizes variations in the spectral line spread function (sometimes called a point

spread function) and produces a series of spectra that are uniform except for photon noise. Although the fibers provide superior illumination of the spectrometer optics, some additional care in the instrument design phase is required to provide excellent flat fielding and sky subtraction. The list of challenges to extreme instrumental precision also includes the optical CCD detectors, with intrapixel quantum efficiency variations, tiny variations in pixel sizes, charge diffusion, and the need for precise controller software to perfectly clock readout of the detector.

In addition to the instrumental precision, another challenge to high Doppler precision is the star itself. Stellar activity, including star spots, p-mode oscillations, and variable granulation, are tied to changes in the strength of stellar magnetic fields. These stellar noise sources are sometimes called stellar jitter and can produce line profile variations that skew the center of mass for a spectral line in a way that is (mis)interpreted by a Doppler code as a velocity change in the star. Although stellar noise signals are subtle, they affect the spectrum in a different way than dynamical velocities. The stellar noise typically has a color dependence and an asymmetric velocity component. In order to reach significantly higher accuracy in velocity measurements, it is likely that we will need to identify and model or decorrelate the stellar noise.

3. INFRARED SPECTROSCOPY

3.1. Doppler Radial Velocities in the Near Infrared

The high fraction of Earth-sized planets estimated to orbit in the habitable zones (HZs) of M dwarfs (*Dressing and Charbonneau,* 2013; *Kopparapu,* 2013; *Bonfils et al.,* 2013) makes the low mass stars very attractive targets for Doppler radial velocity (RV) surveys. The lower stellar mass of the M dwarfs, as well as the short orbital periods of HZ planets, increases the amplitude of the Doppler wobble (and the ease of its detectability) caused by such a terrestrial-mass planet. However, nearly all the stars in current optical RV surveys are earlier in spectral type than ~M5, since later spectral types are difficult targets even on large telescopes due to their intrinsic faintness in the optical: They emit most of their flux in the red optical and near-infrared (NIR) between 0.8 and 1.8 μm (the NIR Y, J, and H bands are 0.98–1.1 μm, 1.1–1.4 μm, and 1.45–1.8 μm). However, it is the low-mass late-type M stars, which are the least luminous, where the velocity amplitude of a terrestrial planet in the habitable zone is highest, making them very desirable targets. Since the flux distribution from M stars peaks sharply in the NIR, stable high-resolution NIR spectrographs capable of delivering high RV precision can observe several hundred of the nearest M dwarfs to examine their planet population.

3.1.1. Fiber-fed near-infrared high-resolution spectrographs. A number of new fiber-fed stabilized spectrographs are now being designed and built for such a purpose: the Habitable Zone Planet finder (*Mahadevan et al.,* 2012) for the 10-m Hobby Eberly Telescope, the Calar Alto high-Resolution search for M dwarfs with Exo-earths with Near-infrared and Visible Echelle Spectrographs (CARMENES) (*Quirrenbach et al.,* 2012) for the 3.6-m Calar Alto Telescope, and SpectroPolarimètre Infra-Rouge (SPIRou) (*Santerne et al.,* 2013) being considered for the Canada-France-Hawaii Telescope (CFHT). The instrumental challenges in the NIR, compared to the optical, are calibration, stable cold operating temperatures of the instrument, and the need to use NIR detectors. The calibration issues seem tractable (see below). Detection of light beyond 1 μm required the use of NIR sensitive detectors like the Hawaii-2 (or 4) RG HgCdTe detectors. These devices are fundamentally different than charge-coupled devices (CCDs) and exhibit effects such as interpixel capacitance and much greater persistence. Initial concerns about the ability to perform precision RV measurements with these devices has largely been retired with laboratory (*Ramsey et al.,* 2008) and on-sky demonstrations (*Ycas et al.,* 2012b) with a Pathfinder spectrograph, although careful attention to ameliorating these effects is still necessary to achieve high RV precision. This upcoming generation of spectrographs, being built to deliver 1–3-m s⁻¹ RV precision in the NIR, will also be able to confirm many of the planets detected with the Transiting Exoplanet Survey Satellite (TESS) and Gaia around low-mass stars. Near-infrared spectroscopy is also an essential tool to be able to discriminate between giant planets and stellar activity in the search for planets around young active stars (*Mahmud et al.,* 2011).

3.1.2. Calibration sources. Unlike iodine in the optical, no single known gas cell simultaneously covers large parts of the NIR z, Y, J, and H bands. Thorium-argon lamps, which are so successfully used in the optical, have very few thorium emission lines in the NIR, making them unsuitable as the calibrator of choice in this wavelength regime. Uranium has been shown to provide a significant increase in the number of lines available for precision wavelength calibration in the NIR. New linelists have been published for uranium lamps (*Redman et al.,* 2011, 2012) and these lamps are now in use in existing and newly commissioned

NIR spectrographs. Laser frequency combs, which offer the prospects of very high precision and accuracy in wavelength calibration, have also been demonstrated with astronomical spectrographs in the NIR (*Ycas et al.,* 2012b), with filtering making them suitable for an astronomical spectrograph. A generation of combs spanning the entire z–H-band region has also been demonstrated in the laboratory (*Ycas et al.,* 2012a). Continuing development efforts are aimed at effectively integrating these combs as calibration sources for M-dwarf Doppler surveys with stabilized NIR spectrographs. Single mode fiber-based Fabry-Pérot cavities fed by supercontinuum light sources have also been demonstrated by *Halverson et al.* (2012). To most astronomical spectrographs the output from these devices looks similar to that of a laser comb, although the frequency of the emission peaks is not known innately to high precision. Such inexpensive and rugged devices may soon be available for most NIR spectrographs, with the superior (and more expensive) laser combs being reserved for the most stable instruments on the larger facilities. While much work remains to be done to refine these calibration sources, the calibration issues in the NIR largely seem to be within reach.

3.1.3. Single-mode fiber-fed spectrographs. The advent of high strehl ratio adaptive optics (AO) systems at most large telescopes makes it possible to seriously consider using a single-mode optical fiber (SMF) to couple the light from the focal plane of the telescope to a spectrograph. Working close to the diffraction limit enables such SMF-fed spectrographs to be very compact while simultaneously capable of providing spectral resolution comparable or superior to natural-seeing spectrographs. A number of groups are pursuing technology development relating to these goals (*Ghasempour et al.,* 2012; *Schwab et al.,* 2012; *Crepp,* 2013). The SMFs provide theoretically perfect scrambling of the input point spread function (PSF), further aiding in the possibility of very-high-precision and compact Doppler spectrometers emerging from such development paths. While subtleties relating to polarization state and its impact on velocity precision remain to be solved, many of the calibration sources discussed above are innately adaptable to use with SMF fiber-fed spectrographs. Since the efficiency of these systems depends steeply on the level of AO correction, it is likely that Doppler RV searches targeting the red optical and NIR wavelengths will benefit the most.

3.2. Spectroscopic Detection of Planetary Companions

Direct spectroscopic detection of the orbit of nontransiting planets has finally yielded successful results this decade. While the traditional Doppler technique relies on detecting the radial velocity of the star only, the direct spectroscopic detection technique relies on observing the star-planet system in the NIR or thermal IR (where the planet-to-star flux ratio is more favorable than the optical) and obtaining high-resolution, very high S/N spectra to be able to spectroscopically measure the radial velocity of both the star and the planet in a manner analogous to the detection of a spectroscopic binary (SB2). The radial velocity observations directly yield the mass ratio of the star-planet system. If the stellar mass is known (or estimated well), the planet mass can be determined with no sin i ambiguity despite the fact that these are not transiting systems. The spectroscopic signature of planets orbiting Tau Boo, 51 Peg, and HD 189733 have recently been detected using the CRyogenic high-resolution InfraRed Echelle Spectrograph (CRIRES) instrument on the Very Large Telescope (VLT) (*Brogi et al.,* 2012, 2013; *de Kok et al.,* 2013; *Rodler et al.,* 2012), and efforts are ongoing by multiple groups to detect other systems using the Near Infrared Echelle Spectrograph (NIRSPEC) instrument at Keck Observatory (*Lockwood et al.,* 2014). The very high S/N required of this technique limits it to the brighter planet hosts and to relatively close-in planets, but yields information about mass and planetary atmospheres that would be difficult to determine otherwise for the nontransiting planets. Such techniques complement the transit detection efforts underway and will increase in sensitivity with telescope aperture, better infrared detectors, and more sophisticated analysis techniques. While we have focused primarily on planet-detection techniques in this review article, high-resolution NIR spectroscopy using large future groundbased telescopes may also be able to detect astrobiologically interesting molecules (e.g., O_2) around Earth-analogs orbiting M dwarfs (*Snellen et al.,* 2013).

4. DOPPLER MEASUREMENTS FROM SPACE

Although there are no current plans to build high-resolution spectrometers for space missions, this environment might offer some advantages for extreme-precision Doppler spectroscopy if the instrument would be in a stable thermal and pressure environment. Without blurring from Earth's atmosphere, the PSF would be very stable and the image size could be small, making it intrinsically easier to obtain high resolution with an extremely compact instrument. Furthermore, the effect of sky subtraction and telluric contamination are currently difficult problems to solve with groundbased instruments, and these issues are eliminated with spacebased instruments.

5. TRANSIT DETECTIONS

At the time of the publication of the *Protostars and Planets IV (PPIV)* volume in 2000 (*Mannings et al.,* 2000), the first transiting extrasolar planet — HD 209458b — had just been found (*Henry et al.,* 2000; *Charbonneau et al.,* 2000). That momentous announcement, however, was too late for the conference volume, and *PPIV*'s single chapter on planet detection was devoted to 14 planets detected by Doppler velocity monitoring, of which only 8 were known prior to the June 1998 meeting. Progress, however, was rapid. In 2007, when the *Protostars and Planets V* volume was published (*Reipurth et al.,* 2007), nearly 200 planets had been found with Doppler radial velocities, and 9 transiting planets were known at that time (*Charbonneau et al.,* 2007).

In the past several years, the field of transit detection has come dramatically into its own. A number of long-running groundbased projects, notably the SuperWASP (*Collier Cameron et al.,* 2007) and HATNet surveys (*Bakos et al.,* 2007), have amassed the discovery of dozens of transiting planets with high-quality light curves in concert with accurate masses determined via precision Doppler velocity measurements. Thousands of additional transiting planetary candidates have been observed from space. Transit timing variations (*Agol et al.,* 2005; *Holman and Murray,* 2005) have progressed from a theoretical exercise to a practiced technique. The Spitzer Space Telescope (along with the Hubble Space Telescope and groundbased assets) has been employed to characterize the atmospheres of dozens of transiting extrasolar planets (*Seager and Deming,* 2010). An entirely new, and astonishingly populous, class of transiting planets in the mass range $R_\oplus < R_P < 4\ R_\oplus$ has been discovered and probed (*Batalha et al.,* 2013). Certainly, with each new iteration of the Protostars and Planets series, the previous edition looks hopelessly quaint and out of date. Is seems certain that progress will ensure that this continues to be the case.

5.1. The Era of Spacebased Transit Discovery

Two space missions, Kepler (*Borucki et al.,* 2010) and Convection, Rotation and planetary Transits (CoRoT) (*Barge et al.,* 2008), have both exhibited excellent productivity, and a third mission, Microvariability and Oscillations of Stars (MOST), has provided photometric transit discoveries of several previously known planets (*Winn et al.,* 2011; *Dragomir et al.,* 2013). Indeed, Fig. 1 shows that during the past six years, transiting planets have come to dominate the roster of new discoveries. Doppler velocimetry, which was overwhelmingly the most productive discovery method through 2006, is rapidly transitioning from a general survey mode to an intensive focus on low-mass planets orbiting very nearby stars (*Mayor et al.,* 2011) and to the characterization of planets discovered in transit via photometry.

The Kepler mission, in particular, has been completely transformative, having generated, at last rapidly evolving count, over 100 planets with mass determinations, as well as hundreds of examples of multiple transiting planets orbiting a single host star, many of which are in highly co-planar, surprisingly crowded systems (*Lissauer et al.,* 2011b). Taken in aggregate, the Kepler candidates indicate that planets with masses $M_P < 30\ M_\oplus$ and orbital periods $P < 100$ d are effectively ubiquitous (*Batalha et al.,* 2011), and as shown in Fig. 3, the distribution of mass ratios and periods of these candidate planets are, in many cases, curiously reminiscent of the regular satellites of the jovian planets within our own solar system.

The CoRoT satellite ceased active data gathering in late 2012, having substantially exceeded its three-year design life. In spring of 2013, just after the end of its nominal mission period, the Kepler satellite experienced a failure of a second reaction wheel, which brought its high-

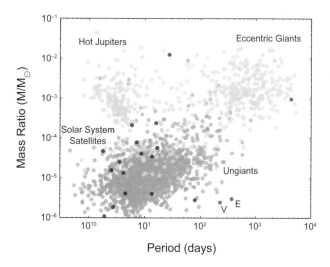

Fig. 3. Medium-gray circles: $\log_{10}$ ($M_{satellite}/M_{primary}$) and $\log_{10}(P)$ for 634 planets securely detected by the radial velocity method (either with or without photometric transits). Light-gray circles: $\log_{10}(M_{satellite}/M_{primary})$ and $\log_{10}(P)$ for the regular satellites of the jovian planets in the solar system. Dark-gray circles: $\log_{10}(M_{satellite}/M_{primary})$ and $\log_{10}(P)$ for 1501 Kepler candidates and objects of interest in which multiple transiting candidate planets are associated with a single primary. Radii for these candidate planets, as reported in *Batalha et al.* (2013), are converted to masses assuming $M/M_\oplus = (R/R_\oplus)^{2.06}$ (*Lissauer et al.*, 2011a), which is obtained by fitting the masses and radii of the solar system planets bounded in mass by Venus and Saturn. Data are from *www. exoplanets.org* as of August 15, 2013.

precision photometric monitoring program to a premature halt. The four years of Kepler data in hand, however, are well-curated, fully public, and are still far from being fully exploited; it is not unreasonable to expect that they will yield additional insight that is equivalent to what has already been gained from the mission to date. *Jenkins et al.* (2010) describe the fiducial Kepler pipeline; steady improvements to the analysis procedures therein have led to large successive increases in the number of planet candidates detected per star (*Batalha et al.*, 2013).

The loss of the Kepler and CoRoT spacecraft has been tempered by the recent approvals of two new space missions. In the spring of 2013, NASA announced selection of the TESS mission for its Small Explorer Program. The TESS mission is currently scheduled for a 2017 launch. It will employ an all-sky strategy to locate transiting planets with periods of weeks to months, and sizes down to $R_p \sim 1$ $R_\oplus$ (for small parent stars) among a sample of 5×10^5 stars brighter than V = 12, including ~1000 red dwarfs (*Ricker et al.*, 2010). TESS is designed to take advantage of the fact that the most heavily studied, and therefore the most scientifically valuable, transiting planets in a given category (hot Jupiters, extremely inflated planets, sub-Neptune-sized planets, etc.) orbit the brightest available parent stars. To date, many of these "fiducial" worlds, such as HD 209458b, HD 149026b,

HD 189733b, and Gliese 436b, have been discovered to transit by photometrically monitoring known Doppler-wobble stars during the time windows when transits are predicted to occur. By surveying all the bright stars, TESS will systematize the discovery of the optimal transiting example planets within every physical category. The CHaracterising ExOPlanet Satellite (CHEOPS) is also scheduled for launch in 2017 (*Broeg et al.*, 2013). It will complement TESS by selectively and intensively searching for transits by candidate planets in the $R_\oplus < R_p < 4$ $R_\oplus$ size range during time windows that have been identified by high-precision Doppler monitoring of the parent stars. It will also perform follow-up observations of interesting TESS candidates.

5.2. Transit Detection

The *a priori* probability that a given planet can be observed in transit is a function of the planetary orbit, and the planetary and stellar radii

$$\mathcal{P}_{tr} = 0.0045 \left(\frac{AU}{a} \right) \left(\frac{R_\star + R_p}{R_\odot} \right) \left[\frac{1 + e \cos(\pi/2 - \omega)}{1 - e^2} \right] \quad (3)$$

where ω is the angle at which orbital periastron occurs, such that $\omega = 90°$ indicates transit, and e is the orbital eccentricity. A typical hot Jupiter with $R_p \gtrsim R_{Jup}$ and P ~ 3 d, orbiting a solar-type star, has a $\tau \sim 3$ hr transit duration, a photometric transit depth, d ~ 1%, and $\mathcal{P} \sim 10\%$. Planets belonging to the ubiquitous super-Earth–sub-Neptune population identified by Kepler (i.e., the gray points in Fig. 3) are typified by $\mathcal{P} \sim 2.5\%$, d ~ 0.1%, and $\tau \sim 6$ hr, whereas Earth-sized planets in an Earth-like orbits around solar-type stars present a challenging combination of $\mathcal{P} \sim 0.5\%$, d ~ 0.01%, and $\tau \sim 15$ hr.

Effective transit search strategies seek the optimal tradeoff between cost, sky coverage, photometric precision, and the median apparent brightness of the stars under observation. For nearly a decade, the community as a whole struggled to implement genuinely productive surveys. For an interesting summary of the early disconnect between expectations and reality, see *Horne* (2003). Starting in the mid-2000s, however, a number of projects began to produce transiting planets (*Konacki et al.*, 2003; *Alonso et al.*, 2004; *McCullough et al.*, 2006), and there are now a range of successful operating surveys. For example, the ongoing Kelt-North project, which has discovered four planets to date (*Collins et al.*, 2013), targets very bright 8 < V < 10 stars throughout a set of 26° × 26° fields that make up ~12% of the full sky. Among nearly 50,000 stars in this survey, 3822 targets have RMS photometric precision better than 1% (for 150-s exposures). A large majority of the known transit-bearing stars, however, are fainter than Kelt's faint limit near V ~ 10. The 10 < V < 12 regime has been repeatedly demonstrated to provide good prospects for Doppler follow-up and detailed physical characterization, along with a large number of actual transiting planets. In this stellar brightness regime, surveys such as HATNet and SuperWASP have led the way. For instance, HAT-South (*Bakos et al.,* 2013), a globally networked extension of the long-running HATNet project, monitors 8.2° × 8.2° fields and reaches 6 millimagnitude (mmag) photometric precision at a four-minute cadence for the brightest nonsaturated stars at r ~ 10.5. SuperWASP's characteristics are roughly similar, and to date, it has been the most productive groundbased transit search program.

To date, the highest-precision groundbased exoplanetary photometry has been obtained with orthogonal phase transfer arrays trained on single, carefully preselected high-value target stars. Using this technique, *Johnson et al.* (2009) obtained 0.47-mmag photometry at 80-s cadency for WASP-10 (V = 12.7). By comparison, with its spaceborne vantage, Kepler obtained a median photometric precision of 29 ppm with 6.5-hour cadence on V = 12 stars. This is ~2× better than the best special-purpose groundbased photometry, and ~20× better than the leading groundbased discovery surveys.

Astrophysical false positives present a serious challenge for wide-field surveys in general and for Kepler in particular, where a majority of the candidate planets lie effectively out of reach of Doppler characterization and confirmation (*Morton and Johnson,* 2011). Stars at the bottom of the main sequence overlap in size with giant planets (*Chabrier and Baraffe,* 2000) and thus present near-identical transit signatures to those of giant planets. Grazing eclipsing binaries can also provide a source of significant confusion for low S/N light curves (*Konacki et al.,* 2003).

Within the Kepler field, pixel "blends" constitute a major channel for false alarms. These occur when an eclipsing binary, either physically related or unrelated, shares line of sight with the target star. Photometry alone can be used to identify many such occurrences (*Batalha et al.,* 2010), whereas in other cases, statistical modeling of the likelihood of blend scenarios (*Torres et al.,* 2004; *Fressin et al.,* 2013) can establish convincingly low false alarm probabilities. High-profile examples of confirmation by statistical validation include the R = 2.2 $R_{\oplus}$ terrestrial candidate planet Kepler 10c by (*Fressin et al.,* 2011), as well as the planets in the Kepler 62 system (*Borucki et al.,* 2013). False alarm probabilities are inferred to be dramatically lower for cases where multiple candidate planets transit the same star. Among the gray points in Fig. 3 there is very likely only a relatively small admixture of false alarms.

5.3. Results and Implications

Aside from the sheer increase in the number of transiting planets that are known, the string of transit discoveries over the past six years have been of fundamentally novel importance. In particular, transit detections have enabled the study of both planets and planetary system architectures for which there are no solar system analogs. A brief tally of significant events logged in order of discovery year might include (1) Gliese 436 b (*Gillon et al.,* 2007), the first tran-

siting Neptune-sized planet and the first planet to transit a low-mass star; (2) HD 17156b, the first transiting planet with a large orbital eccentricity (e = 0.69) and an orbital period (P = 21 d) that is substantially larger than the 2 d < P < 5 d range occupied by a typical hot Jupiter (*Barbieri et al.,* 2007); (3) CoRoT 7 b (*Léger et al.,* 2009) and Gliese 1214b (*Charbonneau et al.,* 2009), the first transiting planets with masses in the so-called "super-Earth" regime 1 $M_{\oplus}$ < M < 10 $M_{\oplus}$; (4) Kepler 9b and 9c (*Holman et al.,* 2010), the first planetary system to show tangible transit timing variations, as well as the first case of transiting planets executing a low-order mean-motion resonance; (5) Kepler 22b, the first transiting planet with a size and an orbital period that could potentially harbor an Earth-like environment (*Borucki et al.,* 2012); and (6) the Kepler 62 system (*Borucki et al.,* 2013), which hosts at least five transiting planets orbiting a K2V primary. The outer two members, Planet "e" with P = 122 d and Planet "f" with P = 267 d, both have 1.25 $R_{\oplus}$ < R_p < 2 $R_{\oplus}$, and receive S = 1.2 ± 0.2 $S_{\odot}$ and S = 0.4 ± 0.05 $S_{\odot}$ of Earth's solar flux respectively.

Bulk densities are measured for transiting planets with parent stars that are bright enough and chromospherically quiet enough to support Doppler measurement of M_P sin i, and can also be obtained by modeling transit timing variations (*Fabrycky et al.,* 2012; *Lithwick et al.,* 2012). More than 100 planetary densities (mostly for hot Jupiters) have been securely measured. These are plotted in Fig. 4, which hints at the broad outlines of an overall distribution. Figure 4 is anticipated to undergo rapid improvement over the next several years as more Kepler candidates receive mass determinations. It appears likely, however, that there exists a very broad range of planetary radii at every mass. For

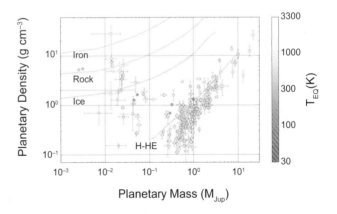

Fig. 4. Density-mass diagram for planets with well-determined masses and radii. Planets are color-coded by the equilibrium temperature, $T_{eq} = (R_*^{1/2}T_*)/((2a)^{1/2}(1-e^2)^{1/8})$ that they would have if they were zero-albedo blackbodies reradiating from the full planetary surface area. The solar system planets more massive than Mars are included in the plotted aggregate. Gray lines show expected $\rho(M_P)$ for planetary models of pure hydrogen-helium, pure water, pure silicate, and pure iron compositions. Planetary data are from *www. exoplanets.org* as of August 15, 2013.

example, to within errors, planets with $M_P \sim 6\ M_\oplus$ appear to range in radius by a factor of at least 3. While a substantial number of short-period giant planets are inflated by unknown energy source(s) (*Batygin and Stevenson,* 2010), compositional variations are at least capable of explaining the observed range of radii for planets with $M_P < 0.2\ M_{Jup}$ (*Fortney et al.,* 2007). The mass-density distribution (and by extension, the composition distribution) of extrasolar planets as a function of stellocentric distance is an important outcome of the planet-formation process. It is still entirely unclear whether planets with $P < 100$ d that have no solar system analogs are the product of migration processes (*Ida and Lin,* 2004) or of *in situ* formation (*Chiang and Laughlin,* 2013). More high-quality measurements of transiting planets will be required to resolve the puzzle.

The large number of candidate multiple transiting planet systems indicate that co-planar architectures are the rule for planets with $P < 100$ d in the size range of $R_p \sim 1.5$–6 $R_\oplus$ (*Moorhead et al.,* 2011). The inclination dispersion of most candidate systems with two or more transiting planets appears to have a median between 1° and 3°. Candidate planets in multiple-transit systems, furthermore, are invariably in dynamically stable configurations when imbued with reasonable mass-radius relations (*Lissauer et al.,* 2011a). Nature has therefore produced a galactic planetary census that is extraordinarily well suited to detection and characterization via the transit method. The advent of the new space missions, in concert with the James Webb Space Telescope's (JWST) potential for atmospheric characterization of low-mass planets (*Deming et al.,* 2009), indicate that transits will remain at the forefront for decades to come.

Finally, transit detection is unique in that it democratizes access to cutting-edge research in exoplanetary science. Nearly all the highly cited groundbased discoveries have been made with small telescopes of aperture d < 1 m. Amateur observers were co-discoverers of the important transits by HD 17156b (Barbieri et al., 2007) and HD 80606b (*Garcia-Melendo and McCullough,* 2009), and citizen scientists have discovered several planets to date in the Kepler data under the auspices of the Planet Hunters project (*Fischer et al.,* 2012; *Lintott et al.,* 2013; *Schwamb et al.,* 2013).

6. DIRECT IMAGING TECHNIQUES

The field of exoplanets is almost unique in astronomical science in that the subjects are almost all studied indirectly, through their effects on more visible objects, rather than being imaged themselves. The study of the dominant constituents of the universe (dark energy and dark matter) through their gravitational effects is of course another example. Direct imaging of the spatially resolved planet is a powerful complement to the other techniques described in this chapter. It is primarily sensitive to planets in wide orbits a > 5 AU, and since photons from the planets are recorded directly, the planets are amenable to spectroscopic or photometric characterization. However, direct detection also represents a staggering technical challenge. If a twin to our solar system

were located at a distance of 10 pc from Earth, the brightest planet would have only ~10^{-9} the flux of the parent star, at an angular separation of 0.5 arcsec.

In spite of this challenge, the field has produced a small number of spectacular successes: the images and spectra of massive (>1000 $M_\oplus$) young self-luminous planets. The advent of the first dedicated exoplanet imaging systems should lead to rapid progress and surveys with statistical power comparable to groundbased Doppler or transit programs. In the next decade, spacebased coronagraphs will bring mature planetary systems into reach, and some day, a dedicated exoplanet telescope may produce an image of an Earth analog orbiting a nearby star.

6.1. Limitations to High-Contrast Imaging

The greatest challenge in direct imaging is separating the light of the planet from residual scattered light from the parent star. This can be done both optically — removing the starlight before it reaches the science detector — and in post-processing, using a feature that distinguishes starlight from planetary light.

6.1.1. High-contrast point spread function, coronagraphs, and adaptive optics. Even in the absence of aberrations, the images created by a telescope will contain features that will swamp any conceivable planet signal. The PSF, as the name implies, is the response of the telescope to an unresolved point source. In the case of an unaberrated telescope, the PSF is the magnitude squared of the Fourier transform of the telescope aperture function. For an unobscured circular aperture, the diffraction pattern is the distinctive Airy rings. (The one-dimensional equivalent would represent the telescope as a top hat function, whose Fourier transform is a sync, giving a central peak and oscillating sidelobes.) More complex apertures will have more complex diffraction patterns.

Removing this diffraction pattern is the task of a coronagraph. Originally developed by *Lyot* (1939) to allow small telescopes to study the coronae of the Sun, chronographs employ optical trickery to remove the light from an on-axis star while allowing some of the flux from the off-axis planet to remain. A wide variety of approches have been developed (*Guyon et al.,* 2006), far too many to enumerate here, although they can be divided into broad families. The classical Lyot coronagraph blocks the on-axis source with a focal plane mask, followed by a pupil-plane Lyot mask that blocks the light diffracted by the focal plane (*Lyot,* 1939; *Sivaramakrishnan et al.,* 2001). Apodizers operate by modifying the transmission of the telescope so that the Fourier transform has substantially less power in the sidelobes; a nonphysical example would be a telescope whose transmission was a smoothly varying Gaussian, which would result in a purely Gaussian PSF. In more practical designs, apodization is implemented through binary "shaped pupil" masks (*Kasdin et al.,* 2003) that sharply reduce diffraction over a target region at a significant cost in throughput. Hybrid Lyot approaches use pupil-plane apodization (*Soummer et al.,* 2011) or complicated focal-plane masks (*Kuchner and Traub,* 2002) to boost the performance of the classic Lyot. Phase-induced amplitude apodization uses complex mirrors to create the tapered beam needed to suppress diffraction without a loss in throughput (*Guyon et al.,* 2005). A particularly promising new technique creates an optical vortex in the focal plane (*Nersisyan et al.,* 2013), removing the diffracted light almost perfectly for an on-axis source in a unobscured aperture. Many more complex coronagraphs exist (see *Guyon et al.,* 2006, for discussion). Typically, the best coronagraphs remove diffraction down to the level of 10^{-10} at separations greater than the inner working angle (IWA), typically $2-4\lambda/D$.

Light is also scattered by optical imperfections — wavefront errors induced by the telescope, camera, or atmospheric turbulence. Even with a perfect coronagraph, atmospheric turbulence, which typically is many waves of phase aberration, produces a PSF that completely overwhelms any planetary signal. Even in the absence of atmospheric turbulence, small wavefront errors from, e.g., polishing marks will still scatter starlight. These can be partially corrected through adaptive optics — using a deformable mirror (DM), controlled by some estimate of the wavefront, to correct the phase of the incoming light. In the case of small phase errors, a Fourier relationship similar to that for diffraction exists between the wavefront and PSF (see *Perrin et al.,* 2003, and *Guyon et al.,* 2006, for discussion and examples). A useful figure of merit for adaptive optics correction is the Strehl ratio, defined as the ratio of the peak intensity of the measured PSF to the theoretical PSF for an equivalent unaberrated telescope. With current-generation adaptive optics systems, Strehl ratios of 0.4–0.8 are common in the K band — meaning that 60–80% of the scattered light remains uncorrected.

The halo of light scattered by wavefront errors is particularly troublesome because it does not form a smooth background, but is broken up into a pattern of speckles. In monochromatic light these speckles resemble the diffraction-limited PSF of the telescope, and hence are easily confused with the signal from a planet. As a result, high-contrast images are usually nowhere near the Poisson limit of photon noise but instead limited by these speckles. Uncorrected atmospheric turbulence produces a halo of speckles that rapidly evolve; static or quasistatic wavefront errors, such as adaptive optics miscalibrations, produce slowly evolving speckles that mask planetary signals.

6.1.2. Post-processing. These speckle patterns can be partially mitigated in post-processing. Such PSF subtraction requires two components. First, there must be some distinction between a planetary signal and the speckle pattern — some diversity. Examples include wavelength diversity, where the wavelength dependence of the speckle pattern differs from that of the planet; rotational diversity, in which the telescope (and associated speckle pattern) rotates with respect to the planet/star combination (*Marois et al.,* 2006); or observations of a completely different target star. Such reference PSFs will never be a perfect match, as the PSF evolves with time, temperature, star brightness, and

wavelength. The second component needed for effective PSF subtraction is an algorithm that can construct the "best" PSF out of a range of possibilities. With a suitable library of PSFs, least-squares fitting (*Lafrenière et al.,* 2007a) or principal components analysis can assemble synthetic PSFs and enhance sensitivity to planets by a factor of 10–100.

6.2. Imaging of Self-Luminous Planets

With these techniques applied to current-generation systems, planets with brightness ~10^{-5} can be seen at angular separations of ~1.0 arcsec. This is far from the level of sensitivity needed to see mature Jupiter-like planets. Fortunately, planets are available that are much easier targets. When a planet forms, significant gravitational potential energy is available. Depending on the details of initial conditions, a newly formed giant planet may have an effective temperature of 1000–2000 K (*Marley et al.,* 2007) and a luminosity of 10^{-5}–10^{-6} $L_\odot$ (Fig. 5). As with the brown dwarfs, a large fraction of this energy could be released in the NIR, bringing the planet into the detectable range. Such planets remain detectable for tens of millions of years. Several surveys have targeted young stars in the solar neighborhood for exoplanet detection (*Liu et al.,* 2010; *Lafrenière et al.,* 2007b; *Chauvin et al.,* 2010), benefitting from the identification of nearby young associations composed of stars with ages 8–50 million years (m.y.) (*Zuckerman and Song,* 2004). Most of these surveys have produced only nondetections, with upper limits on the number of giant planets as a function of semimajor axis that exclude large numbers of very-wide-orbit (50 AU) planets.

A handful of spectacular successes have been obtained. One of the first detections was a 5 M_{Jup} object that was orbiting not a star but a young brown dwarf, 2M1207B (*Chauvin et al.,* 2004). A spectacular example of planetary companions to a main-sequence star is the HR 8799 multiplanet system (Fig. 6). This consists of four objects near a young F0V star, orbiting in counterclockwise directions. The object's luminosities are well constrained by broadband photometry (*Marois et al.,* 2008; *Currie et al.,* 2011). Estimates of the planetary mass depend on knowledge of the stellar age — thought to be 30 m.y. (*Marois et al.,* 2010; *Baines et al.,* 2012) — and initial conditions; for "hot start" planets the masses are 3–7× that of Jupiter. Multi-planet gravitational interactions provide a further constraint on the mass (*Marois et al.,* 2010; *Fabrycky and Murray-Clay,* 2010), excluding massive brown dwarf companions. Other notable examples of directly imaged exoplanets include the very young object 1RXS J1609b (*Lafrenière et al.,* 2010), the cool planet candidate GJ 504B (*Kuzuhara et al.,* 2013), and the planet responsible for clearing the gap inside the β Pictoris disk (*Lagrange et al.,* 2010). A candidate optical HST image of an exoplanet was reported orbiting Fomalhaut (*Kalas et al.,* 2008), but very blue colors and a belt-crossing orbit (*Kalas et al.,* 2013) indicate that what is seen is likely light scattered by a debris cloud or disk (that may still be associated with a planet).

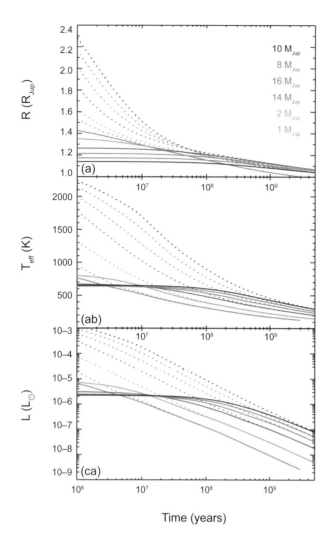

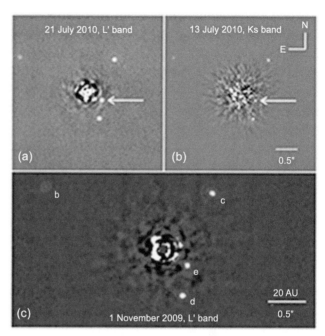

Fig. 6. Near-infrared Keck adaptive optics images of the HR 8799 system from *Marois et al.* (2010). Four giant planets, 3–7× the mass of Jupiter, are visible in near-infrared emission. The residual speckle pattern after PSF subtraction can be seen in the center of each image.

Fig. 5. Reproduction of Fig. 4 of *Marley et al.* (2007) showing the model radius, temperature, and luminosity of young Jupiters as a function of time since the beginning of their formation. Different colors reflect different planetary masses. Dotted lines indicate "hot start" planets, where adiabatic formation retains most of the initial energy and entropy; solid lines indicate "cold start," where accretion through a shock (as in the standard core accretion paradigm) results in loss of entropy. In either case, planets are significantly easier to detect at young ages.

The photometric detections of self-luminous planets have highlighted the complexities of modeling the atmospheres of these objects. Although they are similar to brown dwarfs, many of the directly imaged planets have temperatures that place them in the transitional region between cloud-dominated L dwarfs and methane-dominated T dwarfs — a change that is poorly understood even for the well-studied brown dwarfs. Cloud parameters in particular can make an enormous difference in estimates of properties like effective temperature and radius [see supplementary material in *Marois et al.* (2008) and subsequent discussion in *Barman et al.* (2011), *Marley et al.* (2012), *Currie et al.* (2011),

and discussion in the chapter by Madhusudhan et al. in this volume].

If a planet can be clearly resolved from its parent star, it is accessible not only through imaging but also spectroscopically. Integral field spectrographs are particularly well suited to this (e.g., *Oppenheimer et al.*, 2013; *Konopacky et al.*, 2013), since they also capture the spectrum of neighboring speckle artifacts, which can be used to estimate the speckle contamination of the planet itself. Spectra show that the self-luminous planets do (as expected) have low gravity and distinct atmospheric structure from brown dwarfs. In some cases, spectra have sufficiently high SNR that individual absorption features (e.g., of CO) can be clearly resolved (*Konopacky et al.*, 2013), allowing direct measurements of atmospheric chemistry and abundances (Fig. 7).

6.3. Future Groundbased and Spacebased Facilities

Most direct imaging of exoplanets to date has taken place with traditional instruments attached to general-purpose AO systems, such as the Near InfraRed Camera (NIRC) 2 on the Keck II telescope or Nasmyth Adaptive Optics System (NACO) on the VLT. In fact, for most of these observations, the presence or absence of a coronagraph has had little effect on sensitivity, which is dominated by wavefront errors uncorrected by the AO system. Some sensitivity enhancement has come from dedicated exoplanet imaging cameras, employing techniques like dual-channel imaging,

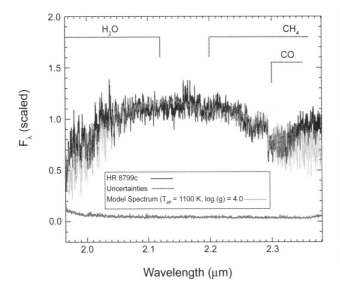

Fig. 7. High-resolution spectrum of the extrasolar planet HR 8799c taken with the OSIRIS spectrograph and the Keck adaptive optics system, reproduced from *Konopacky et al.* (2013). Residual speckle noise changes the overall spectral shape (e.g., the upturn at the long wavelength end) but does not inject narrow features; the CO break is clearly detected, as are many individual CO and H_2O lines, while methane is absent.

in combination with conventional adaptive optics (*Nielsen et al.*, 2013; *Janson et al.*, 2013). The combination of pyramid wavefront sensing and adaptive secondary mirrors on the Large Binocular Telescope (LBT) and Magellan telescopes has shown excellent high-contrast performance (*Skemer et al.*, 2012).

However, to significantly increase the number of imaged exoplanets will require dedicated instruments that combine very-high-performance adaptive optics, suitable corona-graphs, and exoplanet-optimized science instruments such as low-spectral-resolution diffraction-limited integral field spectrographs (IFS). The first such instrument to become operational is the Project 1640 coronagraphic IFS (*Oppenheimer et al.*, 2013), integrated with a 3000-actuator AO system on the 5-m Hale telescope. The Subaru Coronagraphic Extreme AO System (SCExAO) (*Martinache et al.*, 2012) is a 2000-actuator AO system that serves as a test bed for a wide variety of advanced technologies including focal-plane wavefront sensing and pupil-remapping coronagraphs. Finally, two facility-class planet imagers will be operational in 2014 on 8-m class telescopes: the Gemini Planet Imager (GPI) (*Macintosh et al.*, 2012) and the VLT Spectro-Polarimetric High-contrast Exoplanet REsearch (SPHERE) facility (*Beuzit et al.*, 2008; *Petit et al.*, 2012). Both have 1500-actuator AO systems, apodized-pupil Lyot coronagraphs, and IFSs (SPHERE also incorporates a dual-channel IR imager and a high-precision optical polarimeter). Laboratory testing and simulations predict that they will achieve on-sky contrasts of better than 10^6 at angles of

0.2 arcsec, although with the limitation of requiring bright stars (I < 8 mag for GPI, V < 12 mag for SPHERE) to reach full performance. Both instruments will be located in the southern hemisphere, where the majority of young nearby stars are located. Simulated surveys (*McBride et al.*, 2011) predict that GPI could discover 20–50 jovian planets in a 900-hour survey.

Direct detection instruments have also been proposed for the upcoming 20–40-m Extremely Large Telescopes. These instruments exploit the large diameters of the telescope to achieve extremely small inner working angles (0.03 arcsec or less), opening up detection of protoplanets in nearby star-forming regions orbiting at the snow line (*Macintosh et al.*, 2006), or reflected light from mature giant planets close to their parent star (*Kasper et al.*, 2010). At their theoretical performance limits, such telescopes could reach the contrast levels needed to detect rocky planets in the habitable zones of nearby M stars, although reaching that level may present insurmountable technical challenges. (*Guyon et al.*, 2012).

A coronagraphic capability has been proposed for the 2.4-m Astrophysics Focused Telescope Assets (AFTA) Wide-Field Infrared Survey Telescope (WFIRST) mission (*Spergel et al.*, 2013). Due to the obscured aperture and relative thermal stability of the telescope, it would likely be limited to contrasts of 10^{-9} at separations of 0.1 or 0.2 arcsec, but this would still enable a large amount of giant-planet and disk science, including spectral character-ization of mature giant planets.

Direct detection of an Earth-analog planet orbiting a solar-type star, however, will almost certainly require a dedicated space telescope using either an advanced co-ronagraph — still equipped with adaptive optics — or a formation-flying starshade occulter.

7. MICROLENSING

7.1. Planetary Microlensing

7.1.1. Microlensing basics. A microlensing event occurs when two stars at different distances pass within ~1 mas of each other on the plane of the sky (*Gaudi*, 2012). Light from the source star "S" is bent by the lens star "L," so that the observer "O" sees the image "I" instead of the true source (see Fig. 8). If the source and the lens are perfectly aligned along the line of sight, the source is lensed into a ring (*Chwolson*, 1924; *Einstein*, 1936; *Renn et al.*, 1997), called an Einstein ring, whose angular size is given by

$$\theta_E = \sqrt{\kappa M_L \pi_{rel}} \sim 0.3 \text{ mas} \qquad (4)$$

for typical values of the lens mass ($M_L = 0.5\ M_\odot$), lens distance ($D_L = 6$ kpc), and source distance ($D_S = 8$ kpc). In equation (4), $\pi_{rel} = (1\ AU/D_L) - (1\ AU = D_S)$ is the trigo-nometric parallax between the source and the lens, and $\kappa = 8.14$ mas $M_\odot^{-1}$.

If the source is offset from the lens by some small amount, it is lensed into two images that appear in line with

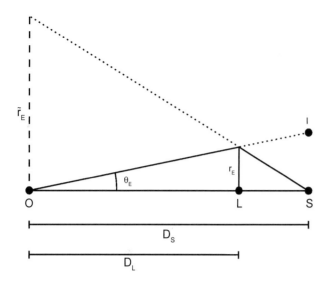

Fig. 8. Basic geometry of microlensing.

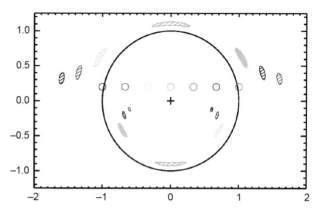

Fig. 9. Images of a lensed source star. The position of the source is indicated by the small circles. The filled voids show the lensed images for each source position. The large black circle shows the Einstein ring. The lens star is at the origin, marked by the plus.

the source and the lens, and close to the Einstein ring as in Fig. 9. Because the size of the Einstein ring is so small, the two images of the source are unresolved and the primary observable is their combined magnification

$$A = \frac{u^2 + 2}{u\sqrt{u^2 + 4}} \qquad (5)$$

where u is the projected separation between the source and the lens as a fraction of the Einstein ring. Since the source and the lens are both moving, u (and therefore A) is a function of time.

7.1.2. Types of planetary perturbations. If planets are gravitationally bound to a lensing star, the planet can be detected if one of the source images passes over or near the position of the planet. This creates a perturbation to the microlensing light curve of the host star. Because the images generally appear close to the Einstein ring, microlensing is most sensitive to planets with projected separations equal to the physical size of the Einstein ring in the lens plane, $r_E = \theta_E D_L$.

Another way to think about this is to consider the magnification map. The magnification of the source by a point lens can be calculated for any position in space using equation (5), giving a radially symmetric magnification map. The source then traces a path across this map creating a microlensing event whose magnification changes as a function of time (and position). The presence of the planet distorts the magnification map of the lens and causes two or more caustics to appear, as shown by the gray curves in Fig. 10a. A perfect point source positioned at a point along the caustic curve will be infinitely magnified. In order to detect the planet, the source trajectory must pass over or near a caustic caused by the planet (*Mao and Paczynski, 1991; Gould and Loeb, 1992; Griest and Safizadeh, 1998*).

There are two kinds of perturbations corresponding to the two sets of caustics produced by the planet. The "planetary caustic" is the larger caustic (or set of caustics) unassociated with the position of the lens star (right side of Fig. 10a). The "central caustic" is much smaller than the planetary caustic and is located at the position of the lens star (left side of Fig. 10a). Figure 10 shows two example source trajectories, their corresponding light curves, and details of the planetary perturbation in a planetary caustic crossing. As the mass ratio, q, decreases, so does the duration of the planetary perturbation. In addition, the detailed shape of the perturbation depends on the size of the source star relative to the size of the Einstein ring, ρ.

7.1.3. Planet masses from higher-order effects. The fundamental observable properties of the planet are the mass ratio between the planet and the lens star, q, and the projected separation between the planet and the lens star as a fraction of the Einstein ring, s. Hence, while $q \lesssim 10^{-3}$ definitively identifies the companion to the lens as a planet, its physical properties cannot be recovered without an estimate of M_L and D_L. However, if θ_E and $\tilde{r}_E$ (the size of the Einstein ring in the observer plane) can be measured, it is possible to obtain measurements of M_L and D_L (see Fig. 8), and hence the physical mass and projected separation of the planet: $m_p = qM_L$ and $a_\perp = s\theta_E D_L$. These variables can be measured from higher-order effects in the microlensing light curve. If finite-source effects are observed (cf. Fig. 10e), θ_E is measured since $\rho = \theta_\star/\theta_E$ and the angular size of the source, $\theta_\star$, can be determined from the color-magnitude diagram (*Yoo et al., 2004*). Finally, as Earth orbits the Sun, the line of sight toward the event changes, giving rise to microlens parallax (*Gould, 1992; Gould et al., 1994*), allowing a measurement of $\tilde{r}_E$

$$\pi_E = \frac{1\mathrm{AU}}{\tilde{r}_E} = \frac{\pi_{rel}}{\theta_E} \qquad (6)$$

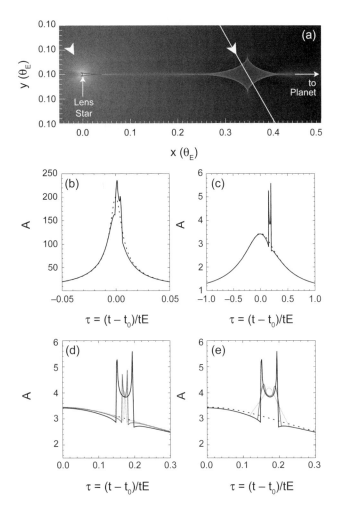

Fig. 10. (a) Magnification map for a planet with q = 0.001 and s = 1.188 and a source size ρ = 0.001. The gray lines indicate the caustics. Two example source trajectories are shown. The scale is such that the Einstein ring is a circle of radius 1.0 centered at (0,0). The planet is located at (1.188,0), just outside the Einstein ring (off the righthand side of the plot). **(b)** Light curve corresponding to the lefthand source trajectory (a central caustic crossing). The dotted line shows the corresponding light curve for a point lens. **(c)** Light curve corresponding to the righthand source trajectory (a planetary caustic crossing). **(d)** Detail of **(c)** showing the variation in the planetary signal for different values of $q = 10^{-3}$, 10^{-4}, and 10^{-5}. **(e)** The variation in the planetary signal for different values of ρ = 0.001, 0.01, and 0.03.

7.1.4. Microlensing degeneracies and false-positives. In microlensing the most common degeneracy is that planets with separation s produce nearly identical central caustics as planets with separation s^{-1} (*Griest and Safizadeh, 1998*). For planetary caustics, this is not a major problem since s (where s is larger than the Einstein ring) produces a "diamond"-shaped caustic, whereas s^{-1} produces a pair of "triangular" caustics (*Gaudi and Gould, 1997*). Additional degeneracies arise when higher-order effects such as parallax and the orbital motion of the lens are significant. In such

cases, the exact orientation of the event on the sky becomes important and can lead to both discrete and continuous degeneracies in the relevant parameters (e.g., *Gould, 2004; Skowron et al., 2011*).

False positives are rare in microlensing events in which the source crosses a caustic. Because the magnification diverges at a caustic, this produces a discontinuity in the slope of the light curve, which is very distinctive (see Fig. 10). However, in events without caustic crossings, planetary signals can be mimicked by a binary source (*Gaudi, 1998; Hwang et al., 2013*), orbital motion of the lens (e.g., *Albrow et al., 2000*), or even starspots (e.g., *Gould et al., 2013*). Often multi-band data can help distinguish these scenarios, as in the case of starspots or lensing of two sources of different colors.

7.2. Microlensing Observations in Practice

The first microlensing searches were undertaken in the late 1980s, primarily as a means to find massive compact halo objects [a dark matter candidate (*Alcock et al., 1992; Aubourg et al., 1993*)]. These searches were quickly expanded to include fields toward the galactic bulge to search for planets and measure the mass function of stars in the inner galaxy (*Paczynski, 1991; Griest et al., 1991*). One million stars must be observed to find one microlensing event, so the first surveys focused on simply detecting microlensing events. These surveys typically observed each field between once and a few times per night. However, the timescale of the planet is much shorter: a day or two for a Jupiter-mass planet down to an hour for an Earth-mass planet. Hence, follow-up groups target the known microlensing events to obtain the higher-cadence observations necessary to detect planets.

In practice, it is not possible to follow up all microlensing events, so the first priority is placed on the high-magnification events (A ≥ 50), i.e., the central caustic crossing events. Not only can the time of peak sensitivity to planets be predicted (around the time of maximum magnification), but these events are much more sensitive to planets than the average events, giving maximal planet-yield for the available resources (*Griest and Safizadeh, 1998*).

To date, almost 20 microlensing planets have been published, most of them found using the survey + follow-up method and in high-magnification events. Currently the main surveys for detecting microlensing events are the Optical Gravitational Lens Experiment (OGLE) (*Udalski, 2003*) and Microlensing Observations in Astrophysics (MOA) (*Bond et al., 2004*). Wise Observatory in Israel is also conducting a microlensing survey toward the bulge (*Gorbikov et al., 2010; Shvartzvald and Maoz, 2012*). These surveys combined now discover more than 2000 microlensing events each year. In addition, several groups are devoted to following up these events: the Microlensing Follow-Up Network (μFUN) (*Gould et al., 2006*), Microlensing Network for the Detection of Small Terrestrial Exoplanets (MiNDSTEp) (*Dominik et al., 2010*), Probing Lensing Anomalies NETwork (PLANET) (*Beaulieu et al., 2006*), and RoboNet (*Tsapras et al., 2009*).

7.3. Microlensing Planet Discoveries

7.3.1. Highlights. 7.3.1.1. *The first microlensing planet:* The first microlensing planet, OGLE-2003-BLG-235/MOA-2003-BLG-53Lb, was a 2.6-M_{Jup} planet discovered in 2003 by the OGLE and MOA surveys (*Bond et al., 2004*). Although it was discovered and characterized by surveys, this planet was found in "follow-up mode" in which the MOA survey changed its observing strategy to follow this event more frequently once the planetary anomaly was detected.

7.3.1.2. *Massive planets around M-dwarfs:* Many of the planets discovered by microlensing have large mass ratios corresponding to jovian planets. At the same time, the microlensing host stars are generally expected to be M dwarfs, since those are the most common stars in the galaxy. Specifically, there are two confirmed examples of events for which the host star has been definitively identified to be an M dwarf hosting a super-Jupiter: OGLE-2005-BLG-071 (*Udalski et al., 2005; Dong et al., 2009*) and MOA-2009-BLG-387 (*Batista et al., 2011*). The existence of such planets is difficult to explain since the core-accretion theory of planet formation predicts that massive, jovian planets should be rare around M dwarfs (*Laughlin et al., 2004; Ida and Lin, 2005*). However, it is possible they formed through gravitational instability and migrated inward (*Boss, 2006*).

7.3.1.3. *Multi-planet systems:* Two of the microlensing events that host planets, OGLE-2006-BLG-109 (*Gaudi et al., 2008*) and OGLE-2012-BLG-0026 (*Han et al., 2013*), have signals from two different planets. The OGLE-2006-BLG-109L system is actually a scale model of our solar system. The planets in this event are a Jupiter and a Saturn analog, with both planets at comparable distances to those planets around the Sun when the difference in the masses of the stars is taken into account.

7.3.1.4. *Free-floating planets:* Because microlensing does not require light to be detected from the lenses, it is uniquely sensitive to detecting free-floating planets. Since θ_E scales as $M^{1/2}$, free-floating planets have extremely small Einstein rings and hence give rise to short-duration events (≤ 1 d). Based on the analysis of several years of MOA survey data, *Sumi et al.* (2011) found that there are two free-floating Jupiters for every star.

7.3.2. The frequency of planets measured with microlensing. Figure 11 compares the sensitivity of microlensing to other techniques, where the semimajor axis has been scaled by the snow line, $a_{snow} = 2.7$ AU ($M_\star/M_\odot$). The "typical" microlensing host is an M dwarf rather than a G dwarf, so from the perspective of the core-accretion theory of planet formation, the relevant scales are all smaller. In this theory, the most important scale for giant planet formation is the location of the snow line, which depends on stellar mass (*Ida and Lin, 2004*). Microlensing is most sensitive to planets at 1 r_E, which is roughly 3 × a_{snow} for an M dwarf (i.e., $a_{snow} \sim 1$ AU and $r_E \sim 3$ AU).

The frequency, or occurrence rate, of planets can be calculated by comparing the sensitivities of individual

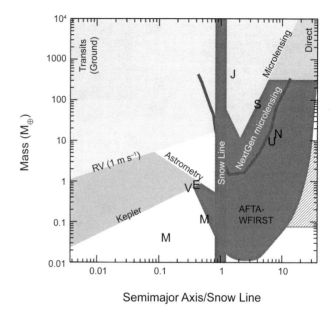

Fig. 11. Sensitivity of microlensing compared to other techniques. Figure courtesy of B. S. Gaudi and M. Penny.

events to the planets detected. *Gould et al.* (2010) analyzed high-magnification microlensing events observed by μFUN from 2005 to 2008 and found dN/(d log q d log s) = 0.31 ± 0.15 planets per dex² normalized at planets with Saturn-mass ratios. *Cassan et al.* (2012) also calculated the frequency of planets using events observed by PLANET, including both high- and low-magnification events. They found a similar planet frequency of dN/(d log a d log m_p) = $10^{-0.62\pm0.22}(m_p/M_{Sat})^{0.73\pm0.17}$ normalized at Saturn masses and flat as a function of semimajor axis. Figures 8 and 9 in *Gould et al.* (2010) compare their result to the results from radial velocity for solar-type stars (*Cumming et al., 2008; Mayor et al., 2009*) and M dwarfs (*Johnson et al., 2010b*).

7.4. The Future of Microlensing

7.4.1. Second-generation microlensing surveys. Advances in camera technology now make it possible to carry out the ideal microlensing survey: one that is simultaneously able to monitor millions of stars while also attaining an ~15-min cadence. Both OGLE and MOA have recently upgraded to larger field-of-view cameras (*Sato et al., 2008; Soszyński et al., 2012*). They have teamed up with the Wise Observatory in Israel to continuously monitor a few of their fields (*Shvartzvald and Maoz, 2012*). In addition, the Korea Microlensing Telescope Network (KMTNet) (*Park et al., 2012*) is currently under construction. This network consists of three identical telescopes in Chile, Australia, and South Africa, which will conduct a high-cadence microlensing survey toward the galactic bulge. As these second-generation surveys get established, they will dominate the microlensing planet detections and the bulk of the detections will shift to planetary caustic crossings. Although high-magnification events are individually more sensitive to planets, they are very rare compared to low-magnification events. Hence, the larger cross-section of the planetary caustics will make low-magnification events the dominant channel for detecting planets in the new surveys.

7.4.2. Spacebased microlensing. The next frontier of microlensing is a spacebased survey, which has the advantages of improved photometric precision, the absence of weather, and better resolution. The improved resolution that can be achieved from space is a major advantage for characterizing the planets found by microlensing. In groundbased searches the stellar density in the bulge is so high that unrelated stars are often blended into the 1″ PSF. This blending makes it impossible to accurately measure the flux from the lens star, and hence unless higher-order microlensing effects are observed, it is difficult to know anything about the lens. In space, it is possible to achieve a much higher resolution that resolves this blending issue, allowing an estimate of the lens mass based on its flux and hence a measurement of true planet masses rather than mass ratios.

The first microlensing survey satellite was proposed in *Bennett and Rhie* (2000, 2002). Currently, a microlensing survey for exoplanets has been proposed as a secondary science project for the Euclid mission (*Penny et al.,* 2012; *Beaulieu et al.,* 2013) and is a major component of the WFIRST mission (*Spergel et al.,* 2013). The WFIRST mission is expected to detect thousands of exoplanets beyond the snow line (*Spergel et al.,* 2013). The parameter space probed by this mission is complementary to that probed by the Kepler mission, which focused on detecting transits from close-in planets (see Fig. 11).

8. ASTROMETRY

8.1. Introduction

Steady advances in the eighteenth century improved the precision of stellar position measurements so that it was possible to measure the proper motions of stars, their parallax displacements due to Earth's motion around the Sun, and orbital motion caused by the gravitational tug of stellar companions (*Perryman,* 2012). While the impact of astrometry on exoplanet detection has so far been limited, the technique has enormous potential and is complementary to other methods (*Gatewood,* 1976; *Black and Scargle,* 1982; *Sozzetti,* 2005). Astrometry is most sensitive to wider orbits, because the center-of-mass displacement amplitude increases with orbital period. As a result, detectable orbital periods are typically several years. The need for measurement stability and precision over such long time baselines has been a challenging requirement for currently available instruments. Fortunately, with the successful launch of the Gaia satellite, the prospects for spacebased astrometric planet searches are good.

8.1.1. Parameterization of orbital motion. The term astrometry refers to the measurement of a star's position relative to the background sky, i.e., an astrometric orbit

corresponds to the barycentric motion of a star caused by an invisible companion. This motion follows Kepler's laws and is parameterized by the period P, the eccentricity e, the time of periastron passage T_0, its inclination relative to the sky plane i, the longitude of periastron ω, the longitude of the ascending node Ω, and the semimajor axis a_1 expressed in angular units (Fig. 12). The Thiele-Innes constants A, B, F, G are commonly used instead of the parameters a_1, ω, Ω, i, because they linearize the orbit term in the general expression for an astrometric signal Λ measured along an axis determined by the angle ψ

$$\Lambda = (\Delta\alpha^* + \mu_{\alpha^*} t)\cos\psi + (\Delta\delta + \mu_\delta t)\sin\psi + \varpi\Pi_\psi$$
$$+ (BX + GY)\cos\psi + (AX + FY)\sin\psi \quad (7)$$

where ϖ is the parallax, Π_ψ is the parallax factor along ψ, X and Y are the rectangular coordinates (*Hilditch,* 2001)

$$X = \cos E - e \quad Y = \sqrt{1 - e^2}\sin E \quad (8)$$

and E is the eccentric anomaly. This relation includes coordinate offsets in the equatorial system ($\Delta\alpha^*$, $\Delta\delta$), proper motions (μ_{α^*}, μ_δ), parallactic motion, and orbital motion. It can be applied to both one- and two-dimensional measurements made by Hipparcos, Gaia, or interferometers.

8.1.2. Signal dependence on mass and distance. The semimajor axis $\bar{a}_1$ of a star's barycentric orbit is related to the stellar mass m_1, the mass of the companion m_2, and the orbital period by Kepler's third law (SI units)

$$4\pi^2 \frac{\bar{a}_1^3}{P^2} = G \frac{M_P^3}{(M_* + M_P)^2} \quad (9)$$

where G is the gravitational constant. The relation between angular and linear semimajor axes is proportional to the parallax, $a_1 \propto \varpi \bar{a}_1$, thus the orbit's apparent angular size

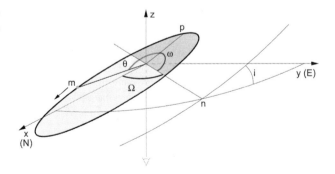

Fig. 12. Illustration of the orbit described by a star (m) about the barycenter located at the origin. The observer sees the sky plane defined by the x-y axes from below along the z axis. The angles i, ω, Ω, Θ, the ascending node n, and the periastron position p are indicated. By convention, x is north and y is east. Figure from *Sahlmann* (2012).

decreases reciprocally with the system's distance from Earth. The value of a_1 determines the semi-amplitude of the periodic signal we intend to detect with astrometric measurements. Figure 13 shows the minimum astrometric signature $a_{1,min}$ derived from equation (9) for planets listed in the *exoplanets.org* database (*Wright et al.,* 2011) on June 1, 2013, that have an entry for distance, star mass, orbital period, and planet mass, where we assumed circular orbits. For radial velocity planets, $a_{1,min}$ is a lower bound because we set sin i = 1. The spread at a given period originates in differing distances, star masses, and planet masses. Figure 13 illustrates the typical signal amplitudes for the known exoplanet population and highlights that only a small fraction of known planets are accessible with a measurement precision of 1mas. It also shows that an improvement by only 1 order of magnitude in precision would set astrometry over the threshold of routine exoplanet detection.

8.1.3. Scientific potential. The motivation for using astrometry to carry out exoplanet searches is founded in the rich and complementary orbital information provided by this technique. Astrometric measurements determine the value of $m_2^3/(m_1 + m_2)^2$, thus if the host star mass is known, then planet mass m_2 can be estimated without the sin i ambiguity of radial velocity measurements. An astrometric study of a statistical sample of exoplanets could therefore accurately determine the planet mass function and help to refine theories of planet formation. Equation (9) implies that any orbital configuration creates an astrometric signal and the amplitude increases with orbital period (see the trend in Fig. 13), making astrometry an ideal technique for the study of planets on long-period orbits. Because the technique measures the photocenter, it is sensitive to the detection of planets around fast-rotating stars with broad spectral lines or around very faint objects like brown dwarfs. There may also be a reduced sensitivity to stellar activity compared to radial velocity or photometric measurements (*Eriksson and Lindegren,* 2007; *Lagrange et al.,* 2011). Since activity is currently hampering the detection of Earth-mass planets (e.g., *Dumusque et al.* 2012), astrometry may hold a distinct advantage for future searches, although the precision needed to detect Earth-like planets around the closest stars is at the level of 1 µarcsec. Astrometry is applicable to planet searches around nearby stars of various masses and ages, with benefits for the study of the planet mass function, of long-period planets, and of planets around active stars.

8.2. Techniques and Instruments

The precision σ of an astrometric measurement is fundamentally limited by the ability to measure an image position on a detector. In the diffraction limit, it is therefore related to the wavelength λ, the aperture size D, and the signal-to-noise S/N, typically limited by photon noise $S/N \sim \sqrt{N_p}$

$$\sigma \propto \frac{1}{S/N} \frac{\lambda}{D} \qquad (10)$$

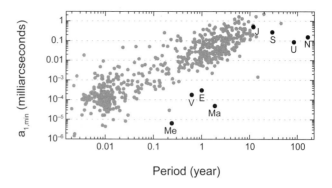

Fig. 13. Minimum astrometric signature of the host star as a function of orbital period for 570 planets (gray circles). For reference, the astrometric signatures of a solar-mass star located at a distance of 10 pc caused by the solar system planets are shown with black circles and labeled with the planet initials.

thus, the achievable astrometric precision improves with the aperture size. For observations from the ground, the turbulence in Earth's atmosphere above the telescope is the dominant error source. It can be mitigated by modeling of seeing-limited observations (*Lazorenko and Lazorenko,* 2004), by the use of adaptive optics (*Cameron et al.,* 2009), and with off-axis fringe tracking in dual-field interferometry (*Shao and Colavita,* 1992). Spaceborne instruments avoid atmospheric perturbations altogether and give access to nearly diffraction-limited observations, thus are ideal for high-precision astrometry work. Regardless of how the data were collected, the number of free astrometric parameters of a system with n planets is 5 + n × 7, i.e., at least 12 (see equation (7)), compared to 1 + n × 5 parameters for a radial velocity orbit adjustment (*Wright and Howard,* 2009). To obtain a robust solution and to minimize parameter correlations, e.g., between proper, parallactic, and orbital motion, a minimum timespan of one year and appropriate sampling of the orbital period are required.

8.2.1. Groundbased astrometry. Repeated imaging of a target and the measurement of its motion relative to background sources is a basic astrometric method, and several planet search surveys use seeing limited optical imaging with intermediate and large telescopes (*Pravdo and Shaklan,* 1996; *Boss et al.,* 2009). Accuracies of better than 0.1 mas have been achieved with this method (*Lazorenko et al.,* 2009), which satisfies the performance improvement necessary for efficient planet detection. Adaptive-optics assisted imaging is also being used, e.g., for a planet search targeting binaries with separations of a few arcseconds (*Röll et al.,* 2011). An optical interferometer realizes a large effective aperture size by combining the light of multiple telescopes that translates into an achievable precision of 0.01 mas in the relative separation measurement of two stars typically less than 1′ apart (*Shao and Colavita,* 1992). Several observatories have implemented the necessary infrastructure and are pursuing astrometric planet search programs (*Launhardt et al.,* 2008; *Muterspaugh et al.,* 2010; *Woillez et al.,* 2010; *Sahlmann et* al., 2013b). Similarly, very long baseline radio interferometry is a promising method for targeting nearby stars sufficiently bright at radio wavelengths (*Bower et al.*, 2009).

8.2.2. Astrometry from space. Space astrometry was firmly established by the Hipparcos mission, which operated in 1989–1992 and resulted in the determination of positions, proper motions, and absolute parallaxes at the 1-mas level for 120,000 stars (*Perryman et al.*, 1997). The satellite's telescope had a diameter of only 29 cm and scanned the entire celestial sphere several times to construct a global and absolute reference frame. However, Hipparcos data do not have the necessary precision to determine the astrometric orbits of the majority of known exoplanets. On a smaller scale but with slightly better precision, the Hubble Space Telescope's fine guidance sensor has made stellar parallax and orbit measurements possible (*Benedict et al.*, 2001). Because the Stellar Interferometry Mission (*Unwin et al.*, 2008) was discontinued, Gaia is the next space astrometry mission capable of detecting extrasolar planets.

8.3. Results from Astrometry

8.3.1. Combination with radial velocities. For a planet detected with RV, five out of seven orbital parameters are constrained. The two remaining parameters, the inclination i and Ω, can be determined by measuring the astrometric orbit. The knowledge of the RV parameters (or the high weight of RV measurements) leads to a significant reduction of the required S/N for a robust astrometric detection. Second, even an astrometric nondetection carries valuable information, e.g., an upper limit to the companion mass. Therefore, this type of combined analysis is so far the most successful application of astrometry in the exoplanet domain. Hipparcos astrometry yielded mass upper limits of RV planets (*Perryman et al.*, 1996; *Torres*, 2007; *Reffert and Quirrenbach*, 2011) and revealed that, in rare cases, brown dwarfs (*Sahlmann et al.*, 2011a) or stellar companions (*Zucker and Mazeh*, 2001) are mistaken for RV planets because their orbital planes are seen with small inclinations. Similarly, the Hubble fine guidance sensor was used to determine the orbits and masses of brown dwarf companions to Sun-like stars initially detected with RV (*Martioli et al.*, 2010; *Benedict et al.*, 2010) and groundbased imaging astrometry yielded a mass upper limit of ~3.6 M_{Jup} to the planet around GJ 317 (*Anglada-Escudé et al.*, 2012). *Sahlmann et al.* (2011b) (Fig. 14) used Hipparcos data to eliminate low-inclination binary systems mimicking brown dwarf companions detected in a large RV survey, revealing a mass range where giant planets and close brown dwarf companions around Sun-like stars are extremely rare.

8.3.2. Independent discoveries. Working toward the goal of exoplanet detection, optical imaging surveys have succeeded in measuring the orbits of low-mass binaries and substellar companions to M dwarfs (*Pravdo et al.*, 2005; *Dahn et al.*, 2008), relying on astrometric measurements only. Interferometric observations revealed the signature of a Jupiter-mass planet around a star in an unresolved

binary (*Muterspaugh et al.*, 2010), which, if confirmed independently, represents the first planet discovered by astrometry. Recent improvements of imaging astrometry techniques toward 0.1-mas precision made the discovery of a 28-M_{Jup} companion to an early L dwarf possible (Fig. 15) and demonstrated that such performance can be realized with a single-dish telescope from the ground.

8.4. The Future

Without a doubt, our expectations are high for the Gaia mission, which was launched on December 19, 2013. Gaia

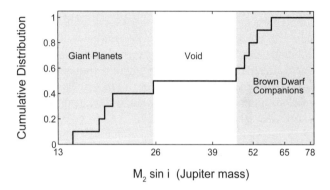

Fig. 14. The minimum mass distribution of substellar companions within 10 AU of Sun-like stars from the Coralie RV survey after constraining the orbital inclinations with Hipparcos astrometry.

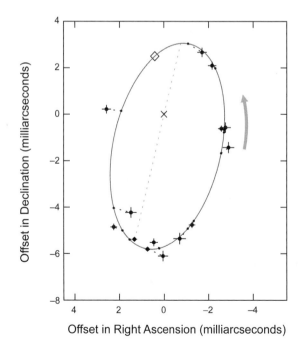

Fig. 15. The barycentric orbit of the L1.5 dwarf DENISP J082303.1-491201 caused by a 28-M_{Jup} companion in a 246-day orbit discovered through groundbased astrometry with an optical camera on an 8-m telescope (*Sahlmann et al.*, 2013a).

is a cornerstone mission of the European Space Agency that will implement an all-sky survey of an estimated billion stellar objects with visible magnitudes of 6–20 (*Perryman et al.*, 2001; *de Bruijne*, 2012). On average, the astrometry of a star will be measured 70 times over the mission lifetime of 5 years with a single measurement precision of ~0.02–0.05 mas for stars brighter than ~14th magnitude. Another look at Fig. 13 shows that hundreds of known exoplanet systems will be detectable, and it is expected that Gaia will discover thousands of new exoplanets (*Casertano et al.*, 2008), yielding a complete census of giant exoplanets in intermediate-period orbits around nearby stars. The sight of astrometric orbits caused by planets around stars will then become just as common as RV curves and dips in light curves are today. Assuming that it will perform as planned, Gaia will therefore add astrometry to the suite of efficient techniques for the study of exoplanet populations and will help us to advance our understanding of (exo)planet formation. It will also pave the way for future space astrometry missions aiming at detecting the Earth-like planets around nearby stars (*Malbet et al.*, 2012).

At the same time, groundbased surveys will remain competitive because they offer long lifetimes, scheduling flexibility, and access to targets not otherwise observable. They are also necessary for technology development and demonstration. The upcoming generation of submillimeter/optical interferometers and telescopes will have larger apertures and wide-field image correction, and hence provide us with even better astrometric performance and new opportunities for exoplanet science.

9. STATISTICAL DISTRIBUTIONS OF EXOPLANET PROPERTIES

In this section, we review and interpret the major statistical properties of extrasolar planets. We focus primarily on results from RV and transit surveys since they have produced the bulk of the discovered planets. Figure 16 shows known planets with measured masses and semimajor axes (projected for microlensing planets). The major archetypes of well-studied planets — cool Jupiters in ~1–5-AU orbits, hot Jupiters in sub-0.1-AU orbits, and sub-Neptune-sized planets orbiting within 1 AU — are all represented, although their relative frequencies are exaggerated due to differing survey sizes and yields. For more thorough reviews of exoplanet properties, the reader is directed to the literature (*Howard*, 2013; *Cumming*, 2011; *Marcy et al.*, 2005; *Udry and Santos*, 2007).

9.1. Abundant, Close-In Small Planets

Planets intermediate in size between Earth and Neptune are surprisingly common in extrasolar systems, but notably absent in our solar system. The planet size and mass distributions (Fig. 17) demonstrate that small planets substantially outnumber large ones, at least for close-in orbits. Doppler surveys using the High Resolution Echelle

Spectrometer (HIRES) at Keck Observatory (*Howard et al.*, 2010) and HARPS (*Lovis et al.*, 2009; *Mayor et al.*, 2011) at the 3.6-m ESO telescope have shown that small planets (Neptune-sized and smaller) significantly outnumber large ones for close-in orbits. Using the detected planets and detection completeness contours, the Eta-Earth Survey at Keck found that the probability of a star hosting a close-in planet scales as (M sin i)$^{-0.48}$: Small planets are more common. In absolute terms, 15% of Sun-like stars host one or more planets with M sin i = 3–30 $M_\oplus$ orbiting within 0.25 AU. The HARPS survey confirmed the rising planet mass function with decreasing mass and extended it to 1–3-$M_\oplus$ planets. It also demonstrated that low-mass planets have small orbital eccentricities and are commonly found in multi-planet systems with two to four small planets orbiting the same star with orbital periods of weeks or months. It found that at least 50% of stars have one or more planets of any mass with P < 100 d.

The distribution of planet sizes (radii) measured by the Kepler mission (Fig. 17) follows the same qualitative trend as the mass distribution, with small planets being more common (*Howard*, 2013; *Petigura et al.*, 2013; *Fressin et al.*, 2013). However, the planet radius distribution extends with small error bars down to 1 $R_\oplus$ for close-in planets, while the mass distribution has 50% uncertainty level near 1 $M_\oplus$. The size distribution is characterized by a power-law rise in occurrence with decreasing size (*Howard*, 2013) down to a critical size of ~2.8 $R_\oplus$, below which planet occurrence plateaus (*Petigura et al.*, 2013). The small planets detected by Kepler (<2 $R_\oplus$) appear to have more circular orbits than

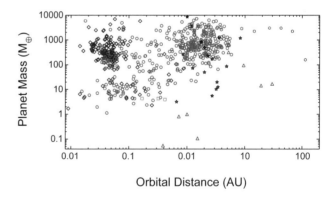

Fig. 16. Masses and orbital distances of planets from the Exoplanet Orbit Database (*Wright et al.*, 2011) (*exoplanets. org*) as of July 2013. The recently discovered Earth-sized planet, Kepler-78b, is also included (*Sanchis- Ojeda et al.*, 2013; *Howard et al.*, 2013; *Pepe et al.*, 2013). Extrasolar planets are coded according to their method of discovery: RV = circles, transit = diamonds, imaging = hexagons, gravitational microlensing = stars, and pulsar timing = squares. Planets in the solar system are triangles. Projected semimajor axis is plotted for microlensing planets while true semimajor axis is plotted for others. The occurrence of some planet types (e.g., hot Jupiters) are exaggerated relative to their true occurrence due to their relative ease of discovery.

larger planets (*Plavchan et al., 2012*), suggesting reduced dynamical interactions.

The high occurrence of small planets with P < 50 d likely extends to more distant orbits. As Kepler accumulates photometric data, it becomes sensitive to planets with smaller sizes and longer orbital periods. Based on 1.5 years of photometry, the small planet occurrence distribution as a function of orbital period is flat to P = 250 d (with higher uncertainty for larger P). Quantitatively, the mean number of planets per star per logarithmic period interval is proportional to $P^{+0.11\pm0.05}$ and $P^{-0.10\pm0.12}$ for 1–2-$R_{\oplus}$ and 2–4-$R_{\oplus}$ planets, respectively (*Dong and Zhu,* 2012).

The Kepler planet distribution also shows that small planets are more abundant around around cool stars (*Howard et al.,* 2012; although see *Fressin et al.,* 2013, for an opposing view). M dwarfs observed by Kepler appear to have a high rate of overall planet occurrence, 0.9 planets per star in the size range 0.5–4 $R_{\oplus}$ in P < 50 d orbits. Earth-sized planets (0.5–1.4 $R_{\oplus}$) are estimated to orbit in the habitable zones (HZ) of 15^{+13}_{-6}% of Kepler's M dwarfs (*Dressing and Charbonneau,* 2013). This estimate depends critically on the orbital bounds of the HZ; using more recent HZ models, the fraction of M dwarfs with Earth-sized planets in the HZ may be 3× higher (*Kopparapu,* 2013).

Of the Kepler planet host stars, 23% show evidence for two or more transiting planets. To be detected, planets in multi-transiting systems likely orbit in nearly the same plane, with mutual inclinations of a few degrees at most. The true number of planets per star (transiting or not) and their mutual inclinations can be estimated from simulated observations constrained by the number of single, double, triple, etc., transiting systems detected by Kepler (*Lissauer et al.,* 2011b). *Fang and Margot* (2012) find an intrinsic multi-planet distribution with 54%, 27%, 13%, 5%, and 2% of systems having 1, 2, 3, 4, and 5 planets with P < 200 d. Nearly all multi-planet systems (85%) have mutual inclinations of less than 3° (*Fang and Margot,* 2013; *Johansen et al.,* 2012). Mutual inclinations of a few degrees are also suggested by comparison between the Kepler and HARPS data (*Figueira et al.,* 2012). This high degree of co-planarity is consistent with planets forming in a protoplanetary disk without significant dynamical perturbations.

The ratios of orbital periods in multi-transiting systems provide additional dynamical constraints. These ratios are largely random (*Fabrycky et al.,* 2012), with a modest excess just outside of period ratios that are consistent with dynamical resonances (ratios of 2.1, 3.2, etc.) and a compensating deficit inside (*Lithwick and Wu,* 2012). The period ratios of adjacent planet pairs demonstrate that >31%, >35%, and >45% of two-planet, three-planet, and four-planet systems are dynamically packed; adding a hypothetical planet would gravitationally perturb the system into instability (*Fang and Margot,* 2013).

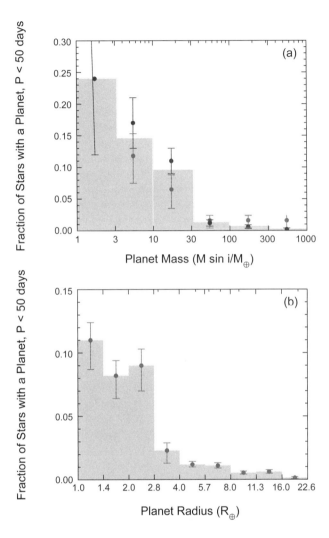

Fig. 17. The **(a)** mass and **(b)** size distributions of planets orbiting close to G and K-type stars. The distributions rise substantially with decreasing size and mass, indicating that small planets are more common than large ones. Planets smaller than 2.8 $R_{\oplus}$ or less massive than 30 $M_{\oplus}$ are found within 0.25 AU of 30–50% of Sun-like stars. In **(a)**, the size distribution is drawn from two studies of Kepler data: *Petigura et al.* (2013) for planets smaller than 4× Earth size and *Howard et al.* (2012) for larger planets. The mass (M sin i) distributions show the fraction of stars having at least one planet with an orbital period shorter than 50 days (orbiting inside ~0.25 AU) are from separate Doppler surveys [darker lines = *Howard et al.* (2010); lighter lines = *Mayor et al.* (2011)], while the histogram shows their average values. Both distributions are corrected for survey incompleteness for small/low-mass planets to show the true occurrence of planets in nature.

9.2. Gas Giant Planets

The orbits of giant planets are the easiest to detect using the Doppler technique and were the first to be studied statistically (e.g., *Udry et al.,* 2003; *Marcy et al.,* 2005). Observations over a decade of a volume-limited sample of ~1000 F-, G-, and K-type dwarf stars at Keck Observatory showed that 10.5% of G- and K-type dwarf stars host one or more giant planets (0.3–10 M_{Jup}) with orbital periods of 2–2000 d (orbital distances of ~0.03–3 AU). Within those parameter ranges, less-massive and more-distant giant planets are more common. Extrapolation of this model suggests that 17–20% of such stars have giant planets orbiting within 20 AU (P = 90 yr) (*Cumming et al.*, 2008). This extrapolation is consistent with a measurement of giant planet occurrence beyond ~2 AU from microlensing surveys (*Gould et al.*, 2010). However, the relatively few planet detections from direct imaging planet searches suggest that the extrapolation is not valid beyond ~65 AU (*Nielsen and Close*, 2010).

These smooth trends in giant planet occurrence mask pile-ups in semimajor axis (*Wright et al.*, 2009). The orbital distances for giant planets show a preference for orbits larger than ~1 AU and to a lesser extent near 0.05 AU ("hot Jupiters") (Fig. 18a). This period valley for apparently single planets is interpreted as a transition region between two categories of jovian planets with different migration histories (*Udry et al.*, 2003). The excess of planets starting at ~1 AU approximately coincides with the location of the ice line, which provides additional solids that may speed the formation of planet cores or act as a migration trap for planets formed farther out (*Ida and Lin*, 2008). The semimajor axis distribution for giant planets in multi-planet systems is more uniform, with hot Jupiters nearly absent and a suppressed peak of planets in >1-AU orbits.

The giant planet eccentricity distribution (Fig. 18b) also differs between single- and multi-planet systems. The eccentricities of single planets can be reproduced by a dynamical model in which initially low eccentricities are excited by planet-planet scattering (*Chatterjee et al.*, 2008). Multiplanet systems with a giant planet likely experienced substantially fewer scattering events. The single-planet systems may represent the survivors of scattering events that ejected other planets in the system.

Metal-rich stars are more likely to host giant planets within 5 AU. This "planet-metallicity correlation" was validated statistically by Doppler surveys of stars with $M_\star = 0.7–1.2\ M_\odot$ and uniformly measured metallicities (*Fischer and Valenti*, 2005; *Santos et al.*, 2004). The probability of a star hosting a giant planet is proportional to the square of the number of iron atoms in the star relative to the Sun, $\mathcal{P}(\text{planet}) \propto N_{Fe}^2$. A later Doppler study spanned a wider range of stellar masses (0.3–2.0 $M_\odot$) and showed that the probability of a star hosting a giant planet correlates with both stellar metal content and stellar mass, $\mathcal{P}(\text{planet}) \propto N_{Fe}^{1.2\pm0.2}\ M_\star^{1.0\pm0.3}$ (*Johnson et al.*, 2010a). Note that the planet-metallicity correlation only applies to gas giant planets. Planets larger than 4 $R_\oplus$ (Neptune-sized) preferentially orbit metal-rich stars, while smaller planets are are nondiscriminating in stellar metallicity (*Buchhave et al.*, 2012). This pattern of host star metallicity can be explained if small planets commonly form in protoplanetary disks, but only a fraction of those small planets grow to a critical size in time to become gas giants.

Although hot Jupiters (giant planets with P ≤ 10 d) are found around only 0.5–1.0% of Sun-like stars (*Wright et al.*, 2012), they are the most well-characterized planets because they are easy to detect and follow up with ground- and spacebased telescopes. However, their origin remains mysterious. In contrast to the commonly multiple sub-Neptune-sized planets, hot Jupiters are usually the only detected

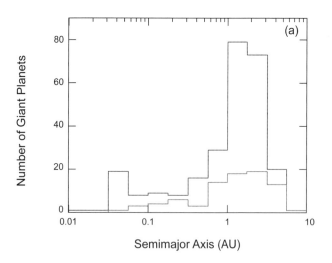

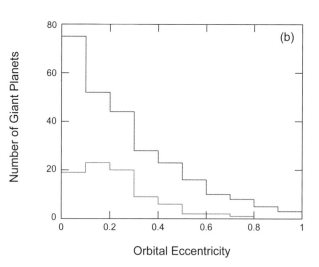

Fig. 18. Orbital characteristics of giant planets ($M_P \sin i > 0.2\ M_{Jup}$) detected by Doppler surveys as cataloged on the Exoplanet Orbit Database (*Wright et al.*, 2011). **(a)** The number distribution of semimajor axes shows that apparently single planets (top line) preferentially orbit at distances of ~0.05 AU and at ~1–3 AU from their host stars. These preferred orbits are diminished in multi-planet systems (bottom line). The decline in number of detected planets for orbits outside ~3 AU is not significant; fewer stars have been searched for such planets compared to the closer orbits. **(b)** The distribution of orbital eccentricities for apparently single planets (top line) span the full range, with low-eccentricity orbits being more common. Giant planets in multi-planet systems (bottom line) have orbits that are more commonly close to circular. The larger eccentricities of single planets suggests that they were dynamically excited from a quiescent, nearly circular origin, perhaps by planet-planet scattering that resulted in the ejection of all but one detectable planet per system.

planet orbiting the host star within observational limits (*Steffen et al.,* 2012). Many hot Jupiters have low eccentricities due to tidal circularization. The measured obliquities of stars hosting hot Jupiters display a peculiar pattern: Obliquities are apparently random above a critical stellar temperature of ~6250 K, but cooler systems are mostly aligned. *In situ* formation is unlikely for hot Jupiters because of insufficient protoplanetary disk mass so close to the star. It is more likely that they formed at several astronomical units, were gravitationally perturbed into orbits with random inclinations and high eccentricities, and were captured at ~0.05 AU by dissipation of orbital energy in tides raised on the planet. For systems with sufficiently strong tides raised by the planet on the star (which depend on a stellar convective zone that is only present below for $T_{eff} \leq 6250$ K), the stellar spin axis aligns to the orbital axis.

9.3. Mass-Radius Relationships

While the mass and size distributions provide valuable information about the relative occurrence of planets of different types, it remains challenging to connect the two. Knowing the mass of a planet only weakly specifies its size, and vice versa. This degeneracy can be lifted for ~200 planets with well-measured masses and radii (Fig. 19), most of which are transiting hot Jupiters. The cloud of points follows a diagonal band from low-mass/small-size to high-mass/large-size. This band of allowable planet mass/size combinations has considerable breadth. Planets less massive than ~30 $M_\oplus$ vary in size by a factor of ~5 and planets larger than ~100 $M_\oplus$ (gas giants) vary by a factor of ~2. For the gas giants, the size dispersion at a given mass is due largely to two effects. First, planets in tight orbits receive higher stellar flux and are more commonly inflated. While higher stellar flux correlates with giant planet inflation (*Weiss et al.,* 2013), it is unclear how the stellar energy is deposited in the planet's interior. Less importantly, the presence of a massive solid core (or distributed heavy elements) increases a planet's surface gravity, causing it to be more compact.

Low-mass planets show an even larger variation in size and composition. Three examples of sub-Neptune-sized planets illustrate the diversity. The planet Kepler-10b has a mass of 4.6 $M_\oplus$ and a density of 9 g cm^{-3}, indicating a rock/iron composition and no atmosphere (*Batalha et al.,* 2011). In contrast, the planet Kepler-11e has a density of 0.5 g cm^{-3} and a mass of 8 $M_\oplus$. A substantial light-element atmosphere (probably hydrogen) is required to explain its mass and radius combination (*Lissauer et al.,* 2011a). The masses and radii of intermediate planets lead to ambiguous conclusions about composition. For example, the bulk physical properties of GJ 1214b [6.5 $M_\oplus$, 2.7 $R_\oplus$, 1.9 g cm^{-3} (*Charbonneau et al.,* 2009)] are consistent with several compositions: a "super-Earth" with a rock/iron core surrounded by ~3% H_2 gas by mass; a water-world planet consisting of a rock/iron core, water ocean, and atmosphere that contribute ~50% of the mass; or a mini-Neptune composed of rock/iron, water, and H/He gas (*Rogers and Seager,* 2010).

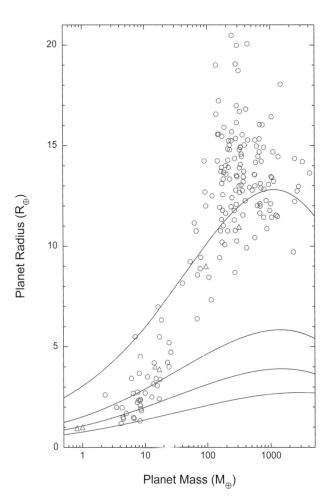

Fig. 19. Masses and radii of well-characterized planets from the Exoplanet Orbit Database (*Wright et al.,* 2011). Extrasolar planets are shown as open circles and solar system planets are open triangles. Lines show model mass-radius relationships for idealized planets consisting of pure hydrogen (*Seager et al.,* 2007), water, rock (Mg$_2$SiO$_4$), or iron (*Fortney et al.,* 2007). Poorly understood heating mechanisms inflate some gas giant planets (larger than ~8 $R_\oplus$) to sizes larger than predicted by the simple hydrogen model. Smaller planets (less massive than ~30 $M_\oplus$) show great diversity in size at a fixed mass, likely due to varying density of solids and atmospheric extent.

Acknowledgments. D.F. acknowledges support from NSF AST1207748 and NASA NNX12AC01C. A.W.H. acknowledges NASA grant NNX12AJ23G. Portions of this work performed under the auspices of the U.S. Department of Energy by Lawrence Livermore National Laboratory under Contract DE-AC52-07NA27344. G.L. acknowledges support from NASA through Spitzer RSA 1438930. S.M. acknowledges support from National Space Foundation (NSF) grants AST-1006676, AST-1126413, AST-1310885, PSARC, the NASA Astrobiology Institute (NNA09DA76A), and the Center for Exoplanets and Habitable Worlds in the pursuit of precise RVs in the NIR. The Center for Exoplanets and Habitable Worlds is supported by the Pennsylvania State University, the Eberly College of Science, and the Pennsylvania Space Grant Consortium. Work by J.Y.

was performed in part by a Distinguished University Fellowship from The Ohio State University and in part under contract with the California Institute of Technology (Caltech)/Jet Propulsion Laboratory (JPL) funded by NASA through the Sagan Fellowship Program executed by the NASA Exoplanet Science Institute.

REFERENCES

Agol E. et al. (2005) *Mon. Not. R. Astron. Soc., 359,* 567.
Albrow M. D. et al. (2000) *Astrophys. J., 534,* 894.
Alcock C. et al. (1992) In *Robotic Telescopes in the 1990s* (A. V. Filippenko, ed.), pp. 193–202. ASP Conf. Ser. 34, Astronomical Society of the Pacific, San Francisco.
Alonso R. et al. (2004) *Astrophys. J. Lett., 613,* L153.
Anglada-Escudé G. et al. (2012) *Astrophys. J., 746,* 37.
Aubourg E. et al. (1993) *The Messenger, 72,* 20.
Baines E. K. et al. (2012) *Astrophys. J., 761,* 57.
Bakos G. Á. et al. (2007) *Astrophys. J., 656,* 552.
Bakos G. Á. et al. (2013) *Publ. Astron. Soc. Pac., 125,* 154.
Barbieri M. et al. (2007) *Astron. Astrophys., 476,* L13.
Barge P. et al. (2008) *Astron. Astrophys., 482,* L17.
Barman T. S. et al. (2011) *Astrophys. J., 733,* 65.
Batalha N. M. et al. (2010) *Astrophys. J. Lett., 713,* L103.
Batalha N. M. et al. (2011) *Astrophys. J., 729,* 27.
Batalha N. M. et al. (2013) *Astrophys. J. Suppl., 204,* 24.
Batista V. et al. (2011) *Astron. Astrophys., 529,* A102.
Batygin K. and Stevenson D. J. (2010) *Astrophys. J. Lett., 714,* L238.
Beaulieu J.-P. et al. (2006) *Nature, 439,* 437.
Beaulieu J. P. et al. (2013) *ArXiv e-prints,* arXiv:1303.6783.
Benedict G. F. et al. (2001) *Astron. J., 121,* 1607.
Benedict G. F. et al. (2010) *Astron. J., 139,* 1844.
Bennett D. P. and Rhie S. H. (2000) In *Bull. Am. Astron. Soc., 32,* 1053.
Bennett D. P. and Rhie S. H. (2002) *Astrophys. J., 574,* 985.
Beuzit J.-L. et al. (2008) In *Ground-Based and Airborne Instrumentation for Astronomy II* (I. S. McLean and M. M. Casali, eds.), SPIE Conf. Series 7014, Bellingham, Washington.
Black D. C. (1997) *Astrophys. J. Lett., 490,* L171.
Black D. C. and Scargle J. D. (1982) *Astrophys. J., 263,* 854.
Bond I. A. et al. (2004) *Astrophys. J. Lett., 606,* L155.
Bonfils X. et al. (2013) *Astron. Astrophys., 549,* A109.
Borucki W. J. et al. (2010) *Science, 327,* 977.
Borucki W. J. et al. (2012) *Astrophys. J., 745,* 120.
Borucki W. J. et al. (2013) *Science, 340,* 587.
Boss A. P. (2006) *Astrophys. J., 643,* 501.
Boss A. P. (2009) *Publ. Astron. Soc. Pac, 121,* 1218.
Bower G. C. et al. (2009) *Astrophys. J., 701,* 1922.
Broeg C. et al. (2013) In *EPJ Web Conf., 47,* 3005.
Brogi M. et al. (2012) *Nature, 486,* 502.
Brogi M. et al. (2013) *Astrophys. J., 767,* 27.
Buchhave L. A. et al. (2012) *Nature, 486,* 375.
Butler R. P. et al. (1999) *Astrophys. J., 526,* 916.
Cameron P. B. et al. (2009) *Astron. J., 137,* 83.
Campbell B. et al. (1988) *Astrophys. J., 331,* 902.
Casertano S. et al. (2008) *Astron. Astrophys., 482,* 699.
Cassan A. et al. (2012) *Nature, 481,* 167.
Chabrier G. and Baraffe I. (2000) *Annu. Rev. Astron. Astrophys., 38,* 337.
Charbonneau D. et al. (2000) *Astrophys. J. Lett., 529,* L45.
Charbonneau D. et al. (2007) In *Protostars and Planets V* (B. Reipurth et al., eds.), pp. 701–716. Univ. of Arizona, Tucson.
Charbonneau D. et al. (2009) *Nature, 462,* 891.
Chatterjee S. et al. (2008) *Astrophys. J., 686,* 580.
Chauvin G. et al. (2004) *Astron. Astrophys., 425,* 2, L29.
Chauvin G. et al. (2010) *Astron. Astrophys., 509,* A52.
Chiang E. and Laughlin G. (2013) *Mon. Not. R. Astron. Soc., 431,* 3444.
Chwolson O. (1924) *Astron. Nachr., 221,* 329.
Collier Cameron A. et al. (2007) *Mon. Not. R. Astron. Soc., 375,* 951.
Collins K. A. et al. (2013) *ArXiv e-prints,* arXiv:1308.2296.
Crepp J. R. (2013) In *AAS Meeting Abstracts, 221,* 149.12.
Cumming A. (2011) In *Exoplanets* (S. Seager., ed.), pp. 191–214. Univ. of Arizona, Tucson.
Cumming A. et al. (2008) *ArXiv e-prints,* arXiv:0803.3357.
Currie T. et al. (2011) *Astrophys. J., 729,* 128.

Dahn C. C. et al. (2008) *Astrophys. J., 686,* 548.
de Bruijne J. H. J. (2012) *Astrophys. Space Sci., 341,* 31.
de Kok R. J. et al. (2013) *Astron. Astrophys., 554,* A82.
Deming D. et al. (2009) *Publ. Astron. Soc. Pac., 121,* 952.
Dominik M. et al. (2010) *Astron. Nachr., 331,* 671.
Dong S. and Zhu Z. (2012) *ArXiv e-prints,* arXiv:1212.4853.
Dong S. et al. (2009) *Astrophys. J., 695,* 970.
Dragomir D. et al. (2013) *Astrophys. J. Lett., 772,* L2.
Dressing C. D. and Charbonneau D. (2013) *Astrophys. J., 767,* 95.
Dumusque X. et al. (2012) *Nature, 491,* 207.
Einstein A. (1936) *Science, 84,* 506.
Eriksson U. and Lindegren L. (2007) *Astron. Astrophys., 476,* 1389.
Fabrycky D. C. and Murray-Clay R. A. (2010) *Astrophys. J., 710,* 1408.
Fabrycky D. C. et al. (2012) *ArXiv e-prints,* arXiv:1201.5415.
Fang J. and Margot J.-L. (2012) *Astrophys. J., 761,* 92.
Fang J. and Margot J.-L. (2013) *Astrophys. J., 767,* 115.
Figueira P. et al. (2012) *Astron. Astrophys., 541,* A139.
Fischer D. A. and Valenti J. (2005) *Astrophys. J., 622,* 1102.
Fischer D. A. et al. (2012) *Mon. Not. R. Astron. Soc., 419,* 2900.
Fischer D. A. et al. (2014) *Astrophys. J., 210,* 5.
Fortney J. J. et al. (2007) *Astrophys. J., 659,* 1661.
Fressin F. et al. (2011) *Astrophys. J. Suppl., 197,* 5.
Fressin F. et al. (2013) *Astrophys. J., 766,* 81.
Garcia-Melendo E. and McCullough P. R. (2009) *Astrophys. J., 698,* 558.
Gatewood G. (1976) *Icarus, 27,* 1.
Gaudi B. S. (1998) *Astrophys. J., 506,* 533.
Gaudi B. S. (2012) *Annu. Rev. Astron. Astrophys., 50,* 411.
Gaudi B. S. and Gould A. (1997) *Astrophys. J., 486,* 85.
Gaudi B. S. et al. (2008) *Science, 319,* 927.
Ghasempour A. et al. (2012) In *Modern Technologies in Space- and Ground-Based Telescopes and Instrumentation II* (R. Navarro et al., eds.), SPIE Conf. Series 8450, Bellingham, Washington.
Gillon M. et al. (2007) *Astron. Astrophys., 472,* L13.
Gorbikov E. et al. (2010) *Astrophys. Space Sci., 326,* 203.
Gould A. (1992) *Astrophys. J., 392,* 442.
Gould A. (2004) *Astrophys. J., 606,* 319.
Gould A. and Loeb A. (1992) *Astrophys. J., 396,* 104.
Gould A. et al. (1994) *Astrophys. J. Lett., 423,* L105.
Gould A. et al. (2006) *Astrophys. J. Lett., 644,* L37.
Gould A. et al. (2010) *Astrophys. J., 720,* 1073.
Gould A. et al. (2013) *Astrophys. J., 763,* 141.
Gray D. F. (1997) *Nature, 385,* 795.
Griest K. and Safizadeh N. (1998) *Astrophys. J., 500,* 37.
Griest K. et al. (1991) *Astrophys. J. Lett., 372,* L79.
Guyon O. et al. (2005) *Astrophy. J., 622,* 1, 744.
Guyon O. et al. (2006) *Astrophys. J. Suppl. Ser., 167,* 81.
Guyon O. et al. (2012) In *Adaptive Optics Systems III* (B. L. Ellerbroek et al., eds.), SPIE Conf. Series 8447, Bellingham, Washington.
Halverson S. et al. (2012) In *Ground-Based and Airborne Instrumentation for Astronomy IV* (I. S. McLean et al., eds.), pp. 84468Q–84468Q-16. SPIE Conf. Series 8446, Bellingham, Washington.
Han C. et al. (2013) *Astrophys. J. Lett., 762,* L28.
Hatzes A. P. et al. (1997) *Astrophys. J., 478,* 374.
Hatzes A. P. et al. (2003) *Astrophys. J., 599,* 1383.
Henry G. W. et al. (2000) *Astrophys. J. Lett., 529,* L41.
Hilditch R.W. (2001) *An Introduction to Close Binary Stars.* Cambridge Univ., Cambridge.
Holman M. J. and Murray N. W. (2005) *Science, 307,* 1288.
Holman M. J. et al. (2010) *Science, 330,* 51.
Horne K. (2003) In *Scientific Frontiers in Research on Extrasolar Planets* (D. Deming and S. Seager, eds.), pp. 361–370. ASP Conf. Ser. 294, Astronomical Society of the Pacific, San Francisco.
Howard A. W. (2013) *Science, 340,* 572.
Howard A. W. et al. (2010) *Science, 330,* 653.
Howard A. W. et al. (2012) *Astrophys. J. Suppl., 201,* 15.
Howard A. W. et al. (2013) *Nature, 503,* 381.
Hunter T. R. and Ramsey L. W. (1992) *Publ. Astron. Soc. Pac., 104,* 1244.
Hwang K.-H. et al. (2013) *Astrophys. J., 778,* 55.
Ida S. and Lin D. N. C. (2004) *Astrophys. J., 604,* 388.
Ida S. and Lin D. N. C. (2005) *Astrophys. J., 626,* 1045.
Ida S. and Lin D. N. C. (2008) *Astrophys. J., 673,* 487.
Janson M. et al. (2013) *Astrophys. J., 773(1),* 73.
Jenkins J. M. et al. (2010) *Astrophys. J. Lett., 713,* L87.
Johansen A. et al. (2012) *Astrophys. J., 758,* 39.
Johnson J. A. et al. (2009) *Astrophys. J. Lett., 692,* L100.

Johnson J. A. et al. (2010a) *Publ. Astron. Soc. Pac, 122*, 905.
Johnson J. A. et al. (2010b) *Publ. Astron. Soc. Pac., 122*, 149.
Kalas P. et al. (2008) *Science, 322*, 1345.
Kalas P. et al. (2013) *Astrophys. J., 775*, 56.
Kasdin N. J. et al. (2003) *Astrophys. J., 582*, 2, 1147.
Kasper M. et al. (2010) In *Ground-Based and Airborne Instrumentation for Astronomy III* (I. S. McLean et al., eds.), SPIE Conf. Series 7735, Bellingham, Washington.
Konacki M. et al. (2003) *Astrophys. J., 597*, 1076.
Konopacky Q. M. et al. (2013) *Science, 339*, 6126, 1398.
Kopparapu R. K. (2013) *Astrophys. J. Lett., 767*, L8.
Kuchner M. J. and Traub W. A. (2002) *Astrophy. J., 570*, 2, 900.
Kuzuhara M. et al. (2013) *Astrophys. J., 774*, 11.
Lafrenière D. et al. (2007a) *Astrophys. J., 660*, 770.
Lafrenière D. et al. (2007b) *Astrophys. J., 670*, 1367.
Lafrenière D. et al. (2010) *Astrophys. J., 719*, 497.
Lagrange A.-M. et al. (2010) *Science, 329*, 57.
Lagrange A.-M. et al. (2011) *Astron. Astrophys., 528*, L9.
Latham D. W. et al. (1989) *Nature, 339*, 38.
Laughlin G. et al. (2004) *Astrophys. J. Lett., 612*, L73.
Launhardt R. et al. (2008) In *Optical and Infrared Interferometry* (M. Schöller et al., eds.), SPIE Conf. Series 7013, Bellingham, Washington.
Lazorenko P. F. and Lazorenko G. A. (2004) *Astron. Astrophys., 427*, 1127.
Lazorenko P. F. et al. (2009) *Astron. Astrophys., 505*, 903.
Léger A. et al. (2009) *Astron. Astrophys., 506*, 287.
Lintott C. J. et al. (2013) *Astron. J., 145*, 151.
Lissauer J. J. et al. (2011a) *Nature, 470*, 53.
Lissauer J. J. et al. (2011b) *Astrophys. J. Suppl., 197*, 8.
Lithwick Y. and Wu Y. (2012) *Astrophys. J. Lett., 756*, L11.
Lithwick Y. et al. (2012) *Astrophys. J., 761*, 122.
Liu M. C. et al. (2010) In *Adaptive Optics Systems II* (B. L. Ellerbroek et al., eds.), SPIE Conf. Series 7736, Bellingham, Washington.
Lockwood A. C. et al. (2014) *ArXiv e-prints*, arXiv:1402.0846.
Lovis C. and Fischer D. (2011) In *Exoplanets* (S. Seager, ed.), pp. 27–53. Univ. of Arizona, Tucson.
Lovis C. et al. (2009) In *Transiting Planets* (F. Pont et al., eds.), pp. 502–505. IAU Symp. 253, Cambridge Univ., Cambridge.
Lyot B. (1939) *Mon. Not. R. Astron. Soc., 99*, 580.
Macintosh B. et al. (2006) In *Advances in Adaptive Optics II* (B. L. Ellerbroek and D. B. Calia, eds.), SPIE Conf. Series 6272, Bellingham, Washington.
Macintosh B. A. et al. (2012) In *Ground-Based and Airborne Instrumentation for Astronomy IV* (I. S. McLean et al., eds.), pp. 84461U–84461U-9. SPIE Conf. Series 8446, Bellingham, Washington.
Mahadevan S. et al. (2012) In *Ground-Based and Airborne Instrumentation for Astronomy IV* (I. S. McLean et al., eds.), SPIE Conf. Series 8446, Bellingham, Washington.
Mahmud N. I. et al. (2011) *Astrophys. J., 736*, 123.
Malbet F. et al. (2012) *Exp. Astron., 34*, 385.
Mannings V. et al., eds. (2000) *Protostars and Planets IV.* Univ. of Arizona, Tucson. 1422 pp.
Mao S. and Paczynski B. (1991) *Astrophys. J. Lett., 374*, L37.
Marcy G. et al. (2005) *Progr. Theor. Phys. Suppl., 158*, 24.
Marley M. S. et al. (2007) *Astrophys. J., 655*, 541.
Marley M. S. et al. (2012) *Astrophys. J., 754*, 135.
Marmier M. et al. (2013) *Astron. Astrophys., 551*, A90.
Marois C. et al. (2006) *Astrophys. J., 641*, 556.
Marois C. et al. (2008) *Science, 322*, 1348.
Marois C. et al. (2010) *Nature, 468*, 1080.
Martinache F. et al. (2012) In *Adaptive Optics Systems III* (B. L. Ellerbroek et al., eds.), pp. 84471Y–84471Y-8. SPIE Conf. Series 8447, Bellingham, Washington.
Martioli E. et al. (2010) *Astrophys. J., 708*, 625.
Mayor M. and Queloz D. (1995) *Nature, 378*, 355.
Mayor M. et al. (2003) *The Messenger, 114*, 20.
Mayor M. et al. (2009) *Astron. Astrophys., 493*, 639.
Mayor M. et al. (2011) *ArXiV e-prints*, arXiv:1109.2497.
McBride J. et al. (2011) *Publ. Astron. Soc. Pac., 123(904)*, 692.
McCullough P. R. et al. (2006) *Astrophys. J., 648*, 1228.
Moorhead A. V. et al. (2011) *Astrophys. J. Suppl., 197*, 1.
Morton T. D. and Johnson J. A. (2011) *Astrophys. J., 738*, 170.
Muterspaugh M. W. et al. (2010) *Astron. J., 140*, 1657.
Nersisyan S. R. et al. (2013) *Optics Express, 21(7)*, 8205.
Nielsen E. L. and Close L. M. (2010) *Astrophys. J., 717*, 878.

Nielsen E. L. et al. (2013) *Astrophys. J., 776(1)*, 4.
Oppenheimer B. R. et al. (2013) *Astrophys. J., 768(1)*, 24.
Paczynski B. (1991) *Astrophys. J. Lett., 371*, L63.
Park B.-G. et al. (2012) In *Ground-Based and Airborne Telescopes IV* (L. M. Stepp et al., eds.), SPIE Conf. Series 8444, Bellingham, Washington.
Penny M. T. et al. (2012) *ArXiv e-prints*, arXiv:1206.5296.
Pepe F. et al. (2011) *Astron. Astrophys., 534*, A58.
Pepe F. et al. (2013) *Nature, 503*, 377.
Perrin M. D. et al. (2003) *Astrophys. J., 596*, 702.
Perryman M. (2012) *EPJ H, 37*, 745.
Perryman M. A. C. et al. (1996) *Astron. Astrophys., 310*, L21.
Perryman M. A. C. et al. (1997) *Astron. Astrophys., 323*, L49.
Perryman M. A. C. et al. (2001) *Astron. Astrophys., 369*, 339.
Petigura E. A. et al. (2013) *Astrophys. J., 770*, 69.
Petit C. et al. (2012) In *Adaptive Optics Systems III* (B. L. Ellerbroek et al., eds.), pp. 84471Z–84471Z-12. SPIE Conf. Series 8447, Bellingham, Washington.
Plavchan P. et al. (2012) *ArXiv e-prints*, arXiv:1203.1887.
Pravdo S. H. and Shaklan S. B. (1996) *Astrophys. J., 465*, 264.
Pravdo S. H. et al. (2005) *Astrophys. J., 630*, 528.
Quirrenbach A. et al. (2012) In *Ground-Based and Airborne Instrumentation for Astronomy IV* (I. S. McLean et al., eds.), SPIE Conf. Series 8446, Bellingham, Washington.
Ramsey L. W. et al. (2008) *Publ. Astron. Soc. Pac., 120*, 887.
Redman S. L. et al. (2011) *Astrophys. J Suppl., 195*, 24.
Redman S. L. et al. (2012) *Astrophys. J. Suppl., 199*, 2.
Reffert S. and Quirrenbach A. (2011) *Astron. Astrophys., 527*, A140.
Reipurth B. et al., eds. (2007) *Protostars and Planets V.* Univ. of Arizona, Tucson. 1024 pp.
Renn J. et al. (1997) *Science, 275*, 184.
Ricker G. R. et al. (2010) *Bull. Am. Astron. Soc., 42*, 450.06.
Rodler F. et al. (2012) *Astrophys. J. Lett., 753*, L25.
Rogers L. A. and Seager S. (2010) *Astrophys. J., 716*, 1208.
Röll T. et al. (2011) In *GAIA: At the Frontiers of Astrometry* (C. Turon et al., eds.), pp. 429–432. EAS Publ. Scr., Vol. 45, Cambridge Univ., Cambridge.
Sahlmann J. (2012) Observing exoplanet populations with high-precision astrometry, Ph.D. thesis, Observatoire de Genève, Université de Genève.
Sahlmann J. et al. (2011a) *Astron. Astrophys., 528*, L8.
Sahlmann J. et al. (2011b) *Astron. Astrophys., 525*, A95.
Sahlmann J. et al. (2013a) *ArXiv e-prints*, arXiv:1306.3225.
Sahlmann J. et al. (2013b) *Astron. Astrophys., 551*, A52.
Sanchis-Ojeda R. et al. (2013) *Astrophys. J., 774*, 54.
Santerne A. et al. (2013) *ArXiv e-prints*, arXiv:1310.0748.
Santos N. C. et al. (2004) *Astron. Astrophys., 415*, 1153.
Sato B. et al. (2008) *Exp. Astron., 22*, 51.
Schwab C. et al. (2012) *ArXiv e-prints*, arXiv:1212.4867.
Schwamb M. E. et al. (2013) *Astrophys. J., 768*, 127.
Seager S. and Deming D. (2010) *Annu. Rev. Astron. Astrophys., 48*, 631.
Seager S. et al. (2007) *Astrophys. J., 669*, 1279.
Shao M. and Colavita M. M. (1992) *Annu. Rev. Astron. Astrophys., 30*, 457.
Shvartzvald Y. and Maoz D. (2012) *Mon. Not. R. Astron. Soc., 419*, 3631.
Sivaramakrishnan A. et al. (2001) *Astrophys. J., 552(1)*, 397.
Skemer A. J. et al. (2012) *Astrophys. J., 753*, 14.
Skowron J. et al. (2011) *Astrophys. J., 738*, 87.
Snellen I. A. G. et al. (2013) *Astrophys. J., 764*, 182.
Soszyński I. et al. (2012) *Acta Astron., 62*, 219.
Soummer R. et al. (2011) *Astrophys. J., 729*, 144.
Sozzetti A. (2005) *Publ. Astron. Soc. Pac., 117*, 1021.
Spanò P. et al. (2012) In *Ground-Based and Airborne Instrumentation for Astronomy IV* (I. S. McLean et al., eds.), SPIE Conf. Series 8446, Bellingham, Washington.
Spergel D. et al. (2013) *ArXiv e-prints*, arXiv:1305.5425.
Steffen J. H. et al. (2012) *Proc. Natl. Acad. Sci., 109*, 7982.
Strassmeier K. G. et al. (2008) In *Ground-Based and Airborne Instrumentation for Astronomy II* (I. S. McLean and M. M. Casali, eds.), SPIE Conf. Series 7014, Bellingham, Washington.
Sumi T. et al. (2011) *Nature, 473*, 349.
Torres G. (2007) *Astrophys. J. Lett., 671*, L65.
Torres G. et al. (2004) *Astrophys. J., 614*, 979.
Tsapras Y. et al. (2009) *Astron. Nachr., 330*, 4.
Udalski A. (2003) *Acta Astron., 53*, 291.

Udalski A. et al. (2005) *Astrophys. J. Lett., 628,* L109.
Udry S. and Santos N. C. (2007) *Annu. Rev. Astron. Astrophys., 45,* 397.
Udry S. et al. (2003) *Astron. Astrophys., 407,* 369.
Unwin S. C. et al. (2008) *Publ. Astron. Soc. Pac., 120,* 38.
Weiss L. M. et al. (2013) *Astrophys. J., 768,* 14.
Winn J. N. et al. (2011) *Astrophys. J. Lett., 737,* L18.
Woillez J. et al. (2010) In *Optical and Infrared Interferometry II*
    (W. C. Danchi et al., eds), p. 7734-12. SPIE Conf. Series 7734,
    Bellingham, Washington.

Wright J. T. and Howard A. W. (2009) *Astrophys. J. Suppl., 182,* 205.
Wright J. T. et al. (2009) *Astrophys. J., 693,* 1084.
Wright J. T. et al. (2011) *Publ. Astron. Soc. Pac., 123,* 412.
Wright J. T. et al. (2012) *Astrophys. J., 753,* 160.
Ycas G. et al. (2012a) *Optics Lett., 37,* 2199.
Ycas G. G. et al. (2012b) *Optics Express, 20,* 6631.
Yoo J. et al. (2004) *Astrophys. J., 603,* 139.
Zucker S. and Mazeh T. (2001) *Astrophys. J., 562,* 549.
Zuckerman B. and Song I. (2004) *Annu. Rev. Astron. Astrophys., 42,* 685.

Madhusudhan N., Knutson H., Fortney J. J., and Barman T. (2014) Exoplanetary atmospheres. In *Protostars and Planets VI* (H. Beuther et al., eds.), pp. 739–762. Univ. of Arizona, Tucson, DOI: 10.2458/azu_uapress_9780816531240-ch032.

Exoplanetary Atmospheres

Nikku Madhusudhan
Yale University and University of Cambridge

Heather Knutson
California Institute of Technology

Jonathan J. Fortney
University of California, Santa Cruz

Travis Barman
Lowell Observatory and The University of Arizona

The study of exoplanetary atmospheres is one of the most exciting and dynamic frontiers in astronomy. Over the past two decades ongoing surveys have revealed an astonishing diversity in the planetary masses, radii, temperatures, orbital parameters, and host stellar properties of exoplanetary systems. We are now moving into an era where we can begin to address fundamental questions concerning the diversity of exoplanetary compositions, atmospheric and interior processes, and formation histories, just as have been pursued for solar system planets over the past century. Exoplanetary atmospheres provide a direct means to address these questions via their observable spectral signatures. In the last decade, and particularly in the last five years, tremendous progress has been made in detecting atmospheric signatures of exoplanets through photometric and spectroscopic methods using a variety of spaceborne and/or groundbased observational facilities. These observations are beginning to provide important constraints on a wide gamut of atmospheric properties, including pressure-temperature profiles, chemical compositions, energy circulation, presence of clouds, and nonequilibrium processes. The latest studies are also beginning to connect the inferred chemical compositions to exoplanetary formation conditions. In the present chapter, we review the most recent developments in the area of exoplanetary atmospheres. Our review covers advances in both observations and theory of exoplanetary atmospheres, and spans a broad range of exoplanet types (gas giants, ice giants, and super-Earths) and detection methods (transiting planets, direct imaging, and radial velocity). A number of upcoming planet-finding surveys will focus on detecting exoplanets orbiting nearby bright stars, which are the best targets for detailed atmospheric characterization. We close with a discussion of the bright prospects for future studies of exoplanetary atmospheres.

1. INTRODUCTION

The study of exoplanetary atmospheres is at the center of the new era of exoplanet science. About 800 confirmed exoplanets, and over 3000 candidates, are now known. The last two decades in exoplanet science have provided exquisite statistics on the census of exoplanets in the solar neighborhood and their macroscopic properties, which include orbital parameters (eccentricities, separations, periods, spin-orbit alignments, multiplicity, etc.), bulk parameters (masses, radii, equilibrium temperatures), and properties of their host stars. The sum total of current knowledge has taught us that exoplanets are extremely diverse in all these macroscopic physical parameters. We are now entering a new era where we begin to understand the diversity of chemical compositions of exoplanets and their atmospheric processes, internal structures, and formation conditions.

The discovery of transiting exoplanets over a decade ago opened a new door for observing exoplanetary atmospheres. The subsequent years saw tremendous growth in the atmospheric observations of a wide range of transiting exoplanets, as well as the detection and characterization of directly imaged planets. Starting with the first detections of atmospheric absorption from a transiting exoplanet (*Charbonneau et al.,* 2002) and thermal emission from a transiting exoplanet (*Charbonneau et al.,* 2005; *Deming et al.,* 2005), atmospheric observations have been reported for more than 50 transiting exoplanets and 5 directly imaged exoplanets to date. These observations, which are generally either broadband photometry or low-resolution spectra, have been obtained using a variety of spaceborne and groundbased observational facilities. Despite the limited spectral resolution of the data and the complexity of the atmospheric models, major advances have been made in our understanding of exoplanetary atmospheres in the past decade.

When combined with their bulk parameters and host stellar properties, atmospheric spectra of exoplanets can provide simultaneous constraints on their atmospheric and interior properties and their formation histories. Figure 1 shows a schematic overview of exoplanet characterization. The possible observables for an exoplanet depend upon its detection method, but the maximal set of observables that are possible for transiting exoplanets comprise the bulk parameters (mass and radius), atmospheric spectra, and host-stellar spectra. Exoplanetary spectra on their own place constraints on the chemical compositions, temperature profiles, and energy distributions in their atmospheres, which in turn allow constraints on myriad atmospheric processes such as equilibrium/nonequilibrium chemical processes, presence or absence of thermal inversions ("stratospheres" in a terrestrial sense), atmospheric dynamics, aerosols, etc. By combining spectra with the mass and radius of the planet, one can begin to constrain the bulk composition of its interior, including the core mass in giant planets and relative fractions of volatiles and rock in rocky planets with implications for interior and surface processes. Furthermore, the elemental abundances of exoplanetary atmospheres, when compared to those of their host stars, provide insights into their formation conditions, planetesimal compositions, and evolutionary histories.

In addition to revealing the diversity of exoplanetary environments, an improved understanding of exoplanetary atmospheres also places the solar system in a cosmic perspective. The exoplanets for which atmospheric characterization is possible now span a large range in mass and effective temperature, including highly irradiated hot Jupiters (T ~ 1300–3000 K), warm distant gas giants (T ~ 500–1500 K), hot Neptunes (T ~ 700–1200 K), and temperate super-Earths (T ~ 500 K). While the majority of these objects have no direct analogs in the solar system in terms of orbital parameters, the distribution of primary chemical elements (e.g., H, C, O) in their atmospheres can be compared with those of solar system planets. As an important example, O/H and C/O ratios are poorly known for the solar system giant planets, but could, in principle, be measured more easily for hot Jupiters. Eventually, our knowledge of chemical abundances in exoplanetary atmospheres could enhance our understanding of how planets form and evolve.

In the present chapter, we review the developments in the area of exoplanetary atmospheres over the past decade. We start with a review in section 2 of the observational methods and available facilities for observing exoplanetary atmospheres. In section 3, we review the theoretical developments that aid in interpreting the observed spectra and characterizing irradiated exoplanetary atmospheres. In section 4, we review observational inferences for hot Jupiter atmospheres, which are the most observed class of exoplanets to date. In sections 5–7, we review the theory and observational inferences for other prominent categories of exoplanets: directly imaged young giant planets,

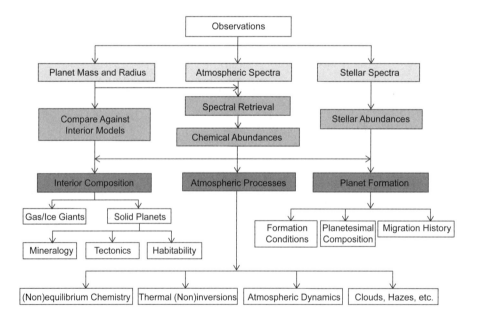

Fig. 1. Schematic diagram of exoplanet characterization. The top two rows show the observables. The third row shows retrieval methods and/or derived parameters. The remaining rows show the various aspects of exoplanetary atmospheres, interiors, and formation that can, in principle, be constrained using all the observables.

transiting hot Neptunes, and super-Earths. We conclude with section 8, where we discuss the future outlook for the field.

2. OBSERVATIONAL METHODS AND FACILITIES

Observational studies of exoplanetary atmospheres can generally be classified into one of two categories: first, measurements of time-varying signals, which include transits, secondary eclipses, and phase curves; and second, spatially resolved imaging and spectroscopy. These techniques allow us to study the properties of the planet's atmosphere, including the relative abundances of common elements such as C, H, O, and N, corresponding atmospheric chemistries, vertical pressure-temperature profiles, and global circulation patterns. The majority of these observations are made at near-infrared (IR) wavelengths, although measurements in the ultraviolet (UV), optical, and mid-IR wavelengths have also been obtained for a subset of planets. The types of planets available for study with each technique directly reflect the biases of the surveys used to detect these objects. Transit and radial velocity (RV) surveys are most sensitive to massive, short-period planets; these planets also display the largest atmospheric signatures, making them popular targets for time-varying characterization studies. Spatially resolved imaging is currently limited to young, massive (≥ 1 M_{Jup}) gas giant planets located at large (tens of astronomical units) separations from their host stars, although this will change as a new generation of adaptive optics (AO) imaging surveys come online. These planets still retain some of the residual heat from their formation, which makes them hot and bright enough to be detectable at the near-IR wavelengths of groundbased surveys.

In order to interpret these observational data, we also need to know the mass and radius of the planet and its host star. For transiting systems the planet-star radius ratio can be obtained from a simple fit to the transit light curve, and for sufficiently bright stars the planet's mass can be determined from the host star's RV semi-amplitude. In systems with multiple gravitationally interacting transiting planets, masses can also be estimated from the measured transit timing variations (*Agol et al.,* 2005; *Holman and Murray,* 2005; *Lithwick et al.,* 2012). For directly imaged planets, masses and radii are typically constrained as part of a global modeling fit to the planet's measured emission spectrum, but the uncertain ages of these systems and the existence of multiple classes of models can result in substantial uncertainties in these fitted parameters. In some systems dynamical sculpting of a nearby debris disk can also provide an independent constraint on the planet's mass.

In the cases where we have good estimates for the planet's mass and/or radius, we can predict the types of atmospheric compositions that would be consistent with the planet's average density and with our current understanding of planet formation processes. All the directly imaged planets detected to date are expected to have hydrogen-dominated atmospheres based on their large estimated masses, while

transiting planets span a much wider range of masses and potential compositions. Although the majority of the transiting planets that have been characterized to date are also hydrogen-dominated gas giants, new discoveries are extending these studies to smaller "super-Earth"-type planets that may be primarily icy or even rocky in nature. In the sections below we focus on four commonly used observational techniques for characterizing the atmospheres of extrasolar planets, discussing their advantages and limitations as well as the types of planets that are best suited for each technique.

2.1. Transmission Spectra

By measuring the depth of the transit when a planet passes in front of its host star we can determine the size of the planet relative to that of the star. A schematic diagram of a planet in transit is shown in Fig. 2. If that planet has an atmosphere, it will appear opaque at some wavelengths and transparent at others, resulting in a wavelength-dependent transit depth. This wavelength-dependent depth is known as a "transmission spectrum," because we are observing the absorption features imprinted on starlight transmitted through the planet's atmosphere (*Seager and Sasselov,* 2000; *Brown,* 2001). This technique is primarily sensitive to the composition of the planet's atmosphere at very low pressures (~0.1−1000 mbar) along the day-night terminator, although it is possible to constrain the local pressure and temperature in this region with high signal-to-noise data (e.g., *Huitson et al.,* 2012). Transmission spectroscopy is also sensitive to high-altitude hazes, which can mask atmospheric absorption features by acting as a gray opacity source (e.g., *Fortney,* 2005). The expected depth of the absorption features in a haze-free atmosphere is proportional to the atmospheric scale height

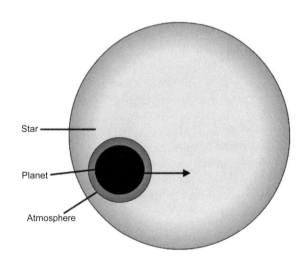

Fig. 2. Schematic illustration of a planet transiting its host star. If the planet has an atmosphere (drawn here as a dark gray annulus) the measured transit depth will vary as a function of wavelength, as the atmosphere will appear opaque at some wavelengths and transparent at others.

$$H = \frac{kT}{\mu g} \qquad (1)$$

where k is Boltzmann's constant, T is the temperature of the planet's atmosphere, μ is the mean molecular weight of the atmosphere, and g is the surface gravity. The size of the absorbing annulus of the planet's atmosphere can be approximated as 5–10 scale heights above the nominal planet radius, resulting in a corresponding change in the measured transit depth of

$$\delta_{depth} \simeq \left(\frac{R_p + 10H}{R_*}\right)^2 - \left(\frac{R_p}{R_*}\right)^2 \qquad (2)$$

where R_p and R_* are the planetary and stellar radii. If we take HD 189733b, a typical hot Jupiter, as an example and assume a globally averaged temperature of 1100 K (*Knutson et al.*, 2009a), a surface gravity of 2140 cm s^{-2} (*Bouchy et al.*, 2005; *Knutson et al.*, 2007), and an atmosphere of H_2, this would correspond to a scale height of 210 km. This planet's nominal 2.5% transit depth would then increase by 0.1% when observed at wavelengths with strong atmospheric absorption features. At visible wavelengths we expect to see sodium and potassium absorption in hot Jupiter atmospheres, while in the near-IR this technique is primarily sensitive to H_2O, CH_4, CO, and CO_2 (e.g., *Seager and Sasselov*, 2000; *Hubbard et al.*, 2001; *Sudarsky et al.*, 2003). These observations constrain the planet's atmospheric chemistry and the presence of hazes, while observations of hydrogen Lyman α absorption and ionized metal lines at ultraviolet wavelengths probe the uppermost layers of the planet's atmosphere and provide estimates of the corresponding mass loss rates for close-in hot Jupiters (e.g., *Lammer et al.*, 2003; *Murray-Clay et al.*, 2009).

This technique has resulted in the detection of multiple molecular and atomic absorption features (see section 4.3), although there is an ongoing debate about the validity of features detected using near-IR transmission spectroscopy with the Near Infrared Camera and Multi-Object Spectrometer (NICMOS) instrument on the Hubble Space Telescope (HST) (e.g., *Swain et al.*, 2008b; *Gibson et al.*, 2011; *Burke et al.*, 2010; *Waldmann et al.*, 2013; *Deming et al.*, 2013). This debate highlights an ongoing challenge in the field, namely that there is often instrumental or telluric variability in the data on the same timescales (minutes to hours) as the timescales of interest for transmission spectroscopy. These time-varying signals must be removed from the data, either by fitting with a functional form that provides a good approximation for the effect (e.g., *Knutson et al.*, 2007), or by assuming that the time-correlated signals are constant across all wavelengths and calculating a differential transmission spectrum (e.g., *Swain et al.*, 2008b; *Deming et al.*, 2013). This naturally leads to debates about whether or not the choice of functional form was appropriate, whether the noise varies as a function of wavelength, and

how these assumptions affect both the inferred planetary transmission spectrum and the corresponding uncertainties on that spectrum. This situation is further complicated when the planet is orbiting an active star, as occulted spots or a time-varying total flux from the star due to rotational spot modulation can both cause wavelength-dependent variations in the measured transit depth (e.g., *Pont et al.*, 2008, 2013; *Sing et al.*, 2011a). Errors in the stellar limb-darkening models used for the fits may also affect the shape of the transmission spectrum; this problem is particularly acute for M stars such as GJ 1214 (*Bean et al.*, 2010, 2011; *Berta et al.*, 2012a). General practice in these cases is to use the model limb-darkening profiles where they provide a good fit to the data, and otherwise to fit for empirical limb-darkening parameters.

2.2. Thermal Spectra

We can characterize the thermal emission spectra of transiting exoplanets by measuring the wavelength-dependent decrease in light when the planet passes behind its host star in an event known as a secondary eclipse. Unlike transmission spectroscopy, which probe the properties of the atmosphere near the day-night terminator, these emission spectra tell us about the global properties of the planet's dayside atmosphere. They are sensitive to both the dayside composition and the vertical pressure-temperature profile, which determine if the molecular absorption features are seen in absorption or emission. Most secondary eclipse observations are made in the near- and mid-IR, where the planets are bright and the star is correspondingly faint. It must be noted that secondary eclipse spectra at shorter wavelengths, e.g., in the visible, could also contain contributions due to reflected or scattered light depending on the planetary albedos, as discussed in sections 2.4 and 4.5. For thermal emission, we can estimate the depth of the secondary eclipse in the Rayleigh-Jeans limit as

$$depth = \left(\frac{R_p}{R_*}\right)^2 \left(\frac{T_p}{T_*}\right) \qquad (3)$$

where R_p and R_* are the planetary and stellar radii, and T_p and T_* are their corresponding temperatures. We can estimate an equilibrium temperature for the planet if we assume that it radiates uniformly as a blackbody across its entire surface

$$T_p = \left(\frac{(1 - A)L_*}{16\pi\sigma d^2}\right)^{1/4} \simeq T_* \sqrt{\frac{R_*}{2d}} \qquad (4)$$

where A is the planet's Bond albedo (fraction of incident light reflected across all wavelengths), σ is the Stefan-Boltzmann constant, and d is the planet-star distance. The righthand expression applies when we assume that the planet's Bond albedo is zero [a reasonable approximation for most hot Jupiters (*Rowe et al.,* 2008; *Burrows et al.,* 2008a; *Cowan and Agol,* 2011b; *Kipping and Spiegel,* 2011; *Demory et al.,* 2011a; *Madhusudhan and Burrows,* 2012)] and treat the star as a blackbody; more generally, A is a free parameter, leading to some uncertainty in T_p. If the planet instead absorbs and reradiates all the incident flux on its dayside alone (see section 2.3), the predicted temperature will increase by a factor of $2^{1/4}$.

The first secondary eclipse detections were reported in 2005 for the hot Jupiters TrES-1 (*Charbonneau et al.,* 2005) and HD 209458b (*Deming et al.,* 2005). More than 50 planets have now been detected in secondary eclipse, of which the majority of observations have come from the Spitzer Space Telescope (for recent reviews, see *Knutson et al.,* 2010; *Seager and Deming,* 2010) and span wavelengths of 3.6–24 µm. There is also a growing body of groundbased near-IR measurements, in the ~0.8–2.1-µm range (*Sing and López-Morales,* 2009; *Anderson et al.,* 2010; *Croll et al.,* 2010, 2011; *Gillon et al.,* 2012b; *Bean et al.,* 2013; *Mancini et al.,* 2013; *Wang et al.,* 2013).

Most of our understanding of the dayside emission spectra of hot Jupiters comes from the combination of broadband Spitzer and groundbased secondary eclipse photometry. The challenge with these observations is that we are often in the position of inferring atmospheric properties from just a few broadband measurements; this can lead to degeneracies in our interpretation of the data (*Madhusudhan and Seager,* 2009, 2010). One solution is to focus on solar-composition models that assume equilibrium chemistry, but this is only a good solution if these assumptions largely hold for this class of planets (see sections 3.3 and 4.3). We can test our assumptions using a few well-studied benchmark cases, such as HD 189733b (*Charbonneau et al.,* 2008) and HD 209458b (*Knutson et al.,* 2008). For these planets we have enough data to resolve most degeneracies in the model fits, while the larger sample of more sparsely sampled planets allows us to fill in the statistical big-picture view as to whether particular atmospheric properties are common or rare. We can also search for correlations between the planet's atmospheric properties and other properties of the system, such as the incident UV flux received by the planet (*Knutson et al.,* 2010), which tells us about the processes that shape planetary atmospheres.

It is also possible to measure emission spectra for non-transiting planets by searching for the RV-shifted signal of the planet's emission in high resolution near-IR spectra obtained using large (~10 m) groundbased telescopes (*Crossfield et al.,* 2012b; *Brogi et al.,* 2013; *de Kok et al.,* 2013), although to date this technique has only been applied successfully to a handful of bright, nearby systems. In order to overcome the low signal-to-noise ratios of individual spectral features, this technique relies on fits using template model atmospheric spectra in order to detect the RV-shifted features in the planet's emission spectrum. This approach is highly complementary to secondary eclipse observations, which are usually obtained at very low spectral resolution.

2.3. Thermal Phase Curves and Atmospheric Dynamics

A majority of the transiting exoplanets known to date have short orbital periods where the timescale for orbital synchronization is much shorter than the age of the system. As a result, we expect these planets to be tidally locked, with permanent daysides and nightsides. One question we might ask for such planets is what fraction of the flux incident on the dayside is transported to the nightside. If this fraction is large, the planet will have a small day-night temperature gradient, whereas if the fraction is small, the planet may have large thermal and chemical gradients between the two hemispheres (see *Showman et al.,* 2008, for a review).

We can estimate the temperature as a function of longitude on these planets and constrain their atmospheric circulation patterns by measuring the changes in the IR brightness of the planet as a function of orbital phase (*Cowan and Agol,* 2008, 2011a). An example of a thermal phase curve is shown in Fig. 3. The amplitude of the flux variations relative to the secondary eclipse tells us the size of the day-night temperature gradient, while the locations of flux maxima and minima indicate the locations of hotter and colder regions in the planet's atmosphere. If we assume that the hottest region of the planet is near the center of its dayside,

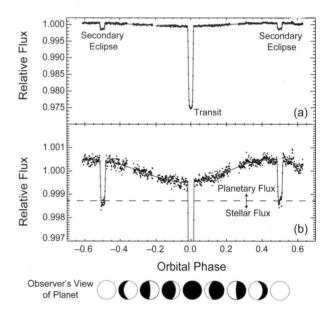

Fig. 3. Phase curve of hot Jupiter HD 189733b measured in the Spitzer 4.5-µm IRAC band. **(a)** Transit and secondary eclipses; **(b)** same data, but with a reduced y axis range in order to better illustrate the change in flux as a function of orbital phase. The horizontal dashed line indicates the measured flux level during secondary eclipse, when only the star is visible. The flux from the planet at any given time can be calculated as the difference between the total measured flux and the dashed line. Beneath the plot we show a schematic diagram indicating the part of the planet that is visible at different orbital phases, with the planet's dayside shown in white and the nightside shown in black.

then we would expect that the amplitude of the measured thermal phase curve should never exceed the depth of the secondary eclipse. In the absence of winds we would expect the hottest and coldest regions of the atmosphere to lie at the substellar and antistellar points, respectively, corresponding to a flux maximum centered on the secondary eclipse and a flux minimum centered on the transit. Atmospheric circulation models predict that hot Jupiters should develop a superrotating equatorial band of wind (*Showman and Polvani, 2011*); such a wind would shift the locations of these hot and cold regions eastward of their original locations, causing the flux minima and maxima to occur just before the transit and secondary eclipse, respectively.

If we only have a phase curve observation in a single wavelength, we can treat the planet as a blackbody and invert our phase curve to make a map of atmospheric temperature vs. longitude (*Cowan and Agol, 2008, 2011a*). However, we know that planetary emission spectra are not blackbodies, and if we are able to obtain phase curve observations at multiple wavelengths, we can map the planetary emission spectrum as a function of longitude on the planet (*Knutson et al., 2009a, 2012*). If we assume that there are no compositional gradients in the planet's atmosphere, the wavelength-dependence of the phase curve shape reflects changes in the atmospheric circulation pattern as a function of depth in the atmosphere, with different wavelengths probing different pressures depending on the atmospheric opacity at that wavelength.

There are two additional observational techniques that can be used to place constraints on the atmospheric circulation patterns and wind speeds of transiting planets. By measuring the wavelength shift of the atmospheric transmission spectrum during ingress, when the leading edge of the planet's atmosphere occults the star, and during egress when the trailing edge occults the star, it is possible to estimate the wind speeds near the dawn and dusk terminators of the planet (*Kempton and Rauscher, 2012; Showman et al., 2012*). A marginal detection of this effect has been reported for HD 209458b (*Snellen et al., 2010b*), but it is a challenging measurement requiring both high spectral resolution and a high signal-to-noise ratio for the spectra. We can also obtain a map of the dayside brightness distributions on these planets by searching for deviations in the shape of the secondary eclipse ingress and egress as compared to the predictions for the occultation of a uniform disk (*Williams et al., 2006; Agol et al., 2010; Majeau et al., 2012; de Wit et al., 2012*). This effect has been successfully measured for HD 189733b at 8 μm and gives a temperature map consistent with that derived from the phase curve observations. Unlike phase curves, which only probe brightness as a function of longitude, eclipse maps also provide some latitudinal information on the dayside brightness distribution.

2.4. Reflected Light

Measurements of the reflected light from transiting extrasolar planets are extremely challenging, as the visible-light planet-star contrast ratio is much smaller than the equivalent value in the IR. The best detections and upper limits on visible-light secondary eclipses to date have all come from space missions, including Microvariability and Oscillations of Stars (MOST) (*Rowe et al., 2008*), Convection, Rotation and planetary Transits (CoRoT) (*Alonso et al., 2009a,b; Snellen et al., 2009, 2010a*), and Kepler (*Borucki et al., 2009, Désert et al., 2011a,b; Kipping and Bakos, 2011; Demory et al., 2011a, 2013; Coughlin and López-Morales, 2012; Morris et al., 2013; Sanchis-Ojeda et al., 2013; Esteves et al., 2013*). Some of these same observations have also detected visible-light phase curves for these planets, which provide information on the planet's albedo as a function of viewing angle. Many of the planets observed are hot enough to have some detectable thermal emission in the visible-light bands, and it is therefore useful to combine the visible-light eclipse with IR observations in order to separate out the thermal and reflected-light components (*Christiansen et al., 2010; Désert et al., 2011b*). One challenge is that many of the transiting planets with measured visible-light secondary eclipses orbit relatively faint stars, making IR eclipse measurements challenging for these targets.

2.5. Polarimetry

There have been some theoretical investigations of polarized light scattered from hot Jupiters (*Seager et al., 2000; Stam et al., 2004, Madhusudhan and Burrows, 2012*) as well as thermally emitted light from young giant planets (*Marley and Sengupta, 2011; de Kok et al., 2011*). In principle, detecting polarized light can constrain the scattering phase function and particle sizes of condensates in a planetary atmosphere, but the data so far for hot Jupiters is unclear. *Berdyugina et al.* (2008) detected a high-amplitude polarization signal in the B band for the hot Jupiter HD 189733b. However, *Wiktorowicz* (2009) was unable to confirm this result. Since the information supplied from polarization measurements would be unique and complementary compared to any other observational technique, further observational work is important.

2.6. Direct Imaging

Finding low-mass companions (planets or brown dwarfs) to stars by direct imaging is extremely challenging and many factors come into play. First and foremost, at the ages of most field stars, even massive planets are prohibitively faint compared to main-sequence host stars. Even at the diffraction limit, planets at separations of less than a few astronomical units will be buried in the point spread function of the star. Fortunately, by taking advantage of planetary evolution and carefully selecting the target stars, direct imaging is not only possible but provides an exquisite means for atmospheric characterization of the planets.

The bolometric luminosity of gas-giant planets is a smoothly varying function of time, is a strong function of mass, and is well approximated by

$$\frac{L_{bol}(t)}{L_{\odot}} \propto \left(\frac{1}{t}\right)^{\alpha} M^{\beta} \kappa^{\gamma} \qquad (5)$$

where t is time, M is mass, and κ is the photospheric Rosseland mean opacity. The exponents α, β, and γ are approximately 5/4, 5/2, and 2/5, respectively (*Stevenson*, 1991). This equation is derived under several simplifying assumptions for the interior equations of state and boundary conditions but, for masses below ~10 M_{Jup}, closely matches the predictions of more detailed numerical simulations (Fig. 4).

The mean opacity in equation (5) is also a function of time but has a smaller impact on the bulk evolution compared to the effect of the monochromatic opacities on the spectral energy distribution (SED), especially as photospheric clouds come and go as a planet cools with time. Moreover, the weak dependence on κ suggests that predictions for L_{bol} are fairly robust against uncertainties in atmospheric opacities. This simple model for planet evolution is often referred to as "hot start" evolution and is very close to an upper limit on L_{bol}. However, it ignores the initial conditions established by the formation process. Figure 4 illustrates the evolution of a planet with initial conditions inspired by core-accretion formation (*Fortney et al.*, 2008b), resulting in a much fainter planet.

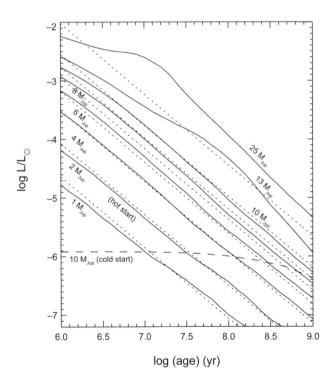

Fig. 4. Evolution of bolometric luminosity for giant planets. Solid lines are hot-start models from *Baraffe et al.* (2003) and dotted lines are based on equation (5) for the same masses (see labels). Large changes in the equation of state make equation (5) invalid above ~8 M_{Jup}. The dashed curve is a cold-start model from *Fortney et al.* (2008b) and illustrates the impact of different initial conditions.

The ideal stars for direct imaging are those that provide maximum contrast between planet light and scattered star light. Equation (5) immediately reveals a key aspect of giant planet evolution relevant for direct imaging searches — young planets are significantly more luminous than older planets of equal mass. Furthermore, effective temperature can be approximated by an equation similar to equation (5) (albeit with different exponents and a slightly weaker mass dependence), indicating that the peak of the SEDs of young planets shift to shorter wavelengths with youth. These two characteristics strongly suggest that the odds of imaging a planet are greatly improved for planets around young stars. Ideal stars are also nearby (for favorable angular separation) and have high proper motion (to easily distinguish bound companions from back/foreground stars).

Even with carefully selected target stars, direct imaging requires extreme high-contrast observations typically using AO and large-aperture telescopes on the ground [e.g., at Keck, the Very Large Telescope (VLT), Subaru, and Gemini] or observations from space (e.g., HST). In practice, candidates are confirmed as companions by establishing common proper motion with the host star over two or more observing seasons.

Generally, direct imaging surveys as yet have had more to say about limits than revealing large samples of planets. For example, using the power-law planet distributions of *Cumming et al.* (2008), *Nielsen et al.* (2010) find that less than 20% of FGKM-type stars have planets >4 M_{Jup} in orbits between 60 and 180 AU (depending on the mass-luminosity-age models used). Examples of discoveries are 2M1207b (*Chauvin et al.*, 2005), Fomalhaut b (*Kalas et al.*, 2008), Beta Pic b (*Lagrange et al.*, 2010), 1RXS J1609 b (*Lafrenière et al.*, 2010) and the quadruple-planetary system HR 8799b, c, d, and e (*Marois et al.*, 2008, 2010). There have been several other recent detections of directly imaged planetary/substellar objects orbiting young stars, e.g., κ And (*Carson et al.*, 2013), GJ 504 (*Kuzuhara et al.*, 2013), and HD 106906 (*Bailey et al.*, 2014), but we note that reliable age estimates are needed in order to obtain accurate masses for these objects (e.g., *Hinkley et al.*, 2013).

3. THEORY OF IRRADIATED ATMOSPHERES

3.1. Atmospheric Models

In parallel to the observational advances, the last decade has also seen substantial progress in the modeling of exoplanetary atmospheres and interpretation of exoplanetary spectra. Models have been reported for exoplanetary atmospheres over a wide range of (1) physical conditions, from highly irradiated giant planets (e.g., hot Jupiters and hot Neptunes) and super-Earths that dominate the transiting planet population to young giant planets on wide orbital separations that have been detected by direct imaging; (2) computational complexity, from one-dimensional plane-parallel models to three-dimensional general circulation models (GCMs); and (3) thermochemical conditions, including models assuming solar-composition in thermochemical equilibrium as well as those with nonsolar abundances and nonequilibrium compositions.

The complexity of models used depend on the nature of data at hand. For low-resolution disk-integrated spectra that are typically observed, one-dimensional models provide adequate means to derive surface-averaged atmospheric temperature and chemical profiles that are generally consistent with those from three-dimensional models. But, when orbital phase curves are observed, three-dimensional models are necessary to accurately model and constrain the atmospheric dynamics.

3.1.1. Self-consistent equilibrium models. Traditionally models of exoplanetary atmospheres have been based on equilibrium considerations. Given the planetary properties (incident stellar irradiation, planetary radius, and gravity) and an assumed set of elemental abundances, equilibrium models compute the emergent stellar spectrum under the assumptions of radiative-convective equilibrium, chemical equilibrium, hydrostatic equilibrium, and local thermodynamic equilibrium (LTE) in a plane-parallel atmosphere (*Seager and Sasselov*, 1998; *Sudarsky et al.*, 2003; *Seager et al.*, 2005; *Barman et al.*, 2005; *Fortney et al.*, 2006; *Burrows et al.*, 2007, 2008b). The constraint of radiative-convective equilibrium allows the determination of a pressure-temperature (P-T) profile consistent with the incident irradiation and the chemical composition (see section 3.2). The constraint of chemical equilibrium allows the determination of atomic and molecular abundances that provide the source of opacity in the atmosphere (see section 3.3) and hydrostatic equilibrium relates pressure to radial distance. Typically, solar abundances are assumed in such models. Additional assumptions in the models include a parametric flux from the interior, parameters representing unknown absorbers (e.g., *Burrows et al.*, 2007, 2008b), and prescriptions for energy redistribution from the dayside to the nightside (e.g., *Fortney et al.*, 2006; *Burrows et al.*, 2007). Yet other models adopt P-T profiles from equilibrium models, or "gray" solutions (e.g., *Hansen*, 2008; *Guillot*, 2010; *Heng et al.*, 2012), but allow the chemical compositions to vary from solar abundances and chemical equilibrium (*Tinetti et al.*, 2007; *Miller-Ricci et al.*, 2009).

Equilibrium models are expected to accurately represent exoplanetary atmospheres in regimes where atmospheric dynamics and nonequilibrium processes do not significantly influence the temperature structure and chemical composition, respectively. Perhaps mostly readily understood are cases where radiative equilibrium is not expected in the visible atmosphere. Such a case is likely found in the atmospheres of hot Jupiters, where extreme dayside forcing and advection should strongly alter the temperature structure from radiative equilibrium (e.g., *Seager et al.*, 2005). When prescribed abundances are used, it can also be difficult to diagnose abundance ratios that are different than assumed, unless a large number of models are run.

3.1.2. Parametric models and atmospheric retrieval. Recent advances in multi-band photometry and/or spectroscopy have motivated the development of parametric models of exoplanetary atmospheres. Parametric models compute radiative transfer in a plane-parallel atmosphere, with constraints of LTE and hydrostatic equilibrium, similar to equilibrium models, but do not assume radiative-convective equilibrium and chemical equilibrium that are assumed in equilibrium models. Instead, in parametric models the P-T profile and chemical abundances are free parameters of the models. For example, in the original models of *Madhusudhan and Seager* (2009) the temperature profile and molecular abundances constitute 10 free parameters — 6 parameters for the temperature profile and 4 for the molecular abundances of H_2O, CO, CH_4, and CO_2 — and the models are constrained by the requirement of global energy balance. On the other hand, *Lee et al.* (2012) and *Line et al.* (2012) consider "level-by-level" P-T profiles where the temperature in each pressure level (of ~100 levels) can be treated as a free parameter. Newer models of carbon-rich atmospheres include additional molecules, e.g., HCN and C_2H_2, as free parameters (*Madhusudhan*, 2012).

Parametric models are used in conjunction with statistical algorithms to simultaneously retrieve the temperature profile and chemical composition of a planetary atmosphere from the observed spectral and/or photometric data. Whereas early retrieval results used grid-based optimization schemes (*Madhusudhan and Seager*, 2009), recent results have utilized more sophisticated statistical methods such as Bayesian Markov chain Monte Carlo methods (*Madhusudhan et al.*, 2011a; *Benneke and Seager*, 2012; *Line et al.*, 2013) and gradient-descent search methods (*Lee et al.*, 2012, 2013; *Line et al.*, 2012).

3.1.3. Opacities. For any atom, molecule, or ion found in a planetary atmosphere, it is important to understand the wavelength-dependent opacity as a function of pressure and temperature. Similarly, for materials that condense to form solid or liquid droplets, one needs tabulated optical properties such that cloud opacity can be calculated from Mie scattering theory. Opacity databases are generally built up from laboratory work and from first-principles quantum mechanical simulation. The rise of modern high-performance computing has allowed for accurate calculations for many molecules over a wide pressure and temperature range.

These databases have generally also been well tested against the spectra of brown dwarfs, which span temperatures of ~300–2400 K, temperatures found in hot Jupiters, directly imaged planets, and cooler gas giants as well. Databases for H_2O, CO, CO_2, and NH_3 have recently improved dramatically (e.g., *Rothman et al.*, 2010; *Tennyson and Yurchenko*, 2012). A remaining requirement among the dominant molecules is a revised database for methane. Recent reviews of opacities used in modeling giant planet and brown dwarf atmospheres can be found in *Sharp and Burrows* (2007) and *Freedman et al.* (2008). For molecules expected to be common in terrestrial exoplanet atmospheres, there is considerable laboratory work done for Earth's atmospheric conditions. However, for very hot rocky planets, databases are lacking at higher temperatures.

3.2. Theory of Thermal Inversions

The atmospheric temperature structure involves the balance of the depth-dependent deposition of stellar energy into the atmosphere, the depth dependent cooling of the atmosphere back to space, along with any additional nonradiative energy sources, such as breaking waves. In terms of radiative transport only, the emergence of thermal inversions can be understood as being controlled by the ratio of opacity at visible wavelengths (which controls the depth to which incident flux penetrates) to the opacity at thermal-IR wavelengths, which controls the cooling of the heating planetary atmosphere. Generally, if the optical opacity is high at low pressure, leading to absorption of stellar flux high in the atmosphere, and the corresponding thermal-IR opacity is low, the upper atmosphere will have slower cooling, leading to elevated temperatures at low pressure. Temperature inversions are common in the solar system, due, for instance, to O_3 in Earth, and absorption by photochemical hazes in the giant planets and Titan.

Early models of hot Jupiters did not feature inversions, because early models had no strong optical absorbers at low pressures. *Hubeny et al.* (2003) pointed out that for the hottest hot Jupiters, TiO and VO gas could lead to absorption of incident flux high in the atmosphere, and drive temperature inversions, as shown in Fig. 5. This was confirmed by *Fortney et al.* (2006) and further refined in *Fortney et al.* (2008a) and *Burrows et al.* (2008b).

At this time there is no compelling evidence for TiO/VO gas in the atmospheres of hot Jupiters. *Burrows et al.* (2008b) in particular has often modeled atmospheres with an *ad hoc* dayside absorber, rather than TiO/VO in particular. Cloud formation is also a ubiquitous process in planetary atmospheres, and cloud formation may actually remove Ti and V from the gas phase, locking these atoms in Ca-Ti-O-bearing condensates below the visible atmosphere. This could keep even the upper atmospheres of the hottest planets free of TiO/VO gas (*Spiegel et al.,* 2009). Cold nightsides could also lead to Ti and V condensing out (*Showman et al.,* 2009a) and being lost from the dayside. It may also be possible that high chromospheric emission from the host star could dissociate inversion-causing compounds in the planetary atmosphere, thereby precluding the formation of thermal inversions (*Knutson et al.,* 2010).

Another possibility is that the chemistry in some atmospheres allows for TiO gas to form in abundance, but in other atmospheres it does not, because there is little available oxygen. *Madhusudhan et al.,* (2011b) showed that the atmospheric C/O ratio can control the abundance of TiO; C/O ≥ 1 can cause substantial depletion of TiO and VO.

3.3. Chemistry in Giant Planetary Atmospheres

As discussed in section 3.1, the chemical composition of a planetary atmosphere strongly influences its temperature structure and spectral signatures, and hence forms a critical component of atmosphere models. Extrasolar giant planets,

similar to Jupiter and Saturn in our solar system, are expected to host primary atmospheres composed largely of hydrogen (H) and helium (He), as evident from their masses and radii (e.g., *Fortney et al.,* 2007; *Seager et al.,* 2007). In addition, the atmospheres are also expected to contain a number of less-abundant molecular and atomic species comprising heavier elements (e.g., O, C, N) in proportions governed by the primordial abundances (*Burrows and Sharp,* 1999). It is these "trace species" that are responsible for the dominant features of giant planet spectra, and are the primary probes of physicochemical processes in their atmospheres and their formation histories.

3.3.1. Equilibrium chemistry. It is customary in exoplanet models to assume that the atmospheres are in chemical equilibrium, and to investigate deviations from equilibrium when data necessitates. The chemical composition of a gas in chemical equilibrium at a given temperature (T), pressure (P), and elemental abundances is governed by the partitioning of species that minimizes the Gibbs free energy of the chemical system (e.g., *Burrows and Sharp,* 1999; *Lodders and Fegley,* 2002; *Seager et al.,* 2005).

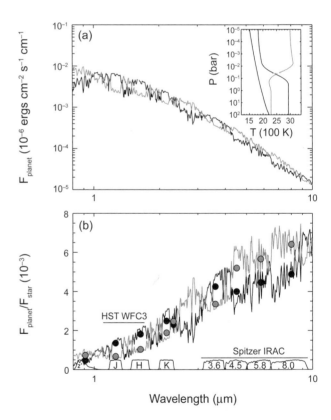

Fig. 5. Models of dayside thermal emission from a typical hot Jupiter atmosphere (see section 3.1). **(a)** Emergent spectra for two models with and without a thermal inversion in the atmosphere (corresponding P-T profiles are shown in inset, along with the TiO condensation curve). The model without (with) a temperature inversion shows molecular bands in absorption (emission). **(b)** Spectrum of the planet-to-star flux ratio. The circles show the model binned in the various photometric bandpasses shown.

Temperatures in atmospheres of currently known irradiated giant exoplanets span a wide range (~500–3000 K). For the elemental abundances, it is customary to adopt solar composition, whereby oxygen is the most dominant element, after H and He, followed by C, the C/O ratio being 0.5 (e.g., *Asplund et al.*, 2009).

Considering solar abundances and chemical equilibrium, H_2O is the dominant carrier of oxygen at all temperatures, whereas carbon is contained in CO and/or CH_4 depending on the temperature (T); at a nominal pressure of 1 bar, CH_4 dominates for T ≤ 1300 K, whereas CO dominates for higher T. Molecules such as CO_2, NH_3, and hydrocarbons other than CH_4 also become prominent depending on the temperature (*Burrows and Sharp*, 1999; *Lodders and Fegley*, 2002; *Madhusudhan and Seager*, 2011; *Visscher and Moses*, 2011; *Moses et al.*, 2011). For higher metallicities than solar, the molar fractions (or "mixing ratios") of all the molecules are systematically enhanced. While solar abundances form a reasonable starting point for atmospheric models, recent studies have shown that the molecular compositions in H-rich atmospheres can be drastically different for different C/O ratios, as shown in Fig. 6. While H_2O is abundant in solar composition atmospheres, it can be significantly depleted for C/O ≥ 1, where C-rich compounds dominate the composition (*Seager et al.*, 2005; *Helling and Lucas*, 2009; *Madhusudhan et al.*, 2011b; *Kopparapu et al.*, 2012; *Madhusudhan*, 2012; *Moses et al.*, 2013).

3.3.2. Nonequilibrium chemistry. The interplay of various physical processes can drive planetary atmospheres out of chemical equilibrium, resulting in deviations in molecular abundances from those discussed above. Nonequilibrium

chemistry in exoplanetary atmospheres have been studied in the context of (1) vertical mixing of species due to turbulent processes on timescales shorter than the equilibrium reaction timescales (also known as "eddy diffusion"), (2) photochemical reactions of species driven by strong incident irradiation, and (3) molecular diffusion or gravitational settling of species. Nonequilibrium chemistry is prevalent in the atmospheres of all solar system giant planets (*Yung and Demore*, 1999) and several brown dwarfs (e.g., *Noll et al.*, 1997), and is similarly expected to be prevalent in the atmospheres of exoplanets (*Zahnle et al.*, 2009; *Moses et al.*, 2011, 2013).

Several recent studies have reported investigations of nonequilibrium chemistry in exoplanetary atmospheres over a wide range of temperatures, metallicities, and C/O ratios (*Cooper and Showman*, 2006; *Zahnle et al.*, 2009; *Line et al.*, 2010; *Visscher and Moses*, 2011; *Madhusudhan and Seager*, 2011; *Moses et al.*, 2011, 2013; *Kopparapu et al.*, 2012; *Venot et al.*, 2012). A general conclusion of these studies is that nonequilibrium chemistry is most prevalent in cooler atmospheres (T_{eq} ≤ 1300), whereas very hot atmospheres (T_{eq} ≥ 2000) tend to be in chemical equilibrium in steady state in regions of exoplanetary atmospheres that are accessible to IR observations, i.e., pressures between ~1 mbar and ~1 bar.

One of the most commonly detectable effects of nonequilibrium chemistry is the CO-CH_4 disequilibrium due to vertical mixing in low-temperature atmospheres. As discussed above, in regions of a planetary atmosphere where T ≤ 1300 K, chemical equilibrium predicts CH_4 to be the dominant carbon-bearing species, whereas CO is expected to be less abundant. However, vertical fluid motions due to a variety of processes can dredge up CO from the lower, hotter regions of the atmosphere into the upper regions, where it becomes accessible to spectroscopic observations. Similar disequilibrium can also take place for the N_2-NH_3 pair in low-temperature atmospheres (T ~ 600–800 K).

Models of nonequilibrium chemistry described above span a wide range in complexity, including photochemistry, kinetics, and eddy diffusion, albeit limited to only the major chemical species as necessitated by available data. The most extensive nonequilibrium models reported to date for H_2-rich atmospheres still consider reaction networks including only the major elements H, He, O, C, N (*Moses et al.*, 2011, 2013), and S (e.g., *Zahnle et al.*, 2009).

3.4. Atmospheric Dynamics

Understanding atmospheric dynamics goes hand in hand with understanding the radiative and chemical properties of atmospheres. For strongly irradiated planets, advection moves energy from day to night and from equator to pole. Additionally, vertical mixing can drive atmospheres away from chemical equilibrium (see section 3.3.2). Several recent studies have reported three-dimensional GCMs of exoplanets, particularly for irradiated giant planets for which phase curve data is available to constrain the models (*Cho et al.*, 2008; *Showman et al.*, 2008, 2009a; *Dobbs-Dixon et*

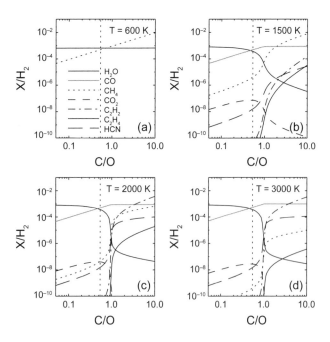

Fig. 6. Dependence of molecular mixing ratios in chemical equilibrium on C/O ratio and temperature (T). Adapted from *Madhusudhan* (2012).

al., 2010; *Heng et al.*, 2011; *Rauscher and Menou*, 2012). A full review of atmospheric dynamics is beyond the scope of this chapter, but recent reviews for hot Jupiters can be found in *Showman et al.* (2008), for exoplanets in general in *Showman et al.* (2009b), and for terrestrial exoplanets in *Showman et al.* (2013).

Here, we will focus on a small number of relatively simple parameters that to first order help us understand the dynamics. The first is the dynamical or advective timescale (τ_{adv}), the characteristic time to move a parcel of gas

$$\tau_{adv} = \frac{R_p}{U} \qquad (6)$$

where U is the characteristic wind speed and R_p the planet radius (*Showman and Guillot*, 2002; *Seager et al.*, 2005).

What effect this may have on the dynamical redistribution of energy in a planetary atmosphere can be considered after estimating the radiative timescale, τ_{rad}. Assuming Newtonian cooling, a temperature disturbance relaxes exponentially toward radiative equilibrium with a characteristic time constant τ_{rad} (e.g., *Goody and Yung*, 1989). At photospheric pressures this value can be approximated by

$$\tau_{rad} \sim \frac{P}{g} \frac{c_P}{4\sigma T^3} \qquad (7)$$

where σ is the Stefan-Boltzmann constant and c_P is the specific heat capacity (*Showman and Guillot*, 2002).

It is the ratio of τ_{rad} to τ_{adv} that helps to understand how efficient temperature homogenization will be in the atmosphere. If $\tau_{rad} \ll \tau_{adv}$, then the atmosphere will readily cool before dynamics can move energy. One would expect a large day-night contrast between the hottest part of the atmosphere at the substellar point and the cooler night side. If $\tau_{rad} \ll \tau_{adv}$, then temperature homogenization is quite efficient and one would expect comparable temperatures on the daysides and nightsides. A toy model demonstrating this effect in hot Jupiters can be found in *Cowan and Agol* (2011a).

For hot Jupiters, which are often assumed to be tidally locked, we are actually in a beneficial position for being able to understand the dynamics observationally. This is due to the flow length scales being planetary-wide in nature. Tidally locked hot Jupiters rotate with periods of days, while Jupiter itself rotates in 10 hours. This significantly slower rotation leads to wider bands of winds (~1–3 for hot Jupiters, rather than ~20 for Jupiter). These flow scales can be understood via the characteristic Rhines length and Rossby deformation radius (*Showman et al.*, 2009b). Both predict that faster rotation will lead to narrower bands and smaller vortices. Furthermore, as discussed in section 2.3, the resulting winds in hot Jupiters cause an eastward shift of the dayside hot spot away from the substellar point.

The study of the atmospheric dynamics of young self-luminous planets (such as Jupiter at a young age) is still in its infancy. *Showman and Kaspi* (2013) stress that the faster rotation periods of these planets (which are not tid-ally locked due to their young ages and large distances from their parent stars) will lead to flow that is rotationally dominated. The forcing of these atmospheres is dominated by the internal heat flux, rather than incident stellar flux. The radiative-convective boundary in such planets is essentially coincident with the visible atmosphere, rather than being down at hundreds of bars as in typical hot Jupiters. Future work will likely aim to better understand time variability of surface fluxes, surface heterogeneity, and vertical mixing.

4. INFERENCES FOR HOT JUPITERS

4.1. Brightness Maps and Atmospheric Dynamics

In order to detect the phase curve of an extrasolar planet, we must measure changes in IR flux with amplitudes less than a few tenths of a percent on timescales of several days. This is challenging for groundbased telescopes, and the only successful phase curve measurements to date have come from Spitzer, CoRoT, and Kepler. The first Spitzer phase curve measurements in the published literature were for the nontransiting planets υ And (*Harrington et al.*, 2006) and HD 179949 (*Cowan et al.*, 2007), but without a radius estimate from the transit and a measurement of the dayside flux from the secondary eclipse depth, these sparsely sampled phase curves are difficult to interpret. *Knutson et al.* (2007) reported the first continuous phase curve measurement for the transiting hot Jupiter HD 189733b at 8 μm using Spitzer. This measurement indicated that the planet had a dayside temperature of ~1200 K and a nightside temperature of ~1000 K, along with a peak flux that occurred just before secondary eclipse, indicating that strong superrotating winds were redistributing energy from the dayside to the nightside, consistent with predictions of GCM models (*Showman et al.*, 2009a).

To date IR Spitzer phase curves have been published for six additional planets, including HD 149206b (*Knutson et al.*, 2009b), HD 209458b (*Crossfield et al.*, 2012a), HD 80606b (*Laughlin et al.*, 2009), HAT-P-2b (*Lewis et al.*, 2013), WASP-12b (*Cowan et al.*, 2012), and WASP-18b (*Maxted et al.*, 2012). Two of these planets, HD 80606b and HAT-P-2b, have eccentric orbits and phase curve shapes that reflect the time-varying heating of their atmospheres. HD 189733b is currently the only planet with published multi-wavelength phase curve observations (*Knutson et al.*, 2009b, 2012), although there are unpublished observations for several additional planets. Unlike previous single-wavelength observations, the multi-wavelength phase curve data for HD 189733b are not well matched by current atmospheric circulation models, which may reflect some combination of excess drag at the bottom of the atmosphere and chemistry that is not in equilibrium.

Snellen et al. (2009) published the first visible-light measurement of a phase curve using CoRoT observations of the hot Jupiter CoRoT-1b. This was followed by a much higher signal-to-noise visible-light phase curve using Kepler observations of the hot Jupiter HAT-P-7b. At these visible wavelengths the measured phase curve is the sum of the thermal and reflected light components, and IR observations are generally required to distinguish between the two. Most recently, *Demory et al.* (2013) reported a Kepler phase curve of the hot Jupiter Kepler-7b and inferred a high geometric albedo (0.35 ± 0.02) and the presence of spatially inhomogeneous clouds in the planetary atmosphere.

We can compare the statistics of atmospheric circulation patterns over the current sample of hot Jupiters. Although our sample of phase curves is limited, it is possible to obtain a rough estimate of the amount of energy that is transported to the planet's nightside by fitting the dayside spectrum with a blackbody and then asking how the estimated temperature compares with the range of expected temperatures corresponding to either no recirculation or full recirculation. *Cowan and Agol* (2011b) published an analysis of the existing body of secondary eclipse data and found that there was a large scatter in the inferred amount of recirculation for planets with predicted dayside temperatures cooler than approximately 2300 K, but that planets hotter than this temperature appeared to have consistently weak recirculation of energy between the two hemispheres. Existing full-orbit phase curve data support this picture, although WASP-12b may be an exception (*Cowan et al.*, 2012).

4.2. Thermal Inversions

4.2.1. Inferences of thermal inversions in hot Jupiters.
The diagnosis of temperature inversions has always been recognized to be a model-dependent process. A clear indication of an inversion would be molecular features seen in emission, rather than absorption, for multiple species across a wide wavelength range. At present, low-resolution emission spectra are available for only a handful of hot Jupiters (e.g., *Grillmair et al.*, 2008; *Swain et al.*, 2008a, 2009a,b), but with no definitive evidence of a thermal inversion.

Progress instead had to rely on wide photometric bandpasses, mostly from Spitzer's Infrared Array Camera (IRAC) from 3 to 10 μm as well as some groundbased measurements, mostly in the K band. These bandpasses average over molecular features, making unique identifications difficult. If the atmosphere is assumed to be in chemical and radiative equilibrium, it becomes easier to identify planets with inversions (*Burrows et al.*, 2008b; *Knutson et al.*, 2008, 2009b), but this assumption may not be accurate for many exoplanetary atmospheres. Even if one assumes a solar C-to-O ratio it is still usually possible to find equally good solutions with and without temperature inversions for the cases where molecular abundances are allowed to vary as free parameters in the fit (*Madhusudhan and Seager*, 2010).

It has long been suggested that there should be a connection between the presence of inversions and a very strong absorber in the optical or near-UV, such as TiO or VO (*Hubeny et al.*, 2003; *Fortney et al.*, 2008a; *Burrows et al.*, 2008b; *Zahnle et al.*, 2009). However, to date there is no compelling observational evidence for such an absorber. The difficulty occurs for at least two reasons. The first is that

at occultation, the planet-to-star flux ratio is often incredibly small, on the order of 10^{-4}–10^{-3} in the near-IR, such that obtaining useful data on the dayside emission/reflection of hot Jupiters is technically quite difficult. A seemingly easier alternative would be to observe optical absorbers in transmission during the transit. However, here we are hampered by the fact that the transmission observations probe the terminator region, rather than the dayside. Since the terminator can be appreciably cooler and receive much less incident flux, the chemical composition of the terminator and daysides can be different. Alternatively, it has been suggested that temperature inversions may also arise from nonradiative processes (e.g., *Menou*, 2012).

4.2.2. Nondetections of thermal inversions and possible interpretations.
For a number of planets there is no compelling evidence for a temperature inversion. This essentially means there is no heating source at low pressure that cannot have this excess energy effectively radiated to space. This likely also negates the possibility of a high-altitude optical absorber, unless this absorber also was opaque in the thermal-IR.

There have been several discussions of how the "TiO hypothesis" could be incorrect. It was originally pointed out by *Hubeny et al.* (2003) that a cold trap could remove TiO gas from the visible atmosphere. If the planetary P-T profile crosses the condensation curve of a Ti-bearing solid at high-pressure (~100–1000) bars in the deep atmosphere, this condensation would effectively remove TiO gas from the upper atmosphere. *Spiegel et al.* (2009) investigated this scenario in some detail and found that vigorous vertical mixing would be required to "break" the cold trap and transport Ti-bearing grains into the visible atmospheres, where they would then have to be vaporized. They also pointed out that TiO itself is a relatively heavy molecule, and K_{zz} values of 10^7 cm^2 s^{-1} are required to even keep the molecule aloft at millibar pressures.

Showman et al. (2009a) discuss the related phenomenon of the nightside code trap, which could also be problematic for TiO gas remaining on the dayside. Superrotation occurs in hot Jupiters, such that gas is transported around the planet from day to night and night to day. On the cooler nightside, Ti could condense into refractory grains and sediment down, thereby leaving the gas that returns to the dayside Ti-free. One might then need a planet hot enough that everywhere on the daysides and nightsides temperatures are hot enough to avoid Ti condensation. Recently *Parmentier et al.* (2013) examined three-dimensional simulations with an eye toward understanding whether Ti-rich grains could be kept aloft by vertical motion. They suggest that temperature inversions could be a transient phenomenon depending on whether a given dayside updraft were rich or poor in Ti-bearing material that could then vaporize to form TiO.

Knutson et al. (2010) demonstrated that the presence or absence of thermal inversions in several systems are correlated with the chromospheric activity of their host stars. In particular, there seems to be a modest correlation between high stellar activity and the absence of thermal inversions.

This could mean that high stellar UV fluxes destroy the molecules responsible for the formation of temperature inversions (*Knutson et al.,* 2010).

Madhusudhan (2012) have suggested that the C/O ratio in hot Jupiter atmospheres could play an important role in controlling the presence or absence of atmospheric TiO, and hence the presence or absence of temperature inversions. In planets with C/O > 1, CO takes up most of the available oxygen, leaving essentially no oxygen for TiO gas. In this framework, the strongly irradiated planets that are hot enough that TiO gas would be expected, but observationally lack temperature inversions, would be carbon-rich gas giants.

4.3. Molecular Detections and Elemental Abundances

Early inferences of molecules in exoplanetary atmospheres were based on few channels of broadband photometry obtained using Spitzer and HST. *Deming et al.* (2005) and *Seager et al.* (2005) reported nondetection of H_2O in the dayside of hot Jupiter HD 209458b, observed in thermal emission. On the other hand, *Barman* (2007) reported an inference of H_2O at the day-night terminator of HD 209458b. *Tinetti et al.* (2007) inferred the presence of H_2O at the day-night terminator of another hot Jupiter, HD 189733b, although the observations have been a subject of debate (see, e.g., *Beaulieu et al.,* 2008; *Désert et al.,* 2009).

Early attempts have also been made to infer molecules using spacebased spectroscopy. The HST NICMOS instrument (1.8–2.3 µm) was used to report inferences of H_2O, CH_4, and/or CO_2 in the atmospheres of hot Jupiters HD 189733b, HD 209458b, and XO-1b (*Swain et al.,* 2008a, 2009a,b; *Tinetti et al.,* 2010). However, there is ongoing debate on the validity of the spectral features derived in these studies (*Gibson et al.,* 2011; *Burke et al.,* 2010; *Crouzet et al.,* 2012) and the interpretation of the data (*Madhusudhan and Seager,* 2009; *Fortney et al.,* 2010). *Grillmair et al.* (2008) inferred H_2O in the dayside atmosphere of HD 189733b using Spitzer Infrared Spectrograph (IRS) spectroscopy (also see *Barman,* 2008), but *Madhusudhan and Seager* (2009) suggest only an upper limit on H_2O based on a broader exploration of the model space.

Recent advances in multi-channel photometry and concomitant theoretical efforts are leading to statistical constraints on molecular compositions. Observations of thermal emission from hot Jupiters in four or more channels of Spitzer photometry, obtained during Spitzer's cryogenic lifetime, are now known for ~20 exoplanets (e.g., *Charbonneau et al.,* 2008; *Knutson et al.,* 2008, 2012; *Machalek et al.,* 2008; *Stevenson et al.,* 2010; *Campo et al.,* 2011; *Todorov et al.,* 2011; *Anderson et al.,* 2013). Parallel efforts in developing inverse modeling, or "retrieval methods," are leading to statistical constraints on molecular abundances and temperature profiles from limited photometric/ spectroscopic data (see section 3.1). Using this approach, *Madhusudhan and Seager* (2009) reported constraints on H_2O, CO, CH_4, and CO_2 in the atmospheres of HD 198733b

and HD 209458b (also see *Madhusudhan and Seager,* 2010, 2011; *Lee et al.,* 2012; *Line et al.,* 2012).

Constraints on molecular compositions of moderately irradiated exoplanets are beginning to indicate departures from chemical equilibrium. As discussed in section 3.3, low-temperature atmospheres provide good probes of nonequilibrium chemistry, especially of CO-CH_4 disequilibrium. *Stevenson et al.* (2010) and *Madhusudhan and Seager* (2011) reported significant nonequilibrium chemistry in the dayside atmosphere of hot Neptune GJ 436b (but see section 6). Nonequilibrium chemistry could also be potentially operating in the atmosphere of hot Jupiter HD 189733b (*Visscher and Moses,* 2011; *Knutson et al.,* 2012). Furthermore, *Swain et al.* (2010) reported a detection of non-local thermodynamic equilibrium (non-LTE) fluorescent CH_4 emission from the hot Jupiter HD 189733b (also see *Waldmann et al.,* 2012; but cf. *Mandell et al.,* 2011; *Birkby et al.,* 2013).

Nominal constraints have also been reported on the elemental abundance ratios in exoplanetary atmospheres. As discussed in section 2, it has now become routinely possible to detect thermal emission from hot Jupiters using groundbased near-IR facilities. The combination of spaceborne and groundbased observations provide a long spectral baseline (0.8–10 µm) facilitating simultaneous constraints on the temperature structure, multiple molecules, and quantities such as the C/O ratio. Using such a dataset (e.g., *Croll et al.,* 2011; *Campo et al.,* 2011), *Madhusudhan et al.* (2011a) reported the first statistical constraint on the C/O ratio in a giant planet atmosphere, inferring C/O ≥ 1 (i.e., carbon-rich) in the dayside atmosphere of hot Jupiter WASP-12b. However, the observations are currently a subject of active debate (see, e.g., *Cowan et al.,* 2012; *Crossfield et al.,* 2012c; *Föhring et al.,* 2013). More recently, *Madhusudhan* (2012) suggested the possibility of both oxygen-rich as well as carbon-rich atmospheres in several other hot Jupiters.

Robust constraints on the elemental abundance ratios such as C/O could help constrain the formation conditions of exoplanets. *Lodders* (2004) suggested the possibility of Jupiter forming by accreting planetesimals dominated by tar rather than water-ice (but cf. *Mousis et al.,* 2012). Following the inference of C/O ≥ 1 in WASP-12b (*Madhusudhan et al.,* 2011a), *Öberg et al.* (2011) suggested that C/O ratios in envelopes of giant planets depend on their formation zones with respect to the various ice lines in the protoplanetary disks. Alternately, inherent inhomogeneities in the C/O ratios of the disk itself may also contribute to high C/O ratios in the giant planets (*Madhusudhan et al.,* 2011b).

We are also beginning to witness early successes in transit spectroscopy using the HST Wide Field Camera 3 (WFC3) spectrograph and groundbased instruments. *Deming et al.* (2013) reported high signal-to-noise transmission spectroscopy of hot Jupiters HD 209458b and XO-1b, and a detection of H_2O in HD 209458b. WFC3 spectra have been reported recently for several other hot Jupiters (*Gibson et al.,* 2012; *Swain et al.,* 2012; *Huitson et al.,* 2013; *Mandell et al.,* 2013). Low-resolution near-IR spectra of hot Jupiters are also being observed from the ground (e.g., *Bean et al.*, 2013).

Recent efforts have also demonstrated the feasibility of constraining molecular abundance ratios of atmospheres of transiting as well as nontransiting hot Jupiters using high-resolution IR spectroscopy (see section 2.2). The technique has now been successfully used to infer the presence of CO and/or H_2O in the nontransiting hot Jupiter τ Bootis (e.g., *Brogi et al.*, 2012; *Rodler et al.*, 2012) and the transiting hot Jupiter HD 189733b (e.g., *de Kok et al.*, 2013; *Rodler et al.*, 2013) (see also *Crossfield et al.*, 2012c).

4.4. Atomic Detections

One of the early predictions of models of hot Jupiters was the presence of atomic alkali absorption in their atmospheres. Spectral features of Na and K, with strong resonance line doublets at 589 and 770 nm, respectively, were predicted to be observable in optical transmission spectra of hot Jupiters (*Seager and Sasselov*, 2000). The strong pressure-broadened lines of these alkalis are well known in brown dwarf atmospheres at these same T_{eff} values. To date, Na has been detected in several planets including HD 209458b (*Charbonneau et al.*, 2002), HD 189733b (*Redfield et al.*, 2008), WASP-17b (*Wood et al.*, 2011), and others, while K has been detected in XO-2b (*Sing et al.*, 2011b). It is not yet clear if the lack of many K detections is significant.

Detailed observations of the well-studied pressure-broadened wings of the Na absorption feature can allow for additional constraints on the planetary temperature structure. Since the line cores of the absorption features are formed at low pressure, and the pressure-broadened wings at higher pressure, models of temperature as a function of height can be tuned to yield the temperature-pressure profile at the terminator of the planet (e.g., *Huitson et al.*, 2012).

The escaping exosphere that has been observed around several hot Jupiters is another area where atomic detections have been made. Most clearly, neutral atomic hydrogen has been observed as a large radius cloud beyond the Roche lobe from two well-studied hot Jupiters, HD 209458b and HD 189733b (*Vidal-Madjar et al.*, 2003; *Lecavelier des Etangs et al.*, 2010; *Linsky et al.*, 2010). A recent review of the theory of exospheres and atmospheric escape from hot Jupiters can be found in *Yelle et al.* (2008). Neutral and ionized metals have been directly observed or inferred in the escaping exospheres of several planets (*Vidal-Madjar et al.*, 2004, *Fossati et al.*, 2010, *Astudillo-Defru and Rojo*, 2013).

4.5. Hazes, Clouds, and Albedos

In the absence of Rayleigh scattering or scattering from clouds, all incident flux from a parent star will be absorbed. Rayleigh scattering, mainly from gaseous H_2, is important in the relatively cloud-free visible atmospheres of Uranus and Neptune. However, warm planets will have much more abundant molecular gaseous opacity sources than solar system ice giants. Thus, the albedos of exoplanets will generally be determined by the relative importance of clouds.

Much of the early theoretical work on giant exoplanet atmospheres focused on albedos and scattered light signatures (*Marley et al.*, 1999, *Seager et al.*, 2000, *Sudarsky et al.*, 2000). In warm planets above 1000 K, clouds of silicates (such as forsterite, Mg_2SiO_4, and enstatite, $MgSiO_3$) were found to dramatically increase albedos for what would otherwise be quite dark planets.

Evidence to date shows that indeed most hot Jupiters are quite dark at optical wavelengths. Geometric albedos have been measured for several hot Jupiters using data from MOST, CoRoT, and especially Kepler. *Rowe et al.* (2008) used MOST data to derive a geometric albedo (A_G) of 0.038 ± 0.045 for HD 209458b. *Snellen et al.* (2010a) measured $A_G = 0.164 ± 0.032$ for CoRoT-2b, but noted that most of this flux likely arose from optical thermal emission, not scattered light. *Barclay et al.* (2012) and *Kipping and Spiegel* (2011) used Kepler data to find $A_G = 0.0136 ± 0.0027$ for TrES-2b, the current "darkest" planet.

The bulk of the evidence is that clouds do not have a dominant role in the absorption/scattering of stellar light for this class of planets. However, there are outliers. In particular, Kepler-7b was found to have $A_G = 0.35 ± 0.02$, likely indicating high silicate clouds (*Demory et al.*, 2011a, 2013).

It is at the time of transit, when the planet's transmission spectrum is observed, that clouds may play a greater role in altering the stellar flux. It was originally noted by *Seager and Sasselov* (2000) that clouds could weaken the spectral features in transmission. *Fortney* (2005) suggested that most such features for nearly all hot Jupiters would be weakened due to minor condensates becoming optically thick at the relevant long slant path lengths through the atmosphere.

The first detection of a hot Jupiter atmosphere was the detection of Na atoms in the atmosphere of HD 209458b (*Charbonneau et al.*, 2002), but the feature was weaker than expected, perhaps due to cloud material. Recent work in the near-IR (*Deming et al.*, 2013) finds weakened water vapor features for both HD 209458b and XO-1b.

The poster child for clouds is HD 189733b, which has an optical spectrum indicating Rayleigh scattering quite high in the planet's atmosphere (*Pont et al.*, 2008; *Sing et al.*, 2011a; *Evans et al.*, 2013). *Lecavelier des Etangs* (2008) suggest a cloud dominated by small Rayleigh scattering silicates.

In no planet is there a definitive determination of the composition of the cloud material. The clouds may originate from condensation of silicates or other refractory "rocky" materials, or from photochemical processes.

5. INFERENCES FOR DIRECTLY IMAGED PLANETS

5.1. Connections with Brown Dwarfs

The search for giant planets and brown dwarfs by direct imaging have gone hand-in-hand over the decades in part because they have overlapping criteria for good stellar targets. In fact, the earliest examples of L and T brown dwarfs,

GD 165B and Gl 229B respectively, were discovered by imaging surveys of white dwarfs and nearby field stars (*Becklin and Zuckerman,* 1988; *Nakajima et al.,* 1995). Numerous brown dwarfs are now known, along with many that are companions to nearby stars.

The atmospheres of brown dwarfs, of all ages, provide the best analogs to the atmospheres of directly imaged planets with the first major similarity being temperature. The coldest brown dwarfs currently known have $T_{eff} \sim 300$ K, while the youngest, most massive, brown dwarfs have $T_{eff} \sim 2500$ K. As illustrated in Fig. 7, this range is expected to encompass the majority of directly imaged planets. The masses of directly imaged planets also overlap with the low-mass end of brown dwarfs, suggesting similar surface gravities (Fig. 7), and convective interiors with thin radiative H + He envelopes. Thus, combinations of radiative, convective, hydrostatic, and chemical equilibrium are likely to be equally as useful baseline assumptions for the atmospheres of directly imaged planets as they are for brown dwarfs. Of course, deviations from equilibrium will occur, just as they do in some brown dwarfs, the planets in our solar system, and short-period exoplanets.

Having characteristics that overlap with brown dwarfs allows us to leverage nearly two decades of observations and modeling that have built detailed understanding of the important opacity sources, how temperature and surface gravity alter SEDs, and how bulk properties (e.g., radius) evolve with time and mass. Brown dwarfs have also been a major impetus for improvements in molecular line data for prominent carbon-, nitrogen-, and oxygen-bearing molecules (H_2O, CO, CH_4, and NH_3), and recent years have seen dramatic improvements (*Tennyson and Yurchenko,* 2012). All these molecules will play an equally important role in understanding directly imaged planets. Of course, having significant overlap with brown dwarfs also implies that directly imaged planets will inherit most of the uncertainty with brown dwarf atmospheres.

5.2. Atmospheric Chemistry

Observationally, the type of data available for brown dwarfs are also available for directly imaged planets — namely traditional spectra and photometry. Directly imaged planets can be compared, side-by-side, with brown dwarfs in color-color and color-magnitude diagrams and within sequences of near-IR spectra. Such comparisons can provide robust identifications of absorption properties and, in some cases, reliable estimates of temperature and gravity.

Directly imaged planets, because of their youth and masses, will have effective temperatures comparable to those of many close-in transiting planets (Fig. 7). However, unlike hot Jupiters, most directly imaged giant planets are far enough from their host star to have minimal stellar irradiation and to not have temperature inversions spanning their near-IR photospheres. The expectation, therefore, is that for most of the atmosphere, temperature decreases outward, resulting in deep molecular absorption features

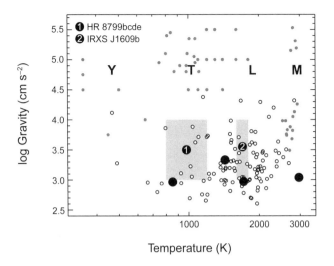

Fig. 7. Surface gravities and temperatures for hot Jupiters (open circles), brown dwarfs (filled gray circles), and approximate locations for five planet-mass companions found by direct imaging (gray regions). Hot Jupiters are plotted with equilibrium temperatures assuming 4π redistribution and brown dwarfs are plotted using values from *Rice et al.* (2010), *Burgasser et al.* (2006), *Cushing et al.* (2011), and *Stephens et al.* (2009). Transiting exoplanets GJ 436b, HD 189733b, HD 209458b, and WASP-12b are indicated with unnumbered filled symbols, from left to right.

comparable to those seen in brown dwarf spectra.

Figure 8 compares low-resolution K-band spectra of two young planets, HR 8799b and HR 8799c (*Barman et al.,* 2011; *Konopacky et al.,* 2013), to the young ~ 5-M_{Jup} brown dwarf companion 2M1207b (*Patience et al.,* 2010) and the old T dwarf HD 3651B. Even at low resolution, water absorption can be identified as responsible for the slope across the blue side of the K band. The spectra of 2M1207b and HR 8799c have the CO absorption band-head at ~ 2.3 μm, characteristic of much hotter brown dwarfs. Individual CO and H_2O lines have been resolved in spectra of HR 8799c, 1RXS J1609, 2M1207b, and many low-mass young brown dwarfs. Methane bands blanket much of the near-IR and IR and, for example, contribute (along with water) to the very strong absorption seen in spectrum of HD 3651b starting around 2.1 μm (Fig. 8). So far, methane absorption has not been unambiguously identified in most young giant planets but likely contributes to the slope across the red wing of the K-band spectrum of HR 8799b (*Barman et al.,* 2011; *Bowler et al.,* 2010). Methane absorption is stronger at longer wavelengths, almost certainly influencing the IR colors across ~ 3 to 5 μm. Recently, *Janson et al.* (2013) reported a detection of methane in the young planet GJ 504b.

As discussed in section 3.3, nonequilibrium chemistry is known to be prevalent in low-temperature planetary atmospheres. Among directly imaged planets and planet-mass companions, HR 8799b, HR 8799c, and 2M1207b stand out as potentially extreme cases of nonequilibrium chemistry. All three objects have effective temperatures below 1200 K

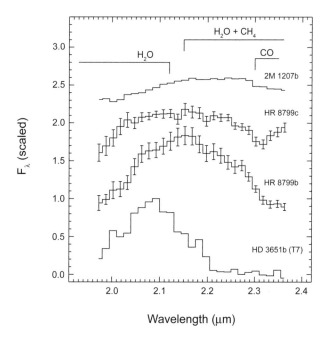

Fig. 8. K-band spectra for 2M1207b (*Patience et al.,* 2010), HR 8799c (*Konopacky et al.,* 2013), HR 8799b (*Barman et al.,* 2011), and HD 3651B, all binned to the same resolution (R ~ 100). Approximate locations of strong absorption bands are indicated for H_2O, CO, and CH_4.

but show only CO absorption and little to no CH_4 in their near-IR spectra (*Bowler et al.,* 2010; *Barman et al.,* 2011; *Konopacky et al.,* 2013). The HR 8799b,c planets are also very cloudy, as inferred from their near-IR broadband colors (as discussed below), and the above-average cloud cover may also be evidence for strong mixing. This situation may very well be common in all young low-mass planets with low surface gravities and suggests that direct imaging may uncover many cloudy, CO-rich, planetary atmospheres.

5.3. Cloud Formation

A major feature of brown dwarf atmospheres is the formation of thick clouds of dust at photospheric depths beginning around $T_{eff} \sim 2500$ K with the condensation of Fe. For cooler and less-massive field L dwarfs, the impact of photospheric dust increases as abundant magnesium-silicate condensates are added (e.g., *Helling et al.,* 2008). The impact of dust opacity is seen as a steady reddening of the SED by about ~0.5 mag in H–K. Mie scattering by small grains dominates the opacity for $\lambda < 1$ µm and is capable of smoothing over all but the strongest absorption features. Field brown dwarfs with $T_{eff} \sim 1400$ K eventually become bluer as the location of atmospheric dust shifts to deeper layers and, eventually, the dust is mostly below the photosphere. Low photospheric dust content is a major characteristic of T-type brown dwarfs. Eventually, for very low masses and effective temperatures below ~500 K, water-ice clouds should form, producing significant changes in the near-IR colors.

The reddening of the SED is often measured in color-color and color-magnitude diagrams (CMD). Figure 9 compares near-IR colors of field brown dwarfs to several young, directly imaged, companions. Objects on the reddest part of the CMD (H–K > 0.5) have cloudy photospheres, while bluer objects are likely cloud-free or have very thin clouds mostly below the photosphere. Interestingly, the first planet-mass companion found by direct imaging, 2M1207b, is one of the reddest objects in near-IR CMDs, indicating that its photosphere is heavily impacted by clouds. The HR 8799 planets are also red and thus similarly impacted by photospheric dust. The physical mechanism that causes the photosphere to transition from cloudy to cloud-free is poorly understood; however, recent observations of young brown dwarfs suggest that the effective temperature at the transition decreases with decreasing surface gravity. Directly imaged planets are not only lower in mass than brown dwarfs, but their surface gravities are also lower because of their youth. That the first set of directly imaged planets are all cloudy is further evidence that the transition temperature decreases with surface gravity and that many young giant planets will have cloudy photospheres even though their effective temperatures are comparable to the much bluer T dwarfs.

5.4. Interpretation of Spectra

Low- to moderate-resolution spectroscopy has been obtained for most planets found so far by direct imaging and discoveries made with the new instruments — Gemini Planet Imager (GPI), Spectro-Polarimetric High-contrast Exoplanet Research (SPHERE), and Project 1640 (P1640) — will be accompanied by low-resolution near-IR spectra. The interpretation of these spectra follows the same techniques used for brown dwarfs. Comparisons can be made to empirical templates spanning the various brown dwarf spectral types to roughly estimate temperature and gravity. However, this method is unreliable for the faintest (lowest mass) and reddest objects in Fig. 9 because "normal" brown dwarfs do not populate the same region of the CMDs. In fact, for 2M1207b, such a comparison would yield a high effective temperature and unphysical radius (*Mohanty et al.,* 2007). In the absence of a large set of well-characterized empirical templates, one is left fitting synthetic model spectra to the observed spectra to infer temperature, gravity, cloud properties and chemical composition.

When analyzing spectra (or broad photometric coverage) of mature brown dwarfs (with known distances), it is common to use the measured bolometric luminosity (L_{bol}) and evolutionary models to establish a plausible range for both effective temperature and surface gravity, then restrict model atmosphere fits to within this range. This greatly reduces the parameter space one must search (especially if vertical mixing and cloud properties are important). This method is fairly robust for old brown dwarfs because

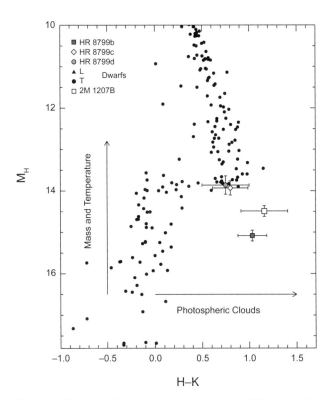

Fig. 9. Absolute H-band magnitudes vs. H-K color for field brown dwarfs (*Dupuy and Liu, 2012*) and the young planet-mass companions HR 8799bcd and 2M1207b. Arrows indicate the basic color and magnitude trends for increasing photospheric cloud coverage/thickness and mass/temperature, respectively.

the evolution is well described by "hot start" models with solar or near-solar abundances (e.g., *Baraffe et al., 2003*). However, the initial conditions set by the formation process will be important at young ages, making it difficult to uniquely connect L_{bol} and age to a narrow range of T_{eff} and log(g) (e.g., *Marley et al., 2007; Spiegel and Burrows, 2012; Bonnefoy et al., 2013*). Furthermore, it is possible that planets formed by core accretion may not have stellar/solar abundances (*Madhusudhan et al., 2011b; Öberg et al., 2011*).

A second approach is to rely exclusively on model atmosphere fits using a large model grid that finely samples the free parameters that define each model (*Currie et al., 2011; Madhusudhan et al., 2011c*). This method has the advantage of being completely independent of the evolution calculations and initial conditions and would allow a mapping of T_{eff} and log(g) as function of age for giant planets. In practice, however, atmosphere-only fitting can be difficult because the number of free parameters is fairly large (especially if basic composition is allowed to vary) and local minima are easily found when minimizing χ^2. Combining photometry across many bandpasses and higher signal-to-noise ratio (SNR) data and higher spectral resolution are often needed to find a reliable match. Of course, giant planet atmospheres are complex

and model inadequacies also contribute to the poor and/or often misleading fits.

5.5. The HR 8799 System

The four planets orbiting the young star HR 8799, between 14 and 70 AU (0.3 to 1.7 arcsec), are the best examples yet of what direct imaging can yield (*Marois et al., 2008, 2010*). This system is providing new insights into the formation and evolution of giant planets and currently serves as the best system for testing planetary models and new instruments. The HR 8799 system is also an excellent early test of our ability to infer basic properties of planets discovered by direct imaging. Unlike transits, when direct imaging is possible, at least one flux measurement is immediately obtained. Thus, planets found in this way immediately lend themselves to atmospheric characterization.

The HR 8799b,c,d discovery included photometry covering J through L′ wavelengths and indicated that all three planets have red IR colors and, therefore traditional cloud-free atmosphere models are inappropriate (*Marois et al., 2008*). Marois et al. found that model atmospheres including condensates suspended at pressures determined only by chemical equilibrium systematically overpredicted T_{eff}, implying radii too small for the well-determined L_{bol} and our current understanding of giant planet formation. Marois et al. also concluded that intermediate cloud models (constrained vertically) could simultaneously match the photometry and match expectations from giant planet evolution tracks. The photometric quality and wavelength coverage continues to improve (*Hinz et al., 2010; Galicher et al., 2011; Currie et al., 2011; Skemer et al., 2012*) and the basic conclusions from these data are that all four planets have $800 < T_{eff} < 1200$ K, have cloud coverage that is comparable to the reddest L dwarfs [or even thicker (*Madhusudhan et al., 2011c*)], and that atmosphere models can easily overestimate effective temperature unless additional, secondary, model parameters are allowed to vary (see *Marley et al., 2012*, for a summary). Most studies agree that CO/CH_4 nonequilibrium chemistry is necessary to fit the photometry. Observations at 3.3 μm, overlapping a broad CH_4 band, are particularly sensitive to nonequilibrium chemistry (*Bowler et al., 2010; Hinz et al., 2010; Skemer et al., 2012*).

In addition to photometry, the HR 8799 planets have been studied spectroscopically. Measurements have been made across two narrow bands, from 3.9–4.1 μm [HR 8799c (*Janson et al., 2010*)] and 2.12–2.23 μm [HR8799b (*Bowler et al., 2010*)], as well as across the full J, H, and K bands [HR 8799b (*Barman et al., 2011*), HR 8799b,c,d (*Oppenheimer et al., 2013*), and HR 8799c (*Konopacky et al., 2013*)]. The H-band spectrum of HR 8799b has a shape indicative of low surface gravity, consistent with its low mass and youth. Low-resolution spectra for HR 8799b and c show only weak or no evidence for CH_4 absorption, best explained by nonequilibrium chemistry, as already suggested by photometry.

Unless initially constrained by evolution model predictions, model atmosphere fits to the low-resolution spectroscopic data (e.g., for HR 8799b) can easily overestimate T_{eff}. This situation is identical to the one encountered when fitting the photometry. The challenge is to produce a cool model atmosphere ($T_{eff} < 1000$ K) that simultaneously has red near-IR colors, weak CH_4 and strong H_2O absorption matching the observed spectra, and the right overall H- and K-band shapes. When minimizing χ^2, local minima are often found because more than one physical property can redden the spectrum, in addition to clouds. As gravity decreases so does collision-induced H_2 opacity (which is strongest in the K band), allowing more flux to escape at longer wavelengths. Also, uniform increases in metals preferentially increase the opacities at shorter wavelengths. So far, the best approach for dealing with potential degeneracies has been to combine photometry covering as much of the SED as possible (influenced mostly by T_{eff}) with as much spectral information as possible (influenced mostly by gravity and nonequilibrium chemistry). Additional constraints on radius from evolution models are also useful to keep the overall parameter space within acceptable/physical boundaries. As atmosphere models and observations improve, some constraints will hopefully be unnecessary.

HR 8799c is nearly a magnitude brighter than HR 8799b, allowing moderate-spectral-resolution observations (R ~ 4000), and has revealed hundreds of molecular lines from H_2O and CO (*Konopacky et al.*, 2013). These data provided additional evidence for low surface gravity and nonequilibrium chemistry. Konopacky et al. detected no CH_4 lines, which strongly supports the nonequilibrium CO/CH_4 hinted at by the photometry and low-resolution spectra. More importantly, access to individual lines allowed the C and O abundances to be estimated, with the best-fitting C/O being larger than the host star, hinting at formation by core accretion (*Konopacky et al.*, 2013; *Madhusudhan et al.*, 2011b; *Öberg et al.*, 2011).

6. ATMOSPHERES OF HOT NEPTUNES

Exo-Neptunes are loosely defined here as planets with masses and radii similar to those of the ice giants in the solar system (M_p between 10 and 30 $M_\oplus$; R_p between 1 and 5 $R_\oplus$). Masses and radii have been measured for six transiting exo-Neptunes to date. Since all these objects have equilibrium temperatures of ~600–1200 K, compared to ~70 K for Neptune, they are referred to as "hot Neptunes." While the interiors of ice giant planets are expected to be substantially enriched in ices and heavy elements, their atmospheres are generally expected to be H_2-rich. Their expected low mean-molecular weight together with their high temperatures make hot Neptunes conducive to atmospheric observations via transit spectroscopy just like hot Jupiters. On the other hand, their temperatures are significantly lower than those of hot Jupiters (~1300–3300 K), making them particularly conducive to studying nonequilibrium processes in their atmospheres (see section 3.3).

To date, atmospheric observations over multiple spectral bands have been reported for only one hot Neptune, GJ 436b. *Stevenson et al.* (2010) reported planet-star flux contrasts of the dayside atmosphere of GJ 436b in six channels of Spitzer broadband photometry, explaining which required a substantial depletion of CH_4 and overabundance of CO and CO_2 compared to equilibrium predictions assuming solar abundances (*Stevenson et al.*, 2010; *Madhusudhan and Seager*, 2011; see also *Spiegel et al.*, 2010). *Knutson et al.* (2011) reported multiple photometric observations of GJ 436b in transit in each of the 3.6- and 4.5-μm Spitzer channels, which indicated variability in the transit depths between the different epochs. For the same set of observations as *Stevenson et al.* (2010) and *Knutson et al.* (2011), *Beaulieu et al.* (2011) used a subset of the data and/or adopted higher uncertainties to suggest the possibility of a methane-rich atmosphere consistent with equilibrium predictions. On the other hand, *Shabram et al.* (2011) reported model fits to the transit data and found their models to be inconsistent with those used by *Beaulieu et al.* (2011).

The inferences of low CH_4 and high CO in GJ 436b suggest extreme departures from equilibrium chemistry. As discussed in section 3.3, in low-temperature atmospheres CH_4 and H_2O are expected to be abundant, whereas CO is expected to be negligible. *Madhusudhan and Seager* (2011) suggested that the inferred CO enhancement in GJ 436b can be explained by a combination of supersolar metallicity ($\geq 10\times$ solar) and nonequilibrium chemistry, via vertical eddy mixing of CO from the hotter lower regions of the atmosphere. The required eddy mixing coefficient (K_{zz}) of ~10^7 cm^2 s^{-1} (*Madhusudhan and Seager*, 2011) is also consistent with that observed in three-dimensional GCMs of GJ 436b (*Lewis et al.*, 2010). However, even though CH_4 can be photochemically depleted by factors of a few in the upper atmospheres of irradiated giant planets (e.g., *Moses et al.*, 2011), the drastic depletion required in the observable lower atmosphere (at the 100-mbar pressure level) of GJ 436b has no explanation to date. Furthermore, *Shabram et al.* (2011) suggested that at the low temperatures of GJ 436b, other higher-order hydrocarbons such as HCN and C_2H_2 can also be abundant in the atmosphere, and contribute spectral features overlapping with those of CH_4. Recently, *Moses et al.* (2013) suggested an alternate explanation that a CH_4-poor and CO-rich composition in GJ 436b could potentially be explained by an extremely high atmospheric metallicity (~$1000\times$ solar). Future observations with better spectral resolution and wider wavelength coverage would be needed to refine the molecular abundances and constrain between the different hypotheses.

7. ATMOSPHERES OF SUPER-EARTHS

"Super-Earths" are nominally defined as planets with masses between 1 and 10 $M_\oplus$. Super-Earths have no analogs in the solar system, their masses being intermediate between those of terrestrial planets and ice giants. As such, it is presently unknown if super-Earths are mini-Neptunes with H/He-rich atmospheres, or are scaled-up terrestrial planets with atmospheres dominated by heavy molecules (H_2O, CO_2, etc.). It is also unclear if the heavier elements beneath the atmospheres are predominantly rock/iron, or if they possess a large mass fraction of volatile ices, as has long been suggested for Uranus and Neptune. As they are the lowest-mass planets whose atmospheres can potentially be observed with existing and upcoming instruments, understanding super-Earths is a major frontier in exoplanetary science today. In the present section, we focus on the atmospheres of volatile-rich super-Earths, such as GJ 1214b.

7.1. Theory

Planetary atmospheres can be classified as either "primary" or "secondary." Primary atmospheres are those that are accreted directly from the nebula, and are therefore dominated by hydrogen and helium gas. Secondary atmospheres are those that are made up of volatiles outgassed from the planet's interior. Within the solar system, Jupiter, Saturn, Uranus, and Neptune have primary atmospheres, while the atmospheres of the smaller bodies (rocky planets and moons) are secondary atmospheres. It is possible that Uranus and Neptune, which are only 10–20% H-He by mass, could have some minor secondary component.

There is a strong expectation that planets along the continuum from 1 to 15 $M_\oplus$ will possess diverse atmospheres that could either be predominantly primary or predominantly secondary or mixed. The accretion of primary atmospheres has long been modeled within the framework of the core-accretion theory for planet formation (see the chapter by Helled et al. in this volume). The outgassing of secondary atmospheres has long been modeled in the solar system, but relatively little work has occurred for exoplanets (*Elkins-Tanton and Seager*, 2008; *Rogers and Seager*, 2010). One particular finding of note is that hydrogen atmospheres up to a few percent of the planet's mass could be outgassed. Unlike primary atmospheres, these would be helium free. A primary future goal of characterizing the atmospheres of planets in the super-Earth mass range is to understand the extremely complex problem of the relative importance of primary and secondary atmospheric origin for planets in this mass range.

The tools to observe super-Earth atmospheres will be the same used for gas giants: initially transmission and emission spectroscopy, and eventually direct imaging. A number of recent studies have reported atmospheric models of super-Earths to aid in the interpretation of observed spectra. These studies investigated theoretical spectra and retrieval methods (*Miller-Ricci et al.*, 2009; *Miller-Ricci and Fortney*, 2010; *Benneke and Seager*, 2012; *Howe and Burrows*, 2012), atmospheric chemistry (e.g., *Miller-Ricci Kempton et al.*, 2012), and clouds and hazes (*Howe and Burrows*, 2012; *Morley et al.*, 2013) for super-Earth atmospheres.

Spectroscopic identification of abundant gases in the planetary atmosphere would be the clearest and most direct path to assessing bulk atmospheric composition. A more indirect route would come from measurements of the scale height of a transiting super-Earth atmosphere in any one molecular band. *Miller-Ricci et al.* (2009) pointed out that a measurement of the atmospheric scale height from a transmission spectrum would be a direct probe of the atmospheric mean molecular weight (MMW) and hence the bulk composition of a super-Earth. Atmospheres with larger scale heights yield more dramatic variation in the transit radius as a function of wavelength. In principle, H-rich atmospheres could be easily separated from those dominated by heavier molecules like steam or carbon dioxide. However, it is important to keep in mind that clouds, which are generally gray absorbers or scatterers, could also obscure transmission spectrum features (*Fortney*, 2005; *Howe and Burrows*, 2012; *Morley et al.*, 2013). However, transmission spectrum observations with a high enough signal-to-noise ratio should readily be able to recover a wealth of atmospheric information (*Benneke and Seager*, 2012).

7.2. GJ 1214b as Prototype

The super-Earth GJ 1214b has been the target of many observational campaigns to better understand the transmission spectrum from the blue to the mid-IR. From a mass and radius measurement alone, the composition is degenerate (*Rogers and Seager*, 2010; *Nettelmann et al.*, 2011; *Valencia et al.*, 2013). If the outer layer of the planet, including the visible atmosphere, can be probed, we will have a much better understanding of the planet's bulk composition (*Miller-Ricci and Fortney*, 2010). A cloud-free atmosphere of solar composition, with a relatively large scale height, was quickly ruled out by *Bean et al.* (2010). Subsequent data from Spitzer (*Désert et al.*, 2011c) and the HST WFC3 (*Berta et al.*, 2012a), along with groundbased data, concurred with this view. With these data, the transmission spectrum was essentially flat (Fig. 10), and was consistent with either a high-MMW atmosphere or an atmosphere blanketed by opaque clouds. Most recently, very high signal-to-noise data with HST WFC3 still show a flat spectrum, ruling out a cloud-free high-MMW atmosphere (*Kreidberg et al.*, 2014).

As has been suggested for GJ 436b, it may be that the atmosphere is *strongly* supersolar in abundances, perhaps hundreds of times solar, or even higher. *Fortney et al.* (2013) suggest from population synthesis models that metal-rich envelopes for low-mass ~5–10-$M_\oplus$ planets may commonly reach values of Z from 0.6 to 0.9. Such atmospheres would naturally appear more featureless in transmission spectra due to a smaller scale height and would have abundant material to form clouds and hazes.

With only one example yet probed, we are not yet sure if the difficulty in obtaining the transmission spectrum of GJ 1214b will be normal, or will be the exception. Photometric detections of transits, and an eclipse, in the Spitzer bandpasses have also been reported for another super-Earth, 55 Cancri e (*Demory et al.*, 2012; *Gillon et al.*, 2012a), but the planet-star radius ratio in this case is significantly smaller compared to GJ 1214b, making spectral observations

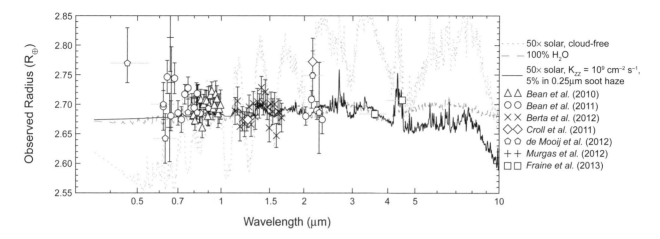

Fig. 10. Model transmission spectra and observations of the super-Earth GJ 1214b. In dotted light gray is a transmission model with a metallicity 50× solar. In dashed dark gray is 100% steam atmosphere. In black is a 50× solar model that also includes a hydrocarbon haze layer from *Morley et al.* (2013). The cloud-free 50× model is clearly ruled out by the data, but with current data the steam and hazy model give very similar quality fits. Future high S/N data in the near-infrared, or especially the mid-infrared, will help to distinguish between models.

challenging. Recently, another transiting planet of a similar mass and radius to GJ 1214b, HD 97658b, was discovered around a bright K star (*Dragomir et al.,* 2013), presenting a promising candidate for spectroscopic follow-up and comparison to GJ 1214b. The ubiquity of 2–3-$R_\oplus$ planets on relatively short period orbits around M stars and Sun-like stars suggests we will eventually have a large comparison sample of these volatile-rich low-mass planets.

8. FUTURE OUTLOOK

Observations with existing and upcoming facilities promise a bright outlook for the characterization of exoplanetary atmospheres. Currently, low-resolution data from spaceborne (HST and Spitzer) and groundbased instruments are already leading to nominal inferences of various atmospheric properties of exoplanets. Future progress in atmospheric observations and theory can advance exoplanetary characterization in three distinct directions, depicted in Fig. 1. First, robust inferences of chemical and thermal properties will enhance our understanding of the various physicochemical processes in giant exoplanetary atmospheres discussed in this chapter. Second, better constraints on the elemental abundance ratios in such atmospheres could begin to inform us about their formation environments. Finally, high-precision observations could help constrain the atmospheric compositions of super-Earths, which in turn could constrain their interior compositions. Several new observational facilities on the horizon will aid in these various aspects of exoplanetary characterization.

8.1. Next-Generation Atmospheric Observations of Transiting Giant Planets

Over the next decade the development of large-aperture telescopes such as the European Extremely Large Telescope

(E-ELT), the Giant Magellan Telescope (GMT), and the Thirty Meter Telescope (TMT) will open up new opportunities for studies of both transiting and nontransiting planets. The increased aperture of these telescopes will provide an immediate advantage for transmission spectroscopy and phase variation studies using visible and near-IR echelle spectroscopy, which are currently photon-noise-limited even for the brightest stars (e.g., *Redfield et al.,* 2008; *Jensen et al.,* 2011, 2012; *Snellen et al.,* 2010b; *Brogi et al.,* 2013; *de Kok et al.,* 2013). Such facilities could also make it feasible to characterize the atmospheres of terrestrial-sized exoplanets (*Snellen et al.,* 2013; *Rodler and López-Morales,* 2014). It is less clear whether or not such large telescopes will provide good opportunities for broadband photometry and low-resolution spectroscopy of transits and secondary eclipses, as this technique requires a large field of view containing multiple comparison stars in order to correct for time-varying telluric absorption and instrumental effects (e.g., *Croll et al.,* 2010, 2011). Most broadband studies using this technique achieve signal-to-noise ratios that are a factor of 2–3 above the photon noise limit, implying that increased aperture may not lead to a corresponding reduction in noise until the limiting systematics are better understood.

The James Webb Space Telescope (JWST) (*Garder et al.,* 2006) currently offers the best future prospects for detailed studies of exoplanetary atmospheres. Given its large aperture and spectral coverage, JWST will be capable of obtaining very high signal-to-noise and high-resolution spectra of transiting giant planets, thereby rigorously constraining their chemical and thermal properties. In particular, the spectroscopic capabilities of the Near Infrared Spectrograph (NIRSpec) (0.6–5.0 μm) and Mid-Infrared Instrument (MIRI) (5–28 μm) onboard JWST will provide a wide wavelength coverage, including spectral features of all the prominent molecules in gas giant atmospheres (see section 3.3). On the other hand, the possibility of dedicated space missions like the Exoplanet Characterisation Observatory (EChO) for atmospheric characterization of exoplanets (*Barstow et al.,* 2013; *Tinetti et al.,* 2013) will significantly increase the sample of giant exoplanets with high-quality spectral data.

8.2. The Small Star Opportunity

If we wish to study the properties of smaller and more Earth-like planets, it is crucial to find these objects orbiting small, nearby stars. The reason for this can be easily understood in light of equations (2) and (3). As the size and temperature of a planet decreases, the corresponding transmission spectrum amplitude and secondary eclipse depth will also drop rapidly. If we evaluate equation (2) for the Earth-Sun system, we find that Earth's atmosphere has a scale height of approximately 8.5 km at a temperature of 290 K, assuming a mean molecular weight of 4.81×10^{-26} kg. The corresponding depth of the absorption features in Earth's transmission spectrum is then 2×10^{-6}, or a factor of ~1000 smaller than the signal from the transiting hot Jupiter HD 189733b. If we carry out the same calculation for a secondary eclipse observation using equation (3), we find a predicted depth of approximately 4×10, also a factor of ~1000 smaller than for HD 189733b. Rather than building a space telescope with a collecting area a million times larger than existing facilities, we can mitigate the challenges inherent in measuring such tiny signals by focusing on observations of small planets transiting small stars.

If we instead place our transiting Earth in front of a late-type M5 dwarf with a radius of 0.27 $R_\odot$ and an effective temperature of 3400 K, we gain a factor of 10 for the transmission spectrum and a factor of 25 for the secondary eclipse depth. These kinds of signals, although more favorable, may still be beyond the reach of the JWST (*Greene et al.,* 2007; *Kaltenegger and Traub,* 2009; *Seager et al.,* 2009). If we are instead willing to consider super-Earth-sized planets, with lower surface gravities and larger scale heights than Earth, it may be possible to measure transmission spectra for these more favorable targets. Similarly, if we extend our secondary eclipse observations to super-Earths, which have larger radii and hence deeper secondary eclipses, such observations could be achievable with JWST.

There are currently three small groundbased transiting planet surveys that are focused exclusively on searching the closest, brightest M dwarfs (*Nutzman and Charbonneau,* 2008; *Berta et al.,* 2012b; *Kóvacs et al.,* 2013; *Sozzetti et al.,* 2013). To date, only one survey, MEarth, has detected a new transiting super-Earth (*Charbonneau et al.,* 2009), but as these surveys mature it is likely that additional systems will be discovered. Another promising avenue is to search for transits of known low-mass planets detected by RV surveys; two transiting super-earths to date have been detected this way (*Winn et al.,* 2011; *Demory et al.,* 2011b; *Dragomir et al.,* 2013). However, neither of these surveys are expected to yield large numbers of low-mass transiting planets orbiting M stars, due to a combination of limited sample sizes and sensitivity limits that are too high to detect very small planets. If we wish to detect large numbers of transiting planets it will require a new space mission that will survey all the brightest stars in the sky with a sensitivity high enough to detect transiting super-Earths. The Transiting Exoplanet Survey Satellite (TESS) mission was recently approved by NASA for launch in 2017, and currently represents our best hope for building up a large sample of small transiting planets suitable for characterization with JWST. In addition, the CHaracterizing ExO-Planet Satellite (CHEOPS) was recently selected by ESA for launch in 2017, which will search for transits of small planets around bright stars that are already known to host planets via RV searches.

8.3. Expectations for New Direct Imaging Instruments

New specialized direct-imaging instruments are, or soon will be, available at groundbased telescopes behind newly designed AO systems. These include the SPHERE instrument for the European Southern Observatory's VLT (*Beuzit et al.,* 2006), P1640 for Palomar Observatory (*Oppenheimer et al.,* 2013), GPI for the Gemini South telescope (*Macintosh et al.,* 2008), the PISCES and LMIRCam facilities for the Large Binocular Telescope (LBT) (e.g., *Skemer et al.,* 2012), and the High-Contrast Coronographic Imager for Adaptive Optics (HiCIAO) instrument for the Subaru telescope for the Strategic Explorations of Exoplanets and Disks with Subaru (SEEDS) Project (e.g., *Tamura,* 2009; *Kuzuhara et al.,* 2013). These instruments are optimized to reduce scattered light within an annulus centered on the star, sacrificing good contrast at very wide separations in favor of a high-contrast field of view optimized for sub-Jupiter-mass planets in the 5–50-AU range. Any candidate within this region will have near-IR fluxes and low-resolution (R ~ 45) spectra measured (e.g., the H band). Surveys with these instruments will determine the frequency of Jupiter-mass planets for a range of stellar types and ages, in wider orbits than are available using RV and transit methods. For a subset of the discoveries, low-resolution spectra, at a high-signal-to-noise ratio, will be obtained across most of the Y–J–H–K band (depending on the instrument). Additional mid-IR observations from the ground will be possible, e.g., with the Magellan AO system providing valuable flux measurements closer to the SED peak for the coldest, lowest-mass, direct imaging discoveries (*Close et al.,* 2012).

The prospects for atmosphere characterization are excellent with access to such a large amount of spectral data. Even at low resolution, broad molecular features from H_2O and CH_4 should be measurable. If clouds are present, the spectra will be noticeably smoother and redder, as discussed in section 5. Having access to Y and J bands will greatly aid in narrowing in on the preferred range of cloud particle sizes and thicknesses. H- and K-band spectra will provide independent estimates of surface gravity. As learned from HR 8799, model atmospheres continue to prove problematic for inferring T_{eff} and surface gravity. However, having spectra for many planets with different ages, masses, and luminosities will quickly lead to improvements in theory.

Giant groundbased telescopes (~30-m-class) and new instrumentation in space promise to greatly increase the number of directly imaged planetary systems, with greater contrasts and down to lower masses and/or older ages. The JWST, with its IR and mid-IR capabilities [Near-Infrared Camera (NIRCam), MIRI, and Tunable Filter Imager (TFI)], will provide much-needed longer-wavelength capability for direct spectroscopy of sub-Jupiter-mass planets.

Acknowledgments. N.M. acknowledges support from Yale University through the YCAA fellowship. T.B. acknowledges support from NASA awards NNX10AH31G and NNH10AO07I and NSF award AST-1405505. J.J.F. acknowledges support from NASA awards NNX09AC22G and NNX12AI43A and NSF award AST-1312545.

REFERENCES

Agol E. et al. (2005) *Astrophys. J., 359,* 567–579.
Agol E. et al. (2010) *Astrophys. J., 721,* 1861–1877.
Alonso R. et al. (2009a) *Astron. Astrophys., 501,* L23–L26.
Alonso R. et al. (2009b) *Astron. Astrophys., 506,* 353–358.
Anderson D. R. et al. (2010) *Astron. Astrophys., 513,* L3.
Anderson D. R. et al. (2013) *Mon. Not. R. Astron. Soc., 430,* 3422–3431.
Asplund M. et al. (2009) *Annu. Rev. Astron. Astrophys., 47,* 481–522.
Astudillo-Defru N. and Rojo P. (2013) *Astron. Astrophys., 557,* A56.
Bailey V. et al. (2014) *Astrophys. J. Lett., 780,* L4.
Baraffe I. et al. (2003) *Astron. Astrophys., 402,* 701–712.
Barclay T. et al. (2012) *Astrophys. J., 761,* 53.
Barman T. S. (2007) *Astrophys. J. Lett., 661,* L191–L194.
Barman T. S. (2008) *Astrophys. J. Lett., 676,* L61–L64.
Barman T. S. et al. (2005) *Astrophys. J., 632,* 1132–1139.
Barman T. S. et al. (2011) *Astrophys. J., 733,* 65.
Barstow J. K. et al. (2013) *Mon. Not. R. Astron. Soc., 430,* 1188–1207.
Bean J. L. et al. (2010) *Nature, 468,* 669–672.
Bean J. L. et al. (2011) *Astrophys. J., 743,* 92–104.
Bean J. L. et al. (2013) *Astrophys. J., 771,* 108–119.
Beaulieu J.-P. et al. (2008) *Astrophys. J., 677,* 1343–1347.
Beaulieu J.-P. et al. (2011) *Astrophys. J., 731,* 16–27.
Becklin E. E. and Zuckerman B. (1988) *Nature, 336,* 656–658.
Benneke B. and Seager S. (2012) *Astrophys. J., 753,* 100.
Berdyugina S. V. et al. (2008) *Astrophys. J. Lett., 673,* L83–L86.
Berta Z. K. et al. (2012a) *Astrophys. J., 747,* 35–52.
Berta Z. K. et al. (2012b) *Astron. J., 144,* 145–165.
Beuzit J.-L. et al. (2006) *The Messenger, 125,* 29–34.
Birkby J. L. et al. (2013) *Mon. Not. R. Astron. Soc. Lett., 436,* L35–L39.
Bonnefoy M. et al. (2013) *Astron. Astrophys., 555,* A107.
Borucki W. J. et al. (2009) *Science, 325,* 709–709.
Bouchy F. et al. (2005) *Astron. Astrophys., 444,* L15–L19.
Bowler B. P. et al. (2010) *Astrophys. J., 723,* 850–868.
Brogi M. et al. (2012) *Nature, 486,* 502–504.
Brogi M. et al. (2013) *Astrophys. J., 767,* 27–37.
Brown T. M. (2001) *Astrophys. J., 553,* 1006–1026.
Burgasser A. et al. (2006) *Astrophys. J., 639,* 1095–1113.
Burrows A. and Sharp C. M. (1999) *Astrophys. J., 512,* 843–863.
Burrows A. et al. (2007) *Astrophys. J. Lett., 668,* L171–L174.
Burrows A. et al. (2008a) *Astrophys. J., 682,* 1277–1282.
Burrows A. et al. (2008b) *Astrophys. J., 678,* 1436–1457.
Burke C. J. et al. (2010) *Astrophys. J., 719,* 1796–1806.
Campo C. et al. (2011) *Astrophys. J., 727,* 125–136.
Carson J. et al. (2013) *Astrophys. J. Lett., 763,* L32–L37.

Charbonneau D. et al. (2002) *Astrophys. J., 568,* 377–384.
Charbonneau D. et al. (2005) *Astrophys. J., 626,* 523–529.
Charbonneau D. et al. (2008) *Astrophys. J., 686,* 1341–1348.
Charbonneau D. et al. (2009) *Nature, 462,* 891–894.
Chauvin G. et al. (2005) *Astron. Astrophys., 438,* L25–L28.
Cho J. et al. (2008) *Astrophys. J., 675,* 817–845.
Christiansen J. L. et al. (2010) *Astrophys. J., 710,* 97–104.
Close L. M. et al. (2012) In *Adaptive Optics Systems III* (B. L. Ellerbroek et al., eds.), p. 84470X. SPIE Conf. Series 8447, Bellingham, Washington.
Cooper C. S. and Showman A. P. (2006) *Astrophys. J., 649,* 1048–1063.
Coughlin J. L. and López-Morales M. (2012) *Astron. J., 143,* 39–61.
Cowan N. and Agol E. (2008) *Astrophys. J. Lett., 678,* L129–L132.
Cowan N. and Agol E. (2011a) *Astrophys. J., 726,* 82–94.
Cowan N. and Agol E. (2011b) *Astrophys. J., 729,* 54–65.
Cowan N. et al. (2007) *Mon. Not. R. Astron. Soc., 379,* 641–646.
Cowan N. et al. (2012) *Astrophys. J., 747,* 82–99.
Croll B. et al. (2010) *Astrophys. J., 717,* 1084–1091.
Croll B. et al. (2011) *Astron. J., 141,* 30–42.
Crossfield I. J. M. et al. (2012a) *Astrophys. J., 752,* 81–94.
Crossfield I. J. M. et al. (2012b) *Astrophys. J., 746,* 46.
Crossfield I. J. M. (2012c) *Astrophys. J., 760,* 140–156.
Crouzet N. et al. (2012) *Astrophys. J., 761,* 7–20.
Cumming A. et al. (2008) *Publ. Astron. Soc. Pac., 120,* 531–554.
Currie T. et al. (2011) *Astrophys. J., 729,* 128–147.
Cushing M. C. et al. (2011) *Astrophys. J., 743,* 50.
de Kok R. et al. (2011) *Astrophys. J., 741,* 59.
de Kok R. et al. (2013) *Astron. Astrophys., 554,* A82–A91.
Deming D. et al. (2005) *Nature, 434,* 738–740.
Deming D. et al. (2013) *Astrophys. J., 774,* 95–111.
Demory B. et al. (2011a) *Astrophys. J. Lett., 735,* L12–L18.
Demory B. et al. (2011b) *Astron. Astrophys., 533,* A114–I21.
Demory B. et al. (2012) *Astrophys. J. Lett., 751,* L28.
de Mooij E. et al. (2012) *Astron. Astrophys., 538,* 46–58.
Demory B. et al. (2013) *Astrophys. J. Lett., 776,* L25.
Désert J.-M. et al. (2009) *Astrophys. J., 699,* 478–485.
Désert J.-M. et al. (2011a) *Astrophys. J., 197,* 14–27.
Désert J.-M. et al. (2011b) *Astrophys. J. Suppl., 197,* 11–21.
Désert J.-M. et al. (2011c) *Astrophys. J., 731,* L40.
de Wit J. et al. (2012) *Astron. Astrophys., 548,* A129–A148.
Dobbs-Dixon I. et al. (2010) *Astrophys. J., 710,* 1395–1407.
Dragomir D. et al. (2013) *Astrophys. J. Lett., 772,* L2.
Dupuy T. and Liu M. (2012) *Astrophys. J. Suppl., 201,* 19.
Elkins-Tanton L. T. and Seager S. (2008) *Astrophys. J., 685,* 1237–1246.
Esteves L. J. et al. (2013) *Astrophys. J., 772,* 51–64.
Evans T. M. et al. (2013) *Astrophys. J. Lett., 772,* L16–L20.
Föhring D. et al. (2013) *Mon. Not. R. Astron. Soc., 435,* 2268–2273.
Fortney J. (2005) *Mon. Not. R. Astron. Soc., 364,* 649–653.
Fortney J. et al. (2006) *Astrophys. J., 642,* 495–504.
Fortney J. et al. (2007) *Astrophys. J., 659,* 1661–1672.
Fortney J. et al. (2008a) *Astrophys. J., 678,* 1419–1435.
Fortney J. et al. (2008b) *Astrophys. J., 683,* 1104–1116.
Fortney J. et al. (2010) *Astrophys. J., 709,* 1396–1406.
Fortney J. et al. (2013) *Astrophys. J., 775,* 80–92.
Fossati L. et al. (2010) *Astrophys. J. Lett., 714,* L222–L227.
Fraine J. D. et al. (2013) *Astrophys. J., 765,* 127–139.
Freedman R. S. et al. (2008) *Astrophys. J. Suppl., 174,* 504–513.
Galicher R. et al. (2011) *Astrophys. J. Lett., 739,* L41.
Gardner J. P. et al. (2006) *Space Sci. Rev., 123,* 485–606.
Gibson N. P. et al. (2011) *Mon. Not. R. Astron. Soc., 411,* 2199–2213.
Gibson N. P. et al. (2012) *Mon. Not. R. Astron. Soc., 422,* 753–760.
Gillon M. et al. (2012a) *Astron. Astrophys., 539,* A28–A34.
Gillon M. et al. (2012b) *Astron. Astrophys., 542,* A4–A18.
Goody R. M. and Yung Y. (1989) *Atmospheric Radiation: Theoretical Basis, 2nd edition.* Oxford Univ., New York.
Greene T. et al. (2007) In *Techniques and Instrumentation for Detection of Exoplanets III* (D. R. Coulter, ed.), pp. 1–10. SPIE Conf. Series 6693, Bellingham, Washington.
Grillmair C. J. et al. (2008) *Nature, 456,* 767–769.
Guillot T. (2010) *Astron. Astrophys., 520,* A27.
Harrington J. et al. (2006) *Science, 314,* 623–626.
Hansen B. (2008) *Astrophys. J. Suppl., 179,* 484–508.
Helling Ch. and Lucas W. (2009) *Mon. Not. R. Astron. Soc., 398,* 985–994.
Helling Ch. et al. (2008) *Astron. Astrophys., 485,* 547–560.

Heng K. et al. (2011) *Mon. Not. R. Astron. Soc., 418,* 2669–2696.
Heng K. et al. (2012) *Mon. Not. R. Astron. Soc., 420,* 20–36.
Hinkley S. et al. (2013) *Astrophys. J., 779,* 153–167.
Hinz P. M. et al. (2010) *Astrophys. J., 716,* 417–426.
Holman M. J. and Murray, N. W. (2005) *Science, 307,* 1288–1291.
Howe A. R. and Burrows A. S. (2012) *Astrophys. J., 756,* 176–189.
Hubbard W. B. et al. (2001) *Astrophys. J., 560,* 413–419.
Hubeny I. et al. (2003) *Astrophys. J., 594,* 1011–1018.
Huitson C. M. et al. (2012) *Mon. Not. R. Astron. Soc., 422,* 2477–2488.
Huitson C. M. et al. (2013) *Mon. Not. R. Astron. Soc., 434,* 3252–3274.
Janson M. et al. (2010) *Astrophys. J. Lett., 710,* L35–L38.
Janson M. et al. (2013) *Astrophys. J. Lett., 778,* L4–L9.
Jensen A. G. et al. (2011) *Astrophys. J., 743,* 203–217.
Jensen A. G. et al. (2012) *Astrophys. J., 751,* 86–102.
Kalas P. et al. (2008) *Science, 322,* 1345.
Kaltenegger L. and Traub W. A. (2009) *Astrophys. J., 698,* 519–527.
Kempton E. M.-R. and Rauscher E. (2012) *Astrophys J., 751,* 117–129.
Kipping D. M. and Bakos G. (2011) *Astrophys. J., 733,* 36–53.
Kipping D. M. and Spiegel D. S. (2011) *Mon. Not. R. Astron. Soc., 417,* L88–L92.
Knutson H. A. et al. (2007) *Nature, 447,* 183–186.
Knutson H. A. et al. (2008) *Astrophys. J., 673,* 526–531.
Knutson H. A. et al. (2009a) *Astrophys. J., 690,* 822–836.
Knutson H. A. et al. (2009b) *Astrophys. J., 703,* 769–784.
Knutson H. A. et al. (2010) *Astrophys. J., 720,* 1569–1576.
Knutson H. A. et al. (2011) *Astrophys. J., 735,* 27–49.
Knutson H. A. et al. (2012) *Astrophys. J., 754,* 22–38.
Konopacky Q. M. et al. (2013) *Science, 339,* 1398–1401.
Kopparapu R. et al. (2012) *Astrophys. J., 745,* 77–86.
Kóvacs et al. (2013) *Mon. Not. R. Astron. Soc., 433,* 889–906.
Kreidberg L. et al. (2014) *Nature, 505(7481),* 69–72.
Kuzuhara M. et al. (2013) *Astrophys. J., 774,* 11–28.
Lafrenière D. et al. (2010) *Astrophys. J. Lett., 719,* 497.
Lagrange A.-M. et al. (2010) *Science, 329,* 57.
Lammer H. et al. (2003) *Astrophys. J. Lett., 598,* L121–L124.
Laughlin G. et al. (2009) *Nature, 457,* 562–564.
Lecavelier des Etangs A. et al. (2008) *Astron. Astrophys., 481,* L83–L86.
Lecavelier des Etangs A. et al. (2010) *Astron. Astrophys., 514,* A72–A82.
Lee J.-M. et al. (2012) *Mon. Not. R. Astron. Soc., 420,* 170–182.
Lee J.-M. et al. (2013) *Astrophys. J., 778,* 97–115.
Lewis N. K. et al. (2010) *Astrophys. J., 720,* 344–356.
Lewis N. K. et al. (2013) *Astrophys. J., 766,* 95–117.
Line M. R. et al. (2010) *Astrophys. J., 717,* 496–502.
Line M. R. et al. (2012) *Astrophys. J., 749,* 93–102.
Line M. R. et al. (2013) *Astrophys. J., 775,* 137.
Linsky J. L. et al. (2010) *Astrophys. J., 717,* 1291–1299.
Lithwick Y. et al. (2012) *Astrophys. J., 761,* 122–133.
Lodders K. (2004) *Astrophys. J., 611,* 587–597.
Lodders K. and Fegley B. (2002) *Icarus, 155,* 393–424.
Machalek P. et al. (2008) *Astrophys. J., 684,* 1427–1432.
Macintosh B. et al. (2008) In *Adaptive Optics Systems* (N. Hubin et al., eds.), p. 701518. SPIE Conf. Series 7015, Bellingham, Washington.
Madhusudhan N. (2012) *Astrophys. J., 758,* 36.
Madhusudhan N. and Burrows A. (2012) *Astrophys. J., 747,* 25–41.
Madhusudhan N. and Seager S. (2009) *Astrophys. J., 707,* 24–39.
Madhusudhan N. and Seager S. (2010) *Astrophys. J., 725,* 261–274.
Madhusudhan N. and Seager S. (2011) *Astrophys. J., 729,* 41–54.
Madhusudhan N. et al. (2011a) *Nature, 469,* 64–67.
Madhusudhan N. et al. (2011b) *Astrophys. J., 743,* 191–202.
Madhusudhan N. et al. (2011c) *Astrophys. J., 737,* 34–48.
Majeau C. et al. (2012) *Astrophys. J. Lett., 747,* L20–L25.
Mancini L. et al. (2013) *Mon. Not. R. Astron. Soc., 436,* 2–18.
Mandell A. et al. (2011) *Astrophys. J., 728,* 18–28.
Mandell A. et al. (2013) *Astrophys. J., 779,* 128–145.
Marley M. S. and Sengupta S. (2011) *Astrophys. J., 417,* 2874–2881.
Marley M. S. et al. (1999) *Astrophys. J., 513,* 879–893.
Marley M. S. et al. (2007) *Astrophys. J., 655,* 541–549.
Marley M. S. et al. (2012) *Science, 754,* 135.
Marois C. et al. (2008) *Science, 322,* 1348–1352.
Marois C. et al. (2010) *Science, 468,* 1080–1083.
Maxted P. F. L. et al. (2012) *Mon. Not. R. Astron. Soc., 428,* 2645–2660.
Menou K. (2012) *Astrophys. J. Lett., 744,* L16.
Miller-Ricci E. and Fortney J. J. (2010) *Astrophys. J. Lett., 716,* L74–L79.
Miller-Ricci E. et al. (2009) *Astrophys. J., 690,* 1056–1067.
Miller-Ricci Kempton E. et al. (2012) *Astrophys. J., 745,* 3.

Mohanty S. et al. (2007) *Astrophys. J., 657,* 1064.
Morley C. V. et al. (2013) *Astrophys. J., 775,* 33–45.
Morris B. M. et al. (2013) *Astrophys. J. Lett., 764,* L22–L27.
Moses J. et al. (2011) *Astrophys. J., 737,* 15–51.
Moses J. et al. (2013) *Astrophys. J., 777,* 34–56.
Mousis O. et al. (2012) *Astrophys. J., 751,* L7.
Murgas F. et al. (2012) *Astron. Astrophys., 544,* A41.
Murray-Clay R. A. et al. (2009) *Astrophys. J., 693,* 23–42.
Nakajima T. et al. (1995) *Nature, 378,* 463–465.
Nettelmann N. et al. (2011) *Astrophys. J., 733,* 2.
Nielsen E. L. et al. (2010) In *Physics and Astrophysics of Planetary Systems* (A.-M. Lagrance et al., eds.), pp. 107–110. EAS Publ. Series, Vol. 41, Cambridge Univ., Cambridge.
Noll K. S. et al. (1997) *Astrophys. J. Lett., 489,* L87–L90.
Nutzman P. and Charbonneau, D. (2008) *Publ. Astron. Soc. Pac., 120,* 317–327.
Öberg K. et al. (2011) *Astrophys. J. Lett., 743,* L16.
Oppenheimer B. R. et al. (2013) *Astrophys. J., 768,* 24.
Parmentier V. et al. (2013) *Astrophys. J., 558,* A91.
Patience J. et al. (2010) *Astron. Astrophys., 517,* 76.
Pont F. et al. (2008) *Mon. Not. R. Astron. Soc. 385,* 109–118.
Pont F. et al. (2013) *Mon. Not. R. Astron. Soc. 432,* 2917–2944.
Rauscher E. and Menou C. (2012) *Astrophys. J., 750,* 96.
Redfield S. et al. (2008) *Astrophys. J. Lett., 673,* L87–L90.
Rice E. L. et al. (2010) *Astrophys. J. Suppl., 186,* 63–84.
Rodler F. and López-Morales M. (2014) *Astrophys. J., 781,* 54.
Rodler F. et al. (2012) *Astrophys. J. Lett., 753,* L25–L29.
Rodler F. et al. (2013) *Mon. Not. R. Astron. Soc. 385(432),* 1980–1988.
Rogers L. A. and Seager S. (2010) *Astrophys. J., 716,* 1208–1216.
Rothman L. S. et al. (2010) *J. Quant. Spectrosc. Radiat. Transfer, 111,* 2139–2150.
Rowe J. F. et al. (2008) *Astrophys. J., 689,* 1345–1353.
Sanchis-Ojeda R. et al. (2013) *Astrophys. J., 774,* 54.
Seager S. and Deming D. (2010) *Annu. Rev. Astron. Astrophys., 48,* 631–672.
Seager S. and Sasselov D. (1998) *Astrophys. J. Lett., 502,* L157–L161.
Seager S. and Sasselov D. (2000) *Astrophys. J., 537,* 916–921.
Seager S. et al. (2000) *Astrophys. J., 540,* 504–520.
Seager S. et al. (2005) *Astrophys. J., 632,* 1122–1131.
Seager S. et al. (2007) *Astrophys. J., 669,* 1279–1297.
Seager S. et al. (2009) In *Astrophysics in the Next Decade* (H. A. Thronson et al., eds.), pp. 123–145. Astrophys. Space Sci. Proc., Springer, The Netherlands.
Shabram M. et al. (2011) *Astrophys. J., 727,* 65–74.
Sharp C. M. and Burrows A. (2007) *Astrophys. J. Suppl., 168,* 140–166.
Showman A. P. and Guillot T. (2002) *Astron. Astrophys., 385,* 166–180.
Showman A. P. and Kaspi Y. (2013) *Astrophys. J., 776,* 85.
Showman A. P. and Polvani L. M. (2011) *Astrophys. J., 738,* 71–95.
Showman A. P. et al. (2008) In *Extreme Solar Systems* (D. Fischer et al., eds.) p. 419. ASP Conf. Ser. 398, Astronomical Society of the Pacific, San Francisco.
Showman A. P. et al. (2009a) *Astrophys. J., 699,* 564–584.
Showman A. P. et al. (2009b) In *Exoplanets* (S. Seager, ed.), pp. 471–516. Univ. of Arizona, Tucson.
Showman A. P. et al. (2012) *Astrophys. J., 762,* 24–44.
Showman A. P. et al. (2013) In *Comparative Climatology of Terrestrial Planets* (S. J. Mackwell et al., eds.), pp. 277–326. Univ. of Arizona, Tucson.
Sing D. K. and López-Morales M. (2009) *Astron. Astrophys., 493,* L31–L34.
Sing D. K. et al. (2011a) *Mon. Not. R. Astron. Soc., 416,* 1443–1455.
Sing D. K. et al. (2011b) *Astron. Astrophys., 527,* A73.
Skemer A. J. et al. (2012) *Astrophys. J., 753,* 14.
Snellen I. et al. (2009) *Nature, 459,* 543–545.
Snellen I. et al. (2010a) *Astron. Astrophys., 513,* A76–A85.
Snellen I. et al. (2010b) *Nature, 465,* 1049–1053.
Snellen I. et al. (2013) *Astrophys. J., 764,* 182–187.
Sozzetti A. et al. (2013) In *Hot Planets and Cool Stars* (R. Saglia, ed.), pp. 3006–3014. EPJ Web Conf. Vol. 47.
Spiegel D. S. and Burrows A. (2012) *Astrophys. J., 745,* 174–188.
Spiegel D. S. et al. (2009) *Astrophys. J., 699,* 1487–1500.
Spiegel D. S. et al. (2010) *Astrophys. J., 709,* 149–158.
Stam D. M. et al. (2004) *Astron. Astrophys., 428,* 663–672.
Stephens D. C. et al. (2009) *Astrophys. J., 702,* 154–170.
Stevenson D. (1991) *Annu. Rev. Astron. Astrophys., 29,* 163–193.
Stevenson K. B. et al. (2010) *Nature, 464,* 1161–1164.

Sudarsky D. et al. (2000) *Astrophys. J., 538,* 885–903.

Sudarsky D. et al. (2003) *Astrophys. J., 588,* 1121–1148.

Swain M. R. et al. (2008a) *Astrophys. J., 674,* 482–497.

Swain M. R. et al. (2008b) *Nature, 452,* 329–331.

Swain M. R. et al. (2009a) *Astrophys. J., 690,* L114–L117.

Swain M. R. et al. (2009b) *Astrophys. J., 704,* 1616–1621.

Swain M. R. et al. (2010) *Nature, 463,* 637–639.

Swain M. R. et al. (2012) *Icarus 225,* 432–445.

Tamura M. (2009) In *Exoplanets and Disks: Their Formation and Diversity* (T. Usuda et al., eds.), p. 11. AIP Conf. Ser. 1158, American Institute of Physics, Melville, New York.

Tennyson J. and Yurchenko S. N. (2012) *Mon. Not. R. Astron. Soc., 425,* 21–33.

Tinetti G. et al. (2007) *Nature, 448,* 169–171.

Tinetti G. et al. (2010) *Astrophys. J. Lett., 712,* L139.

Tinetti G. et al. (2013) *Exp. Astron., 34,* 311–353.

Todorov K. et al. (2010) *Astrophys. J., 708,* 498–504.

Valencia D. et al. (2013) *Astrophys. J., 775,* 10–21.

Venot O. et al. (2012) *Astron. Astrophys., 546,* A43–A61.

Vidal-Madjar A. et al. (2003) *Nature, 422,* 143–146.

Vidal-Madjar A. et al. (2004) *Astrophys. J. Lett., 604,* L69–L72.

Visscher C. and Moses J. (2011) *Astrophys. J., 738,* 72–83.

Waldmann I. P. et al. (2012) *Astrophys. J., 744,* 35–45.

Waldmann I. P. et al. (2013) *Astrophys. J., 766,* 7–16.

Wang W. et al. (2013) *Astrophys. J., 770,* 70.

Wiktorowicz S. (2009) *Astrophys. J., 696,* 1116–1124.

Williams P. et al. (2006) *Astrophys. J., 649,* 1020–1027.

Winn J. N. et al. (2011) *Astrophys. J. Lett., 737,* L18–L24.

Wood P. L. et al. (2011) *Mon. Not. R. Astron. Soc., 412,* 2376–2382.

Yelle R. et al. (2008) *Space Sci. Rev., 139,* 437–451.

Yung Y. and Demore W. B. (1999) *Photochemistry of Planetary Atmospheres.* Oxford Univ., New York.

Zahnle K. et al. (2009) *Astrophys. J. Lett., 701,* L20–L24.

Baraffe I., Chabrier G., Fortney J., and Sotin C. (2014) Planetary internal structures. In *Protostars and Planets VI* (H. Beuther et al., eds.), pp. 763–786. Univ. of Arizona, Tucson, DOI: 10.2458/azu_uapress_9780816531240-ch033.

Planetary Internal Structures

Isabelle Baraffe
University of Exeter

Gilles Chabrier
Ecole Normale Supérieure de Lyon and University of Exeter

Jonathan Fortney
University of California, Santa Cruz

Christophe Sotin
Jet Propulsion Laboratory/California Institute of Technology

This chapter reviews the most recent advancements on the topic of terrestrial and giant planet interiors, including solar system and extrasolar objects. Starting from an observed mass-radius diagram for known planets in the universe, we will discuss the various types of planets appearing in this diagram and describe internal structures for each type. The review will summarize the status of theoretical and experimental works performed in the field of equations of state (EOS) for materials relevant to planetary interiors and will address the main theoretical and experimental uncertainties and challenges. It will discuss the impact of new EOS on interior structures and bulk composition determination. We will discuss important dynamical processes that strongly impact the interior and evolutionary properties of planets (e.g., plate tectonics, semiconvection) and describe nonstandard models recently suggested for our giant planets. We will address the case of short-period, strongly irradiated exoplanets and critically analyze some of the physical mechanisms that have been suggested to explain their anomalously large radius.

1. INTRODUCTION

The 1990s were marked by two historical discoveries: the first planetary system around a star other than the Sun, namely a pulsar (*Wolszczan and Frail,* 1992), and the first Jupiter-mass companion to a solar-type star (*Mayor and Queloz,* 1995). Before that, the development and application of planetary structure theory was restricted to the few planets belonging to our solar system. Planetary interiors provide natural laboratories to study materials under high pressure, complementing experiments that can be done on Earth. This explains why planets have long been of interest to physicists studying the equations of state (EOS) of hydrogen, helium, and other heavy materials made of water, silicates, or iron. Planets also provide laboratories complementary to stellar interiors to study physical processes common to both families of objects, such as semiconvection, tidal dissipation, irradiation, or ohmic dissipation. For solar system planets, space missions and *in situ* explorations have provided a wealth of information on their atmospheric composition, ground compositions for terrestrial planets, and gravitational fields for giant planets, which provide constraints on their interior density distribution. The theory of planetary struc-tures has thus long been built on our knowledge of our own planet, Earth, and of our neighbors in the solar system. The diversity of planetary systems revealed by the discoveries of thousands of exoplanet candidates now tells us that the solar system is not a universal template for planetary struc-tures and system architectures. The information collected on exoplanets will never reach the level of accuracy and details obtained for the solar system planets. Current measurements are mostly limited to gross physical properties, including the mass, radius, orbital properties, and occasionally some information on the atmospheric composition. Nevertheless, the lack of refined information is compensated for by the large number of detected exoplanets, completing the more precise but restricted knowledge provided by our solar sys-tem planets. The general theory describing planetary struc-tures needs now to broaden and account for more physical processes to describe the diversity of exoplanet properties and understand some puzzles. This diversity is illustrated by the mass-radius relation of known planets displayed in Fig. 1. Naively, one would expect a connection between the mass and the radius of a planet. But exoplanets tell us that knowing a planet's mass does not tell us its size, and vice versa. Figure 1 shows that the interpretation of transit radii must be made with caution, since the mass range corresponding to a given radius can span up to 2 orders of magnitudes. Conversely, a planet of the mass of Neptune, for example, could have a variety of sizes, depending on its bulk composition and the mass of its atmosphere. Even more troubling is the existence of ambiguous conclusions about bulk composition, as illustrated by the properties of the exoplanet GJ 1214b, with mass 6.5 $M_\oplus$ and radius 2.5 $R_\oplus$, and which could be explained by at least three very different sets of structures (*Rogers and Seager,* 2010). The study of planetary structures is a giant construction game where progress is at the mercy of advances in both experiment and theory of materials at high pressure (see section 2), improvement in the knowledge of solar system planets paced by exploratory missions (see section 3 and section 4), accumulation of data for a wide variety of exoplanets, and our creativity in filling the shortfall of accurate data and interpreting some amazing properties (see section 5). This

chapter will describe in detail the recent building pieces that elaborate the current theory of planetary internal structures. Descriptions of how models for planets are built and their basic equations and ingredients have already been described in detail in previous reviews and will not be repeated here (see *Guillot,* 1999; *Fortney and Nettelmann,* 2010; *Fortney et al.,* 2011a).

2. EQUATIONS OF STATE OF PLANETARY MATERIALS

The correct determination of the interior structure and evolution of planets depends on the accuracy of the description of the thermodynamic properties of matter under the relevant conditions of temperature and pressure. These latter reach up to about 20,000 K and 70 Mbar (7000 GPa) for Jupiter typical central conditions. While terrestrial planets (or Earth-like planets) are essentially composed of a solid/liquid core of heavy material with a thin atmosphere (see section 3), giant planets have an envelope essentially composed of hydrogen and helium, with some heavier material enrichment, and a core of heavy elements (see section 4). As a new term appearing with the discoveries of exoplanets, super-Earths refer to objects with masses greater than Earth's, regardless of their composition. The heavier elements consist of C, N, and O, often referred to as "ices" under their molecule-bearing volatile forms (H_2O, the most abundant of these elements for solar C/O and N/O ratios, CH_4, NH_3, CO, N_2, and possibly CO_2). The remaining constituents consist of silicates (Mg-, Si-, and O-rich material) and iron (as mixtures of more refractory elements under the form of metal, oxide, sulfide, or substituting for Mg in the silicates). In the pressure-temperature (P-T) domain characteristic of planet interiors, elements go from a molecular or atomic state in the low-density outermost regions to an ionized, metallic one in the dense inner parts, covering the regime of *pressure*-dissociation and ionization. Interactions between molecules, atoms, ions, and electrons are dominant and degeneracy effects for the electrons play a crucial role, making the derivation of an accurate EOS a challenging task. Other phenomena such as phase transition or phase separation may take place in the interior of planets, adding complexity to the problem.

The correct description of the structure and cooling of solar or extrasolar planets thus requires the knowledge of the EOS and the transport properties of various materials under the aforementioned density and temperature conditions. In the section below, we summarize the recent improvements in this field, both on the experimental and theoretical fronts, and show that tremendous improvements have been accomplished on both sides within recent years.

2.1. Hydrogen and Helium Equation of State

A lot of experimental work has been devoted to the exploration of hydrogen (or its isotope deuterium) and helium at high densities, in the regime of pressure ionization.

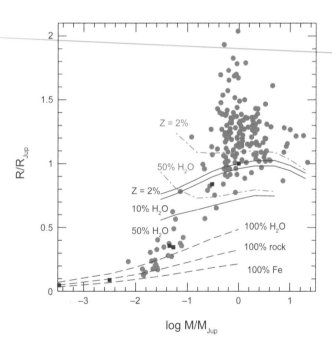

Fig. 1. See Plate 15 for color version. Mass vs. radius of known planets, including solar system planets (blue squares: Jupiter, Saturn, Neptune, Uranus, Earth, and Mars) and transiting exoplanets (magenta dots, from the *Extrasolar Planet Encyclopedia* at *http://exoplanet.eu*). Note that the sample shown in this figure has been cleaned up and only planets where the parameters are reasonably precise are included. The curves correspond to models from 0.1 $M_\oplus$ to 20 M_{Jup} at 4.5 G.y. with various internal chemical compositions. The solid curves correspond to a mixture of H, He, and heavy elements (models from *Baraffe et al.,* 2008). The long dashed lines correspond to models composed of pure water, rock, or iron from *Fortney et al.* (2007). The "rock" composition here is olivine (forsterite Mg_2SiO_4) or dunite. Solid and long-dashed lines (in black) are for non-irradiated models. Dash-dotted (red) curves correspond to irradiated models at 0.045 AU from a Sun.

Modern techniques include laser-driven shock-wave experiments (*Collins et al.,* 1998, 2001; *Mostovych et al.,* 2000), pulse-power compression experiments (*Knudson et al.,* 2004), and convergent spherical shock wave experiments (*Belov et al.,* 2002; *Boriskov et al.,* 2003). They achieve pressures of several megabars in fluid deuterium at high temperature, exploring for the first time the regime of pressure-dissociation. For years, however, the difference between the different experimental results gave rise to a major controversy. While the laser-driven experiments were predicting significant compression of D along the Hugoniot curve, with a maximum compression factor of $\rho/\rho_0 \simeq 6$, where $\rho_0 = 0.17$ g cm^{-3} is the initial density of liquid D at 20 K, the pulse experiments were predicting significantly stiffer EOS, with $\rho/\rho_0 \simeq 4$. The controversy has been resolved recently thanks to a new determination of the quartz EOS up to 16 Mbar (*Knudson and Desjarlais,* 2009). Indeed, evaluation of the properties of high-pressure material is made through comparison with a shock wave standard. Quartz is the most commonly used reference material in the laser experiments. The new experiments have shown that, due to phase changes at high density, the quartz EOS was significantly stiffer than previously assumed. The direct consequence is thus a stiffer D and H EOS, now in good agreement with the pulse-power compression data.

Except for the release of the new quartz EOS (see above), which led to a reanalysis of the existing data, no new result has been obtained for H on the experimental front for more than a decade. Because of the high diffusivity of this material, experimentalists have elected to focus on helium, more easy to lock into precompressed cells. Furthermore, as helium pressure ionization has been predicted to occur directly from the atomic (He) state into the fully ionized (He^{++}) one (*Winisdoerffer and Chabrier,* 2005), this material holds the promise of bringing valuable information on the very nature of pressure metallization. For this reason, even though He by itself represents only 10% of giant planet interior compositions, against 90% for H, all the progress on the high-pressure EOS of light elements since *Protostars and Planets V* (*Reipurth et al.,* 2007) has been obtained for this material, as reviewed below.

Recent high-pressure experiments, using statically precompressed samples in dynamical compression experiments, have achieved up to 2 Mbar for various Hugoniot initial conditions, allowing testing of the EOS over a relatively broad range of T-P conditions (*Eggert et al.,* 2008). These experiments seem to show a larger compressibility than for H, possibly due to electronic excitations, in good agreement with the Saumon-Chabrier-Van Horn (SCVH) EOS (*Saumon et al.,* 1995). These results, however, should be reanalyzed carefully before any robust conclusion can be reached. Indeed, these experiments were still calibrated on the old quartz EOS. It is thus expected that using the recent (stiffer) one will lead to a stiffer EOS for He, as for H, in better agreement with *ab initio* calculations (*Militzer,* 2006).

Tremendous progress has also been accomplished on the theoretical front. Constant increase in computer performanc-

es now allows *ab initio* simulations to be performed over a large enough domain of the T-P diagram to generate appropriate EOS tables. These approaches include essentially quantum molecular dynamics (QMD) simulations, which combine molecular dynamics (MD) to calculate the forces on the (classical) ions and density functional theory (DFT) to take into account the quantum nature of the electrons. We are now in a position where the semi-analytical SCVH EOS for H/He can be supplemented by first-principle EOS (*Caillabet et al.,* 2011; *Nettelmann et al.,* 2012; *Militzer,* 2013; *Soubiran et al.,* in preparation).

A point of importance concerning the H and He EOS is that, while the experimental determination of hydrogen pressure ionization can be envisaged in a foreseeable future, reaching adequate pressures for helium ionization remains presently out of reach. One way to circumvent this problem is to explore the dynamical (conductivity) and optical (reflectivity) properties of helium at high density. Shock compression conductivity measurements had suggested that He should become a conductor at ~1.5 g cm^{-3} (*Fortov et al.,* 2003). The conductivity was indeed found to rise rapidly slightly above ~1 g cm^{-3}, with a very weak dependence upon temperature, a behavior attributed to helium pressure ionization (alternatively the closure of the band gap). Note, however, that the reported measurements are model dependent and that the conductivity determinations imply some underlying EOS model that has not been probed in the domain of interest. Besides the conductivity, reflectivity is another diagnostic of the state of helium at high density. Reflectivity can indeed be related to the optical conductivity through the complex index of refraction, which involves the contribution of both bound and charge carrier electron concentrations. Optical measurements of reflectivity along Hugoniot curves were recently obtained in high-pressure experiments, combining static and shock-wave high-pressure techniques, reaching a range of final densities ~0.7–1.5 g cm^{-3} and temperatures 6×10^4 K (*Celliers et al.,* 2010). Within the regime probed by the experiments, helium reflectivity was found to increase continuously, indicating increasing ionization. A fit to the data, based on a semiconductor Drude-like model, was also predicting a mobility gap closing, and thus helium metallization, around ~1.9 g cm^{-3} for temperatures below $T \lesssim 3 \times 10^4$ K. These results were quite astonishing. It is indeed surprising that helium, with a tightly bound, closed-shell electronic structure and an ionization energy of 24.6 eV, would ionize at such a density, typical of hydrogen ionization.

All these results were questioned by QMD simulations, which were predicting exactly the opposite trend, with a weak dependence of conductivity upon density and a strong dependence on temperature (*Kowalski et al.,* 2007). The QMD calculations found that reflectivity remains very small at a density ~1 g cm^{-3} for $T \lesssim 2$ eV, then rises rapidly at higher temperatures, due to the reduction of the band gap and the increasing occupation of the conduction band arising from the (roughly exponential) thermal excitation of electrons. Accordingly, the calculations were predicting a band gap closure at significantly higher densities. Indeed, the band gap energy found in the QMD simulations ($E_{gap} \sim 15$–20 eV) is found to remain much larger than kT in the region covered by the experiments (around ~1 g cm^{-3} and 1–2 eV).

This controversy between experimental results and theoretical predictions was resolved recently by QMD simulations, similar to the ones mentioned above but exploring a larger density range (*Soubiran et al.,* 2012). These simulations were able to reproduce the experimental data of *Celliers et al.* (2010) with excellent accuracy, both for the reflectivity and the conductivity, but with a drastically different predicted behavior of the gap energy at high density. Based on these results, these authors revisited the experimental ones. They used a similar Drude-like model to fit their data, with a similar functional dependence for the gap energy on density and temperature, but with a major difference. While Celliers et al. ignored the temperature dependence of the gap energy, keeping only a density dependent term, *Soubiran et al.* (2012) kept both terms. As a result, these authors find a much lower dependence on density than assumed in the previous experimental data analysis and a strong dependence on temperature. This leads to a predicted band closure, thus He metallization, in the range ~5–10 g cm^{-3} around 3 eV. Interestingly, this predicted ionization regime is in good agreement with theoretical predictions of the He phase diagram (*Winisdoerffer and Chabrier,* 2005). Importantly enough, the regime $\rho \sim$ 2–4 g cm^{-3}, $T \leq 10^4$ K should be accessed in the near future with precompressed targets and drive laser energies ≥10 kJ. Such experiments should thus be able to confirm or reject the theoretical predictions, providing crucial information on the ionization state of He at high densities, characteristic of giant planet interiors.

2.2. Heavy Elements

2.2.1. Iron. Current diamond anvil cell experiments reach several thousand degrees at a maximum pressure of about 2 Mbar for iron (*Boehler,* 1993), still insufficient to explore the melting curve at Earth's inner core boundary (~3 Mbar and ~5000 K). On the other hand, dynamic experiments yield temperatures too high to explore the relevant P-T domain for Earth but may be useful to probe, e.g., Neptune-like exoplanet interior conditions. So far, one must thus rely on simulations to infer the iron melting curve for Earth, super-Earth, and giant planet conditions. Applying the same type of QMD simulations as mentioned in the previous sections, Morard et al. (in preparation) have determined the melting line of Fe up to T = 14000 K and P =15 Mbar. For Earth, although this melting line lies above the typical temperature at the core mantle boundary, adding up to ~10–30% of impurities would lower the melting temperature enough to cross this boundary. The very nature of iron alloy at Earth's inner core boundary thus remains presently undetermined. In contrast, the melting line, even when considering the possible impact of impurities, lies largely above the T-P conditions characteristic of

the core mantle boundary for super-Earth exoplanets (M ~ 1–10 M$_\oplus$). It is thus rather robust to assert that the iron core of these bodies should be in a solid phase, precluding the generation of a large magnetic field by convection of melted iron inside these objects. These calculations should be extended to higher temperatures and pressures in the near future to explore the melting line of iron under jovian planet conditions.

2.2.2. Water and rocks. The most widely used EOS models for heavy elements are ANEOS (*Thompson and Lauson,* 1972) and SESAME (*Lyon and Johnson,* 1992), which describe the thermodynamic properties of water, "rocks" (olivine (fosterite Mg$_2$SiO$_4$), or dunite in ANEOS, a mixture of silicates and other heavy elements called "drysand" in SESAME, and iron. These EOS consist of interpolations between models calibrated on existing Hugoniot data, with thermal corrections approximated by a Grüneisen parameter ($\gamma = \frac{V}{C_V}(\frac{dP}{dV})_T$), at low to moderately high (≤0.5 Mbar) pressure, and Thomas-Fermi or more sophisticated first-principle calculations at very high density (P ≥ 100 Mbar), where ionized species dominate. Interpolation between these limits, however, provides no insight about the correct structural and electronic properties of the element as a function of pressure, and thus no information about its compressibility, ionization stage (thus conductibility), or even its phase change, solid or liquid. All these properties can have a large impact on the internal structure and the evolution of the planets (see section 5.1).

Recent high-pressure shock compression experiments of unprecedented accuracy for water, however, show significant departures from both the SESAME and ANEOS EOS in the T-P domain characteristic of planetary interiors, revealing a much lower compressibility (*Knudson et al.,* 2012). This is consequential for giant planet interiors, in particular Uranus or Neptune like planets. Indeed, as the calculated amount of H and He in the planet decreases with the stiffness of the water EOS, this new EOS suggests the presence of some H/He fraction in the deep interior of Neptune and Uranus, excluding an inner envelope composed entirely of "water like" material (*Fortney and Nettelmann,* 2010). As H would be metallic, this bears important consequences for the generation of the magnetic field, as addressed below.

The new data, on the other hand, are in excellent agreement with QMD simulations (*French et al.,* 2009). This again reinforces confidence in EOS calculated with first-principle methods, and we are now in a state to be able to use these latter to compute reliable EOS for water or other material in the density regime relevant for planetary interiors (*French et al.,* 2009, 2012; Licari et al., in preparation).

Several major predictions emerge from all the aforementioned first-principle calculations devoted to the EOS of water at high pressure and density. First, all these simulations predict a stable superionic phase for water at high density, characterized by mobile protons in an icy structure (*Redmer et al.,* 2011; *Wilson and Militzer,* 2012b; *Wilson et al.,* 2013). While water is always found to be in a liquid, plasma state under the conditions of Jupiter's core (~20,000 K, 50 Mbar), it should be in the superionic state in a significant fraction of Neptune's and Uranus' inner envelope. For Saturn's, its state is more uncertain, as the uncertainties on the exact T-P core-envelope boundary conditions for this planet encompass the predicted phase transition line. Besides having a significant impact on the determination of the very nature of the core inside solar or extrasolar giant planets, including the so-called "ocean planets," which are predicted to have large water envelopes, these calculations bear major consequences on the generation of the magnetic fields in Uranus and Neptune, with unusual nondipolar components, in contrast to that of Earth.

2.3. Phase Separation

The existence of a phase separation between hydrogen and helium under conditions characteristic of the interiors of Jupiter and Saturn, suggested several decades ago (*Smoluchowski*, 1973; *Salpeter*, 1973), remains an open problem. Such a phase separation is suggested by the measurement of atmospheric He abundance in Jupiter (see section 4.1). Under the action of the planet's gravity field, a density discontinuity yields an extra source of gravitational energy as the dense phase droplets (He-rich ones in the present context) sink toward the planet's center. Conversion of this gravitational energy into heat delays the cooling of the planet, which implies a larger age given the timescale needed to reach a given luminosity compared with a planet with a homogeneous interior. In Saturn's case, such an additional source of energy has traditionally been suggested to explain the planet's bright luminosity at the correct age, i.e., the age of the solar system, ~4.5 × 10⁹ yr (*Stevenson and Salpeter*, 1977b), although an alternative explanation has recently been proposed by *Leconte and Chabrier* (2013) (see section 4).

Recently, two groups have tried to determine the shape of the H/He phase diagram under Jupiter and Saturn interior conditions, based again on QMD simulations for the H/He mixture under appropriate conditions (*Morales et al.*, 2009; *Lorenzen et al.*, 2011). Although basically using the same techniques and reaching globally comparable results, the H/He phase diagram and critical line for H/He for the appropriate He concentration obtained in these two studies differ enough to have significantly different consequences for Jupiter and Saturn's cooling histories. Indeed, it must be kept in mind that, even if H/He phase separation does occur, for instance, in Saturn's interior, not only must it encompass a large enough fraction of the planet interior for the induced gravitational energy release to be significant, but it must occur early enough in the planet's cooling history for the time delay to be consequential today. This implies rather strict conditions on the shape of the phase diagram (see, e.g., *Fortney and Hubbard*, 2004). The critical curve obtained by *Morales et al.* (2009), for instance, does not yield enough energy release to explain Saturn's extra luminosity and excludes H/He in Jupiter's present interior. In contrast, the critical line obtained by

Lorenzen et al. (2011) predicts a phase separation taking place inside both Jupiter and Saturn's present interiors and roughly fulfills the required conditions to explain Saturn's extra luminosity. This does not necessarily imply that one of the two diagrams, if either, is correct and the other one is not. Indeed, as shown by *Leconte and Chabrier* (2013), if layered convection occurs within Saturn's interior, it can explain all or part of the excess luminosity (see section 4.4), making the contribution of a possible phase separation less stringent. Clearly, the issue of the H/He phase diagram under giant planet interior conditions and its exact impact on the planet cooling presently remains an unsettled issue.

To complement this study, recent similar QMD calculations have focused on the possible optical signature of an equimolar or near equimolar H/He mixture undergoing demixtion (*Hamel et al.*, 2011; *Soubiran et al.*, 2013). These studies have calculated the expected change of reflectivity upon demixing. It is predicted that reflectivity exhibits a distinctive signature between the homogeneous and demixed phases potentially observable with current laser-driven experiments (*Soubiran et al.*, 2013).

Of notable interest concerning immiscibility effects in jovian planet interiors are also the recent results of *Wilson and Militzer* (2012b,a). Performing QMD simulations, these authors have shown that for pressures and temperatures characteristic of the core envelope boundary in Jupiter and Saturn, water and MgO (representative of rocky material) are both soluble in metallic hydrogen. This implies that the core material in these planets should dissolve into the fluid envelope, suggesting the possibility of significant core erosion and upward redistribution of heavy material in jovian solar and extrasolar planets. A non-uniform heavy element distribution will limit the rate at which the heat flux can be transported outward, with substantial implications for the thermal evolution and radius contraction of giant planets (see section 4.4 and section 5.2).

3. STRUCTURE OF TERRESTRIAL PLANETS

A terrestrial exoplanet is defined as an exoplanet having characteristics similar to Earth, Mars, and Venus. Earth is the only planet for which the existence of life has been proven. The search for terrestrial exoplanets is driven by the question of life on alien worlds and the belief that it will be on a planet that resembles Earth. The information available for characterizing exoplanets is limited to the distance to its star, the radius, the mass, and very rarely some information on the atmospheric composition (e.g., *Swain et al.*, 2008). This section is devoted to the description of the interior structure and dynamics of terrestrial planets, starting with Earth, which is the best known terrestrial planet. One unique characteristic of Earth is that its surface is divided into several rigid plates that move relative to one another. The motion of the plates is driven by thermal convection in the solid mantle (*Schubert et al.*, 2001). This convection regime is known as the mobile lid regime. Most of Earth's volcanism happens at the plate boundaries. Plate tectonics provide a recycle mechanism that may be important for sustaining life on a planet, although this is not proven. It is also a very efficient way to cool down a planet compared to the so-called "stagnant-lid" regime (no plate tectonics) that characterizes both Venus and Mars. After describing the interior structure of a terrestrial planet (section 3.1), this chapter describes the interior dynamics and addresses the question of the link between convection and plate tectonics (section 3.2). Some small exoplanets with a radius less than 2 $R_\oplus$ (Earth radius) may be terrestrial exoplanets. The last section briefly describes some of them and shows that none of them can have liquid water on their surface.

3.1. Interior Structure of Terrestrial Planets

3.1.1. Elements that form a terrestrial planet. Earth is the terrestrial planet for which we have the most information about its elementary composition. The different layers that form Earth are, from outside to inside, the atmosphere, the hydrosphere, the crust, the mantle, the liquid core, and the solid core (Fig. 2). The formation of the different layers is the result of differentiation processes that started during the accretion at 4.5 Ga and are still operating at the present time.

The atmosphere of Earth is mainly composed of N and O. Its mass is less than 10^{-6} $M_\oplus$, which is quite negligible. The hydrosphere (liquid H_2O) is important for life to form and to develop. Its mass is about 2×10^{-4} $M_\oplus$. Having a stable liquid layer at the surface of a planet is a characteristic shared only with Titan, Saturn's largest moon, where

liquid hydrocarbons form seas and lakes (*Stofan et al.,* 2007). The (P,T) stability domain of the liquid layer is very limited in temperature (Fig. 3), and the conditions for liquid water to be present at the surface of a planet impose strong constraints on the atmospheric processes. Two different kinds of crust are present at the surface of Earth: (1) the oceanic crust that is dense, thin (6 km), and continuously formed at mid-ocean ridges and recycled into the mantle at subduction zones; and (2) the continental crust, which is old (on average), thick, and light. The oceanic crust is formed by melting processes that start deep in the mantle beneath mid-ocean ridges. The crust represents about 0.4% of Earth's mass and will be neglected for determining the total mass of a planet. Thus, the three layers that are most important for life to start and evolve (atmosphere, hydrosphere, and crust) represent a very small fraction of Earth's mass and cannot be detected just by having the mass and radius of a planet.

The two most massive layers are the mantle and the core. They differentiated from one another because an iron alloy is denser and melts at lower temperature than the silicates. This differentiation process may have occurred in the planetesimals by segregation and in protoplanets by Rayleigh-Taylor instabilities (*Chambers,* 2005). When the protoplanets collided to form the terrestrial planets, their iron cores would have merged. Due to Earth's seismically active interior, the location of the different interfaces and the pressure and density profiles are very well known (*Dziewonski*

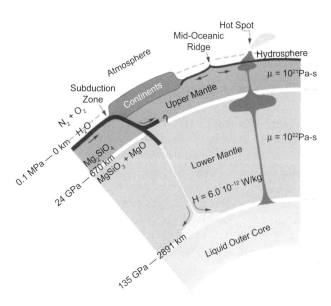

Fig. 2. Structure of Earth. Seismic data, laboratory experiments, and numerical simulations are used in order to simulate the interior structure and dynamics of Earth. Pressure and depth of major interfaces are indicated on the left. The elements and molecules that compose each layer are described in the text. Adapted from *Sotin et al.* (2011).

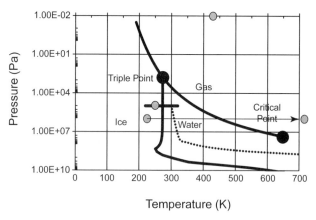

Fig. 3. Phase diagram of H_2O, indicating the small temperature range for the stability of water at the surface of an exoplanet. The horizontal bar represents the (P, T) range at Earth's surface. The gray circles give the effective temperatures for Earth, Venus, and Mercury and are placed at the surface pressure of those planets. For Venus, the arrow links the effective temperature to the surface temperature, stressing the role of the atmosphere. The temperature profile inside the Earth (dotted line) lies within the stability domain of liquid water. If the surface pressure is too small, above the triple point (Mars, Mercury), or if the temperature is too hot, above the critical point (Venus), then the liquid water is not stable at the surface.

and *Anderson*, 1981). The mantle represents approximately two-thirds of the mass and is separated into an upper mantle and a lower mantle. The difference is due to the mineralogical transformation of olivine $(Mg,Fe)_2SiO_4$ at low pressure to perovskite $(Mg,Fe)SiO_3$ and magnesiowüstite $(Mg,Fe)O$ at higher pressure. The rock that composes the upper mantle is known as peridotite and is made of olivine, orthopyroxene $(Mg,Fe)_2Si_2O_6$, clinopyroxene $Ca(Mg,Fe)Si_2O_6$, and garnet, which is an aluminium-rich silicate. The core is composed of a solid inner core of pure iron nestled inside a liquid outer core that contains a light element, presumably sulfur, in addition to iron.

The elementary composition of Earth can be restricted to eight elements that represent more than 99.9% of the total mass (*Javoy,* 1995). Four of these eight elements (O, Fe, Mg, Si) account for more than 95% of the total mass. The four other elements (S, Ca, Al, Ni) add complexities to the model. As described in *Sotin et al.* (2007), Ni is present with iron in the core, sulfur can be present in the liquid outer core and form FeS, calcium behaves like magnesium and forms clinopyroxene in the upper mantle and calcium-perovskite in the lower mantle, and aluminium substitutes to Si and Mg in the silicates. Although CI chondrites have a solar composition, it has been suggested that the composition of Earth may be similar to that of EH enstatite chondrites (*Javoy,* 1995; *Mattern et al.,* 2005). In this case, rocks are enriched in silicon and the ratios of Mg/Si and Fe/Si are lowered to 0.734 and 0.878, respectively (Table 1). However, such variations have minor influences on the values of mass and radius, and it seems appropriate to use the stellar composition as a good first-order approximation of the composition of the planet (*Sotin et al.,* 2007, 2011; *Grasset et al.,* 2009).

The elementary composition of some of the stars hosting exoplanet candidates is known (*Beirão et al.,* 2005; *Gilli et al.,* 2006). *Grasset et al.* (2009) demonstrated that their corrected values of Mg/Si and Fe/Si (adding Ca, Al, and Ni to their closest major element) vary between 1 and 1.6 and between 0.9 and 1.2, respectively. The Sun is on the lower end of these ratios, suggesting that it is enriched in silicon compared to these stars. However, such variability should not significantly affect the radius of an exoplanet for a given mass (*Grasset et al.,* 2009).

For terrestrial planets that are larger than Earth, higher pressures will be reached in the interior and additional transformation to more condensed phases should occur. For example, a post-perovskite phase has been predicted from *ab initio* calculations (*Stamenković et al.,* 2011, and references therein). However, very little is known about this phase and its thermal and transport properties are still debated (*Tackley et al.,* 2013). The elements that compose these very high-pressure minerals shall remain the same.

3.1.2. Equation of state for the different layers. The most accessible information about exoplanets is mass and radius. Following the pioneering work of *Zapolsky and Salpeter* (1969), several studies have looked at the relationships between radius and mass (*Valencia et al.,* 2006; *Sotin*

TABLE 1. Abundances of magnesium and iron relative to silicon for the solar model and the enstatite model of *Javoy* (1995).

	Solar[*]	EH[*]	Solar[†]	EH[†]
Fe/Si	0.977	0.878	0.986	0.909
Mg/Si	1.072	0.734	1.131	0.803

[*] Four elements (O, Fe, Mg, Si).

[†] Eight elements (O, Fe, Si, Mg, Ni, Ca, Al, S).

et al., 2007). The mass of a planet is integrated along the radius assuming that a planet can be described as a one-dimensional sphere, i.e., density depends only on radius and does not vary significantly with longitude or latitude. The density at a given radius depends on pressure and temperature and is computed using EOS, which have been obtained for most of the pressure range relevant for Earth due to recent progress in high-pressure experiments [see, e.g., *Angel et al.* (2009) and section 2]. The pressure is determined assuming hydrostatic equilibrium. The temperature is calculated assuming thermal convection in the mantle and core. The temperature gradient is adiabatic, except in the thermal boundary layers (e.g., *Sotin et al.,* 2011).

Two different approaches are commonly used in Earth sciences for describing the pressure and temperature dependences of materials (e.g., *Jackson,* 1998). One method introduces the effect of temperature in the parameters that describe the mineral's isothermal EOS and is achieved using the third-order Birch-Murnaghan EOS with the thermal effect incorporated using the mineral's thermal expansion coefficient. The second approach dissociates static pressure and thermal pressure by implementing the Mie-Grüneisen-Debye (MGD) formulation. The third-order Birch-Murnaghan (BM) EOS is usually chosen for the upper mantle where the pressure range is limited to less than 25 GPa by the dissociation of ringwoodite into perovskite and periclase. The Mie-Grüneisen-Debye (MGD) formulation is preferred for the lower mantle and core. Other EOS can be used, such as the Vinet EOS, which provides the same result as BM and MGD at low pressure, because the parameters entered into these equations are well constrained from laboratory experiments. Within the temperature range of Earth's lower mantle, the thermal pressure term in the MGD formulation provides estimates close to EOS derived from *ab initio* calculations and shock experiments (*Thompson,* 1990). The composition, i.e., the relative amount of silicon, iron, and magnesium, does not play a major role within the variability of chemical models for Earth. Whether one takes the EH model or the solar/chondritic model (Table 1), the value of the radius remains within the error bars of radius determination for a given mass. The stellar composition variability is on the same order as the variability in composition between the EH model and the chondritic model (*Grasset et al.,* 2009). It seems therefore reasonable to use the stellar composition (Fe/Si and Mg/Si) as a reasonable estimate of the planet elementary composition. Then the relative amount of each mineral can be calculated as described in *Sotin et al.* (2007).

Super-Earths are more massive than Earth and the validity of these equations at much higher pressures and temperatures is questionable (see section 2) (*Grasset et al.,* 2009; *Valencia et al.,* 2009). The Vinet and MGD formulation appear to be valid up to 200 GPa (e.g., *Seager et al.,* 2007). Above 200 GPa, electronic pressure becomes an important component that cannot be neglected. At very high pressure (P > 10 TPa), first principles EOS such as the Thomas-Fermi-Dirac (TFD) formulation can be used (see section 2) (*Fortney et al.,* 2007). The pressure at the core-mantle boundary of terrestrial planets 5 and 10× more massive than Earth is equal to 500 GPa and 1 TPa, respectively (*Sotin et al.,* 2007). In this intermediate pressure range, one possibility is to use the ANEOS EOS mentioned in section 2.2. The study by *Grasset et al.* (2009) compares the density-pressure curves of iron and forsterite using the MGD, TFD, and ANEOS formulations. The TFD formulation predicts values of densities much too small at low pressure. On the other hand, the ANEOS seems to fit the MGD at low pressure and the TFD at very high pressure. Therefore, the ANEOS appears to be a good choice in the intermediate pressure range from 0.2 to 10 TPa. If the density was constant with radius, the radius of a planet would vary as $M^{1/3}$. Since the density increases with increasing pressure and the temperature effect is negligible (*Grasset et al.,* 2009), the exponent is actually lower than 1/3 and equal to 0.274 for planets between 1 and 10 $M_\oplus$. For smaller planets between 0.01 and 1 $M_\oplus$ the coefficient is equal to 0.306, which is closer to 1/3, as expected (Fig. 4).

3.1.3. Observations. Two different observations can be used to test the validity of such models. First, one can compare the mass and radius of the terrestrial planets in our solar system. Second, the density profile of Earth can be compared with the calculated one. The mass and radius of Earth, Mars, and Venus are very well known. The curve describing how a radius depends on the mass of a terrestrial planet is shown on Fig. 4. The fit is very good with the predicted radius of Mars and Mercury being slightly smaller and larger, respectively, than the measured value. For Mercury, it is well known that the ratio Fe/Si is much larger than for the other terrestrial planets. Having more iron reduces the size of the planet. It is therefore not surprising that the radius calculated for a solar-type composition is too large. However, the difference is quite small compared to the uncertainties attached to the determination of the radius and mass of exoplanets. For Mars, the difference is also very small. Mars may have less iron. Future measurements obtained by the Interior Exploration using Seismic Investigations, Geodesy and Heat Transport (InSight) mission that will carry a seismometer on the surface of Mars will help resolve that issue. For Venus and Earth, the calculated value is equal to the observed one at less than 1%. For example, for Earth the calculated value is 43 km larger than the observed value (*Sotin et al.,* 2007). For Venus, whose mass is 0.81 $M_\oplus$, the calculated radius is only

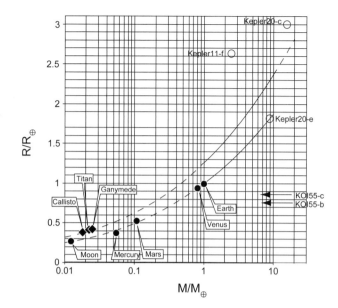

Fig. 4. The mass-radius curve for terrestrial planets very well predicts the values observed for the solar system terrestrial planets. The second curve corresponds to the (R, M) curve for planets having 50% mass of H_2O. The radius is about 25% larger than that of a terrestrial planet. The large icy moons of Jupiter and Saturn fit very well on this curve. The open circles give the values for some small exoplanets (see section 3.3).

5 km larger. The simple model, using solar composition for eight elements, provides a very good approximation of the observed properties (mass, radius) of the terrestrial planets in our solar system.

The earthquakes generated by the motion of the plates have allowed a precise description of the density structure of Earth (*Dziewonski and Anderson,* 1981). This density profile can be compared with the one calculated (Fig. 5). The major difference is in the density of the inner core because the simple model does not model the crystallization state of the core. On Earth, the inner core is solid and made of iron and nickel without any light elements that remain in the liquid phase. Including the growth of an inner core in the model adds complexities that are not required because it does not significantly change the radius for a given mass. The density profile is different in the inner core, but information on the presence of an inner core in exoplanets is not yet available. Earth's inner core only represents one-tenth of Earth's core in mass, which can be translated into about 4% of the total mass of Earth. Besides the inner core, the agreement between the calculated density and the observed density is very good and one can barely see the difference. Another place where the two curves differ is at the transition between the lower and the upper mantle. In the model, the difference between the two mantles is an abrupt change due to the transformation of olivine into perovskite and magnesowüstite. However, we know from seismic observations and laboratory experiments that there is a transition zone

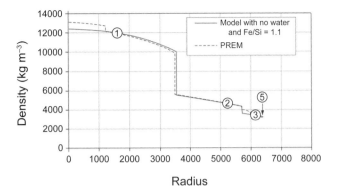

Fig. 5. The density profile simulated by the simple Earth model compares very well with the Preliminary Reference Earth Model (PREM) determined by inversion of seismic waves propagating through Earth. The simple model of the Earth has only three layers: the core (1), the lower mantle (2), and the upper mantle (3). Layer (5) is the hydrosphere, which represents less than 0.1% of the total mass. The model and the observation are somewhat different for the inner core because the inner core is made of pure iron, whereas the model includes the sulfur in both the solid inner iron core and the liquid outer iron-rich layer.

in which the olivine transforms into spinel phases. This complexity is not included in the model since it does not affect the value of the radius for a given mass.

3.2. Internal Dynamics

One unique feature of Earth is the presence of plate tectonics. Although Venus has global properties very similar to Earth, its surface does not present any sign of such tectonics. The reason(s) why two very similar planets have evolved along very different convective regimes is not known (*Moresi and Solomatov*, 1998; *Stevenson*, 2003). The next section provides an overview of mantle convection followed by a summary of some recent numerical simulations on the relationships between convection and plate tectonics.

3.2.1. Subsolidus convection in the silicate mantle. The convection pattern in Earth's mantle is characterized by hot plumes rising from the core-mantle boundary toward the surface. They stop at the cold boundary layer some tens of kilometers below the surface because the upper part of the mantle has a viscosity that is much too high (see below), and eventually form hot spots. A cold sheet of oceanic lithosphere dives into the mantle at subduction zones (Fig. 2). Thermal convection is much more efficient than conduction to remove heat from the interior of a terrestrial planet. Heat sources include radiogenic internal heating due to the decay of the long-lived radioactive elements ^{40}K, ^{232}U, ^{235}U, and ^{242}Th, and the initial heat stored into the planet during the accretion and the differentiation. The convective processes in the mantle control the thermal evolution of the planet (e.g., *Schubert et al.*, 2001). In a fluid that is heated from within, cooled from the top (cold surface temperature),

and heated from below (hot core), cold plumes form at the upper cold thermal boundary layer and hot plumes form at the hot thermal boundary layer, which corresponds to the core-mantle boundary (e.g., *Sun et al.*, 2007). The efficiency of heat transfer is mainly controlled by the mantle viscosity, which depends on a number of parameters, including (but not limited to) the mineral composition of the mantle, the temperature, the pressure, and the grain size. Laboratory experiments (*Davaille and Jaupart*, 1993) and numerical studies (e.g., *Moresi and Solomatov*, 1998; *Grasset and Parmentier*, 1998) have stressed the major role of temperature-dependent viscosity. The vigor of convection is measured by the Rayleigh number, which represents the ratio between the buoyancy forces (density variations induced by temperature) and the viscous force. A small viscosity leads to small viscous forces and large values of the Rayleigh number. Scaling laws have been derived to express the heat flux as a function of the Rayleigh number. A high value of the Rayleigh number leads to a high value of the heat flux. These laws have been employed to predict the thermal evolution of exoplanets (e.g., *Valencia et al.*, 2007). Such scaling laws are valid in the stagnant-lid regime.

One characteristic of the stagnant-lid regime is the presence of a thick conductive layer above the convective mantle. The presence of this layer is due to the very strong temperature-dependence of viscosity (*Davaille and Jaupart*, 1993). At the surface where the temperature is cold, the material has a viscosity that is several orders of magnitude larger than the viscosity of the convective mantle. This layer is known as the lithosphere. The cold thermal boundary layer where cold plumes can form is located under this stagnant lid. Upwelling plumes can deform that layer and provide a topographic signal, as is the case for Venus (*Stofan et al.*, 1995). The adiabatic decompression of the material contained in the uprising plume may lead to partial melt at shallow depths. The partial melt eventually migrates to the surface and causes volcanism that releases gases that were present in the mantle into the atmosphere.

The formation of hot plumes requires the presence of a thermal boundary layer at depth, most likely at the core-mantle boundary. It means that the core is hotter than the mantle, which is possible if the mantle can cool down as quickly as the core. Previous studies based on scaling laws describing the thermal evolution of the core and the mantle (e.g., *Stevenson et al.*, 1983) show that the temperature difference between the core and the mantle decreases quickly. Such a decrease may stop the formation of hot plumes and the convection pattern would be characterized by cold plumes sinking into the mantle and global upwelling, as is the case for a fluid heated from within and/or cooled from above (e.g., *Parmentier et al.*, 1994). Another consequence would be the shutdown of the magnetic dynamo (*Stevenson et al.*, 1983), which has implications on the escape rate of the atmosphere since the presence of a magnetosphere is thought to protect the atmosphere. Mars and Venus do not have a magnetic dynamo at the present time; Mars had one during its early history and the characteristics of this magnetic dynamo are recorded in old crustal rocks that contain a remnant magnetization (*Acuna et al.,* 1999). It is also thought that the presence of a magnetic dynamo on Earth is related to the more efficient cooling rate of the mantle induced by the plate tectonics regime. Indeed, plate tectonics are more efficient than stagnant-lid convection to remove the heat and to cool down the mantle, such that the temperature difference between the mantle and the core enables convective motions in the core.

Mantle convection is an important process that influences the evolution of the surface and the atmosphere. The relationship between convection and plate tectonics is critical to our understanding of the evolution of a terrestrial planet.

3.2.2. Relationships between convection and plate tectonics. Plate tectonics imply that the lithosphere can be broken. As described in *Moresi and Solomatov* (1998), the lithosphere breaks when the stresses induced by convection become larger than the yield stress. This approach was used by *O'Neill and Lenardic* (2007) to simulate the transition from the stagnant-lid regime to the mobile-lid regime, using Earth's case to scale the transition to planets of different sizes. On the other hand, *Valencia et al.* (2007) used scaling laws that provide relationships between the convective stresses and the Rayleigh number (*Schubert et al.,* 2001). The two studies reach opposite conclusions. On the one hand, *O'Neill and Lenardic* (2007) concluded that increasing planetary radius acts to decrease the ratio of driving to resisting stresses, and thus super-sized Earths are likely to be in an episodic or stagnant-lid regime. The episodic regime occurs when the planet experiences episodes of mobile-lid regime. On the other hand, *Valencia et al.* (2007) concluded that as planetary mass increases, the shear stress available to overcome resistance to plate motion increases while the plate thickness decreases, thereby enhancing plate weakness. *Sotin et al.* (2011) pointed out that the two studies can be reconciled if the proper definition of the boundary layer is taken into account. The size of a planet may not be the key parameter for the transition from stagnant-lid regime to mobile lid (plate tectonics) regime. More important is the value of the yield strength.

The yield strength of the lithosphere depends on its history. If the lithosphere has been weakened by impacts, then the yield strength may be much smaller. Also, if the surface temperature is larger, the deformation of the lithosphere may prevent global faulting. *Lenardic and Crowley* (2012) have developed a model of coupled mantle convection and planetary tectonics to demonstrate that history dependence can outweigh the effects of a planet's energy content and material parameters in determining its tectonic state. This conclusion was already mentioned by *Stevenson* (2003) to explain the different paths followed by Earth and Venus. The tectonic mode of the system is then determined by its specific geologic and climatic history. The study by *Lenardic and Crowley* (2012) concludes that models of tectonics and mantle convection will not be able to uniquely determine the tectonic mode of a terrestrial planet without the addition of historical data. It points toward a better

understanding of how climate (surface temperature) can influence the tectonic mode. The coupling between a radiative transfer model of the atmosphere and an internal dynamics model is therefore required. The atmospheric model should include the effects of greenhouse gases such as H_2O and CH_4 and would provide the boundary conditions (pressure, temperature) for the internal dynamics model. The internal dynamics model would determine to what extent gases can be extracted from the mantle into the atmosphere and would provide the surface heat flux. Such models are not yet available.

3.2.3. Melting in the mantle. Melting is a critical process that is responsible for both the formation of the iron-rich core and volcanism that transfers gases dissolved in the mantle to the atmosphere. Inspection of the melting curves of iron, iron alloys (Fe-FeS system), and silicates (see Fig. 3 in *Sotin et al.,* 2007) shows that the melting temperature of an Fe-FeS system is lower than the melting temperature of the silicate mantle. It implies that an iron-rich liquid would form and migrate toward the center of the planet because its density is very large (e.g., *Ricard et al.,* 2009).

The solidus of the silicates is above the horizontally averaged temperature profile in Earth's mantle. The difference between the temperature profile and the solidus decreases with decreasing pressure. Melting of silicates occurs close to the surface (around 100 km) in the hot plumes. This partial melt is less dense and can migrate into magma chambers. Then, the melt contained in the magma chambers eventually reaches the surface (volcanism). Gases play an important role because the solubility of gases decreases with decreasing pressure. Therefore, as the magma migrates toward the surface, the gases exsolve and occupy a larger volume, creating more buoyancy. This runaway effect causes magmatic eruptions, and gases are expelled into the atmosphere. The role of surface pressure is important since it may limit the intensity of the eruptions and the amount of gases released in the atmosphere.

H_2O is an important gas. First, it is a greenhouse gas that will increase the surface temperature. Second, liquid water is thought to be a key ingredient for life. Its presence on the surface is limited to a narrow temperature range (Fig. 3). The evolution of the surface temperature with time as a planet differentiates by volcanism is still poorly understood. The role of plate tectonics is important because it recycles some of the water into the mantle. The water also reacts with the silicates to form hydrated silicates that have properties, including yield strength, very different from those of dry silicates. Modeling the H_2O cycle of an exoplanet is required to understand whether it can be similar to Earth and harbor life.

3.3. Has a Terrestrial Exoplanet Been Found?

If the definition of a terrestrial planet is limited to its size and mass, the answer to this question is probably yes. For example, the characteristics of the planet Kepler-20b lie upon the silicate (mass, radius) curve (*Gautier et al.,* 2012;

Fressin et al., 2012). However, if additional conditions, such as surface temperature and atmospheric composition, are required for defining the terrestrial nature of an exoplanet, then the answer is negative. For Kepler-20e (see below), the equilibrium temperature is on the order of 1200 K, suggesting the presence of molten silicates at shallow depth if not at the surface.

The Kepler mission has provided about a dozen planets with a radius lower than 2 $R_\oplus$. The two smallest ones are in the Kepler-42 system. Also known as the Kepler Object of Interest (KOI) 961 system, the Kepler-42 system is composed of three planets orbiting an M dwarf with orbital periods of less than two days (*Muirhead et al., 2012*). There is no information about their mass, which makes the determination of their density impossible. Although they orbit an M dwarf, their close distance to the star makes the temperature high, with values ranging from 800 K to 500 K. Precise photometric time series obtained by the Kepler spacecraft over a little less than two years have revealed five periodic transit-like signals in the G8 star Kepler 20 (*Gautier et al., 2012; Fressin et al., 2012*). These observations provide the radius of the five planets, which are named, in increasing distance from the star, 20b, 20e, 20c, 20f, and 20d (*Gautier et al., 2012*). Radial velocities measurements provide the mass of Kepler-20b (8.7 $M_\oplus$) and Kepler-20c (16.1 $M_\oplus$). The mass and radius of Kepler-20b are consistent with a terrestrial composition (*Gautier et al., 2012*). The mass and radius of Kepler-20c are more consistent with a sub-Neptune composition. A maximum value of the mass has been inferred for Kepler-20d, which also makes it consistent with a Neptune-like composition. The two planets Kepler-20e and Kepler-20f are not massive enough to provide a measurable radial velocity on the star. Without the radial velocity measurement, the determination of the mass relies on theoretical considerations (*Fressin et al., 2012*) leading to upper and lower bounds: $0.39 <$ $M_{Kepler-20e}/M_\oplus < 1.67$ and $0.66 < M_{Kepler-20f}/M_\oplus < 3.04$. These two planets are located further away from the star than Kepler-20b, and one might therefore expect them to contain more volatiles. Since Kepler-20b lies upon the terrestrial mass-radius curve, the planets Kepler-20e and Kepler-20f are likely above, although that statement needs radial velocity measurements to be validated. Finally, the equilibrium temperature is quite large, equal to 1136 K and 771 K for Kepler-20e and Kepler-20f, respectively.

The Kepler-11 system is composed of six planets for which the masses of the five closer to the star have been estimated by their transit time variations (*Lissauer et al., 2011*). Although transit time variations can only provide upper limits for the mass, the values suggest that the densest planet is the closest to the star. However, even with the maximum mass, these planets are less dense than silicate planets. Finally, two of the smallest known planets are orbiting the post-red-giant, hot B subdwarf star KIC 05807616 at distances of 0.0060 and 0.0076 AU, with orbital periods of 5.7625 and 8.2293 hours, respectively (*Charpinet et al., 2011*). The radius of the two planets KOI-55b and 55c is

equal to 0.759 $R_\oplus$ and 0.867 $R_\oplus$, respectively (see Fig. 4). These planets are smaller than Earth. However, there is no constraint on their mass and it is therefore impossible to conclude that they are terrestrial planets. Finally, their equilibrium temperature is larger than 7000 K. Such high surface temperature values do not fit the canonical model of a terrestrial planet.

3.4. Perspectives

The curve (mass, radius) of terrestrial planets is well determined and has been tested against Earth's characteristics. Varying the elementary composition does not provide significant variations to the terrestrial curve. The Kepler mission has detected more than a dozen planets with a radius smaller than 2 $R_\oplus$. However, only a few of them fall upon the terrestrial curve. For some of them, precise determination of the mass is still lacking. Plate tectonics may play a major role in the development of life. Terrestrial planets that resemble Earth have to be in that regime.

Several research topics must be studied to improve our ability to find Earth-like exoplanets. As discussed in this chapter, understanding the relationships between mantle convection and the tectonic regime is required. Most information on that topic would be achieved by comparing Venus and Earth. A better understanding of Venus' interior structure and dynamics would be achieved by dedicated missions to Venus. The discovery of terrestrial exoplanets also stresses the necessity for a model that combines an atmospheric radiative transfer model with an internal dynamics model. Finally, it is crucial to determine the mass of the small terrestrial exoplanets and to get additional information on their atmospheric composition. Such measurements may be obtained from Earth-based very large telescopes or from space telescopes (see section 6).

4. GIANT PLANETS IN THE SOLAR SYSTEM

Much like the Sun is our reference standard for stars, the solar system's giant planets are our standards for giant planets. They can be observed in great detail from Earth, and space missions such as Pioneer, Voyager 1 and 2, Galileo, and Cassini can provide refined measurements of important quantities, including the planetary gravity field. *In situ* measurement has also taken place, thanks to the Galileo Entry Probe. Our four giant planets show incredible diversity in physical properties, as perhaps one might expect given the factor of 20 difference in mass between Jupiter and the much smaller Uranus and Neptune. Our luck at having four nearby examples of giant planets for study, which could well be a rarity in planetary systems, has allowed us to appreciate the complexity of these planets.

In this section we will first discuss the key observations of our solar system's giant planets and then our "classical" views of their structure. These will be used as jumping off points to understand recent work, which in some cases has dramatically revised our understanding of these planets.

4.1. Classical Inferences for the Solar System Planets

With knowledge of the EOS of hydrogen and helium under high pressure, one can compute a cold-curve (T = 0) mass-radius relation for a solar H-He mix. Such curves, over a wide range in mass, were computed by, e.g., *Zapolsky and Salpeter* (1969). This shows that Jupiter and Saturn are predominantly H/He objects. One can also compute similar curves for adiabatic models of the planets, where interior temperatures reach ~10,000–20,000 K (e.g., *Fortney et al.*, 2007; *Baraffe et al.*, 2008). Both Jupiter and Saturn are found to be smaller and denser than pure H/He adiabatic objects, with Saturn farther from pure H-He composition. Thus one can tell from the planets' masses and radii alone that they are enhanced in metals compared to the Sun. Similar calculations for Uranus and Neptune suggest that the planets are predominately composed of metals, with a minority of their mass in the H-He envelope that makes up the visible outer layers. More sophisticated models can yield additional information, but in the era of exoplanets it is always important to keep in mind that much can be learned from mass and radius alone.

There are actually a reasonably large number of observables beyond mass and radius that can be used to better understand the current structure of giant planets. These include the rotation rate, equatorial radius, polar radius, temperature at a 1-bar reference pressure, total flux emitted by the planet, total flux scattered by the planet, and gravity field. Historically, the oblateness of the planet and the gravity field, when combined with measured rotation rate, have yielded useful constraints on the planetary density as a function of radius. The planets rotate, and how they respond to this rotation via their shape, and via a gravity field that differs from a point mass, yields essential information.

The gravity field is generally parameterized by the gravitational moments J_n, which are the leading coefficients if the external gravitational potential is expanded as a sum of Legendre polynomials [see, e.g., equations (10)–(11) of *Fortney et al.* (2011a)]. These coefficients, "the Js," can be measured by observing the acceleration of spacecraft via Doppler shift of their emitted radio signals. In some cases the coefficients can also be constrained from long-term motions of small moons. A method is then needed to calculate J_n based on an interior structure model that obeys all observational constraints. The state of the art for many decades has been the "theory of figures," as described in full detail in *Zharkov and Trubitsyn* (1978). At this time the method is accurate enough for calculations out to J_6.

The classical view of the planets was well solidified by around 1980, when the core-accretion model of *Mizuno* (1980) suggested the giant planets needed ~10-$M_\oplus$ primordial cores in order to form. Around this same time *Hubbard and Macfarlane* (1980) found that the structure of all four giant planets were consistent with ~10–15-$M_\oplus$ cores at the present day. Even very precise knowledge of the gravity is of limited help when directly constraining the mass of the core. This is because the gravitational moments predomi-

nantly probe the outer planetary layers, and the weighting moves closer to the surface with higher order (see Fig. 1 of *Helled et al.*, 2011a). For Jupiter and Saturn, the region of the core is not directly probed, while for Uranus and Neptune, one has more leverage on core structure.

Models of the thermal evolution of giant planets aim to understand the flux being emitted by the planets. There are two separate components, since our relatively cool giant planets are warmed by the Sun. These components are generally written in terms of corresponding temperatures. T_{therm} characterizes the total thermal flux emitted by the planet. T_{int} characterizes the thermal flux due only to the loss of remnant formation/contraction energy, which is much larger at earlier ages when a planet's interior is hotter. T_{eq} characterizes the component that is due only to absorbed solar flux, which is then reradiated to space. This component can also be time varying (to a lesser degree) due to changes in the solar luminosity, which are readily understood, and changes to the planetary Bond albedo, which are harder to model. With these definitions, at any age $T_{therm}^4 = T_{int}^4 + T_{eq}^4$. At a very young age $T_{int}^4 \gg T_{eq}^4$, while today $T_{int} \ll T_{eq}$ and $T_{therm} \sim T_{eq}$.

Cooling calculations yield the planetary T_{therm} over time, and a correct model would match at least the current T_{therm} and radius of the planet at 4.5 Ga. There can also be other relevant observational constraints. Cooling models originated with Jupiter in *Graboske et al.* (1975), and within a few years (*Hubbard*, 1977; *Bodenheimer et al.*, 1980) it was clear that a Jupiter could have cooled to its present T_{therm} of 125 K from a hot, high-entropy phase in the age of the solar system. However, *Stevenson and Salpeter* (1977a) and *Pollack et al.* (1977) showed that Saturn is currently much more luminous (by ~50%) than one obtains from a similar simple model. It has long been suggested that a key to understanding this model "cooling shortfall" is in an additional energy source within Saturn due to the differentiation of the planet (see section 2.3). *Stevenson* (1975) suggested that, at megabar pressures and temperature below ~10,000 K, H and He could phase separate, leading to helium-rich droplets raining down within a giant planet (*Stevenson and Salpeter*, 1977a). This helium rain would essentially be a change of gravitational potential energy into thermal energy (see section 2.3).

From the Galileo Entry Probe it is known that the atmosphere of Jupiter is modestly depleted in helium, with a mass fraction Y = 0.234 ± 0.005 (*von Zahn et al.*, 1998) compared to the protosolar value of 0.2741 ± 0.0120 (*Lodders*, 2003), signaling that helium phase separation is underway (see section 2.3). Jupiter's atmosphere is also strongly depleted in neon, which seems to be preferentially incorporated in the He-rich phase, which sediments out, as shown by theoretical calculations (e.g., *Wilson and Militzer*, 2010). In Saturn, where He phase separation should be more pronounced, there has been no entry probe, and the helium abundance is much more uncertain. *Conrath and Gautier* (2000) suggest Y = 0.18–0.25. Thermal evolution models with He phase separation have been calculated by *Hubbard et al.* (1999) and *Fortney and Hubbard* (2003) for Saturn, suggesting that if He rains down nearly to the planet's core, then the high planetary T_{therm} and relatively modest atmospheric Y depletion can be simultaneously matched.

Models of the thermal evolution of Uranus and Neptune have had to be generally more exploratory, since the main components are not known with certainty, and the EOS of the components have not been studied in as much detail. However, they have long been an indication that both planets are actually underluminous compared to a simple model with an adiabatic interior (*Podolak et al.*, 1991; *Hubbard et al.*, 1995). However, with recent advances in the water EOS (see section 2.2) (*French et al.*, 2009), evolutionary models are being revisited (*Fortney et al.*, 2011b) (see section 4.5).

4.2. Common Model Assumptions

Modeling the interior structure of giant planets and comparing these models with observations is fraught with degeneracies. This is because planetary density and gravity field measurements are integrated quantities that sample large fractions of the planetary interior. With the acknowledgement that some of these assumptions may not hold in the interiors of planets, workers have traditionally forged ahead while making traditional assumptions. These are a fully convective (and hence adiabatic and isentropic) interior, a central heavy element core that is distinct from the predominantly H-He envelope, a composition in the H-He envelope that is entirely uniform, or one that is broken up into two layers, which may differ in helium or metal content, solid body rotation or rotation on cylinders, and a rotation period for the deep interior that is given by the measured magnetic field rotation rate. Additionally for Uranus and Neptune, the deep interior is broken into two heavy element layers, meaning a rocky core, a water-rich middle envelope, and a predominantly H-He outer envelope.

Therefore most models of giant planets have a three-layer structure. For Jupiter and Saturn, these three layers from the inside out would be layer 1, core material (which could be made of rock and/or water, in a mixture or in two layers); layer 2, an inner envelope made predominantly of liquid metallic hydrogen, which is often enhanced in helium, with a prescribed metal mass fraction; and layer 3, an outer layer of predominantly fluid molecular hydrogen, which is often depleted in helium, with a prescribed metal mass fraction. The boundary between layers 2 and 3 is often left as a free parameter, but is often around 1–2 Mbar, the approximate transition region for hydrogen to change from molecular to metallic form. This is also the region where He is expected to be most immiscible, so placing a He discontinuity here makes some physical sense. If the giant planets are fully convective, it is not as clear that discontinuity in metals should also occur at this *same* pressure, since mixing could homogenize the metals. However, the dredge-up of any heavy element core could lead to metal enhancement in the deeper regions. The values of the mass of any core, and the amount of metals within the H-He

envelope, are iterated until the calculated J_n values agree with observations.

Within Uranus and Neptune, the three-layer structure is manifested in a similar manner. The inner core is assumed to be rock or rock/iron. The middle layer is mostly water or some mixture of fluid icy volatiles (like H_2O, NH_3, and CH_4). The outer envelope is predominantly fluid H_2 and He. Pressures in the outer layer are not high enough for liquid metallic hydrogen to form. However, to match the gravity fields of the planets, the middle layer must often be less dense than an icy mixture (so that H-He is included in the middle layer) and the outer envelope is more dense than H-He, so that an admixture of icy material is included in the H-He layer. Since metals make up most of their mass, one would certainly be interested for more detailed information, in particular the ice-to-rock ratio in Uranus and Neptune. However, one can arrange mixtures of rock, water, and H-He in a variety of reasonable configurations, yielding a variety of degenerate solutions.

4.3. Results from Classic Models

"Classic" models on the interiors of the giant planets remain relevant for a number of reasons. Perhaps most importantly, the relevant input physics have often been poorly known. To put it simply, if one does not understand the EOS of hydrogen well, which is the most important component, there is potentially little to be gained by adding additional complications to a structure model. Even in the era of the advent of more sophisticated models, classic models will still represent an important area of work. But we are now in the position to more properly evaluate if common assumptions are indeed true.

Models of Jupiter and Saturn in particular aim to constrain the total metal mass fraction within the interior, and what fraction of these metals are found in a heavy element core. Figure 6 summarizes the results of modern adiabatic models. These are models computed by several groups, using different EOS (*Militzer et al.*, 2008; *Fortney and Nettelmann*, 2010; *Nettelmann et al.*, 2013b; *Helled and Guillot*, 2013).

For Jupiter, the upper left region in Fig. 6 is from *Militzer et al.* (2008), who used a first-principles H/He EOS. The larger region on the bottom right and down is from *Fortney and Nettelmann* (2010), who use a wide range of possible H/He EOS, and also allow for a possible discontinuity in the heavy element enrichment in the H/He envelope. Models connecting these two regions are therefore also plausible, given the uncertainty in the H/He EOS.

For Saturn, the lefthand region is from *Helled and Guillot* (2013), who considered models with a homogeneous H/He envelope and the SCVH EOS. The larger region on the right is from *Nettelmann et al.* (2013b), who considered a wider range of H/He EOS, and also allow for a discontinuity in heavy elements in the H/He envelope. *Nettelmann et al.* (2013b) also find a range of models that essentially encompass the *Helled and Guillot* (2013) results as a subset.

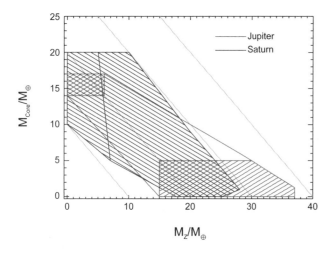

Fig. 6. See Plate 16 for color version. Inferred constraints on the interior of Jupiter and Saturn from fully convective models with a distinct core. The y-axis shows metals in the core, while the x-axis shows metals within the H-He envelope. Gray lines are constant metal mass. See text (section 4.3) for explanations of hatched regions and lines.

Nettelmann et al. (2013b) models with small cores generally have very large enrichments in heavy elements, up to 30× solar, in the deep region of the H/He envelope. Therefore, even with small cores, an appreciable central concentration remains in Saturn.

Note that if Jupiter were precisely of solar composition, it would contain 1.4% metals by mass (*Asplund et al.*, 2009), or 4.5 $M_\oplus$. For Saturn this would be 1.3 $M_\oplus$. On the whole, planets are substantially enhanced in metals, a factor of ~3 to 8 for Jupiter, and ~12 to 21 for Saturn, compared to the solar composition.

For Uranus and Neptune, one would like to constrain the ice-to-rock ratio in the interior, as well as the distribution of these metals. The gravity fields for these planets are less well known. However, the main obstacle is the significant degeneracy in possible composition. These issues were reviewed in detail in *Podolak et al.* (1991) and *Hubbard et al.* (1995). A mixture of high-pressure rock and H-He has an EOS that can closely mimic that of water. Therefore it is possible to construct acceptable models of the planets without water/ice of any kind, if the bulk of the interior is a high-pressure rock-H-He mixture. Also, evidence for a high-density rocky core is not particularly strong. Models with a compressed water-like density throughout most of the deep interior are acceptable.

In Fig. 7 we show calculations from *Nettelmann et al.* (2013a) of current constraints on the interior of Uranus and Neptune. These models constrain the metallicity of the "water-rich" inner envelope with metallicity Z_2, and the H-He rich outer envelope, with metallicity Z_1. As discussed in *Nettelmann et al.* (2013a), the gravity field of Uranus is better constrained than that of Neptune, so *within the framework of three-layer adiabatic models,* its structure is better

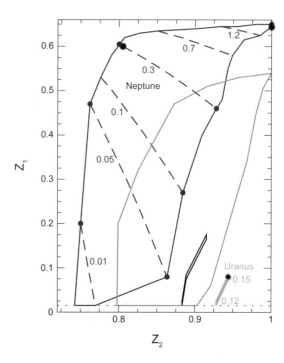

Fig. 7. See Plate 17 for color version. Heavy element mass fraction in the outer envelope (Z_1) and inner envelope (Z_2) of Uranus models (black) and Neptune models (gray) as labeled with the Voyager shape and rotation data. The solid lines frame the full set of solutions for each planet. Dashed lines within the box of Neptune models indicate solutions of same transition pressure between inner and outer envelopes (in M_{bar} as labeled). The dotted line is a guide to the eye for the solar metallicity $Z = 0.015$. The colored areas use the modified shape and rotation data for Uranus (green) and Neptune (blue). Adapted from *Nettelmann et al.* (2013a).

constrained. In these models the fluid ices are modeled using the water EOS of *French et al.* (2009). For both planets, the inner water-rich envelope must have an admixture of a lower-density component (here H-He in solar proportions).

4.4. A More Modern View of Jupiter and Saturn

Over the past few years researchers have begun to question in earnest some of the long-held assumptions of giant planet modeling, in particular for the gas giants Jupiter and Saturn. These assumptions include a fully adiabatic interior and the likelihood that a distinct core exists. Some 20 years ago *Guillot et al.* (1994) investigated whether the giant planets may have radiative windows in the H_2-dominated outer layers. A follow-up study (*Guillot et al.*, 2004) using modern opacities well tested against brown dwarf atmospheres found that adiabatic convection did hold.

Stevenson (1982) and *Stevenson* (1985) suggested that there was no sound reason to believe the primordial cores would remain distinct from the overlying H-He envelope, and *Guillot et al.* (2004) put forth a simple model of how these cores could be dredged up by convection over time.

As described in section 2, with the rise of detailed *ab initio* EOS calculations, it has become feasible to study the miscibility of water, rock, and iron in liquid metallic hydrogen at high pressure. B. Militzer and collaborators have shown that each of these components is miscible at the pressures and temperatures relevant to Jupiter's core (*Wilson and Militzer,* 2012a,b; *Wahl et al.,* 2013). In Saturn, much the same story holds, but the lower temperatures may lead rocky materials to stay unmixed.

If core material is able to be mixed into the overlying H-He envelope, the next natural questions are "Does this occur?" and "How does it affect the planet's evolution?" The redistribution of any core material is not a simple problem. It takes work for the planet to gradually bring up the denser material (be it watery or rocky), and the efficiency of this process is essentially not understood. One could imagine that a core may be gradually diluted into the overlying H/He mixture. The composition gradient may then extend over the interior of the planet, from a highly metal-rich central region, with a smaller metal mass fraction with increasing radius. The process would be an energy sink, since it would alter the gravitational energy of the planet to make the planet less centrally condensed. Much of this redistribution could have potentially happened quite early on, when the planet's interior was hotter and convection more vigorous.

Only very recently have researchers begun examining how such composition gradients would affect the past, current, and future structure and cooling of giant planets. *Leconte and Chabrier* (2012, 2013), extending earlier work by *Chabrier and Baraffe* (2007), have examined the interior structure and thermal evolution of Jupiter and Saturn with heavy element composition gradients throughout most of the interior of the planets. They have used results of extensive three-dimensional calculations of energy transport in the "double-diffusive" regime where both temperature and a dense component may be redistributed (*Rosenblum et al.,* 2011; *Mirouh et al.,* 2012).

Leconte and Chabrier (2012, 2013) investigate how composition gradients, based on parameterizations of the efficiency of energy transport within the interiors of the planets, change the temperature stratification of the interior. For Jupiter and Saturn they find interior structure models that are hotter than homogeneous adiabatic models, and therefore they are more heavy-element-rich to match the planet's gravity field constraints. They also have calculated thermal evolution models for Saturn than can match the planet's anomalously high intrinsic flux without the need for helium phase separation. Composition gradients suppress interior cooling, such that compared to a standard adiabatic cooling model, intrinsic fluxes are lower at young ages, but are in turn higher at old ages. These new models are important and suggestive. To "close the loop" on a modern alternative view of giant planet interiors, the final piece of the puzzle, mentioned above, is an understanding of how planet formation, or rapid core dredge up at young age, could lead to the kinds of composition gradients studied by Leconte and Chabrier.

4.5. A More Modern View of Uranus and Neptune

While in the past 20 years Jupiter has been visited by Galileo (and soon by Juno), and Cassini has been at Saturn for nearly a decade, Uranus and Neptune were only visited via a single flyby by Voyager 2. Progress in our understanding of these planets has been slowed by a lack of data compared to Jupiter and Saturn, a lower precision on measured quantities, and finally, the inherent complexities of the planets themselves.

Recent progress in understanding Uranus and Neptune have come from a new appreciation of past measurement uncertainties, as well as from the application of new EOS for modeling the interior structure and thermal evolution of the planets. *Helled et al.* (2011a) confirmed previous results from *Marley et al.* (1995) that the gravity field of the planets do not require having distinct compositional layers within the planets. *French et al.* (2009) have published a new *ab initio* EOS for water, which has been shown to agree very well with high-pressure shock data (*Knudson et al.,* 2012). Thermal evolution models of Uranus and Neptune that use this new EOS can lead to dramatically revised cooling histories. *Fortney and Nettelmann* (2010) and *Fortney et al.* (2011b) found that they could match the measured T_{therm} of Neptune at 4.5 Ga without having to resort to *ad hoc* cooling histories. Models with fully adiabatic interiors and three distinct layers (the simplest possible model) match the planet's cooling history. However, evolutionary models for Uranus still show a large difference between the predicted intrinsic flux from the planet and the lower measured flux.

Most recently, *Kaspi et al.* (2013), following up on work from *Hubbard et al.* (1991), have shown that the available gravity field data for Uranus and Neptune strongly constrain the depth to which the visible atmospheric differential rotation penetrates into the interior. They show that J_4 constrains the dynamics to the outermost 0.15% of the total mass of Uranus and the outermost 0.2% of the total mass of Neptune. Therefore the dynamics visible in the atmosphere are confined to a thin layer no more than about 1000 km deep on both planets. Similar constraints for Jupiter and Saturn are not yet available, but should be in 2017 from new spacecraft data described below.

Finally, but quite importantly, there has recently been extensive work on discussion on the true rotation periods of all the giant planets, and whether the measured magnetic field rotation periods are consistent with the measured oblateness measurements of the planets and implications for interior models (*Helled et al.,* 2009, 2010).

4.6. Seismology of the Giant Planets

Seismological data has been essential to our current understanding of the internal structure of Earth, the Sun, and other stars. There has also long been interest in measuring the oscillations of the solar system's giant planets, as such data would at a minimum be complementary, at best significantly more constraining, than gravity field data. A definitive detection of oscillations of any jovian planet could in principle serve to accurately determine the core radius and rotation profile of the planet. Since such determinations would remove two important sources of uncertainty surrounding the interior structure, more information could then be gleaned from the traditional interior model constraints. Seismology might also help to constrain more accurately the location of any layer boundaries, such as the transition from molecular to metallic hydrogen or a region of helium enhancement.

After a long history of ambiguous observations, there have recently been two observational breakthroughs. The first was the detection by *Gaulme et al.* (2011) of Jupiter's oscillations with the Seismographic Imaging Interferometer for Monitoring of Planetary Atmospheres (SYMPA) instrument (*Schmider et al.,* 2007), with a measurement of the frequency of maximum amplitude, the mean large frequency spacing between radial harmonics, and the mode maximum amplitude. In particular, the large frequency spacing of 155.3 ± 2.2 µHz was fully consistent with most computed models of Jupiter's interior.

The other avenue into giant planet seismology is "using Saturn's rings as a seismograph," which was discussed in detail by *Marley* (1991). Going back to *Rosen et al.* (1991), it was found that there were spiral structures in Saturn's rings seen with Voyager 2 that could not be explained by any satellite resonances, but could reasonably be explained due to perturbations on the rings from Saturn free oscillation (*Marley and Porco,* 1993). *Hedman and Nicholson* (2013) very recently used superior Cassini data to definitely confirm the presence of features in the rings at locations suggested by *Marley* (1991), and ruled out moon-based resonances based on wave speeds. While the orbital separations of the ring features were as expected, the unexpected feature from the *Hedman and Nicholson* (2013) analysis was that they identified multiple waves with the same number of arms and very similar pattern speeds. This indicates that multiple m = 3 and m = 2 sectoral (l = m) modes may exist within the planet. This peculiar feature was not anticipated from any Saturn model. The methods of modern helioseismology, now brought to bear on giant planets, may allow for important advances (e.g., *Jackiewicz et al.,* 2012).

4.7. Juno at Jupiter and Future Space Missions

In 2016 the NASA Juno spacecraft will enter a low-periapse polar orbit around Jupiter. Many of the major science goals for Juno involve answering major questions about the interior of the planet. First, the mission will accurately map the gravitational field of the planet at high accuracy out to very high order, perhaps J_{12}. There will also be sensitivity to nonzero odd J_n values, such that deviations from hydrostatic equilibrium can be constrained. The degree to which the outer layers of the planet rotate differentially, on cylinders, or via solid-body rotation will be strongly constrained. Tides raised on the planet by Io may be detectable, which would lead to a measurement of the k_2 tidal Love number (*Hubbard et al.,* 1999). The

Lense-Thirring effect is probably observable and will enable a direct measurement of the planet's rotational angular momentum, which could then strongly constrain the planet's moment of inertia (*Helled et al.,* 2011b).

The magnetic field of the planet will also be extensively mapped. While in principle there should be tremendous synergy between interior modeling to better understand composition and structure, and interior models to understand convection and the interior dynamo, in practice these communities do not have enough cross-talk. With outstanding datasets on both gravity and magnetic field, this will hopefully improve. Finally, Juno will have a microwave radiometer that will allow spectroscopy of the planet's thermal emission down to a depth of ~100 bar. This depth is important since the Galileo Entry Probe only reached 22 bar, which apparently was not deep enough to reach below the level where water vapor is partially condensed into clouds, such that the measured water abundance was only a lower limit. With Juno the mixing ratios of H_2O and NH_3 should both be measured, as they are two of the most abundant molecules (along with CH_4) in the H-He envelope. With a refined understanding of the mass fraction of heavy elements within the H-He envelope, the total mass of metals within the planet will be better constrained.

In 2017, the Cassini Solstice extended mission at Saturn is scheduled to end. The current plan for the final stages of the Cassini spacecraft is to bring it into a somewhat Juno-like low-periapse polar orbit, so that precision mapping of the planet's gravity and magnetic fields can be undertaken in gravity up to J_{10}. Much of the same work on understanding differential rotation inside the planet will be enabled. Again in analogy to Juno at Jupiter, it has been suggested by *Helled* (2011) that the Lense-Thirring effect could be observed, which would further constrain the current interior structure. Eventually the craft will be maneuvered down into Saturn's atmosphere, where it will break up. It may be possible for *in situ* sampling of Saturn's atmospheric mixing ratios by the mass spectrometer before breakup.

Given that precise gravity field data will be available to high order, new methods are now being developed to allow for highly accurate calculation of the gravity field from internal structure models. These new methods, which deviate from the theory of figures developed in the 1970s (*Zharkov and Trubitsyn,* 1978), are now just being published (e.g., *Hubbard,* 2012, 2013; *Kong et al.,* 2013), but they are an exciting advance. The generalization of these models beyond merely one-dimensional structure models should also be expected.

There are currently no plans for missions to the Uranus or Neptune systems. Given that Neptune-class planets significantly outnumber true gas giants in exoplanetary systems, it is essential to better understand our lower-mass giant planets. Proposed European medium (M)-class missions that would venture to either Uranus or Neptune in the mid 2030s are the Outer Solar System (OSS) mission to Neptune (*Christophe et al.,* 2012) and the Uranus Pathfinder (*Arridge et al.,* 2012).

5. EXTRASOLAR GIANT PLANETS

5.1. Status of Current Models

There are two key physical ingredients that characterize state-of-the-art models for the internal structure and evolution of exoplanets. The first is the use of modern EOS, as described in section 2, which allow a variation of mixtures and amounts of heavy elements to account for the wide variety of bulk compositions observed for transiting planets. The second is the use of accurate atmospheric boundary conditions, taking into account up-to-date opacities and irradiation from the parent star for close-in planets (see the chapter by Madhusudhan et al. in this volume).

5.1.1. Heavy material enrichment. As mentioned in section 4.1, the impact of heavy material enrichment on planetary radius dates back to the pioneering work by *Zapolsky and Salpeter* (1969), who studied the mass-radius relationship for objects composed of zero-temperature single elements. If the energy transport in the interior is due to efficient convection, the internal temperature profile is quasi-adiabatic. The mechanical structure and thermal internal profile thus strongly depend on the EOS of its chemical constituents.

Figure 1 shows a spread in radius for a given planetary mass, which is partly due to diversity in the bulk chemical composition of planets. Such diversity is expected from our current understanding of giant planet formation processes (see, for example, the chapter by Helled et al. in this volume). Additional physical processes, required to explain abnormally large radii observed for a significant fraction of close-in exoplanets and discussed in section 5.2, are also certainly responsible for a spread in radius. While an increase of the interior heavy element content in a giant planet contributes to a decrease of its radius (see Fig. 1), the additional physical processes mentioned previously work in reverse and contribute to increasing its radius. Since the mechanisms responsible for inflation of close-in exoplanets have not yet been firmly identified, inferring their bulk composition from the knowledge of their mass and radius is still highly uncertain. As discussed recently in *Miller and Fortney* (2011), a relationship between stellar metallicity and planetary heavy elements seems to exist, as originally suggested by *Guillot et al.* (2006). Such a relationship, if confirmed and determined with some confidence, could be used to infer the amount of heavy material in a given inflated close-in planet, based only on the parent star's metallicity and mass. This would provide a powerful tool to determine the amount of additional energy needed to explain a planet's inflated radius.

Currently, the uncertainty resulting from the lack of knowledge of the inflation mechanisms (see section 5.2) adds to the uncertainty in the current EOS used to model planetary interiors. The most widely used EOS to describe the thermodynamic properties of an exoplanet's interiors are the Saumon-Chabrier-Van Horn EOS for H/He mixtures (*Saumon et al.*, 1995), and ANEOS or SESAME for heavy elements, usually water, rock, or iron (see section 2). These EOS have been used to construct models covering a wide range of planetary masses including different amounts and compositions of heavy material (*Guillot et al.*, 2006; *Burrows et al.*, 2007; *Fortney et al.*, 2007; *Baraffe et al.*, 2008). A detailed analysis of the main uncertainties in planetary models that may result from the choice of the EOS and the amount and distribution of heavy material within the planet has been conducted by *Baraffe et al.* (2008). This work shows the sensitivity of the radius to the choice of the EOS (i.e., ANEOS vs. SESAME) and the distribution of heavy material within the planet. The largest difference between the various EOS models, reaching up to ~40–60% in P(ρ) and ~10–15% on the entropy S(P,T), occurs in the T ~ 10^3–10^4 K, P ~ 0.01–1-Mbar interpolated region, the typical domain of Neptune-like planets. The radius of a 20 $M_\oplus$ planet having 50% heavy element enrichment can vary by 25% depending on whether the heavy material is in a central core or distributed over the whole planet. Such high sensitivity of the radius is due to differences in entropy predicted by the different heavy material EOS used and by different impacts of the entropy of heavy material, whether it is in a core or mixed with H/He in the envelope (*Baraffe et al.*, 2008). Improved EOS as described in section 2 are now starting to be used to model the solar system giant planets (see section 4) and specific exoplanets (*Nettelmann et al.*, 2010, 2011). A preliminary revision of the mass-radius relationship for exoplanets based on *ab initio* EOS for H and He shows deviation compared to models based on the Saumon-Chabrier-Van Horn EOS (*Militzer and Hubbard*, 2013). Extensive work is now required to systematically investigate the effects of new *ab initio* EOS on the structure and evolution of exoplanets, promising some exciting results in the near future. It will be of particular interest to know whether improved EOS confirm or not the sensitivity of the radius to the distribution of heavy material within a planet.

5.1.2. Irradiation effects. Because of the high proximity of many newly discovered exoplanets from their parent star, due to current detection techniques, the heating from the incident stellar flux of the planet atmosphere needs to be accounted for. The incident flux can be defined by

$$F_{inc} = \alpha \left(\frac{R_\star}{r(t)} \right)^2 \sigma T_\star^4 \tag{1}$$

where $R_\star$ and $T_\star$ are the host star radius and effective temperature respectively, and $\alpha \in [1/4,1]$ is a scaling factor used to crudely account for day-to-night energy redistribution. The latter represents a proxy for atmospheric circulation, derived by considering whether the absorbed stellar radiation is redistributed evenly throughout the planet's atmosphere ($\alpha = 1/4$), e.g., due to strong winds rapidly redistributing the heat, or whether the stellar radiation is redistributed and reemitted only over the dayside ($\alpha = 1/2$). Since exoplanets are found with a wide range of eccentricities, the general (time-dependent) planet-star separation is given by

$$r(t) = \frac{a\left(1 - e^2\right)}{1 + e \cos \theta(t)} \tag{2}$$

where a is the semimajor axis, e is the eccentricity, and $\theta(t)$ is the angle swept out by the planet during an orbit. As with the case of solar system giant planets (see section 4.2), a useful characteristic atmospheric temperature for irradiated exoplanets is T_{therm}, the total thermal flux emitted by the planet, given by

$$\sigma T_{therm}^4 = \sigma T_{int}^4 + (1 - A)F_{inc} \qquad (3)$$

with T_{int} the internal effective temperature, which is a measure of the energy contribution from the planet interior (see section 4.2), and A the bond albedo. The second righthand term defines the equilibrium temperature $\sigma T_{eq}^4 = (1-A)F_{inc}$ mentioned in section 4 for our giant planets.

For planets with ages of a few gigayears, $T_{int} \ll T_{eq}$ and $T_{therm} \sim T_{eq}$. In this case, as already stressed in *Barman et al.* (2001), an irradiated atmospheric structure characterized by T_{eq} can differ significantly from a non-irradiated structure at the same effective temperature $T_{eff} = T_{eq}$. This point (see also *Seager and Sasselov*, 1998; *Guillot and Showman*, 2002) emphasizes the fact that adopting outer boundary conditions for evolutionary calculations from atmospheric profiles of non-irradiated models with $T_{eff} = T_{eq}$ is incorrect and yields erroneous evolutionary properties for irradiated objects. More correct outer boundary conditions for evolutionary models are based on sophisticated one-dimensional static atmosphere models, including state-of-the-art chemical and radiative transfer calculations, with the impinging radiation field explicitly included in the solution of the radiative transfer equation and in the computation of the atmospheric structure (*Barman et al.*, 2001; *Sudarsky et al.*, 2003; *Barman*, 2007; *Seager et al.*, 2005; *Fortney et al.*, 2006; *Burrows et al.*, 2008). Irradiation produces an extended radiative zone in the planet outer layers and below the photosphere, pushing to deeper layers (i.e., larger pressure) the top of the convective envelope compared to the non-irradiated counterpart. The main effect of irradiation on planet evolution is to slow down the contraction and produce a radius slightly larger than the non-irradiated counterpart at the same age. For known planetary systems, the effect is rather modest for hot Jupiters, with an increase in radius of less than ~10% (*Baraffe et al.*, 2003). For lower-mass planets, however, the effect can be significant, as illustrated by the red curves in Fig. 1. In a recent work, *Batygin and Stevenson* (2013) highlights the fact that the radius of an irradiated super-Earth planet may exceed that of its isolated counterpart by as much as a factor of ~2. Interestingly enough, this brings the radius of such a planet well into the characteristic range of giant planets.

5.1.3. Atmospheric evaporation. It is now well admitted that atmospheric escape may play an important role in irradiated, close-in planets and shape their faces. Planet evaporation can take place through a variety of mechanisms: nonthermal escape, thermal Jean's escape, and hydrodynamic escape. Hydrodynamic evaporation or "blow-off" may be experienced by hydrogen-rich upper atmospheres of close-in giants orbiting at less than 0.1 AU and heated

up to temperatures of more than 10,000 K by X-rays and ultraviolet (UV) radiation from the parent star (*Lammer et al.*, 2003; *Yelle*, 2004). This mechanism has received most attention in the exoplanet literature since it is the only one which may produce high enough mass loss to affect the structure and evolution of close-in planets. Originally, models of extreme ultraviolet (XUV)-driven mass loss were first developed to study water loss from early Venus (*Hunten*, 1982; *Kasting and Pollack*, 1983) and hydrogen loss from the early Earth (e.g., *Watson et al.*, 1981). These kinds of models were extended and applied to the study of hot Jupiters (e.g., *Lammer et al.*, 2003; *Yelle*, 2004; *Tian et al.*, 2005; *Murray-Clay et al.*, 2009; *Koskinen et al.*, 2010; *Ehrenreich and Désert*, 2011; *Owen and Jackson*, 2012). Those objects remain the best studied cases for atmospheric erosion of exoplanets. Mass loss for hot Jupiters typically occurs at a rate $M \sim 10^{10}$–10^{11} g s^{-1}. HD 209458b is the first transiting planet for which HI absorption at or above the planet's Roche lobe was observed, suggesting that hydrogen is escaping from the planet (*Vidal-Madjar et al.*, 2003). Further detection of oxygen and carbon in the extended upper atmosphere of this planet confirmed the hydrodynamical escape scenario, or "blow-off" of the atmosphere, where the hydrodynamical flow of escaping HI is able to drag other species like carbon and oxygen (*Vidal-Madjar et al.*, 2004). Despite many debates, at the time of these observations, regarding their interpretation, there is now a consensus that HD 209458b has an extended exosphere and is losing mass, although details are still not well understood (see, e.g., *Linsky et al.*, 2010).

A common approach to estimate the mass-loss rate and study its effect on a planet's evolution is to assume that some fixed fraction of the incident stellar XUV energy is converted into heat, which works on the atmosphere to remove mass. This is the so-called "energy-limited" approximation (*Watson et al.*, 1981), which provides a simple analytic description of the mass loss rates (*Erkaev et al.*, 2007)

$$\dot{M} = \frac{\varepsilon \pi R_P^3 F_{XUV}}{GM_P K_{tide}} \qquad (4)$$

where F_{XUV} is the stellar XUV flux, M_P and R_P are the planet mass and radius respectively, ε is an efficiency factor (<1) that represents the fraction of the incident XUV flux that is converted into work, and K_{tide} is a correction factor that accounts for the fact that mass only needs to reach the Roche radius in order to escape (see details in *Erkaev et al.*, 2007).

One main uncertainty stems from the determination of the efficiency factor ε. Estimates from observations of the mass loss rate for HD 09458b suggest $\dot{M} \sim 4 \times 10^{10}$ g s^{-1} (e.g., *Yelle et al.*, 2008), which corresponds to an efficiency factor $\varepsilon \sim 0.4$. *Lammer et al.* (2009) suggest that ε is probably less than ~0.6 for hot Jupiters, but should depend, among other things, on the star's activity and the planet atmospheric composition. An analysis of atmospheric escape from Venus suggests that $\varepsilon = 0.15$ may be a more appropriate value (*Chassefière*, 1996). Finally, the value of ε is certainly not constant over the lifetime of the planet (*Owen and Wu*, 2013). Despite these uncertainties and difficulties, equation (4) certainly provides a reasonable estimate of evaporation of planetary masses by XUV flux and has often been used to study the effect of evaporation on the evolution of exoplanets for a variety of planetary masses (*Baraffe et al.*, 2004; *Hubbard et al.*, 2007; *Jackson et al.*, 2010; *Valencia et al.*, 2010).

The order of magnitude for the mass loss rate estimated for HD 209458b is confirmed with recent observations for HD 189733b, another transiting hot Jupiter, for which atmospheric evaporation has also been detected (*Lecavelier Des Etangs et al.*, 2010; *Bourrier et al.*, 2013). With such rates, only a few percent of a hot Jupiter mass is lost over its lifetime (*Yelle*, 2006; *García Muñoz*, 2007; *Murray-Clay et al.*, 2009). This erosion hardly affects a hot Jupiter's internal structure and its evolution, even at very young ages when the high-energy flux of the parent star is expected to be quite large. Indeed, *Murray-Clay et al.* (2009) demonstrated that, in the case of UV evaporation of hot Jupiters, the energy limited approximation is only valid at low fluxes. At high fluxes, the mass loss process is controlled by ionization and recombination balance, with photoionization heating balanced by Lyα cooling. Mass loss under those conditions is "radiation/recombination-limited" with rates that scale as $\dot{M} \propto F_{UV}^{1/2}$, instead of $\dot{M} \propto F_{UV}$ in the energy-limited case, because the input UV power is largely lost to cooling radiation.

If evaporation is not expected to play a key role in the evolution of giant exoplanets, the story may be different for lower-mass exoplanets. Increasing efforts are now devoted to the study of this process for super-Earths and rocky planets. Because of their lower escape velocities, lower-mass planets should have their atmospheres more significantly sculpted by evaporation, like the super-Earths CoRoT-7 b (*Jackson et al.*, 2010) and Kepler-11b (*Lopez et al.*, 2012), which could be stripped entirely by UV evaporation. As highlighted by *Jackson et al.* (2010), for those low-mass planets — which may be strongly affected by evaporation and may not have formed *in situ* — the coupling between tidal evolution, affecting the orbital parameters and thus the evaporation rate, and mass loss, affecting the planetary mass and thus the tidal evolution, needs to be considered in order to have a consistent description of their evolution. If this is indeed the case, evolutionary calculations for these kinds of planets become an extremely complicated and uncertain game. Models are now also being developed for rocky planets with sub-Earth masses. As recently suggested by *Perez-Becker and Chiang* (2013), even the optical photons from the parent star could vaporize close-in, very-low-mass rocky planets. Evaporation is thus important for the interpretation of observed mass distributions of low-mass close-in planets, as provided by Kepler. Improvement in the modeling of evaporation can be done by treating the thermosphere as an integral part of the whole atmosphere,

rather than a separate entity as done in many earlier studies, including detailed photochemistry of heavy atoms and ions and more realistic description of heating efficiencies (e.g., *Koskinen et al.*, 2013). Further developments are expected in the future to better understand this key process, which can strongly affect the original mass of detected close-in, low-mass planets and thus the understanding of their formation process.

5.1.4. Deuterium-burning planets. The International Astronomical Union (IAU) definitions for a planet and a brown dwarf, based on the deuterium-burning minimum mass, is admittedly practical from an observational point of view but is arbitrary with respect to the formation process. Characterizing these two families of objects by their formation processes rather than using the IAU definition is certainly more natural and physical (see the chapter by Chabrier et al. in this volume). If planets with a massive core can form above the aforementioned deuterium-burning minimum mass, a key question is to determine whether or not the presence of the core can prevent deuterium burning to occur in the deepest layers of the H/He envelope. The question is relevant, since it cannot be excluded that objects forming in a protoplanetary disk via core accretion grow beyond the minimum mass for deuterium burning, e.g., 12 M_{Jup} (*Kley and Dirksen*, 2006). The answer to this question was first provided by *Baraffe et al.* (2008), who showed that in a 25-M_{Jup} planet with a 100-$M_\oplus$ core, deuterium-fusion ignition does occur in the layers above the core and deuterium is completely depleted in the convective H/He envelope after ~10 million years (m.y.), independently of the composition of the core material (water or rock). A follow-up study has confirmed these results, combining core-accretion formation models with deuterium-burning calculations (*Mollière and Mordasini*, 2012). These results highlight the utter confusion provided by a definition of a planet based on the deuterium-burning limit (see the chapter by Chabrier et al. in this volume).

5.2. Inflated Exoplanets

One of the most intriguing problems in exoplanet physics is the anomalously large radii of a significant fraction of close-in gas planets. They show radii larger than predicted by standard models of giant planet cooling and contraction. Despite numerous theoretical studies, starting with the work of *Bodenheimer et al.* (2001), and the suggestion of many possible mechanisms (see reviews by *Baraffe et al.*, 2010; *Fortney and Nettelmann*, 2010), no universal mechanism seems to fully account for the observed anomalies. This implies that either we are still missing some important physics in planetary interiors, or that the solution does not reside in a unique mechanism but in a combination of various processes as described below.

One can classify the proposed mechanisms into three categories, as suggested by *Weiss et al.* (2013): incident stellar-flux-driven mechanisms, tidal mechanisms, and delayed contraction. This classification is meaningful in light of increasing evidence for a correlation between the incident stellar flux and the radius anomaly (*Laughlin et al.*, 2011; *Weiss et al.*, 2013). Additionally, increasing statistics, thanks in particular to Kepler data, reveal a lack of inflated radii for giant planets receiving modest stellar irradiation (*Miller and Fortney*, 2011; *Demory and Seager*, 2011). Based on a sample of 70 Kepler planetary candidates, *Demory and Seager* (2011) find that below an incident flux of about 2×10^8 erg s^{-1} cm^{-2}, which represents a subsample of ~30 candidates, the radii is independent of the stellar incident flux. These findings motivate a further search for correlations between planetary radius, mass, and incident flux. Additional future data from Kepler and other surveys should allow an extension of current studies to planets with longer orbits. Despite these trends, it is worth discussing mechanisms falling into the three aforementioned categories. Indeed, even if some of them cannot explain the aforementioned observed correlations, they may still be important physical processes playing a role at any orbital distance or stellar flux, and thus belonging to the modern theory of planetary structures. We will briefly mention below these mechanisms and their current status as viable explanations for the radius anomaly.

5.2.1. Incident stellar-flux-driven mechanisms. An attractive scenario that falls into this category is the idea based on atmospheric circulation by *Showman and Guillot* (2002). A close-in, tidally locked planet constantly receives the stellar flux on the same hemisphere, producing strong temperature contrasts between its day- and nightsides. The resulting fast winds may produce a heating mechanism in the deep interior of the planet, slowing down its evolution and yielding a larger than expected radius. This suggestion was based on numerical simulations of atmospheric circulation by *Showman and Guillot* (2002), which produce a downward flux of kinetic energy of about ~1% of the absorbed stellar flux. This heat flux is supposed to dissipate in the deep interior, producing an extra source of energy during the planet's evolution. The validity of this scenario, however, is still debated and the dissipation mechanism still needs to be found. The substantial transport of kinetic energy found in the simulations of *Showman and Guillot* (2002) strongly depends on the details of atmospheric circulation models and has not been confirmed by further simulations. Important efforts are now devoted to the development of sophisticated three-dimensional dynamical simulations of strongly irradiated atmospheres (*Showman et al.*, 2011). Most numerical studies, however, are based on global circulation models (GCMs) developed for the study of Earth's weather and climate. They usually solve the so-called "primitive" equations of meteorology, involving the approximation of vertical hydrostatic equilibrium and a "shallow atmosphere" (see, e.g., *Vallis*, 2006). These approximations may provide an inaccurate description of the interaction between the upper and deeper atmosphere, crucial to test the *Showman and Guillot* (2002) idea. Developments of state-of-the-art GCM models that solve the full set of hydrodynamical equations, without any of the previ-

ous approximations, and coupled to sophisticated radiative transfer schemes may help solving this problem in the near future (*Dobbs-Dixon et al.*, 2012; *Mayne et al.*, 2014).

Another mechanism that is receiving more and more attention is ohmic dissipation. The idea is that atmospheric winds blowing across the planetary magnetic field produce currents penetrating the interior and giving rise to ohmic heating in the deeper layers, which could affect the radius evolution of the planet (*Batygin and Stevenson*, 2010). At the same time, *Perna et al.* (2010a,b) also suggested the importance of magnetic drag on the dynamics of the atmospheric flows, finding that a significant amount of energy could be dissipated by ohmic heating at depths. This results in energy deposition in the planet given by the ohmic dissipation per unit mass

$$\dot{E} = \frac{J^2}{\rho \sigma} \qquad (5)$$

with J the current given by Ohm's law and σ the electrical conductivity of the gas (*Liu et al.*, 2008; *Batygin and Stevenson*, 2010). At temperatures characteristic of hot Jupiters, the primary source of electrons in their atmosphere stems from thermally ionized alkali metals with low first ionization potential, e.g., Na, Al, and K, with K probably playing a dominant role (*Batygin and Stevenson*, 2010; *Perna et al.*, 2010b). The mechanism has been studied based on approximate models for the atmospheric circulation (*Batygin and Stevenson*, 2010) or using more complex models (*Perna et al.*, 2010b; *Rauscher and Menou*, 2012, 2013). Ohmic dissipation has been incorporated in calculations of planet thermal evolution in order to study its effect on the structure and evolution of hot Jupiters (*Perna et al.*, 2010b; *Batygin et al.*, 2011; *Huang and Cumming*, 2012; *Wu and Lithwick*, 2013). Most studies suggest that for magnetic field strength greater than B = 3–10 G, this dissipation can produce enough heat in the planetary interior to slow down its cooling sufficiently to explain the inflated radii of hot Jupiters (*Perna et al.*, 2010b; *Batygin et al.*, 2011; *Rauscher and Menou*, 2013; *Wu and Lithwick*, 2013). However, a recent study by *Huang and Cumming* (2012) disagrees with these conclusions and finds it difficult for ohmic heating to explain the large radii of hot Jupiters with large masses and large equilibrium temperature (i.e., a large amount of stellar irradiation received). The discrepancy between their results and previous studies stems from the fact that they do not assume a fixed heating efficiency, namely the fraction of the irradiation going into ohmic power, as in *Batygin et al.* (2011) and *Wu and Lithwick* (2013). Instead, *Huang and Cumming* (2012) use a wind zone model to set the induced magnetic field in the wind zone and the magnitude of the heating.

Many uncertainties are inherent to current studies (see, e.g., *Rauscher and Menou*, 2013), with most calculations of the magnetic drag and ohmic heating based on a formalism that assumes axisymmetry in the flow structure and atmospheric resistivities (*Liu et al.*, 2008), which are not realized in highly irradiated planet atmospheres. Simple geometry for the planetary magnetic field, e.g., aligned dipole field, are also usually assumed. Variation of the magnetic geometry, e.g., assuming that the planet's magnetic axis is misaligned from its axis of rotation or that the field is multipolar, which may be the case in reality, could have important impacts on the atmospheric circulation. The magnetic drag is usually applied to the zonal flow, although the meridional flow may also experience drag. Another uncertainty stems from the electrical conductivity σ, which affects the amount of ohmic power, indirectly through its role to determine the current in the low-density wind region and directly since the ohmic dissipation is proportional to $1/\sigma$ (see equation (5)). Finally, the actual wind speed in the layers that may experience the Lorentz drag depends on how deep the wind zone extends, which depends on how well the weather layer couples dynamically to the deeper atmosphere (*Wu and Lithwick,* 2013), a process currently not well understood and modeled, as previously discussed. The final unknown in this process is the strength of hot Jupiter magnetic fields. Based on scaling-law arguments, their magnetic field can range between a few and tens of Gauss (*Reiners and Christensen,* 2010). Direct measure of the planet magnetic field seems difficult since its signal is expected to be buried in the stellar signal. Strategies in the future need to be developed (see discussion in *Rauscher and Menou,* 2013) to provide a better estimate of these planet magnetic fields and consequently the significance of ohmic dissipation. There are thus still many issues and open questions regarding the impact of ohmic dissipation. At the time of this writing, it cannot be concluded whether this mechanism is the universal explanation for hot Jupiter inflated radii or not.

Two other stellar-flux-dependent mechanisms have been suggested, namely the thermal tides (*Arras and Socrates,* 2010) and the mechanical greenhouse (*Youdin and Mitchell,* 2010). They have not received as much attention as the previous mechanisms. It has yet to be demonstrated that they could explain all the observed anomalies, and thus we invite the reader to refer to the original papers for details.

5.2.2. Tidal mechanisms. Tidal dissipation in a giant planet's interior, producing heating that would stop or slow down the planet's contraction, was among the first mechanisms proposed to explain the large observed radius of HD 209458b (*Bodenheimer et al.,* 2001). The idea was that an unseen planetary companion would force nonzero orbital eccentricity, which would be tidally damped. The idea was ruled out for HD 209458b and for other planets for which further observations indicated an eccentricity of zero. The idea of tidal dissipation also evolved with the discovery of more and more inflated planets, indicating the need for a more general mechanism than the accidental presence of a low-mass companion. Since then, this mechanism has received a great deal of attention (see *Fortney and Nettelmann,* 2010; *Baraffe et al.,* 2010; and references therein). With the increasing number of studies gaining in complexity and level of details, the excitement has slowly but

surely subsided. This mechanism should certainly be very important for some systems, but the emerging view now is that it cannot be the universal radius inflation mechanism. Radius inflation by tidal heating is indeed a short-lived phenomenon, whereas the average observed planetary system age is several gigayears. As acknowledged by *Weiss et al.* (2013), this view was advanced in particular by *Leconte et al.* (2010), who used the most detailed tidal evolution equations. This paper demonstrates that quasicircular approximation, with tidal equations truncated on the order of e^2 in eccentricity as usually assumed in tidal calculations of transiting planets, is not valid for exoplanetary systems that have, or were born with, even a modest value of the eccentricity ($e \geq 0.2$). In particular, it is shown that truncating the tidal equations at second-order eccentricity can overestimate the characteristic timescales of the various orbital parameters by several orders of magnitudes. Based on the complete tidal equations, *Leconte et al.* (2010) find that tidal heating can explain moderately inflated planets, but cannot reproduce both the orbital parameters and radius of, e.g., HD 209458b (see Fig. 8) or the radius anomaly of very inflated planets like WASP-12 b, which has a radius of ~1.8 R_{Jup} (*Hebb et al.,* 2009). If tidal dissipation is likely not the universal mechanism responsible for the radius anomaly, it is still an important mechanism to account for since it impacts the evolution of the orbital properties of planetary systems, playing a role in shaping the distribution of orbits of close-in exoplanets and thus having consequences for our understanding of planet formation, migration, and planet-disk interaction (*Jackson et al.,* 2008). Orbits of observed hot Jupiters may still be decaying and end up collapsing with their host star, as suggested by *Levrard et al.* (2009), who show the lack of tidal equilibrium states for many transiting planets, implying that the orbital and rotational parameters evolve over the lifetime of the system.

5.2.3. Delayed contraction. The final class of mechanisms that may contribute to larger radii comprises atmospheric enhanced opacities (*Burrows et al.,* 2007) and reduced interior heat transport (*Chabrier and Baraffe,* 2007). Regarding the first scenario, *Burrows et al.* (2007) suggested that atmospheres with enhanced opacities would slow down the contraction and cooling of planetary interiors, yielding larger radii at a given age. Enhanced atmospheric metallicities or alteration of atmospheric properties (e.g., photochemical processes driven by strong irradiation or cloud formation) could provide the source of opacity enhancement. This effect may play a role in some transiting planets and affect their evolution, but cannot explain the most inflated transiting planets (see discussion in *Baraffe et al.,* 2010). Also, as noted by *Guillot* (2008), an overall increase of the planet metallicity to account for the opacity enhancement also requires an increase of the molecular weight in the planet interior, which would negate the high-opacity effect.

Finally, following the idea of *Stevenson* (1985), *Chabrier and Baraffe* (2007) suggested that heavy element gradients in the planet's interior could decrease heat transport efficiency, slowing down its cooling and contraction, and

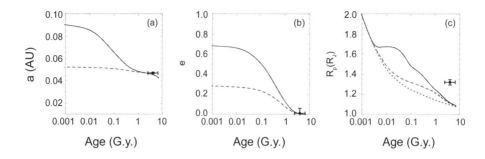

Fig. 8. Tidal/internal evolution of HD 209458b with different initial conditions (solid and dashed) computed with complete tidal equations (from *Leconte et al., 2010*). HD 209458b is a 0.657-M_{Jup} planet orbiting a 1.01-$M_\odot$ star (observed parameters with error bars indicated as solid dots in the panels). These runs assume a tidal quality factor of $Q = 10^6$ for both the planet and the star [see *Leconte et al.* (2010) for details and definitions]. Shown are the evolution of the **(a)** orbital distance, **(b)** eccentricity, and **(c)** radius. In **(c)**, the radius of an isolated planet (no tidal heating) is also shown for comparison (dotted curve).

providing an explanation for hot-Jupiter inflated radii. As previously invoked for solar system giants (section 4.4), if a vertical gradient of heavy elements is present, resulting from, e.g., planetesimal accretion during the formation process or central core erosion, the resulting mean molecular weight gradient can prevent the development of large-scale convection. The combination of heat and chemical species diffusion can drive a hydrodynamical instability and yield so-called "double diffusive" or "layered" convection, also known as "semiconvection" in the stellar community. Because this process does not depend on the stellar incident flux, it is unlikely to be the universal solution to the radius anomaly. But it could be a common property of giant planet interiors, affecting planets at any orbital distance and even our solar system planets, as mentioned in section 4.4. The work of *Leconte and Chabrier* (2013), showing that layered convection could explain Saturn's luminosity anomaly (section 4.4), suggests that the cooling history of giant planets in general, based on the standard assumption of adiabatic homogeneous structure, could be more complex and may need to be revised. The revival of this process for planetary interiors provides a new incentive for the development of three-dimensional numerical simulations (*Rosenblum et al.*, 2011; *Wood et al.*, 2013; *Zaussinger and Spruit*, 2013), which promise advancement in the understanding of semiconvection effects, a long-standing problem in stellar and planetary evolution.

6. CONCLUSIONS

There has been substantial progress done since the previous Protostar and Planets meeting in 2005 on the understanding of planetary structures. Major advancements were performed regarding the development of first-principle numerical methods to study EOS for hydrogen, helium, and various heavy materials relevant for planet interiors. Ongoing and future high-pressure experiments in various national laboratories (Laser Mega-Joule laser projects in France; Livermore and Sandia in the U.S.) promise sub-

stantial progress on this front as well. The perspective to significantly reduce current uncertainties of EOS in the near future, with such combinations of theoretical and experimental works, will allow significant progress in planetary structure modeling.

Regarding our solar system planets, it could be said that the open questions and uncertainties in modeling giant planets are the same as those from ~30 years ago, following the classic review by *Stevenson* (1982). While for the most part this is true, we are now in an excellent position to make real progress on issues that have been around for quite some time. In particular, the questions of the overall metal enrichment of our giant planets, how these metals are distributed, and whether they are predominantly originally made from rocky or icy material will soon be better addressed with the aforementioned progress in EOS and the future space missions to Jupiter and Saturn. Similar progress is also expected for inferring the rotation state of Jupiter and Saturn. To what extent the interiors of our giant planets, and more generally extrasolar gaseous planets, are adiabatic and freely convecting is another long-standing problem in the field. Substantial activities to develop three-dimensional hydrodynamical numerical simulations to tackle this problem will shed important light on the relevance of semiconvection in planetary interiors.

All these questions are also relevant to the modeling of exoplanets. At the time of this writing, hot Jupiters are the most well-characterized of these planets, as they are easy to detect and to follow-up with ground- and spacebased telescopes. Complementary to their bulk composition, information on their atmospheric properties and composition is also obtained based on atmospheric follow-up with, e.g., Spitzer and the Hubble Space Telescope (HST), with more data expected from the James Webb Space Telescope (JWST) and the Exoplanet Characterisation Observatory (EChO). Gaining more and more information on exoplanet properties has opened new challenges that theoretical models now need to address. The anomalous radius observed for a significant fraction of hot Jupiters, related to the stellar incident flux, is one of these challenges. Also, measuring the masses of Neptune-sized and smaller planets that receive high incident flux is now necessary to probe the mass-radius-incident-flux relation for low-mass planets and progress on this front. Future transiting projects such as the Next Generation Transit Survey (NGTS), CHaracterising ExOPlanets Satellite (CHEOPS), and Transiting Exoplanet Survey Satellite (TESS), which will be devoted to super-Earth and Neptune-mass planets, will provide in the near future an important contribution in this context. The detection of Earth-sized planets remains a prominent observational goal and we are now reaping the first harvest. But there is still a long way to go to perform follow-up for Earth-sized planets and characterize their atmospheres.

Direct imaging of giant planets is starting to bear fruit and will become a major detection and characterization technique in the coming decades, with projects like the Gemini Planet Imager (GPI), Very Large Telescope Spectro-Polarimetric High-contrast Exoplanet REsearch (VLT-SPHERE), and the next generation of Extremely Large Telescopes (ELT). As this technique directly probes planet atmospheres, it is on a better footing than indirect methods (secondary eclipse, transmission spectroscopy) for in-depth studies of atmospheric properties (e.g., chemical composition) and inferring global thermal properties of a planet (luminosity, surface temperature). It also completes the sample of planets detected by transit or radial velocity, mostly with young giant planets far from their parent star. The possibility of characterizing planets just after their formation process offers an exciting perspective for better understanding planet formation and thermal evolution.

Acknowledgments. This work is partly supported by the Royal Society Award WM090065 and by the European Research Council under the Union's Seventh Framework (FP7/2007-2013/ERC grant agreement no. 320478).

REFERENCES

Acuna M. H. et al. (1999) *Science, 284,* 790.
Angel R. et al. (2009) *Eur. J. Mineral, 21,* 525.
Arras P. and Socrates A. (2010) *Astrophys. J., 714,* 1.
Arridge C. S. et al. (2012) *Exp. Astron., 33,* 753.
Asplund M. et al. (2009) *Annu. Rev. Astron. Astrophys., 47,* 481.
Baraffe I. et al. (2003) *Astron. Astrophys., 402,* 701.
Baraffe I. et al. (2004) *Astron. Astrophys., 419,* L13.
Baraffe I. et al. (2008) *Astron. Astrophys., 482,* 315.
Baraffe I. et al. (2010) *Rept. Prog. Phys., 73(1),* 016901.
Barman T. (2007) *Astrophys. J. Lett., 661,* L191.
Barman T. S. et al. (2001) *Astrophys. J., 556,* 885.
Batygin K. and Stevenson D. (2010) *Astrophys. J. Lett., 714,* L238.
Batygin K. and Stevenson D. (2013) *Astrophys. J. Lett., 769,* L9.
Batygin K. et al. (2011) *Astrophys. J., 738,* 1.
Beirão P. et al. (2005) *Astron. Astrophys., 438,* 251.
Belov S. I. et al. (2002) *Soviet J. Exp. Theor. Phys. Lett., 76,* 433.
Bodenheimer P. et al. (1980) *Icarus, 41,* 293.
Bodenheimer P. et al. (2001) *Astrophys. J., 548,* 466.
Boehler R. (1993) *Nature, 363,* 534.
Boriskov G. V. et al. (2003) *Sov. Doklady Phys., 48,* 553.
Bourrier V. et al. (2013) *Astron. Astrophys., 551,* A63.
Burrows A. et al. (2007) *Astrophys. J., 661,* 502.
Burrows A. et al. (2008) *Astrophys. J., 678,* 1436.
Caillabet L. et al. (2011) *Phys. Rev. B, 83(9),* 094101.

Celliers P. M. et al. (2010) *Phys. Rev. Lett., 104(18),* 184503.
Chabrier G. and Baraffe I. (2007) *Astrophys. J. Lett., 661,* L81.
Chambers J. E. (2005) In *Treatise on Geochemistry, Vol. 1: Meteorites, Comets and Planets* (A. M. Davis, ed.), p. 461. Elsevier, Amsterdam.
Charpinet S. et al. (2011) *Nature, 480,* 496.
Chassefière E. (1996) *J. Geophys. Res., 101,* 26039.
Christophe B. et al. (2012) *Exp. Astron., 34,* 203.
Collins G. W. et al. (1998) *Science, 281,* 1178.
Collins G. W. et al. (2001) *Phys. Rev. Lett., 87(16),* 165504.
Conrath B. J. and Gautier D. (2000) *Icarus, 144,* 124.
Davaille A. and Jaupart C. (1993) *J. Fluid Mech., 253,* 141.
Demory B.-O. and Seager S. (2011) *Astrophys. J. Suppl., 197,* 12.
Dobbs-Dixon I. et al. (2012) *Astrophys. J., 751,* 87.
Dziewonski A. M. and Anderson D. L. (1981) *Phys. Earth Planet. Inter., 25,* 297.
Eggert J. et al. (2008) *Phys. Rev. Lett., 100(12),* 124503.
Ehrenreich D. and Désert J. (2011) *Astron. Astrophys., 529,* A136.
Erkaev N. V. et al. (2007) *Astron. Astrophys., 472,* 329.
Fortney J. J. and Hubbard W. B. (2003) *Icarus, 164,* 228.
Fortney J. J. and Hubbard W. B. (2004) *Astrophys. J., 608,* 1039.
Fortney J. J. and Nettelmann N. (2010) *Space Sci. Rev., 152,* 423.
Fortney J. J. et al. (2006) *Astrophys. J., 642,* 495.
Fortney J. J. et al. (2007) *Astrophys. J., 659,* 1661.
Fortney J. J. et al. (2011a) In *Exoplanets* (S. Seager, ed.), pp. 397–418. Univ. of Arizona, Tucson.
Fortney J. J. et al. (2011b) *Astrophys. J., 729,* 32.
Fortov V. E. et al. (2003) *Soviet J. Exp. Theor. Phys., 97,* 259.
French M. et al. (2009) *Phys. Rev. B, 79(5),* 054107.
French M. et al. (2012) *Astrophys. J. Suppl., 202,* 5.
Fressin F. et al. (2012) *Nature, 482,* 195.
García Muñoz A. (2007) *Planet. Space Sci., 55,* 1426.
Gaulme P. et al. (2011) *Astron. Astrophys., 531,* A104.
Gautier T. N. III et al. (2012) *Astrophys. J., 749,* 15.
Gilli G. et al. (2006) *Astron. Astrophys., 449,* 723.
Graboske H. C. et al. (1975) *Astrophys. J., 199,* 265.
Grasset O. and Parmentier E. (1998) *J. Geophys. Res., 103,* 18171.
Grasset O. et al. (2009) *Astrophys. J., 693,* 722.
Guillot T. (1999) *Planet. Space Sci., 47,* 1183.
Guillot T. (2008) *Phys. Scr. Vol. T, 130(1),* 014023.
Guillot T. and Showman A. P. (2002) *Astron. Astrophys., 385,* 156.
Guillot T. et al. (1994) *Icarus, 112,* 337.
Guillot T. et al. (2004) In *Jupiter: The Planet, Satellites and Magnetosphere* (F. Bagenal et al., eds.), pp. 35–57. Cambridge Univ., Cambridge.
Guillot T. et al. (2006) *Astron. Astrophys., 453,* L21.
Hamel S. et al. (2011) *Phys. Rev. B, 84(16),* 165110.
Hebb L. et al. (2009) *Astrophys. J., 693,* 1920.
Hedman M. M. and Nicholson P. D. (2013) *Astron. J., 146,* 12.
Helled R. (2011) *Astrophys. J. Lett., 735,* L16.
Helled R. and Guillot T. (2013) *Astrophys. J., 767,* 113.
Helled R. et al. (2009) *Planet. Space Sci., 57,* 1467.
Helled R. et al. (2010) *Icarus, 210,* 446.
Helled R. et al. (2011a) *Astrophys. J., 726,* 15.
Helled R. et al. (2011b) *Icarus, 216,* 440.
Huang X. and Cumming A. (2012) *Astrophys. J., 757,* 47.
Hubbard W. B. (1977) *Icarus, 30,* 305.
Hubbard W. B. (2012) *Astrophys. J. Lett., 756,* L15.
Hubbard W. B. (2013) *Astrophys. J., 768,* 43.
Hubbard W. and Macfarlane J. (1980) *J. Geophys. Res., 85,* 225.
Hubbard W. B. et al. (1991) *Science, 253,* 648.
Hubbard W. B. et al. (1995) In *Neptune and Triton* (D. P. Cruikshank, ed.), pp. 109–140. Univ. of Arizona, Tucson.
Hubbard W. B. et al. (1999) *Planet. Space Sci., 47,* 1175.
Hubbard W. B. et al. (2007) *Icarus, 187,* 358.
Hunten D. M. (1982) *Planet. Space Sci., 30,* 773.
Jackiewicz J. et al. (2012) *Icarus, 220,* 844.
Jackson B. et al. (2008) *Astrophys. J., 678,* 1396.
Jackson B. et al. (2010) *Mon. Not. R. Astron. Soc., 407,* 910.
Jackson I. (1998) *Geophys. J. Intl., 134,* 291.
Javoy M. (1995) *Geophys. Res. Lett., 22,* 2219.
Kaspi Y. et al. (2013) *Nature, 497,* 344.
Kasting J. F. and Pollack J. B. (1983) *Icarus, 53,* 479.
Kley W. and Dirksen G. (2006) *Astron. Astrophys., 447,* 369.
Knudson M. et al. (2004) *Phys. Rev. B, 69(14),* 144209.
Knudson M. et al. (2012) *Phys. Rev. Lett., 108(9),* 091102.

Knudson M. D. and Desjarlais M. P. (2009) *Phys. Rev. Lett.,*
    103(22), 225501.
Kong D. et al. (2013) *Astrophys. J., 764,* 67.
Koskinen T. T. et al. (2010) *Astrophys. J., 722,* 178.
Koskinen T. T. et al. (2013) *Icarus, 226,* 1695.
Kowalski P. M. et al. (2007) *Phys. Rev. B, 76(7),* 075112.
Lammer H. et al. (2003) *Astrophys. J. Lett., 598,* L121.
Lammer H. et al. (2009) *Astron. Astrophys., 506,* 399.
Laughlin G. et al. (2011) *Astrophys. J. Lett., 729,* L7.
Lecavelier Des Etangs A. et al. (2010) *Astron. Astrophys., 514,* A72.
Leconte J. and Chabrier G. (2012) *Astron. Astrophys., 540,* A20.
Leconte J. and Chabrier G. (2013) *Nature Geosci., 6,* 347.
Leconte J. et al. (2010) *Astron. Astrophys., 516,* A64.
Lenardic A. and Crowley J. W. (2012) *Astrophys. J., 755,* 132.
Levrard B. et al. (2009) *Astrophys. J. Lett., 692,* L9.
Linsky J. L. et al. (2010) *Astrophys. J., 717,* 1291.
Lissauer J. J. et al. (2011) *Nature, 470,* 53.
Liu J. et al. (2008) *Icarus, 196,* 653.
Lodders K. (2003) *Astrophys. J., 591,* 1220.
Lopez E. D. et al. (2012) *Astrophys. J., 761,* 59.
Lorenzen W. et al. (2011) *Phys. Rev. B, 84(23),* 235109.
Lyon S. and Johnson J. (1992) *SESAME: The Los Alamos*
    National Laboratory Equation of State Database. LANL Report
    LA-UR-92-3407, Los Alamos National Laboratory, Los Alamos,
    New Mexico.
Marley M. S. (1991) *Icarus, 94,* 420.
Marley M. S. and Porco C. C. (1993) *Icarus, 106,* 508.
Marley M. S. et al. (1995) *J. Geophys. Res., 100,* 23349.
Mattern E. et al. (2005) *Geophys. J. Intl., 160,* 973.
Mayne N. J. et al. (2014) *Astron. Astrophys., 561,* A1.
Mayor M. and Queloz D. (1995) *Nature, 378,* 355.
Militzer B. (2006) *Phys. Rev. Lett., 97(17),* 175501.
Militzer B. (2013) *Phys. Rev. B, 87(1),* 014202.
Militzer B. and Hubbard W. B. (2013) *Astrophys. J., 774,* 148.
Militzer B. et al. (2008) *Astrophys. J. Lett., 688,* L45.
Miller N. and Fortney J. J. (2011) *Astrophys. J. Lett., 736,* L29.
Mirouh G. M. et al. (2012) *Astrophys. J., 750,* 61.
Mizuno H. (1980) *Progr. Theor. Phys., 64,* 544.
Mollière P. and Mordasini C. (2012) *Astron. Astrophys., 547,* 105.
Morales M. A. et al. (2009) *Proc. Natl. Acad. Sci., 106,* 1324.
Moresi L. and Solomatov V. (1998) *Geophys. J. Intl., 133,* 669.
Mostovych A. N. et al. (2000) *Phys. Rev. Lett., 85,* 3870.
Muirhead P. S. et al. (2012) *Astrophys. J., 747,* 144.
Murray-Clay R. A. et al. (2009) *Astrophys. J., 693,* 23.
Nettelmann N. et al. (2008) *Astron. Astrophys., 523,* A26.
Nettelmann N. et al. (2011) *Astrophys. J., 733,* 2.
Nettelmann N. et al. (2012) *Astrophys. J., 750,* 52.
Nettelmann N. et al. (2013a) *Planet. Space Sci., 77,* 143.
Nettelmann N. et al. (2013b) *Icarus, 225,* 548.
O'Neill C. and Lenardic A. (2007) *Geophys. Res. Lett., 34,* L19204.
Owen J. E. and Jackson A. P. (2012) *Mon. Not. R. Astron. Soc.,*
    425, 2931.
Owen J. E. and Wu Y. (2013) *Astrophys. J., 775,* 105.
Parmentier E. M. et al. (1994) *Geophys. J. Intl., 116,* 241.
Perez-Becker D. and Chiang E. (2013) *Mon. Not. R. Astron. Soc.,*
    433(3), 2294.
Perna R. et al. (2010a) *Astrophys. J., 719,* 1421.
Perna R. et al. (2010b) *Astrophys. J., 724,* 313.
Podolak M. et al. (1991) In *Uranus* (J. T. Bergstrahl et al., eds.),
    pp. 29–61. Univ. of Arizona, Tucson.
Pollack J. B. et al. (1977) *Icarus, 30,* 111.
Rauscher E. and Menou K. (2012) *Astrophys. J., 750,* 96.
Rauscher E. and Menou K. (2013) *Astrophys. J., 764,* 103.
Redmer R. et al. (2011) *Icarus, 211,* 798.
Reiners A. and Christensen U. (2010) *Astron. Astrophys., 522,* 13.
Reipurth B. et al., eds. (2007) *Protostars and Planets V.* Univ. of
    Arizona, Tucson. 951 pp.
Ricard Y. et al. (2009) *Earth Planet. Sci. Lett., 284,* 144.
Rogers L. A. and Seager S. (2010) *Astrophys. J., 716,* 1208.
Rosen P. A. et al. (1991) *Icarus, 93,* 25.
Rosenblum E. et al. (2011) *Astrophys. J., 731,* 66.

Salpeter E. E. (1973) *Astrophys. J. Lett., 181,* L83.
Saumon D. et al. (1995) *Astrophys. J. Suppl., 99,* 713.
Schmider F. X. et al. (2007) *Astron. Astrophys., 474,* 1073.
Schubert G. et al. (2001) *Mantle Convection in the Earth and Planets.*
    Cambridge Univ., Cambridge. 940 pp.
Seager S. and Sasselov D. D. (1998) *Astrophys. J. Lett., 502,* L157.
Seager S. et al. (2005) *Astrophys. J., 632,* 1122.
Seager S. et al. (2007) *Astrophys. J., 669,* 1279.
Showman A. P. and Guillot T. (2002) *Astron. Astrophys., 385,* 166.
Showman A. P. et al. (2011) In *Exoplanets* (S. Seager, ed.), pp. 471–
    516. Univ. of Arizona, Tucson.
Smoluchowski R. (1973) *Astrophys. J. Lett., 185,* L95.
Sotin C. et al. (2007) *Icarus, 191,* 337.
Sotin C. et al. (2011) In *Exoplanets* (S. Seager, ed.), pp. 375–395. Univ.
    of Arizona, Tucson.
Soubiran F. et al. (2012) *Phys. Rev. B, 86(11),* 115102.
Soubiran F. et al. (2013) *Phys. Rev. B, 87(16),* 165114.
Stamenković V. et al. (2011) *Icarus, 216,* 572.
Stevenson D. J. (1975) *Phys. Rev. B, 12,* 3999.
Stevenson D. J. (1982) *Annu. Rev. Earth Planet. Sci., 10,* 257.
Stevenson D. J. (1985) *Icarus, 62,* 4.
Stevenson D. J. (2003) *Compt. Rend. Geosci., 335,* 99.
Stevenson D. J. and Salpeter E. E. (1977a) *Astrophys. J. Suppl.,*
    35, 239.
Stevenson D. J. and Salpeter E. E. (1977b) *Astrophys. J. Suppl.,*
    35, 221.
Stevenson D. J. et al. (1983) *Icarus, 54,* 466.
Stofan E. R. et al. (1995) *J. Geophys. Res., 100,* 23317.
Stofan E. R. et al. (2007) *Nature, 445,* 61.
Sudarsky D. et al. (2003) *Astrophys. J., 588,* 1121.
Sun P. et al. (2007) *Proc. Natl. Acad. Sci., 104(22),* 9151–9155.
Swain M. R. et al. (2008) *Nature, 452,* 329.
Tackley P. J. et al. (2013) *Icarus, 225,* 50.
Thompson S. L. (1990) *ANEOS — Analytic Equations of State for*
    Shock Physics Codes, Sandia Natl. Lab. Doc. SAND89-2951,
    Sandia National Laboratories, Albuquerque, New Mexico.
Thompson S. and Lauson H. (1972) *Improvements in the Chart-D*
    Radiation-Hydrodynamic Code III: Revised Analytical Equation
    of State. Technical Report SC-RR-61-0714, Sandia National
    Laboratories, Albuquerque, New Mexico.
Tian F. et al. (2005) *Astrophys. J., 621,* 1049.
Valencia D. et al. (2006) *Icarus, 181,* 545.
Valencia D. et al. (2007) *Astrophys. J. Lett., 670,* L45.
Valencia D. et al. (2009) *Astrophys. Space Sci., 322,* 135.
Valencia D. et al. (2010) *Astron. Astrophys., 516,* A20.
Vallis G. K. (2006) *Atmospheric and Oceanic Fluid Dynamics:*
    Fundamentals and Large-Scale Circulation. Cambridge Univ.,
    Cambridge. 745 pp.
Vidal-Madjar A. et al. (2003) *Nature, 422,* 143.
Vidal-Madjar A. et al. (2004) *Astrophys. J. Lett., 604,* L69.
von Zahn U. et al. (1998) *J. Geophys. Res., 103,* 22815.
Wahl S. M. et al. (2013) *Astrophys. J., 773,* 95.
Watson A. J. et al. (1981) *Icarus, 48,* 150.
Weiss L. M. et al. (2013) *Astrophys. J., 768,* 14.
Wilson H. and Militzer B. (2010) *Phys. Rev. Lett., 104(12),* 121101.
Wilson H. and Militzer B. (2012a) *Phys. Rev. Lett., 108(11),* 111101.
Wilson H. and Militzer B. (2012b) *Astrophys. J., 745,* 54.
Wilson H. F. et al. (2013) *Phys. Rev. Lett., 110,* 15, 151102.
Winisdoerffer C. and Chabrier G. (2005) *Phys. Rev. E, 71(2),* 026402.
Wolszczan A. and Frail D. A. (1992) *Nature, 355,* 145.
Wood T. S. et al. (2013) *Astrophys. J., 768,* 157.
Wu Y. and Lithwick Y. (2013) *Astrophys. J., 763,* 13.
Yelle R. et al. (2008) *Space Sci. Rev., 139,* 437.
Yelle R. V. (2004) *Icarus, 170,* 167.
Yelle R. V. (2006) *Icarus, 183,* 508.
Youdin A. N. and Mitchell J. L. (2010) *Astrophys. J., 721,* 1113.
Zapolsky H. S. and Salpeter E. E. (1969) *Astrophys. J., 158,* 809.
Zaussinger F. and Spruit H. (2013) *Astron. Astrophys., 554,* A119.
Zharkov V. N. and Trubitsyn V. P. (1978) *Physics of Planetary*
    Interiors. Astronomy and Astrophysics Series, Pachart, Tucson.

Davies M. B., Adams F. C., Armitage P., Chambers J., Ford E., Morbidelli A., Raymond S. N., and Veras D. (2014) The long-term dynamical evolution of planetary systems. In *Protostars and Planets VI* (H. Beuther et al., eds.), pp. 787–808. Univ. of Arizona, Tucson, DOI: 10.2458/azu_uapress_9780816531240-ch034.

The Long-Term Dynamical Evolution of Planetary Systems

Melvyn B. Davies
Lund University

Fred C. Adams
University of Michigan

Philip Armitage
University of Colorado, Boulder

John Chambers
Carnegie Institution of Washington

Eric Ford
The Pennsylvania State University and University of Florida

Alessandro Morbidelli
University of Nice

Sean N. Raymond
University of Bordeaux

Dimitri Veras
University of Cambridge

This chapter concerns the long-term dynamical evolution of planetary systems from both theoretical and observational perspectives. We begin by discussing the planet-planet interactions that take place within our own solar system. We then describe such interactions in more tightly packed planetary systems. As planet-planet interactions build up, some systems become dynamically unstable, leading to strong encounters and ultimately either ejections or collisions of planets. After discussing the basic physical processes involved, we consider how these interactions apply to extrasolar planetary systems and explore the constraints provided by observed systems. The presence of a residual planetesimal disk can lead to planetary migration and hence cause instabilities induced by resonance crossing; however, such disks can also stabilize planetary systems. The crowded birth environment of a planetary system can have a significant impact: Close encounters and binary companions can act to destabilize systems, or sculpt their properties. In the case of binaries, the Kozai mechanism can place planets on extremely eccentric orbits that may later circularize to produce hot Jupiters.

1. INTRODUCTION

Currently observed planetary systems have typically evolved between the time when the last gas in the protoplanetary disk was dispersed and today. The clearest evidence for this assertion comes from the distribution of Kuiper belt objects in the outer solar system, and from the eccentricities of massive extrasolar planets, but many other observed properties of planetary systems may also plausibly be the consequence of dynamical evolution. This chapter summarizes the different types of gravitational interactions that lead to long-term evolution of planetary systems, and reviews the application of theoretical models to observations of the solar system and extrasolar planetary systems.

Planetary systems evolve due to the exchange of angular momentum and/or energy among multiple planets, between planets and disks of numerous small bodies ("planetesimals"), between planets and other stars, and via tides with the stellar host. A diverse array of dynamical evolution ensues. In the simplest cases, such as a well-separated two-planet system, the mutual perturbations lead only to periodic oscillations in the planets' eccentricity and inclination. Of greater interest are more complex multiple planet systems where the dynamics are chaotic. In different circumstances the chaos can lead to unpredictable (but bounded) excursions in planetary orbits, to large increases in eccentricity as the system explores the full region of phase space allowed by conservation laws, or to close approaches between planets resulting in collisions or ejections. Qualitative changes to the architecture of planetary systems can likewise be caused by dynamical interactions in binary systems, by stellar encounters in clusters, or by changes to planetary orbits due to interactions with planetesimal disks.

Theoretically, there has been substantial progress since the last Protostars and Planets meeting in understanding the dynamics that can reshape planetary systems. Observational progress has been yet more dramatic. Radial velocity surveys and the Kepler mission have provided extensive catalogs of single- and multiple-planet systems that can be used to constrain the prior dynamical evolution of planetary systems (see the chapter by Fischer et al. in this volume for more details). Routine measurements of the Rossiter-McLaughlin effect for transiting extrasolar planets have shown that a significant fraction of hot Jupiters have orbits that are misaligned with respect to the stellar rotation axis and have prompted new models for how hot Jupiters form. Despite this wealth of data, the relative importance of different dynamical processes in producing what we see remains unclear, and we will discuss in this review what new data is needed to break degeneracies in the predictions of theoretical models. Also uncertain is which observed properties of planetary systems reflect dynamical evolution taking place subsequent to the dispersal of the gas disk (the subject of this chapter), and which involve the coupled dynamics and *hydrodynamics* of planets, planetesimals, and gas within the protoplanetary disk. The chapter by Baruteau et al. in this volume reviews this earlier phase of evolution.

We begin this chapter by considering the long-term stability of the solar system. The solar system is chaotic, but our four giant planets are fundamentally stable, and there is only a small probability that the terrestrial planets will experience instability during the remaining main-sequence lifetime of the Sun. We then compare the current solar system to more tightly packed planetary systems, which are hypothesized progenitors to both the solar system and extrasolar planetary systems. We discuss the conditions, timescales, and outcomes of the dynamical instabilities that can be present in such systems, and compare theoretical models to the observed population of extrasolar planets. We then review how interactions between planets and residual planetesimal disks can lead to planetary migration, which depending on the circumstances can either stabilize or destabilize a planetary system. Finally, we discuss the outcome of dynamical interactions between planetary systems and other stars, whether bound in binaries or interlopers that

perturb planets around stars in stellar clusters. Dynamical evolution driven by inclined stellar mass (and possibly substellar or planetary mass) companions provides a route to the formation of hot Jupiters whose orbits are misaligned to the stellar equator, and we review the status of models for this process (often called the Kozai mechanism). We close with a summary of the key points of this chapter.

2. THE SOLAR SYSTEM TODAY

A quick glance at our system, with the planets moving on quasicircular and almost co-planar orbits, well separated from each other, suggests the idea of a perfect clockwork system, where the orbital frequencies tick the time with unsurpassable precision. But is it really so? In reality, due to their mutual perturbations, the orbits of the planets must vary over time.

To a first approximation, these variations can be described by a secular theory developed by Lagrange and Laplace (see *Murray and Dermott*, 1999) in which the orbital elements that describe a fixed Keplerian orbit change slowly over time. The variations can be found using Hamilton's equations, expanding the Hamiltonian in a power series in terms of the eccentricity e and inclination i of each planet, and neglecting high-frequency terms that depend on the mean longitudes. Only the lowest-order terms are retained since e and i are small for the planetary orbits. The variations for a system of planets j (ranging from 1 to N) can then be expressed as

$$e_j \sin \varpi_j = \sum_{k=1}^{N} e_{kj} \sin\left(g_k t + \beta_k\right)$$

$$e_j \cos \varpi_j = \sum_{k=1}^{N} e_{kj} \cos\left(g_k t + \beta_k\right) \tag{1}$$

with similar expressions for i. Here ϖ is the longitude of perihelion, and the quantities e_{kj}, g_k, and β_k are determined by the planet's masses and initial orbits.

In the Lagrange-Laplace theory, the orbits' semimajor axes a remain constant, while e and i undergo oscillations with periods of hundreds of thousands of years. The orbits change, but the variations are bounded, and there are no long-term trends. Even at peak values, the eccentricities are small enough that the orbits do not come close to intersecting. Therefore, the Lagrange-Laplace theory concludes that the solar system is stable.

The reality, however, is not so simple. The Lagrange-Laplace theory has several drawbacks that limit its usefulness in real planetary systems. It is restricted to small values of e and i; theories based on higher-order expansions exist, but they describe a much more complex time-dependence of eccentricities and inclinations, whose Fourier expansions involve harmonics with argument νt where $\nu = \Sigma_{k=1}^{N} n_k g_k + m_k s_k$ and n_k, m_k are integers, and g and s are secular frequencies associated with e and i, respectively. The coefficients of these harmonics are roughly inversely proportional to v, so that the Fourier Series representation breaks down when v ~ 0, a situation called *secular resonance*.

Moreover, the Lagrange-Laplace theory ignores the effects of mean-motion resonances or near resonances between the orbital periods of the planets. The existence of mean-motion resonances can fundamentally change the dynamics of a planetary system and alter its stability in ways not predicted by the Lagrange-Laplace theory. In particular, the terms dependent on the orbital frequencies, ignored in the Lagrange-Laplace theory, become important when the ratio of two orbital periods is close to the ratio of two integers. This situation arises whenever the critical argument ϕ varies slowly over time, where

$$\phi = k_1\lambda_i + k_2\lambda_j + k_3\varpi_i + k_4\varpi_j + k_5\Omega_i + k_6\Omega_j \qquad (2)$$

for planets i and j, where λ is the mean longitude, Ω is the longitude of the ascending node, and k_{1-6} are integers. The k_{3-6} terms are included because the orbits will precess in general, so a precise resonance is slightly displaced from the case where k_1/k_2 is a ratio of integers.

To leading order, the evolution at a resonance can be described quite well using the equation of motion for a pendulum (*Murray and Dermott*, 1999)

$$\ddot{\phi} = -\omega^2 \sin\phi \qquad (3)$$

At the center of the resonance, ϕ and $\dot{\phi}$ are zero and remain fixed. When ϕ is slightly nonzero initially, ϕ librates about the equilibrium point with a frequency ω that depends on the amplitude of the librations. The semimajor axes and eccentricities of the bodies involved in the resonance undergo oscillations with the same frequency. There is a maximum libration amplitude for which ϕ approaches $\pm\pi$. At larger separations from the equilibrium point, ϕ circulates instead and the system is no longer bound in resonance.

Whereas a single resonance exhibits a well-behaved, pendulum-like motion as described above, when multiple resonances overlap the behavior near the boundary of each resonance becomes erratic, which leads to chaotic evolution. Because the frequencies of the angles $\varpi_{i,j}$ and $\Omega_{i,j}$ are small relative to the orbital frequencies (i.e., the frequencies of $\lambda_{i,j}$), in general resonances with the same k_1, k_2 but different values of $k_{3,...6}$ overlap with each other. Thus, quite generically, a region of chaotic motion can be found associated with each mean-motion resonance.

There are no mean-motion resonances between pairs of solar system planets. Jupiter and Saturn, however, are close to the 2:5 resonance, and Uranus and Neptune are close to the 1:2 resonance. In both cases, the resonant angle ϕ is in circulation and does not exhibit chaotic motion. In general, systems can support N-body resonances, which involve integer combinations of the orbital frequencies of N planets, with N > 2. The forest of possible resonances becomes rapidly dense, and increases with the number N of planets involved. As a result, analytic models become inappropriate for describing precisely the dynamical evolution of a system with many planets.

Fortunately, modern computers allow the long-term evolution of the solar system to be studied numerically using N-body integrations. These calculations can include all relevant gravitational interactions and avoid the approximations inherent in analytic theories. One drawback is that N-body integrations can never prove the stability of a system, only its stability for the finite length of an integration.

N-body integrations can be used to distinguish between regular and chaotic regions, and quantify the strength of chaos, by calculating the system's Lyapunov exponent Γ, given by

$$\Gamma \lim_{t\to\infty} \frac{\ln\left[d(t)/d(0)\right]}{t} \qquad (4)$$

where d is the separation between two initially neighboring orbits. Regular orbits diverge from one another at a rate that is a power of time. Chaotic orbits diverge exponentially over long timespans, although they can be "sticky," mimicking regular motion for extended time intervals. If the solar system is chaotic, even a tiny uncertainty in the current orbits of the planets will make it impossible to predict their future evolution indefinitely.

One of the first indications that the solar system is chaotic came from an 845-million-year (m.y.) integration of the orbits of the outer planets plus Pluto (*Sussman and Wisdom*, 1988). Pluto was still considered a planet at the time due to its grossly overestimated mass. This work showed that Pluto's orbit is chaotic with a Lyapunov timescale ($T_L = 1/\Gamma$) of 20 m.y. Using a 200-m.y. integration of the secular equations of motion, *Laskar* (1989) showed that the eight major planets are chaotic with a Lyapunov time of 5 m.y. This result was later confirmed using full N-body integrations (*Sussman and Wisdom*, 1992). The source of the chaos is due to the existence of two secular resonances, with frequencies $v_1 = 2(g_4-g_3)-(s_4-s_3)$ and $v_2 = (g_1-g_5)-(s_1-s_2)$, i.e., frequencies not appearing in the Lagrange-Laplace theory (*Laskar*, 1990).

The four giant planets by themselves may be chaotic (*Sussman and Wisdom*, 1992), due to a three-body resonance involving Jupiter, Saturn, and Uranus (*Murray and Dermott*, 1999). Assuming that the evolution can be described as a random diffusion through the chaotic phase space within this resonance, *Murray and Dermott* (1999) estimated that it will take ~10^{18} yr for Uranus' orbit to cross those of its neighbors. However, careful examination suggests that chaotic and regular solutions both exist within the current range of uncertainty for the orbits of the outer planets (*Guzzo*, 2005; *Hayes*, 2008), so the lifetime of this subsystem could be longer, and even infinite.

Numerical integrations can also be used to assess the long-term stability of the planetary system, with the caveat that the precise evolution can never be known, so that simulations provide only a statistical measure of the likely behavior. Numerical simulations confirm the expectation that the orbits of the giant planets will not change significantly over the lifetime of the Sun (*Batygin and Laughlin*, 2008). Nonetheless, there remains a remote possibility that the orbits of the inner planets will become crossing on a timescale of a few billion years (*Laskar,* 1994, 2008; *Batygin and Laughlin,* 2008; *Laskar and Gastineau,* 2009).

In principle, if the terrestrial planets were alone in the solar system, they would be stable for all time. (In practice, of course, the solar system architecture will change on a timescale of only ~7 gigayears (G.y.) due to solar evolution and mass loss.) In fact, even if the orbits of the inner planets are free to diffuse through phase space, their evolution would still be constrained by their total energy and angular momentum.

A useful ingredient in orbital evolution is the angular momentum deficit (AMD), given by

$$\text{AMD} \equiv \sum_k \Lambda_k \left(1 - \cos i_k \sqrt{1 - e_k^2}\right) \qquad (5)$$

where $\Lambda_k = M_k M_* / (M_k + M_*) \sqrt{G(M_* + M_k) a_k}$ is the angular momentum of planet k with mass M_k, semimajor axis a_k, eccentricity e_k, and inclination i_k to the invariable plane, and M_* is the mass of the host star.

In the absence of mean-motion resonances, to a good approximation, a remains constant for each planet, so AMD is also constant [e.g., *Laskar* (1997), although this result dates to Laplace], excursions in e and i are constrained by conservation of AMD, and the maximum values attainable by any one planet occur when all the others have e = i = 0. Mercury's low mass means it can acquire a more eccentric and inclined orbit than the other planets. However, even when Mercury absorbs all available AMD, its orbit does not cross Venus. However, when the giant planets are taken into account, the exchange of a small amount of angular momentum between the outer planets and the inner planets can give the latter enough AMD to develop mutual crossing orbits. As a result, long-term stability is not guaranteed.

The possible instability of the terrestrial planets on a timescale of a few billion years shows that the solar system has not yet finished evolving dynamically. Its overall structure can still change in the future, even if marginally within the remaining main-sequence lifetime of the Sun.

This potential change in the orbital structure of the planets would not be the first one in the history of the solar system. In fact, several aspects of the current orbital architecture of the solar system suggest that the planetary orbits changed quite drastically after the epoch of planet formation and the disappearance of the protoplanetary disk of gas. For instance, as described above, the giant planets are not in mean-motion resonance with each other. However, their early interaction with the disk of gas in which they formed could have driven them into mutual mean-motion resonances such as the 1:2, 2:3, or 3:4, where the orbital separations are much narrower than the current ones (*Lee and Peale,* 2002; *Kley et al.,* 2005; *Morbidelli et al.,* 2007). We do indeed observe many resonant configurations among extrasolar planetary systems and in numerical simulations of planetary migration. Also, the current eccentricities and inclinations of the giant planets of the solar system, even if smaller than those of many extrasolar planets, are nevertheless nonzero. In a disk of gas, without any mean-

motion-resonant interaction, the damping effects would have annihilated the eccentricities and inclinations of the giant planets in a few hundred orbits (*Kley and Dirksen,* 2006; *Cresswell et al.,* 2007). So some mechanism must have extracted the giant planets from any original mean-motion resonances and placed them onto their current, partially eccentric and inclined orbits after gas removal.

The populations of small bodies of the solar system also attest to significant changes in the aftermath of gas removal and possibly several 100 m.y. later. The existence and properties of these populations thus provide important constraints on our dynamical history (e.g., see section 6). There are three main reservoirs of small bodies: (1) the asteroid belt between Mars and Jupiter, (2) the Kuiper belt immediately beyond Neptune, and (3) the Oort cloud at the outskirts of the solar system. At the present epoch, both the asteroid belt and the Kuiper belt contain only a small fraction of the mass (0.1% or less) that is thought to exist in these regions when the large objects that we observe today formed (*Kenyon and Bromley,* 2004). As a result, the vast majority of the primordial objects (by number) have been dynamically removed. Some of these bodies were incorporated into the forming planets, some were scattered to other locations within the solar system, and some were ejected. The remaining objects (those that make up the current populations of the asteroid and Kuiper belts) are dynamically excited, in the sense that their eccentricities and inclinations cover the entire range of values allowed by long-term stability constraints; for instance, the inclinations can be as large as 40°, much larger than the nearly co-planar orbits of the planets themselves. This evidence suggests that some dynamical mechanism removed more than 99.9% of the objects and left the survivors on excited orbits, much different from their original circular and co-planar ones. Another constraint is provided by the absence of a correlation between the size of the objects and their orbital excitation. Presumably, the mechanism that altered the orbits of these small bodies acted *after* the removal of the nebular gas; otherwise, gas drag, which is notoriously a size-dependent process, would have imprinted such a correlation. Similarly, the Oort cloud (the source of long-period comets) contains hundreds of billions of kilometer-sized objects, and most trace through orbits with high eccentricities and inclinations (*Wiegert and Tremaine,* 1999; *Kaib and Quinn,* 2009). Since the Oort cloud extends far beyond the expected size of circumstellar disks, the population is thought to have been scattered to such large distances. In the presence of gas, however, these small objects could not have been scattered out by the giant planets, because gas drag would have circularized their orbits just beyond the giant planets (*Brasser et al.,* 2007). Thus, presumably the formation of the Oort cloud postdates gas removal; this timing is consistent with considerations of the solar birth environment (section 6).

Finally, there is evidence for a surge in the impact rates and/or impact velocities on the bodies of the inner solar system, including the terrestrial planets, the Moon, and the asteroid Vesta (*Tera et al.,* 1974; *Ryder,* 2002; *Kring and*

Cohen, 2002; *Marchi et al.,* 2012a,b, 2013). This surge seems to have occurred during a time interval ranging from 4.1 to 3.8 G.y. ago, i.e., starting about 400 m.y. after planet formation. This event is often referred to as the "terminal lunar cataclysm" or the "late heavy bombardment." If such a cataclysm really happened [its existence is still debated today; see, e.g., *Hartmann et al.* (2000)], it suggests that the changes in the structure of the solar system described above did not happen immediately after the removal of gas from the early solar nebula, but only after a significant delay of several 100 m.y.

All this discussion provides hints that the mechanisms of orbital instability discussed in the next sections of this chapter in the framework of extrasolar planet evolution were probably not foreign to the solar system. Specifically, a scenario of possible evolution of the solar system that explains all these aspects will be presented in section 5.

3. INSTABILITIES IN TIGHTLY PACKED SYSTEMS

We will see in this section that the timescale for a planetary system to become unstable is a very sensitive function of planetary separations. Thus the separation of planets within a system is an important quantity. Unfortunately, the typical separation of planets in newly formed planetary systems is unknown observationally and theoretical predictions are also unclear. For terrestrial planets whose final assembly occurs after gas-disk dispersal, the results of *Kokubo and Ida* (1998) suggest that separations of ≃10 mutual Hill radii, $R_{Hill,m}$ are typical. Where

$$R_{Hill,m} \equiv \left(\frac{M_k + M_{k+1}}{3M_*} \right)^{1/3} \frac{a_k + a_{k+1}}{2} \qquad (6)$$

(This is one definition of the mutual Hill radius; there are others in use in the literature.) Here M_* is the stellar mass, M_k is planetary mass, and a_k is semimajor axis. However, terrestrial planets that form more rapidly (i.e., before gas-disk dispersal) may be more tightly packed, as may also be the case for giant planets that necessarily form in a dissipative environment. Giant planets (or their cores) can migrate due to either planetesimal (*Levison et al.,* 2010) or gas-disk interactions (*Kley and Nelson,* 2012). Hydrodynamic simulations of multiple planets interacting with each other and with a surrounding gas disk (*Moeckel et al.,* 2008; *Marzari et al.,* 2010; *Moeckel and Armitage,* 2012; *Lega et al.,* 2013) show that resonant, tightly packed, or well-separated systems can form, but the probabilities for these channels cannot be predicted from first principles. Constraints on the dynamics must currently be derived from comparison of the predicted end states with observed systems.

For giant planets, two types of instability provide dynamical paths that may explain the origin of hot Jupiters and eccentric giant planets. If the planets start on circular, co-planar orbits, *planet-planet* scattering results in a com-

bination of physical collisions, ejection, and generation of eccentricity. If the system instead forms with planets on widely separated eccentric or inclined orbits, *secular chaos* can result in the diffusive evolution of eccentricity to high values even in the absence of close encounters.

3.1. Conditions and Timescales of Instability

The condition for stability can be analytically derived for two-planet systems. *Gladman* (1993), drawing on results from *Marchal and Bozis* (1982) and others, showed that two-planet systems with initially circular and co-planar orbits are Hill stable for separations $\Delta \equiv (a_2-a_1)/R_{Hill,m} \gtrsim 2\sqrt{3}$. Hill stability implies that two planets with at least this separation are analytically guaranteed to never experience a close approach. Numerically, it is found that some systems can be stable at smaller separations within mean-motion resonances, and that the stronger condition of Lagrange stability — which requires that both planets remain bound and ordered for all time — requires only modestly greater spacing than Hill stability (*Barnes and Greenberg,* 2006b; *Veras and Mustill,* 2013).

There is no analytic criterion for the absolute stability of systems with N ≥ 3 planets. The degree of instability can be characterized numerically by evaluating the timescale for orbit crossing, or for the first close encounters between planets, to occur (in practice, different reasonable definitions of "instability" are nearly equivalent). Figure 1 illustrates the median instability timescale as a function of the separation in units of $R_{Hill,m}$ for planetary systems of three equal-mass planets with varying mass ratios $\mu = M_p/M_*$. The behavior is simplest for low-mass planets. In this regime, *Chambers et al.* (1996) found that the time before the first close encounters could be approximated as $\log(t_{close}) = b\Delta + c$, with b and c being constants. The time to a first close encounter is found to vary enormously over a small range of initial planetary separations. Systems with N = 5 were less stable than the N = 3 case, but there was little further decrease in the stability time with further increase in the planet number to N = 10 or N = 20. The scaling of the instability time with separation was found to be mass dependent if measured in units of $R_{Hill,m}$ (with a steeper dependence for larger μ), but approximately *independent* if measured in units $\propto M_p^{1/4}$. *Smith and Lissauer* (2009) extended these results with longer integrations. They found that a single slope provided a good fit to their data for Earth-mass planets for $3.5 \leq \Delta \leq 8$, but that there was a sharp increase in b for $\Delta > 8$. Sufficiently widely separated systems thus rapidly become stable for practical purposes. As in the two-planet case, resonant systems can evade the nonresonant stability scalings, but only for a limited number of planets (*Matsumoto et al.,* 2012).

Similar numerical experiments for more massive planets with $\mu \sim 10^{-3}$ were conducted by *Marzari and Weidenschilling* (2002) and by *Chatterjee et al.* (2008). At these mass ratios, the plot of $t_{close}(\Delta)$ exhibits a great deal of structure associated with the 2:1 (and to a lesser extent 3:1) mean-motion resonance (Fig. 1). A simple exponential fit is no

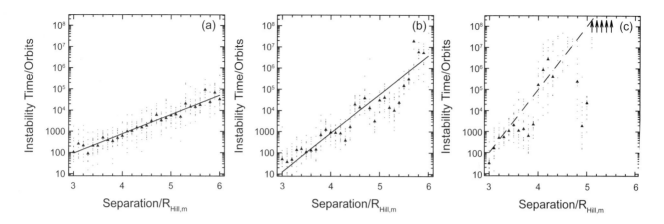

Fig. 1. The median (triangles) instability timescale for initially circular, coplanar three-planet systems, as a function of the separation in units of mutual Hill radii (after *Chambers et al.,* 1996; *Marzari and Weidenschilling,* 2002). The dots show the instability timescale obtained for individual realizations of a planetary system. The instability timescale is defined as the time (in units of the initial orbital period of the inner planet) until the first pair of planets approach within one Hill radius. The panels show mass ratios $\mu = M_p/M_*$ of **(a)** 10^{-6}, **(b)** 3×10^{-5}, and **(c)** 10^{-3}.

longer a good approximation for instability timescales of 10^6–10^8 yr (at a few astronomical units), which might be highly relevant for gas giants emerging from the gas disk. *Chatterjee et al.* (2008) quote an improved fitting formula for the instability time (valid away from resonances), but there is an unavoidable dependence on details of the system architecture such as the radial ordering of the masses in systems with unequal-mass planets (*Raymond et al.,* 2010).

For a single planet, resonance overlap leads to chaos and instability of test particle orbits out to a distance that scales with planet mass as $\Delta a \propto \mu^{2/7}$ for low eccentricities (*Wisdom,* 1980) and $\Delta a \propto \mu^{1/5}$ above a (small) critical eccentricity (*Mustill and Wyatt,* 2012). Resonance overlap is similarly thought to underlie the numerical results for the stability time of multiple-planet systems (*Lissauer,* 1995; *Morbidelli and Froeschlé,* 1996), although the details are not fully understood. *Quillen* (2011) considered the criterion for the overlap of three-body resonances, finding that the density of these resonances was (to within an order of magnitude) sufficient to explain the origin of chaos and instability at the separations probed numerically by *Smith and Lissauer* (2009).

Once stars depart the main sequence, the increase in μ as the star loses mass can destabilize either multiple-planet systems (*Duncan and Lissauer,* 1998; *Debes and Sigurdsson,* 2002; *Veras et al.,* 2013; *Voyatzis et al.,* 2013), or asteroid belts whose members are driven to encounter mean-motion resonances with a single massive planet (*Debes et al.,* 2012). These "late" instabilities may explain the origin of metal-polluted white dwarfs, and white dwarfs with debris disks.

3.2. Outcome of Instability

The outcome of instability in tightly packed planetary systems is a combination of ejections, physical collisions between planets, and planet-star close approaches that may lead to tidal dissipation or direct collision with the star. In more widely spaced systems the same outcomes can occur, but chaotic motion can also persist indefinitely without dramatic dynamical consequences. In the inner solar system, for example, the Lyapunov timescale is very short (≈ 5 m.y.), but the only known pathway to a planetary collision — via the entry of Mercury into a secular resonance with Jupiter — has a probability of only $P \approx 10^{-2}$ within 5 G.y. (*Laskar,* 1989; *Batygin and Laughlin,* 2008; *Laskar and Gastineau,* 2009).

During a close encounter between two planets, scattering is statistically favored over collisions when the escape speed from the planets' surfaces is larger than the escape speed from the planetary system (*Goldreich et al.,* 2004). This can be quantified by the Safronov number Θ

$$\Theta^2 = \left(\frac{M_p}{M_\star} \right) \left(\frac{R_p}{a_p} \right)^{-1} \tag{7}$$

where M_p, R_p, and a_p represent the planet's mass, radius, and orbital distance, and $M_\star$ is the stellar mass (see discussion in *Ford and Rasio,* 2008). We expect scattering to be most important among massive, dense planets and in the outer parts of planetary systems. In the solar system, the giant planets' escape speeds are roughly 2–6× larger than the highest value for the terrestrial planets (Earth's). They are also at larger a_p.

Physical collisions lead to modest eccentricities for the merged remnants (*Ford et al.,* 2001). Scattering at orbital radii beyond the snow line ($a \approx 3$ AU), conversely, results in a broad eccentricity distribution consistent with that observed for massive extrasolar planets (*Chatterjee et al.,* 2008; *Jurić and Tremaine,* 2008). Ejection proceeds via scattering on to highly eccentric orbits, and hence a prediction of planet-planet scattering models is the existence of a population of planets around young stars with very large orbital separation (*Veras et al.,* 2009; *Scharf and Menou,* 2009; *Malmberg et al.,* 2011). The frequency of ejections from scattering is probably too small to explain the large abundance of apparently unbound Jupiter-mass objects discovered by microlensing (*Sumi et al.,* 2011), if those are to be free-floating rather than simply on wide but bound orbits (*Veras and Raymond,* 2012).

The outcome of instabilities that occur within the first few million years (or as the disk disperses) may be modified by gravitational torques or mass accretion from the protoplanetary disk, while instabilities toward the outer edges of planetary systems will result in interactions with outer planetesimal belts. *Matsumura et al.* (2010), using N-body integrations coupled to a one-dimensional gas disk model, and *Moeckel and Armitage* (2012), using two-dimensional hydrodynamic disk models, found that realistic transitions between gas-rich and gas-poor dynamics did not preclude the generation of high eccentricities via scattering. Larger numbers of resonant systems were, however, predicted. *Lega et al.* (2013), instead, found that if a planetary system becomes unstable during the gas-disk phase, it is likely to stabilize (after the ejection or the collision of some planets) in resonant low-eccentricity orbits, and avoid further instabilities after the gas is removed. *Raymond et al.* (2010) ran scattering experiments in which planets at larger radii interacted with massive collisionless planetesimal disks. The disks strongly suppressed the final eccentricities of lower-mass outer planetary systems.

3.3. Secular Chaos

Significantly different evolution is possible in multiple-planet systems if one or more planets possess a substantial eccentricity or inclination from the *beginning*. The departure from circular co-planar orbits can be quantified by the AMD defined earlier.

Wu and Lithwick (2011) proposed secular chaos (combined with tidal effects) as the origin of hot Jupiters. They presented a proof of concept numerical integration of a widely separated (but chaotic) planetary system in which most of the AMD initially resided in an eccentric ($e \approx 0.3$) outer gas giant ($M = 1.5 \, M_{Jup}$, $a = 16$ AU). Diffusion of AMD among the three planets in the system eventually led to the innermost planet attaining $e \simeq 1$ without prior close encounters among the planets. They noted several characteristic features of secular chaos as a mechanism for forming hot Jupiters — it works best for low-mass inner planets, requires the presence of multiple additional planets at large radii, and can result in star-planet tidal interactions that occur very late (in principle after a billion years).

The range of planetary systems for which secular chaos yields dynamically interesting outcomes on short enough timescales remains to be quantified. A prerequisite is a sufficiently large AMD in the initial conditions. This might originate from eccentricity excitation of planets by the gas disk [although this is unlikely to occur for low planet masses

(e.g., *Papaloizou et al.,* 2001; *D'Angelo et al.,* 2006; *Dunhill et al.,* 2013)], from external perturbations such as flybys (*Malmberg et al.,* 2011; *Boley et al.,* 2012), or from a prior epoch of scattering among more tightly packed planets.

4. PLANET-PLANET SCATTERING CONSTRAINED BY DYNAMICS AND OBSERVED EXOPLANETS

The planet-planet scattering model was developed to explain the existence of hot Jupiters (*Rasio and Ford,* 1996; *Weidenschilling and Marzari,* 1996) and giant planets on very eccentric orbits (*Lin and Ida,* 1997; *Papaloizou and Terquem,* 2001; *Ford et al.,* 2001). Given the great successes of exoplanet searches there now exists a database of observations against which to test the planet-planet scattering model. The relevant observational constraints come from the subset of extrasolar systems containing giant planets. In this section we first review the relevant observational constraints. Next, we show how the dynamics of scattering depend on the parameters of the system. We then show that, with simple assumptions, the scattering model can match the observations.

4.1. Constraints from Giant Exoplanets

The sample of extrasolar planets that can directly constrain models of planet-planet scattering now numbers more than 300. These are giant planets with masses larger than Saturn's and smaller than 13 M_{Jup} (*Wright et al.,* 2011; *Schneider et al.,* 2011) with orbital semimajor axes larger than 0.2 AU to avoid contamination from star-planet tidal circularization.

The giant planets have a broad eccentricity distribution with a median of ~0.22 (*Butler et al.,* 2006; *Udry and Santos,* 2007). There are a number of planets with very eccentric orbits: Roughly 16%/6%/1% of the sample has eccentricities larger than 0.5/0.7/0.9.

More massive planets have more eccentric orbits. Giant exoplanets with minimum masses $M_p > M_{Jup}$ have statistically higher eccentricities [as measured by a Kolmogorov-Smirnov (K-S) test] than planets with $M_p < M_{Jup}$ (*Jones et al.,* 2006; *Ribas and Miralda-Escudé,* 2007; *Ford and Rasio,* 2008; *Wright et al.,* 2009). Excluding hot Jupiters, there is no measured correlation between orbital radius and eccentricity (*Ford and Rasio,* 2008).

Additional constraints can be extracted from multiple-planet systems. For example, the known two-planet systems are observed to cluster just beyond the Hill stability limit (*Barnes and Greenberg,* 2006a; *Raymond et al.,* 2009b). However, it is unclear to what extent detection biases contribute to this clustering. In addition, studies of the long-term dynamics of some well-characterized systems can constrain their secular behavior, and several systems have been found very close to the boundary between apsidal libration and circulation (*Ford et al.,* 2005; *Barnes and Greenberg,* 2006b; *Veras and Ford,* 2009).

Often, there are significant uncertainties associated with the observations (*Ford, 2005*). Orbital eccentricities are especially hard to pin down, and the current sample may be modestly biased toward higher eccentricities (*Shen and Turner, 2008; Zakamska et al., 2011*), although it is clear that with their near-circular orbits the solar system's giant planets are unusual in the context of giant exoplanets.

4.2. Scattering Experiments: Effect of System Parameters

We now explore how the dynamics and outcomes of planet-planet scattering are affected by parameters of the giant planets, in particular their masses and mass ratios. Our goal is to understand what initial conditions are able to match all the observed constraints from section 4.1.

During a close gravitational encounter, the magnitude of the gravitational kick that a planet imparts depends on the planet's escape speed. To be more precise, what is important is the Safronov number Θ, as given in equation (7). At a given orbital distance, more massive planets kick harder, i.e., they impart a stronger change in velocity. Within systems with a fixed number of equal-mass planets, more massive systems evolve more quickly because they require fewer close encounters to give kicks equivalent to reaching zero orbital energy and being liberated from the system. The duration of an instability — the time during which the planets' orbits cross — is thus linked to the planet masses. For example, in a set of simulations with three planets at a few to 10 AU, systems with $M_p = 3\ M_{Jup}$ planets underwent a median of 81 scattering events during the instability, but this number increased to 175, 542, and 1871 for $M_p = M_{Jup}$, M_{Sat}, and 30 $M_\oplus$, respectively (*Raymond et al., 2010*). The instabilities lasted 10^4–10^6 yr; longer for lower-mass planets.

Thus, the timescale for orbital instabilities in a system starts off longer than the planet-formation timescale and decreases as planets grow their masses. The natural outcome of giant planet formation is a planetary system with multiple giant planets that undergo repeated instabilities on progressively longer timescales, until the instability time exceeds the age of the system.

The orbits of surviving planets in equal-mass systems also depend on the planet mass. Massive planets end up on orbits with larger eccentricities than less massive planets (*Ford et al., 2003; Raymond et al., 2008, 2009a, 2010*). This is a natural consequence of the stronger kicks delivered by more massive planets. However, the inclinations of surviving massive planets are smaller than for less massive planets in terms of both the inclination with respect to the initial orbital plane (presumably corresponding to the stellar equator) and the mutual inclination between the orbits of multiple surviving planets (*Raymond et al., 2010*). This increase in inclination appears to be intimately linked with the number of encounters the planets have experienced in ejecting other planets rather than their strength.

The planetary mass ratios also play a key role in the scattering process. Scattering between equal-mass planets represents the most energy-intensive scenario for ejection. Since it requires far less energy to eject a less massive planet, the recoil that is felt by the surviving planets is much less. Thus, the planets that survive instabilities between unequal-mass planets have smaller eccentricities and inclinations compared with the planets that survive equal-mass scattering (*Ford et al., 2003; Raymond et al., 2008, 2010, 2012*). This does not appear to be overly sensitive to the planets' ordering, i.e., if the lower-mass planets are located on interior or exterior orbits (*Raymond et al., 2010*).

The timing of the instability may also be important for the outcome because giant planets cool and contract on 10^{7-8}-yr timescales (*Spiegel and Burrows, 2012*). Thus, the planets' escape speeds and Θ values increase over time. Instabilities that occur very early in planetary system histories may thus be less efficient at ejecting planets and may include a higher rate of collisions compared with most scattering calculations to date.

4.3. Scattering Experiments: Matching Observations

Let us consider a simple numerical experiment where all planetary systems containing giant planets are assumed to form three giant planets. The masses of these planets follow the observed mass distribution $dN/dM \propto M^{-1.1}$ (*Butler et al., 2006; Udry and Santos, 2007*) and the masses within a system are not correlated. All systems become unstable and undergo planet-planet scattering. With no fine tuning, the outcome of this experiment matches the observed eccentricity distribution (*Raymond et al., 2008*).

The exoplanet eccentricity distribution can be reproduced with a wide range of initial conditions (*Adams and Laughlin, 2003; Moorhead and Adams, 2005; Jurić and Tremaine, 2008; Chatterjee et al., 2008; Ford and Rasio, 2008; Raymond et al., 2010; Beaugé and Nesvorný, 2012*). The same simulations also reproduce the dynamical quantities that can be inferred from multiple-planet systems: the distribution of parameterized distances of two-planet systems from the Hill stability limit (*Raymond et al., 2009b*) and the secular configuration of two-planet systems (*Timpe et al., 2012*).

Certain observations are difficult to reproduce. Some observed systems show no evidence of having undergone an instability (e.g., they contain multiple giant planets on near-circular orbits or in resonances). The scattering model must thus be able to reproduce the observed distributions, including a fraction of the systems remaining on stable orbits. The fraction of systems that are stable is typically 10–30%, where assumptions are made about the distribution of planetary masses and orbits (*Jurić and Tremaine, 2008; Raymond et al., 2010, 2011*). One should also note that migration is required in order to match the observed distribution of semimajor axes, i.e., scattering alone cannot transport Jupiter-like planets on Jupiter-like orbits to semimajor axes less than 1 AU (*Moorhead and Adams, 2005*).

Perhaps the most difficult observation to match is the observed positive correlation between planet mass and eccentricity. In the simulations from the simple experiment mentioned above, lower-mass surviving planets actually have higher eccentricities than more massive ones (*Raymond et al., 2010*). For massive planets to have higher eccentricities they must form in systems with other massive planets, since scattering among equal-mass massive planets produces the highest eccentricities. This is in agreement with planet formation models: When the conditions are ripe for giant planet formation (e.g., massive disk, high metallicity), one would expect all giants to have high masses, and vice versa.

It is relatively simple to construct a population of systems that can reproduce the observed eccentricity distribution as well as the mass-eccentricity correlation and also the mass distribution. The population consists of equal-mass high-mass systems and a diversity of lower-mass systems that can include unequal-mass systems or equal-mass ones. This population also naturally reproduces the observed distribution of two-planet systems that pile up close to the Hill stability limit (as shown in Fig. 2).

We conclude this section with a note of caution. Given the observational uncertainties and the long list of system parameters, to what degree can our understanding of planet-planet scattering constrain preinstability planetary systems? The scattering mechanism is robust and can reproduce observations for a range of initial conditions. Beyond the limitations of mass, energy, and angular momentum conservation, the details of prescattered systems remain largely hidden from our view. The only constraint that appears to require correlated initial conditions to solve is the mass-eccentricity correlation.

Finally, planet-planet scattering does not happen in isolation but rather affects the other components of planetary systems. Giant planet instabilities are generally destructive to both inner rocky planets (*Veras and Armitage, 2005, 2006; Raymond et al., 2011, 2012; Matsumura et al.,*

2013) and to outer planetesimal disks (*Raymond et al., 2011, 2012; Raymond and Armitage, 2013*). Additional constraints on giant planet dynamics may thus be found in a number of places. Recent discoveries by the Kepler mission of Earth- and super-Earth-sized planets relatively close to their host star reveal only a small fraction with giant planets orbiting nearby (*Lissauer et al., 2011; Ciardi et al., 2013*). Multiple systems also appear to be rather flat and stable for planet-planet interactions (*Lissauer et al., 2011; Tremaine and Dong, 2012; Johansen et al., 2012; Fang and Margot, 2012a*).

Characterizing the relationship between small planets orbiting close to the host star and more massive planets at larger separations could provide insights into the effects of planet scattering on the formation of inner, rocky planets. Similarly, the abundance and mass distribution of planets with large orbital separations (*Malmberg et al., 2011; Boley et al., 2012*) or free-floating planets may provide insights into the planets that are scattered to the outskirts of planetary systems or into the Galaxy as free-floating planets (*Veras and Ford, 2009; Veras et al., 2011; Veras and Raymond, 2012; Veras and Mustill, 2013*). In addition, the fact that planet-planet scattering perturbs both the inner and outer parts of planetary systems may introduce a natural correlation between the presence of debris disks and close-in low-mass planets, as well as an anticorrelation between debris disks and eccentric giant planets (*Raymond et al., 2011, 2012; Raymond and Armitage, 2013*).

5. PLANETS AND PLANETESIMAL DISKS

This section considers the evolution of planetary orbits due to interactions with planetesimal disks. These disks are likely to remain intact after the gaseous portion of the disk has gone away, i.e., for system ages greater than 3–10 m.y.,

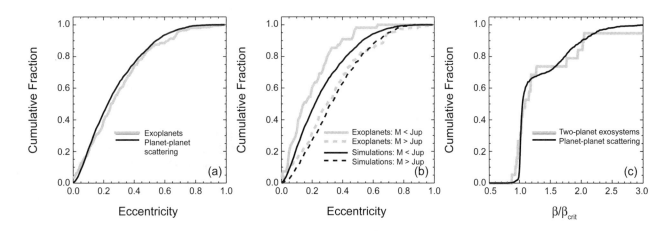

Fig. 2. A comparison between observed properties of giant exoplanets (shown in gray) and planet-planet scattering simulations (in black). **(a)** The eccentricity distribution; **(b)** the eccentricity distribution for low-mass (dashed curves) and high-mass (solid curves) planets; **(c)** the proximity to the Hill stability limit, measured by the quantity β/β_{crit} (*Barnes and Greenberg, 2006b; Raymond et al., 2009b*), where the stability limit occurs at $\beta/\beta_{crit} = 1$ and systems with larger values of β/β_{crit} are stable.

and will continue to evolve in time. Such planetesimal disks are likely to be most effective during the subsequent decade of time, for system ages in the range 10–100 m.y. One of the main effects of a residual planetesimal disk is to drive planetary migration. The subsequent changes in the orbital elements of the planets can cause instabilities (e.g., induced by resonance crossing or by the extraction of the planets from their original resonances), and such action can drive orbital eccentricities to larger values. On the other hand, planetesimal disks can also damp orbital eccentricity and thereby act to stabilize planetary systems.

We begin with a brief overview of the basic properties of planetismal disks. Unfortunately, we cannot observe planetesimal disks directly. Instead, we can piece together an understanding of their properties by considering protoplanetary disks around newly formed stars (see the review of *Williams and Cieza,* 2011) and debris disks (see the reviews of *Zuckerman,* 2001; *Wyatt,* 2008). These latter systems represent the late stages of circumstellar disk evolution, after the gas has been removed, either by photoevaporation or by accretion onto the central star (see the chapter by Alexander et al. in this volume for more details of the photoevaporation process).

Most of the observational information that we have concerning both types of disks is found through their spectral energy distributions (SEDs), especially the radiation emitted at infrared wavelengths. Since these SEDs are primarily sensitive to dust grains, rather than the large planetesimals of interest here, much of our information is indirect. The defining characteristic of debris disks is their fractional luminosity f, essentially the ratio of power emitted at infrared wavelengths to the total power of the star itself. True debris disks are defined to have $f < 10^{-2}$ (*Lagrange et al.,* 2000), whereas systems with larger values of f are considered to be protoplanetary disks. For both types of systems, the observed fluxes can be used to make disk mass estimates. The results show that the disk masses decrease steadily with time. For young systems with ages ~1 m.y., the masses in solid material are typically $M_d \sim 100\ M_\oplus$, albeit with substantial scatter about this value. Note that this amount of solid material is not unlike that of the minimum mass solar/extrasolar nebula (e.g., *Weidenschilling,* 1977; *Kuchner,* 2004; *Chiang and Laughlin,* 2013). By the time these systems reach ages of ~100 m.y., in the inner part of the disk the masses have fallen to only ~0.01 $M_\oplus$ (see Fig. 3 of *Wyatt,* 2008). Note that these mass estimates correspond to the material that is contained in small dust grains; some fraction of the original material is thought to be locked up in larger bodies — the planetesimals of interest here. As a result, the total mass of the planetesimal disk does not necessarily fall as rapidly with time as the SEDs suggest. At large distances from the star, instead, the decay of the mass of the dust population is much slower, suggesting that belts containing tens of Earth masses can exist around billion-year-old stars (*Booth et al.,* 2009).

For a given mass contained in the planetesimal disk, we expect the surface density of solid material to initially fol-low a power-law form so that $\sigma \propto r^{-p}$, where the index p typically falls in the approximate range $1/2 \le p \le 2$ (*Cassen and Moosman,* 1981). The disks are initially expected to have inner edges where the protoplanetary disks are truncated by magnetic fields, where this boundary occurs at r ~ 0.05 AU. Similarly, the outer boundaries are initially set by disk formation considerations. The angular momentum barrier during protostellar collapse implies that disks start with outer radii $r_d \sim 10$–100 AU (*Cassen and Moosman,* 1981; *Adams and Shu,* 1986). Further environmental sculpting of disks (see section 6 below) reinforces this outer boundary. The properties outlined here apply primarily to the starting configurations of the planetesimal disks. As the disks evolve and interact with planets, the surface density must change accordingly.

Given the properties of planetesimal disks, we now consider how they interact with planets and drive planetary migration. The physical mechanism by which planetesimals lead to planet migration can be roughly described as follows (see, e.g., *Levison et al.,* 2007): Within the disk, as the planetesimals gravitationally scatter off the planet, the larger body must recoil and thereby change its trajectory. Since the planet resides within a veritable sea of planetesimals, the smaller bodies approach the larger body from all directions, so that the momentum impulses felt by the planet are randomly oriented. As a result, the orbital elements of the planet will experience a random walk through parameter space. In particular, the semimajor axis of the planetary orbit will undergo a random walk. If the random walk is perfectly symmetric, then the planet has an equal probability of migrating inward or outward; nonetheless, changes will accumulate (proportional to the square root of the number of scattering events). In practice, however, many factors break the symmetry (e.g., the surface density of solids generally decreases with radius) and one direction is preferred.

The details of planetary migration by planetesimal disks depend on the specific properties of the planetary system, including the number of planets, the planet masses, their separations, the radial extent of the disk, and of course the total mass in planetesimals. In spite of these complications, we can identify some general principles that guide the evolution. Some of these are outlined below:

We first note that a disk of planetesimals containing planets often evolves in what can be called a "diffusive regime," where many small scattering events act to make the orbital elements of both the planets and the planetesimals undergo a random walk. As a result, the system has the tendency to spread out. This behavior is often seen in numerical simulations. For a system consisting of a disk of planetesimals and analogs of the four giant planets in our solar system, the scattering events in general lead to Jupiter migrating inward and the remaining three planets migrating outward (*Fernandez and Ip,* 1984; *Hahn and Malhotra,* 1999; *Gomes et al.,* 2004). Similarly, in numerical simulations of systems with two planets embedded in a disk of planetesimals, the inner planet often migrates inward, while the outer planet migrates outward (*Levison et al.*, 2007), whereas a single planet in a planetesimal disk in general migrates inward (*Kirsh et al.*, 2009). This can be understood using some considerations concerning energy conservation. For a system made of a planet and a disk of planetesimals, energy is conserved until planetesimals are lost. Planetesimals can be lost by collisions with the planet, collisions with the Sun, or ejection onto hyperbolic orbit. In the case where the latter is the major loss mechanism, which happens if the planet is massive and the disk is dynamically excited, the removed planetesimals subtract energy from the system of bodies remaining in orbit around the star. Consequently, most of the mass has to move inward and the planet has to follow this trend. Given the expected total masses in planetesimals (see the above discussion), this mode of migration cannot change the semimajor axis of a planetary orbit by a large factor. In particular, this mechanism is unlikely to produce hot Jupiters with periods of about four days and therefore these planets must form by a different mechanism (migration in a gas disk or planet-planet scattering and tidal damping).

Finally, we note that the effects discussed above can be modified in the early phases of evolution by the presence of a gaseous component of the disk (*Capobianco et al.*, 2011). For instance, planetesimals scattered inward by the planet may have their orbits circularized by gas drag, so they cannot be scattered again by the planet. Consequently, the trend is that the planetesimal population loses energy and the planet has to migrate outward.

Planetary migration enforced by the scattering of planetesimals will produce a back reaction on the disk. As outlined above, the disk will tend to spread out, and some planetesimals are lost by being scattered out of the system. Both of these effects reduce the surface density in planetesimals. In addition, the planets can create gaps in the disk of planetesimals. These processes are important because they are potentially observable with the next generation of interferometers [e.g., the Atacama Large Millimeter Array (ALMA)]. We note, however, that submillimeter (and millimeter-wave) observations are primarily sensitive to dust grains rather than planetesimals themselves. The stirring of the planetesimal disk by planets can lead to planetesimal collisions and dust production, thereby allowing these processes to be observed.

Since planetesimal disks are likely to be present in most systems, it is interesting to consider what might have occurred in the early evolution of our own solar system. Here, we follow the description provided by the so-called "Nice model." As described in section 2, it is expected that at the end of the gas-disk phase, the giant planets were in a chain of resonant (or nearly resonant) orbits, with small eccentricities and inclinations and narrow mutual separations. These orbital configurations are those that are found to reach a steady state, and hence are used as the initial condition for the Nice model (*Morbidelli et al.*, 2007). The model also assumes that beyond the orbit of Neptune there was a planetesimal disk, carrying cumulatively ~30–50 $M_\oplus$. A disk in this mass range is necessary to allow the giant

planets to evolve from their original, compact configuration to the orbits they have today. More specifically, the perturbations between the planets and this disk, although weak, accumulated over time and eventually extracted a pair of planets from their resonance. The breaking of the resonance lock makes the planetary system unstable. After leaving resonance, the planets behave as described in sections 3 and 4 and shown in Fig. 3. Their mutual close encounters act to spread out the planetary system and excite the orbital eccentricities and inclinations. In particular, this model suggests that Uranus and Neptune were scattered outward by Jupiter and Saturn, penetrated into the original transneptunian disk, and then dispersed it. The dispersal of the planetesimal disk, in turn, damped the orbital eccentricities of Uranus and Neptune (and to a lesser extent those of Jupiter and Saturn), so that the four giant planets eventually reached orbits analogous to the current ones (*Morbidelli et al.*, 2007; *Batygin and Brown*, 2010; *Batygin et al.*, 2012; *Nesvorný and Morbidelli*, 2012).

In addition to the current orbits of the giant planets, the Nice model accounts for the properties of the small-body populations that, as we described in section 3, suggest that the structure of the solar system experienced dramatic changes after gas dissipation. In fact, the model has been shown to explain the structure of the Kuiper belt (*Levison et al.*, 2008; *Batygin et al.*, 2011), the asteroid belt (*Morbidelli et al.*, 2010), and even the origin of the Oort cloud (*Brasser and Morbidelli*, 2013). Moreover, it has been shown that, with reasonable assumptions concerning the planetesimal disk, the instability of the planetary orbits could occur after hundreds of millions of years of apparent stability (*Tsiganis et al.*, 2005; *Gomes et al.*, 2005; *Levison et al.*, 2011); the subsequent epoch of instability could excite the orbits of the small bodies and thereby produce a shower of projectiles into the inner solar system that quantitatively explains the origin of the late heavy bombardment (*Bottke et al.*, 2012).

At the present time, there are no gross characteristics of the solar system that are at odds with the Nice model. Nevertheless, some aspects of the general picture associated with the model need to be revised or explored. For instance, the cold population (which is a subpopulation of the Kuiper belt characterized by small orbital inclinations) probably formed *in situ* (*Parker and Quanz*, 2012) instead of being implanted from within ~30 AU, as envisioned in *Levison et al.* (2008). Also, *Nesvorný and Morbidelli* (2012) showed that the current orbits of the planets are better reproduced if one postulates the existence of a fifth planet with a mass comparable to those of Uranus and Neptune, eventually ejected from the solar system. However, it has not yet been shown that a five-planet system can become unstable late, as the work of *Levison et al.* (2011) was conducted in the framework of a four-planet system. These issues, however, are unlikely to invalidate the Nice model as a whole.

It may be surprising that the current eccentricities and inclinations of the giant planets of our solar system can be so much smaller than those of many extrasolar planets, especially if they experienced a similar phase of global

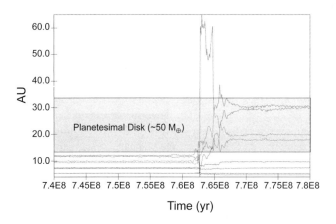

Fig. 3. Orbital evolution of the four giant planets in our solar system according to the Nice model. Here each planet is represented by two curves, denoting perihelion and aphelion distance, hence when the curves overlap, the orbit is circular. The region spanned originally by the planetesimal disk of 50 $M_\oplus$ is shown as a gray area. Notice that the planetary system becomes unstable at t = 762 m.y. in this simulation with the planetary orbits changing radically at this point. Data from *Levison et al.* (2011).

instability (see sections 3 and 4). There are two main reasons for this result. First, the "giant" planets of our solar system have masses that are significantly smaller than those of many extrasolar giant planets. This lower (total) mass made the instability less violent and allowed the orbits of our giant planets to be damped more efficiently by interactions with the planetesimal disk (*Raymond et al.,* 2009a). This trend is particularly applicable for Uranus and Neptune: Their eccentricities could be damped from more than 0.5 to almost 0 by the dispersal of the planetesimal disk, which is assumed to carry about twice the sum of their masses. Second, Jupiter and Saturn fortuitously avoided having close encounters with each other (whereas they both had, according to the Nice model, encounters with a Neptune-mass planet). In fact, in the simulations of solar system instability where Jupiter and Saturn have a close encounter with each other, Jupiter typically ends up on an orbit with eccentricity in the range e = 0.3–0.4 (typical of many extrasolar planets) and recoils to 4.5 AU, while all the other planets are ejected from the system.

6. DYNAMICAL INTERACTIONS OF PLANETARY SYSTEMS WITHIN STELLAR CLUSTERS

In this section, we consider dynamical interactions between the constituent members of young stellar clusters with a focus on the consequent effects on young and forming planetary systems.

Most stars form within some type of cluster or association. In order to quantify the resulting effects on planetary systems, we first consider the basic properties of these cluster environments. About 10% of the stellar population

is born within clusters that are sufficiently robust to become open clusters, which live for 100 m.y. to 1 G.y. The remaining 90% of the stellar population is born within shorter-lived cluster systems that we call embedded clusters. Embedded clusters become unbound and fall apart when residual gas is ejected through the effects of stellar winds and/or supernovae. This dispersal occurs on a timescale of ~10 m.y.

Open clusters span a wide range of masses or, equivalently, number of members. To leading order, the cluster distribution function $f_{cl} \sim 1/N^2$ over a range from N = 1 (single stars) to N = 10^6 [this law requires combining data from different sources (e.g., *Lada and Lada,* 2003; *Chandar et al.,* 1999)]. With this distribution, the probability that a star is born within a cluster of size N scales as P = N $f_{cl} \sim$ 1/N, so that the cumulative probability $\propto$ log N. In other words, stars are equally likely to be born within clusters in each decade of stellar membership size N. For the lower end of the range, clusters have radii on the order of R = 1 pc, and the cluster radius scales as R $\approx$ 1 pc(N/300)$^{1/2}$, so that the clusters have (approximately) constant column density (*Lada and Lada,* 2003; *Adams et al.,* 2006). For the upper end of the range, this law tends to saturate, so that the more massive clusters are denser than expected from this law. Nonetheless, a typical mean density is only about 100 stars/pc^3. Clusters having typical masses and radii have velocity dispersions ~1 km s^{-1}.

Dynamical interactions within clusters are often subject to gravitational focusing. In rough terms, focusing becomes important when the encounter distance is small enough that the speed of one body is affected by the potential well of the other body. For a typical velocity dispersion of 1 km s^{-1}, and for solar-type stars, this critical distance is about 1000 AU. As outlined below, this distance is comparable to the closest expected encounter distance for typical clusters. As a result, gravitational focusing is important — but not dominant — in these systems.

The timescale for a given star to undergo a close encounter (where gravitational focusing is important) with another star within a distance r_{min} can be approximated by (*Binney and Tremaine,* 1987)

$$\tau_{enc} \simeq 3.3 \times 10^7 \text{ yr} \left(\frac{100 \text{pc}^{-3}}{n} \right) \left(\frac{v_\infty}{1 \text{ km s}^{-1}} \right) \times \left(\frac{10^3 \text{AU}}{r_{min}} \right) \left(\frac{M_\odot}{M} \right) \quad (8)$$

Here n is the stellar number density in the cluster, v_∞ is the mean relative speed at infinity of the stars in the cluster, r_{min} is the encounter distance, and M is the total mass of the stars involved in the encounter. The effect of gravitational focusing is included in the above equation.

The estimate suggested above is verified by numerical N-body calculations, which determine a distribution of close encounters (e.g., *Adams et al.,* 2006; *Malmberg et al.,* 2007b; *Proszkow and Adams,* 2009). These studies show that the distribution has a power-law form so that the rate Γ at which a given star encounters other cluster members at a distance of closest approach less than b has the form $\Gamma = \Gamma_0(b/b_0)^\gamma$, where (Γ_0, γ) are constants and b_0 is a fiducial value. The index $\gamma < 2$ is due to gravitational focusing. Typical encounter rates are shown in Fig. 4 for clusters with N = 100, 300, and 1000 members. With this power-law form for the distribution, the expectation value for the closest encounter experienced over a 10-m.y. time span is about $\langle b \rangle \approx 1000$ AU. The initial conditions in a cluster can have an important effect on the subsequent encounter rates. In particular, subvirial clusters that contain significant substructure (i.e., lumps) will have higher encounter rates (*Allison et al.*, 2009; *Parker and Quanz*, 2012) (see also Fig. 4).

Planetary systems are affected by passing stars and binaries in a variety of ways (e.g., *Laughlin and Adams*, 1998; *Adams and Laughlin*, 2001; *Bonnell et al.*, 2001; *Davies and Sigurdsson*, 2001; *Adams et al.*, 2006; *Malmberg et al.*, 2007b, 2011; *Malmberg and Davies*, 2009; *Spurzem et al.*, 2009; *Hao et al.*, 2013). Sufficiently close encounters can eject planets, although the cross sections for direct ejections are relatively small and can be written in the form (from *Adams et al.*, 2006)

$$\Sigma_{ej} \approx 1350 \left(AU \right)^2 \left(\frac{M_*}{1M_\odot} \right)^{-1/2} \left(\frac{a_p}{1AU} \right) \qquad (9)$$

where M_* is the mass of the planet-hosting star and a_p is the (starting) semimajor axis of the planet. Note that the cross section scales as one power of a_p (instead of two) due to gravitational focusing.

Note that equation (9) provides the cross section for direct ejection, where the planet in question leaves its star immediately after (or during) the encounter. Another class of encounters leads to indirect ejection. In this latter case, the flyby event perturbs the orbits of planets in a multiple-planet system, and planetary interactions later lead to the ejection of a planet. Typical instability timescales lie in the range 1–100 m.y. (e.g., *Malmberg et al.*, 2011, see their Fig. 7), although a much wider range is possible. On a similar note, ejection of Earth from our own solar system is more likely to occur indirectly through perturbations of Jupiter's orbit (so that Jupiter eventually drives the ejection of Earth), rather than via direct ejection from a passing star (*Laughlin and Adams*, 2000). Planetary systems residing in wide stellar binaries in the field of the Galaxy are also vulnerable to external perturbations. Passing stars and the Galactic tidal field can change the stellar orbits of wide binaries, making them eccentric, leading to strong interactions with planetary systems (*Kaib et al.*, 2013).

Although direct planetary ejections due to stellar encounters are relatively rare, such interactions can nonetheless perturb planetary orbits, i.e., these encounters can change the orbital elements of planets. Both the eccentricity and the inclination angles can be perturbed substantially, and changes in these quantities are well correlated (*Adams and*

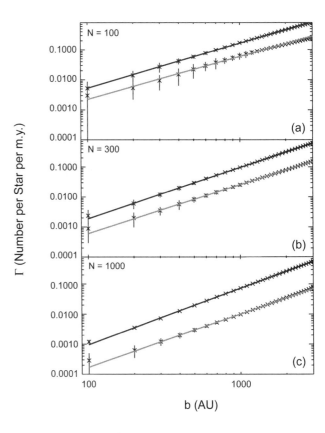

Fig. 4. Distribution of closest approaches for the solar systems in young embedded clusters. Each panel shows the distribution of closest approaches, plotted vs. approach distance b, for clusters with both virial (bottom) and subvirial (top) starting conditions. Results are shown for clusters with N = 100, 300, and 1000 members, as labeled. The error bars shown represent the standard deviation over the compilations. From Fig. 5 of *Adams et al.* (2006), reproduced by permission of the American Astronomical Society.

Laughlin, 2001). As one benchmark, the cross section for doubling the eccentricity of Neptune in our solar system is about $\Sigma \approx (400\ AU)^2$; the cross section for increasing the spread of inclination angles in our solar system to 3.5° has a similar value. Note that these cross sections are much larger than the geometric cross sections of the solar system. The semimajor axes are also altered, but generally suffer smaller changes in a relative sense, i.e., $(\Delta a)/a \ll (\Delta e)/e$.

Sometime after flyby encounters, the orbital elements of planetary systems can be altered significantly due to the subsequent planet-planet interactions (see Fig. 10 of *Malmberg et al.*, 2011). When a planetary system becomes unstable, planets may be ejected via scattering with other planets. However, planets are often ejected as the result of several (many tens to one hundred) scattering events, with each scattering event making the planet slightly less bound to the host star. As a result, snapshots of planetary systems taken some time after the initial flyby sometimes reveal planets on much wider (less bound) orbits with separations in excess of 100 AU. In Fig. 5 we plot the fraction of postflyby systems containing planets on orbits with

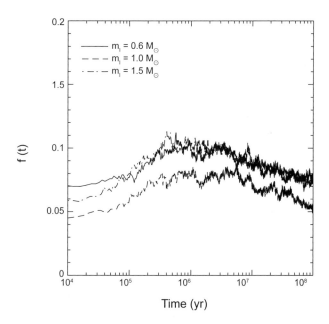

Fig. 5. The fraction of solar systems f(t) containing planets with semimajor axes greater than 100 AU, plotted here as a function of time t after a close encounter for intruder stars of masses 0.6, 1.0, and 1.5 $M_\odot$. These simulations of flybys involve the four gas giants of the solar system with r_{min} < 100 AU. From Fig. 11 of *Malmberg et al.* (2011), reproduced by permission of the Royal Astronomical Society.

semimajor axes a > 100 AU as a function of time after the flyby encounters. This trend is found both in systems made unstable by flybys and those that become unstable without external influence, as described earlier in section 3 (*Scharf and Menou*, 2009; *Veras et al.*, 2009). Planets on such wide orbits should be detectable via direct imaging campaigns; thus the fraction of stars possessing them will place limits on the population of unstable planetary systems.

During flyby encounters, intruding stars can also pick up planets from the planetary system (see Fig. 12 of *Malmberg et al.*, 2011). Clusters thus provide rich environments that can shape the planetary systems forming within them. In particular, if the intruding star already possesses its own planetary system, the addition of the extra planet may destabilize the system.

By combining encounter rates for planetary systems in clusters with cross sections that describe the various channels of disruption, we can estimate the probability (equivalently, the perturbation rate) that a planetary system will suffer, e.g., the ejection of at least one planet over a given time interval (see *Laughlin and Adams*, 1998; *Malmberg et al.*, 2011). By combining the cross section for planetary ejection with the encounter histories found through N-body simulations of stellar clusters, one finds that planets will be ejected in ~5–10% of planetary systems in long-lived clusters (*Malmberg et al.*, 2011). The number of planets ejected directly during flybys will be somewhat lower: For example, only a few planets are expected to

be ejected per (young) embedded cluster. As a result, large numbers of free-floating planets in young clusters point to other mechanisms for ejection, most likely planet-planet interactions in young planetary systems (*Moorhead and Adams*, 2005; *Chatterjee et al.*, 2008). One should also note that dynamical mechanisms are unlikely to be able to explain the population of free-floating planets as inferred by microlensing observations (*Veras and Raymond*, 2012).

Encounters involving binary stars also play an important role in the evolution of planetary systems residing in stellar clusters. A star that hosts a planetary system may exchange into a (wide) binary. If its orbit is sufficiently inclined, the stellar companion can affect planetary orbits via the Kozai mechanism (see also section 7). These interactions force planets onto eccentric orbits that can cross the orbits of other planets, and can thereby result in strong planetary scatterings (*Malmberg et al.*, 2007a). The rate of such encounters depends on the binary population, but Kozai-induced scattering may account for the destruction of at least a few percent of planetary systems in clusters (*Malmberg et al.*, 2007b).

For completeness, we note that dynamical interactions can also affect circumstellar disks, prior to the formation of planetary systems. In this earlier phase of evolution, the disks are subject to truncation by passing stars. As a general rule, the disks are truncated to about one-third of the distance of closest approach (*Ostriker*, 1994; *Heller*, 1995); with the expected distribution of interaction distances, we expect disks to be typically truncated to about 300 AU through this process, although closer encounters will lead to smaller disks around a subset of stars. Since most planet formation takes place at radii smaller than 300 AU, these interactions have only a modest effect. In a similar vein, circumstellar disks can sweep up ambient gas in the clusters through Bondi-Hoyle accretion. Under favorable conditions, a disk can gain a mass equivalent to the minimum mass solar nebula through this mechanism (*Throop and Bally*, 2008).

The orientations of circumstellar disks can also be altered by later accretion of material on the disks and by interactions with other stars within clustered environments (*Bate et al.*, 2010). Alternatively, interactions with a companion star within a binary can also reorient a disk (*Batygin*, 2012). Both processes represent an alternative to the dynamical processes described in sections 7 and 8 as a way to produce hot Jupiters on highly inclined orbits.

Since clusters have significant effects on the solar systems forming within them, and since our own solar system is likely to have formed within a cluster, one can use these ideas to constrain the birth environment of the Sun (*Adams*, 2010). Our solar system has been only moderately perturbed via dynamical interactions, which implies that our birth cluster was not overly destructive. On the other hand, a close encounter with another star (or binary) may be necessary to explain the observed edge of the Kuiper belt and the orbit of the dwarf planet Sedna (*Kobayashi and Ida*, 2001; *Kenyon and Bromley*, 2004; *Morbidelli and*

Levison, 2004), and the need for such an encounter implies an interactive environment. Note that the expected timescale for an encounter (about 2000 yr) is much longer than the orbital period at the edge of the Kuiper belt (350 yr), so that the edge can become well-defined. Adding to the picture, meteoritic evidence suggests that the early solar system was enriched in short-lived radioactive isotopes by a supernova explosion (*Wadhwa et al.,* 2007), an asymptotic giant branch (AGB) star (*Wasserburg et al.,* 2006), or some combination of many supernovae and a second-generation massive star (*Gounelle and Meynet,* 2012; *Gounelle et al.,* 2013). Taken together, these constraints jointly imply that the birth cluster of the solar system was moderately large, with stellar membership size $N = 10^3$–10^4 (e.g., *Hester et al.,* 2004; *Adams,* 2010).

7. THE LIDOV-KOZAI MECHANISM

The perturbing effect of Jupiter on the orbits of asteroids around the Sun was considered by *Kozai* (1962). It was found that for sufficiently highly inclined orbits, the asteroid would undergo large, periodic changes in both eccentricity and inclination. Work by *Lidov* (1962) showed that similar effects could be seen for an artificial satellite orbiting a planet. Here, we will refer to such perturbations as the Lidov-Kozai mechanism when also applied to perturbations of planetary orbits due to an inclined stellar companion. Strong periodic interactions can also occur between two planets, when one is highly inclined.

The Lidov-Kozai mechanism is a possible formation channel for hot Jupiters (*Fabrycky and Tremaine,* 2007; *Nagasawa et al.,* 2008) (as will be discussed in section 8). Concurrently, the Lidov-Kozai mechanism has found wide applicability to other astrophysical problems: binary supermassive black holes (*Blaes et al.,* 2002), binary minor planets (where the Sun is considered the massive outer perturber) (*Perets and Naoz,* 2009; *Fang and Margot,* 2012b), binary millisecond pulsars (*Gopakumar et al.,* 2009); stellar-disk-induced Lidov-Kozai oscillations in the Galactic center (*Chang,* 2009), binary white dwarfs or binary neutron stars (*Thompson,* 2011), and evolving triple star systems with mass loss (*Shappee and Thompson,* 2013).

Lidov-Kozai evolution is approximated analytically by applying Lagrange's planetary equations to a truncated and averaged form of the disturbing function (*Valtonen and Karttunen,* 2006, chapter 9). The truncation is justified because of the hierarchical ordering of the mutual distances of the three bodies. The averaging occurs twice, over a longitude or anomaly of both the lightest body and the outermost body.

Traditionally, the disturbing function is secular, meaning that the planetary semimajor axes remain fixed and the system is free from the influence of mean-motion resonances. The term "Lidov-Kozai resonance" refers simply to the Lidov-Kozai mechanism, which includes large and periodic secular eccentricity and inclination variations. Confusingly, an alteration of this mechanism that introduced nonsecu-

lar contributions (*Kozai,* 1985) has been referred to as a Lidov-Kozai resonance *within* a mean-motion resonance. This formalism has been utilized to help model dynamics in the Kuiper belt (*Gallardo,* 2006; *Wan and Huang,* 2007; *Gallardo et al.,* 2012). The Lidov-Kozai mechanism can also be modified by including the effects of star-planet tides (see section 8), resulting in significant changes in the semimajor axis of a planet (a process sometimes known as "Lidov-Kozai migration") (*Wu and Murray,* 2003).

The complete analytic solution to the secular equations resulting from the truncated and averaged disturbing function may be expressed in terms of elliptic functions (*Vashkovýak,* 1999). However, more commonly an approximate solution is found for small initial values of eccentricities (e) [but where initial values of inclination (i) are sufficiently large for Lidov-Kozai cycles to occur] by truncating the disturbing function to quadrupole order in the mutual distances between the bodies. This solution demonstrates that the argument of pericenter oscillates around 90° or 270°, a dynamical signature of the Lidov-Kozai mechanism. The solution also yields a useful relation: $\sqrt{1-e^2}\cos i \approx$ constant subject to $i > \arcsin(2/5)$ and $e < \sqrt{1-(5/3)\cos^2(i)}$. This relation demonstrates the interplay between eccentricity and inclination due to angular momentum transfer. The period of the eccentricity and inclination oscillations for most observable configurations lie well within a main-sequence-star lifetime, thus at least thousands of such oscillations may occur before the star evolves. The period is on the order of (*Kiseleva et al.,* 1998)

$$\tau_{Kozai} = \frac{2P_{out}^2}{3\pi P_{in}} \frac{M_1 + M_2 + M_3}{M_3} \left(1 - e_{out}^2\right)^{3/2} \qquad (10)$$

where M_1 and M_2 are the masses of the innermost two bodies orbiting each other with period P_{in} and M_3 is the mass of the outermost body orbiting the inner binary with period P_{out} and eccentricity e_{out}.

Recent work has demonstrated that the approximations employed above fail to reproduce important aspects of the true motion, which can be modeled with three-body numerical simulations. By instead retaining the octupole term in the disturbing function, *Ford et al.* (2000) derived more accurate, albeit complex, evolution equations for the true motion. Subsequent relaxation of the assumption of small initial eccentricities has allowed for a wider region of phase space of the true motion to be reproduced by the Lidov-Kozai mechanism (*Katz et al.,* 2011; *Lithwick and Naoz,* 2011; *Naoz et al.,* 2011; *Libert and Delsate,* 2012).

One outstanding consequence of retaining the octupole term is that the effect may flip a planet's orbital evolution from prograde to retrograde, and consequently may explain observations. The projected angle between stellar rotation and planetary orbital angular momentum has been measured for tens of hot Jupiters with the Rossiter-Mclaughlin effect (*Triaud et al.,* 2010). These observations show us that some 20% of hot Jupiters most probably have retrograde orbits, while 50% or so are aligned (*Albrecht et al., 2012*). In at least one case (*Winn et al., 2009*) the true angle is at least 86°, helping to reinforce indications of a subset of planets orbiting in a retrograde fashion with inclinations above 90° (*Albrecht et al., 2012*). *Naoz et al.* (2011) demonstrated how the Lidov-Kozai mechanism can produce these orbits. [They also discovered an important error in the original derivation of the truncated, averaged disturbing function (*Kozai, 1962*), which did not conserve angular momentum due to an erroneous assumption about the longitudes of ascending nodes.] Figure 6 illustrates how this inclination "flipping" can occur naturally in a three-body system, and the consequences of truncating the disturbing function to quadrupole order. In Figs. 6a and 6b, one sees that when the disturbing function is truncated at the quadrupole term, then the spread of inclinations and eccentricities are small, whereas when going to octupole order, the orbit flips (i.e., inclinations above 90°), and eccentricities reach values very close to unity.

Lidov-Kozai oscillations do not work in isolation. Bodies are not point masses: Effects of tides, stellar oblateness, and general relativity may play crucial roles in the evolution. These contributions have been detailed by *Fabrycky and Tremaine* (2007), *Chang* (2009), *Veras and Ford* (2010), and

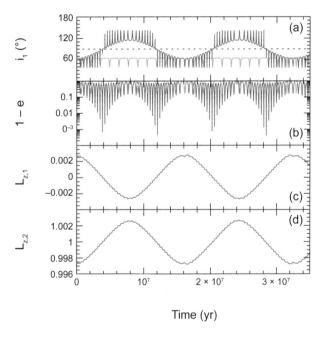

Fig. 6. Inclination flipping due to the Lidov-Kozai mechanism when the disturbing function is truncated to octupole order [upper curve on **(a)** and lower curve on **(b)**] vs. quadrupole order [lower curve on **(a)** and upper curve on **(b)**]. The inner binary consists of a 1-$M_\odot$ star and a 1-M_{Jup} planet separated by 6 AU with e_{in} = 0.001, and the outer body is a brown dwarf of 40 M_{Jup} at a distance of 100 AU from the center of mass of the inner binary with e_{out} = 0.6. The bottom two plots display the normalized vertical components of the inner and outer orbit angular momentum. From Fig. 1 of *Naoz et al.* (2011), reprinted by permission of Macmillan Publishers Ltd.

Beust et al. (2012). A major consequence of these effects, in particular tides, is that planets formed beyond the snow line may become hot Jupiters through orbital shrinkage and tidal circularization (see section 8).

The Lidov-Kozai mechanism may also be affected by the presence of a nascent protoplanetary disk, and hence play a crucial role during the formation of planets around one component of a binary system. If the secondary star is sufficiently inclined to and separated from the orbital plane of the primary, then the Lidov-Kozai mechanism might "turn on," exciting the eccentricity of planetesimals and diminishing prospects for planet formation (*Marzari and Barbieri, 2007*; *Fragner et al., 2011*). However, the inclusion of the effects of gas drag (*Xie et al., 2011*) and protoplanetary disk self-gravity (*Batygin et al., 2011*) assuage the destructive effect of Lidov-Kozai oscillations, helping to provide favorable conditions for planetary growth.

Lidov-Kozai-like oscillations are also active in systems with more than three bodies, which is the focus of the remainder of this section. Examples include quadruple-star systems (*Beust and Dutrey, 2006*), triple-star systems with one planet (*Marzari and Barbieri, 2007*), and multiple-planet systems with or without additional stellar companions.

Given the likelihood of multiple exoplanets in binary systems, how the Lidov-Kozai mechanism operates in these systems is of particular interest. In particular, eccentricity and inclination oscillations produced by a wide-binary stellar companion can induce planet-planet scattering, leading to dynamical instability (*Innanen et al., 1997*). *Malmberg et al.* (2007a) demonstrate that this type of instability can result in planet stripping in the stellar birth cluster, where binaries are formed and disrupted at high inclinations.

Alternatively, multiple-planet systems within a wide binary may remain stable due to Lidov-Kozai oscillations. Understanding the conditions in which stability may occur and the consequences for the orbital system evolution can help explain current observations. *Takeda et al.* (2008) outline how to achieve this characterization by considering the secular evolution of a two-planet system in a wide binary, such that the mutual planet-planet interactions produce no change in semimajor axis. These planet-planet interactions may be coupled analytically to Lidov-Kozai oscillations because the latter are usually considered to result from secular evolution.

Takeda et al. (2008) compare the period of Lidov-Kozai oscillations with the period of the oscillations produced by Laplace-Lagrange secular interactions (see section 2), as shown in Fig. 7. The figure provides an example of where Lidov-Kozai oscillations dominate or become suppressed (by comparing the curves), and the character of the resulting dynamical evolution (identified by the shaded regions). Because the Lidov-Kozai oscillation timescale increases with binary separation (see equation (10)), Lidov-Kozai oscillations are less likely to have an important effect on multi-planet evolution contained in wider stellar binaries. However, for binaries that are sufficiently wide, Galactic tides can cause close pericenter passages every few billion years, perhaps explaining the difference in the observed eccentricity distribution of the population of giant planets in close binaries vs. those in wide binaries (*Kaib et al., 2013*).

8. DYNAMICAL ORIGIN OF HOT JUPITERS

The first exoplanet confirmed to orbit a main-sequence star, 51 Pegasi b (*Mayor and Queloz, 1995*) became a prototype for a class of exoplanets known as hot Jupiters. The transiting subset of the hot Jupiters (here defined as a ≤ 0.1 AU and $M_p \sin i = 0.25$–20 M_{Jup}) provide unique constraints on planetary evolution, and allow observational studies of atmospheric phenomena that are currently not possible for more distant planets. Identifying the dynamical origin of hot Jupiters is a long-standing problem and is the topic of this section.

It is generally accepted that hot Jupiters cannot form *in situ* (*Bodenheimer et al., 2000*). If true, then their origin requires either migration through a massive, and presumably gaseous, disk or dynamical interactions involving multiple stellar or planetary bodies. Although both mechanisms remain possible, recent observations (e.g., *Winn et al., 2010*) indicate that roughly one-fourth of hot Jupiter orbits are substantially misaligned with respect to the stellar rotation axis. These systems (and perhaps others) are naturally explained by dynamical processes, which are the focus of this section. The relevant dynamics may involve (1) Lidov-Kozai evolution of a one-planet system perturbed by a binary stellar companion, (2) Lidov-Kozai evolution in

a multiple-planet system, (3) scattering of multiple planets or secular evolution unrelated to the Kozai resonance, or (4) secular chaos.

All these processes are likely to occur at some level, so the main open question is their relative contribution to forming the observed hot Jupiter population. In every case, tidal interactions — which are inevitable for planets with the orbital period of hot Jupiters — are required in order to shrink and circularize the orbit (*Ivanov and Papaloizou, 2004; Guillochon et al., 2011*). Tides raised on the star are expected to dominate orbital decay, while tides raised on the planet are likely to dominate circularization. However, tides on both bodies can contribute significantly. Quantitative comparisons between theoretical models and observations are limited by considerable uncertainties in tidal physics and planetary structure, as well as observational biases in determining the orbital period distribution and the eccentricity of nearly circular orbits (*Zakamska et al., 2011; Gaidos and Mann, 2013*).

A mechanism for generating high eccentricities, and tidal damping, are minimal ingredients for the dynamical formation of hot Jupiters. Several other processes, however, including general relativity, oblateness (*Correia et al., 2012*), and tides, can lead to precession of short-period orbits. These processes need to be included in models of hot Jupiter formation via secular dynamical effects. *Wu and Murray (2003)* quote formulae for the precession rates

$$\dot{\omega}_{GR} = 3n \frac{GM_*}{a_p c^2 \left(1 - e_p^2\right)}$$

$$\dot{\omega}_{tid} = \frac{15}{2} nk_2 \frac{1 + (3/2)e_p^2 + (1/8)e_p^4}{\left(1 - e_p^2\right)^5} \frac{M_*}{M_p} \left(\frac{R_p}{a_p}\right)^5 \qquad (11)$$

$$\dot{\omega}_{obl} = \frac{1}{2} n \frac{k_2}{\left(1 - e_p^2\right)^2} \left(\frac{\Omega_p}{n}\right)^2 \frac{M_*}{M_p} \left(\frac{R_p}{a_p}\right)^5$$

Here a_p, e_p, M_p, R_p, and Ω_p are the planet's semimajor axis, eccentricity, mass, radius, and spin frequency; n is the mean motion; and k_2 is the tidal Love number. Lidov-Kozai oscillations are generally suppressed if any of these precession rates exceed $\dot{\omega}_{Kozai}$. This can easily occur for a_p < 1 AU. Eccentricity can be enhanced when $\dot{\omega}_{GR} \simeq -\dot{\omega}_{Kozai}$ (*Ford et al., 2000*).

Figure 8, from *Wu and Murray (2003)*, illustrates how hot Jupiters can form in a single-planet system with an inclined stellar companion. Lidov-Kozai oscillations in the planetary eccentricity result in periods where the pericenter distance is close enough for stellar tides to shrink the orbit. As the orbit shrinks, the additional dynamical effects described above become more important, resulting in a decrease in the amplitude of the eccentricity and inclination variations. In this example, after 700 m.y., GR suppresses the Lidov-Kozai oscillations completely. The influence of

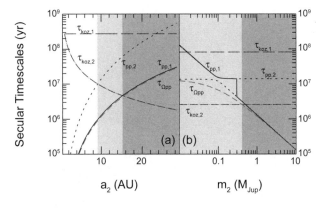

Fig. 7. Comparison of timescales and regimes of motion for a four-body system consisting of two planets secularly orbiting one star of a wide binary. The binary separation and eccentricity are fixed at 500 AU and 0.5 for two 1-$M_\odot$ stars, and the inner 0.3-M_{Jup} planet resides 1 AU away from its parent star. Subscripts "1" and "2" refer to the inner and outer planet, "koz" the Lidov-Kozai characteristic timescale due to the distant star, and "pp" timescales due to Laplace-Lagrange secular theory. The unshaded, lightly shaded, and darkly shaded regions refer to where Lidov-Kozai cycles on the outer planet are suppressed, the planetary orbits precess in concert, and the inner planet's eccentricity grows chaotically. From Fig. 14 of *Takeda et al.* (2008), reproduced by permission of the American Astronomical Society.

the secondary star then becomes negligible as tidal interactions dominate. *Wu et al.* (2007), using a binary population model and estimates for the radial distribution of massive extrasolar planets, estimated that this process could account for 10% or more of the hot Jupiter population.

A combination of gravitational scattering (*Ford and Rasio*, 2006) and Lidov-Kozai oscillations can also lead to the production of hot Jupiters from multiple-planet systems around single stars. Figure 9 provides an example of three-planet scattering in which the outer planet is ejected, triggering Lidov-Kozai cycles between the other two planets. These oscillations are not as regular as those in Fig. 6 because in Fig. 9 the outermost body is comparable in mass to the middle body. Nevertheless, the oscillations of the argument of pericenter about 90° is indicative of the Lidov-Kozai mechanism at work. By about 3 m.y., the semimajor axis of the inner planet has shrunk to a value of 0.07 AU, after which other physical effects dictate the future evolution of the planet.

Both the fraction of scattering systems that yield star-grazing planets, and the fraction of those systems that yield surviving hot Jupiters, are uncertain. *Nagasawa et al.* (2008) and *Nagasawa and Ida* (2011) integrated ensembles of unstable three-planet systems, using a model that included both gravitational and tidal forces. They obtained an extremely high yield (≈30%) of highly eccentric planets that was larger than the yield found in earlier calculations that did not include tides (*Chatterjee et al.*, 2008). One should note that these numbers do not reflect the expected fraction

of scattering systems that would yield *long-lived* hot Jupiters, as many of the highly eccentric planets circularize into orbits with tidal decay times less than the main-sequence lifetime. *Beaugé and Nesvorný* (2012), on the other hand, using a different tidal model, a dispersion in planet masses, and resonant initial conditions, found a yield of surviving hot Jupiters of approximately 10%. Ten percent is also roughly the fraction of hot Jupiters in an unbiased sample of all massive planets with orbital radii less than a few astronomical units. The efficiency of hot Jupiter production from scattering plus tidal circularization is thus high enough for this channel to contribute substantially to the population, if one assumes that scattering occurs in the majority of all such planetary systems.

Hot Jupiters can originate from multi-planet dynamics without the Lidov-Kozai effect. *Wu and Lithwick* (2011) present special configurations of the four-body problem that allow for the innermost planet in a three-planet system to be forced into the tidal circularization radius. These configurations require a significant AMD in the original

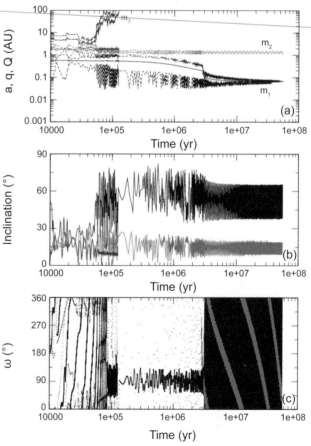

Fig. 9. Hot Jupiter production by Lidov-Kozai forcing from multiple-planet interactions. The inner, middle, and outer planets have masses of 1 M_{Jup}, 2 M_{Jup}, and 1 M_{Jup} and initially nearly circular (e < 0.1) and coplanar (I < 1°) orbits. From Fig. 11 of *Beaugé and Nesvorný* (2012), reproduced by permission of the American Astronomical Society.

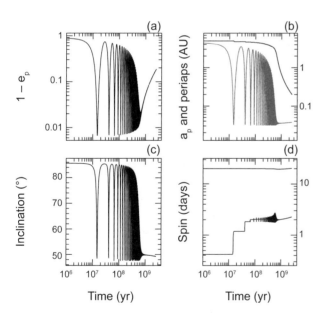

Fig. 8. Hot Jupiter production by Lidov-Kozai forcing from a stellar binary companion. A 7.80-M_{Jup} planet with a_p = 5.0 AU, $e_p(t = 0)$ = 0.1, $I_p(t = 0)$ = 85.6°, and $\omega_p(t = 0)$ = 45° is evolving under the influence of two 1.1-$M_\odot$ stars separated by 1000 AU on a e_B = 0.5 orbit. From Fig. 1 of *Wu and Murray* (2003), reproduced by permission of the American Astronomical Society.

planetary system. Figure 10 presents an example of hot Jupiter production without the Lidov-Kozai effect. Note that all three planets survive the evolution and never cross orbits.

If planet formation produces multiple planets on nearly circular, co-planar orbits, then secular evolution alone is not sufficient to produce hot Jupiters. Planet-disk interactions are not currently thought to be able to generate significant planetary eccentricity (except for high-mass planets), and hence the easiest route toward forming systems with a significant AMD appears to be an initial phase of strong planet-planet scattering. In such a model, hot Jupiters could either be formed early (from highly eccentric scattered

planets) or late (from long-term secular evolution among the remaining planets after scattering). Which channel would dominate is unclear.

The observation of strongly misaligned and retrograde orbits from Rossiter-McLaughlin measurements provides evidence in favor of dynamical formation mechanisms (e.g., *Winn et al.,* 2009), but does not immediately discriminate among different dynamical scenarios. In particular, although pure Lidov-Kozai evolution involving a circular stellar companion cannot create a retrograde planet, the presence of eccentricity in either a stellar or planetary perturber can (*Katz et al.,* 2011; *Lithwick and Naoz,* 2011). The same is true for secular evolution without the Lidov-Kozai effect.

Dawson et al. (2012b) compare the three potential origin channels described above, and suggest that Lidov-Kozai evolution in binary stellar systems is unlikely to dominate, but may explain $15^{+29}_{-11}\%$ of hot Jupiters, corroborating the *Naoz et al.* (2012) result of ≈30%. These estimates rely upon the results of *Nagasawa and Ida* (2011) that find a large fraction of scattering systems (~30%) form (at least initially) a hot Jupiter, a result that they attribute in substantial part to Lidov-Kozai evolution. Similar calculations by *Beaugé and Nesvorný* (2012), using a more realistic tidal model, find less-efficient hot Jupiter formation driven primarily by scattering events. They suggest that higher initial planetary multiplicity results in hot Jupiter populations in better accord with observations. *Morton and Johnson* (2011) provide constraints on the frequency of the two different Lidov-Kozai-induced hot Jupiter formation scenarios from spin-orbit data. They find that the results depend critically on the presence or lack of a population of aligned planetary systems. If this population exists, multiple-planet Lidov-Kozai scattering [specifically from the model of *Nagasawa et al.* (2008)] is the favored formation mechanism. Otherwise, binary-induced Lidov-Kozai evolution [specifically from the model of *Fabrycky and Tremaine* (2007)] is the favored model. The above results also depend upon the assumed tidal physics. *Dawson et al.* (2012a) identified Kepler object of interest (KOI) 1474.01 as a proto-hot Jupiter based on the long transit duration for its orbital period. The presence of large transit timing variations suggests that scattering by a more distant giant planet may explain the origin of KOI 1474.01's high eccentricity. With further analysis, the abundance of transiting proto-hot Jupiters could provide a constraint on the timescale of the high eccentricity migration phase of hot Jupiter's formation.

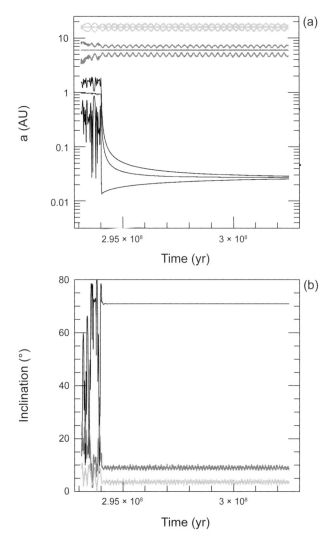

Fig. 10. Hot Jupiter production by multiple planet evolution with no Lidov-Kozai cycles. The inner, middle, and outer planets have masses of 0.5 M_{Jup}, 1.0 M_{Jup}, and 1.5 M_{Jup} and initial semimajor axes of 1 AU, 6 AU, and 16 AU; eccentricities of 0.066, 0.188, and 0.334; and inclinations of 4.5°, 19.9°, and 7.9°. This choice places the AMD primarily in the outer planets. From Fig. 2 of *Wu and Lithwick* (2011), reproduced by permission of the American Astronomical Society.

9. SUMMARY

We have reviewed the long-term dynamical evolution of planetary systems. Our key points are listed below:

1. The giant-planet subsystem of the solar system is stable, although the terrestrial-planet subsystem is marginally unstable with a small chance of planet-planet encounters during the lifetime of the Sun.

2. Planet-planet scattering in tighter planetary systems can lead to close encounters between planets. The timescale

before a system undergoes such encounters is a strong function of the separation of planets.

3. Secular interactions cause the redistribution of angular momentum among planets in a system. In systems with a sufficiently large AMD, such redistribution can lead to close planetary encounters.

4. The outcome of planetary close encounters is a function of the Safronov number. Collisions dominate when the planetary surface escape speeds are smaller than orbital speeds. Planetary scattering will be more common when the surface escape speeds are larger than the planetary orbital speeds in a system.

5. Planets are predicted to pass through a phase of wide orbits within unstable planetary systems as ejections occur only after several scatterings. Imaging surveys will therefore inform us about the frequency of unstable systems.

6. The observed eccentricity distribution is consistent with being an outcome of planet-planet scattering in unstable systems.

7. Interactions with planetesimal disks will cause planets to migrate, which in turn can lead to instabilities within a planetary system. This process probably played an important role in the early history of our own solar system.

8. Aging systems may become unstable when the host star evolves to become a white dwarf and loses mass, as the relative strength of the planet-planet interactions increase compared to the interactions between the planets and host star.

9. Flyby encounters in stellar clusters will occur in dense birth environments. Such encounters may lead to the direct ejection of planets in some cases. In other encounters, perturbations to the planetary orbits lead to instabilities on longer timescales. The intruding star may also pick up a planet from the system.

10. Exchange into binaries can occur in stellar clusters. Planetary systems may be destabilized by the perturbing effect of the companion star through the Lidov-Kozai mechanism where the outer planet suffers periods of higher eccentricity, leading it to have strong encounters with other planets.

11. The Lidov-Kozai mechanism may also operate within primordial binaries or planetary multiple systems, leading to the periodic increase in eccentricity of the planets' orbits and planet-planet encounters in the case of multiple-planet systems.

12. The origin of hot Jupiters through dynamical interactions may involve one of five possible routes: Lidov-Kozai evolution of a one-planet system perturbed by a binary stellar companion; Lidov-Kozai evolution in a multiple-planet system; scattering of multiple planets; secular evolution unrelated to the Kozai resonance; or the reorientation of circumstellar disks before planets form via interaction with a companion star, or via late infall of material or interactions or neighboring stars in clustered birth environments.

Acknowledgments. M.B.D. was supported by the Swedish Research Council (grants 2008-4089 and 2011-3991). P.J.A. acknowledges support from NASA grants NNX11AE12G and NNX13AI58G, and from grant HST-AR-12814 awarded by the Space Telescope Science Institute, which is operated by the Association of Universities for Research in Astronomy, Inc., for NASA, under contract NAS 5-26555. J.C. would like to thank NASA's Origins of Solar Systems program for support. E.B.F. was supported by the National Aeronautics and Space Administration under Origins of Solar Systems grant NNX09AB35G and grants NNX08AR04G and NNX12AF73G issued through the Kepler Participating Scientist Program. We thank the anonymous referee for useful comments.

REFERENCES

Adams F. C. (2010) *Annu. Rev. Astron. Astrophys., 48,* 47.
Adams F. C. and Laughlin G. (2001) *Icarus, 150,* 151.
Adams F. C. and Laughlin G. (2003) *Icarus, 163,* 290.
Adams F. C. and Shu F. H. (1986) *Astrophys. J., 308,* 836.
Adams F. C. et al. (2006) *Astrophys. J., 641,* 504.
Albrecht S. et al. (2012) *Astrophys. J., 757,* 18.
Allison R. J. et al. (2009) *Astrophys. J. Lett., 700,* L99.
Barnes R. and Greenberg R. (2006a) *Astrophys. J. Lett., 652,* L53.
Barnes R. and Greenberg R. (2006b) *Astrophys. J. Lett., 647,* L163.
Bate M. R. et al. (2010) *Mon. Not. R. Astron. Soc., 401,* 1505.
Batygin K. (2012) *Nature, 491,* 418.
Batygin K. and Brown M. E. (2010) *Astrophys. J., 716,* 1323.
Batygin K. and Laughlin G. (2008) *Astrophys. J., 683,* 1207.
Batygin K. et al. (2011) *Astron. Astrophys., 533,* A7.
Batygin K. et al. (2012) *Astrophys. J. Lett., 744,* L3.
Beaugé C. and Nesvorný D. (2012) *Astrophys. J., 751,* 119.
Beust H. and Dutrey A. (2006) *Astron. Astrophys., 446,* 137.
Beust H. et al. (2012) *Astron. Astrophys., 545,* A88.
Binney J. and Tremaine S. (1987) *Galactic Dynamics.* Princeton Univ., Princeton. 735 pp.
Blaes O. et al. (2002) *Astrophys. J., 578,* 775.
Bodenheimer P. et al. (2000) *Icarus, 143,* 2.
Boley A. C. et al. (2012) *Astrophys. J., 750,* L21.
Bonnell I. A. et al. (2001) *Mon. Not. R. Astron. Soc., 322,* 859.
Booth M. et al. (2009) *Mon. Not. R. Astron. Soc., 399,* 385.
Bottke W. F. et al. (2012) *Nature, 485,* 78.
Brasser R. and Morbidelli A. (2013) *ArXiv e-prints,* arXiv:1303.3098.
Brasser R. (2007) *Icarus, 191,* 413.
Butler R. P. et al. (2006) *Astrophys. J., 646,* 505.
Capobianco C. C. et al. (2011) *Icarus, 211,* 819.
Cassen P. and Moosman A. (1981) *Icarus, 48,* 353.
Chambers J. E. et al. (1996) *Icarus, 119,* 261.
Chandar R. et al. (1999) *Astrophys. J. Suppl., 122,* 431.
Chang P. (2009) *Mon. Not. R. Astron. Soc., 393,* 224.
Chatterjee S. et al. (2008) *Astrophys. J., 686,* 580.
Chiang E. and Laughlin G. (2013) *Mon. Not. R. Astron. Soc., 431,* 3444.
Ciardi D. R. et al. (2013) *Astrophys. J., 763,* 41.
Correia A. C. M. et al. (2012) *Astrophys. J. Lett., 744,* L23.
Cresswell P. et al. (2007) *Astron. Astrophys., 473,* 329.
D'Angelo G. et al. (2006) *Astrophys. J., 652,* 1698.
Davies M. B. and Sigurdsson S. (2001) *Mon. Not. R. Astron. Soc., 324,* 612.
Dawson R. I. et al. (2012a) *Astrophys. J., 761,* 163.
Dawson R. I. et al. (2012b) *ArXiv e-prints,* arXiv:1211.0554.
Debes J. H. and Sigurdsson S. (2002) *Astrophys. J., 572,* 556.
Debes J. H. et al. (2012) *Astrophys. J., 747,* 148.
Duncan M. J. and Lissauer J. J. (1998) *Icarus, 134,* 303.
Dunhill A. C. et al. (2013) *Mon. Not. R. Astron. Soc., 428,* 3072.
Fabrycky D. and Tremaine S. (2007) *Astrophys. J., 669,* 1298.
Fang J. and Margot J.-L. (2012a) *Astrophys. J., 761,* 92.
Fang J. and Margot J.-L. (2012b) *Astron. J., 143,* 59.
Fernandez J. A. and Ip W.-H. (1984) *Icarus, 58,* 109.
Ford E. B. (2005) *Astron. J., 129,* 1706.
Ford E. B. and Rasio F. A. (2006) *Astrophys. J. Lett., 638,* L45.
Ford E. B. and Rasio F. A. (2008) *Astrophys. J., 686,* 621.
Ford E. B. et al. (2000) *Astrophys. J., 535,* 385.
Ford E. B. et al. (2001) *Icarus, 150,* 303.
Ford E. B. et al. (2003) In *Scientific Frontiers in Research on Extrasolar Planets* (D. Deming and S. Seager, eds.), pp. 181–188.

ASP Conf. Series 294, Astronomical Society of the Pacific, San Francisco.

Ford E. B. et al. (2005) *Nature, 434,* 873.

Fragner M. M. et al. (2011) *Astron. Astrophys., 528,* A40.

Gaidos E. and Mann A. W. (2013) *Astrophys. J., 762,* 41.

Gallardo T. (2006) *Icarus, 181,* 205.

Gallardo T. et al. (2012) *Icarus, 220,* 392.

Gladman B. (1993) *Icarus, 106,* 247.

Goldreich P. et al. (2004) *Astrophys. J., 614,* 497.

Gomes R. S. et al. (2004) *Icarus, 170,* 492.

Gomes R. et al. (2005) *Nature, 435,* 466.

Gopakumar A. et al. (2009) *Mon. Not. R. Astron. Soc., 399,* L123.

Gounelle M. and Meynet G. (2012) *Astron. Astrophys., 545,* A4.

Gounelle M. et al. (2013) *Astrophys. J. Lett., 763,* L33.

Guillochon J. et al. (2011) *Astrophys. J., 732,* 74.

Guzzo M. (2005) *Icarus, 174,* 273.

Hahn J. M. and Malhotra R. (1999) *Astron. J., 117,* 3041.

Hao W. et al. (2013) *Mon. Not. R. Astron. Soc., 433,* 867.

Hartmann W. K. et al. (2000) In *Origin of the Earth and Moon* (R. M. Canup and K. Righter, eds.), pp. 493–512. Univ. of Arizona, Tucson.

Hayes W. B. (2008) *Mon. Not. R. Astron. Soc., 386,* 295.

Heller C. H. (1995) *Astrophys. J., 455,* 252.

Hester J. J. et al. (2004) *Science, 304,* 1116.

Innanen K. A. et al. (1997) *Astron. J., 113,* 1915.

Ivanov P. B. and Papaloizou J. C. B. (2004) *Mon. Not. R. Astron. Soc., 347,* 437.

Johansen A. et al. (2012) *Astrophys. J., 758,* 39.

Jones H. R. A. et al. (2006) *Mon. Not. R. Astron. Soc., 369,* 249.

Jurić M. and Tremaine S. (2008) *Astrophys. J., 686,* 603.

Kaib N. A. and Quinn T. (2009) *Science, 325,* 1234.

Kaib N. A. et al. (2013) *Nature, 493,* 381.

Katz B. et al. (2011) *Phys. Rev. Lett., 107(18),* 181101.

Kenyon S. J. and Bromley B. C. (2004) *Nature, 432,* 598.

Kirsh D. R. et al. (2009) *Icarus, 199,* 197.

Kiseleva L. G. et al. (1998) *Mon. Not. R. Astron. Soc., 300,* 292.

Kley W. and Dirksen G. (2006) *Astron. Astrophys., 447,* 369.

Kley W. and Nelson R. P. (2012) *Annu. Rev. Astron. Astrophys., 50,* 211.

Kley W. et al. (2005) *Astron. Astrophys., 437,* 727.

Kobayashi H. and Ida S. (2001) *Icarus, 153,* 416.

Kokubo E. and Ida S. (1998) *Icarus, 131,* 171.

Kozai Y. (1962) *Astron. J., 67,* 591.

Kozai Y. (1985) *Cel. Mech., 36,* 47.

Kring D. A. and Cohen B. A. (2002) *J. Geophys. Res.–Planets, 107,* 5009.

Kuchner M. J. (2004) *Astrophys. J., 612,* 1147.

Lada C. J. and Lada E. A. (2003) *Annu. Rev. Astron. Astrophys., 41,* 57.

Lagrange A.-M. et al. (2000) In *Protostars and Planets IV* (V. Mannings et al., eds.), p. 639. Univ. of Arizona, Tucson.

Laskar J. (1989) *Nature, 338,* 237.

Laskar J. (1990) *Icarus, 88,* 266.

Laskar J. (1994) *Astron. Astrophys., 287,* L9.

Laskar J. (1997) *Astron. Astrophys., 317,* L75.

Laskar J. (2008) *Icarus, 196,* 1.

Laskar J. and Gastineau M. (2009) *Nature, 459,* 817.

Laughlin G. and Adams F. C. (1998) *Astrophys. J. Lett., 508,* L171.

Laughlin G. and Adams F. C. (2000) *Icarus, 145,* 614.

Lee M. H. and Peale S. J. (2002) *Astrophys. J., 567,* 596.

Lega E. et al. (2013) *Mon. Not. R. Astron. Soc., 431,* 3494.

Levison H. F. et al. (2007) In *Protostars and Planets V* (B. Reipurth et al., eds.), pp. 669–684. Univ. of Arizona, Tucson.

Levison H. F. et al. (2008) *Icarus, 196,* 258.

Levison H. F. et al. (2010) *Astron. J., 139,* 1297.

Levison H. F. et al. (2011) *Astron. J., 142,* 152.

Libert A.-S. and Delsate N. (2012) *Mon. Not. R. Astron. Soc., 422,* 2725.

Lidov M. L. (1962) *Planet. Space Sci., 9,* 719.

Lin D. N. C. and Ida S. (1997) *Astrophys. J., 477,* 781.

Lissauer J. J. (1995) *Icarus, 114,* 217.

Lissauer J. J. et al. (2011) *Astrophys. J. Suppl., 197,* 8.

Lithwick Y. and Naoz S. (2011) *Astrophys. J., 742,* 94.

Malmberg D. and Davies M. B. (2009) *Mon. Not. R. Astron. Soc., 394,* L26.

Malmberg D. et al. (2007a) *Mon. Not. R. Astron. Soc., 377,* L1.

Malmberg D. et al. (2007b) *Mon. Not. R. Astron. Soc., 378,* 1207.

Malmberg D. et al. (2011) *Mon. Not. R. Astron. Soc., 411,* 859.

Marchal C. and Bozis G. (1982) *Cel. Mech., 26,* 311.

Marchi S. et al. (2012a) *Earth Planet. Sci. Lett., 325,* 27.

Marchi S. et al. (2012b) *Science, 336,* 690.

Marchi S. et al. (2013) *ArXiv e-prints,* arXiv:1305.6679.

Marzari F. and Barbieri M. (2007) *Astron. Astrophys., 472,* 643.

Marzari F. and Weidenschilling S. J. (2002) *Icarus, 156,* 570.

Marzari F. et al. (2010) *Astron. Astrophys., 514,* L4.

Matsumoto Y. et al. (2012) *Icarus, 221,* 624.

Matsumura S. et al. (2010) *Astrophys. J., 714,* 194.

Matsumura S. et al. (2013) *Astrophys. J., 767,* 129.

Mayor M. and Queloz D. (1995) *Nature, 378,* 355.

Moeckel N. and Armitage P. J. (2012) *Mon. Not. R. Astron. Soc., 419,* 366.

Moeckel N. et al. (2008) *Astrophys. J., 688,* 1361.

Moorhead A. V. and Adams F. C. (2005) *Icarus, 178,* 517.

Morbidelli A. and Froeschlé C. (1996) *Cel. Mech. Dynam. Astron., 63,* 227.

Morbidelli A. and Levison H. F. (2004) *Astron. J., 128,* 2564.

Morbidelli A. et al. (2007) *Astron. J., 134,* 1790.

Morbidelli A. et al. (2010) *Astron. J., 140,* 1391.

Morton T. D. and Johnson J. A. (2011) *Astrophys. J., 729,* 138.

Murray C. D. and Dermott S. F. (1999) *Solar System Dynamics.* Cambridge Univ., Cambridge. 592 pp.

Mustill A. J. and Wyatt M. C. (2012) *Mon. Not. R. Astron. Soc., 419,* 3074.

Nagasawa M. and Ida S. (2011) *Astrophys. J., 742,* 72.

Nagasawa M. et al. (2008) *Astrophys. J., 678,* 498.

Naoz S. et al. (2011) *Nature, 473,* 187.

Naoz S. et al. (2012) *Astrophys. J. Lett., 754,* L36.

Nesvorný D. and Morbidelli A. (2012) *Astron. J., 144,* 117.

Ostriker E. C. (1994) *Astrophys. J., 424,* 292.

Papaloizou J. C. B. and Terquem C. (2001) *Mon. Not. R. Astron. Soc., 325,* 221.

Papaloizou J. C. B. et al. (2001) *Astron. Astrophys., 366,* 263.

Parker R. J. and Quanz S. P. (2012) *Mon. Not. R. Astron. Soc., 419,* 2448.

Perets H. B. and Naoz S. (2009) *Astrophys. J. Lett., 699,* L17.

Proszkow E.-M. and Adams F. C. (2009) *Astrophys. J. Suppl., 185,* 486.

Quillen A. C. (2011) *Mon. Not. R. Astron. Soc., 418,* 1043.

Rasio F. A. and Ford E. B. (1996) *Science, 274,* 954.

Raymond S. N. and Armitage P. J. (2013) *Mon. Not. R. Astron. Soc., 429,* L99.

Raymond S. N. et al. (2008) *Astrophys. J. Lett., 687,* L107.

Raymond S. N. et al. (2009a) *Astrophys. J. Lett., 699,* L88.

Raymond S. N. et al. (2009b) *Astrophys. J. Lett., 696,* L98.

Raymond S. N. et al. (2010) *Astrophys. J., 711,* 772.

Raymond S. N. et al. (2011) *Astron. Astrophys., 530,* A62.

Raymond S. N. et al. (2012) *Astron. Astrophys., 541,* A11.

Ribas I. and Miralda-Escudé J. (2007) *Astron. Astrophys., 464,* 779.

Ryder G. (2002) *J. Geophys. Res.–Planets, 107,* 5022.

Scharf C. and Menou K. (2009) *Astrophys. J. Lett., 693,* L113.

Schneider J. et al. (2011) *Astron. Astrophys., 532,* A79.

Shappee B. J. and Thompson T. A. (2013) *Astrophys. J., 766,* 64.

Shen Y. and Turner E. L. (2008) *Astrophys. J., 685,* 553.

Smith A. W. and Lissauer J. J. (2009) *Icarus, 201,* 381.

Spiegel D. S. and Burrows A. (2012) *Astrophys. J., 745,* 174.

Spurzem R. et al. (2009) *Astrophys. J., 697,* 458.

Sumi T. et al. (2011) *Nature, 473,* 349.

Sussman G. J. and Wisdom J. (1988) *Science, 241,* 433.

Sussman G. J. and Wisdom J. (1992) *Science, 257,* 56.

Takeda G. et al. (2008) *Astrophys. J., 683,* 1063.

Tera F. et al. (1974) *Earth Planet. Sci. Lett., 22,* 1.

Thompson T. A. (2011) *Astrophys. J., 741,* 82.

Throop H. B. and Bally J. (2008) *Astron. J., 135,* 2380.

Timpe M. L. et al. (2012) In *Am. Astron. Soc. Meeting Abstracts, 219,* 339.12.

Tremaine S. and Dong S. (2012) *Astron. J., 143,* 94.

Triaud A. H. M. J. et al. (2010) *Astron. Astrophys., 524,* A25.

Tsiganis K. et al. (2005) *Nature, 435,* 459.

Udry S. and Santos N. C. (2007) *Annu. Rev. Astron. Astrophys., 45,* 397.

Valtonen M. and Karttunen H. (2006) *The Three-Body Problem.* Cambridge Univ., Cambridge.

Vashkovýak M. A. (1999) *Astron. Lett., 25,* 476.

Veras D. and Armitage P. J. (2005) *Astrophys. J. Lett., 620,* L111.

Veras D. and Armitage P. J. (2006) *Astrophys. J., 645,* 1509.

Veras D. and Ford E. B. (2009) *Astrophys. J. Lett., 690,* L1.

Veras D. and Ford E. B. (2010) *Astrophys. J., 715,* 803.

Veras D. and Mustill A. J. (2013) *ArXiv e-prints,* arXiv:1305.5540.

Veras D. and Raymond S. N. (2012) *Mon. Not. R. Astron. Soc., 421,* L117.

Veras D. et al. (2009) *Astrophys. J., 696,* 1600.

Veras D. et al. (2011) *Mon. Not. R. Astron. Soc., 417,* 2104.

Veras D. et al. (2013) *Mon. Not. R. Astron. Soc., 431,* 1686.

Voyatzis G. et al. (2013) *Mon. Not. R. Astron. Soc., 430,* 3383.

Wadhwa M. et al. (2007) In *Protostars and Planets V* (B. Reipurth et al., eds.), pp. 835–848. Univ. of Arizona, Tucson.

Wan X.-S. and Huang T.-Y. (2007) *Mon. Not. R. Astron. Soc., 377,* 133.

Wasserburg G. J. et al. (2006) *Nucl. Phys. A, 777,* 5.

Weidenschilling S. J. (1977) *Astrophys. Space Sci., 51,* 153.

Weidenschilling S. J. and Marzari F. (1996) *Nature, 384,* 619.

Wiegert P. and Tremaine S. (1999) *Icarus, 137,* 84.

Williams J. P. and Cieza L. A. (2011) *Annu. Rev. Astron. Astrophys., 49,* 67.

Winn J. N. et al. (2009) *Astrophys. J. Lett., 703,* L99.

Winn J. N. et al. (2010) *Astrophys. J. Lett., 718,* L145.

Wisdom J. (1980) *Astron. J., 85,* 1122.

Wright J. T. et al. (2009) *Astrophys. J., 693,* 1084.

Wright J. T. et al. (2011) *Publ. Astron. Soc. Pac., 123,* 412.

Wu Y. and Lithwick Y. (2011) *Astrophys. J., 735,* 109.

Wu Y. and Murray N. (2003) *Astrophys. J., 589,* 605.

Wu Y. et al. (2007) *Astrophys. J., 670,* 820.

Wyatt M. C. (2008) *Annu. Rev. Astron. Astrophys., 46,* 339.

Xie J.-W. et al. (2011) *Astrophys. J., 735,* 10.

Zakamska N. L., Pan M., and Ford E. B. (2011) *Mon. Not. R. Astron. Soc., 410,* 1895.

Zuckerman B. (2001) *Annu. Rev. Astron. Astrophys., 39,* 549.

Davis A. M., Alexander C. M. O'D., Ciesla F. J., Gounelle M., Krot A. N., Petaev M. I., and Stephan T. (2014) Samples of the solar system: Recent developments. In *Protostars and Planets VI* (H. Beuther et al., eds.), pp. 809–831. Univ. of Arizona, Tucson, DOI: 10.2458/azu_uapress_9780816531240-ch035.

Samples of the Solar System: Recent Developments

Andrew M. Davis
The University of Chicago

Conel M. O'D. Alexander
Carnegie Institution for Science

Fred J. Ciesla
The University of Chicago

Matthieu Gounelle
Museum National d'Histoire Naturelle

Alexander N. Krot
University of Hawai'i

Michail I. Petaev
Harvard University and Harvard-Smithsonian Center for Astrophysics

Thomas Stephan
The University of Chicago

We review some of the major findings in cosmochemistry since the Protostars and Planets V meeting: (1) the results of the sample-return space missions Genesis, Stardust, and Hayabusa, which yielded the oxygen and nitrogen isotopic composition of the Sun, evidence for significant radial transport of solids in the protoplanetary disk and the chondrite-comet connection, and the connection between ordinary chondrites and S-type asteroids, respectively; (2) the deuterium/hydrogen (D/H) ratio of chondritic water as a test of the Nice and Grand Tack dynamical models and implications for the origin of Earth's volatiles; (3) the origin, initial abundances, and distribution of the short-lived radionuclides ^{10}Be, ^{26}Al, ^{36}Cl, ^{41}Ca, ^{53}Mn, and ^{60}Fe in the early solar system; (4) the absolute (U-Pb) and relative (short-lived radionuclide) chronology of early solar system processes; (5) the astrophysical setting of the solar system formation; (6) the formation of chondrules under highly nonsolar conditions; and (7) the origin of enstatite chondrites.

1. INTRODUCTION

The seven years since publication of *Protostars and Planets V* (*Reipurth et al.,* 2007) have seen a number of major discoveries about the chemistry and physics of the solar system and other stars based on the laboratory study of samples returned to Earth by spacecraft and by nature (i.e., meteorites and interplanetary dust particles). In this chapter, we highlight some of the more important discoveries and their implications.

2. THE GENESIS MISSION

The Genesis mission, which returned samples of the solar wind to Earth in 2004, has provided two very surprising results: that the solar wind, and by inference, the Sun, is significantly different from Earth and other sampled rocky bodies of the solar system in its oxygen and nitrogen isotopic compositions (*Burnett and Genesis Science Team,* 2011).

Genesis was launched in 2001 and placed at the L1 point, ~1% of the way to the Sun, where a variety of ultrapure materials were exposed to the solar wind for 27 months. The spacecraft carried an electrostatic concentrator, which concentrated solar wind by ~20× onto a 6-cm-diameter target and increased the implantation depth by about a factor of 3. Genesis returned to Earth on September 8, 2004, and crashed in the Utah desert because a parachute failed to deploy. The Genesis team was able to recover many pieces of collector materials, and the concentrator target survived nearly intact (*Burnett and Genesis Science Team,* 2011).

2.1. Oxygen Isotopes in the Solar System

Oxygen has three stable isotopes, ^{16}O, ^{17}O, and ^{18}O. In most physical and chemical processes occurring on Earth, the $^{18}O/^{16}O$ ratio fractionates by twice as much as the $^{17}O/^{16}O$ ratio. However, *Clayton et al.* (1973) discovered that some components in meteorites are enriched in ^{16}O by about 4%, implying mixing between ^{16}O-rich and ^{16}O-poor components. Oxygen isotopic compositions are now well-established for exploring relationships among planetary materials (*Clayton, 2003; Krot et al., 2014; Scott and Krot, 2014*). The origin of mass-independent oxygen isotopic variations among solar system materials has remained a mystery for 40 years. For much of that time, these variations were thought by most to be relics of incomplete mixing of different nucleosynthetic components in the early solar system, although chemically produced mass-independent fractionation processes were demonstrated in the laboratory and proposed as an alternative mechanism (e.g., *Thiemens and Heidenreich, 1983; Thiemens, 2006*).

Most bulk meteorites, including those from Mars, are within a few permil of average terrestrial oxygen isotopic composition, and it was thought that the Sun would have a similar isotopic composition. Thus, the initial goal of the Genesis mission was to measure the solar wind and infer the solar oxygen isotopic composition with a precision of 1‰ or better. This all changed when a very brief paper by *Clayton* (2002) suggested that the oxygen isotopic variations were caused by photochemical self-shielding of CO in the solar nebula and that the Sun could be significantly ^{16}O-rich compared to most meteorites and planets. Self-shielding was already thought to cause carbon and oxygen isotopic variations in molecular clouds (*Bally and Langer, 1982; van Dishoeck and Black, 1988*). Ultraviolet (UV) photodissociation of CO yields O atoms that may react with ambient H_2 to form H_2O. The path length for UV photodissociation of $C^{16}O$ is much shorter than for those of $C^{17}O$ and $C^{18}O$, because ^{16}O is much more abundant than ^{17}O and ^{18}O. Water made from O from photodissociated CO in the interior of a CO-containing cloud could then be trapped in rock-forming compounds and make silicates enriched in $^{17}O/^{16}O$ and $^{18}O/^{16}O$ compared to the bulk cloud. There are currently three proposals for where CO self-shielding could take place: (1) near the Sun (*Clayton, 2002*); (2) at the surface of the solar nebula, several astronomical units from the Sun (*Lyons and Young, 2005*); or (3) in molecular clouds (*Yurimoto and Kuramoto, 2004*). In the latter proposal, isotopic heterogeneities in molecular cloud material generated by CO self-shielding are preserved in materials formed in the early solar system.

Measurement of the oxygen isotopic composition of solar wind was the highest priority of the Genesis mission. A new instrument combining the techniques of secondary ion mass spectrometry and accelerator mass spectrometry was built at the University of California, Los Angeles, and showed that the Sun is enriched in ^{16}O by about 5% relative to Earth and all known rocky material in the solar system except refrac-

tory inclusions (*McKeegan et al., 2011*). The oxygen isotopic compositions of solar system materials are shown in Fig. 1. There is likely some mass-dependent isotopic fractionation of oxygen during acceleration of the solar wind. The oxygen isotopic composition of the Sun was determined by drawing a line with slope-1/2 through the measured solar-wind composition and assuming that the Sun is at the intersection of that line and a line of slope 0.95 along which lie refractory inclusions and a few unusual chondrules from carbonaceous chondrites (*McKeegan et al., 2011*).

The difference in oxygen isotopic composition between the bulk solar system and rocky material is in the direction predicted by CO self-shielding, but other mechanisms have been proposed to explain oxygen isotopic variations in solar system materials. A difference in oxygen isotopic composition between gas and dust could have arisen in presolar materials by galactic chemical evolution (GCE) (*Clayton, 1988; Krot et al., 2010a*). However, GCE could lead to dust being enriched or depleted in ^{16}O, and GCE does not necessarily increase ^{17}O and ^{18}O equally relative to ^{16}O (*Lugaro et al., 2012a; Nittler and Gaidos, 2012*). The $^{18}O/^{17}O$ ratio of the solar system is 5.2 ± 0.2 (*Young et al., 2011*), but molecular clouds (*Penzias, 1981; Wouterloot et al., 2008*) and young stellar objects (*Smith et al., 2009*) in our region of the Milky Way tend to have lower values of 4.1 ± 0.1 and 4.1–4.4, respectively — there is a hint of a gradient in $^{18}O/^{17}O$ as a function of galactocentric radius (*Wouterloot et al., 2008; Nittler and Gaidos, 2012*), but there is no overlap in error bounds between the solar system ratio and those of objects of similar galactocentric radius. Ultraviolet photodissociation of CO has been shown to produce large, mass-independent oxygen isotopic fractionation (*Chakraborty et al., 2008*), although a vigorous

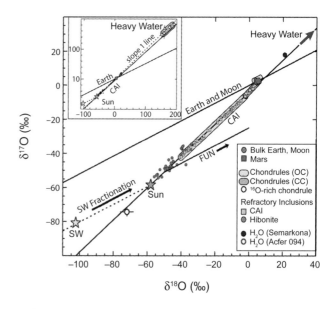

Fig. 1. Oxygen isotopic composition of solar system materials, expressed in δ units: ($\delta^i O = [(^iO/^{16}O)/(^iO/^{16}O)_{SMOW}]-1) \times 1000$, i = 17 or 18; SMOW is standard mean ocean water. From *McKeegan et al.* (2011).

debate about the mechanism ensued (*Federman and Young,* 2009; *Lyons et al.,* 2009a; *Yin et al.,* 2009; *Chakraborty et al.,* 2009). Gas-phase reaction of SiO with O has also produced mass-independent oxygen isotopic fractionation in the laboratory (*Chakraborty et al.,* 2013b).

2.2. Nitrogen Isotopes in the Solar System

Nitrogen has only two stable isotopes, ^{14}N and ^{15}N, so that it can be difficult to disentangle chemical/physical and nucleosynthetic effects. The $^{15}N/^{14}N$ ratio of the solar wind, measured in the Genesis concentrator and corrected for fractionation during acceleration of the solar wind, is $(2.268 \pm 0.028) \times 10^{-3}$ (*Marty et al.,* 2011), ~40% below that of atmospheric N_2, 3.676×10^{-3}. The former value is taken to be representative of the bulk Sun. The solar $^{15}N/^{14}N$ ratio is quite similar to that of Jupiter's atmosphere, $(2.3 \pm 0.3) \times 10^{-3}$ (*Owen et al.,* 2001) and in osbornite, TiN, in an unusual refractory inclusion in the Isheyevo carbonaceous chondrite, $(2.356 \pm 0.0.018) \times 10^{-3}$, which was argued to have condensed in the solar nebula (*Meibom et al.,* 2007). Most meteorites have $^{15}N/^{14}N$ ratios within 10% of the ratio in Earth's atmosphere, although CR (Renazzo-type), CB (Bencubbin-type), and a couple of ungrouped carbonaceous chondrites can have much higher ratios (*Prombo and Clayton,* 1985; *Grady and Pillinger,* 1990; *Alexander et al.,* 2012). Insoluble organic matter (IOM) is the main carrier of nitrogen in carbonaceous chondrites (*Alexander et al.,* 2007) and micrometer-sized hot spots in IOM in carbonaceous chondrites can have large enrichments in ^{15}N (*Busemann et al.,* 2006; *Nakamura-Messenger et al.,* 2006; *Briani et al.,* 2009; *De Gregorio et al.,* 2013). Much remains to be done to understand the causes of nitrogen isotopic variations in the solar system. Large isotopic fractionations via ion-molecule reactions in cold molecular clouds may explain some of the large enrichments in $^{15}N/^{14}N$; UV self-shielding of N_2, a molecule that is isoelectronic with CO, may play a role (*Clayton,* 2002; *Lyons et al.,* 2009b), and UV photodissociation of N_2 has been observed to produce large isotope effects in the laboratory (*Chakraborty et al.,* 2013a), although the mechanism is unclear.

3. WATER IN THE SOLAR SYSTEM

The debate over whether asteroids and/or comets were the major sources of water and other volatiles for the terrestrial planets has been revitalized by the publication of the Grand Tack model (*Walsh et al.,* 2011) and the discovery of a Jupiter-family comet (JFC), Hartley 2, with water ice that has a roughly terrestrial D/H ratio (*Hartogh et al.,* 2011).

The Grand Tack model was originally designed to explain the small mass of Mars, and the scenario it outlines takes place in the first few million years of solar system history while the solar nebula is still present. In it, the still-growing giant planets experience an episode of dramatic radial migration. Jupiter migrates from its formation location at ~3.5 AU from the Sun inward to about 1.5 AU,

at which point Saturn catches up with it and then the two migrate outward again to about 5 and 7 AU, respectively. Uranus and Neptune are also forced to migrate to greater distances (roughly 10 and 13 AU, respectively) as Jupiter and Saturn moved outward. As a result of this migration, not only are the planetesimals at ~1.5–3 AU depleted, explaining Mars' low mass, but the planetesimals between 4 and 13 AU are strongly perturbed, and some of them are scattered into the inner solar system. These presumably icy bodies would have been potent sources of volatiles for the growing terrestrial planets. Some of these outer solar system interlopers also come to dominate the outer part of the asteroid belt.

At the end of the Grand Tack model, the giant planets are not in their current orbits. Their final orbits are established by a second episode of orbital migration ~4 Ga, in a scenario known as the Nice model (*Gomes et al.,* 2005; *Levison et al.,* 2011). The Nice model not only reproduces many features of the Kuiper belt and scattered disk, but also provides an explanation for the late heavy bombardment as planetesimals between ~15 and 30 AU are scattered, primarily by the migration of Neptune. As in the Grand Tack model, some of these scattered objects become trapped in the asteroid belt, particularly the outer part of it (*Levison et al.,* 2009).

There is a radial zonation in the relative abundances of asteroids belonging to the two main spectral classes (*Gradie and Tedesco,* 1982; *Bus and Binzel,* 2002), with the S-complex asteroids dominating the inner asteroid belt and the C-complex asteroids dominating the outer asteroid belt. Thus, the Grand Tack and Nice dynamical models predict that the S-complex asteroids formed in the inner solar system (<4 AU), while the C-complex asteroids formed in the outer solar system (>4 AU).

Most meteorites, except those from the Moon and Mars, are thought to come from the asteroid belt, although it has been suggested that a few come from comets (*Gounelle et al.,* 2006b). Spectroscopically, the ordinary chondrites have been linked to the S-complex asteroids and the carbonaceous chondrites to the C-complex asteroids (*Burbine et al.,* 2002; *Burbine,* 2014). The link between S-complex asteroids and ordinary chondrites has been demonstrated by the Near Earth Asteroid Rendezvous (NEAR) mission to the S-asteroid Eros (*Nittler et al.,* 2001; *Foley et al.,* 2006) and the Hayabusa sample return mission to the S-asteroid 25143 Itokawa (*Nakamura et al.,* 2011; *Yurimoto et al.,* 2011). To date, there have been no missions to C-complex asteroids, but Hayabusa 2 and the Origins Spectral Interpretation Resource Identification and Security-Regolith Explorer (OSIRIS-REx) plan to launch within the next three years and return samples of C-type asteroids to Earth in the 2020s. Confirming a link between carbonaceous chondrites and C-complex asteroids would bolster the Grand Tack and Nice models.

There does seem to be a continuum of petrologic properties between chondritic meteorites and dust from the one object unambiguously from the outer solar system — the JFC 81P/Wild 2, which was sampled by the Stardust mission (section 7) (*Brownlee et al.,* 2006; *Gounelle et al.,* 2008; *Ishii et al.,* 2008a; *Zolensky et al.,* 2008). This continuum need not indicate formation in similar locations. It could simply reflect vigorous radial mixing in the early solar nebula (*Ciesla,* 2007; *Boss,* 2008), and none of the petrologic features are clear indicators of formation distance from the Sun. There is also evidence for two isotopically distinct reservoirs of material in the inner solar system (*Warren,* 2011) — one populated by Earth, the Moon, Mars, ordinary and enstatite chondrites, and differentiated meteorites, and the other composed of the carbonaceous chondrites. While these differences are consistent with the dynamical models, again they are not unambiguous indicators of formation distance from the Sun.

3.1. Deuterium/Hydrogen Ratios and Water

Most outer solar system planetesimals appear to be water-ice-rich, and the D/H ratio of water is a promising indicator of formation distance. Simple models of an already formed and turbulent disk suggest that there would have been a strong radial gradient in the water-ice D/H ratio in the solar nebula due to radial mixing between (1) outer solar system ice that was dominated by very D-rich (D/H $\geq 10^{-3}$) ice inherited from the presolar molecular cloud; and (2) inner solar system water that had been reequilibrated with H_2 gas at high temperatures near the Sun (D/H $\approx 2 \times 10^{-5}$), before being transported out beyond the snow line (e.g., *Drouart et al.,* 1999; *Mousis et al.,* 2000; *Horner et al.,* 2007, 2008; *Jacquet and Robert,* 2013). More realistic models that take into account the initial collapse phase of solar system formation, as well as continuing infall of molecular cloud material onto the disk for some time after it is established, predict more complex temporal and radial variations in water D/H ratios (*Yang et al.,* 2013). For instance, in the simulations, about 1 million year (m.y.) after the start of solar system formation, water D/H ratios are uniformly low (~2×10^{-5}) in the inner disk, rising steeply at ~1 AU to a D-rich plateau (3–4 × 10^{-4}) that extends out to tens of astronomical units before falling again. Nevertheless, these more complex models still predict that outer solar system objects that formed in the same regions as the giant planets should be D-rich compared to the initial solar value and to Earth.

The establishment of a fairly uniform water D/H ratio throughout much of the outer solar system is consistent with the similarly D-enriched (D/H ≈ 2.5–4×10^{-4}) compositions of Saturn's moon Enceladus and most measured comets (*Waite et al.,* 2009; *Hartogh et al.,* 2011) relative to Earth (D/H $\approx 1.5 \times 10^{-4}$). Most of the measured comets are from the Oort cloud. The one JFC that has been measured, Hartley 2, has a more Earth-like water D/H ratio (*Hartogh et al.,* 2011). The model of *Yang et al.* (2013) predicts low-D/H water throughout the solar nebula very early and at all times beyond the D/H-rich plateau, either of which are possible formation locations and times for JFCs.

Irrespective of the details, the general expectations of the water D/H models, which are largely backed up by observations, are that ice in outer solar system objects will be significantly more D-rich than the initial compositions of inner solar system objects (Venus and Mars are now exceptions, but this is because both have lost significant water, which led to enhanced D/H ratios; their original D/H ratios are thought to be similar to that of Earth). Therefore, the Grand Tack and Nice models predict that the parent bodies of the carbonaceous chondrites should have water with elevated D/H ratios relative to Earth. On the other hand, the ordinary, enstatite, and Rumuruti-type (R) chondrites that are thought to have formed in the inner solar system should have water D/H ratios similar to or below that of Earth.

Determining the bulk isotopic composition of water in chondrites is not straightforward because hydrated minerals are generally very fine-grained and intimately mixed with D-rich organic matter. Internal isotopic fractionation and exchange processes may also have operated during aqueous alteration. Lastly, there is the possibility of terrestrial contamination due to weathering and exchange with atmospheric water. To try to overcome these problems, *Alexander et al.* (2012) used the bulk elemental and isotopic compositions of hydrogen and carbon in chondrites to estimate the water D/H ratios for a number of chondrite groups, by extrapolating to C/H = 0 on a plot of δD vs. C/H. They found no evidence for severe terrestrial water contamination and concluded that the Ivuna-type (CI) and Mighei-type (CM) carbonaceous chondrites, as well as the unique carbonaceous chondrite Tagish Lake, have bulk water D/H ratios that are significantly ($\geq$40%) depleted in D compared to Earth. Only the CR chondrites have water D/H ratios that are higher than Earth's, but only by ~15% and not the factor of ~2 seen in most comets. On the other hand, water in the most primitive ordinary chondrite, Semarkona, is enriched in D by roughly a factor of 2 compared to Earth, which is similar to those of Oort cloud comets (OCCs). Previous *in situ* measurements of hydrated silicates in an R chondrite showed that it contains water that is more D-rich than any measured comet (*McCanta et al.,* 2008). The water contents of other carbonaceous chondrite groups [Ornans-type (CO); Vigarano-type (CV); Karoonda-type (CK); Allan Hills 85085-like (CH), and CB] are too low to obtain reliable water D/H ratios.

These results are the reverse of what was anticipated from the Grand Tack models and Nice models. The Tagish Lake results are particularly problematic for the Nice model. Spectroscopically, Tagish Lake resembles D-type asteroids (*Hiroi et al.,* 2001), and it was D-type asteroids that *Levison et al.* (2009) suggested were the most likely to have been implanted into the outer asteroid belt during the Nice planetary migrations. If the OCCs and Enceladus formed between the giant planets, the lower water-D/H ratios in all carbonaceous chondrites suggest that they did not form much beyond the orbit of Jupiter. On the other hand, if the OCCs formed in the same regions as the JFCs (i.e., beyond 15 AU) as some prefer, Enceladus becomes the only constraint on water D/H ratios in the interplanetary region. The high D/H ratio of Enceladus indicates that it formed toward the end of Saturn's growth when accretion rates were low and the saturnian subdisk was cool (*Waite et al.*, 2009). Otherwise, H_2-H_2O reequilibration would have driven down the D/H ratio of the water. In the Grand Tack model, Saturn approaches its final mass as it migrates outward from ~3 AU to ~7 AU. Thus, Enceladus indicates that the carbonaceous chondrites formed <7 AU from the Sun and possibly <3 AU (*Alexander et al.*, 2012). This is in contrast to the predictions of the Grand Tack model that the asteroid belt would have been repopulated with bodies from distances out to ~13 AU following the Grand Tack.

The fact that the ordinary and R chondrites have much higher D enrichments in their water than the carbonaceous chondrites does not necessarily mean that they formed in the outer solar system or accreted water ice that had migrated into the inner solar system. There are potentially very large hydrogen isotopic fractionations associated with the oxidation of metallic iron by water that will leave the remaining water isotopically much heavier (*Alexander et al.*, 2010). The extent of any D-enrichments will have depended on the fraction of water consumed by these reactions and the temperature of reaction. The ordinary chondrites, at least, seem to have had much lower water/metal ratios on their parent bodies than the CI, CM, CR, and Tagish Lake carbonaceous chondrites and, as a result, a much higher fraction of their water was consumed by metal oxidation. Consequently, it is not surprising that the remaining water in Semarkona is much more D-rich. This process should have also affected the carbonaceous chondrites, so that their water D/H ratios at the time of accretion would also have been lower than estimated by *Alexander et al.* (2012), increasing the difference in the water D/H ratios between them and the measured outer solar system objects. At present, it appears that both ordinary and carbonaceous chondrites formed relatively close to the Sun, contradicting the predictions of the two dynamical models.

3.2. Deuterium/Hydrogen Constraints on the Sources of Earth's Volatiles

The high water D/H ratios in the measured OCCs make them unlikely candidates for the major sources of Earth's water. The same conclusion applies to the Moon and Mars, because the hydrogen isotopic compositions of water in their interiors, inferred from measurements of martian meteorites and lunar samples, appear to be similar to that of Earth's (e.g., *Usui et al.*, 2012; *Hallis et al.*, 2012; *Saal et al.*, 2013). As discussed above, it seems unlikely that planetesimals from much of the giant planet region were the sources of Earth's water. The limits on how far from the Sun volatile planetesimals could have come from are largely determined by the formation location of Enceladus, which would have been between 3 and 7 AU in the context of the Grand Tack model.

Based on the Earth-like water D/H ratio of Hartley 2, it has been suggested that JFCs are potential sources for Earth's volatiles (*Hartogh et al.*, 2011). However, there are a number of reasons that make this unlikely. First, the planets would have accreted whole comets and not just their ices. Comets are thought to be organic-rich and, if these organics are similar to those in chondrites and interplanetary dust particles (IDPs), this organic material is very D-rich. As a result, the bulk D/H ratio of Hartley 2 (and other comets) is likely to be significantly higher than Earth's (*Alexander et al.*, 2012). Second, delivery of large numbers of JFCs to Earth from their source region in the Kuiper belt requires a mechanism like the orbital migration envisaged by the Nice model. The Nice model takes place several hundred million years after the earliest evidence for the existence of an atmosphere (*Pepin and Porcelli*, 2006; *Mukhopadhyay*, 2012) and ocean (*Harrison*, 2009) on Earth. Third, the evidence from highly-siderophile-element abundances and isotopic compositions is that the material accreted by Earth after the end of core formation (the so-called late veneer), normally assumed to be the Moon-forming impact [~4.4–4.5 G.y. ago (*Touboul et al.*, 2007; *Borg et al.*, 2011)], was too little (~0.5%) to account for Earth's volatile budget and was not carbonaceous-chondrite-like but closer in composition to the volatile-poor enstatite and ordinary chondrites (*Walker*, 2009; *Burkhardt et al.*, 2011). Indeed, recent models suggest that most of the material accreted since the end of core formation was from the inner asteroid belt (*Bottke et al.*, 2010, 2012).

Thus, asteroids remain the most likely sources of volatiles for Earth and the other terrestrial planets. Assuming that all the major types of asteroid are represented in our meteorite collections, which ones most closely resemble estimates of Earth's volatile budget? The aqueously altered carbonaceous chondrites are the only meteorite groups that contain sufficient volatile contents. In terms of both hydrogen and nitrogen isotopes, Earth's bulk composition most closely resembles those of the CI and CM carbonaceous chondrites (*Alexander et al.*, 2012). Estimates of the relative abundances of hydrogen, carbon, chlorine, bromine, iodine, and the noble gases (except xenon) in Earth are also similar to CI and CM chondrites (*Marty*, 2012; *Marty et al.*, 2013). Addition of about 2 wt.% of CI material (or somewhat more CM material) would explain the absolute abundances of these elements in Earth (*Alexander et al.*, 2012; *Marty*, 2012; *Marty et al.*, 2013). To satisfy the siderophile-element isotopic constraints, accretion of CI/CM material must have predated the end of core formation (i.e., the Moon-forming impact).

4. EARLY SOLAR SYSTEM PROCESSES AND CHRONOLOGY

4.1. Origin of Refractory Inclusions, Chondrules, and Matrices: Current Views

Chondritic meteorites consist of three major components: refractory inclusions [calcium-aluminum-rich inclusions (CAIs) and amoeboid olivine aggregates (AOAs)], chondrules, and fine-grained matrices.

The mineralogy and major-element chemistry of refractory inclusions are similar to those expected for solids in equilibrium with a high-temperature ($\geq$1350 K) gas of solar composition (e.g., *Grossman et al.,* 2002). Oxygen-isotope compositions of refractory inclusions are enriched in ^{16}O relative to the standard mean ocean water (SMOW) [$\delta^{17,18}$O ~ –50‰, Δ^{17}O ~ –25‰, where δ^{17}O and δ^{18}O are defined in the caption to Fig. 1 and Δ^{17}O = δ^{17}O–0.52 × δ^{18}O (*Yurimoto,* 2008; *Gounelle et al.,* 2009a)] and similar to that of the Sun inferred from the analyses of the solar wind returned by the Genesis spacecraft (*McKeegan et al.,* 2011) (Fig. 1). Calcium-aluminum-rich inclusions preserve isotopic records of solar energetic particle irradiation (see section 4.5). It is inferred that CAIs formed by evaporation, condensation, aggregation, and, in some cases, melting in a gas of approximately solar composition in a hot (ambient temperature $\geq$1350 K) part of the protoplanetary disk, possibly near the proto-Sun. They were subsequently transported throughout the disk, possibly by a stellar wind (*Shu et al.,* 1996), by turbulent mixing (*Ciesla,* 2010), and/or by gravitational instabilities following FU-Orionis outbursts (*Boss et al.,* 2012), and accreted into chondrites and comets

Chondrules, up to millimeter-sized spherules of silicate, metal, and sulfide, are the most characteristic features of chondrites, with the exception of the CI chondrites. The chondrules in each chondrite group have a distinct range of physical and chemical properties, as well as abundances [from 20 to 70–80 vol.% (*Brearley and Jones,* 1998; *Jones et al.,* 2005; *Rubin,* 2010)]. The mineralogy, petrology, and bulk chemistry of chondrules indicate that they formed by localized transient heating events in the dust-enriched relatively cold (ambient temperature $\leq$650 K) regions of the protoplanetary disk having ^{16}O-depleted compositions (Δ^{17}O ~ ±5‰) compared to the majority of CAIs (e.g., *Alexander et al.,* 2008; *Krot et al.,* 2009). Most chondrules formed by incomplete melting of isotopically diverse solid precursors, including CAIs, chondrules of previous generations, and, possibly, fragments of thermally processed planetesimals (*Scott and Krot,* 2014). Chondrules in the CB carbonaceous chondrites appear to have formed by complete melting of solid precursors and by gas-liquid condensation, most likely in an impact-generated plume (*Krot et al.,* 2005, 2010b; *Olsen et al.,* 2013). For additional information about the physicochemical conditions and possible mechanisms of chondrule formation, see below.

Fine-grained matrices in primitive (unmetamorphosed and unaltered) chondrites consist of micrometer-sized crystalline magnesian olivine and pyroxene, Fe,Ni-metal and sulfides, and amorphous ferromagnesian silicates (e.g., *Abreu and Brearley,* 2010). Whether matrix and chondrules are chemically complementary and therefore genetically related is still a matter of debate (*Bland et al.,* 2005; *Hezel and Palme,* 2010; *Zanda et al.,* 2012). It has been argued that a significant fraction of matrix materials experienced evaporation and recondensation during chondrule formation (e.g., *Wasson,* 2008), because fine-grained materials

would certainly have evaporated under the conditions of chondrule melting and cooling. However, fine-grained matrices of unmetamorphosed chondrites also contain presolar silicates, silicon carbide, and graphite (*Zinner,* 2014), as well as other highly unequilibrated materials (*Alexander et al.,* 2001; *Alexander,* 2005; *Bland et al.,* 2007), including organic matter (e.g., *Alexander et al.,* 2007). None of these materials can have been heated to high temperature during chondrule formation, as their isotopic and chemical signatures would have been erased by exchange during evaporation, mixing with chondrule gases, and recondensation.

4.2. Astrophysically and Cosmochemically Important Short-Lived and Long-Lived Radionuclides

The elements in the solar system were formed in previous generations of stars. Their relative abundances represent the result of the general chemical evolution of the Galaxy enhanced by local additions from one or more neighboring stellar sources prior to, contemporaneously with, or after the collapse of the protosolar molecular cloud. Some of the elements are radioactive and decay to stable isotopes with different half-lives. The decay products of radioactive isotopes can be measured with high precision using various kinds of mass spectrometry [thermal ionization mass spectrometry (TIMS); secondary ion mass spectrometry (SIMS); and inductively-coupled plasma mass spectrometry (ICPMS)]. These measurements provide the basis for understanding (1) the astrophysical setting of the solar system formation; (2) the irradiation environment in the protoplanetary disk; (3) the chronology of the processes in the early solar system such as the formation of refractory inclusions and chondrules, and aqueous alteration, metamorphism, melting, and differentiation of planetesimals; and (4) the lifetime of the protoplanetary disk. In this section, we summarize recent developments in understanding the origin and distribution of several short-lived radionuclides (SLRs) ($t_{1/2} \leq$ 100 m.y.; Table 1) having important astrophysical and cosmochemical implications (^{10}Be, ^{26}Al, ^{36}Cl, ^{41}Ca, ^{53}Mn, ^{60}Fe, and ^{182}Hf). We also summarize the latest results on absolute chronology of CAIs, chondrules, and differentiated meteorites. For an up-to-date survey of SLRs, see *Davis and McKeegan* (2014).

4.3. Aluminum-26

4.3.1. Origin and distribution of aluminum-26 in the early solar system: Current views. The short half-life of ^{26}Al and its presence in the earliest solids formed in the inner solar system [CAIs, chondrules, chondrites, and achondrites (*Lee et al.,* 1977; *Davis and McKeegan,* 2014, and references therein)] make it an important heat source for early accreted planetesimals. If ^{26}Al was uniformly distributed in the inner solar system, it is also important for high-precision relative chronology of CAI and chondrule formation and melting and differentiation of

TABLE 1. Selected early solar system SLRs.

R	D	S	$t_{1/2}$	R/S
^{10}Be	^{10}B	^{9}Be	1.387	$(0.3-10) \times 10^{-3}$
^{26}Al	^{26}Mg	^{27}Al	0.717	$(5.23 \pm 0.13) \times 10^{-5}$
^{36}Cl	^{36}S and ^{36}Ar	^{35}Cl	0.301	$\gg 2 \times 10^{-5}$
^{41}Ca	^{41}K	^{40}Ca	0.102	4×10^{-9}
^{53}Mn	^{53}Cr	^{55}Mn	3.74	$(6.71 \pm 0.56) \times 10^{-6}$
^{60}Fe	^{60}Ni	^{56}Fe	2.62	$(9.7 \pm 1.9) \times 10^{-9}$
^{182}W	^{182}Hf	^{180}Hf	8.90	$(9.81 \pm 0.41) \times 10^{-6}$

R is the radionuclide, D its daughter isotope, S the reference stable isotope, and $t_{1/2}$ is the half-life in million years [from *www.nndc.bnl.gov*, except for ^{60}Fe, which is from *Rugel et al.* (2009)]. Inferred early solar system ratios, R/S, are from *Davis and McKeegan* (2014), except for ^{60}Fe (see section 4.4), scaled to an absolute solar system age of 4567.30 Ma, which is inferred from absolute ages of CV CAIs (*Connelly et al.*, 2012).

planetesimals. Both the origin (local irradiation vs. external stellar nucleosynthesis) and distribution (homogeneous vs. heterogeneous) of ^{26}Al in the early solar system remain controversial.

4.3.2. Heterogeneous distribution of aluminum-26 in the protoplanetary disk during an epoch of the formation of calcium-aluminum-rich inclusions. Calcium-aluminum-rich inclusions are the oldest solar system solids that have been dated (*Amelin et al.*, 2010; *Bouvier et al.*, 2011; *Connelly et al.*, 2012), and therefore provide important constraints on the origin and distribution of ^{26}Al in the early solar system. On an isochron diagram (^{26}Mg/^{24}Mg vs. ^{27}Al/^{24}Mg), the compositions of whole CAIs from the CV chondrite Allende measured by multicollector ICPMS (MC-ICPMS) yield a well-defined regression line (model isochron) with the initial ^{26}Al/^{27}Al ratio, $(^{26}$Al/^{27}Al$)_0$, of $(5.23 \pm 0.13) \times 10^{-5}$ (*Jacobsen et al.*, 2008), named the "canonical" ratio. A similar value for $(^{26}$Al/^{27}Al$)_0$, $(5.25 \pm 0.02) \times 10^{-5}$, was later inferred from measurements of whole CAIs and AOAs from the less-altered CV chondrite Efremovka (Fig. 2) (*Larsen et al.*, 2011). The previously reported so-called "supracanonical" $(^{26}$Al/^{27}Al$)_0$ values ($\geq 5.6 \times 10^{-5}$) in CV CAIs (*Young et al.*, 2005; *Thrane et al.*, 2006) appear to be either analytical artifacts or the result of late-stage disturbances of the ^{26}Al-^{26}Mg systematics of the CV CAIs in the solar nebula and/or in the CV chondrite parent asteroid.

The canonical ^{26}Al/^{27}Al ratio and the reported homogeneity of this ratio to ~2% in the whole-rock CV CAIs are often interpreted as evidence for (1) uniform distribution of ^{26}Al in the solar system and (2) a very short (≤ 0.04 or ≤ 0.002 m.y.) duration of CAI formation (e.g., *Jacobsen et al.*, 2008; *Larsen et al.*, 2011, respectively). This interpretation, however, has several caveats and is most likely incorrect. (1) The CAI-forming region could have had a different initial ^{26}Al/^{27}Al ratio from the rest of the protoplanetary disk. (2) There are multiple generations of CAIs in different chondrite groups (e.g., *Krot et al.*, 2009, and references therein), but ^{26}Al-^{26}Mg systematics have been measured with high precision only in CV CAIs. (3) Some

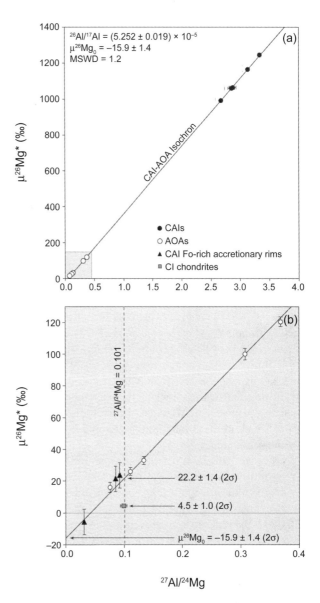

Fig. 2. **(a)** Aluminum-magnesium isochron for bulk AOAs and CAIs from the reduced chondrite Efremovka. **(b)** The bulk Efremovka CAI-AOA isochron intercepts the solar ^{27}Al/^{24}Mg at μ^{26}Mg* = 22.2 ± 1.4 ppm, whereas CI chondrites define a μ^{26}Mg* value of 4.5 ± 1.0 ppm. μ^{26}Mg* = $[(^{26}$Mg/^{24}Mg)/$(^{26}$Mg/^{24}Mg)_\oplus]-1 \times 10^6$, after correction for mass-dependent fractionation. After *Larsen et al.* (2011).

CAIs are known to have formed with $(^{26}$Al/^{27}Al$)_0$ lower than the canonical value, but otherwise have the similar mineralogy and oxygen isotopic compositions as CAIs with the canonical $(^{26}$Al/^{27}Al$)_0$ (e.g., *Simon et al.*, 2002; *Liu et al.*, 2009, 2012a; *Makide et al.*, 2011; *Krot et al.*, 2008, 2012; *Holst et al.*, 2013). Based on these observations, it is inferred that ^{26}Al was heterogeneously distributed in the CAI-forming region during an epoch of CAI formation. Calcium-aluminum-rich inclusions may have recorded injection of ^{26}Al into the protosolar molecular cloud or its core, and its subsequent homogenization in the protoplanetary disk. Therefore, the duration of the CAI-forming epoch cannot be inferred from ^{26}Al-^{26}Mg systematics of CAIs. The canonical $^{26}Al/^{27}Al$ ratio recorded by CV CAIs does not necessarily represent the initial abundance of ^{26}Al in the whole solar system. Instead, it may (*Kita et al.,* 2013) or may not (*Larsen et al.,* 2011) represent the average abundance of ^{26}Al in the protoplanetary disk, which depends on the formation time of the canonical CV CAIs relative to the timing of the completion of delivery of stellar ^{26}Al to the disk and degree of its homogenization in the disk (*Kita et al.,* 2013).

4.3.3. Evidence for and against heterogeneous distribution of aluminum-26 in the disk after the epoch of the formation of calcium-aluminum-rich inclusions. There is an ongoing debate concerning the level of ^{26}Al heterogeneity in the protoplanetary disk after the epoch of CAI formation (e.g., *Larsen et al.,* 2011; *Kita et al.,* 2013). *Larsen et al.* (2011) reported a very precise model Al-Mg isochron of the bulk CV CAIs and AOAs, with an initial $^{26}Al/^{27}Al$ ratio of $(5.25 \pm 0.02) \times 10^{-5}$ and a y-intercept or initial $\mu^{26}Mg^*$ of -15.9 ± 1.4 ppm (Fig. 2). *Larsen et al.* (2011) also demonstrated widespread heterogeneity in the mass-independent magnesium isotopic composition ($\mu^{26}Mg^*$) of bulk chondrites having solar or near-solar Al/Mg ratios (Figs. 2b, 3), and attributed this variability largely to heterogeneity in the initial abundance of ^{26}Al across the protoplanetary disk, although they allowed magnesium isotopic heterogeneity as another possibility.

Kita et al. (2013) argue that the combined (CAIs + AOAs) bulk Al-Mg isochron reported by *Larsen et al.* (2011) is compromised by AOAs, which could have formed later than CAIs. If the regression line is calculated only for bulk CAIs, the bulk ^{26}Mg of chondrites are consistent with uniform distribution of ^{26}Al in the disk at the canonical level (Fig. 4).

A rare type of CAI containing mass-fractionated oxygen, magnesium, and silicon isotopic composition and mass-independent anomalies in a number of elements that are an order of magnitude or more larger than those in normal CAIs was discovered in the late 1970s (*Clayton and Mayeda,* 1977; *Wasserburg et al.,* 1977; *MacPherson,* 2014). These CAIs were named "FUN" inclusions, for their fractionated and unidentified nuclear isotopic anomalies. *Wasserburg et al.* (2012) reported a deficit in bulk ^{26}Mg (-127 ± 32 ppm) in a Allende FUN CAI having a canonical $(^{26}Al/^{27}Al)_0$, $(5.27 \pm 0.18) \times 10^{-5}$, and concluded that the whole-rock magnesium isotopic compositions of chondrites cannot be used as an argument in favor of ^{26}Al heterogeneity in the protoplanetary disk.

The reported mismatch of ~1 m.y. between the absolute uranium-corrected Pb-Pb (see section 4.7) and relative ^{26}Al-^{26}Mg ages of quenched angrites [basaltic achondrites that crystallized rapidly and suffered no subsequent thermal metamorphism and internal reequilibration (*Amelin,* 2008; *Spivak-Birndorf et al.,* 2009; *Larsen et al.,* 2011)] vs. CAIs appears to support heterogeneity in the initial abundance of ^{26}Al of a factor of ~2 across the disk. This conclusion has been recently questioned by *Kruijer et al.* (2014). These

authors compared ^{182}Hf-^{182}W, ^{26}Al-^{26}Mg, and U-corrected Pb-Pb ages of angrites (*Spivak-Birndorf et al.,* 2009; *Schiller et al.,* 2010; *Brennecka and Wadhwa,* 2012; *Kleine et al.,* 2012) and canonical CAIs (*Jacobsen et al.,* 2008; *Connelly et al.,* 2012; *Kruijer et al.,* 2014), and found no evidence for ^{26}Al heterogeneity in the disk regions where the angrite parent body and CAIs formed.

4.3.4. Variations in initial aluminum-26/aluminum-27 ratios among chondrules. The initial $^{26}Al/^{27}Al$ ratios inferred from internal isochrons of chondrules in the least-metamorphosed chondrites (petrologic type 3.0) are systematically lower than the canonical CV CAI value (Fig. 5). If ^{26}Al was uniformly distributed in the protoplanetary disk at the canonical level, these observations suggest that most chondrules formed 1–3 m.y. after CAIs with the canonical $^{26}Al/^{27}Al$ ratio.

Most chondrules within a single chondrite group show a narrow range of $(^{26}Al/^{27}Al)_0$ (Fig. 5), suggesting formation from an isotopically uniform reservoir, possibly within a short period of time (*Kita and Ushikubo,* 2012). There are, however, resolvable differences in $(^{26}Al/^{27}Al)_0$ among some chondrules within individual chondrites as well (*Villeneuve et al.,* 2009) (Fig. 5), suggesting accretion of several generations of chondrules into the same asteroidal body.

4.4. Iron-60

Iron-60 β-decays to ^{60}Ni with a half-life of 2.62 m.y. (*Rugel et al.,* 2009), which is significantly longer than the

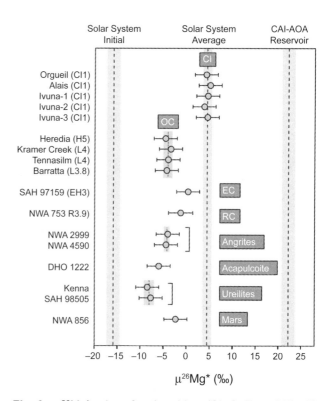

Fig. 3. $\mu^{26}Mg^*$ values for chondrites (CI, O, E, and R), differentiated meteorites (angrites, acapulcoites, and ureilites), and martian meteorites (Mars). From *Larsen et al.* (2011).

value used in earlier publications (1.5 m.y.). Although the presence of live ^{60}Fe in the early solar system was demonstrated almost 20 years ago (*Shukolyukov and Lugmair,* 1993), the initial abundance of ^{60}Fe in the solar system has seen some significant downward revisions. High initial ^{60}Fe/^{56}Fe ratios (up to ~10^{-6}) were originally reported in sulfides and ferromagnesian silicates in unequilibrated ordinary chondrites (UOCs) using SIMS (e.g., *Tachibana and Huss,* 2003; *Mostefaoui et al.,* 2005; *Tachibana et al.,* 2006). During the last several years, the estimates of the initial ^{60}Fe/^{56}Fe ratio in the solar system have varied from 1 × 10^{-8} to 5 × 10^{-7}. Such a broad range of (^{60}Fe/^{56}Fe)$_0$ largely reflects differences between Fe-Ni-isotope measurements by SIMS and MC-ICPMS, which are not fully understood and may reflect (1) the heterogeneous distribution of ^{60}Fe in the early solar system; (2) disturbances of Fe-Ni systematics in some or all of the meteorites studied; or (3) analytical artifacts. Here, we summarize the latest results of Fe-Ni-isotope measurements of chondrules and differentiated meteorites by SIMS, MC-ICPMS, and TIMS.

Telus et al. (2012) demonstrated that much of the previously published SIMS Fe-Ni isotope data were statistically biased due to improper data reduction as discussed in detail by *Ogliore et al.* (2011). After correcting the data, most of the originally reported high initial ^{60}Fe/^{56}Fe ratios (e.g., *Tachibana and Huss,* 2003; *Tachibana et al.,* 2006) became unresolvable or much lower than the original estimates. The most recently reported initial ^{60}Fe/^{56}Fe ratios in ferromagnesian chondrules from UOCs measured by SIMS (*Telus et al.,* 2013) revealed that 16 out of 18 chondrules show no clear evidence for resolvable ^{60}Ni excess with typical upper limits on (^{60}Fe/^{56}Fe)$_0$ of ≤2 × 10^{-7}. Although two other chondrules show resolvable (^{60}Fe/^{56}Fe)$_0$, the isochron

regressions through the data show weak correlations and have reduced χ^2 of 11 and 3.3, suggesting that there has been a disturbance of the Fe-Ni-isotope systematics, most likely during fluid-assisted thermal metamorphism.

Recent MC-ICPMS and TIMS measurements of bulk chondrules in unequilibrated ordinary chondrites give (^{60}Fe/^{56}Fe)$_0$ of ~1 × 10^{-8} (*Tang and Dauphas,* 2012; *Spivak-Birndorf et al.,* 2012; *Chen et al.,* 2013). This value is consistent with the estimated (^{60}Fe/^{56}Fe)$_0$ in the solar system inferred from Fe-Ni-isotope measurements by MC-ICPMS of the whole-rock and mineral separates of quenched angrites with measured absolute ages (*Quitté et al.,* 2010, 2011; *Spivak-Birndorf et al.,* 2011b,a; *Tang and Dauphas,* 2012) and with 4565-Ma troilite (FeS) from an iron meteorite (*Moynier et al.,* 2011). Iron-60 is coproduced with ^{58}Fe in supernovae and in asymptotic giant branch (AGB) stars (*Dauphas et al.,* 2008). Those authors and *Tang and Dauphas* (2012) measured iron and nickel isotopes in their studies of the ^{60}Fe-^{60}Ni system and found no anomalies in ^{58}Fe. They inferred that ^{60}Fe was uniformly distributed in the protoplanetary disk with an initial ^{60}Fe/^{56}Fe ratio of ~10^{-8} (Fig. 6).

4.5. Beryllium-10, Chlorine-36, and Calcium-41: Irradiation Tracers

4.5.1. Beryllium-10. Since its discovery (*McKeegan et al.,* 2000), ^{10}Be has been identified in many CAIs from a variety of meteorite groups. Most CAIs have inferred initial ^{10}Be/^{9}Be ratios of (3–9) × 10^{-4} (*MacPherson et al.,* 2003; *Chaussidon et al.,* 2006; *Liu et al.,* 2009, 2010; *Wielandt et al.,* 2012; *Srinivasan and Chaussidon,* 2013), but one CAI from the Isheyevo CB/CH chondrite has an initial ^{10}Be/^{9}Be ratio of 10^{-2} (*Gounelle et al.,* 2013). This variability contradicts the presolar origin for ^{10}Be via trapping of galactic

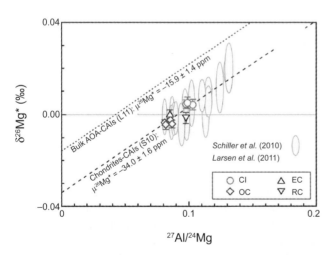

Fig. 4. The ^{26}Al-^{26}Mg system of bulk chondrites; data from *Schiller et al.* (2010) and *Larsen et al.* (2011). Bulk CAI-AOA isochron is from *Larsen et al.* (2011) with intercept at −15.9 ± 1.4 ppm and the (^{26}Al/^{27}Al)$_0$ = (5.25 ± 0.02) × 10^{-5}. The bulk CAI-chondrite isochron is from *Schiller et al.* (2010) with an intercept of −34.0 ± 1.6 ppm and the (^{26}Al/^{27}Al)$_0$ = (5.21 ± 0.06) × 10^{-5}, which combined bulk CAI data of *Jacobsen et al.* (2008). From *Kita et al.* (2013).

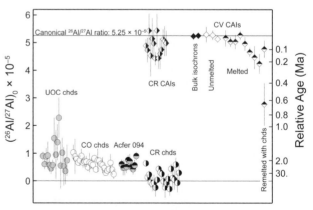

Fig. 5. The inferred initial ^{26}Al/^{27}Al ratios in chondrules (chds) from unequilibrated ordinary chondrites (UOCs), CO and CR carbonaceous chondrites, and the Acfer 094 (ungrouped) carbonaceous chondrite, and CAIs in CR and CV carbonaceous chondrites. Data from *Kita et al.* (2000), *Kita and Ushikubo* (2012), *Mostefaoui et al.* (2002), *Nagashima et al.* (2007, 2008), *Kurahashi et al.* (2008), *MacPherson et al.* (2010, 2012), *Ushikubo et al.* (2013), and *Schrader et al.* (2013).

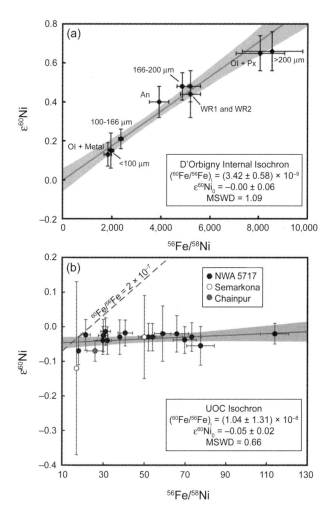

Fig. 6. (a) Iron-60–nickel-60 isochron diagram of the quenched angrite D'Orbigny, which has a uranium-corrected Pb-Pb age of 4563.37 ± 0.25 Ma (*Brennecka and Wadhwa, 2012*). Following correction for ^{60}Fe decay after formation of CV CAIs with a uranium-corrected Pb-Pb absolute age of 4567.30 ± 0.13 Ma (*Connelly et al., 2012*), these measurements give a ^{60}Fe/^{56}Fe ratio at the time of CV CAI formation of (9.7 ± 1.8) × 10^{-9}. **(b)** Iron-60–nickel-60 isochron diagram of chondrules from UOCs Semarkona (LL3.0), NWA 5717 (ungrouped 3.05), and Chainpur (LL3.4). The inferred initial ^{60}Fe/^{56}Fe ratio of (1.04 ± 1.31) × 10^{-8} is inconsistent with the higher values inferred from the *in situ* SIMS measurements of some chondrules in UOCs ~2 × 10^{-7}. From *Tang and Dauphas (2012)*.

cosmic rays suggested by *Desch et al.* (2004), while the production in the solar atmosphere recently proposed by Bricker and Caffee (2010) is not expected to produce the high initial ^{10}B excesses observed in some CAIs (*Chaussidon et al., 2006; Gounelle et al., 2013*). The more likely origin of ^{10}Be is a combination of inherited molecular cloud ^{10}Be generated via enhanced trapping of galactic cosmic rays (*Desch et al., 2004*) and an early irradiation possibly during the infrared classes 0 and I (*Wielandt et al., 2012; Gounelle et al., 2013*). Model calculations indicate that the fluences necessary to have produced ^{10}Be could be as high as a few 10^{20} cm^{-2} (*Duprat and Tatischeff, 2007; Gounelle et al.,*

2013), consistent with X-ray observations of protostars that record intense magnetic activity that can accelerate protons (*Feigelson, 2010*).

4.5.2. Chlorine-36. Although the main decay product of ^{36}Cl is ^{36}Ar (98%), most studies have focused on the ^{36}S decay channel, which is easier to measure by *in situ* SIMS techniques. During the lifetime of the protoplanetary disk, chlorine is present mainly as HCl gas; it will likely condense by reaction with solids below 400 K; the solid HCl hydrate (HCl•3H$_2$O) could form when temperatures fall below ~160 K and adhere to mineral grains and water-ice particles (*Zolotov and Mironenko, 2007*), but equilibrium seems unlikely at such low temperatures. Chlorine is a mobile element that is easily transported in an aqueous solution. It is present almost exclusively in secondary minerals formed during fluid-assisted thermal metamorphism. Sulfur-36 excesses were detected in sodalite and wadalite replacing primary minerals in CAIs from the CV chondrites Allende and Ningqiang (*Lin et al., 2005; Hsu et al., 2006; Nakashima et al., 2008; Jacobsen et al., 2011*). The inferred initial ^{36}Cl/^{35}Cl ratio in sodalite and wadalite varies from 1.8 × 10^{-6} to 1.8 × 10^{-5}. Both minerals formed during late-stage alteration on the CV parent asteroid, after the nearly complete decay of ^{26}Al, ~3 m.y. after CV CAIs (*Brearley and Krot, 2012*). Correction for decay using ^{26}Al/^{27}Al as a chronometer yields inferred inner solar system ^{36}Cl/^{35}Cl values that are far in excess of what can be produced in stars, excluding a stellar origin for ^{36}Cl. Instead, late-stage (during the class III phase) solar-energetic-particle irradiation of either an HCl-rich gas or dust particles mantled by HCl hydrates prior to accretion into the CV parent asteroid may explain the enhanced production of ^{36}Cl (*Gounelle et al., 2006a; Jacobsen et al., 2011; Wasserburg et al., 2011*).

4.5.3. Calcium-41. Calcium-41 has been positively identified in only a few CAIs (*Srinivasan et al., 1996; Sahijpal et al., 2000; Ito et al., 2006; Liu et al., 2012b*). While its initial abundance was originally thought to be given by a ^{41}Ca/^{40}Ca ratio of 1.5 × 10^{-8}, the ratio is now believed to be on the order of 4 × 10^{-9} (*Ito et al., 2006; Liu et al., 2012b*). Given its very short half-life, it is difficult to envision a stellar origin for ^{41}Ca, except in the case of an improbable fine-tuning of the injection parameters (*Liu et al., 2012b*). A more likely origin is early solar system irradiation, although at the moment, models overproduce ^{41}Ca relative to ^{10}Be and their respective initial solar system abundances. Such a discrepancy could be due to erroneous cross sections that have not been experimentally measured for ^{41}Ca. In addition, it is possible that the initial ^{41}Ca/^{40}Ca ratio has been underestimated in meteoritic measurements, due to the volatile nature of potassium (^{41}Ca decays to ^{41}K).

4.6. The Astrophysical Context

Although it is clear that ^{10}Be, ^{36}Cl, and ^{41}Ca record irradiation processes rather than a stellar input, there is not at the moment a unified, coherent model that can account for observed relative abundances in solar system objects.

Furthermore, these SLRs might have recorded different irradiation epochs ranging from class 0 to class III protostars. A more detailed knowledge of the initial content of these SLRs will help us pin down these different contributions, as will cross-section measurements (*Bowers et al.,* 2013) and model improvements. Obviously, given their putative production mode, these SLRs are not suitable for early solar system chronology.

Short-lived radionuclides with mean-lives that are longer than 5 m.y. probably originate from the galactic background, and therefore cannot help us decipher the astrophysical context of our solar system birth (e.g., *Meyer and Clayton,* 2000), nor can the SLRs discussed above that were made by irradiation processes. On the contrary, ^{60}Fe and ^{26}Al are the tools of choice when it comes to understanding the context of formation of our Sun.

For many years, ^{60}Fe was taken as indisputable evidence for the presence of a core-collapse supernova (SN) nearby (~1 pc) the nascent solar system, and injection of SLRs either into the molecular cloud core or into the protoplanetary disk (e.g., *Boss,* 2013; *Hester et al.,* 2004; *Gritschneder et al.,* 2012). It was, however, shown that the presence of a supernova close to a protoplanetary disk (or a molecular core) is highly unlikely (e.g., *Williams and Gaidos,* 2007; *Gounelle and Meibom,* 2008), because by the time a massive star is ready to explode as a SN, it has carved out a ~5-pc large H$_{II}$ region where there are no disks or cores. At the time of a SN explosion, star formation occurs at the boundary of the H$_{II}$ region, which is too far away to receive the correct amount of SLRs. In addition, in the case of the disk, SN injection of SLRs leads to a modification of the disk oxygen isotopes [except for very specific models (see *Gounelle and Meibom,* 2007; *Ellinger et al.,* 2010)], which is not observed. Finally, supernovae produce large excesses of ^{60}Fe relative to ^{26}Al and their respective solar system initial abundances (*Vasileiadis et al.,* 2013). This is especially true since the initial abundance of ^{60}Fe has been revised downward. Therefore, the nearby SN model, first proposed by *Cameron and Truran* (1977), can no longer account for the origin of ^{60}Fe and ^{26}Al in our solar system.

Given the (low) abundance of ^{60}Fe and its relatively short half-life (2.6 m.y.) compared to the typical mixing timescales in the Galaxy (~100 m.y.) (see *de Avillez and Mac Low,* 2002), its presence in the nascent solar system can be accounted for by supernovae that exploded 10–15 m.y. before the formation of the Sun (*Gounelle et al.,* 2009b), a setting that can be seen as an extension of the classical galactic background model.

Any ^{26}Al made with the ^{60}Fe would have essentially fully decayed before formation of the solar system. In addition, the initial abundance of ^{26}Al in the solar system appears to have been low and heterogeneous, but rapidly grew to the canonical value during the epoch of CAI formation (*Liu et al.,* 2012a). This is contrary to the idea of *Jura et al.* (2013), which envisions ubiquitous ^{26}Al at the molecular-cloud scale. These observations require a recent injection of ^{26}Al by a neighboring star into the protosolar molecular core

prior to or contemporaneously with its collapse. Several potential stellar sources of ^{26}Al have been discussed in the literature. An AGB star has been proposed as a source (e.g., *Wasserburg et al.,* 2006; *Lugaro et al.,* 2012b), but it is very unlikely that an AGB star was nearby a low-mass protostar (*Kastner and Myers,* 1994). Wolf-Rayet (WR) stars were first suggested by *Arnould et al.* (1997). Recent models propose either that ^{26}Al was delivered at the bulk-molecular-cloud scale by a set of WR stars (*Gaidos et al.,* 2009), or that a single runaway (v ≥ 20 km s^{-1}) WR star injected ^{26}Al in a shock-induced dense shell (*Tatischeff et al.,* 2010). The problem with the first model is that at the spatial and temporal scale of a molecular cloud, massive stars explode as supernovae at the end of the WR stage, leading to an excess of ^{60}Fe relative to ^{26}Al in the whole cloud. In the second model, given the required velocity of the star (20 pc m.y.$^{-1}$) and gas ambient density (n ~ 100 cm^{-3}), it is likely that the star will have escaped the molecular cloud well before it collects enough gas and enters the WR phase during which ^{26}Al is injected into the cloud. Both models also suffer from the brevity of the WR phase, which means that rare WR stars have to be in just the right place at the right time. To escape that caveat, *Gounelle and Meynet* (2012) have proposed that ^{26}Al originated in the wind-collected shell around a massive star. In this model, the dense shell collected by the massive stellar wind is continuously enriched in ^{26}Al brought at the surface of the star by convection promoted by rotation (*Meynet et al.,* 2008). After several million years, the collected shell reaches a density large enough to make it gravitationally unstable, and a new generation of stars forms in the shell. The Sun would have formed together with hundreds of other stars in such a collected shell. It is therefore a second-generation star, and the massive star having provided ^{26}Al can be seen as its parent star. This mechanism is generic in the sense that this mode of star formation, although not the rule, is relatively common (*Hennebelle et al.,* 2009). The *Gounelle and Meynet* (2012) model for formation of the Sun as one of many stars secondary to a massive star strongly constrains the astrophysical context of our Sun's formation, as it requires that the parent cluster to which the parent star belongs contains roughly 1200 stars. Although WR wind models are attractive for allowing the possibility of formation of new stars with high ^{26}Al/^{60}Fe, the hydrodynamic models of *Boss and Keiser* (2013) show that WR winds are likely to cause molecular cloud cores to shred, rather than collapse. Because this conclusion depends on density, numerical simulations are the next step to explore the origin of wind-delivered ^{26}Al in the solar system. Although considerable progress has been made, there is more to be done to gain a full understanding of the origin of SLRs in the early solar system.

4.7. Chronology of the Early Solar System Processes

4.7.1. Uranium-lead absolute chronology and uranium-235/uranium-238 ratio in the solar system. In contrast to relative SLR chronologies, chronologies based on long-lived radionuclide (LLR) systems (e.g., $^{235,238}U$-$^{207,206}Pb$, ^{147}Sm-^{143}Nd) rely on the knowledge of the present-day abundances of the parent and daughter isotopes in a sample, and are therefore free from assumptions of parent nuclide homogeneity. Of the various LLR systems, the Pb-Pb dating method is the only one precise enough to establish a high-resolution absolute chronology of the first 10 m.y. of the solar system. This chronometer is based on two isotopes of uranium, ^{238}U and ^{235}U, which decay by chains of successive α- and β-emission to the stable lead isotopes ^{206}Pb and ^{207}Pb respectively. This results in $^{207}Pb_R/^{206}Pb_R$ (where R means radiogenic) ratios that correspond to the amount of time passed since the U-Pb system closed. Quantitatively, the relationship is given by the equation $^{207}Pb_R/^{206}Pb_R = (^{235}U/^{238}U) \times [(e^{\lambda_1 t}-1)/(e^{\lambda_2 t}-1)]$, where λ_1 and λ_2 are the decay constants of ^{235}U and ^{238}U, respectively, and t is time. The $^{207}Pb_R/^{206}Pb_R$ ratio of an object (CAI, chondrule) is calculated by extrapolating from an array of measured lead isotopic values that represent varying mixtures of radiogenic lead and initial lead isotopic composition defined by the Nantan iron meteorite (*Blichert-Toft et al.,* 2010). The $^{235}U/^{238}U$ ratio has been traditionally assumed to be constant (137.88) in the solar system. However, it was recently demonstrated to vary in CAIs (*Brennecka et al.,* 2010). Furthermore, the "constant" value of 137.88 has turned out to be slightly off in most other materials (*Brennecka and Wadhwa,* 2012; *Connelly et al.,* 2012). The observed variations in the $^{235}U/^{238}U$ ratio in CAIs correspond to offsets in calculated Pb-Pb ages of up to 5 Ma. As a result, (1) the use of Pb-Pb-isotope systematics for absolute chronology of a sample requires measurements of its $^{235}U/^{238}U$ ratio, and (2) all previously reported Pb-Pb ages of the early solar system materials calculated using a constant $^{235}U/^{238}U$ ratio of 137.88 are potentially incorrect.

To understand the degree of uranium-isotope heterogeneity in the solar system, *Connelly et al.* (2012) measured $^{235}U/^{238}U$ ratios in different solar system materials — CAIs, chondrules, bulk chondrites, and achondrites. They discovered that variations in $^{235}U/^{238}U$ ratios are observed only in CAIs; all other solar system materials are characterized by a constant $^{235}U/^{238}U$ ratio of 137.786 ± 0.013, which, is, however, distinctly different from the previously used value; *Brennecka and Wadhwa* (2012) inferred a statistically indistinguishable value for several angrites, 137.780 ± 0.021. Therefore the previously published Pb-Pb ages of the solar system materials should be recalculated using the new value of the $^{235}U/^{238}U$ ratio.

4.7.2. Uranium-corrected lead-lead ages of calcium-aluminum-rich inclusions and chondrules. Amelin et al. (2010), *Bouvier et al.* (2011), and *Connelly et al.* (2012) reported uranium-corrected Pb-Pb ages of several CAIs with the canonical bulk $^{26}Al/^{27}Al$ ratio from CV chondrites. The data reported by *Amelin et al.* (2010) and *Connelly et al.* (2012) are consistent with each other and suggest formation of CV CAIs at 4567.30 ± 0.16 Ma. This age is slightly younger than the age reported by *Bouvier et al.* (2011), 4567.94 ± 0.31 Ma; the reasons for the discrepancy are

not yet clear. In cosmochemistry, the absolute age of CV CAIs is considered to represent the age of the solar system and the time zero for relative SLR chronology. We note, however, that the relationship between the timing of CAI formation and other stages of the evolution of the solar system is not yet understood.

Connelly et al. (2012) reported uranium-corrected Pb-Pb ages of three chondrules from CV chondrites and two from ordinary chondrites. These ages range from 4567.32 ± 0.42 to 4564.71 ± 0.30 Ma. Chondrules in CB chondrites, which appear to have formed in an impact generated plume (*Krot et al.,* 2005), have a well-defined younger age, 4562.52 ± 0.44 Ma (*Bollard et al.,* 2013). These data indicate that chondrule formation started contemporaneously with CAIs and lasted for at least 3 m.y. The inferred duration of chondrule formation provides the best cosmochemical constraints on the lifetime of the solar protoplanetary disk.

4.7.3. Manganese-53. Manganese-53 appears to have been uniformly distributed in the early solar system and has been extensively used for dating early solar system processes (e.g., *Nyquist et al.,* 2009, and references therein). We note that the initial $^{53}Mn/^{55}Mn$ ratio in CV CAIs, the oldest dated solids formed in the solar system, is not known, due to the presence of isotopically anomalous chromium, the late-stage disturbance of ^{53}Mn-^{53}Cr systematics in the CV CAIs, and the nonrefractory nature of manganese. Therefore the initial $^{53}Mn/^{55}Mn$ ratio in the solar system, (6.71 ± 0.56) × 10^{-6}, is calculated using the initial $^{53}Mn/^{55}Mn$ ratio in the D'Orbigny angrite, (3.24 ± 0.04) × 10^{-6} (*Glavin et al.,* 2004), and the uranium-corrected Pb-Pb ages of D'Orbigny, 4563.37 ± 0.25 Ma (*Brennecka and Wadhwa,* 2012), and CAIs, 4567.30 ± 0.16 Ma (*Connelly et al.,* 2012).

Manganese and chromium have different solubilities in aqueous solutions, and are commonly fractionated during aqueous activity on the chondrite parent asteroids. The aqueously formed secondary minerals in ordinary and carbonaceous chondrites, fayalite (Fe_2SiO_4) and carbonates [(Ca,Mg,Mn,Fe)CO_3], are characterized by high Mn/Cr ratios, and are therefore suitable for relative age dating by *in situ* Mn-Cr-isotope measurements using SIMS instruments. Secondary ion mass spectrometry measurements of ^{53}Mn-^{53}Cr systematics of olivine and carbonates require suitable standards, which have been synthesized only recently (*Fujiya et al.,* 2012, 2013; *McKibbin et al.,* 2013; *Doyle et al.,* 2013). As a result, Mn-Cr relative ages of fayalite and carbonates published prior to 2012 (e.g., *Hutcheon et al.,* 1998; *Hua et al.,* 2005; *Hoppe et al.,* 2007; *Jogo et al.,* 2009; *Petitat et al.,* 2011) are incorrect, but could be corrected using the new sensitivity factors. The recently reported ^{53}Mn-^{53}Cr ages of fayalite in CV and CO chondrites (*Doyle et al.,* 2013), 4–5 m.y. after CV CAIs, are indistinguishable from those of carbonates in CM and CI chondrites (*Fujiya et al.,* 2012, 2013; *Jilly et al.,* 2013), suggesting that aqueous activity on several chondrite parent bodies occurred nearly contemporaneously (Fig. 7).

4.7.4. Hafnium-182. Hafnium is a lithophile (oxygen loving) element and tungsten is siderophile (metal-loving), so that at equilibrium, hafnium partitions into silicate and tungsten into metal. Thus, the ^{182}Hf-^{182}W system is particularly useful for dating core formation in differentiated planets and asteroids. Chronological information can be obtained from this system from two pieces of information: (1) the slope of an isochron on a ^{182}W/^{184}W vs. ^{180}Hf/^{184}W diagram gives the ^{182}Hf/^{180}Hf ratio at the time of formation; and (2) the intercepts of isochrons or the tungsten isotopic compositions of hafnium-free iron meteorites give the tungsten isotopic composition at the time of Hf/W fractionation. Differences in ^{182}Hf/^{180}Hf among objects can be used to construct relative chronologies and the tungsten isotopic composition alone can be used for relative chronology if the Hf/W ratio before differentiation can be inferred. The early solar system ^{182}Hf/^{180}Hf at the time of CAI formation, $(9.71 \pm 0.41) \times 10^{-6}$, is now well known, as is the initial ^{182}W/^{184}W ratio, expressed as $\varepsilon^{182}W = [(^{182}W/^{184}W)_{sample}/(^{182}W/^{184}W)_\oplus - 1] \times 10000 = -3.51 \pm 0.10$ (*Burkhardt et al.*, 2008, 2012).

The difference between the early solar system $\varepsilon^{182}W$ value of -3.51 and the present-day bulk solar system value of -1.9 [the value found in carbonaceous chondrites today (*Kleine et al.*, 2009)] can be used to infer the timing of core-formation events that led to formation of iron meteorites. A complication arises because of the long cosmic exposure ages of iron meteorites, wherein neutron capture effects can affect the tungsten isotopic compositions. Uncorrected iron meteorite data sometimes gave $\varepsilon^{182}W \leq -3.51$, which implied asteroidal differentiation at or before the time of CAI formation, which is counterintuitive. Reliable

methods based on platinum isotopic compositions have now been developed to correct for cosmic-ray exposure in iron meteorites (*Kruijer et al.*, 2013). Data for most iron meteorite groups with either low cosmic-ray-exposure ages or reliably corrected tungsten isotopic compositions imply core-formation events 1 to 2.7 m.y. after CAI formation (*Kruijer et al.*, 2012, 2013), which is still very early and may predate formation of most chondrules in the known groups of chondritic meteorites. It is thought that bodies that formed early melted and differentiated due to heat generated by the decay of ^{26}Al, whereas chondritic bodies, which obviously have never been melted after accreting, formed later when the amount of ^{26}Al was insufficient for melting.

The tungsten isotopic composition of the martian mantle has been known for some time (*Kleine et al.*, 2009, and references therein) from analyses of martian meteorites, but uncertainty in the Hf/W ratio of the mantle prevented precise chronological interpretation of the data. *Dauphas and Pourmand* (2011) were able to constrain the Hf/W ratio of the martian mantle from an excellent correlation they observed between Th/Hf and ^{176}Hf/^{177}Hf ratios in martian meteorite samples. This led them to the remarkable conclusion that Mars accreted very rapidly and reached half its present size only 1.8 m.y. after solar system formation. They suggest that Mars is a stranded planetary embryo, a conclusion consistent with the Grand Tack scenario (*Walsh et al.*, 2011).

5. CHONDRULES

Chondrules appear to have formed as largely molten droplets in brief (hours to days) high-temperature heating events with peak temperatures of 1400°–1800°C and subsequent cooling rates of 1°–1000°C h^{-1} (*Hewins et al.*, 2005; *Miyamoto et al.*, 2009; *Wick and Jones*, 2012). From the abundance of chondrules in chondrites, and their presence even in Comet Wild 2 (*Nakamura et al.*, 2008; *Ogliore et al.*, 2012), they seem to have been the products of one of the more pervasive and energetic processes that operated in the early solar system.

Given such high-formation temperatures and extended formation timescales, along with the low-pressure and hydrogen-rich conditions of the solar nebula, chondrules should have experienced considerable evaporation, not only of volatile elements, such as sodium and potassium, but also of major elements like iron, silicon, and magnesium. Free evaporation (i.e., not limited by diffusion in the melt or by back-reaction with the evaporated gas) should result in large, correlated elemental and isotopic variations among chondrules (*Davis et al.*, 2005; *Richter et al.*, 2011). There are elemental variations among chondrules that appear to be related to volatility (*Jones et al.*, 2005). However, despite numerous searches, no large or systematic isotopic variations have been found in chondrules (*Davis et al.*, 2005; *Georg et al.*, 2007; *Hezel et al.*, 2010).

The favored explanation for the limited isotopic variations found in chondrules is that chondrules formed in dense enough clumps. In dense clumps, there needs to be

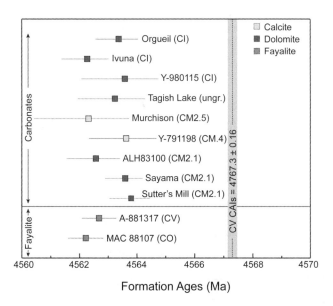

Fig. 7. Manganese-53–chromium-53 relative ages of aqueously formed minerals [fayalite, Fe_2SiO_4; dolomite (Ca,Mg) CO_3; and calcite, $CaCO_3$] in CV, CO, CM, and CI carbonaceous chondrites anchored to the uranium-corrected Pb-Pb age of the D'Orbigny angrite (*Brennecka and Wadhwa*, 2012). The uranium-corrected Pb-Pb age of the CV CAIs is from *Connelly et al.* (2012). Data from *Fujiya et al.* (2012, 2013), *Doyle et al.* (2013), and *Jilly et al.* (2013).

proportionately less evaporation from chondrules before an equilibrium vapor is generated and isotopic exchange between gas and condensed phases can occur. The smaller the amount of evaporation that is required, the faster the chondrules can reequilibrate with the vapor, thereby erasing the isotopic fractionations. Modeling suggests that relatively high pressures (roughly 10^{-4} to 10^{-3} bar) and solid (chondrule + dust)/gas enrichments (~1000× solar) would be needed in the nebula to reduce isotopic fractionations to observed levels on chondrule-formation timescales (*Alexander*, 2004; *Fedkin et al.*, 2012). These conditions are inconsistent with ambient nebular conditions, but could potentially be achieved through shock wave heating/compression (*Desch et al.*, 2005; *Morris and Desch*, 2010; *Hood and Weidenschilling*, 2012; *Morris et al.*, 2012) of regions that had already experienced solid enrichment through turbulent concentration (*Cuzzi et al.*, 2008). However, turbulent concentration produces a spectrum of solid enrichments in the nebula, with most of the material in regions with much lower enrichments. It is puzzling, therefore, that we do not find any evidence for chondrules with large isotopic fractionations that formed in these less-enriched environments.

Even with the above conditions, there should have been complete evaporative loss of volatile elements like sodium and potassium at near peak temperatures. These elements would not have reentered the chondrules until after crystallization of the major minerals olivine and pyroxene were almost complete (*Ebel and Grossman*, 2000; *Fedkin et al.*, 2012). The recent discovery that sodium was present in chondrules at near its current levels throughout crystallization (*Alexander et al.*, 2008; *Borisov et al.*, 2008; *Hewins et al.*, 2012) has led to a dramatic increase in the estimates of the number densities of chondrules in their formation regions (*Alexander et al.*, 2008; *Fedkin and Grossman*, 2013). *Morris et al.* (2012) suggested that the chondrule densities could have been lower if they formed in the bow shock of a planetary embryo with a magma ocean. They suggest that outgassing from a magma ocean would have created a relatively volatile-rich environment in the regions where the chondrules would have formed. However, it is not just volatiles like sodium that require high chondrule number densities. Iron-metal and sulfide also require high densities to prevent them from experiencing extensive evaporation (*Alexander and Ebel*, 2012). Iron and sulfur are less likely to have had high enough vapor pressures above a magma ocean because the high bulk densities of metal and sulfide melts will have caused them to sink to a planetesimal's core. There is also textural evidence that chondrules formed in very-high-density regions where chondrule-chondrule collision and aggregation were common (*Hutchison et al.*, 1980; *Alexander and Ebel*, 2012; *Metzler*, 2012).

The very high chondrule number density estimates during formation cannot, at present, be explained by purely nebula-based models. This, along with the evidence that chondrites may have formed after the parent bodies of many differentiated meteorites (section 4.7.1), has led to a renewed interest in chondrule formation by planetesimal

collisions (*Asphaug et al.*, 2011; *Sanders and Scott*, 2012; *Fedkin and Grossman*, 2013). Formation of impact melts during collisions between cold objects would have been relatively inefficient, with most of the energy going into fragmentation of the impactors (*Davison et al.*, 2010). Relative to chondrules, fragments of preexisting objects are rare in chondrites. However, if the impactors were internally heated by short-lived radionuclides, perhaps even partially or completely molten, droplet formation in the impact ejecta may have been much more efficient. Indeed, *Libourel and Krot* (2007) have suggested on textural grounds that chondrules contained planetesimal material in their precursors.

Nevertheless, many details of the impact scenario for chondrule formation remain to be worked out. For instance, realistic pressure-temperature-time histories of impact ejecta have yet to be calculated for comparison with chondrule thermal histories and compositions. Molten planetesimals are likely to differentiate rapidly, requiring the mantle and core to be remixed in the impact to explain the metal and sulfide in chondrules. There is also no geochemical evidence in chondrules for the elemental fractionations that are likely to have developed during the melting and subsequent crystallization of internally heated planetesimals. Ways of possibly overcoming these and other objections to the impact mechanism have been discussed qualitatively by *Sanders and Scott* (2012), but quantitative models have yet to be developed to test these ideas.

Data for the Pb-Pb and ^{26}Al-^{26}Mg systems suggest time differences of 1–3 m.y. between formation of CAIs and most chondrules (e.g., Fig. 5). This poses problems for all models of chondrule formation, because gas drag on centimeter-sized CAIs would cause them to spiral into the Sun on timescales of as little as 10^5 yr (*Adachi et al.*, 1976; *Weidenschilling*, 1977), yet they accreted onto the same parent bodies as chondrules and compound objects, indicating that CAIs and chondrules that were molten at the same time are quite rare. Furthermore, different types of chondrites contain their own populations of chondrules that are unique in their chemical and isotopic compositions. This implies limited radial mixing of chondrules in the solar nebula or accretion shortly after chondrule formation, but how were CAIs preserved until this happened, and why are some chondrites relatively CAI-rich and others have almost no CAIs? *Ciesla* (2010) has suggested that most of the CAIs that survived to be sampled by chondrite parent bodies were formed in the first 10^5 yr of solar system history. Much remains to be understood.

6. ENSTATITE CHONDRITES

The recent high-precision isotopic studies of terrestrial and extraterrestrial materials (e.g., *Javoy*, 1995; *Clayton*, 2003; *Trinquier et al.*, 2007, 2009) showed that the enstatite chondrites are the only chondrite type with the isotopic compositions of many elements being identical to Earth and Moon, with silicon (*Savage and Moynier*, 2013) and titanium (*Zhang et al.*, 2012) being the exceptions. This, along with the putative discovery of calcium sulfides on the surface of Mercury by the MErcury Surface, Space ENvironment, GEochemistry, and Ranging (MESSENGER) spacecraft (*Nittler et al.,* 2011), revived speculations about formation of enstatite chondrites in the inner solar system and being the building blocks of the terrestrial planets other than Mars. Despite the attractiveness of this idea, building Earth and the Moon from enstatite chondrites runs into significant problems of accounting for the small, but significant differences in silicon and titanium isotopes and the drastic differences in FeO contents, Mg/Si ratios, and volatile contents between these objects.

The enstatite chondrites include two groups with high (EH) and low (EL) iron contents, with both having lower than solar Mg/Si ratios and showing depletions in refractory (e.g., calcium, aluminum, titanium) but not volatile (e.g., sodium, potassium, rubidium, sulfur, fluorine, chlorine, bromine) elements (e.g., *Keil,* 1968; *Sears et al.,* 1982; *Kallemeyn and Wasson,* 1986; *Rubin and Choi,* 2009). There are also closely related enstatite achondrites or aubrites that are believed to be the products of crystallization of melts formed by partial or complete melting of enstatite chondrites (*Keil,* 2010). Despite some differences among these groups (for a detailed review see *Weisberg and Kimura,* 2012), their common feature is the very low-FeO content in silicates and the presence of silicon-bearing Fe, Ni metal and calcium, magnesium, manganese, sodium, and potassium sulfides, indicative of formation in either very reducing or sulfur-rich environments, or both.

The general assumption of the primary condensation of calcium and magnesium sulfides from a nebular gas (analogs of CAIs) requires much more reducing conditions than those in a classic nebula (*Larimer and Bartholomay,* 1979). The most popular model for the enstatite chondrites formation invokes a parcel of nebular gas with an enhanced C/O ratio of ~1 and solar abundances of other elements. Condensation of such a gas results in formation of refractory carbides, silicides, and sulfides, including CaS and MgS, which upon further cooling below ~1100 K would react with the residual gas to form silicates and large amounts of graphite (*Larimer and Bartholomay,* 1979; *Ebel,* 2006). However, neither refractory carbides and abundant graphite nor the predicted substitution of sulfides by silicates has been observed in the enstatite chondrites.

By contrast, the substitution of silicates and oxides by sulfides in silicate chondrules (*Rubin,* 1983, 1984), metal-sulfide nodules (*Ikeda,* 1989), and CAIs (*Lin et al.,* 2003) of the EH chondrites has been described but has not attracted much attention. Recently, *Lehner et al.* (2013) documented in detail the substitution of ferromagnesian silicates by the assemblage of calcium, magnesium, and iron sulfides and silica in a subset of EH chondrite chondrules and interpreted these observations as the evidence for reactions of the primary ferromagnesian silicates with an external sulfur-rich gaseous reservoir. They concluded that the sulfidation reactions of ferromagnesian silicates in chondrules have taken place in a hydrogen-depleted (>2500×) system at 1400–1600 K, when the chondrules were partially molten,

under near-nebular redox conditions and very high S_2 vapor pressures close to those over a partially molten metal-troilite assemblage. They also noticed that the progressive sulfidation of chondrules (defined by increasing sulfide/silicate ratio) results in magnesium loss and an increase of volatile element (manganese, sodium, potassium) contents. The sulfidation of silicates produces a peculiar phase, amorphous porous silica, which so far has only been found associated with sulfidized silicates in the EH chondrites. Overall, the occurrence of porous silica in all components of the EH chondrites (*Lehner et al.,* 2013) strongly suggests that the calcium and magnesium sulfides in enstatite chondrites are the products of secondary alteration of common ferromagnesian silicates rather than primary nebular condensates.

The inferred secondary nature of calcium and magnesium sulfides removes the need for invoking a portion of the solar nebula with C/O > 1 (for which explanations have been difficult) and hints at there having been ordinary FeO-bearing silicates as the precursors of both the enstatite chondrites and the Earth-Moon system.

But the question of where, when, and how such reactions have taken place remains unanswered. A model put forward by *Ebel and Alexander* (2011) considers evaporation of an inner nebula region enriched in the H_2O-depleted dust of chondritic interplanetary dust particle composition by >300× followed by recondensation that initially produces FeO-bearing silicate melts (a.k.a. chondrules), which upon further cooling below ~1400 K would react with the residual S_2-rich gas to form minor calcium and magnesium sulfides. However, further cooling to low temperatures necessary to acquire abundant volatile elements would result in conversion of these sulfides back to silicates. *Lehner et al.* (2013) did not provide a specific model; they just suggested that the sulfidation of chondrules should have taken place in a nebular chondrule-forming region enriched in metal-sulfide assemblage and, to a lesser degree, in organics. While the required depletion in hydrogen (i.e., increased dust/gas ratio) is broadly consistent with the current estimates of the dust/gas ratios in the chondrule-forming regions, a mechanism for the required enrichment in the metal-sulfide component has still to be found.

7. THE STARDUST MISSION

NASA's Stardust spacecraft was launched in February 1999 and returned to Earth in January 2006 (*Brownlee et al.,* 2003, 2004, 2006; *Tsou et al.,* 2003). The prime objective of the mission was to collect samples from the JFC 81P/Wild 2 during passage through its coma in January 2004. Besides cometary dust, Stardust was also supposed to collect contemporary interstellar dust during the cruise phase of the spacecraft before the cometary encounter. This interstellar dust is most likely dust that formed or was reprocessed in the interstellar medium (ISM), but it could also be genuine circumstellar stardust.

In order to decelerate impacting dust particles as gently as possible, low-density silica aerogel with a graded density from 5 to 50 mg cm^{-3} was used as a primary capture medium (*Tsou et al.,* 2003). The cometary and interstellar sample collection trays were mounted back-to-back. Each had 132 aerogel cells with the bars between the cells covered with aluminum foil for collector surface areas of a little over 1000 cm^2 of aerogel and ~150 cm^2 of aluminum foil.

7.1. Cometary Samples

During the Wild 2 flyby, 234 km from the surface of the cometary nucleus, Stardust collected dust particles at a relative speed of 6.1 km s^{-1} (*Brownlee et al.,* 2004). Despite the low-density properties of aerogel, cometary dust particles often disintegrated upon capture. The impacts produced deep cavities with shapes varying from slender, carrot-shaped tracks (type A) to bulbous cavities with (type B) or without (type C) slender tracks emerging (Fig. 8) (*Hörz et al.,* 2006; *Burchell et al.,* 2008). At the end of those slender tracks, compact cometary dust grains survived (terminal particles), while poorly consolidated aggregate particles disintegrated with fragments ending up in the walls of the bulbous parts of the tracks (*Trigo-Rodríguez et al.,* 2008; *Kearsley et al.,* 2012). Track morphology, therefore, already provides information on particle structure, composition, and density (*Kearsley et al.,* 2012). According to *Burchell et al.* (2008), 65%, 33%, and 2% of all observed tracks were of types A, B, and C, respectively. However, for large tracks, bulbous types seem to be more dominant and probably contain the majority of mass collected from Wild 2.

For further analysis of particles residing in aerogel, techniques for extracting small volumes of aerogel (called "keystones") from the capture cells were developed (*Westphal et al.,* 2004). These keystones often contain entire particle tracks and can be subdivided for accessing individual terminal particles or fragments residing in cavity walls. In the initial phase of Wild 2 particle analysis, the work was mainly focused on terminal particles. These are much easier to extract from the aerogel and less intermingled with compressed and melted aerogel. Melting of aerogel and mixing with cometary matter that often partially melted itself upon impact is a severe problem, since it changes the chemical

and mineralogical properties of the cometary dust.

Deceleration of cometary particles upon aluminum foil impact was even more disruptive than aerogel capture. On foils, hypervelocity craters formed that are associated with residual cometary matter, typically as discontinuous layers of impact-produced melts inside crater cavities and on crater rims (*Hörz et al.,* 2006). Although the dust grains were even more severely heated upon aluminum foil impact, the cometary matter associated with the craters is much more easily accessible for analysis and does not require sophisticated sample preparation techniques. Also, these samples provide the sole opportunity to study silicon in Wild 2 material, since, in silica aerogel, it cannot be distinguished from the target material. Similarly, aluminum can only be studied in dust collected by aerogel. As with aerogel, information on particle size, shape, structure, and composition can be inferred from crater morphologies (*Kearsley et al.,* 2008). In some cases, crystalline material from Comet Wild 2 even survived the aluminum foil impact without complete melting, preserving its crystalline structure (*Wozniakiewicz et al.,* 2012).

7.2. Results from Particle Analysis

The elemental abundances for individual cometary grains captured by Stardust were determined by analyzing entire tracks in aerogel using synchrotron X-ray fluorescence (SXRF) (*Flynn et al.,* 2006; *Lanzirotti et al.,* 2008; *Ishii et al.,* 2008b). Particle fragments residing in (*Stephan et al.,* 2008a) or extracted from aerogel (*Stephan et al.,* 2008b) were studied for their elemental composition using time-of-flight (TOF) SIMS. Impact residues associated with impact craters on aluminum foils were analyzed with scanning electron microscopy using energy-dispersive X-ray spectroscopy (SEM-EDS) (*Kearsley et al.,* 2008) and TOF-SIMS (*Leitner et al.,* 2008). The overall chemical composition of Wild 2 dust captured by Stardust seems to be similar to the composition of CI carbonaceous chondrites, although the composition of individual grains often deviates by several orders of magnitude from CI composition (*Stephan,* 2008). For elements heavier than oxygen, with the exception of noble gases, CI chondrite element ratios show excellent agreement with abundances in the solar photosphere (*Lodders et al.,* 2009). Therefore, it can be ascertained that the rock-forming elements in Wild 2 are in the same proportions as in the solar nebula. Some deviations from CI abundances can be explained as a result of volatilization during capture or contamination by the capture medium (*Stephan,* 2008). Unfortunately, there is no reliable data for Wild 2 about the abundance of carbon, hydrogen, oxygen, and nitrogen (CHON) elements, which are, compared to the solar abundances, depleted in CI chondrites and seemed to be enriched, compared to CI, in Comet 1P/Halley, which otherwise also showed CI abundances (*Jessberger et al.,* 1988). Carbon, hydrogen, oxygen, and nitrogen element abundances could not be determined for Wild 2 from Stardust samples due to the nature of the aerogel capture

Fig. 8. Cometary dust tracks in Stardust aerogel. Type A tracks are carrot-shaped, type B are bulbous with slender tracks emerging, and type C are bulbous with no slender tracks. From *Hörz et al.* (2006).

medium, which apart from silicon and oxygen turned out to be contaminated with nitrogen-containing hydrocarbons (*Sandford et al.,* 2010). Also, these elements are at least partially lost upon aerogel capture and, more severely, during aluminum foil impact.

Despite these problems, some information on organic compounds in aerogel as well as associated with aluminum foil craters was obtained by using a variety of spectroscopic and mass spectrometric techniques (*Sandford et al.,* 2006, 2010; *Keller et al.,* 2006; *Leitner et al.,* 2008; *Stephan et al.,* 2008a,b; *Matrajt et al.,* 2008; *Gallien et al.,* 2008; *Cody et al.,* 2008; *Rotundi et al.,* 2008; *Wopenka,* 2012). Organics in Wild 2 are heterogeneously distributed and unequilibrated in composition. There are significant differences from organics found in meteorites, micrometeorites, or IDPs. Wild 2 organics are more labile and richer in oxygen and nitrogen. Infrared spectra of extracted grains and of individual impact tracks are consistent with long-chain aliphatic hydrocarbons. Aromatic hydrocarbons are present but seem to be less abundant than in meteorites or IDPs. Excesses of D and ^{15}N, possibly of interstellar origin, have been found in Stardust although at a lower abundance that in IDPs or micrometeorites. The wide range in organic chemistry found in Wild 2 matter suggests that the comet's organic solids may have multiple origins (*Cody et al.,* 2008).

Wild 2 samples collected by Stardust appear to be weakly consolidated mixtures of nanometer-sized grains, with occasional much larger (>1 μm) ferromagnesian silicates, Fe-Ni sulfides, Fe-Ni metal, and accessory phases (*Zolensky et al.,* 2006). Olivine and low-calcium pyroxene show a wide range of compositions, while iron-nickel sulfides are mostly restricted to a very narrow compositional range, which is similar to the range observed for anhydrous chondritic porous (CP) IDPs, with strong indications for sulfur loss during capture (*Zolensky et al.,* 2006, 2008). Hydrous phases have not been found in Wild 2 material. Furthermore, there is no indication for the presence of any aqueous alteration product like phyllosilicates or indigenous carbonates in Wild 2 particles (*Zolensky et al.,* 2006, 2008). However, cubanite, a copper-iron sulfide usually thought to be formed by hydrothermal alteration, has been found (*Berger et al.,* 2011). Laboratory simulations and collections of IDPs from Earth orbit by the Mir space station show that even substantial heating and structural modification during capture of these grains should not have completely annealed characteristic compositions, grain morphologies, and microscopic crystal structure of phyllosilicates or carbonates had they been present (*Hörz et al.,* 2000; *Burchell et al.,* 2006; *Zolensky et al.,* 2006). Less-abundant Wild 2 materials that were found in the Stardust collection include chondrule fragments (*Zolensky et al.,* 2006; *Nakamura et al.,* 2008; *Ogliore et al.,* 2012) and refractory particles that resemble CAIs found in chondrites (*Zolensky et al.,* 2006; *Simon et al.,* 2008). The Wild 2 samples recovered by Stardust are mixtures of crystalline and amorphous materials (*Zolensky et al.,* 2006, 2008). However, it remains unclear what fraction of the amorphous material is indigenous to the comet

and what fraction was amorphized from crystalline solids upon particle capture (*Leroux et al.,* 2008). The amorphous material known as glass with embedded metal and sulfides (GEMS) is a major constituent of CP IDPs, for which a cometary origin has been suggested (*Bradley,* 1994). Its presence in Wild 2 dust has neither been unambiguously confirmed nor could it be excluded (*Ishii et al.,* 2008a).

7.3. Stardust in Stardust

One of the major expectations for the Stardust mission was that cometary matter should be a mixture of silicates and organics of presolar and nebular origin (*Brownlee et al.,* 2003). Therefore, presolar grains, old stardust, was expected to be highly abundant in Wild 2 material. However, only very few presolar grains have been identified in Stardust samples from Wild 2 (*McKeegan et al.,* 2006; *Stadermann et al.,* 2008; *Floss et al.,* 2013). They seem to be even less abundant in Stardust samples than in other primitive solar system material like carbonaceous chondrites or IDPs (*Floss and Stadermann,* 2012). Recent studies, however, seem to indicate that presolar signatures, which prevail as huge deviations from average solar system isotopic ratios in primitive meteorites or IDPs, were lost during capture of the Wild 2 grains (*Floss et al.,* 2013). The true abundance of presolar grains in Wild 2 dust may be similar to that in primitive IDPs.

7.4. Comparison with Other Solar System Samples

Of all extraterrestrial material that is available for laboratory analysis, Wild 2 samples seem to resemble mostly anhydrous CP IDPs and carbonaceous chondrites (*Zolensky et al.,* 2006, 2008; *Stephan,* 2009). A wide range of olivine and low-calcium pyroxene compositions and the restricted range for iron-nickel sulfides support this contention (*Zolensky et al.,* 2006, 2008). However, the lack of clear evidence for the presence of indigenous amorphous material like GEMS leaves room for other interpretations. Also, Wild 2 matter seems to be more diverse than material found in CP IDPs. The presence of refractory, CAI-like material connects Stardust samples to carbonaceous chondrites. Therefore a petrogenetic continuum between comets and asteroids, as parent bodies of carbonaceous chondrites, has been suggested (*Gounelle et al.,* 2006b; *Ishii et al.,* 2008a). However, the lack of aqueous alteration products in Wild 2 samples precludes a common origin with types 1 or 2 carbonaceous chondrites, and type 3 chondrites generally have narrower olivine and pyroxene compositional ranges (*Zolensky et al.,* 2006). On the other hand, IDPs with refractory inclusions have been found before (*Greshake et al.,* 1996). There is even an entire class of IDPs, called coarse-grained (CG) IDPs, that consists of compact phases like sulfides, olivine, pyroxene, or CAI-like minerals (*Jessberger et al.,* 2001) and seem to resemble some terminal particles found in Stardust aerogel (*Stephan,* 2009). Prior to the return of the Stardust spacecraft, CG IDPs were typically not connected to comets, and therefore attracted much less attention than CP IDPs and are poorly understood (*Zolensky et al.,* 2006; *Stephan,* 2009).

In order to explain the wide compositional range of Wild 2 matter captured by the Stardust spacecraft, different formation locations within the protoplanetary disk are required (*Zolensky et al.,* 2006). Therefore CAI-like material, e.g., could only have formed under high temperatures, presumably close to the young Sun, while Wild 2 itself should have spent its life as a comet beyond the snow line so that thermal metamorphism or aqueous alteration did not take place. It has been suggested that Wild 2 material is a composite of crystalline inner solar system material and amorphous cold molecular cloud material, with more than 50% being processed in the inner solar system (*Ogliore et al.,* 2009). Although the actual fraction of inner solar system material might be less constrained due to the large uncertainty about what fraction of volatile material was lost during capture, it is obvious that radial transport of material across the protoplanetary disk must have taken place in order to explain the observations from Wild 2 samples returned by Stardust.

7.5. Transport and Mixing of Cometary Building Blocks

Prior to the Stardust mission, cometary silicates were known to be present in both crystalline and noncrystalline forms, with the mineralogy determined by both spacecraft and telescopic observations (see review by *Hanner and Bradley,* 2004). The origin of the crystalline component had proven to be a mystery, as such minerals can only be formed at high temperatures. That is, they can form through condensation from a hot solar nebula gas at temperatures ~1200–1500 K or they could form via annealing of amorphous precursors at temperatures of 800–1000 K (*Hallenbeck et al.,* 2000; *Djouadi et al.,* 2005; *Fabian et al.,* 2000; *Nuth et al.,* 2000; *Nuth and Johnson,* 2006), yet comets formed in the cold, outer regions of the solar nebula where temperatures were low enough for water ice to condense (<150 K). Crystalline silicates had also been detected in the outer regions of other protoplanetary disks (e.g., *van Boekel et al.,* 2004), suggesting that the formation of these grains is a common occurrence in protoplanetary disks.

Two possible solutions had been offered to explain the presence of these grains in comets. The first came from recognizing that crystalline grains would have readily formed close to the Sun, where the needed high temperatures were expected (e.g., *Boss,* 1998), and that the grains were transported outward by some dynamic process. *Shu et al.* (1996) suggested that magnetically driven winds would loft grains above the disk and launch them on ballistic trajectories, leading them to fall back onto the solar nebula. Alternatively, *Bockelée-Morvan et al.* (2002) argued that turbulence and the viscous spreading of the solar nebula could carry grains outward with time. In both cases, the crystalline grains would originate in the inner solar nebula, inside 2–4 AU, and be carried outward and mix with the more pristine materials that originated far from the Sun, before being incorporated into comets.

Harker and Desch (2002) offered the second explanation, suggesting that the crystalline silicates were annealed by transient heating events within the solar nebula. Specifically, *Harker and Desch* (2002) argued that if shock waves were present in the solar nebula, silicate particles would be heated through frictional and thermal collisions with the shocked gas. Provided the gas density was high enough, such that the rate of heat deposition would sufficiently exceed the rate of heat loss by radiation, grains could reach temperatures necessary for annealing to take place. Comets would then form from the mix of grains that were processed and avoided processing in the shocks.

The return of the Stardust samples allowed these various models to be tested. Analysis of the rocky materials demonstrated that the high-temperature components collected by the spacecraft were isotopically similar to those found in chondritic meteorites, suggesting a common origin (*McKeegan et al.,* 2006; *Nakamura et al.,* 2008; *Ogliore et al.,* 2012). Had the crystalline silicates been the product of annealing of interstellar dust, extreme isotopic variations would have been expected. Instead, the isotopically "normal" (isotopic ratios approximately equal to those of the bulk solar system) indicated that these high-temperature materials were the product of vaporization and recondensation of dust in the solar nebula. In addition, the Stardust samples contained a number of refractory inclusions whose chemical, mineralogical, and isotopic make-up were found to be similar to the CAIs found in chondritic meteorites (*McKeegan et al.,* 2006; *Simon et al.,* 2008; *Matzel et al.,* 2010). These objects also require high temperatures to form [>1300 K (*Grossman,* 1972; *Davis and Richter,* 2014)], and have unique oxygen isotope signatures in that they are enriched in ^{16}O compared to other solar system rocks (*MacPherson,* 2014, and references therein). This unique signature again pointed to a shared formation environment of cometary and chondritic rocks.

Together, these results indicate that localized annealing in the solar nebula was not the major contributor to the crystalline grains found in comets, and instead, the high-temperature grains found in these objects are the result of dynamical processes carrying these materials outward from the hot, innermost regions of the solar nebula. While some mechanisms for outward transport had been recognized prior to the Stardust mission (e.g., *Shu et al.,* 1996; *Bockelée-Morvan et al.,* 2002), the results from the mission motivated further research into understanding how and when this transport would have occurred. The mechanisms considered to date largely fall into two categories: above-disk transport and transport within the disk.

Above-disk transport could occur via winds launched from the disk via magnetic fields (*Safier,* 1993; *Shu et al.,* 1996, 1997) or photolevitation, where radiation from the star and disk offset the gravitational settling of particles and push the grains outward (*Wurm and Haack,* 2009; *Vinković,* 2009). Within the disk, proposed mechanisms include tapping into the radial spreading of a disk due to mass and angular momentum transport (*Bockelée-Morvan et al.*, 2002; *Ciesla*, 2007; *Boss*, 2008; *Yang and Ciesla*, 2012) or photophoretic forces (*Mousis et al.*, 2007). Each of these mechanisms can, provided conditions in the disk are right, carry grains from the innermost regions of the solar nebula to where comets would form. Distinguishing which of these models was most likely responsible for the transport of grains in the solar nebula is still the subject of ongoing debate. Clues have been taken from the Stardust samples themselves, other properties of solar system materials, and the properties of materials around other protoplanetary disks, with no definitive answer yet emerging.

In the case of those "within-disk" models that tap into the dynamical evolution of the protoplanetary disk, it is generally found that outward transport is most efficient during the earliest stages of disk evolution when mass accretion rates onto the star, and the corresponding outward transport of angular momentum, are highest. Transport in this way allows for observed narrow distribution of ages in the large CAIs found in chondritic meteorites (*Ciesla*, 2010), and is consistent with the observational evidence that the crystallinity fraction of protoplanetary disks was set in the first 10^5 yr of evolution (*Oliveira et al.*, 2011; *Ciesla*, 2011). Furthermore, as solids would not be the only materials transported in these models — gaseous species would be subjected to the same transport — these models predict that gaseous species should also be processed at high temperatures and redistributed through the disk. Evidence for this is currently seen in the D/H ratios of water in solar system objects (*Drouart et al.*, 1999; *Mousis et al.*, 2000; *Yang et al.*, 2013). Indeed, predicting and measuring such correlations may prove to be the best means of testing various disk models in the future.

While there is reason to be optimistic about the matches between the "within-disk" models described above, there are also observations that seem to be at odds with the predictions of these models. Namely, in measuring the ^{26}Al–^{26}Mg systematics of Stardust samples, no evidence of live ^{26}Al was found in one of the refractory grains, Coki (*Matzel et al.*, 2010), or the chondrule fragments (*Ogliore et al.*, 2012). Considering measurement precisions, the upper limits on ^{26}Al/^{26}Al, and homogeneity of nebular ^{26}Al, these data imply that the refractory inclusion must have formed ≥1.7 m.y. after CAIs while the chondrule fragments would have formed ~3 m.y. after CAIs, requiring a mechanism that would allow late transport of grains through the solar nebula. Transport this late could be provided via photophoresis (*Mousis et al.*, 2007), although this requires that fine dust particles are so low in abundance that light from the young Sun could directly reach the particles throughout their transit in the disk. This would imply that fine dust made up only a small fraction of the mass at the time of transport and that comets were likely largely formed late when transport did occur. Alternatively, if ^{26}Al was not homogeneously distributed in the early solar nebula, at least at the time of formation of the cometary particles, then it is

possible such grains formed and were transported outward, having sampled a formation environment that had not yet had ^{26}Al introduced into it (*Yang and Ciesla*, 2012).

To date, the "above-disk" models have not been explored to the same level of detail as the "within-disk" models, making direct comparisons between observations and predictions difficult. It is worth noting, however, that the presence of CAIs in comets was in fact predicted by *Shu et al.* (1996) in the context of their X-wind model. However, *Desch et al.* (2010) have carried out a detailed analysis of the X-wind model and argue against it being an important process for the redistribution of chondritic and cometary materials in the solar nebula.

7.6. Interstellar Samples

Stardust's second major objective was to collect contemporary interstellar dust during the cruise phase of the spacecraft before the cometary encounter (*Brownlee et al.*, 2003; *Tsou et al.*, 2003). For this purpose, a second aerogel collector tray was used, mounted on the backside of the cometary collector. The existence of an interstellar dust stream that is currently passing through the solar system had been discovered by the Ulysses and Galileo spacecrafts (*Grün et al.*, 1993, 1994; *Baguhl et al.*, 1995). Based on these observations, *Landgraf et al.* (1999) predicted that the interstellar side of the Stardust aerogel collector would contain 120 interstellar particles, 40 of which would have sizes greater than 1 μm in radius. These estimates were made assuming a total exposure time for the interstellar dust collector of 290 days, split into three sampling periods, two before and one after the Wild 2 encounter. However, after the comet encounter, it was decided to forgo the third opportunity for collecting interstellar dust in order not to risk the Wild 2 samples by reopening the sample collector again (*Tsou et al.*, 2003). Taking the actual collection time for the interstellar particles of 195 days into account, only ~25 particles greater than 1 μm in radius and ~50 particles smaller than 1 μm in radius were estimated to have impacted the collector (*Westphal et al.*, 2013a).

Compared to the cometary particles returned by Stardust, the interstellar particles are much smaller in size and much lower in abundance. To find such particles in 1037 cm² of aerogel required a totally new approach. With the help of >30,000 volunteers, organized through Stardust@home, an Internet-based platform providing a virtual microscope in order to search through microscope images taken in an automated way, 71 particle tracks in ~250 cm² of aerogel collectors were identified (*Westphal et al.*, 2013b). From their trajectories, 24 of these tracks are consistent with an origin either in the interstellar dust stream or as secondary ejecta from the sample return capsule (*Frank et al.*, 2013). Thirteen of these tracks were eventually extracted, mostly as picokeystones, with the aerogel surrounding the track milled down a thickness of approximately 70 μm (*Frank et al.*, 2013). Elemental analyses by scanning transmission X-ray microscopy (STXM) (*Butterworth et al.*, 2013) and SXRF

(*Brenker et al.,* 2013; *Simionovici et al.,* 2013; *Flynn et al.,* 2013) of track material still inside the aerogel allowed eight tracks to be excluded, as they were generated by secondary ejecta from the sample return capsule. Two tracks showed inconclusive results or were too dense for STXM analysis. Finally, three candidates for interstellar tracks remain. Two tracks showed compositions that seem to point toward a cosmic origin, and one track, although having no detectable residue, was inferred through a combination of optical characterization, simulation comparison, and dynamical modeling to have a likely origin in the interstellar medium (*Frank et al.,* 2013).

The impact directions and speeds of all three interstellar candidate tracks are most consistent with those calculated from interstellar dust models; an interplanetary origin is statistically less likely (*Sterken et al.,* 2013). Laboratory hypervelocity impact experiments with analog material, however, seem to indicate that two of the three interstellar candidate grains were captured at speeds below 10 km s^{-1}, much lower than initially predicted and pointing toward low-density, porous particles, while the track that had no detectable residue was probably generated by an interstellar particle captured at 15–20 km s^{-1} (*Postberg et al.,* 2013). Synchrotron-based X-ray diffraction measurements identified crystalline material, mainly olivine, in both particles found in interstellar track candidates (*Gainsforth et al.,* 2013). From infrared spectroscopy, there is no clear evidence of extraterrestrial organics associated with tracks or terminal particles (*Bechtel et al.,* 2013).

Along with the aerogel, 153 cm^2 of aluminum foil area was also exposed to the interstellar particle flux. Unlike particle tracks in aerogel, impact craters in the aluminum foils do not record a uniquely interpretable morphological signature of the particle trajectory, which could be matched to the interstellar particle flux (*Stroud et al.,* 2013). So far, 13 aluminum foils covering 4.84 cm^2 from the interstellar collector tray have been imaged by SEM. They show abundant impact craters or crater-like features, for which the majority are shown by Auger electron spectroscopy and EDS analyses to be either the result of impacting debris fragments from solar cell panels or intrinsic defects in the foils (*Stroud et al.,* 2013). However, among the 25 crater features analyzed, four craters contained residues that seem to indicate an extraterrestrial origin, either as interplanetary dust particles or interstellar dust particles (*Stroud et al.,* 2013).

So far, all analyses of interstellar candidate tracks or foil craters were performed with minimal destruction. Now, after the end of the preliminary examination phase, more destructive analyses, including mass spectrometry, will occur. Such analyses will enable measurement of isotopic abundances in these samples, in order to ultimately decipher their origin either in the interstellar medium or within the solar system.

8. PRESOLAR GRAINS

In addition to the Stardust mission, genuine stardust, or presolar grains, deserve some discussion. One of the great discoveries in cosmochemistry occurred about 25 years ago, with the identification of isotopically highly anomalous diamond, silicon carbide, and graphite in primitive meteorites (*Lewis et al.,* 1987; *Bernatowicz et al.,* 1987; *Amari et al.,* 1990). These grains, for the most part, are circumstellar grains formed around dying stars like AGB stars and core-collapse supernovae (*Davis,* 2011; *Zinner,* 2013, 2014), and thus are perhaps not relevant to a volume on protostars and planets, but the survival of presolar grains in our solar system provides information on conditions in the solar nebula.

Among the more important discoveries in recent years are that primitive meteorites and IDPs contain presolar silicates (*Messenger et al.,* 2003; *Nguyen and Zinner,* 2004). Presolar silicates are much more easily destroyed by parent-body metamorphism and aqueous alteration than presolar diamond, silicon carbide, and graphite, but the most primitive meteorites can contain ~200 ppm presolar silicates in their matrices (*Floss and Stadermann,* 2012). A recent sample of IDPs collected shortly after Earth passed through the tail of Comet 26P/Grigg-Skjellerup shows that some contain up to 1% presolar silicates (*Busemann et al.,* 2009).

9. THE HAYABUSA MISSION

In May 2003, the Hayabusa spacecraft was launched by the Japan Aerospace Exploration Agency (JAXA) to visit near-Earth asteroid 25143 Itokawa, acquire samples from the asteroid's surface, and return them to Earth (*Fujiwara et al.,* 2006). After some irregularities during sampling at the asteroid in 2005 and problems with the propulsion system that led to a three-year delay of the return of the spacecraft (*Fujiwara et al.,* 2006), Hayabusa successfully returned to Earth in June 2010, carrying more than 1500 rocky particles from Itokawa (*Nakamura et al.,* 2011). Particles range in size between 0.5 and 180 μm, but are mostly smaller than 10 μm (*Nakamura et al.,* 2011; *Tsuchiyama et al.,* 2011).

Near-infrared spectral results obtained by the Hayabusa spacecraft indicated that the surface of Itokawa has an olivine-rich mineral assemblage potentially similar to that of type LL5 or LL6 ordinary chondrites (*Abe et al.,* 2006). The analysis of the samples returned by Hayabusa eventually confirmed these findings. Mineralogy and mineral chemistry as well as oxygen isotopic compositions suggest a strong link to equilibrated LL chondrites (*Nakamura et al.,* 2011; *Yurimoto et al.,* 2011; *Ebihara et al.,* 2011; *Nakashima et al.,* 2013). These findings provide unequivocal evidence that S-type asteroids like Itokawa are one of the sources of ordinary chondrites.

In addition, Itokawa particles allow studies of regolith formation on an asteroid (*Tsuchiyama et al.,* 2011, 2013). Noble gas measurements helped to elucidate the cosmic-ray and solar-wind irradiation history of Itokawa (*Nagao et al.,* 2011). In order to study the alteration of the surfaces of airless bodies exposed to the space environment, known as space weathering, samples from asteroid Itokawa returned by the Hayabusa spacecraft are ideally suited (*Noguchi et al.,* 2011).

Acknowledgments. This work was supported by grants from the NASA Cosmochemistry and Laboratory Analysis of Returned Samples Programs. We thank the anonymous reviewer for a detailed review.

REFERENCES

Abe M. et al. (2006) *Science, 312,* 1334.
Abreu N. M. and Brearley A. J. (2010) *Geochim. Cosmochim. Acta, 74,* 1146.
Adachi I. et al. (1976) *Prog. Theor. Phys., 56,* 1756.
Alexander C. M. O'D. (2004) *Geochim. Cosmochim. Acta, 68,* 3943.
Alexander C. M. O'D. (2005) *Meteoritics & Planet. Sci., 40,* 943.
Alexander C. M. O'D. and Ebel D. S. (2012) *Meteoritics & Planet. Sci., 47,* 1157.
Alexander C. M. O'D. et al. (2001) *Science, 293,* 64.
Alexander C. M. O'D. et al. (2007) *Geochim. Cosmochim. Acta, 71,* 4380.
Alexander C. M. O'D. et al. (2008) *Science, 320,* 1617.
Alexander C. M. O'D. et al. (2010) *Geochim. Cosmochim. Acta, 74,* 4417.
Alexander C. M. O'D. et al. (2012) *Science, 337,* 721.
Amari S. et al. (1990) *Nature, 345,* 238.
Amelin Y. (2008) *Geochim. Cosmochim. Acta, 72,* 221.
Amelin Y. et al. (2010) *Earth Planet. Sci. Lett., 300,* 343.
Arnould M. et al. (1997) *Astron. Astrophys., 321,* 452.
Asphaug E. et al. (2011) *Earth Planet. Sci. Lett., 308,* 369.
Baguhl M. et al. (1995) *Space Sci. Rev., 72,* 471.
Bally J. and Langer W. D. (1982) *Astrophys. J., 255,* 143.
Bechtel H. A. et al. (2013) *Meteoritics & Planet. Sci.,* in press.
Berger E. L. et al. (2011) *Geochim. Cosmochim. Acta, 75,* 3501.
Bernatowicz T. et al. (1987) *Nature, 330,* 728.
Bland P. A. et al. (2005) *Proc. Natl. Acad. Sci., 102,* 13755.
Bland P. A. et al. (2007) *Meteoritics & Planet. Sci., 42,* 1417.
Blichert-Toft J. et al. (2010) *Earth Planet. Sci. Lett., 300,* 152.
Bockelée-Morvan D. et al. (2002) *Astron. Astrophys., 384,* 1107.
Bollard J. et al. (2013) *Mineral. Mag., 77,* 732.
Borg L. E. et al. (2011) *Nature, 477,* 70.
Borisov A. et al. (2008) *Geochim. Cosmochim. Acta, 72,* 5558.
Boss A. P. (1998) *Annu. Rev. Earth Planet. Sci., 26,* 53.
Boss A. P. (2008) *Earth Planet. Sci. Lett., 268,* 102.
Boss A. P. (2013) *Astrophys. J., 773,* 5.
Boss A. P. and Keiser S. A. (2013) *Astrophys. J., 770,* 51.
Boss A. P. et al. (2012) *Earth Planet. Sci. Lett., 345,* 18.
Bottke W. F. et al. (2010) *Science, 330,* 1527.
Bottke W. F. et al. (2012) *Nature, 485,* 78.
Bouvier A. et al. (2011) In *Workshop on the Formation of the First Solids in the Solar System,* Abstract #9054. LPI Contribution No. 1639, Lunar and Planetary Institute, Houston.
Bowers M. et al. (2013) *ArXiv e-prints,* arXiv:1310.3231.
Bradley J. P. (1994) *Science, 265,* 925.
Brearley A. J. and Jones R. H. (1998) *Rev. Mineral. Geochem., 36,* 3.1.
Brearley A. J. and Krot A. N. (2012) In *Metasomatism and the Chemical Transformation of Rock* (D. E. Harlov et. al., eds.), pp. 659–789. Springer-Verlag, Heidelberg.
Brenker F. E. et al. (2013) *Meteoritics & Planet. Sci.,* in press.
Brennecka G. A. and Wadhwa M. (2012) *Proc. Natl. Acad. Sci., 109,* 9299.
Brennecka G. A. et al. (2010) *Science, 327,* 449.
Briani G. et al. (2009) *Proc. Natl. Acad. Sci., 106,* 10522.
Bricker G. E. and Caffee M. W. (2010) *Astrophys. J., 725,* 443.
Brownlee D. E. et al. (2003) *J. Geophys. Res.–Planets, 108,* E8111.
Brownlee D. E. et al. (2004) *Science, 304,* 1764.
Brownlee D. et al. (2006) *Science, 314,* 1711.
Burbine T. H. (2014) In *Treatise on Geochemistry, Second Edition, Vol. 2: Planets, Asteroids, Comets and the Solar System* (A. M. Davis, ed.), pp. 365–415, Elsevier.
Burbine T. H. et al. (2002) In *Asteroids III* (W. F. Bottke Jr. et. al., eds.), pp. 653–667. Univ. of Arizona, Tucson.
Burchell M. J. et al. (2006) *Annu. Rev. Earth Planet. Sci., 34,* 385.
Burchell M. J. et al. (2008) *Meteoritics & Planet. Sci., 43,* 23.
Burkhardt C. et al. (2008) *Geochim. Cosmochim. Acta, 72,* 6177.
Burkhardt C. et al. (2011) *Earth Planet. Sci. Lett., 312,* 390.
Burkhardt C. et al. (2012) *Astrophys. J. Lett., 753,* L6.

Burnett D. S. and the Genesis Science Team (2011) *Proc. Natl. Acad. Sci., 108,* 19147.
Bus S. J. and Binzel R. P. (2002) *Icarus, 158,* 146.
Busemann H. et al. (2006) *Science, 312,* 727.
Busemann H. et al. (2009) *Earth Planet. Sci. Lett., 288,* 44.
Butterworth A. L. et al. (2013) *Meteoritics & Planet. Sci.,* in press.
Cameron A. G. W. and Truran J. W. (1977) *Icarus, 30,* 447.
Chakraborty S. et al. (2008) *Science, 321,* 1328.
Chakraborty S. et al. (2009) *Science, 324,* 1516d.
Chakraborty S. et al. (2013a) In *Lunar and Planetary Science XLIV,* Abstract #1043. Lunar and Planetary Institute, Houston.
Chakraborty S. et al. (2013b) *Science, 342,* 463.
Chaussidon M. et al. (2006) *Geochim. Cosmochim. Acta, 70,* 224.
Chen J. H. et al. (2013) In *Lunar and Planetary Science XLIV,* Abstract #2649. Lunar and Planetary Institute, Houston.
Ciesla F. J. (2007) *Science, 318,* 613.
Ciesla F. J. (2010) *Icarus, 208,* 455.
Ciesla F. J. (2011) *Astrophys. J., 740,* 9.
Clayton D. D. (1988) *Astrophys. J., 334,* 191.
Clayton R. N. (2002) *Nature, 415,* 860.
Clayton R. N. (2003) In *Treatise on Geochemistry, Vol. 1: Meteorites, Comets, and Planets* (A. M. Davis, ed.), pp. 129–142, Elsevier.
Clayton R. N. and Mayeda T. K. (1977) *Geophys. Res. Lett., 4,* 295.
Clayton R. N. et al. (1973) *Science, 182,* 485.
Cody G. D. et al. (2008) *Meteoritics & Planet. Sci., 43,* 353.
Connelly J. N. et al. (2012) *Science, 338,* 651.
Cuzzi J. N. et al. (2008) *Astrophys. J., 687,* 1432.
Dauphas N. and Pourmand A. (2011) *Nature, 473,* 489.
Dauphas N. et al. (2008) *Astrophys. J., 686,* 560.
Davis A. M. (2011) *Proc. Natl. Acad. Sci., 108,* 19142.
Davis A. M. and McKeegan K. D. (2014) In *Treatise on Geochemistry, Second Edition, Vol. 1: Meteorites and Cosmochemical Processes* (A. M. Davis, ed.), pp. 361–395, Elsevier.
Davis A. M. and Richter F. M. (2014) In *Treatise on Geochemistry, Second Edition, Vol. 1: Meteorites and Cosmochemical Processes* (A. M. Davis, ed.), pp. 335–360, Elsevier.
Davis A. M. et al. (2005) In *Chondrites and the Protoplanetary Disk* (A. N. Krot et. al., eds.), pp. 432–455. ASP Conf. Ser. 341, Astronomical Society of the Pacific, San Francisco.
Davison T. M. et al. (2010) *Icarus, 208,* 468.
de Avillez M. A. and Mac Low M.-M. (2002) *Astrophys. J., 581,* 1047.
De Gregorio B. T. et al. (2013) *Meteoritics & Planet. Sci., 48,* 904.
Desch S. J. et al. (2004) *Astrophys. J., 602,* 528.
Desch S. J. et al. (2005) In *Chondrites and the Protoplanetary Disk* (A. N. Krot et. al., eds.), p. 849. ASP Conf. Ser. 341, Astronomical Society of the Pacific, San Francisco.
Desch S. J. et al. (2010) *Astrophys. J., 725,* 692.
Djouadi Z. et al. (2005) *Astron. Astrophys., 440,* 179.
Doyle P. M. et al. (2013) In *Lunar and Planetary Science XLIV,* Abstract #1793. Lunar and Planetary Institute, Houston.
Drouart A. et al. (1999) *Icarus, 140,* 129.
Duprat J. and Tatischeff V. (2007) *Astrophys. J. Lett., 671,* L69.
Ebel D. S. (2006) In *Meteorites and the Early Solar System II* (D. S. Lauretta et al., eds.), pp. 253–277. Univ. of Arizona, Tucson.
Ebel D. S. and Alexander C. M. O'D. (2011) *Planet. Space Sci., 59,* 1888.
Ebel D. S. and Grossman L. (2000) *Geochim. Cosmochim. Acta, 64,* 339.
Ebihara M. et al. (2011) *Science, 333,* 1119.
Ellinger C. I. et al. (2010) *Astrophys. J., 725,* 1495.
Fabian D. et al. (2000) *Astron. Astrophys., 364,* 282.
Federman S. R. and Young E. D. (2009) *Science, 324,* 1516b.
Fedkin A. V. and Grossman L. (2013) *Geochim. Cosmochim. Acta, 112,* 226.
Fedkin A. V. et al. (2012) *Geochim. Cosmochim. Acta, 87,* 81.
Feigelson E. D. (2010) *Proc. Natl. Acad. Sci., 107,* 7153.
Floss C. and Stadermann F. J. (2012) *Meteoritics & Planet. Sci., 47,* 992.
Floss C. et al. (2013) *Astrophys. J., 763,* 140.
Flynn G. J. et al. (2006) *Science, 314,* 1731.
Flynn G. et al. (2013) *Meteoritics & Planet. Sci.,* in press.
Foley C. N. et al. (2006) *Icarus, 184,* 338.
Frank D. R. et al. (2013) *Meteoritics & Planet. Sci.,* in press.
Fujiwara A. et al. (2006) *Science, 312,* 1330.
Fujiya W. et al. (2012) *Nature Comm., 3,* 627.
Fujiya W. et al. (2013) *Earth Planet. Sci. Lett., 362,* 130.
Gaidos E. et al. (2009) *Astrophys. J., 696,* 1854.

Gainsforth Z. et al. (2013) *Meteoritics & Planet. Sci.,* in press.
Gallien J.-P. et al. (2008) *Meteoritics & Planet. Sci., 43,* 335.
Georg R. B. et al. (2007) *Nature, 447,* 1102.
Glavin D. P. et al. (2004) *Meteoritics & Planet. Sci., 39,* 693.
Gomes R. et al. (2005) *Nature, 435,* 466.
Gounelle M. and Meibom A. (2007) *Astrophys. J. Lett., 664,* L123.
Gounelle M. and Meibom A. (2008) *Astrophys. J., 680,* 781.
Gounelle M. and Meynet G. (2012) *Astron. Astrophys., 545,* A4.
Gounelle M. et al. (2006a) *Astrophys. J., 640,* 1163.
Gounelle M. et al. (2006b) *Meteoritics & Planet. Sci., 41,* 135.
Gounelle M. et al. (2008) In *The Solar System Beyond Neptune* (M. A.
    Barucci et al., eds.), pp. 525–541. Univ. of Arizona, Tucson.
Gounelle M. et al. (2009a) *Astrophys. J. Lett., 698,* L18.
Gounelle M. et al. (2009b) *Astrophys. J. Lett., 694,* L1.
Gounelle M. et al. (2013) *Astrophys. J. Lett., 763,* L33.
Gradie J. and Tedesco E. (1982) *Science, 216,* 1405.
Grady M. M. and Pillinger C. T. (1990) *Earth Planet. Sci. Lett., 97,* 29.
Greshake A. et al. (1996) *Meteoritics & Planet. Sci., 31,* 739.
Gritschneder M. et al. (2012) *Astrophys. J., 745,* 22.
Grossman L. (1972) *Geochim. Cosmochim. Acta, 36,* 597.
Grossman L. et al. (2002) *Geochim. Cosmochim. Acta, 66,* 145.
Grün E. et al. (1993) *Nature, 362,* 428.
Grün E. et al. (1994) *Astron. Astrophys., 286,* 915.
Hallenbeck S. L. et al. (2000) *Astrophys. J., 535,* 247.
Hallis L. J. et al. (2012) *Earth Planet. Sci. Lett., 359,* 84.
Hanner M. S. and Bradley J. P. (2004) In *Comets II* (M. C. Festou
    et al., eds.), pp. 555–564. Univ. of Arizona, Tucson.
Harker D. E. and Desch S. J. (2002) *Astrophys. J. Lett., 565,* L109.
Harrison T. M. (2009) *Annu. Rev. Earth Planet. Sci., 37,* 479.
Hartogh P. et al. (2011) *Nature, 478,* 218.
Hennebelle P. et al. (2009) In *Structure Formation in Astrophysics*
    (G. Chabrier, ed.), p. 205. Cambridge Univ., Cambridge.
Hester J. J. et al. (2004) *Science, 304,* 1116.
Hewins R. H. et al. (2005) In *Chondrites and the Protoplanetary
    Disk* (A. N. Krot et al., eds.), pp. 286–316. ASP Conf. Ser. 341,
    Astronomical Society of the Pacific, San Francisco.
Hewins R. H. et al. (2012) *Geochim. Cosmochim. Acta, 78,* 1.
Hezel D. C. and Palme H. (2010) *Earth Planet. Sci. Lett., 294,* 85.
Hezel D. C. et al. (2010) *Earth Planet. Sci. Lett., 296,* 423.
Hiroi T. et al. (2001) *Science, 293,* 2234.
Holst J. C. et al. (2013) *Proc. Natl. Acad. Sci., 110,* 8819.
Hood L. L. and Weidenschilling S. J. (2012) *Meteoritics & Planet. Sci.,
    47,* 1715.
Hoppe P. et al. (2007) *Meteoritics & Planet. Sci., 42,* 1309.
Horner J. et al. (2007) *Earth Moon Planets, 100,* 43.
Horner J. et al. (2008) *Planet. Space Sci., 56,* 1585.
Hörz F. et al. (2000) *Icarus, 147,* 559.
Hörz F. et al. (2006) *Science, 314,* 1716.
Hsu W. et al. (2006) *Astrophys. J., 640,* 525.
Hua X. et al. (2005) *Geochim. Cosmochim. Acta, 69,* 1333.
Hutcheon I. D. et al. (1998) *Science, 282,* 1865.
Hutchison R. et al. (1980) *Nature, 287,* 787.
Ikeda Y. (1989) *Antarct. Meteorite Res., 2,* 75.
Ishii H. A. et al. (2008a) *Science, 319,* 447.
Ishii H. A. et al. (2008b) *Meteoritics & Planet. Sci., 43,* 215.
Ito M. et al. (2006) *Meteoritics & Planet. Sci., 41,* 1871.
Jacobsen B. et al. (2008) *Earth Planet. Sci. Lett., 272,* 353.
Jacobsen B. et al. (2011) *Astrophys. J. Lett., 731,* L28.
Jacquet E. and Robert F. (2013) *Icarus, 223,* 722.
Javoy M. (1995) *Geophys. Res. Lett., 22,* 2219.
Jessberger E. K. et al. (1988) *Nature, 332,* 691.
Jessberger E. K. et al. (2001) In *Interplanetary Dust* (E. Grün et al.,
    eds.), pp. 253–294. Springer-Verlag, Berlin.
Jilly C. E. et al. (2013) *Meteoritics & Planet. Sci. Suppl., 76,* 5303.
Jogo K. et al. (2009) *Earth Planet. Sci. Lett., 287,* 320.
Jones R. H. et al. (2005) In *Chondrites and the Protoplanetary Disk*
    (A. N. Krot et al., eds.), p. 251. ASP Conf. Ser. 341, Astronomical
    Society of the Pacific, San Francisco.
Kallemeyn G. W. and Wasson J. T. (1986) *Geochim. Cosmochim. Acta,
    50,* 2153.
Kastner J. H. and Myers P. C. (1994) *Astrophys. J., 421,* 605.
Kearsley A. T. et al. (2008) *Meteoritics & Planet. Sci., 43,* 41.
Kearsley A. T. et al. (2012) *Meteoritics & Planet. Sci., 47,* 737.
Keil K. (1968) *J. Geophys. Res., 73,* 6945.
Keil K. (2010) *Chem. Erde, 70,* 295.
Keller L. P. et al. (2006) *Science, 314,* 1728.

Kita N. T. and Ushikubo T. (2012) *Meteoritics & Planet. Sci., 47,* 1108.
Kita N. T. et al. (2000) *Geochim. Cosmochim. Acta, 64,* 3913.
Kita N. T. et al. (2013) *Meteoritics & Planet. Sci., 48,* 1383.
Kleine T. et al. (2009) *Geochim. Cosmochim. Acta, 73,* 5150.
Kleine T. et al. (2012) *Geochim. Cosmochim. Acta, 84,* 186.
Krot A. N. et al. (2005) *Nature, 436,* 989.
Krot A. N. et al. (2008) *Astrophys. J., 672,* 713.
Krot A. N. et al. (2009) *Geochim. Cosmochim. Acta, 73,* 4963.
Krot A. N. et al. (2010a) *Astrophys. J., 713,* 1159.
Krot A. N. et al. (2010b) *Geochim. Cosmochim. Acta, 74,* 2190.
Krot A. N. et al. (2012) *Meteoritics & Planet. Sci., 47,* 1948.
Krot A. N. et al. (2014) In *Treatise on Geochemistry, Second Edition,
    Vol. 1: Meteorites and Cosmochemical Processes* (A. M. Davis,
    ed.), pp. 1–63, Elsevier.
Kruijer T. S. et al. (2012) *Geochim. Cosmochim. Acta, 99,* 287.
Kruijer T. S. et al. (2013) *Earth Planet. Sci. Lett., 361,* 162.
Kruijer T. S. et al. (2014) In *Lunar and Planetary Science XLV,*
    Abstract #1786. Lunar and Planetary Institute, Houston.
Kurahashi E. et al. (2008) *Geochim. Cosmochim. Acta, 72,* 3865.
Landgraf M. et al. (1999) *Planet. Space Sci., 47,* 1029.
Lanzirotti A. et al. (2008) *Meteoritics & Planet. Sci., 43,* 187.
Larimer J. W. and Bartholomay M. (1979) *Geochim. Cosmochim. Acta,
    43,* 1455.
Larsen K. K. et al. (2011) *Astrophys. J. Lett., 735,* L37.
Lee T. et al. (1977) *Astrophys. J. Lett., 211,* L107.
Lehner S. W. et al. (2013) *Geochim. Cosmochim. Acta, 101,* 34.
Leitner J. et al. (2008) *Meteoritics & Planet. Sci., 43,* 161.
Leroux H. et al. (2008) *Meteoritics & Planet. Sci., 43,* 97.
Levison H. F. et al. (2009) *Nature, 460,* 364.
Levison H. F. et al. (2011) *Astron. J., 142,* 152.
Lewis R. S. et al. (1987) *Nature, 326,* 160.
Libourel G. and Krot A. N. (2007) *Earth Planet. Sci. Lett., 254,* 1.
Lin Y. et al. (2003) *Geochim. Cosmochim. Acta, 67,* 4935.
Lin Y. et al. (2005) *Proc. Natl. Acad. Sci., 102,* 1306.
Liu M.-C. et al. (2009) *Geochim. Cosmochim. Acta, 73,* 5051.
Liu M.-C. et al. (2010) *Astrophys. J. Lett., 719,* L99.
Liu M.-C. et al. (2012a) *Earth Planet. Sci. Lett., 327,* 75.
Liu M.-C. et al. (2012b) *Astrophys. J., 761,* 137.
Lodders K. et al. (2009) In *The Landolt-Börnstein Database* (J. E.
    Trümper, ed.), pp. 34/1–34/59. Springer-Verlag, Berlin.
Lugaro M. et al. (2012a) *Astrophys. J., 759,* 51.
Lugaro M. et al. (2012b) *Meteoritics & Planet. Sci., 47,* 1998.
Lyons J. R. and Young E. D. (2005) *Nature, 435,* 317.
Lyons J. R. et al. (2009a) *Science, 324,* 1516a.
Lyons J. R. et al. (2009b) *Geochim. Cosmochim. Acta, 73,* 4998.
MacPherson G. J. (2014) In *Treatise on Geochemistry, Second Edition,
    Vol. 1: Meteorites and Cosmochemical Processes* (A. M. Davis,
    ed.), pp. 139–179, Elsevier.
MacPherson G. J. et al. (2003) *Geochim. Cosmochim. Acta, 67,* 3165.
MacPherson G. J. et al. (2010) *Astrophys. J. Lett., 711,* L117.
MacPherson G. J. et al. (2012) *Earth Planet. Sci. Lett., 331,* 43.
Makide K. et al. (2011) *Astrophys. J. Lett., 733,* L31.
Marty B. (2012) *Earth Planet. Sci. Lett., 313,* 56.
Marty B. et al. (2011) *Science, 332,* 1533.
Marty B. et al. (2013) *Rev. Mineral. Geochem., 75,* 149.
Matrajt G. et al. (2008) *Meteoritics & Planet. Sci., 43,* 315.
Matzel J. E. P. et al. (2010) *Science, 328(5977),* 483.
McCanta M. C. et al. (2008) *Geochim. Cosmochim. Acta, 72,* 5757.
McKeegan K. D. et al. (2000) *Science, 289,* 1334.
McKeegan K. D. et al. (2006) *Science, 314,* 1724.
McKeegan K. D. et al. (2011) *Science, 332,* 1528.
McKibbin S. J. et al. (2013) *Geochim. Cosmochim. Acta, 110,* 216.
Meibom A. et al. (2007) *Astrophys. J. Lett., 656,* L33.
Messenger S. et al. (2003) *Science, 300,* 105.
Metzler K. (2012) *Meteoritics & Planet. Sci., 47,* 2193.
Meyer B. S. and Clayton D. D. (2000) *Space Sci. Rev., 92,* 133.
Meynet G. et al. (2008) In *Massive Stars as Cosmic Engines* (F.
    Bresolin et al., eds.), pp. 147–160. IAU Symp. 250, Cambridge
    Univ., Cambridge.
Miyamoto M. et al. (2009) *Meteoritics & Planet. Sci., 44,* 521.
Morris M. A. and Desch S. J. (2010) *Astrophys. J., 722,* 1474.
Morris M. A. et al. (2012) *Astrophys. J., 752,* 27.
Mostefaoui S. et al. (2002) *Meteoritics & Planet. Sci., 37,* 421.
Mostefaoui S. et al. (2005) *Astrophys. J., 625,* 271.
Mousis O. et al. (2000) *Icarus, 148,* 513.
Mousis O. et al. (2007) *Astron. Astrophys., 466,* L9.

Moynier F. et al. (2011) *Astrophys. J., 741,* 71.

Mukhopadhyay S. (2012) *Nature, 486,* 101.

Nagao K. et al. (2011) *Science, 333,* 1128.

Nagashima K. et al. (2007) *Meteoritics & Planet. Sci., 42,* 5291.

Nagashima K. et al. (2008) In *Lunar and Planetary Science XXXIX,* Abstract #2224. Lunar and Planetary Institute, Houston.

Nakamura T. et al. (2008) *Science, 321,* 1664.

Nakamura T. et al. (2011) *Science, 333,* 1113.

Nakamura-Messenger K. et al. (2006) *Science, 314,* 1439.

Nakashima D. et al. (2008) *Geochim. Cosmochim. Acta, 72,* 6141.

Nakashima D. et al. (2013) *Earth Planet. Sci. Lett., 379,* 127.

Nguyen A. N. and Zinner E. (2004) *Science, 303,* 1496.

Nittler L. R. and Gaidos E. (2012) *Meteoritics & Planet. Sci., 47,* 2031.

Nittler L. R. et al. (2001) *Meteoritics & Planet. Sci., 36,* 1673.

Nittler L. R. et al. (2011) *Science, 333,* 1847.

Noguchi T. et al. (2011) *Science, 333,* 1121.

Nuth J. A. III and Johnson N. M. (2006) *Icarus, 180,* 243.

Nuth J. A. III et al. (2000) *Nature, 406,* 275.

Nyquist L. E. et al. (2009) *Geochim. Cosmochim. Acta, 73,* 5115.

Ogliore R. C. et al. (2009) *Meteoritics & Planet. Sci., 44,* 1675.

Ogliore R. C. et al. (2011) *Nucl. Inst. Meth. Phys. Res. B, 269,* 1910.

Ogliore R. C. et al. (2012) *Astrophys. J. Lett., 745,* L19.

Oliveira I. et al. (2011) *Astrophys. J., 734,* 51.

Olsen M. B. et al. (2013) *Astrophys. J. Lett., 776,* L1.

Owen T. et al. (2001) *Astrophys. J. Lett., 553,* L77.

Penzias A. A. (1981) *Astrophys. J., 249,* 518.

Pepin R. O. and Porcelli D. (2006) *Earth Planet. Sci. Lett., 250,* 470.

Petitat M. et al. (2011) *Meteoritics & Planet. Sci., 46,* 275.

Postberg F. et al. (2013) *Meteoritics & Planet. Sci.,* in press.

Prombo C. A. and Clayton R. N. (1985) *Science, 230,* 935.

Quitté G. et al. (2010) *Astrophys. J., 720,* 1215.

Quitté G. et al. (2011) *Geochim. Cosmochim. Acta, 75,* 7698.

Reipurth B. et al., eds. (2007) *Protostars and Planets V.* Univ. of Arizona, Tucson. 951 pp.

Richter F. M. et al. (2011) *Meteoritics & Planet. Sci., 46,* 1152.

Rotundi A. et al. (2008) *Meteoritics & Planet. Sci., 43,* 367.

Rubin A. E. (1983) *Earth Planet. Sci. Lett., 64,* 201.

Rubin A. E. (1984) *Earth Planet. Sci. Lett., 67,* 273.

Rubin A. E. (2010) *Geochim. Cosmochim. Acta, 74,* 4807.

Rubin A. E. and Choi B.-G. (2009) *Earth Moon Planets, 105,* 41.

Rugel G. et al. (2009) *Phys. Rev. Lett., 103(7),* 072502.

Saal A. E. et al. (2013) *Science, 340,* 1317.

Safier P. N. (1993) *Astrophys. J., 408,* 115.

Sahijpal S. et al. (2000) *Geochim. Cosmochim. Acta, 64,* 1989.

Sanders I. S. and Scott E. R. D. (2012) *Meteoritics & Planet. Sci., 47,* 2170.

Sandford S. A. et al. (2006) *Science, 314,* 1720.

Sandford S. A. et al. (2010) *Meteoritics & Planet. Sci., 45,* 406.

Savage P. S. and Moynier F. (2013) *Earth Planet. Sci. Lett., 361,* 487.

Schiller M. et al. (2010) *Geochim. Cosmochim. Acta, 74,* 4844.

Schrader D. L. et al. (2013) *Meteoritics & Planet. Sci., 76,* 5141.

Scott E. R. D. and Krot A. N. (2014) In *Treatise on Geochemistry, Second Edition, Vol. 1: Meteorites and Cosmochemical Processes* (A. M. Davis, ed.), pp. 65–137, Elsevier.

Sears D. W. et al. (1982) *Geochim. Cosmochim. Acta, 46,* 597.

Shu F. H. et al. (1996) *Science, 271,* 1545.

Shu F. H. et al. (1997) *Science, 277,* 1475.

Shukolyukov A. and Lugmair G. W. (1993) *Science, 259,* 1138.

Simionovici A. S. et al. (2013) *Meteoritics & Planet. Sci.,* in press.

Simon S. B. et al. (2002) *Meteoritics & Planet. Sci., 37,* 533.

Simon S. B. et al. (2008) *Meteoritics & Planet. Sci., 43,* 1861.

Smith R. L. et al. (2009) *Astrophys. J., 701,* 163.

Spivak-Birndorf L. et al. (2009) *Geochim. Cosmochim. Acta, 73,* 5202.

Spivak-Birndorf L. J. et al. (2011a) *Workshop on the Formation of the First Solids in the Solar System,* Abstract #9130. LPI Contribution No. 1639, Lunar and Planetary Institute, Houston.

Spivak-Birndorf L. J. et al. (2011b) In *Lunar and Planetary Science XLII,* Abstract #2281. Lunar and Planetary Institute, Houston.

Spivak-Birndorf L. J. et al. (2012) In *Lunar and Planetary Science XLIII,* Abstract #2861. Lunar and Planetary Institute, Houston.

Srinivasan G. and Chaussidon M. (2013) *Earth Planet. Sci. Lett., 374,* 11.

Srinivasan G. et al. (1996) *Geochim. Cosmochim. Acta, 60,* 1823.

Stadermann F. J. et al. (2008) *Meteoritics & Planet. Sci., 43,* 299.

Stephan T. (2008) *Space Sci. Rev., 138,* 247.

Stephan T. (2009) In *Cosmic Dust — Near and Far* (T. Henning et al., eds.), pp. 168–177. ASP Conf. Ser. 414, Astronomical Society of the Pacific, San Francisco.

Stephan T. et al. (2008a) *Meteoritics & Planet. Sci., 43,* 233.

Stephan T. et al. (2008b) *Meteoritics & Planet. Sci., 43,* 285.

Sterken V. J. et al. (2013) *Meteoritics & Planet. Sci.,* in press.

Stroud R. M. et al. (2013) *Meteoritics & Planet. Sci.,* in press.

Tachibana S. and Huss G. R. (2003) *Astrophys. J. Lett., 588,* L41.

Tachibana S. et al. (2006) *Astrophys. J. Lett., 639,* L87.

Tang H. and Dauphas N. (2012) *Earth Planet. Sci. Lett., 359,* 248.

Tatischeff V. et al. (2010) *Astrophys. J. Lett., 714,* L26.

Telus M. et al. (2012) *Meteoritics & Planet. Sci., 47,* 2013.

Telus M. et al. (2013) *Lunar Planet. Sci. Conf., 44,* 2964.

Thiemens M. H. (2006) *Annu. Rev. Earth Planet. Sci., 34,* 217.

Thiemens M. H. and Heidenreich J. E. III (1983) *Science, 219,* 1073.

Thrane K. et al. (2006) *Astrophys. J. Lett., 646,* L159.

Touboul M. et al. (2007) *Nature, 450,* 1206.

Trigo-Rodríguez J. M. et al. (2008) *Meteoritics & Planet. Sci., 43,* 75.

Trinquier A. et al. (2007) *Astrophys. J., 655,* 1179.

Trinquier A. et al. (2009) *Science, 324,* 374.

Tsou P. et al. (2003) *J. Geophys. Res.–Planets, 108,* E8113.

Tsuchiyama A. et al. (2011) *Science, 333,* 1125.

Tsuchiyama A. et al. (2013) *Geochim. Cosmochim. Acta, 116,* 5.

Ushikubo T. et al. (2013) *Geochim. Cosmochim. Acta, 109,* 280.

Usui T. et al. (2012) *Earth Planet. Sci. Lett., 357,* 119.

van Boekel R. et al. (2004) *Nature, 432,* 479.

van Dishoeck E. F. and Black J. H. (1988) *Astrophys. J., 334,* 771.

Vasileiadis A. et al. (2013) *Astrophys. J. Lett., 769,* L8.

Villeneuve J. et al. (2009) *Science, 325,* 985.

Vinković D. (2009) *Nature, 459,* 227.

Waite J. H. Jr. et al. (2009) *Nature, 460,* 487.

Walker R. J. (2009) *Chem. Erde, 69,* 101.

Walsh K. J. et al. (2011) *Nature, 475,* 206.

Warren P. H. (2011) *Earth Planet. Sci. Lett., 311,* 93.

Wasserburg G. J. et al. (1977) *Geophys. Res. Lett., 4,* 299.

Wasserburg G. J. et al. (2006) *Nuclear Phys. A, 777,* 5.

Wasserburg G. J. et al. (2011) *Geochim. Cosmochim. Acta, 75,* 4752.

Wasserburg G. J. et al. (2012) *Meteoritics & Planet. Sci., 47,* 1980.

Wasson J. T. (2008) *Icarus, 195,* 895.

Weidenschilling S. J. (1977) *Mon. Not. R. Astron. Soc., 180,* 57.

Weisberg M. K. and Kimura M. (2012) *Chem. Erde, 72,* 101.

Westphal A. J. et al. (2004) *Meteoritics & Planet. Sci., 39,* 1375.

Westphal A. J. et al. (2013a) *Meteoritics & Planet. Sci.,* in press.

Westphal A. J. et al. (2013b) *Meteoritics & Planet. Sci.,* in press.

Wick M. J. and Jones R. H. (2012) *Geochim. Cosmochim. Acta, 98,* 140.

Wielandt D. et al. (2012) *Astrophys. J. Lett., 748,* L25.

Williams J. P. and Gaidos E. (2007) *Astrophys. J. Lett., 663,* L33.

Wopenka B. (2012) *Meteoritics & Planet. Sci., 47,* 565.

Wouterloot J. G. A. et al. (2008) *Astron. Astrophys., 487,* 237.

Wozniakiewicz P. J. et al. (2012) *Meteoritics & Planet. Sci., 47,* 660.

Wurm G. and Haack H. (2009) *Meteoritics & Planet. Sci., 44,* 689.

Yang L. and Ciesla F. J. (2012) *Meteoritics & Planet. Sci., 47,* 99.

Yang L. et al. (2013) *Icarus, 226,* 256.

Yin Q.-Z. et al. (2009) *Science, 324,* 1516c.

Young E. D. et al. (2005) *Science, 308,* 223.

Young E. D. et al. (2011) *Astrophys. J., 729,* 43.

Yurimoto H. (2008) *Rev. Mineral. Geochem., 68,* 141.

Yurimoto H. and Kuramoto K. (2004) *Science, 305,* 1763.

Yurimoto H. et al. (2011) *Science, 333,* 1116.

Zanda B. et al. (2012) In *Lunar and Planetary Science XLIII,* Abstract #2413. Lunar and Planetary Institute, Houston.

Zhang J. et al. (2012) *Nature Geosci., 5,* 251.

Zinner E. (2013) *Anal. Chem., 85,* 1264.

Zinner E. (2014) In *Treatise on Geochemistry, Second Edition, Vol. 1: Meteorites and Cosmochemical Processes* (A. M. Davis, ed.), pp. 181–213, Elsevier.

Zolensky M. et al. (2008) *Meteoritics & Planet. Sci., 43,* 261.

Zolensky M. E. et al. (2006) *Science, 314,* 1735.

Zolotov M. Y. and Mironenko M. V. (2007) In *Lunar and Planetary Science XXXVIII,* Abstract #2340. Lunar and Planetary Institute, Houston.

Part IV:
Astrophysical Conditions for Life

van Dishoeck E. F., Bergin E. A., Lis D. C., and Lunine J. I. (2014) Water: From clouds to planets. In *Protostars and Planets VI* (H. Beuther et al., eds.), pp. 835–858. Univ. of Arizona, Tucson, DOI: 10.2458/azu_uapress_9780816531240-ch036.

Water: From Clouds to Planets

Ewine F. van Dishoeck
Leiden University and Max Planck Institute for Extraterrestrial Physics

Edwin A. Bergin
University of Michigan

Dariusz C. Lis
California Institute of Technology

Jonathan I. Lunine
Cornell University

Results from recent space missions, in particular Spitzer and Herschel, have led to significant progress in our understanding of the formation and transport of water from clouds to disks, planetesimals, and planets. In this review, we provide the underpinnings for the basic molecular physics and chemistry of water and outline these advances in the context of water formation in space, its transport to a forming disk, its evolution in the disk, and finally the delivery to forming terrestrial worlds and accretion by gas giants. Throughout, we pay close attention to the disposition of water as vapor or solid and whether it might be subject to processing at any stage. The context of the water in the solar system and the isotopic ratios (D/H) in various bodies are discussed as grounding data points for this evolution. Additional advances include growing knowledge of the composition of atmospheres of extrasolar gas giants, which may be influenced by the variable phases of water in the protoplanetary disk. Furthermore, the architecture of extrasolar systems leaves strong hints of dynamical interactions, which are important for the delivery of water and subsequent evolution of planetary systems. We conclude with an exploration of water on Earth and note that all the processes and key parameters identified here should also hold for exoplanetary systems.

1. INTRODUCTION

With nearly 1000 exoplanets discovered to date and statistics indicating that every star hosts at least one planet (*Batalha et al.,* 2013), the next step in our search for life elsewhere in the universe is to characterize these planets. The presence of water on a planet is universally accepted as essential for its potential habitability. Water in gaseous form acts as a coolant that allows interstellar gas clouds to collapse to form stars, whereas water ice facilitates the sticking of small dust particles that eventually must grow to planetesimals and planets. The development of life requires liquid water, and even the most primitive cellular life on Earth consists primarily of water. Water assists many chemical reactions leading to complexity by acting as an effective solvent. It shapes the geology and climate on rocky planets, and is a major or primary constituent of the solid bodies of the outer solar system.

How common are planets that contain water, and how does the water content depend on the planet's formation his-

tory and other properties of the star-planet system? Thanks to a number of recent space missions, culminating with the Herschel Space Observatory, an enormous step forward has been made in our understanding of where water is formed in space, what its abundance is in various physical environments, and how it is transported from collapsing clouds to forming planetary systems. At the same time, new results are emerging on the water content of bodies in our own solar system and in the atmospheres of known exoplanets. This review attempts to synthesize the results from these different fields by summarizing our current understanding of the water trail from clouds to planets.

Speculations about the presence of water on Mars and other planets in our solar system date back many centuries. Water is firmly detected as gas in the atmospheres of all planets, including Mercury, and as ice on the surfaces of the terrestrial planets, the Moon, several moons of giant planets, asteroids, comets, and Kuiper belt objects (see review by *Encrenaz,* 2008). Evidence for past liquid water on Mars has been strengthened by recent data from the Curiosity rover

(*Williams et al.,* 2013). Water has also been detected in spectra of the Sun (*Wallace et al.,* 1995) and those of other cool stars. In interstellar space, gaseous water was detected more than 40 yr ago in the Orion nebula through its masing transition at 22 GHz (1 cm) (*Cheung et al.,* 1969) and water ice was discovered a few years later through its infrared (IR) bands toward protostars (*Gillett and Forrest,* 1973). Water vapor and ice have now been observed in many star and planet-forming regions throughout the galaxy (reviews by *Cernicharo and Crovisier,* 2005; *Boogert et al.,* 2008; *Melnick,* 2009; *Bergin and van Dishoeck,* 2012) and even in external galaxies out to high redshifts (e.g., *Shimonishi et al.,* 2010; *Lis et al.,* 2011; *Weiss et al.,* 2013). Water is indeed ubiquitous throughout the universe.

On their journey from clouds to cores, the water molecules encounter a wide range of conditions, with temperatures ranging from <10 K in cold prestellar cores to ~2000 K in shocks and the inner regions of protoplanetary disks. Densities vary from ~10^4 cm^{-3} in molecular clouds to 10^{13} cm^{-3} in the mid-planes of disks and 10^{19} cm^{-3} in planetary atmospheres. The chemistry naturally responds to these changing conditions. A major question addressed here is to what extent the water molecules produced in interstellar clouds are preserved all the way to exoplanetary atmospheres, or whether water is produced *in situ* in planet-forming regions. Understanding how, where, and when water forms is critical for answering the question of whether water-containing planets are common.

2. WATER PHYSICS AND CHEMISTRY

2.1. Water Phases

Water can exist as a gas (vapor or "steam"), as a solid (ice), or as a liquid. At the low pressures of interstellar space, only water vapor and ice occur, with the temperature at which the transition occurs depending on density. At typical cloud densities of 10^4 particles cm^{-3}, water sublimates around 100 K (*Fraser et al.,* 2001), but at densities of 10^{13} cm^{-3}, corresponding to the mid-planes of protoplanetary disks, the sublimation temperature increases to ~160 K. According to the phase diagram of water, liquid water can exist above the triple point at 273 K and 6.12 mbar (~10^{17} cm^{-3}). Such pressures and temperatures are typically achieved at the surfaces of bodies of the size of Mars or larger and at distances between 0.7 and 1.7 AU for a solar-type star.

Water ice can take many different crystalline and amorphous forms depending on temperature and pressure. At interstellar densities, crystallization of an initially amorphous ice to the cubic configuration, I_c, occurs around 90 K. This phase change is irreversible: Even when the ice is cooled down again, the crystal structure remains and it therefore provides a record of the temperature history of the ice. Below 90 K, interstellar ice is mostly in a compact high-density amorphous (HDA) phase, which does not naturally occur on planetary surfaces (*Jenniskens and*

Blake, 1994). The densities of water ice in the HDA, low-density amorphous (LDA), and I_c phases are 1.17, 0.94, and 0.92 g cm^{-3}, respectively, much lower than those of rocks (3.2–4.4 g cm^{-3} for magnesium-iron silicates).

Clathrate hydrates are crystalline water-based solids in which small nonpolar molecules can be trapped inside "cages" of the hydrogen-bonded water molecules. They can be formed when a gas of water mixed with other species condenses out at high pressure and has enough entropy to form a stable clathrate structure (*Lunine and Stevenson,* 1985; *Mousis et al.,* 2010). (Strictly speaking, the term "condensation" refers to the gas to liquid transition; we adopt here the astronomical parlance where it is also used to denote the gas-to-solid transition.) Clathrate hydrates are found in large quantities on Earth, with methane clathrates on the deep ocean floor and in permafrost as the best known examples. They have been postulated to occur in large quantities on other planets and icy solar system bodies.

2.2. Water Spectroscopy

Except for *in situ* mass spectroscopy in planetary and cometary atmospheres, all information about interstellar and solar system water comes from spectroscopic data obtained with telescopes. Because of the high abundance of water in Earth's atmosphere, the bulk of the data comes from space observatories. Like any molecule, water has electronic, vibrational, and rotational energy levels. Dipole-allowed transitions between electronic states occur at ultraviolet (UV) wavelengths, between vibrational states at near- to mid-IR wavelengths, and between rotational states from mid- to far-IR and submillimeter wavelengths.

Interstellar water vapor observations target mostly the pure rotational transitions. Water is an asymmetric rotor with a highly irregular set of energy levels, characterized by quantum numbers $J_{K_A K_C}$. Because water is a light molecule, the spacing of its rotational energy levels is much larger than that of heavy rotors, such as CO or CS, and the corresponding wavelengths much shorter (0.5 mm vs. 3–7 mm for the lowest transitions). The nuclear spins of the two hydrogen atoms can be either parallel or antiparallel, and this results in a grouping of the H_2O energy levels into ortho ($K_A + K_C$ = odd) and para ($K_A + K_C$ = even) ladders, with a statistical weight ratio of 3:1, respectively. Radiative transitions between these two ladders are forbidden to high order, and only chemical reactions in which an H atom of water is exchanged with an H-atom of a reactant can transform ortho- to para-H_2O and vice versa.

Infrared spectroscopy can reveal both water vapor and ice. Water has three active vibrational modes: the fundamental v=1–0 bands of the v_1 and v_3 symmetric and asymmetric stretches centered at 2.7 μm and 2.65 μm, respectively, and the v_2 bending mode at 6.2 μm. Overtone ($\Delta v = 2$ or larger) and combination (e.g., $v_2 + v_3$) transitions occur in hot gas at shorter wavelengths (see Fig. 1, for example). Gas-phase water therefore has a rich vibration rotation spectrum with many individual lines depending on

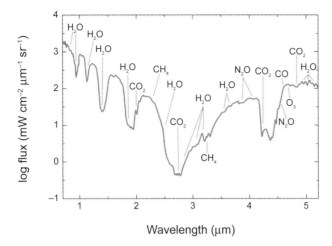

Fig. 1. The near-IR spectrum of Earth showing the many water vibrational bands together with CO_2. The bands below 3 μm are due to overtones and combination bands and are often targeted in exoplanet searches. This spectrum was observed with the NIMS instrument on the Galileo spacecraft during its Earth flyby in December 1990. From *Encrenaz* (2008), with permission from *Annual Reviews*, based on *Drossart et al.* (1993).

the temperature of the gas. In contrast, the vibrational bands of water ice have no rotational substructure and consist of very broad profiles, with the much stronger ν_3 band overwhelming the weak ν_1 band. The ice profile shapes depend on the morphology, temperature, and environment of the water molecules (*Hudgins et al.,* 1993). Crystalline water ice is readily distinguished by a sharp feature around 3.1 μm that is lacking in amorphous water ice. Libration modes of crystalline water ice are found at 45 and 63 μm (*Moore and Hudson,* 1994).

Spectra of hydrous silicates (also known as phyllosilicates, layer-lattice silicates, or "clays") show sharp features at 2.70–2.75 μm due to isolated OH groups and a broader absorption from 2.75 to 3.2 μm caused by interlayered ("bound") water molecules. At longer wavelengths, various peaks can occur depending on the composition; for example, the hydrous silicate montmorillonite has bands at 49 and 100 μm (*Koike et al.,* 1982).

Bound-bound electronic transitions of water occur at far-UV wavelengths around 1240 Å, but have not yet been detected in space.

2.3. Water Excitation

The strength of an emission or absorption line of water depends on the number of molecules in the telescope beam and, for gaseous water, on the populations of the individual energy levels. These populations, in turn, are determined by the balance between the collisional and radiative excitation and deexcitation of the levels. The radiative processes involve both spontaneous emission and stimulated absorption

and emission by a radiation field produced by a nearby star, by warm dust, or by the molecules themselves.

The main collisional partner in interstellar clouds is H_2. Accurate state-to-state collisional rate coefficients, $C_{u\ell}$, of H_2O with both ortho- and para-H_2 over a wide range of temperatures have recently become available thanks to a dedicated chemical physics study (*Daniel et al.,* 2011). Other collision partners such as H, He, and electrons are generally less important. In cometary atmospheres, water itself provides most of the collisional excitation.

Astronomers traditionally analyze molecular observations through a Boltzmann diagram, in which the level populations are plotted vs. the energy of the level involved. The slope of the diagram gives the inverse of the excitation temperature. If collisional processes dominate over radiative processes, the populations are in "local thermodynamic equilibrium" (LTE) and the excitation temperature is equal to the kinetic temperature of the gas, $T_{ex} = T_{kin}$. Generally level populations are far from LTE and molecules are excited by collisions and deexcited by spontaneous emission, leading to $T_{ex} < T_{kin}$. The critical density roughly delineates the transition between these regimes: $n_{cr} = A_{u\ell}/C_{u\ell}$ and therefore scales with $\mu_{u\ell}^2 \nu_{u\ell}^3$, where A is the Einstein spontaneous emission coefficient, μ the electric dipole moment, and ν the frequency of the transition $u \rightarrow \ell$. In the case of water, the combination of a large dipole moment (1.86 Debye) and high frequencies results in high critical densities of 10^8–10^9 cm^{-3} for pure rotational transitions.

Analysis of water lines is much more complex than for simple molecules, such as CO, for a variety of reasons. First, because of the large dipole moment and high frequencies, the rotational transitions of water are usually highly optically thick, even for abundances as low as 10^{-10}. Second, the water transitions couple effectively with mid- and far-IR radiation from warm dust, which can pump higher energy levels. Third, the fact that the "backbone" levels with $K_A = 0$ or 1 have lower radiative decay rates than higher K_A levels can lead to population "inversion," in which the population in the upper state divided by its statistical weight exceeds that for the lower state (i.e., T_{ex} becomes negative). Infrared pumping can also initiate this inversion. The result is the well-known maser phenomenon, which is widely observed in several water transitions in star-forming regions (e.g., *Furuya et al.,* 2003; *Neufeld et al.,* 2013; *Hollenbach et al.,* 2013). The bottom line is that accurate analysis of interstellar water spectra often requires additional independent constraints, for example, from $H_2^{18}O$ or $H_2^{17}O$ isotopologs, whose abundances are reduced by factors of about 550 and 2500, respectively, and whose lines are more optically thin. At IR wavelengths, lines are often spectrally unresolved, which further hinders the interpretation.

2.4. Water Chemistry

2.4.1. Elemental abundances and equilibrium chemistry. The overall abundance of elemental oxygen with respect to total hydrogen nuclei in the interstellar medium is estimated to be 5.75×10^{-4} (*Przybilla et al.,* 2008), of which 16–24% is locked up in refractory silicate material in the diffuse interstellar medium (*Whittet,* 2010). The abundance of volatile oxygen (i.e., not tied up in some refractory form) is measured to be 3.2×10^{-4} in diffuse clouds (*Meyer et al.,* 1998), so this is the maximum amount of oxygen that can cycle between water vapor and ice in dense clouds. Counting up all the forms of detected oxygen in diffuse clouds, the sum is less than the overall elemental oxygen abundance. Thus, a fraction of oxygen is postulated to be in some yet unknown refractory form, called "unknown depleted oxygen" (UDO), whose fraction may increase from 20% in diffuse clouds up to 50% in dense star-forming regions (*Whittet,* 2010). For comparison, the abundances of elemental carbon and nitrogen are 3×10^{-4} and 1×10^{-4}, respectively, with about two-thirds of the carbon thought to be locked up in solid carbonaceous material.

For a gas in thermodynamic equilibrium (TE), the fractional abundance of water is simply determined by the elemental composition of the gas and the stabilities of the molecules and solids that can be produced from it. For standard interstellar abundances with [O]/[C] > 1, there are two molecules in which oxygen can be locked up: CO and H_2O. (The notation [X] indicates the overall abundance of element X in all forms, be it atoms, molecules or solids.) At high pressures in TE, the fraction of CO results from the equilibrium between CO and CH_4, with CO favored at higher temperatures. For the volatile elemental abundances quoted above, this results in an H_2O fractional abundance of $(2–3) \times 10^{-4}$ with respect to total hydrogen, if the CO fractional abundance ranges from $(0–1) \times 10^{-4}$. With respect to H_2, the water abundance would then be $(5–6) \times 10^{-4}$ assuming that the fraction of hydrogen in atomic form is negligible [the density of hydrogen nuclei $n_H = n(H) + 2n(H_2)$]. Equilibrium chemistry is established at densities above roughly 10^{13} cm^{-3}, when three-body processes become significant. Such conditions are found in planetary atmospheres and in the shielded mid-planes of the inner few astronomical units of protoplanetary disks.

Under most conditions in interstellar space, however, the densities are too low for equilibrium chemistry to be established. Also, strong UV irradiation drives the chemistry out of equilibrium, even in high-density environments, such as the upper atmospheres of planets and disks. Under these conditions, the fractional abundances are determined by the kinetics of the two-body reactions between the various species in the gas. Figure 2 summarizes the three routes to water formation that have been identified. Each of these routes dominates in a specific environment.

2.4.2. Low temperature gas-phase chemistry. In diffuse and translucent interstellar clouds with densities less than ~10^4 cm^{-3} and temperatures below 100 K, water is formed largely by a series of ion-molecule reactions (e.g., *Herbst and Klemperer,* 1973). The network starts with the reactions $O + H_3^+$ and $O^+ + H_2$ leading to OH^+. The H_3^+ ion is produced by interactions of energetic cosmic-ray particles with the

gas, producing H_2^+ and H^+, with the subsequent fast reaction of $H_2^+ + H_2$ leading to H_3^+. The cosmic-ray-ionization rate of atomic hydrogen denoted by ζ_H can be as high as 10^{-15} s^{-1} in some diffuse clouds, but drops to 10^{-17} s^{-1} in denser regions (*Indriolo and Mc-Call,* 2012; *Rimmer et al.,* 2012). The ionization rate of H_2 is $\zeta_{H_2} \approx 2\zeta_H$.

A series of rapid reactions of OH^+ and H_2O^+ with H_2 lead to H_3O^+, which can dissociatively recombine to form H_2O and OH with branching ratios of ~0.17 and 0.83, respectively (*Buhr et al.,* 2010). Water is destroyed by photodissociation and by reactions with C^+, H_3^+, and other ions such as HCO^+. Photodissociation of H_2O starts to be effective shortward of 1800 Å and continues down to the ionization threshold at 983 Å (12.61 eV), including Ly α at 1216 Å. Its lifetime in the general interstellar radiation field, as given by *Draine* (1978), is only 40 yr.

2.4.3. High-temperature gas-phase chemistry. At temperatures above 230 K, the energy barriers for reactions with H_2 can be overcome and the reaction $O + H_2 \rightarrow OH + H$ becomes the dominant channel initiating water formation (*Elitzur and Watson,* 1978). Hydroxide subsequently reacts with H_2 to form H_2O, a reaction that is exothermic, but has an energy barrier of ~2100 K (*Atkinson et al.,* 2004). This route drives all the available gas-phase oxygen into H_2O, unless strong UV or a high atomic H abundance converts some water back to OH and O. High-temperature chemistry dominates the formation of water in shocks, in the inner

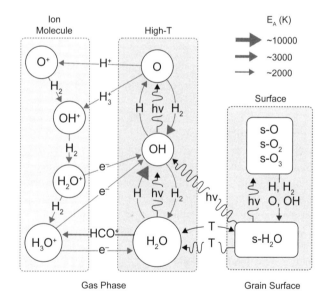

Fig. 2. Summary of the main gas-phase and solid-state chemical reactions leading to the formation and destruction of H_2O under non-equilibrium conditions. Three different chemical regimes can be distinguished: (1) ion-molecule chemistry, which dominates gas-phase chemistry at low T; (2) high-temperature neutral-neutral chemistry; and (3) solid-state chemistry. e stands for electron, ν for photon, and s-X indicates that species X is on the grains. Simplified version of figure by *van Dishoeck et al.* (2011).

envelopes around protostars, and in the warm surface layers of protoplanetary disks.

2.4.4. Ice chemistry. The timescale for an atom or molecule to collide with a grain and stick to it is $t_{fo} = 3 \times 10^9/n_{H_2}$ yr for normal size grains and sticking probabilities close to unity (*Hollenbach et al.,* 2009). Thus, for densities greater than 10^4 cm^{-3}, the timescales for freeze-out are less than a few $\times$ 10^5 yr, generally smaller than the lifetime of dense cores (at least 10^5 yr). Reactions involving dust grains are therefore an integral part of the chemistry. Even weakly bound species, such as atomic H, have a long enough residence time on the grains at temperatures of 10–20 K to react; H_2 also participates in some surface reactions, but remains largely in the gas. *Tielens and Hagen* (1982) postulated that the formation of water from O atoms proceeds through three routes involving hydrogenation of s-O, s-O_2, and s-O_3, respectively, where s-X indicates a species on the surface. All three routes have recently been verified and quantified in the laboratory and detailed networks with simulations have been drawn up (for summaries, see *Cuppen et al.,* 2010; *Oba et al.,* 2012; *Lamberts et al.,* 2013).

Water ice formation is in competition with various desorption processes, which limit the ice build-up. At dust temperatures below the thermal sublimation limit, photodesorption is an effective mechanism to get species back to the gas phase, although only a small fraction of the UV absorptions results in desorption of intact H_2O molecules (*Andersson and van Dishoeck,* 2008). The efficiency is about 10^{-3} per incident photon, as determined through laboratory experiments and theory (*Westley et al.,* 1995; *Öberg et al.,* 2009; *Arasa et al.,* 2010). Only the top few monolayers of the ice contribute. The UV needed to trigger photodesorption can come either from a nearby star, or from the general interstellar radiation field. Deep inside clouds, cosmic rays produce a low level of UV flux, $\sim 10^4$ photons cm^{-2} s^{-1}, through interaction with H_2 (*Prasad and Tarafdar,* 1983). Photodesorption via X-rays is judged to be inefficient, although there are large uncertainties in the transfer of heat within a porous aggregate (*Najita et al.,* 2001). Ultraviolet photodesorption of ice is thought to dominate the production of gaseous water in cold prestellar cores, the cold outer envelopes of protostars, and the outer parts of protoplanetary disks.

Other nonthermal ice-desorption processes include cosmic-ray-induced spot heating (which works for CO, but is generally not efficient for strongly bound molecules like H_2O) and desorption due to the energy liberated by the reaction (called "reactive" or "chemical" desorption). These processes are less well explored than photodesorption, but a recent laboratory study of s-D + s-OD → s-D_2O suggests that as much as 90% of the product can be released into the gas phase (*Dulieu et al.,* 2013). The details of this mechanism, which has not yet been included in models, are not yet understood and may strongly depend on the substrate.

Once the dust temperature rises above ~ 100 K (precise value being pressure dependent), H_2O ice thermally sublimates on timescales of years, leading to initial gas-phase abundances of H_2O as high as the original ice abundances. These simulations use a binding energy of 5600 K for amorphous ice and a slightly higher value of 5770 K for crystalline ice, derived from laboratory experiments (*Fraser et al.,* 2001). Thermal desorption of ices contributes to the gas-phase water abundance in the warm inner protostellar envelopes ("hot cores") and inside the snow line in disks.

2.4.5. Water deuteration. Deuterated water, HDO and D_2O, is formed through the same processes as illustrated in Fig. 2. There are, however, a number of chemical processes that can enhance the HDO/H_2O and D_2O/H_2O ratios by orders of magnitude compared with the overall [D]/[H] ratio of 2.0×10^{-5} in the local interstellar medium (*Prodanović et al.,* 2010). A detailed description is given in the chapter by Ceccarelli et al. in this volume; here, only a brief summary is provided.

In terms of pure gas-phase chemistry, the direct exchange reaction H_2O +HD ↔ HDO + H_2 is often considered in solar system models (*Richet et al.,* 1977). In thermochemical equilibrium this reaction can provide at most a factor of 3 enhancement, and even that may be limited by kinetics (*Lécluse and Robert,* 1994). The exchange reaction D + OH → H + OD, which has a barrier of ~ 100 K (*Sultanov and Balakrishnan,* 2004), is particularly effective in high-temperature gas such as is present in the inner disk (*Thi et al.,* 2010b). Photodissociation of HDO enhances OD compared with OH by a factor of 2–3, which could be a route to further fractionation.

The bulk of the deuterium fractionation in cold clouds comes from gas-grain processes. *Tielens* (1983) pointed out that the fraction of deuterium relative to hydrogen atoms arriving on a grain surface, D/H, is much higher than the overall [D]/[H] ratio, which can be implanted into molecules in the ice. This naturally leads to enhanced formation of OD, HDO, and D_2O ice according to the grain-surface formation routes. The high atomic D/H ratio in the gas arises from the enhanced gaseous H_2D^+, HD_2^+, and D_3^+ abundances at low temperatures (≤ 25 K), when the ortho-H_2 abundance drops and their main destroyer, CO, freezes out on the grains (*Pagani et al.,* 1992; *Roberts et al.,* 2003). Dissociative recombination with electrons then produces enhanced D. The enhanced H_2D^+ also leads to enhanced H_2DO^+ and thus HDO in cold gas, but this is usually a minor route compared with gas-grain processes.

On the grains, tunneling reactions can have the opposite effect, reducing the deuterium fractionation. For example, the OD + H_2 tunneling reaction producing HDO ice is expected to occur slower than the OH + H_2 reaction, leading to H_2O ice. On the other hand, thermal exchange reactions in the ice, such as H_2O + OD → HDO + OH, have been shown to occur rapidly in ices at higher temperatures; these can both enhance and decrease the fractionation. Both thermal desorption at high ice temperatures and photodesorption at low ice temperatures have a negligible effect on the deuterium fractionation, i.e., the gaseous HDO/H_2O and D_2O/H_2O ratios reflect the ice ratios if no other gas-phase processes are involved.

3. CLOUDS AND PRESTELLAR CORES: ONSET OF WATER FORMATION

In this and following sections, our knowledge of the water reservoirs during the various evolutionary stages from clouds to planets will be discussed. The focus is on low-mass protostars (<100 $L_\odot$) and pre-main-sequence stars (spectral type A or later). Unless stated otherwise, fractional abundances are quoted with respect to H_2 and are simply called "abundances." Often the denominator, i.e., the (column) density of H_2, is more uncertain than the numerator.

The bulk of the water in space is formed on the surfaces of dust grains in dense molecular clouds. Although a small amount of water is produced in the gas in diffuse molecular clouds through ion-molecule chemistry, its abundance of ~10^{-8} found by Herschel's Heterodyne Instrument for the Far Infrared (HIFI) (*Flagey et al., 2013*) is negligible compared with that produced in the solid state. In contrast, observations of the 3-μm water-ice band toward numerous IR sources behind molecular clouds, from the ground and from space, show that water-ice formation starts at a threshold extinction of $A_V \approx 3$ mag (*Whittet et al., 2013*). These clouds have densities of at least 1000 cm^{-3}, but are not yet collapsing to form stars. The ice abundance is s-$H_2O/H_2 \approx 5 \times 10^{-5}$, indicating that a significant fraction of the available oxygen has been transformed to water ice even at this early stage (*Whittet et al., 1988; Murakawa et al., 2000; Boogert et al., 2011*). Such high ice abundances are too large to result from freeze-out of gas-phase water produced by ion-molecule reactions.

The densest cold cores just prior to collapse have such high extinctions that direct IR ice observations are not possible. In contrast, the water reservoir (gas plus ice) can be inferred from Herschel-HIFI observations of such cores. Figure 3 presents the detection of the H_2O $1_{10}-1_{01}$ 557-GHz line toward L1544 (*Caselli et al., 2012*). The line shows blue-shifted emission and red-shifted absorption, indicative of inward motions in the core. Because of the high critical density of water, the emission indicates that water vapor must be present in the dense central part. The infalling red-shifted gas originates on the nearside. Because the different parts of the line profile probe different parts of the core, the line shape can be used to reconstruct the water vapor abundance as a function of position throughout the entire core.

The best-fit water abundance profile is obtained with a simple gas-grain model, in which atomic O is converted into water ice on the grains, with only a small fraction returned back into the gas by photodesorption (*Bergin et al., 2000; Roberts and Herbst, 2002; Hollenbach et al., 2009*). The maximum gas-phase water abundance of ~10^{-7} occurs in a ring at the edge of the core around $A_V \approx 4$ mag, where external UV photons can still penetrate to photodesorb the ice, but where they are no longer effective in photodissociating the water vapor. In the central shielded part of the core, cosmic-ray-induced UV photons keep a small, ~10^{-9}, but measurable fraction of water in the gas (*Caselli et al.,*

2012). Quantitatively, the models indicate that the bulk of the available oxygen has been transformed into water ice in the core, with an ice abundance of ~10^{-4} with respect to H_2.

4. PROTOSTARS AND OUTFLOWS

4.1. Outflows

Herschel-HIFI and Photodetector Array Camera and Spectrometer (PACS) data show strong and broad water profiles characteristic of shocks associated with embedded

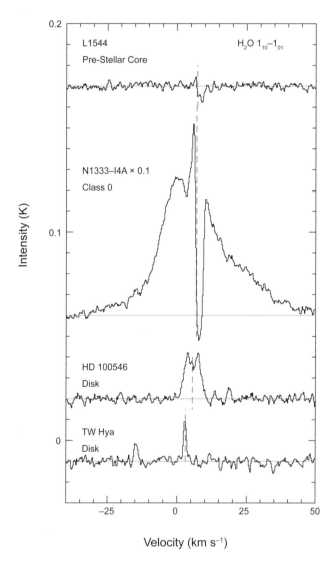

Fig. 3. Herschel-HIFI spectra of the H_2O $1_{10}-1_{01}$ line at 557 GHz in a prestellar core (top), protostellar envelope (middle), and two protoplanetary disks (bottom) (spectra shifted vertically for clarity). The dashed line indicates the rest velocity of the source. Note the different scales: Water vapor emission is strong toward protostars, but very weak in cold cores and disks. The feature at −15 km s^{-1} in the TW Hya spectrum is due to NH_3. Figure by L. Kristensen, adapted from *Caselli et al.* (2012), *Kristensen et al.* (2012), and *Hogerheijde et al.* (2011 and in preparation).

protostars, from low to high mass. In fact, for low-mass protostars this shocked water emission completely overwhelms the narrower lines from the bulk of the collapsing envelope, even though the shocks contain less than 1% of the mass of the system. Maps of the water emission around solar-mass protostars such as L1157 reveal water not only at the protostellar position but also along the outflow at "hot spots" where the precessing jet interacts with the cloud (*Nisini et al., 2010*). Thus, water traces the *currently* shocked gas at positions that are somewhat offset from the bulk of the cooler entrained outflow gas seen in the red- and blue-shifted lobes of low-J CO lines (*Tafalla et al., 2013; Lefloch et al., 2010*).

Determinations of the water abundance in shocks vary from values as low as 10^{-7} to as high as 10^{-4} (for a summary, see *van Dishoeck et al., 2013*). In nondissociative shocks, the temperature reaches values of a few thousand Kelvin and all available oxygen is expected to be driven into water (*Kaufman and Neufeld, 1996*). The low values likely point to the importance of UV radiation in the shock chemistry and shock structure. For the purposes of this chapter, the main point is that even though water is rapidly produced in shocks at potentially high abundances, the amount of water contained in the shocks is small, and, moreover, most of it is lost to space through outflows.

4.2. Protostellar Envelopes: The Cold Outer Reservoir

As the cloud collapses to form a protostar in the center, the water-ice-coated grains created in the natal molecular cloud move inward, feeding the growing star and its surrounding disk (Fig. 4). The water-ice abundance can be measured directly through IR spectroscopy of various water-ice bands toward the protostar itself. Close to a hundred sources have been observed, from very-low-luminosity objects ("proto-brown dwarfs") to the highest-mass protostars (*Gibb et al., 2004; Pontoppidan et al., 2004; Boogert et al., 2008; Zasowski et al., 2009; Öberg et al., 2011*). Inferred ice abundances with respect to H_2 integrated along the line of sight are $(0.5-1) \times 10^{-4}$.

The water vapor abundance in protostellar envelopes is probed through spectrally resolved Herschel-HIFI lines. Because the gaseous water line profiles are dominated by broad outflow emission (Fig. 3), this component needs to be subtracted, or an optically thin water isotopolog needs to be used to determine the quiescent water. Clues to the water vapor abundance structure can be obtained through narrow absorption and emission features in so-called (inverse) P-Cygni profiles (see NGC 1333 IRAS 4A in Fig. 3). The analysis of these data proceeds along the same lines as for prestellar cores. The main difference is that the dust temperature now increases inward, from a low value of 10–20 K at the edge to a high value of several hundred Kelvin in the center of the core (Fig. 4). In the simplest spherically symmetric case, the density follows a power-law $n \propto r^{-p}$ with $p = 1–2$. As for prestellar cores, the data require the

presence of a photodesorption layer at the edge of the core with a decreasing water abundance at smaller radii, where gaseous water is maintained by the cosmic-ray-induced photodesorption of water ice (*Coutens et al., 2012; Mottram et al., 2013*). Analysis of the combined gaseous water and water-ice data for the same source shows that the ice/gas ratio is at least 10^4 (*Boonman and van Dishoeck, 2003*). Thus, the bulk of the water stays in the ice in this cold part, at a high abundance of $\sim 10^{-4}$ as indicated by direct measurements of both the water ice and gas.

4.3. Protostellar Envelopes: The Warm Inner Part

When the infalling parcel enters the radius at which the dust temperature reaches ~ 100 K, the gaseous water abundance jumps from a low value around 10^{-10} to values as high as 10^{-4} (e.g., *Boonman et al., 2003; Herpin et al., 2012; Coutens et al., 2012*). The 100-K radius scales roughly as $2.3 \times 10^{14}\sqrt{(L/L_\odot)}$ cm (*Bisschop et al., 2007*), and is small, <100 AU, for low-mass sources and a few thousand astronomical units for high-mass protostars. The precise abundance of water in the warm gas is still uncertain, however, and can range from 10^{-6} to 10^{-4} depending on the source and analysis (*Emprechtinger et al., 2013; Visser et al., 2013*). A high water abundance would indicate that all water sublimates from the grains in the "hot core" before the material enters the disk; a low abundance indicates the opposite.

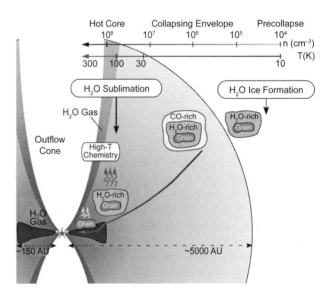

Fig. 4. Schematic representation of a protostellar envelope and embedded disk with key steps in the water chemistry indicated. Water ice is formed in the parent cloud before collapse and stays mostly as ice until the ice sublimation temperature of ~ 100 K close to the protostar is reached. Hot water is formed in high abundances in shocks associated with the outflow, but this water is not incorporated into the planet-forming disk. Figure by R. Visser, adapted from *Herbst and van Dishoeck* (2009).

The fate of water in protostellar envelopes on scales of the size of the embedded disk is currently not well understood, yet it is a crucial step in the water trail from clouds to disks. To probe the inner few hundred astronomical units, a high excitation line of a water isotopolog line not dominated by the outflow or high angular resolution is needed: Groundbased millimeter interferometry of the $H_2^{18}O$ $3_{13}-2_{20}$ ($E_u = 204$ K) line at 203 GHz (*Persson et al.*, 2012) and deep Herschel-HIFI spectra of excited $H_2^{18}O$ or $H_2^{17}O$ lines, such as the $3_{12}-3_{03}$ ($E_u = 249$ K) line at 1095 GHz, have been used. Two main problems need to be faced in the analysis. First, comparison of groundbased and Herschel lines for the same source show that the high-frequency HIFI lines can be optically thick even for $H_2^{18}O$ and $H_2^{17}O$ because of their much higher Einstein A coefficients. Second, the physical structure of the envelope and embedded disk on scales of a few hundred astronomical units is not well understood (*Jørgensen et al.*, 2005), so that abundances are difficult to determine since the column of warm H_2 is poorly constrained. Compact flattened dust structures are not necessarily disks in Keplerian rotation (*Chiang et al.*, 2008) and only a fraction of this material may be at high temperatures.

Jørgensen and van Dishoeck (2010b) and *Persson et al.* (2012) measure water columns and use H_2 columns derived from continuum interferometry data on the same scales (~1″) to determine water abundances of ~$10^{-8}-10^{-5}$ for three low-mass protostars, consistent with the fact that the bulk of the gas on these scales is cold and water is frozen. From a combined analysis of the interferometric and HIFI data, using $C^{18}O$ 9–8 and 10–9 data to determine the warm H_2 column, *Visser et al.* (2013) infer water abundances of $2 \times 10^{-5}-2 \times 10^{-4}$ in the ≥100 K gas, as expected for the larger-scale hot cores.

The important implication of these results is that the bulk of the water stays as ice in the inner few hundred astronomical units and only a few percent of the dust may be at high enough temperatures to thermally sublimate H_2O. This small fraction of gas passing through high-temperature conditions for ice sublimation is consistent with two-dimensional models of collapsing envelope and disk formation, which give fractions of <1−20% depending on initial conditions (*Visser et al.*, 2009, 2011; *Ilee et al.*, 2011; *Harsono et al.*, 2013; *Hincelin et al.*, 2013).

4.4. Entering the Disk: The Accretion Shock and History of Water in Disks

The fact that only a small fraction of the material within a few hundred astronomical units radius is at ≥100 K (section 4.3) implies that most of the water is present as ice and is still moving inward (Fig. 4). At some radius, however, the high-velocity infalling parcels must encounter the low-velocity embedded disk, resulting in a shock at the boundary. This shock results in higher dust temperatures behind the shock front than those achieved by stellar heating (*Neufeld and Hollenbach*, 1994; see *Visser et al.*,

2009, for a simple fitting formula) and can also sputter ices. At early times, accretion velocities are high and all ices would sublimate or experience a shock strong enough to induce sputtering. However, this material normally ends up in the star rather than in the disk, so it is not of interest for the current story. The bulk of the disk is thought to be made up through layered accretion of parcels that fall in later in the collapse process, and which enter the disk at large radii, where the shock is much weaker (*Visser et al.*, 2009). Indeed, the narrow line widths of $H_2^{18}O$ of only 1 km s^{-1} seen in the interferometric data (*Jørgensen and van Dishoeck*, 2010b) argue against earlier suggestions, based on Spitzer data, of large amounts of hot water going through an accretion shock in the embedded phase, or even being created through high-temperature chemistry in such a shock (*Watson et al.*, 2007). This view that accretion shocks do not play a role also contrasts with the traditional view in the solar system community that all ices evaporate and recondense when entering the disk (*Lunine et al.*, 1991; *Owen and Bar-Nun*, 1993).

Figure 5 shows the history of water molecules in disks at the end of the collapse phase at $t_{acc} = 2.5 \times 10^5$ yr for a standard model with an initial core mass of 1 $M_\odot$, angular momentum $\Omega_0 = 10^{-14}$ s^{-1}, and sound speed $c_s = 0.26$ km s^{-1} (*Visser et al.*, 2011). The material ending up in zone 1 is the only water that is completely "pristine," i.e., formed as ice in the cloud and never sublimated, ending up intact in the disk. Material ending up in the other zones contains water that sublimated at some point along the infalling trajectory. In zones 2, 3, and 4, close to the outflow cavity, most of the oxygen is in atomic form due to photodissociation, with varying degrees of subsequent reformation. In zones 5 and 6, most oxygen is in gaseous water. Material in zone 7 enters the disk early and comes close enough to the star to sublimate. This material does not end up in the star, however, but is transported outward in the disk to conserve angular momentum, refreezing when the temperature becomes low enough. The detailed chemistry and fractions of water in each of these zones depend on the adopted physical model and on whether vertical mixing is included (*Semenov and Wiebe*, 2011), but the overall picture is robust.

5. PROTOPLANETARY DISKS

Once accretion stops and the envelope has dissipated, a pre-main-sequence star is left, surrounded by a disk of gas and dust. These protoplanetary disks form the crucial link between material in clouds and that in planetary systems. Thanks to the new observational facilities, combined with sophisticated disk chemistry models, the various water reservoirs in disks are now starting to be mapped out. Throughout this chapter, we will call the disk out of which our own solar system formed the "solar-nebula disk." (Alternative nomenclatures in the literature include "primordial disk," "presolar disk," "protosolar nebula," or "primitive solar nebula.")

5.1. Hot and Cold Water in Disks: Observations

With increasing wavelengths, regions further out and deeper into the disk can be probed. The surface layers of the inner few astronomical units of disks are probed by near- and mid-IR observations. Spitzer's Infrared Spectrograph (IRS) detected a surprising wealth of highly excited pure rotational lines of warm water at 10–30 μm (*Carr and Najita*, 2008; *Salyk et al.*, 2008), and these lines have since been shown to be ubiquitous in disks around low-mass T Tauri stars (*Pontoppidan et al.*, 2010a; *Salyk et al.*, 2011), with line profiles consistent with a disk origin (*Pontoppidan et al.*, 2010b). Typical water excitation temperatures are $T_{ex} \approx 450$ K. Spectrally resolved groundbased near-IR vibration-rotation lines around 3 μm show that in some sources the water originates in both a disk and a slow disk wind (*Salyk et al.*, 2008; *Mandell et al.*, 2012). Abundance ratios are difficult to extract from the observations, because the lines are highly saturated and, in the case of Spitzer data, spectrally unresolved. Also, the IR lines only probe down to moderate height in the disk until the dust becomes optically thick. Nevertheless, within the more than order of magnitude uncertainty, abundance ratios of $H_2O/CO \sim 1$–10 have been inferred for emitting radii up to a few astronomical units (*Salyk et al.*, 2011; *Mandell et al.*, 2012). This indicates that the inner disks have high water abundances on the order of $\sim 10^{-4}$ and are thus not dry, at least not in their surface layers. The IR data show a clear dichotomy in the H_2O detection rate between disks around the lower-mass T Tauri stars and higher-mass, hot-ter A-type stars (*Pontoppidan et al.*, 2010a; *Fedele et al.*, 2011). Also, transition disks with inner dust holes show a lack of water line emission. This is likely due to more rapid photodissociation by stars with higher T_*, and thus stronger UV radiation, in regions where the molecules are not shielded by dust.

Moving to longer wavelengths, Herschel-PACS spectra probe gas at intermediate radii of the disk, out to 100 AU. Far-IR lines from warm water have been detected in a few disks (*Rivière-Marichalar et al.*, 2012; *Meeus et al.*, 2012; *Fedele et al.*, 2012, 2013). As for the inner disk, the abundance ratios derived from these data are highly uncertain. Sources in which both H_2O and CO far-IR lines have been detected (only a few) indicate H_2O/CO column density ratios of 10^{-1}, suggesting a water abundance on the order of 10^{-5} at intermediate layers, but upper limits in other disks suggest values that may be significantly less. Again the disks around T Tauri stars appear to be richer in water than those around A-type stars (Fig. 6).

In principle, the pattern of water lines with wavelength should allow the transition from the gaseous water-rich to the water-poor (the snow line) to be probed. As shown by LTE excitation-disk models, the largest sensitivity to the location of the snow line is provided by lines in the 40–60-μm region, which is exactly the wavelength range without observational facilities except for the Stratospheric Observatory for Infrared Astronomy (SOFIA) (*Meijerink et al.*, 2009). For one disk, that around TW Hya, the available shorter- and longer-wavelength water data have been used to put together a water-abundance profile across the entire disk (*Zhang et al.*, 2013). This disk has a dust hole within 4 AU, within which water is found to be depleted. The water abundance rises sharply to a high abundance at the inner edge of the outer disk at 4 AU, but then drops again to very low values as water freezes out in the cold outer disk.

The cold gaseous water reservoir beyond 100 AU is uniquely probed by Herschel-HIFI data of the ground

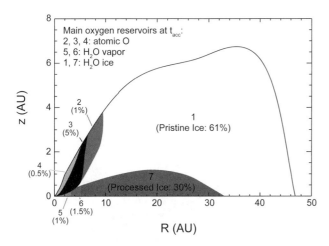

Fig. 5. Schematic view of the history of H_2O gas and ice throughout a young disk at the end of the accretion phase. The main oxygen reservoir is indicated for each zone. The percentages indicate the fraction of disk mass contained in each zone. Zone 1 contains pristine H_2O formed prior to star formation and never altered during the trajectory from cloud to disk. In Zone 7, the ice has sublimated once and recondensed again. Thus, the ice in planet- and planetesimal-forming zones of disks is a mix of pristine and processed ice. From *Visser et al.* (2011).

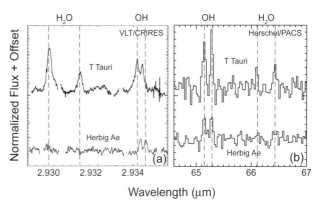

Fig. 6. **(a)** Near-IR and **(b)** far-IR spectra of a T Tau and a Herbig Ae disk, showing OH lines in both but H_2O primarily in disks around cooler T Tau stars. Figure by D. Fedele, based on *Fedele et al.* (2011, 2013).

rotational transitions at 557 and 1113 GHz. Weak, but clear detections of both lines have been obtained in two disks, around the nearby T Tau star TW Hya (*Hogerheijde et al.,* 2011) and the Herbig Ae star HD 100546 (Hogerheijde et al., in preparation) (Fig. 3). These are the deepest integrations obtained with the HIFI instrument, with integration times up to 25 hr per line. Similarly deep integrations on five other disks do not show detections of water at the same level, nor do shallower observations of a dozen other disks of different characteristics. One possible exception, DG Tau (*Podio et al.,* 2013), is a late class I source with a well-known jet and a high X-ray flux. The TW Hya detection implies abundances of gaseous water around 10^{-7} in the intermediate layer of the disk, with the bulk of the oxygen in ice on grains at lower layers. Quantitatively, 0.005 Earth oceans of gaseous water and a few thousand oceans of water ice have been detected (1 Earth ocean = 1.4×10^{24} g = 0.00023 $M_{\oplus}$). While this is plenty of water to seed an Earth-like planet with water, a single jovian-type planet formed in this ice-rich region could lock up the bulk of this water.

Direct detections of water ice are complicated by the fact that IR absorption spectroscopy requires a background light source, and thus a favorable near-edge-on orientation of the disk. In addition, care has to be taken that foreground clouds do not contribute to the water-ice absorption (*Pontoppidan et al.,* 2005). The 3-μm water-ice band has been detected in only a few disks (*Terada et al.,* 2007; *Honda et al.,* 2009). To measure the bulk of the ice, one needs to go to longer wavelengths, where the ice features can be seen in emission. Indeed, the crystalline H_2O features at 45 or 60 μm have been detected in several sources with the Infrared Space Observatory (ISO) Long Wavelength Spectrometer (LWS) (*Malfait et al.,* 1998, 1999; *Chiang et al.,* 2001) and Herschel-PACS (*McClure et al.,* 2012, Bouwman et al., in preparation). Quantitatively, the data are consistent with 25–50% of the oxygen in water ice on grains in the emitting layer.

The ISO-LWS far-IR spectra also suggested a strong signature of hydrated silicates in at least one target (*Malfait et al.,* 1999). Newer Herschel-PACS data show no sign of such a feature in the same target (Bouwman, personal communication). An earlier claim of hydrated silicates at 2.7 μm in diffuse clouds has now also been refuted (*Whittet et al.,* 1998; *Whittet,* 2010). Moreover, there is no convincing detection of any mid-IR feature of hydrated silicates in hundreds of Spitzer spectra of T Tauri (e.g., *Olofsson et al.,* 2009; *Watson et al.,* 2009), Herbig Ae (e.g., *Juhász et al.,* 2010), and warm debris (e.g., *Olofsson et al.,* 2012) disks. Overall, the strong observational consensus is that the silicates prior to planet formation are "dry."

5.2. Chemical Models of Disks

The observations of gaseous water discussed in section 5.1 indicate the presence of both rotationally hot $T_{ex} \approx$ 450 K and cold ($T_{ex} < 50$ K) water vapor, with abundances of ~10^{-4} and much lower values, respectively. Based on

the chemistry of water vapor discussed in section 2.4, we expect it to have a relatively well understood distribution within the framework of the disk thermal structure, potentially modified by motions of the various solid or gaseous reservoirs. This is broadly consistent with the observations.

Traditionally, the snow line plays a critical role in the distribution of water, representing the condensation or sublimation front of water in the disk, where the gas temperatures and pressures allow water to transition between the solid and gaseous states (Fig. 7). For the solar-nebula disk, there is a rich literature on the topic (*Hayashi,* 1981; *Sasselov and Lecar,* 2000; *Podolak and Zucker,* 2004; *Lecar et al.,* 2006; *Davis,* 2007; *Dodson-Robinson et al.,* 2009). Within our modern astrophysical understanding, this dividing line in the mid-plane is altered when viewed within the framework of the entire disk physical structure. There are a number of recent models of the water distribution that elucidate these key issues (*Glassgold et al.,* 2009; *Woitke et al.,* 2009b; *Bethell and Bergin,* 2009; *Willacy and Woods,* 2009; *Gorti et al.,* 2011; *Najita et al.,* 2011; *Vasyunin et al.,* 2011; *Fogel et al.,* 2011; *Walsh et al.,* 2012; *Kamp et al.,* 2013).

5.2.1. General distribution of gaseous water. Figure 8 shows the distribution of water vapor in a typical kinetic chemical disk model with radius R and height z. The disk gas temperature distribution is crucial for the chemistry. It is commonly recognized that dust on the disk surface is warmer than in the mid-plane due to direct stellar photon heating (*Calvet et al.,* 1992; *Chiang and Goldreich,* 1997). Furthermore, the gas temperature is decoupled from the dust in the upper layers due to direct gas heating (e.g., *Kamp and Dullemond,* 2004). There are roughly three areas where water vapor is predicted to be abundant and therefore possibly emissive (see also discussion in *Woitke et al.,* 2009a).

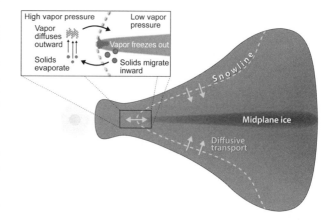

Fig. 7. Cartoon illustrating the snow line as a function of radius and height in a disk and transport of icy planetesimals across the snowline. Diffusion of water vapor from inner to outer disk followed by freeze-out results in pile-up of ice just beyond the snowline (the cold finger effect). Figure by M. Persson, based on *Meijerink et al.* (2009) and *Ciesla and Cuzzi* (2006).

These three areas or "regions" are labeled with coordinates (radial and vertical) that are specific to the physical structure (radiation field, temperature, density, dust properties) of this model. Different models (with similar dust- and gas-rich conditions) find the same general structure, but not at the exact same physical location.

5.2.1.1. Region 1 (R = inner radius to 1.5 AU; z/R < 0.1): This region coincides with the condensation/sublimation front in the mid-plane at the snow line. Inside the snow line water vapor will be abundant. Reaction timescales imposed by chemical kinetics limit the overall abundance depending on the gas temperature. As seen in Fig. 8, if the gas temperature exceeds ~400 K then the mid-plane water will be quite abundant, carrying all available oxygen not locked in CO and refractory grains. If the gas temperature is below this value, but above the sublimation temperature of ~160 K, then chemical kinetics could redistribute the oxygen toward other species. During the early gas- and dust-rich stages up to a few million years, this water-vapor-dominated region will persist and is seen in nearly all models. However, as solids grow, the penetrating power of UV radiation is increased. Since water vapor is sensitive to photodissociation by far-UV, this could lead to gradual decay of this layer, which would be consistent with the nondetection of water vapor inside the gaps of a small sample of transition disks (e.g., *Pontoppidan et al.,* 2010a; *Zhang et al.,* 2013).

5.2.1.2. Region 2 (R > 20 AU; surface layers and outer disk mid-plane): In these disk layers the *dust* temperature is uniformly below the sublimation temperature of water. Furthermore, at these high densities (n > 10^6 cm^{-3}) atoms and molecules freeze out on dust grains on short timescales (section 2.4). Under these circumstances, in the absence of nonthermal desorption mechanisms, models predict strong freeze-out with the majority of available oxygen present on grains as water ice. Much of this may be primordial water ice supplied by the natal cloud (*Visser et al.,* 2011) (Fig. 5).

The detection of rotationally cold water vapor emission in the outer disk of TW Hya demonstrates that a tenuous layer of water vapor is present and that some nonthermal desorption process is active (*Hogerheijde et al.,* 2011). The leading candidate is photodesorption of water ice (*Dominik et al.,* 2005; *Öberg et al.,* 2009), as discussed in section 2.4.4, particularly given the high-UV luminosities of T Tauri stars (*Yang et al.,* 2012a). This UV excess is generated by accretion and dominated by Lyα line emission (*Schindhelm et al.,* 2012).

Once desorbed as OH and H_2O, the UV radiation then also destroys the water vapor molecules, leading to a balance between these processes and a peak abundance near $(1-3) \times 10^{-7}$ (*Dominik et al.,* 2005; *Hollenbach et al.,* 2009). In general, most models exhibit this layer, which is strongly dependent on the location and surface area of ice-coated grains (i.e., less surface area reduces the effectiveness of photodesorption). Direct comparison of models with observations finds that the amount of water vapor predicted to be present exceeds the observed emission (*Bergin et al.,* 2010; *Hogerheijde et al.,* 2011). This led to the suggestion that the process of grain growth and sedimentation could operate to remove water ice from the UV-exposed disk surface layers. This is consistent with spectroscopic data of the TW Hya scattered light disk, which do not show water-ice features in the spectrum originating from this layer (*Debes et al.,* 2013). However, further fine tuning of this settling mechanism is needed (Dominik and Dullemond, in preparation, *Akimkin et al.,* 2013). An alternative explanation may be a smaller dust disk compared with the gas disk (*Qi et al.,* 2013).

5.2.1.3. Region 3 (R < 20 AU; z/R > 0.1): Closer to the exposed disk surface, the gas and dust become thermally decoupled. The density where this occurs depends on the relative amount of dust grains in the upper atmosphere, which may be altered by dust coagulation and settling (*Jonkheid et al.,* 2004; *Nomura et al.,* 2007) and on the thermal accommodation of the dust-gas interaction (*Burke and Hollenbach,* 1983). In these decoupled layers $T_{gas} \gg T_{dust}$, and when the gas temperature exceeds a few hundred Kelvin the neutral-neutral gas-phase pathways for water formation become efficient, leading to water abundances on the order of 10^{-5} (Fig. 8).

More directly, the disk surface is predicted to be water vapor rich at gas temperatures greater than or equal to a

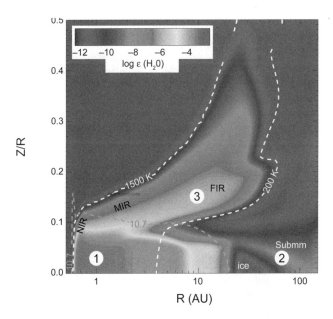

Fig. 8. Abundance of gaseous water relative to total hydrogen as a function of radial distance, R, and relative height above the midplane, z/R, for a disk around an A-type star ($T_* = 8600$ K). Three regions with high H_2O abundance can be distinguished. Regions 1 and 3 involve high-temperature chemistry, whereas region 2 lies beyond the snow line and involves photodesorption of water ice. The white contours indicate gas temperatures of 200 and 1500 K, whereas the gray contour shows the $n_H = 5 \times 10^{10}$ cm^{-3} density contour. From *Woitke et al.* (2009b).

few hundred Kelvin and dust temperatures ~100 K. Indeed, there should exist surface layers at radii where the mid-plane temperature is sufficiently low to freeze water vapor, but where the surface can support water formation via the high-temperature chemistry (i.e., region 3 goes out to larger radii than region 1). Thus the water zone on the disk surface presents the largest surface area and it is this water that is readily detected with current astronomical observations of high-lying transitions of $H_2^{16}O$ with Spitzer and Herschel. The snow line in the mid-plane is thus potentially hidden by the forest of water transitions produced by the hot chemistry on the surface.

There are some key dependences and differences that can be highlighted. One important factor is the shape of the UV radiation field. In general, models that use a scaled interstellar UV radiation field, for example, based on Far Ultraviolet Spectroscopic Explorer (FUSE)/International Ultraviolet Explorer (IUE)/Hubble Space Telescope (HST) observations of the UV excess (*Yang et al.,* 2012a), neglect the fact that some molecules like H_2 and CO require very energetic photons to photodissociate, which are not provided by very cool stars. A better approach is to take the actual stellar continua into account (*van Dishoeck et al.,* 2006), with UV excess due to accretion added where appropriate (*van Zadelhoff et al.,* 2003). A very important factor in this regard is the relative strength of the Lyα line to the overall UV continuum. Observations find that Lyα has nearly an order of magnitude more UV flux than the stellar FUV continuum in accreting sources (*Bergin et al.,* 2003; *Schindhelm et al.,* 2012). In addition, because of the difference in scattering (Lyα isotropic from H atom surface; UV continuum anisotropic from dust grains), Lyα will dominate the radiation field deeper into the disk (*Bethell and Bergin,* 2011).

Most models find the presence of this warm water layer in dust-dominated disks. However, *Bethell and Bergin* (2009) suggest that water can form in such high abundances in the surface layer that it mediates the transport of the energetic UV radiation by becoming self-shielding (Fig. 9). If this is the case, then the surface water would survive for longer timescales, because it is somewhat decoupled from the dust evolution. As a consequence, water and chemistry in the mid-plane might be protected even as the FUV-absorbing dust grains settle to the mid-plane.

Additional factors of importance for the survival of this emissive surface layer are the gas temperature and molecular hydrogen abundance. Although theoretical solutions for the gas temperature are inherently uncertain, it is clear that hot (T_{gas} > few hundred Kelvin) layers exist on disk surfaces (*Bruderer et al.,* 2012). However, as the gas disk dissipates, the accretion rate onto the star decays on timescales of a few million years. Thus the UV luminosity that is associated with this accretion declines and the disk will cool down, cutting off the production of water from the hot ($T \geq 400$ K) gas-phase chemistry on the exposed surface. In addition, as shown by *Glassgold et al.* (2009) and *Ádámkovics et al.* (2013), the formation of surface water

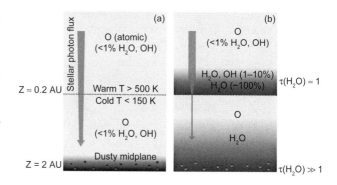

Fig. 9. Cartoon illustrating the water self-shielding mechanism and the resulting vertical stratification of O and H_2O, **(a)** neglecting water self-shielding and **(b)** including water self-shielding. Inclusion of water self-shielding in the upper layers leads to a "wet" warm layer. From *Bethell and Bergin* (2009).

requires the presence of H_2 to power the initiating reaction. Finally, vertical mixing through turbulence or disk winds can bring water ice from the lower to the upper layers where the ice sublimates and adds to the oxygen budget and water emission (*Heinzeller et al.,* 2011).

5.2.2. Planetesimal formation and water-ice transport. The dust particles in disks collide and grow, with water-ice mantles generally thought to help the coagulation processes. The evolution of dust to pebbles, rocks, and planetesimals (1–100-km bodies, the precursors of comets and asteroids) is described in the chapters by Testi et al. and Johansen et al. in this volume. The disk models cited above do not take into account transport of ice-rich planetesimals from the cold outer to the warm inner disk, even though such radial drift is known to be highly effective at a few astronomical units for rocks up to meter size (or millimeter size further out in the disk) (e.g., *Weidenschilling and Cuzzi,* 1993). This drift of icy planetesimals can be a source of water vapor enrichment inside the snow line (*Ciesla and Cuzzi,* 2006). The astrophysical signature of this phenomenon would be the presence of water vapor in the inner disk with an abundance greater than the stellar oxygen abundance, because the planetesimals are hydrogen-poor, i.e., the main volatile species, H_2, is not present in water-rich planetesimal ices. Some of this hot water can diffuse outward again and recondense just outside the snowline (the cold finger effect; Fig. 7), which increases the density of solids by a factor of 2–4 and thereby assists planet formation (*Stevenson and Lunine,* 1988). Alternatively, icy grains can be trapped in pressure bumps where they can grow rapidly to planetesimals before moving inward (e.g., *van der Marel et al.,* 2013).

6. WATER IN THE OUTER SOLAR SYSTEM

In the standard model of the disk out of which our solar system formed, the snow line was at 2.7 AU at the end of the gas-rich phase (*Hayashi,* 1981). This snow line likely moved inward from larger distances in the early embedded phase (*Kennedy and Kenyon*, 2008). Thus, it is no surprise that water ice is a major constituent of all solar system bodies that formed and stayed beyond the snow line. Nevertheless, their water-ice content, as measured by the mass in ice with respect to total ice + rock mass, can differ substantially, from <1% for some asteroids to typically 50% for comets. Also, the observational signatures of water on these icy bodies and its isotopic ratio can differ. In the following sections, we review our knowledge of water ice in the present-day solar system. In section 8, possible mechanisms of supplying water from these reservoirs to the terrestrial planet zone will be discussed.

6.1. Outer Asteroid Belt

Since the outer asteroid belt is located outside the Hayashi snow line, it provides a natural reservoir of icy bodies in the solar system. This part of the belt is dominated by so-called C-, P-, and D-class asteroids with sizes up to a few hundred kilometers at distances of ~3, 4, and ≥4 AU, respectivily, characterized by their particularly red colors and very low albedos, ≤0.1 (*Bus and Binzel*, 2002). Because of their spectroscopic similarities to the chemically primitive carbonaceous chondrites found as meteorites on Earth, C-type asteroids have been regarded as largely unaltered, volatile-rich bodies. The P- and D-type asteroids may be even richer in organics.

The water content in these objects has been studied though IR spectroscopy of the 3-μm band. In today's solar system, any water ice on the surface would rapidly sublimate at the distance of the belt, so only water bonded to the rocky silicate surface is expected to be detected. Hydrated minerals can be formed if the material has been in contact with liquid water. The majority of the C-type asteroids show hydrated silicate absorption, indicating that they indeed underwent heating and aqueous alteration episodes (*Jones et al.*, 1990, and references therein). However, only 10% of the P- and D-type spectra show weak water absorption, suggesting that they have largely escaped this processing and that the abundance of hydrated silicates gradually declines in the outer asteroid belt. Nevertheless, asteroids that do not display water absorption on their surfaces (mostly located beyond 3.5 AU) may still retain ices in their interior. Indeed, water fractions of 5–10% of their total mass have been estimated. This is consistent with models that show that buried ice can persist in the asteroid belt within the top few meters of the surface over billions of years, as long as the mean surface temperature is less than about 145 K (*Schorghofer*, 2008). Water vapor has recently been detected around the dwarf planet Ceres at 2.7 AU in the asteroid belt, with a production rate of at least 10^{26} mol s^{-1}, directly confirming the presence of water (*Küppers et al.*, 2014).

Hsieh and Jewitt (2006) discovered a new population of small objects in the main asteroid belt, displaying cometary characteristics. These so-called main-belt comets (10 are currently known) display clearly asteroidal orbits, yet have been observed to eject dust and thus satisfy the observational definition of a comet. These objects are unlikely to have originated elsewhere in the solar system and to have subsequently been trapped in their current orbits. Instead, they are intrinsically icy bodies, formed and stored at their current locations, with their cometary activity triggered by some recent event.

Since main-belt comets are optically faint, it is not known whether they display hydration spectral features that could point to the presence of water. Activity of main-belt comets is limited to the release of dust, and direct outgassing of volatiles, like for Ceres, has so far not been detected. The most stringent indirect upper limit for the water-production rate derived from CN observations is that in the prototypical main-belt comet 133P/Elst-Pizarro, $<1.3 \times 10^{24}$ mol s^{-1} (*Licandro et al.*, 2011), which is subject to uncertainties in the assumed water-to-CN abundance ratio. For comparison, this is 5 orders of magnitude lower than the water-production rate of Comet Hale-Bopp. Its mean density is 1.3 g cm^{-3}, suggesting a moderately high ice fraction (*Hsieh et al.*, 2004). Herschel provided the most stringent direct upper limits for water outgassing in 176P/LINEAR [$<4 \times 10^{25}$ mol s^{-1}, 3σ (*de Val-Borro et al.*, 2012)] and P/2012 T1 PANSTARRS [$<8 \times 10^{25}$ mol s^{-1} (*O'Rourke et al.*, 2013)].

Another exciting discovery is the direct spectroscopic detection of water ice on the asteroid 24 Themis (*Campins et al.*, 2010; *Rivkin and Emery*, 2010), the largest (198 km diameter) member of the Themis dynamical family at ~3.2 AU, which also includes three main-belt comets. The 3.1-μm spectral feature detected in Themis is significantly different from those in other asteroids, meteorites, and all plausible mineral samples available. *Campins et al.* (2010) argue that the observations can be accurately matched by small ice particles evenly distributed on the surface. A subsurface ice reservoir could also be present if Themis underwent differentiation resulting in a rocky core and an ice mantle. *Jewitt and Guilbert-Lepoutre* (2012) find no direct evidence of outgassing from the surface of Themis or Cybele with a 5σ upper limit for the water-production rate 1.3×10^{28} mol s^{-1}, assuming a cometary water-to-CN mixing ratio. They conclude that any ice that exists on these bodies should be relatively clean and confined to a <10% fraction of the Earth-facing surface.

Altogether, these results suggest that water ice may be common below asteroidal surfaces and widespread in asteroidal interiors down to smaller heliocentric distances than previously expected. Their water contents are clearly higher than those of meteorites that originate from the inner asteroid belt, which have only 0.01% of their mass in water (*Hutson and Ruzicka*, 2000).

6.2. Comets

Comets are small solar system bodies with radii less than 20 km that have formed and remained for most of their lifetimes at large heliocentric distances. Therefore, they likely contain some of the least-processed, pristine ices from the solar-nebula disk. They have often been described as "dirty snowballs," following the model of *Whipple* (1950), in which the nucleus is visualized as a conglomerate of ices, such as water, ammonia, methane, carbon dioxide, and carbon monoxide, combined with meteoritic materials. However, Rosetta observations during the Deep Impact encounter (*Küppers et al.,* 2005) suggest a dust-to-gas ratio in excess of unity in Comet 9P/Tempel 1. Typically, cometary ice/rock ratios are on the order of unity with an implied porosity well over 50% (*A'Hearn,* 2011).

The presence and amount of water in comets is usually quantified through their water-production rates, which are traditionally inferred from radio observations of its photodissociation product, OH, at 18 cm (*Crovisier et al.,* 2002). Measured rates vary from 10^{26} to 10^{29} mol s^{-1}. The first direct detection of gaseous water in Comet 1P/Halley, through its v_3 vibrational band at 2.65 μm, was obtained using the Kuiper Airborne Observatory (KAO) (*Mumma et al.,* 1986). The 557-GHz transition of ortho-water was observed by the Submillimeter Wave Astronomy Satellite (SWAS) (*Neufeld et al.,* 2000) and Odin (*Lecacheux et al.,* 2003), whereas Herschel provided for the first time access to multiple rotational transitions of both ortho- and para-water (*Hartogh et al.,* 2011). These multi-transition mapping observations show that the derived water-production rates are sensitive to the details of the excitation model used, in particular the ill-constrained temperature profile within the coma, with uncertainties up to 50% (*Bockelée-Morvan et al.,* 2012).

Dynamically, comets can be separated into two general groups: short-period, Jupiter-family comets and long-period comets (but see *Horner et al.,* 2003, for a more detailed classification). Short-period comets are thought to originate from the Kuiper belt, or the associated scattered disk, beyond the orbit of Neptune, while long-period comets formed in the Jupiter-Neptune region and were subsequently ejected into the Oort cloud by gravitational interactions with the giant planets. In reality, the picture is significantly more complex due to migration of the giant planets in the early solar system (see below). In addition, recent simulations (*Levison et al.,* 2010) suggest that the Sun may have captured comets from other stars in its birth cluster. In this case, a substantial fraction of the Oort cloud comets, perhaps in excess of 90%, may not even have formed in the Sun's protoplanetary disk. Consequently, there is increasing emphasis on classifying comets based on their chemical and isotopic composition rather than orbital dynamics (*Mumma and Charnley,* 2011). There is even evidence for heterogeneity within a single comet, illustrating that comets may be built up from cometesimals originating at different locations in the disk (*A'Hearn,* 2011).

Traditionally, the ortho-to-para ratio in water and other cometary volatiles has been used to constrain the formation and thermal history of the ices. Recent laboratory experiments suggest that this ratio is modified by the desorption processes, both thermal sublimation and photodesorption,

and may therefore tell astronomers less about the water-formation location than previously thought (see discussions in *van Dishoeck et al.,* 2013; *Tielens,* 2013).

6.3. Water in the Outer Satellites

Water is a significant or major component of almost all moons of the giant planets for which densities or spectral information are available. Jupiter's Galilean moons exhibit a strong gradient from the innermost (Io, essentially all rock, no ice detected) to the outermost (Callisto, an equal mixture of rock and ice). Ganymede has nearly the same composition and hence rock-to-ice ratio as Callisto. Assuming that the outermost of the moons reflects the coldest part of the circumplanetary disk out of which the moons formed, and hence full condensation of water, Callisto's (uncompressed) density matches that of solar-composition material in which the dominant carbon-carrier was methane rather than carbon monoxide (*Wong et al.,* 2008). The fact that Callisto has close to (but not quite) the full complement of water expected based on the solar oxygen abundance implies that the disk around Jupiter had a different chemical composition (CH_4-rich, CO-poor) from that of the solar-nebula disk, which was CO-dominated (*Prinn and Fegley,* 1981).

The saturnian satellites are very different. For satellites large enough to be unaffected by porosity, but excluding massive Titan, the ice-to-rock ratio is higher than for the Galilean moons (*Johnson and Estrada,* 2009). However, Titan — by far the most massive moon — has a bulk density and mass just in between, and closely resembling, Ganymede's and Callisto's. Evidently the saturnian satellite system had a complex collisional history, in which the original ice-rock ratio of the system was not preserved except perhaps in Titan. The moon Enceladus, which exhibits volcanic and geyser activity, offers the unique opportunity to sample saturnian system water directly. Neptune's Triton, like Pluto, has a bulk density and hence ice-rock ratio consistent with what is expected for a solar-nebula disk in which CO dominated over CH_4. Its water fraction is about 15–35%. At this large distance from the Sun, N_2 can also be frozen out and Triton's spectrum is indeed dominated by N_2 ice with traces of CH_4 and CO ices; the water signatures are much weaker than on other satellites (*Cruikshank et al.,* 2000). The smaller transneptunion objects (TNOs) (less than a few hundred kilometer-sized bodies) are usually found to have mean densities around 1 g cm^{-3} and thus a high ice fraction. Larger TNOs such as Quaoar and Haumea have much higher densities (2.6–3.3 g cm^{-3}), suggesting a much lower ice content, even compared with Pluto (2.0 g cm^{-3}) (*Fornasier et al.,* 2013).

In summary, the water-ice content of the outer satellites varies with position and temperature, not only as a function of distance from the Sun, but also from its parent planet. Ice fractions are generally high (≥50%) in the colder parts and consistent with solar abundances depending on the amount of oxygen locked up in CO. However, collisions can have caused a strong reduction of the water-ice content.

6.4. Water in the Giant Planets

The largest reservoirs of what was once water ice in the solar-nebula disk are presumably locked up in the giant planets. If the formation of giant planets started with an initial solid core of 10–15 $M_\oplus$ with subsequent growth from a swarm of planetesimals of ice and rock with solar composition, the giant planets could have had several Earth masses of oxygen, some or much of which may have been in water molecules in the original protoplanetary disk. The core also gravitationally attracts the surrounding gas in the disk, consisting mostly of H and He, with the other elements in solar composition. The resulting gas giant planet has a large mass and diameter, but a low overall density compared with rocky planets. The giant planet atmosphere is expected to have an excess in heavy elements, either due to the vaporization of the icy planetesimals when they entered the envelopes of the growing planet during the heating phase, or due to partial erosion of the original core, or both (*Encrenaz*, 2008; *Mousis et al.*, 2009). These calculations assume that all heavy elements are equally trapped within the ices initially, which is a debatable assumption, and that the ices fully evaporate, with most of the refractory material sedimenting onto the core. In principle, the excesses provide insight into giant planet formation mechanisms and constraints on the composition of their building blocks.

6.4.1. Jupiter and Saturn.

For Jupiter, elemental abundances can be derived from spectroscopic observations and from data collected by the Galileo probe, which descended into the jovian atmosphere. Elements like C, N, S, and the noble gases Ar, Kr, and Xe have measured excesses as expected at 4 ± 2 (*Owen and Encrenaz*, 2006). However, O appears to show significant depletion. Unfortunately, the deep oxygen abundance in Jupiter is not known. The measured abundance of gaseous water — the primary carrier of oxygen in the jovian atmosphere since there is little CO — provides only a lower limit since the troposphere at about 100 mbar is a region of minimum temperature (~110 K for Jupiter) and therefore acts as a cold trap where water can freeze out. Thus, the amount of gaseous water is strongly affected by condensation and rainout associated with large-scale advective motions (*Showman and Dowling*, 2000), meteorological processes (*Lunine and Hunten*, 1987), or both. The Galileo probe fell into a so-called "hot spot" (for the excess brightness observed in such regions at 5-μm wavelengths), with enhanced transparency and hence depleted in water, and is thus not representative of the planet as a whole. The water abundance, less than one-tenth the solar value in the upper atmosphere, was observed to be higher at higher pressures, toward the end of the descent (*Roos-Serote et al.*, 2004). The sparseness of the measurements made it impossible to know whether the water had "leveled out" at a value corresponding to one-third solar or would have increased further had the probe returned data below the final 21-bar level.

As noted above, predictions for standard models of planetesimal accretion — where volatiles are either adsorbed on, or enclathrated in, water ice — give oxygen abundances 3–10× solar in the jovian deep interior. Although atmospheric explanations for the depleted water abundance in Jupiter are attractive, one must not rule out the possibility that water truly is depleted in the jovian interior — i.e., the oxygen-to-hydrogen ratio in Jupiter is less enriched than the carbon value at 4 ± 2× solar. A motivation for making such a case is that at least one planet with C/O > 1 — a "carbon-rich planet" — has been discovered (*Madhusudhan et al.*, 2011a), a companion to the star WASP-12a with a C/O ratio of 0.44, roughly solar. One explanation is that the portion of this system's protoplanetary disk was somehow depleted in water at the time the planet formed and acquired its heavy-element inventory (*Madhusudhan et al.*, 2011b).

Prior to this discovery, the possibility of a carbon-rich Jupiter was considered on the basis of the Galileo results alone by *Lodders* (2004), who proposed that in the early solar system formation the snow line might have been further from the Sun than the point at which Jupiter formed, and volatiles adhering to solid organics rather than water ice were carried into Jupiter.

Mousis et al. (2012) looked at the possibility that Jupiter may have acquired planetesimals from an oxygen-depleted region by examining element-by-element the fit to the Galileo probe data of two contrasting models: one in which the planetesimal building blocks of Jupiter derived from a disk with C/O = 1/2 (roughly, the solar value), and the other in which C/O = 1. Within the current error bars, the two cases cannot be distinguished. However, any determination of the deep oxygen abundance in Jupiter yielding an enrichment of less than 2× solar would be a strong argument in favor of a water depletion, and thus high C/O ratio, at certain places and times in the solar-nebula disk. Are such depletions plausible? A wide range of oxidation states existed in the solar-nebula disk at different times and locations. For example, the driest rocks, the enstatite chondrites, are thought to have come from parent bodies formed inward of all the other parent bodies, and their mineralogy suggests reducing conditions — consistent with a depleted water vapor abundance — in the region of the nebula where they formed (*Krot et al.*, 2000). One recent model of the early evolution of Jupiter and Saturn hypothesizes that these giant planets moved inward significantly during the late stages of their formation, reaching 1.5 AU in the case of Jupiter (*Walsh et al.*, 2011), almost certainly inward of the snow line (see section 8.2). If planetesimals in this region accreted volatiles on refractory organic and silicate surfaces that were then incorporated into Jupiter, the latter would appear carbon-rich and oxygen-depleted. However, temperatures in that region may not have been low enough to provide sufficient amounts of the more volatile phases. A second possibility is that Jupiter's migration scattered these water-poor planetesimals to the colder outer solar system, where they trapped noble gases and carbon- and nitrogen-bearing species at lower temperature — but water ice, already frozen out, was not available.

In order to test such models, one must directly determine the abundance of O-bearing species in Jupiter and, if possible, in Saturn. The case of Saturn is similar to that of Jupiter in the sense that the carbon excess as derived from CH_4 spectroscopy is as expected, but again no reliable oxygen abundance can be determined. Under the conditions present in the jovian envelope at least (if not its core), water will dominate regardless of the initial carbon oxidation state in the solar nebula. NASA's Juno mission to Jupiter will measure the water abundance down to many tens of bars via a microwave radiometer (MWR) (*Janssen et al.*, 2005). Complementary to the MWR is the near-IR spectrometer Jovian Infrared Auroral Mapper (JIRAM) that will obtain the water abundance in the meteorological layer (*Adriani et al.*, 2008). The two instruments together will be able to provide a definitive answer for whether the water abundance is below or above solar, and in the latter case, by how much, when Juno arrives in 2016. Juno will also determine the mass of the heavy-element core of Jupiter, allowing for an interpretation of the significance of the envelope water abundance in terms of total oxygen inventory.

There is no approved mission yet to send a probe into Saturn akin to Galileo. However, the Cassini Saturn Orbiter will make very close flybys of Saturn starting in 2016, similar to what Juno will do at Jupiter. Unfortunately, a microwave radiometer akin to that on Juno is not present on Cassini, but a determination of the heavy-element core mass of Saturn may be obtained from remote sensing. In summary, the fascinating possibility that Jupiter and Saturn may have distinct oxygen abundances because they sampled at different times and to differing extents regions of the solar-nebula disk heterogeneous in oxygen (i.e., water) abundance is testable if the bulk oxygen abundances can be measured.

6.4.2. Uranus and Neptune. The carbon excesses measured from CH_4 near-IR spectroscopy give values of 30–50 for Uranus and Neptune, compared with 9 and 4 for Saturn and Jupiter (*Encrenaz*, 2008). These values are consistent with the assumption of an initial core of 10–15 $M_\oplus$ with heavy elements in solar abundances. Unfortunately, nothing is known about the water or oxygen content in Uranus and Neptune, since water condensation occurs at such deep levels that even tropospheric water vapor cannot be detected. Unexpectedly, ISO detected water vapor in the upper atmospheres of both ice giants, with mixing ratios orders of magnitude higher than the saturation level at the temperature inversion (*Feuchtgruber et al.*, 1997). These observations can only be accounted for by an external flux of water molecules, due to interplanetary dust, or sputtering from rings or satellites. Similarly, Herschel maps of water of Jupiter demonstrate that even 15 yr after the Shoemaker-Levy 9 impact, more than 95% of the stratospheric jovian water comes from the impact (*Cavalié et al.*, 2013). On the other hand, the low D/H ratios in molecular hydrogen measured by Herschel in the atmospheres of Uranus and Neptune may imply a lower ice mass fraction of their cores than previously thought (14–32%) (*Feuchtgruber et al.*, 2013).

It is unlikely that information on the deep oxygen abundance will be available for Uranus and Neptune anytime soon, since reaching below the upper layers to determine the bulk water abundance is very difficult. Thus, data complementary to that for Jupiter and Saturn will likely come first from observations of Neptune-like exoplanets.

7. WATER IN THE INNER SOLAR SYSTEM

Terrestrial planets were formed by accretion of rocky planetesimals, and they have atmospheres that are only a small fraction of their total masses. They are small and have high densities compared with giant planets. Earth and Venus have very similar sizes and masses, whereas Mars has a mass of only 10% of that of Earth and a somewhat lower density (3.9 vs. 5.2 g cm^{-3}). Terrestrial planets undergo different evolutionary processes compared with their gas-rich counterparts. In particular, their atmospheres result mostly from outgassing and from external bombardment. Internal differentiation of the solid material after formation leads to a structure in which the heavier elements sink to the center, resulting in a molten core (consisting of metals like iron and nickel), a mantle (consisting of a viscous hot dense layer of magnesium-rich silicates), and a thin upper crust (consisting of colder rocks).

Water has very different appearances on Venus, Earth, and Mars, which is directly related to their distances from the Sun at 0.7, 1.0, and 1.7 AU, respectively. On Venus, with surface temperatures around 730 K, only gaseous water is found, whereas the surface of Mars has seasonal variations ranging from 150 to 300 K resulting in water freezing and sublimation. Its current mean surface pressure of 6 mbar is too low for liquid water to exist, but there is ample evidence for liquid water on Mars in its early history. Most of the water currently on Mars is likely subsurface in the crust down to 2 km depth. Earth is unique in that its mean surface temperature of 288 K and pressure of 1 bar allow all three forms of water to be present: vapor, liquid, and ice. In section 8, the origin of water on these three planets will be further discussed. It is important to keep in mind that even though these planets have very different atmospheres today, they may well have started out with comparable initial water mass fractions and similar atmospheric compositions dominated by H_2O, CO_2, and N_2 (*Encrenaz*, 2008).

Earth has a current water content that is non-negligible. The mass of the water contained in Earth's crust (including the oceans and the atmosphere) is 2.8×10^{-4} $M_\oplus$, denoted as "one Earth ocean" because almost all this is in the surface waters of Earth. The mass of the water in the present-day mantle is uncertain. *Lécuyer et al.* (1998) estimate it to be in the range of $(0.8–8) \times 10^{-4}$ $M_\oplus$, equivalent to 0.3–3 Earth oceans. More recently, *Marty* (2012) provides arguments in favor of a mantle water content as high as ~7 Earth oceans. However, an even larger quantity of water may have resided in the primitive Earth and been subsequently lost during differentiation and impacts. Thus, the current Earth has a water content of roughly 0.1% by mass, larger than that of enstatite chondrites, and it is possible that the primitive Earth had a water content comparable with or larger than that of ordinary chondrites, definitely larger than the water content of meteorites and material that condensed at 1 AU (Fig. 10).

Remote sensing and space probes have detected water in the lower atmosphere of Venus at a level of a few tens of 16 ppm (*Gurwell et al.,* 2007), confirming that it is a minor component. Radar images show that the surface of Venus is covered with volcanos and is likely very young, only 1 Ga (i.e., formed 1 billion years ago). Direct sampling of the venusian crust is needed to determine the extent of past and present hydration, which will provide an indication of whether Venus once had an amount of water comparable to that on Earth. Hints that there may have been much more water on Venus and Mars come from D/H measurements (see section 8.1).

8. ORIGIN OF WATER TERRESTRIAL PLANETS

Earth, Mars, and perhaps also Venus all have water mass fractions that are higher than those found in meteorites in the inner solar-nebula disk. Where did this water come from, if the local planetesimals were dry? There are two lines of arguments that provide clues to its origin. One clue comes from measuring the D/H ratio of the various water mass reservoirs. The second argument looks at the water mass fractions of various types of asteroids and comets, combined with the mass delivery rates of these objects on the young planets at the time of their formation, based on models of the dynamics of these planetesimals. For example, comets have plenty of water, but the number of comets that enter the inner 2 AU and collide with proto-Earth or -Mars is small. To get one Earth ocean, $\sim 10^8$ impacting comets are needed. It is therefore not *a priori* obvious which of the reservoirs shown in Fig. 10 dominates.

Related to this point is the question of whether the planets formed "wet" or "dry." In the wet scenario, the planets either accreted a water-rich atmosphere or they formed from planetesimals with water bonded to silicate grains at 1 AU, with subsequent outgassing of water as the material is heated up during planet formation. In the dry scenario, the terrestrial planets are initially built up from planetesimals with low water mass fractions and water is delivered to their surfaces by water-rich planetesimals, either late in the building phase, or even later after the planets have formed and differentiated. If after differentiation, this scenario is often called a "late veneer." In the following, we first discuss the D/H ratios and then come back to the mass fraction arguments.

8.1. Deuterium/Hydrogen Ratios in Solar System Water Reservoirs

The D/H ratio of water potentially provides a unique fingerprint of the origin and thermal history of water. Figure 11 summarizes the various measurements of solar system bodies, as well as interstellar ices and protostellar objects.

The primordial [D]/[H] ratio set shortly after the Big Bang is 2.7×10^{-5}. Since then, deuterium has been lost due to nuclear fusion in stars, so the deuterium abundance at the time of our solar system formation, 4.6 Ga (redshift of about $z \approx 1.4$), should be lower than the primordial value, but higher than the current-day interstellar [D]/[H] ratio. The latter value has been measured in the diffuse local

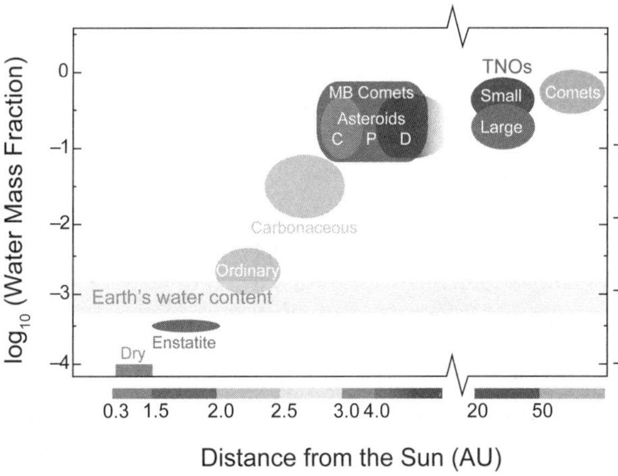

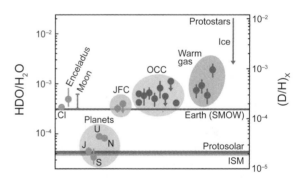

Fig. 10. Water content of various parent bodies at the time of Earth's growth as a function of radial distance from the Sun. Enstatite chondrites originating from asteroids around 1.8 AU in the inner disk are very dry. In contrast, carbonaceous chondrites originating from the outer asteroid belt and beyond have a water content of 5–10%, and main-belt comets, TNOs, and regular comets have even more. Figure by M. Persson, adapted from *Morbidelli et al.* (2012).

Fig. 11. D/H ratio in water in comets and warm protostellar envelopes compared to values in Earth's oceans, giant planets, the solar nebula disk, and the interstellar medium. Values for carbonaceous meteorites (CI), the Moon, and Saturn's moon Enceladus are presented as well. Note that both $(D/H)_X$, the deuterium to hydrogen ratio in molecule X, and HDO/H_2O are plotted; measured HD/H_2 and HDO/H_2O ratios are $2 \times (D/H)_X$. Figure by M. Persson, based on *Bockelée-Morvan et al.* (2012) and subsequent measurements cited in the text. The protostellar data refer to warm gas (*Persson et al.,* 2013, 2014).

interstellar medium from UV absorption lines of atomic D and H, and varies from place to place but can be as high as 2.3×10^{-5} (*Prodanović et al., 2010*). These higher values suggest that relatively little deuterium has been converted in stars, so a [D]/[H] value of the solar-nebula disk only slightly above the current interstellar medium (ISM) value is expected at the time of solar system formation. The solar-nebula [D]/[H] ratio can be measured from the solar-wind composition as well as from HD/H_2 in the atmospheres of Jupiter and Saturn and is found to be $(2.5 \pm 0.5) \times 10^{-5}$ (*Robert et al., 2000*, and references therein), indeed in between the primordial and the current ISM ratios. (Note that measured HD/H_2 and HDO/H_2O ratios are $2 \times (D/H)X$, the D/H ratios in these molecules.)

The D/H ratio of Earth's ocean water is 1.5576×10^{-4} [the "Vienna Standard Mean Ocean Water" (VSMOW, or SMOW)]. Whether this value is representative of the bulk of Earth's water remains unclear, as no measurements exist for the mantle or the core. It is thought that recycling of water in the deep mantle does not significantly change the D/H ratio. In any case, the water D/H ratio is at least a factor of 6 higher than that of the gas out of which our solar system formed. Thus, Earth's water must have undergone fractionation processes that enhance deuterium relative to hydrogen at some stages during its history, of the kind described in section 2.4 for interstellar chemistry at low temperatures.

Which solar system bodies show D/H ratios in water similar to those found in Earth's oceans? The highest [D]/[H] ratios are found in two types of primitive meteorites (*Robert et al., 2000*): LL3, an ordinary chondrite that may have the near-Earth asteroid 433 Eros as its parent body (up to a factor of 44 enhancement compared to the protosolar ratio in H_2), and some carbonaceous chondrites (a factor of 15 to 25 enhancement). Most chondrites show lower enhancements than the most primitive meteorites. Pre-Herschel observations of six Oort cloud or long-period comets give a D/H ratio in water of $\sim 3 \times 10^{-4}$, a factor of 12 higher than the protosolar ratio in H_2 (*Mumma and Charnley, 2011*).

Herschel provided the first measurement of the D/H ratio in a Jupiter-family comet. A low value of $(1.61 \pm 0.24) \times 10^{-4}$, consistent with VSMOW, was measured in Comet 103P/Hartley 2 (*Hartogh et al., 2011*). In addition, a relatively low ratio of $(2.06 \pm 0.22) \times 10^{-4}$ was found in the Oort cloud comet C/2009 P1 Garradd (*Bockelée-Morvan et al., 2012*) and a sensitive upper limit of $<2 \times 10^{-4}$ (3σ) was obtained in another Jupiter-family comet, 45P/Honda-Mrkos-Pajdušáková (*Lis et al., 2013*). The Jupiter-family comets 103P and 45P are thought to originate from the large reservoir of water-rich material in the Kuiper belt or scattered disk at 30–50 AU. In contrast, Oort cloud comets, currently at much larger distances from the Sun, may have formed closer in, near the current orbits of the giant planets at 5–20 AU, although this traditional view has been challenged in recent years (see section 6.2). Another caveat is that isotopic ratios in cometary water may have been altered by the outgassing process (*Brown et al., 2011*). Either way,

the Herschel observations demonstrate that the earlier high D/H values are not representative of all comets.

How do these solar system values compare with interstellar and protostellar D/H ratios? Measured D/H ratios in water in protostellar envelopes vary strongly. In the cold outer envelope, D/H values for gas-phase water are as high as 10^{-2} (*Liu et al., 2011; Coutens et al., 2012, 2013*), larger than the upper limits in ices of $<(2–5) \times 10^{-3}$ obtained from IR spectroscopy (*Dartois et al., 2003; Parise et al., 2003*). In the inner warm envelope, previous discrepancies for gaseous water appear to have been resolved in favor of the lower values, down to $(3–5) \times 10^{-4}$ (*Jørgensen and van Dishoeck, 2010a; Visser et al., 2013; Persson et al., 2013, 2014*.) The latter values are within a factor of 2 of the cometary values.

Based on these data, the jury is still out as to whether the D/H ratio in solar system water was already set by ices in the early (pre-)collapse phase and transported largely unaltered to the comet-forming zone (cf. Fig. 4), or whether further alteration of D/H took place in the solar-nebula disk along the lines described in section 2.4.5 (see also the chapter by Ceccarelli et al. in this volume). The original D/H ratio in ices may even have been reset early in the embedded phase by thermal cycling of material due to accretion events onto the star (see the chapter by Audard et al. in this volume). Is the high value of 10^{-2} found in cold gas preserved in the material entering the disk, or are the lower values of $<10^{-3}$ found in hot cores and ices more representative? Or do the different values reflect different D/H ratios in layered ices (*Taquet et al., 2013*)?

Regardless of the precise initial value, models have shown that vertical and radial mixing within the solar-nebula disk reduces the D/H ratios from initial values as high as 10^{-2} to values as low as 10^{-4} in the comet-forming zones (*Willacy and Woods, 2009; Yang et al., 2012b; Jacquet and Robert, 2013; Furuya et al., 2013; Albertsson et al., 2014*). For water on Earth, the fact that the D enrichment is a factor of ~ 6 above the solar-nebula value rules out the warm thermal exchange reaction of $H_2O + HD \rightarrow HDO + H_2$ in the inner nebula (~ 1 AU) as the sole cause. Mixing with some cold reservoir with enhanced D/H at larger distances is needed.

Figure 11 contains values for several other solar system targets. Enceladus, one of Saturn's moons with volcanic activity, has a high D/H ratio consistent with it being built up from outer solar system planetesimals. Some measurements indicate that lunar water may have a factor of 2 higher hydrogen isotopic ratio than Earth's oceans (*Greenwood et al., 2011*), although this has been refuted (*Saal et al., 2013*).

Mars and Venus have interestingly high D/H ratios. The D/H ratio of water measured in the martian atmosphere is $5.5 \times$ VSMOW, which is interpreted to imply a significant loss of water over martian history and associated enhancement of deuterated water. Because D is heavier than H, it escapes more slowly, therefore over time the atmosphere is enriched in deuterated species like HDO. A time-dependent model for the enrichment of deuterium on Mars assuming a rough outgassing efficiency of 50% suggests that the amount of water that Mars must have accreted is 0.04–0.4 Earth oceans. The amount of outgassing may be tested by Curiosity Mars rover measurements of other isotopic ratios, such as [38]Ar to [36]Ar.

The extremely high D/H ratio for water measured in the atmosphere of Venus, about 100× VSMOW, is almost certainly a result of early loss of substantial amounts of crustal waters, regardless of whether the starting value was equal to VSMOW or twice that value. The amount of water lost is very uncertain because the mechanism and rate of loss affects the deuterium fractionation (*Donahue and Hodges,* 1992). Solar wind stripping, hydrodynamic escape, and thermal (Jeans') escape have very different efficiencies for the enrichment of deuterium per unit amount of water lost. Values ranging from as little as 0.1% to 1 Earth ocean have been proposed.

8.2. Water Delivery to the Terrestrial Planet Zone

8.2.1. Dry scenario. The above summary indicates at least two reservoirs of water-ice-rich material in the present-day solar system with D/H ratios consistent with that in Earth's oceans: the outer asteroid belt and the Kuiper belt or scattered disk (as traced by Comets 103P and 45P). As illustrated in Fig. 10, these same reservoirs also have a high enough water mass fraction to deliver the overall water content of terrestrial planets. The key question is then whether the delivery of water from these reservoirs is consistent with the current understanding of the early solar system dynamics.

Over the last decade, there has been increasing evidence that the giant planets did not form and stay at their current location in the solar system but migrated through the protosolar disk. The Grand Tack scenario (*Walsh et al.,* 2011) invokes movement of Jupiter just after it formed in the early, gas-rich stage of the disk (few million years). The Nice model describes the dynamics and migration of the giant planets in the much later, gas-poor phase of the disk, some 800–900 million years (m.y.) after formation (*Gomes et al.,* 2005; *Morbidelli et al.,* 2005; *Tsiganis et al.,* 2005) (see also the chapter by Raymond et al. in this volume).

The Grand Tack model posits that if Jupiter formed earlier than Saturn just outside the snow line, both planets would have first migrated inward until they got locked in the 3:2 resonance, then they would have migrated outward together until the complete disappearance of the gas disk (see Fig. 12). As summarized by *Morbidelli et al.* (2012), the reversal of Jupiter's migration at 1.5 AU provides a natural explanation for the outer edge of the inner disk of (dry) embryos and planetesimals at 1 AU, which is required to explain the low mass of Mars. The asteroid belt at 2–4 AU is first fully depleted and then repopulated by a small fraction of the planetesimals scattered by the giant planets during their formation. The outer part of this belt around 4 AU would have been mainly populated by planetesimals originally beyond the orbits of the giant planets

(>5 AU), which explains similarities between primitive asteroids (C-type) and comets. Simulations by *Walsh et al.* (2011) show that as the outer asteroid belt is repopulated, $(3-11) \times 10^{-2}\, M_\oplus$ of C-type material enters the terrestrial planet region, which exceeds by a factor of 6–22 the minimum mass required to bring the current amount of water to Earth. In this picture, Kuiper belt objects would at best be only minor contributors to Earth's water budget.

The Nice model identifies Jupiter and Saturn crossing their 1:2 orbital resonance as the next key event in the dynamical evolution of the disk in the gas-poor phase. After the resonance crossing time at ~880 m.y., the orbits of the ice giants, Uranus and Neptune, became unstable. They then disrupted the outer disk and scattered objects throughout the solar system, including into the terrestrial planet region. However, *Gomes et al.* (2005) estimate the amount of cometary material delivered to the Earth to be only about 6% of the current ocean mass. A larger influx of material from the asteroid belt is expected, as resonances between the orbits of asteroids and giant planets can drive objects onto orbits with eccentricities and inclinations large enough to allow them to evolve into the inner solar system. By this time, the terrestrial planets should have formed already, so this material would be part of a 'late delivery.'

The Grand Tack model provides an attractive, although not unchallenged, explanation for the observed morphology of the inner solar system and the delivery of water to the Earth. In the asteroidal scenario, water is accreted during the formation phase of the terrestrial planets and not afterwards through bombardment as a late veneer. Typically, 50% of the water is accreted after the Earth has reached 60–70% of its final mass (*Morbidelli et al.,* 2012). This appears in contradiction with the measurements of the distinct D/H ratio in lunar water mentioned above, approximately twice

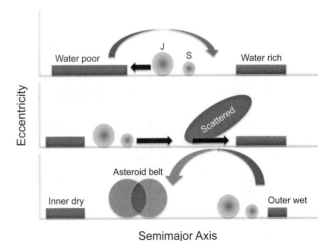

Fig. 12. Cartoon of the delivery of water-rich planetesimals in the outer asteroid belt and terrestrial planet-forming zone based on the Grand Tack scenario, in which Jupiter and Saturn first move inward to 1.5 AU and then back out again. Figure based on *Walsh et al.* (2011).

that in the Earth's oceans (*Greenwood et al.*, 2011). If valid, this would indicate that a significant delivery of cometary water to the Earth-Moon system occurred shortly after the Moon-forming impact. The Earth water would thus be a late addition, resulting from only one, or at most a few collisions with the Earth that missed the Moon (*Robert*, 2011). Expanding the sample of objects with accurate D/H measurements is thus a high priority, long-term science goal for the new submillimeter facilities.

While the case for late delivery of water on Earth is still open, *Lunine et al.* (2003) showed that the abundance of water derived for early Mars is consistent with the general picture of late delivery by comets and bodies from the early asteroid belt. Unlike Earth, however, the small size of Mars dictates that it acquired its water primarily from very small bodies — asteroidal (sizes up to 100 km) — rather than lunar (~1700 km) or larger. The stochastic nature of the accretion process allows this to be one outcome out of many, but the Jupiter Grand Tack scenario provides a specific mechanism for removing larger bodies from the region where Mars formed, thereby resulting in its small size.

8.2.2. Wet scenario. There are two ways for Earth to get its water locally around 1 AU rather than through delivery from the outer solar system. The first option is that local planetesimals have retained some water at high temperatures through chemisorption onto silicate grains (*Drake*, 2005; *de Leeuw et al.*, 2010). There is no evidence for such material today in meteorites nor in the past in interstellar and protoplanetary dust (section 5.1). However, the solid grains were bathed in abundant water vapor during the entire lifetime of gas-rich disks. This mechanism should differentiate between Earth and Venus. Computation and laboratory studies (*Stimpfl et al.*, 2006) suggest that chemisorption may be just marginally able to supply the inventory needed to explain Earth's crustal water, with some mantle water, but moving inward to 0.7 AU the efficiency of chemisorption should be significantly lower. Finding that Venus was significantly drier than Earth early in the solar system history would argue for the local wet source model, although such a model may also overpredict the amount of water acquired by Mars.

The second wet scenario is that Earth accretes a water-rich atmosphere directly from the gas in the inner disk. This would require Earth to have had in the past a massive hydrogen atmosphere (with a molar ratio H_2/H_2O larger than 1) that experienced a slow hydrodynamical escape (*Ikoma and Genda*, 2006). One problem with this scenario is that the timescales for terrestrial planet formation are much longer than the lifetime of the gas disk. Also, the D/H ratio would be too low, unless photochemical processes at the disk surface enhance D/H in water and mix it down to the midplane (*Thi et al.*, 2010a) or over long timescales via mass-dependent atmospheric escape (*Genda and Ikoma*, 2008).

9. EXOPLANETARY ATMOSPHERES

Water is expected to be one of the dominant components in the atmospheres of giant exoplanets, so searches for signatures of water vapor started immediately when the field of transit spectroscopy with Spitzer opened up in 2007. Near-IR spectroscopy with HST and from the ground can be a powerful complement to these data because of the clean spectral features at 1–1.6 μm (Fig. 1). Early detections of water in HD 189733b, HD 209458, and XO1b were not uniformly accepted (see *Seager and Deming*, 2010; *Tinetti et al.*, 2012, for summaries) (see the chapter by Madhusudhan et al. in this volume). Part of the problem may stem from hazes or clouds that can affect the spectra shortward of 1.5 μm and hide water, which is visible at longer wavelengths. Even for the strongest cases for detection, it is difficult to retrieve an accurate water abundance profile from these data, because of the low S/N and low spectral resolution (*Madhusudhan and Seager*, 2010). The advent of a spatial scan mode has improved HST's ability to detect exoplanetary water with the Wide Field Camera 3 (WFC3) (e.g., *Deming et al.*, 2013; *Mandell et al.*, 2013). New groundbased techniques are also yielding improved spectra (*Birkby et al.*, 2013), but dramatic improvements in sensitivity and wavelength range must await the James Webb Space Telescope (JWST). Until that time, only qualitative conclusions can be drawn.

9.1. Gas-Giant Planets

Models of hydrogen-dominated atmospheres of giant exoplanets (T ≈ 700–3000 K; warm Neptune to hot Jupiter) solve for the molecular abundances (i.e., ratios of number densities, also called "mixing ratios") under the assumption of thermochemical equilibrium. Non-equilibrium chemistry is important if transport (e.g., vertical mixing) is rapid enough that material moves out of a given temperature and pressure zone on a timescale short compared with that needed to reach equilibrium. The CH_4/CO ratio is most affected by such "disequilibrium" chemistry, especially in the upper atmospheres. Since CO locks up some of the oxygen, this can also affect the H_2O abundance.

The two main parameters determining the water abundance profiles are the temperature-pressure combination and the C/O ratio. Giant planets often show a temperature inversion in their atmospheres, with temperature increasing with height into the stratosphere, which affects both the abundances and the spectral appearance. However, for the full range of pressures (10^2–10^{-5} bar) and temperatures (700–3000 K), water is present at abundances greater than 10^{-3} with respect to H_2 for a solar C/O abundance ratio of 0.54, i.e., most of the oxygen is locked up in H_2O. In contrast, when the C/O ratio becomes close to unity or higher, the H_2O abundance can drop by orders of magnitude especially at lower pressures and higher temperatures, since the very stable CO molecule now locks up the bulk of the oxygen (*Kuchner and Seager*, 2005; *Madhusudhan et al.*, 2011b). Methane and other hydrocarbons are enhanced as well. As noted previously, observational evidence (not without controversy) for at least one case of such a carbon-rich giant planet, WASP-12b, has been found (*Madhusudhan et al.*, 2011a). Hot Jupiters like WASP-12b are easier targets for measuring C/O ratios, because their elevated temperature profiles allow both water- and carbon-bearing species to be measured without the interference from condensation processes that hamper measurements in Jupiter and Saturn in our own solar system, although cloud-forming species more refractory than water ice can reduce spectral contrast for some temperature ranges. Figure 13 illustrates the changing spectral appearance with C/O ratio for a hot Jupiter (T = 2300 K); for cooler planets (<1000 K) the changes are less obvious.

What can cause exoplanetary atmospheres to have very different C/O ratios from those found in the interstellar medium or even in their parent star? If giant planets indeed form outside the snow line at low temperatures through accretion of planetesimals, then the composition of the ices is a key ingredient in setting the C/O ratio. The volatiles (but not the rocky cores) vaporize when they enter the envelope of the planet and are mixed in the atmosphere during the homogenization process. As the giant planet migrates through the disk, it encounters different conditions and thus different ice compositions as a function of radius (*Mousis et al.*, 2009; *Öberg et al.*, 2011). Specifically, a giant planet formed outside the CO snow line at 20–30 AU around a solar-mass star may have a higher C/O ratio than that formed inside the CO (but outside the H_2O) snow line.

The exact composition of these planetesimals depends on the adopted model. The solar system community traditionally employs a model of the solar-nebula disk, which starts hot and then cools off with time allowing various species to refreeze. In this model, the abundance ratios of the ices are set by their thermodynamic properties ("condensation sequence") and the relative elemental abundances. For solar abundances, water is trapped in various clathrate hydrates like NH_3-H_2O and H_2S-5.75 H_2O between 5 and 20 AU. In a carbon-rich case, no water ice is formed and all the oxygen is CO, CO_2, and CH_3OH ice (*Madhusudhan et al.*, 2011b). The interstellar astrochemistry community, on the other hand, treats the entire disk with non-equilibrium chemistry, both with height and radius. Also, the heritage of gas and ice from the protostellar stage into the disk can be considered (*Aikawa and Herbst*, 1999; *Visser et al.*, 2009; *Aikawa et al.*, 2012; *Hincelin et al.*, 2013) (see Fig. 5). The non-equilibrium models show that indeed the C/O ratio in ices can vary depending on location and differ by factors of 2–3 from the overall (stellar) abundances (*Öberg et al.*, 2011).

9.2. Rocky Planets, Super-Earths

The composition of Earth-like and super-Earth planets (up to 10 $M_\oplus$) is determined by similar thermodynamic arguments, with the difference that most of the material stays in solid form and no hydrogen-rich atmosphere is attracted. Well outside the snow line, at large distances from the parent star, the planets are built up largely from planetesimals that are half rock and half ice. If these planets move inward, the water can become liquid, resulting in "ocean planets" or "water worlds," in which the entire surface of the planet is covered with water (*Kuchner*, 2003; *Léger et al.*, 2004). The oceans on such planets could be hundreds of kilometers deep and their atmospheres are likely thicker and warmer than on Earth because of the greenhouse effect of water vapor. Based on their relatively low bulk densities (from the mass-radius relation), the super-Earths GJ 1214b (*Charbonneau et al.*, 2009) and Kepler 22b (*Borucki et al.*, 2012) are candidate ocean planets.

Further examination of terrestrial planet models shows that there could be two types of water worlds: those that are truly water rich in their bulk composition, and those that are mostly rocky but have a significant fraction of their surface covered with water (Earth is in the latter category) (*Kaltenegger et al.*, 2013). Kepler 62e and f are prototypes of water-rich planets within the habitable zone, a category of planets that does not exist in our own solar system (*Borucki et al.*, 2013). Computing the atmospheric composition of terrestrial exoplanets is significantly more complex than that of giant exoplanets and requires consideration of many additional processes, including even plate tectonics (*Meadows and Seager*, 2011; *Fortney et al.*, 2013).

As for the terrestrial planets in our own solar system, the presence of one or more giant planets can strongly affect the amount of water on exo-(super-)Earths (Fig. 12). A wide variety of dynamical and population-synthesis models on possible outcomes have been explored (e.g., *Raymond et al.*, 2006; *Mandell et al.*, 2007; *Bond et al.*, 2010; *Ida and Lin*, 2004, 2010; *Alibert et al.*, 2011; *Mordasini et al.*, 2009).

Water on a terrestrial planet in the habitable zone may be cycled many times between the liquid oceans and the atmosphere through evaporation and rain-out. However, only a very small fraction of water molecules are destroyed or formed over the lifetime of a planet like Earth. The total Earth hydrogen loss is estimated to be 3 kg s^{-1}. Even if all the hydrogen comes from photodissociation of water, the

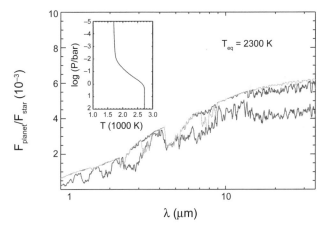

Fig. 13. Spectra of the hot Jupiters with C/O ratios varying from 0.54 (solar, bottom) to 1 (middle) and 3 (top). Note the disappearance of the H_2O features at 1–3 μm as the C/O ratio increases. Figure from *Madhusudhan et al.* (2011b).

water loss would be 27 kg s^{-1}. Given the total mass of the hydrosphere of 1.5×10^{21} kg, it would take 1.8×10^{12} yr to deplete the water reservoir. The vast majority of the water bonds present today were, therefore, formed by the chemistry that led to the bulk of water in interstellar clouds and protoplanetary disks.

10. WATER TRAIL FROM CLOUDS TO PLANETS

Here we summarize the key points and list some outstanding questions.

1. Water is formed on ice in dense molecular clouds. Some water is also formed in hot gas in shocks associated with protostars, but that water is largely lost to space.

2. Water stays mostly as ice during protostellar collapse and infall. Only a small fraction of the gas in the inner envelope is in a "hot core" where the water vapor abundance is high due to ice sublimation.

3. Water enters the disk mostly as ice at large radii and is less affected by the accretion shock than previously thought.

4. Water vapor is found in three different reservoirs in protoplanetary disks: the inner gaseous reservoir, the outer icy belt, and the hot surface layers. The latter two reservoirs have now been observed with Spitzer and Herschel and quantified.

5. Models suggest that the water that ends up in the planet and comet-forming zones of disks is partly pristine ice and partly processed ice, i.e., ice that has at least once sublimated and recondensed when the material comes close to the young star.

6. Water ice promotes grain growth to larger sizes; water-coated grains grow rapidly to planetesimal sizes in dust traps that have now been observed.

7. Planetesimals inside the disk's snow line are expected to be dry, as found in our solar system. However, gaseous water can have very high abundances in the inner astronomical unit in the gas-rich phase. Some grains may have chemically bound water to silicates at higher temperatures, but there is no evidence of hydrated silicates in meteorites nor in interstellar grains.

8. Dynamics affects what type of planetesimals are available for planet formation in a certain location. The presence and migration of giant planets can cause scattering of water-rich planetesimals from the outer disk into the inner dry zone.

9. Several comets have now been found with D/H ratios in water consistent with that of Earth's oceans. This helps to constrain models for the origin of water on the terrestrial planets, but does not yet give an unambiguous answer.

10. Both dry and wet formation scenarios for Earth are still open, although most arguments favor accretion of water-rich planetesimals from the asteroid belt during the late stages of terrestrial planet formation. Delivery of water through bombardment to planetary surfaces in a "late veneer" can contribute as well but may not have been dominant for Earth, in contrast with Mars.

11. Jupiter may be poor in oxygen and water. If confirmed by the Juno mission to Jupiter, this may indicate a changing C/O ratio with disk radius. Evidence for this scenario comes from the detection of at least one carbon-rich exoplanet.

12. All the processes and key parameters identified here should also hold true for exoplanetary systems.

There are a number of open questions that remain. A critical phase is the feeding of material from the collapsing core onto the disk, and the evolution of the young disk during the bulk of the phase of star formation. At present this phase has few observational constraints. In addition, numerous models posit that viscous evolution and the relative movement of the dust to the gas can have an impact on the overall water vapor evolution; this needs an observational basis. Astronomical observations have been confined to water vapor and ice emission from the disk surface but with the mid-plane hidden from view. What does this surface reservoir tells us about forming planets, both terrestrial and giant? Is there a way to detect the mid-plane, perhaps using HCO$^+$ (which is destroyed by water) as a probe of the snowline? However, this needs a source of ionization in the densest parts of the disk that may or may not be present (*Cleeves et al.*, 2013).

The exploration of deuterium enrichments continues to hold promise in the solar system, with need for more information on the D/H values for water in the outer asteroids and (main-belt) comets. It is also clear that we know less about the bulk elemental abundance of the solar system's largest reservoir of planetary material (i.e., Jupiter and Saturn) than one might have assumed. This must have implications for studies of extrasolar systems. In terms of the origin of Earth's oceans, matching the full range of geochemical constraints remains difficult.

Despite the uncertainties, there is a better understanding of the water trail from the clouds to planets. In this light, it is fascinating to consider that most of the water molecules in Earth's oceans and in our bodies may have been formed 4.6 b.y. ago in the cloud out of which our solar system formed.

Acknowledgments. The authors thank many colleagues for collaborations and input, and various funding agencies for support, including NASA/JPL/Caltech. Figures by M. Persson, R. Visser, L. Kristensen, and D. Fedele are much appreciated.

REFERENCES

Ádámkovics M. et al. (2013) *Astrophys. J., 786*, 135.
Adriani A. et al. (2008) *Astrobiology, 8*, 613.
A'Hearn M. F. (2011) *Annu. Rev. Astron. Astrophys., 49*, 281.
Aikawa Y. and Herbst E. (1999) *Astron. Astrophys., 351*, 233.
Aikawa Y. et al. (2012) *Astrophys. J., 760*, 40.
Akimkin V. et al. (2013) *Astrophys. J., 766*, 8.
Albertsson T. et al. (2014) *Astrophys. J., 784*, 39.
Alibert Y. et al. (2011) *Astron. Astrophys., 526*, A63.
Andersson S. and van Dishoeck E. F. (2008) *Astron. Astrophys., 491*, 907.
Arasa C. et al. (2010) *J. Chem. Phys., 132(18)*, 184510.
Atkinson R. et al. (2004) *Atmos. Chem. Phys., 4*, 1461.

Batalha N. M. et al. (2013) *Astrophys. J. Suppl., 204,* 24.

Bergin E. A. and van Dishoeck E. F. (2012) *Philos. Trans. R. Soc. London Ser. A, 370,* 2778.

Bergin E. A. et al. (2000) *Astrophys. J. Lett., 539,* L129.

Bergin E. A. et al. (2003) *Astrophys. J., 582,* 830.

Bergin E. A. et al. (2010) *Astron. Astrophys., 521,* L33.

Bethell T. J. and Bergin E. (2009) *Science, 326,* 1675.

Bethell T. J. and Bergin E. A. (2011) *Astrophys. J., 739,* 78.

Birkby J. L. et al. (2013) *Mon. Not. R. Astron. Soc., 436,* L35.

Bisschop S. E. et al. (2007) *Astron. Astrophys., 465,* 913.

Bockelée-Morvan D. et al. (2012) *Astron. Astrophys., 544,* L15.

Bond J. C. et al. (2010) *Icarus, 205,* 321.

Boogert A. C. A. et al. (2008) *Astrophys. J., 678,* 985.

Boogert A. C. A. et al. (2011) *Astrophys. J., 729,* 92.

Boonman A. M. S. and van Dishoeck E. F. (2003) *Astron. Astrophys., 403,* 1003.

Boonman A. M. S. et al. (2003) *Astron. Astrophys., 406,* 937.

Borucki W. J. et al. (2012) *Astrophys. J., 745,* 120.

Borucki W. J. et al. (2013) *Science, 340,* 587.

Brown J. C. et al. (2011) *Astron. Astrophys., 535,* A71.

Bruderer S. et al. (2012) *Astron. Astrophys., 541,* A91.

Buhr H. et al. (2010) *Phys. Rev. Lett., 105,* 10.

Burke J. R. and Hollenbach D. J. (1983) *Astrophys. J., 265,* 223.

Bus S. J. and Binzel R. P. (2002) *Icarus, 158,* 146.

Calvet N. et al. (1992) *Rev. Mex. Astron. Astrofis., 24,* 27.

Campins H. et al. (2010) *Nature, 464,* 1320.

Carr J. S. and Najita J. R. (2008) *Science, 319,* 1504.

Caselli P. et al. (2012) *Astrophys. J. Lett., 759,* L37.

Cavalié T. et al. (2013) *Astron. Astrophys., 553,* A21.

Cernicharo J. and Crovisier J. (2005) *Space Sci. Rev., 119,* 29.

Charbonneau D. et al. (2009) *Nature, 462,* 891.

Cheung A. C. et al. (1969) *Nature, 221,* 626.

Chiang E. I. and Goldreich P. (1997) *Astrophys. J., 490,* 368.

Chiang E. I. et al. (2001) *Astrophys. J., 547,* 1077.

Chiang H.-F. et al. (2008) *Astrophys. J., 680,* 474.

Ciesla F. J. and Cuzzi J. N. (2006) *Icarus, 181,* 178.

Cleeves L. I. et al. (2013) *Astrophys. J., 777,* 28.

Coutens A. et al. (2012) *Astron. Astrophys., 539,* A132.

Coutens A. et al. (2013) *Astron. Astrophys., 560,* A39.

Crovisier J. et al. (2002) *Astron. Astrophys., 393,* 1053.

Cruikshank D. P. et al. (2000) *Icarus, 147,* 309.

Cuppen H. M. et al. (2010) *Phys. Chem. Chem. Phys., 12(38),* 12077.

Daniel F. et al. (2011) *Astron. Astrophys., 536,* A76.

Dartois E. et al. (2003) *Astron. Astrophys., 399,* 1009.

Davis S. S. (2007) *Astrophys. J., 660,* 1580.

Debes J. H. et al. (2013) *Astrophys. J., 771,* 45.

de Leeuw N. H. et al. (2010) *Chem. Commun., 46,* 8923.

Deming D. et al. (2013) *Astrophys. J., 774,* 95.

de Val-Borro M. et al. (2012) *Astron. Astrophys., 546,* L4.

Dodson-Robinson S. E. et al. (2009) *Icarus, 200,* 672.

Dominik C. et al. (2005) *Astrophys. J. Lett., 635,* L85.

Donahue T. M. and Hodges Jr. R. R. (1992) *J. Geophys. Res., 97,* 6083.

Draine B. T. (1978) *Astrophys. J. Suppl., 36,* 595.

Drake M. J. (2005) *Meteoritics & Planet. Sci., 40,* 519.

Drossart P. et al. (1993) *Planet. Space Sci., 41,* 551.

Dulieu F. et al. (2013) *Nature Rept., 3,* 1338.

Elitzur M. and Watson W. D. (1978) *Astrophys. J. Lett., 222,* L141.

Emprechtinger M. et al. (2013) *Astrophys. J., 765,* 61.

Encrenaz T. (2008) *Annu. Rev. Astron. Astrophys., 46,* 57.

Fedele D. et al. (2011) *Astrophys. J., 732,* 106.

Fedele D. et al. (2012) *Astron. Astrophys., 544,* L9.

Fedele D. et al. (2013) *Astron. Astrophys., 559,* A77.

Feuchtgruber H. et al. (1997) *Nature, 389,* 159.

Feuchtgruber H. et al. (2013) *Astron. Astrophys., 551,* A126.

Flagey N. et al. (2013) *Astrophys. J., 762,* 11.

Fogel J. K. J. et al. (2011) *Astrophys. J., 726,* 29.

Fornasier S. et al. (2013) *Astron. Astrophys., 555,* A15.

Fortney J. J. et al. (2013) *Astrophys. J., 775,* 80.

Fraser H. J. et al. (2001) *Mon. Not. R. Astron. Soc., 327,* 1165.

Furuya K. et al. (2013) *Astrophys. J., 779,* 11.

Furuya R. S. et al. (2003) *Astrophys. J. Suppl., 144,* 71.

Genda H. and Ikoma M. (2008) *Icarus, 194,* 42.

Gibb E. L. et al. (2004) *Astrophys. J. Suppl., 151,* 35.

Gillett F. C. and Forrest W. J. (1973) *Astrophys. J., 179,* 483.

Glassgold A. E. et al. (2009) *Astrophys. J., 701,* 142.

Gomes R. et al. (2005) *Nature, 435,* 466.

Gorti U. et al. (2011) *Astrophys. J., 735,* 90.

Greenwood J. P. et al. (2011) *Nature Geosci., 4,* 79.

Gurwell M. A. et al. (2007) *Icarus, 188,* 288.

Harsono D. et al. (2013) *Astron. Astrophys., 555,* A45.

Hartogh P. et al. (2011) *Nature, 478,* 218.

Hayashi C. (1981) *Progr. Theor. Phys. Suppl., 70,* 35.

Heinzeller D. et al. (2011) *Astrophys. J., 731,* 115.

Herbst E. and Klemperer W. (1973) *Astrophys. J., 185,* 505.

Herbst E. and van Dishoeck E. F. (2009) *Annu. Rev. Astron. Astrophys., 47,* 427.

Herpin F. et al. (2012) *Astron. Astrophys., 542,* A76.

Hincelin U. et al. (2013) *Astrophys. J., 775,* 44.

Hogerheijde M. R. et al. (2011) *Science, 334,* 338.

Hollenbach D. et al. (2009) *Astrophys. J., 690,* 1497.

Hollenbach D. et al. (2013) *Astrophys. J., 773,* 70.

Honda M. et al. (2009) *Astrophys. J. Lett., 690,* L110.

Horner J. et al. (2003) *Mon. Not. R. Astron. Soc., 343,* 1057.

Hsieh H. H. and Jewitt D. (2006) *Science, 312,* 561.

Hsieh H. H. et al. (2004) *Astron. J., 127,* 2997.

Hudgins D. M. et al. (1993) *Astrophys. J. Suppl., 86,* 713.

Hutson M. and Ruzicka A. (2000) *Meteoritics & Planet. Sci., 35,* 601.

Ida S. and Lin D. N. C. (2004) *Astrophys. J., 616,* 567.

Ida S. and Lin D. N. C. (2010) *Astrophys. J., 719,* 810.

Ikoma M. and Genda H. (2006) *Astrophys. J., 648,* 696.

Ilee J. D. et al. (2011) *Mon. Not. R. Astron. Soc., 417,* 2950.

Indriolo N. and McCall B. J. (2012) *Astrophys. J., 745,* 91.

Jacquet E. and Robert F. (2013) *Icarus, 223,* 722.

Janssen M. A. et al. (2005) *Icarus, 173,* 447.

Jenniskens P. and Blake D. (1994) *Science, 265(5173),* 753.

Jewitt D. and Guilbert-Lepoutre A. (2012) *Astron. J., 143,* 21.

Johnson T. V. and Estrada P. R. (2009) In *Saturn from Cassini-Huygens* (M. K. Dougherty et al., eds.), p. 55. Springer-Verlag, Berlin.

Jones T. D. et al. (1990) *Icarus, 88,* 172.

Jonkheid B. et al. (2004) *Astron. Astrophys., 428,* 511.

Jørgensen J. K. and van Dishoeck E. F. (2010a) *Astrophys. J. Lett., 725,* L172.

Jørgensen J. K. and van Dishoeck E. F. (2010b) *Astrophys. J. Lett., 710,* L72.

Jørgensen J. K. et al. (2005) *Astrophys. J., 632,* 973.

Juhász A. et al. (2010) *Astrophys. J., 721,* 431.

Kaltenegger L. et al. (2013) *Astrophys. J. Lett., 775,* L47.

Kamp I. and Dullemond C. P. (2004) *Astrophys. J., 615,* 991.

Kamp I. et al. (2013) *Astron. Astrophys., 559,* A24.

Kaufman M. J. and Neufeld D. A. (1996) *Astrophys. J., 456,* 611.

Kennedy G. M. and Kenyon S. J. (2008) *Astrophys. J., 673,* 502.

Koike C. et al. (1982) *Astrophys. Space Sci., 88,* 89.

Kristensen L. E. et al. (2012) *Astron. Astrophys., 542,* A8.

Krot A. N. et al. (2000) In *Protostars and Planets IV* (V. Mannings et al., eds.), p. 1019. Univ. of Arizona, Tucson.

Kuchner M. J. (2003) *Astrophys. J. Lett., 596,* L105.

Kuchner M. J. and Seager S. (2005) *ArXiv e-prints,* arXiv:0504214.

Küppers M. et al. (2005) *Nature, 437,* 987.

Küppers M. et al. (2014) *Nature, 505,* 525.

Lamberts T. et al. (2013) *Phys. Chem. Chem. Phys., 15,* 8287.

Lecacheux A. et al. (2003) *Astron. Astrophys., 402,* L55.

Lecar M. et al. (2006) *Astrophys. J., 640,* 1115.

Lécluse C. and Robert F. (1994) *Geochim. Cosmochim. Acta, 58,* 2927.

Lécuyer C. et al. (1998) *Chem. Geol., 145,* 249.

Lefloch B. et al. (2010) *Astron. Astrophys., 518,* L113.

Léger A. et al. (2004) *Icarus, 169,* 499.

Levison H. F. et al. (2010) *Science, 329,* 187.

Licandro J. et al. (2011) *Astron. Astrophys., 532,* A65.

Lis D. C. et al. (2011) *Astrophys. J. Lett., 738,* L6.

Liu F.-C. et al. (2011) *Astron. Astrophys., 527,* A19.

Lodders K. (2004) *Astrophys. J., 611,* 587.

Lunine J. I. and Hunten D. M. (1987) *Icarus, 69,* 566.

Lunine J. I. and Stevenson D. J. (1985) *Astrophys. J. Suppl., 58,* 493.

Lunine J. I. (1991) *Icarus, 94,* 333.

Lunine J. I. et al. (2003) *Icarus, 165,* 1.

Madhusudhan N. and Seager S. (2010) *Astrophys. J., 725,* 261.

Madhusudhan N. et al. (2011a) *Nature, 469,* 64.

Madhusudhan N. et al. (2011b) *Astrophys. J., 743,* 191.

Malfait K. et al. (1998) *Astron. Astrophys., 332,* L25.

Malfait K. et al. (1999) *Astron. Astrophys., 345,* 181.

Mandell A. M. et al. (2007) *Astrophys. J., 660,* 823.

Mandell A. M. et al. (2012) *Astrophys. J., 747,* 92.

Mandell A. M. et al. (2013) *Astrophys. J., 779*, 128.
Marty B. (2012) *Earth Planet. Sci. Lett., 313*, 56.
McClure M. K. et al. (2012) *Astrophys. J. Lett., 759*, L10.
Meadows V. and Seager S. (2011) In *Exoplanets* (S. Seager, ed.), p. 441. Univ. of Arizona, Tucson.
Meeus G. et al. (2012) *Astron. Astrophys., 544*, A78.
Meijerink R. et al. (2009) *Astrophys. J., 704*, 1471.
Melnick G. J. (2009) In *Submillimeter Astrophysics and Technology* (D. C. Lis et al., eds.), p. 59. ASP Conf. Ser. 417, Astronomical Society of the Pacific, San Francisco.
Meyer D. M. et al. (1998) *Astrophys. J., 493*, 222.
Moore M. H. and Hudson R. L. (1994) *Astron. Astrophys. Suppl., 103*, 45.
Morbidelli A. et al. (2005) *Nature, 435*, 462.
Morbidelli A. et al. (2012) *Annu. Rev. Earth Planet. Sci., 40*, 251.
Mordasini C. et al. (2009) *Astron. Astrophys., 501*, 1139.
Mottram J. C. et al. (2013) *Astron. Astrophys., 558*, A126.
Mousis O. et al. (2009) *Astrophys. J., 696*, 1348.
Mousis O. et al. (2010) *Faraday Discussions, 147*, 509.
Mousis O. et al. (2012) *Astrophys. J. Lett., 751*, L7.
Mumma M. J. and Charnley S. B. (2011) *Annu. Rev. Astron. Astrophys., 49*, 471.
Mumma M. J. et al. (1986) *Science, 232*, 1523.
Murakawa K. et al. (2000) *Astrophys. J. Suppl., 128*, 603.
Najita J. R. et al. (2001) *Astrophys. J., 561*, 880.
Najita J. R. et al. (2011) *Astrophys. J., 743*, 147.
Neufeld D. A. and Hollenbach D. J. (1994) *Astrophys. J., 428*, 170.
Neufeld D. A. et al. (2000) *Astrophys. J. Lett., 539*, L151.
Neufeld D. A. et al. (2013) *Astrophys. J. Lett., 767*, L3.
Nisini B. et al. (2010) *Astron. Astrophys., 518*, L120.
Nomura H. et al. (2007) *Astrophys. J., 661*, 334.
Oba Y. et al. (2012) *Astrophys. J., 749*, 67.
Öberg K. I. et al. (2009) *Astrophys. J., 693*, 1209.
Öberg K. I. et al. (2011) *Astrophys. J., 740*, 109.
Olofsson J. et al. (2009) *Astron. Astrophys., 507*, 327.
Olofsson J. et al. (2012) *Astron. Astrophys., 542*, A90.
O'Rourke L. et al. (2013) *Astrophys. J. Lett., 774*, L13.
Owen T. and Bar-Nun A. (1993) *Nature, 361*, 693.
Owen T. and Encrenaz T. (2006) *Planet. Space Sci., 54*, 1188.
Pagani L. et al. (1992) *Astron. Astrophys., 258*, 479.
Parise B. et al. (2003) *Astron. Astrophys., 410*, 897.
Persson M. V. et al. (2012) *Astron. Astrophys., 541*, A39.
Persson M. V. et al. (2013) *Astron. Astrophys., 549*, L3.
Persson M. V. et al. (2014) *Astron. Astrophys., 563*, A74.
Podio L. et al. (2013) *Astrophys. J. Lett., 766*, L5.
Podolak M. and Zucker S. (2004) *Meteoritics & Planet. Sci., 39*, 1859.
Pontoppidan K. M. et al. (2004) *Astron. Astrophys., 426*, 925.
Pontoppidan K. M. et al. (2005) *Astrophys. J., 622*, 463.
Pontoppidan K. M. et al. (2010a) *Astrophys. J., 720*, 887.
Pontoppidan K. M. et al. (2010b) *Astrophys. J. Lett., 722*, L173.
Prasad S. S. and Tarafdar S. P. (1983) *Astrophys. J., 267*, 603.
Prinn R. G. and Fegley Jr. B. (1981) *Astrophys. J., 249*, 308.
Prodanović T. et al. (2010) *Mon. Not. R. Astron. Soc., 406*, 1108.
Przybilla N. et al. (2008) *Astrophys. J. Lett., 688*, L103.
Qi C. et al. (2013) *Science, 341*, 630.
Raymond S. N. et al. (2006) *Science, 313*, 1413.
Richet P. et al. (1977) *Annu. Rev. Earth Planet. Sci., 5*, 65.
Rimmer P. B. et al. (2012) *Astron. Astrophys., 537*, A7.
Rivière-Marichalar P. et al. (2012) *Astron. Astrophys., 538*, L3.
Rivkin A. S. and Emery J. P. (2010) *Nature, 464*, 1322.
Robert F. (2011) *Nature Geosci., 4*, 74.
Robert F. et al. (2000) *Space Sci. Rev., 92*, 201.
Roberts H. and Herbst E. (2002) *Astron. Astrophys., 395*, 233.

Roberts H. et al. (2003) *Astrophys. J. Lett., 591*, L41.
Roos-Serote M. et al. (2004) *Planet. Space Sci., 52*, 397.
Saal A. E. et al. (2013) *Science, 340(6138)*, 1317.
Salyk C. et al. (2008) *Astrophys. J. Lett., 676*, L49.
Salyk C. et al. (2011) *Astrophys. J., 731*, 130.
Sasselov D. D. and Lecar M. (2000) *Astrophys. J., 528*, 995.
Schindhelm E. et al. (2012) *Astrophys. J. Lett., 756*, L23.
Schorghofer N. (2008) *Astrophys. J., 682*, 697.
Seager S. and Deming D. (2010) *Annu. Rev. Astron. Astrophys., 48*, 631.
Semenov D. and Wiebe D. (2011) *Astrophys. J. Suppl., 196*, 25.
Shimonishi T. et al. (2010) *Astron. Astrophys., 514*, A12.
Showman A. P. and Dowling T. E. (2000) *Science, 289*, 1737.
Stevenson D. J. and Lunine J. I. (1988) *Icarus, 75*, 146.
Stimpfl M. et al. (2006) *J. Cryst. Growth, 294*, 1, 83.
Sultanov R. and Balakrishnan N. (2004) *J. Chem. Phys., 121(22)*, 11038.
Tafalla M. et al. (2013) *Astron. Astrophys., 551*, A116.
Taquet V. et al. (2013) *Astrophys. J. Lett., 768*, L29.
Terada H. et al. (2007) *Astrophys. J., 667*, 303.
Thi W. et al. (2010a) *Astron. Astrophys., 518*, L125.
Thi W.-F. et al. (2010b) *Mon. Not. R. Astron. Soc., 407*, 232.
Tielens A. G. G. M. (1983) *Astron. Astrophys., 119*, 177.
Tielens A. G. G. M. (2013) *Rev. Mod. Phys., 85*, 1021.
Tielens A. G. G. M. and Hagen W. (1982) *Astron. Astrophys., 114*, 245.
Tinetti G. et al. (2012) *Philos. Trans. R. Soc. London Ser. A, 370(1968)*, 2749.
Tsiganis K. et al. (2005) *Nature, 435*, 459.
van der Marel N. et al. (2013) *Science, 340*, 1199.
van Dishoeck E. F. et al. (2006) *Faraday Discussions, 133*, 231.
van Dishoeck E. F. et al. (2011) *Publ. Astron. Soc. Pac., 123*, 138.
van Dishoeck E. et al. (2013) *Chem. Rev., 113*, 9043.
Vasyunin A. I. et al. (2011) *Astrophys. J., 727*, 76.
van Zadelhoff G.-J. et al. (2003) *Astron. Astrophys., 397*, 789.
Visser R. et al. (2009) *Astron. Astrophys., 495*, 881.
Visser R. et al. (2011) *Astron. Astrophys., 534*, A132.
Visser R. et al. (2013) *Astrophys. J., 769*, 19.
Wallace L. et al. (1995) *Science, 268*, 1155.
Walsh C. et al. (2012) *Astrophys. J., 747*, 114.
Walsh K. J. et al. (2011) *Nature, 475*, 206.
Watson D. M. et al. (2007) *Nature, 448*, 1026.
Watson D. M. et al. (2009) *Astrophys. J. Suppl., 180*, 84.
Weidenschilling S. J. and Cuzzi J. N. (1993) *In Protostars and Planets III* (E. H. Levy and J. I. Lunine, eds.) p. 1031. Univ. of Arizona, Tucson.
Weiss A. et al. (2013) *Astrophys. J., 767*, 88.
Westley M. S. et al. (1995) *Nature, 373*, 405.
Whipple F. L. (1950) *Astrophys. J., 111*, 375.
Whittet D. C. B. (2010) *Astrophys. J., 710*, 1009.
Whittet D. C. B. et al. (1988) *Mon. Not. R. Astron. Soc., 233*, 321.
Whittet D. C. B. et al. (1998) *Astrophys. J. Lett., 498*, L159.
Whittet D. C. B. et al. (2013) *Astrophys. J., 774*, 102.
Willacy K. and Woods P. M. (2009) *Astrophys. J., 703*, 479.
Williams R. M. E. et al. (2013) *Science, 340*, 1068.
Woitke P. et al. (2009a) *Astron. Astrophys., 501*, L5.
Woitke P. et al. (2009b) *Astron. Astrophys., 501*, 383.
Wong M. H. et al. (2008) In *Oxygen in the Solar System* (G. MacPherson et al., eds.), p. 219. Reviews of Mineralogy and Geochemistry, Vol. 68.
Yang H. et al. (2012a) *Astrophys. J., 744*, 121.
Yang L. et al. (2012b) In *Lunar Planet. Sci. XLIII*, Abstract #2023. Lunar and Planetary Institute, Houston.
Zasowski G. et al. (2009) *Astrophys. J., 694*, 459.
Zhang K. et al. (2013) *Astrophys. J., 766*, 82.

Ceccarelli C., Caselli P., Bockelée-Morvan D., Mousis O., Pizzarello S., Robert F., and Semenov D. (2014) Deuterium fractionation: The Ariadne's thread from the precollapse phase to meteorites and comets today. In *Protostars and Planets VI* (H. Beuther et al., eds.), pp. 859–882. Univ. of Arizona, Tucson, DOI: 10.2458/azu_uapress_9780816531240-ch037.

Deuterium Fractionation: The Ariadne's Thread from the Precollapse Phase to Meteorites and Comets Today

Cecilia Ceccarelli
Université J. Fourier de Grenoble

Paola Caselli
University of Leeds and Max Planck Institute for Extraterrestrial Physics

Dominique Bockelée-Morvan
Observatoire de Paris

Olivier Mousis
Université Franche-Comté and Université de Toulouse

Sandra Pizzarello
Arizona State University

Francois Robert
Muséum National d'Histoire Naturelle de Paris

Dmitry Semenov
Max Planck Institute for Astronomy, Heidelberg

The solar system formed about 4.6 billion years (b.y.) ago from condensation of matter inside a molecular cloud. Trying to reconstruct what happened is the goal of this chapter. For that, we combine our understanding of Galactic objects that will eventually form new suns and planetary systems with our knowledge on comets, meteorites, and small bodies of the solar system today. The specific tool we are using is molecular deuteration, namely the amount of deuterium with respect to hydrogen in molecules. This is the Ariadne's thread that helps us to find the way out from a labyrinth of possible histories of our solar system. The chapter reviews the observations and theories of the deuterium fractionation in prestellar cores, protostars, protoplanetary disks (PPDs), comets, interplanetary dust particles (IDPs), and meteorites and links them together to build up a coherent picture of the history of the solar system formation. We emphasize the interdisciplinary nature of this chapter, which gathers together researchers from different communities with the common goal of understanding solar system history.

1. INTRODUCTION

The ancient Greek legend reads that Theseus volunteered to enter the Minotaur's labyrinth to kill the monster and liberate Athens from having to periodically sacrifice young women and men. The task was almost impossible to achieve because killing the Minotaur was not even half the problem; getting out of the labyrinth was even more difficult. But Ariadne, the guardian of the labyrinth and daughter of the king of Crete, provided Theseus with a ball of thread, so he could unroll it when entering the labyrinth and then follow it back to find his way out — which he did.

Our labyrinth here is the history of the formation of the solar system. We are deep inside the labyrinth, with Earth and the planets formed, but we do not know exactly how this happened. There are several paths that go into different directions, but which is the one that will bring us out of this labyrinth, the path that nature followed to form Earth and the other solar system planets and bodies?

Our story reads that once upon a time, there existed an interstellar cloud of gas and dust. Then, about 4.6 billion years (b.y.) ago, one cloud fragment became the solar system. What happened to that primordial condensation? When, why, and how did it happen? Answering these questions involves putting together all the information we have on the present-day solar system bodies and microparticles. But this is not enough, and comparing that information with our understanding of the formation process of solar-type stars in our Galaxy turns out to be indispensable as well.

Our Ariadne's thread for this chapter is deuterium fractionation, namely the process that enriches the amount of deuterium with respect to hydrogen in molecules. Although deuterium atoms are only $\sim 1.6 \times 10^{-5}$ (Table 1) times as abundant as the hydrogen atoms in the universe, its relative abundance in molecules, larger than the elemental deuterium/hydrogen (D/H) abundance in very specific situations, provides a remarkable and almost unique diagnostic tool. Analyzing the deuterium fractionation in different Galactic objects that will eventually form new suns, and in comets, meteorites, and small bodies of the solar system, is like having in our hands a box of old photos with the imprint of memories from the very first steps of the solar system formation. The goal of this chapter is trying to understand the message that these photos bring, using our knowledge of the different objects and, in particular, the Ariadne's thread of the deuterium fractionation to link them together in a sequence that is the one that followed the solar system formation.

The chapter is organized as follows. In section 2, we review the mechanisms of deuterium fractionation in the different environments and establish the basis for understanding the language of the different communities involved in the study of solar system formation. We then

TABLE 1. Definitions of symbols and terms used in the chapter.

Acronym	Definition
aa	Amino acids
A_V	Visual extinction, measured in magnitudes (mag), proportional to the H nuclei column density in the line of sight
ALMA	Atacama Large Millimeter Array
CSO	Caltech Submillimeter Observatory
HSO	Herschel Space Observatory
JCMT	James Clerk Maxwell Telescope
IOM	Insoluble organic matter
ISM	Interstellar medium
NOEMA	Northern Extended Millimeter Array
PSN	Protosolar nebula
SOC	Soluble organic compounds (same as SOM)
SOM	Soluble organic matter (same as SOC)
VLT	Very Large Telescope

Astrophysical Key Objects Mentioned in the Chapter	
PSC	Prestellar Cores: dense and cold condensations; the first step of solar-type star formation
Class 0	Class 0 sources: youngest known solar type protostars
Hot corino	Warm and dense inner regions of the envelopes of class 0 protostars
PPD	Protoplanetary disks: disks of material surrounding young protostars

Origin of Solar System Bodies Mentioned in the Chapter	
JFCs	Jupiter-family comets: ecliptic short-period comets whose reservoir is the Kuiper belt; probably formed in the transneptunian region
OCC	Oort cloud comets: long-period comets whose reservoir is the Oort cloud; probably formed in the Uranus-Neptune region, with some contribution from the Jupiter-Saturn region
CCs	Carbonaceous chondrites: the most primitive meteorites, mostly coming from the main belt (2–4 AU)
IDPs	Interplanetary dust particles: at least 50% are fragments of comets, the rest fragments of main-belt asteroids

| *Measure of Molecular Deuteration* | |

1. In geochemical and biochemical studies, the molecular deuteration is measured by the deuterium enrichment in a sample compared to that of a chosen standard value, in ‰ with respect to the standard ratio.

For deuterium: $\delta D = \dfrac{D/H_{sample} - D/H_{VSMOW}}{D/H_{VSMOW}} \times 10^3.$

2. In astrochemical studies, the molecular deuteration is measured by the heavy/light isotope abundance.

3. In the PSN models, the molecular deuteration is given by the enrichment factor f, defined as the ratio between the molecule D/H and the PSN D/H.

Acronym	Definition	D/H	δD	f	Reference
A_D	Cosmic elemental deuterium abundance	1.6×10^{-5}	–900	0.8	[1]
PSN D/H	Deuterium abundance in the PSN	2.1×10^{-5}	–860	1.0	[2]
VSMOW	Vienna Standard Mean Ocean Water (refers to evaporated ocean waters)	1.5×10^{-4}	0	7.1	[3]

References: [1] *Linsky* (2007); [2] *Geiss and Gloeckler* (1998); [3] *Lécuyer et al.* (1998).

briefly review the major steps of the formation process in section 3. The following sections, from section 4 through section 9, will review in detail observations and theories of deuterium fractionation in the different objects: prestellar cores, protostars, protoplanetary disks (PPDs), comets, and meteorites. In section 10, we will try to follow back the thread, unrolled in the preceding sections, to understand what happened to the solar system, including the formation of the terrestrial oceans. We will conclude with section 11.

2. THE LANGUAGE OF DEUTERIUM FRACTIONATION

2.1. Chemical Processes of Deuterium Fractionation

Deuterium (D) is formed at the birth of the universe with an abundant D/H estimated to be $A_D = 1.6 \times 10^{-5}$ (Table 1) and destroyed in the interiors of the stars. Therefore, its abundance may vary from place to place; for example, it is lower in regions close to the Galactic Center, where the star formation is high, than in the solar system neighborhood (*Lubowich et al.,* 2000). If there were no deuterium fractionation, a species with one H atom, e.g., HCN, would have a relative abundance of D-atom- over H-atom-bearing molecules equal to A_D, namely DCN/HCN $= 1.6 \times 10^{-5}$. As another important example, water would have HDO/H$_2$O $= 2 \times A_D = 3.2 \times 10^{-5}$. Similarly, a species with two hydrogens will have a relative abundance of molecules with two D atoms proportional to A_D^2 (e.g., D$_2$O/H$_2$O $= 2.6 \times 10^{-10}$) and so on. In practice, if there were no deuterium fractionation, the abundance of D-bearing molecules would be ridiculously low.

But in space, certain conditions are present that provide a propitious environment for deuterium fractionation (or molecular deuteration or deuterium enrichment) to occur. This can be summarized in three basic steps, shown in Fig. 1.

1. *Formation of* H_3^+: In cold (~10 K) molecular gas, the fastest reactions are those involving ions, as neutral-neutral reactions have activation barriers and are generally slower. The first formed molecular ion is H_3^+, a product of the cosmic-ray ionization of H_2 and H.

2. *Formation of* H_2D^+, D_2H^+, *and* D_3^+: In cold molecular gas, H_3^+ reacts with HD, the major reservoir of D atoms, and once every three times the D atom is transferred from HD to H_2D^+. The inverse reaction $H_2 + H_2D^+$ that would form HD has a (small) activation barrier so that at low temperatures H_2D^+/H_3^+ becomes larger than A_D. Similarly, D_2H^+ and D_3^+ are formed by reactions with HD.

3. *Formation of other D-bearing molecules:* H_2D^+, D_2H^+, and D_3^+ react with other molecules and atoms, transferring D atoms to all the other species. This can happen directly in the gas phase (step 3a in Fig. 1) or on the grain mantles (step 3b) via the D atoms created by the H_2D^+, D_2H^+, and D_3^+ dissociative recombination with electrons. In both cases, the deuterium fractionation depends on the H_2D^+/H_3^+, D_2H^+/H_3^+, and D_3^+/H_3^+ abundance ratios.

Therefore, generally speaking, the basic molecule for deuterium fractionation is H_2D^+ (and D_2H^+ and D_3^+ in ex-

treme conditions). The cause for the enhancement of H_2D^+ with respect to H_3^+ and, consequently, deuterium fractionation is the larger mass (equivalent to a higher zero energy) of H_2D^+ with respect to H_3^+, which causes the activation barrier in step 2. The quantity that governs whether the barrier can be overcome and, consequently, the deuterium fractionation is the temperature: The lower the temperature, the larger the deuterium fractionation.

Besides, if abundant neutrals and important destruction partners of H_3^+ isotopologs, such as O and CO, deplete from the gas phase (e.g., because of the freeze-out onto dust grains in cold and dense regions; see section 4 and section 6), the deuterium fraction further increases (*Dalgarno and Lepp,* 1984; *Roberts et al.,* 2003). This is due to the fact that the destruction rates of all the H_3^+ isotopologs drop, while the formation rate of the deuterated species increases because of the enhanced H_3^+ abundance.

There is another factor that strongly affects the deuterium fractionation: the ortho-to-para abundance ratio of H_2 molecules. In fact, if this ratio is larger than ~10^{-3}, the internal energy of the ortho-H$_2$ molecules (whose lowest energy level is ~175 K) can be enough to overcome the $H_2D^+ + H_2 \rightarrow HD + H_3^+$ barrier and limit the H_2D^+/H_3^+ ratio (*Gerlich and Schlemmer,* 2002; *Flower et al.,* 2006; *Rist et al.,* 2013). In general, it is believed that ortho- and para-H$_2$ are formed on the surface of dust grains with a statistical ratio of 3:1. Proton-exchange reactions then convert ortho- into para-H$_2$, especially at the low temperatures of dense cloud cores, where the ortho-to-para H$_2$ ratio is predicted to drop below 10^{-3} (*Sipilä et al.,* 2013).

So far we have discussed the deuterium fractionation routes in cold (T ≲ 40–50 K) gas. Different routes occur in

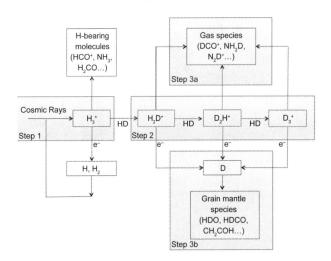

Fig. 1. In cold (≤30 K) gas, deuterium fractionation occurs through three basic steps: (1) formation of H_3^+ ions from the interaction of cosmic rays with H and H_2; (2) formation of H_2D^+ (HD$_2^+$ and D_3^+) from the reaction of H_3^+ (H_2D^+ and HD$_2^+$) with HD; and (3) formation of other D-bearing molecules from reactions with H_2D^+ (HD$_2^+$ and D_3^+) in the gas phase (Step 3a) and on the grain mantles (Step 3b).

warm (30 ≤ T ≤ 100 K) and hot (100 ≤ T ≤ 1000 K) gas. In warm gas (T ≤ 70–80 K) the D atoms can be transferred to molecules by CH_2D^+, whose activation barrier of the reaction with H_2 is larger than that of H_2D^+ (*Roberts and Millar,* 2000). At even higher temperatures, OD transfers D atoms from HD to water molecules (*Thi et al.,* 2010b). In these last two cases, CH_2D^+ and OD play the role of H_2D^+ at lower temperatures. At ≥500 K water can directly exchange D and H atoms with H_2 (*Geiss and Reeves,* 1981; *Lécluse and Robert,* 1994).

Finally, some molecules, notably water, are synthesized on the surfaces of interstellar and interplanetary grains by addition and/or substitution of H and D atoms (section 2.2). In this case, the deuterium fractionation depends on the D/H ratio of the atomic gas. As discussed previously in this section, the enhanced abundance of H_2D^+ (D_2H^+ and D_3^+) also implies an increased atomic D/H ratio in the gas (step 3b in Fig. 1), as D atoms are formed upon dissociative recombination of the H_3^+ deuterated isotopologs, whereas H atoms maintain an about constant density of ≥1 cm^{-3}, determined by the balance between surface formation and cosmic-ray dissociation of H_2 molecules (*Li and Goldsmith,* 2003).

2.2. Deuterium Fractionation of Water

Given the particular role of deuterated water in understanding solar system history, we summarize the three major processes (reported in the literature) that cause water deuteration. They are schematically shown in Fig. 2.

1. *Formation and deuteration on the surfaces of cold grains:* In cold molecular clouds and star-forming regions, water is mostly formed by the addition of H and D atoms to O, O_2, and O_3 on the grain surfaces, as demonstrated by several laboratory experiments (*Hiraoka et al.,* 1998; *Dulieu et al.,* 2010; *Oba et al.,* 2012). In this case, therefore, the key parameter governing water deuteration is the atomic D/H ratio in the gas, which depends on the H_2D^+/H_3^+ ratio as discussed in the previous section.

2. *Hydrogen-deuterium exchange in the gas phase:* As for any other molecule, D atoms can be transferred from H_2D^+ (D_2H^+ and D_3^+) to H_2O in cold gas (section 2.1), and more efficiently through the HD + OH$^+$ → HDO + H and HD + OH$_2^+$ → HDO + H_2 reactions. In warm gas, it is in principle possible to have direct exchange between HD and H_2O to form HDO (*Geiss and Reeves,* 1981; *Lécluse and Robert,* 1994). However, being a neutral-neutral reaction, it possesses an activation barrier (*Richet et al.,* 1977), which makes this route very slow at T ≤ 500 K. On the contrary, for temperatures high enough (≥100 K), the OH + HD and OD + H_2 reactions can form HDO. Based on modeling, *Thi et al.* (2010b) demonstrated that the HD + O → OD + H followed by the OD + H_2 → HDO + H reaction is indeed a major route for the HDO formation in warm gas.

3. *Isotopic exchange between solid H_2O and HDO with other solid species:* Laboratory experiments have shown that D and H atoms can be exchanged between water ice and other molecules trapped in the ice, e.g., CH_3OH (*Ratajczak et al.,* 2009; *Gálvez et al.,* 2011). The exchange

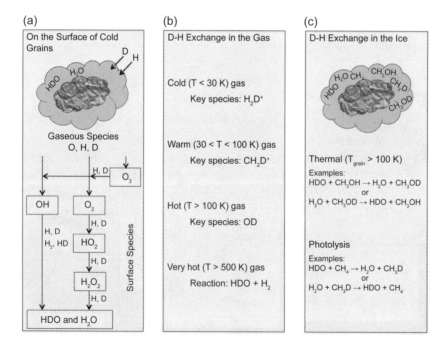

Fig. 2. The three principal routes of water formation and deuteration. **(a)** Gas species as atomic O, H, and D land on the grain surfaces and form water through OH, O_2, and O_3 hydrogenation. **(b)** In the gas, water deuteration proceeds via reactions with H_2D^+, CH_2D^+, OD, and HDO with H_2, depending on the gas temperature. **(c)** D and H atoms can also be exchanged between frozen species and water while they are on the grain surfaces.

very likely occurs during the ice sublimation phase, with the reorganization of the crystal. Similarly, H-D exchange in ice can be promoted by photolysis (*Weber et al.*, 2009). Note that this mechanism not only can alter the HDO/H$_2$O abundance ratio in the ice, but also it can pass D atoms to organic matter trapped in the ice, enriching it in deuterium.

2.3. Toward a Common Language

The ambition of this chapter is to bring together researchers from different communities. One of the disadvantages, which we aim to overcome here, is that these different communities do not always speak the same language. Table 1 is a brief dictionary designed to help readers translate the chapter into their own scientific terminology. In addition, several acronyms used throughout this chapter are also listed in the table. With this, we are ready now to start our voyage through the different objects.

3. A BRIEF HISTORY OF SOLAR SYSTEM FORMATION

According to the widely accepted scenario, there are five major phases of solar-type star formation (Fig. 3):

1. *Prestellar cores.* These are the starting point of solar-type star formation. In these "small clouds" with evidence of contraction motions, contrary to starless cores, matter slowly accumulates toward the center, causing an increase in density while the temperature remains low (≤10 K). Atoms and molecules in the gas phase freeze out onto the cold surfaces of the submicrometer dust grains, forming the so-called icy grain mantles. This is the moment when the deuterium fractionation is most effective; the frozen molecules, including water, acquire a high deuterium fraction.

2. *Protostars.* The collapse starts, gravitational energy is converted into radiation, and the envelope around the central object, the future star, warms up. When and where the temperature reaches the mantle sublimation temperature (~100–120 K), in the so-called hot corinos, the molecules in the mantles are injected into the gas phase, where they are observed via their rotational lines. Complex organic molecules, precursors of prebiotic species, are also detected at this stage.

3. *Protoplanetary disks.* The envelope dissipates with time and eventually only a circumstellar PPD remains. In the hot regions, close to the central object or the disk surface, some molecules, notably water, can be enriched in deuterium via neutral-neutral reactions. In the cold regions, in the mid-plane, where the vast majority of matter resides, the molecules formed in the protostellar phase freeze out again onto the grain mantles, where part of the ice from the prestellar phase may still be present. The deuterium fractionation process again becomes important.

4. *Planetesimal formation.* The process of "conservation and heritage" begins. The submicrometer dust grains coagulate into larger rocks, called planetesimals, the seeds of the future planets, comets, and asteroids. Some of the icy grain mantles are likely preserved while the grains bind together. At least part of the previous chemical history may be conserved in the building blocks of the forming planetary system rocky bodies and eventually passed along as a heritage to the planets. However, migration and diffusion may scramble the original distribution of the D-enriched material.

From a Diffuse Cloud to a Sun-Like + Planetary System:
The Deuterium Fractionation History

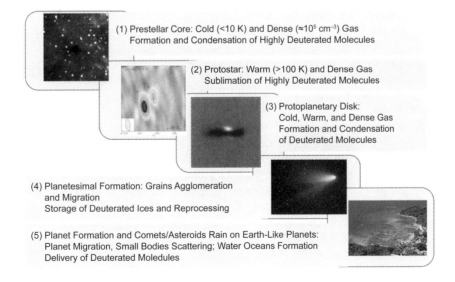

(1) Prestellar Core: Cold (<10 K) and Dense (≈10^5 cm^{-3}) Gas
Formation and Condensation of Highly Deuterated Molecules

(2) Protostar: Warm (>100 K) and Dense Gas
Sublimation of Highly Deuterated Molecules

(3) Protoplanetary Disk:
Cold, Warm, and Dense Gas
Formation and Condensation
of Deuterated Molecules

(4) Planetesimal Formation: Grains Agglomeration
and Migration
Storage of Deuterated Ices and Reprocessing

(5) Planet Formation and Comets/Asteroids Rain on Earth-Like Planets:
Planet Migration, Small Bodies Scattering; Water Oceans Formation
Delivery of Deuterated Moledules

Fig. 3. Schematic summary of the different phases that likely gave birth to the solar system, with the main deuterium processes highlighted (adapted from *Caselli and Ceccarelli, 2012*).

5. *Planet formation.* The last phase of rocky planet formation is characterized by giant impacts between planet embryos, which, in the case of the solar system, resulted in the formation of the Moon and Earth. Giant planets may migrate, inducing a scattering of the small bodies all over the PPD. Oceans are formed on the young Earth and perhaps in other rocky planets. The leftovers of the process become comets and asteroids. In the solar system, their fragments continuously rain on Earth, releasing the heritage stored in the primitive D-enriched ices. Life takes over sometime around 2 b.y. after the formation of Earth and the Moon (*Czaja*, 2010).

In the remainder of this chapter, we will discuss each of these steps, the measured deuterium fractionation, and the processes responsible.

4. PRESTELLAR CORE PHASE

4.1. Structure of Prestellar Cores

Stars form within fragments of molecular clouds, the so-called dense cores (*Benson and Myers,* 1989), produced by the combined action of gravity, magnetic fields, and turbulence (*McKee and Ostriker,* 2007). Some of the starless dense cores can be transient entities and diffuse back into the parent cloud, while others (the prestellar cores) will dynamically evolve until one or more planetary systems are formed. It is therefore important to gather kinematic information to identify prestellar cores, which represent the initial conditions in the process of star and planet formation. Their structure and physical characteristics depend on the density and temperature of the surrounding cloud, i.e., on the external pressure (see the chapter by Tan et al. in this volume). The well-studied prestellar cores in nearby molecular clouds have sizes of $\simeq$10,000 AU, similar to the Oort cloud, masses of a few solar masses, and visual extinctions $\geq$50 mag. They are centrally concentrated (*Ward-Thompson et al.,* 1999) with central densities larger than 10^5 H_2 cm^{-3} (*Bacmann et al.,* 2003; *Crapsi et al.,* 2005; *Keto and Caselli,* 2008), central temperatures close to 7 K (*Crapsi et al.,* 2007; *Pagani et al.,* 2007), and evidence of subsonic gravitational contraction (*Keto and Caselli,* 2010), as well as gas accretion from the surrounding cloud (*Lee and Myers,* 2011). The European Space Agency (ESA) Herschel Space Observatory (HSO) has detected water vapor for the first time toward a prestellar core and unveiled contraction motions in the central $\simeq$1000 AU (*Caselli et al.,* 2012): Large amounts of water (a few Jupiter masses, mainly in ice form) are transported toward the future solar-type star and its potential planetary system.

A schematic summary of the main chemical and physical characteristics of prestellar cores is shown in Fig. 4. Figure 4a shows one of the best-studied objects: L1544 in the Taurus molecular cloud complex, 140 pc away. The largest white contour roughly indicates the size of the dense core, and the outer edge represents the transition region between L1544 and the surrounding molecular cloud, where the ex-

tinction drops below $\simeq$4 mag and photochemistry becomes important. This is where water-ice copiously forms on the surface of dust grains (*Whittet et al.,* 2007; *Hollenbach et al.,* 2009) and low levels of water deuteration are taking place (*Cazaux et al.,* 2011; *Taquet et al.,* 2012a). Within the *dark-cloud zone*, where the carbon is mostly locked in CO, gas-phase chemistry is regulated by ion-molecule reactions (Fig. 4b).

Within the central $\simeq$7000 AU, the volume density becomes higher than a few $\times 10^4$ cm^{-3} (Fig. 4c) and the freeze-out timescale ($\simeq 10^9/n_H$ yr) becomes shorter than a few $\times 10^4$ yr. This is the deuteration zone, where the freeze-out of abundant neutrals such as CO and O, the main destructive partners of H_3^+ isotopologs, favor the formation of deuterated molecules (see section 2.1). Deuteration is one of the main chemical processes at work, making deuterated species the best tools for studying the physical structure and dynamics of the earliest phases of star formation.

4.2. Deuterium Fractionation

Prestellar cores chemically stand out among other starless dense cores, as they show the largest deuterium fractions [>10% (*Bacmann et al.,* 2003; *Crapsi et al.,* 2005; *Pagani et al.,* 2007)] and CO depletions [>90% of CO molecules are frozen onto dust grain surfaces (*Caselli et al.,* 1999; *Bacmann et al.,* 2002)]. The largest deuterium fractions have been observed in N_2H^+ [$N_2D^+/N_2H^+ \simeq$ 0.1–0.7 (*Crapsi et al.,* 2005; *Pagani et al.,* 2007)], ammonia [$NH_2D/NH_3 \simeq$ 0.1–0.4 (*Shah and Wootten,* 2001; *Crapsi et al.,* 2007)], and formaldehyde [$D_2CO/H_2CO \simeq$ 0.01–0.1 (*Bacmann et al.,* 2003)]. HCN, HNC, and HCO$^+$ show somewhat lower deuterations [between 0.01 and 0.1 (*Turner,* 2001; *Hirota et al.,* 2003; *Butner et al.,* 1995; *Caselli et al.,* 2002)]. *Spezzano et al.* (2013) recently detected doubly deuterated cyclopropenylidene (c-C_3D_2), finding a c-C_3D_2/c-C_3H_2 abundance ratio of about 0.02. Unfortunately, no measurement of water deuteration is available yet.

Prestellar cores are the strongest emitters of the ground state rotational transition of ortho-H_2D^+ (*Caselli et al.,* 2003) and the only objects where para-D_2H^+ has been detected (*Vastel et al.,* 2004; *Parise et al.,* 2011). It is interesting to note that the strength of the ortho-H_2D^+ line does not correlate with the amount of deuterium fraction found in other molecules. This is probably due to variations of the H_2D^+ ortho-to-para ratio in different environments (*Caselli et al.,* 2008), an important clue in the investigation of how external conditions affect the chemical and physical properties of prestellar cores (see also *Friesen et al.,* 2013).

4.3. Origin of Deuterium Fractionation

Prestellar cores are "deuterium fractionation factories." The reason for this is twofold: First, they are very cold (with typical gas and dust temperatures between 7 and 13 K). This implies a one-way direction of the reaction $H_3^+ + HD \rightarrow H_2D^+ + H_2$, the starting point of the whole

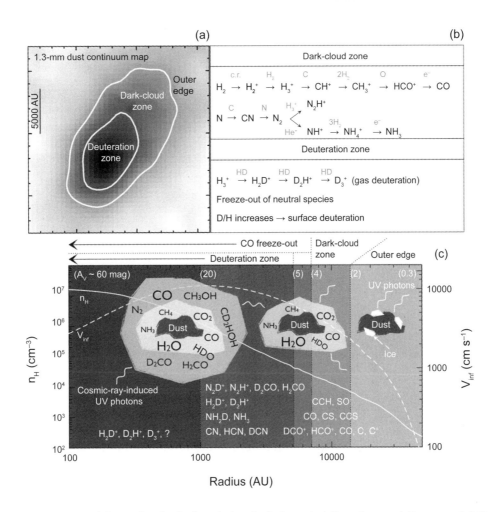

Fig. 4. Schematic summary of the main physical and chemical characteristics of a prestellar core. **(a)** Grayscale is the 1.2-mm dust continuum emission of L1544, the prototypical prestellar core (data from *Ward-Thompson et al.,* 1999). White contours mark the transition from the outer edge to the dense core and the deuteration zone, where CO is highly frozen onto dust grains. **(b)** Main chemical processes in the dark-cloud zone and the deuteration zone. Gray symbols above the arrows are the reaction partners, and the arrow indicates the reaction direction. **(c)** Radial slice of L1544, indicating the number density (n_H), infall velocity (V_{inf}) profiles, visual extinction (A_V), main molecular diagnostic tools in the various zones, and a rough summary of dust grain evolution within the prestellar core (see also *Caselli and Ceccarelli,* 2012).

process of molecular deuteration (see Fig. 1, steps 2 and 3a). Second, a large fraction of neutral heavy species such as CO and O freeze out onto dust grains. As mentioned in section 2.1, the disappearance of neutrals from the gas phase implies less destruction events for H_3^+ and its deuterated forms, with a consequent increase in not just H_2D^+ but also D_2H^+ and D_3^+ (*Dalgarno and Lepp,* 1984; *Roberts et al.,* 2003; *Walmsley et al.,* 2004). This simple combination of low temperatures and the tendency for molecules to stick on icy mantles on top of dust grains can easily explain the observed deuterium fraction measured in prestellar cores (see also *Roueff et al.,* 2005).

The case of formaldehyde (H_2CO) deuteration requires an extra note, as not all the data can be explained by gas-phase models including freeze-out. As discussed in *Bacmann et al.* (2003), another source of deuteration is needed. A promising mechanism is the chemical process-

ing of icy mantles (surface chemistry), coupled with partial desorption of surface molecules upon formation (*Garrod et al.,* 2006). In particular, once CO freezes out onto the surface of dust grains, it can either be stored in the ice mantles, or be "attacked" by reactive elements, in particular atomic hydrogen. In the latter case, CO is first transformed into HCO, then formaldehyde, and eventually methanol (CH_3OH). In prestellar cores, D atoms are also abundant because of the dissociative recombination of the abundant H_2D^+, D_2H^+, and possibly D_3^+ (see Fig. 1, step 3b). Chemical models predict D/H between 0.3 and 0.9 in the inner zones of prestellar cores with large CO freeze-out (*Roberts et al.,* 2003; *Taquet et al.,* 2012a), implying a large deuteration of formaldehyde and methanol on the surface of dust grains (see dust grain cartoons overlaid on Fig. 4c). Thus the measured large deuteration of gas-phase formaldehyde in prestellar cores (and possibly methanol, although this still awaits observational evidence) can be better understood with the contribution of surface chemistry, as a fraction of surface molecules can desorb upon formation (thanks to their formation energy).

5. PROTOSTAR PHASE

5.1. Structure of Class 0 Protostars

Class 0 sources are the youngest protostars. Their luminosity is powered by the gravitational energy, namely the material falling toward the central object, accreting at a rate of $\leq 10^{-5}$ M$_\odot$ yr^{-1}. They last for a short period, $\sim 10^5$ yr (*Evans et al., 2009*) (see also the chapter by Dunham et al. in this volume). The central object, the future star, is totally obscured by the collapsing envelope, whose sizes as the prestellar cores (section 4) are $\sim 10^4$ AU. It is not clear whether a disk exists at this stage (*Maury et al., 2010*), as the original magnetic field frozen on the infalling matter tends to inhibit its formation (see the chapter by Z.-Y. Li et al. in this volume). On the contrary, powerful outflows of supersonic matter are one of the notable characteristics of these objects (see the chapter by Frank et al. in this volume).

In class 0 protostars, the density of the envelope increases with decreasing distance from the center ($n \propto r^{-3/2}$), as well as temperature (*Ceccarelli et al., 1996; Jørgensen et al., 2002*). From a chemical point of view, the envelope is approximatively divided in two regions, delimited by the ice sublimation temperature (100–120 K): a cold outer envelope, where molecules are more or less frozen onto the grain mantles, and an inner envelope, called hot the corino, where the mantles, built up during the prestellar core phase (section 4), sublimate (*Ceccarelli et al., 2000*) [see *Caselli and Ceccarelli (2012)* for a more accurate description of the class 0 protostars]. This transition occurs at distances from the center between 10 and 100 AU, depending on the source luminosity (*Maret et al., 2004, 2005*). Relevant to this chapter, in the hot corinos, the species formed at the prestellar core epoch are injected into the gas phase, carrying the memory of their origin. For different reasons, the outer envelope and the hot corino have molecules highly enriched in deuterium: in the first case because of the low temperatures and CO depletion (section 2.1 and section 4), and in the second case because of the inheritance of the prestellar ices.

5.2. Deuterium Fractionation

Class 0 protostars are the objects where the highest deuterium fractionation has been detected so far and the first where the extreme deuteration, called "superdeuteration" in the literature, has been discovered (*Ceccarelli et al., 1998, 2007*): doubly and even triply deuterated forms with D/H enhancements with respect to the elemental D/H abundance (Table 1) of up to 13 orders of magnitude.

The first and vast majority of measurements were obtained with single-dish observations, so that they cannot

disentangle the outer envelope and the hot corino, if not indirectly, by modeling the line emission in some cases (*Coutens et al., 2012, 2013a*). The following species with more than two atoms of deuterium have been detected [see *Ceccarelli et al. (2007)* for a list of singly-deuterated species]: formaldehyde [D$_2$CO (*Ceccarelli et al., 1998; Parise et al., 2006*)], methanol [CHD$_2$OH and CD$_3$OH (*Parise et al., 2002, 2004, 2006*)], ammonia [NHD$_2$ and ND$_3$ (*Loinard et al., 2001; van der Tak et al., 2002; Lis et al., 2002*)], hydrogen sulfide [D$_2$S (*Vastel et al., 2004*)], thioformaldehyde [D$_2$CS (*Marcelino et al., 2005*)] and water [D$_2$O (*Butner et al., 2007; Vastel et al., 2010*)]. In a few cases, interferometric observations provided us with measurements of water deuterium fractionation in the hot corinos (*Codella et al., 2010; Persson et al., 2012; Taquet et al., 2013a*). Finally, recent observations have detected deuterated species in molecular outflows (*Codella et al., 2010, 2012*). The situation is graphically summarized in Fig. 5.

We note that the methanol deuteration depends on the bond energy of the functional group in which the hydrogen is located. In fact, the abundance ratio CH$_2$DOH/CH$_3$OD is larger than ~ 20 (*Parise et al., 2006; Ratajczak et al., 2011*), whereas it should be 3 if the D atoms were statistically distributed. To explain this, it has been invoked that CH$_3$OD could be selectively destroyed in the gas phase (*Charnley et al., 1997*), that H-D exchange reactions in solid

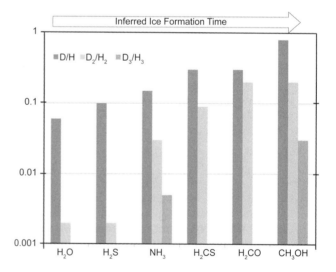

Fig. 5. Measured deuteration ratios of singly, doubly, and triply deuterated isotopologs (adapted from *Caselli and Ceccarelli, 2012*). Modeling the formation of H$_2$O, H$_2$CO, and CH$_3$OH (*Cazaux et al., 2011; Taquet et al., 2013b*) suggests that the increasing deuteration reflects the formation time of the species on the ices. Data sources are H$_2$O: *Liu et al. (2011), Coutens et al. (2012), Taquet et al. (2013a), Butner et al. (2007), Vastel et al. (2010)*; H$_2$S: *Vastel et al. (2003)*; NH$_3$: *Loinard et al. (2001), van der Tak et al. (2002)*; H$_2$CS: *Marcelino et al. (2005)*; H$_2$CO: *Ceccarelli et al. (1998), Parise et al. (2006)*; and CH$_3$OH: *Parise et al. (2002, 2004, 2006)*.

state could contribute to reduce CH_3OD (*Nagaoka et al.,* 2005), or, finally, that the CH_3OD underabundance is due to D-H exchange with water in the solid state (*Ratajczak et al.,* 2009, 2011). The reason for this overdeuteration of the methyl group with respect to the hydroxyl group may help to understand the origin of the deuterium fractionation in the different functional groups of the *insoluble organic matter* (section 8; see Fig. 11). This point will be discussed further in section 10.

The measure of the abundance of doubly deuterated species can also help to understand the formation/destruction routes of the species. Figure 6 shows the D/D_2 abundance ratio of the molecules in Fig. 5. For species forming on the grain surfaces, if the D atoms were purely statistically distributed, namely just proportional to the D/H ratio, then it would hold that D-species/D2-species = 4 (D-species/H-species)$^{-1}$. As shown in Fig. 6, this is not the case for H_2O, NH_3, H_2CS, and H_2CO. A plausible explanation is that the D- and D_2-bearing forms of these species are formed at different times on the grain surfaces: The larger the deuterium fraction, the younger the species (see section 5.3).

In the context of this chapter, the deuteration of water deserves particular attention. As water is very difficult to observe with groundbased telescopes (see the chapter by van Dishoeck et al. in this volume), measurements of HDO/H_2O exist only for four class 0 sources, and they are, unfortunately, in disagreement. Table 2 summarizes the situation. With the Atacama Large Millimeter/submillimeter Array (ALMA) facility it will be possible to settle the issue in the near future. Nonetheless, one thing is already clear from those measurements: The deuteration of water is smaller than that of the other molecules. This is also confirmed by the upper limits on the HDO/H_2O in the solid phase: $\leq 1\%$ (*Dartois et al.,* 2003; *Parise et al.,* 2003).

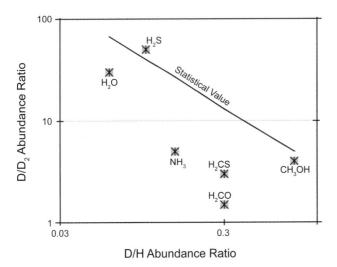

Fig. 6. Measured ratios of singly to doubly deuterated isotopologs of the species marked in the plot. The line shows the statistical value if the D atoms were statistically distributed in the molecules formed on the grain surfaces. References as in Fig. 5. Adapted from *Caselli and Ceccarelli* (2012).

TABLE 2. Measurement of the HDO/H_2O ratio in class 0 sources.

Source	HDO/H_2O (10^{-3})	Note*	Reference
IRAS 16293-2423	3–15	out	[1]
	4–51	hc	[1]
	0.7–1.2	hc	[2]
NGC 1333-IRAS2A	9–180	out	[3]
	~1	hc	[3,4]
	3–80	hc	[5]
NGC 1333-IRAS4A	5–30	hc	[5]
NGC 1333-IRAS4B	≤ 0.6	hc	[6]

*Indicates whether the measure refers to the outer envelope (out) or the hot corino (hc).

References: [1] *Coutens et al.* (2012, 2013b); [2] *Persson et al.* (2012); [3] *Liu et al.* (2011); [4] *Visser et al.* (2013); [5] *Taquet et al.* (2013a); [6] *Jørgensen and van Dishoeck* (2010).

5.3. Origin of Deuterium Fractionation

As discussed in section 2.1 and section 4, the key elements for a large deuterium fractionation are a cold and CO-depleted gas. The large deuterium enrichment observed in class 0 sources, and especially in the hot corinos, is therefore necessarily inherited from the prestellar core phase. The short lifetimes estimated for the class 0 sources support that statement. The different deuterium fractionation then tells us of the past history of the class 0 sources. Following section 4, it is very likely that water, formaldehyde, and methanol were formed on the grain surface at that epoch by hydrogenation of O and CO. The O hydrogenation leading to water occurs first, during the molecular cloud phase, and once formed, water remains frozen on the grains (*Hollenbach et al.,* 2009). CO hydrogenation leading to H_2CO and subsequently to CH_3OH occurs later, in the prestellar core phase (*Taquet et al.,* 2012b). Consequently, the water deuterium enrichment is lower than that of H_2CO and CH_3OH. In other words, the increased deuterium enrichment, linked to the H_2D^+/H_3^+ ratio (section 2.1), corresponds to a later formation of the molecule. Models by *Cazaux et al.* (2011) and *Taquet et al.* (2012a, 2013b) provide quantitative estimates in excellent agreement with the observed values. The same explanation applied to Fig. 6 tells us that H_2CO was synthesized during a large interval of time, so that the singly and doubly deuterated forms did not inherit the same D/H atomic ratio. On the contrary, CH_3OH was formed on a short time interval. Again, this agrees with model predictions.

6. PROTOPLANETARY DISK PHASE

6.1. Structure of Protoplanetary Disks

The protostellar phase has a short lifetime [~0.5 million years (m.y.)], during which the powerful outflows disperse the surrounding envelope, reducing the infall of material from the envelope onto the young disks and the accretion of material from the disk to the protostar. The central few hundred astronomical units are eventually revealed and the "classical" PPD phase starts. This is the final, dynamically active stage, lasting several million years, when the remaining ~0.001–0.1 $M_\odot$ of gas and dust is accreted onto the newly formed central star and assembled into a planetary system.

A summary of the physical and chemical structure of a PPD of a protosolar nebula (PSN) analog (a T Tauri star) is sketched in Fig. 7 (a more detailed description can be found in the chapters by Dutrey et al. and Testi et al. in this volume) (see also *Caselli and Ceccarelli*, 2012; *Henning and Semenov*, 2013, and references therein). Starting from the location of the accreting protostar, undergoing magnetospheric accretion and emitting ultraviolet (UV) photons and X-rays, we note that the first fraction astronomical unit is mainly deprived of dust because of the large temperatures, above the dust melting point (≈1500 K). The dusty disk starts at around 0.1 AU with an inner wall, puffed up by the high temperature (*Dullemond et al.*, 2002). The rest of the disk, up to a few hundred astronomical units, can be approximately divided in three layers: (1) the ionized atmosphere, where the chemistry is dominated by photodissociation due to stellar and interstellar UV photons as well as stellar X-rays (*Thi et al.*, 2010a; *Aresu et al.*, 2012); (2) a warm molecular layer, where molecules survive but photochemistry still plays an important role (*Dominik et al.*, 2005; *Henning et al.*, 2010); and (3) the cooler zones around the mid-plane, where the temperature drops to values close to those measured in prestellar cores (section 4). The densities (up to ~10^{11} cm^{-3}) are significantly higher than those of prestellar cores (up to a few ×10^7 cm^{-3}), so that dust coagulation, freeze-out, and deuterium fractionation are expected to proceed faster (*Ceccarelli and Dominik*, 2005). The temperature increases vertically, away from the

mid-plane, and radially, toward the central object. For a T Tauri star as that in Fig. 7, the inner region (1< radius < 30 AU) has temperatures between ≈30 and 150 K, while lower temperatures are found in the mid-plane at ≥30 AU.

The sketch in Fig. 7 is of course only a rough snapshot. During the several million years of its lifetime, the disk thermal and density structure, X-ray/UV radiation intensities, and grain properties undergo enormous transformation [as traced by infrared (IR) spectroscopy, (sub)millimeter/centimeter interferometry, and advanced disk modeling (*Andrews and Williams*, 2007; *Bouwman et al.*, 2008; *Andrews et al.*, 2009; *Guilloteau et al.*, 2011; *Williams and Cieza*, 2011; *Grady et al.*, 2013)]. The grain growth and sedimentation toward the mid-plane leads to more transparent and thus hotter disk atmosphere and flatter vertical structure, shrinking the zone when gas and dust are collisionally coupled and have similar temperatures. The disk chemical composition changes along with the evolution of the physical conditions (*Aikawa and Nomura*, 2006; *Nomura et al.*, 2007; *Cleeves et al.*, 2011; *Ilee et al.*, 2011; *Fogel et al.*, 2011; *Vasyunin et al.*, 2011; *Akimkin et al.*, 2013). As a result of the increasing X-ray/UV-irradiation with time, the molecularly rich, warm layer shifts closer to the cold mid-plane and the pace of surface chemistry can be delayed by the grain growth.

6.2. Deuterium Fractionation

Observational studies of the detailed chemical structure of PPDs require high sensitivity and high-angular-resolution measurements, which became available only in the past few years. For this reason, measurements of deuterium fractions are very sparse and so far limited to two PSN analog disks (the T Tauri disks TW Hya and DM Tau) and one disk around a 2-$M_\odot$ star [the Herbig Ae star HD 163296 (*Mathews et al.*, 2013), not discussed here].

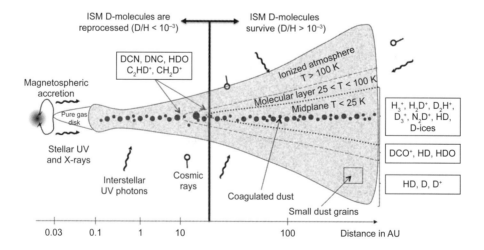

Fig. 7. Schematic vertical cut through a protoplanetary disk around a Sun-like T Tauri star. From the chemical point of view, the disk can be divided onto three layers: (1) the cold mid-plane dominated by ices and grain surface processes; (2) the warmer molecular layer with active gas-grain processes, where many gaseous molecules concentrate; and (3) the hot-irradiated atmosphere dominated by photochemistry and populated by ions and small molecules.

6.2.1. TW Hya. The first deuterated molecule, DCO⁺, was detected by *van Dishoeck et al.* (2003) in the T Tauri star TW Hya. The DCO⁺/HCO⁺ abundance ratio was found to be ≈0.04, similar to that measured for prestellar cores (section 4). The first image of the DCO⁺ emission has been obtained by *Qi et al.* (2008a) using the Submillimeter Array (SMA). They found variations of the DCO⁺/HCO⁺ across the disk, increasing outward from ~1% to ~10% and with a rapid falloff at radii ≥90 AU. DCN/HCN was also measured in the same observing run and was found to be ≈2%. Combining SMA with ALMA data, *Öberg et al.* (2012) have inferred that DCO⁺ and DCN do not trace the same regions: DCN is centrally concentrated and traces inner zones. This is in agreement with the theoretical predictions that DCN can also be produced in gas warmer than 30 K via the intermediate molecular ion CH_2D^+ (see section 2.1), unlike DCO⁺, which is linked to H_2D^+. The CO snow line has been recently imaged using ALMA observations of N_2H^+ (a molecular ion known to increase in abundance when CO starts to freeze out, as CO is one of the destruction partners), finding a radius of ~30 AU (*Qi et al.*, 2013). Deuterated species are expected to thrive close to the CO snow line and more observations should be planned to improve our understanding of the deuterium fractionation around this disk.

6.2.2. DM Tau. *Guilloteau et al.* (2006) found a DCO⁺/HCO⁺ abundance ratio toward DM Tau (≈0.004) about 1 order of magnitude lower than toward TW Hya, indicating interesting differences between apparently similar sources, which need to be explored in more detail.

6.3. Origin of Deuterium Fractionation

The deuterium fractionation in PPDs is expected to occur mainly in the cold (T ≤ 30 K) mid-plane region, where the CO freeze-out implies large abundances of the H_3^+ deuterated isotopologs (*Ceccarelli and Dominik*, 2005). The fractionation route based on CH_2D^+ is also effective in the warm inner disk mid-plane and molecular layer (Fig. 7 and section 2). Modern models of PPDs include a multitude of deuterium fractionation reactions and multiply deuterated species, still without nuclear spin-state processes, however.

Willacy (2007) have investigated deuterium chemistry in the outer disk regions. They found that the observed DCO⁺/HCO⁺ ratios can be reproduced, but, in general, the deuteration of gaseous molecules (in particular double deuteration) can be greatly changed by chemical processing in the disk. In their later study, *Willacy and Woods* (2009) have investigated deuterium chemistry in the inner 30 AU, using the same physical structure and accounting for gas and dust thermal balance. While a good agreement between the model predictions and observations of several nondeuterated gaseous species in a number of PPDs was obtained, the calculated D/H ratios for ices were higher than measured in the solar system comets.

Thi et al. (2010b) have focused on understanding deuterium fractionation in the inner warm disk regions and investigated the water D/H ratio in dense (≥10⁶ cm⁻³) and warm (~100–1000 K) gas, caused by photochemistry and neutral-neutral reactions. Using the T Tau disk structure calculated with "ProDiMo" (*Woitke et al.*, 2009), they predicted that in the terrestrial planet-forming region at ≤3 AU the water D/H ratio may reach ~1%, which is higher than the value of ≈1.5 × 10⁻⁴ measured in Earth's ocean water (Table 1). As mentioned in section 2.2, the formation of HDO at high temperatures (≥100 K) proceeds through the $OH + HD \rightarrow OD + H_2$ reaction (see Fig. 2).

As presented above, our knowledge of deuterium chemistry in PPDs, particularly from the observational perspective, is still rather poor. Only the deuterated molecules DCN and DCO⁺ have been detected so far [plus the recent Herschel detection of HD toward TW Hya (*Bergin et al.*, 2013)]. The analysis of their relatively poorly resolved spectra shows that DCO⁺ emission tends to peak toward outer colder disk regions, while DCN emission appears to be more centrally peaked. Their chemical differentiation seems to be qualitatively well understood from the theoretical point of view, with DCN being produced at warmer temperatures (≤70–80 K) by fractionation via CH_2D^+ and C_2HD^+, and DCO⁺ being produced at cold temperatures (≤10–30 K) by fractionation via H_3^+ isotopologs. However, theoretical models of the inner warm disk regions still tend to overpredict the D/H ratios of molecules observed in comets. The full capabilities of ALMA will provide us with a wealth of valuable new data on deuterated species in many PPDs, which will shed light on the chemical and physical processes during this delicate phase of planet formation.

7. COMETS

7.1. Origin of Comets

Having retained and preserved pristine material from the PSN at the moment of their accretion, comets contain unique clues to the history and evolution of the solar system. Their study provides the natural link between interstellar matter and solar system bodies and their formation. Comets accreted far from the Sun, where ices could condense, and have remained for most of their lifetime outside the orbit of Pluto. According to the reference scenario, the Oort cloud, which is the reservoir of long-period comets [also called Oort cloud comets (OCCs)], was populated by objects gravitationally scattered from the Uranus-Neptune region, with some contribution from the Jupiter-Saturn region (*Dones et al.*, 2004; *Brasser*, 2008). The reservoir of ecliptic short-period comets [the so-called Jupiter-family comets (JFCs)] is the Kuiper belt beyond Neptune, more precisely the scattered disk component of that population characterized by highly eccentric orbits (*Duncan and Levison*, 1997). In the current picture described by the Nice model, the present scattered disk is the remnant of a much larger scattered population produced by the interaction of the primordial planetesimal disk with a migrating Neptune. In the simulations of *Brasser and Morbidelli* (2013) made in the framework of the Nice model, the Oort cloud and the scattered disk formed simultaneously from objects from the same parent population. In summary, either the two comet reservoirs come from the same population of objects formed in the Uranus-Neptune zone, or OCCs formed on average closer to the Sun. Finally, a third option reported in the literature is that OCCs were captured from other stars when the Sun was in its birth cluster (*Levison et al.*, 2010). The molecular and isotopic composition in these two reservoirs can provide key answers in this respect.

7.2. Deuterium Fractionation

Because of the faint signatures of deuterated species, data were only obtained in a handful of bright comets. The first measurements of the D/H ratio in cometary H_2O were obtained in Comet 1P/Halley from mass-resolved ion spectra of H_3O^+ acquired with the *ion mass spectrometer* (*Balsiger et al.*, 1995) and the *neutral mass spectrometer* (*Eberhardt et al.*, 1995) intruments onboard the ESA Giotto spacecraft. These independent data provided precise D/H values of ($3.08^{+0.38}_{-0.53}$) × 10^{-4} and (3.06 ± 0.34) × 10^{-4} (*Balsiger et al.*, 1995; *Eberhardt et al.*, 1995). We note, however, that *Brown et al.* (2012) reexamined the mass spectrometer measurements, reevaluating these values to 2.1 × 10^{-4}.

From observations undertaken with the Caltech Submillimeter Observatory (CSO) and James Clerk Maxwell Telescope (JCMT), HDO was detected in the bright OCCs C/1996 B2 (Hyakutake) and C/1995 O1 (Hale-Bopp) from its $1_{01}-0_{00}$ line at 464.925 GHz (*Bockelée-Morvan et al.*, 1998; *Meier et al.*, 1998a). The derived D/H values [(2.9 ± 1.0) × 10^{-4} and (3.3 ± 0.8) × 10^{-4}, for Comets Hyakutake and Hale-Bopp, respectively] are in agreement with the determinations in Comet Halley (Fig. 8). Finally, observations of the HDO $1_{10}-1_{01}$ transition at 509.292 GHz in the Halley-type Comet 153P/Ikeya-Zhang yielded D/H < 2.5 × 10^{-4} (*Biver et al.*, 2006).

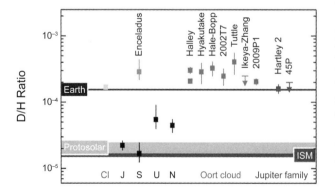

Fig. 8. D/H ratio in the water of comets compared to values in carbonaceous meteorites (CI), Earth's oceans (VSMOW; see Table 1), and Enceladus. Displayed data for planets, the interstellar medium (ISM), and the PSN refer to the value in H_2 (see text for details). Adapted from *Lis et al.* (2013).

Using the new high-resolution Cryogenic Infrared Echelle Spectrograph (CRIRES) of the European Southern Observatory (ESO) Very Large Telescope (VLT), *Villanueva et al.* (2009) observed the HDO rovibrational transitions near 3.7 μm in the Halley-family Comet 8P/Tuttle originating from the Oort cloud. Twenty-three lines were co-added to get a marginal detection of HDO, from which a formal value of D/H of (4.09 ± 1.45) × 10^{-4} was derived.

In cometary atmospheres, water photodissociates into mainly OH and H. The OD/OH ratio was measured to be (2.5 ± 0.7) × 10^{-4} in the OCC C/2002 T7 (LINEAR) through groundbased observations of the OH A $^2\Sigma^+$–X $^2\Pi_i$ UV bands at 310 nm obtained with VLT feeding the Ultraviolet and Visual Echelle Spectrograph (UVES) spectrograph (*Hutsemékers et al.*, 2008). No individual OD line was detected, but a marginal 3σ detection of OD was obtained by co-adding the brightest lines. Atomic deuterium emission was also discovered during UV observations of Comet C/2001 Q4 (NEAT) in April 2004 using the Space Telescope Imaging Spectrograph (STIS) of the Hubble Space Telescope (*Weaver et al.*, 2008). The Lyman-α (Lyα) emission from both deuterium and atomic hydrogen were detected, from which a preliminary value D/H = (4.6 ± 1.4) × 10^{-4} was derived, assuming that H_2O is the dominant source of the observed deuterium and hydrogen, as is likely the case. The strength of the optical method is that both normal and rare isotopologs have lines in the same spectral interval and are observed simultaneously, avoiding problems related to comet variable activity.

The most recent D/H measurements in cometary water were acquired using the ESA Herschel satellite. The HDO $1_{10}-1_{01}$ transition at 509.292 GHz was observed using the Heterodyne Instrument for the Far-Infrared (HIFI) spectrometer in the JFCs 103P/Hartley 2 and 45P/Honda-Mrkos-Pajdušáková (*Hartogh et al.*, 2011; *Lis et al.*, 2013), and in the OCC C/2009 P1 (Garradd) (*Bockelée-Morvan et al.*, 2012). Observations of HDO were interleaved with observations of the H_2O and $H_2^{18}O$ $1_{10}-1_{01}$ lines. Since the H_2O ground state rotational lines in comets are optically thick, optically thin lines of $H_2^{18}O$ provide, in principle, a more reliable reference for the D/H determination. The HDO/$H_2^{18}O$ was measured to be 0.161 ± 0.017 for Comet Hartley 2, i.e., consistent with the Vienna standard mean ocean water (VSMOW) (0.1554 ± 0.0001). The HDO/$H_2^{18}O$ value of 0.215 ± 0.023 for Comet Garradd suggests a significant difference (3σ) in deuterium content between the two comets. *Hartogh et al.* (2011) derived a D/H ratio of (1.61 ± 0.24) × 10^{-4} for Comet Hartley, assuming an $H_2^{16}O$/$H_2^{18}O$ ratio of 500 ± 50, which encompasses the VSMOW value and values measured in cometary water (*Jehin et al.*, 2009). For Comet Garradd, the derived D/H ratio is (2.06 ± 0.22) × 10^{-4} based on the HDO/$H_2^{16}O$ production rate ratio, and (2.15 ± 0.32) × 10^{-4}, using the same method as *Hartogh et al.* (2011). Herschel observations in the JFC 45P/ Honda-Mrkos-Pajdušáková were unsuccessful in detecting the HDO 50-GHz line, but resulted in a sensitive 3σ upper limit for the D/H ratio of 2.0 × 10^{-4}, which is consistent with the value measured in Comet 103P/Hartley 2 and excludes the canonical pre-Herschel value measured in OCCs of $\sim 3 \times 10^{-4}$ at the 4.5σ level (*Lis et al.,* 2013).

Besides H_2O, the D/H ratio was also measured in HCN, from the detection of the J=5–4 rotational transition of DCN in Comet Hale-Bopp with the JCMT (*Meier et al.,* 1998b). The inferred D/H value in HCN is 2.3×10^{-3}, i.e., seven times the value measured in water in the same comet. Finally, D/H upper limits for several molecules were obtained, but most of these upper limits exceed a few percent (*Crovisier et al.,* 2004). For CH_4, observed in the near-IR region, the most stringent D/H upper limits are $\sim 0.5\%$ (*Gibb et al.,* 2012, and references therein).

Table 3 summarizes the situation and Fig. 8 shows the water ice D/H values measured so far in comets, compared to the protosolar and Earth's ocean value of VSMOW (see Table 1). Isotopic diversity is present in the population of OCCs, with members having a deuterium enrichment of up to a factor of 2 with respect to Earth's value. The only two JFCs for which the D/H ratio has been measured present low values consistent with Earth's ocean value.

7.3. Origin of Deuterium Fractionation

The high D/H ratio measured in comets cannot be explained by deuterium exchanges between H_2 and H_2O in the PSN, and is interpreted as resulting from the mixing of D-rich water vapor originating from the presolar cloud or outer cold disk, and material reprocessed in the inner hot PSN (section 9). The discovery of an Earth-like D/H ratio in the JFC 103P/Hartley 2 came as a great surprise. Indeed, JFCs were expected to have D/H ratios higher than or similar to OCCs according to current scenarios (*Kavelaars et al.,* 2011), implying that the source regions of these two populations or the models investigating the gradient of D/H in the PSN should be revisited. Actually, the only dynamical theory that could explain in principle an isotopic dichotomy in D/H between OCCs and JFCs is the one arguing that a substantial fraction of OCCs were captured from other stars when the Sun was in its birth cluster (*Levison et al.,* 2010). Another possible solution is that there was a large-scale movement of planetesimals between the inner and outer PSN. According to the Grand Tack scenario proposed for the early solar system, when the giant planets were still embedded in the nebular gas disk, there was a general radial mixing of the distribution of comets and asteroids (*Walsh et al.,* 2011). Both the similarity of the D/H ratio in Comets 103P/Hartley 2 and 45P/Honda-Mrkos-Pajdušáková with that found in carbonaceous chondrites (CCs), and the isotopic diversity observed in the Oort cloud population, would be in agreement with this scenario, although D/H variations in the JFCs might be expected as well. Finally, another possibility is a nonmonotonic D/H ratio as a function of the heliocentric distance, as predicted by models considering a residual infalling envelope surrounding the PSN disk (*Yang et al.,* 2013). Section 9 will explore these models in greater detail.

TABLE 3. Summary of the measured water D/H values in comets (see text for the references).

Comet	D/H	Type[*]
Halley	$(3.1 \pm 0.5) \times 10^{-4}$	OCC
Hyakutake	$(2.9 \pm 1.0) \times 10^{-4}$	OCC
Hale-Bopp	$(3.3 \pm 0.8) \times 10^{-4}$	OCC
2002 T7	$(2.5 \pm 0.7) \times 10^{-4}$	OCC
Tuttle	$(4.1 \pm 1.5) \times 10^{-4}$	OCC
Ikeya-Zhang	$\leq 2.5 \times 10^{-4}$	OCC
2009 P1	$(2.06 \pm 0.22) \times 10^{-4}$	OCC
2001 Q4	$(4.6 \pm 1.4) \times 10^{-4}$	OCC
Hartley 2	$(1.61 \pm 0.24) \times 10^{-4}$	JFC
45P	$\leq 2.0 \times 10^{-4}$	JFC

[*]OCC: Oort cloud comet; JFC: Jupiter-family comet.

8. CARBONACEOUS CHONDRITES AND INTERSTELLAR DUST PARTICLES

Meteorites have fallen to Earth throughout its history. Interplanetary dust particles are collected in the stratosphere or in the ices of the Earth polar regions. While most IDPs are believed to be cometary fragments, probably from JFCs (*Nesvorný et al.,* 2010), meteorites are fragments of main-belt asteroids [2–4 AU (*Morbidelli et al.,* 2000)], with the few exceptions of martian and lunar meteorites. Both meteorites and IDPs provide us with a huge amount of information on the early phases of the solar system formation. Among the various groups of meteorites (see, e.g., the review by *Alexander,* 2011), the CCs are believed to be the more pristine, little altered by parent bodies, so we will focus this section on this group. Briefly, CCs are stony fragments of primitive asteroids of near-solar bulk composition that were not heated or differentiated to any extent. It is in these mineral matrixes that clay minerals (the so-called phyllosilicates) and diverse soluble and insoluble organic materials are found. Based on the techniques used to study it, the organic matter is classified as insoluble organic matter (IOM) and soluble organic compounds (SOC; sometime this is also referred as soluble organic matter or SOM). The possible interdependence between minerals and organic materials through their solar and presolar formative history is not known and remains one unexplained feature of chemical evolution. Evidence of the exposure of CCs to liquid water also indicates that aqueous processes and partaking minerals contributed to their final organic inventory. In this section, we describe the deuterium fractionation in the clays, IOM, and SOC separately.

8.1. Clay Minerals in Carbonaceous Chondrites and Interplanetary Dust Particles

For more than two decades (*Robert et al.,* 1979), it has been known that CC clay minerals exhibit a systematic enrichment in deuterium relative to the PSN D/H ratio (Table 1). Similarly, clays in IDPs are enriched in deuterium

(*Zinner et al.,* 1983; *Messenger,* 2000; *Aléon et al.,* 2000). Figure 9 shows the distribution of the bulk D/H ratio in the CCs and IDPs. In CCs, the contribution of the D-rich organic matter (see below) to the variations of the bulk D/H ratio is negligible relative to the large domain of isotopic variations, i.e., $\leq 1.5 \times 10^{-5}$. However, for IDPs, the relative contribution of organic H to hydroxyls is not well known, so the tail in the D/H distribution toward the high values may be caused by the contribution of D-rich organic hydrogen. Both CCs and IDPs have D/H ratios ~7× larger than the PSN. Their D/H distribution is peaked around the value of the terrestrial oceans, the VSMOW (Table 1), with a shoulder extending up to about 3.5×10^{-4}. The majority of CCs and IDPs therefore possess a D/H ratio lower than that measured in most comets (section 7).

Alexander et al. (2012) recently discovered a correlation between the D/H ratio and the amount of organic hydrogen relative to the bulk hydrogen (which, as mentioned above, is dominated by the hydroxyl groups of clays), indicating that the water in the meteorite parent body had a D/H ratio — prior to isotopic exchange with D-rich organics — as low as 9.5×10^{-5}, namely smaller than the standard mean ocean water (SMOW) value. In addition, these authors found that the D/H ratio of this pristine water varies according to the meteorite class, indicating a different degree of isotopic enrichment of the PSN water prior to its introduction in the CC parent bodies. Based on these findings, these authors propose that the original D/H ratio in meteoritic water (i.e., before the clay formation) is therefore smaller than what is measured nowadays, with consequences for the terrestrial water origin (section 10.3).

8.2. Insoluble Organic Matter

Insoluble organic matter represents the most abundant (70–90%) form of carbon isolated from CCs (*Bitz and Nagy,* 1966). It is mostly composed of carbon rings and chains, as shown in Fig. 10, and contains several other atoms: O, N, S, and P (*Remusat et al.,* 2005; *Cody and Alexander,* 2005; *Robert and Derenne,* 2006; *Derenne and Robert,* 2010). Note, however, that the molecular structure sketched in the figure should not be regarded as an organic formula of the IOM, but rather a statistical representation of the relative abundances of the organic bonds in the IOM. In addition, a study by *Cody et al.* (2011) showed a molecular relationship between chondritic and cometary organic solids. Remarkably, the possible molecular structure of interstellar refractory organic, namely polyaromatic hydrocarbons (PAHs), seems to be quite different from that of the IOM (*Kwok,* 2004; *Remusat et al.,* 2005). The origin of this difference is still an open issue: It may provide a hint as to a different origin of the two, or to a substantial restructuring of PAHs on the protosolar grains.

Studies of the molecular deuteration in the IOM have been carried out by several groups and with a variety of techniques used in organic chemistry (*Robert and Epstein,* 1982; *Yang and Epstein,* 1983; *Alexander,* 2011). Also, in this case, high values have been measured, as summarized in Fig. 9. However, contrary to the CC clay minerals and IDPs, the D/H distribution is not peaked around the terrestrial value but rather spread from 1.5 to $\sim 3.7 \times 10^{-4}$. Besides, the measured values are larger than those in the bulk of CCs (dominated by clay minerals; see above), IDPs, and Earth and more similar to what is measured in cometary water (Fig. 9). As discussed in section 5, the difference between water and organic deuterium fractionation in protostellar sources, where water is systematically less deuterated, has been interpreted as due to a different history of the water and the organic species formation on the grain surfaces. This may also be the case for the water and organic material in CCs and IDPs, as we will discuss in the following.

Remarkably, *Remusat et al.* (2006, 2009, 2012) have discovered a relation between the D/H ratio measured in the IOM and the energy of the different C-H bonds, namely aromatic (protonated carbon in aromatic moieties), aliphatic (in chains linking aromatic moieties), and benzilic bonds (in the alpha position of the cycles). In addition to these three bonds, the D/H ratio of the benzilic carbons attached to aromatic radicals has been measured through electron paramagnetic resonance (EPR). These radicals are considerably enriched in deuterium, with D/H = $(1.5 \pm 0.5) \times 10^{-2}$ (cf. Fig. 5) (*Gourier et al.,* 2008). This is the largest deuterium enrichment measured in the primitive objects of the solar system. The occurrence of the so-called "deuterium hotspots" observed with the NanoSIMS technique and that

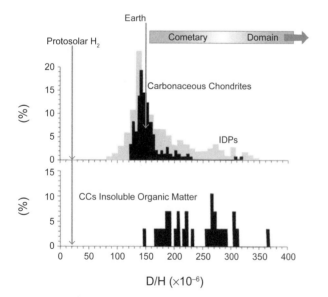

Fig. 9. Histograms of the D/H ratio in CCs (upper panel, black), IDPs (upper panel, gray), and the IOM (lower panel). The figure also shows the PSN and terrestrial water D/H ratios (Table 1). Note that the D/H ratios measured in the IOM "hot spots" are not reported here. Figure based on personal compilation of F. Robert and includes published measurements up to June 2013.

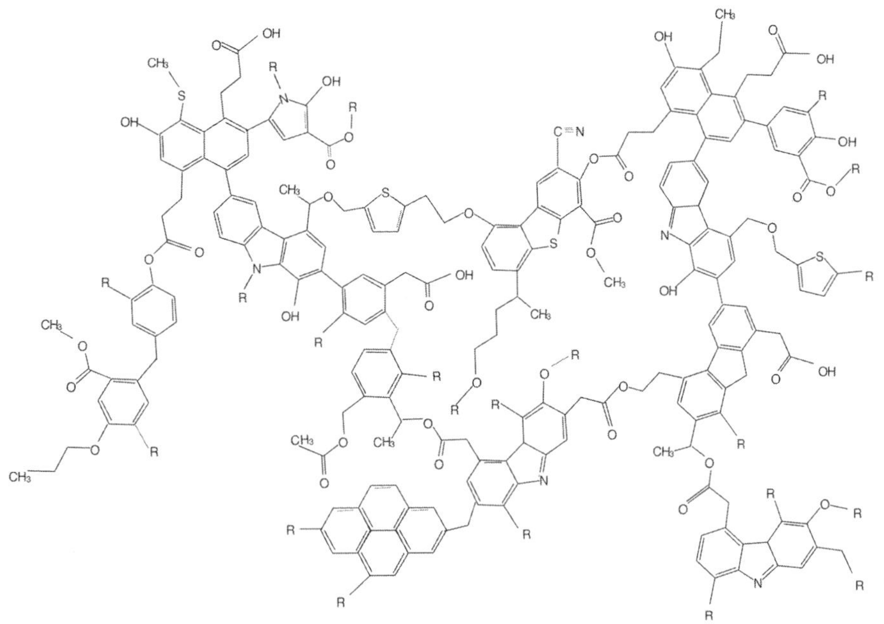

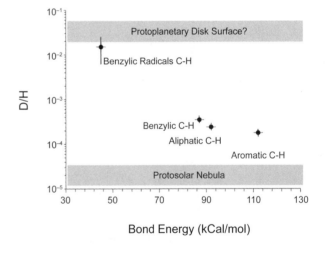

Fig. 10. Statistical model of the molecular structure of the IOM isolated from the Murchison CC. The "R" marks the presence of an unknown organic group, called moiety. Adapted from *Derenne and Robert* (2010).

are dispersed in the IOM is ascribed to a local micrometer-sized concentration of these radicals among the molecular network of the IOM (*Remusat et al.,* 2009).

The relation between the measured D/H and the energy of the bond is shown in Fig. 11: The larger the bond energy, the smaller the measured D/H. This relation between the energy bond and the deuterium fractionation led *Remusat et al.* (2006), *Gourier et al.* (2008), and *Derenne and Robert* (2010) to propose that the deuteration of the IOM occurred during the PPD phase of the solar system rather than during the pre-/protostellar phase or during the parent-body metamorphism (see section 9).

8.3. Soluble Organic Compounds

The distribution of the most abundant species of CC soluble compounds is shown in Fig. 12. Besides the ones shown, more molecules of diverse composition were identified, such as di- and polycarboxylic acids, polyhols, imides, and amides (*Pizzarello et al.,* 2006, and references therein), nucleobases (*Callahan et al.,* 2011), and numerous others are implied by spectroscopy (*Schmitt-Kopplin et al.,* 2010). Their distributions vary significantly between CCs, e.g., ammonia and amino acids are the most abundant compounds in Renazzo-type (CR) but hydrocarbons are in Mighei-type (CM) meteorites.

Amid this complex suite, amino acids (aa) appear to carry a unique prebiotic significance because of their essential role as constituent of proteins in the extant biosphere. Moreover, meteoritic amino acids may display chiral asymmetry in similarity with their terrestrial counterparts (*Pizzarello*

Fig. 11. Measured D/H ratio in functional groups as a function of the bond energy of the group.

and Groy, 2011, and references therein), suggesting that abiotic cosmochemical processes might have provided the early Earth with a primed inventory of exogenous molecules carrying an evolutionary advantage. The amino acids of CCs comprise a numerous and diverse group, having linear or branched alkyl chains and primary as well as secondary amino groups (Fig. 13) in different positions along those chains. These subgroups may require different pathways of formation and/or precursor molecules; all are enriched in ^{13}C, ^{15}N, and D and appear to have their own isotopic signatures. This isotopic diversity was first made evident

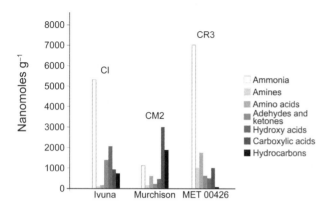

Fig. 12. Abundance distribution of soluble organic compounds in different types of carbonaceous chondrites (CC) (CI: Ivuna-type; CM: Mighei-type; CR: Renazzo-type; MET: Meteorite Hills). In general, the numbers 1–3 after the meteorite type indicate the extent of asteroidal aqueous alteration, with 3 being the least altered.

by the varying $\delta^{13}C$ values detected for Murchison aa (+4.9 to +52.8) and the significant differences found within and between aa subgroups; e.g., the ^{13}C content of 2-aa (Fig. 13) declines with chain length, as seen for carboxylic acids and alkanes, while 2-, 3-, and 4-amino- or dicarboxylic amino acids do not have similar trends. Also, 2-methyl-2-aa are more enriched in ^{13}C than the corresponding 2-H homologs (*Pizzarello et al.,* 2004). The δD values, determined for a larger number of aa in several meteorites (Fig. 13), show that aa subgroups are isotopically distinct in regard to this element as well and reveal additional differences among types of meteorites; again, branched aa display the largest δD values. A higher deuterium content of branched alkyl chains was seen also in all other aa subgroups analyzed, i.e., in the dicarboxylic aa and 3-, 4-, and 5-amino compounds.

The locales, precursor molecules, and synthetic processes that might account for meteoritic aa molecular, isotopic, and chiral properties are still being debated and have been only elucidated in part. As for their synthetic pathways, the first analytical indication came from the finding in meteorites of similar suites of hydroxy- and amino acids, which led *Peltzer and Bada* (1978) to propose a Strecker-type synthesis for both compound groups

The reaction proceeds via the addition of HCN to aldehydes and ketones in the presence of ammonia and water; the larger the ammonia abundance, the larger the expected ratio of NH_2-OH-acids, the latter being the only compounds formed in the absence of ammonia. The scheme agrees with many data from CC analyses but was further corroborated recently by enantiomeric excesses observed in the 6C aa isoleucine of several meteorites (*Pizzarello et al.,* 2012),

which demonstrated the possible origin and survival of enantiomeric excesses from their precursor aldehyde.

A Strecker synthesis readily accounts for meteoritic aa large D/H ratios: All reactants necessary for its first phase have been detected in interstellar clouds and are highly deuterated. The incorporation of little-altered interstellar species into volatile-rich CC parent bodies would then provide the subsequent aqueous phase.

The possible reliance upon minerals by organic materials through their solar and presolar formative history is not known and remains one unexplained feature of chemical evolution. For carbonaceous meteorites, evidence of CC exposure to liquid water indicates that aqueous processes and partaking minerals contributed to their final organic inventory.

9. PROTOSOLAR NEBULA MODELS

In this section we discuss separately the models that have been proposed to explain the measured deuterium fractionation of water and organics in comets, CCs, and IDPs. As in the relevant literature, we will use the enrichment factor f, defined as the ratio between the water D/H and the H_2 D/H of the PSN (Table 1), which for the terrestrial evaporated water (VSMOW) is 7.

9.1. Deuterium Fractionation of Water

The PSN, namely the PPD out of which Earth formed, was initially composed of dust and gas (section 6) with a D/H ratio close to the cosmic one (~2.1 × 10^{-5} vs. ~1.6 × 10^{-5}; Table 1). As described in sections 2 through 6, several reactions in the prestellar to PPD phases lead to the fractionation of deuterium of H-bearing species, and, in particular, of the water ice (section 2.2). The planetesimals that eventually formed comets, the parent bodies of IDPs and meteorites, and possibly Earth (see the discussion in section 10.3) were therefore covered by icy mantles enriched in deuterium. In the warm (≥100–120 K) regions of the PSN, inside the snow line, these ices sublimated. In general, PSN models aiming to model the fate of the ice (*Drouart et al.,* 1999; *Mousis et al.,* 2000; *Mousis,* 2004; *Hersant et al.,* 2001; *Horner et al.,* 2007, 2008; *Kavelaars et al.,* 2011) assumes that the PSN is a "diffusional" disk, in the sense that the turbulence is entirely responsible for its angular momentum transport. The first models used the simplest description for the disk provided by the famous α-viscosity one-dimensional (vertical-averaged) disk (*Shakura and Sunyaev,* 1973). Later versions took into account the two-dimensional structure of the disk (*Hersant et al.,* 2001). The viscosity, in the numerical models set by the value of α, is not only responsible for the mass accretion from the disk to the star (and consequently the disk thermal structure), but also for the dispersion of (part of) the angular momentum by diffusing matter outward. From a chemical point of view, these models consider only the exchange of D atoms between water and molecular hydrogen through the reaction (*Geiss and Reeves,* 1981)

Amino Acid		CM2 (δD or D/H in 10^{-4})	CR2 (δD or D/H in 10^{-4})
glycine		366–399 or 2.12–2.17	868–1070 or 2.89–3.21
DL alanine	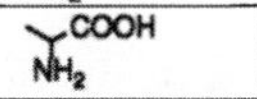	360–765 or 2.11–2.74	1159–1693 or 3.35–4.17
DL-2-a. butyric		1091–1634 or 3.24–4.08	1920–3409 or 4.53–6.63
norvaline	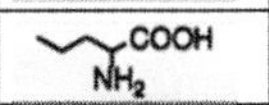	1505 or 3.88	nd
Branched chain compounds			
2-a. isobutyric		2362–3097 or 5.21–6.35	4303–7257 or 8.22–12.80
isovaline	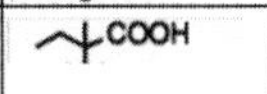	2081–3419 or 4.78–6.85	3813–7050 or 7.46–12.48
DL-valine	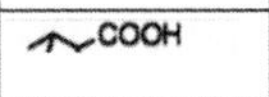	1216–2432 or 3.43–5.32	2086–3307 or 4.78–6.68
2-a. 2,3 methylbutyric	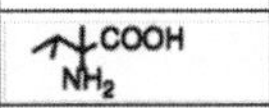	3318–3604 or 6.69–7.14	nd
DL-2 methylnorvaline	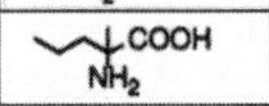	2686–3021 or 5.71–6.23	nd
DL-allo isoleucine	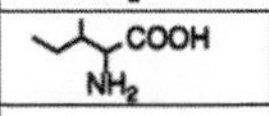	2206–2496 or 4.97–5.42	nd
L-leucine	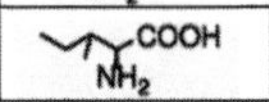	1792–1846 or 4.33–4.41	nd
N substituted amino acids			
Sarcosine	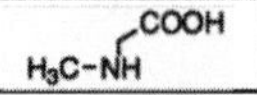	1274–1400 or 3.52–3.72	nd
DL-N-methylalanine	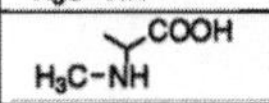	1224–1310 or 3.44–3.58	nd
N-methyl-2am. isobutyric	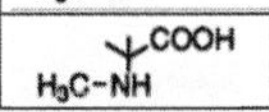	3431–3461 or 6.87–6.91	nd

Fig. 13. Range of δD ‰ values (see Table 1 for definition) and in D/H units of 10^{-4} determined for meteoritic 2-amino, and 2-methylamino acids, by compound-specific isotopic analyses for CM$_2$ (Murchison, Murray, and Lonewolf Nunataks 94101) and CR2 (for Graves Nunataks 95229 and Elephant Moraine 92042). Data are from *Pizzarello and Huang* (2005) and *Elsila et al.* (2012). nd = not determined.

$$H_2O + HD \rightleftharpoons HDO + H_2 \qquad (1)$$

and consider the isotopic exchange proportional to the reaction rate, measured by *Lécluse and Robert* (1994), and the isotopic fractionation at equilibrium, measured by *Richet et al.* (1977). Therefore, in these models, the deuterium fractionation of water across the PSN depends on the deuterium exchange in the gas (where T ≥ 200 K) and the mixing of the outward-diffused D-unenriched water with the inward-diffused D-enriched water. At high temperatures (≥500 K) this exchange leads to f = 1, namely no water deuterium enrichment. At lower temperatures, the exchange leads to f ≤ 3, but, being a neutral-neutral reaction, at temperatures ≤200 K the reaction is too slow to play any role (*Drouart et al.*, 1999). Therefore the deuterium fractionation of water across the PSN in these models depends on the deuterium exchange in the gas (where 200 ≤ T ≤ 500 K) and the mixing of the outward-diffused D-enriched water (sublimated from the ices) with the D-unenriched gas. When the gas reaches a temperature equal to that of water condensation (which is a function of the heliocentric distance and time), the evolution of f stops at that point. It is then assumed that the D-enrichment of the ice is f and that any object formed at that distance will inherit the same enrichment. This implies that transport mechanisms of grains, such as turbulent diffusion or gas drag, are neglected and that there is no evolution afterward [e.g., in cometary nuclei during their orbital evolution (*Brown et al.*, 2012)]. Figure 14 shows a cartoon of the situation.

In summary, these models, based on a strong assumption about the structure and dynamics of the PSN, depend on two basic parameters: the α value (or the turbulent viscosity prescription), which governs the thermal evolution of the disk and the diffusion, and the initial f profile before the disk evolution starts. Figure 15 shows the evolution of f as a function of the heliocentric distance as computed by *Kavelaars et al.* (2011). In this model, the initial f is constant across the disk and equal to the value found in LL3 meteorites [D/H = 0.9–7.3 × 10^{-4} (*Deloule et al.*, 1998)]. The D-enrichment profile matches quite well the HDO/H$_2$O values measured in the OCCs (section 7), if they are formed in the Uranus-Neptune region as predicted by the Nice model (*Levison et al.*, 2008; *Brasser and Morbidelli*, 2013). The computed D-enrichment profile also matches

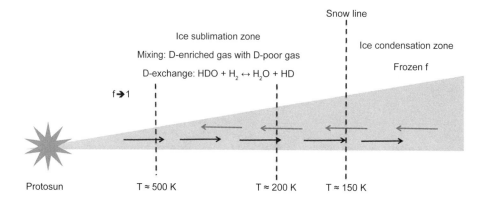

Fig. 14. Cartoon of PSN models. The PSN is described as a turbulent disk where matter diffuses inward and outward. Inside the snow line, which moves outward with increasing time, ices sublimate injecting D-enriched water vapor in the gas. The water vapor then exchanges the D-atoms with H_2 via the reaction $H_2O + HD \rightleftharpoons HDO + H_2$ where $T \geq 200$ K. At $T \geq 500$ K the exchange is very efficient and f tends to unity. When the outgoing gas reaches the condensation zone, it freezes out onto the grain ices. The acquired D-enrichment (f) is then conserved and inherited by the objects formed at that point.

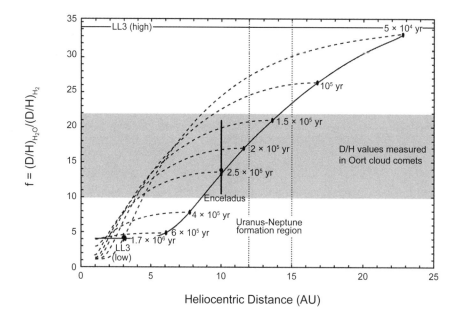

Fig. 15. Enrichment factor f (Table 1) as a function of the heliocentric distance, as predicted by the *Kavelaars et al.* (2011) model. f increases in the gas (dashed curves) during the PSN diffusive phase evolution (Fig. 14; see text). The f increase stops when gas condenses (solid curve), which occurs at different times (marked as dots) for different heliocentric distances. The two horizontal lines marked with LL3 represent the D/H enrichments in LL3 meteorites, low and highly enriched components respectively. The vertical line marked Enceladus gives the range of f measured in this moon. The vertical dotted lines mark the region between Uranus and Neptune, where presumably the Oort cloud comets formed. The gray area corresponds to the f values measured in the Oort cloud comets (section 7).

the D/H value measured by the Cassini spacecraft in the plumes of Saturn's satellite Enceladus at 10 AU, implying that the building blocks of this satellite may have formed at the same location as the OCCs in the outer PSN (*Waite et al.*, 2009; *Kavelaars et al.*, 2011). However, the same model cannot reproduce the more recent HDO/H_2O value observed toward the JFC 103P/Hartley (section 7).

To solve this problem, *Yang et al.* (2013) developed a new model whose major differences with the models mentioned above are (1) the PSN disk is not an isolated system but is fed by the infalling surrounding envelope, and (2) the chemistry is not only governed by the H_2O-H_2 isotopic exchange, but other important reactions are taken into account, following the work by *Thi et al.* (2010b) (see also section 6). The inclusion of fresh material from the envelope outer regions leads to a nonmonotonic f profile (Fig. 16), contrary to what was predicted by the previous studies. This result also stands if the "old" chemical network, namely the isotopic exchange between water and hydrogen, is adopted. This nonmonotonic gradient in deuteration could then explain the HDO/H$_2$O ratio measured toward 103P/Hartley 2 and 45P/Honda-Mrkos-Pajdušáková.

Another important issue, raised by the asymmetric distribution of the water D/H in CCs and IDPs clays, has been recently addressed by *Jacquet and Robert* (2013). In this model, the initial D/H value in the water ice is assumed to be 3×10^{-4}, the average observed value in OCCs (section 7). As in the previous models, iced grains drift inward and vaporize at the snow line. Inside the snow line, the water vapor exchanges its D with H$_2$ and, as a result, the D/H ratio in water vapor decreases. Reaction kinetics compel isotopic exchange between water and hydrogen gas to stop at T $\sim$ 500 K, well inside the snow line ($\sim$200 K). As a consequence, an isotopic gradient is established in the vapor phase between 200 K and 500 K by the mixing between the cometary source and the D/H ratio established at 500 K. The maximum value reached by the D/H ratio around the snow line is 1.2×10^{-4}. A fraction of this water having a D/H ratio of 1.2×10^{-4} can be transported beyond the snow line via turbulent diffusion where it (re)condenses as ice grains that are then mixed with isotopically comet-like water (D/H $\sim 3 \times 10^{-4}$). Thus, the competition between outward diffusion and net inward advection establishes an isotopic gradient that is at the origin of the large isotopic variations in the CCs and their asymmetric distribution. Under this scenario, the calculated probability distribution function of the D/H ratio of ice mimics the observed distribution reported for CCs in Fig. 9. In this interpretation, the minimum D/H ratio of the distribution ($\sim$1.2 $\times$ 10^{-4}) corresponds to the composition of water (re)condensed at the snow line. The distribution function depends on several parameters of the model that are discussed by *Jacquet and Robert* (2013).

Interestingly, the presence of water on Earth, which exhibits a bulk (i.e., including surface water and mantle hydroxyls) D/H value of $1.49 \pm 0.03 \times 10^{-4}$(*Lécuyer et al.,* 1998), can simply be accounted for by formation models advocating that terrestrial planets accreted from embryos sharing a CC composition (*Morbidelli et al.,* 2000). In this case, the contribution of comets to Earth's water budget would be relatively negligible.

9.2. Deuterium Fractionation of Organics

As discussed in section 8, organic matter shows a systematic larger deuterium fractionation with respect to water (Fig. 9). A precious hint of what could be the reason for that is given by the relation between the energy of the functional group bond and the measured D/H value (Fig. 11). To explain this behavior, several authors (*Remusat et al.,* 2006; *Gourier et al.,* 2008; *Derenne and Robert,* 2010) suggested that D atoms are transmitted from HD to the organics via the H$_2$D$^+$ ions present in the disk (*Ceccarelli and Dominik,* 2005; *Qi et al.,* 2008b; *Willacy and Woods,* 2009) (see also section 6), basically following the scheme

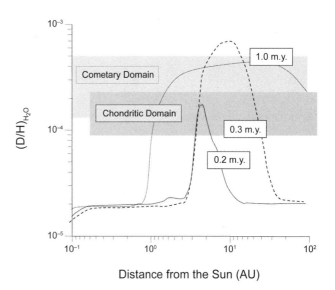

Fig. 16. The radial distribution of D/H in the PSN at different epochs of its evolution (adapted from *Yang et al.,* 2013). Jupiter-family comets, presumably formed in the transneptunian region, are predicted to have a lower D/H ratio than the Oort cloud comets, with the majority presumably formed in the Uranus/Neptune region.

of Fig. 1. Laboratory experiments add support to this hypothesis. *Robert et al.* (2011) experimentally measured the exchange rate of D atoms from H$_2$D$^+$ to organic molecules containing benzylic, aliphatic, and aromatic bonds, finding relative ratios as measured in the IOM. Extrapolating to the typical conditions of PPDs, namely scaling to the theoretical ion molecular ratio (i.e., to the H$_2$D$^+$/H$_2$ ratio) at the surface of the disk, these authors estimated that the exchange is very fast ($10^{4\pm1}$ yr) and therefore a plausible mechanism. Therefore, the fact that the D/H distribution for the IOM is systematically enriched in deuterium compared with that for water (cf. Fig. 9) could result from reactions in the gas phase and are possibly not connected with a fluid circulating in parent body meteorites during a hydrothermal episode. Indeed, all known isotopic reactions of organic molecules with a fluid should have resulted in the opposite distribution, i.e., a systematic depletion in the organic D/H ratio compared with that of clays.

10. FOLLOWING THE UNROLLED DEUTERIUM THREAD

10.1. Overview of Deuterium Fractionation Through Different Phases

Figure 17 graphically summarizes the measured values of the D/H ratio from prestellar cores up to solar system bodies, whereas Table 4 lists the measured D/H values and the main characteristics of the observations of solar system bodies. Overall, Fig. 17 and Table 4, together with Fig. 9,

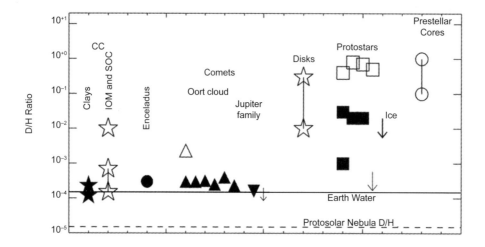

Fig. 17. D/H ratio in the solar system objects and astronomical sources. The filled symbols refer to water measures whereas open symbols refer to organic matter. The different objects are marked.

TABLE 4. Main characteristics of deuteration fractionation in comets and carbonaceous chondrites (CC) (for explanations and references, see sections 7 and 8).

Object	D/H [×10^{-4}]	Note
Oort cloud comets*	~3	Measured in seven comets
Jupiter-family comets*	≤2	Measured in two comets: 1.61 ± 0.24 and ≤2
IDPs*	1.5–3.5	D/H distribution peaked at 1.3, with shoulder from 0.8 to 3.5 (mimics the CC clays D/H distribution)
CC clays*	1.2–2.3	D/H distribution peaked at 1.4, with shoulder from 1.7 to 2.3
CC IOM†	1.5–3.5	D/H distribution does not have a peak
CC IOM functional groups†	1.5–5.5	The larger the functional group bonding energy, the larger the D/H
CC IOM deuterium hot spots†	~100	Micrometer-sized spots in the IOM networks where D/H is highly enhanced
CC SOC†	2.1–7.1	Branched aa show largest D/H

*HDO/H$_2$O.
†Deuterium fractionation in organics.

raise a number of questions, often found in the literature as well, that we address here.

1. *Why is the D/H ratio systematically lower in solar system bodies than in pre- and protostellar objects? When and how did this change occur?* Maybe here the answer is simple after all. All the solar system bodies examined in this chapter (meteorites and comets) were originally at distances less than 20 AU from the Sun. The D/H measurements of all pre- and protostellar objects examined in this chapter (prestellar cores, protostars, and PPDs) refer to distances larger than that, where the temperatures are lower and consequently the deuteration is expected to be much larger (section 2.1). Therefore, this systematic difference may tell us that the PSN did not undergo a global scale remixing of the material from the outer (≥20 AU) to the inner regions. There are exceptions, though, represented by the "hot spots" in meteorites, which, on the contrary, may be the only representatives of this large-scale remixing of the material in the PSN.

2. *Why does organic material have a systematically higher D/H value than water, regardless of the object and evolutionary phase?* The study of the deuterium fractionation in the prestellar cores (section 4) and protostars (section 5) has taught us that the formation of water and organics in different epochs is likely the reason why they have a different D/H ratio (Fig. 5). Water ices form first, when the density is relatively low (~10^4 cm^{-3}) and CO not yet depleted, so the H$_2$D$^+$/H$_3^+$ ratio is moderate (section 2.1). Organics form at a later stage, when the cloud is denser (≥10^5 cm^{-3}) and CO (and other neutrals) condenses on the grain surfaces, making possible a large enhancement of the H$_2$D$^+$/H$_3^+$. Can this also explain, *mutatis mutandis*, the different deuterium fractionation of water and organics measured in CCs and comets? After all, if the PSN cooled down from a hot phase, the condensation of volatiles would follow a similar sequence: oxygen/water first, when the temperature is higher, and carbon/organics later, when the temperature is lower. This would lead to deuterium fractions

lower in water than in organics, regardless of whether the synthesis of water and organic was on the gas phase or on the grain surfaces. Obviously, this is at the moment speculative but a road to explore in our opinion. We emphasize that this "different epoch formation" hypothesis is fully compatible with the theory described in section 9.2 on the origin of the organics deuteration from H_2D^+ (*Robert et al.*, 2011).

3. *Why do most comets exhibit higher D/H water values, about a factor 2, than CCs and IDPs?* A possible answer is that comets are formed at distances ≥5 AU, larger than those where the CCs originate, ≤4 AU (Table 1; see also section 9). A little more puzzling, though, is why the large majority of IDPs have D/H values lower than cometary water, although at least 50% of IDPs are predicted to be cometary fragments (*Nesvorny et al.*, 2010). The new Herschel observations of JFCs indeed indicate lower D/H values than those found in OCCs (Table 3; section 7), so the observed D/H values of IDPs may be consistent with the cometary fragments theory. In conclusion, the D/H distribution of CCs and IDPs is a powerful diagnostic tool for probing the distribution of their origin in the PSN.

4. *Why does the D/H distribution of the IDPs, which are thought to be fragments of comets, mimic the bulk D/H distribution of CCs? And why is it asymmetric, namely with a shoulder toward the large D/H values?* As written above, the D/H ratio distribution depends on the distribution of the original distance from where CCs and IDPs come from or, in other words, their parental bodies. Carbonaceous chondrites are likely fragments of asteroids from the main belt and a good fraction of them are of cometary origin. Interplanetary dust particles are fragments from JFCs and main-belt asteroids. Their similar D/H distribution strongly suggests that the mixture of the two classes of objects is roughly similar, which argues for a difference between the two, more in terms of sizes than in origin. Besides, the asymmetric distribution testifies that a majority of CCs and IDPs originate from parent bodies closer to the Sun (see also *Jacquet and Robert*, 2013).

5. *Why is the D/H ratio distribution of CC organics and water so different?* There are in principle two possibilities: Water and organics formed in different locations at the same time and then were mixed together, or, on the contrary, they were formed at the same heliocentric distance but at (slightly) different epochs. The discussion in point 2 above would favor the latter hypothesis, although this is speculative at the moment. If this is true, the D/H distribution potentially provides us with their history, namely when each of the two components formed. Again, as CC organics are more D-enriched than water, they were formed at later stages.

10.2. Comparison Between Protoplanetary Disk and Protosolar Nebula Models: Need for a Paradigm Shift?

There are several differences between the PPDs and PSN models. The first and major one is that PSN models assume a dense and hot phase for the solar PPD. For example,

temperatures higher than 1000 K persist for $\sim10^5$ yr at 3 AU and $\sim10^6$ yr at 1 AU in the *Yang et al.* (2013) model. Generally, models of PPDs around solar-type stars, on the contrary, never predict such high temperatures at those distances. Second, PPD models consider complex chemical networks with some horizontal and vertical turbulence that modifies, although not substantially, the chemical composition across the disk. On the contrary, in PSN models, chemistry networks are generally very simplified, whereas the turbulence plays a major role in the final D-enrichment across the disk. However, since the density adopted in the PSN models is very high, more complex chemical networks have a limited impact (*Yang et al.*, 2013). Third, the above-mentioned PSN models do not explicitly compute the dust particle coagulation and migration (if not as diffusion) and gas-grain decoupling, whereas (some) disk models do (see the chapters by Turner et al. and Testi et al. in this volume).

In our opinion, the most significant difference is the first one, and the urgent question that needs to be answered is, what is a good description of the PSN? A highly turbulent, hot and diffusive disk, or a cooler and likely calmer disk, like the ones we see around T Tauri stars? Or something else?

The community of those studying PSN models have reasons to believe that the PSN disk was once hot. One of the major reasons is the measurement of the depletion of moderately volatile elements (more volatile than silicon) in chondritic meteorites (*Palme et al.*, 1988), which can be explained by dust coagulation during the cooling of an initially hot (≥1400 K, the sublimation temperature of silicates) disk in the terrestrial planet-formation region [≤2 AU (*Lewis*, 1974; *Cameron and Truran*, 1977; *Cassen and Woolum*, 1996)]. The explanation, though, assumes that the heating of the dust at ≥1400 K was global in extent, ordered, and systematic. Alternatively, it is also possible that it was highly localized and, in this case, the hot initial nebula would be not necessary. For example, the so-called X-wind models assume that the hot processing occurs much closer to the star and the matter is then deposited outward by the early solar wind (*Shu et al.*, 1996, 2000; *Gounelle et al.*, 2001). In addition, the detection of crystalline silicates in comets (*Wooden et al.*, 2004) has been also taken as proof that the solar system passed through a hot phase (*Mousis et al.*, 2007).

Certainly T Tau disks do not show the high temperatures (≥1400 K at r ~ 2 AU) assumed by the PSN model. These temperatures are predicted in the mid-plane very close (≤0.1 AU) to the central star or in the high atmosphere of the disk, but always at distances lower than fractions of astronomical units (*Thi et al.*, 2010b). Even the most recent models of very young and embedded PPDs, with or without gravitational instabilities (see the review by *Caselli and Ceccarelli*, 2012), predict temperatures much lower than ~1400 K. One can ask whether the hot phase of the PSN could in fact be the hot corino phase observed in class 0 sources (section 5). Although the presently available facilities do not allow the probing of regions of a few astronomical units, the very rough extrapolation of the temperature profile predicted for the envelope of IRAS 16293-2422 (the prototype of class 0 sources) by *Crimier et al.* (2010) gives ~1000 K at a few astronomical units. The question is, therefore, should we not compare the PSN with the hot corino models rather than the PPD models? At present, the hot corino models are very much focused only on the ≥10-AU regions, which can be observationally probed, and only consider the gas composition. Considering what happens in the very innermost regions of the class 0 sources, should we not study the dust fate as well? If so, the link with the deuterium fractionation that we observe in the hot corino phase may become much more relevant in the construction of realistic PSN models. Last but not least, the present PSN models are based on the transport and mixing of material with different initial D/H water because of diffusion, which is responsible for the angular momentum dispersion. In class 0 sources, though, the dispersion of the angular momentum is thought to be mainly due to the powerful jets and outflows and not by the diffusion of inward/outward matter. Moreover, during the class 0 phase, material from the cold protostellar envelope continues to rain down onto the central region and the accreting disk (see the chapter by Z.-Y. Li et al. in this volume), replenishing them with highly deuterated material. The resulting D/H gradient across the PSN may therefore be different than that predicted by current theories.

10.3. Where Does the Terrestrial Water Come From?

Several reviews discussing the origin of the terrestrial water are present in the literature (*Horner et al.,* 2009; *Morbidelli et al.,* 2012; *Caselli and Ceccarelli,* 2012) (see also the chapter by van Dishoeck et al. in this volume), so here we will just summarize the points emphasizing the open issues.

Let us first remember that the water budget on Earth is itself the subject of debate. In fact, while the lithosphere budget is relatively easy to measure [~10^{-4} $M_{\oplus}$ (*Lécuyer et al.,* 1998)], the water contained in the mantle, which contains by far the largest mass of our planet, is extremely difficult to measure, and indirect probes, usually noble gases, are used for that (e.g., *Allegre et al.,* 1983), with associated larger error bars. It is even more difficult to evaluate the water content of the early Earth, which was likely more volatile rich than at present (*Kawamoto,* 1996). The most recent estimates give ~2×10^{-3} $M_{\oplus}$ (*Marty,* 2012), namely 20× larger than the value of the lithosphere water. If the water mantle is the dominant water reservoir, as it seems to be, then the D/H value of the terrestrial oceans may be misleading if geochemical processes can alter it. Evidently, measurement of the mantle water D/H is even more difficult. The last estimates suggest a value slightly lower than the terrestrial ocean water. In this chapter, given this uncertainty regarding Earth's bulk water content and D/H, we adopted as reference the evaporated ocean water D/H, the VSMOW (Table 1).

The "problem" of the origin of the terrestrial oceans arises because, if Earth was formed by planetesimals at ~1 AU heliocentric distance, they would have been "dry" and no water should exist on Earth. One theory, called the "late veneer," assumes that water was brought to Earth after its formation by, e.g., comets (*Delsemme,* 1992; *Owen and Bar-Nun,* 1995). This theory is based on the assumption that the D/H cometary water is the same as the Earth water D/H, but, based on observations toward comets (section 7), this assumption is probably wrong. The second theory, based on the work by *Morbidelli et al.* (2000), assumes that a fraction of the planetesimals that built Earth came from more distant (2–4 AU) regions, and were therefore "wet." Dynamical simulations of the early solar system evolution have added support to this theory (*Morbidelli et al.,* 2000; *Morbidelli and Nesvorny,* 2012; *Raymond et al.,* 2009), while at the same time challenging the idea that the flux of late veneer comets and asteroids could have been large enough to make up the amount of water on Earth. Moreover, the D/H value measured in CCs (section 8) adds support to this theory. The recent findings by *Alexander et al.* (2012) would argue for a large contribution of a group of CCs, the CI type.

In summary, the origin of terrestrial water is still a source of intense debate.

11. SUMMARY AND FUTURE PROSPECTS

In this chapter we have established a link between the various phases in the process of solar-type star and planet formation and our solar system. This link has been metaphorically called "Ariadne's thread," and it is represented by the D-fractionation process. Deuterium fractionation is active everywhere in time and space, from prestellar cores, to protostellar envelopes and hot corinos, to PPDs. Its past activity can be witnessed by us today in comets, CCs, and IDPs. Trying to understand and connect this process in the various phases of star and planet formation, while stretching the thread to our solar system, has opened new horizons in the quest of our origins. The ultimate goal is to chemically and physically connect the various phases and identify the particular route taken by our solar system. The steps toward this goal are many, of course, but much will be learned just by starting with the following important points:

1. Bridge the gap between prestellar cores and PPDs and understand how different initial conditions affect the physical and chemical structure and evolution of PPDs.

2. Study the reprocessing of material during the early stages of PPDs; in particular, can self-gravitating accretion disks, thought to be in present in the hot corino/protostellar phase, help in understanding some of the observed chemical and physical features of more evolved PPDs and our solar system?

3. Study the reprocessing of material (including dust coagulation and chemical evolution of trapped ices) throughout PPD evolution; in particular, which conditions favor the production of the organic material observed in pristine bodies of our solar system?

4. Compare PSN models with PPD models and include the main physical and chemical processes, in particular, dust coagulation and ice mantle evolution.

The future is bright thanks to the great instruments available now and in the near future [e.g., ALMA and the Northern Extended Millimeter Array (NOEMA), as well as the ESA Rosetta mission] and the advances in techniques used for the analysis of meteoritic material. Trying to fill the D/H plot shown in Fig. 17 with new observations is of course one priority, but this needs to proceed hand in hand with developments in theoretical chemistry and more laboratory work. Certainly one lesson has been learned from this interdisciplinary work: Our studies of the solar system and of star/planet-forming regions represent two treasures that cannot be kept in two different coffers. It is now time to work together to fully exploit such treasures, with the goal of eventually understanding our astrochemical heritage.

Acknowledgments. C.C. acknowledges financial support from the French Agence Nationale pour la Recherche (ANR) (project FORCOMS, contract ANR-08-BLAN-0225) and the French spatial agency CNES. P.C. acknowledges the financial support of the European Research Council (ERC) (project PALs 320620) and successive rolling grants awarded by the UK Science and Technology Funding Council. O.M. acknowledges support from CNES. S.P. acknowledges support through the years from the NASA Exobiology and Origins of the Solar System Programs. D.S. acknowledges support by the Deutsche Forschungsgemeinschaft through SPP 1385: "The first ten million years of the solar system — a planetary materials approach" (SE 1962/1-1 and 1-2). This research made use of NASA's Astrophysics Data System. We wish to thank A. Morbidelli, C. Alexander, and L. Bonal for a critical reading of the manuscript. We also thank an anonymous referee and the editor, whose comments helped to improve the clarity of the chapter.

REFERENCES

Aikawa Y. and Nomura H. (2006) *Astrophys. J., 642,* 1152.
Akimkin V. et al. (2013) *Astrophys. J., 766,* 8.
Aléon J. et al. (2000) *Meteoritics & Planet. Sci. Suppl., 35,* 19.
Alexander C. M. O. (2011) In *The Molecular Universe* (J. Cernicharo and R. Bachiller, eds.), pp. 288–301. IAU Symp. 280, Cambridge Univ., Cambridge.
Alexander C. M. O. et al. (2012) *Science, 337,* 721.
Allegre C. J. et al. (1983) *Nature, 303,* 762.
Andrews S. M. and Williams J. P. (2007) *Astrophys. J., 659,* 705.
Andrews S. M. et al. (2009) *Astrophys. J., 700,* 1502.
Aresu G. et al. (2012) *Astron. Astrophys., 547,* A69.
Bacmann A. et al. (2002) *Astron. Astrophys., 389,* L6.
Bacmann A. et al. (2003) *Astrophys. J. Lett., 585,* L55.
Balsiger H. et al. (1995) *J. Geophys. Res., 100,* 5827.
Benson P. J. and Myers P. C. (1989) *Astrophys. J. Suppl., 71,* 89.
Bergin E. A. et al. (2013) *Nature, 493,* 644.
Bitz S. M. C. and Nagy B. (1966) *Proc. Natl. Acad. Sci., 56,* 1383.
Biver N. et al. (2006) *Astron. Astrophys., 449,* 1255.
Bockelée-Morvan D. et al. (1998) *Icarus, 133,* 147.
Bockelée-Morvan D. et al. (2012) *Astron. Astrophys., 544,* L15.
Bouwman J. et al. (2008) *Astrophys. J., 683,* 479.
Brasser R. (2008) *Astron. Astrophys., 492,* 251.
Brasser R. and Morbidelli A. (2013) *Icarus, 225,* 40.
Brown R. H. et al. (2012) *Planet. Space Sci., 60,* 166.
Butner H. M. et al. (1995) *Astrophys. J., 448,* 207.
Butner H. M. et al. (2007) *Astrophys. J. Lett., 659,* L137.
Callahan M. P. et al. (2011) *Proc. Natl. Acad. Sci., 108,* 13995.

Cameron A. G. W. and Truran J. W. (1977) *Icarus, 30,* 447.
Caselli P. and Ceccarelli C. (2012) *Astron. Astrophys. Rev., 20,* 56.
Caselli P. et al. (1999) *Astrophys. J. Lett., 523,* L165.
Caselli P. et al. (2002) *Astrophys. J., 565,* 344.
Caselli P. et al. (2003) *Astron. Astrophys., 403,* L37.
Caselli P. et al. (2008) *Astron. Astrophys., 492,* 703.
Caselli P. et al. (2012) *Astrophys. J. Lett., 759,* L37.
Cassen P. and Woolum D. S. (1996) *Astrophys. J., 472,* 789.
Cazaux S. et al. (2011) *Astrophys. J. Lett., 741,* L34.
Ceccarelli C. and Dominik C. (2005) *Astron. Astrophys., 440,* 583.
Ceccarelli C. et al. (1996) *Astrophys. J., 471,* 400.
Ceccarelli C. et al. (1998) *Astron. Astrophys., 338,* L43.
Ceccarelli C. et al. (2000) *Astron. Astrophys., 357,* L9.
Ceccarelli C. et al. (2007) In *Protostars and Planets V* (B. Reipurth et al., eds.), pp. 47–62. Univ. of Arizona, Tucson.
Charnley S. B. et al. (1997) *Astrophys. J. Lett., 482,* L203.
Cleeves L. I. et al. (2011) *Astrophys. J. Lett., 743,* L2.
Codella C. et al. (2010) *Astron. Astrophys., 522,* L1.
Codella C. et al. (2012) *Astrophys. J. Lett., 757,* L9.
Cody G. D. and Alexander C. M. O. (2005) *Geochim. Cosmochim. Acta, 69,* 1085.
Cody G. D. et al. (2011) *Proc. Natl. Acad. Sci., 108,* 19171.
Coutens A. et al. (2012) *Astron. Astrophys., 539,* A132.
Coutens A. et al. (2013a) *Astron. Astrophys., 560,* A39.
Coutens A. et al. (2013b) *Astron. Astrophys., 553,* A75.
Crapsi A. et al. (2005) *Astrophys. J., 619,* 379.
Crapsi A. et al. (2007) *Astron. Astrophys., 470,* 221.
Crimier N. et al. (2010) *Astron. Astrophys., 519,* A65.
Crovisier J. et al. (2004) *Astron. Astrophys., 418,* 1141.
Czaja A. D. (2010) *Nature Geosci., 3,* 522.
Dalgarno A. and Lepp S. (1984) *Astrophys. J. Lett., 287,* L47.
Dartois E. et al. (2003) *Astron. Astrophys., 399,* 1009.
Deloule E. et al. (1998) *Geochim. Cosmochim. Acta, 62,* 3367.
Delsemme A. H. (1992) *Adv. Space Res., 12,* 5.
Derenne S. and Robert F. (2010) *Meteoritics & Planet. Sci., 45,* 1461.
Dominik C. et al. (2005) *Astrophys. J. Lett., 635,* L85.
Dones L. et al. (2004) In *Comets II* (M. C. Festou et al., eds.), pp. 153–174. Univ. of Arizona, Tucson.
Drouart A. et al. (1999) *Icarus, 140,* 129.
Dulieu F. et al. (2010) *Astron. Astrophys., 512,* A30.
Dullemond C. P. et al. (2002) *Astron. Astrophys., 389,* 464.
Duncan M. J. and Levison H. F. (1997) *Science, 276,* 1670.
Eberhardt P. et al. (1995) *Astron. Astrophys., 302,* 301.
Elsila J. E. et al. (2012) *Meteoritics & Planet. Sci., 47,* 1517–1536.
Evans N. J. II et al. (2009) *Astrophys. J. Suppl., 181,* 321.
Flower D. R. et al. (2006) *Astron. Astrophys., 449,* 621.
Fogel J. K. J. et al. (2011) *Astrophys. J., 726,* 29.
Friesen R. K. et al. (2013) *Astrophys. J., 765,* 59.
Gálvez Ó. et al. (2011) *Astrophys. J., 738,* 133.
Garrod R. et al. (2006) *Faraday Discussions, 133,* 51.
Geiss J. and Gloeckler G. (1998) *Space Sci. Rev., 84,* 239.
Geiss J. and Reeves H. (1981) *Astron. Astrophys., 93,* 189.
Gerlich D. and Schlemmer S. (2002) *Planet. Space Sci., 50,* 1287.
Gibb E. L. et al. (2012) *Astrophys. J., 750,* 102.
Gounelle M. et al. (2001) *Astrophys. J., 548,* 1051.
Gourier D. et al. (2008) *Geochim. Cosmochim. Acta, 72,* 1914.
Grady C. A. et al. (2013) *Astrophys. J., 762,* 48.
Guilloteau S. et al. (2006) *Astron. Astrophys., 448,* L5.
Guilloteau S. et al. (2011) *Astron. Astrophys., 529,* A105.
Hartogh P. et al. (2011) *Nature, 478,* 218.
Henning T. and Semenov D. (2013) *Chem. Rev., 113,* 9016.
Henning T. et al. (2010) *Astrophys. J., 714,* 1511.
Hersant F. et al. (2001) *Astrophys. J., 554,* 391.
Hiraoka K. et al. (1998) *Astrophys. J., 498,* 710.
Hirota T. et al. (2003) *Astrophys. J., 594,* 859.
Hollenbach D. et al. (2009) *Astrophys. J., 690,* 1497.
Horner J. et al. (2007) *Earth Moon Planets, 100,* 43.
Horner J. et al. (2008) *Planet. Space Sci., 56,* 1585.
Horner J. et al. (2009) *Planet. Space Sci., 57,* 1338.
Hutsemékers D. et al. (2008) *Astron. Astrophys., 490,* L31.
Ilee J. D. et al. (2011) *Mon. Not. R. Astron. Soc, 417,* 2950.
Jacquet E. and Robert F. (2013) *Icarus, 223,* 722.
Jehin E. et al. (2009) *Earth Moon Planets, 105,* 167.
Jørgensen J. K. and van Dishoeck E. F. (2010) *Astrophys. J. Lett., 725,* L172.
Jørgensen J. K. et al. (2002) *Astron. Astrophys., 389,* 908.

Kavelaars J. J. et al. (2011) *Astrophys. J. Lett., 734,* L30.
Kawamoto T. (1996) *Earth Planet. Sci. Lett., 144,* 577.
Keto E. and Caselli P. (2008) *Astrophys. J., 683,* 238.
Keto E. and Caselli P. (2010) *Mon. Not. R. Astron. Soc., 402,* 1625.
Kwok S. (2004) *Nature, 430,* 985.
Lécluse C. and Robert F. (1994) *Geochim. Cosmochim. Acta, 58,* 2927.
Lécuyer C. et al. (1998) *Chem. Geol., 145,* 249.
Lee C. W. and Myers P. C. (2011) *Astrophys. J., 734,* 60.
Levison H. F. et al. (2008) *Icarus, 196,* 258.
Levison H. F. et al. (2010) *Science, 329,* 187.
Lewis J. S. (1974) *Science, 186,* 440.
Li D. and Goldsmith P. F. (2003) *Astrophys. J., 585,* 823.
Linsky J. L. (2007) *Space Sci. Rev., 130,* 367.
Lis D. C. et al. (2002) *Astrophys. J. Lett., 571,* L55.
Lis D. C. et al. (2013) *Astrophys. J. Lett., 774,* L3.
Liu F.-C. et al. (2011) *Astron. Astrophys., 527,* A19.
Loinard L. et al. (2001) *Astrophys. J. Lett., 552,* L163.
Lubowich D. A. et al. (2000) *Nature, 405,* 1025.
Marcelino N. et al. (2005) *Astrophys. J., 620,* 308.
Maret S. et al. (2004) *Astron. Astrophys., 416,* 577.
Maret S. et al. (2005) *Astron. Astrophys., 442,* 527.
Marty B. (2012) *Earth Planet. Sci. Lett., 313,* 56.
Mathews G. S. et al. (2013) *Astron. Astrophys., 557,* 132.
Maury A. J. et al. (2010) *Astron. Astrophys., 512,* A40.
McKee C. F. and Ostriker E. C. (2007) *Annu. Rev. Astron. Astrophys., 45,* 565.
Meier R. et al. (1998a) *Science, 279,* 842–844.
Meier R. et al. (1998b) *Science, 279,* 1707–1710.
Messenger S. (2000) *Nature, 404,* 968.
Morbidelli A. and Nesvorny D. (2012) *Astron. Astrophys., 546,* A18.
Morbidelli A. et al. (2000) *Meteoritics & Planet. Sci., 35,* 1309.
Morbidelli A. et al. (2012) *Annu. Rev. Earth Planet. Sci., 40,* 251.
Mousis O. (2004) *Astron. Astrophys., 414,* 1165.
Mousis O. et al. (2000) *Icarus, 148,* 513.
Mousis O. et al. (2007) *Astron. Astrophys., 466,* L9.
Nagaoka A. et al. (2005) *Astrophys. J. Lett., 624,* L29.
Nesvorný D. et al. (2010) *Astrophys. J., 713,* 816.
Nomura H. et al. (2007) *Astrophys. J., 661,* 334.
Oba Y. et al. (2012) *Astrophys. J., 749,* 67.
Öberg K. I. et al. (2012) *Astrophys. J., 749,* 162.
Owen T. and Bar-Nun A. (1995) *Icarus, 116,* 215.
Pagani L. et al. (2007) *Astron. Astrophys., 467,* 179.
Palme H. et al. (1988) *Meteoritics & Planet. Sci., 23,* 49.
Parise B. et al. (2002) *Astron. Astrophys., 393,* L49.
Parise B. et al. (2003) *Astron. Astrophys., 410,* 897.
Parise B. et al. (2004) *Astron. Astrophys., 416,* 159.
Parise B. et al. (2006) *Astron. Astrophys., 453,* 949.
Parise B. et al. (2011) *Astron. Astrophys., 526,* A31.
Peltzer E. T. and Bada J. L. (1978) *Nature, 272,* 443.
Persson M. V. et al. (2012) *Astron. Astrophys., 541,* A39.
Pizzarello S. and Groy T. L. (2011) *Geochim. Cosmochim. Acta, 75,* 645.
Pizzarello S. and Huang Y. (2005) *Geochim. Cosmochim. Acta, 69,* 599.
Pizzarello S. et al. (2004) *Geochim. Cosmochim. Acta, 68,* 4963.
Pizzarello S. et al. (2006) In *Meteorites and the Early Solar System II* (D. S. Lauretta and H. Y. McSween Jr., eds.), pp. 625–651. Univ. of Arizona, Tucson.
Pizzarello S. et al. (2012) *Proc. Natl. Acad. Sci., 109,* 11949.
Qi C. et al. (2008a) *Astrophys. J., 681,* 1396.
Qi C. et al. (2008b) *Astrophys. J., 681,* 1396.
Qi C. et al. (2013) *Science, 341,* 630.

Ratajczak A. et al. (2009) *Astron. Astrophys., 496,* L21.
Ratajczak A. et al. (2011) *Astron. Astrophys., 528,* L13.
Raymond S. N. et al. (2009) *Icarus, 203,* 644.
Remusat L. et al. (2005) In *Lunar and Planetary Science XXXVI,* Abstract #1350. Lunar and Planetary Institute, Houston.
Remusat L. et al. (2006) *Earth Planet. Sci. Lett., 243,* 15.
Remusat L. et al. (2009) *Astrophys. J., 698,* 2087.
Remusat L. et al. (2012) *Geochim. Cosmochim. Acta, 96,* 319.
Richet P. et al. (1977) *Annu. Rev. Earth Planet. Sci., 5,* 65.
Rist C. et al. (2013) *J. Phys. Chem. A, 117,* 9800.
Robert F. and Derenne S. (2006) *Meteoritics & Planet. Sci. Suppl., 41,* 5259.
Robert F. and Epstein S. (1982) *Geochim. Cosmochim. Acta, 46,* 81.
Robert F. et al. (1979) *Nature, 282,* 785.
Robert F. et al. (2011) *Geochim. Cosmochim. Acta, 75,* 7522.
Roberts H. and Millar T. J. (2000) *Astron. Astrophys., 361,* 388.
Roberts H. et al. (2003) *Astrophys. J. Lett., 591,* L41.
Roueff E. et al. (2005) *Astron. Astrophys., 438,* 585.
Schmitt-Kopplin P. et al. (2010) *Proc. Natl. Acad. Sci., 107,* 2763.
Shah R. Y. and Wootten A. (2001) *Astrophys. J., 554,* 933.
Shakura N. I. and Sunyaev R. A. (1973) *Astron. Astrophys., 24,* 337.
Shu F. H. et al. (1996) *Science, 271,* 1545.
Shu F. H. et al. (2000) In *Protostars and Planets IV* (V. Mannings et al., eds.), p. 789. Univ. of Arizona, Tucson.
Sipilä O. et al. (2013) *Astron. Astrophys., 554,* A92.
Spezzano S. et al. (2013) *Astrophys. J. Lett., 769,* L19.
Taquet V. et al. (2012a) *Astrophys. J. Lett., 748,* L3.
Taquet V. et al. (2012b) *Astron. Astrophys., 538,* A42.
Taquet V. et al. (2013a) *Astrophys. J. Lett., 768,* L29.
Taquet V. et al. (2013b) *Astron. Astrophys., 550,* A127.
Thi W.-F. et al. (2010a) *Astron. Astrophys., 518,* L125.
Thi W.-F. et al. (2010b) *Mon. Not. R. Astron. Soc., 407,* 232.
Turner B. E. (2001) *Astrophys. J. Suppl., 136,* 579.
van der Tak F. F. S. et al. (2002) *Astron. Astrophys., 388,* L53.
van Dishoeck E. F. et al. (2003) *Astron. Astrophys., 400,* L1.
Vastel C. et al. (2003) *Astrophys. J. Lett., 593,* L97.
Vastel C. et al. (2004) *Astrophys. J. Lett., 606,* L127.
Vastel C. et al. (2010) *Astron. Astrophys., 521,* L31.
Vasyunin A. I. et al. (2011) *Astrophys. J., 727,* 76.
Villanueva G. L. et al. (2009) *Astrophys. J. Lett., 690,* L5.
Visser R. et al. (2013) *Astrophys. J., 769,* 19.
Waite J. H. Jr. et al. (2009) *Nature, 460,* 487.
Walmsley C. M. et al. (2004) *Astron. Astrophys., 418,* 1035.
Walsh K. J. et al. (2011) *Nature, 475,* 206.
Ward-Thompson D. et al. (1999) *Mon. Not. R. Astron. Soc., 305,* 143.
Weaver H. A. et al. (2008) *Asteroids, Comets, and Meteors 2008,* Abstract #8216. LPI Contribution No. 1405, Lunar and Planetary Institute, Houston.
Weber A. S. et al. (2009) *Astrophys. J., 703,* 1030.
Whittet D. C. B. et al. (2007) *Astrophys. J., 655,* 332.
Willacy K. (2007) *Astrophys. J., 660,* 441.
Willacy K. and Woods P. M. (2009) *Astrophys. J., 703,* 479.
Williams J. P. and Cieza L. A. (2011) *Annu. Rev. Astron. Astrophys., 49,* 67.
Woitke P. et al. (2009) *Astron. Astrophys., 501,* 383.
Wooden D. H. et al. (2004) *Astrophys. J. Lett., 612,* L77.
Yang J. and Epstein S. (1983) *Geochim. Cosmochim. Acta, 47,* 2199.
Yang L. et al. (2013) *Icarus, 226,* 256–267.
Zinner E. et al. (1983) *Nature, 305,* 119.

Güdel M., Dvorak R., Erkaev N., Kasting J., Khodachenko M., Lammer H., Pilat-Lohinger E., Rauer H., Ribas I., and Wood B. E. (2014)
Astrophysical conditions for planetary habitability. In *Protostars and Planets VI* (H. Beuther et al., eds.), pp. 883–906.
Univ. of Arizona, Tucson, DOI: 10.2458/azu_uapress_9780816531240-ch038.

Astrophysical Conditions for Planetary Habitability

Manuel Güdel and Rudolf Dvorak
University of Vienna

Nikolai Erkaev
Russian Academy of Sciences

James Kasting
Pennsylvania State University

Maxim Khodachenko and Helmut Lammer
Austrian Academy of Sciences

Elke Pilat-Lohinger
University of Vienna

Heike Rauer
Deutsches Zentrum für Luft- und Raumfahrt (DLR) and Technische Universität Berlin

Ignasi Ribas
Institut d'Estudis Espacials de Catalunya–Consejo Superior de Investigaciones Científicas (CSIC)

Brian E. Wood
Naval Research Laboratory

With the discovery of hundreds of exoplanets and a potentially huge number of Earth-like planets waiting to be discovered, the conditions for their habitability have become a focal point in exoplanetary research. The classical picture of habitable zones primarily relies on the stellar flux allowing liquid water to exist on the surface of an Earth-like planet with a suitable atmosphere. However, numerous further stellar and planetary properties constrain habitability. Apart from "geophysical" processes depending on the internal structure and composition of a planet, a complex array of astrophysical factors additionally determine habitability. Among these, variable stellar ultraviolet (UV), extreme ultraviolet (EUV), and X-ray radiation; stellar and interplanetary magnetic fields; ionized winds; and energetic particles control the constitution of upper planetary atmospheres and their physical and chemical evolution. Short- and long-term stellar variability necessitates full time-dependent studies to understand planetary habitability at any point in time. Furthermore, dynamical effects in planetary systems and transport of water to Earth-like planets set fundamentally important constraints. We will review these astrophysical conditions for habitability under the crucial aspects of the long-term evolution of stellar properties, the consequent extreme conditions in the early evolutionary phase of planetary systems, and the important interplay between properties of the host star and its planets.

1. INTRODUCTION

Discovery of planets around other stars is now well underway. Over 1000 extrasolar planets have been found mainly by groundbased radial velocity (RV) measurements and space- and groundbased photometric transit surveys (*exoplanets.org*). In addition, ≈3500 "planet candidates" have been found by Kepler (*Batalha et al., 2013*) (updates as of November 2013). Due to instrument sensitivity the first planets detected were predominantly gas or ice giants, like Jupiter or Neptune. Recent work by *Howard* (2013) and *Fressin et al.* (2013) show that small transiting planets are more numerous than big ones. Some potentially rocky planets have been identified (*Léger et al., 2009; Batalha et al., 2011; Borucki et al., 2013*). Several planets with minimum masses consistent with rocky planets have been found mostly around cooler stars, including planets in the habitable zones (HZ) (*Pepe et al.,* 2011; *Anglada-Escudé et al.,* 2012; *Bonfils et al.,* 2013), as these stars are less massive and smaller, and hence more easily accelerated by small planets, and show an improved planet/star contrast ratio in the photometric transit method. The search is now on for rocky planets around FGKM stars. A handful of these have already been found (*Buchhave et al.,* 2012) (see also *exoplanets.org*). This search is difficult to do by RV because the induced stellar velocities are small. Kepler can find close-in rocky planets around FGK stars rather easily compared to planets at larger radii. However, small, potentially rocky planets in the HZ are difficult to verify, since most of the solar-like stars are noisier than the Sun, and those identified by Kepler are far away, hence dim (*Gilliland et al.,* 2011). Nevertheless, the first few transiting planets with radii consistent with rocky planets have been recently detected in HZ (*Borucki et al.,* 2013; *Kopparapu,* 2013; *Kaltenegger et al.,* 2013).

In this chapter we are concerned with habitable planets. Below, we expand on what we mean by that term. A fundamental requirement of habitability, though — perhaps the only indisputable requirement — is that the planet must have a solid or liquid surface to provide stable pressure-temperature conditions. In his book *Cosmos,* Carl Sagan imagined hypothetical bag-like creatures that could live on Jupiter by adjusting their height in the atmosphere, just as some fish can adjust their depth in Earth's oceans using air bladders. But one needs to ask how such creatures might possibly evolve. Single-celled organisms could not maintain their altitude on a gas or ice giant planet, and hence would eventually be wafted up to the very cold (unless it was a hot Jupiter) upper atmosphere or down into the hot interior. Here we assume that Sagan's gas giant life forms do not exist, and we focus our attention on rocky planets. Habitability for extrasolar planets is usually defined by liquid water on the surface of a rocky planet. Other liquids can be imagined, as discussed below. Furthermore, water may exist also subsurface, e.g., as expected for Jupiter's icy moon Europa, which would make such a body also potentially habitable. In terms of exoplanets, however, such subsurface life would be very difficult to detect, and we neglect it here.

2. WHAT IS HABITABILITY?

To begin, we must define what constitutes a rocky planet. That definition is easy enough if you can characterize the planet the way we do for objects in our own solar system: It must have a solid or liquid surface. But if one knows only the planet's minimum mass (from RV) or its diameter (from transits), this question becomes more difficult. Astrophysical theory predicts that planets that grow larger than about 10 $M_\oplus$ during the time that the stellar nebula is still around will capture significant amounts of gas and turn into gas or ice giants (*Mizuno et al.,* 1978; *Pollack et al.,* 1996). The theory is not robust, however, because the opacity calculations within the accreting atmosphere are

complex and did not include such phenomena as collision-induced absorption of infrared (IR) radiation by molecular hydrogen, and because it does not exclude higher-mass planets that complete the process of accretion after the nebular gas has disappeared. Similarly, measurement of a planet's diameter via the transit method does not distinguish conclusively between a rocky planet like Earth or an ice giant like Neptune. Neptune's diameter is about 4× that of Earth. Initial data indicate that planets with radii below 2 $R_\oplus$ are rocky. The Kepler team assumes that planets with diameters >2 $R_\oplus$ are ice giants.

Once we are sure that a planet is rocky, the next most fundamental requirement for life as we know it is that it should have access to liquid water. Even though some terrestrial organisms — those that form spores, for example — can persist for long time intervals without water, all terrestrial life forms require liquid water to metabolize and reproduce. There are good biochemical reasons for this dependence. Most importantly, water is a highly polar solvent that can dissolve the polar molecules on which carbon-based life depends. That said, researchers do not completely discount the possibility that alien life forms might have different biochemistries (*Baross et al.,* 2007; *National Academy of Sciences,* 2007). Carbon has a more complex chemistry than any other element, and so most workers agree that alien life would also be carbon-based. Other chain-building elements like silicon and phosphorus exist, but are less energetically favorable than carbon in this respect. Among alternative solvents, even liquid CH_4 is considered by some workers to be a possible medium in which life might exist, hence the interest in Saturn's moon Titan, which has lakes of liquid methane on its surface. Liquid methane is only stable at very low temperatures, though, and so any organic chemistry in it may be too slow to create or to power life. We suspect that liquid water, or some mixture containing liquid water, is needed to support all forms of life.

Confining our interest to planets with liquid water is not all that restrictive. Oxygen is the third most abundant element in the universe, and so many, or most, rocky planets are probably endowed with appreciable amounts of water when they form (e.g., *Raymond et al.,* 2007). In our own solar system, Mercury and Venus lack liquid water, while Earth and Mars have it (although modern Mars has no liquid water in amounts sufficient to develop life). The lack of water on the innermost two planets is easily explained, though, as these planets lie outside the boundaries of the HZ — the region around a star where liquid water can be present on a planet's surface (*Hart,* 1979; *Kasting et al.,* 1993). Mars lacks liquid water on its surface today, but shows evidence that water flowed there earlier in its history. And Mars may have liquid water in its subsurface. Mars' internal heat flow is thought to be about one-third that of Earth (*Montési and Zuber,* 2003), hence, depending on the thermal conductivity of the regolith, Mars could have liquid water within a few kilometers of the surface.

As astronomers, we are interested in discovering life on extrasolar planets. For the foreseeable future, at least, such planets can only be studied remotely, by doing spectroscopy on their atmospheres using big telescopes on the ground or, more likely, in space. For us to detect life on an exoplanet, that life must be able to modify the planet's atmosphere in such a way that we can detect it, as has been done here on Earth. Earth's atmosphere contains 21% O_2 and 1.7 ppmv CH_4, both of which are produced almost entirely by organisms. The maintenance of extreme disequilibrium in a planet's atmosphere, and specifically the coexistence of free O_2 with reduced gases such as CH_4 or N_2O, was suggested many years ago as the best remote evidence for life (*Lederberg,* 1965; *Lovelock,* 1965). If we wish to propose a testable hypothesis about life on other planets, we should therefore restrict our attention to planets like Earth that have liquid water on their surfaces. This allows for the possibility that photosynthetic life might flourish there, greatly increasing productivity (*Kharecha et al.,* 2005), and possibly producing atmospheric biosignatures that we might one day detect.

From a practical standpoint, then, this means that we should design our astronomical searches to look for planets lying within the conventional liquid-water HZ of their parent star. This is a necessary, but not sufficient, condition, as factors such as volatile inventories, high-energy radiation, magnetospheres, and stellar winds may further constrain habitability, as discussed further in this chapter. The boundaries of the HZ were estimated by *Kasting et al.* (1993) based on one-dimensional climate modeling calculations for Earth-like planets. These calculations were done with a cloud-free model (clouds were "painted" on the ground) and they assumed fully saturated tropospheres. Conservatively, planets may lose their water when they lie closer to the star than the "moist greenhouse" limit. At this distance, ≈0.95 AU in these old calculations, a planet's stratosphere becomes wet, and the water is lost by photodissociation followed by escape of hydrogen to space. A stricter limit of the inner edge of the HZ is the so-called runaway greenhouse limit. At this limit the entire water ocean of Earth would reside within the atmosphere due to the self-enhancing water vapor feedback cycle. The outer edge of the HZ is defined by the "maximum greenhouse" limit, beyond which a CO_2-H_2O greenhouse is no longer capable of maintaining a warm surface. This limit is based on the assumption that an Earth-like planet will have volcanos that emit CO_2 and that CO_2 will accumulate in the planet's atmosphere as the surface becomes cold. But, at some point, CO_2 will begin to condense out of the planet's atmosphere, limiting the range of distances over which this feedback process works. In these old calculations, the maximum greenhouse limit occurs at ≈1.67 AU for our Sun. The "first CO_2 condensation" limit of *Kasting et al.* (1993) is no longer considered valid, because CO_2 clouds are now thought to usually warm a planet's surface (*Forget and Pierrehumbert,* 1997; *Mischna et al.,* 2000). *Kitzmann et al.* (2013), however, suggest that the warming effect may have been overestimated in some cases. More optimistic empirical limits on the HZ are derived from

the observation that early Mars appears to lie within it, and Venus could conceivably have been within it prior to ≈1 gigayears (G.y.) ago. An even more optimistic outer edge has been suggested for the HZ by *Seager* (2013), based on a calculation by *Pierrehumbert and Gaidos* (2011) showing that super-Earths with dense captured H_2 atmospheres could remain habitable out to as far as 10 AU in our solar system. Such planets will remain speculative, however, until they are observed, and one would not want to count on their existence while defining the requirements for a telescope to search for extrasolar life (*Kasting et al.,* 2014).

Recently, *Kopparapu et al.* (2013) rederived the HZ boundaries using a new one-dimensional climate model based on the HITEMP database for H_2O. HITEMP differs from the older HITRAN database by including many more weak absorption lines, including some that extend all the way down to near-UV wavelengths. They furthermore included a new formulation of the water vapor continuum by *Paynter and Ramaswamy* (2011). This causes the albedo of an H_2O-rich atmosphere to be substantially lower than previously calculated. Consequently, the moist greenhouse limit for the inner edge moves out to 0.99 AU (Fig. 1). This does not indicate that Earth is actually that close to it, as these calculations continue to assume a fully saturated troposphere and omit cloud feedback (*Kitzmann et al.,* 2010; *Zsom et al.,* 2012). *Abe et al.* (2011) and *Leconte et al.* (2013) performed three-dimensional modeling of rocky planets and show that the inner boundary of the HZ strongly depends on the water reservoir of the planet itself. Water ice may form at the polar caps or on the nightside of a tidally locked planet, allowing for liquid water at some transitional regions. *Von Paris et al.* (2010) and *Wordsworth*

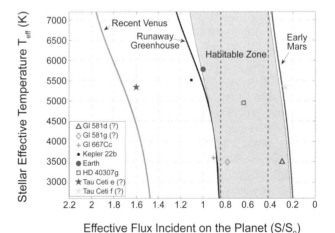

Fig. 1. Habitable-zone boundaries in incident flux as a function of stellar T_{eff}. The gray area is delineated by the (conservative) moist greenhouse (inner border) and the maximum greenhouse (outer border). Somewhat wider HZ are possible, especially related to empirical findings on early Mars and recent Venus. The loci of various exoplanets and Earth are marked. From *Kopparapu et al.* (2013) (copyright AAS, reproduced with permission).

et al. (2010) presented the first (one-dimensional) studies to simulate the climate of Gl 581d, which lies close to the outer edge of the HZ, finding that the planet could be potentially habitable assuming several bars of a CO_2-atmosphere. *Wordsworth et al.* (2011) studied Gl 581d with a three-dimensional climate model, also showing that not only does the amount atmospheric CO_2 determines its habitability, but also the water reservoir and the resonance between the orbital and rotational period of the planet.

The fact that a planet lies within the HZ of its parent star does not guarantee that it will be habitable. Life depends on other elements besides carbon, oxygen, and hydrogen and on other compounds besides water. In particular, the elements nitrogen, phosphorus, and sulfur all play critical roles in terrestrial life. Nitrogen is a constituent of both the amino acids in proteins and the nucleic acids in RNA and DNA. Phosphorus is needed for the phosphate linkages between nucleotides in the latter two compounds, and sulfur is needed for some particular amino acids. The compounds formed by these elements are termed "volatiles" because they are typically gases or liquids at room temperature. Habitable planets therefore require sources of volatiles. Earth's volatiles are mostly thought to have originated from the asteroid belt or beyond (*Morbidelli et al.,* 2000; *Raymond et al.,* 2004), and hence modes of planetary accretion must be considered. To hold onto its water, a planet needs an atmosphere in addition to H_2O vapor. Planets that are too small, like Mars, are thought to lose their atmospheres by various thermal and nonthermal processes, as discussed later in this chapter. Planetary magnetic fields may also be important. Mars, which lacks an intrinsic magnetic field, has had much of its atmosphere stripped by direct interaction with the solar wind. Venus, on the other hand, also lacks an intrinsic magnetic field, yet it has successfully held onto its dense CO_2 atmosphere, suggesting that planetary size is important. Atmospheric composition may also be important, as a less CO_2-rich atmosphere on Venus would have been hotter and more extended in its upper parts and may not have been as well retained. In the following sections of this chapter, we discuss in more detail various factors that contribute to planetary habitability. We focus on *astrophysical* conditions, while geophysical factors (such as internal heat sources, plate tectonics, landmass fraction, outgassing, volcanism, magnetic dynamos), lower-atmosphere issues (climate, winds, clouds), or biological conditions (damaging radiation doses, extreme biological environments) are not considered here.

3. DYNAMICS AND WATER TRANSPORT

3.1. Dynamics and Stability

The evolution of a biosphere is a very gradual process occurring over long time intervals. The long-term stability of a terrestrial planet moving in the HZ is therefore certainly one of the basic requirements for the formation of life on such a planet. The eccentricity of a planetary orbit is a cru-cial factor in orbital dynamics because only a sufficiently small value guarantees that the planet remains within the HZ. However, to assure appropriate conditions for habitability, the dynamics of the whole planetary system matter. For extensive reviews, we refer the reader to the chapters by Raymond et al., Baruteau et al., Benz et al., and Davies et al. in this volume. We focus on selected issues directly relevant for HZ planets here.

Observations of planetary systems and numerical studies indicate that planets are not born where we find them today. Simulations of the late stage of formation of the outer solar system show that the giant planets arrived at their current configuration through a phase of orbital instability. The so-called Nice models (e.g., *Tsiganis et al.,* 2005, *Morbidelli et al.,* 2007) provide the best-matching results for the final configuration of the outer solar system. These simulations start with a more compact system of the four giant planets (in a multi-resonant configuration in the latest version); it then experiences slight changes in the orbits due to planetesimal-driven migration leading to a resonance crossing of Saturn. This crossing causes an unstable phase for the giant planets with faster migration toward the final orbits. While during planetesimal-driven migration dynamical friction dampens orbit eccentricities, the resonance crossing will increase eccentricities and the orbits will diverge from each other (*Chiang,* 2003; *Tsiganis et al.,* 2005).

Since Jupiter and Saturn exert a strong influence on the inner solar system through secular perturbations, a change of the orbital separation of the two giant planets will modify the secular frequencies of their orbits. For example, *Pilat-Lohinger et al.* (2008) studied the secular influence of Jupiter and Saturn on the planetary motion in the HZ of a G2V star for various separations of the two giants. While Jupiter was fixed at its actual orbit of 5.2 AU, Saturn's initial semimajor axis was varied between 8.0 and 11.0 AU.

Figure 2 gives a global view of the perturbations of various Jupiter-Saturn configurations on the HZ of a Sun-type star. We can clearly see that most of the area in the a_{t_p}-a_s plot (semimajor axes for the terrestrial planet and Saturn, respectively) is not perturbed by the two giant planets, but note the arched band showing higher maximum eccentricities for test planets. This is due to the secular frequency associated with the precession of the perihelion of Jupiter. Special attention should be paid to the location of Earth's orbit, where strong variations of the eccentricity (up to ≈ 0.6) can be observed for semimajor axes of Saturn between 8.5 and 8.8 AU. We thus see that the dynamical evolution of giant planets is of utmost relevance for habitability conditions on Earth-like planets.

The huge diversity of planetary systems differing significantly from the architecture of our solar system suggests very different evolutionary tracks where processes like planet migration, planet-planet scattering, resonant trapping, mutual collisions, or hyperbolic ejections influence the final constitution of a planetary system. Observations reveal large eccentricities for many exoplanets, probably indicating phases of strong interactions.

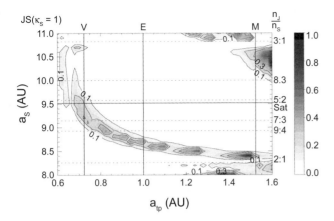

Fig. 2. Maximum-eccentricity map for test planets under the influence of Jupiter and Saturn. The x axis shows the initial semimajor axis of the test planet and the y axis indicates the variation of the semimajor axis of Saturn. Vertical solid lines indicate the positions of Venus (V), Earth (E), and Mars (M), and the horizontal solid line shows the actual position of Saturn (Sat). The dashed horizontal lines represent the mean-motion resonances (MMRs) of Jupiter and Saturn. The gray scale indicates maximum eccentricities of the test planets. From *Pilat-Lohinger et al.* (2008) (copyright AAS, reproduced with permission).

Since all planets in our solar system move in nearly circular orbits, it appears that its instability phase was moderate, probably enabling the successful formation and evolution of life on Earth. It may then be questionable whether planetary systems harboring a giant planet in an eccentric orbit can provide suitable conditions for habitability at all.

On a further note, recalling that most of the stars in our galaxy are members of binary or multiple star systems, additional gravitational perturbations from the binary interactions should also be taken into account for the assessment of habitability, as shown in detail by *Eggl et al.* (2012), *Kane and Hinkel* (2013), *Kaltenegger and Haghighipour* (2013), and *Haghighipour and Kaltenegger* (2013).

3.2. Water Transport

Where did the essential water on Earth in the amount of 0.02% then come from? Simulations show (e.g., *O'Brien et al.,* 2006) that our terrestrial planets likely formed out of material located in a ring around their present locations, i.e., they formed in the region of the HZ. However, both meteoritic evidence as well as evolutionary disk models show that the disk in this region was too warm to allow for the presence of sufficient amounts of water in solids. Therefore, water (ice) delivery should have happened later; the sources were unlikely to be comets, because statistically their collision probability with Earth is too low (*Morbidelli,* 2013). Current thinking is that the chondritic material from the outer asteroid belt is primarily responsible for Earth's water content. Water from chondritic asteroid material also — in contrast to many comets — on average matches

Earth's ocean water deuterium/hydrogen (D/H) ratio. For a summary on these aspects, we also refer to the chapter by van Dishoeck et al. in this volume.

The growth of the planetesimals to protoplanets and finally planets is closely connected with the collisions of the bodies in the early evolution of a planetary system. Such collisions may lead to the enrichment of water, depending on the origin of the colliding bodies. Various simulations (e.g., *Raymond et al.,* 2004, 2009; *O'Brien et al.,* 2006) (see the chapter by Raymond et al. in this volume) produce solar systems akin to ours, but aiming at reproducing properties of our own solar system in detail has repeatedly led to results contradicting one or several properties [masses of inner planets, eccentricities, water content of inner planets, asteroid belt architecture (*Raymond et al.,* 2009)]. The simulations, however, very sensitively depend on the initial conditions and the distribution of the planetesimals. Growth of terrestrial planets via collisions (with bodies from the dry inner regions, and also the water-rich outer regions of the asteroid belt) is often thought to take place when the disk gas has disappeared and only massive solid bodies remain. The mutual perturbations between these planetary embryos lead to orbital crossings and collisions; additionally, giant planets such as Jupiter trigger collisions.

The general assumption now is that about 10% of the mass of Earth accreted from beyond 2.5 AU, suggesting that the composition of the relevant planetary embryos is close to the composition of carbonaceous chondrites, which contain on the order of 5–10% of water (e.g., *Lunine et al.,* 2011).

Walsh et al. (2011) developed a new model that seems to explain several solar system features: A planetesimal disk still embedded in gas is present inside 3 AU (i.e., volatile-poor S-types) as well as between and outside the gas giants (water-rich C-types). Jupiter migrated inward when Saturn was still accumulating mass; Saturn also started to migrate inward because of scattering of rocky planetesimals out from the inner part of the belt. The two planets eventually fell into the 2:3 mean-motion resonance, with Jupiter located at 1.5 AU; meanwhile, the giants had scattered S-type planetesimals outward and confined the inner planetesimal disk to about 1–1.5 AU. Now, at resonance, the inward migration stopped and both giants started to migrate outward again, up to the moment when the disk gas had dissipated. During the outward motion, Jupiter and Saturn scattered predominantly dry respectively wet planetesimals back into the inner solar system, forming the asteroid belt with its characteristic gradient of volatiles. Further scattering and collision of the forming terrestrial planets with the water-rich planetesimals eventually accumulated water on the inner planets. This "Grand Tack" scenario is depicted in Fig. 3 (after *Walsh et al.,* 2011; *Morbidelli,* 2013). Its end point could be the starting point for the Nice model mentioned in section 3.1, organizing the architecture of the present-day outer solar system and finally inducing the late heavy bombardment after some 800 million years (m.y.). The Grand Tack model successfully explains several

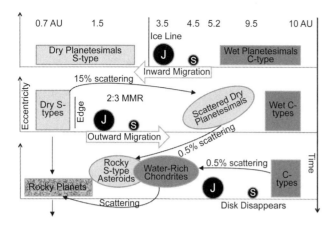

Fig. 3. Schematic view of the formation of planets in the habitable zone of the solar system according to the "Grand Tack" model. After *Walsh et al.* (2011) and *Morbidelli* (2013).

features relevant for habitability of the inner solar system planets (eccentricities, planetary masses, water content).

It is important that the collisions building up the water content on Earth occurred with velocities smaller than the escape velocity from the planetesimals or protoplanets, in order to ensure that complete merging of the bodies occurs (*Stewart and Leinhardt,* 2012; *Dvorak et al.,* 2012, *Maindl and Dvorak,* 2014; *Maindl et al.,* 2013). Obviously, the scenario for the water delivery to Earth is still incomplete and depends highly on the initial conditions one is using.

4. THE HOST STAR IN TIME

The conditions on the surface and in the atmospheres of terrestrial planets are controlled by the radiative and particle output of the host star; as this output varies on timescales of minutes to billions of years, conditions in planetary atmospheres are temporarily or permanently modified.

The photospheric radiative output is a consequence of the nuclear energy production in the star's interior; it dominates the optical and partly the near-ultraviolet (NUV) spectrum. The star's effective temperature and its luminosity follow from the laws of stellar structure and consequent nuclear reactions in the core, which gradually vary in time.

In contrast, radiation in the UV (2100–3500 Å), the far-ultraviolet (FUV) (912–2100 Å), the extreme ultraviolet (EUV) (100–912 Å), and the X-ray (<100 Å) range is a consequence of (often violent) energy release in the stellar atmosphere due to annihilation of unstable magnetic fields. Magnetic activity is expressed in cool photospheric spots, chromospheric plage emitting in the UV, and hot (million-degree) coronal plasma in "magnetic loops" with strong X-ray radiation; also, highly energetic, accelerated particles, shock fronts from mass ejections, short-term energetic flares, and perhaps also the solar wind indicate the operation of magnetic energy release. Magnetic fields are ultimately generated in a magnetic dynamo in the stellar

interior, driven by the interaction between convection and differential rotation. The short wavelength stellar output is therefore a function of stellar rotation (depending on stellar age) and of the stellar spectral type (determining the depth of the convection zone, itself also somewhat dependent on age).

A late-type star's rotation period, P, steadily declines during its main-sequence (MS) life (see the chapter by Bouvier et al. in this volume). This is a consequence of the stellar magnetic field permeating the ionized stellar wind, which transports away angular momentum. The declining rotation period itself feeds back to the dynamo, weakening the production of magnetic activity. As a consequence, P not only declines in time but it converges, after a few hundred million years for a solar analog, to a value dependent on age but independent on the initial zero-age main-sequence (ZAMS) rotation rate. Magnetic activity and the related XUV (FUV, EUV, and X-ray) output therefore also become a function of stellar age. A generalized formula has been derived by *Mamajek and Hillenbrand* (2008), namely, $P = 0.407([B–V]_0–0.495)^{0.325}t_6^{0.566}$[d] where $[B–V]_0$ is the intrinsic stellar B–V color and t_6 is the stellar age in million years. For younger stars, the rotation period also depends on the presence of a circumstellar disk [around pre-main sequence (PMS) stars] leading to "disk-locked," relatively slow rotation, and the history of disk dispersal before the arrival on the main sequence (MS), leading to a wide range of ZAMS rotation periods (*Soderblom et al.,* 1993).

4.1. The Evolution of Stellar Optical Radiation

Before arriving on the MS, a star's luminosity first decreases when it moves down the Hayashi track as a T Tauri star; in that phase, the Sun was several times brighter than today, although its effective temperature was lower (T ≈ 4500 K) and its radius consequently larger (≈2–3 $R_\odot$). Before reaching the ZAMS, solar-like stars reach a minimum bolometric luminosity, L_{bol}, which for a solar-mass star is as low as ≈0.45 $L_{bol,\odot}$ [at an age of ≈13 Ma (*Baraffe et al.,* 1998)].

The Sun's bolometric luminosity has steadily increased since its arrival on the ZAMS, starting from ≈70% of today's luminosity. At the same time, the effective temperature also increased somewhat (*Sackmann and Boothroyd,* 2003). This increase is a consequence of the evolution of nuclear reactions in the core of the star. Similar trends hold for other cool MS stars.

During the later MS evolution, the Sun's brightness will increase up to ≈3 $L_\odot$ at an age of 10 Ga. The significant change in L_{bol} implies that the classical HZ around the star moves outward during this evolution. *Kasting et al.* (1993) derived the HZ width for the entire MS evolution of the Sun (and cooler stars), defining the "continuously habitable zone." The latter becomes narrower as the inner border moves outward while — under the assumption that planets moving into the HZ at a later time do not evolve into habitable planets — the outer border remains the same.

Therefore, the total width of the HZ narrows as the star evolves. Earth will in fact be inside the inner HZ radius at an age of ≈7 Ga, i.e., only 2.4 G.y. in the future (for the most pessimistic models) and so it will lose its water. This effect is less serious for later-type stars for the same time interval because of their slower evolution.

4.2. Evolution of Stellar Ultraviolet Radiation

The stellar FUV-NUV flux ($91.2 \leq \lambda \leq 350$ nm) covers the transition from the emission spectrum of the hot chromospheric/transition region gas, at typical temperatures of 10^4–10^5 K, to the rising photospheric emission, dominating above ~200 nm. This wavelength range contains the strongest emission line, H Lyα, which is responsible for more than 50% of the stellar high-energy (XUV) flux. The UV domain is especially relevant in the context of planetary atmospheres since UV photons trigger a variety of photochemical reactions (e.g., *Hu et al.,* 2012; *Yung and DeMore,* 1999) (section 5.1 below). Important molecules such as H_2O, CO_2, CH_4, CO, and NH_3 have photodissociation cross sections that become significant below 200 nm. Also, the abundance of the biosignature molecule O_3 is mainly controlled by photochemical reactions, hence understanding the interaction of the stellar radiation with the atmospheric chemistry is of great importance for the determination of possible atmospheric biosignatures. A reliable calculation of the photodissociation rates and their time evolution is very complex due to the interplay between the strongly decreasing cross sections and the rapidly rising stellar flux toward longer wavelengths. For example, the flux of solar-like stars increases by more than 2 orders of magnitude from 150 to 200 nm and the photodissociation cross section of, e.g., H_2O, decreases by more than 3 orders of magnitude in the same interval.

The FUV-NUV flux from low-mass stars is variable over many timescales, including hours (flares), days (rotation period), years (activity cycles), and billions of years (rotational spindown). We focus here on the long-term evolution, which is comparable to the nuclear timescale. The UV photospheric flux (i.e., $\lambda \gtrsim 200$ nm) rises with time as the star evolves off the ZAMS and increases its radius and effective temperature. For the case of the Sun, the overall MS UV flux increase ranges within 30% to 200% depending on the wavelength, as illustrated by *Claire et al.* (2012) on the basis of Kurucz atmosphere models.

The chromospheric and transition region emission is much more difficult to describe, as it depends on the evolution of the magnetic properties of the star. A successful technique employed to investigate the long-term evolution of stellar activity is to identify stars that can act as proxies for different ages (e.g., *Ribas et al.,* 2005). In the UV this is especially complicated by the difficulty in finding stars that are close enough to an evolutionary track and by the need to use space facilities [Far Ultraviolet Spectroscopic Explorer (FUSE), Galaxy Evolution Explorer (GALEX), International Ultraviolet Explorer (IUE), Hubble Space

Telescope (HST)] that have only been able to provide high-quality data for a limited number of nearby objects. The best example of the application of the proxy technique is the star κ¹ Cet, which is a close match to the Sun at an age of ~0.6 Ga, as presented by *Ribas et al.* (2010). *Claire et al.* (2012) generalized the study to the entire MS evolution of solar-like stars, resulting in approximate flux multipliers as a function of wavelength that can be applied to the current solar flux to calculate the spectral irradiance at different ages. Figure 4 plots the flux multipliers. Better constraints to the UV flux evolution of solar-like stars will follow from observations with new instrumentation (*Linsky et al.,* 2012).

The flux multipliers of *Claire et al.* (2012) are appropriate for the Sun and solar-like stars but may not be adequate for stars of lower mass. Measurements of UV fluxes of a handful of M dwarfs are available (*France et al.,* 2013, and references therein). France et al. show that the flux in the FUV band relative to the NUV band for most M dwarfs is 3 orders of magnitude larger than the solar value, and this has strong photochemical implications. But, in spite of these measurements, we still lack information on the time evolution of such fluxes, the main reason being the difficulty in estimating stellar ages. The development of new alternative age determination methods for MS K and M dwarfs (*Garcés et al.,* 2011) should help to define the necessary UV flux-time relationships.

The determination of the H Lyα line flux deserves special attention. Precise measurement of the H Lyα line is extremely challenging due to contamination from Earth geocoronal emission and the difficulty of correcting interstellar and astrospheric absorption (*Wood et al.,* 2005). *Linsky et al.* (2013) present a compilation of HST's Space Telescope Imaging Spectrograph (STIS) observations of a relatively large sample of stars from F to M spectral types,

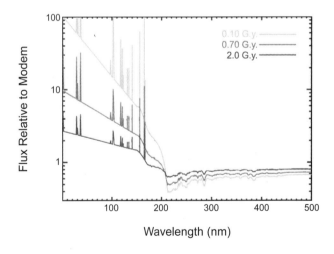

Fig. 4. Flux ratios relative to the modern (present-day) Sun at 1 AU covering the UV range of solar-like stars of different ages. Both continuum and line fluxes are included using the procedure in *Claire et al.* (2012). Adapted from *Claire et al.* (2012) (copyright AAS, reproduced with permission).

with Lyα fluxes and correlations with other emission line fluxes. When normalizing fluxes to the stellar HZ, Linsky et al. show a clear increasing trend of Lyα flux with decreasing stellar T_{eff}. M dwarfs produce, on average, a nearly 1 order of magnitude higher Lyα flux in their HZ than do solar-like stars in their respective HZ. Unfortunately, most of the stars used by Linsky et al. have no well-determined ages and thus a time-evolution determination is not possible. The Lyα flux evolution relationship of *Ribas et al.* (2005) for solar analogs is still up to date

$$F \approx 19.2 t_9^{-0.72} \left[\text{erg s}^{-1} \text{ cm}^{-2} \right] \quad (1)$$

where F is the normalized flux at a distance of 1 AU and t_9 is the stellar age in billion years.

4.3. Evolution of High-Energy Radiation and Activity

The strongest evolutionary change in magnetically induced stellar radiation is seen in the high-energy range including EUV, X-ray, and gamma-ray radiation. The integrated luminosity in the soft X-ray (0.1–10 keV photon energy) and the EUV (0.014–0.1 keV) depends on the stellar rotation period, P, and therefore age; for a solar analog (*Güdel et al.,* 1997; *Sanz-Forcada et al.,* 2011)

$$L_X \approx 10^{31.05 \pm 0.12} P^{-2.64 \pm 0.12} \left[\text{erg s}^{-1} \right] \quad (2)$$

$$L_X \approx (3 \pm 1) \times 10^{28} t_9^{-1.5 \pm 0.3} \left[\text{erg s}^{-1} \right] \quad (3)$$

$$L_{EUV} \approx (1.3 \pm 0.3) \times 10^{29} t_9^{-1.24 \pm 0.15} \left[\text{erg s}^{-1} \right] \quad (4)$$

(t_9 is the stellar age in billion years). However, these dependencies only hold for sufficiently slow rotation. For young, rapid rotators, magnetic activity and therefore the X-ray output reaches a saturation level. Empirically, $L_X \approx 10^{-3} L_{bol}$ for all late-type MS and PMS stars.

These dependencies can be generalized to stars of other spectral class. For example, using the Rossby number Ro = P/τ_c (where τ_c is the spectral-class and age-dependent convective turnover time), one finds a unified description of the L_X–Ro and L_X–P dependencies (*Pizzolato et al.,* 2003), namely $L_X \approx 10^{-3} L_{bol}$ for rapid rotators (young age) and $L_X \propto P^{-2}$ for slow rotators (more evolved stars).

Combining these trends with the mass- or spectral-type-dependent spindown rate of a star has important consequences for the stellar high-energy output in time. Attempts to unify the X-ray decay law were described in *Scalo et al.* (2007), *Guinan et al.* (2009), and *Mamajek and Hillenbrand* (2008). M dwarfs, for example, remain at the saturation level for much longer than solar-like stars, up to on the order of a billion years [as seen, for example, in the Hyades cluster (*Stern et al.,* 1995)], in contrast to a solar analog that leaves this regime typically in the first 100 m.y.

A second evolutionary trend in X-rays is due to the average coronal temperature, resulting in a long-term trend in spectral hardness. X-ray hardness is an important factor determining the penetration depth of X-rays into a planetary atmosphere. While the present-day solar corona shows an average temperature, T_{cor}, of about 2 MK, young, magnetically active stars reveal T_{cor} around 10 MK and even higher, implying a corresponding shift of the dominant spectral radiation to harder photon energies (≈ 1 keV). The empirical relation for solar-mass stars is (*Telleschi et al.,* 2005)

$$L_X = 1.61 \times 10^{26} T_{cor}^{4.05 \pm 0.25} \left[\text{erg s}^{-1} \right] \quad (5)$$

This relation holds similarly for PMS stars.

Short-wavelength (UV to X-ray) emission-line fluxes $F(T_{max},t)$ for maximum line-formation temperatures T_{max} ($4 \leq \log T_{max} \leq 7$) decay in time ($t_9$ in billion years) as

$$F(T_{max}, t) = \alpha t_9^{-\beta} \quad (6)$$

$$\beta = 0.32 \log T_{max} - 0.46 \quad (7)$$

(*Güdel,* 2007) [similar relations hold for the continuum, where T_{max} corresponds to hc/(λk), h is the Planck constant, k the Boltzmann constant, c the speed of light, and λ the continuum wavelength]. In other words, the decay is steeper for shorter wavelengths, implying the largest enhancement factors for young stars for the shortest wavelengths (Fig. 5). The average enhancement factors are summarized in Table 1 for a solar analog throughout its MS life. We emphasize that enhancement factors for individual young stars may

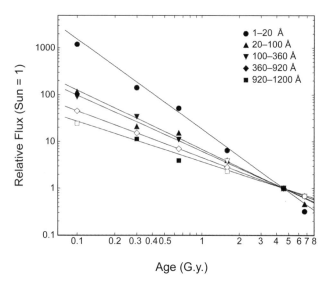

Fig. 5. Power-law decay of spectral output of a solar analog during its MS (age 0.05–10 G.y.) life, normalized to the present-day solar flux. The shortest-wavelength flux decays the fastest. From *Ribas et al.* (2005) (copyright AAS, reproduced with permission).

TABLE 1. Enhancement factors of X-ray/EUV/XUV/FUV fluxes in solar history
(normalized to ZAMS age of 4.6 G.y.).

Solar Age (G.y.)	Time Before Present (G.y.)	X-Rays (1–20 Å)	Soft-X (20–100 Å) EUV (100–360 Å)	FUV (920–1180 Å)
			Enhancement in	
0.1	4.5	1600*	100	25
0.2	4.4	400	50	14
0.7	3.9	40	10	5
1.1	3.5	15	6	3
1.9	2.7	5	3	2
2.6	2.0	3	2	1.6
3.2	1.4	2	1.5	1.4
4.6	0	1	1	1

*Large scatter possible due to unknown initial rotation period of Sun.

scatter around the given values because the rotation period may not have converged to an age-independent value on the MS. We refer to the chapter of Bouvier et al. in this volume for more details on the rotational evolution of stars.

4.4. The Evolution of Stellar Winds and Mass Loss

In addition to being exposed to electromagnetic radiation from their host stars, planets are also exposed to high-speed outflows of particles from the stellar atmosphere. For cool MS stars like the Sun, stellar winds arise in the hot coronae that represent the outermost atmospheres of the stars. Although the mechanisms of coronal heating and coronal wind acceleration remain hot topics of research, *Parker* (1958) demonstrated long ago that once you have a hot corona, a wind much like that of the Sun arises naturally through thermal expansion. Thus, any star known to have a hot corona can be expected to possess a coronal wind something like that of the Sun. Observations from X-ray observatories such as Einstein, the ROentgen SATellite (ROSAT), Chandra, and the X-ray Multi-Mirror Mission (XMM-Newton) have demonstrated that X-ray emitting coronae are ubiquitous among cool MS stars, so coronal winds can be expected to be ubiquitous as well.

Unfortunately, detecting and studying these winds is much harder than detecting and studying the coronae in which they arise. Current observational capabilities do not yet allow us to directly detect solar-like coronal winds emanating from other stars. Bremsstrahlung radio emission may uncover ionized winds. An upper limit for the mass-loss rate of $\dot{M}_w \leq 2 \times 10^{-11}$ $M_\odot$ yr^{-1} was inferred for the F5 IV-V subgiant Procyon (*Drake et al.*, 1993), and $\dot{M}_w \leq 7 \times 10^{-12}$ $M_\odot$ yr^{-1} for the M dwarf Proxima Centauri [using a wind temperature $T_w = 10^4$ K and a wind velocity $v_w = 300$ km s^{-1} (*Lim et al.*, 1996)]. *Van den Oord and Doyle* (1997) also found $\dot{M}_w \leq 10^{-12}$ $M_\odot$ yr^{-1} for observed dMe stars assuming $T_w \approx 1$ MK. For young (a few hundred million year) solar analogs, *Gaidos et al.* (2000) found upper limits of $\dot{M}_w \leq (4–5) \times 10^{-11}$ $M_\odot$ yr^{-1}.

Sensitive upper limits can also be derived from the absence of radio attenuation of coronal (flare or quiescent) radio emission by an overlying wind (*Lim and White*, 1996). This method applies only to spherically symmetric winds. The most sensitive upper limits based on low-frequency radio emission apply to the dMe star YZ CMi, with $\dot{M}_w \leq 5 \times 10^{-14}$ $M_\odot$ yr^{-1}, $\leq 10^{-12}$ $M_\odot$ yr^{-1}, and $\leq 10^{-11}$ $M_\odot$ yr^{-1} for $v_w = 300$ km s^{-1} and $T_w = 10^4$ K, 10^6 K, and 10^7 K, respectively.

Alternatively, neutrals of the interstellar medium flowing into an astrosphere blown by a stellar wind may induce charge exchange that could in principle be detected by the ensuing line emission at X-ray wavelengths. Spatially resolved observations are needed, but again, no detection has been reported yet (see *Wargelin and Drake*, 2001).

So far, the only successful method relies not on detecting the winds themselves, but their collision with the interstellar medium (ISM) (*Linsky and Wood*, 1996). The collision regions are called "astrospheres," analogous to the "heliosphere" that defines the solar wind's realm of influence around the Sun. Because the ISM surrounding the Sun is partially neutral, the solar wind/ISM collision yields populations of hot hydrogen gas throughout the heliosphere created by charge exchange processes. The highest HI densities are in the so-called "hydrogen wall" just beyond the heliopause, hundreds of astronomical units from the Sun. Ultraviolet spectra from the HST have detected Lyα absorption from both the heliospheric hydrogen wall (*Linsky and Wood*, 1996), and astrospheric hydrogen walls surrounding the observed stars (*Gayley et al.*, 1997; *Wood et al.*, 2002, 2005). In the case of astrospheres, the amount of absorption is related to the strength of the stellar winds, so with guidance from hydrodynamic models of the astrospheres, stellar mass loss rates have been inferred from the astrospheric Lyα data. The number of astrospheric wind measurements is only 13, however, and many more measurements are required for further progress to be made in this area.

These observations imply that winds generally correlate with stellar activity (*Wood et al.*, 2005) (Fig. 6). The mass-loss rate $\dot{M}_w$ per unit stellar surface correlates with the stellar X-ray surface flux F_X; correspondingly, the total

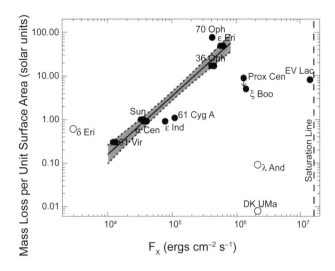

Fig. 6. Mass-loss rates per unit surface area vs. stellar X-ray surface fluxes. Main-sequence stars are shown by filled circles. From *Wood et al.* (2005) (copyright AAS, reproduced with permission).

mass-loss rate $\dot{M}_w$ correlates with the X-ray luminosity L_X and therefore age, t

$$\dot{M}_w \propto L_X^{1.34 \pm 0.18} \qquad (8)$$

$$\dot{M}_w \propto t^{-2.33 \pm 0.55} \qquad (9)$$

where for the second formula the activity-age relation has been used (*Wood et al.*, 2005). Extrapolating the above law up to the X-ray saturation limit at $F_X \approx 2 \times 10^7$ erg cm^{-2} s^{-1}, we would infer $\dot{M}_w$ of the youngest solar analogs to exceed the present-day solar mass loss [$\dot{M}_\odot \approx 2 \times 10^{-14}$ $M_\odot$ yr^{-1} (e.g., *Feldman et al.*, 1977)] by a factor of 1000. However, there is some evidence that the most active stars with $F_X \gtrsim 8 \times 10^5$ erg cm^{-2} s^{-1} are inconsistent with this correlation and may actually have surprisingly weak winds (*Wood et al.*, 2005). Mass loss appears to increase with activity up to about $\dot{M}_w = 100\times$ the current solar wind, but then appears to actually weaken toward even higher stellar activity, down to about $10\times$ the current solar mass-loss rate. Possibly, the appearance of high-latitude active regions (spots) with a magnetic field more akin to a global dipole may inhibit wind escape (*Wood et al.*, 2005).

We emphasize that this method requires the presence of *neutrals* (e.g., at least a partially neutral cloud) around the observed star. If the interstellar medium around the star is fully ionized, no hydrogen walls can form, and the absence of absorption features in Lyα will *not* indicate weak winds.

4.5. Stellar Flares

A stellar flare is the result of short-term (minutes to hours) explosive energy release tapped from nonpotential energy in the coronal magnetic fields, ultimately derived

from convective energy at the stellar surface. During flares, plasma in the stellar atmosphere is rapidly heated, and therefore fluxes increase across the electromagnetic spectrum, in very rare cases exceeding the entire quiescent stellar output (*Osten et al.*, 2010). Flare radiation amplitudes are particularly high in the short-wavelength (XUV, gamma-ray) range, and also in the radio regime.

Flares occur at rates depending on the flare amplitude or its released (radiative or total) energy, E, in the form of a power law, $dN/dE \propto E^{-\alpha}$, where $\alpha \approx 2 \pm 0.4$ has been found by various investigators, i.e., small flares occur at a much higher rate than the rare big flares (*Güdel*, 2004, and references therein).

Very energetic flares or "superflares" may contribute to the ionization of upper planetary atmospheres (ionospheres) and upper atmospheric chemistry, thus potentially changing planetary habitable conditions. *Audard et al.* (2000) found that the rate of large EUV flares exceeding some given radiative energy threshold is proportional to the average, long-term X-ray luminosity of the star, suggesting a very high rate of superflares in young, active stars. This relation at the same time indicates that superflares are not occurring at a higher rate in M dwarfs than in solar analogs; it is the closer-in HZ around M dwarfs combined with the slower decay of M-dwarf activity that subjects habitable planets around M dwarfs to much higher flare-radiation doses for a longer time.

Schrijver et al. (2012) investigated the presence of superflares on the Sun in historical times, using natural archives (nitrate in polar ice cores and various cosmogenic radionuclides) together with flare statistics and the historical record of sunspots; they conclude that flares in the past four centuries are unlikely to have exceeded the largest observed solar flare with total energy of $\approx 10^{33}$ erg. The solar magnetic field (and similarly, stellar magnetic fields) provides an upper bound to the maximum flare energy release simply by the amount of free magnetic energy that could coherently be converted. This limit is indeed about $\approx 10^{33}$ erg, a tenfold-stronger flare requiring a spot coverage of 10%, unobserved on the modern Sun (*Schrijver et al.*, 2012).

However, *Schaefer et al.* (2000) reported, from observations in various wavebands, nine "superflares" with total radiative energies of 10^{33}–10^{38} erg on solar (F- and G-type) analog stars, none of which is exceptionally young or extremely active or a member of a close binary system. Some of the flaring objects were even older than the Sun. They concluded that the average recurrence time for superflares on such stars may be decades to centuries, although this would obviously be an overestimate for the Sun. With the new Kepler satellite data in hand, *Maehara et al.* (2012) surprisingly found many superflares with bolometric energies of 10^{33}–10^{36} erg, with an estimated recurrence time of ≈ 800 yr for 10^{34}-erg flares on old, solar-like stars. None of these stars is known to host close-in exoplanets that could induce such energy release.

The impact of stellar flares on the atmospheres of terrestrial extrasolar planets has been studied by, e.g., *Segura et al.* (2010) and *Grenfell et al.* (2012). The importance of such single events would be their potential to melt ice surfaces on habitable planets, the possible temporary breakup of the ionosphere, and ozone depletion after creation of nitrogen oxides in the irradiated atmosphere; an event of 10^{36} erg (ionizing energy) was estimated to result in a loss of 80% of the total ozone content for more than one year, with the consequent increase in UV irradiation (*Schaefer et al.*, 2000).

4.6. Coronal Mass Ejections

Apart from stellar winds, exoplanets may interact with coronal mass ejections (CMEs), occurring sporadically and propagating within the stellar wind as large-scale plasma-magnetic structures. The high speeds of CMEs (up to thousands of kilometers per second), their intrinsic magnetic field, and their increased density compared to the stellar wind background make CMEs an active factor that strongly influences the planetary environments and magnetospheres. Often, collisions of the close-orbit exoplanets with massive stellar CME plasmas compress planetary magnetospheres much deeper toward the surface of the exoplanet. This would result in much higher ion loss rates than expected during the usual stellar wind conditions (see section 6 below).

Coronal mass ejections can be directly observed only on the Sun. They are associated with flares and prominence eruptions and their sources are usually located in active regions and prominence sites. The likelihood of CME events increases with the size and power of the related flare event. Generally, it is expected that the frequent and powerful flares on magnetically active flaring stars should be accompanied by an increased rate of CME production. Based on estimates of solar CME plasma density n_{CME}, using the *in situ* spacecraft measurements (at distances >0.4 AU) and the analysis of white-light coronagraph images (at distances ≤30 $R_\odot$ ≈ 0.14 AU), *Khodachenko et al.* (2007a) provided general power-law interpolations for the n_{CME} dependence on the orbital distance d (in AU) to a star

$$n_{CME}^{min}(d) = 4.88d^{-2.3}, \quad n_{CME}^{max}(d) = 7.10d^{-3.0} \qquad (10)$$

Equation (10) identifies a typical maximum-minimum range of n_{CME}. The dependence of stellar CME speed v_{CME} on d can be approximated by the formula

$$v_{CME} = v_0 \left(1 - \exp\left[\frac{2.8R_\odot - d}{8.1R_\odot} \right] \right)^{1/2} \qquad (11)$$

proposed in *Sheeley et al.* (1997) on the basis of tracking of several solar wind density enhancements at close distances (d < 0.1 AU). For the approximation of average- and high-speed CMEs, one may take in equation (11) $v_0 = 500$ km s^{-1} and $v_0 = 800$ km s^{-1}, respectively. Besides that, the average mass of CMEs is estimated as 10^{15} g, whereas their average duration at distances of ≈0.05 AU is close to 8 hr.

Because of the relatively short range of propagation of most CMEs, they should strongly impact the planets in close orbits (≤0.3 AU). *Khodachenko et al.* (2007a) have found that for a critical CME production rate f_{CME}^{cr} ≈ 36 CMEs per day (and higher), a close-orbit exoplanet appears to be under continuous action of the stellar CMEs plasma, so that each next CME collides with the planet during the time when the previous CME is still passing over it, acting like an additional type of "wind." Therefore investigations of evolutionary paths of close-orbit exoplanets in potential HZ around young active stars should take into account the effects of such relatively dense magnetic clouds (MCs) and CMEs, apart from the ordinary winds.

5. ATMOSPHERES AND STELLAR RADIATION

The radiation environment during the active period of a young star is of great relevance for the evolution and escape of planetary atmospheres. The visible radiation penetrates through a planetary atmosphere down to its surface unless it is blocked by clouds, aerosols, or dust particles. The average skin or effective temperature, T_{eff}, of a planet can be estimated from the energy balance between the optical/near-IR emission irradiating the planet and the mid-IR thermal radiation that is lost to space

$$T_{eff} = \left[\frac{S(1 - A)}{4\sigma} \right]^{0.25} \qquad (12)$$

with S being the solar radiative energy flux at a particular orbit location of a planet, A the Bond albedo describing the fraction of the radiation reflected from the planet, and σ the Stefan-Boltzmann constant. Most of the IR radiation emitted from Venus or Earth does not come from the surface; rather, it comes from the cloud top level on Venus or the mid-troposphere on Earth and yields T_{eff} ≈ 220 K for Venus and 255 K for Earth. Above the mesopause, radiation with short wavelengths in the X-ray, soft X-ray, and EUV part of the spectrum (the latter two defined as XUV henceforth) is absorbed in the upper atmosphere, leading to dissociation of molecular species, ionization, and heating, so that the thermosphere temperature $T_{th} \gg T_{eff}$.

5.1. Surface Climate and Chemical Processing

The surface climate conditions on terrestrial planets are largely affected by the presence of greenhouse gases (main contributors: H_2O, CO_2, O_3, CH_4, N_2O), absorbing and reemitting IR radiation and thereby heating the lower atmosphere. The greenhouse effect on the surface of Earth is about 33 K and about 460 K for the dense CO_2 atmosphere of Venus. In the troposphere of the terrestrial planets, adiabatic cooling leads to decreasing temperatures with height. In the case of Earth, the ozone (O_3) layer peaking at about 30 km height absorbs stellar radiation around 250 nm, resulting in atmospheric heating and a stratospheric temperature inversion. This is not observed for those terrestrial planets in our solar system that do not have sufficient amounts of ozone, i.e., Mars and Venus. In the context of extrasolar planets, ozone is of special interest not only for its climatic impact, but also as a potential biosignature that can be detected by spectroscopic absorption bands in the IR spectral range.

Ozone is formed in Earth's stratosphere via photolytic processes from O_2 (*Chapman*, 1930) operating at wavelengths <200 nm. The ozone molecule itself is also destroyed photolytically in the UVB and via HO_x and NO_x catalytic cycles (*Crutzen*, 1970), which are also favored by UV radiation. In Earth's troposphere, O_3 is mainly produced by the smog mechanism (*Haagen-Smit et al.*, 1952).

Terrestrial exoplanets orbit host stars of different types, i.e., stars with different effective temperatures and hence spectral energy distributions. Cool dwarf stars, such as M-types, exhibit less flux at optical and near-UV wavelengths than the Sun down to about 200 nm, depending on stellar class. They can, however, be very active in the UV range below 170–190 nm, hence at wavelengths relevant for photochemistry of O_2 and O_3 (section 4.2). Planets with oxygen-bearing atmospheres around M dwarf stars may therefore have weaker or stronger ozone layers, depending on UV fluxes below about 200 nm (e.g., *Segura et al.*, 2005; *Grenfell et al.*, 2013). Cool M dwarfs with reduced UV fluxes allow for less O_2 destruction and therefore less O_3 is produced by the Chapman mechanism. In the case of very cool M dwarfs this production mechanism could be severely reduced and the smog mechanism could take over even in the mid-atmosphere (*Grenfell et al.*, 2013). For Earth-like planets around such cool and quiet M dwarfs, atmospheric ozone abundances can be significantly reduced, whereas the same kind of planet around an active M dwarf with high UV fluxes would show a prominent ozone layer (*Segura et al.*, 2005; *Grenfell et al.*, 2014). The amount of stratospheric ozone on exoplanets around M dwarfs may therefore highly depend on their UV activity. Surface temperatures are somewhat higher on Earth-like planets (i.e., with modern Earth type N_2-O_2-dominated atmospheres) orbiting M dwarfs for the same total stellar insolation, because, e.g., Rayleigh scattering contributes less to the planetary albedo owing to the λ^{-4} dependence of the Rayleigh scattering coefficient. The largest impact of the different stellar energy flux distribution of cool stars on a planet's T-p profile is, however, seen in the mid-atmosphere. Earth-like terrestrial exoplanets around M dwarfs do not show a strong stratospheric temperature inversion, even if a thick ozone layer is present in the mid-atmosphere (*Segura et al.*, 2005; *Rauer et al.*, 2011; *Grenfell et al.*, 2013). For example, as discussed above, some active M dwarfs with very high fluxes below 200 nm can efficiently destroy oxygen to form a strong ozone layer (*Segura et al.*, 2005), but still show no stratospheric T-inversion (Fig. 7). This is primarily caused by the reduced near-UV flux of M dwarf stars around 250 nm, which is relevant for stratospheric heating. In Fig. 7, increasing UV (from an inactive M7 dwarf — "×1" — to 1000× stronger) leads to mid-atmosphere cooling because UV stimulates OH, which reduces CH_4 (an important heater). On increas-

ing UV further, however, O_3 is stimulated, which leads to strong stratospheric heating. Results suggest that a strong T-inversion is less relevant for surface climate but can make the detection of atmospheric absorption bands more difficult (e.g., *Rauer et al.*, 2011; *Kaltenegger et al.*, 2011; *Grenfell et al.*, 2014). In summary, UV can, on the one hand, enhance O_3, hence heat the mid-atmosphere, but on the other hand it can stimulate OH, hence remove CH_4, which has the opposite effect; the effect that is strongest depends on the amount of O_3, CH_4, and the UV intensity.

Other atmospheric species relevant for climate, habitability, and biosignatures, such as water, methane, or nitrous oxide, are also affected by the different host star spectral energy distributions. Water and methane are strong greenhouse gases and therefore relevant for habitability. Water is produced in the stratosphere via methane oxidation (CH_4 + $2O_2 \rightarrow CO_2 + 2H_2O$) and destroyed by photolysis. Earth-like planets around M dwarf stars show somewhat higher atmospheric water abundances for the same incident stellar flux due to higher surface temperatures and reduced stratospheric photolytic destruction. In addition, the production of O^1D is reduced due to the reduced M dwarf UV flux, affecting the production of OH via $H_2O + O^1D \rightarrow 2OH$. This in turn leads to reduced methane destruction, hence increased methane abundances in the atmosphere (e.g., *Segura et al.*, 2005; *Rauer et al.*, 2011). Also, the good biosignature N_2O (good in the sense of attributability of spectral features to biogenic activity) could be significantly increased for planets around cool M dwarfs (*Rauer et al.*, 2011; *Seager*

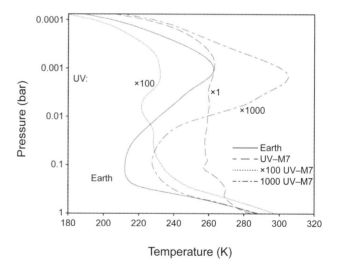

Fig. 7. Response of the atmospheric T-p profile of an Earth-like planet in the HZ of an M7 red dwarf star to UV irradiation. The profiles are for a star with very low UV flux (long dash, "×1"), 100× and 1000× this flux (dotted and long-short dash respectively). The T-p profile for the present Earth is also shown (solid line). The cooling from 1× to 100× is due to the destruction of CH_4 by UV, resulting in less heating. Very high UV levels lead to a temperature inversion in the ozone layer. Adapted from *Grenfell et al.* (2014).

et al., 2013). This is interesting because N_2O is normally not detectable for Earth-like planets around solar-like stars due to its weak spectral features, but may be detectable for planets orbiting M dwarf stars. These examples illustrate how the central star spectral energy distribution affects atmospheric chemistry, abundances, and the strength of potentially observable atmospheric molecular absorption bands. The discussion has just started, and many more parameter studies are needed to identify the most dominant effects on observable signals.

Clearly, the visibility of indicators for habitability (e.g., H_2O, CO_2, CH_4) in spectra obtained from terrestrial exoplanets are significantly affected by the central star spectral type and activity (e.g., *Segura et al.*, 2005, 2010; *Selsis et al.*, 2008; *Rauer et al.*, 2011; *Hedelt et al.*, 2013; *Grenfell et al.*, 2014; *Rugheimer et al.*, 2013). This holds in particular for M dwarfs, which are favorable targets for transit spectroscopy during primary and secondary eclipse because of their improved star/planet contrast. It has become evident that a good characterization of the host star stellar energy distribution and its temporal variability is crucial for the interpretation of absorption spectra and identification of potential false-positive signals.

Clouds are a common phenomenon in the atmospheres of many terrestrial bodies of our solar system (e.g., Venus, Earth, Mars, Titan) and are also expected in atmospheres of terrestrial extrasolar planets. They play a major role in the climate of terrestrial planets by affecting the atmospheric energy budget in several competing ways. Clouds can scatter a fraction of the incident stellar radiation back to space, resulting in atmospheric cooling. On the other hand, trapping thermal radiation within the lower atmosphere clouds can yield a greenhouse effect. This greenhouse effect can occur either by absorption and reemission at their local temperature (classical greenhouse effect) or by scattering thermal radiation back toward the planetary surface (scattering greenhouse effect). In Earth's atmosphere, for example, low-level water clouds usually cool the surface, whereas high-level water ice clouds can lead to a heating effect. The net climatic effect of a cloud is largely determined by the wavelength-dependent optical properties of the cloud particles (e.g., absorption and scattering cross sections, single-scattering albedo, asymmetry parameter, or scattering phase function). These properties can differ considerably for different cloud-forming species, particle sizes, and shapes [see *Marley et al.* (2013) for a broad review on clouds in the atmospheres of extrasolar planets].

The climatic effect of clouds has been discussed, for example, as possible solutions to the faint young Sun paradox (e.g., *Rondanelli and Lindzen*, 2010; *Goldblatt and Zahnle*, 2011a) (see section 7.1 below), for planets at the inner boundary of the HZ (e.g., *Kasting*, 1988), and their effect on exoplanet spectroscopy of planets at different ages (e.g., *Kaltenegger et al.*, 2007). The climatic effect of clouds in atmospheres of Earth-like exoplanets and the resulting effects on the potential masking of molecular absorption bands in the planetary spectra has been studied

again recently by, e.g., *Kitzmann et al.* (2010, 2011a,b, 2013), *Zsom et al.* (2012), *Rugheimer et al.* (2013), and *Vasquez et al.* (2013a,b). The impact of clouds on the climate of particular exoplanets is, however, difficult to evaluate without observational constraints on their cloud cover or all the microphysical details (e.g., condensation nuclei) that determine the cloud particle sizes and shapes (*Marley et al.*, 2013).

Active M dwarfs or even young solar-like stars may exhibit strong stellar cosmic-ray emissions. So-called air shower events lead to secondary electron emissions that can break up strong atmospheric molecules like N_2. This can perturb the atmospheric photochemistry, hence affect biosignature molecules and their detectability [see, e.g., *Segura et al.* (2010) and *Grenfell et al.* (2007, 2012) for a discussion on the effect on ozone and other spectral biosignatures]. For the extreme "flaring case" (see also section 4.5), results suggest that ozone may be strongly removed by cosmic-ray-induced NO_x (*Grenfell et al.*, 2012), whereas the nitrous oxide biosignature may survive. Again, this illustrates the importance of characterizing stellar activity when discussing atmospheric molecular signatures.

5.2. Upper Atmosphere Extreme Ultraviolet Heating and Infrared Cooling

Planets irradiated by high XUV radiation experience heating of the upper atmosphere, related expansion, and thermal escape. Ionizing XUV radiation at short wavelengths ($\lambda \leq 102.7$ nm) and dissociating radiation (UV) as well as chemical heating in exothermic three-body reactions are responsible for the heating of the upper planetary atmospheres (e.g., *Lammer*, 2013). The volume heat production rate q caused by the absorption of the stellar XUV radiation can then be written as (e.g., *Erkaev et al.*, 2013; *Lammer et al.*, 2013b)

$$q_{XUV} = \eta n \sigma_a J e^{-\tau} \qquad (13)$$

with the energy flux related to the corresponding wavelength range outside the atmosphere, J, the absorption cross-section of atmospheric species, σ_a, the optical depth, τ, and the number density, n. In today's Venus or Earth atmospheres, thermospheric heating processes occur near the homopause in the lower thermosphere, where the eddy diffusion coefficient is equal to the molecular diffusion coefficient, at an altitude of ~110 km (e.g., *von Zahn et al.*, 1980). Depending on the availability of the main atmospheric constituents and minor species, the thermospheric temperature can be reduced by IR emission in the vibrational-rotational bands of molecules such as CO_2, O_3, H_3^+, etc.

The fraction of the absorbed stellar XUV radiation transformed into thermal energy is the so-called heating efficiency η; it lies in the range of ~15–60% (*Chassefière*, 1996a; *Yelle*, 2004; *Lammer et al.*, 2009; *Koskinen et al.*, 2013). Recent calculations by *Koskinen et al.* (2013) estimate the total heating rate based on the absorption of stellar XUV radiation, photoelectrons, and photochemistry for the hydrogen-rich "hot Jupiter" HD 209458b with an orbit of 0.047 AU around a solar-like star; they found that the heating efficiency for XUV fluxes, ~450× today's solar flux at 1 AU, is most likely between ~40% and 60%. For hydrogen-rich exoplanets exposed to such high XUV fluxes, most molecules such as H_2 in the thermosphere are dissociated; thus, IR cooling via H_3^+ produced by photochemical reactions with H_2 molecules becomes less important (*Koskinen et al.*, 2007). However, depending on the availability of IR-cooling molecules in planetary atmospheres at larger orbital distances exposed to lower XUV fluxes, the heating efficiency may be closer to the lower value of ~15%.

5.3. Extreme-Ultraviolet- and X-Ray-Induced Thermal Atmospheric Escape

Tian et al. (2008) studied the response of Earth's upper atmosphere to high solar XUV fluxes in more detail and discovered that one can classify the thermosphere response into two thermal regimes. In the first regime, the thermosphere is in hydrostatic equilibrium. In this stage, the bulk atmosphere below the exobase can be considered as static, similar to the thermospheres of the solar system planets today. However, if the star's EUV flux was/is much higher than that of the present Sun, or if the planet's atmosphere is hydrogen-rich (*Watson et al.*, 1981), the thermosphere can enter a second regime. There, the upper atmosphere is heated to such temperatures that the thermosphere becomes nonhydrostatic. In the hydrodynamic flow regime, the major gases in the thermosphere can expand very efficiently to several planetary radii above the planetary surface. As a result, light atoms experience high thermal escape rates and expand to large distances.

In the hydrostatic regime the upper atmosphere experiences classical Jeans escape where particles populating the high-energy tail of a Maxwell distribution at the exobase level have escape energy so that they are lost from the planet. The so-called Jeans parameter, $\lambda_J(r)$, can generally be expressed as (e.g., *Chamberlain*, 1963)

$$\lambda_J(r) = \frac{GM_{pl}m}{rkT(r)} \qquad (14)$$

with the gravitational constant G, the planetary mass M_{pl}, the mass of an atmospheric species m, the Boltzmann constant k, and the upper atmosphere temperature T as a function of planetary distance r. As long as $\lambda(r)$ is >30, an atmosphere is bound to the planet. For lower values, thermal escape occurs and can become very high if $\lambda(r) \sim$ 2–3.5 (*Volkov and Johnson*, 2013), depending on the atmospheric main species, the XUV flux, and planetary parameters.

In the nonhydrostatic regime, it is still possible that not all atoms will reach escape velocity at the exobase level. In such cases, one can expect hydrodynamically expanding thermospheres, where the loss of the upward-flowing

atmosphere results in a strong Jeans-type or controlled hydrodynamic escape, but not in hydrodynamic blow-off, where no control mechanism influences the escaping gas.

Only for $\lambda(r) < 1.5$ does hydrodynamic blow-off occur, and the escape becomes uncontrolled. This happens when the mean thermal energy of the gases at the exobase level exceeds their gravitational energy. Hydrodynamic blow-off can be considered as the most efficient atmospheric escape process. In this extreme condition, the escape is very high because the whole exosphere evaporates and will be refilled by the upward-flowing planetary gas of the dynamically expanding thermosphere as long as the thermosphere can remain in this extreme condition. In such cases the thermosphere, accompanied by adiabatic cooling, starts to dynamically expand to several planetary radii (*Sekiya et al.*, 1980; *Watson et al.*, 1981; *Chassefière*, 1996a; *Tian et al.*, 2005, 2008; *Kulikov et al.*, 2007).

To study the atmospheric structure of an upward-flowing, hydrodynamically expanding thermosphere, one has to solve the set of the hydrodynamic equations. For exoplanets orbiting very close to their host stars, one cannot neglect external forces such as gravitational effects related to the Roche lobe (*Penz et al.*, 2008).

5.4. Extreme-Ultraviolet- and X-Ray-Powered Escape of Hydrogen-Rich Protoatmospheres

The most efficient atmospheric escape period starts after the planetary nebula dissipated and the protoplanet with its nebula-captured hydrogen-dominated envelope is exposed to the extreme radiation and plasma environment of the young star. The time period of this extreme escape process depends generally on the host star's evolving XUV flux, the planet's orbit location, its gravity, and the main atmospheric constituents in the upper atmosphere. Some planets can be in danger of being stripped of their whole atmospheres; on the other hand, if the atmospheric escape processes are too weak, a planet may have problems getting rid of its protoatmosphere (*Lammer et al.*, 2011b,c; *Lammer*, 2013; *Erkaev et al.*, 2013; *Kislyakova et al.*, 2013). Both scenarios will have essential implications for habitability.

That the early atmospheres of terrestrial planets were even hydrogen-dominated or contained more hydrogen than at the present day was considered decades ago by researchers such as *Holland* (1962), *Walker* (1977), *Ringwood* (1979), *Sekiya et al.* (1980, 1981), *Watson et al.* (1981), and more recently by *Ikoma and Genda* (2006). The capture or accumulation of hydrogen envelopes depends on the planetary formation time, nebula dissipation time, depletion factor of dust grains in the nebula, nebula opacity, orbital parameters of other planets in a system, the gravity of the protoplanet, and its orbital location, as well as the host star's radiation and plasma environment (e.g., *Ikoma and Genda*, 2006; *Rafikov*, 2006). Depending on all these parameters, theoretical studies let us expect that fast-growing terrestrial planets from Earth to "super-Earth" type may capture from tens to hundreds or even several thousands of Earth ocean equivalent amounts of hydrogen around their rocky cores (e.g., *Hayashi et al.*, 1979; *Mizuno*, 1980; *Wuchterl*, 1993; *Ikoma et al.*, 2000; *Ikoma and Genda*, 2006; *Rafikov*, 2006), making them really "mini-Neptunes." The recent discoveries of "mini-Neptunes" with known sizes and masses indicate that many of them have rocky cores surrounded by a significant amount of hydrogen envelopes (e.g., *Lissauer et al.*, 2011; *Erkaev et al.*, 2013, 2014; *Kislyakova et al.*, 2013; *Lammer*, 2013).

Adams et al. (2008), *Mordasini et al.* (2012), and *Kipping et al.* (2013) studied the mass-radius relationship of planets with a primordial hydrogen envelope. The hydrogen-envelope mass fraction can be written as

$$f = \frac{M_{atm}}{M_{atm} + M_{core}} \qquad (15)$$

with atmosphere mass M_{atm} and core mass M_{core}. Their results indicate that for a given mass, there is a considerable diversity of radii, mainly due to different bulk compositions, reflecting different formation histories. According to *Mordasini et al.* (2012), a protoplanet inside the HZ at 1 AU, with T_{eff} = 250 K, a core mass of 1 $M_{\oplus}$, and a hydrogen-envelope mass function f of ~0.01 could have a radius of about ~2 $R_{\oplus}$ depending on its age, orbital distance, and activity of its host star.

Additional to such nebula-based hydrogen envelopes, catastrophically outgassed volatile-rich steam atmospheres, which depend on the impact history and the initial volatile content of a planet's interior, could also be formed after a planet finished its accretion (e.g., *Elkins-Tanton and Seager*, 2008; *Elkins-Tanton*, 2011). *Zahnle et al.* (1988) studied the evolution of steam atmospheres as expected on the early Earth and found for surface temperatures close to 2000 K atmospheric temperatures on the order of ~1000 K. Recently, *Lebrun et al.* (2013) studied the lifetime of catastrophically outgassed steam atmospheres of early Mars, Earth, and Venus and found that water vapor condenses to an ocean after 0.1, 1.5, and 10 m.y., respectively. *Hamano et al.* (2013) defined two planet categories: a type I planet formed beyond a certain critical distance from the host star, solidifying, in agreement with *Lebrun et al.* (2013), within several million years; and type II planets formed inside the critical distance where a magma ocean can be sustained for longer time periods of up to 100 m.y., even with a larger initial amount of water.

Figure 8 compares the hydrodynamically expanding hydrogen-dominated thermosphere structure for two cases: an Earth-like planet with T_{eff} of 250 K, located within a G-star HZ at 1 AU, exposed to an XUV flux 100× the present-day Sun's on the one hand, and a similar planet (dashed line) with a more extended nebular-based hydrogen-envelope mass fraction of ~0.01 with a corresponding radius of ~2 $R_{\oplus}$ (*Mordasini et al.*, 2012) on the other hand. For both scenarios shown in Fig. 8 the transonic points are reached below the exobase levels so that the considered

atmospheres are in the controlled hydrodynamic regime.

The thermal escape rate with a heating efficiency of 15% for the case where the base of the thermosphere is close to the planetary radius is ~3.3 × 10³¹ s⁻¹; for the hydrogen-dominated protoatmosphere with f = 0.01 and the corresponding radius at ~2 $R_{\oplus}$, it is ~5.0 × 10³² s⁻¹. The total loss of hydrogen in equivalent amounts of Earth ocean contents (1 EO_H ~ 1.5 × 10²³ g) for the first 100 m.y. would be ~1 EO_H for the case with the smaller lower thermosphere distance and ~15.5 EO_H for the extended hydrogen-dominated protoatmosphere.

One can see from these results that an equivalent amount of only a few Earth oceans may be lost during the first 100 m.y (when the stellar XUV radiation is saturated at high levels); one may expect that terrestrial planets growing rapidly to Earth-mass or "super-Earth-mass" bodies inside the nebula may have a problem with losing their captured hydrogen envelopes. Some planets may keep tens, hundreds, or even thousands of EO_H (*Lammer*, 2013; *Lammer et al.*, 2013b, 2014). It is obvious that early Venus or Earth did not capture such huge amounts of nebula gas and most likely did not accrete so fast by collisions with planetary embryos after the nebula disappeared (*Albarède and Blichert-Toft*, 2007).

For all atmospheric escape processes, the age of the host star and planet as well as the host star's activity are very important. After the planet's origin, the briefly discussed

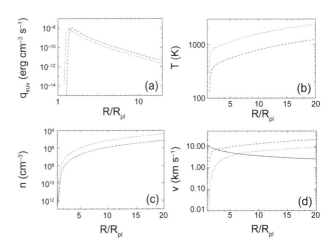

Fig. 8. Profiles of the **(a)** volume heating rate, **(b)** thermosphere temperature, **(c)** atmospheric density, and **(d)** outflow velocity as a function of altitude, for an atomic-hydrogen dominated upper atmosphere of an Earth-like planet with a heating efficiency of 15%, subject to an XUV exposure 100 the present Sun's at 1 AU. The dotted and dashed lines correspond to a temperature T_{eff} of 250 K at the base of the thermosphere with a lower thermosphere near the planetary surface (dotted line) and one for a hydrogen-dominated protoatmosphere with an envelope mass fraction f of 0.01 and a corresponding radius of 2 $R_{\oplus}$ (*Mordasini et al.*, 2012). The solid line in **(d)** corresponds to the escape velocity as a function of distance.

hydrodynamic expansion and outflow due to the extreme stellar and protoplanetary conditions result in the most efficient atmospheric escape process. During this phase no intrinsic planetary magnetic field will protect the expanded upper atmosphere from ion erosion by the stellar wind. The hydrodynamic outflow changes with decreasing XUV flux and cooling of the lower thermosphere from the hydrodynamic to the hydrostatic regime. For this later phase, a recent study by *Kislyakova et al.* (2013, 2014) indicates that stellar-wind-induced ion pick-up of light species such as atomic hydrogen is always less efficient than thermal escape. However, after the XUV flux of a young and active host star decreases to more moderate XUV flux levels <5× that of the present solar value, nonthermal escape processes such as ion pick-up, detached plasma clouds, sputtering, and photochemical losses especially for planets with low gravity become relevant for heavier atmospheric species such as oxygen, carbon, and nitrogen.

6. MAGNETOSPHERIC PROTECTION

As a consequence of upper-atmospheric heating by soft X rays and EUV radiation, the expanding upper atmospheres may reach and even exceed the boundaries of possible planetary magnetospheres. In this case the upper atmosphere will be directly exposed to the plasma flows of the stellar wind and coronal mass ejections (CMEs) with the consequent loss due to ion pick-up, as well as sputtering, and different kinds of photochemical energizing mechanisms, all contributing to *nonthermal* atmospheric mass-loss process (*Lundin et al.*, 2007, and references therein; *Lichtenegger et al.*, 2009; *Lammer*, 2013, and references therein). The size of the planetary magnetosphere appears here as a crucial parameter. Altogether, this makes the planetary magnetic field, as well as the parameters of the stellar wind (mainly its density n_w and speed v_w), very important for the processes of nonthermal atmospheric erosion and mass loss of a planet, eventually affecting the whole evolution of its environment and possible habitability.

6.1. The Challenge of Magnetospheric Protection

The magnetosphere acts as an obstacle that interacts with the stellar wind, deflecting it and protecting planetary ionospheres and upper atmospheres against the direct impact of stellar wind plasmas and energetic particles (e.g., cosmic rays). For an efficient magnetospheric protection of a planet, the size of its magnetosphere characterized by the magnetopause stand-off distance R_s should be much larger than the height of the outer layers of the exosphere. The value of R_s is determined by the balance between the stellar wind ram pressure and the planetary magnetic field pressure at the substellar point (*Griessmeier et al.*, 2004; *Khodachenko et al.*, 2007a). In most studies related to exoplanetary magnetospheric protection, highly simplified planetary dipole-dominated magnetospheres have been assumed. This means that only the intrinsic magnetic dipole moment of an exoplanet, $\mathcal{M}$,

and the corresponding magnetopause electric currents (i.e., the "screened magnetic dipole" case) are considered as the major magnetosphere-forming factors. In this case, i.e., assuming $B(r) \propto \mathcal{M}/r^3$, the value of R_s has been defined by the expression (*Baumjohann and Treumann*, 1997)

$$R_s \equiv R_s^{(dip)} = \left[\frac{\mu_0 f_0^2 \mathcal{M}^2}{8\pi^2 \rho_w \tilde{v}_w^2} \right]^{1/6}$$

(16)

where μ_0 is the diamagnetic permeability of free space, $f_0 \approx 1.22$ is a form-factor of the magnetosphere caused by the account of the magnetopause electric currents, $\rho_w = n_w m$ is the mass density of the stellar wind, and $\tilde{v}_w$ is the relative velocity of the stellar wind plasma, which also includes the planetary orbital velocity.

The planetary magnetic field is generated by the magnetic dynamo. The existence and efficiency of the dynamo are closely related to the type of the planet and its interior structure. Limitations of the existing observational techniques make direct measurements of the magnetic fields of exoplanets impossible. Therefore, an estimate of $\mathcal{M}$ from simple scaling laws found from an assessment of different contributions in the governing equations of planetary magnetic dynamo theory (*Farrell et al.*, 1999; *Sánchez-Lavega*, 2004; *Griessmeier et al.*, 2004; *Christensen*, 2010) is widely used. Most of these scaling laws reveal a connection between the intrinsic magnetic field and rotation of a planet. More recently, *Reiners and Christensen* (2010), based on scaling properties of convection-driven dynamos (*Christensen and Aubert*, 2006), calculated the evolution of average magnetic fields of "hot Jupiters" and found that (1) extrasolar gas giants may start their evolution with rather high intrinsic magnetic fields, which then decrease during the planet lifetime, and (2) the planetary magnetic moment may be independent of planetary rotation (*Reiners and Christensen*, 2010). In the case of *rotation-dependent* dynamo models, the estimations of $\mathcal{M}$ give rather small values for tidally locked close-orbit exoplanets, resulting in small sizes of dipole-dominated magnetospheres, $R_s = R_s^{(dip)}$, compressed by the stellar wind plasma flow. The survival of planets in the extreme conditions close to active stars is often used as an argument in favor of *rotation-independent* models, or in favor of a generalization of the dipole-dominated magnetosphere model.

Khodachenko et al. (2007b) studied the mass loss of the hot Jupiter HD 209458b due to the ion pick-up mechanism caused by stellar CMEs colliding with the planet. In spite of the sporadic character of the CME-planetary collisions for the moderately active host star, it has been shown that the integral action of the stellar CME impacts over the exoplanet's lifetime can produce a significant effect on the planetary mass loss. The estimates of the nonthermal mass loss of a weakly magnetically protected HD 209458b due to stellar wind ion pick-up suggest significant and sometimes unrealistic values — losses up to several tens of planetary masses M_p during a planet's lifetime (*Khodachenko et al.*,

2007b). Because multiple close-in giant exoplanets exist, comparable in mass and size to Jupiter, and because it is unlikely that all of them began their life as much more massive (>10×) objects, one may conclude that additional factors and processes must be considered to avoid full planetary destruction.

6.2. Magnetodisk-Dominated Magnetospheres

The magnetosphere of a close-orbit exoplanet is a complex object whose formation depends on different external and internal factors. These factors may be subdivided into two basic groups: (1) *stellar factors*, e.g., stellar radiation, stellar wind plasma flow, stellar magnetic field; and (2) *planetary factors*, e.g., the type of planet, its orbital characteristics, its escaping material flow, and its magnetic field. The structure of an exoplanetary magnetosphere depends also on the speed regime of the stellar wind plasma relative to the planet (*Erkaev et al.*, 2005; *Ip et al.*, 2004). In particular, for an exoplanet at sufficiently large orbital distance where the stellar wind is supersonic and super-Alfvénic, i.e., where the ram pressure of the stellar wind dominates the magnetic pressure, a Jupiter-type magnetosphere with a bow shock, magnetopause, and magnetotail is formed. At the same time, in the case of an extremely close orbital location of an exoplanet (e.g., d < 0.03 AU for a solar analog), where the stellar wind is still being accelerated and remains submagnetosonic and sub-Alfvénic (*Ip et al.*, 2004; *Preusse et al.*, 2005), an Alfvénic wing-type magnetosphere without a shock in the upstream region is formed. The character of the stellar wind impact on the immediate planetary plasma environment and the atmosphere is different for the super- and sub-Alfvénic types of the magnetosphere. Here, we briefly discuss moderately short-orbit giant planets around solar-type stars subject to a super-Alfvénic stellar wind flow, i.e., the magnetospheres having in general a bow shock, a magnetopause, and a magnetotail, similar to Jupiter. Equivalent configurations are expected for Earth-like planets with evaporating atmospheres.

To explain the obvious survival and sufficient magnetospheric protection of close-orbit hot Jupiters under the extreme conditions of their host stars, *Khodachenko et al.* (2012) proposed a more generic view of a hot Jupiter magnetosphere. A key element in the proposed approach consists of taking into account the upper atmosphere of a planet as an expanding dynamical gas layer heated and ionized by the stellar XUV radiation (*Johansson et al.*, 2009; *Koskinen et al.*, 2010). Interaction of the outflowing plasma with the rotating planetary magnetic dipole field leads to the development of a current-carrying magnetodisk surrounding the exoplanet. The inner edge of the magnetodisk is located at the so-called Alfvénic surface ($r = R_A$) where the kinetic energy density of the moving plasma becomes equal to the energy density of the planetary magnetic field (*Havnes and Goertz*, 1984).

Two major regions, separated by the Alfvénic surface at $r = R_A$, with different topologies of the magnetic field may be distinguished in the magnetosphere of a hot Jupiter driven by the escaping plasma flow (*Mestel*, 1968; *Havnes and Goertz*, 1984). The first region corresponds to the inner magnetosphere, or so-called "dead zone," filled with closed dipole-type magnetic field lines. The magnetic field in the "dead zone" is strong enough to keep plasma locked with the planet. In the second region, the so-called "wind zone," the expanding plasma drags and opens the magnetic field lines (see Fig. 9). The plasma escaping along field lines beyond the Alfvénic surface not only deforms and stretches the original planetary dipole field, but also creates a thin disk-type current sheet in the equatorial region. Altogether, this leads to the development of a new type of magnetodisk-dominated magnetosphere of a hot Jupiter, which has no analog among the solar system planets (*Khodachenko et al.*, 2012). It enlarges the magnetospheric obstacle, thus helping to *suppress* nonthermal escape processes.

6.3. Wind-Exosphere-Magnetosphere Interaction

As discussed before, extended hydrogen exospheres will be produced if hydrogen is the dominating constituent in the upper atmosphere. *Vidal-Madjar et al.* (2003) were the first to observe the transiting exoplanet HD 209458b with the HST/STIS and discovered a 15 ± 4% intensity drop in the stellar Lyα line in the high-velocity part of the spectra. The estimate of the absorption rate has later been carried out independently by *Ben-Jaffel* (2007), who found a lower absorption rate of about 8.9 ± 2.1%, also significantly greater than the transit depth due to the planetary disk alone (*Ben-Jaffel and Sona Hosseini*, 2010). Recently, two other observations of extended upper atmospheres due to Lyα absorption during the transits of the short-periodic gas giant, HD 189733b, have also been reported (*Lecavelier des Etangs et al.*, 2010, 2012). *Holmström et al.* (2008), *Ekenbäck et al.* (2010), and *Lammer et al.* (2011a) have demonstrated that the exosphere-magnetosphere envi-

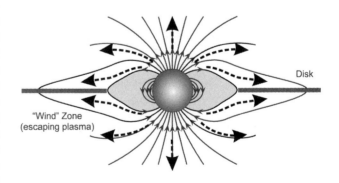

Fig. 9. Schematic view of magnetodisk formation: The expanding plasma flows along the magnetic field lines change the initial dipole topology of the field by the creation of a current disk. In the dead zone (shadowed region), plasma is locked by the strong magnetic field; in the wind zone, plasma escapes along the open field lines in the form of an expanding wind.

ronment of hydrogen-rich gas giants in orbit locations ≤0.05 AU should be strongly affected by the production of energetic neutral hydrogen atoms (ENAs) as well as nonthermal ion escape processes.

Similar processes can be expected for XUV-heated and hydrodynamically expanding upper atmospheres of terrestrial exoplanets, where the upper atmosphere can expand beyond possible magnetopause configurations, so that ENAs will be produced via charge exchange between the charged stellar wind plasma flow and the exospheric particles. From the ionization of exospheric planetary neutral atoms one can probe the stellar-wind-induced ion pick-up loss rate. Recently, *Kislyakova et al.* (2013) applied a stellar wind plasma flow and upper atmosphere interaction model to hydrogen-rich terrestrial planets, exposed to an XUV flux 100× the present solar strength within the HZ of a typical M star at ~0.24 AU. This study found nonthermal escape rates of ~5 × 10^{30} s^{-1}, which is about 1 order of magnitude weaker than the XUV-driven thermal escape rates.

As one can see from Fig. 10, the hydrogen atoms populate a wide area around the planet. The XUV flux is deposited mainly in the lower thermosphere, while the fraction of the ENAs that are directed toward the planet should be deposited in an atmospheric layer below the exobase level where they contribute to thermospheric heating. The production of such huge hydrogen coronae and

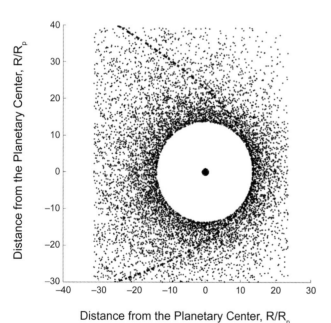

Fig. 10. Modeled extended neutral atomic hydrogen exosphere around a hydrogen-rich "super-Earth" inside an M star HZ at 0.24 AU with the size of 2 R$_\oplus$ and 10 M$_\oplus$; the XUV flux is 50× the present Sun's. The stellar wind plasma density and velocity are ~250 cm^{-3} and 330 km s^{-1}, respectively. The small dots represent the neutral hydrogen atoms, the big dot in the center represents the planet. After *Kislyakova et al.* (2013).

related ENAs was predicted by *Chassefière* (1996a) when he studied the hydrodynamic escape of hydrogen from a hot H$_2$O-rich early Venus thermosphere. This author suggested that the production of ENAs under such extreme conditions should make an important contribution to the heat budget in the upper atmosphere. The expected energy deposition to hydrodynamically expanding, but adiabatically cooling upper atmospheres has to be studied in the future because this important stellar-wind/plasma-induced feedback process can raise the temperature to higher values than the XUV radiation, thus possibly enhancing thermal escape.

7. APPLICATIONS

7.1. The Faint Young Sun Paradox

Stellar evolution theory indicates that the ZAMS Sun was fainter by about 30% than the present-day Sun; the radiative output has increased gradually since then (section 4.1). Estimates of Earth's and Mars' surface temperatures in the presence of their modern atmospheres indicate that both should have been at average temperatures much below the freezing point of liquid water (*Sagan and Mullen*, 1972). Geological evidence, however, clearly suggests warm climates in the first billion years, along with the presence of liquid water (*Kasting and Toon*, 1989; *Valley et al.*, 2002; *Feulner*, 2012). Without some source of additional warming, Earth's surface temperature should not have risen above the freezing point until about 2 b.y ago, long after the first traces of life on Earth appeared (*Kasting and Catling*, 2003). This apparent contradiction is known as the "faint young Sun paradox" (FYSP).

The FYSP has several different possible solutions, and it is not yet agreed which, if any, is correct. These solutions fall into three basic categories described in the subsections below.

7.1.1. Additional greenhouse gases. Higher levels of greenhouse gases could raise the atmospheric temperatures significantly (important also in the present-day Earth atmosphere). For Earth, increased levels of atmospheric CO$_2$ together with water vapor would potentially suffice, and the CO$_2$ level could even self-regulate the mild climate though the carbonate-silicate cycle (*Walker et al.*, 1981). However, geological evidence from paleosols (*Sheldon*, 2006; *Driese et al.*, 2011) argues against massively increased CO$_2$ levels on the young Earth.

Previous paleosol constraints published by *Rye et al.* (1995) have been criticized (*Sheldon*, 2006), as have the very low CO$_2$ estimates from banded iron formations published by *Rosing et al.* (2010) (see also *Reinhard and Planavsky*, 2011; *Dauphas and Kasting*, 2011). The *Sheldon* (2006) and *Driese et al.* (2011) CO$_2$ estimates should be regarded as highly uncertain, as they depend on poorly known factors such as soil porosity and moisture content. Furthermore, recent three-dimensional climate calculations by *Kienert et al.* (2012) suggest that CO$_2$ partial pressures on the early Earth must have been even higher than previously calculated [see also the one-dimensional models in *Haqq-Misra et al.* (2008) and *von Paris et al.* (2008)], because ice albedo feedback tends to amplify the cold temperatures predicted when CO_2 is low. All this suggests that additional greenhouse gases were probably needed.

Other greenhouse gases such as NH_3 may dissociate rapidly (*Kuhn and Atreya,* 1979), and CH_4 may form hazes, increasing the albedo, unless at least an equal amount of CO_2 is present (*Haqq-Misra et al.,* 2008). Ethane (C_2H_6) and carbonyl sulfide (OCS) have been proposed as efficient greenhouse gases in the young terrestrial atmosphere (*Haqq-Misra et al.,* 2008; *Ueno et al.,* 2009). Another candidate is N_2O, which can build up biotically to quite large concentrations and act as a strong greenhouse gas (*Buick,* 2007; *Roberson et al.,* 2011; *Grenfell et al.,* 2011).

Calculations by *Ramirez et al.* (2014) suggest that the early Mars climate problem could instead be resolved by adding 5–20% H_2 to the atmosphere, as has been suggested recently for early Earth (*Wordsworth and Pierrehumbert,* 2013). In addition, higher partial pressures of N_2 might enhance the effectiveness of the CO_2 greenhouse effect on both early Earth and early Mars (*Li et al.,* 2009; *Goldblatt et al.,* 2009; *von Paris et al.,* 2013). Furthermore, SO_2 has been investigated for early Mars climate studies (*Postawko and Kuhn,* 1986; *Tian et al.,* 2010; *Mischna et al.,* 2013).

7.1.2. Lower albedo. A planet's albedo could be changed by lower cloud coverage, in particular in cooler environments (*Rossow et al.,* 1982). Detailed studies of clouds show two competing effects (*Goldblatt and Zahnle,* 2011a; *Zsom et al.,* 2012): Low-lying clouds reflect solar radiation, thus increasing the albedo; in contrast, high clouds, in particular ice-crystal cirrus clouds, add to atmospheric greenhouse warming, potentially sufficient to solve the FYSP (*Rondanelli and Lindzen,* 2010) although requiring full coverage by an unrealistically thick and cold cloud cover (*Goldblatt and Zahnle,* 2011a).

A lower surface albedo could be the result of the suggested smaller continental area in early epochs, although feedback on cloud coverage may result in opposite trends. The absence of cloud condensation nuclei induced biologically (*Charlson et al.,* 1987) also decreases the albedo. Considering these effects and claiming less reflective clouds because of larger cloud droplets (because of fewer biogenically induced condensation nuclei), *Rosing et al.* (2010) inferred a clement climate for the young Earth without the need for high amounts of greenhouse gases (*Rosing et al.,* 2010), although this solution was again questioned (*Goldblatt and Zahnle,* 2011b).

7.1.3. A brighter young Sun. If the Sun had been slightly more massive upon arrival on the ZAMS, then its luminosity would have exceeded standard estimates, owing to the $L \propto M^3$ dependence of a cool star's luminosity, L, on the stellar mass, M. The excess mass could have been shed by a much stronger solar wind in the early phases of solar evolution (*Whitmire et al.,* 1995; *Sackmann and Boothroyd,* 2003). This hypothesis is constrained by the requirement that martian temperatures be supportive of

liquid water, while preventing a runaway greenhouse on Earth. The resulting ZAMS mass of the Sun would then be 1.03–1.07 $M_\odot$ (*Sackmann and Boothroyd,* 2003). Upper limits to the ionized-wind mass-loss rates deduced from radio observations of young solar analogs are marginally compatible with the "bright young Sun" requirement, indicating a maximum ZAMS solar mass of 1.06 $M_\odot$ (*Gaidos et al.,* 2000). Estimates of the young Sun's wind mass loss, however (section 4.4), suggest a total MS mass loss of the Sun of only 0.3% (*Minton and Malhotra,* 2007), too little to explain the FYSP.

7.2. Venus Versus Earth

The atmosphere of present Venus is extremely dry when compared to the terrestrial atmosphere [a total water content of 2×10^{19} g or 0.0014% of the terrestrial ocean, vs. 1.39×10^{24} g on Earth (*Kasting and Pollack,* 1983)]. The similar masses of Venus and Earth and their formation at similar locations in the solar nebula would suggest that both started with similar water reservoirs (*Kasting and Pollack,* 1983). Terrestrial planets such as Venus, Earth, and Mars, with silicate mantles and metallic cores, most likely obtain water and carbon compounds during their accretion (see section 3). The largest part of a planet's initial volatile inventory is outgassed into the atmosphere during the end of magma ocean solidification (e.g., *Zahnle et al.,* 1988, 2007; *Elkins-Tanton,* 2008; *Lebrun et al.,* 2013; *Hamano et al.,* 2013; *Lammer,* 2013). Only a small fraction of the initial volatile inventory will be released into the atmospheres through volcanism. From estimates by considering the possible range of the initial water content in the building blocks of Venus and Earth, it appears possible that both planets may have outgassed hot steam atmospheres with surface pressures up to ~500 bar (*Elkins-Tanton,* 2008).

Recently, *Lebrun et al.* (2013) studied the thermal evolution of early magma oceans in interaction with outgassed steam atmospheres for early Venus, Earth and Mars. Their results indicate that the water vapor of these steam atmospheres condensed to a liquid ocean after ~10, 1.5, and 0.1 m.y. for Venus, Earth, and Mars, respectively, and remained in vapor form for Earth-type planets orbiting closer than 0.66 AU around a solar-like star. However, these authors point out that their results depend on the chosen opacities and it is conceivable that no liquid oceans formed on early Venus. This possibility is also suggested by *Hamano et al.* (2013), who studied the emergence of steam atmosphere produced water oceans on terrestrial planets.

The critical solar flux leading to complete evaporation of a water ocean [a "runaway greenhouse" (*Ingersoll,* 1969)] is $\approx 1.4 \, S_\odot$ nearly independently of the amount of cooling CO_2 in the atmosphere (*Kasting,* 1988), where $S_\odot = 1360$ Wm^{-2} is the present-day solar constant. This critical flux is just about the flux expected from the young Sun at the orbit of Venus (and is about 70% of the present-day value at the same solar distance). New calculations by *Kopparapu et al.* (2013) give a runaway greenhouse limit of 1.06 $S_\odot$. That limit, however, like the *Kasting* (1988) value before it, is almost certainly too low, because it assumes a fully saturated atmosphere and it neglects cloud feedback.

Venus' atmosphere also reveals a surprisingly high deuterium-to-hydrogen (D/H) abundance ratio of $(1.6 \pm 0.2) \times 10^{-2}$ or $100\times$ the terrestrial value of 1.56×10^{-4} (*Kasting and Pollack,* 1983). These observations have motivated a basic model in which water was abundantly available on the primitive Venus. Any water oceans would evaporate, inducing a "runaway greenhouse" in which water vapor would become the major constituent of the atmosphere. This vapor was then photodissociated in the upper atmosphere by *enhanced* solar XUV irradiation (*Kasting and Pollack,* 1983) as well as by frequent meteorite impacts in the lower thermosphere (*Chassefière,* 1996b), followed by the escape of hydrogen into space (*Watson et al.,* 1981; *Kasting and Pollack,* 1983; *Chassefière,* 1996a,b; *Lammer et al.,* 2011c). During the phase of strong thermal escape of atomic hydrogen, remaining oxygen and similar species can be dragged to space by the outward-flowing hydrogen flux (*Hunten et al.,* 1987; *Zahnle and Kasting,* 1986; *Chassefière,* 1996a; *Lammer et al.,* 2011c).

Figure 11 shows the hydrodynamic escape of a hypothetical 500-bar steam atmosphere from early Venus exposed to an EUV flux $100\times$ the present solar value, for assumed heating efficiencies of 15% and 40% (*Lammer et al.,* 2011c; *Lammer,* 2013). In this scenario, it is assumed that early Venus finished its accretion ~50 m.y. after the origin of the Sun and outgassed the maximum expected fraction of its initial water inventory after the magma ocean solidified. One can see that the whole steam atmosphere could have been lost within the condensation timescale of about 10 m.y. modeled by *Lebrun et al.* (2013) for a heating efficiency of 40% as estimated for "hot Jupiters" (*Koskinen et al.,* 2013). For a heating efficiency of 15% (e.g., *Chassefière,* 1996a, 1996b) and assuming that the outgassed steam atmosphere remains in vapor form, the planet would have lost its hydrogen after ~30 m.y., leaving behind some of its oxygen. If the remaining oxygen atoms also experienced high thermal escape rates during the remaining high XUV phase, they would be lost by a combination of thermal and nonthermal escape processes (*Lundin et al.,* 2007). If the high-density CO_2 lower thermosphere was efficiently cooled down to suppress high thermal escape rates of oxygen, the remaining oxygen in the upper atmosphere could have been lost by nonthermal escape processes such as solar-wind-induced ion pick-up and cool ion outflow (e.g., *Kulikov et al.,* 2006; *Lundin et al.,* 2007; *Lammer,* 2013).

If the steam atmosphere cooled within ~10 m.y., so that the remaining water vapor could condense, then a liquid water ocean may have formed on Venus' surface for some period before it again evaporated by the above-mentioned "runaway greenhouse" effect. On the other hand, if early Venus degassed an initial water content that resulted in a steam atmosphere with <250 bar, then the water would most likely have been lost during its vapor phase so that an ocean could never form on the planet's surface.

As discussed in *Lammer et al.* (2011c) and studied in detail in *Lebrun et al.* (2013) and *Hamano et al.* (2013), the outgassed steam atmosphere of the early Earth cooled faster because of its larger distance from the Sun, and this resulted in a shorter condensation timescale compared to the escape. Therefore, a huge fraction of the water vapor condensed and produced Earth's early ocean.

7.3. Mars

The latest research on the formation of Mars indicates that the body formed within ~5 m.y. and remained as a planetary embryo that never accreted toward a larger planet (*Brasser,* 2013). From the building blocks and their water contents expected in the martian orbit, it is estimated that early Mars catastrophically outgassed volatiles amounting to ~50–250 bar of H_2O and ~10–55 bar of CO_2, after the solidification of its magma ocean (*Elkins-Tanton,* 2008). As shown in Fig. 12, due to the high XUV flux of the young Sun, such a steam atmosphere was most likely lost via hydrodynamic escape of atomic hydrogen, dragging heavier atoms such as carbon and oxygen during the first million years after the magma ocean solidified (*Lammer et al.,* 2013c; *Erkaev et al.,* 2014). Although *Lebrun et al.* (2013) found that a degassed steam atmosphere on early Mars would condense after ~0.1 m.y., frequent impacts of large planetesimals and small embryos during the early Noachian most likely kept the protoatmosphere in vapor form (*Lammer et al.,* 2013c). After early Mars lost its out-

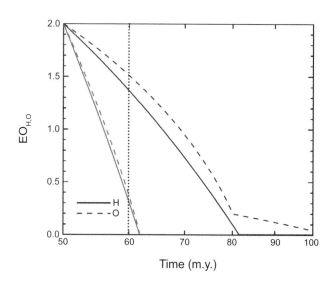

Fig. 11. Hydrodynamic loss of atomic hydrogen and dragged oxygen atoms from a 500-bar outgassed, dissociated steam atmosphere on early Venus during the young Sun's XUV saturation phase ($100\times$ the present XUV flux). The black and gray lines correspond to heating efficiencies of 15% and 40%, respectively (after *Lammer,* 2013). The dotted line marks the time when the steam atmosphere may have condensed (*Lebrun et al.,* 2013) so that an ocean could have formed on the planet's surface.

gassed protoatmosphere, the atmospheric escape rates were most likely balanced by a secondary outgassed atmosphere and delivered volatiles by impacts until the activity of the young Sun decreased so that the atmospheric sources could dominate over the losses (*Tian et al.,* 2009; *Lammer et al.,* 2013c). There is substantial evidence for a warmer and wetter climate on ancient Mars ~3.5–4 G.y. ago (e.g., *Carr and Head,* 2003). Mantle outgassing and impact-related volatiles may have resulted in the buildup of a secondary CO_2 atmosphere of a few tens to a few hundred millibars around the end of the Noachian (*Grott et al.,* 2011; *Lammer et al.,* 2013c). The evolution of the planet's secondary atmosphere and its water content during the past 4 G.y. was most likely determined by an interplay of nonthermal atmospheric escape processes such as direct escape of photochemically hot atoms, atmospheric sputtering, ion pick-up, and cool ion outflow as well as impacts (e.g., *Lundin et al.,* 2007; *Lammer et al.,* 2013c, and references therein), carbonate precipitation, and serpentinization (*Chassefière and Leblanc,* 2011a,b; *Chassefière et al.,* 2013), which finally led to the present-day environmental conditions.

8. THE VARIETY OF HABITABLE CONDITIONS

With the rapidly increasing number of detected exoplanetary systems, the diversity of system architectures, sizes, and radii of the individual planets and the range of average planetary densities have become the focus of *planetary characterization*. It is clear by now that a much wider range of planetary characteristics is realized in exoplanetary systems than available in the solar system. Hot Jupiters, massive planets in the HZ, resonant systems, planets in highly eccentric orbits, and small planets with very low average densities (below that of water) are challenging planetary formation, migration, and evolution theories.

Whether a planet is ultimately habitable depends on a surprisingly large number of astrophysical and geophysical factors, the former of which we have reviewed here. The correct distance from the host star is the most obvious and defining parameter for a HZ, but it by no means provides a sufficient condition for habitability. Migration and resonant processes in the young planetary systems are crucial for the fate of stable orbits of potentially habitable planets. In what looks like a tricky paradox, water is not available in solids in the very region where HZ evolve around cool stars; rather, water needs to be brought there, and gas giant planets may have to do the job of stirring the orbits of the outer planetesimals to scatter them to regions where water is scarce. Water transport through collisions is itself a difficult process, as water needs to survive such collisions to accumulate in the required amounts on the forming planet. Atmospheres outgassed from the planet and trapped by the early protoplanetary disk wrap the growing planet into an extremely dense envelope that itself defies habitability but may be protective of forming secondary atmospheres. Once atmospheres evolve, the harsh environment of the young star begins to act severely. Extreme ultraviolet and X-ray radiation hundreds to thousands of times stronger than the present-day Sun's and stellar winds up to hundreds of times stronger than at present begin to erode upper atmospheres of planets, in particular if no magnetosphere is present that may sufficiently protect the atmosphere. Ultraviolet emission determines the atmospheric temperature profiles and induces a rich chemistry that may become relevant for the formation of life. Here, host star spectral type matters for the formation of an atmospheric temperature inversion. An abiotically forming ozone layer may be one of the important results, especially for active M dwarfs.

The outcome may be an extremely rich variety of potentially habitable environments, but we still lack thorough knowledge of all relevant factors that really determine habitability. While M dwarfs have become favored targets to search for planets in their HZ, such planets may be affected by the strong, long-term high-energy radiation and harsh wind conditions for too long to actually develop habitable planetary surfaces. Their atmospheres may be subject to excessively strong erosion processes.

On the other hand, the development of two habitable planets in the solar system, at least one with thriving life, provides ample evidence that some of the risks, e.g., protoatmospheres that are too heavy or erosion of early nitrogen atmospheres that is too rapid, were successfully overcome (*Lammer et al.,* 2013a). One wonders if some sort of fine-tuning was at work, e.g., between the timing of planet formation and disk dispersal, the outgassing history and the eroding high-energy radiation, the presence of giants like Jupiter or Saturn control-

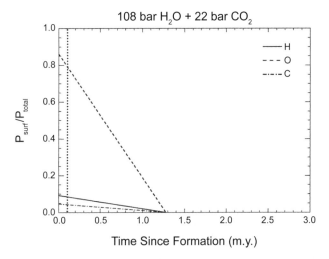

108 bar H_2O + 22 bar CO_2

Fig. 12. Hydrodynamic loss of atomic hydrogen and dragged oxygen and carbon atoms from an outgassed and dissociated steam atmosphere (108 bar H_2O, 22 bar CO_2) on early Mars during the young Sun's XUV saturation phase [100× the present solar XUV level (after *Erkaev et al.,* 2014)]. The dotted line marks the time period when the steam atmosphere may have condensed (*Lebrun et al.,* 2013), so that an ocean could have formed on the planet's surface (neglecting the energy deposited at the planet's surface by frequent planetesimal impacts and small embryos).

ling the architecture and stability of planetary orbits in the HZ, or the mass- and perhaps rotation-dependent formation of a magnetosphere protecting the upper atmosphere of a planet. If such fine-tuning did not succeed, then HZ planets may end up with huge high-pressure hydrogen atmospheres, suggested by the detection of small, low-density exoplanets, strongly eroded atmospheres such as on Mars, or the complete loss of their atmospheres. Other planets may be close to the HZ but may dwell in eccentric orbits, making surface life more problematic.

Apart from the diverse astrophysical conditions, a plethora of *geophysical* conditions matter, such as planetary internal convection and geodynamos, plate tectonics, volcanism, the presence and distribution of land mass, the orientation and stability of the rotation axis, etc. They interact as well with the diverse astrophysical factors discussed in this article, to add further constraints on actual habitability.

Acknowledgments. We thank an anonymous referee for useful comments that helped improve this chapter. M.G.R.D., M.L.K., H.L., and E.P.-L. thank the Austrian Science Fund (FWF) for financial support (research grants related to planetary habitability: S116001-N16 and S116004-N16 for M.G., S11603-N16 for R.D., S116606-N16 for M.L.K., S116607-N16 for H.L., S11608-N16 and P22603-N16 for E.P.-L.). N.V.E. acknowledges support by the RFBR grant No 12-05-00152-a. J.K. thanks the NASA Astrobiology and Exobiology programs. H.R. and H.L. thank the Helmholtz Association for support through the research alliance "Planetary Evolution and Life." I.R. acknowledges financial support from the Spanish Ministry of Economy and Competitiveness (MINECO) and the "Fondo Europeo de Desarrollo Regional" (FEDER) through grant AYA2012-39612-C03-01. M.G., H.L., and M.L.K. acknowledge support by the International Space Science Institute in Bern, Switzerland, and the ISSI Team "Characterizing Stellar and Exoplanetary Environments."

REFERENCES

Abe Y. et al. (2011) *Astrobiology, 11,* 443–460.
Adams E. R. et al. (2008) *Astrophys. J., 673,* 1160–1164.
Albarède F. and Blichert-Toft J. (2007) *Compt. Rend. Geosci., 339,* 917–927.
Anglada-Escudé G. et al. (2012) *Astrophys. J., 751,* L16.
Audard M. et al. (2000) *Astrophys. J., 541,* 396–409.
Baraffe I. et al. (1998) *Astron. Astrophys., 337,* 403–412.
Baross J. A. et al. (2007) In *Planets and Life: The Emerging Science of Astrobiology* (W. T. I. Sullivan et al., eds.), pp. 275–291. Cambridge Univ., Cambridge.
Batalha N. M. et al. (2011) *Astrophys. J., 729,* 27.
Batalha N. M. et al. (2013) *Astrophys. J. Suppl., 204,* 24.
Baumjohann W. and Treumann R. A. (1997) *Basic Space Plasma Physics.* Imperial College, London.
Ben-Jaffel L. (2007) *Astrophys. J. Lett., 671,* L61–L64.
Ben-Jaffel L. and Sona Hosseini S. (2010) *Astrophys. J., 709,* 1284–1296.
Bonfils X., Delfosse X., Udry S. et al. (2013) *Astron. Astrophys., 549,* A109.
Borucki W. J. et al. (2013) *Science, 340,* 587–590.
Brasser R. (2013) *Space Sci. Rev., 174,* 11–25.
Buchhave L. A. et al. (2012) *Nature, 486,* 375–377.
Buick R. (2007) *Geobiol., 5,* 97–100.
Carr M. H. and Head J. W. (2003) *J. Geophys. Res., 108,* Cite ID 5042.
Chamberlain J. W. (1963) *Planet. Space. Sci., 11,* 901–996.
Chapman S. A. (1930) *Mem. R. Meteor. Soc., 3,* 103–125.
Charlson R. J. et al. (1987) *Nature, 326,* 655–661.
Chassefière E. (1996a) *J. Geophys. Res., 101,* 26039–26056.
Chassefière E. (1996b) *Icarus, 124,* 537–552.
Chassefière E. and Leblanc F. (2011a) *Earth Planet. Sci. Lett., 310,* 262–271.
Chassefière E. and Leblanc F. (2011b) *Planet. Space Sci. 59,* 218–226.
Chassefière E. et al. (2013) *J. Geophys. Res., 118,* 1123–1134.
Chiang E. I. (2003) *Astrophys. J., 584,* 465–471.
Christensen U. R. (2010) *Space Sci. Rev., 152,* 565–590.
Christensen U. R. and Aubert J. (2006) *Geophys. J. Intl., 166,* 97–114.
Claire M. W. et al. (2012) *Astrophys. J., 757,* 95.
Crutzen P. J. (1970) *Q. J. R. Meteorol. Soc., 96,* 320–325.
Dauphas N. and Kasting J. F. (2011) *Nature, 474,* E2–E3.
Drake S. A. et al. (1993) *Astrophys. J., 406,* 247–251.
Driese S. G. et al. (2011) *Precambrian Res., 189,* 1–17.
Dvorak R. et al. (2012) In *Let's Face Chaos Through Nonlinear Dynamics* (M. Robnik et al., eds.), pp. 137–147. American Institute of Physics, Melville, New York.
Eggl S. et al. (2012) *Astrophys. J., 752,* 74–85.
Ekenbäck A. et al. (2010) *Astrophys. J., 709,* 670–679.
Elkins-Tanton L. T. (2008) *Earth Planet. Sci. Lett., 271,* 181–191.
Elkins-Tanton L. T. (2011) *Astrophys. Space Sci., 332,* 359–364.
Elkins-Tanton L. and Seager S. (2008) *Astrophys. J., 685,* 1237–1246.
Erkaev N. V. et al. (2005) *Astrophys. J. Suppl., 157,* 396–401.
Erkaev N. V. et al. (2013) *Astrobiology, 13,* 1011–1029.
Erkaev N. V. et al. (2014) *Planet. Space Sci., 98,* 106–119.
Farrell W. M. et al. (1999) *J. Geophys. Res., 104(E6),* 14025–14032.
Feldman W. C. et al. (1977) In *The Solar Output and Its Variation* (O. R. White, ed.), pp. 351–382. Colorado Associated Univ., Boulder.
Feulner G. (2012) *Rev. Geophys., 50,* Cite ID RG2006.
Forget F. and Pierrehumbert R. T. (1997) *Science, 278,* 1273–1276.
France K. et al. (2013) *Astrophys. J., 763,* 149.
Fressin F. et al. (2013) *Astrophys. J., 766,* 81.
Garcés A. et al. (2011) *Astron. Astrophys., 531,* A7.
Gaidos E. J. et al. (2000) *Geophys. Res. Lett., 27,* 501–503.
Gayley K. G. et al. (1997) *Astrophys. J., 487,* 259–270.
Gilliland R. L. et al. (2011) *Astrophys. J. Suppl., 197,* 6.
Goldblatt M. T. and Zahnle K. J. (2011a) *Climate Past, 7,* 203–220.
Goldblatt M. T. and Zahnle K. J. (2011b) *Nature, 474,* E3–E4.
Goldblatt C. et al. (2009) *Nature Geosci., 2,* 891–896.
Grenfell J. L. et al. (2007) *Astrobiology, 7,* 208–221.
Grenfell J. L. et al. (2011) *Icarus, 211,* 81–88.
Grenfell J. L. et al. (2012) *Astrobiology, 12,* 1109–1122.
Grenfell J. L. et al. (2013) *Astrobiology, 13,* 415–438.
Grenfell J. L. et al. (2014) *Planet. Space Sci.,* in press.
Griessmeier J.-M. et al. (2004) *Astron. Astrophys., 425,* 753–762.
Grott M. et al. (2011) *Earth Planet. Sci. Lett., 308,* 391–400.
Güdel M. (2004) *Astron. Astrophys. Rev., 12,* 71–237.
Güdel M. (2007) *Liv. Rev. Solar Phys., 4(3),* available online at *solarphysics.livingreviews.org/Articles/lrsp–2007–3.*
Güdel M. et al. (1997) *Astrophys. J., 483,* 947–960.
Guinan E. F. and Engle S. G. (2009) In *The Ages of Stars* (E. E. Mamajek et al., eds.), pp. 395–408. IAU Symp. 258, Cambridge Univ., Cambridge.
Haagen-Smit A. J. et al. (1952) In *2nd Proc. Natl. Air Symp.,* pp. 54–56.
Haghighipour N. and Kaltenegger L. (2013) *Astrophys. J., 777,* 166.
Hamano K. et al. (2013) *Nature, 497,* 607–610.
Haqq-Misra J. D. et al. (2008) *Astrobiology, 8,* 1127–1137.
Hart M. H. (1979) *Icarus, 37,* 351–357.
Havnes O. and Goertz C. K. (1984) *Astron. Astrophys., 138,* 421–430.
Hayashi C. et al. (1979) *Earth Planet. Sci. Lett., 43,* 22–28.
Hedelt P. et al. (2013) *Astron. Astrophys., 553,* A9.
Holland H. D. (1962) In *Petrologic Studies* (A. E. J. Engel et al., eds.), Geological Society of America, Boulder.
Holmström M. et al. (2008) *Nature, 451,* 970–972.
Howard A. W. (2013) *Science, 340,* 572–576.
Hu R. et al. (2012) *Astrophys. J., 761,* 166.
Hunten D. M. et al. (1987) *Icarus, 69,* 532–549.
Ikoma M. et al. (2000) *Astrophys. J., 537,* 1013–1025.
Ikoma M. and Genda H. (2006) *Astrophys. J., 648,* 696–706.
Ingersoll A. P. (1969) *J. Atmospheric Sci., 26,* 1191–1198.
Ip W.-H. et al. (2004) *Astrophys. J. Lett., 602,* L53–L56.
Johansson E. P. G. et al. (2009) *Astron. Astrophys., 496,* 869–877.
Kaltenegger L. and Haghighipour N. (2013) *Astrophys. J., 777,* 165.
Kaltenegger L. et al. (2007) *Astrophys. J., 658,* 598–616.
Kaltenegger L. et al. (2011) *Astrophys. J., 733,* 35.
Kaltenegger L. et al. (2013) *Astrophys. J. Lett., 755,* L47.

Kane S. R. and Hinkel N. R. (2013) *Astrophys. J., 762*, 7.

Kasting J. F. (1988) *Icarus, 74*, 472–494.

Kasting J. F. and Catling D. (2003) *Annu. Rev. Astron. Astrophys., 41*, 429–463.

Kasting J. F. and Pollack J. B. (1983) *Icarus, 53*, 479–508.

Kasting J. F. and Toon O. B. (1989) In *Origin and Evolution of Planetary and Satellite Atmospheres* (S. K. Atreya et al., eds.), pp. 423–449. Univ. of Arizona, Tucson.

Kasting J. F. et al. (1993) *Icarus, 101*, 108–128.

Kasting J. F. et al. (2014) *Proc. Natl. Acad. Sci.*, in press.

Kharecha P. et al. (2005) *Geobiol., 3*, 53–76.

Khodachenko M. L. et al. (2007a) *Astrobiology, 7*, 167–184.

Khodachenko M. L. et al. (2007b) *Planet. Space Sci., 55*, 631–642.

Khodachenko M. L. et al. (2012) *Astrophys. J., 744*, 70–86.

Kienert H. et al. (2012) *Geophys. Res. Lett., 39(23)*, L23710.

Kipping D. M. et al. (2013) *Mon. Not. Roy. Astron. Soc., 434*, 1883–1888.

Kislyakova G. K. et al. (2013) *Astrobiology, 13*, 1030–1048.

Kislyakova G. K. et al. (2014) *Astron. Astrophys., 562*, A116.

Kitzmann D. et al. (2010) *Astron. Astrophys., 511*, A66.

Kitzmann D. et al. (2011a) *Astron. Astrophys., 531*, A62.

Kitzmann D. et al. (2011b) *Astron. Astrophys., 534*, A63.

Kitzmann D. et al. (2013) *Astron. Astrophys., 557*, A6.

Kopparapu R. K. (2013) *Astrophys. J. Lett., 767*, L8.

Kopparapu R. K. et al. (2013) *Astrophys. J., 765*, 131.

Koskinen T. T. et al. (2007) *Nature, 450*, 845–848.

Koskinen T. et al. (2010) *Astrophys. J., 723*, 116–128.

Koskinen T. T. et al. (2013) *Icarus, 226*, 1678–1694.

Kuhn W. R. and Atreya S. K. (1979) *Icarus, 37*, 207–213.

Kulikov Y. N. et al. (2006) *Planet. Space Sci., 54*, 1425–1444.

Kulikov Yu. N. et al. (2007) *Space Sci Rev., 129*, 207–243.

Lammer H. (2013) *Origin and Evolution of Planetary Atmospheres: Implications for Habitability.* Springer Briefs in Astronomy, Springer, Heidelberg/New York. 98 pp.

Lammer H. et al. (2009) *Astron. Astrophys., 506*, 399–410.

Lammer H. et al. (2011a) *Astrophys. Space Sci., 335*, 9–23.

Lammer H. et al. (2011b) *Astrophys. Space Sci., 335*, 39–50.

Lammer H. et al. (2011c) *Orig. Life Evol. Bisoph., 41*, 503–522.

Lammer H. et al. (2013a) In *The Early Evolution of the Atmospheres of Terrestrial Planets* (J. M. Trigo-Rodriguez et al., eds.), pp. 35–52. Astrophys. Space Sci. Proc., Springer-Verlag, Heidelberg/New York.

Lammer H. et al. (2013b) *Mon. Not. R. Astron. Soc., 430*, 1247–1256.

Lammer H. et al. (2013c) *Space Sci. Rev., 174*, 113–154.

Lammer H. et al. (2014) *Mon. Not. R. Astron. Soc., 138*, 359–391.

Lebrun T. et al. (2013) *J. Geophys. Res., 118*, 1–22.

Lecavelier des Etangs A. et al. (2010) *Astron. Astrophys., 514*, A72.

Lecavelier des Etangs A. et al. (2012) *Astron. Astrophys., 543*, L4.

Leconte J. et al. (2013) *Astron. Astrophys., 554*, A69.

Lederberg J. (1965) *Nature, 207*, 9–13.

Léger A. et al. (2009) *Astron. Astrophys., 506*, 287–302.

Li K. et al. (2009) *Proc. Nat. Academy Sci.*, 9576–9579.

Lichtenegger H. I. M. et al. (2009) *Geophys. Res. Lett., 36*, L10204.

Lim J. and White S. M. (1996) *Astrophys. J., 462*, L91–L94.

Lim J. et al. (1996) *Astrophys. J., 460*, 976–983.

Linsky J. L. and Wood B. E. (1996) *Astrophys. J., 463*, 254–270.

Linsky J. L. et al. (2012) *Astrophys. J., 745*, 25.

Linsky J. L. et al. (2013) *Astrophys. J., 766*, 69.

Lissauer J. J. and the Kepler Team (2011) *Nature, 470*, 53–58.

Lovelock J. E. (1965) *Nature, 207*, 568–570.

Lundin R. et al. (2007) *Space Sci. Rev., 129*, 245–278.

Lunine J. et al. (2011) *Adv. Sci. Lett., 4*, 325–338.

Maehara H. et al. (2012) *Nature, 485*, 478–481.

Maindl T. I. and Dvorak, R. (2014) In *Exploring the Formation and Evolution of Planetary Systems* (M. Booth et al., eds.), in press. IAU Symp. 299, Cambridge Univ., Cambridge.

Maindl T. I. et al. (2013) *Astron. Notes, 334*, 996.

Mamajek E. E. and Hillenbrand L. A. (2008) *Astrophys. J., 687*, 1264–1293.

Marley M. S. et al. (2013) In *Comparative Climatology of Terrestrial Planets* (S. J. Mackwell et al., eds.), pp. 367–391. Univ. of Arizona, Tucson.

Mestel L. (1968), *Mon. Not. R. Astron. Soc., 138*, 359–391.

Minton D. A. and Malhotra R. (2007) *Astrophys. J., 660*, 1700–1706.

Mischna M. M. et al. (2000) *Icarus, 145*, 546–554.

Mischna M. A. et al. (2013) *J. Geophys. Res.–Planets, 118*, 560–576.

Mizuno H. (1980) *Prog. Theor. Phys., 64*, 544–557.

Mizuno H. et al. (1978) *Prog. Theor. Phys., 60*, 699–710.

Montési L. G. J. and Zuber M. T. (2003) *J. Geophys. Res.–Planets, 108*, 5048.

Morbidelli A. (2013) In *Planets, Stars and Stellar Systems* (T. D. Oswalt et al., eds.), pp. 63–109. Springer, Dordrecht.

Morbidelli A. et al. (2000) *Meteoritics & Planet. Sci., 35*, 1309–1320.

Morbidelli A. et al. (2007) *Astron. J., 134*, 1790–1798.

Mordasini C. et al. (2012) *Astron. Astrophys., 547*, A112.

National Academy of Sciences (2007) *The Limits of Organic Life in Planetary Systems.* National Academy of Sciences, Washington, DC.

O'Brien D. P. et al. (2006) *Icarus, 184*, 39–58.

Osten R. A. et al. (2010) *Astrophys. J., 721*, 785–801.

Parker E. N. (1958) *Astrophys. J., 128*, 664–675.

Paynter D. J. and Ramaswamy, V. (2011) *J. Geophys. Res., 116*, D20302.

Penz T. et al. (2008) *Planet. Space Sci., 56*, 1260–1272.

Pepe F. et al. (2011) *Astron. Astrophys., 534*, A58.

Pierrehumbert R. and Gaidos E. (2011) *Astrophys. J., 734*, L13.

Pilat–Lohinger E. et al. (2008) *Astrophys. J., 681*, 1639–1645.

Pizzolato N. et al. (2003) *Astron. Astrophys., 397*, 147–157.

Pollack J. B. et al. (1996) *Icarus, 124*, 62–85.

Postawko S. E. and Kuhn W. R. (1986) *J. Geophys. Res., 91*, 431–438.

Preusse S. et al. (2005) *Astron. Astrophys., 434*, 1191–1200.

Rafikov R. R. (2006) *Astrophys. J., 648*, 666–682.

Ramirez R. M. et al. (2014) *Nature Geosci., 7*, 59–63.

Rauer H. et al. (2011) *Astron. Astrophys., 529*, A8.

Raymond S. N. et al. (2004) *Icarus, 168*, 1–17.

Raymond S. N. et al. (2007) *Astrobiology, 7*, 66–84.

Raymond S. N. et al. (2009) *Icarus, 203*, 644–662.

Reiners A. and Christensen U. R. (2010) *Astron. Astrophys., 522*, A13.

Reinhard C. T. and Planavsky N. J. (2011) *Nature, 474*, E1–E2.

Ribas I. et al. (2005) *Astrophys. J., 622*, 680–694.

Ribas I. et al. (2010) *Astrophys. J., 714*, 384–395.

Ringwood A. E. (1979) *Origin of the Earth and Moon.* Springer, Heidelberg/New York.

Roberson A. et al. (2011) *Geobiol., 9*, 313–320.

Rondanelli R. and Lindzen R. S. (2010) *J. Geophys. Res., 115*, D02108.

Rosing M. T. et al. (2010) *Nature, 464*, 744–747.

Rossow W. B. et al. (1982) *Science, 217*, 1245–1247.

Rugheimer S. et al. (2013) *Astrobiol., 13*, 251–269.

Rye R. et al. (1995) *Nature, 378*, 603–605.

Sackmann I.-J. and Boothroyd A. I. (2003) *Astrophys. J., 583*, 1024–1039.

Sagan C. and Mullen G. (1972) *Science, 177*, 52–56.

Sánchez-Lavega A. (2004) *Astrophys. J. Lett., 609*, L87–90.

Sanz-Forcada J. et al. (2011) *Astron. Astrophys., 532*, A6.

Scalo J. et al. (2007) *Astrobiol., 7*, 85–166.

Schaefer B. E. et al. (2000) *Astrophys. J., 529*, 1026–1030.

Schrijver C. J. et al. (2012) *J. Geophys. Res., 117*, A08103.

Seager S. (2013) *Science, 340*, 577–581.

Seager S. et al. (2013) *Astrophys. J., 777*, 95.

Segura A. et al. (2005) *Astrobiol., 5*, 706–725.

Segura A. et al. (2010) *Astrobiol., 10*, 751–771.

Sekiya M. et al. (1980) *Prog. Theor. Phys., 64*, 1968–1985.

Sekiya M. et al. (1981) *Prog. Theor. Phys., 66*, 1301–1316.

Selsis F. et al. (2008) *Phys. Scr., 130*, 014032.

Sheeley N. R. Jr. et al. (1997) *Astrophys. J., 484*, 472–478.

Sheldon N. D. (2006) *Precambrian Res., 147*, 148–55.

Soderblom D. R. et al. (1993) *Astrophys. J., 409*, 624–634.

Stern R. A. et al. (1995) *Astrophys. J., 448*, 683–704.

Stewart S. T. and Leinhardt Z. M. (2012) *Astrophys. J., 751*, 32.

Telleschi A. et al. (2005) *Astrophys. J., 622*, 653–679.

Tian F. et al. (2005) *Astrophys. J., 621*, 1049–1060.

Tian F. et al. (2008) *J. Geophys. Res., 113*, E05008.

Tian F. et al. (2009) *Geophys. Res. Lett., 36*, L02205.

Tian F. et al. (2010), *Earth Planet. Sci. Lett., 295*, 412–418.

Tsiganis K. et al. (2005) *Nature, 435*, 459–461.

Ueno Y. et al. (2009) *Proc. Natl. Acad. Sci., 106*, 14784–14789.

Valley J. W. et al. (2002) *Geology, 30*, 351–354.

van den Oord G. H. J. and Doyle J. G. (1997) *Astron. Astrophys., 319*, 578–588.

Vasquez M. et al. (2013a) *Astron. Astrophys., 549*, A26.

Vasquez M. et al. (2013b) *Astron. Astrophys., 557*, A46.

Vidal–Madjar A. et al. (2003) *Nature, 422*, 143–146.

Volkov A. N. and Johnson R. E. (2013) *Astrophys. J., 765*, 90.

von Paris P. et al. (2008) *Planet. Space Sci., 45*, 1254–1259.

von Paris P. et al. (2010) *Astron. Astrophys., 522*, A23.

von Paris P. et al. (2013) *Planet. Space Sci., 82,* 149–154.
von Zahn U. et al. (1980) *J. Geophys. Res., 85,* 7829–7840.
Walker J. C. G. (1977) *Evolution of the Atmosphere.* McMillan, New York.
Walker J. C. G. et al. (1981) *J. Geophys. Res., 86,* 9776–9782.
Walsh K. J. et al. (2011) *Nature, 475,* 206–209.
Wargelin B. J. and Drake J. J. (2001) *Astrophys. J. Lett., 546,* L57–L60.
Watson A. J. et al. (1981) *Icarus, 48,* 150–166.
Whitmire D. P. et al. (1995) *J. Geophys. Res., 100,* 5457–5464.
Wood B. E. et al. (2002) *Astrophys. J., 574,* 412–425.
Wood B. E. et al. (2005). *Astrophys. J., 628,* L143–146.

Wordsworth R. and Pierrehumbert R. (2013) *Science, 339,* 64–67
Wordsworth R. D. et al. (2011) *Astrophys. J. Lett., 733,* 48.
Wordsworth R. D. et al. (2010) *Astron. Astrophys., 522,* A22.
Wuchterl G. (1993) *Icarus, 106,* 323–334.
Yelle R. V. (2004) *Icarus, 170,* 167–179.
Yung Y. L. and DeMore W. B. (1999) *Photochemistry of Planetary Atmospheres.* Oxford Univ., Oxford.
Zahnle K. J. and Kasting J. F. (1986) *Icarus, 680,* 462–480.
Zahnle K. J. et al. (1988) *Icarus, 74,* 62–97.
Zahnle K. et al. (2007) *Space Sci. Rev., 129,* 35–78.
Zsom A. et al. (2012) *Icarus, 221,* 603–616.

Index

Page numbers refer to specific pages on which an index term or concept is discussed. "ff" indicates that the term is also discussed on the following pages.